THE STOCKMAN'S HANDBOOK

(Animal Agriculture Series)

Other books by M. E. Ensminger
available from The Interstate:

Animal Science
Beef Cattle Science
Dairy Cattle Science
Sheep and Wool Science
Swine Science
(with R. O. Parker)
Horses and Horsemanship
Poultry Science

Animal Science presents a perspective or panorama of the far-flung livestock industry; whereas each of the other books presents specialized material pertaining to the specific class of farm animals indicated by its respective title.

THE STOCKMAN'S HANDBOOK

(Animal Agriculture Series)

by

M. E. Ensminger, B.S., M.A., Ph.D.

Formerly: Assistant Professor in Animal Science
University of Massachusetts

Chairman, Department of Animal Science
Washington State University

Consultant, General Electric Company
Nucleonics Department (Atomic Energy Commission)

Currently: President
Consultants-Agriservices
Clovis, California

President
Agriservices Foundation

Collaborator
U.S. Department of Agriculture

Adjunct Professor
California State University–Fresno

Adjunct Professor
The University of Arizona–Tucson

Distinguished Professor
University of Wisconsin–River Falls

Sixth Edition

THE INTERSTATE
PRINTERS & PUBLISHERS, INC.

Danville, Illinois

This is my best

TO

my best, Audrey Helen

Editions:

First 1955
Second 1959
Third 1962
Fourth 1970
Fifth 1978
Sixth 1983

Translation: Translated into Spanish under the direction of Dr. Mauricio B. Helman, Professor, Veterinary Sciences, Catholic University of Argentina; and published by El Ateneo, Florida 34-344, Buenos Aires, Argentina.

Library of Congress Catalog Card No. 83-80195

ISBN 0-8134-2295-7

Preface to the Sixth Edition

My guiding philosophy as an author has always been to meet the need, without thought of monetary return. Not only that, I spare neither time nor expense in honing my works. My goal for each title: The best book in the subject matter area. Fortunately, I have been blessed by this philosophy, for my books are used all over the world.

Recently, my longtime publisher friend, Mr. Russell Guin, The Interstate Printers & Publishers, Inc., told me that there is need for a new edition of this book, *The Stockman's Handbook.* So, I invoked my nocturnal habit, pushed a pencil over reams of yellow paper, gave my Executive Assistant a copy of my Missouri hieroglyphics to put through a typewriter, and met my publisher's deadline.

The Stockman's Handbook is what the name implies—it's a handbook for everyone involved in the food chain, from range to range (stove). It's for teachers, students, farmers and ranchers, processors, marketers, and consumers. It's for those who wish to know both the how and the why. It brings together a vast array of information in one source—in concise, quick and easy-to-find form. Moreover, it is for the tomorrow mind, instead of for the yesterday mind.

A ranch couple, whom I have never met, wrote to me as follows relative to the previous edition (the fifth edition) of *The Stockman's Handbook:*

> This past year, we got a copy of *The Stockman's Handbook.* In the preface, you stated that you continually asked yourself: "Is the material helpful; is it clear, concise, and accurate? Does it make for pleasurable reading?" To the first four-pronged question, our answer is *yes, yes, yes, yes.* As to pleasurable reading, more often than not your books are our bedtime reading material.
>
> With all the books written on various aspects of agriculture, there was need for a book that tied it all together. And you did it superbly in *The Stockman's Handbook.*

Because of folks like the couple quoted above, I shall never lay down my pen.

The author gratefully acknowledges the contributions of all those who participated in this revision. Special appreciation is expressed to the following members of my staff for their dedicated efforts: To Audrey Ensminger (Mrs. E), who supervised the revision; to Dr. Richard O. Parker, for his counsel, and for adapting some of the NRC tables; to Greg Bitney for artwork; and to Robert Vann, for ingeniously putting the revision in final form. Also, I am grateful to Doris Sanchez, Program Analyst, Office of the Administrator, Economic Research Service, U.S. Department of Agriculture, who gave unselfishly and liberally of her time and talents in providing much of the recent source material for updating. Additionally, at appropriate places in the book, due acknowledgment and appreciation is expressed to those who responded so liberally to my call for illustrations and information. Most of all, I am grateful to the readers and users of this book, who cause me to keep on keeping on.

M. E. Ensminger

Clovis, California
1983

Contents

Section		Page
1.	Animal Behavior and Environment	1
2.	Business Aspects	31
3.	Breeding	121
4.	Feeding	213
5.	Pasture and Range Forages; Green Chop	587
6.	Hay and Crop Residues	667
7.	Silage and Haylage; High-Moisture Grain	701
8.	Management	723
9.	Buildings and Equipment	795
10.	Animal Health, Disease Prevention, and Parasite Control	873
11.	Selecting and Judging Livestock	1007
12.	Fitting and Showing Livestock	1043
13.	Marketing Livestock and Milk	1061
14.	Meat and Milk	1077
15.	Wool and Mohair	1105
16.	Law on the Livestock Farm	1111
17.	Breed Registry Associations	1123
18.	Agricultural Magazines	1149
19.	Where to Go for Help	1159
20.	Weights and Measures	1167
	Index	1181

SECTION 1

ANIMAL BEHAVIOR AND ENVIRONMENT

Contents **Page**

Animal Behavior 2
Why Animals Behave as They Do 2
Genetic 2
Simple Learning 2
Habituation 2
Conditioning (Operant Conditioning) 3
Imprinting (Socialization) 3
Complex Learning 3
Insight Learning (Reasoning) 4
How Animals Behave—behavioral systems 4
Social Relationships 12
Social Order (Dominance) 12
Leader-Follower 13
Interspecies Relationships 14
Man-Animal Relationships 14
How Animals Communicate 14
Sound 14
Chemicals 15
Visual Displays 15
Normal Animal Behavior 15
Abnormal Animal Behavior 17
Applied Animal Behavior 17
Breeding for Adaptation 17
Quick Adaptation—Early Training 18
Loading Chute and Corral Design 18
Manure Elimination 19
Control of Pests 19
Training Horses 19
Companionship 20
Controlling Animal Behaviors 20
Animal Environment 21
How Environment Affects Animals 21
Feed and Nutrition 22
Weather 22
Environmentally Controlled Buildings 22
Artificial Lighting 23
Health 23
Stress 23
Control Pollution 24
Pollution Laws and Regulations 24
Environmental Effects of Grazing Lands 25
Manure 26
Precautions When Using Manure as a Fertilizer on the Land 27
How Much Manure Can Be Applied to Land 27
Agricultural Chemicals 27
Zoning 28
Conserve Energy 28

Today, there is great interest in animal behavior and environment. Those who grew up around animals and dealt with them in practical ways already have accumulated substantial workaday knowledge about their reaction to certain stimuli or their environment. But those who are less familiar with them may need to acquaint themselves with animal behavior in a man-made environment, better to produce and care for them, and in order to recognize the signs when all is not well in the barnyard. Whether we come from farms or are city-bred, the principles and application of animal behavior and environment depend on understanding.

This section is presented for the purpose of bridging the gap between something old and something new in animal behavior and environment.

The way it used to be done

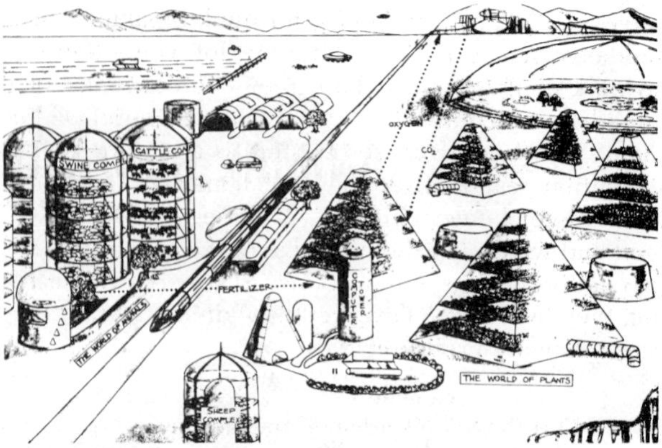

Futuristic farm

Fig. 1-1. The factory-like farm of the future will likely confine convenience animals (the kind that give birth to young without assistance, and which possess many other built-in conveniences), adapted to a man-made environment in high rises, which are fully automated, operated by computer, environmentally controlled, pollution-free, and powered by solar or nuclear energy. (Left picture, courtesy USDA; right picture, by artist, Toby Escola.)

ANIMAL BEHAVIOR

Animal behavior is the reaction of animals to certain stimuli, or the manner in which they react to their environment. Through the years, behavior has received less attention than the quantity and quality of the meat, milk, eggs, fiber, and power produced by animals. But modern breeding, feeding, and management have brought renewed interest in behavior, especially as a factor in obtaining maximum production and efficiency. With the restriction, or confinement, of herds and flocks, many abnormal behaviors evolved to plague those who raise them, including cannibalism, loss of appetite, stereotyped movements, poor parental care, over-aggressiveness, dullness, degenerate sexual behavior, tail biting, cribbing, and a host of other behavior disorders. Confinement has not only limited space, but it has interfered with the habitat and social organization to which the species was adapted and best suited. This has been due to a genetic time lag; man has altered the environment faster than the genetic makeup of animals.

WHY ANIMALS BEHAVE AS THEY DO

Animal behavior is caused by, or is the result of, three forces: (1) genetic, (2) simple learning, and (3) complex learning.

Genetic

The relationship of breeding and selection to behavior becomes obvious in a group of weaning foals of mixed breeding. Upon racing across a field, some amble off in a rhythmic running walk, nodding their heads as they go; others travel high enough to clear the tops of the daisies; still others break away in an easy gallop. Each of these three actions is executed with equal ease and naturalness. The first weanlings described are Tennessee Walking Horses, the second are Hackneys, and the third are Thoroughbreds. In each of these breeds, the distinctive way of going has been accomplished through years of breeding and selection—through heredity.

Simple Learning

No horse—whether he be used for saddle, race, or other purposes—reaches a high degree of proficiency without an education. Thus, if the offspring of Man o' War and six of the fastest mares ever to grace the tracks had merely worked on laundry trucks until six years old, then if they were suddenly—without warning or other preparation—placed upon a racetrack, the immediate results would have been disappointing. Their natural aptitude and conformation in breeding would not have been enough. Schooling and training would still have been necessary in order to bring out their inherent abilities.

In general, the behavior of animals depends upon the particular reaction patterns with which they were born. These are called *instincts* and *reflexes*. They are unlearned forms of behavior. Thus, all horses instinctively like to run. But how well and how fast they run depends upon the training to which they are subjected. They learn by experience. However, the training is only as effective as the inherited neural pathways will permit. Several types of simple learning processes are known; among them, habituation, conditioning (operant conditioning), and imprinting.

HABITUATION

Habituation is getting used to, or ignoring, certain stimuli. Bunk breaking calves is an example. If calves are weaned without prior bunk breaking, then suddenly transferred to a cattle feedlot where there is no milk or grass, and where their feed must be obtained from a bunk, it is a traumatic experience for them. This is so because (1) there is a mother-young separation reaction, (2) they get homesick (and animals do get homesick), and (3) there is a change in feed and water. On the other hand, if they have been preconditioned, as well as bunk broke, prior to weaning, they take to the new feed bunk in the feedlot because they are used to it.

CONDITIONING (Operant Conditioning)

Conditioning learning involves the establishment of a new stimulus/response association. For example, upon hearing the rolling of a barn door a cow may lick her tongue and moo, even though she can see no feed; and upon hearing the rattle of a milk pail, she may "let down" her milk.

Artificial insemination techniques have been developed around the understanding of normal reproductive behaviors and the modification of these behaviors. Semen collection routines are based on behavioral responses that can range from impotence to optimum performance and high-quality semen. Proper stimulation of some bulls, for example, can increase sperm cell output by nearly 40 percent, compared to ejaculates after minimum stimulation.

Another example of conditioning is the use of an electric fence. When an electric fence is installed, the immediate response of animals is to investigate—to touch it with their noses. Upon receiving a shock, they back off and let it alone. Thereafter, the electricity can be shut off for a considerable period of time before some animal again tests it.

Operant learning involves behavior or responses that operate on the environment to produce rewarding and reinforcing effects, such as a food reward immediately following a desired response. It's the learning of an act that has some consequence; i.e., one that operates the environment—like pressing a bar that supplies some feed or turns off a light. An example of natural operant learning is the lifting of the cover of a self-feeder by pigs.

Broadly speaking, training horses is operant conditioning—it's an attempt to modify an animal's behavior. There are two types of training: (1) reinforced training, usually with positive rewards, and (2) forced training in which the animal is compelled to do certain things. Thus, the training of horses is best accomplished through the judicious employment of rewards and punishments. This doesn't mean that the animal is rewarded each time he obeys, or that he is punished when he refuses to do something. But horses are big and strong; hence, it's best that they want to do something, rather than have to be forced. Also, too frequent or improper use of such artificial aids as whips, spurs, reins, and bits makes them less effective; worse yet, it will likely make for a mean horse. However, horses appreciate a pat on the shoulder or a word of praise. Even better results may be obtained by working on an equine's greediness—his fondness for such things as carrots, or a sugar cube. Also, treats may be used effectively as rewards to teach some specific thing such as posing, or to cure a vice like moving while the rider is mounting; but this should not be overdone.

IMPRINTING (Socialization)

Imprinting is a form of early social learning which has been observed in some species. Thus, if a man is present during the critical period of socialization, a puppy may form a lasting association with him. Dog kennels use this principle to produce the most desirable pets, work dogs, or guide dogs for the blind.

Apparently, inheritance controls the time and the length of the critical period when an individual can be imprinted, the type of object to which it can be imprinted, the tendency to respond to the first object to which it is exposed, and the permanence of the attachment to the object following imprinting. For example, goslings can usually be imprinted only within the first 36 hours following hatching.

Although human-socialized animals make adorable pets, they are often a nuisance on the farm. People who have hand reared an orphan lamb, foal, or duck have found that such animals seek human companionship and never fit in well with their own kind. Animals that are socialized to people can also be dangerous, especially if they are large and attempt to respond in the same manner as they do to other members of their own species. For example, upon reaching maturity a deer that was bottle fed as a fawn may react toward humans in the rutting season as if they were other deer, with the result that a stag may attack a human and inflict serious injury.

Complex Learning

Complex learning is the capacity to acquire and apply knowledge–the ability to learn from experience and to solve problems. It is the ability to solve complex problems by something more than simple trial-and-error, habit, or stimulus-response modifications. In man, we recognize this capacity as the ability to develop concepts, to behave according to general principles, and to put together elements from past experience into a new organization.

Animals learn to do some things, whereas they inherit the ability to do others. The latter is often called instinct. Thus, ducks do not have to learn to swim—instinctively, they take to water.

Generally speaking, behavioral scientists are agreed that each species has its own special abilities and capacities, and that it should only be tested on these. For example, the dog, pig, and rat, are more adept at solving a maze test (a pathway complicated by at least one blind alley, used in learning experiments and intelligence tests) than the horse, in order to get food. Hence, solving a maze in order to find food favors the scavengers (and the dog, the pig, and the rat are all scavengers)—they have connived for their food since the beginning of time. However, the horse, whose natural feed was the grass that lay around him, never had to develop this kind of intelligence. He was a plains-living animal, highly specialized for speed as a means of escape from his enemies and with almost no powers of manipulation. Thus, a horse should be good at any problem that can be solved by running, including racing, polo, pole bending, calf roping, etc. Indeed, had equines not been smart and adapted to their particular environment, they would never have made it through 58 million years.

Thus, each species is uniquely adapted to only one ecological niche. Moreover, a niche is filled by the particular species that can solve food finding therein; and that is best adapted under the conditions that prevail. It follows that intelligence comparisons between species are not meaningful, and that it is absurd to say that one species is smarter than another.

Of course, man's intelligence is generally recognized. In fact, were it not for his superior mental faculties, along with his limited muscular force, he might find himself under the saddle or between the shafts, instead of the horse.

INSIGHT LEARNING (Reasoning)

Insight learning is the sudden adaptive reorganization of experience or sudden production of a new adaptive response not arrived at by overt trial-and-error behavior. It replaces trial-and-error. Of course, it is difficult to be certain in such cases that the animal did not have a similar type of problem before. Even so, the immediate application of past experience to a new situation is a noteworthy capacity.

This type of learning is most prevalent in the higher animals. Some examples of insight learning are (1) a detour, or barrier, in which the animal must go away from the feed in order to reach it; (2) a chimpanzee obtaining a banana that is out of reach by stacking boxes beneath it and climbing up; and (3) a chimpanzee fitting two sticks together to pull in a piece of food that is out of reach of either stick alone.

The most important single factor to remember in training animals is that none of them (dogs included) can reason things out. An animal's mind functions by intuition, not logic. Moreover, it has no conscious sense of right and wrong. Thus, it is one of the trainer's tasks to teach an animal the difference between right and wrong—between good and bad. Although the animal cannot utilize pure reason, it can remember; and it has the ability to use the memory of one situation as it applies to another.

HOW ANIMALS BEHAVE—behavioral systems

Man has always had to know something about the behavior of the animals around him. It required knowledge of basic behavior patterns to capture, confine, and herd animals, as did breeding, feeding, watering, and sheltering them. Without this understanding, domestication would have failed and animals would not have survived.

The Chinese have given so many reports on the unusual behavior of animals prior to earthquakes that it is difficult to dismiss them. An illustrated pamphlet, published in 1973 by the Earthquake Office in Tientsin, advises that an earthquake may be imminent when animals exhibit the following erratic behavior:

When cattle, sheep or horses refuse to get into the corral;

when rats run out from their hiding place;

when chickens fly up to the trees and pigs break out from their pens;

when ducks refuse to go to the water and dogs bark for no obvious reason;

when snakes come out from their winter hibernation;

when pigeons are frightened and will not return to their nests;

when rabbits with their ears standing jump up or crash into things; and

when fish jump out of the water as if frightened.

Animals behave differently, according to species. Also, some behavioral systems or patterns are better developed in certain species than in others. Ingestive and sexual behavior systems have been most extensively studied because of their importance commercially. Nevertheless, most animals exhibit the following eight general functions or behavioral systems, each of which is summarized in Table 1-1.

1. Agonistic behavior (combat)
2. Allelomimetic behavior (gregarious behavior)
3. Care-giving and care-seeking (mother-young) behavior
4. Eliminative behavior
5. Ingestive behavior (eating and drinking)
6. Investigative behavior
7. Sexual behavior
8. Shelter-seeking behavior

Fig. 1-2. Agonistic behavior exhibited by a bull. Pawing the ground and bellowing are generally the first stages of combat in cattle.

Fig. 1-3. Courtship of the bull, showing chin-resting, in which the chin and throat are rested on the cow's rump.

Fig. 1-4. Courtship in sheep, showing ram biting ewe's fleece.

Fig. 1-5. Courtship in swine, showing male nudging flank. (The boar also vocalizes with a courting song.)

Fig. 1-6. Stallions on pasture or range approach a mare with care in order to avoid being kicked or struck.

TABLE 1-1

HOW ANIMALS BEHAVE

Behavior	Cattle (Beef and Dairy)	Sheep	Swine	Horses
AGONISTIC BEHAVIOR (Combat): This type of behavior includes fighting, flight, and other related reactions associated with conflict. Among all species of farm mammals, males are more likely to fight than females. Nevertheless, females may exhibit fighting behavior under certain conditions. Castrated males are usually quite passive, which indicates that hormones (especially testosterone) are involved in this type of behavior. Thus, farmers have for centuries used castration as a means of producing docile males, particularly cattle, swine, and horses. Bulls, rams, boars, and stallions that are run together from a very young age seldom fight. Perhaps they have already settled their social rank. On the other hand, bringing together sexually mature strange males of these species almost always results in a fight. The intensity of fighting depends upon the tenacity of the two combatants. Although fighting rarely results in death, it usually continues until one gives up.	In combat, bulls paw the ground and bellow, followed by putting their heads together and butting. Although young bulls raised together will seldom fight, a group of bulls may single out one individual and ride him to death, unless he is removed from the group. Bringing together sexually mature strange bulls almost always results in a fight. Also, it is noteworthy that breeds of cattle differ in their agonistic behavior. There is the hazard that bulls will be stifled as a result of fighting; hence, conditions that result in combat should be minimized. Under range conditions, it is common for large numbers of bulls to be run together with a herd of cows. Even though many different bulls of different ages are included in the herd, fighting among them seldom occurs. Outside of the breeding season, as in the fall of the year, it is not uncommon to see bulls congregated together on the range, away from the cow herd.	Rams fight by backing off and charging at each other headlong. The fight generally continues until one ram gives up, usually after both combatants have bloody noses.	When strange boars are brought together, some fighting ensues. Sows and barrows will also fight, but they do not exhibit the jaw-clicking and saliva-producing (champing) characteristic of fighting boars. A sow will try to bite, whereas a boar will slash his opponent with his tusks. When strange boars are first penned together, they smell one another and begin to circle as they "size up" each other. They frequently strut shoulder to shoulder with the hair on their crest bristled, ears cocked, and head raised in an alert, threatening position. In a serious encounter, the combatants utter deep-throated barking grunts and champ throughout the fight. As the fighting becomes intense, each boar repeatedly thrusts his head and neck sideways and upward, with his jaws open and his teeth bared. If the boars have tusks, slashes are usually inflicted on the shoulders of each other. Fighting boars await the opportunity to discontinue shoulder contact and to nip at the ears or the neck and front legs. Sometimes, they even charge the side of the opponent with their mouth wide open. Fighting may continue for as long as an hour, or it may end very quickly. In any event, it will continue until the dominant boar is satisfied and the loser retreats, with the winner biting and slashing him as he scampers away. A fight on a summer day may end in the death of one or both combatants due to heat exhaustion.	Bringing together sexually mature, strange stallions for the first time almost always results in a vicious fight. Stallions fight by biting, kicking, and striking. Generally they fight head to head and most of the biting is on the neck, shoulders, and front legs. Although fighting rarely results in death, it usually continues until one gives up—battle scarred by teeth and hoof marks. Fighting among mares is less vicious than between stallions. Body biting and kicking are used as a means of establishing social order. Geldings may fight much like mares. Jacks are unusually vicious fighters. They rely on their teeth, rather than kicking. Sometimes wild jacks killed a rival by cutting his windpipe or jugular vein. Also, it is reported that the dominant jack occasionally castrated the weaker jacks with his teeth. Agonistic behavior is of practical importance when strange horses are first put together. One way or another, a social order must be established. Hence, there is always the potential of injury until rank is settled. Also, agonistic behavior may create a potentially dangerous situation to both horses and riders in group riding. To reduce the hazard of such accidents, all horses should be spaced well apart when standing or moving. Wild bands of horses and bands of domestic horses on the range behave very much alike. The stallions have keen sight, hearing, and smell; and each stallion leader is very good at protecting his harem, which usually includes 10 to 20 mares. When frightened or facing danger, the stallion warns his band with snorting and restless movements and takes his place ready for battle if necessary.

(Continued)

TABLE 1-1 (Continued)

Behavior	Cattle (Beef and Dairy)	Sheep	Swine	Horses
ALLELOMIMETIC BEHAVIOR:				
Allelomimetic behavior is mutual mimicking behavior. Thus, when one member of a group does something, another tends to do the same thing; and because others are doing it, the original individual continues. In the wild state, this trait was advantageous in detecting the enemy, and in providing protection therefrom. In wolves and coyotes, this behavior is important in attacks on prey, since a pack working together is much more likely to be successful than an individual working alone. Under domestication, animals are usually protected from predators. Nevertheless, the allelomimetic behavior still has important consequences.	Cows moving across a pasture toward a milking barn often display allelomimetic behavior. One cow starts toward the barn, and the others follow. Because the rest of the herd is following, the first cow proceeds on. Because of stimulating and competing with each other, there is usually higher per steer feed consumption among a group of steers than by one steer alone. Thus one steer penned alone may eat "X" pounds of feed per day. However, when he is placed with other steers, his intake may be "X + Y" pounds. But, of course, the feed consumption advantage can be nullified when the animals are placed together too closely, with the result that the agonistic behavior comes into play.	Sheep walk, run, graze, and bed down together.	Swine exhibit allelomimetic behavior in their eating habits. Thus, when one pig eats, there is a tendency for the rest to join him. As a result, pigs in a group usually average higher feed consumption than one pig alone.	A timid horse will follow behind a pack, in order not to be left behind.
GREGARIOUS BEHAVIOR:				
Gregarious behavior refers to the flocking or herding instinct of certain species. It is closely related to allelomimetic behavior. If animals imitate each other, they must stay together. If they stay together as a mobile group, they must use allelomimetic behavior to do so. All such behavior arises out of the process of social attachment. Gregarious behavior differs among species.	Cattle tend to roam in groups of various sizes when a large herd is placed on a pasture or range. However, there is usually considerable space between the members of the herd. Moreover, on close observation it is evident that there are several small groups within a herd, each ranging from three to five head.	The gregarious, or flocking, instinct is particularly strong in sheep. Moreover, it is more evident in some breeds than in others. The Merino, and animals carrying Merino breeding, are noted for their flocking instinct. This makes it possible to herd them on the range. It is noteworthy that the gregarious instinct of sheep diminishes to some extent when they are placed within fenced holdings, instead of herded. As a result, those who handle western range bands do not try to switch back and forth from fenced range to herding, for the reason that the band becomes unmanageable from the standpoint of herding once they have been in a fenced holding for an extended period of time. Packers use the gregarious instinct of sheep by having an old goat, appropriately called a "Judas," lead sheep to slaughter. A well-trained Judas will lead group after group of sheep to slaughter all day long.	In the wild state, swine roved through the forest in herds. Usually these wild groups consisted of 5 to 10 sows, under the leadership of a boar. The wild boar, with his large and long head and strong tusks, was a formidable match for most any enemy. Under domestication, swine retain their gregarious nature. However, man has altered it a great deal. Today, hogs are usually confined to a very limited area. Also, under domestication, they have lost most of their ferocity and are usually gentle and easily handled.	In the wild state, horses ran in bands; thus, they were gregarious by nature. These bands seldom consisted of more than 40 animals; and always there was a stallion in each group. Under domestication, horses show definite preferences for their herdmates; they will even avoid certain horses in the herd. In the draft horse era, animals that were worked together usually stayed together when they were turned to pasture.

(Continued)

TABLE 1-1 (Continued)

Behavior	Cattle (Beef and Dairy)	Sheep	Swine	Horses
CARE-GIVING AND CARE-SEEKING (Mother-Young) BEHAVIOR The care-giving behavior is largely confined to females among domestic animals, where it is usually described as "maternal"; the care-seeking behavior is normal for young animals. This behavior begins shortly after birth and extends until the young are weaned. Care-giving and care-seeking vary widely among different species of farm animals.	Nature ordained that cows seek isolation at calving time. So, where possible, they'll hide out. Following birth, the care-giving behavior of the new mother becomes evident almost immediately. She gets up and begins to dry her newborn calf by licking it. Simultaneously, some cows "talk" to their newborn. They may become quite concerned and nervous as their "baby" first attempts to stand, takes a few footsteps—and falters. Aided by its mother's licking, and encouraged by her "talking," eventually the calf makes it to its unsteady feet and commences to search for a teat. A newborn calf cannot see too well, but it can smell, touch, and taste. It associates everything that is good and that cares for it with its mother. This is the beginning of herd instinct. If on pasture, the new mother usually hides her calf. During the first day or two, the calf sleeps a great deal, while the mother grazes nearby. But a mother takes great pains not to disclose the hiding place of her calf. At intervals, she returns to feed it. If it is necessary for her to leave her calf in order to get water or supplemental feed, she does not tarry much along the way. Frequently, where there are a number of newborn calves, the cows "baby-sit" for each other. Part of the cows will leave for feed or water, but one or two will remain behind and guard all the calves. Then, when the first cows to leave have returned, the "baby-sitters" will take their turn and depart. In this manner, there are older cows with the calves at all times. When a calf in hiding is approached by a human, it will usually lie as close to the ground as possible, without any movement except for its eyes. If picked up, and if scared, it may bawl (cry) for its mother. If the mother hears the call, she will come running—often ready to fight. Frequently, other cows in the vicinity, especially if they have calves of their own, may join in the response. If the disturbed calf runs away, it will return to the area after the danger has passed. By the time the calf is two days old, the mother wanders more extensively, with the calf at her side. Soon, they rejoin the herd.	After parturition, the ewe licks the newborn lamb, removing moisture and placental membranes. The lamb soon staggers to its feet and makes awkward efforts to find a teat to nurse. Quite often a very weak lamb will have to be held to the teat. Normally, lambs suckle in a standing or kneeling position. While suckling, they wiggle their tails from side to side. The mother-young bond in sheep is very strong; the ewe becomes attached to her offspring, and the lamb develops an attachment to its mother. Although ewes are normally timid and easily frightened, they will defend their young even if the attacker is formidable. It is noteworthy, too, that sheep will accept and suckle orphan goats (kids), and vice versa (interspecies rearing).	The sow is very protective of her pigs, especially if they squeal. She goes toward the intruder with mouth open and emits a series of sharp, barking grunts in rapid succession. She continues to mother her pigs until they are weaned, but after 2 to 3 days' separation she loses interest in them. If pigs are left with the sow for 3 or 4 months, she will usually wean them herself. Sows will readily accept pigs from another litter, provided the transfer is made the first day or two following farrowing. Exchanging of pigs among sows, in order to even out the size litters, is a common practice in herds where many sows are farrowing about the same time. Some nervous sows eat their pigs during or immediately after farrowing. If this trait is observed, all pigs, both live and dead, along with the placental membranes should be removed as soon as possible, before the sow has an opportunity to eat them. Once the sow has acquired a taste for flesh, she may develop a permanent pig-eating habit. Usually, such nervous sows calm down following farrowing, after which their pigs may be returned to them and they will express normal protective behavior.	Mares show much the same maternal behavior toward their young as is exhibited by females of other species of farm animals. Thus, a mare calls her foal with a neigh or whinny and exhibits nervousness and distress when her young is disturbed. When mares are separated from their foals, such as sometimes happens when they are worked or taken away for rebreeding, there is usually a noisy exchange of whinnying between mother and foal when they are put back together again and the foal is allowed to nurse. It is noteworthy that a mare will devote as much attention and affection to a mule colt—a hybrid (ass X horse)—as she will to a horse foal.

Recognition between mother and calf is by smell (olfactory), sight (visual), and sound (auditory). Cows usually sniff their calves after being away for a time; and the calf recognizes its mother's call. The attachment of the mother to her calf is very strong. However, the calf accepts separation with less stress.

If a calf is stillborn, or dies soon after birth, some cows will leave the place where the fetus lies, never to return. Others may return to their dead calves at frequent intervals over a period of several days, smelling it and mooing gently.

Beef calves are normally weaned at about 7 months of age. The bond between cows and calves is very considerable, with the result that the separation is a traumatic experience. Thus, both mothers and calves bawl, often in unison, for 2 to 3 days. In all cases, however, the weaning separation should be complete and final, preferably with no opportunity for the calf to see or hear its dam again. In no case should the cows and calves be turned together once the separation has been made, for it will only prolong the weaning process, and it may cause digestive disorders in the calf.

Dairy calves are normally removed from their mothers when they are from 1 to 4 days of age, with the result that the tie between the mothers and offspring is soon severed.

(Continued)

TABLE 1-1 (Continued)

Behavior	Cattle (Beef and Dairy)	Sheep	Swine	Horses
ELIMINATIVE BEHAVIOR: In recent years, elimination has become a most important phenomenon, and pollution has become a dirty word. Nevertheless, nature ordained that if animals eat, they must eliminate. A full understanding of the eliminative behavior will make for improved animal building design and give a big assist in handling manure. Right off, it should be recognized that the eliminative behavior in farm animals tends to follow the general pattern of their wild ancestors; but it can be influenced by the method of management.	Cattle deposit their feces in a random fashion. Although cows can defecate while walking, with the result that their feces are scattered, generally they manage to deposit their "chips" in neat piles. Most cows hump up to urinate, whereas bulls are inclined to stand squarely on all "fours."	The eliminative behavior in sheep is very similar to that of cattle. However, ewes usually assume a squat position when they urinate.	If given an opportunity, pigs are of very clean habits. They like to keep their bedding area clean and dry. Hence, they usually deposit their feces in a corner of the pen, away from the sleeping quarters. Modern methods of raising pigs in restricted quarters, which are often overcrowded, has disturbed their natural eliminative patterns.	Horses tend to deposit their feces in certain locations, such as along well-traveled paths, like those leading to waterholes. Hence, if given the opportunity, they often return to these locations.
INGESTIVE BEHAVIOR (Eating and Drinking): This type of behavior includes eating and drinking; hence, it is characteristic of animals of all species and all ages. It is very important because animals cannot live without feed and water. *Rumination is the act of chewing the cud, characteristic of herbivorous animals with split hoofs–cattle, sheep, and goats.* It involves regurgitation of ingesta from the reticulo-rumen, swallowing of regurgitated liquids, remastication of the solids accompanied by reinsalivation, and reswallowing of the bolus. The first ingestive behavior trait, common to all young mammals, is suckling. Each species has its own particular method of ingesting food.	Cattle wrap their tongues around grass, then jerk their heads forward so that the vegetation is cut by the lower teeth. Rumination occupies about 8 hours of the cow's time each day. (In addition, the harvesting or grazing time may take another 8 hours. This means that a cow may work a 16-hour day.)	Sheep graze very much like cattle, but their cleft upper lip allows them to graze vegetation closer to the ground. Goats can graze like cattle and sheep, but they are very fond of browse—the young shoots of shrubs and trees.	Swine possess teeth in the upper and lower jaws; hence, they bite off grass or take a mouthful of grain, then chew and swallow it. Pigs have a single stomach, whereas ruminants have a four-compartment stomach. By nature, pigs love to root. If given the opportunity, they will stick their noses into the ground and lift forward and upward, moving earth out of the way and exposing earthworms, grubs, and roots.	The horse is somewhat between ruminants and monogastrics, primarily due to the large "blind gut," which is the seat of considerable bacterial action.

(Continued)

TABLE 1-1 (Continued)

Behavior	Cattle (Beef and Dairy)	Sheep	Swine	Horses
INVESTIGATIVE BEHAVIOR: All animals are curious and have a tendency to explore their environment. Investigation takes place through seeing, hearing, smelling, tasting, and touching. Whenever an animal is introduced into a new area, its first reaction is to explore it. Experienced stockmen recognize that it is important to allow animals time for investigation before attempting to work them, either when they are placed in new quarters or when new animals are introduced into the herd.	If they are not afraid, cattle investigate a strange object at close range. They proceed toward it with their ears pointed forward and their eyes focused directly upon it. As they approach the object, they sniff and their nostrils quiver. When they reach the object, sniffing is replaced by licking; and if the object is small and pliable, they may chew it or even swallow it. Cattle exhibit investigative behavior when placed in a new pasture or in a new barn. As a result, if there is an open gate in a pasture or a hole in the fence, they usually find it, then proceed to explore the new area. Calves are generally more curious than older cattle. Perhaps this is due to the fact that older animals have seen more objects, with the result that fewer things are new or strange to them.	Sheep investigate strange objects and quarters much like cattle. They also approach objects in the same heads-up, ears forward, and eyes-fixed manner. However, sheep are much more timid than cattle, with the result that they usually turn and run if the object moves or if something frightens them.	Pigs are curious. When a strange person approaches a herd of hogs, an alarm, or "woof," is sounded and the animals scatter—scampering as fast as they can for a short distance. In the meantime, if the intruder remains stationary, either standing or sitting, the pigs invariably return to investigate by smelling, rooting, and nibbling. Of course, when pigs are placed in confinement, they have little area to investigate.	Foals are more curious than older horses. Young equines spend much of their time looking at and sniffing objects in their pastures or stalls. As the foal grows older, it may exhibit fear of certain objects. At this stage, it may even move away from its caretaker. When this happens, the horseman should never run after the foal. Rather, stand still; very soon, the foal's curiosity will get the best of it, and it will return. A mare frequently becomes very nervous as she watches her offspring investigate, fearful that it may get hurt in the process.
SEXUAL BEHAVIOR: Reproduction is the first and most important requisite of livestock breeding. Without young being born and born alive, the other economic traits are of academic interest only. Thus, it is important that all those who breed animals should have a working knowledge of sexual behavior. Sexual behavior involves courtship and mating. It is largely controlled by hormones, although males that are castrated after reaching sexual maturity (which, among farm animals, are known as stags) usually retain considerable sex drive and exhibit sexual behavior. This suggests that psychological, or learned, as well as hormonal factors may be involved in sexual behavior. Each animal species has a special pattern of sexual behavior. As a result, interspecies matings do not often occur. There Cont. next page	Experienced cattlemen can usually detect in-heat cows through one or more of the following characteristic symptoms: (1) restlessness; (2) mounting other cows, and standing to be mounted by another cow (standing heat appears to be the best single indicator of the proper time to breed); (3) a noticeable swelling of the labia of the vulva; (4) an inflamed appearance about the lips of the vulva; (5) frequent urination; (6) switching and raising the tail; and (7) a mucous discharge. A day or two following estrus, a bloody discharge is sometimes seen. Dry cows and heifers usually show a noticeable swelling or enlargement of the udder during estrus, whereas in lactating cows a rather sharp decrease in milk production is often encountered. When kept alone, some cows become restless, walk the fence, and Cont. next page	Unlike other farm animals, the ewe shows few visible external indications of heat. The acceptance of the ram (or of a teaser with an apron) is the best method of detection. Ovulation seems to occur late in the heat period usually from about 24 to 30 hours after the onset of estrus. In sheep, the display of sexual behavior of the male is more elaborate than that of the female. Typically, the ram responds to the urination of a ewe in estrus by sniffing the urine, then extending his head with lips upcurled. He sticks his tongue in and out of his mouth as he follows the ewe, noses her external genitalia, and rubs along her side biting her wool. A characteristic part of the sexual display, or teasing, by the ram is the raising and lowering of one front leg in a stiff-legged striking motion.	The external signs of heat in the sow are restless activity; swelling or enlargement of, and discharge from, the vulva (although these signs are not always present); mounting of other sows; frequent urination; and occasional loud grunting. The boar often nudges the sow or gilt around the head or in the flanks with his head and nose and emits a courting song. He will then attempt to mount her.	The signs of estrus in the mare are (1) the relaxation of the external genitalia; (2) frequent urination in small quantities; (3) the teasing of other mares; (4) the apparent desire for company; (5) a slight mucous discharge from the vulva; (6) allowing the stallion to smell and bite her; (7) spreading the hind legs; and (8) lifting the tail sideways. But many mares are shy breeders. Thus, when there is any question about a mare's being in season, she should be tried with the stallion. When possible, it is usually good business regularly to present mares to the teaser every day or every other day as the breeding season approaches. A systematic plan of this sort will save much time and trouble. The courtship of the stallion is characterized by neighing and smelling the external genitalia of Cont. next page

TABLE 1-1 (Continued)

Behavior	Cattle (Beef and Dairy)	Sheep	Swine	Horses
are two notable exceptions, however: The best-known cross between animal species is the mule, a hybrid, which is a cross between the horse and the ass. Also, when sheep and goats are confined, they readily mate with each other, although such matings are never fertile. Males in most species of farm animals detect females in heat by sight or smell. Also, it is noteworthy that courtship is more intense on pasture or range than under confinement, and that captivity has the effect of producing many distortions of sexual behavior compared to wild animals. Perhaps this explains the high percentage foal crop of wild bands of mares, where conception and foaling rates of 90 percent or better were commonplace, in comparison with the average 50 percent foaling rate under domestication. Today, man is attempting to control the sex life of animals, by bringing about ovulation at the time of choice of the owner, rather than of the female.	bawl when they are in heat. Some may even jump the fence, or go through it, as they attempt to find a bull. A bull can often detect a cow that's coming in heat 24 to 48 hours before she will mate, at which time he will remain in her company. Courtship of the bull consists of following the in-heat cow; licking and smelling the external genitalia, with the head extended horizontally and the lip upcurled; and chin-resting, with the chin and throat resting on the cow's rump.			the mare, followed by extended head and upcurled upper lip; and pinching the mare in the croup area with his teeth.
SHELTER-SEEKING BEHAVIOR: All species of animals seek shelter—protection from the sun, wind, rain and snow, insects, and predators.	Cattle are not as sensitive to extremes in temperature—heat and cold, as are swine. Nevertheless, they do seek shelter under natural conditions—this may consist of hills, valleys, timber, and other natural windbreaks; or they may even group closely together. Cattle seem to be able to sense the coming of a storm, at which time they may race about and "act up." During a severe rain or snow storm, they turn their rear ends to the storm and tend to drift away from the direction of the wind. By contrast, bison (buffalo) face a storm head on. During the hot summer months, cattle seek either shade or a waterhole during the heat of the day. Then, they graze in the cool of the evening or early morning. There are well known breed differences in tolerance to heat. Brahman cattle can withstand more heat than the European breeds, whereas the heat tolerance of the Santa Gertrudis is intermediate.	Sheep seek shelter by moving into barns or under trees, by huddling together to keep off flies, by crowding together in extremely cold weather, and by pawing the ground and lying down. Like cattle, during a severe storm they turn their rear ends towards the wind. When there is no shelter, there is danger of sheep massing together and smothering during a very severe storm.	Hogs are very sensitive to extremes of heat and cold; hence, shelter seeking is a very important trait with them. It is particularly important that swine be provided with shade during hot weather so that they may avoid the direct rays of the sun, because they do not possess an adequate cooling mechanism. In hot weather, hogs will wallow in water if given the opportunity. When they are hot, hogs pant rapidly and sleep stretched out full length—so as to expose the maximum body surface to the air. During cold weather, swine sleep curled up and huddled together, thereby exposing minimal body surface to the air.	Horses are not very sensitive to either heat or cold. When not confined in cold areas, they develop a shaggy coat of hair in the wintertime and seek shelter from storms under trees and in the valleys. They paw to get their feed supply when the ground is covered with snow. Like cattle, horses face away from the direction of a severe storm.

SOCIAL RELATIONSHIPS

Social behavior may be defined as any behavior caused by or affecting another animal, usually of the same species, but also, in some cases, of another species.

Social organization may be defined as an aggregation of individuals into a fairly well integrated and self-consistent group in which the unity is based upon the interdependence of the separate organisms and upon their responses to one another.

The social structure and infrastructure in the herd are of great practical importance. Some of the ideas on peck order ("bunt," or "hooking" order) have had to undergo changes as a result of increased understanding of the social organization within the herd.

It is now obvious that there is no simple hierarchial organization; i.e., stepladder or peck order in numerical progression. It is much more complex. For example, in a cow herd the old theory was that A was dominant over B, B over C, C over D, and so on. However, in studying a herd on the range (one not in restricted quarters, or confinement) with ample feed and water, as well as protection, it becomes obvious that the ladder-type hierarchy exists only within small groups of cattle. For example, there might be a group of 5 cows in which there is a ladder-type hierarchy—A, B, C, D, and E. But a group will then be found to be dominant over another little group of perhaps 2 or 3 cows, within which there is again a ladder-type hierarchy. This grouping becomes very important when we see exactly what happens within a herd under normal conditions. Geographically, the herd spreads out with a conglomeration of neutral (neither dominant nor subservient) cows in the middle of the group, with both the dominant cows and the subservient cows spread around the periphery; usually, the dominant cows and the subservient cows are on opposite sides of the circle or herd. Careful study reveals that there are certain lines of dominance throughout the entire herd. For example, one line of dominance might advance from the left-hand side of the herd and continue right down through the center of the herd, group by group, with each group varying in size, from small groups at the top of the dominancy scale to larger groups within the major mass of the herd. As this line of dominance exists at the other, or subservient, end of the scale, the groups get smaller and smaller and have a greater distance between them. It is important to understand that there might be another line extending through the herd in a similar way, but starting on the right-hand side of the herd. In a very big herd, there might be up to 5 or 6 of these group lines throughout a herd. As the dominant groups are approached by subordinate (or more subordinate) groups, the individual distance between them is very carefully established. First, it is established between individuals within the group; next, the intergroup distance is clearly defined. The individual distance is that distance to which a dominant will allow a subservient to approach her. This distance is remarkably constant. So, by studying a herd carefully, one can work out the average distance. This becomes of great practical importance as the area is restricted and the herd is confined. Under extensive conditions, the intercow distance is of the order of 12 to 14 feet for European cattle (*Bos taurus*), and slightly less than this for Brahman cattle (*Bos indicus*). This does not mean that there is a distance of 12 to 14 feet between every cow; rather, it means that the average throughout the herd will be 12 to 14 feet. There will be dominants that will not allow subservients closer to them than 30 to 40 feet, and in extreme cases, not nearer than 60 feet. Also, there are dominants that will run right through the herd to attack a particular subservient.

When we restrict or confine animals and force them into spaces that bring them within the individual distance that has been established, we immediately create stress throughout the herd. Thereupon, the dominants have to pay more attention to maintaining their dominance and to protecting their own field or territory. They will have to be more aggressive in their reactions. The subservients become far more nervous, and their nervousness spreads throughout the herd.

Social Order (Dominance)

Fig. 1-7. Dominance. This shows a dominant cow attacking the neck of a subordinate. The latter submits and avoids a fight.

Within most groups of farm animals of the same species, there is a well-organized social rank. Animals observe this order in their relationships just as carefully as protocol demands that it be observed at a "State Dinner." In chickens, in which it was first observed, the social rank order is called the "peck order."

Thus, in most species of farm animals, the alpha animal in the herd or flock will be dominant over all other individuals and the omega animal will be subordinate to all. In between, some animals will be subordinate in some relationships and dominant in others. Moreover, once these relationships are established, they seldom change. The social order is usually important only in females, because mature male animals are seldom run together in groups. It is unimportant in domestic ewes.

Once the social rank order is established, it results in a peaceful coexistence of the herd or flock. Thereafter, when the dominant one merely threatens, the subordinate animal submits and avoids conflict. Of course, there are some pairs that fight every time they chance to meet. Also, if strange animals are introduced into such a group, social disorganization results in the outbreak of new fighting, as a new social rank order is established.

Among wild animals, social rank order is nature's way of giving mating priority to the top ranking males. Hence, they leave behind more of their progeny than do the less dominant males. Also, dominance establishes priority in feeding.

Social rank among farm animals is of little consequence as long as they are on pasture or range, and if there is plenty of feed and water. But it becomes of very great importance when animals are placed in confinement. When cows are moved into winter quarters, social dominance decrees that replacement heifers be sorted out and fed separately, that young bulls be cared for in separate quarters, and that old cows with poor mouths be fed separately; otherwise, these animals will not get enough feed.

Of course, social rank becomes of importance when a group of animals is fed in confinement; and it becomes doubly important if limited feeding is practiced. Under such circumstances, the dominant individuals crowd the subordinate ones away from the feed bunk, with the result that they may go hungry. This happens both in feedlot cattle and in breeding cattle being wintered.

Several factors influence social rank; among them, (1) age—both young animals and those that are senile rank toward the bottom; (2) early experience—once a subordinate in a particular herd, usually always a subordinate; (3) weight and size; and (4) aggressiveness or timidity. Also, it is noteworthy that social rank is influenced by hormones; for example, a capon (castrated male chicken) automatically goes to the bottom of the totem pole, whereas the injection of roosters and hens with the male sex hormone, testosterone, increases their social rank.

In feedlots and other confinement operations, social facilitation is of great practical importance. Dominants should be sorted out, and, if possible, grouped together. Of course, they will fight it out until a new social order is established. In the meantime, both feed efficiency and gains will suffer. But, as a result of removing the dominants, the feed intake of the rest of the animals will be improved, followed by greater feed efficiency and profit. Among the more settled animals, social facilitation will become more evident. After the dominants have been removed, the rest of the animals will settle down into a new hierarchy, but within the limits of their dominance. Their interaction or social facilitation will be far more likely to have a calming effect on this group, to both the economic and practical advantages of the operator.

Dominance and subordination are not inherited as such, for these relations are developed by experience. Rather, the capacity to fight (agonistic behavior) is inherited, and, in turn, this determines dominance and subordination. Hence, when combat has been bred into the herd, such herds never have the same settled appearance and docility that is desired of high-production and intensive animals.

Leader-Follower

Leader-follower relationships are important in cattle, sheep, hogs, and horses. In each case, the

Fig. 1-8. Horses on the move—Indian file (not abreast), with the lead mare in front and the stallion bringing up the rear.

young follow their mothers; hence, they continue to follow their elders. Leader-follower relationships are particulary strong in sheep, where lambs follow their mothers from birth.

Right off, it is important to distinguish leader-follower relationships from dominance; in the latter, the herd is driven, rather than led. After the dominants have been removed from the herd, the leader-follower phenomenon usually becomes more evident. It is well known that the dominant animal is not necessarily the leader; in fact, it is very rarely the leader. It pays too much attention to other matters of dominance in its relationship within the herd, with the result that it does not develop the qualities of leadership.

In a naturally formed flock of sheep, the oldest ewes lead, followed immediately by their young lambs. Each is follwed less closely by her descendants, with the females followed by their own lambs. Thus, the leader in the flock is usually the oldest ewe with the largest number of descendants. This type of leadership is broken up in flocks where unrelated animals are brought together.

Taking a page from history and people, it is noteworthy that one of the world's greatest leaders, Napoleon, was small in stature. So it is with animals. The leader may be small, but it is always intelligent. In horses, the leader is usually a mare with a well-developed investigatory sense.

Interspecies Relationships

Social relationships are normally formed between members of the same species. However, they can be developed between two different species. In domestication this tendency is important (1) because it permits several species to be kept together in the same pasture or corral, and (2) because of the close relationship between man and animals. Such interspecies relationships can be produced artificially, generally by taking advantage of the maternal instinct of females and using them as foster mothers. It's not unusual, for example, to set duck eggs under a hen. All goes well until the young ducklings take to water for their first swim. Thereupon, the mother hen becomes quite excited, fearful that her babies will drown (little realizing that swimming comes naturally for young ducks).

All sorts of bizarre interspecies relationships have been arranged—including cows raising pigs; bitches (dogs) raising pigs, rabbits, and cats; and cats raising mice.

Man-Animal Relationships

Social relationships can also be transferred to human beings. Thus, an animal caretaker usually forms a care-dependency relationship with the animals under his care. This is particularly true with pets—horses, dogs, and cats. Also, this close relationship was notable in many old-time herdsmen; for example, it was said of Amos Cruckshank, the beloved bachelor herdsman of Aberdeenshire, Scotland, who contributed so richly to the improvement of Shorthorn cattle, that "he was married only to his animals."

Without doubt, one of the best-known stories of a man-animal relationship pertained to groom Will Harbut and the great Thoroughbred, Man o' War (affectionately known as Big Red). When training, Man o' War's morning came early. Will Harbut gave him his first meal at 3:30 a.m.; at 7:30 a.m., he was groomed. Big Red was very fond of his caretaker; he liked to snatch his hat and carry it around as he showed off for visitors. Will Harbut, who had quite a way with words as well as with horses, never tired of telling the thousands of visitors who came to see Man o' War that, "He was the mostest horse that ever was."

HOW ANIMALS COMMUNICATE

Communication between animals involves giving off by one individual of some chemical or physical signal, which, on being received by another, influences its behavior. Communication between animals need not be visual or auditory, nor must it be confined to the present.

Without doubt, this trait accounts, in part at least, for the foundation horse stock of the American Indians and the hardy bands of Mustangs—the feral horses of the Great Plains. In some mysterious manner, the abandoned and stray horses of the expeditions of de Soto and Coronado communicated and found each other; otherwise, they would not have reproduced.

Sound

Fig. 1-9. Communication by sound—a bellowing bull.

Sound communication is of special interest because it forms the fundamental basis of human language. The gift of language alone sets man apart from the rest of the animals and gives him enormous advantages in his adaptation to his environment and in his social organization.

Sound is also an important means of communication among animals. They use sounds in many ways; among them, (1) feeding, in sounds of hunger by young, or food finding, and of hunting cries; (2) distress calls, which announce the approach or presence of an enemy, and the all-clear signal following the departure of a predator; (3) sexual behavior, courting songs, and related fighting; (4) mother-young interrelations to establish contact and evoke care behavior; and (5) maintaining the group in its movements and assembly.

Animals have a very acute sense of hearing, perceiving higher and fainter noises than the human ear.

Chemicals

In mammals, females in estrus secrete a substance that attracts males. Hence, males locate females that are in heat by the sense of smell.

On the range, it has been observed that each stallion usually stakes out a territory for himself and his harem of mares with the outside boundary thereof marked by his feces.

Visual Displays

Birds are noted for their sexual behavior in the act of courtship. Visual displays during courtship are less evident among four-footed animals, but they do occur to some extent.

Most animals will strike a hostile stance when they are excited or prior to fighting. Also, boars bristle—the hair on the top of their necks rises up; this serves to make them look larger and more formidable.

NORMAL ANIMAL BEHAVIOR

The producer needs to be familiar with behavioral norms of animals in order to detect and treat abnormal situations—especially illness. Many sicknesses are first suspected because of some change in behavior—loss of appetite (anorexia); listlessness; labored breathing; posture; reluctance or unusual movement; persistent rubbing or licking; and altered social behavior, such as one animal leaving the herd or flock and going off by itself—these are among the useful diagnostic tools.

Also, it is important to know how animals see. Their orbital vision, rather than binocular vision as in man, explains why they will go through a curved chute more easily than a straight one.

Normal behavior in sleep should be recognized, especially since it differs widely between species.

Table 1-2 is a summary of normal animal behavior.

TABLE 1-2
NORMAL ANIMAL BEHAVIOR

Behavior Norms	Cattle (Beef and Dairy)	Sheep	Swine	Horses
HEALTH:				
Some of the signs of good health are: 1. Contentment. 2. Alertness. 3. Eating with relish and cudding by ruminants. 4. Sleek coat and pliable and elastic skin. 5. Bright eyes and pink membranes. 6. Normal feces and urine. 7. Normal temperature, pulse rate, and breathing rate.	*Normal rectal temperature:* Average, 101.5° F Range, 100.4 - 102.8° F *Normal pulse rate:* 60-70/min. *Normal breathing rate:* 10-30/min.	*Normal rectal temperature:* Average, 102.3° F Range, 100.9 - 103.8° F *Normal pulse rate:* 70-80/min. *Normal breathing rate:* 12-20/min.	*Normal rectal temperature:* Average, 102.6° F Range, 102 - 103.6° F *Normal pulse rate:* 60-80/min. *Normal breathing rate:* 8-13/min.	*Normal rectal temperature:* Average, 100.5° F Range, 99 - 100.8° F *Normal pulse rate:* 32-44/min. *Normal breathing rate:* 8-16/min.

(Continued)

TABLE 1-2 (Continued)

Behavior Norms	Cattle (Beef and Dairy)	Sheep	Swine	Horses
SIGHT: The eyes of most animals are on the side of the head (the cat is an exception). This gives them an orbital, or panoramic, view—to the front, to the side, and to the back—virtually at the same time. Also, this is a rounded, or globular, type of vision. This leads to a different interpretation than that of the binocular type of vision of man.	The wide-set eyes of cattle enable them to have a large panoramic field of vision, even to the extent of seeing everything around them, with slight head movements. Only what is immediately behind their hindquarters is outside their field of view.	Members of the flock maintain contact with each other largely through vision. As a flock grazes, each individual throws up its head at intervals, presumably to respond to the position of other members.	Swine, with their large snouts, have more efficient scent direction than sight.	In its natural habitat, the adult horse keeps a sharp lookout for its enemies, even while grazing. It is rare to see all members of a herd lying down together; one horse is almost always on the lookout. Horses have monocular vision; that is, each eye is independent of the other and can see different pictures. This gives them a panoramic view—to the sides, the front, and the back—virtually at the same time. When a horse wants to see an object very clearly, it will face the object and use both eyes in a binocular manner. By contrast, humans have binocular vision and see the same picture with both eyes. The lens of the horse's eyes is nonelastic; but the retina is arranged on a slope, the bottom part being nearer the lens than the top part. Thus, in order to focus on objects at different distances, the horse has to raise or lower his head so that the image is brought on to that part of the retina at the correct distance to achieve a sharp image. Because of its monocular vision, it is difficult for a horse to judge distance accurately. In its evolution, it was more important that the horse see a wide area around him as he watched for predators than to judge distance. With domesticated horses, however, being able to judge distances is very important in certain types of performance. Thus, a rope horse must accurately judge the distance between himself and the animal he is following; a barrel racing horse must accurately judge the distance to the barrel as he prepares for the turn; and a jumping horse must accurately determine distance to the jump in order to select the take-off point, and he must determine the height and spread of the jump. Of course, top performing horses used for these purposes possess the ability to learn to judge distances; and they receive expert training. Also, it is noteworthy that the horse has good vision in darkness. It's not as good as a cat's night vision, but it is considerably better than that of man. Thus, a horse may be ridden at night with reasonable safety, particularly if he is familiar with the area.
SLEEP: Normal behavior in sleep should be recognized, especially since it differs widely between species.	Cattle typically lie on their stomach or tilt to one side, with the fore limbs folded under the body; one hind limb extends forward, while the other protrudes toward the outside. Although cattle rest in this manner, they do not sleep in the sense that the term usually connotes. While lying down, they do shut their eyes for short periods of time. Beef bulls sometimes assume a sitting position. Calves commonly spend up to one-half hour at a time with their heads turned back in the flank position.	The normal sleeping posture of sheep is on the stomach but tilted to the side with one front leg folded under the body and the other extended forward. Usually the head is turned to one side and the eyes are closed. Although sheep are usually inactive about half of the day, as with cattle there is considerable debate as to whether they actually sleep. Certainly, sheep do not enter the state of deep sleep that exists in horses, dogs, and cats.	The resting position of swine varies according to temperature—in the summer, they sleep stretched full length; in cold weather, they sleep curled up. In any event, pigs sleep soundly—they even snore.	The horse rests and sleeps standing up. This is made possible by a system of ligaments, which do not get tired like muscles, and which take the weight off muscles during rest. Sometimes horses lie down in the sun, apparently to expose the body to warmth. In contrast to cattle and sheep which sleep very little, the horse may sleep soundly for as much as 7 hours out of each 24 hours, mostly during the warmest part of the day. But not all of the 7 hours of sleep are taken at one time; rather, it is short and irregular, depending on the degree of hunger and the climatic conditions.

ABNORMAL ANIMAL BEHAVIOR

Abnormal behaviors of domestic animals are not fully understood. As with human behavior disorders, more work is needed. However, we have learned from studies of captured wild animals that when the amount and quality, including variability, of the surroundings of an animal are reduced, there is increased probability that abnormal behaviors will develop. Also, it is recognized that confinement of animals makes for lack of space; this often leads to unfavorable changes in habitat and social interactions for which the species have become adapted and best suited over thousands of years of evolution. Among the abnormal behaviors that frequently develop with domestic animals in confinement are those summarized in Table 1-3.

Fig. 1-10. A cribber (wind sucker, or stump sucker) in action. This is the vice of biting or setting the teeth against some object, such as a post or manger, while sucking in air.

TABLE 1-3
ABNORMAL ANIMAL BEHAVIOR

Abnormal Behavior	Cattle (Beef and Dairy)	Sheep	Swine	Horses
Abnormal behavior in animals develops where there is a combination of confinement, excess stimulation and forced production with a lack of opportunity to adapt to the situation. Homosexual behavior is common among all species where adult mammals of one sex are confined together.	There are inherited differences in the temperaments of cattle. Nevertheless, constant stress can change the temperament of an animal, just as it can in people. Thus, when a bull is kept for hand mating in a corral by which the cow herd passes each day, cows in heat, or coming in heat, stimulate his sexual behavior. Since he cannot respond naturally through coitus, he becomes a mean bull. Thus, the "mean bull" complex is an example of abnormal behavior in cattle.	As a consultant for the Atomic Energy Commission, the author had an opportunity to observe sheep that were kept in confinement, generation after generation, for 20 years. These animals developed a "wool-eating" habit. They didn't inflict special harm, for they only took small nibbles of wool from each other. Since these sheep were getting the most complete diet that science knew how to formulate, the only conclusion was that the wool-eating habit came about as a result of the unnatural confinement.	Tail biting accompanies close confinement. It results when pigs are prevented from rooting, nibbling, and chewing—it follows when the pig's normal behavior pattern is disturbed. Docking is the best way in which to stop pigs from tail biting. Swine producers have tried all sorts of things to prevent tail biting. Some have substituted other materials for the pigs to bite, such as rubber tires or chains hung near the pigpen. Others have tried spraying the tails with distasteful chemicals. But none of these methods work very well.	Few animals have undergone such drastic change through evolution as equines. Little Eohippus (the dawn horse of 58 million years ago) was a denizen of the swamp. Later, through evolution, the horse became a creature of the prairie. Even though his natural habitat shifted during this long predomestication period, until man confined him he gleaned the feeds provided by nature. Inevitably, this occupied his time and provided exercise. But domestication and confinement to stalls wrought many changes—changes which spawned abnormal behaviors, including balking, bolting feed, cribbing, halter pulling, kicking, tail rubbing, weaving, wood chewing (pica), backing, rearing, shying, striking with the front feet, a tendency to run away, and objection to harnessing, saddling, and grooming. Many of these vices originate with incompetent handling; nevertheless, they may be difficult to cope with or to correct. This is especially true in older animals.

APPLIED ANIMAL BEHAVIOR

In the beginning of this chapter, it was stated that the application of animal behavior depends upon understanding. The presentation to this point has been for the purpose of understanding. Let us next turn to some practical applications of animal behavior—its application in the barnyard. The following sections pertain thereto.

Breeding for Adaptation

The wide variety of livestock in different parts of the world reflects a continuous process of natural and artificial selection which has resulted in the survival of animals well adapted to climate and other environmental factors. Among the examples are: haired sheep (devoid of wool) in desert areas, fat-tailed sheep in arid zones, *Bos indicus* (Zebu) types of cattle in tropical areas, and *Bos taurus* cattle in temperate zones. Such adaptations relate to survival of the animals, but they do not necessarily entail maximum productivity of food for man. European cattle usually have much higher yields of milk and propensities for rapid growth than have the breeds native to Africa or India. It is understandable, therefore, why there have been many attempts to introduce improved European livestock into countries in which the productivity of native stock is low. But there are many problems in breed replacement, with the result that a large number of experimental introductions of new breeds have not been successful. Tropical Africa provides an example. Because of disease problems, poor resistance to high temperatures, and limited feed supplies, many of the attempts made by former colonial powers to improve the output of native stock by replacing them with the European breeds failed. Breed replacement or a crossbreeding system might seem to be a simple panacea for low productivity. However, unless associated with special provisions for subsequent importation of breeding stock and simultaneous improvement of the nutritional, parasitological, disease, and husbandry environments of the crossbreds, it is not likely to succeed.

Selection should be from among animals kept under an environment similar to that under which it is expected that their offspring shall perform—this requisite applies to animals brought in from another herd, either foundation or replacement animals. For example, animals that are going into a range herd should be selected from among animals handled under range conditions, rather than from among stall-fed animals. This recommendation is based on the results of a long-time experiment conducted by the author and his colleagues at Washington State University.[1]

Animals can be changed through heredity and selection. For example, in Israel, which has one of the highest average milk yields per cow of any nation in the world, the flight distance between cows approaches contiguousness in some herds; this is due to Israel's having selected intensively for docility for 25 years. In other words, the animals are literally touching each other, with no antagonistic or dominant-type response. This allows them to concentrate their animals even more than they had previously, thereby giving them a higher productivity per unit area. The only problem reported by Israel is that estrus, or heat, in animals in close proximity is difficult to detect.

Quick Adaptation—Early Training

We need to breed and select animals that adapt quickly to man-made environment—animals that not only survive, but that thrive, under the conditions that man imposes upon them. Actually, they may be less intelligent than their wild ancestors.

Also, early training and experience are extremely important. In general, young animals learn more quickly and easily than adults; hence, advance preparation for adult life will pay handsome dividends. The optimum time for such training varies according to species. Thus, the ideal time to introduce a puppy to its life work is between 8 and 12 weeks of age, whereas the establishment of feed preference in chicks can be traced back to their first meal. Also, stress can be reduced or avoided entirely if animals proceed through a graduated sequence of events leading to an otherwise noxious experience. Preconditioning of cattle is an application of the latter principle to production practices. If calves are properly preconditioned (started on feed, vaccinated, treated for parasites, etc.) prior to weaning, the stress of subsequent weaning and movement to a feedlot is minimized.

Loading Chute and Corral Design

Fig. 1-11. Cattle entering a curved loading chute.

With knowledge of basic behavior patterns and of social habits within the herd, we have at our disposal the necessary tools for designing facilities and housing which will enable us to make animals do what we want them to do, when we want them to do it—and with a saving in both labor and tempers. As an example, let us consider the matter of putting cattle through a chute on their own accord without interference or any extra driving from the handlers.

At the outset, it is recognized that cattle will follow the leader, and that they will automatically try to escape through a gap, or opening. Hence, they will

[1]*Wash. Agr. Exp. Sta. Bull. 34*, 1961.

follow the leader through a curved chute more readily than through a straight one. As the lead animal approaches the chute and realizes that there is an opening either to the left or right, it goes forward with the idea of going through this gap—and escaping. Thus, if one can get the leader-type cow into the chute first, it's a simple matter to get the rest of the herd to follow through a curved chute. So, the practical application of cattle behavior and social habits to handling facilities calls for a curved chute, with a curved entrance on the outer portion of the corral and the normal straight side on the funnel (inside) portion of the corral.

The corral should always match the work requirements, labor, and herd size. The chute should always be curved; and the corral should be designed so that there is always a gap to the left or right. Then, if the animals have been selected for their lack of dominance, they will automatically enter the funnel portion, thence the chute, thence proceed at their own pace throughout the whole of the facility.

• **The circular corral**—By designing corral facilities in an entirely different fashion from the traditional—by abandoning the straight-sided corral—a facility can be designed that will lend itself to very rapid and adequate handling of stock, with a minimum of interference from the handlers. This has the advantages of (1) cutting down stress, (2) speeding up the work, and (3) distributing the cattle after an examination for a particular series of operations or cutouts. This takes the shape of a circular corral, with the diameter determined by the number of animals to be handled.

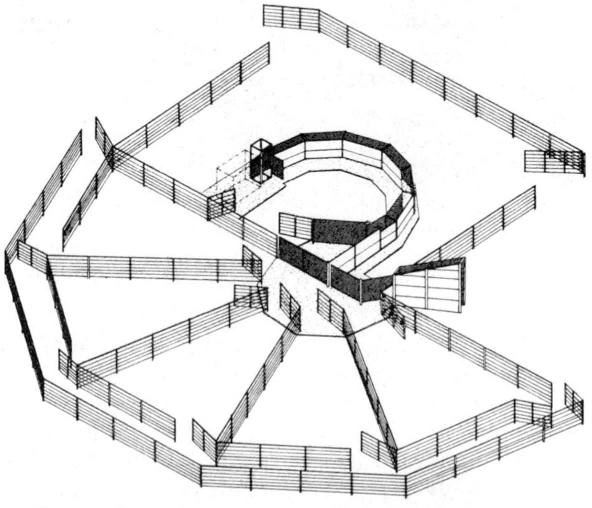

Fig. 1-12. Circular corral. A good corral is more than just a fence. It's a working area. It should have the following facilities: (1) crowding area, (2) working chute, (3) squeeze chute or head gate, and (4) loading chute. As shown above, it can be built initially with only part of the holding pens, then expanded as the herd gets bigger. Note the outside circular passageway and the long, curved working chute which will hold five or six cows at one time. (Adapted by the author from *Plans and Building Information*, Extension Agricultural Engineering, Oklahoma State University, Stillwater, Okla.)

Manure Elimination

Waste is a major concern; it is expensive, time-consuming, and a major pollution problem. But manure handling can be facilitated by an understanding and application of eliminative behavior.

Cattle are indiscriminate eliminators. Even so, this trait can be used effectively. For example, if cattle are fed at the same time each day, feed is released from the rumen into the true stomach regularly; and the moment the latter happens, there is a gastro-colic reflex. When this happens and cattle are put under slight stress, they defecate. Knowing this, cattle can be moved to the defecating area at the right time.

By applying this technique to one of the big dairies in South Africa, it is reported that they were able to get up to 90 percent of the manure dropped on concrete within 1 to 1¼ hours. From there, it was flushed to a sump, thence pumped onto the land. This procedure dramatically reduced the manure handling labor cost by $32 per day on an 800-cow dairy.[2]

Control of Pests

Behavioral studies are involved in pest control. This is so because successful pest control is based on two approaches: (1) finding the weakness of the pest, and (2) developing animals more resistant to, or able to cope with, pests. Pesticide development must consider feeding habits, motility and perception of the organisms, and pollution. Parasite control must often consider the behaviors of the host organisms and of the various life forms of the parasite itself. Often, more knowledge is needed relative to the lives of the afflicted animals. Screwworm control is the classic example of applied behavior in parasite control.

Training Horses

Fig. 1-13. Training a yearling on "how to be a racehorse." Learning to gallop beside a lead pony is one of the first lessons.

---

[2]McFarlane, I. S., "Rationale in Design of Housing and Handling Facilities," *Dairy Science Handbook*, Vol. 9, Agriservices Foundation, Clovis, Calif., 1976.

There are as many successful ways to train horses as there are to train children. The author has observed several top professional trainers. Each used a different technique, yet all ended up with the same result—a champion. Most of them apply the basic principles of behavior given herein.

The good horseman who has followed a program of training and educating the foal from the time it was a few days old has already eliminated the word "breaking." To him, the saddling and/or harnessing of the young horse is merely another step in the training program, which is done with apparent ease and satisfaction.

Each animal species has characteristic ways of performing certain functions, and rarely departs therefrom. The horse is no exception. A good understanding of horse behavior enhances horse training. The faculties of the horse which must be understood and played upon to obtain skillful training and control at all times are: memory, confidence and fear, association of ideas, and willingness.

Companionship

Companionship in animals is of great practical importance. Except for the cat, all domestic animals are highly social and have constant need for companionship.

If not too crowded, placing animals together sometimes accomplishes two things: (1) greater feed consumption, due to the competition between them (social facilitation); and (2) a quieting effect. For example, nervous high-strung boars are frequently provided with a barrow as a companion.

Among horses, the two strongest sources of motivation are (1) the desire to be with other horses (companionship), and (2) the desire to go home. Thus, a pack train will always go better if the leader rides a fast horse. Then the slower ones will try to keep up, rather than be left behind. When away from companions, a horse will try to go home.

The best-known animal companionship pertains to high-strung racehorses and stallions, in which all sorts of companions are used—a goat, a sheep, a chicken, a duck, or a pony. Such companions are commonly referred to as "mascots."

The expression "to get his goat" was born of the common custom of having goats for mascots. Back in the days when skulduggery was as important as form in winning races, the men of one stable sometimes plotted to kidnap the goat mascot of a rival's horse. By "getting the goat" of a favorite, they cleaned up by betting against a horse that was odds-on to win, but too upset to run at his best.

Controlling Animal Behaviors

Many abnormal behaviors can be controlled rather easily, even though the cause is not removed; among them, those which follow.

• **Tail biting**—Tail biting accompanies close confinement. It results when pigs are prevented from rooting, nibbling, and chewing—from disturbing the pig's normal behavior pattern.

Docking is the best way in which to stop pigs from tail biting. Swine producers have tried all sorts of things to keep pigs from tail biting. Some have substituted other materials for the pigs to bite, such as rubber tires or chains hung near the pigpens. Others have tried spraying the tails with distasteful chemicals. But none of these methods works very well.

It is recommended that tail docking be a part of the regular management program, with the tails docked at the same time that the needle teeth are cut, when the pigs are about 3 days old. The side-cutting type pliers will work for both jobs, but tails will bleed less when they're cut with a dull blade. Emasculators and poultry debeakers also work well. To dock the tail, clean it first, then cut it ½ to ¾ inch from its base, lifting it gently so as not to stretch the skin. The skin won't heal over the end bone as rapidly if you pull the tail away from the body. Don't cut the tail shorter than ½ inch because it will make for excess bleeding and slow healing.

• **Bolting feed**—Horses that eat too rapidly are said to be bolting their feed. It can be lessened by spreading the concentrate thinly over the bottom of a large grain box, so that the horse cannot get a large mouthful; or by placing in the grain box a few smooth stones about the size of baseballs, so that the horse has to work to get feed.

• **Eating bedding**—Sometimes gluttonous animals eat their bedding. This is undesirable because (1) most bedding materials are low in nutritional value, and (2) feces-soiled bedding adds to the parasite problem. The problem can be alleviated by muzzling the horse.

• **Wood chewing**—Wood chewing is a common abnormal behavior in horses. In the final analysis, there is only one foolproof way in which to prevent wood chewing; to have no wood on which they can chew—to use metal, or other similar materials, for fences and barns. Of course, this isn't always practical. But wood chewing can be lessened, although it cannot be entirely prevented, by one or more of the following practices:

1. Stepping up the exercise.
2. Feeding three times a day, rather than twice a day, even though the total daily feed allowance remains the same.
3. Spreading out the feed in a larger feed con-

tainer, and/or placing a few large stones about the size of a baseball in the feed container.

4. Providing 2 to 4 pounds of straw or coarse grass hay per animal per day, thereby giving the horse something to nibble on during his spare time.

ANIMAL ENVIRONMENT

Environment may be defined as all the conditions, circumstances, and influences surrounding and affecting the growth, development, and production of animals.

Stockmen were not concerned with the effect of environment on animals so long as they grazed on pastures or ranges. But rising feed and land costs, along with the concentration of animals into smaller spaces, changed all this.

Man achieves environmental control through clothing, vacationing in resort areas, and air-conditioned homes and cars. In animals, environmental control involves space requirements, light, air temperature, relative humidity, air velocity, wet bedding, ammonia buildup, dust, odors, and manure disposal. Control or modification of these factors offers possibilities for improving animal performance. Although there is still much to be learned about environmental control, the gap between awareness and application is becoming smaller.

In the present era, pollution control is the first and most important requisite in locating a new livestock establishment, or in continuing an old one. The location should be such as to avoid (1) the neighbors complaining about odors, insects, and dust; and (2) pollution of surface and underground water. Without knowledge of animal behavior, or without pollution control, no amount of capital, native intelligence, and sweat will make for a successful livestock enterprise.

Presently, the global use of fossil fuels—the stored photosynthates of previous millenia—is in negative balance. But it took the energy crisis of the early 1970s to presage the need to face up to this situation. We can, and we must, conserve energy.

How Environment Affects Animals

The effect of environment on dairy cattle was clearly demonstrated in an experiment in New Zealand. It involved the selection of 20 calves from low-producing herds and 20 calves from high-producing herds. All of them were sired artificially by outstanding bulls. The 40 head were assembled at the Ruakura Experiment Station, raised and milked together for the first lactation. Under these conditions, no significant difference between the production of the two groups was observed. Then, they were sent back to the respective herds from whence they came, whereupon their production was comparable to that of the cows with which they were being milked. Then, for a second time, they were returned to the Ruakura Experiment Station, where again there was no significant difference in their production. The Ruakura Station then went one step further; they confirmed these results by using identical twins, with both twins milked at the Ruakura Station, and then later divided between high- and low-producing herds for subsequent lactation.

The New Zealand experiment underscores the importance of environment. No matter how good the genetics, a good environment is essential to obtain high production.

Heredity has already made its contribution at the time of fertilization, but environment works ceaselessly away until death. Among the environmental factors affecting animals are the following:

1. Feed and nutrition
2. Weather
   a. Environmentally controlled buildings
   b. Artificial lighting
3. Health
4. Stress

We now know that controlled environment must embrace far more than an air-conditioned chamber, along with ample feed and water. The producer needs to concern himself more with the natural habitat of animals. Nature ordained that they do more than eat, sleep, and reproduce. For example, studies on the behavior of swine show that they spend much of their day in active investigative behavior, primarily rooting and manipulating their movement. When free ranging, pigs may spend 40% of their day resting, 35% investigating novel surroundings, 15% eating, and 10% in other activities. What happens when pigs are confined in a building on slotted floors? How is the neural energy dissipated that would normally be used to satisfy the drives for investigating and rooting? Evidently, environmental deficiencies are manifested by tail biting, gastric ulcers, poor maternal care and loss of young, or other physiological functions resulting in a sudden death syndrome or tissue degeneration.

Preventing disorders by merely cutting off the tails of pigs to prevent tail biting, debeaking poultry to alleviate cannibalism, and using choke collars on horses to inhibit cribbing, is not unlike trying to control malaria fever in humans by the use of drugs without getting rid of mosquitoes. Rather, we need to recognize these disorders for what they are—warning signals that conditions are not right. Correcting the cause of the disorder is the best solution. Unfortunately, this is not usually the easiest. Correcting the cause may involve trying to emulate the natural conditions of the species, such as altering space per animal and group size, providing training and experience at opportune times, promoting exercise, and gradually changing rations. Over the long pull, selection provides a major answer to correcting confinement and other behavioral problems; we need to breed animals adapted to man-made environments.

FEED AND NUTRITION

The most important influence in the environment is the feed. Animals may be affected by (1) too little feed, (2) rations that are too low in one or more nutrients, (3) an imbalance between certain nutrients, or (4) objection to the physical form of the ration—for example, it may be ground too finely.

Forced production (such as growth, milk production, and racing two-year-olds) and the feeding of forages and grains which are often produced on leached and depleted soils have created many problems in nutrition. These conditions have been further aggravated through the increased confinement of animals, many animals being confined to stalls or lots all or a large part of the year. Under these unnatural conditions, nutritional diseases and ailments have become increasingly common.

Also, nutritional reproductive failures plague livestock operations. Generally speaking, energy is more important than protein in reproduction. The level and kind of feed before and after parturition will determine how many females will show heat—and conceive. After giving birth, feed requirements increase tremendously because of milk production; hence, a female suckling young needs approximately 50 percent greater feed allowance than during the pregnancy period. Otherwise, she will suffer a serious loss in weight, and she may fail to come in heat and conceive.

The following additional feed-environmental factors are pertinent:

1. **Regularity of feeding**—Animals are creatures of habit; hence, they should be fed at regular times each day, by the clock.
2. **Underfeeding**—Too little feed results in slow and stunted growth of young stock; in loss of weight, poor condition, and excessive fatigue of mature animals; and in poor reproduction, failure of some females to show heat, more services per conception, lowered young crop, and light birth weights.
3. **Overfeeding**—Too much feed is wasteful. Besides, it creates a health hazard; there is usually lowered reproduction in breeding animals, and a higher incidence of digestive disturbances (bloat, founder, and scours)—and even death. Animals that suffer from mild digestive disturbances are commonly referred to as "off feed."

WEATHER

Weather affects the maintenance requirements of animals. For example, under ideal October weather conditions in Missouri, a horse may require 14 lb of a 60 percent TDN ration daily, whereas in the same area and doing the same work, the same horse may require 16 lb daily of the same feed in July and August and 20 lb in the winter. A good horseman senses this situation and changes the feed allowance accordingly.

During hot weather, feedlot cattle "peak" their eating during early morning and again during the evening hours—when it is cool. In cool weather, they eat more during midday than when it's hot. The feeder should sense these changes in cattle eating habits and program their feeding accordingly. Also, cattle eat more following a bad storm or a hot spell. At such times, the bunk may be "slick" for two or three hours and the cattle may line up waiting to be fed. When this happens, the ration should be increased. Also, by going to a higher roughage ration at these times, acidosis and laminitis can be minimized. The ability to recognize the "sign language of animals" and to change the feeding program accordingly is responsible for the oft-quoted statement that "the eye of the master fattens his cattle."

The maintenance requirements of animals increase as temperature, humidity, and air movements depart from the comfort zone. Likewise, the heat loss from animals is affected by these three items.

Environmentally Controlled Buildings

With the shift to confinement structures and high-density production operations, building design became more critical, with consideration given to air temperature, relative humidity, air velocity, wet bedding, dust, light, ammonia buildup, odors, and space requirements.

Environmentally controlled buildings are costly to construct, but they make for the ultimate in animal comfort, health, and efficiency of feed utilization. Also, they lend themselves to automation, which results in a saving in labor; and, because of minimizing space requirements, they effect a saving in land cost. Today, environmental control is rather common in poultry and swine housing, and it is on the increase with other classes of livestock—especially dairy cattle.

In hot climates, increased use is being made of shades for the purpose of enhancing animal comfort and minimizing the maintenance requirements. Also, studies with lactating dairy cows reveal that putting only the head in an air-conditioned chamber, with the rest of the body left exposed to the heat, will increase production and feed efficiency.

Before an environmental system can be designed for animals, it is important to know their (1) heat production, (2) vapor production, and (3) space requirements. This information is presented in Section 9, Buildings and Equipment, of this book; hence, the reader is referred thereto.

Artificial Lighting

The number of hours of light in the day affects the initiation of the normal breeding season of ewes and mares, both of which are seasonal breeders. It is noteworthy, too, that the reproductive function in poultry and migratory fowl is regulated by the length of daylight.

The ratio of hours of daylight to darkness throughout the year acts on nerves in the region of the pituitary gland, and stimulates or inhibits the release of the follicle-stimulating hormone (FSH). Lengthening the daylight hours activates the pituitary, and causes it to release increasing amounts of the FSH which stimulates ovarian function. Thus, sometime after the daylight period increases, the estrous cycle begins in ewes and mares.

Artificial lighting will accomplish the same thing as daylight; hence, it may be used to alter the estrous cycle in both ewes and mares.

- **Sheep**—Normally, ewes come in heat during the late summer or early fall, though there is both an area and a breed difference. The breeding season is usually restricted to about four months.

Ewes generally begin cycling when the number of daylight hours drops below 14. This is the reason that most breeds of sheep come into heat during the fall months. To initiate estrus, however, it appears that the shorter days must be preceded by longer days.

- **Horses**—Normally, the natural breeding season of mares begins in March and extends to late July or August.

Artificial lighting of broodmares enables breeders to bring mares in season about six weeks earlier than normal. By the use of the artificial light technique, a mare that would normally conceive on March 15 may get in foal sometime in January. By avoiding the necessity of skipping a year due to late breeding, this technique may actually result in obtaining two additional foals during the lifetime of a mare.

The procedure consists in using a 200-watt light bulb in a box stall so as to extend the hours of light to 16 hours daily. By beginning the light treatment of mares about December 1, they may be bred the latter part of January.

Slight adjustments in the schedule will need to be made in different locations, depending upon the sunrise and sunset times of the particular area.

HEALTH

Diseases and parasites (external and internal) are ever present animal enviromental factors. Death takes a tremendous toll. Even greater economic losses result from retarded growth and poor feed efficiency, carcass condemnations, decreases in meat quality, and in labor and drug costs. The signs of good health are summarized in Table 1-2 of this section; hence, the reader is referred thereto.

Any departure from the signs of good health constitutes a warning of trouble. Most sicknesses are ushered in by one or more signs of poor health—by indicators that tell the expert caretaker that all is not well—that tell him that his animals will go off feed tomorrow, and that prompt him to do something about it today.

Among the signs of animal ill health are: lack of appetite—the animal does not eat or graze normally; listlessness; droopy ears; sunken eyes; humped-up appearance; abnormal dung—either very hard or watery dung suggests an upset in the water balance or some intestinal disturbance following infection; abnormal urine—repeated attempts to urinate without success or off-colored urine should be cause for suspicion; abnormal discharges from the nose, mouth, and eyes, or a swelling under the jaw; unusual posture—such as standing with the head down or extreme nervousness; persistent rubbing or licking; dull hair coat and dry, scurfy, hidebound skin; pale, red, or purple mucous membranes lining the eyes and gums; reluctance to move or unusual movements; higher than normal temperature; labored breathing—increased rate and depth; altered social behavior such as leaving the herd and going off alone; and sudden drop in production—weight gains, milk, wool, or work.

STRESS

Stress of any kind affects animals. Among the external forces which may stress animals are excitement, presence of strangers, fatigue, number of animals together, space, changing corral and corral mates, previous training, previous nutrition, and management.

Race and show horses are always under stress; and the greater the speed and the more tired they become, the greater the stress. Also, the greater the stress, the more exacting the nutritive requirements. Thus, the ration of race and show horses should be scientifically formulated, rather than based on fads, foibles, and trade secrets.

Animals can be prepared in such a manner as to reduce stress. For example, if calves are properly preconditioned (started on feed, vaccinated, treated for parasites, etc.) prior to weaning, the stress of subsequent weaning and movement to a feedlot will be minimized.

CONTROL POLLUTION

Pollution is the issue of the decade. It matters little whether pollution is due to agriculture or factories. Everything that defiles, desecrates, or makes impure or unclean streams or atmosphere must be controlled.

We must ever be mindful that life, beauty, wealth, and progress depend upon how wisely man uses nature's gifts—the soil, the water, the air, the minerals, and the plant and animal life.

In agriculture, we need to give particular attention to any pollution that may be caused by manure, fertilizer, insecticides, herbicides, and growth promotants.

In recent years, there has been a worldwide awakening to the problem of pollution of the environment (air, water, and soil) and its effect on human health and on other forms of life. Much of this concern stemmed from the amount of manure produced by the sudden increase of animals in confinement. Certainly, there have been abuses of the environment (and it hasn't been limited to agriculture). There is no argument that such neglect should be rectified in a sound, orderly manner, but it should be done with a minimum disruption of the economy and lowering of the standard of living.

In altogether too many cases extreme environmentalists advocate policy changes and legislation that may in the end be detrimental to agriculture, to our food production potential, and to society in general. Frequently, these new messiahs have only used the data that support their theories about ecological doom. One of their favorite comparisons deals with the relative magnitude of the effect on the environment caused by animal manure, industrial waste, and municipal waste. Then, they add the "scare" to their story by citing the number of blue babies and suffocated fish caused by runoff from manure. In particular, they have incriminated cattle feedlots as major culprits. Unfortunately, many of their facts and figures have been in error. In order to set the records straight, and to assist stockmen and others in controlling pollution to the maximum, this section is presented.

Invoking an old law (the Refuse Act of 1899, which gave the Corps of Engineers control over runoff or seepage into any stream which flows into navigable waters), the Federal Environmental Protection Agency (EPA) launched the program to control water pollution by requiring that all cattle feedlots which had 1,000 head or more the previous year must apply for a permit by July 1, 1971. The states followed suit; although differing in their regulations, all of them increased legal pressures for clean water and air. Then followed the Federal Water Pollution Control Act Amendments, enacted by Congress in 1971, charging the EPA with developing a broad national program to eliminate water pollution.

Pollution Laws and Regulations

Both open lot and confinement livestock systems come under pollution regulations. Open lots present drainage and runoff problems. Confinement systems must be coordinated with the disposal areas in order that pollution will not be created when storage pits are emptied.

Registration of facilities and a permit to operate are the primary requirements that the Federal Government and the various states are using to insure that livestock wastes are properly handled. Since state regulations vary somewhat, it is recommended that the stockman check into the regulations of the state in which he is operating or plans to operate. The Federal guidelines, which are rather broad, follow:

1. **Who must apply**—The basic provisions of the Federal regulations are (In some states two permits are needed—one state, the other Federal.):

   a. Feedlots with 1,000 or more animal units (1,000 cattle, 700 mature dairy cattle, 2,500 swine weighing over 55 lb, 12,000 sheep or lambs, 55,000 turkeys, 180,000 laying hens, or 290,000 broilers) must obtain a permit.

   b. Feedlots with fewer than 1,000 animal units, but more than 300 animal units (more than 300 cattle, 200 mature dairy cattle, 750 swine weighing over 55 lb, or 3,000 sheep), must obtain a permit if the facility discharges pollutants either in (1) a man-made conveyance constructed for the purpose, or (2) waters that pass through the confined area.

   c. Feedlots with fewer than 300 animal units are not subject to the permit requirements except where so designated on a case-by-case basis.

   d. Livestock confinement facilities include open feedlots, confined operations, stockyards, livestock auction barns, and buying stations. Also, the regulations apply to any combination of species in the same feedlot.

   e. Cropland runoff and manure spreading are not specifically regulated under the Federal Act, although they may be in the future.

2. **How to apply**—Forms may be secured from the offices of EPA and state environmental agencies, the county agent, or the SCS district offices. Fill out a *Short Form B* and send it, along with a $10 filing fee, to the EPA regional office. Then, either the Federal EPA or the state agency will make an on-the-site inspection. They will draft a proposed permit, put it on the public notice, and give the applicant and the public 30 days to comment on it. Then, if there are no protests, the Federal discharge permit will be issued.

3. **Cost-sharing help**—In 1973-74, the Rural Environmental Assistance Program (REAP) provided cost-share payments on certain livestock waste storage

and diversion facilities, with a limitation of 50 percent of the cost and a maximum of $2,500. The stockman should check on the availability of such funds in the future.

Environmental Effects of Grazing Lands

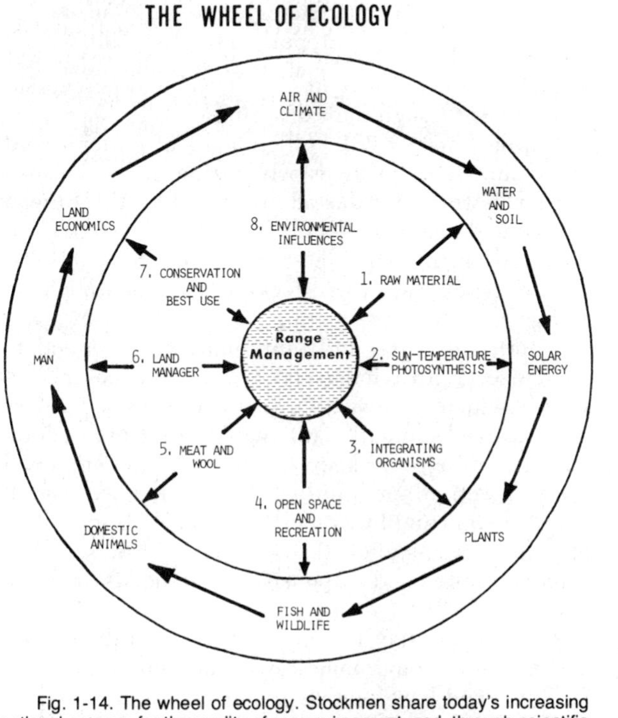

Fig. 1-14. The wheel of ecology. Stockmen share today's increasing national concern for the quality of our environment, and, through scientific range management, they are doing much to improve it.

Little pollution potential exists from pasture systems with low animal densities or numbers, or where pastures are rotated. So, except for high-density pasture systems involving a number of animals, pollution is no problem. Nevertheless, some environmentalists have centered their attack on the grazing of public lands.

Grazing influences the environment on Federal lands. Under poor range management, the environment is affected adversely; under good range management, such as exists on most ranges today, grazing actually improves the environment.

Eating of plant materials by animals is a natural process in earthly and aquatic systems. Thus, the coming of the white man to what is now the United States, along with the introduction of domestic animals, did not constitute an entirely new component in the environment. Rather, domestic animals replaced, or added to, the wild animals that were already there.

Mistakes in grazing practices have occurred in the United States in the past, the most significant of which was the exploitative grazing practiced between 1865 and the 1930s. The effects were almost catastrophic. Nevertheless, they were not the result of grazing ranges that had never been grazed before. Rather, they resulted from several decades of grazing the western ranges with too many animals for too long, and often at the wrong season of the year. Most range livestock operators of that period were not aware of the benefits that could accrue to them from improved range management.

Scientific management of rangeland began at the turn of the century. Range managers and livestock operators found that controlling grazing improved both range conditions and livestock production. Development of this new concept marked the beginning of the end of the exploitative period of grazing and the introduction of managed grazing on the western ranges.

The environmental effects of grazing depend upon the kind of range, the intensity of grazing, and the kind of management employed to control livestock on the range. It is generally recognized that unregulated heavy grazing results in loss of desirable forage plants, increased runoff and erosion, and other indications of range deterioration. On the other hand, planned seasonal grazing and controlled animal distribution foster rapid vegetational growth. Most grazing experiments have shown that ranges may be improved more rapidly under proper grazing management than with no grazing at all.

There is no evidence that well-managed grazing of domestic livestock is incompatible with a high-quality environment. But there is ample evidence that managed grazing by livestock enhances certain uses and that poor management detracts from them. Properly managed grazing is a reasonable and beneficial use of the range.

Ecologists tell us that good range management will support more wildlife than the wilderness. This explains why big game numbers on Federal lands have increased during recent years, and why wildlife production is an increasingly important use of rangelands.

Indeed, ranges actually improve while being properly utilized by domestic livestock. The benefits which accrue to the range include increased vegetation cover, improved plant species composition, improved soil fertility and soil structure, and greater yield of high-quality water. When sheep and cattle go, rank underbrush takes over, and fire becomes a real hazard.

Both upland game birds and big game animals are benefited by grazing that promotes good cover for mating sites and enhances food supply and other habitat requirements.

On ranges with mixed types of vegetation, herbaceous species increase and browse species decline when grazed only by game. The converse is true when cattle graze the land. The combined grazing by two groups of animals maintains a better balance of browse species—preferred by game animals, and of herbaceous species, preferred by cattle and sheep.

Heavy livestock grazing is beneficial to irrigated pastures used by geese and other migratory waterfowl. Unless the vegetation is closely cropped, these areas are unattractive to the birds.

Thus, livestock grazing of the public lands is contributing to improved wildlife habitat conditions and increased numbers of game animals. Range development programs, particularly livestock water developments, have made more public land usable by game animals and is partly responsible for the vast increase in game numbers over the years.

On many grass-shrub ranges, livestock grazing reduces the danger of fire by preventing a buildup of dry grass, which is highly inflammable.

Grazing systems and manipulation of vegetation can create contrast in vegetation color and pattern, thereby improving the aesthetical value of the landscape. Also, the livestock industry is traditional to the West; hence, a well-managed range with its cattle herd and roundup, or with its sheep camp, has recreational values. Indeed, cattle and sheep on the landscape are pleasing to tourists who come to view the "old West."

Ranges properly grazed by hoofed animals produce safe water. Counts of fecal coliform organisms, as indicators of water pollution by warm-blooded animals, relate more closely to the quantity of the fecal material than to the kind of animal. Investigations have shown that the count of harmful bacteria in streams is no greater in areas grazed by livestock than in areas grazed by wild animals alone, and that modern livestock grazing has little effect upon the chemical and physical quality of the water.

It is noteworthy, too, that few western ranges are ever in a stable, natural condition, whether or not they are grazed by domestic animals. Rather, most of them are in a stage of vegetational development following disturbances by such phenomena as drought, flood, avalanche, frost, or fire. Also, cyclic phenomena, such as large numbers of deer, rodent epidemics, or insect plagues, temporarily change the natural ecosystems. Thus, an absolutely stable rangeland is seldom attained or maintained.

Significantly, the greatest diversity of animal and plant species and the highest rates of reproduction occur when the landscape supports many stages of ecosystem development. Fire, grazing, and drought stimulate plants and animals to new growth. Each stage of vegetational development is more productive of certain animal species than of others.

Finally, in an era of world food shortages, the contribution of properly managed Federal lands in terms of food and fiber production needs to be recognized. More and more grains will be used for direct human consumption. As a result, there will be an increased reliance on ranges for meat and wool production. It just makes sense to preserve all the natural food and fiber that we can. Remember that petroleum is not needed to make wool. Remember, too, that cattle and sheep are completely recyclable. It takes thousands of years to create coal, oil, and natural gas; and when they're gone, they're gone forever. But animals produce a new crop each year and perpetuate themselves through their offspring.

Approximately 261 million acres of Federal land are administered for livestock grazing. In 1980, lands in the 11 western states administered by the Bureau of Land Management and the U.S. Forest Service provided grazing all or part of the year for 5,981,980 head of all classes of livestock—cattle, horses, sheep, and goats.

Both stockmen and environmentalists need to recognize (1) that forage is a renewable natural resource, which regrows each year and is wasted unless it is utilized annually; (2) that grazing on Federal rangelands helps to keep the natural environmental systems active and productive; (3) that we cannot allow overgrazing by domestic livestock, bison, deer, or wild horses; and (4) that grazing must be scientifically controlled and responsive to the needs of all users.

Indeed, it may be said that man's influence on and use of the environment will determine how well we live—and how long we live.

Manure

No doubt, the manure pollution problem, suspicioned or real, will persist. However, the energy crisis, accompanied by high chemical fertilizer prices, has caused manure to be looked upon as a resource and not a waste that presents a disposal problem. At current fertilizer prices—(per pound nitrogen (N) = 25¢; phosphorus (P_2O_5) = 20¢; and potassium (K_2O) = 10¢)—one ton of average cow manure is worth more than $4.92 per ton. As a result, a growing number of American farmers are returning to organic farming—they're using more manure—the unwanted barnyard centerpiece of years gone-by. They are discovering that they are just as good reapers of the land and far better stewards of the soil.

In the future, as fertilizer and feed become increasingly scarce and expensive, the economic value of manure will increase.

PRECAUTIONS WHEN USING MANURE AS A FERTILIZER ON THE LAND

The following precautions should be observed when using manure as a fertilizer:

1. Avoid applying closer than 100 feet to waterways, streams, lakes, wells, springs, or ponds.
2. Do not apply where downward movement of water is not good, or where irrigation water is very salty or inadequate to move salts down.
3. Incorporate (preferably by plowing or discing) manure into the soil as quickly as possible after application. This will maximize nutrient conservation, reduce odors, and minimize runoff pollution.
4. Distribute the waste as uniformly as possible on the area to be covered.
5. Irrigate thoroughly to leach excess salts below the root zone.
6. Allow about a month after irrigation before planting, to enable soil microorganisms to begin decomposition of manure.
7. Minimize odor problems by—
   a. Spreading raw manure frequently, especially during the summer.
   b. Spreading early in the day as the air is warming up, rather than late in the day when the air is cooling.
   c. Spreading only on days when the wind is not blowing toward populated areas.

HOW MUCH MANURE CAN BE APPLIED TO LAND

With today's heavy animal concentration in one location, the question is being asked: How many tons of manure can be applied per acre without depressing crop yield, making for salt problems in the soil, making for nitrate problems in feed, or contributing excess nitrate to groundwater or surface streams?

Based on earlier studies in the Midwest, before the rise of commercial fertilizers, it would appear that one can apply from 5 to 20 tons of manure per acre, year after year with benefit.

Heavier applications can be made, but probably should not be repeated every year. With higher rates than 20 tons per year annually, there may be excess salt and nitrate buildup. Excess nitrate from manure can pollute streams and groundwater and result in toxic levels of nitrate in crops. Without doubt the maximum rate at which manure can be applied to the land will vary widely according to soil type, rainfall, and temperature.

State regulations differ in limiting the rate of manure application. Missouri draws the line at 30 tons per acre on pasture, and 40 tons per acre on cropland. Indiana limits manure application according to the amount of nitrogen applied, with the maximum limit set at 225 pounds per acre per year. Nebraska requires only one-half acre of land for liquid manure disposal per acre of feedlot, which appears to be the least acreage for manure disposal required by any state.

When a farmer has sufficient land, he should use rates of manure which supply only the nutrients needed by the crop rather than the maximum possible amounts suggested for pollution control.

Agricultural Chemicals

In the everyday pursuit of modern agriculture, more and more chemicals and drugs are being used. Hand in hand with this development, there has been increased public concern over the use of these products, for fear of poisoning human food.

Chemicals and drugs must be used with discretion, especially those designed to kill some living organism. But sometimes choices must be made; for example, between malaria-carrying mosquitoes and some fish, or between hordes of locusts and grasshoppers and the crops that they devour. This merely underscores the need for (1) careful testing through properly designed experiments of all products prior to use, (2) conforming with federal and state laws, and (3) accurate labeling and use of products. Additionally, food producers need to relate the miracle of agriculture. They need to tell that back of food and clothing are agricultural chemicals and drugs—herbicides, insecticides, pesticides, disease control materials, feed additives, fertilizers, and many others. They need to show how these products are as indispensable to modern food production as tractors, trucks, hybrid seeds, and improved livestock.

In an era of food shortages, losses of feed, food, and fiber will increasingly concern all people, producers and consumers alike. If the world is to "waste not, want not," it must use more agricultural chemicals and drugs in the future, not fewer.

Through the proper use of agricultural chemicals—herbicides, insecticides, pesticides, and other chemicals—food and fiber losses could be reduced substantially—perhaps by as much as 30 to 50 percent. The net result would be an increase of 10 to 15 percent in the world food supply, with no new land required. In no other way can the hungry gap be filled so quickly at so little cost.

Farmers have an obligation to produce more foods and fibers efficiently, to reduce production costs, and to increase their income. In turn, consumers—all—benefit from agricultural chemicals and drugs which produce more products at less cost.

When properly used, agricultural chemicals are an important adjunct in providing feed for animals and food for people. However, improper use can result in toxicity. Moreover, certain chemicals can accumulate in the body fat of animals, and be found in the meat or secreted in the milk.

The vast majority of agricultural chemicals and drugs have been properly used. Of course, it shouldn't be too surprising that a few have been improperly used when it is realized that there are approximately 300,000 trade name products on the market.

When chemical poisoning or drug misuse happens, it can be both devastating and perplexing. Usually, the causative agent can be diagnosed after an investigation of the environment and the feed. However, few poisons can be diagnosed with certainty by clinical symptoms alone. When trouble is encountered, the producer should promptly call a veterinarian if animals are involved and a medical doctor if people are involved.

A voluminous amount of information is available on the deleterious effects of poisonous chemicals and drugs—both artificial and natural.

Farmers know that unless they follow state and federal regulations they risk having their products condemned and seized, or refused by food processors. Nevertheless, the economics dictate that new products be used as soon as they prove useful. On the other hand, food faddists may feel that they are being poisoned; wildlife conservationists may be concerned over possible damage to songbirds and other animals; beekeepers become unhappy if insecticides kill honeybees; and public health agencies are concerned about contamination of soil, water, and food supplies.

Zoning

In recent years, some of the concern over the problem of pollution of the environment (air, water, and soil) and its effect on human health has stemmed from horses kept in the suburbs, with some environmentalists protesting on the basis that they make for more flies, dust, and odors. This has resulted in zoning.

Thus, the owner who desires to keep a horse, or other animals, in his backyard, on a plot of land within corporate limits of a municipality or within an urban area, should first consult the zoning ordinance to determine if there are any restrictions against it.

Most zoning regulations are local ordinances, covering a municipality, a county, or certain zoning districts. Also, there are rapid and basic changes in zoning ordinances; many cities and counties are rewriting and making changes in them. Thus, a prospective backyard animal keeper should check with local zoning authorities before purchasing an animal or animals.

CONSERVE ENERGY

Fossil fuels—the stored photosynthates of previous millenia—are like a bank account. There is nothing wrong with drawing upon either of them, but neither is inexhaustible. It is highly imprudent not to be aware of big withdrawals and not to cover them. Within a short span of a few years, the world made the transition from a positive energy balance based upon the capture of the energy of the sun via green plants, crops, and forests to an imbalance, or even a negative balance, by resorting primarily to the bank of trapped sun energy of fossil fuels that had accumulated over millions of years.

Fig. 1-15. An Oriental wet rice peasant, using animal power (water buffalo), expends only 1 calorie of energy to produce each 50 calories of food. By comparison, the average U.S. farmer, using mechanical power (tractors), expends 2.5 calories of fuel energy to produce 1 calorie of food.

Currently, the global use of resources is increasing at an alarming rate—far faster than population; at the present pace, energy use will double in 14 years as compared to population doubling in 33 years. But it took the energy crisis of the early 1970s to cause the world to face up to this dilemma.

Everyone knows that cars and airplanes are powered by fuel. But how many people realize that modern, mechanized food production in the developed countries requires an extra input of fuel, which is mostly of fossil origin? How many people know that the developed nations have used fossil fuels to supplement the natural energy that comes directly from the sun on a day-to-day basis, and that they have grown dependent upon them?

Of course, the direct input of fuel into food production is of rather recent origin. It all began in a very small way about 1840, when fuel-powered ships transported fertilizer (guano, and later bone meal) from South America to Europe. Then, after 1910, transportation vehicles relied almost exclusively on fossil fuels. But the direct use of fossil fuels in agriculture started with the manufacturing of chemical fertilizer beginning about 1922—little more than 60 years ago. Following closely in period of time, the tractor was substituted for horses, mules, and oxen—eventually almost completely replacing them.

In addition to food production on the farm, there are two other important steps in the food line as it moves from the producer to the consumer; namely, processing and marketing, both of which require higher energy inputs than to produce the food on the farm (see Table 1-4).

Table 1-4 points up the increasing drain that modern food production is putting on the energy supply. In 1980, U.S. farms put in 2.8 calories of fuel per calorie of food grown, 3.1 times more than the on-farm energy input in 1940.

TABLE 1-4

MODERN FOOD PRODUCTION IS INEFFICIENT IN ENERGY UTILIZATION—THE STORY FROM PRODUCER TO CONSUMER[1]

Year	On the Farm	Food Processing	Marketing and Home Cooking	Total/Person/ Year
1940[2]				
Million kcal . . .	0.9	2.2	2.1	5.2
Percent	18.0	42.0	40.0	100.0
1980[3]				
Million kcal . . .	2.8	4.7	4.6	12.1[4]
Percent	23.0	39.0	38.0	100.0
Increase, times 1940-1980	3.1	2.1	2.2	2.3

[1]Energy in million kcal used per capita to produce 1 million kcal of food in the U.S.

[2]Values from Borgstrom, G., "The Price of a Tractor," *Ceres*, FAO of the U.N., Rome, Italy, Nov.-Dec. 1974, p. 18, Table 3.

[3]Author's estimate based on several reports detailing trends in energy usage.

[4]This means that in 1980 it required 12.1 million kcal to produce 1 million kcal of food for each person, a daily consumption of 2,740 kcal (1,000,000 ÷ 365 = 2,740).

Table 1-4 also shows that, in the United States in 1980, a total of 12.1 calories were used in the production, food processing, and marketing-cooking for every calorie of food consumed, with a percentage distribution of the total cost of energy at each step from producer to consumer as follows: on the farm, 23%; food processing, 39%; and marketing and home cooking, 38%. In 1940, it took only 5.2 calories—slighlty less than half the 1980 figure—to get 1 calorie of food on the table. It's noteworthy, too, that more energy is required for food processing and marketing-home cooking than for growing the product; and that, from 1940 to 1980, the on-the-farm energy requirement increased by 3.1 times, in comparison with an increase of 2.1 and 2.2 times for each of the other steps—processing and marketing-home cooking.

Prior to the advent of machines and fuel in crop production, 1 calorie of energy input on the farm produced about 16 calories of food energy. Today, on the average, U.S. farms put in about 2.8 calories of fuel per calorie of food grown; hence, to produce a daily intake of 3,000 calories of edible food from cultivated crops may require 8,400 calories of energy from fossil fuels—an exhaustible source. It's more surprising yet—and thought-provoking—to know that, even today in the poorer or developing countries, it takes only 1 calorie to produce each 10 calories of food consumed. The Oriental wet rice peasant uses only 1 unit of energy to produce 50 units of food energy. This gives the Orientals a favorable position among the major powers as the energy crisis worsens.

The following additional points are pertinent to any energy conservation program:

1. **Photosynthesis fixes energy**—Photosynthesis is by far the most important energy-producing process. But currently only about 1% of the solar energy falling on an area is fixed by photosynthesis; and only 5% of this captured energy is fixed in a form suitable as food for man. Thus, (a) man's manipulation of plants for increased efficiency of solar energy conversion, and (b) converting a greater percentage of total energy fixed as chemical energy in plants (the other 95%) into a form available to man would appear to hold great promise in solving the future food problems of the world.

2. **Animals step up energy**—The increase in the energy level through animal products—through animals consuming the photosynthetic energy in crops—almost equals the energy subsidies at each of the two steps after the product leaves the farm (in food processing and marketing-home cooking—see Table 1-4).

3. **Crop residues contain energy**—Crop residues left in the field, above or below the soil surface, may well constitute four to five times more energy than is harvested. Increasingly, this potential source of added feed, organic fertilizer, and energy will be utilized in the future.

4. **Increased yields; the law of diminishing returns in energy**—Modern intensive farming has markedly increased crop yields per acre and per manhour—by as much as 50- to 100-fold. But this has been done at the cost of large inputs of fuel (including electricity).

For a surprising number of modern cropping systems, a 10- to 50-fold increase in the energy output merely doubles or triples the food energy. Substantial expenditures fail to produce corresponding increases in yields. Thus, the law of diminishing return prevails.

High petroleum costs have spurred a search for other energy sources and for means of conserving energy. Higher productivity of the agriculture of tomorrow must be achieved through ingenious approaches in order to reverse the present lopsided energy balance. In obtaining increased food yields, we must consider the use of energy to produce energy. We must remember that photosynthesis is by far the most important energy-producing process; indeed, that it is the only basic food-manufacturing process in the world. We must remember, too, that grazing animals do not require fuel outside of their own body use to harvest the energy and other nutrients of grass (solar energy converted into chemical energy by grass), a renewable source.

SECTION 2

BUSINESS ASPECTS

Contents **Page**

Business Organization 32
Types of Business Organization 33
Proprietorship (Individual) 33
Partnership (General Partnership) 34
Limited Partnership 34
Corporation 35
Acquiring a Farm or Ranch 36
Ways to Acquire a Farm or Ranch 36
What's a Farm or Ranch Worth? 38
Financing the Farm or Ranch 39
Other Considerations When Buying a Farm or Ranch 43
Choosing the Class of Animals 46
Capital Needs 46
Cow-Calf Capital Needs 46
Cattle Feedlot Capital Needs 49
Guidelines Relative to Facility and Equipment Costs 51
Credit in the Livestock Business 52
Types of Credit or Loans 52
Credit Sources 52
Public Offerings 54
Central Money Markets 54
Credit Factors Considered and Evaluated by Lenders 55
Credit Factors Considered by Borrowers 56
Helpful Hints for Building and Maintaining a Good Credit Rating 56
Borrow Money to Make Money 57
Calculating Interest 57
Manager 58
Traits of a Good Livestock Manager 58
Organization Chart and Job Description 59
An Incentive Basis for the Help 59
Farm Records and Accounts 65
Analyzing a Livestock Business—Is It Profitable? 69
Cow-Calf Profit Indicators 70
Cattle Feedlot Profit Indicators 70
Budgets in the Livestock Business 71
Computers in the Livestock Business 73
Business Aspects of Cattle Feeding 75
Computing Break-Even Prices for Cattle 76
Custom (Contract) Feeding 79
Co-op Owned Feedlots 83
Condominium Cattle Feedlots 85
Record Forms 86
Stocker and Grower Contracts 87
Dairy Cattle Contracts 88
Horse Contracts 91
Syndicated Horses 91
Stallion Breeding Contract 94
Boarding Agreement 94
Futures Trading 97
Livestock Futures Trading 97
Futures Trading in Finished Cattle, Feeder Cattle, and Feed 97
What Constitutes a Futures Cattle Contract? 98
Glossary of Futures Market Terms 104
Tax Management and Reporting 106
Tax Pointers for Stockmen 106
Tax Planning 113
Tax Sheltered Livestock Investments 114
Estate Planning 115
Wills 118
Trusts 118
Partnership Contract 118
Livestock Insurance 119

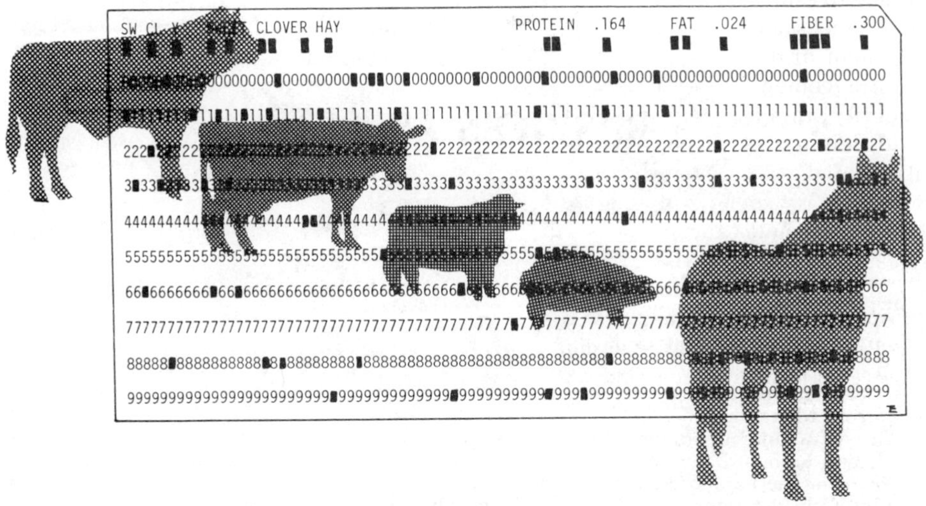

Fig. 2-1. Farming has gone modern! Today's successful farmers and ranchers must have, and use, as complete records as any other business. Also, records must be kept current. To meet this need, many big and complex agricultural enterprises have replaced hand record keeping with computers.

Agriculture, with assets totaling $1,090.3 billion in 1981, (1) ranks as the nation's biggest single industry, and (2) has a value equivalent to three-fifths of all corporations in the United States. Moreover, it is destined to get bigger and more complicated.

From 1935 to 1981, within a span of 46 years, the number of farms decreased from 6.8 million to 2.4 million and the size of farms increased from 154.8 acres to 431 acres.[1] Thus, within 46 years, nearly 65 percent of the farms disappeared from American agriculture and the average size of farms more than doubled. With this transition, herds, flocks, and feedlots became bigger.

In 1952, each farm worker supplied enough food and fiber for 17 persons, including himself; by 1981, each farm worker supplied enough food and fiber for 76 persons, including himself.

The above trends to bigness will continue.

The business and management aspects of animal production will be increasingly important in the future. Changes in the type of business organization and in financial management will come. More capital will be required, more money will be borrowed, competent managers will be in demand, better and more complete records will be necessary, futures trading will increase, and stockmen will become more knowledgeable relative to tax management, estate planning, and liability. The net result will be that those engaged in the business of agriculture must treat it as the big business that it is and become more sophisticated and efficient; otherwise, they won't be in business very long.

[1]The Census definition of a farm is as follows: "A place that sells $1,000 a year in ag products."

The stockman of the future will be a good businessman, as well as a good stockman. He also will need an operation that is large enough to provide his family an adequate standard of living and generate enough capital to keep expanding. Since profit margins will likely decline still further, there will be greater stress on business and financial management skills.

Generating both equity and debt capital, or risk and borrowed capital, will be one of the main concerns of the future stockman. The large investment, plus the need to keep competitive by utilizing new and usually expensive technological advances, will cause capital to be very important.

To obtain capital several things will be necessary. The stockman-businessman will have to prepare (1) profit and loss statements to show that his operation is profitable, (2) financial statements to show that progress is being made, and (3) cash-flow projections to show loan repayability. Then, and then only, will he be ready to go looking for money.

Skill in capital budgeting and analyzing alternative investment opportunities will be needed to see that the limited capital is invested where payoff will be the greatest. Stockmen will also have to exercise budget and cost controls of their business. Skill in building sound credit will be needed.

The greatest payoffs in the future are likely to accrue to those stockmen who improve their skills in business and financial management.

BUSINESS ORGANIZATION

Big land-livestock operations will get bigger, demanding more and more capital and top management. In the years ahead, many investors will become part owners in land and livestock, much as they now do through corporate stocks and bonds; and they will leave the management of the holdings to the professionals. Such an arrangement will also make it possible to (1) diversify in countries and types of investments—in different areas of the United States, and in Australia, Canada, South America, and other areas where there are vast acreages of rangeland with great potential for improvement that can be secured at reasonable prices; (2) minimize risks of loss from droughts and local depressions; (3) obtain for investors, big and little, the benefits that accrue to bigness, such as lower investment per animal unit, and lower feed and labor costs; (4) furnish recreational and vacation areas on farms and ranches of which they are part owner; and (5) provide know-how, continuity, and able management.

Part ownership in land and livestock investment companies affords a modern way in which to spread investments and minimize risks—as is done with stocks and bonds, grain and livestock futures, and syndicated sires.

Types of Business Organization

The success of today's farming is very dependent on the type of business organization. No one type of organization is superior under all circumstances; rather, each situation must be considered individually. The size of the operation, the family situation, the enterprises, the objectives—all these, and more, are important in determining the best way in which to organize the farming business.

Three major types of business organizations are commonly found among farming enterprises: (1) the sole proprietorship, (2) the partnership, and (3) the corporation. Among the factors which should be considered when deciding which business form best fits a given set of circumstances are the following:

1. Which type of organization is most likely to be looked upon favorably from the standpoint of more credit and capital?
2. How much capital will be required of each individual involved?
3. Are there tax advantages to be gained from the business organization?
4. Is expansion of the business feasible and facilitated?
5. Which type of organization reduces risks and liability most?
6. Which type of organization can be terminated most easily and readily?
7. Which type of ownership provides for the most continuity and ease of transfer?
8. What costs for legal and accounting fees are involved, in setting up the organization and in the preparation of the annual reports required by law?
9. Who will manage the business?

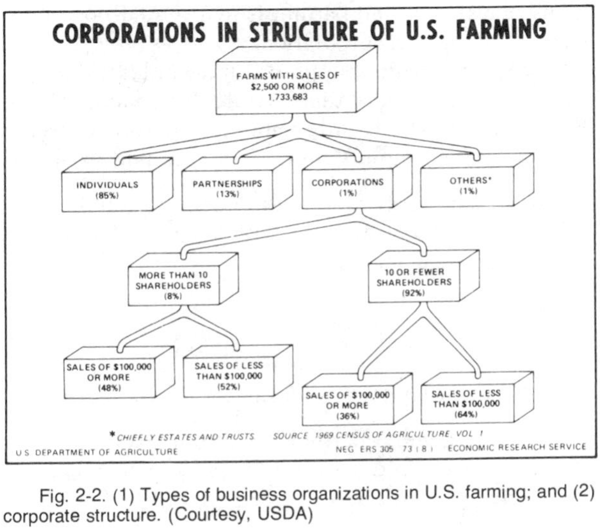

Fig. 2-2. (1) Types of business organizations in U.S. farming; and (2) corporate structure. (Courtesy, USDA)

Most agricultural enterprises are operated as sole proprietorships, not necessarily because this is the best type of organization, but with no effort to form some other type of organization it naturally results. Both the partnership and the corporation, which require special planning and effort to bring about, are well suited to the operation of large livestock establishments.

PROPRIETORSHIP (Individual)

This is the most common type of business organization in U.S. farming—90 percent of the nation's farms are individually or family owned. Under the sole proprietorship, or individual (or family) ownership, one man (or family) controls the business. He may not provide all the capital used in the business; in fact, he usually does not. However, he has sole management and control of the operation, although this may be modified and delegated somewhat through lease agreements, contracts, etc. Basically, the sole proprietor gets all the profits of the business; likewise, he must absorb all the losses.

In comparison with other forms of organization, the sole proprietorship has two major limitations: (1) It may be more difficult to acquire new capital for expansion; and (2) not much can be done to provide for continuity and to keep the present business going as a unit, with the result that it usually goes out of existence with the passing of the owner.

PARTNERSHIP (General Partnership)

A partnership is an association of two or more persons who, as co-owners, operate the business. About 13 percent of U.S. farms are partnerships.

The basic idea of two or more persons joining together to carry out a business venture can be traced back to the syndicates that were used in major trading centers in western Europe in the Middle Ages. Many of the early efforts to colonize the New World were also partnerships, or "companies" which provided venture capital, ships, provisions, and trade goods to induce settlement of large land grants.

Most farm partnerships involve family members who have pooled land, machinery, working capital, and often their labor and management to operate a larger business than would be possible if each member limited his operation to his own resources. It is a good way in which to bring a son, who is usually short on capital, into the business, yet keep the father in active participation. Although there are financial risks to each member of such a partnership, and potential conflicts in management decisions, the existence of family ties tends to minimize such problems.

In order for a partnership to be successful, the enterprise must be sufficiently large to utilize the abilities and skills of the partners and to compensate each adequately in keeping with his contribution to the business.

A partnership has the following *advantages:*

1. **Combining resources**—A partnership often increases returns from the operation due to combining resources. For example, one partner may contribute his labor and management skills, whereas another may provide the capital. Under such an arrangement, it is very important that the partners agree on the value of each person's contribution to the business, and that this be clearly spelled out in the partnership agreement.
2. **Equitable management**—Unless otherwise agreed upon, all partners have equal rights, regardless of financial interest. Any limitations, such as voting rights proportionate to investments, should be a written part of the agreement.
3. **Tax savings**—A partnership does not pay any tax on its income, but it must file an informational return. The tax is paid as part of the individual tax returns of the respective partners, usually at lower tax rates.
4. **Flexibility**—Usually, the partnership does not need outside approval to change its structure or operation—the vote of the partners suffices.

Partnerships may have the following *disadvantages:*

1. **Liability for debts and obligations of the partnership**—In a partnership, each partner is liable for all the debts and obligations of the partnership.
2. **Uncertainty of length of agreement**—A partnership ceases with the death or withdrawal of any partner, unless the agreement provides for continuation by the remaining partners.
3. **Difficulty of determining value of partner's interest**—Since a partner owns a share of every individual item involved in the partnership, it is often very difficult to judge value. This tends to make transfer of a partnership difficult. This disadvantage may be lessened by determining market values regularly.

The above is what is known as a partnership or general partnership. It is characterized by (1) management of the business being shared by the partners, and (2) each partner being responsible for the activities and liabilities of all of the partners, in addition to his own activities within the partnership.

LIMITED PARTNERSHIP

A limited partnership is an arrangement in which two or more parties supply the capital, but only one partner is involved in the management. This is a special type of partnership with one or more "general partners" and one or more "limited partners."

The limited partnership avoids many of the problems inherent in a general partnership and has become the chief legal device for attracting outside investor capital into farm and ranch ventures. Although this device has been widely used in the oil and gas industry, and for acquiring income-producing urban real estate for a number of years, its application to agricultural ventures on a national scale is quite new. As the term implies, the financial liability of each partner is limited to his original investment, and the partnership does not require, and in fact prohibits, direct involvement of the limited partners in management. In many ways, a limited partner is in a similar position to a stockholder in a corporation.

A limited partnership must have at least one general partner who is responsible for managing the business and who is fully liable for all obligations.

The *advantages* of a limited partnership are:

1. It facilitates bringing in outside capital.
2. It need not dissolve with the loss of a partner.
3. Interests may be sold or transferred.
4. The business is taxed as a partnership.
5. Liability is limited.
6. It may be used as a tax shelter.

The *disadvantages* of a limited partnership are:

1. The general partner has unlimited liability.
2. The limited partners have no voice in management.

CORPORATION

A corporation is a device for carrying out a farming or ranching enterprise as an entity entirely distinct from the persons who are interested in and control it. Each state authorizes the existence of corporations. As long as the corporation complies with the provisions of the law, it continues to exist—irrespective of changes in its membership.

Until about 1960, few farms and ranches were operated as corporations. In recent years, however, there has been increased interest in the use of corporations for the conducting of farm and ranch business. Even so, only about two percent of U.S. farms use the corporate structure.

From an operational standpoint, a corporation possesses many of the privileges and responsibilities of a real person. It can own property; it can hire labor; it can sue and be sued; and it pays taxes.

Separation of ownership and management is a unique feature of corporations. The owners' interest in a corporation is represented by shares of stock. The shareholders elect the board of directors who, in turn, elect the officers. The officers are responsible for the day-to-day operation of the business. Of course, in a close family corporation, shareholders, directors, and officers can be the same persons.

The major *advantages* of a corporate structure are:

1. It provides continuity despite the death of a stockholder.
2. It facilitates transfer of ownership.
3. It limits the liability of shareholders to the value of their stock.
4. It may make for some savings in income taxes.

The major *disadvantages* of a corporation are:

1. It is restricted to doing only what is specified in its charter.
2. It must register in each state.
3. It must comply with stipulated regulations which involve considerable paperwork and expense.
4. It is subject to the hazard of higher taxes.
5. It is possible to lose control.

FAMILY OWNED (PRIVATELY OWNED) CORPORATION

Still another type of corporation is family owned (privately owned). It enjoys most of the advantages of its generally larger outside investor counterpart, with few of the disadvantages. The chief *advantages* of the family owned corporation over a partnership arrangement are:

1. **It alleviates unlimited liability**—For this reason, a lawsuit cannot destroy the entire business and all the individual partners with it.
2. **It facilitates estate planning and ownership transfer**—It makes it possible to handle the estate and keep the business in the family and going if one of the partners should die. Each of the heirs can be given shares of stock—which are easy to sell or transfer and can be used as collateral to borrow money—while leaving the management of the enterprise to those heirs interested in operating it, or even to outsiders.

TAX-OPTION CORPORATION (SUBCHAPTER S CORPORATION)

Instead of paying a corporate tax, a corporation with no more than 35 stockholders may elect to be taxed as a partnership, with the income or losses passed directly to the shareholders, each of whom pays taxes on his share of the profits. This special type of corporation is variously referred to as a "tax-option" corporation, "subchapter S" corporation, pseudocorporation, or elective corporation.

For income tax purposes, the owners of a tax-option corporation are taxed as if they were a partnership. That is, income earned by the corporation passes through the corporation to the personal income tax returns of the individual shareholders. Thus, the corporation does not pay any income tax. Instead, each shareholder pays tax on his share of corporate income at his individual tax rate; and each shareholder reports his share of long-term capital gains and receives his deductions therefor. Although each shareholder's portion of any corporate losses from current operations is deducted from his personal return, capital losses incurred by the corporation cannot be passed through to the shareholders.

Thus, there are some very real advantages to be gained from a "subchapter S" or "tax-option" corporation. However, in order to qualify as a subchapter S corporation, the following requisites must be met:

1. There cannot be more than 35 stockholders.
2. All stockholders must agree to be taxed as a partnership.
3. Nonresident aliens cannot own stock.
4. There can be only one class of stock.
5. Not more than 20 percent of the gross receipts of the corporation can be from royalties, rents, dividends, interest, or annuities plus gains from sale or exchange of stock and securities; and not more than 80 percent of gross receipts can be from sources outside the United States.

ADVANTAGES OF LIMITED PARTNERSHIPS AND CORPORATIONS

In addition to the advantages peculiar to (1) limited partnerships, and (2) corporations, and covered under each, limited partnerships and corporations have the following advantages over individual ownership in the acquisition of capital:

1. They make it possible for several producers to pool their resources and develop an economically sized operation, which might be too large for any one of them to finance individually.
2. They make it possible for persons outside agriculture to invest through purchase of shares of stock in the business.
3. They can generally borrow money easier because the strength of the loan is not dependent on the financial and managerial capability of one person.
4. They give assurance that the business will continue, even if one of the owners should die or decide to sell his interest.
5. They provide built-in management, with continuity; and, generally speaking, they attract very able management.

Thus, stockmen can and do use either of these two business organizations—a limited partnership or a corporation—to develop and maintain an economically sound operation. Actually, no one type of business organization is best suited for all purposes. Rather, each case must be analyzed, with the assistance of qualified specialists, to determine whether there is an advantage to using one of these types of organizations, and, if so, which organization is best suited to the proposed business.

AGENCY SERVICES

As an alternative to entering into the limited partnership arrangement or owning stock in a corporation, nonfarm investors wishing to engage in cattle feeding (and in certain other agricultural enterprises) can utilize the services of several firms which specialize in purchasing feeders, contracting with feedlots, and selling market cattle. Also, similar services are available for acquiring and managing commercial breeding herds. Under the agency arrangement, the investor establishes a drawing account for the agent, arranges his own financing if he wishes, and can withdraw any profits realized. Because he obtains legal title to specific lots of cattle, he may use them as collateral for loans.

Firms offering such services charge a flat fee per head, or a percentage of gross sales. They usually do not have financial interests in the feedlots or ranches with whom they contract.

ACQUIRING A FARM OR RANCH

Acquiring a suitable farm is a big problem, particularly for the beginner. He must compete for available farms with established farmers as well as with other beginners. Many established farmers need more land to enlarge their operations. Others move during the year, getting a better or more suitable farm. Some simply move to a new locality for personal reasons.

Whatever the motive for acquiring a farm or ranch, some buyers will be happy with their purchases. Others will rue the day that they made the decision. How well the farm or ranch was bought, and how it was bought, may make the difference. Since most stockmen buy only one farm and operate it for a lifetime, it follows that the vast majority of them lack expertise relative to the procedures to follow when acquiring a farm or ranch. The guidelines that follow have been prepared to fill this need.

Ways to Acquire a Farm or Ranch

Farms or ranches may be acquired through gift or inheritance, by marriage, by renting or leasing, or by purchasing.

1. **Gift or inheritance**—Many farms and ranches are inherited and stay in the family for several generations. Certainly, inheriting a farm or ranch is a real advantage to anyone desiring to stay in the business and having the know-how and ability to operate the enterprise. Farm and ranch land and livestock (as well as corporate stock, U.S. savings bonds, mutual funds, money, whole life and annuity life insurance, and commercial real estate) can be transferred to relatives and friends, with certain tax savings, provided certain well-established rules are observed.

The basic rule in giving land or animals to another member of the family is that the donor (person giving the property) must give up control. Of course, if it actually or constructively passes through the hands of the donor, it is taxed to him.

You can give relatives or friends income-producing property and the income will be taxed to the person receiving the gift (donee). But the property must actually be transferred. Any strings attached to the ownership whereby the donor can have control over the property or get it back at some future time will not meet the requirements of the law. The basic elements are:

a. There must be intention to make a gift.
b. Transfer of legal title and control.
c. The donee accepts the gift.
d. No consideration (money or property) to be exchanged for the property.

If a person is expecting to inherit a farm or ranch, he should become fully cognizant of the inheritance tax laws. Also, it is most important that he have clear title to the land. If the title of ownership of a farm or ranch is left unsettled through two or three generations, the value of the land may almost be expended in settling the estate.

Where gift or inheritance money or property are involved, always seek the advice of a tax accountant and/or tax attorney.

2. **Marry it**—Although young men wanting to enter farming or ranching are frequently admonished, facetiously, that they should "marry for love, but love a woman with plenty of money," there is more than a little bit of truth in the advice.

In some countries, the social orders call for the parents to arrange the marriages of sons and daughters, primarily to keep them within the same class strata, thereby not dividing property with those who "have not." However, this method is fast giving way to the new social order, which, like that in the United States, results in marriages between individuals of vastly different amounts of wealth.

3. **Rent or lease**—The main advantage in renting over buying is that less capital is required and less financial risk is involved. The main disadvantages are insecurity of tenure and that the farming enterprise may be limited in size or kind because the landowner is reluctant to make needed additional investments in buildings and facilities. These disadvantages can be minimized, and sometimes eliminated, by a suitable lease—an agreement between landlord and tenant under which the farm or ranch is rented and operated. Such a lease should always be in writing.

The most common types of farm and ranch leases are:

a. **Cash lease**—This is a good type of lease for (1) the small farm or where the landlord lives at a distance, and (2) a tenant who has adequate livestock, equipment, and working capital. It encourages livestock farming because all of the crop can easily be fed on the farm. Also, it is simple, with little chance for controversy.

There are two types of cash leases: (1) that type in which a fixed rent per acre is agreed upon when the lease is drawn, and (2) that type in which the rent is adjusted to prices of farm products which prevail during the lease year. Under the second plan, the landlord bears part of the risk of price changes; however, it is difficult to keep cash rent in line with farm product prices. If product prices are used as a basis for rent changes, the products, markets, and dates should be specified.

The landlord may prefer a cash lease because (1) the amount paid is definite, and (2) it requires less supervision by the owner. On the other hand, it may not always be desirable from the standpoint of the landlord because (1) it generally makes for lower income, (2) it gives him less control of the farm, and (3) it is difficult to collect rent if crops fail.

The tenant may prefer a cash lease because (1) it will make for more profit if he is a successful manager, (2) it makes for more independence in the operation, and (3) it makes for more profit in the good years.

b. **Livestock share lease**—Livestock share leases vary considerably. But most of them provide for 50-50 ownership of the herd or flock; 50-50 sharing of the costs of production—especially feed and veterinary expenses; and 50-50 division of the income from the sale of animals. Buildings are generally a cost borne by the landowner. Labor is the responsibility of the operator.

A livestock share lease fits the tenant who wants to raise livestock, but cannot finance a program. It is especially suited where tenant and landlord get along well and where the landlord can make a good contribution in management.

In order for this type of lease to work best, the landlord should live close to the farm, and either give it his personal attention or arrange for adequate management help such as can be provided through a professional farm management service.

The landlord may prefer a livestock share lease because (1) it encourages more livestock and more manure, (2) low-quality crops can be utilized more easily, (3) he retains an active interest in management, and (4) it generally makes for more profit.

The tenant may prefer a livestock share lease because (1) the risk is less since rent is based on net income on the farm, (2) it requires less tenant capital, (3) the landlord is more willing to make improvements, and (4) he can gain experience from the guidance of a successful owner.

A careful determination of lease provisions and putting them in writing will result in a lease that is more equitable to both tenant and landowner, and will avert later misunderstandings and friction between the two parties. Standard lease forms are available, so the detailed provisions of a lease need not be spelled out in this book.

Renting or leasing might be a desirable way to start in the livestock business even if funds are available for purchase. This is particularly true where there is an option-to-buy clause. This gives the renter an opportunity to study the ranch more carefully and gain additional management experience without committing his entire assets.

4. **Purchase**—Purchase of a livestock farm or ranch has the advantages of security of tenure and freedom to make management decisions. Earnings from the operator's equity capital may be added to labor and management earnings for living expenses, reinvestment in the business, or other uses. Also, the value of the land may rise over a period of time. On

the other hand, ownership may involve substantial indebtedness. Also, risks of financial loss are greater than in renting.

Of course, ownership brings with it financial responsibility that is both greater and longer lasting than the financial responsibility which renting entails. Few persons buy more than one farm in a lifetime; moves are time-consuming and expensive.

Some part-time farmers, working at nonfarm jobs, use their off-farm income to move gradually into full-time farming. They use their initial savings to make a down payment on a small farm and to buy enough livestock and equipment to permit limited farming or ranching operations for the first few years. The off-farm income makes them better credit risks for lenders than if they were wholly dependent on farm earnings. They can continue to borrow to build up their farm business to a point where it will support their family and pay off previous loans. Such a gradual shift into full-time farming can usually be made with less sacrifice in family living standards and better chances of eventual success than an abrupt change to full-time farming.

Generally speaking, purchase of land is either by (a) land purchase contract, or (b) mortgage contract.

Purchase of land by use of land purchase contract has become much more important in recent years. These contracts allow the use of lower buyer down payments (usually from nothing to 29%) with the balance paid over a long period of years in annual payments. For the buyer, this offers a way in which to acquire land without having to make a big down payment. For the seller, it has certain capital gain tax advantages and it usually attracts more prospective buyers and makes for a higher sale price. In order to qualify for special treatment on capital gains for federal income tax purposes, the seller must not receive more than 30 percent of the purchase price in the year of sale.

Mortgage contracts differ from land purchase contracts in the following ways: (1) They are of longer duration—usually 20 to 30 years, or up to 40 years, whereas land purchase contracts are commonly for 10 years or less; (2) the law provides for specified grace periods after default in payments before the seller can foreclose the mortgage; and (3) larger down payments are normally required—frequently 40 or 50 percent of the purchase price.

What's a Farm or Ranch Worth?

The above question is best answered by still another question: How much will it make? The most logical answer to the latter question is that the farm or ranch should be expected to return to the prospective buyer as much on his investment as he could earn were his money invested in the best alternative enterprise of comparable risk. Thus, if an alternative enterprise of comparable risk will return 10% on investment, the farm or ranch should do likewise.

The above income-productivity approach does not take into consideration a possible tax shelter or increase in land values. However, neither of these should be counted upon. Although the prospective land buyer should have no objection to Uncle Sam's being lenient on his taxes, or to striking oil, or to having a fashionable summer resort go up on the adjacent property, none of these possibilities can be counted upon. The tax structure may change; and land values may not increase. Besides, the stockman living upon the land and depending upon his earnings from the operation to pay interest on borrowed money and buy groceries for his family cannot rely on "paper profits" (unrealized profits).

The income-productivity approach, which calls for calculating receipts and expenses, should be based on projected longtime productivity levels, prices, and costs, without either undue optimism or pessimism.

In plain simple terms, the income-productivity approach is based on the capacity of the farm or ranch to make money—the more money it will make, the more it is worth. Step by step, it is determined as follows:

1. Record expected receipts and expenses from the operation of the farm or ranch.
2. Determine the net returns to the land by subtracting all expenses from gross receipts, including a return to the operator for all his labor and management.
3. Divide the dollar returns to land by the rate of interest that you could get were your money invested in the best alternative enterprise of comparable risk to arrive at the "income-productivity value" of the farm or ranch. This value is the maximum that a buyer can pay for a farm or ranch based on its "productivity value." If the price is below this figure, it's a good buy. If it's above, watch out.

The above procedure is given in Table 2-1, based on an actual ranch.

Thus, based on income productivity, a buyer who wishes to allow himself $25,000 per year for labor and management, and who wishes to realize 10% on his investment, could consider the Bar-None Ranch—a 500-cow, 5,000-acre ranch—a good buy at up to $1,000,000, or $200 per acre. On the basis of a 500-cow carrying capacity, that's $2,000 per cow for the land and improvements.

Historically, the "income productivity" value of farms and ranches has been below current market prices. This means that either land is too high or animals are too cheap. Moreover, a prospective purchaser must recognize that he will likely have to pay the going market price for a farm or ranch. That is, he

TABLE 2-1
INCOME PRODUCTIVITY VALUE OF
BAR-NONE RANCH
A 500-COW OPERATION ON A 5,000-ACRE RANCH[1]

	Amount
Receipts:	
Cattle sales	$171,250
Expenses:	
Cash expenses:	
Feed purchases	12,115
Hired labor	5,000
Machinery	7,785
Property tax	4,025
Other	1,900
Noncash expenses:	
Depreciation	11,425
Interest on operating expenses	4,000
Operator labor and management	25,000
Total expenses	71,250
Net return to land (ranch):	
Income productivity value at 10% rate of interest ($100,000 ÷ .10)	$1,000,000

[1]This is an abbreviated form. A competent appraiser will detail this. For example, under "cattle sales," there should be a breakdown as to number, weight, and price of steer calves; number, weight, and price of heifer calves; number, weight, and price of yearlings, by sex; number, weight, and price of cows; and number, weight, and price of bulls. Also, hay sales, crop sales, and pasture rents would be included—if income from these sources is expected.

Similarly, under expenses there should be a more detailed breakdown of costs.

will have to pay a price close to that for which comparable land is selling. Except when buying from relatives, there are precious few really good buys. However, there are two exceptions: (1) when an existing stockman enlarges his present holdings by acquiring nearby land, thereby lowering the cost per animal unit expenses—primarily through more efficient use of labor, management, and equipment; and (2) when unimproved land can be developed to where its "income productivity" exceeds its cost plus development.

Fig. 2-3 shows farmland value 1960-1982, based on actual sales.

U.S. FARMLAND VALUE $ PER ACRE-MARCH 1

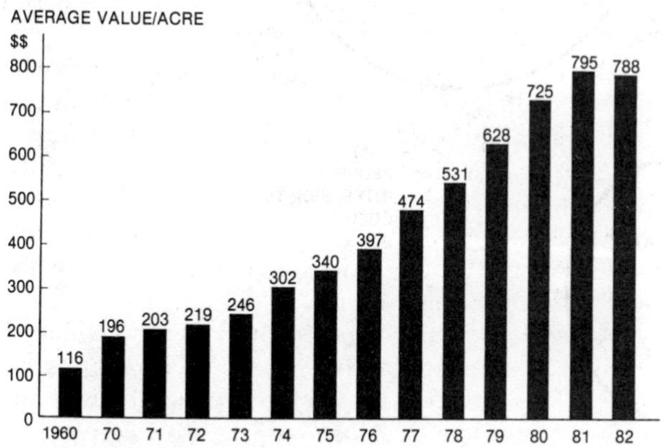

Fig. 2-3. U.S. farmland value in dollars per acre, 1960-1982. (Source: USDA, ERS, National Economic Division)

OTHER PLUS VALUES TO LAND AND LIVESTOCK OWNERSHIP

Some buy land to balance operations and cut costs; some to keep the children down on the farm; others because it adjoins their present property; and still others because of pride of ownership. There are also other important plus values to land and cattle ownership. Among them:

1. Appreciation in land values. From 1960 to 1982, U.S. farmland values rose from $116 per acre to $788 per acre, an increase of 6.8 times. All indications point to continued rises in land price in the years ahead.
2. As an effective hedge against inflation.
3. As a desirable alternative to a jittery stock market.
4. As a tax shelter.
5. A boom down on the farm in the years ahead brought about by (a) expanded export opportunities for U.S. farm products which have shifted the expectations of many farmland buyers and their lenders to a new plateau regarding the future well-being of U.S. agriculture, with the result that more people are interested in buying farmland; and (b) spreading roads and suburbs, and precious little new land that can be brought under production.

There is little doubt that farm real estate values will continue to climb. The only question is "how much?"

Financing the Farm or Ranch

The average U.S. farmer or rancher has more than $300,000 invested in land, machinery, livestock, working capital, and farm buildings other than his house. Many stockmen have much larger amounts invested. Investments of over $1,000,000 per farm or ranch are not uncommon. Hence, a big cattle spread necessitates both money and knowledge of financing.

• **Credit**—*Credit may be defined as "belief in the truth of a statement, or in the sincerity of a person."* In farming and ranching, or in any other business transaction, credit means confidence that men will take care of their future obligations. Credit is the lifeblood of the livestock business. Without it, few large operations would be possible; for not many people are able to provide all of the capital that they need.

Most commercial lenders have guides and standards that set upper limits on the amount they will lend. Usually, to get credit on a mortgage for buying a farm, the borrower is expected to make a down payment of 40 to 50 percent of the purchase price. Lenders usually will make loans on livestock and on new machinery for up to 80 percent of the purchase price.

Total farm investment in land, buildings, livestock, and equipment has increased more than six fold (6.4 times) in 25 years, rising from $170 billion in 1956 to $1090 billion on January 1, 1981. Farm debts have increased even more—they are 9.3 times larger than 25 years ago. The amount of debt owed by farmers and ranchers has risen from $18.8 billion in 1956 to $174.5 billion in 1981, and the trend shows no signs of letting up.

• **Sources of loans**—Hand in hand with getting the right kind of loan, it is important that the best available source of the loan be secured. Table 2-2 shows the primary sources of the three main kinds of loans. It is noteworthy that banks, merchants, and individuals provide 80 percent of the farm credit.

In seeking a loan, it usually pays to "shop around" in advance of actual need to see which source is best under the circumstances. Compare cost of credit, length of loan, loan fees, repayment privileges, and security required. Also, for long-term loans check the reputation of the lenders for staying with worthy borrowers in times of adversity.

In comparing costs of credit, look at the total dollar amounts, not just interest rates. Lenders figure charges in different ways. For example, if you buy feed or farm machinery on the time-purchase plan, you may find that you will pay more in interest than you would if you borrowed the money to pay for the purchase outright.

TABLE 2-2
PRINCIPAL SOURCES OF THREE MAIN KINDS OF FARM LOANS

Credit Source	Kind of Loan: Long Term	Intermediate Term	Short Term
Commercial banks	X	X	X
Dealers and merchants		X	X
Farm mortgage companies	X		
Farmers Home Administration	X	X	X
Federal Land Bank Associations	X	X	
Individual lenders	X	X	X
Insurance companies	X		
Production Credit Associations		X	X

To obtain loans or information about loans from commercial banks, dealers and merchants, farm mortgage companies, individuals, and insurance companies, apply directly to these sources or to their local representatives. Local banks and farm real estate dealers often serve as loan correspondents for life insurance or farm mortgage companies; they can tell you about loan requirements, terms, conditions, and interest rates, and arrange for loans.

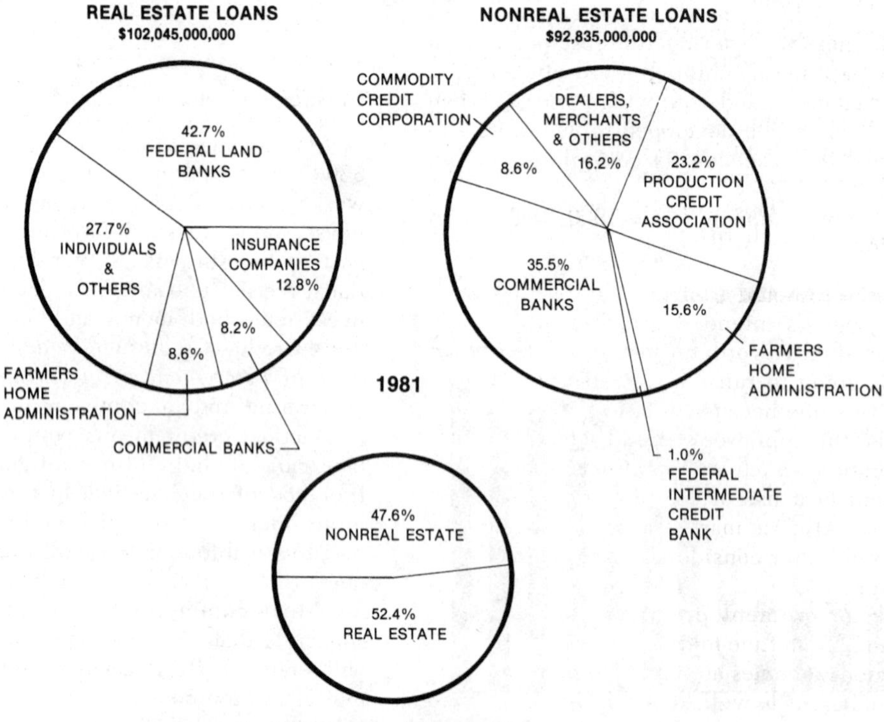

Fig. 2-4. Where farmers borrow (1981). (Source: Governor, Farm Credit Administration, Washington, D.C.)

Things to Do When Buying a Farm or Ranch

After the prospective purchaser has found the farm or ranch that he likes and has determined that the price is right, the following things should be done:

1. **Have it appraised**—Buying a farm or ranch is a big financial transaction. Thus, it is good business to have an accredited rural appraiser make a detailed appraisal of the property. The appraiser's fee will vary depending upon the size of the ranch to be inspected and the time required to document income and expense items, search out and view comparable sales, and prepare a confidential written appraisal report. Appraisal fees normally run from $2.00 per acre on 1,000-acre ranches down to $0.75 per acre on much larger ranches.

There are three basic approaches to appraisal of land, or estimating its value. These are (a) market value, or what similar land has sold for recently; (b) productive value, or net income the land will produce; and (c) present value of useful improvements.

The appraisal should show the fair market value of the property. This is frequently defined as "the price at which a willing seller would sell and a willing buyer would buy, neither being under abnormal pressure." This definition assumes that both buyer and seller are fully informed as to the property and as to the state of the market for that type of property, and that the property has been exposed in the open market for a reasonable time.

The appraisal should include maps of the farm, showing the physical features and the various soils. There should be a summary sheet listing all the buildings and their size and description. The appraiser should also allocate a value to buildings, fences, tiling, wells, pipelines, and other depreciable items for income tax purposes. This will give the buyer a reliable value from which to set up a depreciation schedule.

The appraiser should evaluate improvements in terms of the owner's intended use. For example, a $20,000 turkey shed is worth only scrap lumber to an owner who is going to run cows. Worse yet, it may have a negative value because of taxes.

Preferably, the appraiser should be familiar with livestock operations. Such an appraiser can point out the highest and best use of the property for the intended purpose. Also, he may be able to indicate factors which would alter considerably the plans of the potential buyer.

2. **Check government programs**—Government programs change from time to time, but usually one or more government agencies are involved in most farms and ranches. Thus, it is well to check into the situation.

The county Agricultural Stabilization and Conservation Service (ASCS) offices advise on and administer commodity programs, including allotments and marketing quotas for the basic commodities. These offices can also supply information about the soil, water, timber, and wildlife conservation practices that the Agricultural Conservation Program helps carry out on individual farms. ASCS offices also are charged with the local administration of price support commodity loans made available through the Commodity Credit Corporation, certain emergency programs in designated areas affected by drought or floods, the feed grain program, and other farm programs.

The Soil Conservation Service (SCS) has offices in nearly every county. It provides technical assistance and information on soil and water conservation, land-use alternatives, soil surveys, and resource use.

3. **Check courthouse records**—A courthouse check is standard procedure for most appraisers. However, the prospective owner should make sure that it is not overlooked.

The courthouse records should show what the property tax has been running on the farm or ranch in question, and if the property is subject to special levies for drainage or irrigation districts. The plat books should be checked to make certain of the boundaries of the property and how many acres are actually involved. What mortgages are on record against the farm? Check the recorder's office also for any special agreements in regard to property-line fences. The latter is especially important if there is a "water gap" where fences must frequently be rebuilt. Also, important water and mineral rights should be checked.

4. **Check mortgages**—The prospective purchaser should check the mortgage situation. Is there a mortgage on the property? If so, how much, at what interest rate; and can it be assumed without penalty? These questions are particularly important during times of scarce money and high interest rates.

Things to Avoid When Buying a Farm or Ranch

In the purchase of a farm or ranch, there are certain pitfalls which should be avoided; among them, the following:

1. **Avoid legal problems**—Regardless of whether a farm is purchased directly from the seller on a firsthand negotiated basis or through a real estate broker, the buyer should have an attorney check the details, thereby lessening the hazard of legal problems.

2. **Beware of the glamour states**—Much land in California, Florida, and Arizona is priced so high that it is difficult to show a profit from the operation of a stock farm or ranch. In these states, either a higher

and better use must be considered, or the land must be purchased on the basis of speculation—its potential for recreational development, housing, etc. Beautiful mountains, trees, streams, and sunsets all make for a heap of living and enjoyment, but they don't feed an animal. Thus, they should be secondary in the selection of a farm or ranch.

3. **Avoid city suburbs and high taxes**—Where rapid-growth cities are involved, it is generally wise to stay at least 50 miles away if one wishes to develop a stock farm or ranch. Of course, there are many small towns or cities that are not subject to rapid growth where it is possible to be closer in without the hazard of subdivisions or high taxes.

4. **Avoid overelaborate improvements**—Improvements are always expensive to maintain and they are subject to taxes. Hence, they should have utility value. No matter how attractive they may be, unless improvements contribute to the income of the farm or ranch, they have a negative value.

Purchase Contract

After a prospective buyer has found the particular property that he wants and has agreed upon a price with the owner, he should have an attorney draft an agreement covering the terms of the purchase. Then he should sign it and submit it as an offer to buy. It does not become a binding contract until the seller signs it, also. After both parties sign the contract, there is little bargaining power left; hence, the buyer should get all stipulations covered in the original contract.

The purchase contract should be relatively simple, amounting to a mere memorandum signed by buyer and seller, but it should clearly specify the following:

1. Amount and method of paying purchase price.
2. The amount of deposit or down payment, and the method of handling it. Is it to be applied to the total purchase price; when is it to be forfeited or returned; and is payment made to a responsible person?
3. Method of financing the purchase. Does purchaser assume and agree to pay existing mortgage? Will purchaser obtain a new mortgage loan?
4. Is seller to furnish an abstract of title brought up to date or a good and clear title that can be insured?
5. Date that possession can be taken.
6. A list of items that go with the property.
7. Who pays accrued and current taxes.
8. The legal rights of any tenant on the property.

It is important to remember that the buyer assumes risk of loss as soon as the contract is signed, even though the deed has not been delivered. If a barn burns and the seller has no insurance, the buyer could be forced to pay the full purchase price agreed upon, even though the barn has burned. So it is important that the buyer make certain that he has insurance on the buildings during the interim period.

It is customary to prorate annual taxes, with the buyer and seller assuming responsibility for the number of months that each has actual possession of the property.

The purchase contract should require the seller to deliver to the buyer an abstract of title for the property, certified to the date of sale. If the seller cannot produce a clear title or satisfactory title insurance by a certain date, the contract should call for a refund of the buyer's down payment.

Of course, the purchase contract should include the price being paid for the property and the date on which the seller guarantees possession. There should be a complete legal description of the holdings. If it is understood that certain portable buildings go with the farm, they should be itemized in the purchase contract.

The contract should detail how payment is to be made and the time of settlement. If the seller is to carry a mortgage on the farm or ranch for part of the purchase price, the buyer will need to work out the usual arrangement as to interest rates, prepayment privileges, etc.

If the buyer wishes to go on the place to do certain work in advance of taking actual possession—like repairing buildings, or reseeding a pasture—this should be spelled out in the contract.

Nothing should be left to oral agreements, because most such agreements are unenforceable. Also, the contract should be binding upon the heirs and assigns of the seller.

In summary, when completing the purchase of a farm or ranch, give attention to the following details:

1. Do you have satisfactory evidence that the seller has complete title to the property and can convey it to the purchaser by deed?
2. Examine the seller's deed. Does the wife release her dower; does seller warrant free and clear of all encumbrances; is the description of boundaries and acreage correct; are the easements or right-of-way for or against the farm; are mineral rights or water rights reserved; and has seller attached U.S. transfer stamps?
3. Examine mortgage and note before signing.
4. Immediately record deed with the Register of Deeds office in the county where the property is located.
5. Insure all uninsured buildings.
6. Make sure that expenses incurred in the last 60 days have been paid for materials or work done on buildings, for land clearing or leveling, and for wells or pipelines.

Other Considerations When Buying a Farm or Ranch

Many factors in addition to those already mentioned should be considered in the selection of a farm or ranch; among them, the following, each of which is discussed in a section which follows: (1) deeded vs government leased land; (2) water and water rights; (3) mineral rights; (4) oil and gas leases; (5) timber; (6) easements and property lines; and (7) risks. Although space limitations will not permit a discussion of them, the following factors should also be considered: (1) area and climate; (2) soil and topography; (3) improvements; (4) wind direction, windbreaks, and natural shelters; (5) service facilities, community and markets; (6) expansion possibilities; and (7) nearness to factories or cities, with the possible advantages of (a) off-farm employment, and (b) big-city attractions; and the likely disadvantages of (c) air and noise pollution, and (d) limited expansion possibilities.

DEEDED VS GOVERNMENT LEASED LAND

It should be noted that ranches made up principally of deeded land sell for substantially more than those made up mainly of government leased land. A rule of thumb is that, on a per animal unit basis, all deeded land is worth twice as much as comparable land of which only one-half is deeded. Of course, the type of government lease, as well as the way the leased area lies in relationship to the deeded land, can make a tremendous difference in value. Yet, no matter what the history of the lease or what the old-timers say, a government lease can be cancelled quickly and without compensation.

WATER; WATER RIGHTS

Only seeing people without water is more disturbing than seeing animals dying of thirst. Both people and animals can survive longer without food than without water. Rainfall, wells, rivers, snow, streams, creeks, springs, lakes, ponds, or any other source of water should be considered. Because of the importance of water, the official weather rainfall records, by months and over a period of years, should be obtained.

In most states, domestic users have the first right to water. Agricultural uses usually rank second. In some instances, where industry is highly centralized and promotes the welfare of the public, manufacturing use is given priority to water over agriculture.

Plenty of good drinking water for human use is a high-priority item. Generally, this involves well water. Wells should be checked to make sure that they meet county health department standards, and that there is adequate water supply, for drilling is costly. The only accurate test of a well is to have it pumped dry and see how fast it fills up, or to see how much water can be pumped out of it and how fast, without substantially changing the water level. Of course, there is variation from season to season, and from year to year. In most western states, the general underlying water table is going down every year; hence, it is impossible to predict what it might be 10 years from date of purchase. Artesian wells are the most risky of all; nobody can predict when they're going to stop flowing.

From the standpoint of the water supply for animals, consideration should be given to the distribution of rainfall, to snowfall in the northern areas and to river frontage wherever streams are found. Where running water is not available, man-made lakes, ponds, developed springs, and wells may be relied upon, but all these cost money. Hence, they should be considered at the time of purchase.

Where water is to be used for irrigation, both availability and cost must be considered. A stockman with free riparian water rights has considerable advantage over his neighbor who must pay an average of $20 per acre per season for the same amount of water.

Water rights have been of prime importance to the development of civilization. Nearly every society had its own system of regulating water. According to the Bedouin "code of the desert," a traveler might drink of a well, but "should the well bear the *wasm* (camel brand) of a local tribe, and should the traveler, without permission, water his flocks and camels not bearing this brand, then should he be slain, and his body left to be devoured by the birds of the air and the beasts of the field."

Water rights have always been essential in arid countries where there is limited water supply. In western and southwestern United States, where the use of water is essential to the productivity of the land and to all living things on the land, the use of waterways is usually written into the deed of the farm or ranch. This water right becomes part and parcel of the land, meaning that it cannot be separated from the land. This is because much of the value of the property depends on accessibility to water. The rights are vested, and thus are considered private property.

Basically, there are two types of recognized doctrines, or water rights—the riparian right, based on English law, and the appropriative right.

- **Riparian right**—A riparian right is the right of an owner who owns land adjacent to a body of water. Under riparian law, the property owner who has land lying next to a stream, or having a stream running through it, is entitled to water that is required for his domestic consumption and for his livestock. The English Common Law, from which the riparian doctrine stems, further states that the owner is entitled to have this stream flow undiminished in quantity or quality.

The Americanized version of the riparian right has a reasonable use clause which allows the riparian owner to make beneficial use of water so long as the quantity and quality of stream flow are not materially reduced. In other words, a downstream riparian landowner enjoys the same rights as an upstream owner. If, through unreasonable use, an upstream owner infringes upon the rights of a downstream owner and deprives him of water, the latter could sue and possibly collect damages and halt the excessive use.

Groundwater may be a different story. Straight riparian rights apply to well-defined underground streams. But percolating groundwater (water moving downward through the soil) is defined as real property in some states. As such, it is owned by the overlying landowner. Use of such water, even when it deprives a neighbor, cannot be contested.

There are many state variations and interpretations of riparian water rights. Beyond the right to use water for domestic consumption, riparian rules are usually vague. Ordinarily, this does not create a problem in the normally high-rainfall areas of eastern United States. Yet, the irrigator does not really have adequate protection. Because of the irrigation boom, most riparian states have a special legislative committee studying water rights. They have established, or they will establish, priority among users.

In new irrigation areas, a landowner should consult an attorney on how to protect his rights and investment. Records of the date irrigation started, the amount of water used, the acreage irrigated, and the return from the crops grown are information that can be valuable later if beneficial use must be proved.

When a person diverts water that other landowners have legal right to, without being stopped, he gains the rights to continued use. This is known as prescriptive rights. To gain prescriptive rights, the diversion must have been for a period of time designated by state statute, and the water must have been used openly.

• **Appropriative rights**—An appropriative right is the right for a certain amount of water at a given place. It is not necessary for the water source to lie next to the land where the water is to be used. Rather, it is transported by such means as an irrigation ditch.

The appropriative rights for regulating water use are found mainly in the arid western states, although certain other states, including Minnesota and Mississippi, have adopted this type of law. Where states recognize both riparian and appropriative rights, the appropriative rights are dominant. Under the appropriative right, the individual may acquire the right to use water for beneficial purposes on a given tract of land by fulfilling certain requirements of written law. The appropriative right fully recognizes the public ownership of water. A landowner must apply to the controlling state agency to obtain the right to use water; and the right can be lost through nonuse after a stated period of time. In the appropriation states, rules vary on groundwater.

In appropriative rights, the right to use water depends upon prior claims made against the water sources. Anyone who filed to use water before you is entitled to his water needs before yours can be filled.

A vested right protects the rights of persons putting water to beneficial use before the appropriative law was passed. Usually, vested rights also apply to water applied beneficially within three years after passage of the law. Although a vested right has three years' priority, it can be lost through nonuse.

• **Where to go for water right help**—It is very important that both prospective and present landowners obtain authoritative information relative to water rights.

In some riparian states, there are agencies that are studying water legislation and are in charge of regulating present water laws.

In states that use the appropriative doctrine, there is an authoritative agency, usually a chief engineer, to whom applications must be placed to obtain a right to use water beneficially. This official can give information concerning what must be done to abide by the statutes of his state.

You should request a copy of the water laws from the state agency, your state senator or representative, or your state agricultural college. Where water rights may be an individual legal question, a competent attorney should be consulted.

MINERAL RIGHTS

Mineral rights are usually involved in the buying and selling of land, especially in oil- and gas-producing areas. Buyers of land should always have the title checked to see if the mineral rights have been severed. Generally, they are broken down into two broad classifications—surface and subsurface.

In the United States, the following two major theories prevail concerning the actual ownership of any subsurface wealth:

1. The ownership-in-place theory generally recognizes that any mineral deposit is actually a part of the land and is owned by the individual holding title to the land.
2. The nonownership theory in which the landowner does not have outright title to underlying mineral deposits but has the right to explore and retain any deposit developed.

Regardless of which theory is recognized in a particular state, the landowner has the following mineral rights:

1. Rights may be transferred to others.

2. Mineral deposits are part of the land.

3. Landowner has right to withdraw minerals, and, in the case of gas and oil, be free of liability for drainage (the pumping of oil or gas from under adjoining, nonowned property).

• **Separating mineral rights**—Surface and subsurface rights can be separated. In "ownership-in-place" states, you dispose of the minerals, while in "nonownership-in-place" states, it is simply the right to explore and retain any minerals recovered.

It should be noted that the term "minerals" refers to gas and oil, unless specifically stated otherwise. Moreover, any conveyance of a named mineral does not include other minerals unless so stated.

Surface and subsurface rights may be severed in any one of the following six basic ways:

1. By deed conveying all or part interest in the minerals.
2. By deed conveying land but retaining mineral rights.
3. By land contract excepting the minerals.
4. By mineral lease in an ownership-in-place jurisdiction (conveys present undivided seven-eighths interest to the lessee on any mineral developed, except in Kansas).
5. Court judgments setting aside mineral rights or part interest in a lawsuit.
6. Formation of a mining partnership.

Separating mineral interests can create problems, especially if interest is divided among many parties. For example, if the property has been through three or four different hands, possibly one-eighth of the minerals might be left for the new purchaser. Not only is such a small fraction of the mineral interest unattractive to the buyer, but widespread breakdown of the interest runs up cost of bringing land abstracts up to date and often discourages leasing and well development. Separation of mineral interest reduces loan values and usually increases difficulty in obtaining credit.

As a general rule, it is not considered good business for a surface rights owner to dispose of over 50 percent of the subsurface rights.

OIL AND GAS LEASES

Oil and gas leases have become fairly standard. Nevertheless, one should check any proposed lease with an attorney before signing.

A general division on any developed oil and gas is ⅛ for the lessor (the mineral owner) and ⅞ for the lessee (the persons taking a lease on the land). Most leases are perpetual as long as certain qualifications are met. Leases are generally subject to termination by the lessee anytime within the base period by failure to begin a well or pay the delay rental by a stated date.

In some areas, mineral rights are more valuable than the surface rights.

TIMBER

Trees are pretty, but they may or may not have monetary value on a farm or ranch. Evaluation of timber is a job for an expert. The value of timber delivered to market may have little relationship to its quality as it stands. In some cases, the cost of cutting and transporting trees is exorbitant.

EASEMENTS; PROPERTY LINES

An easement is the right to go on and use the land of another in a particular manner. Two common types of easements are: (1) the grant of one landowner to another of the right to build or use a roadway across the land to provide access to another tract; and (2) where a power company purchases an easement to string an electric line across land. Before purchasing a farm or ranch, all easements should be checked because an easement (1) limits the landowner's use of his property, and (2) is valid against the purchaser. The author recalls one near-sale of a ranch that was being bought primarily because of a beautiful site on which to build the new owner's dream house. Just as the deal was about to be closed, it was discovered that the county had a permanent easement for a 25-yard strip for a road right through the intended house location.

Three types of legal descriptions of farms and ranches are commonly used: (1) the rectangular survey, based on meridians and parallels; (2) metes and bounds (metes are measures of length—feet, inches, or perches—a perch equals 16½ feet; bounds are artificial boundaries such as roads, streams, adjoining farms); or (3) monuments (iron pin, blazed tree, lake, stream). Disagreements over property lines have led to feuds and lawsuits. Accordingly, before purchasing land, the property lines should be determined. Usually, it is wise to engage the services of a professional surveyor. The author knows of one case where failure to do this resulted in the stockman building a new ranch home only to discover, some years later, that the house was on an adjacent property—not his own.

RISKS

The ownership of a farm or ranch does involve some risks, which are not pleasantly recalled by most current owners, and of which all prospective owners should be aware. Among them are the following:

1. **Drought**—A long dry period will shorten the

grazing season; cut down on crops to be harvested for winter feed; result in loss in weight of cows; make for a smaller calf crop percentage; lower weaning weights; necessitate the purchase of feed; and/or even cause liquidation of all or part of the herd at unfavorable prices.

2. **Floods, storms, blizzards, fires**—Historically, some cattle are lost as a result of sudden rainstorms and blizzards. Also, some are lost by fire.

3. **Disease outbreaks**—Traditionally, a stockman vaccinates against the diseases most common to his area, provided such a preventive exists. Then, he takes a calculated risk relative to a long list of other diseases. But costly disease outbreaks do occur, resulting in death losses and inefficiency among the living.

CHOOSING THE CLASS OF ANIMALS

Under some conditions a hog farmer may have one neighbor who is a cattle feeder, a second who operates a dairy, a third who keeps a sizable farm flock of sheep, a fourth who produces light horses for recreation and sport, and a fifth whose chief source of income is from poultry. All may be successful and satisfied with their respective livestock enterprises. This indicates, therefore, that several types of livestock farming may be nearly equally well adapted to an area or region. This means that the selection of the dominant type of livestock enterprise should be analyzed from the standpoint of the individual farm or ranch.

Usually a combination of several factors suggests the livestock enterprise or enterprises best adapted to a particular farm or ranch. One of these factors is the labor requirement. Available feeds and market outlets for animals and animal products are also very important. Table 2-3 may be of assistance in arriving at a decision as to the kind or kinds of livestock best suited to the individual farm or ranch.

Fig. 2-5. Proper choice of the class or classes of animals makes for more profits.

CAPITAL NEEDS

In 1981, farm assets—investments in land, improvements, machinery, equipment, animals, feed, and supplies—totaled $1,090.3 billion, while farm debt totaled $174.5 billion. Thus, in the aggregate, farmers had 84 percent equity in their business and 16 percent borrowed money (debts). The balance sheet of U.S. farming from 1945 through 1980 is shown in Fig. 2-6.

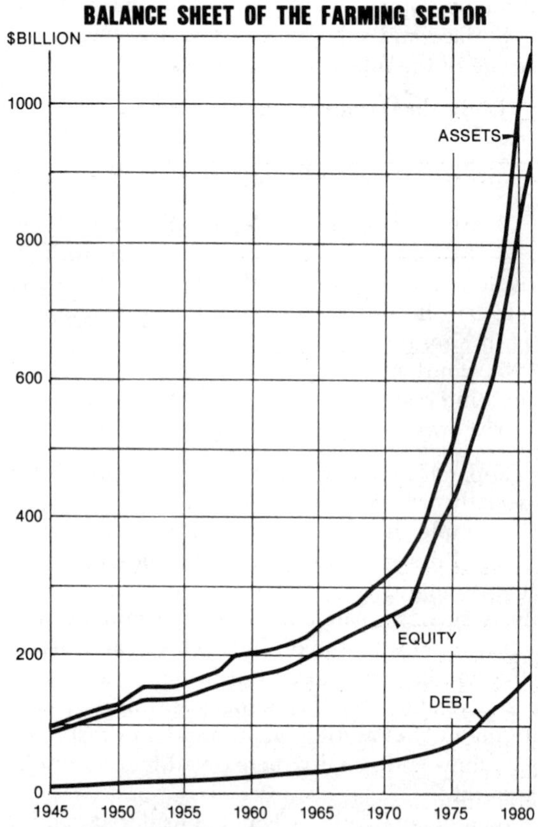

Fig. 2-6. Balance sheet of U.S. farming, showing (1) assets, (2) debts (liabilities), and (3) equities. (Courtesy, Farm Credit Administration)

Perhaps agriculturalists have been too conservative, for it is estimated that ¼ to ⅓ of American farmers could profit from the use of more credit in their operations.

Another statistic which points up the enormity of capital needs is that it takes about $33 in farm assets to produce $1 of net farm income.

Cow-Calf Capital Needs

Those thinking of becoming cow-calf operators inevitably ask, "How much money will it take, and what can I make?" Unfortunately, there is a paucity of information on this subject on which to base an an-

TABLE 2-3

FACTORS TO CONSIDER IN CHOOSING THE CLASS OF ANIMALS

Animal or Product	Man-hours/Cwt Production[1] 1935-1939	1976-1980	No. Head Cared for in 1982 by One Man in the Most Efficient Operations[2]		Characteristics, Requisites, and/or Conditions Under Which They Fit
	(hours)	(hours)			
Beef cattle	4.2	1.3	Cow-calf (brood cows)	300-500	Plenty of roughages (pasture and hay); moderate capital available on intermediate and long-term basis.
			Feedlot cattle	2,500	Plenty of grain or cheap by-product feeds; strong finances; adequate know-how in buying, feeding, and marketing cattle.
Milk cows	3.4	0.4	Milk cows	150	Plenty of labor; intensive farming; suitable outlet for milk.
Sheep	6.1	2.0[2]	Farm flock ewes Range ewes	1,000 1,000-2,000	Plenty of roughages (pasture and hay); moderate capital.
			Feedlot lambs	7,500-10,000	Plenty of grain or cheap by-product feeds; strong finances; adequate know-how of buying, feeding, and marketing.
Hogs	3.2	0.5	Sows	200	Feed grains abundant; good sanitation; where capital is limited.
Layers (eggs)	1.7/100 eggs	0.2/100 eggs	Laying hens (cage) Laying hens (floor)	40,000[3] 15,000[3]	Suitable market outlet; considerable short-term and intermediate credit; good sanitation and disease prevention; adequate size, automation, and know-how.
Broilers	8.5	0.1		75,000	Suitable market outlet; considerable short-term and intermediate credit; good sanitation and disease prevention; adequate size, automation, and know-how.
Turkeys	23.7	0.4		40,000	Suitable market outlet; considerable short-term and intermediate credit; good sanitation and disease prevention; adequate size, automation, and know-how.

[1]Source: Data on "man-hours/cwt production," except sheep, from *Agricultural Statistics*, 1981, USDA, p. 441, Table 637.
[2]Estimates by Dr. M. E. Ensminger.
[3]Does not include time devoted to egg processing.

swer, despite the fact that it is sorely needed by both investors and lenders. To fill this gap, the author made a nationwide survey, the results of which are reported in Table 2-4. This reveals that, on the average, (1) it requires an investment of $1,000 to $1,700 per mature cow for land, improvements, machinery and equipment; (2) physical plant and cattle are the main costs; (3) the larger the herd, the lower the per animal unit investment and the feed and labor costs; and (4) purebred operations are much more costly than commercial. (See page 48 for Table 2-4.)

WHAT RETURN?

All cow-calf enterprises are owned by people, and most people want to be paid for their investment and work. This is as it should be. The only question is—how much return?

The following figures, taken from Table 2-4, for a commercial herd within the range of 300 to 499 cows, show (1) the cost (exclusive of depreciation) to produce one calf to weaning age, and (2) the possible net income.

Note well: The example that follows and Table 2-4 are presented for the purpose of illustrating principles, and not as indicators of actual costs and returns. Each cow-calf operator should insert his own current figures.

1. Cash expense—	
Feed	$ 40.00
Labor	12.00
Other	4.00
Total	$ 56.00
2. Investment in—	
Land	850.00
Improvements	100.00
Machinery and equipment	125.00
Cattle	320.00
Total	$1,395.00
@8% interest	.08
Yearly interest	$ 111.60
3. Total cost—	
Cash expense for feed, labor, and other	$ 56.00
Interest	111.60
Total cost	$ 167.60

4. Gross income, basis 450-pound calf, at 50¢—450 × .50 = $225.00

5. Net income—
$225 − 167.60 = $57.40

TABLE 2-4

AVERAGE INVESTMENT IN U.S. COW-CALF ENTERPRISES[1]

Investment—	In a *Commercial* cow-calf operation with the following number of cows of breeding age				In a *Purebred* cow-calf operation with the following number of cows of breeding age			
	Under 99	100-299	300-499	500 or more	Under 99	100-299	300-499	500 or more
	($)	($)	($)	($)	($)	($)	($)	($)
1. Investment/animal unit (one mature cow):								
a. In land and improvements (real estate)	1,100	1,025	950	900	1,400	1,325	1,250	1,200
b. In machinery and equipment	200	175	125	100	300	275	225	200
Total	1,300	1,200	1,075	1,000	1,700	1,600	1,475	1,400
2. Investment to produce one beef calf to weaning age (with provision made for barren cows):								
a. Land (investment/cow unit × going rate of interest)[2]	68.00	68.00	68.00	88.00	88.00	88.00	88.00	88.00
b. Improvements (buildings, etc.; depreciation plus interest on a per calf basis)[3]	30.00	21.00	12.00	6.00	36.00	27.00	18.00	12.00
c. Machinery and equipment (depreciation plus interest on a per calf basis)[4]	36.00	31.50	22.50	18.00	54.00	49.50	40.50	36.00
d. Cattle (depreciation plus interest on a per calf basis)[5]	49.90	49.90	49.90	49.90	72.15	72.15	72.15	72.15
e. Feed (including feed for dam)	50.00	45.00	40.00	35.00	75.00	70.00	65.00	60.00
f. Labor (including labor for dam)	18.00	15.00	12.00	10.00	36.00	30.00	24.00	20.00
g. Other	6.00	5.00	4.00	3.00	12.00	10.00	8.00	6.00
Total	257.90	235.40	208.40	209.90	373.15	346.65	315.65	294.15

[1]Survey made by Dr. M. E. Ensminger.
[2]Land for commercial cattle at $850/animal unit; land for purebred cattle at $1,100/animal unit; interest, 8%.
[3]Buildings depreciated 25 years; interest, 8%.
[4]Machinery and equipment depreciated 10 years; interest, 8%.
[5]Commercial cows, $320; purebred cows, $420; cattle depreciated 7 years; salvage value, $150/head; interest, 8%.

Thus, based on the above figures, a commercial cattleman would need a 300-cow herd to make $17,000 per year (300 × $57.40 = $17,220). Of course, efficient operators will do much better. High-quality feeder calves will bring $1.00/cwt more than the average kind—that's $4.50 per head, or $1,215 per year on a 300-cow herd with a 90% calf crop. A 5% increase in the calf crop means 15 more calves, or $3,375 in a 300-cow herd, with calves at 50¢ per pound. Fifty pounds more weaning weight per calf can mean $6,750 more in a 300-cow herd, with a 90% calf crop and 50¢ calves.

Of course, some land is overpriced, particularly in the more populous areas. Yet, there are compensating factors in many of the latter regions—their greater appreciation as potential building sites and recreational areas.

Goodsell and Belfield reported returns from cattle ranch capital for the period 1960 to 1972 (see Table 2-5).

TABLE 2-5

RETURNS TO CATTLE RANCH CAPITAL, 1960-72[1]

Year	Cattle Ranches		
	Southwest	Northern Plains	Rocky Mountain
	(%)[2]	(%)[2]	(%)[2]
1960	NA[3]	2.8	3.1
1961	NA[3]	2.3	4.1
1962	NA[3]	4.5	5.7
1963	NA[3]	3.7	5.0
1964	NA[3]	2.0	2.5
1965	1.6	2.1	3.5
1966	1.9	3.3	4.8
1967	1.2	3.2	4.9
1968	1.4	3.5	5.5
1969	1.6	4.1	7.0
1970	1.1	4.7	6.5
1971	1.0	5.4	7.4
1972	3.6	7.6	10.8

[1]Goodsell, W. D., and M. J. Belfield, *Cost and Returns, Northwest Cattle Ranches*, ERS-525, USDA, 1972, p. 9, Table 8.
[2]Net ranch income less a nominal charge (annual wage to year-round hands × 1.25) for operator's labor and management divided by total ranch investment.
[3]NA = Not available.

As shown in Table 2-5, during the years for which data are available on southwest cattle ranches, returns were much lower than on cattle ranches in the Northern Plains and Rocky Mountains. There are 2 reasons for this: (1) Investment runs high on southwest ranches; and (2) the Southwest was plagued with droughts during much of the time reported. For the 3 areas, the returns on ranch capital averaged 3.92 percent for the period 1960 to 1972, which is not enough. Hence, it prompts the following 3 questions, to which the author gives his answers:

1. Q. *Would an investor do better in the stock market?*

A. Not likely. The 1960-69 composite return on common stocks reported by Standard and Poor was 3.19%. For 1970-72, it was 3.27%. As noted above, the average return on cattle ranches was 3.92% for the 13-year period, which was slightly higher than the return from common stocks.

2. Q. *Would an investor make more money in some agricultural enterprise other than cattle?*

A. Not likely. In the 10-year period 1963 to 1972, farmers as a whole averaged only a 3.9 percent return on the equity of their capital investment in farming—the same as cattlemen averaged during this period.

3. Q. *Would an investor put money in a cattle ranch if he knew in advance that he would get an average return of only 3.92 percent?*

A. What's better or more sure than an investment in the good earth? Marshall Field put it this way: "Buying real estate is not only the best way, the quickest way, and the safest way, but the only way to become wealthy." Andrew Carnegie said: "Ninety percent of all millionaires became so through owning real estate. More money has been made in real estate than in all industrial investments. The wise young man or wage earner of today invests his money in real estate."

Cattle Feedlot Capital Needs

Cattle feeders need to know the size investment that they need, and can justify, in cattle feeding facilities. A nomograph may be used for this purpose. It can give a quick, preliminary idea of cost, gross profit, returns, and investment relationships. But a nomograph should not replace more detailed figuring which should precede all major investment decisions. Also, one should realize that a nomograph will give erroneous and misleading information unless based upon accurate and realistic cost and return data from the problem at hand.

Fig. 2-7 is a nomograph for use in making quick calculations of justifiable investments in beef feeding systems. The section that follows gives an example and step by step instructions for using the nomograph. By working through an example, it will soon be discovered how quickly the nomograph can help in evaluating capital investments for cattle feeding enterprises. (See page 50 for Fig. 2-7.)

INSTRUCTIONS FOR USING NOMOGRAPH

This nomograph provides a quick method of computing the investment in cattle feeding facilities one can justify on the basis of three factors: (1) the percent of the total investment to "charge off" as annual costs each year; (2) the gross profit (G.P.) expected per head; and (3) the return per head (R/C & L) desired for labor, management, and interest on the capital required for cattle, feed, and miscellaneous equipment (*excluding* the investment in the feedlot facilities under consideration).

• **Determining "percent annual costs"**—Annual fixed costs include such items as interest, insurance, and taxes (usually from 4 to 6%) plus an allowance for depreciation or annual principal repayments required, if making a cash-flow analysis (usually from 5 to 15%). These 2 percents should be combined to get the proper "total annual costs, percent of investment" for this analysis.

Data Needed for Computations	Example	Your Farm
1. Percent (%) annual costs	15%	________
2. Gross profit (return over variable costs) per head	$20.00	________
3. Return to labor, management, and capital (R/C & L)	$14.00	________
4. Investment per head justified	$40.00	________

• **Procedure for determining the investment justified in cattle feeding facilities—**

1. Locate the *vertical line* representing the "total annual costs, % of investment" (15% in the example).

2. Follow this line straight down to the point where it intersects with the diagonal line representing the expected gross profit (returns over variable costs) per head ($20 in the example)—*Mark this point.*

3. Locate the point on the left-hand vertical scale which represents the desired return ($/head) for labor, management, and capital invested in cattle, feed, and miscellaneous equipment ($14/head in example)—*Mark this point.*

4. Draw a straight line to connect the two points located above. The point of intersection with the right-hand vertical scale indicates the justified dollar-per-head investment in facilities ($40/head in example).

• **Procedure for determining the probable return ($/head) with a known, or contemplated, investment per head in facilities**—Follow the above procedure through point 2. Then, locate the dollar investment per head on the right-hand vertical scale ($40/head in example). Connect a straight line from this point through the one located in point 2. The point of intersection with the left-hand vertical scale gives the probable $ return/head ($14 in example).

Other variations in the use of the nomograph are obvious. For example, if the investment per head in facilities is $50, a return of $15/head is desired, and the gross profit per head is $20—10 percent annual costs would be indicated.

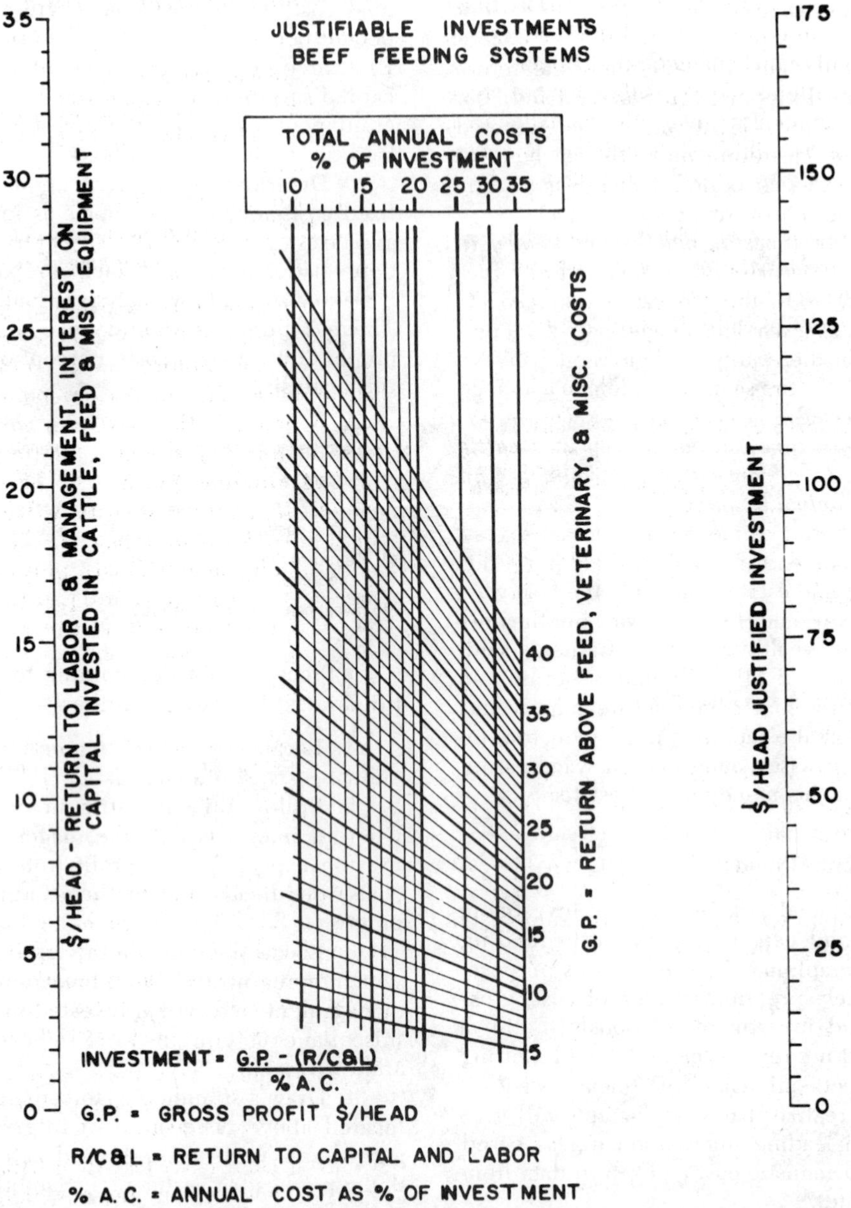

Fig. 2-7. Nomograph for computing "justifiable investments in beef feeding systems." (Nomograph, and instructions for using it, prepared by Robert M. George, University of Missouri)

COST OF FEEDLOT

Before constructing a feedlot, cost must be considered for two reasons: (1) capital must be secured; and (2) cost must be amortized. The usual basis of computing cost is on a "per animal unit capacity." This will run about the same whether calves or yearlings are involved, because per unit capacity must consider carrying the animals to market time. Table 2-15, Cattle Finishing Profit Indicators, gives the estimated initial (new) land, feedlot, and equipment cost on the basis of per animal unit capacity.

The area affects cost from the standpoint of shelter requirements and land values. Thus, because of the necessity for winter protection and shelters, feedlot costs are higher in the northern tier of states than in the South. Land values are higher in California than most areas of the United States, with the result that land costs become a factor.

Size of feedlot affects per animal cost. Most studies reveal that investment savings do accrue to the larger feedlots. Thus, the cost per animal usually decreases up to about 10,000-head capacity, then it increases slightly with larger lots. The slightly higher cost per head capacity of the larger lots appears to be due to duplication in equipment and the tendency to become more highly mechanized and elaborate.

An open lot without shelter is the cheapest type of feedlot construction. In the Southern Plains area, where the weather is mild and shelters are unnecessary, investment costs range from \$30 to \$50 per head of capacity.[2]

Housing increases costs, and the more elaborate the housing the greater the cost. University of Minnesota studies showed the costs given in Table 2-6 relative to three types of cattle feedlot facilities.[3]

TABLE 2-6

COSTS OF THREE TYPES OF CATTLE FEEDLOT FACILITIES

Type of Facility	Sq Ft per Animal	Cost per Animal Capacity
		(\$)
Open shed	20	70
Cold confinement	17	115
Warm confinement (heated)	17	170

Guidelines Relative to Facility and Equipment Costs

Overinvestment in facilities is a mistake. Some stockmen are prone to invest more in facilities and equipment than reasonably can be expected to make a satisfactory return; others invest too much in feed mills. Sometimes small cattle feeders fail to recognize that it may cost half as much to mechanize to feed 500 head as it costs to mechanize to feed 2,000 head.

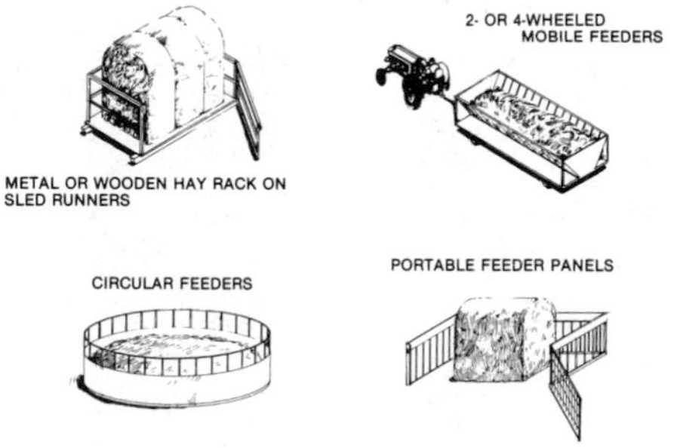

Fig. 2-8. Feeding has been automated.

In order to lessen overinvestment by the uninformed, guidelines are useful. Here are some:

1. **Guideline No. 1**—*The break-even point on how much you can afford to invest in equipment to replace hired labor can be arrived at by the following formula:*

$$\frac{\text{Annual saving in hired labor from new equipment}}{\text{(divide by) .15}} = \text{amount you can afford to invest.}$$

Example:

If hired labor costs \$9,600 per year, this becomes—

$$\frac{\$9{,}600}{.15} = \$64{,}000\text{, the break-even point on new equipment}$$

Since labor costs are going up faster than machinery and equipment costs, it may be good business to exceed this limitation under some circumstances. Nevertheless, the break-even point, \$64,000 in this case, is probably the maximum expenditure that can be economically justified at the time.

2. **Guideline No. 2**—*The break-even point on new facility-equipment costs is five times the annual salary of each person replaced.*

Assuming an annual cost plus operation of power machinery and equipment equal to 20% of new cost, the break-even point to justify replacement of one hired man is as follows:

Example:

If annual cost of one hired man is[4]	The break-even point on new investment is
\$7,000 (20%) × 5	\$35,000
8,000 (20%) × 5	40,000
9,000 (20%) × 5	45,000

Assume that the new cost of added equipment comes to \$15,000, that the annual cost is 20% of this amount, and that the new equipment would save one hour of labor per day for 6 months of the year. Here's how to figure the value of labor to justify an expenditure of \$15,000 for this item:

\$15,000 (new cost) × 20% = \$3,000

3,000 (annual ownership use cost) ÷ 180 hours (labor saved) = \$16.67 per hour.

So, if labor costs less than \$16.67 per hour, you probably shouldn't buy the new item.

3. **Guideline No. 3**—*Do not spend over 4.5% of the annual product sold for annual cattle feedlot and equipment costs.*

Example No. 1:

1,000 steers sold @ \$675 each, for a total of \$675,000
\$675,000 × 4.5% (av.) = \$30,375
\$30,375 ÷ 1,000 = \$30.37/head/year, the maximum that may be spent on feedlot and equipment.

Example No. 2:

2,000 steers sold @ \$675 each, for a total of \$1,350,000
\$1,350,000 × 4.5% (av.) = \$60,750
\$60,750 ÷ 2,000 = \$30.37/head/year.

[2]Gill, D., and M. D. Paine, Oklahoma State University, "Feedlot Design and Construction," *Feedlot Management,* Nov. 1972, p. 84.

[3]*Feedlot Management,* Nov. 1972, p. 44.

[4]This is assuming that the productivity of men at different salaries is the same, which may or may not be the case.

CREDIT IN THE LIVESTOCK BUSINESS

Total farm assets are estimated at $1090.3 billion, while farm debt is about $174.5 billion. This means that, in the aggregate, farmers have nearly an 84% equity in their business, and 16% borrowed capital. Perhaps they have been too conservative, for it is estimated that ¼ to ⅓ of American farmers could profit from the use of more credit in their operations.

Credit is an integral part of today's livestock business. Wise use of it can be profitable, but unwise use of it can be disastrous. Accordingly, stockmen should know more about it. They need to know something about the lending agencies available to them, the types of credit, and how to go about obtaining a loan.

Types of Credit or Loans

Getting the needed credit through the right kind of loan is an important part of sound financial farm management. The following three general types of agricultural credit are available, based on length of life and type of collateral needed:

- **Short-term loans**—This type of loan is made for operating expenses and is usually for one year or less. It is used for the purchase of feeders or birds, feed, seed, fertilizer, gasoline, and family living expenses. Security, such as a chattel mortgage on the feeders, birds, or crop, may be required by the lender; and the loan is repaid when the animals or crop are sold.
- **Intermediate-term loans**—These loans are used to buy equipment and breeding stock, for making land improvements, and for remodeling existing buildings. They are paid back in one to seven years. Generally, they are secured by a chattel mortgage on livestock and machinery.
- **Long-term loans**—These loans are secured by mortgage on real estate and are used to buy land or make major improvements to farmland and buildings or to finance construction of new buildings. They may be for as long as 40 years. Usually they are paid off in regular annual or semiannual payments. The best sources for long-term loans are: an insurance company, the Federal Land Bank, the Farm Home Administration, or an individual.

Credit Sources

Table 2-7 shows where farmers borrow, the amount of loans from each source, and the percent of the total held by each type of lender. Also, it shows borrowing trends (see last column).

But, agricultural financing is changing, and it will continue to change even more in the years ahead. Today, farmers are tapping the vast supply of farm equity or risk capital that is constantly seeking investment opportunities—nonfarm equity capital is being used in agriculture.

Some time or other most farmers and ranchers find it necessary to borrow money to buy land; to construct buildings and other improvements; to purchase equipment, seed, and livestock; and/or to pay for seasonal labor. They should know something, therefore, about the lending organizations available to them in order that they may determine which one will best serve their needs. The leading sources of farm credit are summarized in Table 2-8.

TABLE 2-7

WHERE FARMERS BORROW (1982)[1]

Type and Source of Loan	Amount of Loan	Percent of Total	Percent of Change from '81
	(million $)	*(%)*	*(%)*
Real estate mortgage loans:			
Federal Land Banks	43, 564	42.7	+ 21.2
Individuals and others	28,250	27.7	+ 5.8
Insurance companies	13,100	12.8	+ 1.3
Farmers Home Administration	8,744	8.6	+ 13.3
Commercial banks	8,387	8.2	− 4.1
Total	102,045	100.0	+ 10.9
Nonreal estate loans:			
Commercial banks	32,948	35.5	+ 4.4
Production Credit Association	21,513	23.2	+ 7.3
Individuals and others	15,000	16.2	+ 7.1
Farmers Home Administration	14,452	15.6	+ 22.9
Commodity Credit Corp	8,008	8.6	+ 60.9
Federal Intermediate Credit Banks	914	1.0	+ 12.7
Total	92,835	100.0	+ 11.6
Total loans	194,880		
Percent real estate	52.4		
Percent nonreal estate	47.6		

[1]Data provided in a personal communication to the author from the Governor, Farm Credit Administration, Washington, D.C.

TABLE 2-8

MAJOR SOURCES OF CREDIT, AND THE CHARACTERISTICS OF EACH

Lenders	Sources of Funds	Limitations on Agricultural Lending	Loans Offered to Farmers	Comments
Commercial banks	Demand and time deposits, bank stock, retained earnings. Funds from correspondent banks, the Federal Reserve and participation with PCAs or the FmHA.	Commercial banks may prefer more profitable alternatives, such as installment loans. Also, legal reserve laws limit the volume of deposits available for loans.	(1) Short-term operating loans repayable within 1 year. (2) Intermediate loans, repayable in 1 to 7 years. (3) Some banks make long-term or real estate loans (7 to 25 years).	A financial statement is required by bank examiners. Some commercial banks have special agricultural representatives who are qualified to assist the borrower in many ways.
Farm Credit System, which embraces the following 3 federal lending units, all under the supervision of the Farm Credit Administration, an independent federal agency:	Sale of its bonds and discount notes in the private money market.			The Federal Land Banks and PCAs supply nearly ⅓ of the credit used by farmers, and the banks for cooperatives provide nearly ⅔ of the borrowed capital used by farmer cooperatives.
1. Federal Land Banks (FLB)	Sell bonds publicly on the national money markets. FLBs also draw on some money available from capital stock and retained earnings.	To borrow, you must buy stock in the local FLB equal to 5% of the requested loan amount. Congress currently limits the loan to 85% or less of the appraised property value.	Real estate loans, amortized over periods ranging up to 40 years. FLBs secure loans for improvements by mortgaging the improved real estate.	The Federal Land Bank is actually a farmer cooperative. Loans are on first mortgage only. When making loans, consideration is given to market value of the real estate, plus the income and the management ability of the borrower.
2. Production Credit Associations (PCA)	PCAs borrow from the Federal Intermediate Credit Bank, which gets funds by selling short-maturity debentures on the national money market. Also capital stock, and retained earnings.	Borrower must buy stock in the local PCA equal to 5% (sometimes 10%) of the loan. You can also get loans for farm related services, such as rural home construction.	Only short-term and intermediate-term loans. PCAs do not offer long-term and real estate loans.	These are local cooperative lending organizations. The loan limit varies with individual cases, but it can be up to 100% of cost. However, 70-30 is most common, with the borrower providing 30% margin.
3. Banks for Cooperatives				Banks for Cooperatives provide the majority of financing for the nation's farm supply, marketing, and business service cooperatives.
Farmers Home Administration (FmHA)	Congressional appropriations. FmHA also secures insured loans from other lenders, and from emergency or revolving funds.	Eligible only to farmers who can't "reasonably" borrow elsewhere. Legal maximums: $100,000 for operating loan, $200,000 for real estate loan (with participation, up to $300,000).	Short-, intermediate-, and long-term. FmHA provides supervision for its loans.	Applicants who are veterans and have farm experience receive preference. Where a natural disaster has occurred, under the Emergency Loan Program a borrower may borrow up to 80% of the loss, but not to exceed $500,000.
Individuals	Personal loans, made by one individual to another.	Compared to the other sources, individuals frequently offer lower interest rates and down payments. However, the repayment period may be shorter.	There's an infinite variety of conditions and interest rates. Range all the way from real estate contracts or mortgages to personal unsecured loans.	One disadvantage of a loan from an individual is that the arrangement may be complicated by the lender's death unless adequate provision has been made for this eventuality.
Insurance companies	Premiums received on insurance policies. They also draw funds from reserves held to pay insurance claims, and from capital and retained earnings.	Prefer long-term loans. Higher yielding, more secure investment opportunities lure money away from agricultural lending. May limit to selected geographic areas.	Typically, insurance companies limit themselves to real estate loans up to 30 to 40 years, and improvement loans which they secure through real estate mortgages.	Generally insurance companies will make loans up to 60% of the appraised value of the farm or up to 50% of the sale value.
Merchants and dealers	Borrow from lending institutions, capital and retained earnings. Their supplier, distributor, or manufacturer may extend similiar credit to them.	Limited credit, because they must keep a certain level of liquidity. Rates are higher than other lenders. Also limited by their supplier and other sources of capital.	Open account or sales contract. Short-term loans on equipment and machinery, intermediate loans on equipment. May charge add-on interest. Some offer discounts if you pay cash.	It must be realized that merchants and dealers extend credit to farmers and ranchers primarily for the purpose of promoting the sale of products and services, and that their profits come from both sales and interest.

PUBLIC OFFERINGS

Three different kinds of offerings are commonly used: (1) SEC registered offerings, (2) regulation A offerings, and (3) intrastate offerings.

SEC Registered Offering

Where a public offering of limited partner interests (such as participation in a cattle feeding fund) in excess of $500,000 is to be marketed, a registration statement providing full and fair disclosure of the character of the securities must be filed with the Securities and Exchange Commission (SEC), in keeping with the Securities Act of 1933. The SEC considers selling interests in an agricultural enterprise, such as in a cattle feeding fund, similar to selling stock in a corporation. Thus, a prospectus, which reveals all pertinent facts of the securities offered, must be printed. The latter is used as informational matter to explain partnership operations to potential investors. Also, the services of a lawyer knowledgeable in the area of public securities offerings is necessary in preparing the prospectus, filing the registration statement, and negotiating with the underwriter. Generally speaking, an investment banker, or underwriter, is needed to market the offering—as a middleman—to bring buyers and sellers together, for which he charges a commission. Although there is nothing to keep the person or persons offering securities from marketing them privately, they usually do not have the necessary time, staff, or expertise. In selling limited partnership interests, underwriters generally work on a best effort basis. This means that they are not obliged to market all the securities. The limited partnership interests are sold to the public at a price previously established by the agricultural company. The time involved and the cost of registering a fund are very considerable. Normally, registration will require 4 to 8 months; and the total cost for legal, accounting, and printing, will run anywhere from $50,000 to $150,000, depending upon the size of the offering, legal fees, and underwriter fees. Other costs of organizing a limited partnership include registering with the state in which business will be conducted. While these costs may be passed on to the purchasers of the limited partners interests, such intention must be stated in the prospectus. Moreover, there is no assurance that the offering will sell, in which case the general partner must stand all costs.

The question of whether or not a registration fee must be filed with the SEC should be answered by an attorney as there are a number of exemptions for which provision is made in the SEC regulations. It is possible that the sale of the securities may be exempt from the registration provisions of the SEC if the offering does not involve a "public offering," if the aggregate amount of the offering to the public does not exceed $500,000, or if the issue is to be sold on an intrastate basis.

Regulation A Offering

Regulation A, issued by the Commission, provides for the exemption of certain classes of domestic and Canadian securities where the aggregate offering to the public does not exceed $500,000. While a registration statement need not be filed with the SEC under Regulation A, notification and reports are required. Also, the regulation requires that offering circulars containing information prescribed by the Commission must be furnished to buyers. Although a filing under Regulation A is not as difficult, time-consuming, or expensive as a registration, it still involves expense and labor.

Intrastate Offering

Some persons or companies may wish to sell their stock within the confines of a state since costs, fees, and time can be saved. Quite often the savings in filing fees, printing, etc., may bring the cost down to 25 percent of the expenditures that are incurred in a SEC registration. Certain states are in a position to qualify an issue within a week if it is presented properly, as compared to the six weeks or longer that are normally required in filing with the SEC.

It is necessary to determine whether an issue is exempt from registration with the SEC under the provisions of Section 3A (11) of the Securities Exchange Act of 1933 which provides:

> Any security which is a part of an issue offered and sold only to persons resident within a single State or Territory, where the issuer of such security is a person resident and doing business within, or, if a corporation, incorporated by and doing business within such State and Territory.

State laws differ as to registration of intrastate issues. Thus, if the offering can be sold within the confines of a state, it would be well to consider with an attorney the advantages and disadvantages of this type of registration.

CENTRAL MONEY MARKETS

This refers to financing operations, such as cattle feeding, by using central money markets through the medium of commercial paper in competition with other lending institutions. This concept was initiated by the Central National Bank of Chicago, through the bank's Central Ag-Finance Corporation, in 1972. According to bank officials, their use of Central Money Markets was prompted because of what they saw as a developing situation in which the public agencies—Federal Intermediate Credit Banks, Production Credit Associations, and other federal and quasi-federal agencies—may not be able to provide the necessary financing for the burgeoning needs of cattle feeders. Also, commercial banks have been somewhat stymied in the past by the limit of their deposits. The answer, according to this school of thought, is to put agricultural financing on a par with commercial financing—into the Central Money Marts with commercial paper as the basis for securing money.

Central Ag-Finance Corporation (a subsidiary of Central National Bank of Chicago) is selling commercial paper in the open market, at rates competitive to the market, and then using the proceeds to make loans generally of one year or less maturity to agricultural producers. This is a new concept in agricultural financing. It may well prove to be the most desirable method of financing large commercial cattle feedlots in the future.

Credit Factors Considered and Evaluated by Lenders

Potential money borrowers sometimes make their first big mistake by going in "cold" to see a lender, without adequate facts and figures, with the result that they have two strikes against getting the loan to begin with.

When you go in to talk over a loan, take along a financial statement, records of past years' operations, tax forms, and full records of production. Also, point out the things that you have done to minimize risk, such as insurance, forward contracting, estate planning, and some cash cushion.

When considering and reviewing loan requests, the lender tries to arrive at the repayment ability of the potential borrower. Likewise, the borrower has no reason to obtain money unless it will make money.

Lenders need certain basic information in order to evaluate the soundness of a loan request. To this end, the following information should be submitted:

1. **Analysis and feasibility study**—Lenders are impressed with a borrower who has a written-down program showing where he is now, where he's going, and how he expects to get there. In addition to spelling out the goals, this should give assurance of the necessary management skills to achieve them. Such an analysis of the present and projection into the future is imperative in big operations.

2. **The applicant, farm or ranch, and financial statement**—It is the borrower's obligation, and in his best interest, to present the following information to the lender.

a. **The applicant:**

(1) Name of applicant and wife; age of applicant.

(2) Number of children (minors; legal age).

(3) Partners in business, if any.

(4) Years in area.

(5) References.

b. **The farm or ranch:**

(1) Owner or tenant.

(2) Location; legal description and county, and direction and distance from nearest town.

(3) Type of enterprise—cow-calf, feedlot, etc.

c. **Financial statement**—This document indicates the borrower's financial record and current financial position; his potential ahead; and his liability to others. The borrower should always have sufficient slack to absorb reasonable losses due to such unforeseen happenstances as storms, droughts, diseases, and poor markets, thereby permitting the lender to stay with him in adversity and to give him a chance to recoup his losses in the future.

The financial statement should include the following:

(1) Current assets:

(a) Livestock.

(b) Feed.

(c) Machinery.

(d) Cash—There should be reasonable cash reserves, to cut interest costs, and to provide a cushion against emergencies.

(e) Bonds or other investments.

(f) Cash value of life insurance.

(2) Fixed assets:

(a) Real property, with estimated value:

i. Farm or ranch property.

ii. City property.

iii. Long term contracts.

(3) Current liabilities:

(a) Mortgages.

(b) Contracts.

(c) Open account—to whom owed.

(d) Co-signer or guarantor on notes.

(e) Any taxes due.

(f) Current portion of real estate indebtedness due.

(4) Fixed liabilities—amount and nature of real estate debt:

(a) Date due.

(b) Interest rate.

(c) To whom payable.

(d) Contract or mortgage.

3. **Other factors**—Shrewd lenders usually ferret out many things; among them:

a. **The potential borrower**—Most lenders recognize that the potential borrower is the most important part of the loan. Lenders consider his—

(1) Character.
(2) Honesty and integrity.
(3) Experience and ability.
(4) Moral and credit rating.
(5) Age and health.
(6) Family cooperation.
(7) Continuity, or line of succession.

Lenders are quick to sense the "high-liver"—the fellow who lives beyond his means; the poor manager—the kind who would have made it except for hard luck, and to whom the hard luck happened many times; and the dishonest, lazy, and incompetent.

In recognition of the importance of the man back of the loan, "key man" insurance on the owner or manager should be considered by both the lender and the borrower.

b. **Production records**—This refers to a good set of records showing efficiency of production. Such records should show weight and price of livestock sold; milk and fat production of lactating cows; calf, lamb, pig and/or foal crop percentages and weaning weights; efficiency of feed utilization and rate of gain on animals finished for market; age of livestock; female replacement program; depreciation schedule; average crop yield; and other pertinent information. Lenders will increasingly insist on good records.

c. **Progress with previous loans**—Has the borrower paid back previous loans plus interest; has he reduced the amount of the loan, thereby giving evidence of progress?

d. **Profit and Loss (P & L) statement**—This serves as a valuable guide to the potential ahead. Preferably, this should cover the previous three years. Also, most lenders prefer that this be on an accrual basis (even if the farmer or rancher is on a cash basis in reporting to the Internal Revenue Service).

e. **Physical plant:**

(1) Is it an economic unit?

(2) Does it have adequate water, and is it well balanced in feed and livestock?

(3) Is there adequate diversification?

(4) Is the right kind of livestock being produced?

(5) Are the right crops and varieties grown, and are approved methods of tillage and fertilizer practices being followed?

(6) Is the farmstead neat and well kept?

f. **Collateral (or security):**

(1) Adequate to cover loan, with margin.

(2) Quality of security:

(a) Grade and age of livestock.

(b) Type and condition of machinery.

(c) If grain storage is involved, adequate protection from moisture and rodents.

(d) Government participation.

(3) Identification of security:

(a) Brands, ear tags, tattoo marks of livestock.

(b) Serial numbers on machinery.

4. **The loan request**—Farmers and ranchers are in competition for money from urban businessmen. Hence, it is important that their request for a loan be well presented and supported. The potential borrower should tell the purpose of the loan; how much money is needed, when it's needed, and what it's needed for; the soundness of the venture; and the repayment schedule.

Credit Factors Considered by Borrowers

Credit is a two-way street; it must be good for both the borrower and the lender. If a borrower is the right kind of person and on a sound basis, more than one lender will want his business. Thus, it is usually well that a borrower shop around a bit—that he be familiar with several sources of credit and see what they have to offer. There are basic differences in length and type of loan, repayment schedules, services provided with the loan, interest rate, and the ability and willingness of lenders to stick by the borrower in emergencies and times of adversity. Thus, interest rates and willingness to loan are only two of the several factors to consider. Also, if at all possible, all borrowing should be done from one source; a one-source lender will know more about the borrower's operations and be in a better position to help him.

Helpful Hints for Building and Maintaining a Good Credit Rating

Stockmen who wish to build up and maintain good credit are admonished to do the following:

1. **Keep credit in one place, or in few places**—Generally, lenders frown upon "split financing." Shop around for a creditor who (a) is able, willing and interested in extending the kind and amount of credit needed, and (b) will lend at a reasonable rate of interest, then stay with you.

2. **Get the right kind of credit**—Don't use short-term credit to finance long-term improvements or

other capital investments. Also, use the credit for the purpose intended.

3. **Be frank with the lender**—Be completely open and aboveboard. Mutual confidence and esteem should prevail between borrower and lender.

4. **Keep complete and accurate records**—Complete and accurate records should be kept by enterprises. By knowing the cost of doing business, decision making can be on a sound basis.

5. **Keep annual inventory**—Take an annual inventory for the purpose of showing progress made during the year.

6. **Repay loans when due**—Borrowers should work out a repayment schedule on each loan, then meet payments when due. Sale proceeds should be promptly applied on loans.

7. **Plan ahead**—Analyze the next year's operation and project ahead.

Borrow Money to Make Money

Stockmen should never borrow money unless they are reasonably certain that it will make or save money. With this in mind, borrowers should ask, "How much should I borrow?" rather than, "How much will you lend me?"

Calculating Interest

The charge for the use of money is called interest. The basic charge is strongly influenced by the following:

1. The *basic cost* of money in the money market.
2. The *servicing costs* of making, handling, collecting, and keeping necessary records on loans.
3. The *risk* of loss.

Interest rates vary among lenders and can be quoted and applied in several different ways. The quoted rate is not always the basis for proper comparison and analysis of credit costs. Even though several lenders may quote the same interest rate, the effective or simple annual rate of interest may vary widely. The more common procedures for determining the actual annual interest rate, or the equivalent of simple interest on the unpaid balance, follow.

1. **Simple or true annual interest on the unpaid balance**—A $1,200 note payable at maturity (12 months) with 12% interest:

Interest paid .... .12 × $1,200 = $144
Average use of the money .... $1,200 for the entire year
Actual rate of interest .... $\frac{\$144 \text{ (interest)}}{\$1{,}200 \text{ (used for one year)}} = 12\%$

2. **Installment loan (with interest on unpaid balance)**[5]—A $1,200 note payable in 12 monthly installments with 12% interest on the unpaid balance:

Interest paid ranges from:

First month .... $\frac{.12 \times \$1{,}200}{12} = \$12$

to

Twelfth month .... $\frac{.12 \times 100}{12} = \$1$

Total for 12 months is $78

Average use of the money ranges from $1,200 for the first month down to $100 for the twelfth month, an average of $650 for 12 months.

Effective rate of interest ...... $\frac{\$78}{\$650} = 12\%$

3. **Add-on installment loan (with interest on face amount)**—A $1,200 note payable in 12 monthly installments with 12% interest on face amount of loan:

Interest paid .... .12 × $1,200 = $144

Average use of the money ranges from $1,200 for the first month down to $100 for the twelfth month, an average of $650 for 12 months.

Effective rate of interest .... $\frac{\$144}{\$650} = 22.15\%$

4. **Points and interest**—Some lenders now charge "points." A point is 1% of the face value of the loan. Thus, if 4 points are being charged on a $1,200 loan, $48 will be deducted and the borrower will receive only $1,152. But he will have to repay the full $1,200. Obviously, this means that the actual interest rate will be more than the stated rate. But how much more?

Assume that a $1,200 loan is for 1 year and the annual rate of interest is 12%. Then the payment by the borrower of 4 points would make the actual interest rate as follows:

Interest .... .12 × $1,200 = $144
Average use of money .... $1,152 for one year
Effective rate of interest .... $\frac{\$144 \text{ (interest)}}{\$1{,}152 \text{ (used for one year)}} = 12.5\%$

[5] This method is used for amortized loans.

5. **If interest is not stated, use this formula to determine the effective annual interest rate:**

Effective rate of interest=

$$\frac{2 \times \text{Number of payment periods in one year}^{6} \times \text{Finance charges}^{7}}{\text{Balance owed}^{8} \times \text{Number of payments in contract plus one}}$$

For example, a store advertises a refrigerator for \$500. It can be purchased on the installment plan for \$80 down and monthly payments of \$35 for 12 months. What is the actual rate of interest if you buy on the time payment plan?

Effective rate of interest=

$$\frac{2 \times 12 \times \$35}{\$420 \times (12 \text{ plus } 1)} = \frac{\$840}{\$5{,}460} = 15.4\%$$

MANAGER

According to Webster, *a manager is one who conducts business affairs with economy; and management is the act, or art, of managing, handling, controlling or directing.*

Three major ingredients are essential to success in the livestock business: (1) good livestock, (2) good feeding, and (3) good management. A manager can make or break any livestock enterprise. Unfortunately, this fact is often overlooked, primarily because the accent is on scientific findings, automation, and new products.

Management gives point and purpose to everything else. The skill of the manager materially affects how well animals are bought and sold, the quality of the animals, the health of the animals, the results of the ration, the stress of the stock, efficiency of production, the performance of labor, the public relations of the establishment, and even the expression of the genetic potential. Indeed, a livestock manager must wear many hats—and he must wear each of them well.

The bigger and the more complicated the operation, the more competent the management required. This point merits emphasis because, currently, (1) bigness is a sign of the times, and (2) the most common method of attempting to "bail out" of an unprofitable business venture is to increase its size. Although it's easier to achieve efficiency of equipment, labor, purchases, and marketing in big operations, bigness alone will not make for greater efficiency, as some owners have discovered to their sorrow, and others will experience. Management is still the key to success. When in financial trouble, owners should have no illusions on this point.

In manufacturing and commerce, the importance and scarcity of top managers are generally recognized and reflected in the salaries paid to persons in such positions. Unfortunately, agriculture as a whole has lagged; and altogether too many owners still subscribe to the philosophy that the way to make money out of the livestock business is to hire a manager cheaply, with the result that they usually get what they pay for—a "cheap" manager.

Traits of a Good Livestock Manager

There are established bases for evaluating many articles of trade including animals, hay, and grain. They are graded according to well-defined standards. Additionally, feeds are chemically analyzed and feeding trials conducted with them. But no such standard or system of evaluation has evolved for livestock managers, despite their acknowledged importance.

The author has prepared the Livestock Manager Checklist, given in Table 2-9, which (1) employers may find useful when selecting or evaluating a manager, (2) managers may apply to themselves for self-improvement purposes, and (3) students may use for guidance as they prepare themselves for management positions. No attempt has been made to assign a per-

TABLE 2-9
LIVESTOCK MANAGER CHECKLIST

☐ *Character*—
Absolute sincerity, honesty, integrity, and loyalty; ethical.

☐ *Industry*—
Work, work, work; enthusiasm, initiative, and agressiveness.

☐ *Ability*—
Livestock know-how and experience, business acumen—including ability systematically to arrive at the financial aspects and convert this information into sound and timely management decisions; knowledge of how to automate and cut costs; common sense; organized; imagination; growth potential.

☐ *Plans*—
Sets goals; prepares organization chart and job description; plans work and works plans.

☐ *Analyzes*—
Identifies the problem, determines pros and cons, then comes to a decision.

☐ *Courage*—
To accept responsibility, to innovate, and to keep on keeping on.

☐ *Promptness and dependability*—
A self-starter; has "T.N.T.," which means that he does it today, not tomorrow.

☐ *Leadership*—
Stimulates subordinates and delegates responsibility.

☐ *Personality*—
Cheerful; not a complainer.

[6]Regardless of the total number of payments to be made, use 12 if the payments are monthly, use 6 if payments are every other month, or use 2 if payments are semiannual.

[7]Use either the time payment price less the cash price, or the amount you pay the lender less the amount you received if negotiating for a loan.

[8]Use cash price less down payment or, if negotiating for a loan, the amount you receive.

centage score to each trait, because this will vary among livestock establishments. Rather, it is hoped that this checklist will serve as a useful guide (1) to the traits of a good manager, and (2) to what the boss wants.

Organization Chart and Job Description

It is important that every worker know to whom he is responsible and for what he is responsible; and the bigger and the more complex the operation, the more important this becomes. This should be written down in an organization chart and a job description. A sample Organization Chart is given in Fig. 2-9, and a sample Job Description is given in Fig. 2-10.

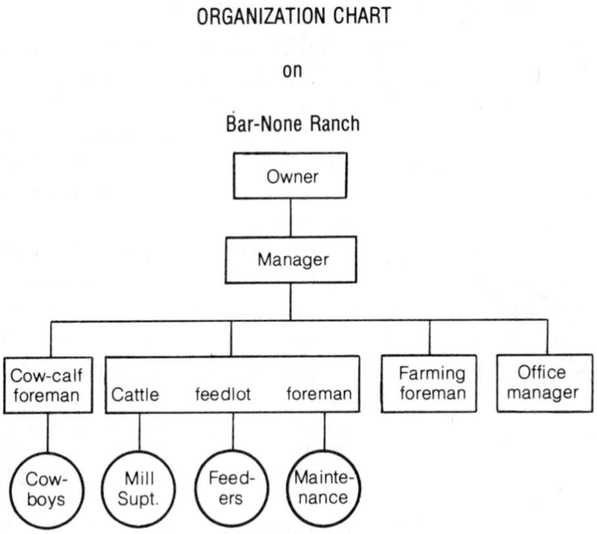

Fig. 2-9. A suggested farm-ranch Organization Chart.

An Incentive Basis for the Help

Big farms and ranches must rely on hired labor, all or in part. Good help—the kind that everyone wants—is hard to come by; it's scarce, in strong demand, and difficult to keep. Moreover, the farm manpower situation is going to become more difficult in the years ahead. There is need, therefore, for some system that will (1) give a big assist in getting and holding top-flight help, and (2) cut costs and boost profits. An incentive basis that makes hired help partners in profit is the answer.

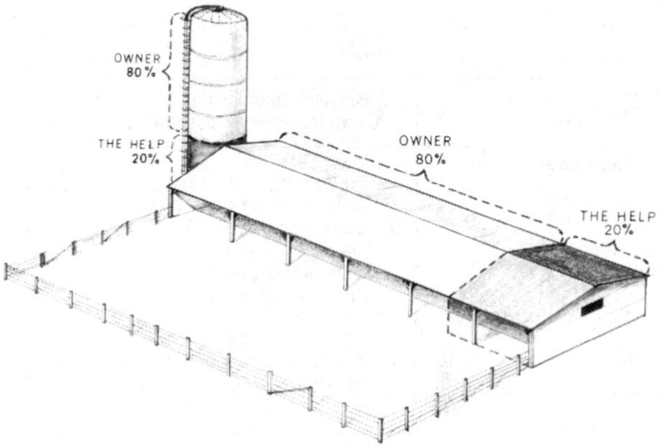

Fig. 2-11. A good incentive basis makes hired help partners in profit.

Many manufacturers have long had an incentive basis. Executives are frequently accorded stock option privileges, through which they prosper as the business prospers. Common laborers may receive bonuses based on piecework or quotas (number of units or pounds produced). Also, most factory workers get overtime pay and have group insurance and a retirement plan. A few industries have a true profit-sharing

JOB DESCRIPTIONS
on
Bar-None Ranch

Owner	Manager	Cow-calf foreman	Cattle feedlot foreman
Responsible for:	*Responsible for:*	*Responsible for:*	*Responsible for:*
1. Selecting management. 2. Making policy decisions. 3. Borrowing capital. 4. (List others.)	1. Supervising all staff. 2. Preparing proposed longtime plan. 3. Budgets. 4. (List others.)	1. Directing cow-calf staff. 2. Selecting and culling. 3. Breeding cows, including A.I. work. 4. Feeding the herd. 5. Calving. 6. Branding (marking), dehorning, castrating, vaccinating. 7. Herd health. 8. Preconditioning. 9. Marketing calf crop. 10. (List others.)	1. Directing feedlot staff. 2. Buying and selling cattle. 3. Processing incomeing cattle. 4. Animal health. 5. Feedlot rations. 6. Feeding. 7. (List others.)

Fig. 2-10. Job description.

arrangement based on net profit as such, a specified percentage of which is divided among employees. No two systems are alike. Yet, each is designed to pay more for labor, provided labor improves production and efficiency. In this way, both owners and laborers benefit from better performance.

Family owned and family operated farms have a built-in incentive basis; there is pride of ownership, and all members of the family are fully cognizant that they prosper as the business prospers.

Many different incentive plans can be, and are, used. There is no best one for all operations.

Note well: The various plans given in the tables and narrative that follow are for the purpose of showing how different incentives work; and *not* as indicators of amounts. The type and amount of incentive chosen, along with the provisions, should be tailored to fit the specific operation, with consideration given to the kind and size of the operation, the extent of the owner's supervision, the present and projected productivity levels, mechanization, and other factors.

For most livestock operations, the author favors a "production sharing and prevailing price" type of incentive.

TABLE 2-10

INCENTIVE PLANS FOR LIVESTOCK ESTABLISHMENTS

Types of Incentives	Pertinent Provisions of Some Known Incentive Systems in Use	Advantages	Disadvantages	Comments
1. **Bonuses**	A flat, arbitrary bonus; at Christmastime, year-end, or quarterly or other intervals. A tenure bonus such as (1) 5 to 10% of the base wage or 2 to 4 weeks' additional salary paid at Christmastime or year-end; (2) 2 to 4 weeks' vacation with pay, depending on length and quality of service; or (3) $10 to $20/week set aside and to be paid if employee stays on the job a specified time.	It's simple and direct.	Not very effective in increasing production and profits.	
2. **Equity-building plan**	Employee is allowed to own a certain number of animals. In breeding operations, these are usually fed without charge.	It imparts pride of ownership to the employee.	The hazard that the owner may feel that employee accords his animals preferential treatment; suspicioned if not proved.	
3. **Production sharing**	$2 to $6/calf weaned, $1/cwt on gain of feeder cattle; 50¢ to $2/head on fed cattle marketed; $1 to $2/pig marketed above 7 pigs/litter; 50¢ to $1/lamb weaned; $1/cwt of gain on lambs fed; 20¢ to 50¢/head on fed lambs marketed; so much per day for meeting certain levels of milk production/cow (for example, a 20¢ bonus/cow/month for 45 to 51 lb milk/cow/day; and 70¢ for 52 lb or above, with the bonus graduated upward with higher production in order to reflect the difficulty of increasing milk production in a herd where milk yield is already high.	It's an effective way to achieve higher production.	Net returns may suffer. For example, a higher rate of gain than is economical may be achieved by feeding stockers more concentrated and expensive feeds than are practical. This can be alleviated by (1) specifying the ration, and (2) setting an upper limit on the gains to which the incentive will apply. If a high performance level already exists, further gains or improvements may be hard to come by.	Incentive payments for production above certain levels—for example, above 450 lb calf weaned/cow bred—are more effective than paying for all units produced.
4. **Profit sharing**		Net income sharing works better for managers, supervisors, and foremen than for common laborers because fewer hazards are involved to opening up the books to them. It's an effective way to get hired help to cut costs. It's a good plan for a hustler.	Percent of gross does not impart cost of production consciousness. Both (1) percent of gross income, and (2) percent of net income expose the books and accounts to workers, who may not understand accounting principles. This can lead to suspicion and distrust. Controversy may arise (1) over accounting procedures—for example, from the standpoint of the owner a fast tax write-off may be desirable on new equipment but this reduces the net shared with the worker; and (2) because some owners are prone to overbuild and overequip, thereby decreasing net.	There must be prior agreement on what constitutes gross or net receipts, as the case may be, and how it is figured.
a. Percent of gross income	1% to 2% of the gross.			
b. Percent of net income	10% to 20% of the net after deducting all costs.			

(Continued)

TABLE 2-10 (Continued)

Types of Incentives	Pertinent Provisions of Some Known Incentive Systems in Use	Advantages	Disadvantages	Comments
5. **Production sharing and prevailing price**	*Cow-calf, ewe-lamb, or sow operation*—Basis (1) percent offspring weaned, and (2) weaning weight (which means pounds offspring weaned/female bred). *Finishing cattle, or finishing hogs*—Basis (1) pounds feed/lb gain, and (2) daily rate of gain. *Dairy*—Basis (1) udder health (for example, 20¢/cow/month for a somatic cell count of 500,000 to 600,000, graduated upward to 50¢/cow/month for a somatic cell count of 400,000 or less); or (2) calving interval (for example, \$20/month for a herd calving interval of 13.0 to 13.4 months, graduated upward to \$50 bonus/month for a calving interval of 12.5 months or under). *Horse breeding establishment*—Basis (1) percent foal crop weaned, and (2) price of yearlings. In each of the above, establish break-even point(s), then split profit(s) beyond this point(s) basis (1) 80% (owner) and 20% (help); or (2) use escalator arrangement, giving help greater percentage as profits rise.	It embraces the best features of both production sharing and profit sharing, without the major disadvantages of each. It (1) encourages high productivity and likely profits, (2) is tied in with prevailing prices, (3) does not necessitate opening the books, and (4) is flexible—it can be split between owner and employee on any basis desired, and the production part can be adapted to a sliding scale or escalator arrangement—for example, the incentive basis can be higher for the quarter pound of feedlot gain made in excess of 2¾ lb than for a quarter-pound gain in excess of 2¼ lb.	It is a bit more complicated than some other plans, and it requires more complete records.	When properly done, and all factors considered, this is the most satisfactory incentive basis for a livestock enterprise.

HOW MUCH INCENTIVE PAY?

After (1) reaching a decision to go on an incentive basis, and (2) deciding on the kind of incentive, it is necessary to arrive at how much to pay. Here are some guidelines that may be helpful in determining this:

1. Pay the going base, or guaranteed, salary; then add the incentive pay above this.
2. Determine the total stipend (the base salary plus incentive) to which you are willing to go.
3. Before making any offers, always check the plan on paper to see (a) how it would have worked out in past years based on your records, and (b) how it will work out as you achieve the future projected production.

Let's take the following example:

> A foreman of a 500-cow herd is now producing an average of 400 pounds of calf weaned per cow bred. He is receiving a base salary of \$1,000 per month, plus house, garden, and 600 pounds of dressed beef per year. The owner prefers a "production sharing and prevailing price" type of incentive.

Step by step, here is the procedure for arriving at an incentive arrangement based on increased production:

1. By checking with local sources, it is determined that the present salary of \$1,000 per month plus extras is the going wage; and, of course, the foreman receives this regardless of what the year's calf production or price turns out to be—it's guaranteed.
2. A study of the cow-calf records reveals that with a little extra care on the part of the foreman—particularly in pregnancy testing, at calving time, and in rotating pastures—the average weaning weight of calf per cow bred can be boosted enough to permit paying him \$1,300 per month, or \$300 per month more than he's now getting. That's \$3,600 per year. This can be fitted into the incentive plan.
3. An average increase of 50 pounds of calf weaned per cow bred at 60 cents per pound would mean \$30 per cow, or \$15,000, on a 500-cow herd. With an 80:20 split between owner and manager, the foreman would get \$3,000, or \$300 per month.

REQUISITES OF AN INCENTIVE BASIS

Owners who have not previously had experience with an incentive basis are admonished not to start with any plan until they are sure of both their plan and their help. Also, it is well to start with a simple plan; then a change can be made to a more inclusive and sophisticated plan after experience is acquired.

Regardless of the incentive plan adopted for a specific operation, it should encompass the following essential features:

1. A good owner (or manager) and good workers. No incentive basis can overcome a poor manager. He must be a good supervisor and fair to his help. Also, on big establishments, he must prepare a written organization chart and job description so the employees know (a) to whom they are responsible, and (b) for what they are responsible. Likewise, no incentive basis can spur employees who are not able, interested, and/or willing. This necessitates that they be selected with special care where they will be on an

incentive basis. Hence, the three—good owner (manager), good employees, and good incentive—go hand in hand.

2. It must be fair to both employer and employees.

3. It must be based on and make for mutual trust and esteem.

4. It must compensate for extra performance, rather than substitute for a reasonable base salary and other considerations (house, utilities, and certain provisions).

5. It must be as simple, direct, and easily understood as possible.

6. It should compensate all members of the team—from cowboys to manager on a cow-calf outfit, from feeders and feed processors to manager in a cattle feedlot, and from foreman to milkers in a dairy.

7. It must be put in writing, so that there will be no misunderstanding. For example, if some production-sharing plan is used in a cattle feedlot, it should stipulate the ration (or who is responsible for ration formulation), the maximum gain of stocker cattle, and the grade to which finishing cattle are to be carried. On a cow-calf outfit, it should stipulate the ration, the culling of cows, and other pertinent factors.

8. It is preferable, although not essential, that workers receive incentive payments (a) at rather frequent intervals, rather than annually, and (b) immediately after accomplishing the extra performance.

9. It should give the hired help a certain amount of responsibility, from the wise exercise of which they will benefit through the incentive arrangement.

10. It must be backed up by good records; otherwise, there is nothing on which to base incentive payments.

11. It should be a two-way street. If employees are compensated for superior performance, they should be penalized (or, under most circumstances, fired) for poor performance. It serves no useful purpose to reward the unwilling, the incompetent, and the stupid. For example, no overtime pay should be given to an employee who must work longer because of slowness or in order to correct mistakes of his own making. Likewise, if the reasonable break-even point on a cow-calf operation is an average of a 400-pound calf weaned per cow bred, and this production level is not reached because of obvious neglect (for example not being on the job at calving time), the employee(s) should be penalized (or fired).

INDIRECT INCENTIVES

Normally, we think of incentives as monetary in nature—as direct payments or bonuses for extra production or efficiency. However, there are other ways of encouraging employees to do a better job. The latter are known as indirect incentives. Among them are (1) good wages; (2) good labor relations; (3) adequate house plus such privileges as the use of the farm truck or car, payment of electric bill, use of a swimming pool, hunting and fishing, use of a horse, and furnishing meat, milk, and eggs; (4) good buildings and equipment; (5) vacation time with pay, time off, sick leave; (6) group health; (7) security; (8) the opportunity for self-improvement that can accrue from working for a top man; (9) the right to invest in the business; (10) an all-expense-paid trip to a short course, show, or convention; and (11) year-end bonus for staying all year. These indirect incentives will be accorded to the help of more and more establishments, especially the big ones.

EXAMPLES OF INCENTIVE BASIS

Space limitations will not permit presentation in this book of all conceivable incentive bases for all classes of livestock and all kinds and sizes of operations. Hence, only two—an incentive basis for cow-calf operators, and an incentive basis for cattle feedlots—will follow. However, the same principles can be adapted and applied to any class of livestock or to any kind of operation. Also, a scorecard for evaluation of employees is included. It may be used in any type of business, big or little.

An Incentive Basis for Cow-Calf Operators

On cow-calf enterprises, there is need for some system which will encourage caretakers to be good nursemaids to newborn calves, though it may mean the loss of sleep, and working with cold, numb fingers. Additionally, there is need to do all those things which make for the maximum percent calf crop weaned at a heavy weight.

From the standpoint of the owner of a cow-calf enterprise, production expenses remain practically unchanged regardless of the efficiency of the operation. Thus, the investment in land, buildings and equipment, cows, feed, and labor differs very little with a change (up or down) in the percent calf crop or the weaning weight of calves; and income above a certain break-even point is largely net profit. Yet, it must be remembered that owners take all the risks; hence, they should benefit most from profits.

On a cow-calf operation, the writer recommends that profits beyond the break-even point (after deducting all expenses, but not including returns to labor and management) be split on an 80:20 basis. This means that every dollar made above a certain level is split, with the owner taking 80 cents and the employees getting 20 cents. Also, there is merit in an escalator arrangement—with the split changed to 70:30, for example, when a certain plateau of efficiency is reached. Moreover, that which goes to the employees should be divided on the basis of their respective contributions, all the way down the line; for example, 25% of it might go to the manager, 25% might be divided among the foremen, and 50% of it divided among the rest of the help.

A true profit-sharing system on a cow-calf outfit based on net profit has the disadvantages of (1) employees not benefiting when there are losses, as frequently happens in the cattle business; and (2) management opening up the books, which may lead to gossip, misinterpretation, and misunderstanding. An incentive system based on major profit factors alleviates these disadvantages.

Gross income in cow-calf operations is determined primarily by (1) percent calf crop weaned, (2) weaning weight of calves, and (3) price. The first two factors can easily be determined. Usually, enough calves are sold to establish price; otherwise, the going price can be used.

The incentive basis proposed in Table 2-11 for cow-calf operation is simple, direct, and easily applied. As noted, it is based on average pounds of calf weaned per cow, which factor encompasses both percent calf crop and weaning weight.

TABLE 2-11

A PROPOSED INCENTIVE BASIS FOR COW-CALF OPERATIONS

Average Pounds of Calf Weaned/ Cow Bred	Here's How It Works
(lb) 350 375 400 (break-even point)	On this particular operation, the break-even point is assumed to be an average of 400 lb of calf weaned per cow bred; and, of course, this is arrived at after all cost of production factors have been included.
425 450 475 500 525 550 575 600	Higher poundage of calf per cow bred can be achieved through (1) increased calf crop percentage, (2) increased weaning weight of calf, and/or (3) a combination of the two. Pounds of calf weaned per cow bred in excess of the break-even point (in this case 400 lb) are sold or evaluated at the going price. If an average of 450 lb of calf per cow bred is weaned, and if mixed steers and heifers of this quality are worth 60¢ per pound, that's $30 more net profit per cow. In a 500-cow herd, that's $15,000. With an 80:20 division, $12,000 would go to the owner, and $3,000 would be distributed among the employees. Or, if desired, and if there is an escalator arrangement, there might be an 80:20 split at 425 lb, a 70:30 split at 450 lb, and 65:35 split at 475 lb.

An Incentive Basis for Cattle Feedlots

Table 2-12 is a cattle feedlot incentive basis.

TABLE 2-12

A PROPOSED INCENTIVE BASIS FOR CATTLE FEEDLOTS

Feed/Lb Gain		Daily Rate of Gain
(lb)		*(lb)*
12.0		0.5
11.5		0.75
11.0		1.00
10.5		1.25
10.0		1.50
9.5		1.75
9.0		2.00
8.5	(Break-even point)	2.25
8.0		2.50
7.5		2.75
7.0		3.00
6.5		3.25
6.0		3.50
5.5		3.75

Here's How It Works

On this particular cattle finishing operation, the break-even points are assumed to be (1) 8.5 lb of feed/lb gain, and (2) 2.25 lb daily gain. Of course, for cattle on grower rations, the break-even points would be different; for example, feed efficiency might be 10.0 and the daily rate of gain 1.00 lb; also, on grower rations an upper limit on daily gains could be set, beyond which there would be no bonus benefits.

Feed saved and gains made in excess of the break-even points are computed at going prices.

If the feed efficiency drops to 8.0 and the gain increases to 2.5, and if feed costs $120 per ton and cattle are worth 60¢ per pound, then these feed savings and increased gains are worth—

Feed saved, lb	Cost of feed/lb	Value of feed saved/ lb gain	Value of feed saved on 2.5 lb gain
0.5	6.0¢	3.0¢	7.5¢

Gains made, lb	Per lb mkt. value of gains	Value of incr. daily gain
0.25	60¢	15.0¢

Increased profit/head/day:

	Cents
Feed	7.5
Gain	15.0
Total	22.5

On steers fed for 140 days, that's $31.50/head. With 500 steers, that's $15,750 total. When divided on an 80:20 basis, that's $12,600 for the owner and $3,150 for the employees.

An incentive basis for cattle feedlot help is needed for motivation purposes, just as it is in cow-calf operations. It is the most effective way in which to lessen absenteeism, poor processing and mixing of feeds, irregular and careless feeding, unsanitary troughs and water, sickness, shrinkage, and other profit-sapping factors.

Whenever possible, the break-even points—(1) pounds feed/pound gain, and (2) daily rate of gain—should be arrived at from actual records accumulated by the specific feedlot, preferably over a period of years. Perhaps, too, they should be moving averages, based on 5 to 10 years, with older years dropped out and more recent years added from time to time—thereby reflecting improvements in efficiency and rate of gain due primarily to changing technology, rather than to the efforts of the caretakers.

With a new feedlot, on which there are no historical records from which to arrive at break-even points of feed efficiency and rate of gain, it is recommended that the figures of other similar feedlots be used at the outset. These can be revised as actual records on the specific feedlot become available. It is important, however, that the new feedlot start on an incentive basis, even though the break-even points must be arbitrarily assumed at the time.

Because of the high correlation between feed efficiency and rate of gain, this incentive basis results in an overlapping of measures. Nevertheless, both efficiency and rate of gain are important profit indicators to cattlemen. Because of the overlapping, however, some may prefer to choose one or the other of the measures, rather than both.

Another incentive basis followed in a few large feedlots consists of the following: a certain percent (say 15%) of the net earnings set aside in a trust account, which is divided among and applied to the account of each employee according to salary and/or length of service, and paid to employees upon retirement or after a specified period of years. The main disadvantages to this incentive basis are that there may not be any net some years, that some employees do not want to wait that long for their added compensation, and that it opens up the books of the business.

THE SCORECARD INCENTIVE BASIS

A scorecard which the author developed and has used is herewith presented as Fig. 2-12.

PERSONAL SCORE CARD For ______________________	*Pts. or %*	*1981 June 30*	*1981 Dec. 31*	*1982 June 30*	*1982 Dec. 31*	*1983 June 30*	*1983 Dec. 31*
USE: This scorecard is applied by the owner-manager to each staff member every 6 months, then the results are discussed with each person in a private conference. PURPOSE: To provide staff with an evaluation which they may use as a basis for self-improvement, and to recognize and reward superior preformance.							
CHARACTER: Absolute sincerity, honesty, integrity, loyalty; ethical	10						
INDUSTRY: Works hard; has enthusiasm, initiative, and aggressiveness; a desire to get the job done, and a willingness to let the boss worry about raises; not afraid of long hours when necessary, not a clock watcher	15						
ABILITY AND PERFORMANCE: Skilled and competent in area of work; is neat and accurate; turns out adequate work; has know-how, clarity, common sense, good judgment, and maturity; is organized; shows growth potential and self-improvement; not a know-it-all; accepts responsibility, masters the job assignments without being told "what to do next"; plans work, organizes, and keeps on top; efficient and cuts costs; recognizes that the "boss" must make decisions and give directions	40						
INTEREST: Genuine interest in work—not just payday and 5 o'clock	10						
COURAGE: To innovate—to try the new, and to keep on keeping on	5						
PROMPTNESS AND DEPENDABILITY: A self-starter, does it today and not tomorrow	10						
PERSONALITY AND APPEARANCE: Cheerful; not a complainer; a member of the team	10						
	100						

Fig. 2-12. Scorecard and incentive basis for staff.

As the name implies, the scorecard method involves the preparation of a scorecard. In the scorecard, each major desired trait and performance is given a numerical value according to its relative importance.

Based on the average of two scores per year (Fig. 2-12), the following considerations are suggested for staff:

1. **Christmas bonus**—

Score	Grade	Employee Should:	Christmas Bonus[9]
59 or under	Poor	Improve, work longer hours, and/or look for a job elsewhere	None
60-74	Good	Improve and work longer hours	None
75-79	Good	Keep Improving	$1.00
80-84	Good	Keep Improving	$2.00
85-89	Good+	Keep Improving	$3.00
90-94	Excellent	Keep Improving	$4.00
95-100	Superior	Keep on keeping on	$5.00

2. **Health insurance; extra vacation time**—Full-time employees on appointment (not on hourly basis) exceeding (and maintaining) a minimum score of 75 after 6 months' service or 80 after 2 years' service, will be accorded the following added considerations:

After Years' Service	Consideration
½ year (6 mo)	Health insurance[10] for employee (not family), with employer and employee each paying 50% of premiums. (Employer will terminate health insurance upon termination of employment of employee.)
2 years	Health insurance for employee (not family), with employer paying premiums. (Employer will terminate health insurance upon termination of employment of employee.)
3 years	Added workday of paid vacation.
4 years	Added workday of paid vacation.
5 years	Added workday of paid vacation.
6 years	Added workday of paid vacation.
7 years	Added workday of paid vacation.[11]

3. **Merit increases**—Salary increases on a merit basis only, and determined by (a) average annual score of the employee, (b) how well the business is doing, and (c) going wages of the area for the particular assignment. Whatever salary increase is accorded will be on a January 1 basis only, following the second six months' review.

[9]Based on full-time employees and 12 months' prior service; regular half-time employees are accorded half these amounts. Those not on the job 12 months are accorded consideration on a proportion to 12 months' basis.

[10]Where the spouse already has health insurance, the insurance of each party should be coordinated by the respective insurance companies.

[11]Thus, at the end of 7 years, employees meeting the stipulated score requisite will have health insurance and 3 weeks' (15 workdays) vacation with pay.

FARM RECORDS AND ACCOUNTS[12]

Modern farming is more than a job; it is a business. Therefore, it should be conducted in a businesslike manner. This means that there should be adequate records and accounts.

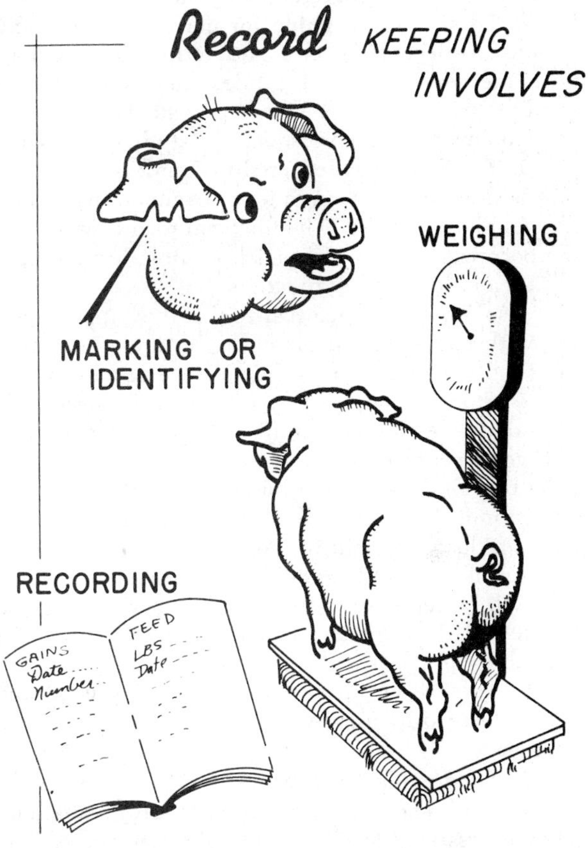

Fig. 2-13. Livestock records and accounts involve three primary essentials: (1) marking or identifying each animal, (2) weighing over the scales, and (3) keeping a farm record and account book. (Drawing by R. F. Johnson)

Why Keep Records?

There are many reasons for keeping good records, the most important of which follow:

1. **To provide profit and progress indicators**—Production records on livestock and crops are profit indicators, and a way to measure progress.

2. **To guide changes in enterprises**—Farm and ranch records should provide information from which the farm business may be analyzed, with its strong and its weak points ascertained. From the facts thus determined, the manager may adjust current operations and develop a more profitable organization. The

[12]In the preparation of this section, the author had the benefit of the authoritative review and suggestions of Prof. A. H. Harrington, Department of Agricultural Economics, Washington State University.

enterprise should be above average before the owner borrows to expand it. Is the ranch too small? Is it more profitable to sell weaners or yearlings? Should the farm or ranch produce or buy hay?

3. **To provide a net worth statement**—Farm and ranch records should provide a net worth statement, showing financial progress during the year.

4. **To serve as a guide for current income and expenses**—Records of cost and returns of previous years are very valuable as guides, and as a means of spotting trouble. Items which deviate substantially from the historical record should be studied with care.

5. **To obtain needed credit**—Lenders need certain basic information in order to evaluate the soundness of a loan request. The financial record will show the borrower's current financial position, his potential ahead, and his liability to others.

Also, stockmen must realize that they are competing with many other users of credit, including retail merchants, manufacturers, home buyers, and professional people. Many of these borrowers can and do provide the lender with profit and loss statements and net worth statements, prepared by a CPA. Also, they usually submit (a) annual budgets, and (b) cash flow projections, showing when money is needed, and showing loan repayability. As stockmen increase the amount of borrowed capital in their operations, they, too, will be required to furnish adequate records and budgets if they are to compete successfully for the available capital.

6. **To save on income taxes**—Keeping good and adequate records usually makes for a savings in income taxes.

7. **To provide for continuity of the business**—"Barn door" and "memory" records are insufficient. They mitigate against continuity of the business. The sudden passing of a manager places a severe stress on a business even under the most favorable circumstances. However, a good set of records gives a big assist to those who must take over during such times.

8. **To keep a historical record**—Good and complete historical records are needed for future reference purposes.

Types of Records

In general, the functions enumerated under the earlier section headed "Why Keep Records?" can be met by the following types of records:

1. **Annual inventory**—The annual inventory is the most valuable record that a farmer can keep. It should include a list and value of real estate, livestock, equipment, feed, supplies, and all other property, including cash on hand, notes, bills receivable, and growing crops. Also, it should include a list of mortgages, notes, and bills payable. It shows the farmer what he owns and what he owes; whether he is getting ahead or going behind. The following pointers may be helpful relative to the annual inventory.

Fig. 2-14. Annual inventory, the most valuable record on the farm or ranch.

a. **Time to take inventory**—The inventory should be taken at the beginning of the account year; usually this means December 31 or January 1.

b. **Proper and complete listing**—It is important that each item be properly and separately listed.

c. **Method of arriving at inventory values**—It is difficult to set up any hard and fast rule to follow in estimating values when taking inventories. Perhaps the following guides are as good as any:

(1) **Real estate**—Estimating the value of farm real estate is, without doubt, the most difficult of all. It is suggested that the farmer use either (a) the cost of the farm, (b) the present sale value of the farm, or (c) the capitalized rent value according to its productive ability with an average operator.

(2) **Buildings**—Buildings are generally inventoried on the basis of cost less observed depreciation and obsolescence. Once the original value of a building is arrived at, it is usu-

ally best to take depreciation on a straight line basis by dividing the original value by the estimated life in terms of years. Usually 4 percent or more depreciation is charged off each year for income tax purposes.

(3) **Livestock**—Animals are usually not too difficult to inventory because there are generally sufficient current sales to serve as a reliable estimate of value.

(4) **Machinery**—The inventory value of machinery is usually arrived at by one of two methods: (a) the original cost less a reasonable allowance for depreciation each year, or (b) the probable price that it would bring at a well-attended auction.

Under conditions of ordinary wear and reasonable care, it can be assumed that the general run of farm machinery (except trucks and autos) will last about 10 years. Thus, with new machinery, the annual depreciation will be the original cost divided by 10.

(5) **Feed and supplies**—The value of feed and supplies can be based on market price.

Two further points are important. Whatever method is used in arriving at inventory value (a) should be followed at both the beginning and the end of the year, and (b) should reflect the operator's opinion of the value of the property involved.

2. **Daily and weekly report**—A farm or ranch manager should keep a daily record, like the daily-weekly report shown in Fig. 2-15. It takes little time to keep it. Nevertheless, a certain time should be set aside daily for this purpose, thereby assuring that it

"BAR-NONE RANCH," FOR WEEK BEGINNING ____________ (month, day, year), Prepared by ____________

	Monday	Tuesday	Wednesday	Thursday	Friday	Saturday	Sunday	Comments for Day or Week
WEATHER Rain or snow, inch .... Temp.—high .... low								
LABOR Accident or sick .... Who, what, how .... Changes								
EQUIPMENT BREAKDOWN Kind and make .... Cause .... Cost to repair								
ANIMAL HEALTH PROGRAM (indicate cow-calf/cattle feedlot) Vaccination .... Treatment								
ANIMAL LOSSES (indicate cow-calf/cattle feedlot) No. .... Kind .... Cause								
ANIMALS RECEIVED (indicate cow-calf/cattle feedlot) No. .... Kind .... From .... Price or custom feed								
ANIMALS SOLD (indicate cow-calf/cattle feedlot) No. .... Kind .... To .... Price								
CROP HARVESTED Kind .... Field No. .... Acreage .... Yield/acre								
VISITORS Name and address								

Fig. 2-15. A good type of daily and weekly report.

will be kept—and that it will be kept accurately. For the manager, such a report provides an invaluable record of the day-by-day operations. For the owner, it's a quick and easy way to keep informed. This record should be filed, where it can be referred to as needed.

3. **Record of livestock and crop production**—A record of the production and sale of animals and their products, and of the yield of crops, is most important, for the success of the farm depends upon production. Such records are important profit indicators and help in analyzing the farm business. They may be few or many, depending upon the wishes of the operator and the class of livestock. The production records of a cow-calf operator, for example, will usually include percent calf crop dropped, percent calf crop weaned, weaning weight of calves at seven months, the pounds of beef per cow or per acre, the pounds of feed required to produce a pound of beef, and death losses.

4. **Record of receipts and expenses**—Such a record is essential to any type of well-managed business. To be most useful, these entries should not only record the amount of the transaction, but should give the source of the income or the purpose of the expense, as the case may be. In other words, they should show the farmer from what sources the income is derived and for what it is spent.

The following kinds and arrangements of farm record books are commonly used for recording receipts and expenditures:

a. Those that devote a separate page to each enterprise; that is, a separate page is used for the beef cattle finishing enterprise, another for swine, still another for sheep, and so on.

b. Those that provide for a record of receipts and expenses on the same page, using one column for receipts and another for expenses. This type is easy to keep, but very difficult to analyze from the standpoint of any particular enterprise.

c. Those that combine the features of both 1 and 2 above. The latter are more difficult to keep than the others, and may be confusing to the person keeping the record.

Household and personal accounts should be kept, but should be handled entirely separate from farm accounts because they are not farming expenses as such.

5. **Financial records**—The name of the game is "profit." Thus, it is necessary that production be translated into dollars and weighed against costs involved in achieving that production.

Many different kinds of financial records are used by stockmen; among the most common ones are the following: annual inventory, budgets, income, expenses, cash flow, depreciation schedule, annual net income, enterprise accounts, profit and loss (P&L) statement, and net worth statement.

Kind of Record and Account Book

A farmer can make his own record book by simply ruling off the pages of a bound notebook to fit his specific needs, but the saving is negligible. Instead, it is recommended that he obtain a copy of a farm record book prepared for and adapted to his area. Such a book may usually be obtained at a nominal cost from the agricultural economics department of each state college of agriculture. Also, certain commercial companies distribute very acceptable farm record and account books at no cost.

At the outset, it should be recognized that a farm record should be easy to keep and should give the information desired to make a valuable analysis of the business.

Most farm record and account books contain simple and specific instructions relative to their use.

Who Shall Keep the Records?

The records may be kept either by someone in the farm or ranch business, or by someone hired to perform this service—a professional.

• **Farm or ranch help**—Very frequently, farm or ranch records are kept by the wife. If she has adequate time, there is no reason why she cannot keep good records.

Most farmers are not good record keepers, primarily because the operation of an ordinary farm requires large amounts of physical labor. As a result, most farmers and ranchers are physically exhausted at the end of the day's work and have neither the time nor the ambition to record in at least four different places each transaction which occurred during the day, as would be necessary of the usual double-entry bookkeeping system.

• **A professional**—There are, of course, individuals and firms who make a business of keeping farm or ranch records. Usually, they are accountants or farm management specialists.

How Shall Records Be Kept?

Records may be kept either by hand or by computer. Accurate and up-to-the-minute records and controls have taken on increasing importance in all agriculture, including the livestock business, as the investment required to engage in farming and ranching has risen and profit margins have narrowed. Today's successful farmers and ranchers must have, and use, as complete records as any other business. Also, records must be kept current; it no longer suffices merely to know the bank balance at the end of the year.

• **By hand**—The hand system can be used, but is slow and tedious. This service is usually performed by a member of the family or by an accountant living in the community. Nevertheless, after learning what is wanted, a first-rate accounting system can be kept by hand.

• **By computer**—Big and complex agricultural enterprises have outgrown hand record keeping. It's too time-consuming, with the result that it doesn't allow management enough time for planning and decision making. Additionally, it does not permit an all-at-once consideration of the complex interrelationships which affect the economic success of the business. This has prompted a new computer technique known as linear programming.

Whether a stockman should keep his own records or hire a professional, and whether records should be kept by hand or by computer, each individual must decide for himself. For the most part, the decision should be based on weighing the usefulness of the information each provides against the cost of obtaining it.

Summarizing and Analyzing the Records

At the end of the year, the second or closing inventory should be taken, using the same method as was followed in taking the initial inventory. The final summary should then be made, following which the records should be analyzed. In the latter connection, the farmer should remember that the purpose of the analysis is not to prove that he has or has not been prosperous. He probably knows the answer to this question already. Rather, the analysis should show actual conditions on the farm and point out ways in which these conditions may be improved.

Although the farmer or rancher can summarize and analyze his own records, there are many advantages in having the services of a specialist for this purpose. Such a specialist is in a better position to make a "cold" appraisal without prejudice, and to compare enterprises with those of other similar operators. Thus, the specialist may discover that, in comparison with other operators, the hogs on a given farm are requiring too much feed to make a hundred pounds of gain, or that the steer feeding enterprise is much less profitable than others have experienced. The local county agent can either render or recommend such specialized assistance. In some areas, it may consist in joining a cooperative farm record group or engaging the services of a consultant; in some states, such service is provided by the state agricultural college.

ANALYZING A LIVESTOCK BUSINESS—IS IT PROFITABLE?

Most people are in business to make money—and stockmen are people. In some areas, particularly near cities and where the population is dense, land values may appreciate so as to be a very considerable profit factor. Also, a tax angle may be important. But neither of these should be counted upon. The livestock operation should make a reasonable return on the investment; otherwise, the owner should not be in the business.

For land and livestock to be profitable, they should yield a return sufficient to the owner to (1) meet the interest payment on the investment, (2) retire a reasonable portion of the loan, and (3) provide satisfactory management return. But there is no more reason why large land and livestock holdings should be debt-free than there is for General Motors, or any other big corporation, to be debt-free.

A livestock owner or manager needs to analyze his business—to determine how well he's doing. With big operations, it's no longer possible to base such an analysis on the bank balance statement at the end of the year. In the first place, once per year is not frequent enough, for it is possible to go broke, without really knowing it, in that period of time. Secondly, a balance statement gives no basis for analyzing an operation—for ferreting out its strengths and weaknesses. In large cattle, lamb, and hog feeding operations, it is strongly recommended that progress be charted by means of monthly or quarterly closings of financial records.

Also, a stockman must not only compete with other stockmen down the road, but he must compete with himself—with his record last year and the year before. He must work ceaselessly at making progress, of improving the end product, and lowering costs of production.

To analyze a livestock business, two things are essential: (1) good records; and (2) yardsticks, or profit indicators, with which to measure an operation.

Profit indicators are gauges for measuring the primary factors contributing to profit. In order for a stockman to determine how well he's doing, he must be able to compare his own operations with something else; for example, (1) his own historical five-year average, (2) the average for the U. S. or for his particular area, or (3) the top 5 percent. The author favors the latter, for high goals have a tendency to spur superior achievement.

Space limitations will not permit presentation in this book of all conceivable profit indicators for all classes of livestock. Hence, only two—cow-calf profit indicators and cattle feedlot profit indicators—will follow. However, these can readily be adapted to sheep, swine, horse, dairy, and poultry operations.

Like most profit indicators, the ones given in Table 2-13 and 2-14 are not perfect. But they will serve as useful guides. Also, on some establishments, there may be reason for adding or deleting some of the indicators; and this can be done. The important thing is that each operation have adequate profit indicators, and that these be applied as frequently as possible; in a cattle feedlot, for example, this may be done monthly with some indicators.

Cow-Calf Profit Indicators

Many factors determine the profitableness of a cow-calf enterprise. Certainly, a favorable per animal unit capital investment in land and improvements is a first requisite. Additionally, percent calf crop weaned and weaning weight are exceedingly important, as shown in Table 2-13.

TABLE 2-13

SIZE AND WEIGHT OF CALF CROP WEANED ARE IMPORTANT

	Yearly Operating Cost—$300 per Cow-Calf Weights				
Calf Crop	450 Lb	425 Lb	400 Lb	375 Lb	350 Lb
(%)					
90	405 74.1¢	382 78.5¢	360 83.3¢	337 89.0¢	315 95.2¢
85	382 78.5¢	361 83.1¢	340 88.2¢	319 94.0¢	298 $1.01
80	360 83.3¢	340 88.2¢	320 93.7¢	300 $1.00	280 $1.07
75	337 89.0¢	319 94.0¢	300 $1.00	281 $1.07	263 $1.14
	Break-even point @ 88-89 cents/lb.				

In Table 2-13, it is assumed that the yearly operating cost is $300 per cow. Then, the effect of size and weight of calf crop is computed. As shown, with a 90% calf crop and an average weaning weight per cow bred of 405 lb (450 × 90% = 405), a selling price of 74.1¢/lb will meet the break-even cost of $300. With a 75% calf crop, because of fewer pounds per cow bred (450 × 75% = 337 lb), the calves would have to bring 89.0¢/lb in order to break even.

Table 2-14 will serve as a yardstick for determining (1) how you stack up with the nation's (a) average, and (b) top five percent of cow-calf operators; and (2) where you are falling down in your cow-calf enterprise.

It is noteworthy that Table 2-14 reveals that the top five percent operators have a higher investment/animal unit in land and improvements than their average counterparts. Obviously, better operators have better land and improvements; their savings are made in the handling of the herd. It's not unlike selecting a ration—where it's net returns, rather than cost per ton, that counts.

No claim is made that Table 2-14 is perfect. Admittedly, there are wide area differences, and no two farms or ranches are alike. Also, there are seasonal differences; for example, a drought will materially affect weaning weight of calves. Yet, Table 2-14 will serve as a useful guide.

Cattle Feedlot Profit Indicators

Cattle feedlot operators need to keep good records and make frequent analyses (at least once monthly) to determine how well they are doing. A determination of assets and liabilities at the end of the fiscal year is not good enough, primarily because it is available only once per year. Cattle feeding requires much capital; hence, records should be kept as current as possible at all times.

TABLE 2-14

COW-CALF PROFIT INDICATORS[1]

	Average for U.S. Cow-Calf Operations			
	Commercial		Purebred	
	Average	Top 5%	Average	Top 5%
Investment/animal unit (one mature cow) in land and improvements (real estate) .... ($)	2,000	2,250	2,500	3,000
Percent calf crop dropped (based on no. cows bred) .... (%)	88	93	89	94
Percent of calf crop weaned (based on no. cows bred) .... (%)	81	89	82	91
Weaning weight of calf at 7 mo. .... (lb)	425	525	450	550
Age and longevity of cows:				
Age when removed from herd .... (yr)	9	10	9.5	10.5
No. calves produced in lifetime of cow	6.0	8.0	7.0	8.5
Labor/cow/year .... (hr)	10.0	9.5	16.0	13.5
Net return per cow to management[2] .... ($)	27.25	41.50	38.00	95.00

[1]This is a consensus (or judgment) arrived at from the following knowledgeable sources: National Cattlemen's Association (*Cattle-Fax*), heads of university animal science departments, and agricultural consultants. A consensus was resorted to for the reasons that (1) no extensive, nationwide, scientific sampling of cattlemen on these matters has ever been made; (2) there was considerable variation in the figures obtained from the various sources; and (3) this information is much needed by cattlemen and those who counsel with them. No claim is made relative to the scientific accuracy of the data; rather, it is presented (1) because it is the best information of its kind presently available on a nationwide basis, and (2) with the hope that it will stimulate needed research along these lines.

[2]Net return to management after deducting from gross receipts all costs, including depreciation on machinery, buildings and cattle, and interest on investment.

Operators are primarily interested in 2 questions; namely, (1) how well am I doing—profitwise, and (2) how do I compare with other feedlots? Table 2-15 gives some guidelines for answering these questions. It may be used in making an analysis of a specific feedlot, for determining (1) the strengths and weaknesses within the feedlot, and (2) how it stacks up with the top 5 percent of the nation's feedlots. It is noteworthy that the top 5 percent operators invest more in land and equipment than their average counterparts. Obviously, they effect savings in operation rather than in physical plant.

Admittedly, profit indicators, such as those given in Table 2-15, are not perfect, simply because no two feedlots are the same. Nationally, there are wide differences in climate, feeds, land costs, salaries and wages, and other factors. Nevertheless, indicators per se serve as a valuable yardstick. Through them, it is possible to measure how well a given feedlot is doing— to ascertain if it is out of line in any one category, and, if so, the extent of same.

After a few years of operation, it is desirable that a feedlot evolve with its own yardstick and profit indicators, based on its own historical records and averages. Even with this, there will be year to year fluctuations due to seasonal differences, cattle and feed price changes, disease outbreaks, changes in managers, wars and inflation, and other happenstances.

TABLE 2-15

CATTLE FINISHING PROFIT INDICATORS[1]

	Calves		Yearlings	
	Av. for U.S. Feedlots	Top 5% of U.S. Feedlots	Av. for U.S. Feedlots	Top 5% of U.S. Feedlots
Initial (new) land, open feedlot, and equipment cost basis/animal capacity ($)	60.00	63.00	65.00	68.00
Feedlot and equipment cost charged off/animal finished out ($)	4.50	4.00	3.50	2.50
Daily nonfeed costs/animal[2] (¢)	12.0	10.0	10.0	9.0
Salaries and wages/head/day (¢)	3.5	3.25	3.25	3.0
Death, losses (%)	2.0	1.5	1.0	0.8
Vet fees and medicine/head/day (¢)	5.5	5.0	4.5	4.0
Feed/lb gain (feed as fed basis):				
Steers (lb)	7.2	6.9	8.0	7.6
Heifers (lb)	7.3	7.0	8.6	8.0
Daily rate of gain:				
Steers (lb)	2.3	2.5	2.8	3.0
Heifers (lb)	2.0	2.15	2.5	2.7
Net return per head to management[3] ($)	14.00	17.00	15.00	18.00

[1]This is a consensus (or judgment) arrived at from the following knowledgeable sources: National Cattlemen's Association (*Cattle-Fax*), heads of university animal science departments, and agricultural consultants. A consensus was resorted to for the reasons that (1) no extensive, nationwide, scientific sampling of cattlemen on these matters has ever been made, (2) there was considerable variation in the figures obtained from the various sources; and (3) this information is much needed by cattlemen and those who counsel with them. No claim is made relative to the scientific accuracy of the data; rather, it is presented (1) because it is the best information of its kind presently available on a nationwide basis, and (2) with the hope that it will stimulate needed research along these lines.

[2]This embraces all costs other than cattle and feed. It includes salaries and wages; taxes, interest, insurance; utilities; gasoline, oil, grease; depreciation; repairs; vet and medical; consultant and legal service; trucking; promotion; and other costs.

[3]This means just what it says; net return to management after deducting from gross receipts all costs, including depreciation on machinery and buildings, and interest on investment.

BUDGETS IN THE LIVESTOCK BUSINESS

A budget is a projection of records and accounts and a plan for organizing and operating ahead for a specified period of time. A short-time budget is usually for one year, whereas a long-time budget is for a period of years. The principal value of a farm budget is that it provides a working plan through which the operation can be coordinated. Changes in prices, droughts, and other factors make adjustments necessary. But these adjustments are more simply and wisely made if there is a written budget to use as a reference.

How to Set Up a Budget

It's unimportant whether a printed form (of which there are many good ones) is used or one made up on an ordinary ruled 8½" × 11" sheet placed sidewise. The important things are: (1) that a budget is kept, (2) that it be on a monthly basis, and (3) that the operator be "comfortable" with whatever forms or system is used.

No budget is perfect. But it should be as good an estimate as can be made—despite the fact that it will be affected by such things as droughts, diseases, markets, and many other unpredictables.

A simple, easily kept, and adequate budget can be prepared by using the following three types of forms:

1. Annual cash expense budget (see Table 2-16).
2. Annual cash income budget (see Table 2-17).
3. Annual cash expense and income budget—cash flow (see Table 2-18).

(See pages 72 and 73 for Tables 2-16, 2-17, and 2-18.)

TABLE 2-16
ANNUAL CASH EXPENSE BUDGET[1]

______________________________ for 19 ________
(name of farm or ranch)

Item	Total	Jan.	Feb.	Mar.	Apr.	May	June	July	Aug.	Sept.	Oct.	Nov.	Dec.
Labor hired													
Feed purchased													
Gas, fuel, grease													
Taxes													
Insurance													
Interest													
Utilities													
etc.													
etc.													
etc.													
Total													

[1]The Annual Cash Expense Budget should show the monthly breakdown of various recurring items—everything except the initial loan and capital improvements. It includes labor, feed, supplies, fertilizer, taxes, interest, utilities, etc.

The annual cash expense budget (Table 2-16) should show the monthly breakdown of various recurring items—everything except the initial loan and capital improvements. It includes labor, feed, supplies, fertilizer, taxes, interest, utilities, etc.

The annual cash income budget (Table 2-17) is just what the name implies—an estimated cash income by months.

The annual cash expense and income budget (Table 2-18) is a cash flow budget, obtained from the first two forms. It's a money "flow" summary by months. From this, it can be ascertained when, and how much, money will need to be borrowed, and the length of the loan along with a repayment schedule. It makes it possible to avoid tying up capital unnecessarily, and to avoid unnecessary interest.

How to Figure Net Income

Table 2-18 shows a gross income statement. There are other expenses that must be taken care of before net profit is determined; namely:

1. **Depreciation on buildings and equipment**—It is suggested that the "useful life" of buildings and equipment be as follows, with depreciation accordingly: buildings, 15 years; and machinery and equip-

TABLE 2-17
ANNUAL CASH INCOME BUDGET[1]

______________________________ 19 ________
(name of farm or ranch)

Item	Total	Jan.	Feb.	Mar.	Apr.	May	June	July	Aug.	Sept.	Oct.	Nov.	Dec.
500 steers													
430 bu oats													
etc.													
etc.													
etc.													
Total													

[1]The Annual Cash Income Budget is just what the name implies—an estimated cash income by months.

TABLE 2-18

ANNUAL CASH EXPENSE AND INCOME BUDGET (cash flow)[1]

______________________________ for 19 ________

(name of farm or ranch)

Item	Total	Jan.	Feb.	Mar.	Apr.	May	June	July	Aug.	Sept.	Oct.	Nov.	Dec.
Gross income	25,670					1,000	1,000						
Gross expense	13,910					575	2,405						
Difference	11,760					425	1,405						
Surplus (+) or Deficit (−)	+					+	−						

[1]The Annual Cash Expense and Income Budget is a cash flow budget, obtained from the first two forms. It's a money "flow" summary by months. From this can be ascertained when, and how much, money will need to be borrowed, the length of the loan and a suitable repayment schedule. It makes it possible to avoid tying up capital unnecessarily, and to avoid unnecessary interest.

ment, 5 years. These depreciation figures are in compliance with the Economic Recovery Tax Act of 1981, Accelerated Cost Recovery System (ACRS).

2. **Interest on owner's money invested in farm and equipment**—This should be computed at the going rate in the area, say 12%.

Here's an example of how the above works: Let's assume that on a given farm there was a gross income of $200,000 and a gross expense of $125,000, or a surplus of $75,000. Let's further assume that there are $60,000 worth of machinery, $60,000 worth of buildings, and $200,000 of the owner's money invested in farm and equipment. Let's further assume that buildings are being depreciated in 15 years and machinery in 5 years. Here is the result:

Gross profit		$75,000
Depreciation—		
Machinery: $ 60,000 @ 20% =	$12,000	
Buildings: $ 60,000 @ 6.67% =	4,002	
	$16,002	
Interest: $200,000 @ 12% =	24,000	
		40,002
Return to labor and management		$34,998

Some people prefer to measure management in terms of return on invested capital, and not wages. This approach may be accomplished by paying management wages first, then figuring return on investment.

Enterprise Accounts

Where a cattle enterprise is diversified (for example, a farm or ranch having a cow-calf operation, a feedlot, and crops), enterprise accounts should be kept—in this case three different accounts for three different enterprises. The reasons for keeping enterprise accounts are:

1. It makes it possible to determine which enterprises have been most profitable, and which least profitable.
2. It makes it possible to compare a given enterprise with competing enterprises of like kind, from the standpoint of ascertaining comparative performance.
3. It makes it possible to determine the profitableness of an enterprise at the margin (the last unit of production). This will give an indication as to whether to increase the size of a certain enterprise at the expense of an alternative existing enterprise when both enterprises are profitable in total.

COMPUTERS IN THE LIVESTOCK BUSINESS

Accurate and up-to-the-minute records and controls have taken on increasing importance in all agriculture, including the livestock business, as the investment required to engage in farming and ranching has risen and profit margins have narrowed. Today's successful farmers and ranchers must have, and use, as complete records as any other business. Also, records must be kept current; it no longer suffices merely to know the bank balance at the end of the year.

The computer technique known as linear programming is similar to budgeting, in that it compares several plans simultaneously and chooses from among them the one likely to yield the highest returns. It is a way in which to analyze a great mass of data and consider many alternatives. It is not a managerial genie, nor will it replace decision-making managers. However, it is a modern and effective tool in the present age, when just a few dollars per head or per acre can spell the difference between profit and loss.

There is hardly any limit to what computers can do if fed the proper information. Among the difficult questions that they can answer for a specific farm or ranch are:

1. **How is the entire operation doing so far?**—It is possible to obtain quarterly or monthly progress reports, often making it possible to spot trouble before it's too late.

2. **What farm enterprises are making money; which ones are freeloading or losing?**—By keeping records by enterprises— cow-calf, cattle feedlot, hogs, wheat, corn, etc.—it is possible to determine strengths and weaknesses, then either to rectify the situation or to shift labor and capital to a more profitable operation. Through "enterprise analysis," some operators have discovered that one part of the farm business may earn $10 or more per hour for labor and management, whereas another may earn only $2 per hour, and still another may lose money.

3. **Is each enterprise yielding maximum returns?**—By having profit, or performance, indicators in each enterprise (see Tables 2-14 and 2-15), it is possible to compare these (a) with the historical average of the same farm or ranch, or (b) with the same indicators of other farms or ranches.

4. **How does this ranch stack up with its competition?**—Without revealing names, the computing center (local, state, area, or national) can determine how a given ranch compares with others—either the average, or the top (say 5%).

5. **How to plan ahead?**—By using projected prices and costs, computers can show what moves to make for the future; they can be a powerful planning tool. They can be used in determining when to plant, when to schedule farm machine use, etc.

6. **How can income taxes be cut to the legal minimum?**—By keeping an accurate record of expenses and figuring depreciations accurately, computers make for a saving in income taxes on most farms and ranches.

7. **What are the "least cost" and "highest net returns" rations?**—Instruction on how to balance a ration by computer is given in this book in Section 4, Balanced Ration, How to Balance Rations, Computer Method. Hence, the reader is referred thereto.

For providing answers to the above questions and many more, computer accounting costs an average of about 1 percent of the gross farm income. By comparison, it is noteworthy that city businesses pay double this amount.

There are three requisites for linear programming a farm or ranch; namely:

1. Access to a computer.
2. Computer know-how, so as to set the program up properly and be able to analyze and interpret the results.
3. Good records.

The pioneering computer services available to farmers and ranchers were operated by universities, trade associations, and the government, with most of them being on an experimental basis. Subsequently, others have entered the field, including commercial data processing firms, banks, machinery companies, feed and fertilizer companies, and farm suppliers. They are using it as a "service sell," as a replacement for the days of the "hard sell."

Programmed farming is here to stay, and it will increase. Space limitations will not permit elucidation in this book of the role of computers in each kind of livestock operation. Only two kinds—cow-calf, and cattle feedlots—will follow. However, the same principles can be adapted to any other livestock operation.

Computers in Cow-Calf Operations

In the past, the biggest deterrent to Production Testing on cow-calf operations, both purebred and commercial, has been the voluminous and time-consuming record keeping involved. Keeping records as such does not change what an animal will transmit, but records must be used to locate and propagate the genetically superior animals if genetic improvement is to be accomplished.

Production Testing has been covered elsewhere in this book (see Section 3); thus, repetition at this point is unnecessary. As has been pointed out, two factors contribute to optimum beef production; namely, (1) heredity, and (2) environment. Records of traits must be adjusted for such well-known sources of variation as age of dam, age of calf, and sex. This is tedious and time-consuming when records are kept and analyzed by hand. However, punched card equipment, or a computer, can handle this assignment efficiently, and with less risk of errors and omissions than the hand method. Several processing centers—universities, state associations, performance registries, breed associations, cooperatives, and private firms—now offer this service. Their goals are similar, but their methods of accomplishing them differ slightly. The important thing is to use one of them; generally speaking, this should be the one in use in the area which best fits the individual needs.

In addition to their use in production testing, computerized records can be, and are, used for herd record purposes—as a means of keeping management up to date and serving as an alert on problems or work to be done. Each animal must be individually identified. Reports can be obtained at such intervals as desired, usually monthly or every two weeks. Also, the owner can keep as complete or as few records as desired, and the system may be adapted to either purebred or commercial herds. Of course, commercial operators do not need pedigrees. Here are several of the records that can be kept by computer:

1. Pedigrees.
2. Animals that need attention such as:

   a. Animals 4 months old that are unregistered.
   b. Females 6 months old that are unvaccinated.
   c. Bulls that are 6 months old, and which should either be marked to keep or be castrated.
   d. Heifers 18 months old that haven't been bred.
   e. Cows that have been bred 2 consecutive times.
   f. Cows not rebred 3 months after calving.
   g. Cows that have reached 9 years of age, and that may be getting shelly and should be culled.
   h. Cows due to calve in 30 days.
   i. Calves 7 months of age that haven't been weaned.
   j. Calves 7 months of age that haven't been scored and weighed.

3. A running, or cumulative, inventory of the herd, by sex, including calves dropped, calves due, and purchases and sales—in number of animals and dollars.
4. The depreciation of purchased animals according to the accounting method of choice.

It is predicted that "cardpokes" will come to be almost as well known in the future as "cowpokes" are at present.

Computers in Cattle Feedlots

Ration formulation is only one use of computers in cattle feedlots. As every cattleman knows, many management decisions are involved in obtaining the highest net returns from a given lot of cattle; and the bigger and the more complex the feedlot, the more important it is that the decisions be right.

Problems can be solved without the aid of a computer. But the machine has the distinct advantages of (1) speed, (2) coming up with answers to each of several problems simultaneously, and (3) offering the best single alternative, all factors considered. Among the possible uses of computers in cattle feedlots, particularly the larger operations, are:

1. Ration formulation.
2. How to determine the best ingredient buy.
3. The most profitable ration—the one that costs the least per pound of gain produced. In many cases, the cheapest ration will actually increase cost of gain. Also, rations that produce the most rapid daily gains may be too costly; for example, silage rations will not produce as high gains as an all-concentrate ration, yet the net profit from their use may be greater.
4. The most profitable kind of cattle—age, weight, sex, and grade—to feed in relation to available feeds and feed prices.
5. Seasonal differences in performance of cattle in a given feedlot.
6. As a means of forecasting profits or losses; with all-at-once consideration given to feeder prices, slaughter cattle outlook, probable rate of gain, and interest and overhead costs.
7. As a means of keeping and updating the voluminous daily feed transfers from feed inventories, to mixed rations, to records for each lot of cattle. Accurate and current feed inventories are necessary for wise ingredient buying and for financing feed inventories; and accurate and current feed records by lots are important for both privately owned and custom fed cattle.
8. When to sell cattle for maximum profit.

BUSINESS ASPECTS OF CATTLE FEEDING

Cattle feeders are businessmen and, like other businessmen, they hope to obtain a reasonably good return for the use of their capital, labor, and management. To this end, their business aspects must become more sophisticated and efficient; they must—

1. Compute break-even prices prior to buying feeder cattle, especially if they do not buy and sell each week.
2. Buy feeder cattle of the right size, quality, and price.
3. Sell the cattle to the best advantage.
4. Integrate when possible.
5. Feed cattle to weight and grade.
6. Evaluate performance.
7. Obtain economies with size.
8. Finance the feedlot and cattle properly and adequately.

Of course, the above eight points represent a great oversimplification of a complex business, but they do clearly set forth the main requisites for profitable cattle feeding. Anyone who wishes to make money feeding cattle must have expertise in these eight areas.

Business aspects outweigh all other factors—feed additives, crossbreds, pollution control, etc.—producing change in cattle feeding. It is important, therefore, that feeders and those who counsel with them be thoroughly grounded in each of these areas.

Other factors than the cost of feed, amount of gain, and price of slaughter cattle affect the price that a feeder can afford to pay for feeder cattle; among them, are the following:

1. Condition of the cattle. Thin cattle, if in good health, will make faster gains than fleshy cattle.

2. Growthy cattle—cattle that are big framed and on the rangy order—make better gains than the little, compact kind; and they may be carried to heavier weights. If the feeder cannot obtain cattle backed by production records, "eyeballing" will help.

3. Younger, lighter weight cattle tend to make more efficient gains.

4. Cattle of known, superior ancestry with gaining ability are worth more.

5. Higher grade cattle are worth more. This is so because better grades generally bring a higher selling price, and, therefore, a higher price is obtained on their gains made in the feedlot.

6. The higher the cost of feed, the greater the necessary margin between the cost of feeder cattle and the selling price of finished animals. This is so because of the high cost of gains as compared to their selling price.

7. The longer the feeding period and the greater the gains necessary to get the cattle in a finished condition, the greater the necessary margin. This is also due to the fact that gains made in the feedlot are expensive, sometimes costing more to produce than can be realized in selling on the market.

8. Feeder steers are generally worth approximately $3 to $4 per cwt more than heifers. This is because they gain about 10 percent faster, require 5 to 10 percent less feed, and bring from $.50 to $1.50 per cwt more than heifers when finished. Additionally, there is no pregnancy problem.

9. Good crossbreds will make 2 to 4 percent more rapid and efficient gains than the average of the parent breeds.

Computing Break-Even Prices for Cattle

Those who feed on a large scale and on a continuous basis try to build in some insurance against the consequences of price changes through their buying programs. When finished cattle are sold, these cattle feeders try to replace them with feeders bought at a price which would allow a suitable profit if they were sold at the same time as the finished cattle they replace. To the extent that prices of finished cattle and feeder cattle move together (both in direction and in magnitude), this works reasonably well. But they don't always move together.

Except for big cattle feeders who buy and sell cattle each week as a means of hedging, cattle feeders should compute break-even prices before buying feeder cattle. Even the big operators who buy and sell weekly do not keep the same number of cattle on feed from year to year, or throughout the year. As a result, the demand for feeders tends to be buoyant following a period of good profits from cattle feeding and depressed following a period of low returns or losses.

A nomograph (a graph or chart) may be used in computing break-even prices for cattle. It can give a quick, preliminary idea of cost, price, and investment relationships. But a nomograph should not replace more detailed budgeting which should precede all major buying, selling, and investment decisions. Also, one should realize that a nomograph will give erroneous and misleading information unless based upon accurate and realistic cost and return data from the problem at hand.

Fig. 2-16 is a nomograph for use in making quick calculation of break-even buying and selling prices for cattle.

Note well: The example that follows is presented for the purpose of showing how to use the nomograph; and *not* as an indicator of today's prices. The cattle feeder should insert his own figures.

INSTRUCTIONS FOR USING NOMOGRAPH

This nomograph provides a quick method of computing the "break-even prices" a feeder can afford to pay for incoming cattle, based upon assumptions relative to costs, amount of gain, and the estimated selling price for finished cattle. Likewise, the break-even selling price for finished cattle can be determined, based upon assumptions concerning costs, amount of gain, and the prices of feeder cattle.

The following step-by-step procedure illustrates the use of the nomograph:

Data Needed		Example	Your Farm
1. Purchase weight of feeder cattle	(W_1)	400#	________
2. Selling weight of finished cattle	(W_2)	1,000#	________
3. Feed costs per cwt of gain	(L)	$32.00	________
4. Price per cwt of feeder cattle	(P_1)	$50.00	________
5. Price per cwt of finished cattle	(P_2)	$39.00	________

• **Procedure for determining the break-even price for feeder cattle—**

Step 1—On Chart 1, locate the purchased weight of feeder cattle on scale W_1 and the final weight on scale W_2 and connect the two with a straight line.

Step 2—Read the W_1/W_2 weight-gain ratio on the center scale.

Step 3—On Chart 2, locate the same weight-gain ratio *vertical line,* W_1/W_2, on the scale to the immediate right.

Step 4—Follow this line straight down to the point where it intersects the *diagonal line* which represents the cost per cwt of gain—*Mark this point.*

Step 5—On the vertical P_2 (break-even price), locate the expected selling price per cwt of finished cattle—*Mark this point.*

Step 6—Draw a straight line to connect the two points located above. The point of intersection with the vertical scale P_1 on the left indicates the break-even price per unit for feeder cattle—the price which will allow recovery of costs.

NOTE: In Step 4, a "diagonal cost line" may be

BEEF CATTLE BREAK-EVEN PRICES: DETERMINED BY PURCHASED PRICE, COST ($/CWT.) AND WEIGHT-GAIN RATIO (INITIAL WT./FINAL WT.)

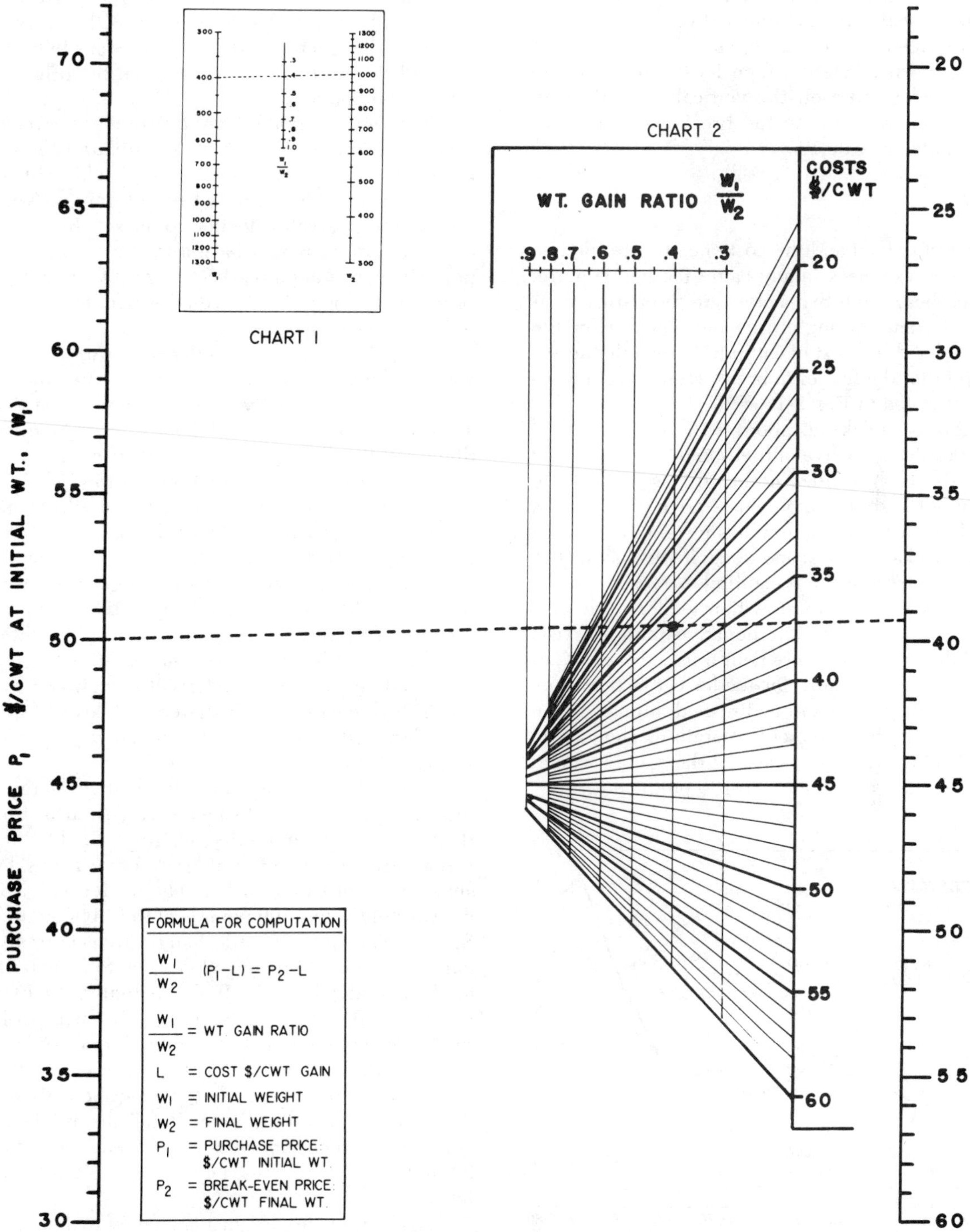

Fig. 2-16. Nomograph for computing the "break-even prices" for cattle. Beef cattle break-even price is determined by purchase price, feed cost ($/cwt), and weight gain ratio. (This ingenious nomograph, and instructions on how to use it, were prepared by Robert M. George and Albert R. Hagan, University of Missouri.)

selected to represent any cost per cwt to be recovered—feed costs only, total variable costs, total variable plus fixed costs, or total costs plus some desired profit per cwt of gain.

• **Procedure for determining the break-even price for finished cattle**—To determine the break-even price one must receive for finished cattle, to recover the costs of feed, feeder cattle, etc., follow the same procedure down to Step 5, then locate the purchase price for feeder cattle on the vertical scale P_1, and, with a straight line, locate the break-even price for finished cattle on scale P_2.

MARGIN

The cost of feed and the cost of cattle are the two major capital expenses in any cattle feeding venture. Likewise, these same two factors are the most important ones in determining profits and losses. How the cost of feed and the cost of cattle influence the necessary margin for the feeder to break even on his operations is indicated in Fig. 2-16.

Generally speaking, about 80% of the cost of finishing cattle (exclusive of the initial purchase price for animals) is for feed. Another 6% is usually absorbed by interest on the purchase price of the cattle. Then labor costs, taxes, purchasing and marketing charges, shrinkage losses, and death losses (about 1 to 2%) make up most of the remaining expenses.

A positive margin exists when feeder cattle cost less than finished cattle. A negative margin exists when feeder cattle cost more than finished cattle. Cattle feeders will pay more for feeder cattle than they expect to receive for them as finished animals when there appears to be a favorable margin on the gain in weight. That is, when they can sell the gain in weight for considerably more than it cost to produce it.

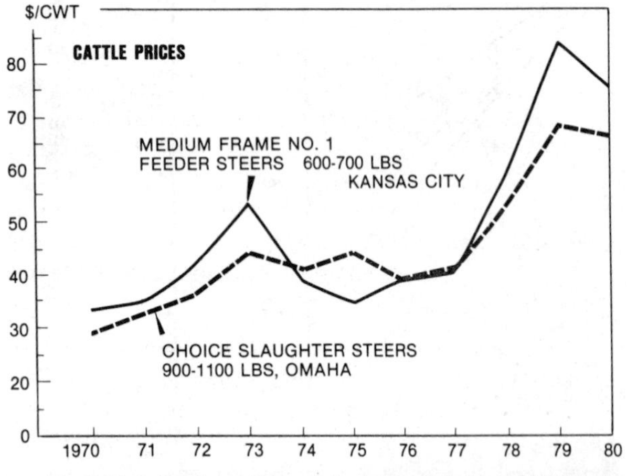

Fig. 2-17. A negative cattle feeder's margin existed until 1974; that is, feeder steers cost more than the price of slaughter steers. (Source: *Livestock and Meat Statistics, Supplement for 1980,* Statistical Bulletin No. 522, USDA)

Profits in cattle feeding come from two different kinds of margins—price margin and feeding margin.

Price margin is the difference between the cost per cwt of the feeder animal and the selling price per cwt of the same animal when finished.

For example, if a feeder pays $66 per cwt for a 600-pound steer and sells him for $63 per cwt, the price margin is a negative $3. This means that the cattle feeder would take an $18 loss on the original 600 pounds he bought.

Feeding margin is the difference between the cost of putting on 100 pounds of gain and the selling price per cwt of the same animal when finished. Thus, if it costs $50 per cwt to put gain on yearling steers, and if a cattle feeder could sell his cattle for $63 per cwt, he would have a feeding margin of $13 per cwt. Assuming a market weight of 1,000 pounds, or of 400 pounds gain, he could expect to make about $52 on feeding margin.

The amount a cattle feeder makes as a result of a good feeding margin can more than offset the losses accruing from a negative price margin, but it doesn't always work that way. It depends on many different things—the selling price, the cost of gain, the price paid for feeder animals, and other factors.

In the example just cited, the feeding margin amounted to $52 per animal. This is not to suggest, however, that cattle feeders should always put more gain on yearling steers. How much gain a cattle feeder should put on depends upon the kind of feeding program being followed, the kind and condition of the feeder cattle when they go into the lot, rate of gain, and several other factors. Research and experiences have clearly demonstrated that costs of gain go up pretty fast if cattle are fed much beyond Choice slaughter grade.

The principles of profits from price margin and feeding margin apply to feeder calves, also. But the relative importance of price margin vs feeding margin is not quite the same for calves as for yearlings. Let us analyze the situation further: The feeder who is feeding yearlings buys 600 pounds of the 1,000 pounds he finally sells. Getting cattle bought right is pretty important to him. If he pays $1 per cwt too much for his feeders, that takes $6 off the potential profit from price margin and from total profits. On the other hand, the feeder who buys calves is more interested in costs of gain and feeding margin than in price margin because about 60 percent of the weight he sells is from the gain he puts on in the feedlot. Thus, if he pays $1 per cwt too much for a 400-pound feeder calf, it hurts, but not quite so much—$4 compared to $6 per cwt per head.

If a farmer-feeder just manages to balance his gains from feeding margin with his losses from price margin, this does not necessarily mean that he should not feed cattle. Actually, he isn't in too bad shape.

He's getting paid market price for his feed, a going wage for his labor, around eight percent on his own capital, and he's getting enough to cover all fixed costs like depreciation, taxes, etc., on his lot and equipment. With the commercial feeder, it's another story. He usually buys most of his feed, and he operates on borrowed capital. Thus, to stay in business, he must turn in a profit over and above these costs.

Custom (Contract) Feeding[13]

Custom cattle feeding is the feeding of cattle for a fee, usually without taking ownership of the animals.

Contract feeding is not new. It made rapid development after 1929, and there was much of it during the severe drought of 1934. From this time to World War II, contract feeding decreased in importance—a decline attributed to improved feed conditions on the western range, higher prices for feeder animals, and the availability of more credit through federal and private loan agencies.

Custom cattle feeding as we know it today paralleled the development of commercial feedlots. California pioneered in it; thence it spread to Arizona, the Northwest, and other areas of the West, Nebraska, Texas, the Oklahoma Panhandle, and western Kansas. Even today, these are the principal custom feeding areas. Custom feeding provided a means of financing the rapid growth of cattle numbers needed to utilize the highly mechanized, large volume feeding operations. Individuals who did not have sufficient capital to build and operate a feedlot large enough to perform economically could acquire the necessary capital and volume by custom feeding cattle for others. In this way, part of the burden of providing capital for efficient operation of a large feedlot was shifted to outside interests.

Capital requirements, periods of severe economic conditions (like scarce money and high interest), times of depressed feeder cattle prices, and adverse pasture conditions caused custom feeding to grow following World War II. These same forces, along with the need for high occupancy (full feedlots) and increased integration, have resulted in further expansion of custom feeding.

Most custom feeders have developed large, highly mechanized, and very efficient plants. Usually, they have on their staffs highly trained nutritionists who are charged with the responsibility of formulating rations and of obtaining maximum gains and feed efficiency at the lowest possible cost. Through custom feeding, they sell the use of their facilities, services, and know-how to cattle owners, usually with profit to each party.

The Packers and Stockyards Administration, of the U.S. Department of Agriculture, ruled that, effective July 1, 1974, (1) packers could not own, operate, or control *custom* feedlots; and (2) *custom* feedlot owners could not own, operate, or control packing plants. This action was taken in order to avoid monopolistic conditions. This ruling does not prohibit packers from feeding their own cattle for their own slaughter needs.

The proportion of custom fed cattle to cattle owned by the feedlot varies (1) in period of time—it increases in times of financial stress (when cattle feeding is not profitable, money is scarce, and interest is high); (2) according to area—for example, there is more custom feeding in California than Colorado; (3) according to size of feedlot—generally speaking, the larger the feedlot, the greater the percentage of custom feeding. Some feedlots do not do any custom feeding whatsoever; others are almost wholly on a custom basis; but most lots have part of each. Feedlots that do both—those in the dual role of custom feeding and owning cattle—vary in the proportion of cattle in each category, but most of them seem to prefer about ⅔ custom fed cattle and ⅓ ownership. It's a good bread-and-butter division; in times when fed cattle lose money, such a feedlot has sufficient assured income to pay its bills.

The ownership of custom fed cattle is diverse. It includes (1) cow-calf men (farmers and ranchers) who wish to retain ownership of the cattle that they produce through the feedlot phase; (2) packers; and (3) investors, including limited partnerships, corporations, cattle buyers, cattle dealers, and others.

PROVISIONS OF A CUSTOM FEEDING CONTRACT

Custom feeding contracts should always be detailed and in writing, for a good understanding is the best way to avoid a misunderstanding. Also, contracts should be fair to both parties—to both the feedlot owner and the cattle owner.

The experience of feedlot owners and cattle owners, and the difficulties encountered, suggest that custom feeding contracts should include provision for the following:

- **Ration**—Some basics about the rations—such as the different rations to be used when getting the cattle on full feed, proportion of concentration to roughage, and energy and protein content—should be spelled out in the contract. Yet, the feedlot operator should be permitted flexibility, so that he can take advantage of price changes in ingredients, etc.

[13]This section was authoritatively reviewed by, and helpful suggestions were received from, the following: Dr. Willard F. Williams, Department of Agricultural Economics, Texas Tech University, Lubbock, Tex.; and Mr. Ronald R. Baker, President, C & B Livestock, Inc., Hermiston, Ore.

• **Veterinary expenses**—Veterinary expenses, medication, dehorning, castrating, and dipping charges should be specified. They will vary with the age of the cattle (calves require more attention than older cattle), the preconditioning, if any, etc. Usually, the assessment for veterinary expenses runs from $5.50 to $7.00 per head.

• **Responsibility for death losses**—The contract should specify responsibility for death losses. The cattle owner normally assumes all losses prior to the arrival of the cattle at the feedlot. After a reasonable time at the feedlot, losses may be the fault of the feedlot; hence, it may logically be expected to assume a share of the loss. Usually the partitioning of death losses is accomplished by (1) the cattle owner's standing loss of the initial cost of the feeder cattle delivered to the lot; and (2) the feedlot operator's canceling out all or part of his charges for yardage, feed, and/or cost of gain from the time the custom fed cattle arrive in the feedlot until the time any death losses may occur.

• **Buying and selling services**—The buying and selling services provided by the feedlot are important to the success of the operation. In most cases, the feedlot is in the best position to sell the finished cattle. But this responsibility should be spelled out in the contract, along with any fee to be charged for the services.

• **Right to reject cattle that are doing poorly**—A feedlot using a payment-for-gain contract should reserve the right to reject cattle that are doing poorly. Of course, where the feedlot purchases the cattle for the owner, he accepts the responsibility to get cattle that are doing well.

• **Power of attorney for feed yard operator**—From a legal standpoint, feed yards should have power of attorney for buying, selling, and borrowing for customers.

• **Arbitration**—In cases of dispute and disagreement over a custom feeding contract that cannot be resolved by the owner of the cattle and the feedlot, the contract should provide for arbitration to be conducted by a committee of three—each party to the contract choosing a representative and these two then choosing a third party, to study the case and recommend settlement.

TYPES OF CUSTOM FEEDING CONTRACTS

The services rendered vary from feedlot to feedlot and according to the type of contract. In some instances, the services may be so complete that the customer never sees the cattle. The feedlot operator may buy the feeder cattle, feed them, market them, and send the customer (his client) a check for the balance, after deducting input costs, interest charges, and custom feeding charges. Less complete services are usually available to suit the customer.

Both the feedlot owner and the cattle owner should analyze different types of contracts and determine which best fits their respective circumstances. Some feedlots offer several types of contracts, thereby according the cattle owner a choice.

Competition may dictate the type of contract and the charges made. But by knowing the variables and managing them correctly, the feedlot owner can write and carry out a contract that will be fair to himself and his customer.

Generally speaking, contracts with fixed charges are the most satisfactory and the most common, primarily because there is less room for misunderstanding.

Although there are many types of custom cattle feeding contracts, and many variations of each kind exist, most of them can be classified under one of the following types:

• **Feed cost plus daily yardage fee per head**—This type of contract is based on the cost of feed plus an additional 15 to 21 cents per head per day to cover handling, yardage, feed grinding, and similar expenses. Generally, an additional charge of $5.50 to $7.50 per head is made to cover medication, vaccination, branding, dehorning, and dipping. The customer finances purchase of the cattle.

With this type of contract, the feedlot does not assume any risk whatsoever. It is merely its intent to sell feed, facilities, and services at an agreed price. However, it does not permit the feedlot owner to participate in the greater revenue generated by high performing animals as does a margin based on markup on feed costs.

• **Feed cost plus markup**—This type of contract calls for reimbursement on cost of feed plus a feed markup on (1) a flat rate, (2) a percent of cost, or (3) a percent of moisture added in steam processing.

With a flat rate markup, a $15 to $21 charge per ton above feed cost is made to cover feed handling, grinding, and labor costs. An additional assessment is made to cover veterinary services and medication; and the customer finances the purchase of cattle. Since actual feed milling costs (for labor, power, insurance, mill maintenance, etc.) run $9 to $15 per ton, profit to the feedlot accrues from having a higher markup than the milling cost. It should be noted that markup on such high-moisture feeds as silage and wet beet pulp is generally computed on a dry basis. On a wet basis, the markup on silage is usually figured at about one-third the dry matter basis. A flat markup per ton of feed favors the feedlot when prices fall, and the cattle owner when feed prices rise.

Also, markup on feed may be on a percentage of cost basis. With this arrangement, higher feed costs favor the feedlot, whereas lower feed costs reduce the actual return to the feedlot.

Markup on feed is sometimes through the addition of steam in processing. Where this method is used, the feedlot must either add enough steam to generate returns comparable to other markup systems or increase the daily yardage fee.

Any system of feed markup will be more profitable to the feedlot with heavy, high performing cattle than with light, slow gaining cattle, simply because the former eat more.

With the "feed cost plus markup contract," the feedlot is essentially a feed manufacturer processing and delivering feed to its customers—the owners of the cattle.

- **Feed cost plus (1) daily yardage fee per head, and (2) markup per ton of feed**—This is a combination of the first two types. Those feedlots that charge the higher yardage rates per head daily add a smaller markup per ton of feed above actual ingredient prices; conversely, those that charge the lower yardage rates per head per day make a higher feed charge over and above actual ingredient cost. As a rule of thumb, feed markup is generally lowered by $1 per ton for each 1¢ per head per day of yardage charged. At the present time, it appears that the lower yardage cost and the higher feed markup is the favored basis, primarily because (1) the owner of the cattle is less inclined to object to such charges, and (2) increased competition has driven custom feeders to make their charges on the least conspicuous basis.
- **Agreement to purchase contract**—In this plan the cattle feedlot operator buys the feeder cattle in his own name, usually with the client required to make a down payment of 20 to 30 percent of the purchase price. The client then executes an agreement to buy the cattle when they are ready for slaughter, including the original purchase price of the feeder cattle (less any down payment made) and all feeding, handling, and interest charges. (Interest charges are tax deductible.) There are several variations of this type of contract. But all of them are much like buying commodities.
- **Payment for weight gained**—This plan is based on a charge per pound of gain. In this arrangement, the feeder is reimbursed on the basis of the gain in weight put on the cattle, at an agreed price of so many dollars per hundred. This type of contract has decreased in importance in recent years, because it frequently results in poor owner-feeder relations—due primarily to the following reasons: (1) It is impossible in advance of feeding to detect those lots of animals that will be "poor doers" because of such factors as nervousness, nutritional deficiencies, diseases, and/or parasites; (2) the longer the feeding period, the greater the cost of gains, and the length of the feeding period is seldom stipulated in such contracts; (3) because weather and disease, which are uncontrollable, affect rate of gain; and (4) the amount of fill or shrinkage when weighing animals in and out the feedlot is of great importance to both the owner and the feeder, and a source of argument. Also, a major disadvantage of using a payment-for-gain system is that it does not adjust for feed costs; the feeder must absorb any increase in feed cost. However, it does allow a feedlot to take advantage of opportunity feeds—such as down corn, or grass which cannot be delivered to the feedlot. Also, the small feedlot operator who does not have scales may use this type of custom contract.

Payment for weight gained may be used in growing and backgrounding operations, where it may be desirable to specify both the minimum and the maximum rate of gain. Also, it may be used on cattle that are pastured for a time before being sent to the feedlot.

- **The incentive basis contract**—This is another system of charging on a payment-for-gain basis that some cattle owners like, because it gives an incentive for the feedlot to produce rapid daily gains. It consists in paying the feedlot for all feed plus a charge arrived at by "multiplying the average daily gain times itself, or times some factor." Thus, if the cattle being finished should average 3 pounds gain daily over the entire feeding period, the per head per day basis of payment to the feeder would be as follows:

3 (gain) × 6[14] = 18¢/head/day

If the average daily gain per head over the entire period is 2 pounds, then the basis of payment would be as follows:

2 (gain) × 6[15] = 12¢/head/day

The incentive basis isn't desirable from the standpoint of the feedlot owner if the cattle are "poor doers" or if, because of disease or other factors beyond his control, poor gains are obtained. For understandable reasons, however, this type of contract does have a very strong appeal to the cattle owner.

- **An investment-type contract**—This refers to an arrangement wherein an investor contracts for cattle, feed, and services, much as he would were he buying stocks or bonds on the New York Stock Exchange. There are several variations of investment contracts.

The essential provisions of the investment-type contract are:

The cattle owner—

1. Puts up $60 per head (only about 10% value at market, compared to 70% required with stocks)—more if their machine projections indicate losses.

---

[14]This factor might be varied in keeping with economic conditions. There is no reason why it must be identical to the average daily gains. It may be higher or lower according to economic conditions.

[15]Ibid.

2. Pays freight, taxes, medicines, and feed bill (all on credit).

3. Assumes risk (average of 1 to 2% death loss).

4. Pays interest (tax deductible).

The custom feeder—

1. Provides balance of capital, and profits on interest rate.

2. Processes feed at a profit.

3. Provides know-how, facilities, and veterinary services.

4. Spreads risks.

OPERATION PROCEDURES

The operation procedures followed by feedlots that custom feed vary. In particular there should be an understanding between the feedlot and the customer relative to billing and payment of feed costs, selling arrangement, and payment for slaughter cattle.

• **Billing for feed or gain costs**—Most feedlots bill their clients for feed or gain costs on the first and fifteenth of each month, at the end of the month, or at the end of the feeding period. The more frequent the billing, the smaller the amount of short-term capital required by the feedlot. Where feedlots carry feed costs longer than a month, they usually charge interest.

The vast majority of custom feedlots worthy of the name are equipped with platform or hopper scales and/or scale trucks with electric or mechanical scales mounted under the feeding box; hence, billings are based on actual weights. Occasionally, a farmer without scales will feed a few cattle on a custom basis. In the latter case, the feedlot owner and the cattle owner may agree to (1) compute feed costs on the basis of 3 percent of the incoming body weight of the cattle plus 7 lb per day (for example, the daily feed consumption of a steer weighing 700 lb at the time of delivery to the yard would be estimated at 28 lb); or (2) feed on a cost-of-gain basis, with payment delayed until the cattle are weighed and marketed at the end of the feeding period.

• **Selling arrangement**—Feedlot managers generally handle the selling of custom fed cattle in the same manner, and through the same channels, as cattle which the feedlot owns. Feedlot managers, or their representatives, are usually in a better position to estimate the weights and grades of cattle on feed than are their clients. Also, they are more familiar with the type and quality of fed cattle desired by various packer buyers.

• **Payment for custom fed cattle**—Market payments for custom finished cattle may be made either directly to the owner of the cattle or directly to the feedlot, depending on prior arrangements between the feedlot and the client. However, commercial banks and other lending institutions generally retain a first lien on the client's cattle; and this must be satisfied. Likewise, finance agencies generally provide the necessary financing for feed and other custom feeding charges; hence, the feedlot operator ordinarily is assured of receiving full payment for feed bills and other services. In the event the client has outstanding bills with the custom feeder, feedlots handling payments for their clients are permitted to retain sufficient funds to satisfy these debts, after satisfying the first mortgage holder.

WHAT CATTLE OWNERS EXPECT OF CUSTOM FEEDLOTS

The owner who entrusts his valuable cattle to a feedlot for custom feeding rightfully expects certain things of the feedlot, chief of which are:

• **Profits**—Cattle owners assume considerable risk; hence, they expect profits commensurate therewith. They will contract with the feedlot which consistently returns the most on their investment.

• **Cattle feeding know-how, honesty, and integrity**—Cattle owners will not, knowingly, place cattle in a custom feedlot under inexperienced management. Also, because cattle owners are not involved in day-to-day management, the honesty and integrity of the custom feedlot management are very important to them.

• **Progress reports**—Cattle owners wish to keep informed of how well their investment is doing. To this end, they expect monthly progress reports giving such pertinent information as (1) rate of gain, (2) feed efficiency, (3) cost of gains, and (4) sickness and death losses.

• **Courteous customer treatment**—Like any other customer, a custom cattle owner likes to feel that the feedlot wants and appreciates his business.

• **Competitiveness**—The cattle owner expects the feedlot in which he places cattle for custom feeding to be competitive in charges and performance with other feedlots of the area.

• **Satisfactory financial position**—The customer expects that the feedlot with which he does business be in sufficiently strong financial position, that it buys feeds when it is most advantageous (usually at harvest), and that it pays its bills regularly.

• **Knowledge of the financial position of packer buyers**—The cattle owner expects that the feedlot owner know the financial position of the packer buyer to whom his cattle are sold, so that there can be no concern relative to payment. A few cattle owners who custom feed even require that the feedlot handling the sale of the finished cattle guarantee payment for them.

HOW TO ATTRACT CUSTOM FEEDERS

Success in attracting customers depends on reputation and performance. Reputation will attract new customers, but only performance will hold them.

If a feedlot relies on custom fed cattle for a certain percentage of its capacity it's important that there be enough clients with sufficient cattle to keep the lots filled, or nearly so, at all times.

- **Be successful**—The best way to recruit and retain customers is to be successful. After a customer has used the services of a given feedlot, he will know what it can do. Until then, the feedlot manager will have to convince the prospective customer of his ability. Among the tools which the manager may use in proving his ability are records and computers.

- **A complete set of records**—A good set of records will allow the feedlot manager to predict with confidence what he can do for the prospective custom feeding client in rate, efficiency, and cost of gain; in disease and death losses; and in buying and selling cattle.

- **A computerized system**—Prospective clients will be impressed with the sophistication of a computerized accounting system and a computerized closeout statement. It indicates the feedlot's capability of complete, accurate, and prompt records. In addition to being a good means of keeping customers well informed, a computerized system may be used to provide them with an accurate and complete set of records for accounting purposes. Also, records are useful to the feedlot, because it provides periodic analysis of progress and allows the manager to monitor the feeding program for needed changes.

- **Financing cattle and feed fed on a custom basis**—Financing helps to attract clients. Since feedlot operators are under pressure to keep their feedlots filled, financing of cattle and feed by custom operators will probably continue to be an important source of funds to those who place cattle in these lots.

Commercial banks are the primary source of financing for cattle fed on a custom basis. They generally require a margin equivalent to 25 to 30 percent of the value of the feeder cattle. In addition, banks make loans to cover feeding charges. It is not uncommon for banks to finance 80 percent of the feeder cattle price plus all of the feeding charges. Depending on the reputation of the feedlot operator and the custom feeder, banks and other lending institutions may secure only the cattle as collateral for the loan. They may also specify that feeder cattle be hedged on a futures market before negotiating loans, although this has not been general practice to date. Some lending agencies will even advance funds for the margin on cattle futures contracts, as part of a feeder cattle loan.

Also, some custom feedlots finance cattle purchases in their own names, with the client executing an "agreement to purchase" the cattle when they are sold at a cost equal to the initial purchase price of the feeder plus all feeding, handling, and interest charges.

- **Bring prospective feeding customers together**—The feedlot manager may attract some customers by bringing them together, especially those who have insufficient funds to carry on a continuous feeding program by themselves. The feedlot manager can provide a real service by assisting them in the formation of clubs or corporations whose pooled resources are large enough to feed cattle continuously.

Some custom feeders honor more than one type of contract, thereby according the cattlemen a selection.

SAMPLE CUSTOM CATTLE FEEDING CONTRACT

A sample custom cattle feeding contract is shown in Fig. 2-18. Variations can be made in this form so as to adapt it to a specific owner and feeder. (See page 84 for Fig. 2-18.)

Co-op Owned Feedlots

Most large-scale cattle feedlots are owned by individuals, partnerships, or corporations. It is expected that this will continue to be the dominant type of ownership. However, there is increasing interest in cooperative cattle feedlots, as a means of (1) accommodating relatively small operators who wish to feed out their own cattle, without necessitating that they have cattle feedlot managerial ability or large capital investment in a feedlot and equipment, and (2) attracting outside investment capital.

ALTERNATE WAYS TO ORGANIZE A CO-OP FEEDLOT

Two alternatives generally exist in setting up a cooperative feedlot; namely, (1) organize it as a separate department of an existing cooperative, or (2) organize a new and distinct cooperative feedlot—start from scratch.

The first method generally reduces the capital outlay, particularly when existing feed milling and storage facilities may be used. On the other hand, such facilities may be so far removed from the feedlot that higher feed transportation costs offset any capital outlay advantage.

The organization of an entirely new cooperative feedlot allows (1) more flexibility, and (2) the opportunity to design modern feed milling and storage facilities tailor-made for the particular lot.

MEMBERSHIP

Certain requisites for membership should be established; among them, (1) that the member provide a stipulated number of cattle, thereby assuring that the feedlot will be kept full, and (2) that the member have a relatively high initial investment, thereby assuring continued support.

CUSTOM CATTLE FEEDING CONTRACT

THIS CONTRACT made and entered into at ______________________, this ________ day of

__________, 19 ______, by and between

______________________________ of ______________________________
(name) (address)

hereinafter designated Owner, and

______________________________ of ______________________________
(name) (address)

hereinafter designated Feeder.

THIS AGREEMENT shall extend ________ days following delivery of cattle to Feeder's yard and cover the following cattle which shall be delivered to Feeder within the period ____________ (month), 19 ________, to ____________ (month), 19 ________. (If part, or all, of the cattle covered by this agreement have not been purchased at time of its execution, the approximate number of head shall be indicated; balance of information shall be filled in at time of purchase or delivery of cattle to Feeder's yard and shall become a part of this contract.)

Number of Head	Sex	Age	Breed	Brand or Identification

The Feeder will—

1. Provide suitable facilities in his feedyard and feed and care for cattle belonging to Owner in a good and husbandlike manner.
2. Brand, vaccinate (except where the services of a veterinarian are needed), spray, dehorn, sort, weigh, hospitalize and care for all sick cattle, and pay for all drugs, except for implants, and for medication prescribed or used by a veterinarian.
3. Reserve the right to salvage or sell for the highest possible price any sick or unthrifty animal when time is of the essence to avoid total loss. In such cases, Feeder shall retain such salvage returns as are due him under this contract and return balance, if any, to Owner.
4. Retain and dispose of the manure.
5. Be free from any and all liability of any kind except for damage or injury resulting from his (Feeder's) gross negligence and willful neglect.

The Owner—

1. Warrants that he is the legal owner of all cattle covered in this contract.
2. Agrees to pay Feeder for all feed ingredients on the basis of cost (feeds produced on ____________ ranch and fed to Owner's cattle to be computed at prevailing prices for their respective kinds and grades) plus milling costs of $ ________ per ton of feed and/or yardage at ________ cents per head per day, plus added cost of implants and veterinarian-prescribed drugs and Three Dollars ($3.00) per head on bulls castrated. These charges will be billed at the end of each month and will be due by the 10th of the following month, after which an interest charge at the rate of ________% per annum will be made until paid.
3. Will bear any losses of cattle (a) while in the feedyard of the Feeder or (b) after purchase but prior to delivery.
4. Will pay for any veterinary services deemed necessary by Feeder, covering such services as the Feeder should not render.
5. Will be responsible for the removal or sale of his cattle, unless he wishes to delegate this assignment to the Feeder or his representative, in which case such service by the Feeder will be without charge to the Owner.
6. Understands and agrees (a) that all loans and charges provided for herein shall be paid (in full if all cattle are removed simultaneously, or on a proportionate basis if part of the animals are removed) prior to withdrawal of Owner's cattle from Feeder's yard; (b) that Feeder has a lien on Owner's cattle in his (Feeder's) possession for any and all unpaid bills plus interest; and (c) that if any sums remain unpaid after the duration of this contract, Feeder may at his option any time thereafter sell such livestock as necessary to pay and discharge the amount due him. Any such sales as may be mandatory under the latter clause may be (a) at either public or private sale and (b) in one or more lots, without demand or notice to owner—demand and notice being hereby waived. From the proceeds of any such mandatory sale or sales, the Feeder shall (a) retain all amounts due him from Owner hereunder, plus expenses incurred in sale including any legal charges or attorney's fees; and (b) pay the excess, if any, to Owner.

______________________________ ______________________________
(Owner) (Feeder)

______________________________ ______________________________
(Date) (Date)

Fig. 2-18. Custom Cattle Feeding Contract.

OPERATING POLICIES

The general principles and practices of buying, handling, and marketing cattle are much the same in both individually and cooperatively owned feedlots; hence, only the operating policies unique to co-ops will be covered in the points that follow:

1. **Management**—In addition to meeting the traits of a good manager, as set forth earlier in this section, a successful co-op manager must (a) be able to work harmoniously with his members and board, and (b) be a strong leader, relying on his board for broad general policy decisions, but avoiding the hazardous pitfall of committee action, compromise, and weakness.

2. **Pooling vs individual ownership of cattle**—A decision must be reached on whether to operate the yard by pooling the cattle or by maintaining individual ownership.

The advantages of operating the feedlot on a pool basis are: (a) it requires fewer pens and eliminates under- and over-utilization of pens; (b) it lessens record keeping; (c) it makes it easier to group and handle small incoming lots of cattle; (d) it facilitates grouping of cattle according to weight, quality, and grade for marketing; (e) it spreads risks for members in that they become partners in a year-around marketing program, rather than seasonal; and (f) it provides greater potential bargaining power when marketing. However, pooling (a) requires that co-ops paying cash for pooled cattle have more capital than where individual ownership is retained; (b) may make for disgruntled members, because cattlemen are prone to feel that their cattle are better than the evaluation that the manager is willing to give them; and (c) takes away the pride of ownership.

For a pooling arrangement to work at its best, the cattle owned by the members must be fairly uniform in quality, and the co-op must be in a strong financial position if cash is to be paid at delivery time (in contrast to the book entry method). Also, some "refusal" arrangement is important. Usually, the latter works as follows: When the cattle are delivered to the feedyard, they are evaluated according to the current price. The owner is then given a choice: (a) If he accepts the price, the cattle are pooled and he receives either cash or a book entry; or (b) if he refuses the price, his cattle are fed separately. Usually, a higher charge is levied where individual ownership is retained than where animals are pooled, so as to cover less efficient utilization of pen space, and other slightly higher costs in care and marketing.

3. **Charges**—Co-ops charge their members on different bases, just as custom feeders do. Basically, the methods of making charges are (a) on a straight tonnage of feed markup above feed costs, (b) daily yardage, (c) a combination of feed markup and yardage, and (d) per pound of gain.

METHOD OF FINANCING

Three basic sources of capital for financing a cooperative feedlot are available; namely, (1) co-op members, (2) nonmember investors, and (3) lending agencies. Sometimes funds are obtained from all three sources.

Membership fees set at a high level raise capital and encourage membership participation in the enterprise. Certificates of indebtedness and preferred stock are often attractive investments in a cooperative for both members and nonmembers. Other means of membership financing include common stock, deferred patronage refunds, and revolving funds. Local banks and banks for cooperatives are also potential sources for financing cooperatives.

Condominium Cattle Feedlots

The condominium concept in feedlots is relatively new. It refers to a feedlot in which there is separate ownership and management of lots, but a joint ownership of part of the facilities (perhaps the feed mill, feed trucks, etc.).

Property taxes are leveled on the individual condominium units. By-laws are required in which procedures are specified for purposes of the overall administration and maintenance of a condominium feedlot.

The *advantages* of condominium cattle feedlots over individually owned and smaller lots are:

1. **Lower feeding costs**—Iowa State University studies show that feeding 600 head of cattle instead of 100 head lowers the feeding costs $3.25 per 100 pounds of gain. Feeding 1,000 head instead of 600 lowers the cost another 75¢ per 100 pounds' gain.

2. **More efficient waste management**—Waste products can be managed more efficiently in larger units.

3. **Easier to adopt new feeding technology**—A larger lot can adopt new cattle feeding technology more easily than a smaller operation.

4. **They attract more buyers**—Cattle buyers will come regularly to a large feedlot, which means more competition from buyers—and could mean a higher price.

But there are *disadvantages* to a condominium. Among them, are:

1. **Capable management difficult to come by**—It takes a more able manager to conduct a big feedlot than a small one.

2. **Disease problems increased**—Disease problems are increased by having different groups of cattle in nearby pens.

3. **Group decision making can be troublesome**—Group decisions sometimes make for com-

promise and weakness and cause other problems among condominium feedlot owners.

More time and research is needed in order to evaluate the condominium concept. Such feedlots will not be the answer for all feeders, but large feedlots can spell extra profits if the necessary conditions exist.

Record Forms

There is no limit to the number of kinds of record

DAILY RECORD FEEDLOT:

Pen No. ______________ Date Started ______________________
Month Day Year

Day of Month	Head In						Death Losses	Head Out			Daily Feed		Sold			Comments
	No.	Origin	Pur. Price	Total Pur. Wt.	Total Wt. at Lot	Av. Wt. at Lot	(cause)	No.	Total Wt.	Av. Wt.	Total Lb	Lb/ Head/ Day	To	Price/ Cwt	Carcass Grade	
1																
2																
3																
4																
5																
6																
7																
8																
9																
10																
11																
12																
13																
14																
15																
16																
17																
18																
19																
20																
21																
22																
23																
24																
25																
26																
27																
28																
29																
30																
31																

Fig. 2-19. Daily Cattle Feedlot Record Form.

forms that can be kept in a given feedlot. Also, there is little similarity in record forms between lots, due to differences between people—primarily managers and bookkeepers. The important things are (1) that record forms be so designed as to facilitate record keeping, with as much ease, efficiency, and accuracy as possible; and (2) that records be kept.

Figs. 2-19 and 2-20 show two basic record forms; Fig. 2-19 is a Daily Record, whereas Fig. 2-20 is a Monthly, Cumulative, and Final Feed Summary. Many variations of these can be made.

STOCKER AND GROWER CONTRACTS

Hand in hand with the development of big feedlots and year-round feeding came the need for an assured supply of feeder cattle of the desired kind on a continuous basis. To meet this need, more and more feedlots have turned to contractual arrangements with stocker growers, with numerous kinds of contracts. Usually, the cattle are owned by the feedlot, most of which are large and in a stronger financial position than the majority of stocker growers. The two most common kinds of contracts are based on either (1) a fixed cost for the gain, or (2) an agreed feed cost plus an extra charge for labor and lot rental. Usually, there is provision for adjusting for death loss. Such contracts should always be in writing, with all provisions, including weighing conditions, spelled out.

Although the use of stocker and grower contracts has increased in recent years, the concept is not new. Many of the Kansas bluestem pasture owners have long grown out yearlings owned by Iowa and other Corn Belt feeders.

Today, many corn farmers in the fertile irrigated area in the vicinity of Greeley, Colorado, make corn silage and feed cattle on a contract basis to stockers owned by one of several large feedlots in the vicinity. Stocker cattle are also being grown under contract on the wheat pastures of Kansas, Oklahoma, and Texas; on hay and other roughages in the irrigated valleys of the West; and on sorghum silage and stalk fields throughout the Southwest.

CATTLE RECORD—Monthly, Cumulative, and Final Feed Summary

__________ (date-month-year) __________ (reported by)

Pen No. ________ No. Head ________ Purchase Price: cents/lb. ________ Date Started ________ Date Closed ________

RATION:	**Days on Each Ration**	**Total**	**Price**	**Total Cost**
	(days)	*(lb)*	*($/av. 2,000 lb ton)*	*($)*
Starter:				
Intermediate:				
Finishing:				
Totals:				

FEED ANALYSIS:

Total Feed fed ________ lb

Feed days (no. head × days) ________ no.

Pen (total) weight out .... (lb) ________
Pen (total) weight in .... (lb) ________
Net gain .... (lb) ________

Av./head weight out .... (lb) ________
Av./head weight in .... (lb) ________

Feed/head/day .... (lb) ________
Gain/head/day .... (lb) ________

Feed conversion: lb feed/lb gain .... (lb) ________

Feed cost/head/day .... ($) ________

Feed cost/lb of gain .... ($) ________

COST ANALYSIS: **($)**

Gross receipts from sale of cattle .... ________

Costs:
Purchase price of cattle .... ________
Total cost of feed .... ________
Other costs:
Milling charges .... ________
Mineral charges .... ________
Medication .... ________
Management .... ________
Labor .... ________
Physical plant (other than milling) .... ________
Total costs .... ________

Net return:
Pen of cattle .... ________
Per head .... ________

Comments:

Fig. 2-20. Monthly, cumulative, and final cattle feedlot record form.

DAIRY CATTLE CONTRACTS

With the increase in specialization and size of enterprises the business aspects of dairy production have become more important. With this transition, the following types of contracts have evolved: (1) heifer replacement contracts, (2) cow rental contracts, and (3) cow pools.

Heifer Replacement Contracts

Contracting for heifer replacements evolved because some dairymen found that it was best to have specialists do this job for them. More specifically, here are the forces that are generally back of the movement:

1. Farmers who either do not wish or cannot afford to make the very considerable capital investment required for market milk production can, at a lower cost, utilize their land, facilities, and time for growing replacement heifers for other dairymen.
2. Often those who grow replacement heifers under contract can arrange for the financing of the operation from the person for whom the heifers are being grown.
3. The desire of the dairyman to expand his milking herd, thereby crowding out the heifer replacement program.
4. High feed and land costs where the milking herd is maintained forcing a shift of heifer replacement production to more isolated and less expensive areas.
5. The desire of milk producers to obtain replacement heifers from genetically superior animals that they have developed in their herds. This can be achieved by placing their own calves out under a contractual arrangement.
6. The contract arrangement may be used as a means of lessening risks to each party, by limiting or apportioning losses.

The following types of heifer raising contracts are in use:

1. **The gain in weight or flat fee contract**—According to this plan, the contractor (the dairyman who is having his heifers grown) pays the contractee (the heifer raiser) on the basis of (a) so many cents per pound of gain, or (b) a flat fee of so many dollars per month. Many variations of this plan exist.
2. **Option to purchase contract**—Under this type of contract, the grower purchases the calves, at anywhere from 4 days to 6 months of age, depending on the terms of the contract. But the original owner of the calves has the option to buy them back after they are bred. Because of changing economic conditions, the pricing arrangement is usually somewhat flexible, with prices of feed and milk being primary determining factors. If the original owner does not wish to exercise his option, the grower may sell them to the highest bidder.

   Under this plan, the grower takes ownership of the calves, thereby providing a built-in incentive for him to do a good job.
3. **Co-op contracting heifers**—This type of co-op grows out surplus heifers belonging to members of the cooperative. Dairymen-members producing market milk transfer their calves, under contract, to farmers who are interested in growing them out cooperatively. The grower is paid on either a cost per pound of gain basis or so many dollars per month.

Naturally, the cost of raising dairy heifer replacements varies widely from area to area, regardless of whether they are raised on the farm where dropped or under contract. The primary factors accounting for these wide area variations are (1) cost of feed and labor, and (2) climate.

For some market milk producers, contracting for replacement heifers is advantageous. For others, there are advantages to on-the-farm raising, including the following:

1. It lessens the chance of introducing disease in the lactating herd.
2. Unless the milk producer places his own heifers out under contract, the source of replacements may be questionable and not up to the genetic standards of the herd in which they are to be used as replacements.
3. Contracting for heifers usually costs more than when they are home produced.
4. Contracting instead of raising heifer replacements alleviates a means of diversifying on the farm—the production of both market milk, and the raising of replacement heifers.
5. Home raised replacements generally have less udder and breeding trouble than contract raised heifers, with the result that they have a longer productive life in the herd.

Cow Rental Contracts

Renting cows is a method of getting control of additional resources, which dairymen may wish to consider.

Dairies are getting bigger and are requiring more capital. But farmers are finding it increasingly difficult, if not impossible, to borrow money. Thus, many businessmen-farmers are casting about for methods through which they may obtain additional capital. Cow rental is a possibility.

A young dairyman can get started with little capital, or an established dairyman can expand, through a rental agreement because there is no investment in the herd. Also, it takes some of the risk out of dairying.

An elderly dairyman who wishes to retire and yet retain ownership of his herd can rent out his cows. Also, it provides a way in which he can turn his capital invested in cows into retirement income; actually, it's a method of herd dispersal and tax management.

Once a dairyman has decided to consider cow renting as an alternative method of doing business, the question of how to calculate rent arises. Since little precedent exists, it is suggested that the several cost factors be calculated as a means of arriving at rental. These are shown in Table 2-19.[16]

Note well: The figures used in Table 2-19, Table 2-20, and the narrative that follows are for the purpose of illustrating the different cost factors involved in cow rentals, and *not* as indicators of actual costs. Each dairyman will need to evolve with his own current figures.

TABLE 2-19

HOW TO CALCULATE ANNUAL COW RENTAL

Line	Item	Example A 3-Yr Dep.	Example B 4-Yr Dep.	Your Situation
1	Depreciation	$ 51.00	$38.25	$ ______
2	Interest	$ 25.88	$25.88	$ ______
3	Property taxes & insurance	$ 6.00	$ 6.00	$ ______
4	Death losses	$ 8.00	$ 8.00	$ ______
5	Trucking	$ 4.00	$ 3.00	$ ______
6	Management & miscellaneous	$ 15.00	$15.00	$ ______
7	Total annual rent	$109.88	$96.13	$ ______

An explanation of Table 2-19 follows:

Line 1-A—Purchase cost $400, salvage value $247 (1,300 lb times 19¢), depreciable balance $153 ($400 minus $247). Annual depreciation $51 ($153 divided by 3 years).

Line 1-B—Same as 1-A. Only $153 divided by 4 equals $38.25.

Line 2-A-B—Multiply the interest rate (8% times the average of the original cost and salvage value.)

$$\frac{(\$400 + \$247)}{2} \times 8\% = \$25.88$$

Line 3-A-B—1½% of cow purchase cost.

Line 4-A-B—2% of cow purchase cost.

Line 5-A—Cost of hauling cow is $12. This assumes one man keeps cow 3 years.

Line 5-B—Cost of hauling cow is $12. This assumes one man keeps cow 4 years.

Line 6-A-B—Charge for owner's time, effort, and cash costs to manage the rental business.

Line 7-A-B—This assumes the dairy farmer renting the cow keeps the calf, rebreeds the cow, and keeps the cow at least 12 months. The total divided by 12 will equal the monthly payment rate per cow.

[16]*Hoard's Dairyman*, June 25, 1970, p. 689, article by R. A. Luening, Farm Records Specialist, University of Wisconsin.

Of course, the figures used in Table 2-19 are for illustrative purposes only. The situation will vary. As noted, depreciation is the largest single cost factor. It will depend on (1) the initial purchase price of the cow, (2) the length of the productive life of the cow, and (3) the salvage or market value of the cow when disposed from the herd.

Next, the renter will need to decide if renting will be a profitable venture on his farm. One way in which to do this is to compare renting vs owning. Table 2-20 presents such a comparison.[17] Preliminary to detailing Table 2-20, the following assumptions were made: (See page 90 for Table 2-20.)

> A cow costing $400 and producing 12,000 pounds of milk annually; milk figures at $4.75/cwt net after hauling; 4 years' productive life in the herd; a 90 percent calf crop, with 25 percent of the calves retained for replacements and the rest sold at $30/head; the salvage value of the cull cow computed at $247, with one-fourth this value sold each year.

Table 2-20 shows that it is more profitable to own a cow than to rent, based on the figures used. Of course, ownership may not be open to the dairyman, simply because he cannot borrow money. In the latter case, it is important to note that a return of $98.29 for labor and management can accrue from rental. With a 100-cow herd, that's $9,829 per year. Hence, when there are fixed costs, such as for buildings and equipment, it would be profitable to rent cows so long as labor and management costs would not exceed $98.29, or about $2.00 per hour, if one allows 50 hours per cow.

In evolving with a cow lease arrangement, it is important that the following factors be agreed upon by both parties and spelled out in the lease: length of lease; who rebreeds the cow; who gets the calf; feeding and management practices; the culling program; testing, vaccinations, and veterinary services; and insurance. An attorney should be consulted. He should draw up the agreement, which should be signed by both parties.

Cow Pools

A cow pool is defined as a business organization or cooperative which cares for and milks cows in a centralized location.

Cow pools are not new. The Walker-Gordon pool, located at Plainsboro, New Jersey, was established in 1891; and it has operated continuously and successfully ever since.

Among the *advantages* ascribed to pools are (1)

[17]Ibid.

TABLE 2-20

INCOME AND COSTS OF OWNING VS RENTING A COW

	Owning		Renting
Income:			
Milk sales	$570.00	(12,00 lb × $4.75)	$570.00
Calf value	27.00	($30 × 90%)	27.00
Cull cow sales	61.75	(247 × 25%)	0.00
Total receipts	$658.75		$597.00
Costs:			
Feed	$284.58	(From Wisconsin Dairy Budget)	$243.58
Operating costs	87.25	(From Wisconsin Dairy Budget)	74.00
Interest costs on cow	25.88	($400 + $247) / 2	0.00
Ownership cost of replacement	8.70	(From Wisconsin Dairy Budget)	0.00
Building & equipment costs	85.00	(From Wisconsin Dairy Budget)	85.00
Cow rental costs	0.00	(From Example B of rental cost)	96.13
Total costs excluding labor and management	$491.41		$498.71
Return to labor and management	$167.34		$ 98.29
Difference in favor of owning		$69.05	

young farmers with limited capital can get into the business by buying cows and putting them into pools, without having to own land and buildings; (2) reduced cost for overhead and equipment; (3) lower purchase price for milled dairy feeds; (4) greater labor efficiency is possible; (5) dairymen who pool their cows may use part of their labor more profitably in other enterprises; (6) farmers have more reasonable hours of work, and are not "tied down" to their milking chores every day of the year; (7) better production records are kept, with the result that the herd is upgraded; (8) a large supply of quality milk is provided, which can be readily marketed; (9) dairymen can qualify for a Grade A market without large capital outlay; and (10) lower milk hauling costs.

But pools have some *disadvantages*; among them are (1) the incidence of disease may be high, because of the concentration of animals; (2) the farmer may lose control over the management of his herd; (3) nonagricultural interests may control the pool (and they may be impractical); (4) the commingling of cows belonging to different dairymen may create management problems; (5) inability of dairymen who go into the pool to use their present on-farm buildings and equipment for anything else; (6) established marketing procedures may be disrupted; (7) if a large number of producers of manufacturing milk are suddenly converted to the production of Grade A milk, a surplus of Grade A milk and depressed prices may follow for a period of time.

A careful weighing of both the advantages and the disadvantages would indicate that cow pools best serve either young or elderly dairymen. Dairymen who are already selling Grade A milk may gain little from going into a cow pool.

The following types of cow pools are in operation:

1. **The Walker-Gordon type pool**—The cow owners in this pool care for and manage their herds in keeping with the rigid health and sanitation requirements specified by Walker-Gordon, in facilities owned by the latter. Owners remove cows to their home farms during the dry period.

Walker-Gordon provides the buildings and equipment; furnishes feed to the cow owners, milks the cows, and purchases, processes, and markets the certified milk. Manure is dehydrated and sold as a garden fertilizer under the trade name of "Bovung."

2. **Custom cow pools**—This refers to pools that care for cows on a custom or contract basis. Two types of cow owners are involved: (a) dairy farmers who wish to be relieved of caring for their cows, and (b) nonfarmers, who buy cows as an investment. In comparison with small, individually operated dairies, cow pools generally provide more mechanization and know-how. Additionally, they appeal to nonfarm investors because they provide a "tax shelter." There has been a trend toward investor-ownership of cows in pools, rather than farmer-ownership. It is expected that this trend will continue.

Some of the larger pools have on their staffs veterinarians who are responsible for the health of the herd, and nutritionists who are charged with the responsibility of formulating rations and obtaining maximum efficiency of production. Through custom or contract arrangement, pools sell the use of their facilities and services to cow owners, usually with a profit to each party.

Several different kinds of custom contracts exist;

hence, there is no standard fee. One rather common arrangement consists in the cow owner and the pool entering into a 3-year contract with (a) the pool operator charging a certain fee (for example, $60) per cow per year, payable in advance, for housing and equipment plus a management fee of 5 percent of that which remains after deducting all operating costs; and (b) the cow owner paying for all feed, labor, veterinary care, DHIA testing, breeding service, and marketing.

Where nonfarmers or investors are involved, a professional management firm may act as the representative of the owner. The usual charge for a 10-year period is 1/9 the value of each cow, payable in advance. Thus, if a cow is valued at $1,200, 1/9 would be $133.33 per cow, or $11.11 per year per cow. Some management firms guarantee the investor-dairyman a return of 1% per month on his investment. Where a guaranteed return exists, the management firm usually (a) gets all income in excess of 1% per month, plus the calves, and (b) agrees to replace death losses up to 25% of the herd owned by the investor.

3. **"Cowtel"**—"Cowtels" differ from cow pools in that their herds are lotted and milked separately by ownership, whereas in a cow pool they are commingled. The main disadvantage of the cowtel is that it is more expensive to operate where each owner's herd is segregated. This problem can be lessened by specifying that the minimum-sized herd for separate penning shall be 30 cows (or whatever the number decided upon).

4. **Co-op cow pools**—Most cow pools are owned by individuals, partnerships, or corporations. However, there is increasing interest in cooperative cow pools, as a means of accommodating co-op members. The following two types of co-op cow pools exist:

a. **The Utah-type milking pool**—This type of pool operates in central and southern Utah. The cows are milked at a central location, according to one of the following three methods:

(1) All cows are kept on the producer's place and driven to the milking barn twice daily.

(2) All cows are corralled and housed on land belonging to the cooperative. The cooperative generally hires the milkers.

(3) A combination of the first two methods is followed; some cows are maintained on the producer's premises and driven back and forth, whereas others are maintained on the premises of the cooperative. In the Utah-type pools, an initial membership fee of so much per cow is charged, plus a service fee per head per day.

b. **Co-op custom cow pools**—These are similar to "custom cow pools" (see point 2 under "Cow Pools"). However, in the case of the co-op custom cow pool, the net profits are savings from the pool operation; hence, they are distributed to the dairymen in proportion to the number of cows placed in the pool by each member.

The three main factors affecting cow pool returns are (1) production per cow and milk price, (2) cow pool charges or fees, and (3) length of the cow's productive life, or depreciation rate. Studies indicate that a cow which does not produce 10,000 pounds or more of milk per year is not likely to return much money in a pool!

HORSE CONTRACTS

Today, many horse enterprises are owned and operated as businesses, rather than as hobbies. In the conduct of horse businesses, the following types of contracts may be involved: (1) syndicated horses, (2) stallion breeding contracts, and (3) boarding agreements.

Syndicated Horses

Reduced to simple terms, a syndicated horse is one that is owned by several people. Most commonly, it's a stallion, although an expensive yearling or broodmare is sometimes syndicated. Also, any number of people can form a syndicate. However, there is a tendency to use the term "partnership" where two to four owners are involved, and to confine the word "syndicate" to a larger group of owners.

Each member of the syndicate owns a certain number of "shares," depending on how much he purchased or contributed. It's much like a stock market investor, who may own one or several shares in General Electric, IBM, or some other company. Sometimes one person may own as much as a half interest in a horse. Occasionally, half shares are sold.

Generally speaking, the number of shares in a stallion is limited to the number of mares that may reasonably be bred to him in one season—usually 30 to 35, with Thoroughbred stallions.

WHY AND HOW HORSE OWNERS SYNDICATE

The owner of a stallion that has raced successfully usually has the opportunity to choose between (1) continuing as sole owner of the horse, and standing him for service privately or publicly, or (2) syndicating him. In recent years, more and more owners of top stallions have elected to syndicate. The most common reasons for so doing are:

1. The stallion owner does not have a breeding farm or an extensive band of broodmares.

2. The owner believes that the stallion under consideration may not nick well with many of his mares; or perhaps the stallion is closely related to the mares.

3. The owner has need for immediate income. Moreover, the profit, according to a tax court ruling, is subject to the frequently advantageous capital gains treatment on income tax. By contrast, if sole ownership is retained, considerable promotional and advertising expenses will be involved for approximately three years—until the stallion's get make their debut on the tracks; and, in the meantime, practically no income can be expected until about a year from entering stud, at which time the usual "live foal" guarantee is met. Until this condition is fulfilled, any stud fees that are collected are generally held in escrow, as protection if they should have to be returned.

4. Syndicating spreads the risk, should the stallion get injured or die, or prove unsuccessful as a sire.

The owner may arrange the syndication himself, usually with competent legal advice; or, if preferred, the syndication can be turned over to a professional manager, who will generally take a free share as his organization fee.

The following pointers are pertinent to successful syndication of stallions:

1. **Check fertility**—Before syndicating, it is a good idea to check the fertility by test-mating to a coldblood (draft) mare. Of course, if the stallion is still racing, and has not been retired to stud, this is impossible.

2. **Establish a stud fee**—A common rule of thumb is that each syndicate share is worth 4 times the stud fee. Hence, if it is decided that the stallion under consideration will command a $10,000-stud fee, each share would be worth $40,000. If 30 shares are involved, the horse would have a value of $1,200,000 for syndication purposes.

3. **Determine time of payment**—In most cases, payment is due upon the signing of the syndicate contract, although some contracts (a) allow 30, 60, or 90 days; or (b) provide that the price of a share may be paid on the installment plan over a 2- or 3-year period.

4. **Put it in writing**—Syndication agreements should be clear, detailed, and in writing. In addition to identifying the horse, the agreement should state (a) the shareholders' proportionate interest (say 1/32); (b) the breeding rights of a shareholder (for example, the right to breed one mare per season to the horse, so long as he is in good health and able to breed); (c) the method of distributing services by lot, should it be necessary to limit the number of mares bred during any given season; (d) the method of disposing of, and the price to charge for, any extra services (over and above one per share, for example) during a given season; (e) the place where the horse shall stand, or how such determination will be made (usually by majority vote of the shareholders); and (f) how other policy matters not covered in the agreement will be determined (usually by majority vote).

Generally, such routine matters as the feed, care, and health of the stallion, and the scheduling of mares are left to the discretion of the syndicate manager, at a stated fee per month, with each shareholder billed proportionate to his number of shares. The manager also handles the promotion and advertising, insurance, and unusual veterinary expenses, as stipulated by the syndicate, with the costs prorated among its members.

Normally, a shareholder can barter his breeding service to another stallion. However, he cannot sell his share without prior approval of the manager and giving the other shareholders the right to buy it at the price offered; and, normally, this same stipulation applies to the sale of a service during any season.

Also, provision is usually made for sale of the horse should the majority of the shareholders so desire, with them also determining, at the time of sale, the price and whether sale shall be at private treaty or auction. Further, the contract usually provides for "pensioning," or otherwise disposing of, a sire should he become sterile or be overtaken by old age before dying.

In short, a syndicate agreement, like any good legal contract, attempts to spell out every foreseeable contingency that may arise during the stud's career, and to arrange for majority vote of the shareholders to settle any unforeseen contingencies.

The Nashua Syndicate Agreement, used by Mr. Leslie B. Combs II, Spendthrift Farm, Lexington, Kentucky, who has probably contributed more than any other person to stallion syndication, follows. It is also noteworthy that, when Mr. Combs syndicated Nashua, he sold all but one share of a total of 32, each at $39,100 (for a total of $1,251,200), over the telephone in one afternoon; and the only reason that the one share was not sold until the next morning was that he couldn't reach one of his regular clients on the telephone that afternoon.

SAMPLE SYNDICATE AGREEMENT

The syndicate agreement for the noted Thoroughbred stallion, Nashua, is given in Fig. 2-21.[18]

[18]This agreement was provided through the courtesy of Mr. Leslie B. Combs II and published by Agriservices Foundation in *Stud Managers' Handbook*, Vol. 1, beginning on p. 45.

SYNDICATION AGREEMENT

(prepared for Thoroughbred stallion, Nashua)

THIS AGREEMENT, made as of ________________ between the several persons whose names and addresses are set out in the Schedule hereto attached as the original Subscribers, and being referred to collectively as "the Shareholders."

WITNESSETH:

WHEREAS, Leslie Combs II, Spendthrift Farm, Ironworks Pike, Lexington, Kentucky, has purchased the Thoroughbred horse NASHUA (B. C., 1952), by *NASRULLAH-SEGULA, by JOHNSTOWN from the Estate of William Woodward, Jr. and has formed a Syndicate to acquire the ownership thereof upon the following terms and conditions:

1. The ownership of NASHUA shall be divided into thirty-two (32) shares, and the purchasers of said thirty-two (32) shares have paid the total purchase price of One Million, Two Hundred and Fifty-One Thousand, Two Hundred Dollars, ($1,251,200.00), or the sum of Thirty-Nine Thousand, One Hundred Dollars, ($39,100.00) per share.

2. Each of the thirty-two (32) shares shall be on an equal basis with the others and shall be indivisible, and only a full share shall have any of the rights hereunder; provided, however, that there is expressly reserved, and the within sale is made subject to one (1) free nomination to NASHUA each year during his life for each of the following named persons, their heirs and assigns: John W. Hanes, 460 Park Avenue, New York City, C. J. Devine, 48 Wall Street, New York City, and Leslie Combs II, Spendthrift Farm, Lexington, Kentucky.

3. NASHUA shall be returned to training as soon as practicable and shall race under the personal management and supervision of a Committee consisting of John W. Hanes, C. J. Devine, and Leslie Combs II, as agents for the shareholders. The Committee shall have full charge of and complete control over the future racing career of NASHUA, including but not limited to (a) the employment of a trainer, (b) the selection of tracks at which he will be trained and raced, (c) the races to which he will be nominated and in which he will be actually started, (d) the selection and employment of a jockey or jockeys, (e) the name and colors under which he will be raced, and (f) how long NASHUA shall race and when he shall be retired from racing, and the actions, decisions and judgments of the Committee with respect to any and all of the foregoing matters shall be final, conclusive and binding upon all of the Shareholders and shall not give rise to any liability upon the Committee or the individual members thereof so long as they act in good faith.

All expenses incurred by the Committee in training and racing NASHUA shall be paid by the Shareholders in proportion to the number of shares owned by each of them, and the earnings of NASHUA shall likewise be divided amongst the Shareholders proportionately. The Committee shall furnish each Shareholder periodically with a statement showing the expenses and earnings.

The Committee is authorized to execute such leases or other instruments as may be required under the rules of The Jockey Club and/or the various Racing Commissions and other governmental bodies having jurisdiction of the premises to qualify NASHUA to race.

If a member of the Committee should die, resign or be unable to serve for any reason, then the remaining members of the Committee shall select his successor from amongst the Shareholders.

4. Upon retirement to the stud, NASHUA shall stand and shall be kept and maintained at Spendthrift Farm, Ironworks Pike, in Fayette County, Kentucky, under the sole personal managment and supervision of Leslie Combs II, and he shall be entitled to charge and receive the prevailing rates for stallion keep. Leslie Combs II shall have complete charge of advertising the stallion and shall have the authority to select a veterinarian. Owners of the shares shall pay all charges, costs and expenses incurred in connection with said stallion in the proportion that their respective shares bear to the whole number of shares.

5. Each Shareholder in each breeding season shall be entitled to one (1) free nomination to said stallion for each share owned by him, subject to the payment of his share of the Syndicate expenses and the provisions of Paragraph 6; provided, however, that in NASHUA's first full season in the stud he shall be limited to a book of twenty-five (25) mares, and the owners of the thirty-two (32) nominations shall be determined by lot at a drawing to be held at such time and place as the aforesaid Committee may determine, and notice of which shall be sent by registered mail or by telegram to each Shareholder at least five (5) days prior thereto. Each share and each free nomination shall be regarded as if it were the subject of separate ownership and shall be on an equal basis, the one with the other.

Thereafter, if the veterinarian attending said stallion and the Syndicate Manager, Leslie Combs II, shall certify that in their opinion NASHUA's book may be increased without injury to him, then additional yearly nominations may be sold by the Syndicate Manager at the regular stud fee and the yearly proceeds thereof shall be divided among the Shareholders in proportion to the number of shares owned by each.

Each mare bred to NASHUA must be in sound breeding condition and free from infection or disease, and no mare shall be covered more than six (6) times in any breeding season.

6. If Leslie Combs II, with the advice and approval of the veterinarian, shall determine that NASHUA shall be bred to less than thirty-five (35) mares in any stud season, then the Shareholders and those persons holding the three (3) free nominations in each year (as provided in Paragraph 2 herein) who collectively shall be entitled to such reduced number of nominations shall be determined by lot, and any Shareholder or holder of said free nominations who has suffered by reason of the drawing of lots in any season shall not be submitted to the risk of the drawing of lots in any subsequent season unless and until all other Shareholders and holders of said free nominations, have suffered as the result thereof; and for the purpose of this clause each share and free nomination shall be regarded as if it were the subject of separate ownership and shall be on an equal basis, the one with the other. Notice of the decision to reduce NASHUA's book to less than thirty-five (35) nominations and of the time and place of the drawing shall be sent by the Syndicate Manager to each Shareholder and holder of a free nomination by registered mail or by telegram at least five (5) days prior to said drawing.

7. Leslie Combs II shall employ the usual care customarily employed in Fayette County, Kentucky, in the management of NASHUA, but shall not be responsible for any injury, disease or death of said stallion, nor for any injury, disease of any mare resulting from breeding or attempted breeding to said stallion.

8. Leslie Combs II shall have and is hereby granted the right and option to purchase any share or shares which any owner desires to sell, and such owner shall first offer the same to Leslie Combs II with the price requested for the same. If Leslie Combs II is unwilling to pay the price requested by the owner, then such owner may secure a written offer elsewhere for such share or shares, and if the owner is willing to accept such written offer he shall present the same to Leslie Combs II, who shall have the right to purchase, within forty-eight hours

(Continued)

Fig. 2-21. Syndication agreement.

thereafter, such share or shares for price so offered in writing and which the owner was willing to accept. In the event Leslie Combs II fails to purchase such share or shares within the time specified, then such owner may accept such written offer. This option shall apply in the same manner and under the same conditions to such share or shares in the new ownership. This option shall apply to and have priority over any hypothecation, distraint or other alienation of said share or shares or any interest therein, and any and all transfers of any share or shares are expressly subject to said option.

9. All notices required hereunder shall be effective and binding if sent by prepaid registed mail, telegram, cable, or delivered in person to the address of the respective Shareholders set out in the Schedule attached or such address as shall hereafter be designated in writing to the Syndicate Manager.

10. The Shareholders accept delivery of NASHUA without examination as to his fertility and breeding soundness, as no veterinary examination with reference thereto has been made or will be made prior to his retirement from racing.

11. The undersigned hereby subscribes for ________________ shares in the Syndicate for the total sum of $________________, payable in cash upon the execution of this Agreement, and in consideration thereof Leslie Combs II has sold and conveyed ________________ Shares to undersigned, subject to all of the terms and conditions herein.

This Agreement may be executed in several counterparts, and when executed by the Shareholders the several counterparts shall constitute the agreement between the parties as if all signatures were appended to one original instrument.

WITNESS the hand of the undersigned as of the day and date first above written.

____
Name

____

____
Address

Approved:

____
Syndicate Manager

Stallion Breeding Contract

Stallion breeding contracts should always be in writing; and the higher the stud fee, the more important it is that good business methods prevail. Neither "gentlemen's agreements" nor barn door records will suffice.

From a legal standpoint, a stallion breeding contract is binding to the parties whose signatures are affixed thereto. Thus, it is important that the contract be carefully read and fully understood before signing.

A sample stallion breeding contract is presented in Fig. 2-22.[19]

In addition to the provisions made in the sample stallion breeding contract presented here, and most other similar contracts, the author suggests that the following matters be covered in the stallion breeding contract:

1. **Facts about the mare**—There should be record of the mare's temperament; thereby lessening danger to her, to the stallion, and to the personnel. Also, historical information should be included about the mare's breeding record and peculiarities, and her health—preferably with the health record provided by the veterinarian who has looked after her.

2. **Some management understanding**—The parties to the contract should reach an understanding relative to the mare's veterinary care, parasite control, seasonal injections, foot trimming, etc., and then put it in writing.

3. **An incentive basis**—Generally, stallion owners guarantee a live foal, which means that the foal must stand and nurse; otherwise, the stud fee is either refunded or not collected, according to the stipulations. Of course, it is in the best interests of both parties that a strong, healthy foal be born. One well-known Quarter Horse establishment reports that their records reveal that of all mares settled during a particular 3-year period, 19% of them subsequently either resorbed or aborted feti, or the foal or mare died. Further, their investigation of these situations showed that the vast majority of these losses could have been averted by better care and management. They found many things wrong—ranging from racing mares in foal to turning them to pastures where there was insufficient feed. To alleviate many, if not most, of these losses—losses that accrue after the mare has been examined and pronounced safe in foal, then taken away from the stallion owner's premises—the author suggests that an incentive basis be incorporated in the stallion breeding contract. For example, the stallion owner might agree to reduce the stud fee (1) by 10, 15, or 20% (state which), provided a live foal is born; or (2) by 25 to 33⅓% provided the mare owner's veterinarian certifies that the mare is safe in foal 30 days after being removed from the place where bred, with payment made at that time and based on conception rather than on birth of a live foal.

Boarding Agreement

Today's tough zoning laws and antipollution campaigns are making it increasingly difficult to keep horses in towns and suburban areas. As a result, more

[19]Sample prepared by the author of this book.

STALLION CONTRACT

(To be executed in duplicate for each mare; one copy to be retained by each party.)
This Contract for the breeding season of ________ made and entered into by and between
(year)

______________________________, ______________________________
(owner of stallion) (address)

herinafter designated Stallion Owner, and

______________________________, ______________________________
(owner of mare) (address)

herinafter designated Mare Owner.
This contract covers __________

The stallion, ______________________, whose service fee is $ ______________;
(name of stallion)

$________ of which is paid with this contract, and the balance of

$________ will be paid before the mare leaves __________

(name of farm or ranch)

and
The mare, ______________, Reg. No. ________, by ______________
(sire)

out of ________, age ________, color ________.
(dam)

I. *The Mare Owner agrees that—*

Upon arrival, the mare will (a) be halter-broken, (b) have the hind shoes removed, and (c) be accompanied by a health certificate signed by a veterinarian, certifying that she is healthy and in sound breeding condition.

Stallion Owner will not be responsible for accident, disease, or death to the mare, or to her foal (if she has a foal).

Stallion Owner may, at his discretion, have his veterinarian (a) check and treat the mare for breeding condition or diseases, and (b) treat her for parasites if needed; with the expenses of such services charged to Mare Owner's account and paid when the mare leaves the farm or ranch.

He will pay the following board on his mare at the time the mare leaves the farm or ranch: Feed and facilites $ ________/day.

Should the mare prove barren, or should the foal die at birth, he will send notice of same, signed by a licensed veterinarian, within five days of such barren determination or death.

Should he fail to deliver the above mare to Stallion Owner's premises on or before ______________,
(date)
Stallion Owner shall be under no further obligation with respect to any matter herein set forth.

This contract shall not be assigned or transferred. In the event the mare is sold, any remaining unpaid fee shall immediately become due and payable and no refund shall be due anyone under any circumstances.

II. *The Stallion Owner agrees that—*

He will provide suitable facilities for the mare and feed and care for her in a good and husbandlike manner.
Mare owner will not be responsible for any disease, accident, or injury to Stallion Owner's horses.
A live foal is guaranteed—meaning a foal that can stand up alone and nurse.

III. *The Stallion Owner and Mare Owner mutually agree that—*

This contract is not valid unless completed in full.

Should the above-named stallion die or become unfit for service, or should the above-named mare die or become unfit to breed, this contract shall become null and void and money paid as part of this contract shall be refunded to Mare Owner.

Should the mare prove barren, or should the foal die at birth, with certification of same provided to Stallion Owner within the time specified, Stallion Owner has the option either to (a) rebreed the mare the following year, or (b) refund the $________ portion of the breeding fee, thereby cancelling this entire contract.

The mare will not receive more than ________ covers during the breeding season; and she will not be
(no.)
bred before ________ 19____
(date)
or after ________ 19____
(date)

__________ ______________________ ______________
(date) (signature; Mare Owner or Rep.) (address)

__________ ______________________ ______________
(date) (signature; Stallion Owner or Rep.) (address)

Fig. 2-22. Stallion breeding contract.

and more horses are being stabled and cared for in boarding establishments out in the country, to which owners commute. This prompts the need for an agreement.

Boarding agreements should always be in writing, rather than verbal, "gentlemen's agreements." From a legal standpoint, a boarding agreement is binding to the parties whose signatures are affixed thereto. Thus, it is important that the agreement be carefully filled out, read, and fully understood before signing. A sample boarding agreement is given in Fig. 2-23.

BOARDING AGREEMENT

(To be executed in duplicate; one copy to be retained by each party.)

This agreement made and entered into by and between ________________ (owner of horse), ________________ (address), hereinafter designated "Horse Owner," and ________________ (owner of stable), ________________ (address), hereinafter designated, "Stable Owner." This agreement covers the horse described as follows:

________________ (name) ________ (sex) ________ (age) ________ (color)

I. *Stable Owner agrees that—*

1. He will keep the horse in a stall and/or paddock described as follows:

2. He will feed, water, and care for the horse in a good and husbandlike manner; feeding horse as follows:

Kind of Feed	Amount of Feed: Morning	Noon	Night
	(lb)	*(lb)*	*(lb)*

3. He will perform the following additional services:

   a. *Grooming (Specify):* ________________

   b. *Exercising (specify):* ________________

   c. *Parasite treatments (specify):* ________________

   d. *Others (list):* ________________

II. *Horse Owner agrees that—*

1. He will make all arrangements for the periodic shoeing of the horse, and assume the cost thereof. Any exception to this shoeing arrangement shall be given in the space that follows:

2. He will pay Stable Owner (a) for the foregoing facilities, feed, and services the sum of $ ________ per month, payable on the ________ day of each month in advance; and (b) for drugs and medications, at cost, the first of each month following invoicing.

3. Stable Owner shall be entitled to a lien against the boarded horse for the value of services rendered, and shall be entitled to enforce said lien according to the appropriate laws of the state, *provided* (a) Stable Owner performs the services herein specified, and (b) Horse Owner fails to make a scheduled payment.

III. *Horse Owner and Stable Owner mutually agree that—*

1. In the event the horse shall require the services of a Veterinarian, Stable Owner will immediately contact Horse Owner. In the event Horse Owner cannot be reached, Stable Owner is hereby authorized, as agent for Horse Owner, (a) to call Dr. ________________, DVM; and, should he be unavailable, (b) to call any other licensed Veterinarian of his choice. All fees charged by said Veterinarian shall be the sole and exclusive responsibility of the Horse Owner, with no liability whatsoever on the part of Stable Owner for such fees.

2. This document constitutes the entire agreement between the parties and there are no other agreements between them except as noted below.

________________ (Signature of Horse Owner) ________ (date)

________________ (Signature of Stable Owner) ________ (date)

Fig. 2-23. Boarding agreement.

FUTURES TRADING[20]

Fig. 2-24. Futures trading—a way to hedge against uncertainties.

Futures trading is not new. It is a well-accepted, century-old procedure used in many commodities, for managing risk, protecting profits, stabilizing prices, and smoothing out the flow of merchandise. For example, it has long been an integral part of the grain industry; grain elevators, flour millers, feed manufacturers, and others have used it to protect themselves against losses due to price fluctuations. Also, a number of livestock products—hides, tallow, frozen pork bellies, and hams—were traded on the futures market before the advent of beef futures. Many of these operators prefer to forego the possibility of making a high speculative profit in favor of earning a normal margin or service charge through efficient operation of their business. They look to futures markets to provide (1) an insurance medium in the marketing field, and (2) the facilities and machinery for underwriting price risks.

A commodity exchange is a place where buyers and sellers meet on an organized market and transact business on paper, without the physical presence of the commodity. The exchange neither buys nor sells; rather, it provides the facilities, establishes rules, serves as a clearinghouse, holds the margin money deposited by both buyers and sellers, and guarantees delivery on all contracts. Buyers and sellers are represented by brokerage firms.

The unique characteristics of futures markets is that trading is in terms of contracts to deliver or to take delivery, rather than on the immediate transfer of the physical commodity. In practice, however, very few contracts are held until the delivery date. The vast majority of them are cancelled by offsetting transactions made before the delivery date.

Many cow-calf men have long forward contracted their calves for future delivery without the medium of an exchange. They contract to sell and deliver to a buyer a certain number and kind of calves at an agreed upon price and place. Hence, the risk of loss from a decrease in price after the contract is shifted to the buyer; and, by the same token, the seller foregoes the possibility of a price rise. In reality, such contracting is a form of futures trading. Unlike futures trading on an exchange, however, actual delivery of the cattle is a must. Also, such privately arranged contracts are not always available, the terms may not be acceptable, and the only recourse to default on the contract is a lawsuit. By contrast, futures contracts are readily available and easily offset.

Livestock Futures Trading

Livestock futures trading is relatively new. It was not until November 1964 that trading in live cattle futures opened on the Chicago Mercantile Exchange. Trading in live hogs began 15 months later. Since then, livestock futures trading has grown enormously. Between 1965 and 1981, the number of cattle futures contracts increased from 59,219 to 4,282,293. Between 1966 and 1981, the number of hog futures contracts increased from 8,063 to 2,258,083.

Futures Trading in Finished Cattle, Feeder Cattle, and Feed

The three big uncertainties in the cattle feeding business, any one of which can cause a cattle feeder to suffer heavy losses, are prices of (1) feeder cattle, (2) feed, and (3) finished cattle. Through futures contracts, the cattle feeder can now hedge all three. In advance of feeding, he can lock in his price of feeder cattle, feed, and finished cattle.

This discussion is devoted primarily to live (slaughter) beef cattle futures as they apply to cattle feedlot operators, because it is the highest risk phase of the cattle business, as well as the least flexible. Unless a feeder contracts ahead, he has no assurance of what his finished cattle will bring when they are

[20]This entire section on beef futures was authoritatively reviewed by and helpful suggestions were received from John Gaines, Ag Lender Marketing Specialist, Chicago Mercantile Exchange, Chicago, Ill.

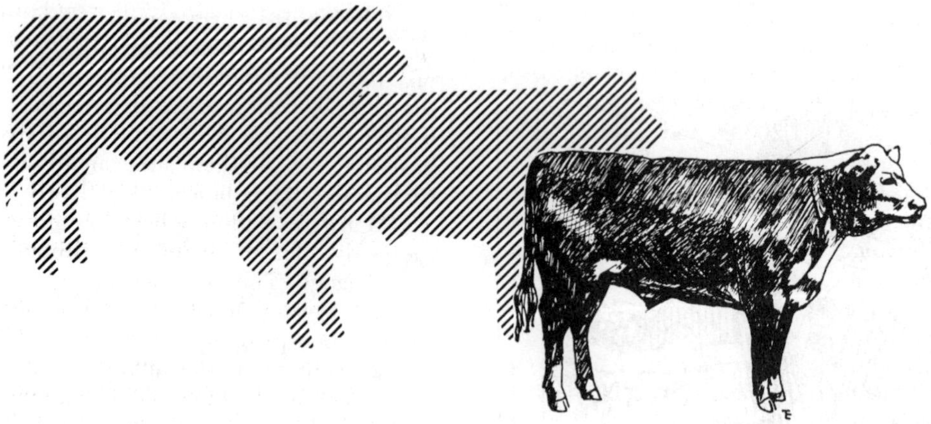

Fig. 2-25. Beef cattle futures—a marketing management tool.

ready to go. Moreover, there is little flexibility in market time, for the reason that excess finish is costly and unwanted by the consumer. As a result of this uncertainty of market price, and in realization of the high risks involved, sleepless nights are rather commonplace among cattle feeders; they find it difficult to concentrate on the business at hand—the efficient feeding and management of cattle. Live (slaughter) beef cattle futures provide a means through which a cattle feeder can fix his selling price before the cattle are ready to be marketed.

The second major item of the triumvirate making for uncertainties in cattle feeding is the price of feeder cattle. Only by contracting ahead, can the cattle feeder be sure of the price that he will have to pay when he is ready to lay in feeder cattle. For many years, a fairly effective, albeit unorganized, cash contracting system has been operating relative to feeder cattle. Feeder cattle futures now offer, on an organized basis, a method for cattle feeders to lock in the price of feeder cattle well ahead of taking delivery, thereby alleviating possible heavy losses due to sharp price rises of feeder cattle. Without feeder cattle, a feedlot is not in business. Yet, much of the overhead cost for facilities and staff continues. Hence, a full feedlot is important. The cow-calf man—the producer of feeder cattle—has more flexibility, and is less dependent on contracting ahead, than the cattle feeder. If the feeder cattle market isn't good, he can hold his calf crop for a time; he may even carry them over for another year—to the yearling stage. Also, rather than accept what he considers to be an unfavorable price for his stockers, he can have them custom fed, or he can feed them out himself. By retaining ownership for a longer period of time, he increases the probability of being able to price his cattle at a profit. Certainly, there are risks in the cow-calf business, but, in comparison with cattle feeding, there is more flexibility, and the timing is not so exacting.

Since feed represents such a large proportion of the cost of feeding cattle (amounting to approximately 80% of the costs exclusive of the purchase price of the feeder cattle), it is wise to set the price months in advance whenever possible. Usually, feed can be bought most advantageously at harvest time. Thus, the cattle feedlot owner who has adequate storage and finances generally buys his main feed ingredients at that time. By so doing, he can project with reasonable accuracy what it will cost him to feed cattle. Corn and soybean meal futures permit the cattle feeder to accomplish the same thing without actually taking delivery on the feed and incurring storage costs and risks of physical deterioration. The cattle feeder can use such futures to protect against increases in feed prices.

What Constitutes a Futures Cattle Contract?

A futures contract is a standardized, legally binding transaction in which the seller promises to make delivery of a specified quantity and type of a commodity at a specified location(s) during a specified future month. The buying and selling are done through a third party (the exchange clearing member) so that the buyer and seller remain anonymous; the validity of the contract is guaranteed by a reputable and well-financed exchange clearing member; and either buyer or seller can readily liquidate his position by simply offsetting sale or purchase.

The Chicago Mercantile Exchange specifications of finished cattle and feeder cattle contracts follow:

- **Specifications for a live (slaughter) cattle con-**

tract are—Delivery and acceptance of 40,000 lb of USDA yield grade 1, 2, 3, or 4 Choice grade steers (approximately 37 head), within the weight range of 1,050 to 1,125.5 lb, and yielding 62%, or within the weight range of 1,125.6 to 1,200 lb, and yielding 63%; stated discounts and tolerances including substitutions in estimated grade, weight, yield, fat thickness, and other details; and delivery to: Peoria, Illinois; Joliet, Illinois; Omaha, Nebraska; Sioux City, Iowa; Guymon, Oklahoma; and Greeley, Colorado.

- **Specifications for a feeder cattle contract are**—44,000 lb of feeder steers averaging 575-700 lb (approximately 65 head) consisting of the USDA No. one (1) muscle thickness and not more than thirteen (13) head of the top one-third (⅓) of the USDA No. two (2) muscle thickness. Par delivery of feeder cattle may be made from approved livestock yards at: Omaha, Nebraska; Oklahoma City, Oklahoma; or Sioux City, Iowa. Deliveries also may be made from approved livestock yards at: Kansas City, Missouri; and St. Joseph, Missouri, at a discount of 25¢ per hundredweight; at St. Paul, Minnesota; Greeley, Colorado; Dodge City, Kansas; and Amarillo, Texas, at a discount of 50¢ per hundredweight; at Billings, Montana, at a discount of 75¢ per hundredweight; and at Montgomery, Alabama, at a discount of $6 per hundredweight.

COMMISSION FEES AND MARGIN REQUIREMENTS ON CATTLE CONTRACTS

The commission fee on all futures contracts covering both purchase and sale (called a round turn) is negotiable between the brokerage firm and customer. In 1982, the minimum hedge margin on live cattle was $700, and the speculative margin was $900. On feeders, the speculative margin was $900 and the hedge margin was $700. The margin deposit may be increased by the broker if the value of the contract should change unfavorably.

HEDGING; SPECULATING

Traditionally, futures contracts have been used for two purposes: (1) hedging, and (2) speculating.

The risks accepted by farmers take different forms; among them, loss of animals by disease, fire, lightning, and theft. Protection from most of these losses can be provided by livestock insurance. But insurance companies do not have policies that cover the loss in animal values that may occur due to price change. Fortunately, through the practice of hedging on commodity futures exchanges, it is possible to reduce risk and uncertainty due to falling prices.

Hedging is the purchase or sale of a futures contract as a temporary substitute for a merchandising transaction which will be made at a later date. Usually this involves opposite transactions in the futures market from those made, or to be made, in the cash market. Since the price movements in the two markets are related, it is anticipated that any loss in one market will be at least partially offset by a gain in the other, with the result that loss through price change will be reduced.

The purpose of the hedge is to protect the merchandising profit anticipated by a handler of a commodity. Merchandising profit is distinguished from speculative profit in that the merchandising profit results from producing and marketing the actual commodity (like finishing and marketing cattle), whereas speculative profit results solely from changes in price and is not the result of a producing or marketing function. Many cattle feeders prefer to leave the assumption of price risk to some other person who is interested in a product in the hope of reselling it at a profit without changing the nature of the product. This other person is referred to as a speculator. *A speculator is a person who is willing to accept the risks associated with price changes in the hope of profiting from increases or decreases in futures prices.* By assuming the risks of price change, the speculator provides many valuable economic functions. His presence in the market gives it both liquidity and continuity.

BASIS

Fig. 2-26. The basis.

The essence of profit and loss in hedging of a commodity, like live cattle and feeder cattle, is the accurate calculation of the *basis* for a particular delivery month. *The difference between cash and futures prices is called the basis.* The difference between your own local cash price and the future price on sale day is *your basis*. The basis is the most important single factor in hedging regardless of the particular commodity involved.

With futures contracts such as those for live cattle and feeder cattle, which are continuously produced, nonstorable commodities, the relationship of the cash price to the futures price has relatively little meaning except during the contract month. Hence, *accurate estimation of the basis for a particular delivery month* is most important in effective hedging of live cattle and feeder cattle. If the estimated basis turns out to be the actual basis on sale day, a perfect price-protecting hedge is the result. However, if the estimated basis is incorrect, there will be a slight gain or loss from the expected results estimated earlier.

Basis variation usually represents an identifiable pattern which repeats itself from year to year and is mostly explainable by economic factors. Hence, accurate estimation of the basis for some future point in time, even if the basis varies during that time, is the key to successful hedging.

In actual practice, *the basis for the par delivery area represented in the futures contract tends to narrow toward zero as the delivery time for the futures contract approaches.* The reason for this is simple. If on April 1, the cash price of a commodity was $1 below the April futures price, merchants would buy the actual commodity, sell the commodity futures for April and make a profit. As they did this, the two prices would rapidly converge until the basis had narrowed and the profit opportunity had disappeared. The convergence of the cash and futures prices during the delivery month means the futures price reflects actual values in the cash market.

There are two methods of determining the basis for any local market: (1) historic price relationships, and (2) cost of delivery.

To calculate the basis with the first method, one must obtain past futures price data and compare those prices to one's local cash market prices. Hence, if one were calculating the basis for live cattle at Kansas City, Missouri, he might find that Kansas City cash prices have normally been 50¢ per cwt below the futures prices at the Chicago Mercantile Exchange in the delivery month. The basis would be—50¢.

To calculate the basis with the second method, one must obtain the actual cost of transporting the cattle from a local market to the delivery point designated by the futures contract. Hence, to calculate the basis between Kansas City and Omaha, via the cost method, one should estimate the transportation cost (including shrink), interest charges, insurance charges, and the like.

There are many factors that cause the basis for any local market to vary over a period of time. These include such things as changes in local supply-demand factors, changes in local production costs, the size of a future crop, government programs, and local market receipts.

In established markets, *basis patterns between markets tend to repeat themselves from one year to the next.* Hence, experienced traders know that their local basis tends to be at a certain level during particular times of the year. The repetition of these patterns from one year to the next makes basis prediction more reliable than price prediction.

To operate successfully, each hedger must develop his own series of data. There is no shortcut to success. Only by keeping his own data will the hedger be familiar (day by day) with the basis situation, with the factors affecting basis, and with comparable basis situations at other times. Only in this way will he have the hedging information pertinent to his hedging location.

The basis reflects the market. When the basis widens, it means that everybody is selling. When the basis narrows, the market is telling you to move your commodity to market.

The stockman (a cattle feeder, or a hog producer) should be knowledgeable relative to the basis for his area before he begins figuring what price he can lock in by using futures.

EXAMPLES OF FUTURES CATTLE CONTRACTS

Examples of futures hedging by each (1) a cattle feeder, (2) a cow-calf man, (3) a packer, and (4) a cattle feeder using a long hedge to protect the price of feeder cattle replacements at the time he forward contracts finished cattle follow. These illustrate hedging procedures, although it must be borne in mind that in actual application the hedges may not work out as perfectly as these.

• **Example 1: A cattle feeder hedging to lock in price (see Table 2-21)**—It is now November, and the cattle feeder has just purchased his feeder cattle to place in the feedlot. Based on past experience, he is quite confident that these cattle should be ready for market the following April. Through good record keeping, he is also quite confident that his production (including labor) and marketing costs should be about $57.50/cwt.

He decides to hedge his cattle with the April futures contract which at the time is selling for $62.45/cwt. He has also estimated his basis will be about $2.10/cwt in April. He subtracts this figure from the April futures price and gets his localized futures price of $60.35/cwt or an estimated $2.85/cwt profit; hence, he sells April futures.

TABLE 2-21

EXAMPLE OF A CATTLE FEEDER USING A SHORT HEDGE TO LOCK IN A PRICE

Cash Market		Futures Market		Basis
	Per Cwt		*Per Cwt*	*Per Cwt*
Nov. 15: Expects to receive in April	$60.35	Sells April futures at	$62.45	–$2.10 Expected
April 10: Sells cattle on cash market at	$57.35	Buys April futures at	$58.85	–$1.50 Actual
Futures gain	$ 3.60			
Realized price	$60.95	Gain	$ 3.60	Gain $0.60

The cattle feeder sold his finished cattle on the cash market for $57.35/cwt which, after subtracting his production costs of $57.50/cwt gives him a loss of $0.15/cwt. However, he realized a profit of $3.60/cwt on his futures transaction, so that his total profit was $3.45/cwt.

This example illustrates what could happen on a declining market. The feeder still showed a profit, even though he had to sell his cattle in the cash market for a price lower than his production costs because this loss was offset by a larger profit in the futures market. This is true because, as the cash price declined, the futures prices also declined.

If, however, the cash and futures prices had risen, he still could have made a profit, this time in the cash market. But because of a loss in the futures market, his total profit would have been less than had he not hedged. Nevertheless, he still received the price protection he desired, which was his main purpose in hedging.

• **Example 2: A cow-calf man using a short hedge (see Table 2-22)**—During April, as his calves are being born, a rancher decides to hedge these calves on a feeder cattle contract. Most of the calves will be sold as feeders during October. Through experience he has estimated that it costs him $65.25/cwt to produce these feeders. The futures market is showing October feeder cattle at $71.25/cwt or at a localized price of $68/cwt. The rancher feels that this assures him of a reasonable profit; hence, he sells October futures.

Even though the cash market was not as strong as the rancher had hoped, he was still able to realize the profit he wanted because of his hedge.

• **Example 3: A meat-packer using a long hedge (see Table 2-23)**—The above examples are illustrations of short hedges. The following example will be of a long hedge, where the futures contract is bought.

The meat-packer has determined, from his basis charts, that the February cattle futures are normally $1/cwt above the local cash price in February. This then assures him of the maximum amount that he will have to pay for cattle in February—that is, if the basis does narrow to $1 during February, the cost of the slaughter cattle will be $1 below the futures price in February.

The meat-packer has also determined that the most he can pay for the slaughter cattle and still make a profit is $60/cwt. He is confident that the basis will be $1 in February.

The profit in the futures market of $7.20/cwt, when applied to the higher than expected cash prices,

TABLE 2-22

EXAMPLE OF A COW-CALF MAN HEDGING TO LOCK IN A PRICE

Cash Market		Futures Market		Basis
	Per Cwt		*Per Cwt*	*Per Cwt*
April 25: Expects to receive in Oct.	$68.00	Sells Oct. futures at	$71.25	–$3.25 Expected
Oct. 10: Sells feeder cattle on cash market at	$68.50	Buys Oct. futures at	$69.00	$0.50 Actual
Futures gain	$ 2.25			
Realized price	$70.75		$ 2.25	Gain $2.75

TABLE 2-23

EXAMPLE OF A PACKER USING A LONG HEDGE

Cash Market		Futures Market		Basis
	Per Cwt		*Per Cwt*	*Per Cwt*
Sept. 27:				
Expects to pay in Feb.	$58.80	Buys amount needed of Feb. futures at	$59.80	−$1.00 Expected
Feb. 20:				
Buys slaughter cattle at	$66.00	Sells Feb. futures at	$67.00	−$1.00 Actual
Futures gain	−$7.20			
Realized purchase price	$58.80	Gain	$ 7.20	

assured the meat-packer that he could purchase the slaughter cattle at a price that allowed him to protect his profit margin.

• **Example 4: A cattle feeder hedging on feeder replacements (see Table 2-24)**—It is not uncommon for a feeder to contract slaughter cattle for future delivery at a set price to a packer before he has acquired the necessary feeder cattle. If by the time of purchase the price of feeders increases beyond the feeder's expectation, his feeding margin may be substantially reduced or he may even suffer a loss on the contract.

The futures contract in feeder cattle can be used in a long hedge to reduce or eliminate the risk involved in an adverse movement of feeder cattle prices. At the time that he negotiates a forward contract with the packer, the feeder would buy feeder futures contracts, preferably for the month in which he actually planned to buy the feeders. If the price of the futures followed local market prices, the feeder would not care what happened to the level of feeder cattle prices because any loss or gain in the cash market would be offset by an opposite outcome in futures.

The arithmetic of this particular long hedge is illustrated in Table 2-24.

If the feeder had not hedged, he would have had to pay $1.90/cwt more for his feeder cattle than he had planned. By hedging he paid $67.10, plus hedging costs, or about what he based his contract price on in January.

It is advisable to place such a hedge in the futures contract month in which the cattle will be purchased because the cash-futures relationship is more predictable at this time. Also, in the event that local feeder cattle prices advanced substantially more than futures, the feeder could stand for delivery of the feeder cattle at the Chicago futures price quotation.

A long hedge to fix the price of feeder cattle may be used in lieu of forward contracting to fix or cheapen the price of feeder cattle needed for replacements several weeks or months in advance of actual purchase. However, the same businesslike procedures are required to get a good buy in feeder futures as in forward contracting. The most likely time to buy futures is when the particular contract in which the feeder is interested is favorably priced relative to local feeder prices after allowing for costs of actually taking delivery.

If futures and cash prices move together closely, the feeder can buy cattle locally when he needs them at about the price prevailing when he placed the long hedge. If futures advance relative to local cash, the feeder can buy cattle as needed and cheapen their

TABLE 2-24

A LONG HEDGE TO PROTECT FEEDING MARGIN ON FORWARD CONTRACTED FED CATTLE

Cash Market		Futures Market		Basis
	Per Cwt		*Per Cwt*	*Per Cwt*
Jan. 15:				
Expects to pay in Feb.	$67.60	Buys Feb. futures at	$66.75	+$0.85 Expected
Feb. 15:				
Buys feeder cattle at	$69.50	Sells Feb. futures at	$69.15	+$0.35 Actual
Futures gain	−$2.40			
Realized purchase price	$67.10	Gain	$ 2.40	Gain $0.50

cost by the gain in the futures transaction. If the futures decline relative to cash, the feeder has the option of taking delivery of his cattle under the futures contract.

DELIVERY AGAINST THE CONTRACT

Although very few contracts, usually fewer than three percent, are consummated by actual delivery of the commodity, a hedger should consider delivery as one of his alternatives, particularly when the cash and futures prices are out of line with each other. However, due consideration must be given to the costs of delivering or receiving delivery, since such costs may be of such magnitude as to offset the differences between the cash and futures prices.

It is not the function of the futures market to provide an alternative source of supply nor an alternate means of disposal of surplus commodities. The purpose of delivery is merely to serve as a safeguard to be used when all else fails.

FACTS ABOUT FUTURES CONTRACTS

A cardinal feature of any workable futures contract—whether it be steers, grain, or any other commodity—is that there shall be maintained a solid connection with the commodity; that is, cash and futures must be tied together.

Any contract held until maturity must be delivered. This keeps the futures price in line with the cash price at the livestock market.

During the delivery month the cash and futures markets tend to come together at the point of delivery. If this were not so, traders would quickly take advantage of the situation. For example, if prior to the termination of trading on August futures, the price of U.S. slaughter steers on the terminal market was $5/cwt below August futures, traders could buy cattle and sell futures, then deliver on the contract for a profit of $5/cwt (less marketing and brokerage fees).

ADVANTAGES AND LIMITATIONS OF LIVE (Slaughter) CATTLE FUTURES

In this section, only live (slaughter) cattle futures will be discussed, simply because they constitute the greatest uncertainty, or risk, in the cattle business; hence, they will always dominate the futures market insofar as cattle feeding is concerned. Nevertheless, many of the same advantages and limitations apply to feeder cattle and feeds.

Live (slaughter) cattle futures are serving a useful purpose; and they are here to stay. Before using them, however, a cattleman should understand what they will and will not do for him.

Among the *advantages* of beef cattle futures are:

1. They serve as a price barometer for several months ahead, thereby increasing the range of information and judgments brought to bear on finished cattle prices and making it easier for feeders to choose a preferred course of action.
2. Through hedging, they can provide price protection or insurance to cattlemen against major breaks in the market.
3. They permit prices to be "locked in" anytime during the feeding period. Thus, they allow selectivity of the market time over the entire feeding period, rather than limit it to the one day that cattle are ready to go to market.
4. They make it possible for cattlemen to obtain credit more easily and to increase financial leverage. For example, let's assume that without hedging, a particular cattleman is able to borrow 70% of the cost of feeder cattle. If he has $90,000 of his own capital to invest, this will enable him to purchase $300,000 worth of feeder cattle. However, if he hedges the cattle he buys, the lender may be willing to lend up to 90% of their cost. His $90,000 of capital will then permit him to purchase $900,000 worth of feeder cattle. In this case, therefore, hedging tripled the number of cattle he could purchase and likewise his profit potential. Knowledgeable lenders will advance funds for the margin on cattle futures contracts.

   Some lenders will advance funds for the margin on cattle futures contracts.
5. They make for a more stable market, with fewer peaks and valleys of price movements.
6. They make it possible for a meat-packer to protect himself when he contracts with a feeder for delivery of finished cattle, (a) for a few months ahead, and (b) at the futures market price at the date of specified delivery. Thereupon, the packer initiates a hedge by selling futures to offset his purchase contract.

Like many good things in life, live beef cattle futures are not perfect. They will not solve all the cattlemen's price problems, they will not raise longtime price levels, nor will they cause people to eat more beef. But these are not disadvantages, they're facts.

Among the *limitations* of live beef cattle futures are the following:

1. During an extended period of rising finished cattle prices, the cattle feeder is disadvantaged when he fixes a price for his cattle in advance. One study revealed that over a 6-year period the consistent hedger would have sacrificed 23% in profits to attain a 74% reduction in profit variability.[21]

[21]Curtis, C. E., Economic Research Service, USDA, "Beef Futures Trading Hit $16 Billion; No Let-Up Is in Sight," *Livestock Breeder Journal*, Aug. 1973, p. 12.

2. No provision for heifers is available; only steers may be delivered, however, heifers can be hedged if the basis, or differences, between steer prices and heifer prices can be determined.

3. Some delivery months may not move exactly as the cash market does.

4. A change in the basis (the spread between the cash price and the price of the futures) can mean a hedging loss as well as a hedging profit.

5. There is a relatively narrow range of time during which it is practical to hold slaughter cattle while waiting for a change in the basis.

6. The feeder must not forget to offset by purchase of another contract at the proper time; otherwise, he may find it necessary to deliver.

7. If the feeder sells futures for a greater amount than the finished weight of his cattle, he is engaged in speculation for the amount of the excess.

8. There are some costs in futures which must be considered; namely, commission and interest on margin capital. These should be considered as costs of doing business; for the protection secured, the cattleman must pay a commission—much as he does for a life insurance policy.

9. Unless a feeder has maintained good and accurate records, and can project his costs with reasonable accuracy, he cannot intelligently determine if a futures price is favorable for placing a hedge.

HOW TO GO ABOUT HEDGING BEEF FUTURES

Here is the "how and where" that a cattleman interested in hedging must follow:

1. Have good and accurate records of costs.
2. Contact a brokerage house that holds a membership in the commodity exchange.
3. Open up a trading account with the broker, by signing an agreement with him authorizing him to execute trades.
4. Deposit with the broker the necessary margin money for each contract desired. He will then maintain a separate account for the cattleman. The commission fee is due when the contract is fulfilled by delivery, offsetting purchase, or sale of another contract.
5. Maintain basis charts, showing the relationship between (a) local prices of feeders and slaughter cattle, and (b) live beef futures.

Glossary of Futures Market Terms

Futures markets have a jargon and language of their own. It is not necessary that cattlemen dealing in futures master all of them, but it will facilitate matters if they at least have a working knowledge of the following:

Basis—The difference or spread between the cash price at a particular market and the price of a futures contract. This spread differs from one market to another and changes with time.

Basis movement—The change which occurs in a particular cash-futures price relationship. It is the change in basis that determines the success or failure of a hedge, rather than changes in market price. One should always hedge according to basis rather than price.

Bear market—A downward moving or lower market is considered "bearish," because the bear strikes down its victim.

Bid—A bid subject to immediate acceptance made on the floor of an exchange to buy a definite quantity of a commodity future at a specified price.

Break—A more or less sharp price decline.

Broker—An agent who handles the execution of all trades. He may also represent a clearinghouse member.

Bull market—An upward moving or higher market is considered "bullish," because the bull tosses his victim upward on impaled horns.

Cash (spot)—The cash price refers to the price of live animals and not futures contract. Also known as spot commodity.

Cash market—Cattle bought and sold for immediate delivery. Also known as spot market.

Chicago Board of Trade—It was founded in 1848. The Chicago Board of Trade handles futures trading in such commodities as wheat, corn, oats, rye, soybeans, and soybean oil and meal.

CFTC—The Commodity Futures Trading Commission, the independent federal agency created by Congress to regulate commodity futures trading. The CFTC Act of 1974 became effective April 21, 1975. Previously, futures trading had been regulated by the Commodity Exchange Authority of the USDA.

Commission—The charge made by a broker for buying or selling a futures contract.

Commission house—A firm which buys and sells actual commodities or futures contracts for the accounts of its customers.

Commitment—A trader is said to have a commitment, when he assumes the obligation to accept or make delivery on a futures contract.

Deferred futures—The futures, of those currently traded, that expire during the most distant months. (See Nearbys.)

Delivery—The tender and receipt of the actual commodity, or warehouse receipts covering such commodity, in settlement of a futures contract.

Delivery points—Those points designated by futures exchanges at which the physical commodity covered by futures contract may be delivered in fulfillment of such a contract.

Discount to futures—When the cash price is under the futures price.

Forward contract—A forward contract calls for delivery at sometime in the future. In a forward contract, a cattleman might make a deal with a buyer during the summer months that calls for delivery of cattle in the fall at the price agreed upon in the contract.

Futures—A term used to designate any and all contracts which are made or established subject to the rules for delivery at a later date.

Hedge—The purchase or sale of a futures contract as a temporary substitute for a merchandising transaction to be made at a later date. Usually it involves opposite positions in the cash market and the futures market at the same time.

Hedgers—Persons who desire to avoid risks, and who try to increase their normal profit margins through buying and selling futures contracts. They are feeders, packers, and others actually involved in production, processing, or marketing of beef. Their primary objective is to establish future prices and costs so that operational decisions can be made on the basis of known relationships.

Limit order—Placing price limitations on orders given the brokerage firm.

Long—The buying side of an open futures contract. A trader whose net position in the futures market shows an excess of open purchases over open sales is said to be "long."

Long hedge—Buying on the futures market contracts against anticipated need in the future in order to protect against a rise in the marketprice. Thus, futures contracts in feeder cattle can be used in a long hedge to reduce or eliminate the risk involved in a rise of feeder cattle prices. At the time the feeder negotiates a forward contract with the packer, he would buy feeder futures contracts, preferably for the month in which he actually planned to buy the feeders.

Margin—Cash or equivalent posted as guarantee of fulfillment of a futures contract (not a payment or purchase).

Margin call—If the market price of a futures contract changes after the cattleman has sold or purchased a futures contract, he will either make a profit or lose money. If the price moves in such a direction so that he loses money, his broker will deduct the losses from his original "margin" and call for additional funds in order to bring the "margin" back up to the original amount.

For example, a feeder might have his broker sell a live cattle futures contract at $60/cwt. He would deposit $700 margin with the broker. If the price of futures were to advance to $61, the feeder would have lost $1/cwt or $400. His broker would deduct the $400 from his original margin of $700, leaving $300. The broker would issue a "margin call" for an additional $400.

Nearbys—The nearest active trading month of a futures market. (See Deferred futures.)

Offer—Indicates a willingness to sell a futures contract at a given price. (See Bid.)

Open interest—Number of open futures contracts. Refers to unliquidated purchases or sales but never to their combined total. (See Commitment.)

Pit—An octagonal platform on the trading floor of an exchange consisting of steps upon which traders and brokers stand while executing futures trades.

Premium—When the cash price is above the futures.

Rally—Quick advance in prices following a decline.

Ring—A circular platform on the trading floor of an exchange, consisting of steps on which traders and brokers stand while executing futures trades.

Round turn—A purchase and its liquidating sale, or a sale and its liquidating purchase.

Security deposit (initial)—Synonymous with the term "margin," a cash amount of funds which must be deposited with the broker for each contract as a guarantee of fulfillment of the futures contract. It is not considered as part payment of purchase.

Security deposit (maintenance)—A sum, usually smaller than, but part of, the original deposit or margin which must be maintained on deposit at all times. If a customer's equity in any futures position drops to or under the maintenance level, the broker must issue a call for the amount of money required to restore the customer's equity in the account to the original margin level.

Settlement price—The daily price at which the clearinghouse clears all trades. The settlement price of each day's trading is based upon the closing range of that day's trading. Settlement prices are used to determine both margin calls and invoice prices for deliveries.

Short—The selling of an open futures contract. A trader whose net position in the futures market shows an excess of open sales over open purchases is said to be "short."

Short hedge—When one owns an inventory of a commodity and hedges by selling an equivalent amount of futures contracts, he has sold short or is short futures and has what is called a short hedge. An example of a short hedge is selling on the futures markets contracts of live cattle which represent cattle that are on feed in the feedlot in order to protect the enterprise against a severe decline in the market.

Speculators—Persons who are willing to accept the risks associated with price changes in the hope of profiting from increases or decreases in futures prices.

Spot commodity—The actual physical commodity such as live cattle as distinguished from the futures. Also known as cash commodity.

Spread—A market position that is simultaneously long and short equivalent amounts of the same or related commodities. In some markets, the term "straddle" is used synonymously.

Ticker—A teletype machine which sends and receives futures market and cash market information.

Trend—The direction prices are taking.

Volume—The number of purchases or sales of a commodity futures contract made during a specified period of time.

TAX MANAGEMENT AND REPORTING[22]

Fig. 2-27. Good tax management and reporting consists in complying with the law, but in paying no more tax than is required. To this end, many farmers and ranchers burn midnight oil! (Drawing by Steve Allured)

Good tax management and reporting consists in complying with the law, but in paying no more tax than is required. It is the duty of revenue agents to see that taxpayers pay the correct amount, and it is the business of taxpayers to make sure that they do not pay more than is required. From both standpoints, it is important that farmers and ranchers should familiarize themselves with as many of the tax regulations as possible.

The cardinal principles of good tax management are: (1) maintenance of adequate records so as to assure payment of taxes in amounts no less or no more than required by law, and (2) conduct of business affairs to the end that the tax required by law is no greater than necessary.

Also, farmers and ranchers need to recognize that good tax management and good farm management do not necessarily go hand in hand. In fact, they may be in conflict. When the latter condition prevails, the advantages of one must be balanced against the disadvantages of the other to the end that there shall be the greatest net return.

Tax Pointers for Stockmen

It is recognized that tax matters constitute a highly specialized and complex field, and that there are many individual farm and ranch differences. Therefore, it is not intended that the pointers which follow should substitute either for careful reading of the official instructions or for consultation with a professional tax specialist. The latter procedure is especially recommended when tax problems are a bit out of the ordinary, and, like a visit to the family doctor, can be most effective when aid is sought before it is too late.

Also, it is recognized that tax laws, regulations, and rulings are frequently revised and that courts are constantly rendering decisions interpreting these laws, rules, and regulations, thus making it important that farmers and ranchers keep abreast of such changes and modernize their tax management accordingly. It is impossible for the average stockman to do this without conferring regularly with a tax consultant who is familiar with the general field of taxation, especially with the taxation of income received by farmers and ranchers.

Some tax pointers of particular interest to stockmen follow:

1. **File an estimate or file your current return**—Remember, if at least two-thirds of your gross income is from the business of farming, and if your tax year begins on January 1, that you may elect to:

   a. File an estimate of your tax and pay this amount by January 15 of the year following the close of your current tax year, then file your current return and pay any balance by April 15; or

[22]This section refers to federal income taxes only.

The author gratefully acknowledges the helpful suggestions of the following noted tax authorities who reviewed this section, along with the sections on Estate Planning and Inheritance and Partnership: Mr. Neal Harding and Ms. Marthea V. Noell, Arthur Andersen & Co., San Jose, Calif.; Mr. James G. Clements, DeForest & Deur, 20 Exchange Place, New York, NY; and Mr. S. P. Kurth and Ms. Laura Lee, Kurth Law Firm, Tax Attorneys, Billings, Mont.

b. File your current return and pay any tax due on or before March 1 of the following year.

2. **Keep adequate and accurate farm records and accounts**—Frequently farmers and ranchers pay more taxes than necessary or have their returns questioned by the Internal Revenue Service simply because they failed to keep adequate and accurate farm records. Farmers and ranchers are not required by law to keep an elaborate set of records and accounts, such as would be necessary to reflect the income of a large automobile manufacturing company, for example. On the other hand, where the accrual method is used, careful inventories must be kept, and farmers and ranchers should have records which will support the figures which appear in these inventories. Even the cash basis farmer must have evidence to support his income and expenditure figures. This is particularly important where a farmer or rancher is claiming capital gains on the sale of livestock held for draft, dairy, or breeding purposes.

Taxpayers should not forget that the burden of proof is on them, and not on the government. There is nothing objectionable, so far as the law is concerned, to managing business affairs to the end that the lawful income tax liability is kept to a minimum. Perspective must be maintained, however, lest the immediate benefit of tax-savings management be overshadowed by a greater long-range tax burden or a greater economic detriment than the actual amount of taxes saved. If the taxpayer cannot prove his income and expenditures, he is at the mercy of the government. In extreme cases where records are very poor, the government is liable to recompute the taxpayer's income on the so-called "net worth" method. Thereupon, if a large discrepancy appears which cannot be explained by the taxpayer, he will find himself in serious trouble with the government.

From the standpoint of good tax reporting, the following farm records and accounts should be kept:

a. A summary of all farm business receipts and expenditures supported by deposit slips and checks, respectively, which deposit slips and checks should carry information identifying the nature of the deposit and withdrawal.

b. A record of profits or losses on the sale of purchased livestock.

c. A depreciation schedule for all farm buildings, machinery, purchased livestock held for draft, dairy, or breeding purposes (unless these animals are included in inventory), and other depreciable property.

d. All income and expenses of a personal nature that should be taken into account.

e. An animal record which will make it possible to take advantage of the capital gains law as it applies to draft, dairy or breeding livestock. Where the livestock enterprise is relatively small, or where a purebred herd is involved, this should include the following:

(1) The date born or purchased.
(2) Ancestry.
(3) The purchase price.
(4) The designation that it is either "held for breeding" or "held for sale."
(5) A record of the use of the animal; e.g., dates bred, etc.
(6) The amount of depreciation, if any, which was deducted for each animal on previous tax returns.
(7) The date sold, and reasons for sale.
(8) The sale price.
(9) Salvage value of breeding animals.
(10) Ratio of ordinary sales to capital sales.

Where a large range operation is involved, it is recognized that it would be difficult to keep an individual animal record on raised animals. Nevertheless, on both the cash and the accrual basis, the stockman should keep an inventory (1) which will show the reason for differences between opening and closing inventories, and (2) from which it is possible to trace the animals from birth to disposition by age groups. Sales should be backed up with sales slips showing the sex and approximate age of the animals sold, and death and disappearance losses should be reported by notes which give the sex and approximate age of the animals lost. Also, all brand inspection reports should be kept.

Since a tax return is no more accurate than the information that goes on it, it is recommended that the above information be recorded in a suitable farm record and account book as the year progresses rather than trust to memory or meager records at tax reporting time.

3. **Separate the farm home from the farm business**—The farmhouse, and such expenses as fuel, insurance on the dwelling and contents, and expenses for groceries, clothing, etc., are not deductible. Likewise, only that part of an automobile that is chargeable to the farm business is deductible as a farm expense.

4. **Keep year-to-year income as steady as possible**—Although it is recognized that climate and prices affect year-to-year income on the farm or ranch, good tax management consists in minimizing these fluctuations. Thus, a farmer with an income of $6,000 one year and $1,000 the next will usually pay more total tax for the two-year period than one who has an average annual income of $3,500.

The Revenue Act of 1964 made provision, beginning in 1964, for "income averaging." Although this does not alleviate the importance of keeping year-to-year income as steady as possible, it can be of substantial benefit to many stockmen whose incomes fluctuate widely from year to year.

The accrual basis of reporting tends to even up differences in year-to-year farm income. For example, livestock breeders and feeders may sometimes find it desirable to withhold the sale of animals one year longer than normal. Under such circumstances, they face the possibility of having to pay tax on income from the sale of 2 years' production in 1 year unless they use the accrual basis of reporting. On the other hand, the farmer on the cash basis is also able to avoid undue fluctuations in income by offsetting unusually high income by purchasing in the year of that high income supplies for use during the following year, and by paying off in the year of high income expenses which he would not normally pay off until the following year. (For example, in some states it is possible to pay 2 years' state taxes in 1 year.) But caution should be exercised in prepaying expenses because this type of tax planning may be challenged by the Internal Revenue Service if it causes a material distortion of income in a particular year.

5. **Select either the "Cash Basis" or the "Accrual Basis" of reporting**—Farmers and ranchers may report on either the "Cash Receipts and Disbursements Basis" (Cash Basis) or the "Accrual Basis." Also, stockmen can use a combination of the two bases, which is known as a "hybrid basis": (a) the accrual basis for livestock, and (b) the cash basis for all other items. Those who have never filed a return before have the option of selecting either basis. However, those who have previously filed have established a basis for reporting and cannot change to the other method without the written consent of the Secretary of the Treasury or his delegate.[23] Most tax consultants favor the cash basis for stockmen. A description of each system follows:

a. **Cash basis**—Under this system, farm income includes all cash or value of merchandise or other property received during the taxable year. It includes all available receipts from the sale of all items produced on the farm and profits from the sales of items which have been purchased, exclusive, generally speaking, of one-half of the profits received from the sale of property used by the farmer in his trade or business, such as breeding stock and farm machinery. It does not include the value of products sold or services performed for which payment was not actually available during the taxable year.

Under the cash system, allowable deductions include those business expenses previously incurred that were actually paid during the year, and depreciation on depreciable items.

[23]Application to make such change is required to be filed within 90 days after the beginning of the taxable year to be covered by the return, and the change will not be granted unless the government and the taxpayer agree to the terms and conditions under which the change will be effected.

b. **Accrual basis**—This system necessitates that complete annual inventories be kept. On the accrual basis, tax is paid on all income earned during the taxable year regardless of whether payment is actually received, and on increases of inventory values of livestock, crops, feed, produce, etc., at the end of the year as compared with the beginning of the year. All expenses incurred during the year's business are deducted from gross income regardless of whether payment is actually made, and deductions are made for any decrease in inventory values of livestock, etc., during the year.

Four methods of inventorying are available to the accrual basis farmer and rancher; namely, (1) cost, (2) cost or market, whichever is lower, (3) farm price, and (4) unit livestock price. The latter two systems are unique systems of inventorying which have been authorized by the government for use specifically by farmers and ranchers. It is recommended that the stockman seek the counsel of a tax advisor in determining which of these four systems he should use.

In summary form, the advantages and the disadvantages of each system—the cash basis, and the accrual basis—are herewith presented.

a. **Some of the advantages of the cash basis are:**

(1) It calls for only the simplest kind of bookkeeping, since no inventories are involved.

(2) Income can be controlled by delaying or advancing sales, purchases and payment of expenses.

(3) A stockman gets a much greater tax advantage than can be obtained on the accrual basis when he sells breeding animals which he has raised (see point No. 6, entitled "Report the maximum income as capital gain, and the minimum as ordinary income," which follows).

b. **Some of the disadvantages of the cash basis are:**

(1) In a bad market year, a taxpayer may be forced to make sales in order to have income against which to deduct his expenses.

(2) Market conditions or other reasons may force the taxpayer to sell two crops in one year, thus doubling his income for that year, but he may be unable to offset the doubled income with advance purchases and payment of expenses.[24]

c. **Some of the advantages of the accrual basis are:**

[24]The net operating loss carry-back, carry-over provisions of the law may overcome this. Also, "income averaging," provided by The Revenue Act of 1964, may help.

(1) The use of inventories tends to even out rises and falls in income more or less automatically and avoids forced sales or bunching up of income.

(2) Expenses may be deducted in the year incurred, whether or not they have been actually paid, thus avoiding any need of borrowing the cash in order to be able to pay expenses prior to the end of the year.

(3) It makes for more accurate bookkeeping and financial statements, and it makes the taxpayer face up to the facts of his operation.

d. **Some of the disadvantages of the accrual basis are**:

(1) The taxpayer is unable to take full advantage of the capital gains law when he sells breeding, dairy or draft livestock (see point No. 6, entitled "Report the maximum income as capital gain, and the minimum as ordinary income," which follows).

(2) It requires the use of inventories, and, hence, more complex record keeping and income reporting.

(3) Tax must be paid on inventory increases, which might prove to be only "paper profits."

6. **Report the maximum income as capital gain, and the minimum as ordinary income**—Since capital gains are not taxed nearly so heavily as ordinary income, it is wise to report the maximum of the former. For most farmers and ranchers, the tax rate on capital gains is based on only 40% of the gain reported. Even for those in the highest tax bracket, the tax on the long-term capital gain will not exceed 20% of such entire gain.

In 1969, the capital gains law was revised to accord capital gains treatment to livestock providing they met the following conditions: (a) cattle and horses were owned for 24 months or more, and other livestock were owned for 12 months or more; and (b) they were held for draft, dairy, breeding, or sporting purposes, and not primarily for sale in the ordinary course of the business. This provision makes it possible to report a greater portion of the profit from livestock as capital gain rather than as ordinary income.

Money spent purchasing work, breeding, or dairy animals is regarded as an investment of capital and shall be depreciated unless (a) such animals are included in the inventory, and (b) the farmer or rancher is using an accrual method of accounting.

From the standpoint of capital gains, the words "livestock held for breeding" refer to animals which the farmer or rancher used or intended to use to produce offspring in his own herd. This includes cattle, hogs, horses, mules, donkeys, sheep, goats, furbearing animals, and other mammals; but does not include poultry (chickens, ducks, turkeys, pigeons, or geese) or other birds, fish, frogs, reptiles, etc. It is not necessary that they have produced offspring prior to their sale in order to qualify, although it is easier to prove that they were held for breeding if offspring were produced. The purpose for which the animal is held is ordinarily shown by the taxpayer's actual use of the animal. However, a draft, breeding, or dairy purpose may be present if an animal is disposed of within a reasonable time after the intended use for such purpose is prevented or made undesirable by reason of accident, disease, drought, unfitness of the animal for such purpose, or a similar factual circumstance.

The determination of whether or not livestock are held for breeding purposes depends upon all the facts and circumstances in each particular case. Some examples follow:

a. The following animals may be classed as intended for breeding purposes:

(1) Animals retained for breeding purposes, but which subsequently proved to be sterile, diseased, or otherwise unsatisfactory as a permanent component part of the breeding herd, and were sold within a reasonable time thereafter.

(2) Young animals which normally would have been used for replacement purposes, but which were sold (a) when an entire herd was dispersed due to the retirement of the owner, or (b) when a herd was reduced because of drought, economic circumstances or other causes.

(3) Gilts that farrowed one litter (on a one-litter system) and that were marketed within a reasonable time thereafter.

b. The following animals cannot be classed as intended for breeding purposes:

(1) Females intended for normal sale, but which were retained for a time and bred prior to disposal.

(2) Bred cows or heifers which were bought for fattening purposes and which calved before marketing.

Where the cash basis of reporting is followed, the gain is computed as follows: If the animal was produced on the farm or ranch, the actual sale price is the gain. If the animal was purchased, the gain is the sale price plus depreciation minus the purchase cost.

Where the accrual basis of reporting is followed for animals produced on the farm or ranch, the gain is the difference between the sale price and the last inventory value of the animal. If the animal were purchased and carried in the depreciation schedule, the gain would be the sale price plus depreciation minus cost. If the animal were purchased and carried in the inventory, the gain would be the sale price minus the last inventory value.

Sales of animals eligible for capital gains treatment are entered on Form 4797. This is true even for sales of those animals which due to the depreciation recapture provisions of the law, or the failure to meet required holding periods, has caused all gains upon sale to be categorized as ordinary income as opposed to capital gains.

Receipts of animals raised primarily for sale, and which cannot qualify as capital gains, are considered as ordinary income, and are reportable in full on Form 1040 F.

7. **Distinguish between capital expenditures and ordinary operating expenses**—The line between capital expenditures and ordinary operating expenses is often difficult to draw. Generally speaking, if the improvement substantially lengthens the life of the building or changes the use of the structure, it should be considered a capital expenditure, and its cost recovered over a period of years; whereas money spent on property to maintain it in its present condition and to prevent deterioration, without adding to the value of the property, is classed as repairs or ordinary operating expenses.

Capital expenditures include such things as new buildings, new equipment, major improvements to old buildings (such as a new roof, or a new foundation), and a major overhaul of machinery. However, the purchase of a new tire for the tractor, or the patching of the roof on the barn, should be charged off as an operating expense in the year the repair is made.

8. **Take advantage of investment credit**—If a taxpayer acquires new or used depreciable property for use in his business and places that property in service during the tax year, he may qualify for an investment credit of 10% of the adjusted cost, provided the property has a useful life of at least 5 years and a 6% credit if the property has a life of at least 3 years. After 1980, the determination of the life of a piece of property is determined by reference to guidelines outlined in the "accelerated Cost Recovery System." This system is discussed in greater detail under point number 10.

9. **Know the major items that constitute gross income and operating expense**—Those on the cash basis will have the following sources of income:

a. Sale of livestock raised.
b. Sale of produce raised.
c. Other farm income.
d. Profit on sale of livestock purchased and other items purchased.

Those on the accrual basis will have the following sources of income:

a. Change in inventory value of livestock, crops and products at the end of the year.
b. Sale of livestock, crops and products during the year.
c. Miscellaneous income receipts.

Fig. 2-28. Depreciation is a means of recovering an investment in property. It represents that loss in value of a physical asset that cannot be made good by repairs and is occasioned by normal wear and tear, obsolescence, or old age.

On both the cash basis and the accrual basis, generally speaking, all expenses necessary to the operation of the farm as a business enterprise, except those for capital investments, are allowable as deductions in arriving at the net farm profit. These expenses include such items as hired labor, feed, fertilizer, seed, soil and water conservation expenditures, gas and oil used in the farm business, taxes and insurance paid on farm property (but not on the farm dwelling), interest paid on mortgages and notes arising out of the farm business, cash rent paid, milk hauling, etc. (see page 1 of Form 1040 F).

10. **Set up depreciation schedules properly**—Depreciation is an estimated operating expense covering wear, tear, exhaustion, and obsolescence of property used in the farm business.

Depreciation may be taken on all farm buildings (except the farm home that is owned and occupied by the taxpayer), and on everything from grain elevators to horse clippers, including tile drains, water systems, fences, machinery, and equipment.

Those who file returns on a cash basis may also take depreciation on dairy cattle, breeding, and work stock which were purchased, but they cannot take depreciation on any livestock which they raise because all costs of raising have been deducted as operating expenses. On the accrual basis, depreciation may be taken on purchased animals if not included in the inventory on page 2 of Form 1040 F.

Taxpayers should list each building, and each piece of machinery on which depreciation is to be computed in the depreciation schedule. Such items as cows and small implements may be grouped on Form 1040 F, but such groupings should be derived from the totaling of a detailed individual list kept current in a permanent farm record book.

The Economic Recovery Tax Act (E.R.T.A.) of 1981 completely revolutionized Federal law relating to the calculation of depreciation allowances by instituting the Accelerated Cost Recovery System (ACRS). The new system is mandatory for all depreciable property placed in service after December 31, 1980, and essentially abandons traditional depreciation concepts of estimated useful lives, salvage value and used property rules. Instead, ACRS requires that depreciation allowances be computed on the unadjusted basis of property in accordance with tables provided in the law. "Unadjusted basis" is defined as the basis of property without adjustment for prior depreciation. For assets placed in service after December 31, 1982, this amount must then be reduced by one-half of the amount of the investment tax credit claimed on the property.

The new law created five property classes into which all eligible property must be categorized. Three of these classes apply to farm related assets. A description of these classes and examples of assets which would typically fall into each class follows:

a. **Three-year property**—This class includes all personal depreciable property which had a mid-point life of four years or less under the Treasury Department's old "asset depreciation ranges." Typically this category will include light farm trucks, automobiles, breeding hogs, and horses which are more than 12 years old when placed in service.

b. **Five-year property**—Five-year property is all property which does not fall into either the 3-year or the 15-year category. Assets included here are farm equipment and machinery (including grain bins and fences); structures used solely for the housing, raising and feeding of livestock and their produce; breeding or dairy cattle; horses not described as three-year property; and breeding sheep and goats.

c. **Fifteen-year property**—This class, which is defined as real property, had a mid-point life of 12.5 years under the old Treasury Department ranges. This category will usually only include farm buildings (other than the livestock facilities described above as five-year property).

As previously mentioned, ACRS depreciation is computed according to tables. All three-year and five-year property placed in service during a given tax year receive the same depreciation provision as calculated by reference to the tables. Hence, an asset placed in service during the first month of a tax year will receive no more depreciation allowance than an asset placed in service during the twelfth month of that year. Depreciation on property in the 15-year class will differ, depending on the month in which the property is placed in service.

Table 2-25 shows the recovery percentages for 3-, 5-, and 15-year property placed in service after December 31, 1980.

While the ACRS system is a much simpler method of computing tax depreciation, it also has drawbacks in that it provides far less flexibility to the taxpayer in calculating each year's depreciation provision. The only flexibility built into the system is the ability for the taxpayer to elect to use the straight-line method (as opposed to the table method) over either the assigned ACRS class life or an extended class life. The optional straight-line lives available by class category are:

3-year property	3, 5, or 12 years
5-year property	5, 12, or 25 years
15-year real property	15, 35, or 45 years

For taxable years beginning after 1981, taxpayers are allowed to elect to deduct currently the cost of a limited amount of property as an expense, rather than capitalizing and depreciating the property. The annual maximum deductible amount under this provision is as follows:

Taxable Years Beginning In	Amount
1981	$ —
1982 or 1983	5,000
1984 or 1985	7,500
1985 and after	10,000

To qualify for this election, the ACRS property must be tangible personal property (which includes livestock, but excludes trees and vines) which was

TABLE 2-25

DEPRECIATION ALLOWANCES, SHOWING 3-, 5-, AND 15-YEAR PROPERTY CLASSES APPLICABLE TO FARMS AND RANCHES

3-Year and 5-Year Property

Ownership Year	Class: 3-year	Class: 5-year
	(%)	(%)
1	25	15
2	38	22
3	37	21
4	—	21
5	—	21
	100	100

15-Year Property

(Use the column for the month in the first year the property is placed in service)

Ownership Year	1	2	3	4	5	6	7	8	9	10	11	12
	(%)											
1	12	11	10	9	8	7	6	5	4	3	2	1
2	10	10	11	11	11	11	11	11	11	11	11	12
3	9	9	9	9	10	10	10	10	10	10	10	10
4	8	8	8	8	8	8	9	9	9	9	9	9
5	7	7	7	7	7	7	8	8	8	8	8	8
6	6	6	6	6	7	7	7	7	7	7	7	7
7	6	6	6	6	6	6	6	6	6	6	6	6
8	6	6	6	6	6	6	5	6	6	6	6	6
9	6	6	6	6	5	6	5	5	5	6	6	6
10	5	6	5	6	5	5	5	5	5	5	6	5
11	5	5	5	5	5	5	5	5	5	5	5	5
12	5	5	5	5	5	5	5	5	5	5	5	5
13	5	5	5	5	5	5	5	5	5	5	5	5
14	5	5	5	5	5	5	5	5	5	5	5	5
15	5	5	5	5	5	5	5	5	5	5	5	5
16	—	—	1	1	2	2	3	3	4	4	4	5
	100	100	100	100	100	100	100	100	100	100	100	100

acquired by purchase and is used in the taxpayer's trade or business. If the taxpayer elects to expense the cost of property, no investment credit can be claimed on the costs that are expensed. Hence, the taxpayer must determine if the present value of the taxes saved by electing to expense currently exceeds the value of the taxes saved by claiming investment credit in the year the asset is placed in service together with depreciating the asset under ACRS.

The new ACRS rules apply only to property placed in service after December 31, 1980. All methods of computing depreciation which the taxpayer elected to use on pre-1981 property remain intact and unchanged by the new law.

11. **Handle livestock death losses as prescribed**—On the cash basis, no death deduction can be made for an animal that was born and raised on the farm, because the cost of raising the animal has been deducted already as operating expense. If the animal was purchased, however, death loss can be deducted. Where the animal lost is not being held for draft, breeding or dairy purposes, but is held primarily for sale, the loss, regardless of the cause of death, should be reported on line 53, Part II of Form 1040 F, as "other deductions," with an explanation. In the case of a purchased animal held for draft, breeding or dairy purposes, and thus considered as a capital asset, the same treatment is nevertheless accorded if the animal dies of old age or disease. However, if such an animal is destroyed (as by lightning, or flood, or by order of government authorities because of a contagious disease), the loss has to be offset against gains, if any, from the sale of other property used in the taxpayer's business. If there is a net loss, it is handled in the same manner as the other losses described above. In any event, the amount of the loss is the cost of the animal, less depreciation and any insurance received.

On the accrual basis, when the value of an animal appears in the beginning-of-year inventory but not in the end-of-year inventory, the loss is automatically accounted for in the change in inventory value. Any money received from insurance or indemnity is entered as other farm income.

12. **Secure tax consideration in handling net operating losses**—The law provides that net operating losses, such as may be encountered from declining

livestock prices or crop failure, can be carried back three years, and, if necessary, any remaining excess losses may be carried forward 15 years. Since the procedure for reporting these losses is very complicated and requires special forms, it is recommended that the aid of an income tax consultant be obtained under such conditions.

13. **Protect forced sales resulting in profit**—From time to time, a farmer or rancher is forced to sell livestock because of disease or drought and realizes a profit on such sales either because (a) he is on a cash basis and the animal stands at zero on his books, or (b) he uses inventories and the animal was carried at a very low inventory value. In such cases, under the involuntary conversion provisions of the code, he will not be taxed on the profit if he uses all of the proceeds of the sale to purchase replacement stock within two years (one year for property involuntarily converted before 1970) from the close of the first tax year in which any part of the gain is realized, or within such longer time as the government may allow upon an application for extension. With respect to drought sales, only animals held for draft, breeding, or dairy sales will qualify.

If, because of soil or other environmental contamination, it is not feasible for a farmer to reinvest the proceeds from compulsory or involuntarily converted livestock in replacement stock; the farmer may invest the proceeds in other property, including real property, used for farming. This replacement property will also allow the farmer to defer recognition of any of the profits realized from the disposition of his livestock due to disease or drought.

14. **Secure gas tax refunds**—For information relative to this subject, secure the publication entitled *Federal Fuel Tax Credit or Refund for Nonhighway and Transit Users*, No. 378, at your local Internal Revenue Service office.

15. **Secure competent help when needed**—A farmer or rancher can get help in making out his federal income tax return from the following sources:

a. Local representatives of the Director of Internal Revenue.

b. Lawyers, accountants, and others who specialize in tax matters. It is noteworthy that fees paid for tax assistance are deductible on the following year's income tax return.

16. **Resort to available recourse if necessary**—In the event the return is audited, the Internal Revenue Agent conducting the examination may propose that certain changes be made to the income as reported with the result that the taxpayer will owe more tax. If, after his explanation and discussion of the matters, the taxpayer is not in agreement, he may proceed as follows:

a. Indicate to the Internal Revenue Agent that he does not agree. The taxpayer will then receive from the Revenue Agent a "30-day letter," reporting the agent's findings and the proposed tax deficiency.

b. Request a conference with the District Conferee. If agreement is not reached at this point, the taxpayer may file a written protest.

c. Have a conference with the Appellate Division. If the differences cannot be resolved at this point, the taxpayer will receive a 90-day letter, or notice of assessment. He may then, within 90 days, file a petition with the Tax Court of the United States, which will consider the case on its merits.

d. At any point following the aforementioned 30-day letter, but prior to filing a petition with the Tax Court, the taxpayer may decline any of the procedures listed, pay the tax when billed, and file a claim for refund. If the claim is not acted upon within six months, or is rejected, the taxpayer may file suit for refund in the U.S. District Court. In some cases, it is also possible to proceed in the U.S. Court of Claims.

e. At any point the taxpayer may sign a waiver (which in effect provides for the assessment of the tax) and pay the tax upon receipt of notice of assessment. This is usually done when the taxpayer is in agreement with the decision at any point of the procedures and limits the amount of interest to be paid on the deficiency should he be unsuccessful in a claim for refund.

Tax Planning

There are some things that a stockman can do to lessen the tax bite. Some of these will be discussed.

CLAIM AS MUCH DEPRECIATION AS POSSIBLE

Generally, this means utilizing the ACRS table percentages for depreciating all property. However, as ACRS utilizes accelerated depreciation, all depreciation taken over the amount of depreciation which would have been allowable using the straight-line method over certain prescribed periods is subject to an "alternative minimum" tax. A tax advisor should be consulted to determine if the advantage of utilizing the accelerated depreciation methods under ACRS outweighs the creation of this depreciation tax preference.

MAKE USE OF INVESTMENT CREDIT

Investment credit can be a massive tax saver. It permits the farmer or rancher to deduct 6 or 10% of the cost of purchased property off the tax he would otherwise have to pay. The effect is to reduce his tax by 6 or 10% on most items.

On eligible new or used property that the stockman purchases, he can take 6 or 10% of the "qualified investment" off the tax he would otherwise have to pay. This means that if he bought a tractor for $20,000, an investment credit of 10% could save him $2,000 in taxes. That's worth as much as $4,000 in added depreciation deduction for a taxpayer in the 25% bracket.

For property acquired after December 31, 1982, a taxpayer must reduce the depreciable basis of an asset by one-half of the amount of investment tax credit claimed on such asset. Hence, a 10% investment credit will allow only 95% of the asset on which the credit was claimed to be depreciated over current and future periods. Some flexibilities exist within the law which allow the taxpayer to claim a lower amount of investment tax credit in exchange for a greater depreciable base. A tax advisor should be consulted to determine which alternative produces the greatest overall tax benefits.

A lot of different kinds of farm property are eligible for the investment credit. Along with the usual array of machinery and equipment, breeding stock now qualifies. Even tile drains, fences, feeding floors, outside electrical installations, and water systems are eligible.

Buildings don't usually qualify, but the credit can be claimed on silos and grain bins. In recent years, structures and equipment used in livestock confinement systems have qualified. The test is whether the structure houses property "used as an integral part of production, manufacturing, or extractive activity." Also, whether the structure is so closely related to use of the equipment that it can be expected to be replaced when the equipment is changed.

OTHER TAX SAVERS

Other tax-saving ideas that stockmen may apply are:

1. **Buy ahead on certain inputs**—The stockman might consider buying ahead feed, fertilizer, seed, chemicals, or other items. In doing so, however, he should pay for it in the year for which claim is being made, and, if possible, take delivery. A mere deposit for future delivery on verbal contract isn't enough to back up a deduction as far as IRS is concerned.

2. **Cull the herd**—The stockman may sell additional breeding stock rather than replacement stock. Remember that only 40% of the income from the sale of home-raised breeding animals is taxable. Remember, too, that cattle and horses must be kept two years and other livestock for one year to qualify as capital gain breeding stock.

3. **Pay children for farm work**—Payments made to children for farm work are business expenses. But the stockman must be able to show that a true employer-employee relationship exists. To do so, children should be assigned definite jobs at agreed-upon wages and paid regularly.

Wages paid to children under 21 years of age are not subject to Social Security tax. However, if a child nets more than $400 on a 4-H or FFA cattle project, he is liable for Social Security tax as a self-employed person.

There are other ways to save on taxes. But there are also some "traps" into which the stockman may fall. That's why it's important to consult with a good farm or ranch tax advisor—one who has had considerable experience in working out tax plans involving breeding animals and other unique aspects of farm taxes.

Tax Sheltered Livestock Investments

Since federal income tax was introduced in 1913, numerous changes have been made in the law, usually in the form of special deductions, credits, exclusions, exemptions, and special rates. Occasionally, these changes allow some taxpayers to escape paying as much tax as their critics think they should. Such provisions are popularly called "tax loopholes" or "tax shelters."

Special farm tax rules, when combined with high nonfarm income, can permit the deferral of income tax on nonfarm income. In addition, some farming activities allow the conversion of ordinary income into capital gains income, which has a lower tax rate. Also, if a farming investment produces a "real" loss, there is little likelihood that the loss will be disallowed for tax purposes. These are the characteristics of an ideal tax shelter.

Some high tax bracket nonagricultural men and women are investing in livestock programs to shelter nonfarm earnings. Such "tax loss" stockmen are not "typical" in U.S. agriculture. But neither are they a rarity.

Today, practically every brokerage house has a tax shelter department which offers cattle feeding, along with oil and real estate, investment programs. Also, investment advisory firms have sprung up, who, for a fee, will advise a potential investor and assist him in setting up a cattle feeding, oil and gas, or real estate tax shelter program.

What, then, is the investment and tax appeal of livestock enterprises? How do they work?

Special treatment for "farmers" under provisions of the Internal Revenue Code arises primarily from the following three sources:

1. **Cash accounting**—Farmers are permitted to

choose between the accrual method and the cash receipts and disbursement method (the cash method). As a result, the cash method may be used as a means of minimizing and postponing income tax payments. For example, an investor may buy feeder cattle near the end of his tax year, and prepay feed and interest costs, which would be deductible as a business expense and cause a loss on farm operations, which could be offset against his ordinary income.

2. **Current deduction of capital expenditures**—Investment spending (the cost of acquiring and developing capital assets) is generally not deductible from income as a current expense for income tax purposes. Instead, these costs are required to be capitalized and recovered through depreciation over the useful life of the asset. Livestock breeders are allowed to deviate from this general rule. In a breeding program, they are allowed to use their expenses to offset ordinary income while the herd is being built up and treated as a capital asset.

Additional provisions allow taxpayers reporting farming operations to deduct certain expenses which, if incurred in other businesses, would be capitalized. These include (a) soil and water conservation expenditures (not to exceed 25% of gross income from farming for the year), (b) land clearing expenses or expenditures incurred in making land suitable for farming (limited to the lesser of $5,000, or 25% of taxable income from farming during the year), and (c) expenses for fertilizer applied on land used in farming.

3. **Livestock as a capital asset**—Livestock held for draft, breeding, dairy, or sporting purposes is treated as property used in a trade or business. This means that livestock (except poultry) held for the above purposes is entitled to capital gain treatment upon sale provided holding period requirements are met. These holding periods are 24 months for cattle and horses and 12 months for all other qualifying livestock.

When cash accounting is used, the benefit of this provision is multiplied. Expenses of raising the animal are deductible currently and the entire sales price is taxed as a capital gain.

Of course, purchased livestock held for draft, breeding, dairy, or sporting purposes can be depreciated just as any other purchased capital asset. But there is recapture of depreciation upon sale.

LIMITATIONS ON LOSSES

The basic rule that hobby or pleasure losses cannot be deducted from nonfarm income beyond the income generated from the livestock business prevails, and the general rule that a person must enter into the livestock business "to make a profit" before deductions for his expenses beyond income from the livestock business was left intact by the 1976 Tax Reform Law.

Under the prior law, the amount that could be written off was without limitation until the losses exceeded $25,000, and the nonfarm income or nonlivestock business income exceeded $50,000. This was referred to as the "farm loss recapture" rules or the "excess deductions account" (EDA). After 1975, no further additions to EDA were required or permitted, but the rules applied to existing accounts.

Under the 1976 Tax Reform Law, the $25,000 limitation was eliminated and the amount of loss that a stockman can deduct from nonlivestock income is limited to the amount that he has "at risk." A taxpayer is generally considered "at risk" with respect to an activity to the extent of his cash and the adjusted basis of the property contributed to the activity, plus any amounts borrowed for use in the activity with respect to which the taxpayer has personal liability for payment from his personal assets. Further, his net fair market value of personal assets which secure nonrecourse borrowings is included in the definition of "at risk." "Nonrecourse" means the taxpayer has no personal liability for the payment of a debt.

These "at risk" provisions apply to losses attributable to amounts paid, incurred, depreciated, or amortized in taxable years after December 31, 1975.

In summary, the stockman will be limited to those dollars he has put out of his pocket, plus contracts or loans for monies or property put into the livestock business that carries his personal liability or responsibility.

If the loss in any year is less than the amount "at risk," the full amount of the loss is deductible and the "at risk" amount is reduced by the loss deducted. The reduced "at risk" amount is then carried over to the next year to determine any limit deductions lost under that year.

If the loss is greater than the amount "at risk," the deductible loss is limited to the amount "at risk" at the end of the year. The amount "at risk" is reduced to "0." In this case, the nondeductible portion of the loss is carried over to the next year and is available for deduction then, if not prevented by application by the "at risk" rule.

If the risk amount has been reduced to "0," no further losses may be deducted until such time as the taxpayer places additional amounts of investment through borrowings or additional cash investment.

ESTATE PLANNING

Human nature being what it is, most farmers shy away from suggestions that someone help plan the disposition of their property and other assets after they are gone. Also, they have a long-standing distrust of lawyers, legal terms, and trusts; and to them the

Fig. 2-29. Estate planning is a way in which to preserve the farm or ranch for use by your chosen successors.

subject of taxes seldom makes for pleasant conversation.

If a farmer has prepared a will or placed his property in joint tenancy, his estate will be distributed as he has specified. If not, it goes to his heirs, according to the laws governing intestate (without a will) succession of property. His heirs are those persons whom the law appoints to succeed to his property in the event of intestacy, and are not necessarily the persons the farmer would want to have his property.

If no plans are made, estate taxes and settlement costs often run considerably higher than if proper estate planning is done and a will is made to carry out these plans. Today, the livestock business is big business; many stockmen have well over $1,000,000 invested in land, animals, and equipment. Thus, it is not a satisfying thought to one who has worked hard to build and maintain a good livestock establishment during his lifetime to feel that his heirs will have to sell the facilities and animals to raise enough cash to pay federal estate and inheritance taxes. By using a good estate planning service, a stockman can generally save thousands of dollars for his family in estate and inheritance taxes and in estate and settlement costs. For assistance, stockmen should go to an estate planning specialist—an individual or company specializing in this work, or to the trust department of a commercial bank.

A limited discussion of some of the revisions in the estate and gift tax laws resulting from the Tax Reform Act of 1976 and the Economic Recovery Tax Act of 1981 follows.

- **Farm valuation is on basis of present use**—One very important aspect of the new laws for the benefit of the stockman's estate is the formula to be used for valuing farms. Many heirs of farmers have been faced with the problem of a farm being valued at its highest and best use, which in many cases was substantially higher than its farming use value and resulted in substantial estate taxes.

Under the 1976 and 1981 Acts, if certain conditions are met, real property used for farming may be valued on the basis of its existing use instead of the highest and best use as under the old law. In no case, however, may the alternate method reduce the gross estate by more than $700,000 for persons dying in 1982 and $750,000 for persons dying in 1983 and thereafter.

In general, the formula for valuing farms is the actual value based on the average annual gross cash rental for comparable farm purpose land located in the same locality, less the average annual state and local real estate taxes for such comparable land, divided by the average annual effective rate for all new federal land bank loans. The general result of the law change is expected to reduce the amount of estate taxes due on farms.

- **Longer time to pay estate taxes**—To lessen the need for forced sales of farms or ranches in order to pay estate taxes, the estate will be able to defer taxes up to 15 years, provided the livestock business is a closely held business and the business value exceeds 35% of the value of the gross estate. In such case, only annual interest payments need be made during the first five years. Thereafter the tax due and interest thereon is payable in 10 annual installments. A special 4% interest rate is allowed on the deferred estate taxes attributable to the first $1 million of the farm, or other closely held business property.

An estate may dispose up to 50% of the closely held business without causing an acceleration of the installment payments. A disposition of more than 50% will cause all unpaid installments to become due currently.

The longer time to pay estate taxes is a great benefit to a farmer's estate that is short on cash, and it alleviates a forced sale of animals or ranch in order to pay taxes.

If more than 65 percent of a decedent's adjusted gross estate is an interest in a closely held business, an executor may elect to pay all or part of the estate taxes, in up to 10 equal annual installments. Moreover, he may elect, for the first installment for a period up to five years, with interest limited to a special 4% rate on the estate taxes attributable to the first $1 million of the farm, or other closely held business property.

However, if there is a disposition of the livestock business or aggregate withdrawals of one-third of its money or property, there will be an acceleration of the taxes that would have been due.

It should be noted that these new liberalized provisions apply not only to closely held animal businesses, but to any type of business.

• **Higher exemptions**—One of the significant revisions in the estate and gift tax law made under the 1976 Act was the implementation of a unified transfer tax credit which replaced the old separate lifetime exemptions for estate and gift taxes. The 1976 Act allowed the new unified credit to increase to a maximum of $47,000 in 1981. The Economic Recovery Act now allows for the credit to increase steadily, phasing in at higher levels over six years according to the schedule in Table 2-26.

TABLE 2-26

UNIFIED CREDIT FOR FEDERAL AND GIFT TAXES

Year	Unified Credit	Equivalent Exemption
	($)	($)
1982	62,800	225,000
1983	79,300	275,000
1984	96,300	325,000
1985	121,800	400,000
1986	155,800	500,000
1987 and later	192,800	600,000

As shown in Table 2-26, the unified credit increased from $47,000 to $62,800 in 1982. This was equivalent to a tax-free estate of $225,000. By the time the full credit of $192,800 is reached in 1987, the equivalent tax-free estate value will be $600,000. Thus, with a unified credit of $192,800, there is no estate or gift tax on taxable transfers aggregating $600,000 or less.

• **Increase in gift tax exclusion**—The nontaxable gift exclusion is increased from $3,000 to $10,000 per person per year. A husband and wife who elect gift-splitting may jointly give $20,000 per recipient per year. And these gifts may be in the form of interest in the livestock operation. Nonsplit gifts not exceeding $10,000 per person per year will not have to be reported even though made within three years of death. Lifetime marital gifts, and gifts less than $10,000, are completely nontaxable and they are not grossed up in the donor's estate for federal estate tax purposes.

Taxable gifts such as nonmarital gifts and gifts in excess of $10,000 per person per year must be reported on gift tax returns. These tax returns must be reported on a calendar year basis. The data due for the return is April 15 of the subsequent year.

The $10,000 a year given to each individual will not in any way increase the donee's estate taxes in the future.

For many years people have been aware of the benefits of getting assets out of the estate to avoid estate taxes. However, up to this point the $3,000 annual exclusion for gift taxes has severely limited people's ability to do this. Now, with the larger $10,000 exclusion and with gift-splitting, giving away the estate before death to avoid taxes is a lot easier.

• **Tax rate reductions**—Under the previous law, the marginal tax rates ranged from 18% for taxable estate values of $10,000 or less to 70% for estates exceeding $5 million in value. Under the 1981 law, the top rates will be reduced yearly until the maximum rate of 50% is obtained in 1985. Thereafter, the 50% bracket will be for taxable estates over $2.5 million. The rates below the 50% level will not change (Table 2-27).

TABLE 2-27

FEDERAL ESTATE AND GIFT TAX RATE SCHEDULE

	Taxable estate and lifetime gifts		Tax		Of excess
	From:	To:	$ +	%	over:
1	$ 0	$ 10,000	$ 0	18	$ 0
2	10,000	20,000	1,800	20	10,000
3	20,000	40,000	3,800	22	20,000
4	40,000	60,000	8,200	24	40,000
5	60,000	80,000	13,000	26	60,000
6	80,000	100,000	18,200	28	80,000
7	100,000	150,000	23,800	30	100,000
8	150,000	250,000	38,800	32	150,000
9	250,000	500,000	70,800	34	250,000
10	500,000	750,000	155,800	37	500,000
11	750,000	1,000,000	248,300	39	750,000
12	1,000,000	1,250,000	345,800	41	1,000,000
13	1,250,000	1,500,000	448,300	43	1,250,000
14	1,500,000	2,000,000	555,800	45	1,500,000
15	2,000,000	2,500,000	780,800	49	2,000,000
16	2,500,000	—	1,025,800	50	2,500,000

Notice in Table 2-27 that the tax rate is graduated. The tentative tax rate on the first $10,000 of the taxable estate or lifetime gifts is 18%. The taxable estate or lifetime gifts from $10,000 to $20,000 will result in a tax of $1,800 on the first $10,000 of property transferred and a 20% tax rate for a transfer of more than the first $10,000, but less than $20,000. For example, for a taxable estate of $110,000, line 7 in the table (taxable estate from $100,000 to $150,000) shows the tentative tax on the first $100,000 to be $23,800 and the tax on the next $10,000 would be $3,000 ($10,000 × 30% = $3,000). The total tentative tax, therefore, on a $110,000 transfer would be $23,800 plus $3,000 or $26,800.

• **The unlimited marital deduction**—By far the most significant change in the gift and estate tax laws made by the Economic Recovery Tax Act of 1981 has been to increase from 50 to 100% the deduction allowable for lifetime and deathtime transfers to one's surviving spouse. This new unlimited marital deduction will permit a farmer to pass his entire estate to his spouse completely free of tax. This, of course, will significantly reduce the need for forced sales of property to raise cash to pay taxes. It may also affect the farmer's need for life insurance.

Use of the unlimited marital deduction is not always the best tax planning. Consideration must be given to the potential estate tax liability of the estate of the surviving spouse and state estate and inheritance taxes. For these reasons, it is always prudent to seek professional advice on these matters.

• **Summary**—The 1976 and 1981 tax reforms provided substantial benefits to the stockman in many areas. It made for more advantages than disadvantages to the stockman.

The new laws cover many areas of taxes and are very complex. With the significant changes affecting the stockman, he should engage the services of a qualified lawyer to assist him in estate planning.

Wills

A will is a set of instructions drawn up by or for an individual which details how he wishes his estate to be handled after his passing. Despite the importance of a will in distributing property in keeping with the individual's wishes, about 50 percent of farmers and ranchers pass away without having written a will. This means that state law determines property distribution in such cases.

Every stockman should have a will. By so doing, (1) the property will be distributed in keeping with his wishes, (2) he can name the executor of the estate, and (3) sizable tax savings can be made by the way in which the property is distributed. Because technical and legal rules govern the preparation, validity, and execution of a will, it should be drawn up by an attorney. Wills can and should be changed and updated from time to time. This can be done either by (1) a properly drawn-up codicil (formal amendment to a will), or (2) a completely new will which revokes the old one.

The same attorney should prepare both the husband's and wife's wills so that a common disaster clause can be incorporated and the estate planning of each can be coordinated.

Trusts

A trust is a written agreement by which an owner of property (the trustor) transfers his title to a trustee for the benefit of persons called beneficiaries. Both real and personal property may be placed in trust.

The trustee may be an individual(s), bank, or corporation, or a combination of two or three of these. Management skill should be considered carefully in choosing a trustee.

A trust can continue for any period of time set by the owner—for a lifetime, until the youngest child reaches age 21, etc. If the trust extends beyond a lifetime, there are limitations which should be explained by an attorney.

KINDS OF TRUSTS

Basically, there are two kinds of trusts, the *living* and the *testamentary*. The living or *inter vivos* trust is in essence an agreement between the trustor and the trustee and may be revocable or irrevocable.

The *revocable trust* can be terminated or altered; under it the trustor is concerned about the here and now, rather than only the hereafter. The trustor continues to make decisions, and he can call off the whole arrangement (it's revocable) if it doesn't work out as expected. The revocable trust offers no special estate tax advantage; the assets of a revocable trust are included in the estate of the deceased creating the trust. However, it can be written in such a manner as to reduce substantially the estate taxes of the beneficiaries. Also, the revocable trust will eliminate the cost of probate—costs which may include executor's fees, attorney's fees, court costs, and appraisal fees.

The *irrevocable trust* cannot be amended, altered, revoked, or terminated. Under an irrevocable trust, the trustor must be willing to part with his trust property forever (irrevocably) and have nothing further to do with it and its administration. However, the irrevocable trust has many favorable aspects in estate planning; it will reduce estate taxes in both the estate of the trustor and the estate(s) of the life beneficiaries, and it avoids probate.

The *testamentary trust* is so-called because it is established under the provisions of the trustor's last will and testament. The testamentary trust does not become effective until after death of the trustor, followed by probate. There is no tax saving in the trustor's estate. However, the trust may be drafted to save estate taxes in the estates of the beneficiaries. A testamentary trust is useful when the heirs are minors or inexperienced in money matters.

Partnership Contract

Another logical step in the transfer of property is a partnership contract between the parents and their heir(s) recorded in accordance with State law. Appropriate counsel should be consulted in the preparation of such an agreement. Where the partnership contract is between the father and the heir, a provision should be included permitting the heir to purchase the father's share of the partnership for a fixed amount, through a buy-sell agreement. Funding for such a buy-sell agreement is commonly done with life insurance. This can provide for proper and uninterrupted operation of the livestock enterprise, because at the father's death, the heir will acquire the father's interest in the partnership.

LIVESTOCK INSURANCE[25]

The ownership of a fine animal constitutes a risk; which means that there is a chance of financial loss. Unless the owner is in such strong financial position that he alone can assume this risk, the animal should be insured.

The rates of American Live Stock Insurance Company, Geneva, Illinois, the only U.S. insurer that covers all classes of livestock, follow:

1. *Cattle—beef or dairy:*

Conditions	Rate
Age limits, 3 mo—7 yr	
15-day term	$1.50/$100
(A 15-day policy may be endorsed "15-day cover to have its effective starting date at time of actual shipment.")	
1-mo term	$2.50/$100
2-mo term	$3.00/$100
3-mo term	$3.50/$100
6-mo term	$4.00/$100
1-yr term	$6.00/$100
Age exceptions—note added premium:	
Calves, 2-7 wk old	$4.00/$100
Calves, 7 wk-3 mo old	$2.00/$100
(This is additional premium to be added to period coverage and considered earned in entirety when written.)	
Bulls past 6, up to eighth birthday, eligible for insurance after amount of cover has been confirmed by company. One dollar per hundred additional premium charged for each year or part over seventh birthday. *At ninth birthday, insurance not available.*	
Cows past 7 may be covered in same manner. *At tenth birthday annual insurance not available.*	

Generally, special stipulations and rates apply to (1) group (herd) insurance for cattle, and (2) 4-H and FFA calves. For information relative to these, or other special types of coverage, the owner should make inquiry of a livestock insurance agent.

2. *Hogs, sheep, and goats:*

Conditions	Rate
Age limits 3 mo—2 yr	
15-day term	$3.00/$100
30-day term	$4.00/$100
60-day term	$5.00/$100
90-day term	$6.00/$100

[25]This section was accorded the authoritative review of Mr. Frank Harding, American Live Stock Insurance Company, 200 South Fourth Street, Geneva, Ill.

3. *Horses:*

Conditions	Rate
• *Arabians*	
Ages 24 hr through 5 mo	6%
Ages 6 mo to 15 yr	3.5%
• *American Saddlebreds, Morgans, Tennessee Walking Horses, and all other breeds of Show, Saddle Horses, and Ponies*	
Rates to apply up to 15 yr of age *Breeding Animals* (permanently retired and used for breeding only)	
Stallions	4.5%
Broodmares	4.5%
Foals	
Inception at age 24 hr	6%
Inception at age 30 days	5%
Inception at age 90 days	4.5%
Yearlings	4.5%
Pleasure and Show 2-15 yr of age	
Colts and Fillies	4.5%
Geldings	5%
• *Quarter Horses and Appaloosas*	
Rates to apply through 12 yr except where noted	
Breeding Animals (permanently retired and used for breeding only)	
Stallions	4.5%
Broodmares	4.5%
Foals	
Inception at age 24 hr	6%
Inception at 30 days	5.5%
Inception at age 60 days	5%
Yearlings	5%
All Other Uses	5%
(Applies to racing animals through 7 years of age only—racing animals subject to claiming race inclusion clause)	
• *Standardbreds (Trotters and Pacers)*	
Breeding Animals Through 12 yr	
Stallions	4%
Broodmares	4.5%
Foals	
Age 24 hours through 7 days	6%
8 days through 29 days	5.5%
30 days through 89 days	5%
90 days and older	4.5%
Yearlings and all racing animals (all subject to claiming race inclusion clause)	4%
• *Thoroughbreds*	
Racing Animals (All subject to claiming race inclusion clause)	
Flat Racers, excluding Geldings	
Values up to $25,000	7.5%
Values over $25,000	6.5%
(Subject to minimum premium of $1,875)	
Flat Racers—Geldings	
All values	8.5%
The above rates apply to Flat Racers only through the age of 7.	
Yearlings, insured prior to November 1	
Values up to $14,999	5%
Values of $15,000 and over	4%
Yearlings insured on and after November 1, rates same as for Flat Racers above.	
Foals	
Age 24 hours through 7 days	7%
8 days through 29 days	6.5%
30 days through 89 days	6%
90 days and older	5.5%
Breeding Animals Through 12 years	
Stallions	4.5%
Broodmares	4.5%
• *Hunters, Jumpers, Polo Ponies*	5.5%
• *Hurdlers* 12-year age limit	8.5%
• *Steeplechasers* 12-year age limit	9%
• *Livestock Specified Peril Coverage*	
Including coverage at racetracks and shows	1%
Excluding coverage at racetracks and shows	0.6%

- *Livestock Special Accident Coverage*

  To cover loss due to death resulting from, caused by, or made necessary by any and all external and visible accident—apply 60% of normal mortality rate for appropriate class of horses.

- *Short-Term Transportation Coverage*

  Available on all classes of horses—normally 1% excluding racing risk.

- *Prospective Foal Insurance*

  Can be arranged that insures the usual risks of an expected foal being slipped, cast, aborted, born dead, or dying during pregnancy and within a specified time after foaling. The complete produce record and history of the mare must be submitted. Each request for this type of insurance is reviewed separately by American Live Stock before a premium is quoted.

For overage animals and stable discount and/or group rating on all horses, submit to the company for rating.

Special stipulations and rates apply to (a) fire, lightning, windstorm and transportation losses only, (b) castration of colts and setting tails, (c) air and ocean transportation, and (d) group insurance. For information relative to these, the owner should see a livestock insurance agent.

In order to obtain insurance, the following information is generally required: Name, registry number, ear tag or tattoo number (markings on horse), breed, sex, date of birth, amount to be insured for and period of insurance required, and a statement of health examination (made not more than five days prior to insuring) by an approved federal or state veterinarian to the effect "that the animal(s) (referring to it by name) is at the time of applying for insurance in a state of good physical health and condition."

SECTION 3

BREEDING

Contents **Page**

- Puberty, Heat, and Gestation Periods of Animals 122
- Flushing and Conditioning 122
- Mating Table 122
- Gestation Table 122
- Color Inheritance in Shorthorns 122
- Trihybrid Cross 123
- Dehorning with Polled Bulls 123
- Dwarfism in Cattle 123
- Double Muscling (Muscular Hypertrophy) 127
- Sex Determination 130
- Pregnancy Testing 131
- Multiple Births in Cattle 132
- Freemartin Heifers 133
- Production Testing 134
  - Production Testing Beef Cattle 134
    - Economically Important Traits and Their Heritability 135
    - Record Forms 135
    - Most Probable Producing Ability (MPPA) 135
    - Ratios 135
  - Production Testing Dairy Cattle 145
    - Economically Important Traits and Their Heritability 145
    - Alternate DHI Testing Plans 145
      - Official DHI Testing 146
        - Dairy Herd Improvement Association (DHIA) 146
        - Dairy Herd Improvement Registry (DHIR) 146
      - Unofficial DHI Testing 146
    - Adjustment Factors 147
      - Length of Lactation 147
      - Number of Milkings Per Day 148
      - Age and Month of Calving 148
      - Fat Content of Milk (FCM) 148
    - Evaluating Dairy Sires (Progeny Testing) 148
      - Proved Sires 149
      - Young Sires 149
      - USDA-DHIA Sire Summaries 149
      - Predicted Difference (PD) 150
      - Repeatability 150
      - Selecting the Dairy Bull 150
  - Production Testing Sheep 151
    - Economically Important Traits and Their Heritability 152
    - Record Forms 152
  - Production Testing Swine 157
    - Economically Important Traits and Their Heritability 157
    - Record Forms 158
    - Meatiness; Measuring Backfat 159
  - Production Testing Horses 163
    - Heritability of Performance 163
    - Record Forms 164
- Crossbreeding 164
  - Hybrid Vigor or Heterosis 164
  - Complementary 167
  - Not All Hybrids Excel Purebreds 168
  - Crossbreeding Beef Cattle 168
    - Buffalo X Cattle Hybrids 171
  - Crossbreeding Dairy Cattle 172
  - Crossbreeding Sheep 172
  - Crossbreeding Swine 173
  - Crossbreeding Horses 174
- Systems of Selection 175
- Cow Heat Detection Methods and Devices 176
- Hormonal Control of Heat 177
- Superovulation and Ova Transplantation 179
- Induced Calving 180
- Artificial Insemination 180
- Blood Typing Cattle and Horses 189
- Breeds of Livestock and Their Characteristics 189

This section contains in summary and illustrated form some of the facts pertaining to genetics and breeding which a modern livestock operator needs to know and to apply in the propagation of farm animals.

PUBERTY, HEAT, AND GESTATION PERIODS OF ANIMALS

The normal age of puberty, and the normal heat and gestation periods of animals are given in Table 3-1.

TABLE 3-1
PUBERTY, HEAT, AND GESTATION PERIODS OF ANIMALS

Class of Animal	Age of Puberty	Duration of Heat		Interval of Heat		Gestation Period	
		Range	Average	Range	Average	Range	Average
	(mos.)			*(days)*	*(days)*	*(days)*	*(days)*
Cattle	8-12	6-30 hr	16-20 hr	19-23	21	278-288	283
Sheep	5-7	20-42 hr	30 hr	14-20	16-17	144-152	148
Goats	4-8	20-80 hr	36-48 hr	12-25	20	140-160	151
Swine	4-7	1-5 days	2-3 days	18-24	21	98-124	114
Horses	12-15	1-37 days	4-6 days	10-37	21	310-370	336

FLUSHING AND CONDITIONING

Flushing is that practice of feeding cows, ewes, and sows more generously 2 to 3 weeks before breeding. This may be accomplished by grain feeding, or cows and ewes may be turned on more lush pasture or range. Under most circumstances, the following amounts per head daily of a suitable concentrate are added to the ration that the females were receiving prior to flushing: cows, 2 to 5 lb; ewes, 1 to 2 lb; and sows, about 2 lb. Immediately after breeding, females should be returned to normal rations.

Although it is not likely that all the benefits ascribed to flushing will be fully realized under all conditions, the general feeling persists that the practice will result in (1) more eggs being shed, (2) the females coming in heat more promptly, (3) more certain and prompt conception—with the young arriving more nearly at the same time, and (4) a 15 to 30 percent increase in lamb and pig crops.

Fat cows, ewes, and sows can best be conditioned for breeding by increasing the exercise.

In mares, preparation for breeding is known as "conditioning." Usually, it involves exercising plus emulating spring conditions through blanketing and exercising.

MATING TABLE

Table 3-2, Mating Table, gives pertinent information relative to the use of sires of different classes of livestock, including considerations that should be given to age and method of mating.

GESTATION TABLE

The stockman who has information relative to breeding dates can easily estimate parturition dates from Table 3-3, Gestation Table (page 124).

COLOR INHERITANCE IN SHORTHORNS

Fig. 3-1 shows how color in Shorthorn cattle is inherited.

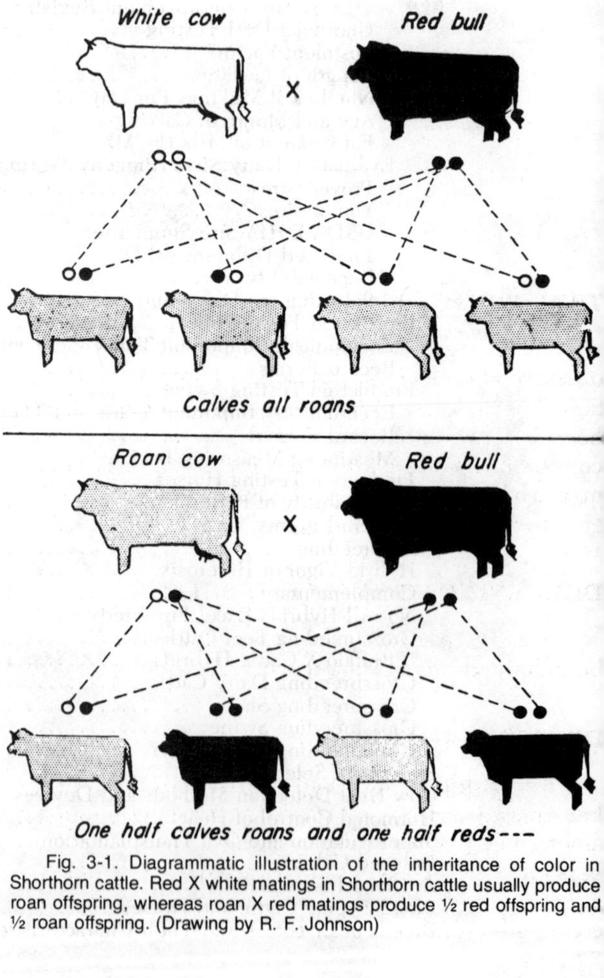

Fig. 3-1. Diagrammatic illustration of the inheritance of color in Shorthorn cattle. Red X white matings in Shorthorn cattle usually produce roan offspring, whereas roan X red matings produce ½ red offspring and ½ roan offspring. (Drawing by R. F. Johnson)

TABLE 3-2

MATING TABLE

Class of Animal	Age	No. of Females Bred/yr: Hand-mating	No. of Females Bred/yr: Pasture-mating	Comments
Bull	Yearling Two-yr-old Three-yr-old or over	10-12 25-30 40-50	8-10 20-25 25-40	Most western ranchers use 1 bull to about 25 cows. A bull should remain a vigorous and reliable breeder up to 10 yr or older; up to 6 to 7 yr under range conditions.
Ram	Lamb	20-25		Lambs should be used in hand-mating only. Unless well grown, and under close supervision of an experienced sheepman, they should not be used at all.
	Yearling or older	50-75	35-60	Most range operators use 1 ram to 25 to 35 ewes. A ram should remain a vigorous and reliable breeder up to 6 to 8 yr of age.
Boar	8 to 12 mo of age Yearling or older	24 50	12 35-40	Boar pigs should be limited to one service/day; older boars to two services/day. A boar should remain a vigorous and reliable breeder up to 6 to 8 yr of age. Under hand-mating, 2 services are recommended; the first mating on gilts should be on the first day of estrus and the first mating on sows on the second day of estrus, with a second mating following the first by 24 hr in each case.
Stallion[1]	Two-yr-old Three-yr-old Four-yr-old Mature horse Over 18-yr-old	10-15 20-40 30-60 80-100 20-40	Preferably no pasture-mating	Limit the 2-yr-old to 2 to 3 services/wk; the 3-yr-old to 1 service/day, and the 4-yr-old or over to 2 service/day. A stallion should remain a vigorous and reliable breeder up to 20 to 25 yr of age.

[1]There are breed differences. Thus, when first entering stud duty, the average 3-year-old Thoroughbred should be limited to 15 to 20 mares per season, whereas a Standardbred of the same age may breed 20 to 30 mares; and the 4- or 5-year-old Thoroughbred should be limited to 25 to 30 mares, whereas a Standardbred of the same age may breed 30 to 40 mares. Mature stallions of the draft breeds may and do breed up to 100 mares in a season.

TRIHYBRID CROSS

Sometimes cattlemen like to cross breeds. The outcome of such crosses cannot be predicted with certainty in all cases. With Angus X Hereford crosses, however, the inheritance of the color and horned condition is well established. Fig. 3-2 shows what one may expect to secure, on the average, from such a crossbreeding program.

DEHORNING WITH POLLED BULLS

Fig. 3-3 (page 125) shows how horned cattle may be dehorned through the use of polled bulls.

DWARFISM IN CATTLE

Beginning about 1940, a disturbing condition known as dwarfism appeared in increasing frequency among beef cattle, probably in all breeds. There are several different types of dwarfism, of which the short-headed, short-legged, pot-bellied dwarf—commonly referred to as the snorter dwarf—is the most frequent. The discussion which follows applies specifically to snorter dwarfism.

There is complete agreement among scientists (1)

Fig. 3-2. Diagram showing a trihybrid cross; the inheritance of polled, white face, and black body characteristics in an Angus X Hereford cross. Note that all first cross (F_1) animals are polled, black bodied, and white faced; whereas, on the average, the F_1 X F_1 cross results in the 27:9:9:9:3:3:3:1 ratio shown. (Drawing by R. F. Johnson)

TABLE 3-3

GESTATION TABLE

Date Bred	Cow 283 days (date due)	Ewe 148 days (date due)	Sow 114 days (date due)	Mare 336 days (date due)
Jan. 1	Oct. 11	May 29	Apr. 25	Dec. 3
Jan. 6	Oct. 16	June 3	Apr. 30	Dec. 8
Jan. 11	Oct. 21	June 8	May 5	Dec. 13
Jan. 16	Oct. 26	June 13	May 10	Dec. 18
Jan. 21	Oct. 31	June 18	May 15	Dec. 23
Jan. 26	Nov. 5	June 23	May 20	Dec. 28
Jan. 31	Nov. 10	June 28	May 25	Jan. 2
Feb. 5	Nov. 15	July 3	May 30	Jan. 7
Feb. 10	Nov. 20	July 8	June 4	Jan. 12
Feb. 15	Nov. 25	July 13	June 9	Jan. 17
Feb. 20	Nov. 30	July 18	June 14	Jan. 22
Feb. 25	Dec. 5	July 23	June 19	Jan. 27
Mar. 2	Dec. 10	July 28	June 24	Feb. 1
Mar. 7	Dec. 15	Aug. 2	June 29	Feb. 6
Mar. 12	Dec. 20	Aug. 7	July 4	Feb. 11
Mar. 17	Dec. 25	Aug. 12	July 9	Feb. 16
Mar. 22	Dec. 30	Aug. 17	July 14	Feb. 21
Mar. 27	Jan. 4	Aug. 22	July 19	Feb. 26
Apr. 1	Jan. 9	Aug. 27	July 24	Mar. 3
Apr. 6	Jan. 14	Sept. 1	July 29	Mar. 8
Apr. 11	Jan. 19	Sept. 6	Aug. 3	Mar. 13
Apr. 16	Jan. 24	Sept. 11	Aug. 8	Mar. 18
Apr. 21	Jan. 29	Sept. 14	Aug. 13	Mar. 23
Apr. 26	Feb. 3	Sept. 21	Aug. 18	Mar. 28
May 1	Feb. 8	Sept. 26	Aug. 23	Apr. 2
May 6	Feb. 13	Oct. 1	Aug. 28	Apr. 7
May 11	Feb. 18	Oct. 6	Sept. 2	Apr. 12
May 16	Feb. 23	Oct. 11	Sept. 7	Apr. 17
May 21	Feb. 28	Oct. 16	Sept. 12	Apr. 22
May 26	Mar. 5	Oct. 21	Sept. 17	Apr. 27
May 31	Mar. 10	Oct. 26	Sept. 22	May 2
June 5	Mar. 15	Oct. 31	Sept. 27	May 7
June 10	Mar. 20	Nov. 5	Oct. 2	May 12
June 15	Mar. 25	Nov. 10	Oct. 7	May 17
June 20	Mar. 30	Nov. 15	Oct. 12	May 22
June 25	Apr. 4	Nov. 20	Oct. 17	May 27
June 30	Apr. 9	Nov. 25	Oct. 22	June 1

Date Bred	Cow 283 days (date due)	Ewe 148 days (date due)	Sow 114 days (date due)	Mare 336 days (date due)
July 5	Apr. 14	Nov. 30	Oct. 27	June 6
July 10	Apr. 19	Dec. 5	Nov. 1	June 11
July 15	Apr. 24	Dec. 10	Nov. 6	June 16
July 20	Apr. 29	Dec. 15	Nov. 11	June 21
July 25	May 4	Dec. 20	Nov. 16	June 26
July 30	May 9	Dec. 25	Nov. 21	July 1
Aug. 4	May 14	Dec. 30	Nov. 26	July 6
Aug. 9	May 19	Jan. 4	Nov. 31	July 11
Aug. 14	May 24	Jan. 9	Dec. 6	July 16
Aug. 19	May 29	Jan. 14	Dec. 11	July 21
Aug. 24	June 3	Jan. 19	Dec. 16	July 26
Aug. 29	June 8	Jan. 24	Dec. 21	July 31
Sept. 3	June 13	Jan. 29	Dec. 26	Aug. 5
Sept. 8	June 18	Feb. 3	Dec. 31	Aug. 10
Sept. 13	June 23	Feb. 8	Jan. 5	Aug. 15
Sept. 18	June 28	Feb. 13	Jan. 10	Aug. 20
Sept. 23	July 3	Feb. 18	Jan. 15	Aug. 25
Sept. 28	July 8	Feb. 23	Jan. 20	Aug. 30
Oct. 3	July 13	Feb. 28	Jan. 25	Sept. 4
Oct. 8	July 18	Mar. 5	Jan. 30	Sept. 9
Oct. 13	July 23	Mar. 10	Feb. 4	Sept. 14
Oct. 18	July 28	Mar. 15	Feb. 9	Sept. 19
Oct. 23	Aug. 2	Mar. 20	Feb. 14	Sept. 24
Oct. 28	Aug. 7	Mar. 25	Feb. 19	Sept. 29
Nov. 2	Aug. 12	Mar. 30	Feb. 24	Oct. 4
Nov. 7	Aug. 17	Apr. 4	Mar. 1	Oct. 9
Nov. 12	Aug. 22	Apr. 9	Mar. 6	Oct. 14
Nov. 17	Aug. 27	Apr. 14	Mar. 11	Oct. 19
Nov. 22	Sept. 1	Apr. 19	Mar. 16	Oct. 24
Nov. 27	Sept. 6	Apr. 24	Mar. 21	Oct. 29
Dec. 2	Sept. 11	Apr. 29	Mar. 26	Nov. 3
Dec. 7	Sept. 16	May 4	Mar. 31	Nov. 8
Dec. 12	Sept. 21	May 9	Apr. 5	Nov. 13
Dec. 17	Sept. 26	May 14	Apr. 10	Nov. 18
Dec. 22	Oct. 1	May 19	Apr. 15	Nov. 23
Dec. 27	Oct. 6	May 24	Apr. 20	Nov. 28

that the dwarf condition is of genetic origin, and (2) that it is inherited as a simple autosomal recessive (the word "autosomal" merely means that it is not carried on the sex chromosomes), and conditioned by at least two pairs of modifying genes. Thus, the birth of a dwarf calf identifies both the sire and the dam as carriers of the dwarf gene.

Conditions Prevailing in Dwarf-Afflicted Herds

One or the other of the conditions (or perhaps both conditions) shown in Figs. 3-4 and 3-5 prevail in any herd of cattle in which dwarf-carrying animals are being used. Thus, 100 offspring from matings of carrier bulls X noncarrier cows will, on the average, possess the following genetic picture from the standpoint of dwarfism:[1]

50—carriers, although not dwarfs
50—noncarriers and nondwarfs
100—total

[1]All ratios are averages based on large numbers; thus, they may not apply to any given herd.

Likewise, 100 offspring from matings of carrier bulls X carrier cows will, on the average, possess the following genetic picture from the standpoint of dwarfism:[2]

25—dwarfs
50—carriers, although not dwarfs
25—noncarriers and nondwarfs
100—total

On the basis of these facts, it may be concluded that the following dwarfism genetic picture applies to any given calf having (1) one carrier parent, or (2) both carrier parents:

1. A calf out of parents, one of which is a known carrier and the other a noncarrier, has a 50 percent chance of being free of the dwarf factor.
2. A calf, both of whose parents are carriers, has only one chance in four of being free of the dwarf factor.

[2]Ibid.

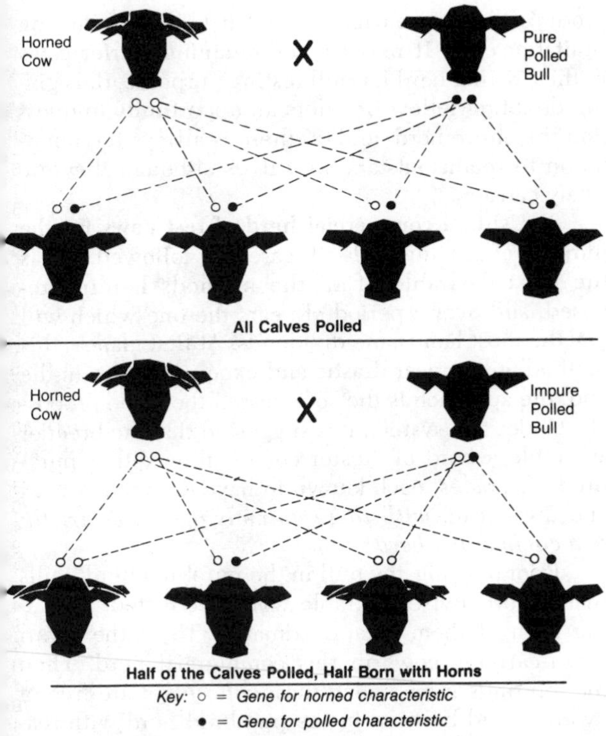

Fig. 3-3. Diagrammatic illustration of the inheritance of horns in cattle. If a bull that is pure or homozygous for the polled character is mated with a number of horned females, all of the calves will be polled; whereas if a bull that is impure or heterozygous for the polled character is mated with a number of horned females, only half of the calves will, on the average, be polled. (Drawing by R. F. Johnson)

It is recognized that the percentage of carrier females in any given herd will vary. Obviously, where dwarf calves have appeared, there are both carrier bulls and carrier cows. Some breeders may remove carrier animals, especially cows, once they have dropped a dwarf calf, thus selecting away from the trait. Others may unwittingly select for, rather than against, animals of the carrier type, if such a carrier type exists and if it is associated with some much sought characteristic, such as a markedly dished face.

From Figs. 3-4 and 3-5, the following deductions of value to practical cattlemen may be made: (1) Where a carrier bull is mated to noncarrier cows, no dwarfs will be produced, but, on the average, ½ the calves will be carriers; and (2) the carrier heifers from this first cross can, and likely will, produce ¼ dwarfs if they are mated back to a carrier bull. In other words, although the use of a carrier bull in a clean herd will not produce any dwarf calves, the seed for dwarfism is sown and it will crop out providing a second carrier bull is used in the herd.

Figs. 3-4 and 3-5 also indicate the futility of continuing the use of carrier bulls or females. Such practice should be continued only (1) if the breeder wishes to accept the eventual economic loss from producing dwarf calves in about the proportion indicated, or (2) if the breeder feels that an animal is sufficiently valuable otherwise to warrant its use, despite propagating dwarf factor genes (and there are those who argue that they would rather get rid of one bad trait—such as dwarfism—than a lot of bad traits such as may exist in some old-fashioned animals that are free of dwarfism). Also it should be recognized that any animal producing a dwarf is a carrier, regardless of the number of dwarfs produced (one or several).

How to Purge the Herd of Dwarfs

The breeding program followed to remove or minimize the dwarf condition will depend somewhat on the type of herd involved—especially on whether it is a commercial or purebred herd.

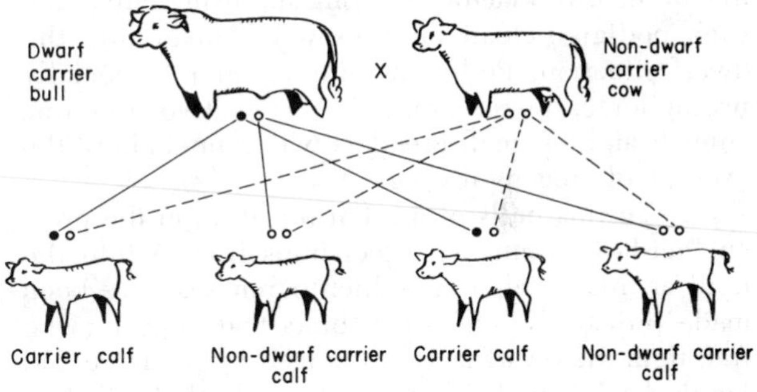

Fig. 3-4. Diagrammatic illustration of the inheritance of the most common kind (short-headed or snorter type) of dwarfism, showing what to expect when a carrier (heterozygous) bull(s) is mated to a noncarrier (homozygous normal) cow(s); or the sexes may be reversed. As shown, carrier X noncarrier matings will, *on the average*, produce calves of which (1) 50% are carriers, although not dwarfs, and (2) 50% are noncarriers, and nondwarfs. Unfortunately, the two groups look alike and cannot be detected by sight. (Drawing by R. F. Johnson)

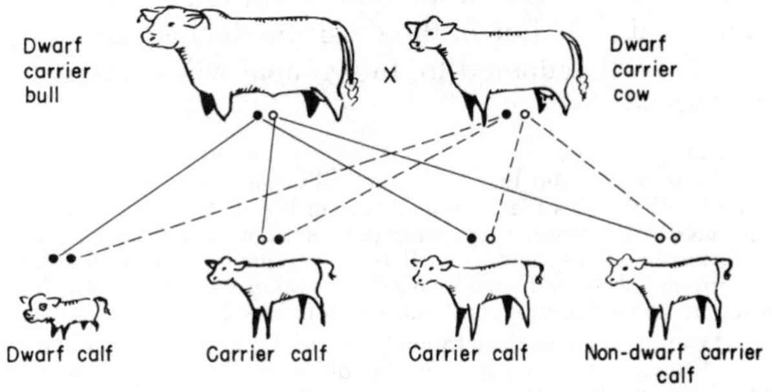

Fig. 3-5. Diagrammatic illustration of the inheritance of the most common kind (short-headed or snorter type) of dwarfism, showing what to expect when a carrier (heterozygous) bull(s) is mated to a carrier (heterozygous) cow(s). As shown, carrier X carrier matings, will, *on the average*, produce calves of which (1) 25% are dwarfs, (2) 50% are carriers, although not dwarfs, and (3) 25% are noncarriers, and nondwarfs. Unfortunately, only the dwarfs can be detected by sight; the 2 nondwarf groups look alike and cannot be distinguished by sight. (Drawing by R. F. Johnson)

In a commercial herd, the breeder may lessen the chances of obtaining dwarfs by using an outcross (unrelated) sire within the same breed or by crossbreeding with a sire from another breed. With this system, the dwarf-carrying cows will remain, but—because of the recessive condition of the dwarf factor—it will be covered up.

In a purebred herd, the action taken in handling the dwarf situation should be more drastic. A reputable purebred breeder has an obligation, not only to himself, but to his customers among both the purebred and commercial herds. Purebred herds should be purged of the undesirable dwarf genes. This can be done through pursuing any one of the following three breeding systems:[3]

1. **Using sires of families free of the dwarf factor; pedigree-clean animals**—Within each breed where dwarfs have appeared, certain families exist that are free of the dwarf factor. Securing and using bulls from such pedigree-clean families will "cover up" the dwarf situation. Pedigree information is especially useful for early screening of prospective breeding animals and for small breeders who cannot afford the expense of progeny testing.

2. **Testing bulls of present breeding in the present herd**—Continue to select bulls from within the herd or from herds from which purchases have been made previously, but select bulls that appear to be free from the dwarf factor as judged by pedigree and family background. Next, test these bulls by mating each of them to cows in the present herd that have produced dwarf calves. If each bull tested is mated to 15 known carrier cows and all the progeny are normal, there would be only 1.3 chances out of a hundred (1.3%), or 1 chance in 75, that the bull is a dwarf-carrier and yet passed the test undetected.[4] A test of about the same validity can be secured by breeding a sire to 30 of his daughters. Accordingly, such a tested bull could be used in any herd with reasonable certainty that he is free of the dwarf-producing factor.

If rigidly adhered to, this system will eventually produce the desired results, but it has the following limitations: (1) It necessitates retaining carrier cows in the existing herd for bull testing purposes, thus giving doubting fellow breeders an opportunity to question the entire herd; and (2) there is always the temptation to retain outstanding calves although they are likely carriers.

3. **Using a commercial herd of test cows for the purpose of proving bulls**—If carefully followed, this is the most desirable of all the methods herein proposed, and over a period of years the one which will pay the most handsome dividends. At first glance, this method will appear drastic and expensive, but, in the end, the approach is the soundest of the three proposals. Under this system, it is suggested that the breeder assemble a herd of "tester cows" (from either purebreds or grades) each known to have dropped at least one dwarf calf, *with these animals operated strictly as a commercial herd.*

Prior to using any bull in the purebred herd, bulls that are otherwise desirable would be tested by mating each of them to approximately 15 of the dwarf factor-carrying cows in this commercial herd. Then the top bulls from among those whose get are free of dwarfs could be used in the purebred herd with reasonable certainty.[5]

Carrier cows mated to bulls as indicated would, on the average, not produce in excess of 25 percent dwarfs if all the bulls were dwarf-carriers (see section relative to "Conditions Prevailing in Dwarf-Afflicted Herds"). Of course, fewer dwarfs would be produced if some of the bulls were dwarf-free, as expected. Thus, there would be considerable remuneration from the sale of calves in the operation of such a commercial herd. Further, and most important, the merits of young sires, from the standpoint of type and efficiency of production, could be determined in the commercial herd prior to using them in the purebred herd—thus making it possible to select sires by modern record of performance methods.

Under any of the three systems herein proposed, it would be wise to eliminate those bulls and cows that are known to have produced dwarf calves as soon as desirable replacement animals proved to be free of the factor are available.

Providing one does not select a larger than normal percentage of carrier heifers as replacements for the herd, each generation of calves sired by dwarf-free bulls would halve (lessen by 50%) the number of carriers in the herd. True enough, there would always be some of the dwarf carriers present, for the incidence of carriers is halved with each generation of such matings, but not eliminated. Yet, after two gen-

[3]It is recognized that it is practically impossible to eliminate completely the dwarf factor from a herd or breed of cattle once it has appeared. With a proper breeding program, however, the incidence of dwarfism will become so small as to be unimportant—much as the occurrence of the red color (caused by another recessive factor) has been minimized in the Angus and Holstein breeds of cattle.

[4]A lesser number than 15 might be used, but the breeder could not be so sure of the results. For example, with 10 or 5 such matings, the chances of failing to detect a carrier bull are increased from 1.3% to 5.6% and 23.7%, respectively. Thus, with 10 such matings and all normal calves, there is only 1 chance in 18 that the bull is a dwarf-carrier and yet passed the test undetected.

If only a limited number of carrier cows are available, it may be desirable to breed each prospective herd sire to 4 to 6 carrier cows initially, followed by more thorough testing of those passing the initial screening.

If the factor becomes so rare that it is impossible to secure sufficient carrier cows for bull testing, the incidence of dwarfism will be so small as to be unimportant.

[5]The chances of avoiding a dwarf factor carrying bull through such testing on 15 carrier cows is covered in the section immediately preceding.

erations of such matings, the incidence of dwarf carriers would be small. Also, it is noteworthy that the use of proved dwarf factor-free bulls would give assurance that no more dwarf calves would be produced in the herd.

It is perfectly obvious that the elimination of the dwarf-producing factor is both slow and costly. Yet it is the only way in which cattle can be freed from dwarfs. Also it may require real courage to recognize openly the situation and discard outstanding animals that are known carriers.

Because dwarfs represent an almost complete economic loss, the problem deserves careful attention.

DOUBLE MUSCLING (Muscular Hypertrophy)

Double muscling refers to cattle characterized by bulging muscles of the shoulder and thigh, a very rounded rear end (as viewed from the side), a wide but shallow body throughout, appearance of intermuscular grooves, and fine bones.

In Germany, the trait is known as doppellender (double rump); in Italy, it's doppia (horse rump); in France it's culard; and in the United States, it's double muscling.

Double muscling is really a misnomer. Likewise, the scientific name, muscular hypertrophy, is incorrect because it implies increased size of fibers in each muscle, which is not the case. Rather, it has been shown that double muscled cattle have more fibers, not larger fibers.

Since beef cattle are produced primarily for their muscle, it's logical that selection should be centered around muscularity. It follows that cattle with "double muscles," or these tendencies, have appeal and have increased in frequency in the United States during recent years. But there are disadvantages as well as advantages to double muscling. Hence, cattlemen should be familiar with the characteristics and genetics of double muscling, and its side effects.

Double muscling is a genetically controlled character. It appears to be caused by a single recessive gene, which tends to be "masked" by the dominant gene in the heterozygous carriers. Other examples of a character controlled by one pair of genes are: polledness and hornedness, and dwarfism. The genetics, therefore, are relatively simple. Since each animal has two genes for such characters, all cattle can be classified as follows:

DM DM—Homozygous normal; two dominant normal genes—a normal animal.

DM dm—Heterozygous; one dominant normal gene (DM) which tends to cover up the one recessive gene (dm)—these are called carriers. This coverup is not complete; hence, there is a tendency toward double muscling.

dm dm—Homozygous recessive; two recessive double muscle genes—a double muscled animal.

The progeny from a sire and dam of all these three genotypes are predictable, on the average, but not necessarily for any one offspring. The possible matings and progeny are shown in Table 3-4.

TABLE 3-4

POSSIBLE MATINGS AND PROGENY OF DOUBLE MUSCLED CATTLE

Sire	Dam	Progeny
DM DM	DM DM	DM DM. All normal.
DM DM DM dm	DM dm DM DM	½ DM DM, ½ DM dm. All normal, but ½ carriers.
DM dm	DM dm	¼ DM DM, ½ DM dm, ¼ dm dm. Of the ¾ normal, 2 out of 3 are carriers; the remaining ¼ are double muscled.
DM DM dm dm	dm dm DM DM	DM dm. ALL carriers.
DM dm dm dm	dm dm DM dm	½ DM dm, ½ dm dm. One-half carriers, ½ double muscled.
dm dm	dm dm	dm dm. ALL double muscled.

Appearance of Homozygous Double Muscled (dm dm) Cattle

Obviously, the problem is to determine if an animal is DM dm (a carrier), rather than DM DM (a normal animal). There are two ways to do this: (1) appearance, and (2) breeding tests. Detection by appearance is not 100 percent sure, but the experienced observer doesn't make many mistakes.

The appearance of homozygous double muscle (dm dm) cattle is summarized in Table 3-5. But remember that the double muscle character is really a syndrome of many characteristics. Remember, too, that all of these traits may not be present to the same degree in any one animal. (See page 128 for Table 3-5.)

Appearance of Carrier (DM dm) Cattle

Generally speaking, homozygous, double muscled animals can be identified. But it isn't easy to pick out the heterozygotes—the carriers of the double muscle gene, due to the wide variation in expression. Some of the carriers look quite normal, others look like homozygous double muscled animals, and still others are intermediate between these two extremes. Also, identity is further complicated because few, if any, double muscled animals show all of the charac-

TABLE 3-5
TRAITS OF DOUBLE MUSCLED CATTLE

Body Part: Trait	Appearance in Double Muscled Cattle
Rump and round	Protruding and rounding; definite grooves, or creases, between the thigh muscles.
Tail	Short. Attached far forward. Prominent tailhead.
Middle; heart girth; flank	Shallow bodied, light heart girth, tucked up flank; animal appears leggy and cylindrical.
Head	Small, long, carried lower than top of shoulders.
Shoulder	Large and bulging; grooves, or creases, evident in arm and forearm.
Cannon bone	Short and fine.
Stance	Animal stands camped out; forelegs extended to front and hindlegs stretched.
Vulva	The vulva of females is small, high, and far forward.
Testicles	Small and carried close to the stomach.
Lying down	Double muscled cattle spend a lot of time lying down; they're "muscle laden"; with a high proportion of flesh to bone.
Age	Double muscling is most conspicuous in young animals. It becomes less apparent with advancing age.
Sex	General lack of masculinity, other than muscularity, in bulls; and lack of femininity in heifers and cows. At breeding age, double muscling is more marked in males than females.
Birth; early growth	Heavy birth weight. Calves often have enlarged tongues and crooked legs, and are weak. Good early growth. But growth markedly slower by one year of age; and small mature size.
Environment	Double muscling is more marked with superior environment; it shows up more in well-fed animals.

teristics. Nevertheless, carriers are characterized by general overall trim appearance, thicker quarter with bulging, thicker round, and a higher tailhead setting than normal animals.

Double Muscled Cattle—Good or Bad?

Double muscled Piemontese (Piedmont) cattle mean to the Italian cattle industry what broad breasted turkeys and Cornish cross broilers mean to the U. S. poultry industry. All are meat producers par excellence.

Italian authorities estimate that 80 percent of the bulls of the Piemontese breed of cattle are double muscled to some degree. Moreover, producers select for the trait; and both feeders and slaughterers vie for double muscled bulls. (In Italy, they feed bulls; not steers.) The reason: All of them make more money from double muscled cattle than from normal cattle.

Piemontese cattle are the most popular breed in Italy. In 1964 (latest census available), there were 674,000 of them. The fact that 80 percent of the bulls are double muscled to some degree indicates that the character responds to selection. Knowledgeable Piemontese cattle breeders in Italy told the author that they expect the following results in their Piemontese breeding programs (see Fig. 3-6):

1. Phenotypically normal heifers mated to double muscled bulls will produce 80 percent double muscled bull calves.
2. By culling out the first calf heifers whose bull calves from the above mating were not double muscled, then mating only proved heterozygotes (or carriers) to double muscled bulls, 95 percent double muscled bull progeny will be produced.

If double muscling is caused by a single recessive gene, as seems to be the case in the British breeds (Angus, Hereford, and Shorthorn), (1) mating phenotypically normal (noncarrier) cows to a homozygous, double muscled bull would produce 100% heterozygotes (carriers), none of which would be double muscled, and (2) mating known carrier cows (heterozygotes) to a homozygous, double muscled bull would produce 50% homozygous, double muscled calves and 50% heterozygous carriers. However, the Italians report breeding results from their Piemontese breed which suggest that modifier genes common in that breed tend to endow the double muscling gene with dominance (partial dominance). As a result of these modifier genes, the 100% heterozygotes (carriers) referred to could, in the Piemontese breed, easily be classified as 80% double muscled and 20% normal, thereby explaining Fig. 3-6, alternate breeding program No. 1. Likewise, in the Piemontese breed, mating known carrier cows to double muscled bulls could result in 95% of the offspring being classed as double muscled (Fig. 3-6, alternate breeding program No. 2), as a result of the action of modifier genes causing a large part of the 50% heterozygotes to be classed as double muscled.

Also, it would seem reasonable to suspect that, if 80% of the bulls of the Piemontese breed are double muscled to some degree, a large part of the heifers are, also. Hence, many of the heifers classed as "phenotypically normal" in Fig. 3-6, alternate breeding program No. 1, would actually be carriers of the double muscled gene. However, due to the presence or absence of various modifier genes, perhaps they may appear completely normal or only moderately

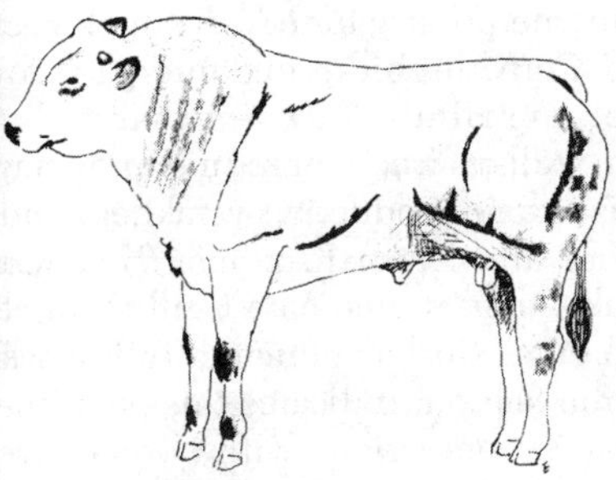

Fig. 3-6 A

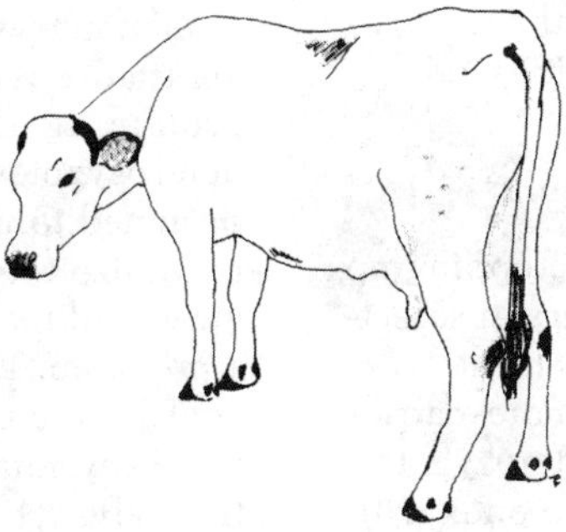

Fig. 3-6 B

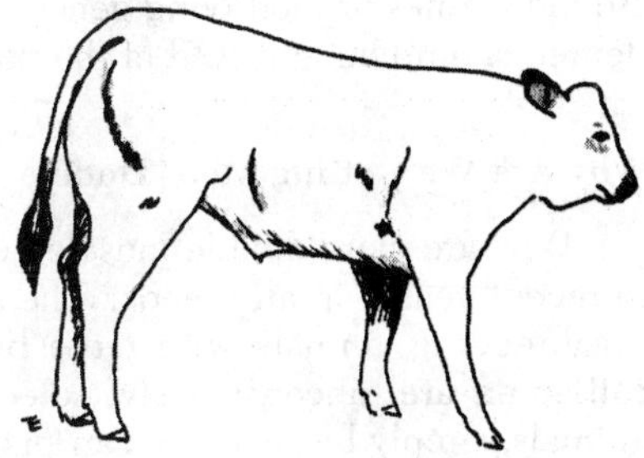

Fig. 3-6 C

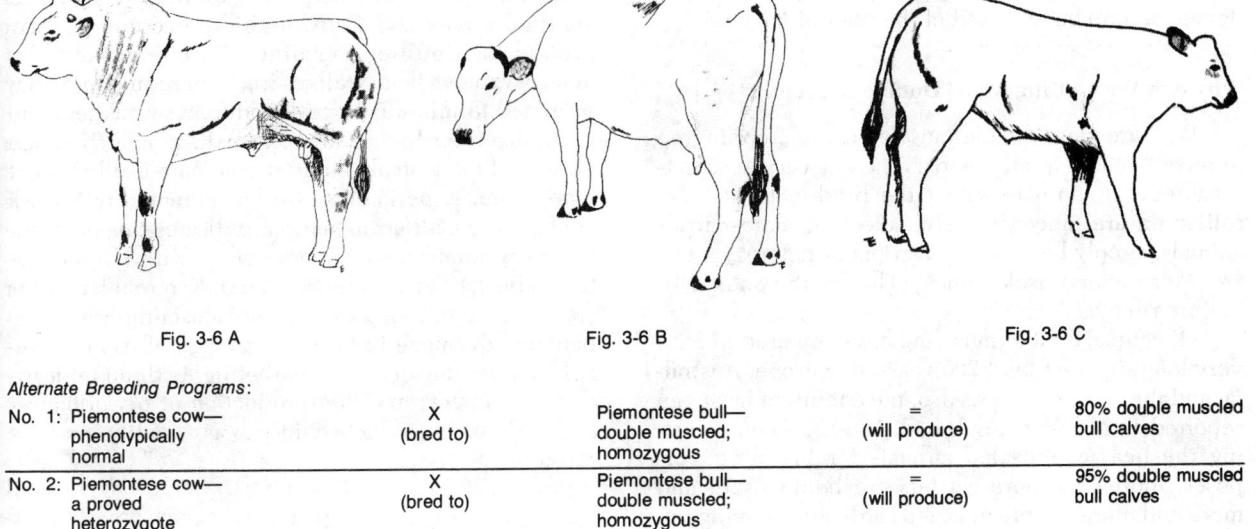

Alternate Breeding Programs:

No. 1: Piemontese cow—phenotypically normal	X (bred to)	Piemontese bull—double muscled; homozygous	= (will produce)	80% double muscled bull calves
No. 2: Piemontese cow—a proved heterozygote (or carrier)	X (bred to)	Piemontese bull—double muscled; homozygous	= (will produce)	95% double muscled bull calves

Fig. 3-6 A-B-C. Double muscling predictability in Piemontese (Piedmont) cattle, in Italy.

double muscled, with the result that they are classed as "normals."

ADVANTAGES OF DOUBLE MUSCLED CATTLE

In Italy, all members of the beef team—cattle producers, feeders, slaughterers, and retailers—are profiting from double muscled cattle. But they are quick to point out that there are other advantages, too. Here are some of the plusses that they list in favor of double muscled over normal cattle:

1. The calves grow more rapidly up to one year of age.
2. They convert feed more efficiently; it requires fewer pounds of feed to produce a pound of beef.
3. They produce a superior carcass. Double muscled cattle dress 72% (vs 63% normal); and, in comparison with normal cattle, they have a larger rib eye, and produce less brisket, plate, flank, and kidney and pelvic fat. Also, double muscled cattle yield a higher proportion of the more desirable cuts—more steaks. The Italian retailer cuts 80% steaks from double muscled carcasses vs 40% from normal carcasses—they literally cut steaks from end to end of a double muscled carcass.

In summary: Double muscled cattle are superior to normal cattle in (1) rate and efficiency of gain to one year of age, and (2) general carcass desirability.

DISADVANTAGES OF DOUBLE MUSCLED CATTLE

It is recognized that there may be breed differences when it comes to the advantages and disadvantages of double muscled cattle. Nevertheless, here are some of the disadvantages to double muscling that have been reported in different countries:

1. The conception rate is lower, due to (a) the infantile reproductive tracts or slow sexual maturity of some animals, and (b) the flat vulva, which makes copulation difficult.
2. The gestation period is about 10 days longer.
3. There is more calving difficulty (caesarian section; pulling calves; stillborn), due to heavier calves at birth (Piemontese double muscled calves average 108 pounds at birth vs 99 pounds for normal calves), along with the enlarged rump and round regions.
4. Double muscled calves are more difficult to raise; due to such things as (a) enlarged tongues (Macroglossia), and (b) greater susceptibility to disease.
5. Double muscled cows are poor milkers; they produce 30 to 50 percent less milk than normal cows.
6. Double muscled cattle must be fed a higher proportion of concentrate to roughage, simply because they cannot utilize roughage effectively.
7. There is less marbling; hence, carcasses of double muscled cattle may be penalized.

It's unlikely that all of the above disadvantages will occur in any one herd at any one time. Moreover, the degree to which they occur among breeds, and within breeds, will vary according to the "background" genes or modifying genes. Nevertheless, cattlemen should be apprised of the possibilities.

Why Are We Getting More Double Muscling?

Why are more double muscled cattle cropping up in recent years—in all breeds? The answer: In selecting breeding animals with more bred-in meat type, cattlemen are, unconsciously, selecting more carrier animals, simply because the carriers or heterozygotes, are the heavier muscled ones. (They're the ones with the big rib eyes.)

Of course, double muscling has been around for a very long time—at least 200 years, in Europe, Australia, and the United States. Also, the condition has been reported in almost every breed. Hence, when selecting the heavier muscled animals for breeding purposes, more and more carriers are being used; and more and more double muscled cattle are showing up, and will continue to show up.

Summary Relative to Double Muscling

Double muscled cattle—good or bad? Obviously, in Italy it's good—very good. If it were not so, no breed that produced 80% double muscled bull calves could survive. The main reason that double muscling in Italy is so good is that their slaughterers pay a premium of $29/cwt on foot for double muscled cattle at market time. With a 1,100-pound animal, that's a premium of $319 per head. The reason for the premium: the reputed 9% higher dressing percentage (72 vs 63%) and twice the steaks (80 vs 40%) of double muscled cattle over normal cattle. Also, the Italians give the impression (without scientific proof) that many of the disadvantages that U.S. cattlemen attribute to double muscling have been minimized—they're less nettlesome—in the Piemontese breed and to Italian cattlemen. For example, they don't seem to complain too much about breeding or calving problems. Maybe they have overcome this through selection. Obviously, the expression of the trait in many approved heterozygote (carrier) Piemontese cows is nil. Moreover, their system of handling early weaned calves—nurse cowing (2 to 5 calves/cow) plus a starter ration, until 4 to 6 weeks of age—makes for a good start in life; even if double muscled calves are less vigorous at birth.

Indeed, Italian cattlemen have something going for them. In the United States, it's another story. Until, and unless, a premium is paid for double muscled beef over the butcher's block, there's no incentive; there's insufficient reason to risk the disadvantages of double muscled cattle—even if they could be minimized through selection. Yet, there are some aspects of double muscling in cattle that might be used in improving efficiency of beef production. Also, cattlemen may well emulate poultry breeders—the broad breasted turkey and Cornish chicken counterpart, by producing double muscled cattle (perhaps the heterozygotes) in a well-planned breeding program designed to minimize their production weakness and maximize their higher dressing percentage, more red meat, and more steaks; then promote and sell them at a premium. Experimental work conducted by Rollins et al., of the California Station, indicates this possibility. They found that, in comparison with normal cattle, calves from double muscled X normal parents (calves heterozygous for double muscling) had a 10 percent advantage in terms of pounds of trimmed retail cuts per day of age at marketing, with no undesirable side effects in either production or performance, and little or no reduction in carcass quality grade at marketing.[6]

SEX DETERMINATION

The most widely accepted theory of sex determination at the present time is that sex is determined by the chromosomal makeup of the individual. Fig. 3-7 shows how sex is determined in farm animals.

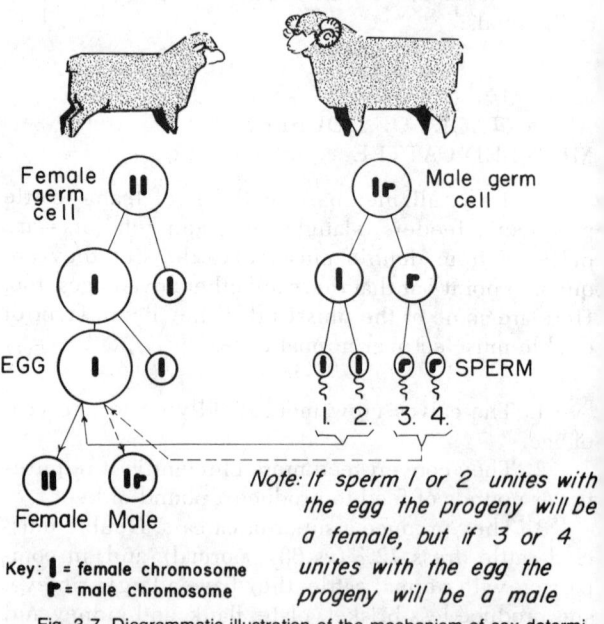

Fig. 3-7. Diagrammatic illustration of the mechanism of sex determination in farm animals, showing how sex is determined by the chromosomal makeup of the individual. The female has a pair of like sex chromosomes, whereas the male has a pair of unlike sex chromosomes. Thus, if an egg and a sperm of like sex chromosomal makeup unite, the offspring will be a female; whereas if an egg and sperm of unlike sex chromosomal makeup unite, the offspring will be a male.

The scientist's symbols for the male and female, respectively, are: ♂ (the sacred shield and spear of Mars, the Roman god of war), and ♀ (the looking glass of Venus, the Roman goddess of love and beauty). (Drawing by R. F. Johnson)

[6]Rollins, W. C., R. B. Thiessen, and M. Tanaka, "Usefulness of Market Calves Heterozygous for Double Muscling Gene," *California Agriculture*, Vol. 28, No. 3, March 1974, p.8.

Many unsuccessful attempts have been made to control sex, including the electrophoretic, mechanical, and chemical methods of separation of the two types of sperm cells. Obviously, some method of controlling sex of offspring would have tremendous significance in the livestock field. However, to date, no practical solution has been found. Until there is adequate experimental evidence, any method or theory that purports to control sex should be regarded with skepticism. Of course, research is being continued because the stakes are high if any workable method can be found.

PREGNANCY TESTING

The absence of heat is not always a sign of pregnancy, but a positive diagnosis can be made. Among the advantages of early pregnancy detection in all species are the following:

1. It gives early warning of breeding troubles, such as infertile males and cystic ovaries of females, and makes it possible to accord special care or treatment.
2. It makes it possible to detect shy breeders and females that show signs of heat even when well advanced in gestation.
3. It makes it possible to rebreed nonpregnant, feed-wasting females.
4. It allows for the separation and grouping of females—as pregnant, and nonpregnant, which is requisite to proper nutrition and husbandry.
5. It makes for more effective use of facilities, including providing adequate facilities for parturition.
6. It makes it possible to guarantee pregnancy on females that are for sale.

Several methods for pregnancy determination have been developed and used experimentally or semipractically. The most common test for each species follows.

Pregnancy Testing Cows

Cows are commonly pregnancy tested by "rectal palpation." By about the second month in heifers and the third month in cows, the uterus becomes enlarged, especially in the pregnant horn, and drops into the abdominal cavity. An experienced technician can ascertain this sign of pregnancy by *feeling with the gloved hand through the rectum wall*. Application of this method depends upon the recognition of changes in tone, size, and location of the uterine horns and changes in the uterine arteries.

This cow pregnancy test is popular because it affords early diagnosis, and there is little hazard when performed by experienced operators. It is recommended that cows be pregnancy tested, by this method, about 2 months after the bulls have been removed. Generally, a veterinarian will charge $1 to $5 per head, depending on the size of the herd. With convenient facilities—corrals and squeeze—an experienced operator can pregnancy test 800, or more, cows per day.

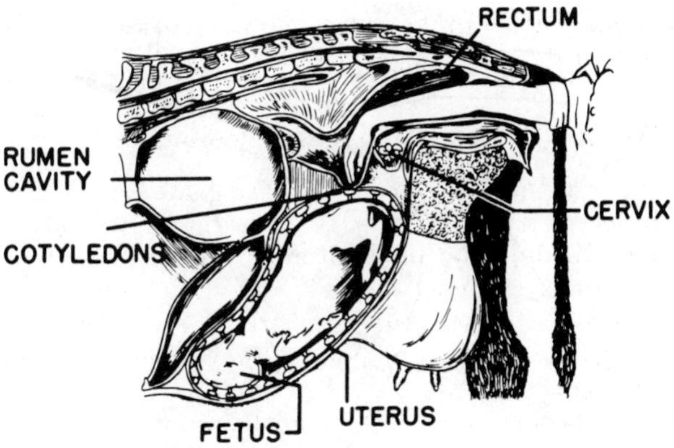

Fig. 3-8. Rectal method for determining pregnancy in the cow.

Pregnancy Testing Ewes

The "rectal abdominal palpation" test, developed by the U.S. Sheep Experiment Station, Dubois, Idaho, is the ewe pregnancy test of choice. The only equipment required is a 16-inch hollow, plastic rod (called a palpation rod), some lubricating material, and a device (such as a cradle) for holding a ewe on her back. This method of pregnancy detection is illustrated in Fig. 3-9 (page 132).

In the hands of an experienced operator, detection of ewes at midpregnancy (65 to 70 days postbreeding) with the pregnancy rod is virtually 100 percent accurate. With proper handling equipment, a technician and 3 assistants can pregnancy test 200 ewes per hour.

Pregnancy Testing Sows

Sows are pregnancy tested by "ultrasonic echo." In 1973, the first pregnancy-testing machine for sows, called the "pregnosticator," became available. (Other similar pregnancy detector machines have subsequently come on the market.) It was developed by Dr. Philip Dziuk, University of Illinois. By ultrasonic echo from the fluid in the uterus, this machine can detect pregnancy accurately in swine as early as 30 days after conception and up to 80 days. It actually measures the amount of fluid in the uterus. This fluid increases rapidly following conception and reaches detectable levels 25 to 30 days after breeding. It remains detectable for 80 to 90 days after breeding, fol-

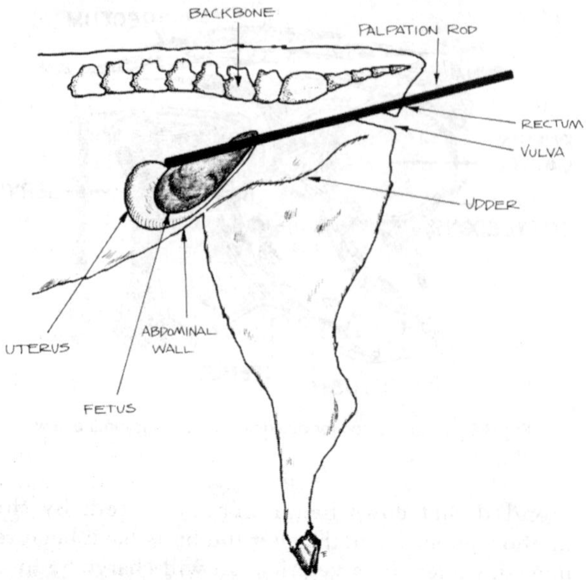

Fig. 3-9. The palpation rod is inserted in the rectum, close to the spine. The objective is to (1) position the rod dorsal to the fetus, and (2) elevate and hold the fetus against the abdominal wall. Then, the free hand is used to feel the fetus through the abdominal wall. *Note*: To facilitate comprehension, the above diagram shows the ewe in an upright position. However, in a pregnancy examination, the ewe is placed on her back in a comfortable horizontal position.

lowing which the mass of pigs in the uterus exceeds the fluid content. When properly used, the detector is almost 100 percent accurate.

Pregnancy Testing Mares

The "manual rectal palpation" test is by far the most used equine pregnancy test, followed to a much lesser extent by the antigen-antibody test known as the "MIP Test."

1. **Manual (rectal) palpation**—An experienced technician can determine pregnancy (or barrenness) of mares at 98 to 100 percent accuracy by feeling with the hand through the rectal wall. Normally, the test is made 43 to 45 days after breeding, but pregnancy in maiden mares often can be detected within 35 to 40 days following conception. When performed by an experienced person, the manual test is quite reliable.

The following procedure is employed by most experienced technicians in making the manual test:

a. The examination is made in surroundings familiar to the mare and in an unhurried manner, as this (1) makes for a minimum of restraint, and (2) avoids roughness.

b. Two helpers are used; one to twitch the mare, and the other to hold the tail to one side. If the mare objects to the twitch, it is left off.

c. The latex obstetrical sleeve with glove attached is slipped on and lubricated. The rectum is entered and evacuated; the arm is inserted nearly to the shoulder, reaching forward and downward until the ovaries are located (the left ovary is most accessible for right-handed operators; the right ovary for left-handed operators); and the uterus is *gently* palpated and massaged. If the mare is 43 to 45 days pregnant, an enlargement approximately the size of a large orange can be located along the bottom of one of the uterine horns.

2. **MIP Test**—There are a number of immunological tests which utilize the antigen-antibody reaction, but the most common one is the "MIP Test" (Mare Immunological Pregnancy Test). This test utilizes the principle whereby pregnant mare serum (gonadotropin) inhibits the agglutination of gonadotropin-coated erythrocytes in the presence of gonadotropin antiserum. The result is the formation of a ring at the bottom of a test tube. The procedure is performed with 2 test tubes—a control test tube and a sample test tube. Mare serum is placed in both, along with erythrocytes and reagents. However, antiserum is placed in the sample test tube, whereas a control solution is used in the other test tube. In the control tube, the pattern of a doughnutlike ring will show at the bottom, whether or not the mare is pregnant. But the sample tube with the antiserum will show the pattern only if gonadotropin is present in the mare's serum—signifying that the mare is pregnant. The MIP Test, which can be run in 2 hours' time, can determine with virtually 100 percent accuracy equine pregnancy from a blood sample taken 41 to 63 days after the mare is serviced. The MIP Test Kit contains all the supplies needed for running the test.[7]

MULTIPLE BIRTHS IN CATTLE

Multiple births among cattle have been observed since their domestication.

A review of the literature reveals that, on the average, such multiple births occur at the frequencies shown in Table 3-6.

TABLE 3-6
FREQUENCY OF TWINS IN CATTLE

Breed	Total Number of Births	Percent of Twin Births
All beef cattle breeds		0.44
Angus	1,111	0.81
Grade Angus	586	1.71
Brown Swiss	14,111	2.70
Holstein	18,736	3.08
Jersey	87,926	1.02
Simmental	12,625	4.61

[7]The MIP Test Kit may be secured from the Denver Chemical Manufacturing Co., Stamford, Conn.

Selection for natural twinning does not appear to hold much promise because the heritability of twinning is low. A herd in which selection for twinning was practiced for more than 20 years showed a twinning frequency of only 1.71 percent during the last 10 years.

The repeatability of twinning is estimated to be three to four times higher than the average of the population, once a cow has given birth to the first set of twins.

Twins may be produced in any of the following five ways:

1. By two eggs being produced at the same heat period, with both fertilized and carried to term.
2. By two eggs being shed at the same heat period, but the cow being bred to two different bulls with a sperm from each of the bulls uniting with an egg.
3. By a cow coming in heat and being bred, then three weeks later coming in heat again and being rebred; with both matings resulting in viable offspring.
4. By a single fertilized ovum splitting during the early stage of development.
5. By the use of hormones to induce superovulation.

Twins may be either fraternal (dizygotic) or identical (monozygotic). Fraternal twins are produced from 2 separate ova that were fertilized by 2 different sperm. Identical twins result when a single fertilized egg divides very early in its embryology, into 2 separate individuals.

In humans, nearly half of the like-sexed twins are identical, whereas in cattle only 5 to 12 percent of such births are identical. Such twins are always of the same sex, a pair of males or a pair of females, and alike genetically—their chromosomes and genes are alike; they are 100 percent related. When identical twins are not entirely separate, they are known as Siamese twins.

Genetically, fraternal twins are no more alike than full brothers and sisters born at different times; they are only 50 percent related. They usually resemble each other more, however, because they were subjected to the same intrauterine environment before birth and generally they are reared under much the same environment. Also, fraternal twins may be of different sexes.

Distinguishing between identical and fraternal twin calves is not easy, but the following characteristics of identical twins will be helpful:

1. Identical twins are usually born in rapid succession, and frequently there is only one placenta.
2. The calves are necessarily of the same sex.
3. The coat colors are identical; i.e., if there is a broken color, there must be a strong degree of resemblance in this respect.
4. There is little variation in birth weights, general conformation and, more particularly, the shape of the head, position of the horns and occurrence of skin pigmentation, rudimentary teats, etc.
5. Muzzle prints show a degree of resemblance.
6. The shape, twisting and position of the horns and behavior of the twins can be observed at a later stage. Identical twins are inclined to keep together when grazing, walking, lying down, or ruminating.
7. Identical twins have the same blood group.

Most cattlemen prefer single births to twins, for the following reasons:

1. The high incidence of stillbirths in twins. Herefords on the range show 3.6% stillbirths among singles vs 15.7% stillbirths among twins. Despite this fact, twinning would result in more live calves per 100 cows calving; 96 live calves from singles ($100 \times 3.6\% = 3.6$; then, $100 - 3.6 = 96$) vs 168 from twins ($200 \times 15.7\% = 31.4$; then, $200 - 31.4 = 168$).
2. About 85 percent of all heifers born twin with a bull are apt to be freemartins (sterile heifers).
3. Twin calves average 20 to 30 percent lighter weights at birth than singles.
4. The tendency of cows that have produced twins to have a lowered conception rate following twinning.

FREEMARTIN HEIFERS

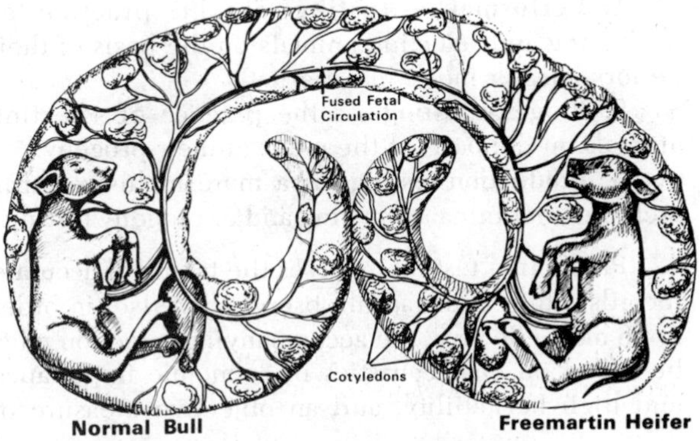

Fig. 3-10. Diagram showing fused fetal circulation of twin calves of opposite sex. Note (1) the fetal circulation of the male fused with that of the female, (2) fetal cotyledon free yolk sac, and (3) normal bull on the left and freemartin heifer on the right. (Source: *Physiology of Reproduction*, by Marshall; courtesy, the publisher, Longmans, Green and Co., Ltd., London, England)

Sterile heifers that are born twin with a bull are known as freemartins. This condition prevails in about 85 percent twin births when a calf of each sex is involved. The fetal circulations fuse, and the male hormones get into the circulation of the unborn female where they interfere with the normal development of sex and modify the female embryo in the direction of the male. In approximately 15 percent of twin births of unlike sexes, fusion of the circulation does not occur, and the animal is normal and fertile.

Since only about 15 percent of such heifers are fertile, it is usually best to assume that they are sterile and market them, unless (1) an experienced person determined at the time of birth that their circulatory systems were not fused, (2) an examination of the vagina reveals that the animal is normal (in freemartin heifers, the vagina is usually about one-third normal length), or (3) skin-grafting[8] or blood-typing[9] techniques show that they are not freemartins and that they may, therefore, be regarded as reproductively normal.

PRODUCTION TESTING[10]

Four bases of selection are available to the livestock breeder; namely, (1) selection based on type of individual, (2) selection based on pedigree, (3) selection based on show-ring winnings, and (4) selection based on production testing. Stockmen have always followed one or all of the first three bases, and, therefore, need no introduction to them; now they are placing increasing emphasis on production testing.

Production testing embraces both (1) performance testing (sometimes called individual merit testing), and (2) progeny testing. The distinction between and the relationship of these terms are set forth in the following definitions:

1. **Performance testing**—is the practice of evaluating and selecting animals on the basis of their performance or individual merit.
2. **Progeny testing**—is the practice of selecting animals on the basis of the merit of their progeny.
3. **Production testing**—is a more inclusive term, including performance testing and/or progeny testing.

Production testing involves the taking of accurate records rather than casual observation. Also, in order to be most effective, the accompanying selection must be based on characteristics of economic importance and high heritability, and an objective measure or "yardstick" (such as pounds, inches, etc.) should be placed upon each of the traits to be measured. Finally, those breeding animals that fail to meet the high standards set forth must be removed from the herd promptly and unflinchingly.

Production Testing Beef Cattle

Production testing is now an accepted beef cattle improvement tool, which may be used by purebred and commercial cattlemen alike.

The traits of economic value include those which contribute to either productive efficiency or desirability of product, the major ones of which are: fertility (reproductive efficiency), mothering and nursing ability, rate of gain, efficiency of feed utilization, longevity, and carcass merit (meatiness and quality).

The performance of cattle on the farm or ranch or in the feedlot, and their value on the rail (in the cooler), varies widely. For example, in the same feedlot and on the same ration, some cattle gain only 1 pound daily while others exceed 3 pounds. Although this difference is disturbing, the fact that such variations do exist, and the further fact that such traits are, to a considerable extent, inherited differences, makes it possible to select and improve. This is possible because research has shown that when cattle are kept under nearly uniform conditions and their performance records are adjusted for known environmental differences, genetically superior animals can be identified.

Thus, production testing is the systematic measurement of differences in traits, the recording of the measurements, and the use of records in selection. It is a selection tool that will increase the rate of genetic improvement in individual herds, and eventually in the breed and in the total cattle population.

The rate of improvement in a herd, breed, and population is dependent on (1) the percentage of observed differences between animals that is due to heredity (heritability), (2) the difference between selected individuals and the average of the herd or group from which they come (selection differential), (3) the genetic association among traits upon which selection is based (genetic correlations), and (4) the average age of parents when the offspring are born (generation interval).

It should be noted that production testing will not increase the production of the animal that is tested. That is, keeping records does nothing to change the genetics of the test animal.

Production records are primarily useful for comparing cattle that are handled alike in a herd, and are not reliable for estimating differences between herds or between groups treated differently within a herd. This is so because large environmental differences due to location, management, and nutrition are likely

[8]Billingham, R. E., and G. H. Lampkin, *J. Embryol. Exp. Morph.*, Vol. 5, Part 4, Dec. 1957, pp. 351-367.

[9]Stormont, C., *Journal of Animal Sciences*, Vol. 13, No. 1, Feb. 1954, pp. 94-98.

[10]The following authorities reviewed this section and made many helpful suggestions for its improvement: Dr. E. J. Warwick, Staff Scientist, Livestock and Veterinary Services; Dr. Clair E. Terrill, Staff Scientist—both of the Agricultural Research Service of the USDA, Beltsville, Md.; and Dr. S. H. Fowler, Texas A & M University, Agricultural Research and Extension Center, Uvalde, Tex.

to exist between herds or between management groups within a herd. It is not possible to adjust accurately for these differences.

ECONOMICALLY IMPORTANT TRAITS AND THEIR HERITABILITY

Research has revealed a high degree of heritability for several economically important traits in beef cattle. By heritability is meant the amount of variation in a trait, such as weaning weight, which is inherited or bred-in. This is usually expressed in a percentage; for example, birth weight has a heritability of 40 percent. The balance, or 60 percent (100 − 40 = 60) is due to environment—such things as feed, climate, and disease. The important thing to remember is that the heritable portion is passed on by parents to their offspring—the environmental portion is not.

This means that if the heritability of a trait is high, marked progress can be made through selection. On the other hand, if most of the improvement in an economically important trait is due to environment, the heritability of that character is low and little progress toward improvement can be made through selection. Nevertheless, the economic value of some traits may be great enough for them to receive emphasis in a breeding program even though their heritabilities may be low; fertility is an example of such a trait. Even though the heritability of many characters is disappointingly small, it is gratifying to know that progress from selection is cumulative and permanent.

Table 3-7, p. 136, lists the economically important traits in beef cattle and gives their estimated heritability.

RECORD FORMS

Production records must be written down. The record forms should be relatively simple, and they should be in a form that will permit easy summarization—for example, the record of one cow should be on one sheet if possible. Suggested record forms are shown in Figs. 3-11, 3-12a, and 3-12b (pages 137, 138, and 139).

MOST PROBABLE PRODUCING ABILITY (MPPA)

It is recommended that MPPA be included on Produce of Dam summaries and that ranking of dams be based on MPPA for the 205-day weaning weight ratio. This is needed to compare dams which do not have the same number of calf records in their averages. For example, suppose six cows have the following records of production:

Cow	No. Calves	Average Weaning Weight Ratio	MPPA
A	1	85	94.0
B	2	88	93.2
C	4	90	92.7
D	3	110	106.7
E	4	112	108.8
F	1	115	106.0

MPPA is most helpful for identifying the lowest producing cows to be culled. In the example, cow A has the lowest lifetime average. However, this is for only a single calf for which environmental conditions or the calf's genetic potential for growth might have been below the average of what the cow would normally produce. One or more calves from cows B or C could also have had a record of 85 or less. All three cows are probably low producers, but MPPA enables more accurate culling and in this example indicates that cows B and C are slightly lower producing cows than A.

MPPA for weaning weight ratio is computed by the following formula:

$$MPPA = H + \frac{NR}{1 + (N\text{-}1)\,R}(C - H)$$

Where H = 100, the herd average weaning weight ratio,
N = the number of calves included in the cow's average,
R = .4, the repeatability factor for weaning weight ratio, and
C = average for weaning weight ratio for all calves the cow has produced.

RATIOS

In selecting within a herd, today's sophisticated cattlemen are not satisfied with merely selecting an individual with a good record, but they wish to go further. They want to know how a bull's contemporaries performed. It may mean very little to have a bull with an adjusted weaning weight of 550 lb if his contemporaries within the same herd average 570 lb. However, the selection of a bull with a 550-lb weaning weight from among contemporaries averaging 470 lb is a different story. It is for this reason that ratios are used to indicate an animal's ranking in relation to the rest of the herd.

"Weight ratio," "gain ratio," and "conformation score ratio" are used to refer to the performance of an individual relative to the average of all animals of the same group. It is calculated as follows:

$$\frac{\text{Individual record}}{\text{Average of animals in group}} \times 100$$

Ratios are a useful device for quickly visualizing the relative rankings of individuals in a group. For example, if the average gain on a group of bulls was 3.0 pounds per day, the gain ratio of a bull gaining 3.3 pounds per day would be 110. A ratio of 100 is average for a particular group. Thus, ratios above 100 indicate animals above average, whereas ratios below 100 indicate animals below average.

TABLE 3-7
ECONOMICALLY IMPORTANT TRAITS IN BEEF CATTLE, AND THEIR HERITABILITY

Economically Important Characters	Approximate Heritability of Characters[1]	Comments
	(%)	
1. Calving interval (fertility)	10	Fertility is economically the most important trait in beef cattle. Without a calf being born, and born alive, cattle are self-eliminating.
2. Birth weight	40	Birth weight is associated with calving survival. Also, it has a positive correlation of .39 with growth rate. Selecting for increased birth weight is generally avoided because of likely increased calving difficulty.
3. Weaning weight	30	Heavy weaning weight is important because: 1. It is indicative of the milking ability of the cow. 2. Gains made before weaning are cheaper than those made after weaning. 3. Those who sell calves at weaning usually make more profit due to the heavier weight available to sell.
4. Cow maternal ability	40	Mothering ability is important in beef cows, because it contributes to calf survival and weaning weight.
5. Feedlot gain	45	Daily rate of gain is important because: 1. It is highly correlated with efficiency of gain. 2. It makes for a shorter time in reaching market weight and condition, thereby effecting a saving in labor and making for a more rapid turnover in capital.
6. Pasture gain	30	Most beef animals spend a good part of their lives on grass; hence, pasture gain is important.
7. Efficiency of gain	40	Efficiency of feed conversion is expressed as pounds of feed intake per 100 pounds of gain. It is seldom measured in performance and progeny tests because of the difficulty in securing feed intake records on individual animals. However, with the development of various types of electronic devices to facilitate such measurements, a number of central testing stations are now securing data on feed efficiency. A positive relationship exists between rate and efficiency of gain, so selection for rapid rate of gain does give some automatic selection for efficiency of gains. However, more rapid genetic progress can be made in improving efficiency of gain by selecting directly for efficiency of feed utilization, because some genes related to efficiency of gain are not related to rapid gains and would be "missed" in selecting only for correlated trait of rapid gain. With beef production becoming so highly competitive, cost factors have become so important to profit and success that an increasing number of stockmen will be measuring and selecting directly for feed efficiency in the years ahead.

(Continued)

TABLE 3-7 (Continued)

Economically Important Characters	Approximate Heritability of Characters[1]	Comments
8. Final feedlot weight	60	Final feedlot weight is usually referred to as *weight per day of age*. It is generally computed at one year of age or at the end of the performance test. It is probably the most important measurement of the estimated value of a beef bull. It is composed of birth weight, weaning weight, and postweaning gain.
9. Conformation score:		This score should be based on skeletal soundness and indications of carcass desirability. Structural soundness, especially of the feet and legs, is most important in breeding animals.
a. Weaning	25	
b. Slaughter	40	
10. Carcass traits		Quality of product and quantity of edible portion are the basic factors of carcass merit. Where breeding animals are involved, and are not to be slaughtered, carcass quality may be evaluated by either (1) ultrasonic measurements, or (2) the K^{40} counter. Ultrasonic can be used to measure rib eye area and outside fat cover. The K^{40} counter evaluates the entire animal; it provides an effective method of measuring the total content of the live animal.
a. Carcass grade	40	High carcass grade is important because it helps determine selling price and is related to the juiciness and palatability of the meat.
b. Rib eye area	70	The rib eye (the large muscle which lies in the angle of the rib and vertebra) is indicative of the bred-in muscling of the entire carcass. Thus, a large area of rib eye is much sought.
c. Tenderness	60	Warner-Bratzler shear test and taste panel test are recommended as methods of measuring tenderness.
d. Fat thickness	45	Fat thickness is taken at the twelfth rib.
11. Cancer eye susceptibility	30	There is indication that susceptibility to cancer eye is hereditary.

[1]Gregory, K. E., *Beef Cattle Breeding*, Ag. Info. Bull. No. 286, Agricultural Research Service, USDA, 1969.

GET OF SIRE RECORD

Calf Crop for Year of ______________________ Sire's Name ______________ Reg. No. ______________

Sex: ______________________ Date of Birth ______________________

Owner and Address ______________________

Herd No. of Calf	Date of Birth	Calf Data							Yearling Data					Dam Data				Remarks
		Weaning Date	Weaning Age in Days	Weight in Lb	Daily Gain from Birth Weight,Lb	Adj. 205-Day Weaning Weight,Lb	Weaning Weight Ratio[1]	Confor-mation Score[2]	Date Weighed	Weight, Lb	Weight Adj. to Days	Year-Weight Ratio[1]	Confor-mation Score[2]	Herd No.	Age This Year	Mature Weight, Lb	Confor-mation Score[2]	
Totals																		
Averages																		

[1]Ratio calculated as follows:

$$\frac{\text{Individual record}}{\text{Av. of all calves on same farm and same season}} \times 100$$

[2]It is suggested that conformation be scored on a numbering system, ranging from 3 to 17, as follows:

17-16-15—Correct in skeletal structure and muscular development. Near perfect.
14-13-12—Skeletal structure basically sound and muscular development average to superior. Minor faults only.
11-10-9—May have moderate to severe faults in skeletal structure or muscular development. Animals suitable for commercial herds.
8-7-6—Animals lacking in beef character. May have serious structural defects and lack muscling.
5-4-3—Thin fleshed cattle. The kind of which no self-respecting stockman would be proud.

Fig 3-11. Get of Sire Record.

INDIVIDUAL COW RECORD

Tattoo ____________________

Name ____________ Reg. No. ____________ Bred by ____________ Birth Date ____________

Purchased from ____________ Birth Wt., Lb ____________

Address ____________ Weaning Wt., Lb ______ Age ______ Conf. Score ______

Sire ____________

Purchase Date ______ Price, $ ______ Yearling Wt., Lb ______ Age ______ Conf. Score ______

Disposition ______ Price, $ ______ Two Year Wt., Lb ______ Age ______ Conf. Score ______

Av. Daily Gain Weaning to 1 yr., Lb ____________

Reason for Disposal ____________ Feed Efficiency ______ lb feed/100 lb gain

Dam ____________

Date ______ Temperament ____________

Faults & Abnormalities ____________

PRODUCE OF DAM RECORD

Calf Data												Yearling Data					Production Testing				
							Weaning Wt., Lb						Yr., Wt., Lb								
Birth Date	Sex	Tattoo	Sire	Birth Wt., Lb	Vigor at Birth[1]	Weaning Age Days	Act.	205-day Adj.	**Weaning Weight Ratio[2]**	Weaning Cond.	Conf. Score[3]	Date	Adj.	Days	**Yearling Weight Ratio[2]**	Conf. Score[3]	Days on Feed	Av. Daily Gain, Lb	Gain Ratio[2]	Lb Feed /100 Lb Gain	Disposition; Price; Remarks

[1]0=dead at birth; 1=definitely undersized at birth; 2=unthrifty, definite indications of disorders; 3=moderately thrifty, slight indications of disorders; 4=thrifty, no signs of disorders, dry hair coat; 5=thrifty, no signs of disorders, sleek hair coat; 6=very large, healthy, and vigorous

[2]Ratio calculated as follows:

$$\frac{\text{Individual record}}{\text{Av. of all calves on same farm and same season}} \times 100$$

[3]It is suggested that conformation be scored on a numbering system, ranging from 3 to 17, as follows:

17-16-15—Correct in skeletal structure and muscular development. Near perfect.

14-13-12—Skeletal structure basically sound and muscular development average to superior. Minor faults only.

11-10-9—May have moderate to severe faults in skeletal structure or muscular development. Animals suitable for commercial herds.

8-7-6—Animals lacking in beef character. May have serious structural defects and lack muscling.

5-4-3—Thin fleshed cattle. The kind of which no self-respecting stockman would be proud.

Fig. 3-12a. Individual Cow Record (see next page for reverse side of record form).

IMMUNIZATION AND TEST RECORD

Immunizations					Health Tests							Remarks
Date[1]	Blklg.	M. Edema	Bangs	Misc.	TB-Bangs	Johnes	Lepto.	Anaplas.	Vib.	Trich.	Misc.	

[1]Indicate vaccinations by check in appropriate column opposite date given; indicate test results by P (positive), N (negative), or S (suspect) opposite date of test.

GENERAL INFORMATION

Record all facts pertinent to the history of this cow, viz.: veterinary treatment (except immunizations), udder condition, mothering instinct, calving peculiarities, etc.

Date	Remarks

Fig. 3-12b. Individual Cow Record. This is the reverse side of the record form shown on the previous page.

PERFORMANCE TESTING ON THE FARM

Basically, beef cattle performance testing is for the purposes of improving growth rate, or rate of gain (which is 45% heritable), and feed efficiency (which is 40% heritable). Ideally, it is conducted as follows: Each animal is individually identified by means of an ear tattoo, ear tag, ear notches, brand, or neck strap. Soon after weaning, and following an adjustment period of 2 to 3 weeks, animals are individually weighed and individually full-fed on weighed amounts of a high energy feed for a period of 140 days, followed by individual weighing at the end of this period; thereby obtaining an individual record of both (1) rate of gain and (2) feed efficiency—the pounds of feed required to make 100 pounds of gain.

Under practical conditions, performance testing of cattle is usually limited to bulls for the following reasons: (1) A bull produces in his lifetime many more offspring than a female, which means that his influence on the total genetic progress of the herd is greater; (2) many more replacement females than herd bulls are selected, with the result that the facilities, labor, and cost of individually testing them in the ideal manner would be overwhelming; and (3) if heifers are production tested in the ideal manner by full-feeding on a high energy ration, they may become so fat that it will impair their reproductive performance. Yet, where possible, heifers should be performance tested, with some modification of the above procedure, because greater genetic progress can be made thereby.

From the above, it may be concluded that the three major problems encountered in performance testing heifers in the ideal manner are:

1. The very considerable labor and expense involved in individually feeding large numbers of replacement heifers in order to measure feed efficiency.
2. The hazard of lowered reproductive performance as a result of heifers becoming too fat when full-fed on a high energy ration.
3. At the close of such a test, fat heifers must be placed on reducing rations, which means that much of the feed that went into making gains during the test period is lost.

To alleviate these three problems, it is recommended that heifers be performance tested somewhat differently than bulls; that their performance test be modified as follows:

1. **Measure rate of gain only**—Heifers should be performance tested by obtaining a record of their (a) weaning weight, and (b) yearling weight, taken over the scales. Heavy weaning weight is indicative of the milking ability of the dam. Heavy yearling weight is indicative of size and growth.

Fortunately, there is a significant correlation between rate and efficiency of gain (r = .40 to .60); hence, selection for rapid gains should also improve efficiency of gains. Thus, it is recommended that heifers be tested only for rate of gain, which can be accomplished by individually weighing at the beginning and end of the test period.

By measuring rate of gain only and eliminating feed efficiency, heifers may be group fed; thereby alleviating the very considerable labor and facilities of individual feeding. Heifers gaining the fastest may then be selected for replacements.

2. **Full-feed a high roughage ration**—In order that heifers will not become too fat when full-fed during the test period, they should be fed a high roughage—low energy ration. They may even be performance tested on pasture, as pasture gain has an estimated heritability of 30 percent (in comparison with 45% for feedlot gain).

By a high roughage, drylot ration is meant one in which the ratio of roughage to concentrate is somewhere between (a) one part of roughage to one part of concentrate, and (b) two parts of roughage to one part of concentrate.

In order to avoid selective eating, high roughage rations should be fed as a complete mixed feed, pelleted or unpelleted. Also, the ration should be full-fed, allowing each heifer to eat according to her appetite, because a good appetite is inherited and conducive to faster gains—a desirable characteristic, especially in feedlot cattle.

3. **Test for 140 days**—The performance test should cover a minimum of 140 days. A period of this length allows each heifer to express her potential growth rate and tends to average some of the environmental effects on growth rate during the feeding period—such as the condition (fatness or thinness) of each heifer when she is placed on test.

Usually, animals are performance tested soon after weaning, primarily for convenience reasons. It is noteworthy, however, that selection for faster growth rate during one period of life should improve growth rate in another period of life.

4. **Either 365-day or 550-day weights may be used**—Often it is not practical to performance test heifers by full-feeding them in a drylot for 140 days or more, with weights taken at the beginning and the end of the period. It is generally more practical to feed them under normal conditions, then obtain either a 365-day or 550-day weight. Heifers gaining the fastest may be selected for replacements.

The 365-day and 550-day weight may be calculated as follows:

a. adjusted 365-day weight =

$$\frac{\text{actual final weight} - \text{(minus) actual 205-day weaning weight}}{\text{number of days between weights}} \times 160$$

plus 205-day weaning weight adjusted for age of dam

b. adjusted 550-day weight =

$$\frac{\text{actual final weight} - \text{(minus) actual 205-day weaning weight}}{\text{number of days between weights}} \times 345$$

plus 205-day weaning weight adjusted for age of dam

Yearling (365-day) weights or 550-day weights are 30 to 35 percent heritable, whereas rate of gain on full feed in a drylot is 45 percent heritable. Nevertheless, selection of heifers on the basis of either 365-day or 550-day weights should result in some progress and is generally much more practical than performance testing on a full feed.

SECURE INFORMATION ON CLOSE RELATIVES

Information on the productivity of *close relatives* (the sire and the dam and the brothers and sisters) can supplement that on the animal itself and thus be a distinct aid in selection. The production records of more distant relatives are of little significance, because, individually, due to the sampling nature of inheritance, they contribute only a few genes to an animal many generations removed.

PRODUCTION TESTING REQUISITES

The following requisites are pertinent to production testing:

1. **Assistance**—Nearly every state now has an approved Beef Cattle Improvement Association (BCIA) program. Also, most breed registry associations have established programs. The Beef Improvement Federation (BIF) is a composite organization, responsible for standardization and uniformity in program systems. But a cattleman can set up and conduct a production program on his own, unassisted.

2. **Identification**—In order to evaluate individual production, it is essential that each animal be posi-

tively identified—by means of ear notches, ear tags, tattoos, or brands. For purebred breeders, who must use a system of identification anyway, this does not constitute an additional detail.

Fig. 3-13. Like mother, like calf! By having identical numbers, a calf and its mother can be readily paired. By having color coded tags for calves, the calf's sire can be readily identified.

3. **Record ages of cows**—Cow ages are needed in order to adjust the weaning weights of calves, because very young and very old cows wean off calves that are lighter than cows in the prime of life. Thus, the age of each cow should be recorded, even if it is necessary to mouth them to get it.

4. **Environmental conditions**—Beef cattle should be tested under much the same conditions as those under which their offspring will be expected to produce.

5. **Rely on milk and grass in preweaning performance test**—In order to evaluate the milking ability of the cow, it is best that neither creep feeding nor nurse cows be used.

Where part of the calves have been creep fed for six weeks or longer, a factor of .97 should be applied to creep-fed calves if they are to be compared to calves not exposed to the creep.

Where calves on nurse cows are to be compared to calves not on nurse cows, the total gain of calves on nurse cows should be decreased by .3 pound per day for the number of days that the calf was on a nurse cow, with a maximum of 240 days.

6. **Ration**—It is recommended that test animals be self-fed, preferably on a complete ration (with the grain and forage combined).

7. **Take conformation scores**—Conformation scores should be taken at two stages—(1) at weaning (approximately 205 days of age), and (2) at the end of the postweaning test; using the 17-point system (see footnotes of Fig. 3-11, Get of Sire Record, and of Fig. 3-12a Individual Cow Record).

Conformation should be reported using a ratio computed separately for each sex-management code group. Both conformation score and score ratio should be reported in sire, dam, and group summaries. MPPA can also be computed for conformation score ratios using the same formula previously shown (see section on "Most Probable Producing Ability [MPPA]"), but with R = .3.

Type scoring should be done by someone qualified to pass judgment in the field of cattle evaluation. Scores at weaning should be used as indicators. Post-weaning grades (grades placed on cattle at one year or 18 months) will give a better indication of type for replacement animals going into the herd.

8. **Adjust the data for age at weaning, age of dam, and sex of calf as follows—**

a. **Adjustment for weaning weight**—It is recommended that the weaning weight be standardized to 205 days (with a maximum range of 160 to 250 days) of age. To this end, the 205-day weights should be computed on the basis of average daily gains from birth to weaning; and the following formula used to provide an estimated 205-day weight, *unadjusted* for age of dam or sex of calf:

$$\text{Adjusted 205-day wt. (lb)} = \frac{\text{Actual weaning wt.} - 70}{\text{Weaning age in days}} \times 205 + 70$$

(The 70 is a constant; use actual birth wt. if available)

b. **Adjustment for age of dam**—To adjust for age of dam, the following adjustment factors are recommended:

2-year-old—multiply computed 205-day wt. by 1.15

3-year-old—multiply computed 205-day wt. by 1.10

4-year-old—multiply computed 205-day wt. by 1.05

5- through 10-year-old—no adjustment

11-year-old and older—multiply computed 205-day wt. by 1.05

c. **Adjustment for sex of calf**—To adjust to a bull equivalent, the following adjustment factors are recommended:

Heifer wt.—multiply by 1.10
Steer wt.—multiply by 1.05

In a commercial herd, where the majority of male calves are steers, records of heifer calves should be adjusted upward to a steer basis by multiplying by 1.05, and records of any bull calves should be adjusted to a steer basis by subtracting 5 percent or multiplying by 95.

9. **Apply the results**—A production test is worthless unless it is used. So, a cattleman should study his records carefully and put them to work.

Use the preweaning record to (a) cull cows producing the lightest calves and/or calves of undesirable type—cows deficient in mothering, milking, and growth potential; (b) sift out bulls that sire light or lower-grading calves; (c) select the heaviest, highest grading heifers for replacements; and (d) improve the management of the cow herd.

Use the postweaning records to (a) select herd sires, (b) select replacement heifers, (c) cull poor producers, (d) improve rate and efficiency of gain, (e) produce more muscular carcasses of higher quality and cutability, and (f) improve the management of the herd.

CENTRAL TESTING STATIONS

Central testing stations are locations where animals are assembled from several herds to evaluate differences in some performance traits under uniform conditions. Central testing stations are used for (1) comparing individual performance of potential seed stock herd sires to similar animals from other herds; (2) comparing bulls being readied for sale to commercial producers; (3) finishing steers or heifers scheduled for slaughter as part of progeny test programs for growth and carcass traits; (4) as an educational tool to acquaint breeders with record of performance; and (5) estimating genetic differences between herds or between sire progenies in gaining ability, feed conversion, conformation, and carcass characteristics.

A bull buyer must decide (1) which herd(s) to buy bulls from, and (2) which bull or bulls to buy within a herd. If the bulls are raised and fed entirely on the farm or ranch where dropped, the buyer has the nearly impossible task of deciding how much of the apparent superiority or inferiority of bulls in a specific herd is due to feeding and herdsmanship. Having them handled for part of their lives (the postweaning test period) under standard conditions minimizes these effects and makes the task of the buyer easier, whether he is buying herd sires for a purebred herd or commercial bulls.

If progeny test groups of steers or heifers from different herds are being fed out to determine the transmitting ability of the sires for growth rate, efficiency, and carcass traits, sire comparisons are more accurate if all progeny are fed under standard conditions for the final feeding period.

Central tests are of limited usefulness for estimating genetic differences between herds. When so used, at least 5 to 10 head per herd should be tested annually for a minimum of 3 years. The larger the herd size, the greater the number of test animals necessary to sample the herd adequately. The precision of the tests may be improved if 5 to 8 progeny of each of 2 or more sires from each herd are tested each year. This permits assessment of within herd differences to compare with between herd differences. Every effort should be made to get a representative sample of animals from each herd on test; otherwise, little real information on herd differences will be obtained.

Central Testing Station Procedure

The following procedure is recommended for central testing of bulls:

1. At the time of delivery to test stations, calves should be at least 180 days of age, and not more than 305 days of age.

2. Herds from which bulls are consigned should be on herd testing programs for preweaning and postweaning performance. Calves should have completed the weaning phase of the performance records program and the following information should be submitted to the test station:

   Sire, dam, birth date, actual weaning weight and date, adjusted 205-day weight, within herd weaning weight ratio (based on average of all bull calves in same weaning season and management group) and the number of calves making up this average.

3. There should be an adjustment or warm-up period of 21 days or more immediately prior to the test period.

4. The length of test should be 140 days or more.

5. Initial and final test weights should be an average of two full weights taken on different days.

6. All bulls sold in a test sale should be examined by a competent veterinarian for reproductive and structural soundness.

7. Test rations will vary according to locally available feeds and test objectives. A complete ration is preferred. Feeding should be ad lib. Rations containing between 60 and 70 percent total digestible nutrients (TDN) should be adequate for the expression of genetic differences in growth. The lower end of this range should result in few health problems and less excessive fattening.

8. Sire group testing of bulls is more desirable than individual testing because it provides more information to the breeder and to the prospective buyers.

Measurements Recommended for Test Stations

Record forms similar to Fig. 3-14 and Fig. 3-15 are recommended for use by test stations, with records required in Fig. 3-14, and records optional in Fig. 3-15.

MEASUREMENTS RECOMMENDED FOR TEST STATIONS

						Weaning			Gain Test					Yearling	
Name and Address of Owner	Breed	Sire	Lot No.	Birth Date	Act-ual Wt.	Wean-ing Date	Adj. 205-Day Wt.	W.W. Ratio w/in Herd and No.	(Date) Initial Test Wt.	(Date) Final Test Wt.	Age in Days	ADG	Test Gain Ratio	Adj. 365-Day Wt.	365-Day Wt. Ratio
			(1)	*(2)*	*(3)*	*(4)*	*(5)*	*(6)*	*(7)*	*(8)*	*(9)*	*(10)*	*(11)*	*(12)*	*(13)*

Note: Each test group (i.e., breed and age group) should be listed together on the report and averaged. (Age range in each group should not exceed 90 days and breeds should be averaged separately within each age group.)

Sire group averages should be shown for 3 or more progeny of same sire.

If sire groups include calves from different age groups, data may be listed together by sires, but only the average of ratios shown.

The explanation that follows is pertinent to the columns designated by the corresponding column numbers:

(1) Ear tag and tattoo numbers.
(2) Month - day - year of birth.
(3) Actual weight used to compute 205-day adj. weight.
(4) Month - day - year when weights were taken to compute 205-day adj. weight.
(5) Weaning weight adjusted to 205 days and for age of dam according to BIF. If creep fed, add C after weight.
(6) Adj. 205-day wt. divided by average of all bull calves in same herd in same weaning season group and same management code. Minimum entrance requirement is optional with test management. The number of calves making the average is listed in parentheses.
(7) Actual weight.
(8) Average of at least 2 full weights taken on different days. May be more than one day apart if desired.
(9) Age at end of test.
(10) Final weight - initial weight ÷ length of test in days. Minimum length 140 days, no maximum.
(11) Average daily gain ÷ test group average of average daily gain. (Breed within age group average.)
(12) $\frac{\text{Final test wt.} - \text{actual weaning wt.}}{\text{Days between wts.}} \times 160$ + adj. 205-day wt. (adj. for dam's age)
(13) Adj. 365-day wt. ÷ test group average of adj. 365-day weights. (Breed within age group average.)

Fig. 3-14. Measurements recommended for all test stations.

OPTIONAL MEASUREMENTS FOR TEST STATIONS

Yearling						
Wt. per Day of Age	Conf. Score	Index	Fat Thick.	Est. Yield Grade	Adj. Feed Conv.	Initial Cond. Score
(14)	*(15)*	*(16)*	*(17)*	*(18)*	*(19)*	*(20)*

(14) Test wt. ÷ days of age when weighed.
(15) Based on structural soundness and estimated potential for carcass desirability (including carcass weight and cutability).
(16) Indices will vary with individual test objectives. They should all be based on ratios to the group average of a trait multiplied by some percentage figure, thus resulting in values ranging below and above a mean of 100.
(17) Fat thickness may be measured by ultrasonics and expressed in hundredths of inches.
(18) Cutability estimates based on ultrasonic readings of rib eye area and fat thickness may be classified into the market yield grades of 1, 2, 3, 4, or 5.
(19) Feed conversion of any group fed together in one pen should be expressed as pounds of feed per 100 pounds of gain. The actual amount of feed should be adjusted to a common body weight to eliminate differences in maintenance requirements.
(20) Initial degree of fatness may be visually estimated and scored on a scale of 1 to 5, with 1 being very thin; 5, excessively fat; and 3, average in condition.

Fig. 3-15. Optional measurements for test stations.

BEEF CARCASS EVALUATION

Beef over the block is the ultimate objective in producing cattle. It is the final objective of all production testing.

The value of a beef carcass depends chiefly upon two factors—the quality of the meat and the amount of salable meat the carcass will yield, particularly the yield of the high value, preferred, retail cuts. To reflect these factors, all USDA graded carcasses are graded for both quality grade and yield grade.

Quality refers to the palatability-indicating characteristics of the lean and is evaluated by considering the marbling (flecks of fat within the lean) and firmness of the lean as observed in a cut surface in relation to the apparent maturity of the animal from which the carcass was produced. Superior quality implies firm, well-muscled lean that is fine in texture and has a light red, youthful color. The USDA quality grades for beef are: Prime, Choice, Good, Standard, Commercial, Utility, Cutter, and Canner.

The yield grade of a beef carcass is determined by considering four characteristics: (1) the amount of external fat; (2) the amount of kidney, pelvic, and heart fat; (3) the area of rib eye muscle; and (4) the carcass weight. The USDA yield grades of beef provide a uniform method of identifying cutability differences among carcasses. There are five USDA yield grades, numbered 1 through 5. Yield Grade 1 carcasses have the highest yields of retail cuts; yield Grade 5 the lowest.

Carcass Evaluation Procedure

The following procedure, step by step, is involved in arriving at the carcass evaluation of a sire:

Step 1: Number of carcasses—8 to 12 carcasses will give a reasonably good evaluation of a sire; as few as 5 to 6 carcasses will indicate the probable transmitting ability of a sire.

Step 2: Kind of carcass sample—Bulls to be compared should be bred to random samples of cows, with the calves born at approximately the same time of the year and reared under similar conditions.

Offspring on which carcass data will be obtained should be unselected and preferably of the same sex. If they are not of the same sex, the proportion of sexes should be equalized as far as possible and sex differences should be taken into account in comparing sire groups.

Step 3: Carrying out carcass evaluation program—Carrying out a carcass evaluation program involves the following procedures:

a. **Feed under standard conditions**—Feed the intended slaughter animals under standard condition. At the close of the feeding period, proceed with slaughter.

b. **Identification**—Animal must be identified on foot and on the rail.

c. **Locate packer**—A cooperative packer should be located in advance.

d. **Contact USDA meat grading service**—Well in advance of slaughtering, make arrangements to have the Federal Meat Grading Service evaluate the carcasses, on a regular charge basis.

e. **Federal grader records evaluation**—After the carcass is chilled, the federal grader records on a standard USDA form the carcass evaluation and forwards this record to the producer (Fig. 3-16).

FORM LS-106
(3-1-66)

BEEF CARCASS EVALUATION REPORT

U. S. DEPARTMENT OF AGRICULTURE
CONSUMER AND MARKETING SERVICE
LIVESTOCK DIVISION

USDA NO.	OTHER IDENTIFICATION	BREED *(As supplied by owner)*	MEAT GRADING CERTIFICATE NO.

NAME OF PRODUCER	NAME OF PACKER

1 QUALITY GRADE ______ BY THIRDS

A. CONFORMATION, MARBLING, AND MATURITY FACTORS

CONFORMATION	DEGREE OF MARBLING	MATURITY (APPROXIMATE AGE SHOWN) *(Circle one)*
		A *(Under 30 mos.)* **B** *(30 to 48 mos.)* **C D E** *(Over 48 mos.)*

B. OTHER FACTORS

TEXTURE OF MARBLING *(Check one)*
☐ FINE ☐ MEDIUM ☐ COARSE

COLOR OF LEAN *(Check one)*
☐ VERY LIGHT CHERRY RED ☐ CHERRY RED ☐ SLIGHTLY DARK RED ☐ MODERATELY DARK RED ☐ DARK RED ☐ VERY DARK RED ☐ BLACK

FIRMNESS OF LEAN *(Check one)*
☐ VERY FIRM ☐ FIRM ☐ MODERATELY FIRM ☐ SLIGHTLY SOFT ☐ SOFT ☐ VERY SOFT ☐ EXTREMELY SOFT

TEXTURE OF LEAN *(Check one)*
☐ VERY FINE ☐ FINE ☐ MODERATELY FINE ☐ SLIGHTLY FINE ☐ SLIGHTLY COARSE ☐ COARSE ☐ VERY COARSE

2 YIELD GRADE ______ BY TENTHS

YIELD FACTORS

CARCASS WEIGHT *(From packer's hot wt. tag)*	FAT THICKNESS *(Inches, nearest 1/10 in.)*		RIB EYE AREA *(from Grid)*	KIDNEY, PELVIC AND HEART FAT *(As per-cent of carcass weight)*
LB.	IN. ACTUAL	IN. ADJUSTED	SQ. IN. BY TENTHS	PCT. ESTIMATED

______ (DATE) ______ (SIGNATURE OF GRADER)

Fig. 3-16. U.S. Department of Agriculture standard Beef Cattle Evaluation Report.

Step 4: Select and use meat sires—Select and use extensively those bulls producing superior beef carcasses, for they will be the ones with bred-in beef quality that will satisfy the consumer.

PERFORMANCE RECORDS IN COMMERCIAL HERDS

Choosing and keeping performance records in a commercial cattle herd is not easy. Yet, records are necessary. Specification of the product offered for sale (calves, in the case of the commercial cattleman) is becoming the rule. Thus, complete records, adequately analyzed and utilized, will help any commercial operation. But records cost money. A large commercial cow-calf operation (of 300 cows or more) may not be able to justify more than a simple feeder calf program involving the sampling of the product that it is offering for sale. On an every other year basis, this might involve a random sampling of calves which are fed out and slaughtered. The gain and carcass data are then used by the producer (1) in the development of his performance reputation (the production of reputation feeder cattle), and (2) in the selection of herd sires to improve his performance. Small commercial operations (with 50 to 300 cows) can well afford to keep more detailed records on their cow herds since this is a means by which they can more effectively compete. Their produce of dam records can be an aid in developing a high-producing cow herd, and in adjusting the management to optimize production. Most state Beef Cattle Improvement Associations have such programs available at nominal cost. The use of such programs, even with multisire pastures, allows the producer at least to evaluate groups of sires purchased. In this manner, he can study the sources of breeding stock supply and be more critical in his future selection.

Production Testing Dairy Cattle[11]

Production records are the most important management tool in dairying. Almost every decision regarding the dairy herd is based on them—how much to feed, when to turn cows dry, which cows to cull, the level of herd health, and which bull to use. Records necessitate that each cow be individually identified, and that milk and butterfat production records be kept.

[11]This entire section on Production Testing Dairy Cattle was authoritatively reviewed by the following: Dr. Stanley N. Gaunt, Department of Veterinary and Animal Sciences, University of Massachusetts, Amherst, Mass.; and Dr. Clinton E. Meadows, Extension Dairy Specialist, Michigan State University, East Lansing, Mich.

ECONOMICALLY IMPORTANT TRAITS AND THEIR HERITABILITY

The economically important characters in dairy cattle are those which contribute to the production of milk, more abundantly and efficiently. Variation and heritability provide the potential for improvement; and the various testing programs provide means of measuring traits.

Heritability is a measure of additive genetic variation. It is a measure of accuracy when choosing parents. To the dairyman, it determines "how much he gets of that for which he selects." When heritability is low, there will be many errors when choosing parents and progress will be slow. When heritability is high, progress will be rapid. Heritability of milk production is ± 25 percent, which is low. But progress will be much faster if dairymen performance test all cows (DHI) and progeny test all bulls (DHI).

The following example will serve to illustrate the importance and application of heritability estimates: If heifers from a herd of cows averaging 13,000 lb of milk were selected from 16,000-lb dams, and by sires equal to these dams, it would be important to know how much of this 3,000-lb superiority in the parents would show up in the offspring. The heritability of milk production is 25%, which indicates that only 25% of the superiority of the selected parents is genetic and will appear in the daughters; 25% of 3,000 is 750 lb of milk. Adding this 750 to 13,000, the estimate of 13,750-lb daughters results from mating 16,000-lb parents that originated in a 13,000-lb herd. Heritability estimates for several traits in dairy cattle are given in Table 3-8.

TABLE 3-8
HERITABILITY ESTIMATES

Trait	Heritability
Production Traits	
Milk Production	.25
Percent Fat	.50
Percent Protein	.50
Percent SNF	.50
Feedlot Gain	.45
Physical Traits	
Stature	.40
Udder Support	.20
Legs and Feet	.15
Management Traits	
Milking Speed	.25
Birth Weight	.40
Temperament	.40
Fertility	.05

ALTERNATE DHI TESTING PLANS

The National Cooperative Dairy Herd Improvement (DHI) Program is a voluntary cooperative effort to improve the level and efficiency of milk production and to increase dairy profits. It involves dairymen, local and state DHI organizations, extension services of land grant colleges and universities, and the U.S. Department of Agriculture.

The U.S. Department of Agriculture (Animal Science Research Division) aids in conducting and distributing results of the sire evaluation phase of the DHI Program. It also coordinates, furnishes materials, provides statistical information, analyzes data, and researches various aspects of the program.

State and local Dairy Herd Improvement Associations(DHIA) conduct the program among dairymen, working through the Coooperative Extension Service in cooperation with the Federal Extension Service and Animal Science Research Service of the USDA.

Dairy cattle record-keeping plans may be either *official* or *unofficial*, with alternate choices under each grouping.

Official DHI Testing

This includes the Standard Dairy Herd Improvement Association (DHIA) and the Dairy Herd Improvement Registry (DHIR). Since both programs are official, a supervisor tests herds at least ten times annually. Records from both programs are used in proving dairy sires.

Diary Herd Improvement Association (DHIA)

This program, first adopted in 1926, is the most complete of all dairy production and record plans. More than half of the cows in the United States on production test are on this program. Both registered and grade cows can be enrolled.

In this program, a supervisor or tester, employed by the local or state testing association, visits the herd one day each month. He identifies all cows in the herd, and he weighs and takes representative samples of the milk from all animals in the herd for 2 consecutive milkings (3 milkings on herds 3-times-daily milkings). He then combines the milk samples and tests them for butterfat or sends them to a central testing laboratory. Records are obtained on an individual cow basis on monthly and accumulative records for milk and fat (the latter in pounds and percentage); amount and cost of feed, and income over feed cost; breeding dates, calving dates, dry dates, and other factors affecting productivity; and in some testing associations, somatic cells or the California Mastitis Test (CMT) is made as an aid in monitoring udder health.

The above information is fed into a computer, programmed to provide monthly summaries on (1) individual cows, and (2) the herd; and this information is sent to each dairyman (Fig. 3-17).

Additional reports are provided in most states. For example, in addition to the Monthly Report (Fig. 3-17) the Pennsylvania Dairy Herd Improvement Program provides the following: Lactation Record Report (a twice annual report of lactation records completed by the members of the herd during the previous six-month period), Reproduction Management Report, Individual Cow Record, Meritorious Lifetime Production Certificate, Herd Production Certificate, and Calf Record Report.

Dairy Herd Improvement Registry (DHIR)

This is the Standard DHIA record *plus* added requirements to satisfy the needs of breed associations. Among the latter, are *surprise tests*, made when the milk production of certain cows exceeds the breed average or another specified amount. Only registered dairy cows are eligible for DHIR records. The production records of herds enrolled in DHIR are mailed to the respective breed registries for official recording.

Unofficial DHI Testing

These unofficial record plans are designed to aid within herd management at minimal cost. Among such plans are: Owner-Sampler (OS), Weigh-A-Day-A-Month (WADAM), Milk Only Record (MOR), and Alternate AM-PM.

Owner-Sampler Records (OS)

Under the owner-sampler plan, the owner himself, rather than a supervisor, weighs and samples the milk. The samples are then tested at a central laboratory.

Weigh-A-Day-A-Month (WADAM)

In this program, the owner weighs the milk from each cow one day each month, and enters the weight and feeding information on the form provided. He mails the information and forms to the supervisor, or a central office, where calculations are completed, following which summaries are returned to him.

Milk Only Record (MOR)

This testing plan, which originated in North Carolina, is receiving considerable attention, especially in the southeastern states. These records involve milk weights recorded by the DHIA supervisor, without fat determination.

Alternate AM-PM Test

This plan, which is now available in some states, requires supervision of only one milking per month

OWNER
MONTHLY REPORT

THE PENNSYLVANIA STATE UNIVERSITY, UNITED STATES DEPARTMENT OF AGRICULTURE, AGRICULTURAL RESEARCH SERVICE, DAIRY CATTLE RESEARCH BRANCH and STATE AGRICULTURAL EXTENSION SERVICES COOPERATING.

PENNSYLVANIA DAIRY HERD IMPROVEMENT PROGRAM

Please examine this report carefully. Report errors to your DHIA Supervisor.
Herbert C. Gilmore
Herbert C. Gilmore
Extension Dairy Specialist

Herd Code CO	Herd	Sampled Mo. Yr.	Cent Day	Page	
H 18	44	9--74	30	1	GEORGE H HAYES
Type of Test Milking Hrs. DHI		P 520	A 535		R D 1 BOX 109
8-30 (From)	10-04 (Thru)	36 (Days in Period)	10-07 (Date Received)	10-10 (Date Processed)	MILL HALL PA 17751

Test Period (Inclusive)

FORAGE: Feeding Schedule, Prices, Analyses (A)

	Milking Cows: Hay	Corn Sil.	Other For.	Suppl't	Past.	Dry Cows: Hay	Corn Sil.	Other For.	Suppl't	Past.
LBS	8	38				8	38			
$	2.00	.60				2.00	.60			
TDN / MOIS	50 / 12	21 / 70				50 / 12	21 / 70			

Milk Price (B) $/CWT		DIFF
7.95	35	077
6.22	SUP 141	
Grain $ A/B	OMC	1 = Protein 2 = S.N.F.

Index No.	Eartag or Registry No.	Breed BW WT CWT	Current Test Day Data: Milk lbs OMC %	% Fat	Fat OMC lbs	% Change Prev. Mo.	Barn Name	Note	Average per Day: Days in Milk	Grain Fed	Grain Rec	Feed Cost $ ¢	Value +/- FC $ ¢	Remarks: Code	Mo.	Day	Date Due: Mo	Day	Lactation Total to Date: Date Calved Mo.	Day	Age Yr.-Mo.	Days in Milk	Milk lbs	% Fat	Fat OMC lbs	Feed Cost $	Value +/- FC $	Sire of Calf / Calf I.D. Estimated 305: Milk lbs	% Fat	Fat lbs
93	5746976	314	44.8	33	1.48	-16	LOIS		36	16	19	1.38	2.11				5	6	5	20	11-09	135	7383	34	252	343	235	13216		451
108	6069034	315	48.8	43	2.10	09	PIEBE		36	30B	24	2.25	1.93						8	6	10-07	57	2618	40	106	255	-40	10289		401
113N	23AHP3062	315					SUSIE	P	11	6	8	.76	-.76	10	8	19	N	23	8	29	9-10	352	17013	38	654	610	870	158263.9		614
115	23AHP3545	314	26.8	39	1.05	-32	BETTY		36	10	14	1.01	1.20				5	18	N	23	10-00	313	14384	41	594	523	752	141704.1		586
117	6054161	314	28.3	49	1.39	-42	CANDY	B	36	28	21	2.13	.42						7	26	10-04	68	2964	33	97	249	-22	9811		312
133R	*23AJC3989	313					RUTH		0	6				59	X	1	9	11	8	30	7-09					82	-82			
200	7725751	313	44.5	36	1.60	-07	BRIDGY	P	(D)	16	20	1.38	2.19				1	20	2	18	3-04	22[illegible]	[illegible]437 (E)	37	462	457	589	14800		554
201	7725749	312	51.1	35	1.79	-15	PAMELA		36	24	22	1.88	2.18				6	6	6	18	3-05	106	6087	39	238	304	187	1339[illegible]	(F)	521
202	7725757	315	50.6	39	1.97	10	TRUDY		36	22B	25	1.76	2.42	61	9	29	7	8	6	15	3-09	109	5856	38	224	283	186	1259[illegible]		479
203	8033319	312	63.6	31	1.97	03	GAIL 1		36	26	26	2.01	2.85	61	9	22	7	1	7	9	3-01	85	5285	31	164	252	143	14217		435
204F	7725759	311	65.6	28	1.84		FRECKLE		21	30	26	2.25	2.61	22	9	11			9	11	3-08	21	1378	28	39	128	-26	1466179		
205N	23AVS6278	313					POLLY		34	6	10	.76	-.76	10	X	3	N	2	7	25	2-06	432	14576	41	603	552	714	112764.1		461
207	23AVS6383	310	26.8	33	.88	-20	BEAUTY		36	10	10	1.01	1.08				3	20	X	3	2-07	364	14965	34	502	523	716	131053.4		440
209N	7725753	312					LUCINDA	P	32	6	8	.76	-.76	10	X	1	N	30	D	14	3-02	288	12316	37	460	437	616	123163.7		460
210	8021097	312	38.3	37	1.42	-15	LORIE 2	P	36	16	17	1.38	1.72				2	7	1	15	2-04	260	13726	34	469	452	678	15373		530
211	8021098	3 9	24.3	50	1.22	-25	FLOSIE2	P	36	14	11	1.26	.95				2	7	2	4	2-01	240	8700	41	361	362	399	10353		433
212	8021093	312	42.3	44	1.86	-04	JAN 2	P	36	16	20	1.38	2.28				2	10	2	22	2-09	222	10945	42	456	382	575	13791		584
213	7725752	313	55.5	44	2.44	04	FAIRY		36	20	30	1.63	3.17	61	9	20	6	29	3	23	3-02	193	10476	43	453	356	561	14038		612
214	8021096	311	55.6	31	1.72	07	FLO 2		36	14	22	1.26	2.99	61	9	19	6	28	5	10	2-06	145	6965	37	260	243	319	12676		481
215	8021099	311	39.6	40	1.58	-01	FAWN 2		36	20	17	1.63	1.67				4	11	5	1	2-08	154	5649	42	239	276	207	9773		418
216	8289820	310	47.6	38	1.81	08	HATTIE		36	20	19	1.63	2.26				6	6	5	24	2-03	131	6371	37	233	247	261	12678		468
217F	8021091	312	63.6	32	2.04		LOIS 2		29	30	29	2.25	2.66	21	9	3			9	3	2-08	29	1844	32	59	65	77	1631889		7812
(C)																														

(* CREDIT NOT COMPLETE FROM CALVING DATE) (* ——— MILKED ONE OR MORE DAYS 3X ———*) (% FAT INDICATES RECORD IS ACTUAL ——) (H)

Herd Totals	Days in Period	Cow Days: In Herd	In Milk	Actual Production: Milk lbs	Fat lbs	OMC lbs	Value $: Milk	+/- FC	Feed Fed (CWT): Grain	Hay	Other	Days Pasture	Feed Cost ($): Grain	Hay	Other	Past.	Total	Herd Test Day: Milk lbs	% Fat
Period	36	2274	2051	88616	3279		7183	399[illegible] (G)	[illegible]371	182	864		2307	364	518		3189	2417	3.8
Rolling	365	22365	18915	861095	33723		74656	42490	3771	926	14750	3731	19393	2035	10086	652	32166		Tank Wt lbs

Per Cow Average: BW	BRD		No. Cows	% Cow Days in Milk	Milk lbs	% Fat	Fat lbs	OMC lbs	Value $: Milk	+/- FC	Feed Fed (lbs): Grain	Hay	Other	Past. Days	Feed Cost ($): Grain	Hay	Other	Past	Total	per Cwt Milk	Milking Cows: Milk	Fat	% All Cows During Year: 1st Calf Heifers	Purchased	Removed	No. of Individual Cows
		Test Period Daily	63.2	90.2	39.0	3.7	1.44		3.16	1.76	16	[illegible]8 (I)	38		1.01	.16	.23		1.40	3.59	43.2	1.60	(J)			
13	83	365 Day Rolling	61.3	84.6	14047	3.9	550		1217	693	6152	1511	24062	61	316	33	164	10	524	3.74	No. Tests 11	Herd Status ►	13	0	9	68

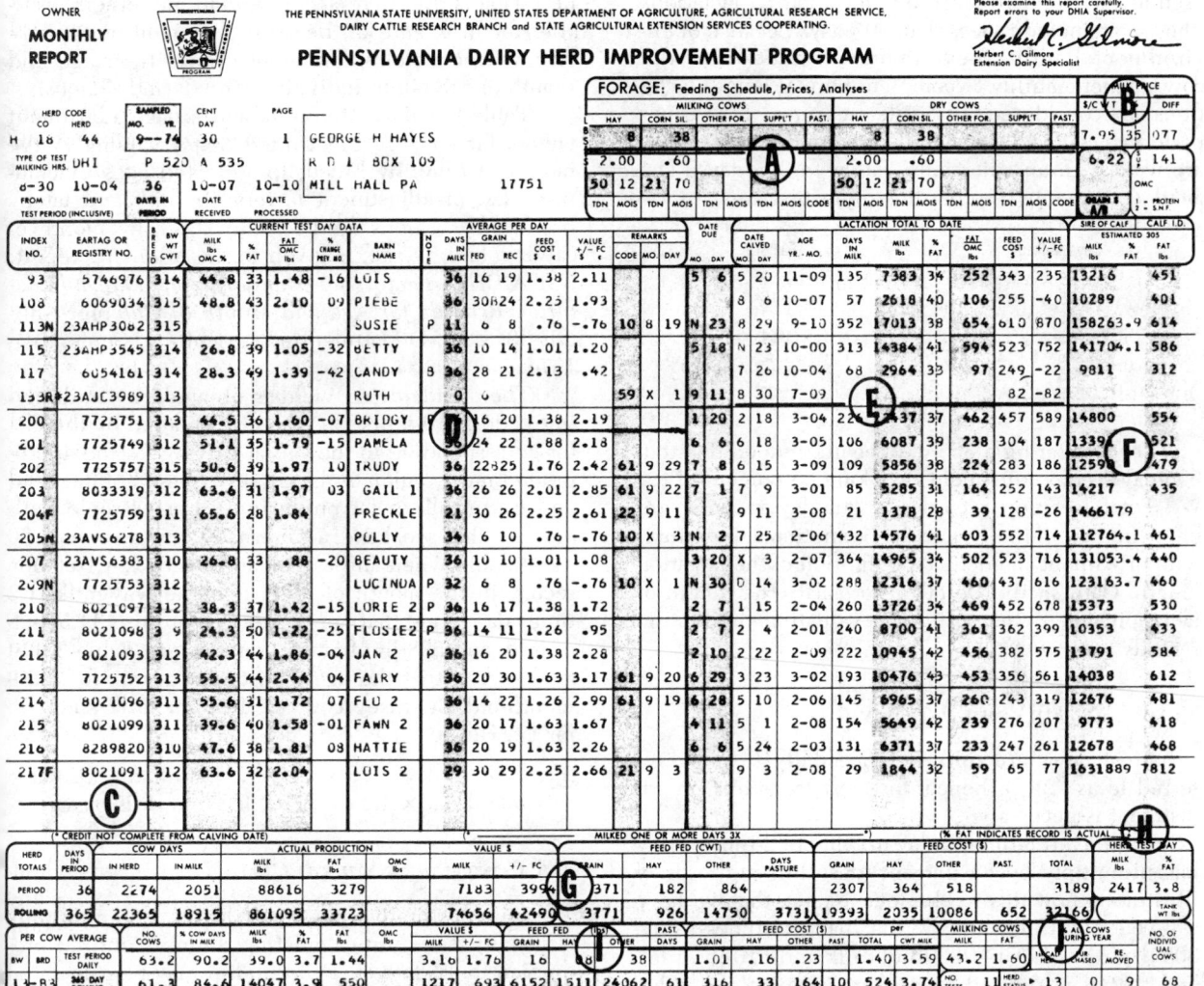

Fig. 3-17. Monthly Report in use by the Pennsylvania Dairy Herd Improvement Program. This report shows how each cow and the entire herd has been doing. To facilitate comparisons, each cow is always listed on the same relative position on each monthly report. An up-to-date analysis of the entire herd is always listed at the bottom of the page. Each cow's production information is divided into three parts: test day, lactation to date, and estimated 305-day.

(instead of two, as in DHIA)—AM in one month, PM in the next. The plan needs further refinement, but it holds considerable promise. Its chief virtue is that it permits the supervisor to enroll more herds, thereby lowering cost. However, AM-PM records have shown greater variation from a cow's true yield than Standard DHI records, with the result that they are not used in genetic evaluation studies of sires and dams.

ADJUSTMENT FACTORS

It is frequently desirable to compare the performance of individuals or groups of animals. To do so, it is necessary to adjust all records to a comparable basis. For this purpose, adjustment factors have been developed for each breed for (1) length of lactation, (2) the number of milkings per day, (3) age and month of calving, and (4) fat content of milk. These four adjustments are important for comparing milk and fat of cows in different environmental conditions. Each of these factors will be discussed. At the outset, however, the following point is pertinent: Although adjustment factors are usually necessary in order to get two or more records to a common basis, it is recognized that records that are comparable without factors are more reliable.

Length of Lactation

The most generally accepted standard length of lactation records is 305 days. When a cow is milked longer than 305 days, her yield for the first 305 days is used as the standard lactation yield. Partial lactations (those terminated in less than 305 days because of environmental influences having no relation to the cow's genetic ability to complete normal length lactations) are considered legitimate measures of the cow's performance up to the time they were terminated and are used with an adjustment factor to 305 days. The factors commonly used for this projection are:

Days Milked	Factor
95	2.82
125	2.16
155	1.77
185	1.51
215	1.32
245	1.18
275	1.08

For comparing a 365-day record, reduce this to a 305-day record equivalent by taking 85 percent of it.

Total lactation records, or 365-day records, are often quoted verbally and in promotional literature, with or without an adequate definition of the lactation length. Care should be taken to clarify the length of lactation when comparing or evaluating production records.

Number of Milkings per Day

Most cows are milked twice daily (usually referred to as "2×"); hence, for most lactations no adjustment is necessary.

To convert 3-times-a-day milking to 2-times-a-day basis, multiply by 83 percent (.83). For purposes of illustrating how this works, let's assume that we have a 4-year-old Holstein cow that has a 3-times-a-day, 305-day record of 14,000 lb of milk and 610 lb of fat, and that it is desired to convert it to 2-times-a-day basis. Simply multiply the cow's record by 0.83. Hence—

14,000 lb milk × 0.83 = 11,620 lb milk on 2 × basis
610 lb fat × 0.83 = 506.3 lb fat on 2 × basis

Age and Month of Calving

The age of a cow is always based on her age when she calved, which is when her record begins. It is estimated, on a rule-of-thumb basis, that at 2 years of age a cow produces approximately 70-80% of her mature production; at 3 years, 80-91%; at 4 years, 91-96%; 5 years, 97-100%; and at 6 years her mature record.

New "age adjustment factors" have been developed to standardize 305-day lactation records to a mature equivalent basis and to minimize environmental variation due to the month of the year in which the record began. These new age and month-of-calving factors are based on a total of 4,452,332 Official DHI and DHIR lactations between 1964 and 1968. These new factors are more accurate than any others available for milk and fat, because they more accurately remove recent environmental effects from age and month of calving in individual breeds and regions.

Table 3-9 shows the milk and fat age adjustment factors for cows in the United States calving in the month of May, by breed, for selected ages. A complete list of adjustment factors for milk and fat by breeds, by regions (and for the U.S.), by month of calving, and by age, is given in the following report: *USDA-DHIA Factors for Standardizing 305-Day Lactation Records for Age and Month of Calving*, ARS-NE-40, U. S. Department of Agriculture, September, 1974.

The standardized yield is obtained by multiplying yield for the first 305 days of lactation by the factor corresponding to the age at calving for the appropriate breed, region of the country, season of the year, and trait (milk or fat production). For example, let's assume that we have a Guernsey cow that was 20 months old when she calved and began her lactation record in the month of May; that she was milked 2 times daily; and that her 305-day record was 11,510 lb of milk and 508 lb of fat. By referring to Table 3-9, it is observed that the age adjustment factors for a 20-month-old Guernsey cow are 1.29 for milk and 1.28 for fat. Hence—

11,510 lb milk × 1.29 = 14,847.9 lb milk on ME basis
508 lb fat × 1.28 = 650.2 lb fat on ME basis

Fat Content of Milk (FCM)

For comparative purposes, the fat content of milk is usually based on calculating the milk and fat production to 4% fat (4% FCM), but it may be calculated to any desired fat basis. The formula for 4% FCM is:

$$4.0\%\ \text{FCM} = (0.4 \times \text{milk weight}) + (15 \times \text{fat weight})$$

EVALUATING DAIRY SIRES (Progeny Testing)

Genetically, the sire and the dam contribute equally to their offspring. However, the breeding value of a dairy sire can be more accurately determined than that of a cow because he will have many more offspring. Thus, it is through the selection of sires that the major portion of progress in genetic improvement is made.

Generally speaking, the dairyman has three sources of herd sires: (1) artificial insemination service, (2) purchase of a herd sire, or (3) raising a herd sire.

Sires should be selected largely on a production basis. They may also be selected by pedigree, type, and family or bloodlines. The dairyman must also decide between a proved sire and a young sire.

TABLE 3-9

MILK AND FAT AGE ADJUSTMENT FACTORS FOR COWS IN THE UNITED STATES CALVING IN THE MONTH OF MAY, BY AGE AND BY BREED[1]

	Ayrshire		Brown Swiss Red Poll		Guernsey		Holstein & Red Dane		Jersey		Milking Shorthorn	
Age	Milk	Fat	Milk	Fat	Milk	Fat	Milk	Fat	Milk	Fat	Milk	Fat
(months)												
20	1.33	1.31	1.54	1.51	1.29	1.28	1.40	1.39	1.37	1.36	1.49	1.47
24	1.23	1.21	1.40	1.37	1.21	1.21	1.30	1.29	1.27	1.26	1.31	1.29
30	1.16	1.13	1.29	1.27	1.13	1.12	1.21	1.20	1.17	1.16	1.19	1.18
36	1.13	1.12	1.20	1.18	1.09	1.08	1.15	1.15	1.12	1.11	1.16	1.16
42	1.09	1.08	1.14	1.13	1.06	1.05	1.10	1.10	1.07	1.06	1.14	1.15
48	1.06	1.06	1.10	1.09	1.04	1.04	1.07	1.07	1.04	1.04	1.12	1.12
54	1.04	1.03	1.06	1.06	1.02	1.02	1.04	1.04	1.02	1.02	1.09	1.10
60	1.02	1.02	1.04	1.04	1.01	1.02	1.02	1.03	1.00	1.01	1.06	1.07
66	1.01	1.02	1.03	1.03	1.01	1.02	1.01	1.02	.99	1.00	1.04	1.05
72	1.01	1.02	1.02	1.03	1.01	1.02	1.01	1.01	.98	1.00	1.02	1.03
84	1.00	1.02	1.01	1.02	1.01	1.03	1.01	1.02	.98	1.00	1.00	1.01
95	1.00	1.02	1.00	1.02	1.01	1.04	1.01	1.02	.98	1.01	.98	1.00
110	1.02	1.04	1.01	1.03	1.03	1.06	1.03	1.05	1.00	1.03	.98	1.01
120	1.03	1.05	1.02	1.05	1.04	1.08	1.05	1.07	1.01	1.04	1.00	1.02
135	1.04	1.07	1.04	1.08	1.06	1.11	1.08	1.10	1.03	1.07	1.01	1.04
140	1.06	1.09	1.05	1.09	1.06	1.11	1.10	1.12	1.04	1.08	1.02	1.05
150	1.08	1.11	1.07	1.11	1.08	1.14	1.12	1.15	1.06	1.10	1.03	1.06
160	1.10	1.14	1.09	1.13	1.10	1.16	1.15	1.18	1.08	1.12	1.04	1.07

[1]*USDA-DHIA Factors for Standardizing 305-Day Lactation Records for Age and Month of Calving*, ARS-NE-40, Agricultural Research Service, USDA, Sept. 1974, pp. 80-91.

Proved Sires

A proved sire is a bull that has a certain number of daughters with milk production records. The summaries of such bulls are the most reliable indicators available as to the performance of future daughters of these sires. The more daughters and the more herds in which the daughters make records, the more accurate the proofs will be.

The four common methods of evaluating proved sires, based on production records, are (1) daughter average, (2) daughter-dam comparison, (3) herdmate comparison, and (4) Modified Contemporary Comparison (MCC). Today, the MCC is the most widely used and most reliable of all methods.

Young Sires

A young sire is a young bull that is selected on the basis of pedigree, but with consideration given to type and freedom from defects.

Until the young sire is proved, it is good procedure to use him on selected cows or heifers for one season, then let him remain idle until his first daughters are in production and tested. But an individual breeder is limited in the genetic accuracy that he can attain by testing a young bull in his herd only. Such proofs are only about 20% accurate (even with as many as 25 daughters) in indicating performances in other herds or in artificial insemination (A.I.). By contrast, a proof on 10 daughters in A.I. (where usually there are only 1 or 2 daughters per herd) has twice the accuracy, or 40%; with 25 daughters, the accuracy is 57%; 50 daughters, 77%; 100 daughters, 89%; and 500 daughters, 97%. As noted, the accuracy goes up very little with an increase in numbers of daughters of a sire within one herd, whereas with A.I. the accuracy increases sharply with increased numbers of daughters. So, a young bull should be tested in several herds.

USDA-DHIA Sire Summaries

The U.S. Department of Agriculture-Dairy Herd Improvement Association sire summary list is published three times yearly by, and free copies may be obtained from, the Dairy Herd Improvement Investigations Unit, Building 263, Agricultural Research Center, Beltsville, Md. This list contains the most recent estimates of transmitting ability of sires of six dairy breeds for milk, fat percent, fat yield, and income. These genetic evaluations are based on the records of the bulls' daughters in herds participating in official production testing programs (DHIA and DHIR).

A sire summary is compiled when 10 or more daughters of a bull have lactation data and herdmate records reported. (Some sire summaries report only sires having 20 or more production tested daughters.) The production records used in USDA-DHIA sire summaries consist of lactation records of 305 days or less, standardized to twice daily (2×) milking, mature equivalent (ME) basis.

The dairy registry associations also publish sire summaries. Some of these are useful for type (conformation) information, which is not included in USDA-DHIA summaries.

Predicted Difference (PD)

Predicted Difference (PD) indicates the amount by which each bull's daughters should, on the average, outproduce (or underproduce) their breed average herdmates on a mature equivalent basis. It is the most accurate available measure of a sire's ability to transmit production potential. For purposes of illustrating, the following Predicted Difference for a bull owned by American Breeders Service is taken from the May 1974, USDA Sire Summary:

Name of Bull	Predicted Difference			
	Milk	*%*	*Fat*	*$*
Westside AB Seaman	+1,572	−.09	+43	+119

Thus, the daughters of Westside AB Seaman may be expected to produce an average of 1,572 pounds more milk and 43 pounds more butterfat than herdmates in breed average herds.

As noted above, the USDA-DHIA sire summary also gives a predicted difference for income (PDI) for each bull listed. This economic index is added to each bull's summary to predict gross income in relation to breed average and is based on the bull's predicted difference for milk production. The Predicted Difference for income considers PD for milk and fat (in lb), the price paid for a unit of milk at base test (percent fat), the fat percentage, and fat differential. The formula uses the average U.S. milk price for 3.5 percent milk, and the average fat differential.

Thus, the level of production of the daughters of an A.I. sire is not very helpful in deciding which bull to use in a herd. The important criterion is the level of improvement shown over herdmates that can be expected in the production of future daughters; i.e., *the predicted difference.* This becomes apparent in Table 3-10.

TABLE 3-10
LEVEL OF PRODUCTION AND PREDICTED DIFFERENCE FOR SEVERAL BULLS

Number of Daughters of Each Bull	Av. Milk Production of Daughters (M.E.-305-2X)	Predicted Difference
	(lb)	*(lb of milk)*
230	15,926	+1,626
756	15,472	+1,421
378	16,007	+1,215
430	16,307	+1,121
585	16,035	+1,010
278	16,037	+ 991
237	16,041	+ 829
343	15,421	+ 732
242	15,930	+ 569
446	16,439	+ 327

If the Table 3-10 bull with the highest production (16,439 lb of milk) were selected, the expected production of his future daughters would only average about 327 lb of milk above breed average herdmates. However, if the second sire in the list were used (production of daughters = 15,472 lb of milk), the expected production of his future daughters would average about 1,421 lb above breed average herdmates. This is almost 1,000 lb more milk per lactation for each daughter over the sire whose daughters average 16,439 lb of milk per lactation.

Repeatability

Repeatability is the value that indicates how sure we are that the Predicted Difference reflects an individual bull's true transmitting ability for future matings.

Repeatability has nothing to do with the level of transmitting ability of a bull for production traits. It simply indicates the reliability, based on available proof, with which we can expect a bull to repeat his Predicted Difference, on the average, in future matings.

Table 3-11 can be used to illustrate the value of repeatability when selecting between sires which have essentially the same predicted difference. Future daughters of the first sire listed in Table 3-11 stand a 99 percent chance of having an average production of 683 pounds of milk above their herdmates. The future daughters of the last bull stand a 28 percent chance of being 671 pounds of milk better than their herdmates. Thus, the first bull's future daughters' production can be predicted with about 3.5 times more confidence than those of the last bull.

TABLE 3-11
VARIATIONS IN "REPEATABILITY" REPORTED WITH APPROXIMATELY IDENTICAL PREDICTED DIFFERENCE MILK VALUES

No. of Daughters	No. of Herds	Av. Milk Production of Daughters (M.E.-305-24)	Repeat-ability of Sire Summary	Predicted Difference Pounds of Milk
		(lb)	*(%)*	
1,515	999	15,385	99	+683
582	423	14,557	97	+696
271	169	14,718	94	+681
44	32	15,164	66	+695
20	12	15,638	49	+689
25	3	18,837	35	+696
27	4	17,779	31	+670
14	3	16,360	28	+671

The higher the repeatability, the more confidence one can place in sire summaries. Repeatability increases with the number of daughters, the number of lactations per daughter, and the number of herds in which daughters are located.

Selecting the Dairy Bull

Predicted difference is an excellent management tool to use in ranking bulls on the basis of their breeding value. It tells what production the dairyman can expect from the daughters of the sires he plans to use in his breeding program. The repeatability value indicates the confidence with which it may be expected that these daughters will attain the predicted level of production. It should be realized that every daughter of a particular bull will not have the same production, but that the average of the group will be as indicated.

For example, if a bull is used with a predicted difference of +1,000 pounds of milk, some daughters will produce below this level and others will produce above this level. Yet as a group they will exceed the breed average production by 1,000 pounds of milk per lactation.

The choice of bulls is almost unlimited with A.I. Predicted difference can range in value from +1,600 or +1,700 pounds of milk to a −1,100 or −1,200 pounds. Since the semen of a bull with the highest predicted difference usually costs a premium, the predicted difference income (PDI) can be used as a tool in determining if expected daughter productivity will more than compensate for the extra semen cost. Thus, what is the dollars and cents production difference between the two bulls, one with a PD of +1,150 pounds of milk, and the other with a PD of +250—with each bull having the same repeatability?

First, it must be realized that it will be about 3 to 3½ years before any difference is evident. So, thinking 3 to 4 years ahead, the dairyman would realize an average of about 900 pounds more milk if he used the bull with the +1,150 proof. At today's milk prices, this would mean over $45 per daughter per lactation. In terms of a 50-cow herd, this would mean $2,250 per year added income.

The most desirable sire is one with high predicted difference of high repeatability.

For herd improvement, a dairyman should choose—

1. Bulls with the highest PD values.
2. Bulls with higher PD values that also have higher repeatabilities (narrower Confidence Ranges).
3. Bulls that are least +600 milk if they have low repeatabilities.

• *Best semen buy*—A simple and easy-to-use formula for ranking sires based on their net return over investment follows:

$$\$ \text{ net return} = \frac{\text{PD}}{10} - (6 \times \$ \text{ cost per breeding unit})$$ [12]

This equation is easy to use by hand or with a small pocket calculator. If one wishes to reflect the economic effect of both milk and fat, PD fat-corrected milk can be used in place of PD milk in the above equation. PD fat-corrected milk can be calculated from PD $ value by dividing the PD $ value of each bull by the price of milk per pound which the U.S. Department of Agriculture used to calculate PD $.

Production Testing Sheep

In sheep, as with other classes of livestock, production testing (performance and progeny testing) is used for the following purposes:

1. To aid in the selection of both male and female herd or flock replacements.
2. To provide an accurate basis for culling herds or flocks.
3. To aid in the promotion and sale of breeding stock.

The amount of improvement in sheep that can be achieved through production testing depends on the following:

1. The accuracy in measuring a trait, and the use made of the record obtained.
2. The selection pressure applied. This is limited by (a) reproductive rate, (b) the number or percentage of animals that need to be saved for flock replacements, and (c) the number of traits being selected for simultaneously. Most of the selection pressure and improvement will result from the selection of rams.
3. The variability in the trait or traits being selected. The greater the variability, the more rapid the improvement.
4. The heritability of the trait for which selection is being made.

ON-THE-FARM VS CENTRAL TEST STATION

From a practical standpoint, some traits can be evaluated only on the farm; among them, fertility, prolificacy, longevity, and all traits of lambs measured before weaning.

Postweaning traits may be measured either on the farm or at a central test station. The *advantages* of central testing include:

1. It facilitates the accurate measurement of some traits that are difficult or impossible to measure on the farm, such as feed efficiency and ultrasonic estimate of fat thickness.
2. Comparisons of rams between flocks are possible, *provided* pretest management differences are minimal so that such comparisons are valid.
3. Greater standardization of the test environment may be possible than on the farm.
4. There is greater confidence in the results of a test conducted by an independent agency.

---

[12]From paper entitled, "Economics of Dairy Cattle Breeding," by R. W. Everett and R. E. Pearson, Cornell University and ARS, USDA.

The *disadvantages* of central testing include:

1. The impossibility of completely alleviating pretest differences in flock management, rations, disease and parasite control, etc.

2. The relatively high cost.

ECONOMICALLY IMPORTANT TRAITS AND THEIR HERITABILITY

Two basic principles of animal breeding are:

1. The more traits included as criteria for selection the slower the progress for any one of them.

2. A trait will not respond to selection unless variation for the particular trait exists in the flock, and unless the trait is heritable.

To the above may be added the observation that more measurements mean more time and cost. Thus, only those traits should be considered which contribute to net income, which display variation in the flock, and which are heritable.

Table 3-12 lists the economically important traits in sheep and gives their heritability.

RECORD FORMS

A carefully planned and executed on-the-farm recording scheme for lamb production is the first step in a sheep production testing program. Records should be in understandable form; and, most important, proper use of the records should be made in selection and culling.

Fig. 3-18 shows an individual ewe or ram record form which will meet the needs of most herds and flocks. (See pages 154 and 155 for Figs. 3-18 and 3-18a.)

PREWEANING LAMB PERFORMANCE TEST

The following preweaning lamb performance tes program is recommended for use in all purebred o commercial farm flocks:

Step 1: Record data—*Minimum data* should in

TABLE 3-12

ECONOMICALLY IMPORTANT TRAITS IN SHEEP AND THEIR HERITABILITY

Economically Important Characters	Approximate Heritability of Character (%)	Comments
1. Multiple birth	15	Where adequate feeds are available, twin lambs are desirable because (1) they greatl increase the weight of lambs sold per ewe, and (2) the annual maintenance require ments of ewes is not far different, whether they are producing twins or singles.
2. Birth weight of lambs	30	The larger lambs at birth are generally more vigorous and make faster gains. Australia workers have increased twinning rate in Merino sheep by 2.3% per year by selectior In New Zealand, the Romney has responded to selection for twinning by increasin 1.1% per year.
3. Weaning weight:		Heavy weaning weights are especially important in those areas where cost of produc tion is largely on a per head rather than on a per pound basis, such as the wester range.
a. 60 days of age	10	
b. 100 days of age	30	
4. Rate of gain	30	Preweaning rate of gain, or growth rate, is largely a reflection of the milk production the ewe. It is affected by twinning, sex of lamb, and age of ewe. Postweaning rate of gain is a reflection of inherent growth potential of the individual. is also positively correlated with mature size. Growth rate is economically important for three reasons: (1) It is highly associated wi feed efficiency—rapid growth is efficient growth; (2) rapid growth allows for the sa of a larger amount of product; and (3) it makes for a shorter time in reaching mark weight and condition, thus effecting a saving in labor, making for less exposure risk and disease, and allowing for more rapid turnover in capital. Postweaning growth rate can be measured effectively by average daily gain, either the farm or in the central station.
5. Type score:		Type can include any or all of the following: (1) characteristics that influence an ar mal's ability to live and perform in its environment—such as feet and legs, teeth, ar udder; (2) traits that indicate meatiness; and/or (3) breed type. Type is a factor in determining today's market values. Yet, type within itself unsupported by performance records for other traits—will not likely be sufficient ensure high selling prices in the future. The determination of optimum type, the evaluation of it, and the use made of the inf mation, should remain the responsiblity of the individual breeder.
a. Weaning	10	
b. Yearling	40	
6. Finish or condition at weaning	17	Finish at weaning is largely determined by available feed and is not highly heritab Yet it is most important because milk-fat lambs suitable for slaughter at weaning tir

(Continue

TABLE 3-12 (Continued)

Economically Important Characters	Approximate Heritability of Character	Comments
		almost always bring more per pound than thinner lambs that are sold as feeders. For the range area as a whole, about 25% of the lambs lack sufficient finish for slaughter at weaning time.
7. Wrinkles or skin folds:		
a. Neck folds (weaning)	39	Sheep with smooth bodies are preferred. Wrinkled sheep are difficult to shear, and lack fiber uniformity.
b. Body folds (yearling)	40	
8. Face covering	56	Wool-blind ewes do not graze well, require more labor if they are clipped around the eyes, and wean fewer pounds of lamb. At the Western Sheep Breeding laboratory, ewes with open faces produced 11 lb more lamb per ewe bred than those with covered faces.
9. Fleece weight:		
a. Grease weight	38	Clean fleece weight is most important, for the fiber is far more important than the materials scoured from grease wool. However, scouring a whole fleece, or even a sample, requires much time and equipment; hence, it likely can only be justified in the selection of stud rams in purebred flocks of fine- and medium-wool sheep. Since there is a close correlation between clean fleece weight and grease fleece weight, grease fleece weight will suffice under most circumstances.
b. Clean weight	40	
10. Staple length:		
a. Weaning	39	Fiber length is important because it is a major factor in determing fleece weight and grade.
b. Yearling	47	
11. Fleece grade	35	The grade of a fleece—which is based primarily on fiber diameter, but with consideration given to length, also—is important because it determines the use and price of wool.
12. Fat thickness over loin eye	23	Fat thickness is a measure of meatiness; excess fat results in an increase in fat trim and a decrease in percent lean cuts.
13. Loin eye area	53	Loin eye area is a good indicator of muscling.
14. Carcass weight/day of age	22	This trait is moderately heritable.
15. Carcass grade	12	High carcass grade is important because it determines eating and selling qualities.
16. Carcass length	31	Long carcasses are usually meaty carcasses.

clude (a) identification—lamb ear tag number; (b) sire number; (c) dam number; (d) age of dam, in years, at lambing time; (e) birth date of lamb; (f) sex of lamb; (g) type of birth—single, twin, triplet; and (h) how reared—single, twin, triplet.

Optional data may include (a) whether or not lamb was creep fed; (b) slaughter grade; (c) 200-day yearling body weight; and (d) grease fleece weight and staple length to nearest 1/10 inch.

Step 2: Wean and weigh—In advance, decide on weaning age—usually 90, 120, or 140 days. Wean and weigh as near to the intended age as possible.

Step 3: Type score—Even though the heritability of type, or conformation, is low at weaning (10%), it is a factor in determining today's market values; hence, all performance records should be augmented by type scores at weaning time. Also, and even more important because of the higher heritability (40%), all animals retained for breeding purposes should be type scored as yearlings.

Type can include any or all of the following: (1) characteristics that influence an animal's ability to live and perform in its environment—such as feet and legs, teeth, udder, and lethals and sublethals; (2) traits that indicate meatiness; and/or (3) breed type.

Also type score may include face cover score, wrinkle score, and record of the presence or absence of scurs or horns.

The determination of optimum type, the evaluation of it, and use made of it, should remain the responsibility of the individual breeder.

Step 4: Adjust for certain environmental factors—Adjust records for certain environmental factors such as age, sex, type of birth and rearing, and age of dam.

Weaning weights of lambs within a flock may be adjusted to 90, 120, or 140 days of age by finding the weight per day of age, then multiplying by the standardized age desired. Thus, the following formula may be used to provide the estimated 120-day weight of lambs:

$$\text{Adjusted 120-day weight (lb)} = \frac{\text{Actual weaning weight}}{\text{Actual days of age}} \times 120$$

Additional adjustment factors are given in Table 3-13 (page 156).

SHEEP

INDIVIDUAL EWE OR RAM RECORD **FLOCK NO.** ....

Breed .... Reg. No. .... Ear Nick .... Tattoo .... Birth Date ....

Sire ....

Dam ....

Type of birth (Single, Twin) ....
Bred by ....
Bought from ....
Address ....
Date Purchased ....
Type, Weaned[1] .... Date....................
Type, Yearling[1] .... Date....................
Back[2]........ Rump[3]........ Leg[4]........
Defects & Abnormalities[5] ....

Date ....
Temperament ....
(Gentle, nervous)
Face Covering[6] ....
(As a lamb)
Face Covering[6] ....
(As yearling)
Disposed to .... Date....................
Why Disposed[11] ....

LAMBS (Use one line for each lamb for ewe's offspring; use one line for the average of a ram's progeny for each year).

Date of birth	Ear nick and No.	Vigor at birth	Type of birth[7]	Type of rearing[8]	Sex	Birth Wt.	Defects and abnormalities[5]	Sire	Milking ability - ewe[9]	Weaning age, days	120 Day Weight	Weaning condition[10]	Weaning type[1]	Disposition[11] or remarks

[1] Trueness to breed appearance and desired mutton conformation:
"1" Excellent; "2" Good; "3" Medium; "4" Fair; "5" Poor.
[2] Straightness, strength, and spring of rib; width.
"1" Excellent; "2" Good; "3" Medium; "4" Fair; "5" Poor.
[3] Width and levelness: 1-2-3-4-5 as above.
[4] Plumpness of thigh: 1-2-3-4-5 as above.
[5] Including overshot or undershot jaw, scurs, black fiber, etc.
[6] "1" Not covered beyond poll; "2" Covered to eyes; "3" Covered slightly below eyes, but open faced; "4" Covered partially below eyes, but not subject to wool blindness; "5" Face covered and subject to wool blindness.
[7] S—Single; T—Twin; Tr—Triplet.
[8] S—Single; T—Twin; Tr—Triplet; Gr—Grafted on foster mother and give her number.
[9] Good, medium, poor.
[10] Condition or degree of fatness:
"1" Excellent; "2" Good; "3" Medium; "4" Fair; "5" Poor.
[11] Cause of death, reason for disposal, kept for breeding purposes, whom sold to.

Fig. 3-18. Individual Ewe or Ram Record Form. (See Fig. 3-18a for reverse side of record form.)

WEIGHT RECORD OF EWE OR RAM

Date	Age	Weight	Condition[1]	Remarks[2]

REMARKS

(For example: bad udder, poor mother, aborted, veterinary treatment and nature of ailment.)

Date:	Remarks:

FLEECE

(Use one line for each year)

Length Side[3]	Fineness[4]			Date of Shearing	Days Growth	Grease Weight	Per cent of Yield	Clean Wt.	Color of Skin	Purity[5]	Remarks About Fleece
	Shoulders	Side	Thigh								

1. Condition or degree of fatness: "1" Excellent; "2" Good; "3" Medium; "4" Fair, "5" Poor.
2. Factors affecting weight: e.g., just shorn, soon lamb, etc.
3. Length of staple, middle at side, to nearest 0.2 cm, just before shearing.
4. Numerical grade as determined by USDA samples, just before shearing.
5. Kemp, black fibers, etc.

Fig. 3-18a. Individual Ewe or Ram Record Form (reverse side of Fig. 3-18).

TABLE 3-13

ADJUSTMENT FACTORS[1]

	Age of Dam		
	1 Year Old	2 Years Old or Over 6 Years Old	3 to 6 Years Old
Ewe Lamb:			
Single	1.22	1.09	1.00
Twin - raised as twin	1.33	1.20	1.11
Twin - raised as single	1.28	1.14	1.05
Triplet - raised as triplet	1.46	1.33	1.22
Triplet - raised as twin	1.42	1.28	1.17
Triplet - raised as single	1.36	1.21	1.11
Wether:			
Single	1.19	1.06	.97
Twin - raised as twin	1.30	1.17	1.08
Twin - raised as single	1.25	1.11	1.02
Triplet - raised as triplet	1.43	1.30	1.19
Triplet - raised as twin	1.39	1.25	1.14
Triplet - raised as single	1.33	1.18	1.08
Ram lamb:			
Single	1.11	.98	.89
Twin - raised as twin	1.22	1.09	1.00
Twin - raised as single	1.17	1.03	.94
Triplet - raised as triplet	1.35	1.22	1.11
Triplet - raised as twin	1.31	1.17	1.06
Triplet - raised as single	1.25	1.10	1.00

Multiply 90-, 120- or 140-day weight by the appropriate factor.

Example: To find the adjusted 120-day weight of a twin-born and reared ram lamb from a 2-year-old ewe that weighed 90 lb at 110 days of age, make the following calculations:

$$\frac{90 \text{ lb}}{110 \text{ days of age}} = .82 \text{ lb} \times 120 = 98 \text{ lb} \times 1.09 \text{ (adjustment factor)} = 107 \text{ lb}$$

The adjusted 120-day weight of the lamb would be 107 lb.

[1]Scott, G., *The Sheepman's Production Handbook*, 2nd ed., Sheep Industry Development Program, May 1975, p. 27.

Step 5: Cull—Cull the lambs that fail to measure up in the preweaning performance test.

POSTWEANING LAMB PERFORMANCE TEST

The postweaning lamb performance test may be conducted either on the farm or at a central test station, as follows:

Step 1: Postweaning feed test—Determine the growth rate by placing weaned lambs on a uniform feeding test for approximately 90 days. Make selection on the basis of growth rate during the feeding test.

Step 2: Cull—Cull the lambs that were poor gainers in the postweaning lamb performance test.

EWE PRODUCTION TEST

Reproductive traits in ewes should be measured as part of the on-the-farm performance record scheme. For this purpose, the record form shown in Figs. 3-18 and 3-18a will suffice for most herds and flocks.

The following ewe performance test program is recommended for use in all purebred flocks:

Step 1: Record data—*Minimum data* should include (a) ewe number; (b) sire number; (c) dam number; (d) age of dam in years; (e) birth date of ewe (f) type of birth of ewe—single, twin, triplet; (g) how reared—single, twin, triplet, artificially; (h) number of lambs born; (i) number of lambs weaned; and (j) adjusted weight of lambs weaned.

Step 2: Type score yearlings—At the yearling stage, type score and cull rigidly (for type score as a yearling is 40% heritable).

Step 3: Evaluate fleece—Record the shearing date, grease fleece weight, staple length to nearest 1/1 inch, and fleece grade.

Step 4: Compute productivity for each ewe in the flock as—

a. Lamb productivity per ewe, in total adjusted weight of lambs produced. It is intended that this should put strong emphasis on twinning.

b. *Ewe combined productivity score,* by combining both lamb and wool as follows:

Ewe combined productivity score = Total pounds of lamb produced (using adjusted weights) + 3 times the 12-month wool total

c. *Ewe productivity weight ratio or index,* to show the performance of an individual in relation to the average of all animals of the same group. Thus, the average productivity (total lamb and/or wool production) of the flock would be considered 100 percent. Individual ewes can then be rated based on their production above or below the flock average. Thus, a ewe producing 20 percent more lamb and/or wool than the average would have a productivity index of 120.

PROGENY TESTING RAMS

Progeny testing of rams can be very reliable carefully planned and executed. For a valid progeny test, ewes must be assigned at random to the rams being tested, and sufficient ewes must be allotted to each ram to allow an accurate test. Progeny testing for rate of gain and carcass merit, for example, requires minimum of 10 ewes per ram.

Progeny tests can be slow and expensive. Thus, rams are tested as yearlings, they will be two and half years old when the test is complete. Of course testing ram lambs would speed up the process. Fortunately, research shows that selecting sires on the basis of their own growth rate (selecting them on the basis of their own performance test) will result in more than half as much economic gain as selecting them on the basis of progeny test. It would appear, therefore, that the added gain from progeny testing, in comparison with performance testing, is not sufficient to justify progeny testing of most rams that are to be used in natural service. If artificial insemination with froze semen becomes practicable in sheep, progeny testing will become more important.

The following procedure is recommended for progeny testing rams:

Step 1: Record individual data—Record the following for each ram being progeny tested: (a) ram number and breed; (b) sire number; (c) dam number; (d) age of dam in years; (e) birth date of ram; (f) type of birth—single, twin, triplet; (g) how reared—single, twin, triplet; (h) lambs born from dam per ewe year; and (i) adjusted 90-, 120-, or 140-day weight of lambs from dam per ewe year (weight per day of age to 200 days of age, or to 12 to 16 months of age, is desirable; but, of course, this won't be available where ram lambs are being tested).

Step 2: Postweaning feed test—Wean prospective stud rams at 60 to 90 days of age and place them on uniform feed test for 90 days. The heritability of growth rate is increased as maternal influence is decreased; hence, selection for growth should be based on the growth rate during the postweaning feeding test.

Step 3: Select top ram lambs—Select ram lambs that have excelled in both preweaning and postweaning performance.

If further testing—progeny testing—is not to be made, select future sires with high postweaning growth rate, that are well muscled and have a minimum of fat, and that are out of ewes producing multiple births with maximum growth rate.

Where elite stud rams are desired, they should be progeny tested—proceed to steps 4, 5, and 6.

Step 4: Mate to randomly chosen ewes—Mate each ram lamb to a minimum group of 10 randomly chosen ewes.

Step 5: Test lambs—Wean the lambs early (60 to 90 days), feed on uniform test for approximately 90 days, and slaughter.

Step 6: Use best rams as yearlings—The progeny test for gain and carcass merit can be computed in time to select and use the top progeny tested rams in their yearling breeding season.

Step 7: Evaluate fleece—Record the shearing date, grease fleece weight, staple length to nearest 1/10 inch, and fleece grade.

Step 8: Type score—At the yearling stage, type score and cull rigidly (for type score as a yearling is 40% heritable).

Step 9: Cull—Cull those rams that fail to measure up in the progeny test.

PRODUCTION TESTING BY SHEEP REGISTRY ASSOCIATIONS

Some of the sheep registry associations have evolved with excellent production testing programs for their respective breeds and members. Literature pertaining to these may be obtained by writing directly to the breed registry association of the breed of special interest. Among such production testing programs are the following:

1. The "Ram Certification Program" of the American Cheviot Sheep Society.
2. The "Production Testing" and "Ram Certification" programs of the American Hampshire Sheep Association.
3. The "Certified Rams" and "Registry of Merit Rams" of the American Rambouillet Sheep Breeders Association.
4. The "Performance Registry" and "Ram Certification" programs of the American Shropshire Registry Association, Inc.
5. The "Performance Record" of the Montadale Sheep Breeders' Association, Inc.
6. The "Production Registry" and "Ram Certification" programs of the National Suffolk Sheep Association.

PRODUCTION TESTING COMMERCIAL SHEEP

Commercial sheep breeders should select and use production tested rams. Also, they should use records for selecting ewe lamb replacements and for culling low producing ewes.

Production Testing Swine

The effectiveness of swine selection can be increased, provided it is based upon carefully taken records rather than upon casual observation. Naturally, it would be illogical to expect upstanding, narrow-bodied, shallow sows and boars to beget meaty barrows that would be market toppers. Breeding animals of acceptable meat type can only transmit these qualities unfailingly to all their offspring when they themselves have been rendered relatively homozygous or pure for the necessary genes—a process that can be gradually accomplished through judgment by the eye method, but which can be made more rapid and certain through securing and intelligently using production records.

ECONOMICALLY IMPORTANT TRAITS AND THEIR HERITABILITY

That swine show variations in economically important traits is generally recognized. The problem is to measure these differences from the standpoint of discovering the most desirable genes and then increasing their concentration and, at the same time, to purge the herd of the less desirable characters.

Production testing begins with the birth of the litter. Females should be selected and culled continually during the growing period. Structurally sound, fast-growing gilts with reasonable fat cover should be chosen as potential breeding animals. At about 180 to 200 pounds, these gilts should be weighed and probed for backfat thickness, with both values adjusted to a 220-pound basis. If desired, backfat thickness may be determined by use of the lean meter or ultrasonic equipment. Details relative to the 3 common methods of making backfat thickness determinations are given in a later section headed, "Meatiness; Measuring Backfat."

Except for backfat thickness, carcass traits can be measured and appraised only after slaughter. So, after selecting the replacement gilts, a representative group of the remaining animals not used for breeding purposes should be slaughtered, with carcass measurements made. The latter data should be considered when making final decisions in selecting herd replacements.

Table 3-14 lists the economically important traits in swine and gives their estimated heritability.

RECORD FORMS

A prerequisite for any swine production test data is that each animal be positively identified—by means of ear notches. For purebred breeders, who must use a system of animal identification anyway, this does not constitute an additional detail. But the taking of weights and grades does require additional time and labor—an expenditure which is highly worthwhile, however.

In order not to be burdensome, the record form should be relatively simple. Figs. 3-19 and 3-20 are recommended forms (pages 159 and 160).

TABLE 3-14

ECONOMICALLY IMPORTANT TRAITS IN SWINE AND THEIR HERITABILITY[1]

Economically Important Characters	Approximate Heritability of Characters[2]	Comments
	(%)	
1. Litter size at birth	15	On the average, a sow will have consumed a total of ¾ to 1 ton of feed during the period between breeding and the date her litter is weaned. Thus, if this quantity of feed must be charged against a litter of 4 or 5 pigs, the chance of eventual profit is small.
2. Litter size at weaning	12	Although greatly influenced by herdsmanship, litter survival to weaning is a measure of the mothering ability of the sow.
3. Birth weight of pigs	5	Very light pigs usually lack vigor.
4. Litter weight at weaning	17	Weaning weight is important, for it has been shown that the pigs that are heaviest at weaning time reach market weight more quickly. The low heritability of this factor indicates that it is largely a function of the nursing ability of the sow rather than genetic.
5. Daily rate of gain from weaning to marketing	30	Daily rate of gain from weaning to marketing is important because (1) it is highly correlated with efficiency of gain, and (2) it makes for a shorter time in reaching market weight and condition, thus effecting a saving in labor, making for less exposure to risk and disease, and allowing for a more rapid turnover in capital. Rate of gain and lardiness may be correlated to some degree. Thus, one should not let this be the only factor upon which selection is based.
6. Efficiency of feed utilization	30	Where convenient, accurate litter feed records should be kept, for the most profitable animals generally require less feed to make 100 lb of gain.
7. Conformation score	29	This heritability figure is likely to be considerably higher in a herd of low quality.
8. Carcass characteristics:		
a. Length	60	Carcass length is perhaps the most highly hereditary trait in hogs. This accounts for the rapid shifts that frequently have been observed; for example, in changing from chuffy to rangy hogs.
b. Backfat thickness	50	The probe, lean meter, or ultrasonic equipment can be used to measure backfat thickness on prospective breeding animals.
c. Loin lean area	50	Loin area is an indication of muscling or red meat.
d. Percent ham, based on carcass weight	58	Ham is a high-priced cut; hence, the aim is to get as large a ham as possible.
e. Percent lean cuts, based on carcass weight	50	A high yield of lean cuts means trimarble fat and more edible meat.

[1]These heritability estimates apply to within herd and within breed variations. Variations between breeds are much higher in heritability than the variations within.

[2]The rest is due to environment. The heritability figures given herein are averages based on large numbers; thus, some variation from these may be expected in individual herds.

MEATINESS; MEASURING BACKFAT

The importance of backfat as a measure of meatiness becomes apparent when it is realized that each additional 0.1 inch of fat in a 140-lb carcass results in a 1.5 percent (2 lb) increase in fat trim and a decrease of about 5 percent (7 lb) in percent lean cuts. This is reflected in the marketplace. Also, it requires less

SWINE

Individual Sow Record

Breed ______________________ Name and registration no. ______________________

Date farrowed ______________________ Identification ______________________
(ear notch, tattoo)

Bred by ______________________
(Name and address)

Sow's pedigree: ______________________ { ______________________ / ______________________
(Sire)

______________________ { ______________________ / ______________________
(Dam)

Record of litter of which the sow was a member:

No. in litter ______________________ No. of pigs weaned ______________________

Weaning wt. at ______________________ days of age:
(fill in)

Her own wt. ______________________ Av. wt. of litter ______________________

Litter mate carcass record, if any:

No. carcasses ____________; av. back fat ____________; loin eye ____________; length ____________.
(in.) (sq in.) (in.)

Number of teats ______________________

Production Record of Sow

	1	2	3	4	5	6	7	8
Litter no.								
Sire								
No. services								
Farrowing data:								
Date								
Temperament of sow: (Gentle; nervous; cross)								
No. pigs born: Alive								
Dead								
Mummies								
Total								
Av. birth weight								
No. functioning teats								
Weaning data: Age								
No. weaned								
Av. weaning wt.								
Offspring saved for breeding: No. gilts								
No. boars								

DISPOSAL OF SOW

Date ______________________ Reasons ______________________

Sold to ______________________
(Name and address)

Price $ ______________________

Fig. 3-19. Individual Sow Record Form.

SWINE

Litter Record

Breed____________________ Litter No.____________________
(notch, tattoo)

Data on Dam:

Pedigree:____________________ { __________ (Sire) __________ (Dam)
(name, reg. no., and ear notch)

Birth date____________________
(date and year)

Litter mate carcass data, if any:

No. carcasses__________; Av. back fat__________ (in.); loin eye__________ (sq in.); length__________ (in.).

Sow's__________ Litter.
(1st, 2nd, etc.)

Data on Sire:

Pedigree:____________________ { __________ (Sire) __________ (Dam)
(name, reg. no., and ear notch)

Birth date____________________

Litter mate carcass data, if any:

No. carcasses__________; Av. back fat__________ (in.); loin eye__________ (sq in.); length__________ (in.).

Date of birth____________________

No. pigs born:

Alive____________________

Dead____________________

Mummies____________________

Total____________________

No. pigs weaned__________

Health Services:

Date cholera vaccinated____________________

Date erysipelas vaccinated____________________

Date wormed____________________

Other, including iron pills or shots (list)__________

____

____

Individual Pig Record

Pig's No.	Sex	No. Teats	Birth Wt.	Off Color Markings	Defects & Abnormalities	Weaning Wt. ____days (fill in)	Date Castrated	Date & Cause of Death	Disposal Date & To Whom	Remarks

Fig. 3-20. Litter Record Form.

feed to produce a pound of lean gain than a pound of fat gain.

Thickness of backfat, which has a heritability of 50 percent, has long been recognized as an important measure of meatiness in hogs. For many years, visual appraisal was the only method of estimating backfat thickness on live animals. However, even the most skilled are oftentimes wrong in their visual measurements.

Today, three mechanical methods are available and may be used by producers in determining backfat on live hogs; namely, the probe, the lean meter, and the sonoray. Each of these methods requires hog restraint. The probe and the lean meter may be used with the hog restrained by a snare, but for the sonoray the animal must be restrained with all four feet off the ground.

The probe and the lean meter were developed for the purpose of obtaining objective measures of backfat. The sonoray is used to determine loin eye area as well as backfat.

• **Probing**—Fig. 3-21 shows the probing sites on the live hog. These three locations correspond to the three locations where backfat determinations are made on the carcass.

The only equipment needed for probing is a

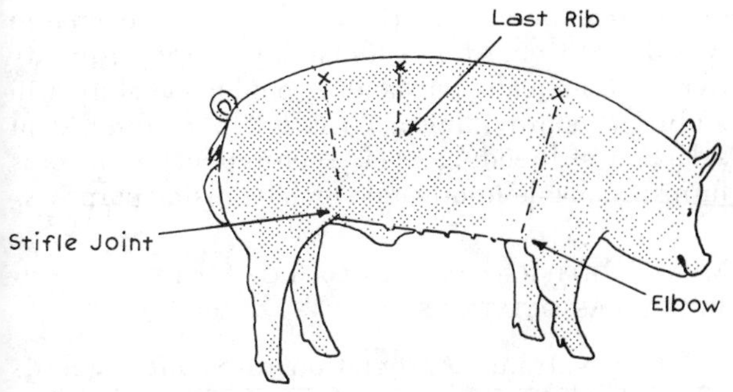

Fig. 3-21. Probe each hog at three locations: (1) midpoint of shoulder above elbow; (2) middle of back where last rib joins the vertebrae; and (3) rump, straight above the stifle joint.

snare to restrain the hog, a sharp knife or scalpel blade, and a narrow 6-inch metal ruler with 1/10-inch gradations. The steps and technique in probing are:

1. Wrap the knife or scalpel with tape about ⅜ inch from the tip (in order to keep the blade from going too deep).
2. Restrain the hog with a nose snare.
3. Jab the knife through the skin at a right angle to the hog's body.
4. Insert the probe in the cut and slant it so that it points toward the center of the hog's body.
5. Force the probe through the fat down to the loin muscle. When the probe reaches the loin muscle, a firm resistance will be noted.
6. Push the clip on the probe down to the skin line. Remove the ruler and read the measurement.

The backfat thickness should be adjusted to a 220-pound basis (see Table 3-16).

Where pinpoint accuracy is not considered essential, measuring backfat at a single probe site is suggested. The recommended site where one probe only is taken is the seventh rib, located at a distance approximately four fingers wide behind the shoulder-probing location and one inch off the midline of the back.

• **Lean meter**—This tool is more sophisticated than the probe, but also more expensive. The method is based on the difference in electrical conductivity of fat and muscle; fat is a relatively poor conductor, whereas muscle and blood are good conductors. The locations for fat determinations with the lean meter are the same as for the probe (see Fig. 3-21); and, like the probe, fat depth is read in tenths of inches.

• **Sonoray (ultrasonic)**—The sonoray machine employs the "pulse echo" technique. Basically, this is the generation of very short bursts of high-frequency (nondestructive, inaudible) sound into the animal, detecting the reflection of the pulses, and measuring the elapsed time between introduction of the sound pulse into the animal and the return of the reflected pulse. When the machine is properly calibrated, the fat depth can be read directly from the scale.

The sonoray backfat reading is made at the last rib, usually at the midline of the back. However, the reading may also be made 2 inches from the midline.

The sonoray may also be used to estimate loin eye area at the last rib.

Any one of these three mechanical methods for determining backfat thickness of the live animal is a valuable adjunct to visual appraisal and scales in the selection of meaty-type breeding animals and the production of higher quality pork carcasses. Hence, backfat measurements should be used in the selection program of both purebreds and commercial producers.

CENTRAL TESTING STATIONS

Several states and local associations operate central swine testing stations, to assist producers in evaluating their breeding stock and to improve the performance and quality of market hogs. These evaluation stations differ somewhat. But the following rules and regulations of the Oklahoma Swine Evaluation Station are rather typical:

1. **Entries**—An entry shall consist of three purebred boars—eligible for registration, which are the progeny of one sire.
2. **Weight of pigs**—The minimum weight for acceptance is 32 pounds; the maximum weight is 1.1 pounds per day of age.
3. **Size litter**—All pigs must be from litters of at least eight live pigs farrowed.
4. **Teats**—Each pig entered must have a minimum of 12 teats, 6 on each side.
5. **Health inspection**—All pigs must pass a health inspection by a veterinarian upon arrival.
6. **Testing procedure**—The pen of three pigs will be tested as follows:
   a. **Start of test**—The test will start when the largest pig in the pen weighs 80 pounds.
   b. **Individual performance**—Individual performance data will be obtained as pigs reach 220 pounds on a weekly basis.
7. **Ration**—An 18 percent crude protein, pelleted ration will be fed
8. **To qualify**—Each boar must meet the following qualifications:
   a. **Backfat**—Less than 1.20 inches of backfat adjusted to 220 pounds weight.
   b. **Loin eye**—A scanogram loin eye estimate of at least 4.5 square inches adjusted to 220 pounds.
   c. **Daily gain**—A minumum daily gain of 1.70 pounds.
   d. **Sound**—Be physically sound, as determined by a test station committee.
   e. **Negative to brucellosis test**—Found negative to a blood test for brucellosis.

9. **Published results**—The following performance information will be published:

a. **Days to 220 pounds**—A relatively short period to reach market weight is desired because it effects a saving in labor, makes for less exposure to risk and disease, and allows for more rapid turnover in capital.

b. **Pen feed efficiency**—The most profitable animals generally require less feed to make 100 pounds of gain.

c. **Individual boar index**—200 + 80 (ADG, or average daily gain) − 60 (BF, or backfat, probe) − 40 (FE, feed efficiency).

PRODUCTION TESTING BY SWINE RECORD ASSOCIATIONS

Several of the swine registry associations of the United States have a system of production testing known as Production Registry (PR). The rules governing these registries are very similar in the breeds and generally include the following:

1. The registries are available only to registered purebreds not possessing any outstanding faults.
2. Litters must be ear-notched at farrowing.
3. Production Registry (PR) litters consist of eight or more (some specify a minimum of eight pigs for a gilt and nine pigs for a sow) live pigs farrowed.
4. To qualify, a sow must produce a PR litter; some registry associations require that a sow produce two PR litters in order to qualify.
5. A boar becomes a "PR Boar" when he has sired a specified number of PR qualifying litters (anywhere from 5 to 15, depending on the registry association); or a specified number of PR daughters (anywhere from 2 to 10, depending on the registry association); or a combination of the two.
6. Breed associations urge testing of whole herds of sows rather than a selected few.

Thus, in addition to pedigreeing, protecting, and promoting, the swine record associations of America now recognize Production Testing. Swine Production Testing as now followed in this country is relatively simple, and, where it has been tried, the breeders are convinced of its merits. It merely involves marking the pigs in each litter—a practice followed by purebred breeders, anyway—then weighing them at a specified age and recording the weights. If the pigs in a litter do not measure up to a certain standard, they and their parents should be discarded from the breeding herd. In a purebred herd, the pigs must also measure up to the standards of the breed. Weights at 154 to 175 days of age plus records of feed consumption would be desirable, but these are more difficult to obtain.

The purpose of Production Registry in swine is to emphasize practical utility points and to enable breeders to coordinate outstanding individuality (type) with equally outstanding production ability. In addition to providing a basis for more intelligent selection of breeding stock, such production records furnish valuable information for advertising purposes.

MEAT CERTIFICATION PROGRAM OF SWINE RECORD ASSOCIATIONS

The National Association of Swine Records adopted a uniform program relative to *Certified Meat Hogs*. This program adds carcass evaluation to Production Registry.

A Certified Litter (CL) must meet the following standard in order to qualify:

1. It must first qualify for Production Registry in its own breed registry association.
2. At 175 days of age, and at a recommended weight of 190 to 240 pounds (when weighed off the truck for slaughter), two gilts or barrows from the litter must be delivered to a cooperating slaughter station. At the time of delivery, they will be individually tattooed.
3. The carcasses from these two pigs must meet the standards given in Table 3-15. These statistics must be reported officially by the slaughtering station to the registry association of the breed represented.[13]

TABLE 3-15

CARCASS STANDARDS FOR MEAT CERTIFICATION

Weight	Days	Length[1] (minimum)	Backfat Thickness (maximum)	Loin Eye (minimum)
(lb)		*(in.)*	*(in.)*	*(sq in.)*
220	175	29.50	1.50	4.50

[1]Some breeds have a minimum length of 0.5 inch less than given.

4. Standards and conversion factors are given in Table 3-16. Thus, equivalent 175-day weights may be calculated by adding 2 pounds for each day under 175 days, or by deducting 2 pounds for each day over 175 days.

TABLE 3-16

THE STANDARDS AND CONVERSION FACTORS

220-Pound Standard		Conversion Factors
Days to 220 lb	175 max.	2 lb/day
Length to 220 lb	29.50 in. min.	0.025 in./lb
Backfat at 220 lb	1.50 in. max.	0.004 in./lb
Loin area at 220 lb	4.50 sq in. min.	0.015 sq in./lb

5. The carcass measurements are obtained as follows:

[13]Some breed registry associations give additional and special recognition for meeting higher standards.

a. The loin is broken at the tenth rib, and the loin area is calculated by means of a planimeter from tracings of loin eye made on parchment paper.

b. The carcass length is calculated from the front of the first rib, where it joins the vertebra, to the front of the aitch bone.

c. The backfat is an average of three measurements taken opposite (1) the first rib, (2) the last rib, and (3) the last lumbar vertebra.

Actual backfat thickness is measured to the outside of the skin and at a right angle to the back.

A Certified Meat Sire (CMS) is one that has sired five Certified Litters; each litter of which is out of a different sow, not more than two of which are full sisters or dam and daughter.

A Certified Mating is the repeat mating of a boar and sow that have produced a Certified Litter.

Production Testing Horses

The breeders of racehorses have always followed a program of mating animals of proved performance on the track. For example, it is noteworthy that the first breed register which appeared in 1791—known as "An Introduction to the General Stud Book"—recorded the pedigrees of all the Thoroughbred horses winning important races. In a similar way, the Standardbred horse—which is an American creation—takes its name from the fact that, in its early history, animals were required to trot a mile in 2 minutes and 30 seconds, or to pace a mile in 2 minutes and 25 seconds, before they could be considered as eligible for registry. The chief aim, therefore, of the early-day breeders of racehorses was to record the pedigree of outstanding performers rather than all members of the breed.

The simplest type of progeny testing in horses consists of the average record of merit of an individual stallion's or mare's offspring. Thus, the offspring of Thoroughbred or Standardbred animals bred for racing may be tested by timing on the track. Less satisfactory tests for saddle horses and harness horses have been devised. However, it is conceivable that actual exhibiting on the tanbark in the great horse shows of the country may be an acceptable criterion for saddle- and harness-bred animals. Also, the dynamometer might conceivably be used for testing animals of draft horse breeding, although it has not been so used in the past.

HERITABILITY OF PERFORMANCE

Relatively little scientific work has been done on the heritability of performance of horses—on the genetics of working ability, racing ability, cutting ability, jumping ability, etc. As a result, the horse is the last of the farm animals to which the science of genetics has been added to the art of breeding. Nevertheless, horsemen have selected for performance. For example, the Thoroughbred horse has been selected and bred for speed and stamina for 300 years. Because more often than not the "best" horse wins, the breeding of the best to the best has resulted in improvement in the track performance of the Thoroughbred horse over the centuries.

The underlying genetic principle which determines the success of mating the best to the best is based upon the assumption that phenotypes of the best for a given trait, such as speed, are due to simple *additive*-type genes without regard to family relationships. On the other hand, when a breeder plans his matings on the basis of a nick, family or pedigree relationships receive careful consideration.

The underlying genetic principle in making an outcross is that the members of the unrelated strains of families will bring together genes which will act in a complementary fashion to produce hybrid vigor in the offspring for the traits desired.

Differences in the performance ability (working, racing, cutting, jumping) are due to two major forces—heredity and environment. Success in selecting superior breeding animals for each of these traits depends entirely upon how accurately we are able to partition the differences in performance capacity of horses into causes due to the environment and causes due to heredity.

The important environmental factors in determining the overall performance of horses are nutrition (both prenatal and postnatal), health care, quality of training, ability of the horseman (teamster, rider), and injuries.

An important genetic principle is that traits as such are not inherited. Rather, what is inherited is the ability to respond to a given set of environmental conditions in order to produce a trait with a measureable effect.

The key to continued genetic improvement in the performance of a horse, such as the racing capacity of the Thoroughbred, rests essentially on two factors: (1) the magnitude of the heritable component (additive genes) of performance (racing) capacity, and (2) the accuracy with which the breeder can identify those individuals which are truly genetically superior to their contemporaries.

Essentially, the breeding value of a horse is the fraction of the differences that will be transmitted to the progeny. The most straightforward measure of this is the heritability of the trait. It follows that an estimate of heritability of a trait is one of the most important considerations in formulating an effective program of improvement through breeding. Reliable estimates on the heritability of performance traits in horses are limited in comparison with other species. Nevertheless, further and important knowledge has been accumulated in recent years. Some heritability estimates follow:

• **Working ability**—In most countries, work horses, as distinct from light horses (sporting breeds), still make up the bulk of the population. In France, for example, only 15 percent of the horse population consists of the sporting breeds.

The main measure of the working ability in a horse is pulling power. This performance trait has been estimated to have a heritability of 26 percent.[14]

• **Racing ability**—Racing performance can be measured in different ways: by purses earned, time per unit distance, handicap weight, or *Timeform* ratings or other year-end handicaps. In a 1971 study, the Texas Agricultural Experiment Station[15] determined the racing ability of individual horses through a computer comparison of the number of lengths (one length = 8 ft) the horse would win or lose to other horses in a typical race. For this unique study, each horse was given a rating called the "Performance Rate." Theoretically, it was assumed that the average horse would have a Performance Rate of zero. Then, in an average race, a horse with a Performance Rate of +12 would, theoretically, finish 12 lengths in front of the average horse. Likewise, a horse whose Performance Rate was −12 would, theoretically, finish 12 lengths behind the average horse and 24 lengths behind one with a +12 Performance Rate.

The Texas Station study included all 3-year-olds which raced on North American tracks in 1971. It involved 6,458 fillies and 7,113 colts and geldings, which were sired by 3,228 different stallions. Statistical analysis of the data showed that racing ability is about 40% heritable. This means that, on the average, about 40% of the difference in racing superiority of one horse over another is due to differences in heredity. The remaining 60% is due to difference in environment—nutrition, state of health, and abilities of trainers and jockeys.

So, after nearly three centuries of selection for speed and stamina, it should still be possible to improve the racing performance of Thoroughbred horses through selection of superior stock for future parents.

• **Cutting ability**—Based on a study made by the Texas Station, the cutting ability of horses is less than 10 percent heritable.[16] Obviously, training is most important in determining cutting ability.

[14]Cunningham, Professor E. P., Head of Animal Breeding and Genetics, Dublin University, Ireland, "Equine Genetics," *The Blood-Horse*, Oct. 6, 1975, p. 4,210.

[15]Kieffer, Dr. N. M., Geneticist, Texas A & M University, College Station, Texas, "Heritability of Racing Ability," *The Blood-Horse*, Oct. 13, 1975, p. 4,292.

[16]Kieffer, Dr. N. M., Geneticist, Texas A & M University, College Station, Texas, *Research Report*, Texas A & M Experiment Station, Dec. 3, 1975.

• **Jumping ability**—Based on a study of steeplechase results in France, involving 3,500 progeny of 326 stallions, the heritability of jumping was estimated to be 18 percent.[17]

Although the heritability estimates of performance traits in horses reported above are disturbingly low, genes are a permanent, transmissible investment, whereas environmental factors are not. When buying horses, therefore, it is important to know whether you're buying desirable genes or superior environment.

RECORD FORMS

Figs. 3-22 (page 165) and 3-23 (page 166) show record forms that will be useful on most breeding establishments.

CROSSBREEDING

Crossbreeding is the mating of animals of different breeds. In a broad sense, crossbreeding also includes the mating of purebred sires of one breed with high-grade females of another breed.

Today, there is great interest in crossbreeding, and increased research is under way on the subject. Crossbreeding is being used by stockmen to (1) increase productivity over straightbreds because of the resulting hybrid vigor or heterosis, just as is being done by commercial corn and poultry producers; (2) produce commercial animals with a desired combination of traits not available in any one breed; and (3) produce foundation stock for developing new breeds.

The motivating forces back of increased crossbreeding in farm animals are (1) more artificial insemination, thereby simplifying the rotation of sires of different breeds; and (2) the necessity for stockmen to become more efficient in order to meet their competition, both from within their respective industries and from without.

Crossbreeding will play an increasing role in the production of market animals in the future, because it offers the several advantages discussed in the sections which follow.

Hybrid Vigor or Heterosis

Heterosis, or hybrid vigor, is the name given to the biological phenomenon which causes crossbreds to outproduce the average of their parents. For numerous traits, the performance of the cross is superior to the average of the parental breeds. This phenomenon has been well known for years and has been used in many breeding programs. The produc-

[17]Cunningham, Professor E. P., Head of Animal Breeding and Genetics, Dublin University, Ireland, "Equine Genetics," *The Blood-Horse*, Oct. 6, 1975, p. 4,210.

HORSE

INDIVIDUAL LIFETIME BROODMARE RECORD

Name of mare ______________________

Number or other identity ______________________

Birth date ______________________

Show or performance record ______________________

Temperament ______________________
(gentle, nervous, cross)

PHOTO

Bred by ______________________
(name and address)

Purchased: from ______________________
(name and address)

Date Price

Disposal: Sold to ______________________
(name and address)

Date Price

Reasons

Production Record of Mares

Year	Sire of foal	Birth date of foal	Temperament of mare at foaling (gentle, nervous, cross)	Foaling (normal, requiring assistance, ret. placenta)	Vigor foal at birth (deformities)	Sex of foal	Identity of foal	Date foal was weaned	Score of foal: suckling or weanling	Score of foal: yearling	Score of foal: 2-year-old	Score of foal: 3-year-old	Disposal of foal: Sold to: (name and address)	Disposal of foal: Date	Disposal of foal: Price	Disposal of foal: Reasons	Disposal of foal: Remarks

Fig. 3-22a. Individual Lifetime Broodmare Record. (See Fig. 3-22b. for reverse side of record form.)

HEALTH RECORD

Immunization: Encephalomyelitis	Immunization: Tetanus	Immunization: Abortion	Immunization	Immunization	Parasite treatment (what)	Other veterinary treatment	Remarks

Fig. 3-22b. Individual Lifetime Broodmare Record. (reverse side of Fig. 3-22a.)

HORSE

INDIVIDUAL YEARLY STALLION BREEDING RECORD

Name of stallion ______________________

Number or other identity ______________________

Birth date ______________________

Show or performance record ______________________

____

PHOTO

For breeding year of ______________________

For foaling year of ______________________

Total number of services ______________________

No. services/conception ______________________

MARES IN FOAL TO STALLION

Name of mare	Dates mare was bred				Date foaled	Vigor of foal at birth	Sex of foal	Disposal of foal				Remarks
								Sold to (name and address)	Date	Price	Reasons	

Fig. 3-23a. Individual Yearly Stallion Breeding Record. (See Fig. 3-23b. for reverse side of record form.)

HEALTH RECORD

Date	Immunization							Parasite treatment (what)	Semen test	Veterinary treatment	Remarks
	Encephalomyelitis	Tetanus	Other								

Fig. 3-23b. Individual Yearly Stallion Breeding Record (reverse side of Fig. 3-23a.)

tion of hybrid seed corn by developing inbred lines and then crossing them is probably the most important attempt by man to take advantage of hybrid vigor. Also, heterosis is being used extensively in commercial swine, sheep, layer, and broiler production today; an estimated 80 percent of market hogs, market lambs, and layers are crossbreds, and 95 percent of broilers are crosses.

The genetic explanation for the hybrid's extra vigor is basically the same, whether it be cattle, hogs, sheep, layers, broilers, hybrid corn, hybrid sorghum, or whatnot. Heterosis is produced by the fact that the dominant gene of a parent is usually more favorable than its recessive partner. When the genetic groups differ in the frequency of genes they have and dominance exists, then heterosis will be produced.

Heterosis is measured by the amount the crossbred offspring exceeds the average of the two parent breeds or inbred lines for a particular trait, using the following formula for any one trait:

$$\frac{\text{Crossbred average} \ (\text{minus}) \ \text{Purebred average}}{\text{Purebred average}} \times 100 = \text{Percent hybrid vigor}$$

Thus, if the average of the 2 parent populations for weaning weight of calves at 205 days of age is 400 lb and the average of their crossbred offspring is 420 lb, application of the above formula shows that the amount of heterosis is 20 lb, or 5 percent.

Traits high in heritability—like tenderness of rib eye in cattle—respond consistently to selection, but show little response in hybrid vigor. Traits low in heritability—like mothering ability, calving interval, and conception rate—usually show good response in hybrid vigor.

The level of hybrid vigor for all traits depends on the breeds crossed. The greater the genetic difference between two breeds, the greater the hybrid vigor expected. The genetic difference between a British breed (*Bos taurus*) and a breed of Indian origin (*Bos indicus*) is greater than the difference between one British breed and another British breed.

Complementary

Complementary refers to the advantage of a cross over another cross or over a purebred, resulting from the manner in which two or more characters combine or complement each other. It is a matching of breeds so that they compensate each other, the objective being to get the desirable traits of each. Thus, in a crossbreeding program, breeds that complement each other should be selected, thereby maximizing the desirable traits and minimizing the undesirable traits. Since breeds which are selected because they tend to express a maximum of some trait (e.g., high daily gain) will have some undesirable traits (e.g., large mature cow size and high maintenance cost), different breeds must be selected for different purposes. A well-known example of breed complementation for improving overall carcass desirability in the market animal is the Angus X Charolais cross, combining the higher carcass grade of the Angus with the higher cutability of the Charolais.

Introduce New Genes Quickly

Crossbreeding provides a way in which to introduce new and desired genes quickly—at a faster rate than can be achieved by selection within a breed. A practical example of this sort is the introduction of new genes for milk production in a beef herd by crossing a dairy breed with a beef breed, then selecting females from within the crossbred foundation for the future cow herd.

Get Hybrid Vigor Expressed in the Female

Except for a two-breed cross, crossbreeding offers an opportunity to have hybrid vigor expressed in breeding females. This is most important in the cow herd where it results in increased fertility, survivability of the calves, milk production, growth rate of calves, and longevity of the cows—all factors that mean more profit to the cowman.

Factors Affecting Magnitude of Advantages from Crossbreeding

Many other examples of each of the advantages of crossbreeding could be cited. It should be noted, however, that the total magnitude of the advantage of these factors—achieving the 15 to 25 percent potential immediate increase in yield per female unit through continuous crossbreeding compared to continuous straight breeding—depends upon the following:

1. **Making wide crosses**—The wider the cross, the greater the heterosis.
2. **Selecting breeds that are complementary**—A crossbreeding program should involve breeds that possess the favorable expression of traits desired in the crossbred offspring that will be produced.
3. **Using high-performing stock**—Once a crossbreeding program is initiated, further genetic improvement is primarily dependent upon the use of superior production tested males.
4. **Following a sound crossbreeding system**—For a continuous high expression of heterosis and maximum output per female, a sound system of crossbreeding must be followed. This should include the use of crossbred females, for research clearly indicates that over one-half the higher profits from a crossbreeding program results therefrom.

5. **Tapping purebreds constantly**—Purebreds must be constantly tapped to renew the vigor of crossbreds; otherwise, the vigor is dissipated.

Not All Hybrids Excel Purebreds

Stockmen can learn from poultrymen when it comes to breeding, for the breeding of chickens has passed through the total presently known systems. In fact, each method has been, and still is, used successfully.

The vast majority of chickens in America today are hybrids of one form or another—they're either strain crosses, breed crosses, or crosses between inbred lines. But they're exceptions!

Despite the fact that hybrids are widely used as commercial layers, it is noteworthy that egg-laying tests show that purebred Single Comb White Leghorns compete on even terms with hybrids under test conditions. Certainly, the hybrids are equal to the purebreds, but the point is that they do not excel them. The same would apply to purbred vs crossbred four-footed animals, *provided* the purebreds reached the pinnacle enjoyed by White Leghorns.

In broiler production, the main objective is the improvement of growth rate to eight weeks of age, although improvement in other economic factors is sought. Generally, growth rate and hybrid vigor are obtained by systematic matings that may involve crossing different breeds, different strains of the same breed, or the crossing of inbred lines. Most of the strains used as sires trace their ancestry to the broad-breasted Cornish breed. But there is some question whether heterosis, as obtained through hybrid breeding, contributes substantially to broiler weight. It is noteworthy, for example, that the best purebred New Hampshire strains generally equal the most rapid gaining crosses. The latter point is of great significance to purebred breeders of four-footed animals.

Body conformation is especially important in turkeys, because they are marketed at heavier weights than broilers and their carcass is usually left whole rather than cut up. Since conformation, size, and color of turkeys are highly heritable, they have responded well to simple methods of breeding and selection. As a result, most turkeys are bred as purebreds, rather than crossbreds. Also, it should be of more than passing interest to cattlemen to know that in turkey breeding programs (1) selections are largely based on physical appearances (phenotype), and (2) mass matings (in which a number of males are allowed to run with the entire flock of hens) are the common practice.

Crossbreeding Beef Cattle

Without a planned breeding program, crossbreeding will almost inevitably end up with (1) a motley collection of females and progeny varying in type and color, and (2) minimum benefits from hybrid vigor or heterosis.

"Where do I go from here?" This is the question that many cattlemen frequently ask, almost frantically, after having heifers of breeding age sired by exotic bulls. Others get worried when they notice that calves out of their crossbred cows aren't doing so well as the first-cross calves. Of course, what these cattlemen really want to know is how they can maintain satisfactory hybrid vigor (heterosis) in animals when a herd is on a continuous crossbreeding program. They want to know how they can maintain 15 to 25 percent greater total efficiency in the crossbreds than the average of their parents; in production rate, calf livability, growth rate, and feed conversion.

Several different systems of crossbreeding may be used. Among them are the following:

1. **Two-breed cross**—This consists of mating purebred bulls to purebred or high-grade cows of another breed. An example would be using Angus bulls on Hereford cows, to give crossbred Angus X Hereford offspring—black baldies. This system of crossing has been used with success by cattlemen for many years.

In the 2-breed cross, only the calves are crossbred—the breeding of the sires and dams remains the same. Hence, the 2-breed cross imparts hybrid vigor only in the calf. On the average, it gives about an 8 to 10 percent increase in pounds of calf weaned per cow bred, plus another 2 to 3 percent advantage in rate of gain in the feedlot. In order to follow the 2-breed cross indefinitely, the purebred females must be replaced with other purebreds sooner or later. They may either be purchased from another breeder or the breeder may want to produce his own purebred heifers within his own herd.

The two-breed cross is relatively simple. However, it has one major deficiency; it does not make use of the crossbred cow.

2. **Two-breed backcross or crisscross**—This system involves the use of bulls of breed A on cows of breed B, then backcrossing the progeny to bulls of either breed A or B. The rotation is accomplished by using bulls of the breed least related to the particular set of cows. For example, if Charolais bulls are mated to Hereford cows, the crossbred Charolais X Hereford heifers could be retained and bred to either a Charolais or a Hereford bull. If Hereford bulls were used, the calves produced would be ¼ Charolais and ¾ Hereford. Later, if the heifers of this breeding are saved, they should be bred to a Charolais bull. The 2-breed backcross results in about 67 percent of the maximum heterosis being attained in the crossbred calves. But since crossbred cows are used, overall performance should be a little better in pounds of calf weaned per cow bred than in the 2-breed cross.

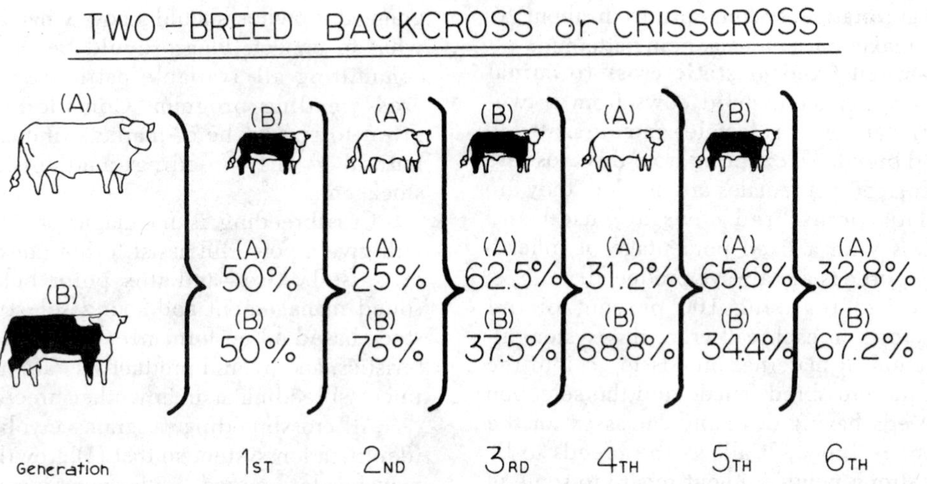

Fig. 3-24. Two-breed backcross or crisscross: Charolais bull X Hereford cow, thence female offspring bred to Hereford bull, thence female offspring bred to Charolais bull.

3. **Three-breed rotation cross**—This system calls for the selection of 3 breeds (e.g., breeds A, B, and C, which might represent Herefords, Brahmans, and Charolais), possessing the combination of maternal, growth, and carcass traits desired in the crossbred cows and the slaughter cattle produced. Crossbred females, selected for growth rate, are retained for breeding and bred to a purebred bull of one of the 3 breeds. Each new generation of crossbred females is retained for breeding and mated to a purebred bull until bulls of all 3 breeds have been used in rotation. Thus, such a system would operate as follows:

Mate the existing B cow herd continuously to bulls of breed A; select crossbred heifers for growth rate and mate them continuously to bulls of breed C; mate the selected C (AB) females to bulls of breed B. After the rotation of bulls from the 3 breeds is completed, the rotation of purebred sires begins all over again. Thus, mate the selected B X (ABC) females to bulls of breed A.

Continue the same system indefinitely, always selecting the best performing crossbred females to be mated to the breed of sire in the program to which they are least related.

In addition to the genetic advantages of this system, commercial cattlemen select their own replacements; hence, the only outside cattle purchases are production tested bulls. The major disadvantage is that after the first four years it is necessary to maintain bulls of all three breeds simultaneously (unless A.I. is used).

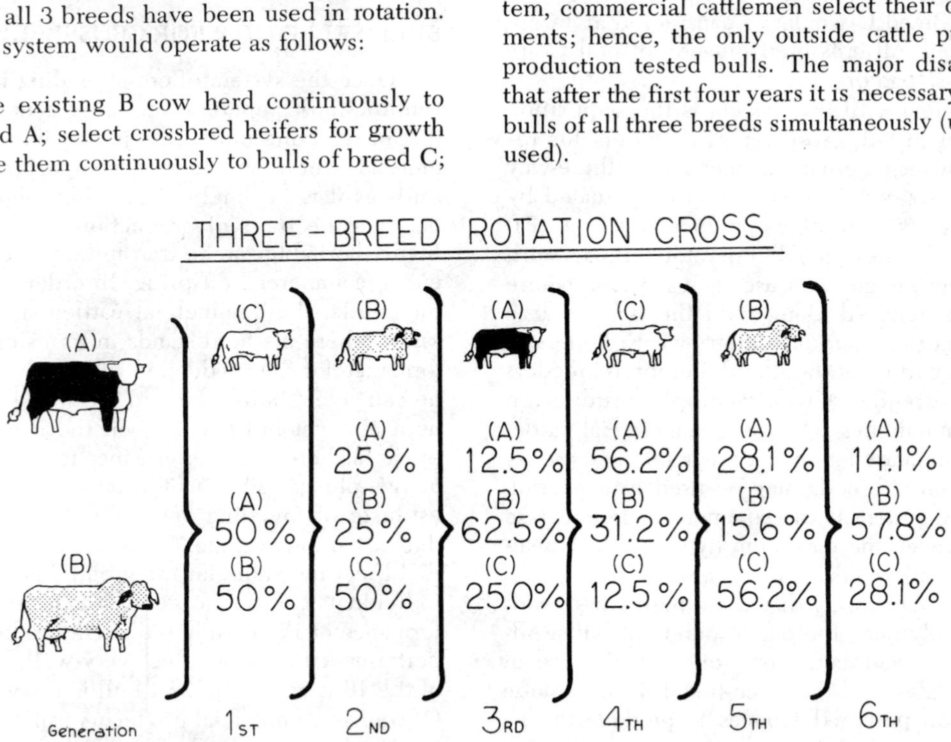

Fig. 3-25. Three-breed rotation cross; Hereford bull X Brahman female, thence female offspring bred to Charolais bull, thence female offspring bred to Brahman bull.

A three-way rotation system results in about 87 percent of the maximum heterosis being attained.

4. **Three-breed fixed or static cross (terminal cross)**—In this system, crossbred cows from a two-breed cross (F_1s) are used as females and are mated to a bull of a third breed. All offspring from this cross are sold. When replacement females are needed, they are purchased. Thus, crossbred cows are used and crossbred calves with a fixed percentage of inheritance from three breeds are always produced.

In addition to realizing 100 percent of the maximum heterosis in each calf crop, this system allows the selection of maternal breeds to go into the production of the crossbred female and the selection of growthy breeds having desirable carcasses for the terminal cross sire breed. It allows the breeds to be used for their strong points without regard to some of their weaker points. A breeder can tailor-make the crossbred market animal, putting together in one animal desirable traits of several breeds. Such specification is not possible in the rotational system because all breeds contribute to maternal performance and calf performance.

The mechanics of this system consist in selecting three breeds for crossbreeding—two breeds (A and B) that will produce crossbred cows with outstanding maternal characteristics for fertility, milking ability, mothering ability, and adaptation. Select a third breed (breed C) with rapid, efficient, postweaning muscle growth rate. Breed C would be considered a "terminal" sire breed. All crossbred progeny of bull C are marketed for slaughter.

The problem with this system is the acquisition of production tested, crossbred (F_1) heifers for replacements in such a program, since all the three-way crosses are marketed. The system is perpetuated by having specialized multipliers produce crossbred (F_1) replacement females. Small operators (those with under 100 cows) might well use such a system where heifers are purchased along with the bulls. Large operators might well produce their own F_1 heifers in a specialized portion of their herd. Purebred breeders (seed stock breeders) would supply production selected terminal sires which the commercial cattleman would purchase for such a program.

Four or more breeds may be used in a rotation crossbreeding system if the commercial producer so desires. However, the maximum hybrid vigor is usually realized with the three-breed cross.

Also, it is noteworthy that all of these crossbreeding systems rely upon the use of purebred bulls. Additionally, the two-breed cross relies on the use of purebred females, and the three-breed fixed or static cross relies on purebred females to produce the F_1 heifers necessary for the program.

Before going into a long-range crossbreeding program, the owner should know what is involved and what to expect. Plans should be developed before committing all available cattle and resources to a crossbreeding program. Consideration should be given to size of herd, markets, number of pastures, natural vs A.I. breeding, availability of breeding stock, etc.

Crossbreeding is no magic or "cure-all," but it will give a powerful assist to the pocketbook if properly used. Also—and this point bears emphasis—sound management and sound selection of breeding stock based on performance, potential carcass characteristics, and overall productivity are just as important in crossbreeding as in any other breeding program.

All crossbreeding programs involve some animal identification system so that (1) growth rate of heifers may be determined, with selection of replacements made on this basis; and (2) where more than one breed is involved, the cow herd can be sorted for assignment to specific sire breeds for mating. Unless A.I. is used, it is necessary to maintain bulls of whatever breeds are involved, along with separate breeding pastures for each sire breed. Also, it should be recognized that where bulls of two or three different breeds are used, the crossbred slaughter progeny will vary considerably in performance and carcass traits, because they will be sired by bulls of different breeds and be produced by cows of divergent breed backgrounds.

BULL SELECTION FOR CROSSBREEDING

Once the system of crossbreeding has been chosen, the most important recurring genetic decision is that of selecting bulls. It is just as important in commercial production to select superior performance bulls as it is in purebred breeding herds. This is so because the traits of production and product being highly heritable can be transmitted directly from parent to commercial offspring. In order intelligently to buy bulls, the commercial cattleman must at least know where his herd stands in terms of average performance for the production and product traits. Then, he can select bulls strong in the weaker points of his herd and get bulls which, on the average, will improve his herd. The performance test of bulls to a year of age, plus possibly a sib carcass test, is an adequate estimate of breeding value for the commercial producer. On the average, the performance test will predict breeding value for the group, but for a particular individual it is not so good. Thus, the average performance of 10 yearling bulls will predict the average performance of their calves very well, but just which of the 10 is really the best bull is not well estimated. Of course, commercial producers using A.I. can afford to use progeny tested bulls available to them.

The selection of breeding herds from which to

select commercial bulls is made difficult since only a fraction of herd differences are genetic—that is, the genetic differences between herds is probably less than 40 percent. On the other hand, research evidence suggests that for the production and product traits, the heritability of within herd differences among animals treated alike is between 40 and 50 percent. Thus, as long as the better herds are being patronized, selection of the best performing bulls within a herd is a safer bet than spending too much time deciding on which herd from which to buy bulls.

CROSSBRED BULLS

Cattlemen are sometimes tempted to use hybrid bulls as herd sires, particularly hybrids sired by high-priced exotics. This inclination is understandable, since the performance of hybrid bulls, as individuals, is often superior to either parent breed. But, before putting a hybrid bull with cows, some cautions are in order.

Unlike improved production resulting from selection, advantages from hybrid vigor are not transmitted from parent to progeny. Hence, to the extent to which superior performance is due to hybrid vigor, a hybrid bull will not breed true. If the hybrid bull is out of purebreds which have been selected for superior performance, this portion of his inherited superiority may be transmitted to his progeny. However, individual performance of hybrids is a less accurate indicator of their breeding value than is the performance of purebreds.

Work at the Experiment Station, Miles City, Montana, indicates that there may be some gain from the use of crossbred bulls over straightbreds, in increased fertility, vigor, and livability of the calves. But the disadvantages of using a crossbred bull outweigh the advantages. Among the disadvantages are the following:

1. **Hybrid vigor may mask the true breeding worth of a hybrid bull**—The sire's effect on profitability is basically indirect through the performance of his progeny. Thus, it is important that sires be accurately selected for the characters desired on their progeny. Hybrid vigor is not transmitted; hence, it may mask the true breeding worth of a hybrid bull. In other words, selection of purebreds is expected to be more effective.

2. **The progeny of hybrid sires tend to be more variable**—Since hybrid sires are less prepotent, their progeny will tend to be more variable in all measures of performance. Also, variation in color and conformation will tend to be more evident. As a result of this lack of uniformity, the market price of their progeny will be lower, especially when they are sold as feeders.

3. **Crossbred bulls have less effect on performance than crossbred cows**—Cows affect offspring through milk production and mothering ability; hence, they affect the performance of offspring more than the sire.

4. **The likelihood that crossbred bulls will be produced by breeders who haven't the best cattle**—A major problem in considering crossbred bulls is that there is a strong likelihood that they will be produced by breeders who haven't the best cattle from which to produce bulls.

Despite the above, there are special situations in which crossbred bulls may be considered. Two of these circumstances are:

1. **For the creation of a new breed**—The use of crossbred bulls is necessary in the creation of new breeds especially adapted to certain conditions. For example, the Santa Gertrudis breed of cattle, a breed derived from ⅝ Shorthorn and ⅜ Brahman, was developed to meet a need in the hot, dry, insect infested area of the Southwest. Experienced cattlemen of the area will vouch for the fact that this is a practical example of a planned system of crossbreeding which has utility value under the environmental conditions common to the country. Still other examples of crossbreeding in the creation of breeds may be cited, including the breeding up to one of the new exotics in which purebred sires may be too expensive or scarce.

2. **For coping with harsh environmental conditions**—Under certain conditions, it may be desirable to incorporate hybrid vigor and adaptability in the bull, as well as in the brood cow, in order to cope with harsh environmental conditions. For this reason, Gulf Coast cattlemen often prefer bulls that have ⅛ to ¼ Brahman breeding.

From the above, it may be concluded that, except for special circumstances, crossbred bulls should not be used. They cannot be counted on to be herd improvers like comparable purebreds.

BUFFALO X CATTLE HYBRIDS[18]

From time to time, American buffalo (*Bos bison*) and domestic beef cattle (*Bos taurus*) have been crossed, in Canada and the United States. Out of such crosses have evolved cattalo (cattle of less than ½ bison parentage), beefalo (⅜ buffalo, ⅜ Charolais, and ¼ Hereford), and the American Breed (⅛ buffalo, ½ Brahman, ¼ Charolais, 1/16 Durham, and 1/16 Hereford). These breeds are variously extolled, on the basis of their adaptability to cold, snowy climates; ability to thrive on weeds, shrubs, and other vegeta-

[18]Buffalo breeders are banded together in the National Buffalo Assn., Box 995, Pierre, S.D. 57501.

tion which domestic cattle pass up; small birth weights (straight buffalo calves weigh only about 25 lb at birth); and leaner and more flavorful meat.

Pertinent information relative to the reproductive ability of the American buffalo (*Bos bison*) X domestic cattle (*Bos taurus*) hybrids follows:

1. Bison and domestic cattle interbreed.
2. Fewer maternal calving losses occur when domestic bulls are used on bison cows, although the reciprocal mating may be made.
3. Half-buffalo bull calves (F_1 hybrids) show normal sexual behavior, but they are always sterile. The scrotum is held close to the body cavity, as in the bison.
4. The half-buffalo heifers (F_1 hybrids) are fertile.
5. A few backcross bull hybrids have produced semen containing some sperm.
6. Reproductive ability improves in both sexes of further generations as the percentage of domestic blood increases.

It is possible that animals carrying a small percentage of buffalo breeding may have a place under certain conditions. However, more scientific research on the subject is needed.

Crossbreeding Dairy Cattle

The crossing of dairy breeds produces hybrid vigor or heterosis, just as it does in meat animals. It results in more live calves, improved livability, and increased early growth. Yet, crossbreeding for heterosis has been little used in dairy cattle breeding programs in the United States. This prompts the question: Why has crossbreeding been more widely used in beef cattle, sheep, and swine than in dairy cattle? The answer: There is one great difference between a dairy cow and females of meat species. A dairy cow is both a breeding animal and a producing animal—she must produce both calves and milk. In meat animals, however, it is practical to keep females primarily for breeding purposes—to produce feeder calves, feeder lambs, or feeder pigs, Thus, it is expected that most U.S. dairymen will continue to rely upon selection based on genetic variation to make genetic improvement in their herds.

The limitation presented above does not preclude the use of crossbreeding to create new dairy breeds. Thus, dairy bulls are being crossed on "native" breeds to create new breeds better adapted to the tropics. This special adaptability to environment is not due primarily to heterosis. Rather, any heterotic effects are incidental to the primary purpose of this type of crossbreeding—to utilize the additive variation present in the parent breeds. Cows that are well adapted and that produce more milk can be developed for many areas of the tropics by crossing dairy bulls of the European breeds on the native cows.

Crossbreeding Sheep

Although the common systems used in breeding sheep are not unlike those applied to other classes of farm animals, there appears to be more crossbreeding among sheep because of (1) the fact that sheep are called upon to produce two products, lamb and wool; (2) the many diverse conditions under which they are produced; and (3) the conviction on the part of many sheepmen that the hybrid vigor of crossbreeding accounts for increased vigor and livability in the lamb crop. Crossbreeding, therefore, is extensively followed in commercial sheep production on the western range. The ewe bands are predominantly of Rambouillet extraction; whereas, for market lamb production, Suffolk or Hampshire rams are generally used. The Rambouillet ewe bands are desired because of their (1) gregarious or flocking instinct, (2) great hardiness, and (3) superior shearing qualities. On the other hand, lambs of this breeding are not so desirable for market lambs. Thus, mutton-type rams are used in order to get large, fast growing lambs that will attain a good market finish on milk and range vegetation or that can be readily sold to go into feedlots for further finishing. As black-faced crossbred lambs of this type are not suitable as flock replacements, both ewe and wether lambs are marketed. Replacement females are obtained by (1) outright purchase from a sheepman who has used white-faced rams (Rambouillets, Columbias, Targhees, or Panamas) for purposes of raising animals for sale as replacements; (2) using white-faced rams on the band every third year and retaining the ewe lambs (some sheepmen with several bands simply use certain bands for producing lambs for replacement purposes); or (3) using both white-faced and black-faced rams simultaneously on the same ewe band. In the last type of program, the better white-faced ewe lambs—which are easily recognized as the offspring of the white-faced rams—are selected out for breeding purposes.

As can be readily surmised, crossbreeding in sheep does make for a considerable problem from the standpoint of producing or purchasing replacement animals. Also, it often makes the ram problem a difficult one. This practice, however, was born of necessity, there being few or no existing breeds or types possessing all the desirable features needed. In recent years, considerable effort has been made toward developing breeds of sheep better adapted to the needs, with the hope of alleviating the necessity of crossbreeding. The Columbia, Targhee, and Panama breeds evolved out of this need.

In addition to the crossbreeding common to the western range, most hothouse lambs are produced through using this system of breeding. Usually grade Merino or Dorset ewes are topped with a Southdown ram. Ewes of this extraction will breed out of season, and they are excellent milkers; and Southdown rams impart to their progeny the ultimate in early maturity and mutton type. Crossbreeding has also gained in popularity in Kentucky where crossbred Hampshire-Rambouillet ewes are frequently bred to Southdown rams for the production of grass-fat lambs.

Studies by the U.S. Department of Agriculture show that purebred ewes crossed with purebred rams of another breed raised 2 more lambs per 100 ewes than purebred ewes bred to rams of the same breed. Also, the lambs averaged 6 pounds heavier at weaning.

From the above cross, the first cross ewe lambs bred to purebred rams of a third breed raised 14 more lambs per 100 ewes that were 10 pounds heavier at weaning than those of the purebred breeds in the cross.

The ewe lambs from the above (containing blood of 3 breeds) crossed with a purebred ram of a fourth breed raised 27 more lambs per 100 ewes that were 7 pounds heavier at weaning than those of the purebred breeds in the cross.

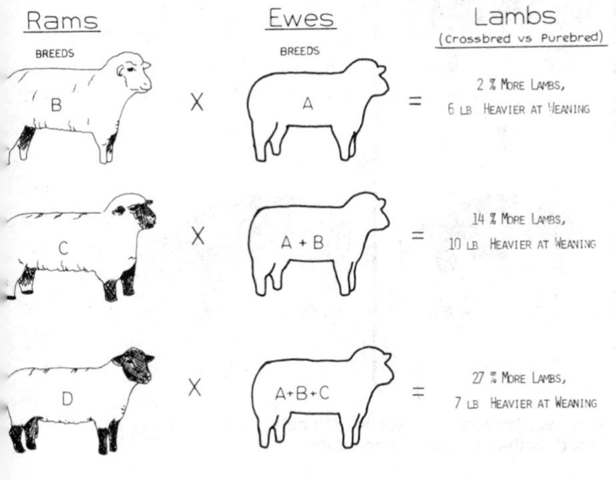

Fig. 3-26. What the USDA sheep crossbreeding studies showed.

In addition to more and heavier lambs per ewe, crossbreeding enables the commercial sheep raiser to benefit from the desirable characteristics of one breed—such as long life, flocking instinct, and wool production—and excellent body conformation and rapid growth of another breed.

Thus, systematic crossbreeding as a mating system for sheep production will continue to provide heterosis, or hybrid vigor, for some very important traits with which sheep producers are concerned. For example, it appears that an adapted crossbred ewe is more fertile than a straightbred ewe. But if such crossbred ewes are used and mated to rams of an unrelated breed for maximum commercial production, a minimum of three breeds must be involved. The breeds that are crossed to produce the crossbred ewe (the F_1) that will be retained as the producer in the flock should possess those characteristics which contribute to making a highly productive female—such as reproductive efficiency, milk production, maternal instincts, and wool quantity and quality. In turn, these crossbred ewes should be bred to one of the ram breeds. The ram breeds would be mated to the crossbred ewes as a terminal cross, to produce market lambs. Such ram breeds should possess growthiness, carcass quality, sexual aggressiveness, and fertility. An example of a crossbreeding program of this type would be:

1. **Foundation ewe breeds**—Rambouillet, Merino, Columbia, Corriedale, Targhee. Two of these breeds to be selected and crossed to produce the F_1 females.

2. **F_1 females**—The F_1 females resulting from the above cross to be bred to Suffolk or Hampshire rams as a terminal cross. All lambs to be marketed.

Crossbreeding Swine

Crossbreeding has been widely used in swine because it makes for increased production and profit. At this time, about 80 to 85 percent of all commercial hogs are crossbreds.

The mating of individuals from different breeds produces heterosis or hybrid vigor, a condition in which the offspring are superior in certain traits to the average of their parents. In a swine crossbreeding program, heterosis results in larger litters farrowed and weaned, stronger pigs at birth, and faster growth. However, heterosis gives little or no improvement in feed efficiency and carcass merit. The latter traits must be obtained through the selection of the parents.

In order to obtain the full benefits of crossbreeding, the producer should follow a systematic program in selecting crossbred replacement gilts and purebred boars. Superior crossbred gilts should be mated to rapid-growing, muscular boars from families with proven performance and carcass quality.

The three common crossbreeding systems followed in swine are:

1. **Two-breed cross**—This consists in mating a purebred boar to purebred or high-grade sows of another breed; for example, a Yorkshire boar and Hampshire sows.

Where this system is held to first crosses only, the

breeder is faced with the problem of sooner or later breeding the females back to a purebred boar of the same breed in order to secure replacement females. Under these conditions, he is prone to make little or no selection and to keep all of the females for replacement purposes. In such a program, it is usually found that the producer does well to maintain the quality of the female herd.

2. **Two-breed backcross or crisscross**—This system, which uses 2 breeds alternately, is recommended where good individuals of only 2 breeds are available. Boars of 2 different breeds (breeds that complement each other) are used in alternate generations. Selected crossbred gilts, produced by mating sows of breed A to a boar of breed B, are bred back to a boar belonging to one of the parent breeds (boar of Breed A, for example). Then, selected offspring from this mating are next bred back to a boar of breed B. Crossbred gilts and sows are always mated to the boar of the breed farthest away in their pedigree.

5-breed cross merely retains that level in later generations. Contrary to the belief of some swine producers, using a 3-breed or 4-breed rotation as described above does not cause a decline in the level of heterosis after several generations *provided* purebred boars are always used.

Before starting a crossbreeding program, the swine producer should know what is involved and what to expect. Also, he should realize that sound management and the selection of superior breeding stock on performance, potential carcass characteristics, and overall productivity are requisites for success. Moreover, the breeds used should complement each other.

Crossbreeding Horses

The greatest use of equine crossbreeding in this country was in the production of mules, prior to the mechanization of American agriculture. The mule,

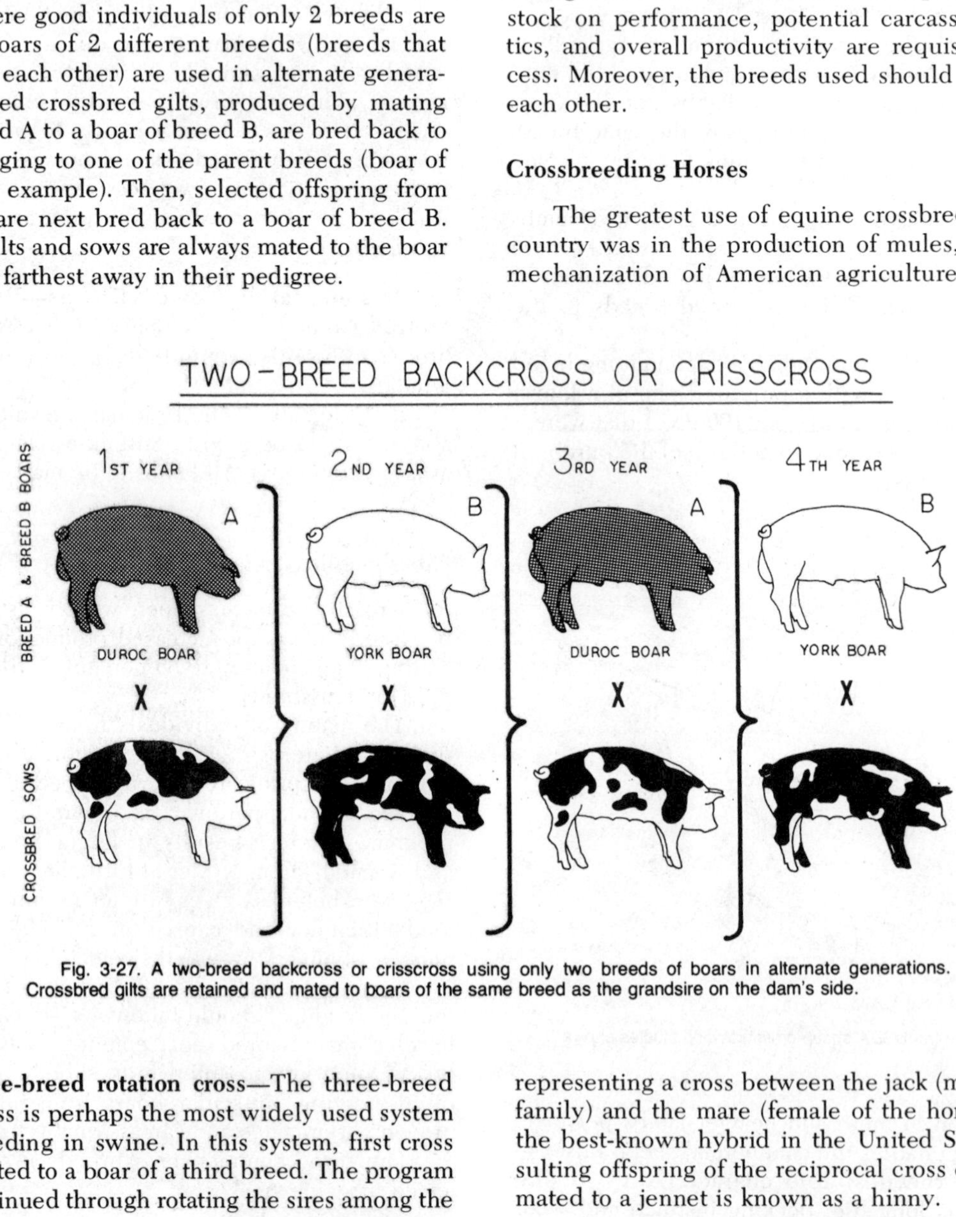

Fig. 3-27. A two-breed backcross or crisscross using only two breeds of boars in alternate generations. Crossbred gilts are retained and mated to boars of the same breed as the grandsire on the dam's side.

3. **Three-breed rotation cross**—The three-breed rotation cross is perhaps the most widely used system of crossbreeding in swine. In this system, first cross gilts are mated to a boar of a third breed. The program is then continued through rotating the sires among the three breeds.

Some swine producers follow a 4-breed, or even a 5-breed, system of crossbreeding. It is noteworthy, however, that the optimum amount of hybrid vigor is attained with the 3-breed cross, and that a 4-breed or

representing a cross between the jack (male of the as family) and the mare (female of the horse family), i the best-known hybrid in the United States. The re sulting offspring of the reciprocal cross of the stallio mated to a jennet is known as a hinny.

Horsemen have crossed certain breeds to obtai desired colors or color patterns. Arabians are some times crossed on other breeds to produce horses wit great endurance and stamina. The infusion of som coldblood (draft breeding) is sometimes relied upo

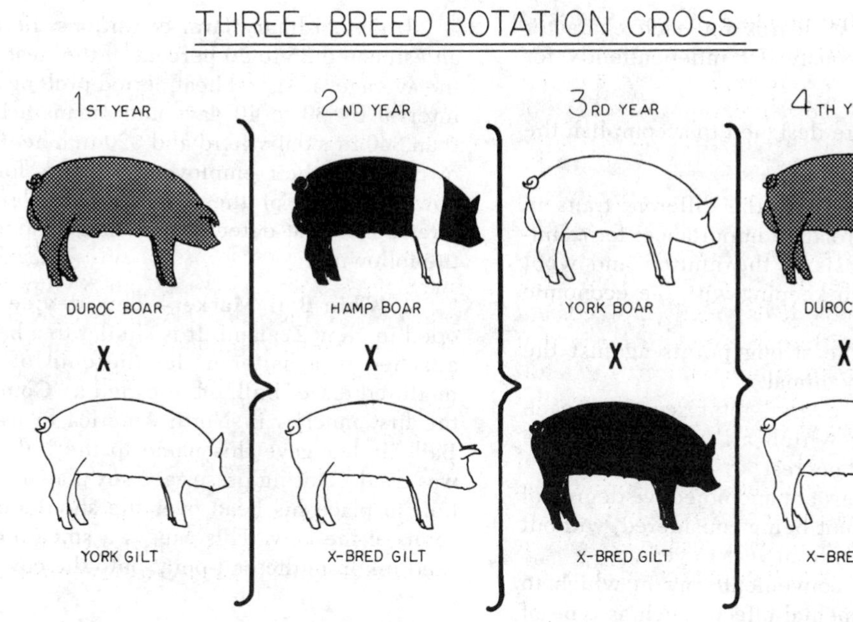

Fig. 3-28. A three-breed rotation cross, using Duroc, Hampshire, and Yorkshire breeds.

to secure hunters of greater size—of the weight carrying variety.

According to the rules of the American Horse Shows Association, harness show ponies may be of any breed or combination of breeds; the only requisite is that they must be under 12-2 hands in height. Three breeds produce animals that qualify under this category—Shetland, Welsh, and Hackney. Heavy harness ponies are, as the name indicates, miniature heavy harness horses—they're under 14-2 hands. Generally they are either purebred Hackneys, or predominantly of Hackney breeding.

Although crossbreeding of horses has a place for the purposes indicated above, purebreeding will continue to control the destiny of further improvement in horses and furnish the desired homozygosity and uniformity which many horsemen insist is a part of the art of breeding better horses.

SYSTEMS OF SELECTION

Hand in hand with the breeding system—with production testing, or crossbreeding—the stockman needs to follow a system of selection which will result in maximum total progress over a period of several years or animal generations. The three common systems are:

1. **Tandem selection**—This refers to that system in which there is selection for only one trait at a time until the desired improvement in that particular trait is reached, following which selection is made for another trait, etc. This system makes it possible to achieve rapid improvement in the trait for which selection is being practiced, but it has two major disadvantages: (a) Usually it is not possible to select for one trait only, and (b) generally income is dependent on several traits.

Tandem selection is recommended only in those rare herds and flocks where one character only is primarily in need of improvement; for example, where a certain flock of fine-wool sheep needs improving primarily in staple length.

2. **Establishing minimum standards for each character, and selecting simultaneously but independently for each character**—This system, in which several of the most important characters are selected for simultaneously, is without doubt the most common system of selection. It involves establishing minimum standards for each character and culling animals which fall below these standards. For example, it might be decided to cull all beef calves having a gain ratio under 110. Of course, the minimum standards may have to vary from year to year if environmental factors change markedly (for example, if calves average light at weaning time due to a severe drought and poor pasture).

The chief weakness of this system is that an individual may be culled because of being faulty in one character only, even though it is well nigh ideal otherwise.

3. **Selection index**—Selection indexes combine all important traits into one overall value or index. Theoretically, a selection index provides a more desirable way in which to select for several traits than either (a) the tandem method, or (b) the method of

establishing minimum standards for each character and selecting simultaneously but independently for each character.

Selection indexes are designed to accomplish the following:

a. To give emphasis to the different traits in keeping with their relative importance; for example, to give emphasis to the mutton and wool qualities of sheep in keeping with the economic importance of each.

b. To balance the strong points against the weak points of each animal.

c. To obtain an overall total score for each animal, following which all animals can be ranked from best to poorest.

d. To assure a constant and objective degree of emphasis on each trait being considered, without any shifting of ideals from year to year.

e. To provide a convenient way in which to correct for environmental effects, such as type of birth (single or twin), age of dam, etc.

Despite their acknowledged virtues, selection indexes are not perfect. Among their weaknesses are the following:

a. Their use may result in covering up or masking certain bad faults or defects, when they are overbalanced by strong points.

b. They may not adequately allow for year to year differences as genetic changes may be confounded or confused with environmental changes.

c. Their accuracy is dependent upon (1) the correct evaluation of the net worth of the economic traits considered, (2) the correctness of the estimate of heritability of the traits, and (3) the genetic correlation between the traits; and these estimates are often difficult to make.

In practice, a well designed selection index, including items measuring both performance and quality or market type, will serve as a good guide in making selections. It should, however, be supplemented with careful observation on individual animals in order to eliminate those which have severe defects not adequately covered in the index. Caution should be used, however, in departing from the index unless there is overwhelming reason for doing so.

COW HEAT DETECTION METHODS AND DEVICES

The problem of heat detection becomes more important as herds get larger, good hired help is more difficult to come by, cows produce more milk, and animal value increases.

Under ordinary farm conditions, herdsmen miss an estimated 25 to 50 percent of the heat periods. On the average, a missed heat period prolongs the calving interval by 30 to 40 days and means a loss of more than $40 in a dairy herd and $20 in a beef herd. Some owners pay their employees a bonus for catching a cow in heat. For these reasons, cattlemen are interested in heat detection methods. Among them are the following:

1. **Chin-Ball Marker**—This device was developed in New Zealand. It is similar to a ball-point pen attached to a halter under the chin of a surgically modified teaser bull, often called a "Gomer." (One of the first ranches in North America to use the Chin-Ball Marker gave this name to the bull on which it was used.) During preservice sex play, it is usual for a bull to place his head over the shoulders, back, and rump of the cow. This causes a smearing of the colored ink from the ball-point onto the cow.

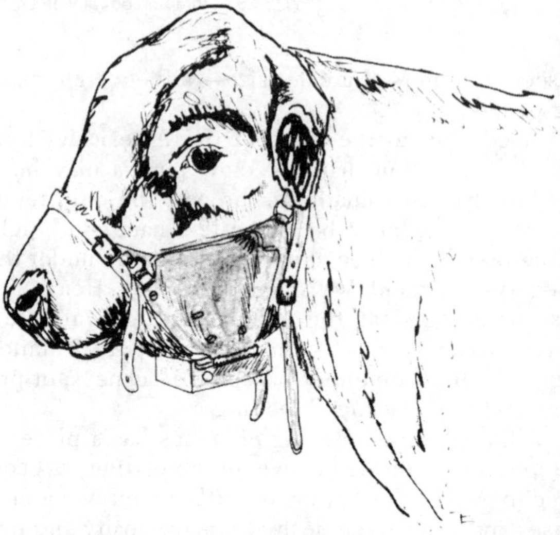

Fig. 3-29. Chin-Ball marking device. (Courtesy, American Breeders Service, De Forest, Wisc.)

One filling of the stainless steel container is sufficient to mark 15 to 25 cows. Experience indicates that one Gomer bull can work approximately 80 cows. In large pastures and in larger sized herds, it is best to have two bulls.

This method of heat detection is a most dependable management tool.

2. **The KaMaR Heat-Mount Detector**—The heat-mount detector is a 2 × 4½-inch fabric base to which is attached a white plastic capsule. Inside the capsule is a small plastic tube containing red dye. The tube is constructed so the dye is released slowly by moderate pressure. When enough dye is released

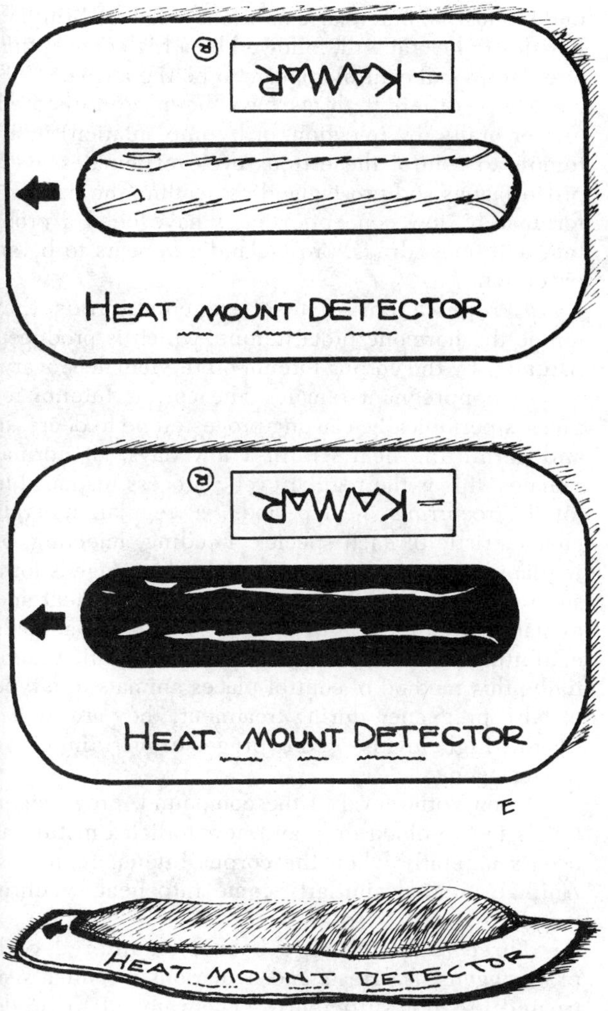

Fig. 3-30. Device for heat detection, as an aid in the artificial insemination of beef and dairy cows. At the top, the KaMaR Heat-Mount Detector is shown before activation. Center shows detector bright red after activation, indicating that cow is in heat. Lower view shows side or profile view of the device, which is applied to cow by an adhesive. (Courtesy, KaMaR, Inc., Steamboat Springs, Colo.)

from the tube (after about 4 to 5 seconds of pressure), it spreads over the inner lining of the capsule, causing it to turn red.

The detector relies on the natural bovine instinct of "bulling" or mounting during estrus. The pressure from the brisket of a mounting animal causes the dye to be released and the detector to turn red. If the cow does not stand for the mounting animal, there will not be enough pressure to release the dye and turn the detector red. This device has resulted in catching 95 percent of the heat periods.

3. **Pen-O-Block**—The Pen-O-Block is a plastic tube placed within the bull's sheath and held in place with a stainless steel pin. The bull can detect cows in heat and mount them in a normal way, but the device

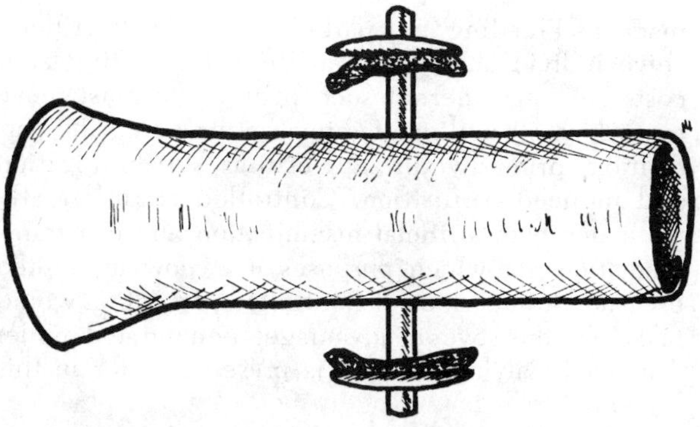

Fig. 3-31. Pen-O-Block marking device. (Courtesy, American Breeders Service, De Forest, Wisc.)

mechanically prevents him from making contact with the cow.

The Pen-O-Block consists of a white plastic tube, the pin or cannula, two washers, and a cotter pin. The device is inserted within the bull's sheath and held in place by the cannula. The procedure is best carried out by a veterinarian, as it requires skill.

Properly used, these three aids will improve heat detection. They are by no means replacements for visual heat detection; nor will they solve all the problems in breeding a beef herd artificially. Other factors that need attention are:

1. **Nutrition**—Cows must have adequate nutrition to cycle at a satisfactory rate for successful breeding.

2. **Rest interval**—This is very important, as cows must have calved at least 50 to 60 days prior to breeding for satisfactory performance.

3. **A.I. facilities**—Facilities should be adequate for handling and breeding the cow herd. Locate them where the cows tend to gather, such as the watering hole.

4. **Personnel**—Trained personnel are needed to do heat detection, gather the in-heat cows, and inseminate the herd.

HORMONAL CONTROL OF HEAT[19]

Today, hormonal control of heat and induced parturition are much sought as management tools. Most phases of animal management have shifted from individual care to herd or flock care, primarily to save labor. Thus, few cattle, sheep, and hogs are individually fed anymore. Instead, they are group fed in a self-feeder or bunk. But no such progress has been

[19]In the preparation of this section, the author benefited from the reviews and suggestions of the following scientists, with expertise in this field: Dr. R. E. Erb, Professor, Department of Animal Sciences, Purdue University, West Lafayette, Ind.; and Dr. Louis J. Boyd, Chairman, Animal Science Division, The University of Georgia, Athens, Ga.

made in breeding and parturition. Animals still receive individual attention at these times. But labor costs no longer permit such luxury. We must move toward mass handling of animals at breeding and parturition, primarily through estrous synchronization and induced parturition. Controlled estrus greatly facilitates both artificial insemination and ova transplantation, for which purposes it is now in limited use—albeit unperfected. Controlled parturition would make for the several advantages enumerated under "Induced Calving," which is presented later in this section.

Planned parenthood is not new. It has long been practiced among females of all species, women included. Stockmen have long "tampered with" the breeding and parturition season that was common in the wild state. Prior to domestication, animals brought forth their young in the fields and glens, inhibited only by age and feed, and influenced somewhat by seasons. But man changed all this—even without the use of hormones. Sly controls have been exercised over breeding for a very long time. For example, farm flock owners controlled reproduction in chickens by the simple act of putting eggs under an old setting hen—unless she hid out. Modern poultry producers regulate chick hatchings by controlling when, and how many, eggs go into the incubator. It is more difficult to accomplish the same thing in four-footed animals.

Horsemen, especially those who race or show, want their mares to foal as soon after January 1 as possible, because a horse's age is computed on a January 1 basis, regardless of how late in the year he may have been born. Also, horsemen are interested in regulating estrus as a means of getting more clinically anestrous mares (mares with functional ovaries that are not detected in estrus during the normal breeding season) bred. Sheepmen and hogmen strive for two crops of offspring per year, and for multiple births. Purebred cattlemen who show plan their breeding programs to take maximum advantage of show classifications; commercial cattlemen are concerned with weather and feed supply; and dairymen want the largest flow of milk at a time when the product is likely to bring the highest price.

Stockmen have altered nature's way in farm animals (1) by confining the male at certain times, or hand mating; (2) by emulating spring conditions—through providing better feed and shelter (and/or blankets) when breeding at other times of the year; (3) by flushing—through feeding females more liberally two to three weeks ahead of the breeding season; and (4) by artificially controlling the hours of light per day—through use of ordinary electric lights, which activate hormone production. Each of these methods has been used with varying degrees of success. All have fallen short of achieving the hoped-for goal under most commercial conditions—that of bringing females in heat at will, followed by a high conception rate. Hormonal control appears to be the answer.

Many different drugs have been administered (either orally, by injection, or by implantation) in attempts to control the estrous cycle of females, with progestagens and prostaglandins heading the list. Unfortunately, low conception rates have been a problem with most drugs; prostaglandin appears to be an exception.

• **Progestagens**—These are compounds that mimic the hormone progesterone, which is produced naturally by the corpus luteum on the female's ovary.

In nonpregnant females, the corpus luteum regresses periodically, causing progesterone to decrease and permitting heat within a few days. In normal females, this is the regular cyclic process responsible for the recurring "heat periods" at regular intervals characteristic of each species. Feeding, injecting, or implanting nonpregnant cows with progestagens for a sustained period of 14 to 20 days prevents heat and ovulation. Withdrawal of the dose results in animals exhibiting heat 2 to 8 days after withdrawal. Essentially, this method of control places animals in a type of false pregnancy during treatment; they are fooled by the high levels of the progesterone-mimicking progestagen.

Upon withdrawal of the compound, progestagen levels in the blood drop and new follicles mature as occurs naturally when the corpus luteum regresses. Animals treated similarly come into heat synchronously.

Three problems have been encountered with progestagens: (1) Heats and ovulations in animals so treated are not sufficiently synchronized to breed them successfully at one time; (2) conception at the first synchronized heat is subnormal (10 to 30% below that of normal untreated animals); and (3) the progestagens must be administered for 14 to 20 days, which involves much labor.

• **Prostaglandins**—With the discovery in the early 1970s that prostaglandins affected reproduction, a completely new era of research got underway. These are hormonelike substances (parahormones), found in almost every body cell and tissue, that are believed to play a key role in regulating cellular metabolism, as well as regulating a variety of body functions, including reproduction, blood pressure, gastric secretion, inflammatory response to injury, and blood clotting. The name "prostaglandin," which is a misnomer, was given to these substances because they were believed to have originated in the male's prostate gland.

These highly potent substances have been called local hormones or tissue hormones because they do their work near the area in which they are produced, as distinguished from most circulating hormones which aim at more distant targets. A synthetic prostag landin is being used to regress the corpus luteum (the growth on the ovary that prevents heat and ovulation). This allows the natural estrous cycle to begin again. Research indicates that the time interval between injection with a single dose of prostaglandin and onset of estrus and ovulation (1) is very short (in cows, about 90 hours after injection), and (2) is predictable.

When mares are injected intramuscularly with prostaglandins, they usually come in heat 3 to 4 days later, with ovulation occurring about 7 days after injection. Mares can be bred during the induced estrus. In one study involving 73 treated mares,[20] 73 percent were detected in estrus as defined by standing or showing signs to the stallion. Pregnancy data were available on only 54 of the 73 treated mares, since some mares were moved prior to evaluation of pregnancy and other mares were not inseminated or bred. Of the 54 mares, 56 percent of them became pregnant from breeding at the estrus which occurred following administration of the prostaglandin. Mares were bred an average of 1.7 times during the estrus.

Because prostaglandin is luteolytic (causes regression of the corpus luteum), an animal must have a corpus luteum in order to respond. Since a high percentage of clinically anestrous mares do have a corpus luteum, prostaglandin is effective. This allows the clinically anestrous mare either to return to estrus or to be bred, as determined by ovarian follicular development and ovulation. Treatment of clinically anestrous mares with prostaglandin has resulted in acceptable pregnancy rates.

Studies to date indicate that females treated with a single dose of prostaglandin may be bred at a predetermined time, over a short period, and still have satisfactory fertility. Although more research needs to be done on this promising new drug, it may be the answer to future controlled breeding.

The major criteria for measuring the success of hormone induced estrus are (1) percentage of females that come in heat, and (2) percentage conceptions when bred at synchronized heat.

Researchers in both colleges and industries are in general agreement that hormone controlled estrous synchronization will work, and that it offers promise of good returns when properly used in a well-managed herd. Scientists also realize that we don't know all the answers—that further research work is necessary. Nevertheless, it appears that planned parenthood is here to stay—that its wide use only awaits getting the technique perfected and lowering cost, both of which will come. In the meantime, stockmen are admonished to keep abreast of developments and to rely on well-informed advisers.

[20]Lauderdale, J. W., "Prostaglandins F2 Alpha: Effective Tool for Control of Estrus of Cows and Mares," *Stud Managers' Handbook*, Vol. 12, Agriservices Foundation, Clovis, Calif., 1976, p. 102.

SUPEROVULATION AND OVA TRANSPLANTATION[21]

Because the only practical use of superovulation at the present time is in conjunction with ova transplantation, both superovulation and ova transplantation are presented in this section; and since the commercial employment of superovulation and ova transplantation is limited primarily to cattle at the present time, the discussion pertains to this class of animals. However, with proper adaptation, the same principles and techniques will work with females of all species.

The bull is capable of producing several billions of sperm daily, whereas the cow normally produces an ovum (occasionally 2 ova) every 17 to 24 days. Now it is possible, through the administration of hormones, to obtain several ova (5 to 50) from a cow at estrous cycle. It is also possible to obtain a large number of eggs from young heifer calves, by injection of hormones.

The basic principle of superovulation is to stimulate extensive follicular development through the use of a hormone preparation, given intramuscularly or subcutaneously, with follicle-stimulating hormone (FSH) activity. The most common sources of such a hormone are pregnant mares' serum (PMSG) and FSH extracts from pituitaries of slaughtered animals. Many animals so treated will come into estrus about five days after initiation of treatment and ovulate, through release of their own luteinizing hormone (LH). However, to help assure that multiple ovulations occur, the ovulating LH from pituitaries or in human chorionic gonadotrophin (HCG) is injected.

Early studies have confirmed that FSH should be administered twice daily over a period of about five days. PMSG has a longer biological life and a single subcutaneous injection is normally used. Five or six days after the original FSH or PMSG "shot," LH or HCG is given intravenously.

Cows or heifers should ovulate by the seventh day after starting hormone treatment.

Since ovulation occurs over a period of time, not all the eggs are fertilized unless the donor (superovulated animal) is inseminated repeatedly. A yield of four or five good fertilized eggs per donor is average.

Of course, the real economic value of superovulation in cattle lies in twinning and in the successful tranfer of excess eggs (ova transplantation) from more valuable donor cows to less valuable recipient cows. As a result of this technique, someday stockmen may refer to litter-bearing cows.

[21]In the preparation of this section, the author benefited from the reviews and suggestions of the following scientists, with expertise in this field: Dr. R. E. Erb, Professor, Department of Animal Sciences, Purdue University, West Lafayette, Ind.; and Dr. Louis J. Boyd, Chairman, Animal Science Division, The University of Georgia, Athens, Ga.

Artificial insemination has given a means for the widespread distribution of desirable genes via the sperm. Similar genetic selection through high-quality females has, however, been limited since, normally, one cow will produce one calf per year and the average number of offspring per female will seldom exceed 5 in a lifetime. Out of the latter arose the idea that a marked increase in the production of offspring from desirable cows might be affected by superovulation, followed by transfer of the fertilized ova to less desirable cows, with the latter serving as host-mothers or foster-mothers to the developing embryo. Ova transplantation is a 7-step process as follows: (1) Synchronize heat cycles of donor and recipient cows; (2) administer a drug to the donor cow so that she superovulates; (3) breed donor cow (A.I. or natural); (4) collect ova from donor cow, 5 days after breeding; (5) examine eggs, making sure that they are normal and fertilized; (6) prepare foster-mothers, by synchronizing (usually by hormone control) their ovulation with the donor; and (7) transfer a fertilized egg to each recipient.

Pregnancy in the recipients can be diagnosed in about 35 days. Full-term pregnancies result in full sibs (brothers and sisters) with the genetic traits of the donor cow and the bull to which she was bred. Recipients have no genetic influence on the calves they carry—they merely serve as "incubators."

The following advantages would accrue from extensive use of ova transplantation:

1. A dozen calves might be obtained from a valuable cow during a year's time.
2. The rate of progress in genetic improvement would be speeded up, because of the increased number of progeny from valuable cows.
3. Valuable cows that produce normal ova but fail to conceive due to some hormonal or anatomical defects would not need to be culled because of sterility; such animals could be used as donors for supplying ova for transplantation.
4. Heifers could be effectively progeny tested at an early age. If large numbers of fertilized eggs could be procured from calves and transplanted to sexually mature recipients, the generation time of cattle could be reduced by one year or more.
5. Genetically inferior cows could be made more productive by using them as incubator cows.

In summary, it may be said that egg transfer can be done successfully by skilled teams. But the application of present techniques is limited to the most elite animals for three reasons: (1) the high cost, (2) the delicate technique required to harvest the fertilized ova, and (3) the likely possibility that females used as donors repeatedly may become incapable of producing young. With more research, techniques will become more efficient, simple, and economical and fertilized egg transfer will be widely used.

INDUCED CALVING[22]

Instead of letting "nature take its course," scientists are now synchronizing parturition. Currently, the technique is being used to a limited extent in beef cattle. The objectives of induced early calving are (1) lowering birth weight of calves, thereby lessening parturition difficulty; (2) predicting calving dates in order to pool labor and concentrate watching; (3) gaining a longer period from calving until rebreeding; and (4) shortening the calving interval and obtaining more offspring in the lifetime of the animal.

Females that have passed the 269th day of pregnancy will calve within 24 to 72 hours following intramuscular injection with an adrenal steroid. Experimental work indicates that such induced calving will result in (1) 5 to 8 days earlier than normal calving, and (2) calves 6 to 8 pounds lighter in birth weight than those carried to term. However, a higher incidence of retained placentas and temporarily lowered milk production accompany early calving. Antibiotics are indicated where the fetal membrane remains attached to the uterus longer than normal. Failure to expel the membranes after induced calving appears to have no effect on fertility as cows suffering this problem usually have no trouble breeding back.

Induced calving should be limited to healthy cows, free from disease, because the steroid "knocks out" the animal's immune-body system for several days. Also, until the problem of retained placentas is solved, it should not be used on dairy cows in commercial milk production, because of (1) restrictions in the use of antibiotics in dairy cows in lactation; (2) anesthetic reasons, where cows are housed and milked in the usual manner; and (3) the high incidence of metritis following retained placentas, which will likely infect herd mates and make for difficult breeding.

ARTIFICIAL INSEMINATION[23]

Artificial insemination is, by definition, *the deposition of spermatozoa in the female genitalia by artificial rather than by natural means.* Legend has

[22]In the preparation of this section, the author benefited from the reviews and suggestions of the following scientists, with expertise in this field: Dr. R. E. Erb, Professor, Department of Animal Science, Purdue University, West Lafayette, Ind.; and Dr. Louis J. Boyd, Chairman, Animal Science Division, College of Agriculture, The University of Georgia, Athens, Ga.

[23]This section was authoritatively reviewed by, and helpful suggestions were received from, the following A.I. specialists: Mr. William M. Durfey, Executive Secretary, National Association of Animal Breeders, Columbia, Mo.; Mr. Miles R. McCarry, Curtiss Breeding Service, Cary, Ill.; and Mr. Richard C. Newman, American Breeders Service, De Forest, Wisc.

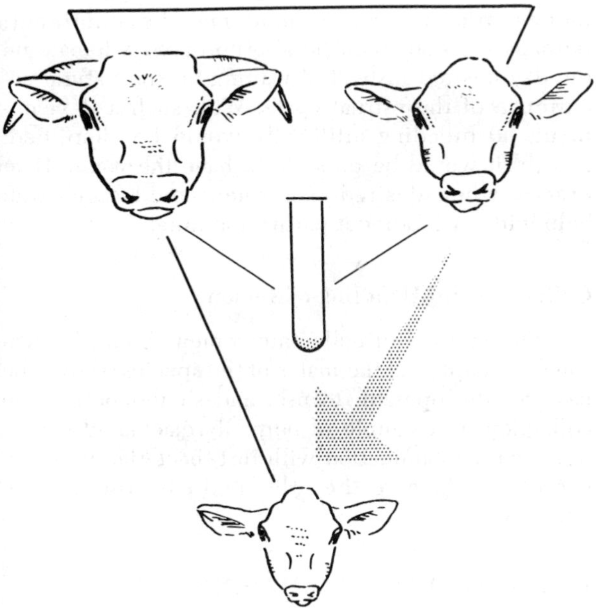

Fig. 3-32. More "test tube babies" may be expected in the future. (Drawing by R. F. Johnson)

that this method had its origin in 1322, at which time an Arab chieftain used artificial methods to impregnate a prized mare with semen stealthily collected by night from the sheath of a beautiful stallion belonging to an enemy tribe. However, the first scientific research relative to the artificial insemination of domestic animals was conducted with dogs by the Italian physiologist, Lazarro Spallanzani, in 1780.

Currently, artificial insemination is most widely practiced with dairy and beef cattle. Accordingly, many of the techniques and much of the application which follow are based on experiments and experiences with this particular species.

Advantages of Artificial Insemination

Some of the advantages of artificial insemination are:

1. **It increases the use of outstanding sires**—Through artificial insemination, many breeders can avail themselves of the use of outstanding sires, whereas the services of such males were formerly limited to a relatively few females of one owner, or, at the most, a small group of owners.

2. **It alleviates the danger and bother of keeping a sire**—Some hazard and bother are usually involved in keeping a sire, especially a bull or a stallion. Usually, the stockman may choose from the breeding programs of one or more established artificial insemination organizations and eliminate the necessity of maintaining a sire.

3. **It makes it possible to overcome certain physical handicaps to mating**—Artificial insemination is of value in (a) mating animals of greatly different sizes—for example, in using heavy, mature sires on young females, and (b) using stifled or otherwise crippled sires that are unable to perform natural service.

4. **It lessens sire costs**—In most herds, artificial insemination is usually less expensive than the ownership of a worthwhile sire together with the accompanying housing, feed, and labor costs.

In beef herds, it reduces bull numbers by three-fourths, thereby lessening sire costs and freeing range for more cows.

5. **It reduces the likelihood of costly delays through using infertile sires**—Because the breeding efficiency of sires used artificially is constantly checked, it reduces the likelihood of breeding females to a sire that is of low fertility or even sterile for an extended period of time.

6. **It helps control diseases**—Since no sire is present to make sexual contact, artificial insemination reduces the spread of venereal diseases, such as vibriosis and trichomoniasis in cattle.

Of course, to gain the benefit of disease control through artificial insemination, it is essential that the sires from which semen originates be free from infectious diseases.

7. **It makes it feasible to prove more sires**—Because of the small size of the herds in which they are used, many sires used in natural service are never proved. Still others are destroyed before their true breeding worth is known. Through artificial insemination, it is possible to determine the genetic worth of a sire at an earlier age and with more certainty than in natural service. The best of the sires proved at an early age are put into heavy use and have a longer period of usefulness than is possible under natural breeding methods.

8. **It creates large families of animals**—The use of artificial insemination makes possible the development of large numbers of animals within a superior family, thus providing uniformity and giving a better basis for a constructive breeding program.

9. **It increases pride of ownership**—The ownership of progeny of outstanding sires inevitably makes for pride of ownership, with accompanying improved feeding and management.

10. **It alleviates distance and time as limiting factors**—The male and the female may be separated by thousands of miles, and, with frozen semen, years may pass between the time of collection of the semen and insemination of the female.

11. **It increases profits**—The offspring of outstanding sires are usually higher and more efficient producers, and thus more profitable. A.I. provides a means of using such sires more widely.

Limitations of Artificial Insemination

Like many other wonderful techniques, artificial insemination is not without its limitations. A full understanding of such limitations, however, will merely accentuate and extend its usefulness. Some of the limitations of artificial insemination are:

1. **It requires trained technicians**—To be successful, artificial insemination must be carried out by skilled technicians. This means training, preferably training augmented by experience. While the cow breeding process is not complicated, it has been found that a small percentage of people who attempt to learn it never succeed in doing so.

2. **It may accentuate the damage of a poor sire**—It must be realized that when a male sires the wrong type of offspring his damage is merely accentuated because of the increased number of progeny possible. For this reason, untried or untested males are seldom used extensively in a stud. Fortunately, suitable standards for evaluating sires of meat and dairy animals have evolved through performance and progeny testing. Thus, it is noteworthy that 60 percent of the dairy sires are proved, and that these sires account for about 80 percent of the matings made. This precautionary measure virtually eliminates the possibility of using a genetically inferior dairy sire.

3. **It restricts the sire market**—It has been argued that the widespread adoption of artificial insemination has greatly decreased the market for poor or average sires. Such an argument is shortsighted. The principal thrust of artificial insemination is in the direction of maximum improvement through maximum use of superior sires. Obviously, this eliminates the necessity for using poor or average sires; hence, it must be regarded as an attribute, rather than a limitation.

4. **It may be subject to certain abuses**—If semen is transported from farm to farm, the character of the technician must be above reproach. Trained workers can detect differences in the spermatozoa of the bull, ram, boar, stallion, or cock; but even the most skilled scientist is unable to differentiate between the semen of a Thoroughbred and a Morgan, to say nothing of the difference between two stallions of the same breed. However, it appears that such abuse is more suspicioned than real. In a blood type study with cattle, Rendel of Sweden found 4.2 percent family records in error out of 615 animals by natural service, compared to 4.0 percent family records in error out of 199 sired by artificial insemination.[24]

5. **It is not yet fully practical to bring females in "true heat" at will**—Many advantages would accrue from bringing females of all species in heat and ovulation when desired and with certainty. By using hormones, planned parenthood may be imminent; perhaps we shall soon be able to breed a female on the day desired instead of waiting for the natural occurrence of the estrual cycle. With such a development, (a) breeding artificially would be simplified; and (b) it would be possible to have the young born exactly when desired—stockmen could then swap help with each other at parturition time.

[24]Rendel, J., "Studies of Cattle Blood Groups. II. Parentage Tests," *Acta. Agric. Scand.*, Vol. 8, No. 131, 1958, p. 140.

Collection and Handling of Semen

The method of collecting semen should be reasonably adapted to the males of the species, should be easy for the operator to use, and should permit the collection of a sample of normally ejaculated semen free from contamination with dirt, bacteria, or excess secretions from either the male or the female genitalia.

PREPARATION OF EQUIPMENT

The success of artificial insemination is directly proportionate to the cleanliness of equipment, operators, and animals. This is because spermatozoa are highly sensitive to, and quickly killed by, dirt, water, urine, excess heat, cold shock, or light, and by just about any other substance foreign to their natural environment.

It is recommended that all equipment that comes in contact with semen during the collection process be washed in a mild detergent after use. It should then be thoroughly rinsed in tap water to remove all detergent residues. The next step is rinsing with distilled water to remove any and all impurities from the tap water.

Rubber equipment should be boiled in distilled water, then rinsed in ethanol. Better yet, disposable sterile plastic should be substituted for rubber wherever possible.

At present, the inseminating tubes, or catheters, used in breeding cows are disposable plastic. So are the gloves used by inseminating technicians. This means that all such equipment can be thrown away after one use. The costly, time-consuming sterilization techniques needed with the glass equipment used in the early days of artificial insemination has been eliminated.

If a mare is to be inseminated, she should be hobbled; her tail should be bandaged; and the region about the vulva should be washed and prepared exactly as if she were to be bred in natural service. The place of insemination should be free from dust and should be sprinkled with water if necessary. Above all, cleanliness is the most important factor; dirty animals, utensils, or operators will seriously decrease the results of artificial insemination. If soap or detergent is used, rinse away all traces.

ARTIFICIAL VAGINA

Many techniques have been developed for the collection of semen, but the most satisfactory one consists of using the artificial vagina, a Russian invention. Males usually respond to this method with little previous training.

The artificial vagina consists of an outer tube or casing—which is usually constructed of heavy rubber, metal, or one of the plastics—and an inner tube, or lining, of thin rubber. The space between the two tubes is usually filled with warm water and a little air. One end of the apparatus is open to allow the entrance of the penis, and the other end is attached to a glass tube or beaker to receive the ejaculated semen.

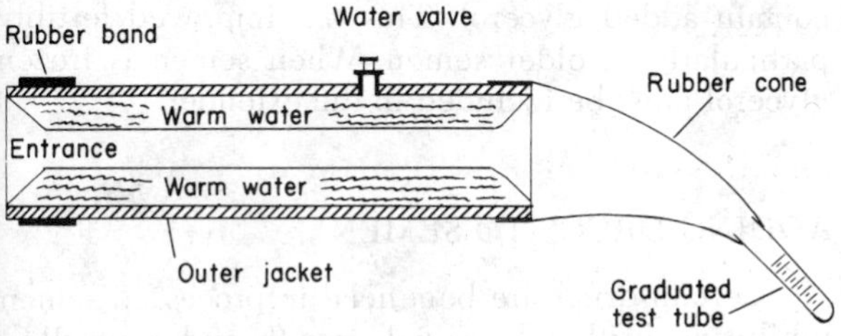

Fig. 3-33. Diagrammatic artificial vagina for the collection of bull semen. (Drawing by R. F. Johnson)

Most males, especially young ones, can be trained to use dummies. Training consists in exposing the male to an in-heat female, without permitting service. After two or more experiences of this kind, the male is introduced to the dummy at the same location; thereby being taught to anticipate service when brought to the breeding quarters. The dummy should be strongly built and firmly anchored.

Many commercial semen-producing businesses have replaced dummies with live "jump stock"—steers as well as cows. It has long since been demonstrated that bulls will mount cows not in heat and stand to be collected with an artificial vagina. More recently, it was discovered that most bulls could be collected in the same manner if a steer replaced the cow as the "jump animal." The advantage is that steers are bigger and stronger than cows—better able to bear the weight of the bull from which collection is being made.

OTHER METHODS OF COLLECTING SEMEN

Other, but less popular, methods of collecting semen are as follows:

1. **Electric stimulation**—This method first found limited use with rams. It has since been adapted to, and widely used in, collecting semen from bulls that will not or cannot mount jump other animals or that lack in libido. It is also useful on farms and ranches where there are no facilities for normal semen collection and in cases where time does not permit training bulls to the artificial vagina.

Electrical stimulation can be, and is, adapted to nearly all farm animals except boars; and electroejaculator apparatuses are available commercially. The simplest equipment introduces a weak alternating current to the sacral and pelvic nerves via electrodes placed in the rectum on a probe or by hand. For bulls, a single electrode, delivering 5 to 30 volts, is placed in the rectum; then a series of stimulations is applied until erection and ejaculation occur.

2. **Collection of semen following rectal massage of ampullae**—This system has been used successfully in bulls and stallions. It consists of the rectal massage of the ampullae and the collection of semen in a funnel leading to a glass tube. This method has the disadvantages of (a) requiring considerable experience on the part of the operator, (b) necessitating two persons for collection, and (c) resulting in abnormal semen. Hence, it should be resorted to only in males that are unable or unwilling to mount and ejaculate.

VOLUME OF SEMEN; CONCENTRATION OF SPERM

The volume of semen ejaculated at one service and the concentration of sperm vary according to species and individuals. Table 3-17 gives average figures by classes of farm animals.

TABLE 3-17

SEMEN VOLUME AND SPERM CONCENTRATION OF FARM ANIMALS

Class of Animal	Av. Volume of Semen per Ejaculate	Av. Concentration of Sperm	No. Females That Can Be Bred per Ejaculate
	(ml)	*(millions/ml)*	
Bull	4-7	1,500	100-600
Ram	1	800-4,000	40-100
Boar	100-500 (av. 240-250)	160-200	10-12
Stallion	90	30-800	8-12

As a rule, the smaller ejaculates of high sperm concentration, such as found in bull and ram semen, are the most suitable for artificial insemination; for they can be diluted into a large volume with special diluters that retain the life of the sperm to a higher degree than the natural seminal fluid. It is possible, however, to fractionate boar ejaculate during collection so as to obtain the sperm-rich portion apart from the large volume of aspermic seminal fluids.

In cattle, sperm numbers are of more importance than semen volume. In swine, however, any volume below 50 ml seems to decrease conception rate.

STORAGE AND SHIPMENT OF SEMEN

If semen is to be used within one or two hours following collection, it may be kept at room temperature. For longer storage as liquid semen, and delayed use, it should be properly diluted and gradually cooled (avoiding temperature shock) and stored at a temperature of 35 to 40°F.

In 1952, British scientists reported that the addition of glycerine to semen diluters permitted them to freeze certain semen at temperatures much below zero (they used dry ice to freeze at a temperature of −79°C or −110°F), and still retain a high degree of fertility following thawing.

Frozen semen is now being used for longtime storage; and this basic discovery is being further investigated and perfected by scientists in several countries. More specialized equipment and improved procedures are being developed.

Today, most semen shipped long distances is frozen in either glass ampules or in straws. Both ampules and straws are shipped attached to metal canes, or holders, placed in an especially designed steel drum-like container partially filled with liquid nitrogen (−320°F), and transported. This is the most convenient and practical method to transport semen and is used almost universally.

Although frozen semen cannot be said to have revolutionized A.I. completely, it has made a highly significant impact. Its principal advantages are (1) reducing semen wastage; (2) making for a wider selection of sires; (3) facilitating planned matings, because semen from a given sire is constantly available; (4) extending the usefulness of valuable sires and making possible the use of their semen long after death (calves have been obtained from semen stored as long as 27 years; hence, it appears that frozen semen can be stored almost indefinitely); and (5) making completely feasible the shipment of semen long distances into locations previously considered inaccessible.

In 1975, frozen semen for dairy cattle was used exclusively by nearly all organizations. Approximately 100 percent of the inseminations made that year were with frozen semen. Without doubt, frozen semen will continue to replace fluid (unfrozen) semen; the primary limitation being its greater cost to small organizations.

In the future, semen may be preserved in the dry state at low temperatures. Progress along these lines is encouraging.

SEMEN EXTENDERS

Addition of extenders to freshly collected semen is almost imperative because (1) they provide needed volume; and (2) they exert a beneficial effect on the sperm.

Table 3-18 summarizes the pertinent facts relative to the four most widely used semen extenders in the United States. Other extenders are (1) glucose plus tartrate, sulfate, or phosphate salts (developed by Milovanov, a Russian); (2) modified Ringer solution (developed by Bonnier and Trulsson, which is considered the best of the extenders for rooster semen); (3) skim milk; (4) dry skim milk powder; (5) homogenized whole milk; (6) synthetic pabulum (developed by Phillips and Spitzer); (7) Krebs solution (developed by Lardy and Phillips, for boar semen); (8) IVT (developed by Illinois); (9) CUE (developed by Cornell); and (10) certain commercially prepared extenders.

Most fluid semen extenders (listed in Table 3-18) presently used in the United States for cattle breeding contain added glycerol. This has improved fertility, particularly in older semen. When semen is frozen, glycerol must be included in the extender.

ADDING DRUGS TO SEMEN

Certain drugs are beneficial in processed semen, inhibiting further bacterial growth and controlling certain pathogenic organisms, if present, and making for definitely higher conception rates, especially when added to semen from males whose low fertility is due to infections—particularly vibriosis. The recommended drugs and levels to add to semen are as follows:

1. **Streptomycin**—Add 500 to 1,000 micrograms of streptomycin per ml of diluted semen.
2. **Penicillin**—Add 500 to 1,000 units of penicillin per ml of diluted semen.
3. **Linco-Spectin**—This added so that raw semen is incubated for 15 minutes with 300/600 mcg/ml of raw semen and the final concentration is 150/300 mcg/ml of processed semen.
4. **Polymyxim B Sulfate**—This has been used in conjunction with dihydrostreptomycin and penicillin for control of *Vibrio fetus*. The procedure consists in mixing 2,000 mcg streptomycin and 1,000 mcg polymyxin with raw semen followed by dilution with extender containing the same antibiotics plus 500 mcg of penicillin. The antibiotics were added at these levels/ml of semen or extender.

For control of *Vibrio fetus*, at least six hours must elapse between addition of antibiotics (streptomycin or penicillin) to the semen and insemination of the cow; and the rate of semen to the diluter must be at least 1:25. Frequently, this procedure is followed as a routine precaution in *Vibrio fetus* control.

Streptomycin or penicillin, and certain other antibiotics, may be added to the extenders listed in Table 3-18.

TABLE 3-18
COMMON SEMEN EXTENDERS

Type of Extender	Formula of Extender	Classes of Animals for Which It Is Satisfactory	Comments
1. Egg yolk-phosphate (developed by Lardy and Phillips)	1. To 100 cc of boiling glass-distilled water add: 0.2 g KH_2PO_4 (chemically pure) 2.0 g $Na_2HPO_4.12H_2O$ (chemically pure). Cool to room temperature. 2. Add an equal volume of *fresh egg* yolk which has been carefully separated from the whites. Solution No. 1 can be stored, but a fresh mixture (Nos. 1 and 2) should be made up each time.	Bull Ram Stallion	When using this extender for stallion semen, add 10 g dextrose or glucose (chemically pure) per 100 cc of boiling distilled water. Currently, this extender is little used in cattle breeding in the U.S.
2. Egg yolk-citrate (developed by Salisbury, Fuller and Willett)	To 100 cc of boiling glass-distilled water add: 3.0 g of $Na_3C_6H_5O_7.2H_2O$ Add 30 to 50% volume of *fresh egg* yolk before use.	Bull Ram	The main advantage of this extender is that it disperses the fat globules of egg yolk, thus facilitating the microscopic examination of extended semen. To date, no experimental work has been reported on the use of egg yolk-citrate as an extender for stallion semen. However, there is no reason to believe that it would not be as satisfactory as yolk-phosphate.
3. Homogenized whole milk (developed by Almquist and Thacker)	Buy fresh homogenized whole milk. Put in double boiler and boil for 5 to 10 minutes. A safe minimum temperature is 200° F. Remove film with a sterile glass rod. Cool to body temperature and dilute at same rate as if egg yolk-phosphate or egg yolk-citrate were used.	Bull Ram Boar	To date, no experimental work has been reported on the use of homogenized whole milk as an extender for stallion semen.
4. Glycine-containing diluents	1. 50% of a 4% glycine solution. 42.5% skim milk, heated to 92°C and then cooled. 7.5% egg yolk. —or— 2. 15% of a 4% glycine solution. 72.25% skim milk, heated to 92° C and then cooled. 12.75% egg yolk.	Boar	These extenders will maintain motility in boar semen for 1 to 2 weeks. However, conception beyond 24 hours' storage is low despite good sperm motility. As a swine semen extender, Aamdal recommends[1] 3% sodium citrate—30% egg yolk, with added antibiotics.

[1]*The Artificial Insemination of Farm Animals*, 3rd ed., ed. by E. J. Perry, Rutgers University Press, New Brunswick, N.J., 1960.

When to Breed

A female is fertile only when an egg is present which can be fertilized. Moreover, an egg can live for only a short time after being shed from an ovary unless it is fertilized. The optimal time for insemination is in advance of the time of ovulation, which varies according to species.

A cow doesn't shed her egg from the ovary until about 10 hours after the close of standing heat, and the egg lives only 6 to 10 hours. Thus, for optimal results from insemination, cows should be bred in the latter two-thirds of heat or within a few hours after having gone out of heat; roughly, this means that a cow should be bred within a 24-hour period after she is first noticed in standing heat (see Fig. 3-34). To accomplish this, it is recommended that cows first observed in standing heat in the morning be bred during the afternoon or evening of the same day, and that those observed in heat in the evening be bred the next morning.

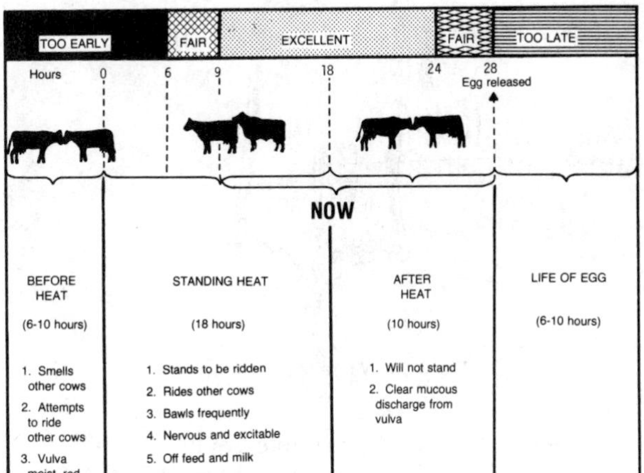

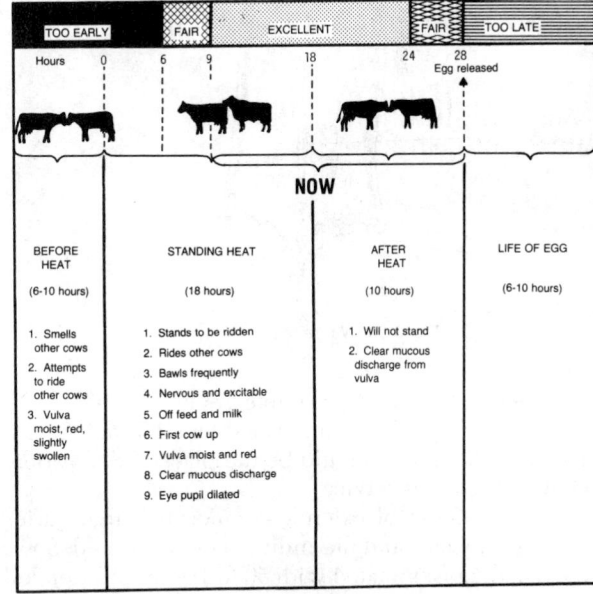

Fig. 3-34. Breeding time guide for the average cow.

Mares normally ovulate 24 to 48 hours before the termination of the heat period, ewes near the end of heat, and sows about 30 to 40 hours after the beginning of the heat period. Thus, a mare with a 5-day heat period (4 to 6 days is considered normal for a mare) should be inseminated on the third day and at least every other day thereafter for the duration of the heat. Ewes should be inseminated during the last day of heat, and sows 12 to 24 hours after the beginning of heat. In sows, a second service 12 to 24 hours later increases conception and litter size.

Insemination of the Female

Cleanliness during all insemination manipulations is essential and is a crucial point for success or failure. This applies to the instruments used, to the hands of the operator, and to the animals.

The inseminating tube should be passed just through the cervix, stopping at the location where the cervix ends and the uterus begins. Precisely at this point the semen should be deposited—slowly (see

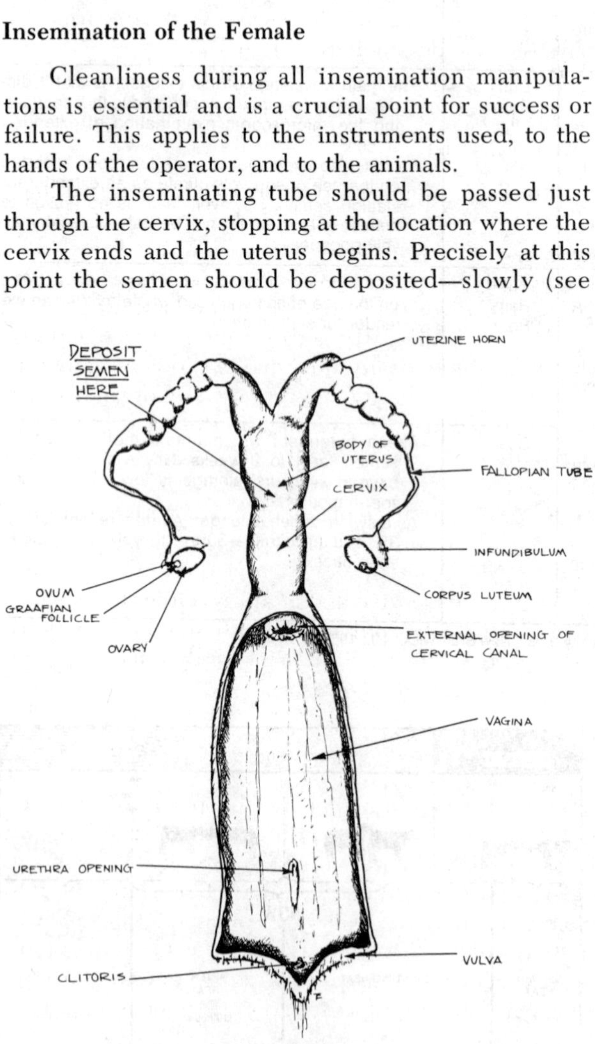

Fig. 3-35. Where to deposit semen.

Fig. 3-35). The most desirable region of deposition has not yet been fully established for all farm animals. In sows, the uterus should be the place of deposition, as it is in natural service.

The amount of extended semen required varies with the species and the individuals concerned. Sows are usually inseminated with 50 to 100 ml of extended semen, mares 20 to 30, cows 0.5 to 1.5, ewes 0.1 to 0.2, and hens 0.1 ml.

Practical Application

Today, artificial insemination is more extensively practiced with dairy cattle and beef cattle than with any other class of farm animals. In 1938, when the program first began in America, only 7,359 cows and 646 herds were bred by this means in organized groups in the United States; whereas, in 1974, an estimated 11,503,560 dairy units and 4,545,502 beef units were using A.I.[25] In 1974, artificial insemination programs involved 60 percent of milking cattle and 90 percent of the dairy herds in the United States.

In the future, there will be a marked increase in the use of A.I. in beef cattle, and to a lesser extent in sheep, swine, horses, dogs, turkeys, and milk goats (a special section devoted to each of the first four classes of farm animals follows). For a list of A.I. organizations, see Section 19 of this book.

Based on present knowledge, gained through research and practical observation, it may be concluded that stockmen can make artificial insemination more successful through the following:

1. Give the female a reasonable rest following parturition and before rebreeding; in cows this should be 50 to 60 days, and in sows 35 to 49 days.
2. Keep records of heat periods and note irregularities.
3. Watch carefully for heat signs, especially at the approximate time.
4. Where an association is involved, notify the insemination technician promptly when an animal comes in heat.
5. Avoid breeding diseased females or females showing cloudy mucus. The latter condition indicates an infection somewhere in the reproductive tract.
6. Have the veterinarian examine females that have been bred three times without conception or that show other reproductive abnormalities.
7. Follow a proper nutrition program at all times.

SEMEN PACKAGING

For many years, frozen semen was packaged in glass ampules with each ampule containing enough semen to breed one cow. Ampules are still widely used, but there is a trend toward the use of straws.

Straws, while still comparatively new in North America, have been used successfully throughout much of Europe for many years. Generally speaking semen in straws will not stand as much abuse or mistreatment as semen in ampules. But this is not considered an important factor because successful artificial insemination involves handling semen carefully

[25]Currently, most reporting is in units of semen, not first service as was the custom for many years. It takes 1½ to 2 units per cow.

nd according to directions without shortcuts and ough handling.

Straws are small, as compared to ampules. They have a higher surface-to-volume ratio. This means hey freeze and thaw more rapidly.

Typically a ½ cc straw can be thawed in less than one minute, as compared to 10 minutes' thawing time for a 1 cc glass ampule. When there are several cows o be bred in one barn or corral, there is a natural in-lination to dump the required number of ampules nto the thaw box at the same time. This saves waiting 10 minutes between cows. It also means that much of he semen will not be used at its peak, or anywhere near 10 minutes after entering the thaw box.

There's no need for such mass thawing with traws. The fact that they thaw in less than one min-ute means that the rancher or dairyman can thaw hem one by one. They'll be ready to use as soon as he inseminator has put on a fresh plastic glove and is eady to breed the next cow.

Another factor is that semen goes directly into the cow from the straw as opposed to going (1) from am-pule, (2) to breeding tube, (3) to cow. This totally eliminates loss of sperm in semen which clings to the mpule and the breeding tube. Just about all sperm in the straw goes into the cow.

The fact that straws are smaller than ampules also nters the picture in the form of storage space re-quirements. A liguid nitrogen refrigerator, or tank, will hold almost twice as much strawed semen as am-puled semen. This means that, with straws, the tech-nician or herd owner can keep a wider selection of semen on hand, and/or take full advantage of semen specials" and "bargain sales." The other alternative s a smaller refrigerator with smaller capital invest-ment and lower operating costs.

THAWING FROZEN SEMEN

With both straws and ampules, there is a wide ariation in thawing techniques. In all cases, it is best o follow the recommendations and instructions of the A.I. organization that actually processed and packaged the semen. Processing techniques vary from stud to tud. Instructions for thawing are drawn up to fit the particular brand of semen involved.

ARTIFICIAL INSEMINATION OF DAIRY CATTLE

In 1937, the North Central School of Agriculture and Experiment Station, Grand Rapids, Minnesota, conducted the first large-scale demonstration of artifi-cial insemination of dairy cattle in the United States. Subsequent growth was rapid. By 1974, artificial in-semination was practiced in 60% of the dairy cattle and, to some extent at least, in 90% of the dairy herds in the United States. It is noteworthy, too, that the most populous U.S. breed, the Holstein, led all purebred dairy breeds in percent of registrations from A.I. matings—with 68.2%.

Artificial insemination has played a significant role in the genetic improvement of U.S. dairy cows. Also, A.I. "barn charts" focused attention on effi-ciency of reproduction. In turn, this had an impact upon fertility of the national herd and the control of certain diseases, and it stimulated research in physiology and pathology of reproduction.

ARTIFICIAL INSEMINATION OF BEEF CATTLE

In 1965, about 14 percent of all inseminations were with beef semen. For the most part, this con-sisted of crossing beef bulls on dairy females—particularly dairy heifers. But there were an estimated 600,000 to 750,000 strictly beef cattle matings.

By 1965, all dairy cattle registry organizations recognized genetically qualified A.I. offspring as eli-gible for registry. But most beef registry organizations had rules stringently limiting, and often preventing, registration of purebred offspring of bulls owned by more than three or four individuals.

This situation changed radically with the intro-duction of continental European or "exotic" beef breeds. Because of very limited numbers, they needed A.I. to survive and multiply. Other breeds fol-lowed suit. Today, only one or two breed associations still resist A.I. Over 20 percent of all semen sold in 1974 was beef semen.

ARTIFICIAL INSEMINATION OF SHEEP

Artificial insemination in sheep has been prac-ticed in The Soviet Union, Eastern and Central Europe, and some areas of South America. It has been used experimentally in the United States, but the fol-lowing factors have made the commercial use of A.I. impractical in this country to date: (1) high labor costs, (2) few rams have been production or progeny tested and identified as superior, (3) conception rates from a single insemination are not high.

ARTIFICIAL INSEMINATION OF SWINE

Artificial insemination of swine in the United States is still in the experimental stage, but there is great interest in its potential. Successful freezing of boar semen, which defied scientists for many years, has been accomplished. This makes it possible to store semen until the desired season, then breed more sows to one boar than was ever before possible. The remaining bottlenecks are:

1. **We need to be able to detect when sows are ready for breeding**—It is sometimes difficult to determine exactly when a sow or gilt is ready for breeding, primarily because most swine producers are unskilled in heat detection—they have always relied upon the boar. If a sow is not bred at the proper time (during the first 12 to 24 hours of the heat period), litter size and conception rate are likely to be reduced.

2. **We need estrus control in gilts so as to synchronize farrowing**—Some control may be obtained in sows through manipulating the pattern of weaning among a group of sows. But there is need for estrus control in gilts so as to synchronize farrowings.

Without doubt, in due time the barriers listed above will be overcome, and artificial insemination in swine will expand just as it has in the American dairy industry.

In Finland, an estimated 35 percent of the sows are bred A.I.

ARTIFICIAL INSEMINATION OF HORSES

To date, there has been less interest in, and more resistance to, artificial insemination in horses than in any other class of livestock. As a result, (1) there is a paucity of experimental work in the field, and (2) several of the breed associations frown upon or forbid the practice.

Before wide scale use can be made of artificial insemination of horses, the following problems need to be solved:

1. **We need to be able to breed more mares per stallion**—At the present time, too few mares can be bred per stallion in any one breeding period or season. For example, it is possible to breed up to 600 cows from one collection of a bull, compared to perhaps 8 to 15 for the stallion.

2. **We need to be able to store stallion semen longer**—Until recently, stallion semen was viable only one to two days and had to be used in the liquid state. However, stallion semen is now being frozen successfully and its use is growing. This development may write a new chapter in horse breeding, but there has been very little progress to date.

3. **We need to be able to detect when mares are ready for breeding**—It is sometimes difficult to determine exactly when a mare should be serviced. If a mare is not bred at the proper time (within 20 to 24 hours before ovulation), the conception rate is very low.

The following report is noteworthy: In 1961, Professor Cheng Pi-liu, of China, reported that he bred artificially 2,798 mares to one stallion in one year, that an average of 15 mares were bred per collection, and that there was a 73.9 percent foal crop.

Bull Costs Vs A.I. Costs

With the increase in artificial insemination in recent years, a frequently asked question is: What's the cost of A.I. vs natural service?

Many cost figures seen in various publications show bull costs by natural service ranging from $6 to $12 per cow. The Nebraska Station reported that the cost of keeping a bull is much higher than this, even if the bull is depreciated over a 4-year period (see Table 3-19).

TABLE 3-19

ANNUAL COSTS OF OWNING AND MAINTAINING A BULL[1]

	1981-82 Prices			
	If Depreciated Over 2 Years		If Depreciated Over 4 Years	
	Total Costs	Direct Costs	Total Costs	Direct Costs
Hay, 1.5 tons	$ 75.00	$ 75.00	$ 75.00	$ 75.00
Winter pasture	25.00	—	25.00	—
Summer pasture	108.00	—	108.00	—
Salt and mineral	4.00	4.00	4.00	4.00
Veterinary and medicine	14.00	14.00	14.00	14.00
Death loss	6.00	6.00	6.00	6.00
Depreciation ($1,500 purchase price, $840 selling price)	330.00	330.00	165.00	165.00
Interest on bull (12%)	140.00	140.00	140.00	140.00
Interest on feed and operating expense (12%)	6.00	6.00	6.00	6.00
Labor (10 hr)	55.00	—	55.00	—
Use of buildings and equipment	9.00	3.00	9.00	3.00
Miscellaneous expense	6.00	6.00	6.00	6.00
TOTAL	$778.00	$584.00	$613.00	$419.00
30 cows	26.00	19.50	20.50	14.00

[1]Estimates submitted by NE Beef and Ag Economics Specialists: Guyer & Jose

As shown in Table 3-19, one of the largest bull cost items is depreciation. These figures are based on a purchase price of $1,500 and a selling price of $840. On this basis, bull depreciation alone would amount to more than $10 per cow. Direct costs, excluding pasture, amount to $13.97 to $19.47 per cow (depending on how long the bull is kept), and $14.87 to $21.27 per cow if the charge for summer pasture is included.

No allowance was made for investment credit in the bull costs shown. According to tax laws, this would be a consideration and would reduce the bull costs slightly. In most cases, it would amount to no more than $3 per cow in the herd, however.

Artificial insemination eliminates the expense and problems associated with maintaining bulls. But of course, it involves some different expenses, primarily for the cost of semen and the added labor.

Based on a study of 37 commercial ranches in Wyoming, the Wyoming Experiment Station reported $1.87 greater cost per calf from A.I. than from natural service.[26] However, the A.I.-sired calves had a $7.05 per head greater value than calves sired by natural service, leaving $5.18 net per calf in favor of A.I. This comparison included all charges related to breeding the cow herds.

BLOOD TYPING CATTLE AND HORSES

Cattle blood typing was developed by the University of Wisconsin during the decade 1940-50, and horse blood typing was developed by the University of California, at Davis, during the period 1958-64. It involves a study of the components of the blood, which are inherited according to strict genetic rules that have been established in the research laboratory. By determining the genetic "markers" in each sample and then applying the rules of inheritance, parentage can be determined. To qualify as the offspring of a given female or male, an animal must not possess any genetic markers not present in its alleged parents. If it does, it constitutes grounds for illegitimacy.

Blood typing is used for the following purposes:

- **To verify parentage**—The test is used in instances where the offspring may bear some unusual color or markings or carry some undesirable recessive characteristic. It may also be used to verify a registration certificate. Since an estimated 5% of all registered animals in the United States are illegitimate, there is need to use blood typing as a bulwark of breed integrity. Through blood typing, parentage can be verified with 90% accuracy.[27] Although this means that 10% of the cases can't be settled, it's not possible to do any better than that in human blood typing.
- **To determine which of two sires**—When a female has been served by two or more males during one breeding season, blood typing can exclude the incorrect male and include the correct male in over 90 percent of the cases.
- **To provide a permanent blood type record for identification purposes**—Two samples of blood are required for each animal to be studied; and the samples must be taken in tubes and in keeping with detailed instructions provided by the laboratory. In parentage cases, this calls for blood samples from the offspring and both parents; in paternity cases, samples must be taken from the offspring, the dam, and all the likely sires.
- **To substitute for fingerprinting**—Much attention is now being given to the idea of utilizing blood typing as a positive means of identification of stolen animals, through proving their parentage.
- **To detect fertile heifers born co-twin with bulls**—About 15 percent of all heifers born twin with a bull are potentially fertile; the other 85 percent are sterile, or freemartins. Blood typing alleviates the need to wait until such heifers reach breeding age in order to ascertain their breeding potentialities. Instead, a blood sample from each of the twins (the bull and the heifer) can be submitted to a service-typing laboratory and a diagnosis made. If the bull and heifer have *like* blood types (except possible differences in the J system), the heifer is diagnosed as a freemartin and nonbreeder. If the bull and the heifer have unlike blood types (except possible differences in the J system alone), the heifer is diagnosed as potentially fertile.

The basis for this remarkable method of diagnosing the breeding potentialities of a heifer born twin with a bull goes back to the early events in the embryology of cattle twins. In about 85 percent of the cattle embryos, some of the chorionic blood vessels become anastomosed or joined together. This results in a communal blood vascular system. Hence, the twins come to share each other's blood forming tissues. As a result, they have like blood types.

- **Blood typing laboratories**—The following laboratories are capable of determining bull parentage:

1. Serology Laboratory, Department of Reproduction, School of Veterinary Medicine, University of California, Davis, California 95612.
2. Immunogenetics Laboratory, Department of Animal Science, Texas A & M University, College Station, Texas 77843.
3. Cattle Blood Typing Laboratory, Department of Dairy Science, 625 Stadium Drive, Columbus, Ohio 43210.
4. Animal Disease Research Institute (E), Health of Animals Branch, Agriculture Canada, P.O. Box 11300, Postal Station "H," Ottawa, Ontario, Canada K2H 8P9.

The Serology Laboratory at the University of California is also capable of determining equine parentage.

BREEDS OF LIVESTOCK AND THEIR CHARACTERISTICS

A purebred animal may be defined as a member of a breed, the animals of which possess a common ancestry and distinctive characteristics, which is either registered or eligible for registry in the herd book of that breed.

Tables 3-20 to 3-25 give in summary form the

---

[26]Stevens, D. M., and T. Mohr, *Artificial Insemination of Range Cattle in Wyoming: An Economic Analysis*, Wyo. Bull. 496, 1969.

[27]In a personal communication to the author, Dr. Clyde Stormont, Professor of Immunogenetics, Department of Reproduction, School of Veterinary Medicine, University of California, Davis, reported that in the California Laboratory they have been able to solve approximately 91% of all cattle and horse parentage cases.

common U.S. Breeds of Livestock and their characteristics.

Sometimes people construe the write-up of a breed of livestock in a book or in a U.S. Department of Agriculture bulletin as an official recognition of the breed. Nothing could be further from the truth, for no person or office has authority to approve a breed. The only legal basis for recognizing a breed is contained in the Tariff Act of 1930, which provides for the duty-free admission of purebred breeding stock provided they are registered in the country of origin. But this stipulation applies to imported animals only.

In this book, no *official* recognition of any breed is intended or implied. Rather, the author has tried earnestly and without favoritism to present the factual story of the breeds. In particular, such information relative to the new and/or less widely distributed breeds is needed, and often difficult to come by.

TABLE 3-20

BREEDS OF BEEF AND DUAL-PURPOSE CATTLE AND THEIR CHARACTERISTICS

Breed	Place of Origin	Color	Distinctive Head Characteristics	Other Distinguishing Characteristics	Disqualifications; Comments
Angus	Scotland; in the northeastern counties of Aberdeen, Angus, Kincardine, and Forfar.	Black.	Polled.	Comparatively smooth coat of hair. Somewhat cylindrical body.	Horns, scurs, or buttons. Red color. A noticeable amount of white above the underline, or in front of the navel, or on one or more legs. Calves from females less than 18 mo. of age when calf was dropped, or from bulls less than 6 mo. of age at the time of service.
Barzona	U.S.A.; on the Bard Ranches of Kirkland, Ariz.; hence, the name Barzona (a contraction of Bard and Arizona). The foundation of the breed, which was laid in 1942, consisted of Africander, Hereford, Santa Gertrudis, and Angus.	Red.		Well adapted to arid and semiarid ranges (the Bard Ranch, where the breed was developed, has an average rainfall of 12.76 in.).	The breeding program followed by Bard Ranches to create the Barzona breed consisted of (1) forming a large genetic pool by crossing breeds, then breeding within the herd, and (2) using records to eliminate undesirable genes and retain desirable genes.
Beef Friesian	U.S.A.; based on dual-purpose Friesians brought from Europe, beginning with an importation from Ireland in 1972. In Europe, the Friesian has always been a dual-purpose animal, whereas the American descendant, the Holstein-Friesian, has been developed exclusively as a dairy breed.	Black and white. Beef Friesian X Angus are generally black.	Broad muzzle, open nostrils, strong jaw, broad and moderately dished forehead, straight bridged nose.	Rate and efficiency of gains comparable to the exotics; little calving difficulty; good milking ability.	Beef Friesians are being developed by 3 approaches: (1) from purebred Beef Friesians, based on European stock; (2) through crossing Beef Friesian bulls on Holstein females, thence grading up; and (3) from a Beef Friesian X Angus cross.
Beefmaster (approx. ½ Brahman, and ¼ each Shorthorn and Hereford)	U.S.A.; Lasater Ranch, Falfurrias, Tex., beginning in 1908.	Red is the dominant color, but color is variable and is disregarded in selection.	The majority are horned, although a few are naturally polled.	Good milk producers under range conditions; heavy weaning and mature weights.	In order that each Beefmaster may be permanently identified with the breeder thereof, the breeder must use a prefix name such as "Jones Beefmaster," "Smith Beefmaster," etc., to designate his cattle. Thus, in a unique way, the responsibility for the continued improvement of the breed is placed squarely upon the individual breeder.

(Continued)

TABLE 3-20 (Continued)

Breed	Place of Origin	Color	Distinctive Head Characteristics	Other Distinguishing Characteristics	Disqualifications; Comments
Belted Galloway	Scotland; in the southwestern district of Galloway. First imported to the U.S. in 1948.	Black with a brownish tinge, or dun; with a white belt completely encircling the body between the shoulders and the hooks.	Polled.	Striking white belt; heavy coat of hair.	Red color, incomplete belt, other white marks, or scurs.
Blonde d'Aquitaine	Southwest France; in 1961, when three French strains of similar background—Garonne, Quercy, and Pyreneenee—combined.	Yellow, brown, fawn, or wheat colored.		The breed is relatively fine boned. There is little calving difficulty, due to the width and shape of the pelvis.	In France, Blonde d'Aquitaine are usually performance and progeny tested. Generally, the top third of the bulls in a performance test are subsequently progeny tested.
Braford (approx. 5/8 Hereford and 3/8 Brahman)	U.S.A.; on Adams Ranches, Fort Pierce, Fla., beginning about 1948. Breed registry not formed until 1973. Evolved from crossing Brahmans and Herefords.	Red or brindle, with white markings on the head and pigmentation around the eyes.	Horned.	Short haired; heat tolerant; only a slight hump; fertile; good milk production. Mature bulls weigh 1,500 to 2,000 lb, and cows 1,000 to 1,500 lb.	Offspring cannot be registered if they are from cows that have not calved annually, that required veterinary assistance at calving or that have bad udders. For registration, the Association requires a pedigree, performance records, and that the animal pass an inspection.
Brahman	India (but a distinct American breed has been created through the amalgamation of several Indian types, probably with a small infusion of European breeding).	Gray or red preferred; either solid color, or a gradual blending of the 2. However, there are brown, black, white, and spotted Brahmans.	Drooping ears. A long face.	Prominent hump over the shoulders. An abundance of loose, pendulous skin under the throat and along the dewlap. A voice that resembles a grunt rather than a low.	Brindle, grulla (a smutty or blackish red), or albino color; muzzle, hoofs, and switch of a color other than black, if all occur in same animal. Cryptorchid bull. Freemartin heifer. Inherited lameness. Dwarf or midget characteristics. Brahman are well adapted to hot, insect-infested areas, and to sparse vegetation.
Brangus (3/8 Brahman X 5/8 Angus)	U.S.A.; on Clear Creek Ranch, Welch, Okla., owned by Frank Buttram, beginning in 1942.	Black.	Polled, with evidence of Brahman influence.	Slight crest over the neck. Smooth, sleek coat.	Horns; any color other than black; white ahead of navel; small for age; extremely nervous; too fine boned.
Charbray (3/4 Charolais X 1/4 Brahman to 7/8 Charolais X 1/8 Brahman)	U.S.A.; in the Rio Grande Valley of Texas, beginning in the late 1930s. From Charolais X Brahman crosses.	Light tan at birth, but usually change to a cream white in a few weeks.	Horned.	A slight hint of the Brahman dewlap remains. The Charbray has the growth thrust of the Charolais and the heat-insect tolerance of the Brahman.	To qualify for registration, Charbray cattle must have at least 1/4 Brahman. Charolais-Brahman of lesser percentages are recorded but not considered registered.
Charolais (usually spelled Charollais in France)	France; in the province of Charolles, in central France. Later, in the province of Nivernais. Breed society founded in France in 1887.	Light tan at birth. Changes to cream white in few weeks.	Horned.	Pink skin and mucous membranes. Noted for large size, growth thrust, and bred-in red meat.	The association disqualifies any animal that (1) has a black nose, (2) is spotted, or (3) has excessive dark skin pigmentation.

(Continued)

TABLE 3-20 (Continued)

Breed	Place of Origin	Color	Distinctive Head Characteristics	Other Distinguishing Characteristics	Disqualifications; Comments
Chianina	Central Italy; in the Chiana Valley (from which they take their name), in the province of Tuscany. Of very ancient origin, going back to the days of the Roman Empire, when they were used for draft.	Porcelain white hair, black switch, and dark skin. Calves are born tan colored, which gradually turns to white at about 60 days of age.	Horned. Narrow head with black pigmentation around eyes, black tongue, and black nose and palate.	The largest breed of cattle in the world. Mature bulls stand about 6 ft high (18 hands) at the withers and weigh up to 4,000 lb. Mature cows weigh up to 2,400 lb. The breed is also noted for trimness of middle; fineness of head, horn, and bone; absence of excessive dewlap and brisket; and poor milkers.	Calving difficulties are infrequent, perhaps due to the rather small heads and long, narrow bodies of the newborn. The growth rate and leanness of the breed give Chianina bulls an important role as a terminal cross in a crossbreeding program.
Devon	England; in the counties of Devon and Somerset.	Red. A rich, dark red is preferred; hence, the name "Ruby Red."	Creamy white horns with black tips. Also, there are polled strains.	Switch varies from whitish red to nearly white at tip. Skin is orange yellow with pigment especially noticeable around eyes and muzzle.	Double muscling. Dwarfism. Excessive white color.
Dexter	Ireland; in the southern and southwestern parts. They were named after their founder, a Mr. Dexter.	Black or red.	Horned. Head is rather long.	Small size and short legs, with smallness accentuated by shortness of legs from knees and hocks down. Mature bulls should not exceed 1,000 lb and mature cows 800 lb. Some mature animals are less than 40 in. high.	Animals having white other than on the belly, switch, udder, or scrotum are disqualified for registry. Bulldog calves, a lethal condition, occurs in some animals of the breed.
Fleckvieh (German Simmental)	Southern Germany; where it has been bred since 1895. Evolved from Simmental cattle, which originally came from Switzerland.	Generally red and white spotted, with a white face. The red varies from dark to a more common diluted, almost yellow shade.	Horned.	In Germany, it is considered a dual-purpose breed, with emphasis on beef. Mature bulls average about 2,550 lb, and cows about 1,550 lb.	The progeny testing and selection program of the Fleckvieh breed in Bavaria is, without doubt, one of the best in the world. To be selected for licensing as a dam of a herd sire, a cow must be in the top 8% of the breed on milk production; classified by a committee for size and conformation; and must meet rigid standards for calving intervals, calving ease, milking ease, disposition, and pedigree.
Galloway	Scotland; in the southwestern province of Galloway.	Black; sometimes with a brownish or reddish tint; or dun. Also, they may be white or belted.	Polled.	Long curly hair; hardiness and ability to rustle in cold weather.	Scurs or horns. White markings on feet or legs, in front of the navel, or above the underline.
Gelbvieh (German Yellow)	Germany; in Bavaria. Descended from red-brown Keltic-German Landrace, on which Simmental and Shorthorn were crossed in early 1800s. Actually came into being when 4 breeds of German cattle—Franconian, Glan-Donnersberg, Lahn, and Limpurg—amalgamated around 1920. First imported to North America in 1972.	Golden red to rust. Solid color.	Horned.	Large, long-bodied, well-muscled, fast-gaining, and high-quality carcasses. Mature bulls weigh from 2,300 to 2,800 lb (average 2,500), and cows 1,400 to 1,800 lb (average 1,500).	In Germany, no Gelbvieh bull is put into general A.I. service until he is 6 years old and his progeny have proven him superior to his contemporaries.

(Continued)

TABLE 3-20 (Continued)

Breed	Place of Origin	Color	Distinctive Head Characteristics	Other Distinguishing Characteristics	Disqualifications; Comments
Hays Converter	In Canada; by the former Minister of Agriculture, Harry Hays, beginning in 1957. Foundation breeds were Hereford, Brown Swiss, and Holstein. Claimed that the breed converts feed into profit; hence, the name "Converter."	Predominant color is black with a white face, white feet, and a white tail. A few are red with white faces. Color is not a factor in selection.		Traits upon which Sen. Hays built the breed are (1) growth; (2) fertility; (3) minimum calving problems; (4) well-attached udders; (5) abundant milk; (6) sound feet and legs; and (7) pigmentation. The Hays Converter is a "dual-purpose beef breed." Mature bulls weigh about 2,200 lb and cows 1,400 lb.	The system and steps used to produce the breed were (1) selected from different breeds the important characteristics needed; (2) combined the genes into one large breeding population; (3) selected intensely for important characteristics and culled ruthlessly for several generations; and (4) measured the resulting animals after hybrid vigor was no longer important, to determine the transmissible genetic superiority.
Hereford	England; in the county of Hereford. The first importation of Hereford cattle into the U.S. was made by Henry Clay of Kentucky, in 1817.	Red with white markings; white face and white on the underline, flank, crest, switch, breast, and below the knees and hocks. White back of the crops, high on the flanks, or too high on the legs is objectionable. Likewise, dark or smutty noses and red necks are frowned upon.	Horned. The white face is the distinct trademark of the breed.	A thick coat of hair.	Calves from females less than 21 mo. of age when calf was dropped, or from bulls less than 12 mo. of age when service producing the calf occurred, cannot be registered.
Indu Brazil (Zebu)	Brazil.	Light grey to silver grey; dun to red.	Prominent forehead and long drooping ears. Symmetrical horns drawing upward and to the rear.	Prominent hump over the shoulders. An abundance of loose, pendulous skin under the throat and along the dewlap. A voice that resembles a grunt rather than a low.	Brindle color combinations. White markings on the nose or switch. Absence of loose, thick, mellow skin. Weak and improperly formed hump.
Limousin	Southwestern France; in the 19th Century. Presently, found in largest numbers around Limoges. The breed takes its name from the Limousin Mountains. First imported to North America in 1967.	Wheat to rust red.	Horned.	Modern meat-type cattle—long, relatively shallow, with moderate to heavy muscling. Mature bulls average about 2,400 lb and cows about 1,300 lb. Noted for ease of calving and high carcass quality.	The Limousin is one of the new European breeds raised primarily for meat production, rather than the dual-purpose of meat and milk.
Lincoln Red	England; in Lincolnshire—the rugged east coast in England. They became the Lincoln Red in 1960.	Deep cherry red, with occasional white markings.	There are both horned and polled strains, with the polled predominating in their native land.	A long body; light birth weights and ease of calving; pigmentation; excellent milk production; fast growth rate; and good fertility. In England, they are a dual-purpose breed, with separate strains for beef and dairy.	It appears that the main use for the breed in America will be in crossbreeding to produce F_1 females, since they make excellent brood cows.

(Continued)

TABLE 3-20 (Continued)

Breed	Place of Origin	Color	Distinctive Head Characteristics	Other Distinguishing Characteristics	Disqualifications; Comments
Maine-Anjou	Western France; in the provinces of Maine and Anjou, from which it takes its name. The French Herd Book was established in 1919.	Dark red with white underline, often with small white patches on the body. Also, dark roans are found.	Most heads are either red, or the eyes are surrounded by red.	The Maine-Anjou is the largest of the French breeds. Mature bulls weigh 2,500 lb or above, and cows 2,000 lb or more. Considered a dual-purpose breed, with emphasis on beef. Cows average about 5,000 lb of milk per lactation, testing 3.7%. They are long, rather up-standing, have a particularly long rump, and are noted for rapid growth.	The logical place for Maine-Anjou in American crossbreeding systems is as maternal sires, although use as terminal sires may occur to some extent.
Marchigiana (pronounced Mar-key-jahna)	Italy; in the Marche region, around Rome. With the fall of the Roman Empire in the 5th Century, nomadic cattle were crossed with the 2 native Italian breeds of the time—the Chianina and the Romagnola. Out of these crosses evolved the basic foundation stock for the Marchigiana. The Herd Book was established in Italy in 1930.	Grayish white, although bulls may be darker. Dark skin pigmentation, and dark muzzle, switch, and below or around the eyes. Calves are born tan but turn white at about 2 months of age.	Horns that appear small in proportion to the size of the cattle.	Ability to do well under adverse conditions. Mature bulls weigh 2,650 to 3,100 lb. Mature cows weigh from 1,400 to 1,800 lb.	In Italy, Marchigianas have been very popular in crossbreeding programs with dairy cattle.
Murray Grey	Australia; from a mating, first made by the Sutherlands on "Thologolong," in the Murray Valley, near Wodonga, Victoria, Australia, in 1905, of a very light roan (almost white) Shorthorn cow and an Angus bull. Because of the use of Angus bulls following the first cross, the Murray Grey is predominantly Angus. The Murray Grey Beef Cattle Society, of Australia, was formed in 1962; and the American Murray Grey Association, Inc., was organized in 1970.	Silver-grey color, which adapts them to sunny areas, as well as colder areas.	Polled.	Ease of calving, because of small calves at birth; dark skin pigmentation, which lessens cancer eye; superior carcass; good dispositions. Bulls weigh around 2,000 lb at maturity, and females from 1,100 to 1,300 lb.	In the American Murray Grey Assn., females with 7/8 Murray Grey blood can be registered. Bulls are eligible for registry with 15/16 Murray Grey blood. In addition, Recordation Certificates can be obtained on any crosses of 1/2 or more Murray Grey breeding. A ranch prefix (the owner's last name, the ranch name, or whatnot) is required of each breeder.
Normande	France; in the areas of Normandy, Brittany, and Maine. The breed registry was established in 1883.	Primarily dark red and white. Colored patches around the eyes give them a "bespectacled" appearance and resistance to cancer eye and pinkeye; and dark pigmentation on the udder prevents sunburn.	Bespectacled eyes, due to dark coloring.	The Normande is known as a dual-purpose breed in France. Mature bulls in good condition average about 2,425 lb, although weights up to 2,850 lb have been reported. Cows weigh 1,550 to 1,750 lb and produce an average of about 8,800 lb of milk per year.	

(Continued)

TABLE 3-20 (Continued)

Breed	Place of Origin	Color	Distinctive Head Characteristics	Other Distinguishing Characteristics	Disqualifications; Comments
Norwegian Red	Norway, where they are known as Norwegian Red-and-White. The first importation of the breed into the U.S. was made by the Southern Cattle Corporation, Memphis, Tenn., in 1973.	Red; red and white.	Horned.	Abundant milk production; excellent feed conversion; and good carcasses. Mature bulls weigh from 2,200 to 2,640 lb, and mature cows from 1,210 to 1,430 lb.	In Norway, the Norwegian Red-and-White is a dual-purpose breed, kept for both milk and beef. Today, the Norwegian Red-and-White is the dominant breed of Norway; there are only 200 cattle of other breeds. It is noteworthy, too, that 58% of the cattle of Norway are registered purebreds.
Piedmont (Piemontese)	Italy; where they are the most popular breed.	White or pale grey with black points.		About 80% of the bulls of the Piedmont breed are double muscled to some degree. In Italy, they report 9% higher dressing percentage and twice the steaks from double-muscled cattle over normal cattle. Piedmont cattle command a very considerable premium on the Italian market. Originally, the Piedmont was considered to be a dual-purpose breed in Italy, but today it is selected and bred for beef qualities.	In Italy, under their system of intensive care, breeding and calving problems of double-muscled Piedmont cattle do not appear to be serious. Double-muscled Piedmont cattle mean to the cattle industry of Italy what broad-breasted turkeys and Cornish cross broilers mean to the U.S. poultry industry; all are meat producers par excellence.
Pinzgauer (Pinzgau)	Austria and adjacent areas of Italy and Germany; in the Alpine region.	Chestnut brown sides, with a white top-line and underline, and usually white feet. Deep orange pigment around eyes and on udder.	Horned.	A "beefy" breed; more so than most of the exotics, although it is classed as a dual-purpose breed in its native land. Mature bulls weigh 2,200 to 2,900 lb; and mature cows from 1,300 to 1,650 lb. Breed is noted for hardiness, longevity (the oldest cows and bulls reach 17 to 18 years of age), fertility, and foraging ability.	In Austria, all animals are subjected to and must pass a rigid conformation test before they can be registered. Additionally, performance of dams and daughters in milk production and butterfat content is a criterion in the selection of breeding bulls.
Polled Hereford	U.S.A.; in Iowa, from a mutation (polled) that appeared in the horned Hereford breed.	Red with white markings, white face and white on underline, flank, crest, switch, breast, and below the knees and hocks. White back of the crops, high on the flanks, or too high on the legs is objectionable. Likewise, dark or smutty noses are frowned upon.	Polled, with a white face.	A thick coat of hair.	Horned animals are disqualified. No calf is eligible for registration unless its sire was at least 12 mo. of age at the time of conception, and its dam at least 21 mo. of age at the time of calving. Polled Herefords recorded in both the American Hereford Breeders Assn. and the American Polled Hereford Breeders Assn. are known as double standard. Those that can be recorded only in the American Polled Hereford Breeders Assn. are called single standard.

(Continued)

TABLE 3-20 (Continued)

Breed	Place of Origin	Color	Distinctive Head Characteristics	Other Distinguishing Characteristics	Disqualifications; Comments
Polled Shorthorn	U.S.A.; in the north-central states, chiefly Ohio and Indiana.	Red, white, or any combination of red and white. A "smutty nose" or dark nose is objectionable.	Polled.	Other than being polled, they resemble Shorthorns, except there are more spotted animals among them.	Horned animals are disqualified.
Ranger	U.S.A.; beginning in 1950, on the following three ranches: Barnes Livestock Company, Riverton, Wyo.; W. W. Ritchie and Family, Buffalo, Wyo.; and Watson Cattle Company, Cedarville, Calif. The following breeds were used in developing the Ranger: Hereford, Milking Shorthorn, Red Angus, Shorthorn, Beefmaster, Scotch Highland, and Brahman. The name "Ranger" was selected because the breed was developed on, and is adapted to, the range areas of the West.	They run the gamut of cattle colors, including both solid and broken colors.		Medium size; hardy; fertile—animals have been selected to calve at an early age, at yearly intervals or less, and without assistance; adequate and persistent milk production; heavy weaning weight; good carcass quality.	The developers of the breed refer to it as, "a 'cow' breed for the cowman who must have a profitable commercial operation."
Red Angus	Scotland.	Red.	Polled.	Similar to black Angus, except for recessive red color. In England and Scotland, both reds and blacks are registered in the same association, without distinction. In the U.S., however, red-colored animals have been barred from registry in the American Angus Assn. since 1917. The Red Angus Assn. of America was organized in 1954.	White any place other than on underline; dwarf, double muscling or other genetic defects.
Red Brangus	U.S., from Brahman X Angus cross, made in 1946. Registry chartered in 1956.	Red.	Broad head with slightly curved forehead and straight profile; with medium sized, moderately drooping ears.	Males have crest immediately forward of the shoulders. Smooth, sleek coat.	White spotting other than on the underline, brindling, or roan on the body, or black skin or mucous membrane. Long hair, or tight hide. Undersized; too rangy or too compact. Mature females with underdeveloped teats or udders. Mature males with an excessive or pendulous sheath, or the absence of a sheath.
Red Poll	England; in the eastern middle coastal counties of Norfolk and Suffolk.	Red, varying from light to dark red. Any white except in the switch is discriminated against. Also, a smoky nose or dark spots on the nose are objectionable.	Polled.		White above underline, above switch of tail, or on legs. Bulls with white on underline forward of the navel region; or with only one testicle. Solid black or blue nose. Scurs or any horny growth. Total blindness.
Salers	France; in the south-central area—a mountainous region. The name "Salers" was first applied to the cattle of this area in 1840, after a small town located in the center of the province. The Salers Herd Book was started in 1906.	Solid, deep cherry red, with a white switch and sometimes white spots under the belly.	Horned, although there are polled strains.	In their native land, the breed is noted for rapid gain, hardiness, and adaptability. Mature bulls have an average weight of 2,530 lb; mature cows average about 1,540 lb. The Salers was founded as a dual-purpose breed.	Salers International, the breed registry, makes the claim that no genetical defect and no double muscling has ever been reported in the Salers breed.

(Continued)

TABLE 3-20 (Continued)

Breed	Place of Origin	Color	Distinctive Head Characteristics	Other Distinguishing Characteristics	Disqualifications; Comments
Santa Gertrudis (5/8 Shorthorn and 3/8 Brahman)	U.S.A.; on the King Ranch in Texas, based on a Shorthorn X Brahman cross. The foundation sire, "Monkey," was born in 1920. Named from the Santa Gertrudis Land Grant, granted by the Crown of Spain, on which the breed evolved, now the headquarters division of King Ranch.	Red or cherry red.	Generally horned, but there are polled strains.	Hair should be short, straight, and slick. Hide should be loose, with surface area increased by neck folds and sheath or navel flap. But neither should be excessive.	White spots out of underline; fawn or cream color, brindling or roan condition; solid black. Heredity deformities, such as hernia, cryptorchid, wry nose, wry tail, double muscling, malformed genitalia, undershot and overshot jaw.
Scotch Highland (or Highland)	Scotland.	Red, yellow, silver, white, dun, black, or brindle.	Short head; long widespread horns; and heavy foretop.	Long, shaggy hair; short legs. Hardiness and ability to rustle in cold weather.	Polled and spotted animals are disqualified. They should be solid color except for an occasional white tip on switch, or white on the underline or udder.
Shorthorn	England; in the northeastern counties of Durham, Northumberland, York, and Lincoln. The Shorthorn was the first breed of cattle to have a Herd Book—the Coates Herd Book, founded by George Coates in 1822. The name "Shorthorn" stems from the fact that, through breeding and selection, the early improvers of the breed shortened the horns of the native cattle.	Red, white or any combination of red and white. A "smutty nose" or dark nose is objectionable.	Rather short, refined, incurving horns.		No calf is eligible for registration unless its sire and dam were each at least 18 mo. of age at the birth date of the calf.
Simmental	Western Switzerland; in the Simme Valley, from which it derives its name. It is much older than the herd register, which was set up in Bern, Switzerland, in 1806. The first Simmental bull was brought to North America in 1967, by a group of southern Alberta cattlemen headed by Travers Smith of Cardston, Alberta, Canada.	Generally red-and-white spotted, although some are nearly solid in color. The red varies from dark to a more common diluted, almost yellow, shade. A white face, which, like the Hereford, appears to be dominant in inheritance.	Horned.	The Simmental was first developed as a dual-purpose breed. They combine meat and milk to an unusually high degree, along with rapid growth rate. Mature bulls average about 2,300 to 2,400 lb, and cows about 1,600 to 1,700 lb. The breed milk production average is about 8,000 lb, with a 4% butterfat test.	Genetic unsoundness. To qualify for registration, animal must be at least ½ Simmental. Any animal whose sire cannot be determined through blood typing cannot be registered.
South Devon	England; originated in southern Devonshire, through infusion of Guernsey blood into the Devon breed. The South Devon has had its own Herd Book in England since 1891. Henry Wallace, former U.S. Vice President, brought the first shipment of South Devons to the U.S. in 1936.	Medium light red color.		The South Devon is a dual-purpose breed. Mature bulls weigh 2,000 to 2,800 lb; and mature cows weigh 1,200 to 1,700 lb. Cows are heavy milkers; they average about 6,550 lb of milk per lactation, with 4.2% fat.	The South Devon is the only breed in England that both (1) receives a Milk Marketing Board premium for rich milk, and (2) qualifies for the British Beef Subsidy. The breed is extolled as a "dam breed," superior in maternal traits.

(Continued)

TABLE 3-20 (Continued)

Breed	Place of Origin	Color	Distinctive Head Characteristics	Other Distinguishing Characteristics	Disqualifications; Comments
Sussex	England; descended from red cattle that inhabited Sussex and Kent Counties at the time of the Norman Conquest (1066). First registered in England in 1840. First imported into the U.S. in 1883.	Deep mahogany red.	A high percentage of Sussex cattle are polled.	In England, the breed has earned the reputation as the "butcher's beast," because of the evenness of fleshing, predominance of lean meat, and high dressing percentage.	In the U.S., the Sussex Cattl Assn. of America register cattle in the English Her Book. American entries i the English Herd Boo commenced in 1967. Sussex cattle were immor talized by Rudyard Kiplin in the poem, "Alnasche and the Oxen."
Tarentaise (Tarine)	France; in the Alps, where the cattle were known as Tarentaise since 1863. The Herd Book was started in 1888.	Solid wheat-colored, ranging from a light cherry to dark blond. Bulls tend to darken around the neck and shoulders with maturity. Frequently, bulls have a darker dorsal stripe.	Black pigmentation of the muzzle and around the eyes.	Noted for easy calving, due to adequate pelvic capacity and small calves; vigorous calves at birth; hardiness; black hair around the eyes and pigmented udders and teats, thereby making cancer eye and sunburned teats rarities; good fertility; and milking ability, with cows averaging about 8,000 lb per lactation. Smaller than most of the exotics. Mature bulls average about 1,800 lb and cows about 1,150 lb.	In France, the breed registr advocates eliminating (a though it does not "disqua ify") widespread patches c white hairs or badger gre coloring, bright red c mahogany overall color, stripe on the back lighte than the general colorinc very dark or black parts c the coat (cheeks, dewlar shoulders, etc.); total ab sence of black pigmenta tion on mucous membrane and extremities; poor gen eral conformation, particu larly a crest-shaped tail. Calves are small and vigorou at birth; hence, Tarentais bulls are well suited for us on virgin heifers.
Texas Longhorn	U.S.A.: from cattle of Spanish extraction. On his second voyage in 1493, Columbus brought Spanish cattle to Santo Domingo. By 1900, the Texas Longhorn was driven to near extinction, replaced by the European breeds—the Shorthorn, Hereford, and Angus. The Texas Longhorn Breeders Assn. of America was organized in 1964. At that time, there were only approximately 1,500 head of genuine Texas Longhorn cattle in existence.	Texas Longhorns are characterized by a great array of colors, in all degrees of richness, and in all possible combinations and patterns.	Large, spreading horns that curve upward. Long head, with small ears. Longer hair between the horns than on the body.	Long legs and high shoulders. Noted for fertility, ease of calving, hardiness, resistance to many common diseases, rustling ability, good feet and legs, longevity, and adaptation to a wide variety of environmental conditions.	
Welsh Black	Wales; where they have long been bred as dual-purpose cattle. The Canadian Welsh Black Cattle Society was formed in 1970.	Black.	Horned, although there is a polled strain in Wales.	High fertility; little calving difficulty; good milk production; adapted to harsh conditons of climate and forage; longevity; relative freedom from sunburned udders and cancer eye. Mature bulls weigh 1,800 to 2,000 lb, and cows from 1,000 to 1,300 lb. Cows give 6,000 to 7,700 lb of milk per lactation.	The major impact of the bree is expected to be on broo cows—as maternal sires i a crossbreeding program. In Wales, white except o udder or scrotal area is disqualification.

TABLE 3-21

BREEDS OF DAIRY CATTLE AND THEIR CHARACTERISTICS

Breed	Place of Origin	Color	Distinctive Head Characteristics	Other Distinguishing Characteristics	Disqualifications; Comments
Ayrshire	County of Ayrshire in southwestern Scotland.	Light to deep cherry red, mahogany, brown, or a combination of these colors, with white or white alone. Black or brindle are objectionable.	Horns are widespread and tend to curve upward and outward. However, there is a polled strain.	The udders are especially symmetrical and well attached to the body. The breed is noted for its style and animation, good feet and legs, and grazing ability.	
Brown Swiss	The Alps of Switzerland. Brought to America in 1869.	Solid brown, varying from very light to dark. White markings are objectionable.	The nose and tongue are black, and there is a characteristic light-colored band around the muzzle. Medium-length horns.	Strong and rugged, with some tendency toward the heavy muscling characteristic of the beef breeds. Calm and unexcitable.	Spotting is undesirable. In 1971, The Brown Swiss Cattle Breeders Association formed the Brown Swiss Beef International, Inc., for registration of beef-type Brown Swiss.
Dutch Belted	Holland, prior to 17th Century.	Black and white.	Horns. Head somewhat long and dished.	White belt extending entirely around the body, from a little back of the shoulder to just in front of the hips.	No belt.
Guernsey	Isle of Guernsey.	Fawn with white markings clearly defined; preferably a clear (buff) muzzle.	Good length of head; horns incline forward, are refined and medium in length, and taper toward the tips.	The milk is especially yellow in color; golden yellow skin pigmentation; the unhaired portions of the body are light or pinkish in color (whereas in the Jersey they are near black); calves are relatively small at birth.	
Holstein-Friesian	Netherlands and northern Germany.	Black and white or red and white.	Clean-cut, broad muzzle, open nostrils, strong jaw, broad and moderately dished forehead, straight bridged nose.	Large angular animal; females should weigh 1,500 lb (mature); males in breeding condition 2,200 lb.	Black and white animals are disqualified if (1) solid black, (2) solid white, (3) black in switch, (4) solid black belly, (5) one or more legs circled with black that touches the hoof, (6) black on one or more legs from the hoof to above knees or hocks, or (7) grayish color. Suffix "Red" is added to name of red and white animals.
Illawarra	Australia; from Bates Shorthorns, with infusion of Ayrshire and Devon breeding. In the U.S., the International Illawarra Association was formed in 1974.	Predominantly red. Also, red and white.	Horns.	Cows flesh up quickly when not lactating. Mature cows weigh 1,300 to 1,600 lb.	The Illawarra is the first new dairy breed in 100 years. Illawarras are not closely related to U.S. Milking Shorthorns, at least in the last 100 years.
Jersey	Island of Jersey.	Jerseys vary greatly in color, but the characteristic color is some shade of fawn, with or without white markings.	Forehead, broad and moderately dished with large, bright eyes. Head clean-cut and proportionate to body.	Jerseys are especially known for their well-shaped udders, strong udder attachments, and ease of calving. They are also very angular and refined.	Total blindness, permanent lameness that interferes with normal function, blind quarter, freemartin heifers, and animals showing signs of being operated upon or tampered with.
Milking Shorthorn	England; the breed traces to a milking strain of Shorthorns developed by Thomas Bates.	Red, white, or any combination of red and white.	Fine horns that are rather short.	Good milk production.	No calf is eligible for registration unless its sire and dam were each at least 18 mo. of age at the birth date of the calf. In 1949, the Milking Shorthorn breed split off from the American Shorthorn Assn. and formed a separate breed registry—the American Milking Shorthorn Society. In 1973, the American Shorthorn Assn. made provision to register Milking Shorthorns in its Herd Book. But there is no reciprocal arrangement for acceptance of Shorthorn (beef) blood in the American Milking Shorthorn Herd Book.

TABLE 3-22

BREEDS OF SHEEP AND THEIR CHARACTERISTICS

Breed	Place of Origin	Color	Distinctive Head Characteristics	Other Distinguishing Characteristics	Disqualifications; Comments
(Classified by type of wool produced)[1] *FINE-WOOL BREEDS:* **American Merino**	Spain.	White. Reddish-brown spots may occasionally appear on lips, ears, and pasterns.	Most rams have horns, but there are some polled strains.	Distinguished from the Delaine Merinos by more skin wrinkles; the more wrinkled American Merinos being the "A" and "B" types. Strong flocking instinct. Ewes will breed out of season.	
Debouillet	U.S.A.; on the Amos Dee Jones Ranches of Roswell and Tatum, N.M., beginning in 1927-30. Association was organized in 1954.	White.	Rams may have horns, but there are also polled strains; open face.	Comparatively smooth body; long staple.	Overshot or undershot jaw; broken-down pasterns; undersized; brown hair on face, ears, or legs; black spots in fleece; too light fleece.
Delaine Merino (the "C" Type, or Texas Delaine)	Spain. The Delaine Merino and Texas Delaine are of similar origin and appearance.	White. Reddish-brown spots may occasionally appear on lips, ears, and pasterns.	Most rams have horns, but there are some polled strains. The females should be free of horns or scurs.	Comparatively smooth bodied; of the "C" type. Strong flocking instinct. Ewes will breed out of season.	*Texas Delaine Sheep Assn. disqualifications* Abnormal testicles; swayback; close horns black spots in the fleece or body; overshot or undershot jaw; weak pasterns.
Rambouillet	France; from Spanish Merino parent stock imported from Spain.	Cream to white.	Most rams have horns, but there are some polled strains. Ewes are hornless.	Largest fine-wool breed. Strong flocking instinct. Ewes will breed out of season.	Abnormal development of testicles or only one testicle descended in scrotum; unsound udder or inverted teats; overshot or undershot jaws black spot in the fleece
MEDIUM-WOOL BREEDS: **Cheviot**	Scotland; in the Cheviot Hills between Scotland and England.	White face with a black nose. Often black spots are on the ears.	Both sexes are polled.	Stylish, alert, and active. Head and legs free from wool.	Black spots other than ears. Overshot or undershot jaw.
Dorset	England; especially in the southern counties of Dorset and Somerset. Polled strain was developed by North Carolina State University in the early 50s.	White and practically free from wool.	There are horned and polled strains, both of which are registered by The Continental Dorset Club. Except for the presence or absence of horns, the two strains are identical.	Ewes will breed out of season.	Incisor teeth not meeting the dental pad; abnormal testes; inverted eyelids; general off-type appearance.
Finnsheep (Finnish Landrace)	Finland. Finnsheep were brought to the U.S. in 1968. An earlier importation from Finland was made to the University of Manitoba, in Canada.	White.	Head is free of wool. Usually, both sexes are hornless, but a few rams have light horns.	Prolificacy. In Finland, ewes (all ages included) average 2.41 lambs per litter.	The wool is usually 50 to 54 spinning count with a fleece weight of about 9.5 lb. Finnsheep are being used in breeding programs to increase litter size (multiple births).
Hampshire	England; in the south-central county of Hampshire.	Rich deep brown, approaching black.	Both sexes are hornless, although rams sometimes have scurs.	Large size; early maturity.	Undesirable traits are crooked legs and poor feet; inverted eyelids abnormal sex organs black fibers; wool blindness; horns; abnormal teeth or jaw development.

Footnotes on last page of table.

(Continued)

TABLE 3-22 (Continued)

Breed	Place of Origin	Color	Distinctive Head Characteristics	Other Distinguishing Characteristics	Disqualifications; Comments
Montadale (Columbia X Cheviot)	U.S.A.; by E. H. Mattingly, St. Louis, Mo.	White.	Both sexes are polled. Head free of wool.	Black hoofs and nose; black spots in ears.	Horns. Black spots in wool. Pink nose.
North Country Cheviot	Scotland; from the old Long Hill sheep, but with infusion of Merino, Ryeland, and Southdown blood in formative period.	White.	Nose straight to slightly Roman. Rams are sometimes horned.	Wool grades 50s to 56s; mature rams weigh up to 300 lb and mature ewes up to 200 lb.	
Oxford	England; in the south-central county of Oxford. The Oxford stems from a Hampshire X Cotswold cross, made in the 1830s.	Variable, from gray to brown.	Both sexes are polled. Topknot of wool.	Largest of the Down breeds.	Black fiber; stub horns.
Shropshire	England; in the central western counties of Shropshire and Stafford.	Dark face, but a gray nose is not objectionable.	Both sexes are polled, although rams frequently have scurs.	Covering of dense wool well over the poll.	Such lack of type as to render the identify of the breed doubtful; horns or stubs (not scurs); overshot or undershot jaws.
Southdown	England; in the southeastern county of Sussex.	Light or mouse brown color preferred.	Both sexes are polled, although rams sometimes have scurs.	Superior conformation and quality of carcass.	Horns; dark poll; speckled markings on face, ears, and legs; white or approaching black face and legs; open or coarse wool; one or both testicles not descended in scrotum; too dark colored or black skin; black or brown fleece; incisor teeth not meeting dental pad; black spot on face, legs, or body.
Suffolk	England; in the southeastern counties of Suffolk, Essex, and Norfolk.	Very black head, ears, and legs.	Both sexes are polled, although rams frequently have scurs.	Head and ears are entirely free from wool.	
Tunis (or American Tunis)	Northern Africa; from the province of Tunis. First imported into U.S. in 1799.	Reddish brown to bright tan.	Both sexes are polled; long, drooping ears; head free from wool.	Originally, it was a fat-tailed sheep, which means that the tail was distinctly broad and fat. However, breeders have selected away from this trait. Pendulous ears. Will mate almost any season of the year.	Horns; red or black wool; one testicle; undershot or overshot jaw. Newborn lambs are red or tan, but they gradually turn white.
LONG-WOOL BREEDS:					
Cotswold	England; in the Cotswold hills of Gloucestershire.	White, although grayish specks and bluish tinge are common.	Both sexes are polled, although rams frequently have scurs.	Natural wavy ringlets or curls in which the fleece hangs all over the body. Tuft of wool on the forehead. Second only to the Lincoln in size.	Unsound animals.

(Continued)

TABLE 3-22 (Continued)

Breed	Place of Origin	Color	Distinctive Head Characteristics	Other Distinguishing Characteristics	Disqualifications; Comments
Leicester	England; in the central county of Leicester.	White, but may have bluish tinge or black spots.	Both sexes are polled.		The Border Leicester is a strain of Leiceste sheep. It is distin guished from the Eng lish Leicester by being open faced and bare legged and having a shorter fleece.
Lincoln	England; along the eastern coast of England and bordering the North Sea, in Lincolnshire.	White. Black spots may be present but are discriminated against.	Both sexes are polled.	Largest of all breeds of sheep. Rams weigh 250 to 375 lb; ewes 200 to 275 lb. Produces the heaviest fleece of any mutton breed.	
Romney	England; in the Romney Marsh region of the County of Kent.	White.	Both sexes are polled. Open face.	In comparison with other long-wool breeds: the Romney is shorter legged, more rugged, and its fleece is shorter, finer, and less open.	Black fleece, or black spots.
CROSSBRED-WOOL BREEDS:[2] **Columbia** (Lincoln rams × Rambouillet ewes)	U.S.A.; in Wyoming and Idaho.	White.	Both sexes are polled.	Open-faced, with no tendency to wool blindness.	Horns or scurs; woo blindness; uneven o light fleece; overshot o undershot jaw; colored wool; excessive folds.
Corriedale (Lincoln and Leicester rams X Merino ewes)	New Zealand.	White, with dark points. Black spots are sometimes present.	Both sexes are polled.		Black or brown spots. Wool blindness. Malformed mouth. Horns or scurs.
Panama (Rambouillet rams X Lincoln ewes)	U.S.A.; by Laidlaw and Brockie of Muldoon, Ida.	White.	Both sexes are polled.		Horns, scurs, or knobs overshot or undersho mouth; excessive folds or wrinkles; colored wool; colored spots larger than ¾ in. in diameter on clear areas any unsound hereditary factor.
Targhee (Rambouillet rams, Lincoln-Rambouillet-Corriedale ewes)	U.S.A.; by the USDA at Dubois, Ida.	White.	Both sexes are polled. Open-faced.	Moderately low set. Long productive life.	Black or brown color in the fleece. Horns or scurs.
CARPET-WOOL BREED: **Black-faced Highland** (or Scottish Blackface)	Scotland; in the highland country.	Black or mottled.	Both sexes have horns.	Striking stylish appearance. Fleece consists of long coarse outer coat and a finer inner coat.	
FUR-SHEEP BREED: **Karakul**	Asia; in the county of Bokhara (U.S.S.R.).	Black or brown.	Rams have horns, but ewes are hornless.	Drooping ears. A fat-tailed sheep. Lamb pelts suitable for fur production.	

[1]Breeds of sheep may be and are classified on several different bases including (1) their degree of suitability for mutton or wool production (mutton or wool type), (2) color c face (white or black), (3) presence or absence of horns (horned or polled), (4) topography of the area in which they originated (mountains, upland or lowland), and (5) type c wool produced. Each system of classification has its special merits, but perhaps a classification based on type of wool produced is as good as any.

[2]The listing of the crosses which produced each of the "crossbred wool breeds" is given for breed history purposes only, and does not imply any lack of purity of the respective breeds. Nor does it indicate that all of them are new breeds: for example, the Corriedale, which is an old breed, was originated in New Zealand about 1880.

TABLE 3-23

BREEDS OF GOATS AND THEIR CHARACTERISTICS[1]

Breed	Place of Origin	Color	Distinctive Head Characteristics	Other Distinguishing Characteristics	Disqualifications; Comments
MOHAIR-BEARING GOAT BREED: **Angora**	Turkey; in the province of Angora.	White face, ears, legs, and mohair.	Both sexes are usually horned, but polled individuals occur.	Outer coat consists of long locks or strands of mohair. Rather thin, long, and pendulous ears.	Any deformed or unsound animal; or any animal having black horns or colored hair, or body not entirely covered with mohair.
MILK GOAT BREEDS:[1] **American La Mancha**	U.S.A., from a short-eared Spanish breed crossed on leading purebred breeds.	Any color or combination of colors.	Short ears or no ears; straight nose; hornless or neatly disbudded.	The hair is short, fine, and glossy. The different-type ears are known as "gopher" or "cookie."	Anything other than gopher ears in males. Ears other than true La Mancha type in females.
French-Alpine	France; originally from the French Alps.	Multicolored coats, with no standard markings.	Some have horns at birth and are disbudded, but others are hornless; erect ears; straight nose.	Large and rangy, yet deerlike. Erect ears; straight face.	Pendulous ears.
Nubian	The Nubian in the U.S. is of mixed origin. It evolved out of crossing Indian Jumna Pari and Egyptian Zariby types on British dairy goats.	They may be any color or colors, solid or patterned. Common colors are: black, gray, cream, white, shades of tan, brown, and rich reddish-brown. Common markings include lighter ears, facial stripes, muzzle, crown, and/or undertrim; over-all light- or dark-colored spots or patches of any size are often found.	Some born with horns and disbudded; others are hornless. Long drooping ears. Roman nose and prominent forehead. Does are beardless.		Upright ears. Dished face.
Rock Alpine	U.S.A.	Multicolored coats, with no standard markings.	Some have horns at birth and are disbudded, but others are hornless; erect ears; straight nose.		
Saanen	Switzerland, in the Saanen Valley.	Pure white or creamy white. The cream color may vary from light to dark fawn.	Hornless animals preferred; straight nose; erect ears.	Largest of the Swiss breeds.	Large (1½" diameter or more) dark spot in hair; pendulous ears.
Swiss Alpine	Switzerland.	Chamoise; solid brown, ranging from light to a deep red bay. Black points.	Hornless or neatly disbudded. Erect ears.		
Toggenburg	Switzerland, in the Toggenburg Valley; but they originally came from the Swiss Alps.	Light fawn to dark chocolate, with 2 white stripes on the face and white on the legs below the knees.	Hornless or debudded; straight or dished nose; erect ears.	Medium size, sturdy, and vigorous.	Tricolor or piebald; large (1½" or more) white spot in males; pendulous ears.

[1]In addition to the specific breed disqualifications given in the right-hand column, the American Dairy Goat Association lists the following disqualifications in any breed: total blindness; permanent lameness or difficulty in walking; blind or nonfunctioning half of udder; blind teat; double teats, extra teats that interfere with milking; hermaphrodism; navel hernia; crooked face in bucks; and extra teats, teats cut off or double orifice in bucks.

TABLE 3-24

BREEDS OF SWINE AND THEIR CHARACTERISTICS

Breed	Place of Origin	Color	Distinctive Head Characteristics	Other Distinguishing Characteristics	Disqualifications; Comments
American Landrace	Denmark	White, although small black skin spots are common.	Medium lop ears, straight snout, and trim jowl.	Very long side. Frequently weak pasterns.	Black in the hair coat. Fewer than 6 teats on either side. Erect ears, with no forward break. Swirls on topline (head to tail).
Berkshire	England; chiefly in the south central counties of Berkshire and Wiltshire.	Black with 6 white points, 4 white feet, some white on the face, and a white switch on the tail. Any or all white points may be missing.	Medium short nose and erect ears.	Striking style and carriage.	A swirl on upper half of body; more than 10% white; any color other than black or white; hernia, cryptorchid, absence of anal opening, rectal or uteral prolapse. Extreme pug nose is objectionable.
Chester White	U.S.A.; chiefly in Chester and Delaware counties of Pennsylvania.	White. Small bluish spots are sometimes found on the skin, but are discriminated against.			Not ⅔ big enough for age, upright ears, off-colored hair; spots on hide larger than a silver dollar, cryptorchidism in males; hernia in males or females; or swirls on body above flanks.
Conner Prairie line C line E line F	Conner Prairie Swine, Inc., Noblesville, Ind., beginning in 1946. Conner Prairie sells about 900 boars each year.	All colors, but usually spotted. Color is not a factor in selection.	Fairly long head; trim jowl; and medium-sized ears, usually turning.	Litter size: 10 to 10.5. Weight at 5 mo.: 198 lb. Backfat thickness: 1.27″. Feed conversion: 284 lb.	These 3 lines represent an amalgamation of (Minnesota No. 1, Minnesota No. 2, Montana No. 1, Maryland No. 1, Beltsville No. 1, Beltsville No. 2, and Minnesota No. 3), along with the infusion of some purebreds of the foundation breeds from which they evolved.
Duroc	U.S.A.; chiefly in New York and New Jersey.	Red, varying from light to dark.	Medium-size ear, tipping forward.		White feet or white spots on any part of body; any white on end of nose; black spots larger than 2″ in diameter; swirls on upper half of the body or neck; ridgeling (1 testicle) boars; or fewer than 6 udder sections on either side.
Hampshire	U.S.A.; in Boone County, Ky.	Black, with a white belt around the shoulders and body, including the front legs.	Longer and straighter in the face than most breeds; ears carried erect.		Too much white on ham, hind legs, or head; or incomplete belt.
Hereford	U.S.A.; by R. U. Webber of La Plata, Mo.	Red body color, with white face, legs, and switch similar to Hereford cattle.	White face.		A white belt extending over shoulders, back or rump; more than ⅓ white markings; no white markings on face; less than 2 white feet; swirl on any part of body; no marks of identification; ridgeling boar; permanent deformities of any kind; unsound underlines of less than 6 teats on each side.

(Continued)

TABLE 3-24 (Continued)

Breed	Place of Origin	Color	Distinctive Head Characteristics	Other Distinguishing Characteristics	Disqualifications Comments
Lacombe (55% Landrace, 23% Berkshire, and 22% Chester White)	Canada; at the Experimental Farm, Lacombe, Alberta, beginning in 1947.	White.	Medium-sized flop ears and a medium-length, slightly dished face.	Of the 3 parent breeds, it resembles the Landrace most closely.	
Managra (45% Swedish Landrace, 20% Wessex Saddleback, 15% Welsh, and 20% from a mixture of Minnesota No. 1, Berkshire, Yorkshire, and Tamworth)	Canada; by University of Manitoba, beginning in 1958.	White.	Lop-eared.	The breed is noted for litter size, maternal ability, and temperament.	
OIC (Ohio Improved Chester)	U.S.A.; in Ohio, by L. B. Silver of Salem, Ohio.	White.	Wide, short head and smooth dished face. Ears droop slightly.		Swirls on upper half of body, hernia, cryptorchidism, spots on skin with other than white hair, or inverted nipples.
Poland China	U.S.A.; in Ohio, in the Miami Valley of Warren and Butler counties.	Black or black with white spots, with 6 white points—the feet, face, and tip of the tail.	Drooping ears.		Fewer than 6 teats on a side, a swirl on upper half of body, hernia, or cryptorchidism.
Spotted	U.S.A.; chiefly in Indiana.	Spotted black and white, about 50% each.		Females must have at least 6 prominent teats on each side to be eligible for show or sale.	Brown or sandy spots; less than 20% or more than 80% white on body; boar with a swirl; small upright ears; not over half normal size; cramped or deformed feet; seriously diseased, barren or blind; or if scoring fewer than 60 points.
Tamworth	England; in the central counties of Stafford, Leicester, Warwick, and Northampton.	Red, varying from light to dark. Black spots may occur, but are objectionable.	Wide between the ears; snout moderately long and straight; neat jowl; and medium-size, erect ears.		Swirls. More than 5% black.
Wessex Saddleback	Hampshire, England	Black, with a white belt around the shoulders and body including the front legs.	Fairly long snout, medium-sized ears; with forward pitch; trim jowl.		
Yorkshire (known as the Large White in England)	England	White, although black "freckles" appear.	Slightly dished face, and erect ears.	Long bodied; prolific.	Swirls on upper third of body, hernia, hair color other than white, cryptorchidism, one testicle or any pronounced abnormal condition of the testicles, hermaphrodite, blind or inverted teats, total blindness, fewer than 6 teats on each side, permanent lameness, or excessive amount of black or dark pigment of the skin.

TABLE 3-25

BREEDS OF LIGHT HORSES, PONIES, DRAFT HORSES, AND ASSES AND MULES, AND THEIR CHARACTERISTICS

Breed	Place of Origin	Color	Other Distinguishing Characteristics	Primary Uses	Disqualifications; Comments
LIGHT HORSES AND PONIES:					
American Bashkir Curly	In Bashkiria, on the eastern slopes of the Ural Mountains, in the U.S.S.R.; hence, the name Bashkir. Modern history of curly horses in America began in 1898, when Peter Damele of Ely, Nev., cut three curly animals from a herd of wild horses in the Peter Hanson Mountain Range. Most of today's curly horses trace to Damele ranch breeding.	All colors are accepted.	Curly coat, with corkscrew mane and wavy tail. In build, Curlies are medium size and chunky, somewhat resembling the early-day Morgan in conformation. The breed is noted for small nostrils, a gentle disposition, and heavy milking. Many of the animals have a natural fox-trot gait.	As a pleasure horse; for utility purposes—including light draft work; family trail horses; and children's mounts.	Weight in excess c 1,350 lb; faulty con formation.
American Creme Horse[1]	U.S.A. Pale cream horses have existed for a very long time, primarily in Oregon and Washington. They were not given breed status until 1970, when the American Albino Assn., Inc., established a separate American Creme Horse Division for their registration.	The following color classifications of American Creme Horses are registered: A—Body ivory white, mane white (lighter than body), eyes blue, skin pink. B—Body cream, mane darker than body, cinnamon buff to ridgeway, eyes dark. C—Body and mane of the same color, pale cream; eyes blue; skin pink. D—Body and mane of same color, sooty cream; eyes blue; skin pink. Combinations of the above classifications are also accepted.		Pleasure and stock horses; for exhibition purposes; as parade and flag-bearer horses.	Pink eyes. Any color other tha ivory white or cream.
American Gotland Horse	Baltic Island of Gotland, a part of Sweden.	Bay, brown, black, dun, chestnut, palomino, roan, and some leopard and blanket markings.	Average about 51″ high, with a range of 11 to 14 hands.	Harness racing, pleasure horses (trotting), and jumpers; for children and moderate-sized adults.	Pintos and animals wi large white marking are disqualified.
American Mustang	Along the Barbary Coast of North Africa. From here, they were taken to Spain by the conquering Moors, propagated in Andalusia, and brought to America by the conquistadores. The American Mustang Assn. was formed in 1962.	Any color.		Pleasure riding, show, trail riding, endurance trails, stock horses, and jumping.	They must be betwee 13-2 and 15 hanc high.
American Saddle Horse	U.S.A.; in Fayette County, Ky.	Bay, brown, chestnut, gray, roan, black, or golden. Gaudy white markings are frowned upon.	Ability to furnish an easy ride, with great style and animation. Long, graceful neck and proud action.	Three- and five-gaited saddle horses. Fine harness horses. Pleasure horses. Stock horses.	
American Walking Pony	U.S.A.; near Macon, Ga., in 1968.	No color coat stipulation. Since it is a cross between Welsh Pony and Tennessee Walking Horse, the colors of both parent breeds occur.	They range in height from 13 to 14-2 hands. Ponies that perform the running walk gait.	Pleasure riding; children's or small adults' mounts.	Standing over 14 hands; not being ab to perform at the ru ning walk. Appaloosa color is n accepted.

(Continue

TABLE 3-25 (Continued)

Breed	Place of Origin	Color	Other Distinguishing Characteristics	Primary Uses	Disqualifications; Comments
American White Horse[1]	U.S.A.; White Horse Ranch, Naper, Neb.	Snow-white hair, as white as clean snow; pink skin; light blue, dark blue (near black), brown, or hazel eyes.		Pleasure horses; trained horses for exhibition purposes; parade and flag-bearer horses.	Pink eyes.
Andalusian	Spain. From desert-bred Barbs (introduced by the invading Moors) crossed on the light, agile horses of southern Spain.	White, grays, and bays most common. Also, there are a few blacks, roans, and chestnuts.	Andalusians stand 14-2 to 16 hands and weigh from 1,000 to 1,200 lb.	Bullfighting, parade horses, dressage, jumping, and pleasure riding.	Animals not tracing to the Spanish Registry, which is supervised by the Army in Spain, are not eligible for registry.
Appaloosa	U.S.A.; in Oregon, Washington, and Idaho; from animals originating in Fergana, Central Asia.	Variable, but usually white over the loin and hips, with dark round or egg-shaped spots thereon.	The eye is encircled by white, the skin is mottled, and the hoofs are striped vertically black and white.	Stock horses. Pleasure horses. Parade horses. Race horses.	Animals not having Appaloosa characteristics, and animals of draft horse or pony; Albino or Pinto breeding; gray or non-Appaloosa roan, or the progeny of a gray or non-Appaloosa roan; cryptorchid or monorchid, or sired by a cryptorchid or monorchid; paint or pinto markings; and animals under 14 hands high after 5 years of age.
Arabian	Arabia.	Bay, gray, and chestnut with an occasional white or black. White marks on the head and legs are common. The skin is always dark.	A beautiful head, short coupling, docility, great endurance, and gay way of going.	Saddle horses. Show horses. Pleasure horses. Stock horses. Racing.	
Buckskin[2]	U.S.A.	Buckskin, red dun, grulla (mouse dun). *International Buckskin Horse Assn.:* Dorsal stripe, leg barring, shoulder stripe or shadowing; black ear tips; hazel eyes; cobwebbing on face; frosted mane and tail.	*American Buckskin Registry Assn., Inc.*: All types accepted. Horses must be over 54″, ponies under 54″.	Stock horses, pleasure horses, and show purposes.	*American Buckskin Registry Assn., Inc., disqualifications*: Palominos, chestnuts, sorrels, or bays with dorsal stripe; draft type; blue or glass eyes; white spots on body (indicating Pinto or Appaloosa blood) or white markings above knees or hocks. *International Buckskin Horse Assn. disqualifications*: Excessive white; showing Paint, Pinto, or Appaloosa characteristics.
Chickasaw	U.S.A. Developed by the Chickasaw Indians (hence, the name of the breed) of Tennessee, North Carolina, and Oklahoma, from horses of Spanish extraction.	Bay, black, chestnut, gray, roan, sorrel, or palomino.	Short head and ears; a short back; a short neck; square, stocky hips; a low-set tail; a wide chest; and great width between the eyes.	Cow ponies.	Preferred height: 53″ to 59″.
Cleveland Bay	England; in the Cleveland district of Yorkshire.	Always solid bay with black legs.	Larger than most light horse breeds; weight from 1,150 to 1,400 lb.	Today, it is used chiefly as a general utility horse; for riding, driving, and doing all kinds of farm work. Also, used in crossbreeding to produce heavy weight hunters.	Any color other than bay, although a few white hairs on the forehead are permissible.

Footnotes on last page of table.

(Continued)

TABLE 3-25 (Continued)

Breed	Place of Origin	Color	Other Distinguishing Characteristics	Primary Uses	Disqualifications; Comments
Connemara Pony	Ireland; along the west coast.	Gray, black, bay, brown, dun, cream, with occasional roans and chestnuts.	The American Connemara Society registers in 2 sections: *Section 1*, "pony," 13 to 14-2 hands; *Section 2*, "small horse," over 14-2 hands.	Jumpers, showing under saddle and in harness; for medium-sized adults and children.	Piebalds, skewbalds and cream with blue eyes not accepted.
Galiceno	Galicia, a province in northwestern Spain.	Solid colors prevail. Bay, black, chestnut (sorrel), dun (buckskin), gray, brown and palomino are most common.	Intermediate in size; at maturity they stand 12 to 13 hands and weigh 625 to 700 lb.	Riding horses.	Albinos, pintos, and paints are ineligible for registry. Cryptorchids or monorchid unless gelded.
Hackney	England; on the eastern coast, in Norfolk and adjoining counties.	Chestnut, bay, and brown are most common colors, although roans and blacks are seen. White marks are common and are desired.	In the show-ring, custom decrees that Hackney horses and ponies be docked and have their manes pulled. High natural action.	Heavy harness or carriage horses and ponies. For crossbreeding purposes to produce hunters and jumpers.	Piebald or skewbald color not accepted.
Hanoverian	Germany, in the Hanover section, beginning in 1732. They were developed for the purpose of providing a superior horse for military use, with emphasis on size, intelligence, and temperament. In 1973, the American Hanoverian Society was formed, with headquarters in Carmel, Ind.	Variable.	Hanoverian horses are big and powerful. Many of them stand 16½ hands or better and weigh 1,200 lb or more. They combine nobility, size, and strength in a unique way.	In Europe, they are used for riding, driving (carriage horses), hunting, jumping, dressage, and utility purposes. In the U.S., the breed is used for all light horse purposes, especially for hunting, jumping, and dressage.	Without doubt, the breeding and selection program followed with Hanoverian horses in Germany is the finest equine production-testing program in the world.
Hungarian Horse	Hungary.	All colors; either solid or broken.	Unique combination of style and beauty with ruggedness.	Stock horses, cutting horses, pleasure horses, trail riding, hunters, and jumpers.	Cryptorchids; glass eyed.
Lipizzan	In Lipizza, Yugoslavia, the town from which the breed takes its name.	Most mature animals are white. But foals are born dark brown or gray, then turn white at 4 to 6 years of age. About 1 in 600 remains black or brown throughout life. When the latter happens, it is considered good luck.	An elastic walk, with considerable knee action.	Dressage (for which purpose they are without a peer), harness horses, pleasure horses, hunters, jumpers, and parade horses.	Glass eyes, extreme Roman nose, or deformed or crooked limbs.
Missouri Fox Trotting Horse	U.S.A.; in the Ozark Hills of Missouri and Arkansas.	Sorrels predominate, but any color is accepted.	The fox-trot gait.	Pleasure horses. Stock horses. Trail riding.	If animal cannot fox-trot
Morab	U.S.A.; in Fresno County, Calif., from a foundation of Morgan mares bred to Arabian stallions to produce stock horses, beginning 1955. However, the breed was not established until 1973, when the American Morab Horse Assn., Inc., was founded.	Bay, black, brown, buckskin, chestnut, dun, gray, grulla, palomino, or roan.	The breed possesses the muscular strength and ruggedness of the Morgan and the refinement and beauty of the Arabian.	Show purposes, pleasure riding, endurance rides, and ranch work.	Animals are disqualified if they have breed characteristics other than Morgan, Arabian or Thoroughbred; they carry Saddle Horse or Quarter Horse breeding; they have albino, appaloosa, paint, or pinto color; or if they are under 14 hands at maturity.

Footnotes on last page of table.

(Continued)

TABLE 3-25 (Continued)

Breed	Place of Origin	Color	Other Distinguishing Characteristics	Primary Uses	Disqualifications; Comments
Morgan	U.S.A.; in the New England states.	Bay, brown, black and chestnut; extensive white markings are uncommon.	Easy keeping qualities, endurance, and docility.	Saddle horses. Stock horses.	Wall-eye (lack of pigmentation of the iris), or natural white markings above the knee or hock except on the face.
Morocco Spotted Horse	U.S.A.	Spotted. The secondary color, white, must comprise not less than 10%, not including white on the legs or face.		Parade horses, saddle horses, stock horses, pleasure horses, and harness horses.	Animals that are under 14-2 hands, or that are of draft or pony breeding, or that show such characteristics, are not eligible for registration.
National Appaloosa Pony	U.S.A.; near Rochester, Ind.	Vari-colored; but leopard, blanket-type, snowflake, or roan are most popular. The skin, nose, and area around the eyes are mottled. White sclera encircles the eyes.	Appaloosa color, and standing under 14-2 hands.	Working ponies; show ponies; and for trail riding, jumping, and racing.	Albino-, pinto-, or paint-colored animals not eligible for registration.
Paint Horse[3]	U.S.A.	White plus any other color. Must be a recognizable paint, as distinguished by 2 color patterns—tobiano or overo. Tobiano is the most numerous, there being about 4 tobianos to 1 overo.	No discrimination is made against glass, blue, or light-colored eyes.	Stock horses. Pleasure horses. Show purposes. Racing.	Lack of white markings above the knees or hocks except on the face; appaloosa color or blood; adult horses under 14 hands; 5-gaited horses.
Palomino	U.S.A.; from animals of Spanish extraction.	Golden (the color of a newly minted gold coin, or 3 shades lighter or darker), with a light-colored mane and tail (white, silver, or ivory, with not more than 15% dark or chestnut hair in either). White markings on the face or below the knees are acceptable. Dark skin.		Stock horses. Parade horses. Pleasure horses. Saddle horses. Fine harness horses.	*The Palomino Horse Breeders of America lists the following disqualifications*: Patches of stained or blackened discoloration; dorsal stripe along the spine; zebra stripes or lighter or darker color running around legs or across shoulder or withers; patches of white hair except if caused by an injury or saddle. *The Palomino Horse Assn., Inc.*, lists the following disqualifications: blue, moon, or pink-eyes; white or dark spots on the body (known as pinto markings); or an animal whose sire or dam was an Albino, pinto, or appaloosa. Horses under 14-1 hands at maturity (5 years of age) are ineligible for registry.
Paso Fino[4]	In the Caribbean area; mostly Puerto Rico, Columbia, Dominican Republic, and Cuba.	Any color, although solid colors are preferred. Bay, chestnut, or black with white markings are most common. Occasionally, palominos and pintos appear.	The paso gait, essentially a broken pace—a lateral (not diagonal) gait. The sequence of the movement of the hooves is: right rear, right fore, left rear, and left fore; with the hind foot touching the ground a fraction of a second before the front foot, producing a 4-beat gait. The paso gait is performed at 3 speeds: (1) paso fino, which is the classic showring gait, performed with the horse fully balanced and collected; (2) paso corto, which is a more relaxed form of the gait and is commonly referred to as the natural paso gait; and (3) paso largo, which is the speed form of the gait. Additionally, Paso Finos walk and canter; hence, they are 3-gaited horses.	Pleasure, cutting, and parade horses; and for endurance riding and drill team work.	Animals that do not possess the paso fino gait or do not trace directly to the purebred Paso Fino ancestry are not eligible for registration.
Peruvian Paso Horse[4]	Peru.	Any color, although solid colors are preferred.	The paso, or gait, a natural (inborn), smooth, 4-beat, lateral gait—in essence, a broken pace. In executing his gait, the Peruvian Paso moves with a "termino"—a graceful, flowing movement in which the forelegs are rolled toward the outside as the horse strides forward, much like the arm motions of a swimmer. Termino is a spectacular and beautiful natural action. Breed enthusiasts insist that the Peruvian Paso does only 2 acceptable gaits, the "paso llano" and the "sobreandando"; and they do not admit to a canter. At the annual National Show in Lima, Peru, the Peruvian Paso is asked to do only the paso llano and the sobreandando gaits.	Pleasure horses, parade horses, and endurance horses.	*Peruvian Paso Horse Registry of North America*: Light forequarters, coarseness, extreme height. *American Assn. of Owners & Breeders of Peruvian Paso Horses* disqualifies paint, pinto, or albino coloration.

Footnotes on last page of table.

(Continued)

TABLE 3-25 (Continued)

Breed	Place of Origin	Color	Other Distinguishing Characteristics	Primary Uses	Disqualifications; Comments
Pinto[3]	U.S.A.; from horses brought in by the Spanish conquistadores.	Preferably half color or colors and half white, with many spots well placed. The two distinct pattern markings are: overo or tobiano.	Glass eyes are not discounted. Association has separate registry for ponies and/or horses under 14 hands.	Any light horse purpose, but especially for show, parade, novice, pleasure purposes, stock horses.	Animals with Appaloosa ancestry or color, or o known draft hors breeding, are ineligi ble for registry.
Pony of the Americas	U.S.A.; Mason City, Iowa.	Similar to appaloosa; white over the loin and hips, with dark, round, or egg-shaped spots.	Happy medium of Arabian and Quarter Horse in miniature, ranging in height from 46″ to 54″, with Appaloosa color.	Children's western type using pony.	Not having appaloosa color; exceeding 54 or under 46″ at maturity (6 yr); pinto, albino, or roan color, or whos sires and/or dams were pinto or albino colore white stockings above either knee and/or eithe hock, or a bald face that covers any part of th sides of the head; or cryptorchids or monor chids.
Quarter Horse	U.S.A.	Chestnut, sorrel, bay, and dun are most common, although they may be palomino, black, brown, roan, or copper-colored.	Well-muscled. Small alert ear; sometimes heavily muscled cheeks and jaw.	Stock horses. Racing. Pleasure horses.	Pinto, appaloosa, an albino colors are in eligible for registry also white marking on the underline, or underlying light skin beyon the following locations: (1) white above a knee o hock, (2) white back or a line from the ear to th corner of the mouth, or (3) white on the lower li above a line connecting the 2 corners of th mouth.
Rangerbred (Colorado Rangers)	U.S.A.; in Colorado. The Assn. was formed in 1937. The name "Colorado Rangers" was selected to signify that they were Colorado-bred, and that they were bred on the range.	All colors accepted; but many Rangerbred horses are spotted.	Similar to Appaloosa.	Stock horses.	The 2 foundation stal lions of the Rangerbred—Leopard, an Arabian and Linden Tree, a pure Barb—were imported from Turkey by Ulys ses S. Grant, retired general and former presi dent of the U.S., in 1878. They were a gift to th General from the Sultan of Turkey. Horses of draft or pony blood are ineligible for reg istration.
Shetland Pony	Shetland Isles.	All colors; either solid or broken.	Small size. Two class sizes recognized by breed registry: (1) 43″ and under, and (2) 43″ to 45″. Good disposition.	Children's mounts. Harness. Roadster. Racing.	Over 46″ in height. Glass eyes are undesi able.
Spanish-Barb	Barb horses were taken from Africa to Spain with the conquest of Spain by the Moors in 711 A.D. From Spain, they were taken to Cuba in 1511, to Mexico in 1519, to southwestern U.S. in 1540, and to Florida in 1565. The Spanish-Barb Breeders Assn. was organized in 1972.	All colors are represented in the breed; but dun, grulla, sorrel, and roan are most common. Most animals are solid colored. A dorsal stripe and zebra markings occur in all duns and grullas and in some sorrels.	Spanish-Barbs are small horses (the standard height is 13-3 to 14-1 hands), with short coupling (usually 5 lumbar vertebrae; at times 17 thoracic vertebrae), deep bodies, good action, and without extreme muscling.	Cow ponies, western riding, English riding, and pack-horses.	
Spanish-Mustang	U.S.A. Trace to feral and semiferal (Indian-owned) horses of Barb and Andalusian ancestry brought to America by the Spanish in the early 1500s and 1600s. Beginning about 1925, Robert E. Brislawn, Sr., and Ferdinand L. Brislawn began gathering pure Spanish Mustangs. Breed registry was founded in 1957, at Sundance, Wyo.	The whole gamut of colors, including all the solid colors and all the broken colors.	Some have 5 to 5½ lumbar vertebrae. Stand 13 to 14½ hands.	Cow ponies, western riding, English riding, and pack-horses.	

[3] Footnotes on last page of table.

(Continued

TABLE 3-25 (Continued)

Breed	Place of Origin	Color	Other Distinguishing Characteristics	Primary Uses	Disqualifications; Comments
Standardbred	U.S.A.	Bay, brown, chestnut, and black are most common; but grays, roans, and duns are found.	Smaller, less leggy and with more substance and ruggedness than the Thoroughbred.	Harness racing, either trotting or pacing. Harness horses in horse shows.	
Tennessee Walking Horse	U.S.A.; in the Middle Basin of Tennessee.	Sorrel, chestnut, black, roan, white, bay, brown, gray, or golden. White markings on the face and legs are common.	The running walk gait.	Plantation Walking Horses, pleasure horses, and show horses.	
Thoroughbred	England. The first edition of the General (English) Stud Book was published in 1793.	Bay, brown, chestnut, and black; less frequently, roan and gray. White markings on the face and legs are common.	Fineness of conformation. Long, straight and well-muscled legs.	Running races. Stock horses. Saddle horses. Polo mounts. Hunters.	All Thoroughbreds in the world trace to 3 stallions: The Darley Arabian, the Byerly Turk, or the Godolphin Barb.
Trakehner	In Trakehnen, East Prussia, in 1732. It evolved from the blending of the indigenous Prussian horses, Thoroughbred, and Arabians. *The American Trakehner Assn., Inc.*, was formed in September, 1974.		Horses with the size of the Thoroughbred, but more rugged, and possessing the elegance of the Arabian.		
Welsh Pony	Wales.	Any color except piebald and skewbald. Gaudy white markings are not popular.	Small size; intermediate between Shetland ponies and other light horse breeds. The American Welsh Stud Book height stipulations are: "A" Div.—cannot exceed 12-2 hands. "B" Div.—over 12-2 and not more than 14 hands.	Mounts for children and small adults. Racing. Roadsters. Trail riding. Parade. Stock cutting. Hunting.	Piebald or skewbald.
Isabella	U.S.A.; on McKenzie Rancho, Williamsport, Ind. The foundation animals were American Saddlers.	Gold, white, or chestnut, with flaxen, silver, or white mane and tail. There may be white markings on the face and legs.		Pleasure riding and as exhibition horses.	Animals having bay color, spots, or black mane and tail are not eligible for registration.
DRAFT HORSES:					
Belgian	Belgium.	Bay, chestnut, and roan are most common, but browns, grays, and blacks are occasionally seen. Many Belgians have a flaxen mane and tail and a white-blazed face.	Lowest set and most massive of all draft breeds.	Farm work horses. Exhibition purposes.	Side bones or curbs disqualify an animal for registry.

Footnotes on last page of table.

(Continued)

TABLE 3-25 (Continued)

Breed	Place of Origin	Color	Other Distinguishing Characteristics	Primary Uses	Disqualifications; Comments
Clydesdale	Scotland; along the River Clyde.	Bay and brown with white markings are most common; but blacks, grays, chestnuts, and roans are occasionally seen.	Superior style and action. Feather or hair on the legs.	Farm work horses. Exhibition purposes.	
Percheron	France; in the northwestern district of La Perche.	Mostly black or gray; but bays, browns, chestnuts, and roans are seen.	In comparison with other draft breeds, noted for its handsome clean-cut head.	Farm work horses. Exhibition purposes.	
Shire	England; primarily in the east central counties of Lincolnshire and Cambridgeshire.	Common colors are bay, brown, and black with white markings; although grays, chestnuts, and roans are occasionally seen.	Taller than any other draft breed. Feather or hair on the legs.	Farm work horses. Exhibition purposes.	
Suffolk	England; in the eastern county of Suffolk.	Chestnut only.	They are the smallest of the draft breeds. Close-to-the-ground and chunky build.	Farm work horses. Exhibition purposes.	Any color other tha chestnut.
ASSES AND MULES: **American Jack (Jack; *Equus asinus*)**	All foundation jack stock introduced into the U.S. came from southern Europe, mostly bordering on the Mediterranean sea. Various breeds and strains were imported and blended into one breed in the U.S. known as the American Jack.	Black, with a white nose, often with light-colored underline and white on inside of legs. Red or sorrel. Gray.	Long and large ears, short hair on mane and tail; absence of "chestnuts" on inside of hind legs; a thick hide; small hoofs; a loud and harsh voice, called a bray; and a 12 months' gestation period for jennets (vs 11 months for mares).	Production of mules.	In the heyday of jack and jennets, to quali for registry, they ha to meet minimur standards for heigh heart girth, and ci cumference of for cannon.
Donkeys (small asses)	Same as above.	Same as above.	Same as above, but smaller.	Driving, riding, working, and as pets.	The term "burro" refe to an unimprove member of the as family.
Miniature Mediterranean Donkeys	Sardinia and Sicily.	Mouse color to almost black.	Dorsal stripe, forming a cross with stripe over withers and down shoulders.	Children's pet.	Over 38" high. Without cross.
Mules (hinnies)	The mule is a hybrid, a cross between the horse and ass.	Variable color, but sorrell is the preferred color.	The mule resembles his sire, the jack, more than the mare. But the desired conformation is identical to that of a horse for similar use, with more stress placed upon size, and set and quality of ear.	Driving, riding, packhorses, working, and show purposes.	

[1]Both the American Creme Horse and the American White Horse are registered by the American Albino Assn., Inc., Crabtree, Ore., with separate divisions provided for eac

[2]Known, officially, as Buckskin or American Buckskin by the American Buckskin Registry Assn., Inc.,; and as I.B.H.A. or International Buckskin by the International Bucksk Horse Assn.

[3]Two different associations have evolved for the registration of these varicolored horses. In the American Paint Horse Assn., the breed is known as the American Paint Hors whereas in the Pinto Horse Assn. of America, Inc., it is known as the Pinto. Both groups of horses are of similar background and color.

[4]In the U.S., 4 different breed associations have evolved for the registration and promotion of horses which come under the general heading of "Paso" horses. But each the registries has slightly different standards. The American Paso Fino Horse Assn., Inc., and the Paso Fino Owners and Breeders Assn., Inc., call their breed "Paso Fino"; th American Assn. of Owners and Breeders of Peruvian Paso Horses and the Peruvian Paso Horse Registry of North America call their breed "Peruvian Paso."

SECTION 4

FEEDING

Contents **Page**

PART I—GENERAL LIVESTOCK FEEDING

Types of Digestive Systems and Kinds of Feeds Eaten 215
Feeds 215
Nutrient Needs 226
  Energy 226
  Protein 226
  Minerals 227
  Vitamins 230
  Water 233
How to Evaluate Feedstuffs 234
Feed Preparation 248
Feeding Standards 250
Balanced Rations 252
  How to Balance Rations 253
Rations for Farm Animals 258
Feed Substitutions 258
Home Mixed Vs Commercial Feeds 259

PART II—FEEDING BEEF CATTLE

Nutrient Requirements 260
  Energy 271
  Protein 272
  Minerals 272
    Beef Cattle Mineral Chart 274
  Vitamins 280
    Beef Cattle Vitamin Chart 282
  Water 281
Feed Additives and Implants 281
Rations for Beef Cattle 287
Feed Substitution Table for Cattle (Beef and Dairy) 288
Feeding Breeding Beef Cattle 295
  Feeding Brood Cows 295
  Feeding Calves 301
  Feeding Replacement Heifers 304
  Feeding Herd Bulls 306
Feeding Stockers 306
  Compensatory Growth 308
Feeding Finishing (Fattening) Cattle 308
  Feed Additives and Implants 314
  Ration Formulation Using Net Energy 315
  Managing Feedlot Cattle 316
  Pasture Finishing Cattle 320
Feeding Show and Sale Beef Cattle 322

PART III—FEEDING DAIRY CATTLE

Nutrient Requirements 324
  Energy 324
  Protein 324
  Minerals 329
    Dairy Cattle Mineral Chart 330
  Vitamins 329
    Dairy Cattle Vitamin Chart 338
  Water 342
Feeds for Dairy Cattle 342
Feeding Lactating Cows 343
  Rations for Lactating Cows 343

Contents **Page**

Challenge or Lead Feeding 346
Feeding on Pasture 347
Feed Substitutions 347
Feeding Dry Cows 347
Feeding Dairy Calves 348
Feeding Replacement Heifers 349
Feeding Dairy Bulls 350
Feeding Show and Sale Dairy Cattle 350

PART IV—FEEDING SHEEP

Nutrient Requirements 351
Energy 351
Protein 351
Minerals 355
Sheep Mineral Chart 356
Vitamins 355
Sheep Vitamin Chart 362
Water 364
Feed Additives and Implants 364
Rations for Sheep 367
Feed Substitution Table 368
Feeding Breeding Sheep 371
Feeding Growing-Finishing Lambs 373
Feeding Finishing (Fattening) Lambs 375
Feeding Show and Sale Sheep 376

PART V—FEEDING SWINE

Nutrient Requirements 377
Recommended Nutrient Allowances 377
Energy 377
Protein 380
Minerals 383
Swine Mineral Chart 384
Vitamins 388
Swine Vitamin Chart 389
Water 392
Feeds for Swine 392
Feed Additives 394
Rations 396
Feed Substitution Table 401
Feeding Breeding Swine 401
Feeding Growing-Finishing Hogs 406
Corn-Hog Ratio 408
Soft Pork 408
Feeding Show and Sale Swine 408

PART VI—FEEDING HORSES

Nutrient Requirements Vs Allowances 409
Nutrient Requirements 409
Recommended Nutrient Allowances 409
Energy 413
Protein 414
Minerals 415
Horse Mineral Chart 414
Vitamins 418
Horse Vitamin Chart 418
Water 418
Rations for Horses 428
Feed Substitution Table 428
Feeding Breeding Horses 428
Feeding Young Equines 428
Feeding Pleasure Horses 433
Feeding Horses in Training 434
Feeding Racehorses 434
Feeding Show and Sale Horses 435

PART VII—NUTRITIONAL DISEASES AND AILMENTS

PART VIII—FEEDS AND THEIR COMPOSITION

Feed Composition Tables 455
Table 4-101—Composition of Feeds 456
Table 4-102—Amino Acid Composition of Feeds 580
Table 4-103—Iodine Content of Feeds 584
Table 4-104—Selenium Content of Feeds 584
Table 4-105—Vitamin D Content of Feeds 585

PART I—GENERAL LIVESTOCK FEEDING[1]

Animals inherit certain genetic possibilities, but how well these potentialities develop depends upon the environment to which they are subjected; and the most important influence in the environment is the feed. In turn, all feeds come directly or indirectly from plants which have their roots in the soil. Thus, we have the cycle as a whole—from the soil, through the plant, thence to the animal and back to the soil again.

The primary purpose of the 597 million (1981 figures) U.S. animals, including poultry, is to convert inedible feeds into food, clothing, power, and recreation. About two-thirds of the feed consumed by U.S. livestock is not suited for human consumption. In this category are hay, pasture, coarse forages (such as straws, fodders, etc.); certain grains; such by-products as are obtained from mills, packinghouses, and food-processing plants; and damaged grains and foods; and garbage. These are converted into meat, milk, wool, and eggs.

Also, feeding is important from an economic standpoint; it is the major item of expense in producing livestock. For example, feed accounts for approximately the following proportions of the cost of livestock production: finishing cattle, 70%; milk production, 55%; feedlot lambs, 50%; and pork, 65 to 75%.

This section is a capsule presentation of livestock feeding. The first part of the section is devoted to general feeding information and recommendations pertinent to all classes of livestock. This is followed by a section devoted to the feeding of each class of livestock—beef cattle (including finishing cattle), dairy cattle, sheep, swine, and horses.

TYPES OF DIGESTIVE SYSTEMS AND KINDS OF FEEDS EATEN

There are major differences in the anatomy and physiology of the organs of the digestive tract of different animal species. These differences are of great nutritional significance, as they affect both the nature of the digestion processes and the kind of feed that can be utilized. Based on type of digestive system, animals may be grouped as (1) monogastric (simple-stomached), (2) polygastric (ruminants), (3) pseudo-ruminants (those with functional ceca), or (4) avian (poultry).

Based on the kind of feed eaten, animals are classed as follows:

1. *Herbivores*—The vegetarians; they depend entirely upon plants for their feed supply.
2. *Carnivores*—The flesh eaters; they feed almost entirely upon the flesh of other animals.
3. *Omnivores*—The consumers of both plants and flesh.

An understanding of specific differences in types of digestive systems and kind of feed eaten is essential to intelligent feeding. This information is presented in Fig. 4-1, page 216.

Monogastric animals, like hogs, must eat a large percentage of grains and other concentrates and depend almost entirely on digestive enzymes to break down these compounds. Ruminants, such as cattle and sheep, with their four stomach compartments and the help of microorganisms, can subsist largely, or entirely, on bulky, high-fiber forages which, because of their low energy per unit weight of dry matter, must be consumed in large quantities to supply their nutrient needs. The horse, because of his greatly enlarged cecum and large intestine, can utilize quantities of roughage intermediate between simple-stomached and ruminant animals.

FEEDS

Feed (or feedstuff) is any ingredient, or material, fed to animals for purposes of sustaining them. Most feedstuffs provide one or more nutrients, but nonnutritive products may be fed for such purposes as increasing production, increasing feed efficiency, providing flavor, adding color, or for other reasons related to palatability, bulk, or preserving feeds.

A wide variety of feedstuffs can be used for animal feeding throughout the world. More than 2,000 different products have been classified as animal feeds, not counting varietal, grade, and stage of maturity differences. However, as shown in Table 4-1, page 217, relatively few of these products make up the great bulk of the U.S. feed supply.

Classes of Feeds

The number of feedstuffs is so great that it is impossible to cover each of them in this book. Rather, at this point they are merely classified as (1) roughages, (2) concentrates, (3) protein supplements, (4) by-product feeds and crop residues, (5) special feeds, (6) minerals, (7) vitamins, and (8) feed additives and implants.

---

[1]Most NRC nutritive requirement and feed composition tables are on a moisture-free basis. Yet, the majority of practical operators prefer an "as-fed" basis. Hence, in this book, where "as-fed" values were not available, NRC's moisture-free requirements were converted to an "as-fed" basis by assuming an average 90% dry matter for the latter. Correctly speaking, the values obtained as a result of this conversion are "air dry," rather than "as-fed." But, since these values approximate "as-fed" values, as closely as can be obtained, the author elected to designate them "as-fed" throughout this section.

Group 1—Monogastric (simple stomach):

Animal	*Class of Food*
Pig	omnivore
Dog	carnivore
Monkey	omnivore
Man	omnivore

Group 2—Polygastric (ruminants):

Animal	*Class of Food*
Cow	herbivore
Sheep	herbivore
Goat	herbivore

Group 3—Pseudoruminants (functional cecum):

Animal	*Class of Food*
Horse	herbivore
Rabbit	herbivore
Guinea pig	herbivore
Hamster	omnivore

Group 4—Avian (poultry):

Animal	*Class of Food*
Chicken	omnivore
Turkey	omnivore
Duck	omnivore
Goose	omnivore

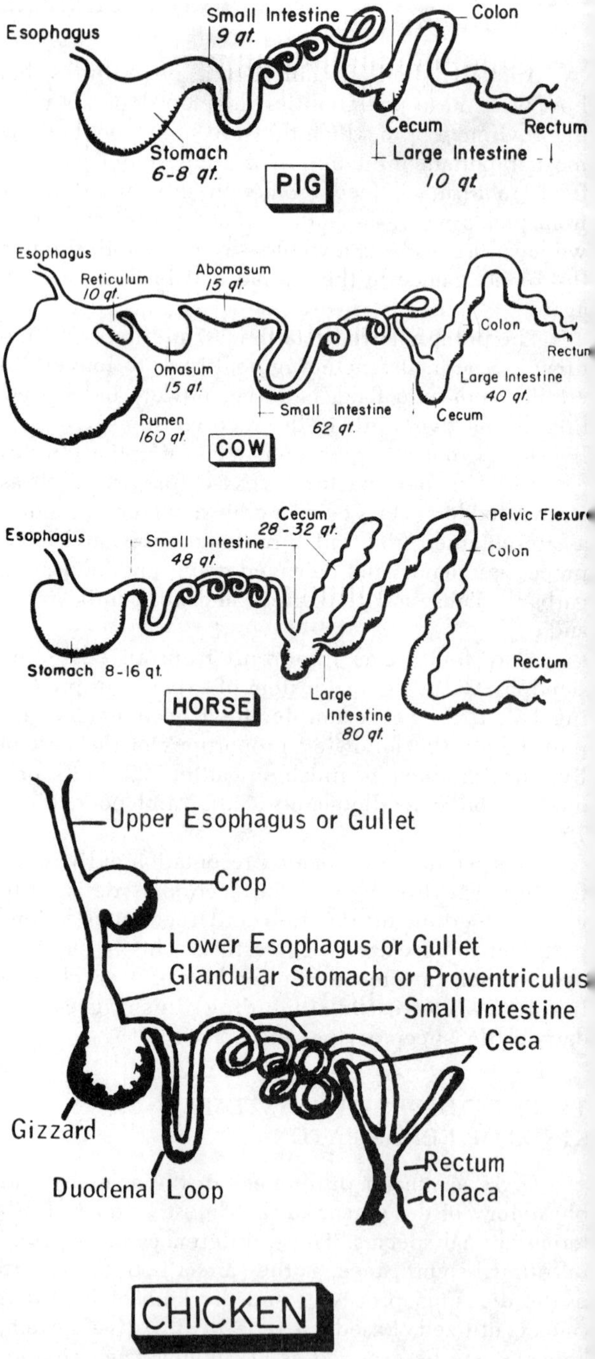

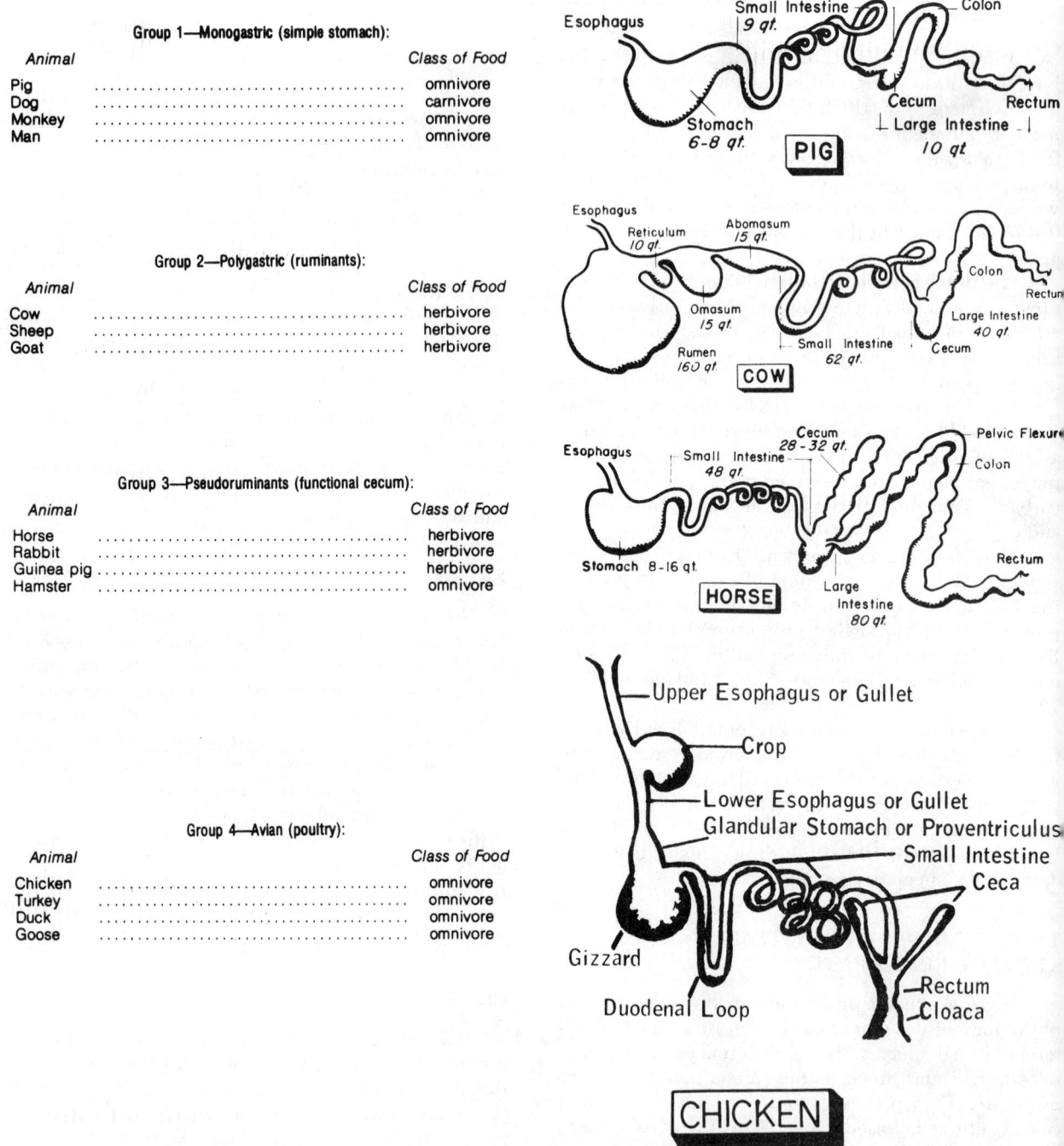

Fig. 4-1. Classification of animals according to type of digestive system and kind of feed eaten, with schematic diagram of digestive tracts of pig, cow, horse, and chicken.

ROUGHAGES

Roughages are bulky feeds that are low in weight per unit volume, contain more than 18 percent crude fiber, and are low in energy. They are the natural feeds of all herbivorous animals, including ruminants and horses. Although swine can survive solely on roughages, productivity is generally too low to be economical.

Roughages include pasture, dry forages, and silages. Fig. 4-2 shows that the three forages—pasture grazing, hay and silage-stover—account for 64% of al

TABLE 4-1

ANIMAL FEEDS CONSUMED IN THE UNITED STATES—1980

	Acreage Harvested	Estimated Use for Feed		Yield per Acre
	(1,000 acres)	*(1,000 tons)*		*(tons)*
Hay[1]				
Alfalfa	26,269	81,261		3.04
All other hay	33,168	57,639		1.54
Silage[1]				
Corn	9,261	111,093		12.0
Sorghum	744	7,109		9.6
Grains[2]			*(mil. bu.)*	*(bu.)*
Corn	73,061	115,892	4,139	91.0
Sorghum	12,722	8,607	307	46.2
Oats	8,640	7,232	452	53.0
Barley	7,233	4,176	174	49.6
High-protein				
Oilseed meal[3]				
Soybean		17,385		
Cottonseed		1,712		
Peanut		400		
Linseed		129		
Sunflower meal		125		
Animal proteins[3]				
Tankage and meat meal		1,800		
Fish meal and solubles		475		
Commercial dried milk products		300		
Noncommercial milk products		288		
Urea		395		
Grain protein feeds[3]				
Gluten feed and meal		1,310		
Distillers' dried grains		450		
Brewers' dried grains		390		
Miscellaneous feeds[3]				
Wheat millfeeds		4,000		
Molasses, inedible		5,463		
Alfalfa meal		1,400		
Dried & molasses beet pulp		1,300		
By-product feeds from hominy, oats, etc.		1,100		
Fats & oils		695		
Rice millfeeds		500		

[1]*Crop Production*, 1980 Annual Summary, USDA, Jan. 15, 1981, p. A-3.
[2]*Feed Situation*, USDA, Aug. 1982.
[3]*Feed Situation*, USDA, Aug. 1982, p. 27, Table 16.

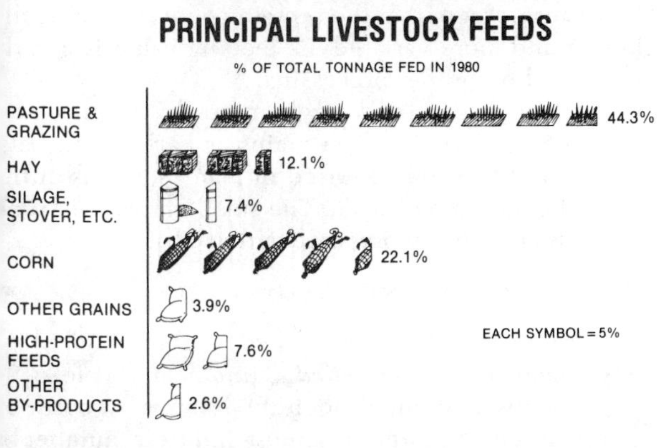

Fig. 4-2. These principal livestock feeds are converted into meat, eggs, milk, and wool. About two-thirds of the feed consumed by animals is not suited for human consumption. (Based on data from the following USDA sources: *Crop Production,* 1980 annual summary, January 15, 1981; and *Feed Situation,* August 1982.)

U.S. livestock feeds. Of course, the proportion of roughage to concentrate consumption varies widely according to class of animal. As shown in Table 4-2, sheep and goats head the list of "roughage burners,"

TABLE 4-2

PERCENTAGE OF FEED FOR DIFFERENT CLASSES OF U.S. LIVESTOCK DERIVED FROM (1) CONCENTRATES, AND (2) ROUGHAGES, INCLUDING PASTURE, 1980-81[1]

Class of Animal	Concentrates	Roughages
	(%)	(%)
Sheep and goats	9.9	90.1
Beef cattle	13.3	86.7
Horses and mules	43.6	56.4
Dairy cattle	37.9	62.1
Swine	86.5	13.5
Poultry (chickens and turkeys)	100.0	
All livestock	363	63.7

[1]Unpublished data provided by Mr. George C. Allen, Agricultural Economist, Commodity Economics Division, Economic Research Service, USDA.

with 90% of their total feed coming therefrom. Beef cattle rank second, with 87% of their feed coming from roughages, including pasture. Horses-mules and dairy cattle rank third and fourth, respectively, in roughage consumption. Swine and poultry consume negligible amounts of roughage.

From a feeding standpoint, the following general characteristics of roughages are pertinent, although some well-known roughage can be cited as an exception to each characteristic (for example, on a dry basis well-eared corn silage runs 18% crude fiber, but the TDN is high—about 70%):

1. *Bulk*—They are bulky feeds with a low weight per unit of volume.
2. *Fiber and energy*—They contain more than 18% crude fiber, and they are lower in energy than the concentrates.
3. *Digestibility*—They are generally lower in digestibility than concentrates, due to lignin content.
4. *Minerals*—They are generally higher in calcium, potassium, and trace minerals than most concentrates, but phosphorus content is apt to be moderate to low.
5. *Vitamins*—They are higher in fat-soluble vitamins than most concentrates. Legumes are good sources of B vitamins.
6. *Protein*—They are variable in protein content. Legumes may run 20% or more crude protein, whereas other roughages, such as straws, may have only 3 to 4% crude protein.

From an overall nutrition standpoint, roughages may range from very good nutrient sources (such as lush young grass, legumes, and high-quality silage) to very poor feeds (such as straws, hulls, and some browse). Nevertheless, all of them can be used advantageously, provided (1) they are properly prepared and supplemented, and (2) the feeder uses judgment in selecting the species and class of animal to which the particular roughage is fed.

CONCENTRATES

Concentrate feeds (not including protein supplements) are those which are high in energy and low in fiber (under 18%), and which contain less than 20% protein. Many different kinds of concentrate feeds can be used as animal feeds. Availability and price are the two most important factors determining the choice of concentrates.

Corn is the most common grain fed to livestock. More than 10 times as much corn as sorghum is fed in the United States. It is palatable and rich in the energy-producing carbohydrates and fats (80% TDN), and low in fiber. Also, corn is easily stored, with only moisture and carotene being lost over a period of time. Corn-and-cob meal (consisting of 20 to 25% cob) is excellent for finishing cattle. However, corn has

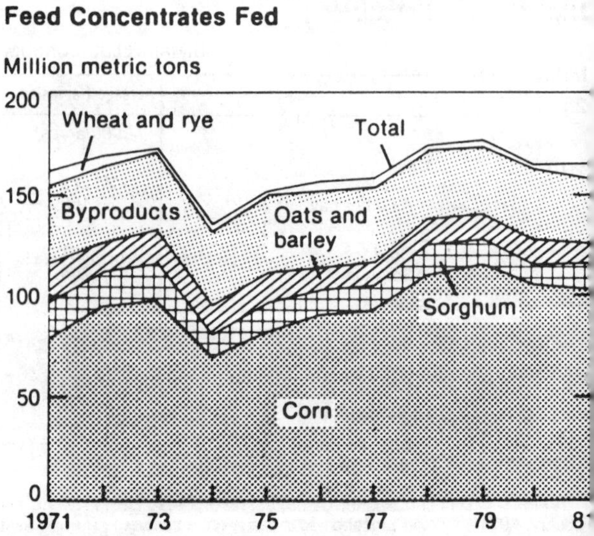

Feed fed to livestock and poultry. Year beginning October 1. 1980 preliminary; 1981 projected.

Fig. 4-3. This shows the concentrates, including by-product feec (oilseed meals, animal protein feeds, and milk by-products only), fed U.S. livestock and poultry, 1971-81. This figure points up (1) that corn by far the most important livestock feed, (2) that by-products rank secon and (3) that wheat is of minor importance as a livestock feed. Some vit statistics back of this figure are: In 1980, in million short tons, the follov ing quantities were fed: total feed grains, 144.6; total by-products, 41. and total concentrates (including by-products), 190.6. (From: *1981 Hanc book of Agricultural Charts*, USDA, p. 79)

certain very definite limitations—it is low in protei (and in the essential amino acids lysine, methionine and tryptophan) and calcium.

The grain sorghums are assuming an increasingl important role in livestock feeding, particularly in th fringe areas of the Corn Belt, and in the South an Southwest where moisture conditions are less favora ble. New and high-yielding varieties have been de veloped and have become popular. As a result, mor and more grain sorghums are being fed to livestock The chemical composition of sorghum (milo) is simi lar to corn except that the protein content is generall higher and more variable. Its feeding value is greatl enhanced by proper processing.

Although corn and sorghum are by far the mos common feed grains, such grains as barley, rye, oats wheat, and triticale, are used in many sections of th United States and Canada. The small grains are excel lent when properly prepared and used.

Protein Supplements

Protein supplements are feeds that are high in ni trogenous compounds called amino acids. At least 2 amino acids have been identified and may occur i combinations to form an almost limitless number c proteins.

High-protein feeds (20%, or more, crude protein are usually named and classified according to thei origin and method of processing. On the basis of ori

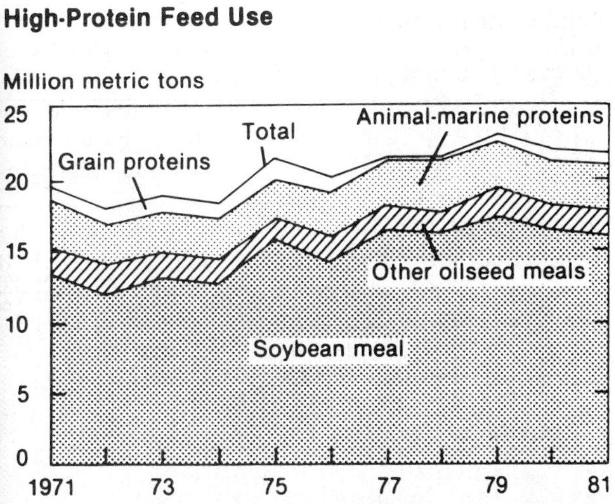

Fig. 4-4. This shows the high-protein feeds fed to U.S. livestock and poultry, 1971-81, in 44% protein soybean meal equivalent. The following explanation is pertinent: Grain proteins include gluten feed and meal, and brewer and distiller dried grains; animal-marine proteins include tankage, meat meal, marine by-products, and milk products; other oilseed meals include cottonseed, linseed, peanut, sunflower, and copra. This figure shows that soybean meal is by far the most important high-protein feed. Some vital statistics back of this figure are: In 1980, in million short tons, the following quantities were fed: soybean meal, 18.5; total of all oil meals, 20.2; animal protein, 2.8; grain protein, 2.9; and total of all high-protein feed, 25.9. (From *1981 Handbook of Agricultural Charts,* USDA, p. 180)

gin, they are usually grouped into two general categories as follows:

1. *Animal proteins*—Animal protein supplements are derived from inedible tissues from meat-packing or rendering plants, from surplus milk or milk products, and from marine sources. They include proteins from meat, fish, poultry, eggs, milk, and their products. With hogs and chickens, one of these protein sources was formerly a must. With the discovery and general availability of vitamin B_{12}, high-protein feeds of animal origin became less essential for swine and poultry.

Not all animal proteins are of high quality. For example, feather meal, a by-product from poultry processing, runs about 85 percent protein, but the protein is very poorly digested by monogastrics and must be hydrolyzed for good utilization. Even then, not more than 3 to 5 percent should be used in swine rations.

2. *Plant proteins*—This group includes the common oilseed by-products—soybean meal, cottonseed meal, linseed meal, peanut meal, safflower meal, sunflower meal, rapeseed meal, and coconut (or copra) meal. They vary in protein content and feeding value, depending on the seed from which they are produced, the amount of hull and/or seed coat included, and the method of oil extraction used.

In practical feeding operations, hogs and chickens are usually provided with some protein feeds of animal origin in order to supplement the proteins found in grains and forages. Protein quality is less important with ruminants and pseudoruminants, because of microbial synthesis. In feeding mature cattle, sheep, and horses, a safe plan to follow is to provide a liberal supply of high-quality legume hay or lush young pasture along with the concentrates. Also, the quality of the proteins in a ration is likely to be higher if a variety of feeds is combined.

In addition to the oilseed meals, numerous good commercially manufactured protein supplements are available. Usually, they are prepared for a particular class of livestock. They are generally blends of animal and vegetable protein ingredients, with urea added for ruminants. They may also include minerals, vitamins, and/or antibiotics.

BY-PRODUCT FEEDS AND CROP RESIDUES

By-product feeds include innumerable roughages and concentrates obtained from plant and animal processing, and from industrial manufacturing. Many of these are standard and valuable livestock feeds; among them, milling by-products from the cereal grains and oilseeds, root crops (cull potatoes and by-products of potato processing, turnips, mangels, swedes, fodder beets, carrots, and parsnips), dried beet pulp, and beet tops (from sugar beet processing), distillery and brewing by-products, unused bakery products, and by-products from numerous fruits and nuts.

As is true of any ration ingredient, the requisites to effective and profitable use of each by-product feed are (1) that it be bought at a favorable price, nutritive composition considered; (2) that its proximate composition be known, and that it be incorporated in a balanced ration; (3) that it be palatable and consumed in adequate quantity; and (4) that it not adversely affect carcass quality, particularly from the standpoint of harmful chemical residues from pesticides applied to crops. Generally speaking, the use of by-product feeds calls for ingenuity and experience in handling them, special knowledge relative to their nutritive qualities and use in balanced rations, and relatively high labor costs. As a result, many feeders are not interested in using them, whereas others find it a lucrative business.

Crop residues are the forages that remain after harvesting a grain crop. Among such crop residues are: corn stalks and husklage, sorghum stalks, soybean refuse, small grain straws and chaff, and legume and grass seed straws. Crop residues must be fed to the right class of animals, and they must be properly supplemented.

Special Feeds

Among the special feeds used for more than one

class of animals are (1) colostrum, (2) fats and oils, (3) high-lysine corn (opaque-2, or O_2), (4) high-moisture (early harvested) grain, (5) milk replacers (synthetic milk), (6) molasses, (7) nonprotein nitrogen (NPN) sources, and (8) single-celled protein. Pertinent information relative to each of these feeds follows. Additional special feeds used primarily by one species are covered in the sections devoted to each class of animals.

COLOSTRUM

Colostrum is the first milk secreted by mammalian females following parturition.

Newborn animals receive immunity to certain diseases from immunoglobulins in the colostrum which they consume during their first 24 hours of life. These antibodies are absorbed across the intestinal wall, but absorption in calves is reduced as early as 24 hours following birth. After that time, blood immunoglobulin levels change very little.

The discussion that follows pertains to calves, because most colostrum studies have been with cattle. But there is reason to believe that the same principles apply to other mammalian species.

Calves should consume 2 to 4 pounds of colostrum per feeding for the first 3 days after birth. Although the immunoglobulin content of colostrum falls rapidly after the first milking, the milk from the next 5 milkings has a higher than normal content of protein, fat, vitamins, and minerals. All of these are important in the nutrition of young calves. Colostrum is particularly high in Vitamins A and B_{12}, and in iron which helps keep up the hemoglobin level of the blood of the young calf.

• *Excess production*—Most cows produce enough colostrum to feed the calf nearly through weaning, provided it can be stored. It is not uncommon for a Holstein cow to produce 100 pounds of colostrum before her milk can be marketed. This is sufficient to feed her calf 5 to 6 pounds daily for 3 weeks. If only heifers are raised, this would extend beyond 5 weeks. Most calves can be weaned after 4 weeks.

• *How to store?*—Freezing is perhaps the simplest and preferred method of storing excess colostrum. The colostrum can be put in gallon or half-gallon containers (such as milk cartons). At feeding time, they can be thawed by adding one part hot water to two parts of colostrum.

Fermented (or pickled) colostrum has gained popularity in recent years. This involves storage at room temperature (50° to 75° F) in an easily cleaned can such as a clean plastic garbage can, lined with a plastic garbage bag. The colostrum will ferment and sour. Milk from cows which have been treated for mastitis should not be used since the antibiotics will inhibit fermentation, especially for the first 24 hour after treatment. Fermented colostrum should b stirred thoroughly before each feeding.

It may be fed as is, or it may be diluted with on part of water to two parts of colostrum. Experiment indicate that it is not necessary to dilute the colos trum.

Since the solids content of colostrum is highe than whole milk, care must be taken to control con sumption. A typical feeding schedule with a choic between whole milk or fermented colostrum follows

Birth Weight (lb)	Whole Milk -OR- (lb per day)	Fermented Colostrum (lb per day)
less than 70	6	4
70-100	7	5
more than 100	8	6

Note that the amount of colostrum or whole mil to feed varies according to the size of calves.

Experiments have shown equal performance o calves fed once or twice daily on either feeding sys tem. Calf scours were not different for the groups. I may be concluded that either whole milk or fe mented colostrum will grow calves to satisfactory siz to weaning.

FATS AND OILS

Animal and vegetable fats seem to be equally e fective additions to rations; thus, selection should b determined solely by comparative price. Ordinarily animal fats are much cheaper than such vegetable oil as soybean oil or cottonseed oil. Vegetable oils ar generally priced out of the animal feed market, for us in margarine and in paint and other industrial uses.

Several different fat products are used as anim feed; among them, acidulated soap stock (foots), ta lows, greases (white and yellow), blended feeding fa house grease, brown grease, sewer grease, and mod ified yellow grease. Each of them should be bough by specifications and guarantees.

The energy value of fat is much higher than th of grains. For example, the TDN of animal fat is 232. whereas the TDN of No. 2 corn is 82.3. For this rea son, it is possible to increase the energy content wit little increase of the bulk of the ration. Thus, with th same feed intake, energy intake is higher.

If the price is favorable, fat may be added to r tions at the following levels: for swine and poultry, to 10%; and for cattle, 2 to 6%. Higher levels of f usually result in drastically lowered feed consum tion. When corn is the major source of grain, fat add tions can be expected to be less useful than with th small grains. This is understandable when it realized that corn contains approximately 4% fat a compared to 1 to 1½ for the other feed grains.

Higher levels of fat than indicated above may be used for young ruminants in milk replacers; depending on the purpose, replacers may contain 15 to 30 percent added fat.

HIGH-LYSINE CORN (Opaque-2, or O_2)

It has been known for many years that corn, the world's third most important human food after rice and wheat, is nutritionally inadequate. In 1914, researchers at the Connecticut Agricultural Experiment Station induced starvation in laboratory rats by feeding them generous helpings of corn. Further, it was found that rats could be restored to health by supplementing the high-corn diet with two protein fractions—the amino acids lysine and tryptophan.

Although normal corn contains about 10 percent protein, half of the protein consists of zein, which is especially poor in lysine and tryptophan, essential amino acids that the nonruminant cannot manufacture and must get from feed.

This deficiency of corn shows up in people wherever corn is a major source—if not the only source—of protein in the diet. Known by the exotic name, kwashiorkor, this nutritional deficiency disease is the leading cause of mortality among infants and children in many parts of the world.

For years, plant scientists assayed the world's corn varieties one by one, looking for a strain with more nutritionally balanced protein. Finally, in 1963, workers at Purdue University analyzed an odd group of corns characterized by soft, floury endosperm inside an opaque, chalk-white kernel. The Purdue scientists found that the opaque characteristic of corn, which had been noted for years without exciting much scientific interest, is associated with a recessive gene that replaces some of the kernel's humanly useless zein with needed lysine and tryptophan. The mutant—routinely labeled opaque-2, or O_2 for short—had a lysine content of 3.4 percent, compared to 2.0 percent for normal corn. Additionally, opaque-2 showed higher levels of tryptophan and other amino acids.

Although the nutritional value of the high-lysine corn is recognized, two major hurdles between research discovery and application must yet be overcome: (1) the mutant gene is linked to opaque-2's soft, floury kernel, which is both light in weight and vulnerable to pest attacks, producing lower yields for farmers; and (2) opaque-2 has not been accepted by the majority of consumers, who are accustomed to the harder "flint" or "dent" kernels with a deeper, translucent color. But the need is great—human lives are at stake. So, plant breeders have set about crossing the opaque-2 gene on corn varieties that better meet the demands of farmers and consumers.

HIGH-MOISTURE (Early Harvested) GRAIN

High-moisture grain, especially corn and sorghum, is grain that is harvested at 24 to 30% moisture content. Experimental tests show that the feed efficiency of a high-moisture milo is improved by 8 to 15%, although there is little increase in daily gain. There is less improvement from high-moisture processing of shelled corn, and the results have been most variable. High-moisture milo and corn should be ground or rolled before it is fed. It is questionable, however, if it pays to process high-moisture corn in rations that have less than 15% roughage.

• *Moisture is important when buying feeds*—When buying grains, a feeder should never lose sight of how much water he may be purchasing. Table 4-3 illustrates the relative value (dry matter purchased) when paying for corn on a 15.5% moisture basis while actually receiving corn of another moisture content. Thus, if the feeder were receiving 19% moisture corn and paying for 15.5% moisture, he would receive only 95.86% of the dry matter for which he paid. On the other hand, if corn is delivered with 7% moisture, while paying on a 15.5% moisture basis, the feeder would receive 110.06% of that for which he paid.

TABLE 4-3

RELATIVE VALUE OF U.S. NO. 2 CORN (15.5% MOISTURE) AS AFFECTED BY CHANGES IN MOISTURE[1]

Moisture	DM Basis Multiplier	Moisture	DM Basis Multiplier
(%)		(%)	
0	1.1834		
1	1.1716	19	.9586
2	1.1598	20	.9467
3	1.1479	21	.9349
4	1.1361	22	.9231
5	1.1243	23	.9112
6	1.1124	24	.8994
7	1.1006	25	.8876
8	1.0888	26	.8757
9	1.0769	27	.8639
10	1.0651	28	.8521
11	1.0533	29	.8402
12	1.0414	30	.8284
13	1.0296	31	.8166
14	1.0178	32	.8047
15	1.0059	33	.7929
16	.9941	34	.7811
17	.9822	35	.7691
18	.9704	36	.7574

[1]If 15.5% moisture corn is the purchase basis, it will require 1.1834 units of purchase basis corn to make 1 unit of 100 dry matter basis corn.

• *Moisture is important in formulating rations*—A careful feeder must constantly watch the moisture content of the feeds he buys, and the effect of moisture on his nutritional quality control. Most good feeders will readjust feeding formulas whenever moisture in a leading ingredient changes over one percent.

The way in which moisture changes cause imbalances is pointed up in the following example:

Let's assume that a feeder is using a ration which has as one of its main ingredients corn silage with 68% moisture content, and that this ration requires 1.9% supplement on an "as fed" basis. Now assume that the moisture of the silage suddenly decreased to 55%, and with it the necessary supplement to balance the ration increased to 2.62%. Obviously, if the feeder did not adjust the feeding formula, a serious shortage of protein could result. In this case, the cattle would receive only 72.5% as much supplement as they should have since the mixing formula was not recalculated.

The multipliers in Table 4-4 may be used to determine the price per unit of dry matter simply by multiplying price times the appropriate factor for the indicated moisture.

TABLE 4-4

CORRECTION FACTORS TO USE WHEN CONVERTING FEEDS OF VARIOUS MOISTURE CONTENTS TO A 100% DRY MATTER BASIS (0% MOISTURE)

Moisture	100% DM Basis Multiplier	Moisture	100% DM Basis Multiplier	Moisture	100% DM Basis Multiplier
0	1.0000	29	1.4084	58	2.3809
1	1.0101	30	1.4285	59	2.4390
2	1.0204	31	1.4492	60	2.5000
3	1.0309	32	1.4705	61	2.5641
4	1.0416	33	1.4925	62	2.6315
5	1.0526	34	1.5151	63	2.7020
6	1.0638	35	1.5384	64	2.7777
7	1.0752	36	1.5625	65	2.8571
8	1.0869	37	1.5873	66	2.9411
9	1.0989	38	1.6129	67	3.0303
10	1.1111	39	1.6393	68	3.1250
11	1.1235	40	1.6666	69	3.2258
12	1.1363	41	1.6949	70	3.3333
13	1.1494	42	1.7241	71	3.4482
14	1.1627	43	1.7543	72	3.5714
15	1.1765	44	1.7857	73	3.7037
16	1.1904	45	1.8181	74	3.8461
17	1.2048	46	1.8518	75	4.0000
18	1.2195	47	1.8867	76	4.1666
19	1.2345	48	1.9231	77	4.3478
20	1.2500	49	1.9607	78	4.5454
21	1.2658	50	2.0000	79	4.7619
22	1.2820	51	2.0408	80	5.0000
23	1.2987	52	2.0833	81	5.2631
24	1.3157	53	2.1276	82	5.5555
25	1.3333	54	2.1739	83	5.8824
26	1.3513	55	2.2222	84	6.2500
27	1.3698	56	2.2727	85	6.6666
28	1.3889	57	2.3255		

• *Acid treated high-moisture grain*—In the past, high-moisture grain has been either (1) artificially dried, or (2) stored in an airtight silo, to prevent spoilage. Now there is a third alternative—the use of naturally occurring acetic and propionic acids for reducing mold growth and other deterioration of high-moisture grain. Tests with acid treated (sprayed on at the time of storage) high-moisture grain suggests that rate o gain and dry matter conversion are at least equal, i not superior, to artificially dried or airtight store grain.

MILK REPLACERS (Synthetic Milk)

Several reputable commercial companies now produce and sell milk replacers or synthetic milk.

Although scientists have not yet learned how t compound a synthetic product that will alleviate th necessity of colostrum, in certain other respects the have been able to improve upon nature's product milk. For example, it has long been known that mil is deficient in iron and copper, thus resulting i anemia in suckling young if proper precautions ar not taken. In addition to correcting these deficiencie milk replacers are fortified with vitamins, mineral and antibiotics.

From the standpoint of livestock producers synthetic milk is of interest in raising orphaned o early weaned animals of each class of livestock. Als it is a valuable adjunct in certain disease control pr grams, especially those diseases that may be transmi ted from dam to offspring; and, in some cases, i makes it practical to retain in production those valu able females which, due to injury or disease to the udder, cannot suckle their young.

The raising of orphaned young of each class of farm animals will be simplified if they have first received colostrum. For such orphans, the milk replacer should be mixed according to the directions of the manufacturer. During the first few days, it is generally best to feed the orphan from a bottle with a rubber nipple. Later, it should be taught to drink from a suitable receptacle. It is important that all feeding utensils be kept absolutely clean and sanitary (cleaned and sterilized each time) and that feeding be at regular intervals. Also, orphaned animals should be given grain (and hay in the case of ruminants and foals) at the earliest possible time.

MOLASSES

Molasses (including cane or blackstrap, beet, citrus, and wood molasses) is extensively used as a livestock feed; almost 5½ million tons are used annually in the United States (see Table 4-1, page 217).

Cane and beet molasses are by-products of the manufacture of sugar from sugarcane and sugar beets, respectively. Citrus molasses is produced from the juice of citrus wastes. Wood molasses is a by-product of the manufacture of paper, fiberboard, and pure cellulose from wood; it's an extract from the more soluble carbohydrates and minerals of the wood material. Cane or blackstrap is by far the most extensively used type.

When used at levels of 10 to 15 percent of the ration, molasses has about three-fourths the energy value of corn. However, molasses has added values as an appetizer, to reduce dustiness of a ration, as a binder for pelleting, to stimulate rumen microbial activity, and as a source of unidentified factors. Also, cane molasses is a good source of certain trace minerals.

Brix is a term used to express molasses quality, as reflected by the relative level of sugar present. It is arrived at by first determining specific gravity. Then by use of conversion tables, the degrees Brix, or level of sucrose present, is obtained.

The different types of molasses may also be available in dehydrated form.

NONPROTEIN NITROGEN (NPN) SOURCES

Certain nonprotein nitrogen sources may be substituted for all or much of the supplemental protein required in most ruminant rations, provided such rations are adequate in minerals and readily available carbohydrates. Among such products are urea, ammoniated molasses, ammoniated beet pulp, ammoniated cottonseed meal, ammoniated citrus pulp, and ammoniated rice hulls. The possibility exists that other products will be forthcoming. Each such product should be evaluated by controlled feeding trials.

UREA

Urea is a white, crystalline, odorless, nonprotein nitrogen compound of the formula N_2H_4CO. It is manufactured in chemical plants that produce anhydrous ammonia by fixing some of the nitrogen of the air; some of the ammonia gas is combined with gaseous carbon dioxide to produce the white crystalline solid urea which is quite stable. In addition to feeds, urea is used as a fertilizer, either dry or in solutions, and in making plastics. Also, it is noteworthy that urea occurs as the principal end product of nitrogen metabolism in nearly all mammals; it is found in the urine of all farm animals and man.

Approximately 395,000 tons of urea are fed annually in the United States, as a source of protein for cattle, sheep, and goats.[2] In recent years there has been increased interest in feeding urea to ruminants, due primarily to the following circumstances:

1. *Shortage of oil meal proteins*—The scarcity and high price of normal supplies of oil meal protein feeds is well known.
2. *Progress in fundamental ruminant nutrition*—Through basic studies, scientists have established many of the nutrient requirements of rumen microorganisms, thereby permitting the preparation of balanced supplements designed to enable animals to get the most out of the roughages they consume. This knowledge has led to the extensive use of such low-grade roughages as corncobs, straws, and poor-grade hays—many of which had been wasted previously.

These factors, plus meeting the needs of a rapidly expanding human population, are likely to continue to accentuate the interest of feed manufacturers and stockmen in utilizing urea and other nonprotein nitrogen sources.

Urea may constitute up to one-third of the total protein of the ration of ruminants, provided additional energy is added in the form of molasses or grain to compensate for the lack of energy in the urea, in order to feed properly the rumen bacteria. By total protein is meant the protein intake of the entire ration—including forage, grain, and protein supplements.

Common guidelines relative to the use of urea for beef cattle are given in Table 4-5, page 224.

[2]Source: USDA. See Table 4-1.

TABLE 4-5

COMMON GUIDELINES TO THE USE OF UREA FOR CATTLE[1]

	For Finishing Cattle	For Grower (stocker) Cattle	For Wintering Pregnant and Lactating Cows
Percent of total protein in ration from urea .... (%)	33 ⅓	25.0	25.0
Maximum urea/animal/day ....(lb)	0.22 (100 g)	0.15 (68 g)	—
Percent of urea, by weight of total air-dry feed consumed ....(%)	1.0	1.0	1.0
Percent of urea, by weight, of concentrate mix (grain plus protein supplement)[2] ....(%)	2.0-3.0	3.0	3.0
Percent of urea, by weight, of the protein supplement ....(%)	20-30[3]	10.0[4]	10.0
Percent of supplemental nitrogen in high-protein supplement from urea[5] .... (%)	60-90[6]	30.0	30.0
Pounds of urea added/ton of corn silage at ensiling time[7] .... (lb)	10.0 (4.5 kg)	10.0 (4.5 kg)	10.0 (4.5 kg)

[1]In the preparation of this table, the author had the authoritative help of Dr. W. M. Beeson, Department of Animal Sciences, Purdue University, Lafayette, Ind.; Dr. William H. Hale, Department of Animal Science, The University of Arizona, Tucson, Ariz.; and Dr. W. E. Dinusson, Department of Animal Science, North Dakota State University, Fargo, N.D.

[2]Feed intake may be depressed if over 1% is used. Yet, many beef men are successfully using 2%.

[3]This means that as much as 60 to 90% of the protein value of the supplement may come from nonprotein sources. However, since such a supplement will constitute only 2 to 5% of the total ration fed, the first rule of thumb given in Table 4-5 still applies; namely, only ¼ to ⅓ of the total protein in the ration will be supplied from a nonprotein source.

[4]A protein supplement containing 10% urea provides 28.1% of protein equivalent (281% × .10) from nonprotein nitrogen.

[5]High urea supplements are best fed in complete mixed rations, which are *thoroughly* mixed. *Supplements containing 20-30% urea require extreme caution when being hand fed.*

[6]In a feedlot ration, this may be equivalent to 25 to 40% of the total nitrogen from all sources.

[7]On a dry matter basis, corn silage ensiled at the well-dented stage runs about 8% protein. The addition of 10 lb of urea per ton (or 5 kg/1,000 kg) of silage increases the protein content from 8 to 13%. However, there is loss of flexibility in feeding such a ration, and the rate of gain will be less than can be secured from higher energy, more dense rations. Also, it is extremely important that the urea be well mixed in the silage, otherwise there is hazard of toxicity.

There is no difference in the nutritional value of liquid and dry supplements built around urea if the supplements contain the same basic nutrients. Thus, it is a matter of personal choice, convenience, and ingredient costs as to which is used by feeders.

Slow-Release Nonprotein Nitrogen

Although products such as urea have many advantages, compounds that are nonprotein in nature possess the ability to liberate free ammonia. When fed to animals in excess of their ability to utilize this free ammonia, elevated blood ammonia levels will occur. Should the levels be sufficiently high that the normal metabolic processes cannot detoxify and eliminate by way of the kidneys (urine), death may result. This has spurred researchers in an effort to improve nonprotein nitrogen sources. Much of this research has centered around slowing down the release of ammonia from these nonprotein nitrogen sources so that the animal's metabolic system is not overworked or overloaded with ammonia at any one given point in time. This lower, more constant supply of ammonia can thereby be more efficiently utilized by the rumen microflora and subsequently the animal. Thus, the protection of protein and the coating of amino acids can improve growth and feed efficiency.

Some of the slow-release nonprotein nitrogen products presently on the market are: Starea, a combination of urea and gelatinized starch; and Golde Pro, a gelatinized corn combined with urea.

SINGLE-CELLED PROTEIN (SCP)

Some single-celled protein types such as yeas algae, and bacteria can be useful sources of protei and vitamins for human and animal feeding. Th safety of these foods depends on the organism selected, the quality of substrate used, and the cond tions of growth. Of course, yeast and bacteria hav been used for centuries in the baking, brewing, an distilling industries, in making cheese and other fe mented foods, and in the storage and preservation foods.

Dried brewers' yeast, a residue from the brewin industry, and Torula yeast, resulting from the fermen tation of wood residues and other cellulose source have been marketed as animal feeds for years. Wit proper processing, they are also suitable for huma foods.

Bacteria grow faster than yeasts under favorabl conditions, doubling their mass in a matter of min utes, rather than hours. Dried bacterial cells contai at least 55 percent protein.

Various bacteria and yeasts can be selected an cultured to grow on organic wastes. These includ animal wastes, sewage, many different chemical res dues from industrial plants, petroleum by-products awdust, and other fibrous residues. Petroleum companies in several countries have built factories to produce bacterial protein for the animal feed market. Also, considerable research is in progress to convert manure from poultry and other animals through bacterial fermentation into animal protein feed. This recycling process could produce much protein and help solve a pollution problem.

Algae are single-celled plants which may contain 20 to 60 percent protein on a dry basis. They synthesize proteins by the use of solar energy. In northern Nigeria, these plants are dried and eaten for human food. A noted University of California scientist conjectures that with the sunshine available in southern California, one acre of algae would feed 500 people for one year. Preliminary results with cultivated freshwater algae indicate that they will produce about 10 times as much protein per unit of land area as soybeans. Although algae grow widely on the earth's water surfaces, problems of harvesting and processing them into acceptable food products remain unsolved.

Feed Additives and Implants

Feed additives and implants are nonnutritive products that improve the rate and/or efficiency of gain of animals, prevent certain diseases, or preserve feeds. But there is no evidence of a nutritional deficiency when they are omitted from a ration.

Most animal scientists and stockmen agree that antibiotics and hormones stand out as the two nutritional discoveries of recent years that have had the greatest impact on the livestock industry. Today, it is estimated that 75 percent of the nation's growing-finishing cattle, lambs, and pigs receive feed additives or implants.

Some glowing reports to the contrary, there is no evidence to indicate that the use of these additives or implants can or will alleviate the need for vigilant sanitation, improved nutrition, and superior management. Instead, with the unfolding and applying of scientific information relative to these promotants, the producer will be able to achieve still greater efficiency of production. Also, practical producers will weigh the benefits of each one against its cost.

ANTIBIOTICS

Antibiotics are substances which are produced by living organisms (molds, bacteria, or green plants) and which have bacteriostatic or bactericidal properties.

Today, hundreds of antibiotics are known—including commercially produced bacitracin (and zinc bacitracin), chlortetracycline (Aureomycin), erythromycin, hygromycin, neomycin, oleandomycin, oxytetracycline (Terramycin), penicillin, streptomycine, and tylosin (Tylan).

Evidence indicates that antibiotics improve rate and efficiency of gain through the following actions:

1. *By reducing the incidence of subclinical levels of bacterial infections in the digestive and respiratory tracts*—This theory is based on the fact that antibiotics fail to show any measurable effect on animals maintained under germ-free conditions.
2. *By stimulating appetite and having a nutrient-sparing effect*—Most antibiotic-fed animals eat more than control animals receiving the same diet without antibiotics. This may largely account for the improved growth and efficiency.

   Also, there is evidence that antibiotics may have a sparing effect on dietary needs of some amino acids and B complex vitamins in young chicks, pigs, and rats, with this beneficial response being greater when diets contain submarginal or minimal levels of those nutrients.
3. *By stimulating certain enzyme systems*—Growth may be enhanced as a result of the stimulation of various enzyme systems.

In addition to improvement in rate and efficiency of gains, antibiotics usually (1) reduce the incidence of diarrhea in young animals, especially in young mammals deprived of colostrum; (2) lessen proneness of animals to go off feed; (3) reduce enterotoxemia and death loss of lambs on high-grain rations; and (4) lower the incidence of abscessed livers of cattle on high-grain rations.

The Food and Drug Administration has ruled that certain antibiotics must be withdrawn at a stipulated period of time prior to slaughter. Thus, those using antibiotics should always familiarize themselves with and follow these regulations.

Some folks object to the continued use of low levels of antibiotics in feeds on the ground that resistant pathogenic strains of microorganisms might develop which could be harmful to humans. This concern has caused the Food and Drug Administration to threaten to ban the use of certain antibiotics in livestock feeds. Although it is true that microbial resistance to antibiotics does occur, there is no scientific evidence that more virulent pathogens have evolved as a result of feeding low levels of antibiotics to animals. As a matter of fact, the evidence shows that resistant strains are nearly always less virulent.

HORMONES (or Hormonelike Products)

Most stockmen are familiar with, or have used, one or more of the hormones or hormonelike products.

The following presently available and approved products have been shown to improve gain and feed efficiency of feedlot cattle and calves: melengestrol acetate (MGA), used orally for heifers; zeranol (Ralgro), implanted; Synovex S (for steers), implanted; Synovex H (for heifers), implanted; and Rumensin. Most of these products appear to work well for lambs, although much lower dosages are required.

In ruminants, the hormones result in increased feed intake and increased nitrogen retention. The usual result is an increase in growth rate and feed efficiency and less finish when animals are fed to the same weight as untreated animals.

Hormones have not shown any consistent benefit for swine.

OTHER ADDITIVES

A number of other additives are used from time to time and for specific purposes; among them—

1. *Anthelmintic (worming) agents*, used to control stomach and intestinal worms.
2. *Antibacterial products (chembiotics)*, which exert much the same effects as the antibiotics.
3. *Antioxidants*, designed to prevent rancidity in feeds.
4. *Bloat control products*, such as poloxalene, which is marketed under the trade name, "Bloat Guard."
5. *Drugs*, incorporated in rations for the purpose of preventing certain diseases. For example, coccidiostats designed to prevent coccidiosis.
6. *Sodium bentonite (clay)*, used as a pellet binder, which also shows promise of improving the nitrogen utilization of ruminants.
7. *Sodium bicarbonate*, which functions as a buffer and pH agent, maintaining sufficient alkaline reserves (buffering capacity) in the animal to ensure normal physiological and metabolical functions.
8. *Tranquilizers*, used to calm animals under stress.

NUTRIENT NEEDS

Nutrients are the chemical substances found in feed materials that can be used, and are necessary, for the maintenance, production, and health of animals. The chief classes of nutrient substances are carbohydrates, fats, proteins, minerals, vitamins, and water. Nutrients are needed by the animal in definite amounts, with the quantities varying according to the class and age of animal, and the purpose for which it is being fed. A deficiency in a nutrient can be, and often is, a limiting factor in animal production.

The economical production of animal products is dependent upon (1) meeting the total nutritional requirement of the animal, and (2) knowing both the nu tritional requirements of the animal and the nutritiv value of the feeds.

Energy

Energy is required for practically all lif processes—for the action of the heart, maintenance blood pressure and muscle tone, transmission nerve impulses, ion transport across membrane reabsorption in the kidneys, protein and fat synthesi the secretion of milk, and the production of eggs an wool.

A deficiency of energy is manifested by slow stunted growth, body tissue losses, and/or lowere production of meat, milk, eggs, or fiber, rather than b specific signs such as those which characterize man mineral and vitamin deficiencies. For this reaso energy deficiencies often go undetected and unrect fied for extended periods of time.

It is common knowledge that a ration must co tain carbohydrates, fats, and proteins. Although eac of these has specific functions in maintaining a no mal body, they can all be used to provide energy f maintenance, for work, or for finishing. From th standpoint of supplying the normal energy needs animals, however, the carbohydrates are by far th most important, more of them being consumed tha any other compound, whereas the fats are next in im portance for energy purposes. Carbohydrates are usu ally more abundant and cheaper, and they are ve easily digested, absorbed, and transformed into bod fat. Also, carbohydrate feeds may be more easil stored than fats in warm weather and for longe periods of time. Feeds high in fat content are likely t become rancid, and rancid feed is unpalatable, if n actually injurious in some instances.

Protein

Proteins are complex organic compounds mad up chiefly of amino acids, which are present i characteristic proportions for each specific protei This nutrient always contains carbon, hydrogen, ox gen, and nitrogen, and in addition it usually contai sulfur and frequently phosphorus. Proteins are esse tial in all plant and animal life as components of th active protoplasm of each living cell.

Crude protein refers to all the nitrogenous com pounds in a feed. It is determined by finding the n trogen content and multiplying the result by 6.25. Th nitrogen content of protein averages about 16 perce (100 ÷ 16 = 6.25).

Animals of all ages and kinds require adequa amounts of protein of suitable quality—for mainte nance, growth, finishing, reproduction, work, an wool production. Of course, the protein requiremen

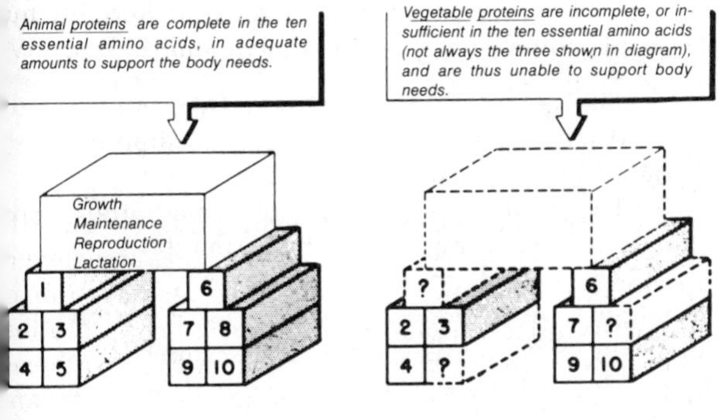
Animal proteins are complete in the ten essential amino acids, in adequate amounts to support the body needs.
Vegetable proteins are incomplete, or insufficient in the ten essential amino acids (not always the three shown in diagram), and are thus unable to support body needs.
Growth
Maintenance
Reproduction
Lactation
1
2 3
4 5
6
7 8
9 10
?
2 3
4 ?
6
7 ?
9 10

The Ten Essential Amino Acids:

1. Phenylalanine
2. Tryptophan
3. Leucine
4. Isoleucine
5. Lysine
6. Methionine
7. Valine
8. Histidine
9. Threonine
10. Arginine

Fig. 4-5. The amino acids are sometimes referred to as the building stones of proteins. Rations that furnish an insufficient amount of the essential building stones (amino acids) are said to have proteins of poor quality. In general, proteins of animal origin are of good quality, whereas proteins of plant origin are of poor quality. (Drawing by R. F. Johnson)

for growth, reproduction, and lactation are the greatest and most critical.

In general, animal proteins are superior to plant proteins for monogastric animals (including man) because they are better balanced in the essential amino acids. For example, zein (a corn protein) is an incomplete plant protein. It is deficient in the essential amino acids lysine and tryptophan. On the other hand, animal proteins are excellent sources of lysine, and many of them (especially milk and eggs) are abundant in tryptophan.

The necessity of each amino acid in the diet of the experimental rat has been thoroughly tested, but less is known about the requirements of large animals or even the human. According to our present knowledge, based largely on work with the rat, the following division of amino acids as essential and nonessential seems proper:

Essential (Indispensable)	Nonessential (Dispensable)
Arginine	Alanine
Histidine	Aspartic acid
Isoleucine	Citrulline
Leucine	Cysteine
Lysine	Cystine
Methionine (may be replaced in part by cystine)	Glutamic acid
Phenylalanine	Glycine
Threonine	Hydroxyglutamic acid
Tryptophan	Hydroxyproline
Valine	Norleucine
	Proline
	Serine
	Tyrosine

Fortunately, the amino acid content of proteins from various sources varies. Thus, the deficiencies of one protein may be improved by combining it with another, and the mixture of the two proteins often will have a higher feeding value than either one alone. It is for this reason that a considerable variety of feeds in the ration is usually recommended.

The feed proteins are broken down into amino acids by digestion. They are then absorbed and distributed by the bloodstream to the body cells, which rebuild these amino acids into body proteins.

Minerals

Minerals are the inorganic elements of animals and plants, determined by burning off the organic matter and weighing the residue, which is called ash.

Eighteen mineral elements are known to be required by at least some animal species. They can be divided into the following two groups based on the relative amounts needed in the ration:

Major or Macro Minerals	Trace or Micro Minerals
Sodium (Na)	Chromium (Cr)
Chlorine (Cl)	Cobalt (Co)
Calcium (Ca)	Copper (Cu)
Phosphorus (P)	Fluorine (F)
Magnesium (Mg)	Iodine (I)
Potassium (K)	Iron (Fe)
Sulfur (S)	Manganese (Mn)
	Molybdenum (Mo)
	Selenium (Se)
	Silicon (Si)
	Zinc (Zn)

The general functions of minerals are as follows:

1. Give rigidity and strength to the skeletal structure.
2. Serve as constituents of the organic compounds, such as protein and lipid, which make up the muscles, organs, blood cells, and other soft tissues of the body.
3. Activate enzyme systems.
4. Control fluid balance—osmotic pressure and excretion.
5. Regulate acid-base balance.
6. Exert characteristic effects on the irritability of muscles and nerves.
7. In mineral-vitamin relationships.

In addition to the general functions in which several minerals may be involved, each essential mineral has one specific role and sometimes more.

In order to balance rations properly and to prevent some of the deficiency diseases and ailments, a stockman needs to be aware of any soil mineral deficiencies or of excesses of certain minerals that may cause toxicity, characteristic of the area in which his farm or ranch is located (see Fig. 4-6, page 228).

To avoid deficiencies, minerals are usually provided, either free-choice or added to the ration. Such supplements should supply only the specific minerals that are deficient—and in the quantities necessary. Excesses and mineral imbalances should be avoided.

A summary of individual mineral functions, defi-

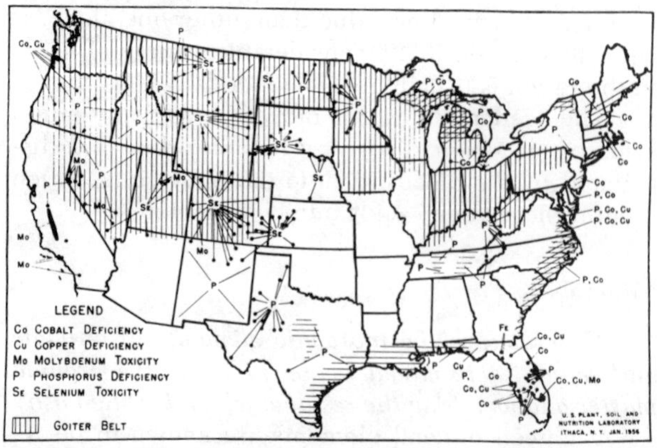

Fig. 4-6. Known areas in the U.S. where mineral nutritional diseases of animals occur. The dots indicate approximate locations where troubles occur. The lines not terminating in dots indicate a generalized area or areas where specific locations have not been reported. (Courtesy, USDA)

ciency symptoms, interrelationships and toxicities, and sources is given under the respective animal species later in this section.

SALT (Sodium Chloride)

Salt, which serves as both a condiment and a nutrient, is needed by all classes of animals, but more especially by herbivora (grass-eating animals). It may be provided in the form of granulated, rock, or block salt. In general, the form selected is determined by price and availability. It is to be pointed out, however, that very hard block and rock salts are difficult for stock to eat, often resulting in sore tongues and inadequate consumption. Also, if there is much competition for the salt block, the more timid animals may not get their requirements.

The salt requirements are greatly increased under conditions which cause heavy sweating, thereby resulting in large losses of this mineral from the body. Unless replaced, fatigue will result. For this reason, when engaged in hard work and perspiring profusely, both horses and humans should receive liberal allowances of salt.

CALCIUM AND PHOSPHORUS

Farm animals are more likely to suffer from a lack of calcium and phosphorus than from any of the other minerals except common salt. These 2 minerals comprise about 70% of the mineral content of the animal body and from ⅓ to ½ of the minerals of milk. About 99% of the calcium and over 80% of the phosphorus are found in the bones and teeth.

Liberal allowances of calcium and phosphorus are especially important for young, growing animals, for those that are pregnant, and for those that are producing milk.

The following general characteristics of feeds in regard to calcium and phosphorus are noteworthy:

1. The cereal grains and their by-products and straws, dried mature grasses, and protein supplements of plant origin are low in calcium.
2. The protein supplements of animal origin and legume forage and rape are rich in calcium.
3. The cereal grains and their by-products are fairly high or even rich in phosphorus, but a large portion of the phosphorus is not readily available.
4. Most all protein-rich supplements are high in phosphorus. But, here again, plant sources of phosphorus contain much of this element in a bound form.
5. Beet by-products and dried, mature, nonleguminous forages (such as grass hays and fodders) are likely to be low in phosphorus.
6. The calcium and phosphorus content of plants can be increased through fertilizing the soil upon which they are grown.

Calcium-Phosphorus Ratio and Vitamin D

In considering the calcium and phosphorus requirements of animals, it is important to realize that the proper utilization of these minerals by the body is dependent upon three factors: (1) an adequate supply of calcium and phosphorus in an available form, (2) a suitable ratio between them (somewhere between 1 to 2 parts of calcium to 1 part of phosphorus), and (3) sufficient vitamin D to make possible the assimilation and utilization of the calcium and phosphorus.

Although nutritionists have generally advocated a calcium-phosphorus ratio somewhere between 1 to 2 parts of calcium to 1 part of phosphorus, there is much evidence that a calcium-phosphorus ratio of 1:1 to 2:1 for nonruminants (hogs and horses) and 1:1 to 7:1 for ruminants are equally satisfactory, but ratios below 1:1 are disastrous.

If plenty of vitamin D is present (as provided either by sunlight or through the ration), the ratio of calcium to phosphorus becomes less important. Also, less vitamin D is needed when there is a desirable calcium-phosphorus ratio.

Recommended Calcium and Phosphorus Supplements

Table 4-6 gives several sources of calcium and phosphorus and the typical analysis of each.

Precautions Relative to Calcium and Phosphorus Supplements

Earlier experiments cast considerable doubt on the availability of phosphorus when it was largely in the form of phytin. Although wheat bran is very high

TABLE 4-6

TYPICAL ANALYSIS OF CALCIUM AND PHOSPHORUS SUPPLEMENTS[1]

Compound	Calcium Content	Phosphorus Content	Sodium Content	Nitrogen Content	Fluoride Content
	(%)	(%)	(%)	(%)	(%)
Calcium compounds:					
Oystershells, ground	38.00	—			
Limestone, ground	34.00	—			
Defluorinated phosphates manufactured from defluorinated phosphoric acid:					
Monocalcium phosphate	16.00	21.00	—	—	0.16
Dicalcium phosphate	21.00	18.50	—	—	0.14
Defluorinated phosphate	32.00	18.00	—	—	0.16
Monoammonium phosphate	0.50	24.00	—	11.00	0.18
Diammonium phosphate	0.50	20.00	—	18.00	0.16
Ammonium polyphosphate solution	0.10	14.50	—	10.00	0.12
Defluorinated wet-process phosphoric acid	0.20	23.70	—	—	0.18
Defluorinated phosphates manufactured from furnace phosphoric acid:					
Monocalcium phosphate	22.00	23.00	—	—	0.03
Dicalcium phosphate	26.00	18.50	—	—	0.05
Tricalcium phosphate	38.00	19.50	—	—	0.05
Monosodium phosphate, anhy	—	25.50	19.00	—	0.03
Disodium phosphate	—	21.50	32.00	—	0.03
Sodium tripolyphosphate	—	25.00	30.00	—	0.03
Ammonium polyphosphate solution	—	16.00	—	11.00	0.02
Feed-grade phosphoric acid	—	23.70	—	—	0.03
High-fluoride phosphates:					
Soft rock phosphate	17.00	9.00	—	—	1.20
Ground rock phosphate	35.00	13.00	—	—	3.70
Ground low-fluorine rock phosphate	36.00	14.00	—	—	0.45

[1]Except for calcium compounds, figures from *Effects of Fluorides in Animals*, NRC-National Academy of Sciences, Washington, D.C., 1974, pp. 8-9, Table 1.

in phosphorus, containing 1.32%, there was some question as to its availability due to the high phytin content of this product. More recent studies, however, indicate that cattle, and perhaps mature swine, can partially utilize phytin phosphorus. Cattle can utilize about 60% of the total phosphorus from most plant sources, whereas swine can utilize only about 50%. It must be emphasized, however, that phosphorus availability depends to a large extent on phosphorus sources, dietary supplies of calcium, and adequate vitamin D. Recent work indicates that high calcium levels enhance the formation of the insoluble phytic acid, whereas high vitamin D levels aid materially in the utilization of phosphorus in the form of phytin. On the other hand, in the case of man and poultry, the evidence seems clear that phytin phosphorus is a less satisfactory source of phosphorus.

During World War II, the shortage of phosphorus feed supplements led to the development of defluorinated phosphates for feeding purposes. Raw, unprocessed rock phosphate usually contains from 3.25 to 4.0 percent fluorine, whereas steamed bone meal normally contains only 0.05 to 0.10 percent. Fortunately, through heating at high temperatures under conditions suitable for elimination of fluorine, the excess fluorine of raw rock phosphate can be removed. Such a product is known as defluorinated rock phosphate.

The Association of American Feed Control Officials has established maximum fluorine content for (1) mineral substances, and (2) total ration (see Table 4-7).

TABLE 4-7

MAXIMUM FLUORINE CONTENT FOR (1) MINERAL SUBSTANCES, AND (2) TOTAL RATION[1]

Class of Animal	Maximum fluorine content of any mineral or mineral mixture which is to be used directly for the feeding of animals shall not exceed—	Fluorine content of rock phosphate (or other ingredients) shall be such that the maximum fluorine content of the total ration shall not exceed—
	(%)	(%)
Cattle	0.30	0.009
Sheep	0.35	0.01
Swine	0.45	0.014
Poultry	0.60	0.035

[1]*Feed Control*, Association of American Feed Control Officials, 1976, pp. 52-53.

TRACE MINERALS

Special attention needs to be given to trace minerals in areas where there is a deficiency of one or more of them, when poor quality roughage is fed, or when high- or all-concentrate rations are fed.

CHELATED TRACE MINERALS

Those selling chelated minerals generally recommend a smaller quantity of them (but at a higher price per pound) and extoll their "fenced-in" properties.

When it comes to synthetic chelating agents, much needs to be learned about their selectivity toward minerals, the kind and quantity most effective, their mode of action, and their behavior with different species of animals and with varying rations.

MINERAL SUPPLEMENTS

Only the specific minerals that are deficient, and in the quantities necessary, should be provided. Excesses and mineral imbalances should be avoided. It follows that needed supplementary minerals will vary according to the animal species, age, production, ration, and the mineral content of the soils and crops in the area where grown. Most feeds provide minerals in addition to basic organic nutrients, although fat and urea are marked exceptions. Nevertheless, most rations require more concentrated sources of one or more mineral elements.

Generally, the macrominerals of concern are sodium chloride (common salt), calcium, phosphorus, magnesium, and sometimes sulfur; and the trace elements that may be deficient are copper, iodine, iron, manganese, and zinc, and in some places cobalt and selenium.

Needed mineral mixes may be either home mixed or provided by a commercial product. Commercial mineral mixes are minerals mixed by manufacturers who specialize in the commercial mineral business, either handling minerals alone or in combination with a feed business. Because mineral mixes have become more complicated with the recognition of the importance of trace elements and interrelationships, and because most farmers and ranchers do not have the equipment with which to mix minerals properly, commercial minerals are finding a place of increasing importance in all livestock feeding.

Minerals may be either incorporated in the ration or self-fed.

When livestock are fed a mixed feed, totally or in part, the needed minerals are generally incorporated in the ration in keeping with known requirements. This is usually accomplished by adding 0.25 to 0.50 percent trace mineralized salt to the total ration, plus calcium and phosphorus (and any other minerals that are in short supply) as needed to balance the ration. In addition, where the lower level of salt (0.25%) is incorporated in the ration, trace mineralized salt is usually provided free-choice.

Where animals are fed an unmixed ration or are on pasture, minerals may be provided as follows:

1. *Where animals are on liberal grain feeding*—Provide free access to a 2-compartment mineral box, with (a) trace mineralized salt in one side; and (b) in the other side, a mixture of ⅓ trace mineralized salt (salt included for purposes of palatability), ⅓ defluorinated phosphate or steamed bone meal, and ⅓ ground limestone or oystershell flour.
2. *Where animals are primarily on roughage (pasture, hay, and/or silage)*—Provide free access to a 2-compartment mineral box, with (a) trace mineralized salt in one side (salt included for purposes of palatability); and (b) in the other side, a mixture of ⅓ trace mineralized salt and ⅔ defluorinated phosphate or steamed bone meal.

As noted, no limestone or oystershell flour (sources of calcium only) are needed in the latter mix because forages are generally more deficient in phosphorus than in calcium.

Vitamins

Vitamins are complex organic compounds that are required in minute amounts by one or more animal species for normal growth, production, reproduction, and/or health.

The omission of a single vitamin from the diet of a species that requires it will produce specific deficiency symptoms. Many of the vitamins function as coenzymes (metabolic catalysts); others have no such role, but perform certain essential functions.

Many phenomena of vitamin nutrition are related to solubility—vitamins are soluble in either fat or water. Consequently, it is important that both nutritionists and producers be well informed about solubility differences in vitamins and make use of such differences in programs and practices. Based on solubility, vitamins may be grouped as follows:

Fat-Soluble Vitamins	Water-Soluble Vitamins (all except vitamin C are known as B vitamins)
A (carotene)	B_{12}
D	Biotin
E	Choline
K	Folic acid (folacin)
	Inositol
	Niacin (nicotinic acid)
	Pantothenic acid
	Para-aminobenzoic acid
	Pyridoxine (B_6)
	Riboflavin (B_2)
	Thiamin (B_1)
	C (ascorbic acid)

The fat-soluble vitamins are stored in appreciable quantities in the body, whereas the water-soluble vitamins are not. Thus, vitamin A and/or carotene may be stored by an animal in its liver and fatty tissue in sufficient quantities to meet its requirements for a period of six months or longer. By contrast, the large amounts of water which pass through most animals daily tend to carry out the water-soluble vitamins of the body, thereby depleting the supply. Thus, they must be supplied in the diet on a day-to-day basis for those animals having a simple stomach in which microbial synthesis is limited.

Each of the vitamins is as much a distinct chemical compound as is cane sugar, for example. All of them contain carbon, hydrogen, and oxygen. In addition, all of the B vitamins except inositol contain nitrogen. Certain of the B vitamins also contain one or more of the mineral elements in their molecules. Even when added to the diet in very small amounts, vitamins are extraordinarily potent.

Pertinent information pertaining to the vitamin needs of each species is contained in the vitamin section devoted to each class of animals, and a summary of each of the vitamin deficiency diseases is contained in Table 4-100, page 438. In reviewing these tables, it must be remembered that single, uncomplicated vitamin deficiencies are the exception rather than the rule. Multiple deficiencies are altogether too common, making diagnosis difficult even to the trained observer.

VITAMIN A (Carotene)

Vitamin A, which is required of all farm animals and man, is strictly a product of animal metabolism, no vitamin A being found in plants. The counterpart in plants is known as carotene, which is the precursor of vitamin A. Because the animal body can transform carotene into vitamin A, this compound is often spoken of as "provitamin A."

The degree of greenness in a roughage is a good index of its carotene content, *provided* it has not been stored too long. Early cut, leafy green hays are very high in carotene.

Aside from yellow corn, practically all of the cereal grains used in livestock feeding have little carotene or vitamin A value. Even yellow corn has only about one-tenth as much carotene as well-cured hay. Dried peas of the green and yellow varieties and carrots are also valuable sources of carotene.

When vitamin A deficiency symptoms appear, stockmen should add a stabilized vitamin A product to the ration. But, it is wasteful to feed more vitamin A than is needed. Also, feeding exceedingly high levels of vitamin A over an extended period of time may cause bone fragility, hyperostosis, and exfoliated epithelium. When fed as directed, the vast majority of livestock feeds won't provide excesses of vitamin A. But where higher product levels than needed are used as a sales gimmick, there is hazard that the caretaker will administer the "stuff" like the Irishman's medicine—"If a little will do some good, a larger dose will do more good."

Other facts pertinent to vitamin A for animals follow:

1. *Circumstances conducive to vitamin A deficiencies*—The circumstances most conducive to vitamin A deficiencies are (a) extended periods of drought, resulting in the pastures becoming dry and bleached; (b) a long winter feeding period on bleached hay or straws, especially overripe cereal hays and straws; and (c) using feeds which have lost their vitamin A potency as a result of either heat or extended storage (for example, it has been found that alfalfa may lose nine-tenths of its vitamin A value in a year's storage). There is reason to believe that mild deficiencies of vitamin A, especially in the winter and early spring, are fairly common.

Fortunately, animals are able to store vitamin A, primarily in the liver, during periods of abundance to tide them through periods of scarcity. Thus, animals on green pasture store reserves to help meet their needs during the winter feeding period when their rations may be deficient. Mature animals may be able to store sufficient vitamin A in the liver to last six months; young animals store much less.

It is generally believed that stressed animals have a higher vitamin A requirement than those not under stress. Among such stress factors are: racing, showing, fatigue, hot weather, confinement, excitement, and number of animals run together.

2. *Measurement of vitamin A potency*—The vitamin A potency (whether due to the vitamin itself, to carotene, or to both) of feeds is usually reported in terms of IU or USP units. These two units of measurement are the same. They are based on the growth response of rats, in which several different levels of the test product are fed to different groups of rats, as a supplement to a vitamin A-free diet which has caused growth to cease. A USP or IU is the vitamin A value for rats of 0.30 microgram of pure vitamin A alcohol, or of 0.60 microgram of pure beta-carotene. The carotene or vitamin A content of feeds is commonly determined by calorimetric or spectroscopic methods.

VITAMIN D

Like vitamin A, vitamin D is required by all farm animals and man.

For four-footed animals (cattle, sheep, swine, and horses), both D_2 (the plant form) and D_3 (the animal form) are equally effective, so there is no need to use some of each. With poultry, however, vitamin D_3 (the animal form) is more active than vitamin D_2 (the plant form), and, therefore, should be used.

Young animals sometimes develop rickets because of insufficient vitamin D, calcium, or phosphorus. Rickets is characterized by reduced bone calcification, stiff and swollen joints, stiffness of gait, irritability, and reduction in serum calcium and phosphorus. It can be prevented by exposing the animal to direct sunlight as much as possible, by allowing free access to a suitable mineral mixture, or by providing good quality sun-cured hay or luxuriant pasture grown on well-fertilized soil. In confinement operations, and in northern areas that do not have adequate sunshine, stockmen should provide young stock with a vitamin D supplement.

With vitamin D, as with vitamin A, there is need for adequacy without harmful excesses. Too much vitamin D may harm an animal. Vitamin D toxicity is characterized by calcification of the blood vessels, heart, and other soft tissues, and by bone abnormalities. Also, there is general weakness and loss of body weight. In general, the levels of vitamin D intake that might prove harmful to farm animals are so far above the amounts fed for nutritional purposes that extensive studies on toxicity have not been made. Also, there appears to be wide specie and individual difference in the tolerance to high levels of vitamin D. As a result, evidence of vitamin D toxicity in farm animals is fragmentary and inadequate. It appears, however, that it takes massive doses of vitamin D to produce toxic symptoms in farm animals. By contrast, human infants and pregnant women have shown a low tolerance for excess vitamin D. For most age groups of people, the NRC recommended intake of vitamin D is a total not to exceed 400 IU per day; however, above 2,000 IU per day (only 5 times the recommended intake) can lead to hypercalcemia (toxicity caused by elevated blood levels of calcium).

The vitamin D requirement is less when a proper balance of calcium and phosphorus exists in the ration.

Other facts pertinent to vitamin D follow:

1. *Vitamin D, and cholesterol and ergosterol*—Most of the commonly used feeds contain little or no vitamin D, yet there is no widespread need for special supplements containing this factor. Fortunately, the skins of animals and many feeds contain provitamins in certain forms of cholesterol and ergosterol, respectively, which, through the action of ultraviolet light (light of such short wavelength that it is invisible) from the sun, are converted into vitamin D. These certain forms of cholesterol and ergosterol themselves have no antirachitic effect.

2. *Vitamin D limited in feeds*—Of all the known vitamins, vitamin D has the most limited distribution in common feeds. Very little of this factor is contained in the cereal grains and their by-products, in roots and tubers, in feeds of animal origin, or in growing pasture grasses. The only important natural sources of vitamin D are sun-cured hay and other roughages. The chief vitamin D-rich concentrates include sun-cured hay, cod-liver and other fish oils, irradiated cholesterol and ergosterol, and irradiated yeast.

3. *Effectiveness of sunlight in producing vitamin D*—The effectiveness of sunlight is determined by the lengths and intensity of the ultraviolet rays which reach the body. It is more potent in the tropics than elsewhere, more potent at noon than earlier or later in the day, more potent in the summer than in the winter, and more potent at high altitudes. The ultraviolet rays are largely screened out by clothing, window glass, clouds, smoke, or dust. Also, some biochemists theorize that the color of the skin of humans is nature's way of regulating the manufacture of vitamin D—that the dark skin of races near the equator filters out excess ultraviolet light.

VITAMIN E (Tocopherols)

There is experimental evidence that vitamin E improves the fertility of both males and females, and that it prevents and corrects anhidrosis (dry, dull hair coat) in horses.

Most practical rations contain liberal quantities of vitamin E, perhaps enough except under conditions of work, stress, or reproduction, or where there is interference with its utilization. Rather than buy and use costly vitamin E concentrates indiscriminately, the stockman should only add them to the ration on the advice of a nutritionist or veterinarian.

The requirements for vitamin E are influenced by interrelationships with other essential nutrients—increased by the presence of interfering substances, and spared by the presence of other substances that may be protective or that may assume part of its functions.

VITAMIN K

When vitamin K is deficient, the coagulation time of the blood is increased and the prothrombin level is decreased. This is the main justification for adding this vitamin to the ration. However, it appears that vitamin K is synthesized in adequate amounts by the intestinal microflora of ruminants and the horse.

B VITAMINS

Some members of this group are referred to by subscript numbers as vitamin B_1, B_2, etc.; others are known by their chemical names; still others have both a number and a chemical designation.

The last new vitamin discovery was made in 1948, when Merck and Company isolated the antianemia factor in liver, which became vitamin B_{12}.

Under normal circumstances, most, if not all, of the B vitamins (and vitamin K) are synthesized by ruminants (cattle and sheep) in sufficient quantities to meet their requirements. They are formed in the rumen and lower gastrointestinal tract as metabolic by-products of microbial fermentations. In pseudoruminants, such as the horse and rabbit, similar microbial synthesis occurs in the cecum and colon.

Unlike ruminants, however, pigs and poultry have one stomach (and no large cecum like the horse). As a result, they do not synthesize enough of certain B vitamins. Consequently, these factors must be provided regularly in the ration in adequate amounts if deficiencies are to be averted.

This means that the nutritionist and the caretaker should provide a dietary source of water-soluble B vitamins on a daily basis for simple-stomached animals—pigs, poultry, and man, but they need not be too concerned about providing the B vitamins to cattle, sheep, goats, horses, and rabbits.

It is also to be emphasized that subacute deficiencies can exist although the actual deficiency does not appear. In fact, borderline deficiencies are both the most costly and the most difficult with which to cope, going unnoticed and unrectified. Such borderline deficiencies result in poor and expensive gains.

Also, under farm conditions one will usually not find a vitamin deficiency which involves only a single vitamin. Instead, deficiencies usually represent a combination of factors, and usually the deficiency symptoms will not be clear-cut.

VITAMIN C (Ascorbic Acid)

A dietary need for vitamin C is limited to man, the guinea pig, and the monkey. Hence, there is no need to add this vitamin to animal rations.

VITAMIN IMBALANCES

Experiments have shown that the amounts needed of certain vitamins may be affected by the supply of another vitamin or of some other nutritive essential. Also, it is known that excess fortification of the animal's diet with certain vitamins may prove more detrimental than helpful. Thus, the stockman should avoid harmful imbalances; he should provide vitamins on the basis of recommended allowances. Also, when fortifying with vitamins, consideration should be given to the vitamins provided by the ingredients of the normal ration, for it is the total composition of the feed that counts.

UNIDENTIFIED FACTORS

In addition to the vitamins as such, certain unidentified or unknown factors are important in animal nutrition. They are referred to as "unidentified" or "unknown" because they have not yet been isolated or synthesized in the laboratory. Nevertheless, rich sources of these factors and their effects have been well established. A diet that supplies the specific levels of all the known nutrients but which does not supply the unidentified factors is inadequate for best performance. There is evidence that the growth factors exist in dried whey, marine and packinghouse by-products, distillers' solubles, antibiotic fermentation residues, alfalfa meal, and certain green forages. There is also evidence that at least one unknown hatchability factor is in fish solubles and green forage. Most of the unidentified factor sources are added to the diet at a level of one to three percent.

VITAMIN SUPPLEMENTS

Formerly, a wide variety of feed ingredients were added to livestock rations for their vitamin content. But it was found that the vitamin concentration of feedstuffs varied tremendously, being affected by plant species and part (leaf, stalk, or seed), harvesting, storing, and processing. Generally speaking, vitamins are easily destroyed by heat, sunlight, oxidation, and mold growth. So, today, nutritionists rely on vitamin supplements, which in many cases are chemically pure sources that need to be used only in very minute amounts. In modern feed formulation, premixes often represent the commonsense approach to providing vitamins.

For adult ruminants, vitamins A, D, and E are of concern, with A being the one most likely to be deficient. Under ordinary circumstances, ruminants synthesize adequate B vitamins, and vitamins C and K. Unless they are kept indoors, they usually receive sufficient exposure from direct sunlight to meet their needs for vitamin D.

Because of the greater prevalence of confinement feeding, along with limited gastrointestinal synthesis, swine are more apt to suffer from vitamin deficiencies than ruminants. Under practical conditions, special consideration should be given to the need for supplementing swine rations with the following vitamins: A, D, E, B_{12}, choline, niacin (nicotinic acid), pantothenic acid, and riboflavin (B_2).

Water

Water is one of the most vital of all nutrients. In fact, animals can survive for a longer period without feed than they can without water. Fortunately, under most conditions, it can be readily provided in abundance and at little cost. In addition to what animals drink, water is found in all feeds, ranging from about 10 percent in air-dry feeds to over 80 percent in fresh green forage.

Water is one of the largest single constituents of the animal body, varying in amount from 40% in fat hogs to 80% in newborn pigs, 50% in a 1,000-pound steer to 70% in a newborn calf, and 50% in a fat lamb to 80% in a newborn lamb. In general, the percentage of water in the bodies of animals varies with their species, condition, and age. The younger the animal, the more water it contains. Also, the fatter the animal, the lower the water content. Thus, as an animal matures, it requires proportionately less water on a weight basis, because it consumes less feed per unit of weight and the water content of the body is being replaced by fat. This accounts for the fact that gains in older animals are more costly than those in younger animals.

The specific water requirements of each class of animals will receive further consideration in the sections devoted to the respective species. In general, however, under practical conditions, the needs for water can best be taken care of by allowing the animals free access to plenty of clean, fresh water at all times.

HOW TO EVALUATE FEEDSTUFFS

Some feeds are more valuable than others; hence, measures of their relative usefulness are important. Among such methods of evaluating feeds are the following:

1. Physical evaluation
2. Best buy:
   a. Cost per unit of nutrients
   b. Tabular method
   c. Chart method
3. Chemical analysis
4. Digestion (or metabolism) trial
5. Measuring and expressing energy value of feedstuffs:
   a. Total digestible nutrients (TDN)
   b. Calorie system
6. Feeding trials
7. Other evaluations

Physical Evaluation of Feedstuffs

In order to produce or buy superior feeds, stockmen need to know what constitutes feed quality, and how to recognize it. They need to be familiar with those recognizable characteristics of feeds which indicate high palatability and nutrient content. If in doubt, the animals will tell them, for they like and thrive on high-quality feed.

The physical evaluation of feedstuffs, especially forages, is based largely on eye and smell appeal. Does it look good and smell good? The easily recognizable characteristics of hay of high feeding value are:

1. It is made from plants cut at an early stage of maturity, thus assuring the maximum content of protein, minerals and vitamins, and the highest digestibility.
2. It is leafy, thus giving assurance of high protein content.
3. It is bright green in color, thus indicating proper curing, a high carotene or provitamin A content, and palatability.
4. It is free from foreign material, such as weeds and stubble.
5. It is free from must or mold and dust.
6. It is fine stemmed and pliable—not coarse, stiff and woody.
7. It has a pleasing, fragrant aroma; it "smells" good enough to eat.

With grains, physical evaluation is largely based on test weight per bushel, moisture content, and amount of foreign material.

Best Buy in Feeds

Feed prices vary widely. For profitable production, therefore, feeds with similar nutritive properties should be interchanged as price relationships warrant.

In buying feeds, the stockman should check prices against values received. Three different methods of arriving at the best buys in feeds are (1) cost per unit of nutrients, (2) tabular method, and (3) chart method.

COST PER UNIT OF NUTRIENTS

One method of arriving at the best buy in feeds is to compute and compare the cost per unit of nutrients. The use of this method can best be illustrated by the examples that follow:

• *Cost per pound of protein and TDN*—If 44 percent protein (crude) soybean meal is selling at $6.00 per 100 pounds whereas 35 percent protein (crude) linseed meal sells for $5.00 per 100 pounds, which is the better buy? Divide $6.00 by 44 to get 13.6¢ per pound of crude protein for the soybean meal. Then divide $5.00 by 35 and get 14.3¢ per pound of crude protein for the linseed meal. Thus, at these prices soybean meal is the better buy—by 0.7¢ (14.3 − 13.6 = 0.7) per pound of crude protein.

When buying energy feed, one can compare the cost per pound of total digestible nutrients (TDN). For example, if corn is priced at $4.00 per 100 pounds and has a TDN of 91 percent, divide $4.00 by 91 and the result is 4.44¢ per pound of TDN.

If barley with 83 percent TDN sells for $3.85 per 100 pounds, divide $3.85 by 83, and the price is 4.64¢ per pound of TDN. Thus, corn would be the better buy by 0.2¢ (4.64 − 4.44 = 0.2) per pound of TDN.

• *Cost per pound of phosphorus*—When buying a mineral, the stockman may also check price against value received. For example, let's assume that the main need is for phosphorus, and that we wish to compare two minerals, which we shall call brands "X" and "Y." Brand "X" contains 12 percent phosphorus and sells at $340 per ton or $17/cwt; whereas, brand "Y" contains 10 percent phosphorus and sells at $320 per ton or $16/cwt. Which is the better buy?

Comparative Value of Brands "X" and "Y"
(based on phosphorus content alone)

Brand	Phosphorus	Price/cwt	Cost/lb phosphorus
	(%)	($)	($)
"X"	12	17.00	1.42
"Y"	10	16.00	1.60

Hence, brand "X" is the better buy, even though it costs $1.00 more per hundred, or $20.00 more per ton.

One other thing is important when buying minerals. As a usual thing, the more scientifically formulated mineral mixes will have plus values in terms of trace mineral needs and balance.

Of course, it is recognized that many other factors affect the actual feeding value of each feed, such as (1) palatability, (2) grade of feed, (3) preparation of feed, (4) ingredients with which each feed is combined, and (5) quantities of each feed fed. It follows that, from the standpoint of the stockman, the most important measurement of a feed's usefulness is in terms of "net returns," rather than cost per bag or cost per ton. To a swine producer, for example, cost per pound or per ton of feed, and pounds of feed required to produce a pound of pork are important only as they reflect or affect the cost per unit of pork produced. Thus, if the cost of a growing-finishing ration is 6¢ a pound and 4 pounds of the ration are required to produce 1 pound body weight, then the feed cost per pound of body weight can be arrived at by multiplying the above figures (6 × 4), which gives a feed cost of 24¢ per pound. Obviously when rations are compared, the ration that produces a pound of pork at the lowest total feed cost is the most desirable from an economic point of view.

TABULAR METHOD[3]

Table 4-8 is a tabular method for determining the best buys in feeds, with feeds classified as follows:

I. Dry Forages and Silages.
II. Energy Feeds.
III. Protein Feeds.

Table 4-8, page 236, shows the value of several feeds compared to two widely used feeds: (1) corn, a high-energy feed; and (2) soybean meal, a high-protein feed. In using this tabular method at any given time, the stockman should first replace the price figures in column 5 with those current for the particular area at the time purchases are to be made.

To use the form, select the feed in question; as an example, barley (see II. Energy Feeds—Including Roots, Tubers, Fruits, Nuts). The first column gives the digestible protein per 100 lb; the next column gives the nonprotein digestible nutrients per 100 lb; the third and fourth columns are the constants for corn and soybean meal (the two base feeds). The fifth column gives the actual price per ton of the feed. The sixth column gives the calculated value of the feed in relation to the selling price of corn and soybean meal. The seventh column gives the difference in the price per ton of the feed and its actual value. In the case of barley (48 lb/bu), the selling price is $107.00 per ton while the actual value is $96.27 per ton. This means that when corn is selling at $105.00 per ton and soybean meal at $135.00, barley at $107.00 per ton is not a good buy; it is overpriced by $10.73 per ton, based on digestible protein content and net energy value.

Most of the values given in columns 1 and 2 of Table 4-8 were obtained from various National Academy of Sciences publications; thus, they closely approximate, but are not necessarily identical to, the values given in Table 4-101, Feed Composition, of this book. Where available, values based on the actual analysis of the feed being evaluated should be used in these columns.

The calculations for Table 4-8 are made as follows:

1. Using barley as the example:

a. Let us assume that the price of No. 2 corn is $105.00 per ton.

b. The barley-corn constant is 0.823.

c. The barley-corn constant times the price of corn (0.823 × $105.00) is equal to $86.42.

d. Let us assume that the selling price of soybean meal is $135.00.

---

[3]The tabular and chart methods of feed evaluation presented in the sections headed Tabular Method and Chart Method were meticulously prepared by A. H. Ensminger and are based on the ingenious formula developed by Dr. William E. Petersen, University of Minnesota, and published in *Journal of Dairy Science*, 15, 1932, pp. 293-297.

TABLE 4-8
TABULAR METHOD OF DETERMINING BEST BUYS IN FEEDS

	(1) Digestible Protein/100 Lb	(2) Nonprotein Digestible Nutrients/ 100 Lb	(3) Constant for Corn (which shows the extent to which the price of corn affects the value)	(4) Constant for Soybean Meal (which shows the extent to which the price of soybean meal affects the value)	(5) Market Price per Ton	(6) Value (when corn is $105/ton, soybean meal is $135/ton)	(7) Difference
					($)	($)	($)
Corn No. 2	6.7	77.2	1.000	0.000	105.00	105.00	—
Soybean meal, sol-extd 45% protein	42.0	31.3	0.000	1.000	135.00	135.00	—
I. **Dry Forages and Silages**							
Alfalfa hay, all analyses	10.8	39.5	0.436	0.188	95.00	71.16	− 23.84
Alfalfa meal, dehy, min 22% protein	17.5	44.6	0.438	0.346	115.00	92.70	− 22.30
Alfalfa and brome hay	9.2	36.7	0.412	0.154	25.00	64.05	+ 39.05
Alfalfa and brome silage, all analyses	2.4	12.3	0.146	0.034	13.00	19.92	+ 6.92
Alfalfa and orchardgrass hay	10.2	41.2	0.466	0.168	75.00	71.61	− 3.39
Alfalfa and orchardgrass silage, min 50% DM	5.5	28.0	0.330	0.078	13.00	45.18	+ 32.18
Alfalfa silage, all analyses	3.4	12.1	0.134	0.060	16.00	22.07	+ 6.17
Almond hulls	0.5	67.6	0.926	−0.136	76.00	78.87	+ 2.87
Alsike clover hay	8.2	42.6	0.504	0.114	75.00	68.31	− 6.69
Bagasse, sugarcane, pulp, dehy	0.0	34.2	0.472	−0.074	40.00	39.57	− 0.43
Bermuda-grass hay, sun-cured	3.6	41.2	0.532	0.000	52.00	55.86	+ 3.86
Corncobs, ground	0.0	44.8	0.618	−0.098	28.00	51.66	+ 23.66
Corn (fodder), all analyses	3.4	47.4	0.620	−0.620	50.00	62.67	+ 12.67
Corn silage ears w/husks (snap ear corn)	2.2	29.7	0.388	−0.008	15.00	39.66	+ 24.66
Corn stover	2.6	49.0	0.650	−0.042	30.00	65.58	+ 32.58
Corn stover silage	1.1	18.3	0.242	−0.012	10.00	23.79	+ 13.79
Cottonseed hulls	0.0	43.1	0.596	−0.094	40.00	49.89	+ 9.89
Grass hay, all analyses	5.2	44.8	0.566	0.034	100.00	64.02	− 35.98
Grass and legume silage, all analyses	1.8	15.7	0.198	0.010	13.00	22.14	+ 9.14
Lespedeza hay, all analyses	8.3	44.7	0.532	0.112	24.00	70.98	+ 46.98
Milo silage (fodder)	0.9	18.4	0.246	−0.018	13.00	23.40	+ 10.40
Oat hay, all analyses	4.0	46.9	0.606	−0.002	89.00	63.36	− 25.64
Oat silage, dough stage	1.9	23.5	0.306	−0.004	13.00	31.59	+ 18.59
Paper waste	0.5	26.8	0.364	−0.048	15.00	31.74	+ 16.74
Prairie hay, all analyses	2.0	44.7	0.598	−0.046	24.00	56.58	+ 32.58
Red clover hay, all analyses	7.3	38.0	0.450	0.102	75.00	61.02	− 13.98
Rice straw	0.8	38.3	0.520	−0.064	48.00	45.96	− 2.04
Sorghum, Atlas, stover	1.2	44.7	0.606	−0.068	30.00	54.45	+ 24.45
Sorghum silage (fodder), dough stage	1.2	14.5	0.189	−0.001	13.00	19.71	+ 6.71
Sorghum silage, stover	0.3	19.9	0.272	−0.036	10.00	23.70	+ 13.70
Soybean straw	1.3	36.5	0.490	−0.046	35.00	45.24	+ 10.24
Sunflower silage, milk stage	1.1	9.9	0.125	0.006	12.00	13.94	+ 1.94
Sweet corn silage, cannery residue	1.4	19.8	0.260	−0.008	12.00	26.22	+ 14.22
Timothy hay, all analyses	2.5	46.0	0.610	−0.036	120.00	59.19	− 60.81
Wheat straw	0.0	38.7	0.534	−0.086	40.00	44.46	+ 4.46
II. **Energy Feeds (Including Roots, Tubers, Fruits, Nuts)**							
Almond hulls	0.5	67.6	0.928	−0.136	74.00	79.08	+ 5.08
Animal fat, hydrolyzed	0.0	232.6	3.210	−0.512	300.00	267.93	− 32.07
Barley grain, 48 lb/bu	8.6	66.0	0.823	0.073	107.00	96.27	− 10.73
Barley grain, Pacific Coast	5.2	66.0	0.858	−0.011	107.00	88.61	− 18.40
Beet mangels	1.1	10.0	0.127	0.006	12.00	14.15	+ 2.15
Citrus pulp, dried	3.0	70.1	0.937	−0.078	80.00	87.86	+ 7.86
Corn and cob meal (ear corn)	4.4	65.5	0.859	−0.033	103.00	85.74	− 17.26
Corn gluten meal	33.5	42.7	0.247	0.757	220.00	128.13	− 91.87
Corn grits (hominy feed)	7.1	75.4	0.968	0.015	117.00	103.67	− 13.34
Cottonseed whole, ground	14.5	74.2	0.876	0.205	149.00	119.66	− 29.35
Distillers' dried grains w/o solubles	18.9	57.3	0.598	0.354	109.00	110.58	+ 1.58
Grain screenings	8.6	48.7	0.585	0.111	77.00	76.41	− 0.59
Lard	0.0	238.2	3.287	−0.524	330.00	274.40	− 55.61
Milo grain	7.4	70.2	0.893	0.033	103.50	98.22	− 5.28
Molasses, citrus	2.9	48.2	0.635	−0.033	58.00	62.22	+ 4.22

(Continued)

TABLE 4-8 (Continued)

	(1) Digestible Protein/100 Lb	(2) Nonprotein Digestible Nutrients/ 100 Lb	(3) Constant for Corn (which shows the extent to which the price of corn affects the value)	(4) Constant for Soybean Meal (which shows the extent to which the price of soybean meal affects the value)	(5) Market Price per Ton	(6) Value (when corn is $105/ton, soybean meal is $135/ ton)	(7) Difference
Molasses, Sugar beet	3.8	58.1	0.763	−0.032	48.00	75.80	+ 27.80
Molasses, Sugarcane (blackstrap)	1.9	62.4	0.842	−0.089	51.00	76.40	+ 25.40
Oats, grain, 32 lb/bu	8.5	60.0	0.742	0.084	96.88	89.25	− 7.63
Potato tubers	1.3	17.4	0.227	−0.006	15.00	23.03	+ 8.03
Rye grain	7.9	65.1	0.817	0.057	95.00	93.48	− 1.52
Sorghum grain	5.6	68.1	0.883	−0.007	92.00	91.77	+ 0.23
Soybean seeds	34.1	49.1	0.330	0.758	220.00	136.98	− 83.02
Sugar beet pulp, dehy	4.7	61.3	0.798	−0.015	92.00	81.77	− 10.24
Triticale grain	14.6	61.4	0.698	0.236	100.00	105.15	+ 5.15
Wheat bran	9.6	62.4	0.763	0.106	121.00	94.43	− 26.58
Wheat grain, min 60 lb/bu	12.0	48.9	0.552	0.197	105.00	84.56	− 20.45
Wheat middlings, standard	13.4	58.2	0.667	0.213	83.00	98.79	+ 15.79
Wood molasses	0.0	54.2	0.748	−0.119	49.00	62.48	+ 13.48
II. **Protein Feeds**							
Ammonium polyphosphate	62.5	0.0	−0.637	1.588	220.00	147.50	− 72.51
Blood meal	56.9	1.9	−0.554	1.441	225.00	136.37	− 88.63
Brewers' grains, dried	19.1	42.6	0.393	0.392	87.00	94.19	+ 7.19
Buttermilk, dried	28.6	53.1	0.441	0.609	760.00	128.52	−631.48
Coconut (copra) meal, sol-extd, 21% protein	16.5	50.7	0.531	0.307	125.00	97.20	− 27.80
Cottonseed meal, prepress, sol-extd, 41% protein	32.5	30.8	0.93	0.757	152.00	111.96	− 40.04
Cottonseed, whole, ground	14.5	74.2	0.876	0.205	165.00	119.66	− 45.35
Distillers' dried grains, w/o solubles	20.0	57.8	0.594	0.381	109.00	113.81	+ 4.81
Feather meal, hydrolyzed	70.1	5.9	−0.634	1.767	185.00	171.98	− 13.03
Fish meal, menhaden	50.3	20.8	−0.226	1.232	275.00	142.59	−132.41
Linseed meal, mech-extd, 33% protein	31.5	38.3	0.207	0.716	160.00	118.40	− 41.61
Liver meal, animal	54.3	28.2	−0.165	1.317	800.00	160.47	−639.53
Meat meal (meat scraps)	47.5	14.0	−0.291	1.176	230.00	128.21	−101.80
Meat and bone meal	42.8	17.4	−0.196	1.049	230.00	121.04	−108.97
Milk, skimmed, dried	30.6	48.5	0.357	0.671	880.00	128.08	−751.92
Peanut meal, sol-extd, 47% protein	42.9	30.1	−0.023	1.023	152.00	135.69	− 16.31
Poultry viscera, w/feet and heads, rendered	47.1	23.5	−0.156	1.145	197.50	138.20	− 59.31
Rapeseed meal, sol-extd, 34% protein	28.5	31.4	0.143	0.655	120.00	103.44	− 16.56
Safflower meal, w/o hulls, sol-extd, 42% protein	37.4	31.6	0.054	0.880	144.00	124.47	− 19.53
Sesame seed meal, mech-extd, 44% protein	37.8	34.1	0.085	0.885	185.00	128.40	− 56.60
Sunflower meal, sol-extd, 44% protein	42.1	22.7	−0.117	1.019	150.00	125.28	− 24.72

e. The barley-soybean meal constant is 0.073.

f. The barley-soybean meal constant times the price of soybean meal (0.073 × $135.00) is equal to $9.86.

g. The value of barley based on corn (a high-energy feed) is $86.42. The value of barley based on soybean (a high-protein feed) is $9.86.

h. The value of barley based on the combined values of corn and soybean meal is $86.42 plus $9.86 or $96.27.

i. Barley is not a good buy, for it is worth $10.73 per ton less ($107.00 − $96.27) than its actual selling price.

2. Using dried skim milk as the example (see III. Protein Feeds, which shows that the market price of skim milk is $880 per ton):

a. The selling price of corn is $105.00.

b. The skim milk-corn constant is 0.357.

c. The skim milk-corn constant times the price of corn (0.357 × $105.00) is equal to $37.49.

d. The selling price of soybean meal is $135.00 (assumed).

e. The skim milk-soybean meal constant is 0.671.

f. The skim milk-soybean meal constant times the price of soybean meal (0.671 × $135.00) is equal to $90.59.

g. The value of skim milk based on corn (a

high-energy feed) is $37.49. The value of skim milk based on soybean meal (a high-protein feed) is $90.59.

h. The value of skim milk based on the combined values of corn and soybean meal is $37.49 plus $90.59 or $128.08.

i. Obviously, dried skim milk is not a good buy based on digestible protein content and net energy value, for it is overpriced by $751.92 per ton ($880.00 – $128.08). Nevertheless, dried skim milk is a valuable feed for young animals because of its high-quality proteins, vitamins, mineral balance, and the beneficial effect of the milk sugar, lactose. In addition, it is palatable and highly digestible.

CHART METHOD

Figs. 4-7, 4-8, and 4-9 are Feed Evaluation Charts for determining the best buy in feeds.

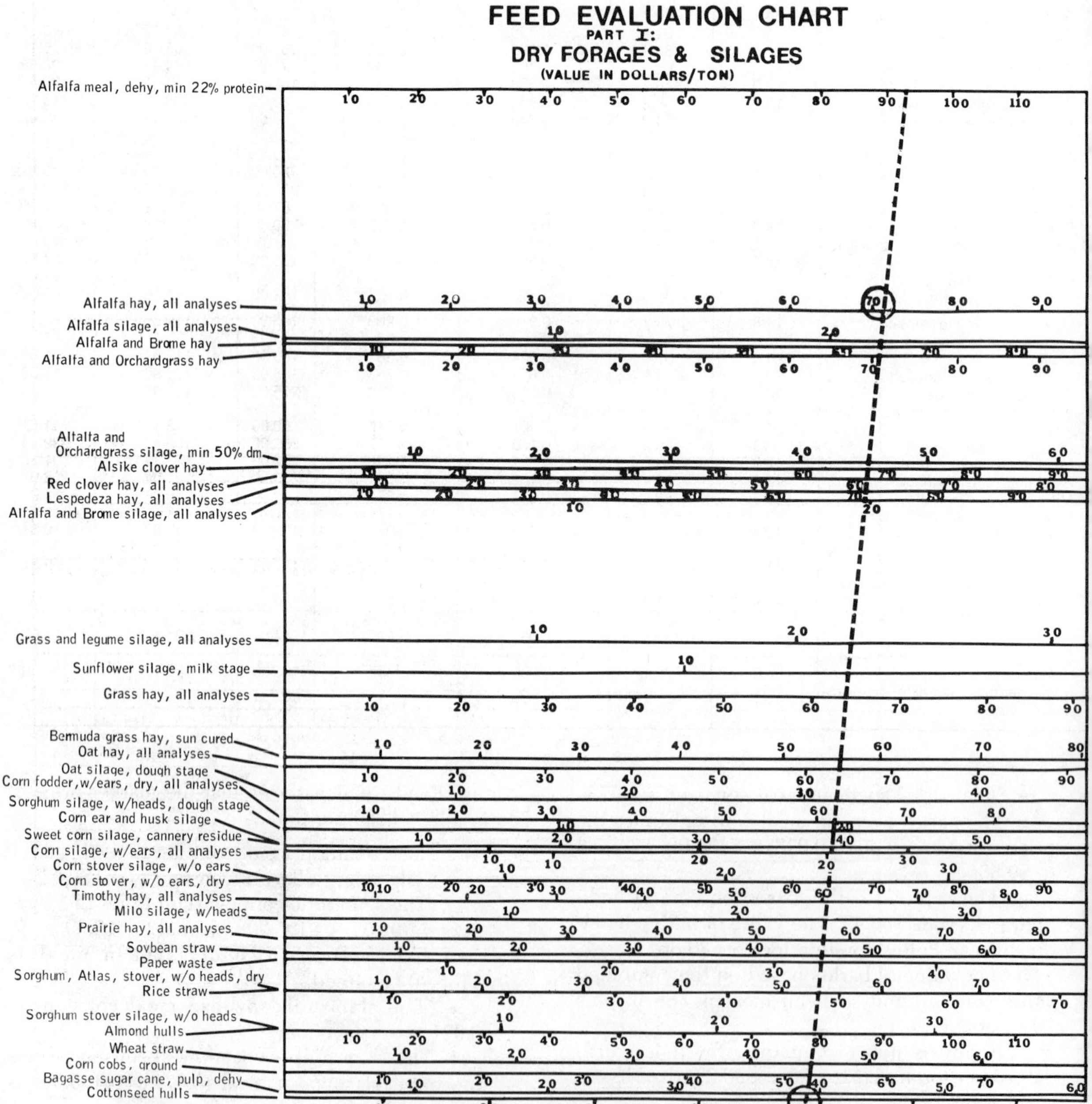

Fig. 4-7. Feed Evaluation Chart for determining the best buy in *dry forages and silages*. (Prepared by A. H. Ensminger)

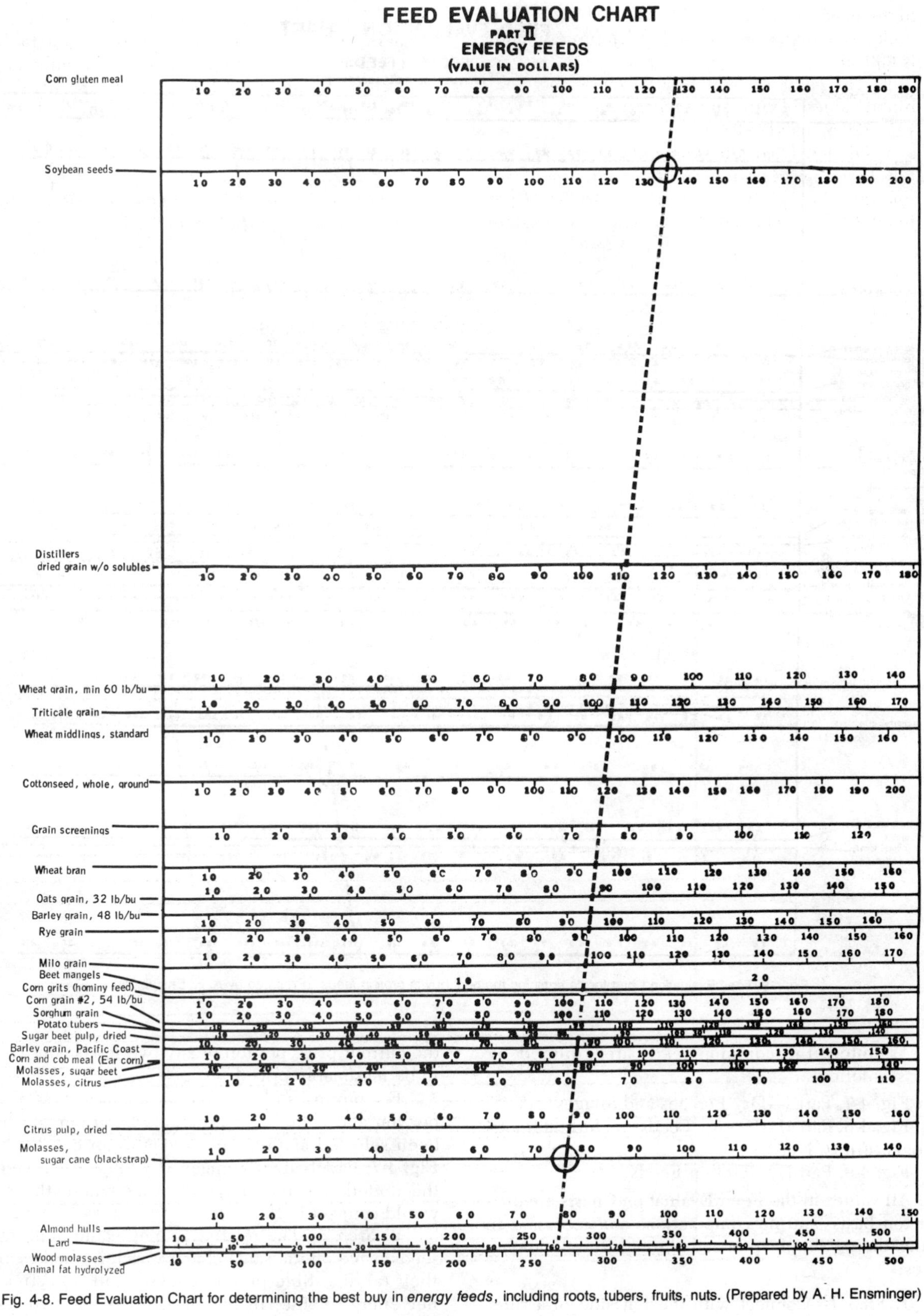

Fig. 4-8. Feed Evaluation Chart for determining the best buy in *energy feeds*, including roots, tubers, fruits, nuts. (Prepared by A. H. Ensminger)

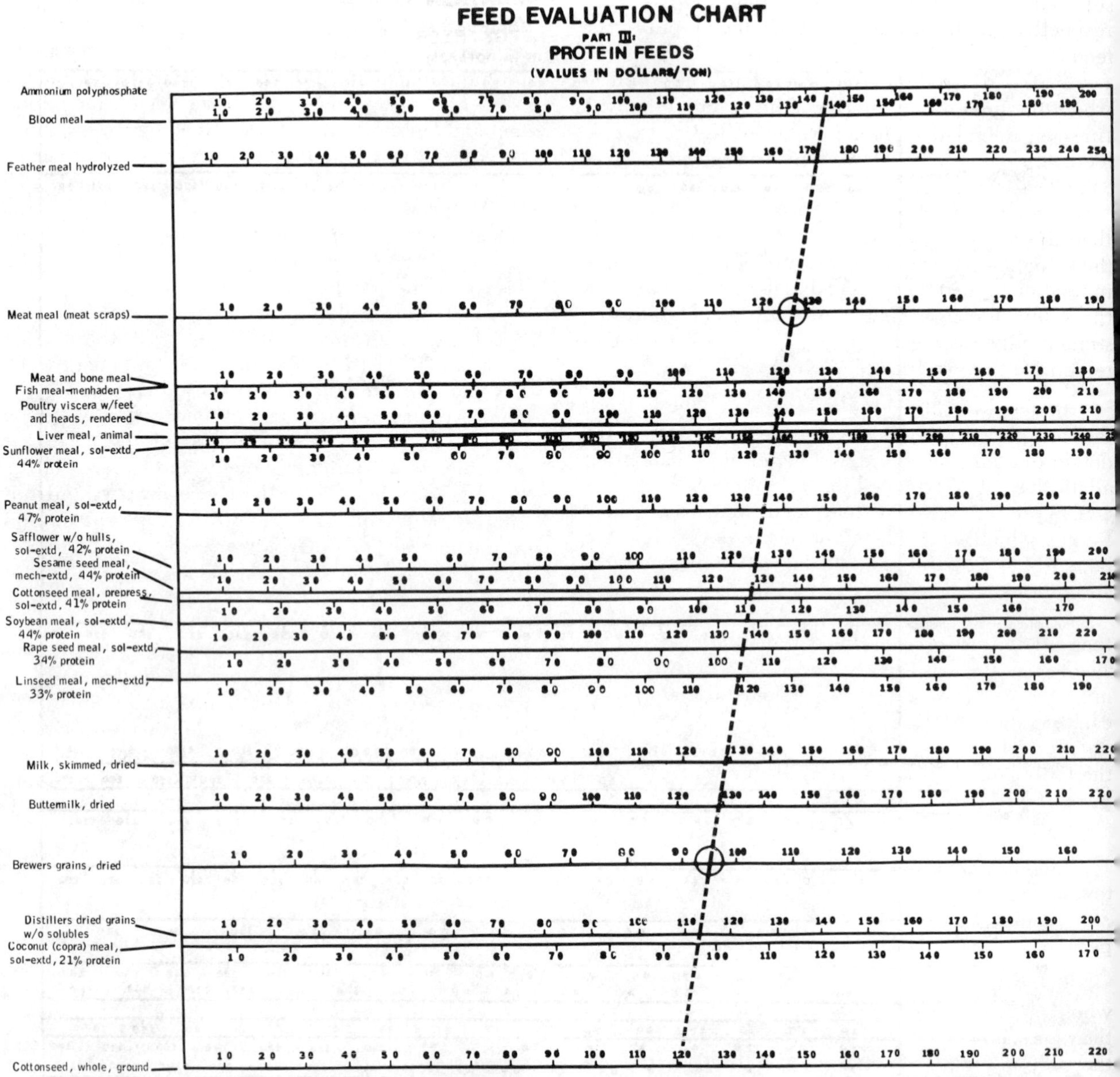

Fig. 4-9. Feed Evaluation Chart for determining the best buy in *protein feeds*. (Prepared by A. H. Ensminger)

As noted, the common feeds are grouped by charts as follows:

Fig. 4-7, Part I—Dry Forages and Silages

Fig. 4-8, Part II—Energy Feeds (Including Roots, Tubers, Fruits, Nuts)

Fig. 4-9, Part III—Protein Feeds

All values in the Feed Evaluation Charts are on a per ton basis, and no computations are necessary in using them. Simply proceed through the following steps:

1. *Step 1*—Connect with the left side of a ruler the current price per ton of any two feeds considered to be reasonably priced.

For purposes of illustration, alfalfa hay at $71.16 per ton and cottonseed hulls at $49.89 per ton have been selected as the two reasonably priced feeds in Fig. 4-7, Part I—Dry Forages and Silages. Therefore, the dotted line on Fig. 4-7 shows where the ruler would connect in this case.

2. *Step 2*—The relative money value of all other feeds listed on the same chart (as determined by both their (a) digestible protein content and (b) estimated net energy value, in comparison to the two feeds

selected) is the point at which the ruler intersects the respective graphs or lines, each of which represents a feed.

In the illustration used, therefore, the point at which the dotted line intersects the several graphs or lines is the relative money value of the respective feeds to corn at $105.00 per ton and soybean meal at $135.00 per ton.

3. *Step 3*—If the actual price of any feed is less than the value shown (to the left of the point where the straight edge intersects the graph), it is a good buy in relation to the two base feeds selected; if more than the value shown (to the right of the point where the straight edge intersects the graph), it is a poor buy in relation to the two base feeds selected.

In the example selected, therefore, those feeds that are priced at figures which lie to the left of the dotted line may be considered better buys than either alfalfa hay at $71.14 per ton or cottonseed hulls at $49.90 per ton; whereas those feeds that are priced at figures which lie to the right of the dotted line are poorer buys than either alfalfa hay or cottonseed hulls at the prices indicated.

If some other products are decidedly cheaper than alfalfa hay or cottonseed hulls, they should be selected as the base feeds. Then, in order to determine the best buy in available feeds, proceed as indicated in steps 1, 2, and 3.

For purposes of illustrating the other two charts, the following feeds have been selected as reasonably priced, and connected by dotted lines in order to show where the ruler would connect:

Fig. 4-8, Part II—Soybean seeds at $136.98 per ton, and sugarcane molasses at $76.40 per ton.

Fig. 4-9, Part III—Brewers' dried grains at $94.19 per ton, and meat meal at $128.21 per ton.

In those occasional but abnormal price periods when protein-rich feeds are as cheap as or cheaper than grains, feeds should be selected on the basis of the economy with which energy is furnished. Then Figs. 4-7, 4-8, and 4-9 may be used to compare feeds on an energy basis alone by the following process:

1. Select a base feed for which the price seems to be reasonable.

2. Place the left side of a straight edge on the price per ton of the base feed, and keep the straight edge parallel to the side of the chart.

3. The money value (as determined by estimated net energy value alone, in comparison with the base feed selected) of all other feeds is the point at which the straight edge intersects the respective graphs.

4. If the actual price of any feed is less than the value shown (to the left of the point where the straight edge intersects the graph), it is a good buy in relation to the base feed selected.

In using the Feed Evaluation Charts, Figs. 4-7, -8, and -9, it should be recognized that these values are based on two factors only, namely (1) the digestible protein content, and (2) the estimated net energy value of the respective feeds. These factors are important and are indicative of actual feeding values, but it is further recognized that there are many other factors that affect the actual feeding value of each feed, such as (1) species of animals (for example, many of the values given are not applicable to poultry, and there are differences between beef cattle, dairy cattle, sheep, swine, and horses); (2) palatability; (3) grade of feed; (4) preparation of feed; (5) ingredients with which each feed is combined; and quantities of each feed fed; (6) relative need for protein and energy; etc. For these reasons, many times the values given in the Feed Evaluation Charts and in the Feed Substitution Tables given in this section under each species do not coincide. This is primarily due to the fact that in the Feed Evaluation Charts both the protein and net energy values of each feed are always considered simultaneously, whereas in the Feed Substitution Tables the protein and/or energy values of each feed are considered separately.

Naturally, it is impossible to list all feeds in such charts as Figs. 4-7, 4-8, and 4-9. When it is desired to determine the value of feeds not listed, the reader is referred to the Feed Substitution Tables (Table 4-27, Feed Substitution Table for Cattle [Beef and Dairy], page 290; Table 4-63, Feed Substitution Table for Sheep, page 368; Table 4-85, Feed Substitution Table for Swine, page 402; and Table 4-97, Feed Substitution Table for Horses, page 430).

Chemical Analysis

A chemical analysis gives a solid foundation on which to start in evaluating feeds. It is a rough indicator of the value of a feedstuff or ration with regard to specific nutrient substances. Thus, feed composition tables serve as a basis for ration formulation and for feed purchasing and merchandising. Commercially prepared feeds are required by state law to be labeled with a list of ingredients and a guaranteed analysis. Although state laws vary slightly, most of them require that the feed label (tag) show in percent the minimum crude protein and fat; maximum crude fiber and ash; minimum calcium and phosphorus; maximum salt; and minimum TDN. These figures are the buyer's assurance that the feed contains the minimal amounts of the higher cost items—protein and fat; and not more than the stipulated amounts of the lower cost, and less valuable, items—the crude fiber and ash.

The proximate analysis, developed by workers at the Weende Experiment Station in Germany, attempts to isolate the least digestible fraction by treat-

TABLE 4-9

THE FRACTIONS OF PROXIMATE ANALYSIS

Procedure[1]	Fraction	Major Components
1. Heat sample to constant weight at temperature just above boiling point of water. Loss in weight equals water.	Moisture (dry matter)	Water and any volatile compounds (100% − H_2O = D.M.%)
2. Burn at 500 to 600° C for 2 hrs.	Ash (mineral matter)	Mineral elements
3. Determine nitrogen by Kjeldahl sulfuric-acid digestion.	Crude protein (protein averages 16% N; hence, N × 6.25 = crude protein)	Proteins, amino acids, nonprotein nitrogen
4. Extraction with ether.	Ether extract (fat)	Fats, oils, waxes, resins, coloring matter
5. Residue after boiling in weak acid and weak alkali.	Crude fiber (CF)[2]	Cellulose, hemicellulose, lignin
6. Remainder; i.e., 100 minus sum of the other fractions.	Nitrogen-free extract (NFE)[2]	Starch, sugars, some cellulose, hemicellulose and lignin

[1]Each procedure is applied to a separate sample, of standard weight, of the feedstuff to be analyzed.
[2]Carbohydrates (CHO = CF + NFE).

ing a feed sample with a weak acid, followed by a weak alkali, to obtain a residue termed crude fiber. It is the most generally used chemical scheme for evaluating feedstuffs, despite the fact that the information it gives may often be of uncertain nutritional significance or even misleading. According to it, a feedstuff is partitioned into the six fractions shown in Table 4-9.

In addition to proximate analysis, methods are available for assaying individual vitamins—biological assays for some, chemical determinations for others.

Despite the recognized value of a chemical analysis, it is not the total answer because of the following reasons:

1. *Feeds vary widely*—Because of wide variations in the composition of feeds, feed composition tables ("book values") should be considered as guides only. For example, the protein and moisture content of hay is quite variable, and the phosphorus and iodine content of soils affect plant composition. So, whenever possible, especially in large operations, it is best to take a representative sample of each major feed ingredient and have a chemical analysis made of it.

2. *It does not tell how much indigestible material there is in a feed*—Unfortunately, the acid-alkali treatment dissolves much of the plant lignin, a plant skeletal substance that no animal can digest, making it impossible to predict accurately how much indigestible matter there is in the feed. The method tends to overestimate the nutritive value of some feeds, underestimate that of others, and to indicate how the constituents of the indigestible residue are related to each other or what function some of them perform in digestion.

3. *It does not go far enough*—A chemical or proximate analysis does not provide any information relative to palatability, texture, toxicity, digestive disturbances, the associated effect of feedstuffs, or nutritional adequacy. Neither does it tell anything about the soil on which the feed was grown, despite the fact that soils high in molybdenum and selenium affect the composition of the feeds produced. Other similar soil-plant-animal relationships exist and are important. Thus, further steps need to be taken to evaluate a feed.

Digestion (Metabolism) Trial

Animals are not able to extract all the nutrients present in feeds. The actual value of ingested nutrients is dependent upon the use which the body is able to make of them. The first consideration here is digestibility, since undigested nutrients do not get into the body proper.

A digestion trial is made by determining the percentage of each nutrient in the feed through chemical analysis; giving the feed to the test animal for a preliminary period (usually 7 to 10 days for ruminants) so that all residues of former feeds will pass out of the digestive tract; giving weighed amounts of the feed during the test period (7 to 10 days for ruminants), collecting, weighing, and analyzing the feces; determining the difference between the amount of the nutrient fed and the amount found in the feces; and computing the percentage of each nutrient digested. The latter figure is known as the *digestion coefficient* for that nutrient in the feed.

Various techniques and equipment may be used to make the fecal collections; among them are: a specially designed digestion stall; collection harness and bag; markers (such as carmine, ferric oxide, chromic oxide, or soot), fed with the ration at the beginning and the end of the collection period; and indicators of an inert reference subject.

The digestibility of a feedstuff is affected by four factors: the species of animal (type of digestive tract), the chemical composition of the feed, the way in which the feed was processed, and the individuality of the animal (some animals have more efficient digestive systems than others).

Measuring and Expressing Energy Value of Feedstuffs

One nutrient cannot be considered as more important than another, because all nutrients must be present in adequate amounts if efficient production is to be maintained. Yet, historically, feedstuffs have been compared or evaluated primarily on their ability to supply energy to animals. This is understandable because (1) energy is required in larger amounts than any other nutrient, (2) energy is most often the limiting factor in livestock production, and (3) energy is the major cost associated with feeding animals.

Cereal grains are higher than roughages in energy. Although grains usually cost more on a weight basis than roughages, they are often a cheaper source of energy.

Our understanding of energy metabolism has increased through the years. With this added knowledge, changes have come in both the methods and terms used to express the energy value of feeds.

Broadly speaking, two methods of measuring energy are employed in this country—the total digestible nutrient system (TDN), and the calorie system. Each system has its advantages and advocates. But, more and more feedstuffs are being evaluated in calories.

TOTAL DIGESTIBLE NUTRIENTS (TDN)

Total Digestible Nutrients (TDN) is the sum of the digestible protein, fiber, nitrogen-free extract, and fat × 2.25. It has been the most extensively used measure for energy in the United States.

Back of TDN values are the following steps:

1. *Digestibility*—The digestibility of a particular feed for a specific species is determined by a digestion trial.

2. *Computation of digestible nutrients*—Digestible nutrients are computed by multiplying the percentage of each nutrient in the feed (protein, fiber, nitrogen-free extract [NFE], and fat) by its digestion coefficient. The result is expressed as digestible protein, digestible fiber, digestible NFE, and digestible fat. For example, if No. 2 corn contains 8.9 percent protein of which 77 percent is digestible, the percent of digestible protein is 6.9.

3. *Computation of total digestible nutrients (TDN)*—The TDN is computed by use of the following formula:

$$\% \text{ TDN} = \frac{\text{DCP} + \text{DCF} + \text{DNFE} + (\text{DEE} \times 2.25) \times 100}{\text{feed consumed}}$$

where DCP = digestible crude protein; DCF = digestible crude fiber; DNFE = digestible nitrogen-free extract; and DEE = digestible ether extract.

TDN is ordinarily expressed as a percent of the ration or in units of weight (lb or kg), not as a caloric figure.

The main *advantage* of the TDN system is that it has been used for a very long time and many people are acquainted with it.

The main *disadvantages* of the TDN system are:

1. It is really a misnomer, because TDN is not an actual total of the digestible nutrients in a feed. It does not include the digestible mineral matter (such as salt, limestone, and defluorinated phosphate—all of which are digestible); and the digestible fat is multiplied by the factor 2.25 before being included in the TDN figure, because its energy value is higher than carbohydrates and protein. As a result of multiplying fat by the factor 2.25, feeds high in fat will sometimes exceed 100 in percentage TDN (a pure fat with a coefficient of digestibility of 100% would have a theoretical TDN value of 225%- - -100% × 2.25).

2. It is an empirical formula based upon chemical determinations that are not related to actual metabolism of the animal.

3. It is expressed as a percent or in weight (lb or kg), whereas energy is expressed in calories.

4. It takes into consideration only digestive losses; it does not take into account other important losses, such as losses in the urine, gasses, and increased heat production (heat increment).

5. It overevaluates roughages in relation to concentrates when fed for high rates of production, due to the higher heat loss per pound of TDN in high-fiber feeds.

Because of these several limitations, in the United States the TDN system is gradually being replaced by other energy evaluation systems, particularly net energy. However, due to the voluminous TDN data on many feeds and long-standing tradition, it will continue to be used by many people for a long time to come.

How to Calculate the TDN of a Mixed Feed

Frequently, stockmen have a chemical analysis made of a mixed feed. Normally, the chemical results are reported in terms of crude protein, crude fat (ether extract), crude fiber, ash, moisture, and NFE.

In finishing rations in particular, the chemical analysis needs to be augmented by the TDN, or energy, value. This poses a difficult question because digestibility is affected by many things—level of feed intake, particle size, condition of animal, and individuality.

If the formulation of the mixed feed is known, the TDN value can be calculated as follows:

1. Obtain the digestible nutrients of each ingredient by multiplying the percentage of each nutrient by the digestion coefficient (see section headed, "Total Digestible Nutrients—TDN," point No. 2).
2. Then, the TDN is the sum of all the digestible organic nutrients—protein, fiber, nitrogen-free extract, and fat (× 2.25).

The above procedure is rather tedious and laborious. A simple and quick rule-of-thumb method for arriving at the TDN of a mixed cattle ration follows:

70% of the crude protein, plus fat times 2¼, plus NFE

The above rule-of-thumb method for determining TDN is close enough for most purposes. However, it is not recommended where considerable amounts of such by-product feeds as almond hulls, raisin stems, or grape seeds are included in the ration.

CALORIE SYSTEM

Calories are used to express the energy value of feedstuffs. One calorie (always written with a small c) is the amount of heat required to raise the temperature of 1 gram of water 1° C.

To measure this heat, an instrument known as the bomb calorimeter is used, in which the feed (or other substance) tested is placed and burned in the presence of oxygen (see Fig. 4-10).

It is noteworthy that the determination of the heat of combustion with a bomb calorimeter is not as difficult or time-consuming as the chemical analyses used in arriving at TDN values. Briefly stated, the procedure is as follows: An electric wire is attached to the material being tested, so that it can be ignited by remote control; 2,000 grams of water are poured around the bomb; 25 to 30 atmospheres of oxygen are added to the bomb; the material is ignited; the heat given off from the burned material warms the water; and a thermometer registers the change in temperature of the water. For example, if 1 gram of material is burned and the temperature of the water is raised one degree centigrade, 2,000 cal are given off. Hence, the

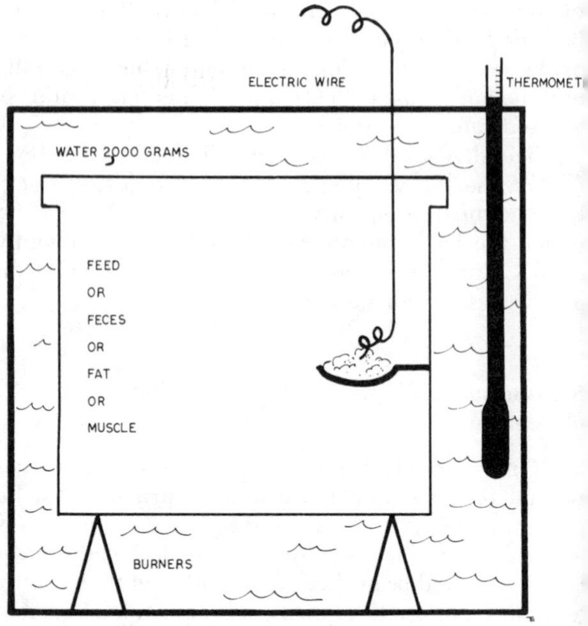

Fig. 4-10. Diagrammatic sketch of a bomb calorimeter used for th determination of the gross energy value (caloric content) of various mat rials.

material contains 2,000 cal per gram, or 907,200 ca per pound (2 million cal per kg), or 907 kilocalorie (kcal) per pound, or 0.907 megacalories (Mcal) pe pound. This value is known as the gross energy (GE content of the material. Thus, 1 kcal is equivalent t 1,000 cal while 1 Mcal is equivalent to 1,000,000 ca or 1,000 kcal.

Through various digestive and metabolic proc esses, much of the energy in feed is dissipated as i passes through the animal's digestive system. Abou 60 percent of the total combustible energy in grai and about 80 percent of the total combustible energ in roughage is lost as feces, urine, gasses, and hea These losses are illustrated in Fig. 4-11.

As shown in Fig. 4-11, energy losses occur in th digestion and metabolism of feed. Measures that ar used to express animal requirements and the energ content of feeds differ primarily in the digestive an metabolic losses that are included in their determina tion. Thus, the following terms are used to expres energy value of feeds:

- *Gross energy (GE)*—Gross energy represent the total combustible energy in a feedstuff. It does no differ greatly between feeds, except for those high i fat. For example, 1 pound of corncobs contains abou the same amount of GE as 1 pound of shelled corn Therefore, GE does little to describe the usefu energy in feeds for finishing animals.
- *Digestible energy (DE)*—Digestible energy i that portion of the GE in a feed that is not excreted i the feces.

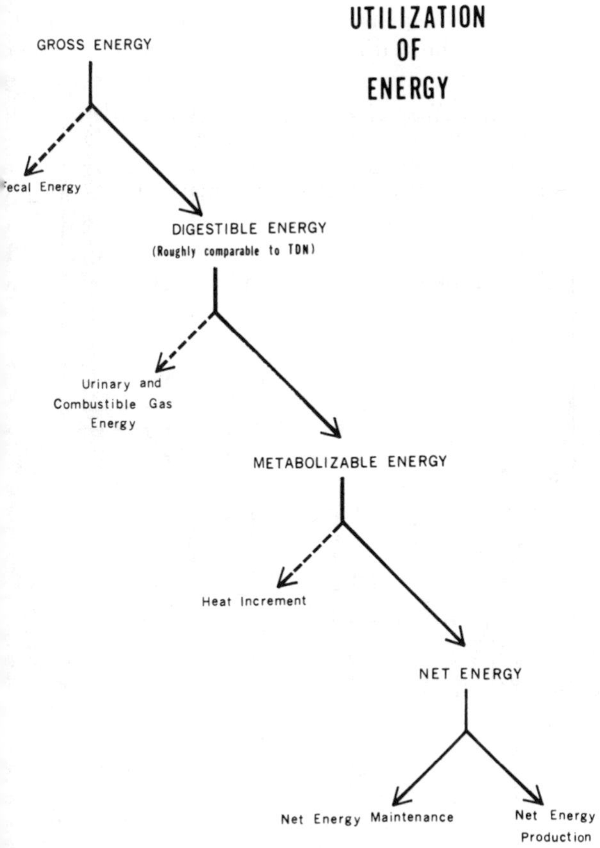

Fig. 4-11. Utilization of energy.

• *Metabolizable energy (ME)*—Metabolizable energy represents that portion of the GE that is not lost in the feces, urine, and gas. Although ME more accurately describes the useful energy in the feed than does GE or DE, it does not take into account the energy lost as heat.

• *Net energy (NE)*—Net energy represents the energy fraction in a feed that is left after the fecal, urinary, gas, and heat losses are deducted from the GE. The net energy, because of its greater accuracy, is being used increasingly in ration formulations, especially in computerized formulations for large operations.

Although net energy is a more precise measure of the real value of the feed than other energy values, it is much more difficult to determine. Heat increment must be determined with the whole body or respiration calorimeter. California workers have recently made effective use of a comparative slaughter technique and balance trials to calculate net energy values of feeds and determine animal requirements.

The net energy values of feeds are different for maintenance and production. Roughages compare more favorably with grain for maintenance than for production.

• *Net energy for maintenance (NE_m)*—This is the fraction of the net energy that keeps the animal in energy equilibrium.

• *Net energy for gain (NE_g)*—This is the fraction of the net energy available for gain in weight.

Balance Methods of Determining Net Energy

Measurements of animal heat began 200 years ago, when Lavoisier and LaPlace, the great French scientists, enclosed a guinea pig in a chamber surrounded by ice. The amount of ice melted by the heat of the animal multiplied by the latent heat of ice indicated the heat given off by the guinea pig. Since that time, more sophisticated types of equipment for measuring body heat have been developed. All of them are adaptions of the following two general types:

1. *Direct calorimetry*—In direct calorimetry, the animal is confined to a well-insulated chamber and the heat losses (by radiation, convection, and conduction from the body surface; by evaporation of water from the skin and lungs; and by excretion of urine and feces) are measured either (a) by the increase in temperature of a known volume of water, or (b) by the electrical current generated as heat passes across thermocouples. These types of calorimeters are very expensive to build and operate.
2. *Indirect calorimetry*—In indirect calorimetry, the heat of production is calculated from measurement of the respiratory exchange—the O_2 consumption, and usually the CO_2 production—of the animal. This method is based on the fact that O_2 consumption and CO_2 production are closely related to heat production. The ratio of carbon dioxide produced to oxygen consumed, which is known as the respiratory quotient (RQ), is distinctive for each compound; hence, it serves to indicate the type of nutrient being metabolized. The RQ for glucose is 1:1 or 1.00; for fat, it is 0.7; and for protein, it is 0.8. The RQ for fat is lower than for glucose or protein because the relatively larger amounts of hydrogen in fat require more oxygen for complete combustion.

The total heat production of the animal may be computed by either (a) measuring RQ along with oxygen, then making readings from tables; or (b) using a single equation relating heat production to the respiratory exchange.

Comparative measurements of direct and indirect calorimetry reveal that the two methods give results that are in close agreement. Since direct calorimeters are more complex and costly, as well as more expensive to operate, than indirect calorimeters, most nutritional studies are now conducted by means of indirect calorimetry.

The actual measurement of the respiratory exchange of animals may be accomplished by several different types of apparatus; among them, (1) the open circuit respiration chamber, (2) the closed circuit respiration chamber, (3) the confinement method, and (4) masks (spirometer) or tracheotomized animals.

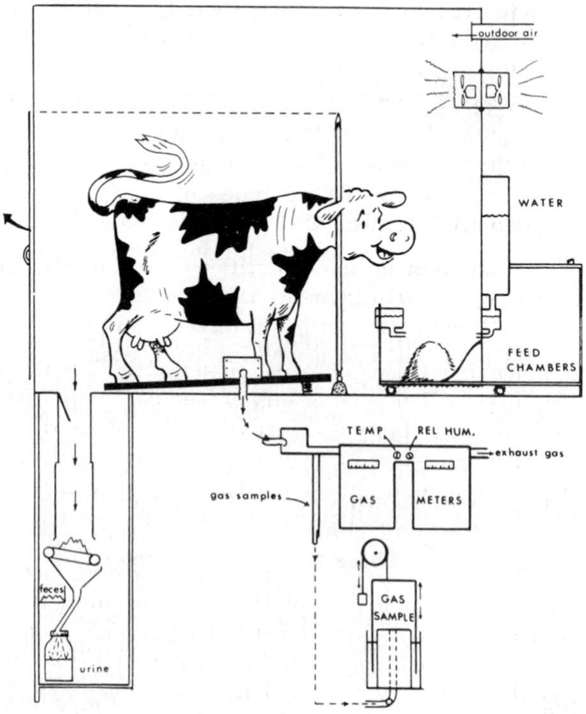

OPEN CIRCUIT RESPIRATION CHAMBER

Fig. 4-12. Illustration of an open circuit respiration chamber for large animals. The gas meters (shown in lower part of illustration) are used to measure the respiratory exchange of the cow. These data, plus the gas composition, provide the information needed to calculate the Heat Production (HP) of the cow. The HP of an animal consuming feed in a thermoneutral environment is composed of the heat increment (heat of fermentation plus heat of nutrient metabolism) plus heat used for maintenance (basal metabolism plus voluntary activity). The use of the open circuit respiration chamber for determining energy provides exact information as to metabolic processes and as to the effect of a specific nutrient or ration. However, it is an expensive and laborious procedure, which must be limited to a few animals and for short periods. (Courtesy, Dairy Cattle Research Branch, USDA)

Comparative Slaughter Method of Determining Net Energy

The comparative slaughter method of determining net energy (energy storage and heat production) is an old technique with a new look. It was first employed by Mitchell et al., of the Illinois Station, in 1926. But, for the most part, it was discontinued, because chemically analyzing the body (carcass) is slow, tedious, and expensive. Then, in 1959, Lofgreen an Garrett, of the California Station, reported an ingen ious modification of the comparative slaughte technique. By making use of established relationship between carcass density and the composition of th animal, they were able to estimate the energy conten of the carcass without analyzing the body chemically The density of the carcass is determined by weighin in water—using a dipping procedure (see Fig. 4-13) which can be done quickly and without affecting th sale value of the carcass.

Fig. 4-13. This shows the dipping procedure being used to get ca cass density from which specific gravity is computed, which, in turn, used to estimate carcass fat content. By use of an initial and a fin slaughter group of animals in each feeding trial, the energy gain can b measured, giving a more accurate measure of the true feed value tha does just liveweight gain. Further, the method provides a measure energy content without grinding and chemical analysis of the carcass.

The modified comparative slaughter method i especially well suited to studies involving growin and fattening animals—cattle, lambs, hogs, and broi ers, in which the amount of energy stored in the ca cass can be measured. However, it is not adapted t use in dairy animals.

The comparative slaughter method requires rela

ively large numbers of animals. A random sample, or heck group, is selected and slaughtered at the be- ;inning of an experiment to determine the initial ody composition. Then, at the close of the experi- nent, the remaining animals are slaughtered and nalyzed. The difference in the calorie content of the wo groups represents the energy storage or gain, vhich is a far more accurate measure of the true value f feed than liveweight gain.

The comparative slaughter method of feed evalu- tion is unique in the following respects:

1. It provides a relatively inexpensive way in vhich to determine net energy values.
2. The animals can be kept under more natural onditions, similar to those found in a commercial en- erprise.
3. Feeds can be assigned NE_m (net energy for naintenance) and NE_g (net energy for gain) values in eeping with the efficiency of metabolizable energy tilization for these different physiological processes.

The modified comparative slaughter technique as had a major impact on feed evaluation and ration ormulation throughout the United States.

'EEDING TRIALS

Each method of evaluating feedstuffs, discussed arlier in this chapter, has a place and is valuable. But one of them takes into consideration all the factors which determine the true value of any feed for a par- icular class of livestock. The "court of last judgment" or determining the true value of a feedstuff is the nimal. How well do animals eat the feed? How does t affect their health and well-being? How are they producing? Answers to these questions call for feed- ng the ingredient or ration under controlled condi- ions to the particular class of livestock.

The U.S. Department of Agriculture and the state experiment stations have conducted numerous exper- iments to determine the feeding value of specific feeds and of different rations, for each class of live- tock. The results of these studies have paid hand- ome dividends.

Generally speaking, it is in the best interest of stockmen that experiment stations continue to assume responsibility for the majority of research, including the evaluation of feedstuffs. In comparison with pri- vate industry, they have more trained research per- sonnel; generally their studies involve greater accu- racy and more controls; and their results are unbiased and unquestioned. But even experiment stations need to bear in mind that adequate animal numbers are important; that there are species and individual ani- mal differences; that feeds differ widely; and that the results must be repeatable.

Conducting Applied Tests on the Farm

When carefully conducted and properly interpreted and used, feeding trials can be a valuable adjunct in many of today's large livestock operations. Among their virtues, the operator can study area and feed differences. Among their limitations, usually there is less accuracy and there are fewer controls than in most university conducted experiments. For the latter reason, most of them should be looked upon as applied tests or demonstrations per se, rather than carefully controlled, basic experiments; terminology which doesn't detract from their value, but which does place them in proper perspective.

OTHER EVALUATIONS

In addition to being nutritionally complete, the following feed requirements are important:

1. *Palatability*—If they don't eat it, they won't produce; and if they don't eat enough, feed efficiency will be poor. The relationship of feed consumption to feed efficiency becomes clear when it is realized that the maintenance requirement of an animal producing at a low rate represents a much greater percent of the total feed required than for an animal producing at a more rapid rate.

Palatability is the result of the following factors sensed by the animal in locating and consuming feed: appearance, odor, taste, texture, temperature, and, in some cases, auditory properties of the feed (like the sound of pigs eating corn). These factors are affected by the physical and chemical nature of the feed.

2. *Variety*—Some variety in the ration is desirable, particularly from the standpoints of assuring (a) increased palatability, and (b) balance of nutrients—for example, all the essential amino acids.
3. *Digestive disturbances*—Bloat, colic, scours, and constipation are the bane of all feeders. The choice of feeds can give a big assist in minimizing such disturbances. For example, bloat in cattle and colic in horses can be lessened by avoiding lush or frosted pastures; scours can be lessened by proper feeding; and constipation can be corrected by feeding alfalfa, wheat bran, linseed meal, or molasses.
4. *Bulk*—The amount of bulk in the ration will vary. Ruminants can consume bulkier feeds than monogastric animals; the younger the animal, the less bulk; and the higher the production desired, the less bulky the ration. Also, the relative cost of feeds—concentrates vs roughages—will influence the relative amount of bulk in the ration.
5. *Cost*—Cost is important. But even more important is net returns; hence, it may well be said that it is net returns rather than cost per ton, or per bag, that counts.

6. *Poisonous plants and feeds*—Poisonous plants and feeds should be avoided. The stockman should know the poisonous plants common to the area, and avoid them. Also, the following poisons should be avoided: arsenic, botulism, ergot, fluorine, lead poisoning, mercury poisoning, mycotoxins, nitrate poisoning, pitch poisoning, scabbed grain, selenium poisoning, smut on grain, and spoiled or moldy feed (see Sec. 10, Table 10-12, Some Potentially Poisonous Elements).

FEED PREPARATION

Feed is the major cost in animal production. Hence, it is economically important that it be processed in such manner as to make for maximum efficiency (1) in handling, from a mechanical and automation standpoint; and (2) in utilization, from the standpoint of the animal.

Feed preparation can influence the nutritive value of a feed. For example, fine grinding and pelleting of forages tend to increase rate of passage through the gut, which lowers fiber digestibility. However, overall animal response to pelleted forages is usually increased over the same forage fed in long or chopped form, because the slightly lower digestibility is more than offset by increased feed consumption.

Generally speaking, the higher the level of feeding and the greater the production desired, the more important proper feed preparation becomes. This is so because (1) the higher the level of feeding, the more selective animals become in their eating habits; and (2) in ruminants, digestibility decreases as level of feeding increases, primarily because the feed does not remain in the digestive tract long enough for maximal effect of the various digestive processes.

Most of the recent technology in feed preparation has been with feedlot cattle. It came in with the advent of large commercial feedlots. But much of it is applicable to all ruminants. Feed preparation for swine and poultry has remained relatively simple as compared with the variety of methods available and in use for ruminant feeds. The major change in horse feed preparation has been the increased use of all-pelleted rations (hay and grain combined).

Processing Grain

Grain is processed in order to increase feed efficiency through improvement in palatability and digestibility.

This results from physical and/or chemical changes in the grain. Physical changes include moisture level, heat, pressure, and particle size. Chemical changes may include structural changes in the starch, protein, and fat of grains resulting in changes in digestibility and metabolic end products. In some cases, so-called physiochemical changes occur in that bot physical and chemical alterations are simultaneousl apparent. Rate of ingesta passage and site of digestio within the gastrointestinal tract are both likely end re sults of physiochemical changes in processed grains.

Grain processing gives greater returns when fee intake of grains is high. Animals on maintenance ra tions are not normally fed much grain; hence, any in crease in feed efficiency may not return the adde processing cost.

Based on moisture, processing may be grouped a follows:

Dry Processing	Wet Processing
Crumbling	Acid preservation of high-moisture grain (fatty acid treated grain)
Dry rolling (cracking, or crimping)	Cooking
Grinding	Extruding (gelantinization)
Jet-sploded grain	High-moisture (early harvested) grain
Micronizing	Reconstituted grain
Pelleting or cubing	Soaking and slopping
Popping	Sprouting grain (germinating grain; hydroponics)
Roasting	Steam rolling; flaking
Whole shelled corn	

Because of space limitations, only two of thes grain-processing methods are presented here.

ACID PRESERVATION OF HIGH-MOISTURE GRAIN

The acid preservation of high-moisture grai (fatty acid treated grain) involves mixing 1 to 1½ pe cent propionic acid (or a mixture of propionic aci with either acetic acid or formic acid) with high moisture cereal grain to inhibit mold or spoilag thereby alleviating artificial drying or the necessity o storage in an airtight silo.

HYDROPONICS (Sprouted Grain)

Hydroponics (or sprouted grain) is the growin of plants with their roots immersed in an aqueous so lution containing the essential mineral nutrient salt instead of in soil. This means that sprouted grain fo feed is produced with water and chemicals, withou dirt.

The Wisconsin Alumni Research Foundatio chemically analyzed and compared the composition o oat grain and 5-day oat grass on a dry matter basis (se Table 4-10).

As shown in Table 4-10, the 5-day oat grass is better source than oat grain of calcium, phosphoru carotene, vitamin E, the B vitamins—riboflavir thiamin, and niacin, and of vitamin C.

It is noteworthy, however, that, based on studie conducted by the different universities, sprouting re sults in an average loss of 83% of the dry matter of th oat grain. One study showed a reduction in TDN fro

TABLE 4-10

OMPOSITION OF OAT GRAIN AND 5-DAY OAT GRASS, MOISTURE-FREE BASIS[1]

Constituent	Oat Grain	Oat Grass
ry matter (%)	100.00	100.00
rotein (%)	15.00	21.00
ther extract (fat) (%)	4.21	5.20
itrogen-free extract (%)	65.86	42.79
iber (%)	11.71	26.11
sh (%)	3.22	3.90
alcium (%)	0.063	0.238
hosphorus (%)	0.360	0.509
	(mg/kg)	*(mg/kg)*
arotene[2]	0	39.067
itamin E	17.95	48.87
iacin	7.18	103.96
iboflavin	1.96	22.29
hiamin	3.14	12.86
itamin C	0	218.3

[1]Analyses by Wisconsin Alumni Research Foundation.

[2]Each mg of beta carotene was considered to be equivalent to 1,556 IU of itamin A.

'5.7% in the oat grain to 70.2% for the sprouted oats. lso, the digestibility of dry matter, energy, protein, ther extract, and nitrogen-free extract was lower for he sprouted oats than for the oat grain.

In arriving at a decision whether or not to proluce feeds hydroponically, consideration should be iven to (1) the needs of different classes of livestock or each nutrient, and (2) the cost of supplying these utrients in the form of sprouted grain.

It should also be recognized that supplemental uantities of calcium and phosphorus can be provided n many forms at a relatively low cost.

Sprouting greatly increases the carotene content see Table 4-10). Thus, if carotene, or vitamin A, is leficient, sprouted grains are a good supplemental ource, although supplemental vitamin A can be provided in a dry, stabilized form at a low cost.

Most rations are adequate in vitamin E. The B viamins listed in Table 4-10 (riboflavin, thiamin, and iacin) are produced by the microorganisms in cattle, heep, and horses; hence, supplemental quantities of hem are not normally needed. Vitamin C is not reuired in the diet of farm animals.

Although sprouted grains are high-quality feeds rom the standpoint of certain minerals and vitamins, he need for supplemental quantities of such nutrients n common rations for livestock is questionable.

In one cattle finishing study, 5 pounds of dry rushed oats was compared to the green oat plants roduced from 5 pounds of oats. The remainder of the ation was ground ear corn and lespedeza hay. The ains of both lots of steers were equal but the cost per ound of gain was 1.9 times greater for the green oat lants.

Based on percent increase in cost of TDN from prouted oats compared with unsprouted oats, the cost of sprouted oats is over four times that of the original oats.

Without doubt, sprouted grains will give an assist when added to poor rations—and the poorer the ration, the bigger the boost. However, with our present knowledge of nutrition, balanced rations can be arranged without the added labor and expense of sprouting grain.

Processing Roughage

The preparation of roughages has received less attention than the processing of grains. However, with increasing world food shortages, roughages will become more important; hence, their processing will assume greater importance.

In processing roughages, one should avoid those (1) with high moisture, which may heat and produce spontaneous combustion, and (2) in which there are foreign objects (wire, etc.) which the animals may not be able to select out, and which may ignite a fire when being processed.

The common methods of roughage preparation are: chopping (shredding), grinding, pelleting, and cubing. Because of space limitations, only cubing, pelleted complete rations, and treatments of high cellulose feeds will be covered in the sections that follow.

PELLETING OR CUBING

Pelleting involves compressing forage, preceded by grinding. The two biggest deterrents to pelleting are (1) fine grinding, and (2) cost.

Cubing refers to the practice of compressing long or coarsely cut hay into cubes about 1¼ inch square and 2 inches long, with a bulk density of 30 to 32 pounds per cubic foot. Cubes offer most of the advantages of pelleted forages, with few of the disadvantages. They alleviate fine grinding, and they facilitate automation in both haymaking and feeding; and they lower milk fat percentage only slightly, if at all. Cubing costs about $5 more per ton than baling.

Both pelleting and cubing will (1) simplify haymaking, (2) lessen transportation costs and storage space, (3) reduce labor, (4) make automatic hay feeding feasible, (5) decrease nutrient losses, and (6) eliminate dust.

With pelleting or cubing, the spread between high- and low-quality roughage is narrowed; that is, the poorer the quality of the roughage, the greater the advantage from pelleting or cubing. This is so because such preparation assures complete consumption of the roughage. Also, pelleting or cubing, especially pelleting, usually speeds up the passage of roughage through the digestive system.

PELLETED COMPLETE RATIONS

Complete pelleted rations—in which the hay and grain are combined, then pelleted—are finding an increasing place for horses, and perhaps swine. Among the virtues ascribed to all-pelleted rations are (1) they prevent selective eating—if properly formulated, each mouthful is a balanced diet; (2) they alleviate waste; (3) they eliminate dust (thereby lessening heaves in horses); (4) they lessen labor and equipment; and (5) they lessen storage.

TREATMENTS OF HIGH-CELLULOSE FEEDS

High feed prices and more stringent burning regulations have spurred research to find a practical method of improving the feeding value of several high-cellulose products, such as rice, wheat and barley straws, corncobs, cottonseed hulls, newspaper, sawdust, and gin trash.

In their natural state, these products make poor feedstuffs because lignin or silica, or a combination of the two, (1) encrust the energy-rich carbohydrates, cellulose, and hemicellulose; and (2) keep the microbes in the ruminant's stomach from breaking them down to release energy.

The answer to this problem lies in some treatment that opens up the fibers enough to permit increased digestion in the rumen. Several methods of chemical treatment are being investigated; among them, the use of sodium, potassium, or ammonium hydroxide; ammonia; and high-pressure steam.

It is noteworthy that the alkali soaking process was first used on straw in Germany in 1919, when there was a critical shortage of livestock feeds.

In addition to increasing the digestibility of high-cellulose products, the goal of modern investigations is to perfect a treatment which will be both economical and nonpolluting.

Liquid Supplements

Liquid supplements (many of which contain molasses and urea, usually with added vitamins and trace minerals) are available and widely used. This is a convenient way of feeding supplements to cattle on pasture. Also, liquid supplements are sometimes added to complete ration mixes, either as part of the mix or as a top dressing.

There is no difference in the nutritional value of liquid and dry supplements if the supplements contain the same basic nutrients. Thus, the choice shoul be made on the basis of cost and convenience.

Preparation of Feeds for Each Class of Livestock

Table 4-11 is a summary of pertinent informatio relative to the preparation of feeds for each class livestock.

FEEDING STANDARDS

Feeding standards are tables showing th amounts of one or more nutrients needed by differe species of animals for different purposes, such growth, finishing, and lactation. They serve as guid in balancing rations and feeding practices. Most fee ing standards are expressed in (1) quantities of nut ents required per day, and/or (2) percent of the ratio the first type being used for animals given exact qua tities of a ration, and the second type used when an mals are fed free-choice (ad libitum).

Today, the most up-to-date feeding standards the United States are those published by the Nation Research Council (NRC) of the National Academy Sciences, based on TDN. In this country, the TD system is gradually giving way to other energy evalu tion systems, particularly net energy. England us metabolizable energy (ME), adjusted according to t efficiency with which a feedstuff or diet is used for particular purpose. Other European standards a based on starch equivalents, Scandinavian Fe Units, and other methods of evaluation.

Although feeding standards are excellent a needed guides, there are still many situations whe nutrient needs cannot be specified with great acc racy for animals. Also, in practical feeding operatior economy must be considered; for example, dairym are interested in obtaining that level of milk produ tion which will make for the largest net returns light of current feed costs and the market price milk. Moreover, feeding standards tell nothing abo the palatability, physical nature, or possible digesti disturbances of a ration. Neither do they give consi eration to individual animal differences, manageme differences, effects of such stresses as weather, d ease, parasitism, surgery (dehorning, castrating, etc Thus, there are many variables that alter the nutrie needs and utilization of animals—variables that a difficult to include quantitatively in feedi standards, even when feed quality is well known.

TABLE 4-11
PREPARATION OF FEEDS

Class of Animal	Grain	Roughage	Comments
BEEF CATTLE	Flaking (or jet-sploded, micronizing, popping, roasting, extruding, or high-moisture grain—with choice determined by cost) is preferable, especially for full-fed animals on a high-grain ration. But flaking equipment (and other similar equipment) is costly to purchase and operate; hence, a large-volume operation is required to cover fixed costs. Dry roll or grind coarsely for most beef cattle, especially those not full fed on high-grain ration and those in smaller operations. On high-concentrate rations (those with 80% or more concentrate), whole corn need not be processed. Grain need not be processed for calves under 6 months of age, for young calves masticate feed thoroughly. Cubes (large pellets) preferred for feeding on pasture or range. Professional herdsmen often cook feed (especially barley) for show cattle to increase palatability.	Long hay satisfactory for most cattle other than commercial feedlot operations, but chopping should be considered from standpoints of ease of handling and lessening wastage (especially of poor quality hay). Chop (2" in length), pellet, or cube. Chopping or shredding fodders and stovers (corn or milo) makes them easier to handle and lessens waste.	Fine grinding grain increases incidence of hyperkeratosis (ruminal parakeratosis) in feedlot cattle. Dry rolling and coarse grinding of grains are of about equal value for most beef cattle. Either method is just as satisfactory as more expensive methods (like flaking) when grain intake is relatively low. Chopping low-quality hay is more advantageous than chopping high-quality hay. Finely ground hay not recommended, as it decreases digestibility.
DAIRY CATTLE	Flaking or coarse grinding preferable. Grain may be fed whole to young calves under 6 months of age.	Long hay or cubes. Cubes lend themselves to mechanization; and they lower milk fat percentage only slightly, if at all.	Finely ground or pelleted roughage will result in reduced rumen acetate production and lower milk fat percentage.
SHEEP	Processing grains not necessary unless seeds are hard (like sorghum or millet) or the teeth are poor. Hard seeds (like sorghum and millet) may be flaked, dry rolled, or ground coarsely, with cost determining the choice. Pellets are increasingly being used by lamb feeders. Cubes or pellets preferred for feeding on pasture or range. Professional shepherds prefer flaked grain for show sheep, as the ration is lighter and there are fewer digestive disturbances.	Chop (2" in length), pellet, or cube. Many lamb feeders are using all-pelleted rations (hay and grain combined).	Sheep masticate grain more thoroughly than cattle, with the result that feed preparation for sheep is of less value than for cattle. A high incidence of parakeratosis—a degeneration of the rumen papilla—appears to result from feeding pellets, especially when low forage-high concentrate pellets are used. Hence, breeding sheep should not be fed for extended periods on pellets without any long or chopped forage.

(Continued)

TABLE 4-11 (Continued)

Class of Animal	Grain	Roughage	Comments
SWINE	Corn, barley, grain sorghum, and oats should be finely ground for swine. Medium to coarse grinding is best for wheat, because fine grinding makes it pasty and less palatable. Pelleting corn-soybean rations generally improves feed utilization and increases rate of gain by at least 4 to 5%. Cook Irish potatoes, beans, soybeans, and garbage. Cooking (except for the feeds listed above), soaking, or fermenting are not of value when swine are on full feed. Liquid and paste feeding give inconsistent results in feed consumption and rate of gain; hence, they should be evaluated on the basis of a mechanical means of dispensing feed. However, slop (slurry or gruel) is desirable for early-weaned pigs, and perhaps for pigs being fitted for show or sale. High-moisture corn does not result in any improvement of efficiency for swine; hence, the value of high-moisture corn as compared to regular corn should be computed on a dry matter basis.	Hay that is to be incorporated in mixed feeds should be ground. Rations containing considerable amounts of fiber are improved by pelleting because of increased consumption, improved carbohydrate digestibility, and reduced sorting and wastage compared to meal rations.	Fine grinding will cause some bridging in self-feeders. Also, finely ground feed is associated with increased incidence of stomach ulcers in swine.
HORSES	Flaking is the preferred method of grain preparation for horses; it makes for a light ration and few digestive disturbances. For horses with good teeth, the value of oats is increased only 5% by processing.	Either feed long hay or an all-pelleted ration (grain and hay combined).	

BALANCED RATIONS

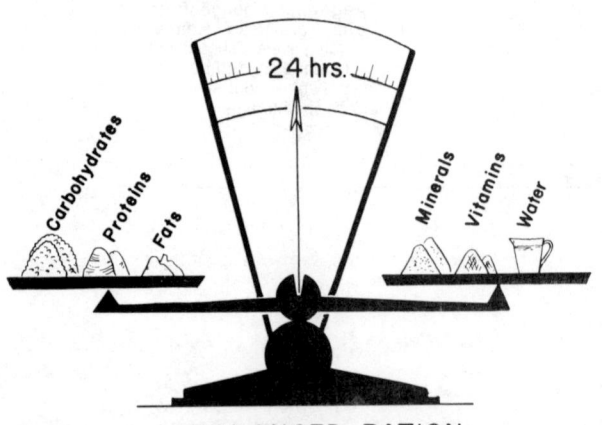

Fig. 4-14. Successful livestock producers feed balanced rations. (Drawing by R. F. Johnson)

To supply all the needs—for maintenanc growth, finishing, reproduction, lactation, work (running), and/or wool—the different classes of ar mals must receive sufficient feed to furnish the nece sary quantity of energy (carbohydrates and fats), pr teins, minerals, vitamins, and water. Perhaps und certain conditions feed additives may be desirable, a though it is not likely that they are essential. A rati that meets all these needs is said to be balance More specifically, by definition, *a balanced ration one which provides an animal the proper proportio and amounts of all the required nutrients for a peri of 24 hours.*[4]

[4]Although Webster defines the noun "ration" as "the amount food (feed) supplied to an animal for a definite period, usually fo day," to most stockmen the word implies the feeds fed to an anir or animals, without limitation to the time in which they are co sumed. In this and other sections of *The Stockman's Handbook*, t author accedes to the common usage of the word, rather than to d tionary correctness.

When in confinement, animals have access only to the feeds provided by the caretaker. This points up the importance of balanced rations.

Several suggested rations for different classes of livestock are given later in this section, under the discussion of the respective classes of livestock. Generally these rations will suffice, but it is recognized that rations should vary with conditions, and that many times they should be formulated to meet the conditions of a specific farm or ranch, or to meet the practices common to an area.

Also, a good stockman should know how to balance a ration. Then, if the occasion demands, he can do it. Perhaps of even greater importance, he will then be able more intelligently to select and buy rations with informed appraisal; to check on how well his manufacturer, dealer, or consultant is meeting his needs; and to evaluate the results.

How to Balance Rations

Ration formulation consists in combining feeds to make a ration that will be eaten in the amount needed to supply the daily nutrient requirements of the animal. This may be accomplished by the methods presented later in this section, but first the following pointers are necessary:

1. In computing rations, more than simple arithmetic should be considered, for no set of figures can substitute for experience. Compounding rations is both an art and a science—the art comes from animal know-how and experience, and keen observation; the science is largely founded on chemistry, physiology, and bacteriology. Both are essential for success.

2. Before attempting to balance a ration, the following major points should be considered:

a. *Availability and cost of the different feed ingredients*—Preferably, cost of ingredients should be based on delivery after processing—because delivery and processing costs are quite variable.

b. *Moisture content*—When considering costs and balancing rations, feeds should be placed on a comparable moisture basis; usually, an air-dry basis, or 10 percent moisture content, is used. This is especially important in the case of high-moisture grain or silage.

c. *Composition of the feeds under consideration*—Feed composition tables ("book values"), or average analysis, should be considered only as guides, because of wide variations in the composition of feeds. For example, the protein and moisture contents of sorghum, hay, and silages are quite variable. Whenever possible, especially with large operations, it is best to take a representative sample of each major feed ingredient and have a chemical analysis made of it for the more common constituents—protein, fat, fiber, nitrogen-free extract, and moisture; and often calcium, phosphorus, and carotene. Such ingredients as oil meals and prepared supplements, which must meet specific standards, need not be analyzed so often, except as quality-control measures.

Despite the recognized value of a chemical analysis, it is not the total answer. It does not provide information on the availability of nutrients to the animal; it does not tell anything about the associated effect of feedstuffs—for example, the apparent way in which beet pulp enhances the value of ground milo; and it does not tell anything about taste, palatability, texture, or undesirable physiological effects such as bloat and laxativeness. Nevertheless, a chemical analysis does give a basis on which to start the evaluation of feeds. Also, with chemical analysis at hand, and bearing in mind that it's the composition of the total feed (the finished ration) that counts, the person formulating the ration can more intelligently determine the quantity of protein to buy, and the kind and amounts of minerals and vitamins to add.

d. *Soil analysis*—If the origin of a given feed ingredient is known, a soil analysis or knowledge of the soils of the area can be very helpful; for example, (1) the phosphorus content of soils affects plant composition, (2) soils high in molybdenum and selenium affect the composition of the feeds produced, (3) iodine- and cobalt-deficient areas are important in animal nutrition, and (4) other similar soil-plant-animal relationships exist.

e. *The nutrient allowances*—This should be known for the particular class of animals for which a ration is to be formulated. Also, it must be recognized that nutrient requirements and allowances must be changed from time to time, as a result of new experimental findings.

3. In addition to providing a proper quantity of feed and to meeting the protein and energy requirements, a well-balanced and satisfactory ration should be:

a. Palatable and digestible.

b. Economical. Generally speaking, this calls for the maximum use of feeds available in the area, especially forages.

c. Adequate in protein content, but not higher than is actually needed. Generally speaking, medium- and high-protein feeds are in scarcer supply and higher in price than high-energy feeds.

d. Well fortified with the needed minerals, or

free access to suitable minerals should be provided; but mineral imbalances should be avoided.

e. Well fortified with the needed vitamins.

f. So formulated, where ruminants are involved, as to nourish the billions of bacteria in the paunch in order that there will be satisfactory (1) digestion of roughages, (2) utilization of lower quality and cheaper proteins and other nitrogenous products (thus, it is possible to use urea to constitute up to one-third of the total protein of the ration of ruminants, provided care is taken to supply enough carbohydrates and other nutrients to assure adequate nutrition for rumen bacteria), and (3) synthesis of B vitamins.

This means that rumen microorganisms must be supplied adequate (1) energy, including small amounts of readily available energy such as sugars or starches; (2) ammonia-bearing ingredients such as proteins, urea, and ammonium salts; (3) major minerals, especially sodium, potassium, and phosphorus; (4) cobalt and possibly other trace minerals; and (5) unidentified factors found in certain natural feeds rich in protein or nonprotein nitrogenous constituents.

g. One that will enhance, rather than impair, the quality of the product (meat, milk, wool, or eggs) produced.

4. In addition to considering changes in availability of feeds and feed prices, ration formulation should be altered at stages to correspond to changes in weight and productivity of animals.

The above points are pertinent to the balancing of rations, regardless of the mechanics of computation used. In the sections that follow, four different methods of ration formulation are presented: (1) the square method; (2) the trial-and-error method; (3) the net energy method; and (4) the computer method. Despite the sometimes confusing mechanics of each system, if done properly, the end result of all four methods is the same—a ration that provides the desired allowance of nutrients in correct proportions economically (or at least cost), but, more important, so as to achieve the greatest net returns—for it's net profit rather than cost per bag that counts. Since feed represents by far the greatest cost item in livestock production, the importance of balanced rations is evident.

An exercise in ration formulation follows for purposes of illustrating the application of each of these four methods:

1. *Square method*, applied to a swine ration.
2. *Trial-and-error method*, applied to a lactating cow ration.
3. *Net energy method*, applied to a cattle finishing ration.
4. *Computer method*, applied to a beef cattle ra tion.

It is emphasized, however, that each method o ration formulation may be used in balancing ration for all classes of livestock.

SQUARE (or Pearson Square) METHOD

The square method is simple, direct, and easy Also, it permits quick substitution of feed ingredient in keeping with market fluctuations, without disturb ing the protein content. The latter virtue is of particu lar value to the feed manufacturer.

In balancing rations by the square method, it i recognized that protein content alone receives majo consideration. Correctly speaking, therefore, it is method of balancing the protein requirement, wit only incidental consideration given to the vitamir mineral, and other nutritive requirements.

To compute balanced rations by the squar method, or by any other method, it is first necessary t have available both feeding standards (see the Nutr ent Requirement tables in this section, under the di cussion of the respective classes of livestock) and fee composition tables (see Part VIII, Feed Compositio Tables, 4-101 to 4-105, pages 456 to 585).

The following example will show how to use th square method in formulating a swine ration:

Example:

A swine producer has 40-pound pigs to which h desires to feed a 16% protein ration until they reac 120 pounds weight. He has on hand corn containin 9.5% protein. He can buy a 36% protein supplemen which is reinforced with minerals and vitamin What percent of the ration should consist of eac corn and the 36% protein supplement?

Step by step, the procedure in balancing this r tion is as follows:

1. Draw a square, and place the number 16 in th center thereof.

2. At the upper left-hand corner of the squar write *concentrate* and its protein content (36); at th lower left-hand corner, write *corn* and its protein co tent (9.5).

3. Subtract diagonally across the square (th smaller number from the larger number), and wri the difference at the corners on the right-hand sic (36 − 16 = 20; 16 − 9.5 = 6.5). The number at th upper right-hand corner gives the parts of concentra by weight, and the number at the lower right-har corner gives the parts of corn by weight to make ration with 16 percent protein.

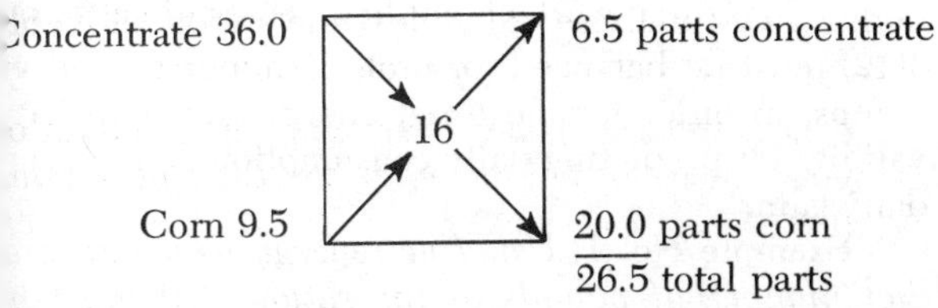

4. To determine what percent of the ration would e corn, divide the parts of corn by the total parts: 20 ÷ 26.5 = 75% corn. The remainder, 25%, would be upplement.

TRIAL-AND-ERROR METHOD

In the example that follows, the trial-and-error nethod is used, with consideration given to energy nd protein. Also, crude protein rather than digestible rotein is used because (1) this is what the feed man-facturer wants to know as he plans a feed formula, nd (2) this is what the dairyman sees on the feed tag vhen he purchases feed. In most mixed dairy feeds pproximately 80 percent of the total protein is diges-ible.

Example:

Let's assume that we have a 1,430-lb cow produc-ng 60 lb of milk testing 4.0 percent fat. The dairy-nan is feeding 15 lb of alfalfa hay and 45 lb of corn ilage per day. Corn, oats, and soybean meal are vailable. What concentrate mix shall he use to meet he needs of this lactating cow, from the standpoint f energy and protein?

Before proceeding further, here are some general ules and assumptions that we shall follow:

1. The TDN of the complete ration of lactating ows should be 70% or better, preferably 74 to 75%.
2. One percent salt and 1% of a low-calcium igh-phosphorus mineral (including trace minerals) re to be added to the grain ration. Also, salt and a nineral mix will be self-fed in a 2-compartment min-ral box.
3. Vitamin A and vitamin D will be added to the ation at a level of 1,000 IU of vitamin A and 150 IU of itamin D per pound of concentrate, respectively.
4. It is assumed that the available feeds have ap-roximately the following composition (as-fed basis):

	TDN (%)	Crude Protein (%)
Alfalfa hay (all analyses)	50.3	15.5
Corn (grain)	80.3	9.5
Corn silage (all analyses)	16.3	2.0
Oats (grain)	67.2	11.7
Soybean meal (solv-extd)	71.6	45.8

[5]From: Table 4-44, "Daily Nutrient Requirements of Lactating airy Cattle," in this section of *The Stockman's Handbook.*

Here are the steps in balancing this ration:

	TDN (lb)	Crude Protein (lb)
Step 1—The daily TDN and crude protein requirements of this cow (1,430 lb body weight, 60 lb of 4% milk) are:[5]		
Requirements of cow for—		
Maintenance	9.9	1.7
Milk production	19.8	4.7
Total	29.7	6.4
Step 2—The roughage (15 lb alfalfa hay, 45 lb corn silage) is supplying:		
Alfalfa hay, 15 lb	7.5	2.3
Corn silage, 45 lb	7.3	0.9
Total from forage	14.8	3.2
Step 3—Remainder, to be supplied by concentrate:	14.9	3.2
Step 4—Let's try out (that's why it is called the "trial-and-error method") a grain mix of 700 lb corn, 280 lb oats, 10 lb monosodium phosphate, and 10 lb salt, and see how much TDN and crude protein in 1,000 lb of the grain mix:		
Corn, 700 lb	562.0	66.5
Oats, 280 lb	188.2	32.8
Monosodium phosphate, 10 lb	—	—
Salt, 10 lb	—	—
Total	750.2	99.3
or in percent	75.0%	9.9%

Step 5—Divide the TDN needed from concentrate (14.9 lb) by the percent TDN in the mixture (75.0%). Thus, feeding 19.9 lb of the concentrate will meet the energy needs.

Step 6—Will this level of grain mix (19.9 lb) also meet the crude protein needs? By multiplying the pounds of concentrate mixture by the percent crude protein (19.9 × 9.9%), we find that the proposed concentrate would supply 1.97 lb of crude protein, whereas 3.2 lb are needed. Therefore, a high-protein supplement must be substituted for some of the homegrown grain.

	TDN (lb)	Crude Protein (lb)
Step 7—Let's substitute 150 lb of soybean meal for 150 lb of corn. Hence, the ration as now proposed will consist of:		
Corn, 550 lb	441.7	52.3
Oats, 280 lb	188.2	32.8
Soybean meal, 150 lb	107.4	68.7
Monosodium phosphate, 10 lb	—	—
Salt, 10 lb	—	—
Total	737.3	153.8
or in percent	73.7%	15.4%
Step 8—By referring back to Step No. 3, we can divide the pounds of TDN and crude protein needed from the concentrate, by the percentage of TDN and crude protein found in the grain mix in Step No. 7. We find that 14.9 ÷ .737 = 20.2 lb needed to supply	14.9	
and 3.2 ÷ .154 = 20.8 lb needed to supply		3.2

Thus, we find that the following ration will supply the needed TDN and crude protein for a 1,430-lb lactating cow producing 60 lb of milk testing 4% fat:

	TDN (lb)	Crude Protein (lb)
Alfalfa hay, 15 lb	7.5	2.3
Corn silage, 45 lb	7.3	0.9
Concentrate mix (Step Nos. 7 & 8) 20.8 lb	15.3	3.2
Total	30.1	6.4

In many sections of the country, especially in grain-deficit areas and on highly specialized dairies where little or no grain is grown, the dairyman may find it most economical to purchase a commercial dairy feed to balance out the roughage that is being fed.

NET ENERGY METHOD

Correctly speaking, the net energy system is not a method of balancing rations. Rather, it is a means of predicting daily feed consumption and average daily gain when using a ration that is balanced for protein, minerals, and vitamins.

In order to use the net energy system in this manner, the following values must be available:

1. A table showing the net energy requirements of the particular class of animals. Table 4-19, page 271, shows the net energy requirements for growing-finishing beef cattle (in megacalories per animal per day), with a breakdown into steers and heifers.
2. A table showing the nutrient composition of feeds, with the net energy of each feed partitioned into energy used for body maintenance and for gain; thus, the net energy values in megacalories (Mcal) per unit (lb or kg) are needed for each feed for maintenance (NE_m) and for gain (NE_g) (see Table 4-12).

TABLE 4-12

RATION FOR FINISHING CATTLE

Ration Ingredient	Lb	Composition of Ingredients (as-fed basis) NEm[1] (Mcal/lb)[3]	Ration Supplies NEm[1] (Mcal)[3]	Composition of Ingredients (as-fed basis) NEg[2] (Mcal/lb)[3]	Ration Supplies NEg[2] (Mcal)[3]
Shelled corn No. 2	68.60	0.92	63.11[4]	0.60	41.16[5]
Soybean meal (solvent)	4.00	0.78	3.12	0.52	2.08
Alfalfa hay (mid-bloom) ...	27.00	0.50	13.50	0.24	6.48
Salt	0.40	—	—	—	—
Total	100.00	—	79.73	—	49.72

[1] NE_m = net energy for maintenance.
[2] NE_g = net energy for gain.
[3] Mcal stands for megacalorie.
[4] 68.60 lb × 0.92 = 63.11.
[5] 68.60 lb × 0.60 Mcal = 41.16.

The two examples that follow will show how to use the net energy method. In the first example, net energy values of feeds are used to predict the number of pounds of a given ration that a steer would need to consume to make a specified daily gain. In the second example, net energy is used to predict average daily gain based on consuming a certain number of pounds of a specified ration. Bear in mind that the ration in both cases (in these examples, the ration in Table 4-12) must be balanced for protein, minerals, and vitamins, in order for these net energy values to have validity for predicting daily consumption and average daily gain.

Example No. 1: *Using net energy values to predict number of pounds of the ration that must be consumed to produce a specific gain—How many calories would a 775-pound yearling steer need to consume to gain 2.9 pounds daily?*

Step 1—Calculate the net energy for maintenance (NE_m) and gain (NE_g) values for a pound of the ration shown in Table 4-12.

Use a table like Table 4-101, at the end of this section, which gives composition values of feeds. One pound of the Table 4-12 ration supplies 0.7973 megacalories of net energy for maintenance (Mcal NE_m) and 0.4972 megacalories of net energy for gain (Mcal NE_g).

Step 2—From Table 4-19 find the requirement for a 775-lb (see 772-lb requirements) yearling steer to gain 2.9 lb daily. This follows:

	Mcal/day
NE_m	6.24
NE_g	6.48

Step 3—Pounds of feed to meet the daily maintenance requirement:

$$6.24 \text{ Mcal} \div .7973 \text{ Mcal} = 7.83 \text{ lb}$$

Step 4—Pounds of feed to meet the requirement for 2.9 lb daily gain:

$$6.48 \text{ Mcal} \div .4972 \text{ Mcal} = 13.03 \text{ lb}$$

Step 5—Total pounds of feed steer must eat daily to gain 2.9 lb:

7.83 lb + 13.03 lb = 20.86 lb

Example No. 2: *Using net energy to predict the average daily gain of a 775-pound steer that is consuming a certain number of pounds of a specified ration—Let's assume that we have a 775-pound steer that is consuming 18 pounds of the ration shown in Table 4-12. What daily gain should be expected?*

Step 1—Pounds of feed to meet the daily maintenance requirement = 7.83 lb (see prior example).
Step 2—Pounds of feed left for gain:

18 lb − 7.83 lb = 10.17 lb

Step 3—Mcal of NEg supplied by remaining feed:

10.17 lb × .4972 Mcal = 5.06 Mcal

Step 4—Daily gain expected from 5.06 Mcal of NE_g (Table 4-19).

5.36 Mcal produces 2.4 lb of gain

Therefore, 5.06 Mcal will produce 2.3 lb daily gain.

COMPUTER METHOD

Many large livestock establishments, and most feed companies, now use computers for ration formulation as well as for other purposes; and their use will increase.

Despite their sophistication, there is nothing magical or mysterious about balancing rations by computer. Although they can alleviate many human errors in calculations, the data which come out of a machine are no better than those which went into it; without a man, they don't know the difference between a Doberman and a Hereford. The men back of the computer—the stockman and his nutritionist who prepare the data that go into it, and who evaluate and apply the results that come out of it—become more important than ever. This is so because an electronic computer doesn't know anything about (1) feed palatability, (2) bloat prevention, (3) limitations that must be imposed on certain feeds to obtain maximum utilization, (4) the goals in the feeding program—such as growing or finishing, (5) homegrown feeds for which there may not be a suitable market, (6) feed processing and storage facilities, (7) the health, environment, and stress of the animals, and (8) the men responsible for actual feed preparation and feeding. Additionally, it must be recognized that a computer may even reflect, without challenge, the prejudices and whims of those who prepare the data for it.

Hand in hand with the use of computers in balancing rations, the term "least-cost ration formulation" evolved. In some respects this designation was unfortunate, for the use of least-cost rations does not necessarily assure the highest net returns—and net profit is more important than cost per ton. For example, the least-cost ration may not produce the desired daily gain or carcass quality.

An electronic computer can do little more than a good mathematician can do, but it can do it a lot faster and it can check all possible combinations. It alleviates the endless calculations and many hours of time required for hand calculations. For example, it is estimated that there may be as many as 500 practical solutions when 6 quality specifications and 10 feedstuffs are considered for a ration.

Generally speaking, electronic feed formulation (1) effects a greater saving when first applied to a ration than in subsequent applications, and (2) is of most use where a wide selection of feed ingredients is available and/or prices shift rather rapidly.

The information needed and the procedure followed in formulating rations by computer are exactly the same as in the hand method of ration formulation; namely, (1) the nutritive requirements for the particular class and kind of animal, (2) nutritive content of the feeds, and (3) ingredient costs. Sometimes this simple fact is overlooked because of the awesomeness of the computer, and the jargon used by those who wish to impress fellow scientists. Step by step, the procedure in formulating rations by computer is:

1. *List available feed ingredients, and the cost of each*—It is necessary that all of the available feeds be listed along with the unit cost (usually per ton) of each; preferably, ingredient cost should be based on market price plus delivery, storage, and processing cost.

2. *Record quality of feed*—The more that is known about the quality of feed the better. This is so because of the wide variation in composition and feeding value within ingredients; for example, between two samples of alfalfa hay.

Whenever possible, an actual chemical analysis of a representative sample of each ingredient under consideration should be available and used. However, the imperfections of a chemical analysis of a feedstuff should be recognized; chiefly, (a) it does not provide information on the availability of nutrients to animals, and (b) there are variations between samples.

3. *Establish ration specifications*—Set down the ration specifications—the nutrients and the levels of each that are to be met. This is exactly the same procedure as is followed in the hand method. For example, in arriving at ration specifications for feedlot cattle, the nutritionist considers (a) age, weight, and grade of cattle; (b) length of feeding period; (c) the

probable market; (d) season of year; (e) background and stress of animals; and (f) other similar factors.

4. *Give restrictions*—Usually it is necessary to establish certain limitations on the use of ingredients; for example, with feedlot cattle the restrictions would likely show (a) the maximum amount of roughage, (b) the maximum amount of urea, (c) the minimum and maximum amounts of fat, (d) the proportion of cottonseed hulls to alfalfa hay, (e) the proportion of one grain to another—such as 60% barley and 40% milo, (f) an upper limit of some ingredients—such as 20% rye, (g) the exact amount of the premix, and (h) the lower and upper limits of molasses, as between 5 and 10%.

It must be recognized that the narrower the limitations imposed on the computer, the less the choice it will have in ration formulation and the higher the cost.

5. *Stipulate feed additives*—Generally speaking, the nutritionist makes rigid stipulations as to amounts of these ingredients, much as he does with added vitamins and minerals. All of them cost money, and many of them must be used in compliance with the Food and Drug Administration regulations.

6. *Obtain program*—Take data, ration specifications, and restrictions to an experienced computer programmer or systems analyst. He will either (a) "tailor-make" a program for a given situation, or (b) suggest one of the "canned" linear programs available from the larger computer companies. The canned programs are, by necessity, general in nature, because they are written for a wide variety of applications; but they cost less than a tailor-made program.

7. *Put data on cards*—The data obtained from steps 1 through 5 (above) must be punched into standard punch cards (unless key-to-tape or key-to-disk equipment is available), and verified by competent keypunchers or verifiers.

8. *"Feed" the punched cards into the computer*—When the punched cards or other input media are fed into the computer, it treats the data as one gigantic algebra problem and arrives at the ration formulation in a matter of minutes. Based on available feeds, analysis, and price, the computer evolves with the mix that will meet the desired nutritive allowances at the least possible cost.

9. *Formulate as necessary*—All rations should be reviewed at frequent intervals, and reformulated when there are shifts in (a) availability of ingredients (certain ingredients may no longer be available, but new ones may have evolved), (b) price, and/or (c) chemical composition.

RATIONS FOR FARM ANIMALS

Suggested rations for different classes and ages of beef cattle, dairy cattle, sheep, swine, and horses are given under the discussion of the respective species. Also, some suggested fitting rations are included therewith. All of these are merely intended as general guides. Variations can and should be made in the rations used. The feeder should give consideration to (1) the supply of homegrown feeds, (2) the availability and price of purchased feeds, (3) the class and age of animals, (4) the health and condition of the animals, (5) the length of the grazing season, and (6) the net returns.

Also, it should be recognized that feeds of similar nutritive properties can and should be interchanged as price relationships warrant. Thus, (1) the cereal grains may consist of corn, barley, wheat, oats, and/or milo; (2) the animal protein supplements may consist of tankage, meat meal, fish meal or other marine by-products, and/or milk by-products; (3) the plant protein supplements may consist of soybean, cottonseed, linseed, peanut, and/or sunflower meals; (4) the roughage may include many varieties of hays and silages; and (5) a vast array of by-product feeds may be utilized.

Although certain principles are usually followed by all good feeders, no book of knowledge or set of instructions can substitute for experience and born livestock intuition. Skill and good judgment are always essential, for the feed requirements of animals do not necessarily remain the same from day to day or from period to period. Thus, the age and size of the animal, the kind and degree of activity, the climatic conditions, the kind, quality and amount of feed, the system of management, and the health, condition, and temperament of the animal are all continually exerting a powerful influence in determining the nutritive needs. How well the feeder understands, anticipates, interprets, and meets these requirements usually determines the success or failure of the ration and the results obtained.

FEED SUBSTITUTIONS

The successful stockman is a keen student of values. He recognizes that feeds of similar nutritive properties can and should be interchanged in the ration as price relationships warrant, thereby making it possible at all times to obtain a balanced ration at the lowest cost.

In arriving at feed substitutions, two primary factors besides cost, chemical composition, and feeding value should be considered—namely, palatability and quality of product produced. Also, when substituting feed ingredients, the following facts should be kept in mind:

1. Feeds differ widely in feeding value. Barley, for example, varies widely in feeding value according to the hull content and the test weight per bushel

and the same can be said relative to oats. There is a wide range in the protein content of sorghum. The forages vary widely in feeding value according to the stage of maturity at which they are cut and how well they are cured and stored.

2. Nonlegume forages may have a higher than normal value relative to legumes when the chief need of the animal is for additional energy rather than for supplemented protein.

3. Based primarily on available supply and price, certain feeds—especially those of medium protein content, such as brewers' dried grains, corn gluten feed (gluten feed), distillers' dried grains, distillers' dried solubles, peanuts, and peas (dried)—may be used interchangeably as (a) grains and by-product feeds, and/or (b) protein supplements.

4. The feeding value of certain feeds is materially affected by preparation. For example, wheat must be coarsely ground or rolled for cattle.

For the reasons noted above, no comparative value of feeds can be absolute. Rather, feed substitutions should be based on the class and age of animal and the quality of feed, together with experiences and experiments. Feed substitution tables for each species are included in this section, under the discussion of the respective classes of livestock.

HOME MIXED VS COMMERCIAL FEEDS

The stockman has the following options from which to choose for home mixing feeds:

1. Purchase of a commercially prepared protein supplement (likely reinforced with vitamins and minerals), which may be blended with local or homegrown grain.
2. Purchase of a commercially prepared vitamin-mineral premix which may be mixed with an oil meal, and then blended with local or homegrown grain.
3. Purchase of individual ingredients (including vitamins and minerals) and mixing the feed from the ground up.

Commercial feeds are just what the term implies—instead of being farm mixed, these feeds are mixed by commercial feed manufacturers who specialize in the business. In 1981, a total of 81.9 million tons of primary feeds (complete feeds) were manufactured in the United States, and an additional 32 million tons of secondary feeds (supplements) were produced; making for a total of 113.9 million tons of commercial feeds. Primary feed is defined as that which is mixed with individual ingredients, sometimes with the addition of a premix at a rate of less

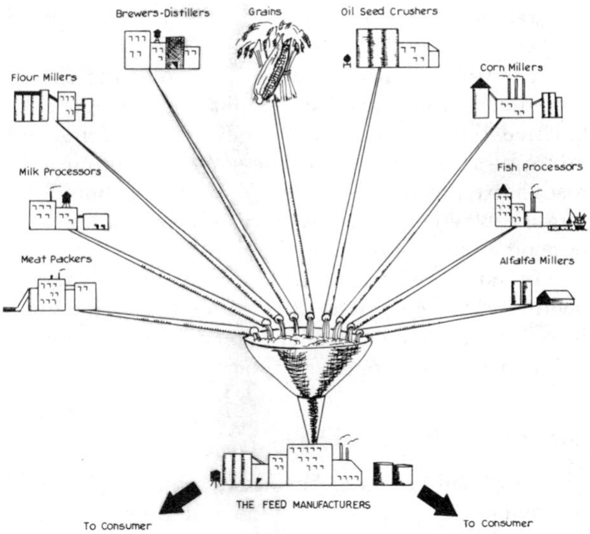

Fig. 4-15. Commercial feed companies obtain their raw materials for feeds from many sources. Over 100 different ingredients are processed into various (1) complete feeds, or (2) concentrates that are fed with homegrown grains.

than 100 pounds per ton of finished feed. Secondary feed is that which is mixed with one or more ingredients and a formula feed supplement, the breakdown, percentagewise, by classes of livestock for which primary (complete) commercial feeds were used in 1981 follows: poultry, 41.8%; dairy, 18.2%; beef and sheep, 18.1%; hogs, 15.6%; and all other, 6.3%.[6]

The commercial feed manufacturer has the distinct advantages of (1) purchasing feed in quantity lots, making possible price advantages; (2) economical and controlled mixing; (3) the hiring of scientifically trained personnel for use in determining the rations; and (4) quality control. Most stockmen have neither the know-how nor the quantity of business to provide these services on their own. Because of these several advantages, commercial feeds are finding a place of increasing importance in livestock feeding.

Numerous types of commercial feeds, ranging from additives to complete rations, are on the market, with most of them designed for a specific species, age, or need. Among them, are complete rations (including hay for ruminants and pseudoruminants), concentrates, pelleted or cubed forages, protein supplements (with or without reinforcements of vitamins and/or minerals), vitamin and/or mineral supplements, additives (antibiotics, hormones, etc.), milk replacers, starters, young stock rations, fitting rations, rations for different levels of production for the idle (like dry

[6]Source: American Feed Manufacturers Assn.

cow rations) to the forced producers, and medicated feeds.

The value of farm-grown grains—plus the cost of ingredients which need to be purchased in order to balance the ration, and the cost of grinding and mixing—as compared to the cost of commercial ready-mixed feeds laid down on the farm, should determine whether it is best to mix feeds at home or depend on ready-mixed feeds.

In summary, it may be said that there exist two good alternative sources of most feeds and rations—home mixed or commercial—and the able manager will choose wisely between them.

State Commercial Feed Laws

Nearly all the states have laws regulating the sale of commercial feeds. These benefit both stockmen and reputable feed manufacturers. In most states the laws require that every brand of commercial feed sold in the state be licensed, and that the chemical composition be guaranteed.

Samples of each commercial feed are taken each year, and analyzed chemically in the state's laboratory to determine if the manufacturer lived up to his guarantee. Additionally, skilled microscopists examine the sample to ascertain that the ingredients present are the same as those guaranteed. Flagrant violations on the latter point may be prosecuted.

Results of these examinations are generally published, annually, by the state department in charge of such regulatory work. Usually, the publication of the guarantee alongside any "short-changing" is sufficient to cause the manufacturer promptly to rectify the situation, for such public information soon becomes known to both users and competitors.

PART II—FEEDING BEEF CATTLE

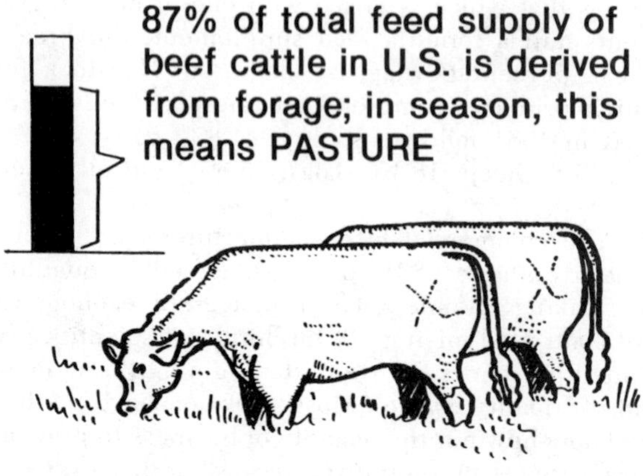

87% of total feed supply of beef cattle in U.S. is derived from forage; in season, this means PASTURE

Fig. 4-16. Forages are the foundation of successful beef production.

The feeding of beef cattle constitutes the greatest single cost item of their production. It is important, therefore, that the feeding practices be as satisfactory and economical as possible.

Pastures and other roughages, preferably with a maximum of the former, are the very foundation of successful beef cattle production. In fact, it may be said that the principal function of beef cattle is to harvest vast acreages of forages, and, with or without supplementation, to convert these feeds into more nutritious and palatable products for human consumption. It is estimated (1) that 87 percent of the total feed of beef cattle is derived from roughages (see Table 4-2, page 217), and (2) that 43.9 percent of the land area of continental United States is pastured all or part of the year, with much of this area utilized by beef cattle. If produced on well-fertilized soils, green grass and well-cured, green, leafy hay can supply all of the nutrient requirements of beef cattle, except the need for common salt and whatever energy-rich feeds may be necessary for additional conditioning or drylot finishing.

NUTRIENT REQUIREMENTS

In recent years, the introduction of crossbreeding and the exotic breeds has produced faster gaining calves, later maturing cattle, and heavier milking cows. Also, more and more heifers are being bred to calve as two-year-olds. Thus, provision is herein made for the nutritive needs created by these changes.

As feeds represent by far the greatest cost item in beef production, it is important that there be a basic understanding of the nutrient requirements. For convenience, these needs will be discussed under the following groups: (1) energy, (2) protein, (3) mineral, (4) vitamin, and (5) water.

The nutrient requirements of beef cattle are given in Tables 4-13 to 4-19, which follow. These figures are, for the most part, requirements (rather than allowances); hence, they do not provide for margins of safety to compensate for variations in feed composition, environment, and possible losses of nutrients during storage or processing.

TABLE 4-13

DAILY NUTRIENT REQUIREMENTS OF GROWING-FINISHING STEER CALVES AND YEARLINGS (per Animal)[1]

Weight[2]		Daily Gain		Minimum Dry Matter Consumption[3]		Roughage[3]	Total Protein		Digestible Protein		NEm	NEg	ME[3]	TDN[3,4]		Ca	P	Vitamin A
(lb)	(kg)	(lb)	(kg)	(lb)	(kg)	(%)	(lb)	(kg)	(lb)	(kg)	(Mcal)	(Mcal)	(Mcal)	(lb)	(kg)	(g)	(g)	(1,000 IU)
220	100	0	0	4.6	2.1	100	0.40	0.18	0.22	0.10	2.43	0	4.2	2.6	1.2	4	4	5
		1.1	0.5	6.4	2.9	70-80	0.79	0.36	0.53	0.24	2.43	0.89	6.6	4.0	1.8	14	11	6
		1.5	0.7	6.0	2.7	50-60	0.88	0.40	0.62	0.28	2.43	1.27	7.1	4.4	2.0	19	13	6
		2.0	0.9	6.2	2.8	25-30	1.01	0.46	0.73	0.33	2.43	1.68	7.7	4.6	2.1	24	16	7
		2.4	1.1	6.0	2.7	<15	1.08	0.49	0.79	0.36	2.43	2.10	8.4	5.1	2.3	28	19	7
331	150	0	0	6.2	2.8	100	0.51	0.23	0.29	0.13	3.30	0	5.6	3.5	1.6	5	5	6
		1.1	0.5	8.8	4.0	70-80	0.97	0.44	0.62	0.28	3.30	1.20	9.0	5.5	2.5	14	12	9
		1.5	0.7	8.6	3.9	50-60	1.08	0.49	0.73	0.33	3.30	1.73	9.6	6.0	2.7	18	14	9
		2.0	0.9	8.4	3.8	25-30	1.19	0.54	0.82	0.37	3.30	2.27	10.7	6.6	3.0	23	17	9
		2.4	1.1	8.2	3.7	<15	1.28	0.58	0.90	0.41	3.30	2.84	11.3	6.8	3.1	28	20	9
441	200	0	0	7.7	3.5	100	0.66	0.30	0.37	0.17	4.10	0	7.0	4.2	1.9	6	6	8
		1.1	0.5	12.8	5.8	80-90	1.26	0.57	0.77	0.35	4.10	1.49	12.1	7.5	3.4	14	13	12
		1.5	0.7	12.6	5.7	70-80	1.34	0.61	0.86	0.39	4.10	2.14	13.0	7.9	3.6	18	16	13
		2.0	0.9	10.8	4.9	35-45	1.34	0.61	0.88	0.40	4.10	2.82	13.3	8.2	3.7	23	18	13
		2.4	1.1	10.1	4.6	<15	1.39	0.63	0.95	0.43	4.10	3.52	14.1	8.6	3.9	27	20	13
551	250	0	0	9.7	4.4	100	0.77	0.35	0.44	0.20	4.84	0	8.2	5.1	2.3	8	8	9
		1.5	0.7	12.8	5.8	55-65	1.37	0.62	0.86	0.39	4.84	2.53	14.4	8.8	4.0	18	16	14
		2.0	0.9	13.7	6.2	45-50	1.52	0.69	0.97	0.44	4.84	3.33	16.2	9.9	4.5	22	19	14
		2.4	1.1	13.2	6.0	20-25	1.61	0.73	1.06	0.48	4.84	4.17	17.0	10.4	4.7	26	21	14
		2.9	1.3	13.2	6.0	<15	1.68	0.76	1.12	0.51	4.84	5.04	18.6	11.5	5.2	30	23	14
661	300	0	0	10.4	4.7	100	0.88	0.40	0.51	0.23	5.55	0	9.4	5.7	2.6	9	9	10
		2.0	0.9	17.9	8.1	55-65	1.79	0.81	1.10	0.50	5.55	3.82	19.5	11.9	5.4	22	19	16
		2.4	1.1	16.8	7.6	20-25	1.81	0.82	1.15	0.52	5.55	4.78	20.4	12.3	5.6	25	22	16
		2.9	1.3	15.6	7.1	<15	1.83	0.83	1.19	0.54	5.55	5.77	21.6	13.2	6.0	29	23	16
		3.1[5]	1.4[5]	16.1	7.3	<15	1.92	0.87	1.26	0.57	5.55	6.29	22.5	13.7	6.2	31	25	16
772	350	0	0	11.7	5.3	100	1.01	0.46	0.57	0.26	6.24	0	10.6	6.4	2.9	10	10	12
		2.0	0.9	17.6	8.0	45-55	1.76	0.80	1.08	0.49	6.24	4.29	20.8	12.8	5.8	20	18	18
		2.4	1.1	17.6	8.0	20-25	1.83	0.83	1.15	0.52	6.24	5.36	22.4	13.7	6.2	23	20	18
		2.9	1.3	17.6	8.0	<15	1.92	0.87	1.21	0.55	6.24	6.48	24.2	15.0	6.8	26	22	18
		3.1[5]	1.4[5]	18.1	8.2	<15	1.98	0.90	1.26	0.57	6.24	7.06	25.3	15.4	7.0	28	24	18
882	400	0	0	13.0	5.9	100	1.12	0.51	0.64	0.29	6.89	0	11.8	7.3	3.3	11	11	13
		2.2	1.0	20.7	9.4	45-55	1.92	0.87	1.19	0.54	6.89	5.33	24.5	15.0	6.8	21	20	19
		2.6	1.2	18.7	8.5	20-25	1.92	0.87	1.19	0.54	6.89	6.54	25.4	15.4	7.0	23	21	19
		2.9	1.3	19.0	8.6	<15	1.98	0.90	1.23	0.56	6.89	7.16	26.5	16.1	7.3	25	22	19
		3.1[5]	1.4[5]	19.8	9.0	<15	2.07	0.94	1.30	0.59	6.89	7.80	28.0	17.0	7.7	26	23	19
992	450	0	0	14.1	6.4	100	1.19	0.54	0.68	0.31	7.52	0	12.8	7.9	3.6	12	12	14
		2.2	1.0	22.7	10.3	45-55	2.12	0.96	1.26	0.57	7.52	5.82	26.7	16.3	7.4	20	20	20
		2.6	1.2	22.5	10.2	20-25	2.14	0.97	1.28	0.58	7.52	7.14	28.6	17.4	7.9	23	22	20
		2.9	1.3	20.5	9.3	<15	2.14	0.97	1.30	0.59	7.52	7.83	29.0	17.6	8.0	24	23	20
		3.1[5]	1.4[5]	21.6	9.8	<15	2.16	0.98	1.32	0.60	7.52	8.52	30.5	18.5	8.4	25	23	20
,102	500	0	0	15.4	7.0	100	1.32	0.60	0.75	0.34	8.14	0	13.9	8.4	3.8	13	13	15
		2.0	0.9	23.1	10.5	45-55	2.09	0.95	1.23	0.56	8.14	5.60	27.1	16.5	7.5	19	19	23
		2.4	1.1	22.9	10.4	20-25	2.12	0.96	1.26	0.57	8.14	7.01	29.2	17.8	8.1	20	20	23
		2.6	1.2	21.2	9.6	<15	2.12	0.96	1.28	0.58	8.14	7.73	29.7	18.1	8.2	21	21	23
		2.9[5]	1.3[5]	22.0	10.0	<15	2.14	0.97	1.32	0.60	8.14	8.47	31.4	19.2	8.7	22	22	23

[1]Adapted by the author from *Nutrient Requirements of Beef Cattle*, No. 4, 5th rev. ed., NRC-National Academy of Sciences, 1976, pp. 22-23.

[2]Average weight for a feeding period.

[3]Dry matter consumption, ME and TDN allowances are based on NE requirements and the general type of diet indicated in the roughage column. Most roughages will ontain 1.9 to 2.2 Mcal of ME/kg dry matter and 90-100% concentrate diets are expected to contain 3.1 to 3.3 Mcal of ME/kg.

[4]TDN was calculated by assuming 3.6155 Mcal of ME per kilogram of TDN.

[5]Most steers of the weight indicated, and not exhibiting compensatory growth, will fail to sustain the energy intake necessary to maintain this rate of gain for an extended eriod.

TABLE 4-14

DAILY NUTRIENT REQUIREMENTS OF GROWING-FINISHING HEIFER CALVES AND YEARLINGS (per Animal)[1]

Weight[2]		Daily Gain		Minimum Dry Matter Consumption[3]		Roughage[3]	Total Protein		Digestible Protein		NEm	NEg	ME[3]	TDN[3,4]		Ca	P	Vitamin A
(lb)	*(kg)*	*(lb)*	*(kg)*	*(lb)*	*(kg)*	*(%)*	*(lb)*	*(kg)*	*(lb)*	*(kg)*	*(Mcal)*	*(Mcal)*	*(Mcal)*	*(lb)*	*(kg)*	*(g)*	*(g)*	*(1,000 IU*
220	*100*	0	*0*	4.6	*2.1*	100	0.40	*0.18*	0.22	*0.10*	2.43	0	4.2	2.6	*1.2*	4	4	5
		1.1	*0.5*	6.6	*3.0*	70-80	0.82	*0.37*	0.55	*0.25*	2.43	0.99	6.9	4.2	*1.9*	14	11	6
		1.5	*0.7*	6.4	*2.9*	50-60	0.93	*0.42*	0.64	*0.29*	2.43	1.44	7.5	4.6	*2.1*	19	14	6
		2.0	*0.9*	6.6	*3.0*	25-30	1.06	*0.48*	0.75	*0.34*	2.43	1.92	8.3	5.1	*2.3*	24	17	7
		2.4	*1.1*	6.6	*3.0*	<15	1.17	*0.53*	0.86	*0.39*	2.43	2.43	9.2	5.5	*2.5*	29	19	7
331	*150*	0	*0*	6.2	*2.8*	100	0.53	*0.24*	0.31	*0.14*	3.30	0	5.6	3.5	*1.6*	5	5	6
		1.1	*0.5*	9.0	*4.1*	70-80	0.99	*0.45*	0.64	*0.29*	3.30	1.34	9.4	5.7	*2.6*	14	12	9
		1.5	*0.7*	8.8	*4.0*	50-60	1.10	*0.50*	0.73	*0.33*	3.30	1.95	10.4	6.2	*2.8*	18	14	9
		2.0	*0.9*	8.8	*4.0*	25-30	1.19	*0.54*	0.82	*0.37*	3.30	2.60	11.3	6.8	*3.1*	23	17	9
		2.4	*1.1*	8.8	*4.0*	<15	1.32	*0.60*	0.93	*0.42*	3.30	3.30	12.4	7.5	*3.4*	28	20	9
441	*200*	0	*0*	7.7	*3.5*	100	0.66	*0.30*	0.37	*0.17*	4.10	0	7.0	4.2	*1.9*	6	6	8
		0.7	*0.3*	11.9	*5.4*	100	1.08	*0.49*	0.64	*0.29*	4.10	0.95	10.8	6.6	*3.0*	10	10	12
		1.1	*0.5*	13.2	*6.0*	80-90	1.28	*0.58*	0.77	*0.35*	4.10	1.66	12.7	7.7	*3.5*	14	13	13
		1.5	*0.7*	13.2	*6.0*	70-80	1.34	*0.61*	0.86	*0.39*	4.10	2.42	13.8	8.4	*3.8*	18	16	13
		2.0	*0.9*	11.7	*5.3*	35-45	1.37	*0.62*	0.88	*0.40*	4.10	3.23	14.3	8.8	*4.0*	22	17	13
		2.4	*1.1*	11.0	*5.0*	<15	1.41	*0.64*	0.95	*0.43*	4.10	4.09	15.4	9.5	*4.3*	25	19	13
551	*250*	0	*0*	9.0	*4.1*	100	0.77	*0.35*	0.44	*0.20*	4.84	0	8.3	5.1	*2.3*	7	7	9
		0.7	*0.3*	14.1	*6.4*	100	1.26	*0.57*	0.73	*0.33*	4.84	1.13	12.8	7.8	*3.5*	12	12	14
		1.1	*0.5*	14.3	*6.5*	80-90	1.37	*0.62*	0.82	*0.37*	4.84	1.96	14.2	8.6	*3.9*	13	13	14
		1.5	*0.7*	12.8	*5.8*	55-65	1.37	*0.62*	0.84	*0.38*	4.84	2.86	15.0	9.1	*4.1*	17	15	14
		2.0	*0.9*	13.0	*5.9*	35-45	1.43	*0.65*	0.93	*0.42*	4.84	3.81	16.5	10.1	*4.6*	21	17	14
		2.4	*1.1*	14.3	*6.5*	20-25	1.63	*0.74*	1.06	*0.48*	4.84	4.84	18.7	11.5	*5.2*	25	20	14
		2.6	*1.2*	13.9	*6.3*	<15	1.65	*0.75*	1.08	*0.49*	4.84	5.37	19.4	11.9	*5.4*	27	21	14
661	*300*	0	*0*	10.4	*4.7*	100	0.88	*0.40*	0.51	*0.23*	5.55	0	9.5	5.7	*2.6*	9	9	10
		0.7	*0.3*	16.3	*7.4*	100	1.39	*0.63*	0.79	*0.36*	5.55	1.29	14.5	8.4	*4.0*	13	13	16
		1.1	*0.5*	16.3	*7.4*	80-90	1.48	*0.67*	0.88	*0.40*	5.55	2.25	16.3	9.9	*4.5*	14	14	16
		1.5	*0.7*	14.6	*6.6*	55-65	1.48	*0.67*	0.88	*0.40*	5.55	3.37	17.1	10.4	*4.7*	16	15	16
		2.0	*0.9*	15.0	*6.8*	35-45	1.54	*0.70*	0.97	*0.44*	5.55	4.37	19.0	11.5	*5.2*	19	17	16
		2.4	*1.1*	16.5	*7.5*	20-25	1.72	*0.78*	1.08	*0.49*	5.55	5.55	21.5	13.2	*6.0*	23	20	16
		2.6	*1.2*	15.9	*7.2*	<15	1.74	*0.79*	1.10	*0.50*	5.55	6.16	22.3	13.7	*6.2*	24	20	16
772	*350*	0	*0*	11.7	*5.3*	100	1.01	*0.46*	0.57	*0.26*	6.24	0	10.6	6.4	*2.9*	10	10	12
		0.7	*0.3*	18.1	*8.2*	100	1.52	*0.69*	0.86	*0.39*	6.24	1.45	16.5	10.0	*4.6*	15	15	18
		1.1	*0.5*	18.3	*8.3*	80-90	1.61	*0.73*	0.93	*0.42*	6.24	2.52	18.3	11.2	*5.1*	15	15	18
		1.5	*0.7*	17.4	*7.9*	55-65	1.61	*0.73*	0.95	*0.43*	6.24	3.68	19.7	11.9	*5.4*	15	15	18
		2.0	*0.9*	17.9	*8.1*	35-45	1.70	*0.77*	1.01	*0.46*	6.24	4.91	21.8	13.2	*6.0*	17	17	18
		2.4	*1.1*	18.3	*8.3*	20-25	1.79	*0.81*	1.10	*0.50*	6.24	6.23	24.0	14.5	*6.6*	20	19	18
		2.6[5]	*1.2*[5]	17.9	*8.1*	<15	1.79	*0.81*	1.10	*0.50*	6.24	6.91	25.0	15.2	*6.9*	21	20	18
882	*400*	0	*0*	13.0	*5.9*	100	1.12	*0.51*	0.64	*0.29*	6.89	0	11.8	7.3	*3.3*	11	11	13
		0.7	*0.3*	20.0	*9.1*	100	1.68	*0.76*	0.95	*0.43*	6.89	1.61	18.2	11.1	*5.0*	16	16	19
		1.1	*0.5*	18.7	*8.5*	70-80	1.72	*0.78*	0.95	*0.43*	6.89	2.79	19.5	11.9	*5.4*	15	15	19
		1.5	*0.7*	19.2	*8.7*	55-65	1.74	*0.79*	1.01	*0.46*	6.89	4.06	21.7	13.2	*6.0*	16	16	19
		2.0	*0.9*	18.5	*8.4*	20-25	1.74	*0.79*	1.04	*0.47*	6.89	5.43	23.5	14.3	*6.5*	17	17	19
		2.4[5]	*1.1*[5]	18.3	*8.3*	<15	1.79	*0.81*	1.08	*0.49*	6.89	6.88	25.9	15.9	*7.2*	19	18	19
992	*450*	0	*0*	14.1	*6.4*	100	1.21	*0.55*	0.68	*0.31*	7.52	0	12.9	7.9	*3.6*	12	12	14
		0.4	*0.2*	19.2	*8.7*	100	1.63	*0.74*	0.90	*0.41*	7.52	1.14	17.4	10.6	*4.8*	16	16	19
		1.1	*0.5*	20.5	*9.3*	70-80	1.76	*0.80*	1.01	*0.46*	7.52	3.05	21.3	13.0	*5.9*	17	17	20
		1.8	*0.8*	20.1	*9.1*	35-45	1.81	*0.82*	1.06	*0.48*	7.52	5.17	24.5	15.0	*6.8*	16	16	20
		2.2[5]	*1.0*[5]	18.7	*8.5*	<15	1.83	*0.83*	1.06	*0.48*	7.52	6.71	26.8	16.3	*7.4*	19	19	20

[1]Adapted by the author from *Nutrient Requirements of Beef Cattle*, No. 4, 5th rev. ed., NRC-National Academy of Sciences, 1976, pp. 24-25.

[2]Average weight for a feeding period.

[3]Dry matter consumption, ME and TDN allowances are based on NE requirements and the general type of diet indicated in the roughage column. Most roughages w contain 1.9 to 2.2 Mcal of ME/kg dry matter and 90-100% concentrate diets are expected to have 3.1 to 3.3 Mcal of ME/kg.

[4]TDN was calculated by assuming 3.6155 kcal of ME per gram of TDN.

[5]Most heifers of the weight indicated, and not exhibiting compensatory growth, will fail to sustain the energy intake necessary to maintain this rate of gain for an extende period.

TABLE 4-15

DAILY NUTRIENT REQUIREMENTS OF BEEF CATTLE BREEDING HERD (per Animal)[1]

Weight[2]		Daily Gain		Minimum Dry Matter Consumption[3]		Roughage[3]	Total Protein		Digestible Protein		NE_m	NE_g	ME[3]	TDN[3,4]		Ca	P	Vitamin A
(lb)	*(kg)*	*(lb)*	*(kg)*	*(lb)*	*(kg)*	*(%)*	*(lb)*	*(kg)*	*(lb)*	*(kg)*	*(Mcal)*	*(Mcal)*	*(Mcal)*	*(lb)*	*(kg)*	*(g)*	*(g)*	*(1,000 IU)*
Pregnant yearling heifers—Last 3-4 months of pregnancy																		
716	*325*	0.9	*0.4*[4]	14.5	*6.6*	100[5]	1.28	*0.58*	0.75	*0.34*	5.89	0.62	12.6	7.7	*3.5*	15	15	19
		1.3	*0.6*	18.7	*8.5*	100	1.65	*0.75*	0.93	*0.42*	5.89	1.52	16.2	9.9	*4.5*	18	18	23
		1.8	*0.8*	20.7	*9.4*	85-100	1.87	*0.85*	1.10	*0.50*	5.89	2.49	20.1	12.3	*5.6*	22	20	26
772	*350*	0.9	*0.4*[4]	15.2	*6.9*	100	1.34	*0.61*	0.77	*0.35*	6.23	0.65	13.2	8.1	*3.7*	15	15	19
		1.3	*0.6*	19.6	*8.9*	100	1.72	*0.78*	0.99	*0.45*	6.23	1.60	16.9	10.3	*4.7*	19	19	25
		1.8	*0.8*	22.0	*10.0*	85-100	1.94	*0.88*	1.12	*0.51*	6.24	2.63	21.1	12.9	*5.8*	22	21	28
827	*375*	0.9	*0.4*[4]	15.9	*7.2*	100	1.39	*0.63*	0.79	*0.36*	6.56	0.68	13.7	8.4	*3.8*	15	15	20
		1.3	*0.6*	20.5	*9.3*	100	1.79	*0.81*	1.01	*0.46*	6.56	1.68	17.7	10.8	*4.9*	19	19	26
		1.8	*0.8*	24.2	*11.0*	85-100	2.12	*0.96*	1.21	*0.55*	6.56	2.76	22.1	13.5	*6.1*	22	22	31
882	*400*	0.9	*0.4*[4]	16.5	*7.5*	100	1.43	*0.65*	0.84	*0.38*	6.89	0.71	14.2	8.7	*4.0*	16	16	21
		1.3	*0.6*	21.4	*9.7*	100	1.85	*0.84*	1.06	*0.48*	6.89	1.76	18.5	11.3	*5.1*	19	19	27
		1.8	*0.8*	25.6	*11.6*	85-100	2.23	*1.01*	1.26	*0.57*	6.89	2.90	23.0	14.0	*6.4*	22	22	33
937	*425*	0.9	*0.4*[4]	17.2	*7.8*	100	1.52	*0.69*	0.88	*0.40*	7.21	0.74	14.8	9.0	*4.1*	16	16	22
		1.3	*0.6*	22.3	*10.1*	100	1.94	*0.88*	1.10	*0.50*	7.21	1.84	19.2	11.7	*5.3*	19	19	28
		1.8	*0.8*	26.7	*12.1*	85-100	2.31	*1.05*	1.32	*0.60*	7.21	3.03	24.0	14.6	*6.6*	22	22	34
Dry pregnant mature cows—Middle third of pregnancy																		
772	*350*			12.2	*5.5*	100[5]	0.71	*0.32*	0.33	*0.15*	6.23		10.8	6.6	*3.0*	10	10	15
882	*400*			13.4	*6.1*	100	0.79	*0.36*	0.37	*0.17*	6.89		11.9	7.3	*3.3*	11	11	17
992	*450*			14.8	*6.7*	100	0.86	*0.39*	0.42	*0.19*	7.52		13.0	7.9	*3.6*	12	12	19
1,102	*500*			15.9	*7.2*	100	0.93	*0.42*	0.44	*0.20*	8.14		14.1	8.6	*3.9*	13	13	20
1,213	*550*			17.0	*7.7*	100	0.99	*0.45*	0.49	*0.22*	8.75		15.1	9.2	*4.2*	14	14	22
1,323	*600*			18.3	*8.3*	100	1.08	*0.49*	0.51	*0.23*	9.33		16.1	9.8	*4.4*	15	15	23
1,433	*650*			19.4	*8.8*	100	1.15	*0.52*	0.55	*0.25*	9.91		17.1	10.4	*4.7*	16	16	25
Dry pregnant mature cows—Last third of pregnancy																		
772	*350*	0.9	*0.4*[4]	15.2	*6.9*	100[5]	0.90	*0.41*	0.42	*0.19*	7.8		13.2	8.0	*3.6*	12	12	19
882	*400*	0.9	*0.4*	16.5	*7.5*	100	0.97	*0.44*	0.46	*0.21*	8.4		14.3	8.7	*4.0*	14	14	21
992	*450*	0.9	*0.4*	17.9	*8.1*	100	1.06	*0.48*	0.51	*0.23*	9.1		15.4	9.4	*4.2*	15	15	23
1,102	*500*	0.9	*0.4*	19.0	*8.6*	100	1.12	*0.51*	0.53	*0.24*	9.7		16.4	10.0	*4.5*	15	15	24
1,213	*550*	0.9	*0.4*	20.1	*9.1*	100	1.19	*0.54*	0.55	*0.25*	10.3		17.5	10.7	*4.8*	16	16	26
1,323	*600*	0.9	*0.4*	21.4	*9.7*	100	1.26	*0.57*	0.60	*0.27*	10.9		18.5	11.2	*5.1*	17	17	27
1,433	*650*	0.9	*0.4*	22.5	*10.2*	100	1.32	*0.60*	0.64	*0.29*	11.5		19.6	11.9	*5.4*	18	18	29
Cows nursing calves—Average milking ability[6]—First 3-4 months postpartum																		
772	*350*			18.1	*8.2*	100[5]	1.65	*0.75*	0.97	*0.44*	9.2		15.9	9.7	*4.4*	24	24	19
882	*400*			19.4	*8.8*	100	1.79	*0.81*	1.06	*0.48*	9.9		17.0	10.4	*4.7*	25	25	21
992	*450*			20.5	*9.3*	100	1.90	*0.86*	1.10	*0.50*	10.5		18.1	11.0	*5.0*	26	26	23
1,102	*500*			21.6	*9.8*	100	1.98	*0.90*	1.17	*0.53*	11.1		19.2	11.7	*5.3*	27	27	24
1,213	*550*			23.1	*10.5*	100	2.14	*0.97*	1.26	*0.57*	11.9		20.3	12.3	*5.6*	28	28	26
1,323	*600*			24.2	*11.0*	100	2.23	*1.01*	1.30	*0.59*	12.3		21.3	13.0	*5.9*	28	28	27
1,433	*650*			25.1	*11.4*	100	2.31	*1.05*	1.37	*0.62*	12.9		22.3	13.7	*6.2*	29	29	29
Cows nursing calves—Superior milking ability[7]—First 3-4 months postpartum																		
772	*350*			22.4	*10.2*	100[8]	2.45	*1.11*	1.43	*0.65*	12.3		21.0	12.8	*5.8*	45	40	32
882	*400*			23.8	*10.8*	100	2.58	*1.17*	1.52	*0.69*	13.0		22.1	13.5	*6.1*	45	41	34
992	*450*			24.9	*11.3*	100	2.71	*1.23*	1.59	*0.72*	13.6		23.2	14.1	*6.4*	45	42	36
1,102	*500*			26.0	*11.8*	100	2.84	*1.29*	1.68	*0.76*	14.2		24.3	14.8	*6.7*	46	43	38
1,213	*550*			27.3	*12.4*	100	2.98	*1.35*	1.74	*0.79*	14.9		25.3	15.4	*7.0*	46	44	41
1,323	*600*			28.4	*12.9*	100	3.11	*1.41*	1.83	*0.83*	15.5		26.4	16.1	*7.3*	46	44	43
1,433	*650*			29.5	*13.4*	100	3.22	*1.46*	1.90	*0.86*	16.2		27.5	16.8	*7.6*	47	45	45
Bulls, growth and maintenance (moderate activity)																		
661	*300*	2.2	*0.90*	19.4	*8.8*	70-75	1.98	*0.90*	1.21	*0.55*	5.6	3.8	20.4	12.3	*5.6*	27	23	34
882	*400*	2.0	*0.70*	24.2	*11.0*	70-75	2.27	*1.03*	1.37	*0.62*	6.9	4.1	25.2	15.4	*7.0*	23	23	43
1,102	*500*	1.5	*0.50*	26.9	*12.2*	80-85	2.36	*1.07*	1.37	*0.62*	8.5	3.7	27.0	16.5	*7.5*	22	22	48
1,323	*600*	1.1	*0.30*	26.4	*12.0*	80-85	2.25	*1.02*	1.32	*0.60*	9.8	3.0	26.4	16.1	*7.3*	22	22	48
1,543	*700*	0.7	*0*	28.4	*12.9*	90-100[8]	2.38	*1.08*	1.32	*0.60*	11.0	2.0	27.7	17.0	*7.7*	23	23	50
1,764	*800*	0	*0*	23.1	*10.5*	100[8]	1.96	*0.89*	1.10	*0.50*	12.2	0	21.0	12.8	*5.8*	19	19	41
1,984	*900*	0	*0*	25.1	*11.4*	100[8]	2.18	*0.99*	1.21	*0.55*	13.3	0	22.8	13.9	*6.3*	21	21	44
2,205	*1000*	0	*0*	27.3	*12.4*	100[8]	2.31	*1.05*	1.32	*0.60*	14.4	0	24.8	15.2	*6.9*	22	22	48

[1]Adapted by the author from *Nutrient Requirements of Beef Cattle*, No. 4, 5th rev. ed., NRC-National Academy of Sciences, 1976, pp. 26-27.
[2]Average weight for a feeding period.
[3]Dry matter consumption, ME and TDN requirements are based on the general type of diet indicated in the roughage column.
[4]Approximately 0.4 ± 0.1 kg of weight gain/day over the last third of pregnancy is accounted for by the products of conception. These nutrients and energy requirements include the quantities estimated as necessary for conceptus development.
[5]Average quality roughage containing about 1.9 to 2.0 Mcal ME/kg dry matter.
[6]5.0 ± 0.5 kg of milk/day. Nutrients and energy for maintenance of the cow and for milk production are included in these requirements.
[7]10 ± 0.5 kg of milk/day. Nutrients and energy for maintenance of the cow and for milk production are included in these requirements.
[8]Good quality roughage containing at least 2.0 Mcal ME/kg dry matter.

TABLE 4-16

NUTRIENT REQUIREMENTS OF RATIONS FOR GROWING-FINISHING STEER CALVES AND YEARLINGS[1,2]
(In Percentage or Amount per Pound or Kilogram of Ration)

Weight[3]		Daily Gain		Moisture Basis (As-fed = est. 90% dry matter. M-F = moisture-free)	Daily Feed Consumption[4]		Roughage[4]	Total Protein	Digestible Protein	TDN[5,6]	ME[5] Mcal per		NEm[5] Mcal per		NEg[5] Mcal per		Ca	P
(lb)	*(kg)*	*(lb)*	*(kg)*		*(lb)*	*(kg)*	*(%)*	*(%)*	*(%)*	*(%)*	*(lb)*	*(kg)*	*(lb)*	*(kg)*	*(lb)*	*(kg)*	*(%)*	*(%)*
220	100	0	0	As-fed	5.1	2.3	100	7.8	4.5	50	0.82	1.8	0.48	1.05	—	—	0.16	0.16
				M-F	*4.6*	*2.1*	*100*	*8.7*	*5.0*	*55*	*0.91*	*2.0*	*0.53*	*1.17*	—	—	*0.18*	*0.18*
		1.1	0.5	As-fed	7.1	3.2	70-80	11.2	7.5	56	0.90	2.0	0.54	1.22	0.21	0.46	0.43	0.34
				M-F	*6.4*	*2.9*	*70-80*	*12.4*	*8.3*	*62*	*1.00*	*2.2*	*0.60*	*1.35*	*0.23*	*0.51*	*0.48*	*0.38*
		1.5	0.7	As-fed	6.7	3.0	50-60	13.3	9.6	63	1.02	2.3	0.64	1.44	0.39	0.86	0.63	0.43
				M-F	*6.0*	*2.7*	*50-60*	*14.8*	*10.7*	*70*	*1.13*	*2.5*	*0.71*	*1.60*	*0.43*	*0.95*	*0.70*	*0.48*
		2.0	0.9	As-fed	6.9	3.1	25-30	14.8	10.6	69	1.14	2.5	0.74	1.63	0.49	1.06	0.77	0.51
				M-F	*6.2*	*2.8*	*25-30*	*16.4*	*11.8*	*77*	*1.27*	*2.8*	*0.82*	*1.81*	*0.54*	*1.18*	*0.86*	*0.57*
		2.4	1.1	As-fed	6.7	3.0	<15	16.4	12.0	77	1.27	2.8	0.85	1.86	0.56	1.23	0.94	0.63
				M-F	*6.0*	*2.7*	*<15*	*18.2*	*13.3*	*86*	*1.41*	*3.1*	*0.94*	*2.07*	*0.62*	*1.37*	*1.04*	*0.70*
331	150	0	0	As-fed	6.9	3.1	100	7.8	4.5	50	0.82	1.8	0.48	1.05	—	—	0.16	0.16
				M-F	*6.2*	*2.8*	*100*	*8.7*	*5.0*	*55*	*0.91*	*2.0*	*0.53*	*1.17*	—	—	*0.18*	*0.18*
		1.1	0.5	As-fed	9.8	4.4	70-80	9.9	6.3	56	0.90	2.0	0.54	1.22	0.21	0.46	0.32	0.29
				M-F	*8.8*	*4.0*	*70-80*	*11.0*	*7.0*	*62*	*1.00*	*2.2*	*0.60*	*1.35*	*0.23*	*0.51*	*0.35*	*0.32*
		1.5	0.7	As-fed	9.6	4.3	50-60	11.3	7.6	63	1.02	2.3	0.64	1.44	0.39	0.86	0.41	0.32
				M-F	*8.6*	*3.9*	*50-60*	*12.6*	*8.5*	*70*	*1.13*	*2.5*	*0.71*	*1.60*	*0.43*	*0.95*	*0.46*	*0.36*
		2.0	0.9	As-fed	9.0	4.2	25-30	12.7	8.7	69	1.14	2.5	0.74	1.63	0.49	1.06	0.55	0.40
				M-F	*8.4*	*3.8*	*25-30*	*14.1*	*9.7*	*77*	*1.27*	*2.8*	*0.82*	*1.81*	*0.54*	*1.18*	*0.61*	*0.45*
		2.4	1.1	As-fed	9.1	4.1	<15	14.0	10.0	77	1.27	2.8	0.85	1.86	0.56	1.23	0.68	0.49
				M-F	*8.2*	*3.7*	*<15*	*15.6*	*11.1*	*86*	*1.41*	*3.1*	*0.94*	*2.07*	*0.62*	*1.37*	*0.76*	*0.54*
441	200	0	0	As-fed	8.6	3.9	100	7.6	4.3	50	0.82	1.8	0.48	1.05	—	—	0.16	0.16
				M-F	*7.7*	*3.5*	*100*	*8.5*	*4.8*	*55*	*0.91*	*2.0*	*0.53*	*1.17*	—	—	*0.18*	*0.18*
		1.1	0.5	As-fed	14.2	6.4	80-90	8.9	5.4	52	0.86	1.9	0.50	1.13	0.24	0.54	0.22	0.20
				M-F	*12.8*	*5.8*	*80-90*	*9.9*	*6.0*	*58*	*0.95*	*2.1*	*0.56*	*1.25*	*0.27*	*0.60*	*0.24*	*0.22*
		1.5	0.7	As-fed	14.0	6.3	70-80	9.7	6.1	58	0.94	2.1	0.58	1.26	0.32	0.70	0.29	0.25
				M-F	*12.6*	*5.7*	*70-80*	*10.8*	*6.8*	*64*	*1.04*	*2.3*	*0.64*	*1.40*	*0.35*	*0.78*	*0.32*	*0.28*
		2.0	0.9	As-fed	12.0	5.4	35-45	11.1	7.4	68	1.10	2.4	0.70	1.53	0.45	0.99	0.42	0.33
				M-F	*10.8*	*4.0*	*35-45*	*12.3*	*8.2*	*75*	*1.22*	*2.7*	*0.78*	*1.70*	*0.50*	*1.10*	*0.47*	*0.37*
		2.4	1.1	As-fed	11.2	5.1	<15	12.2	8.4	77	1.27	2.8	0.85	1.86	0.56	1.23	0.53	0.39
				M-F	*10.1*	*4.6*	*<15*	*13.6*	*9.3*	*86*	*1.41*	*3.1*	*0.94*	*2.07*	*0.62*	*1.37*	*0.59*	*0.43*
551	250	0	0	As-fed	10.8	4.6	100	7.6	4.3	50	0.82	1.8	0.48	1.05	—	—	0.16	0.16
				M-F	*9.7*	*4.1*	*100*	*8.5*	*4.8*	*55*	*0.91*	*2.0*	*0.53*	*1.17*	—	—	*0.18*	*0.18*
		1.5	0.7	As-fed	14.2	6.4	55-65	9.6	6.0	63	1.02	2.3	0.64	1.40	0.39	0.86	0.28	0.25
				M-F	*12.8*	*5.8*	*55-65*	*10.7*	*6.7*	*70*	*1.13*	*2.5*	*0.71*	*1.56*	*0.43*	*0.95*	*0.31*	*0.28*
		2.0	0.9	As-fed	15.2	6.9	45-50	10.0	6.2	65	1.06	2.3	0.67	1.48	0.41	0.92	0.32	0.28
				M-F	*13.7*	*6.2*	*45-50*	*11.1*	*7.1*	*72*	*1.18*	*2.6*	*0.74*	*1.64*	*0.46*	*1.02*	*0.35*	*0.31*
		2.4	1.1	As-fed	14.7	6.7	20-25	10.9	5.6	69	1.14	2.5	0.74	1.63	0.49	1.06	0.39	0.32
				M-F	*13.2*	*6.0*	*20-25*	*12.1*	*8.0*	*77*	*1.27*	*2.8*	*0.82*	*1.81*	*0.54*	*1.18*	*0.43*	*0.35*
		2.9	1.3	As-fed	14.7	6.7	<15	11.4	7.6	77	1.27	2.8	0.85	1.86	0.56	1.23	0.45	0.34
				M-F	*13.2*	*6.0*	*<15*	*12.7*	*8.5*	*86*	*1.41*	*3.1*	*0.94*	*2.07*	*0.62*	*1.37*	*0.50*	*0.38*
661	300	0	0	As-fed	11.6	5.2	100	7.7	4.3	50	0.82	1.8	0.48	1.05	—	—	0.16	0.16
				M-F	*10.4*	*4.7*	*100*	*8.6*	*4.8*	*55*	*0.91*	*2.0*	*0.53*	*1.17*	—	—	*0.18*	*0.18*
		2.0	0.9	As-fed	19.9	9.0	55-65	9.0	5.6	63	1.06	2.3	0.64	1.40	0.39	0.86	0.24	0.21
				M-F	*17.9*	*8.1*	*55-65*	*10.0*	*6.2*	*70*	*1.18*	*2.5*	*0.71*	*1.56*	*0.43*	*0.95*	*0.27*	*0.23*
		2.4	1.1	As-fed	18.7	8.4	20-25	9.7	6.1	69	1.14	2.5	0.74	1.63	0.49	1.06	0.30	0.26
				M-F	*16.8*	*7.6*	*20-25*	*10.8*	*6.8*	*77*	*1.27*	*2.8*	*0.82*	*1.81*	*0.54*	*1.18*	*0.33*	*0.29*

Footnotes on last page of table.

(Continued)

TABLE 4-16 (Continued)

Weight[3]		Daily Gain		Moisture Basis (As-fed = est. 90% dry matter. M-F = moisture-free)	Daily Feed Consumption[4]		Roughage[4]	Total Protein	Digestible Protein	TDN[5,6]	ME[5] Mcal per		NEm[5] Mcal per		NEg[5] Mcal per		Ca	P
(lb)	(kg)	(lb)	(kg)		(lb)	(kg)	(%)	(%)	(%)	(%)	(lb)	(kg)	(lb)	(kg)	(lb)	(kg)	(%)	(%)
		2.9	1.3	As-fed	17.3	7.9	<15	10.5	6.8	75	1.22	2.7	0.81	1.78	0.53	1.18	0.37	0.29
				M-F	*15.6*	*7.1*	*<15*	*11.7*	*7.6*	*83*	*1.36*	*3.0*	*0.90*	*1.98*	*0.59*	*1.31*	*0.41*	*0.32*
		3.1[7]	1.4[7]	As-fed	17.9	8.1	<15	10.7	7.0	77	1.27	2.8	0.85	1.86	0.56	1.23	0.38	0.31
				M-F	*16.1*	*7.3*	*<15*	*11.9*	*7.8*	*86*	*1.41*	*3.1*	*0.94*	*2.07*	*0.62*	*1.37*	*0.42*	*0.34*
772	350	0	0	As-fed	13.0	5.9	100	7.6	4.3	50	0.82	1.8	0.48	1.05	—	—	0.16	0.16
				M-F	*11.7*	*5.3*	*100*	*8.5*	*4.8*	*55*	*0.91*	*2.0*	*0.53*	*1.17*	—	—	*0.18*	*0.18*
		2.0	0.9	As-fed	19.6	8.9	45-55	9.0	5.5	65	1.06	2.3	0.67	1.48	0.41	0.92	0.22	0.20
				M-F	*17.6*	*8.0*	*45-55*	*10.0*	*6.1*	*72*	*1.18*	*2.6*	*0.74*	*1.64*	*0.46*	*1.02*	*0.25*	*0.22*
		2.4	1.1	As-fed	19.6	8.9	20-25	9.4	5.8	72	1.14	2.5	0.74	1.63	0.49	1.06	0.26	0.22
				M-F	*17.6*	*8.0*	*20-25*	*10.4*	*6.5*	*80*	*1.27*	*2.8*	*0.82*	*1.81*	*0.54*	*1.18*	*0.29*	*0.25*
		2.9	1.3	As-fed	19.6	8.9	<15	9.7	6.2	75	1.22	2.7	0.81	1.78	0.53	1.18	0.29	0.25
				M-F	*17.6*	*8.0*	*<15*	*10.8*	*6.9*	*83*	*1.36*	*3.0*	*0.90*	*1.98*	*0.59*	*1.31*	*0.32*	*0.28*
		3.1[7]	1.4[7]	As-fed	20.1	9.1	<15	9.8	6.3	77	1.27	2.8	0.88	1.86	0.56	1.23	0.31	0.26
				M-F	*18.1*	*8.2*	*<15*	*10.9*	*7.0*	*86*	*1.41*	*3.1*	*0.98*	*2.07*	*0.62*	*1.37*	*0.34*	*0.29*
882	400	0	0	As-fed	14.4	6.6	100	7.6	4.3	50	0.82	1.8	0.48	1.05	—	—	0.16	0.16
				M-F	*13.0*	*5.9*	*100*	*8.5*	*4.8*	*55*	*0.91*	*2.0*	*0.53*	*1.17*	—	—	*0.18*	*0.18*
		2.2	1.0	As-fed	23.0	10.4	45-55	8.5	5.1	65	1.06	2.3	0.67	1.48	0.41	0.92	0.20	0.19
				M-F	*20.7*	*9.4*	*45-55*	*9.4*	*5.7*	*72*	*1.18*	*2.6*	*0.74*	*1.64*	*0.46*	*1.02*	*0.22*	*0.21*
		2.6	1.2	As-fed	20.8	9.4	20-25	9.2	5.7	72	1.14	2.5	0.74	1.63	0.49	1.06	0.24	0.22
				M-F	*18.7*	*8.5*	*20-25*	*10.2*	*6.3*	*80*	*1.27*	*2.8*	*0.82*	*1.81*	*0.54*	*1.18*	*0.27*	*0.25*
		2.9	1.3	As-fed	21.1	9.6	<15	9.4	5.8	77	1.27	2.8	0.88	1.86	0.56	1.23	0.26	0.23
				M-F	*19.0*	*8.6*	*<15*	*10.4*	*6.5*	*86*	*1.41*	*3.1*	*0.98*	*2.07*	*0.62*	*1.37*	*0.29*	*0.26*
		3.1[7]	1.4[7]	As-fed	22.0	10.0	<15	9.4	5.9	77	1.27	2.8	0.88	1.86	0.56	1.23	0.26	0.23
				M-F	*19.8*	*9.0*	*<15*	*10.5*	*6.6*	*86*	*1.41*	*3.1*	*0.98*	*2.07*	*0.62*	*1.37*	*0.29*	*0.26*
992	450	0	0	As-fed	15.7	7.1	100	7.6	4.3	50	0.82	1.8	0.48	1.05	—	—	0.16	0.16
				M-F	*14.1*	*6.4*	*100*	*8.5*	*4.8*	*55*	*0.91*	*2.0*	*0.53*	*1.17*	—	—	*0.18*	*0.18*
		2.2	1.0	As-fed	25.2	11.4	45-55	8.4	5.0	65	1.06	2.3	0.67	1.48	0.41	0.92	0.17	0.17
				M-F	*22.7*	*10.3*	*45-55*	*9.3*	*5.5*	*72*	*1.18*	*2.6*	*0.74*	*1.64*	*0.46*	*1.02*	*0.19*	*0.19*
		2.6	1.2	As-fed	25.0	11.3	20-25	8.6	5.1	72	1.14	2.5	0.74	1.63	0.49	1.06	0.21	0.20
				M-F	*22.5*	*10.2*	*20-25*	*9.5*	*5.7*	*80*	*1.27*	*2.8*	*0.82*	*1.81*	*0.54*	*1.18*	*0.23*	*0.22*
		2.9	1.3	As-fed	22.8	10.3	<15	9.4	5.7	77	1.27	2.8	0.88	1.86	0.56	1.18	0.23	0.22
				M-F	*20.5*	*9.3*	*<15*	*10.4*	*6.3*	*86*	*1.41*	*3.1*	*0.98*	*2.07*	*0.62*	*1.31*	*0.26*	*0.25*
		3.1[7]	1.4[7]	As-fed	24.0	10.9	<15	9.0	5.5	77	1.27	2.8	0.88	1.86	0.56	1.23	0.23	0.21
				M-F	*21.6*	*9.8*	*<15*	*10.0*	*6.1*	*86*	*1.41*	*3.1*	*0.98*	*2.07*	*0.62*	*1.37*	*0.26*	*0.23*
1102	500	0	0	As-fed	17.1	7.8	100	7.6	4.3	50	0.82	1.8	0.48	1.05	—	—	0.16	0.16
				M-F	*15.4*	*7.0*	*100*	*8.5*	*4.8*	*55*	*0.91*	*2.0*	*0.53*	*1.17*	—	—	*0.18*	*0.18*
		2.0	0.9	As-fed	25.7	11.7	45-55	8.2	4.8	65	1.06	2.3	0.67	1.48	0.41	0.92	0.16	0.16
				M-F	*23.1*	*10.5*	*45-55*	*9.1*	*5.3*	*72*	*1.18*	*2.6*	*0.74*	*1.64*	*0.46*	*1.02*	*0.18*	*0.18*
		2.4	1.1	As-fed	25.4	11.6	20-25	8.3	5.0	72	1.14	2.5	0.74	1.63	0.49	1.06	0.17	0.17
				M-F	*22.9*	*10.4*	*20-25*	*9.2*	*5.5*	*80*	*1.27*	*2.8*	*0.82*	*1.81*	*0.54*	*1.18*	*0.19*	*0.19*
		2.6	1.2	As-fed	23.6	10.7	<15	9.0	5.4	77	1.27	2.8	0.88	1.86	0.56	1.18	0.20	0.20
				M-F	*21.2*	*9.6*	*<15*	*10.0*	*6.0*	*86*	*1.41*	*3.1*	*0.98*	*2.07*	*0.62*	*1.31*	*0.22*	*0.22*
		2.9[7]	1.3[7]	As-fed	24.4	11.1	<15	8.7	5.4	77	1.27	2.8	0.88	1.86	0.56	1.23	0.20	0.20
				M-F	*22.0*	*10.0*	*<15*	*9.7*	*6.0*	*86*	*1.41*	*3.1*	*0.98*	*2.07*	*0.62*	*1.37*	*0.22*	*0.22*

[1]Adapted by the author from *Nutrient Requirements of Beef Cattle*, No. 4, 5th rev. ed., NRC-National Academy of Sciences, 1976, pp. 28-29.
[2]The concentration of vitamin A in all diets for finishing steers is 2,200 IU/kg of dry diet.
[3]Average weight for a feeding period.
[4]Dry matter consumption, ME and TDN allowances are based on NE requirements and the general type of diet indicated in the roughage column. Most roughages will contain 1.9 to 2.2 Mcal of ME/kg dry matter, and 90-100% concentrate diets are expected to contain 3.1 to 3.3 Mcal of ME/kg.
[5]Due to conversion and rounding variation, the figures in these columns may not be in exact agreement with a similar energy concentration figure calculated from the data of Table 4-13.
[6]TDN was calculated by assuming 3.6155 Mcal of ME per kilogram of TDN.
[7]Most steers of the weight indicated, and not exhibiting compensatory growth, will fail to sustain an energy intake necessary to maintain this rate of gain for an extended period.

TABLE 4-17

NUTRIENT REQUIREMENTS OF RATIONS FOR GROWING-FINISHING HEIFER CALVES AND YEARLINGS[1,2]
(In Percentage or Amount per Pound or Kilogram of Ration)

Weight[3]		Daily Gain		Moisture Basis (As-fed = est. 90% dry matter. M-F = moisture-free)	Daily Feed Consumption[4]		Roughage[4]	Total Protein	Digestible Protein	TDN[5,6]	ME[5] Mcal per		NEm[5] Mcal per		NEg[5] Mcal per		Ca	P
(lb)	(kg)	(lb)	(kg)		(lb)	(kg)	(%)	(%)	(%)	(%)	(lb)	(kg)	(lb)	(kg)	(lb)	(kg)	(%)	(%)
220	100	0	0	As-fed	5.1	2.3	100	7.8	4.5	50	0.82	1.8	0.48	1.05	—	—	0.16	0.16
				M-F	*4.6*	*2.1*	*100*	*8.7*	*5.0*	*55*	*0.91*	*2.0*	*0.53*	*1.17*	—	—	*0.18*	*0.18*
		1.1	0.5	As-fed	7.3	3.3	70-80	11.2	7.5	55	0.90	2.0	0.54	1.22	0.29	0.63	0.42	0.33
				M-F	*6.6*	*3.0*	*70-80*	*12.4*	*8.3*	*61*	*1.00*	*2.2*	*0.60*	*1.32*	*0.32*	*0.70*	*0.47*	*0.37*
		1.5	0.7	As-Fed	7.1	3.2	50-60	13.0	9.0	62	1.02	2.3	0.64	1.40	0.39	0.86	0.59	0.43
				M-F	*6.4*	*2.9*	*50-60*	*14.4*	*10.0*	*69*	*1.13*	*2.5*	*0.71*	*1.56*	*0.43*	*0.95*	*0.66*	*0.48*
		2.0	0.9	As-fed	7.3	3.3	25-30	14.3	10.2	69	1.14	2.5	0.74	1.63	0.49	1.06	0.72	0.51
				M-F	*6.6*	*3.0*	*25-30*	*15.9*	*11.3*	*77*	*1.27*	*2.8*	*0.82*	*1.81*	*0.54*	*1.18*	*0.80*	*0.57*
		2.4	1.1	As-fed	7.3	3.3	<15	16.0	11.7	77	1.27	2.8	0.85	1.86	0.56	1.23	0.87	0.57
				M-F	*6.6*	*3.0*	*<15*	*17.8*	*13.0*	*86*	*1.41*	*3.1*	*0.94*	*2.07*	*0.62*	*1.37*	*0.97*	*0.63*
331	150	0	0	As-fed	6.9	3.1	100	7.8	4.5	50	0.82	1.8	0.48	1.05	—	—	0.16	0.16
				M-F	*6.2*	*2.8*	*100*	*8.7*	*5.0*	*55*	*0.91*	*2.0*	*0.53*	*1.17*	—	—	*0.18*	*0.18*
		1.1	0.5	As-fed	10.0	4.5	70-80	9.9	6.4	55	0.90	2.0	0.54	1.22	0.29	0.63	0.31	0.26
				M-F	*9.0*	*4.1*	*70-80*	*11.0*	*7.1*	*61*	*1.00*	*2.2*	*0.60*	*1.32*	*0.32*	*0.70*	*0.34*	*0.29*
		1.5	0.7	As-fed	9.8	4.4	50-60	11.2	7.4	62	1.02	2.3	0.64	1.40	0.39	0.86	0.41	0.32
				M-F	*8.8*	*4.0*	*50-60*	*12.4*	*8.2*	*69*	*1.13*	*2.5*	*0.71*	*1.56*	*0.43*	*0.95*	*0.45*	*0.35*
		2.0	0.9	As-fed	9.8	4.4	25-30	12.2	8.3	69	1.14	2.5	0.74	1.63	0.49	1.06	0.51	0.38
				M-F	*8.8*	*4.0*	*25-30*	*13.5*	*9.2*	*77*	*1.27*	*2.8*	*0.82*	*1.81*	*0.54*	*1.18*	*0.57*	*0.42*
		2.4	1.1	As-fed	9.8	4.4	<15	13.5	9.4	77	1.27	2.8	0.85	1.86	0.56	1.23	0.63	0.45
				M-F	*8.8*	*4.0*	*<15*	*15.0*	*10.5*	*86*	*1.41*	*3.1*	*0.94*	*2.07*	*0.62*	*1.37*	*0.70*	*0.50*
441	200	0	0	As-fed	8.6	3.9	100	7.6	4.4	50	0.82	1.8	0.48	1.05	—	—	0.16	0.16
				M-F	*7.7*	*3.5*	*100*	*8.5*	*4.9*	*55*	*0.91*	*2.0*	*0.53*	*1.17*	—	—	*0.18*	*0.18*
		0.7	0.3	As-fed	13.2	6.0	100	8.2	4.9	50	0.82	1.8	0.48	1.05	0.21	0.45	0.16	0.16
				M-F	*11.9*	*5.4*	*100*	*9.1*	*5.4*	*55*	*0.91*	*2.0*	*0.53*	*1.17*	*0.23*	*0.50*	*0.18*	*0.18*
		1.1	0.5	As-fed	14.7	6.7	80-90	8.6	5.2	52	0.86	1.9	0.50	1.12	0.24	0.54	0.21	0.20
				M-F	*13.2*	*6.0*	*80-90*	*9.6*	*5.8*	*58*	*0.95*	*2.1*	*0.56*	*1.24*	*0.27*	*0.60*	*0.23*	*0.22*
		1.5	0.7	As-fed	14.7	6.7	70-80	9.2	5.8	58	0.94	2.1	0.58	1.26	0.35	0.78	0.27	0.24
				M-F	*13.2*	*6.0*	*70-80*	*10.2*	*6.5*	*64*	*1.04*	*2.3*	*0.64*	*1.40*	*0.39*	*0.86*	*0.30*	*0.27*
		2.0	0.9	As-fed	13.0	5.9	35-45	10.5	6.8	68	1.10	2.4	0.70	1.55	0.45	0.99	0.37	0.29
				M-F	*11.7*	*5.3*	*35-45*	*11.7*	*7.5*	*75*	*1.22*	*2.7*	*0.78*	*1.72*	*0.50*	*1.10*	*0.41*	*0.32*
		2.4	1.1	As-fed	12.2	5.6	<15	11.5	7.7	77	1.27	2.8	0.85	1.86	0.56	1.23	0.45	0.34
				M-F	*11.0*	*5.0*	*<15*	*12.8*	*8.6*	*86*	*1.41*	*3.1*	*0.94*	*2.07*	*0.62*	*1.37*	*0.50*	*0.38*
551	250	0	0	As-fed	10.0	4.5	100	7.6	4.4	50	0.82	1.8	0.48	1.05	—	—	0.16	0.16
				M-F	*9.0*	*4.1*	*100*	*8.5*	*4.9*	*55*	*0.91*	*2.0*	*0.53*	*1.17*	—	—	*0.18*	*0.18*
		0.7	0.3	As-fed	15.7	7.1	100	8.0	4.7	50	0.82	1.8	0.48	1.05	0.21	0.45	0.16	0.16
				M-F	*14.1*	*6.4*	*100*	*8.9*	*5.2*	*55*	*0.91*	*2.0*	*0.53*	*1.17*	*0.23*	*0.50*	*0.18*	*0.18*
		1.1	0.5	As-fed	15.9	7.2	80-90	8.6	5.1	52	0.86	1.9	0.50	1.12	0.24	0.54	0.18	0.18
				M-F	*14.3*	*6.5*	*80-90*	*9.5*	*5.7*	*58*	*0.95*	*2.1*	*0.56*	*1.24*	*0.27*	*0.60*	*0.20*	*0.20*
		1.5	0.7	As-fed	14.2	6.4	55-65	9.4	5.8	65	1.06	2.3	0.67	1.48	0.41	0.92	0.26	0.23
				M-F	*12.8*	*5.8*	*55-65*	*10.5*	*6.5*	*72*	*1.18*	*2.6*	*0.74*	*1.64*	*0.46*	*1.02*	*0.29*	*0.26*
		2.0	0.9	As-fed	14.4	6.6	35-45	10.1	6.4	69	1.14	2.5	0.74	1.63	0.49	1.06	0.32	0.26
				M-F	*13.0*	*5.9*	*35-45*	*11.1*	*7.1*	*77*	*1.27*	*2.8*	*0.82*	*1.81*	*0.54*	*1.18*	*0.36*	*0.29*
		2.4	1.1	As-fed	15.9	7.2	20-25	10.3	6.7	72	1.18	2.6	0.77	1.70	0.51	1.13	0.34	0.28
				M-F	*14.3*	*6.5*	*20-25*	*11.4*	*7.4*	*80*	*1.31*	*2.9*	*0.86*	*1.89*	*0.57*	*1.25*	*0.39*	*0.31*
		2.6	1.2	As-fed	15.4	7.0	<15	10.7	7.0	77	1.27	2.8	0.85	1.86	0.56	1.23	0.39	0.30
				M-F	*13.9*	*6.3*	*<15*	*11.9*	*7.8*	*86*	*1.41*	*3.1*	*0.94*	*2.07*	*0.62*	*1.37*	*0.43*	*0.33*
661	300	0	0	As-fed	11.6	5.2	100	7.7	4.4	50	0.82	1.8	0.48	1.05	—	—	0.16	0.16
				M-F	*10.4*	*4.7*	*100*	*8.6*	*4.9*	*55*	*0.91*	*2.0*	*0.53*	*1.17*	—	—	*0.18*	*0.18*
		0.7	0.3	As-fed	18.1	8.2	100	7.9	4.4	50	0.82	1.8	0.48	1.05	0.21	0.45	0.16	0.16
				M-F	*16.3*	*7.4*	*100*	*8.8*	*4.9*	*55*	*0.91*	*2.0*	*0.53*	*1.17*	*0.23*	*0.50*	*0.18*	*0.18*

Footnotes on last page of table.

(Continued)

TABLE 4-17 (Continued)

Weight[3]		Daily Gain		Moisture Basis (As-fed = est. 90% dry matter. M-F = moisture-free)	Daily Feed Consumption[4]		Roughage[4]	Total Protein	Digestible Protein	TDN[5, 6]	ME[5] Mcal per		NEm[5] Mcal per		NEg[5] Mcal per		Ca	P
(lb)	*(kg)*	*(lb)*	*(kg)*		*(lb)*	*(kg)*	*(%)*	*(%)*	*(%)*	*(%)*	*(lb)*	*(kg)*	*(lb)*	*(kg)*	*(lb)*	*(kg)*	*(%)*	*(%)*
		1.1	0.5	As-fed	18.1	8.2	80-90	8.3	4.9	55	0.90	2.0	0.54	1.22	0.29	0.63	0.17	0.17
				M-F	*16.3*	*7.4*	*80-90*	*9.2*	*5.4*	*61*	*1.00*	*2.2*	*0.60*	*1.32*	*0.32*	*0.70*	*0.19*	*0.19*
		1.5	0.7	As-fed	16.2	7.3	55-65	9.1	5.5	65	1.06	2.3	0.67	1.48	0.41	0.92	0.22	0.21
				M-F	*14.6*	*6.6*	*55-65*	*10.1*	*6.1*	*72*	*1.18*	*2.6*	*0.74*	*1.64*	*0.46*	*1.02*	*0.24*	*0.23*
		2.0	0.9	As-fed	16.7	7.6	35-45	9.4	5.8	69	1.14	2.5	0.74	1.63	0.49	1.06	0.25	0.22
				M-F	*15.0*	*6.8*	*35-45*	*10.4*	*6.5*	*77*	*1.27*	*2.8*	*0.82*	*1.81*	*0.54*	*1.18*	*0.28*	*0.25*
		2.4	1.1	As-fed	18.3	8.3	20-25	9.4	5.8	72	1.18	2.6	0.77	1.70	0.51	1.13	0.28	0.24
				M-F	*16.5*	*7.5*	*20-25*	*10.4*	*6.5*	*80*	*1.31*	*2.9*	*0.86*	*1.89*	*0.57*	*1.25*	*0.31*	*0.27*
		2.6	1.2	As-fed	17.7	8.0	<15	9.8	6.2	77	1.27	2.8	0.85	1.86	0.56	1.23	0.30	0.25
				M-F	*15.9*	*7.2*	*<15*	*10.9*	*6.9*	*86*	*1.41*	*3.1*	*0.94*	*2.07*	*0.62*	*1.37*	*0.33*	*0.28*
772	350	0	0	As-fed	13.0	5.9	100	7.6	4.3	50	0.82	1.8	0.48	1.05	—	—	0.16	0.16
				M-F	*11.7*	*5.3*	*100*	*8.5*	*4.8*	*55*	*0.91*	*2.0*	*0.53*	*1.17*	—	—	*0.18*	*0.18*
		0.7	0.3	As-fed	20.1	9.1	100	7.6	4.3	50	0.82	1.8	0.48	1.05	0.21	0.45	0.16	0.16
				M-F	*18.1*	*8.2*	*100*	*8.5*	*4.8*	*55*	*0.91*	*2.0*	*0.53*	*1.17*	*0.23*	*0.50*	*0.18*	*0.18*
		1.1	0.5	As-fed	20.3	9.2	80-90	7.8	4.6	55	0.90	2.0	0.54	1.22	0.29	0.63	0.16	0.16
				M-F	*18.3*	*8.3*	*80-90*	*8.7*	*5.1*	*61*	*1.00*	*2.2*	*0.60*	*1.32*	*0.32*	*0.70*	*0.18*	*0.18*
		1.5	0.7	As-fed	19.3	8.8	55-65	8.3	4.9	62	1.02	2.3	0.64	1.40	0.39	0.86	0.17	0.17
				M-F	*17.4*	*7.9*	*55-65*	*9.2*	*5.4*	*69*	*1.13*	*2.5*	*0.71*	*1.56*	*0.43*	*0.95*	*0.19*	*0.19*
		2.0	0.9	As-fed	19.9	9.0	35-45	8.6	5.1	68	1.10	2.4	0.70	1.55	0.45	0.99	0.19	0.19
				M-F	*17.9*	*8.1*	*35-45*	*9.5*	*5.7*	*75*	*1.22*	*2.7*	*0.78*	*1.72*	*0.50*	*1.10*	*0.21*	*0.21*
		2.4	1.1	As-fed	20.3	9.2	20-25	8.9	5.4	72	1.18	2.6	0.77	1.70	0.51	1.13	0.22	0.21
				M-F	*18.3*	*8.3*	*20-25*	*9.9*	*6.0*	*80*	*1.31*	*2.9*	*0.86*	*1.89*	*0.57*	*1.25*	*0.24*	*0.23*
		2.6[7]	1.2[7]	As-fed	19.9	9.0	<15	9.0	5.6	77	1.27	2.8	0.85	1.86	0.56	1.23	0.23	0.22
				M-F	*17.9*	*8.1*	*<15*	*10.0*	*6.2*	*86*	*1.41*	*3.1*	*0.94*	*2.07*	*0.62*	*1.37*	*0.26*	*0.25*
882	400	0	0	As-fed	14.4	6.6	100	7.6	4.3	50	0.82	1.8	0.48	1.05	—	—	0.16	0.16
				M-F	*13.0*	*5.9*	*100*	*8.5*	*4.8*	*55*	*0.91*	*2.0*	*0.53*	*1.17*	—	—	*0.18*	*0.18*
		0.7	0.3	As-fed	22.2	10.1	100	7.6	4.3	50	0.82	1.8	0.48	1.05	0.21	0.45	0.16	0.16
				M-F	*20.0*	*9.1*	*100*	*8.5*	*4.8*	*55*	*0.91*	*2.0*	*0.53*	*1.17*	*0.23*	*0.50*	*0.18*	*0.18*
		1.1	0.5	As-fed	20.8	9.4	70-80	7.9	4.6	58	0.94	2.1	0.58	1.26	0.32	0.70	0.16	0.16
				M-F	*18.7*	*8.5*	*70-80*	*8.8*	*5.1*	*64*	*1.04*	*2.3*	*0.64*	*1.40*	*0.35*	*0.78*	*0.18*	*0.18*
		1.5	0.7	As-fed	21.3	9.7	55-65	8.1	4.8	59	0.98	2.3	0.64	1.40	0.39	0.86	0.16	0.16
				M-F	*19.2*	*8.7*	*55-65*	*9.0*	*5.3*	*66*	*1.09*	*2.5*	*0.71*	*1.56*	*0.43*	*0.95*	*0.18*	*0.18*
		2.0	0.9	As-fed	20.6	9.3	20-25	8.5	5.0	69	1.14	2.5	0.74	1.63	0.49	1.06	0.18	0.18
				M-F	*18.5*	*8.4*	*20-25*	*9.4*	*5.6*	*77*	*1.27*	*2.8*	*0.82*	*1.81*	*0.54*	*1.18*	*0.20*	*0.20*
		2.4[7]	1.1[7]	As-fed	20.3	9.2	<15	8.7	5.3	77	1.27	2.8	0.85	1.86	0.56	1.23	0.21	0.20
				M-F	*18.3*	*8.3*	*<15*	*9.7*	*5.9*	*86*	*1.41*	*3.1*	*0.94*	*2.07*	*0.62*	*1.37*	*0.23*	*0.22*
992	450	0	0	As-fed	15.7	7.1	100	7.6	4.3	50	0.82	1.8	0.48	1.05	—	—	0.16	0.16
				M-F	*14.1*	*6.4*	*100*	*8.5*	*4.8*	*55*	*0.91*	*2.0*	*0.53*	*1.17*	—	—	*0.18*	*0.18*
		0.4	0.2	As-fed	21.3	9.7	100	7.6	4.2	50	0.82	1.8	0.48	1.05	0.21	0.45	0.16	0.16
				M-F	*19.2*	*8.7*	*100*	*8.5*	*4.7*	*55*	*0.91*	*2.0*	*0.53*	*1.17*	*0.23*	*0.50*	*0.18*	*0.18*
		1.1	0.5	As-fed	22.8	10.3	70-80	7.7	4.4	58	0.94	2.1	0.58	1.26	0.32	0.70	0.16	0.16
				M-F	*20.5*	*9.3*	*70-80*	*8.6*	*4.9*	*64*	*1.04*	*2.3*	*0.64*	*1.40*	*0.35*	*0.78*	*0.18*	*0.18*
		1.8	0.8	As-fed	22.3	10.1	35-45	8.1	4.8	68	1.10	2.4	0.70	1.55	0.45	0.99	0.16	0.16
				M-F	*20.1*	*9.1*	*35-45*	*9.0*	*5.3*	*75*	*1.22*	*2.7*	*0.78*	*1.72*	*0.50*	*1.10*	*0.18*	*0.18*
		2.2[7]	1.0[7]	As-fed	20.8	9.4	<15	8.6	5.0	77	1.27	2.8	0.85	1.86	0.56	1.23	0.20	0.20
				M-F	*18.7*	*8.5*	*<15*	*9.5*	*5.6*	*86*	*1.41*	*3.1*	*0.94*	*2.07*	*0.62*	*1.37*	*0.22*	*0.22*

[1]Adapted by the author from *Nutrient Requirements of Beef Cattle*, No. 4, 5th rev. ed., NRC-National Academy of Sciences, 1976, pp. 30-31.

[2]The concentration of vitamin A in all diets for finishing heifers is 2,200 IU/kg of dry diet.

[3]Average weight for a feeding period.

[4]Dry matter consumption, ME and TDN allowances are based on NE requirements and the general type of diet indicated in the roughage column. Most roughages will contain 1.9 to 2.2 Mcal of ME/kg dry matter, and 90-100% concentrate diets are expected to have 3.1 to 3.3 Mcal of ME/kg.

[5]Due to conversion and rounding variation, the figures in these columns may not be in exact agreement with a similar energy concentration figure calculated from the data of Table 4-14.

[6]TDN was calculated by assuming 3.6155 kcal of ME per gram of TDN.

[7]Most heifers of the weight indicated, and not exhibiting compensatory growth, will fail to sustain the energy intake necessary to maintain this rate of gain for an extended period.

TABLE 4-18

NUTRIENT REQUIREMENTS OF RATIONS FOR BEEF CATTLE BREEDING HERD[1,2]
(In Percentage or Amount per Pound or Kilogram of Ration)

Weight[3]		Daily Gain		Moisture Basis (As-fed = est. 90% dry matter. M-F = moisture-free)	Daily Feed Consumption[4]		Roughage[4]	Total Protein	Digestible Protein	TDN[5]	ME[5] Mcal per		NEm[5] Mcal per		NEg[5] Mcal per		Ca	P
(lb)	(kg)	(lb)	(kg)		(lb)	(kg)	(%)	(%)	(%)	(%)	(lb)	(kg)	(lb)	(kg)	(lb)	(kg)	(%)	(%)
Pregnant yearling heifers—Last third of pregnancy																		
716	325	0.9	0.4[6]	As-fed	16.1	7.3	100[7]	7.9	4.6	47	0.77	1.7	0.44	0.98	0.15	0.34	0.21	0.21
				M-F	*14.5*	*6.6*	*100[7]*	*8.8*	*5.1*	*52*	*0.86*	*1.9*	*0.49*	*1.09*	*0.17*	*0.38*	*0.23*	*0.23*
		1.3	0.6	As-fed	20.8	9.4	100	7.9	4.6	47	0.77	1.7	0.44	0.98	0.15	0.34	0.19	0.19
				M-F	*18.7*	*8.5*	*100*	*8.8*	*5.1*	*52*	*0.86*	*1.9*	*0.49*	*1.09*	*0.17*	*0.38*	*0.21*	*0.21*
		1.8	0.8	As-fed	23.0	10.4	85-100	8.1	4.8	52	0.86	1.9	0.50	1.12	0.24	0.54	0.21	0.19
				M-F	*20.7*	*9.4*	*85-100*	*9.0*	*5.3*	*58*	*0.95*	*2.1*	*0.56*	*1.24*	*0.27*	*0.60*	*0.23*	*0.21*
772	350	0.9	0.4[6]	As-fed	16.9	7.7	100	7.9	4.6	47	0.77	1.7	0.44	0.98	0.15	0.34	0.20	0.20
				M-F	*15.2*	*6.9*	*100*	*8.8*	*5.1*	*52*	*0.86*	*1.9*	*0.49*	*1.09*	*0.17*	*0.38*	*0.22*	*0.22*
		1.3	0.6	As-fed	21.8	9.9	100	7.9	4.6	47	0.77	1.7	0.44	0.98	0.15	0.34	0.19	0.19
				M-F	*19.6*	*8.9*	*100*	*8.8*	*5.1*	*52*	*0.86*	*1.9*	*0.49*	*1.09*	*0.17*	*0.38*	*0.21*	*0.21*
		1.8	0.8	As-fed	24.4	11.1	85-100	7.9	4.6	52	0.86	1.9	0.50	1.12	0.24	0.54	0.20	0.19
				M-F	*22.0*	*10.0*	*85-100*	*8.8*	*5.1*	*58*	*0.95*	*2.1*	*0.56*	*1.24*	*0.27*	*0.60*	*0.22*	*0.21*
827	375	0.9	0.4[6]	As-fed	17.7	8.0	100	7.8	4.5	47	0.77	1.7	0.44	0.98	0.15	0.34	0.19	0.19
				M-F	*15.9*	*7.2*	*100*	*8.7*	*5.0*	*52*	*0.86*	*1.9*	*0.49*	*1.09*	*0.17*	*0.38*	*0.21*	*0.21*
		1.3	0.6	As-fed	22.8	10.3	100	7.8	4.5	47	0.77	1.7	0.44	0.98	0.15	0.34	0.18	0.18
				M-F	*20.5*	*9.3*	*100*	*8.7*	*5.0*	*52*	*0.86*	*1.9*	*0.49*	*1.09*	*0.17*	*0.38*	*0.20*	*0.20*
		1.8	0.8	As-fed	26.9	12.2	85-100	7.8	4.5	50	0.82	1.8	0.48	1.05	0.21	0.45	0.18	0.18
				M-F	*24.2*	*11.0*	*85-100*	*8.7*	*5.0*	*55*	*0.91*	*2.0*	*0.53*	*1.17*	*0.23*	*0.50*	*0.20*	*0.20*
882	400	0.9	0.4[6]	As-fed	18.3	8.3	100	7.8	4.5	47	0.77	1.7	0.44	0.98	0.15	0.34	0.19	0.19
				M-F	*16.5*	*7.5*	*100*	*8.7*	*5.0*	*52*	*0.86*	*1.9*	*0.49*	*1.09*	*0.17*	*0.38*	*0.21*	*0.21*
		1.3	0.6	As-fed	23.8	10.8	100	7.8	4.5	47	0.77	1.7	0.44	0.98	0.15	0.34	0.18	0.18
				M-F	*21.4*	*9.7*	*100*	*8.7*	*5.0*	*52*	*0.86*	*1.9*	*0.49*	*1.09*	*0.17*	*0.38*	*0.20*	*0.20*
		1.8	0.8	As-fed	28.4	12.9	85-100	7.8	4.5	50	0.82	1.8	0.48	1.05	0.21	0.45	0.17	0.17
				M-F	*25.6*	*11.6*	*85-100*	*8.7*	*5.0*	*55*	*0.91*	*2.0*	*0.53*	*1.17*	*0.23*	*0.50*	*0.19*	*0.19*
937	425	0.9	0.4[6]	As-fed	19.1	8.7	100	7.9	4.6	47	0.77	1.7	0.44	0.98	0.15	0.34	0.18	0.18
				M-F	*17.2*	*7.8*	*100*	*8.8*	*5.1*	*52*	*0.86*	*1.9*	*0.49*	*1.09*	*0.17*	*0.38*	*0.20*	*0.20*
		1.3	0.6	As-fed	24.8	11.2	100	7.8	4.5	47	0.77	1.7	0.44	0.98	0.15	0.34	0.17	0.17
				M-F	*22.3*	*10.1*	*100*	*8.7*	*5.0*	*52*	*0.86*	*1.9*	*0.49*	*1.09*	*0.17*	*0.38*	*0.19*	*0.19*
		1.8	0.8	As-fed	29.7	13.4	85-100	7.8	4.5	50	0.82	1.8	0.48	1.05	0.21	0.45	0.16	0.16
				M-F	*26.7*	*12.1*	*85-100*	*8.7*	*5.0*	*55*	*0.91*	*2.0*	*0.53*	*1.17*	*0.23*	*0.50*	*0.18*	*0.18*
Dry pregnant mature cows—Middle third of pregnancy																		
772	350	0	0	As-fed	13.6	6.1	100[7]	5.3	2.5	47	0.77	1.7	0.44	0.98	—	—	0.16	0.16
				M-F	*12.2*	*5.5*	*100[7]*	*5.9*	*2.8*	*52*	*0.86*	*1.9*	*0.49*	*1.09*	—	—	*0.18*	*0.18*
882	400	0	0	As-fed	14.9	6.8	100	5.3	2.5	47	0.77	1.7	0.44	0.98	—	—	0.16	0.16
				M-F	*13.4*	*6.1*	*100*	*5.9*	*2.8*	*52*	*0.86*	*1.9*	*0.49*	*1.09*	—	—	*0.18*	*0.18*
992	450	0	0	As-fed	16.4	7.4	100	5.3	2.5	47	0.77	1.7	0.44	0.98	—	—	0.16	0.16
				M-F	*14.8*	*6.7*	*100*	*5.9*	*2.8*	*52*	*0.86*	*1.9*	*0.49*	*1.09*	—	—	*0.18*	*0.18*
1,102	500	0	0	As-fed	17.7	8.0	100	5.3	2.5	47	0.77	1.7	0.44	0.98	—	—	0.16	0.16
				M-F	*15.9*	*7.2*	*100*	*5.9*	*2.8*	*52*	*0.86*	*1.9*	*0.49*	*1.09*	—	—	*0.18*	*0.18*
1,213	550	0	0	As-fed	18.9	8.6	100	5.3	2.5	47	0.77	1.7	0.44	0.98	—	—	0.16	0.16
				M-F	*17.0*	*7.7*	*100*	*5.9*	*2.8*	*52*	*0.86*	*1.9*	*0.49*	*1.09*	—	—	*0.18*	*0.18*
1,323	600	0	0	As-fed	20.3	9.2	100	5.3	2.5	47	0.77	1.7	0.44	0.98	—	—	0.16	0.16
				M-F	*18.3*	*8.3*	*100*	*5.9*	*2.8*	*52*	*0.86*	*1.9*	*0.49*	*1.09*	—	—	*0.18*	*0.18*
1,433	650	0	0	As-fed	21.6	9.8	100	5.3	2.5	47	0.77	1.7	0.44	0.98	—	—	0.16	0.16
				M-F	*19.4*	*8.8*	*100*	*5.9*	*2.8*	*52*	*0.86*	*1.9*	*0.49*	*1.09*	—	—	*0.18*	*0.18*

Footnotes on last page of table.

(Continued)

TABLE 4-18 (Continued)

Weight[3]		Daily Gain		Moisture Basis (As-fed = est. 90% dry matter. M-F = moisture-free)	Daily Feed Consumption[4]		Roughage[4]	Total Protein	Digestible Protein	TDN[5]	ME[5] Mcal per		NE_m[5] Mcal per		NE_g[5] Mcal per		Ca	P
(lb)	(kg)	(lb)	(kg)		(lb)	(kg)	(%)	(%)	(%)	(%)	(lb)	(kg)	(lb)	(kg)	(lb)	(kg)	(%)	(%)
Dry pregnant mature cows—Last third of pregnancy																		
772	350	0.9[6]	0.4[6]	As-fed	16.9	7.7	100[7]	5.3	2.5	47	0.77	1.7	0.44	0.98	—	—	0.16	0.16
				M-F	*15.2*	*6.9*	*100*[7]	*5.9*	*2.8*	*52*	*0.86*	*1.9*	*0.49*	*1.09*	—	—	*0.18*	*0.18*
882	400	0.9	0.4	As-fed	18.3	8.3	100	5.3	2.5	47	0.77	1.7	0.44	0.98	—	—	0.16	0.16
				M-F	*16.5*	*7.5*	*100*	*5.9*	*2.8*	*52*	*0.86*	*1.9*	*0.49*	*1.09*	—	—	*0.18*	*0.18*
992	450	0.9	0.4	As-fed	19.8	9.0	100	5.3	2.5	47	0.77	1.7	0.44	0.98	—	—	0.16	0.16
				M-F	*17.8*	*8.1*	*100*	*5.9*	*2.8*	*52*	*0.86*	*1.9*	*0.49*	*1.09*	—	—	*0.18*	*0.18*
1,102	500	0.9	0.4	As-fed	21.0	9.6	100	5.3	2.5	47	0.77	1.7	0.44	0.98	—	—	0.16	0.16
				M-F	*18.9*	*8.6*	*100*	*5.9*	*2.8*	*52*	*0.86*	*1.9*	*0.49*	*1.09*	—	—	*0.18*	*0.18*
1,213	550	0.9	0.4	As-fed	22.3	10.1	100	5.3	2.5	47	0.77	1.7	0.44	0.98	—	—	0.16	0.16
				M-F	*20.1*	*9.1*	*100*	*5.9*	*2.8*	*52*	*0.86*	*1.9*	*0.49*	*1.09*	—	—	*0.18*	*0.18*
1,323	600	0.9	0.4	As-fed	23.8	10.8	100	5.3	2.5	47	0.77	1.7	0.44	0.98	—	—	0.16	0.16
				M-F	*21.4*	*9.7*	*100*	*5.9*	*2.8*	*52*	*0.86*	*1.9*	*0.49*	*1.09*	—	—	*0.18*	*0.18*
1,433	650	0.9	0.4	As-fed	25.0	11.3	100	5.3	2.5	47	0.77	1.7	0.44	0.98	—	—	0.16	0.16
				M-F	*22.5*	*10.2*	*100*	*5.9*	*2.8*	*52*	*0.86*	*1.9*	*0.49*	*1.09*	—	-	*0.18*	*0.18*
Cows nursing calves—Average milking ability[8]—First 3-4 months postpartum																		
772	350	—	—	As-fed	20.1	9.1	100[7]	8.3	4.9	47	0.77	1.7	0.44	0.98	—	—	0.26	0.26
				M-F	*18.1*	*8.2*	*100*[7]	*9.2*	*5.4*	*52*	*0.86*	*1.9*	*0.49*	*1.09*	—	—	*0.29*	*0.29*
882	400	—	—	As-fed	21.6	9.8	100	8.3	4.9	47	0.77	1.7	0.44	0.98	—	—	0.25	0.25
				M-F	*19.4*	*8.8*	*100*	*9.2*	*5.4*	*52*	*0.86*	*1.9*	*0.49*	*1.09*	—	—	*0.28*	*0.28*
992	450	—	—	As-fed	22.8	10.3	100	8.3	4.9	47	0.77	1.7	0.44	0.98	—	—	0.25	0.25
				M-F	*20.5*	*9.3*	*100*	*9.2*	*5.4*	*52*	*0.86*	*1.9*	*0.49*	*1.09*	—	—	*0.28*	*0.28*
1,102	500	—	—	As-fed	24.0	10.9	100	8.3	4.9	47	0.77	1.7	0.44	0.98	—	—	0.25	0.25
				M-F	*21.6*	*9.8*	*100*	*9.2*	*5.4*	*52*	*0.86*	*1.9*	*0.49*	*1.09*	—	—	*0.28*	*0.28*
1,213	550	—	—	As-fed	25.7	11.7	100	8.3	4.9	47	0.77	1.7	0.44	0.98	—	—	0.24	0.24
				M-F	*23.1*	*10.5*	*100*	*9.2*	*5.4*	*52*	*0.86*	*1.9*	*0.49*	*1.09*	—	—	*0.27*	*0.27*
1,323	600	—	—	As-fed	26.9	12.2	100	8.3	4.9	47	0.77	1.7	0.44	0.98	—	—	0.22	0.22
				M-F	*24.2*	*11.0*	*100*	*9.2*	*5.4*	*52*	*0.86*	*1.9*	*0.49*	*1.09*	—	—	*0.25*	*0.25*
1,433	650	—	—	As-fed	27.9	12.7	100	8.3	4.9	47	0.77	1.7	0.44	0.98	—	—	0.22	0.22
				M-F	*25.1*	*11.4*	*100*	*9.2*	*5.4*	*52*	*0.86*	*1.9*	*0.49*	*1.09*	—	—	*0.25*	*0.25*
Cows nursing calves—Superior milking ability[9]—First 3-4 months postpartum																		
772	350	—	—	As-fed	24.9	11.3	100[10]	9.8	5.8	50	0.82	1.8	0.48	1.05	—	—	0.40	0.35
				M-F	*22.4*	*10.2*	*100*[10]	*10.9*	*6.4*	*55*	*0.91*	*2.0*	*0.53*	*1.17*	—	—	*0.44*	*0.39*
882	400	—	—	As-fed	26.4	12.0	100	9.8	5.8	50	0.82	1.8	0.48	1.05	—	—	0.38	0.34
				M-F	*23.8*	*10.8*	*100*	*10.9*	*6.4*	*55*	*0.91*	*2.0*	*0.53*	*1.17*	—	—	*0.42*	*0.38*
992	450	—	—	As-fed	27.7	12.6	100	9.8	5.8	50	0.82	1.8	0.48	1.05	—	—	0.36	0.33
				M-F	*24.9*	*11.3*	*100*	*10.9*	*6.4*	*55*	*0.91*	*2.0*	*0.53*	*1.17*	—	—	*0.40*	*0.37*
1,102	500	—	—	As-fed	28.9	13.1	100	9.8	5.8	50	0.82	1.8	0.48	1.05	—	—	0.35	0.32
				M-F	*26.0*	*11.8*	*100*	*10.9*	*6.4*	*55*	*0.91*	*2.0*	*0.53*	*1.17*	—	—	*0.39*	*0.36*
1,213	550	—	—	As-fed	30.3	13.8	100	9.8	5.8	50	0.82	1.8	0.48	1.05	—	—	0.33	0.32
				M-F	*27.3*	*12.4*	*100*	*10.9*	*6.4*	*55*	*0.91*	*2.0*	*0.53*	*1.17*	—	—	*0.37*	*0.35*
1,323	600	—	—	As-fed	31.6	14.3	100	9.8	5.8	50	0.82	1.8	0.48	1.05	—	—	0.32	0.31
				M-F	*28.4*	*12.9*	*100*	*10.9*	*6.4*	*55*	*0.91*	*2.0*	*0.53*	*1.17*	—	—	*0.36*	*0.34*
1,433	650	—	—	As-fed	32.8	14.9	100	9.8	5.8	50	0.82	1.8	0.48	1.05	—	—	0.32	0.30
				M-F	*29.5*	*13.4*	*100*	*10.9*	*6.4*	*55*	*0.91*	*2.0*	*0.53*	*1.17*	—	—	*0.35*	*0.33*

Footnotes on last page of table.

(Continued)

TABLE 4-18 (Continued)

Weight[3]		Daily Gain		Moisture Basis (As-fed = est. 90% dry matter. M-F = moisture-free)	Daily Feed Consumption[4]		Roughage[4]	Total Protein	Digestible Protein	TDN[5]	ME[5] Mcal per		NEm[5] Mcal per		NEg[5] Mcal per		Ca	P
(lb)	*(kg)*	*(lb)*	*(kg)*		*(lb)*	*(kg)*	*(%)*	*(%)*	*(%)*	*(%)*	*(lb)*	*(kg)*	*(lb)*	*(kg)*	*(lb)*	*(kg)*	*(%)*	*(%)*
Bulls, growth and maintenance (moderate activity)																		
661	300	2.2	1.0	As-fed	21.6	9.8	70-75	9.1	5.7	58	0.94	2.1	0.58	1.26	0.32	0.70	0.28	0.23
				M-F	*19.4*	*8.8*	*70-75*	*10.2*	*6.3*	*64*	*1.04*	*2.3*	*0.64*	*1.40*	*0.35*	*0.78*	*0.31*	*0.26*
882	400	2.0	0.9	As-fed	26.9	12.2	70-75	8.5	5.0	58	0.94	2.1	0.58	1.26	0.32	0.70	0.19	0.19
				M-F	*24.2*	*11.0*	*70-75*	*9.4*	*5.6*	*64*	*1.04*	*2.3*	*0.64*	*1.40*	*0.35*	*0.78*	*0.21*	*0.21*
1,102	500	1.5	0.7	As-fed	29.9	13.6	80-85	7.9	4.6	55	0.90	2.0	0.54	1.22	0.29	0.63	0.16	0.16
				M-F	*26.9*	*12.2*	*80-85*	*8.8*	*5.1*	*61*	*1.00*	*2.2*	*0.60*	*1.32*	*0.32*	*0.70*	*0.18*	*0.18*
1,323	600	1.1	0.5	As-fed	29.3	13.3	80-85	7.9	4.5	55	0.90	2.0	0.54	1.22	0.29	0.63	0.16	0.16
				M-F	*26.4*	*12.0*	*80-85*	*8.8*	*5.0*	*61*	*1.00*	*2.2*	*0.60*	*1.32*	*0.32*	*0.70*	*0.18*	*0.18*
1,543	700	0.7	0.3	As-fed	31.6	14.3	90-100[10]	7.7	4.3	50	0.82	1.8	0.48	1.05	0.15	0.45	0.16	0.16
				M-F	*28.4*	*12.9*	*90-100*[10]	*8.5*	*4.8*	*55*	*0.91*	*2.0*	*0.53*	*1.17*	*0.17*	*0.50*	*0.18*	*0.18*
1,764	800	0.0	0.0	As-fed	25.7	11.7	100[10]	7.7	4.3	50	0.82	1.8	0.48	1.05	—	—	0.16	0.16
				M-F	*23.1*	*10.5*	*100*[10]	*8.5*	*4.8*	*55*	*0.91*	*2.0*	*0.53*	*1.17*	—	—	*0.18*	*0.18*
1,984	900	0.0	0.0	As-fed	27.9	12.7	100[10]	7.7	4.3	50	0.82	1.8	0.48	1.05	—	—	0.16	0.16
				M-F	*25.1*	*11.4*	*100*[10]	*8.5*	*4.8*	*55*	*0.91*	*2.0*	*0.53*	*1.17*	—	—	*0.18*	*0.18*
2,205	1,000	0.0	0.0	As-fed	30.3	13.8	100[10]	7.7	4.3	50	0.82	1.8	0.48	1.05	—	—	0.16	0.16
				M-F	*27.3*	*12.4*	*100*[10]	*8.5*	*4.8*	*55*	*0.91*	*2.0*	*0.53*	*1.17*	—	—	*0.18*	*0.18*

[1]Adapted by the author from *Nutrient Requirements of Beef Cattle*, No. 4, 5th rev. ed., NRC-National Academy of Sciences, 1976, pp. 32-33.
[2]The concentration of vitamin A in all diets for pregnant heifers and cows is 2,800 IU/kg dry diet; for lactating cows and breeding bulls, 3,900 IU/kg.
[3]Average weight for a feeding period.
[4]Dry matter consumption, ME and TDN requirements are based on the general type of diet indicated in the roughage column.
[5]Due to conversion and rounding variation, the figures in these columns may not be in exact agreement with a similar figure calculated from the data in Table 4-15.
[6]Approximately 0.4 ± 0.1 kg of weight gain/day over the last third of pregnancy is accounted for by the products of conception.
[7]Average quality roughage containing about 1.9-2.0 Mcal ME/kg dry matter.
[8]5.0 ± 0.5 kg of milk/day.
[9]10 ± 1 kg of milk/day.
[10]Good quality roughage containing 2.0 Mcal ME/kg dry matter.

TABLE 4-19

NET ENERGY REQUIREMENTS OF GROWING AND FINISHING BEEF CATTLE
(Megacalories per Animal per Day)[1]

Body Weight .... (lb) / (kg)		220 / 100	331 / 150	441 / 200	551 / 250	661 / 300	772 / 350	882 / 400	992 / 450	1,102 / 500
NE_m Required		2.43	3.30	4.10	4.84	5.55	6.24	6.89	7.52	8.14
Daily Gain		NE_g Required								
(lb)	*(kg)*									
Steers										
0.2	0.1	0.17	0.23	0.28	0.34	0.39	0.43	0.48	0.52	0.56
0.4	0.2	0.34	0.46	0.57	0.68	0.78	0.88	0.97	1.06	1.14
0.7	0.3	0.52	0.70	0.87	1.03	1.18	1.33	1.47	1.61	1.74
0.9	0.4	0.70	0.95	1.18	1.40	1.60	1.80	1.99	2.17	2.34
1.1	0.5	0.89	1.20	1.49	1.77	2.02	2.27	2.51	2.74	2.97
1.3	0.6	1.08	1.46	1.81	2.15	2.46	2.76	3.05	3.33	3.60
1.5	0.7	1.27	1.73	2.14	2.53	2.90	3.26	3.60	3.93	4.25
1.8	0.8	1.47	2.00	2.47	2.93	3.36	3.77	4.17	4.55	4.92
2.0	0.9	1.68	2.27	2.82	3.33	3.82	4.29	4.74	5.18	5.60
2.2	1.0	1.88	2.55	3.16	3.75	4.29	4.82	5.33	5.82	6.29
2.4	1.1	2.10	2.84	3.52	4.17	4.78	5.36	5.93	6.47	7.01
2.6	1.2	2.31	3.13	3.88	4.60	5.27	5.92	6.54	7.14	7.73
2.9	1.3	2.53	3.43	4.26	5.04	5.77	6.48	7.16	7.83	8.47
3.1	1.4	2.76	3.74	4.63	5.49	6.29	7.06	7.80	8.52	9.22
3.3	1.5	2.99	4.05	5.02	5.95	6.81	7.65	8.46	9.23	9.98
Heifers										
0.2	0.1	0.18	0.25	0.30	0.36	0.41	0.46	0.51	0.56	0.61
0.4	0.2	0.37	0.50	0.62	0.74	0.84	0.95	1.05	1.14	1.24
0.7	0.3	0.57	0.77	0.95	1.13	1.29	1.45	1.61	1.75	1.90
0.9	0.4	0.77	1.05	1.30	1.54	1.76	1.98	2.18	2.39	2.58
1.1	0.5	0.99	1.34	1.66	1.96	2.25	2.52	2.79	3.05	3.30
1.3	0.6	1.21	1.64	2.03	2.40	2.75	3.09	3.41	3.73	4.03
1.5	0.7	1.44	1.95	2.42	2.86	3.27	3.68	4.06	4.44	4.80
1.8	0.8	1.67	2.28	2.81	3.33	3.82	4.28	4.73	5.17	5.59
2.0	0.9	1.92	2.60	3.23	3.81	4.37	4.91	5.43	5.93	6.41
2.2	1.0	2.17	2.94	3.65	4.32	4.95	5.56	6.14	6.71	7.26
2.4	1.1	2.43	3.30	4.09	4.84	5.55	6.23	6.88	7.52	8.13
2.6	1.2	2.70	3.66	4.55	5.37	6.16	6.91	7.64	8.35	9.03
2.9	1.3	2.98	4.04	5.01	5.92	6.79	7.63	8.42	9.21	9.96
3.1	1.4	3.26	4.42	5.49	6.49	7.44	8.36	9.23	10.09	10.91
3.3	1.5	3.56	4.82	5.98	7.07	8.11	9.11	10.06	11.00	11.90

[1]Adapted by the author from *Nutrient Requirements of Beef Cattle*, No. 4, 5th rev. ed., NRC-National Academy of Sciences, 1976, p. 34.

Energy

Carbohydrates, which constitute about 75 percent of all the dry matter of plants, are the chief sources of energy of cattle feeds. Next to carbohydrates, fats are important as energy sources.

The first and most important function of feeds is that of meeting the maintenance needs. If there is not sufficient feed, as is frequently true during periods of drought or when winter rations are skimpy, the energy needs of the body are met by the breakdown of tissue. This results in loss of condition and body weight.

After the energy needs for body maintenance have been met, any surplus energy may be used for growth, finishing, reproduction, or lactation. When cattle are finished at early ages, growth and finishing are in most instances simultaneous, and, therefore, not easily separated.

In the finishing process, the percentage of protein, ash, and water steadily decreases as the animal matures and fattens, whereas the percentage of fat increases. Thus, the body of a calf at birth may contain about 70% water and 4% fat; whereas the body of a fat 2-year-old steer may contain only 45 to 50% water but from 30 to 35% fat. This storage of fat requires a liberal allowance of energy feeds.

Through bacterial action in the rumen, cattle are able to utilize a considerable portion of roughages as sources of energy. Yet it must be realized that with extremely bulky rations, the animal cannot consume sufficient quantities to produce the maximum amount of fat. For this reason, finishing rations contain a considerable proportion of concentrated feeds, mostly cereal grains. On the other hand, when the energy requirements are primarily for maintenance, roughages are usually the most economical sources of energy for beef cattle.

SYMPTOMS OF ENERGY DEFICIENCY (Underfeeding)

Lack of sufficient total feed is probably the most common deficiency suffered by beef cattle, although it is recognized that underfeeding is frequently complicated by a concomitant shortage of protein and other nutrients. Restricted rations often occur during periods of drought, when pastures or ranges are overstocked, or when winter rations are skimpy. Also, many range cattlemen regularly plan that cows in good flesh should lose some condition during the winter months; they feel that it is uneconomical to feed sufficiently to retain the fleshy condition. Fortunately, during such times of restricted feed intake, animals have nutritive reserves upon which they can draw. Although they may survive for a considerable period of time under these conditions, there is an inevitable loss in body weight and condition; and, varying with the degree of underfeeding, there may be a slowing or cessation of growth (including skeletal growth), failure to conceive, and increased mortality. Low feed intake also commonly results in increased deaths from toxic plants and from lowered resistance to parasites and diseases.

Protein

The protein allowance for beef cattle, regardless of age or system of production, should be ample to replace the daily breakdown of the tissues of the body including the growth of hair, horns, and hoofs. In general, the protein needs are greatest for the growth of the young calf and for the gestating-lactating cow.

With stocker cattle, or in the maintenance of the beef breeding herd, it usually does not pay to add a protein supplement when a legume hay is fed. With feedlot cattle on high-concentrate rations, or when the breeding herd is being wintered on a nonlegume roughage, sufficient protein supplement—usually 1 to 2 pounds daily—should be added to the ration.

SYMPTOMS OF PROTEIN DEFICIENCY

Depressed appetite is the primary symptom of protein deficiency in beef cattle rations. Depressed appetite may, in turn, lead to an inadequate intake of energy; hence, protein deficiency and energy deficiency often occur together.

Other symptoms of protein deficiency are loss of weight, poor growth, irregular or delayed estrus, and reduced milk production.

Minerals

Beef cattle are liable to the usual inefficiencies and ailments when exposed to (1) prolonged and severe mineral deficiencies, or (2) excesses of fluorine, selenium, or molybdenum (see Table 4-100, later in Section 4, for a summary of Nutritional Diseases and Ailments of Animals).

Needed minerals may be incorporated in beef cattle rations or in the water. In addition, it is recommended that all classes and ages of cattle be allowed free access to suitable minerals. Free-choice feeding is in the nature of cheap insurance, with the animals consuming the minerals if they are needed.

MINERAL REQUIREMENTS

The National Research Council mineral requirements of beef cattle are summarized in Table 4-20.

BEEF CATTLE MINERAL CHART

Table 4-21, Beef Cattle Mineral Chart, page 274, gives, in summary form, the following pertinent information relative to each mineral listed: (1) Conditions usually prevailing where deficiencies are reported, (2) function, (3) deficiency symptoms, (4) nutrient requirements, (5) recommended allowances, and (6) practical sources.

MINERAL SUPPLEMENTATION

When buying and home mixing minerals, or when buying commercial mineral mixes, the cattleman should first determine his needs, based on (1) available feeds, (2) area (for example, the Northern Great Plains and the Southwest are phosphorus-deficient areas), and (3) the age and reproduction status (pregnancy and lactation make a difference) of the animals for which the mineral mix is intended. Excesses and mineral imbalances should be avoided.

The mineral recommendations for all classes and

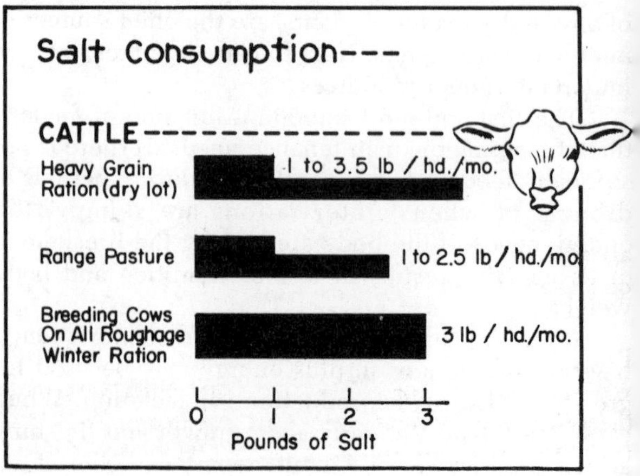

Fig. 4-17. The average salt consumption of cattle.

TABLE 4-20

MINERAL REQUIREMENTS OF RATIONS FOR BEEF CATTLE[1]
(In Percentage or Amount per Pound or Kilogram of Ration)

Mineral	Moisture Basis (As-fed = est. 90% dry matter. M-F = moisture-free)	Growing and Finishing Steers and Heifers		Dry Pregnant Cows		Breeding Bulls and Lactating Cows		Possible Toxic Levels[2]	
Major or macro minerals:									
Sodium (Na) .... (%)	As-fed *M-F*	0.05 *0.06*		0.05 *0.06*		0.05 *0.06*		— —	
Calcium (Ca)[3] .... (%)	As-fed *M-F*	0.16-0.94 *0.18-1.04*		0.16 *0.18*		0.16-0.40 *0.18-0.44*		—	
Phosphorus (P)[3] .... (%)	As-fed *M-F*	0.16-0.63 *0.18-0.70*		0.16 *0.18*		0.16-0.35 *0.18-0.39*		— —	
Magnesium (Mg) .... (%)	As-fed *M-F*	0.04-0.09 *0.04-0.10*		— —[4]		0.16 *0.18*		—	
Potassium (K) .... (%)	As-fed *M-F*	0.45-0.72 *0.6 -0.8*		— —[4]		— —[4]		—	
Sulfur (S) .... (%)	As-fed *M-F*	0.09 *0.1*		— —[4]		— —[4]		—	
Trace or micro minerals:		*(lb)*	*(kg)*	*(lb)*	*(kg)*	*(lb)*	*(kg)*	*(lb)*	*(kg)*
Cobalt .... (mg)	As-fed *M-F*	0.02-0.05 *0.02-0.05*	0.05-0.09 *0.05-0.10*	0.02-0.05 *0.02-0.05*	0.05-0.09 *0.05-0.10*	0.02-0.05 *0.02-0.05*	0.05-0.09 *0.05-0.10*	4.5-6.3 *5-7*	9-14 *10-15*
Copper (Cu) .... (mg)	As-fed *M-F*	1.8 *2*	3.6 *4*	— —[4]		— —[4]		46.8 *52*	104 *115*
Iodine (I) .... (mcg)	As-fed *M-F*	— —[5]		20.7-40.5 *23-45*	45-90 *50-100*	20.7-40.5 *23-45*	45-90 *50-100*	40.5 *45*	90 *100*
Iron (Fe) .... (mg)	As-fed *M-F*	4.5 *5*	9 *10*	— —[4]		— —[4]		164 *182*	360 *400*
Manganese (Mn) ...... (mg)	As-fed *M-F*	0.45-4.5 *0.50-5.0*	0.9-9.0 *1.0-10.0*	8.2 *9.1*	18.0 *20.0*	— —[4]		61 *68*	135 *150*
Selenium (Se) .... (mg)	As-fed *M-F*	0.05 *0.05*	0.09 *0.10*	0.02-0.05 *0.02-0.05*	0.05-0.09 *0.05-0.10*	0.02-0.05 *0.02-0.05*	0.05-0.09 *0.05-0.10*	1.8 *2*	4.5 *5*
Zinc (Zn) .... (mg)	As-fed *M-F*	8.1-12.6 *9-14*	18-27 *20-30*	— —[4]		— —[4]		368 *409*	811 *900*

[1]Adapted by the author from *Nutrient Requirements of Beef Cattle*, No. 4, 5th rev. ed., NRC-National Academy of Sciences, Washington, D.C., 1976, p. 35.
[2]The level of mineral that is toxic is at best an estimate and is dependent upon such factors as length of intake, availability of the mineral in the feedstuff or compound, and other mineral levels.
[3]See Tables 4-13 to 4-18 for more detailed data on requirements.
[4]Unknown. It is suggested that the level for the growing and finishing animal be used.
[5]Very small, but unknown.

ages of cattle, especially those fed unmixed rations or on pasture, are:

1. *Where animals are on liberal grain feeding*—Provide free access to a 2-compartment mineral box, with (a) trace mineralized salt in one side, and (b) in the other side, a mixture of ⅓ trace mineralized salt (salt included for purposes of palatability), ⅓ defluorinated phosphate or steamed bone meal, and ⅓ ground limestone or oystershell flour.

2. *Where animals are primarily on roughage (pasture, hay, and/or silage)*—Provide free access to a 2-compartment mineral box, with (a) trace mineralized salt in one side (salt included for purposes of palatability), and (b) in the other side, a mixture of ⅓ trace mineralized salt and ⅔ defluorinated phosphate or steamed bone meal.

Salt should always be available on a free-choice basis in addition to whatever mineral mix is provided.

TABLE 4-21—BEEF CATTLE

Minerals Which May Be Deficient Under Normal Conditions	Conditions Usually Prevailing Where Deficiencies Are Reported	Function of Mineral	Some Deficiency Symptoms
MAJOR OR MACRO MINERALS:			
Salt (sodium and chlorine—NaCl). The requirements for sodium and chlorine are commonly expressed as salt requirements because salt is an effective, economical way of supplementing diets with these elements.	Negligence; for salt is inexpensive.	Sodium chloride helps maintain osmotic pressure in body cells, upon which depends the transfer of nutrients to the cells and the removal of waste materials and the maintenance of water balance among the tissues. Also sodium is important in making bile, which aids in the digestion of fats and carbohydrates, and chlorine is required for the formation of hydrochloric acid in the gastric juice so vital to protein digestion. It is noteworthy that when salt is omitted, sodium expresses its deficiency first.	Intensive craving of salt, manifested by the animals chewing and licking various objects. Prolonged deficiency results in lack of appetite, unthrifty appearance, and decreased production. High producing milk cows may collapse and die when salt deficiency has been of long duration. Excessive salt intake can result in toxicity. But as much as 3 lb can be consumed per cow daily without harm provided animals have free access to plenty of water.
Calcium (Ca)	When finishing cattle are fed heavily on concentrates and limited quantities of legume roughage, especially young cattle on a long feed. Adding calcium to such a ration increases the rate of gain, improves feed utilization, results in heavier, stronger bones, and enhances market grades. When the diet consists chiefly of dried mature grasses or cereal straws. When cows are in heavy lactation.	Essential for development and maintenance of normal bones and teeth. Important in blood coagulation and lactation. Enables heart, nerves, and muscles to function. Regulates permeability of tissue cells. Affects availability of phosphorus and zinc.	Calcium deficiency in beef cattle is rare and mild; the symptoms are inconspicuous. With severe privation, there may be fractured bones, poor gains and bone development, and lower market grade.
Phosphorus (P)	Semiarid regions are commonly associated with soils deficient in phosphorus. The phosphorus content of plants generally decreases markedly with maturity, with the result that deficiencies often occur in cattle subsisting for long periods on mature dried forage.	Essential for sound bones and teeth, and for the assimilation of carbohydrates and fats. A vital ingredient of the proteins in all body cells. Necessary for enzyme activation. Acts as a buffer in blood and tissue. Occupies a key position in biologic oxidation, and reactions requiring energy.	Loss of appetite, poor gains, decreased milk production, decreased feed efficiency, depraved appetite—with special craving for chewing bones and eating soil. Lameness and stiffness of joints, broken bones. Rickets in young animals and osteomalacia, osteoporosis, and osteitis fibrosa in mature animals. Breeding problems, milk fever, retained afterbirth, and blindness. (Phosphorus is necessary to convert carotene to vitamin A.)

Footnotes on last page of table.

MINERAL CHART

Nutrient Requirements[1]				
Daily Nutrients/ Animal	Percentage of Rations	Recommended Allowances[1]	Practical Sources of the Mineral	Comments
Sodium requirement of young, growing calves is 2 to 3 g per head daily.	*As-fed*[2] *M-F* **NaCl:* 0.09 *0.10* **Na for growing calves:* 0.05 *0.05*	Cows on pasture or on high-roughage winter rations will consume from 1 to 3 lb (.45 to 1.36 kg) salt per head per month; finishing steers on heavy grain rations in drylot will consume 1 to 3.5 lb (.45 to 1.59 kg) per head per month; a wide range due to differences in age, rations, form of salt (rock vs block), and weather losses. Most ranchers compute the yearly salt requirements on the basis of 25 lb (11 kg) per cow. The careful location of the salt supply is an important adjunct in range management.	Salt should be available at all times. It should be both (1) self-fed, free-choice, and (2) mixed with other ration ingredients. Free access to salt in the form of loose, rock, or block salt. Cattle prefer loose salt to block salt, since it can be eaten more rapidly and with less effort. However, experiments with growing dairy heifers and lactating cows have shown fully as good results with block salt as with loose salt even though smaller quantities were consumed. This means that the additional intake of loose salt over block salt does not appear to benefit cattle. Commercial mineral mixes (in block, or loose form) may contain ⅓ or more salt.	The salt requirements of cattle differ (1) between individuals; (2) according to whether milk is produced (being higher for lactating cows than for dry cows, because of the salt in the milk); (3) from season to season; (4) according to the weathering losses to which the salt is subjected (being higher on pasture than in the drylot; exposed block salt loses about 15% per month); (5) between block and loose salt (animals often consuming twice as much easy-to-get loose salt as block salt); and (6) according to the salt content of the soil, feed, and water (being higher when vegetable proteins are fed, than when animal proteins are fed, higher on predominantly forage rations than on predominantly concentrate rations, and higher on lush early pasture than on more mature grasses). These are some of the reasons why free-choice feeding of salt is advocated.
Variable, according to class, age, and weight of cattle (see Tables 4-13 to 4-15).	*Variable according to class, age, and weight of cattle, but should be at least 0.18% (see Tables 4-16 to 4-18).	Free access to a calcium supplement, or 0.1 lb (45 g) of a calcium supplement added to the daily ration. Calf rations should contain a minimum of .4% calcium, .3% phosphorus, and 200 IU of vitamin D per pound (440 IU per kg).	Ground limestone, steamed bone meal, oystershell flour, dicalcium phosphate, or defluorinated phosphate; free-choice, or 0.1 lb (45 g) per head daily added to the ration. Where both calcium and phosphorus need to be supplemented, they should be provided in a readily available and palatable form such as dicalcium phosphate, defluorinated phosphate, or bone meal.	In addition to an adequate supply of calcium, proper utilization is dependent upon (1) a highly available source of the mineral, (2) a suitable ratio between calcium and phosphorus (somewhere between 1 to 2 parts of calcium to 1 part of phosphorus). Calcium-phosphorus ratios of 2:1 have been shown to be beneficial in reducing urinary calculi. When calculi problems are encountered, even higher levels of calcium may be advisable. Ratios between calcium and phosphorus of 7:1 have been reported to be satisfactory for cattle. Generally when cattle receive at least ⅓ of a legume forage, ample calcium will be provided. But even nonlegume forages contain more calcium than cereal grains. Plants grown on calcium-rich soils are high in calcium. Calcium availability of 70% is generally assumed for all feedstuffs.
*Variable, according to class, age, and weight of cattle (see Tables 4-13 to 4-15).	*High-energy rations should contain at least 0.22% phosphorus; other rations should contain at least 0.18% (see Tables 4-16 to 4-18).	Free access to a phosphorus supplement, or 0.1 lb (45 g) of a phosphorus supplement added to the daily ration. Where phosphorus is added to water, either of the following methods may be employed: 1. Added by hand at rate of ¼ oz of monosodium phosphate/8 gal water, or ¼ oz/head/day. 2. Added by dispenser, using stock solution of 2½ lb of monosodium phosphate/gal water (or 100 lb/40 gal water).	Dicalcium phosphate, defluorinated phosphate, monosodium phosphate, diammonium phosphate, steamed bone meal; free-choice, or 0.1 lb (45 g) per head daily added to the ration.	Grains, grain by-products, and high-protein supplements are fairly high in phosphorus; hence, rations high in such ingredients require little or no phosphorus supplementation. Calcium-phosphorus ratios of 2:1 are beneficial in reducing urinary calculi; and even higher levels of calcium may be necessary when urinary calculi is encountered. Ratios between calcium and phosphorus of 7:1 have been reported to be satisfactory for cattle.

(Continued)

TABLE 4-2

Minerals Which May Be Deficient Under Normal Conditions	Conditions Usually Prevailing Where Deficiencies Are Reported	Function of Mineral	Some Deficiency Symptoms
Magnesium (Mg)	When milk feeding of calves is prolonged without grain or hay. (Milk is rather low in magnesium.) Certain pastures early in the spring. Lactating cows are most commonly affected.	Essential for the bones and teeth, and required for various body processes. Aids in maintaining acid base equilibrium and in activating many enzyme systems. In cells, magnesium is present in far greater concentration than calcium.	Grass tetany or grass staggers, chara terized by anorexia, hyperemi hyperirritability, convulsions, ar death.
Potassium (K)	When drylot finishing cattle receive high or all-concentrate rations.	Essential for proper enzyme, muscle and nerve function, rumen microorganism activity, and appetite.	Poor appetite and feed conversion, slo growth, stiffness, and emaciation.
Sulfur (S)	In high-urea rations.	Essential for the synthesis of methionine.	Depressed appetite, loss of weight, po growth, irregular or delayed estrus, ar reduced milk production.
TRACE OR MICRO MINERALS: **Cobalt** (Co)	In cobalt-deficient areas (soils) where this element is not provided (in Fla., Mich., Wisc., Mass., N.H., Penn., and N.Y.).	Cobalt is an integral component of the vitamin B_{12} molecule, and vitamin B_{12} is synthesized by microorganisms in the reticulum.	Affected animals become wea emaciated, and eventually die. Oth symptoms include loss of appetit craving for hair and wood, scaliness skin, and sometimes diarrhea.
Copper (Cu)	In copper-deficient areas (soils), as in Florida and the Coastal Plain region. On peat and muck soils. Deficiencies have occurred in calves kept on an exclusive milk diet for long periods.	Copper, along with iron and vitamin B_{12}, is necessary for hemoglobin formation, although it forms no part of the hemoglobin molecule (or red blood cells). Copper is essential in enzyme systems, hair development and pigmentation, bone development, reproduction, and lactation.	Emaciation, depigmentation (cattle tu yellowish) and loss of hair, stunte growth, anemia, and brittle and ma formed bones. Also heat periods a suppressed, and there may be de praved appetite and diarrhea. Youn calves may have straight pasterns an stand forward on their toes.
Fluorine (F)	Feeding rock phosphate which has not been defluorinated and which may contain 3.5 to 4.0% fluorine, a toxic level.	No essential known function in beef cattle. But excessive levels of fluorine are harmful to cattle.	Teeth may erode and the enamel ma become mottled; the bones becom thickened and soft, and their breakin strength decreases; appetite is de creased, and slow growth results.

Footnotes on last page of table.

(Continued)

Nutrient Requirements[1]				
Daily Nutrients/ Animal	Percentage of Rations	Recommended Allowances[1]	Practical Sources of the Mineral	Comments
*5.5 to 13.6 mg/lb (12 to 30 mg/kg) body weight per day. In problem areas, up to 20 g of supplemental magnesium per head daily may be required to prevent grass tetany.	*For lactating cows, 0.16% as fed or 0.18% of moisture-free ration.		Commonly fed roughages and concentrates usually contain ample magnesium, but it may not be present in an available form. Magnesium sulfate or oxide may be used as a supplement.	Although grass tetany is attributed to magnesium deficiency, uncomplicated cases have been produced only on purified diets or by prolonged feeding of calves on milk. However, supplemental feeding of magnesium (6 to 20 g/day) reduces the incidence of grass tetany in many outbreaks.
	As-fed[2] *M-F* *0.5 to 0.7 *0.6 to 0.8*	0.8 to 1.0% of the total moisture-free ration.	Roughages usually contain ample potassium. Potassium chloride is the supplement of choice.	Grains often contain less than 0.5% potassium. Excessive levels of potassium have been found to interfere with magnesium absorption. Also, excessive levels of potassium, along with high levels of phosphorus, increase the incidence of phosphatic urinary calculi.
	*0.09 *0.1*	*3 g of inorganic sulfur to 100 g urea, or 1 part of inorganic sulfur to 15 parts of nonprotein nitrogen.	Organic forms of sulfur are most readily utilized, elemental sulfur is least so, and sulfates are intermediate in this respect.	Sulfur requirements are primarily those involving amino acid nutrition.
0.07 to 0.10 mg cobalt/100 lb (45.4 kg) body weight.	*Between 0.023 and 0.045 mg/lb *(0.05 and 0.10 mg/kg)* of moisture-free ration.	Free access to a cobaltized mineral mixture in cobalt-deficient areas.	A cobaltized mineral mixture may be prepared by adding cobalt at the rate of 0.2 oz per 100 lb (1.25 mg/kg) of salt as cobalt chloride or cobalt sulfate, cobalt carbonate, cobalt oxide; or a good commercial mineral mixture or sait product may be used.	Several good commercial cobalt-containing minerals are on the market. A vitamin B_{12} injection will relieve a cobalt deficiency. *Toxicity has been produced in calves by feeding 0.5 mg of cobalt per pound (1.2 mg/kg) body weight.
	*1.8 mg/lb *(4 mg/kg)* of total moisture-free ration when rations contain low levels of molybdenum and sulfate. When rations contain high levels of molybdenum and sulfate, copper requirement may be increased 2- or 3-fold.	*Copper deficiency can be prevented by adding 0.5% copper sulfate to salt fed free-choice. Copper (Cu) added to total feed (moisture-free basis) 8.0 ppm. Copper may also be injected as glycinate to meet the nutritional needs for the mineral.	Salt containing 0.25 to 0.5% copper sulfate or copper carbonate.	Copper-deficient cattle can be returned to normal by feeding 3 g of copper sulfate or blue vitriol every 10 days. An interesting interrelation exists between copper and molybdenum. An excess of molybdenum (in the presence of sulfate) causes a condition which can be cured only by administering copper. Excess copper is toxic; it accumulates in the liver, and death may result.
	**Safe fluorine levels: No more than* 45 mg/lb *(100 mg/kg)* of moisture-free ration for finishing cattle and *no more than* 18.2 mg/lb *(40 mg/kg)* for animals to be kept for breeding.			Fluorine is a cumulative poison; hence, the toxic effects may not be noticed for some time.

(Continued)

TABLE 4-2

Minerals Which May Be Deficient Under Normal Conditions	Conditions Usually Prevailing Where Deficiencies Are Reported	Function of Mineral	Some Deficiency Symptoms
Iodine (I)	In iodine-deficient areas (soils) where iodized salt is not fed (in northwestern U.S. and in the Great Lakes region). Where feeds come from iodine-deficient areas.	Iodine is needed by the thyroid gland in making thyroxin (an iodine-containing hormone which controls the rate of body metabolism or heat production).	Production of weak, goitrous, or dea[d] calves. Occasional borderline cases may sur[vive]; in these, the moderate thyroid en[largement] disappears in a few weeks.
Iron (Fe)	Calves on an exclusive milk diet.	Necessary constituent of hemoglobin (oxygen carrying system of the blood). Deficiencies result in anemia. Up to 20 weeks of age, supplemental iron contributes to improved weight gain in calves fed milk diets.	Anemia and decreased growth in calves on exclusive milk diet. Excessive amounts of iron are toxic.
Manganese (Mn)	In northwestern U.S. All-concentrate diets based on corn supplemented with nonprotein nitrogen.		Delayed estrus, reduced fertility, abortions, and deformed young. Calves born to manganese-deficient cows may exhibit deformed legs (enlarged joints, stiffness, twisted legs, "overknuckling"), weak and shortened bones, and poor growth.
Molybdenum (Mo)	Molybdenum toxicity occurs only occasionally in cattle and appears to be an area problem.	Specific symptoms of molybdenum deficiency in cattle have not been described.	Molybdenum toxicity results in severe scours and loss of condition.
Selenium (Se)	Low-selenium forage and low vitamin E. It is an area problem which occurs in many parts of the U.S.	Vitamin E fills a vital role in the functioning of every cell of the body. It serves as a powerful biological antioxidant in the body cells and in the digestive tract. It has a sparing effect on the selenium requirement. Toxic levels reported in S.D., N.D., Mont., Wyo., Utah, Neb., Kan., and Colo.	White muscle disease characterized by white muscle, heart failure, and paralysis. Hollow or swayed back. Often a dystrophic tongue.
Zinc (Zn)		Essential in skin, hair, and bone development.	Parakeratosis in young calves, evidenced by inflamed nose and mouth, unthrifty appearance, roughened haircoat, and stiffness of joints.

[1]As used herein, the distinction between "nutrient requirements" and "recommended allowances" is as follows: In nutrient requirements, no margins of safety are included intentionally; whereas in recommended allowances, margins of safety are provided to compensate for variations in feed composition, environment, and possible losses during storage or processing.

Where preceded by an asterisk, the nutrient requirements, recommended allowances, and other facts presented herein are taken from *Nutrient Requirements of Beef Cattle*, No. 4, 5th rev. ed., NRC-National Academy of Sciences, Washington, D.C., 1976.

[2]Estimated 90% dry matter.

Continued)

Nutrient Requirements[1]				
Daily Nutrients/ Animal	Percentage of Rations	Recommended Allowances[1]	Practical Sources of the Mineral	Comments
400 to 800 micrograms iodine per day for pregnant and lactating beef cows.	*As-fed*[2] *M-F* **NaCl:* 0.09 *0.10*	Free access to stabilized iodized salt containing 0.01% potassium iodide (0.0076% iodine). Iodine (I) added to total feed (moisture-free basis) 0.5 ppm.	Stabilized iodized salt containing 0.01% potassium iodide. Calcium iodate. Ethylenediamine dihydriodide (EDDI).	The enlargement of the thyroid gland (goiter) is nature's way of trying to make enough thyroxin, when there is insufficient iodine in the feed. Toxicity can occur, resulting in depressed appetite, dull listless appearance, difficulty in swallowing, hacking cough, and weepy eyes.
30 mg daily during first 4 to 8 weeks or injections of 500 mg at birth and at 8 weeks of age.	*4.5 mg/lb *(10 mg/kg)* of moisture-free ration.	Iron (Fe) 40 mg daily.	Levels of iron in feed believed to be ample, since feeds contain 36 to 45 mg/lb (80-100 mg/kg) in most regions. Iron sulfate.	After calves are past 20 weeks of age, iron does not seem to be beneficial. About 30% of all calves are affected by prenatal iron deficiencies.
	*0.45 to 4.5 mg/lb *(1 to 10 mg/kg)* of moisture-free ration.	*An intake of 20 ppm will prevent deformities in the fetus.	Most roughages contain over 13.6 mg of manganese per pound *(30 mg/kg)* of dry matter, and most grains contain about half this amount. Manganous oxide, sulfate, and carbonate.	A deficiency of manganese exists in northwestern U.S., where it has been shown to cause "crooked calves."
	*.0045 mg/lb *(0.01 mg/kg)* of moisture-free ration.	As a feed additive, molybdenum is not cleared by Food and Drug Administration.	Many feeds contain 6.8 to 13.6 mg/lb of ration dry matter.	Toxic levels of molybdenum interfere with copper metabolism and thus increase copper requirements. Increasing copper level in diet to 1 g/head daily is effective in overcoming molybdenum toxicity in beef cattle. Phosphorus, manganese, potassium, zinc, and sulfur have also been reported to affect the degree of molybdenum toxicity.
	*0.045 mg/lb *(0.1 mg/kg)* of moisture-free ration. Cows grazing on pastures with less than this quantity of selenium produce calves with white muscle disease.		Any of following 4 procedures are effective against white muscle disease: (1) selenium as a drench; (2) subcutaneous or intramuscular injections; (3) selenium as a feed additive; or (4) adding selenium to fertilizer applied to pasture.	*Selenium toxicity may occur when cattle consume feeds containing 10-30 ppm of selenium on a dry matter basis for an extended period. It has been shown that the performance of feedlot cattle fed a selenium-deficient ration is improved by injecting selenium, the only approved method of selenium supplementation.
	*9.1 to 13.6 mg/lb *(20 to 30 mg/kg)* of moisture-free ration.	50-90 ppm zinc in the total feed (air-dry basis).	Many feeds contain 6.8 to 13.6 mg/lb *(15-30 mg/kg)* of ration dry matter. Adding zinc to fertilizer applied to pasture.	Mild zinc deficiency in feedlot cattle results in lowered weight gains without the development of a specific syndrome.

Note: Mineral recommendations for all classes and ages of cattle, especially those fed unmixed rations or on pasture, are—

1. *Where animals are on liberal grain feeding*—Provide free access to a 2-compartment mineral box, with (a) trace mineralized salt in one side, and (b) in the other side, a nixture of ⅓ trace mineralized salt (salt included for purposes of palatability), ⅓ defluorinated phosphate or steamed bone meal, and ⅓ ground limestone or oystershell flour.

2. *Where animals are primarily on roughage (pasture, hay, and/or silage)*—Provide free access to a 2-compartment mineral box, with (a) trace mineralized salt in one side salt included for purposes of palatability), and (b) in the other side, a mixture of ⅓ trace mineralized salt and ⅔ defluorinated phosphate or steamed bone meal.

Vitamins

Vitamin deficiencies in cattle may occur as a result of lack of availability of vitamins or because of the presence of antimetabolites. Both are important concepts. For example, analyses show corn to be adequate in niacin. Yet, due either to an antimetabolite or unavailability, there may be niacin deficiencies when corn is fed—deficiencies which can be remedied by niacin supplementation.

VITAMIN REQUIREMENTS

The National Research Council vitamin requirements of beef cattle are summarized in Table 4-22.

TABLE 4-22

VITAMIN REQUIREMENTS OF BEEF CATTLE PER UNIT OF RATION[1]

Nutrient	Moisture Basis (As-fed = est. 90% dry matter. Moisture-free = 100% DM)	Growing and Finishing Steers and Heifers		Dry Pregnant Cows		Breeding Bulls and Lactating Cows	
		(IU/lb)	*(IU/kg)*	*(IU/lb)*	*(IU/kg)*	*(IU/lb)*	*(IU/kg)*
Vitamin A activity[2]	As-fed *Moisture-free*	900 *1,000*	1,980 *2,200*	1,146 *1,273*	2,520 *2,800*	1,598 *1,773*	3,510 *3,900*
Vitamin D	As-fed *Moisture-free*	112.5 *125*	247.5 *275*	112.5 *125*	247.5 *275*	112.5 *125*	247.5 *275*
Vitamin E	As-fed *Moisture-free*	.6-2.4 *.7-2.7*	13.5-54 *15-60*	—[3]	—[3]	.6-2.4 *.7-2.7*	13.5-54 *15-60*

[1]Adapted by the author from *Nutrient Requirements of Beef Cattle*, No. 4, 5th rev. ed., NRC-National Academy of Sciences, Washington, D.C., 1976, p. 35.
[2]May be vitamin A or provitamin A equivalent. See Tables 4-13 to 4-15 for more detailed data on requirements.
[3]Unknown. It is suggested that the level for the growing and finishing animal be used.

BEEF CATTLE VITAMIN CHART

Table 4-23, Beef Cattle Vitamin Chart, page 282, gives, in summary form, the following pertinent information relative to each vitamin listed: (1) Conditions usually prevailing where deficiencies are reported, (2) function, (3) deficiency symptoms, (4) nutrient requirements, (5) recommended allowances, and (6) practical sources.

VITAMIN SUPPLEMENTATION

The absence of one or more vitamins in the ration may lead to a failure in growth or reproduction, or to characteristic disorders known as deficiency diseases. In severe cases, death itself may follow. Although the occasional deficiency symptoms are the most striking result of vitamin deficiencies, it must be emphasized that in practice mild deficiencies probably caus higher total economic losses than do severe deficien cies. It is relatively uncommon for a ration, or diet, t contain so little of a vitamin that obvious symptoms o a deficiency occur. When one such case does appea it is reasonable to suppose that there must be severa cases that are too mild to produce characteristi symptoms but which are sufficiently severe to lowe the state of health and the efficiency of production.

• *Vitamin A (carotene)*—The vitamin most likel to be deficient in beef cattle rations is vitamin A. Tru vitamin A is a chemically formed compound, whic does not occur in plants. It is furnished in most bee cattle rations in the form of its precursor, carotene However, plants are a variable and often undependa ble source of carotene, due to oxidation. Also, cattl are relatively inefficient converters of carotene to vi tamin A. The latter fact was taken into consideratio in the development of international standards for vi tamin A, which are based on the rate at which the ra converts beta-carotene to vitamin A. The conversio rate for the rat is 1 mg of beta-carotene to 1,667 IU o vitamin A, whereas it is estimated that 1 mg of beta carotene is equal to 400 IU of vitamin A in cattle Moreover, the conversion rate for cattle varies unde different conditions; it is influenced by type o carotenoid, breed, individual differences in animals and level of carotene intake. Stress conditions—suc as extremely hot weather, viral infections, and altere thyroid function—have also been suggested as cause for reduced conversion.

Under practical feeding conditions, cattlemen should consider (1) previous feeding as it influence ody stores of vitamin A; (2) vitamin A destruction during processing or when mixed with oxidizing materials; and (3) carotene destruction in feeds during storage.

• *Vitamin D*—When exposed to enough direct sunlight, beef cattle normally acquire their vitamin D needs, for the ultraviolet rays in sunlight penetrate the skin and produce vitamin D from traces of sterols in the tissues. Also, cattle obtain vitamin D from sun-cured roughages. However, the addition of vitamin D to the ration is important where cattle, especially calves, are kept in a barn most of the day, where there is limited sunshine, where the calcium-phosphorus ratio leaves much to be desired, and/or where little or no sun-cured hay is fed. Vitamin D helps build strong and sturdy frames. It is usually added at a level of about one-seventh the level of added vitamin A.

• *Vitamin E*—Added vitamin E may be necessary under certain conditions because of its relationship to vitamin A utilization and the prevention of white muscle disease.

• *Vitamin K*—Under normal conditions, adequate vitamin K is synthesized in the rumen of cattle. However, symptoms of inadequacy (a bleeding syndrome known as "sweetclover disease") occur when moldy sweetclover hay, high in dicoumarol content, is fed.

• *B vitamins*—Dietary requirements for the B vitamins—B_{12}, biotin, niacin (nicotinic acid), pantothenic acid, pyridoxine (B_6), riboflavin (B_2), and thiamin (B_1)—have been demonstrated experimentally for the young calf during the first eight weeks of life, prior to the development of the functioning rumen. At this stage in life, these requirements are usually met by the milk of the dam. Later, the B vitamins appear to be synthesized in sufficient quantities by rumen bacterial fermentation. However, inadequacy of protein or other nutrients in the ration may impair rumen fermentation, with the result that sufficient quantities of the B vitamins will not be synthesized.

Water

Water is the cheapest feed! Hence, beef cattle should have an abundant supply of it before them at all times. Mature cattle will consume an average of about 11 gallons of water per head daily, with younger animals requiring proportionately less. The water requirement is influenced by several factors, including rate and composition of gain, pregnancy, lactation, activity, type of ration, feed intake, and environmental temperature.

Saline water containing one percent soluble salts may be toxic. Excessive nitrates or alkalinity may make water unsatisfactory for cattle.

In the northern latitudes, heaters must be provided to make the water available, but they are not needed to warm the water further.

FEED ADDITIVES AND IMPLANTS[7]

Table 4-24, page 284, summarizes the growth stimulants that are presently available and can be used. All of these products have been shown to improve gain and feed efficiency of cattle significantly. The information presented in Table 4-24 is the most recent available. But feed additives and implants do change from time to time; new products are developed, and sometimes old products are banned by the Food and Drug Administration. So, those using additives should always confer with local authorities and read and follow manufacturer's label directions for more complete details on the use of a specific drug or combination of drugs.

In considering the additives listed in Table 4-24, it should be noted that there is no evidence to indicate that the use of these products can or will alleviate the need for vigilant sanitation, improved nutrition, and superior management. Also, the benefits of each one must be weighed against its cost.

FEED PREPARATION

The preparation of feeds for cattle is fully covered earlier in this section under the heading "Feed Preparation," including Table 4-11 therein; hence, the reader is referred thereto.

BEEF CATTLE FEEDING GUIDE

Table 4-25, Beef Cattle Feeding Guide, page 285, will serve as a useful guide for the cattleman who wishes to formulate his own rations. In using this table, the following points should be noted:[8]

1. Under "Description of Cattle"—column 1 of Table 4-25—are sufficient groups to cover the vast majority of cattle found on the nation's farms and ranches.

2. Columns 2 to 8 give pertinent recommendations relative to both forages and concentrates. These recommendations are in keeping with those advocated by scientists, and with the actual practices followed by successful operators.

---

[7]This section and Table 4-24 were authoritatively reviewed by the following: Dr. T. W. Perry, Department of Animal Sciences, Purdue University, Lafayette, Ind.; Dr. Wise Burroughs, Department of Animal Science, Iowa State University, Ames, Iowa; Dr. Wilton W. Heinemann, Animal Scientist, Irrigated Agriculture Research and Extension Center, Washington State University, Prosser, Wash.; Dr. A. T. Ralston, Department of Animal Science, Oregon State University, Corvallis, Ore.; Dr. Dean E. Hodge, Director, Beef Cattle Research Division, Ralston Purina Company, St. Louis, Mo.; Dr. Jack E. Martin, Sterling Nutritional Service, Inc., Sterling Colo.; Dr. T. M. Means, International Animal Research Coordinator, Eli Lilly and Company, Greenfield, Ind.; Dr. Calvin Drake, Aid, Inc., Wichita, Kan.; and Dr. Aaron L. Andrews, General Manager, Hess and Clark, Ashland, Ohio.

[8]In addition, see pertinent footnotes which accompany Table 4-25.

TABLE 4-23—BEEF CATTL

Vitamin Which May Be Deficient Under Normal Conditions	Conditions Usually Prevailing Where Deficiencies Are Reported	Function of Vitamin	Some Deficiency Symptoms
A (Vitamin A is found only in animals; plants contain the precursor—carotene)	Vitamin A deficiencies may occur when—(1) extended drought results in dry, bleached pastures; (2) winter feeding on bleached hays (especially overripe cereal hays or straws) with little or no green hay or silage; (3) drylot finishing on rations with little or no green forage or yellow corn, especially for feeding periods longer than 2 to 3 months; and (4) there is high nitrate intake, in either water or feed.	Vitamin A—(1) promotes growth and stimulates appetite; (2) assists in reproduction and lactation; (3) helps keep the mucous membranes of respiratory and other tracts in healthy condition; and (4) makes for normal vision.	*Mild deficiency:* Lowered feed consumption and weig gains. *Severe deficiency:* Night blindness, muscular incoordin tion, staggering gait, and convulsiv seizures. Total and permanent blin ness in young animals. Other localize paralysis may occur. Excessive wate ing of the eyes (rather tha xerophthalmia) usually occurs; the co neas of the eyes become keratinize and may, upon infection, develop u ceration. Severe and intermitte diarrhea at advanced stages of de ciency is characteristic. *Finishing cattle:* Generalized edema or anasarca may occur, with symptoms of lameness in th hock and knee joints and swelling in the brisket area. Pulmonary complicatior culminating in pneumonia have been reported. *Bulls of breeding age:* Decline in sexual activity. Spermatozoa decrease in numbers and motility, ar there is a marked increase in abnormal forms. *Breeding cows:* Estrus may continue, but conception rate may be low. Pregnant cows may abc or give birth at term of dead, weak, or blind calves. Retained placentas a common. *Vitamin A deficiency can be detected by carotene and vitamin A analysis blood and liver tissue of cattle.*
D	Young calves kept indoors, especially in the winter-time. Finishing cattle in northern U.S. on high-silage and grain rations and a minimum of sun-cured hay.	Aids in assimilation and utilization of calcium and phosphorus, and necessary in the normal bone development of animals—including the bones of the fetus.	Rickets in young calves, the symptoms which are: decreased appetite, lo ered growth rate, digestive di turbances, stiffness in gait, labore breathing, irritability, weakness, an occasionally, tetany and convulsion Later, enlargement of the joints, slig arching of the back, bowing of the leg and the erosion of the joint surface cause difficulty in locomotion. Posteri paralysis may follow fracture of vert brae. Symptoms develop more slow in older animals. Vitamin D deficiency in the pregna animal may result in dead, weak or d formed calves at birth.
E (also see Table 4-100)	Abnormally high levels of nitrites may produce vitamin E deficiencies. Where soils are very low in selenium.	Serves as a physiological antioxidant, facilitating the absorption and storage of vitamin A. Its other biochemical roles in the animal body appear to be related to its antioxidant capability, including the protection of vitamin A.	Muscular dystrophy (commonly calle white muscle disease) in calves 2 to weeks of age, characterized by hea failure and paralysis varying in sever from slight lameness to inability stand. Also, a dystrophic tongue often seen in affected animals.
B Vitamins: B_{12} biotin niacin (nicotinic acid) pantothenic acid pyridoxine (B_6) riboflavin (B_2) thiamin (B_1)	Severe inadequacy of protein or other nutrients in the diet may impair rumen fermentation to such an extent that sufficient quantities of B vitamins will not be synthesized.		
K	When moldy sweetclover hay high in dicoumarol content is fed, resulting in a bleeding syndrome called sweetclover poisoning or bleeding disease.		

[1]As used herein, the distinction between "nutrient requirements" and "recommended allowances" is as follows: In nutrient requirements, no margins of safety are include intentionally; whereas in nutrient allowances, margins of safety are provided in order to compensate for variations in feed composition, environment, and possible losses durir

VITAMIN CHART

Nutrient Requirements[1]		Recommended Allowances[1]	Practical Sources of the Vitamin	Comments
Daily Nutrients/ Animal	Percentage of Rations			
*Variable according to class, age, and weight of cattle (see Tables 4-13 to 4-15). *Injection of 1 million IU of vitamin A intramuscularly will prevent deficiency symptoms for 2-4 months in growing or breeding cattle.	*Variable according to class, age, and weight of cattle. On a moisture-free basis, the vitamin A requirements are about as follows: *1. Growing-finishing steers and heifers, 1,000 IU/lb *(2,200 IU/kg)*. *2. Pregnant heifers and cows, 1,270 IU/lb *(2,800 IU/kg)*. *3. Lactating cows and breeding bulls, 1,770 IU/lb *(3,900 IU/kg)*.		Stabilized vitamin A. Green pasture. Grass or legume silages. Yellow corn. Green hay not over 1 yr. old. The average carotene content of some common feeds is as follows: **mg Carotene/ lb or kg** Legume hays (including alfalfa) average quality 9-14/lb *20-31/kg* Nonlegume hays, average quality 4-8/lb *9-18/kg* Dehydrated alfalfa meal, average quality .... 50-70/lb *110-154/kg* Yellow corn ..... 0.8-1.0/lb *1.8-2.2/kg* Silages, corn or sorghum 2-10/lb *4-22/kg*	Hay over 1 yr. old, regardless of green color, is usually not an adequate source of carotene or vitamin A activity. The younger the animal, the quicker vitamin A deficiencies will show up. Mature animals may store sufficient vitamin A to last 6 months. When deficiency symptoms appear, it is recommended that there be added to the ration either (1) a stabilized vitamin A product, or (2) dehydrated alfalfa or grass. Corn and sorghum silage may contain a substance which destroys carotene and/or vitamin A.
	**As-fed:* 112 IU/lb *(247 IU/kg)* *Moisture-free:* 125 IU/lb *(275 IU/kg)*	Normally, beef cattle receive sufficient vitamin D from exposure to direct sunlight or from sun-cured hay.	Exposure to direct sunlight. Sun-cured hay. Irradiated yeast.	Sun-cured alfalfa hay contains 300 to 1,000 IU/lb (661-2,204/kg). Vitamin D is usually added at level of about 1/7 level of vitamin A.
	*For young calves, *dl*-alpha-tocopherol acetate added to *as-fed ration:* 6.1 to 24.6 IU/lb *(13.5 to 54 IU/kg); moisture-free ration:* 6.8 to 27.3 IU/lb *(15 to 60 IU/kg)*.	Generally natural feeds supply adequate quantities of alpha-tocopherol for mature cattle, although muscular dystrophy in calves occurs in certain areas.	Alpha-tocopherol, added to the diet or injected intramuscularly. Commercial vitamin E supplements. Grains contain 6-15 mg vitamin E/lb (13-33 mg/kg).	The incidence of white muscle disease appears to be lower where the cows receive 2 to 3 lb (.91-1.36 kg) of grain during last 60 days of pregnancy. Where supplemental vitamin E is needed, it may be added to the ration or injected intramuscularly.
		Usually, no dietary B vitamins need be supplied to cattle.	Milk supplied by the cow during early lactation.	During the first 8 weeks of life of the calf, the dietary requirements for the B vitamins are usually adequately met by milk from the dam; after this, these vitamins are usually synthesized by the rumen bacteria.
				Except when the dicoumarol content of hay is excessively high (as in moldy sweetclover hay) sufficient vitamin K is synthesized in the rumen of cattle.

storage or processing. Where preceded by an asterisk, the nutrient requirements, recommended allowances, and other facts presented herein were taken from *Nutrient Requirements of Beef Cattle*, No. 4, 5th rev. ed., NRC-National Academy of Sciences, Washington, D.C., 1976.

TABLE 4-24

IMPLANTS AND GROWTH STIMULANTS FOR FINISHING CATTLE

Cattle	Additive	Method of Administering	Dosage	Increase in Daily Rate of Gain	Increase in Feed Efficiency	Effect on Carcass Quality	Other Comments	Withdrawal Period Prior to Slaughter
Finishing Steers	1. Antibiotics[1]	Oral	10 mg/100 lb body wt. daily; or 70-75 mg/head daily	6%	4%	Improves carcass quality slightly; more fat deposition and marbling.	Antibiotics will also reduce the disease level. More effective on high-roughage rations than on high-concentrate rations.	None
	2. Compudose	Implant	24 mg estradiol	17%	7%	No effect.	It gives improved performance for 200 days.	No withdrawal required.
	3. Ralgro (Zeranol)	Implant	36 mg resorcyclic acid lactone	10%	5-10%	No effect.	Nonestrogenic.	65 days
	4. Rumensin (monensin sodium)	Oral	50-360 mg/head/day	Gain not affected. Saves feed.	10%	No effect.	Not a hormone. It results in more propionic acid and less butyric and acetic acids; hence, more energy.	No withdrawal required. Do not allow horses or other equines access to formulations containing Rumensin, for it has been fatal to equines.
	5. Synovex S (for steers)	Implant	200 mg progesterone 20 mg estradiol benzoate	10-15%	5-10%	No effect.		
Finishing Heifers	1. Antibiotics[1]	Oral	10 mg/100 lb body wt. daily; or 70-75 mg/head daily	6%	4%	Improves carcass quality slightly; more fat deposition and marbling.	Antibiotics will also reduce the disease level. More effective on high-roughage than on high-concentrate rations.	None
	2. MGA (Malengestrol acetate)	Oral	0.25-0.50 mg daily melengestrol acetate	11%	8%	MGA will lower the incidence of estrus in heifers and increase rate and efficiency of gain. It is not effective with pregnant heifers.	MGA is effective for heifers, but not for steers.	48 hours
	3. Ralgro (Zeranol)	Implant	36 mg resorcyclic acid lactone	10%	5-10%	No effect.	Nonestrogenic.	65 days
	4. Rumensin (monensin sodium)	Oral	50-360 mg/head/day	Gain not affected. Saves feed.	10%	No effect.	Not a hormone. It results in more propionic acid and less butyric and acetic acids; hence, more energy.	No withdrawal required. Do not allow horses or other equines access to formulations containing Rumensin, for it has been fatal to equines.
	5. Synovex H (for heifers)	Implant	200 mg testosterone propionate 20 mg estradiol benzoate	10%	5-10%	No effect.	Recommended for use in heifers during last 60-150 days of the finishing period.	
Suckling Calves	1. Antibiotics[1]	Oral (in creep feed)	15 to 20 mg/100 lb body wt. daily	6%	4%		Antibiotics will also reduce the disease level.	None
	2. Compudose	Implant	24 mg estradiol	5%		No effect.	It is long lasting.	No withdrawal time.
	3. Ralgro (Zeranol)	Implant	36 mg resorcyclic acid lactone	10%	5-10%	No effect.	Nonestrogenic.	65 days

[1]Bacitracin, Bacitracin Zinc, Chlortetracycline, Erythromycin, Oxytetracycline.

TABLE 4-25

BEEF CATTLE FEEDING GUIDE[1]

(As-fed Basis)

Description of Cattle (1)	Recommendations[2] (2)	In drylot, with following types of forages: Legume and/or Legume-Nonlegume Mixed Forages of High Quality; Consisting of Dry Forages and/or Silage (high-protein forages) (3)		Legume and Nonlegume Forages Mixed; Consisting of Dry Forages and/or Silage (medium-protein forages) (4)		Nonlegume Forage; Consisting of Dry Forages and/or Silage (low-protein forages) (5)		On pasture of the following grades: Excellent (6)		Fair to Good (7)		Poor, Including Winter Pasture Consisting of Dry Grass Cured on the Stalk[3] (8)	
		(lb)	*(kg)*	*(lb)*	*(kg)*	*(lb)*	*(kg)*	*(lb)*	*(kg)*	*(lb)*	*(kg)*	*(lb)*	*(kg)*
ature pregnant beef breeding cows (av. wt. 1,100 lb; *500 kg*). Medium- and low-protein forages may be used for pregnant cows.	Forage per head daily, in lb or *kg*.	18-20	*8.2-9.1*	18-20	*8.2-9.1*	18-20	*8.2-9.1*						
	Concentrate: (1) Supplement allowance of soybean meal (or equivalent 41-45% crude protein) per head daily, in lb or *kg*.[4]					½-1½	*.23-.68*					½-1½	*.23-.68*
ature lactating beef breeding cows (av. wt. 1,100 lb; *500 kg*). When possible, use high-quality, high-protein forage for nursing cows.	Forage per head daily, in lb or *kg*.	26	*11.8*	24	*10.9*	22	*10.0*						
	Concentrate: (1) Total concentrate allowance per head daily, including protein supplement, in lb or *kg*.			2½	*1.1*	5	*2.3*			2½	*1.1*	5	*2.3*
	(2) Supplement allowance of soybean meal (or equivalent 41-45% crude protein) per head daily, in lb or *kg*.[4,5]			1½	*.68*	3	*1.4*			1½	*.68*	3	*1.4*
	(3) Crude protein composition of total concentrate, in %.	10-14		14-18		18-20		10-14		14-18		18-20	
eplacement heifers (weighing 400-500 lb; *181-227 kg*); to be bred to calve as 2-year-olds. Heifers bred to calve as 3-year-olds can be wintered at a lower level.	Forage per head daily, in lb or *kg*.	12-18	*5.4-8.2*	12-18	*5.4-8.2*	12-18	*5.4-8.2*						
	Concentrate: (1) Total concentrate allowance per head daily, including protein supplement, in lb or *kg*.	2-4	*.9-1.8*	2½-4	*1.1-1.8*	2½-4½	*1.1-2.0*					2½-4½	*1.1-2.0*
	(2) Supplement allowance of soybean meal (or equivalent 41-45% crude protein) per head daily, in lb or *kg*.[4,5]			½-1	*.23-.45*	1¼-1½	*.57-.68*					1¼-1½	*.57-.68*
	(3) Crude protein composition of total concentrate, in %.	9-13 (Cereal grains only will suffice)		14-18		17-22						17-22	
tocker calves: roughed through the winter and generally grazed the following summer. Fed for winter gains of ¾ to 1 lb *(0.34-0.45 kg)* per head daily (weighing 400-500 lb; *181-227 kg*, start of period).	Forage per head daily, in lb or *kg*.	12-18	*5.4-8.2*	12-18	*5.4-8.2*	12-18	*5.4-8.2*						
	Concentrate: (1) Supplement allowance of soybean meal (or equivalent 41-45% crude protein) per head daily, in lb or *kg*.[4]			¼-1	*.1-.45*	1¼-1½	*.57-.68*					1¼-1½	*.57-.68*
inishing calves (weighing 400-500 lb [or *181-227 kg*] start of feeding, and 750-850 lb [or *340-386 kg*] at marketing).	Forage per head daily.	2-6	*.9-2.7*	2-6	*.9-2.7*	2-5	*.9-2.3*						
	Concentrate: (1) Total concentrate allowance per head daily, including protein supplement, in lb or *kg*.	12-15	*5.4-6.8*	12-15	*5.4-6.8*	12-15	*5.4-6.8*	10-12	*4.5-5.4*	11-13	*5-5.9*	12-14	*5.4-6.4*
	(2) Supplement allowance of soybean meal (or equivalent 41-45% crude protein) per head daily, lb or *kg*.[4,5]	1-1½	*.45-.68*	1½-1¾	*.68-.8*	1¾-2¼	*.8-1.0*			1½-1¾	*.68-.8*	1¾-2¼	*.8-1.0*
	(3) Crude protein composition of total concentrate, in %.	9-11 (Cereal grains only will suffice)		12-13		13-15		9-11 (Cereal grains only will suffice)		12-13		13-15	

(Continued)

TABLE 4-25 (Continued)

Description of Cattle (1)	Recommendations[2] (2)	In drylot, with following types of forages: Legume and/or Legume-Nonlegume Mixed Forages of High Quality; Consisting of Dry Forages and/or Silage (high-protein forages) (3)	Legume and Nonlegume Forages Mixed; Consisting of Dry Forages and/or Silage (medium-protein forages) (4)	Nonlegume Forage; Consisting of Dry Forages and/or Silage (low-protein forages) (5)	On pasture of the following grades: Excellent (6)	Fair to Good (7)	Poor, Including Winter Pasture Consisting of Dr Grass Cured on th Stalk[3] (8)
Yearlings: roughed through the winter, and pasture finished the following summer. Fed for winter gains of 1 to 1¼ lb *(0.45-0.57 kg)* per head daily (weighing about 600 lb [or *272.8 kg*] start of wintering).	Forage per head daily, in lb or *kg*.	16-24 *7.3-10.9*	16-24 *7.3-10.9*	16-24 *7.3-10.9*			
	Concentrate: (1) Supplement allowance of soybean meal (or equivalent 41-45% crude protein) per head daily, in lb or *kg*.[4,5]		1-1½ *.45-.68*	1½-1¾ *.68-.8*			1½-1¾ *.68-.8*
Finishing yearlings (weighing about 600 lb [or *272 kg*] start of feeding, and 850 to 1,050 lb [or *386-476 kg*] at marketing).	Forage per head daily, in lb or *kg*.	2-8 *.9-3.6*	2-8 *.9-3.6*	2-8 *.9-3.6*			
	Concentrate: (1) Total concentrate allowance per head daily, including protein supplement, in lb or *kg*.	15-19½ *6.8-8.9*	15-19¾ *6.8-9.0*	15-20 *6.8-9.1*	12-18 *5.4-8.2*	13-19 *5.9-8.6*	14-20 *8.4-9.*
	(2) Supplement allowance of soybean meal (or equivalent 41-45% crude protein) per head daily, in lb or *kg*.[4,5]	1-1½ *.45-.68*	1¼-1¾ *.57-.8*	1½-2½ *.68-1.1*		1¼-1¾ *.57-.8*	1½-2½ *.68-1.*
	(3) Crude protein composition of total concentrate, in %.	8-10 (Cereal grains only will suffice)	11-12	12-13	8-10 (Cereal grains only will suffice)	11-12	12-13
Finishing long-yearling steers (weighing about 800 lb [or *363 kg*] start of feeding and 1,000 to 1,100 lb [or *454-499 kg*] at marketing).	Forage per head daily, in lb or *kg*.	2-12 *.9-5.4*	2-12 *.9-5.9*	2-12 *.9-5.4*			
	Concentrate: (1) Total concentrate allowance per head daily, including protein supplement, in lb or *kg*.	16-22 *7.3-10*	16-22 *7.3-10*	16½-22¾ *7.5-10.3*	13-19 *5.9-8.6*	14-20 *6.4-9.1*	15-21 *6.8-9.*
	(2) Supplement allowance of soybean meal (or equivalent 41-45% crude protein) per head daily, in lb or *kg*.[4,5]		½-¾ *.23-.34*	1½-1¾ *.68-.8*		½-¾ *.23-.34*	1½-1¾ *.68-.8*
	(3) Crude protein composition of total concentrate, in %.	9-12 (Cereal grains only will suffice)	10-11	11-12	9-10 (Cereal grains only will suffice)	10-11	11-12

[1]This table was authoritatively reviewed by Dr. Robert Totusek, Department of Animal Sciences and Industry, Oklahoma State University, Stillwater, Okla.; Dr. W. M. Warr Head, Department of Animal and Dairy Sciences, Auburn University, Auburn, Ala.; Dr. A. T. Ralston, Department of Animal Science, Oregon State University, Corvallis, Ore.; E. R. Barrick, Department of Animal Science, North Carolina State University, Raleigh, N.C.; Dr. H. B. Geurin, Director of Feed Research, W. R. Grace and Co., St. Louis, Mo.; R. I. Pick, Walnut Grove Products, Atlantic, Iowa; and Dr. W. P. Lehrer, Jr., Director, Nutrition and Research, Albers Milling Company, Los Angeles, Calif.

[2]The daily forage recommendations are based on dry forage. When silage is included in the ration, figure 3 lb of silage equivalent to 1 lb of dry forage, due to the hig moisture content of silage. Many cattlemen do not winter-feed as liberally as herein recommended. In general, these operators feel that it is more profitable (1) to let cattle "h their own" or even lose in condition during the winter months (so long as they remain healthy), to keep winter feed and labor costs at a minimum, and (2) to make all or mos the gains on grass.

[3]On a dry basis, the crude protein content of mature, weathered grasses may be 3% or less. The upper limit of the concentrate allowance recommended in column 8 sho be fed on winter range when (1) the grass is less abundant, and/or (2) the grass is relatively low in protein.

[4]Soybean meal, which usually ranges from 41 to 45% protein content, is herein used as a standard merely because it is the leading U.S. protein supplement. It is to emphasized, however, (1) that other protein supplements, including numerous commercial products, may be used; (2) that, in general, those supplements should be purchas which provide a unit of protein at the lowest cost, and those feeds which are highest in protein content are usually the most economical; and (3) that where other protein fee are substituted for the soybean meal recommended herein (41-45% protein), an equivalent amount of crude protein should be provided—for example, approximately 2 lb c 20% crude protein supplement should be provided to replace each 1 lb of soybean meal (although it is recognized that 2 lb of a 20% protein feed will generally provide m energy, and may supply more of certain other important nutrients, than 1 lb of soybean meal).

[5]The recommended supplement allowance is based on the assumptions (1) that cereal grains, averaging 9 to 13% crude protein content, comprise the major part of concentrate mix, and (2) that the forage is not comprised entirely or predominantly of nonlegume silage. Naturally, less protein supplement will need to be added where feeds higher protein content than the cereal grains predominate. Also less protein supplement is required to balance a ration consisting predominantly of barley (of 12.7% cru protein content) than one mostly of corn (of 8.7% crude protein content). Likewise, the upper limit of protein supplement recommended herein (or even a higher figure) required to balance a ration where the forage is comprised entirely or largely of very low-protein forages such as those that are mature and weathered.

In particular, it should be noted that all protein ecommendations are in terms of *crude protein* con-ent,[9] rather than digestible protein. This was decided ipon because (a) this is what the feed manufacturer vants to know as he plans a feed formula, and (b) this s what the stockman sees on the feed tag when he purchases feed.

3. It is recognized that most farmers and ranchers generally grow their own forages, and purchase part or all of the concentrates. Thus, they generally wish to know what crude protein content of concentrate alone (including grains, by-product feeds, and/or protein supplements) they need to feed to balance out the forage which is available. Likewise, feed manufactur-ers have need for this information in compounding mixes. For these reasons, harvested forages in Table 4-25 are classified as (a) high-protein forages, (b) medium-protein forages, and (c) low-protein forages; and specific recommendations are made for each. Similar classifications and recommendations are made for (1) excellent, (2) fair to good, and (3) poor pastures.

4. It is often hazardous to formulate rations for excellent pastures that are different from those for poor pastures, because (a) cattlemen may be in error in appraising the quality of their pastures, and (b) pas-tures are generally excellent in the early spring, but become progressively poorer as the season advances unless they are irrigated and fertilized.

For purposes of illustration, let us refer to Table 4-25. Under column 5, it is noted that a mature beef breeding cow (av. wt. 1,100 lb) that is being fed a daily ration of somewhere between 18 to 20 lb of grass hay or other nonlegume dry roughage should receive, in addition, ½ to 1½ lb daily of a protein supplement of soybean meal (or some other protein supplement which will provide an amount equivalent to 41 to 45% crude protein). To be sure, it is entirely proper to meet this recommended crude protein content of concentrate by feeding double the allowance of some protein supplement with approximately 20 percent crude protein content. Many times the latter may be more economical, and even advisable—for example, when the forage is of poor quality and added energy feed is needed. In general, however, those feeds should be purchased which furnish a unit of protein at the lowest cost, and those feeds which supply the protein in the most concentrated form are usually the most economical.

Under column 2 of Table 4-25, additional information, of value to both the feed manufacturer and the stockman who mixes his own rations, is given. For example, in Table 4-25 under "Finishing long yearling steers . . . ," recommendations are given relative to the following:

"(3) Crude protein composition of total concentrate, in %."

RATIONS FOR BEEF CATTLE[10]

Table 4-26 gives some suggested rations for different classes and ages of cattle. All of these are merely intended as general guides. Variations can and should be made in the rations used. The feeder should give consideration to (1) the supply of home-grown feeds, (2) the availability and price of purchased feeds, (3) the class and age of cattle, (4) the health and condition of the animals, and (5) the length of the grazing season.

[9]Also, it is recognized (1) that beef cattle consume a large propor-tion of forage, and (2) that the percentage digestibility of protein of forages differs tremendously—for example, the percent digestibility of protein of wheat straw is 11, whereas for alfalfa hay it is 71. On the other hand, the grains do not differ greatly in percent digestibility of protein. The National Research Council expresses digestible protein as 77.5% of total protein.

[10]Insofar as possible, these rations are based on the requirements of the National Research Council, as applied by the author.

TABLE 4-26—DAILY RATION
(As-fe

Suggested Rations With all rations and for all classes and ages of cattle, provide free access in separate containers to (1) salt (iodized salt in iodine-deficient areas), and (2) a suitable mineral mixture.	Wintering mature pregnant beef breeding cows (av. wt. 1,100 lb, or *499 kg*)		Wintering mature lactating beef breeding cows (av. wt. 1,100 lb, or *499 kg*)		Wintering replacement heifer (weighing 400-500 lb, o *181-227 kg*, start of winterin	
	(lb)	*(kg)*	*(lb)*	*(kg)*	*(lb)*	*(kg)*
1. Legume hay or grass-legume mixed hay, good quality	18-20	*8.2 - 9.1*	30	*13.6*	13-15[4]	*5.9 - 6.8*[4]
Grain	-	-	-	-	2-3	*.91- 1.36*
Protein supplement	-	-	-	-	-	-
2. Grass hay or other nonlegume dry roughage	18-20	*8.2 - 9.1*	24-26	*10.9-11.8*	12-18[4]	*5.4 - 8.2*[4]
Grain	-	-	2	*.91*	2½-4½	*1.13- 2.04*
Protein supplement	½-1	*.23- .45*	3	*1.36*	1¼-1½	*.57- .68*
3. Legume hay or grass-legume mixed hay, good quality	7-11	*3.2 - 5.0*	26-28	*11.8-12.7*	8-12[4]	*3.6 - 5.4*[4]
Grass hay or other nonlegume dry roughage	9-11	*4.1 - 5.0*	-	-	4-6	*1.8 - 2.7*
Grain	-	-	1	*.45*	2½-4	*1.13- 1.81*
Protein supplement	-	-	1	*.45*	½-1	*.23- .54*
4. Corn or sorghum silage	50-55	*22.7 -25*	55	*25.0*	25-40	*11.3 -18.1*
Grain	-	-	2	*.91*	-	-
Protein supplement	0-½	*0 - .23*	3	*1.36*	1½-1¾	*.68- .79*
5. Grass silage, half or more legume	50	*22.7*	50	*22.7*	25-40	*11.3 -18.2*
Grain	-	-	4	*1.81*	3-4	*1.36- 1.81*
Protein supplement	-	-	-	-	½	*.23*
6. Silage (corn or sorghum silage fed with legume hay or legume silage fed with grass hay)	35	*15.9*	40	*18.1*	15-30	*6.8 -13.6*
Hay	5-6	*2.3 - 2.7*	10	*4.5*	3-4	*1.4 - 1.8*
Grain	-	-	-	-	1-2	*.45- .91*
Protein supplement	0-½	*.23*	-	-	½-1	*.22- .45*

[1]This table was authoritatively reviewed by Dr. Robert Totusek, Department of Animal Sciences and Industry, Oklahoma State University, Stillwater, Okla.; Dr. W. M. Warrer Head, Department of Animal and Dairy Sciences, Auburn University, Auburn, Ala.; Dr. A. T. Ralston, Department of Animal Science, Oregon State University, Corvallis, Ore.; D E. R. Barrick, Department of Animal Science, North Carolina State University, Raleigh, N.C.; Dr. H. B. Geurin, Director of Feed Research, W. R. Grace and Co., St. Louis, Mo.; D R. I. Pick, Walnut Grove Products, Atlantic, Iowa; and Dr. W. P. Lehrer, Jr., Director, Nutrition and Research, Albers Milling Company, Los Angeles, Calif.

[2]If stocker calves are late or the roughage is fair to poor in quality, it may be desirable to add 2-4 lb (*.91-1.81 kg*) of grain per head daily. If farm scales are availabl monthly weights may be used as the criterion for grain feeding. Keep in mind that the calves should gain ¾-1 lb (*.34-.45 kg*) daily.

FEED SUBSTITUTION TABLE FOR CATTLE (Beef and Dairy)

The successful cattleman is a keen student of values. He recognizes that feeds of similar nutritive properties can and should be interchanged in the ration as price relationships warrant, thereby making it possible at all times to obtain a balanced ration at the lowest cost. Thus, (1) the cereal grains may consist of corn, barley, wheat, oats, and/or sorghum; (2) the protein supplement may consist of soybean, cottonseed, peanut, sunflower, and/or linseed meal; (3) the roughage may include many varieties of hays and silages; and (4) a vast array of by-product feeds may be utilized.

Table 4-27, Feed Substitution Table for Cattle (Beef and Dairy), page 290, is a summary of the com parative values of the most common U.S. feeds. In ar riving at these values, two primary factors beside chemical composition and feeding value have bee considered—namely, palatability and carcass quality.

In using this feed substitution table, the follow ing facts should be recognized:

1. That, for best results, different ages and group of animals within classes should be fed differently.

2. That individual feeds differ widely in feedin value. Barley and oats, for example, vary widely i feeding value according to the hull content and th test weight per bushel, and forages vary widely ac cording to the stage of maturity at which they are cu and how well they are cured and stored.

OR BEEF CATTLE[1]
Basis)

Wintering stocker calves roughed hrough winter and grazed the fol- owing summer. Fed for winter gain of ¾-1 lb (*.34-.45 kg*) per head daily (weighing 400-500 lb, or *181-227 kg*, start of wintering)[2]		Finishing calves in drylot, generally in winter (weighing 400-500 lb, or *181-227 kg*, start of feeding and 750-850 lb, or *340-386 kg*, at marketing)[3]		Wintering yearlings; roughed through the winter, and generally pasture finished the following summer. Fed for winter gains of 1-1¼ lb, or *.45-.57 kg*, per head daily (weighing about 600 lb, or *272 kg*, start of wintering)		Finishing yearlings in drylot, generally in winter (weighing about 600 lb, or *272 kg*, start of feeding, and 900-1,050 lb, or *409-477 kg*, at marketing)[3]		Finishing long-yearling steers in drylot generally in winter (weighing about 850 lb, or *386 kg*, start of feeding and 1,000-1,100 lb, or *454-499 kg*, at marketing)[3]	
(lb)	*(kg)*	*(lb)*	*(kg)*	*(lb)*	*(kg)*	*(lb)*	*(kg)*	*(lb)*	*(kg)*
12-18[4] - -	*5.4 - 8.2* - -	4-6 12-15 1-1½	*1.8 - 2.7* *5.4 - 6.8* *.45- .68*	16-24 - -	*7.2 -10.9* - -	4-8 15-19½ 1-1½	*1.8 - 3.6* *6.8 - 8.8* *.45- .68*	6-12 16-22 -	*2.7 - 5.4* *7.2 -10.0* -
12-18[4] - 1¼-1½	*5.4 - 8.2* - *.57- .68*	4-5 12-15 1¾-2	*1.8 - 2.3* *5.4 - 6.8* *.79- .91*	16-24 - 1½-1¾	*7.2 -10.9* - *.68- .79*	4-8 15-20 1½-2½	*1.8 - 3.6* *6.8 - 9.1* *.68- 1.1*	6-12 16½-22¾ 1½-1¾	*2.7 - 5.4* *7.5 -10.3* *.68- .79*
12-18[4] 4-6 - ¼-1	*5.4 - 8.2* *1.8 - 2.7* - *.11- .45*	2-3 2-3 12-15 1½-1¾	*.91- 1.36* *.91- 1.36* *5.4 - 6.8* *.68- .79*	6-8 10-16 - 1-1½	*2.7 - 3.6* *4.5 - 7.3* - *.45- .68*	2-4 2-4 15-19¾ 1¼-1¾	*.91- 1.81* *.91- 1.81* *6.8 - 9.0* *.57- .79*	3-6 3-6 16-22 ½-¾	*1.4 - 2.7* *1.4 - 2.7* *7.2 -10.0* *.23- .34*
25-40 - 1-1¼	*11.3 -18.1* - *.45- .57*	6-16 8-12 2	*2.7 - 7.3* *3.6 - 5.4* *.91*	40-55 - 1¼-1½	*18.1 -24.9* - *.57- .68*	6-25 11-16 2	*2.7 -11.3* *5.0 - 7.3* *.91*	6-35 15-21 1¼-1½	*2.7 -15.9* *6.8 - 9.5* *.57- .68*
25-40 2-3 ½	*11.3 -18.1* *.91- 1.36* *.23*	6-16 8-12 1-2	*2.7 - 7.3* *3.6 - 5.4* *.45- .91*	40-55 4-5 ½	*18.1 -24.9* *1.8 - 2.3* *.23*	6-25 11-16 1-1½	*2.7 11.3* *5.0 - 7.3* *.45- .68*	6-35 15-21 1	*2.7 -15.9* *6.8 - 9.5* *.45*
15-30 3-4 1-2 ½	*6.8 -13.6* *1.4 - 1.8* *.45- .91* *.23*	3-8 1-3 8-12 1-2	*1.4 - 3.6* *.45- 1.4* *2.6 - 5.4* *.45- .91*	20-35 7 - ½-¾	*9.1 -15.9* *3.2* - *.23- .34*	3-15 1-4 11-16 1-1	*1.4 - 6.8* *.45- 1.8* *5.0 - 7.3* *.45- .79*	3-15 1-7 15-21 1-1¼	*1.4 - 6.8* *.45- 3.2* *6.8 - 9.5* *.45- .57*

[3]In general, the experienced feeder plans that cattle on full feed shall consume (1) feeds in amounts (daily: air-dry basis) equal to about 2.5-3.0% of their liveweight, (2) 70-90% concentrates, and (3) a minimum of 2-4 lb *(.91-1.81 kg)* roughage for each 100 lb *(45 kg)* liveweight. In areas where roughage is more abundant and comparatively cheaper than grain, the proportions of roughage to grain should be somewhat higher than indicated. In computing roughage consumption, 3 lb *(1.36 kg)* of silage are considered equivalent to one lb *(.45 kg)* of hay.

[4]With calves (both replacement heifers and stockers) an extra 2 lb *(.91 kg)* of hay daily, over and above requirements, are herewith indicated to allow for wastage. Practical operators generally feed stemmy or other hay left over by calves to the cow herd.

3. That nonlegume forages may have a higher relative value to legumes than herein indicated provided the chief need of the animal is for additional energy rather than for supplemented protein. Thus, the nonlegume forages of low value can be used to better advantage for wintering mature, dry beef cows than for young calves.

On the other hand, legumes may actually have a higher value relative to nonlegumes than herein indicated provided the chief need is for additional protein rather than for added energy. Thus, no protein supplement is necessary for breeding beef cows provided a good quality legume forage is fed.

4. That, based primarily on available supply and price, certain feeds—especially those of medium protein content, such as brewers' dried grains, corn gluten feed (gluten feed), distillers' dried grains, distillers' dried solubles, peanuts, and peas (dried)—may be used interchangeably as (a) energy feeds, and/or (b) protein supplements.

5. That the feeding value of certain feeds is materially affected by preparation. Thus, wheat must be coarsely ground or rolled for cattle. The values herein reported are based on proper feed preparation in each case.

For the reasons noted above, the comparative values of feeds shown in the feed substitution table are not absolute. Rather, they are reasonably accurate approximations based on average quality feeds, together with experiences and experiments.

TABLE 4-27

FEED SUBSTITUTION TABLE FOR CATTLE (Beef and Dairy)
(As-fed Basis)

Feedstuff	Relative Feeding Value (lb for lb) in comparison with the designated (underlined) base feed which = 100	Maximum Percentage of Base Feed (or comparable feed or feeds) which it can replace for best results	Remarks
GRAINS, BY-PRODUCT FEEDS, ROOTS AND TUBERS:[1] (Low- and Medium-Protein Feeds)			
Corn, No. 2	100	100	The most important concentrate for finishing cattle in the U.S. Grind coarsely or roll unless pigs follow cattle.
Almond hulls, dried, no shells	70-75	15-30	
Almond hulls and shell meal	35	15-20	
Apple pomace, air-dry	78	33⅓	
Bakery products, dried	100	15-30	
Bakery waste, not dried (30% water)	75	15-30	
Barley	90	25-100	The heavier the barley and the smaller the proportion of hulls, the higher the feeding value. Pigs following barley-fed cattle produce less pork than where corn is fed. Grind coarsely or roll for cattle. In Canada, where considerable barley is fed, it is often used as the only basal feed in the ration once animals are accustomed to it.
Beans (cull)	80	10	Best when cooked, but can also be fed raw. Beans should be ground. When cooked, 3 to 4 lb/head daily; when raw, 1 to 2 lb. Scouring may occur if they constitute more than 15% of total ration. Use in finishing grain ration.
Beet pulp, dried	90	50	
Beet pulp, molasses, dried	90-95	50	
Beet pulp, wet	25	40	50% the value of corn silage. May compose 40% of ration on dry matter basis.
Brewers' dried grains	80	33⅓	Not very palatable. Fed chiefly to dairy cattle. Too bulky and usually too costly to be used in finishing rations.
Brewers' grains (wet)	13-15	33⅓	Grains usually come from barley. Best to haul and feed directly. Can be stored in silo if salt is added at rate of 25 lb per ton of grains.
Buckwheat	55-75	33⅓	Should be ground and mixed with other grains.
Carrots (cull)	10-15	20-25	Store 3 to 4 weeks before using; fresh carrots cause scouring. Feed whole or sliced.
Citrus pulp, dried	80-88	25-50	
Corn and cob meal	85-90	100	
Corn gluten feed (gluten feed)	85-90	50	
Distillers' dried grains	73-90	33⅓	Rye distillers' dried grains are of lower value than similar products made from corn or wheat. Distillers' dried grains are used chiefly for dairy cattle.
Distillers' dried solubles	73-90	33⅓	The chief difference between distillers' dried grains and distillers' dried solubles is the higher B vitamin content of the latter. Normally this is not important for cattle or sheep.
Fat (animal or vegetable)	225	5	Fat has 203 megacalories energy/100 lb for maintenance and 127 megacalories for weight gain, as compared to 92 and 60, respectively, for corn.
Hominy feed	100	50	
Manure, cattle	75	50	Approximately 80% of the total nutrients of feeds is excreted as animal manure. However, the feeding value of manure will vary according to (1) the nutritive value of the feeds initially fed, (2) the class, age, and individuality of the animal to which the feeds were initially fed, and (3) the handling and processing of the manure.
Manure, poultry (see poultry house litter)			
Molasses, beet	75	10-40	Value is highest when used as an appetizer. May be laxative if fed at levels above 6 lb daily.
Molasses, cane	75	10-40	Value is highest when used as an appetizer.
Molasses, citrus	65-90	10-40	
Molasses, wood	26-30	10-20	Rather unpalatable.

Footnotes on last page of table.

(Continued)

TABLE 4-27 (Continued)

Feedstuff	Relative Feeding Value (lb for lb) in comparison with the designated (underlined) base feed which = 100	Maximum Percentage of Base Feed (or comparable feed or feeds) which it can replace for best results	Remarks
ats	70-90	10-100	Valuable for young stock, for breeding stock and for getting animals on feed. Oats have lowest value for finishing cattle and should be limited to ⅓ of such rations. Also, the feeding value of oats varies according to the test weight per bushel. Grind or roll for cattle.
aunch, dried (also see "Paunch-blood" under Protein Supplements of this table)	90	5-10	Dried paunch is not too palatable, with the result that it depresses appetite. Rate of gain is not affected, but feed efficiency is slightly lowered.
eas (cull), dried	88	40	Peas appear to be unpalatable to certain individuals. Also, there is bloat hazard if they exceed 40% of the ration.
ear waste, air-dry	75	40	
otatoes (Irish), wet	20-25	85	When fed with alfalfa hay, they are worth about 80% as much per ton as corn silage. Do not feed frozen. Sunburned, decomposed, or sprouted potatoes should not make up more than 10% of potatoes fed. Keep steers' heads down while eating to prevent choking.
otatoes (Irish), dehydrated	88	50	Excellent source of energy, but deficient in protein, minerals, and vitamins.
otatoes (Sweet)	25	85	
otatoes (Sweet), dehydrated	95-100	50	Dehydrated sweet potatoes are more palatable than dehydrated Irish potatoes.
oultry house litter	10-40	15-25	Poultry house litter may also be used as a protein source (see Protein Supplement, this table).
unes	62	15	Because of the laxative quality of prunes, they should be limited to 7% of the total ration.
aisins (cull)	70	33⅓	
aisin pulp	53	25	
ce (rough rice)	80	100	
ce bran	66⅔-75	33⅓	
ce polishings	88	25	
ye	100	33⅓	Not palatable when fed in larger amounts.
creenings, refuse	62-70	25-35	Should be finely ground in order to kill noxious weed seeds. Quality varies; good quality screenings are equal to oats, whereas poor quality screenings resemble straw.
orghum (milo, kafir), grain	90-95	100	Varieties vary in protein content. Grind or roll for cattle.
pelt and emmer	70-90	30-100	Similar to oats.
heat	100-105	50	Grind coarsely, or roll.
heat bran	70-90	25-33⅓	Because of its bulk and fiber, bran is not desirable for finishing rations. Bran is valuable for young animals, for breeding animals, and for starting animals on feed.
heat-mixed feed (mill run)	95	33⅓	Sometimes fed to the breeding herd, to young calves, and to finishing cattle being started on feed.
heat screenings	85	50	
ood (cooked)	75-80	70	Wood products, which are largely cellulose and lignin must be cooked before animals can digest them.

(Continued)

TABLE 4-27 (Continued)

Feedstuff	Relative Feeding Value (lb for lb) in comparison with the designated (underlined) base feed which = 100	Maximum Percentage of Base Feed (or comparable feed or feeds) which it can replace for best results	Remarks
PROTEIN SUPPLEMENTS:			
Soybean meal (41%)	100	100	Slightly laxative effect.
Alfalfa or clover screenings	70-75	50	Grind finely to destroy weed seeds.
Brewers' dried grains	55-65	50	Not very palatable. Fed chiefly to dairy cattle.
Copra meal (coconut oil meal), 21%	90-100	50	
Corn gluten feed (gluten feed)	65-75	50-100	
Corn gluten meal (gluten meal)	90-100	50	Somewhat unpalatable.
Cottonseed meal (41%)	100	100	Among practical cattlemen, the feeling persists tha cottonseed meal has a constipating effect; som experimental work to the contrary. Although it may be fed as the only protein supple ment, best results are secured when it is fed wit linseed meal for finishing cattle.
Distillers' dried grains	65-70	100	Rye distillers' dried grains are about 10% lower i protein than similar products made from corn c wheat.
Distillers' dried solubles	70	100	Low in palatability.
Feather meal (hydrolyzed; 84% protein)	175	50	Feather meal is unpalatable; hence, cattle must b accustomed to it gradually and it must be limite in quantity. It is best used for wintering brood cow and stocker cattle.
Legume screenings	75	75	Satisfactory, but less palatable than soybean or co tonseed meal.
Linseed meal (35%)	(For other than finishing cattle) 95	100	Linseed meal has a laxative effect. Some cattle wi not tolerate more than 5 to 8% linseed meal in th ration.
	(For finishing cattle) 115	100	Higher value for finishing cattle due to both greate efficiency and higher selling price of the cattle be cause of the increased bloom.
Paunch-blood feed (also see "Paunch, dried" under Grains section of this table)	100	100	At slaughter, each cow yields about 20 lb of paunc and 20 lb of blood. Dried paunch runs about 10% protein, dried blood around 80%, and a 50-50 mix ture of the two products, around 45%.
Peanut meal (45%)	100	100	Peanut meal may become rancid if stored too long especially in warm, moist climates.
Peas (cull), dried	65-75	50	
Poultry house litter	50-55	25	Poultry house litter may also be used as an energ source (see Grains section of this table).
Rapeseed meal (37%)	88	75	Rapeseed meal should be limited to not more than lb per cow.
Safflower meal, well hulled (42%)	92	100	
Safflower meal, with hulls (20%)	40-45	100	Safflower meal with hulls is unpalatable. Thus, i should be mixed with more palatable feeds.
Sesame meal	90-95	25	
Soybeans, whole	95-100	100	Not so satisfactory for finishing calves. Soybean allowance should be limited to amour necessary to balance the ration. Larger amount may be unduly laxative and throw cattle "off feed."
Sunflower meal (39%)	95-100	100	If poorly hulled and lower protein content than 39% feeding value will be lowered accordingly. It i well liked by cattle and keeps well in storage.

Footnotes on last page of table.

(Continued)

TABLE 4-27 (Continued)

Feedstuff	Relative Feeding Value (lb for lb) in comparison with the designated (underlined) base feed which = 100	Maximum Percentage of Base Feed (or comparable feed or feeds) which it can replace for best results	Remarks
DRY FORAGES AND SILAGES:[2]			All the dry nonlegume forages listed herein are satisfactory when needed minerals and either a limited amount of legume hay or a protein supplement are supplied to balance the ration.
Alfalfa hay, all analysis	100	100	Does away with or lessens protein supplement requirements.
Alfalfa silage	33⅓-50	50-85	When alfalfa silage replaces corn silage, more energy feed must be provided but less protein.
Alfalfa straw	37	50	Feed with good hay.
Apple pomace silage	17-25	50-85	Usually fed as a substitute for corn or grass silage. 50% the value of corn silage. Sometimes fed out of a stack or trench silo.
Apples	17-25	50-85	Do not feed more than 25 lb/cow. Not recommended for finishing cattle. Danger of choking when fed whole. Relatively high handling cost.
Bagasse, dried; sugarcane or sorghum	10-20	5-10	Has negative protein value.
Barley hay	70	100	Avoid bearded varieties.
Barley straw	63	70	Of the cereal straws, barley ranks next to oat straw in feeding value. Use for dry pregnant cows. Supplement daily with 5-6 lb alfalfa hay or 1-2 lb of 30-40% protein supplement.
Bean straw	34	50	Feed with good hay.
Beet tops, fresh	20	33⅓-50	In the West, large acreages of fresh beet tops are pastured off by cattle and sheep. Bloat may be problem when tops are frozen. Tops are laxative. Add 2½ lb of ground limestone/ton of feed.
Beet top silage, sugar	17-25	33⅓-50	Feed 2 oz of finely ground limestone or chalk with each 100 lb of tops, as calcium changes the oxalic acid to insoluble calcium oxalate.
Clover hay, crimson	90-100	100	Crimson clover hay has a considerably lower value if not cut at an early stage.
Clover hay, red	90-100	100	If the rest of the ration is adequate in protein, clover hay will be equal to alfalfa in feeding value; otherwise, it will be lower.
Clover-timothy hay	80-90	100	Value of clover-timothy mixed hay depends on the proportion of clover present and the stage of maturity at which it is cut.
Corncobs, ground	70	90	Ground corncobs can be used as the only roughage for beef cattle if properly supplemented with proteins, minerals, and vitamins.
Corn fodder	75	80-90	
Corn husklage (shucklage)	50	80-90	Highest and best use is for dry pregnant cows. It is slightly higher in energy and more palatable than corn stover.
Corn silage	33⅓-50	50-85	
Corn (sweet) silage, cannery waste	26-40	50-85	
Corn stover	45	70-90	Corn stover will meet the energy needs of dry pregnant cows, but is deficient in protein and low in phosphorus and vitamin A. Two acres of cornstalks will carry a cow 100-120 days.
Corn (sweet) stover	50	80-90	
Cottonseed hulls	66⅔	75	Use for dry pregnant cows. Supplement daily with 4-6 lb of good legume hay or 1-2 lb of 30-40% protein supplement.
Cowpea hay	90-100	100	
Gin trash, cotton	75	75	
Grape pomace or meal	15-30	10-25	Add molasses to improve palatability; low in TDN and protein; high in fiber.
Grass-legume mixed hay	80-90	100	Value depends on the proportion of legume present and the stage of maturity at which it is cut.
Grass-legume silage	32-47	50-85	Unless grain is added as a preservative, grass silage requires more energy feed, but less protein supplement than corn silage when fed to finishing cattle.
Grass silage	30-45	50-85	For finishing cattle, grass silage must be supplemented with additional energy feeds, such as cereal grain or molasses, to be of the same value as corn silage.

(Continued)

TABLE 4-27 (Continued)

Feedstuff	Relative Feeding Value (lb for lb) in comparison with the designated (underlined) base feed which = 100	Maximum Percentage of Base Feed (or comparable feed or feeds) which it can replace for best results	Remarks
Hop vine silage	20	50-75	It should be chopped when placed in the silo.
Hops, spent, dehydrated	80	50-65	Devoid of carotene; feed with legume hay.
Johnsongrass hay	70	100	
Lespedeza hay	80-100	100	Feeding value of lespedeza hay varies considerably with stage of maturity at which it is cut.
Mint hay	70-80	75	Cattle tire of mint hay when it is fed as the only roughage for extended periods.
Oat hay	75	100	
Oat silage	32-47	50-85	Must be chopped finely to exclude air from silo.
Oat straw	66⅔	75	Oat straw is the best of the cereal straws. Use for dry pregnant cows. Supplement daily with 4-6 lb of good legume hay or 1-2 lb of 30-40% protein supplement.
Paper (newspaper; waste paper)	66⅔	50	Paper varies in feeding value in proportion to the cellulose (most paper is 60-90% cellulose) and lignin content. Magazine and bookstock paper are higher in cellulose and lower in lignin than newspapers; hence, of higher feeding value. Pelleting or cubing may increase the value of paper. *Caution*: Some newspapers contain heavy metals (boron, lead, barium, and antimony), sometimes used as a dye carrier in printer's ink, which may be toxic to animals. This is especially true of "funny" papers because of the quantity of heavy metals carried on the colored ink of the comics.
Pea straw	45-75	60-75	
Pea-vine hay	100-110	75-90	Can constitute the only roughage for finishing cattle.
Pea-vine silage	33⅓-50	50-85	Unless grain is added as a preservative, pea-vine silage requires more energy feed, but less protein supplement than corn silage when fed finishing cattle.
Potato silage	25-30	50-75	About 75% the value of corn silage.
Prairie hay	65-70	100	
Reed canarygrass hay	70	100	
Rice straw	47	70	High levels of rice straw can be used for wintering cattle if the straw is properly fortified.
Sawdust	75-80	70	Digestibility is increased by cooking and other treatments. There are indications that the presence of sawdust will reduce liver abscesses in feedlot cattle.
Sorghum fodder	70	100	
Sorghum silage (grain varieties)	32-47	50-85	For finishing cattle, 85 to 90% as valuable as corn silage and must be supplemented in the same manner as corn silage.
Sorghum silage (sweet varieties)	25-30	50-85	Nearly equal to grain varieties in value per acre because of greater yield.
Sorghum (milo) stover	35	70-90	Can be grazed or harvested and stored either as dry feed or silage. About 2% higher in protein, but less palatable, than corn stover.
Soybean hay	85-90	50-75	Lower value than alfalfa hay, largely due to greater wastage in feeding. It may cause scouring when fed alone.
Sudangrass hay	70	100	
Sunflower silage	25-35	50-85	65 to 75% value of corn silage. Somewhat unpalatable and may cause constipation. Harvest for silage when ½ to ⅔ of heads are in bloom.
Sweetclover hay	100	100	Value of sweetclover hay varies widely. Second year sweetclover hay is less desirable than first year sweetclover hay and is more apt to cause sweetclover disease.
Timothy hay	70	100	
Vetch-oat hay	80-90	100	The higher the proportion of vetch, the higher the value.
Wheat hay	70	100	
Wheat straw	60	65	Of the cereal straws, wheat ranks third in nutritive value, behind oat straw and barley straw. Highest and best use is for dry pregnant cows. Supplement daily with 6 lb of alfalfa or 2 lb of a 30-40% protein supplement.

[1]Roots and tubers are of lower value than the grain and by-product feeds due to their higher moisture content.
[2]Silages are of lower value than dry forages due to their higher moisture content.

FEEDING BREEDING BEEF CATTLE

The beef breeding herd must be properly fed if a good calf crop is to be obtained. The size of the calf crop, the vigor and size the calves attain by market time, and the feed efficiency of the herd largely determine the profit realized.

Heavy grain feeding is uneconomical and unnecessary for the beef breeding herd. The nutrient requirements should be adequate merely to provide for maintenance, growth (if the animals are immature), and reproduction. Fortunately, these requirements can largely be met through feeding roughages—pasture in season, and dry forages and silages during the winter months.

Feeding Brood Cows

Feed affects total profit and cow productivity. It accounts for 65 to 75% of the total cost of keeping cows, and it exerts a powerful influence on cow fertility and calf weaning weight—the two biggest success factors in the cattle business.

NUTRITIONAL REQUIREMENTS

Experiments and practical observations reveal that the period during which calf crop percentage is affected most by nutrition extends from 30 days before calving until 70 days after calving, until after rebreeding—a period of approximately 100 days. This, then, is the most critical period in the cow-calf business. It's when life begins—that period within which one calf is born and another is conceived. The needs for the cow during this most critical production period are approximately equal to her needs for the remainder of the year.

A second important requisite of a sound beef cattle nutrition program is to feed animals according to their requirements. It is impossible to feed the herd properly where calving occurs the year around, or when dry pregnant cows, replacement heifers, and cows nursing calves are run together. This point becomes apparent from the following nutritional differences of (1) cows nursing calves, and milking well; (2) yearling heifers, last 3 to 4 months of pregnancy; and (3) dry mature cows, last 2 to 3 months of pregnancy:

1. The requirements of cows nursing calves are higher and more critical than those (a) of yearling heifers the last 3 to 4 months of pregnancy, or (b) of dry mature cows the last 2 to 3 months of pregnancy in total feed consumed, in energy and protein of the ration, and in calcium and phosphorus. After a cow calves, her energy needs jump about 50 percent, her protein needs double, and her calcium and phosphorus needs triple.

2. The requirements of yearling heifers the last 3 to 4 months of pregnancy are higher than those of dry mature cows the last 2 to 3 months of pregnancy in energy and protein, and in calcium and phosphorus.

Weight also makes a difference, as shown in Table 4-28, which gives the daily nutrient requirements at various weights of (1) dry pregnant cows, and (2) cows nursing calves.

TABLE 4-28

DAILY NUTRIENT REQUIREMENTS OF BEEF COWS (per animal)[1]

Body Weight		Crude Protein		TDN		Calcium	Phosphorus
(lb)	*(kg)*	*(lb)*	*(kg)*	*(lb)*	*(kg)*	*(g)*	*(g)*
Dry pregnant mature cows (middle third of pregnancy)							
772	*350*	.71	*.32*	6.6	*3.0*	10	10
882	*400*	.79	*.36*	7.3	*3.3*	11	11
992	*450*	.86	*.39*	7.9	*3.6*	12	12
1,102	*500*	.93	*.42*	8.6	*3.9*	13	13
1,213	*550*	.99	*.45*	9.2	*4.2*	14	14
1,323	*600*	1.08	*.49*	9.8	*4.4*	15	15
1,433	*650*	1.15	*.52*	10.4	*4.7*	16	16
Cows nursing calves, first 3 to 4 months after calving (superior milking ability)							
772	*350*	2.45	*1.11*	12.8	*5.8*	45	40
882	*400*	2.58	*1.17*	13.5	*6.1*	45	41
992	*450*	2.71	*1.23*	14.1	*6.4*	45	42
1,102	*500*	2.84	*1.29*	14.8	*6.7*	46	43
1,213	*550*	2.98	*1.35*	15.4	*7.0*	46	44
1,323	*600*	3.11	*1.41*	16.1	*7.3*	46	44
1,433	*650*	3.22	*1.46*	16.8	*7.6*	47	45

[1]Adapted by the author from *Nutrient Requirements of Beef Cattle*, No. 4, 5th rev. ed., NRC-National Academy of Sciences, Washington, D.C., 1976, pp. 26, 27; with U.S. Customary added by the author (also see Table 4-15 in this section).

Nutritional Reproductive Failure

Since cattlemen largely determine their own destiny when it comes to feeding, it is important that they know the causes of nutritional reproductive failure and how to rectify them. The following points are pertinent thereto:

1. Energy is more important than protein in reproduction.
2. Beef cows receiving inadequate energy reproduce at a low level.
3. Phosphorus supplementation of cows on range areas deficient in phosphorus increases the calf crop.
4. Administering additional vitamin A to heifers grazing dry forage increases the calf crop.
5. The level and kind of feed before and after calving will determine how many cows will show heat—and conceive. After calving, feed requirements increase tremendously because of milk production; hence, when a cow is suckling a calf, she needs approximately 50 percent greater feed allowance than during the pregnancy period (see Table 4-28). Otherwise, she will suffer a serious loss in weight and fail to come in heat and conceive.
6. Cows in average condition should gain a minimum of 100 lb during the pregnancy period, followed by a gain of ½ to ¾ lb daily after calving and extending through the breeding season. If they are on the thin side at calving time, they should gain 1½ to 2 lb daily after they drop calves. This calls for 7 to 11 lb of TDN daily before calving (which can be provided by feeding 14 to 22 lb of average quality hay), and 10 to 17 lb of TDN after calving (which can be provided by feeding 14 to 28 lb of hay plus 4 lb grain), with the lactating requirement dependent on both cow weight and milking ability. Additionally, there must be adequate protein, minerals, and vitamins.

WINTER FEEDING

Winter feeding is the most expensive time in cow-calf operations. From an economic standpoint, therefore, it is important that wintering practices be both knowledgeable and wise. The cheaper home-grown roughages should constitute the bulk of the winter ration for dry pregnant cows. A practical ration may consist of silage and/or dry roughages (legume or grass hays) combined with a small quantity of protein-rich concentrates (such as soybean meal or cottonseed meal). With the use of a leguminous roughage, the protein-rich concentrate may be omitted. Dusty or moldy feed and frozen silage should be avoided in feeding all cattle—especially in the case of the pregnant cow, for such feed may produce complications and possible abortion.

Except during the winter months, pastures constitute most of the feed of beef cattle. By fall, however, grass is usually in short supply and relatively poor as a source of protein, certain minerals (especially phosphorus), and carotene (provitamin A). To overcome these deficiencies, the cattleman must resort to either (1) supplemental feeding on pasture, or (2) drylot feeding. At no other time in the operations is a possible profit so likely to be dissipated and replaced by loss.

Fall feeding should not be delayed so long that animals begin to lose weight. The reason cattle often eat and get poor on dry, weathered grass is that it is low in energy, protein, carotene, phosphorus, and perhaps certain other minerals. These deficiencies become more acute and increase in severity as winter advances. Cattle simply cannot consume sufficient quantities of such bulky, low-quality roughage to meet their needs; and the younger the animal the more acute the problem. Under such circumstances the maintenance needs are met by the breakdown of body tissues, accompanied by the observed loss in weight and condition. Young animals fail to grow; it makes for lightweight calves. Also, reproduction is affected adversely; serious underfeeding results in lowered calf crops. Supplementing fall grass with a concentrated type of supplement is the practical and ideal way in which to alleviate such nutritive deficiencies.

Rations for Dry Pregnant Cows

Dry pregnant cows in average condition should make sufficient gain in weight to account for the growth of the fetus (60 to 90 lb), plus sufficient increase in weight and condition to carry them through the suckling period. In total, they should gain 100 to 150 lb during the pregnancy period, or at the rate of approximately ½ lb daily. Of course, the size and condition of the cow is the best gauge as to the feed allowance and desired gain. As previously noted, dry cows require less supplementation than cows suckling calves. Nevertheless, they should not be permitted to lose too much flesh, unless, of course, they are overfat. Also, it is recognized that it requires less feed to keep cattle from losing flesh than it does to restore them to proper condition after they have become thin. Thus, it is good economy to start feeding before cows show any signs of malnutrition. Unless a good quality legume roughage is fed, the concentrate should provide protein, energy, and needed minerals and vitamins.

When winter grazing is not possible, one of the rations in Table 4-29 may be used to meet the daily needs for energy and protein of a 1,100-pound, dry pregnant cow. A combination of legume roughage with lower quality roughage (such as stalklage, straw corncobs, or cottonseed hulls) will meet both the energy and protein requirements without the use of a supplement.

TABLE 4-29

WINTERING RATIONS FOR A 1,100-POUND, DRY PREGNANT COW
(As-fed Basis)

	Rations				
	1	2	3	4	5
	(lb/day)				
Legume-grass hay	18				10
Legume-grass haylage[1]		30			
Corn or grain sorghum silage			35		
Stalklage or husklage				45	
Straw, cobs, or cottonseed hulls					10
Supplement[2]			.5	1	

[1]Haylage figured at 55% dry matter, corn or grain sorghum silage at 35% dry matter, stalklage-husklage at 45% dry matter.
[2]Supplement figured at 48% crude protein. Quantity to be adjusted in keeping with the protein content of the supplement. For example, if a 24% crude protein supplement is fed, the quantity of supplement should be doubled.

Rations for Cows Nursing Calves

Cows with calves at side should be fed for the production of milk, for which the requirements are more rigorous than those during pregnancy. This is important because, until weaning time, the growth of the calf is determined chiefly by the nourishment available through the milk of its dam. The principal part of the calf's ration, therefore, may be cheaply and safely provided by giving its mother the proper feed for the production of milk. To stimulate milk flow, most beef cows need a concentrate during the winter months; and the poorer or the more limited the roughage, the higher the supplement requirement. On the average, cows that calve in the fall should be fed a minimum of 4 to 6 pounds concentrate daily throughout the winter.

The energy requirement of a cow nursing a calf is about 50 percent higher than that of a dry pregnant cow; and the protein, calcium, and phosphorus requirements are about double. Since the vast majority of the nation's cows with calves at side are on pasture most, if not all, of the lactation period, the only supplemental need is for salt and other minerals, unless the pasture is insufficient in quantity or quality of feed to support adequate milk production. The rations in Table 4-30 may be used for drylot feeding of beef cows nursing calves. Of course, the daily levels shown in this table should be approached gradually so that nutritional scours will not develop in baby calves.

TABLE 4-30

WINTERING RATION FOR A 1,100-POUND COW NURSING A CALF
(As-fed Basis)

	Rations				
	1	2	3	4	5
	(lb/day)				
Legume-grass hay	30			20	10
Legume-grass haylage[1]		50			
Corn or grain sorghum silage[1]			60		40
Grain				5	
Supplement[2]			1.5		

[1]Haylage figured at 55% dry matter; corn or grain sorghum silage figured at 35% dry matter.
[2]Supplement figured at 48% crude protein. Quantity to be adjusted in keeping with the protein content of the supplement. For example, if a 24% crude protein supplement is fed, the quantity of the supplement should be doubled.

Crop Residues and Winter Pastures

Two requisites are important in wintering the cow herd: (1) bringing them through the winter in proper condition for calving, and (2) keeping feeding costs to the minimum consistent with nutritional demands. Meeting these requirements has prompted increased use of crop residues and winter pastures for brood cows. As the ever-increasing human population of the world consumes a higher proportion of grains and seeds, and their by-products, directly, cattle will utilize a maximum of crop residues and pastures and a minimum of products suitable for human consumption. Thus, more and more farmers with crops will include a beef herd in their operations and realize a fair return from feeds which would otherwise be wasted.

Generally speaking, crop residues may be grazed, processed as dry feed, or made into silage. The important thing to remember is that their relatively low value, in comparison with grains, necessitates low-cost harvesting, storing, and feeding. Also, they must be fed to the right class of animals, and they must be properly supplemented. Remember, too, that there is a marked difference between economical wintering and deficient wintering.

Corn Residues

Of all crop residues, the residue of corn is produced in greatest abundance and offers the greatest potential for expansion in cow numbers. In 1980, 73,061,000 acres of corn, yielding 91 bushels per acre, were harvested in this country. For the most part, over and above the grain, 2½ to 3 tons of dry matter produced per acre (40 to 50% of the energy value of the total corn plant) were left to rot in the field. That's 200 million tons of potential cow feed wasted, enough to winter 151.5 million dry pregnant cows consuming an average of 22 pounds of corn refuse per head per day during a 4-month period. Moreover, mature cows are physiologically well adapted to utilizing such roughage. And that's not all! When corn residue is used to the maximum as cow feed, acreage which would otherwise be used to pasture the herd is liberated to produce more corn and other crops. Also, there are many other crop residues, which, if properly utilized, could increase the 151.5 million figure.

Fig. 4-18. Cows grazing cornstalks.

Although corn refuse offers tremendous potential as a cow feed, there are difficulties in harvesting and storing it. But science and technology have teamed up and are working ceaselessly away at solving these problems.

Broadly speaking, three alternate methods of salvaging corn refuse are being used: (1) grazing, (2) harvesting and dry feeding, and (3) ensiling; with different ways of accomplishing each.

Other Crop Residues

A host of crop residues, other than corn residue, can be used for feeding cows; among them, the following: sorghum (milo), soybean refuse, small grain refuse, legume and grass seed straws, and cottonseed hulls.

Winter Pasture

Where feasible, winter pasture offers cattlemen a means of reducing costs. By accumulating the feed in the field, rather than harvesting, storing, and handling the forage, the cost and labor of winter feeding can be substantially reduced. Also, costs of bedding and manure hauling can be eliminated.

Tall fescue is used as a winter pasture in the area to which it is adapted—Missouri, Illinois, Indiana and Ohio. Usually, the new regrowth is baled in late June into round bales and left in the field. The round bales shed rain and snow and, together with the regrowth, make excellent late fall and winter grazing. Experience shows that field stored forage has adequate quality to maintain beef cows in good condition.

Fig. 4-19. Cattle winter grazing (continuous access) on round bales and tall fescue regrowth.

PASTURES AND RANGES

Good pasture or range is the cornerstone of successful beef cattle production, as is attested by the following facts:

1. A total of 26 percent of the total land area of the United States (50 states) is used solely for grassland.

2. A total of 87 percent of the total feed supply of all U.S. beef cattle is derived from forage; in season this means pasture.

3. Good pasture alone will produce 200 to 400 pounds of beef per acre annually (in weight of calves weaned, or in added weight of older cattle); superior pastures will do much better.

4. No method of harvesting has yet been devised which is as cheap as that which can be accomplished by animals.

5. Pasture gains are generally cheaper than drylot gains because (a) less labor is required, (b) grass is the cheapest of all roughages, (c) less expensive protein supplement is required, (d) the animals scatter their own droppings, thus alleviating hauling manure, and (e) fewer buildings and less equipment are necessary.

Western pastures that receive less than 20 inches of rainfall annually are classified as the western range. The carrying capacity of much of the range is low, and little of it provides yearlong grazing. Moreover, variation in vegetative types, climate, and topography in the range country is accompanied by great diversity in the seasonal use made of it. As a result, rangelands are usually grazed during different parts of the year, and the herds migrate with the season, moving to the mountains and higher elevations in summer and returning to the lower ranges in winter.

From the standpoint of vegetation and utilization by livestock, ranges differ from tame, or cultivated, pastures as follows:

1. They are less productive.
2. They are more likely to progress to less palatable plants.
3. They are more difficult to restore when depleted.
4. They often serve multiple use—for wildlife production, recreation, timber production, and mineral production, as well as use by cattle and other domestic animals.

RANGE NUTRIENT DEFICIENCIES

Every cattleman worthy of the name forces his young stock for an early market; most soils are deficient in certain nutrients, which, in turn, affect the plants and the animals feeding thereon; during droughts and early and late in the season, feed may be in short supply (thereby limiting energy and other nutrients); early spring pastures are washy and lacking in energy; during droughts and late in the season, grasses become mature, leached, and bleached—they increase in fiber and decrease in protein, phosphorus, and carotene. To meet these conditions, a supplemental source of energy, protein, phosphorus, and vitamin A is necessary.

PASTURE AND RANGE SUPPLEMENTS

Improved pasture or range should be the first goal of cattlemen, without using supplemental feeding as a substitute for good grass or as a crutch for poor range. Instead, the two—good range and proper supplemental feeding—go hand in hand.

Where dried grass cured on the stalk is grazed, or where insufficient pasture is available—perhaps due to drought or overstocking—supplemental feeding is necessary. Also, supplemental feeding is a way in which to extend the grazing season, both early and late.

There is no one best and most practical pasture or range supplement for any and all conditions. Many different feeds may be used; among them, (1) farm or ranch (or local) produced hay, (2) alfalfa pellets or cubes, with or without fortification, and (3) supplements of various kinds.

Also, cattlemen can lessen the labor attendant to the daily feeding of a pasture or range supplement by (1) hand feeding cubes at intervals, rather than daily, (2) use of protein blocks, (3) use of liquid protein supplements, or (4) self-feeding salt-feed mixtures. Where these feeding systems do not result in the neglect of the herd, there is no effect upon the health and weight of the cows, percent calf crop, or weaning weight of calves.

• *Range cubes or pellets*—Traditionally, cattle have been supplemented either once or twice daily on pasture or range, with the cubes scattered on the ground.

Urea-containing supplements, particularly those containing high levels of urea, should not be fed at intervals on the range because (1) range forages are relatively low in energy, and (2) urea is extremely soluble and its nitrogen becomes available very quickly in the rumen. Where nonprotein nitrogen is used in a range cube or pellet, a slow-release product is safest. Two suggested formulations for pasture-range cubes or pellets are given herein: one without urea (Table 4-31), and the other with urea (Table 4-32, page 300).

TABLE 4-31

RANGE CUBE OR PELLET, WITHOUT UREA
(As-fed Basis)

Ingredient	Percent	Per Ton
	(%)	*(lb)*
Soybean meal, 44% (or cottonseed meal)	72.7	1,454
Alfalfa meal, 15%	15.0	300
Molasses (sugarcane)	8.5	170
Dical., or equivalent	2.0	40
Salt	1.0	20
Trace minerals	.5	10
Vitamin A[1] (30,000 IU/gram potency)	.3	6
TOTAL	100.0	2,000

Proximate Analysis:

	(%)
Crude protein	35.9
Fat	1.2
Fiber	8.3
Calcium	1.01
Phosphorus	.9
TDN	68.7

[1]In low-sunshine areas, also add 6 million IU of vitamin D/ton of finished feed.

• *Hand feeding at intervals, rather than daily*—Based on a four-year study done at the Texas Station, plus observations and experiences, the author recommends feeding nonurea range supplement twice weekly, allocating in each of the two feedings one-half as much supplement as would have been fed in a week on a daily feeding basis.

TABLE 4-32

RANGE CUBE OR PELLET, WITH UREA
(Preferably a Slow-Release Product)
(As-fed Basis)

Ingredient	Percent	Per Ton
	(%)	(lb)
Corn #2 (barley, wheat, oats, and/or milo)	34.7	694
Soybean meal, 44% (cottonseed, linseed[1] and/or peanut meal)	32.5	650
Alfalfa meal, 15%	15.0	300
Molasses, Sugarcane	10.0	200
Urea, 45% grade (use slow-release product)	4.0	80
Dical., or equivalent	2.0	40
Salt	1.0	20
Trace minerals	.5	10
Vitamin A[2] (30,000 IU/gram potency)	.3	6
TOTAL	100.0	2,000

Proximate Analysis:

	(%)
Crude protein[3]	31.8
Fat	2.2
Fiber	6.6
Calcium	.9
Phosphorus	.7
TDN	67.5

[1]If linseed is used, limit to 6% of the ration.
[2]In low-sunshine areas, also add 6 million IU of vitamin D/ton of finished feed.
[3]This includes not more than 11.24% equivalent protein from nonprotein nitrogen; 34.9% of the total protein is furnished by urea (use a slow-release product).

Protein cubes may be scattered on the ground, two or three times a week. This offers a method of checking the animals because they are attracted by the sight or sound of the vehicle when they know that there is something to eat.

Twice-weekly feeding has two distinct advantages over the use of salt-feed mixes: (1) It alleviates the cost of using excess salt, which has no nutritive value when so used; and (2) it forces inspection of the herd two times per week, which is as infrequent as is desirable.

• *Protein blocks*—Protein blocks are just what the designation implies. They are compressed protein blocks, generally weighing from 30 to 50 pounds each.

Blocks may be placed in grazing areas where cattle have frequent access to them, with one block provided to 15 cows. Intake will vary with the feed supply and the type of block. Generally, it is planned to limit feed consumption to about 2 pounds per head per day by hardness of block and salt and/or fat content.

• *Liquid protein supplements*—Liquid supplements in a "lick" tank can be offered free-choice. This is a convenient and satisfactory way in which to supply protein, energy, and other nutrients, so long as the cattle do not consume more than they need.

• *Self-feeding salt-feed mixture*—The practice of using salt as a governor to limit feed consumption on pasture or range has been around a very long time. It was ushered in as a laborsaving device for cattle and sheep in inaccessible and rough areas.

The proportion of salt to feed may vary anywhere from 5 to 40% (with 30 to 33⅓% salt content being most common), with the actual intake of feed supplement limited to 1 to 2½ pounds daily. By varying the proportion of salt in the mixture, it is possible to hold the consumption of feed supplement to any level desired. In some range areas, a reduction of the salt level from 33⅓ to 24% will increase consumption by about 50%. When a liberal feeding of grain on pasture is desired, 4% salt may be sufficient.

Two suggested salt-meal supplements follow. Either 41 percent cottonseed or soybean meal may be used. Neither mix should be pelleted.

Ingredient	Salt-Meal Mix No. 1	Salt-Meal Mix No. 2
	(lb)	(lb)
Salt	665	499
Meal, either 41% cottonseed or soybean meal	1,331	1,497
Vitamin A (30,000 IU/g)	4	4
	2,000	2,000
Guarantee:		
Crude protein	27%	30%
Salt	max. 35%	max. 27%
Vitamin A	18,000 IU/lb	13,000 IU/lb
Consumption level	approx. 1½ lb daily	approx 2 lb daily

RATIONS FOR CONFINEMENT COWS

Confining (drylotting) beef cows refers to the practice of confining beef cows to small quarters—to drylots, all or part of the year.

From a feeding standpoint, the following points are pertinent in a drylot beef cow operation:

1. All feed must be mechanically harvested and moved to the feedlot, rather than being harvested directly by the cows.

2. An assured, adequate, and economic feed supply must be available. The capital tied up in stored feeds may be quite large.

3. More knowledge of beef cow nutrition and ration formulation is needed than for unconfined cows.

Rations for drylot cows generally consist of cheap roughages—such as crop refuse, straw, cottonseed hulls, and gin trash—supplemented with protein, grain, vitamins, and minerals as required. Where available, higher quality roughages—such as silages, hays, and haylages—may be used, especially (1) during the critical 100 days, beginning 30 days before calving and extending to 70 days after calving; and (2) for heifers calving as 2-year-olds. Also, during the summer and fall, green chop is frequently fed.

The mineral needs of confinement cows may be met either by incorporating the needed minerals in the supplement which is fed, or by feeding the required minerals free-choice.

Vitamin A supplementation is extremely important for drylot cows. The carotene content of the dry forage should be disregarded and the total vitamin A requirement met by supplementation. This can be done by feeding a supplement of 2 pounds of mill feed containing 1 million IU of vitamin A per animal—feeding this vitamin A supplement once a month to heifers, and every other month to older cows. With older cows receiving high levels of dry forages containing normal amounts of carotene, it is probable that the vitamin A requirements are being met. However, it has been demonstrated under range conditions that (1) percent calf crop is markedly increased by supplementing with vitamin A during drought years, and (2) calves respond to vitamin A treatments given their dams 90 days prior to calving.

RATIONS FOR SEMICONFINEMENT (or Partial Confinement) COWS

A semiconfinement (or partial confinement) operation is one which takes advantage of grazing during part of the year, such as winter grazing of corn or sorghum stalks or seasonal grazing of pastures. In addition to providing low-cost feed and allowing the animals to do their own harvesting, breeding may be timed so that the calves will be dropped on clean pasture as a means of (1) preventing calf scours, and (2) stimulating milk flow.

Feeding Calves

Cattlemen, as a whole, have lagged in applying much of what we know about feeding and managing calves. They're inclined to let mother cows and mother nature fend for the calves. Indeed, more good proven practices, based on both successful operations and research, need to be put to use in feeding and handling calves.

FEEDING ORPHAN AND MULTIPLE BIRTH CALVES

Occasionally a cow dies during or immediately after parturition, leaving an orphan calf to be raised. Also, there are times when cows fail to give a sufficient quantity of milk for the newborn calf. Sometimes, there are multiple births.

Fortunately, orphan calves can now be raised successfully on a milk replacer and calf starter ration, using them as directed. The milk replacer may be fed by using a bottle or pail equipped with a rubber nipple, or the calf may be taught to drink from a pail. It is important that all receptacles be kept absolutely clean and sanitary (clean and scald each time) and that feeding be at regular intervals. Dry feed should be started at the earliest possible time, not later than one week of age. With proper management, healthy calves may be switched entirely to a suitable dry feed at four to five weeks of age.

Basically, calves are fed according to one of three systems: (1) the whole milk system, (2) the combination whole milk-milk replacer system, or (3) the combination whole milk-calf starter system. Of course, various combinations of these three systems are used, also. A suggested schedule for each of these three systems and two suggested starter rations are given in Part III, Feeding Dairy Cattle, "Feeding Dairy Calves" (see Table 4-50, page 348, for schedule, and see Table 4-51, page 349, for starter rations).

FEEDING EARLY WEANED CALVES

Early weaning refers to the practice of weaning calves earlier than the usual weaning age of about 7 months, usually within the range of 45 days to 5 months of age. Although it is not common practice among U.S. beef cattlemen, dairymen have been weaning 3-day-old calves for years. Also, early weaning has long been an integral part of many of the beef programs of Europe.

Currently, there is much interest in early weaning because (1) it fits into a drylot cow-calf management system; and (2) it can give a big assist in getting females, especially two-year-old heifers, to rebreed in a short period of time.

Considering the low efficiency involved in converting supplemental energy to milk and in converting milk to meat, it is apparent that a more efficient use of feed could be achieved by giving the supplemental feed directly to the calf. A lactating cow requires about 50 percent more feed than a dry cow. So, rather than give her that additional feed, it is more efficient to give feed directly to the calf, thereby favoring early weaning.

Where early weaning is successful, the only responsibility of the beef cow is to produce a calf and give it a good start in life for a brief period, then go on a maintenance ration the rest of the year.

Like many good things in life, early weaning does have some disadvantages. To be successful, superior nutrition and management are essential; and the earlier the weaning age, the more exacting these requirements.

From 45 days of age on, early-weaned calves can be fed any good starter ration (see Table 4-51), which generally contains dry skim milk. Most commercial feed companies manufacture a starter ration. Of course, the starter ration should be made available to the calves well ahead of weaning in order that they will be accustomed to it, thereby avoiding any setback.

CREEP FEEDING

Creep feeding is the supplementation of calves while they are nursing their dams. It increases weaning weight.

Fig. 4-20 shows why creep feeding is important. From birth to weaning, the protein and energy requirements of a growing calf increase well beyond the ability of most beef cows to meet those needs. For example, to meet the protein and energy requirements for growth, a 100-lb calf needs 10 lb of milk, whereas a 500-lb calf needs 50 lb of milk. Since the average beef cow gives only 13 lb of milk per day throughout a 7-month suckling period, a 500-lb calf lacks 40 lb of getting enough milk from its dam at this stage of lactation to meet its needs—that's the "hungry-calf gap."

To fill the "hungry-calf gap"—the nutrient requirements over and above that provided by 13 lb of milk—would require the consumption of 50 lb of green grass daily. Of course, that's a physical impossibility, because a 500-lb calf simply cannot hold that much bulk. So, the best way to fill the "hungry-calf gap" is to creep feed.

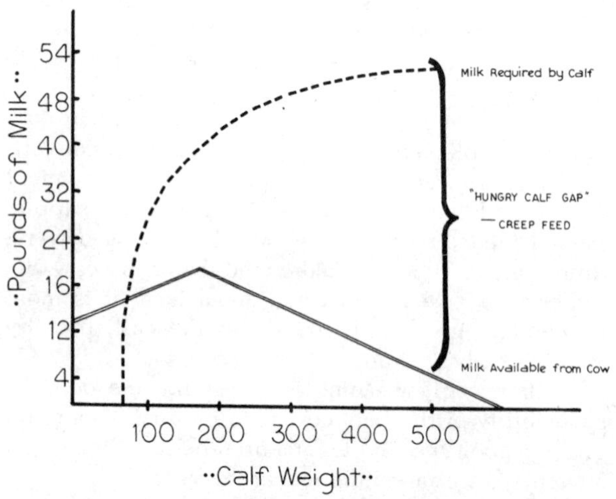

Fig. 4-20. The hungry-calf gap—the difference between (1) the milk required by the calf, and (2) the milk available from the cow.

THE CREEP

A creep is an enclosure or feeder for feeding purposes which is accessible to the calves but through which the cows cannot pass. It allows for the feeding of the calves but not their dams. The enclosing fence may be of board, pole, or metal construction, with an entrance 16 to 20 inches wide and 3 to 3½ feet high.

Fig. 4-21. A calf creep.

CREEP RATIONS; FEEDING DIRECTIONS

Creep-fed calves need special rations. They require a feed that is high in protein; rich in readily available energy; fortified with vitamins, minerals, and unidentified factors; and with all the nutrients in proper balance. Also, the ration must be very palatable. This calls for an exacting ration. To meet these needs, more and more cattlemen are finding it practical to buy a commercially prepared complete creep feed, or a well-fortified and highly concentrated supplement to add to locally available feeds, rather than purchase individual ingredients and mix from the ground up.

Table 4-33 and 4-34 show two creep rations, formulated by the author, that have been widely and successfully used.

When 3 to 4 weeks of age, calves should be started on feed very gradually. For the first 3 to 5 days, only about ¼ pound of feed per calf should be placed in the container(s) each day, and any leftover feed should be removed and given to the cows. In this manner, the feed will be kept clean and fresh. When calves are on lush pasture and their mothers are milking well, difficulty may be experienced in getting them to eat; but time and patience will pay off, and the results will become evident in 2 to 3 months.

After 5 to 7 days of hand feeding, the creep ration can be left before the calves safely. Once they are on full feed, never let the feeder become empty; and avoid sudden changes. During the first 30 days, they will consume about a pound per head daily. By the end of the fifth month, they should be up to 8 pounds daily. Of course, with good pastures and plenty of

TABLE 4-33
CALF CREEP RATION #1[1]
(As-fed Basis)

Ingredient	Percent	Per Ton
	(%)	(lb)
Oats	39.6	800
Corn #2	14.8	300
Barley	9.9	200
Wheat bran	9.9	200
Dried molasses beet pulp	9.9	200
Soybean meal, 44%	9.9	200
Molasses	4.9	100
Salt	.5	10
Dicalcium phosphate	.5	10
Trace minerals	.04	1
Vitamin A (30,000 IU/g)	.06	1.5
TOTAL	100.0	2,022.5

Proximate Analysis:

	(%)
Crude protein	14.3
Fat	3.2
Fiber	8.3
Calcium	.32
Phosphorus	.51
TDN	69.6

[1]*Feed Preparation*: Preferably 1/8- or 3/16-inch pellets. Otherwise, steam roll and flake grains, or grind grains coarsely.

milk, these consumption figures may be halved. Calves will consume approximately 500 pounds of creep feed per head from one month of age to weaning. In years of lush pasture, it will be less; in dry years more.

TABLE 4-34
CALF CREEP RATION #2
(As-fed Basis)

Ingredient	Percent	Per Ton
	(%)	(lb)
Corn #2	24.25	485
Alfalfa meal, 15%	22.50	450
Oats	20.00	400
Alfalfa hay (all analyses)	10.00	200
Soybean meal, 44%	6.20	124
Bran	5.00	100
Linseed meal, 35%	5.00	100
Molasses	5.00	100
Dicalcium phosphate	2.00	40
Trace minerals	.05	1
Vitamin A (325,000 IU/g)[1]	—	84 g
TOTAL	100.0	2,000

Proximate Analysis:

	(%)
Crude protein	15.1
Fat	3.0
Fiber	12.7
Calcium	1.04
Phosphorus	.73
TDN	64.9

[1]When 4 lb/head/day of the calf creep ration is consumed, 54,600 IU of vitamin A will be obtained in the feed.

WHY CREEP FEED?

The three major reasons for creep feeding are:

1. *It provides a way to fill the "hungry-calf gap"*—Creep feeding provides a logical and practical way to compensate for insufficient milk.
2. *It makes for heavier weaning weights*—Creep feeding results in 50 to 70 pounds heavier weaning weight per calf, at no extra cost for the capital investment in land and cows.
3. *It usually pays*—The potential profitability of creep feeding depends upon (a) the price of cattle, and (b) the price of feed.

The following rule of thumb may be used to determine whether or not it will pay to creep feed: It pays to creep feed when the selling price per hundred pounds of calf is greater than the cost of ¼ ton (500 lb) of feed.

LIMITATIONS OF CREEP FEEDING

Like many good things, creep feeding does have its limitations; among them, the following: It isn't always profitable; it lowers subsequent feedlot gains and efficiency; it makes for less desirable stockers, because the latter are normally placed on less nutritious growing rations consisting predominantly of roughages; it mitigates against selecting cows for milk production; it is difficult to do in remote areas; and it cannot be done where there are hogs, sheep, or goats.

GROWTH STIMULANTS FOR CALVES

Growth-promoting implants generally increase rate of gain and weaning weight of creep-fed calves. But there is little or no benefit from implanting noncreep-fed calves because the function of a growth stimulant is to improve utilization of energy, especially concentrates. The growth stimulants that are presently available and approved for suckling calves are listed in Table 4-24, Implants and Growth Stimulants for Finishing Cattle, page 284.

PRECONDITIONING

Preconditioning is a way of preparing the calf to withstand the stress and rigors of leaving its mother, learning to eat new kinds of feed, and shipping from the farm or ranch to the feedlot. To the cow-calf producer, it is a program of management, nutrition, and immunization. It, along with improved breeding based on production testing, is the trademark of the producer of reputation feeder calves. To the feedlot operator, preconditioning is a way in which to prepare calves to fit into his program and to minimize costly and unnecessary procedures. (Also see Section 8, Management.)

Although preconditioning embraces many steps and practices, none is more important than adjusting to feed bunks and water troughs and starting on a ration similar to that which they will get in the feedlot. For the first 3 days following weaning, calves should have access to loose grass hay. Additionally, they should be started on a ration of about the following composition:

Crude protein, minimum %	12.0
Calcium, %	.5
Phosphorus, %	.3
Vitamin A, IU/lb	5,000
Net energy for production (NE_p), Mcal	38
Roughage: concentrate ratio, approx.	40:60

Fig. 4-22. Preconditioning is the schedule of practices followed in preparing calves for separation from their mothers.

If weaning is totally impractical, calves should be started on a creep feed similar to the above ration. This type of ration will be very similar to the starting ration that they will receive when they arrive in the feedlot.

Feeding Replacement Heifers

The feeding program of replacement heifers will have a lifelong effect on their productivity. It will determine how young they may be bred, whether they calve early or late, whether they are good milkers or poor milkers, the weaning weight of their calves, and how long they remain in the herd. Also, feed accounts for 40 to 70 percent of the cost of raising replacement heifers.

The pregnancy requirements of replacement heifers are really not too great. The body of an 80-lb, newborn calf contains only about 12 lb of protein, 3. lb of fat, and 3.6 lb of mineral matter. But the lactatio requirements are much more rigorous. If a 2-year-ol heifer gives her calf an average of 1¾ gallons of mil per day over a 7-month suckling period, she will pro duce in that milk a total of 93 lb of protein, 107 lb o fat, 133 lb of sugar, and 20 lb of minerals. Hence, th comparison: 12 lb of protein in the fetus vs 93 lb i the milk during the suckling period. This means tha nearly 8 times more protein is required for 7 month of lactation than for 9 months of pregnancy.

Also, when breeding yearlings to calve as two year-olds, cattlemen should be aware that nature ha ordained that the growth of the fetus, and the lactatio which follows, shall take priority over the materna requirements. Hence, when there is a nutritive defi ciency, the young mother's body will be deprived, o stunted, before the developing fetus or milk produc tion will be materially affected.

NUTRIENT REQUIREMENTS

Meeting the nutrient requirements of heifer from weaning to first calving is of great importance The requirements of heifers of different body weight and growth rates are given in Table 4-35.

SUGGESTED RATIONS

In season, good pasture plus mineral supple ments fed free-choice will meet the nutrient require ments for proper growth and development of heifers

On winter pasture, when dry forage is of lo quality, and sometimes not too abundant, 1 to pounds of a protein supplement should be provide in the form of cubes, blocks, meal-salt, or liquid When consumed at the intended level, the supple ment should contain sufficient vitamin A to meet th requirements. Mineral supplements should also b provided, preferably free-choice.

Where winter grazing is not available, heifer must be drylotted and fed a complete ration. Suff cient nutrients should be provided to meet the re quirements and to keep heifers in a thrifty conditio neither too fat nor too thin.

The wintering rations in Table 4-36 for 500 pound heifer calves should result in a rate of gain of to 1.5 pounds per day.

The wintering rations in Table 4-37 for 800- t 900-pound bred yearling heifers should allow a gai of 0.75 to 1 pound per day during the winterin period prior to calving.

Replacement heifers should be fed rathe liberally—more so than stocker cattle which are bein grown for the feedlot, to the end that they will acquir most of their growth and development before calving With limited feeding, they will not have enoug

TABLE 4-35

DAILY NUTRIENT REQUIREMENTS OF GROWING HEIFERS (per animal)[1]

Body Weight		Daily Gain		Crude Protein		TDN		Calcium	Phosphorus
(lb)	*(kg)*	*(lb)*	*(kg)*	*(lb)*	*(kg)*	*(lb)*	*(kg)*	*(g)*	*(g)*
331	*150*	1.50	*0.7*	1.10	*0.5*	6.2	*2.8*	18	14
441	*200*	1.50	*0.7*	1.34	*0.6*	8.4	*3.8*	18	16
551	*250*	1.50	*0.7*	1.37	*0.6*	9.1	*4.1*	17	15
661	*300*	2.00	*0.9*	1.54	*0.7*	11.5	*5.2*	19	17
772	*350*	0.7	*0.3*	1.52	*0.7*	10.0	*4.5*	15	15
882	*400*	0.7	*0.3*	1.68	*0.8*	11.1	*5.0*	16	16

[1]Adapted by the author from *Nutrient Requirements of Beef Cattle*, No. 4, NRC-National Academy of Sciences, Washington, D.C., 1976, pp. 24, 25, Table 1A (also see Table 4-14 of this section).

TABLE 4-36

DAILY RATIONS FOR HEIFER CALVES (500 LB)
(As-fed Basis)

	Rations									
	1		2		3		4		5	
	(lb)	*(kg)*	*(lb)*	*(kg)*	*(lb)*	*(kg)*	*(lb)*	*(kg)*	*(lb)*	*(kg)*
Legume-grass haylage	25	*11*								
Legume-grass hay			10	*4.5*	10	*4.5*			5	*2.3*
Corn or sorghum silage							30	*13.6*	20	*9.1*
Ground ear corn			4	*1.8*						
Corn, grain sorghum, or barley					3	*1.4*				
Supplement[1]							1	*.45*		

[1]Supplement contains 48% crude protein. Quantity to be adjusted in keeping with the protein content of the supplement.

TABLE 4-37

RATIONS FOR BRED YEARLING HEIFERS (800 to 900 LB)
(As-fed Basis)

	Rations									
	1		2		3		4		5	
	(lb)	*(kg)*	*(lb)*	*(kg)*	*(lb)*	*(kg)*	*(lb)*	*(kg)*	*(lb)*	*(kg)*
Corn or sorghum silage	45	*20.5*	25	*11*						
Legume-grass hay			10	*4.5*	20	*9.1*			15	*6.8*
Legume-grass haylage							35	*15.9*		
Corn, grain sorghum, or barley									3	*1.4*
Supplement[1]	1.5	*0.7*								

[1]Supplement contains 48% crude protein. Quantity to be adjusted in keeping with the protein content of the supplement.

weight for age to breed when they are 15 months old; and it is best not to have them calve until they are 30 months of age. Slower development and delayed breeding may be practical where forage is abundant and cheap, while concentrates are scarce and high.

Occasionally, a replacement animal is injured by overfeeding or by fitting for the show, but such losses are insignificant compared with those resulting from the thousands of undersized, poorly developed animals that are grossly underfed.

SEPARATE HEIFERS BY AGES

The nutritive requirements for heifers differ according to body weight and expected daily gain (Table 4-35). Consequently, the recommended ration for a 500-pound heifer calf (Table 4-36) differs from that of an 800- to 900-pound bred heifer (Table 4-37). It is important, therefore, that replacement heifers be separated by ages for wintering, with coming yearlings in one group and coming 2s in another.

SUMMARY RELATIVE TO CALVING TWO-YEAR-OLDS

From the above, it may be concluded that unless forage is abundant and cheap cattlemen can, and should, breed yearling heifers to calve as two-year-olds. (Also see Section 8, Management.)

Feeding Herd Bulls

Frequently, little thought is given to the feeding and management of bulls except during the breeding season. Instead, the feeding program for herd bulls should be such as to keep them in a thrifty, vigorous condition at all times. They should neither be overfitted nor in thin, run-down condition. Also, exercise is necessary for the normal well-being of the bull.

Winter is the proper time to condition bulls for the next breeding season. Bulls that have been running on pasture with the cows are likely to be thin; thus, they require sufficient concentrate to put them in proper flesh. Mature bulls will consume daily amounts of feeds equal to 1½ to 3 percent of their liveweight, depending upon condition and individuality.

Mature bulls should be fed all the legume hay they will eat plus 3 to 5 pounds of ground or rolled grain and 1 pound of a 32 percent protein supplement (or equivalent) per head per day. Also, free access to a suitable mineral mixture should be provided. About 60 days before the bulls are turned with the cows, the concentrate allowance should be increased by 25 to 50 percent, with the amount of the increase determined by the condition of the bulls.

Mature herd bulls need no additional feed when running with the cow herd on good summer pasture.

FEEDING STOCKERS

Stockers are calves and yearlings, both steers and heifers, that are intended for eventual finishing and slaughtering and which are being fed and cared for in such manner that growth rather than finishing will be realized. They are generally younger and thinner than feeder cattle.

Today, the stocker stage is changing as a result of forces working in opposite directions, with one force favoring lengthening of the stocker stage and the other favoring shortening it:

1. Scarce and high-priced grains favor more roughage feeding and less grain feeding, resulting in carrying stockers to older ages and heavier weights, followed by a shorter feedlot period.

2. Heavier milking cows and heavier weaning weights, coupled with high-priced land, favor shortening the stocker stage, or even eliminating it, as 600-pound, or heavier, weaning weights are achieved.

In the future, both types of stocker operations will prevail, with the choice determined primarily by the price of grain and the weaning weight of the calves. Heavy weaned calves will likely go directly into the feedlot or for slaughter. Calves with light to average weaning weights will likely be carried as stockers to 700- to 800-pound weights, thereby shortening the feedlot period and lessening grain feeding.

Today, stockers are grown according to two systems: (1) Calves or light yearlings are either (a) roughed through the winter, followed by grazing, or (b) grazed only, then sold as feeders in late summer and fall; or (2) calves or yearlings are fed harvested roughage and grain in drylot, and then transferred to another location for finishing.

Backgrounding is an old practice with a new emphasis and a new name. Actually, backgrounding and the stocker stage are one and the same. Both refer to that period in the life of a calf from weaning to around an 800-pound weight, to the stage when calves are ready to go on a high-energy finishing ration. However, the term "backgrounding," which was ushered in with the development of large feedlots, indicates a shift in emphasis. The term "stocker stage" connotes emphasis on marketing roughages through thin cattle whereas "backgrounding" connotes emphasis on growing out feeder calves ready to go on a high-energy finishing ration. Backgrounding may be done on pasture or in the drylot, or some combination of both. At its best, the animals should be in good health, bunk broke, and ready to go on full feed.

For a stocker operation to be profitable, the grower must be ever aware of the following reasons back of it, then feed stockers accordingly: (1) to provide a supply of the kind of cattle desired by finishing lots at the time needed; (2) to utilize roughages and other low-cost feeds; and (3) to "cheapen down" the cattle.

Because of the very nature of the operation, the successful feeding of stockers requires the maximum of economy consistent with normal growth and development. This necessitates cheap feed—either pasture or range grazing or such cheap harvested roughage as hay, straw, fodder, and silage. In general, the winter feeds for stockers consist of the less desirable and less marketable roughages. It is important, therefore, that the high-roughage rations of young stockers be properly supplemented from the standpoints of proteins, minerals, and vitamins.

Of course, too small gains may be unprofitable to the grower. Besides, young animals can be stunted. To make maximum growth without fattening—just to maintain condition—calves of the British breeds and crossbreds should gain 1.25 pounds daily, and yearlings should gain 0.9 pound daily.

TABLE 4-38

DAILY RATION FOR STOCKER CALVES (400-500 LB)[1]
(As-fed Basis)

	Rations (fed for gains of 1.25 lb/head/day)																			
	1		2		3		4		5		6		7		8		9		10	
	(lb)	*(kg)*	*(lb)*	*(kg)*	*(lb)*	*(kg)*	*(lb)*	*(kg)*	*(lb)*	*(kg)*	*(lb)*	*(kg)*	*(lb)*	*(kg)*	*(lb)*	*(kg)*	*(lb)*	*(kg)*	*(lb)*	*(kg)*
Legume hay or grass-legume mixed hay	12-18	*5.4-8.2*			8-12	*3.6-5.4*			2-4	*0.9-1.8*			8-10	*3.6-4.5*						
Grass hay			12-18	*5.4-8.2*	4-6	*1.8-2.7*					2-4	*0.9-1.8*					10-12	*4.5-5.4*		
Straw, corncobs, cornstalks, stalklage, or cottonseed hulls													2-4	*0.9-1.8*	2-3	*0.9-1.4*			2	0.9
Corn or sorghum silage							25-40	*11.4-18.1*	20-30	*9.1-13.6*	20-30	*9.1-13.6*								
Legume-grass silage, or oat silage															20-25	*9.1-11.4*				
Legume-grass haylage, or oat haylage																			20-25	*9.1-11.4*
Grain (corn, sorghum, barley, or oats)													4-5	*1.8-2.3*			4-5	*1.8-2.3*	4-5	*1.8-2.3*
Protein supplement (41% or equivalent)			1¼-1½	*0.6-0.7*	¼-1	*0.1-0.5*	1-1¼	*0.5-0.6*	¾-1	*0.3-0.5*	1¼-1½	*0.6-0.7*			1-1½	*0.5-0.7*	1-1½	*0.5-0.7*		

[1]With all rations, provide suitable minerals (see Tables 4-20 and 4-21).

TABLE 4-39

DAILY RATION FOR YEARLING STOCKER (600-700 LB)[1]
(As-fed Basis)

	Rations (fed for gains of 0.9 lb/head/day)																			
	1		2		3		4		5		6		7		8		9		10	
	(lb)	*(kg)*	*(lb)*	*(kg)*	*(lb)*	*(kg)*	*(lb)*	*(kg)*	*(lb)*	*(kg)*	*(lb)*	*(kg)*	*(lb)*	*(kg)*	*(lb)*	*(kg)*	*(lb)*	*(kg)*	*(lb)*	*(kg)*
Legume hay or grass-legume mixed hay	16-24	*7.3-10.9*			6-8	*2.7-3.6*			2-4	*0.9-1.8*			6-8	*2.7-3.6*						
Grass hay			16-24	*7.3-10.9*	10-16	*4.5-7.3*					2-4	*0.9-1.8*					16-20	*7.3-9.1*		
Straw, corncobs, cornstalks, stalklage, or cottonseed hulls													12-15	*5.4-6.8*	10-12	*4.5-5.4*			2	0.9
Corn or sorghum silage							45-55	*20.4-25.0*	40-50	*18.2-22.7*	40-50	*18.2-22.7*								
Legume-grass silage, or oat silage															20	*9.1*				
Legume-grass haylage, or oat haylage																			35-40	*15.9-18.2*
Grain (corn, sorghum, barley, or oats)													5-6	*2.3-2.7*			5-6	*2.3-2.7*	5-6	*2.3-2.7*
Protein supplement (41% or equivalent)			1½-1¾	*0.7-0.8*	1-1½	*0.5-0.7*	1¼-1½	*0.6-0.7*	¾-1	*0.3-0.5*	1¼-1½	*0.6-0.7*			1	*0.5*	1-1½	*0.5-0.7*		

[1]With all rations, provide suitable minerals (see Tables 4-20 and 4-21).

Nutrient Allowances

Where grower rations are formulated on the basis of percentage of nutrients in the ration, the following are recommended:

Protein

For under 1.5 lb daily gain ....10.0 %
For 1.5 lb or more daily gain ....10.5 %

Calcium and Phosphorus

For under 500 lb liveweight .... .30%
For 500 lb or over liveweight .... .25%

Vitamin A

Air-dry feed
(10% moisture) .... 800 to 1,000 IU/lb/day
.... 10,000 IU/head/day

Implant

With gains of more than
1.5 lb/head/day .... Include additive or implant

Rations for Stockers

Tables 4-38 and 4-39 contain some recommended rations for stocker cattle.

Variations can and should be made in the rations used. The grower should give consideration to (1) the supply of homegrown feeds, (2) the availability and price of purchased feeds, (3) the class and age of cattle, (4) the health and condition of animals, and (5) the kind of feeder cattle in demand by feedlots.

Level of Wintering

The level of wintering stockers affects the gains in the next stage. Thus, calves gaining the most during the winter make the least gains on pasture the following summer

Calves wintered to gain 1.0 pound daily make satisfactory summer pasture gains. This level is recommended for calves to be grazed season-long the following summer, provided the same ownership is retained all the way through. One to 2 pounds' daily gain during the winter is usually desirable if calves (1) are to be sold in the spring, (2) will be on full feed 2 to 3 months after going to grass, (3) will be receiving a limited feed of grain on grass, or (4) are replacement heifers that are to be bred at 13 to 15 months of age.

Since yearlings are not growing as rapidly as calves, they may be fed for smaller gains than calves, and yet show comparable condition. Thus, for maximum growth without fattening (for just holding their condition) calves should gain approximately 1.25 pounds daily, whereas yearlings need to gain only 0.9 pound daily.

Winter Pastures

Wherever possible, stocker calf operations are planned around a winter pasture program. Weanling calves or lightweight, thin yearlings are purchased in the fall. In some cases, homegrown calves are retained and developed under this system for sale as yearlings. As would be expected, winter pasturing of stockers is largely limited to the southern part of the United States, with the kind of pasture varying from area to area.

- *Winter wheat pastures*—Winter wheat pastures are widely used for stocker cattle in Kansas, Oklahoma, and Texas. When such pastures are good, cattle make very acceptable gains on them. However wet weather or droughts make winter wheat pastures unreliable, with the result that it is important that there be flexibility in the stocker program, both in numbers and season of use.
- *Other cool-season pastures*—In the southern states, extensive use is made of oats, rye, ryegrass vetch, and fescue—a perennial grass that remains green throughout the winter. This area is turning more and more to winter grazing, as a means of making profitable year-round use of their land and labor and providing 600- to 800-pound feeder cattle in greatest demand by feedlots.

Compensatory Growth

It is common practice for stocker cattle to be "roughed through" the winter as cheaply as possible with limited daily gains. Then, in the spring, the animals are turned to lush spring pasture or put in a feedlot on a high-energy ration. Animals so managed exhibit the phenomenon of "compensatory growth"; that is, on the high-energy diet they gain faster and more efficiently than similar cattle which were fed more liberally during the wintering period. Feedlot operators were quick to sense this situation, and to take advantage of it. This is the chief reason for the popularity of Okie-type cattle. Usually, they are animals whose growth has been held back to less than their genetic potential. When fed more liberally, they exhibit a surge in growth rate and feed efficiency. Large compensatory growth usually indicates that someone (the stocker operator) has lost money while someone else (the feeder) has made money. It is noteworthy that Holsteins and the larger exotics should never be handled so as to exhibit compensatory gains. If they're held back in the winter, they're too heavy when they finish.

FEEDING FINISHING (Fattening) CATTLE

The finishing of cattle is what the name implies: the laying on of fat.

Fig. 4-23. Feeding finishing (fattening) cattle.

In a general way, there are two methods of finish-ng cattle for market: (1) cattle feedlots, including con-finement (sheltered) finishing; and (2) pasture finish-ng. The growth of cattle feeding and composition of :attle slaughter are shown in Fig. 4-24, page 310.

Cattle feeders are commonly classed as either (1) ommercial feeders, or (2) farmer-feeders, based argely on numbers. From the standpoint of statistical eporting, the U.S. Department of Agriculture com-nonly draws the line at 1,000 head. A commercial attle-feeding operation is defined as one having a apacity of 1,000 head or more, at any one time. Commercial feeders differ from farmer-feeders in the ollowing respects: (1) They usually feed cattle on a ear-round basis, rather than during the winter nonths only; (2) they may grow little, or none, of their eed; (3) they are highly mechanized; and (4) they are nowledgeable of costs and returns, skillful buyers nd sellers, and aware of market trends. Today, com-nercial feedlots with more than 1,000-head capacity ominate cattle feeding.

Feedlots exceeding 100,000 capacity are now in peration in California, Colorado, Arizona, and the orthern Texas Panhandle. Even in the Corn Belt, eedlots are getting bigger.

eeds

For convenience, the commonly used beef cattle nishing feeds are herein classified as (1) concen-trates, (2) by-product feeds, (3) roughages, (4) protein supplements, (5) minerals, (6) vitamins, and (7) water. Also, feed additives and implants are a part of modern cattle feeding.

CONCENTRATES

Concentrate feeds are those which are high in energy and low in fiber. Many different kinds of concentrate feeds can be, and are, used in beef cattle finishing. Availability and price are the two most important factors determining the choice of concentrates.

Corn is the most common desirable grain used in finishing cattle. It is palatable and rich in the energy-producing carbohydrates and fats, and low in fiber. Also, corn is easily stored, only moisture and carotene being lost over a period of time. However, corn has certain very definite limitations—it is low in protein and calcium.

The grain sorghums are assuming an increasingly important role in cattle feeding, particularly in the fringe areas of the Corn Belt, and in the South and Southwest where moisture conditions are less favorable. New and high-yielding varieties have been developed and become popular. As a result, more and more grain sorghums are being fed to cattle. The chemical composition of sorghum (milo) is similar to corn except that the protein content is generally higher and more variable. Its feeding value is greatly enhanced by steam processing and other similar methods of preparation.

Although corn and sorghum are by far the most common grains used in finishing steers, such grains as barley, rye, oats, and wheat are used in many sections of the United States and Canada. The small grains are excellent for finishing cattle when properly used. In comparison with corn feeding: (1) barley-fed cattle are more susceptible to bloat (for this reason, it is best not to use a straight legume hay along with a grain ration high in barley; a mixture of barley and dried beet pulp is commonly used in the West); (2) barley- or wheat-fed animals are more apt to tire of their ration during a long feeding period; (3) rye should not constitute more than ⅓ of the grain ration because it is unpalatable; (4) more care is necessary to prevent wheat-fed cattle from going off feed; and (5) oats should not constitute more than ½ of the ration, and preferably not more than ⅓ because of its bulk. Fortunately, these limitations can be lessened considerably by mixing these feeds together, or by mixing the cereal grain with beet pulp, silage, or chopped hay. Also, it is important that the small grains be coarsely ground or properly rolled. It is recognized that wheat and oats are frequently too expensive to include in cattle finishing rations.

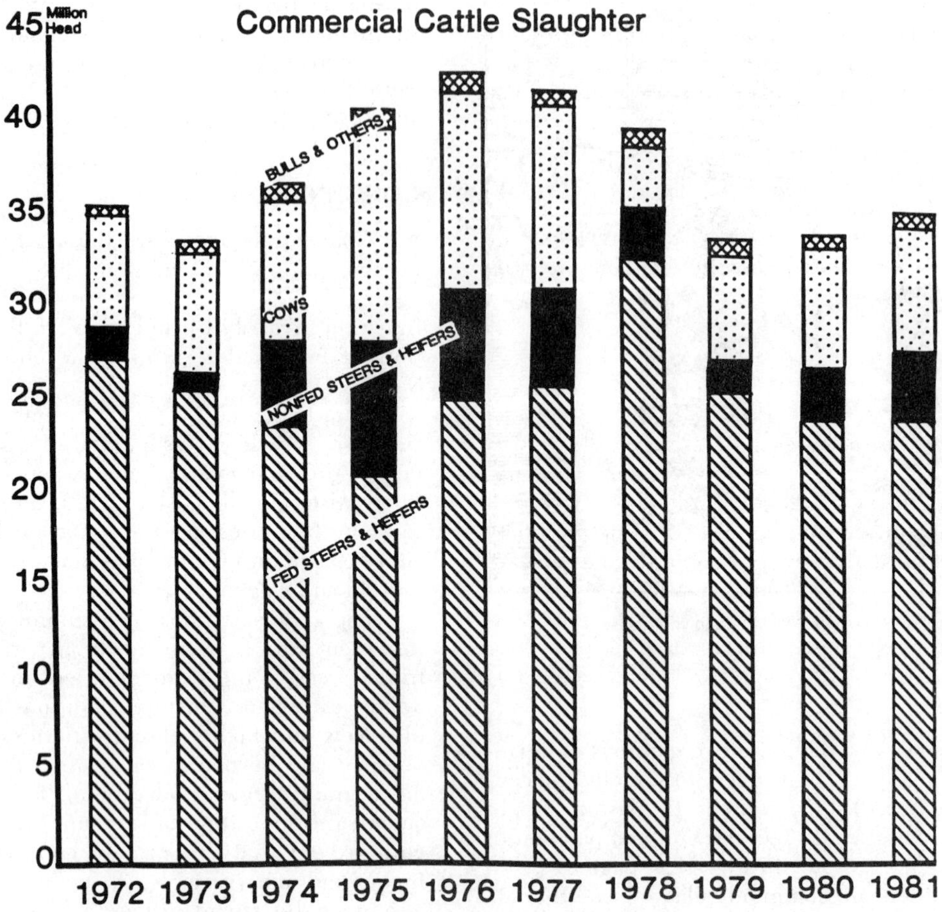

Fig. 4-24. Growth of cattle feeding and composition of cattle slaughter, 1972 to 1981. (Courtesy, National Cattlemen's Assn., *Cattle-Fax*, Denver, Colo.)

BY-PRODUCT FEEDS

Innumerable by-products—both roughages and concentrates—from plant and animal processing, and from industrial manufacturing, are available and used as cattle feeds, in different areas, including the following: potatoes and potato pulp, pea vines and corn refuse silage from the canning industry, and by-products from numerous fruits and nuts.

As is true of any ration ingredient, the requisit to effective and profitable use of each by-product fee in cattle feeding are: (1) that it be bought at a favor ble price, nutritive composition considered; (2) th its proximate composition be known, and that it l incorporated in a balanced ration; (3) that it be palat ble and consumed in adequate quantity; and (4) that not adversely affect carcass quality, particularly fro the standpoint of harmful chemical residues from pe ticides applied to crops. Generally speaking, the use of by-product feeds calls for ingenuity and experience in handling them, special knowledge relative to their nutritive qualities and use in balanced rations, and relatively high labor costs. As a result, many cattle feeders are not interested in using them, whereas others find it a lucrative business.

The feeding value and the maximum amount that can be fed to cattle of several by-product feeds are given in Table 4-27, Feed Substitution Table for Cattle (Beef and Dairy), page 290.

ROUGHAGES

Roughages are used in feedlot rations to supply bulk, physical properties, energy, protein, minerals, and vitamins. They contain considerable fiber (cellulose, hemicellulose, and lignin); consequently, they have a lower available energy content than concentrates. For this reason, only limited amounts of roughages are incorporated in finishing rations, particularly toward the end of the feeding period. They are, however, used extensively in growing programs and in warm-up rations.

The amount of roughage in feedlot rations varies over a wide range—from roughage alone in some grower rations to all-concentrate rations in some finishing rations, and many roughage proportions between these two extremes. Each of these roughage to concentrate ratios may be highly successful and very practical under certain conditions. In the final analysis, therefore, the roughage to concentrate ratio for a given feedlot should be determined by (1) the available feeds and comparative prices, (2) feed processing facilities, (3) the feed handling charges, (4) the age and quality of cattle, (5) the stage in the feeding period (i.e., starting vs finishing period), (6) temperature (decrease roughage and increase concentrate in hot summer months, because high-roughage rations produce more body heat), (7) the feeder, (8) the troubles encountered (off feed, founder, scours, bloat), and (9) the results obtained by previous experience.

Also, step-wise reductions in the amount of roughage are usually made at least three times during the finishing period; for example—

Starter ration—70% roughage, 30% concentrate

Intermediate ration—30% roughage, 70% concentrate

Final ration—10% roughage, 90% concentrate

Such step changes in roughage to concentrate ratios should be made gradually, by blending the 2 mixes for 2 to 4 days.

In drylot finishing, the kind of roughage fed varies from area to area. This is so because, normally, it is not practical to move roughages great distances. Thus, generally speaking, cattle feeders utilize those roughages that are most readily available and lowest in price.

PROTEIN SUPPLEMENTS

The daily protein requirements of growing-finishing cattle are given in Tables 4-13 and 4-14, pages 261 and 262, and the percentage of protein needed in the ration is given in Tables 4-16 and 4-17, pages 264 to 267.

Although excess protein in the ration can be partly utilized for energy, each one percent increase in protein above the required level may increase the cost of gain ¼ to ½ cent per pound. However, underfeeding protein can cost much more than overfeeding protein due to the slow gains and poor feed efficiency.

KINDS OF PROTEIN

Quality of protein, or balance of essential amino acids, is not a critical factor in most beef cattle finishing rations, because bacteria in the rumen "manufacture" proteins that are used by cattle.

The choice of a protein supplement should usually be determined by the comparative price of a pound of protein in the available supplements (see earlier section on "Cost per Unit of Nutrients"). The leading protein supplements for finishing cattle are soybean meal, cottonseed meal, linseed meal, and urea or slow-released nonprotein nitrogen products.

Urea

Urea is not a protein. It is a simple nitrogen compound, with the following structural formula:

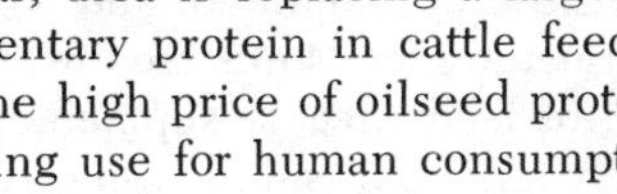

From urea, microorganisms can obtain nitrogen, synthesize amino acids, and finally bacterial protein—provided that all the nutrients essential for protein synthesis are present.

Each year, urea is replacing a larger percent of the supplementary protein in cattle feedlot rations. Because of the high price of oilseed proteins, due to their increasing use for human consumption and for monogastric animals, eventually urea and/or other nonprotein nitrogen compounds will be used as a major source of supplementary protein for feedlot cattle.

Urea or other nonprotein nitrogen compounds can furnish 33 percent of the total protein requirement of feedlot cattle; higher levels cause a depression in gain and feed efficiency. It is noteworthy, however, that, even in supplements in which 90 percent of the protein equivalent is from urea, only about one-third of the total protein in the ration is supplied from nonprotein nitrogen—the remainder is supplied from grain and roughages. Guidelines relative to the use of urea for cattle are given in Table 4-5, page 224.

The following conditions are essential for proper urea utilization by feedlot cattle:

1. High level of bacteria population in the rumen. Cattle off feed or sick cattle do not utilize urea very effectively.
2. Slow release of ammonia from urea.
3. Two to four weeks' adjustment period.
4. Low level of natural protein in the diet. Urea utilization is depressed when used with increasing levels of natural protein. Microorganisms prefer the nitrogen from natural protein to nonprotein nitrogen because natural proteins, such as soybean meal, furnish other nutrients which are beneficial to bacterial and protozoan life.
5. High-quality ingredients are required in high-urea supplements. For this reason, high-fiber filler feeds (such as ground corncobs, oat hulls, rice hulls, cottonseed hulls, cellulose, paper, and sawdust) should not be used with urea.
6. Response to urea is greater on high energy-high roughage rations than on low energy-limited roughage diets.
7. Urea supplements should be well mixed (homogenous).
8. High-urea dry supplements should be protected from rain and kept as dry as possible, because urea is hygroscopic. (It picks up moisture.)
9. Essential nutritional factors including (a) readily available source of energy (such as grain or molasses); (b) adequate levels of calcium and phosphorus; (c) required level of trace mineral elements; (d) a nitrogen-sulfur ratio which is not wider than 15:1; (e) unidentified urea-protein synthesis factors (dehydrated alfalfa meal in dry urea mixes, and distillers' solubles in liquid supplements); (f) iodized salt, to improve the palatability and mask the taste of urea; (g) fortified with proper levels of synthetic vitamin A, to furnish a minimum of 20,000 IU of vitamin A daily for growing and finishing cattle; (h) fortified with vitamin D if cattle are confined, and (i) fortified with vitamin E if natural feedstuffs are low.
10. Do not feed urea or supplements containing urea to newly arrived or shipped-in cattle for a period of 21 to 28 days.
11. Do not feed urea to cattle that have been starved or off feed for 36 hours until they have had a chance to fill the rumen with feed.
12. For growing cattle, feed a maximum of 0.15 pound (68 grams) of urea daily.
13. For finishing steers or heifers on grain and roughage, do not feed more than 0.22 pound (10 grams) of urea per head daily.
14. Formulate complete cattle rations so that n more than 33% of the crude protein or nitrogen is derived from urea. Protein supplements may contain 8 to 90% of the protein from urea; but when blende with natural feedstuffs like grain and roughage, th total contribution of protein from urea is usually les than 33%.
15. Do not feed urea over and above the protei requirement; add only enough properly balanced ure supplement to meet the protein needs.
16. Urea should be either thoroughly mixed in properly balanced supplement or incorporated in complete ration.
17. If supplements containing urea are self-fed the intake should be controlled by using a "lic wheel" for liquid supplements or incorporating hig levels of salt and dry supplements.

AMOUNT OF PROTEIN SUPPLEMENT TO FEED

The percent protein supplement to add to the ra tion will depend upon the age of the cattle, the kin and amount of roughage, and the protein content o the grain(s), or other carbonaceous concentrate bein fed. Also, more protein is needed in rations wit higher energy density. Thus, the amount of supple ment should be determined for each lot. On a pe centage of total ration basis, it decreases as the cattl grow older. Here are the recommendations for crud protein in the total ration, on an air dry basis (which i about 10% moisture):

	Stage in Feedlot	Crude Protein (%)
Calves	First 60-90 days	11.0-12.0
	Next 100-200 days	10.5-11.0
	200 days to market	10.0-10.5
Yearlings	First 60-90 days	10-5-11.0
	100 days to market	10.0-10.5

Based on recent studies, the Ohio Station worker recommend that protein supplements be deleted fro feedlot rations after cattle are past 750 pounds weight After removing the protein supplement, their test ra tion contained from 8.2 to 8.6 percent protein, a goo 25 percent below the usual recommendations. The do warn that when lowering the protein the feede must not decrease the levels of minerals and vitamin in the ration.

For cattle on relatively high concentrate ration in the drylot, the following rules of thumb may b used:

1. Where no legume hay is fed, add 2 pounds o oilseed protein supplement per head daily.

2. Where half the roughage consists of legume ay, add 1 pound of oilseed protein supplement per ead daily.

Where cattle are getting a full feed of good quality legume hay, haylage, or silage, and limited grain, t is not necessary to add a protein supplement.

Because protein supplements are usually expenive, normally one should not add more than is required to balance the ration. Neither should they be horted, for digestion of roughage is lowered if there s a lack of protein in the ration.

PROTEIN POINTERS

The following points should be taken into consideration to assure proper protein utilization by inishing beef cattle:

1. The protein content of the major feed ingrelient(s) should be known so that the ration formulaion can be precise.
2. Consideration should be given to the protein ontent and digestibility of the grain, because the rain is the most economical source of protein and upplies the largest percent of the ration protein.
3. Higher levels of protein are needed in alloncentrate rations.
4. When possible, use a roughage high in protein o as to lessen the amount of supplemental protein ecessary.
5. Urea can be successfully used at a level up to ne percent of the total finishing ration to replace atural protein, provided the ration formulation perits optimum utilization of the urea by the rumen mi-roorganisms.
6. When excessive protein levels are fed, the verage is wasted and performance may actually be educed with high-performing animals.
7. Absolute protein requirements depend upon he age of the animal, and probably the energy conent of the ration.
8. When per head consumption of corn reaches a evel of 18 pounds per day, no supplemental protein s required, although other additives such as minerals nd vitamins are necessary. Thus, protein supplenents are not needed in the late finishing stages of earling cattle.
9. The production of single-cell protein (SCP) rom solid waste as a protein supplement for feedlot attle will likely increase in importance.

MINERALS

Minerals play an important role in the nutrition of eedlot cattle. The amount of each mineral in the raion is important; both deficiencies and excesses are o be avoided. Therefore, an analysis should be made, particularly where new feeds and feeds from new areas are involved.

Where a complete mixed feed (roughage and concentrate combined) is fed to finishing cattle, it should contain 0.25 to 0.5 percent salt. Also, in the larger feedlots, the other needed minerals are usually incorporated in the ration as a special mineral supplement or in the protein supplement. For recommended kinds and allowances of minerals, see Tables 4-20 and 4-21, pages 273 and 274. Special attention needs to be given to trace minerals in areas where there is a deficiency of one or more of them, when poor quality roughage is fed, or when high- or all-concentrate rations are fed.

Even when minerals are added to the ration of finishing cattle, the author favors self-feeding them in addition. For this purpose, the mineral recommendations are:

1. *Where animals are on liberal grain feeding*—Provide free access to a 2-compartment mineral box, with (a) trace mineralized salt in one side, and (b) in the other side, a mixture of ⅓ trace mineralized salt (salt included for purposes of palatability), ⅓ defluorinated phosphate or steamed bone meal, and ⅓ ground limestone or oystershell flour.
2. *Where animals are primarily on roughage (pasture, hay, and/or silage)*—Provide free access to a 2-compartment mineral box, with (a) trace mineralized salt in one side (salt included for purposes of palatability), and (b) in the other side, a mixture of ⅓ trace mineralized salt and ⅔ defluorinated phosphate or steamed bone meal.

With this arrangement, if cattle need added minerals, they will consume them; if they don't need them, they'll pass them up. In particular, incoming feedlot cattle frequently crave minerals, due to deficiencies in their previous feeding. Such cattle, however, should be given only limited quantities of minerals until the danger of overeating has passed.

(Also see Part I section on "Minerals.")

VITAMINS

Vitamins A, E, and in some cases vitamin D, should be added to feedlot rations.

The rumen organisms synthesize adequate B vitamins and vitamin K, and nothing is gained by adding these to feedlot rations of healthy cattle. Likewise, no benefit has been reported from supplemental vitamin C. Vitamin D is produced in the skin of animals in direct sunlight, but during cloudy, winter weather, or when cattle are confined, it should be added to the ration.

Feeders should watch for the following symptoms of vitamin A deficiency in feedlot cattle: rough hair

TABLE 4-40
RECOMMENDED LEVELS OF VITAMIN A FOR FEEDLOT CATTLE

	Vitamin A/ Head/Day	(As-fed Basis) Vitamin A/Ton of Supplement When It Is Fed at Level of	
		1 lb/Head/day	2 lb/Head/day
	(IU)	*(IU)*	*(IU)*
Cattle on growing ration in winter	10,000	20,000,000	10,000,000
Cattle on full-fed finishing ration in winter	20,000	40,000,000	20,000,000
Cattle full-fed grain on pasture	20,000	40,000,000	20,000,000
Cattle fed in drylot in summer	30,000	60,000,000	30,000,000

coat, watery eyes, loose and watery droppings, edema (stovepipe legs), and night blindness.

Table 4-22, page 280, gives the vitamin requirements of beef cattle, whereas Table 4-40 gives the recommended levels of vitamin A for feedlot cattle, with overage for safety.

A common guideline on the level of vitamin A for feedlot cattle is to use 3,000 IU per 100 pounds body weight, or 1,000 IU for each pound of total feed. One million IU of vitamin A will cost about 3 cents as a feed additive, or about 7 cents as an injectable.

When vitamin D is needed and added to the ration, it is recommended that 4,000 to 6,000 IU of it be given per head per day. This is approximately 1/6 to 1/7 the recommended level of vitamin A.

Where grains are heat processed for feedlot cattle, some research shows that it may be advisable to provide supplemental vitamin E. The National Research Council indicates that the requirement for vitamin E is about 7 to 27 IU per pound of ration dry matter for growing and finishing cattle; hence, feeding this level may be advisable where grain is subjected to heat processing.

WATER

Feedlot cattle should have access to plenty of clean, fresh water at all times. They will consume 7 to 12 gallons per head per day. In cold climates, waterers should be equipped with heaters. Where the water supply is not limited by cost or volume, continuous-flow waterers are excellent. In order to keep the pathogen and algae content at a minimum, water tanks should be cleaned at least once a week in the winter and twice a week in the summer. In sick pens and pens of new cattle, the water tanks should be cleaned daily.

Feed Additives and Implants

Feed additives first made headlines in 1952 when Iowa State University researchers announced the results of cattle feeding trials indicating a major breakthrough in lowering feed usage and increasing weight gains by feeding the compound diethylstilbestrol (DES). Other additives and implants followed.

Note well: The Food and Drug Administration (FDA) banned the use of diethylstilbestrol for cattle and sheep effective November 1, 1959.

Table 4-24, Implants and Growth Stimulants for Finishing Cattle, which appears earlier in this section (see page 284), summarizes the growth stimulants that are presently available and can be used; hence, the reader is referred thereto. All of these products have been shown to improve gain and feed efficiency of feedlot cattle significantly.

Feed Preparation

Prior to 1960, very little attention was given to feed processing for commercial cattle production other than grinding or crushing grain and chopping forage. But in recent years great progress has been made and many new techniques have been developed.

GRAIN PREPARATION

Modern day fattening rations usually contain from 75 to 95 percent concentrate. Moreover, grains supply up to 90 percent of the usable energy of the ration. Thus, any improvement in the efficiency of utilization of grain will be reflected in improved performance and feed requirement of fattening cattle.

The primary reasons for processing grains for feedlot cattle are:

- To increase digestibility.
- To increase palatability.
- To increase surface area for greater microbial activity.
- To give rumen microorganisms and digestive enzymes easier access to the starches and readily utilizable nutrients.
- To affect the rate of passage of feed through the digestive tract, or to affect rumen mobility by increasing the bulk through certain processing methods.
- To increase feed efficiency through a combination of the above factors.

Among the factors to consider when deciding on the grain processing method are the size of kernel, percentage of moisture, and percentage of concentrate in the ration.

When any of the dry processing methods are used, it is important that the kernel be broken, but that there be coarseness and relative freedom from fines.

Among the methods of preparing grain for finishing (fattening) cattle are the following: dry rolling, fatty acid-treated grain, flaking, gelatinization, grinding, high-moisture (early harvested) grain, jet-sploded grain, micronizing, pelleting concentrates, popping, reconstituted grain, roasting, and whole shelled corn. Further elucidation relative to preparing grain for cattle is contained earlier in Section 4, under the heading "Processing Grain" and in Table 4-11, page 251; hence, the reader is referred thereto.

ROUGHAGE PREPARATION

The three common methods of roughage preparation are chopping, cubing, and pelleting. Each of these methods is discussed earlier in this section under the heading "Processing Roughage" and in Table 4-11, page 251; hence, the reader is referred thereto.

MIXED RATIONS VS FEEDING ROUGHAGES AND CONCENTRATES SEPARATELY

Most experiments and experiences have not shown any difference between mixed rations and the feeding of roughage and concentrates separately insofar as rate and efficiency of gain are concerned. However, a mixed ration has the following advantages:

1. It makes for greater efficiency in feeding and lessens the sorting at the feed bunk.
2. Where the roughage is relatively unpalatable, a mixed ration forces consumption.
3. Where it is desired to limit concentrate consumption, mixing with the roughage is desirable.
4. After cattle have become adjusted to the feedlot, a mixed ration makes it easier to get them on full feed.

Thus, each feeder must make his own decision on the matter of mixed vs feeding roughage and concentrate separately, with relative costs and other factors considered. Most large feedlots use completely mixed rations.

ALL-CONCENTRATE AND HIGH-CONCENTRATE RATIONS

Based on experiments and experiences, the following conclusions relative to all-concentrate and high-concentrate rations appear to be justified:

1. Ruminants need some "roughness factor" or "scratch factor" to stimulate the rumen papillae for normal functioning. In high- or all-concentrate rations, this can be achieved partially by rolling or coarse grinding.
2. With the possible exception of whole shelled corn, and such high fiber feeds as oats and barley, a high level of management is needed to make an all-concentrate system work under feedlot conditions. Problems associated with high-concentrate rations include acidosis, founder, and liver abscesses.
3. Some research indicates that continued high levels of performance can be better maintained by including 10 to 15 percent roughage in high-concentrate rations.
4. Rations having a concentrate content of 90 percent or more should be self-fed, and a liberal amount of feed should be available at all times.
5. Feed efficiency is improved in high- and all-concentrate rations, due to their high energy; but rate of gain is not materially affected.
6. The energy of a ration may be increased without eliminating much of the roughage by adding 4 to 5 percent fat.
7. Feed formulation and balance of nutrients become more critical on high- and all-concentrate rations; specifically—
    a. Higher levels of vitamin A must be added—50,000 to 100,000 IU/head/day.
    b. The ration must be fortified with the calcium, phosphorus, and trace elements that are normally provided through the roughage.
    c. The unidentified growth factors are usually reduced with a reduction of the roughage. To compensate therefor, 5 percent dehydrated alfalfa meal may be added to the ration.
8. Cattle on high- and all-concentrate rations stall and go off feed more frequently.
9. Pelleting high- or all-concentrate rations lowers daily gains.
10. In the final analysis, the comparative price of concentrates and roughages—the economics of the situation—along with management practices, will be the major determining factors.

Ration Formulation Using Net Energy

Ration formulation for finishing (fattening) cattle, using net energy, consists of combining feeds to make a ration that will be eaten in the amount needed to supply the daily nutrient requirements of the animals. This necessitates that the following net energy values be available:

1. The net energy requirements for finishing cattle (in megacalories per animal per day), with a breakdown for steers and heifers (see Table 4-19, page 271).
2. The nutrient composition of feeds, with the net energy of each feed partitioned into energy used for body maintenance and for gain; thus, the net energy values in megacalories (Mcal) per unit (lb or kg) are needed for each feed for maintenance (NE_m) and for gain (NE_g) (see Part VIII, Table 4-101, page 456).

With the above information available, net energy

values may be used to formulate rations. When using the net energy method, it is important to bear in mind that rations must also be balanced for protein, minerals, and vitamins, in order for net energy values to express their maximum potential.

An exercise in ration formulation illustrating the application of the net energy method to a cattle finishing ration is given earlier under the heading "How to Balance Rations"; hence, the reader is referred thereto (see page 253).

Some suggested rations for finishing cattle that may serve as useful guides are given in Table 4-26, Daily Rations for Beef Cattle, page 288.

Managing Feedlot Cattle

Although it is not possible to arrive at any overall, certain formula for success in operating a cattle feedlot, those operators who have made money have paid close attention to the details of management.

There are many facets of cattle management. Only those involving some aspect of feeding will be covered in the discussion that follows.

BACKGROUNDING

Backgrounding is the preparation of cattle from weaning until placing on finishing rations. It involves maximum roughage consumption and moderate gains.

The growing of calves from weaning until placing on finishing rations is not new. Only the term "backgrounding" is new. Likewise, some new "wrinkles" have been added to the method of conducting it.

Growing of calves to the yearling stage for placement in feedlots was, and still is, known as growing stockers. Farmers have, for many years, fed high-roughage rations to calves prior to marketing. Some ranch operators have, historically, retained calves for a second grazing season. Also, wintering cattle on small grain pasture is a well-established practice in the South and Lower Plains.

The term "backgrounding" came in with the development of large commercial cattle feedlots—outfits that usually had limited amounts of available roughage and other cheap feeds, and that had need for, on a year-round basis, growthy, but unfinished, cattle of a certain weight, usually within the range of 600 to 750 pounds. Today, there is renewed interest in backgrounded cattle, due to high grain prices and the need to produce more beef from roughage. Also, it isn't particularly efficient for large feedlots to tie up capital for feeding cattle where limited gains are involved.

KINDS OF BACKGROUNDING

Backgrounding of stockers and feeders can be di vided into two systems: (1) backgrounding on pasture in which calves or light yearlings are wintered an grazed, or grazed only, and sold as feeders in the lat summer or fall; and (2) backgrounding in the drylo in which the cattle are fed harvested roughage an grain and then transferred to another lot for finishing

KINDS OF CATTLE TO BACKGROUND

Generally speaking, the English beef breeds ar best suited for backgrounding purposes. This is be cause they should be grown to approximately 600 t 750 pounds before placing on finishing, or high energy, rations. Holsteins and some of the large growthier exotics are not well suited to backgroun ing, unless heavy finishing weights are planned. The need to be placed on high-energy rations at weanin time; otherwise, they will not finish out at desirabl weights of 1,050 to 1,100 pounds—instead, they wi be too heavy at market time.

RATE OF GAIN OF BACKGROUNDED CATTLE

Properly backgrounded cattle should gain fro 0.75 to 1.50 pounds per head per day. Cattle finishe object to cattle that have made higher gains, becaus it lessens, or eliminates, compensatory growth. Th is, when put on high-energy rations, animals that hav been backgrounded so as to make minimal daily gai usually gain better than similar cattle that have bee fed more liberally during the backgrounding perio For the latter reason, when contracting for bac grounding calves, feedlots commonly specify the kin of ration and the range in gains.

HANDLING NEWLY ARRIVED CATTLE

The most critical period for feeder cattle is th first 21 to 28 days in the feedlot. The following re ommendations pertaining to incoming cattle wi minimize death losses and maximize performance:

• *Provide clean, dry, comfortable quarters*— Whether it be an open lot or a building, incoming ca tle should be provided with clean, dry, comfortab quarters. A dry and comfortable bed for resting is ve essential because cattle are tired and have a low r sistance to respiratory diseases.

• *Process upon arrival*—The relative merits processing calves (1) at point of origin, (2) upon ar val at destination, or (3) two to three weeks after ar val are often debated.

In a well-designed experiment, involving 3

alves that originated in Texas, the University of :alifornia provided the answer to this question. ased on (1) rate of gain, (2) disease resistance, and 3) cost per pound of gain for feed, processing, and ıedication, the California study showed that process-ıg at arrival is best, and that processing at point of rigin is preferable to delayed processing.

• *Provide clean fresh water*—Give the cattle easy ccess to clean, fresh water because they are usually ehydrated and thirsty upon arrival and will drink ater before they eat feed. Open water tanks are referable to automatic water bowls because most ırm and ranch cattle are accustomed to drinking from ınks or ponds.

• *Provide a palatable ration*—Feeding a palatable ation—one that cattle will start eating soon after they re unloaded in the feedlot—will reduce the inci-.ence of shipping fever and make the cattle recover heir weight loss more rapidly.

1. *Roughage*—The best roughage for newly ar-ived feedlot cattle is *long grass* hay, because it is ery similar in composition and taste to the grass to which most feedlot cattle have been accustomed. 'hus, cattle will usually eat long grass hay more uickly than any other roughage. In areas where grass ıays are not available, or are too expensive to feed, ny other nonlegume roughage can be fed, such as orn silage, sorghum silage, cottonseed hulls, orncobs, or grass-legume hay that contains more rass than legumes. Above all, do not feed high-uality alfalfa hay because it is too laxative, and it will ause scouring which will trigger shipping fever. The ame may be said relative to alfalfa haylage or alfalfa ilage.

Corn silage of approximately 65 percent moisture ontent is an excellent feed for new cattle. If cattle do ıot eat the corn silage too well at the outset, the eeder should sprinkle a little grass hay on the top of t to encourage them to start eating.

2. *Concentrate*—Incoming cattle may be fed ap-ıroximately 4 lb of concentrate per head daily, con-isting of 2 lb of grain and 2 lb of protein supplement. 'he protein supplement should be fortified so as to ırovide 50,000 IU of vitamin A daily. For heavily tressed cattle, the protein supplement should also ontain a high level of antibiotic, or a combination of ıntibiotic and a bactericidal agent such as sul-amethazine. The following level of antibiotic-ulfamethazine is recommended:

> Feed 350 mg of antibiotic plus 350 mg of sulfamethazine per head daily to newly arrived cattle for a period of 28 days. With the antibiotic-sulfamethazine treatment, shipping fever is practically alleviated.

Do not feed urea for the first 28 days after the cattle arrive. Starvation destroys the ability of the rumen to utilize urea or other nonprotein nitrogen and makes cattle more sensitive to urea toxicity. Therefore, it is not wise to put extra stress on cattle by using urea during this adjustment period.

• *Satisfy mineral hunger*—Incoming cattle are usually hungry for minerals, especially if they have been on dry range forage. Thus, they should have access either to a mineral mixture consisting of two parts of dicalcium phosphate and one part of salt, or to a good commercial mineral.

• *Observe, isolate, and treat sick animals*—Newly arrived cattle should be observed at least twice daily. Sick animals should be removed and treated. Treating sick animals promptly, rather than waiting until tomorrow, may mean the difference between life and death. Animals that show clinical signs of shipping fever—sunken eyes, runny nose, drooling at the mouth, labored breathing, and/or weaving (unsteady gait)—should be isolated in a separate "sick pen" or "hospital."

Rest, fresh water, good feed, proper medication, and TLC (tender loving care) are the cardinal essentials for preventing shipping fever and death losses.

SCHEDULE FOR GETTING CATTLE ON FEED

When new cattle arrive at the feedlot, the objective is to get them on full feed as rapidly as possible, without throwing them off feed. This is not easily accomplished because many factors influence the difficulties experienced in starting new cattle on feed, among them: (1) the length of time that the cattle have been without feed; (2) the kind of feed to which the cattle were accustomed prior to shipment; (3) the age of the cattle—young cattle adapt to a change in feed more easily than old cattle; (4) whether or not the cattle have been fed and watered out of troughs before; (5) the weather conditions; and (6) existing nutritional deficiencies.

• *Traditional procedures in getting older cattle (not calves) on feed*—when first brought into the feedlot, cattle that are not accustomed to grain may be started on feed by either of the following procedures:

1. Self-fed long grass hay (and/or corn or sorghum silage), and hand-fed the concentrate according to the following schedule (with the cattle automatically lessening their self-fed hay consumption as the grain is increased):

First day—Feed 4 lb of concentrate/head/day, consisting of 2 lb of grain and 2 lb of protein supplement.

Daily increase—Step up the grain by one lb/head/day until cattle are receiving one lb/cwt body weight.

Increase every third day—After a level of one lb daily/cwt body weight is reached, make increases every third day as follows:

Calves—¼ lb
Yearlings—½ lb
2-year-olds—1 lb

2. Hand-fed a mixed ration of chopped grass hay (and/or corn or sorghum silage) and concentrate, with the proportion of roughage decreased and the grain increased according to the following schedule:

Day	Kind of Feed	Roughage (%)
1	Grass hay and/or nonlegume silage	100
2-4	Grass hay plus starter	60-90
5-14	Starter ration	40-60
15-21	Transition ration	15-40
22-to market	Finisher ration	5-15

Although one of the above procedures may serve as a useful guide, it is recognized that no set of instructions can replace the cattle intuition and good judgment of an experienced feeder.

After cattle are on full feed, they may either be self-fed or hand-fed. Most large feedlots feed twice daily, barely letting the cattle clean up the previous feed before the next feeding.

• *Schedule and ration for getting calves on feed*—Most cattle feeders follow the procedure and type of ration given above for getting cattle on feed; they start them on a high roughage ration, then work them over to a high concentrate ration as they progress through the feeding program. However, based on University of California studies, it appears that for calves (not older cattle) a starting ration consisting of 28% hay (roughage) and 72% concentrate is best. This is similar to the procedure outlined in point 2 above for older cattle, except that the calves are immediately started on a lower roughage (28%)—higher concentrate (72%) ration.

Among the problems encountered in new cattle are feed and water refusal, lactic acidosis, bloat, and diarrhea. Refusal of feed and water is generally due to the fact that the animals are not used to conventional troughs and/or the feed is so different.

Lactic acidosis generally results from feeding hungry cattle excessive levels of rapidly fermentable feeds. The condition is characterized by an accumulation of lactic acid in the rumen and a lowering of the pH in the blood and urine. The problem can be minimized by starting cattle on a high-roughage ration and shifting them gradually to a high-concentrate ration.

Bloat occasionally occurs in new cattle, although it is more frequent during the later stages of feeding. Bloat and diarrhea in new cattle can generally be prevented by feeding generous quantities of such roughages as straw, grass hay, cottonseed hulls, o corncobs.

FREQUENCY OF FEEDING

Experiments and experiences show that in creased frequency of feeding from 1 to 3 times pe day improves performance above increased costs. Fo this reason, most commercial feedlot cattle are fed times daily. Some experiments indicate that eve more frequent feeding—more than 3 times daily—wil produce slightly more rapid gains and result i greater feed efficiency. However, the improved per formance is not enough to warrant the increased costs

FEED REGULARLY

Feedlot cattle should be fed at regular times eac day, by the clock. This means that in the larger lot cattle in each alley should be fed in the same orde each day. Be prompt—remember that cattle are crea tures of habit.

AMOUNT TO FEED; FULL VS LIMITED FEEDING

Feed intake is one of the key factors affectin feedlot performance. Perhaps no other factor has suc overriding importance in determining rate and effi ciency of gain, and, ultimately, the profit derived fror feeding cattle. Of course, the reason for emphasis o high feed intake is that once a sufficient amount of th ration is consumed to meet the maintenance needs c a finishing animal, the remainder is converted to gai with remarkable efficiency. Thus, as shown in Fig 4-25, by adding 4 lb to the daily feed intake of 600-lb steer, rate of gain may be increased by $1^{1}/_{10}$ l per day. Conversely, poor feed intake results in to high a percentage of the total nutrients being ex pended for maintenance.

Thus, finishing cattle should receive a maximum ration over and above the maintenance requirements In general, they will consume daily an amount (on a air-dry basis) equal to 2.5 to 3.0 percent of thei liveweight. Feed intake will vary according to th condition of the cattle, the palatability of the feeds the energy of the ration (in general, animals eat t meet their energy needs), the weather conditions, an the management practices. For example, older an more fleshy cattle consume less feed per hundred weight than do younger animals carrying less condi tion; thus, mature, overfinished steers will consum feeds in amounts equal to about 1.5 percent of thei liveweight, whereas thin steers under 2 years of age will consume fully twice as much feed per uni liveweight.

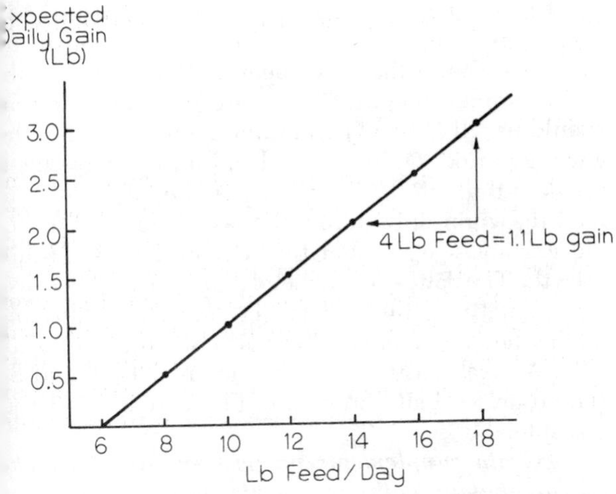

Fig. 4-25. Relationship of daily feed intake to rate of gain; 600-lb teers fed 85% concentrate ration.

Limited feeding means just what the name ndicates—not giving the animals all they want. Limted feeding generally decreases the rate of gain, adersely affects feed conversion, and increases cost of ;ains. Under most conditions, cattle should be full-fed hroughout the finishing period.

OVERFINISHING

Overfeeding is undesirable, being wasteful of eeds and creating a health hazard. When overfeeding xists, there is usually considerable leftover feed and vastage, and there is a high incidence of bloat, founler, scours, and even death. Animals that suffer from nild digestive disturbances are commonly referred to s "off feed."

Experienced cattle feeders are fully aware of the act that to carry finishing cattle to an unnecessarily high finish is usually prohibitive from a profit standpoint. This is true because the gains in weight then consist chiefly of fat but little water. In addition, a very fat animal eats less heartily, with the result that a small proportion of the nutrients, over and above the maintenance requirement, is available for making body tissue.

Fig. 4-26 and Table 4-41 show that the heavier the cattle, the more expensive the gains. Also, these figures point up (1) the importance of topping out finished cattle, rather than waiting until the entire lot is ready; and (2) the reason why it is generally wise to sell cattle when they are ready to go, rather than to hold for a higher market.

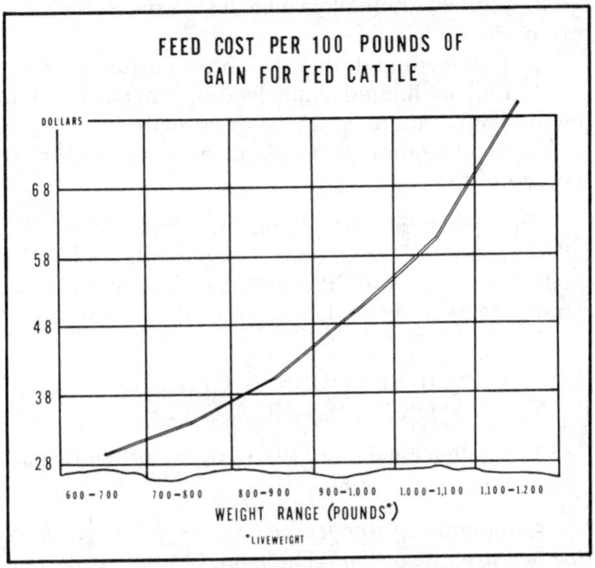

Fig. 4-26. This graph illustrates changes in feed conversion efficiency for cattle from normal feeder weights to slaughter weights. Note that feed costs per 100 lb gain more than double from 600-700 lb to 1,000-1,100 lb, and that the conversion efficiency ratio changes even more sharply when cattle pass 1,100 lb.

TABLE 4-41
APPROXIMATE COST OF ADDING 100 POUNDS OF BEEF ON STEERS AT VARIOUS WEIGHTS

Steer Weight	If Cost of No. 2 Corn per Bushel Is				
	$2.50	$2.75	$3.00	$3.25	$3.50
(lb)					
400	22.78	25.16	27.55	29.52	31.77
500	23.89	26.40	28.77	33.83	33.38
600	25.97	28.74	31.36	36.64	36.21
700	29.03	32.18	35.32	41.07	40.52
800	33.89	37.54	40.91	47.86	47.25
900	40.69	45.10	49.09	52.68	56.81
1,000	50.42	55.69	60.82	65.27	70.27
1,100	64.03	70.81	77.32	82.98	89.38
1,200	83.33	92.26	103.36	108.06	116.44

Pasture Finishing Cattle

When grains are scarce and high in price, more cattle are grass finished. But, because young cattle grow and do not reach market finish under usual pasture conditions, it is impossible to finish them at early ages and light weights without either supplemental feeding on pasture and/or lot finishing at the end of the grazing season.

SYSTEMS OF PASTURE FINISHING

When cattle are finished on pasture, any one of the following systems may be employed:

1. Finishing on pastures alone—no concentrates being fed.
2. Limited grain allowance during the entire pasture period.
3. Full feeding during the entire pasture period.
4. Full or limited grain feeding on pasture following the period of peak pasture growth.
5. Short feeding (60 to 120 days) in the feedlot at the end of the pasture period.

The system of pasture finishing that will be decided upon will depend upon the age of the cattle, the quality of the pasture, the price of concentrates, the rapidity of gains desired, and the market conditions.

BASIC CONSIDERATIONS IN UTILIZING PASTURES FOR FINISHING CATTLE

The following points are basic in utilizing pastures for finishing cattle:

- *Moderate winter feeding makes for most effective pasture utilization*—The more liberally cattle are fed during the winter, the less will be their effective utilization of pasture the following summer—the less the compensatory gains. Generally speaking, for maximum utilization of pasture, stocker calves should be fed for winter gains not in excess of 1.25 pounds per head daily, and yearlings not in excess of 0.9 pound.
- *Early pastures are "washy" but high in protein*—Cattle should not be turned to pasture too early. The first growth is extremely "washy," possessing little energy. However, the crude protein content of the forages is high during the early stages of growth and rapidly decreases as the forages mature. This would indicate the importance of pasturing rather heavily during the period of maximum growth in the spring and early summer.
- *Sudden changes are to be avoided*—Changes from drylot to pastures or from less succulent to more succulent pastures should be made with care; for grass is a laxative, and the cattle may shrink severely. Also, bloat may occur.
- *Time of starting grain feeding on pastures is determined by condition of cattle and quality of pastures*—Cattle that have been fed grain rather liberally through the winter and are in good condition should usually be fed grain from the beginning of the grazing period. On the other hand, if they have been roughed through the winter, it may be just as well to feed the grain only during the last 80 to 120 days of the grazing season, after the season of peak pasture growth. The latter recommendation is made because it is sometimes difficult to get animals to consume grain when an abundance of palatable forage is available. At peak pasture growth, the animals should be started on feed and brought to full feed as rapidly as possible.
- *Grain supplements on pastures usually make for larger daily gains and earlier marketing*—Young cattle (calves and yearlings) on summer pasture usually do not grow at their maximum potential due to energy and protein deficiencies in the feed at various times of the season. Thus, the addition of a grain supplement for cattle on pasture makes for larger daily gains and earlier marketing—either directly off grass or with a shorter drylot finishing period. The owner thus avoids late fall competition and lower prices on strictly grass cattle. Also, because cattle that are grain fed on pasture can be marketed over a wider period of time, there is greater flexibility in the operations. However, cattle that are supplemented on summer pasture often sell for less to go into feedlots because feedlot operators fear that they may not gain as rapidly as cattle that are not supplemented on pasture; they feel that more rapid gains on pasture make for less rapid and efficient gains in the feedlot—for less of compensatory growth.
- *Whole corn preferred to rolled corn*—When self-feeding steers on pasture, whole corn is preferred to rolled corn for the following reasons: (1) Slightly less feed is required per 100 pounds gain; (2) it alleviates processing cost; and (3) it results in less incidence of founder and rumen parakeratosis because whole corn supplies some "roughness factor" in the ration to stimulate the rumen.
- *Protein supplement not needed on good pasture*—As long as pasture is green and growing, no supplemental protein is required. During drought periods and late fall when the grass matures, extra protein is needed. At such times, it is good business to add 1 pound of protein supplement to each 8 to 12 pounds of grain. Usually this will increase the rate and efficiency of gain.
- *Carrying capacity of pastures will vary*—The carrying capacity of pastures will vary with the amount of grain supplement, the quality of pasture and the age and condition of the cattle. Because of these factors, the acreage per steer will vary all the way from 1 to 10.

• *Age is a factor*—Young cattle (yearlings) tend to grow as well as to fatten. Thus, older cattle (two years or older) will reach a high degree of finish on pastures alone. As good as the pastures are, it must be remembered that grass is still a roughage.

• *Minerals for cattle on pasture*—Salt is especially necessary when grass is being utilized. Finishing steers consume from ¾ to 1½ ounces of salt per head daily. Also, cattle on pasture should have free access to a 2-compartment mineral box, with (1) trace mineralized salt in one side (salt included for purposes of palatability), and (2) in the other side, a mixture of ⅓ trace mineralized salt and ⅔ defluorinated phosphate or steamed bone meal.

• *Species of grasses or legumes will vary*—The most desirable species of grasses or legumes or grass-legume mixtures to be seeded will vary according to the area, especially according to the soil and climatic conditions (see Section 5, "Pasture and Range Forages; Green Chop," for recommended grass and/or legume species).

Temporary or supplemental pastures, such as Sudangrass or millet, are used for a short period and are usually more productive and palatable than permanent pastures. They are seeded for the purpose of providing supplemental grazing during the season when the regular permanent or rotation pastures are relatively unproductive.

• *Grass vs grass-legume mixtures should be considered*—In general, where adapted legumes can be successfully grown—either alone or with grass mixtures—the results are superior to yields obtained from pure stands of the grasses.

• *Grain feeding will lengthen the grazing season*—At the Washington Agricultural Experiment Station, in experiments conducted by the author, grain feeding cattle on pasture lengthened the grazing season by an average of 57 days.

• *Self-feeding vs hand feeding on pasture*—Self-feeding grain on pasture has generally proved superior to hand feeding, as the animals consume more feed, make more rapid gains, and return more profit.

• *Economy of grain feeding on pasture*—Whether or not it will be profitable to feed grain on pasture will depend primarily upon the price of grain, the premium paid for cattle of higher finish and grade, the season in which it is desired to market, and the area and quality of pasture.

• *Pasture bloat can be controlled*—The following practices will be helpful in reducing the bloat hazard:

1. Give a full feed of hay or other dry roughage before the animals are turned to legume pastures, to prevent the animals from filling too rapidly on the green material.
2. After the animals are once turned to pasture, they should be left there continuously. If they must be removed overnight or for longer periods, they should be filled with dry roughage before they are returned to pasture.
3. Mixtures that contain approximately half grasses and half legumes should be used.
4. Water and salt should be conveniently accessible at all times.
5. The animals should not be allowed to become empty when they congregate in a drylot for shade or insect protection and then be allowed to gorge themselves suddenly on the green forage.
6. Many practical cattlemen feel that the bloat hazard is reduced by mowing alternate strips through the pasture, thus allowing the animals to consume the dry forage along with the pasture. Others keep in the pasture a rack well filled with dry hay or straw.
7. Because of the many serrations on the leaves, Sudan hay appears especially effective in preventing bloat when fed to cattle on legume pastures.
8. Consider the use of poloxalene, a nonionic surfactant, developed through research at Kansas State University, for the control of legume bloat in cattle. One such product is marketed under the trade name Bloat Guard. Always use such products according to the manufacturer's directions.

Other Methods of Improving Rate and Efficiency of Gain

Several other methods, in addition to additives, can be used to increase the rate and efficiency of gain of feedlot cattle. Among them are the following:

1. Feed young bulls (uncastrated males) instead of steers. The male hormones secreted by the testicles are excellent growth stimulants and will improve gain and feed efficiency by 10 to 15 percent. Alternatives to bulls that merit consideration are short scrotum bulls (induced cryptorchidism), and Russian castrates. With these methods, testosterone is produced, yet, in comparison with bulls, the animals are easier to handle and may be carried to advanced ages without being labeled "bull beef."
2. Take advantage of the genetic improvement of beef cattle by crossbreeding and the introduction of genes from the exotic breeds. This offers one of the most permanent ways of increasing the weaning weight of calves and improving performance in feedlot cattle. The selection of fast gaining and efficient cattle within the straightbreds, and by crossbreeding, may improve the efficiency of performance of feedlot cattle by 10 percent or more.
3. Reduce the cost of producing beef by improving the quality of cattle rations through grain processing and nutritionally balanced protein supplements.
4. Eliminate internal and external parasites, and

protect cattle against the common diseases. This will save millions of dollars for both cattle feeders and consumers.

5. Keep abreast of new developments, including the discovery of new growth stimulants.

FEEDING SHOW AND SALE BEEF CATTLE[11]

All animals intended for show purposes, including both breeding animals and steers, must be placed in proper condition. To accomplish this, a suitable ration must be selected and the animal or animals must be fed with care over a sufficiently long period. The rations listed in Table 4-42 have been used by successful showmen. They are higher in protein content than rations normally used in commercial finishing operations, but most experienced herdsmen feel that by such means they get more bloom. In general, when show animals are being force fed on any one of these concentrate mixtures, experienced herdsmen prefer to feed a grass hay or a grass-legume mixed hay to a straight legume, because of the laxative effect and possible bloat hazard of the latter.

Ration 11, which the author has used extensively in fitting show steers, is prepared as follows: The whole barley is processed by (1) adding water in the proportion of 2 to 2½ gallons to each gallon of dry barley, and (2) cooking until the kernels are thoroughly swelled and can be easily squashed between the thumb and forefinger. Then it is mixed with the balance of the ration in about the proportions (on a dry basis) indicated. Each steer also receives 4 lb daily of a milk replacer (a nurse cow replacer). As the animal approaches show finish, the ration is changed by decreasing the rolled barley by 7 lb and increasing the rolled oats by 5 lb and the wheat bran by 2 lb.

The selection of the fitting ration should be made largely on the basis of availability and price of feeds and the results obtained. Other important points in compounding the show ration and feeding the anima are:

1. Use care in getting the animal on feed. Avoi digestive disturbances and setbacks. A safe plan con sists in feeding not more than 1 lb of grain at the firs feed, or 2 lb for the day. This may be increased b approximately ¼ to ½ lb daily until the animal is o full feed 3 to 4 weeks later.

From the beginning, it is safe to full feed gras hay or the hay to which the animal is accustome Oats are the best concentrate for the beginning ratio As the grain feed is increased according to the direc tions given above, gradually (a) replace the oats wit the mixed ration selected, and (b) decrease the ha until the animal is eating only 3 to 6 pounds of ha daily at the end of the feeding period.

2. When on full feed, the average animal will ea from 1½ to 2½ pounds of grain for each 100 pounds o liveweight. Feed only as much grain as the anima will clean up in ½ to 1 hour's time.

3. Most herdsmen prefer flaked grains in fittin rations. But they may be coarsely ground or crushed.

4. Provide needed minerals.

5. If the droppings are too thin or there is scour ing, (a) cut down on the grain allowance, and (b) clea up the quarters. If trouble still persists, cut down o the legume roughage and the protein supplemen (especially linseed meal). Many experience herdsmen prefer feeding grass or grass-legume mixe hay to a straight legume hay, because of some diffi culty in keeping force-fed animals on feed when legume is fed.

6. The palatability of the ration may be enhance by adding blackstrap molasses. Make it by diluting ½ to 1 pint of molasses with an equal volume of wate and mixing it with each grain ration just before feed ing. Although blackstrap molasses is preferable, bee molasses is satisfactory.

7. Satisfactory milk replacers, which alleviate the need for nurse cows, are now on the market. These products should be used in keeping with manufac turer's directions.

[11]Further information on selecting, fitting, and showing beef cattle is contained in Section 12 of this book.

TABLE 4-42

FITTING RATIONS FOR SHOW AND SALE CATTLE
(As-fed Basis)

Rations 1 to 5 are bulky. They are recommended for use (1) by the inexperienced feeder, and (2) in starting prospective show animals on feed.

Rations 6 to 11 are less bulky and higher in energy. They are recommended for use (1) by the experienced feeder, and (2) during the latter part of the fitting period.

	(lb)	*(kg)*
Ration No. 1		
Rolled barley	50	*22.7*
Crushed oats	20	*9.1*
Wheat bran	20	*9.1*
Protein supplement[1]	10	*4.5*
Ration No. 2		
Rolled barley	30	*13.6*
Flaked corn	20	*9.1*
Crushed oats	20	*9.1*
Wheat bran	20	*9.1*
Protein supplement[1]	10	*4.5*
Ration No. 3		
Flaked corn	40	*18.1*
Crushed oats	30	*13.6*
Wheat bran	20	*9.1*
Protein supplement[1]	10	*4.5*
Ration No. 4		
Crushed oats	30	*13.6*
Rolled barley	30	*13.6*
Wheat bran	20	*9.1*
Flaked corn	10	*4.5*
Protein supplement[1]	10	*4.5*
Ration No. 5		
Flaked corn	60	*27.2*
Crushed oats	30	*13.6*
Protein supplement[1]	10	*4.5*
Ration No. 6		
Flaked corn or sorghum	50	*22.7*
Rolled barley	40	*18.1*
Protein supplement[1]	10	*4.5*
Ration No. 7		
Flaked corn	60	*27.2*
Crushed oats	20	*9.1*
Dry beet pulp	10	*4.5*
Protein supplement[1]	10	*4.5*
Ration No. 8		
Flaked corn	40	*18.1*
Rolled barley	20	*9.1*
Crushed oats	10	*4.5*
Dried beet pulp	10	*4.5*
Wheat bran	10	*4.5*
Protein supplement[1]	10	*4.5*
Ration No. 9		
Crushed oats	25	*11.3*
Rolled barley	20	*9.1*
Rolled wheat	20	*9.1*
Flaked corn	20	*9.1*
Wheat bran	10	*4.5*
Protein supplement[1]	5	*2.3*
Ration No. 10		
Rolled barley	35	*15.9*
Crushed oats	20	*9.1*
Rolled wheat	20	*9.1*
Dry beet pulp	15	*6.8*
Protein supplement[1]	10	*4.5*
Ration No. 11		
Rolled barley	20	*9.1*
Flaked corn	20	*9.1*
Crushed oats	20	*9.1*
Whole barley (dry wt basis but cooked before feeding)	13	*5.9*
Commercial supplement	8	*3.6*
Linseed meal	8	*3.6*
Wheat bran	6	*2.7*
Beet pulp, dried molasses	4	*1.8*
Salt	1	*.5*

[1]The protein supplement may consist of linseed, soybean, cottonseed, or peanut meal. With most herdsmen, linseed meal is the preferred protein supplement. It gives the animal a sleek hair coat and a pliable hide. Because it is a laxative feed, however, caution should be used in feeding it. Although it is true that an animal getting good clover or alfalfa hay needs less protein supplement than does one eating nonleguminous roughage, it is not possible to supply all the needed protein with hay and still get enough grain into young animals to finish them quickly.

PART III—FEEDING DAIRY CATTLE

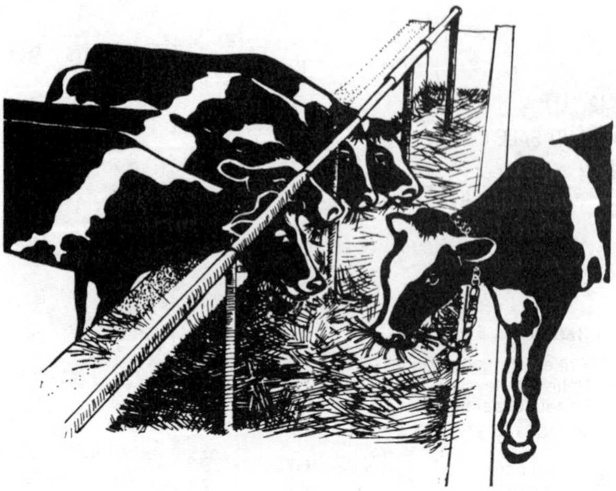

Fig. 4-27. Feeding dairy cattle.

Feed, more than any other one factor, determines the productivity and profitably of dairy cows. Within a herd, approximately 25% of the difference in milk production between cows is due to heredity; the remaining 75% is determined by environmental factors with feed making up the largest portion. Feed accounts for about 55% (with a range of 45 to 65%) of the cost of milk production. Therefore, a good feeding program is necessary for profitable milk production.

It costs slightly more to feed high producers than low producers. But high producers generally return more net income over feed cost than low producers. Fig. 4-28 shows how income over feed cost improves as production per cow increases.

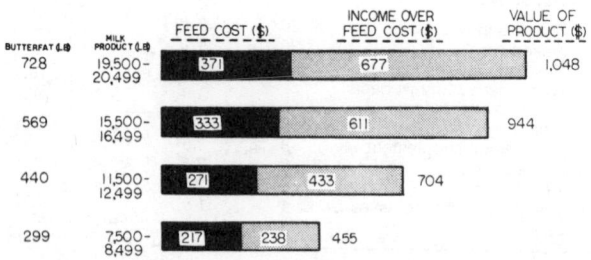

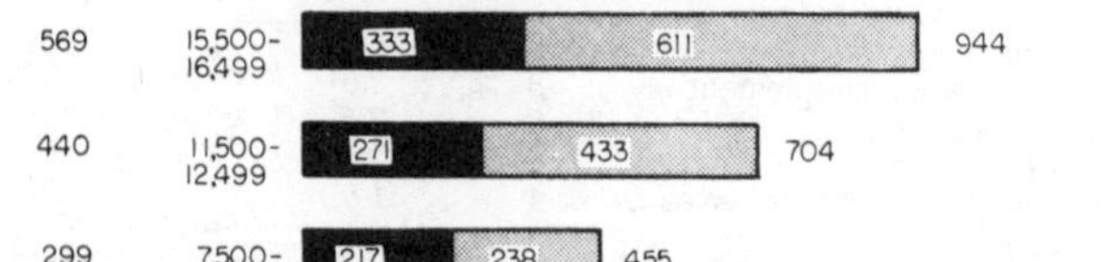

Fig. 4-28. It costs more to feed high producing cows—but it pays. The reason: Feed and overhead costs for maintenance are practically the same, regardless of level of production.

NUTRIENT REQUIREMENTS

The nutrient requirements for dairy cattle have been established by the National Academy of Sciences. These requirements are herein reproduced in Tables 4-43, 4-44, 4-45, and 4-45a. In using these ta bles, cognizance should be taken of the fact that th nutrient requirements given in them do not allow fo any margin of safety; that is, they do not provide fo animal differences, feed differences, and losses of cer tain nutrients in storage. Accordingly, in the formula tion of rations, certain margins of safety should b provided.

Energy

Lack of energy is the most common deficiency o dairy rations. Cows cannot produce milk at pea levels if their rations are too low in energy.

Most of the energy required is supplied by ca bohydrates and fats in forage and grain. All cows, ex cept low producing ones—those producing less tha 15 to 20 pounds of milk per day, need some grain they are to yield at top levels.

The net energy (NE) value of a feed depends o whether it is being used for maintenance, fattening growth, or milk production. Thus, two systems of ex pressing net energy are used. The first system ex presses the requirements in net energy for mainte nance (NE_m) and net energy for gain (NE_g); these tw values are given in Table 4-43. The second system expresses energy in one value ($NE_{lactating\ cows}$) tha can be used for maintenance, pregnancy, and mil production; this value is given in Table 4-44. Thus three net energy values are included in the NRC dairy feed composition tables: (1) (NE_m) for th maintenance of nonlactating animals, (2) (NE_g) for th deposition of body gain in nonlactating animals, an (3) ($NE_{lactating\ cows}$) for maintenance, gain, pregnancy and milk production of lactating cows.

Since energy is used with different degrees of ef ficiency for maintenance and body gain in nonlactat ing animals, both NE_m and NE_g are used to expres the total energy needs of growing heifers and bulls Since energy is used with similar degrees of effi ciency for maintenance and milk production in lactat ing animals, a single net energy value of feeds ($NE_{lac-tating\ cows}$) is adequate to calculate rations for both maintenance and milk production.

Protein

Proteins are essential for growth, repair of old tis sue, milk production, and development of the unborn calf.

For dairy cattle, it's the total amount of protein that counts, not the source. This is so because cattle have microorganisms in the rumen which build up most of the amino acids needed.

TABLE 4-43

DAILY NUTRIENT REQUIREMENTS OF DAIRY CATTLE (per Animal)[1]

Body Weight		Daily Gain		Dry Feed		Energy						Total Crude Protein		Minerals		Vitamins	
						NE_m	NE_g	ME	DE	TDN				Ca	P	A	D
(lb)	(kg)	(lb)	(kg)	(lb)	(kg)	(Mcal)	(Mcal)	(Mcal)	(Mcal)	(lb)	(kg)	(lb)	(g)	(g)	(g)	(1,000 IU)	(IU)
Growing heifers (large breeds)																	
93	42	.88	.4	1.39	.6	1.25	.70	2.98	3.31	1.65	.8	.33	148	8	5	1.8	280
110	50	1.10	.5	3.20	1.5	1.45	.96	4.82	5.42	2.71	1.2	.44	198	10	6	2.1	330
165	75	1.54	.7	4.63	2.1	1.96	1.37	6.71	7.67	3.75	1.7	.70	318	15	8	3.2	495
220	100	1.54	.7	6.17	2.8	2.43	1.47	8.09	9.26	4.63	2.1	.89	402	18	9	4.2	660
331	150	1.54	.7	8.82	4.0	3.30	1.68	10.49	12.17	6.17	2.8	1.12	510	19	12	6.4	990
441	200	1.54	.7	11.46	5.2	4.10	1.96	13.01	15.20	7.72	3.5	1.37	620	21	14	8.5	1,320
551	250	1.54	.7	13.89	6.3	4.84	2.17	15.20	17.86	9.04	4.1	1.55	704	23	17	10.6	1,650
661	300	1.54	.7	15.87	7.2	5.55	2.38	17.07	20.11	10.05	4.6	1.70	771	24	18	12.7	1,980
772	350	1.54	.7	17.63	8.0	6.24	2.52	18.88	22.26	11.13	5.1	1.82	826	25	19	14.8	2,310
882	400	1.54	.7	18.96	8.6	6.89	2.66	20.40	24.03	12.01	5.5	1.90	864	25	20	17.0	2,640
992	450	1.54	.7	20.06	9.1	7.52	2.80	21.82	25.66	12.83	5.8	1.97	892	27	21	19.1	2,970
1,102	500	1.32	.6	20.94	9.5	8.14	2.52	22.26	26.28	13.18	6.0	1.99	903	27	21	21.2	3,300
1,213	550	.88	.4	21.60	9.8	8.75	1.76	21.33	25.48	12.74	5.8	2.01	913	27	20	23.3	3,630
1,323	600	.44	.2	21.16	9.6	9.33	.90	19.60	23.68	11.84	5.4	1.94	879	25	18	25.4	3,960
Growing heifers (small breeds)																	
55	25	.66	.3	.99	.5	.85	.53	2.14	2.38	1.19	.5	.24	111	6	4	1.1	165
66	30	.77	.4	1.15	.5	.95	.63	2.49	2.77	1.39	.6	.28	128	7	4	1.3	200
110	50	.88	.4	3.09	1.4	1.45	.76	4.36	4.94	2.47	1.1	.39	176	9	6	2.1	330
165	75	1.10	.5	4.62	2.1	1.96	.98	5.96	6.94	3.42	1.6	.61	275	13	7	3.2	495
220	100	1.10	.5	6.17	2.8	2.43	1.05	7.17	8.35	4.17	1.9	.79	360	16	8	4.2	660
331	150	1.10	.5	8.82	4.0	3.30	1.20	9.42	11.11	5.56	2.5	1.04	474	17	11	6.4	990
441	200	1.10	.5	11.46	5.2	4.10	1.40	11.86	14.06	7.03	3.2	1.29	586	20	13	8.5	1,320
551	250	1.10	.5	13.89	6.3	4.84	1.55	13.81	16.49	8.25	3.7	1.49	678	22	16	10.6	1,650
661	300	1.10	.5	15.87	7.2	5.55	1.70	15.69	18.74	9.37	4.3	1.64	746	23	17	12.7	1,980
772	350	.88	.4	16.31	7.4	6.24	1.44	15.99	19.14	9.57	4.3	1.63	738	23	17	14.8	2,310
882	400	.44	.2	16.09	7.3	6.89	.76	14.85	17.94	8.97	4.1	1.53	692	21	16	17.0	2,640
Growing bulls (large breeds)																	
93	42	.88	.4	1.39	.6	1.25	.70	2.98	3.31	1.65	.8	.33	148	8	5	1.8	280
110	50	1.10	.5	3.20	1.5	1.45	.96	4.82	5.42	2.71	1.2	.44	198	10	6	2.1	330
165	75	1.54	.7	4.63	2.1	1.96	1.37	6.71	7.67	3.75	1.7	.70	318	15	8	3.2	495
220	100	1.76	.8	6.17	2.8	2.43	1.68	8.47	9.63	4.81	2.2	.94	427	19	10	4.2	660
331	150	2.20	1.0	8.82	4.0	3.30	2.30	11.73	13.40	6.70	3.0	1.29	583	22	13	6.4	990
441	200	2.20	1.0	11.46	5.2	4.10	2.50	14.05	16.23	8.11	3.7	1.55	702	23	16	8.5	1,320
551	250	2.20	1.0	13.89	6.3	4.84	2.70	16.13	18.78	9.39	4.3	1.72	778	25	18	10.6	1,650
661	300	2.20	1.0	16.31	7.4	5.69	2.95	18.67	21.78	10.89	4.9	1.90	862	27	20	12.7	1,980
772	350	2.20	1.0	18.30	8.3	6.54	3.20	20.89	24.38	12.19	5.5	2.02	917	28	21	14.8	2,310
882	400	2.20	1.0	19.84	9.0	7.41	3.50	22.93	26.72	13.36	6.1	2.09	947	29	23	17.0	2,640
992	450	2.20	1.0	20.94	9.5	8.27	3.80	25.08	29.07	14.53	6.6	2.06	934	29	23	19.1	2,970
1,102	500	1.98	.9	22.05	10.0	8.95	3.60	25.56	29.76	14.88	6.8	2.15	973	29	23	21.2	3,300
1,213	550	1.54	.7	23.15	10.5	9.62	2.91	25.51	29.94	14.97	6.8	2.15	976	29	22	23.3	3,630
1,323	600	1.54	.7	23.81	10.8	10.27	3.01	26.58	31.13	15.56	7.1	2.18	988	29	23	25.4	3,960
1,433	650	1.10	.5	24.47	11.1	10.90	2.20	25.75	30.44	15.21	6.9	2.19	992	29	23	27.6	4,290
1,543	700	1.10	.5	25.13	11.4	11.53	2.25	26.94	31.75	15.87	7.2	2.20	998	30	23	29.7	4,620
1,653	750	.66	.3	25.79	11.7	12.14	1.35	25.48	30.44	15.21	6.9	2.26	1,024	30	23	31.8	4,950
1,764	800	.66	.3	26.46	12.0	12.74	1.35	26.35	31.44	15.72	7.1	2.29	1,040	30	23	33.9	5,280
Growing bulls (small breeds)																	
55	25	.66	.3	.99	.5	.85	.53	2.14	2.38	1.19	.5	.24	111	6	4	1.1	165
66	30	.77	.4	1.15	.5	.95	.63	2.49	2.77	1.39	.6	.28	128	7	4	1.3	200
110	50	.88	.4	3.09	1.4	1.45	.76	4.36	4.94	2.47	1.1	.39	176	9	6	2.1	330
165	75	1.10	.5	4.62	2.1	1.96	.98	5.96	6.94	3.42	1.6	.61	275	13	7	3.2	495
220	100	1.32	.6	6.17	2.8	2.43	1.26	7.64	8.81	4.41	2.0	.84	381	17	9	4.2	660
331	150	1.54	.7	8.82	4.0	3.30	1.61	10.30	11.98	5.95	2.7	1.15	520	20	12	6.4	990
441	200	1.54	.7	11.46	5.2	4.10	1.75	12.59	14.78	7.39	3.4	1.41	640	21	14	8.5	1,320
551	250	1.54	.7	13.89	6.3	4.84	1.89	14.62	17.28	8.64	3.9	1.58	718	23	17	10.6	1,650
661	300	1.54	.7	16.31	7.4	5.69	2.07	16.89	20.02	10.00	4.5	1.79	811	26	19	12.7	1,980
772	350	1.54	.7	18.30	8.3	6.54	2.24	18.93	22.44	11.22	5.1	1.92	873	27	20	14.8	2,310
882	400	1.54	.7	19.84	9.0	7.41	2.45	20.84	24.64	12.32	5.6	2.01	910	28	22	17.0	2,640
992	450	1.32	.6	20.94	9.5	8.27	2.28	21.83	25.84	12.92	5.8	1.98	898	28	22	19.1	2,970
1,102	500	1.10	.5	22.05	10.0	8.95	2.00	22.22	26.45	13.23	6.0	2.07	941	28	23	21.2	3,300
1,213	550	.66	.3	22.49	10.2	9.62	1.25	21.29	25.62	12.81	5.8	2.06	935	28	22	23.3	3,630
1,323	600	.22	.1	20.77	9.4	10.27	.43	19.27	23.28	11.64	5.3	1.84	833	25	19	25.4	3,960

(Continued)

Footnote on last page of table.

TABLE 4-43 (Continued)

Body Weight		Daily Gain		Dry Feed		Energy						Total Crude Protein		Minerals		Vitamins	
						NE_m	NE_g	ME	DE	TDN				Ca	P	A	D
(lb)	*(kg)*	*(lb)*	*(kg)*	*(lb)*	*(kg)*	*(Mcal)*	*(Mcal)*	*(Mcal)*	*(Mcal)*	*(lb)*	*(kg)*	*(lb)*	*(g)*	*(g)*	*(g)*	*(1,000 IU)*	*(IU)*
Veal calves																	
77	35	1.10	.5	1.48	.7	.98	.90	3.17	3.52	1.76	.8	.38	173	7	4	1.5	231
99	45	1.76	.8	2.34	1.1	1.36	1.52	5.04	5.60	2.80	1.3	.57	259	8	5	1.9	297
121	55	1.98	.9	2.65	1.2	1.55	1.73	5.74	6.38	3.20	1.5	.64	292	11	7	2.3	363
143	65	2.20	1.0	3.00	1.4	1.76	1.95	6.48	7.20	3.59	1.6	.71	324	13	8	2.8	429
165	75	2.31	1.1	3.26	1.5	1.96	2.10	7.05	7.83	3.92	1.8	.74	334	15	9	3.2	495
220	100	2.42	1.1	3.73	1.7	2.43	2.31	8.05	8.94	4.48	2.0	.79	357	17	10	4.2	660
276	125	2.65	1.2	4.30	2.0	2.88	2.64	9.30	10.33	5.16	2.3	.86	392	19	11	5.3	825
331	150	2.87	1.3	4.89	2.2	3.30	2.99	10.58	11.75	5.86	2.7	.94	428	20	12	6.4	990
Maintenance of mature breeding bulls																	
1,102	500	—	—	17.20	7.8	9.36	—	15.95	19.27	9.63	4.4	1.48	673	20	15	21	—
1,323	600	—	—	19.73	9.0	10.74	—	18.29	22.09	11.04	5.0	1.69	766	23	17	25	—
1,543	700	—	—	22.13	10.0	12.05	—	20.52	24.78	12.39	5.6	1.88	852	26	19	30	—
1,764	800	—	—	24.47	11.1	13.32	—	22.52	27.20	13.60	6.2	2.08	942	29	21	34	—
1,984	900	—	—	26.74	12.1	14.55	—	24.79	29.94	14.97	6.8	2.24	1,017	31	23	38	—
2,205	1,000	—	—	28.92	13.1	15.75	—	26.83	32.41	16.02	7.4	2.41	1,093	34	25	42	—
2,425	1,100	—	—	31.08	14.1	16.91	—	28.84	34.83	17.42	7.9	2.58	1,169	36	27	47	—
2,646	1,200	—	—	33.18	15.1	18.05	—	30.77	37.17	18.58	8.4	2.74	1,244	39	29	51	—
2,866	1,300	—	—	35.23	16.0	19.17	—	32.67	39.46	19.84	9.0	2.90	1,316	41	31	55	—
3,086	1,400	—	—	37.21	16.9	20.27	—	34.49	41.66	20.83	9.5	3.06	1,386	43	33	59	—

[1]Adapted by the author from *Nutrient Requirements of Dairy Cattle*, No. 3, 5th rev. ed., NRC-National Academy of Sciences, Washington, D.C., 1978, pp. 32-35.

TABLE 4-44

DAILY NUTRIENT REQUIREMENTS OF LACTATING DAIRY CATTLE (per Animal)[1]

Body Weight		Energy					Total Crude Protein		Minerals		Vitamin A
		$NE_{lactating\ cows}$	ME	DE	TDN				Ca	P	
(lb)	*(kg)*	*(Mcal)*	*(Mcal)*	*(Mcal)*	*(lb)*	*(kg)*	*(lb)*	*(g)*	*(g)*	*(g)*	*(1,000 IU)*
Maintenance of mature lactating cows[2]											
770	350	6.47	10.76	12.54	6.28	2.9	.75	341	14	11	27
880	400	7.16	11.90	13.86	6.94	3.2	.82	373	15	13	30
990	450	7.82	12.99	15.14	7.58	3.4	.89	403	17	14	34
1,100	500	8.46	14.06	16.39	8.20	3.7	.95	432	18	15	38
1,210	550	9.09	15.11	17.60	8.82	4.0	1.02	461	20	16	42
1,320	600	9.70	16.12	18.79	9.41	4.3	1.08	489	21	17	46
1,430	650	10.30	17.12	19.95	9.99	4.5	1.14	515	22	18	50
1,540	700	10.89	18.10	21.09	10.56	4.8	1.19	542	24	19	53
1,650	750	11.47	19.06	22.21	11.11	5.0	1.25	567	25	20	57
1,760	800	12.03	20.01	23.32	11.66	5.3	1.31	592	27	21	61
Maintenance and pregnancy (last 2 months of gestation)											
770	350	8.42	14.00	16.26	8.18	3.7	1.42	642	23	16	27
880	400	9.30	15.47	17.98	9.04	4.1	1.55	702	26	18	30
990	450	10.16	16.90	19.64	9.85	4.5	1.68	763	29	20	34
1,100	500	11.00	18.29	21.25	10.67	4.8	1.81	821	31	22	38
1,210	550	11.81	19.65	22.83	11.46	5.2	1.93	877	34	24	42
1,320	600	12.61	20.97	24.37	12.24	5.6	2.05	931	37	26	46
1,430	650	13.39	22.27	25.87	13.01	5.9	2.17	984	39	28	50
1,540	700	14.15	23.54	27.35	13.73	6.2	2.28	1,035	42	30	53
1,650	750	14.90	24.79	28.81	14.46	6.6	2.39	1,086	45	32	57
1,760	800	15.60	26.02	30.24	15.19	6.9	2.50	1,136	47	34	61
Milk production (nutrients required per 2.2 lb or 1 kg of milk)											
% Fat											
2.5		.59	.99	1.15	.57	.26	.16	72	2.4	1.65	
3.0		.64	1.07	1.24	.62	.28	.17	77	2.5	1.70	
3.5		.69	1.16	1.34	.66	.30	.18	82	2.6	1.75	
4.0		.74	1.24	1.44	.73	.33	.19	87	2.7	1.80	
4.5		.78	1.31	1.52	.75	.34	.20	92	2.8	1.85	
5.0		.83	1.39	1.61	.82	.37	.22	98	2.9	1.90	
5.5		.88	1.48	1.71	.86	.39	.23	103	3.0	2.00	
6.0		.93	1.56	1.81	.90	.41	.24	108	3.1	2.05	

[1]Adapted by the author from *Nutrient Requirements of Dairy Cattle*, No. 3, 5th rev. ed., NRC-National Academy of Sciences, Washington, D.C., 1978, p. 35.

[2]To allow for growth of young lactating cows, increase the maintenance allowances for all nutrients except vitamin A by 20% during the first lactation and 10% during the second lactation.

TABLE 4-45

NUTRIENT REQUIREMENTS OF RATIONS FOR DAIRY CATTLE
(In As-fed Ration. See Table 4-45a for Moisture-free Basis)[1]

Nutrients	Unit	Calf Milk Replacer[2]		Calf Starter Concentrate Mix		Growing Bulls and Heifers		Dry Pregnant Cow Ration		Lactating Cow Rations When Daily Milk Production Is— 40-57 lb (18-26 kg)		57-77 lb (26-35 kg)		More than 77 lb (35 kg)		Mature Bull Ration	
		(per lb)	(per kg)	(per lb)	(per kg)	(per lb)	(per kg)	(per lb)	(per kg)	(per lb)	(per kg)	(per lb)	(per kg)	(per lb)	(per kg)	(per lb)	(per kg)
Energy																	
Digestible (DE)	Mcal	1.71	*3.77*	1.44	*3.18*	1.08	*2.39*	1.08	*2.39*	1.21	*2.66*	1.28	*2.82*	1.35	*2.98*	1.01	*2.22*
Metabolizable (ME)	Mcal	1.54	*3.40*	1.27	*2.81*	.91	*2.01*	.91	*2.01*	1.04	*2.28*	1.11	*2.44*	1.18	*2.60*	.84	*1.84*
NE_m	Mcal	.98	*2.16*	.77	*1.71*	.51	*1.13*	-	-	-	-	-	-	-	-	.49	*1.08*
NE_{gain}	Mcal	.63	*1.40*	.49	*1.08*	.24	*.54*	-	-	-	-	-	-	-	-	-	-
$NE_{lactation}$	Mcal	-	-	-	-	-	-	.55	*1.22*	.62	*1.37*	.66	*1.46*	.70	*1.55*	-	-
TDN	%	85.5		72.0		54.0		54.0		60.3		63.9		67.5		50.4	
Crude protein	%	19.8		14.4		10.8		9.9		12.6		13.5		14.4		7.7	
Crude fiber	%	-		-		13.5		15.3		15.3		15.3		15.3		13.5	
Ether extract	%	9.0		1.8		1.8		1.8		1.8		1.8		1.8		1.8	
Major or macro minerals:[3]																	
Calcium (Ca)	%	.63		.54		.36		.33		.43		.49		.54		.22	
Phosphorus (P)	%	.45		.38		.23		.23		.31		.34		.36		.16	
Sodium (Na)	%	.09		.09		.09		.09		.16		.16		.16		.09	
Sodium chloride (NaCl)	%	.23		.23		.23		.23		.42		.42		.42		.23	
Magnesium (Mg)[4]	%	.06		.06		.14		.14		.18		.18		.18		.14	
Potassium (K)	%	.72		.72		.72		.72		.72		.72		.72		.72	
Sulfur (S)	%	.26		.19		.14		.15		.18		.18		.18		.10	
Trace or micro minerals:[3]																	
Cobalt (Co)	ppm	.09		.09		.09		.09		.09		.09		.09		.09	
Copper (Cu)[5]	ppm	9		9		9		9		9		9		9		9	
Fluorine (F)	ppm	27		27		27		27		27		27		27		27	
Iodine (I)	ppm	.23		.23		.23		.45		.45		.45		.45		.45	
Iron (Fe)[6]	ppm	90		90		45		45		45		45		45		45	
Manganese (Mn)[5]	ppm	36		36		36		36		36		36		36		36	
Molybdenum (Mo)	ppm	5.4		5.4		5.4		5.4		5.4		5.4		5.4		5.4	
Selenium (Se)	ppm	.09		.09		.09		.09		.09		.09		.09		.09	
Zinc (Zn)	ppm	36		36		36		36		36		36		36		36	
Fat-soluble vitamins:																	
Vitamin A	IU	1,551	*3,420*	898	*1,980*	898	*1,980*	1,306	*2,880*	1,306	*2,880*	1,306	*2,880*	1,306	*2,880*	1,306	*2,880*
Vitamin D	IU	245	*540*	122	*270*	122	*270*	122	*270*	122	*270*	122	*270*	122	*270*	122	*270*
Vitamin E	ppm	270		-		-		-		-		-		-		-	

[1]Adapted by the author from *Nutrient Requirements of Dairy Cattle*, No. 3, 5th rev. ed., NRC-National Academy of Sciences, Washington, D.C., 1978, p. 36.

[2]The following minimum quantities of B-complex vitamins are suggested per unit of milk replacer: niacin, 2.6 ppm; pantothenic acid, 13 ppm; riboflavin, 6.5 ppm; pyridoxine, 6.5 ppm; thiamin, 6.5 ppm; folic acid, 0.5 ppm; biotin, 0.1 ppm; vitamin B-12, 0.07 ppm; choline, 0.26%. It appears that adequate amounts of these vitamins are furnished when calves have functional rumens (usually at 6 weeks of age) by a combination of rumen synthesis and natural feedstuffs.

[3]The maximum safe levels for the same minerals are as follows: sodium chloride, 4.5%; sulfur, 0.32%; cobalt, 9 ppm; copper, 72 ppm; fluorine, 27 ppm; iodine, 45 ppm; iron, 900 ppm; manganese, 900 ppm; molybdenum, 5 ppm; selenium, 5 ppm; and zinc, 450 ppm. These values are based on very limited data and safe levels may be substantially affected by specific feeding conditions.

[4]Under conditions conducive to grass tetany should be increased to 0.23% or higher.

[5]Values are the maximum concentration for all classes. Minimum requirements not yet established.

[6]The maximum safe level of supplemental iron in some forms is materially lower than 900 ppm.

TABLE 4-45a

NUTRIENT REQUIREMENTS OF RATIONS FOR DAIRY CATTLE
(In Moisture-free Ration. See Table 4-45 for As-fed Basis)[1]

Nutrients	Unit	Calf Milk Replacer[2]		Calf Starter Concentrate Mix		Growing Bulls and Heifers		Dry Pregnant Cow Ration		Lactating Cow Rations When Daily Milk Production Is— 40-57 lb (18-26 kg)		57-77 lb (26-35 kg)		More than 77 lb (35 kg)		Mature Bull Ration	
		(per lb)	*(per kg)*	*(per lb)*	*(per kg)*	*(per lb)*	*per kg)*	*(per lb)*	*(per lb)*	*(per kg)*	*(per kg)*	*(per lb)*	*(per kg)*	*(per lb)*	*(per kg)*	*(per lb)*	*(per kg)*
Energy																	
Digestible (DE)	Mcal	1.90	*4.19*	1.60	*3.53*	1.20	*2.65*	1.20	*2.65*	1.34	*2.95*	1.42	*3.13*	1.50	*3.31*	1.12	*2.47*
Metabolizable (ME)	Mcal	1.71	*3.78*	1.41	*3.12*	1.01	*2.23*	1.01	*2.23*	1.15	*2.53*	1.23	*2.71*	1.31	*2.89*	.93	*2.04*
NE_m	Mcal	1.09	*2.40*	.86	*1.90*	.57	*1.26*	-	-	-	-	-	-	-	-	.54	*1.20*
NE_{gain}	Mcal	.70	*1.55*	.54	*1.20*	.27	*.60*	-	-	-	-	-	-	-	-	-	-
$NE_{lactation}$	Mcal	-	-	-	-	-	-	.61	*1.35*	.69	*1.52*	.73	*1.62*	.78	*1.72*	-	-
TDN	%	95		80		60		60		67		71		75		56	
Crude protein	%	22.0		16.0		12.0		11.0		14		15.0		16.0		8.5	
Crude fiber	%	-		-		15		17		17		17		17		15	
Ether extract	%	10		2		2		2		2		2		2		2	
Major or macro minerals:[3]																	
Calcium (Ca)	%	.70		.60		.40		.37		.48		.54		.60		.24	
Phosphorus (P)	%	.50		.42		.26		.26		.34		.38		.40		.18	
Sodium (Na)	%	.10		.10		.10		.10		.18		.18		.18		.10	
Sodium chloride (NaCl)	%	.25		.25		.25		.25		.46		.46		.46		.25	
Magnesium (Mg)[4]	%	.07		.07		.16		.16		.20		.20		.20		.16	
Potassium (K)	%	.80		.80		.80		.80		.80		.80		.80		.80	
Sulfur (S)	%	.29		.21		.16		.17		.20		.20		.20		.11	
Trace or micro minerals:[3]																	
Cobalt (Co)	ppm	.10		.10		.10		.10		.10		.10		.10		.10	
Copper (Cu)	ppm	10		10		10		10		10		10		10		10	
Fluorine (F)[5]	ppm	30		30		30		30		30		30		30		30	
Iodine (I)	ppm	.25		.25		.25		.50		.50		.50		.50		.50	
Iron (Fe)[6]	ppm	100		100		50		50		50		50		50		50	
Manganese (MN)	ppm	40		40		40		40		40		40		40		40	
Molybdenum (Mo)[5]	ppm	6		6		6		6		6		6		6		6	
Selenium (Se)	ppm	.10		.10		.10		.10		.10		.10		.10		.10	
Zinc (Zn)	ppm	40		40		40		40		40		40		40		40	
Fat-soluble vitamins:																	
Vitamin A	IU	1,723	*3,800*	998	*2,200*	998	*2,200*	1,451	*3,200*	1,451	*3,200*	1,451	*3,200*	1,451	*3,200*	1,451	*3,200*
Vitamin D	IU	272	*600*	136	*300*	136	*300*	136	*300*	136	*300*	136	*300*	136	*300*	136	*300*
Vitamin E	ppm	300		-		-		-		-		-		-		-	

[1]Adapted by the author from *Nutrient Requirements of Dairy Cattle*, No. 3, 5th rev. ed., NRC-National Academy of Sciences, Washington, D.C., 1978, p. 36.

[2]The following minimum quantities of B-complex vitamins are suggested per unit of milk replacer: niacin, 2.6 ppm; pantothenic acid, 13 ppm; riboflavin, 6.5 ppm; pyridoxine, 6.5 ppm; thiamin, 6.5 ppm; folic acid, 0.5 ppm; biotin, 0.1 ppm; vitamin B-12, 0.07 ppm; choline 0.26%. It appears that adequate amounts of these vitamins are furnished when calves have functional rumens (usually at 6 weeks of age) by a combination of rumen synthesis and natural feedstuffs.

[3]The maximum safe levels for the same minerals are as follows: sodium chloride, 5%; sulfur, 0.35%; cobalt, 10 ppm; copper, 80 ppm; fluorine, 30 ppm; iodine, 50 ppm; iron, 1,000 ppm; manganese, 1,000 ppm; molybdenum, 6 ppm; selenium, 5 ppm; and zinc, 500 ppm. These values are based on very limited data and safe levels may be substantially affected by specific feeding conditions.

[4]Under conditions conducive to grass tetany should be increased to 0.25% or higher.

[5]Values are the maximum concentration for all classes. Minimum requirements not yet established.

[6]The maximum safe level of supplemental iron in some forms is materially lower than 1,000 ppm.

In the feed industry, protein levels are usually shown in terms of total crude protein. To convert the figure on the tag to approximate digestible protein, take 80% of it (multiply by 4/5). Thus, a 16% crude protein contains 12.8% digestible protein.

The amount of protein needed in the grain mix depends on the kind and quality of roughage. As the amount of legume increases, the percentage protein in the grain mix can be lowered.

When more protein is fed than needed, it is used as a source of energy. Because protein concentrates are more expensive than carbohydrate feeds, it usually is more economical to feed only the amount needed.

Minerals

The mineral elements required by dairy cattle are: sodium chloride, calcium, phosphorus, magnesium, potassium, sulfur, cobalt, copper, iodine, iron, manganese, molybdenum, selenium, and zinc.

The lactating cow has additional mineral requirements, over and above those needed for her own body or for the developing fetus. Milk contains about 0.7 percent minerals. Thus, one cow producing 15,000 lb of milk gives 105 lb of minerals per year. By way of comparison, it is noteworthy that 3 steers produce only 120 lb of minerals by the time they reach 1,000 lb in weight (at 18 months of age). This means that in one lactation a cow produces in her milk nearly as much minerals as 3 steers store in their bodies in 54 months (3 × 18). Additionally, a milk cow needs minerals for body maintenance (which requirements are about the same as those of a steer), for development of the unborn calf, and for growth if she is a young cow.

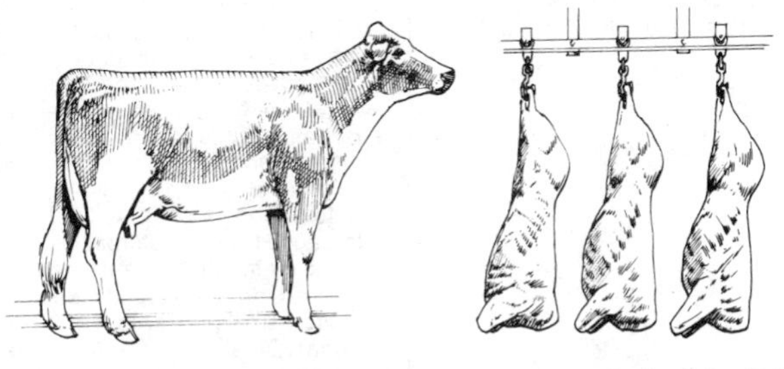

Fig. 4-29. It takes 3 steers 18 months' time (for each steer) to store as much minerals in their bodies as one cow produces in milk in one year. And the cow remains alive to do it all over again!

Dairy cattle of all ages and stages of production are more apt to suffer from a lack of phosphorus in their feed than from a deficiency of any other mineral element. Major changes in dairy cattle feeding have accentuated the phosphorus deficiency in recent years. Among these changes are (1) increased crop yields as a result of improved varieties and heavy nitrogen fertilization, which have depleted the phosphorus of the soil; (2) alfalfa constituting more of the roughage component, and alfalfa is always a rich source of calcium (in alfalfa, calcium-phosphorus ratios of 6 to 8:1 are not uncommon); (3) urea is being fed (and urea does not contain any minerals), whereas protein-rich oil meal supplements are a valuable source of phosphorus; and (4) high-level feeding and increased milk production, which carries with it a built in stress factor that tends to emphasize any difficulty that may be encountered if any nutrient is deficient or supplied in excess. Generally speaking, the calcium-phosphorus ratio of the total ration should not be wider than 2:1.

It is also good business to guard against any trace mineral deficiencies by providing cobalt, copper, iodine, manganese, and zinc. These trace minerals may be provided in the mineral mix, in trace mineralized salt, or in the ration itself.

Salt and other minerals may be added to the concentrate mix, usually at the rate of about one percent salt and one percent other minerals. Even so, they should always be available free-choice.

DAIRY CATTLE MINERAL CHART

Table 4-46, Dairy Cattle Mineral Chart, page 330, lists the minerals required by dairy cattle and gives pertinent information pertaining to each.

Vitamins

Dairy cattle, like other animals, require vitamins. Fortunately, under normal conditions, natural feeds furnish most vitamins or their precursors in adequate amounts. Furthermore, members of the B vitamin group and vitamin K are synthesized in the rumen, and vitamin C is synthesized in the tissues. However, the adequacy of vitamin intakes should be verified under certain conditions, such as (1) when forage is fed in limited amounts or is of low quality, (2) when sun-cured hay or exposure of animals to sunlight is limited, or (3) when milk replacers for young calves are relied on extensively. When it is impractical to supply adequate amounts of vitamins from natural sources, they can be furnished from commercially prepared supplements.

DAIRY CATTLE VITAMIN CHART

Table 4-47, Dairy Cattle Vitamin Chart, page 338, gives, in summary form, the following pertinent information relative to each vitamin listed: (1) conditions usually prevailing where deficiencies are reported, (2) function, (3) deficiency symptoms, (4) nutrient requirements, (5) recommended allowances, and (6) practical sources.

TABL
DAIRY CATTL

Minerals Which May Be Deficient Under Normal Conditions	Conditions Usually Prevailing Where Deficiencies Are Reported	Function of Mineral	Some Deficiency Symptoms
MAJOR OR MACRO MINERALS:			
Salt (sodium and chlorine—NaCl) The requirements for sodium and chlorine are commonly expressed as salt requirements because salt is an effective, economical way of supplementing diets with these elements.	Negligence; for salt is inexpensive.	Sodium chloride helps maintain osmotic pressure in body cells, upon which depends the transfer of nutrients to the cells and the removal of waste materials and the maintenance of water balance among the tissues. Also sodium is important in making bile, which aids in the digestion of fats and carbohydrates; and chlorine is required for the formation of hydrochloric acid in the gastric juice so vital to protein digestion. It is noteworthy that when salt is omitted, sodium expresses its deficiency first.	Intense craving for salt, lack of appe tite, haggard appearance, du eyes, and a rough hair coat. In milking cows, there is a rapid los of weight, failure of appetite, and decline in milk production. Terminal symptoms include shive ing, incoordination, weakness, ca diac arrythmia, and death. Cows recover quickly and complete when rations are supplemente with salt. Excessive salt intake can result in to icity, but as much as 3 lb *(1.36 kg* can be consumed per cow dail without harm provided animal have free access to plenty of water
Calcium (Ca)	When the diet consists chiefly of dried mature grasses or cereal straws. When cows are in heavy lactation.	Essential for development and maintenance of normal bones and teeth. Important in blood coagulation and lactation. Enables heart, nerves, and muscles to function. Regulates permeability of tissue cells. Affects availability of phosphorus and zinc.	In young calves, a calcium deficienc prevents normal bone growth an retards general growth and de velopment. Their bones are low i calcium and phosphorus and frac ture easily. In mature cows, feeding rations low i calcium over an extended perio may cause depletion of the calciu and phosphorus in the bones an result in fragile, easily fracture bones and in reduced milk, bu there is no reduction in the calciu concentration in the milk. Milk fever (parturient paresis) in cow is caused by a disturbance in ca cium metabolism manifested by marked drop in blood serum ca cium. (Also see Table 4-100, Nutri tional Diseases and Ailments c Animals, for a more complete dis cussion of "Milk fever.")

Footnotes on last page of table.

4-46
MINERAL CHART

Nutrient Requirements[1]: Daily Nutrients/ Animal	Nutrient Requirements[1]: Percentage of Rations	Recommended Allowances[1]	Practical Sources of the Mineral	Comments
For young, growing animals: 2 to 3 g of sodium. *Lactating cows:* 30 g of supplemental salt/head/day. *Lactating cows:* 21.3 g of sodium/cow/day.	**Lactating cows:* 0.16 as-fed or 0.18% of sodium (equivalent to 0.45% sodium chloride) in the moisture-free ration.	Cows on pasture or on high-roughage rations will consume from 1 to 3 lb *(.45-1.36 kg)* salt per head per month.	Salt should be available at all times. It should be both (1) self-fed, free-choice; and (2) mixed with other ration ingredients. Free access to salt in the form of loose, rock, or block salt. Cattle prefer loose salt to block salt, since it can be eaten more rapidly and with less effort. However, experiments with growing dairy heifers and lactating cows have shown fully as good results with block salt as with loose salt even though smaller quantities were consumed. This means that the additional intake of loose salt over block salt does not appear to benefit cattle. Commercial mineral mixes (in block, or loose form) may contain ⅓ or more salt.	The salt requirements of cattle differ (1) between individuals; (2) according to whether milk is produced (being higher for lactating cows than for dry cows, because of the salt in the milk); (3) from season to season; (4) according to the weathering losses to which the salt is subjected (being higher on pasture than in the drylot; exposed block salt loses about 15% per month); (5) between block and loose salt (animals often consuming twice as much easy-to-get loose salt as block salt); and (6) according to the salt content of the soil, feed, and water (being higher when vegetable proteins are fed, than when animal proteins are fed, higher on predominantly forage rations than on predominantly concentrate rations, and higher on lush early pasture than on more mature grasses). These are some of the reasons why free-choice feeding of salt is advocated.
Variable, according to class, age, and weight of dairy cattle (see Tables 4-43 and 4-44). *Dairy heifers 3 to 6 mo. of age:* 9 g of calcium/head/day.	*Variable, according to class, age, and weight of dairy cattle (see Table 4-45). **Dairy heifers:* 0.13 as-fed or 0.14% of the moisture-free ration **Lactating cows:* 0.14 as-fed or 0.16% of the moisture-free ration.	Free access to a calcium supplement, or 0.1 lb *(45 g)* of a calcium supplement added to the daily ration. Calf rations should contain a minimum of .4% calcium, .3% phosphorus, and 200 IU of vitamin D/lb *(440 IU/kg)*.	*Studies on availability of calcium indicate that calves absorb 95% of the calcium in milk. In older animals, true absorption is quite variable—ranging from 5 to 55%. *Each 2.2 lb *(1 kg)* of 4% milk contains 1.23 g of calcium. Ground limestone, steamed bone meal, oystershell flour, dicalcium phosphate, or defluorinated phosphate; free-choice, or 0.1 lb *(45 g)* per head daily added to the ration. Where both calcium and phosphorus need to be supplemented, they should be provided in a readily available and palatable form such as dicalcium phosphate, defluorinated phosphate, or bone meal.	Cows may be in negative calcium balance during early lactation, but the deficit is made up in late lactation and in the dry period. Many lactating cows, especially those fed legume forages, consume considerably more calcium than necessary without harmful effects. In addition to an adequate supply of calcium, proper utilization is dependent upon (1) a highly available source of the mineral, (2) a suitable ratio between calcium and phosphorus (somewhere between 1 to 2 parts of calcium to 1 part of phosphorus). Ratios between calcium and phosphorus of 7:1 have been reported to be satisfactory for cattle. Calcium availability of 70% is generally assumed for all feedstuffs.

(Continued)

TABLE 4-46

Minerals Which May Be Deficient Under Normal Conditions	Conditions Usually Prevailing Where Deficiencies Are Reported	Function of Mineral	Some Deficiency Symptoms
Phosphorus (P)	Semiarid regions are commonly associated with soils deficient in phosphorus. The phosphorus content of plants generally decreases markedly with maturity, with the result that deficiencies often occur in cattle subsisting for long periods on mature dried forage.	Essential for sound bones and teeth, and for the assimilation of carbohydrates and fats. A vital ingredient of the proteins in all body cells. Necessary for enzyme activation. Acts as a buffer in blood and tissue. Occupies a key position in biologic oxidation, and reactions requiring energy.	Phosphorus deficiency symptoms are loss of appetite and retarded growth. Often, depraved appetite—chewing bones, wood, and hair—is observed. However, cows may suffer from extreme phosphorus deficiency without manifesting depraved appetite. In chronic phosphorus deficiency sometimes animals become stiff in the joints. Anestrus and low conception rates may be manifested in females of breeding age with inadequate phosphorus intakes, but the phosphorus content of the milk does not decrease. Rickets in young animals and osteomalacia, osteoporosis, and osteitis fibrosa in mature animals.
Magnesium (Mg)	Hypomagnesemic tetany appears to arise from a shortage of magnesium in blood serum because of dietary deficiency, low availability of feed magnesium, and failure to mobilize skeletal reserves. When milk feeding of calves is prolonged without grain or hay (milk is rather low in magnesium). Certain pastures early in the spring. Lactating cows are most commonly affected.	Essential for the bones and teeth, and required for various body processes. Aids in maintaining acid base equilibrium and in activating many enzyme systems. In cells, magnesium is present in far greater concentration than calcium.	Grass tetany or grass staggers characterized by loss of appetite, hyperemia, hyperirritability, convulsions, and death.
Potassium (K)	When cattle receive high-concentrate rations.	Essential for proper enzyme, muscle and nerve function, rumen microorganism activity, and appetite.	Marked decrease in feed intake, pica, loss of hair bloom, decreased pliability of hide, significantly lower blood plasma and milk potassium and higher hematocrit readings.
Sulfur (S)	In high-urea rations.	Essential for the synthesis of methionine. Affects cellulose digestion.	Depressed appetite, loss of weight, poor growth, irregular or delayed estrus, and reduced milk production.

Footnotes on last page of table.

(Continued)

Nutrient Requirements[1]		Recommended Allowances[1]	Practical Sources of the Mineral	Comments
Daily Nutrients/ Animal	Percentage of Rations			
*Variable, according to class, age, and weight of dairy cattle (see Tables 4-43 and 4-44).	*Variable, according to class, age, and weight of dairy cattle (see Table 4-45). **Dairy calves:* 0.27% as-fed or 0.30% of the moisture-free ration.	Free access to a phosphorus supplement, or 0.1 lb *(45 g)* of a phosphorus supplement added to the daily ration.	Dicalcium phosphate, defluorinated phosphate, monosodium phosphate, diammonium phosphate, steamed bone meal; free-choice, or 0.1 lb *(45 g)* per head daily added to the ration.	*The NRC phosphorus requirements in Tables 4-43 and 4-44 are based on a decline in true digestibility of phosphorus from about 94% in calves to 55% in animals over 880 lb *(400 kg)* liveweight. The ratio of calcium to phosphorus in bone is about 2:1 in older animals and somewhat lower in young animals. In milk, the ratio is approximately 1.3:1. Ratios between calcium and phosphorus of 7:1 have been reported to be satisfactory for calves. Hence, the ratio of calcium to phosphorus in the diet is far less critical for ruminants than it is for monogastric animals.
**Calves fed milk:* 5.5 to 7.3 mg/lb *(12-16 mg/kg)* body weight per day. **Lactating cows:* 2.0 to 2.5 g of available magnesium plus 0.05 g for each lb *(0.12 g for each kg)* of milk produced. In problem areas, up to 20 g of supplemental magnesium per head daily may be required to prevent grass tetany.	*0.05% as-fed or 0.06% in the moisture-free ration of young calves, increasing to 0.18% as-fed or 0.20% in the moisture-free ration of lactating cows.		Commonly fed roughages and concentrates usually contain ample magnesium, but it may not be present in an available form. Magnesium sulfate or oxide may be used as a supplement.	Although grass tetany is attributed to magnesium deficiency, uncomplicated cases have been produced only on purified diets or by prolonged feeding of calves on milk. However, supplemental feeding of magnesium (6 to 20 g/day) reduces the incidence of grass tetany in many outbreaks. *While 60% of the body magnesium is stored in the bones, this reserve is not easily mobilized. Consequently, an abrupt change from a normal diet to a magnesium-deficient diet can result in hypomagnesia within 2 to 18 days.
	*0.5 to 0.7% as-fed or 0.5 to 0.8% of the total moisture-free ration.	0.8 to 1.0% of the total ration dry matter.	Roughages usually contain ample potassium. Potassium chloride is the supplement of choice.	Grains often contain less than 0.5% potassium. Excessive levels of potassium have been found to interfere with magnesium absorption. Also, excessive levels of potassium along with high levels of phosphorus, increase the incidence of phosphatic urinary calculi.
	*As low as 0.1% of the moisture-free ration.	*Suggested dietary level for dairy cattle is 0.2% sulfur and a nitrogen-sulfur ratio of 10:1 in urea-containing rations.	Organic forms of sulfur are most readily utilized, elemental sulfur is least so, and sulfates are intermediate in this respect.	Sulfur requirements are primarily those involving amino acid nutrition.

(Continued)

TABLE 4-46

Minerals Which May Be Deficient Under Normal Conditions	Conditions Usually Prevailing Where Deficiencies Are Reported	Function of Mineral	Some Deficiency Symptoms
TRACE OR MICRO MINERALS:			
Cobalt (Co)	In cobalt-deficient areas (soils) where this element is not provided (in Fla., Mich., Wisc., Mass., N.H., Penn., and N.Y.).	Cobalt is an integral component of the vitamin B_{12} molecule, and vitamin B_{12} is synthesized by microorganisms in the reticulum.	Loss of appetite, listlessness, retarded growth or loss of weight, development of anemia, pale mucous membranes, muscular incoordination, a stumbling gait, rough hair coat, decline in milk production and high mortality rate among calves. Also, there may be craving for hair and wood.
Copper (Cu)	In copper-deficient areas (soils), as in Florida and the Coastal Plain region. On peat and muck soils. Deficiencies have occurred in calves kept on an exclusive milk diet for long periods.	Copper, along with iron and vitamin B_{12}, is necessary for hemoglobin formation, although it forms no part of the hemoglobin molecule (or red blood cells). Copper is essential in enzyme systems, hair development and pigmentation, bone development, reproduction, and lactation.	Most cases of copper deficiency start with severe diarrhea, followed by rapid loss of weight. Other symptoms are: cessation of growth; abnormal appetite; rough, coarse, bleached or graying hair coat; anemia; swelling of the ends of the leg bones, especially above the pasterns; fragile bones; and affected animals may develop a pacing gait. Osteomalacia develops in mature cows. Copper depleted cows may fail to conceive, may have difficulty at calving (retained placenta), or may give birth to calves with congenital rickets.
Fluorine (F)	Feeding rock phosphate which has not been defluorinated and which may contain 3.5 to 4.0% fluorine, a toxic level.	The essentiality of fluorine for animals has not been established. But excessive levels of fluorine are harmful to cattle.	Excess fluorine is characterized by severe reduction in feed intake, reduced production, stiffness in legs, enlarged bones, rapid decline in health, and death. Teeth may erode and the enamel may become mottled.
Iodine (I)	In iodine-deficient areas (soils) where iodized salt is not fed (in northwestern U.S. and in the Great Lakes region). Where feeds come from iodine-deficient areas.	Iodine is needed by the thyroid gland in making thyroxin (an iodine-containing hormone which controls the rate of body metabolism or heat production).	Production of weak, goitrous, or dead calves. Occasional borderline cases may survive; in these, the moderate thyroid enlargement disappears in a few weeks.

Footnotes on last page of table.

(Continued)

Nutrient Requirements[1]				
Daily Nutrients/ Animal	Percentage of Rations	Recommended Allowances[1]	Practical Sources of the Mineral	Comments
	*0.03 to 0.045 mg/lb *(0.07-0.10 mg/kg)* of the moisture-free ration.	Free access to cobaltized mineral mixture in cobalt-deficient areas.	A cobalt supplement made by mixing 60 g of cobalt sulfate, or 40 to 50 g of cobalt carbonate, with 220 lb *(100 kg)* of salt, or a good commercial mineral mixture or salt product may be used.	A vitamin B_{12} injection will relieve a cobalt deficiency. The level of cobalt toxicity is about 100 times that normally supplied. *Growing dairy calves can consume 40 to 50 mg of cobalt daily per 100 kg of liveweight without toxic effects. Calves fed excessive amounts of cobalt show decreased appetite and growth rate, decreased water consumption, rough hair coat, and lack of muscular coordination.
	*4.5 mg/lb or 10 mg of copper/kg of feed, except where molybdenum levels are high, in which case the copper requirement increases.	*Copper deficiency can be prevented by adding 0.5% copper sulfate to salt fed free-choice. Copper added to total feed (moisture-free basis) 8.0 ppm. Copper may also be injected as glycinate to meet the nutritional needs for the minerals.	*The addition of 0.5% copper sulfate to salt is usually recommended in copper-deficient areas.	Copper-deficient cattle can be returned to normal by feeding 3 g of copper sulfate or blue vitriol every 10 days. An interesting interrelationship exists between copper and molybdenum. An excess of molybdenum (in the presence of sulfate) causes a condition which can be cured only by administering copper. Excess copper is toxic; it accumulates in the liver, and death may result. Excessively high copper intakes may also increase the copper content of the milk and increase its susceptibility to oxidized flavor.
	**Safe fluorine levels:* For dairy heifers, the ration should not contain more than 13.6 mg fluorine/lb *(30 mg/kg)*; and for adult breeding and lactating cattle, not more than 18.2 mg fluorine/lb *(40 mg/kg)*.			Under normal conditions, fluorine accumulates in the skeleton of the animal throughout life. An intake of 0.64 mg fluorine/lb *(1.4 mg/kg)* of body weight produces marginal toxicity. Larger intakes cause extensive metabolic changes and even death. Toxicity may be counteracted to some extent by green forages, adequate nutrition, dietary aluminum salts, high calcium levels, and liberal grain feeding.
00 to 800 micrograms iodine/day for a 1,000-lb *(454 kg)* cow producing 40 lb *(18.1 kg)* of milk daily.	*Iodized salt at rate of 0.09% as-fed or 0.10% of moisture-free ration.	Free access to stabilized iodized salt containing 0.01% potassium iodide (0.0076% iodine).	Stabilized iodized salt containing 0.01% potassium iodide. Calcium iodate. Ethylenediamine dihydriodide (EDDI).	The enlargement of the thyroid gland (goiter) is nature's way of trying to make enough thyroxin when there is insufficient iodine in the feed. Lactation and gestation increase the demand for iodine. Iodine is always present in milk, but in variable quantities. It is higher in colostrum than in normal milk. Adding iodine to feed increases the iodine content of milk. Toxicity can occur, resulting in depressed appetite, dull listless appearance, difficulty in swallowing, hacking cough, and weepy eyes.

(Continued)

TABLE 4-46

Minerals Which May Be Deficient Under Normal Conditions	Conditions Usually Prevailing Where Deficiencies Are Reported	Function of Mineral	Some Deficiency Symptoms
Iron (Fe)	Calves on an exclusive milk diet. Where there is a severe loss of blood because of parasitic infestations or disease.	Necessary constituent of hemoglobin (oxygen carrying system of the blood). Deficiencies result in anemia. Up to 20 weeks of age, supplemental iron contributes to improved weight gain in calves fed milk diets.	Anemia and decreased growth i calves on exclusive milk diet. Excessive amounts of iron are toxic.
Manganese (Mn)	In northwestern U.S., where a deficiency of manganese exists. All-concentrate diets based on corn supplemented with nonprotein nitrogen.		Delayed estrus, reduced fertility abortions, and deformed young Calves born to manganese deficient cows may exhibi deformed legs (enlarged joints stiffness, twisted legs, "overknuckling"—commonly calle crooked calf disease), weak an shortened bones, and poor growth.
Molybdenum (Mo)	Molybdenum toxicity occurs only occasionally in cattle and appears to be an area problem.		Specific symptoms of molybdenum deficiency in cattle have not bee described. Molybdenum toxicity results in sever scours, unthriftiness, rough hai coat, loss of hair color, dehydration, arching of the back, listlessness and weakness, brittle bones emaciation, and in extreme cases death. Over an extended period there is a disturbance of phosphorus metabolism, with lameness joint abnormalities, and osteoporosis.
Selenium (Se)	Low-selenium forage and low vitamin E. It is an area problem, but it occurs in many parts of the U.S.	The biological functions of selenium have not been clarified, but they are closely connected with the function of vitamin E. Vitamin E has a sparing effect on the selenium requirement. Toxic levels reported in S.D., N.D., Mont., Wyo., Utah, Neb., Kan., and Colo.	White muscle disease—characterize by white muscle, heart failure, an paralysis. Hollow or swayed back Often a dystrophic tongue.
Zinc (Zn)		Essential in skin, hair, and bone development.	Parakeratosis in young calves, evidenced by inflamed nose an mouth, unthrifty appearance roughened hair coat, and stiffnes of joints.

[1]As used herein, the distinction between "nutrient requirements" and "recommended allowances" is as follows: In nutrient requirements, no margins of safety are include intentionally; whereas in "recommended allowances," margins of safety are provided to compensate for variations in feed composition, environment, and possible losses durin storage or processing.

Where preceded by an asterisk, the nutrient requirements, recommended allowances, and other facts presented herein were taken from *Nutrient Requirements of Dair Cattle*, No. 3, 4th rev. ed., NRC-National Academy of Sciences, Washington, D.C., 1971.

Continued)

Nutrient Requirements[1]		Recommended Allowances[1]	Practical Sources of the Mineral	Comments
Daily Nutrients/ Animal	Percentage of Rations			
30 to 60 mg iron/day to calves gaining approximately 2 lb *(0.9 kg)* daily.		Iron, 40 mg daily.	Levels of iron in feed believed to be ample, since feeds contain 36 to 45 mg/lb *(80-100 mg/kg)* in most regions. The iron in ferric oxide is much less available than iron in ferrous sulfate, ferrous carbonate, or ferric chloride.	After calves are past 20 weeks of age, iron does not seem to be beneficial. About 30% of all calves are affected by prenatal iron deficiency.
	*9 mg/lb *(20 mg/kg)* of feed.	An intake of 20 ppm will prevent deformities in the fetus.	Most roughages contain over 13.6 mg of manganese/lb *(30 mg/kg)* of dry matter, and most grains contain about half this amount. Manganous oxide, sulfate, and carbonate.	The dietary requirement for manganese is increased by high intakes of calcium and phosphorus.
ꞮRequirements for molybdenum by cattle have not been established.		As a feed additive, molybdenum is not cleared by the Food and Drug Administration.	Many feeds contain 6.8 to 13.6 mg/lb of ration dry matter.	*Levels above 9.1 mg/lb *(20 mg/kg)* of forage are often associated with occurrence of teart pasture and toxicity. Toxic levels of molybdenum interfere with copper metabolism and thus increase copper requirements. Increasing copper level in diet to 1 g/head daily is effective in overcoming molybdenum toxicity in beef cattle. Phosphorus, manganese, potassium, zinc, and sulfur have also been reported to affect the degree of molybdenum toxicity.
	*0.045 mg/lb *(0.1 mg/kg)* of moisture-free ration, depending on the presence of enhancing or interfering substances in the ration. Cows grazing on pastures with less than this quantity of selenium produce calves with white muscle disease. Selenium is not cleared by the Food and Drug Administration as a feed additive for cattle.		Any of following four procedures are effective against white muscle disease: (1) selenium as a drench; (2) subcutaneous or intramuscular injections; (3) selenium as a feed additive; or (4) adding selenium to fertilizer applied to pasture.	Selenium toxicity may occur when cattle consume feeds containing 10-30 ppm of selenium on a dry matter basis for an extended period. It has been shown that the performance of cattle fed a selenium-deficient ration is improved by injecting selenium, the only approved method of selenium supplementation for cattle.
	*18.2 mg/lb *(40 mg/kg)* of moisture-free ration.	50-90 ppm zinc in the total feed (air-dry basis).	Many feeds contain 6.8 to 13.6 mg/lb *(15-30 mg/kg)* on moisture-free basis; adding zinc to fertilizer applied to pasture.	Mild zinc deficiency in cattle results in lowered weight gains without the development of a specific syndrome.

Note: Mineral recommendations for all classes and ages of dairy cattle, especially those fed unmixed rations or on pasture, are—

1. *Where dairy cattle are on liberal grain feeding*—Provide free access to a 2-compartment mineral box, with (a) trace mineralized salt in one side, and (b) in the other side, a mixture of ⅓ trace mineralized salt (salt included for purposes of palatability), ⅓ defluorinated phosphate or steamed bone meal, and ⅓ ground limestone or oystershell flour.
2. *Where dairy cattle are primarily on roughage (pasture, hay, and/or silage)*—Provide free access to a 2-compartment mineral box, with (a) trace mineralized salt in one side (salt included for purposes of palatability), and (b) in the other side, a mixture of ⅓ trace mineralized salt and ⅔ defluorinated phosphate or steamed bone meal.

TABLE

DAIRY CATTLE

Vitamins Which May Be Deficient Under Normal Conditions	Conditions Usually Prevailing Where Deficiencies Are Reported	Function of Vitamin	Some Deficiency Symptoms
FAT-SOLUBLE VITAMINS: **A** (carotene)	Where poor quality forage or low levels of forage are fed. When calves are fed only limited amounts of colostrum or whole milk.	Promotes growth and stimulates appetite. Assists in reproduction and lactation. Helps keep the mucous membranes of respiratory and other tracts in healthy condition. Makes for normal vision.	Stratified keratinization of epthelial tissue is characteristic of vitamin A deficiency. In cattle, this may be observed by degeneration of the mucosa of the respiratory tract mouth, salivary glands, eyes, tear glands intestinal tract, urethra, kidneys, and vagina. Tissues so affected are highly susceptible to infection; and colds and pneumonia often occur. Diarrhea, loss of appetite, and emaciation are commonly observed at this stage of deficiency. In later stages, characteristic changes in the eye may take place: excessive lacrimation keratitis, softening of the cornea xerophthalmia, cloudiness of the cornea and sometimes permanent blindness resulting from infection. As vitamin A deficiency develops, night blindness occurs which can readily be detected by driving animals among unfamiliar obstacles in dim light. Staggering gait, convulsive seizures and papilledema are usually present, resulting from elevated cerebrospinal fluid pressure. *Bulls of breeding age:* Decline in sexual activity. Spermatozoa decrease in numbers and motility, and there is a marked increase in abnormal forms. *Breeding cows:* Shortened gestation periods a high incidence of retained placenta, and birth of dead, incoordinated, or blind calves (such blindnesses are usually permanent). *Vitamin A deficiency can be detected by carotene and vitamin A analyses of blood and liver tissue of cattle.*
D	Young calves kept indoors, especially in the wintertime.	Aids in assimilation and utilization of calcium and phosphorus, and necessary in the normal bone development of animals—including the bone of the fetus.	Rickets in young calves, the symptoms of which are: decreased appetite, lowered growth rate, digestive disturbances, stiffness in gait, labored breathing, irritability weakness, and, occasionally, tetany and convulsions. Later, enlargement of the joints, slight arching of the back, bowing of the legs, and the erosion of the joint surfaces cause difficulty in locomotion. Posterior paralysis may follow fracture of vertebrae. Symptoms develop more slowly in older animals. Vitamin D deficiency in the pregnant animal may result in dead, weak or deformed calves at birth.
E (Also see Table 4-100.)	Abnormally high levels of nitrates may produce vitamin E deficiencies. Where soils are very low in selenium.	Serves as a physiological antioxidant, facilitating the absorption and storage of vitamin A. Its other biochemical roles in the animal body appear to be related to its antioxidant capability, including the protection of vitamin A.	Muscular dystrophy (commonly called white muscle disease) in calves 2 to 12 weeks of age; characterized by heart failure and paralysis varying in severity from slight lameness to inability to stand. Also, a dystrophic tongue is often seen in affected animals.
K	When moldy sweetclover hay high in dicoumarol content is fed, resulting in a bleeding syndrome called "sweetclover poisoning" or bleeding disease. This disease can be effectively treated with vitamin K.		

Footnote on last page of table.

4-47

VITAMIN CHART

Nutrient Requirements[1]				
Daily Nutrients/ Animal (or injection)	Amount/lb (or kg) of Feed	Recommended Allowances[1]	Practical Sources of the Vitamin	Comments
Variable according to class, age, and weight of animal (see Tables 4-43 and 4-44). **For growing calves:* 4.8 mg carotene/100 lb *(10.6 mg/100 kg)* of liveweight. In the absence of more definitive data, it is assumed that a similar level is required for maintenance.	Variable according to class, age, and weight of animal (see Table 4-45).		Stabilized vitamin A. Green pasture. Grass or legume silages. Yellow corn. Green hay 1 yr. old. The average carotene content of some common feeds is as follows: **mg Carotene/ lb or *kg*** Legume hays (including alfalfa), average quality 9-14/lb *20-31/kg* Nonlegume hays, average quality 4-8/lb *9-18/kg* Dehydrated alfalfa meal, average quality .... 50-70/lb *110-154/kg* Yellow corn ..... 0.8-1.0/lb *1.8-2.2/kg* Silages, corn or sorghum 2-10/lb *4-22/kg*	Vitamin A is found only in animals; plants contain the precursor, carotene. With practical dairy cattle rations, consider 1.0 mg of carotene equivalent to only 400 IU of vitamin A. Hay over 1 yr. old, regardless of green color, is usually not an adequate source of carotene or vitamin A activity. The younger the animal the quicker vitamin A deficiencies will show up. Mature animals may store sufficient vitamin A to last 6 months. When deficiency symptoms appear, it is recommended that there be added to the ration either (1) a stabilized vitamin A product, or (2) dehydrated alfalfa or grass. Corn and sorghum silage may contain a substance which destroys carotene and/or vitamin A.
*Variable according to class, age, and weight of animal (see Tables 4-43 and 4-44). *Calves:* 300 IU/100 lb *(660 IU/100 kg)* of liveweight. **Mature cows—maintenance, reproduction, lactation:* 5,000 to 6,000 IU/cow/day.	*Variable according to class, age, and weight of animal (see Table 4-45).	*Normally, dairy cattle receive sufficient vitamin D from exposure to direct sunlight or from sun-cured hay. Massive doses (20,000,000 IU of vitamin D/day), starting 5 days before the expected calving date and continuing through the first day postpartum, with a maximum dosage period of 7 days, have been helpful in controlling milk fever.	Exposure to direct sunlight. Sun-cured hay. Irradiated yeast.	There is no danger of vitamin D deficiency when animals receive sun-cured forage or are exposed to ultraviolet light or sunlight. Sun-cured alfalfa hay contains 300 to 1,000 IU/lb *(661-2,204 IU/kg)*. Vitamin D is usually added at level of about 1/7 level of vitamin A.
	**Calves:* 136.4 mg *dl*-alpha-tocopherol acetate/lb *(300 mg/kg)* of moisture-free ration.	Generally natural feeds supply adequate quantities of alpha-tocopherol for mature cattle, although muscular dystrophy in calves occurs in certain areas.	Alpha-tocopherol, added to the diet or injected intramuscularly. Commercial vitamin E supplements. Grains contain 6 to 15 mg vitamin E/lb *(13-33 mg/kg)*.	The incidence of white muscle disease appears to be lower where the cows receive 2 to 3 lb *(.91-1.36 kg)* of grain during last 60 days of pregnancy. Where supplemental vitamin E is needed, it may be added to the ration or injected intramuscularly.
				Except when the dicoumarol content of hay is excessively high (as in moldy sweet clover hay) sufficient vitamin K is synthesized in the rumen of cattle.

(Continued)

TABLE 4-47

Vitamins Which May Be Deficient Under Normal Conditions	Conditions Usually Prevailing Where Deficiencies Are Reported	Function of Vitamin	Some Deficiency Symptoms
B VITAMINS:	Severe inadequacy of protein or other nutrients in the diet may impair rumen fermentation to such an extent that sufficient quantities of B vitamins will not be synthesized.	*The deficiency signs of each of the B vitamins that follow were produced in young calves fed purified and semi-purified diets.*	
B_{12}			Signs of a B_{12} deficiency include poor appetite and growth, muscular weakness, and poor general condition.
Biotin			Deficiency of biotin is characterized by paralysis of the hindquarters.
Choline			Within 6 to 8 days on a choline-deficient diet calves developed extreme weakness and labored breathing and were unable to stand.
Folic acid (folacin)			A deficiency of folic acid has not been described in the calf.
Niacin (nicotinic acid)			Deficiency signs are sudden loss of appetite, severe diarrhea, and dehydration, followed by sudden death.
Pantothenic acid			The most characteristic sign of a deficiency in the calf is scaly dermatitis around the eyes and muzzle. Loss of appetite and diarrhea follow after 11 to 20 weeks on a deficient diet and calves become weak and are unable to stand and may develop convulsions. Calves are also susceptible to mucosal infection, especially of the respiratory tract.
Pyridoxine (B_6)			The signs of deficiency are loss of appetite and cessation of growth at 3½ to 10 weeks. Some calves exhibited severe epileptoid fits, with wild thrashing of the legs and head and grinding of the teeth.
Riboflavin (B_2)			Hyperemia of the mucosa of the mouth and along the edge of the lips, loss of hair—especially bilaterally on the abdomen around the navel, copious salivation, and excess tears. Other less specific signs are loss of appetite, diarrhea, and retarded growth.
Thiamin (B_1)			Polyneuritis, characterized by poor coordination of the legs, especially the forelimbs and by inability to rise and stand. Frequently, the head is retracted along the shoulder. Usually, there is loss of appetite and severe diarrhea, followed by dehydration and death.

[1]As used herein, the distinction between "nutrient requirements" and "recommended allowances" is as follows: In nutrient requirements, no margins of safety are included intentionally; whereas in nutrient allowances, margins of safety are provided in order to compensate for variations in feed composition, environment, and possible losses during storage or processing. Where preceded by an asterisk, the nutrient requirements, recommended allowances, and other facts presented herein were taken from *Nutrient Requirements of Dairy Cattle*, No. 3, 4th rev. ed., NRC-National Academy of Sciences, Washington, D.C., 1971.

Continued)

Nutrient Requirements[1]				
Daily Nutrients/ Animal (or injection)	Amount/lb (or kg) of Feed	Recommended Allowances[1]	Practical Sources of the Vitamin	Comments
		Usually, no dietary B vitamins need be supplied to cattle. *The recommended B vitamin allowances that follow prevented deficiency signs from developing.*	Milk supplied by the cow during early lactation.	During the first 8 weeks of life of the calf, the dietary requirements for the B vitamins are usually adequately met by milk from the dam; after this, these vitamins are usually synthesized by the rumen bacteria.
		*The recommended allowance of vitamin B_{12} is between 0.15 and 0.31 mcg/lb *(0.34 and 0.68 mcg/kg)* of liveweight.		
		*No symptoms developed when synthetic milk was supplemented with 10 mcg of biotin/kg and fed at 10% *(1.0 mcg/kg)* of liveweight.		
		*Supplementation with 246 mg of choline/qt *(260 mg/liter)* of synthetic milk prevented the deficiency signs.		
		*Supplementation with 2.5 mg of nicotinic acid/qt *(2.6 mg/liter)* of milk fed free-choice twice daily prevented the deficiency.		
		*No pantothenic acid deficiency signs were observed with 0.6 mg of calcium pantothenate/lb *(1.30 mg/kg)* of liquid diet or 60 mcg/lb liveweight *(130 mcg/kg)*.		
		*29.5 mcg/lb *(65 mcg/kg)* liveweight prevented pyridoxine deficiency symptoms.		
		*The requirement is thought to be less than 30 mcg/lb *(65 mcg/kg)* liveweight.		
		*Clinical signs in calves weighing less than 110 lb *(50 kg)* prevented with 0.30 mg thiamin-HCl/lb *(0.65 mg/kg)* of liquid diet fed at 10% of liveweight.		

Water

Large amounts of water are essential if a lactating cow is to produce to her maximum capacity. Cows drink an average of 100 to 200 pounds of water per day, with heavy producers drinking up to 300 pounds per day. The amount of water a cow will drink depends on her size and milk yield, the temperature and relative humidity of the air, the temperature of the water, and the amount of moisture in her feed.

In extremely cold weather, it is a good idea to have a tank heater to keep the water from freezing. Also, frequency of watering is important. Cows stabled in a stanchion-type barn produce 3½ to 4 percent more milk if they have drinking cups available than if they are watered twice daily. Contrary to some opinions, cows do not produce more milk from softened than from normal hard water.

FEEDS FOR DAIRY CATTLE

For convenience, the commonly used dairy feeds are herewith classified as (1) roughages, (2) concentrates, (3) special feeds, and (4) commercial feeds.

Roughages

Cows can produce up to 70 percent of their ability when fed good quality roughage alone. But with high-producing cows, a greater percentage of the total feed intake must be in the form of concentrates. Even so, large amounts of high-quality roughage should be the basis for feeding on most dairy farms.

In using any kind of roughage, three important points should be kept in mind: (1) To obtain the most nutrients from roughage, it must be of good quality; (2) the better the roughage, the smaller the requirement for grains; and (3) the cow is, by nature, a good consumer of forage.

The common rules of thumb for forage consumption of dairy cows are as follows:

1. A cow will eat 2 to 3 lb of hay per 100 lb of body weight if fed hay alone.
2. It takes three pounds of silage to equal (or replace) 1 lb of hay, the lower feeding value of silage being due to its high moisture content.
3. It takes about 3 lb of good hay to supply the same amount of usable energy as 2 lb of grain.
4. Cows will consume 100 to 200 lb of pasture per day; since pasture normally contains 70 to 85 percent moisture, that's 15 to 60 lb of dry matter per day.

If roughage is of very high quality, cows will eat more of it, with the result that the grain requirement will be lessened. However, over and above meeting the minimum roughage requirement, the proportion of roughage to concentrate should be determined primarily by the economics of the situation—that is, i should be decided on the basis of the relative price o available roughage and concentrate, the milk produc tion, and the net returns.

Concentrates

Concentrate feeds are those which are high i energy and low in fiber. Many different kinds of con centrate feeds can be, and are, used in dairy cattl feeding. They are usually classed according to tota crude protein content as (1) low-protein, (2 medium-protein, and (3) high-protein feeds. Th chemical analysis of feeds can be obtained from fee composition tables (see Tables 4-101 to 4-105, page 456 to 585).

Three factors besides chemical composition ar important in evaluating concentrates for milk cows— palatability, quality of milk produced, and cost. Th most infallible way in which to appraise the first tw factors is through actual feeding trials. Consideratio of the third factor necessitates that the dairyman be keen student of values. He must change the formula tions of his ration(s) in keeping with comparativ prices.

Special Feeds

Among the special feeds used by dairymen ar the following:

1. *Urea*—When added to the dairy cattle ration urea must be mixed thoroughly to ensure even dis tribution, and used according to directions. It may b toxic when improperly used, or when fed in too larg amounts.

Normally, dairymen limit urea to 1.5 to 2.0 per cent of the concentrate ration. Higher levels (up t 2.75% of the concentrate) are unpalatable and depres appetite. However, the unpalatability may be al leviated by pelleting the urea with alfalfa.

2. *Antibiotics*—Antibiotics lessen the incidenc of diarrhea in calves and result in increased rate o gain in dairy calves. For this reason, they are usuall included in milk replacers and calf starters.

3. *Thyroprotein (iodinated casein)*—Whe treated in a special way with iodine, dried milk pro tein yields a product that has a similar activity in th body as the thyroid hormone. This drug has foun limited use with dairy cows.

When added to the feed at certain levels (usuall about 15 g/head/day) and under certain conditions thyroprotein stimulates the cow to produce more mil (up to 20% more) of a higher fat (up to 30% more content. However, to be effective, this product mus be added at a specific time during the lactation perio and it must be withdrawn very gradually before th

›nd of lactation. Also, cows must be given additional eed to produce the extra milk. Not all cows will re- pond to thyroprotein. Because of the problems in- ʼolved in feeding the product, it is not widely used in eeding practice. Also, since thyroprotein is classified ıs a drug, milk and butterfat production records of ›ows fed thyroprotein are not acceptable under DHIA ules.

4. *Sodium propionate, sodium and calcium lac- ate, and propylene glycol*—Sodium propionate and odium and calcium lactate powder have some value n preventing ketosis in dairy cows when fed at levels ›f ¼ to ½ pound per cow per day, beginning one veek before calving and extending 6 weeks after calv- ng. Propylene glycol, a liquid, fed at levels of ¼ to ½ ›int per cow per day, can be used in a similar manner. ›ince these products are not palatable, they should be .dded to the ration gradually or used as a drench.

5. *Sodium bicarbonate ($NaHCO_3$)*—Sodium ›icarbonate is known to function as a buffer and pH .gent, maintaining sufficient alkaline reserves (buffer- ng capacity) in the animal to ensure normal ›hysiological and metabolic functions. In limited ex- ›erimental work, the University of Hawaii found that odium bicarbonate fed as 3.84% of the concentrate or ıs 0.75 lb per cow daily significantly increased 4% fat ›orrected milk, daily fat production, and acetic- ›ropionic ratio with all levels of roughage fed. Fur- her, it was found that the amount of roughage in the liet influences the response to sodium bicarbonate.

6. *Irradiated yeast*—Yeast contains considerable ›rgosterol, which, when exposed to ultraviolet light, ›roduces vitamin D. This high-potency vitamin)-containing yeast is helpful in preventing milk fever n dairy cows, when fed at a level of 20 million units ›f vitamin D daily, in the form of irradiated yeast, for ʼ days just before calving. It should not be fed longer han the 7-day period, as it may result in detriment to he cow's health.

Commercial Dairy Feeds

Several different types of commercial feeds are .vailable for dairy cattle; among them, (1) complete lairy concentrates, (2) dry cow rations, (3) fitting ra- ions, (4) growing or young stock rations, (5) calf start- rs, (6) milk replacer feeds, and (7) protein supple- nents.

An enlightened dairyman will know how to de- ermine what constitutes the best in commercial feeds or his specific needs. He will not rely solely on how he feed looks and smells.

FEED PREPARATION

Pertinent information relative to the preparation ›f feeds for all classes of livestock is presented earlier in this section under the heading "Feed Preparation," and in Table 4-11, page 251; hence, the reader is referred thereto.

FEEDING LACTATING COWS

Dairymen realize greatest profit from feeding when cows convert the maximum proportion of their feed into milk. The nutrient requirements for production depend primarily on the amount and composition of the milk. These needs for cows of all sizes and levels of production are given in Tables 4-43 to 4-45a, pages 325 to 328. Rations that fulfill these requirements, plus a margin of safety, can be formulated based on composition of feeds listed in Tables 4-101 to 4-105. The primary concern in feeding lactating cows is to provide a ration adequate in energy, protein, salt, calcium, phosphorus, and vitamin A (or carotene). When allowances for these nutrients are met, other minerals and vitamins usually are present in sufficient amounts, also.

Additional considerations in feeding lactating cows include palatability of the ration; physical form; fiber level; protein and mineral content of concentrates, proportion of concentrate to roughage; relative prices of ingredients; voluntary feed intake; and frequency and regularity of feeding. Thus, the proper feeding of lactating cows necessitates that the dairyman have sufficient knowledge relative to basic nutrient requirements and principles to plan an efficient feeding program and the experience and management ability to apply it.

Rations for Lactating Cows

Cows fed an all-forage ration produce about 30 percent less milk than cows fed concentrates in average amounts. This is because a cow's stomach simply isn't big enough to hold all the roughage necessary to get the amount of energy needed. To provide the needed energy, grain must be added. Grain provides energy in concentrated form. For example, 5 lb of barley contains as much energy as 8 lb of hay or 25 lb of silage.

Grain rations also supply protein. The percentage total protein needed in the grain ration depends upon the type of roughage fed. Some guides are given in Table 4-48.

TABLE 4-48
PERCENT PROTEIN NEEDED IN GRAIN MIX

Roughage Fed	Percent Protein in Grain Mix
All legume	10-14
Mixed (part legume and part grass)	14-18
All grass	18-20

As shown in Table 4-48 with high-quality legume roughage, a 10 percent protein grain mix will suffice. With a low-quality grass roughage, however, the protein level of the grain mix should be much higher, up to 18 to 20 percent.

Since proteins are always the most expensive part of a ration, for practical reasons the dairyman should not feed more of them than necessary. For this reason, the protein content of the grain mix should always be balanced out on the basis of the roughage being fed.

The concentrate ration needed to supplement the available roughage on the dairy farm may either be home mixed or a commercial concentrate. About 3 percent of all concentrates fed to dairy cattle in the United States are commercially mixed. Commercially mixed dairy concentrates have largely taken over, and replaced home mixed grains, in dairy herds operating in grain-deficit areas and among those highly specialized dairy enterprises on which little or no grains are produced. This trend will continue.

On the smaller dairies, and particularly where grain is homegrown or abundantly available locally, home mixing of dairy concentrates will continue.

Table 4-49, Lactating Cow Feeding Guide, will

TABLE 4-49
LACTATING COW FEEDING GUIDE
(As-fed Basis)

Suggested Grain Mix, Based on Kind of Roughage Available	Low-Protein (Under 12%) Concentrates	Low-Medium Protein (12% to 18%) Concentrates	Medium-High Protein (18% to 28%) Concentrates	High-Protein (Over 32%) Concentrates
	(% protein)	*(% protein)*	*(% protein)*	*(% protein)*
FEEDS:	Molasses* ..... 3.2 Corn & cob meal ..... 8.1 Corn #2 ..... 8.9 Dried beet pulp ..... 9.1 Hominy feed ..... 10.8 Sorghum (milo) ..... 11.0 Barley ..... 11.6 Oats ..... 11.7 Rye* ..... 11.9 12% dairy feed ..... 12.0 Wheat* 12.7	16% dairy feed ..... 16.0 Wheat bran ..... 16.0 Wheat middlings 17.2	18-24% dairy feed ..... 18-24 Copra (coconut) meal ..... 21.3 Peas* ..... 23.4 Corn gluten feed ..... 25.8 Brewers' dried grains* ..... 25.9 Malt sprouts ..... 26.4 Distillers' dried grains 27.3	32-34% dairy feed ..... 32-34 Linseed meal ..... 35 Cottonseed meal* ..... 41 Corn gluten meal ..... 42 Soybean meal ..... 45 Peanut meal 47
	(lb)	*(lb)*	*(lb)*	*(lb)*
Excellent Roughage—High-protein forage: (1) legume or (2) legume and non-legume mixed forages of *high quality*; consisting of dry forages and/or silage. Mix No. 1-e	1,000			
Mix No. 2-e	900			100
Mix No. 3-e	800		200	
Mix No. 4-e	850	100		50
Medium Roughage—Medium-protein forage: (1) legume or (2) legume and nonlegume mixed forages of *medium quality*; consisting of dry forages and/or silage. Mix No. 1-m	800			200
Mix No. 2-m	650		350	
Mix No. 3-m	700	100	100	100
Mix No. 4-m	Straight 16% dairy feed, or ½ Mix No. 1-p & ½ 16% dairy feed			
Poor Roughage—Low-protein forage: nonlegume forage; consisting of dry forages and/or silage. Mix No. 1-p	700	300		
Mix No. 2-p	600		200	200
Mix No. 3-p	600	100	100	200
Mix No. 4-p	500	and 500 lb 32% Dairy Feed		

Comments:
Add—To all rations (1) 1% iodized or trace-mineralized salt; (2) 1% steamed bone meal, dicalcium phosphate, or the equivalent (use monosodium phosphate or a high phosphorus commercial mineral where alfalfa is fed liberally); (3) 1,000 IU of vitamin A/lb of concentrate and, unless cows are in sunlight, add 150 IU of vitamin D/lb of concentrate.

**Limitations*—Wheat, not more than 50% of the ration; dried molasses beet pulp, 20%; molasses, 15%; peas and brewers' dried grains, 30%; rye, 10%; and cottonseed meal, 20% of the mix for calves, but as needed for mature cows.

serve as a useful guide for the dairyman who wishes to mix his own concentrate, with the ingredients therein based on the kind of roughage available.

Here is how to use Table 4-49: Let's assume that a dairyman has (1) a medium-quality forage, and (2) both low- and medium-high protein concentrates from which to choose. How many pounds each of the low- and medium-high protein concentrates will be required in a 1,000-pound concentrate mix? Step by step, here is the answer:

1. Look under "medium-roughage–medium-protein forage" (column to the left).

2. Mix No. 2-m, containing 650 lb of low-protein concentrates (under 12% protein) and 350 lb of medium-high protein concentrates (18 to 28% protein), will meet the needs. The concentrates may be chosen from among those listed at the top of the respective columns of Table 4-49—the low-protein concentrates from column 2 (under 12%) and the medium-high protein concentrates from column 4 (18 to 28%).

How to Balance a Dairy Ration

Generally speaking, the rations given in Table 4-49 will suffice. But it is recognized that rations should vary with conditions, and that many times they should be formulated to meet the conditions of a specific dairy farm. Also, a good dairyman should know how to balance a ration. Complete instructions on how to balance a ration (including an example of balancing a dairy ration) are given under the heading "How to Balance Rations" (see page 253); hence, the reader is referred thereto. By (1) following these instructions, and (2) using nutrient requirement tables similar to Tables 4-43 to 4-45, it is possible to balance rations for specific weights of animals and levels of production.

Amount of Concentrate to Feed

In the past, a number of rule-of-thumb feeding guides were used to estimate the amount of concentrates to feed to dairy cows. A very common one was to feed 1 lb of concentrate for each 4 lb of milk produced by Holstein, Ayrshire, and Brown Swiss cows, and 1 lb of concentrate for each 3 lb of milk produced by Guernsey and Jersey cows.

Today, rule-of-thumb concentrate feeding guides are in disrepute, because they're actually stumbling blocks to wise concentrate feeding. With cows not capable of high production, or those on very good roughage, some rules can lead to wasteful overfeeding. On the other hand, some rules can lead to underfeeding the high-producing cows. It can be concluded, therefore, that no simple set of rules of thumb can replace experience and dairy intuition. The rule of the successful feeder is to increase concentrates as long as the cow responds with extra milk at a profit. The commonly used profit indicator is the "milk-feed price ratio," which is the pounds of concentrate equal in value to 1 pound of milk.

Generally the cost of concentrate per pound is somewhere between ½ and ¾ the price received for milk; of course, there are yearly, seasonal, and area variations. For the United States as a whole, the milk-feed price ratio was 1.35 in 1974 and 1.44 in 1981.[12] Thus, the cost per pound of concentrate was 74% the price received per pound of milk in 1974, and 69% in 1981. As a result of the high feed prices in

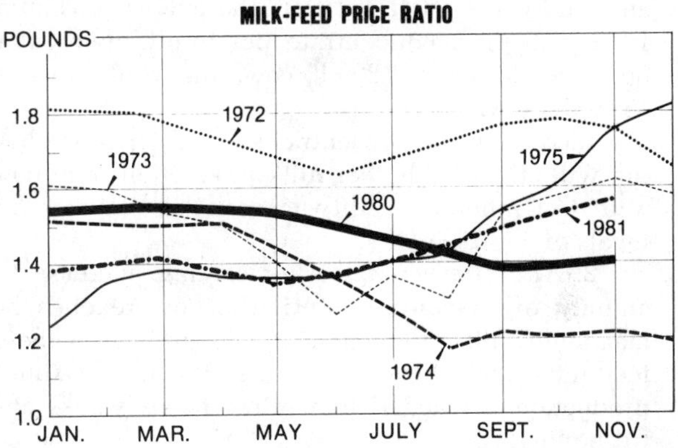

Fig. 4-30. Pounds of concentrate ration equal in value to 1 pound of milk sold to plants, 1972-81. (Adapted by the author from *Dairy Situation*, USDA, March 1976, p. 11, Fig. 8; and from *Dairy Outlook & Situation*, USDA, June 1982, p. 5).

1974 (pound for pound, the concentrate cost was 74% the price received for milk), many of the nation's dairymen either went broke or went out of business.

The amount of grain that it will pay to feed milking cows depends upon several factors: (1) quality of roughage, (2) price of roughage, (3) price and quality of concentrates, (4) price of milk, and (5) inherent producing ability of cows. Of these, the most important is the inherent milk-producing ability of the cow. This necessitates (1) using milk and fat production records of each cow, and (2) weighing and feeding the concentrate to each individual cow.

With loose housing and milking parlors, cows generally do not have time to consume over 10 to 15 pounds of grain per day while in the milking parlor. Thus, high-producing cows should be fed additional concentrate away from the parlor.

The corral system of feeding is being used more and more, primarily as a means of saving labor. It involves segregating the lactating cows into two or more

[12]*Dairy Situation*, DS-359, ERS, USDA, March 1976, p. 11; and *Dairy Outlook & Situation*, USDA, June 1982, p. 5.

production groups, with the grain allowance of each group in keeping with its production. Generally, fence-line feeding with a self-unloading truck is used.

Challenge or Lead Feeding

Challenge feeding, or lead feeding, refers to feeding the lactating cow so that she is challenged to reach her maximum production level early in lactation, without being limited by lack of available energy for milk production at that time. Here is how it works:

1. *Last 2 to 3 weeks of the dry period*—Beginning 2 to 3 weeks before freshening, feed the cow about 4 lb of concentrate per day. Increase this amount by 1 lb each day until the cow is consuming 1.0 to 1.5 lb of concentrate per hundredweight of body weight. For a 1,200-lb cow, this would be 12 to 18 lb per day.

Increasing the concentrates just prior to freshening will stimulate higher milk production from cows with the inherent ability to respond to increased levels of energy intake.

2. *After freshening*—After calving, increase the amount of concentrate until the cow reaches her maximum milk production or maximum free-choice feed intake, whichever comes first. Usually, maximum production is reached from three to six weeks after freshening.

3. *Test, then adjust concentrate to production*—Following the first test in the dairy record-keeping program after the cow has been fresh for at least two weeks, adjust the concentrate to the amount justified by her production.

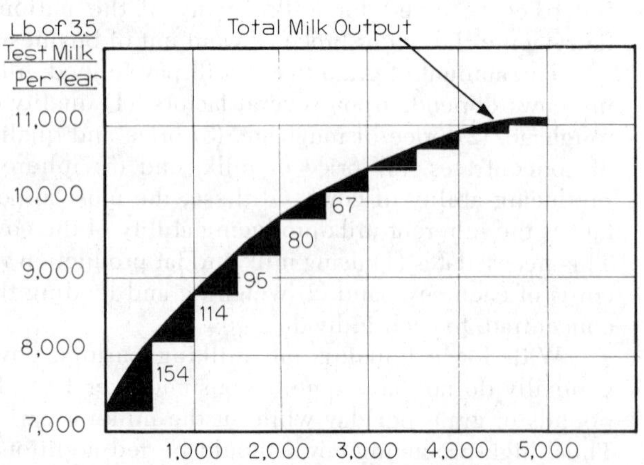

Fig. 4-31. Challenge feed (stair-step feed) "for production," rather than "according to production." The cow charted above has the ability to produce 10,000 lb of milk per year when fed 1 lb grain for every 4 lb of milk produced. Production on roughage alone is around 7,500 lb of milk per year. With added grain, milk production increases. The first 500 lb of grain give 154 lb of milk for each 100 lb of grain. The second 500 lb result in somewhat less milk and so on until very little additional milk is obtained with the last 500 lb.

4. *Adjust concentrate on basis monthly mi tests*—For the rest of the lactation period, adjust th amount of concentrate to the production of the cow indicated by monthly milk tests. Keep adding grain l 1-pound increases until the value of the added mi produced no longer pays for the cost of the grain. Th "stair-step" method of feeding, as it is sometim called, is illustrated in Fig. 4-31.

Cows respond best to challenge feeding early the lactation period when the lactation drive greatest. As the lactation period progresses, the co centrate feeding schedule should be checked wi care, at least once per month, and adjusted to the mo economical feeding level. As would be expecte some cows respond to extra concentrate feeding mo than others. Those which do not respond should l cut back until the grain feeding level is such that th intake of feed pays off in added milk.

Some dairymen report that they have increase milk production by 2,000 pounds per cow by cha lenge feeding—by using less roughage and more co centrates properly fortified, fed to the right animals the right time; and increases of over 5,000 pounds milk per lactation have been achieved.

The challenge feeding method has several *adva tages* over the old, conventional method of feeding pound of concentrate for each 3 to 4 pounds of mi produced; among them—

1. It allows the cow's ruminal microorganisms adjust to high-concentrate levels before calving.
2. It makes for a fairly high level of concentra intake prior to freshening, which is likely to contin after calving.
3. It provides abundant energy to the cow at th precise time when she needs it most—early in lact tion, thereby challenging her to reach her maximu production level.
4. It allows maximum production because th decrease in amount of concentrates fed follows th drop in milk production rather than precedes it.
5. It results in a higher yield of milk at the pea of lactation, which tends to persist throughout th production cycle.
6. It lessens ketosis, because high-producin cows do not need to depend on the breakdown body fat for additional energy for milk production.
7. It helps maintain body weight.
8. It prevents uneconomical feeding since th input of energy is determined by the ability of th cow to respond.

Among the *disadvantages*, or objections, voice by some dairymen to challenge feeding are the fo lowing:

1. Feeding high-concentrate levels prior freshening results in a higher incidence of udd

edema (caked udder). However, controlled experiments have not substantiated this claim; they have shown no difference in the incidence of udder edema in cows fed high- and low-concentrate levels just prior to freshening.

2. High production resulting from a high-concentrate ration will cause a higher incidence of mastitis. It is true that a high-concentrate ration will put a greater strain on the udder, and that chronic cases of mastitis may flare up because of the extra strain. However, the higher level of concentrates does not cause the mastitis; it merely accentuates the appearance of the mastitis which was already in the udder.

3. Not all cows respond to extra concentrate feeding with increased production. So, some cows will be overfed from three to six weeks before their maximum production is determined on test day and the concentrate level reduced accordingly. However, in high-producing herds, only a few cows will not respond; and these should likely be culled.

4. Challenge feeding is more profitable when concentrates are more favorably priced than roughages, comparatively speaking. This is a valid criticism, because the concept of challenge feeding is based on maximum grain feeding—at a profit. There are times when feeding more high-priced grain to increase production may be uneconomical, when it may be well to rely on more high-quality forage to produce milk at a profit. In short, the important thing is to feed for high profit—not top production.

Feeding on Pasture

Problems of milk production are at a minimum during the early pasture season, when plant growth is lush. However, when the weather gets hot, it is a different story; this is the period known as the "summer slump." High temperatures actually affect pasture growth more than the well-being of the cows.

The following summer feeding program is recommended for most dairy cows:

1. Have good pastures and follow good pasture management.
2. Consider ensiling the early pasture growth, thereby avoiding wasted grass early in the season.
3. Bunk feed hay with pasture, regardless of pasture quality. If the quality of the pasture is comparable to the roughage that was used in the winter ration, continue with the concentrate mix that was used in the winter. On the other hand, if summer pastures are considerably better or considerably poorer than the quality of the winter roughages, the grain mix should be changed accordingly. In other words, the grain mix used when cows are on pasture should be formulated so as to balance out the deficiencies of the grass, just the same as the winter concentrate balances out the deficiencies of the winter roughages.
4. When pastures are poor and/or the weather is very hot, feed more grain and limit the hay, thereby providing needed added energy and avoiding the excess heat of high-fiber rations. But do not restrict roughage to the point that butterfat test drops.
5. Provide adequate shade for cows on summer pasture.
6. Consider supplementing summer pastures with silage.

Feed Substitutions

Feed substitutions for beef and dairy animals are similar. Hence, the reader is referred to Table 4-27, Feed Substitution Table for Cattle (Beef and Dairy), presented in Part II—Feeding Beef Cattle, page 290.

Feed Terms—Conversions

The following terms are generally used in analyzing dairy feed costs and practices:

1. *Animal Unit Month (AUM)*—is the amount of feed required for one mature cow for one month. It is equivalent in nutrients to 0.4 ton of average hay, or 320 Mcal of energy.
2. *Fat Corrected Milk (FCM)*—is a term used to compare milks of different composition on a standard energy basis. Four percent FCM is calculated by multiplying the actual milk yield by 0.4, then adding to this product the actual fat yield multiplied by 15. For example, if a cow produces 60 lb of 3.5% milk (60 × 3.5%) or 2.1 lb fat, her Fat Corrected Milk production is (60 × .4) plus (2.1 × 15), or 55.5 lb of 4% milk.

FEEDING DRY COWS

Dry cows have three important jobs: (1) recovering from a heavy milk producing period and resting the mammary glands, (2) developing the unborn calf (more than half the fetal growth occurs during the last two months of lactation), and (3) storing up body reserves for the next milking period. This necessitates that they be properly fed.

The following routine is recommended for dry cows:

1. Turn first- and second-lactation heifers dry 60 to 65 days before expected calving. Turn cows more than 4 years of age dry 50 to 60 days before freshening.
2. Stop grain feeding and milking abruptly to hasten drying off. Examine the udder at intervals, and rub oil (such as camphorated oil or a mixture of lard

and spirits of camphor) on it at intervals, but do not milk it out. At the end of five to seven days when the udder is soft and flabby, milk out what little secretion remains.

3. Feed only good quality roughage during the first 2 to 3 weeks of the dry period. Beginning 2 to 6 weeks before expected calving, start feeding concentrates twice daily. The amount of concentrate to feed should be determined primarily by the cow's condition and the quality of the roughage. It may vary from 4 to 20 pounds per head daily. A guide to desirable gains follows:

Weight of Cows	Pounds Gain
800-1,000	75-150
1,000-1,200	100-200
1,200-1,400	125-250

4. A special dry cow concentrate mix can be fed during this period. The dry cow mix may contain 2 to 4 percent less protein than the grain mixture normally fed to the milking herd. Also, if milk fever is a problem, as it is in most herds, it is desirable that the dry cow ration (a) be balanced out for calcium and phosphorus (a 1:1 ratio preferably, and not to exceed a 2:1 calcium-phosphorus ratio), and (b) contain added v tamin D.

Many successful dairymen follow a program challenge or lead feeding. They reach a feeding lev of 1 to 1½ pounds of grain to each 100 pounds liveweight about one week before freshening, a continuing at this rate up to freshening. This prec feeding gets the rumen, and the cow's appetite a eating habits, adjusted to liberal feeding befo freshening. Also, a cow freshening in good conditi starts off better and maintains a higher level of pr duction; her milk is usually higher in total solids; a the incidence of milk fever and ketosis is usually r duced.

FEEDING DAIRY CALVES

One of the most important phases of dairy pr duction is that of feeding and managing the dai calves raised for replacement purposes. Statistics r veal that more than 20 percent of the dairy calves d of sickness or disease before reaching maturity. Ma of these deaths are caused by faulty nutrition.

TABLE 4-50

SCHEDULE FOR FEEDING DAIRY CALVES BY THREE DIFFERENT SYSTEMS

Age of Calf	Whole Milk System	Whole Milk-Milk Replacer System	Whole Milk-Calf Starter System
0 to 3 days	Calf should receive colostrum during first 3 days.	Calf should receive colostrum during first 3 days.	Calf should receive colostrum during first 3 days.
3 days	Start feeding whole milk at the rate of 1 lb milk to 10 lb body weight.[1]	Start feeding whole milk at the rate of 1 lb milk to 10 lb body weight.[1]	Start feeding whole milk at the rate of 1 lb milk to 10 lb body weight.[1]
7 days	Make grain available in box in pen.	Make calf starter available in box in pen.	Make calf starter available in box in pen.
7 to 10 days		Start replacing whole milk with fluid milk replacer. Replace 1 to 2 lb milk daily with fluid milk replacer until change is completed.	
14 days		Transition to milk replacer should be completed.	
21 days	Make good quality hay available in rack in pen.	Make good quality hay available in rack in pen.	Make good quality hay available in rack in pen.
60 days		Discontinue milk replacer.	Discontinue feeding whole milk. Larger, more vigorous calves may have whole milk stopped as early as 42 days.
60 to 120 days	Permit calves to consume grain free-choice up to 4 or 5 lb daily. Rest of nourishment should be obtained from hay.	Permit calves to consume calf starter free-choice, up to 4 to 5 lb daily. Rest of nourishment should be obtained from hay.	Permit calves to consume calf starter free-choice up to 4 or 5 lb daily. Rest of nourishment should be obtained from hay.
90 days	Discontinue whole milk.		

[1]For economic reasons, it is never advisable to feed calves more than 12 lb whole milk daily during the entire whole milk feeding period.

Regardless of the calf-feeding system followed later, all calves should receive the first milk—colostrum. It contains antibodies which protect calves against disease. Also, colostrum is high in protein and in vitamins and minerals. Surplus colostrum can be frozen and stored for a period of one year or longer. It may then be thawed and warmed to about 100°F and fed as needed. Colostrum is a valuable feed for calves, regardless of their age; hence, it can be fed to them at any time with no ill effects.

All calf-feeding systems make use of colostrum for the first two to three days following birth. At this time, cleanliness and sanitation are especially important.

Basically, calves are fed according to one of three systems: (1) the whole milk system, (2) the combination whole milk-milk replacer system, or (3) the combination whole milk-calf starter system. Of course, various combinations of these three systems are used, also. A suggested schedule for each of these three systems is given in Table 4-50.

The whole milk method costs the most, but it produces the fastest gains, the best-appearing calves, and requires the least skill of any system.

Milk replacers are composed of sizable amounts of milk by-products, such as dry skim milk, buttermilk, or whey; and they are generally fortified by antibiotics, vitamins, and minerals. They can be fed as the only feed immediately following the colostrum period; or, as shown in Table 4-50, they may replace whole milk beginning on about the seventh day.

There is hardly any limit to the number of calf starters on the market. Most of them are mixed commercially. It might well be added that because of the difficulty in formulating a home mixed calf ration, the purchase of a good commercial feed usually represents a wise investment.

Two suggested calf starter rations are given in Table 4-51.

Starter Ration A, of Table 4-51, is designed for feeding anytime after the first day following birth. Starter Ration B is designed for feeding beginning about 45 days of age. As is true in any ration change, the transition from Ration A to Ration B should be made gradually by blending the feeds over a period of 2 to 3 days.

Good quality hay for young calves is essential; it provides an economical source of nutrients, helps maintain rate of gain, and speeds up the development of the rumen.

Many dairymen make the mistake of placing calves on pasture at too early an age. Unless pastures are properly supplemented, young calves simply cannot hold enough grass, or other pasturage, to obtain sufficient nutrients for their growing bodies. Accordingly, growth will be retarded.

Veal calf production is also becoming an important part of the dairy industry. Whether or not it is profitable, will depend on the price of veal, the price of feeds used to produce the veal, and the skill of the operator to maintain healthy calves and a low death loss.

TABLE 4-51

CALF STARTER RATIONS
(As-fed Basis)

Ingredients	Starter Ration A[1] (for feeding first 45 days, along with liquid skim milk)	Starter Ration B[1] (for feeding after first 45 days, with dry skim milk therein)
	(lb)	*(lb)*
Barley	1,000	750
Soybean or cottonseed meal (41%)	560	450
Dried skim milk		400
Molasses	200	200
Wheat bran	200	150
Dicalcium phosphate	20	20
Trace mineralized salt	20	20
Antibiotic (follow mfg.'s directions)	10	10
Vitamin A	10,000 IU/lb	2,000 IU/lb
Vitamin D (not needed if calf is in sunlight)	2,000 IU/lb	400 IU/lb
TOTAL	2,010	2,000
Proximate Analysis:	*(%)*	*(%)*
Crude protein	19.9	22.3
Fiber	5.3	4.1
Fat	2.6	2.2
Calcium	.47	.70
Phosphorus	.68	.78
TDN	69.4	71.2

[1]In 1/8- or 3/16-in. pellets.

FEEDING REPLACEMENT HEIFERS

Without doubt, dairy heifers are more commonly neglected than any other animals on the dairy farm. This is especially true during the period from six months of age until freshening. Since growing heifers are the nucleus of the future dairy herd, it is important that they be given every opportunity to grow and develop into large, deep-bodied two-year-olds at the time of freshening.

When the feeding of milk or other special calf meals is discontinued, heifers should have an abundance of other feeds, especially of good quality roughage, so that their growth will not be retarded. Whether receiving dry roughage or on pasture, young heifers need a supplemental feeding of concentrates. The amount of grain fed will vary according to the age of the animal and the quality of the forage. Generally it will range from 3 to 5 pounds per head per day, but, with poor roughage, 6 to 8 pounds may be necessary. In any event, replacement heifers should neither be overfed nor underfed. Heifers that are underfed do not come in heat at an early date and they may be too small to breed. On the other hand, heifers that are too fat do not conceive easily, and they do not produce as well as heifers that are fed properly.

Well-grown heifers can be bred at 14 to 16 months of age. Table 4-52 can be used as a guide to determine if the heifers have obtained sufficient growth to permit breeding at the age indicated.

Normal Growth of Calves and Heifers

Table 4-52 shows the weight and heart girth measurements of dairy calves or heifers at monthly intervals up to two years of age. The author suggests that the figures given therein for Holsteins also be used for Brown Swiss, and that the figures given therein for Ayrshires also be used for Milking Shorthorn and Red Poll heifers. If the dairyman does not have scales, he can measure the heart girth with tape to estimate (within 95% accuracy) weight. In any event, weight for age is important from the standpoint of determining the growth progress made in herd replacements.

FEEDING DAIRY BULLS

Bull calves raised for breeding purposes shoulc be fed and handled much the same as heifers. Olde bulls should be kept in thrifty, vigorous condition, bu they should not be permitted to get too fat. Mature bulls can be fed the same grain ration as the lactating cows. Depending on the quality of the roughage, usu ally about ½ pound of grain per 100 pounds of body weight will suffice for the mature bull. Also, indi vidual differences must be considered, for some bull are easier keepers than others.

In addition to the grain and roughage ration, the bull should have free access to a double compartmen mineral box, with ground salt in one side and a suita ble mineral mixture in the other.

FEEDING SHOW AND SALE DAIRY CATTLE

Dairy animals intended for show or sale shoulc be fed so as to achieve a certain amount of finish o bloom, but they should not be too fat. Linseed meal beet pulp, oats, barley, and wheat bran are popula feeds in a fitting and showing ration. Likewise, gooc roughages are always very important.

TABLE 4-52

NORMAL HEART GIRTH MEASUREMENTS AND WEIGHT OF CALVES AND HEIFERS DURING THE GROWING PERIOD[1]

Age in Months	Holstein (also Brown Swiss)		Ayrshire (also Milking Shorthorn and Red Poll)		Guernsey		Jersey	
	(in.)	(lb)	(in.)	(lb)	(in.)	(lb)	(in.)	(lb)
Birth	31	96	29½	72	29	66	24½	56
1	33½	118	32	98	31½	90	29½	72
2	37	161	35½	132	34½	122	32½	102
3	40¼	213	38¾	179	38	164	35½	138
4	43½	272	42¾	236	41¼	217	38¼	181
5	47	335	45½	291	44¼	265	41½	228
6	50	396	48¼	340	47	304	44½	277
7	52½	455	51¼	408	49¾	362	47¼	325
8	54¾	508	53	447	51¾	410	49¾	369
9	57	559	55	485	53¾	448	51¾	409
10	58¾	609	57	526	55	486	53¼	446
11	60½	658	58	563	56¾	521	55	481
12	62½	714	59	583	58¼	549	56½	520
13	63¼	740	60¾	630	59¼	587	57½	540
14	64¼	774	62	666	60½	615	58½	565
15	65¼	805	63	703	61¾	640	59	585
16	66¼	841	64	731	62½	674	59¾	611
17	67¼	874	65¼	758	63½	696	60½	635
18	68½	912	66	781	65	727	61½	660
19	69¼	946	66½	813	65½	752	62½	687
20	70½	985	67½	841	66¼	780	63	712
21	71½	1,025	68½	885	67½	816	64	740

[1]Body weights for Holsteins and Jerseys from USDA Technical Bulletins 1098 and 1099. Heart girth measurements for these weights taken from Neb. Ag. Exp. Sta. Res. Bull. 194 (1960). Weights and heart measurements for Ayrshires and Guernseys calculated from data furnished by Professor H. P. Davis, University of Nebraska, Lincoln, Neb.

PART IV—FEEDING SHEEP

Fig. 4-32. Feeding sheep.

Sheep consume a higher proportion of forages than any other class of livestock, it being estimated that 90 percent of the total feed supply of U.S. sheep production is derived from roughages. They are naturally adapted to grazing on pastures and ranges which supply a variety of forage plants, and they thrive best on forage that is short and fine rather than high and coarse. Although sheep will eat considerable quantities of weeds and brush, they prefer choice grasses and legumes.

NUTRIENT REQUIREMENTS

The nutrient requirements of sheep are given in Tables 4-53 and 4-54, pages 352 and 353.

Energy

Lack of energy—hunger—is probably the most common nutritional deficiency of sheep. It may result from lack of feed or from the consumption of poor quality feed.

Inadequate amounts of feed may result from overgrazing, drought, snow covering the feed, or from a low dry matter content of lush, washy feeds. Also, poorly digested low-quality forage leads to reduced feed intake.

The energy needs of sheep are largely met through the consumption and digestion of roughages—pasture, hay, and silage. Grains, such as corn, barley, milo, wheat, and oats, are used to raise the energy level of the ration during periods when supplementation is necessary. In general, sheep subsist on an even higher proportion of roughages to concentrates than do beef cattle, and this applies to finishing lambs. The bacterial action in the paunch of the sheep efficiently converts roughages into suitable sources of energy.

It is generally recognized that the energy requirements of sheep are affected by size, age, pregnancy, lactation, growth, and protein content of the ration. It is also affected by environment, shearing, and sex.

An energy deficiency is characterized by slowing and cessation of growth, loss of weight, reduced fertility or reproductive failure, lowered milk production and shortened lactation period, reduced quantity and quality of wool (including breaks in the fiber), lowered resistance to infection with internal parasites, and increased mortality.

Protein

Sheep need protein, as do other classes of animals, for maintenance, growth, reproduction, and finishing. Additionally, sheep need protein for the production of wool—a protein product. Wool is especially rich in the sulfur-containing amino acid, cystine, but this requirement is usually amply met by the cystine of feeds or by methionine, another amino acid which is also rather widely distributed in natural sources and which is derived from rumen synthesis.

Green pastures and legume hays (alfalfa, clover, soybeans, lespedeza, etc.) are excellent and practical sources of proteins for sheep in most areas. Where the ranges are bleached and dry for an extended period, or where legume hays cannot be produced for winter feeding, however, it may be desirable to provide sheep with such protein-rich supplements as soybean meal, cottonseed meal, linseed meal, peanut meal, sunflower meal, or a commercial protein supplement, at the rate of about ¼ to ⅓ pound per ewe per day.

The protein requirements of sheep are affected by growth, pregnancy, lactation, mature size, weight for age, body condition, rate of gain, and protein-energy ratio. Though correspondingly less because of their smaller body size and lower milk production, the protein requirement of ewes nursing lambs are much like those of lactating cows.

A protein deficiency is characterized by reduced appetite, lowered feed intake, and poor feed efficiency. In turn, this makes for poor growth, poor muscular development, loss of weight, reduced reproductive efficiency, and reduced wool production. Under extreme conditions, there are severe digestive disturbances, nutritional anemia, and edema.

TABLE 4-53
DAILY NUTRIENT REQUIREMENTS OF SHEEP (per Animal)[1]

Body Weight		Gain or Loss		Dry Matter[2] Per Animal		Dry Matter[2] % Live-Weight	Nutrients per Animal: Energy TDN		DE[3]	ME	Total Protein		DP[4]		Grams DP per Mcal DE	Ca	P	Vita-min A	Caro-tene	Vita-min
(lb)	(kg)	(lb)	(g)	(lb)	(kg)	(%)	(lb)	(kg)	(Mcal)	(Mcal)	(lb)	(g)	(lb)	(g)	(g)	(g)	(g)	(IU)	(mg)	(IU)
EWES[5]																				
Maintenance																				
110	50	.02	10	2.2	1.0	2.0	1.21	.55	2.42	1.98	.20	89	.11	48	20	3.0	2.8	1,275	1.9	278
132	60	.02	10	2.4	1.1	1.8	1.34	.61	2.68	2.20	.22	98	.12	53	20	3.1	2.9	1,530	2.2	333
154	70	.02	10	2.6	1.2	1.7	1.46	.66	2.90	2.38	.24	107	.13	58	20	3.2	3.0	1,785	2.6	388
176	80	.02	10	2.9	1.3	1.6	1.59	.72	3.17	2.60	.26	116	.14	63	20	3.3	3.1	2,040	3.0	444
Nonlactating and first 15 weeks of gestation																				
110	50	.07	30	2.4	1.1	2.2	1.32	.60	2.64	2.16	.22	99	.12	54	20	3.0	2.8	1,275	1.9	278
132	60	.07	30	2.9	1.3	2.1	1.59	.72	3.17	2.60	.26	117	.14	64	20	3.1	2.9	1,530	2.2	333
154	70	.07	30	3.1	1.4	2.0	1.70	.77	3.39	2.78	.28	126	.15	69	20	3.2	3.0	1,785	2.6	388
176	80	.07	30	3.3	1.5	1.9	1.81	.82	3.61	2.96	.30	135	.16	74	20	3.3	3.1	2,040	3.0	444
Last 6 weeks of gestation or last 8 weeks of lactation suckling singles[6]																				
110	50	.39	175(+45)	3.7	1.7	3.3	2.18	.99	4.36	3.58	.35	158	.19	88	20	4.1	3.9	4,250	6.2	278
132	60	.40	180(+45)	4.2	1.9	3.2	2.43	1.10	4.84	3.97	.39	177	.22	99	20	4.4	4.1	5,100	7.5	333
154	70	.41	185(+45)	4.6	2.1	3.0	2.69	1.22	5.37	4.40	.43	195	.24	109	20	4.5	4.3	5,950	8.8	388
176	80	.42	190(+45)	4.8	2.2	2.8	2.82	1.28	5.63	4.62	.45	205	.25	114	20	4.8	4.5	6,800	10.0	444
First 8 weeks of lactation suckling singles or last 8 weeks of lactation suckling twins[7]																				
110	50	−.06	−25(+80)	4.6	2.1	4.2	3.00	1.36	5.98	4.90	.48	218	.29	130	22	10.9	7.8	4,250	6.2	278
132	60	−.06	−25(+80)	5.1	2.3	3.9	3.31	1.50	6.60	5.41	.53	239	.32	143	22	11.5	8.2	5,100	7.5	333
154	70	−.06	−25(+80)	5.5	2.5	3.6	3.59	1.63	7.17	5.88	.57	260	.34	155	22	12.0	8.6	5,950	8.8	388
176	80	−.06	−25(+80)	5.7	2.6	3.2	3.73	1.69	7.44	6.10	.60	270	.35	161	22	12.6	9.0	6,800	10.0	444
First 8 weeks of lactation suckling twins																				
110	50	−.13	−60	5.3	2.4	4.8	3.44	1.56	6.86	5.63	.61	276	.38	173	25	12.5	8.9	4,250	6.2	278
132	60	−.13	−60	5.7	2.6	4.3	3.73	1.69	7.44	6.10	.66	299	.41	187	25	13.0	9.4	5,100	7.5	333
154	70	−.13	−60	6.2	2.8	4.0	4.01	1.82	8.01	6.57	.71	322	.45	202	25	13.4	9.5	5,950	8.8	388
176	80	−.13	−60	6.6	3.0	3.7	4.30	1.95	8.58	7.04	.76	345	.48	216	25	14.4	10.2	6,800	10.0	444
Replacement lambs and yearlings[8]																				
66	30	.40	180	2.9	1.3	4.3	1.78	.81	3.56	2.92	.29	130	.16	75	21	5.9	3.3	1,275	1.9	166
88	40	.26	120	3.1	1.4	3.5	1.81	.82	3.61	2.96	.29	133	.16	74	20	6.1	3.4	1,700	2.5	222
110	50	.18	80	3.3	1.5	3.0	1.83	.83	3.65	2.99	.29	133	.16	73	20	6.3	3.5	2,125	3.1	278
132	60	.09	40	3.3	1.5	2.5	1.81	.82	3.61	2.96	.29	133	.16	72	20	6.5	3.6	2,550	3.8	333
RAMS																				
Replacement lambs and yearlings[8]																				
88	40	.55	250	4.0	1.8	4.5	2.58	1.17	5.15	4.22	.41	184	.24	108	21	6.3	3.5	1,700	2.5	222
132	60	.44	200	5.1	2.3	3.8	3.04	1.38	6.07	4.98	.48	219	.27	122	20	7.2	4.0	2,550	3.8	333
176	80	.33	150	6.2	2.8	3.5	3.40	1.54	6.78	5.56	.55	249	.30	134	20	7.9	4.4	3,400	5.0	444
220	100	.22	100	6.2	2.8	2.8	3.40	1.54	6.78	5.56	.55	249	.30	134	20	8.3	4.6	4,250	6.2	555
265	120	.11	50	5.7	2.6	2.2	3.15	1.43	6.29	5.16	.51	231	.28	125	20	8.5	4.7	5,100	7.5	666
LAMBS																				
Finishing[9]																				
66	30	.44	200	2.9	1.3	4.3	1.83	.83	3.65	2.99	.32	143	.19	87	24	4.8	3.0	765	1.1	166
77	35	.48	220	3.1	1.4	4.0	2.07	.94	4.14	3.39	.34	154	.21	94	23	4.8	3.0	892	1.3	194
88	40	.55	250	3.5	1.6	4.0	2.47	1.12	4.93	4.04	.39	176	.24	107	22	5.0	3.1	1,020	1.5	222
99	45	.55	250	3.7	1.7	3.8	2.62	1.19	5.24	4.30	.41	187	.25	114	22	5.0	3.1	1,148	1.7	250
110	50	.48	220	4.0	1.8	3.6	2.78	1.26	5.54	4.54	.44	198	.27	121	22	5.0	3.1	1,275	1.9	278
121	55	.44	200	4.2	1.9	3.5	2.93	1.33	5.85	4.80	.46	209	.28	127	22	5.0	3.1	1,402	2.1	305
Early weaned[10]																				
22	10	.55	250	1.3	.6	6.0	.97	.44	1.94	1.59	.21	96	.15	69	36	2.4	1.6	850	1.2	67
44	20	.60	275	2.2	1.0	5.0	1.61	.73	3.21	2.63	.35	160	.25	115	36	3.6	2.4	1,700	2.5	133
66	30	.66	300	3.1	1.4	4.7	2.25	1.02	4.49	3.68	.43	196	.29	133	30	5.0	3.3	2,550	3.8	200

[1]Adapted by the author from *Nutrient Requirements of Sheep*, No. 5, 5th rev. ed., NRC-National Academy of Sciences, Washington, D.C., 1975, pp. 42-43.
[2]To convert dry matter to an as-fed basis, divide dry matter by percentage of dry matter.
[3]1 kg TDN = 4.4 Mcal DE (digestible energy). DE may be converted to ME (metabolizable energy) by multiplying by 82%.
[4]DP = digestible protein.
[5]Values are for ewes in moderate condition, not excessively fat or thin. Fat ewes should be fed at the next lower weight, thin ewes at the next higher weight. Onc maintenance weight is established, such weight would follow through all production phases.
[6]Values in parentheses are for ewes suckling singles last 8 weeks of lactation.
[7]Value in parentheses are for ewes suckling twins last 8 weeks of lactation.
[8]Requirements for replacement lambs (ewe and ram) start when the lambs are weaned.
[9]Maximum gains expected. If lambs are held for later market, they should be fed as replacement ewe lambs are fed. Lambs capable of gaining faster than indicated shoul be fed at a higher level. Lambs finish at the maximum rate if they are self-fed.
[10]An 88-lb *(40-kg)* early weaned lamb should be fed the same as a finishing lamb of the same weight.

TABLE 4-54

NUTRIENT REQUIREMENTS OF RATIONS FOR SHEEP[1]
(In Percentage or Amount per Pound or Kilogram of Ration)

Weight		Daily Gain or Loss		Moisture Basis (As-fed = est. 90% dry matter. M-F = moisture-free)	Daily Feed Consumption[2]		Total Protein	Digestible Protein[3]	TDN	DE[4]		ME		Ca	P	Vitamin A		Carotene		Vitamin D	
(lb)	(kg)	(lb)	(g)		(lb)	(kg)	(%)	(%)	(%)	Mcal per (lb)	(kg)	Mcal per (lb)	(kg)	(%)	(%)	IU per (lb)	(kg)	Mg per (lb)	(kg)	IU per (lb)	(kg)
WES[5]																					
laintenance																					
110	50	.02	10	As-fed	2.4	1.1	8.0	4.3	50	1.0	2.2	.8	1.8	.27	.25	520	1,148	.8	1.7	113	250
				M-F	*2.2*	*1.0*	*8.9*	*4.8*	*55*	*1.1*	*2.4*	*.9*	*2.0*	*.30*	*.28*	*578*	*1,275*	*.9*	*1.9*	*126*	*278*
132	60	.02	10	As-fed	2.7	1.2	8.0	4.3	50	1.0	2.2	.8	1.8	.25	.23	568	1,252	.8	1.8	124	273
				M-F	*2.4*	*1.1*	*8.9*	*4.8*	*55*	*1.1*	*2.4*	*.9*	*2.0*	*.28*	*.26*	*631*	*1,391*	*1.0*	*2.0*	*137*	*303*
154	70	.02	10	As-fed	2.9	1.3	8.0	4.3	50	1.0	2.2	.8	1.8	.24	.22	607	1,339	.9	2.0	132	291
				M-F	*2.6*	*1.2*	*8.9*	*4.8*	*55*	*1.1*	*2.4*	*.9*	*2.0*	*.27*	*.25*	*675*	*1,488*	*1.0*	*2.2*	*147*	*323*
176	80	.02	10	As-fed	3.2	1.4	8.0	4.3	50	1.0	2.2	.8	1.8	.22	.22	641	1,412	.9	2.1	140	308
				M-F	*2.9*	*1.3*	*8.9*	*4.8*	*55*	*1.1*	*2.4*	*.9*	*2.0*	*.25*	*.24*	*712*	*1,569*	*1.0*	*2.3*	*155*	*342*
lonlactating and first 15 weeks of gestation																					
110	50	.07	30	As-fed	2.7	1.2	8.1	4.4	50	1.0	2.2	.8	1.8	.24	.22	473	1,043	.7	1.5	103	228
				M-F	*2.4*	*1.1*	*9.0*	*4.9*	*55*	*1.1*	*2.4*	*.9*	*2.0*	*.27*	*.25*	*526*	*1,159*	*.8*	*1.7*	*115*	*253*
132	60	.07	30	As-fed	3.2	1.4	8.1	4.4	50	1.0	2.2	.8	1.8	.22	.20	480	1,059	.7	1.5	105	230
				M-F	*2.9*	*1.3*	*9.0*	*4.9*	*55*	*1.1*	*2.4*	*.9*	*2.0*	*.24*	*.22*	*534*	*1,177*	*.8*	*1.7*	*116*	*256*
154	70	.07	30	As-fed	3.4	1.6	8.1	4.4	50	1.0	2.2	.8	1.8	.21	.19	520	1,148	.8	1.7	113	249
				M-F	*3.1*	*1.4*	*9.0*	*4.9*	*55*	*1.1*	*2.4*	*.9*	*2.0*	*.23*	*.21*	*578*	*1,275*	*.9*	*1.9*	*126*	*277*
176	80	.07	30	As-fed	3.7	1.7	8.1	4.4	50	1.0	2.2	.8	1.8	.20	.19	555	1,224	.8	1.8	121	266
				M-F	*3.3*	*1.5*	*9.0*	*4.9*	*55*	*1.1*	*2.4*	*.9*	*2.0*	*.22*	*.21*	*617*	*1,360*	*.9*	*2.0*	*134*	*296*
.ast 6 weeks of gestation or last 8 weeks of lactation suckling singles[6]																					
110	50	.39	175	As-fed	4.1	1.9	8.4	4.7	52	1.1	2.3	0.9	1.9	.22	.21	1,020	2,250	1.5	3.2	67	148
				M-F	*3.7*	*1.7*	*9.3*	*5.2*	*58*	*1.2*	*2.6*	*1.0*	*2.1*	*.24*	*.23*	*1,134*	*2,500*	*1.6*	*3.6*	*74*	*164*
132	60	.40	180	As-fed	4.7	2.1	8.4	4.7	52	1.1	2.3	0.9	1.9	.21	.20	1,096	2,416	1.6	3.5	71	158
				M-F	*4.2*	*1.9*	*9.3*	*5.2*	*58*	*1.2*	*2.6*	*1.0*	*2.1*	*.23*	*.22*	*1,217*	*2,684*	*1.8*	*3.9*	*79*	*175*
154	70	.41	185	As-fed	5.1	2.3	8.4	4.7	52	1.1	2.3	0.9	1.9	.19	.18	1,157	2,550	1.7	3.8	76	167
				M-F	*4.6*	*2.1*	*9.3*	*5.2*	*58*	*1.2*	*2.6*	*1.0*	*2.1*	*.21*	*.20*	*1,285*	*2,833*	*1.9*	*4.2*	*84*	*185*
176	80	.42	190	As-fed	5.3	2.4	8.4	4.7	52	1.1	2.3	0.9	1.9	.19	.18	1,262	2,782	1.8	4.1	82	182
				M-F	*4.8*	*2.2*	*9.3*	*5.2*	*58*	*1.2*	*2.6*	*1.0*	*2.1*	*.21*	*.20*	*1,402*	*3,091*	*2.0*	*4.5*	*92*	*202*
'irst 8 weeks of lactation suckling singles or last 8 weeks of lactation suckling twins[7]																					
110	50	−.06	−25	As-fed	5.1	2.3	9.4	5.6	58	1.2	2.6	1.1	2.2	.47	.33	826	1,822	1.2	2.7	54	119
				M-F	*4.6*	*2.1*	*10.4*	*6.2*	*65*	*1.3*	*2.9*	*1.2*	*2.4*	*.52*	*.37*	*918*	*2,024*	*1.4*	*3.0*	*60*	*132*
132	60	−.06	−25	As-fed	5.7	2.6	9.4	5.6	58	1.2	2.6	1.1	2.2	.45	.32	905	1,995	1.3	3.0	59	131
				M-F	*5.1*	*2.3*	*10.4*	*6.2*	*65*	*1.3*	*2.9*	*1.2*	*2.4*	*.50*	*.36*	*1,006*	*2,217*	*1.5*	*3.3*	*66*	*145*
154	70	−.06	−25	As-fed	6.1	2.8	9.4	5.6	58	1.2	2.6	1.1	2.2	.43	.31	972	2,142	1.4	3.2	63	140
				M-f	*5.5*	*2.5*	*10.4*	*6.2*	*65*	*1.3*	*2.9*	*1.2*	*2.4*	*.48*	*.34*	*1,080*	*2,380*	*1.6*	*3.5*	*70*	*155*
176	80	−.06	−25	As-fed	6.3	2.9	9.4	5.6	58	1.2	2.6	1.1	2.2	.43	.31	1,068	2,354	1.6	3.4	70	154
				M-F	*5.7*	*2.6*	*10.4*	*6.2*	*65*	*1.3*	*2.9*	*1.2*	*2.4*	*.48*	*.34*	*1,186*	*2,615*	*1.7*	*3.8*	*78*	*171*
First 8 weeks of lactation suckling twins																					
110	50	−.13	−60	As-fed	5.9	2.7	10.4	6.5	58	1.2	2.6	1.1	2.2	.47	.33	723	1,594	1.1	2.3	47	104
				M-F	*5.3*	*2.4*	*11.5*	*7.2*	*65*	*1.3*	*2.9*	*1.2*	*2.4*	*.52*	*.37*	*803*	*1,771*	*1.2*	*2.6*	*53*	*116*
132	60	−.13	−60	As-fed	6.3	2.9	10.4	6.5	58	1.2	2.6	1.1	2.2	.45	.32	801	1,768	1.2	2.9	52	115
				M-F	*5.7*	*2.6*	*11.5*	*7.2*	*65*	*1.3*	*2.9*	*1.2*	*2.4*	*.50*	*.36*	*890*	*1,962*	*1.3*	*2.9*	*58*	*128*
154	70	−.13	−60	As-fed	6.9	3.1	10.4	6.5	58	1.2	2.6	1.1	2.2	.43	.31	868	1,913	1.3	2.8	57	125
				M-F	*6.2*	*2.8*	*11.5*	*7.2*	*65*	*1.3*	*2.9*	*1.2*	*2.4*	*.48*	*.34*	*964*	*2,125*	*1.4*	*3.1*	*63*	*139*
176	80	−.13	−60	As-fed	7.3	3.3	10.4	6.5	58	1.2	2.6	1.1	2.2	.43	.31	925	2,040	1.3	3.0	60	133
				M-F	*6.6*	*3.0*	*11.5*	*7.2*	*65*	*1.3*	*2.9*	*1.2*	*2.4*	*.48*	*.34*	*1,028*	*2,267*	*1.5*	*3.3*	*67*	*148*

(Continued)

Footnotes on last page of table.

TABLE 4-54 (Continued)

Weight		Daily Gain or Loss		Moisture Basis (As-fed = est. 90% dry matter. M-F = moisture-free)	Daily Feed Consumption[2]		Total Protein	Digestible Protein[3]	TDN	DE[4]		ME		Ca	P	Vitamin A		Carotene		Vitamin D	
(lb)	*(kg)*	*(lb)*	*(g)*		*(lb)*	*(kg)*	*(%)*	*(%)*	*(%)*	*Mcal per* (lb)	(kg)	*Mcal per* (lb)	(kg)	*(%)*	*(%)*	*IU per* (lb)	(kg)	*Mg per* (lb)	(kg)	*IU per* (lb)	(kg)
Replacement lambs and yearlings[8]																					
66	30	.40	180	As-fed	3.2	1.4	9.0	5.2	56	1.1	2.4	.9	2.0	.40	.22	400	883	.6	1.4	52	115
				M-F	*2.9*	*1.3*	*10.0*	*5.8*	*62*	*1.2*	*2.7*	*1.0*	*2.2*	*.45*	*.25*	*445*	*981*	*.7*	*1.5*	*58*	*128*
88	40	.26	120	As-fed	3.4	1.6	8.6	4.8	54	1.1	2.3	.9	1.9	.40	.22	496	1,093	.7	1.6	65	143
				M-F	*3.1*	*1.4*	*9.5*	*5.3*	*60*	*1.2*	*2.6*	*1.0*	*2.1*	*.44*	*.24*	*551*	*1,214*	*.8*	*1.8*	*72*	*159*
110	50	.18	80	As-fed	3.7	1.7	8.0	4.3	50	1.1	2.2	.8	1.8	.38	.21	578	1,275	.9	1.9	76	167
				M-F	*3.3*	*1.5*	*8.9*	*4.8*	*55*	*1.2*	*2.4*	*.9*	*2.0*	*.42*	*.23*	*643*	*1,417*	*1.0*	*2.1*	*84*	*185*
132	60	.09	40	As-fed	3.7	1.7	8.0	4.3	50	1.1	2.2	.8	1.8	.39	.22	694	1,530	1.0	2.3	91	200
				M-F	*3.3*	*1.5*	*8.9*	*4.8*	*55*	*1.2*	*2.4*	*.9*	*2.0*	*.43*	*.24*	*771*	*1,700*	*1.1*	*2.5*	*100*	*222*
RAMS																					
Replacement lambs and yearlings[8]																					
88	40	.55	250	As-fed	4.4	2.0	9.1	5.4	58	1.2	2.6	1.1	2.2	.32	.17	385	850	.6	1.3	50	111
				M-F	*4.0*	*1.8*	*10.2*	*6.0*	*65*	*1.6*	*2.9*	*1.2*	*2.4*	*.35*	*.19*	*428*	*944*	*.6*	*1.4*	*56*	*123*
132	60	.44	200	As-fed	5.7	2.6	8.6	4.8	54	1.1	2.3	.9	1.9	.28	.15	453	998	.7	1.5	60	131
				M-F	*5.1*	*2.3*	*9.5*	*5.3*	*60*	*1.2*	*2.6*	*1.0*	*2.1*	*.31*	*.17*	*503*	*1,109*	*.8*	*1.7*	*66*	*145*
176	80	.33	150	As-fed	6.9	3.1	8.0	4.3	50	1.1	2.2	.8	1.8	.25	.14	496	1,093	.7	1.6	65	143
				M-F	*6.2*	*2.8*	*8.9*	*4.8*	*55*	*1.2*	*2.4*	*.9*	*2.0*	*.28*	*.16*	*551*	*1,214*	*.8*	*1.8*	*72*	*159*
220	100	.22	100	As-fed	6.9	3.1	8.0	4.3	50	1.1	2.2	.8	1.8	.27	.15	620	1,366	.9	2.0	80	178
				M-F	*6.2*	*2.8*	*8.9*	*4.8*	*55*	*1.2*	*2.4*	*.9*	*2.0*	*.30*	*.17*	*689*	*1,518*	*1.0*	*2.2*	*90*	*198*
265	120	.11	50	As-fed	6.3	2.9	8.0	4.3	50	1.1	2.2	.8	1.8	.30	.16	801	1,766	1.2	2.6	105	230
				M-F	*5.7*	*2.6*	*8.9*	*4.8*	*55*	*1.2*	*2.4*	*.9*	*2.0*	*.33*	*.18*	*890*	*1,962*	*1.3*	*2.9*	*116*	*256*
LAMBS																					
Finishing[9]																					
66	30	.44	200	As-fed	3.2	1.4	9.9	6.0	58	1.1	2.5	.9	2.1	.33	.21	240	529	.3	.7	52	115
				M-F	*2.9*	*1.3*	*11.0*	*6.7*	*64*	*1.3*	*2.8*	*1.0*	*2.3*	*.37*	*.23*	*267*	*588*	*.4*	*.8*	*58*	*128*
77	35	.48	220	As-fed	3.4	1.6	9.9	6.0	60	1.2	2.7	1.0	2.2	.31	.19	260	573	.4	.8	57	125
				M-F	*3.1*	*1.4*	*11.0*	*6.7*	*67*	*1.4*	*3.0*	*1.1*	*2.4*	*.34*	*.21*	*289*	*637*	*.4*	*.9*	*63*	*139*
88	40	.55	250	As-fed	3.9	1.8	9.9	6.0	63	1.3	2.8	1.0	2.3	.28	.17	260	574	.4	.8	57	125
				M-F	*3.5*	*1.6*	*11.0*	*6.7*	*70*	*1.4*	*3.1*	*1.1*	*2.5*	*.31*	*.19*	*289*	*638*	*.4*	*.9*	*63*	*139*
99	45	.55	250	As-fed	4.1	1.9	9.9	6.0	63	1.3	2.8	1.0	2.3	.26	.16	276	608	.4	.9	60	132
				M-F	*3.7*	*1.7*	*11.0*	*6.7*	*70*	*1.4*	*3.1*	*1.1*	*2.5*	*.29*	*.18*	*306*	*675*	*.5*	*1.0*	*67*	*147*
110	50	.48	220	As-fed	4.4	2.0	9.9	6.0	63	1.3	2.8	1.0	2.3	.25	.15	289	637	.4	1.0	63	139
				M-F	*4.0*	*1.8*	*11.0*	*6.7*	*70*	*1.4*	*3.1*	*1.1*	*2.5*	*.28*	*.17*	*321*	*708*	*.5*	*1.1*	*70*	*154*
121	55	.44	200	As-fed	4.7	2.1	9.9	6.0	63	1.3	2.8	1.0	2.3	.23	.14	301	664	.4	1.0	66	145
				M-F	*4.2*	*1.9*	*11.0*	*6.7*	*70*	*1.4*	*3.1*	*1.3*	*2.5*	*.26*	*.16*	*335*	*738*	*.5*	*1.1*	*73*	*161*
Early weaned[10]																					
22	10	.55	250	As-fed	1.4	.7	14.4	10.4	66	1.3	2.9	1.1	2.3	.36	.24	578	1,275	.8	1.8	46	101
				M-F	*1.3*	*.6*	*16.0*	*11.5*	*73*	*1.5*	*3.2*	*1.2*	*2.6*	*.40*	*.27*	*643*	*1,417*	*.9*	*2.0*	*51*	*112*
44	20	.60	275	As-fed	2.4	1.1	14.4	10.4	66	1.3	2.9	1.1	2.3	.32	.22	694	1,530	1.0	2.3	54	120
				M-F	*2.2*	*1.0*	*16.0*	*11.5*	*73*	*1.5*	*3.2*	*1.2*	*2.6*	*.36*	*.24*	*771*	*1,700*	*1.1*	*2.5*	*60*	*133*
66	30	.66	300	As-fed	3.4	1.6	12.6	8.6	66	1.3	2.9	1.1	2.3	.32	.22	743	1,639	1.1	2.4	58	129
				M-F	*3.1*	*1.4*	*14.0*	*9.5*	*73*	*1.5*	*3.2*	*1.2*	*2.6*	*.36*	*.24*	*826*	*1,821*	*1.2*	*2.7*	*65*	*143*

[1]Adapted by the author from *Nutrient Requirements of Sheep*, No. 5, 5th rev. ed., NRC-National Academy of Sciences, 1975, pp. 44-45.
[2]To convert dry matter to an as-fed basis, divide dry matter by percentage of dry matter.
[3]DP = digestible protein.
[4]1 kg TDN = 4.4 Mcal DE (digestible energy). DE may be converted to ME (metabolizable energy) by multiplying by 82%. Because of rounding errors, calculations between Table 4-53 and Table 4-54 may not give the same values.
[5]Values are for ewes in moderate condition, not excessively fat or thin. Fat ewes should be fed at the next lower weight, thin ewes at the next higher weight. Once maintenance weight is established, such weight would follow through all production phases.
[6]For ewes suckling singles last 8 weeks of lactation, daily gain should be an additional .10 lb (45 g).
[7]For ewes suckling twins last 8 weeks of lactation, daily gain should be .12 lb (55 g).
[8]Requirements for replacement lambs (ewe and ram) start when the lambs are weaned.
[9]Maximum gains expected. If lambs are held for later market, they should be fed as replacement ewe lambs are fed. Lambs capable of gaining faster than indicated should be fed at a higher level. Lambs finish at the maximum rate if they are self-fed.
[10]An 88-lb *(40 kg)* early weaned lamb should be fed the same as a finishing lamb of the same weight.

Minerals

Although the body contains many mineral elements, only 15 have been demonstrated to be essential for sheep—7 major mineral constituents, and 8 trace elements. These minerals, along with the requirements of each, are presented in Tables 4-55 and 4-56. Where known, the toxic levels of the micro minerals are given, also. (Note that fluorine is not required.)

TABLE 4-55

MACRO MINERAL REQUIREMENTS OF SHEEP[1]
(Percentage of Ration)

Nutrient	Requirement[2]	
	As-fed[3]	Moisture-free
	(%)	(%)
Sodium (Na)	0.04-0.09	0.04-0.10
Chlorine (Cl)	-	-
Calcium (Ca)	0.19-0.47	0.21-0.52
Phosphorus (P)	0.14-0.33	0.16-0.37
Magnesium (Mg)	0.04-0.07	0.04-0.08
Potassium (K)	0.45	0.50
Sulfur (S)	0.13-0.23	0.14-0.26

[1]From: *Nutrient Requirements of Sheep*, No. 5, 5th rev. ed., NRC-National Academy of Sciences, Washington, D.C., 1975, p. 47, Table 10.
[2]Values are estimates based on experimental data.
[3]Estimated 90% dry matter.

SHEEP MINERAL CHART

Table 4-57, Sheep Mineral Chart, page 356, gives, in summary form, the following pertinent information relative to each mineral listed: (1) conditions usually prevailing where deficiencies are reported, (2) function, (3) deficiency symptoms, (4) nutrient requirements, (5) recommended allowances, and (6) practical sources.

Vitamins

Mature sheep require the fat-soluble vitamins A, D, E, and K, but they do not need added sources of the B vitamins, since the latter are synthesized in adequate amounts by rumen microorganisms.

Normal ewe rations are adequate in all the fat-soluble vitamins with the exception of the low carotene and/or vitamin A content of dry winter ranges. However, ewes can build up liver stores of vitamin A adequate to maintain production for 3 to 4 months. Vitamin D deficiency is not a problem unless ewes are maintained on vitamin D-deficient diets in environments devoid of sunlight. In some coastal areas of the United States, the latter condition may prevail.

The vitamin A and vitamin D requirements of sheep are given in Table 4-58, page 360.

SHEEP VITAMIN CHART

Table 4-59, Sheep Vitamin Chart, page 362, gives, in summary form, the following pertinent information relative to each vitamin listed: (1) conditions usually prevailing where deficiencies are reported, (2) function, (3) deficiency symptoms, (4) nutrient requirements, (5) recommended allowances, and (6) practical sources.

TABLE 4-56

MICRO MINERAL REQUIREMENTS OF SHEEP AND TOXIC LEVELS[1]
(Parts per Million or Mg/kg, of Ration)

Nutrient	Requirement[2]		Toxic Level[2]	
	As-fed[3]	Moisture-free	As-fed[3]	Moisture-free
	(ppm)	(ppm)	(ppm)	(ppm)
Cobalt (Co)	0.09	0.1	90-180	100-200
Copper (Cu)	4.5	5	7-23	8-25
Fluorine (F)	-	-	55-180	60-200
Iodine (I)	0.09-0.73	0.1-0.80[4]	7.3+	8+
Iron (Fe)	27-45	30-50	-	-
Manganese (Mn)	18-36	20-40	-	-
Molybdenum (Mo)	0.45	0.5	4.5-18	5-20
Selenium (Se)	0.09	0.1	1.8	2
Zinc (Zn)	32-45	35-50	900	1,000

[1]From: *Nutrient Requirements of Sheep*, No. 5, 5th rev. ed., NRC-National Academy of Sciences, Washington, D.C., 1975, p. 47, Table 11.
[2]Values are estimates based on experimental data.
[3]Estimated 90% dry matter.
[4]High level for pregnancy and lactation in diets not containing goitrogens; should be increased if diets contain goitrogens.

TABLE 4-57—SHEEP

Minerals Which May Be Deficient Under Normal Conditions	Conditions Usually Prevailing Where Deficiencies Are Reported	Function of Mineral	Some Deficiency Symptoms
MAJOR OR MACRO MINERALS:			
Salt (Sodium and chlorine—NaCl)	Negligence; for salt is inexpensive.	Salt is known to have many regulatory functions in the body. When deprived of salt, feed consumption and water intake are decreased. Also, with lack of salt, milk production and growth rate may be reduced.	A deficiency of salt may result in a depraved appetite, with the sheep trying to satisfy their craving by chewing wood, licking dirt, or eating toxic amounts of poisonous plants; decreased feed consumption; and decreased efficiency in the utilization of nutrients.
Calcium (Ca)	Lack of vitamin D. When finishing lambs are fed heavily on concentrates and limited quantities of legume roughage. When the feed consists largely of dried mature grasses or corn silage. Calcium-deficient areas (where pasture and range forages are deficient in Ca) are: Fla., La., Neb., Va., and W.Va.	Essential for development and maintenance of normal bones and teeth. Important in blood coagulation and lactation. Enables heart, nerves, and muscles to function. Regulates permeability of tissue cells. Affects availability of phosphorus and zinc.	Subnormal development of bone; rickets in young animals, and osteomalacia in adults.
Phosphorus (P)	Lack of vitamin D. When sheep subsist for long periods on mature forages (such as dry range or grass or cereal hays). When the ration consists of a high proportion of beet by-products. When sheep subsist on pastures in phosphorus-deficient areas.	Essential for sound bones and teeth, and for the assimilation of carbohydrates and fats. A vital ingredient of the proteins in all body cells, Necessary for enzyme activation. Acts as a buffer in blood and tissue. Occupies a key position in biologic oxidation, and reactions requiring energy.	Slow growth, depraved appetite, unthrifty appearance, listlessness, low level of phosphorus in the blood (less than 4 mg/100 ml of plasma), and development of knock-knees.
Magnesium (Mg)		Necessary for many enzyme systems and for proper functioning of the nervous system. Closely associated with the metabolism of calcium and phosphorus.	Hypomagnesemic tetany, a hyperirritability of the neuromuscular system. Sometimes this condition is accompanied by hypocalcemia. Acute tetany may occur and as a result of insufficient dietary magnesium or inability to mobilize skeletal magnesium.
Potassium (K)	When finishing lambs are fed high-concentrate and urea rations and limited amounts of dry roughage.	Essential for proper enzyme, muscle and nerve function, rumen microorganism activity, and appetite.	Poor appetite and feed conversion, progressive stiffness from front to rear, and dry wool.
Sulfur (S)	When finishing lambs are fed high-concentrate and urea rations and limited amounts of roughage.	Functions in synthesis of sulfur-containing amino acids in the rumen and various compounds of the body. Wool is high in sulfur; hence, sulfur is closely related to wool production.	Poor weight gains, feed efficiency and wool growth. Also, excessive salivation, lacrimation, and shedding of wool.

Footnotes on last page of table.

INERAL CHART

Nutrient Requirements[1]					
Daily Nutrients/ Animal	Percentage of Ration		Recommended Allowances[1]	Practical Sources of the Mineral	Comments
	As-fed[2]	*M-F*			
.ambs in drylot consume about 9 g of salt daily. Mature sheep in drylot may consume more.	**Salt for growing lambs:* 0.36	0.40	**Salt for mature sheep:* 0.5% of the complete feed, or 1.0% to the concentrate portion.	Free access to salt. Loose salt, rather than block salt, should be provided, for the reason that sheep bite at salt blocks, rather than lick, with the result that their teeth may be broken. In iodine-deficient areas, stabilized iodized salt should always be provided.	Sheep consume about 5 times more salt/100 lb body wt. than cattle. This is thought to be due to their consumption of a higher proportion of forages to concentrates than other classes of livestock.
	**Na requirement of sheep:* 0.04 to 0.09	0.04 to 0.10	*Range operators commonly provide ½ to ¾ lb (¼-⅓ *kg*) salt/ewe/month. Mature sheep in drylot may consume more.		In alkaline areas, the water may contain enough salt to meet the requirements, and supplemental salt may not be needed.
ariable, according to class, age, and weight of sheep (see Table 4-53, p. 352).	* 0.19 to 0.47	0.21 to 0.52	Self-feed suitable mineral, or add calcium to the ration as required to bring level of total ration slightly above requirements.	Ground limestone, or oyster-shell flour. Where both calcium and phosphorus are needed, use bone meal, dicalcium phosphate, or defluorinated phosphate.	Most pasture and range forage contains adequate amounts of calcium. Forage containing from 0.24 to 0.32% calcium is considered adequate. Calcium requirements are usually met when sheep receive at least ⅓ of a legume forage. Normal blood levels of calcium are from 9 to 12 mg per 100 ml of serum.
ariable, according to class, age, and weight of sheep (see Table 4-53, p. 352).	* 0.14 to 0.33	0.16 to 0.37	Self-feed suitable mineral, or add phosphorus to the ration as required to bring level of total ration slightly above requirements.	Monosodium phosphate, or diammonium phosphate. Where both calcium and phosphorus are needed, use bone meal, dicalcium phosphate, or defluorinated phosphate.	The proper calcium-phosphorus ratio should be kept in mind. Forage containing below 0.16% phosphorus is usually considered deficient for ewes during gestation, and 0.20% borderline during lactation. *A phosphorus deficiency may be manifested when the blood phosphorus level falls below 4 mg/100 ml of plasma.
	*0.04 to 0.07	0.04 to 0.08			*Blood serum normally contains about 2.5 mg/100 ml.
	*Rations or forages containing 0.06% are considered adequate for the adult range ewe.				
	*0.45	0.5	0.7 to 1.0% of total air-dry ration.	Roughages usually contain adequate potassium, with the possible exception of nonlegume silage. Potassium chloride is the supplement of choice.	The feeding of potassium chloride appears to reduce the incidence of urinary calculi in feedlot lambs. This is especially true with high-milo rations.
	**Mature ewes:* 0.13 to 0.16	0.14 to 0.18	*It is recommended that a dietary nitrogen-sulfur ratio of 10:1 be maintained.	Sulfate sulfur, elemental sulfur, or sulfur-containing proteins or amino acids. Inorganic compounds are generally more convenient and economical for supplemental feeding.	*Practically all common feedstuffs contain more than 0.1% sulfur. However, mature grass and grass hays are sometimes low in sulfur. Where forages are low in sulfur or high in urea, increased weight gains and wool growth can be obtained by feeding sulfur.
	**Young lambs:* 0.16 to 0.23	0.18 to 0.26			

(Continued)

TABLE 4-5

Minerals Which May Be Deficient Under Normal Conditions	Conditions Usually Prevailing Where Deficiencies Are Reported	Function of Mineral	Some Deficiency Symptoms
TRACE OR MICRO MINERALS:			
Cobalt (Co)	Cobalt-deficient areas or soils (in Fla., and in parts of Mich., Wisc., N.H., Mass., Penn., N.Y., and Alberta, Canada).	Promote synthesis of B_{12} in the rumen.	Lack of appetite, lack of thrift, sever emaciation, weakness, anemia, de creased fertility, and decrease milk and wool production.
Copper (Cu)	In copper-deficient areas (soils), as in Florida and in the coastal plains region of the Southeast.	Anemia is associated with copper deficiency. Animals suffering from inadequate copper intake appear unable to absorb iron at a normal rate, and a deficiency in hemoglobin synthesis results.	Signs in suckling lambs includ muscular incoordination, partia paralysis of the hindquarters, an degeneration of the myelin sheat of the nerve fibers. Lambs may b born weak and may die because c their inability to nurse. Sheep suffering from a copper defi ciency may produce "steely" o "stringy" wool, which is lacking i crimp, tensile strength, affinity fo dyes, and elasticity. Depigmenta tion of the wool of black sheep ha been noted as a sign of severe de ficiency.
Fluorine (F)	*Conditions which may result in fluorine toxicity:* High fluorine in the water supply. Use of rock phosphate that contains 3 to 4% fluorine.		Fluorine deficiency not reported Rather, the hazard is fluorine toxic ity.
Iodine (I)	Iodine-deficient areas or soils (in northwestern U.S. and in the Great Lakes region) where iodized salt is not fed. Feeds from iodine-deficient areas.	Formation of thyroxine, a hormone of the thyroid gland.	Lambs born with goiter; usuall stillborn or die soon after birth Usually, such lambs have very littl wool. In mature sheep an iodine deficienc may result in reduced wool yiel and reduced rate of conception.
Iron (Fe)	Iron-deficiency anemia sometimes occurs in lambs raised on slotted floors.		
Manganese (Mn)	Lambs on a purified diet containing less than 1 ppm of manganese over a 5-month period.	For skeletal development.	Bone changes. In goats, low manganese has resulte in a reduction of birth weight o kids.
Molybdenum (Mo)	Excess molybdenum in the soil such as is found in areas of California, Nevada, and England. Increased incidence of copper toxicity in recent years seems to be associated chiefly with drylot feeding, a tendency noted in New Zealand, Great Britain, Canada, and the U.S.	It is believed that the molybdenum binds and inactivates the copper in the intestine. Aids digestion and makes for more rapid gains.	A low intake of molybdenum causes excess copper to accumulate in the tissues, especially the liver, even when the copper intake is moderate, thus producing fatal jaundice (easily detected in the eyes). This disease can be prevented by increasing the molybdenum intake. *Excess molybdenum causes a scouring disease in sheep; a conditior which is controlled by increasing the copper level in the diet to 5 ppm.

Footnotes on last page of table.

-(Continued)

Nutrient Requirements[1]				
Daily Nutrients/ Animal	Percentage of Ration	Recommended Allowances[1]	Practical Sources of the Mineral	Comments
.1 mg daily for a 120-lb *(54-kg)* sheep. obalt should be ingested frequently, even daily.		*Feed cobalt at the rate of 12g/220 lb *(12 g/100 kg)* of salt as cobalt chloride or cobalt sulfate.	A cobalt mineral mixture. Other effective methods of providing cobalt are (1) to add cobalt to the soil, or (2) to place cobalt pellets into the rumen.	Several good commercial cobalt-containing minerals are on the market, in either the block or loose form. Cobalt is much more effective when given by mouth than when given intravenously. *Feed or forage containing more than 0.07 ppm of cobalt on a dry matter basis has been shown to prevent a deficiency.
	*4.5 ppm *5 ppm* —if molybdenum and sulfate levels are normal. Merino sheep are less efficient than British breeds in absorbing copper from feeds; hence, they need an additional 1-2 ppm in their ration.	*Add copper sulfate to the salt at rate of 0.5%.	Salt containing 0.5% of copper sulfate. Or copper carbonate may be used.	Copper deficiencies may exist alone or along with deficiencies of cobalt and iron. An interesting interrelation exists between copper, molybdenum, and sulfate. An excess of molybdenum causes a pathological condition which can be cured only by administering copper. Stores of copper in the liver, kidney, heart, lungs, pancreas, and spleen serve as a reserve for as long as 4-6 months when animals are grazing copper-deficient forage. Sheep are much more susceptible to copper toxicity than cattle. As much as 25 mg of copper in the daily ration of sheep is considered toxic; and about 9 mg/day is considered the safe tolerance level.
	*Breeding sheep should not be fed a ration containing more than 55 ppm (as-fed) or 60 ppm of fluorine on a moisture-free basis. *Finishing lambs can tolerate up to 135 ppm (as-fed) or 150 ppm of fluorine in the ration on a moisture-free basis. Acute toxicity can occur at 200 ppm.			Symptoms of fluorine toxicity are loss of appetite; the nomal ivory color of bones changes to chalky white; bones thicken; and the teeth, especially the incisors, may become pitted and eroded to such an extent that the nerves are exposed.
		0.5 ppm of total feed (moisture-free basis), or .454 g/ton of feed. *Free access to stabilized iodized salt containing 0.0078% iodine.	Stabilized iodized salt containing 0.0078% iodine. Calcium iodate.	Do not use iodized salt in a mixture with a concentrate to limit feed intake, as the animals may consume an excessive amount of iodine.
		*Intramuscular injections of iron-dextran; 2 injections, 150 mg of iron in each, given 3 weeks apart.		
	*The exact manganese requirements of sheep are unknown.			
	*The Food and Drug Administration does not recognize molybdenum as safe; hence, the law prohibits adding it to feed for sheep.	The two contrasting situations—(1) high molybdenum and copper deficiency, or (2) low molybdenum and excess copper accumulation—make it very difficult to define nutrient requirements of molybdenum and copper.		A high molybdenum intake induces a copper deficiency even when the copper content of pasture is quite high; the scouring effect can be prevented by providing an increased copper intake. Sheep are less affected than cattle by high molybdenum intakes. *In treating copper toxicity, both molybdenum and sulfate should be administered. Drench each lamb daily for 3 weeks with 100 mg of ammonium molybdate and 1 g of sodium sulfate in 20 ml of water.

(Continued)

TABLE 4-57

Minerals Which May Be Deficient Under Normal Conditions	Conditions Usually Prevailing Where Deficiencies Are Reported	Function of Mineral	Some Deficiency Symptoms
Selenium (Se)	Areas where selenium content of crops is below 0.1 ppm, such as northwestern, northeastern, and southeastern U.S. Parts of S.D., Wyo., and Utah produce forage containing excess selenium which causes toxicity in farm animals.	Related to vitamin E metabolism.	Reduced growth and white muscl disease, which affects lambs 2- weeks of age. Impaired fertility has been reported i New Zealand.
Zinc (Zn)			Ram lambs show 2 signs of zinc defi ciency: (1) impaired growth of th testes, and (2) cessation of sper matogenesis. Lambs on zinc deficient diet show slipping o wool, swelling and lesions aroun hooves and periobital regions o eyes, excessive salivation, loss o appetite, eating of wool, listless- ness, and reduced growth.

[1]As used herein, the distinction between "nutrient requirements" and "recommended allowances" is as follows: In nutrient requirements, no margins of safety are includec intentionally; whereas in recommended allowances, margins of safety are provided to compensate for variations in feed composition, environment, and possible losses durinc storage or processing.

Where preceded by an asterisk, the requirements, recommended allowances, and other facts presented herein were taken from *Nutrient Requirements of Sheep*, No. 5, 5th rev. ed., NRC-National Academy of Sciences, Washington, D.C., 1975.

[2]Estimated 90% dry matter.

TABLE 4-58—DAILY VITAMIN A AND D

	Vitamin A (IU)[2] for Body Weights Indicated												
	Body Weight, Lb												
	265	220	176	154	132	121	110	99	88	77	66	44	22
Ewes													
Maintenance			2,040	1,785	1,530		1,275						
Nonlactating and first 15 weeks of gestation			2,040	1,785	1,530		1,275						
Last 6 weeks of gestation or last 8 weeks of lactation suckling singles ..			6,800	5,950	5,100		4,250						
First 8 weeks of lactation suckling singles or last 8 weeks of lactation suckling twins			6,800	5,950	5,100		4,250						
First 8 weeks of lactation suckling twins			6,800	5,950	5,100		4,250						
Replacement lambs and yearlings[3]					2,550		2,125		1,700		1,275		
Rams													
Replacement lambs and yearlings[3]	5,100	4,250	3,400		2,550				1,700				
Lambs													
Finishing						1,402	1,275	1,148	1,020	892	765		
Early-weaned									1,020		2,550	1,700	850

[1]From *Nutrient Requirements of Sheep*, No. 5, 5th rev. ed., NRC-National Academy of Sciences, Washington, D.C., 1975, pp. 42-43, Table 1.
[2]May be vitamin A or provitamin A equivalent.

—(Continued)

Nutrient Requirements[1]				
Daily Nutrients/ Animal	Percentage of Ration	Recommended Allowances[1]	Practical Sources of the Mineral	Comments
	Feeding selenium is not approved by the Food and Drug Administration. (In 1974, FDA approved the addition of selenium to the diets of chickens, turkeys, and swine; but not ruminants or horses.) Injection may be used.	*Inject a commercial pharmacological product containing selenium and vitamin E. Give 2 muscular injections—the first at birth, and the second 2 weeks later. The first injection contains 0.25 mg of selenium as sodium selenite and 68 IU of vitamin E as *d*-alpha-tocopherol. The second injection contains 1.0 mg of selenium from the same source and 68 IU of vitamin E.		*Chronic selenium toxicity occurs when sheep consume feeds containing more than 3 ppm of selenium on a dry basis over a prolonged period. Toxicity signs include loss of wool, soreness and sloughing of the hooves, and marked reduction in reproductive performance. Plants grown on the same seleniferous soils vary greatly in their uptake of selenium, with a range of 1,000 ppm to only 10-25 ppm. The most practical way to prevent livestock losses from selenium poisoning is to manage the grazing so that animals alternate between selenium-bearing and other areas.
Zinc is essential, but the requirement is not known.				*Zinc deficiency symptoms have been alleviated by feeding 100 ppm of zinc.

Note: Mineral recommendations for all classes and ages of sheep, especially those fed unmixed rations or on pasture, are–

1. *Where sheep are on liberal grain feeding*—Provide free access to a 2-compartment mineral box, with (a) trace mineralized salt in one side, and (b) in the other side, a mixture of ⅓ trace mineralized salt (salt included for purposes of palatability), ⅓ defluorinated phosphate or steamed bone meal, and ⅓ ground limestone or oystershell flour.

2. *Where sheep are primarily on roughage (pasture, hay, and/or silage)*—Provide free access to a 2-compartment mineral box, with (a) trace mineralized salt in one side (salt included for purposes of palability), and (b) in the other side, a mixture of ⅓ trace mineralized salt and ⅔ defluorinated phosphate or steamed bone meal.

Additionally, in those areas where cobalt and/or copper deficiencies exist in the soil (and plants), add cobalt and/or copper sulfate to either the salt or salt-phosphorus mixture in the proportions indicated. If desired, the mineral supplement may be incorporated in the ration in keeping with the recommended allowances given in this table.

REQUIREMENTS OF SHEEP (per Animal)[1]

Vitamin D (IU) for Body Weights Indicated												
Body Weight, Lb												
265	220	176	154	132	121	110	99	88	77	66	44	22
		444	388	333		278						
		444	388	333		278						
		444	388	333		278						
		444	388	333		278						
		444	388	333		278						
				333		278		222		166		
666	555	444		333				222				
					305	278	250	222	194	166		
								222		200	133	67

[3]Requirements for replacement lambs (ewe and ram) start when the lambs are weaned.

TABLE 4-59—SHEEP

Vitamin Which May Be Deficient Under Normal Conditions	Conditions Usually Prevailing Where Deficiencies Are Reported	Function of Vitamin	Some Deficiency Symptoms
A	Vitamin A deficiencies may occur when—(1) extended drought results in dry, bleached pastures; (2) winter feeding on bleached hays (especially overripe cereal hays or straws) with little or no green hay or silage; (3) drylot finishing on rations with little or no green forage or yellow corn, especially for feeding periods longer than 2 to 3 months; and (4) there is high nitrate intake, in either water or feed.	Necessary for maintaining normal epithelial tissue.	Keratinization of the respiratory, alimentary, reproductive, urinary, and ocular epithelia; lowered resistance to infection; abnormal development of bone; birth of lambs that are weak, malformed, or dead; and night blindness.
D	When ration consists predominantly of dehydrated hays, green feeds, and seeds and their by-products. Prolonged cloudy weather or when kept inside, especially in fast-growing young lambs.	Prevention of rickets.	Rickets in young lambs. Congenital malformations in newborn lambs from extreme deficiencies.
E	Under drylot conditions when red kidney beans are fed. When lambs are making rapid growth, although this isn't always the case. When old hay is fed, as oxidation destroys vitamin E.	Prevention of white muscle disease (also called nutritional muscular dystrophy, or stiff lamb disease).	Stiff lamb disease, or white muscle disease; characterized by a stiff, stilted way of moving and a "roached" back. Sometimes a paralysis of hind legs. However, not all stiff lambs are the result of Vitamin E deficiency.
Other Vitamins	B vitamin deficiencies may be evident in poorly fed and unhealthy animals. Vitamin K deficiency may occur when the dicoumarol content of hay is excessively high, as when moldy sweet clover hay is fed.	Vitamin K or K_2 is necessary in the blood clotting mechanism.	

[1]As used herein, the distinction between "nutrient requirements" and "recommended allowances" is as follows: In nutrient requirements, no margins of safety are included intentionally; whereas in nutrient allowances, margins of safety are provided in order to compensate for variations in feed composition, environment, and possible losses during storage or processing.

VITAMIN CHART

Nutrient Requirements[1]		Recommended Allowances[1]	Practical Sources of the Vitamin	Comments
Daily Nutrients/ Animal	Amount/Lb (or Kg) of Feed			
Variable, according to class, age, and weight of sheep (see Table 4-53, p. 352).	Variable, according to class, age, and weight of sheep (see Table 4-54, p. 353).	*Sheep that are deficient in vitamin A and weigh 70 lb *(32 kg)* or more should recieve 100,000 IU of vitamin A by injection, and their diets should be adjusted to provide recommended levels of vitamin A or carotene. *Ewes deficient in vitamin A should be given vitamin A either orally or by injection prior to breeding.	Stabilized vitamin A. Green pasture. Grass or legume silages. Yellow corn. Green hay not over 1 yr. old. The average carotene content of some common feeds is as follows: **Carotene *(mg)*** Legume hays (including alfalfa), average quality .... 9-14/lb *20-31/kg* Nonlegume hays, average quality .... 4-8/lb *9-18/kg* Dehydrated alfalfa meal, average quality .... 50-70/lb *110-154/kg* Yellow corn ..... 0.8-1.0/lb *1.8-2.2/kg* Silages, corn or sorghum 2-10/lb *4-22/kg*	*Sheep do not convert carotene to vitamin A as efficiently as rats. For sheep, 1 mg of feed carotene is equivalent to 400-700 IU of vitamin A. *It requires 200 days entirely to deplete liver storage of ewe lambs previously pastured on green feed. Because of this storage, animals that normally graze on green forage during the growing season are able to do reasonably well on a low-carotene diet of dry feed for periods of 4 to 6 months.
**For all sheep except early weaned lambs:* 555 IU per 200 lb *(100 kg)* body weight. **For early weaned lambs:* 666 IU per 220 lb *(100 kg)* body weight.	Variable, according to class, age, and weight of sheep (see Table 4-54).	Breeding sheep, 500 to 800 IU/head/day. Feeder lambs, 500 IU/ head/day.	Exposure to sunlight, through irradiation. Sun-cured hays. Irradiated yeast.	Newborn lambs are provided with enough vitamin D from their dams to prevent rickets for 4 to 6 weeks if the ewes have adequate storage. Sheep with white skin and short wool receive more vitamin D activity from irradiation by sunlight than do animals with dark skin or long wool.
		*Good response is obtained when lambs receive 5 mg/lb *(11 mg/kg)* body weight of *dl*-alpha-tocopherol weekly in rations containing 0.1-1.0 ppm of selenium.	Vitamin E is widely distributed in natural feeds, but oxidation rapidly destroys it. Thus, old hay or ground feeds may be poor sources. White muscle disease in lambs is prevented by adding alpha-tocopherol (either *dl* or *d* forms) and selenium to the diet.	Experiments have failed to relate vitamin E deficiency with reproductive failure in sheep. The need for vitamin E in the diet of young nursing lambs is related to the selenium level in the diet. Selenium has a sparing effect on the vitamin E requirement; the higher the selenium level in the diet, the lower the vitamin E requirement, and vice versa.
		*The B vitamins are not required in the diet of sheep with functioning rumens, because the microorganisms synthesize these vitamins in adequate amounts. *Vitamin K_2 is normally synthesized in large amounts in the rumen; no need for dietary supplementation has been established.	Green leafy materials of any kind, fresh or dry, are good sources of K_1.	Addition of B vitamins has not been shown to be beneficial to mature sheep. However, young lambs (to about 2 months of age) with undeveloped rumens have been shown to have a dietary need for vitamin B_{12}, thiamin, pyridoxine, riboflavin, niacin, folic acid, and possibly some of the other B vitamins, since they will not be receiving these in the milk from their dams. Cobalt is necessary for the synthesis of vitamin B_{12} in the rumen. *No supplementary dietary need for vitamin C has been shown.

Where preceded by an asterisk, the nutrient requirements, recommended allowances, and other facts presented herein were taken from *Nutrient Requirements of Sheep*, No. 5, 5th rev. ed., NRC-National Academy of Sciences, Washington, D.C., 1975.

Water

Contrary to frequent, but unfounded, opinions, sheep do require considerable water. Mature animals will consume an average of approximately a gallon per day; whereas feeder lambs require about half this amount. Like other classes of animals, the water consumption of sheep varies with the climate and type of feed. They consume more water in the summer than in the winter and more when on dry feeds than when eating considerable roots or other succulent feeds. Sheep will go for weeks without water when foraging on grasses and other feeds of high moisture content. The latter conditions often prevail on desert ranges in the early spring and on many of the mountain ranges during the summer months.

FEED ADDITIVES AND IMPLANTS[13]

Table 4-60 summarizes the growth stimulants that are presently available and can be used. All of these products have been shown to improve gain and feed efficiency of sheep. The information presented in Table 4-60 is the most recent available. But feed additives and implants do change from time to time; new products are developed, and sometimes old products are banned by the Food and Drug Administration. So, those using additives should always confer with local

[13]This section and Table 4-60 were authoritatively reviewed by the following: Dr. Robert M. Jordan, Department of Animal Science, University of Minnesota, St. Paul, Minn.; and Dr. T. W. Perry and Dr. J. B. Outhouse, Department of Animal Sciences, Purdue University, West Lafayette, Ind.

TABLE 4-60

SHEEP FEED ADDITIVES AND IMPLANTS

Type of Additive	Method of Administering	Dosage	Effect on: Daily Rate of Gain	Effect on: Feed Efficiency	Effect on: Carcass Quality	Comments
			(% Incr.)	*(% Incr.)*		
Antibiotics (Chlortetracycline and Oxytetracycline)	Feeding (oral)	Aureomycin (chlortetracycline) 20 to 50 gm/ton of feed. Terramycin (oxytetracycline) 10 to 20 gm/ton of feed.	Range: 0-31 Av.: 11	Range: 4-27 Av.: 10	No effect to slight improvement.	Antibiotics (especially chlortetracycline and oxytetracycline) may improve performance when added to creep and lamb finishing rations. Response to antibiotics varies markedly according to differences in management and degree of stress to which lambs are subjected. There is some evidence that antibiotics reduce the incidence of enterotoxemia.
Ralgro (Zeranol)	Implant	12 mg/head	Range: 0-25 Av.: 10	Av.: 6		Do not implant animals within 40 days of slaughter.

uthorities and read and follow manufacturer's label lirections for more complete details on the use of a pecific drug or combination of drugs.

'EED PREPARATION

The preparation of feeds for sheep is fully covred earlier in Section 4, under the heading, "Feed 'reparation," including Table 4-11, page 251, therein; ence, the reader is referred thereto.

HEEP FEEDING GUIDE

In order to compute balanced rations, it is first ecessary to have available both feeding standards nd feed composition tables. The latter are given in 'ables 4-101 to 4-105, Feed Composition Tables, ages 456 to 585.

For purposes of convenience and simplification, he author prepared Table 4-61, page 366, which acilitates the formulation of sheep rations. In using his table, the following points should be noted:[14]

1. Under "Description of Animals"—column No. —are sufficient groups to cover the vast majority of heep found on the nation's farms and ranches.
2. Columns 2 to 8 give pertinent recommendaions relative to both forages and concentrates. These ecommendations are in keeping with those advoated by scientists, and with the actual practices folowed by successful operators.

In particular, it should be noted that all protein ecommendations are stated in terms of *crude protein* ontent, rather than digestible protein. This was done ecause (a) this is what the feed manufacturer wants o know as he plans a feed formula, and (b) this is vhat the sheepman sees on the feed tag when he purhases feed.

3. It is recognized that most sheepmen generally row their own forages, and purchase part or all of the concentrates. Thus, they generally wish to know what crude protein content of concentrate alone (including grains, by-product feeds, and/or protein supplements) they need to feed in order to balance out the forage which is available. Likewise, feed manufacturers have need for this information in compounding mixes. For these reasons, harvested forages are classified as (a) high-protein forages, (b) medium-protein forages, and (c) low-protein forages; and specific recommendations are made for each. Similar classifications and recommendations are made for (d) excellent pastures, (e) fair to good pastures, and (f) poor pastures.
4. It is often hazardous to formulate rations for "excellent" pastures that are different from those for "poor" pastures, because (a) sheepmen may be in error in appraising the quality of their pastures, and (b) pastures are generally excellent in the early spring, but become progressively poorer as the season advances unless they are irrigated and fertilized.

For purposes of illustration, let us refer to Table 4-61. Under column 5, it is noted that during the first 100 days of gestation, ewes (average weight 100 to 150 lb) that are being fed a daily ration of somewhere between 3 and 5 lb of a forage of grass hay or other nonlegume dry roughage should receive, in addition, ¼ to ⅓ lb daily of a protein supplement of soybean meal (or some other protein supplement which will provide an amount equivalent to 41 to 45% crude protein). To be sure, it is entirely proper to meet this recommended crude protein content of concentrate by feeding double the allowance of some protein supplement with approximately 20 percent crude protein content. Many times the latter may be more economical, and even advisable—for example, when the forage is of poor quality and added energy feed is needed. In general, however, those feeds should be purchased which furnish a unit of protein at the lowest cost, and those feeds which supply the protein in the most concentrated form are the most economical.

Under column No. 2 of Table 4-61, additional information, of value to both the feed manufacturer and the stockman who mixes his own rations, is given.

[14]In addition, see pertinent footnotes which accompany Table -61.

TABLE 4-61
SHEEP FEEDING GUIDE
(As-fed Basis)

Description of Animals (1)	Recommendations[1] (2)	Recommendations Relative to Concentrate Only (including grain, by-product feeds, and/or protein supplements)										
		In Drylot, with Following Types of Forages						On Pasture of the Following Grades:				
		(1) Legume and/or (2) legume and nonlegume mixed forages of high quality; consisting of dry forages and/or silage (high-protein forages) (3)		Legume and nonlegume forages mixed; consisting of dry forages and/or silage (medium-protein forages) (4)		Nonlegume forage; consisting of dry forages and/or silage (low-protein forages) (5)		Excellent (6)	Fair to good (7)		Poor; including winter pasture consisting of dry grass cured on the stalk[2] (8)	
		(lb)	*(kg)*	*(lb)*	*(kg)*	*(lb)*	*(kg)*		*(lb)*	*(kg)*	*(lb)*	*(kg)*
Ewes first 100 days of gestation (weighing 100-150 lb, or *45-68 kg*)	Forage per head daily.	3-5	*1.36-2.27*	3-5	*1.36-2.27*	3-5	*1.36-2.27*	—	—	—	—	—
	Concentrate: (1) Supplement allowance of soybean meal (or equivalent 41-45% crude protein) per head daily.[3]	—	—	—	—	¼-⅓	*.11-.15*	—	—	—	¼-⅓	*.11-.15*
Ewes last 6 weeks of gestation (weighing 115-165 lb, or *52-75 kg*)	Forage per head daily.	3-5	*1.36-2.27*	3-5	*1.36-2.27*	3-5	*1.36-2.27*	—	—	—	—	—
	Concentrate: (1) Total concentrate allowance per head daily.	½-¾	*.23-.34*	½-¾	*.23-.34*	¾-1	*.34-.45*	—	—	—	¾-1	*.34-.45*
	(2) Supplement allowance of soybean meal (or equivalent 41-45% crude protein) per head daily.[3,4]	—	—	—	—	¼-⅓	*.11-.15*	—	—	—	¼-⅓	*.11-.15*
	(3) Crude protein composition of total concentrate, in %.	9-12 (Cereal grains only will suffice)		11-14 (Cereal grains only will suffice)		17-21		—	—	—	17-21	
Ewes in lactation (weighing 100-150 lb, or *45-68 kg*)	Forage per head daily.	3-5	*1.36-2.27*	3-5	*1.36-2.27*	3-5	*1.36-2.27*	—	—	—	—	—
	Concentrate: (1)Total concentrate allowance per head daily.	¾-1½	*.34-.68*	¾-1½	*.34-.68*	1-1¾	*.45-.79*	—	—	—	1-1¾	*.45-.79*
	(2) Supplement allowance of soybean meal (or equivalent 41-45% crude protein) per head daily.[3,4]	—	—	1/10	*0.045*	⅓	*.15*	—	—	—	⅓	*.15*
	(3) Crude protein composition of total concentrate, in %.	11-13 (Cereal grains only will suffice)		14-17		22-24		—	—	—	22-24	
Finishing lambs of full feed (weighing 55-90 lb, or *25-41 kg*)	Forage per head daily.	1¼-1¾	*.57-.79*	1¼-1¾	*.57-.79*	1¼-1¾	*.57-.79*	—	—	—	—	—
	Concentrate: (1) Total concentrate allowance per head daily.	1¼-1¾	*.57-.79*	1¼-1¾	*.57-.79*	1½-2	*.68-.91*	—	1-1½	*.45-.68*	1½-2	*.68-.91*
	(2) Supplement allowance of soybean meal (or equivalent 41-45% crude protein) per head daily.[3,4]	—	—	0.1-0.2	*.05-.09*	0.2-0.3	*.09-.14*	—	0.1-0.2	*.05-.09*	0.2-0.3	*.09-.14*
	(3) Crude protein composition of total concentrate, in %.	9-13 (Cereal grains only will suffice)		11-14		13-15		—	12-13		13-14	

[1] The daily forage recommendations given herein are based on dry forage. When silage is included in the ration, figure 3 lb of silage equivalent to 1 lb of dry forage, due to the higher moisture content of silage.

[2] On a dry basis, the crude protein content of mature, weathered grasses may be 3% or less.

The upper limit of the concentrate allowance recommended in column No. 8 should be fed on winter range when (1) the grass is less abundant, and/or (2) the grass is relatively low in protein.

[3] Soybean meal, which ranges from 41 to 45% protein content, is herein used as a standard merely because it is the leading U.S. protein supplement. It is to be emphasized, however, (1) that other protein supplements, including numerous commercial products, may be used, (2) that, in general, those supplements should be purchased which provide a unit of protein at the lowest cost, and those feeds which are highest in protein content are usually the most economical, and (3) that where other protein feeds are substituted for the soybean meal recommended herein (41-45% protein), an equivalent amount of crude protein should be provided—for example, approximately 2 lb of a 20% crude protein supplement should be provided in order to replace each 1 lb of soybean meal (although it is recognized that 2 lb of a 20% protein feed will generally provide more energy, and may supply more of certain other important nutrients, than 1 lb of soybean meal).

[4] The recommended supplement allowance is based on the assumptions (1) that cereal grains, averaging 9 to 13% crude protein content, comprise the major part of the concentrate mix, and (2) that the forage is not comprised entirely or predominantly of nonlegume silage. Naturally, less protein supplement will need to be added where feeds of higher protein content than the cereal grains predominate. Also less protein supplement is required to balance out a ration consisting predominantly of barley (of 11.6% crude protein content) than one consisting mostly of corn (of 9.0% crude protein content). Likewise, the upper limit of protein supplement recommended herein (or even a higher figure) is required to balance out a ration where the forage is comprised entirely or largely of nonlegume silage (for corn silage has a crude protein content of 2.1%, whereas timothy hay, for example, has a crude protein composition of 6.8%).

RATIONS FOR SHEEP

Sheep rations vary with the section of the coun-ry, depending chiefly on available local feeds. Ex-ept at lambing time or when emergencies occur as a esult of drought or inclement weather, western bands eceive little supplemental feed. Even with farm locks, a minimum of grain is fed to breeding animals. Grain feeding usually is limited to the latter part of gestation and to the lactation period prior to turning to pasture.

Table 4-62 contains some Rations for Ewes based on what successful sheep operators in various sections of the country use.

The following points are also pertinent in feeding heep of all classes and ages:

1. They should be provided needed minerals; ee Table 4-57, Sheep Mineral Chart, page 356, for ecommendations.

2. Unless grains are unusually hard, they need not be ground for sheep. The animals prefer to do heir own grinding, and the feeds are no more effec-ively utilized when ground.

Some sheepmen, especially purebred and farm flock breeders, prefer to feed grain mixtures rather than one grain only to ewes during the gestation and lactation periods. Then any one of the following six mixtures may be used (as-fed basis):

Ration No. 1[1]	(lb)	(kg)
Oats	50	22.7
Barley	25	11.4
Wheat bran	10	4.5
Dry beet pulp	10	4.5
Linseed meal	5	2.3

Ration No. 2	(lb)	(kg)
Corn or Sorghum	50	22.7
Oats	20	9.1
Wheat bran	20	9.1
Protein supplement[2]	10	4.5

Ration No. 3	(lb)	(kg)
Corn	60	27.2
Oats or wheat bran	30	13.6
Protein supplement[2]	10	4.5

Ration No. 4	(lb)	(kg)
Oats	30	13.6
Corn	20	9.1
Wheat bran	30	13.6
Protein supplement[2]	20	9.1

Ration No. 5	(lb)	(kg)
Oats	60	27.2
Wheat bran	25	11.4
Protein supplement[2]	15	6.8

Ration No. 6	(lb)	(kg)
Corn	90	40.9
Protein supplement[2]	10	4.5

[1]Ration 1 has been used extensively by the author of this book. If the lactating ewes lose considerable flesh prior to turning to spring pasture, the shepherd adds a little corn to the ration.

[2]Linseed, cottonseed, and/or soybean meal.

TABLE 4-62

DAILY RATIONS FOR EWES[1]

(As-fed Basis)

(Rams may be fed any of the rations listed for ewes, but they should receive slightly more liberal allowances)

Type of Ration	First 100 Days of Gestation[2] (weighing 100-150 lb, or 45-68 kg) (lb)	(kg)	Last 6 Weeks of Gestation[2] (weighing 115-165 lb, or 52-75 kg)	Ewes in Lactation (weighing 100-150 lb, or 45-68 kg)
1. Legume hay[3] or grass-legume mixed hay, good quality	3-5	1.4-2.3	To each ration listed in the first column add ½ to ¾ lb (.23-.34 kg) grain[6] daily.	To each ration listed in the first column add ¾ to 1½ lb (.34-.68 kg) grain[6] daily, plus ¼ lb (.1 kg) protein supplement to each ration having less than 2 lb (.9 kg) legume.
2. Legume hay[3] or grass-legume mixed hay, good quality	1½-2½	.7-1.1		
Grass hay or other nonlegume dry roughage	1½-2½	.7-1.1		
3. Legume hay[3] or grass-legume mixed hay, good quality	1½-2	.7- .9		
Corn or sorghum silage	4-6	1.8-2.7		
4. Grass hay or other nonlegume dry roughage	3-5	1.4-2.3		
Protein supplement[4]	¼-⅓	.1- .2		
5. Corn or other nonlegume silage	8-11	3.6-5.0		
Protein supplement[4]	¼-⅓	.1- .2		
6. Roots[5]	5-6	2.3-2.7		
Legume hay[3] or grass-legume mixed hay, good quality	2¼-3¼	1.0-1.5		
7. Grass hay or other nonlegume dry roughage	2-2½	.9-1.1		
Corn or sorghum silage	3-4	1.4-1.8		
Protein supplement[4]	¼-⅓	.1- .2		

[1]The upper limits of hay given herein are higher than required because it is realized that ewes will refuse up to 30% of their forage allotment—the amount of waste varying according to the quality of the forage.

[2]Ewes should gain in weight during the entire pregnancy period, making a total gain of 15 to 25 lb.

[3]The legume hay may consist of alfalfa, clover, soybean, lespedeza, etc.

[4]The protein supplement may consist of linseed, cottonseed, and/or soybean meal—with nutted (pea-size) products preferred.

[5]The important root crops for sheep are: mangels (stock beets), rutabagas (swedes), turnips, and carrots.

[6]The grain usually consists of whole corn, barley, wheat, oats, and/or sorghum; although other grains are used. Grain feeding the last 6 weeks of pregnancy will lessen pregnancy disease, increase the livability of lambs, and increase milk production.

Pound for pound, any of these mixed rations can replace the grain in any of the 7 suggested rations of Table 4-62, but they are (1) slightly higher in protein and dry matter, and (2) slightly lower in TDN.

FEED SUBSTITUTION TABLE

The successful sheepman is a keen student of values. He recognizes that feeds of similar nutritive properties can and should be interchanged in the ration as price relationships warrant, thus making it possible at all times to obtain a balanced ration at the lowest cost.

Table 4-63, Feed Substitution Table for Sheep, is a summary of the comparative values of the most common U.S. feeds. In arriving at these values, two primary factors besides chemical composition and feeding value have been considered; namely, palatability and carcass quality.

In using this feed substitution table, the following facts should be recognized:

1. That, for best results, different ages and groups of sheep within classes should be fed differently.

2. That individual feeds differ widely in feeding value. Barley and oats, for example, vary widely in feeding value according to the hull content and the test weight per bushel, and forages vary widely according to the stage of maturity at which they are cut and how well they are cured and stored.

3. That nonlegume forages may have a higher relative value to legumes than herein indicated provided the chief need of the animal is for additional energy rather than for supplemented protein. Thus, the nonlegume forages of low value can be used to better advantage for wintering mature ewes than for young lambs.

On the other hand, legumes may actually have a higher value relative to nonlegumes than herein indicated provided the chief need is for additional protein rather than for added energy. Thus, no protein supplement is necessary for breeding ewes provided a good quality legume forage is fed.

4. That, based primarily on available supply and price, certain feeds—especially those of medium protein content, such as brewers' dried grains, corn gluten feed (gluten feed), distillers' dried grains, distillers' dried solubles, and peas (dried)—are used interchangeably as (a) grains and by-products feeds, and/or (b) protein supplements.

5. That the feeding value of certain feeds is materially affected by preparation. The values herein reported are based on proper feed preparation in each case.

For the reasons noted above, the comparative values of feeds shown in the feed substitution table (Table 4-63) are not absolute. Rather, they are reasonably accurate approximations based on average-quality feeds.

TABLE 4-63

FEED SUBSTITUTION TABLE FOR SHEEP
(As-fed Basis)

Feedstuff	Relative Feeding Value (lb for lb) in comparison with the designated (underlined) base feed which = 100	Maximum Percentage of Base Feed (or comparable feed or feeds) which it can replace for best results	Remarks
GRAINS, BY-PRODUCT FEEDS, ROOTS AND TUBERS:[1] (Low- and Medium-Protein Feeds)			
<u>Corn, No. 2</u>	<u>100</u>	<u>100</u>	Grinding not necessary unless (1) for old ewes with poor teeth, (2) for lambs under 5-6 wks., (3) for incorporation in a mixed ration.
Apple pomace, dehydrated	82-86	33⅓	
Barley	85-100	100	It does not pay to grind barley for sheep.
Beet pulp, dried	95	33⅓-50	Value of about 80% when used as the only concentrate for finishing lambs.
Beet pulp, molasses, dried	95	33⅓-50	Value of about 80% when used as the only concentrate for finishing lambs.
Beet pulp, wet	25	33⅓-50	
Brewers' dried grains	80-95	33⅓	Not very palatable. Fed chiefly to dairy cattle.
Citrus pulp, dried	95	25-50	
Corn gluten feed (gluten feed)	85-90	50	
Distillers' dried grains	95-100	33⅓-50	
Distillers' dried solubles	95-100	33⅓-50	
Hominy feed	100	100	
Molasses, beet	75	20	Actual value may be higher as an appetizer.
Molasses, cane	75	25	Actual value may be higher as an appetizer.

Footnotes on last page of table.

(Continued)

TABLE 4-63 (Continued)

Feedstuff	Relative Feeding Value (lb for lb) in comparison with the designated (underlined) base feed which = 100	Maximum Percentage of Base Feed (or comparable feed or feeds) which it can replace for best results	Remarks
Molasses, citrus	80-90	20	
Oats	75-100	10-100	Lower value when used as the only grain for finishing lambs. Highest value for young lambs, for breeding animals and for starting lambs on feed. Need not be ground for sheep. Should not constitute more than ⅓ of finishing rations. Feeding value varies according to the test weight per bushel.
Peas, dried	100	40	
Rice (rough rice)	55-75	100	
Rice bran	66⅔-75	33⅓	
Rice polishings	85-90	25	
Potatoes (Irish)	25-35	85	Contrary to popular belief, potatoes can be fed successfully through the pregnancy and lactation periods.
Roots (chiefly mangels or stock beets, rutabagas or swedes, turnips, and carrots)	25-35	50	Some sheepmen believe that the feeding of high levels of roots over a long period will produce urinary calculi. Therefore, caution should be exercised in feeding them to rams and wethers (females not affected). Many shepherds add roots to the ration of show sheep, for conditioning purposes.
Rye	83-87	50-100	Apparently rye is more palatable to sheep than to other classes of animals. Rye may be fed whole to sheep.
Sorghum, milo	100	100	All varieties have about the same feeding value. There is no advantage in grinding sorghum for sheep.
Wheat	90-95	100	May be fed as the only grain, but it is improved by mixing with another grain. Wheat may be fed whole. Wheat-fed sheep appear to be especially susceptible to founder.
Wheat bran	90	10-33⅓	Because of its bulk and fiber, wheat bran should not constitute more than 10-15% of a finishing ration. Bran is valuable for young animals, for breeding animals, and for starting animals on feed.
Wheat mixed feed (mill-run)	90-95	10-33⅓	Can be used in about the same way and in the same quantities as wheat bran for sheep.
PROTEIN SUPPLEMENTS:			
Soybean meal (41½)	100	100	
Alfalfa or clover screenings	70-75	50	Grind finely to destroy weed seeds.
Brewers' dried grains	75	100	
Copra meal (coconut meal) (21%)	90-100	50	
Corn gluten feed (gluten feed)	60-70	50-100	
Corn gluten meal (gluten meal)	100	50	
Cottonseed meal (41%)	100	100	Unlike the situation with finishing cattle, cottonseed meal is about equal to linseed meal for finishing lambs.
Distillers' dried grains	90	100	Rye distillers' dried grains are about 10% lower in protein than similar products made from corn or wheat.
Distillers' dried solubles	90	100	
Linseed meal (35%)	100	100	
Peanut meal (45%)	100	100	
Peas, dried	65-75	50	
Safflower meal, with hulls (42%)	40-45	100	
Soybeans	95-100	100	It does not pay to grind soybeans for sheep.
DRY FORAGES AND SILAGES:[2]			
Alfalfa hay, all analyses	100	100	
Alfalfa silage	33⅓-50	50-85	When alfalfa silage replaces corn silage, more energy feed must be provided but less protein, unless grain is used as a preservative.
Barley hay	70	50	The beards may be harmful, especially to woolly faced sheep.

(Continued)

TABLE 4-63 (Continued)

Feedstuff	Relative Feeding Value (lb for lb) in comparison with the designated (underlined) base feed which = 100	Maximum Percentage of Base Feed (or comparable feed or feeds) which it can replace for best results	Remarks
Beet tops, fresh	16-25	33⅓-50	In the West, large acreages of fresh beet tops a pastured off by sheep and cattle.
Beet tops, dry	70	50	
Beet top silage, sugar	17-25	33⅓-50	Either provide some dry forage or feed 2 oz of fine ground limestone to each 100 lb of silage.
Bromegrass hay	75	100	
Clover hay, crimson	90-100	100	Crimson clover hay has a considerably lower value not cut at an early stage.
Clover hay, red	90-100	100	If the rest of the ration is adequate in protein, clov hay will be equal to alfalfa in feeding value; othe wise, it will be lower.
Clover-timothy hay	80-90	100	
Corn fodder	75	100	Should be chopped.
Corn silage	33⅓-50	50-85	Although a ration in which corn silage is the only fo age is sometimes fed to sheep, most feeders pr fer to limit the silage and use some hay.
Corn stover	35	50	Unsatisfactory for finishing lambs, but cut or shre ded stover may be used as a part of the roughag for breeding ewes if fed along with a good legum
Cowpea hay	95-100	100	
Grass-legume mixed hay	80-90	100	Value depends on the proportion of legumes prese and the stage of maturity at which it is cut.
Grass-legume silage	32-45	50-85	Although a ration in which grass silage is the on forage is sometimes fed to sheep, most feede prefer to limit the silage and use some hay.
Grass silage	30-45	50-85	
Johnsongrass hay	70	100	
Lespedeza hay	80-100	100	Feeding value varies considerably with stage of m turity at which it is cut.
Mint hay	80-95	75	
Oat hay	75	50	
Pea-vine silage	33⅓-50	50-85	Unless grain is added as a preservative, pea-vine s lage requires more energy feed, but less prote supplement than corn silage when fed to finishir lambs.
Pea-vine hay	100-110	75	
Prairie hay	65-70	100	
Reed canarygrass	70	100	
Sorghum fodder	70	100	
Sorghum silage (grain varieties)	32-47	50-85	Although a ration in which sorghum silage is the on forage is sometimes fed to sheep, most feede prefer to limit the silage and use some hay.
Sorghum silage (sweet varieties)	25-30	50-85	Nearly equal to grain varieties in value per acre b cause of greater yield.
Sorghum stover	35	50	Unsatisfactory for finishing lambs, but cut or shre ded stover may be used as a part of the roughag for breeding ewes if fed along with a good legum
Soybean hay	85-100	100	The lower value is for finishing lambs. For oth classes of sheep, it is equal to alfalfa hay.
Sudangrass hay	50-75	50	
Sweetclover hay	100	100	Value of sweetclover hay varies widely. Second ye sweetclover hay is less desirable than first ye clover hay and is more apt to cause sweetclov disease.
Timothy hay	70	50	
Vetch-oat hay	80-90	100	The higher the proportion of vetch, the higher th value.
Wheat hay	70	50	

[1]Roots and tubers are of lower value than the grain and by-product feeds due to their higher moisture content.
[2]Silages are of lower value than dry forages due to their higher moisture content.

FEEDING BREEDING SHEEP

Success in the sheep business is largely measured by the percent lamb crop raised and the pounds of lamb marketed per ewe. This calls for proper feeding of the breeding flock.

The nutrient requirements should be adequate to provide for maintenance, growth (if the animals are immature), and reproduction. For the most part, these requirements can be met through feeding roughages—pasture in season, and dry forages and silages during the winter months.

Feeding Ewes

The most important factor affecting percent lamb crop and pounds of lamb marketed is the feed of the ewe, which generally involves the following distinct periods: (1) flushing ewes, (2) feeding pregnant ewes, (3) feeding at lambing time, and (4) feeding lactating ewes.

- *Flushing ewes—Flushing is the practice of conditioning or having thin ewes gain in weight just prior to breeding.* This special feeding usually begins 2 to 3 weeks prior to breeding and continues into the breeding season. It may be accomplished by turning the ewes to a fresh, luxuriant pasture 2 to 3 weeks before breeding time; or if such a pasture is not available, satisfactory results may be brought about by feeding a grain allowance of ½ to ¾ of a pound daily over a like period of time.

Although it is not likely that all of the benefits ascribed to flushing will be fully realized under all conditions, the general feeling persists that the practice will result in a 15 to 20 percent increase in the lamb crop, and that the ewes will breed both earlier and more nearly at the same time. Hence, it follows that the lamb crop will be earlier and more uniform in age and size.

- *Feeding pregnant ewes*—If a strong, healthy crop of lambs is to be expected, the ewes must be properly fed and cared for throughout the period of pregnancy. In general, this means the feeding of a suitable and well-balanced ration, together with the necessary minerals and vitamins as required for maintenance (and growth, if the ewe is not fully mature), growth of the fleece, and development of the fetus.

After the ewes are bred, they should have access to pastures as long as they are available and open. When the ground is firm, winter pasture or range, stalk, or stubble fields may be pastured to advantage.

Where winter pastures are either unavailable or inadequate, supplemental feeds must be provided. The most satisfactory forage is a good-quality legume hay, such as alfalfa, clover, lespedeza, or soybeans. A 150-pound ewe will eat about 4 pounds of hay daily.

Ewes should gain in weight during the entire period of pregnancy, making a total gain of 20 to 30 pounds for the period. They should enter the nursing period with some reserve flesh, because the lactation requirements are much more rigorous than those of the gestation period.

About ⅔ of the birth weight of the developing fetus is attained during the last 6 weeks of pregnancy. As a result, the protein requirement of a 140-pound ewe is ⅓ higher in late pregnancy than in early pregnancy, and the TDN requirements are over 40 percent higher. To meet these increased last-of-pregnancy needs, a concentrate allowance of ½ to 1 pound daily—the amount depending upon the quality of roughage available and the condition of the ewes—should be fed beginning a month to 6 weeks before lambing time.

On the western ranges, ewes are normally maintained on winter grazing areas, with or without supplemental feeds, as long as possible. When the vegetation is sparse or covered by deep snow, supplemental feeds of hays, preferably alfalfa or some other legume, or concentrates are provided. Often protein supplements in the form of pellets or cubes are used, for these may be scattered about the feeding grounds, neither being blown away nor difficult for the sheep to find.

- *Feeding at lambing time*—Usually, some five to seven days should elapse before ewes are placed on full feed following parturition. In general, feeds of a bulky and laxative nature should be provided during the first few days. A mixture of equal parts of oats and wheat bran is excellent. Soon after lambing, the ewe should be given water with the chill removed but should not be allowed to gorge.

- *Feeding lactating ewes*—Following lambing, the feed allowance of the ewe should be increased according to her capacity and needs.

The importance of adequate nutrition for the ewe during lactation cannot be overemphasized. Several research workers report that early lamb growth (at least up to 30 days of age) and milk production are highly correlated (.90). Milk secretion studies in New Zealand indicate that the more important factors influencing milk yield are age of ewe, time of lambing, ewe health, number of lambs suckled, genetic factors, and, most striking of all, plane of nutrition.

Pastures should be provided as soon as possible after lambing, but in the meantime a high-quality legume hay or, better yet, a combination of hay and silage will take care of the roughage needs. Though varying somewhat according to the size and condition of the ewe and whether there are twins or merely a single, an adequate ration for lactating ewes may consist of approximately 4 to 5 lb of high-quality alfalfa hay plus 1 to 2 lb of grain daily. With silage, the daily ration may consist of 1 and 1½ to 2 lb of hay, 4 to 6 lb of silage, and 1 to 2 lb of grain. If neither a legume hay nor legume silage is available, a protein supplement should be included in the grain ration.

Feeding the Farm Flock

In the northern latitudes, farm flock ewes are frequently given from ½ to 1 pound daily of a grain ration in addition to the roughage allowance from about a month before lambing to the time that they are turned to spring pasture. Higher levels of grain are fed during the suckling period than during gestation. Many of the farm flocks of the South, however, are kept in good thrifty condition, and the lambs are raised to the marketing stages, without the feeding of any grain. In still other areas, the ewes are fed only during periods of deep snows or extended droughts.

In general, for practical reasons, the ration of ewes should consist of as nearly year-around pastures as possible, with well-cured hay and other forages available the balance of the year, plus a limited grain allowance under certain conditions. Good-quality, sun-cured hay and lush pastures not only will provide most of the necessary proteins but also are excellent sources of calcium and the vitamins, especially vitamins A and D.

Feeding the Range Band

Today, the practical and successful range sheepman winter feeds. The progressive rancher is also equipped to meet emergency feeding periods, of which droughts are the most common.

Ewes are normally maintained on winter grazing areas, with or without supplemental feeds, as long as possible. Usually these ranges are located at the lower altitudes and the vegetation consists of rather mature and bleached grasses or brush and browse. When the vegetation is sparse or covered by deep snow, supplemental feeds of hays, preferably alfalfa, some other legume, or concentrates are provided. Often protein supplements in the form of pellets or cubes are used, for these may be scattered about the feeding grounds, neither being blown away nor difficult for the sheep to find. Usually such expensive protein supplements are fed only when native grass hays are being utilized, high-quality alfalfa not requiring a protein supplement.

Because of the magnitude of the range sheep industry and the fact that it is a highly specialized type of operation, in the sections that follow special discussion is devoted to the feeding and management of sheep on the range.

NUTRIENT DEFICIENCIES OF RANGE FORAGE

Hunger, due to lack of feed, is the most common deficiency on the western range. In particular, ther may be a shortage of energy during droughts, late i the season, or early in the spring when grass is wash Under such energy-deficient conditions, sheep los weight and condition, and lambs fail to grow. Als reproduction is adversely affected.

Mature, weathered native range grass is almo always deficient in protein—being as low as 3 pe cent, or less. Protein leaching losses due to fall an winter rains may range from 37 to 73 percent.

Phosphorus deficiencies are rather commo among range sheep, but calcium deficiency is seldo encountered.

Of the vitamins, vitamin A is most likely to b deficient in range forage, because dry, bleached rang grass is very low in carotene (the precursor of vitami A).

RANGE SUPPLEMENTS

Three suggested range supplements, varyin from high to low protein, are given in Table 4-6 Note that the energy concentrations of these suppl ments are in inverse order to their protein contents.

Sheep on range grass may be supplemented b feeding the high-protein formulation in Table 4-6 where there is need to correct both protein and pho phorus deficiencies. Of course, the supplements i Table 4-64 may be modified in keeping with avail bility and cost of feeds, and yet meet known deficie cies. For example, if phosphorus is the only def ciency, it may be corrected by feeding a phosphor supplement free-choice.

There is no one best and most practical rang supplement for any and all conditions. Many differe feeds may be used; among them, (1) ranch or local produced hay, (2) alfalfa pellets or cubes, with without fortification, and (3) supplements of vario kinds.

Also, sheepmen can lessen the labor attendant t the daily feeding of a pasture or range supplement b (1) using protein blocks, or (2) self-feeding salt-fee mixtures.

Where salt is used for the purpose of governin consumption, the proportion of salt to feed may va anywhere from 5 to 40 percent (with 30 to 33⅓% sa content being most common).

RATE OF SUPPLEMENTAL FEEDING

The time and rate of supplemental feeding is d termined by the reason for feeding supplement Supplements are fed for two primary purposes: (1) balance diets by adding small quantities of a nutrie (such as protein, a mineral, or a vitamin) or a combin tion of nutrients; and (2) to provide nutrients durin short-term emergencies. As an example of the latter, supplement may be needed to prevent sheep fro

TABLE 4-64
FORMULAS FOR RANGE SHEEP SUPPLEMENTS[1]

Feed[2]	Recommended Level of Protein					
	High		Medium		Low	
	(%)		(%)		(%)	
Cottonseed meal, 41%	62.5		32.0		0	
Soybean meal, 41%	10.0		10.0		0	
Barley	0		33.0		67.0	
Corn 2	5.0		10.0		15.0	
Alfalfa meal, min. 17%	12.5		6.0		5.0	
Molasses, sugarcane	5.0		5.0		10.0	
Dicalcium phosphate	4.0		3.0		2.0	
Salt or trace mineralized salt	1.0		1.0		1.0	
	100.0		100.0		100.0	
Composition	**As-fed[3]**	**M-F**	**As-fed[3]**	**M-F**	**As-fed[3]**	**M-F**
Crude protein (%)	32.9	*36.5*	23.8	*26.4*	10.8	*12.0*
Digestible protein (%)	26.7	*29.6*	19.3	*21.4*	8.2	*9.1*
Metabolizable energy (Mcal/lb)	1.0	*1.1*	1.1	*1.2*	1.3	*1.4*
Phosphorus (%)	1.5	*1.7*	1.2	*1.3*	0.7	*0.8*
Carotene (mg/lb)	9.0	*10.0*	4.5	*5.0*	3.7	*4.1*
Rate of Feeding (lb/day)	0.22-0.45	*0.25-0.5*	0.22-0.45	*0.25-0.5*	0.22-0.45	*0.25-[4] 0.5*

[1]Adapted by the author from *Nutrient Requirements of Sheep*, No. 5, 5th rev. ed., NRC-National Academy of Sciences, Washington, D.C., 1975, p. 48, Table 14.
[2]Feeds to be mixed and fed in meal or pellet form.
[3]Estimated 90% dry matter.
[4]In emergency situations, up to 1.0 lb may be fed.

eating poisonous plants during periods when they are on the trail or when forage is covered with snow.

Supplemental feeding should be timed to start when it is needed. If phosphorus supplementation is required, it should be provided continuously, perhaps by free-choice feeding. Where energy, protein, and/or vitamin A supplementation are involved, it takes a unique skill to recognize the nutritional state of the sheep, the range condition, and the need for supplement—both in kind and amount. The successful manager develops a grazing plan that minimizes the need for supplements, yet he provides the proper supplement at the proper time and in the proper amounts.

The normal range of supplementation for sheep is ¼ to ½ lb per head per day. Rates above ½ lb approach a level that will result in reduced intake of range forage. Where range vegetation is so short as to require supplementation in excess of ½ lb per head per day, consideration should be given either to moving the sheep into a drylot or to moving them to a better grazing area.

Some managers divide their sheep according to age, condition, and twins vs single lambs. Of course, this is facilitated where there are several bands. By so doing, it is possible (1) to give the animals that require the highest level of nutrition the best pasture or range, and/or (2) to supplement according to need.

Feeding Rams

Rams should be fed so as to remain in vigorous, active breeding condition. In general, they should be fed the same kind of feeds as ewes but in slightly larger quantities. They need a generous allowance of relatively high-quality feed just before and during the breeding season, when pasture is not available. During the balance of the year, pasture is usually adequate when available; otherwise, the ration may be comparable to that of the ewes.

FEEDING GROWING-FINISHING LAMBS

The growing-finishing stage of lambs refers to that period extending from birth to weaning at four to six months of age. At no other period in the life of the sheep is the promotion of growth and the prevention of disease so important.

Where succulent pastures are available, most practical sheepmen, including producers with both farm flocks and range bands, consider that a combination of such green forage plus the ewe's milk is ample. In fact, lambs are unique among farm animals, inasmuch as they may be marketed at top prices off grass. Although young cattle may be sold off grass without having any other feed, they will usually fail to get sufficiently fat to bring top prices.

Early Weaning

Early weaning refers to the practice of weaning lambs earlier than usual—weaning at five to eight weeks of age or earlier. There is much interest in early weaning because—

1. Of lambing out of season, multiple births, and more than one lamb crop per year.

2. Lactating ewes usually reach a peak in milk production 3 to 4 weeks after lambing, then decline thereafter. By 3 to 4 months after lambing, many ewes will be producing very little milk.

3. Fewer parasite problems accompany an early weaning program.

4. Increased knowledge of nutrition now makes it possible for scientists to improve upon milk (except for colostrum), chiefly by reinforcing it with certain vitamins and minerals.

5. Young gains are cheap gains, due to (a) the higher water and lower fat content of young animals in comparison with older animals, and (b) the higher feed consumption per unit weight of young animals.

6. Following weaning, ewes can be maintained on a limited feed allowance, thereby effecting a saving in cost.

For successful early weaning, superior nutrition and management are essential; and the earlier the weaning age the more exacting these requirements.

Early weaning of lambs is, to a considerable extent, a matter of preparation, rather than the abrupt separation of lambs from their mothers. Lambs that are to be early weaned should be creep fed from the time they are old enough to eat. At weaning time, the separation should be made by removing the ewes from the lambs, rather than vice versa. By keeping the lambs in familiar surroundings, stress is minimized.

An early-weaned lamb ration should meet the following specifications: contain a minimum of 16 percent crude protein; be fortified with supplemental iron if the lambs are raised on slotted floors; and have a calcium-phosphorus ratio of at least 1:1 (2:1 if urinary calculi has been experienced).

Milk replacers containing approximately 30% fat and 24% protein have been used successfully in feeding lambs receiving colostrum and weaned at one da of age. Replacers with reduced lactose content (fror 27 to 42% on a dry matter basis) give improved per formance. The milk is fed (1) cold at 36 to 40° F rather than warm to reduce overeating and bacteria contamination, and (2) free-choice. From the begir ning, lambs are offered a very palatable solid feed i addition to the milk. The milk replacer is discor tinued when the lambs are eating sufficient quantitie of the dry feed, usually at 21 to 35 days of age.

Creep Feeding

The practice of supplemental feeding of nursin lambs in a separate enclosure away from their dam is known as creep feeding. Lambs will usually cor sume some creep feed at 10 to 14 days of age.

Creep rations can either be hand-fed or self-fec Many sheepmen hand-feed until the lambs begin t eat regularly, then self-feed from this point on.

The amount of creep feed consumed is inversel proportional to the ewe's milk production. For thi reason, (1) twin lambs usually consume more tha single lambs, and (2) significant amounts of cree feed are consumed at six to eight weeks of age, a which time the ewe's milk production usually drops.

Until lambs are six weeks old, the grain should b crimped, cracked, or rolled, unless a pelleted ration i used. After this age, whole grain may be fed unless i is extremely hard (like millet).

It is important that the creep ration be very pala able. For this reason, rolled oats, wheat bran, soybea meal, and molasses are important ingredients in creep ration. Even then, if lambs have access to lus pasture, they may prefer it to the creep feed.

Four suggested creep rations are given in Tabl 4-65.

TABLE 4-65

CREEP RATIONS FOR LAMBS UNDER VARIED CONDITIONS[1]

(As-fed Basis)

Ingredient[2]	Ration No. 1 (grind at first; feed whole later)	Ration No. 2 (hand- or self-fed, ground or pelleted)	Ration No. 3 (hand- or self-fed, ground or pelleted)	Ration No. 4 (ground diet may be hand- or self-fed)
	(%)	(%)	(%)	(%)
Alfalfa, hay, s-c, early bloom	free-choice	65.0	30.0	
Corn, dent yellow	58.5	12.0		84.0
Corn, ears, ground			55.0	
Oats or barley	20.0	9.0		
Wheat bran	10.0			
Soybean meal or linseed meal	10.0	10.0	10.0	15.0
Molasses, sugarcane		3.0	5.0	
Limestone, ground, min. 33% calcium	1.0			1.0
Sodium phosphate, monobasic		1.0		
Trace mineralized salt	0.5			
	100.0	100.0	100.0	100.0
Antibiotic (chlortetracycline or oxytetracycline)		7-12 mg/lb 15-26 mg/kg	7-12 mg/lb 15-26 mg/kg	
Vitamin A supplement		250 IU/lb 550 IU/kg		
Vitamin D supplement		25 IU/lb 55 IU/kg		

[1]Adapted by the author from *Nutrient Requirements of Sheep*, No. 5, 5th rev. ed., NRC-National Academy of Sciences, Washington, D.C., 1975, p. 49, Table 15.
[2]Feed the highest quality hay in a separate rack. Feed hay and grain twice daily to keep them fresh. Offer trace mineralized salt free-choice for all rations.

FEEDING FINISHING (Fattening) LAMBS

The primary objective of the sheepman is that of producing milk-fat lambs suitable for slaughter at weaning time. Only when the pasture is inadequate are lambs sold via the feeder route. Almost all feeder lambs come from the range area. Some range areas produce only a small percentage of lambs which are classed as feeders, whereas in other areas almost all the lambs must be sold as feeders because the vegetation is not sufficient to promote rapid growth and finishing. It is estimated that, for the range area as a whole, an average of at least 50 percent of all lambs produced in any one year receive additional feed after they are removed from the range and prior to slaughter.

Colorado is the leading lamb-feeding state of the nation, finishing out about one-sixth of the sheep and lambs fed in the United States. Here, locally grown alfalfa, sugar beet by-products, and barley are used extensively, along with considerable corn and protein supplements which are shipped in from outside areas. Texas ranks second in lamb feeding, followed by Wyoming, Nebraska, and California.

Numerous feeding practices and a great variety of feeds are used in lamb-finishing operations. In general, however, all methods may be classified as (1) drylot feeding, or (2) field finishing.

Drylot Feeding

TABLE 4-66

RATIONS FOR FINISHING (FATTENING) LAMBS[1]
(As-fed Basis)

(Finishing lambs on full feed, weighing 55-105 lb, or 25-48 kg; many lamb feeders prefer to chop the hay and mix and/or pellet it with the grain ration)

Type of Ration	Lb/Day	Kg/Day
1. Legume hay	1¼-1¾	*.57-.79*
Grain[2]	1¼-1¾	*.57-.79*
2. Grass-legume mixed hay, good quality	1¼-1¾	*.57-.79*
Grain[2]	1¼-1¾	*.57-.79*
Protein supplement[3]	0.1-0.2	*.05-.09*
3. Legume hay[4]	1¼-1¾	*.57-.79*
Corn or sorghum silage	3¾-5¼	*1.70-2.38*
Protein supplement[3]	.15-0.2	*.07-.09*
4. Legume hay	¾-1	*.34-.45*
Corn or sorghum silage	1½-2½	*.68-1.14*
Grain[2]	1¼-1¾	*.57-.79*
Protein supplement[3]	0.1	*.05*
5. Corn or sorghum silage[4,5]	2¾-4½	*1.25-2.04*
Grain[2]	1¼-1¾	*.57-.79*
Protein supplement[3]	0.2-0.3	*.09-.14*
6. Legume hay	1½-2½	*.68-1.14*
Grain[2]	¾	*.34*
Beet pulp, wet	2-3	*.91-1.36*

[1]Rations 3 and 6 are especially suited to starting lightweight lambs weighing 55 to 70 lb, or *25-32 kg*, following which they should be switched to one of the other rations.

[2]Whole corn, barley, wheat, heavy oats, and/or sorghum.

[3]Linseed, cottonseed, and/or soybean meal.

[4]Although these rations are occasionally used, they are controversial. Most experienced sheep feeders prefer (1) a ration with some grain to ration No. 3, and (2) not to use silage for the only roughage as in ration No. 5.

[5]Ration No. 5 will be considerably improved by the addition of (1) ½ lb of either dehydrated alfalfa or good legume hay, and (2) 0.01 lb of calcium carbonate.

Drylot feeding is, as the name indicates, feeding under restricted conditions. This may either be (1) shelter or barn feeding, or (2) open yard feeding. Table 4-66 gives some suggested rations for finishing (fattening) lambs in drylot.

Field Finishing

This method of finishing lambs is somewhat comparable to the pasture finishing of cattle, except that a greater variety of feeds is used by lamb feeders.

The kind of field feeding varies from area to area and even between farms within the same locality. Most of the feeder lambs are shipped to these feeding areas in August and September at the time the lambs are normally weaned from range ewes. Usually these field-fed lambs are ready for market in November and early December.

Throughout the Corn Belt, feeder lambs are usually used as scavengers during the early part of the field feeding process. Frequently, the lambs are pastured in the stubble fields or on the meadows until all these feeds are consumed, after which they are turned into cornfields.

In Kansas, Oklahoma, Nebraska, and Texas, thousands of lambs are finished primarily by fall pasturing of the wheat fields. In the Pacific Northwest, a limited number of lambs are finished by gleaning pea stubble.

Basic Considerations in Finishing Lambs

Although no rules of success are applicable to any and all conditions, the following basic considerations in finishing lambs are worth noting:

1. In lamb feeding operations, the purchase price of the lambs represents 60-70 percent of all costs. This indicates the importance of keeping death losses to a minimum.

2. Experienced feeders normally expect to lose about 2.5 percent of lambs on feed. This is about twice the loss that occurs in commercial cattle feeding operations.

3. Lamb feeding is seasonal in nature, usually extending from August to about the following May. This seasonal condition is due to the fact that (a) suitable feeder lambs are not available until the late summer and fall months, and (b) following the growing and harvesting seasons, the feeders have available quantities of marketable and unmarketable feeds which may be utilized by lambs.

4. As in cattle feeding, feedlot gains are expensive, usually costing more per pound than the selling price on the market. Thus, a reasonable margin or difference between the cost and selling price per hundredweight is necessary.

5. Feed accounts for approximately 50 percent of

the cost of finishing feedlot lambs, exclusive of the initial purchase price of the feeder lambs.

6. Though the situation varies according to the kind of feed and the age of animal, it requires about 800 pounds of feed to produce 100 pounds of on-foot lamb. Thus, lambs utilize feeds more efficiently than cattle.

7. In a 250-mile shipment lambs will shrink about 5 percent. If properly fed, watered, and cared for en route, lambs may be shipped 1,500 to 2,000 miles without much greater shrinkage than this.

8. Most feeder lambs weigh between 55 and 70 pounds when placed on feed and from 85 to 105 pounds following a 90- to 120-day feeding period.

9. Wool is of importance in selecting feeder lambs because it has a bearing on their market value, the pelt being the most valuable slaughter by-product.

10. Range feeder lambs are more plentiful than native feeders, thus allowing for greater selection; and usually they are more uniform and are less heavily infected with parasites.

11. Lambs are frequently fed on a contract basis, with many and varied agreements being used.

12. Wether lambs appear to make slightly more rapid gains than ewe lambs but they do not finish quite so early as ewe lambs.

13. Where western lambs have undergone a long shipment immediately after being taken from their mothers, special care is necessary in starting them on feed. After rest following shipment, lambs are usually started on grain by feeding about ¼ pound per head daily. Gradually this allowance is increased so that the lambs are getting a full feed of about 2 pounds of grain per head daily and about the same amount of hay when on full feed 3 to 4 weeks later.

14. A great variety of feeds can be used in lamb feeding. In general, the successful feeder balances out the ration by selecting those feeds which are most readily available at the lowest possible price.

15. Unless such extremely hard seeds as millet are included in the ration, it does not pay to grind feeds for finishing lambs.

FEEDING SHOW AND SALE SHEEP

In addition to being reasonably economical (mostly homegrown) and well-balanced, the fitting ration for show sheep should be palatable. Many feed combinations meet these specifications. The ration selected is usually determined by (1) the availability and price of the feed in the area, and (2) the preference and judgment of the shepherd.

Some suggested grain fitting rations follow. To each of these grain rations should be added (1) good quality roughage—usually homegrown—in about the proportion indicated in Table 4-66, Rations for Finishing (Fattening) Lambs, and (2) salt and other minerals on a free-choice basis as previously advocated.

1. *Rations for lambs* (either creep-fed or weanec lambs that are being fitted for show. As-fed basis)— Show lambs on full feed will eat about 2½ pounds o grain per head daily:

Ration No. 1[1]	*(lb)*	*(kg)*
Oats	65	29.5
Wheat bran	17.5	7.9
Linseed meal	17.5	7.9

Ration No. 2	*(lb)*	*(kg)*
Corn	40	18.2
Oats	40	18.2
Wheat bran	10	4.5
Protein supplement[2]	10	4.5

Ration No. 3	*(lb)*	*(kg)*
Oats	30	13.9
Barley	20	9.1
Corn	20	9.1
Wheat bran	20	9.1
Protein supplement[2]	10	4.5

Ration No. 4	*(lb)*	*(kg)*
Oats	70	31.8
Wheat bran	20	9.1
Protein supplement[2]	10	4.5

Ration No. 5	*(lb)*	*(kg)*
Oats	80	36.3
Wheat bran	20	9.1

Ration No. 6	*(lb)*	*(kg)*
Corn	45	20.4
Oats	45	20.4
Protein supplement[2]	10	4.5

Ration No. 7	*(lb)*	*(kg)*
Corn, cracked	85	38.6
Soybean meal	15	6.8

[1]The author has used this ration extensively. Near the end of the fitting perioc the shepherd adds 50 lb of barley and 50 lb of peas to each 100 lb of Ration 1.
[2]Linseed, cottonseed, and/or soybean meal.

2. *Rations for fitting yearlings and mature sheep* (as-fed basis)—Show yearlings on full feed will ea about 3 pounds of grain per head daily, whereas ma ture sheep will eat about 3½ pounds of grain per heac daily:

Ration No. 1	*(lb)*	*(kg)*
Oats	50	22.7
Peas (split)	40	18.2
Wheat bran	10	4.5

Ration No. 2	*(lb)*	*(kg)*
Oats	50	22.7
Barley	40	18.2
Wheat bran	10	4.5

Ration No. 3	*(lb)*	*(kg)*
Corn	40	18.2
Oats	40	18.2
Wheat bran	10	4.5
Protein supplement[1]	10	4.5

Ration No. 4	*(lb)*	*(kg)*
Oats	60	27.2
Peas	10	4.5
Barley	10	4.5
Wheat bran	10	4.5
Protein supplement[1]	10	4.5

[1]Linseed, cottonseed, and/or soybean meal.

When feeding small grains (oats, barley, o wheat), shepherds prefer that they be steam rollec and they prefer nutted (pea-sized) linseed meal. Cor is usually cracked or coarsely ground, and peas ar split or cracked. When pastures are not available, a falfa is the most popular hay, but any good legume i quite satisfactory. The lighter types of lamb ration (see "1. *Rations for Lambs,*" 1-1, 4, and 5) are usuall used for summer feeding, especially when anima are being fitted for the late shows.

In fitting animals for show or sale, most succes ful shepherds feed a limited quantity of cabbag sliced carrots, mangels (stock beets), rutabaga (swedes), or turnips. These succulent feeds are highl relished by sheep and appear to help their digestio and general thrift.

PART V—FEEDING SWINE[15]

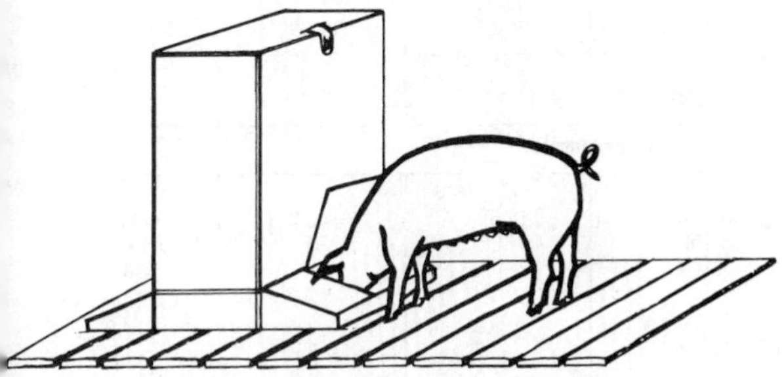

Fig. 4-33. Feeding swine.

Domestic swine have less choice in their selection of feed than any other class of four-footed animals. For the most part they are able to consume only what the caretaker provides. This consists largely of concentrated feeds with only a small proportion of roughage. These conditions are made more critical because hogs grow much faster in proportion to their body weight than the larger farm animals, and they produce young at an earlier age. Thus, a knowledge of the nutritional needs of swine is especially important.

Extensive surveys indicate that about 25 to 30 percent of all pigs farrowed fail to live to weaning age. Although these heavy losses are due to many and variable factors, certainly nutritional deficiencies play a major role.

Knowledge of feeding swine is also important from an economic standpoint, because feed accounts for approximately 65 to 75 percent of the total cost of producing pork.

NUTRIENT REQUIREMENTS

The nutrient requirements of swine are influenced by age, function, disease level, nutrient interaction, environment, etc. It has been established that the pig has a requirement for over 40 different nutrients. Fortunately, not all of them are of practical concern.

No one nutrient is more important than another—and all are essential. Each nutrient has one or more particular and specific functions to perform in the body. If the nutrient is not supplied by the ration in proper amounts, the functions (growth, reproduction, lactation, etc.) will obviously be impaired. Since the modern swine producer is interested in maximum performance, it follows that the input of nutrients must be ample to bring this about.

The nutrient requirements of swine are given in Tables 4-67, 4-68, 4-69, and 4-70, which follow. These figures are, for the most part, requirements (rather than allowances); hence, they do not provide for margins of safety to compensate for variations in feed composition, environment, and possible losses of nutrients during storage or processing.

Sometimes nutritional requirements, or standards, like those in Tables 4-67 to 4-70, pages 378 to 381, impart the erroneous impression that such figures are absolute, final, and unchangeable. Nothing could be further from the truth. Rather, the figures used in these tables are guides based on research.

In using Tables 4-67 to 4-70, the following pertinent points should be recognized:

1. Feedstuffs produced in various parts of the country vary in nutritive value.
2. The environment in which pigs are produced can modify the requirements.
3. Animals bred for high performance have nutritional needs that are quite different from average performers.

RECOMMENDED NUTRIENT ALLOWANCES

Presently available information indicates that the nutrient allowances recommended in Table 4-71, page 382, will meet the minimum requirements for swine and provide reasonable margins of safety.

Energy

Energy is the body's fuel supply. Every movement and activity of the pig's life involves the expenditure of fuel—energy for breathing, heart action, digestion, muscular movement, as well as heat to keep the body warm. If more energy is consumed than necessary to carry on vital functions, the excess is stored as body fat. In fact, this is what is done in finishing hogs. More energy is eaten than is needed for growth and body maintenance, with the result that the animal lays down fatty tissue with the excess.

The main nutrients supplying energy are carbohydrates. There are several forms of carbohydrates in plants. In feed analysis these forms are identified as either nitrogen-free-extract (NFE) or crude fiber. The NFE fraction includes the more soluble carbohydrates—sugars, starch, and some hemicellulose. All but hemicellulose are very digestible. Crude fiber, however, contains cellulose, hemicellulose, and lignin, all of which are highly indigestible by the pig.

---

[15]The recommended nutrient allowances and suggested rations given in this section are based on National Research Council (NRC) standards.

TABLE 4-67

DAILY NUTRIENT REQUIREMENTS OF GROWING-FINISHING SWINE FED *AD LIBITUM* (per Animal)[1]

Liveweight (lb) *(kg)*	2-11[2] *1-5*	11-22 *5-10*	22-44 *10-20*	44-77 *20-35*	77-132 *35-60*	132-220 *60-100*
Feed intake (air-dry) (lb) *(g)*	0.6 *250*	1.1 *500*	2.2 *1,000*	3.3 *1,500*	4.4 *2,000*	6.6 *3,000*
Nutrients	**Requirements**					
Energy and protein:						
Digestible energy[3] (kcal)	925	1,750	3,370	5,055	6,740	10,110
Metabolizable energy[3] (kcal)	900	1,700	3,160	4,740	6,320	9,480
Crude protein[4] (g)	67.5	100	180	240	280	390
Amino acids:						
Arginine (g)	0.8	1.3	2.3	3.0	3.6	4.8
Histidine (g)	0.8	1.2	2.0	2.7	3.2	4.5
Isoleucine (g)	2.1	3.2	5.6	7.5	8.8	12.3
Leucine (g)	2.5	3.8	6.8	9.0	10.4	14.4
Lysine (g)	3.2	4.8	7.9	10.5	12.2	17.1
Methionine + cystine[5] (g)	1.9	2.8	5.1	6.8	8.0	9.0
Phenylalanine + tyrosine[6] (g)	3.0	4.4	7.9	10.5	12.2	17.1
Threonine (g)	1.9	2.8	5.1	6.8	7.8	11.1
Tryptophan[7] (g)	0.5	0.8	1.3	1.8	2.2	3.0
Valine (g)	2.1	3.2	5.6	7.5	8.8	12.3
Major or macro minerals:						
Calcium (g)	2.3	4.0	6.5	9.0	11.0	15.0
Phosphorus[8] (g)	1.8	3.0	5.5	7.5	9.0	12.0
Sodium (g)	0.25	0.5	1.0	1.5	2.0	3.0
Chlorine (g)	0.33	0.7	1.3	2.0	2.6	3.9
Magnesium (g)	0.10	0.2	0.4	0.6	0.8	1.2
Potassium (g)	0.75	1.3	2.6	3.5	4.0	5.1
Trace or micro minerals:						
Copper (mg)	1.5	3	5	6	6	9
Iodine (mg)	0.04	0.07	0.14	0.21	0.28	0.42
Iron (mg)	38	70	80	90	100	120
Manganese (mg)	1.0	2	3	3	4	6
Selenium (mg)	0.04	0.08	0.15	0.22	0.30	0.30
Zinc (mg)	25	50	80	90	100	150
Fat-soluble vitamins:						
Vitamin A (IU)	550	1,100	1,750	1,950	2,600	3,900
or β-carotene (mg)	2.2	4.4	7.0	7.8	10.4	15.6
Vitamin D (IU)	55	110	200	300	300	375
Vitamin E (IU)	2.8	5.5	11	17	22	33
Vitamin K (menadione) (mg)	0.50	1.1	2.2	3.3	4.4	6
Water-soluble vitamins:						
Biotin[9] (mg)	0.03	0.05	0.10	0.15	0.20	0.30
Choline[10] (mg)	275	550	900	1,050	1,100	1,200
Folacin[9] (mg)	0.15	0.30	0.60	0.90	1.2	1.8
Niacin[11] (mg)	5.5	11	18	21	24	30
Pantothenic acid (mg)	3.3	6.5	11	17	22	33
Riboflavin (mg)	0.75	1.5	3.0	3.9	4.4	7
Thiamin (mg)	0.33	0.65	1.1	1.7	2.2	3.3
Vitamin B-6 (mg)	0.38	0.75	1.5	1.7	2.2	3.3
Vitamin B-12 (μg)	5.5	11	15	17	22	33

[1]Adapted by the authors from *Nutrient Requirements of Swine*, No. 2, 8th rev. ed., NRC-National Academy of Sciences, 1979, p. 23.

[2]Requirements reflect the estimated levels of each nutrient needed for optimal performance when a fortified grain-soybean meal diet is fed, except that a substantial level of milk products should be included in the diet of the 2-11 lb *(1-5 kg)* pig. Concentrations are based upon amounts per unit of air-dry diet (i.e., 90% dry matter).

[3]These are not absolute requirements, but are suggested energy levels derived from diets containing corn and soybean meal (44% crude protein). When lower energy grains are fed, these energy levels will not be met; consequently, feed efficiency would be lowered.

[4]Approximate protein levels required to meet the need for indispensable amino acids when a fortified grain-soybean meal diet is fed to pigs weighing more than 11 lb *(5 kg)*.

[5]Methionine can fulfill the total requirements; cystine can meet at least 50% of the total requirement.

[6]Phenylalanine can fulfill the total requirement; tyrosine can meet at least 50% of the total requirement.

[7]It is assumed that usable tryptophan content of corn does not exceed 0.05%.

[8]At least 30% of the phosphorus requirement should be provided by inorganic and/or animal product sources.

[9]These levels are suggested. No requirements have been established.

[10]In excess of its requirement for protein synthesis, methionine can spare dietary choline (4.3 mg methionine is equal in methylating capacity to 1 mg choline).

[11]It is assumed that most of the niacin present in cereal grains and their by-products is in bound form and thus unavailable to swine. The niacin contributed by these sources is not included in the requirement listed. In excess of its requirement for protein synthesis, tryptophan can be converted to niacin (50 mg tryptophan yields 1 mg niacin).

The kind of carbohydrate a feed contains determines its value as a source of energy for the pig. Cereal grains are widely used in swine feeding because of their very high NFE (60-70%) and low crude fiber content.

Another group of energy nutrients is the fats and oils. Fat, which is abundant in such common hog feeds as peanuts and soybeans, is a very concentrated source of fuel. It supplies approximately 2.25 times as much metabolizable energy as an equal weight of carbohydrates. Therefore, a feed high in fat, or a ra tion containing added fat, is much higher in energ value than a feed or ration low in fat. It is em phasized, however, that liberal quantities of eithe soybeans or peanuts will produce soft pork.

Although roughages are a good source of energ for ruminants, because of their bulky nature and th restricted size of the digestive tract of hogs in com parison with ruminants, only limited quantities o them are contained in normal swine rations

TABLE 4-68

DAILY NUTRIENT REQUIREMENTS OF BREEDING SWINE (per Animal)[1]

Production Period	Bred Gilts and Sows; Young and Adult Boars	Lactating Gilts and Sows		
Feed intake (air-dry) (lb)	4[2]	8.8	10.5	12.1
(g)	*1,800*	*4,000*	*4,750*	*5,500*
Nutrients	**Requirements**			
Energy and protein:				
Digestible energy (kcal)	6,120[3]	13,580	16,130	18,670
Metabolizable energy (kcal)	5,760[3]	12,780	15,180	17,570
Crude protein (g)	216	520	618	715
Amino acids:				
Arginine (g)	0	16.0	19.0	22.0
Histidine (g)	2.7	10.0	11.9	13.8
Isoleucine (g)	6.7	15.6	18.5	21.4
Leucine (g)	7.6	28.0	33.2	38.5
Lysine (g)	7.7	23.2	27.6	31.9
Methionine + cystine[4] (g)	4.1	14.4	17.1	19.8
Phenylalanine + tyrosine[5] (g)	9.4	34.0	40.4	46.8
Threonine (g)	6.1	17.2	20.4	23.6
Tryptophan[6] (g)	1.6	4.8	5.7	6.6
Valine (g)	8.3	22.0	26.1	30.2
Major or macro minerals:				
Calcium (g)	13.5	30.0	35.6	41.2
Phosphorus[7] (g)	10.8	20.0	23.8	27.5
Sodium (g)	2.7	8.0	9.5	11.0
Chlorine (g)	4.5	12.0	14.2	16.5
Magnesium (g)	0.7	1.6	1.9	2.2
Potassium (g)	3.6	8.0	9.5	11.0
Trace or micro minerals:				
Copper (mg)	9	20	24	28
Iodine (mg)	0.25	0.56	0.66	0.77
Iron (mg)	144	320	380	440
Manganese (mg)	18	40	48	55
Selenium (mg)	0.27	0.40	0.48	0.55
Zinc (mg)	90	200	238	275
Fat-soluble vitamins:				
Vitamin A (IU)	7,200	8,000	9,500	11,000
or β-carotene (mg)	28.8	32.0	38.0	44.0
Vitamin D (IU)	360	800	950	1,100
Vitamin E (IU)	18.0	40.0	47.5	55.0
Vitamin K (mg)	3.6	8.0	9.5	11.0
Water-soluble vitamins:				
Biotin[8] (mg)	0.18	0.4	0.48	0.55
Choline (mg)	2,250	5,000	5,940	6,875
Folacin[8] (mg)	1.08	2.4	2.8	3.3
Niacin[9] (mg)	18.0	40.0	47.5	55.0
Pantothenic acid (mg)	21.6	48.0	57.0	66.0
Riboflavin (mg)	5.4	12.0	14.2	16.5
Thiamin (mg)	1.8	4.0	4.8	5.5
Vitamin B-6 (mg)	1.8	4.0	4.8	5.5
Vitamin B-12 (μg)	27.0	60.0	71.2	82.5

[1]Adapted by the authors from *Nutrient Requirements of Swine*, No. 2, 8th rev. ed., NRC-National Academy of Sciences, 1979, p. 25. Requirements reflect the estimated levels of each nutrient needed for optimal performance when a fortified grain-soybean meal diet is fed. Concentrations are based upon amounts per unit of air-dry diet (i.e., 90% dry matter).

[2]An additional 25% should be fed to working boars.

[3]Individual feeding and moderate climatic conditions are assumed. An energy reduction of about 10% is possible when gilts and sows are tethered or individually penned in a stall in environmentally controlled housing. An energy increase of about 25% is suggested for cold climatic (winter) conditions.

[4]Methionine can fulfill the total requirement; cystine can meet at least 50% of the total requirement.

[5]Phenylalanine can fulfill the total requirement; tyrosine can meet at least 50% of the total requirement.

[6]It is assumed that usable tryptophan content of corn does not exceed 0.05%.

[7]At least 30% of the phosphorus requirement should be provided by inorganic and/or animal product sources.

[8]These levels are suggested. No requirements have been established.

[9]It is assumed that most of the niacin present in cereal grains and their by-products is in bound form and thus unavailable to swine. The niacin contributed by these sources is not included in the requirement listed. In excess of its requirement for protein synthesis, tryptophan can be converted to niacin (50 mg tryptophan yields 1 mg niacin).

Roughages (pastures and ground legume hays) are added to swine rations because of their protein, minerals, and vitamins, rather than for energy purposes.

Energy values are generally expressed in feed tables as total digestible nutrients (TDN), digestible energy (DE), metabolizable energy (ME), and/or net energy (see Table 4-101, page 456). Since energy expenditure can be measured as heat, modern nutritionists measure the energy needs of the animal in calories (a unit of heat). TDN values may be converted to DE by assuming that 1 pound of TDN is equivalent to 2,000 kilocalories of DE. ME values are calculated from the formula—

$$ME = DE \frac{(96 - 0.2 \times \%\ \text{crude protein})}{100}$$

The latter formula indicates that ME values are somewhat less than 96 percent of DE values.

TABLE 4-69

NUTRIENT REQUIREMENTS OF GROWING-FINISHING SWINE FED *AD LIBITUM*[1]
(Percentage or Amount per Pound and per Kilogram of Diet)

Liveweight (lb) / *(kg)*	2-11[2] *1-5*		11-22 *5-10*		22-44 *10-20*		44-77 *20-35*		77-132 *35-60*		132-220 *60-100*	
Daily gain (expected) (lb) / *(g)*	0.4 *200*		0.7 *300*		1.1 *500*		1.3 *600*		1.5 *700*		1.8 *800*	
Nutrients	**Requirements**											
Energy and protein:												
Digestible energy[3] (kcal)	1,678	3,700	1,587	3,500	1,528	3,370	1,533	3,380	1,537	3,390	1,540	3,395
Metabolizable energy[3] (kcal)	1,633	3,600	1,542	3,400	1,433	3,160	1,440	3,175	1,447	3,190	1,449	3,195
Crude protein[4] (%)	27		20		18		16		14		13	
Amino acids:												
Arginine (%)	0.33		0.25		0.23		0.20		0.18		0.16	
Histidine (%)	0.31		0.23		0.20		0.18		0.16		0.15	
Isoleucine (%)	0.85		0.63		0.56		0.50		0.44		0.41	
Leucine (%)	1.01		0.75		0.68		0.60		0.52		0.48	
Lysine (%)	1.28		0.95		0.79		0.70		0.61		0.57	
Methionine + cystine[5] (%)	0.76		0.56		0.51		0.45		0.40		0.30	
Phenylalanine + tyrosine[6] (%)	1.18		0.88		0.79		0.70		0.61		0.57	
Threonine (%)	0.76		0.56		0.51		0.45		0.39		0.37	
Tryptophan[7] (%)	0.20		0.15		0.13		0.12		0.11		0.10	
Valine (%)	0.85		0.63		0.56		0.50		0.44		0.41	
Major or macro minerals:												
Calcium (%)	0.90		0.80		0.65		0.60		0.55		0.50	
Phosphorus[8] (%)	0.70		0.60		0.55		0.50		0.45		0.40	
Sodium (%)	0.10		0.10		0.10		0.10		0.10		0.10	
Chlorine (%)	0.13		0.13		0.13		0.13		0.13		0.13	
Magnesium (%)	0.04		0.04		0.04		0.04		0.04		0.04	
Potassium (%)	0.30		0.26		0.26		0.23		0.20		0.17	
	(per lb)	*(per kg)*	*(per lb)*	*(per kg)*	*(per lb)*	*(per kg)*	*(per lb)*	*(per kg)*	*(per lb)*	*(per kg)*	*(per lb)*	*(per kg)*
Trace or micro minerals:												
Copper (mg)	2.7	*6.0*	2.7	*6.0*	2.3	*5.0*	1.8	*4.0*	1.4	*3.0*	1.4	*3.*
Iodine (mg)	0.07	*0.14*	0.07	*0.14*	0.07	*0.14*	0.07	*0.14*	0.07	*0.14*	0.07	*0.*
Iron (mg)	68	*150*	64	*140*	37	*80*	28	*60*	23	*50*	18	*40*
Manganese (mg)	1.8	*4.0*	1.8	*4.0*	1.4	*3.0*	0.9	*2.0*	0.9	*2.0*	0.9	*2.*
Selenium (mg)	0.07	*0.15*	0.07	*0.15*	0.07	*0.15*	0.07	*0.15*	0.07	*0.15*	*0.07*	*0.*
Zinc (mg)	46	*100*	46	*100*	37	*80*	28	*60*	23	*50*	23	*50*
Fat-soluble vitamins:												
Vitamin A (IU)	998	*2,200*	998	*2,200*	794	*1,750*	590	*1,300*	590	*1,300*	590	*1,300*
or β-carotene (mg)	4.0	*8.8*	4.0	*8.8*	3.2	*7.0*	2.4	*5.2*	2.4	*5.2*	2.4	*5.*
Vitamin D (IU)	100	*220*	100	*220*	91	*200*	91	*200*	68	*150*	57	*125*
Vitamin E (IU)	5	*11*	5	*11*	5	*11*	5	*11*	5	*11*	5	*11*
Vitamin K (menadione) (mg)	0.9	*2.0*	0.9	*2.0*	0.9	*2.0*	0.9	*2.0*	0.9	*2.0*	0.9	*2.*
Water-soluble vitamins:												
Biotin[9] (mg)	0.05	*0.10*	0.05	*0.10*	0.05	*0.10*	0.05	*0.10*	0.05	*0.10*	0.05	*0.*
Choline[10] (mg)	499	*1,100*	499	*1,100*	408	*900*	317	*700*	249	*550*	181	*400*
Folacin[9] (mg)	0.27	*0.60*	0.27	*0.60*	0.27	*0.60*	0.27	*0.60*	0.27	*0.60*	0.27	*0.*
Niacin[11] (mg)	10	*22*	10	*22*	8.2	*18*	6.4	*14*	5.4	*12*	4.5	*10*
Pantothenic acid (mg)	6	*13*	6	*13*	5	*11*	5	*11*	5	*11*	5	*11*
Riboflavin (mg)	1.4	*3.0*	1.4	*3.0*	1.4	*3.0*	1.2	*2.6*	1.0	*2.2*	1.0	*2.*
Thiamin (mg)	0.6	*1.3*	0.6	*1.3*	0.5	*1.1*	0.5	*1.1*	0.5	*1.1*	0.5	*1.*
Vitamin B-6 (mg)	0.7	*1.5*	0.7	*1.5*	0.7	*1.5*	0.5	*1.1*	0.5	*1.1*	0.5	*1.*
Vitamin B-12 (μg)	10	*22*	10	*22*	7	*15*	5	*11*	5	*11*	5	*11*

[1]Adapted by the authors from *Nutrient Requirements of Swine,* No. 2, 8th rev. ed., NRC-National Academy of Sciences, 1979, p. 22.

[2]Requirements reflect the estimated levels of each nutrient needed for optimal performance when a fortified grain-soybean meal diet is fed, except that a substantial level of milk products should be included in the diet of the 2-11 lb *(1-5 kg)* pig. Concentrations are based upon amounts per unit of air-dry diet (i.e., 90% dry matter).

[3]These are not absolute requirements, but are suggested energy levels derived from diets containing corn and soybean meal (44% crude protein). When lower energy grains are fed, these energy levels will not be met; consequently, feed efficiency would be lowered.

[4]Approximate protein levels required to meet the need for indispensable amino acids when a fortified grain-soybean meal diet is fed to pigs weighing more than 11 lb *(5 kg).*

[5]Methionine can fulfill the total requirement; cystine can meet at least 50% of the total requirement.

[6]Phenylalanine can fulfill the total requirement; tyrosine can meet at least 50% of the total requirement.

[7]It is assumed that usable tryptophan content of corn does not exceed 0.05%.

[8]At least 30% of the phosphorus requirement should be provided by inorganic and/or animal product sources.

[9]These levels are suggested. No requirements have been established.

[10]In excess of its requirement for protein synthesis, methionine can spare dietary choline (4.3 mg methionine is equal in methylating capacity to 1 mg choline).

[11]It is assumed that most of the niacin present in cereal grains and their by-products is in bound form and thus unavailable to swine. The niacin contributed by these sources is not included in the requirement listed. In excess of its requirement for protein synthesis, tryptophan can be converted to niacin (50 mg tryptophan yields 1 mg niacin).

Symptoms of energy deficiency are: slow or interrupted growth, lowered reproduction, and offspring dead or weak at birth.

In addition to supplying energy, certain dietary fats supply essential fatty acids, most commonly linoleic acid. Practical swine rations contain adequate amounts of essential fatty acids. After the essential fatty acid requirement has been met, additions of fat increase the energy in the ration.

Protein

Proteins supply the building materials fro which body tissue and many body regulators, such enzymes and hormones, are made. Each protein made up of several nitrogen compounds called ami acids.

The pig has a specific requirement for each of t essential amino acids. (See Part I, "Protein," for a li

TABLE 4-70

NUTRIENT REQUIREMENTS OF BREEDING SWINE[1]

(Percentage or Amount per Pound and per Kilogram of Diet)

Nutrient	Bred Gilts and Sows; Young and Adult Boars[2]		Lactating Gilts and Sows	
Energy and protein:				
Digestible energy .... (kcal)	1,542	3,400	1,540	3,395
Metabolizable energy ..... (kcal)	1,451	3,200	1,449	3,195
Crude protein[3] .... (%)	12		13	
Amino acids:				
Arginine .... (%)	0		0.40	
Histidine .... (%)	0.15		0.25	
Isoleucine .... (%)	0.37		0.39	
Leucine .... (%)	0.42		0.70	
Lysine .... (%)	0.43		0.58	
Methionine + cystine[4] .... (%)	0.23		0.36	
Phenylalanine + tyrosine[5] ... (%)	0.52		0.85	
Threonine .... (%)	0.34		0.43	
Tryptophan[6] .... (%)	0.09		0.12	
Valine .... (%)	0.46		0.55	
Major or macro minerals:				
Calcium .... (%)	0.75		0.75	
Phosphorus[7] .... (%)	0.60		0.50	
Sodium .... (%)	0.15		0.20	
Chlorine .... (%)	0.25		0.30	
Magnesium .... (%)	0.04		0.04	
Potassium .... (%)	0.20		0.20	
	(per lb)	*(per kg)*	*(per lb)*	*(per kg)*
Trace or micro minerals:				
Copper .... (mg)	2.3	*5*	2.3	*5*
Iodine .... (mg)	0.06	*0.14*	0.06	*0.14*
Iron .... (mg)	36.3	*80*	36.3	*80*
Manganese .... (mg)	4.5	*10*	4.5	*10*
Selenium .... (mg)	0.07	*0.15*	0.07	*0.15*
Zinc .... (mg)	22.7	*50*	22.7	*50*
Fat-soluble vitamins:				
Vitamin A .... (IU)	1,814	*4,000*	907	*2,000*
or β-carotene .... (mg)	7.3	*16*	3.6	*8*
Vitamin D .... (IU)	90.7	*200*	90.7	*200*
Vitamin E .... (IU)	4.5	*10*	4.5	*10*
Vitamin K (menadione) ..... (mg)	0.9	*2*	0.9	*2*
Water-soluble vitamins:				
Biotin[8] .... (mg)	0.05	0.1	0.05	0.1
Choline .... (mg)	566.9	1,250	566.9	1,250
Folacin[8] .... (mg)	0.3	*0.6*	0.3	*0.6*
Niacin[9] .... (mg)	4.5	*10*	4.5	*10*
Pantothenic acid .... (mg)	5.4	*12*	5.4	*12*
Riboflavin .... (mg)	1.4	*3*	1.4	*3*
Thiamin .... (mg)	0.5	*1*	0.5	*1*
Vitamin B-6 .... (mg)	0.5	*1*	0.5	*1*
Vitamin B-12 .... (μg)	6.8	*15*	6.8	*15*

[1]Adapted by the authors from *Nutrient Requirements of Swine*, No. 2, 8th rev. ed., NRC-National Academy of Sciences, 1979, p. 24. Requirements reflect the estimated levels of each nutrient needed for optimal performance when a fortified grain-soybean meal diet is fed. Concentrations are based upon amounts per unit of air-dry diet (i.e., 90% dry matter).

[2]Requirements for boars of breeding age have not been established. It is suggested that the requirements will not differ significantly from that of bred gilts and sows.

[3]Approximate protein levels required to meet the need for indispensable amino acids when a fortified grain-soybean meal diet is fed. The true digestibilities of the amino acids were assumed to be 90%.

[4]Methionine can fulfill the total requirement, cystine can meet at least 50% of the total requirement.

[5]Phenylalanine can fulfill the total requirement, tyrosine can meet at least 50% of the total requirement.

[6]It is assumed that usable tryptophan content of corn does not exceed 0.05%.

[7]At least 30% of the phosphorus requirement should be provided by inorganic and/or animal product sources.

[8]These levels are suggested. No requirements have been established.

[9]It is assumed that most of the niacin present in cereal grains and their by-products is in bound form and thus unavailable to swine. The niacin contributed by these sources is not included in the requirement listed. In excess of its requirement for protein synthesis, tryptophan can be converted to niacin (50 mg tryptophan yields 1 mg niacin).

of amino acids.) Since they are needed for the formation of every new cell, the need is most critical when growth is rapid. This makes ration formulation for the young pig very important, because the protein provided at this time must supply the amino acids for muscle growth (lean meat), internal organs, blood, bone, and all other parts associated with growth and development. The usefulness of a protein source depends upon its amino acid composition, because the real need of the pig is for amino acids and not for protein as such.

Although it is common practice to refer to "percent protein" in a ration, this term has little significance in swine nutrition unless there is information about the amino acids present. For swine, quality is just as important as quantity. It is possible for pigs to perform better on a 12 percent protein ration, well-balanced for amino acids, than on a 16 percent protein ration having a poor amino acid balance.

From a practical standpoint, the problem of building a balanced ration for swine is centered around correcting the deficiencies of the cereal grains. Although corn, wheat, and barley may contain from 8 to 12 percent protein, their protein is seriously deficient

TABLE 4-71

RECOMMENDED NUTRIENT ALLOWANCE FOR SWINE[1]

(in Percentage or Amount per Pound or Kilogram of Ration, As-fed Basis)

Item	Animal Production Period															
	Growing-Finishing						Sow Reproduction				Boars					
	Starter (10-40 lb, or *4.5-18 kg*)		Grower (40-120 lb, or *18-55 kg*)		Finisher (120 lb, or *55 kg*, to mkt. wt.)		Gestation		Lactation		(40-120 lb, or *18-55 kg*)		(120-250 lb, or *55-114 kg*)		Adult	
	(lb)	(kg)	(lb)	(kg)	(lb)	(kg)	(lb)	(kg)	(lb)	(kg)	(lb)	(kg)	(lb)	(kg)	(lb)	(kg)
Expected performance (from beginning to end of period)																
Daily feed intake	0.6-2.0	*0.3-0.9*	2.0-6.0	*0.9-2.7*	6.0-9.0	*2.7-4.1*	4.0-4.5[2]	*1.8-2.0*	10.0-15.0	*4.5-6.8*	2.0-6.0	*0.9-2.7*	6.0-9.0	*2.7-4.1*	5.0-7.0[3]	*2.3-3.2*
Daily gain	0.5-1.2	*0.2-0.5*	1.2-1.6	*0.5-0.7*	1.6-2.0	*0.7-0.9*	0.7-0.9	*0.3-0.4*	—	—	1.2-1.7	*0.5-0.8*	1.7-2.4	*0.8-1.1*	—	—
	Nutrient Requirements per Pound or Kilogram of Ration[4]															
Protein:		*18*		*16*		*13*		*12*[5]		*16*		*18*		*16*		*16*
Amino acids:																
Arginine (%)		*0.37*		*0.25*		*0.15*		*0.25*		*0.34*						
Histidine (%)		*0.34*		*0.23*		*0.14*		*0.17*		*0.26*						
Isoleucine (%)		*0.76*		*0.52*		*0.31*		*0.37*		*0.67*						
Leucine (%)		*0.84*		*0.67*		*0.40*		*0.56*		*0.99*						
Lysine (%)		*0.90*		*0.76*		*0.57*		*0.42*		*0.76*						
Methionine & cystine[6] (%)		*0.56*		*0.35*		*0.20*		*0.20*		*0.32*						
Phenylalanine & tyrosine[7] (%)		*0.79*		*0.54*		*0.32*		*0.33*		*1.00*						
Threonine (%)		*0.66*		*0.45*		*0.27*		*0.34*		*0.51*						
Tryptophan (%)		*0.18*		*0.12*		*0.07*		*0.07*		*0.13*						
Valine (%)		*0.67*		*0.46*		*0.28*		*0.46*		*0.68*						
Energy:																
Digestible energy (kcal)	1,590	*3,505*	1,500	*3,307*	1,500	*3,307*	1,500	*3,307*	1,500	*3,307*	1,500	*3,307*	1,500	*3,307*	1,500	*3,307*
Metabolizable energy (kcal)	1,520	*3,351*	1,440	*3,175*	1,440	*3,175*	1,440	*3,175*	1,440	*3,175*	1,440	*3,175*	1,440	*3,175*	1,440	*3,175*
Major or macro minerals:																
Salt[8] (%)		*0.35*		*0.35*		*0.35*		*0.35*		*0.35*		*0.35*		*0.35*		*0.35*
Calcium[9] (%)		*0.80*		*0.65*		*0.60*		*0.80*		*0.80*		*0.90*		*0.75*		*0.80*
Phosphorus[9] (%)		*0.60*		*0.50*		*0.45*		*0.60*		*0.60*		*0.70*		*0.60*		*0.60*
Trace or micro minerals:																
Copper (ppm)		*6*		*6*		*6*		*6*		*6*		*6*		*6*		*6*
Iodine (ppm)		*0.2*		*0.2*		*0.2*		*0.3*		*0.2*		*0.2*		*0.2*		*0.2*
Iron (ppm)		*80*		*80*		*80*		*80*		*80*		*80*		*80*		*80*
Manganese (ppm)		*30*		*30*		*30*		*40*		*30*		*30*		*30*		*30*
Selenium (ppm)		*0.2*		*0.15*		*0.15*		*0.15*		*0.15*		*0.15*		*0.15*		*0.15*
Zinc (ppm)		*100*		*75*		*50*		*100*		*75*		*125*		*100*		*75*
Fat-soluble vitamins:																
Vitamin A (carotene) (IU)	1,500	*3,300*	1,000	*2,200*	1,000	*2,200*	2,000	*4,400*	1,500	*3,300*	2,000	*4,400*	2,000	*4,400*	2,000	*4,400*
Vitamin D (IU)	300	*660*	150	*330*	100	*220*	150	*330*	150	*330*	—	–	—	–	150	*330*
Vitamin E (IU)	10	*22*	10	*22*	10	*22*	10	*22*	10	*22*	10	*22*	10	*22*	10	*22*
Vitamin K (mg)	4	*9*	4	*9*	4	*9*	4	*9*	4	*9*	4	*9*	4	*9*	4	*9*
Water-soluble vitamins:																
Vitamin B_{12} (mcg)	9	*20*	6	*13*	4	*9*	7	*15*	5	*11*	7	*15*	7	*15*	5	*11*
Choline (mg)	600	*1,320*	400	*880*	350	*770*	660	*1,320*	400	*880*	400	*880*	400	*880*	400	*880*
Niacin (nicotinic acid) (mg)	12	*26*	8	*18*	6	*13*	10	*22*	8	*18*	10	*22*	10	*22*	8	*18*
Pantothenic acid (mg)	6	*13*	5	*11*	5	*11*	7.5	*17*	6	*13*	7.5	*17*	7.5	*17*	6	*13*
Riboflavin (B_2) (mg)	1.5	*3.3*	1.2	*2.6*	1	*2.2*	2	*4.4*	2	*4.4*	2	*4.4*	2.2	*4.4*	2	*4.4*

[1]These recommended nutrient allowances, which were prepared by The Ohio State University, are rather typical of Corn Belt swine formulations.

[2]The quantity will vary depending on environmental conditions. Animals should be fed the quantity of diet to achieve 90 lb gain in gilts and 70 lb gain in sows during the period from day of mating to immediately postfarrowing (this would correspond to about 0.90 and 0.79 lb of gain per day, respectively, during gestation).

[3]The adult boar should be fed an adequate quantity to keep in good breeding condition, but not overly fat.

[4]The nutrient requirements for the boar have largely been estimated.

[5]This value is conservative. Recent research suggests that this value can safely be lowered to 9% crude protein, but further experimentation is needed.

[6]Cystine can fulfill up to 50% of the suggested allowance.

[7]Tyrosine can fulfill up to 30% of the suggested allowance.

[8]Trace mineral salt is recommended to be added to the diet at 0.50%.

[9]The values are total quantites (not available). Recent research suggests that the boar may need the value indicated for maximum bone development.

in the essential amino acid, lysine; corn is also deficient in tryptophan. Moreover, the digestive tract of the pig is not adapted to extensive synthesis of proteins by microorganisms like the paunch of ruminants.

Previously, it was stated that when an excess of energy is consumed by the pig it is stored in the form of fat. Protein is not stored in the body in appreciable amounts. If an excess of protein is fed, the unused nitrogen portion is discarded as urea in the urine and the carbon fraction is used as a source of energy. From an economic standpoint, it is unprofitable to feed more protein than needed to meet the nutritional requirements of the pig.

Symptoms of protein (amino acid) deficiency are: reduced feed intake, stunted growth, poor hair and skin condition, and lowered reproduction.

Minerals

Of all common farm animals, the pig is most likely to suffer from mineral deficiencies. This is due to the following peculiarities of swine husbandry.

1. Hogs are fed principally upon cereal grains and their by-products, all of which are relatively low in mineral matter, particularly in calcium.

2. The skeleton of the pig supports greater weight in proportion to its size than that of any other farm animal.

3. Hogs do not normally consume great amounts of roughage (pasturage or dry forage), which would tend to balance the mineral deficiencies of grains.

4. Hogs are fed to grow at a maximum rate for an early market, before they are mature.

5. Hogs reproduce when less mature than other classes of livestock.

Salt (sodium and chlorine), calcium, and phosphorus are the supplemental minerals needed in largest quantities by the pig. Other minerals are required in small amounts and are known as "trace minerals." The latter include copper, iodine, iron, manganese, selenium, and zinc. Although minerals constitute a small percentage of the swine ration, their importance to the health and well-being of the pig cannot be minimized.

The mineral requirements of swine are presented in Tables 4-67 to 4-70. Because of the importance of trace minerals (micro minerals) in swine nutrition, the following special mineral tables are presented herein:

Table 4-72, Micro mineral Requirements of Swine
Table 4-73, Toxic Levels and Symptoms of Excesses of Certain Minerals

TABLE 4-72

MICRO MINERAL REQUIREMENTS OF SWINE
(Amount per Pound or Kilogram of Ration, As-fed Basis)[1]

Mineral Element	Requirement in Ration	
	(mg/lb)	*(mg/kg)*
Copper	2.7	6[2]
Iodine	0.09	0.2
Iron	36	80[2]
Manganese	9	20
Selenium	0.04	0.1
Zinc	23	50[3]

[1]Adapted by the author from *Nutrient Requirements of Swine*, No. 2, 7th rev. ed., NRC-National Academy of Sciences, 1973, p. 33.
[2]Baby pig requirement.
[3]Higher levels may be needed if excessive calcium is fed.

TABLE 4-73

TOXIC LEVELS AND SYMPTOMS OF EXCESSES OF CERTAIN MINERALS[1]

Element	Toxic Level	Age	Symptoms
Salt (NaCl)	6-8% (if limited water is available)	All ages	Nervousness, staggering, weakness, paralysis, and death.
Copper (Cu)	136-227 mg/lb *(300-500 mg/kg)*, in absence of higher levels of dietary iron and zinc.[2]	Immature	Reduced growth, lower hemoglobin, icterus, and death.[3]
Iodine (I)	364 mg/lb *(800 mg/kg)*	Immature	Depressed feed intake and rate of gain; lowered hemoglobin and eye lesions.
Iron (Fe)	2,273 mg/lb *(5,000 mg/kg)*	Immature	Depressed feed intake and rate of gain; reduced serum inorganic phosphorus and femur ash; rickets.
Manganese (Mn)	1,818 mg/lb *(4,000 mg/kg)*	Immature	Depressed feed intake; reduced growth rate; stiffness and stilted gait.
Selenium (Se)	2.3-3.6 mg/lb *(5-8 mg/kg)*	Immature	Hoofs separate from coronary bands; emaciation; loss of hair; cirrhosis and atrophy of liver.
	4.5 mg/lb *(10 mg/kg)*	Breeding (sows)	Reduced conception; pigs small, weak, or dead at birth.
Zinc (Zn)	909 mg/lb *(2,000 mg/kg)*	Immature	Reduced performance; arthritis; extensive hemorrhage and gastritis.

[1]Adapted by the author from *Nutrient Requirements of Swine*, No. 2, 7th rev. ed., NRC-National Academy of Sciences, 1973, p. 35.
[2]In a few instances, a dietary level of 114 mg/lb (250 mg/kg) has resulted in symptoms of excess.
[3]In some instances, 227 mg/lb (500 mg/kg) of copper has been fed without icterus or death occurring.

SWINE MINERAL CHART

Table 4-74, Swine Mineral Chart, gives, in summary form, the following pertinent information relative to each mineral listed: (1) conditions usually prevailing where deficiencies are reported, (2) functio (3) deficiency symptoms, (4) nutrient requirement (5) recommended allowances, and (6) practica sources.

TABLE 4-74—SWIN

Minerals Which May Be Deficient Under Normal Conditions	Conditions Usually Prevailing Where Deficiencies Are Reported	Function of Mineral	Some Deficiency Symptoms
MAJOR OR MACRO MINERALS:			
Salt (sodium and chlorine—NaCl)	Salt deficiencies may exist when the protein supplement is all or chiefly of plant origin, although herbivorous animals require more salt than swine.	Improves appetite, promotes growth, helps regulate body pH, and is essential for hydrochloric acid formation in the stomach.	Poor and depraved appetite, unthrif condition, and failure to grow.
Calcium (Ca)	When the protein supplements are chiefly of plant origin and little forage is used. When swine are raised in confinement without vitamin D added to the ration. When feed intake is restricted during gestation. When there is poor calcium-phosphorus ratio. Retention of calcium is affected by source of dietary protein (or phytic acid content) and the level of magnesium.	Bone and teeth formation; nerve function; muscle contraction; blood coagulation; cell permeability. Essential for milk production.	Loss of appetite and poor growt lack of thrift, lameness and sti ness, weakened bone structur and impaired reproduction. Sever cases may show reduced seru calcium and tetany. Rickets ma develop in young pigs, or os teomalacia in older animals.
Phosphorus (P)	Rations containing only plant ingredients; late gestation; lactation; high calcium rations; swine in confinement without vitamin D added to the ration; poor calcium to phosphorus ratio. Retention of phosphorus is affected by source of dietary protein (or phytic acid content) and the level of magnesium.	Bone and teeth formation; a component of phospholipids which are important in lipid transport and metabolism and cell-membrane structure. In energy metabolism. A component of RNA and DNA, the vital cellular constituents required for protein synthesis. A constituent of several enzyme systems.	Loss of appetite and poor growt lameness and stiffness, weakene bone structure, reduced inorgani blood phosphorus, depraved appe tite, breeding difficulties, and rick ets in young pigs, or osteomalaci in older animals.
Magnesium (Mg)	Essential in many enzyme systems.	Essential for normal skeletal development, as a constituent of bone; also essential as an enzyme activator, primarily in glycolytic system.	Hyperirritability, muscular twitching reluctance to stand, stepping syn drome, weak pasterns, loss o equilibrium, and tetany, followe by death.
Potassium (K)		Major cation of intracellular fluid where it is involved is osmotic pressure and acid-base balance. Muscle activity. Required in enzyme reaction involving phosphorylation of creatine. Influences carbohydrate metabolism.	Loss of appetite, slow growth, poo hair and skin condition, decrease feed efficiency, and cardiac im pairment.

Footnotes on last page of table.

INERAL CHART

Nutrient Requirements[1]				
Daily Nutrients/ Animal	Percentage of Rations	Recommended Allowances[1]	Practical Sources of the Mineral	Comments
Variable, according to class, age, and weight of swine (see Tables 4-67 and 4-68).	Variable, according to class, age, and weight of swine (see Tables 4-69 and 4-70). *Pigs weighing 28.7 to 77.2 lb *(13-35 kg)* require 0.08 to 0.10% sodium and 0.12 to 0.13% chlorine in the as-fed ration.	*0.25 to 0.5% in the as-fed ration, or give hogs free access to salt alone or in a mineral mixture.	Salt in loose form.	In iodine-deficient areas, stabilized iodized salt should be used. When pigs are salt starved, precaution should be taken to prevent overeating of it. The salt-poisoning syndrome associated with feeding brine or salted fish meal to swine can be produced by adding 6-8% salt (on a dry matter basis) to the regular diet of pigs and giving a limited amount of water.
Variable, according to class, age, and weight of swine (see Tables 4-67 and 4-68).	Variable, according to class, age, and weight of swine (see Tables 4-69 and 4-70).	Self-feed suitable mineral, or add Ca to the as-fed ration as required to bring level of total ration slightly above requirements. *0.75% Ca in the as-fed ration is adequate for both male and female breeding swine.	Ground limestone, or oyster-shell flour. Where both Ca and P are needed, use monocalcium phosphate, dicalcium phosphate, tricalcium phosphate, defluorinated phosphate, or bone meal.	Because cereal grains (which largely form the diet of swine) are low in Ca, swine are more apt to suffer from Ca deficiencies than from any of the other minerals except salt. *Most favorable Ca:P ratio is between 1:1 and 1.5:1. Sow's milk contains a Ca:P ratio of 1.3:1.
Variable, according to class, age, and weight of swine (see Tables 4-67 and 4-68).	Variable, according to class, age, and weight of swine (see Tables 4-69 and 4-70).	Self-feed suitable mineral, or add P to the ration as required to bring level of total ration slightly above requirements.	Monosodium phosphate, disodium phosphate, sodium tripolyphosphate, ammonium phosphate solution, or feed-grade phosphoric acid. Where both Ca and P are needed, use monocalcium phosphate, dicalcium phosphate, tricalcium phosphate, defluorinated phosphate, or bone meal.	One-half to ⅔ of P in grains is in phytate form, of which 20 to 50% is not available to swine; although fairly good utilization of phytate P is achieved through action of enzyme phytase(s) in the intestine. *Most favorable Ca:P ratio is between 1:1 and 1.5:1. Sow's milk contains a Ca:P ratio of 1.3:1.
Exact requirement is not known.		*181.8 mg/lb *(400 mg/kg)* as-fed ration.	Magnesium oxide or magnesium sulfate.	Practical rations adequate in magnesium.
*Between 2.5 and 5.0 daily for 100-lb *(45-kg)* pig.	*0.26 as-fed ration for 10-lb *(4.5-kg)* pig. *0.23 to 0.28% as-fed ration for 35-lb *(16-kg)* pig.		Corn contains 0.27% potassium, and other cereals contain 0.42% to 0.49% potassium.	Deficiency of potassium not observed in practical rations.

(Continued)

Minerals Which May Be Deficient Under Normal Conditions	Conditions Usually Prevailing Where Deficiencies Are Reported	Function of Mineral	Some Deficiency Symptoms
TRACE OR MICRO MINERALS:			
Cobalt (Co)	If vitamin B_{12} is limited.	An essential component of vitamin B_{12}.	
Copper (Cu)	Suckling pigs kept off soil.	Essential element in a number of enzyme systems and necessary for synthesizing hemoglobin and preventing nutritional anemia.	Slow growth, poor hair and skin condition, lameness and stiffness, weakened bone structure, weak and crooked legs, and anemia.
Iodine (I)	Iodine-deficient areas or soils (in northwestern U.S. and in the Great Lakes region) when iodized salt is not fed. Where feeds come from iodine-deficient areas.	Needed by the thyroid gland for making thyroxin, an iodine-containing hormone which controls the rate of body metabolism or heat production.	Loss of appetite, slow growth, poor hair and skin condition, impaired breeding or gestation, offspring dead or weak at birth, pigs hairless at birth, and/or goiter.
Iron (Fe)	Suckling pigs kept off soil.	Necessary for formation of hemoglobin, an iron-containing compound which enables the blood to carry oxygen. Iron is also important to certain enzyme systems.	Loss of appetite, slow growth, poor hair and skin condition, high mortality in young pigs, susceptibility to disease, thumps, and anemia. The number of grams of hemoglobin per 100 ml of blood is a rapid, reliable indicator of the iron status of the pig.
Manganese (Mn)		Necessary for growth, bone structure, and reproduction.	Lameness or stiffness, weakened bone structure, impaired reproduction, pigs dead or weak at birth, reduced skeletal growth, increased backfat, and irregular estrus.
Selenium (Se)	A selenium-deficient diet.	Not completely known. But involved in vitamin E absorption and/or retention. Also, a required nutrient in its own right.	Loss of appetite, slow growth, marked necrosis of the liver, a yellowish-brown discoloration of body fat, and sudden death.
Zinc (Zn)	High levels of calcium in relation to zinc levels.	Zinc is a component of several enzyme systems, including peptidases and carbonic anhydrase. Also, zinc is required for normal protein synthesis and metabolism and is a component of insulin.	Parakeratosis or swine dermatitis, pigs have a mangy look, reduced appetite, unthriftiness, poor growth rate, and diarrhea, and there may be vomiting. It affects swine of all ages.

[1]As used herein, the distinction between "nutrient requirements" and "recommended allowances" is as follows: In nutrient requirements, no margins of safety are included intentionally; whereas in nutrient allowances, margins of safety are provided in order to compensate for variations in feed composition, environment, and possible losses during storage or processing.

Where preceded by an asterisk, the nutrient requirements, recommended allowances, and other facts presented herein were taken from *Nutrient Requirements of Swine*, No. 2, 7th rev. ed., NRC-National Academy of Sciences, Washington, D.C., 1973.

—(Continued)

Nutrient Requirements[1]				
Daily Nutrients/ Animal	Percentage of Rations	Recommended Allowances[1]	Practical Sources of the Mineral	Comments
No requirements for cobalt have been established.		*Cobalt levels of about 0.045 mg/lb *(0.1 mg/kg)* are often added to swine feeds (as-fed basis).	Cobalt chloride, cobalt sulfate, cobalt oxide, or cobalt carbonate. Also, several good commercial minerals are on the market.	
0.045 to 0.068 mg/lb *(0.1-0.15 mg/kg)* body weight.	*2.7 mg/lb *(6 mg/kg)* as-fed ration for baby pigs.		Copper sulfate, copper carbonate, and copper oxide are about equally effective.	Beyond the suckling period, natural feedstuffs usually contain enough copper. *Apart from the role of copper as an essential trace element, much higher levels (56.8-113.6 mg/lb, or *125-250 mg/kg*) in the diet have been shown to support increased rate and efficiency of gains of pigs to breeding age.
For pregnant sows: 2.0 mcg/lb *(4.4 mcg/kg)* body wt. daily, and somewhat less for growing swine.	*0.09 mg/lb *(0.2 mg/kg)* as-fed ration.	*Use stabilized, iodized salt containing 0.007% iodine incorporated at 0.5% of grain ration or fed free-choice. 0.09 mg/lb *(0.2 mg/kg)* of as-fed ration.	Stabilized iodized salt containing 0.007% iodine.	
Newborn pigs require 7 mg of absorbed iron daily for normal growth.	*36.36 mg/lb *(80 mg/kg)* of as-fed ration for baby pigs.	*36.4 mg/lb *(80 mg/kg)* as-fed ration for baby pigs. Suitable iron preparations, injected at levels of 150 to 200 mg into baby pigs at 1 to 3 days of age, will prevent anemia due to iron deficiency.	Ferrous sulfate or ferric ammonium citrate. For the prevention or treatment of anemia in young pigs, either (1) place a little uncontaminated sod (topsoil, from an area where hogs have not run for years) in the corner of the pen daily, (2) inject a suitable iron preparation at a level of 150 to 200 mg into baby pigs at 1 to 3 days of age, (3) swab the sow's udder with iron solution, (4) give an iron-copper pill, or (5) allow access to oral iron preparations. In addition, the pigs should be encouraged to eat a grain ration as soon as they are old enough.	Newborn pigs contain an average of 47 mg of iron. Iron has a detoxifying effect when added to gossypol-containing diets. Add iron from soluble source to free gossypol at a weight ratio of 1:1. Milk is deficient in iron (sow's milk contains an average of 1 mg of iron/liter) and copper. Pigs should be encouraged to eat grain ration as soon as old enough. Iron levels of 2,273 mg/lb *(5,000 mg/kg)* of diet are considered toxic.
Minimum requirements for manganese not well defined.		*9.1 mg/lb *(20 mg/kg)* as-fed ration.	Manganous oxide.	
	*0.045 mg/lb *(0.10 mg/kg)* of as-fed ration.	Selenium in either sodium selenite or sodium selenate at rate of 0.1 ppm of complete as-fed ration. *Injection of 5 mg sodium selenite or barium selenate every 28 days will prevent selenium deficiency.	Sodium selenite or sodium selenate.	Selenium is related to vitamin E absorption. *Caution:* Toxic level of selenium is in range of 2.27-3.63 mg/lb *(5-8 mg/kg)* selenium in the feed. The U.S. Food and Drug Administration approved the addition of selenium to swine diets in 1974.
	*22.7 mg/lb *(50 mg/kg)* of as-fed ration containing soybean protein. *When calcium level of ration is 1½ to 2%, double the zinc allowance.		Zinc carbonate or zinc sulfate.	It has been shown that parakeratosis is caused by zinc and calcium forming an unavailable complex. Zinc toxicosis has been produced by zinc or zinc carbonate at level of 909 mg/lb *(2,000 mg/kg)* of corn-soybean meal diet.

Note: *Mineral recommendations for all classes and ages of swine, especially those fed unmixed rations or on pasture, are—*

1. *Where animals are on liberal grain feeding*—Provide free access to a 2-compartment mineral box, with (a) trace mineralized salt in one side, and (b) in the other side, a mixture of ⅓ trace mineralized salt (salt included for purposes of palatability), ⅓ defluorinated phosphate or steamed bone meal, and ⅓ ground limestone or oystershell flour.
2. *Where animals are primarily on roughage (pasture, hay, and/or silage)*—Provide free access to a 2-compartment mineral box, with (a) trace mineralized salt in one side (salt included for purposes of palatability), and (b) in the other side, a mixture of ⅓ trace mineralized salt and ⅔ defluorinated phosphate or steamed bone meal.

Vitamins

Vitamins are complex organic compounds needed in minute amounts, which are essential for health and normal body functions. Like amino acids, each vitamin has a specific function to perform. Vitamins are classified into two groups—fat-soluble, and water-soluble. The primary fat-soluble vitamins of practical concern are vitamins A, D, and E. Vitamin K may be of concern under some circumstances. Riboflavin, niacin, pantothenic acid, and vitamin B_{12} are the water-soluble vitamins most likely to be deficient in swine rations. Choline, which is usually listed with the water-soluble vitamins, is also frequently added to practical swine rations. The body can store reserves of the fat-soluble vitamins for a considerable period of time. But stores of the water-soluble vitamins are depleted quite rapidly.

Because of the greater prevalence of confinement feeding, swine are more likely to suffer from vitamin deficiencies than any other class of four-footed animals.

TABLE 4-75—SWINE

Vitamins Which May Be Deficient Under Normal Conditions	Conditions Usually Prevailing Where Deficiencies Are Reported	Function of Vitamin	Some Deficiency Symptoms
FAT-SOLUBLE VITAMINS:			
A	Absence of green forages, either pasture or green hay—especially under drylot conditions. Where the ration consists chiefly of white corn, milo, barley, wheat, oats, or rye; or by-products of these grains; or yellow corn that has been stored in excess of a year.	Essential for normal maintenance and functioning of the epithelial tissues, particularly of the eye and the respiratory, digestive, reproductive, nerve, and urinary systems.	Night and day blindness, very irritable, poor appetite and slow growth lameness and impaired reproduction with dead or weak offspring Low resistance to respiratory infections.
D	Limited sunlight and/or limited quantities of sun-cured hay in drylot rations.	Aids in assimilation and utilization of calcium and phosphorus, and necessary in the normal bone development of animals—including the bones of the fetus.	Rickets in young pigs, or osteomalacia in mature hogs. Both conditions result in large joints and weak bones.
E (tocopherol)	Diets containing excessive amounts of highly unsaturated fatty acids or oxidized fats.	Antioxidant. Muscle structure. Reproduction.	Loss of appetite and slow growth. Increased embryonic mortality and muscular incoordination in suckling pigs from sows fed vitamin E deficient diets during gestation and lactation.
K	Moldy feed.	Essential for prothrombin formation and blood clotting.	Bleeding condition in young pigs, which responds to injection or oral administration of vitamin K.

Footnote on last page of table.

SWINE VITAMIN CHART

Table 4-75, Swine Vitamin Chart, gives, in summary form, the following pertinent information relative to each vitamin listed: (1) conditions usually prevailing where deficiencies are reported, (2) function, (3) deficiency symptoms, (4) nutrient requirements, (5) recommended allowances, and (6) practical sources.

VITAMIN CHART

Nutrient Requirements[1]				
Daily Nutrients/ Animal	Percentage of Rations	Recommended Allowances	Practical Sources of the Vitamin	Comments
Variable, according to class, age, and weight of swine (see Tables 4-67 and 4-68).	Variable, according to class, age, and weight of swine (see Tables 4-69 and 4-70).	Add vitamin A to the ration to bring level of total ration slightly above requirements.	Stabilized vitamin A.	One mg of beta-carotene from natural feedstuffs is equal to approximately 500 IU of vitamin A activity for swine. Meals from artificially dehydrated forages are much higher in carotene than sun-cured products. Taken together, liver storage levels of plasma vitamin A and pressure of cerebrospinal fluid give reliable estimates of the vitamin A status of the pig.
Variable, according to class, age, and weight of swine (see Tables 4-67 and 4-68).	Variable, according to class, age, and weight of swine (see Tables 4-69 and 4-70).	Add vitamin D to the ration to bring level of total ration slightly above requirements.	Vitamin D_2 (irradiated ergosterol) and vitamin D_3 (irradiated 7-dehydrocholesterol) are similar in biological activity for swine. Exposure to sunlight. Sun-cured hay (10% alfalfa in the total ration will normally supply sufficient vitamin D).	Grains, grain by-products, and high-protein feedstuffs are practically devoid of vitamin D; therefore, unless swine are exposed daily to the ultraviolet rays of the sun, the diet should be fortified with vitamin D. When animals are exposed to direct sun-light, the ultraviolet light produces vitamin D from traces of cholesterol in the skin. The vitamin D requirement is less when a proper balance of calcium and phosphorus exists in the ration. One international unit vitamin D is defined as the biological activity of 0.025 micrograms of crystalline vitamin D_3.
Unknown		*5 IU/lb *(11 IU/kg)* of diet.	High-quality green feeds, whole cereal grains, and the germ of cereal grains.	Tocopherols differ in their biological activity, with *d*-alpha-tocopherol being the most active. One IU of vitamin E is the equivalent in biopotency of 1 mg *dl*-alpha-tocopherol acetate.
		*If there is evidence of a vitamin K deficiency, supplement the as-fed ration with menadione at levels of 1.0 mg/lb *(2.2 mg/kg)*.	Under practical conditions, the vitamin K requirement is met by vitamin K in feedstuffs and by intestinal synthesis.	

(Continued)

TABLE 4-75

Vitamins Which May Be Deficient Under Normal Conditions	Conditions Usually Prevailing Where Deficiencies Are Reported	Function of Vitamin	Some Deficiency Symptoms
WATER-SOLUBLE VITAMINS:			
B_{12}		Numerous metabolic functions, and essential for normal growth and reproduction in swine.	Poor growth, lowered reproduction and anemia.
Biotin	When pigs are fed (1) dried, raw egg white, or (2) sulfathalidine. When young pigs are fed a diet devoid of biotin.		Alopecia, spasticity of the hind legs, cracks in the feet, and a dermatosis.
Choline	Baby pigs fed a synthetic milk diet containing not more than 0.8% methionine.	Involved in nerve impulses. A component of phospholipids. Donor of methyl groups.	Unthriftiness, lack of coordination, spraddled hind legs at birth, fatty infiltration of liver, poor reproduction, poor lactation, and decreased survival of the young.
Folacin (folic acid)		Related to B_{12} metabolism. Metabolic reactions involving incorporation of single carbon units into larger molecules.	Poor growth. Macrocytic anemia.
Niacin (nicotinic acid)		Required by all living cells, and an essential component of important metabolic enzyme systems involved in glycolysis and tissue respiration.	Loss of appetite and decreased gain, followed by diarrhea, occasional vomiting, dermatitis, and loss of hair.
Pantothenic acid	Long period of inadequate pantothenic acid intake.	Part of coenzyme A, a necessary factor for intermediary metabolism.	A goose-stepping gait, loss of appetite, poor growth, diarrhea, reduced fertility, and breeding failure.
Pyridoxine (B_6)		As coenzyme in protein and nitrogen metabolism. Involved in red blood cell formation. Important in endocrine systems.	Loss of appetite and poor growth, unsteady gait, anemia, and epileptic-like fits (convulsions).
Riboflavin (B_2)		A component of enzyme systems essential to normal metabolic processes.	Loss of appetite, poor growth, rough hair coat, diarrhea, reproductive failure in the sow, pigs dead or weak at birth, and crooked legs and incoordination.

Footnote on last page of table.

—(Continued)

Nutrient Requirements[1] Daily Nutrients/ Animal	Percentage of Rations	Recommended Allowances	Practical Sources of the Vitamin	Comments
Variable, according to class, age, and weight of swine (see Tables 4-67 and 4-68).	Variable, according to class, age, and weight of swine (see Tables 4-69 and 4-70).		Synthetic B_{12}. Protein supplements of animal origin. Fermentation products.	Vitamin B_{12} is apt to be lacking in swine rations. Synthesis of vitamin B_{12} intestinal flora may supplement dietary sources. B_{12} contains the trace element cobalt; hence, the synthesis of B_{12} in the intestines is dependent on the presence of cobalt in the feed. This may be the major, if not the only, function of cobalt as an essential nutrient.
				Very young pigs (under about 3 weeks of age) do not produce enough biotin until they develop an intestinal flora capable of synthesizing it. The protein avidin in raw egg white makes biotin unavailable to pigs. Heat treatment inactivates avidin and makes egg white safe for feeding to pigs.
Variable, according to class, age, and weight of swine (see Tables 4-67 and 4-68). Choline at level of 9.09 mg/lb *(20 mg/kg)* body weight has prevented symptoms of deficiency in sows.	Variable, according to class, age, and weight of swine (see Tables 4-69 and 4-70). Choline at level of 0.1% of the as-fed ration has prevented symptoms of deficiency in baby pigs.		Choline chlorides or choline dihydrogen. Choline content of normal feed is sufficient.	Choline content of normal feeds is usually sufficient.
Requirements have not been determined.			Practical swine rations are believed to be adequate in folacin. Synthetic folacin.	
Variable, according to class, age, and weight of swine (see Tables 4-67 and 4-68).	Variable, according to class, age, and weight of swine (see Tables 4-69 and 4-70).			Niacin occurs in corn, wheat, and milo in bound form; hence, it may be unavailable to the pig. Also, the tryptophan level affects the niacin requirement, because of the conversion of tryptophan to niacin.
Variable, according to class, age, and weight of swine (see Tables 4-67 and 4-68).	Variable, according to class, age, and weight of swine (see Tables 4-69 and 4-70).		Calcium pantothenate (only the D isomer has vitamin activity). Fish solubles.	Widely distributed and occurs in practically all feedstuffs. However, the quantity present may not always be sufficient to meet the needs of the pig.
Variable, according to class, age, and weight of swine (see Tables 4-67 and 4-68).	Variable, according to class, age, and weight of swine (see Tables 4-69 and 4-70).		Vitamin B_6. Cereal grains and their by-products. Rice bran and polished rice. Green pastures. Well-cured alfalfa hay. Yeast.	Pyridoxine content of normal feeds is usually sufficient.
Variable, according to class, age, and weight of swine (see Tables 4-67 and 4-68).	Variable, according to class, age, and weight of swine (see Tables 4-69 and 4-70).		Synthetic riboflavin. Green pastures. Milk and milk products. Meat scraps and fish meal.	Riboflavin is apt to be lacking in swine rations.

(Continued)

TABLE 4-75

Vitamins Which May Be Deficient Under Normal Conditions	Conditions Usually Prevailing Where Deficiencies Are Reported	Function of Vitamin	Some Deficiency Symptoms
Thiamin (B_1)		As a coenzyme in energy metabolism. Promotes appetite and growth, required for normal carbohydrate metabolism, and aids reproduction.	Loss of appetite and poor growth, diarrhea, dead or weak offspring, slow pulse, low body temperature, and flabby heart.
Vitamin C (ascorbic acid)			
Either there is sufficient intestinal synthesis or the pig does not need the following B vitamins: inositol and para-aminobenzoic acid.			
Unidentified factors, both organic and inorganic.		Contribute factor or factors, or correct imbalances.	

[1]As used herein, the distinction between "nutrient requirements" and "recommended allowances" is as follows: In nutrient requirements, no margins of safety are included intentionally; whereas in nutrient allowances, margins of safety are provided in order to compensate for variations in feed composition, environment, and possible losses during storage or processing.

Water

Water is so common that it is seldom thought of as a nutrient. However, it is the largest single part of nearly all living things. The body of a baby pig is about three-fourths water. Also, it is noteworthy that swine can live longer without feed than without water.

The daily water requirements of hogs vary from ½ gallon to 1½ gallons per 100 pounds liveweight. The higher requirements are for young pigs and lactating sows. Also, the higher the temperature, the greater the water consumption. It is preferable that swine have access to automatic waterers, with water available at all times. Otherwise they should be hand watered at least twice daily. During winter, the drinking water should not be permitted to fall below 40°F.

FEEDS FOR SWINE

Throughout the world, swine are raised on a great variety of feeds, including numerous by-products. Except when on pasture or when ground dry forages are incorporated in the ration, they eat relatively little roughage; only 15 percent of the total feed consumed by swine in the United States is derived from roughages.

Although corn is the chief concentrate fed to swine, the agriculture of the 50 states is very diverse, and the diet of the pig is readily adapted to the feeds produced locally. A similar adaptation in feeding practices is found in other countries. Thus, in most sections of the world, swine are fed predominantly on homegrown feeds. Ireland depends largely upon potatoes and dairy by-products; the swine industry of Denmark has been built up to augment the dairy industry, with milk and whey supplementing homegrown and imported cereals (mostly barley); in Germany, the pig is fed on such crops as potatoes, sugar beets, and green forage; and in other countries, the pig is often a scavenger, competing very little for grains suitable for human consumption.

Concentrates

Because of their simple monogastric stomach, swine consume more concentrates and less roughages than any other class of large farm animals. This characteristic gives pigs limited opportunity to consume large quantities of calcium and of vitamin-rich and better quality protein roughages, with the result that they suffer from more nutritional deficiencies than any other species except poultry.

—(Continued)

Nutrient Requirements[1]				
Daily Nutrients/ Animal	Percentage of Rations	Recommended Allowances	Practical Sources of the Vitamin	Comments
Variable, according to class, age, and weight of swine (see Tables 4-67 and 4-68).	Variable, according to class, age, and weight of swine (see Tables 4-69 and 4-70).		Thiamin hydrochloride. Green pastures. Well cured, green leafy hays. Cereal grains. Peas. Brewers' yeast.	Thiamin content of normal feeds is usually sufficient.
			Crystalline ascorbic acid.	Normally, pigs are able to synthesize vitamin C in amounts sufficient to meet their requirements. However, there is limited evidence that dietary ascorbic acid is beneficial under some conditions.
			Distillers' dried solubles, fish solubles, dried whey, grass juice concentrate, green pasture, high-quality grass silage, soil, alfalfa meal, brewers' dried yeast, and liver.	Unidentified factors contribute other than known nutrients of benefit to growing pigs and gestating-lactating sows.

Where preceded by an asterisk, the nutrient requirements, recommended allowances, and other facts presented herein were taken from *Nutrient Requirements of Swine*, No. 2, 7th rev. ed., NRC-National Academy of Sciences, Washington, D.C., 1973.

Although most concentrate feeds are not suitable as the sole ration for hogs, it must be realized that swine can utilize a larger variety of feeds to greater advantage than other farm animals. In general, the grain crops—corn, barley, wheat, oats, rye, and the sorghums—constitute the major component of the swine ration. However, sweet potatoes and peanuts are successfully and extensively used in the South, soybeans in the central states, and peas in the Northwest. In those districts where they are grown, potatoes (cull) also are utilized in considerable quantities in feeding hogs. In addition, in almost every section of the country one or more by-product feeds are fed to hogs—including the by-products of the fishing, meat-packing, milling, and dairy industries. Human food wastes, such as refuse or garbage, are also fed extensively.

The protein and vitamin requirements of the monogastric pig differ very greatly from those of the ruminant, for the latter improves the quality of proteins and creates certain vitamins through bacterial synthesis.

Despite all this, it is possible to meet the nutritive needs of the pig on concentrated feeds by keeping in mind the following facts when balancing the ration:

1. The cereal grains and their by-products are relatively good in phosphorus, but low in calcium and the other minerals.
2. Except for the carotene content of yellow corn and green peas, the grains are very poor sources of the vitamins.
3. Most cereal grains supply proteins of poor quality.
4. Protein supplements of animal origin and soybean meal generally supply proteins of high quality, whereas proteins of plant origin, other than soybean meal, generally supply proteins of low quality.
5. Because of the inadequacies of most concentrates, it is usually necessary to rely on fortifications with minerals and vitamins.

Dry Roughages

The favorable results obtained from feeding a corn-soybean meal ration, properly fortified with minerals and vitamins, in an era of relatively cheap grain, caused many researchers and swine producers to question the wisdom of using alfalfa meal in the ration. There was never any doubt about alfalfa meal being an excellent source of quality protein, carotene (vitamin A), B vitamins (riboflavin, pantothenic acid, and niacin), vitamin D if sun-cured, calcium, and unidentified factors. But these nutrients could be purchased more cheaply from other sources. Besides, it has not been proven that alfalfa meal contains some unknown nutrient essential for improved performance; and its high fiber (25 to 30%) and low palatability limit the amounts of it that can be used in baby pig and in growing-finishing rations. As a result, the swing was away from the use of alfalfa meal in swine rations.

But scarce and high-priced grains have caused alfalfa to return in favor as a feedstuff, particularly in rations for gestating sows. If the price of alfalfa meal is right—if it is a cheaper source of protein and energy than corn and other grains—15 to 35 percent (or even higher levels) alfalfa meal may, to advantage, be incorporated in gestating-lactating sow rations and 2 to 5 percent may be used in growing-finishing rations. These levels serve as a safety factor to ensure the presence of certain vitamins, minerals, and unidentified factors.

Feed Additives[16]

Certain feed additives have become somewhat standard ingredients of swine rations, especially for pigs from birth to market weight. They are not nutrients as such; hence, they should not be considered as dietary essentials.

Antibiotics and other antimicrobial compounds, alone or in combination, should be used only at approved levels and for the specific purpose for which they are authorized. From time to time, new products are added and old products are banned. Also, regulations concerning levels and withdrawal prior to marketing are subject to change. Hence, when using additives, the swine producer should always seek the counsel of local authorities and/or read and follow manufacturer's labels.

Additives used in swine rations include (1) antibiotics, (2) arsenicals, (3) copper compounds, and (4) combinations of additives.[17]

Continuous feeding of the same additive may decrease its effectiveness. Hence, periodic changing of drugs seems to be logical. The best time to rotate additives is when major changes are made in the ration, such as when going from a grower to a finishing ration.

• **Antibiotics**—Antibiotics are the most common additive in swine rations. They are used to stimulate growth, improve feed efficiency, secure uniformity of performance, and control infections. The response secured from their use depends on (1) the age of the pig, (2) the sanitary conditions, (3) the level fed, (4) the health and environment of the animal, (5) the type of ration, and (6) the season of the year. When fed to young, unthrifty pigs, antibiotics have increased growth rate by over 200%. For growing-finishing hogs under good sanitary conditions, antibiotics generally result in about 10% faster gains on 5% less feed. Pigs up to 100 pounds weight give the greatest response to antibiotic feeding. Experimental results have been inconsistent relative to the value of antibiotics in brood sow rations, but it appears that breeding herds with a high disease level may show a favorable response.

Regardless of the standard of sanitation practiced, within each swine facility there is an "environmental disease level." It may or may not be obvious to the casual observer, depending upon its degree of severity. Just how antibiotics counter these unfavorable conditions and bring about improved performance in growing swine is not exactly known. However, they apparently do have an influence on the intestinal bacteria and improve the health of the animal, as unthrifty pigs respond more to antibiotic feeding than do healthy ones. Antibiotics may function as follows:

1. They favor bacteria that synthesize known or unknown nutrients needed by the animal.
2. They inhibit the growth of nutrient-destroying microorganisms.
3. They improve the availability and/or absorption of certain nutrients.
4. They inhibit the growth of organisms that produce excessive amounts of ammonia and other toxic waste products in the intestine.
5. They prevent or control certain diseases in the intestinal tract and other parts of the body.

Despite what has been said, in no case should antibiotics be used to replace good sanitation and management.

Among the antibiotics commonly used in swine rations are bacitracin, chlortetracycline, oleandomycin, oxytetracycline, penicillin, tylosin, and zinc bacitracin. Most of these antibiotics or combinations of antibiotics are sold under different trade names.

Table 4-76 gives the recommended levels for antibiotics and arsenicals.

[16]This section was authoritatively reviewed by the following persons: Dr. Robert H. Grummer, Professor of Animal Science, University of Wisconsin, Madison, Wisc.; Dr. Richard F. Wilson, Faculty Member in Charge of Swine, Animal Science Department, The Ohio State University, Columbus, Ohio; and Dr. E. R. Barrick, Professor of Animal Science, North Carolina State University, Raleigh, N.C.

[17]Only a partial list of feed additives and combinations is given herein. *The Feed Additive Compendium*, published annually by the Miller Publishing Company, Minneapolis, Minn., contains a more complete listing.

TABLE 4-76

ANTIBIOTIC AND ARSENICAL SUPPLEMENTATION OF SWINE RATIONS
(As-fed Basis)

Ration[2]	Supplementary Level, Grams per Ton of Feed[1]		
	Antibiotic	Arsanilic Acid[3]	3-Nitro-Hydroxy-phenylarsonic Acid[3]
	(g)	*(g)*	*(g)*
Complete Feed			
Breeder[4]	0 to 20-30		
Creep and starter (5-30 lb, or *2.3-13.6 kg*)	40	90	22
Grower (30-100 lb, or *13.6-45.5 kg*)	10-20	90	22
Finisher (100 lb, or *45.5 kg*, to market weight)	10	90	22
Supplement (35-40% protein)			
Pig (up to 100 lb, or *45.5 kg*)	50-100	450	100
Hog (100 lb, or *45.5 kg*, to market weight)	0-50	450	100
Sow[4]	0 to 100-150		
Therapeutic	100-200	90	22

[1]*Conversion factors:* grams to pounds, divide by 454; percent to grams/ton, divide by 11 and move decimal 5 places to right; ppm to percent, move decimal 4 places to left.
[2]Feeds containing an arsenical alone, or containing an antibiotic at 50 grams or more per ton, are designated as therapeutic or medicated feeds.
[3]Never use both arsenicals in a single ration at the levels indicated.
[4]The effect of antibiotics on breeding stock is not conclusive, but there is some evidence to indicate that the feeding of antibiotics to pregnant sows may increase the birth weight, livability, and weaning weight of pigs.

• **Arsenicals**—Arsenic compounds, such as arsanilic acid or sodium arsanilate, are also growth stimulants that can be fed to growing-finishing swine.

• **Copper compounds**—Copper compounds have growth-stimulating value equal to antibiotics. They are also effective as a therapeutic treatment for intestinal disorders that do not respond satisfactorily to antibiotics or arsenicals. Copper is toxic when fed in excessive amounts (250 ppm or more) for prolonged periods. Copper compounds should not be added to hog rations if a lagoon manure disposal system is being used, as the copper will seriously interfere with the bacterial action in the lagoon.

Table 4-77 gives the recommended levels for copper compounds.

TABLE 4-77

RECOMMENDED LEVELS OF COPPER COMPOUNDS FOR GROWING-FINISHING SWINE[1]

Copper Compounds	Percent of Copper	Grams per Ton
	(%)	*(g)*
Cupric carbonate ($CuCO_3$)	51	222
Cupric oxide (CuO)	80	142
Cupric sulfate ($CuSO_4.5H_2O$)	25	454

[1]Amount to add per ton of complete ration (as-fed basis) to furnish 125 ppm of copper.

• **Additive combinations**—Many hog producers use combinations of additives. Experience as well as research trials would indicate that frequently these give a greater response than a single additive. Several combinations have been approved for use in swine feeds; among them, those listed in Table 4-78.

APPROVED LEVELS AND WITHDRAWAL TIME OF FEED ADDITIVES

In order to protect consumers from drug residues that may be harmful to their health, some antibiotics or chemotherapeutics, and all arsenic and sulfa compounds, are required by federal law to be withdrawn from rations within a specified time prior to slaughter. The Food and Drug Administration is responsible for establishing the tolerances, or the maximum quantity of a chemical that can be present in a food. So, regardless of the material used, the following safety precautions should be observed:

1. Use only those materials specifically authorized for use on swine. Regulations change frequently, so check them often.
2. Carefully follow label directions.
3. Do not mix unauthorized combinations.

4. Use minimal effective amounts.
5. Add or apply precise quantities.
6. Observe the proper interval between use and slaughter.

Table 4-78 gives the approved levels and withdrawal time (the interval between use and slaughter) of some common swine additives.

TABLE 4-78

APPROVED LEVELS AND WITHDRAWAL PERIODS FOR COMMONLY USED FEED ADDITIVES

Feed Additive	Approved Growth Promotion Level, (grams/ton)	Pre-slaughter Withdrawal Period[1]
Antibiotics		
Bacitracin	10-50	None
Bacitracin, M.D.	10-50	None
Bacitracin, Zinc	10-50	None
Bambermycins	2	None
Chlortetracycline	10-50	None
Erythromycin	10-70	None
Oleandomycin	5-11.25	None
Oxytetracycline	7.5-50	None
Penicillin	10-50	None
Tylosin	10-100	None
Virginiamycin	5-10	None
Chemotherapeutics		
Arsanilic acid	45-90	5 days
Carbadox	10-25	10 wk (75 lb)
Furazolidone	—	5 days
Roxarsone	22.7-68.1	5 days
Sodium arsanilate	45-90	5 days
Combinations[2]		
Chlortetracycline	100	15 days
+ sulfamethazine	100	
+ penicillin	50	
Chlortetracycline	100	7 days
+ sulfathiazole	100	
+ penicillin	50	
Penicillin +	1.5-8.5	None
streptomycin	7.5-41.5	
Tylosin +	100	15 days
sulfamethazine	100	

[1]Period of time the drug must be removed from the diet before slaughter.
[2]Not a complete list of approved combinations; for further information consult the Feed Additive Compendium, Miller Publishing Co., Minneapolis, Minn.

FEED PREPARATION

The preparation of feeds for swine is fully covered earlier in this section under the heading, "Feed Preparation," including Table 4-11 therein, page 251; hence, the reader is referred thereto.

RATIONS

The nutritive requirements of swine vary according to age, weight, sex, stage of gestation or lactation, and environment. This calls for a feed for every need. Table 4-79 gives, in summary form, the recommended types of rations and feeding programs for all classes of swine.

Hogs can and do use many different concentrates. This makes it possible to choose as chief ingredient those most readily available at the lowest price.

The rations given in Table 4-80 will serve as useful guides. As noted, there are suggested rations for each of the following: (1) starter, 10-40 lb; (2) growing-finishing, 40-120 lb; (3) growing-finishing, 120-220 lb; (4) gestation; (5) lactation; (6) boars-grower, 40-120 lb; (7) boars-finisher, 120-250 lb; and (8) boars-adult.

The suggested rations given in Table 4-80 meet the nutrient requirements set forth in Tables 4-67 to 4-70.

In addition to the Table 4-80 rations, a milk replacer ration should be used for pigs weaned prior to 3 weeks of age and for orphan pigs. For this purpose any of the formulations shown in Table 4-81, page 398, may be used. Because milk replacer rations are both complex and used in limited quantity, it is usually best to rely on a good commercial product, rather than attempt to home mix.

In addition to the complete rations given in Table 4-80, Tables 4-82, page 398, and 4-83, page 399, give suggested formulas for 40% and 35% protein supplements, respectively. These may be either purchased commercially or mixed on the farm. Table 4-84, page 400, shows how these protein supplements (40% and 35%, respectively) may be combined with different cereal grains to obtain a ration of the desired level of protein.

TABLE 4-79

TYPES OF RATIONS AND FEEDING PROGRAMS FOR SWINE

Ration	Source	Age or Size of Pig	Level of Protein
			(%)
Milk replacer (prestarter)	Commercial	Early weaned or orphan pigs under 3 weeks of age.	20-23
Starter	Commercial	3 weeks of age to 40 lb body weight.	18
Growing-finishing	Commercial or farm mixed	40-120 lb.	16
Growing-finishing	Commercial or farm mixed	120 lb to market weight (220 lb).	13
Pregestation and gestation	Limit-feed gilts and sows about 4 lb per head per day. In extreme cold weather and/or for females in poor condition a level of 5 to 6 lb is suggested. Flush gilts by full feeding or hand feeding 6 to 8 lb per day 2 weeks prior to breeding. Two to 3 weeks prior to farrowing, feeding level should be increased 1 lb per head per day.		12
Farrowing	Reduce feed intake slightly 4 to 5 days before and after farrowing. Laxative rations, those containing 10-15% beet pulp, should be fed starting 10 days before farrowing and during the first week of lactation.		14-16
Lactation	Self-feed during lactation or hand-feed to appetite.		16
Boars	Limit-feed young growing boars 5 to 5.5 lb and adult boars 4.5 lb per head per day. Protein content: young boars, 18%; boars 120-250 lb, 16%; mature boars, 14%.		14-18

TABLE 4-80

BALANCED RATIONS FOR SWINE
(As-fed Basis)

	Diet Percentage Composition and Period of Animal Production							
	Growing-Finishing			Sow Reproduction		Boars		
Ingredient	Starter 10-40 lb *(4.5-18 kg)*	Grower 40-120 lb *(18-55 kg)*	Finisher 120-220 lb *(55-100 kg)*	Gestation	Lactation	Grower 40-120 lb *(18-55 kg)*	Finisher 120-250 lb *(55-114 kg)*	Adult
Corn #2	45.30	74.25	83.05	84.20	73.85	66.35	73.95	80.30
Soybean meal, 44%	25.05	22.10	13.50	11.05	22.10	28.30	22.00	16.00
Rolled oats	25.00	—	—	—	—	—	—	—
Dicalcium phosphate	1.10	0.85	0.80	1.65	1.40	2.10	1.50	1.05
Limestone	1.25	1.00	1.00	1.05	1.10	0.95	0.90	1.10
Trace mineral salt[1]	0.50	0.50	0.50	0.50	0.50	0.50	0.50	0.50
Selenium premix	0.05	0.05	0.05	0.05	0.05	0.05	0.05	0.05
Vitamin premix[2]	1.50	1.00	1.00	1.50	1.00	1.50	1.00	1.00
Antibiotic	0.25	0.25	0.10	—	—	0.25	0.10	—
Total	100.00	100.00	100.00	100.00	100.00	100.00	100.00	100.00
Proximate Analysis:								
Crude protein (%)	18.4	16.7	13.6	12.6	16.7	18.9	16.7	14.5
Calcium (%)	0.83	0.65	0.61	0.84	0.82	0.97	0.78	0.71
Phosphorus (%)	0.60	0.54	0.50	0.64	0.64	0.78	0.65	0.55

[1]Make sure that the product used meets the recommended trace mineral allowance given in Tables 4-71 and 4-72.

[2]The percentage (or amount) of this material to be added will depend on the manufacturer. Follow the label or instructions by the manufacturer: but make sure that the product used meets the recommended vitamin allowances given in Tables 4-71 and 4-75.

TABLE 4-81

MILK REPLACERS FOR BABY PIGS BEFORE 3 WEEKS OF AGE[1]
(As-fed Basis)

Percent Protein	Ingredient	1	2	3	4
		(lb)			
8.9	Ground yellow corn #2	881	652	551	530
50.9	Soybean meal, 49%	455	500	600	500
16.7	Rolled oat groats	—	—	—	200
33.5	Skim milk, dried	200	400	200	200
13.8	Whey, dried	200	200	400	400
30.3	Fish solubles, condensed	—	50	50	—
	Sugar	200	100	100	100
	Stabilized animal fat	—	50	50	20
	Oystershells, ground (38% Ca)	14	10	10	10
	Dicalcium phosphate (26% Ca, 18.5% P)	22	10	10	10
	Salt	5	5	5	5
	Trace mineral premix	3	3	3	3
	Vitamin premix	20	20	20	20
	DL-methionine	—	—	1	2
	Feed additives[2] (g/ton)	100-300	100-300	100-300	100-300
	Total	2,000	2,000	2,000	2,000
	Proximate Analysis:				
	Protein (%)	19.45	23.61	23.48	21.69
	Calcium (%)	0.98	0.72	0.69	0.69
	Phosphorus (%)	0.81	0.62	0.61	0.61
	Lysine (%)	1.23	1.58	1.56	1.40
	Methionine (%)	0.33	0.43	0.44	0.45
	Cystine (%)	0.37	0.35	0.36	0.34
	Tryptophan (%)	0.24	0.30	0.30	0.28
	Metabolizable energy (kcal/lb)	1,380	1,429	1,418	1,405

Feeding Directions

[1]The milk replacer ration is normally fed in only limited amounts. It should be used for pigs weaned prior to 3 weeks of age until they reach approximately 12 lb. Then they can be switched to a starter ration. It is a good ration to feed (a) orphan pigs when the sow dies, (b) when extreme disease outbreak (TGE) occurs, or (c) if the sow fails to produce milk.

[2]The feed additive may be part of the vitamin premix, or if a separate premix, it should replace an equal amount of corn.

TABLE 4-82

40 PERCENT SUPPLEMENTS[1,2]
(As-fed Basis)

Percent Protein	Ingredient	1	2	3	4
		(lb)			
50.9	Soybean meal, 49%	1,628	—	1,238	1,458
45.8	Soybean meal, 44%	—	1,223	—	—
15.3	Alfalfa meal, dehydrated	—	—	100	—
49.5	Meat and bone meal[3]	—	550	400	—
60.4	Fish meal, menhaden	—	—	—	200
	Oystershells, ground (38% Ca)	90	45	55	80
	Dicalcium phosphate (26% Ca, 18.5% P)	160	60	85	140
	Iodized salt	50	50	50	50
	Trace mineral premix	12	12	12	12
	Vitamin premix	60	60	60	60
	Feed additives[4]	—	—	—	—
	Total	2,000	2,000	2,000	2,000
	Proximate Analysis:				
	Protein (%)	39.48	40.66	40.87	41.46
	Calcium (%)	3.95	4.02	3.96	3.98
	Phosphorus (%)	2.01	2.05	2.02	2.05
	Salt added (%)	2.50	2.50	2.50	2.50
	Lysine (%)	2.67	2.55	2.60	2.86
	Methionine (%)	0.55	0.56	0.56	0.66
	Cystine (%)	0.59	0.48	0.52	0.58
	Tryptophan (%)	0.55	0.46	0.49	0.56
	Metabolizable energy (kcal/lb)	1,123	968	1,047	1,120

Feeding Directions

[1]These supplements can be used to make growing-finishing, gestation, or lactation rations. See Table 4-84 footnote for mixing directions.

[2]Supplements 2 and 3 may be self-fed free-choice with shelled corn.

[3]The meat and bone meal was considered to have 8.1% calcium and 4.1% phosphorus. If meat and bone meal with a higher concentration of calcium and phosphorus is used, the amount of dicalcium phosphate should be reduced accordingly.

[4]The level of feed additives will depend on the type of ration in which the supplement is going to be used, but should be 4 to 6 times higher than desired in the complete ration.

TABLE 4-83

35 PERCENT SUPPLEMENTS[1,2]
(As-fed Basis)

Percent Protein	Ingredient	1	2	3	4
		(lb)			
45.8	Soybean meal, 44%	1,670	1,575	1,100	1,050
15.3	Alfalfa meal, dehydrated	—	100	—	200
49.5	Meat and bone meal[3]	—	—	400	400
8.9	Corn #2[4]	—	—	290	145
	Oystershells, ground (38% Ca)	80	75	50	45
	Dicalcium phosphate (26% Ca, 18.5% P)	140	140	60	60
	Iodized salt	50	50	40	40
	Trace mineral premix	10	10	10	10
	Vitamin premix	50	50	50	50
	Feed additives[5]	—	—	—	—
	Total	2,000	2,000	2,000	2,000
	Proximate Analysis:				
	Protein (%)	36.74	35.50	35.49	35.45
	Calcium (%)	3.55	3.51	3.49	3.52
	Phosphorus (%)	1.80	1.78	1.74	1.74
	Salt added (%)	2.50	2.50	2.00	2.00
	Lysine (%)	2.51	2.40	2.21	2.20
	Methionine (%)	0.53	0.51	0.50	0.49
	Cystine (%)	0.56	0.54	0.44	0.45
	Trypotophan (%)	0.53	0.52	0.41	0.43
	Metabolizable energy (kcal/lb)	1,022	991	1,035	958

Feeding Directions

[1]These supplements can be used to make growing-finishing, gestation, or lactation rations. See Table 4-84 footnote for mixing directions.

[2]Supplements 3 and 4 may be self-fed free-choice with shelled corn.

[3]The meat and bone meal was considered to have 8.1% calcium and 4.1% phosphorus. If meat and bone meal with a higher concentration of calcium and phosphorus is used, the amount of dicalcium phosphate should be reduced accordingly.

[4]The corn can be replaced by wheat middlings, corn distiller grains with solubles, or other grain by-products.

[5]The level of feed additives will depend on the type of ration in which the supplement is going to be used, but should be 3 to 5 times higher than desired in the complete ration.

In formulating the several rations given in Tables 4-80 to 4-84 cognizance was taken of the fact that two ystems of swine feeding predominate: (1) the use of omplete rations, either commercially or homemixed, 's (2) the grain and the supplement fed separately. Also, both hand-feeding and self-feeding methods are ollowed, with the latter being more common except or gestating sows.

Where commercial protein supplements are ought (usually a combined protein-mineral-vitamin-ntibiotic supplement—similar to Table 4-82 or 4-83 ormulations) to use with farm-grown grains, they are tilized in any of the following ways:

1. Mixed with ground, farm-grown grain in appropriate amounts to make a complete ration. (See Table 4-84. The footnote for this table gives the directions for its use.)

2. Self-fed in separate self-feeders, with the ground or whole grain also being self-fed in separate self-feeders.

3. Hand fed; the supplement and the grain each being hand fed in the proportions recommended (see Table 4-84).

In formulating the suggested rations herein presented, the following facts were also considered:

1. The most critical periods, nutritionally, in the life of the pig are (a) birth to 40 pounds weight, (b) gestation, and (c) lactation; hence, the proteins, minerals, and vitamins are especially important during these periods. Accordingly, in the rations suggested in this section, a considerable margin of safety in the most essential nutrients is provided over and above minimum requirements for these periods.

2. Complete mixed rations are much preferable for pigs up to 40 pounds in weight. Because of the difficulty in formulating and home mixing satisfactory milk replacer and starter rations, the purchase of a good commercial feed usually represents a wise investment.

TABLE 4-84

RATIO OF GRAIN TO PROTEIN SUPPLEMENTS NEEDED TO OBTAIN THE DESIRED LEVEL OF PROTEIN IN A RATION, WITH AND WITHOUT ALFALFA MEAL[1]
(As-fed Basis)

% Protein Desired	40% Supplement		35% Supplement		Corn (8.9% crude protein)		Ground Barley (11.6% crude protein)		Ground Oats (11.7% crude protein)		Ground Wheat (12.7% crude protein)		Alfalfa Meal (15.3% crude protein)	
	(lb)	*(kg)*	*(lb)*	*(kg)*	*(lb)*	*(kg)*	*(lb)*	*(kg)*	*(lb)*	*(kg)*	*(lb)*	*(kg)*	*(lb)*	*(kg)*
18	575	*261*	—	—	1,325	*602*	—	—	—	—	—	—	100	*45*
18	500	*227*	—	—	750	*341*	650	295	—	—	—	—	100	*45*
18	525	*238*	—	—	1,000	*454*	—	—	375	*170*	—	—	100	*45*
18	450	*204*	—	—	—	—	925	*420*	—	—	525	*238*	100	*45*
18	375	*170*	—	—	—	—	—	—	550	*250*	975	*443*	100	*45*
18	—	—	675	*307*	1,225	*556*	—	—	—	—	—	—	100	*45*
18	—	—	575	*261*	675	*307*	650	*295*	—	—	—	—	100	*45*
18	—	—	650	*295*	1,000	*454*	—	—	—	—	250	*114*	100	*45*
18	—	—	500	*227*	—	—	1,000	*454*	—	—	400	*182*	100	*45*
18	—	—	500	*227*	—	—	—	—	500	*227*	900	*409*	100	*45*
18	600	*272*	—	—	1,400	*636*	—	—	—	—	—	—	—	—
18	525	*238*	—	—	875	*397*	600	*272*	—	—	—	—	—	—
18	550	*250*	—	—	1,000	*454*	—	—	450	*204*	—	—	—	—
18	425	*193*	—	—	—	—	925	*420*	—	—	650	*295*	—	—
18	400	*182*	—	—	—	—	—	—	700	*317*	900	*409*	—	—
18	—	—	725	*329*	1,275	*579*	—	—	—	—	—	—	—	—
18	—	—	650	*295*	750	*341*	600	*272*	—	—	—	—	—	—
18	—	—	675	*307*	1,000	*454*	—	—	325	*148*	—	—	—	—
18	—	—	500	*227*	—	—	850	*386*	—	—	650	*295*	—	—
18	—	—	500	*227*	—	—	—	—	550	*250*	950	*431*	—	—
16	450	*204*	—	—	1,450	*658*	—	—	—	—	—	—	100	*45*
16	350	*159*	—	—	850	*386*	700	*317*	—	—	—	—	100	*45*
16	300	*136*	—	—	—	—	950	*431*	—	—	650	*295*	100	*45*
16	400	*182*	—	—	1,150	*522*	—	—	350	*159*	—	—	100	*45*
16	275	*125*	—	—	—	—	—	—	625	*284*	1,000	*454*	100	*45*
16	—	—	550	*250*	1,350	*613*	—	—	—	—	—	—	100	*45*
16	—	—	425	*193*	775	*352*	700	*317*	—	—	—	—	100	*45*
16	—	—	375	*170*	—	—	875	*397*	—	—	650	*295*	100	*45*
16	—	—	500	*227*	1,050	*477*	—	—	350	*159*	—	—	100	*45*
16	—	—	325	*148*	—	—	—	—	575	*261*	1,000	*454*	100	*45*
16	375	*170*	—	—	925	*420*	700	*317*	—	—	—	—	—	—
16	475	*216*	—	—	1,525	*692*	—	—	—	—	—	—	—	—
16	425	*193*	—	—	1,225	*556*	—	—	350	*159*	—	—	—	—
16	250	*114*	—	—	1,150	*522*	—	—	—	—	700	*318*	—	—
16	250	*114*	—	—	—	—	—	—	750	*341*	1,000	*454*	—	—
16	—	—	450	*204*	850	*386*	700	*317*	—	—	—	—	—	—
16	—	—	575	*261*	1,425	*647*	—	—	—	—	—	—	—	—
16	—	—	525	*238*	1,125	*511*	—	—	350	*159*	—	—	—	—
16	—	—	200	*91*	1,100	*499*	—	—	—	—	700	318	—	—
16	—	—	175	*80*	—	—	—	—	725	329	1,100	*499*	—	—
14	325	*148*	—	—	1,575	*715*	—	—	—	—	—	—	100	*45*
14	225	*102*	—	—	875	*397*	800	*363*	—	—	—	—	100	*45*
14	275	*125*	—	—	1,325	*602*	—	—	300	*136*	—	—	100	*45*
14	—	—	375	*170*	1,525	*692*	—	—	—	—	—	—	100	*45*
14	—	—	250	*114*	850	*386*	800	*363*	—	—	—	—	100	*45*
14	—	—	350	*159*	1,250	*568*	—	—	300	*136*	—	—	100	*45*
14	350	*159*	—	—	1,650	*749*	—	—	—	—	—	—	—	—
14	250	*114*	—	—	950	*431*	800	*363*	—	—	—	—	—	—
14	300	*136*	—	—	1,300	*590*	—	—	400	*182*	—	—	—	—
14	—	—	425	*193*	1,575	*715*	—	—	—	—	—	—	—	—
14	—	—	300	*136*	900	*409*	800	*363*	—	—	—	—	—	—
14	—	—	350	*159*	1,250	*568*	—	—	400	*182*	—	—	—	—
12	200	*91*	—	—	1,700	*772*	—	—	—	—	—	—	100	*45*
12	150	*68*	—	—	1,350	*613*	400	*182*	—	—	—	—	100	*45*
12	150	*68*	—	—	1,450	*658*	—	—	300	*136*	—	—	100	*45*
12	—	—	250	*114*	1,650	*749*	—	—	—	—	—	—	100	*45*
12	—	—	175	*80*	1,325	*602*	400	*182*	—	—	—	—	100	*45*
12	—	—	200	*91*	1,400	*636*	—	—	300	*136*	—	—	100	*45*
12	225	*102*	—	—	1,775	*806*	—	—	—	—	—	—	—	—
12	150	*68*	—	—	1,250	*568*	600	*272*	—	—	—	—	—	—
12	200	*91*	—	—	1,500	*681*	—	—	300	*136*	—	—	—	—
12	—	—	275	*125*	1,725	*783*	—	—	—	—	—	—	—	—
12	—	—	200	*91*	1,300	*590*	500	*227*	—	—	—	—	—	—
12	—	—	225	*102*	1,475	*670*	—	—	300	*136*	—	—	—	—

[1]In order to obtain an 18% protein feed, one could mix 575 lb of 40% supplement, 1,325 lb of corn, and 100 lb of alfalfa meal. Likewise, a 14% supplement without alfal meal (for use on pasture) could be obtained by mixing 300 lb of 35% supplement, 900 lb of corn, and 800 lb of barley.

Pointers in Formulating Rations and Feeding Swine

In formulating rations and in feeding swine, the following points are noteworthy:

1. Feeds of similar nutritive properties can be interchanged in the ration as price relationships warrant. See Table 4-85 for some energy feeds that may be substituted for corn, and for some protein feeds that may be substituted for soybean meal.
2. If wheat, barley, oats, or grain sorghum is used instead of corn as the grain in a ration, the protein supplement may be slightly reduced, because these grains have a higher protein content than corn.
3. Pacific Coast grains are generally lower in protein content than grains produced elsewhere.
4. When proteins of animal origin predominate, adequate mineral protection can be obtained by allowing hogs free access to a 2-compartment box or self-feeder with (a) salt (trace mineralized) in one side, and (b) a mixture of ⅓ salt (salt added for purposes of palatability) and ⅔ monosodium phosphate or other phosphorus supplement, in the other side. When supplements of plant origin constitute most of the source of proteins, add a third compartment to the mineral box and place in it a mixture of ⅓ salt (trace mineralized) and ⅔ ground limestone or oystershell flour.
5. Where there is insufficient sunlight or where dehydrated alfalfa meal is fed, vitamin D should be added in keeping with the recommended allowances (see Tables 4-71 and 4-75).
6. Where the ration consists chiefly of white corn, barley, wheat, oats, rye, kafir, or by-products of these grains, there may be a deficiency of vitamin A (see Tables 4-71 and 4-75 for recommended allowances).
7. Except for gestating sows and boars of breeding age, hogs are generally self-fed. All of the ingredients may be mixed together and placed in the same self-feeder, or the grain may be placed in one self-feeder (or compartment) and the protein supplements (including any ground alfalfa) in another. If the (a) cereal grains, and (b) protein supplements (including ground alfalfa) are hand fed, the grain and supplement should be fed separately, in the proportions indicated in the suggested rations.
8. An exception should be made to the cafeteria-style feeding when the grain ration consists of barley, oats, rye, or kafir. These feeds are higher in protein content than corn, and for this reason are generally fed as a mixed ration. Otherwise, the pigs will often eat more protein supplement than is necessary to balance the ration. Likewise, when corn is fed as the grain, sometimes such protein supplements as (a) roasted soybeans, (b) soybean meal, and (c) peanut meal are too palatable to be fed separately from the corn.
9. Full-fed finishing hogs will consume 4 to 5 lb of feed daily per 100 lb liveweight until they weigh 100 lb. They will eat 3 to 4 lb daily per 100 lb weight from this stage until marketing.

FEED SUBSTITUTION TABLE

The successful swine producer is a keen student of values. He recognizes that feeds of similar nutritive properties can and should be interchanged in the ration as price relationships warrant, thus making it possible at all times to obtain a balanced ration at the lowest cost.

Table 4-85, Feed Substitution Table for Swine, page 402, is a summary of the comparative values of the most common U.S. and Canadian feeds. In arriving at these values, two primary factors besides chemical composition and feeding value have been considered; namely, palatability and carcass quality.

In using this feed substitution table, the following facts should be recognized:

1. That, for best results, different ages and groups of animals within classes should be fed differently.
2. That individual feeds differ widely in feeding value. Barley and oats, for example, vary widely in feeding value according to the hull content and the test weight per bushel.
3. That, based primarily on available supply and price, certain feeds—especially those of medium protein content, such as peanuts and peas (dried)—are used interchangeably as (a) grains and by-product feeds, and/or (b) protein supplements.
4. That the feeding value of certain feeds is materially affected by preparation; thus, potatoes and beans should always be cooked for hogs. The values herein reported are based on proper feed preparation in each case.

For these reasons, the comparative values of feeds shown in the feed substitution table which follows are not absolute. Rather, they are reasonably accurate approximations based on average quality feeds.

FEEDING BREEDING SWINE

Profit or loss in swine breeding is largely determined by the number of pigs weaned and marketed per sow. In turn, feeding the herd at breeding time and during gestation greatly influences the number of pigs farrowed and weaned. Proper feeding will result in (1) more pigs farrowed per sow, (2) larger and healthier pigs at birth, (3) fewer dead pigs, runts, and abnormal pigs per litter, (4) more milk produced by the sows, and (5) more and heavier pigs weaned per litter.

TABLE 4-85

FEED SUBSTITUTION TABLE FOR SWINE
(As-fed Basis)

Feedstuff	Relative Feeding Value (lb for lb) in comparison with the designated (underlined) base feed which = 100	Maximum Percentage of Base Feed (or comparable feed or feeds) which it can replace for best results	Remarks
GRAINS, BY-PRODUCT FEEDS, ROOTS AND TUBERS:[1] (Low- and Medium-Protein Feeds)			
Corn, No. 2	100	100	Corn is the leading U.S. swine feed, about 50% the total production being fed to hogs. It does not pay to grind corn for growing-finishin pigs, but it should be ground for older hogs.
Bakery waste	95	50	Bakery wastes average about 10% protein. The should be supplemented with protein, mineral and vitamins about the same as the cereal grain
Barley	90-95	100	Of variable feeding value due to wide spread in te wt./bu. Should be ground or rolled. In Canada, where high-quality bacon is produce barley is considered preferable to corn for finis ing hogs.
Beans (cull)	90	33-66	Cook thoroughly; on a dry weight basis, limit to ½ th grain ration for pigs under 100 lb and ⅔ of th grain ration for pigs above 100 lb; supplement wi animal protein.
Carrots (or beets, mangels, or turnips)	12-20	25	
Cassava, dried meal	85	33⅓	
Corn meal	100	20	
Hominy feed	95	50	Hominy feed will produce soft pork if it constitute more than ½ the grain ration.
Lard	230	5	
Millet (hog millet)	85-90	50	
Molasses, beet	70	20	
Molasses, cane	70	20	In Cuba, where cane molasses is abundant an cheap, it is sometimes self-fed to market hog with 150 lb hogs consuming up to 4½ lb/head/da Such high levels do make for soft, watery feces.
Molasses, citrus	70	10-20	It takes pigs 5 to 7 days to get used to the bitter tas of citrus molasses.
Oats	70-80	33⅓-100	Grind for swine. Feeding value varies according test wt./bu.
Oats, rolled or dehulled	107	100	Due to bulk, oats are a better feed for breeding swir than for young pigs or finishing hogs. For growing-finishing pigs, oats is equal to cor when limited to ⅓ of the ration.
Peanuts	120-125	100	Peanuts are usually fed by hogging-off.
Peas, dried	90-100	100	Normally peas should be fed to swine as a prote supplement. Two tons of peas = 1 ton of grain + ton of soybean meal.
Potatoes (Irish)	25-28	25-50	Potatoes contain only 23% as much dry matter a shelled corn. Not palatable in the raw state; must be cooked. When cooked and fed in a ratio of 3 lb of potatoes 1 lb of grain, they are worth 25-28% as much a corn.
Potatoes (Irish), dehydrated	100	33⅓	
Potatoes (sweet)	20-25	33⅓-50	Cooking also improves the feeding value of swe potatoes.
Potatoes (sweet), dehydrated	90	33⅓	
Rice (rough rice)	80-85	50	Rice should be ground.
Rice bran	100	33⅓	If more than ⅓ of the grain consists of rice bran, sc pork will result.
Rice polishings	100-120	33⅓	Limited because feed becomes rancid in storag and soft pork will be produced.
Rice screenings	95	50	
Rye	90	30	Should be limited because it is unpalatable. Grind for swine.
Sorghum, grain	95	100	All varieties have about the same feeding value. Check protein content and add supplement a necessary. Both very dry grain and bird-resistant varieti should be ground.

Footnote on last page of table.

(Continue

TABLE 4-85 (Continued)

Feedstuff	Relative Feeding Value (lb for lb) in comparison with the designated (underlined) base feed which = 100	Maximum Percentage of Base Feed (or comparable feed or feeds) which it can replace for best results	Remarks
Spelt	65-80	25	Value varies according to the amount of hulls.
Sunflower seed	100	50	
Tallow	240	5	
Triticale	80-90	50	Higher levels not palatable.
Wheat	100-105	100	Feed whole if self-fed. Otherwise, grind coarsely; fine grinding makes it pasty and unpalatable. Wheat-corn mixtures are more efficient than wheat alone.
Wheat bran	75	15-25	Bran is particularly valuable at farrowing time. In Canada, where high-quality bacon is produced, 15 to 25% wheat bran is sometimes incorporated in the finishing ration in order to obtain a lean carcass.
Wheat flour middlings	103	20	
Wheat standard middlings	85-100	25-50	Combine with animal protein and limit to 1 lb/head/day.
Wheat red dog and wheat white shorts	115-120	25	
PROTEIN SUPPLEMENTS:			
Soybean meal (41%)	100	100	Well balanced in amino acids. Best quality of all plant protein supplements. Very palatable.
Blood meal (80%)	123	20	High in protein (above 80%), but protein is lower in digestibility and quality than most other protein feeds. Not very palatable.
Buttermilk, dry	90-105	100	
Buttermilk, liquid	15	100	Pound for pound, worth 1/10 as much as dried buttermilk.
Buttermilk, semisolid	33⅓-50	100	Pound for pound, worth ⅓ as much as dried buttermilk.
Copra meal (coconut meal) (21%)	50	25	
Corn gluten meal (gluten meal)	50-75	50	
Cottonseed meal (41%)	85	33⅓	Except where new glandless cottonseed meal is used, high levels may produce gossypol poisoning and the level of cottonseed meal in swine rations should not exceed 8 to 9% of the total ration. Cottonseed meal is low in lysine.
Fish meal, Menhaden (60%)	115	100	Expensive. Excellent balance of amino acids, and good source of calcium and phosphorus.
Linseed meal (35%)	80	25-50	Low in lysine; slightly laxative.
Malt sprouts	100	10	Malt sprouts contain a growth factor(s). They result in increased feed intake and gain.
Meat and bone meal (50%)	100	100	Low in tryptophan; good source of calcium and phosphorus.
Meat scraps (50-55%)	100	100	
Peanut meal (45%)	95	50	Becomes rancid when stored too long. Low in lysine; very palatable.
Peanuts	60-70	50	Peanuts are usually fed by hogging-off. High levels will produce soft pork.
Peas, dried	50	50	
Shrimp meal	90-100	50	
Skim milk, dried	90-120	100	Excellent quality protein; very palatable; expensive. Especially good in prestarter and starter rations.
Skim milk, liquid	15	100	Pound for pound, worth 1/10 as much as dried skim milk.
Soybeans	70-75	50	
Tankage (60%)	110	100	Good source of calcium and phosphorus. Low in tryptophan.
Whey, dried	45	100	
Whey, liquid	30	50	Worth ½ as much as skim milk.

(Continued)

TABLE 4-85 (Continued)

Feedstuff	Relative Feeding Value (lb for lb) in comparison with the designated (underlined) base feed which = 100	Maximum Percentage of Base Feed (or comparable feed or feeds) which it can replace for best results	Remarks
PASTURES AND DRY LEGUMES:			
Pasture, good		5-20% of grain, and 20-50% of protein supplement.	Pasture and dry legumes are sources of good quali proteins, of minerals, and of vitamins. Thus, swir should have access to either pasture or groun legumes.
Alfalfa meal		It can replace all of pasture, in drylot rations.	For drylot rations, include 5-10% alfalfa in ration growing-finishing pigs, and 15-35% in ration gestating-lactating sows. In Canada, where high-quality bacon is produce up to 25% alfalfa meal is sometimes included the finishing ration in order to obtain a lean ca cass.

[1]Roots and tubers are of lower value than the grain and by-product feeds due to their higher moisture content.

Feeding Prospective Breeding Gilts

Prospective breeding gilts should be kept from getting too fat. Meat-type animals can usually be left on a high-energy ration until they reach 175 to 200 pounds without becoming too fat. It is neither necessary nor desirable that females intended for breeding purposes carry the same degree of finish as market animals. After selecting replacement gilts, they should be fed as follows:

1. Give about 5 pounds per day through their second heat period.
2. Flush—full feed—after the second heat period until breeding on the third heat period.
3. After breeding, limit the feed intake to 3 to 5 pounds per day. Overfeeding during gestation can cause embryonic death and thus decrease litter size.

Feeding Brood Sows

The nutrition of brood sows is critical, for it may materially affect conception, reproduction, and lactation. Proper feeding of sows should begin with replacement gilts and continue through each stage of the breeding cycle—flushing, gestation, farrowing, and lactation.

• **Flushing sows**—*The practice of conditioning or having the sows gain in weight just prior to breeding is known as flushing.* The purpose of flushing is to increase the number of ova shed during estrus. About 10 to 14 days prior to expected breeding, the female should be fed a ration that will make for gains of 1 to 1¼ pounds per day. Generally 6 to 8 pounds per day of a high-energy, 14 to 16 percent protein feed that is well balanced in minerals and vitamins, is adequate. Immediately after breeding, the females should b put back on limited feeding. Continuation of a hig level of feeding after breeding will result in a highe embryo mortality.

• **Gestation period**—The nutrients fed the preg nant gilt or sow must first take care of the usua maintenance needs. If the gilt is not fully mature, nu trients are required for both maternal growth an growth of the fetus. Quality and quantity of proteins minerals, and vitamins become particularly importan in the ration of young pregnant gilts, for their re quirements are much greater and more exacting tha those of the mature sow.

Approximately two-thirds of the growth of th fetus is made during the last month of the gestatio period. It may be said, therefore, that the demand resulting from pregnancy are particularly accelerate during the latter third of the gestation period. Again the increased needs are primarily for proteins, vita mins, and minerals.

During gestation, it is also necessary that bod reserves be stored for subsequent use during lacta tion. With a large litter and a sow that is a heav milker, the demands for milk production are generall greater than can be supplied by the ration fed at th time of lactation. Although desired gains will var somewhat according to the initial condition, matur sows are generally fed to gain about 70 lb during th pregnancy period, and first litter pregnant gilts are fe to gain about 90 lb. This means that from the day o mating until farrowing time gilts should be fed to gai about 0.9 lb per day and mature sows about 0.7 lb pe day. This calls for a daily feed allowance of approxi mately 4 to 4.5 lb per head, with variation accordin to environmental conditions.

It is important that the condition of dry sows should be regulated so that they are neither too fat nor too thin at farrowing time. Overly fat sows may have difficulty in farrowing and give birth to weak or dead pigs. Sows that are too thin at farrowing tend to become suckled down during lactation. Thus, one way or another, limited feeding is a must for gestating gilts and sows. This may be accomplished by any one of the following feeding systems:

1. By adding sufficient bulk.
2. By interval feeding.
3. By group hand feeding.
4. By individual feeding.

In addition to the above limited feeding systems, the use of pasture should be considered. Where available, a leguminous pasture is the ideal way in which to limit-feed gestating gilts and sows. Dry sows on good legume pasture are usually fed ½ pound less supplement and 2 pounds less grain per day. In addition to limiting the feed intake, the pasture system provides valuable quality protein, minerals, vitamins, and exercise.

• **Farrowing time**—It is considered good practice to feed lightly and with bulky laxative feeds from four to five days before and after farrowing. Wheat bran or oats may constitute half of the limited ration, and a small amount of linseed meal may be added.

The sow may be watered at frequent intervals before or after farrowing, but in no event should she be allowed to gorge. It is also a good plan to take the chill off the water in the wintertime.

• **Lactation period**—The nutritive requirements of a lactating sow are more rigorous than those during gestation. They are very similar to those of a milk cow, except they are more exacting relative to quality proteins and the B vitamins because of the absence of rumen synthesis in the pig. A good lactating sow will produce an average of about 1 gallon of milk daily during the suckling period. A sow's milk is also richer than cow's milk in all nutrients, especially in fat. Thus, sows suckling litters need a liberal allowance of concentrates rich in protein, calcium, phosphorus, and vitamins.

It is essential that suckling pigs receive a generous supply of milk, for at no other stage in life will they make such economical gains. The gains made by pigs from birth to weaning are largely determined by the milk production of the sows; and this in turn is dependent upon the ration fed and the sow's inherent ability to produce milk. The lactating sow should be provided with a liberal feed allowance—ranging from 2½ to 4½ pounds daily for each 100 pounds weight. Generous feeding during lactation, with a small shrinkage in weight, is more economical than a stingy allowance of feed, for the nutrients in milk must come either from the feed or from the sow's back. Lactating sows are commonly self-fed, because even when hand-fed they are practically on full feed.

A suggested lactation ration is given in Table 4-80, page 397.

Feeding Orphan Pigs

There is no replacement for the sow's colostrum. If the newborn pig does not receive colostrum, it has a lesser chance for survival. An orphan pig can obtain colostrum by being placed with another sow (a foster sow) that has just farrowed. If no such sow is available, the orphan can be fed a commercial milk replacer, several good ones of which are on the market. A homemade milk replacer can be prepared by mixing the following ingredients:

1 quart milk
1 pint half-and-half
1 raw egg

Portions of this mixture should be warmed to 98 to 100°F and fed about every three hours. The use of a shallow pan for feeding is recommended. Immersing the baby pig's nose in the milk a few times will result in its drinking readily. It is extremely important that all feeding utensils be kept clean and sanitary; otherwise, scouring will occur.

The orphan pig can be fed a dry 20 to 23 percent milk replacer from 5 to 7 days of age until about 2 to 3 weeks of age (see Table 4-81, page 398). At this time, it can be switched to an 18 to 20 percent pig starter (see Table 4-80, page 397).

Creep Feeding Pigs

Baby pigs should have access to a creep feed beginning at 7 to 10 days of age. Commercial milk replacers (prestarters) and starters are readily available, or farm mixed creep rations can be used (see Table 4-81 for suggested milk replacers, and see Table 4-80 for a suggested starter ration). Pigs should receive a total of 3 to 5 pounds of milk replacer, after which they should be switched to starter until they weigh about 40 pounds. Early availability of a quality creep feed to young pigs will result in:

1. More uniform pigs with fewer runts.
2. Heavier weaning weight.
3. Less mortality of baby pigs.
4. Decrease in incidence and severity of baby pig scours.
5. Less setback to young pigs when weaned from the sow. The earlier pigs are to be weaned, the more important it is that they be eating dry feed at an early age.
6. Less weight loss by the sow.

For successful creep feeding, the following pointers are pertinent:

1. Begin by giving the baby pigs a mere handful of creep feed, replenished daily. The creep feed should not be allowed to become stale or contaminated. Place feed in flat pans.

2. Once the pigs have started to eat readily, place the ration in a creep feeder so that they have access to the ration at all times. One linear foot of feeder space should be provided for each five pigs. The edge of the feeder trough should not be more than four inches above the floor. For maximum consumption of the creep ration, the feeder should be located close to the waterer for the baby pigs.

3. Make clean, fresh water available to the young pigs in a waterer. It is not sufficient to rely on the waterer used by the sow to furnish water to the baby pigs.

4. The creep area should be light, warm, dry, and draft-free. It should be located in an area where the pigs are the least disturbed. Excitement, noise, and a change in feeding routine affect eating habits and subsequent feed consumption. Having the creep area near the sleeping area encourages more frequent eating. Arrange the creep area in such a way that it can be easily cleaned, and so that feed and water can be supplied conveniently without the producer getting into the area.

5. Individual litter creep areas are preferable. Where several litters have access to one creep area, it is advisable to limit the number of pigs to about 40 per creep.

Feeding Boars

The feed allowance of young boars should vary according to condition of the animals, the climatic condition, and the individuality. If the animals are inclined to get too fat, which is likely to happen in self-feeding, the ration may well contain a considerable amount of bulky feeds; otherwise, limited feeding may be necessary.

The feed requirements of the herd boar are about the same as those of a female of equal weight. He should always be kept in thrifty, vigorous condition and virile. In no case should boars be overfat, nor should they be in a thin run-down condition. Normally, the following feed allowances will suffice: for boars weighing 120 to 150 lb, 6 to 9 lb of feed daily; for mature boars, 5 to 7 lb of feed daily. A more liberal ration must be provided in the wintertime and when the sire is in heavy service. The feed allowance should be varied with the age, development, temperament, breeding demands, and roughage consumed.

Three suggested boar rations are given in Table 4-80, page 397; one for young boars weighing 40 to 120 pounds, another for boars weighing from 120 to 250 pounds, and a third for adult boars. Where only a few boars are involved, they may be fed the same ration as is provided for the sows.

FEEDING GROWING-FINISHING HOGS

In the practical swine enterprise, growing-finishing generally refers to that period from weaning to market weight of about 220 pounds. Because hogs are finished at an early age, the process really consists of both growing and finishing. In a general way, there are 2 methods of finishing hogs for market: (1) full feeding all the time until the animals attain a market weight, and (2) limited feeding early in the period with full feeding the last 60 to 75 days of the period before marketing.

For the production of lean (bacon) carcasses, the rate of gain should be restricted to about 1½ pounds daily after a liveweight of 100 to 125 pounds. This is easily accomplished by using a lighter, bulkier finishing ration (made by including 10 to 20% bran, oats, alfalfa, or other suitable bulky feed in the ration). Level of protein has no direct effect on carcass excellence, though it does affect the growth of the pig.

Neither system, full feeding nor limited feeding, can be recommended as being best for any and all conditions. The plan to follow should be determined by (1) market conditions, (2) type and breeding of the pigs, (3) price of feeds, (4) feeds available on the farm, (5) kind and extent of pastures available, (6) available labor, etc. Self-feeders are well adapted to a system of full feeding, but hand-feeding or interval feeding are necessary in any plan for limiting the ration.

Research indicates that there is little difference in the feed efficiency of group-fed pigs between (1) limited feeding of 5 pounds per day from 125 pounds to market, and (2) self-feeding. However, limited feeding results in slower gains, increased labor or mechanization, variable performance, and increased supervision. For these reasons, the practice can be justified only when sufficient premium is paid for the modest increase achieved in the lean-to-fat ratio. It should be added that the selection of hogs with bred in meat-type carcasses has largely alleviated the need for restricting rate of gain and using bulky rations as a means of getting leaner carcasses.

When on full feed, finishing pigs will consume 5 to 6 lb of feed daily per 100 lb liveweight up to 100 lb in weight. From 100 lb to a finished weight of 220 lb pigs on full feed consume about 4 lb of feed daily for each 100 lb of liveweight. With good feed and management, about 350 lb of feed are required to produce 100 lb of gain during the growing-finishing period from 40 lb to 220 lb weight. This means that about 630 lb of grower-finishing ration are required to make 180 lb of gain from 40 lb to 220 lb weight, but the amount varies with the inherent ability of the animals, thrift, and the kind and amount of pasture utilized. Right off, this appears to be high, particularly when compared with the conversion rates obtained in boar testing stations. But it must be remembered that boars are more efficient than barrows or gilts, that the use of pelleted rations and small number of pigs per pen improve feed efficiency in test stations, and that test stations represent near ideal conditions. Of course, the feed consumption of the breeding herd and that of the pigs during the suckling period must be added to the growing-finishing feed consumption figures in order to arrive at the total feed requirements. Studies conducted by the University of Illinois show that 27 percent of the total feed required to produce market hogs is consumed by the breeding herd (including the pigs to weaning) and 73 percent by the pigs after weaning. Hence, in producing a 200-lb market hog, the total feed requirements are:

	Pounds Feed	Percent of Total Feed
Breeding herd, including pigs to weaning ...	233	27
Pigs from 40 to 220 lb	630	73
Total	863	100

The protein requirements of the pig are greatest early in life. For this reason, decreasing percentages of protein supplement should be incorporated in mixed rations as the finishing process progresses.

However, ample protein should always be provided in the ration, otherwise growth will be retarded. It is also important that the mineral and vitamin needs of growing-finishing pigs be met.

As previously indicated, pigs can utilize a great variety of concentrates. The chief ingredients of a growing-finishing ration, therefore, are usually, for practical reasons, those most readily available at the lowest possible price.

Suggested growing-finishing rations are given in Table 4-80, page 397. Note that there are 3 different suggested rations, designed to meet the changing needs at different stages of growth—10 to 40 lb, 40 to 120 lb, and 120 lb to market weight.

In addition to the subject matter covered above, there are other feed and management aspects of great importance in feeding growing-finishing hogs, including (1) feed required to produce a pound of market hog, (2) effect of sex, (3) corn-hog ratio, and (4) soft pork. Each of these will be discussed in the sections that follow:

Feed Required to Produce a Pound of Market Hog

The Table 4-86 figures are realistic goals of the feed requirement to produce a pound of market hog. As noted, including the feed required by the sow and the boar, 3.9 pounds of feed per 1 pound of market hog is a reasonable goal. This means that 390 lb of feed are required to produce 100 lb of market hog, when including the feed consumed by the breeding herd.

TABLE 4-86

ESTIMATED FEED REQUIRED TO PRODUCE 220-POUND MARKET HOG

Stage of Production	Feed Required per 220-Lb Market Hog
	(lb)
Sow gestation ration (includes pregestation and breeding)	124
Boar ration	4
Lactation ration	49
Starter ration (creep to 40 lb)	50
Grower-finisher ration (40 to 220 lb)	630
Total	857
Per 100 lb of on-foot hog produced $\left(\frac{857}{220} \times 100\right) = 390$	
Per 1 lb of on-foot hog produced $(390 \div 100) = 3.9$	

The values given in Table 4-86 are estimates based on standards for apportioning the quantities of sow and boar feed to each pig and the feed conversion normally attained during the starter and grower-finisher periods. Although this table does not consider pig deaths after weaning, normal milling losses, and feed wastage, it is assumed that these losses would likely be offset by sow weight gains which are not considered in the pounds of pork produced. To achieve this goal—3.9 pounds of feed per 1 pound of on-foot hog produced—it is of prime importance that the swine operation be well managed, including limited feeding of pregnant sows, high conception rates, large litters weaned, early weaning and rebreeding, low death losses, minimal disease problems, balanced rations, and minimal feed wastage.

Effect of Sex on Performance of Growing-Finishing Hogs

When full fed, boars consume 10 to 15% less feed daily than barrows or gilts and are 10 to 15% more efficient in feed conversion. Also, boars gain faster than barrows and gilts. Barrows gain approximately .10 pound faster per day than gilts, which reduces their age at slaughter by 10 days. Feed per pound of gain is similar for barrows and gilts. Gilts yield carcasses having .11 inch less backfat, .52 square inch larger loin eye area, and 1.8% more lean cuts than barrows. Dressing percentage usually favors barrows, which is consistent with their greater depth of backfat.

Corn-Hog Ratio

The corn-hog ratio refers to the number of bushels of corn required to be equivalent in value to 100 pounds of live hogs at local markets, based on average prices received by farmers for corn and hogs. During the 10-year period 1972-1981, the corn-hog ratio averaged 17.5. This means that the price relationship was such that 17.5 bushels of corn equaled in value 100 pounds of hogs.

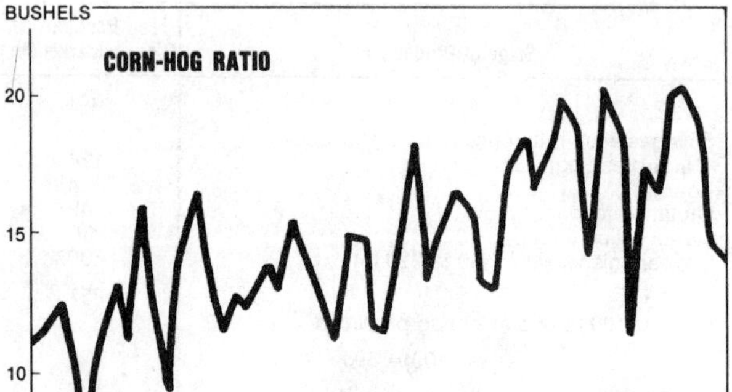

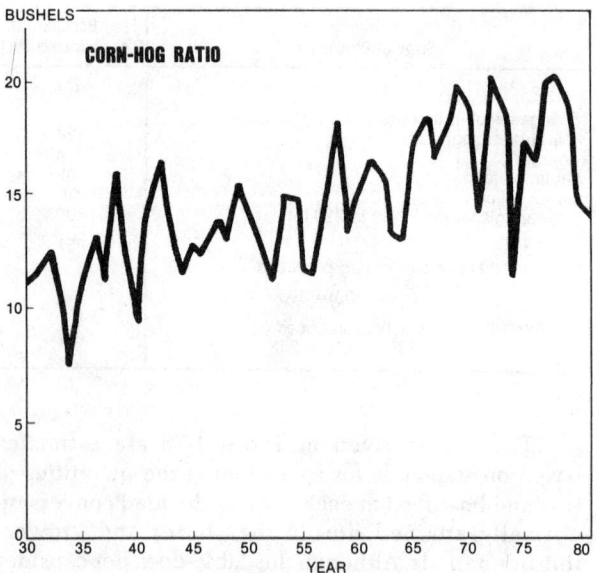

Fig. 4-34. The corn-hog ratio, 1930 to 1981. which averaged 12.6 for the entire period. (Based on data from *Agricultural Statistics*, USDA)

A high corn-hog ratio—one above 17.5 in recent years—indicates cheap corn and high-priced hogs and likely profit to the producer—conditions that stimulate more breeding and more feeding to heavier weights. On the other hand, a low ratio, one which is below 17.5 means high-riced corn and low-priced hogs—conditions that result in less breeding and feeding of swine.

Soft Pork

Feed fats are laid down in the body without undergoing much change. Thus, when finishing hogs are liberally fed on high fat content feeds in which the fat is liquid at ordinary temperature, soft pork results. This condition prevails when hogs are liberally fed such feeds as soybeans, peanuts, mast, or garbage.

Soft pork is undesirable from the standpoint of both the processor and the consumer. It remains flabby and oily even under refrigeration. In soft pork, there is a higher shrinkage in processing; the cuts do not stand up and are unattractive in the showcase; it is difficult to slice the bacon; and the cooking losses are higher through loss of fat. For these reasons, hogs that are liberally fed on those feeds known to produce sc pork are heavily discounted on the market.

The firmness of pork carcasses may be judged b (1) grasping the flank below the ham, (2) lifting or end of the cut while permitting the other end to re on the table (a firm pork cut will not bend readily), (3) applying a slight pressure of the thumb (not gou ing) on a cut surface. Experimentally, either th iodine number or the refractive index is used in de termining the degree of softness; this is a measure the degree of unsaturation.

Unless the producer is willing to take the norm reduction in price (about $1.00 per cwt), it is recom mended that feeds which normally produce soft po be fed liberally only to pigs under 85 pounds i weight and to the breeding herd. For growin finishing pigs over 85 pounds in weight, soybeans an peanuts should not constitute more than 10 percent the ration if a serious soft pork problem is to b averted.

Experimental evidence and practical observatio have shown, however, that when a ration producin hard fat is given following a period of feeds rich i unsaturated fats, the body fat gradually becom harder. It has also been found that this process take place more rapidly if the animals are first fasted for period before the change in ration is made. This prac tice is called "hardening off." Thus, many hogs th are, for practical reasons, finished primarily on suc feeds as soybeans, peanuts, or garbage, are hardene off with a ration of corn or some other suitable grain

FEEDING SHOW AND SALE SWINE

Any of the rations listed in Table 4-80, page 39 for the respective classes and ages of swine are suita ble for use in fitting show animals of similar classific tion. Because of the high cost of labor, the rece trend has been toward self-feeding both young bree ing animals and market barrows and gilts that a being fitted for show. Many of them are left on sel feeders right up to show time, others are hand fe during the last month or two only of the fitting perio However, most experienced herdsmen feel that the can get superior bloom and condition by either (1 hand feeding, or (2) using a combination of hand fee ing and self-feeding (hand feeding twice daily and a lowing free access to a self-feeder). When hand fee ing they also prefer mixing the ration with skim mil buttermilk, or condensed buttermilk, and feeding th entire ration in the form of a slop.

Adding milk to a ration that is already proper balanced does make for a higher protein content tha necessary. On the other hand, most experience herdsmen prefer using rations of higher protein co tent for fitting purposes. They feel they get mor bloom that way. In general, however, when skim mil r buttermilk is used in slop feeding, the protein eeds of the ration may be reduced by one-half with-ut harm to the animal.

In fitting show barrows, it may be necessary to lecrease or discontinue slop-feeding two to four veeks before the show to avoid paunchiness and low-ring of the dressing percentage.

When oatmeal (oat groats, rolled hulled oats) is not too high priced, many successful hog showmen replace up to 50 percent of the grain (corn, wheat, bar-ley, oats, and/or sorghum) in the ration with oatmeal. They do this especially when fitting hogs—both breeding animals and barrows—in the younger age groups. Oatmeal is highly palatable, lighter, and less fattening than corn.

Suitable minerals should always be provided.

PART VI—FEEDING HORSES

Fig. 4-35. Feeding horses.

Feed is the most important influence in the envi-onment of horses. Unless they are fed properly, their naximum potential in reproduction, growth, body orm, speed, endurance, style, and attractiveness can-not be achieved. Also, feed constitutes the greatest single cost item in the horse business.

Horses differ from other farm animals in that they have greater value; are kept for recreation, sport, and work; are fed for a longer life of usefulness; have a smaller digestive tract; should not carry surplus weight; and are fed for nerve, mettle, animation, and character of muscle.

NUTRIENT REQUIREMENTS VS ALLOWANCES

In ration formulation, two words are commonly used—"requirements" and "allowances." Require-ments do not provide for margins of safety. Thus, to feed a horse on the basis of meeting the bare re-quirements would not be unlike building a bridge without providing margins of safety for heavier than average loads or for floods. No competent engineer would be so foolish as to design such a bridge. Likewise, knowledgeable horse nutritionists provide for margins of safety—they provide for the necessary nutritive allowances. They allow for variations in feed composition; possible losses during storage and proc-essing; day-to-day, and period-to-period, differences in needs of animals; age and size of animal; stage of gestation and lactation; the kind and degree of activ-ity; the amount of stress; the system of management; the health, condition, and temperament of the animal; and the kind, quality, and amount of feed—all of which exert a powerful influence in determining nu-tritive needs.

Meeting the nutrient needs of horses is a major factor in determining their efficiency and years of service. In the discussion that follows, both require-ments and allowances will be covered; and the rec-ommended nutritive allowances of the horse will be discussed under the following headings: (1) energy (carbohydrates and fats), (2) protein, (3) minerals, (4) vitamins, and (5) water.

NUTRIENT REQUIREMENTS

The nutrient requirements of horses are given in Tables 4-87, 4-88, 4-89, and 4-90, pages 410 to 412, which follow.

RECOMMENDED NUTRIENT ALLOWANCES

Presently available information indicates that the nutrient allowances recommended in Tables 4-91 and 4-92, page 413, will meet the minimum requirements for horses and provide reasonable margins of safety.

In using the recommended allowances given in Tables 4-91 and 4-92, page 413, as guides in horse ra-tion formulations, consideration should be given to the nutrients provided by the ingredients of the ra-tion, for it's the total composition of the finished feed that counts.

• *Stress affects nutritive needs*—Stress may be caused by excitement, temperament, fatigue, number of horses together, previous nutrition, breed, age, and

TABLE 4-87

DAILY NUTRIENT REQUIREMENTS OF MATURE HORSES, PREGNANT MARES, AND LACTATING MARES[1] (per Animal)

Body Weight		Daily Feed[2]		Digestible Energy	Protein		Digestible Protein		Ca	P	Vitamin A[3]
(lb)	*(kg)*	*(lb)*	*(kg)*	*(Mcal)*	*(lb)*	*(g)*	*(lb)*	*(g)*	*(g)*	*(g)*	*(1,000 IU)*
Mature horses at rest (maintenance)											
441	*200*	8.2	*3.75*	8.24	.70	*320*	.31	*140*	*9*	*6*	5.0
882	*400*	13.9	*6.30*	13.86	1.19	*540*	.53	*240*	*18*	*11*	10.0
1,102	*500*	16.4	*7.45*	16.39	1.39	*630*	.64	*290*	*23*	*14*	12.5
1,323	*600*	18.8	*8.50*	18.79	1.61	*730*	.73	*330*	*27*	*17*	15.0
Mares, last 90 days of pregnancy											
441	*200*	8.1	*3.70*	9.23	.86	*390*	.44	*200*	*14*	*9*	10.0
882	*400*	13.7	*6.20*	15.52	1.41	*640*	.75	*340*	*27*	*19*	20.0
1,102	*500*	16.2	*7.35*	18.36	1.65	*750*	.86	*390*	*34*	*23*	25.0
1,323	*600*	18.5	*8.40*	21.04	1.91	*870*	1.01	*460*	*40*	*27*	30.0
Mares, first 3 months of lactation											
441	*200*	11.5	*5.20*	14.58	1.56	*710*	1.19	*540*	*24*	*16*	13.0
882	*400*	18.4	*8.35*	23.36	2.46	*1,120*	1.50	*680*	*40*	*27*	22.0
1,102	*500*	22.2	*10.10*	28.27	2.99	*1,360*	1.85	*840*	*50*	*34*	27.5
1,323	*600*	26.0	*11.80*	33.05	3.52	*1,600*	2.18	*990*	*60*	*40*	33.0
Mares, lactation from 3 months to weaning											
441	*200*	11.0	*5.00*	12.99	1.32	*600*	.75	*340*	*20*	*13*	11.0
882	*400*	17.1	*7.75*	20.20	2.00	*910*	1.12	*510*	*33*	*22*	18.0
1,102	*500*	20.6	*9.35*	24.31	2.42	*1,100*	1.36	*620*	*41*	*27*	22.5
1,323	*600*	23.9	*10.90*	28.29	2.84	*1,290*	1.61	*730*	*49*	*30*	27.0

[1]Adapted by the author from *Nutrient Requirements of Horses*, No. 6, 4th rev. ed., NRC-National Academy of Sciences, 1978, p. 17-20.
[2]Assume 2.75 Mcal of digestible energy per kilogram of 100% dry feed.
[3]One milligram of beta-carotene equals 400 IU of vitamin A.

TABLE 4-88

DAILY NUTRIENT REQUIREMENTS OF GROWING HORSES (per Animal)[1]

Age	Body Weight		Percentage of Mature Weight	Daily Gain		Daily Feed[2]		Digestible Energy	Protein		Digestible Protein		Ca	P	Vitamin A[3]
(mo)	*(lb)*	*(kg)*		*(lb)*	*(kg)*	*(lb)*	*(kg)*	*(Mcal)*	*(lb)*	*(g)*	*(lb)*	*(g)*	*(g)*	*(g)*	*(1,000 IU)*
441-lb mature weight (200 kg)															
3	132	*60*	30.0	1.54	*.70*	5.0	*2.25*	7.35	.90	*410*	.84	*380*	*18*	*11*	2.4
6	209	*95*	47.5	1.10	*.50*	6.3	*2.85*	8.80	1.03	*470*	.68	*310*	*19*	*14*	3.8
12	308	*140*	70.0	.44	*.20*	6.4	*2.90*	8.15	.77	*350*	.44	*200*	*12*	*9*	5.5
18	374	*170*	85.0	.22	*.10*	6.8	*3.10*	8.10	.70	*320*	.37	*170*	*11*	*7*	6.0
24	407	*185*	92.5	.11	*.05*	6.8	*3.10*	8.10	.66	*300*	.33	*150*	*10*	*7*	5.5
882-lb mature weight (400 kg)															
3	275	*125*	31.3	2.20	*1.00*	7.8	*3.55*	11.51	1.43	*650*	1.10	*500*	*27*	*17*	5.0
6	407	*185*	46.3	1.43	*.65*	9.2	*4.20*	13.03	1.45	*660*	.95	*430*	*27*	*20*	7.4
12	583	*265*	66.3	.88	*.40*	10.9	*4.95*	13.80	1.32	*600*	.77	*350*	*24*	*17*	10.0
18	726	*330*	82.5	.55	*.25*	12.2	*5.50*	14.36	1.30	*590*	.70	*320*	*22*	*15*	11.5
24	803	*365*	91.3	.22	*.10*	11.8	*5.35*	13.89	1.14	*520*	.59	*270*	*20*	*13*	11.0
1,102-lb mature weight (500 kg)															
3	341	*155*	31.0	2.64	*1.20*	9.2	*4.20*	13.66	1.65	*750*	1.19	*540*	*33*	*20*	6.2
6	506	*203*	40.6	1.76	*.80*	11.0	*5.00*	15.60	1.74	*790*	1.14	*520*	*34*	*25*	9.2
12	715	*325*	65.0	1.21	*.55*	13.2	*6.00*	16.81	1.67	*760*	.99	*450*	*31*	*22*	12.0
18	880	*400*	80.0	.77	*.35*	14.3	*6.50*	17.00	1.56	*710*	.86	*390*	*28*	*19*	14.0
24	990	*450*	90.0	.33	*.15*	14.5	*6.60*	16.45	1.39	*630*	.72	*330*	*25*	*17*	13.0
1,323-lb mature weight (600 kg)															
3	374	*170*	28.3	3.08	*1.40*	10.2	*4.65*	15.05	1.85	*840*	1.72	*780*	*36*	*23*	6.8
6	583	*265*	44.2	1.87	*.85*	12.0	*5.45*	16.92	1.89	*860*	1.25	*570*	*37*	*27*	10.6
12	847	*385*	64.2	1.32	*.60*	14.8	*6.75*	18.85	1.98	*900*	1.10	*500*	*35*	*25*	14.0
18	1,045	*475*	79.2	.77	*.35*	16.2	*7.35*	19.06	1.65	*750*	.95	*430*	*32*	*22*	13.5
24	1,188	*540*	90.0	.44	*.20*	16.3	*7.40*	19.26	1.63	*740*	.86	*390*	*31*	*20*	13.0

[1]Adapted by the author from *Nutrient Requirements of Horses*, No. 6, 4th rev. ed., NRC-National Academy of Sciences, 1978, p. 17-20.
[2]Assume 2.75 Mcal of digestible energy per kilogram of 100% dry feed.
[3]One milligram of beta-carotene equals 400 IU of vitamin A.

TABLE 4-89

NUTRIENT REQUIREMENTS OF RATIONS FOR HORSES AND PONIES EXPRESSED ON AN AS-FED BASIS[1]
(In Percentage or Amount per Pound or Kilogram of Ration)

	Digestible Energy		Crude Protein	Calcium	Phosphorus	Vitamin A	
	(Mcal/lb)	*(Mcal/kg)*	*(%)*	*(%)*	*(%)*	*(IU/lb)*	*(IU/kg)*
Mature horses and ponies at maintenance	0.9	*2.0*	7.7	0.27	0.18	650	*1,450*
Mares, last 90 days of gestation	1.0	*2.25*	10.0	0.45	0.30	1,400	*3,000*
Lactating mare, first 3 months	1.2	*2.6*	12.5	0.45	0.30	1,150	*2,500*
Lactating mare, 3 months to weaning	1.1	*2.3*	11.0	0.40	0.25	1,000	*2,200*
Creep feed	1.4	*3.15*	16.0	0.80	0.55	—	—
Foal (3 months of age)	1.35	*2.9*	16.0	0.80	0.55	800	*1,800*
Weanling (6 months of age)	1.25	*2.8*	14.5	0.60	0.45	800	*1,800*
Yearling (12 months of age)	1.2	*2.6*	12.0	0.50	0.35	800	*1,800*
Long yearling (18 months of age)	1.1	*2.3*	10.0	0.40	0.30	800	*1,800*
Two-year old (light training)	1.2	*2.6*	9.0	0.40	0.30	800	*1,800*
Mature working horses:							
light work[2]	1.0	*2.25*	7.7	0.27	0.18	650	*1,450*
moderate work[3]	1.2	*2.6*	7.7	0.27	0.18	650	*1,450*
intense work[4]	1.25	*2.8*	7.7	0.27	0.18	650	*1,450*

[1]Adapted by the author from *Nutrient Requirements of Horses*, No. 6, 4th rev. ed., NRC-National Academy of Sciences, 1978, p. 21. As-fed is estimated to be 90% dry matter.
[2]Examples are horses used in western pleasure, bridle path hack, equitation, etc.
[3]Examples are ranch work, roping, cutting, barrel racing, jumping, etc.
[4]Examples are race training, polo, etc.

TABLE 4-90

NUTRIENT REQUIREMENTS OF RATIONS FOR HORSES AND PONIES EXPRESSED ON A MOISTURE-FREE BASIS[1]
(In Percentage or Amount per Pound or Kilogram of Ration)

	Digestible Energy		Crude Protein	Calcium	Phosphorus	Vitamin A	
	(Mcal/kg)	*(Mcal/kg)*	*(%)*	*(%)*	*(%)*	*(IU/lb)*	*(IU/kg)*
Mature horses and ponies at maintenance	1.0	*2.2*	8.5	0.30	0.20	725	*1,600*
Mares, last 90 days of gestation	1.1	*2.5*	11.0	0.50	0.35	1,550	*3,400*
Lactating mare, first 3 months	1.3	*2.8*	14.0	0.50	0.35	1,275	*2,800*
Lactating mare, 3 months to weanling	1.2	*2.6*	12.0	0.45	0.30	1,150	*2,450*
Creep feed	1.6	*3.5*	18.0	0.85	0.60	—	—
Foal (3 months of age)	1.5	*3.25*	18.0	0.85	0.60	900	*2,000*
Weanling (6 months of age)	1.4	*3.1*	16.0	0.70	0.50	900	*2,000*
Yearling (12 months of age)	1.3	*2.8*	13.5	0.55	0.40	900	*2,000*
Long yearling (18 months of age)	1.2	*2.6*	11.0	0.45	0.35	900	*2,000*
Two-year old (light training)	1.3	*2.6*	10.0	0.45	0.35	900	*2,000*
Mature working horses:							
light work[2]	1.1	*2.5*	8.5	0.30	0.20	725	*1,600*
moderate work[3]	1.3	*2.9*	8.5	0.30	0.20	725	*1,600*
intense work[4]	1.4	*3.1*	8.5	0.30	0.20	725	*1,600*

[1]Adapted by the author from *Nutrient Requirements of Horses,* No. 6, 4th rev. ed., NRC-National Academy of Sciences, 1978, p. 21. Moisture-free is 100% dry matter.
[2]Examples are horses used in western pleasure, bridle path hack, equitation, etc.
[3]Examples are ranch work, roping, cutting, barrel racing, jumping, etc.
[4]Examples are race training, polo, etc.

TABLE 4-91

RECOMMENDED ALLOWANCES OF PROTEIN, FIBER, AND TOTAL DIGESTIBLE NUTRIENTS (TDN)

Type of Horse	Minimum Crude Protein	Maximum Crude Fiber	Minimum TDN
	(%)	(%)	(%)
Most mature horses used for race, show, or pleasure	12	25	53 to 70[1]
Broodmares	13	25	50 to 60
Stallions	14	25	50 to 68[2]
Young equines:			
Foals, 2 weeks to 10 months old	21	8	68 to 74
Weanlings to 18 months old	14	20	60
18 months to 3 years old	13	25	50 to 60

[1]The heavier the work, the more energy required.
[2]Increase the energy immediately before and during the breeding season.

TABLE 4-92

RECOMMENDED ALLOWANCES OF MINERALS AND VITAMINS

Kind of Mineral or Vitamin	Daily Allowance per 1,000-Lb *(455-kg)* Horse[1]	Allowances per Ton of Finished Feed (hay and grain; As-fed basis)[2]
Major or macro minerals:		
Salt	2 oz	10.0 lb
Calcium	70.0 g	12.33 lb
Phosphorus	60.0 g	10.57 lb
Magnesium	6.4 g	1.3 lb
Potassium	68.1 g	12.0 lb
Trace or micro minerals:		
Cobalt	1.5 mg	.12 g
Copper	90 mg	7.2 g
Iodine	2.6 mg	.21 g
Iron	640 mg	51.2 g
Manganese	340 mg	27.2 g
Zinc	400 mg	32.0 g
Fat-soluble vitamins:		
Vitamin A (USP)	50,000	4,000,000
Vitamin D_2 (USP)	7,000	560,000
Vitamin E (IU)	200	16,000
Vitamin K (mg)	8	640
Water-soluble vitamins:		
Vitamin B_{12} (mcg)[3]	125	10,000
Choline (mg)	400	32,000
Folic acid (mg)	2.5	200
Niacin (nicotinic acid) (mg)	50	4,000
Pantothenic acid (mg)	60	4,800
Riboflavin (B_2) (mg)	40	3,200
Thiamin (B_1) (mg)	25	2,000

[1]This is based on an allowance of 25 lb *(11 kg)* of feed per 1,000-lb *(455-kg)* horse per day, or 2.5 lb *(1.1 kg)* of feed per 100 lb *(45 kg)* of body weight.
[2]Where hay is fed separately, double this amount should be added to the concentrate.
[3]Micrograms.

management. Race and show horses are always under stress; and the greater the speed, and the more tired they become, the greater the stress. Thus, the ration for race and show horses should be scientifically formulated. The greater the stress, the more exacting the nutritive requirements.

• *Other factors affect nutritive needs*—The nutrient requirements of horses vary with the individuality, age, and size of the animal; the kind, amount, and severity of work performed; the condition and training of the animal; the environmental temperature; the kind, quality, and amount of feed; the system of management; and the health and temperament of the animal; the stage of gestation or lactation of a mare; and the ability of the rider or driver. All these forces, and more, are continually exerting a powerful influence in determining nutritive needs. Also, the nutrient needs do not remain the same from day to day or from period to period. How well the horseman understands, anticipates, interprets, and meets these requirements usually determines the success or failure of the ration.

Energy

The National Academy of Sciences reports the following energy requirements for various activities of light horses:[18]

Activity	Requirement
	(kcal/hour/ kg of mass)
Walking	0.5
Slow trotting, some cantering	5.1
Fast trotting, cantering, some jumping	12.5
Cantering, galloping, jumping	24.0
Strenuous effort	39.0

Note that the energy requirements during "strenuous effort" are 78 times greater than in "walking."

Fig. 4-36. In racing, horses may use up to 100 times the energy utilized at rest.

Generally, increased energy for horses is met by increasing the grain and decreasing the roughage.

A lack of energy may cause slow and stunted growth in foals and loss of weight, poor condition, and excessive fatigue in mature horses.

[18]*Nutrient Requirements of Horses*, No. 6, 3rd rev. ed., NRC-National Academy of Sciences, Washington, D.C., 1973.

Protein

Horses of all ages and kinds require adequate amounts of protein of suitable quality for maintenance, growth, finishing, reproduction, and work. Of course, the protein requirements for growth and reproduction are the greatest and most critical.

Since the vast majority of protein requirements given in feeding standards meet minimum needs only, the allowances for race, show, breeding, and young animals should be higher.

The extent to which the horse's ration is supplemented with proteins depends primarily on the age of the horse and on the quality of the forage fed. Growing or lactating animals require somewhat more protein than horses that are idle, gestating, or working. Also, grass hays are generally low in quality and quantity of proteins and require more supplementation than legumes.

A deficiency of proteins in the horse may result in the following deficiency symptoms: depressed appetite, poor growth, loss of weight, reduced milk pr duction, irregular estrus, lowered foal crops, loss condition, and lack of stamina.

PROTEIN POISONING

Some opinions to the contrary, protein poisonin as such has never been documented. There is n proof that heavy feeding of high-protein feeds horses is harmful, provided (1) the ration is balance out in all other respects, (2) the animal's kidneys a normal and healthy (a large excess of protein in term of body needs increases the work of the kidneys f the excretion of the urea), (3) any ration change high-protein feed is made gradually, as is recom mended in any change in feed, and (4) there adequate exercise and normal metabolism.

Some horses do appear to be allergic to certa proteins or to excesses of specific amino acids, as result of which they may develop "protein bumps."

TABLE 4-93—HOR

Minerals Which May Be Deficient Under Normal Conditions	Conditions Usually Prevailing Where Deficiencies Are Reported	Function of Mineral	Some Deficiency Symptoms
MAJOR OR MACRO MINERALS:			
Salt (sodium and chlorine, NaCl)	Negligence, for salt is cheap. Horses sweating excessively, as with vigorous exercise and during warm weather.	Sodium and chlorine help maintain osmotic pressure in body cells, upon which depends the transfer of nutrients to the cells and the removal of waste materials. Sodium is associated with muscle contraction and is important in making bile, which aids in the digestion of fats and carbohydrates. Chlorine is required for the formation of hydrochloric acid in the gastric juice so vital to protein digestion.	In warm or hot weather, wor horses show heat stress. Long-term symptoms of sodiu deficiency are: depraved a petite, rough hair coat, reduce growth of young animals, ar decreased milk production.
Calcium (Ca)	The typical horse ration of grass hay and farm grains—usually deficient in calcium.	Builds strong bones and sound teeth. Very important during lactation. Affects availability of phosphorus.	Rickets in young horses; o teomalacia in mature horses.
Phosphorus (P)	Horses pastured on phosphorus-deficient areas or fed for a long period on mature, weathered forage.	Important in the development of bones and teeth. Essential to metabolism of carbohydrates and fats, and enzyme activation.	Rickets in young horses; o teomalacia in mature horses.
Magnesium (Mg)	Horses on high grain-low forage ration.	Reduces stress and irritability.	Horses under stress are keyed u high-strung, and jumpy.

Footnotes on last page of table.

It is recognized that protein in excess of what the body can use tends to be wasted insofar as its specific functions are concerned, since it cannot be stored in any but very limited amounts and must be catabolized. Nevertheless, some wastage of protein in terms of its known functions may be both physiologically and economically desirable in order to (1) maintain the protein reserves, (2) provide an adequate protein-calorie ratio for efficient energy utilization, and (3) assure that protein quality needs are met, despite the marked difference of quality among commonly fed rations. Generally speaking, high-protein feeds are more expensive than high-energy feeds (feeds high in carbohydrates and fats), with the result that there is the temptation to feed too little of them.

Minerals

The classical horse ration of grass hay and farm grains is usually deficient in calcium, but adequate in phosphorus. Also, salt is almost always deficient; and many horse rations do not contain sufficient iodine and certain other trace elements. Thus, horses usually need special mineral supplements. But they should not be fed either more or less minerals than needed.

Although acute mineral deficiency diseases and actual death losses are relatively rare, inadequate supplies of any one of the essential mineral elements may result in lack of thrift, poor gains, inefficient feed utilization, lowered reproduction, and decreased performance in racing, showing, riding, or whatnot. This does not mean that all mineral elements known to be required by at least one animal species must always be included in horse mineral supplements. Rather, only the specific minerals that are deficient in the ration—and in the quantities necessary—should be supplied. *Excesses and mineral imbalances are to be avoided.*

Table 4-93, Horse Mineral Chart, lists the minerals required by horses and gives pertinent information pertaining to each.

MINERAL CHART

Recommended Allowances[1]		Practical Sources of the Mineral	Comments
Daily per 1,000-lb *(455-kg)* Horse	Per Ton or Percent of Total Ration[2]		
2.0 oz (56 g)	10 lb *(4.5 kg)* per ton (as-fed) 0.5 to 1.0% of ration.	Salt provided free choice, preferably in loose form, or 0.5 to 1.0% salt added to the ration.	Horses require both sodium and chlorine, but the requirement for chlorine is approximately half that of sodium. Generally, the chlorine requirements will be met if the sodium needs are adequate. Sodium and chlorine are low in feeds of plant origin. There is little danger of overfeeding salt unless a salt starved animal is suddenly exposed to too much salt, or if liberal amounts of water are not available.
70 g	12.3 lb *(5.6 kg)* per ton (as-fed) 0.6 to 0.7% of ration.	Ground limestone or oystershell flour. Where both calcium and phosphorus are needed, use bone meal or dicalcium phosphate (see Table 4-6, p. 229).	The calcium-phosphorus ratio should be maintained close to 1.1:1 although 2:1 is acceptable when the higher calcium content is due to the presence of legume. Narrower ratios may cause osteomalacia in mature horses. Where there is a shortage of calcium in the ration, it is withdrawn from the bones.
60 g	10.6 lb *(4.8 kg)* per ton (as-fed) 0.5% of ration.	Monosodium phosphate, disodium phosphate, or sodium tripolyphosphate. Where both calcium and phosphorus are needed, use bone meal or dicalcium phosphate. (see Table 4-6, p. 229).	Same as stated for calcium under "Comments" above. If plenty of vitamin D is present, the ratio of calcium to phosphorus becomes less important. Apparently phosphorus cannot be withdrawn from the bones.
6.4 g	1.3 lb *(0.6 kg)* per ton (as-fed).	Magnesium sulfate. Magnesium oxide.	Excess of magnesium upsets calcium and phosphorus metabolism.

(Continued)

TABLE 4-

Minerals Which May Be Deficient Under Normal Conditions	Conditions Usually Prevailing Where Deficiencies Are Reported	Function of Mineral	Some Deficiency Symptoms
Potassium (K)	Where stabled horses are on high-concentrate rations.	Major cation of intracellular fluid where it is involved in osmotic pressure and acid-base balance. Muscle activity. Required in enzyme reaction involving phosphorylation of creatine. Influences carbohydrate metabolism.	Growth retardation, unsteady ga general muscle weaknes pica, diarrhea, distended a domen, emaciation, followed death. Abnormal electrocardiograms.
TRACE OR MICRO MINERALS:			
Cobalt (Co)	Animals pastured in cobalt-deficient areas, such as Australia, western Canada, and in following states of U.S.: Fla., Mich., Wisc., N.H., Penn., and N.Y.	Cobalt is required for the synthesis of vitamin B_{12} in the intestinal tract of the horse.	Anemia
Copper (Cu)	Suckling foals. Mare's milk, along with milk from other species, is low in copper. Horses grazing pastures low in copper, as in Australia and in Florida and the Coastal Plain regions of the U.S.	Copper, along with iron and vitamin B_{12}, is necessary for hemoglobin formation, although it forms no part of the hemoglobin molecule (or red blood cells). Closely associated with normal bone development in young growing animals.	Anemia, characterized by few than normal red cells and le than normal amount of hem globin. Abnormal bone development foals.
Iodine (I)	Iodine-deficient areas or soils (in northwestern U.S. and in the Great Lakes region) when iodized salt is not fed. Use of feeds that come from iodine-deficient areas.	Iodine is needed by the thyroid gland in making thyroxin, an iodine-containing compound which controls the rate of body metabolism or heat production.	Foals born dead, or very wea and unable to stand or nurse. Higher than normal incidence navel ill.
Iron (Fe)	Suckling foals kept away from soil and feed other than milk.	Necessary for formation of hemoglobin, an iron-containing compound which enables the blood to carry oxygen. Also, important to certain enzyme systems.	Iron-deficiency anemia, charac terized by fewer than norm red cells and less than norm amount of hemoglobin.
Manganese (Mn)	Excess calcium and phosphorus which decreases absorption of manganese.	Essential for normal bone formation (as a component of the organic matrix). Thought to be an activator of enzyme systems. Growth and reproduction.	Poor growth. Lameness, shortening and bow ing of legs, and enlarged joints Impaired reproduction (testicula degeneration of males; defec tive ovulation of females).
Zinc (Zn)	Feeds low in zinc. Excess calcium may reduce the absorption and utilization of zinc.	Required for normal protein synthesis and metabolism. Imparts gloss or "bloom" to the hair coat.	Rough, dull hair coat. Loss of appetite.

[1]These are recommended allowances, and not requirements. The author's position on recommended allowances vs requirements for horses is clearly stated in the narrative of this book under the heading entitled "Nutrient Requirements Vs Allowances."

[2]Where hay is fed separately, double these amounts should be added to the concentrate.

—(Continued)

Recommended Allowances[1]		Practical Sources of the Mineral	Comments
Daily per 1,000-lb *(455-kg)* Horse	Per Ton or Percent of Total Ration[2]		
3.1 g	12 lb *(5.4 kg)* per ton (as-fed).	Potassium chloride. Roughages usually contain ample potassium.	Potassium deficiency may occur in stabled horses on a high-concentrate ration.
.5 mg	0.12 g/ton (as-fed).	Cobaltized mineral mix made by adding cobalt at the rate of 0.2 oz/100 lb of salt as cobalt chloride, cobalt sulfate, cobalt oxide, or cobalt carbonate. Also, several good commercial cobalt-containing minerals are on the market.	The disease called "salt sick" in Florida is due to a cobalt deficiency associated with a copper deficiency.
90 mg	7.2 g/ton (as-fed).	Trace mineralized salt containing copper sulfate or copper carbonate.	A copper deficiency in horses has been reported in Australia. High molybdenum in forages does not appear to affect horses as much as ruminants. However, in high-molybdenum areas, more copper may be added to horse rations; but excesses and toxicity should be avoided.
2.6 mg	0.21 g/ton (as-fed).	Stabilized iodized salt containing 0.01% potassium iodide (0.0076% iodine). Calcium iodate. Ethylenediamine dihydriodide (EDDI).	Enlargement of the thyroid gland (goiter) is nature's way of trying to make enough thyroxin (an iodine-containing hormone) when there is insufficient iodine in the feed. Feeding excess iodine continuously will also produce goiter in foals.
640 mg	51.2 g/ton (as-fed).	Ferrous sulfate administered orally. Trace mineralized salt. Cane molasses.	The horse's body contains about 0.004% iron. Milk is deficient in iron, and the iron content of the mother cannot be increased through feeding iron. Thus, foals should be individually or creep fed as soon as they are old enough. A variable store of both iron and copper is located in liver and spleen, and some iron is found in the kidneys. Too much iron may be harmful.
340 mg	27.2 g/ton (as-fed).	Trace mineralized salt containing 0.25% manganese (or more).	Most natural feedstuffs are rich in manganese.
400 mg	32 g/ton (as-fed).	Zinc carbonate. Zinc sulfate.	If zinc in the feed is on the low side, the addition of zinc should improve the hair coat.

Note: Mineral recommendations for all classes and ages of horses, especially those fed unmixed rations or on pasture, are—

1. *Where animals are on liberal grain feeding*—Provide free access to a 2-compartment mineral box, with (a) trace mineralized salt in one side, and (b) in the other side, a mixture of ⅓ trace mineralized salt (salt included for purposes of palatability), ⅓ defluorinated phosphate or steamed bone meal, and ⅓ ground limestone or oystershell flour.
2. *Where animals are primarily on roughage (pasture, hay, and/or silage)*—Provide free access to a 2-compartment mineral box, with (a) trace mineralized salt in one side (salt included for purposes of palatability), and (b) in the other side, a mixture of ⅓ trace mineralized salt and ⅔ defluorinated phosphate or steamed bone meal.

Vitamins

Certain vitamins are necessary for the growth, development, health, and reproduction of horses. Deficiencies of vitamins A and D are sometimes encountered. Also, indications are that vitamin E and some of the B vitamins are required by horses. Further, it is recognized that single, uncomplicated vitamin deficiencies are the exception rather than the rule.

High-quality, leafy, green forages plus plenty of sunshine generally give horses most of the vitamins they need. Horses get carotene (which they can convert to vitamin A) and riboflavin from green pasture and green hay not over a year old, and they get vitamin D from sunlight and sun-cured hay. If plenty of green forage and sunlight are not available, the horseman should get the advice of a nutritionist or veterinarian on the use of vitamin additives to the feed.

Table 4-94, Horse Vitamin Chart, lists the vitamins required by horses and gives pertinent information pertaining to each.

Water

Horses should have access to ample quantities of clean, fresh water at all times. They will drink 10 to 12 gallons daily; the amount depends on weather, amount of work done (sweating), rations fed, and size of horse.

Free access to water is desirable. When this is not possible, horses should be watered at approximately the same times daily. Opinions vary among horsemen as to the proper times and method of watering horses. All agree, however, that regularity and frequency are desirable. Most horsemen agree that water may be given before, during, or after feeding.

Frequent, small waterings between feedings are desirable during warm weather or when the animal is being put to hard use. Do not allow a horse to drink heavily when he is hot, because he may founder; and do not allow a horse to drink heavily just before being put to work.

Water should be available in both stalls and corrals. All waterers should have drains for easy clean-

TABLE 4-94—HORSE

Vitamins Which May Be Deficient Under Normal Conditions	Conditions Usually Prevailing Where Deficiencies Are Reported	Functions of Vitamins	Some Deficiency Symptoms
FAT-SOLUBLE VITAMINS:			
A	Extended drought. Bleached hays. Stall feeding where there is little or no green forage or yellow corn. Following great stress, as when race or show horses are put in training.	Promotes growth and stimulates appetite. Assists in reproduction and lactation. Keeps the mucous membranes of respiratory and other tracts in healthy condition. Makes for normal vision. Prevents night blindness.	Reproductive failure, nerve degeneration, night blindness, uneven and poor hoof development, a predisposition to respiratory infection, lacrimation (tears), incoordination, keratinization of the cornea, progressive weakness, certain bone disorders, and finicky appetite.
D	Limited sunlight, and or limited sun-cured hay, especially when horse is kept inside most of the time.	Assimilation and utilization of calcium and phosphorus, necessary in normal bone development—including the bones of the fetus.	Rickets in foals, osteomalacia in mature horses. Both conditions result in large joints and weak bones.
E	It is possible that more vitamin E is destroyed or used up by horses during times of stress or strain than can be obtained through normal feeds.	Serves as insurance against destruction of vitamin A. Makes for improved reproduction. Prevents anhidrosis.	Lowered breeding performance in both mares and stallions. Anhidrosis—a dry, dull hair coat.
K	Following intestinal disorders.	Concerned with blood coagulation.	Increased clotting time of the blood.

Footnotes on last page of table.

ng, and should be heated to 40° to 45° F during the winter months in cold regions.

EEDS FOR HORSES

More than one kind of hay makes for appetite appeal. In season, any good pasture can replace part or ll of the hay unless work or training conditions make ubstitution impractical.

Good quality oats and timothy hay always have een considered standard feeds for light horses. Iowever, feeds of similar nutritive properties can be nterchanged in the ration as price relationships warant; among them, the grains—oats, corn, barley, vheat, and sorghum; the protein supplements—inseed meal, soybean meal, and cottonseed meal; nd hays of many varieties. Feed substitution makes it ossible to obtain a balanced ration at lowest cost.

During the winter months, it is well to add a few liced carrots to the ration, an occasional bran mash, or a small amount of linseed meal. Also a bran mash or linseed meal may be used to regulate the bowels.

The proportion of concentrates must be increased and the roughages decreased as energy needs rise with the greater amount, severity, or speed of work. A horse that works at a trot needs considerably more feed than one that works at a walk. For this reason, riding horses in medium to light use require somewhat less grain and more hay in proportion to body weight than light horses that are racing. Also, from an esthetic standpoint, large, paunchy stomachs are objectionable on horses that are used for recreation and sport.

In addition to making for a nutritionally complete ration, the following factors should be considered when choosing horse feeds: cost, palatability, preparation, variety, bulk, and laxativeness.

For purposes of convenience in the discussion that follows, the author has classed feeds as (1) pasture, (2) hay, (3) concentrates, (4) protein supplements, and (5) special feeds and additives.

ITAMIN CHART

Recommended Allowances[1]		Practical Sources of the Vitamin	Comments
Daily per 1,000-lb *(455-kg)* Horse	Per Ton of Total Ration[2]		
0,000 USP	4,000,000 USP per ton (as-fed)	Stabilized vitamin A. Green grass. Green hay not over 1 year old. Grass or legume silage.	A considerable margin of safety in vitamin A and carotene is provided in the recommended allowances due to the oxidative destruction of these materials in feeds during storage. Hay over 1 year old, regardless of green color, is usually not an adequate source of carotene or vitamin A activity. The younger the animal, the quicker vitamin A deficiencies will show up. Mature animals may store sufficient vitamin A to last 6 months. When deficiency symptoms appear, add stabilized vitamin A to the ration.
,000 USP	560,000 USP per ton (as-fed)	Either vitamin D_2 (the plant form) or D_3 (the animal form) is equally effective for the horse. Exposure to sunlight. Sun-cured hays.	The vitamin D requirement is less when a proper balance of calcium and phosphorus exists in the ration. When animals are exposed to direct sunlight the ultraviolet light produces vitamin D from traces of cholesterol in the skin. Stabled horses, exercised in the early morning, will not get sufficient vitamin D in this manner.
200 IU	16,000 IU per ton (as-fed)	Alpha tocopherol, a stable form of vitamin E. Germ or germ oil of plants. Green plants. Green hays.	Most rations contain ample vitamin E. Before adding it, the horseman should seek the advice of a competent authority. Utilization of vitamin E is dependent on adequate selenium.
8 mg	640 mg per ton (as-fed)	Menadione (vitamin K_3). Green pasture. Well-cured hays. Fish meal.	High levels of vitamin K will overcome bleeding due to dicoumarol. Vitamin K is generally (1) widely distributed in normal feeds, and/or (2) synthesized in adequate amounts by the intestinal microflora of the horse.

(Continued)

TABLE 4-9

Vitamins Which May Be Deficient Under Normal Conditions	Conditions Usually Prevailing Where Deficiencies Are Reported	Functions of Vitamins	Some Deficiency Symptoms
WATER-SOLUBLE VITAMINS:			
B$_{12}$	When few, or no feeds of animal origin are fed.	Coenzyme in several enzyme systems. Closely linked with folic acid.	Loss of appetite and poor growt
Choline	Ration low in methionine, an amino acid.	Essential in building and maintaining cell structure and in the transmission of nerve impulses.	Slow growth.
Folic acid (folacin)		Related to B$_{12}$ metabolism. Metabolic reactions involving incorporation of single-carbon units into larger molecules.	Poor growth. Anemia.
Niacin (nicotinic acid)		Constituent of coenzymes. Hydrogen transport.	Reduced growth and appetite. Skin rashes, diarrhea, nerve dis orders.
Pantothenic acid		Part of coenzyme A, a necessary factor for life processes.	Poor growth, skin rashes, poc appetite, nerve disorders.
Riboflavin (B$_2$)	When green feeds (pasture, hay, or silage) are not available.	Probably for synthesis of ocular vitamin C or its protecting substance. Important in protein metabolism.	Periodic ophthalmia (or moo blindness). Decreased rate of growth an feed efficiency. Porous and weak bones; liga ments and joints impaired.
Thiamin (B$_1$)	Poor quality hay and grain. When sulfa drugs or antibiotics are given to the horse, the synthesis of B vitamins is impaired.	Required for normal carbohydrate metabolism. Promotes appetite and growth.	Decreased feed consumptio (loss of weight), incoordinatio (especially in the hindquarters), lowered blood thiamin, ele vated blood pyruvic acid, en larged heart, and nervou symptoms.
Unidentified factors	Since the U.S. foal crop is only around 50%, it is obvious that there is room for improvement somewhere alon the line; and perhaps unidentified factors are involved. Also, optimal results with horses during the critica periods (growth, gestation-lactation and when under stress as in racing or showing) appear to be dependen upon providing unidentified factors through such ingredients as distillers' dried solubles, dehydrated alfalfa meal, condensed fish solubles, brewers' dried yeast, antibiotic fermentation residues, dried whey, and corr fermentation solubles.		

[1]These are recommended allowances, and not requirements. The author's position on recommended allowances vs requirements for horses is clearly stated in the narrativ of this book under the heading entitled "Nutrient Requirements Vs Allowances."

—(Continued)

Recommended Allowances[1]		Practical Sources of the Vitamin	Comments
Daily per 1,000-lb *(455-kg)* Horse	Per Ton of Total Ration[2]		
25 mcg	10,000 mcg per ton (as-fed)	Synthetic B_{12}. Protein supplements of animal origin. Fermentation products.	
400 mg	32,000 mg per ton (as-fed)	Choline chloride. Choline dihydrogen.	Choline content of normal feeds is sufficient. All naturally occurring fats contain some choline.
2.5 mg	200 mg per ton (as-fed)	Synthetic folacin. Green, leafy plants.	Folic acid is widely distributed in horse feeds. Also, folic acid is synthesized in the lower gut.
50 mg	4,000 mg per ton (as-fed)	Synthetic niacin. Animal by-products. Green alfalfa.	The horse can convert the essential amino acid tryptophan into niacin. Hence, it is important to make certain that the ration is adequate in niacin; otherwise, the horse will use tryptophan to supply niacin needs.
60 mg	4,800 mg per ton (as-fed)	Calcium pantothenate. Fish solubles.	Grain is very deficient in pantothenic acid. Intestinal synthesis of pantothenic acid likely meets the body needs of the horse.
40 mg	3,200 mg per ton (as-fed)	Synthetic riboflavin. Green pasture. Green hay. Silages. Milk and milk products.	Lack of vitamin B_2 is not the only cause of moon blindness. Sometimes, moon blindness follows leptospirosis, and it may be caused by an allergic reaction.
35 mg	2,800 mg per ton (as-fed)	Thiamin hydrochloride. Green pastures. Well-cured, green, leafy hays. Cereal grains. Brewers' yeast.	Thiamin is synthesized in the lower gut of the horse by bacterial action, but there is some doubt as to its sufficiency. When neither green pasture nor high-quality roughage is available, thiamin hydrochloride should be added to the ration. Since carbohydrate metabolism is increased during physical exertion, it is important that B_1 be available in quantity at such times.

[2]Where hay is fed separately—that is, where an all-in-one pellet is not fed, for example—double these amounts should be added to the concentrate.

Pastures

Good pastures are excellent for horses—especially for idle horses, broodmares, and young stock. Horses in use may be turned to pasture at nights or over the weekend. Yet, it is becoming increasingly difficult to provide good pastures for many horses, especially those in suburban areas. The subject of pastures is fully covered in Section 5, "Pasture and Range Forages; Green Chop," of this book; hence, the reader is referred thereto.

Hays

Under most conditions, the hay requirement of horses ranges from 0.5 percent to 1.0 percent of body weight, or from 5 to 10 pounds of roughage daily for a 1,000-pound horse.

Usually, young horses and idle horses can be provided with an unlimited allowance of hay. In fact, much good will result from feeding young and idle horses more roughage and less grain. But one should gradually increase the grain and decrease the hay as work or training begins.

Racehorses should receive a minimum of roughage, since they need a maximum of energy. Sometimes it is necessary to muzzle greedy horses to keep them from eating bedding when their roughage allowance is limited.

The hay should be early cut, leafy, green, well cured, and free from dust and mold. Hay native to the locality is usually fed. However, horsemen everywhere prefer good quality timothy. With young stock and breeding animals especially, it is desirable that a sweet grass-legume mixture of alfalfa hay be fed. The legume provides a source of high-quality proteins and certain minerals and vitamins.

Horses like variety. Therefore, if at all possible, it is wise to have more than one kind of hay in the stable. For example, timothy may be provided at one feeding and a grass-legume mixed hay at the other feeding. Good horsemen often vary the amount of alfalfa fed, for increased amounts of alfalfa in the ration will increase urination and give a softer consistency to the bowel movements. This means that elimination from kidneys and bowels can be carefully regulated by the amount and frequency of alfalfa feedings. Naturally, such regulation becomes more necessary with irregular use and idleness. On the other hand, in some areas alfalfa is fed as the sole roughage with good results.

Concentrates

Of all the concentrates, heavy oats most nearly meet the needs of horses; and, because of the uniformly good results obtained from their use, they have always been recognized as the leading grain for horses. Corn is also widely used as a horse feed, particularly in the central states. Despite occasional prejudice to the contrary, barley is a good horse feed. As proof of the latter assertion, it is noteworthy that the Arab—who was a good horseman—fed barley almost exclusively. Also, wheat, wheat bran, molasses, and commercial mixed feeds are extensively used. Milk by-products and milk replacers are commonly fed to young stock. It is to be emphasized, therefore, that careful attention should be given to the prevailing price of feeds available locally, for many feeds are well suited to horses. Often substitutions can be made that will result in a marked saving without affecting the nutritive value of the ration. When corn or other heavy grains are fed, it is important that a little linseed meal or wheat bran be used, in order to regulate the bowels.

Protein Supplements

Grass hays and farm grains are low in quality and quantity of protein. Hence, they should be supplemented with other sources of protein. The following oil meals are most commonly used as protein supplements for horses: linseed meal, soybean meal, cottonseed meal, and sunflower meal.

UREA FOR HORSES

It is recognized that horses frequently consume urea-containing cubes and blocks intended for cattle and sheep, particularly on the western range. Moreover, it appears that mature horses are able to do so without untoward effects. The latter observation was confirmed in one limited experiment[19] in which 4 horses consumed an average of 4.57 lb per day of a urea-containing supplement, or 0.55 lb/head/day of feed urea (262%), for 5 months. Also, the Louisiana Station[20] did not find urea detrimental or toxic to horses when it constituted up to 5 percent of the grain ration, with up to 0.5 lb per day of urea consumed. There are reports, however, of urea toxicity in foals, in which bacterial action is more limited than in older horses.

Thus, there is some evidence that nonprotein nitrogen (urea) can be substituted for protein in the diet of the horse, but the conversion to protein is inefficient. Up to five percent of urea in the total ration does not appear to be harmful to mature horses. Nevertheless, in recognition of the more limited bacterial action in the horse and the hazard of toxicity—especially to young equines, most state laws

[19]*Veterinary Medicine*, Vol. 58, No. 12, Dec. 1963, pp. 945-946.

[20]"Non-Toxicity of Urea Feeding to Horses," *Veterinary Medicine/Small Animal Clinician*, Nov. 1965.

ɔrbid the use of such nonprotein nitrogen sources as rea in horse rations.

pecial Feeds and Additives

Special feeds may be needed from time to time ɔr regulating the bowels, imparting bloom or gloss to ıe hair, promoting growth of young stock, or as re-ʼards.

ɪRAN MASH

Feeding a bran mash is the traditional way of egulating the bowels of horses on idle days and at uch other times as required.

The mash is prepared by filling a 2- to 2½-gallon ucket with wheat bran, pouring enough boiling hot vater over it to make it the consistency of breakfast atmeal, covering the bucket with a blanket and al-owing it to steam until cool, then feeding it to the ɪorse.

Occasionally, when a horse is offered a bran mash or the first time, he may refuse to eat it. When this ɔccurs, the animal may be enticed to eat the mash by ither (1) introducing him to a little of it by hand, or 2) sprinkling some sugar, or some other well-liked eed, over it.

ʼEEDS THAT IMPART BLOOM

Bloom or gloss is important in horses. But some-imes they lack this desired quality—their hair is dull ɪnd dry. Feeding a well-balanced ration will usually ectify this situation. Also, feeding the following ɔroducts will make for an attractive, shiny coat:

1. *Corn oil or safflower oil*—Feed at the rate of 2 ɔunces (4 tablespoons) per horse twice per day.
2. *Whole flaxseed soaked*—Put a handful of vhole flaxseed in a teacup, cover it with water, let it stand overnight, then pour it over the morning feed. Repeat twice each week.

Unless the horse is afflicted with lice, mange, or some other ailment, either of the above treatments will impart bloom or gloss to the coat.

LYSINE

Cornell University reported that the addition of lysine to the diet of growing horses increased weight gains, feed consumption, and feed efficiency. But, the experimental diet used contained linseed meal as the major source of protein. Normally, it is much more practical to supply a source of good quality protein, such as milk protein or soybean meal (perhaps along with some linseed meal), rather than add lysine to lin-seed meal.

It is not recommended that horsemen spend money on lysine for horses. Instead, a well-balanced ration should be fed to horses of all ages, and espe-cially high-quality proteins should be incorporated in the ration of young equines. There is no experimental evidence that the addition of lysine will improve a good ration.

ANTIBIOTICS

Antibiotics are not nutrients; they're drugs. They are a chemical substance, produced by molds or bac-teria, which has the ability to inhibit the growth of or to destroy other microorganisms.

Certain antibiotics, at stipulated levels, are ap-proved by the FDA for growth promotion and for the improvement of feed efficiency of young equines up to one year of age. Unless there is a disease level, however, there is no evidence to warrant the continu-ous feeding of antibiotics to mature horses. Such prac-tice may even be harmful. Hence, where antibiotics are needed for therapeutic purposes, it is best to seek the advice of a veterinarian.

It appears that antibiotics may be especially help-ful for young foals which suffer setbacks from infec-tions, digestive disturbances, inclement weather, and other stress factors. Also, horses may benefit from an-tibiotics (1) when being transported from one location to another—for example, when being moved to a new show or track; (2) when there is a low disease level in the herd; or (3) when mares are foaling.

The poorer the feed, the greater the response from antibiotics; and the poorer the management, the greater the response from antibiotics. It follows, therefore, that there is a temptation to use antibiotics as a "crutch," rather than improve the regimen.

When used in feed, the level of antiobitcs should be in keeping with the directions of the manufacturer and with the Food and Drug Administration regula-tions.

TREATS

Horses are fed a great variety of treats. Their menus may include a choice of carrots or other roots,

Fig. 4-37. The carrot and the stick.

apples and other fruits, pumpkins, squashes, melons, molasses, sugar, honey, or innumerable other goodies.

Within moderation, treats are fine—especially as rewards. But they can be overdone, with the result that the *prima donnas* of the equine world don't "eat like a horse"; they eat like people—and often they are just as finicky. Hence, a horse should not be permitted to eat too much of any treat, simply because he likes it.

FEED PREPARATION

The physical preparation of cereal grains for horses has been practiced by horsemen for a very long time. Basically, grain is either soaked, cooked, ground, or rolled (wet or dry), and hay is either fed long, or pelleted, or cubed.

For horses, flaking is the preferred method of grain preparation; it makes for a light ration and fewer digestive disturbances. For animals with good teeth, the value of oats is increased only five percent by processing.

Hay for horses is usually fed long or incorporated in an all-pelleted ration (with the grain and hay combined).

Further discussion of feed preparation for horses is contained under the heading, "Feed Preparation," including Table 4-11, page 251, therein, hence, the reader is referred thereto.

HORSE FEEDING GUIDE

In order to compute balanced rations, it is first necessary to have available both feeding standards and feed composition tables. The latter are given in Table 4-101, Feed Composition Table, at the end of this section.

For purposes of convenience and simplification the author prepared Table 4-95, which facilitates the formulation of horse rations. In using this table, the following points should be noted:

1. That, for best results, different ages of animals should be fed differently.

2. That individual feeds differ widely in feeding value. Barley and oats, for example, vary widely in feeding value according to the hull content and the test weight per bushel, and forages vary widely according to the stage of maturity at which they are cut and how well they are cured and stored.

3. That nonlegume forages may have a higher relative value to legumes than herein indicated provided the chief need of the animal is for additional energy rather than for supplemented protein. Thus, the nonlegume forages of low value can be used to better advantage for wintering mature horses than for young foals.

On the other hand, legumes may have a higher actual value relative to nonlegumes than herein indicated provided the chief need is for additional protein rather than for added energy. Thus, no protein supplement is necessary for broodmares provided a good quality legume forage is fed.

4. That, based primarily on available supply and price, certain feeds—especially those of medium protein content, such as brewers' dried grains, distillers' dried solubles, and peas (dried)—are used interchangeably as (a) grains and by-product feeds, and/or (b) protein supplements.

5. That the feeding value of certain feeds is materially affected by preparation. The values herein reported are based on proper feed preparation in each case.

TABLE 4-95
HORSE FEEDING GUIDE
(As-fed Basis)

Description of Animals (1)	Recommendations[1] (2)	Recommendations Relative to Concentrate Only (Including grain, by-product feeds, and/or protein supplements)											
		In corral, with following types of forages:						On pasture of the following grades:					
		(1) Legume and/ or (2) legume and nonlegume mixed forages of high quality; consisting of dry forages and/or silage (high-protein forages) (3)		Legume and non-legume forages mixed; consisting of dry forages and/or silage (medium-protein forages) (4)		Nonlegume forage; consisting of dry forages and/or silage (low-protein forages) (5)		Excellent (6)		Fair to good (7)		Poor; including winter pasture consisting of dry grass cured on the stalk[2] (8)	
		(lb)	*(kg)*	*(lb)*	*(kg)*	*(lb)*	*(kg)*	*(lb)*	*(kg)*	*(lb)*	*(kg)*	*(lb)*	*(kg)*
Stallions in breeding season (weighing 900-1,400 lb, or *409-636 kg*)	Forage per 100 lb liveweight daily.[3]	¾-1½	*.34-.68*	¾-1½	*.34-.68*	¾-1½	*.34-.68*	¾-1½	*.34-.68*	¾-1½	*.34-.68*	¾-1½	*.34-.68*
	Concentrate: (1) Total concentrate allowance per 100 lb liveweight daily.	¾-1½	*.34-.68*	¾-1½	*.34-.68*	¾-1½	*.34-.68*	¾-1½	*.34-.68*	¾-1½	*.34-.68*	¾-1½	*.34-.68*
	(2) Amount of linseed meal (or equivalent 35% crude protein) to incorporate in total concentrate mix, in % of mix.[4,5]	Oats or equivalent will suffice		2½-8		5-13		Oats or equivalent will suffice		2½-8		5-13	
	(3) Crude protein composition of total concentrate, in %.	11-13		12-14		13-15		11-13		12-14		13-15	
Pregnant mares (weighing 900-1,400 lb, or *409-636 kg*)	Forage per 100 lb liveweight daily.[6]	¾-1½	*.34-.68*	¾-1½	*.34-.68*	¾-1½	*.34-.68*	—		—		—	
	Concentrate: (1) Total concentrate allowance per 100 lb liveweight daily.	¾-1½	*.34-.68*	¾-1½	*.34-.68*	¾-1½	*.34-.68*	—		—		¾-1½	*.34-.68*
	(2) Amount of linseed meal (or equivalent 35% crude protein) to incorporate in total concentrate mix, in % of mix.[4,5]	Oats or equivalent will suffice		2½		5-8		—		—		5-8	
	(3) Crude protein composition of total concentrate, in %.	12		13		13-14		—		—		13-14	
Lactating mares (weighing 900-1,400 lb, or *409-636 kg*)	Forage per 100 lb liveweight daily.	¾-1½	*.34-.68*	¾-1½	*.34-.68*	¾-1½	*.34-.68*	—		—		—	
	Concentrate: (1) Total concentrate allowance per 100 lb liveweight daily.	¾-1½	*.34-.68*	¾-1½	*.34-.68*	¾-1½	*.34-.68*	—		—		¾-1½	*.34-.68*
	(2) Amount of linseed meal (or equivalent 35% crude protein) to incorporate in total concentrate mix, in % of mix.[4,5]	5		7½		13-18		—		—		13-18	
	(3) Crude protein composition of total concentrate, in %.	13		14		15-16		—		—		15-16	
Foals before weaning (weighing 100-350 lb, or *45-159 kg*) with projected mature wts. of 900-1,400 lb, or *409-636 kg*)	Forage per 100 lb liveweight daily.[7]	½-¾	*.23-.34*	½-¾	*.23-.34*	½-¾	*.23-.34*	—		—		½-¾	*.23-.34*
	Concentrate: (1) Total concentrate allowance per 100 lb liveweight daily.	½-¾	*.23-.34*	½-¾	*.23-.34*	½-¾	*.23-.34*	½-¾	*.23-.34*	½-¾	*.23-.34*	½-¾	*.23-.34*
	(2) Amount of linseed meal (or equivalent 35% crude protein) to incorporate in total concentrate mix, in % of mix.[4,5]	8-13		13-18		18-24		8-13		13-18		18-24	
	(3) Crude protein composition of total concentrate, in %.	14-15		15-16		16-17		14-15		15-16		16-17	

Footnotes on last page of table.

(Continued)

TABLE 4-95 (Continued)

Description of Animals (1)	Recommendations[1] (2)	Recommendations Relative to Concentrate Only (Including grain, by-product feeds, and/or protein supplements)											
		In corral, with following types of forages:						On pasture of the following grades:					
		(1) Legume and/or (2) legume and nonlegume mixed forages of high quality; consisting of dry forages and/or silage (high-protein forages) (3)		Legume and nonlegume forages mixed; consisting of dry forages and/or silage (medium-protein forages) (4)		Nonlegume forage; consisting of dry forages and/or silage (low-protein forages) (5)		Excellent (6)		Fair to good (7)		Poor; including winter pasture consisting of dry grass cured on the stalk[2] (8)	
		(lb)	*(kg)*	*(lb)*	*(kg)*	*(lb)*	*(kg)*	*(lb)*	*(kg)*	*(lb)*	*(kg)*	*(lb)*	*(kg)*
The weanling (weighing 350-450 lb, or *159-205 kg*)	Forage per 100 lb liveweight daily.[8]	1½-2	*.68-.91*	1½-2	*.68-.91*	1½-2	*.68-.91*	—		—		1½-2	*.68-.91*
	Concentrate: (1) Total concentrate allowance per 100 lb liveweight daily.	1-1½	*.45-.68*	1-1½	*.45-.68*	1-1½	*.45-.68*	1-1½	*.45-.68*	1-1½	*.45-.68*	1-1½	*.45-.68*
	(2) Amount of linseed meal (or equivalent 35% crude protein) to incorporate in total concentrate mix, in % of mix.[4,5]	5-12½		10-13		13-18		5-12½		10-13		13-18	
	(3) Crude protein composition of total concentrate, in %.	13-15		14-15		15-17		13-15		14-15		15-17	
The yearling, 2nd summer (weighing 450-700 lb, or *205-318 kg*)	Forage per 100 lb liveweight daily.	1¼-1¾	*.57-.80*	1¼-1¾	*.57-.80*	1¼-1¾	*.57-.80*	—		—		—	
	Concentrate: (1) Total concentrate allowance per 100 lb liveweight daily.	¾-1¼	*.34-.57*	¾-1¼	*.34-.57*	¾-1¼	*.34-.57*	—		—		¾-1¼	*.34-.57*
	(2) Amount of linseed meal (or equivalent 35% crude protein) to incorporate in total concentrate mix, in % of mix.[4,5]	Oats or equivalent will suffice		0-8		8-18		—		—		8-18	
	(3) Crude protein composition of total concentrate, in %.	9-11		12-14		14-16		—		—		14-16	
The yearling, or rising 2-yr-old, 2nd winter (weighing 700-1,000 lb, or *318-455 kg*)	Forage per 100 lb liveweight daily.[9]	1-1½	*.45-.68*	1-1½	*.45-.68*	1-1½	*.45-.68*	—		—		1-1½	*.45-.68*
	Concentrate: (1) Total concentrate allowance per 100 lb liveweight daily.	½-1	*.23-.45*	½-1	*.23-.45*	½-1	*.23-.45*	—		—		½-1	*.23-.45*
	(2) Amount of linseed meal (or equivalent 35% crude protein) to incorporate in total concentrate mix, in % of mix.[4,5]	Oats or equivalent will suffice		Oats or equivalent will suffice		4-8		—		—		4-8	
	(3) Crude protein composition of total concentrate, in %.	9-10		11-13		13-14		—		—		13-14	

Footnotes on last page of table.

(Continued)

TABLE 4-95 (Continued)

Description of Animals (1)	Recommendations[1] (2)	Recommendations Relative to Concentrate Only (Including grain, by-product feeds, and/or protein supplements)											
		In corral, with following types of forages:						On pasture of the following grades:					
		(1) Legume and/or (2) legume and nonlegume mixed forages of high quality; consisting of dry forages and/or silage (high-protein forages) (3)		Legume and nonlegume forages mixed; consisting of dry forages and/or silage (medium-protein forages) (4)		Nonlegume forage; consisting of dry forages and/or silage (low-protein forages) (5)		Excellent (6)		Fair to good (7)		Poor; including winter pasture consisting of dry grass cured on the stalk[2] (8)	
		(lb)	*(kg)*	*(lb)*	*(kg)*	*(lb)*	*(kg)*	*(lb)*	*(kg)*	*(lb)*	*(kg)*	*(lb)*	*(kg)*
Light horses at work; in riding, in driving or in racing.	Forage per 100 lb liveweight daily.[10]												
	Hard use	1-1¼	*.45-.57*	1-1¼	*.45-.57*	1-1¼	*.45-.57*	¾-1	*.34-.45*	¾-1	*.34-.45*	1-1¼	*.45-.57*
	Med. use	1-1¼	*.45-.57*	1-1¼	*.45-.57*	1-1¼	*.45-.57*	½-¾	*.23-.34*	½-¾	*.23-.34*	¾-1	*.34-.45*
	Light use	1¼-1½	*.57-.68*	1¼-1½	*.57-.68*	1¼-1½	*.57-.68*	—	—	—	—	½-¾	*.23-.34*
	Concentrate:[10]												
	(1) Total concentrate allowance per 100 lb liveweight daily.												
	Hard use	1¼-1⅓	*.57-.61*	1¼-1⅓	*.57-.61*	1¼-1⅓	*.57-.61*	1¼-1⅓	*.57-.61*	1¼-1⅓	*.57-.61*	1¼-1¾	*.57-.79*
	Med. use	¾-1	*.34-.45*	¾-1	*.34-.45*	¾-1	*.34-.45*	¾-1	*.34-.45*	¾-1	*.34-.45*	¾-1	*.34-.45*
	Light use	2/5-½	*.18-.23*	2/5-½	*.18-.23*	2/5-½	*.18-.23*	2/5-½	*.18-.23*	2/5-½	*.18-.23*	2/5-½	*.18-.23*
	(2) Amount of linseed meal (or equivalent 35% crude protein) to incorporate in total concentrate mix, in % of mix.[4,5]												
	Hard use	Oats or equivalent will suffice		Oats or equivalent will suffice		Hard use = 0-8		Oats or equivalent will suffice		Oats or equivalent will suffice		Hard use = 0-8	
	Med. use	Oats or equivalent will suffice		Oats or equivalent will suffice		Med. and light use: Oats or equivalent will suffice		Oats or equivalent will suffice		Oats or equivalent will suffice		Med. and light use: Oats or equivalent will suffice	
	Light use	Oats or equivalent will suffice		Oats or equivalent will suffice		Med. and light use: Oats or equivalent will suffice		Oats or equivalent will suffice		Oats or equivalent will suffice		Med. and light use: Oats or equivalent will suffice	
	(3) Crude protein composition of total concentrate, in %.												
	Hard use	10-12		11-13		12-14		10-12		11-13		12-14	
	Med. use	9-11		10-12		11-13		9-11		10-12		11-13	
	Light use	8-10		9-11		10-12		8-10		9-11		10-12	
Mature idle horses; stallions, mares, and geldings (weighing 900-1,400 lb, or *409-636 kg*)	Forage (no concentrate necessary) per 100 lb liveweight daily.[11]	1½-1¾	*.57-.80*	1½-1¾	*.57-.80*	1½-1¾	*.57-.80*	—		—		1½-1¾	*.57-.80*

[1]The forage recommendations given herein are based on dry forage. When silage is included in the ration, figure 3 lb of silage equivalent to 1 lb of dry forage, due to the higher moisture content of silage.

[2]On a dry basis, the crude protein content of mature, weathered grasses may be 3% or less.

[3]In breeding season, a stallion should always have some legume hay.

[4]Linseed meal, which averages about 35% protein content, is herein used as a standard merely because it is the leading U.S. high-protein supplement for horses. It is to be emphasized, however, (a) that other protein supplements, including numerous commercial products, can and should be used; (b) that, in general, those supplements should be purchased which provide a unit of protein at the lowest cost, and those feeds which are highest in protein content are usually the most economical; and (c) that where other protein feeds are substituted for the linseed meal recommended herein, an equivalent amount of crude protein should be provided—for example, approximately 1¾ lb of 20% crude protein supplement should be provided in order to replace each 1 lb of linseed meal (although it is recognized that 1¾ lb of a 20% protein feed will generally provide more energy than 1 lb of linseed meal). Because linseed meal has a laxative effect, with some animals and rations it may have to be limited to 8 to 10% of the total concentrate ration.

[5]The recommended supplement allowance is based on the assumption that oats, averaging 11.8% crude protein content, comprise the major part of the concentrate mix. Naturally, less protein supplement will need to be added where feeds of higher protein content than oats predominate.

[6]During the last half of pregnancy, mares should always have some legume hay.

[7]Foals before weaning should have legume hay only.

[8]Weanlings should always have some legume hay.

[9]Yearlings are usually fed grass hay or other nonlegume forage.

[10]When on pasture, light horses at work should receive a concentrate allowance similar to that which is provided when not on pasture, but the forage allowance may be lessened.

[11]Idle horses are usually fed grass hay or other nonlegume forage.

RATIONS FOR HORSES

Several suggested rations are given in Table 4-96, page 429.

FEED SUBSTITUTION TABLE

The successful horseman is a keen student of values. He recognizes that feeds of similar nutritive properties can and should be interchanged in the ration as price relationships warrant, thereby making it possible at all times to obtain a balanced ration at the lowest cost.

Table 4-97, Feed Substitution Table for Horses, page 430, is a summary of the comparative values of the most common U.S. horse feeds. In arriving at these values, chemical composition, feeding value, and palatability have been considered. The comparative values shown are not absolute. Rather, they are reasonably accurate approximations based on average quality feeds.

FEEDING BREEDING HORSES

Regular and normal reproduction is the basis for profit on any horse breeding establishment. Despite this undeniable fact, it has been estimated that only 40 to 60 percent of all mares bred actually produce foals. Certainly, there are many causes of reproductive failure, but most scientists are agreed that inadequate nutrition is a major one.

As with all species, most of the growth of the fetus of the mare occurs during the last third of pregnancy, thus making the reproductive requirements most critical during this period.

The nutritive requirements for moderate to heavy milk production are much more rigorous than the pregnancy requirements. There is special need for a rather liberal protein, mineral, and vitamin allowance.

In the case of young, growing, pregnant females, additional protein, minerals, and vitamins, above the ordinary requirements, must be provided; otherwise, the fetus will not develop properly or milk will be produced at the expense of the tissues of the dam.

It is also known that the ration exerts a powerful effect on sperm production and semen quality. Too fat a condition can even lead to temporary or permanent sterility. Moreover, there is abundant evidence that greater fertility of stallions exists under conditions where a well-balanced ration and plenty of exercise are provided.

Feeding Broodmares

The following pointers are pertinent to feeding broodmares properly:

1. Condition the mare for breeding by providing adequate and proper feed and the right amount of exercise prior to the breeding season.
2. See that adequate proteins, minerals, and vitamins are available during the last third of pregnancy when the fetus grows most rapidly.
3. Feed and water with care immediately before and after foaling. For the first 24 hours after parturition, the mare may have a little hay and a limited amount of water from which the chill has been taken. A light feed of bran or a wet bran mash is suitable for the first feed and the following meal may consist of oats or a mixture of oats and bran. A reasonably generous allowance of good quality hay is permissible after the first day. If confined to the stable, as may be necessary in inclement weather, the mare should be kept on a limited and light grain and hay ration for about 10 days after foaling. Feeding too much grain at this time is likely to produce digestive disturbances in the mare and, even more hazardous, it may produce too much milk, which may cause indigestion in the foal. If weather conditions are favorable and it is possible to allow the mare to foal on a clean, lush pasture she will regulate her own feed needs most admirably.
4. Provide adequate nutrition during lactation because the requirements during this period are more rigorous than the requirements during pregnancy.
5. Make sure that young growing mares receive adequate nutrients; otherwise, the fetus will not develop properly or the dam will not produce milk except at the expense of her body tissues.

Feeding Stallions

The ration exerts a powerful effect on sperm production and semen quality. Successful breeders adhere to the following stallion feeding rules:

1. Feed a balanced ration, giving particular attention to proteins, minerals, and vitamins.
2. Regulate the feed allowance because the stallion can become infertile if he gets too fat. Also, increase the exercise when the stallion is not a sure breeder.
3. Provide pasture in season as a source of both nutrients and exercise.

Feeding Young Equines

Growth is the very foundation of horse production. This is so because horses cannot perform properly or possess the necessary speed and endurance if their growth has been stunted or their skeletons have been injured by inadequate rations during early age. Naturally, these requirements become increasingly acute when horses are forced for early use, such as the training and racing of the two-year-old.

Young equines should be fed according to age,

TABLE 4-96

RATIONS FOR HORSES[1]

(As-fed Basis)

Age, Sex, and Use	Daily Allowance	Kind of Hay (More than one kind of hay makes for variety and appetite appeal. In season, any good pasture can replace part or all of the hay except for horses at work or in training.)	Suggested Grain Rations (With all rations, and for all classes and ages of horses, provide suitable minerals; see Table 4-93, Horse Mineral Chart, p. 414, for recommendations.) Ration No. 1	(lb)	(kg)	Ration No. 2	(lb)	(kg)	Ration No. 3	(lb)	(kg)
Stallions in breeding season (weighing 900-1,400 lb; or *409-636 kg*)	¾ to 1½ lb *(.3-.7 kg)* grain per 100 lb *(45 kg)* body weight, together with a quantity of hay within same range.	Grass-legume mixed (or ⅓ to ½ legume hay, with balance grass hay)	Oats Wheat Wheat bran Linseed meal	55 20 20 5	*25* *9* *9* *2*	Corn Oats Wheat Wheat bran	35 35 15 15	*16* *16* *7* *7*	Oats	100	*45*
Pregnant mares (weighing 900-1,400 lb; or *409-636 kg*)	¾ to 1½ lb *(.3-.7 kg)* grain per 100 lb *(45 kg)* body weight, together with a quantity of hay within same range.	Grass-legume mixed; or ⅓ to ½ legume hay, with balance grass hay (straight grass hay may be used first half of pregnancy)	Oats Wheat bran	80 20	*36* *9*	Barley Oats Wheat bran	45 45 10	*20* *20* *5*	Oats Linseed meal	95 5	*43* *2*
Foals before weaning (weighing 100-350 lb, or *45-159 kg*; with projected mature weights of 900-1,400 lb; or *409-636 kg*)	½ to ¾ lb *(.2-.3 kg)* grain per 100 lb *(45 kg)* body weight, together with a quantity of hay within same range.	Legume hay	Oats Wheat bran Linseed meal	50 40 10	*23* *18* *5*	Oats Barley Wheat bran Linseed meal	30 30 30 10	*14* *14* *14* *5*	Oats Wheat bran	80 20	*36* *9*
			(Rations balanced basis of following assumptions: Mares of mature weights of 600, 800, 1,000, and 1,200 lb [or *273, 364, 455, and 545 kg*] may produce 36, 42, 44, and 49 lb [or *16, 19, 20, and 22 kg*] of milk daily.)								
Weanlings (weighing 350-450 lb; or *159-204 kg*)	1 to 1½ lb *(.5-.7 kg)* grain and 1½ to 2 lb *(.7-.9 kg)* hay per 100 lb *(45 kg)* body weight.	Grass-legume mixed (or ½ legume hay, with balance grass hay)	Oats Barley Wheat bran Linseed meal	30 30 30 10	*14* *14* *14* *5*	Oats Wheat bran Linseed meal	70 15 15	*32* *7* *7*	Oats Linseed meal	80 20	*36* *9*
Yearlings; 2nd summer (weighing 450-700 lb; or *204-317 kg*)	Good luxuriant pastures. (If in training or for other reasons without access to pastures, the ration should be intermediate between the adjacent upper and lower groups.)										
Yearling or rising 2-yr-old; 2nd winter (weighing 700-1,000 lb; or *317-454 kg*)	½ to 1 lb *(.2-.5 kg)* of grain and 1 to 1½ lb *(.5-.7 kg)* hay per 100 lb *(45 kg)* body weight.	Grass-legume mixed; or ⅓ to ½ legume hay, with remainder grass hay.	Oats Wheat bran	80 20	*36* *9*	Barley Oats Wheat bran Linseed meal	35 35 15 15	*16* *16* *7* *7*	Oats	100	*45*
Light horses at work, in riding, driving, and racing (weighing 900-1,400 lb; or *409-636 kg*)	**Hard Use**—1¼ to 1⅓ lb *(.57-.6 kg)* grain and 1 to 1¼ lb *(.5-.57 kg)* hay per 100 lb *(45 kg)* body weight. **Med. Use**—¾ to 1 lb *(.3-.5 kg)* grain and 1 to 1¼ lb *(.5-.6 kg)* hay per 100 lb *(45 kg)* body weight. **Light Use**—2/5 to ½ lb *(.18-.2 kg)* grain and 1¼ to 1½ lb *(.6-.7 kg)* hay per 100 lb *(45 kg)* body weight.	Grass hay.	Oats	100	*45*	Oats Corn	70 30	*32* *14*	Oats Barley	70 30	*32* *14*
Mature idle horses; stallions, mares, and geldings (weighing 900-1,400 lb; or *409-636 kg*)	1½ to 1¾ lb *(.7-.8 kg)* of hay per 100 lb *(45 kg)* body weight.	Pasture in season; or grass-legume mixed hay.	(With grass hay, add ¾ lb *(.34 kg)* daily of a high-protein supplement.)								

[1]Mineral recommendations for all classes and ages of horses, especially those fed unmixed rations or on pasture:

a. *Where animals are on liberal grain feeding*—Provide free access to a 2-compartment mineral box, with (1) trace mineralized salt in one side; and (2) in the other side, a mixture of ⅓ trace mineralized salt (salt included for purposes of palatability), ⅓ defluorinated phosphate or steamed bone meal, and ⅓ ground limestone or oystershell flour.

b. *Where animals are primarily on roughage (pasture, hay, and/or silage)*—Provide free access to a 2-compartment mineral box, with (1) trace mineralized salt in one side (salt included for purposes of palatability); and (2) in the other side, a mixture of ⅓ trace mineralized salt and ⅔ defluorinated phosphate or steamed bone meal.

TABLE 4-97

FEED SUBSTITUTION TABLE FOR HORSES
(As-fed basis)

Feedstuffs	Relative Feeding Value (lb for lb) in comparison with the designated (underlined) base feed which = 100	Maximum Percentage of Base Feed (or comparable feed or feeds) which it can replace for best results	Remarks
GRAINS, BY-PRODUCT FEEDS, ROOTS AND TUBERS:[1] (Low- and Medium-Protein Feeds)			
Oats	100	100	The leading horse feed. The feeding value of oats varies according to the hull content and test weight per bushel. Need not be ground.
Barley	110	100	Most horsemen feel that it is preferable to feed barley along with more bulky feeds; for example, 25% oats or 15% wheat bran. Crush for horses.
Beet pulp, dried	100	33⅓	Not palatable to horses.
Beet pulp, molasses, dried	100	33⅓	Not palatable to horses.
Brewers' dried grains	100	50	
Carrots	15-25	10	Horses are very fond of carrots.
Corn No. 2	115	100	Ranks second to oats as a light horse feed.
Corn gluten feed (gluten feed)	100	50	It has a lower value than indicated when forage is of low protein content.
Distillers' dried grain	90-100	25	
Distillers' dried solubles	90-100	25	
Hominy feed	115	100	
Molasses, beet	80-95	10	In hot, humid areas, molasses should be limited to 5%; otherwise, mold may develop unless an inhibitor is used. Cane molasses is slightly preferred to beet molasses.
Molasses, cane	80-95	10	In hot, humid areas, molasses should be limited to 5%; otherwise, mold may develop unless an inhibitor is used.
Peas, dried	100	40	
Rice (rough rice)	115	50	Grind for horses.
Rye	115	33⅓	Higher levels, or abrupt changes to rye may cause digestive disturbances. Not palatable.
Sorghum, grain	110-115	85	All varieties have about the same feeding value. Crush for horses.
Wheat	115	50	Wheat should be mixed with a more bulky feed in order to prevent colic.
Wheat bran	100	20	Valuable for horses because of its bulky nature and laxative properties.
Wheat-mixed feed (mill run)	105	20	Excessive quantities will cause colic or other digestive upsets.
PROTEIN SUPPLEMENTS:			
Linseed meal (35%)	100	100	Linseed meal (old process) is the preferred protein supplement for horses. It is valued because of its laxative properties and because of the sleek hair coat which it imparts.
Brewers' dried grains	65-70	50	
Buttermilk, dried	100	100	May be used in place of dried skimmed milk for foals.
Copra meal (coconut meal) (21%)	90-100	50	
Corn gluten feed (gluten feed)	70	100	
Corn gluten meal (gluten meal)	100	50	Somewhat unpalatable to horses.
Cottonseed meal (41%)	100	100	Satisfactory if limited to amounts necessary to balance ordinary rations.
Peanut meal (45%)	100	100	
Peas, dried	75	50	
Skimmed milk, dried	100	100	Especially valuable for young equines; for creep feeding until past weaning.
Soybean meal (41%)	100	100	
Soybeans	100	100	Soybeans should be limited to ⅓ of the concentrate ration.
Sunflower meal (41%)	100	33⅓	Sunflower meal should not constitute more than ⅓ of the protein supplement for palatability reasons.
Whey, dried	50	50	Whey may be laxative.

Footnotes on last page of table.

(Continued)

TABLE 4-97 (Continued)

Feedstuffs	Relative Feeding Value (lb for lb) in comparison with the designated (underlined) base feed which = 100	Maximum Percentage of Base Feed (or comparable feed or feeds) which it can replace for best results	Remarks
RY FORAGES AND SILAGES:[2]			
mothy hay	100	100	The preferred hay of horsemen.
falfa hay, all analyses	133⅓	100	Good-quality alfalfa is excellent for horses. Alfalfa may be ground and pelleted. It provides high-quality proteins, and certain minerals and vitamins. It is somewhat laxative. Contrary to some "old wives' tales," it will not damage the kidneys.
arley hay	100	100	Lower value if not cut at the early dough stage.
omegrass hay	100	100	
lover hay, crimson	125	100	Crimson clover hay has considerably lower value if not cut at an early stage.
lover hay, red	125	100	Clover hay should be well cured and free from dust and mold.
lover-timothy hay	110-115	100	Value of clover-timothy mixed hay depends on the proportion of clover present and the stage of maturity at which it is cut.
orn fodder	100	50	Preferably fed along with a good legume hay. It is best to shred the fodder.
orn silage	45-55	33⅓-50	Preferably fed along with a good legume hay.
orn stover	60	50	It is best to shred the stover.
owpea hay	110	100	
rass-legume mixed hay	110-115	100	
rass-legume silage	45-50	33⅓-50	
rass silage	40-45	33⅓-50	
ohnsongrass hay	90-95	100	
espedeza hay	115	100	
at hay	100	100	Lower value if not cut at the early dough stage.
rchardgrass	100	100	Should be cut before maturity. It is a safe feed for horses.
rairie hay	100	100	
eed canarygrass	90-95	100	
orghum fodder	100	50	Preferably fed along with a good legume hay. It is best to shred the fodder.
orghum silage	40-45	33⅓-50	Preferably fed along with a good legume hay.
orghum stover	60	50	It is best to shred the stover.
oybean hay	110	100	
udangrass hay	90-95	100	
etch-oat hay	110-115	100	The higher the proportion of vetch, the higher the value.
Vheat hay	100	100	

[1]Roots and tubers are of lower value than the grain and by-product feeds due to their higher moisture content.
[2]Well preserved silage of good quality, free from mold and not frozen, affords a highly nutritious succulent forage for horses during the winter months—especially for idle orses, brood mares, and growing colts. Silages are of lower value than dry forages due to their higher moisture content.

with differentiation made for foals, weanlings, yearlings, and two- and three-year-olds.

FEEDING FOALS

As with all young mammals, milk from the dam gives the foal a good start in life. Within 30 minutes to 2 hours after birth, the foal should be up on its feet and getting the colostrum. Increased growth, durability, and soundness of foals may be obtained by feeding them apart from their dams; either (1) by tying the mare while the foal eats, or (2) by providing a creep for the foals.

The need for a foal feeding program, starting early in life, is due to the decline in mare's milk in both quantity and nutrients following foaling.

When the foal is between 10 days and 3 weeks of age, it will begin to nibble on grain and hay. In order to promote thrift and early development and to avoid setback at weaning time, it is important to encourage the foal to eat supplementary feed as early as possible. For this purpose, a low-built grain box should be provided especially for the foal; or, if on pasture, the foal may be creep fed. The choice between individual feeding and creep feeding may be left to the horseman; the important thing is that foals receive supplemental feed.

A creep is an enclosure for feeding purposes, made accessible to the foal(s), but through which the dam cannot pass. For best results, the creep should be built at a spot where the mares are inclined to loiter. The ideal location is on high ground, well drained, in the shade, and near the place of watering. Keeping the salt supply nearby will be helpful in holding mares near the creep.

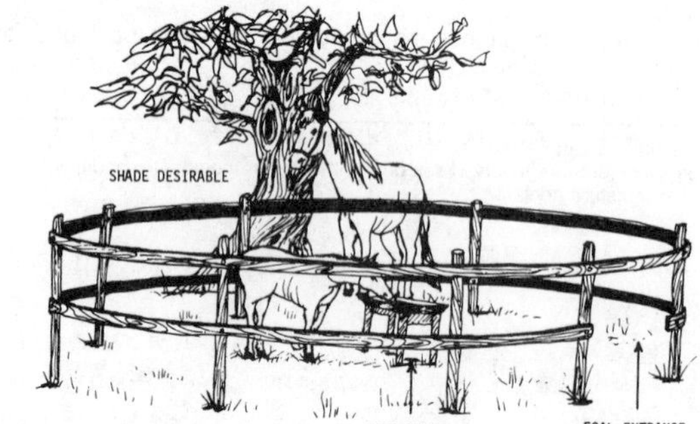

Fig. 4-38. A foal creep.

It is important that foals be started on feed carefully, and at an early age. At first only a small amount of feed should be placed in the trough each day, any surplus being removed and given to other horses. In this manner, the feed will be kept clean and fresh, and the foals will not be consuming any moldy or sour feed.

Rolled oats and wheat bran, to which a little brown sugar has been added, is especially palatable as a starting ration.

Table 4-98 gives the formulation of an excellent foal ration, which may be either individually fed or creep fed.

TABLE 4-98
FOAL RATION
(As-fed Basis)

Ingredients	Percent	Amount in 500-lb (227-kg) Mix	
	(%)	(lb)	(kg)
Corn (flaked)	37.4	187.0	85.0
Soybean meal (41%)	33.0	165.0	75.0
Oats (rolled)	23.0	115.0	52.3
Brewers' yeast	0.5	2.5	1.1
Molasses	3.0	15.0	6.8
Dicalcium phosphate	1.0	5.0	2.3
Limestone	1.0	5.0	2.3
Salt (trace mineralized)	1.0	5.0	2.3
Vitamins A and D	0.1	0.5	0.2
Total	100.0	500.0	227.3

Because of the difficulty in formulating and home mixing a foal ration, the purchase of a good commercial feed usually represents a wise investment.

In addition to its grain ration, the foal should be given good quality hay (preferably a legume), unless it is on good pasture.

Free access to salt and a suitable mineral mixture should be provided. The mineral will be consumed to best advantage if placed in a convenient place and under shelter; or it may be incorporated in the ration. Plenty of fresh water must be available at all times.

At 4 to 5 weeks of age, the normal healthy foal should be consuming ½ lb of grain daily per 100 lb of liveweight. By weaning time, this should be increased to about ¾ lb or more per 100 lb liveweight (or 6 to 8 lb of feed/head/day), the exact amount varying with the individual, the type of feed, and the development desired.

Under such a system of care and management the foal will become less dependent upon its dam and the weaning process will be facilitated. If properly cared for, foals will normally attain one-half of their mature weight during the first year. Most Thoroughbred and Standardbred breeders plan to have the animals attain full height by the time they are two years of age. However, such results require liberal feeding from the beginning.

It is well recognized that the forced development of race, show, and sale horses must be done expertly if the animals are to remain durable and sound. This calls for particular emphasis on the kind of ration, feed allowance, and exercise.

Raising the Orphan Foal

Occasionally a mare dies during or immediately after parturition, leaving an orphan foal to be raised. Also, there are times when mares fail to give a sufficient quantity of milk for the newborn foal. Sometimes there are twins. In such cases, it is necessary to resort to other milk supplies. The problems will be simplified if the foal has at least received the colostrum from the dam, for it does play a very important part in the well-being of the newborn young.

If at all possible, the foal should be shifted to another mare. Some breeding establishments regularly follow the plan of breeding a mare that is a good milk producer but whose foal is expected to be of little value. Her own foal is either destroyed or raised on a bottle, and the mare is used as a foster mother or nurse mare.

The larger nurseries usually keep a supply of colostrum on hand. They remove colostrum from mares that (1) have had dead foals, or (2) produce excess milk, then store it in a freezer for future use for foals that do not receive colostrum from their dams. When needed, it can be removed from the freezer, heated, and fed. This is an excellent practice.

If no colostrum is available, the foal should be placed on either (1) cow's milk made as nearly as possible of the same composition as mare's milk (Mare's milk is higher in percentage of water and sugar than cow's milk and is lower in other components.); or (2) a synthetic milk replacer.

For best results in raising the orphan foal, milk from a fresh cow, low in butterfat, should be used. To about a pint of milk, add a tablespoon of sugar and from 3 to 5 tablespoonsful of lime water. Warm to body temperature, and for the first few days feed about ¼ pint every hour. After 3 or 4 weeks the sugar can be stopped, and at 5 or 6 weeks skimmed milk can be used entirely.

Orphan foals may also be raised on milk replacer, fed according to the directions of the manufacturer. Here again the situation is simplified if the foal has first received colostrum.

For the first few days, the milk (either cow's milk or milk replacer) may be fed by using a bottle and a rubber nipple. Later, the foal should be taught to drink from a pail. It is important that all receptacles be kept absolutely clean and sanitary (clean and scald each time), and that feeding be at regular intervals. Grain feeding should be started at the earliest possible time with the orphan foal.

FEEDING WEANLINGS

The most critical period in the entire life of a horse is that interval from weaning time (about six months of age) until one year of age. Foals suckling their dams and receiving no grain may develop very satisfactorily up to weaning time. However, lack of preparation prior to weaning and neglect following the separation from the dam may prevent the animal from developing properly.

No great setback or disturbances will be encountered at weaning time provided that the foals have developed a certain independence from proper grain feedings during the suckling period. Generally, weanlings should receive 1 to 1½ lb of grain and 1½ to 2 lb of hay daily per each 100 lb of liveweight. The amount of feed will vary somewhat with the individuality of the animal, the quality of roughage, available pastures, the price of feeds, and whether the weanling is being developed for show, race, or sale. Naturally, animals being developed for early use or sale should be fed more liberally, although it is equally important to retain clean, sound joints, legs, and feet—a condition which cannot be obtained so easily in heavily fitted animals.

Because of the rapid development of bone and muscle in weanlings, it is important that, in addition to ample quantity of feed, the ration also provides quality proteins, and adequate minerals and vitamins.

FEEDING YEARLINGS

If young animals have been fed and cared for so that they are well grown and thrifty as yearlings, usually little difficulty will be experienced at any later date.

When on pasture, yearlings that are being grown for show or sale should receive grain in addition to grass.

The winter feeding program for the rising 2-year-olds should be such as to produce plenty of bone and muscle rather than fat. From ½ to 1 lb of grain and 1 to 1½ lb of hay should be fed for each 100 lb of liveweight. The quantity will vary with the quality of the roughage, the individuality of the animal, and the use for which the animal is produced.

FEEDING TWO- AND THREE-YEAR-OLDS

Except for the fact that the two- and three-year-olds will be larger, and, therefore, will require more feed, a description of their proper care and management would be merely a repetition of the principles that have already been discussed for the yearling.

FEEDING PLEASURE HORSES

Keeping pleasure horses—horses used for recreation and sport—in peak condition makes for greater satisfaction when they're used.

It is difficult to feed pleasure horses properly because their use is often irregular. Sometimes they're used moderately; at other times they're idle; at still other times they're worked hard over the weekend or on a trail ride.

Most horses used for pleasure are worked lightly, perhaps 1 to 3 hours of riding per day. Others are worked medium hard, as when ridden 3 to 5 hours per day. Still others are worked very hard, as when raced or when ridden 5 to 8 hours per day. The recommended daily feed allowance per 100 pounds body weight of pleasure horses in light, medium, and hard use is given in Table 4-99.

FEEDING HORSES IN TRAINING

Horses in heavy training for specific purposes-such as training for racing, cutting, roping, jumpin or hunting—have a higher nutritional requireme than most pleasure horses. And the younger the an mal in training, the higher the level of nutritio needed in order to develop and maintain sound leg and build a strong frame and body. Therefore, th level of work, the temperament of the individual, an the age of the horse determine the nutritional need For this reason, horses in training should be fed a individuals.

TABLE 4-99
FEED ALLOWANCE OF PLEASURE HORSES
(As-fed Basis)

	Lb Daily/100 Lb *(45 kg)* Weight of Horse					
	Light Use		Medium Use		Hard Use	
	(lb)	*(kg)*	*(lb)*	*(kg)*	*(lb)*	*(kg)*
Hay	1.25-1.5	0.57-0.68	1.0 -1.25	0.45-0.57	1.0 -1.25	0.45-0.57
Grain	0.4 -0.5	0.18-0.23	0.75-1.0	0.34-0.45	1.25-1.33	0.57-0.60

As shown above, the roughage content of the ration decreases and the concentrate content increases as the amount of work increases. This is because the digestibility and the efficiency of conversion are greater for high-energy concentrates than for roughages.

Of course, horses differ in temperament and in ease of keeping. Also, no two horses will perform the same amount of work with an equal expenditure of energy, and no two horsemen will get the same amount of work out of the same horse. So, the feed allowance should be increased if the horse fails to maintain condition, and it should be decreased if the animal becomes too fat.

In season, pasture may replace hay, all or in part, according to the quality of the pasture. But the concentrate allowance of the working horse should remain about the same on pasture as in the stable or dry corral. There is a tendency of the pastured working horse to sweat and tire more easily (be "soft"), probably due to the high water content of green forage.

In addition to forage and grain, pleasure horses should have access to salt and a suitable mineral mix, free-choice. The mineral requirements of the working horse differ from those of the idle horse mainly in the salt requirements, due to the loss of salt in perspiration.

The vitamin requirements of working horses are approximately the same as those of idle horses, except for the increase in the B complex requirements due to the greater carbohydrate metabolism of the working horse.

Horses in training will eat about 1½ lb of grai and 1 lb of hay per 100 lb liveweight.

FEEDING RACEHORSES

Racehorses are equine athletes whose nutritiv requirements are the most exacting, but the mos poorly met, of all animals.

High-strung and highly stressed racehorses nee special rations just as human athletes do—and for th same reasons; and, the younger the age, the mor acute the need. This calls for rations high in protein rich in readily available energy, fortified with vita mins, minerals, and unidentified factors—and with al nutrients in proper balance.

A racehorse is asked to develop a large amount o horsepower in a period of one to three minutes. Th oxidations that occur in a racehorse's body are at higher pitch than in an idle horse, and, therefore more vitamins are required.

Also, racehorses are the *prima donnas* of th equine world; most of them are temperamental, an no two of them can be fed alike. They vary in rapidit of eating, in the quantity of feed that they will con sume, in the proportion of concentrate to roughag that they will take, and in response to differen caretakers. Thus, for best results, they must be fed a individuals.

Most racehorse rations are deplorably deficient i protein, simply because they are based on th minimum requirements of little-stressed, slow, plod ding draft horses.

During the racing season, the hay of a racehorse should be limited to 7 or 8 pounds, whereas the concentrate allowance may range up to 16 pounds. Heavy roughage eaters may have to be muzzled, to keep them from eating their bedding. A bran mash is commonly fed once a week.

FEEDING SHOW AND SALE HORSES

Each year, many horses are fitted for shows or sales. In both cases, a fattening process is involved, but exercise is doubly essential.

For horses that are being fitted for shows, the conditioning process is also a matter of hardening, and the horses are used daily in harness or under saddle. Regardless of whether a sale or a show is the major objective, fleshing should be obtained without sacrificing action or soundness or without causing filling of the legs and hocks.

In fattening horses, the animals should be brought to full feed rather gradually, until the ration reaches a maximum of about 2 lb of grain daily for each 100 lb of liveweight. When on full feed, horses make surprising gains. Daily weight gains of 4 to 5 lb are not uncommon. Such animals soon become fat, sleek, and attractive. This is probably the basis for the statement that "fat will cover up a multitude of sins in a horse."

Although exercise is desirable from the standpoint of keeping the animals sound, it is estimated that such activity decreases the daily rate of gains by as much as 20 percent. Because of the greater cost of gains and the expense involved in bringing about forced exercise, most feeders of sale horses limit the exercise to that obtained naturally from running in a paddock.

In comparison with finishing cattle or sheep, there is more risk in fattening horses. Heavily fed horses kept in idleness are likely to become blemished and injured through playfulness, and there are more sicknesses among liberally fed horses than in other classes of stock handled in a similar manner.

In fitting show horses, the finish must remain firm and hard, the action superb, and the soundness unquestioned. Thus, they must be carefully fed, groomed, and exercised to bring them to proper bloom.

Horsemen who fit and sell yearlings or younger animals may feed a palatable milk replacer or commercial feed to advantage.

PART VII—NUTRITIONAL DISEASES AND AILMENTS

More animals (and people) throughout the world suffer from hunger—from just plain lack of sufficient feed—than from the lack of a specific nutrient (or nutrients); therefore, it is recognized that nutritional deficiencies may be brought about either by (1) too little feed, or (2) rations that are too low in one or more nutrients.

Also, forced production (such as very high milk yields and finishing animals at early ages) and the feeding of forages and grains which are often produced on leached and depleted soils have created many problems in nutrition. This condition has been further aggravated through the increased confinement of stock, many animals being confined to lots or buildings all or a large part of the year. Under these unnatural conditions, nutritional diseases and ailments have become increasingly common.

Although the cause, prevention, and treatment of most of these nutritional diseases and ailments are known, they continue to reduce profits in the livestock industry simply because the available knowledge is not put into practice. Moreover, those widespread nutritional deficiencies which are not of sufficient proportions to produce clear-cut deficiency symptoms cause even greater economic losses because they go unnoticed and unrectified. Table 4-100 contains a summary of the important nutritional diseases and ailments affecting animals.

Manganese deficiency

Pantothenic Acid deficiency

Goiter

Fig. 4-39. Some nutritional diseases and ailments.

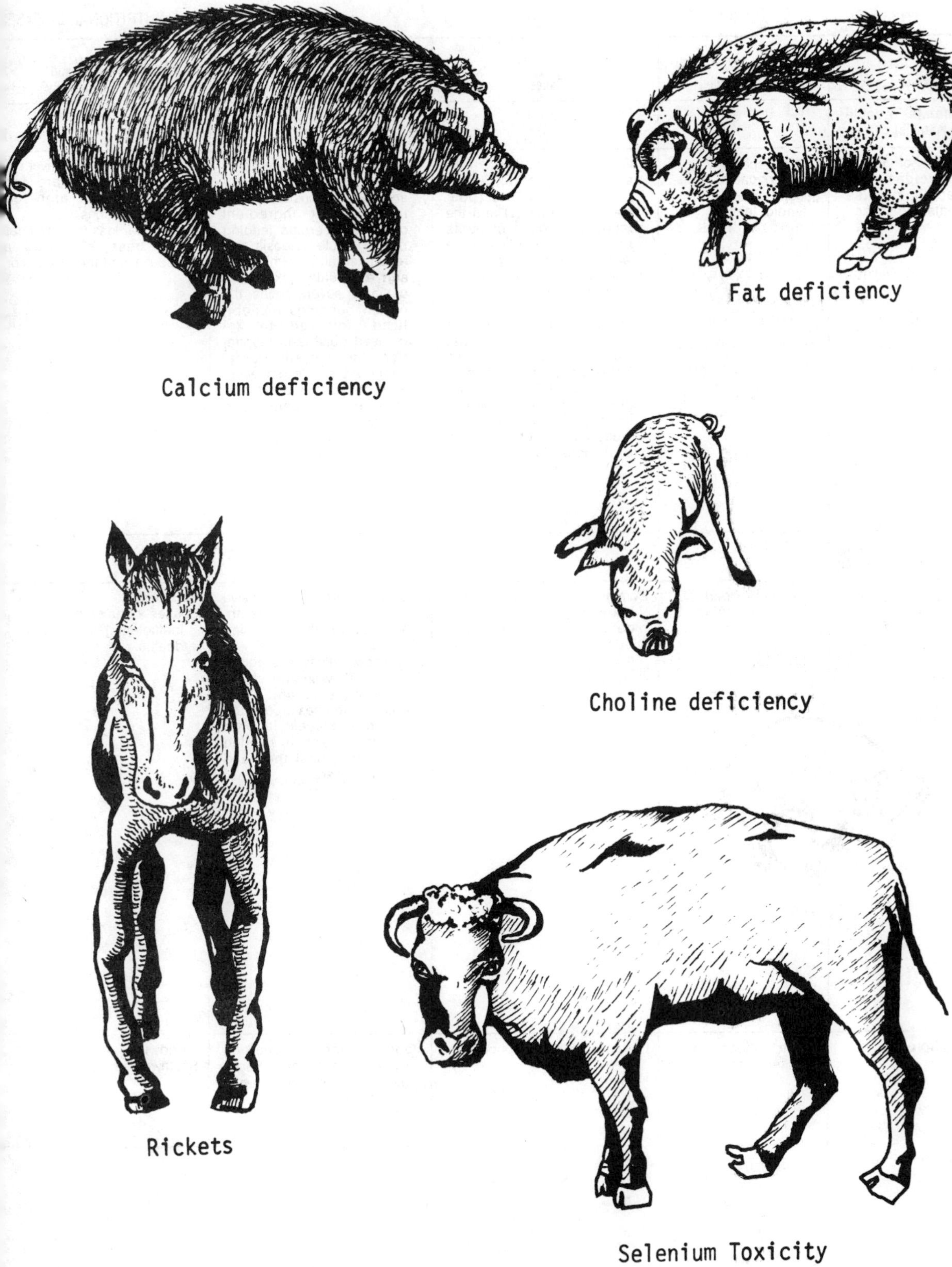

Fig. 4-39 (continued)

TABLE 4-100—NUTRITIONAL DISEASES

Disease	Species Affected	Cause	Symptoms and Signs (or age group most affected)	Distribution and Losses Caused By[2]
Acetonemia in cattle (see Ketosis)				
Acidosis (or lactic acid acidosis)—a metabolic disease of cattle and sheep	Cattle, especially feedlot cattle. Sheep, especially feedlot lambs.	Acidosis is caused by an increase in lactic acid-producing bacteria (both the d- and l-forms) and the rapid production of lactic acid. It commonly occurs when there is a sudden shift from a high-roughage to a high-concentrate ration. However, cattle maintained on high-energy rations are constantly in a marginal state of acidosis due to the formation of lactic acid in the rumen flora. Thus, ingredient changes, poor mixing of grain in the ration, or faulty feeding can promote acute acidosis.	Marginal acidosis is characterized by poor performance and inconsistent feed ingestion. If ingredient changes or erratic feeding persist, acute acidosis may result, creating laminitis—and eventually "ski shoe" cattle. In severe cases, the rumen becomes immobilized, followed by increased pulse and respiration rate, variable rectal temperature, sunken eyes, loss of dermal elasticity, staggering, coma, and death.	Acidosis occurs wherever cattle and lambs are fed, especially on high-concentrate rations. The annual loss from acidosis has been estimated at about 1% of the production
Alkali disease (see Selenium poisoning)				
Anemia, nutritional  Fig. 4-40	All warm-blooded animals, including man.	Commonly an iron deficiency, but it may be caused by a deficiency of copper, cobalt, and/or certain vitamins.	Loss of appetite, progressive emaciation, and death. Most prevalent in suckling young. Pigs show listlessness, rough hair coat, wrinkled skins, drooping ears and tails, pale membranes around the mouth and eyes, labored breathing, and a swollen condition about the head and shoulders.	Worldwide. Losses consist of slow and inefficient gains, and deaths.
Aphosphorosis	Cattle; sheep to a lesser extent.	Low available phosphorus in feed.	Depraved appetite; chewing bones, wood, hair, rags, etc. Stiff joints and fragile bones. Breeding problems and a high incidence of milk fever in dairy cattle.	Worldwide. Southwestern U.S.

Footnotes on last page of table.

AND AILMENTS OF ANIMALS[1]

Treatment	Control and Eradication	Prevention	Remarks
Different treatments have been used with varying degrees of success; among them: (1) removal of rumen contents and replacement by contents of an animal on a normal ration; (2) feeding a high level of penicillin (12-20 million units) to suppress lactic acid-producing bacteria; (3) drenching (or intravenous injection) with a solution of sodium bicarbonate to restore the acid base balance; (4) daily intramuscular administration of antihistamines and cortical steroids for each of several days to help prevent intoxication and laminitis; or (5) backing the cattle down on both amount and kind of feed (lessening the ration, and returning to the mix that was being used before trouble was encountered).	Acidosis is best controlled by (1) avoiding accidental access of cattle to large amounts of concentrates, and (2) changing gradually and stepwise from a low to a high proportion of concentrate in the ration.	Prevention consists in avoiding erratic feeding and abrupt ration changes.	
Provide dietary sources of the nutrient or nutrients, the deficiency of which is known to cause the condition.	When nutritional anemia is encountered, it can usually be brought under control by supplying dietary sources of the nutrient or nutrients, the deficiency of which is known to cause the condition.	Supply dietary sources of iron, copper, cobalt, and certain vitamins. Keep suckling animals confined to a minimum and provide feeds at an early age. Anemia in pigs can be prevented by providing supplemental iron in one of following forms:	Anemia is a condition in which the blood is either deficient in quality or quantity. (A deficient quality refers to a deficiency in hemoglobin and/or red cells.) Levels of iron in feed believed to be ample, since feeds contain 40 to 400 mg/lb.
	1. Two injections into the ham muscle of iron dextran (100 and 50 mg iron, respectively) at 3 and 21 days of age. 2. Place clean soil in the farrowing pen daily. Soil should not be contaminated with parasite eggs and other disease organisms. Iron sulfate can be sprinkled over the soil. 3. Give the pigs iron tablets or paste at 2 to 3 days of age. Repeat the treatment every 7 to 10 days until the pigs are eating creep ration adequately. If pills are given, it is important to see that the pigs swallow them and not spit them out. 4. Swab sow's udder daily with a solution of 1 lb iron sulfate dissolved in 1 gal of warm water. 5. Give the baby pigs a solution made by dissolving 1 lb of iron sulfate and 1 oz of copper sulfate in 1 qt warm water. One teaspoonful a week should be given to the young pigs until they are eating creep feed adequately. 6. Provide pigs with access to a creep feed by the time they are 10 days old.		
Intravenous drench with suitable phosphorus solution. Add phosphorus to the ration.	Controlled by feeding phosphorus, either free-choice or added to the ration.	Feed phosphorus in feed and/or as mineral supplement (free-choice). Keep the calcium-phosphorus ratio within the range 2:1 or 1:1.	Generally caused by lack of phosphorus in the pasture. Phosphorus fertilizing may help.

(Continued)

TABLE 4-10

Disease	Species Affected	Cause	Symptoms and Signs (or age group most affected)	Distribution and Losses Caused By[2]
Azoturia (hemoglobinuria, Monday morning disease, blackwater)	Horses.	Sudden exercise, following a day or two of rest during which time the horse has been on full feed, resulting in partial spasm or "tie-up." Thought to be caused by an abnormal amount of glycogen stored in the muscle. As the glycogen breaks down, lactic acid is formed. The lactic acid builds up in the muscle, causing a myocitis which manifests itself as partial spasm, or "tie-up."	Profuse sweating, abdominal distress, wine-colored urine, stiff gait, reluctance to move, and lameness. Finally, animal assumes a sitting position, and eventually falls prostrate on the side.	Worldwide, but the disease seldom seen in horses a pasture and rarely in horse at constant work.
Baby pig shakes (see Hypoglycemia)				
Bloat—Feedlot	All ruminants.	High-concentrate rations increase numbers of slime-producing bacteria in rumen. Slime traps fermentation gas and produces bloat.	Symptoms same as pasture bloat (see "Bloat—Pasture" which follows). Occurs when cattle or sheep have been fed high-concentrate low-roughage rations for approximately 60 days or longer.	A survey of Kansas feedlot showed the following losse from bloat: 0.1% died c bloat; 0.2% bloated se verely; and 0.6% bloate mildly to moderately, wit animal performance a fected adversely.[3]
Bloat—Pasture  Fig. 4-41	All ruminants.	Most common on lush legume pastures. Incidence on wheat pasture has been increasing in recent years. Pasture bloat is a frothy bloat caused by interaction of several factors—plant, animal, and microbial. Soluble plant proteins play a prominent role in permitting stable froth formation.	First observed as distention of paunch on left side in front of hipbone. This is followed by distention of right side, protrusion of anus, respiratory distress, cyanosis of tongue, struggling, and death if not treated.	Widespread, although some areas appear to have mor bloat than others. Often results in death. 36% of all mortality due t nutritional diseases an ailments is attributed t bloat.[4] Causes average annua losses in beef and dair cattle (including milk) o $104,904,000.
Colic	Horses.	Improper feeding, working, or watering.	Excruciating pain; and, depending on the type of colic, other symptoms are: distended abdomen, increased intestinal rumbling, violent rolling and kicking, profuse sweating, constipation, and refusal of feed and water.	Worldwide.
Crooked calves	Cattle.	Manganese deficiency.	Calves born with crooked necks and legs.	Northwestern U.S.
Fluorine poisoning (fluorosis)	All farm animals, poultry, fish, and man.	Ingesting excessive quantities of fluorine through either the feed, air, water, or some combination of these.	Abnormal teeth (especially mottled enamel) and bones (bones become thickened and softened), stiffness of joints, loss of appetite, emaciation, reduction in milk flow, diarrhea, and salt hunger.	The water in parts of Ark Calif., S.C., and Tex. ha been reported to contai excess fluoride. Occasior ally, throughout the U.S high-fluorine phosphate are used in mineral mixe Areas near certain indus tries which heat earth materials or burn high fluoride coal may be problem.

Footnotes on last page of table.

—(Continued)

Treatment	Control and Eradication	Prevention	Remarks
Absolute rest and quiet. While awaiting the veterinarian, apply heated cloths or blankets, or hot-water bottles to the swollen and hardened muscles. The veterinarian should determine treatment. In mild cases, he may use a tranquilizer or sedative. In severe cases, he may use muscle relaxers or sodium bicarbonate in solution to readjust the acid balance in the muscles.	When trouble is encountered, decrease the ration and increase the exercise on idle days.	Restrict the ration and provide daily exercise when the animal is idle. Give a wet bran mash the evening before an idle day or turn the idle horses to pasture. Some believe that a diuretic (a drug which will increase the flow of urine) will prevent the tie-up syndrome. This is a common treatment of racehorses. Others feel that increased B vitamins will prevent the lactic acid buildup.	The chances of recovery are good for horses that remain standing, are not forced to move after the signs are noticed, and whose pulse returns to normal within 24 hours.
Drench cattle with 1 to 2 oz poloxalene (Therabloat®), and then relieve free gas with stomach tube 10 minutes after treatment.	If feasible, increase proportion of roughage in ration. However, good quality legume hay may increase incidence of feedlot bloat.	No effective preventive drug available.	Feedlot bloat may occur during any month of year; however, more common during hot, humid weather.
Time permitting, severe cases of bloat should be treated by a veterinarian. Puncturing of the paunch should be a last resort. Mild cases may be home-treated by (1) keeping the animal on its feet and moving, and (2) drenching cattle either with (a) ½ to 1 qt mineral oil or (b) 1 to 2 oz poloxalene (Therabloat®). Mineral oil will cause cattle to go off feed whereas poloxalene will not.	When there is high incidence of bloat, it may be desirable to change the feed. Where legume bloat is encountered, use poloxalene (Bloat Guard®) according to manufacturer's directions.	The incidence is lessened by (1) avoiding straight legume pastures, (2) feeding dry forage along with pasture, (3) avoiding a rapid fill from an empty start, (4) keeping animals continuously on pasture after they are once turned out, (5) keeping salt and water conveniently accessible at all times, and (6) avoiding frosted pastures. Use poloxalene (Bloat Guard®), a nonionic surfactant, according to manufacturer's directions, for the control of legume bloat.	Legume pastures, alfalfa hay, and barley appear to be associated with a higher incidence of bloat than many other feeds. Legume pastures are particularly hazardous when moist, after a light rain or dew.
Call a veterinarian. To avoid danger of inflicting self-injury, (1) place the animal in a large, well-bedded stable, or (2) take it for a slow walk. Depending on diagnosis, veterinarian may use one or more of following: sedatives; laxatives, such as mineral oil; drugs; or surgery.	Follow a good management program, including parasite control.	Proper feeding, working, watering. Control parasites.	Colic is also a symptom of abdominal pain that can be caused by a number of different conditions. For example, bloodworms cause a colic due to damage in the wall of blood vessels. This results in poor circulation to the intestine.
		Feed manganese; 30 ppm of total feed.	The Utah station has produced crooked calves by feeding lupine.
Any damage may be permanent, but animals which have not developed severe symptoms may be helped to some extent, if the sources of excess fluorine are eliminated.	Discontinue the use of feeds, water, or mineral supplements containing excessive fluorine.	Avoid the use of feeds, water, or mineral supplements containing excessive fluorine. (See Table 4-7, p. 229, "Maximum Fluorine Content for (1) Mineral Substances, and (2) Total Ration.") 100 ppm (0.01%) fluorine of the total dry ration is the borderline in toxicity for cattle. At levels of 25-100 ppm, some mottling of the teeth may occur over periods of 3-5 years. In breeding animals, therefore, the permissible level is 30 ppm of the total dry ration. Not more than 65-100 ppm fluorine should be present in dry matter of rations when rock phosphate is fed.	Fluorine is a cumulative poison. Underfluorinated rock phosphate often contains 3.5 to 4.0% (35,000-40,000 ppm) of fluorine. Phosphate clays (soft phosphates) are usually too high in fluorine to be used safely unless defluorinated.

(Continued)

TABLE 4-1C

Disease	Species Affected	Cause	Symptoms and Signs (or age group most affected)	Distribution and Losses Caused By[2]
Founder (laminitis)  Fig. 4-42	Horses. Cattle. Sheep. Goats.	Overeating (grain; or lush legume or grass—known as "grass founder"), overdrinking, or from inflammation of the uterus following parturition. Also intestinal inflammation. Too rapid change in the ration.	Extreme pain, fever (103° to 106°F), and reluctance to move. If neglected, chronic laminitis will develop, resulting in a dropping of the hoof soles and a turning up of the toe walls.	Worldwide. Actual death losses fro founder are not very grea but usefulness may be a fected.
Goiter (see Iodine deficiency)				
Grass tetany (grass staggers)  Fig. 4-43	Cattle. Sheep.	Magnesium deficiency.	Generally occurs during first 2 weeks of pasture season. Nervousness, twitching of muscles (usually of head and neck), head held high, accelerated respiration, high temperature, gnashing of the teeth, and abundant salivation. Slight stimulus may precipitate a crash to the ground, and finally death.	Reported in Neb., Ky., Mc Iowa, Wash., and othe states. Also found in Ne Zealand, England and Ho land. Highly fatal if not treate quickly. Metabolic diseases of catt (which include grass tetan ketosis, milk fever, lame ness, and reduced resis ance to infectious agent cause estimated losses c $15 million annually.[5]
Heaves	Horses. Mules.	Exact cause unknown, but it is known that the condition is often associated with the feeding of damaged, dusty, or moldy hay. It often follows severe respiratory infection such as strangles. Probably an allergy.	Difficulty in forcing air out of the lungs, resulting in a jerking of flanks (double flank action) and coughing. The nostrils are often slightly dilated and there is a nasal discharge.	Worldwide. Losses are negligible.
Hypoglycemia (or Baby pig shakes)  Fig. 4-44	Swine.	Low blood-sugar level accompanies the trouble, but cause of the low blood sugar is unknown. The hog cholera virus can also cause this disease.	Shivering, weakness, failure to nurse, with no evidence of scouring. If disturbed, the pigs emit a weak, crying squeal. Hair becomes erect and rough, and the heart action slow and feeble. Without treatment, death usually comes in 24 to 36 hours after the first symptoms appear. Confined to baby pigs only.	Throughout the U.S.; mortalit may be high.
Iodine deficiency (goiter) 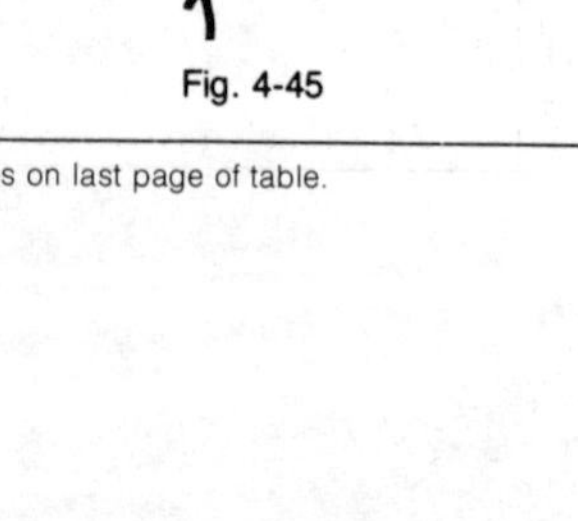 Fig. 4-45	All farm animals and man.	A failure of the body to obtain sufficient iodine from which the thyroid gland can form thyroxine (an iodine-containing compound).	Goiter (big neck) is the most characteristic symptom in humans, calves, lambs, and kids. Also, there may be reproductive failures and weak offspring that fail to survive. Pigs may be born hairless, and show edema of shoulders and neck. Foals may be weak.	Northwestern U.S. from th Great Lakes region t Washington. Also reporte in California and Texas.

Footnotes on last page of table.

(Continued)

Treatment	Control and Eradication	Prevention	Remarks
nding arrival of the veterinar-an, the attendant should stand he animal's feet in a cold-water bath. tihistamines, restricting the diet, use of diuretics, and anti-nflammatory agents such as corticosteroids or phenyl-butazone, may speed recovery and alleviate serious aftereffects.	Alleviate the causes.	Avoid (1) overeating, (2) over-drinking (especially when hot), and/or (3) inflammation of the uterus following parturition. Veterinary attention should be given if mares retain the after-birth longer than 12 hours.	Unless foundered animals are quite valuable, it is usually desirable to dispose of them following a case of severe founder.
ravenous injection of a solution of calcium and/or magnesium salt by a veterinarian.	(See Prevention.)	Grass tetany can be prevented by not turning animals to pasture, but this is not practical. Feeding hay at night during the first 2 weeks of the pasture season is helpful. A salt lick of 10 parts each of magnesium sulfate and calcium diphosphate with 80 parts of salt will aid in prevention. Also, a mixture of 2 parts magnesium oxide to 1 part salt as the only source of salt is effective.	Affected animals show low blood magnesium, often low serum calcium. Treated cattle may be aggressive on arising; so watch out!
tihistamine granules can be administered in feed to control coughing due to lung congestion. fected animals are less bothered if turned to pasture, if used only at light work, if fed an all-pelleted ration, or if the hay s sprinkled lightly with water.		Avoid the use of damaged feeds. Feed an all-pelleted ration, thereby alleviating dust.	Unlike a man, a horse cannot breathe through his mouth. Basically, heaves is a rupture of some of the alveoli in the lungs, of which the specific cause is unknown.
ovide heat lamps for pigs. earliest symptoms either (1) force feed at frequent intervals a mixture of 1 part corn syrup diluted with 2 parts of water or (2) give intraperitoneal injections of 5% glucose solution every 4 to 6 hours. Consult veterinarian.	Not contagious	Adequate rations and good care and management of the gestating sows may lessen the incidence of the disease. Be sure there is adequate milk for baby pigs during first days of life.	One of the hazards of hypoglycemia is that the milk flow of the sow will not be stimulated or may even cease due to the inactivity of the affected pigs. In the latter case, the pigs may have to be either transferred to a foster mother or hand fed.
casionally borderline cases may survive; in these the moderate thyroid enlargement disappears in a few weeks. ce the iodine deficiency symptoms appear, no treatment is very effective.	At the first signs of iodine deficiency, iodized salt should be fed to all farm animals.	In iodine-deficient areas, feed iodized salt to all farm animals throughout the year. Stabilized iodized salt containing 0.01% potassium iodide is recommended. Organic iodide is also a suitable source of iodine, but is usually more costly.	The enlarged thyroid gland (goiter) is nature's way of attempting to make sufficient thyroxine under conditions where an iodine deficiency exists. Large excesses of iodine systemically may cause abortions.

(Continued)

TABLE 4-1(

Disease	Species Affected	Cause	Symptoms and Signs (or age group most affected)	Distribution and Losses Caused By[2]
Ketosis (also known as acetonemia in cattle and pregnancy disease in sheep)  Fig. 4-46	Cattle. Sheep. Goats.	A metabolic disorder, thought to be a disturbance in the carbohydrate metabolism. May involve adrenal insufficiency.	In cows, ketosis or acetonemia is usually observed within first 1-6 weeks after calving. Affected animals show loss in appetite and condition, a marked decline in milk production, and the production of a peculiar sweetish chloroformlike odor of acetone that may be present in the milk and pervade the barn. In ewes and goats, ketosis or pregnancy disease generally strikes during last 2 weeks of pregnancy. Usually affected ewes are carrying twins or triplets. Symptoms include grinding of teeth, dullness, weakness, frequent urination and trembling when exercised—with the final stage being complete collapse, followed by death in 90% of the cases.	Worldwide. Ketosis or acetonemia affec dairy cattle throughout th U.S. Metabolic diseases of catt (which include ketosis, mi fever, grass tetany, lam ness, and reduced resis ance to infectious agent cause estimated losses $15 million annually.[5] Ketosis or pregnancy diseas of sheep affects farm flock more than range bands, th losses in the former som times being as high a 25%.
Manganese deficiency (See Crooked calf disease and Grass tetany.)	All farm animals and poultry.	Deficiency of manganese.	Young born with stiff, curved, or crooked necks and backs, and permanently bent forward legs caused by contracted tendons.	Reported in Wash., Mont., an Utah.
Milk fever (parturient paresis; hypocalcemia) 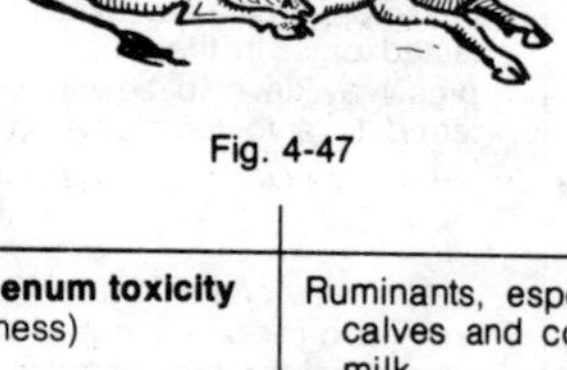 Fig. 4-47	Cattle. Sheep. Goats (occasionally).	Low blood calcium concentration. Too much calcium in the ration can cause this condition. In milking cows, the calcium-phosphorus ratio should not exceed 2:1 for cows with a history of milk fever.	Commonly occurs soon after calving and in high-producing cows. Rarely occurs at first calving. First symptoms are loss of appetite, constipation, and general depression. This is followed by nervousness and finally collapse and complete loss of consciousness. The head is usually turned back.	A common, widespread dis ease of dairy cows. Losses are not great, althoug untreated animals are likel to die. Metabolic diseases of cattl (which include milk feve ketosis, grass tetany, lame ness, and reduced resis ance to infectious agents cause estimated losses c $15 million annually.[5]
Molybdenum toxicity (teartness)	Ruminants, especially calves and cows in milk.	As little as 10 to 20 ppm in forages result in toxic symptoms.	Toxic levels of molybdenum interfere with copper metabolism, thus increasing the copper requirement and producing typical copper deficiency symptoms. The physical symptoms are anemia and extreme diarrhea, with consequent loss in weight and milk yield. Black hair may turn brown.	England, and in Florida California, and Manitoba.

Footnotes on last page of table.

—(Continued)

Treatment	Control and Eradication	Prevention	Remarks
Cattle: ½ to 1 lb of either propylene glycol or sodium propionate daily, with the dose divided into 2 treatments for 5 to 10 days. Put treatment in grain if cow is eating; otherwise, give as drench. Intravenous injection of glucose (500 ml of 50% glucose solution) is rapid way of getting sugar in blood. *Sheep and goats:* 4 oz of propylene glycol, given orally twice daily.	*Cows:* Maintain relatively high energy intake before calving; increase energy intake substantially after calving. *Ewes:* Avoid obesity in early pregnancy. Feed rather liberally last 6 weeks of pregnancy.	The incidence of ketosis can be lessened by (1) not allowing cows to be excessively fat at calving; (2) increasing the level of concentrates rapidly after calving; (3) feeding good quality roughage after calving, and avoiding abrupt changes in roughage; (4) feeding adequate proteins, minerals, and vitamins, and (5) providing comfort, exercise, and ventilation. In problem herds, feeding ¼ lb daily of propylene glycol or sodium propionate may be helpful.	The clinical findings are similar in the case of affected cattle and sheep, but it usually strikes ewes just before lambing, whereas cows are usually affected within the first 1 to 6 weeks after calving.
	(See Prevention.)	Feed a mineral containing manganese; 30 ppm of total feed, or 27.24 g/ton feed.	The Utah station has also produced crooked calves by feeding lupine. Alkali can tie up the manganese in water, soils, or plants.
Have veterinarian give injection of a calcium salt ($CaCl_2$, Ca lactate, Ca Gluconate, or other Ca salts) to elevate blood serum Ca above the concentration of the 5 or 6 mg/ml that is associated with the onset of tetany (normal level is 9-10).	(See Prevention.)	Each of the following measures will lessen the incidence of milk fever: 1. *Calcium-phosphorus ratio and amounts*—Approximately a 2.3:1 Ca:P ratio. Feed a ration that contains 0.5 to 0.7% Ca and 0.3 to 0.4% P. 2. *Calcium shock treatment*—10 to 14 days before calving, feed a Ca-deficient ration with a Ca:P ratio of 1:2. This activates the cow's calcium-mobilizing mechanism for drawing calcium from the bones, with the result that it is functioning before calving and, milk fever is avoided. 3. *High vitamin D*—This consists in feeding 20 million units of vitamin D/cow/day starting about 5 days before calving and continuing through the first day postpartum, with a maximum dosage period of 7 days.	The name "milk fever" is a misnomer, because the disease is not accompanied by fever, the temperature really being below normal.
One gram of copper sulfate per head daily will cure symptoms of molybdenum toxicity.		One gram of copper sulfate per head daily will prevent molybdenum toxicity. It is usually provided as part of a commercial mineral or concentrate mix.	When feeds are high in sulfate, toxic symptoms will be produced on lower levels of molybdenum and, conversely, higher levels of molybdenum can be tolerated with low levels of sulfate.

(Continued)

TABLE 4-100

Disease	Species Affected	Cause	Symptoms and Signs (or age group most affected)	Distribution and Losses Caused By[2]
Nitrate poisoning (oat hay poisoning, corn stalk poisoning)	Primarily cattle. Sheep. Horses.	Forages (vegetative part) of most grain crops (oats, wheat, barley, rye, corn, sorghum), Sudangrass, and numerous weeds, especially (1) when under stress such as drought, insufficient sunlight, or after spraying with weed killer (herbicide); or (2) following heavy nitrate fertilization of soils (commercial, green manure crop, barnyard manure). Some nitrate may be formed after forage is stacked. Inorganic nitrate or nitrite salts, or fertilizers left where animals have access to them, or where they may be mistaken for salt. Pond or shallow well into which surface runoff from barnyard or well-fertilized soil might drain.	Accelerated respiration and pulse rate; diarrhea; frequent urination; loss of appetite; general weakness; trembling and staggering gait; frothing from mouth; lowered milk production; abortion; blue color of the mucous membrane, muzzle, and udder due to lack of oxygen in blood; death within 4½ to 9 hrs. after consuming nitrates. A rapid and accurate diagnosis of nitrate poisoning may be made by examining blood. Normal blood is red and becomes brighter when exposed to air, whereas blood from cows toxic with nitrates is a brown color due to formation of methemoglobin. Nitrates oxidize ferrous hemoglobin (oxyhemoglobin) to ferric hemoglobin (methemoglobin) which cannot transport oxygen. The animal essentially suffocates for lack of oxygen in tissues. When ¾ of the oxyhemoglobin is converted to methemoglobin, the animal will die.	Excessive nitrate content of feeds is an increasingly important cause of poisoning in farm animals, due primarily to more and more high nitrogen fertilization. But nitrate toxicity is not new, having been reported as early as 1850, and having occurred in semiarid regions of this and other countries for years.
Oat hay poisoning (see Nitrate poisoning)				
Osteomalacia 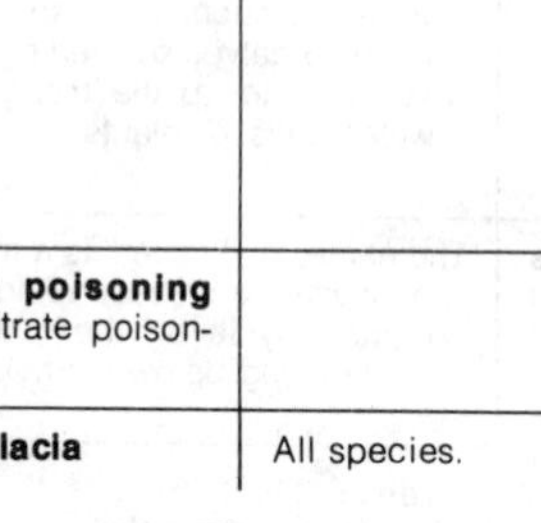 Fig. 4-48	All species.	Inadequate phosphorus (sometimes inadequate calcium). Lack of vitamin D in confined animals. Incorrect ratio of calcium to phosphorus.	Phosphorus deficiency symptoms are: depraved appetite (gnawing on bones, wood, or other objects, or eating dirt); lack of appetite, stiffness of joints, failure to breed regularly, decreased milk production, and an emaciated appearance. Calcium deficiency symptoms are: fragile bones, reproductive failures, and lowered lactations. Mature animals most affected. Most of the acute cases occur during pregnancy and lactation.	Southwestern U.S. is classed as a phosphorus-deficient area, whereas calcium-deficient areas have been reported in parts of Florida, Louisiana, Nebraska, Virginia, and West Virginia.
Parakeratosis (greasy skin disease)	Swine.	High calcium levels in the diet—above 0.8%.	Pigs have mangy look, reduced appetite and growth rate, diarrhea, and vomiting. It affects pigs 1-5 months of age.	Mortality is not high; economic loss is mainly in reduced gains and lowered feed efficiency.
Periodic ophthalmia (moon blindness)	Horses. Mules. Asses.	It may be caused by (1) leptospirosis, (2) a localized hypersensitivity or allergic reaction or (3) lack of riboflavin.	Periods of cloudy vision, in one or both eyes, which may last for a few days to a week or two and then clear up; but it recurs at intervals, eventually culminating in blindness in one or both eyes.	In many parts of the world. In the U.S., it occurs most frequently in the states east of the Missouri River.

Footnotes on last page of table.

—(Continued)

Treatment	Control and Eradication	Prevention	Remarks
4% solution of methylene blue (in a 5% glucose or a 1.8% sodium sulfate solution) administered by a veterinarian intravenously at the rate of 100 cc/1,000 lb liveweight.	(See Prevention.)	More than 0.5% nitrate nitrogen (dry basis) may be considered as potentially toxic. Feed should be analyzed when in question, by using a simple test to detect presence of nitrates (qualitative); if present, follow with a quantitative test to determine how much is present. Nitrate poisoning may be reduced by (1) feeding high levels of grains and other high-energy feeds (molasses) and vitamin A, (2) limiting the amount of high-nitrate feeds, (3) ensiling forages which are high in nitrates, (fermentation reduces some nitrates to gas, but care must be taken to avoid nitric oxide and nitrogen dioxide released in early stages of fermentation) and avoid feeding until 3 to 4 weeks in storage.	Nitrate form of nitrogen does not appear to cause the actual toxicity. During digestion, the nitrate is reduced to nitrite, a far more toxic form (10 to 15 times more toxic than nitrates). In cows and sheep, this conversion takes place in the rumen (paunch); in horses in the cecum. Lethal dose varies with (1) nutritional state, size and type of animal; and (2) the consumption of feed other than nitrate-containing material. (Nitrate over 5% of total ration is a potential source of trouble; 0.75% content nitrate forages must be fed with caution, and milk production will be lowered; and 1.5% death will likely occur.) Where nitrate troubles are suspected, consult the local veterinarian or county agent.
Select natural feeds that contain sufficient quantities of calcium and phosphorus. Feed a special mineral supplement or supplements. If this disease is far advanced, treatment will not be successful.	(See Treatment.)	Feed balanced rations, and allow animals free access to a suitable phosphorus and calcium supplement. Increase the calcium and phosphorus content of feed through fertilizing the soils.	Calcium deficiencies are much more rare than phosphorus deficiencies in cattle, sheep, and horses. Calcium deficiencies are fairly common in swine because grains, which are their chief feed, are low in this mineral.
Add 0.4 lb of zinc carbonate or 0.9 lb of zinc sulfate heptahydrate/ton of feed.	It is not contagious.	Add 0.4 lb of zinc carbonate or 0.9 lb of zinc sulfate heptahydrate/ton of feed where the disease is encountered.	Excess calcium reduces the absorption and utilization of zinc. In swine, this causes parakeratosis.
Antibiotics administered promptly are helpful in some cases. Immediately (1) change to greener hay or grass, and (2) add riboflavin at the rate of 40 mg/day/animal.		Feed green grass, or well-cured green, leafy hay; or add riboflavin to the ration at the rate of 40 mg/animal/day.	This disease has been known to exist for at least 2,000 years.

(Continued)

TABLE 4-1(

Disease	Species Affected	Cause	Symptoms and Signs (or age group most affected)	Distribution and Losses Caused By[2]
Polioencephalomalacia (cerebrocortical necrosis, or forage poisoning)	Cattle. Sheep.	Believed to be due to a thiamin (B_1) deficiency. It is noninfectious.	Sudden deaths in feedlot cattle. Sick animals are excitable, incoordinated, and have impaired vision. On driving, these animals go down into convulsions.	Most common in feedlot ani mals. Causes severe econom losses—in reduced feed e ficiency, prolonged finis ing, expensive treatment and deaths.
Pregnancy disease in sheep (see Ketosis)				
Protein poisoning	Horses in particular.	High levels of protein incriminated. Some horses do appear to be allergic to certain proteins or to excesses of specific amino acids.	"Protein bumps" over the body—an allergic reaction.	
Rickets  Fig. 4-49	All farm animals and man.	Lack of either calcium, phosphorus, or vitamin D; or an incorrect ratio of the 2 minerals.	Enlargement of the knee and hock joints, and the animal may exhibit great pain when moving about. Irregular bulges (beaded ribs) at juncture of ribs with breastbone, and bowed legs. Rickets is a disease of young animals—calves, foals, pigs, lambs, kids, pups, and chicks.	Worldwide. It is seldom fatal.
Salt deficiency (sodium chloride)	All farm animals and man.	Lack of salt (sodium chloride).	Loss of appetite, retarded growth, loss of weight, a rough coat, lowered production of milk, and a ravenous appetite for salt.	Worldwide, especially amon grass-eating animals.
Salt poisoning (sodium chloride)	All farm animals, but swine and sheep most frequently affected.	Brine from cured meats; wet salt. Where large amounts of brine or salt have been mixed in hog slop. When excess salt is fed following salt starvation. When salt is improperly used to govern self-feeding of concentrate.	Sudden onset—1 to 2 hours after ingesting salt; extreme nervousness; muscle twitching and fine tremors; much weaving, wobbling, staggering, and circling; blindness; weakness; normal temperature, rapid but weak pulse, and very rapid and shallow breathing; diarrhea; death from a few hours up to 48 hours. Convulsions seldom occur, except in pigs.	Salt poisoning is relatively rare.
Salt sick (cobalt deficiency)	Cattle. Sheep. Goats.	Cobalt deficiency, associated with copper and perhaps iron deficiencies.	Loss of appetite, depraved appetite, scaliness of skin, listlessness, and lack of thrift.	Australia, Western Canada Fla., N.H., Mich., Wisc. N.Y., and N.C. On sandy soils.

Footnotes on last page of table.

-(Continued)

Treatment	Control and Eradication	Prevention	Remarks
tramuscular or intravenous thiamin injections should be administered to sick animals. Supplementary fluids should be given by way of the stomach tube.		Until the cause is discovered, little can be done to prevent the disease, except to provide a good ration.	
ower protein content of the ration.	Economics generally control the level of protein feeding. High protein feeds are more expensive than high energy feeds, with the result that there is temptation to feed too little of them.	Do not feed excessive levels of protein.	There is no proof that heavy feeding of high-protein feeds to horses is harmful, provided (1) the ration is balanced out in other respects, (2) the animal's kidneys are normal and healthy (a large excess of protein in terms of body needs increases the work of the kidneys for the excretion of the urea), (3) any ration change to high-protein feed is made gradually, as is recommended in any change in feed, and (4) there is adequate exercise and normal metabolism.
the disease has not advanced too far, treatment may be successful by supplying adequate amounts of vitamin D, calcium, and phosphorus, and/or adjusting the ratio of calcium to phosphorus.	(See Prevention.)	Provide (1) sufficient calcium, phosphorus, and vitamin D, and (2) a correct ratio of the 2 minerals.	Rickets is characterized by a failure of growing bone to ossify, or harden, properly.
Salt starved animals should be gradually accustomed to salt, slowly increasing the hand-fed allowance until the animals may be safely allowed free access to it.	(See Treatment and Prevention.)	Provide plenty of salt at all times, preferably by free-choice feeding.	Common salt is one of the most essential minerals for grass-eating animals, and one of the easiest and cheapest to provide. Excessive salt intake can result in toxicity if animals are deprived of water (see Salt poisoning).
Provide large quantities of fresh water to affected animals. Those that can and do drink seldom need additional treatment. Those unable to drink should be given water via stomach tube, by the veterinarian. The vet may also give (I.V. or intraperitoneally) calcium gluconate to severely affected animals.	(See Prevention.)	If animals have not had salt for a long time, they should first be hand-fed salt, gradually increasing daily allowance until they leave a little in the mineral box; then self-feed.	Indians and pioneers handed down many legendary stories about huge numbers of wild animals that killed themselves simply by gorging at a newly found salt lick after having been salt-starved for long periods of time.
Provide 0.2 to 0.5 oz cobalt salt/100 lb of salt—or feed a suitable trace mineral supplement.		Mix 0.2 to 0.5 oz of cobalt chloride, cobalt sulfate, or cobalt carbonate/100 lb of either (1) salt, or (2) whatever mineral mix is being used.	Cobalt is needed especially for microbial synthesis of vitamin B_{12}. Nonruminants must be fed preformed vitamin B_{12}.

(Continued)

TABLE 4-10

Disease	Species Affected	Cause	Symptoms and Signs (or age group most affected)	Distribution and Losses Caused By[2]
Selenium poisoning (alkali disease) 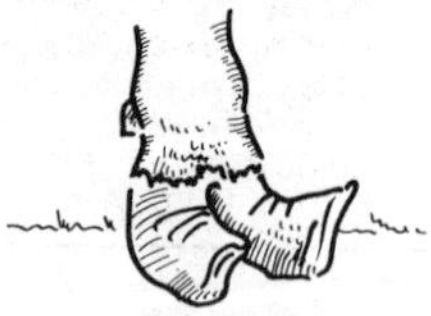 Fig. 4-50	All farm animals and man.	Consumption of plants grown on soils containing selenium.	Loss of hair from the tail in cattle, a general loss of hair in swine, and a loss of hair from the mane and tail in horses. In severe cases, the hoofs slough off, lameness occurs, feed consumption decreases, and death may occur by starvation.	In certain regions of the west ern U.S.—especially certain areas in S.D., Mont., Wyo. Neb., Kan., and perhaps areas in other states in the Great Plains and Rocky Mountains. Also in Canada.
Stiff-lamb disease (white muscle disease, muscular dystrophy)  Fig. 4-51	Lambs. (Also, a similar condition, white muscle disease, appears in calves.)	Selenium deficiency and lack of vitamin E; or perhaps for some as yet unknown reason the vitamin E of the ration is not available to the animal due to an inhibitor, or for other reasons.	A stiff, stilted way of moving, chiefly in hind legs, although front legs and shoulders may be involved. Back usually humped or "roached." Lambs that live are usually stunted. Young, rapidly growing lambs 1-5 weeks of age especially susceptible.	Throughout the U.S., but incidence is highest in intermountain area, between the Rocky and Cascade Mountains. Affected lambs often die.
Sweet clover disease  Fig. 4-52	Cattle; rarely affects sheep or horses.	Usually produced only by moldy or spoiled sweet clover, hay or silage. Caused by presence of dicoumarol which interferes with vitamin K in blood clotting.	Loss of clotting power of the blood. As a result, blood forms soft swellings beneath skin on different parts of body. Serious or fatal bleeding may occur at time of dehorning, castration, parturition, or following injury. All ages affected. A newborn animal may also have the condition at birth.	Wherever sweet clover is grown and cured for live stock feed.
Urinary calculi (gravel, stones, water belly, urolithiasis)	Cattle. Sheep. Horses. Man.	Unknown, but it does seem to be nutritional. Experiments and experiences have shown a higher incidence of urinary calculi when there is (1) a high potassium intake, (2) a high phosphorus-low calcium ratio (from the standpoint of preventing urinary calculi, the Ca:P ratio should be about 2:1), (3) a high silica content in the ration, or a high proportion of high-silica grains and forages, such as native grasses, wheat straw, sugar beet leaves or pulp, sorghums, and cottonseed meal. A deficiency of vitamin A may be contributing factors.	Frequent attempts to urinate, dribbling or stoppage of the urine, pain and renal colic. Usually only males affected, the females being able to pass the concretions. Bladder may rupture, with death following. Otherwise, uremic poisoning may set in.	Worldwide. Affected animals seldom recover completely. Causes estimated average annual loss in beef cattle of $4,052,000.

Footnotes on last page of table.

—(Continued)

Treatment	Control and Eradication	Prevention	Remarks
The use of salt containing 37.5 ppm of arsenic may reduce the incidence of chronic selenium poisoning on seleniferous range. Pasture rotation and use of supplemental feeds from nonseleniferous areas are practical solutions to the problem. There is no known treatment for acute selenium poisoning.	(Control measures based on Prevention.)	Abandon areas where soils contain excess selenium, because crops produced on such soils constitute a menace to both animals and man.	Chronic cases of selenium poisoning occur when cattle consume feeds containing 8.5 ppm of selenium over an extended period; acute cases occur on 500 to 1,000 ppm. The toxic level of selenium is in the range of 2.27-4.54 mg/lb (5-10 ppm) of feed.
Injections of selenium and tocopherol give dramatic results.	If stiff-lamb disease occurs in the flock, give prompt attention to any possible improvements in the ration. In season, turn animals to good pasture.	Inject lambs with a solution of vitamin E and selenium (1) at docking time (3-4 days) and (2) 2-4 weeks of age. Since several concentrations are available, follow the directions on the label.	Most natural rations, even corncobs, contain an abundance of vitamin E. At Cornell, a high incidence of stiff-lamb disease occurred when a ration high in cull red kidney beans was fed. Stiff-lamb disease is most common (1) in rapidly growing lambs and (2) on lush pastures. Linseed meal is probably effective in preventing stiff-lamb disease because it contains selenium.
Remove the offending materials and administer menadione (vitamin K_3). The veterinarian usually gives the affected animal an injection of plasma or whole blood from a normal animal that was not fed on the same feed.	When a case of sweet clover disease is observed in the herd, either (1) discontinue feeding the damaged product, or (2) alternate it with a better-quality hay, especially alfalfa.	Properly cure any sweet clover hay or ensilage.	The disease has also been produced from feeding moldy lespedeza hay and from sweet clover pasture.
When calculi develop, it may be advisable to dispose of the animal, since treatments have limited success. Treatment: (1) add ammonium chloride at the rate of 1 oz (lambs) or 1¼ to 1½ oz (cattle) per head daily—or 50 to 60% more ammonium sulfate; (2) increase salt content of ration to 1½% if the animals are not ready for market—too much may lower feed intake; (3) incorporate 20% alfalfa in the ration; (4) administer muscle relaxants to help the passage of calculi from the bladder; or (5) surgically remove the calculi; however, males will become nonbreeders after such an operation.	If severe outbreaks of urinary calculi occur in finishing steers or lambs it is usually well to dispose of them if they are carrying acceptable finish. For feedlot wether lambs, add to the ration 0.5% ammonium chloride (1 oz/head/day) or 0.9% ammonium sulfate.	Good feed and management appear to lessen the incidence. Delayed castration (castration of bull calves at 4-5 mo. of age) and high-salt diets of feedlot cattle (1-3% salt in the grain ration, using the upper limits in the winter months) in order to induce more water consumption are effective preventive measures. Avoid high phosphorus and low calcium. Provide adequate vitamin A, salt, and water.	Calculi are stonelike concretions in the urinary tract which almost always originate in the kidneys. These stones block the passage of urine, resulting in the condition commonly referred to as "water belly." The mineral deposits may be of variable sizes, shapes, and composition. In cattle, the phosphatic type predominates under feedlot conditions and the silicerous type occurs most frequently in range cattle. Ammonium chloride (see Control and Eradication) appears to be the product of choice. However, ammonium sulfate may be used, at the rate of 1.7 to 2.0 oz/head/day. Add it to the ration when an outbreak occurs.

(Continued)

TABLE 4-10

Disease	Species Affected	Cause	Symptoms and Signs (or age group most affected)	Distribution and Losses Caused By[2]
Vitamin A deficiency (night blindness and xerophthalmia)	All farm animals and man.	Vitamin A intake too low. High levels of nitrate intake from hay, silage, and/or water.	Night blindness, the first symptom of vitamin A deficiency, is characterized by faulty vision, especially noticeable when the affected animal is forced to move about in twilight in strange surroundings. Rough hair coat. Reduced fertility. Xerophthalmia in cattle develops in the advanced stages of vitamin A deficiency. The eyes become severely affected, and blindness may follow. Severe diarrhea in young calves and intermittent diarrhea in advanced stages in adults. In finishing cattle, generalized edema or anasarca with lameness in hock and knee joints and swelling in the brisket area. Young animals fail to grow.	Worldwide. Especially preva lent where one of the follow ing conditions frequentl prevails: (1) extende drought, and (2) winte feeding on bleached gras cured on the stalk or o bleached hay.
White muscle disease (Muscular dystrophy; in sheep, stiff-lamb disease)	Calves. Lambs.	Selenium and vitamin E deficiency; it may be lack of availability of vitamin E, or presence of an inhibitor.	Symptoms range from mild "founderlike" stiffness to sudden death. Calves continue to nurse as long as they can reach the cow's teats. Many calves stand or lie with protruded tongue, fighting for breath against severe pulmonary edema. It seems that more calves than lambs develop fatal heart damage. Affected calves show pathological lesions similar to those of "stiff lambs" (white muscle disease in lambs); namely, whitish areas or streaks in the heart and other muscles. Affects calves from birth to 3 months of age.	Throughout the U.S., but the incidence appears to be highest in the intermountain area, between the Rocky and Cascade Mountains.

[1]This table was authoritatively reviewed by the following: Dr. Erle E. Bartley, Department of Dairy and Poultry Science, Kansas University, Manhattan, Kan.; Dr. C. Brent Theurer, Department of Animal Science, The University of Arizona, Tucson, Ariz.; Dr. J. T. Huber, Department of Dairy Science, Michigan State University, East Lansing, Mich.; and Dr. T. H. Blosser, Agricultural Research Service, USDA, Washington, D.C.

[2]Unless otherwise indicated, the estimated average annual loss figures (in dollars) given in column 4 of this table were taken from *Losses in Agriculture*, Ag. Hdbk. No. 291, USDA, Washington, D.C., 1965.

—(Continued)

Treatment	Control and Eradication	Prevention	Remarks
reatment consists in correcting the dietary deficiencies and (1) adding vitamin A to the ration, or (2) injecting cattle intramuscularly or intraruminal 500,000 to 1,000,000 IU of vitamin A.	(See Prevention and Treatment.)	Provide good sources of carotene (provitamin A) through green, leafy hays, silage, lush green pasture, yellow corn. Add stabilized vitamin A to ration or inject slow-release vitamin A intramuscularly.	High levels of nitrates interfere with the conversion of carotene to vitamin A. Sheep will not develop xerophthalmia on a vitamin A deficiency.
jection of selenium and tocopherol (vitamin E). Confine affected animals to a stall and give plenty of rest.	(See Prevention and Treatment.)	Feed 1¼ lb linseed meal per cow during last 2 months of pregnancy, for its selenium content. Alpha tocopherol (vitamin E) added to the diet or injected intramuscularly shortly after birth and again at 2 to 4 weeks of age. Use commercial vitamin E according to directions.	Most natural rations, even corncobs, contain an abundance of vitamin E. At Cornell University, stiff-lamb disease developed on a ration high in cull red kidney beans.

[3]Meyer, R. M., *27th Kansas Formula Feed Conference Proceedings*, P. H1, Kansas State University, Manhattan, Kan., 1972.
[4]Ensminger, M. E., M. W. Galgan, and W. L. Slocum, *Problems and Practices of American Cattlemen*, Wash. Ag. Exp. Sta. Bull. 562, 1955, p. 18.
[5]From a report prepared for the Council of Deans, Association of American Veterinary Medical Colleges, based on information available on disease losses as of Feb. 1, 981.

PART VIII—FEEDS AND THEIR COMPOSITION

A feed (or feedstuff) is any ingredient, or material, fed to animals for the purpose of sustaining them. A wide variety of feeds is being used for animal feeding throughout the world. More than 2,000 different products have been classified as animal feeds, not counting varietal, grade, and stage of maturity differences.

Nutrient compositions of feedstuffs are necessary for intelligent ration preparation. These are presented in the following tables:

Table 4-101—Composition of Feeds
Table 4-102—Amino Acid Composition of Feeds
Table 4-103—Iodine Content of Feeds
Table 4-104—Selenium Content of Feeds
Table 4-105—Vitamin D Content of Feeds

Table 4-101 contains the compositions of the great array of feeds. Relatively few feeds have been analyzed for amino acid, iodine, selenium, or vitamin D content; nevertheless, such values as are available are reported in Tables 4-102, 4-103, 4-104, and 4-105.

For convenience, the feeds listed in Tables 4-101 and 4-102 have been classified as (1) energy feeds, (2) protein feeds, (3) dry forages and roughages, (4) silages and haylages, (5) pasture and range plants, (6) mineral supplements, and (7) vitamin supplements. It is emphasized, however, that these partitions are only approximate and that there is some overlapping—with certain feeds listed in more than one category; for example, whole soybean seed may be used either as an energy feed (83.2% TDN, as fed) and/or as a protein feed (37.9%, as fed). Definitions of each of these feed classes follow:

1. *Energy feeds*—Feeds that are high in energy, low in fiber (under 18%), and contain less than 20% protein.

2. *Protein feeds*—Feeds that contain more than 20% protein or protein equivalent.

3. *Dry forages and roughages*—Feeds in the dry state that are bulky and low in weight per unit volume, contain more than 18% crude fiber, and are relatively low in energy.

4. *Silages and haylages*—

a. Silages are fermented, high-moisture forages stored under anaerobic conditions in a silo. Usually, they are green crops, or dry crops to which moisture has been added, which are chopped when stored, and which contain 65 to 70% moisture.

b. *Haylage* is low-moisture silage, which usually contains 35 to 55% moisture.

5. *Pasture and range plants*—Grass, browse, and other forages that are harvested by grazing animals.

6. *Mineral supplements*—Rich sources of one o more of the inorganic elements needed to perfor certain essential body functions.

7. *Vitamin supplements*—Rich synthetic o natural feed sources of one or more of the comple organic compounds, called vitamins, that are require in minute amounts by animals for normal growt production, reproduction, and/or health.

To the extent available, the feed compositions i Tables 4-101, 4-102, 4-103, 4-104, and 4-105 wer taken from the various National Academy of Science National Research Council publications. Additiona feeds and compositions were provided by the autho with compositions obtained from experimental re ports, industries, and other reliable sources.

MOISTURE CONTENT OF FEEDS

All feeds (except oven-dry) contain some mois ture, with the amount depending upon the manne and length of time in which they are stored and th amount of moisture in the air. Most dry forages an grains contain 8 to 12% moisture, with an average o about 10%. Silage generally runs 65 to 70% water haylage is in the 35 to 55% moisture range, and pas ture varies from 50 to 83% water, depending on th stage of maturity, season, and soil moisture.

It is necessary to know the amount of water i feeds if comparisons of compositions are to be mean ingful. Obviously, when grain is bought or fed, it value with 14% moisture is not the same as with 10% moisture.

The composition of a feed is usually expressed o one or more of the following bases:

1. *As-fed (wet, fresh)*—This refers to feed a normally fed to animals. It may range from 0% t 100% dry matter.

2. *Air dry (approximately 90% dry matter)*—Thi refers to feed that is dried by means of natural ai movement, usually in the open. It may be either a actual or an assumed dry matter content; the latter i approximately 90%. Most feeds are fed in an air dr state.

3. *Moisture-free (M-F, oven-dry, 100% dr matter)*—This refers to a sample of feed that has bee dried in an oven at 221° F (105° C) until all the mois ture has been removed.

Dry Matter Calculations

The percent water is determined by drying a finely ground sample of feed in an oven until a con stant weight is attained. The dry matter is determine by finding the percent water in a feed and subtracting this value from 100 percent. The calculations involved follow:

$$\% \text{ water} = \frac{(\text{wt. of feed sample before drying}) - (\text{wt. of feed sample after drying})}{\text{Wt. of feed sample before drying}} \times 100$$

% DM = 100 minus percent water

Adjusting Moisture Content

The significance of water content of feeds becomes obvious in the following examples. When using Total Digestible Nutrients (TDN) as a measure of energy value, some of the high-moisture tubers show almost the same feeding value per unit of their dry matter content as the cereal grains:

Feed	Water (%)	Dry Matter (%)	Energy Value (TDN) As-Fed	Energy Value (TDN) Dry-Matter Basis
Corn, grain	10	90	80	90
Barley, grain	10	90	77	85
Melons, whole ...	94	6	5	80
Potatoes, tubers .	79	21	18	85
Apples, fruit	82	18	13	74

As shown, dry matter becomes a common denominator for the comparison of feeds, particularly as to energy value; but this applies to other nutrients also.

The feed compositions in Tables 4-101 to 4-105 are on both (1) as-fed and (2) moisture-free bases. The formulas that follow may be used for adjusting moisture content from (1) moisture-free to as-fed, or (2) as-fed to moisture-free.

- *From moisture-free (100% dry matter) to as-fed (90% dry matter)*—Moisture-free (100% DM) can be converted to as-fed (90% DM) by using the following formula:

$$\times\ (100\%\ \text{DM}) \times .9 = \text{y}\ (90\%\ \text{DM})$$

- *From as-fed (90% dry matter) to moisture-free (100% dry matter)*—As-fed (90% DM) can be converted to moisture-free (100% DM) by using the following formula:

$$\frac{\times\ (90\%\ \text{DM})}{.9} = \text{y}\ (100\%\ \text{DM})$$

FEED COMPOSITION TABLES

The Feed Composition Tables are organized as follows:

1. *Table 4-101, Composition of Feeds, pages 456-567*—Each energy feed, protein feed, dry forage and roughage, silage and haylage, and pasture and range plant is presented in tabular form on four pages (two double-page spreads), with one page devoted to each of the following categories:
   a. First left-hand page, chemical analysis
   b. First right-hand page, energy
   c. Second left-hand page, minerals
   d. Second right-hand page, vitamins
2. *Table 4-101, Composition of Feeds, pages 568-571*—Each mineral supplement is presented in a two-page spread.
3. *Table 4-101, Composition of Feeds, pages 572-579*—Each vitamin supplement is presented on four pages (two double-page spreads), with one page devoted to each of the following categories:
   a. First left-hand page, chemical analysis
   b. First right-hand page, energy
   c. Second left-hand page, minerals
   d. Second right-hand page, vitamins
4. *Table 4-102, Amino Acid Composition of Feeds, pages 580-583*—Each feed is presented on one page.
5. *Table 4-103, Iodine Content of Feeds, page 584.*
6. *Table 4-104, Selenium Content of Feeds, page 584.*
7. *Table 4-105, Vitamin D Content of Feeds, page 585.*

TABLE 4-101 COMPOSITION OF FEEDS—

Feed #	Feed Name—Description	Moisture Basis: As Fed or M-F (moisture-free)	Proximate Analysis						Digestible Protein		
			Dry Matter	Ash	Crude Fiber	Ether Extract (Fat)	N-Free Extract	Crude Protein	Ruminant	Non-Ruminant	Horse Rabbit
ENERGY FEEDS			(%)	(%)	(%)	(%)	(%)	(%)	(%)	(%)	(%)
1	Almond hulls *(Prunus amygdalus)*	As Fed	91.0	5.9	13.8	2.2	65.2	3.9	.5		1.4
		M-F	*100.0*	*6.5*	*15.2*	*2.4*	*71.7*	*4.2*	*.5*		*1.5*
2	Animal fat, hydrolized	As Fed	95.0			95.0					
		M-F	*100.0*			*100.0*					
3	Animal fat, heat rendered	As Fed	99.5			99.4					
		M-F	*100.0*			*99.9*					
4	Animal fat, lard *(Sus scrofa)*	As Fed									
		M-F	*100.0*			*100.0*					
5	Animal, tallow *(Bos spp)*	As Fed	96.0	.1		94.0		1.5			
		M-F	*100.0*	*.1*		*97.9*		*1.6*			
6	Bakery residue, dried	As Fed	91.6	3.5	.7	13.7	62.8	10.9	7.1		7.5
		M-F	*100.0*	*3.8*	*.8*	*15.0*	*68.6*	*11.9*	*7.7*		*8.1*
7	Barley grain *(Hordeum vulgare)*	As Fed	89.0	3.0	5.3	1.7	67.4	11.6	8.5	8.7	9.5
		M-F	*100.0*	*3.4*	*6.0*	*1.9*	*75.7*	*13.0*	*9.6*	*9.8*	*10.7*
8	Barley grain, 48 lb/bu	As Fed	89.0	2.2	6.0	1.9	66.8	12.1	7.7		8.1
		M-F	*100.0*	*2.5*	*6.7*	*2.1*	*75.1*	*13.6*	*8.7*		*9.1*
9	Barley grain, Pacific Coast	As Fed	89.0	2.3	6.2	2.2	68.5	9.8	7.1		5.5
		M-F	*100.0*	*2.6*	*7.0*	*2.5*	*77.0*	*10.9*	*8.0*		*6.2*
10	Beans, kidney *(Phaseolus vulgaris)*	As Fed	88.8	3.7	4.2	1.3	57.7	21.9	14.7		
		M-F	*100.0*	*4.2*	*4.7*	*1.5*	*64.9*	*24.7*	*16.5*		
11	Beans, mung *(Phaseolus aureus)*	As Fed	89.8	3.7	3.9	1.3	57.0	23.9			
		M-F	*100.0*	*4.2*	*4.3*	*1.4*	*63.5*	*26.6*			
12	Beans, navy *(Phaseolus vulgaris)*	As Fed	90.0	4.1	4.2	1.4	57.4	22.9			
		M-F	*100.0*	*4.6*	*4.7*	*1.6*	*63.7*	*25.4*			
13	Beans, pinto *(Phaseolus spp)*	As Fed	90.3	4.3	4.0	1.3	57.9	22.8	19.8		
		M-F	*100.0*	*4.8*	*4.5*	*1.4*	*64.1*	*25.2*	*21.9*		
14	Beet, mangels *(Beta spp)*	As Fed	10.6	1.0	.9	.1	7.2	1.4	1.0	1.0	1.0
		M-F	*100.0*	*9.7*	*8.3*	*.6*	*68.2*	*13.2*	*7.5*	*7.5*	*7.6*
15	Beet, sugar, pulp, wet *(Beta saccharifera)*	As Fed	10.0	.7	2.0	.2	6.2	.9	.4		
		M-F	*100.0*	*6.7*	*20.0*	*2.1*	*62.2*	*9.0*	*4.0*		
16	Beet, sugar, pulp dried	As Fed	91.0	3.6	19.0	.6	58.7	9.1	4.3	3.5	6.0
		M-F	*100.0*	*3.9*	*20.9*	*.7*	*64.5*	*10.0*	*4.7*	*3.9*	*6.6*
17	Beet, sugar, pulp, w/ molasses, dried	As Fed	92.0	5.7	16.0	.6	60.6	9.1	6.0	2.2	5.7
		M-F	*100.0*	*6.2*	*17.4*	*.7*	*65.8*	*9.9*	*6.5*	*2.4*	*6.2*
18	Beet, sugar, molasses *(Beta saccharifera)*	As Fed	77.0	8.2	0	.2	61.9	6.7	3.1		3.9
		M-F	*100.0*	*10.6*	*0*	*.3*	*80.4*	*8.7*	*4.0*		*5.1*
19	Blackstrap molasses *(Saccharum officinarum)*	As Fed	75.0	8.1		.1	63.6	3.2	1.4		2.0
		M-F	*100.0*	*10.8*		*.1*	*84.8*	*4.3*	*1.8*		*2.6*
20	Bread, dehy.	As Fed	86.0	1.5	.5	1.0	72.0	11.0	7.4		7.7
		M-F	*100.0*	*1.7*	*.6*	*1.2*	*83.7*	*12.8*	*8.6*		*9.0*
21	Brewers' grains, wet	As Fed	23.8	1.2	3.8	1.5	11.8	5.5	4.0		
		M-F	*100.0*	*4.8*	*16.1*	*6.5*	*49.6*	*23.0*	*16.8*		

NERGY FEEDS

Feed #	TDN Ruminant (%)	TDN Non-Ruminant (%)	TDN Horse Rabbit (%)	DE Ruminant Mcal per lb	DE Ruminant Mcal per kg	DE Non-Ruminant kcal per lb	DE Non-Ruminant kcal per kg	DE Horse Rabbit Mcal per lb	DE Horse Rabbit Mcal per kg	ME Ruminant Mcal per lb	ME Ruminant Mcal per kg	ME Non-Ruminant ME kcal per lb	ME Non-Ruminant ME kcal per kg	ME Non-Ruminant Chicken ME_n kcal per lb	ME Non-Ruminant Chicken ME_n kcal per kg	ME Horse Rabbit Mcal per lb	ME Horse Rabbit Mcal per kg	NE_m Mcal per lb	NE_m Mcal per kg	NE_g Mcal per lb	NE_g Mcal per kg	NE_{lc} Mcal per lb	NE_{lc} Mcal per kg
1	68.1			1.4	3.0					1.1	2.5												
	74.8			*1.5*	*3.3*					*1.2*	*2.7*												
2	232.6					3,695	8,130			3.8	8.4	3,591	7,900	3,207	7,055								
	244.8					*3,890*	*8,558*			*4.0*	*8.8*	*3,780*	*8,316*	*3,375*	*7,426*								
3	225.3									3.9	8.5			3,223	7,090			2.0	4.5	1.2	2.6	1.0	2.2
	237.2									*3.9*	*8.6*			*3,239*	*7,126*			*2.1*	*4.6*	*1.2*	*2.6*	*1.0*	*2.2*
4																							
						3,527	*7,760*					*3,500*	*7,700*										
5																							
6	82.0	86.1		1.6	3.6	1,726	3,798			1.4	3.0	1,616	3,555	1,803	3,966								
	89.6	*94.1*		*1.8*	*4.0*	*1,885*	*4,148*			*1.5*	*3.2*	*1,765*	*3,883*	*1,969*	*4,332*								
7	72.9	70.8	72.5	1.5	3.2	1,419	3,122		3.2	1.2	2.6	1,325	2,915	1,198	2,635	1.2	2.6	.9	1.9	.6	1.3	1.0	2.1
	81.9	*79.5*	*81.5*	*1.6*	*3.6*	*1,593*	*3,505*		*3.6*	*1.4*	*3.0*	*1,488*	*3,274*	*1,345*	*2,960*	*1.4*	*3.0*	*1.0*	*2.1*	*.6*	*1.4*	*1.0*	*2.3*
8	75.0	72.2		1.5	3.3	1,447	3,183			1.2	2.7	1,351	2,973										
	84.3	*81.1*		*1.7*	*3.7*	*1,626*	*3,577*			*1.4*	*3.1*	*1,518*	*3,340*										
9	71.6	63.0		1.5	3.2	1,263	2,778			1.2	2.6	1,035	2,278					.8	1.7	.5	1.1	.9	1.9
	80.5	*70.8*		*1.6*	*3.6*	*1,419*	*3,122*			*1.4*	*3.0*	*1,164*	*2,560*					*.9*	*1.9*	*.6*	*1.3*	*1.0*	*2.1*
10	76.0	84.0		1.5	3.4	1,684	3,705			1.3	2.8	1,533	3,372										
	85.6	*94.6*		*1.7*	*3.8*	*1,897*	*4,173*			*1.4*	*3.1*	*1,726*	*3,798*										
11	76.3	84.4		1.5	3.4	1,692	3,722			1.3	2.8	1,533	3,373										
	85.0	*94.0*		*1.7*	*3.8*	*1,884*	*4,144*			*1.4*	*3.1*	*1,707*	*3,756*										
12	76.5	84.8		1.5	3.4	1,699	3,738			1.3	2.8	1,545	3,398	1,063	2,338			.8	1.8	.5	1.2	.9	1.9
	85.0	*94.2*		*1.7*	*3.8*	*1,888*	*4,153*			*1.4*	*3.1*	*1,716*	*3,775*	*1,181*	*2,598*			*.9*	*2.0*	*.6*	*1.3*	*1.0*	*2.1*
13	77.4	84.8		1.5	3.4	1,700	3,739			1.3	2.8	1,700	3,399										
	85.7	*93.9*		*1.7*	*3.8*	*1,882*	*4,140*			*1.4*	*3.1*	*1,711*	*3,764*										
14	8.7	8.3		.2	.4	166	365			.1	.3	155	342					.1	.2	.05	.1	.1	.2
	82.0	*78.0*		*1.6*	*3.6*	*1,564*	*3,440*			*1.3*	*2.9*	*1,465*	*3,224*					*.8*	*1.8*	*.5*	*1.2*	*.9*	*2.0*
15	7.2		8.0	.1	.3			.2	.4	.1	.3					.1	.3	.1	.2	.05	.1	.4	.2
	71.5		*80.0*	*1.5*	*3.2*			*1.6*	*3.5*	*1.2*	*2.6*					*1.3*	*2.9*	*.7*	*1.5*	*.5*	*1.0*	*.7*	*1.6*
16	65.5	68.3	74.3	1.3	2.9	1,369	3,012	1.5	3.3	1.1	2.4	1,288	2,833	255	561	1.2	2.7	.7	1.5	.4	.9	.7	1.6
	72.0	*75.1*	*81.6*	*1.5*	*3.2*	*1,505*	*3,310*	*1.6*	*3.6*	*1.2*	*2.6*	*1,415*	*3,113*	*280*	*617*	*1.4*	*3.0*	*.7*	*1.6*	*.5*	*1.0*	*.8*	*1.8*
17	70.4	69.5		1.4	3.1	1,393	3,064			1.2	2.6	1,310	2,881	299	658			.9	1.9	.5	1.2	.8	1.7
	76.5	*75.6*		*1.5*	*3.4*	*1,505*	*3,331*			*1.3*	*2.8*	*1,424*	*3,132*	*325*	*715*			*.9*	*2.0*	*.6*	*1.3*	*.8*	*1.8*
18	61.6	55.5		1.1	2.4	1,112	2,447			1.0	2.2			871	1,917			.7	1.5	.5	1.0	.7	1.5
	80.0	*72.1*		*1.5*	*3.2*	*1,445*	*3,178*			*1.3*	*2.9*			*1.132*	*2,490*			*.9*	*1.9*	*.6*	*1.3*	*.9*	*1.9*
19	61.1	55.7		1.2	2.7	1,117	2,458			1.0	2.3	1,073	2,360	889	1,955			.7	1.6	.5	1.0	.8	1.8
	81.5	*74.3*		*1.6*	*3.6*	*1,490*	*3,277*			*1.4*	*3.0*	*1,430*	*3,146*	*1,185*	*2,607*			*1.0*	*2.1*	*.6*	*1.3*	*1.1*	*2.4*
20	75.5	82.3		1.5	3.3	1,650	3,630			1.2	2.7	1,541	3,391	1,603	3,527								
	87.7	*95.7*		*1.8*	*3.9*	*1,919*	*4,221*			*1.5*	*3.2*	*1,792*	*3,943*	*1,864*	*4,101*								
21	16.4	21.7		.3	.7	435	956			.3	.6	397	874										
	69.0	*91.3*		*1.4*	*3.0*	*1,830*	*4,026*			*1.1*	*2.5*	*1,672*	*3,678*										

TABLE 4-101 COMPOSITION OF FEEDS—

Feed #	Feed Name—Description	Moisture Basis: As Fed or M-F (moisture-free)	Dry Matter	Macrominerals: Calcium (Ca)	Phosphorus (P)	Sodium (Na)	Chlorine (Cl)	Magnesium (Mg)	Potassium (K)	Sulfur (S)	Microminerals: Cobalt (Co)	Copper (Cu)	Iron (Fe)	Manganese (Mn)	Zinc (Zn)
			(%)	(%)	(%)	(%)	(%)	(%)	(%)	(%)	(ppm or mg/kg)	(ppm or mg/kg)	(%)	(ppm or mg/kg)	(ppm or mg/kg)
ENERGY FEEDS															
1	Almond hulls	As Fed	91.0												
		M-F	100.0												
2	Animal fat, hydrolized	As Fed	95.0												
		M-F	100.0												
3	Animal fat, heat rendered	As Fed	99.5												
		M-F	100.0												
4	Animal fat, lard	As Fed													
		M-F	100.0												
5	Animal, tallow	As Fed	96.0												
		M-F	100.0												
6	Bakery residue, dried	As Fed	91.6	.06	.47			.32	.83	.02	.0	.0	.006	.0	
		M-F	100.0	.06	.51			.35	.91	.02	.0	.0	.007	.0	
7	Barley grain	As Fed	89.0	.07	.40	.06		.13	.49	.15		5.8	.008	8.0	
		M-F	100.0	.08	.45	.07		.15	.55	.17		6.5	.009	8.9	
8	Barley grain, 48 lb/bu	As Fed	89.0	.24	.36										
		M-F	100.0	.27	.41										
9	Barley grain, Pacific Coast	As Fed	89.0	.06	.40										
		M-F	100.0	.07	.45										
10	Beans, kidney	As Fed	88.8	.11	.40	.01			.98				.007		
		M-F	100.0	.12	.45	.01			1.10				.008		
11	Beans, mung	As Fed	89.8	.13	.34	.01			1.04				.008		
		M-F	100.0	.14	.38	.01			1.15				.009		
12	Beans, navy	As Fed	90.0	.15	.57	.05	.04	.17	1.70	.23		9.9	.009	18.5	
		M-F	100.0	.17	.63	.06	.04	.19	1.89	.26		11.0	.011	20.6	
13	Beans, pinto	As Fed	90.3	.14	.35										
		M-F	100.0	.16	.39										
14	Beet, mangels	As Fed	10.6	.02	.02	.07	.13	.02	.21	.02			.002		
		M-F	100.0	.19	.19	.66	1.23	.19	1.98	.19			.019		
15	Beet, sugar, pulp, wet	As Fed	10.0	.09	.01				.02						
		M-F	100.0	.90	.10				.20						
16	Beet, sugar, pulp, dried	As Fed	91.0	.68	.10	.17	.04	.27	.21	.20	.091	12.5	.030	35.0	.7
		M-F	100.0	.75	.11	.19	.04	.30	.23	.22	.100	13.7	.033	38.5	.8
17	Beet, sugar, pulp, w/molasses, dried	As Fed	92.0	.56	.10			.13	1.64	.39					
		M-F	100.0	.61	.11			.14	1.78	.42					
18	Beet, sugar, molasses	As Fed	77.0	.16	.03	1.17	1.25	.23	4.77		.400	17.6	.008	4.6	
		M-F	100.0	.21	.04	1.52	1.62	.30	6.20		.500	22.9	.010	6.0	
19	Blackstrap molasses	As Fed	75.0	.89	.08	.15	2.79	.35	2.38	.35	.912	59.6	.019	42.2	
		M-F	100.0	1.19	.11	.20	3.72	.47	3.17	.46	1.216	79.4	.025	56.3	
20	Bread, dehy.	As Fed	86.0	.05	.10										
		M-F	100.0	.06	.12										
21	Brewers' grains, wet	As Fed	23.8	.07	.12				.02						
		M-F	100.0	.30	.51				.08						

ENERGY FEEDS

Feed #	A (1 mg car. = 1,667 IU-A)	Carotene (Pro-vitamin A)	E (α-tocopherol)	K	B_{12}	Biotin	Choline	Folic Acid (Folacin)	Niacin (Nicotinic Acid)	Pantothenic Acid	Pyridoxine (B_6)	Riboflavin (B_2)	Thiamin (B_1)
	Fat-Soluble Vitamins				Water-Soluble Vitamins								
		(ppm or mg/kg)	*(ppm or mg/kg)*	*(ppm or mg/kg)*	*(ppm or mg/kg)*	*(ppm or mg/kg)*	*(ppm or mg/kg)*	*(ppm or mg/kg)*	*(ppm or mg/kg)*	*(ppm or mg/kg)*	*(ppm or mg/kg)*	*(ppm or mg/kg)*	*(ppm or mg/kg)*
1													
2													
3													
4													
5													
6	7,668	4.6	410.3				1,235		19.0	14.7	26.01	1.9	1.5
	8,335	*5.0*	*448.1*				*1,349*		*20.8*	*16.0*	*28.42*	*2.0*	*1.6*
7			36.0			.18	988	.53	52.9	8.1	2.94	1.8	5.1
			40.4			*.20*	*1,110*	*.60*	*59.4*	*9.1*	*3.30*	*2.0*	*5.7*
8													4.3
													4.8
9			36.0				937		44.1	7.3		1.3	4.0
			40.5				*1,054*		*49.6*	*8.2*		*1.5*	*4.5*
10	295	.2							22.8			2.0	5.1
	333	*.2*							*25.7*			*2.0*	*5.7*
11	748	.4							26.1			2.1	3.8
	834	*.5*							*29.1*			*2.4*	*4.3*
12							1,764		24.5	3.1		2.0	6.6
							1,960		*27.2*	*3.4*		*2.2*	*7.3*
13													
14													
15													
16	303	.2					829		16.3	1.5		.7	.4
	333	*.2*					*912*		*17.9*	*1.6*		*.8*	*.4*
17	303	.2					823		16.3	1.6		.6	
	333	*.2*					*895*		*17.7*	*1.7*		*.7*	
18							862		42.2	4.6		2.4	
							1,119		*54.8*	*6.0*		*3.1*	
19							876		34.3	38.3		3.3	.9
							1,167		*45.7*	*51.1*		*4.4*	*1.2*
20							882		28.7	11.0		2.0	
							1,025		*33.3*	*12.8*		*2.3*	
21													

TABLE 4-101 COMPOSITION OF FEEDS—

Feed #	Feed Name—Description	Moisture Basis: As Fed or M-F (moisture-free)	Proximate Analysis: Dry Matter (%)	Ash (%)	Crude Fiber (%)	Ether Extract (Fat) (%)	N-Free Extract (%)	Crude Protein (%)	Digestible Protein: Ruminant (%)	Non-Ruminant (%)	Horse Rabbit (%)
ENERGY FEEDS											
22	Brewers' grains, dried	As Fed	92.0	3.6	15.0	6.2	41.3	25.9	19.1	19.1	20.7
		M-F	100.0	3.8	16.3	6.7	45.0	28.2	20.8	20.8	22.5
23	Carrot roots *(Daucus* spp)	As Fed	12.9	1.2	1.2	.2	9.0	1.3	.9	.9	1.2
		M-F	100.0	9.7	9.0	1.4	69.6	10.3	6.5	7.3	9.1
24	Cattle, tallow, raw *(Bos* spp)	As Fed	96.0	.1		94.0		1.5			
		M-F	100.0	.1		97.9		1.6			
25	Citrus molasses (syrup) *(Citrus* spp)	As Fed	65.0	5.2	0	.2	52.5	7.1	3.6		3.2
		(M-F)	100.0	8.0	0	.3	80.8	10.9	5.6		4.9
26	Citrus pulp, dried *(Citrus* spp)	As Fed	90.0	6.3	13.0	3.4	60.7	6.6	3.5		3.3
		M-F	100.0	7.0	14.4	3.8	67.5	7.3	3.9		3.7
27	Citrus pulp, wet	As Fed	18.3	1.4	2.3	.6	12.8	1.2	.5		.6
		M-F	100.0	7.7	12.6	3.3	69.8	6.6	2.6		3.2
28	Coconut (Copra) meal, mech-extd *(Cocos nucifera)*	As Fed	93.0	6.7	12.0	6.7	47.2	20.4	17.1		
		M-F	100.0	7.2	12.9	7.2	50.8	21.9	18.4		
29	Coconut (Copra) meal, sol-extd	As Fed	92.0	6.3	15.0	2.1	47.3	21.3	16.7		
		M-F	100.0	6.9	16.3	2.3	51.3	23.2	18.1		
30	Corn & cob meal (ear corn) *(Zea mays)*	As Fed	87.0	1.5	8.0	3.5	65.9	8.1	4.3	5.8	4.0
		M-F	100.0	1.7	9.2	4.0	75.8	9.3	4.9	6.7	4.6
31	Corn germ meal	As Fed	93.0	3.1	12.0	2.0	57.9	18.0	13.6		
		M-F	100.0	3.3	12.9	2.2	62.2	19.4	14.6		
32	Corn gluten feed	As Fed	90.4	6.6	7.3	2.7	48.0	25.8	22.2		
		M-F	100.0	7.3	8.0	2.9	53.2	28.6	24.6		
33	Corn gluten meal	As Fed	91.0	2.4	4.0	2.3	39.4	42.9	35.7		42.9
		M-F	100.0	2.6	4.4	2.5	43.4	47.1	39.2		47.1
34	Corn grain *(Zea mays, indentata)*	As Fed	87.0	1.4	2.1	4.1	69.9	9.5	6.6	6.3	7.4
		M-F	100.0	1.6	2.4	4.7	80.3	10.9	7.6	7.2	8.5
35	Corn grain #1, 56 lb/bu	As Fed	86.3	1.3	2.0	4.0	70.2	8.8	6.4	7.1	5.7
		M-F	100.0	1.5	2.3	4.6	81.4	10.2	7.4	8.2	6.6
36	Corn grain #2, 54 lb/bu	As Fed	89.0	1.3	2.0	4.1	72.7	8.9	6.7		6.1
		M-F	100.0	1.5	2.2	4.6	81.7	10.0	7.5		6.8
37	Corn grain, high lysine	As Fed	92.0	1.8	3.7	4.0	71.7	10.8			
		M-F	100.0	2.0	4.0	4.3	78.0	11.7			
38	Corn grain, high moisture	As Fed	69.8	1.3	2.1	2.8	56.3	7.5	5.4	6.0	4.9
		M-F	100.0	1.8	3.0	4.0	80.5	10.7	7.8	8.6	7.0
39	Corn grits (see Hominy feeds)										
40	Corn molasses	As Fed	72.5	8.0	0	0	64.2	.3	0		0
		M-F	100.0	11.0	0	0	88.6	.4	0		0
41	Corn oil	As Fed	99.0			96.0					
		M-F	100.0			97.0					
42	Corn starch	As Fed	90.4	.2	.1	.2	89.3	.6	0		0
		M-F	100.0	.2	.2	.2	98.7	.7	0		0

NERGY FEEDS

Feed #	TDN Ruminant (%)	TDN Non-Ruminant (%)	TDN Horse Rabbit (%)	DE Ruminant Mcal per lb	DE Ruminant Mcal per kg	DE Non-Ruminant kcal per lb	DE Non-Ruminant kcal per kg	DE Horse Rabbit Mcal per lb	DE Horse Rabbit Mcal per kg	ME Ruminant Mcal per lb	ME Ruminant Mcal per kg	ME Non-Ruminant ME kcal per lb	ME Non-Ruminant ME kcal per kg	ME Non-Ruminant Chicken ME_n kcal per lb	ME Non-Ruminant Chicken ME_n kcal per kg	ME Horse Rabbit Mcal per lb	ME Horse Rabbit Mcal per kg	NE_m Mcal per lb	NE_m Mcal per kg	NE_g Mcal per lb	NE_g Mcal per kg	NE_{lc} Mcal per lb	NE_{lc} Mcal per kg
22	62.1	40.4	46.9	1.2	2.7	809	1,779	.9	2.1	1.0	2.2	730	1,606	1,096	2,412	.8	1.7	.6	1.3	.4	.8	.6	1.4
	67.5	*43.9*	*51.0*	*1.4*	*3.0*	*879*	*1,934*	*1.0*	*2.3*	*1.1*	*2.4*	*794*	*1,746*	*1,192*	*2,622*	*.8*	*1.8*	*.6*	*1.4*	*.4*	*.8*	*.7*	*1.6*
23	10.9	10.6	11.0	.2	.5	213	469	.2	.5	.2	.4	200	440			.2	.4						
	84.7	*82.5*	*85.6*	*1.7*	*3.7*	*1,654*	*3,638*	*1.7*	*3.8*	*1.4*	*3.1*	*1,553*	*3,417*			*1.4*	*3.1*						
24																							
25	50.1	52.3		1.0	2.2	1,048	2,306			.8	1.8	988	2,174					.6	1.3	.4	.9	.6	1.3
	77.0	*80.4*		*1.5*	*3.4*	*1,611*	*3,545*			*1.3*	*2.8*	*1,520*	*3,344*					*.9*	*2.0*	*.6*	*1.3*	*.9*	*1.9*
26	69.3	47.2	63.0	1.4	3.1			1.3	2.8	1.1	2.5	776	1,708					.8	1.8	.5	1.2	.8	1.7
	77.0	*52.5*	*70.0*	*1.5*	*3.4*			*1.4*	*3.1*	*1.3*	*2.8*	*863*	*1,898*					*.9*	*2.0*	*.6*	*1.3*	*.9*	*1.9*
27	14.6	9.2		.3	.6	188	414			.2	.5	281	618										
	80.1	*51.3*		*1.6*	*3.5*	*1,028*	*2,261*			*1.3*	*2.9*	*1,535*	*3,377*										
28	76.7	87.8		1.5	3.4	1,752	3,855			1.3	2.8	1,603	3,527	809	1,779			.8	1.7	.5	1.2	1.0	2.1
	82.5	*94.4*		*1.6*	*3.6*	*1,884*	*4,145*			*1.4*	*3.0*	*1,724*	*3,792*	*870*	*1,913*			*.9*	*1.9*	*.6*	*1.3*	*1.0*	*2.2*
29	67.8	76.5		1.4	3.0	1,519	3,341			1.1	2.5	1,386	3,050					.7	1.5	.5	1.0	.8	1.8
	73.7	*83.1*		*1.5*	*3.2*	*1,651*	*3,632*			*1.2*	*2.7*	*1,507*	*3,315*					*.7*	*1.6*	*.5*	*1.1*	*.9*	*2.0*
30	73.3	69.0	67.9	1.4	3.1	1,384	3,044	1.4	3.0	1.2	2.7	1,302	2,865	1,283	2,822	1.1	2.5						
	84.3	*79.4*	*78.0*	*1.6*	*3.6*	*1,590*	*3,499*	*1.5*	*3.4*	*1.4*	*3.1*	*1,497*	*3,293*	*1,475*	*3,244*	*1.3*	*2.8*						
31	74.9	65.9		1.5	3.3	1,320	2,905			1.2	2.7	1,302	2,864	773	1,700								
	80.6	*70.9*		*1.6*	*3.6*	*1,421*	*3,126*			*1.3*	*2.9*	*1,400*	*3,080*	*831*	*1,828*								
32	74.7	80.1		1.5	3.3	1,604	3,528			1.2	2.7	1,448	3,186	755	1,661			.8	1.7	.5	1.2	1.0	2.1
	82.6	*88.5*		*1.6*	*3.6*	*1,772*	*3,898*			*1.4*	*3.0*	*1,601*	*3,523*	*835*	*1,837*			*.9*	*1.9*	*.6*	*1.3*	*1.0*	*2.3*
33	76.4	79.8		1.5	3.4	1,597	3,514	1.5	3.4	1.3	2.8	1,533	3,373	1,503	3,307	1.3	2.8	.8	1.8	.5	1.2	.9	2.0
	84.0	*87.7*		*1.7*	*3.7*	*1,755*	*3,862*	*1.7*	*3.7*	*1.4*	*3.0*	*1,685*	*3,707*	*1,652*	*3,634*	*1.4*	*3.0*	*.9*	*2.0*	*.6*	*1.3*	*1.0*	*2.2*
34	80.3	67.8	79.2	1.6	3.5	1,359	2,989	1.6	3.5	1.3	2.9	1,274	2,803			1.3	2.9						
	92.3	*77.9*	*91.9*	*1.9*	*4.1*	*1,561*	*3,435*	*1.9*	*4.1*	*1.5*	*3.3*	*1,465*	*3,223*			*1.5*	*3.3*						
35	81.7	79.6		1.6	3.6	1,595	3,510			1.4	3.0	1,498	3,296										
	94.7	*92.2*		*1.9*	*4.2*	*1,848*	*4,065*			*1.5*	*3.4*	*1,735*	*3,818*										
36	82.3	82.1		1.6	3.6	1,645	3,620			1.4	3.0	1,544	3,397	1,388	3,053			.9	2.0	.6	1.3	1.0	2.2
	92.5	*92.2*		*1.9*	*4.1*	*1,848*	*4,065*			*1.5*	*3.3*	*1,735*	*3,817*	*1,559*	*3,430*			*1.0*	*2.3*	*.7*	*1.5*	*1.1*	*2.4*
37	82.0			1.6	3.6					1.4	3.0												
	89.0			*1.8*	*3.9*					*1.5*	*3.2*												
38	65.0	63.5		1.3	2.9	1,273	2,800			1.1	2.4	1,195	2,629										
	93.1	*91.0*		*1.9*	*4.1*	*1,824*	*4,012*			*1.5*	*3.4*	*1,712*	*3,766*										
39																							
40	60.9	50.8		1.2	2.7	1,017	2,238			1.0	2.2	977	2,150										
	84.0	*70.1*		*1.7*	*3.7*	*1,404*	*3,088*			*1.4*	*3.0*	*1,348*	*2,965*										
41		171.1				3,429	7,544					3,303	7,267										
		172.8				*3,464*	*7,620*					*3,336*	*7,340*										
42	84.3	86.7		1.7	3.7	1,738	3,823			1.4	3.0	1,673	3,680	1,659	3,650								
	93.2	*95.8*		*1.9*	*4.1*	*1,920*	*4,224*			*1.5*	*3.4*	*1,850*	*4,070*	*1,835*	*4,038*								

TABLE 4-101 COMPOSITION OF FEEDS [cut off]

Feed #	Feed Name—Description	Moisture Basis: As Fed or M-F (moisture-free)	Dry Matter	Macrominerals							Microminerals				
				Calcium (Ca)	Phosphorus (P)	Sodium (Na)	Chlorine (Cl)	Magnesium (Mg)	Potassium (K)	Sulfur (S)	Cobalt (Co)	Copper (Cu)	Iron (Fe)	Manganese (Mn)	Zinc (Zn)
ENERGY FEEDS			(%)	(%)	(%)	(%)	(%)	(%)	(%)	(%)	(ppm or mg/kg)	(ppm or mg/kg)	(%)	(ppm or mg/kg)	(ppm o[cut off] mg/kg)
22	Brewers' grains, dried	As Fed	92.0	.27	.50	.26	.17	.14	.08	.30	.092	21.3	.025	37.6	
		M-F	*100.0*	*.29*	*.54*	*.28*	*.19*	*.15*	*.09*	*.33*	*.100*	*23.1*	*.027*	*40.9*	
23	Carrot roots	As Fed	12.9	.04	.04	.12		.02	.30	.02		1.3	.001	3.7	
		M-F	*100.0*	*.37*	*.32*	*1.00*		*.17*	*2.50*	*.17*		*11.1*	*.011*	*31.5*	
24	Cattle, tallow, raw	As Fed	96.0												
		M-F	*100.0*												
25	Citrus molasses	As Fed	65.0	1.31	.16	.26	.07	.14	.09		.103	72.8	.033	26.0	88.9
		M-F	*100.0*	*2.01*	*.25*	*.40*	*.10*	*.22*	*.14*		*.159*	*112.0*	*.050*	*40.0*	*136.7*
26	Citrus pulp, dried	As Fed	90.0	1.96	.12			.16	.62			5.7	.016	6.8	14.5
		M-F	*100.0*	*2.18*	*.13*			*.18*	*.69*			*6.3*	*.018*	*7.6*	*16.1*
27	Citrus pulp, wet	As Fed	18.3												
		M-F	*100.0*												
28	Coconut (Copra) meal, mech-extd	As Fed	93.0	.21	.61	.04		.26	1.12	.37	.138	18.7	.196	55.4	
		M-F	*100.0*	*.23*	*.66*	*.04*		*.28*	*1.20*	*.40*	*.148*	*20.1*	*.211*	*59.6*	
29	Coconut (Copra) meal, sol-extd	As Fed	92.0	.17	.61	.04	.03							55.0	
		M-F	*100.0*	*.18*	*.66*	*.04*	*.03*							*59.7*	
30	Corn & cob meal (ear corn)	As Fed	87.0	.04	.27	.01		.15	.53		.261	7.7	.007	13.1	9.0
		M-F	*100.0*	*.05*	*.31*	*.01*		*.17*	*.61*		*.300*	*8.8*	*.008*	*15.0*	*10.3*
31	Corn germ meal	As Fed	93.0	.10	.40				.20					16.0	
		M-F	*100.0*	*.11*	*.43*				*.22*					*17.2*	
32	Corn gluten feed	As Fed	90.4	.44	.78	.95	.22	.29	.57	.22	.088	47.9	.046	23.8	
		M-F	*100.0*	*.49*	*.86*	*1.05*	*.24*	*.32*	*.63*	*.24*	*.098*	*52.9*	*.051*	*26.4*	
33	Corn gluten meal	As Fed	91.0	.16	.40	.09		.04	.03		.091	28.2	.040	7.3	
		M-F	*100.0*	*.18*	*.44*	*.10*		*.05*	*.03*		*.100*	*31.0*	*.044*	*8.0*	
34	Corn grain	As Fed	87.0	.04	.30	.07		.11	.30	.08		3.0	.004	6.1	
		M-F	*100.0*	*.05*	*.35*	*.08*		*.13*	*.35*	*.09*		*3.5*	*.005*	*7.0*	
35	Corn grain θ1, 56 lb/bu	As Fed	86.3	.02	.28	.01		.10	.28	.12		4.1	.002	5.6	
		M-F	*100.0*	*.02*	*.32*	*.01*		*.11*	*.33*	*.14*		*4.7*	*.003*	*6.5*	
36	Corn grain θ2, 54 lb/bu	As Fed	89.0	.02	.31						.018		.002		
		M-F	*100.0*	*.02*	*.35*						*.020*		*.003*		
37	Corn grain, high lysine	As Fed	92.0	.02	.20										
		M-F	*100.0*	*.02*	*.20*										
38	Corn grain, high moisture	As Fed	69.8	.01	.25				.27						
		M-F	*100.0*	*.02*	*.36*				*.39*						
39	Corn grits (see Hominy feeds)														
40	Corn molasses	As Fed	72.5												
		M-F	*100.0*												
41	Corn oil	As Fed	99.0												
		M-F	*100.0*												
42	Corn starch	As Fed	90.4												
		M-F	*100.0*												

ERGY FEEDS

Feed #	Fat-Soluble Vitamins				Water-Soluble Vitamins								
	A (1 mg car. = 1,667 IU-A)	Carotene (Pro-vitamin A)	E (α-tocopherol)	K	B_{12}	Biotin	Choline	Folic Acid (Folacin)	Niacin (Nicotinic Acid)	Pantothenic Acid	Pyridoxine (B_6)	Riboflavin (B_2)	Thiamin (B_1)
		(ppm or mg/kg)	*(ppm or mg/kg)*	*(ppm or mg/kg)*	*(ppm or mg/kg)*	*(ppm or mg/kg)*	*(ppm or mg/kg)*	*(ppm or mg/kg)*	*(ppm or mg/kg)*	*(ppm or mg/kg)*	*(ppm or mg/kg)*	*(ppm or mg/kg)*	*(ppm or mg/kg)*
22						0	1,587	.22	43.4	8.6	.66	1.5	.7
						0	*1,725*	*.24*	*47.2*	*9.3*	*.72*	*1.6*	*.8*
23	167	.1							6.0			.5	.6
	1,000	*.6*							*50.8*			*4.2*	*5.1*
24													
25									26.6	12.5		6.2	
									41.0	*19.3*		*9.5*	
26							769		21.5	13.0		2.4	1.5
							854		*23.9*	*14.4*		*2.7*	*1.7*
27													
28							920	1.30	24.9	6.6		3.1	.7
							989	*1.40*	*26.8*	*7.1*		*3.3*	*.8*
29							1,100	.30	24.0	6.6	4.40	13.2	.9
							1,196	*.33*	*26.0*	*7.2*	*4.78*	*14.3*	*1.0*
30			20.0			.04	550	.30	20.0	5.0	5.00	1.1	
			23.0			*.05*	*632*	*.34*	*23.0*	*5.7*	*5.75*	*1.3*	
31			86.9			2.99	1,800	.19	35.1	4.1		4.1	1.0
			93.5			*3.22*	*1,936*	*.21*	*37.7*	*4.4*		*4.4*	*1.1*
32	14,003	8.4				.33	1,514	.31	71.8	17.1	15.00	2.4	2.0
	15,503	*9.3*				*.37*	*1,674*	*.34*	*79.4*	*18.9*	*16.58*	*2.7*	*2.2*
33	27,339	16.4	24.0			.15	330	.19	49.9	10.3	7.99	1.5	.2
	30,006	*18.0*	*26.4*			*.16*	*363*	*.21*	*54.8*	*11.3*	*8.79*	*1.6*	*.2*
34							590						
							678						
35						.06		.21					
						.07		*.24*					
36	3,001	1.8	22.0						26.3	3.9		1.3	3.6
	3,334	*2.0*	*24.7*						*29.5*	*4.4*		*1.5*	*4.0*
37							528		19.8	4.8		1.1	
							574		*21.5*	*5.2*		*1.2*	
38													
39													
40													
41													
42													

TABLE 4-101 COMPOSITION OF FEEDS—

Feed #	Feed Name—Description	Moisture Basis: As Fed or M-F (moisture-free)	Proximate Analysis: Dry Matter	Ash	Crude Fiber	Ether Extract (Fat)	N-Free Extract	Crude Protein	Digestible Protein: Ruminant	Non-Ruminant	Horse Rabbit
			(%)	(%)	(%)	(%)	(%)	(%)	(%)	(%)	(%)
ENERGY FEEDS											
43	Corn, sweet *(Zea mays saccharata)*	As Fed	90.9	1.9	2.5	7.9	66.9	11.7	8.6	9.3	8.1
		M-F	*100.0*	*2.1*	*2.8*	*8.7*	*73.6*	*12.8*	*9.5*	*10.2*	*9.0*
44	Cottonseed, whole, grnd *(Gossypium spp)*	As Fed	92.7	3.5	16.9	22.9	26.3	23.1	16.4		
		M-F	*100.0*	*3.8*	*18.2*	*24.7*	*28.4*	*24.9*	*17.7*		
45	Distillers' dried grains, w/o solubles	As Fed	92.5	1.6	12.8	7.4	43.4	27.3	18.9		
		M-F	*100.0*	*1.7*	*13.8*	*8.0*	*46.9*	*29.6*	*20.4*		
46	Distillers' dried corn grain w/solubles *(Zea mays)*	As Fed	93.0	7.4	4.0	9.8	44.9	26.9	21.0		
		M-F	*100.0*	*8.0*	*4.3*	*10.5*	*48.3*	*28.9*	*22.6*		
47	Distillers' dried rye grain *(Secale cereale)*	As Fed	93.0	1.9	14.0	7.4	47.3	22.4	9.7		
		M-F	*100.0*	*2.0*	*15.1*	*8.0*	*50.8*	*24.1*	*10.4*		
48	Distillers' dried solubles	As Fed	92.1	6.2	3.4	8.9	44.8	28.8			
		M-F	*100.0*	*6.8*	*3.7*	*9.6*	*48.7*	*31.2*			
49	Distillers' dried sorghum grain, w/o solubles *(Sorghum vulgare)*	As Fed	94.0	4.0	12.0	8.5	38.3	31.2	25.9		
		M-F	*100.0*	*4.3*	*12.8*	*9.0*	*40.7*	*33.2*	*27.6*		
50	Distillers' dried sorghum, w/solubles	As Fed	94.8	4.2	10.1	9.4	38.0	33.1			
		M-F	*100.0*	*4.4*	*10.7*	*9.9*	*40.1*	*34.9*			
51	Distillers' rye grain w/solubles, dehy *(Secale cereale)*	As Fed	90.5	6.4	8.1	4.1	44.7	27.2			
		M-F	*100.0*	*7.1*	*9.0*	*4.5*	*49.3*	*30.1*			
52	Emmer grain *(Triticum dicoccum)*	As Fed	91.0	3.5	10.0	2.0	62.6	12.9	8.7		8.5
		M-F	*100.0*	*3.9*	*11.0*	*2.2*	*68.7*	*14.2*	*9.6*		*9.3*
53	Flax (see Linseed meal)										
54	Flax seed, whole *(Linum usitatissimum)*	As Fed	93.2	4.9	6.1	35.0	24.1	23.1	19.4		
		M-F	*100.0*	*5.2*	*6.6*	*37.5*	*25.8*	*24.9*	*20.9*		
55	Garbage, commercial, cooked, wet	As Fed	26.3	1.4	.7	5.9	14.0	4.3	3.0		3.2
		M-F	*100.0*	*5.3*	*2.7*	*22.4*	*53.3*	*16.3*	*11.4*		*12.2*
56	Garbage, commercial, cooked, dried	As Fed	53.6	3.5	1.6	14.5	24.5	9.5	7.0		7.2
		M-F	*100.0*	*6.6*	*2.9*	*27.1*	*45.7*	*17.7*	*13.0*		*13.4*
57	Grain screening	As Fed	90.3	9.1	28.3	4.8	33.7	14.4	10.3		8.8
		M-F	*100.0*	*10.1*	*31.3*	*5.3*	*37.3*	*16.0*	*11.5*		*9.8*
58	Grass seed screenings	As Fed	88.0		17.6			7.0	5.7		
		M-F	*100.0*		*20.0*			*8.0*	*6.5*		
59	Grits, corn (see Hominy feed)										
60	Hominy feed (corn grits), white *(Zea mays)*	As Fed	89.9	3.0	4.7	5.7	65.7	10.8	6.6		7.2
		M-F	*100.0*	*3.3*	*5.2*	*6.3*	*73.2*	*12.0*	*7.4*		*8.0*
61	Hominy feed (corn grits), yellow	As Fed	90.6	2.7	5.0	7.0	65.2	10.7	7.1		7.3
		M-F	*100.0*	*3.0*	*5.5*	*7.7*	*72.0*	*11.8*	*7.8*		*8.1*
62	Kafir grains *(Sorghum vulgare)*	As Fed	90.0	1.6	2.0	2.9	71.7	11.8	8.2	8.4	7.5
		M-F	*100.0*	*1.8*	*2.2*	*3.3*	*79.6*	*13.1*	*9.1*	*9.3*	*8.3*
63	Kafir head chops	As Fed	86.9	2.9	7.3	2.5	64.8	9.4	4.3		6.2
		M-F	*100.0*	*3.4*	*8.4*	*2.9*	*74.5*	*10.8*	*5.0*		*7.2*

ENERGY FEEDS

Feed #	TDN			Digestible Energy						Metabolizable Energy								Net Energy-Ruminant					
	Ruminant	Non-Ruminant	Horse Rabbit	Ruminant		Non-Ruminant		Horse Rabbit		Ruminant		Non-Ruminant ME		Non-Ruminant Chicken ME_n		Horse Rabbit		NE_m		NE_g		NE_{lc}	
	(%)	(%)	(%)	Mcal per		kcal per		Mcal per		Mcal per		kcal per		kcal per		Mcal per		Mcal per		Mcal per		Mcal per	
				lb	kg	lb	kg	lb	kg	lb	kg	lb	kg	lb	kg	lb	kg	lb	kg	lb	kg	lb	kg
43	88.6	85.2		1.8	3.9	1,707	3,756			1.5	3.2	1,595	3,508										
	97.4	*93.7*		*2.0*	*4.3*	*1,878*	*4,131*			*1.6*	*3.5*	*1,755*	*3,860*										
44	88.7	109.7		1.8	3.9	2,196	4,832			1.5	3.2	2,109	4,639					.9	1.9	.5	1.1	1.1	2.4
	95.7	*118.4*		*1.9*	*4.2*	*2,370*	*5,215*			*1.6*	*3.5*	*2,275*	*5,006*					*.9*	*2.0*	*.5*	*1.2*	*1.2*	*2.6*
45	76.2	88.8		1.5	3.4	1,780	3,915			1.3	2.8	1,602	3,525										
	82.4	*95.9*		*1.6*	*3.6*	*1,922*	*4,228*			*1.4*	*3.0*	*1,731*	*3,809*										
46	81.8			1.6	3.6					1.4	3.0							.9	2.0	.6	1.3		
	88.0			*1.8*	*3.9*					*1.5*	*3.2*							*1.0*	*2.2*	*.6*	*1.4*		
47	42.8	62.6		1.0	2.3	1,254	2,759			.7	1.5	1,148	2,525					.4	.9	.05	.1		
	46.0	*67.3*		*1.1*	*2.5*	*1,349*	*2,967*			*.8*	*1.7*	*1,234*	*2,715*					*.5*	*1.0*	*.05*	*.1*		
48	80.1	83.4		1.6	3.5	1,672	3,679			1.3	2.9	1,500	3,301	1,386	3,049								
	87.0	*90.6*		*1.7*	*3.8*	*1,816*	*3,995*			*1.4*	*3.1*	*1,629*	*3,584*	*1,505*	*3,311*								
49	78.0	88.3		1.5	3.4	1,769	3,892			1.3	2.8	1,578	3,474					.8	1.8	.5	1.2	.9	2.0
	83.0	*93.9*		*1.7*	*3.7*	*1,882*	*4,140*			*1.4*	*3.0*	*1,680*	*3,696*					*.9*	*1.9*	*.6*	*1.3*	*1.0*	*2.1*
50	80.6	91.8		1.6	3.6	1,840	4,047			1.3	2.9	1,636	3,599										
	85.1	*96.8*		*1.7*	*3.8*	*1,940*	*4,269*			*1.4*	*3.1*	*1,726*	*3,797*										
51	57.4	60.1		1.1	2.5	1,205	2,650			1.0	2.1	988	2,173										
	63.5	*66.4*		*1.3*	*2.8*	*1,331*	*2,928*			*1.0*	*2.3*	*1,091*	*2,401*										
52	70.3	61.2	63.7	1.4	3.1	1,227	2,699	1,278	2,812	1.1	2.5	1,145	2,518			1.0	2.3						
	77.3	*67.3*	*70.0*	*1.5*	*3.4*	*1,348*	*2,966*	*1,405*	*3,090*	*1.3*	*2.8*	*1,258*	*2,767*			*1.1*	*2.5*						
53																							
54	105.1	111.2		2.1	4.6	2,229	4,903			1.7	3.8	2,028	4,461										
	112.8	*119.3*		*2.3*	*5.0*	*2,391*	*5,260*			*1.9*	*4.1*	*2,176*	*4,787*										
55	22.7	23.6		.5	1.0	473	1,041			.4	.8	439	965										
	86.1	*89.8*		*1.7*	*3.8*	*1,799*	*3,957*			*1.4*	*3.1*	*1,667*	*3,668*										
56	46.6	47.1		.9	2.0	943	2,075			.8	1.7	873	1,920										
	87.1	*88.0*		*1.7*	*3.8*	*1,762*	*3,876*			*1.5*	*3.2*	*1,630*	*3,585*										
57	47.4	18.7		1.0	2.1	375	824			.8	1.7	360	792					.5	1.0	.1	.3	.4	.9
	52.5	*20.7*		*1.0*	*2.3*	*415*	*912*			*.9*	*1.9*	*398*	*876*					*.5*	*1.1*	*.1*	*.3*	*.5*	*1.0*
58	57.2			1.1	2.5					1.0	2.1												
	65.0			*1.3*	*2.9*					*1.1*	*2.4*												
59																							
60	76.8	73.4		1.5	3.4	1,470	3,234			1.3	2.8	1,376	3,028										
	85.4	*81.6*		*1.7*	*3.8*	*1,635*	*3,597*			*1.4*	*3.1*	*1,531*	*3,368*										
61	84.3	67.8		1.7	3.7	1,357	2,986			1.4	3.1	1,270	2,795	1,320	2,905			1.0	2.2	.6	1.4	1.0	2.3
	93.0	*74.8*		*1.9*	*4.1*	*1,498*	*3,296*			*1.5*	*3.4*	*1,402*	*3,085*	*1,457*	*3,206*			*1.1*	*2.5*	*.7*	*1.6*	*1.2*	*2.6*
62	63.9	83.5		1.3	2.8	1,674	3,683			1.0	2.3	1,566	3,446					.6	1.4	.4	.9		
	71.0	*92.8*		*1.4*	*3.1*	*1,860*	*4,092*			*1.2*	*2.6*	*1,740*	*3,829*					*.7*	*1.6*	*.5*	*1.0*		
63	45.5	63.3		.9	2.0	1,269	2,791			.7	1.6	1,190	2,618										
	52.4	*72.8*		*1.0*	*2.3*	*1,460*	*3,211*			*.9*	*1.9*	*1,369*	*3,012*										

TABLE 4-101 COMPOSITION OF FEEDS—

Feed #	Feed Name—Description	Moisture Basis: As Fed or M-F (moisture-free)	Dry Matter	Macrominerals: Calcium (Ca)	Phosphorus (P)	Sodium (Na)	Chlorine (Cl)	Magnesium (Mg)	Potassium (K)	Sulfur (S)	Microminerals: Cobalt (Co)	Copper (Cu)	Iron (Fe)	Manganese (Mn)	Zinc (Zn)
			(%)	(%)	(%)	(%)	(%)	(%)	(%)	(%)	(ppm or mg/kg)	(ppm or mg/kg)	(%)	(ppm or mg/kg)	(ppm or mg/kg)
ENERGY FEEDS															
43	Corn, sweet	As Fed	90.9	.01	.41							5.4	.005		
		M-F	*100.0*	*.01*	*.45*							*6.0*	*.005*		
44	Cottonseed, whole, grnd	As Fed	92.7	.14	.68	.29		.32	1.11	.24		50.0	.014	12.1	
		M-F	*100.0*	*.15*	*.73*	*.31*		*.35*	*1.20*	*.26*		*54.0*	*.015*	*13.1*	
45	Distillers' dried grains, w/o solubles	As Fed	92.5	.16	1.06	.04		.14	1.19	.45	.091	45.2	.032	53.6	194.3
		M-F	*100.0*	*.17*	*1.15*	*.04*		*.15*	*1.28*	*.49*	*.099*	*48.9*	*.035*	*57.9*	*210.0*
46	Distillers' dried corn grain w/solubles	As Fed	93.0	.35	1.37	.24	.26	.64	1.74	.37	.196	82.7	.055	73.0	
		M-F	*100.0*	*.38*	*1.47*	*.26*	*.28*	*.69*	*1.87*	*.40*	*.211*	*88.9*	*.059*	*79.0*	
47	Distillers' dried rye grain	As Fed	93.0	.13	.41	.17	.05	.17	.11	.44				18.5	
		M-F	*100.0*	*.14*	*.44*	*.18*	*.05*	*.18*	*.12*	*.47*				*19.9*	
48	Distillers' dried solubles	As Fed	92.1	.21	1.23	.15		.46	1.84		.196	71.7	.030	64.2	138.1
		M-F	*100.0*	*.22*	*1.33*	*.16*		*.50*	*2.00*		*.213*	*77.9*	*.033*	*69.7*	*150.0*
49	Distillers' dried sorghum grain, w/o solubles	As Fed	94.0	.14	.59										
		M-F	*100.0*	*.15*	*.63*										
50	Distillers' dried sorghum, w/solubles	As Fed	94.8	.17	.92									104.5	
		M-F	*100.0*	*.18*	*.97*									*110.2*	
51	Distillers' rye grain w/solubles, dehy	As Fed	90.5												
		M-F	*100.0*												
52	Emmer grain	As Fed	91.0		.43				.47			34.5	.005	85.8	
		M-F	*100.0*		*.47*				*.52*			*37.9*	*.006*	*94.3*	
53	Flax (see Linseed meal)														
54	Flax seed, whole	As Fed	93.2	.22	.52			.40	.78	.23			.009	60.7	
		M-F	*100.0*	*.23*	*.55*			*.43*	*.84*	*.25*			*.010*	*65.1*	
55	Garbage, commercial, cooked, wet	As Fed	26.3	.11	.07										
		M-F	*100.0*	*.42*	*.27*										
56	Garbage, commercial, cooked, dried	As Fed	53.6	.32	.21			.17				9.6	.013	4.8	
		M-F	*100.0*	*.60*	*.40*			*.32*				*18.0*	*.024*	*9.0*	
57	Grain screenings	As Fed	90.3	.20	.20										
		M-F	*100.0*	*.22*	*.22*										
58	Grass seed screenings	As Fed	88.0												
		M-F	*100.0*												
59	Grits, corn (see Hominy feed)														
60	Hominy feed (corn grits), white	As Fed	89.9	.05	1.00						.020	13.3	.007	13.9	
		M-F	*100.0*	*.06*	*1.10*						*.022*	*14.7*	*.008*	*15.5*	
61	Hominy feed (corn grits), yellow	As Fed	90.6	.05	.53	.40	.05	.24	.67	.03	.060	14.6	.006	14.6	
		M-F	*100.0*	*.06*	*.58*	*.56*	*.07*	*.26*	*.74*	*.03*	*.066*	*16.1*	*.007*	*16.1*	
62	Kafir grains	As Fed	90.0	.04	.33	.06	.10	.15	.34	.16	.391	6.3	.009	15.8	
		M-F	*100.0*	*.04*	*.37*	*.07*	*.11*	*.17*	*.38*	*.18*	*.434*	*7.0*	*.010*	*17.6*	
63	Kafir head chops	As Fed	86.9	.08	.24			.23							
		M-F	*100.0*	*.09*	*.28*			*.27*							

NERGY FEEDS

Feed #	Fat-Soluble Vitamins				Water-Soluble Vitamins								
	A (1 mg car. = 1,667 IU-A)	Carotene (Pro-vitamin A)	E (α-tocopherol)	K	B_{12}	Biotin	Choline	Folic Acid (Folacin)	Niacin (Nicotinic Acid)	Pantothenic Acid	Pyridoxine (B_6)	Riboflavin (B_2)	Thiamin (B_1)
		(ppm or mg/kg)	*(ppm or mg/kg)*	*(ppm or mg/kg)*	*(ppm or mg/kg)*	*(ppm or mg/kg)*	*(ppm or mg/kg)*	*(ppm or mg/kg)*	*(ppm or mg/kg)*	*(ppm or mg/kg)*	*(ppm or mg/kg)*	*(ppm or mg/kg)*	*(ppm or mg/kg)*
43													
44													
45	13,003	7.8							46.8	11.6		3.8	2.5
	14,003	*8.4*							*50.5*	*12.5*		*4.1*	*2.6*
46	1,167	.7	54.8			1.50	4,826	1.13	115.4	20.9		16.9	6.8
	1,334	*.8*	*58.9*			*1.61*	*5,189*	*1.21*	*124.1*	*22.5*		*18.2*	*7.3*
47									17.0	5.3		3.3	1.3
									18.3	*5.7*		*3.6*	*1.4*
48	1,834	1.1			.0029	2.85	4,254		143.2	25.4	8.67	11.3	6.9
	2,000	*1.2*			*.0031*	*3.09*	*4,619*		*155.5*	*27.5*	*9.42*	*12.3*	*7.5*
49													
50							844		61.1	12.3		4.2	1.3
							891		*64.4*	*13.0*		*4.4*	*1.4*
51									62.8	17.4		8.2	3.1
									69.4	*19.2*		*9.0*	*3.4*
52	1,000	.6							50.2	14.0		1.6	
	1,167	*.7*							*56.0*	*15.7*		*1.8*	
53													
54													
55													
56													
57									47.1	22.8	2.37	.6	.4
									52.2	*25.3*	*2.63*	*.7*	*.5*
58													
59													
60						.13			55.3	6.7	13.24	2.2	13.1
						.15			*61.5*	*7.5*	*14.73*	*2.4*	*14.6*
61	15,336	9.2				.13	1,000	.28	51.1	7.5	11.00	2.0	7.9
	16,837	*10.1*				*.14*	*1,104*	*.31*	*56.4*	*8.3*	*12.14*	*2.2*	*8.7*
62						.23		.20					
						.26		*.22*					
63													

TABLE 4-101 COMPOSITION OF FEEDS—

Feed #	Feed Name—Description	Moisture Basis: As Fed or M-F (moisture-free)	Proximate Analysis: Dry Matter	Ash	Crude Fiber	Ether Extract (Fat)	N-Free Extract	Crude Protein	Digestible Protein: Ruminant	Non-Ruminant	Horse Rabbit
			(%)	(%)	(%)	(%)	(%)	(%)	(%)	(%)	(%)
ENERGY FEEDS											
64	Kelp, dried (seaweed) *(Laminariales, Fucales)*	As Fed	91.3	35.2	6.5	.5	42.6	6.5	2.8		3.5
		M-F	*100.0*	*38.6*	*7.1*	*.5*	*46.7*	*7.1*	*3.1*		*3.9*
65	Lard *(Sus scrofa)*	As Fed	100.0			100.0					
		M-F	*100.0*			*100.0*					
66	Linseed meal, mech- or sol-extd, 35% protein *(Linum usitatissimum)*	As Fed	90.6	5.7	8.8	3.4	36.7	36.0	29.5	32.1	
		M-F	*100.0*	*6.3*	*9.7*	*3.8*	*40.5*	*39.7*	*32.6*	*35.4*	
67	Linseed meal, mech- or sol-extd, 37% protein	As Fed	91.6	5.4	8.1	5.8	34.9	37.4	32.5		
		M-F	*100.0*	*5.9*	*8.8*	*6.3*	*38.2*	*40.8*	*35.5*		
68	Mangels, roots	As Fed	10.6	1.0	.9	.1	7.2	1.4	1.0	1.0	1.0
		M-F	*100.0*	*9.7*	*8.3*	*.6*	*68.2*	*13.2*	*7.5*	*7.5*	*7.6*
69	Manure, cattle	As Fed	93.5	17.9	26.6	2.7	34.1	12.2	66.0		
		M-F	*100.0*	*19.1*	*28.4*	*2.9*	*36.6*	*13.0*	*70.5*		
70	Manure, poultry, cage, dried	As Fed	88.6	26.5	13.5	1.7	35.5	28.7	12.8		
		M-F	*100.0*	*29.9*	*15.2*	*1.9*	*40.2*	*32.4*	*14.4*		
71	Manure, poultry, floor, dried	As Fed	84.5	14.1	15.7	2.3	27.1	25.3	19.7		
		M-F	*100.0*	*16.7*	*18.5*	*2.7*	*32.2*	*29.9*	*23.3*		
72	Millet grain *(Setaria spp)*	As Fed	89.9	2.9	6.4	4.0	64.3	12.2	7.5		8.7
		M-F	*100.0*	*3.3*	*7.2*	*4.5*	*71.5*	*13.5*	*8.3*		*9.6*
73	Milo heads, chopped *(Sorghum vulgare)*	As Fed	90.0	4.8	7.0	2.5	65.8	9.9	6.7		6.0
		M-F	*100.0*	*5.3*	*7.8*	*2.7*	*73.1*	*11.1*	*7.4*		*6.6*
74	Milo grain *(Sorghum vulgare)*	As Fed	89.0	1.7	2.0	2.8	71.6	11.0	7.4	7.7	6.8
		M-F	*100.0*	*1.9*	*2.2*	*3.1*	*80.4*	*12.4*	*8.3*	*8.7*	*7.6*
75	Molasses, citrus	As Fed	65.0	5.2	0	.2	52.5	7.1	3.6		3.2
		M-F	*100.0*	*8.0*	*0*	*.3*	*80.8*	*10.9*	*5.6*		*4.9*
76	Molasses, pear	As Fed	76.4	5.3	0	0	69.9	1.2			
		M-F	*100.0*	*6.9*	*0*	*0*	*91.5*	*1.6*			
77	Molasses, sugar beet *(Beta saccharifera)*	As Fed	77.0	8.2	0	.2	61.9	6.7	3.1		3.9
		M-F	*100.0*	*10.6*	*0*	*.3*	*80.4*	*8.7*	*4.0*		*5.1*
78	Molasses, sugarcane (blackstrap) *(Saccharum officinarum)*	As Fed	75.0	8.1	0	.1	63.6	3.2	1.4		2.0
		M-F	*100.0*	*10.8*	*0*	*.1*	*84.8*	*4.3*	*1.8*		*2.6*
79	Molasses, sugarcane, dried	As Fed	96.0	13.8	5.0	.6	66.3	10.3	6.4		6.8
		M-F	*100.0*	*14.4*	*5.2*	*.6*	*69.1*	*10.7*	*6.7*		*7.1*
80	Molasses, wood	As Fed	62.4	3.2	0	.1	58.5	.6			
		M-F	*100.0*	*5.0*	*0*	*.2*	*93.8*	*1.0*			
81	Oats, grain *(Avena sativa)*	As Fed	89.0	3.4	11.0	4.5	58.4	11.7	9.0	9.5	7.4
		M-F	*100.0*	*3.8*	*12.4*	*5.1*	*65.5*	*13.2*	*10.1*	*10.7*	*8.3*
82	Oats, grain, 32 lb/bu	As Fed	89.0	3.3	11.0	4.2	59.2	11.3	8.4		7.0
		M-F	*100.0*	*3.7*	*12.4*	*4.7*	*66.5*	*12.7*	*9.5*		*7.9*
83	Oats, grain, Pacific Coast	As Fed	91.2	3.7	11.0	5.4	62.1	9.0	6.9		4.7
		M-F	*100.0*	*4.1*	*12.1*	*5.9*	*68.0*	*9.9*	*7.6*		*5.2*
84	Oat groats (hulled oats)	As Fed	91.0	2.2	3.0	6.0	63.1	16.7	13.5		12.1
		M-F	*100.0*	*2.4*	*3.3*	*6.6*	*69.3*	*18.4*	*14.8*		*13.3*

ENERGY FEEDS

	TDN			Digestible Energy						Metabolizable Energy										Net Energy-Ruminant					
												Non-Ruminant													
Feed #	Ruminant	Non-Ruminant	Horse Rabbit	Ruminant		Non-Ruminant		Horse Rabbit		Ruminant		ME		Chicken ME_n		Horse Rabbit		NE_m		NE_g		NE_{lc}	
	(%)	(%)	(%)	Mcal per		kcal per		Mcal per		Mcal per		kcal per		kcal per		Mcal per		Mcal per		Mcal per		Mcal per	
				lb	kg	lb	kg	lb	kg	lb	kg	lb	kg	lb	kg	lb	kg	lb	kg	lb	kg	lb	kg
64	28.9			.6	1.3					.5	1.0												
	31.6			*.6*	*1.4*					*.5*	*1.1*												
65	238.2	212.9		4.8	10.5	3,527	7,760			3.9	8.6	3,500	7,700	3,909	8,600								
	238.2	*212.9*		*4.8*	*10.5*	*3,527*	*7,760*			*3.9*	*8.6*	*3,500*	*7,700*	*3,909*	*8,600*								
66	71.6	71.2		1.5	3.2	1,426	3,137			1.2	2.6	1,255	2,760	665	1,463			.7	1.6	.5	1.1	.9	1.9
	79.0	*78.6*		*1.6*	*3.5*	*1,574*	*3,463*			*1.3*	*2.9*	*1,385*	*3,046*	*734*	*1,615*			*.8*	*1.8*	*.5*	*1.2*	*1.0*	*2.1*
67	77.9			1.5	3.4					1.3	2.8												
	85.0			*1.7*	*3.7*					*1.4*	*3.1*												
68	8.7	8.3		.2	.4	166	365			.1	.3	155	342					.1	.2	.05	.1	.1	.2
	82.0	*78.0*		*1.6*	*3.6*	*1,564*	*3,440*			*1.3*	*2.9*	*1,465*	*3,224*					*.8*	*1.8*	*.5*	*1.2*	*.9*	*2.0*
69	70.0			1.4	3.1					1.1	2.5												
	74.9			*1.5*	*3.3*					*1.2*	*2.7*												
70	46.3			.9	2.0					.8	1.7			386	850								
	52.3			*1.0*	*2.3*					*.9*	*1.9*			*450*	*990*								
71	61.3			1.2	2.7					.8	1.8												
	72.5			*1.5*	*3.2*					*1.0*	*2.2*												
72	57.7	69.7		1.1	2.5	1,397	3,073			1.0	2.1	1,303	2,866										
	64.2	*77.6*		*1.3*	*2.8*	*1,555*	*3,421*			*1.0*	*2.3*	*1,450*	*3,190*										
73	72.5	62.5		1.5	3.2	1,253	2,756			1.2	2.6	1,202	2,645					.7	1.6	.5	1.0		
	80.5	*69.4*		*1.6*	*3.5*	*1,391*	*3,060*			*1.3*	*2.9*	*1,335*	*2,937*					*.8*	*1.7*	*.5*	*1.1*		
74	77.7	77.2	71.2	1.5	3.4	1,547	3,404	1.4	3.1	1.3	2.8	1,447	3,183	1,479	3,254			.7	1.6	.5	1.1		
	87.3	*86.7*	*80.0*	*1.7*	*3.8*	*1,738*	*3,823*	*1.6*	*3.5*	*1.5*	*3.2*	*1,626*	*3,577*	*1,662*	*3,656*			*.9*	*1.9*	*.5*	*1.2*		
75	50.1	52.3		1.0	2.2	1,048	2,306			.8	1.8	988	2,174					.6	1.3	.4	.9	.6	1.3
	77.0	*80.4*		*1.5*	*3.4*	*1,611*	*3,545*			*1.3*	*2.8*	*1,520*	*3,344*					*.9*	*2.0*	*.6*	*1.3*	*.9*	*1.9*
76	62.9	61.4		1.3	2.8	1,230	2,707			1.0	2.3	1,177	2,590										
	82.3	*80.3*		*1.6*	*3.6*	*1,610*	*3,543*			*1.4*	*3.0*	*1,541*	*3,390*										
77	61.6	55.5		1.1	2.4	1,112	2,447			1.0	2.2			871	1,917			.7	1.5	.5	1.0	.7	1.5
	80.0	*72.1*		*1.5*	*3.2*	*1,445*	*3,178*			*1.3*	*2.9*			*1.132*	*2,490*			*.9*	*1.9*	*.6*	*1.3*	*.9*	*1.9*
78	61.1	55.7		1.2	2.7	1,117	2,458			1.0	2.3	1,073	2,360	889	1,955			.7	1.6	.5	1.0	.8	1.8
	81.5	*74.3*		*1.6*	*3.6*	*1,490*	*3,277*			*1.4*	*3.0*	*1,430*	*3,146*	*1,185*	*2,607*			*1.0*	*2.1*	*.6*	*1.3*	*1.1*	*2.4*
79	75.8	69.8		1.5	3.3	1,399	3,077			1.2	2.7	1,312	2,886	1,312	2,408			.8	1.7	.5	1.1		
	79.0	*72.7*		*1.6*	*3.5*	*1,457*	*3,205*			*1.3*	*2.9*	*1,366*	*3,006*	*1,140*	*2,508*			*.8*	*1.8*	*.5*	*1.2*		
80	54.3	52.8		1.1	2.4	1,059	2,329			.9	2.0	1,014	2,231										
	87.0	*84.7*		*1.7*	*3.8*	*1,698*	*3,735*			*1.4*	*3.1*	*1,627*	*3,579*										
81	67.3	62.0	62.3	1.4	3.0	1,241	2,731	1.3	2.8	1.1	2.5	1,160	2,553	1,137	2,501	1.0	2.3	.7	1.5	.5	1.0	.8	1.7
	75.6	*69.7*	*70.0*	*1.5*	*3.3*	*1,395*	*3,070*	*1.4*	*3.1*	*1.3*	*2.8*	*1,304*	*2,868*	*1,277*	*2,810*	*1.1*	*2.5*	*.8*	*1.7*	*.5*	*1.1*	*.9*	*1.9*
82	68.6	57.2	62.3	1.4	3.0	1,146	2,522	1.3	2.8	1.1	2.5	1,071	2,357			1.0	2.3						
	77.1	*64.3*	*70.0*	*1.5*	*3.4*	*1,287*	*2,832*	*1.4*	*3.1*	*1.3*	*2.8*	*1,204*	*2,648*			*1.1*	*2.5*						
83	71.1	66.7	63.8	1.5	3.2	1,337	2,941	1.3	2.8	1.2	2.6	1,256	2,763			1.0	2.3	.7	1.6	.5	1.1	.8	1.8
	78.0	*73.1*	*70.0*	*1.6*	*3.5*	*1,465*	*3,223*	*1.4*	*3.1*	*1.3*	*2.8*	*1,377*	*3,030*			*1.1*	*2.5*	*.8*	*1.8*	*.5*	*1.2*	*.9*	*1.9*
84	87.1	83.7	76.4	1.8	3.9	1,677	3,690	1.5	3.4	1.5	3.2	1,549	3,407	1,541	3,390			1.0	2.1	.6	1.4	1.0	2.3
	95.7	*92.0*	*84.0*	*2.0*	*4.3*	*1,843*	*4,055*	*1.7*	*3.7*	*1.6*	*3.5*	*1,702*	*3,744*	*1,693*	*3,725*			*1.1*	*2.4*	*.7*	*1.5*	*1.1*	*2.5*

TABLE 4-101 COMPOSITION OF FEEDS—

Feed #	Feed Name—Description	Moisture Basis: As Fed or M-F (moisture-free)	Dry Matter	Macrominerals: Calcium (Ca)	Phosphorus (P)	Sodium (Na)	Chlorine (Cl)	Magnesium (Mg)	Potassium (K)	Sulfur (S)	Microminerals: Cobalt (Co)	Copper (Cu)	Iron (Fe)	Manganese (Mn)	Zinc (Zn)
			(%)	(%)	(%)	(%)	(%)	(%)	(%)	(%)	(ppm or mg/kg)	(ppm or mg/kg)	(%)	(ppm or mg/kg)	(ppm or mg/kg)
ENERGY FEEDS															
64	Kelp, dried (seaweed)	As Fed	91.3	2.48	.28			.85							
		M-F	*100.0*	*2.72*	*.31*			*.93*							
65	Lard	As Fed	100.0												
		M-F	*100.0*												
66	Linseed meal, mech- or sol-extd, 35% protein	As Fed	90.6	.40	.84	.12	.04	.58	1.30	.39	.298	33.8	.024	34.9	
		M-F	*100.0*	*.44*	*.93*	*.13*	*.04*	*.64*	*1.43*	*.43*	*.329*	*37.3*	*.027*	*38.5*	
67	Linseed meal, mech- or sol extd, 37% protein	As Fed	91.6	.38	.85				1.09						
		M-F	*100.0*	*.41*	*.93*				*1.19*						
68	Mangels, roots	As Fed	10.6	.02	.02	.07	.13	.02	.21	.02			.002		
		M-F	*100.0*	*.19*	*.19*	*.66*	*1.23*	*.19*	*1.98*	*.19*			*.019*		
69	Manure, cattle	As Fed	93.5	1.89	.66	.36		.27	.72			15.0	.008		9.9
		M-F	*100.0*	*2.02*	*.71*	*.39*		*.29*	*.77*			*16.0*	*.009*		*10.6*
70	Manure, poultry, cage, dried	As Fed	88.6	7.80	2.2	.42		.63	1.37		.001	61.0	.002	291.0	325.0
		M-F	*100.0*	*8.80*	*2.5*	*.47*		*.71*	*1.55*		*.001*	*69.0*	*.003*	*328.0*	*367.0*
71	Manure, poultry, floor, dried	As Fed	84.5	2.50	1.6	.42		.35	1.77			23.0	.038	190.0	343.0
		M-F	*100.0*	*3.00*	*1.9*	*.50*		*.41*	*2.09*			*27.0*	*.045*	*225.0*	*406.0*
72	Millet grain	As Fed	89.9	.05	.28	.04	.14	.16	.43	.13	.044	21.6	.004	29.1	13.9
		M-F	*100.0*	*.06*	*.31*	*.04*	*.16*	*.18*	*.48*	*.14*	*.049*	*24.0*	*.005*	*32.4*	*15.4*
73	Milo heads, chopped	As Fed	90.0	.12	.25										
		M-F	*100.0*	*.13*	*.28*										
74	Milo grain	As Fed	89.0	.04	.29	.01	.08	.20	.35		.089	14.1	.004	12.9	
		M-F	*100.0*	*.04*	*.33*	*.01*	*.09*	*.22*	*.39*		*.100*	*15.8*	*.005*	*14.5*	
75	Molasses, citrus	As Fed	65.0	1.31	.16	.16	.07	.14	.09		.103	72.8	.033	26.0	88.9
		M-F	*100.0*	*2.01*	*.25*	*.40*	*.10*	*.22*	*.14*		*.159*	*112.0*	*.050*	*40.0*	*136.7*
76	Molasses, pear	As Fed	76.4												
		M-F	*100.0*												
77	Molasses, sugar beet	As Fed	77.0	.16	.03	1.17	1.25	.23	4.77		.400	17.6	.008	4.6	
		M-F	*100.0*	*.21*	*.04*	*1.52*	*1.62*	*.30*	*6.20*		*.500*	*22.9*	*.010*	*6.0*	
78	Molasses, sugarcane (blackstrap)	As Fed	75.0	.89	.08	.15	2.79	.35	2.38	.35	.912	59.6	.019	42.2	
		M-F	*100.0*	*1.19*	*.11*	*.20*	*3.72*	*.47*	*3.17*	*.46*	*1.216*	*79.4*	*.025*	*56.3*	
79	Molasses, sugarcane, dried	As Fed	96.0	1.18	.14										
		M-F	*100.0*	*1.23*	*.15*										
80	Molasses, wood	As Fed	62.4	1.45	.03	.03	.12	.07	.04	.03				12.6	
		M-F	*100.0*	*2.33*	*.05*	*.05*	*.20*	*.11*	*.06*	*.05*				*20.3*	
81	Oats, grain	As Fed	89.0	.10	.35	.06	.11	.17	.37		.062	5.9	.007	38.2	
		M-F	*100.0*	*.11*	*.39*	*.07*	*.12*	*.19*	*.42*		*.070*	*6.6*	*.008*	*42.9*	
82	Oats, grain, 32 lb/bu	As Fed	89.0	.06	.27										
		M-F	*100.0*	*.07*	*.30*										
83	Oats, grain, Pacific Coast	As Fed	91.2	.09	.33										
		M-F	*100.0*	*.10*	*.36*										
84	Oat groats (hulled oats)	As Fed	91.0	.07	.43	.05	.09	.09	.34	.20		6.4	.007	28.6	
		M-F	*100.0*	*.08*	*.47*	*.06*	*.10*	*.10*	*.3*	*.22*		*7.0*	*.008*	*31.4*	

ENERGY FEEDS

Feed #	Fat-Soluble Vitamins				Water-Soluble Vitamins									
	A (1 mg car. = 1,667 IU-A)	Carotene (Pro-vitamin A)	E (α-tocopherol)	K	B_{12}	Biotin	Choline	Folic Acid (Folacin)	Niacin (Nicotinic Acid)	Pantothenic Acid	Pyridoxine (B_6)	Riboflavin (B_2)	Thiamin (B_1)
		(ppm or mg/kg)	*(ppm or mg/kg)*	*(ppm or mg/kg)*	*(ppm or mg/kg)*	*(ppm or mg/kg)*	*(ppm or mg/kg)*	*(ppm or mg/kg)*	*(ppm or mg/kg)*	*(ppm or mg/kg)*	*(ppm or mg/kg)*	*(ppm or mg/kg)*	*(ppm or mg/kg)*
64													
65													
66	167	.1					1,488		34.1	13.6		3.2	7.2
	167	*.1*					*1,643*		*37.6*	*15.0*		*3.5*	*8.0*
67													
68													
69													
70													
71													
72							788		52.5	7.3		1.6	6.5
							877		*58.4*	*8.2*		*1.8*	*7.3*
73	1,000	.6										2.0	
	1,167	*.7*										*2.2*	
74			12.0			.25	678	.22	42.7	11.4	4.10	1.2	3.9
			13.5			*.29*	*761*	*.24*	*48.0*	*12.8*	*4.60*	*1.3*	*4.4*
75									26.6	12.5		6.2	
									41.0	*19.3*		*9.5*	
76													
77							862		42.2	4.6		2.4	
							1,119		*54.8*	*6.0*		*3.1*	
78							876		34.3	38.3		3.3	.9
							1,167		*45.7*	*51.1*		*4.4*	*1.2*
79													
80													
81			32.6				946		13.7	13.1		1.6	6.2
			36.6				*1,063*		*15.4*	*14.7*		*1.8*	*7.0*
82													
83			20.0				959		13.9	13.2		1.1	
			21.9				*1,052*		*15.2*	*14.5*		*1.2*	
84									8.1	14.7	1.10	1.3	6.8
									8.9	*16.2*	*1.21*	*1.4*	*7.5*

TABLE 4-101 COMPOSITION OF FEEDS-

Feed #	Feed Name—Description	Moisture Basis: As Fed or M-F (moisture-free)	Proximate Analysis: Dry Matter	Ash	Crude Fiber	Ether Extract (Fat)	N-Free Extract	Crude Protein	Digestible Protein: Ruminant	Non-Ruminant	Horse Rabbit
			(%)	(%)	(%)	(%)	(%)	(%)	(%)	(%)	(%)
ENERGY FEEDS											
85	Pea feed or pea meal *(Pisum spp)*	As Fed	90.0	3.5	23.7	1.4	43.7	17.7	12.9		12.7
		M-F	*100.0*	*3.9*	*26.3*	*1.6*	*48.6*	*19.7*	*14.4*		*14.1*
86	Pea seed, field *(Pisum sativum arvense)*	As Fed	90.7	3.0	6.1	1.2	57.0	23.4			
		M-F	*100.0*	*3.3*	*6.7*	*1.3*	*62.8*	*25.9*			
87	Peanut meal, mech-extd, 45% protein *(Arachis hypogaea)*	As Fed	92.0	2.2	11.0	7.5	25.5	45.8	41.4	43.7	
		M-F	*100.0*	*2.3*	*12.0*	*8.2*	*27.7*	*49.8*	*45.0*	*47.5*	
88	Peanut meal, sol-extd, 47% protein	As Fed	92.0	4.5	13.0	1.2	25.9	47.4	42.3		
		M-F	*100.0*	*4.9*	*14.1*	*1.3*	*28.2*	*51.5*	*46.0*		
89	Pearlmillet *(Pennisetum glaucum)*	As Fed	89.1	1.7	2.5	4.3	68.3	12.3	8.2	7.5	8.8
		M-F	*100.0*	*2.0*	*2.8*	*4.8*	*76.6*	*13.8*	*9.1*	*8.4*	*9.8*
90	Pineapple, cannery residue *(Ananas comosus)*	As Fed	87.1	3.0	17.0	1.4	61.7	4.0	.9		1.2
		M-F	*100.0*	*3.4*	*19.6*	*1.6*	*70.8*	*4.6*	*1.0*		*1.4*
91	Pineapple, cannery residue, w/molasses, dehy	As Fed	87.4	3.2	15.9	1.0	63.4	3.9	.8		1.2
		M-F	*100.0*	*3.7*	*18.2*	*1.1*	*72.5*	*4.5*	*.9*		*1.3*
92	Potato peelings *(Solanum tuberosum)*	As Fed	23.0	1.4	.8	.1	18.4	2.3	1.4	1.7	1.5
		M-F	*100.0*	*6.0*	*3.4*	*.4*	*80.3*	*9.9*	*5.9*	*7.3*	*6.3*
93	Potato processing residue, wet	As Fed	13.1	.4	1.6	.1	9.9	1.0	.4	.3	.6
		M-F	*100.0*	*2.9*	*12.4*	*.9*	*75.9*	*7.9*	*3.2*	*2.1*	*4.4*
94	Potato processing residue, dried	As Fed	88.6	4.2	6.1	.4	70.2	7.7	3.4	2.0	4.6
		M-F	*100.0*	*4.7*	*6.9*	*.5*	*79.2*	*8.7*	*3.8*	*2.3*	*5.2*
95	Potato tubers *(Solanum tuberosum)*	As Fed	24.6	1.2	.5	.1	20.6	2.2	1.5	.8	1.5
		M-F	*100.0*	*4.8*	*2.1*	*.3*	*83.8*	*9.0*	*6.1*	*3.2*	*6.0*
96	Potato tubers, cooked	As Fed	24.3	1.3	.7	.1	20.0	2.2	1.2	1.5	1.4
		M-F	*100.0*	*5.3*	*3.0*	*.3*	*82.3*	*9.1*	*4.7*	*6.0*	*5.6*
97	Potato tubers, dried (meal)	As Fed	91.4	4.3	2.1	.3	75.0	9.7	6.0		6.4
		M-F	*100.0*	*4.7*	*2.3*	*.3*	*82.1*	*10.6*	*6.6*		*7.0*
98	Rice bran *(Oryza sativa)*	As Fed	91.0	10.9	11.0	15.1	40.5	13.5	9.0		9.4
		M-F	*100.0*	*12.0*	*12.1*	*16.6*	*44.5*	*14.8*	*9.9*		*10.3*
99	Rice grain w/hulls, grnd (paddyrice) *(Oryza sativa)*	As Fed	88.1	5.2	7.8	1.5	66.6	7.0	4.1		4.0
		M-F	*100.0*	*5.8*	*8.8*	*1.7*	*75.6*	*8.1*	*4.7*		*4.6*
100	Rice groats (brown rice)	As Fed	88.2	1.0	.9	1.7	76.3	8.3	4.9		5.2
		M-F	*100.0*	*1.2*	*1.0*	*1.9*	*86.5*	*9.4*	*5.5*		*5.9*
101	Rice groats, polished (polished rice)	As Fed	89.0	.5	.4	.4	81.8	5.9	4.4		4.1
		M-F	*100.0*	*.6*	*.4*	*.5*	*91.9*	*6.6*	*4.9*		*4.6*
102	Rice polishings, dried	As Fed	90.4	6.9	3.2	11.8	56.1	12.6	8.9	10.3	8.9
		M-F	*100.0*	*7.6*	*3.6*	*13.1*	*62.0*	*13.7*	*9.9*	*11.4*	*9.9*
103	Rye, distillers' grains w/solubles, dehy *(Secale cereale)*	As Fed	90.5	6.4	8.1	4.1	44.7	27.2			
		M-F	*100.0*	*7.1*	*9.0*	*4.5*	*49.3*	*30.1*			
104	Rye grain *(Secale cereale)*	As Fed	89.0	1.8	2.0	1.5	71.8	11.9	9.4	9.2	7.9
		M-F	*100.0*	*2.0*	*2.2*	*1.7*	*80.7*	*13.4*	*10.6*	*10.3*	*8.9*
105	Safflower seeds *(Carthamus tinctorius)*	As Fed	93.1	2.9	26.6	29.8	17.5	16.3	13.0		12.4
		M-F	*100.0*	*3.1*	*28.6*	*32.0*	*18.8*	*17.5*	*14.0*		*13.3*

ENERGY FEEDS

Feed #	TDN			Digestible Energy						Metabolizable Energy								Net Energy-Ruminant					
	Ruminant	Non-Ruminant	Horse Rabbit	Ruminant		Non-Ruminant		Horse Rabbit		Ruminant		Non-Ruminant ME		Non-Ruminant Chicken ME_n		Horse Rabbit		NE_m		NE_g		NE_{lc}	
	(%)	(%)	(%)	Mcal per		kcal per		Mcal per		Mcal per		kcal per		kcal per		Mcal per		Mcal per		Mcal per		Mcal per	
				lb	kg	lb	kg	lb	kg	lb	kg	lb	kg	lb	kg	lb	kg	lb	kg	lb	kg	lb	kg
85	76.5			1.5	3.4					1.3	2.8												
	85.0			1.7	3.7					1.4	3.1												
86	71.4	84.1		1.4	3.1	1,685	3,707			1.2	2.6	1,618	3,559	1,000	2,200								
	78.8	92.7		1.6	3.5	1,858	4,087			1.3	2.8	1,784	3,924	1,103	2,426								
87	78.7	84.3		1.6	3.5	1,689	3,715			1.3	2.8	1,449	3,187	1,132	2,491			.8	1.8	.5	1.2	.9	2.0
	85.5	91.6		1.7	3.8	1,835	4,038			1.4	3.1	1,575	3,464	1,231	2,708			.9	2.0	.6	1.3	1.0	2.1
88	73.6	64.9		1.5	3.2	1,293	2,845	1.4	3.1	1.2	2.7	1,065	2,344	1,002	2,205			.7	1.6	.5	1.1	.9	1.9
	80.0	70.5		1.6	3.5	1,455	3,201	1.5	3.4	1.3	2.9	1,209	2,659	1,090	2,397			.8	1.8	.5	1.2	1.0	2.1
89	69.7	56.8		1.4	3.1	1,139	2,506			1.1	2.5	1,062	2,336										
	78.2	63.8		1.5	3.4	1,278	2,812			1.3	2.8	1,191	2,621										
90	63.1			1.3	2.8					1.0	2.3												
	72.5			1.5	3.2					1.2	2.6												
91	63.8			1.3	2.8					1.0	2.3												
	73.0			1.5	3.2					1.2	2.6												
92	19.0	20.0		.4	.8	401	882			.3	.7	376	827										
	82.4	86.8		1.6	3.6	1,740	3,827			1.4	3.0	1,635	3,598										
93	8.8	8.2		.2	.4	164	361			.1	.3	155	340										
	67.5	62.6		1.4	3.0	1,254	2,758			1.1	2.4	1,185	2,608										
94	70.4	79.3		1.4	3.1	1,589	3,496			1.1	2.5	1,497	3,293										
	79.5	89.5		1.6	3.5	1,794	3,946			1.3	2.9	1,690	3,717										
95	20.0	20.7		.4	.9	415	913			.3	.7	391	860					.2	.5	.1	.3	.2	.5
	81.5	84.2		1.6	3.6	1,687	3,712			1.4	3.0	1,588	3,494					.9	2.0	.6	1.3	.9	2.0
96	16.4	21.3		.3	.7	427	939			.3	.6	403	886										
	67.8	88.0		1.4	3.0	1,764	3,880			1.1	2.5	1,661	3,654										
97	73.8	76.3		1.5	3.3	1,529	3,364			1.2	2.7	1,436	3,159	1,603	3,527								
	80.8	83.5		1.6	3.6	1,674	3,682			1.3	2.9	1,571	3,456	1,754	3,859								
98	64.2	66.2		1.3	2.8	1,326	2,917			1.0	2.3	1,273	2,800	741	1,630			.6	1.3	.4	.8	.6	1.4
	70.6	72.7		1.4	3.1	1,457	3,205			1.2	2.6	1,400	3,077	814	1,791			.6	1.4	.4	.9	.7	1.6
99	72.7	74.5		1.5	3.2	1,493	3,285			1.2	2.6	1,410	3,102										
	82.5	84.6		1.6	3.6	1,695	3,730			1.4	3.0	1,600	3,521										
100	78.8	84.0		1.6	3.5	1,684	3,704			1.3	2.9	1,584	3,485										
	89.4	95.3		1.8	3.9	1,910	4,202			1.5	3.2	1,796	3,952										
101	76.3	86.3		1.5	3.4	1,730	3,805			1.3	2.8	1,633	3,592	1,417	3,118			.8	1.8	.5	1.2	.9	1.9
	85.7	97.0		1.7	3.8	1,944	4,277			1.4	3.1	1,835	4,036	1,592	3,503			.9	2.0	.6	1.3	1.0	2.2
102	81.8	87.0		1.6	3.6	1.744	3,836			1.4	3.0	1,626	3,577	1,300	2,860								
	90.4	96.2		1.8	4.0	1,928	4,241			1.5	3.3	1,798	3,955	1,438	3,164								
103	57.4	60.1		1.1	2.5	1,205	2,650			1.0	2.1	988	2,173										
	63.5	66.4		1.3	2.8	1,331	2,928			1.0	2.3	1,091	2,401										
104	75.7	77.9		1.5	3.3	1,561	3,434			1.2	2.7	1,458	3,208	1,169	2,572			.8	1.8	.5	1.2		
	85.0	87.5		1.7	3.8	1,572	3,858			1.4	3.1	1,638	3,604	1,314	2,890			.9	2.0	.6	1.4		
105	82.4	80.5		1.6	3.6	1,613	3,549			1.4	3.0	1,549	3,407					.9	2.0	.6	1.3	1.0	2.2
	88.5	86.5		1.8	3.9	1,734	3,814			1.5	3.3	1,664	3,661					1.0	2.2	.6	1.4	1.0	2.3

TABLE 4-101 COMPOSITION OF FEEDS-

Feed #	Feed Name—Description	Moisture Basis: As Fed or M-F (moisture-free)	Dry Matter	Macrominerals: Calcium (Ca)	Phosphorus (P)	Sodium (Na)	Chlorine (Cl)	Magnesium (Mg)	Potassium (K)	Sulfur (S)	Microminerals: Cobalt (Co)	Copper (Cu)	Iron (Fe)	Manganese (Mn)	Zinc (Zn)
			(%)	(%)	(%)	(%)	(%)	(%)	(%)	(%)	(ppm or mg/kg)	(ppm or mg/kg)	(%)	(ppm or mg/kg)	(ppm or mg/kg)
ENERGY FEEDS															
85	Pea feed or pea meal	As Fed	90.0												
		M-F	100.0												
86	Pea seed, field	As Fed	90.7	.17	.50				1.03						
		M-F	100.0	.19	.55				1.14						
87	Peanut meal, mech-extd, 45% protein	As Fed	92.0	.16	.57			.33	1.15	.29				25.5	
		M-F	100.0	.18	.62			.36	1.25	.32				27.7	
88	Peanut meal, sol-extd, 47% protein	As Fed	92.0	.20	.65	.07	.03	.04	1.20					29.0	20.0
		M-F	100.0	.22	.71	.08	.03	.04	1.30					31.5	21.7
89	Pearlmillet	As Fed	89.1	.05	.38								.001		
		M-F	100.0	.06	.43								.001		
90	Pineapple, cannery residue	As Fed	87.1	.21	.10								.049		
		M-F	100.0	.24	.12								.056		
91	Pineapple, cannery residue, w/molasses, dehy	As Fed	87.4												
		M-F	100.0												
92	Potato peelings	As Fed	23.0	.03	.04										
		M-F	100.0	.14	.19										
93	Potato processing residue, wet	As Fed	13.1	.03	.03				.17						
		M-f	100.0	.23	.24				1.33						
94	Potato processing residue, dried	As Fed	88.6												
		M-F	100.0												
95	Potato tubers	As Fed	24.6	.01	.06	.02	.07	.03	.56	.02		4.3	.002	10.2	
		M-F	100.0	.05	.24	.09	.28	.14	2.26	.09		17.7	.009	41.6	
96	Potato tubers, cooked	As Fed	24.3												
		M-F	100.0												
97	Potato tubers, dried (meal)	As Fed	91.4	.07	.20				1.97					2.9	
		M-F	100.0	.08	.22				2.16					3.1	
98	Rice bran	As Fed	91.0	.06	1.82	.07	.07	.95	1.74	.18		13.0	.019	417.9	29.9
		M-F	100.0	.07	2.00	.08	.08	1.04	1.91	.20		14.3	.021	459.2	32.9
99	Rice grain w/hulls, grnd (paddyrice)	As Fed	88.1	.05	.39		.05	.10	.22	.05					
		M-F	100.0	.06	.45		.06	.11	.25	.06					
100	Rice groats (brown rice)	As Fed	88.2	.04	.25	.03	.13	.09	.21	.02		3.4	.003	13.3	
		M-F	100.0	.05	.28	.04	.15	.10	.23	.03		3.9	.003	15.1	
101	Rice groats, polished (polished rice)	As Fed	89.0	.03	.12	.01	.02	.02	.13	.08		2.9	.002	10.9	1.8
		M-F	100.0	.03	.14	.01	.02	.02	.15	.09		3.3	.002	12.3	2.0
102	Rice polishings, dried	As Fed	90.4	.05	1.24	.11	.13	.65	1.02	.17			.016		
		M-F	100.0	.06	1.37	.12	.14	.72	1.13	.19			.018		
103	Rye, distillers' grains w/solubles, dehy	As Fed	90.5												
		M-F	100.0												
104	Rye grain	As Fed	89.0	.06	.34	.02	.03	.12	.46	.15		7.8	.008	66.9	30.5
		M-F	100.0	.07	.38	.02	.03	.13	.52	.17		8.8	.009	75.2	34.3
105	Safflower seeds	As Fed	93.1												
		M-F	100.0												

NERGY FEEDS

Feed #	Fat-Soluble Vitamins: A (1 mg car. = 1,667 IU-A)	Carotene (Pro-vitamin A) (ppm or mg/kg)	E (α-tocopherol) (ppm or mg/kg)	K (ppm or mg/kg)	Water-Soluble Vitamins: B_{12} (ppm or mg/kg)	Biotin (ppm or mg/kg)	Choline (ppm or mg/kg)	Folic Acid (Folacin) (ppm or mg/kg)	Niacin (Nicotinic Acid) (ppm or mg/kg)	Pantothenic Acid (ppm or mg/kg)	Pyridoxine (B_6) (ppm or mg/kg)	Riboflavin (B_2) (ppm or mg/kg)	Thiamin (B_1) (ppm or mg/kg)
85													
86													
87	306	.2					1,683		169.0	48.2		5.3	7.3
	333	*.2*					*1,829*		*183.7*	*52.4*		*5.8*	*7.9*
88			3.0			.39	2,000	.36	170.1	53.0	10.00	11.0	7.3
			3.3			*.42*	*2,174*	*.39*	*184.9*	*57.6*	*10.87*	*12.0*	*7.9*
89													
90													
91													
92													
93													
94													
95													
96													
97													
98			60.0			.42	1,254		303.2	23.5	29.23	2.6	22.4
			65.9			*.46*	*1,378*		*333.2*	*25.8*	*32.12*	*2.9*	*24.6*
99	0		14.0			.08		.29	49.0				
	0		*15.9*			*.09*		*.33*	*55.6*				
100			8.7			.09		.19	45.3	10.7	7.00	.6	3.2
			9.9			*.10*		*.21*	*51.4*	*12.1*	*7.94*	*.7*	*3.6*
101	0		3.6				909		16.6	3.3	.39	.4	.8
	0		*4.0*				*1,021*		*18.7*	*3.7*	*.44*	*.5*	*.9*
102			90.0			.62	1,317		409.4	58.8	27.95	1.8	19.6
			99.6			*.69*	*1,457*		*452.7*	*65.0*	*30.90*	*2.0*	*21.7*
103									62.8	17.4		8.2	3.1
									69.4	*19.2*		*9.0*	*3.4*
104		0	15.0			.06	439	6.0	1.2	6.9		1.6	3.9
		0	*16.9*			*.07*	*493*	*6.7*	*1.3*	*7.8*		*1.8*	*4.4*
105			1.0										
			1.1										

TABLE 4-101 COMPOSITION OF FEEDS-

Feed #	Feed Name—Description	Moisture Basis: As Fed or M-F (moisture-free)	Proximate Analysis						Digestible Protein		
			Dry Matter	Ash	Crude Fiber	Ether Extract (Fat)	N-Free Extract	Crude Protein	Ruminant	Non-Ruminant	Horse Rabbit
			(%)	(%)	(%)	(%)	(%)	(%)	(%)	(%)	(%)
ENERGY FEEDS											
106	Screenings, grain	As Fed	90.3	9.1	28.3	4.8	33.7	14.4	10.3		8.8
		M-F	*100.0*	*10.1*	*31.3*	*5.3*	*37.3*	*16.0*	*11.5*		*9.8*
107	Sorghum, grain, all analyses *(Sorghum vulgare)*	As Fed	89.0	2.1	2.0	3.1	70.7	11.1	6.3		5.7
		M-F	*100.0*	*2.4*	*2.2*	*3.5*	*79.4*	*12.5*	*7.1*		*6.4*
108	Sorghum, kafir, grain *(Sorghum vulgare)*	As Fed	90.0	1.6	2.0	2.9	71.7	11.8	8.2	8.4	7.5
		M-F	*100.0*	*1.8*	*2.2*	*3.3*	*79.6*	*13.1*	*9.1*	*9.3*	*8.3*
109	Sorghum, milo, head chops	As Fed	90.0	4.8	7.0	2.5	65.8	9.9	6.7		6.0
		M-F	*100.0*	*5.3*	*7.8*	*2.7*	*73.1*	*11.1*	*7.4*		*6.6*
110	Sorghum, milo, grain *(Sorghum vulgare)*	As Fed	89.0	1.7	2.0	2.8	71.6	11.0	7.4	7.7	6.8
		M-F	*100.0*	*1.9*	*2.2*	*3.1*	*80.4*	*12.4*	*8.3*	*8.7*	*7.6*
111	Soybean oil *(Glycine max)*	As Fed	100.0			100.0					
		M-F	*100.0*			*100.0*					
112	Soybean seeds *(Glycine max)*	As Fed	90.0	4.9	5.0	17.3	24.9	37.9	34.1	30.8	
		M-F	*100.0*	*5.4*	*5.6*	*19.2*	*27.7*	*42.1*	*37.9*	*34.2*	
113	Starch, corn	As Fed	90.4	.2	.1	.2	89.3	.6	0		0
		M-F	*100.0*	*.2*	*.2*	*.2*	*98.7*	*.7*	*0*		*0*
114	Spelt grain *(Triticum spelta)*	As Fed	89.6	3.5	8.2	2.0	64.0	11.9	8.9		8.4
		M-F	*100.0*	*3.9*	*9.1*	*2.2*	*71.5*	*13.3*	*9.9*		*9.4*
115	Sugar beet molasses *(Beta saccharifera)*	As Fed	77.0	8.2	0	.2	61.9	6.7	3.1		3.9
		M-F	*100.0*	*10.6*	*0*	*.3*	*80.4*	*8.7*	*4.0*		*5.1*
116	Sugar beet pulp, wet	As Fed	10.0	.7	2.0	.2	6.2	.9	.4		
		M-F	*100.0*	*6.7*	*20.0*	*2.1*	*62.2*	*9.0*	*4.0*		
117	Sugar beet pulp, dried	As Fed	91.0	3.6	19.0	.6	58.7	9.1	4.3	3.5	6.0
		M-F	*100.0*	*3.9*	*20.9*	*.7*	*64.5*	*10.0*	*4.7*	*3.9*	*6.6*
118	Sugar beet pulp w/ molasses, dried	As Fed	92.0	5.7	16.0	.6	60.6	9.1	6.0	2.2	5.7
		M-F	*100.0*	*6.2*	*17.4*	*.7*	*65.8*	*9.9*	*6.5*	*2.4*	*6.2*
119	Sugarcane molasses *(Saccharum officinarum)*	As Fed	75.0	8.1	0	.1	63.6	3.2	1.4		2.0
		M-F	*100.0*	*10.8*	*0*	*.1*	*84.8*	*4.3*	*1.8*		*2.6*
120	Sunflower seeds *(Helianthus spp)*	As Fed	93.6	3.1	29.0	25.9	18.8	16.8			
		M-F	*100.0*	*3.3*	*31.0*	*27.7*	*20.1*	*17.9*			
121	Sweet potato roots, dried *(Ipomoea batatas)*	As Fed	30.6	1.1	1.3	.4	26.1	1.7	.6		.7
		M-F	*100.0*	*3.6*	*4.2*	*1.3*	*85.5*	*5.4*	*1.8*		*2.2*
122	Tallow *(Bos spp)*	As Fed	96.0	.1		94.0		1.5			
		M-F	*100.0*	*.1*		*97.9*		*1.6*			
123	Triticale grain *(Triticum spp x secale cereale)*	As Fed	90.0	1.8	4.0	1.1	67.6	15.5	14.6		
		M-F	*100.0*	*2.0*	*4.4*	*1.2*	*75.2*	*17.2*	*16.2*		
124	Wheat bran *(Triticum spp)*	As Fed	89.0	6.1	10.0	4.1	52.8	16.0	12.2	12.0	11.5
		M-F	*100.0*	*6.9*	*11.2*	*4.6*	*59.3*	*18.0*	*13.7*	*13.5*	*12.9*
125	Wheat grain *(Triticum spp)*	As Fed	89.0	1.6	3.0	1.7	70.0	12.7	10.0		8.4
		M-F	*100.0*	*1.8*	*3.4*	*1.9*	*78.6*	*14.3*	*11.2*		*9.4*
126	Wheat grain, min 60 lb/bu	As Fed	88.0	1.8	2.2	1.4	69.0	13.6	10.6		10.0
		M-F	*100.0*	*2.0*	*2.5*	*1.6*	*78.5*	*15.4*	*12.0*		*11.4*

NERGY FEEDS

Feed #	TDN			Digestible Energy						Metabolizable Energy								Net Energy-Ruminant					
	Ruminant	Non-Ruminant	Horse Rabbit	Ruminant		Non-Ruminant		Horse Rabbit		Ruminant		Non-Ruminant ME		Non-Ruminant Chicken ME_n		Horse Rabbit		NE_m		NE_g		NE_{lc}	
	(%)	(%)	(%)	Mcal per		kcal per		Mcal per		Mcal per		kcal per		kcal per		Mcal per		Mcal per		Mcal per		Mcal per	
				lb	kg	lb	kg	lb	kg	lb	kg	lb	kg	lb	kg	lb	kg	lb	kg	lb	kg	lb	kg
106	47.4	18.7		1.0	2.1	375	824			.8	1.7	360	792					.5	1.0	.1	.3	.4	.9
	52.5	*20.7*		*1.0*	*2.3*	*415*	*912*			*.9*	*1.9*	*398*	*876*					*.5*	*1.1*	*.1*	*.3*	*.5*	*1.0*
107	73.9	79.1		1.5	3.3	1,575	3,466			1.2	2.7	1,480	3,257	1,582	3,480			.8	1.7	.5	1.2	.9	1.9
	83.0	*88.3*		*1.7*	*3.7*	*1,770*	*3,894*			*1.4*	*3.1*	*1,664*	*3,660*	*1,732*	*3,810*			*.9*	*2.0*	*.6*	*1.3*	*1.0*	*2.1*
108	63.9	83.5		1.3	2.8	1,674	3,683			1.0	2.3	1,566	3,446					.6	1.4	.4	.9		
	71.0	*92.8*		*1.4*	*3.1*	*1,860*	*4,092*			*1.2*	*2.6*	*1,740*	*3,829*					*.7*	*1.6*	*.5*	*1.0*		
109	72.5	62.5		1.5	3.2	1,253	2,756			1.2	2.6	1,202	2,645					.7	1.6	.5	1.0		
	80.5	*69.4*		*1.6*	*3.5*	*1,391*	*3,060*			*1.3*	*2.9*	*1,335*	*2,937*					*.8*	*1.7*	*.5*	*1.1*		
110	77.7	77.2	71.2	1.5	3.4	1,547	3,404	1.4	3.1	1.3	2.8	1,447	3,183	1,479	3,254			.7	1.6	.5	1.1		
	87.3	*86.7*	*80.0*	*1.7*	*3.8*	*1,738*	*3,823*	*1.6*	*3.5*	*1.5*	*3.2*	*1,626*	*3,577*	*1,662*	*3,656*			*.9*	*1.9*	*.5*	*1.2*		
111																							
112	84.1	90.5		1.7	3.7	1,815	3,992			1.4	3.1	1,487	3,272	1,391	3,060			1.0	2.2	.6	1.4	1.0	2.3
	93.5	*100.6*		*1.9*	*4.1*	*2,016*	*4,436*			*1.6*	*3.5*	*1,654*	*3,638*	*1,545*	*3,400*			*1.1*	*2.4*	*.7*	*1.5*	*1.1*	*2.5*
113	84.3	86.7		1.7	3.7	1,738	3,823			1.4	3.0	1,673	3,680	1,659	3,650								
	93.2	*95.8*		*1.9*	*4.1*	*1,920*	*4,224*			*1.5*	*3.4*	*1,850*	*4,070*	*1,835*	*4,038*								
114	64.8	63.3		1.3	2.9	1,269	2,791			1.0	2.3	1,184	2,605										
	72.3	*70.7*		*1.5*	*3.2*	*1,417*	*3,117*			*1.2*	*2.6*	*1,321*	*2,907*										
115	61.6	55.5		1.1	2.4	1,112	2,447			1.0	2.2	1,068	2,349	871	1,917			.7	1.5	.5	1.0	.7	1.5
	80.0	*72.1*		*1.5*	*3.2*	*1,445*	*3,178*			*1.3*	*2.9*	*1,387*	*3,051*	*1,132*	*2,490*			*.9*	*1.9*	*.6*	*1.3*	*.9*	*1.9*
116	7.2		8.0	.1	.3			.2	.4	.1	.3					.1	.3	.1	.2	.05	.1	.1	.2
	71.5		*80.0*	*1.5*	*3.2*			*1.6*	*3.5*	*1.2*	*2.6*					*1.3*	*2.9*	*.7*	*1.5*	*.5*	*1.0*	*.7*	*1.6*
117	65.5	68.3	74.3	1.3	2.9	1,369	3,012	1.5	3.3	1.1	2.4	1,288	2,833	255	561	1.2	2.7	.6	1.5	.4	.9	.7	1.6
	72.0	*75.1*	*81.6*	*1.5*	*3.2*	*1,505*	*3,310*	*1.6*	*3.6*	*1.2*	*2.6*	*1,415*	*3,113*	*280*	*617*	*1.4*	*3.0*	*.7*	*1.6*	*.5*	*1.0*	*.8*	*1.8*
118	70.4	69.5		1.4	3.1	1,393	3,064			1.2	2.6	1,310	2,881	299	658			.8	1.9	.5	1.2	.7	1.7
	76.5	*75.6*		*1.5*	*3.4*	*1,514*	*3,331*			*1.3*	*2.8*	*1,424*	*3,132*	*325*	*715*			*.9*	*2.0*	*.6*	*1.3*	*.8*	*1.8*
119	61.1	55.7		1.2	2.7	1,117	2,458			1.0	2.3			889	1,955			.7	1.6	.5	1.0	.8	1.8
	81.5	*74.3*		*1.6*	*3.6*	*1,490*	*3,277*			*1.4*	*3.0*			*1,185*	*2,607*			*1.0*	*2.1*	*.6*	*1.3*	*1.1*	*2.4*
120	75.9	76.3		1.5	3.3	1,529	3,364			1.2	2.7	1,413	3,108										
	81.0	*81.5*		*1.6*	*3.6*	*1,633*	*3,593*			*1.3*	*2.9*	*1,509*	*3,320*										
121	25.7	24.5		.5	1.1	491	1,080			.4	.9	466	1,026										
	84.1	*80.1*		*1.7*	*3.7*	*1,605*	*3,532*			*1.4*	*3.0*	*1,524*	*3,353*										
122																							
123	76.0			1.5	3.4					1.5	3.2												
	84.0			*1.7*	*3.7*					*1.6*	*3.5*												
124	60.5	55.7	57.9	1.2	2.7	1,141	2,511	1.0	2.3	1.0	2.2	1,055	2,320	521	1,146			.6	1.4	.4	.9	.7	1.5
	68.0	*64.0*	*65.0*	*1.4*	*3.0*	*1,282*	*2,821*	*1.2*	*2.6*	*1.1*	*2.5*	*1,185*	*2,607*	*585*	*1,288*			*.7*	*1.5*	*.5*	*1.0*	*.8*	*1.7*
125	78.3	81.8	78.3	1.6	3.5	1,640	3,607	1.6	3.5	1.3	2.9	1,535	3,378	1,396	3,071			.9	1.9	.5	1.3	.9	2.1
	88.0	*91.9*	*88.0*	*1.8*	*3.9*	*1,842*	*4,053*	*1.8*	*3.9*	*1.5*	*3.2*	*1,725*	*3,796*	*1,569*	*3,451*			*1.0*	*2.2*	*.6*	*1.4*	*1.0*	*2.3*
126	77.4	80.4		1.5	3.4	1,611	3,545			1.3	2.9	1,498	3,295					.9	1.9	.5	1.3	.9	2.0
	88.0	*91.4*		*1.8*	*3.9*	*1,832*	*4,030*			*1.5*	*3.2*	*1,702*	*3,744*					*1.0*	*2.2*	*.6*	*1.4*	*1.0*	*2.3*

TABLE 4-101 COMPOSITION OF FEEDS—

Feed #	Feed Name—Description	Moisture Basis: As Fed or M-F (moisture-free)	Dry Matter	Macrominerals: Calcium (Ca)	Phosphorus (P)	Sodium (Na)	Chlorine (Cl)	Magnesium (Mg)	Potassium (K)	Sulfur (S)	Microminerals: Cobalt (Co)	Copper (Cu)	Iron (Fe)	Manganese (Mn)	Zinc (Zn)
			(%)	(%)	(%)	(%)	(%)	(%)	(%)	(%)	(ppm or mg/kg)	(ppm or mg/kg)	(%)	(ppm or mg/kg)	(ppm or mg/kg)
ENERGY FEEDS															
106	Screenings, grain	As Fed	90.3	.20	.20										
		M-F	*100.0*	*.22*	*.22*										
107	Sorghum, grain, all analyses	As Fed	89.0	.04	.31	.04	.09	.17	.34	.16	.123	9.6		14.5	13.7
		M-F	*100.0*	*.05*	*.35*	*.05*	*.10*	*.19*	*.38*	*.18*	*.138*	*10.8*		*16.3*	*15.4*
108	Sorghum, kafir, grain	As Fed	90.0	.04	.33	.06	.10	.15	.34	.16	.391	6.3	.009	15.8	
		M-F	*100.0*	*.04*	*.37*	*.07*	*.11*	*.17*	*.38*	*.18*	*.434*	*7.0*	*.010*	*17.6*	
109	Sorghum, milo, head chops	As Fed	90.0	.12	.25										
		M-F	*100.0*	*.13*	*.28*										
110	Sorghum, milo, grain	As Fed	89.0	.04	.29	.01	.08	.20	.35		.089	14.1	.004	12.9	
		M-F	*100.0*	*.04*	*.33*	*.01*	*.09*	*.22*	*.39*		*.100*	*15.8*	*.005*	*14.5*	
111	Soybean oil	As Fed	100.0												
		M-F	*100.0*												
112	Soybean seeds	As Fed	90.0	.25	.59	.12	.03	.28	1.59	.22		15.7	.008	29.5	
		M-F	*100.0*	*.28*	*.66*	*.13*	*.03*	*.31*	*1.77*	*.24*		*17.4*	*.009*	*32.8*	
113	Starch, corn	As Fed	90.4												
		M-F	*100.0*												
114	Spelt grain	As Fed	89.6												
		M-F	*100.0*												
115	Sugar beet molasses	As Fed	77.0	.16	.03	1.17	1.25	.23	4.77		.400	17.6	.008	4.6	
		M-F	*100.0*	*.21*	*.04*	*1.52*	*1.62*	*.30*	*6.20*		*.500*	*22.9*	*.010*	*6.0*	
116	Sugar beet pulp, wet	As Fed	10.0	.09	.01				.02						
		M-F	*100.0*	*.90*	*.10*				*.20*						
117	Sugar beet pulp, dried	As Fed	91.0	.68	.10	.17	.04	.27	.21	.20	.091	12.5	.030	35.0	.7
		M-F	*100.0*	*.75*	*.11*	*.19*	*.04*	*.30*	*.23*	*.22*	*.100*	*13.7*	*.033*	*38.5*	*.8*
118	Sugar beet pulp w/ molasses, dried	As Fed	92.0	.56	.10			.13	1.64	.39					
		M-F	*100.0*	*.61*	*.11*			*.14*	*1.78*	*.42*					
119	Sugarcane molasses	As Fed	75.0	.89	.08	.15	2.79	.35	2.38	.35	.912	59.6	.019	42.2	
		M-F	*100.0*	*1.19*	*.11*	*.20*	*3.72*	*.47*	*3.17*	*.46*	*1.216*	*79.4*	*.025*	*56.3*	
120	Sunflower seeds	As Fed	93.6	.17	.52				.66				.003	21.6	
		M-F	*100.0*	*.18*	*.56*				*.71*				*.003*	*23.1*	
121	Sweet potato roots, dried	As Fed	30.6	.03	.05	.01		.05	.31	.04		1.3	.001	3.4	
		M-F	*100.0*	*.10*	*.15*	*.05*		*.16*	*1.01*	*.13*		*4.2*	*.005*	*11.1*	
122	Tallow	As Fed	96.0												
		M-F	*100.0*												
123	Triticale grain	As Fed	90.0	.05	.27	.01		.11	.48						
		M-F	*100.0*	*.06*	*.30*	*.01*		*.12*	*.53*						
124	Wheat bran	As Fed	89.0	.14	1.17	.06	.06	.55	1.24			12.3	.017	115.7	
		M-F	*100.0*	*.16*	*1.32*	*.07*	*.07*	*.62*	*1.39*			*13.8*	*.019*	*130.0*	
125	Wheat grain	As Fed	89.0	.05	.36	.09	.08	.16	.52		.080	7.2	.005	48.8	13.7
		M-F	*100.0*	*.06*	*.41*	*.10*	*.09*	*.18*	*.58*		*.090*	*8.1*	*.006*	*54.7*	*15.4*
126	Wheat grain, min 60 lb/bu	As Fed	88.0												
		M-F	*100.0*												

NERGY FEEDS

Feed #	Fat-Soluble Vitamins				Water-Soluble Vitamins								
	A (1 mg car. = 1,667 IU-A)	Carotene (Pro-vitamin A)	E (α-tocopherol)	K	B_{12}	Biotin	Choline	Folic Acid (Folacin)	Niacin (Nicotinic Acid)	Pantothenic Acid	Pyridoxine (B_6)	Riboflavin (B_2)	Thiamin (B_1)
		(ppm or mg/kg)	*(ppm or mg/kg)*	*(ppm or mg/kg)*	*(ppm or mg/kg)*	*(ppm or mg/kg)*	*(ppm or mg/kg)*	*(ppm or mg/kg)*	*(ppm or mg/kg)*	*(ppm or mg/kg)*	*(ppm or mg/kg)*	*(ppm or mg/kg)*	*(ppm or mg/kg)*
106									47.1	22.8	2.37	.6	.4
									52.2	*25.3*	*2.63*	*.7*	*.5*
107						2.60	678	.20	43.1	11.1	5.30	1.3	4.1
						2.92	*762*	*.22*	*48.4*	*12.5*	*5.96*	*1.5*	*4.6*
108						.23		.20					
						.26		*.22*					
109	1,000	.6										2.0	
	1,167	*.7*										*2.2*	
110			12.0			.25	678	.22	42.7	11.4	4.10	1.2	3.9
			13.5			*.29*	*761*	*.24*	*48.0*	*12.8*	*4.60*	*1.3*	*4.4*
111													
112	751	.5											
	834	*.5*											
113													
114									47.6				
									53.1				
115							862		42.2	4.6		2.4	
							1,119		*54.8*	*6.0*		*3.1*	
116													
117	303	.2					829		16.3	1.5		.7	.4
	333	*.2*					*912*		*17.9*	*1.6*		*.8*	*.4*
118	306	.2					823		16.3	1.6		.6	
	333	*.2*					*895*		*17.7*	*1.7*		*.7*	
119							876		34.3	38.3		3.3	.9
							1,167		*45.7*	*51.1*		*4.4*	*1.2*
120													
121	91,518	54.9							6.2			.6	1.0
	299,226	*179.5*							*20.4*			*2.0*	*3.4*
122													
123							462					.4	
							513					*.5*	
124			10.8			.48	988	1.80	209.2	29.0	10.00	3.1	7.9
			12.1			*.54*	*1,110*	*2.00*	*235.1*	*32.6*	*11.24*	*3.5*	*8.9*
125			15.5			.10	830	.40	56.6	12.1		1.2	4.9
			17.4			*.11*	*933*	*.45*	*63.6*	*13.6*		*1.3*	*5.5*
126													

TABLE 4-101 COMPOSITION OF FEEDS

Feed #	Feed Name—Description	Moisture Basis: As Fed or M-F (moisture-free)	Proximate Analysis: Dry Matter	Ash	Crude Fiber	Ether Extract (Fat)	N-Free Extract	Crude Protein	Digestible Protein: Ruminant	Non-Ruminant	Horse Rabbit
			(%)	(%)	(%)	(%)	(%)	(%)	(%)	(%)	(%)
ENERGY FEEDS											
127	Wheat grain, Pacific Coast	As Fed	89.2	1.9	2.7	2.0	72.8	9.1	7.7		
		M-F	*100.0*	*2.1*	*3.0*	*2.2*	*81.6*	*11.1*	*8.6*		
128	Wheat, durum, grain *(Triticum durum)*	As Fed	89.5	1.8	2.2	2.0	70.1	13.4	9.3		9.3
		M-F	*100.0*	*2.0*	*2.5*	*2.2*	*78.3*	*15.0*	*10.3*		*10.4*
129	Wheat middlings (red dog flour)	As Fed	89.0	2.5	2.0	3.6	62.9	18.0	13.3	14.1	12.0
		M-F	*100.0*	*2.8*	*2.2*	*4.0*	*70.8*	*20.2*	*15.0*	*15.8*	*13.5*
130	Wheat middlings, standard	As Fed	90.0	4.3	8.0	4.9	55.6	17.2	12.9	14.4	13.7
		M-F	*100.0*	*4.8*	*8.9*	*5.4*	*61.8*	*19.1*	*14.3*	*16.0*	*15.2*
131	Wheat millrun	As Fed	90.0	5.1	8.0	4.2	57.4	15.3	10.4		11.8
		M-F	*100.0*	*5.7*	*8.9*	*4.7*	*63.7*	*17.0*	*11.6*		*13.1*
132	Wheat shorts	As Fed	90.0	4.1	5.0	5.0	57.5	18.4	14.4		12.4
		M-F	*100.0*	*4.5*	*5.6*	*5.5*	*64.0*	*20.4*	*16.0*		*13.8*
133	Wood molasses	As Fed	62.4	3.2	0	.1	58.5	.6	0		0
		M-F	*100.0*	*5.0*	*0*	*.2*	*93.8*	*1.0*	*0*		*0*

NERGY FEEDS

Feed #	TDN			Digestible Energy						Metabolizable Energy								Net Energy-Ruminant					
												Non-Ruminant											
	Ruminant	Non-Ruminant	Horse Rabbit	Ruminant		Non-Ruminant		Horse Rabbit		Ruminant		ME		Chicken ME_n		Horse Rabbit		NE_m		NE_g		NE_{lc}	
	(%)	(%)	(%)	Mcal per		kcal per		Mcal per		Mcal per		kcal per		kcal per		Mcal per		Mcal per		Mcal per		Mcal per	
				lb	kg	lb	kg	lb	kg	lb	kg	lb	kg	lb	kg	lb	kg	lb	kg	lb	kg	lb	kg
27	78.0			1.6	3.5					1.3	2.9							.9	1.9	.5	1.3	.9	2.1
	88.0			*1.8*	*3.9*					*1.5*	*3.2*							*1.0*	*2.2*	*.6*	*1.4*	*1.0*	*2.3*
28	77.3	81.5		1.5	3.4	1,633	3,593			1.3	2.8	1,521	3,346										
	86.4	*91.1*		*1.7*	*3.8*	*1,826*	*4,017*			*1.4*	*3.1*	*1,700*	*3,739*										
29	80.7	74.1	75.7	1.6	3.6	1,458	3,208	1.5	3.3	1.3	2.9	1,348	2,965	1,253	2,756								
	90.7	*83.3*	*85.0*	*1.8*	*4.0*	*1,638*	*3,604*	*1.7*	*3.8*	*1.5*	*3.3*	*1,514*	*3,331*	*1,408*	*3,097*								
30	72.3	69.6		1.4	3.1	1,395	3,069			1.2	2.6	1,283	2,822	838	1,843			.8	1.8	.5	1.2	.9	1.9
	80.3	*77.3*		*1.6*	*3.5*	*1,549*	*3,408*			*1.3*	*2.9*	*1,425*	*3,135*	*931*	*2,048*			*.9*	*2.0*	*.6*	*1.3*	*1.0*	*2.1*
31	72.9	62.0		1.5	3.2	1,243	2,734			1.2	2.7	1,150	2,529	802	1,764			.8	1.7	.5	1.1	.9	1.9
	81.0	*68.9*		*1.6*	*3.6*	*1,381*	*3,038*			*1.4*	*3.0*	*1,277*	*2,810*	*891*	*1,960*			*.9*	*1.9*	*.6*	*1.3*	*1.0*	*2.1*
32	77.1	68.6		1.5	3.4	1,329	2,924			1.3	2.8	1,227	2,700	1,203	2,646			.9	1.9	.5	1.2	.9	2.0
	85.7	*76.2*		*1.7*	*3.8*	*1,477*	*3,249*			*1.4*	*3.1*	*1,364*	*3,000*	*1,336*	*2,940*			*1.0*	*2.1*	*.6*	*1.4*	*1.0*	*2.2*
33	54.3	52.8		1.1	2.4	1,059	2,329			.9	2.0	1,014	2,231										
	87.0	*84.7*		*1.7*	*3.8*	*1,698*	*3,735*			*1.4*	*3.1*	*1,627*	*3,579*										

TABLE 4-101 COMPOSITION OF FEEDS—

Feed #	Feed Name—Description	Moisture Basis: As Fed or M-F (moisture-free)	Dry Matter	Macrominerals							Microminerals				
				Calcium (Ca)	Phosphorus (P)	Sodium (Na)	Chlorine (Cl)	Magnesium (Mg)	Potassium (K)	Sulfur (S)	Cobalt (Co)	Copper (Cu)	Iron (Fe)	Manganese (Mn)	Zinc (Zn)
ENERGY FEEDS			*(%)*	*(%)*	*(%)*	*(%)*	*(%)*	*(%)*	*(%)*	*(%)*	*(ppm or mg/kg)*	*(ppm or mg/kg)*	*(%)*	*(ppm or mg/kg)*	*(ppm or mg/kg)*
127	Wheat grain, Pacific Coast	As Fed	89.2	.12	.30										
		M-F	*100.0*	*.14*	*.34*										
128	Wheat durum, grain	As Fed	89.5	.15	.40	0			.46			7.7	.004	28.7	
		M-F	*100.0*	*.17*	*.45*	*0*			*.51*			*8.6*	*.004*	*32.1*	
129	Wheat middlings, (red dog flour)	As Fed	89.0	.08	.52	.66	.03	.29	.60	.24	.114	4.4	.006	37.6	64.9
		M-F	*100.0*	*.09*	*.58*	*.74*	*.03*	*.33*	*.67*	*.27*	*.128*	*4.9*	*.007*	*42.3*	*72.9*
130	Wheat middlings, standard	As Fed	90.0	.14	.91	.22	.03	.37	.97	.17	.090	22.0	.009	118.3	
		M-F	*100.0*	*.16*	*1.01*	*.24*	*.03*	*.41*	*1.08*	*.19*	*.100*	*24.4*	*.010*	*131.5*	
131	Wheat millrun	As Fed	90.0	.09	1.02	.22		.51	1.28		.180	18.7	.009	102.7	
		M-F	*100.0*	*.10*	*1.13*	*.24*		*.57*	*1.42*		*.200*	*20.8*	*.010*	*114.1*	
132	Wheat shorts	As Fed	90.0	.11	.76	.07	.07	.26	.85		.090	9.2	.010	104.5	111.0
		M-F	*100.0*	*.12*	*.84*	*.08*	*.08*	*.29*	*.94*		*.100*	*10.3*	*.011*	*116.1*	*123.3*
133	Wood molasses	As Fed	62.4	1.45	.03	.03	.12	.07	.04	.03				12.6	
		M-F	*100.0*	*2.33*	*.05*	*.05*	*.20*	*.11*	*.06*	*.05*				*20.3*	

ENERGY FEEDS

Feed #	Fat-Soluble Vitamins				Water-Soluble Vitamins								
	A (1 mg car. = 1,667 IU-A)	Carotene (Pro-vitamin A)	E (α-tocopherol)	K	B_{12}	Biotin	Choline	Folic Acid (Folacin)	Niacin (Nicotinic Acid)	Pantothenic Acid	Pyridoxine (B_6)	Riboflavin (B_2)	Thiamin (B_1)
		(ppm or mg/kg)	*(ppm or mg/kg)*	*(ppm or mg/kg)*	*(ppm or mg/kg)*	*(ppm or mg/kg)*	*(ppm or mg/kg)*	*(ppm or mg/kg)*	*(ppm or mg/kg)*	*(ppm or mg/kg)*	*(ppm or mg/kg)*	*(ppm or mg/kg)*	*(ppm or mg/kg)*
127													
128								.39	45.3			1.3	6.3
								.44	*50.6*			*1.4*	*7.0*
129			57.6				1,100	1.10	52.6	13.6	11.00	1.5	18.9
			64.7				*1,247*	*1.25*	*59.1*	*15.3*	*12.47*	*1.7*	*21.2*
130	5,168	3.1	21.1				1,168	.99	99.8	16.7		2.0	16.7
	5,668	*3.4*	*23.4*				*1,298*	*1.10*	*110.9*	*18.6*		*2.2*	*18.6*
131							981		112.0	13.2		2.4	15.2
							1,090		*124.4*	*14.7*		*2.7*	*16.9*
132			29.9			.37	928	1.10	94.6	17.6	11.00	2.0	15.8
			31.7			*.41*	*1,031*	*1.22*	*105.1*	*19.6*	*12.22*	*2.2*	*17.6*
133													

TABLE 4-101 COMPOSITION OF FEEDS—

Feed #	Feed Name—Description	Moisture Basis: As Fed or M-F (moisture-free)	Proximate Analysis: Dry Matter	Ash	Crude Fiber	Ether Extract (Fat)	N-Free Extract	Crude Protein	Digestible Protein: Ruminant	Non-Ruminant	Horse Rabbit
PROTEIN FEEDS			(%)	(%)	(%)	(%)	(%)	(%)	(%)	(%)	(%)
134	Ammoniated citrus molasses *(Citrus spp)*	As Fed	60.7	4.7	0	2.1	32.5	21.4			
		M-F	*100.0*	*7.8*	*0*	*3.5*	*53.5*	*35.2*			
135	Ammoniated citrus pulp	As Fed	87.2	4.6	13.2	5.5	51.8	12.1	8.3		8.6
		M-F	*100.0*	*5.3*	*15.1*	*6.4*	*59.4*	*13.8*	*9.5*		*9.9*
136	Ammoniated cottonseed meal *(Gossypium spp)*	As Fed	92.0	6.7	14.3	3.8		44.9			
		M-F	*100.0*	*7.3*	*15.5*	*4.2*		*48.9*			
137	Ammoniated furfural *(Zea mays)*	As Fed	94.3	4.9	51.7	.4	2.4	34.9			
		M-F	*100.0*	*5.3*	*54.8*	*.4*	*2.5*	*37.0*			
138	Ammoniated molasses *(Saccharum officinarum)*	As Fed	66.4	6.0	0	0	35.8	24.6	14.5		
		M-F	*100.0*	*9.1*	*0*	*0*	*53.8*	*37.1*	*21.9*		
139	Ammoniated rice hulls *(Oryza sativa)*	As Fed	92.0	19.1	44.7	.9	16.9	10.4	4.4		
		M-F	*100.0*	*20.7*	*48.6*	*1.0*	*18.4*	*11.3*	*4.8*		
140	Ammoniated sugar beet pulp *(Beta saccharifera)*	As Fed	93.0	11.4	15.2	.4	48.9	17.4	14.1		
		M-F	*100.0*	*12.2*	*16.3*	*.5*	*52.4*	*18.7*	*15.1*		
141	Ammoniated polyphosphate	As Fed	60.0					62.5			
		M-F	*100.0*					*104.2*			
142	Ammonium phosphate, dibasic	As Fed	97.0	34.5				112.5			
		M-F	*100.0*	*35.6*				*115.9*			
143	Ammonium phosphate, monobasic	As Fed	97.0	34.5				68.8			
		M-F	*100.0*	*35.6*				*70.9*			
144	Barley malt sprouts *(Hordeum vulgare)*	As Fed	91.9	6.2	14.4	1.4	44.2	25.7	23.4		
		M-F	*100.0*	*6.7*	*15.6*	*1.6*	*48.1*	*28.0*	*25.5*		
145	Beans, kidney *(Phaseolus vulgaris)*	As Fed	88.8	3.7	4.2	1.3	57.7	21.9	14.7		
		M-F	*100.0*	*4.2*	*4.7*	*1.5*	*64.9*	*24.7*	*16.5*		
146	Beans, mung *(Phaseolus aureus)*	As Fed	89.8	3.7	3.9	1.3	57.0	23.9			
		M-F	*100.0*	*4.2*	*4.3*	*1.4*	*63.5*	*26.6*			
147	Beans, navy *(Phaseolus vulgaris)*	As Fed	90.0	4.1	4.2	1.4	57.4	22.9			
		M-F	*100.0*	*4.6*	*4.7*	*1.6*	*63.7*	*25.4*			
148	Beans, pinto *(Phaseolus spp)*	As Fed	90.3	4.3	4.0	1.3	57.9	22.8	19.8		
		M-F	*100.0*	*4.8*	*4.5*	*1.4*	*64.1*	*25.2*	*21.9*		
149	Blood meal	As Fed	91.0	4.9	1.0	1.4	3.8	79.9	58.0	63.7	
		M-F	*100.0*	*5.4*	*1.1*	*1.5*	*4.2*	*87.8*	*63.7*	*70.0*	
150	Brewers' grains, wet	As Fed	23.8	1.1	3.8	1.5	11.9	5.5	4.0		
		M-F	*100.0*	*4.8*	*16.1*	*6.5*	*49.6*	*23.0*	*16.8*		
151	Brewers' grains, dried	As Fed	92.0	3.6	15.0	6.2	41.3	25.9	19.1	19.1	20.7
		M-F	*100.0*	*3.8*	*16.3*	*6.7*	*45.0*	*28.2*	*20.8*	*20.8*	*22.5*
152	Buttermilk, condensed	As Fed	29.2	3.6	.1	2.4	12.3	10.8	9.7		
		M-F	*100.0*	*12.4*	*.2*	*8.1*	*42.4*	*36.9*	*33.2*		
153	Buttermilk, dried	As Fed	93.0	9.0	.4	5.2	46.6	31.8	28.6	29.5	
		M-F	*100.0*	*9.6*	*.4*	*5.6*	*50.2*	*34.2*	*30.7*	*31.8*	
154	Casein, dried	As Fed	90.7	2.9	.2	.8	5.2	81.6	79.1	75.9	
		M-F	*100.0*	*3.2*	*.2*	*.9*	*5.7*	*90.0*	*87.3*	*83.7*	

ROTEIN FEEDS

Feed #	TDN			Digestible Energy						Metabolizable Energy								Net Energy-Ruminant					
	Rumi-nant	Non-Rumi-nant	Horse Rabbit	Rumi-nant		Non-Rumi-nant		Horse Rabbit		Rumi-nant		Non-Ruminant ME		Non-Ruminant Chicken ME_n		Horse Rabbit		NE_m		NE_g		NE_{lc}	
	(%)	(%)	(%)	Mcal per		kcal per		Mcal per		Mcal per		kcal per		kcal per		Mcal per		Mcal per		Mcal per		Mcal per	
				lb	kg	lb	kg	lb	kg	lb	kg	lb	kg	lb	kg	lb	kg	lb	kg	lb	kg	lb	kg
34	53.5	57.0		1.1	2.4	1,142	2,513			.9	1.9	937	2,061										
	88.2	93.9		1.8	3.9	1,882	4,141			1.5	3.2	1,543	3,395										
35	68.8			1.4	3.0					1.1	2.5												
	78.9			1.6	3.5					1.3	2.9												
36	66.0	67.0		1.3	2.9	1,340	2,955			1.1	2.4	1,181	2,605	892	1,966			.7	1.5	.4	.9	.7	1.6
	72.0	73.0		1.4	3.2	1,457	3,212			1.2	2.6	1,284	2,831	969	2,137			.7	1.6	.5	1.0	.8	1.7
37	39.8	33.2		.8	1.8	665	1,463			.6	1.4	546	1,201										
	42.2	35.2		.9	1.9	705	1,551			.7	1.5	579	1,273										
38	49.0	50.5		1.0	2.2	1,011	2,225			.8	1.8	830	1,826										
	73.9	76.0		1.5	3.3	1,523	3,351			1.2	2.7	1,249	2,748										
39	12.4			.2	.5					.2	.4												
	13.5			.3	.6					.2	.5												
40	71.0																						
	77.0																						
41																							
42																							
43																							
44	62.9	73.1		1.3	2.8	1,465	3,222			1.0	2.3	1,323	2,911	649	1,428								
	68.5	79.5		1.4	3.0	1,594	3,507			1.1	2.5	1,440	3,168	706	1,554								
45	76.0	84.0		1.5	3.4	1,684	3,705			1.3	2.8	1,533	3,372										
	85.6	94.6		1.7	3.8	1,897	4,173			1.4	3.1	1,726	3,798										
46	76.3	84.4		1.5	3.4	1,692	3,722			1.3	2.8	1,533	3,373										
	85.0	94.0		1.7	3.8	1,884	4,144			1.4	3.1	1,707	3,756										
47	76.5	84.8		1.5	3.4	1,699	3,738			1.3	2.8	1,545	3,398	1,063	2,338			.8	1.8	.5	1.2	.9	1.9
	85.0	94.2		1.7	3.8	1,888	4,153			1.4	3.1	1,716	3,775	1,181	2,598			.9	2.0	.6	1.3	1.0	2.1
148	77.4	84.8		1.5	3.4	1,700	3,739			1.3	2.8	1,545	3,399										
	85.7	93.9		1.7	3.8	1,882	4,140			1.4	3.1	1,711	3,764										
149	60.0	68.4		1.2	2.6	1,125	2,475							1,293	2,844								
	65.9	75.2		1.3	2.9	1,236	2,719							1,420	3,125								
150	16.4	21.7		.3	.7	435	956			.3	.6	397	874										
	69.0	91.3		1.4	3.0	1,828	4,022			1.1	2.5	1,672	3,678										
151	62.1	40.4	46.9	1.2	2.7	809	1,779	.9	2.1	1.0	2.2	730	1,606	1,096	2,412	.8	1.7	.6	1.3	.4	.7	.6	1.4
	67.5	43.9	51.0	1.4	3.0	879	1,934	1.0	2.3	1.1	2.4	794	1,746	1,192	2,622	.8	1.8	.6	1.4	.4	.8	.7	1.6
152	25.6	27.8		.5	1.1	557	1,225			.4	.9	493	1,085										
	87.8	95.3		1.8	3.9	1,910	4,202			1.5	3.2	1,691	3,721										
153	81.7	77.3		1.6	3.6	1,550	3,410			1.4	3.0	1,381	3,039	1,252	2,754							1.0	2.2
	87.9	83.2		1.8	3.9	1,667	3,668			1.5	3.2	1,486	3,269	1,346	2,962							1.1	2.4
154	80.1	78.9		1.6	3.5	1,581	3,479			1.3	2.9	1,230	2,706	1,841	4,051								
	88.3	87.0		1.8	3.9	1,744	3,836			1.5	3.2	1,356	2,984	2,030	4,466								

TABLE 4-101 COMPOSITION OF FEEDS–

Feed #	Feed Name—Description	Moisture Basis: As Fed or M-F (moisture-free)	Dry Matter	Macrominerals							Microminerals				
				Calcium (Ca)	Phosphorus (P)	Sodium (Na)	Chlorine (Cl)	Magnesium (Mg)	Potassium (K)	Sulfur (S)	Cobalt (Co)	Copper (Cu)	Iron (Fe)	Manganese (Mn)	Zinc (Zn)
PROTEIN FEEDS			(%)	(%)	(%)	(%)	(%)	(%)	(%)	(%)	(ppm or mg/kg)	(ppm or mg/kg)	(%)	(ppm or mg/kg)	(ppm or mg/kg)
134	Ammoniated citrus molasses	As Fed	60.7	1.02	.16			.08							
		M-F	100.0	1.68	.26			.13							
135	Ammoniated citrus pulp	As Fed	87.2	1.66	.12			.07							
		M-F	100.0	1.90	.14			.08							
136	Ammoniated cottonseed meal	As Fed	92.0	.18	.94	.03		.55	1.31		.152	18.8	.020	22.6	
		M-F	100.0	.19	1.02	.04		.60	1.43		.165	20.4	.022	24.5	
137	Ammoniated furfural	As Fed													
		M-F	100.0												
138	Ammoniated molasses	As Fed	66.4	.81	.13										
		M-F	100.0	1.22	.20										
139	Ammoniated rice hulls	As Fed	92.0	.15	.19										
		M-F	100.0	.16	.21										
140	Ammoniated sugar beet pulp	As Fed	93.0	.30	.07	.29		.10	.62	.41	.200	.9	0.016	23.5	1.7
		M-F	100.0	.32	.08	.31		.11	.66	.44	.300	.9	0.017	25.2	1.8
141	Ammoniated polyphosphate	As Fed	60.0	.10	14.50										
		M-F	100.0	.17	24.20										
142	Ammonium phosphate, dibasic	As Fed	97.0	.50	20.00	.04	0	.45	.01	2.5		91	1.200	400	342
		M-F	100.0	.59	20.60	.04	0	.46	.01	2.6		94	1.237	412	353
143	Ammonium phosphate, monobasic	As Fed	97.0	.50	24.00	.06		.45		.7		80	1.200	400	300
		M-F	100.0	.52	24.74	.06		.46		.7		82	1.237	412	309
144	Barley malt sprouts	As Fed	91.9	.24	.77	1.34		.18	.21	.79				31.4	
		M-F	100.0	.26	.84	1.46		.19	.23	.86				34.2	
145	Beans, kidney	As Fed	88.8	.11	.40	.01			.98				.007		
		M-F	100.0	.12	.45	.01			1.10				.008		
146	Beans, mung	As Fed	89.8	.13	.34	.01			1.04				.008		
		M-F	100.0	.14	.38	.01			1.15				.009		
147	Beans, navy	As Fed	90.0	.15	.57	.05	.04	.17	1.70	.23		9.9	.009	18.5	
		M-F	100.0	.17	.63	.06	.04	.19	1.89	.26		11.0	.011	20.6	
148	Beans, pinto	As Fed	90.3	.14	.35										
		M-F	100.0	.16	.39										
149	Blood meal	As Fed	91.0	.28	.22	.32	.27	.22	.90	.33		9.9	.376	5.3	
		M-F	100.0	.31	.24	.35	.30	.24	.99	.36		10.9	.413	5.8	
150	Brewers' grains, wet	As Fed	23.8	.07	.12				.02						
		M-F	100.0	.30	.51				.08						
151	Brewers' grains, dried	As Fed	92.0	.27	.50	.26	.17	.14	.08	.30	.092	21.3	.025	37.6	
		M-F	100.0	.29	.54	.28	.19	.15	.09	.33	.100	23.1	.027	40.9	
152	Buttermilk, condensed	As Fed	29.2	.44	.26	.31	.12	.19	.23	.03					
		M-F	100.0	1.51	.89	1.06	.41	.65	.79	.10					
153	Buttermilk, dried	As Fed	93.0	1.33	.95	.80		.48	.99	.08			.001	3.6	
		M-F	100.0	1.43	1.02	.86		.52	1.06	.09			.001	3.8	
154	Casein, dried	As Fed	90.7	.61	.99									4.4	
		M-F	100.0	.67	1.10									4.9	

ROTEIN FEEDS

Feed #	Fat-Soluble Vitamins				Water-Soluble Vitamins									
	A (1 mg car. = 1,667 IU-A)	Carotene (Pro-vitamin A)	E (α-tocopherol)	K	B_{12}	Biotin	Choline	Folic Acid (Folacin)	Niacin (Nicotinic Acid)	Pantothenic Acid	Pyridoxine (B_6)	Riboflavin (B_2)	Thiamin (B_1)
		(ppm or mg/kg)	*(ppm or mg/kg)*	*(ppm or mg/kg)*	*(ppm or mg/kg)*	*(ppm or mg/kg)*	*(ppm or mg/kg)*	*(ppm or mg/kg)*	*(ppm or mg/kg)*	*(ppm or mg/kg)*	*(ppm or mg/kg)*	*(ppm or mg/kg)*	*(ppm or mg/kg)*
34													
35													
36			9.3			1.12	2,742	3.76	29.0	9.7		5.3	4.5
			10.2			*1.21*	*2,980*	*4.08*	*32.0*	*10.6*		*5.7*	*4.9*
37													
38													
39													
40	1,834	1.1											
	2,167	*1.3*											
141													
142													
143													
144									57.7			10.3	9.1
									62.8			*11.3*	*9.9*
145									22.8			2.0	5.1
									25.7			*2.2*	*5.7*
146									26.1			2.1	3.8
									29.1			*2.4*	*4.3*
147							1,764		24.5	3.1		2.0	6.6
							1,960		*27.2*	*3.4*		*2.2*	*7.3*
148													
149			0				757		31.5	1.1	1.46	1.5	
			0				*831*		*34.6*	*1.2*	*1.60*	*1.7*	
150													
151						0	1,587	.22	43.4	8.6	.66	1.5	.7
						0	*1,725*	*.24*	*47.2*	*9.3*	*.72*	*1.6*	*.8*
152												14.5	
												49.8	
153	2,167	1.3			.0184	.29	1,822	.40	8.6	30.4	2.44	26.3	3.0
	2,334	*1.4*			*.0198*	*.31*	*1,960*	*.43*	*9.3*	*32.7*	*2.62*	*28.3*	*3.2*
154						.04	210	.51	1.3	2.7	.44	1.5	.4
						.05	*232*	*.56*	*1.5*	*2.9*	*.49*	*1.7*	*.5*

TABLE 4-101 COMPOSITION OF FEEDS—

Feed #	Feed Name—Description	Moisture Basis: As Fed or M-F (moisture-free)	Proximate Analysis: Dry Matter	Ash	Crude Fiber	Ether Extract (Fat)	N-Free Extract	Crude Protein	Digestible Protein: Ruminant	Non-Ruminant	Horse Rabbit
			(%)	(%)	(%)	(%)	(%)	(%)	(%)	(%)	(%)
PROTEIN FEEDS											
155	Cheese rind	As Fed	82.8	7.5	.2	19.3	11.4	44.4			
		M-F	*100.0*	*9.0*	*.3*	*23.3*	*13.7*	*53.7*			
156	Citrus molasses, ammoniated *(Citrus spp)*	As Fed	60.7	4.7	0	2.1	32.5	21.4			
		M-F	*100.0*	*7.8*	*0*	*3.5*	*53.5*	*35.2*			
157	Citrus pulp, ammoniated	As Fed	87.2	4.6	13.2	5.5	51.8	12.1	8.3		8.6
		M-F	*100.0*	*5.3*	*15.1*	*6.4*	*59.4*	*13.8*	*9.5*		*9.9*
158	Coconut (Copra) meal, mech-extd, 20% protein *(Cocos nucifera)*	As Fed	92.7	6.6	11.3	6.7	47.4	20.7	17.1		
		M-F	*100.0*	*7.2*	*12.2*	*7.2*	*51.1*	*22.3*	*18.4*		
159	Coconut (Copra) meal, sol-extd, 21% protein	As Fed	91.2	6.3	14.0	2.1	47.5	21.3	16.5		
		M-F	*100.0*	*6.9*	*15.4*	*2.3*	*52.0*	*23.4*	*18.1*		
160	Corn distillers' grains, dried *(Zea mays)*	As Fed	92.5	1.5	12.8	7.4	43.4	27.4	18.9		
		M-F	*100.0*	*1.7*	*13.8*	*8.0*	*46.9*	*29.6*	*20.4*		
161	Corn germ meal	As Fed	93.0		12.0	2.0		18.0			
		M-F	*100.0*		*12.9*	*2.2*		*19.4*			
162	Corn gluten feed	As Fed	90.4	6.6	7.3	2.7	48.0	25.8	22.2		
		M-F	*100.0*	*7.3*	*8.0*	*2.9*	*53.2*	*28.6*	*24.6*		
163	Corn gluten meal	As Fed	91.0	2.4	4.0	2.3	39.4	42.9	35.7		42.9
		M-F	*100.0*	*2.6*	*4.4*	*2.5*	*43.4*	*47.1*	*39.2*		*47.1*
164	Cottonseed meal, mech-extd, 36% protein *(Gossypium spp)*	As Fed	93.5	6.1	15.7	6.5	25.6	39.6	32.4		
		M-F	*100.0*	*6.5*	*16.8*	*7.0*	*27.3*	*42.4*	*34.7*		
165	Cottonseed meal, mech-extd, 41% protein	As Fed	94.0	6.3	12.0	5.6	29.1	41.0	33.2		
		M-F	*100.0*	*6.7*	*12.8*	*6.0*	*30.9*	*43.6*	*35.3*		
166	Cottonseed meal, pre-press, sol-extd, 41% protein	As Fed	92.5	6.8	12.0	.9	31.8	41.0			
		M-F	*100.0*	*7.4*	*13.0*	*1.0*	*34.3*	*44.3*			
167	Cottonseed meal, sol-extd, 41% protein	As Fed	91.1	6.2	11.4	2.1	31.8	41.9	33.9		
		M-F	*100.0*	*6.8*	*12.5*	*2.3*	*34.9*	*46.0*	*37.3*		
168	Cottonseed meal, 45% protein, min.	As Fed									
		M-F	*100.0*								
169	Cottonseed meal, prepress, sol-extd, 48% protein	As Fed	92.5	7.2	8.5	1.1	26.7	50.0	40.4		
		M-F	*100.0*	*7.8*	*9.2*	*1.2*	*28.9*	*54.0*	*43.7*		
170	Cottonseed meal, glandless	As Fed	92.7	5.8	12.7	1.5	31.3	41.5			
		M-F	*100.0*	*6.3*	*13.7*	*1.6*	*33.7*	*44.8*			
171	Cottonseed meal, ammoniated	As Fed	92.0	6.7	14.3	3.8		44.9			
		M-F	*100.0*	*7.3*	*15.5*	*4.2*		*48.9*			
172	Cottonseed, whole, grnd *(Gossypium spp)*	As Fed	92.7	3.5	16.9	22.9	26.3	23.1	16.4		
		M-F	*100.0*	*3.8*	*18.2*	*24.7*	*28.4*	*24.9*	*17.7*		
173	Diammonium phosphate	As Fed	97.0	34.5				112.45			
		M-F	*100.0*	*35.6*				*115.9*			
174	Distillers' dried grain, w/o solubles	As Fed	92.5	1.6	12.8	7.4	43.4	27.3	18.9		
		M-F	*100.0*	*1.7*	*13.8*	*8.0*	*46.9*	*29.6*	*20.4*		
175	Distillers' dried corn grain, w/solubles	As Fed	93.0	7.4	4.0	9.8	44.9	26.9	21.0		
		M-F	*100.0*	*8.0*	*4.3*	*10.5*	*48.3*	*28.9*	*22.6*		

ROTEIN FEEDS

Feed #	TDN			Digestible Energy						Metabolizable Energy								Net Energy-Ruminant					
	Ruminant	Non-Ruminant	Horse Rabbit	Ruminant		Non-Ruminant		Horse Rabbit		Ruminant		Non-Ruminant ME		Non-Ruminant Chicken ME_n		Horse Rabbit		NE_m		NE_g		NE_{lc}	
	(%)	(%)	(%)	Mcal per		kcal per		Mcal per		Mcal per		kcal per		kcal per		Mcal per		Mcal per		Mcal per		Mcal per	
				lb	kg	lb	kg	lb	kg	lb	kg	lb	kg	lb	kg	lb	kg	lb	kg	lb	kg	lb	kg
155	75.6	78.4		1.5	3.3	1,571	3,457			1.2	2.7	1,339	2,945										
	91.4	*94.7*		*1.8*	*4.0*	*1,898*	*4,175*			*1.5*	*3.3*	*1,617*	*3,557*										
156	53.5	57.0		1.1	2.4	1,142	2,513			.9	1.9	937	2,061										
	88.2	*93.9*		*1.8*	*3.9*	*1,882*	*4,141*			*1.5*	*3.2*	*1,543*	*3,395*										
157	68.8			1.4	3.0					1.1	2.5												
	78.9			*1.6*	*3.5*					*1.3*	*2.9*												
158	76.4	87.4		1.5	3.4	1,752	3,855			1.3	2.8	1,603	3,527	806	1,773			.8	1.7	.5	1.2	.9	2.1
	82.5	*94.4*		*1.6*	*3.6*	*1,891*	*4,160*			*1.4*	*3.0*	*1,730*	*3,807*	*870*	*1,913*			*.9*	*1.9*	*.6*	*1.3*	*1.0*	*2.2*
159	67.2	75.8		1.4	3.0	1,519	3,341			1.1	2.4	1,386	3,050					.7	1.5	.5	1.0	.8	1.8
	73.7	*83.1*		*1.5*	*3.3*	*1,666*	*3,665*			*1.2*	*2.7*	*1,520*	*3,345*					*.8*	*1.7*	*.5*	*1.1*	*.9*	*2.0*
160	76.2	88.8		1.5	3.4	1,778	3,912			1.3	2.8	1,602	3,525										
	82.4	*95.9*		*1.6*	*3.6*	*1,920*	*4,224*			*1.4*	*3.0*	*1,731*	*3,809*										
161														773	1,700								
														831	*1,828*								
162	74.1	80.1		1.5	3.3	1,604	3,528			1.2	2.7	1,448	3,186	755	1,661			.8	1.7	.5	1.2	.9	2.1
	82.6	*88.5*		*1.6*	*3.6*	*1,772*	*3,898*			*1.4*	*3.0*	*1,601*	*3,523*	*835*	*1,837*			*.9*	*1.9*	*.6*	*1.3*	*1.0*	*2.3*
163	76.4	79.8	76.4	1.5	3.4	1,597	3,514	1.5	3.4	1.3	2.8	1,395	3,069	1,438	3,307	1.3	2.8	.8	1.8	.5	1.2	.9	2.0
	84.0	*87.7*	*84.0*	*1.7*	*3.7*	*1,755*	*3,862*	*1.7*	*3.7*	*1.4*	*3.1*	*1,533*	*3,373*	*1,652*	*3,634*	*1.4*	*3.0*	*.9*	*2.0*	*.6*	*1.3*	*1.0*	*2.2*
164	75.3	72.9		1.4	3.0	1,461	3,215			1.1	2.4	1,282	2,820					.7	1.6	.5	1.0	1.1	2.5
	80.5	*78.0*		*1.5*	*3.2*	*1,563*	*3,439*			*1.2*	*2.6*	*1,371*	*3,016*					*.8*	*1.7*	*.5*	*1.1*	*1.2*	*2.6*
165	73.3	81.0		1.5	3.3	1,528	3,361			1.2	2.7	1,331	2,929	1,068	2,349			.8	1.7	.5	1.1	.9	1.9
	78.0	*86.2*		*1.5*	*3.4*	*1,625*	*3,575*			*1.3*	*2.8*	*1,416*	*3,116*	*1,136*	*2,499*			*.8*	*1.8*	*.5*	*1.2*	*.9*	*2.0*
166	77.1			1.5	3.4	1,224	2,692			1.3	2.8	1,065	2,342	902	1,984								
	83.3			*1.7*	*3.7*	*1,323*	*2,910*			*1.4*	*3.0*	*1,151*	*2,532*	*975*	*2,145*								
167	65.5			1.5	3.4					1.2	2.6			954	2,098			.7	1.6	.5	1.0	.9	1.9
	72.0			*1.7*	*3.7*					*1.3*	*2.8*			*1,047*	*2,303*			*.8*	*1.7*	*.5*	*1.1*	*.9*	*2.0*
168																							
169	69.4			1.4	3.1					1.1	2.5							.7	1.6	.5	1.0	.8	1.7
	75.0			*1.5*	*3.3*					*1.2*	*2.7*							*.8*	*1.7*	*.5*	*1.1*	*.9*	*1.9*
170	66.8	68.4		1.3	2.9	1,370	3,014			1.1	2.4	1,191	2,621										
	72.0	*73.7*		*1.5*	*3.2*	*1,478*	*3,252*			*1.2*	*2.6*	*1,285*	*2,828*										
171	66.0	67.0		1.3	2.9	1,340	2,955			1.1	2.4	1,181	2,605	292	1,966			1.7	1.5	.4	.9	.7	1.6
	72.0	*73.0*		*1.4*	*3.2*	*1,457*	*3,212*			*1.2*	*2.6*	*1,284*	*2,831*	*969*	*2,137*			*.7*	*1.6*	*.5*	*1.0*	*.8*	*1.7*
172	88.7	109.7		1.8	3.9	2,196	4,832			1.5	3.2	1,376	4,402					.9	1.9	.5	1.1	1.1	2.4
	95.7	*118.4*		*1.9*	*4.2*	*2,370*	*5,215*			*1.6*	*3.5*	*2,159*	*4,749*					*.9*	*2.0*	*.5*	*1.2*	*1.2*	*2.6*
173																							
174	76.2	88.8		1.5	3.4	1,780	3,915			1.3	2.8	1,602	3,525										
	82.4	*95.9*		*1.6*	*3.6*	*1,922*	*4,228*			*1.4*	*3.0*	*1,731*	*3,809*										
175	81.8			1.6	3.6					1.4	3.0							.9	2.0	.6	1.3		
	88.0			*1.8*	*3.9*					*1.5*	*3.2*							*1.0*	*2.2*	*.6*	*1.4*		

TABLE 4-101 COMPOSITION OF FEEDS—

Feed #	Feed Name—Description	Moisture Basis: As Fed or M-F (moisture-free)	Dry Matter	Macrominerals							Microminerals				
				Calcium (Ca)	Phosphorus (P)	Sodium (Na)	Chlorine (Cl)	Magnesium (Mg)	Potassium (K)	Sulfur (S)	Cobalt (Co)	Copper (Cu)	Iron (Fe)	Manganese (Mn)	Zinc (Zn)
PROTEIN FEEDS			(%)	(%)	(%)	(%)	(%)	(%)	(%)	(%)	*(ppm or mg/kg)*	*(ppm or mg/kg)*	(%)	*(ppm or mg/kg)*	*(ppm or mg/kg)*
155	Cheese rind	As Fed	82.8	.96	.54	.79	.59	.02	.27						
		M-F	*100.0*	*1.15*	*.66*	*.95*	*.71*	*.03*	*.32*						
156	Citrus molasses, ammoniated	As Fed	60.7	1.02	.16			.08							
		M-F	*100.0*	*1.68*	*.26*			*.13*							
157	Citrus pulp, ammoniated	As Fed	87.2	1.66	.12			.07							
		M-F	*100.0*	*1.90*	*.14*			*.08*							
158	Coconut (Copra) meal, mech-extd. 20% protein	As Fed	92.7	.21	.62	.04		.31	1.53	.34	.128	14.1	.132	65.4	53.0
		M-F	*100.0*	*.23*	*.67*	*.04*		*.33*	*1.65*	*.37*	*.138*	*15.2*	*.142*	*70.6*	*59.0*
159	Coconut (Copra) meal, sol-extd, 21% protein	As Fed	91.2	.17	.61		.03								
		M-F	*100.0*	*.19*	*.67*		*.03*								
160	Corn distillers' grains, dried	As Fed	92.5	.16	1.06	.04	.07	.14	1.19	.45	.091	45.2	.032	53.6	194.3
		M-F	*100.0*	*.17*	*1.15*	*.04*	*.08*	*.15*	*1.28*	*.49*	*.099*	*48.9*	*.035*	*57.9*	*210.0*
161	Corn germ meal	As Fed	93.0	.10	.40				.20					16.0	
		M-F	*100.0*	*.11*	*.43*				*.22*					*17.2*	
162	Corn gluten feed	As Fed	90.4	.44	.78	.95	.22	.29	.57	.22	.088	47.9	.046	23.8	
		M-F	*100.0*	*.49*	*.86*	*1.05*	*.24*	*.32*	*.63*	*.24*	*.098*	*52.9*	*.051*	*26.4*	
163	Corn gluten meal	As Fed	91.0	.16	.40	.09		.04	.03		.091	28.2	.040	7.3	
		M-F	*100.0*	*.18*	*.44*	*.10*		*.05*	*.03*		*.100*	*31.0*	*.044*	*8.0*	
164	Cottonseed meal, mech-extd, 36% protein	As Fed	93.5	.18	1.02	.06	.02	.49	1.43	.26					
		M-F	*100.0*	*.20*	*1.09*	*.07*	*.02*	*.53*	*1.53*	*.28*					
165	Cottonseed meal, mech-extd, 41% protein	As Fed	94.0	.16	1.20	.04	.05	.56	1.40	.40	.150	19.4	.030	21.5	
		M-F	*100.0*	*.17*	*1.28*	*.04*	*.05*	*.60*	*1.49*	*.43*	*.160*	*20.7*	*.032*	*22.9*	
166	Cottonseed meal, prepress, sol-extd, 41% protein	As Fed	92.5	.16	1.20	.04		.56	1.40		1.931	19.5	.030	21.5	93.4
		M-F	*100.0*	*.17*	*1.30*	*.04*		*.61*	*1.51*		*2.088*	*21.1*	*.032*	*23.2*	*101.0*
167	Cottonseed meal, sol-extd, 41% protein	As Fed	91.1	.16	1.06	.06	.04	.47	1.26	.21	1.90	18.1	.010	21.8	60.0
		M-F	*100.0*	*.18*	*1.16*	*.07*	*.04*	*.52*	*1.38*	*.23*	*2.10*	*19.9*	*.010*	*23.9*	*66.0*
168	Cottonseed meal, 45% protein, min.	As Fed		.22	1.11			.54				11.0	.018	9.2	
		M-F	*100.0*												
169	Cottonseed meal, prepress, sol-extd, 48% protein	As Fed	92.5	.16	1.01	.05		.46	1.26		.093	17.9	.011	22.8	73.3
		M-F	*100.0*	*.17*	*1.09*	*.05*		*.50*	*1.36*		*.100*	*19.4*	*.012*	*24.6*	*79.2*
170	Cottonseed meal, glandless	As Fed	92.7												
		M-F	*100.0*												
171	Cottonseed meal, ammoniated	As Fed	92.0	.18	.94	.03		.55	1.31		.152	18.8	.020	22.6	
		M-F	*100.0*	*.19*	*1.02*	*.04*		*.60*	*1.43*		*.165*	*20.4*	*.022*	*24.5*	
172	Cottonseed, whole, grnd	As Fed	92.7	.14	.68	.29		.32	1.11	.24		50.0	.014	12.1	
		M-F	*100.0*	*.15*	*.73*	*.31*		*.35*	*1.20*	*.26*		*54.0*	*.015*	*13.1*	
173	Diammonium phosphate	As Fed	97.0	.57	20.60	.04		.45	.01	2.5		91.3	1.200	400	342
		M-F	*100.0*	*.59*	*20.60*	*.04*		*.46*	*.01*	*2.6*		*94.1*	*1.237*	*412*	*353.0*
174	Distillers' dried grain, w/o solubles	As Fed	92.5	.16	1.06	.04		.14	1.19	.45	.091	45.2	.032	53.6	194.3
		M-F	*100.0*	*.17*	*1.15*	*.04*		*.15*	*1.28*	*.49*	*.099*	*48.9*	*.035*	*57.9*	*210.0*
175	Distillers' dried corn grain, w/solubles	As Fed	93.0	.35	1.37	.24	.26	.64	1.74	.37	.196	82.7	.055	73.0	
		M-F	*100.0*	*.38*	*1.47*	*.26*	*.28*	*.69*	*1.87*	*.40*	*.211*	*88.9*	*.059*	*79.0*	

'ROTEIN FEEDS

Feed #	Fat-Soluble Vitamins				Water-Soluble Vitamins								
	A (1 mg car. = 1,667 IU-A)	Carotene (Pro-vitamin A)	E (α-tocopherol)	K	B_{12}	Biotin	Choline	Folic Acid (Folacin)	Niacin (Nicotinic Acid)	Pantothenic Acid	Pyridoxine (B_6)	Riboflavin (B_2)	Thiamin (B_1)
	(ppm or mg/kg)	*(ppm or mg/kg)*	*(ppm or mg/kg)*	*(ppm or mg/kg)*	*(ppm or mg/kg)*	*(ppm or mg/kg)*	*(ppm or mg/kg)*	*(ppm or mg/kg)*	*(ppm or mg/kg)*	*(ppm or mg/kg)*	*(ppm or mg/kg)*	*(ppm or mg/kg)*	*(ppm or mg/kg)*
155													
156													
157													
158							1,092	1.39	27.9	6.2		3.3	.7
							1,178	*1.50*	*30.2*	*6.7*		*3.6*	*.7*
159							1,263		31.0	5.8		3.5	
							1,385		*34.0*	*6.3*		*3.9*	
160	13,003	7.8							46.8	11.6		3.8	2.5
	14,003	*8.4*							*50.5*	*12.5*		*4.1*	*2.6*
161			87.0			3.00	1,800	.20	35.1	4.1		4.1	1.0
			93.5			*3.22*	*1,936*	*.21*	*37.7*	*4.4*		*4.4*	*1.1*
162	14,003	8.4				.33	1,514	.31	71.8	17.1	15.00	2.4	2.0
	15,503	*9.3*				*.37*	*1,674*	*.34*	*79.4*	*18.9*	*16.58*	*2.7*	*2.2*
163	27,339	16.4	24.0			.15	330	.19	49.9	10.3	7.99	1.5	.2
	30,006	*18.0*	*26.4*			*.16*	*363*	*.21*	*54.8*	*11.3*	*8.79*	*1.6*	*.2*
164													
165			40.0				2,780	2.30	39.5	14.0	5.30	5.0	6.5
			42.5				*2,957*	*2.45*	*42.0*	*14.9*	*5.64*	*5.3*	*6.9*
166						.53	2,860	2.30	39.5	14.0		5.0	6.5
						.57	*3,092*	*2.49*	*42.7*	*15.1*		*5.4*	*7.0*
167			15.0			.54	2,784	2.79	42.4	13.8	4.8	4.4	7.6
			16.3			*.59*	*3,056*	*3.06*	*46.5*	*15.1*	*5.3*	*4.8*	*8.3*
168													
169													
170													20.3
													21.9
171			9.3			1.12	2,742	3.76	29.0	9.7		5.3	4.5
			10.2			*1.21*	*2,980*	*4.08*	*32.0*	*10.6*		*5.7*	*4.9*
172													
173													
174	13,003	7.8							46.8	11.6		3.8	2.5
	14,003	*8.4*							*50.5*	*12.5*		*4.1*	*2.6*
175	1,167	.7	54.8			1.50	4,826	1.13	115.4	20.9		16.9	6.8
	1,337	*.8*	*58.9*			*1.61*	*5,189*	*1.21*	*124.1*	*22.5*		*18.2*	*7.3*

TABLE 4-101 COMPOSITION OF FEEDS—

Feed #	Feed Name—Description	Moisture Basis: As Fed or M-F (moisture-free)	Proximate Analysis: Dry Matter (%)	Ash (%)	Crude Fiber (%)	Ether Extract (Fat) (%)	N-Free Extract (%)	Crude Protein (%)	Digestible Protein: Ruminant (%)	Non-Ruminant (%)	Horse Rabbit (%)
PROTEIN FEEDS											
176	Distillers' dried rye grain *(Secale cereale)*	As Fed	93.0	1.9	14.0	7.4	47.3	22.4	9.7		
		M-F	*100.0*	*2.0*	*15.1*	*8.0*	*50.8*	*24.1*	*10.4*		
177	Distillers' dried solubles	As Fed	92.1	6.2	3.4	8.9	44.8	28.8			
		M-F	*100.0*	*6.8*	*3.7*	*9.6*	*48.7*	*31.2*			
178	Distillers' dried sorghum grain, w/o solubles *(Sorghum vulgare)*	As Fed	94.0	4.0	12.0	8.5	38.3	31.2	25.9		
		M-F	*100.0*	*4.3*	*12.8*	*9.0*	*40.7*	*33.2*	*27.6*		
179	Distillers' dried sorghum grain w/ solubles	As Fed	94.8	4.2	10.1	9.4	38.0	33.1			
		M-F	*100.0*	*4.4*	*10.7*	*9.9*	*40.1*	*34.9*			
180	Distillers', rye grains w/solubles, dehy *(Secale cereale)*	As Fed	90.5	6.4	8.1	4.1	44.7	27.2			
		M-F	*100.0*	*7.1*	*9.0*	*4.5*	*49.3*	*30.1*			
181	Feather meal, hydrolized	As Fed	93.2	4.2	1.5	2.5		85.0	70.1		
		M-F	*100.0*	*4.5*	*1.6*	*2.7*		*91.2*	*75.2*		
182	Fish liver meal	As Fed	92.8	6.1	1.2	17.3	5.4	62.8			
		M-F	*100.0*	*6.6*	*1.3*	*18.6*	*5.8*	*67.7*			
183	Fish meal, Anchovy	As Fed	92.0	15.0		5.3		65.6			
		M-F	*100.0*	*16.3*		*5.8*		*71.3*			
184	Fish meal, mech-extd	As Fed	88.4	20.2	1.0	5.6	2.2	59.4	52.9	54.7	
		M-F	*100.0*	*22.8*	*1.1*	*6.3*	*2.6*	*67.2*	*59.8*	*61.8*	
185	Fish meal, Menhaden	As Fed	91.4	19.0	.06	9.8	1.6	60.4	48.9		
		M-F	*100.0*	*20.8*	*.06*	*10.7*	*1.8*	*66.1*	*53.5*		
186	Fish meal, Sardine	As Fed	92.9	15.7	.9	5.4	5.7	65.2		63.2	
		M-F	*100.0*	*16.9*	*.9*	*5.8*	*6.3*	*70.1*		*68.0*	
187	Fish meal, Tuna	As Fed	93.2	22.8	.7	7.0	3.3	59.4	45.1	57.6	
		M-F	*100.0*	*24.5*	*.8*	*7.5*	*3.5*	*63.7*	*48.4*	*61.8*	
188	Fish meal, white	As Fed	90.6	24.0	.5	4.2	1.0	60.9	56.6	59.0	
		M-F	*100.0*	*26.5*	*.6*	*4.6*	*1.1*	*67.2*	*62.5*	*65.2*	
189	Fish solubles, condensed	As Fed	50.1	9.4	.1	7.7	2.6	30.3		29.1	
		M-F	*100.0*	*18.7*	*.2*	*15.4*	*5.2*	*60.5*		*58.1*	
190	Fish solubles, dried	As Fed	93.7	14.9	.5	9.3	3.5	65.5			
		M-F	*100.0*	*15.9*	*.5*	*9.9*	*3.8*	*69.9*			
191	Flax (see Linseed meal)										
192	Flax seed, whole *(Linum usitatissimum)*	As Fed	95.7					17.8			
		M-F	*100.0*					*18.6*			
193	Furfural, ammoniated *(Zea mays)*	As Fed	94.3	4.9	51.7	.4	2.4	34.9			
		M-F	*100.0*	*5.3*	*54.8*	*.4*	*2.5*	*37.0*			
194	Horse bean, seeds *(Vicia faba equina)*	As Fed	87.3	3.5	7.7	1.3	49.3	25.5			19.4
		M-F	*100.0*	*4.0*	*8.8*	*1.5*	*56.5*	*29.2*			*22.2*
195	Linseed meal, mech-extd, 33% protein *(Linum usitatissimum)*	As Fed	89.9	6.2	8.1	8.4	33.6	33.6	31.5		
		M-F	*100.0*	*6.9*	*8.9*	*9.4*	*37.4*	*37.4*	*35.0*		
196	Linseed meal, mech- or sol-extd, 35% protein	As Fed	90.6	5.7	8.8	3.4	36.8	35.9	31.2	32.3	
		M-F	*100.0*	*6.2*	*9.7*	*3.7*	*40.6*	*39.8*	*34.4*	*35.4*	

ROTEIN FEEDS

Feed #	TDN Ruminant (%)	TDN Non-Ruminant (%)	TDN Horse Rabbit (%)	DE Ruminant Mcal per lb	DE Ruminant Mcal per kg	DE Non-Ruminant kcal per lb	DE Non-Ruminant kcal per kg	DE Horse Rabbit Mcal per lb	DE Horse Rabbit Mcal per kg	ME Ruminant Mcal per lb	ME Ruminant Mcal per kg	ME Non-Ruminant ME kcal per lb	ME Non-Ruminant ME kcal per kg	ME Non-Ruminant Chicken ME_n kcal per lb	ME Non-Ruminant Chicken ME_n kcal per kg	ME Horse Rabbit Mcal per lb	ME Horse Rabbit Mcal per kg	NE_m Mcal per lb	NE_m Mcal per kg	NE_g Mcal per lb	NE_g Mcal per kg	NE_{lc} Mcal per lb	NE_{lc} Mcal per kg
176	42.8	62.6		1.0	2.3	1,254	2,759			.7	1.5	1,148	2,525					.4	.9	.05	.1		
	46.0	67.3		1.1	2.5	1,349	2,967			.8	1.7	1,234	2,715					.5	1.0	.05	.1		
177	80.1	83.4		1.6	3.5	1,672	3,679			1.3	2.9	1,500	3,301	1,386	3,049								
	87.0	90.6		1.7	3.8	1,816	3,995			1.4	3.1	1,629	3,584	1,505	3,311								
178	78.0	88.3		1.5	3.4	1,769	3,892			1.3	2.8	1,579	3,474					.8	1.8	.5	1.2	.9	2.0
	83.0	93.9		1.7	3.7	1,882	4,140			1.4	3.0	1,680	3,696					.9	1.9	.6	1.3	1.0	2.1
179	80.6	91.8		1.6	3.6	1,840	4,047			1.3	2.9	1,636	3,599										
	85.1	96.8		1.7	3.8	1,940	4,269			1.4	3.1	1,726	3,797										
180	57.4	60.1		1.1	2.5	1,205	2,650			1.0	2.1	988	2,173										
	63.5	66.4		1.3	2.8	1,331	2,928			1.0	2.3	1,091	2,401										
181	76.0	63.0		1.5	3.3	1,261	2,775			1.1	2.4	1,032	2,270	1,070	2,354								
	81.5	68.0		1.6	3.6	1,361	2,995			1.1	2.5	1,107	2,436	1,148	2,526								
182	80.2	94.5		1.6	3.5	1,894	4,167			1.3	2.9	1,560	3,431										
	86.5	101.8		1.7	3.8	2,041	4,490			1.4	3.1	1,680	3,697										
183														1,200	2,640								
														1,305	2,870								
184	61.9	63.5		1.2	2.7	1,274	2,802			1.0	2.2	1,050	2,309	1,072	2,359								
	70.1	71.9		1.4	3.1	1,441	3,170			1.1	2.5	1,188	2,613	1,213	2,669								
185	63.0	69.7		1.3	2.8	1,492	3,282			1.0	2.3	1,292	2,843	1,317	2,898								
	68.9	76.2		1.4	3.0	1,632	3,590			1.1	2.5	1,414	3,110	1,441	3,171								
186	66.4	65.0		1.3	2.9	1,304	2,868			1.1	2.4	1,067	2,347	1,322	2,908								
	71.5	70.0		1.5	3.2	1,403	3,086			1.2	2.6	1,148	2,525	1,422	3,129								
187	62.0	63.8		1.2	2.7	1,279	2,813			1.0	2.2	1,063	2,339	884	1,944								
	66.5	68.4		1.3	2.9	1,371	3,017			1.1	2.4	1,140	2,508	948	2,086								
188	64.8	63.6		1.3	2.9	1,275	2,804			1.0	2.3	1,050	2,310	1,200	2,640								
	71.1	70.2		1.5	3.2	1,407	3,095			1.2	2.6	1,159	2,550	1,325	2,914								
189	39.7	44.6		.8	1.8	893	1,965			.6	1.4	748	1,646	671	1,477								
	79.2	89.0		1.6	3.5	1,783	3,922			1.3	2.9	1,494	3,286	1,340	2,948								
190	71.2	73.3		1.4	3.1	1,469	3,232			1.2	2.6	1,203	2,646	1,295	2,850								
	76.0	78.2		1.5	3.4	1,568	3,449			1.3	2.8	1,284	2,824	1,382	3,041								
191																							
192																							
193	39.8	33.2		.8	1.8	665	1,463			.6	1.4	589	1,295										
	42.2	35.2		.9	1.9	705	1,551			.7	1.5	624	1,373										
194	72.2	77.2	69.9	1.5	3.2	1,546	3,402	1.4	3.1	1.2	2.6	1,393	3,065			1.1	2.5						
	82.8	88.4	80.1	1.7	3.7	1,772	3,899	1.6	3.5	1.4	3.0	1,597	3,513			1.3	2.9						
195	69.8	84.4		1.5	3.3	1,690	3,719			1.2	2.7	1,495	3,290										
	77.6	93.9		1.6	3.6	1,881	4,139			1.4	3.0	1,664	3,661										
196	71.6	71.1		1.5	3.2	1,426	3,137			1.2	2.6	1,254	2,759	665	1,463			.7	1.6	.5	1.1	.9	1.9
	79.0	78.5		1.6	3.5	1,574	3,463			1.3	2.9	1,385	3,046	734	1,614			.8	1.8	.5	1.2	1.0	2.1

TABLE 4-101 COMPOSITION OF FEEDS–

Feed #	Feed Name— Description	Moisture Basis: As Fed or M-F (moisture-free)	Dry Matter	Macrominerals: Calcium (Ca)	Phosphorus (P)	Sodium (Na)	Chlorine (Cl)	Magnesium (Mg)	Potassium (K)	Sulfur (S)	Microminerals: Cobalt (Co)	Copper (Cu)	Iron (Fe)	Manganese (Mn)	Zinc (Zn)
			(%)	(%)	(%)	(%)	(%)	(%)	(%)	(%)	(ppm or mg/kg)	(ppm or mg/kg)	(%)	(ppm or mg/kg)	(ppm or mg/kg)
PROTEIN FEEDS															
176	Distillers' dried rye grain	As Fed	93.0	.13	.41	.17	.05	.17	.11	.44				18.5	
		M-F	*100.0*	*.14*	*.44*	*.18*	*.05*	*.18*	*.12*	*.47*				*19.9*	
177	Distillers' dried solubles	As Fed	92.1	.21	1.23	.15		.46	1.84		.196	71.7	.030	64.2	138.1
		M-F	*100.0*	*.22*	*1.33*	*.16*		*.50*	*2.00*		*.213*	*77.9*	*.033*	*69.7*	*150.0*
178	Distillers' dried sorghum grain, w/o solubles	As Fed	94.0	.14	.59										
		M-F	*100.0*	*.15*	*.63*										
179	Distillers' dried sorghum grain w/solubles	As Fed	94.8	.17	.92									104.5	
		M-F	*100.0*	*.18*	*.97*									*110.2*	
180	Distillers' rye grains w/soluble, dehy	As Fed	90.5												
		M-F	*100.0*												
181	Feather meal, hydrolized	As Fed	93.2	.20	.70	.70		.20	.30						
		M-F	*100.0*	*.21*	*.75*	*.75*		*.21*	*.32*						
182	Fish liver meal	As Fed	92.8												
		M-F	*100.0*												
183	Fish meal, Anchovy	As Fed	92.0	3.74	2.46	.95		.25	.74			9.7	.023	9.3	105.5
		M-F	*100.0*	*4.07*	*2.67*	*1.04*		*.27*	*.81*			*10.5*	*.025*	*10.1*	*114.7*
184	Fish meal, mech-extd	As Fed	88.4	5.48	3.33	1.07	1.21	.21	.39	.24	.106	14.6	.036	22.8	
		M-F	*100.0*	*6.20*	*3.77*	*1.21*	*1.37*	*.24*	*.44*	*.27*	*.119*	*16.5*	*.041*	*25.8*	
185	Fish meal, Menhaden	As Fed	91.4	5.14	2.91	.34	.32	.14	.73		.197	11.0	.045	34.1	150.6
		M-F	*100.0*	*5.62*	*3.18*	*.37*	*.34*	*.16*	*.80*		*.215*	*12.0*	*.050*	*37.3*	*164.7*
186	Fish meal, Sardine	As Fed	92.9	4.50	2.61	.18	.41	.10	.33		.183	20.1	.030	22.1	
		M-F	*100.0*	*4.84*	*2.81*	*.19*	*.44*	*.11*	*.35*		*.196*	*21.7*	*.032*	*23.8*	
187	Fish meal, Tuna	As Fed	93.2	8.03	4.33	.73		.23	.73			10.7	.037	8.6	211.3
		M-F	*100.0*	*8.61*	*4.65*	*.78*		*.25*	*.78*			*11.5*	*.040*	*9.2*	*226.7*
188	Fish meal, white	As Fed	90.6	7.27	3.63							2.8	.016	14.2	
		M-F	*100.0*	*8.02*	*4.00*							*3.0*	*.018*	*15.7*	
189	Fish solubles, condensed	As Fed	50.1	.17	.82	2.13	2.62	.02		.12		41.7	.003	16.8	
		M-F	*100.0*	*.34*	*1.64*	*4.25*	*5.23*	*.04*		*.24*		*83.2*	*.006*	*33.4*	
190	Fish solubles, dried	As Fed	93.7	1.27	1.69										
		M-F	*100.0*	*1.36*	*1.80*										
191	Flax (see Linseed meal)														
192	Flax seed, whole	As Fed	95.7	.28	.55										
		M-F	*100.0*	*.29*	*.57*										
193	Furfural, ammoniated	As Fed	94.3												
		M-F	*100.0*												
194	Horse bean, seeds	As Fed	87.3	.13	.54				1.16						
		M-F	*100.0*	*.15*	*.62*				*1.33*						
195	Linseed meal, mech-extd, 33% protein	As Fed	89.9												
		M-F	*100.0*												
196	Linseed meal, mech- or sol-extd, 35% protein	As Fed	90.6	.39	.84	.12	.04	.57	1.29	.39	.298	33.8	.025	34.9	
		M-F	*100.0*	*.43*	*.93*	*.13*	*.04*	*.64*	*1.43*	*.43*	*.329*	*37.3*	*.027*	*38.5*	

ROTEIN FEEDS

Feed #	Fat-Soluble Vitamins				Water-Soluble Vitamins								
	A (1 mg car. = 1,667 IU-A)	Carotene (Pro-vita-min A)	E (α-toco-pherol)	K	B_{12}	Biotin	Choline	Folic Acid (Folacin)	Niacin (Nicotinic Acid)	Panto-thenic Acid	Pyridoxine (B_6)	Riboflavin (B_2)	Thiamin (B_1)
		(ppm or mg/kg)	*(ppm or mg/kg)*	*(ppm or mg/kg)*	*(ppm or mg/kg)*	*(ppm or mg/kg)*	*(ppm or mg/kg)*	*(ppm or mg/kg)*	*(ppm or mg/kg)*	*(ppm or mg/kg)*	*(ppm or mg/kg)*	*(ppm or mg/kg)*	*(ppm or mg/kg)*
76									17.0	5.3		3.3	1.3
									18.3	*5.7*		*3.6*	*1.4*
77	1,834	1.1			.0029	2.85	4,254		143.2	25.4	8.67	11.3	6.9
	2,000	*1.2*			*.0031*	*3.09*	*4,619*		*155.5*	*27.5*	*9.42*	*12.3*	*7.5*
78													
79							844		61.1	12.3		4.2	1.3
							891		*64.4*	*13.0*		*4.4*	*1.4*
80									62.8	17.4		8.2	3.1
									69.4	*19.2*		*9.0*	*3.4*
81					.0746	.04	880	.22	30.8	11.0		2.0	.1
					.0800	*.05*	*044*	*.24*	*33.0*	*11.8*		*2.1*	*.1*
82													
83			3.4										
			3.7										
84			18.5		.2495		3,510		60.8	8.7	14.14	6.5	1.3
			20.9		*.2823*		*3,972*		*68.8*	*9.8*	*16.00*	*7.4*	*1.4*
85			9.0		.0761		2,826		55.4	8.7		4.8	.7
			9.8		*.0832*		*3,092*		*60.6*	*9.5*		*5.2*	*.7*
86					.1719		2,983		62.5	9.3		6.0	.4
					.1850		*3,209*		*67.3*	*10.0*		*6.4*	*.5*
87					.3007		2,738		141.4	7.1		6.8	
					.3225		*2,936*		*151.7*	*7.6*		*7.3*	
88			9.0		.0983		8,838		69.1	8.7	8.50	8.9	1.7
			9.9		*.1085*		*9,755*		*76.3*	*9.6*	*9.39*	*9.9*	*1.9*
89							2,998		169.1	35.5		14.6	
							5,985		*337.5*	*70.8*		*29.0*	
90							5,535		280.1	56.8		17.8	
							5,907		*298.9*	*60.6*		*19.0*	
91													
92													
93													
94													
95													
96	151	.1					1,490		34.1	13.5		3.1	7.2
	167	*.1*					*1,643*		*37.6*	*14.9*		*3.4*	*8.0*

TABLE 4-101 COMPOSITION OF FEEDS—

Feed #	Feed Name—Description	Moisture Basis: As Fed or M-F (moisture-free)	Proximate Analysis						Digestible Protein		
			Dry Matter	Ash	Crude Fiber	Ether Extract (Fat)	N-Free Extract	Crude Protein	Ruminant	Non-Ruminant	Horse Rabbit
PROTEIN FEEDS			(%)	(%)	(%)	(%)	(%)	(%)	(%)	(%)	(%)
197	Linseed meal, mech- or sol-extd, 37% protein	As Fed	91.6	5.6	8.3	5.8	34.5	37.4	32.5		
		M-F	*100.0*	*6.1*	*9.1*	*6.3*	*37.8*	*40.7*	*35.5*		
198	Liver meal, animal	As Fed	92.7	6.3	1.4	15.7	2.8	66.5			
		M-F	*100.0*	*6.8*	*1.5*	*17.0*	*3.0*	*71.7*			
199	Meat & bone meal	As Fed	93.6	28.6	1.8	11.0	2.7	49.5	42.8	44.0	
		M-F	*100.0*	*30.5*	*1.9*	*11.8*	*2.9*	*52.9*	*45.7*	*47.1*	
200	Meat & bone meal (tankage)	As Fed	94.1	28.0	2.4	12.4	4.2	47.1			
		M-F	*100.0*	*29.8*	*2.6*	*13.2*	*4.4*	*50.0*			
201	Meat meal (meat scraps)	As Fed	93.5	24.7	2.3	9.3	3.7	53.5	47.1		
		M-F	*100.0*	*26.4*	*2.5*	*10.0*	*4.0*	*57.1*	*50.4*		
202	Milk, whole, fluid *(Bos spp)*	As Fed	12.0	.8	0	3.4	4.7	3.1	3.0	3.2	
		M-F	*100.0*	*6.3*	*0*	*28.5*	*39.4*	*25.8*	*24.8*	*26.9*	
203	Milk, whole, dried	As Fed	93.7	5.3	.2	26.0	36.9	25.3			
		M-F	*100.0*	*5.7*	*.2*	*27.8*	*39.4*	*26.9*			
204	Milk, skimmed, fluid	As Fed	9.6	.8	0	.1	6.0	2.7	2.6	3.3	
		M-F	*100.0*	*8.5*	*0*	*.8*	*62.2*	*28.5*	*27.4*	*34.4*	
205	Milk, skimmed, dried	As Fed	94.0	7.6	.2	.9	51.8	33.5	30.4	33.2	
		M-F	*100.0*	*8.1*	*.2*	*1.0*	*55.1*	*35.6*	*32.4*	*35.3*	
206	Milk, cow's *(Bos spp)*	As Fed	12.0	.8	0	3.4	4.7	3.1	3.0	3.2	
		M-F	*100.0*	*6.3*	*0*	*28.5*	*39.4*	*25.8*	*24.8*	*26.9*	
207	Milk, ewe's *(Ovis avies)*	As Fed	19.2	.9	0	6.9	4.9	6.5			
		M-F	*100.0*	*4.7*	*0*	*35.9*	*25.5*	*33.9*			
208	Milk, goat's *(Capra hircus)*	As Fed	13.2	.8	0	4.1	4.7	3.6			
		M-F	*100.0*	*6.1*	*0*	*31.0*	*35.6*	*27.3*			
209	Milk, mare's *(Equus caballus)*	As Fed	9.4	.4	0	1.1	5.9	2.0			
		M-F	*100.0*	*4.2*	*0*	*11.7*	*62.8*	*21.3*			
210	Milk, sow's *(Sus scrofa)*	As Fed	20.1	1.0	0	6.7	5.1	7.3			
		M-F	*100.0*	*5.0*	*0*	*33.3*	*25.4*	*36.3*			
211	Molasses, ammoniated	As Fed	66.4	6.0	0	0	35.8	24.6	14.5		
		M-F	*100.0*	*9.1*	*0*	*0*	*53.8*	*37.1*	*21.9*		
212	Monoammonium phosphate	As Fed	97.0	34.5				68.8			
		M-F	*100.0*	*35.6*				*70.9*			
213	Navy beans *(Phaseolus vulgaris)*	As Fed	90.0	4.1	4.2	1.4	57.4	22.9			
		M-F	*100.0*	*4.6*	*4.7*	*1.6*	*63.7*	*25.4*			
214	Pea feed or pea meal *(Pisum spp)*	As Fed	87.7	3.0	6.8	1.4	53.3	23.2	20.0	20.3	19.3
		M-F	*100.0*	*3.4*	*7.7*	*1.6*	*60.8*	*26.5*	*22.8*	*23.2*	*22.0*
215	Pea seeds, cull *(Pisum sativum arvense)*	As Fed	91.6	2.8	5.9	1.1	59.8	22.0			
		M-F	*100.0*	*3.1*	*6.4*	*1.2*	*65.3*	*24.0*			
216	Peanut meal, mech-extd, 45% protein *(Arachis hypogaea)*	As Fed	92.0	2.2	11.0	7.5	25.5	45.8	41.4	43.7	
		M-F	*100.0*	*2.3*	*12.0*	*8.2*	*27.7*	*49.8*	*45.0*	*47.5*	
217	Peanut meal, sol-extd, 47% protein	As Fed	92.0	4.5	13.0	1.2	25.9	47.4	42.3		
		M-F	*100.0*	*4.9*	*14.1*	*1.3*	*28.2*	*51.5*	*46.0*		

PROTEIN FEEDS

Feed #	TDN Ruminant	TDN Non-Ruminant	TDN Horse Rabbit	DE Ruminant		DE Non-Ruminant		DE Horse Rabbit		ME Ruminant		ME Non-Ruminant ME		ME Non-Ruminant Chicken ME_n		ME Horse Rabbit		NE_m		NE_g		NE_{lc}	
	(%)	(%)	(%)	Mcal per lb	Mcal per kg	kcal per lb	kcal per kg	Mcal per lb	Mcal per kg	Mcal per lb	Mcal per kg	kcal per lb	kcal per kg	kcal per lb	kcal per kg	Mcal per lb	Mcal per kg	Mcal per lb	Mcal per kg	Mcal per lb	Mcal per kg	Mcal per lb	Mcal per kg
197	78.1			1.5	3.4					1.3	2.8												
	85.3			1.7	3.8					1.4	3.1												
198	82.5	89.7		1.6	3.6	1,798	3,956			1.4	3.0	1,465	3,224										
	89.0	96.8		1.8	3.9	1,940	4,269			1.5	3.2	1,581	3,479										
199	60.2	81.6		1.2	2.7	897	1,974			1.0	2.2	749	1,647	875	1,924			.7	1.5	.5	1.0	.8	1.8
	64.4	87.2		1.3	2.8	959	2,110			1.0	2.3	800	1,760	935	2,057			.7	1.6	.5	1.0	.9	1.9
200	62.1	72.6		1.2	2.7	1,455	3,200			1.0	2.3	1,249	2,748										
	66.0	77.1		1.3	2.9	1,545	3,399			1.1	2.4	1,327	2,920										
201	65.4	66.6		1.3	2.9	1,334	2,935			1.1	2.4	1,124	2,472	895	1,969			.7	1.6	.5	1.1	.9	1.9
	70.0	71.2		1.4	3.1	1,427	3,139			1.1	2.5	1,202	2,644	957	2,106			.8	1.7	.5	1.1	.9	2.0
202	15.6	14.9		.3	.7	298	656			.3	.6	270	594					.2	.5	.1	.2	.2	.5
	130.0	124.2		2.5	5.5	2,489	5,476			2.1	4.7	2,250	4,950					2.1	4.6	.9	2.0	1.8	4.0
203	121.8	130.1		2.5	5.4	2,605	5,731			2.0	4.4	2,362	5,197					2.0	4.3	.9	1.9		
	130.0	138.8		2.6	5.7	2,779	6,114			2.1	4.7	2,521	5,547					2.1	4.6	.9	2.0		
204	8.9	9.4								.1	.3	168	369					.1	.2	.05	.1	.1	.3
	93.0	98.2								1.5	3.4	1,750	3,849					1.0	2.3	.7	1.5	1.2	2.7
205	80.8	84.8		1.6	3.6	1,698	3,735			1.3	2.9	1,507	3,316	1,146	2,521			.9	1.9	.6	1.3		
	86.0	90.2		1.7	3.8	1,806	3,973			1.4	3.1	1,604	3,528	1,219	2,682			1.0	2.1	.6	1.4		
206	15.6	14.9		.3	.7	298	656			.3	.6	270	594					.3	.6	.1	.2	.2	.5
	130.0	124.2		2.5	5.5	2,489	5,476			2.1	4.7	2,250	4,950					2.1	4.6	.9	2.0	1.8	4.0
207	21.9	28.7		.5	1.0	575	1,265			.4	.8	513	1,128										
	114.5	149.4		2.3	5.1	2,995	6,589			1.9	4.1	2,670	5,875										
208	15.4	18.9		.3	.7	380	835			.3	.6	344	756										
	117.3	143.5		2.4	5.2	2,876	6,328			1.9	4.2	2,603	5,726										
209	9.6	11.0		.2	.4	221	486			.1	.3	203	446										
	102.2	117.4		2.0	4.5	2,352	5,175			1.7	3.7	2,157	4,745										
210	22.5	28.9		.5	1.0	579	1,274			.4	.8	514	1,130										
	111.8	143.8		2.2	4.9	2,881	6,339			1.8	4.0	2,555	5,620										
211	49.0	50.5		1.0	2.2	1,011	2,225			.8	1.8	895	1,969										
	73.9	76.0		1.5	3.3	1,523	3,351			1.2	2.7	1,348	2,966										
212																							
213	76.5	84.8		1.5	3.4	1,699	3,738			1.3	2.8	1,545	3,398	1,063	2,338			.8	1.8	.5	1.2	.9	1.9
	85.0	94.2		1.7	3.8	1,888	4,153			1.4	3.1	1,716	3,775	1,181	2,598			.9	2.0	.6	1.3	1.0	2.1
214	73.1	74.7	68.0	1.5	3.2	1,498	3,295			1.2	2.6	1,357	2,986			1.1	2.5						
	83.3	85.2	77.5	1.6	3.6	1,707	3,756			1.4	3.0	1,548	3,405			1.3	2.8						
215	72.9	86.0		1.5	3.2	1,725	3,794			1.2	2.6	1,572	3,458										
	79.6	93.9		1.6	3.5	1,882	4,141			1.3	2.9	1,716	3,775										
216	78.7	84.3		1.6	3.5	1,689	3,715			1.3	2.8	1,449	3,187	1,132	2,491			.8	1.8	.5	1.2	.9	2.0
	85.5	91.6		1.7	3.8	1,835	4,038			1.4	3.1	1,575	3,464	1,231	2,708			.9	2.0	.6	1.3	1.0	2.1
217	73.6	64.9		1.5	3.2	1,293	2,845	1.4	3.1	1.2	2.7	1,065	2,344	1,002	2,205			.7	1.6	.5	1.1	.9	1.9
	80.0	70.5		1.6	3.5	1,455	3,201	1.5	3.4	1.3	2.9	1,209	2,659	1,090	2,397			.8	1.8	.5	1.2	1.0	2.1

TABLE 4-101 COMPOSITION OF FEEDS—

Feed #	Feed Name—Description	Moisture Basis: As Fed or M-F (moisture-free)	Dry Matter	Macrominerals							Microminerals				
				Calcium (Ca)	Phosphorus (P)	Sodium (Na)	Chlorine (Cl)	Magnesium (Mg)	Potassium (K)	Sulfur (S)	Cobalt (Co)	Copper (Cu)	Iron (Fe)	Manganese (Mn)	Zinc (Zn)
PROTEIN FEEDS			(%)	(%)	(%)	(%)	(%)	(%)	(%)	(%)	(ppm or mg/kg)	(ppm or mg/kg)	(%)	(ppm or mg/kg)	(ppm or mg/kg)
197	Linseed meal, mech- or sol-extd, 37% protein	As Fed	91.6	.38	.85				1.11						
		M-F	*100.0*	*.41*	*.93*				*1.21*						
198	Liver meal, animal	As Fed	92.7	.56	1.26						.135	89.4	.063	8.8	
		M-F	*100.0*	*.61*	*1.36*						*.145*	*96.5*	*.068*	*9.5*	
199	Meat & bone meal	As Fed	93.6	11.42	5.69	.73	.75	1.13	1.46		.182	1.5	.050	12.3	
		M-F	*100.0*	*12.20*	*6.08*	*.78*	*.80*	*1.20*	*1.56*		*.195*	*1.6*	*.053*	*13.2*	
200	Meat & bone meal (tankage)	As Fed	94.1	11.47	5.25										
		M-F	*100.0*	*12.19*	*5.58*										
201	Meat meal (meat scraps)	As Fed	93.5	7.94	4.03	1.68	1.31	.27	.55	.50	.128	9.7	.044	9.5	
		M-F	*100.0*	*8.49*	*4.31*	*1.80*	*1.40*	*.29*	*.59*	*.53*	*.137*	*10.4*	*.047*	*10.2*	
202	Milk, whole, fluid	As Fed	12.0	.11	.09	.05	.19		.13			0			
		M-F	*100.0*	*.93*	*.75*	*.39*	*1.56*		*1.11*			*.3*			
203	Milk, whole, dried	As Fed	93.7	.89	.67	.36		1.01	1.11			.8	.017	.4	
		M-F	*100.0*	*.95*	*.72*	*.39*		*1.08*	*1.18*			*.9*	*.018*	*.4*	
204	Milk, skimmed, fluid	As Fed	9.6	.12	.10			.01	.10	.03	.011	.1	.002	.2	
		M-F	*100.0*	*1.26*	*1.03*			*.11*	*1.01*	*.32*	*.110*	*.9*	*.017*	*2.3*	
205	Milk, skimmed, dried	As Fed	94.0	1.26	1.03	.52		.11	1.67	.32	.110	11.5	.005	2.2	
		M-F	*100.0*	*1.34*	*1.10*	*.55*		*.12*	*1.78*	*.34*	*.117*	*12.2*	*.005*	*2.3*	
206	Milk, cow's	As Fed	12.0	.11	.09	.05	.19		.13			0			
		M-F	*100.0*	*.93*	*.75*	*.39*	*1.56*		*1.11*			*.3*			
207	Milk, ewe's	As Fed	19.2	.21	.12				.19						
		M-F	*100.0*	*1.09*	*.63*				*.99*						
208	Milk, goat's	As Fed	13.2	.13	.11				.18						
		M-F	*100.0*	*.98*	*.83*				*1.36*						
209	Milk, mare's	As Fed	9.4	.08	.05				.08						
		M-F	*100.0*	*.85*	*.53*				*.85*						
210	Milk, sow's	As Fed	20.1												
		M-F	*100.0*												
211	Molasses, ammoniated	As Fed	66.4	.81	.13										
		M-F	*100.0*	*1.22*	*.20*										
212	Monoammonium phosphate	As Fed	97.0	.50	24.00	.06		.45		.7		80	1.200	400	300
		M-F	*100.0*	*.52*	*24.74*	*.06*		*.46*		*.7*		*82*	*1.237*	*412*	*309*
213	Navy beans	As Fed	90.0	.15	.57	.05	.04	.17	1.70	.23		9.9	.009	18.5	
		M-F	*100.0*	*.17*	*.63*	*.06*	*.04*	*.19*	*1.89*	*.26*		*11.0*	*.011*	*20.6*	
214	Pea feed or pea meal	As Fed	87.7												
		M-F	*100.0*												
215	Pea seeds, cull	As Fed	91.6	.17	.32										
		M-F	*100.0*	*.19*	*.35*										
216	Peanut meal, mech-extd, 45% protein	As Fed	92.0	.16	.57			.33	1.15	.29				25.5	
		M-F	*100.0*	*.18*	*.62*			*.36*	*1.25*	*.32*				*27.7*	
217	Peanut meal, sol-extd, 47% protein	As Fed	92.0	.20	.65	.07	.03	.04	1.20					29.0	20.0
		M-F	*100.0*	*.22*	*.71*	*.08*	*.03*	*.04*	*1.30*					*31.5*	*21.7*

ROTEIN FEEDS

Feed #	Fat-Soluble Vitamins				Water-Soluble Vitamins								
	A (1 mg car. = 1,667 IU-A)	Carotene (Pro-vitamin A)	E (α-tocopherol)	K	B_{12}	Biotin	Choline	Folic Acid (Folacin)	Niacin (Nicotinic Acid)	Pantothenic Acid	Pyridoxine (B_6)	Riboflavin (B_2)	Thiamin (B_1)
		(ppm or mg/kg)	*(ppm or mg/kg)*	*(ppm or mg/kg)*	*(ppm or mg/kg)*	*(ppm or mg/kg)*	*(ppm or mg/kg)*	*(ppm or mg/kg)*	*(ppm or mg/kg)*	*(ppm or mg/kg)*	*(ppm or mg/kg)*	*(ppm or mg/kg)*	*(ppm or mg/kg)*
197													
198					.5019	.02		5.56	205.0	31.0		46.3	.2
					.5416	*.02*		*6.00*	*221.2*	*33.4*		*50.0*	*.2*
199			1.0		.0983		1,980		46.4	3.6		4.2	1.1
			1.1		*.1051*		*2,116*		*49.4*	*3.8*		*4.5*	*1.2*
200													
201			1.0				1,948		56.5	4.8		5.2	
			1.1				*2,083*		*60.4*	*5.1*		*5.6*	
202	1,400	.8							.9			1.6	.3
	11,400	*6.8*							*7.9*			*13.4*	*2.4*
203	11,800	7.0				3.7			7.5	22.7	4.63	16.8	3.3
	12,600	*7.5*				*4.0*			*8.0*	*24.2*	*4.94*	*17.9*	*3.5*
204													
205	300	.2	9.2		.0419	.33	1,426	.62	11.5	33.7	3.97	20.1	3.5
	300	*.2*	*9.8*		*.0446*	*.35*	*1,517*	*.68*	*12.2*	*35.8*	*4.22*	*21.4*	*3.7*
206	1,400	.8							.9			1.6	.3
	11,400	*6.8*							*7.9*			*13.4*	*2.4*
207													
208													
209													
210													
211													
212													
213							1,764		24.5	3.1		2.0	6.6
							1,960		*27.2*	*3.4*		*2.2*	*7.3*
214							624						193.4
							712						*220.5*
215													
216	400	.2					1,683		169.0	48.2		5.3	7.3
	400	*.2*					*1,829*		*183.7*	*52.4*		*5.8*	*7.9*
217			3.0			.39	2,000	.36	170.1	53.0	10.00	11.0	7.3
			3.3			*.42*	*2,174*	*.39*	*184.9*	*57.6*	*10.87*	*12.0*	*7.9*

TABLE 4-101 COMPOSITION OF FEEDS—

Feed #	Feed Name—Description	Moisture Basis: As Fed or M-F (moisture-free)	Proximate Analysis: Dry Matter	Ash	Crude Fiber	Ether Extract (Fat)	N-Free Extract	Crude Protein	Digestible Protein: Ruminant	Non-Ruminant	Horse Rabbit
			(%)	(%)	(%)	(%)	(%)	(%)	(%)	(%)	(%)
PROTEIN FEEDS											
218	Poultry manure, cage, dried	As Fed	88.6	26.5	13.5	1.7	35.5	28.7	12.8		
		M-F	*100.0*	*29.9*	*15.2*	*1.9*	*40.2*	*32.4*	*14.4*		
219	Poultry manure, floor, dried	As Fed	84.5	14.1	15.7	2.3	27.1	25.3	19.7		
		M-F	*100.0*	*16.7*	*18.5*	*2.7*	*32.3*	*29.9*	*23.3*		
220	Poultry viscera w/ feet & heads rendered	As Fed	93.4	18.7	1.6	13.1	4.6	55.4			
		M-F	*100.0*	*20.0*	*1.8*	*14.0*	*4.9*	*59.3*			
221	Rape, Canada, seeds, cooked, sol-extd, 40% protein *(Brassica napus var)*	As Fed	92.0	7.2	9.3	1.1	33.9	40.5			
		M-F	*100.0*	*7.8*	*10.1*	*1.2*	*36.9*	*44.0*			
222	Rape seed meal, sol-extd, 34% protein, *(Brassica* spp)	As Fed	90.3	7.0	13.8	1.5	28.5	39.5	33.0		
		M-F	*100.0*	*7.8*	*15.3*	*1.7*	*31.6*	*43.6*	*36.6*		
223	Rape seed meal, mech-extd, 37% protein	As Fed	93.6	7.3	13.7	8.7	26.8	37.1	31.1		
		M-F	*100.0*	*7.8*	*14.6*	*9.3*	*28.7*	*39.6*	*33.2*		
224	Rye distillers' grains, dried	As Fed	93.0	1.9	14.0	7.4	47.3	22.4	9.7		
		M-F	*100.0*	*2.0*	*15.1*	*8.0*	*50.8*	*24.1*	*10.4*		
225	Rye distillers' grains, w/solubles, dehy *(Secale cereale)*	As Fed	90.5	6.4	8.1	4.1	44.7	27.2			
		M-F	*100.0*	*7.1*	*9.0*	*4.5*	*49.3*	*30.1*			
226	Safflower meal, mech-extd, 20% protein *(Carthamus tinctorius)*	As Fed	91.0	3.7	31.0	6.5	30.1	19.7	15.7		
		M-F	*100.0*	*4.1*	*34.1*	*7.2*	*32.9*	*21.7*	*17.3*		
227	Safflower meal w/o hulls, mech-extd, 42% protein	As Fed	88.5	6.4	8.5	6.7	26.5	40.4			
		M-F	*100.0*	*7.2*	*9.6*	*7.6*	*29.9*	*45.7*			
228	Safflower meal, sol-extd, 20% protein	As Fed	91.8	4.7	32.3	3.9	29.5	21.4	17.2		
		M-F	*100.0*	*5.1*	*35.2*	*4.2*	*32.2*	*23.3*	*18.7*		
229	Safflower meal w/o hulls, sol-extd, 42% protein	As Fed	90.5	6.4	8.5	1.7	29.4	44.5	37.4		
		M-F	*100.0*	*7.1*	*9.4*	*1.9*	*32.5*	*49.1*	*41.3*		
230	Sesame seeds *(Sesamum indicum)*	As Fed	92.0	5.6	10.3	42.9	10.9	22.3	37.9		
		M-F	*100.0*	*6.2*	*11.2*	*46.6*	*11.8*	*24.2*	*41.0*		
231	Sesame seed meal, mech-extd, 44% protein	As Fed	93.0	10.4	5.0	8.7	21.0	47.9	38.3		
		M-F	*100.0*	*11.2*	*5.4*	*9.4*	*22.5*	*51.5*	*41.2*		
232	Shrimp meal	As Fed	89.9	27.4	11.2	2.9	1.5	46.9			
		M-F	*100.0*	*30.5*	*12.4*	*3.2*	*1.7*	*52.2*			
233	Sorghum distillers' grains, dried	As Fed	94.0	4.0	12.0	8.5	38.3	31.2	25.9		
		M-F	*100.0*	*4.3*	*12.8*	*9.0*	*40.7*	*33.2*	*27.6*		
234	Soybean flour *(Glycine max)*	As Fed	92.3	6.1	2.3	.8	32.9	50.2	46.2	45.7	
		M-F	*100.0*	*6.6*	*2.4*	*.9*	*35.5*	*54.5*	*50.1*	*49.6*	
235	Soybean meal, mech-extd, 41% protein	As Fed	90.0	6.0	6.0	4.7	31.1	43.8	38.3		
		M-F	*100.0*	*6.7*	*6.7*	*5.2*	*34.6*	*48.7*	*42.6*		
236	Soybean meal, sol-extd, 44% protein	As Fed	89.0	5.9	6.0	1.2	30.1	45.8	41.4		
		M-F	*100.0*	*6.6*	*6.7*	*1.3*	*33.9*	*51.5*	*46.5*		
237	Soybean meal, sol-extd, 49% protein	As Fed	89.8	5.9	2.8	1.1	30.5	50.9	47.7		44.6
		M-F	*100.0*	*6.6*	*3.1*	*1.2*	*34.0*	*56.7*	*53.1*		*49.7*
238	Soybean seeds	As Fed	90.0	4.9	5.0	17.3	24.9	37.9	34.1	30.8	
		M-F	*100.0*	*5.4*	*5.6*	*19.2*	*27.7*	*42.1*	*37.9*	*34.2*	

PROTEIN FEEDS

Feed #	TDN			Digestible Energy						Metabolizable Energy								Net Energy-Ruminant					
												Non-Ruminant											
	Ruminant	Non-Ruminant	Horse Rabbit	Ruminant		Non-Ruminant		Horse Rabbit		Ruminant		ME		Chicken ME_n		Horse Rabbit		NE_m		NE_g		NE_{lc}	
	(%)	(%)	(%)	Mcal per lb	Mcal per kg	kcal per lb	kcal per kg	Mcal per lb	Mcal per kg	Mcal per lb	Mcal per kg	kcal per lb	kcal per kg	kcal per lb	kcal per kg	Mcal per lb	Mcal per kg	Mcal per lb	Mcal per kg	Mcal per lb	Mcal per kg	Mcal per lb	Mcal per kg
218	46.3			.9	2.0					.8	1.7			386	850								
	52.3			*1.0*	*2.3*					*.9*	*1.9*			*450*	*990*								
219	61.3			1.2	2.7					.8	1.8												
	72.5			*1.5*	*3.2*					*1.0*	*2.2*												
220	70.6	79.0		1.4	3.1	1,583	3,483			1.2	2.6	1,330	2,926	1,213	2,668								
	75.6	*84.6*		*1.5*	*3.3*	*1,695*	*3,729*			*1.2*	*2.7*	*1,424*	*3,133*	*1,298*	*2,856*								
221	68.0	69.5		1.4	3.0	1,392	3,062			1.2	2.7	1,212	2,667										
	73.9	*75.5*		*1.5*	*3.3*	*1,513*	*3,328*			*1.3*	*2.9*	*1,318*	*2,899*										
222	63.7	67.8		1.3	2.8	1,357	2,986			1.0	2.3	1,191	2,621					.6	1.4	.4	.8	.7	1.5
	70.5	*75.1*		*1.4*	*3.1*	*1,504*	*3,308*			*1.1*	*2.5*	*1,320*	*2,903*					*.7*	*1.5*	*.4*	*.9*	*.8*	*1.7*
223	69.3	84.5		1.4	3.1	1,694	3,726			1.2	2.6	1,505	3,319					.7	1.6	.5	1.0	.8	1.7
	74.0	*90.3*		*1.5*	*3.3*	*1,810*	*3,981*			*1.2*	*2.7*	*1,612*	*3,546*					*.8*	*1.7*	*.5*	*1.1*	*.8*	*1.8*
224	42.8	62.6		1.0	2.3	1,254	2,759			.7	1.5	1,148	2,525					.4	.8	.05	.1		
	46.0	*67.3*		*1.1*	*2.5*	*1,349*	*2,967*			*.8*	*1.7*	*1,234*	*2,715*					*.4*	*.9*	*.05*	*.1*		
225	92.7	100.3		1.9	4.1	2,011	4,424			1.5	3.4	1,800	3,959										
	102.0	*110.4*		*2.0*	*4.5*	*2,212*	*4,867*			*1.7*	*3.7*	*1,980*	*4,356*										
226	47.3	54.7		1.0	2.1	1,096	2,411			.8	1.8	899	1,977					.5	1.0	.05	.1	.5	1.0
	52.0	*60.1*		*1.0*	*2.3*	*1,205*	*2,650*			*.9*	*1.9*	*988*	*2,173*					*.5*	*1.1*	*.2*	*.5*	*.5*	*1.1*
227	70.1	76.0		1.4	3.1	1,524	3,352			1.1	2.5	1,322	2,908	768	1,690								
	79.2	*85.9*		*1.6*	*3.5*	*1,723*	*3,790*			*1.3*	*2.9*	*1,495*	*3,288*	*868*	*1,910*								
228	50.2			1.0	2.2					.8	1.8							.5	1.1	.2	.4	.5	1.0
	54.7			*1.1*	*2.4*					*.9*	*2.0*							*.5*	*1.2*	*.2*	*.5*	*.5*	*1.2*
229	69.0			1.4	3.0					1.1	2.5							.6	1.4	.4	.9	.8	1.7
	76.3			*1.5*	*3.4*					*1.3*	*2.8*							*.7*	*1.6*	*.5*	*1.0*	*.9*	*1.9*
230	99.4	106.3		2.0	4.4	2,130	4,685			1.6	3.6	1,746	3,841										
	108.0	*115.5*		*2.2*	*4.8*	*2,315*	*5,092*			*1.8*	*3.9*	*1,898*	*4,175*										
231	69.7	81.4		1.4	3.1	1,631	3,589			1.2	2.6	1,407	3,096	1,194	2,626			.7	1.6	.5	1.0	.8	1.7
	75.0	*87.5*		*1.5*	*3.3*	*1,754*	*3,858*			*1.2*	*2.7*	*1,513*	*3,329*	*1,284*	*2,824*			*.8*	*1.7*	*.5*	*1.1*	*.9*	*1.9*
232	42.7	44.4		.9	1.9	891	1,960			.7	1.5	761	1,674	763	1,678								
	47.5	*49.5*		*1.0*	*2.1*	*991*	*2,180*			*.8*	*1.7*	*847*	*1,863*	*846*	*1,862*								
233	78.0	88.3		1.5	3.4	1,769	3,892			1.3	2.8	1,579	3,474					.8	1.8	.5	1.2	.9	2.0
	83.0	*93.9*		*1.7*	*3.7*	*1,882*	*4,140*			*1.4*	*3.0*	*1,680*	*3,696*					*.9*	*1.9*	*.6*	*1.3*	*1.0*	*2.1*
234	79.6	78.5		1.6	3.5	1,574	3,462			1.3	2.9	1,338	2,943										
	86.2	*85.1*		*1.7*	*3.8*	*1,706*	*3,753*			*1.4*	*3.1*	*1,450*	*3,190*										
235	75.6	76.6		1.5	3.4	1,535	3,378			1.2	2.7	1,241	2,731	1,114	2,450			.9	1.9	.5	1.2	1.0	2.1
	84.0	*85.1*		*1.7*	*3.7*	*1,706*	*3,753*			*1.4*	*3.0*	*1,477*	*3,249*	*1,237*	*2,722*			*1.0*	*2.1*	*.6*	*1.4*	*1.1*	*2.4*
236	76.3	78.9	71.2	1.5	3.2	1,579	3,474	1.4	3.1	1.2	2.6	1,296	2,851	1,022	2,249	1.2	2.6	.8	1.7	.5	1.1		
	85.7	*88.6*	*80.0*	*1.6*	*3.5*	*1,774*	*3,903*	*1.6*	*3.5*	*1.3*	*2.9*	*1,456*	*3,203*	*1,149*	*2,527*	*1.3*	*2.9*	*.9*	*1.9*	*.6*	*1.3*		
237	78.8	71.2	75.4	1.5	3.3	1,792	3,942	1.5	3.3	1.3	2.9	1,610	3,542	1,102	2,425	1.2	2.7	1.1	2.4				
	87.8	*79.3*	*84.0*	*1.7*	*3.7*	*1,995*	*4,390*	*1.7*	*3.7*	*1.5*	*3.2*	*1,793*	*3,944*	*1,227*	*2,700*	*1.4*	*3.0*	*1.2*	*2.7*				
238	84.1	90.5		1.7	3.7	1,815	3,992			1.4	3.1	1,487	3,272	1,391	3,060			1.0	2.2	.6	1.4	1.0	2.3
	93.5	*100.6*		*1.9*	*4.1*	*2,016*	*4,436*			*1.6*	*3.5*	*1,654*	*3,638*	*1,545*	*3,400*			*1.1*	*2.4*	*.7*	*1.5*	*1.1*	*2.5*

TABLE 4-101 COMPOSITION OF FEEDS—

Feed #	Feed Name—Description	Moisture Basis: As Fed or M-F (moisture-free)	Dry Matter	Macrominerals							Microminerals				
				Calcium (Ca)	Phosphorus (P)	Sodium (Na)	Chlorine (Cl)	Magnesium (Mg)	Potassium (K)	Sulfur (S)	Cobalt (Co)	Copper (Cu)	Iron (Fe)	Manganese (Mn)	Zinc (Zn)
PROTEIN FEEDS			(%)	(%)	(%)	(%)	(%)	(%)	(%)	(%)	(ppm or mg/kg)	(ppm or mg/kg)	(%)	(ppm or mg/kg)	(ppm or mg/kg)
218	Poultry manure, cage, dried	As Fed	88.6	7.80	2.2	.42		.63	1.37		.001	61.0	.002	291.0	325.0
		M-F	*100.0*	*8.80*	*2.5*	*.47*		*.71*	*1.55*		*.001*	*69.0*	*.003*	*328.0*	*367.0*
219	Poultry manure, floor, dried	As Fed	84.5	2.50	1.6	.42		.35	1.77			23.0	.038	190.0	343.0
		M-F	*100.0*	*3.00*	*1.9*	*.50*		*.41*	*2.09*			*27.0*	*.045*	*225.0*	*406.0*
220	Poultry viscera w/feet & heads rendered	As Fed	93.4	3.00	1.70										
		M-F	*100.0*	*3.21*	*1.82*										
221	Rape, Canada, seeds, cooked, sol-extd, 40% protein	As Fed	92.0	.66	.93										
		M-F	*100.0*	*.72*	*1.01*										
222	Rape seed meal, sol-extd, 34% protein	As Fed	90.3	.40	.90										
		M-F	*100.0*	*.44*	*1.00*										
223	Rape seed meal, mech-extd, 37% protein	As Fed	93.6	.60	.97									61.9	
		M-F	*100.0*	*.64*	*1.04*									*66.1*	
224	Rye distillers' grains, dried	As Fed	93.0	.13	.41	.17	.05	.17	.11	.44				18.5	
		M-F	*100.0*	*.14*	*.44*	*.18*	*.05*	*.18*	*.12*	*.47*				*19.9*	
225	Rye distillers' grains, w/solubles, dehy	As Fed	90.5												
		M-F	*100.0*												
226	Safflower meal, mech-extd, 20% protein	As Fed	91.0	.23	.71	.05	.04	.33	.72	.05		9.7	.045	17.8	39.8
		M-F	*100.0*	*.25*	*.78*	*.06*	*.05*	*.36*	*.79*	*.06*		*10.7*	*.050*	*19.6*	*43.7*
227	Safflower meal w/o hulls, mech-extd, 42% protein	As Fed	88.5	.32	.59										
		M-F	*100.0*	*.36*	*.67*										
228	Safflower meal, sol-extd, 20% protein	As Fed	91.8	.34	.84										
		M-F	*100.0*	*.37*	*.92*										
229	Safflower meal w/o hulls, sol-extd, 42% protein	As Fed	90.5	.24	1.66										
		M-F	*100.0*	*.26*	*1.83*										
230	Sesame seeds	As Fed	92.0	.94	.70										
		M-F	*100.0*	*1.02*	*.76*										
231	Sesame seed meal, mech extd, 44 % protein	As Fed	93.0	2.03	1.29									48.0	
		M-F	*100.0*	*2.18*	*1.39*									*51.6*	
232	Shrimp meal	As Fed	89.9	7.34	1.59			.54					.011	30.1	
		M-F	*100.0*	*8.17*	*1.77*			*.60*					*.012*	*33.5*	
233	Sorghum distillers' grains, dried	As Fed	94.0	.14	.59										
		M-F	*100.0*	*.15*	*.63*										
234	Soybean flour	As Fed	92.3	.33	.62	.34	.20	.20	1.89	.41		16.0	.020	31.7	19.9
		M-F	*100.0*	*.35*	*.67*	*.37*	*.22*	*.22*	*2.05*	*.44*		*17.3*	*.022*	*34.4*	*21.6*
235	Soybean meal, mech-extd, 41% protein	As Fed	90.0	.27	.63	.24	.07	.25	1.71	.33	.180	18.0	.016	32.3	
		M-F	*100.0*	*.30*	*.70*	*.27*	*.08*	*.28*	*1.90*	*.37*	*.200*	*20.0*	*.018*	*35.9*	
236	Soybean meal, sol-extd, 44% protein	As Fed	89.0	.32	.67	.34	0	.27	1.97	.43	.089	36.3	.012	27.5	27.0
		M-F	*100.0*	*.36*	*.75*	*.38*	*0*	*.30*	*2.21*	*.48*	*.100*	*40.8*	*.013*	*30.9*	*30.3*
237	Soybean meal, sol-extd, 49% protein	As Fed	89.8	.26	.62	.01	.07		2.02					45.5	45.0
		M-F	*100.0*	*.29*	*.69*	*.01*	*.08*		*2.25*					*50.7*	*50.1*
238	Soybean seeds	As Fed	90.0	.25	.59	.12	.03	.28	1.59	.22		15.7	.008	29.5	
		M-F	*100.0*	*.28*	*.66*	*.13*	*.03*	*.31*	*1.77*	*.24*		*17.4*	*.009*	*32.8*	

ROTEIN FEEDS

Feed #	Fat-Soluble Vitamins				Water-Soluble Vitamins								
	A (1 mg car. = 1,667 IU-A)	Carotene (Pro-vitamin A)	E (α-tocopherol)	K	B_{12}	Biotin	Choline	Folic Acid (Folacin)	Niacin (Nicotinic Acid)	Pantothenic Acid	Pyridoxine (B_6)	Riboflavin (B_2)	Thiamin (B_1)
		(ppm or mg/kg)	*(ppm or mg/kg)*	*(ppm or mg/kg)*	*(ppm or mg/kg)*	*(ppm or mg/kg)*	*(ppm or mg/kg)*	*(ppm or mg/kg)*	*(ppm or mg/kg)*	*(ppm or mg/kg)*	*(ppm or mg/kg)*	*(ppm or mg/kg)*	*(ppm or mg/kg)*
218													
219													
220			2.0				5,952		39.7	8.8		11.0	
			2.1				*6,373*		*42.5*	*9.4*		*11.8*	
221													
222													
223			18.9										
			20.2										
224									17.0	5.3		3.3	1.3
									18.3	*5.7*		*3.6*	*1.4*
225									62.8	17.4		8.2	3.1
									69.4	*19.2*		*9.0*	*3.4*
226							1,499		12.8	51.4		2.4	
							1,647		*14.1*	*56.5*		*2.6*	
227							2,641		22.6	88.0		4.1	
							2,985		*25.5*	*99.5*		*4.6*	
228													
229													
230													
231							1,533			6.3		3.7	
							1,649			*6.8*		*4.0*	
232							5,836					4.0	
							6,494					*4.4*	
233													
234	700	.4				.72	2,233		59.5	14.0		3.9	1.5
	800	*.5*				*.78*	*2,420*		*64.5*	*15.2*		*4.3*	*1.7*
235	300	.2	6.6			.30	2,673	6.60	30.4	14.9		3.5	4.0
	333	*.2*	*7.3*			*.33*	*2,940*	*7.33*	*33.8*	*16.6*		*3.9*	*4.4*
236			3.0			.32	2,743	.70	26.8	14.5	8.00	3.3	6.6
			3.4			*.36*	*3,082*	*.79*	*30.1*	*16.3*	*9.00*	*3.7*	*7.4*
237			3.3			.32	2,761	3.60	21.6	14.5	8.00	3.1	2.4
			3.7			*.36*	*3,075*	*4.01*	*24.1*	*16.1*	*8.91*	*3.5*	*2.7*
238	800	.5							22.0			3.1	11.0
	900	*.5*							*24.4*			*3.4*	*12.2*

TABLE 4-101 COMPOSITION OF FEEDS-

Feed #	Feed Name—Description	Moisture Basis: As Fed or M-F (moisture-free)	Proximate Analysis						Digestible Protein		
			Dry Matter	Ash	Crude Fiber	Ether Extract (Fat)	N-Free Extract	Crude Protein	Ruminant	Non-Ruminant	Horse Rabbit
PROTEIN FEEDS			(%)	(%)	(%)	(%)	(%)	(%)	(%)	(%)	(%)
239	Sunflower meal, mech extd, 41% protein *(Helianthus annuus)*	As Fed	92.9	6.8	13.3	7.6	24.2	41.0	36.9		
		M-F	*100.0*	*7.3*	*14.3*	*8.2*	*26.0*	*44.2*	*39.7*		
240	Sunflower meal, sol-extd, 44% protein	As Fed	93.0	7.7	11.0	2.9	24.6	46.8	42.1		
		M-F	*100.0*	*8.3*	*11.8*	*3.1*	*26.5*	*50.3*	*45.3*		
241	Tankage (digester tankage)	As Fed	92.0	21.4	2.0	8.2	.6	59.8	50.6		
		M-F	*100.0*	*23.3*	*2.2*	*8.8*	*.7*	*65.0*	*55.0*		
242	Urea (Feed grade)	As Fed							281.0		
		M-F	*100.0*								
243	Whey *(Bos spp)*	As Fed	6.9	.7	0	.3	5.0	.9	.6		.6
		M-F	*100.0*	*9.4*	*0*	*4.3*	*73.3*	*13.0*	*8.6*		*9.2*
244	Whey, condensed	As Fed	63.6	6.4	.2	.6	47.7	8.7	5.8		6.2
		M-F	*100.0*	*10.0*	*.4*	*.9*	*75.0*	*13.7*	*9.2*		*9.8*
245	Whey, dried	As Fed	94.0	9.7	0	.8	69.7	13.8	10.6		11.2
		M-F	*100.0*	*10.3*	*0*	*.9*	*74.1*	*14.7*	*11.3*		*11.9*
246	Yeast, brewers' dried *(Saccharomyces cervisiae)*	As Fed	93.0	6.9	1.5	1.0	39.3	44.3			
		M-F	*100.0*	*7.4*	*1.6*	*1.1*	*42.3*	*47.6*			
247	Yeast, irradiated, dried	As Fed	93.9	6.2	6.1	1.1	32.1	48.4	42.8	42.6	
		M-F	*100.0*	*6.6*	*6.5*	*1.2*	*34.2*	*51.5*	*45.6*	*45.4*	
248	Yeast, torula, dried *(Torulopsis utilis)*	As Fed	93.0	7.8	2.0	2.5	32.4	48.3	42.7	38.7	
		M-F	*100.0*	*8.4*	*2.2*	*2.7*	*34.8*	*51.9*	*45.9*	*41.6*	

ROTEIN FEEDS

Feed #	TDN Rumi-nant	TDN Non-Rumi-nant	TDN Horse Rabbit	DE Rumi-nant		DE Non-Rumi-nant		DE Horse Rabbit		ME Rumi-nant		ME Non-Ruminant ME		ME Non-Ruminant Chicken ME_n		ME Horse Rabbit		NE_m		NE_g		NE_{lc}	
	(%)	(%)	(%)	Mcal per		kcal per		Mcal per		Mcal per		kcal per		kcal per		Mcal per		Mcal per		Mcal per		Mcal per	
				lb	kg	lb	kg	lb	kg	lb	kg	lb	kg	lb	kg	lb	kg	lb	kg	lb	kg	lb	kg
39	69.1	77.0		1.4	3.1	1,543	3,394			1.1	2.5	1,343	2,955					.6	1.4	.4	.9	.8	1.7
	74.4	*82.9*		*1.5*	*3.3*	*1,660*	*3,653*			*1.2*	*2.7*	*1,446*	*3,181*					*.7*	*1.5*	*.5*	*1.0*	*.8*	*1.8*
40	64.8	68.3		1.2	2.7	1,369	3,012			1.0	2.2	1,175	2,586	920	2,024			.6	1.3	.4	.8	.6	1.4
	69.7	*73.5*		*1.3*	*2.9*	*1,472*	*3,239*			*1.1*	*2.4*	*1,264*	*2,780*	*990*	*2,178*			*.6*	*1.4*	*.4*	*.8*	*.7*	*1.5*
41	64.3	66.9		1.3	2.8	1,340	2,949			1.0	2.3	958	2,107	1,187	2,612								
	69.9	*72.7*		*1.4*	*3.1*	*1,457*	*3,206*			*1.1*	*2.5*	*1,041*	*2,290*	*1,290*	*2,839*								
42																							
43	5.8	5.5		.1	.3	110	243			.1	.2	103	227										
	84.6	*79.8*		*1.7*	*3.7*	*1,599*	*3,518*			*1.4*	*3.1*	*1,493*	*3,284*										
44	51.5	48.8		1.0	2.3	978	2,152			.9	1.9	912	2,006										
	81.1	*76.7*		*1.6*	*3.6*	*1,537*	*3,382*			*1.3*	*2.9*	*1,433*	*3,153*										
45	82.2	70.5		1.6	3.6	1,414	3,111			1.4	3.0	1,363	2,999	842	1,852							1.0	2.2
	87.5	*75.1*		*1.8*	*3.9*	*1,505*	*3,310*			*1.5*	*3.2*	*1,450*	*3,190*	*895*	*1,970*							*1.0*	*2.3*
46	79.7	76.7		1.6	3.5	1,538	3,384			1.3	2.9	1,329	2,923	949	2,087								
	85.7	*82.5*		*1.7*	*3.8*	*1,654*	*3,639*			*1.4*	*3.1*	*1,429*	*3,143*	*1,020*	*2,244*								
47	66.9	67.7		1.4	3.0	1,358	2,987			1.1	2.4	1,162	2,556										
	71.3	*72.1*		*1.4*	*3.1*	*1,446*	*3,181*			*1.2*	*2.6*	*1,237*	*2,722*										
48	68.3	65.6		1.4	3.0	1,314	2,890			1.2	2.6	957	2,106	1,102	2,425			.8	1.7	.5	1.2	.9	1.9
	73.5	*70.5*		*1.5*	*3.2*	*1,413*	*3,108*			*1.3*	*2.8*	*1,030*	*2,265*	*1,185*	*2,608*			*.9*	*1.9*	*.5*	*1.2*	*1.0*	*2.1*

TABLE 4-101 COMPOSITION OF FEEDS

Feed #	Feed Name—Description	Moisture Basis: As Fed or M-F (moisture-free)	Dry Matter	Macrominerals							Microminerals				
				Calcium (Ca)	Phosphorus (P)	Sodium (Na)	Chlorine (Cl)	Magnesium (Mg)	Potassium (K)	Sulfur (S)	Cobalt (Co)	Copper (Cu)	Iron (Fe)	Manganese (Mn)	Zinc (Zn)
PROTEIN FEEDS			(%)	(%)	(%)	(%)	(%)	(%)	(%)	(%)	(ppm or mg/kg)	(ppm or mg/kg)	(%)	(ppm or mg/kg)	(ppm c mg/kg,
239	Sunflower meal, mech-extd, 41% protein	As Fed	92.9	.43	1.04				1.08					22.9	
		M-F	*100.0*	*.46*	*1.12*				*1.16*					*24.7*	
240	Sunflower meal, sol-extd, 44% protein	As Fed	93.0	.40	1.00		.10		1.00					23.0	
		M-F	*100.0*	*.43*	*1.07*		*.10*		*1.08*					*24.7*	
241	Tankage (digester tankage)	As Fed	92.0	5.95	3.17	1.67		.16	.56	.70	.184	38.7	.190	19.1	
		M-F	*100.0*	*6.46*	*3.45*	*1.82*		*.17*	*.61*	*.76*	*.200*	*42.1*	*.207*	*20.8*	
242	Urea (Feed grade)	As Fed													
		M-F	*100.0*												
243	Whey	As Fed	6.9	.05	.04				.19				.001		
		M-F	*100.0*	*.72*	*.65*				*2.75*				*.015*		
244	Whey, condensed	As Fed	63.6	.38	.58										
		M-F	*100.0*	*.60*	*.91*										
245	Whey, dried	As Fed	94.0	.87	.79	.48	.70	.13	1.20	1.05	.109	43.1	.016	4.6	
		M-F	*100.0*	*.94*	*.84*	*.51*	*.74*	*.14*	*1.28*	*1.12*	*.116*	*45.9*	*.017*	*4.9*	
246	Yeast, brewers' dried	As Fed	93.0	.15	1.48	.10		.25	1.48	.49	1.476	24.6	.005	5.9	37.4
		M-F	*100.0*	*.16*	*1.59*	*.11*		*.26*	*1.59*	*.53*	*1.587*	*26.5*	*.005*	*6.3*	*40.2*
247	Yeast, irradiated, dried	As Fed	93.9	.77	1.41				2.14						
		M-F	*100.0*	*.83*	*1.51*				*2.28*						
248	Yeast, torula, dried	As Fed	93.0	.57	1.68	.01		.13	1.88			13.4	.009	12.7	99.2
		M-F	*100.0*	*.61*	*1.81*	*.01*		*.14*	*2.02*			*14.4*	*.010*	*13.7*	*106.7*

ROTEIN FEEDS

Feed #	Fat-Soluble Vitamins				Water-Soluble Vitamins								
	A (1 mg car. = 1,667 IU-A)	Carotene (Pro-vitamin A)	E (α-tocopherol)	K	B_{12}	Biotin	Choline	Folic Acid (Folacin)	Niacin (Nicotinic Acid)	Pantothenic Acid	Pyridoxine (B_6)	Riboflavin (B_2)	Thiamin (B_1)
		(ppm or mg/kg)	*(ppm or mg/kg)*	*(ppm or mg/kg)*	*(ppm or mg/kg)*	*(ppm or mg/kg)*	*(ppm or mg/kg)*	*(ppm or mg/kg)*	*(ppm or mg/kg)*	*(ppm or mg/kg)*	*(ppm or mg/kg)*	*(ppm or mg/kg)*	*(ppm or mg/kg)*
39													
:40			11.0				2,900		220.0	10.0	16.00	3.1	
			11.8				*3,118*		*236.5*	*10.7*	*17.20*	*3.3*	
41							2,169	1.50	39.2	2.4		2.4	.4
							2,358	*1.60*	*42.6*	*2.6*		*2.6*	*.5*
'42													
'43	100	.1							1.0			1.4	.3
	1400	*.8*							*14.5*			*20.3*	*4.3*
:44										14.6		16.0	3.3
										23.0		*25.2*	*5.2*
'45	460	.3			.0300	.40	1,698	.90	11.2	47.7	2.50	29.9	3.7
	500	*.3*			*.0319*	*.42*	*1,806*	*.96*	*11.9*	*50.7*	*2.66*	*31.8*	*3.9*
:46						1.08	4,183	14.76	467.5	110.7	46.75	32.0	110.7
						1.16	*4,497*	*15.87*	*502.6*	*119.0*	*50.26*	*34.4*	*119.0*
:47												18.5	
												19.7	
:48						1.12	2,909	23.25	500.1	82.9	29.48	44.4	6.2
						1.20	*3,129*	*25.00*	*537.8*	*89.1*	*31.70*	*47.7*	*6.7*

TABLE 4-101 COMPOSITION OF FEEDS

Feed #	Feed Name—Description	Moisture Basis: As Fed or M-F (moisture-free)	Proximate Analysis: Dry Matter	Ash	Crude Fiber	Ether Extract (Fat)	N-Free Extract	Crude Protein	Digestible Protein: Ruminant	Non-Ruminant	Horse Rabbit
			(%)	(%)	(%)	(%)	(%)	(%)	(%)	(%)	(%)
DRY FORAGES & ROUGHAGES											
249	Alfalfa hay, all analyses *(Medicago sativa)*	As Fed	91.4	9.0	28.0	1.7	37.2	15.5	10.8	7.3	11.1
		M-F	*100.0*	*9.9*	*30.6*	*1.9*	*40.6*	*17.0*	*11.8*	*8.0*	*12.2*
250	Alfalfa hay, prebloom	As Fed	84.5	7.0	23.7	2.5	34.6	16.6	10.6		12.1
		M-F	*100.0*	*8.3*	*28.1*	*3.0*	*41.0*	*19.6*	*12.6*		*14.3*
251	Alfalfa hay, early bloom	As Fed	90.0	7.8	26.5	2.0	36.9	16.8	11.9		11.1
		M-F	*100.0*	*8.7*	*29.4*	*2.2*	*41.0*	*18.7*	*13.2*		*12.3*
252	Alfalfa hay, midbloom	As Fed	89.2	7.3	27.5	1.4	37.6	15.3	10.9		9.8
		M-F	*100.0*	*8.2*	*30.8*	*1.6*	*42.2*	*17.2*	*12.2*		*11.0*
253	Alfalfa hay, full bloom	As Fed	87.7	7.8	29.4	1.7	34.7	14.1	10.0		8.7
		M-F	*100.0*	*8.9*	*33.5*	*1.9*	*39.6*	*16.1*	*11.4*		*9.9*
254	Alfalfa hay, mature	As Fed	91.2	6.7	36.2	1.0	34.8	12.5	8.5		7.0
		M-F	*100.0*	*7.4*	*39.7*	*1.1*	*38.2*	*13.6*	*9.3*		*7.7*
255	Alfalfa hay, very leafy	As Fed	90.9	8.6	22.4	2.5	39.2	18.2	13.4		13.0
		M-F	*100.0*	*9.5*	*24.6*	*2.8*	*43.1*	*20.0*	*14.6*		*14.3*
256	Alfalfa leaf meal, dehy	As Fed	92.3	11.3	18.5	3.1	38.6	20.8	15.7		15.1
		M-F	*100.0*	*12.2*	*20.1*	*3.4*	*41.7*	*22.6*	*17.0*		*16.4*
257	Alfalfa meal, dehy, min. 15% protein	As Fed	92.8	9.1	26.2	2.3	39.9	15.3	11.3		9.6
		M-F	*100.0*	*9.8*	*28.2*	*2.5*	*43.0*	*16.5*	*12.2*		*10.3*
258	Alfalfa meal, dehy, min. 22% protein	As Fed	92.9	7.7	18.5	3.7	40.5	22.5	17.5		16.7
		M-F	*100.0*	*8.3*	*19.9*	*4.0*	*43.6*	*24.2*	*19.0*		*18.2*
259	Alfalfa & bromegrass hay *(Medicago sativa, Bromus inermis)*	As Fed	82.5	6.4	27.8	2.2	32.8	13.3	9.2		9.2
		M-F	*100.0*	*7.8*	*33.7*	*2.7*	*39.6*	*16.2*	*11.2*		*11.2*
260	Alfalfa & grass hay	As Fed	89.6	6.2	30.4	1.5	39.5	12.0	7.8		8.0
		M-F	*100.0*	*6.9*	*33.9*	*1.7*	*44.1*	*13.4*	*8.7*		*8.9*
261	Alfalfa & orchardgrass hay *(Medicago sativa, Dactylis glomerata)*	As Fed	91.2	7.7	29.5	2.0	37.2	14.8	10.2		10.2
		M-F	*100.0*	*8.4*	*32.4*	*2.2*	*40.8*	*16.2*	*11.3*		*11.2*
262	Alfalfa & timothy hay *(Medicago sativa, Phleum pratense)*	As Fed	92.9	5.0	30.6	1.3	46.2	9.8	5.7		6.1
		M-F	*100.0*	*5.4*	*32.9*	*1.4*	*49.8*	*10.5*	*6.1*		*6.6*
263	Almond hulls *(Prunus amygdalus)*	As Fed	91.0	5.9	13.8	2.2	65.2	3.9	.5		1.4
		M-F	*100.0*	*6.5*	*15.2*	*2.4*	*71.7*	*4.2*	*.5*		*1.5*
264	Almond shells *(Prunus amygdalus)*	As Fed	92.5	3.4	41.8	4.8					
		M-F	*100.0*	*3.7*	*45.2*	*5.2*					
265	Alsike clover hay (Trifolium hybridum)	As Fed	87.9	7.6	25.8	2.4	39.2	12.9	8.6		7.6
		M-F	*100.0*	*8.7*	*29.4*	*2.7*	*44.5*	*14.7*	*9.8*		*8.7*
266	Bagasse, sugarcane, pulp, dehy *(Saccharum officinarum)*	As Fed	91.5	2.8	44.5	.7	42.0	1.5	0		0
		M-F	*100.0*	*3.1*	*48.6*	*.7*	*45.9*	*1.7*	*0*		*0*
267	Barley straw *(Hordeum vulgare)*	As Fed	88.2	6.1	37.4	1.7	39.4	3.6	.5		.4
		M-F	*100.0*	*6.9*	*42.4*	*1.9*	*44.7*	*4.1*	*.6*		*.5*
268	Beet, sugar (see Sugar beet)										
269	Bermudagrass hay, sun cured *(Cynodon dactylon)*	As Fed	91.1	7.5	27.0	1.6	46.9	8.1	4.3		4.1
		M-F	*100.0*	*8.2*	*29.6*	*1.8*	*51.5*	*8.9*	*4.7*		*4.5*

RY FORAGES AND ROUGHAGES

	TDN			Digestible Energy						Metabolizable Energy								Net Energy-Ruminant					
												Non-Ruminant											
Feed #	Ruminant	Non-Ruminant	Horse Rabbit	Ruminant		Non-Ruminant		Horse Rabbit		Ruminant		ME		Chicken ME_n		Horse Rabbit		NE_m		NE_g		NE_{lc}	
	(%)	(%)	(%)	Mcal per		kcal per		Mcal per		Mcal per		kcal per		kcal per		Mcal per		Mcal per		Mcal per		Mcal per	
				lb	kg	lb	kg	lb	kg	lb	kg	lb	kg	lb	kg	lb	kg	lb	kg	lb	kg	lb	kg
49	50.3	32.2	45.4	1.0	2.2	645	1,419	.9	2.0	.8	1.8	597	1,313			.7	1.6						
	55.1	*35.2*	*49.7*	*1.1*	*2.4*	*706*	*1,553*	*1.0*	*2.2*	*.9*	*2.0*	*654*	*1,438*			*.8*	*1.8*						
50	50.4			1.0	2.2					.8	1.8							.5	1.1	.3	.6	.5	1.2
	59.7			*1.2*	*2.6*					*1.0*	*2.2*							*.6*	*1.3*	*.3*	*.7*	*.7*	*1.5*
51	51.0		53.1	1.0	2.3			1.0	2.1	.8	1.8							.5	1.2	.2	.5	.5	1.1
	56.7		*59.0*	*1.1*	*2.5*			*1.0*	*2.3*	*1.0*	*2.1*							*.6*	*1.3*	*.2*	*.5*	*.6*	*1.3*
52	49.1		50.8	1.0	2.2			.9	2.0	.8	1.8							.5	1.1	.2	.4	.5	1.1
	55.0		*57.0*	*1.1*	*2.4*			*1.0*	*2.3*	*.9*	*2.0*							*.5*	*1.2*	*.2*	*.5*	*.5*	*1.2*
53	47.4		46.5	1.0	2.1			.8	1.8	.8	1.7							.5	1.0	.2	.4	.5	1.1
	54.0		*53.0*	*1.1*	*2.4*			*1.0*	*2.1*	*.9*	*2.0*							*.5*	*1.2*	*.2*	*.5*	*.6*	*1.3*
54	47.2		44.7	1.0	2.1			.8	1.7	.8	1.7							.5	1.0	.1	.3	.5	1.1
	51.7		*49.0*	*1.0*	*2.3*			*.9*	*1.9*	*.9*	*1.9*							*.5*	*1.1*	*.2*	*.4*	*.5*	*1.2*
55	55.9			1.1	2.5					.9	2.0												
	61.5			*1.2*	*2.7*					*1.0*	*2.2*												
56	58.2			1.2	2.6					1.0	2.1			718	1,580								
	63.1			*1.3*	*2.8*					*1.0*	*2.3*			*778*	*1,712*								
57	53.5	44.1	53.8	1.1	2.4	882	1,941	1.0	2.1	.9	1.9	724	1,592	305	670			.5	1.2	.3	.6	.6	1.4
	57.7	*47.5*	*58.0*	*1.1*	*2.5*	*951*	*2,092*	*1.0*	*2.3*	*1.0*	*2.1*	*780*	*1,716*	*328*	*722*			*.6*	*1.3*	*.3*	*.7*	*.7*	*1.5*
58	62.1			1.2	2.7					1.0	2.3			802	1,764								
	67.3			*1.4*	*3.0*					*1.1*	*2.4*			*863*	*1,899*								
59	45.9			.9	2.0					.8	1.7							.5	1.0	.2	.4	.5	1.0
	55.6			*1.1*	*2.5*					*.9*	*2.0*							*.5*	*1.2*	*.2*	*.5*	*.5*	*1.2*
60	50.4			1.0	2.2					.8	1.8												
	56.3			*1.1*	*2.5*					*.9*	*2.0*												
61	51.4			1.0	2.3					.9	1.9												
	56.4			*1.1*	*2.5*					*.9*	*2.0*												
62	52.1			1.0	2.3					.9	1.9												
	56.1			*1.1*	*2.5*					*.9*	*2.0*												
63	68.1			1.4	3.0					1.1	2.5												
	74.8			*1.5*	*3.3*					*1.2*	*2.7*												
64																							
65	51.0		49.2	1.0	2.3			.9	1.9	.8	1.8							.5	1.1	.3	.6	.6	1.3
	58.0		*56.0*	*1.2*	*2.6*			*1.0*	*2.2*	*1.0*	*2.1*							*.6*	*1.3*	*.3*	*.7*	*.7*	*1.5*
66	34.2			.7	1.5					.5	1.2												
	37.4			*.8*	*1.7*					*.6*	*1.4*												
67	38.5		33.5	.8	1.7			.6	1.4	.6	1.4							.4	.9	.1	.2	.4	.9
	43.7		*38.0*	*.9*	*1.9*			*.7*	*1.6*	*.7*	*1.6*							*.5*	*1.0*	*.1*	*.2*	*.5*	*1.0*
68																							
69	44.2			.9	1.9					.7	1.6							.4	.9	.1	.2		
	48.5			*1.0*	*2.1*					*.8*	*1.8*							*.5*	*1.0*	*.1*	*.2*		

TABLE 4-101 COMPOSITION OF FEEDS

Feed #	Feed Name—Description	Moisture Basis: As Fed or M-F (moisture-free)	Dry Matter	Macrominerals: Calcium (Ca)	Phosphorus (P)	Sodium (Na)	Chlorine (Cl)	Magnesium (Mg)	Potassium (K)	Sulfur (S)	Microminerals: Cobalt (Co)	Copper (Cu)	Iron (Fe)	Manganese (Mn)	Zinc (Zn)
DRY FORAGES & ROUGHAGES			(%)	(%)	(%)	(%)	(%)	(%)	(%)	(%)	(ppm or mg/kg)	(ppm or mg/kg)	(%)	(ppm or mg/kg)	(ppm o mg/kg)
249	Alfalfa hay, all analyses	As Fed	91.4	1.29	.21	.15	.28	.31	1.99	.29	.205	18.5	.017	56.5	
		M-F	*100.0*	*1.41*	*.24*	*.17*	*.31*	*.34*	*2.18*	*.32*	*.225*	*20.2*	*.019*	*61.9*	
250	Alfalfa hay, prebloom	As Fed	84.5	1.06	.19	.18	.29	.21	1.99	.53			.021	29.0	
		M-F	*100.0*	*1.25*	*.23*	*.22*	*.34*	*.25*	*2.36*	*.63*			*.025*	*34.3*	
251	Alfalfa hay, early bloom	As Fed	90.0	1.13	.21	.14	.34	.27	1.87	.27	.081	12.1	.018	28.4	
		M-F	*100.0*	*1.25*	*.23*	*.15*	*.38*	*.30*	*2.08*	*.30*	*.090*	*13.4*	*.020*	*31.5*	
252	Alfalfa hay, midbloom	As Fed	89.2	1.20	.20	.13	.34	.31	1.30	.27		13.7	.009	14.7	
		M-F	*100.0*	*1.35*	*.22*	*.15*	*.38*	*.35*	*1.34*	*.30*		*15.4*	*.010*	*16.5*	
253	Alfalfa hay, full bloom	As Fed	87.7	1.12	.18			.31	.48		.109	11.6	.018	29.6	
		M-F	*100.0*	*1.28*	*.20*			*.35*	*.55*		*.124*	*13.4*	*.020*	*33.7*	
254	Alfalfa hay, mature	As Fed	91.2	.95	.15			.33	2.18				.014		
		M-F	*100.0*	*1.04*	*.16*			*.36*	*2.39*				*.015*		
255	Alfalfa hay, very leafy	As Fed	90.9	1.62	.24				1.96						
		M-F	*100.0*	*1.78*	*.27*				*2.15*						
256	Alfalfa leaf meal, dehy	As Fed	92.3	1.74	.28	.06					.200				
		M-F	*100.0*	*1.88*	*.30*	*.07*					*.216*				
257	Alfalfa meal, dehy, min. 15% protein	As Fed	92.8	1.22	.22	.07		.29	2.32	.17	.18	10.4	.031	29.0	20.0
		M-F	*100.0*	*1.32*	*.24*	*.08*		*.30*	*2.50*	*.18*	*.19*	*11.2*	*.033*	*31.1*	*21.5*
258	Alfalfa meal, dehy, min. 22% protein	As Fed	92.9	1.48	.28	.11	.52	.34	2.51	.30	.30	11.1	.045	37.0	20.0
		M-F	*100.0*	*1.59*	*.30*	*.12*	*.56*	*.36*	*2.70*	*.32*	*.32*	*11.9*	*.048*	*39.8*	*21.5*
259	Alfalfa & bromegrass hay	As Fed	82.5	.85	.25	.35	.39	.45	1.53	.19	.071	13.5	.011	35.1	
		M-F	*100.0*	*1.03*	*.30*	*.42*	*.47*	*.54*	*1.85*	*.23*	*.086*	*16.3*	*.013*	*42.5*	
260	Alfalfa & grass hay	As Fed	89.6	1.18	.24									30.9	
		M-F	*100.0*	*1.32*	*.27*									*34.4*	
261	Alfalfa & orchardgrass hay	As Fed	91.2												
		M-F	*100.0*												
262	Alfalfa & timothy hay	As Fed	92.9												
		M-F	*100.0*												
263	Almond hulls	As Fed	91.0												
		M-F	*100.0*												
264	Almond shells	As Fed	92.5												
		M-F	*100.0*												
265	Alsike clover hay	As Fed	87.9	1.15	.22	.41	.69	.40	1.49	.18		5.3	.026	60.7	
		M-F	*100.0*	*1.31*	*.25*	*.46*	*.78*	*.45*	*1.70*	*.21*		*6.0*	*.030*	*69.0*	
266	Bagasse, sugarcane, pulp, dehy	As Fed	91.5												
		M-F	*100.0*												
267	Barley straw	As Fed	88.2	.30	.08	.12	.59	.17	2.01	.15			.026	15.2	
		M-F	*100.0*	*.34*	*.09*	*.14*	*.68*	*.19*	*2.28*	*.17*			*.030*	*17.2*	
268	Beet, sugar (see Sugar beet)														
269	Bermudagrass hay, sun cured	As Fed	91.1	.42	.18			.15	1.34				.026		
		M-F	*100.0*	*.46*	*.20*			*.17*	*1.47*				*.029*		

RY FORAGES AND ROUGHAGES

Feed #	Fat-Soluble Vitamins				Water-Soluble Vitamins								
	A (1 mg car. = 1,667 IU-A)	Carotene (Pro-vitamin A)	E (α-tocopherol)	K	B_{12}	Biotin	Choline	Folic Acid (Folacin)	Niacin (Nicotinic Acid)	Pantothenic Acid	Pyridoxine (B_6)	Riboflavin (B_2)	Thiamin (B_1)
		(ppm or mg/kg)	*(ppm or mg/kg)*	*(ppm or mg/kg)*	*(ppm or mg/kg)*	*(ppm or mg/kg)*	*(ppm or mg/kg)*	*(ppm or mg/kg)*	*(ppm or mg/kg)*	*(ppm or mg/kg)*	*(ppm or mg/kg)*	*(ppm or mg/kg)*	*(ppm or mg/kg)*
249	108,688	65.2	101.6		0	.16		3.10					
	119,023	*71.4*	*111.2*		*0*	*.18*		*3.40*					
250													
251	190,871	114.5	23.4										
	212,042	*127.2*	*26.0*										
252	49,510	29.7										9.5	
	55,511	*33.3*										*10.6*	
253	54,010	32.4											
	61,679	*37.0*											
254	24,005	14.4											
	26,339	*15.8*											
255													
256													
257	106,188	63.7	98.6	9.86			894	1.55	38.9	18.1	6.53	10.6	3.0
	114,356	*68.6*	*106.2*	*10.63*			*963*	*1.67*	*41.9*	*19.5*	*7.04*	*11.4*	*3.2*
258	420,917	252.5	151.0	8.40			1,853	3.00	58.8	33.0	7.80	17.4	4.2
	453,091	*271.8*	*162.5*	*9.04*			*1,995*	*3.23*	*63.3*	*35.5*	*8.40*	*18.7*	*4.5*
259	35,840	21.5							22.3	19.4		5.5	
	43,342	*26.0*							*27.0*	*23.5*		*6.7*	
260													
261													
262													
263													
264													
265	273,388	164.0											
	311,729	*187.0*											
266													
267	3,334	2.0											
	3,834	*2.3*											
268													
269	195,372	117.2											
	214,543	*128.7*											

TABLE 4-101 COMPOSITION OF FEEDS

Feed #	Feed Name—Description	Moisture Basis: As Fed or M-F (moisture-free)	Proximate Analysis						Digestible Protein		
			Dry Matter	Ash	Crude Fiber	Ether Extract (Fat)	N-Free Extract	Crude Protein	Ruminant	Non-Ruminant	Horse Rabbit
DRY FORAGES & ROUGHAGES			(%)	(%)	(%)	(%)	(%)	(%)	(%)	(%)	(%)
270	Bromegrass hay, all analyses *(Bromus spp)*	As Fed	89.7	7.1	28.7	2.4	40.9	10.6	5.1		5.6
		M-F	*100.0*	*7.9*	*32.0*	*2.7*	*45.6*	*11.8*	*5.7*		*6.2*
271	Canarygrass, reed, hay *(Phalaris arundinacea)*	As Fed	91.3	7.5	31.6	2.8	40.0	9.4	4.8		5.5
		M-F	*100.0*	*8.2*	*34.6*	*3.1*	*43.8*	*10.3*	*5.3*		*6.0*
272	Cereals, immature, dehy	As Fed	92.9	14.5	16.2	4.8	32.8	24.6	19.2		
		M-F	*100.0*	*15.6*	*17.4*	*5.2*	*35.3*	*26.5*	*20.7*		
273	Clover, alsike, hay *(Trifolium hybridum)*	As Fed	87.9	7.6	25.8	2.4	39.2	12.9	8.6		7.6
		M-F	*100.0*	*8.7*	*29.4*	*2.7*	*44.5*	*14.7*	*9.8*		*8.7*
274	Clover, crimson, hay *(Trifolium incarnatum)*	As Fed	87.4	8.0	28.1	2.4	34.1	14.8	10.1		9.5
		M-F	*100.0*	*9.2*	*32.2*	*2.7*	*39.0*	*16.9*	*11.6*		*10.9*
275	Clover & grass, hay	As Fed	88.7	6.3	29.3	2.4	40.2	10.5	6.5		6.8
		M-F	*100.0*	*7.1*	*33.0*	*2.7*	*45.4*	*11.8*	*7.3*		*7.7*
276	Clover, Ladino, hay *(Trifolium repens)*	As Fed	91.2	8.5	17.5	1.7	42.2	21.0	13.1		13.5
		M-F	*100.0*	*9.3*	*19.2*	*1.9*	*46.6*	*23.0*	*14.4*		*14.8*
277	Clover, red, hay, all analyses *(Trifolium pratense)*	As Fed	88.3	7.0	26.5	2.3	39.5	13.0	7.9		7.3
		M-F	*100.0*	*7.9*	*30.0*	*2.6*	*44.8*	*14.7*	*9.0*		*8.3*
278	Clover, red & grass hay	As Fed	88.7	5.9	29.9	1.6	37.3	14.0	9.6		9.7
		M-F	*100.0*	*6.6*	*33.7*	*1.8*	*42.1*	*15.8*	*10.8*		*10.9*
279	Clover & timothy hay *(Trifolium spp)*	As Fed	88.5	5.6	31.7	3.0	38.9	9.3	5.4		5.8
		m-F	*100.0*	*6.3*	*35.8*	*3.4*	*44.0*	*10.5*	*6.1*		*6.6*
280	Corn cobs, grnd *(Zea mays)*	As Fed	90.4	1.6	32.4	.6	53.3	2.5	.2		.7
		M-F	*100.0*	*1.8*	*35.8*	*.7*	*58.9*	*2.8*	*.2*		*.8*
281	Corn fodder, dry, all analyses *(Zea mays)*	As Fed	82.4	5.4	21.3	2.0	46.4	7.3	3.6		4.1
		M-F	*100.0*	*6.6*	*25.9*	*2.4*	*56.2*	*8.9*	*4.4*		*5.0*
282	Corn husks	As Fed	88.6	3.2	29.3	.8	52.0	3.3	.2		1.0
		M-F	*100.0*	*3.6*	*33.0*	*.9*	*58.7*	*3.8*	*.3*		*1.1*
283	Corn stover, w/o ears	As Fed	87.2	6.4	32.4	1.1	42.2	5.1	2.2		3.1
		M-F	*100.0*	*7.3*	*37.1*	*1.3*	*48.4*	*5.9*	*2.5*		*3.6*
284	Cotton gin trash *(Gossypium spp)*	As Fed	90.3	7.7	33.1	1.5	41.3	6.7	3.0		3.7
		M-F	*100.0*	*8.5*	*36.7*	*1.7*	*45.7*	*7.4*	*3.4*		*4.1*
285	Cottonseed hulls	As Fed	90.3	2.6	42.9	1.4	39.5	3.9	.3		1.4
		M-F	*100.0*	*2.9*	*47.5*	*1.6*	*43.7*	*4.3*	*.3*		*1.6*
286	Crested wheatgrass hay *(Agropyron cristatum)*	As Fed	92.0	6.3	30.0	2.2	43.6	9.9	4.7		4.1
		M-F	*100.0*	*6.8*	*32.6*	*2.4*	*47.4*	*10.8*	*5.1*		*4.5*
287	Crimson clover hay *(Trifolium incarnatum)*	As Fed	87.4	8.0	28.1	2.4	34.1	14.8	10.1		9.5
		M-F	*100.0*	*9.2*	*32.2*	*2.7*	*39.0*	*16.9*	*11.6*		*10.9*
288	Fescue, alta, hay *(Festuca arundinacia)*	As Fed	89.0	5.9	33.7	2.0	39.7	7.7	4.0		3.3
		M-F	*100.0*	*6.6*	*37.9*	*2.2*	*44.6*	*8.7*	*4.5*		*3.7*
289	Fescue, meadow, hay *(Festuca elatior)*	As Fed	86.5	8.0	28.1	2.2	40.0	8.2	4.7		5.0
		M-F	*100.0*	*9.2*	*32.5*	*2.6*	*46.2*	*9.5*	*5.5*		*5.8*
290	Foxtail, meadow, hay *(Alopecurus pratensis)*	As Fed	88.1	8.2	25.1	1.9	40.0	12.9	8.7		8.8
		M-F	*100.0*	*9.4*	*28.5*	*2.2*	*45.2*	*14.7*	*9.9*		*10.0*

RY FORAGES AND ROUGHAGES

Feed #	TDN Ruminant (%)	TDN Non-Ruminant (%)	TDN Horse Rabbit (%)	DE Ruminant Mcal per lb	DE Ruminant Mcal per kg	DE Non-Ruminant kcal per lb	DE Non-Ruminant kcal per kg	DE Horse Rabbit Mcal per lb	DE Horse Rabbit Mcal per kg	ME Ruminant Mcal per lb	ME Ruminant Mcal per kg	ME Non-Ruminant ME kcal per lb	ME Non-Ruminant ME kcal per kg	ME Non-Ruminant Chicken ME_n kcal per lb	ME Non-Ruminant Chicken ME_n kcal per kg	ME Horse Rabbit Mcal per lb	ME Horse Rabbit Mcal per kg	NE_m Mcal per lb	NE_m Mcal per kg	NE_g Mcal per lb	NE_g Mcal per kg	NE_{lc} Mcal per lb	NE_{lc} Mcal per kg
270	47.3		39.5	1.0	2.1			.8	1.7	.8	1.7							.5	1.0	.2	.4	.5	1.0
	52.7		*44.0*	*1.0*	*2.3*			*.9*	*1.9*	*.9*	*1.9*							*.5*	*1.1*	*.2*	*.4*	*.5*	*1.1*
271	46.1		42.0	.9	2.0			.8	1.8	.8	1.7					.7	1.5	.5	1.0	.1	.3		
	50.5		*46.0*	*1.0*	*2.2*			*.9*	*1.9*	*.8*	*1.8*					*.8*	*1.7*	*.5*	*1.1*	*.1*	*.3*		
272	56.2			1.1	2.5					.9	2.0												
	60.5			*1.2*	*2.7*					*1.0*	*2.2*												
273	51.0		49.2	1.0	2.3			.9	1.9	.8	1.8							.5	1.1	.3	.6	.6	1.3
	58.0		*56.0*	*1.2*	*2.6*			*1.0*	*2.2*	*1.0*	*2.1*							*.6*	*1.3*	*.3*	*.7*	*.7*	*1.5*
274	50.1		48.1	1.0	2.2			.9	1.9	.8	1.8							.5	1.1	.3	.6	.5	1.2
	57.3		*55.0*	*1.1*	*2.5*			*1.0*	*2.2*	*.9*	*2.0*							*.6*	*1.3*	*.3*	*.7*	*.6*	*1.4*
275	50.8			1.0	2.2					.8	1.8												
	57.2			*1.1*	*2.5*					*1.0*	*2.1*												
276	54.4			1.1	2.4					.9	2.0												
	59.7			*1.2*	*2.6*					*1.0*	*2.2*												
277	51.8		48.6	1.0	2.3			.9	1.9	.9	1.9					.7	1.5	.5	1.1	.2	.5	.5	1.2
	58.7		*55.0*	*1.2*	*2.6*			*1.0*	*2.2*	*1.0*	*2.1*					*.8*	*1.7*	*.6*	*1.3*	*.3*	*.6*	*.6*	*1.3*
278	49.4			1.0	2.2					.8	1.8												
	55.7			*1.1*	*2.5*					*.9*	*2.0*												
279	47.9			1.0	2.1																		
	54.1			*1.1*	*2.4*																		
280	43.1		26.2	.9	1.9			.5	1.1	.7	1.6							.5	1.0	.1	.2	.4	.8
	47.7		*29.0*	*1.0*	*2.1*			*.5*	*1.2*	*.8*	*1.7*							*.5*	*1.1*	*.1*	*.3*	*.4*	*.9*
281	53.3			1.1	2.4					.9	1.9							.5	1.2	.2	.5	.6	1.3
	64.7			*1.3*	*2.9*					*1.0*	*2.3*							*.6*	*1.4*	*.3*	*.6*	*.7*	*1.5*
282	50.2			1.0	2.2					.8	1.8												
	56.7			*1.1*	*2.5*					*1.0*	*2.1*												
283	52.3			1.0	2.3					.9	1.9							.5	1.1	.2	.5	.5	1.1
	60.0			*1.2*	*2.6*					*1.0*	*2.2*							*.5*	*1.2*	*.3*	*.6*	*.6*	*1.3*
284	42.4			.9	1.9					.7	1.5												
	45.9			*.9*	*2.0*					*.8*	*1.7*												
285	42.4			.9	1.9					.7	1.5							.4	.9	.1	.2	.3	.7
	47.0			*1.0*	*2.1*					*.8*	*1.7*							*.5*	*1.0*	*.1*	*.2*	*.4*	*.8*
286	50.6			1.0	2.2					.8	1.8							.4	1.1	.2	.5		
	55.0			*1.1*	*2.4*					*.9*	*2.0*							*.5*	*1.2*	*.3*	*.6*		
287	50.1		48.1	1.0	2.2			.9	1.9	.8	1.8							.5	1.1	.3	.6	.5	1.2
	57.3		*55.0*	*1.1*	*2.5*			*1.0*	*2.2*	*1.0*	*2.1*							*.6*	*1.3*	*.3*	*.7*	*.6*	*1.4*
288	47.0		40.1	1.0	2.1			.7	1.6	.8	1.7												
	52.7		*45.0*	*1.0*	*2.3*			*.8*	*1.8*	*.9*	*1.9*												
289	49.6			1.0	2.2					.8	1.8												
	57.3			*1.3*	*2.5*					*1.0*	*2.1*												
290	54.7			1.1	2.4					.9	2.0												
	62.1			*1.2*	*2.7*					*1.0*	*2.2*												

TABLE 4-101 COMPOSITION OF FEEDS

Feed #	Feed Name—Description	Moisture Basis: As Fed or M-F (moisture-free)	Dry Matter	Macrominerals: Calcium (Ca)	Phosphorus (P)	Sodium (Na)	Chlorine (Cl)	Magnesium (Mg)	Potassium (K)	Sulfur (S)	Microminerals: Cobalt (Co)	Copper (Cu)	Iron (Fe)	Manganese (Mn)	Zinc (Zn)
DRY FORAGES & ROUGHAGES			(%)	(%)	(%)	(%)	(%)	(%)	(%)	(%)	(ppm or mg/kg)	(ppm or mg/kg)	(%)	(ppm or mg/kg)	(ppm or mg/kg)
270	Bromegrass hay, all analyses	As Fed	89.7	.36	.18	.57	.48	.20	2.26	.17		7.7	.011	52.0	
		M-F	100.0	.40	.20	.63	.54	.22	2.52	.19		8.6	.012	58.0	
271	Canarygrass, reed, hay	As Fed	91.3	.31	.23			.24	2.15			10.9	.018	84.4	
		M-F	100.0	.34	.25			.26	2.35			11.9	.020	92.4	
272	Cereals, immature, dehy	As Fed	92.9	.66	.46										
		M-F	100.0	.71	.50										
273	Clover, alsike, hay	As Fed	87.9	1.15	.22	.41	.69	.40	1.49	.18		5.3	.026	60.7	
		M-F	100.0	1.31	.25	.46	.78	.45	1.70	.21		6.0	.030	69.0	
274	Clover, crimson, hay	As Fed	87.4	1.24	.16	.34	.55	.24	1.35	.24			.061	149.7	
		M-F	100.0	1.42	.18	.39	.63	.27	1.54	.28			.070	171.3	
275	Clover & grass, hay	As Fed	88.7	.87	.20	.17	.63	.25	1.43	.13		7.0	.022	91.8	
		M-F	100.0	.99	.23	.19	.71	.28	1.62	.14		7.9	.025	103.5	
276	Clover, Ladino, hay	As Fed	91.2	1.26	.36	.12	.26	.46	1.98	.20	.137	8.0	.055	120.8	15.5
		M-F	100.0	1.38	.40	.13	.28	.50	2.17	.22	.150	8.8	.060	132.5	17.0
277	Clover, red, hay, all analyses	As Fed	88.3	1.39	.19	.14	.23	.42	1.76	.15	.132	9.8	.009	60.5	
		M-F	100.0	1.57	.22	.16	.26	.47	1.99	.17	.150	11.1	.010	68.5	
278	Clover, red & grass hay	As Fed	88.7								.137				
		M-F	100.0								.154				
279	Clover & timothy hay	As Fed	88.5												
		M-F	100.0												
280	Corn cobs, grnd	As Fed	90.4	.11	.04			.06	.76	.42	.118	6.6	.021	5.6	
		M-F	100.0	.12	.04			.07	.84	.47	.130	7.3	.023	6.2	
281	Corn fodder, dry, all analyses	As Fed	82.4	.42	.21	.02	.16	.24	.77	.12		4.0	.008	56.1	
		M-F	100.0	.50	.25	.03	.19	.29	.93	.14		4.8	.010	68.1	
282	Corn husks	As Fed	88.6	.16	.13				.57						
		M-F	100.0	.18	.14				.65						
283	Corn stover, w/o ears	As Fed	87.2	.43	.08	.06		.39	1.43	.15		4.4	.019	118.4	
		M-F	100.0	.49	.09	.07		.45	1.64	.17		5.1	.022	135.8	
284	Cotton gin trash	As Fed	90.3												
		M-F	100.0												
285	Cottonseed hulls	As Fed	90.3	.14	.09	.02	.02	.13	.76		.018	14.2	.009	106.3	19.9
		M-F	100.0	.16	.10	.02	.02	.14	.84		.020	15.7	.010	117.7	22.0
286	Crested wheatgrass hay	As Fed	92.0	.30	.19						.221				
		M-F	100.0	.33	.21						.240				
287	Crimson clover hay	As Fed	87.4	1.24	.16	.34	.55	.24	1.35	.24			.061	149.7	
		M-F	100.0	1.42	.18	.39	.63	.27	1.54	.28			.070	171.3	
288	Fescue, alta, hay	As Fed	89.0	.34	.21			.22	2.12						
		M-F	100.0	.39	.24			.24	2.38						
289	Fescue, meadow, hay	As Fed	86.5	.43	.19			.43	1.39					21.2	
		M-F	100.0	.50	.22			.50	1.60					24.5	
290	Foxtail, meadow, hay	As Fed	88.1												
		M-F	100.0												

RY FORAGES AND ROUGHAGES

Feed #	Fat-Soluble Vitamins				Water-Soluble Vitamins									
	A (1 mg car. = 1,667 IU-A)	Carotene (Pro-vitamin A)	E (α-tocopherol)	K	B_{12}	Biotin	Choline	Folic Acid (Folacin)	Niacin (Nicotinic Acid)	Pantothenic Acid	Pyridoxine (B_6)	Riboflavin (B_2)	Thiamin (B_1)
		(ppm or mg/kg)	*(ppm or mg/kg)*	*(ppm or mg/kg)*	*(ppm or mg/kg)*	*(ppm or mg/kg)*	*(ppm or mg/kg)*	*(ppm or mg/kg)*	*(ppm or mg/kg)*	*(ppm or mg/kg)*	*(ppm or mg/kg)*	*(ppm or mg/kg)*	*(ppm or mg/kg)*
270	18,004	10.8											
	20,004	*12.0*											
271	10,688	6.4											
	11,669	*7.0*											
272													
273	273,388	164.0											
	311,729	*187.0*											
274	51,010	30.6											
	58,345	*35.0*											
275	24,172	14.5											
	27,172	*16.3*											
276	245,049	147.0											
	268,720	*161.2*											
277	53,510	32.1				.09							
	60,512	*36.3*				*.11*							
278	16,670	10.0											
	18,670	*11.2*											
279													
280	1,000	.6											
	1,167	*.7*											
281	6,001	3.6											
	7,335	*4.4*											
282													
283													
284													
285													
286	34,174	20.5											
	37,174	*22.3*											
287	51,010	30.6											
	58,345	*35.0*											
288	30,840	18.5											
	34,507	*20.7*											
289	103,021	61.8	117.3										
	119,024	*71.4*	*135.6*										
290													

TABLE 4-101 COMPOSITION OF FEEDS-

Feed #	Feed Name—Description	Moisture Basis: As Fed or M-F (moisture-free)	Proximate Analysis: Dry Matter	Ash	Crude Fiber	Ether Extract (Fat)	N-Free Extract	Crude Protein	Digestible Protein: Ruminant	Non-Ruminant	Horse Rabbit
DRY FORAGES & ROUGHAGES			(%)	(%)	(%)	(%)	(%)	(%)	(%)	(%)	(%)
291	Foxtail, millet, hay (*Setaria italica*)	As Fed	85.8	7.0	25.0	2.8	42.5	8.5	4.9		5.3
		M-F	*100.0*	*8.1*	*29.2*	*3.3*	*49.5*	*9.9*	*5.7*		*6.1*
292	Gamagrass hay (*Tripsacum* spp)	As Fed	89.3	6.3	29.3	1.8	44.3	7.6	3.8		4.5
		M-F	*100.0*	*7.0*	*32.8*	*2.0*	*49.7*	*8.5*	*4.3*		*5.0*
293	Grama hay (*Bouteloua* spp)	As Fed	89.2	8.4	29.1	1.5	44.6	5.6	2.0		2.8
		M-F	*100.0*	*9.5*	*32.6*	*1.7*	*50.0*	*6.2*	*2.3*		*3.1*
294	Grass, dehy	As Fed	89.0	8.5	16.6	3.5	39.8	20.6	15.4		15.0
		M-F	*100.0*	*9.5*	*18.7*	*3.9*	*44.8*	*23.1*	*17.3*		*16.8*
295	Grass hay, all analyses	As Fed	89.3	6.9	29.6	2.3	41.9	8.6	5.2		5.5
		M-F	*100.0*	*7.7*	*33.1*	*2.6*	*47.0*	*9.6*	*5.8*		*6.1*
296	Grass & legume mixed hay (30% legume)	As Fed	88.9	6.8	28.4	3.5	37.3	12.9	8.6		8.8
		M-F	*100.0*	*7.6*	*32.0*	*3.9*	*42.0*	*14.5*	*9.7*		*9.9*
297	Johnsongrass hay (*Sorghum halepense*)	As Fed	90.7	8.1	30.2	1.8	43.6	7.0	3.1		3.9
		M-F	*100.0*	*8.9*	*33.3*	*2.0*	*48.1*	*7.7*	*3.4*		*4.3*
298	Kafir fodder, very dry (*Sorghum vulgare*)	As Fed	92.0	7.9	19.6	1.0	56.8	6.7	2.5		3.7
		M-F	*100.0*	*8.6*	*21.3*	*1.1*	*61.7*	*7.3*	*2.7*		*4.0*
299	Kafir stover, very dry	As Fed	91.0	11.2	23.8	2.1	44.3	10.0	6.3	6.2	
		M-F	*100.0*	*12.3*	*26.0*	*2.3*	*48.5*	*10.9*	*6.9*	*6.8*	
300	Kelp, seaweed (*Laminariales* [order], *Fucales* [order])	As Fed	91.3	35.2	6.5	.5	42.6	6.5	2.8		3.5
		M-F	*100.0*	*38.6*	*7.1*	*.5*	*46.7*	*7.1*	*3.1*		*3.9*
301	Ladino clover hay (*Trifolium repens*)	As Fed	91.2	8.5	17.5	1.7	42.2	21.0	13.1		13.5
		M-F	*100.0*	*9.3*	*19.2*	*1.9*	*46.6*	*23.0*	*14.4*		*14.8*
302	Ladino clover & grass hay	As Fed	89.2	7.3	19.9	2.2	43.3	16.5	11.9		11.7
		M-F	*100.0*	*8.2*	*22.3*	*2.5*	*48.5*	*18.5*	*13.4*		*13.1*
303	Lespedeza (annual) hay, all analyses (*Lespedeza* spp)	As Fed	90.0	5.3	26.9	2.5	42.2	13.1	8.8		8.9
		M-F	*100.0*	*5.9*	*29.9*	*2.8*	*46.8*	*14.6*	*9.7*		*9.9*
304	Lespedeza, sericea, hay, all analyses (*Lespedeza cuneata*)	As Fed	90.8	4.9	28.4	1.8	43.1	12.6	8.4		8.5
		M-F	*100.0*	*5.4*	*31.3*	*2.0*	*47.4*	*13.9*	*9.2*		*9.4*
305	Lucerne (see Alfalfa)										
306	Manure, poultry, cage, dried	As Fed	88.6	26.5	13.5	1.7	35.5	28.7	12.8		
		M-F	*100.0*	*29.9*	*15.2*	*1.9*	*40.2*	*32.4*	*14.4*		
307	Manure, poultry, floor, dried	As Fed	84.5	14.1	15.7	2.3	27.1	25.3	19.7		
		M-F	*100.0*	*16.7*	*18.5*	*2.7*	*32.2*	*29.9*	*23.3*		
308	Meadow hay	As Fed	92.9	8.0	28.0	2.3	46.2	8.5	2.7		3.9
		M-F	*100.0*	*8.6*	*30.1*	*2.5*	*49.7*	*9.1*	*2.9*		*4.2*
309	Meadow fescue hay (*Festuca elatior*)	As Fed	86.5	8.0	28.1	2.2	40.0	8.2	4.7		5.0
		M-F	*100.0*	*9.2*	*32.5*	*2.6*	*46.2*	*9.5*	*5.5*		*5.8*
310	Meadow foxtail hay (*Alopecurus pratensis*)	As Fed	88.1	8.2	25.1	1.9	40.0	12.9	8.7		8.8
		M-F	*100.0*	*9.4*	*28.5*	*2.2*	*45.2*	*14.7*	*9.9*		*10.0*
311	Millet hay (*Setaria* spp)	As Fed	90.6	9.0	27.9	1.6	47.8	4.3	1.0		1.7
		M-F	*100.0*	*9.9*	*30.8*	*1.8*	*52.8*	*4.7*	*1.1*		*1.9*

RY FORAGES AND ROUGHAGES

Feed #	TDN Ruminant (%)	TDN Non-Ruminant (%)	TDN Horse Rabbit (%)	Digestible Energy Ruminant Mcal per lb	Digestible Energy Ruminant Mcal per kg	Digestible Energy Non-Ruminant kcal per lb	Digestible Energy Non-Ruminant kcal per kg	Digestible Energy Horse Rabbit Mcal per lb	Digestible Energy Horse Rabbit Mcal per kg	Metabolizable Energy Ruminant Mcal per lb	Metabolizable Energy Ruminant Mcal per kg	Metabolizable Energy Non-Ruminant ME kcal per lb	Metabolizable Energy Non-Ruminant ME kcal per kg	Metabolizable Energy Non-Ruminant Chicken ME_n kcal per lb	Metabolizable Energy Non-Ruminant Chicken ME_n kcal per kg	Metabolizable Energy Horse Rabbit Mcal per lb	Metabolizable Energy Horse Rabbit Mcal per kg	Net Energy-Ruminant NE_m Mcal per lb	Net Energy-Ruminant NE_m Mcal per kg	Net Energy-Ruminant NE_g Mcal per lb	Net Energy-Ruminant NE_g Mcal per kg	Net Energy-Ruminant NE_{lc} Mcal per lb	Net Energy-Ruminant NE_{lc} Mcal per kg
291	51.8			1.0	2.3					.9	1.9												
	60.4			*1.2*	*2.7*					*1.0*	*2.2*												
292	41.9			.8	1.8					.7	1.5												
	46.9			*1.0*	*2.1*					*.8*	*1.7*												
293	39.1			.8	1.7					.6	1.4												
	43.8			*.9*	*1.9*					*.7*	*1.6*												
294	59.7			1.2	2.6					1.0	2.2												
	67.0			*1.4*	*3.0*					*1.1*	*2.4*												
295	50.0		38.1	1.0	2.2			.8	1.7	.8	1.8					.6	1.4						
	56.0		*42.7*	*1.1*	*2.5*			*.9*	*1.9*	*.9*	*2.0*					*.7*	*1.5*						
296	50.7			1.0	2.2					.8	1.8												
	57.0			*1.1*	*2.5*					*1.0*	*2.1*												
297	50.8			1.0	2.2					.8	1.8							.5	1.1	.2	.5	.5	1.1
	56.0			*1.1*	*2.5*					*.9*	*2.0*							*.5*	*1.2*	*.2*	*.5*	*.5*	*1.2*
298	43.0			.9	1.9					.7	1.6												
	46.8			*1.0*	*2.1*					*.8*	*1.7*												
299	55.0		44.0	1.1	2.4			.8	1.8	.9	2.0					.7	1.4	.5	1.2	.3	.6	.5	1.1
	66.0		*48.0*	*1.2*	*2.6*			*.9*	*1.9*	*1.0*	*2.2*					*.7*	*1.6*	*.6*	*1.3*	*.3*	*.7*	*.6*	*1.2*
300	28.9			.6	1.3					.5	1.0												
	31.6			*.6*	*1.4*					*.5*	*1.1*												
301	54.4			1.1	2.4					.9	2.0							.5	1.2	.3	.6	.6	1.3
	59.7			*1.2*	*2.6*					*1.0*	*2.2*							*.6*	*1.3*	*.3*	*.7*	*.6*	*1.4*
302	56.3			1.1	2.5					.9	2.0												
	63.1			*1.3*	*2.8*					*1.0*	*2.3*												
303	45.4			.9	2.0					.7	1.6												
	50.5			*1.0*	*2.2*					*.8*	*1.8*												
304	42.9			.9	1.9					.7	1.6												
	47.2			*1.0*	*2.1*					*.8*	*1.7*												
305																							
306	46.3			.9	2.0					.8	1.7			386	850								
	52.3			*1.0*	*2.3*					*.9*	*1.9*			*450*	*990*								
307	61.3			1 2	2.7					.8	1.8												
	72.5			*1.5*	*3.2*					*1.0*	*2.2*												
308	48.3		37.2	1.0	2.1			.7	1.5	.8	1.7							.5	1.0	1.3	2.9		
	52.0		*40.0*	*1.0*	*2.3*			*.7*	*1.6*	*.9*	*1.9*							*.5*	*1.1*	*1.4*	*3.1*		
309	49.6			1.0	2.2					.8	1.8												
	57.3			*1.1*	*2.5*					*1.0*	*2.1*												
310	54.7			1.1	2.4					.9	2.0												
	62.1			*1.2*	*2.7*					*1.0*	*2.2*												
311	49.6			1.0	2.2					.8	1.8												
	54.7			*1.1*	*2.4*					*.9*	*2.0*												

TABLE 4-101 COMPOSITION OF FEEDS-

Feed #	Feed Name—Description	Moisture Basis: As Fed or M-F (moisture-free)	Dry Matter	Macrominerals: Calcium (Ca)	Phosphorus (P)	Sodium (Na)	Chlorine (Cl)	Magnesium (Mg)	Potassium (K)	Sulfur (S)	Microminerals: Cobalt (Co)	Copper (Cu)	Iron (Fe)	Manganese (Mn)	Zinc (Zn)
			(%)	(%)	(%)	(%)	(%)	(%)	(%)	(%)	(ppm or mg/kg)	(ppm or mg/kg)	(%)	(ppm or mg/kg)	(ppm or mg/kg)
DRY FORAGES & ROUGHAGES															
291	Foxtail, millet, hay	As Fed	85.8	.28	.16	.09	.11	.20	1.66	.14				118.5	
		M-F	100.0	.33	.18	.10	.13	.23	1.94	.16				138.2	
292	Gamagrass hay	As Fed	89.3												
		M-F	100.0												
293	Grama hay	As Fed	89.2	.34	.18										
		M-F	100.0	.38	.20										
294	Grass, dehy	As Fed	89.0												
		M-F	100.0												
295	Grass hay, all analyses	As Fed	89.3	.48	.21			.16	1.20	.15			.056	66.2	
		M-F	100.0	.53	.24			.18	1.35	.17			.063	74.1	
296	Grass & legume mixed hay (30% legume)	As Fed	88.9	1.02	.34									128.2	
		M-F	100.0	1.15	.38									144.2	
297	Johnsongrass hay	As Fed	90.7	.73	.28			.32	1.22				.054		
		M-F	100.0	.81	.31			.35	1.35				.060		
298	Kafir fodder, very dry	As Fed	92.0												
		M-F	100.0												
299	Kafir stover, very dry	As Fed	91.0												
		M-F	100.0												
300	Kelp, seaweed	As Fed	91.3												
		M-F	100.0												
301	Ladino clover hay	As Fed	91.2	1.26	.36	.12	.26	.46	1.98	.20	.137	8.0	.055	120.8	15.5
		M-F	100.0	1.38	.40	.13	.28	.50	2.17	.22	.150	8.8	.060	132.5	17.0
302	Ladino clover & grass hay	As Fed	89.2	.95	.22			.30	1.79			14.4	.015	39.7	
		M-F	100.0	1.07	.24			.34	2.00			16.1	.017	44.5	
303	Lespedeza (annual) hay, all analyses	As Fed	90.0	.96	.18			.22	.94				.026	131.0	
		M-F	100.0	1.07	.20			.24	1.04				.029	145.5	
304	Lespedeza, sericea, hay, all analyses	As Fed	90.8	.94	.22			.20	1.00				.026	91.6	
		M-F	100.0	1.03	.25			.22	1.10				.029	100.8	
305	Lucerne (see Alfalfa)														
306	Manure, poultry, cage dried	As Fed	88.6	7.80	2.2	.42		.63	1.37		.001	61.0	.002	291.0	325.0
		M-F	100.0	8.80	2.5	.47		.71	1.55		.001	69.0	.003	328.0	367.0
307	Manure, poultry, floor, dried	As Fed	84.5	2.50	1.6	.42		.35	1.77			23.0	.038	190.0	343.0
		M-F	100.0	3.00	1.9	.50		.41	2.09			27.0	.045	225.0	406.0
308	Meadow hay	As Fed	92.9	.53	.16										
		M-F	100.0	.57	.17										
309	Meadow fescue hay	As Fed	86.5	.43	.19			.43	1.39					21.2	
		M-F	100.0	.50	.22			.50	1.60					24.5	
310	Meadow foxtail hay	As Fed	88.1												
		M-F	100.0												
311	Millet hay	As Fed	90.6	.29	.18	.09		.20	1.72	.14				118.7	
		M-F	100.0	.33	.20	.10		.23	1.94	.16				134.0	

RY FORAGES AND ROUGHAGES

Feed #	Fat-Soluble Vitamins				Water-Soluble Vitamins								
	A (1 mg car. = 1,667 IU-A)	Carotene (Pro-vita-min A)	E (α-toco-pherol)	K	B_{12}	Biotin	Choline	Folic Acid (Folacin)	Niacin (Nicotinic Acid)	Panto-thenic Acid	Pyridoxine (B_6)	Riboflavin (B_2)	Thiamin (B_1)
		(ppm or mg/kg)	*(ppm or mg/kg)*	*(ppm or mg/kg)*	*(ppm or mg/kg)*	*(ppm or mg/kg)*	*(ppm or mg/kg)*	*(ppm or mg/kg)*	*(ppm or mg/kg)*	*(ppm or mg/kg)*	*(ppm or mg/kg)*	*(ppm or mg/kg)*	*(ppm or mg/kg)*
291													
292													
293													
294													
295	33,840	20.3											
	37,841	*22.7*											
296	138,194	82.9											
	155,531	*93.3*											
297													
298													
299													
300	94,686	56.8											
	106,188	*63.7*											
301	67,180	40.3											
	73,515	*44.1*											
302	245,049	147.0											
	268,720	*161.2*											
303													
304	59,678	35.8										8.8	
	65,846	*39.5*										*9.7*	
305													
306													
307													
308	66,347	39.8											
	71,348	*42.8*											
309	103,021	61.8	117.3										
	119,024	*71,4*	*135.6*										
310													
311													

TABLE 4-101 COMPOSITION OF FEEDS-

Feed #	Feed Name—Description	Moisture Basis: As Fed or M-F (moisture-free)	Proximate Analysis: Dry Matter	Ash	Crude Fiber	Ether Extract (Fat)	N-Free Extract	Crude Protein	Digestible Protein: Ruminant	Non-Ruminant	Horse Rabbit
			(%)	(%)	(%)	(%)	(%)	(%)	(%)	(%)	(%)
DRY FORAGES & ROUGHAGES											
312	Milo fodder *(Sorghum vulgare)*	As Fed	89.0	7.2	21.9	2.5	50.9	6.5	2.7		3.6
		M-F	*100.0*	*8.1*	*24.7*	*2.8*	*57.1*	*7.3*	*3.0*		*4.0*
313	Milo stover	As Fed	92.5	9.9	31.7	1.4	46.2	3.3	1.1		.9
		M-F	*100.0*	*10.7*	*34.2*	*1.5*	*50.0*	*3.6*	*1.2*		*1.0*
314	Native grass hay, intermountain (Meadow hay)	As Fed	92.9	8.0	28.0	2.3	46.2	8.5	2.7		3.9
		M-F	*100.0*	*8.6*	*30.1*	*2.5*	*49.7*	*9.1*	*2.9*		*4.2*
315	Native grass hay, midwest (Prairie hay)	As Fed	91.0	7.2	30.7	2.1	45.2	5.8	2.0		3.0
		M-F	*100.0*	*8.0*	*33.7*	*2.3*	*49.6*	*6.4*	*2.2*		*3.2*
316	Oat hay, all analyses *(Arena sativa)*	As Fed	88.2	7.2	27.3	1.9	43.7	8.1	4.1		3.8
		M-F	*100.0*	*8.2*	*31.0*	*2.1*	*49.5*	*9.2*	*4.6*		*4.3*
317	Oat straw	As Fed	90.1	6.8	36.9	2.1	40.3	4.0	1.0		2.0
		M-F	*100.0*	*7.6*	*41.0*	*2.3*	*44.7*	*4.4*	*1.1*		*2.2*
318	Orchardgrass hay *(Dactylis glomerata)*	As Fed	88.3	6.5	30.1	2.9	40.4	8.4	4.9		3.7
		M-F	*100.0*	*7.4*	*34.1*	*3.3*	*45.7*	*9.5*	*5.5*		*4.2*
319	Paper, waste	As Fed									
		M-F	*100.0*	*.9*	*68.9*	*3.7*	*25.8*	*.7*	*0*		*0*
320	Pasture grasses, western plains, all analyses	As Fed	90.0	7.3	28.5	2.3	43.4	8.5	4.7		
		M-F	*100.0*	*8.1*	*31.7*	*2.6*	*48.2*	*9.4*	*5.2*		
321	Pea & oat hay	As Fed	89.2	7.7	27.3	3.0	39.0	12.2	8.7		
		M-F	*100.0*	*8.6*	*30.6*	*3.4*	*43.7*	*13.7*	*9.8*		
322	Pea straw *(Pisum spp)*	As Fed	87.3	5.7	34.0	1.6	39.4	6.6	4.1		3.5
		M-F	*100.0*	*6.5*	*38.9*	*1.8*	*45.2*	*7.6*	*4.7*		*4.0*
323	Peanut hay *(Arachis hypogaea)*	As Fed	91.2	9.7	23.7	5.1	42.1	10.6	6.7		6.8
		M-F	*100.0*	*10.6*	*26.0*	*5.6*	*46.2*	*11.6*	*7.3*		*7.5*
324	Pearlmillet *(Pennisetum glaucum)*	As Fed	87.6	8.9	32.3	1.8	37.3	7.3	3.7		4.3
		M-F	*100.0*	*10.2*	*36.9*	*2.0*	*42.5*	*8.4*	*4.2*		*4.9*
325	Poultry manure (see Manure, poultry)										
326	Prairie hay, all analyses	As Fed	91.0	7.2	30.7	2.1	45.2	5.8	2.0		3.0
		M-F	*100.0*	*8.0*	*33.7*	*2.3*	*49.6*	*6.4*	*2.2*		*3.2*
327	Quackgrass hay *(Agropyron repens)*	As Fed	88.8	6.2	30.1	2.4	40.3	9.8	5.8		6.2
		M-F	*100.0*	*7.0*	*33.9*	*2.8*	*45.2*	*11.1*	*6.6*		*7.0*
328	Red clover hay, all analyses *(Trifolium pratense)*	As Fed	88.3	7.0	26.5	2.3	39.5	13.0	7.9		7.3
		M-F	*100.0*	*7.9*	*30.0*	*2.6*	*44.8*	*14.7*	*9.0*		*8.3*
329	Red clover & grass hay	As Fed	88.7	5.9	29.9	1.6	37.3	14.0	9.6		9.7
		M-F	*100.0*	*6.6*	*33.7*	*1.8*	*42.1*	*15.8*	*10.8*		*10.9*
330	Reed canarygrass hay *(Phalaris arundinacea)*	As Fed	91.3	7.5	31.6	2.8	40.0	9.4	4.8		5.5
		M-F	*100.0*	*8.2*	*34.6*	*3.1*	*43.8*	*10.3*	*5.3*		*6.0*
331	Rice hulls *(Oryza sativa)*	As Fed	92.4	18.4	41.1	.8	29.3	2.8	.2		.5
		M-F	*100.0*	*19.9*	*44.5*	*.9*	*31.6*	*3.1*	*.2*		*.5*
332	Rice straws	As Fed	90.5	15.4	31.8	1.3	38.0	4.0	.8		1.5
		M-F	*100.0*	*17.0*	*35.1*	*1.4*	*42.0*	*4.5*	*.9*		*1.7*

RY FORAGES AND ROUGHAGES

	TDN			Digestible Energy						Metabolizable Energy								Net Energy-Ruminant					
												Non-Ruminant											
Feed #	Ruminant	Non-Ruminant	Horse Rabbit	Ruminant		Non-Ruminant		Horse Rabbit		Ruminant		ME		Chicken ME_n		Horse Rabbit		NE_m		NE_g		NE_{lc}	
	(%)	(%)	(%)	Mcal per		kcal per		Mcal per		Mcal per		kcal per		kcal per		Mcal per		Mcal per		Mcal per		Mcal per	
				lb	kg	lb	kg	lb	kg	lb	kg	lb	kg	lb	kg	lb	kg	lb	kg	lb	kg	lb	kg
312	57.3			1.1	2.5					1.0	2.1												
	64.4			1.3	2.8					1.0	2.3												
313	48.6			1.0	2.1					.8	1.8												
	52.6			1.0	2.3					.9	1.9												
314	48.3		37.2	1.0	2.1			.7	1.5	.8	1.7							.5	1.0	1.3	2.9		
	52.0		40.0	1.0	2.3			.7	1.6	.9	1.9							.5	1.1	1.4	3.1		
315	46.7			1.0	2.1					.8	1.7												
	51.3			1.0	2.3					.9	1.9												
316	51.8		43.2	1.0	2.3			.8	1.7	.9	1.9							.5	1.2	.3	.6	.5	1.2
	58.7		49.0	1.2	2.6			.9	2.0	1.0	2.1							.6	1.3	.3	.7	.6	1.4
317	44.1		43.3	.9	1.9					.7	1.6					.7	1.6	.5	1.0	.1	.3	.5	1.0
	49.0		48.1	1.0	2.2					.8	1.8					.8	1.7	.5	1.1	.2	.4	.5	1.1
318	49.4		40.6	1.0	2.2			.8	1.6	.8	1.8							.5	1.1	.2	.5	.5	1.1
	56.0		46.0	1.1	2.5			.9	1.9	.9	2.0							.5	1.2	.3	.6	.6	1.3
319																							
	26.8			.5	1.2					.5	1.0												
320	48.8			1.0	2.2					.8	1.8												
	54.2			1.1	2.4					.9	2.0												
321	53.0			1.0	2.3					.9	1.9												
	59.4			1.2	2.6					1.0	2.1												
322	42.4		38.4	.9	1.9			.7	1.6	.7	1.5												
	48.6		44.0	1.0	2.1			.8	1.8	.8	1.8												
323	58.4			1.2	2.6					1.0	2.1												
	64.0			1.3	2.8					1.0	2.3												
324	47.2			1.0	2.1					.8	1.7												
	53.8			1.1	2.4					.9	1.9												
325																							
326	46.7			1.0	2.1					.8	1.7												
	51.3			1.0	2.3					.9	1.9												
327	51.9			1.0	2.3					.9	1.9												
	58.5			1.2	2.6					1.0	2.1												
328	51.8		48.6	1.0	2.3			.9	1.9	.9	1.9					.7	1.5	.5	1.1	.2	.5	.5	1.2
	58.7		55.0	1.2	2.6			1.0	2.2	1.0	2.1					.8	1.7	.6	1.3	.3	.6	.6	1.3
329	49.4			1.0	2.2					.8	1.8												
	55.7			1.1	2.5					.9	2.0												
330	46.1		42.0	.9	2.0			.8	1.8	.8	1.7					.7	1.5	.5	1.0	.1	.3		
	50.5		46.0	1.0	2.2			.9	1.9	.8	1.8					.8	1.7	.5	1.1	.1	.3		
331	12.1			.2	.5					.2	.4												
	13.1			.3	.6					.2	.5												
332	39.1			.8	1.7					.6	1.4												
	43.2			.9	1.9					.7	1.6												

TABLE 4-101 COMPOSITION OF FEEDS—

Feed #	Feed Name—Description	Moisture Basis: As Fed or M-F (moisture-free)	Dry Matter	Macrominerals: Calcium (Ca)	Phosphorus (P)	Sodium (Na)	Chlorine (Cl)	Magnesium (Mg)	Potassium (K)	Sulfur (S)	Microminerals: Cobalt (Co)	Copper (Cu)	Iron (Fe)	Manganese (Mn)	Zinc (Zn)
			(%)	(%)	(%)	(%)	(%)	(%)	(%)	(%)	(ppm or mg/kg)	(ppm or mg/kg)	(%)	(ppm or mg/kg)	(ppm or mg/kg)
DRY FORAGES & ROUGHAGES															
312	Milo fodder	As Feα	89.0	.35	.18										
		M-F	*100.0*	*.40*	*.20*										
313	Milo stover	As Fed	92.5	.59	.11										
		M-F	*100.0*	*.64*	*.12*										
314	Native grass hay, intermountain (Meadow hay)	As Fed	92.9	.53	.16										
		M-F	*100.0*	*.57*	*.17*										
315	Native grass hay, midwest (Prairie hay)	As Fed	91.0	.32	.12			.22	.98					.008	
		M-F	*100.0*	*.35*	*.14*			*.24*	*1.08*					*.009*	
316	Oat hay, all analyses	As Fed	88.2	.23	.21	.15	.46	.26	.86		.06	3.9	.044	65.9	
		M-F	*100.0*	*.26*	*.24*	*.17*	*.52*	*.29*	*.97*		*.07*	*4.4*	*.050*	*74.7*	
317	Oat straw	As Fed	90.1	.30	.09	.33	.70	.16	2.20	.22		9.1	.018	35.3	
		M-F	*100.0*	*.33*	*.10*	*.37*	*.78*	*.18*	*2.44*	*.24*		*10.1*	*.020*	*39.2*	
318	Orchardgrass hay	As Fed	88.3	.40	.29		.36	.26	1.85	.23	.02	12.1	.009	220.4	16.0
		M-F	*100.0*	*.45*	*.33*		*.41*	*.30*	*2.10*	*.26*	*.02*	*13.7*	*.010*	*249.6*	*18.1*
319	Paper, waste	As Fed													
		M-F	*100.0*	*.10*	*.07*			*.63*	*.30*			*12.0*	*.021*	*10.0*	*138.0*
320	Pasture grasses, western plains, all analyses	As Fed	90.0	.36	.15				.74						
		M-F	*100.0*	*.40*	*.17*				*.82*						
321	Pea & oat hay	As Fed	89.2	.73	.23				1.05						
		M-F	*100.0*	*.82*	*.26*				*1.18*						
322	Pea straw	As Fed													
		M-F	*100.0*												
323	Peanut hay	As Fed	91.2	1.12	.24			.37	.74		.072				
		M-F	*100.0*	*1.23*	*.26*			*.40*	*.81*		*.079*				
324	Pearlmillet	As Fed													
		M-F	*100.0*												
325	Poultry manure (see Manure, poultry)														
326	Prairie hay, all analyses	As Fed	91.0	.32	.12			.22	.98				.008		
		M-F	*100.0*	*.35*	*.14*			*.24*	*1.08*				*.009*		
327	Quackgrass hay	As Fed	88.8	.31	.11									40.9	
		M-F	*100.0*	*.33*	*.12*									*46.1*	
328	Red clover hay, all analyses	As Fed	88.3	1.39	.19	.14	.23	.42	1.76	.15	.132	9.8	.009	60.5	
		M-F	*100.0*	*1.57*	*.22*	*.16*	*.26*	*.47*	*1.99*	*.17*	*.150*	*11.1*	*.010*	*68.5*	
329	Red clover & grass hay	As Fed	88.7								.137				
		M-F	*100.0*								*.154*				
330	Reed canarygrass hay	As Fed	91.3	.31	.23			.24	2.15			10.9	.018	84.4	
		M-F	*100.0*	*.34*	*.25*			*.26*	*2.35*			*11.9*	*.020*	*92.4*	
331	Rice hulls	As Fed	92.4	.08	.07				.31					308.0	
		M-F	*100.0*	*.09*	*.08*				*.34*					*333.3*	
332	Rice straws	As Fed	90.5	.19	.07	.28		.10	1.19					313.1	
		M-F	*100.0*	*.21*	*.08*	*.31*		*.11*	*1.32*					*345.8*	

DRY FORAGES AND ROUGHAGES

Feed #	Fat-Soluble Vitamins				Water-Soluble Vitamins								
	A (1 mg car. = 1,667 IU-A)	Carotene (Pro-vitamin A)	E (α-tocopherol)	K	B_{12}	Biotin	Choline	Folic Acid (Folacin)	Niacin (Nicotinic Acid)	Pantothenic Acid	Pyridoxine (B_6)	Riboflavin (B_2)	Thiamin (B_1)
		(ppm or mg/kg)	*(ppm or mg/kg)*	*(ppm or mg/kg)*	*(ppm or mg/kg)*	*(ppm or mg/kg)*	*(ppm or mg/kg)*	*(ppm or mg/kg)*	*(ppm or mg/kg)*	*(ppm or mg/kg)*	*(ppm or mg/kg)*	*(ppm or mg/kg)*	*(ppm or mg/kg)*
312													
313													
314	66,347	39.8											
	71,348	*42.8*											
315													
316	148,530	89.1											
	168,367	*101.0*											
317													
318	43,342	26.0	218.8									6.0	2.6
	49,010	*29.4*	*247.8*									*6.8*	*2.9*
319													
320													
321													
322													
323	76,349	45.8										8.8	
	83,850	*50.3*										*9.7*	
324													
325													
326													
327	57,512	34.5											
	62,012	*37.2*											
328	53,511	32.1				.09							
	60,512	*36.3*				*.11*							
329	16,670	10.0											
	18,670	*11.2*											
330													
331									36.5			.6	2.2
									39.5			*.7*	*2.4*
332													

TABLE 4-101 COMPOSITION OF FEEDS—

Feed #	Feed Name—Description	Moisture Basis: As Fed or M-F (moisture-free)	Proximate Analysis: Dry Matter	Ash	Crude Fiber	Ether Extract (Fat)	N-Free Extract	Crude Protein	Digestible Protein: Ruminant	Non-Ruminant	Horse Rabbit
			(%)	(%)	(%)	(%)	(%)	(%)	(%)	(%)	(%)
DRY FORAGES & ROUGHAGES											
333	Russian-thistle hay (Tumbleweed) *(Salsola kali tenuifolia)*	As Fed	88.4	13.6	25.0	1.9	37.4	10.5	6.8		7.2
		M-F	*100.0*	*15.4*	*28.3*	*2.1*	*42.3*	*11.9*	*7.7*		*8.1*
334	Safflower hulls *(Carthamus tinctorius)*	As Fed	91.3	1.6	53.1	3.4	29.9	3.3	0		.9
		M-F	*100.0*	*1.8*	*58.2*	*3.7*	*32.7*	*3.6*	*0*		*1.0*
335	Sainfoin hay *(Onobrychis* spp)	As Fed	85.7	6.5	22.7	3.2	39.1	14.2	10.0		9.9
		M-F	*100.0*	*7.6*	*26.5*	*3.7*	*45.6*	*16.6*	*11.6*		*11.6*
336	Sawdust, wood	As Fed	84.0	.5	62.4	.6	20.2	.3	0		
		M-F	*100.0*	*.6*	*74.3*	*.7*	*24.0*	*.4*	*0*		
337	Seaweed meal (kelp) *(Laminariales [order] Fucales [order])*	As Fed	91.3	35.2	6.5	.5	42.6	6.5	2.8		3.5
		M-F	*100.0*	*38.6*	*7.1*	*.5*	*46.7*	*7.1*	*3.1*		*3.9*
338	Sorghum fodder, sweet, dry *(Sorghum vulgare saccharatum)*	As Fed	82.3	6.0	23.4	2.3	44.9	5.7	2.3		3.1
		M-F	*100.0*	*7.3*	*28.4*	*2.8*	*54.6*	*6.9*	*2.8*		*3.7*
339	Sorghum grain fodder *(Sorghum vulgare)*	As Fed	90.2	8.9	24.8	1.7	48.6	6.2	2.6		3.3
		M-F	*100.0*	*9.9*	*27.5*	*1.9*	*53.8*	*6.9*	*2.8*		*3.7*
340	Sorghum grain stover	As Fed	85.1	8.2	27.7	1.8	42.9	4.5	1.2		
		M-F	*100.0*	*9.6*	*32.6*	*2.1*	*50.4*	*5.3*	*1.4*		
341	Soybean hay *(Glycine max)*	As Fed	89.2	7.2	28.6	2.0	36.9	14.5	9.4		8.9
		M-F	*100.0*	*8.0*	*32.1*	*2.2*	*41.4*	*16.3*	*10.5*		*10.0*
342	Soybean straw	As Fed	87.6	5.6	38.6	1.2	37.4	4.8	1.5		2.0
		M-F	*100.0*	*6.4*	*44.1*	*1.4*	*42.6*	*5.5*	*1.7*		*2.3*
343	Stargrass *(Cynodon plectostachyun)*	As Fed	90.1	8.3	37.2	1.6	37.5	5.4	1.9		2.7
		M-F	*100.0*	*9.2*	*41.3*	*1.8*	*41.6*	*6.1*	*2.2*		*3.1*
344	Sudangrass hay *(Sorghum vulgare sudanense)*	As Fed	88.9	8.5	25.7	1.6	41.8	11.3	5.2		
		M-F	*100.0*	*9.6*	*28.9*	*1.8*	*47.0*	*12.7*	*5.8*		
345	Sugar beet leaves w/crowns, dehy *(Beta saccharifera)*	As Fed	84.5	15.9	11.5	1.0	45.6	10.5	6.6	4.6	6.9
		M-F	*100.0*	*18.9*	*13.6*	*1.2*	*53.9*	*12.4*	*7.8*	*5.5*	*8.2*
346	Sweet potato vines, dried *(Ipomoea batatas)*	As Fed	89.2	10.1	22.2	3.2	42.2	11.5	6.4		7.6
		M-F	*100.0*	*11.3*	*24.9*	*3.6*	*47.4*	*12.8*	*7.2*		*8.5*
347	Timothy hay, all analyses *(Phleum pratense)*	As Fed	88.6	4.5	30.2	2.3	45.3	6.3	2.5		3.1
		M-F	*100.0*	*5.1*	*34.0*	*2.5*	*51.3*	*7.1*	*2.9*		*3.4*
348	Timothy & clover hay *(Phleum pratense, Trifolium* spp)	As Fed	88.9	5.5	31.2	2.2	41.8	8.3	4.5		5.0
		M-F	*100.0*	*6.1*	*35.1*	*2.5*	*47.0*	*9.3*	*5.0*		*5.6*
349	Tumbleweed (Russian-thistle), hay *(Salsola kali tenuifolia)*	As Fed	88.4	13.6	25.0	1.9	37.4	10.5	6.8		7.2
		M-F	*100.0*	*15.4*	*28.3*	*2.1*	*42.3*	*11.9*	*7.7*		*8.1*
350	Velvet bean hay *(Stizolobium* spp)	As Fed	92.8	7.4	27.5	3.1	38.4	16.4	11.7		11.5
		M-F	*100.0*	*8.0*	*29.6*	*3.3*	*41.4*	*17.7*	*12.6*		*12.4*
351	Vetch & oat hay *(Vicia* spp, *Avena sativa)*	As Fed	87.6	8.2	27.3	2.7	37.5	11.9	7.8		8.0
		M-F	*100.0*	*9.4*	*31.2*	*3.1*	*42.7*	*13.6*	*8.9*		*9.2*
352	Vetch & wheat hay *(Vicia* spp, *Triticum aestivum)*	As Fed	90.0	7.2	28.8	2.2	36.4	15.4	10.9		10.8
		M-F	*100.0*	*8.0*	*32.0*	*2.4*	*40.5*	*17.1*	*12.1*		*12.0*
353	Vetch hay *(Vicia* spp)	As Fed	88.2	3.4	25.1	1.2	40.9	17.6	12.3		10.8
		M-F	*100.0*	*3.8*	*28.5*	*1.4*	*46.3*	*20.0*	*14.0*		*12.3*

DRY FORAGES AND ROUGHAGES

	TDN			Digestible Energy						Metabolizable Energy								Net Energy-Ruminant					
												Non-Ruminant											
Feed #	Ruminant	Non-Ruminant	Horse Rabbit	Ruminant		Non-Ruminant		Horse Rabbit		Ruminant		ME		Chicken ME_n		Horse Rabbit		NE_m		NE_g		NE_{lc}	
	(%)	(%)	(%)	Mcal per		kcal per		Mcal per		Mcal per		kcal per		kcal per		Mcal per		Mcal per		Mcal per		Mcal per	
				lb	kg	lb	kg	lb	kg	lb	kg	lb	kg	lb	kg	lb	kg	lb	kg	lb	kg	lb	kg
333	40.7			.9	1.8					.7	1.5							.4	.9	.1	.2		
	46.0			1.0	2.0					.8	1.7							.5	1.1	.1	.2		
334	14.0			.3	.6					.2	.5			420	924								
	15.4			.3	.7					.3	.6			460	1,012								
335	52.0			1.0	2.3					.9	1.9												
	60.7			1.2	2.7					1.0	2.2												
336	.9			.02	.04					.01	.03												
	1.1			.02	.05					.02	.04												
337	28.9			.6	1.3					.5	1.0												
	31.6			.6	1.4					.5	1.1												
338	48.8			1.0	2.2					.8	1.8												
	59.3			1.2	2.6					1.0	2.1												
339	50.6			1.0	2.2					.8	1.8							.5	1.1	.2	.5	.5	1.2
	56.2			1.1	2.5					.9	2.0							.5	1.2	.3	.6	.6	1.4
340	45.9			.9	2.0					.8	1.7							.5	1.0	.2	.5	.5	1.1
	54.0			1.1	2.4					.9	1.9							.5	1.2	.3	.6	.5	1.2
341	47.0			1.0	2.1					.8	1.7							.5	1.0	.1	.3	.5	1.0
	52.1			1.0	2.3					.9	1.9							.5	1.1	.2	.4	.5	1.1
342	34.8			.7	1.5					.6	1.3							.3	.7			.2	.5
	39.7			.8	1.8					.6	1.4							.4	.9			.3	.6
343	43.9		35.3	.9	1.9			.6	1.4	.7	1.6					.5	1.2	.4	.9	.1	.2	.4	.9
	48.7		39.2	1.0	2.2			.7	1.6	.8	1.8					.6	1.3	.5	1.0	.1	.2	.5	1.0
344	50.4			1.0	2.2					.8	1.8							.5	1.1	.3	.6	.5	1.2
	56.7			1.1	2.5					1.0	2.1							.6	1.3	.3	.6	.6	1.3
345	51.3	46.4		1.0	2.3	.9	2.0			.9	1.9	893	1,964										
	60.7	55.0		1.2	2.7	1.1	2.4			1.0	2.2	1,058	2,328										
346	49.4			1.0	2.2					.8	1.8												
	55.4			1.1	2.4					.9	2.0												
347	48.5		40.4	1.0	2.1					.8	1.8					.7	1.5						
	54.8		45.6	1.1	2.4					.9	2.0					.8	1.7						
348	48.4			1.0	2.1					.8	1.8												
	54.4			1.1	2.4					.9	2.0												
349	40.7			.8	1.8					.7	1.5							.4	.9	.1	.2		
	46.0			.9	2.0					.8	1.7							.5	1.1	.1	.2		
350	54.3			1.1	2.4					.9	2.0												
	58.5			1.2	2.6					1.0	2.1												
351	50.4			1.0	2.2					.8	1.8												
	57.6			1.0	2.5					1.0	2.1												
352	57.9			1.2	2.6			1.0	2.1														
	64.3			1.3	2.8			1.0	2.3														
353	54.4			1.1	2.4					.9	2.0							.5	1.1	.3	.6	.5	1.2
	61.7			1.2	2.7					1.0	2.2							.6	1.3	.3	.7	.6	1.4

TABLE 4-101 COMPOSITION OF FEEDS—

Feed #	Feed Name—Description	Moisture Basis: As Fed or M-F (moisture-free)	Dry Matter	Macrominerals							Microminerals				
				Calcium (Ca)	Phosphorus (P)	Sodium (Na)	Chlorine (Cl)	Magnesium (Mg)	Potassium (K)	Sulfur (S)	Cobalt (Co)	Copper (Cu)	Iron (Fe)	Manganese (Mn)	Zinc (Zn)
			(%)	(%)	(%)	(%)	(%)	(%)	(%)	(%)	(ppm or mg/kg)	(ppm or mg/kg)	(%)	(ppm or mg/kg)	(ppm or mg/kg)
DRY FORAGES & ROUGHAGES															
333	Russian-thistle hay (Tumbleweed)	As Fed	88.4	1.63	.26			1.05	6.06						
		M-F	*100.0*	*1.84*	*.29*			*1.19*	*6.85*						
334	Safflower hulls	As Fed	91.3												
		M-F	*100.0*												
335	Sainfoin hay	As Fed	85.7												
		M-F	*100.0*												
336	Sawdust, wood	As Fed	84.0	.18	.01			.03	.06			6.0	.004	10.0	19.0
		M-F	*100.0*	*.21*	*.01*			*.03*	*.06*			*7.1*	*.005*	*11.9*	*22.6*
337	Seaweed meal (kelp)	As Fed	91.3												
		M-F	*100.0*												
338	Sorghum fodder, sweet, dry	As Fed	82.3	.31	.13		.32	.29	1.20					107.6	
		M-F	*100.0*	*.38*	*.16*		*.39*	*.35*	*1.46*					*130.7*	
339	Sorghum grain fodder	As Fed	90.2	.56	.17										
		M-F	*100.0*	*.62*	*.19*										
340	Sorghum grain stover	As Fed	85.1	.34	.09										
		M-F	*100.0*	*.40*	*.11*										
341	Soybean hay	As Fed	89.2	1.15	.21	.11	.13	.70	.87	.23	.080	8.0	.027	82.6	21.4
		M-F	*100.0*	*1.29*	*.23*	*.12*	*.15*	*.79*	*.97*	*.26*	*.090*	*9.0*	*.030*	*92.6*	*24.0*
342	Soybean straw	As Fed	87.6	1.39	.05			.81	.46					44.9	
		M-F	*100.0*	*1.59*	*.06*			*.92*	*.53*					*51.2*	
343	Stargrass	As Fed	90.1												
		M-F	*100.0*												
344	Sudangrass hay	As Fed	88.9	.50	.28	.02		.36	1.37	.05	.116	32.7	.018	82.9	
		M-F	*100.0*	*.56*	*.31*	*.02*		*.40*	*1.54*	*.06*	*.130*	*36.8*	*.020*	*93.3*	
345	Sugar beet leaves w/crowns, dehy	As Fed	84.5												
		M-F	*100.0*												
346	Sweet potato vines, dried	As Fed	89.2												
		M-F	*100.0*												
347	Timothy hay, all analyses	As Fed	88.6	.36	.15	.16		.16	1.41	.12		4.4	.012	41.8	
		M-F	*100.0*	*.40*	*.16*	*.18*		*.18*	*1.59*	*.13*		*5.0*	*.014*	*47.1*	
348	Timothy & clover hay	As Fed	88.9	.62	.15	.17		.23	1.61	.13		6.2	.014	47.4	
		M-F	*100.0*	*.70*	*.17*	*.19*		*.26*	*1.81*	*.14*		*7.0*	*.015*	*53.3*	
349	Tumbleweed (Russian-thistle), hay	As Fed	88.4	1.63	.26			1.05	6.06						
		M-F	*100.0*	*1.84*	*.29*			*1.19*	*6.85*						
350	Velvet bean hay	As Fed	92.8		.24				2.20						
		M-F	*100.0*		*.26*				*2.37*						
351	Vetch & oat hay	As Fed	87.6	.76	.27				1.51						
		M-F	*100.0*	*.87*	*.31*				*1.72*						
352	Vetch & wheat hay	As Fed	90.0												
		M-F	*100.0*												
353	Vetch hay	As Fed	88.2	1.20	.30	.46		.24	1.87	.13	.309	8.7	.044	53.7	
		M-F	*100.0*	*1.36*	*.34*	*.52*		*.27*	*2.12*	*.15*	*.350*	*9.9*	*.050*	*60.9*	

RY FORAGES AND ROUGHAGES

Feed #	A (1 mg car. = 1,667 IU-A)	Carotene (Pro-vitamin A)	E (α-tocopherol)	K	B_{12}	Biotin	Choline	Folic Acid (Folacin)	Niacin (Nicotinic Acid)	Pantothenic Acid	Pyridoxine (B_6)	Riboflavin (B_2)	Thiamin (B_1)
	Fat-Soluble Vitamins				Water-Soluble Vitamins								
		(ppm or mg/kg)	*(ppm or mg/kg)*	*(ppm or mg/kg)*	*(ppm or mg/kg)*	*(ppm or mg/kg)*	*(ppm or mg/kg)*	*(ppm or mg/kg)*	*(ppm or mg/kg)*	*(ppm or mg/kg)*	*(ppm or mg/kg)*	*(ppm or mg/kg)*	*(ppm or mg/kg)*
333													
334													
335													
336													
337	94,686	56.8											
	106,188	*63.7*											
338	3,667	2.2											
	4,334	*2.6*											
339													
340													
341	53,011	31.8	23.8										
	59,512	*35.7*	*26.7*										
342													
343													
344	7,835	4.7											
	8,835	*5.3*											
345													
346													
347			55.9			.06		2.03					
			63.1			*.07*		*2.29*					
348	39,508	23.7											
	44,509	*26.7*											
349													
350													
351													
352													
353													

TABLE 4-101 COMPOSITION OF FEEDS–

Feed #	Feed Name—Description	Moisture Basis: As Fed or M-F (moisture-free)	Proximate Analysis: Dry Matter	Ash	Crude Fiber	Ether Extract (Fat)	N-Free Extract	Crude Protein	Digestible Protein: Ruminant	Non-Ruminant	Horse Rabbit
			(%)	(%)	(%)	(%)	(%)	(%)	(%)	(%)	(%)
DRY FORAGES & ROUGHAGES											
354	Wheat straw *(Triticum* spp)	As Fed	90.1	6.5	37.4	1.4	41.6	3.2	0		.2
		M-F	*100.0*	*7.2*	*41.5*	*1.5*	*46.2*	*3.6*	*0*		*.2*
355	Wheatgrass, crested, hay *(Agropyron cristatum)*	As Fed	92.0	6.3	30.0	2.2	43.6	9.9	4.7		4.1
		M-F	*100.0*	*6.8*	*32.6*	*2.4*	*47.4*	*10.8*	*5.1*		*4.5*
356	Wood, sawdust	As Fed	84.0	.5	62.4	.6	20.2	.3	0		
		M-F	*100.0*	*.6*	*74.3*	*.7*	*24.0*	*.4*	*0*		

DRY FORAGES AND ROUGHAGES

Feed #	TDN Ruminant (%)	TDN Non-Ruminant (%)	TDN Horse Rabbit (%)	DE Ruminant Mcal per lb	DE Ruminant Mcal per kg	DE Non-Ruminant kcal per lb	DE Non-Ruminant kcal per kg	DE Horse Rabbit Mcal per lb	DE Horse Rabbit Mcal per kg	ME Ruminant Mcal per lb	ME Ruminant Mcal per kg	ME Non-Ruminant ME kcal per lb	ME Non-Ruminant ME kcal per kg	ME Non-Ruminant Chicken ME_n kcal per lb	ME Non-Ruminant Chicken ME_n kcal per kg	ME Horse Rabbit Mcal per lb	ME Horse Rabbit Mcal per kg	NE_m Mcal per lb	NE_m Mcal per kg	NE_g Mcal per lb	NE_g Mcal per kg	NE_{lc} Mcal per lb	NE_{lc} Mcal per kg
354	39.7		31.5	.8	1.8			.6	1.3	.6	1.4							.4	.9	.1	.2	.4	.9
	44.1		*35.0*	*.9*	*1.9*			*.7*	*1.5*	*.7*	*1.6*							*.5*	*1.0*	*.1*	*.2*	*.5*	*1.0*
355	50.6			1.0	2.2					.8	1.8							.5	1.1	.2	.5		
	55.0			*1.1*	*2.4*					*.9*	*2.0*							*.5*	*1.2*	*.3*	*.6*		
356	.9			.02	.04					.01	.03												
	1.1			*.02*	*.05*					*.02*	*.04*												

TABLE 4-101 COMPOSITION OF FEEDS-

Feed #	Feed Name—Description	Moisture Basis: As Fed or M-F (moisture-free)	Dry Matter	Macrominerals: Calcium (Ca)	Phosphorus (P)	Sodium (Na)	Chlorine (Cl)	Magnesium (Mg)	Potassium (K)	Sulfur (S)	Microminerals: Cobalt (Co)	Copper (Cu)	Iron (Fe)	Manganese (Mn)	Zinc (Zn)
			(%)	(%)	(%)	(%)	(%)	(%)	(%)	(%)	(ppm or mg/kg)	(ppm or mg/kg)	(%)	(ppm or mg/kg)	(ppm or mg/kg)
DRY FORAGES & ROUGHAGES															
354	Wheat straw	As Fed	90.1	.15	.07	.13		.11	1.00	.17	.036	3.0	.018	36.4	
		M-F	*100.0*	*.17*	*.08*	*.14*		*.12*	*1.11*	*.19*	*.040*	*3.3*	*.020*	*40.4*	
355	Wheatgrass, crested, hay	As Fed	92.0	.30	.19						.221				
		M-F	*100.0*	*.33*	*.21*						*.240*				
356	Wood, sawdust	As Fed	84.0	.18	.01			.03	.06			6.0	.004	10.0	19.0
		M-F	*100.0*	*.21*	*.01*			*.03*	*.06*			*7.1*	*.005*	*11.9*	*22.6*

DRY FORAGES AND ROUGHAGES

Feed #	Fat-Soluble Vitamins				Water-Soluble Vitamins								
	A (1 mg car. = 1,667 IU-A)	Carotene (Pro-vitamin A)	E (α-tocopherol)	K	B_{12}	Biotin	Choline	Folic Acid (Folacin)	Niacin (Nicotinic Acid)	Pantothenic Acid	Pyridoxine (B_6)	Riboflavin (B_2)	Thiamin (B_1)
		(ppm or mg/kg)	*(ppm or mg/kg)*	*(ppm or mg/kg)*	*(ppm or mg/kg)*	*(ppm or mg/kg)*	*(ppm or mg/kg)*	*(ppm or mg/kg)*	*(ppm or mg/kg)*	*(ppm or mg/kg)*	*(ppm or mg/kg)*	*(ppm or mg/kg)*	*(ppm or mg/kg)*
354	3,334	2.0											
	3,667	*2.2*											
355	34,173	20.5											
	37,174	*22.3*											
356													

TABLE 4-101 COMPOSITION OF FEEDS–…

Feed #	Feed Name—Description	Moisture Basis: As Fed or M-F (moisture-free)	Proximate Analysis: Dry Matter	Ash	Crude Fiber	Ether Extract (Fat)	N-Free Extract	Crude Protein	Digestible Protein: Ruminant	Non-Ruminant	Horse Rabbit
			(%)	(%)	(%)	(%)	(%)	(%)	(%)	(%)	(%)
SILAGES AND HAYLAGES											
357	Alfalfa silage, all analyses *(Medicago sativa)*	As Fed	29.7	2.8	9.0	1.0	11.6	5.3	3.6		3.7
		M-F	*100.0*	*9.3*	*30.4*	*3.3*	*39.2*	*17.8*	*12.1*		*12.5*
358	Alfalfa, wilted	As Fed	36.2	3.1	10.9	1.3	14.4	6.4	4.3		4.5
		M-F	*100.0*	*8.6*	*30.2*	*3.6*	*39.8*	*17.8*	*11.9*		*12.4*
359	Alfalfa silage w/ corn grain	As Fed	25.5	2.0	6.3	1.2	11.1	4.9	3.4		3.4
		M-F	*100.0*	*7.8*	*24.8*	*4.7*	*43.7*	*19.0*	*13.5*		*13.5*
360	Alfalfa silage w/ molasses	As Fed	31.1	2.6	9.0	1.2	12.9	5.4	3.9		3.6
		M-F	*100.0*	*8.5*	*28.8*	*4.0*	*41.3*	*17.4*	*12.4*		*11.9*
361	Alfalfa & brome, all analyses *(Medicago sativa, Bromus inermis)*	As Fed	27.5	2.0	9.8	1.1	10.8	3.8	2.4		2.4
		M-F	*100.0*	*7.4*	*35.6*	*4.1*	*39.3*	*13.6*	*8.6*		*8.6*
362	Alfalfa & brome, wilted	As Fed	46.1	4.1	15.2	1.1	18.6	7.1	4.8		4.8
		M-F	*100.0*	*8.8*	*33.0*	*2.3*	*40.4*	*15.5*	*10.3*		*10.3*
363	Alfalfa & orchardgrass, min 50% D-M *(Medicago sativa, Dactylis glomerata)*	As Fed	61.0		18.6			9.9	5.5		
		M-F	*100.0*		*30.5*			*16.2*	*9.1*		
364	Atlas sorghum stover silage *(Sorghum vulgare)*	As Fed	61.5	3.9	15.4	1.1	37.6	3.5	.8		.8
		M-F	*100.0*	*6.3*	*25.0*	*1.9*	*61.2*	*5.6*	*1.3*		*1.3*
365	Beet pulp silage *(Beta saccharifera)*	As Fed	12.2	.6	3.9	.3	5.9	1.5	.9		.9
		M-F	*100.0*	*5.1*	*32.0*	*1.9*	*48.5*	*12.5*	*7.7*		*7.7*
366	Beet, sugar (see Sugar beet)										
367	Bermudagrass, coastal, wilted *(Cynodon dactylon)*	As Fed	39.3	3.2	12.5	1.3	16.6	5.7	3.7		3.7
		M-F	*100.0*	*8.1*	*31.9*	*3.4*	*42.2*	*14.4*	*9.3*		*9.3*
368	Canarygrass, reed *(Phalaris canariensis)*	As Fed	27.5	2.0	9.1	.7	13.2	2.5	1.2		1.2
		M-F	*100.0*	*7.3*	*33.1*	*2.5*	*48.2*	*8.9*	*4.2*		*4.2*
369	Citrus pulp *(Citrus spp)*	As Fed	19.5	1.0	3.1	1.7	12.3	1.4	.4		.5
		M-F	*100.0*	*5.3*	*15.9*	*8.8*	*62.9*	*7.1*	*1.8*		*2.7*
370	Clover, Ladino *(Trifolium repens)*	As Fed	24.9	2.4	5.3	1.0	10.3	5.9	4.3		4.4
		M-F	*100.0*	*9.8*	*21.2*	*4.1*	*41.5*	*23.4*	*17.4*		*17.6*
371	Clover, red *(Trifolium pratense)*	As Fed	26.2	2.5	8.5	1.0	10.4	3.8	2.4		2.5
		M-F	*100.0*	*9.6*	*32.5*	*3.6*	*39.8*	*14.5*	*9.2*		*9.4*
372	Clover, red & grass silage	As Fed	35.5	3.0	11.1	1.1	15.9	4.4	2.6		2.6
		M-F	*100.0*	*8.5*	*31.3*	*3.1*	*44.8*	*12.3*	*7.4*		*7.4*
373	Clover, red w/molasses	As Fed	33.9	2.8	10.3	1.0	15.3	4.5	2.9		2.8
		M-F	*100.0*	*8.4*	*30.3*	*2.9*	*45.1*	*13.3*	*8.5*		*8.4*
374	Corn silage, w/ears, all analyses *(Zea mays)*	As Fed	24.1	1.5	5.9	.7	14.0	2.0	1.0		.9
		M-F	*100.0*	*6.3*	*24.5*	*2.9*	*58.1*	*8.2*	*4.0*		*3.6*
375	Corn silage, ears	As Fed	70.0	1.6	7.3	2.5	51.6	7.0	3.8		3.8
		M-F	*100.0*	*2.3*	*10.4*	*3.6*	*73.6*	*10.1*	*5.4*		*5.4*
376	Corn ear & husk (snap ear corn)	As Fed	43.4	.8	5.1	1.5	32.2	3.8	2.1		2.2
		M-F	*100.0*	*1.8*	*11.8*	*3.5*	*74.1*	*8.8*	*4.8*		*5.1*
377	Corn husks (husklage)	As Fed	78.0	2.6	26.8	.7	45.0	2.9			
		M-F	*100.0*	*3.4*	*34.3*	.9	*57.7*	*3.7*			

SILAGES AND HAYLAGES

Feed #	TDN			Digestible Energy						Metabolizable Energy								Net Energy-Ruminant					
												Non-Ruminant											
	Ruminant	Non-Ruminant	Horse Rabbit	Ruminant		Non-Ruminant		Horse Rabbit		Ruminant		ME		Chicken ME_n		Horse Rabbit		NE_m		NE_g		NE_{lc}	
	(%)	(%)	(%)	Mcal per		kcal per		Mcal per		Mcal per		kcal per		kcal per		Mcal per		Mcal per		Mcal per		Mcal per	
				lb	kg	lb	kg	lb	kg	lb	kg	lb	kg	lb	kg	lb	kg	lb	kg	lb	kg	lb	kg
357	16.2			.3	.7					.3	.6							.1	.3	.05	.1	.2	.4
	54.7			1.1	2.4					.9	2.0							.5	1.2	.2	.5	.5	1.2
358	21.0			.4	.9					.4	.8							.2	.5	.1	.3	.2	.5
	58.0			1.2	2.6					1.0	2.1							.6	1.3	.3	.7	.6	1.3
359	14.5			.4	.8					.2	.5												
	56.8			1.1	2.5					1.0	2.1												
360	17.6			.4	.8					.3	.6							.2	.4	.1	.2	.2	.4
	56.5			1.1	2.5					.9	2.0							.6	1.3	.3	.6	.6	1.3
361	14.7			.3	.6					.2	.5												
	53.6			1.1	2.4					.9	1.9												
362	25.1			.5	1.1					.4	.9												
	54.4			1.1	2.4					.9	2.0												
363	33.5			.7	1.5					.5	1.2							.3	.7	.1	.3	.3	.7
	55.0			1.1	2.4					.9	2.0							.5	1.2	.2	.4	.5	1.1
364	36.9			.7	1.6					.6	1.3												
	60.1			1.2	2.7					1.0	2.2												
365	8.3			.2	.4					.1	.3												
	68.0			1.4	3.0					1.1	2.5												
366																							
367	23.4			.5	1.0					.4	.9												
	59.4			1.2	2.6					1.0	2.2												
368	13.5			.3	.6					.2	.5												
	49.2			1.0	2.2					.8	1.8												
369	17.2			.4	.8					.3	.6							.2	.4	.1	.3	.2	.5
	88.0			1.8	3.9					1.5	3.2							1.0	2.2	.6	1.4	1.0	2.3
370	17.1			.4	.8					.3	.6												
	68.9			1.4	3.0					1.1	2.5												
371	14.9			.3	.7					.2	.5												
	56.8			1.1	2.5					1.0	2.1												
372	21.8			.5	1.0					.4	.8												
	61.3			1.2	2.7					1.0	2.2												
373	20.4			.4	.9					.3	.7												
	60.1			1.2	2.7					1.0	2.2												
374	16.3			.3	.7					.3	.6												
	67.7			1.4	3.0					1.1	2.4												
375	51.0			1.0	2.2					.8	1.8												
	72.9			1.5	3.2					1.2	2.6												
376	31.2			.6	1.4					.5	1.2							.3	.7	.2	.4	.4	.8
	72.0			1.5	3.2					1.2	2.7							.7	1.6	.5	1.0	.8	1.8
377	46.7			1.0	2.1					.8	1.7												
	59.9			1.2	2.6					1.0	2.2												

TABLE 4-101 COMPOSITION OF FEEDS-

Feed #	Feed Name—Description	Moisture Basis: As Fed or M-F (moisture-free)	Dry Matter	Macrominerals							Microminerals				
				Calcium (Ca)	Phosphorus (P)	Sodium (Na)	Chlorine (Cl)	Magnesium (Mg)	Potassium (K)	Sulfur (S)	Cobalt (Co)	Copper (Cu)	Iron (Fe)	Manganese (Mn)	Zinc (Zn)
SILAGES AND HAYLAGES			*(%)*	*(%)*	*(%)*	*(%)*	*(%)*	*(%)*	*(%)*	*(%)*	*(ppm or mg/kg)*	*(ppm or mg/kg)*	*(%)*	*(ppm or mg/kg)*	*(ppm or mg/kg)*
357	Alfalfa silage, all analyses	As Fed	29.7	.48	.11	.05	.15	.10	.71	.11	.045	.29	.009	14.9	
		M-F	*100.0*	*1.61*	*.38*	*.16*	*.50*	*.34*	*2.40*	*.36*	*.150*	*.97*	*.030*	*50.3*	
358	Alfalfa, wilted	As Fed	36.2	.51	.12	.06	.15	.12	.85	.13		3.4	.011	18.8	
		M-F	*100.0*	*1.40*	*.32*	*.14*	*.41*	*.33*	*2.36*	*.36*		*9.3*	*.030*	*52.0*	
359	Alfalfa silage w/ corn grain	As Fed	25.5												
		M-F	*100.0*												
360	Alfalfa silage w/ molasses	As Fed	31.1	.54	.10			.11	.80			3.9	.009	13.2	
		M-F	*100.0*	*1.74*	*.31*			*.34*	*2.56*			*12.6*	*.030*	*42.6*	
361	Alfalfa & brome, all analyses	As Fed	27.5	.18	.05			.05	.51		.031	3.1	.004	8.3	
		M-F	*100.0*	*.64*	*.20*			*.20*	*1.86*		*.112*	*11.2*	*.015*	*30.2*	
362	Alfalfa & brome, wilted	As Fed	46.1	.72	.12										
		M-F	*100.0*	*1.56*	*.26*										
363	Alfalfa & orchard-grass, min. 50% D-M	As Fed	61.0												
		M-F	*100.0*												
364	Atlas sorghum stover silage	As Fed	61.5	.24	.07										
		M-F	*100.0*	*.39*	*.11*										
365	Beet pulp silage	As Fed	12.2												
		M-F	*100.0*												
366	Beet, sugar (see Sugar beet)														
367	Bermudagrass, coastal, wilted	As Fed	39.3												
		M-F	*100.0*												
368	Canarygrass, reed	As Fed	27.5												
		M-F	*100.0*												
369	Citrus pulp	As Fed	19.5	.40	.03			.03	.12				.003		
		M-F	*100.0*	*2.04*	*.15*			*.16*	*.62*				*.016*		
370	Clover, Ladino	As Fed	24.9												
		M-F	*100.0*												
371	Clover, red	As Fed	26.2	.42	.06	.06		.10	.46	.04		2.9	.009	34.1	
		M-F	*100.0*	*1.61*	*.23*	*.23*		*.40*	*1.76*	*.16*		*11.0*	*.033*	*130.0*	
372	Clover, red & grass silage	As Fed	35.5	.39	.07										
		M-F	*100.0*	*1.11*	*.20*										
373	Clover, red w/ molasses	As Fed	33.9												
		M-F	*100.0*												
374	Corn silage, w/ears, all analyses	As Fed	24.1	.12	.05				.20						
		M-F	*100.0*	*.50*	*.20*				*.88*						
375	Corn silage, ears	As Fed	70.0												
		M-F	*100.0*												
376	Corn ear & husk (snap ear corn)	As Fed	43.4	.03	.12										
		M-F	*100.0*	*.06*	*.27*										
377	Corn husks (husklage)	As Fed	78.0	.12	.06	.06		.17	.70			11.0	.001	12.0	19.0
		M-F	*100.0*	*.16*	*.08*	*.08*		*.22*	*.90*			*14.0*	*.001*	*16.0*	*24.0*

SILAGES AND HAYLAGES

Feed #	Fat-Soluble Vitamins				Water-Soluble Vitamins								
	A (1 mg car. = 1,667 IU-A)	Carotene (Pro-vitamin A)	E (α-tocopherol)	K	B_{12}	Biotin	Choline	Folic Acid (Folacin)	Niacin (Nicotinic Acid)	Pantothenic Acid	Pyridoxine (B_6)	Riboflavin (B_2)	Thiamin (B_1)
		(ppm or mg/kg)	*(ppm or mg/kg)*	*(ppm or mg/kg)*	*(ppm or mg/kg)*	*(ppm or mg/kg)*	*(ppm or mg/kg)*	*(ppm or mg/kg)*	*(ppm or mg/kg)*	*(ppm or mg/kg)*	*(ppm or mg/kg)*	*(ppm or mg/kg)*	*(ppm or mg/kg)*
357	44,342	26.6											
	149,530	*89.7*											
358	31,173	18.7											
	86,017	*51.6*											
359													
360	50,343	30.2											
	162,032	*97.2*											
361	47,510	28.5											
	173,035	*103.8*											
362													
363													
364	23,338	14.0											
	37,841	*22.7*											
365													
366													
367													
368													
369													
370													
371	90,185	54.1											
	344,069	*206.4*											
372													
373													
374	18,337	11.0							10.4				
	76,015	*45.6*							*43.0*				
375													
376													
377													

TABLE 4-101 COMPOSITION OF FEEDS-

Feed #	Feed Name—Description	Moisture Basis: As Fed or M-F (moisture-free)	Proximate Analysis: Dry Matter	Ash	Crude Fiber	Ether Extract (Fat)	N-Free Extract	Crude Protein	Digestible Protein: Ruminant	Non-Ruminant	Horse Rabbit
SILAGES AND HAYLAGES			(%)	(%)	(%)	(%)	(%)	(%)	(%)	(%)	(%)
378	Corn stalks (stalkage)	As Fed	66.0	7.0				2.8			
		M-F	*100.0*	*10.6*				*4.2*			
379	Corn stover, w/o ears	As Fed	27.2	2.3	8.7	.7	13.5	2.0	.8		.8
		M-F	*100.0*	*8.6*	*32.1*	*2.4*	*49.7*	*7.2*	*2.8*		*3.0*
380	Corn stover, stack silage, w/o ears	As Fed	35.0	2.5	9.9	1.1	19.1	2.4	.7		.9
		M-F	*100.0*	*7.0*	*28.3*	*3.0*	*54.8*	*6.9*	*2.0*		*2.5*
381	Corn, sweet, cannery residue *(Zea mays saccharata)*	As Fed	29.4	1.0	7.9	.7	17.2	2.6	1.4		
		M-F	*100.0*	*3.3*	*26.8*	*2.3*	*58.8*	*8.8*	*4.9*		
382	Grass silage, immature	As Fed	28.3	2.5	9.0	1.2	11.9	3.7	2.3		2.3
		M-F	*100.0*	*8.7*	*31.9*	*4.0*	*42.2*	*13.2*	*8.2*		*8.2*
383	Grass, early bloom	As Fed	23.4	2.1	7.3	.6	10.6	2.8	1.7		1.7
		M-F	*100.0*	*9.1*	*31.0*	*2.7*	*45.1*	*12.1*	*7.2*		*7.2*
384	Grass, w/molasses	As Fed	25.8	3.8	6.1	1.4	10.2	4.3	2.9		2.9
		M-F	*100.0*	*14.6*	*23.8*	*5.3*	*39.8*	*16.5*	*11.4*		*11.2*
385	Grass & legume, all analyses	As Fed	29.3	2.2	9.2	.9	13.5	3.5	1.8		1.8
		M-F	*100.0*	*7.6*	*31.4*	*3.2*	*46.0*	*11.8*	*6.2*		*6.2*
386	Grass & legume, wilted	As Fed	33.0	3.0	9.6	1.2	14.4	4.8	3.0		3.0
		M-F	*100.0*	*9.2*	*29.2*	*3.7*	*43.6*	*14.3*	*9.2*		*9.2*
387	Grass & legume w/ barley grain	As Fed	34.0	2.2	8.6	1.4	16.6	5.2	3.4		3.4
		M-F	*100.0*	*6.4*	*25.4*	*4.0*	*49.0*	*15.2*	*10.0*		*10.0*
388	Grass & legume w/ molasses	As Fed	30.0	2.2	9.3	1.0	14.1	3.4	1.7		1.9
		M-F	*100.0*	*7.4*	*31.1*	*3.3*	*47.0*	*11.2*	*5.7*		*6.4*
389	Kafir silage *(Sorghum vulgare)*	As Fed	29.5	2.2	8.0	1.0	16.2	2.1	.8		.8
		M-F	*100.0*	*7.6*	*27.3*	*3.3*	*54.6*	*7.2*	*2.8*		*2.8*
390	Ladino clover & grass *(Trifolium repens)*	As Fed	29.9	2.6	7.5	1.5	12.9	5.4	3.8		3.8
		M-F	*100.0*	*8.7*	*25.1*	*5.0*	*43.1*	*18.1*	*12.6*		*12.6*
391	Lespedeza, annual, silage *(Lespedeza spp)*	As Fed	30.2	1.8	9.5	.8	13.8	4.3	2.8		2.8
		M-F	*100.0*	*6.0*	*31.5*	*2.6*	*45.7*	*14.2*	*9.2*		*9.2*
392	Lespedeza, sericea *(Lespedeza cuneata)*	As Fed	30.4	1.7	9.5	.9	14.0	4.3	2.7		2.7
		M-F	*100.0*	*5.5*	*31.3*	*3.1*	*46.1*	*14.0*	*8.9*		*8.9*
393	Lucerne (see Alfalfa)										
394	Millet, Japanese & soybean *(Echinochloa crusgalli, Glycine max)*	As Fed	21.1	2.9	7.3	1.1	6.9	2.9	1.7		
		M-F	*100.0*	*13.7*	*34.6*	*5.2*	*32.8*	*13.7*	*8.1*		
395	Milo silage, w/heads *(Sorghum vulgare)*	As Fed	32.3	2.0	6.5	.5	21.0	2.3	.9		.9
		M-F	*100.0*	*6.2*	*20.1*	*1.5*	*65.1*	*7.1*	*2.7*		*2.7*
396	Oat, dough stage *(Avena sativa)*	As Fed	36.6	2.7	12.6	1.3	16.5	3.5	1.9		1.9
		M-F	*100.0*	*7.3*	*34.4*	*3.6*	*45.0*	*9.7*	*5.2*		*5.1*
397	Oat w/molasses	As Fed	33.0	2.5	10.5	1.2	15.9	2.9	1.4		1.4
		M-F	*100.0*	*7.4*	*31.7*	*3.8*	*48.3*	*8.8*	*4.2*		*4.2*
398	Oat & vetch *(Avena sativa, Vicia spp)*	As Fed	27.3	2.3	8.0	1.2	12.4	3.4	2.1		2.1
		M-F	*100.0*	*8.1*	*29.4*	*4.3*	*45.6*	*12.6*	*7.7*		*7.7*

LAGES AND HAYLAGES

Feed #	TDN Ruminant (%)	TDN Non-Ruminant (%)	TDN Horse Rabbit (%)	Digestible Energy Ruminant Mcal per lb	Digestible Energy Ruminant Mcal per kg	Digestible Energy Non-Ruminant kcal per lb	Digestible Energy Non-Ruminant kcal per kg	Digestible Energy Horse Rabbit Mcal per lb	Digestible Energy Horse Rabbit Mcal per kg	Metabolizable Energy Ruminant Mcal per lb	Metabolizable Energy Ruminant Mcal per kg	Non-Ruminant ME kcal per lb	Non-Ruminant ME kcal per kg	Non-Ruminant Chicken ME_n kcal per lb	Non-Ruminant Chicken ME_n kcal per kg	Horse Rabbit Mcal per lb	Horse Rabbit Mcal per kg	Net Energy-Ruminant NE_m Mcal per lb	NE_m Mcal per kg	NE_g Mcal per lb	NE_g Mcal per kg	NE_{lc} Mcal per lb	NE_{lc} Mcal per kg
78																							
79	15.4			.3	.7					.3	.6							.1	.3	.1	.2	.1	.3
	56.7			1.1	2.5					1.0	2.1							.5	1.2	.3	.6	.6	1.3
80	19.2			.4	.8					.3	.7												
	54.8			1.1	2.4					.9	2.0												
81	21.2			.4	.9					.4	.8							.2	.5	.1	.3	.2	.5
	72.0			1.5	3.2					1.2	2.6							.7	1.6	.5	1.0	.8	1.8
82	17.1			.4	.8					.3	.6												
	60.7			1.2	2.7					1.0	2.2												
83	14.4			.3	.6					.2	.5												
	61.6			1.2	2.7					1.0	2.2												
84	18.0			.4	.8					.3	.7												
	70.0			1.4	3.1					1.1	2.5												
85	16.4			.3	.7					.3	.6							.1	.3	.05	.1	.2	.4
	56.0			1.1	2.5					.9	2.0							.5	1.2	.2	.5	.5	1.2
86	20.3			.4	.9					.3	.7												
	61.6			1.2	2.7					1.0	2.2												
87	20.2			.4	.9					.3	.7												
	59.5			1.2	2.6					1.0	2.2												
88	17.1			.4	.8					.3	.6							.2	.4	.1	.2	.2	.4
	57.0			1.1	2.5					1.0	2.1							.5	1.2	.3	.6	.6	1.3
89	14.7			.3	.6					.2	.5												
	56.7			1.1	2.5					1.0													
90	18.5			.4	.8					.3	.7												
	63.3			1.3	2.8					1.0	2.3												
91	15.5			.3	.7					.3	.6												
	51.2			1.0	2.3					.9	1.9												
92	16.5			.3	.7					.3	.6												
	54.2			1.1	2.4					.9	2.0												
93																							
94				.3	.6					.2	.5												
				1.2	2.6					1.0	2.1												
95	19.3			.4	.9					.3	.7												
	59.8			1.2	2.6					1.0	2.2												
96	25.4			.5	1.1					.3	.7												
	55.7			1.1	2.5					.9	2.0												
97	17.6			.4	.8					.3	.6												
	53.2			1.0	2.3					.9	1.9												
98	17.0			.4	.8					.3	.6												
	62.4			1.3	2.8					1.0	2.3												

TABLE 4-101 COMPOSITION OF FEEDS-

Feed #	Feed Name—Description	Moisture Basis: As Fed or M-F (moisture-free)	Dry Matter	Macrominerals							Microminerals				
				Calcium (Ca)	Phosphorus (P)	Sodium (Na)	Chlorine (Cl)	Magnesium (Mg)	Potassium (K)	Sulfur (S)	Cobalt (Co)	Copper (Cu)	Iron (Fe)	Manganese (Mn)	Zinc (Zn)
SILAGES AND HAYLAGES			(%)	(%)	(%)	(%)	(%)	(%)	(%)	(%)	*(ppm or mg/kg)*	*(ppm or mg/kg)*	(%)	*(ppm or mg/kg)*	*(ppm or mg/kg)*
378	Corn stalks (stalkage)	As Fed	66.0	.24	.08	.07		.16	.61	.08		8.0	.005	34.0	16.0
		M-F	*100.0*	*.37*	*.12*	*.10*		*.24*	*.92*	*.12*		*12.0*	*.008*	*52.0*	*24.0*
379	Corn stover, w/o ears	As Fed	27.2	.10	.05			.08	.39						
		M-F	*100.0*	*.38*	*.19*			*.31*	*1.43*						
380	Corn stover, stack silage, w/o ears	As Fed	35.0												
		M-F	*100.0*												
381	Corn, sweet, cannery residue	As Fed	29.4												
		M-F	*100.0*												
382	Grass silage, immature	As Fed	28.3												
		M-F	*100.0*												
383	Grass, early bloom	As Fed	23.4												
		M-F	*100.0*												
384	Grass, w/molasses	As Fed	25.8	.27	.07										
		M-F	*100.0*	*1.04*	*.28*										
385	Grass & legume all analyses	As Fed	29.3	.23	.08		.31			.20					
		M-F	*100.0*	*.78*	*.28*		*1.06*			*.69*					
386	Grass & legume, wilted	As Fed	33.0	.24	.12										
		M-F	*100.0*	*.73*	*.35*										
387	Grass & legume w/ barley grain	As Fed	34.0	.26	.12										
		M-F	*100.0*	*.75*	*.35*										
388	Grass & legume w/ molasses	As Fed	30.0	.31	.08										
		M-F	*100.0*	*1.04*	*.28*										
389	Kafir silage	As Fed	29.5	.07	.05			.08	.50						
		M-F	*100.0*	*.24*	*.17*			*.27*	*1.68*						
390	Ladino clover & grass	As Fed	29.9	.31	.07			.09							
		M-F	*100.0*	*1.04*	*.23*			*.30*							
391	Lespedeza, annual, silage	As Fed	30.2												
		M-F	*100.0*												
392	Lespedeza, sericea	As Fed	30.4												
		M-F	*100.0*												
393	Lucerne (see Alfalfa)														
394	Millet, Japanese & soybean	As Fed	21.1												
		M-F	*100.0*												
395	Milo silage, w/heads	As Fed	32.3												
		M-F	*100.0*												
396	Oat, dough stage	As Fed	36.6	.17	.12										
		M-F	*100.0*	*.47*	*.33*										
397	Oat w/molasses	As Fed	33.0	.10	.09										
		M-F	*100.0*	*.31*	*.28*										
398	Oat & vetch	As Fed	27.3												
		M-F	*100.0*												

ILAGES AND HAYLAGES

Feed #	Fat-Soluble Vitamins				Water-Soluble Vitamins								
	A (1 mg car. = 1,667 IU-A)	Carotene (Pro-vitamin A)	E (α-tocopherol)	K	B_{12}	Biotin	Choline	Folic Acid (Folacin)	Niacin (Nicotinic Acid)	Pantothenic Acid	Pyridoxine (B_6)	Riboflavin (B_2)	Thiamin (B_1)
		(ppm or mg/kg)	*(ppm or mg/kg)*	*(ppm or mg/kg)*	*(ppm or mg/kg)*	*(ppm or mg/kg)*	*(ppm or mg/kg)*	*(ppm or mg/kg)*	*(ppm or mg/kg)*	*(ppm or mg/kg)*	*(ppm or mg/kg)*	*(ppm or mg/kg)*	*(ppm or mg/kg)*
378													
379													
380													
381													
382													
383													
384													
385	112,356	67.4							13.4				
	383,243	*229.9*							*45.6*				
386													
387													
388													
389	5,334	3.2											
	18,004	*10.8*											
390													
391													
392													
393													
394													
395													
396													
397													
398													

TABLE 4-101 COMPOSITION OF FEEDS-

Feed #	Feed Name—Description	Moisture Basis: As Fed or M-F (moisture-free)	Proximate Analysis: Dry Matter	Ash	Crude Fiber	Ether Extract (Fat)	N-Free Extract	Crude Protein	Digestible Protein: Ruminant	Non-Ruminant	Horse Rabbit
SILAGES AND HAYLAGES			(%)	(%)	(%)	(%)	(%)	(%)	(%)	(%)	(%)
399	Orchardgrass *(Dactylis glomerata)*	As Fed	29.5	2.7	10.5	1.2	10.9	4.2	2.8		2.7
		M-F	*100.0*	*9.1*	*35.4*	*4.2*	*37.0*	*14.3*	*9.8*		*9.2*
400	Pea, field *(Pisum spp)*	As Fed	27.0	2.5	7.4	1.2	12.1	3.8	2.3		2.4
		M-F	*100.0*	*9.3*	*27.5*	*4.3*	*44.8*	*14.1*	*8.8*		*9.0*
401	Pea & oat	As Fed	28.5	2.8	9.1	1.1	12.1	3.4	2.1		
		M-F	*100.0*	*9.8*	*31.9*	*3.9*	*42.5*	*11.9*	*7.4*		
402	Potato & mixed hay silage	As Fed	33.8	2.1	6.1	.9	20.8	3.9	2.3		
		M-F	*100.0*	*6.2*	*18.0*	*2.7*	*38.4*	*11.5*	*6.8*		
403	Potato & corn meal silage	As Fed	31.8	1.1	1.0	.4	27.2	2.1	1.1		
		M-F	*100.0*	*3.5*	*3.1*	*1.3*	*85.5*	*6.6*	*3.5*		
404	Potato tuber *(Solanum tuberosum)*	As Fed	25.1	1.4	2.1	.1	20.2	2.5	1.6	1.3	.7
		M-F	*100.0*	*5.5*	*8.5*	*.4*	*80.6*	*10.0*	*6.3*	*5.3*	*2.7*
405	Red clover silage *(Trifolium pratense)*	As Fed	26.2	2.5	8.5	1.0	10.4	3.8	2.4		2.5
		M-F	*100.0*	*9.6*	*32.5*	*3.6*	*39.8*	*14.5*	*9.2*		*9.4*
406	Reed canarygrass *(Phalaris canariensis)*	As Fed	27.5	2.0	9.1	.7	13.2	2.5	1.2		1.2
		M-F	*100.0*	*7.3*	*33.1*	*2.5*	*48.2*	*8.9*	*4.2*		*4.2*
407	Russian-thistle *(Salsola spp)*	As Fed	34.4	6.2	10.3	.9	14.5	2.5	1.0		1.0
		M-F	*100.0*	*17.9*	*29.8*	*2.8*	*42.1*	*7.4*	*2.9*		*2.9*
408	Sorghum, dough stage *(Sorghum vulgare)*	As Fed	26.9	1.8	7.5	.7	14.4	2.5	1.2		1.2
		M-F	*100.0*	*6.7*	*28.1*	*2.7*	*53.3*	*9.2*	*4.6*		*4.6*
409	Sorghum stover silage	As Fed	35.1	2.9	10.8	.8	19.0	1.6	.3		.7
		M-F	*100.0*	*7.9*	*30.9*	*2.3*	*54.2*	*4.7*	*.9*		*1.9*
410	Sorghum, sweet *(Sorghum vulgare saccharatum)*	As Fed	26.0	2.0	7.0	.7	14.7	1.6	.4		.6
		M-F	*100.0*	*7.7*	*26.8*	*2.5*	*56.7*	*6.3*	*1.6*		*2.4*
411	Sorghum, kafir *(Sorghum vulgare)*	As Fed	29.5	2.2	8.0	1.0	16.2	2.1	.8		.8
		M-F	*100.0*	*7.6*	*27.3*	*3.3*	*54.6*	*7.2*	*2.8*		*2.8*
412	Sorghum, milo *(Sorghum vulgare)*	As Fed	32.3	2.0	6.5	.5	21.0	2.3	.9		.9
		M-F	*100.0*	*6.2*	*20.1*	*1.5*	*65.1*	*7.1*	*2.7*		*2.7*
413	Soybean silage (fodder) *(Glycine max)*	As Fed	28.0	2.8	8.7	.7	11.6	4.2	2.8		3.4
		M-F	*100.0*	*10.0*	*31.1*	*2.6*	*41.6*	*14.7*	*10.1*		*12.1*
414	Soybean & Sudangrass *(Glycine max, Sorghum vulgare sudanense)*	As Fed	25.4	2.2	8.7	.5	10.6	3.4	2.1		2.1
		M-F	*100.0*	*8.6*	*34.1*	*2.1*	*41.9*	*13.3*	*8.3*		*8.3*
415	Sudangrass *(Sorghum vulgare sudanense)*	As Fed	23.3	2.3	8.0	.7	9.9	2.4	1.4		1.5
		M-F	*100.0*	*9.7*	*34.4*	*3.1*	*42.6*	*10.2*	*6.1*		*6.5*
416	Sugar beet pulp *(Beta saccharifera)*	As Fed	12.2	.6	3.9	.3	5.9	1.5	.9		.9
		M-F	*100.0*	*5.1*	*32.0*	*1.9*	*48.5*	*12.5*	*7.7*		*7.7*
417	Sugar beet tops w/ crowns	As Fed	20.7	6.8	2.8	.6	7.9	2.6	2.1		1.7
		M-F	*100.0*	*33.0*	*13.3*	*2.6*	*38.4*	*12.7*	*10.0*		*8.0*
418	Sunflower, milk stage *(Helianthus annuus)*	As Fed	21.2	2.1	6.2	1.3	9.5	2.1	1.1		1.1
		M-F	*100.0*	*10.0*	*29.4*	*5.9*	*45.0*	*9.7*	*4.9*		*5.0*
419	Sunflower, mature	As Fed	22.8	2.2	8.9	1.1	8.9	1.6	.6		.6
		M-F	*100.0*	*9.7*	*39.1*	*4.9*	*39.1*	*7.2*	*2.5*		*2.8*

SILAGES AND HAYLAGES

Feed #	TDN Ruminant (%)	TDN Non-Ruminant (%)	TDN Horse Rabbit (%)	Digestible Energy Ruminant Mcal per lb	Digestible Energy Ruminant Mcal per kg	Digestible Energy Non-Ruminant kcal per lb	Digestible Energy Non-Ruminant kcal per kg	Digestible Energy Horse Rabbit Mcal per lb	Digestible Energy Horse Rabbit Mcal per kg	Metabolizable Energy Ruminant Mcal per lb	Metabolizable Energy Ruminant Mcal per kg	Metabolizable Energy Non-Ruminant ME kcal per lb	Metabolizable Energy Non-Ruminant ME kcal per kg	Metabolizable Energy Non-Ruminant Chicken ME_n kcal per lb	Metabolizable Energy Non-Ruminant Chicken ME_n kcal per kg	Metabolizable Energy Horse Rabbit Mcal per lb	Metabolizable Energy Horse Rabbit Mcal per kg	Net Energy-Ruminant NE_m Mcal per lb	Net Energy-Ruminant NE_m Mcal per kg	Net Energy-Ruminant NE_g Mcal per lb	Net Energy-Ruminant NE_g Mcal per kg	Net Energy-Ruminant NE_{lc} Mcal per lb	Net Energy-Ruminant NE_{lc} Mcal per kg
399	18.8			.4	.8					.3	.6												
	63.7			*1.3*	*2.8*					*1.0*	*2.3*												
400	16.4			.3	.7					.3	.6												
	60.8			*1.2*	*2.7*					*1.0*	*2.2*												
401	17.6			.4	.8					.3	.6												
	61.8			*1.2*	*2.7*					*1.0*	*2.2*												
402	21.7			.5	1.0					.4	.8												
	64.2			*1.3*	*2.8*					*1.0*	*2.3*												
403	27.1			.5	1.2					.3	1.0												
	85.2			*1.7*	*3.8*					*1.4*	*3.1*												
404	20.0	22.1		.4	.9	443	974			.3	.7	419	922					.2	.5	.1	.3	.2	.5
	79.0	*88.1*		*1.6*	*3.5*	*1,765*	*3,884*			*1.3*	*2.9*	*1,670*	*3,674*					*.8*	*1.8*	*.5*	*1.2*	*.9*	*2.0*
405	14.9			.3	.7					.2	.5												
	56.8			*1.1*	*2.5*					*1.0*	*2.1*												
406	13.5			.3	.6					.2	.5												
	49.2			*1.0*	*2.2*					*.8*	*1.8*												
407	14.0			.3	.6					.2	.5												
	40.8			*.8*	*1.8*					*.7*	*1.5*												
408	15.7			.3	.7					.3	.6												
	58.6			*1.2*	*2.6*					*1.0*	*2.1*												
409	20.2			.4	.9					.3	.7												
	57.5			*1.1*	*2.5*					*1.0*	*2.1*												
410	15.2			.3	.7					.2	.5							.1	.3	.1	.2	.1	.3
	58.3			*1.2*	*2.6*					*1.0*	*2.1*							*.6*	*1.3*	*.3*	*.6*	*.6*	*1.3*
411	14.7			.3	.6					.2	.5												
	56.7			*1.1*	*2.5*					*1.0*	*2.1*												
412	19.3			.4	.9					.3	.7												
	59.8			*1.2*	*2.6*					*1.0*	*2.2*												
413	15.2			.3	.7					.3	.6							.1	.3	.05	.1	.1	.3
	54.3			*1.1*	*2.4*					*.9*	*2.0*							*.5*	*1.2*	*.2*	*.5*	*.5*	*1.1*
414	14.8			.3	.7					.2	.5												
	58.5			*1.2*	*2.6*					*1.0*	*2.1*												
415	13.5			.3	.6					.2	.5							.1	.3	.05	.1	.1	.3
	58.0			*1.2*	*2.6*					*1.0*	*2.1*							*.6*	*1.3*	*.3*	*.6*	*.6*	*1.3*
416	8.3			.2	.4					.1	.3												
	68.0			*1.4*	*3.0*					*1.1*	*2.5*												
417	11.2			.2	.5					.2	.4							.1	.2	.04	.09	.1	.2
	54.0			*1.1*	*2.4*					*.9*	*2.0*							*.5*	*1.1*	*.2*	*.4*	*.5*	*1.2*
418	11.0			.2	.5					.2	.4												
	52.0			*1.0*	*2.3*					*.9*	*1.9*												
419	10.8			.2	.5					.2	.4												
	47.5			*1.0*	*2.1*					*.8*	*1.7*												

TABLE 4-101 COMPOSITION OF FEEDS-

Feed #	Feed Name—Description	Moisture Basis: As Fed or M-F (moisture-free)	Dry Matter	Macrominerals: Calcium (Ca)	Phosphorus (P)	Sodium (Na)	Chlorine (Cl)	Magnesium (Mg)	Potassium (K)	Sulfur (S)	Microminerals: Cobalt (Co)	Copper (Cu)	Iron (Fe)	Manganese (Mn)	Zinc (Zn)
			(%)	(%)	(%)	(%)	(%)	(%)	(%)	(%)	(ppm or mg/kg)	(ppm or mg/kg)	(%)	(ppm or mg/kg)	(ppm or mg/kg)
SILAGES AND HAYLAGES															
399	Orchardgrass	As Fed	29.5												
		M-F	*100.0*												
400	Pea, field	As Fed	27.0	.37	.08			.11	.38	.07					
		M-F	*100.0*	*1.36*	*.29*			*.39*	*1.40*	*.25*					
401	Pea & oat	As Fed	28.5	.17	.09										
		M-F	*100.0*	*.60*	*.32*										
402	Potato & mixed hay silage	As Fed	33.8												
		M-F	*100.0*												
403	Potato & corn meal silage	As Fed	31.8												
		M-F	*100.0*												
404	Potato tuber	As Fed	25.1	.01	.06										
		M-F	*100.0*	*.04*	*.23*										
405	Red clover silage	As Fed	26.2	.42	.06	.06		.10	.46	.04		2.9	.009	34.1	
		M-F	*100.0*	*1.61*	*.23*	*.23*		*.40*	*1.76*	*.16*		*11.0*	*.033*	*130.0*	
406	Reed canarygrass	As Fed	27.5												
		M-F	*100.0*												
407	Russian-thistle	As Fed	34.4												
		M-F	*100.0*												
408	Sorghum, dough stage	As Fed	26.9												
		M-F	*100.0*												
409	Sorghum stover silage	As Fed	35.1	.14	.04									51.9	
		M-F	*100.0*	*.40*	*.11*									*147.9*	
410	Sorghum, sweet	As Fed	26.0	.09	.05	.01	.01	.07	.32			8.1	.005	17.5	
		M-F	*100.0*	*.35*	*.20*	*.03*	*.06*	*.27*	*1.22*			*31.3*	*.020*	*67.3*	
411	Sorghum, kafir	As Fed	29.5	.07	.05			.08	.50						
		M-F	*100.0*	*.24*	*.17*			*.27*	*1.68*						
412	Sorghum, milo	As Fed	32.3												
		M-F	*100.0*												
413	Soybean silage (fodder)	As Fed	28.0	.35	.14			.11	.26			2.6	.011	31.8	
		M-F	*100.0*	*1.25*	*.49*			*.38*	*.93*			*9.3*	*.040*	*113.5*	
414	Soybean & Sudangrass	As Fed	25.4												
		M-F	*100.0*												
415	Sudangrass	As Fed	23.3	.15	.05			.11	.72			8.5	.002	23.0	
		M-F	*100.0*	*.64*	*.23*			*.49*	*3.07*			*36.6*	*.010*	*98.8*	
416	Sugar beet pulp	As Fed	12.2												
		M-F	*100.0*												
417	Sugar beet tops w/ crowns	As Fed	20.7	.48	.04				1.18						
		M-F	*100.0*	*2.32*	*.20*				*5.70*						
418	Sunflower, milk stage	As Fed	21.2												
		M-F	*100.0*												
419	Sunflower, mature	As Fed	22.8												
		M-F	*100.0*												

LAGES AND HAYLAGES

Feed #	Fat-Soluble Vitamins: A (1 mg car. = 1,667 IU-A)	Carotene (Pro-vita-min A)	E (α-toco-pherol)	K	Water-Soluble Vitamins: B_{12}	Biotin	Choline	Folic Acid (Folacin)	Niacin (Nicotinic Acid)	Panto-thenic Acid	Pyridoxine (B_6)	Riboflavin (B_2)	Thiamin (B_1)
		(ppm or mg/kg)	(ppm or mg/kg)	(ppm or mg/kg)	(ppm or mg/kg)	(ppm or mg/kg)	(ppm or mg/kg)	(ppm or mg/kg)	(ppm or mg/kg)	(ppm or mg/kg)	(ppm or mg/kg)	(ppm or mg/kg)	(ppm or mg/kg)
399	83,183	49.9											
	281,890	*169.1*											
400													
401													
402													
403													
404													
405	90,185	54.1											
	344,069	*206.4*											
406													
407													
408													
409	4,001	2.4											
	11,336	*6.8*											
410													
411	5,334	3.2											
	18,004	*10.8*											
412													
413													
414	15,836	9.5											
	62,512	*37.5*											
415	30,173	18.1											
	129,359	*77.6*											
416													
417													
418													
419													

TABLE 4-101 COMPOSITION OF FEEDS–

Feed #	Feed Name—Description	Moisture Basis: As Fed or M-F (moisture-free)	Proximate Analysis						Digestible Protein		
			Dry Matter	Ash	Crude Fiber	Ether Extract (Fat)	N-Free Extract	Crude Protein	Ruminant	Non-Ruminant	Horse Rabbit
SILAGES AND HAYLAGES			(%)	(%)	(%)	(%)	(%)	(%)	(%)	(%)	(%)
420	Sweet corn, cannery residue *(Zea mays saccharata)*	As Fed	29.4	1.0	7.9	.7	17.2	2.6	1.4		
		M-F	*100.0*	*3.3*	*26.8*	*2.3*	*58.8*	*8.8*	*4.9*		
421	Sweet potato vines *(Ipomoea batatas)*	As Fed	12.1	1.4	3.5	.5	5.1	1.6	.9		1.0
		M-F	*100.0*	*11.5*	*29.0*	*4.3*	*42.0*	*13.2*	*7.2*		*8.2*
422	Vetch *(Vicia spp)*	As Fed	30.1	2.4	9.8	1.0	13.4	3.5	2.0		2.0
		M-F	*100.0*	*7.9*	*32.7*	*3.3*	*44.4*	*11.7*	*6.7*		*6.8*
423	Vetch & oat *(Vicia spp, Avena sativa)*	As Fed	26.4	1.9	8.8	.6	12.9	2.2	1.0		1.0
		M-F	*100.0*	*7.2*	*33.3*	*2.3*	*48.9*	*8.3*	*3.8*		*3.8*

ILAGES AND HAYLAGES

Feed #	TDN Ruminant	TDN Non-Ruminant	TDN Horse Rabbit	DE Ruminant		DE Non-Ruminant		DE Horse Rabbit		ME Ruminant		ME Non-Ruminant ME		ME Non-Ruminant Chicken ME_n		ME Horse Rabbit		NE_m		NE_g		NE_{lc}	
	(%)	(%)	(%)	Mcal per lb	Mcal per kg	kcal per lb	kcal per kg	Mcal per lb	Mcal per kg	Mcal per lb	Mcal per kg	kcal per lb	kcal per kg	kcal per lb	kcal per kg	Mcal per lb	Mcal per kg	Mcal per lb	Mcal per kg	Mcal per lb	Mcal per kg	Mcal per lb	Mcal per kg
420	21.2			.4	.9					.4	.8							.2	.5	.1	.3	.2	.5
	72.0			*1.5*	*3.2*					*1.2*	*2.6*							*.7*	*1.6*	*.5*	*1.0*	*.8*	*1.8*
421	6.0			.1	.3					.1	.2												
	50.2			*1.0*	*2.2*					*.8*	*1.8*												
422	19.0			.4	.8					.3	.7												
	62.9			*1.3*	*2.8*					*1.0*	*2.3*												
423	16.4			.3	.7					.3	.6												
	62.0			*1.2*	*2.7*					*1.0*	*2.2*												

TABLE 4-101 COMPOSITION OF FEEDS-

Feed #	Feed Name—Description	Moisture Basis: As Fed or M-F (moisture-free)	Dry Matter	Macrominerals: Calcium (Ca)	Phosphorus (P)	Sodium (Na)	Chlorine (Cl)	Magnesium (Mg)	Potassium (K)	Sulfur (S)	Microminerals: Cobalt (Co)	Copper (Cu)	Iron (Fe)	Manganese (Mn)	Zinc (Zn)
			(%)	(%)	(%)	(%)	(%)	(%)	(%)	(%)	(ppm or mg/kg)	(ppm or mg/kg)	(%)	(ppm or mg/kg)	(ppm or mg/kg)
SILAGES AND HAYLAGES															
420	Sweet corn, cannery residue	As Fed	29.4												
		M-F	*100.0*												
421	Sweet potato vines	As Fed	12.1												
		M-F	*100.0*												
422	Vetch	As Fed	30.1												
		M-F	*100.0*												
423	Vetch & oat	As Fed	26.4												
		M-F	*100.0*												

SILAGES AND HAYLAGES

Feed #	Fat-Soluble Vitamins				Water-Soluble Vitamins								
	A (1 mg car. = 1,667 IU-A)	Carotene (Pro-vitamin A)	E (α-tocopherol)	K	B_{12}	Biotin	Choline	Folic Acid (Folacin)	Niacin (Nicotinic Acid)	Pantothenic Acid	Pyridoxine (B_6)	Riboflavin (B_2)	Thiamin (B_1)
		(ppm or mg/kg)	*(ppm or mg/kg)*	*(ppm or mg/kg)*	*(ppm or mg/kg)*	*(ppm or mg/kg)*	*(ppm or mg/kg)*	*(ppm or mg/kg)*	*(ppm or mg/kg)*	*(ppm or mg/kg)*	*(ppm or mg/kg)*	*(ppm or mg/kg)*	*(ppm or mg/kg)*
420													
421													
422													
423													

TABLE 4-101 COMPOSITION OF FEEDS—

Feed #	Feed Name—Description	Moisture Basis: As Fed or M-F (moisture-free)	Proximate Analysis: Dry Matter	Ash	Crude Fiber	Ether Extract (Fat)	N-Free Extract	Crude Protein	Digestible Protein: Ruminant	Non-Ruminant	Horse Rabbit
PASTURE & RANGE PLANTS			(%)	(%)	(%)	(%)	(%)	(%)	(%)	(%)	(%)
424	Alfalfa, all analyses *(Medicago sativa)*	As Fed	27.2	2.5	6.9	1.1	11.1	5.6	4.1		
		M-F	*100.0*	*9.1*	*25.3*	*4.1*	*40.8*	*20.7*	*15.4*		
425	Alfalfa, midbloom	As Fed	24.2	2.1	6.7	.7	9.7	5.0	3.8		3.5
		M-F	*100.0*	*8.8*	*27.7*	*2.8*	*40.3*	*20.4*	*15.6*		*14.7*
426	Alfalfa, full bloom	As Fed	25.3	2.5	8.0	.7	9.9	4.2	3.4		2.8
		M-F	*100.0*	*9.8*	*31.7*	*2.6*	*39.0*	*16.9*	*13.3*		*10.9*
427	Alfalfa & bromegrass *(Medicago sativa, Bromus* spp)	As Fed	22.5	2.2	5.3	.8	9.4	4.8	3.7		3.5
		M-F	*100.0*	*9.8*	*23.6*	*3.6*	*41.8*	*21.2*	*16.5*		*15.4*
428	Alfalfa & orchardgrass *(Medicago sativa, Dactylis glomerata*	As Fed	25.0	2.5	7.6	1.1	9.8	4.0	2.7		2.7
		M-F	*100.0*	*10.1*	*30.5*	*4.3*	*38.9*	*16.2*	*10.9*		*10.9*
429	Alfalfa & timothy *(Medicago sativa, Phleum pratense)*	As Fed	21.8	2.3	4.6	.8	9.6	4.5	3.4		3.1
		M-F	*100.0*	*10.6*	*21.1*	*3.7*	*44.2*	*20.4*	*15.6*		*14.6*
430	Alsike clover *(Trifolium hybridum)*	As Fed	22.5	2.1	5.2	.8	10.3	4.1	3.0		2.8
		M-F	*100.0*	*9.3*	*23.3*	*3.6*	*45.7*	*18.1*	*13.3*		*12.7*
431	Alta (tall) fescue *(Festuca arundinacea)*	As Fed	23.9	2.0	7.1	.8	11.3	2.7	1.9		1.8
		M-F	*100.0*	*8.3*	*29.6*	*3.3*	*47.2*	*11.6*	*7.7*		*7.5*
432	Beardgrass (see Bluestem)										
433	Beet, sugar, tops w/crowns *(Beta saccharifera)*	As Fed	15.9	3.3	1.7	.4	7.7	2.8	2.1		1.9
		M-F	*100.0*	*20.6*	*10.7*	*2.5*	*48.7*	*17.5*	*13.2*		*12.2*
434	Bermudagrass *(Cynodon dactylon)*	As Fed	36.7	4.6	9.5	.7	17.7	4.2	2.7		2.5
		M-F	*100.0*	*12.4*	*25.9*	*2.0*	*48.1*	*11.6*	*7.3*		*6.8*
435	Bermudagrass, coastal	As Fed	28.8	1.8	8.2	1.1	13.4	4.3	3.1		3.0
		M-F	*100.0*	*6.3*	*28.4*	*3.8*	*46.6*	*14.9*	*10.8*		*10.3*
436	Bluegrass, Canada, all analyses *(Poa compressa)*	As Fed	30.6	2.6	8.1	1.1	13.6	5.2	3.8		2.6
		M-F	*100.0*	*8.4*	*26.4*	*3.7*	*44.5*	*17.0*	*12.3*		*8.6*
437	Bluegrass, Kentucky, immature *(Poa pratensis)*	As Fed	30.5	2.9	7.7	1.1	13.5	5.3	3.9		
		M-F	*100.0*	*9.4*	*25.1*	*3.5*	*44.7*	*17.3*	*12.9*		
438	Bluegrass, Kentucky, & white clover pasture *(Poa pratensis, Trifolium repens)*	As Fed	24.4	2.6	4.5	.9	11.4	5.0	3.9		3.7
		M-F	*100.0*	*11.0*	*18.3*	*3.5*	*46.6*	*20.6*	*15.8*		*14.8*
439	Bluestem, immature *(Andropogon spp)*	As Fed	31.6	2.7	9.1	.9	15.4	3.5	2.3		1.8
		M-F	*100.0*	*8.5*	*28.9*	*2.8*	*48.8*	*11.0*	*7.2*		*5.6*
440	Bromegrass, immature *(Bromus spp)*	As Fed	32.5	3.4	7.8	1.5	13.2	6.6	5.1		3.2
		M-F	*100.0*	*10.7*	*23.9*	*4.5*	*40.6*	*20.3*	*15.6*		*9.9*
441	Buffalograss *(Buchloe dactyloides)*	As Fed	47.7	6.0	13.2	.9	23.2	4.4	2.7		2.8
		M-F	*100.0*	*12.5*	*27.7*	*1.9*	*48.7*	*9.2*	*5.7*		*5.8*
442	Buffelgrass, young pasture *(Cenchrus ciliaris)*	As Fed	41.4	4.1	15.9	2.2	15.1	4.1			
		M-F	*100.0*	*9.8*	*38.4*	*5.4*	*36.6*	*9.8*			
443	Bur-clover *(Medicago denticulata)*	As Fed	20.8	2.3	3.9	1.7	7.8	5.1	4.0		3.8
		M-F	*100.0*	*11.1*	*18.8*	*8.2*	*37.5*	*24.4*	*19.3*		*18.0*
444	Cactus, prickly-pear *(Opuntia spp)*	As Fed	17.1	3.2	2.3	.4	10.3	.9	.5		.3
		M-F	*100.0*	*18.9*	*13.3*	*2.3*	*60.5*	*5.0*	*2.7*		*1.9*

ASTURE AND RANGE PLANTS

Feed #	TDN Ruminant (%)	TDN Non-Ruminant (%)	TDN Horse Rabbit (%)	DE Ruminant Mcal per lb	DE Ruminant Mcal per kg	DE Non-Ruminant kcal per lb	DE Non-Ruminant kcal per kg	DE Horse Rabbit Mcal per lb	DE Horse Rabbit Mcal per kg	ME Ruminant Mcal per lb	ME Ruminant Mcal per kg	ME Non-Ruminant ME kcal per lb	ME Non-Ruminant ME kcal per kg	Chicken ME_n kcal per lb	Chicken ME_n kcal per kg	ME Horse Rabbit Mcal per lb	ME Horse Rabbit Mcal per kg	NE_m Mcal per lb	NE_m Mcal per kg	NE_g Mcal per lb	NE_g Mcal per kg	NE_{lc} Mcal per lb	NE_{lc} Mcal per kg
424	16.2			.3	.7					.3	.6							.2	.4	.1	.2	.2	.4
	59.7			*1.2*	*2.6*					*1.0*	*2.2*							*.6*	*1.3*	*.3*	*.6*	*.6*	*1.4*
425	14.5			.3	.6					.2	.5												
	60.2			*1.2*	*2.7*					*1.0*	*2.2*												
426	14.4		13.7	.3	.6			.2	.5	.2	.5												
	57.1		*54.0*	*1.1*	*2.5*			*1.0*	*2.1*	*1.0*	*2.1*												
427	14.2			.3	.6					.2	.5												
	63.2			*1.3*	*2.8*					*1.0*	*2.3*												
428	14.5			.3	.6					.2	.5												
	58.0			*1.2*	*2.6*					*1.0*	*2.1*												
429	14.9			.3	.7					.2	.5												
	68.6			*1.4*	*3.0*					*1.1*	*2.5*												
430	15.0			.3	.7					.2	.5												
	66.6			*1.3*	*2.9*					*1.1*	*2.4*												
431	15.6			.3	.7					.3	.6												
	65.2			*1.3*	*2.9*					*1.1*	*2.4*												
432																							
433	10.1			.2	.4					.2	.4												
	63.7			*1.3*	*2.8*					*1.0*	*2.3*												
434	23.2		20.6	.5	1.0			.4	.8	.4	.8												
	63.2		*56.0*	*1.3*	*2.8*			*1.0*	*2.2*	*1.0*	*2.3*												
435	18.8			.4	.8					.3	.7												
	65.1			*1.3*	*2.9*					*1.1*	*2.4*												
436	19.0			.4	.8					.3	.7							.2	.4	.1	.2	.2	.5
	62.0			*1.2*	*2.7*					*1.0*	*2.2*							*.6*	*1.4*	*.4*	*.8*	*.8*	*1.7*
437	20.1			.4	.9					.3	.7							.2	.5	.1	.3	.2	.5
	66.0			*1.3*	*2.9*					*1.1*	*2.4*							*.7*	*1.5*	*.4*	*.9*	*.8*	*1.8*
438	17.0			.4	.8					.3	.6												
	69.6			*1.4*	*3.1*					*1.1*	*2.5*												
439	18.0			.4	.8					.3	.7							.2	.4	.1	.2		
	57.0			*1.1*	*2.5*					*1.0*	*2.1*							*.5*	*1.2*	*.3*	*.6*		
440	21.0			.4	.9					.4	.8							.2	.5	.1	.3	.2	.5
	64.7			*1.3*	*2.9*					*1.1*	*2.4*							*.6*	*1.4*	*.4*	*.9*	*.7*	*1.6*
441	26.7			.5	1.2					.5	1.0							.3	.6	.09	.2		
	56.0			*1.1*	*2.5*					*.9*	*2.0*							*.5*	*1.2*	*.2*	*.5*		
442																							
443	14.5			.3	.6					.2	.5												
	69.4			*1.4*	*3.1*					*1.1*	*2.5*												
444	9.4			.2	.4					.1	.3							.1	.2	.05	.1		
	55.5			*1.1*	*2.5*					*.9*	*2.0*							*.6*	*1.3*	*.3*	*.7*		

TABLE 4-101 COMPOSITION OF FEEDS–

Feed #	Feed Name—Description	Moisture Basis: As Fed or M-F (moisture-free)	Dry Matter	Macrominerals: Calcium (Ca)	Phosphorus (P)	Sodium (Na)	Chlorine (Cl)	Magnesium (Mg)	Potassium (K)	Sulfur (S)	Microminerals: Cobalt (Co)	Copper (Cu)	Iron (Fe)	Manganese (Mn)	Zinc (Zn)
			(%)	(%)	(%)	(%)	(%)	(%)	(%)	(%)	(ppm or mg/kg)	(ppm or mg/kg)	(%)	(ppm or mg/kg)	(ppm or mg/kg)
PASTURE AND RANGE PLANTS															
424	Alfalfa, all analyses	As Fed	27.2	.47	.08	.05	.12	.07	.55	.11	.024	2.7	.008	13.7	4.8
		M-F	*100.0*	*1.72*	*.31*	*.20*	*.45*	*.27*	*2.03*	*.39*	*.090*	*9.9*	*.030*	*50.5*	*17.6*
425	Alfalfa, midbloom	As Fed	24.2	.49	.07	.04		.06	.50	.07					5.7
		M-F	*100,0*	*2.01*	*.28*	*.16*		*.26*	*2.06*	*.29*					*23.6*
426	Alfalfa, full bloom	As Fed	25.3	.39	.07	.04		.07	.54	.08			.010	39.3	3.6
		M-F	*100.0*	*1.53*	*.27*	*.15*		*.27*	*2.13*	*.31*			*.040*	*155.2*	*14.1*
427	Alfalfa & bromegrass	As Fed	22.5	.28	.07				.63						
		M-F	*100.0*	*1.24*	*.31*				*2.80*						
428	Alfalfa & orchard-grass	As Fed	25.0	.10	.13			.06							
		M-F	*100.0*	*.40*	*.52*			*.24*							
429	Alfalfa & timothy	As Fed	21.8												
		M-F	*100.0*												
430	Alsike clover	As Fed	22.5	.31	.06	.10	.17	.07	.61	.05		1.3	.010	26.3	13.5
		M-F	*100.0*	*1.36*	*.29*	*.45*	*.77*	*.32*	*2.70*	*.22*		*6.0*	*.043*	*117.1*	*60.2*
431	Alta (tall) fescue	As Fed	23.9												
		M-F	*100.0*												
432	Beardgrass (see Bluestem)														
433	Beet, sugar, tops w/crowns	As Fed	15.9	.16	.04	.09		.17	.92	.09		2.2	.003	7.7	
		M-F	*100.0*	*1.01*	*.22*	*.54*		*1.07*	*5.79*	*.57*		*13.6*	*.017*	*48.4*	
434	Bermudagrass	As Fed	36.7	.19	.08	.16		.08	.60		.026	2.1	.040	36.7	
		M-F	*100.0*	*.53*	*.22*	*.44*		*.23*	*1.63*		*.070*	*5.7*	*.110*	*100.1*	
435	Bermudagrass, coastal	As Fed	28.8	.14	.08										
		M-F	*100.0*	*.49*	*.27*										
436	Bluegrass, Canada, all analyses	As Fed	30.6	.12	.12			.05	.62					24.2	
		M-F	*100.0*	*.39*	*.39*			*.16*	*2.04*					*79.2*	
437	Bluegrass, Kentucky, immature	As Fed	30.5	.17	.14			.05	.70	.20		4.3	.009	27.5	52.0
		M-F	*100.0*	*.56*	*.47*			*.18*	*2.28*	*.66*		*14.1*	*.030*	*80.3*	*170.4*
438	Bluegrass, Kentucky, & white clover pasture	As Fed	24.4	.31	.11										
		M-F	*100.0*	*1.29*	*.46*										
439	Bluestem, immature	As Fed	31.6	.20	.05			.02	.43			11.6	.022	26.3	
		M-F	*100.0*	*.63*	*.17*			*.07*	*1.35*			*36.8*	*.070*	*83.3*	
440	Bromegrass, immature	As Fed	32.5	.19	.12			.06	1.40						
		M-F	*100.0*	*.59*	*.37*			*.18*	*4.30*						
441	Buffalograss	As Fed	47.7	.25	.08			.07	.34						
		M-F	*100.0*	*.52*	*.16*			*.14*	*.71*						
442	Buffelgrass, young pasture	As Fed	41.4												
		M-F	*100.0*												
443	Bur-clover	As Fed	20.8												
		M-F	*100.0*												
444	Cactus, prickly-pear	As Fed	17.1	1.08	.01		.04	.28	.21	.04			.015		
		M-F	*100.0*	*6.29*	*.08*		*.21*	*1.65*	*1.21*	*.23*			*.090*		

PASTURE AND RANGE PLANTS

	Fat-Soluble Vitamins				Water-Soluble Vitamins								
Feed #	A (1 mg car. = 1,667 IU-A)	Carotene (Pro-vita-min A)	E (α-toco-pherol)	K	B_{12}	Biotin	Choline	Folic Acid (Folacin)	Niacin (Nicotinic Acid)	Panto-thenic Acid	Pyridoxine (B_6)	Riboflavin (B_2)	Thiamin (B_1)
		(ppm or mg/kg)	*(ppm or mg/kg)*	*(ppm or mg/kg)*	*(ppm or mg/kg)*	*(ppm or mg/kg)*	*(ppm or mg/kg)*	*(ppm or mg/kg)*	*(ppm or mg/kg)*	*(ppm or mg/kg)*	*(ppm or mg/kg)*	*(ppm or mg/kg)*	*(ppm or mg/kg)*
424	90,185	54.1	41.4			.13		.67					
	331,566	*198.9*	*152.1*			*.49*		*2.47*					
425													
426	15,670	9.4								7.9			
	53,344	*32.0*								*31.3*			
427													
428													
429													
430												4.4	2.0
												19.6	*8.8*
431													
432													
433	9,168	5.5										1.1	
	58,012	*34.8*										*6.6*	
434	172,034	103.2											
	468,594	*281.1*											
435	158,865	95.3											
	550,943	*330.5*											
436													
437	194,706	116.8											
	638,461	*383.0*											
438													
439	115,523	69.3											
	365,406	*219.2*											
440	248,883	149.3											
	765,986	*459.5*											
441													
442													
443													
444													

TABLE 4-101 COMPOSITION OF FEEDS-

Feed #	Feed Name—Description	Moisture Basis: As Fed or M-F (moisture-free)	Proximate Analysis: Dry Matter (%)	Ash (%)	Crude Fiber (%)	Ether Extract (Fat) (%)	N-Free Extract (%)	Crude Protein (%)	Digestible Protein: Ruminant (%)	Non-Ruminant (%)	Horse Rabbit (%)
PASTURE & RANGE PLANTS											
445	Canada bluegrass (see Bluegrass, Canada)										
446	Canarygrass, reed *(Phalaris arundinacea)*	As Fed	25.8	2.6	6.9	1.1	11.7	3.4	2.4		3.2
		M-F	*100.0*	*10.0*	*26.8*	*4.4*	*45.6*	*13.2*	*9.2*		*12.7*
447	Clover, alsike *(Trifolium hybridum)*	As Fed	22.5	2.1	5.2	.8	10.3	4.1	3.0		2.8
		M-F	*100.0*	*9.3*	*23.3*	*3.6*	*45.7*	*18.1*	*13.3*		*12.7*
448	Clover, crimson *(Trifolium incarnatum)*	As Fed	17.6	1.7	4.9	.6	7.4	3.0	2.1		2.1
		M-F	*100.0*	*9.5*	*27.7*	*3.3*	*42.5*	*17.0*	*11.9*		*11.8*
449	Clover, Ladino *(Trifolium repens)*	As Fed	18.0	1.9	2.5	.9	8.2	4.5	3.5		3.2
		M-F	*100.0*	*10.8*	*14.1*	*4.8*	*45.6*	*24.7*	*19.5*		*18.1*
450	Clover, red, early bloom *(Trifolium pratense)*	As Fed	19.6	2.0	3.7	1.0	8.8	4.1	2.9		2.7
		M-F	*100.0*	*10.2*	*19.0*	*5.0*	*44.7*	*21.1*	*14.8*		*13.8*
451	Clover, subterranean (sub) *(Trifolium subterraneum)*	As Fed	17.4	1.7	5.0	.8	7.7	2.2			
		M-F	*100.0*	*9.8*	*28.7*	*4.4*	*44.6*	*12.5*			
452	Clover, white *(Trifolium repens)*	As Fed	17.7	2.1	2.8	.6	7.2	5.0	3.4		3.7
		M-F	*100.0*	*11.9*	*15.7*	*3.3*	*40.9*	*28.2*	*19.6*		*21.0*
453	Comfrey, prickly *(Symphytum asperrimum)*	As Fed	13.2	2.4	1.8	.3	6.1	2.6	2.0		1.8
		M-F	*100.0*	*18.2*	*13.8*	*2.4*	*45.9*	*19.7*	*14.9*		*14.0*
454	Corn, sweet, ears & husks *(Zea mays saccharata)*	As Fed	37.8	.9	4.3	2.6	26.2	3.8	2.3		2.4
		M-F	*100.0*	*2.4*	*11.4*	*6.9*	*69.3*	*10.0*	*6.0*		*6.4*
455	Corn stover (ears removed), green, field corn	As Fed	22.7	1.4	6.0	.4	13.6	1.3	.5		.6
		M-F	*100.0*	*6.2*	*26.4*	*1.8*	*59.9*	*5.7*	*2.3*		*2.8*
456	Corn stover (ears removed), sweet *(Zea mays saccharata)*	As Fed	22.0	1.4	5.7	.4	12.9	1.6	.8		.9
		M-F	*100.0*	*6.2*	*26.0*	*1.8*	*58.7*	*7.3*	*3.8*		*4.0*
457	Crested wheatgrass, early bloom *(Agropyron cristatum)*	As Fed	42.5	3.1	12.5	.7	21.5	4.7	3.1		1.7
		M-F	*100.0*	*7.3*	*29.5*	*1.6*	*50.6*	*11.0*	*7.2*		*4.0*
458	Crimson clover & ryegrass *(Trifolium incarnatum, Lolium spp)*	As Fed	18.3	1.9	3.4	1.1	8.0	3.9	3.0		2.8
		M-F	*100.0*	*10.4*	*18.6*	*6.0*	*43.7*	*21.3*	*16.4*		*15.3*
459	Curly mesquite, browse *(Hilaria belangeri)*	As Fed									
		M-F	*100.0*	*17.1*	*28.8*	*2.2*	*45.9*	*6.0*	*2.6*		*3.0*
460	Fescue, meadow *(Festuca elatior)*	As Fed	27.6	2.4	7.5	1.2	12.3	4.2	2.3		2.8
		M-F	*100.0*	*8.6*	*27.1*	*4.2*	*45.0*	*15.1*	*8.4*		*10.3*
461	Fescue, tall *(Festuca arundinacea)*	As Fed	23.9	2.0	7.1	.8	11.3	2.7	1.9		1.8
		M-F	*100.0*	*8.3*	*29.6*	*3.3*	*47.2*	*11.6*	*7.7*		*7.5*
462	Foxtail, meadow *(Alopecurus pratensis)*	As Fed	26.1	2.8	5.6	1.2	12.0	4.5	3.3		3.1
		M-F	*100.0*	*10.7*	*21.5*	*4.6*	*46.0*	*17.2*	*12.7*		*12.0*
463	Gamagrass *(Tripsacum dactyloides)*	As Fed									
		M-F	*100.0*	*10.2*	*30.2*	*2.0*	*49.1*	*8.5*	*4.8*		*5.0*
464	Grama grass, mature *(Bouteloua spp)*	As Fed	63.4	6.3	20.7	1.1	31.2	4.1	2.0		
		M-F	*100.0*	*9.9*	*32.7*	*1.7*	*49.2*	*6.5*	*3.2*		
465	Grass, early bloom	As Fed	30.8	1.8	10.6	1.3	14.1	3.0	1.7		1.8
		M-F	*100.0*	*5.8*	*34.4*	*4.2*	*45.8*	*9.8*	*5.4*		*6.0*

ASTURE AND RANGE PLANTS

Feed #	TDN			Digestible Energy						Metabolizable Energy								Net Energy-Ruminant					
												Non-Ruminant											
	Rumi-nant	Non-Rumi-nant	Horse Rabbit	Rumi-nant		Non-Rumi-nant		Horse Rabbit		Rumi-nant		ME		Chicken ME_n		Horse Rabbit		NE_m		NE_g		NE_{lc}	
	(%)	(%)	(%)	Mcal per		kcal per		Mcal per		Mcal per		kcal per		kcal per		Mcal per		Mcal per		Mcal per		Mcal per	
				lb	kg	lb	kg	lb	kg	lb	kg	lb	kg	lb	kg	lb	kg	lb	kg	lb	kg	lb	kg
445																							
446	14.8			.3	.7					.2	.5							.2	.4	.1	.2		
	57.5			*1.1*	*2.5*					*1.0*	*2.1*							*.7*	*1.5*	*.4*	*.9*		
447	15.0			.3	.7					.2	.5												
	66.6			*1.3*	*2.9*					*1.1*	*2.4*												
448	11.2			.2	.5					.2	.4												
	64.0			*1.3*	*2.8*					*1.0*	*2.3*												
449	13.2			.3	.6					.2	.5												
	73.4			*1.5*	*3.2*					*1.2*	*2.7*												
450	13.3			.3	.6					.2	.5							.1	.3	.1	.2	.1	.3
	67.7			*1.4*	*3.0*					*1.1*	*2.5*							*.7*	*1.6*	*.5*	*1.0*	*.8*	*1.7*
451																							
452	11.1			.2	.5					.2	.4												
	62.4			*1.3*	*2.8*					*1.0*	*2.3*												
453	8.2									.1	.3							.2	.4	.1	.2		
	61.8									*1.0*	*2.2*							*1.4*	*3.0*	*.8*	*1.8*		
454	30.3	26.1		.6	1.3	523	1,151			.5	1.1	492	1,082										
	80.2	*69.1*		*1.6*	*3.5*	*1,385*	*3,046*			*1.3*	*2.9*	*1,300*	*2,862*										
455	13.0			.3	.6					.2	.5												
	57.3			*1.1*	*2.5*					*1.0*	*2.1*												
456	13.9			.3	.6					.2	.5												
	63.3			*1.3*	*2.8*					*1.0*	*2.3*												
457	24.7			.5	1.1					.4	.9							.2	.5	.1	.3		
	58.0			*1.2*	*2.6*					*1.0*	*2.1*							*.5*	*1.2*	*.3*	*.6*		
458	12.6			.3	.6					.2	.5												
	69.1			*1.4*	*3.0*					*1.1*	*2.5*												
459																							
	48.8			*1.0*	*2.2*					*.8*	*1.8*												
460	17.8		13.8	.4	.8			.3	.6	.3	.6												
	64.4		*50.0*	*1.3*	*2.8*			*.9*	*2.0*	*1.0*	*2.3*												
461	15.6			.3	.7					.3	.6												
	65.2			*1.3*	*2.9*					*1.1*	*2.4*												
462	17.4			.4	.8					.3	.6												
	66.7			*1.3*	*2.9*					*1.1*	*2.4*												
463																							
	64.2			*1.3*	*2.8*					*1.0*	*2.3*												
464	34.2			.7	1.5					.5	1.2							.4	.8	.2	.4		
	54.0			*1.1*	*2.4*					*.9*	*2.0*							*.5*	*1.2*	*.3*	*.6*		
465	21.1			.4	.9					.4	.8												
	68.4			*1.4*	*3.0*					*1.1*	*2.5*												

TABLE 4-101 COMPOSITION OF FEEDS–

Feed #	Feed Name—Description	Moisture Basis: As Fed or M-F (moisture-free)	Dry Matter	Macrominerals: Calcium (Ca)	Phosphorus (P)	Sodium (Na)	Chlorine (Cl)	Magnesium (Mg)	Potassium (K)	Sulfur (S)	Microminerals: Cobalt (Co)	Copper (Cu)	Iron (Fe)	Manganese (Mn)	Zinc (Zn)
	PASTURE AND RANGE PLANTS		(%)	(%)	(%)	(%)	(%)	(%)	(%)	(%)	(ppm or mg/kg)	(ppm or mg/kg)	(%)	(ppm or mg/kg)	(ppm or mg/kg)
445	Canada bluegrass (see Bluegrass, Canada)														
446	Canarygrass, reed	As Fed	25.8	.10	.08				.94						
		M-F	100.0	.40	.30				3.64						
447	Clover, alsike	As Fed	22.5	.31	.06	.10	.17	.07	.61	.05		1.3	.010	26.3	13.5
		M-F	100.0	1.36	.29	.45	.77	.32	2.70	.22		6.0	.043	117.1	60.2
448	Clover, crimson	As Fed	17.6	.24	.05	.07	.11	.05	.55	.05					
		M-F	100.0	1.38	.29	.40	.61	.29	3.10	.28					
449	Clover, Ladino	As Fed	18.0	.23	.08	.02		.09	.34	.02			.006	12.9	
		M-F	100.0	1.27	.42	.12		.48	1.87	.12			.036	71.7	
450	Clover, red, early bloom	As Fed	19.6	.44	.07				.49						
		M-F	100.0	2.26	.38				2.49						
451	Clover, subterranean (sub)	As Fed	17.4	.26	.06										
		M-F	100.0	1.49	.33										
452	Clover, white	As Fed	17.7	.25	.09	.07	.11	.08	.38	.06			.006	54.3	
		M-F	100.0	1.40	.51	.39	.61	.45	2.13	.33			.034	307.2	
453	Comfrey, prickly	As Fed	13.2	.15	.07	.01		.05	.60	.03	.22	1.3	.011	12.1	3.1
		M-F	100.0	1.12	.55	.10		.38	4.53	.24	1.66	9.8	.080	92.0	23.6
454	Corn, sweet, ears & husks	As Fed	37.8												
		M-F	100.0												
455	Corn stover (ears removed), green field corn	As Fed	22.7	.14	.02	.01	.07	.05	.37	.04		1.1	.005	30.9	
		M-F	100.0	.62	.09	.04	.31	.22	1.63	.18		4.9	.022	136.0	
456	Corn stover (ears removed), sweet	As Fed	22.0												
		M-F	100.0												
457	Crested wheatgrass, early bloom	As Fed	42.5												
		M-F	100.0												
458	Crimson clover & ryegrass	As Fed	18.3	.12	.12										
		M-F	100.0	.66	.66										
459	Curly mesquite, browse	As Fed													
		M-F	100.0	.55	.09			.15	.39						
460	Fescue, meadow	As Fed	27.6	.14	.10			.10	.55		.037	1.1			
		M-F	100.0	.51	.38			.37	2.00		.135	4.0			
461	Fescue, tall	As Fed	23.9												
		M-F	100.0												
462	Foxtail, meadow	As Fed	26.1	.15	.12										
		M-F	100.0	.57	.46										
463	Gamagrass	As Fed													
		M-F	100.0	.62	.31										
464	Grama grass, mature	As Fed	63.4	.22	.08			.08	.22		.114	8.1		30.0	
		M-F	100.0	.34	.12			.13	.35		.180	12.8		47.4	
465	Grass, early bloom	As Fed	30.8	.17	.07				.41						
		M-F	100.0	.55	.23				1.33						

ASTURE AND RANGE PLANTS

Feed #	Fat-Soluble Vitamins: A (1 mg car. = 1,667 IU-A)	Carotene (Pro-vitamin A)	E (α-tocopherol)	K	Water-Soluble Vitamins: B_{12}	Biotin	Choline	Folic Acid (Folacin)	Niacin (Nicotinic Acid)	Pantothenic Acid	Pyridoxine (B_6)	Riboflavin (B_2)	Thiamin (B_1)
		(ppm or mg/kg)	*(ppm or mg/kg)*	*(ppm or mg/kg)*	*(ppm or mg/kg)*	*(ppm or mg/kg)*	*(ppm or mg/kg)*	*(ppm or mg/kg)*	*(ppm or mg/kg)*	*(ppm or mg/kg)*	*(ppm or mg/kg)*	*(ppm or mg/kg)*	*(ppm or mg/kg)*
445													
446													
447												4.4	2.0
												19.6	*8.8*
448													
449	95,852	57.5											
	532,440	*319.4*											
450													
451													
452	43,842	26.3	54.6						11.1			15.9	2.5
	248,383	*149.0*	*308.6*						*62.8*			*90.2*	*14.1*
453													
454													
455													
456													
457													
458													
459													
460	155,364	93.2	45.6									2.4	3.3
	562,613	*337.5*	*165.1*									*8.6*	*11.9*
461													
462													
463													
464	32,173	19.3											
	50,677	*30.4*											
465													

TABLE 4-101 COMPOSITION OF FEEDS—

Feed #	Feed Name—Description	Moisture Basis: As Fed or M-F (moisture-free)	Proximate Analysis: Dry Matter	Ash	Crude Fiber	Ether Extract (Fat)	N-Free Extract	Crude Protein	Digestible Protein: Ruminant	Non-Ruminant	Horse Rabbit
PASTURE & RANGE PLANTS			(%)	(%)	(%)	(%)	(%)	(%)	(%)	(%)	(%)
466	Grass, mature	As Fed	53.6	3.1	19.4	1.0	27.2	2.9	1.1		1.3
		M-F	*100.0*	*5.8*	*36.2*	*1.8*	*50.7*	*5.5*	*2.1*		*2.5*
467	Grass & legume (30% legume)	As Fed	23.5	2.5	5.9	.9	10.7	3.5	2.5		2.4
		M-F	*100.0*	*10.7*	*25.0*	*3.7*	*45.5*	*15.1*	*10.8*		*10.3*
468	Johnsongrass, immature *(Sorghum halepense)*	As Fed	19.8	2.1	5.6	.6	8.4	3.1	2.2		2.1
		M-F	*100.0*	*10.6*	*28.5*	*3.2*	*42.2*	*15.5*	*11.2*		*10.6*
469	Junegrass, immature *(Koeleria cristata)*	As Fed									
		M-F	*100.0*	*7.8*	*25.8*	*2.3*	*40.3*	*23.8*	*18.7*		*17.3*
470	Kafir fodder, all analyses *(Sorghum vulgare)*	As Fed	57.1	4.6	16.0	1.7	29.0	5.8	3.6		3.6
		M-F	*100.0*	*8.1*	*28.0*	*3.0*	*50.7*	*10.2*	*6.3*		*6.4*
471	Kentucky bluegrass (see Bluegrass, Kentucky)										
472	Koa haole (lead tree) *(Leucaena glauca)*	As Fed	29.4	1.8	9.7	.6	12.0	5.3	4.0		3.8
		M-F	*100.0*	*6.2*	*33.0*	*2.0*	*40.8*	*18.0*	*13.5*		*12.7*
473	Ladino clover & grass, early bloom	As Fed	20.0	2.2	4.8	.6	9.3	3.1	2.2		2.1
		M-F	*100.0*	*11.2*	*23.9*	*2.8*	*46.8*	*15.3*	*11.0*		*10.5*
474	Lespedeza (annual), immature *(Lespedeza spp)*	As Fed	31.1	3.3	8.5	.8	12.7	5.8	4.4		4.1
		M-F	*100.0*	*10.6*	*27.3*	*2.7*	*40.7*	*18.7*	*14.1*		*13.2*
475	Lespedeza (annual), early bloom	As Fed	25.0		8.0			4.1	3.0		
		M-F	*100.0*		*32.0*			*16.4*	*12.1*		
476	Lespedeza, sericea, all analyses *(Lespedeza cuneata)*	As Fed	32.8	2.0	7.4	1.2	16.3	5.9	4.4		4.2
		M-F	*100.0*	*6.2*	*22.7*	*3.8*	*49.3*	*18.0*	*13.5*		*12.7*
477	Lucerne (see Alfalfa)										
478	Mangels tops w/ crowns *(Beta spp)*	As Fed	12.6	2.4	1.4	.5	6.2	2.1	1.7		1.5
		M-F	*100.0*	*19.2*	*11.4*	*4.2*	*48.2*	*17.0*	*12.9*		*11.9*
479	Meadow fescue *(Festuca elatior)*	As Fed	27.6	2.4	7.5	1.2	12.3	4.2	2.3		2.8
		M-F	*100.0*	*8.6*	*27.1*	*4.2*	*45.0*	*15.1*	*8.4*		*10.3*
480	Meadow foxtail *(Alopecurus pratensis)*	As Fed	26.1	2.8	5.6	1.2	12.0	4.5	3.3		3.1
		M-F	*100.0*	*10.7*	*21.5*	*4.6*	*46.0*	*17.2*	*12.7*		*12.0*
481	Mesquite, common, browse *(Prosopis juliflora)*	As Fed									
		M-F	*100.0*	*5.9*	*27.3*	*3.4*	*42.3*	*21.1*	*16.2*		*15.2*
482	Millet, Japanese *(Echinochloa crusgalli)*	As Fed	21.5	2.0	6.2	.5	10.5	2.3	1.3		1.5
		M-F	*100.0*	*9.3*	*28.8*	*2.6*	*48.8*	*10.5*	*6.1*		*6.6*
483	Millet, Pearl *(Pennisetum glaucum)*	As Fed	20.7	1.9	6.4	.6	9.7	2.1	1.3		1.3
		M-F	*100.0*	*9.2*	*31.1*	*2.9*	*46.8*	*10.1*	*6.2*		*6.3*
484	Milo *(Sorghum vulgare)*	As Fed	67.0	4.1	20.7	1.2	35.7	5.3	2.9		3.0
		M-F	*100.0*	*6.2*	*30.8*	*1.8*	*53.3*	*7.9*	*4.3*		*4.5*
485	Napiergrass, prebloom *(Pennisetum purpureum)*	As Fed	14.9	1.3	4.7	.4	6.8	1.7	1.0		.5
		M-F	*100.0*	*8.6*	*31.5*	*3.0*	*45.9*	*11.0*	*6.7*		*3.3*
486	Oatgrass, tall *(Arrhenatherum elatius)*	As Fed	30.3	2.0	10.5	.9	14.3	2.6	1.5		1.5
		M-F	*100.0*	*6.6*	*34.7*	*3.0*	*47.1*	*8.6*	*4.9*		*5.0*

ASTURE AND RANGE PLANTS

Feed #	TDN Ruminant (%)	TDN Non-Ruminant (%)	TDN Horse Rabbit (%)	DE Ruminant Mcal per lb	DE Ruminant Mcal per kg	DE Non-Ruminant kcal per lb	DE Non-Ruminant kcal per kg	DE Horse Rabbit Mcal per lb	DE Horse Rabbit Mcal per kg	ME Ruminant Mcal per lb	ME Ruminant Mcal per kg	ME Non-Ruminant ME kcal per lb	ME Non-Ruminant ME kcal per kg	ME Non-Ruminant Chicken ME_n kcal per lb	ME Non-Ruminant Chicken ME_n kcal per kg	ME Horse Rabbit Mcal per lb	ME Horse Rabbit Mcal per kg	NE_m Mcal per lb	NE_m Mcal per kg	NE_g Mcal per lb	NE_g Mcal per kg	NE_{lc} Mcal per lb	NE_{lc} Mcal per kg
466	26.1			.5	1.2					.4	.9												
	48.7			1.0	2.1					.8	1.8												
467	15.1			.3	.7					.2	.5												
	64.0			1.3	2.8					1.0	2.3												
468	11.9			.2	.5					.2	.4												
	60.0			1.2	2.6					1.0	2.2												
469																							
	57.7			1.1	2.5					1.0	2.1												
470	35.9			.7	1.6					.6	1.3												
	62.8			1.3	2.8					1.0	2.3												
471																							
472	17.3			.4	.8					.3	.6												
	58.8			1.2	2.6					1.0	2.1												
473	12.9			.3	.6					.2	.5												
	64.5			1.3	2.8					1.0	2.3												
474																							
475	14.6			.3	.6					.2	.5												
	58.3			1.2	2.6					1.0	2.1												
476	20.1			.4	.9					.3	.7												
	61.3			1.2	2.7					1.0	2.2												
477																							
478	7.9			.1	.3					.1	.3												
	62.9			1.3	2.8					1.0	2.3												
479	17.8		13.8	.4	.8			.3	.6	.3	.6												
	64.4		50.0	1.3	2.8			.9	2.0	1.0	2.3												
480	17.4			.4	.8					.3	.6												
	66.7			1.3	2.9					1.1	2.4												
481																							
482	13.3			.3	.6					.2	.5												
	61.8			1.2	2.7					1.0	2.2												
483	13.1			.3	.6					.2	.5												
	63.3			1.3	2.8					1.0	2.3												
484	38.6			.8	1.7					.6	1.4												
	57.7			1.1	2.5					1.0	2.1												
485	8.6			.2	.4					.1	.3							.1	.2	.05	.1	.1	.2
	57.7			1.1	2.5					1.0	2.1							.6	1.3	.3	.7	.6	1.3
486	17.3			.4	.8					.3	.6												
	57.3			1.1	2.5					1.0	2.1												

TABLE 4-101 COMPOSITION OF FEEDS–

Feed #	Feed Name—Description	Moisture Basis: As Fed or M-F (moisture-free)	Dry Matter	Macrominerals: Calcium (Ca)	Phosphorus (P)	Sodium (Na)	Chlorine (Cl)	Magnesium (Mg)	Potassium (K)	Sulfur (S)	Microminerals: Cobalt (Co)	Copper (Cu)	Iron (Fe)	Manganese (Mn)	Zinc (Zn)
			(%)	(%)	(%)	(%)	(%)	(%)	(%)	(%)	(ppm or mg/kg)	(ppm or mg/kg)	(%)	(ppm or mg/kg)	(ppm or mg/kg)
PASTURE AND RANGE PLANTS															
466	Grass, mature	As Fed	53.6	.16	.02										
		M-F	*100.0*	*.31*	*.05*										
467	Grass & legume (30% legume)	As Fed	23.5	.13	.07				.32						
		M-F	*100.0*	*.57*	*.32*				*1.36*						
468	Johnsongrass, immature	As Fed	19.8	.18	.06										
		M-F	*100.0*	*.93*	*.31*										
469	Junegrass, immature	As Fed													
		M-F	*100.0*												
470	Kafir fodder, all analyses	As Fed	57.1	.22	.10			.17	.97						
		M-F	*100.0*	*.38*	*.17*			*.30*	*1.69*						
471	Kentucky bluegrass (see Bluegrass, Kentucky)														
472	Koa haole (lead tree)	As Fed	29.4												
		M-F	*100.0*												
473	Ladino clover & grass, early bloom	As Fed	20.0												
		M-F	*100.0*												
474	Lespedeza (annual), immature	As Fed	31.1												
		M-F	*100.0*												
475	Lespedeza (annual), early bloom	As Fed	25.0	.34	.05			.08	.28				.006	52.2	
		M-F	*100.0*	*1.35*	*.21*			*.27*	*1.12*				*.025*	*208.6*	
476	Lespedeza, sericea, all analyses	As Fed	32.8	.42	.10			.07	.39		.023		.008	34.1	
		M-F	*100.0*	*1.27*	*.29*			*.22*	*1.20*		*.071*		*.024*	*103.8*	
477	Lucerne (see Alfalfa)														
478	Mangels tops w/ crowns	As Fed	12.6												
		M-F	*100.0*												
479	Meadow fescue	As Fed	27.6	.14	.10			.10	.55		.037	1.1			
		M-F	*100.0*	*.51*	*.38*			*.37*	*2.00*		*.135*	*4.0*			
480	Meadow foxtail	As Fed	26.1	.15	.12										
		M-F	*100.0*	*.57*	*.46*										
481	Mesquite, common, browse	As Fed													
		M-F	*100.0*	*1.94*	*.19*			*.23*	*1.41*						
482	Millet, Japanese	As Fed	21.5	.11	.07				.52						
		M-F	*100.0*	*.51*	*.32*				*2.40*						
483	Millet, Pearl	As Fed	20.7												
		M-F	*100.0*												
484	Milo	As Fed	67.0	.27	.14				1.83						
		M-F	*100.0*	*.40*	*.21*				*2.73*						
485	Napiergrass, prebloom	As Fed	14.9	.09	.06										
		M-F	*100.0*	*.60*	*.41*										
486	Oatgrass, tall	As Fed	30.3	.12	.14				.91						
		M-F	*100.0*	*.40*	*.46*				*3.00*						

ASTURE AND RANGE PLANTS

Feed #	Fat-Soluble Vitamins				Water-Soluble Vitamins								
	A (1 mg car. = 1,667 IU-A)	Carotene (Pro-vitamin A)	E (α-tocopherol)	K	B_{12}	Biotin	Choline	Folic Acid (Folacin)	Niacin (Nicotinic Acid)	Pantothenic Acid	Pyridoxine (B_6)	Riboflavin (B_2)	Thiamin (B_1)
		(ppm or mg/kg)	*(ppm or mg/kg)*	*(ppm or mg/kg)*	*(ppm or mg/kg)*	*(ppm or mg/kg)*	*(ppm or mg/kg)*	*(ppm or mg/kg)*	*(ppm or mg/kg)*	*(ppm or mg/kg)*	*(ppm or mg/kg)*	*(ppm or mg/kg)*	*(ppm or mg/kg)*
466													
467													
468													
469													
	16,837	10.1							22.4	8.0	3.40	2.4	
470	*29,339*	*17.6*							*39.2*	*14.1*	*5.95*	*4.2*	
471													
472													
473													
474													
475													
476													
477													
478													
	155,364	93.2	45.6									2.4	3.3
479	*562,612*	*337.5*	*165.1*									*8.6*	*11.9*
480													
481													
482													
483													
	2,167	1.3											
484	*3,334*	*2.0*											
485													
486													

TABLE 4-101 COMPOSITION OF FEEDS-

Feed #	Feed Name—Description	Moisture Basis: As Fed or M-F (moisture-free)	Proximate Analysis: Dry Matter	Ash	Crude Fiber	Ether Extract (Fat)	N-Free Extract	Crude Protein	Digestible Protein: Ruminant	Non-Ruminant	Horse Rabbit
PASTURE & RANGE PLANTS			(%)	(%)	(%)	(%)	(%)	(%)	(%)	(%)	(%)
487	Oats, immature *(Avena sativa)*	As Fed	14.2	1.5	3.5	.4	6.6	2.2			
		M-F	*100.0*	*10.6*	*24.9*	*2.6*	*46.3*	*15.6*			
488	Oats, dough stage	As Fed	25.6	2.1	8.1	.9	11.9	2.6	1.6		1.6
		M-F	*100.0*	*8.0*	*31.7*	*3.6*	*46.6*	*10.1*	*6.4*		*6.3*
489	Oats & peas, early stage *(Avena sativa, Vigna spp)*	As Fed	22.9	2.0	7.0	.9	9.5	3.5	2.5		2.4
		M-F	*100.0*	*8.8*	*30.5*	*3.9*	*41.5*	*15.3*	*11.0*		*10.5*
490	Oats & vetch, milk stage *(Avena sativa, Vicia spp)*	As Fed	32.5	2.5	9.1	1.0	16.4	3.5	2.3		2.2
		M-F	*100.0*	*7.8*	*28.1*	*3.0*	*50.3*	*10.8*	*6.9*		*6.8*
491	Orchardgrass, immature *(Dactylis glomerata)*	As Fed	23.8	2.7	5.6	1.2	9.9	4.4	3.3		3.1
		M-F	*100.0*	*11.3*	*23.6*	*5.0*	*41.7*	*18.4*	*13.8*		*13.1*
492	Pangolagrass *(Digitaria decumbens)*	As Fed	20.0	2.0	6.0	.4	9.4	2.2	1.4		1.3
		M-F	*100.0*	*9.8*	*29.8*	*2.0*	*47.6*	*10.8*	*6.9*		*6.8*
493	Pasture grasses, western plains, all analyses	As Fed	60.2	4.4	21.2	1.2	30.5	2.9	1.3		
		M-F	*100.0*	*7.3*	*35.2*	*2.0*	*50.7*	*4.8*	*2.2*		
494	Pearlmillet *(Pennisetum glaucum)*	As Fed	20.7	1.9	6.4	.6	9.7	2.1	1.3		1.3
		M-F	*100.0*	*9.2*	*31.1*	*2.9*	*46.8*	*10.1*	*6.2*		*6.3*
495	Pea, field *(Pisum sativum arvense)*	As Fed	18.4	1.7	4.7	.6	7.7	3.7	2.8		2.7
		M-F	*100.0*	*9.1*	*25.4*	*3.4*	*42.1*	*20.0*	*15.2*		*14.3*
496	Prickly-pear cactus *(Opuntia spp)*	As Fed	17.1	3.2	2.3	.4	10.3	.9	.5		.3
		M-F	*100.0*	*18.9*	*13.3*	*2.3*	*60.5*	*5.0*	*2.7*		*1.9*
497	Rape *(Brassica spp)*	As Fed	16.9	2.1	2.5	.6	8.6	3.1	2.3		2.1
		M-F	*100.0*	*12.6*	*14.7*	*3.8*	*51.2*	*17.7*	*13.5*		*12.4*
498	Red clover, early bloom *(Trifolium pratense)*	As Fed	19.6	2.0	3.7	1.0	8.8	4.1	2.9		2.7
		M-F	*100.0*	*10.2*	*19.0*	*5.0*	*44.7*	*21.1*	*14.8*		*13.8*
499	Redtop, full bloom *(Agrostis alba)*	As Fed	26.3		6.6			2.1	1.2		
		M-F	*100.0*		*25.1*			*8.1*	*4.6*		
500	Reed canarygrass *(Phalaris arundinacea)*	As Fed	25.8	2.6	6.9	1.1	11.7	3.4	2.4		3.2
		M-F	*100.0*	*10.0*	*26.8*	*4.4*	*45.6*	*13.2*	*9.2*		*12.7*
501	Rescuegrass *(Bromus catharticus)*	As Fed	28.9	4.0	6.7	1.0	12.2	5.0	3.7		3.5
		M-F	*100.0*	*13.8*	*23.2*	*3.5*	*42.2*	*17.3*	*12.8*		*12.1*
502	Russian-thistle (tumbleweed) *(Salsola spp)*	As Fed	80.0					11.8	7.8		
		M-F	*100.0*					*14.7*	*9.7*		
503	Rye pasture *(Secale cereale)*	As Fed	20.6	2.1	5.2	.9	8.2	4.2	3.4		3.0
		M-F	*100.0*	*10.3*	*25.2*	*4.1*	*39.7*	*20.7*	*16.2*		*14.8*
504	Ryegrass, all analvses *(Lolium spp)*	As Fed	21.5	1.6	6.6	.6	10.6	2.1	1.3		1.3
		M-F	*100.0*	*7.6*	*30.6*	*2.8*	*49.5*	*9.5*	*5.8*		*5.8*
505	Sainfoin *(Onobrychis spp)*	As Fed	22.7	1.9	5.8	.7	10.4	3.9	3.0		2.8
		M-F	*100.0*	*8.4*	*25.5*	*3.3*	*45.4*	*17.4*	*12.9*		*12.2*
506	Sorghum, milo *(Sorghum vulgare)*	As Fed	67.0	4.1	20.7	1.2	35.7	5.3	2.9		3.0
		M-F	*100.0*	*6.2*	*30.8*	*2.8*	*53.3*	*7.9*	*4.3*		*4.5*
507	Soybean forage, all analyses *(Glycine max)*	As Fed	22.7	2.4	6.2	.9	9.1	4.1	3.1		2.8
		M-F	*100.0*	*10.5*	*27.3*	*4.0*	*40.3*	*17.9*	*13.5*		*12.6*

STURE AND RANGE PLANTS

Feed #	TDN Rumi-nant	TDN Non-Rumi-nant	TDN Horse Rabbit	DE Rumi-nant		DE Non-Rumi-nant		DE Horse Rabbit		ME Rumi-nant		ME Non-Ruminant ME		ME Non-Ruminant Chicken ME_n		ME Horse Rabbit		Net Energy-Ruminant NE_m		Net Energy-Ruminant NE_g		Net Energy-Ruminant NE_{lc}	
	(%)	(%)	(%)	Mcal per lb	Mcal per kg	kcal per lb	kcal per kg	Mcal per lb	Mcal per kg	Mcal per lb	Mcal per kg	kcal per lb	kcal per kg	kcal per lb	kcal per kg	Mcal per lb	Mcal per kg	Mcal per lb	Mcal per kg	Mcal per lb	Mcal per kg	Mcal per lb	Mcal per kg
87	8.6			.2	.4					.1	.3												
	60.7			1.2	2.7					1.0	2.2												
88	16.5			.3	.7					.3	.6												
	64.7			1.3	2.9					1.0	2.1												
89	14.8			.3	.7					.2	.5												
	64.7			1.3	2.9					1.0	2.3												
90	21.5			.5	1.0					.4	.8												
	66.0			1.3	2.9					1.1	2.4												
91	15.8		14.3	.3	.7			.3	.6	.3	.6							.1	.3	.1	.2	.2	.4
	66.3		60.0	1.3	2.9			1.1	2.4	1.1	2.4							.6	1.4	.4	.9	.7	1.6
92																							
93	29.3			.6	1.3					.5	1.1												
	48.7			1.0	2.1					.8	1.8												
94	13.1			.3	.6					.2	.5												
	63.3			1.3	2.8					1.0	2.3												
95	13.1			.3	.6					.2	.5												
	70.6			1.4	3.1					1.2	2.6												
96	9.4			.2	.4					.1	.3							.1	.2	.05	.1		
	55.5			1.1	2.4					.9	2.0							.6	1.3	.3	.7		
97	13.1			.3	.6					.2	.5												
	77.8			1.5	3.4					1.3	2.8												
98	13.3			.3	.6					.2	.5							.1	.3	.1	.2	.1	.3
	67.7			1.4	3.0					1.1	2.4							.7	1.6	.5	1.0	.8	1.7
99	15.9			.3	.7					.3	.6							.2	.4	.1	.2	.2	.4
	60.3			1.2	2.7					1.0	2.2							.6	1.3	.3	.7	.6	1.4
00	14.8			.3	.7					.2	.5												
	57.5			1.1	2.5					1.0	2.1												
01	18.9			.4	.8					.3	.7												
	65.5			1.3	2.9					1.1	2.4												
02	40.0			.8	1.8					.6	1.4							.4	.8	.1	.2		
	50.0			1.0	2.2					.8	1.8							.5	1.1	.1	.2		
03	14.4			.3	.6					.2	.5												
	70.1			1.4	3.1					1.1	2.5												
04	14.4			.3	.6					.2	.5												
	66.9			1.4	3.0					1.1	2.4												
05	14.5			.3	.6					.2	.5												
	63.8			1.3	2.8					1.0	2.3												
06	38.6			.8	1.7					.6	1.4												
	57.7			1.1	2.5					1.0	2.1												
07	14.9			.3	.7					.2	.5												
	65.5			1.3	2.9					1.1	2.4												

TABLE 4-101 COMPOSITION OF FEEDS

Feed #	Feed Name—Description	Moisture Basis: As Fed or M-F (moisture-free)	Dry Matter	Macrominerals: Calcium (Ca)	Phosphorus (P)	Sodium (Na)	Chlorine (Cl)	Magnesium (Mg)	Potassium (K)	Sulfur (S)	Microminerals: Cobalt (Co)	Copper (Cu)	Iron (Fe)	Manganese (Mn)	Zinc (Zn)
			(%)	(%)	(%)	(%)	(%)	(%)	(%)	(%)	(ppm or mg/kg)	(ppm or mg/kg)	(%)	(ppm or mg/kg)	(ppm or mg/kg)
PASTURE AND RANGE PLANTS															
487	Oats, immature	As Fed	14.2			.02				.01					
		M-F	*100.0*			*.11*				*.08*					
488	Oats, dough stage	As Fed	25.6	.08	.08				.60						
		M-F	*100.0*	*.30*	*.30*				*2.34*						
489	Oats & peas, early stage	As Fed	22.9												
		M-F	*100.0*												
490	Oats & vetch, milk stage	As Fed	32.5												
		M-F	*100.0*												
491	Orchardgrass, immature	As Fed	23.8	.14	.13	.01		.07	.80	.05			.005	32.0	
		M-F	*100.0*	*.58*	*.55*	*.04*		*.31*	*3.38*	*.21*			*.020*	*134.3*	
492	Pangolagrass	As Fed	20.0	.09	.07			.03							
		M-F	*100.0*	*.45*	*.35*			*.14*							
493	Pasture grasses, western plains, all analyses	As Fed	60.2	.19	.04										
		M-F	*100.0*	*.32*	*.07*										
494	Pearlmillet	As Fed	20.7												
		M-F	*100.0*												
495	Pea, field	As Fed	18.4	.22	.04			.04	.28				.007	15.6	
		M-F	*100.0*	*1.21*	*.23*			*.22*	*1.50*				*.040*	*85.1*	
496	Prickly-pear cactus	As Fed	17.1	1.08	.01		.04	.28	.21	.04			.015		
		M-F	*100.0*	*6.29*	*.08*		*.21*	*1.65*	*1.21*	*.23*			*.090*		
497	Rape	As Fed	16.9	.25	.07			.01	.57	.11		1.4	.003	7.7	
		M-F	*100.0*	*1.47*	*.43*			*.06*	*3.37*	*.67*		*8.1*	*.018*	*46.0*	
498	Red clover, early bloom	As Fed	19.6	.44	.07				.49						
		M-F	*100.0*	*2.26*	*.38*				*2.49*						
499	Redtop, full bloom	As Fed	26.3												
		M-F	*100.0*												
500	Reed canarygrass	As Fed	25.8	.10	.08				.94						
		M-F	*100.0*	*.40*	*.30*				*3.64*						
501	Rescuegrass	As Fed	28.9	.15	.08										
		M-F	*100.0*	*.52*	*.28*										
502	Russian-thistle (tumbleweed)	As Fed	80.0	2.64	.13										
		M-F	*100.0*	*3.30*	*.16*										
503	Rye pasture	As Fed	20.6	.11	.09			.07							
		M-F	*100.0*	*.51*	*.41*			*.36*							
504	Ryegrass, all analyses	As Fed	21.5												
		M-F	*100.0*												
505	Sainfoin	As Fed	22.7												
		M-F	*100.0*												
506	Sorghum, milo	As Fed	67.0	.27	.14				1.83						
		M-F	*100.0*	*.40*	*.21*				*2.73*						
507	Soybean forage, all analyses	As Fed	22.7	.25	.07			.12	.21			2.1	.005	27.1	
		M-F	*100.0*	*1.08*	*.29*			*.54*	*.92*			*9.2*	*.021*	*119.4*	

ASTURE AND RANGE PLANTS

Feed #	Fat-Soluble Vitamins				Water-Soluble Vitamins								
	A (1 mg car. = 1,667 IU-A)	Carotene (Pro-vita-min A)	E (α-toco-pherol)	K	B_{12}	Biotin	Choline	Folic Acid (Folacin)	Niacin (Nicotinic Acid)	Panto-thenic Acid	Pyridoxine (B_6)	Riboflavin (B_2)	Thiamin (B_1)
		(ppm or mg/kg)	(ppm or mg/kg)	(ppm or mg/kg)	(ppm or mg/kg)	(ppm or mg/kg)	(ppm or mg/kg)	(ppm or mg/kg)	(ppm or mg/kg)	(ppm or mg/kg)	(ppm or mg/kg)	(ppm or mg/kg)	(ppm or mg/kg)
487	132,693	79.6											
	934,520	*560.6*											
488													
489													
490													
491	133,860	80.3											
	562,446	*337.4*											
492													
493													
494													
495													
496													
497													
498													
499													
500													
501	45,176	27.1											
	156,198	*93.7*											
502	12,002	7.2											
	15,003	*9.0*											
503	117,690	70.6											
	571,114	*342.6*											
504													
505													
506	2,167	1.3											
	3,334	*2.0*											
507	109,688	65.8	63.6										
	483,597	*290.1*	*280.4*										

TABLE 4-101 COMPOSITION OF FEEDS

Feed #	Feed Name—Description	Moisture Basis: As Fed or M-F (moisture-free)	Proximate Analysis: Dry Matter	Ash	Crude Fiber	Ether Extract (Fat)	N-Free Extract	Crude Protein	Digestible Protein: Ruminant	Non-Ruminant	Horse Rabbit
			(%)	(%)	(%)	(%)	(%)	(%)	(%)	(%)	(%)
PASTURE & RANGE PLANTS											
508	Soybean & Sudangrass *(Glycine max, Sorghum vulgare sudanese)*	As Fed	24.2	1.6	8.3	.5	11.1	2.7	1.8		1.8
		M-F	*100.0*	*6.6*	*34.3*	*2.1*	*45.8*	*11.2*	*7.3*		*7.2*
509	Soybean & millet *(Glycine max, Setaria spp)*	As Fed	23.5	1.8	7.5	.5	11.4	2.3	1.4		1.5
		M-F	*100.0*	*7.7*	*31.9*	*2.1*	*48.5*	*9.8*	*6.0*		*6.0*
510	Stargrass *(Cynodon plectostachyum)*	As Fed	28.1	3.0	9.5	.9	12.0	2.8	1.8		1.7
		M-F	*100.0*	*10.6*	*33.6*	*3.1*	*42.6*	*10.1*	*6.5*		*6.3*
511	St. Augustine grass *(Stenotaphrum secundatum)*	As Fed	18.1	1.3	5.4	.5	8.2	2.7	1.9		1.8
		M-F	*100.0*	*7.2*	*29.8*	*2.8*	*45.3*	*14.9*	*10.7*		*10.2*
512	Subterranean (sub) clover *(Trifolium subterraneum)*	As Fed	17.4	1.7	5.0	.8	7.7	2.2			
		M-F	*100.0*	*9.8*	*28.7*	*4.4*	*44.6*	*12.5*			
513	Sudangrass, all analyses *(Sorghum vulgare)*	As Fed	20.8	1.6	5.7	.7	9.9	2.9	2.1		2.0
		M-F	*100.0*	*7.8*	*27.5*	*3.3*	*47.3*	*14.1*	*9.9*		*9.6*
514	Sugar beet, tops w/ crowns *(Beta saccharifera)*	As Fed	15.9	3.3	1.7	.4	7.7	2.8	2.1		1.9
		M-F	*100.0*	*20.6*	*10.7*	*2.5*	*48.7*	*17.5*	*13.2*		*12.2*
515	Sweet clover *(Melilotus spp)*	As Fed	25.0	2.1	7.4	.7	10.3	4.5	3.4		3.1
		M-F	*100.0*	*8.2*	*29.4*	*2.8*	*41.8*	*17.8*	*13.8*		*12.5*
516	Sweet corn, cannery residue *(Zea mays saccharata)*	As Fed	77.0	2.5	17.0	1.8	48.9	6.8	3.9		4.1
		M-F	*100.0*	*3.3*	*22.1*	*2.3*	*63.5*	*8.8*	*5.0*		*5.3*
517	Tall oatgrass (see Oatgrass, tall)										
518	Thistle, Russian *(Salsola spp)*	As Fed	80.0					11.8	7.8		
		M-F	*100.0*					*14.7*	*9.7*		
519	Timothy, all analyses *(Phleum pratense)*	As Fed	27.6	2.2	7.4	1.2	13.3	3.5	2.2		2.4
		M-F	*100.0*	*8.0*	*26.8*	*4.2*	*48.2*	*12.8*	*7.9*		*8.5*
520	Tumbleweed *(Salsola spp)*	As Fed	80.0					11.8	7.8		
		M-F	*100.0*					*14.7*	*9.7*		
521	Velvet bean, dough stage *(Stizolobium spp)*	As Fed	22.5	2.4	5.3	.5	10.8	3.5	2.5		2.4
		M-F	*100.0*	*10.6*	*23.6*	*2.3*	*47.7*	*15.8*	*11.1*		*10.8*
522	Vetch, common *(Vicia sativa)*	As Fed	20.4	2.1	5.5	.5	8.5	3.8	2.8		2.7
		M-F	*100.0*	*10.3*	*27.0*	*2.5*	*41.6*	*18.6*	*13.9*		*13.2*
523	Vetch, hairy *(Vicia villosa)*	As Fed	18.5	2.0	5.2	.6	6.2	4.4	3.5		3.2
		M-F	*100.0*	*10.7*	*28.3*	*3.5*	*33.7*	*23.8*	*18.9*		*17.4*
524	Vetch & oats *(Vicia spp, Avena sativa)*	As Fed	25.9	3.2	6.3	.9	10.8	4.7	3.5		3.4
		M-F	*100.0*	*12.3*	*24.3*	*3.5*	*41.8*	*18.1*	*13.6*		*12.8*
525	Wheat, pasture *(Triticum spp)*	As Fed	22.9	2.2	5.2	.7	11.1	3.7			
		M-F	*100.0*	*9.5*	*22.9*	*3.3*	*48.3*	*16.0*			
526	Wheatgrass, crested (see Crested wheatgrass)										
527	Wheatgrass, Bluebunch, prebloom *(Agropyron spicatum)*	As Fed									
		M-F	*100.0*	*6.3*	*31.5*	*3.1*	*45.6*	*13.5*	*9.4*		*9.1*
528	White clover *(Trifolium repens)*	As Fed	17.7	2.1	2.8	.6	7.2	5.0	3.4		3.7
		M-F	*100.0*	*11.9*	*15.7*	*3.3*	*40.9*	*28.2*	*19.6*		*21.0*
529	Yellow trefoil (black medic) *(Medica lupulins)*	As Fed	22.7	2.3	5.6	.8	9.1	4.9	3.8		3.6
		M-F	*100.0*	*10.2*	*24.7*	*3.4*	*40.1*	*21.6*	*16.7*		*15.6*

ASTURE AND RANGE PLANTS

Feed #	TDN Ruminant (%)	TDN Non-Ruminant (%)	TDN Horse Rabbit (%)	DE Ruminant Mcal per lb	DE Ruminant Mcal per kg	DE Non-Ruminant kcal per lb	DE Non-Ruminant kcal per kg	DE Horse Rabbit Mcal per lb	DE Horse Rabbit Mcal per kg	ME Ruminant Mcal per lb	ME Ruminant Mcal per kg	ME Non-Ruminant ME kcal per lb	ME Non-Ruminant ME kcal per kg	Chicken ME_n kcal per lb	Chicken ME_n kcal per kg	ME Horse Rabbit Mcal per lb	ME Horse Rabbit Mcal per kg	NE_m Mcal per lb	NE_m Mcal per kg	NE_g Mcal per lb	NE_g Mcal per kg	NE_{lc} Mcal per lb	NE_{lc} Mcal per kg
508	15.9			.3	.7					.3	.6												
	65.5			*1.3*	*2.9*					*1.1*	*2.4*												
509	15.6			.3	.7					.3	.6												
	66.2			*1.3*	*2.9*					*1.1*	*2.4*												
510	15.7			.3	.7					.2	.5							.1	.3	.04	.1	.1	.3
	55.4			*1.1*	*2.5*					*.9*	*1.9*							*.5*	*1.1*	*.2*	*.4*	*.5*	*1.1*
511	11.5			.2	.5					.2	.4												
	63.8			*1.3*	*2.8*					*1.0*	*2.3*												
512																							
513	14.1			.3	.6					.2	.5												
	67.9			*1.4*	*3.0*					*1.1*	*2.5*												
514	10.1			.2	.4					.2	.4												
	63.7			*1.3*	*2.8*					*1.0*	*2.3*												
515	16.0			.3	.7					.3	.6												
	64.1			*1.3*	*2.8*					*1.0*	*2.3*												
516	54.4			1.1	2.4					.9	2.0												
	70.7			*1.4*	*3.1*					*1.2*	*2.6*												
517																							
518	40.0			.8	1.8					.6	1.4							.4	.8	.1	.2		
	50.0			*1.0*	*2.2*					*.8*	*1.8*							*.5*	*1.1*	*.1*	*.2*		
519	18.1			.4	.8					.3	.7												
	65.4			*1.3*	*2.9*					*1.1*	*2.4*												
520	40.0			.8	1.8					.6	1.4							.4	.8	.1	.2		
	50.0			*1.0*	*2.2*					*.8*	*1.8*							*.5*	*1.1*	*.1*	*.2*		
521	15.2			.3	.7					.3	.6												
	67.6			*1.4*	*3.0*					*1.1*	*2.5*												
522	12.2			.2	.5					.2	.4												
	59.8			*1.2*	*2.6*					*1.0*	*2.2*												
523	12.6			.3	.6					.2	.5												
	68.3			*1.4*	*3.0*					*1.1*	*2.5*												
524	17.1			1.8	.8					.3	.6												
	66.1			*1.3*	*2.9*					*1.1*	*2.4*												
525	15.6			.3	.7					.3	.6							.2	.4	.1	.3	.2	.4
	68.3			*1.4*	*3.0*					*1.1*	*2.5*							*.7*	*1.6*	*.5*	*1.1*	*.9*	*1.9*
526																							
527																							
528	11.1			.2	.5					.2	.4												
	62.4			*1.3*	*2.8*					*1.0*	*2.3*												
529	13.9			.3	.6					.2	.5												
	61.0			*1.2*	*2.7*					*1.0*	*2.2*												

TABLE 4-101 COMPOSITION OF FEEDS

Feed #	Feed Name—Description	Moisture Basis: As Fed or M-F (moisture-free)	Dry Matter	Macrominerals: Calcium (Ca)	Phosphorus (P)	Sodium (Na)	Chlorine (Cl)	Magnesium (Mg)	Potassium (K)	Sulfur (S)	Microminerals: Cobalt (Co)	Copper (Cu)	Iron (Fe)	Manganese (Mn)	Zinc (Zn)
PASTURE AND RANGE PLANTS			(%)	(%)	(%)	(%)	(%)	(%)	(%)	(%)	(ppm or mg/kg)	(ppm or mg/kg)	(%)	(ppm or mg/kg)	(ppm o mg/kg
508	Soybean & Sudangrass	As Fed	24.2												
		M-F	*100.0*												
509	Soybean & millet	As Fed	23.5												
		M-F	*100.0*												
510	Stargrass	As Fed	28.1	.04	.05										
		M-F	*100.0*	*.15*	*.18*										
511	St. Augustine grass	As Fed	18.1												
		M-F	*100.0*												
512	Subterranean (sub) clover	As Fed	17.4	.26	.06										
		M-F	*100.0*	*1.49*	*.33*										
513	Sudangrass, all analyses	As Fed	20.8	.10	.09			.07	.44	.02	.027	7.5	.004	16.9	
		M-F	*100.0*	*.49*	*.44*			*.35*	*2.14*	*.11*	*.132*	*35.9*	*.021*	*81.3*	
514	Sugar beet, tops w/ crowns	As Fed	15.9	.16	.04	.09		.17	.92	.09		2.2	.003	7.7	
		M-F	*100.0*	*1.01*	*.22*	*.54*		*1.07*	*5.79*	*.57*		*13.6*	*.017*	*48.4*	
515	Sweet clover	As Fed	25.0	.33	.07	.03		.08	.41	.12		2.5	.004	31.4	12.5
		M-F	*100.0*	*1.32*	*.27*	*.10*		*.33*	*1.65*	*.49*		*9.9*	*.014*	*125.4*	*50.0*
516	Sweet corn, cannery residue	As Fed	77.0		.69										
		M-F	*100.0*		*.90*										
517	Tall oatgrass (see Oatgrass, tall)														
518	Thistle, Russian	As Fed	80.0	2.64	.13										
		M-F	*100.0*	*3.30*	*.16*										
519	Timothy all analyses	As Fed	27.6	.16	.10	.03	.14	.07	.58	.04	.011				
		M-F	*100.0*	*.59*	*.38*	*.11*	*.51*	*.25*	*2.09*	*.13*	*.040*				
520	Tumbleweed	As Fed	80.0	2.64	.13										
		M-F	*100.0*	*3.30*	*.16*										
521	Velvet bean, dough stage	As Fed	22.5												
		M-F	*100.0*												
522	Vetch, common	As Fed	20.4	.27	.07			.04	.51	.02		2.0	.008	24.5	
		M-F	*100.0*	*1.32*	*.34*			*.20*	*2.50*	*.10*		*9.7*	*.039*	*120.0*	
523	Vetch, hairy	As Fed	18.5	.20	.06				.42						
		M-F	*100.0*	*1.10*	*.33*				*2.25*						
524	Vetch & oats	As Fed	25.9	.14	.11			.06	.44						
		M-F	*100.0*	*.54*	*.41*			*.24*	*1.70*						
525	Wheat, pasture	As Fed	22.9												
		M-F	*100.0*												
526	Wheatgrass, crested (see Crested wheatgrass)														
527	Wheatgrass, Bluebunch, prebloom	As Fed													
		M-F	*100.0*	*.41*	*.30*										
528	White clover	As Fed	17.7	.25	.09	.07	.11	.08	.38	.06			.006	54.3	
		M-F	*100.0*	*1.40*	*.51*	*.39*	*.61*	*.45*	*2.13*	*.33*			*.034*	*307.2*	
529	Yellow trefoil (black medic)	As Fed	22.7												
		M-F	*100.0*												

ASTURE AND RANGE PLANTS

Feed #	Fat-Soluble Vitamins				Water-Soluble Vitamins								
	A (1 mg car. = 1,667 IU-A)	Carotene (Pro-vitamin A)	E (α-tocopherol)	K	B_{12}	Biotin	Choline	Folic Acid (Folacin)	Niacin (Nicotinic Acid)	Pantothenic Acid	Pyridoxine (B_6)	Riboflavin (B_2)	Thiamin (B_1)
		(ppm or mg/kg)	*(ppm or mg/kg)*	*(ppm or mg/kg)*	*(ppm or mg/kg)*	*(ppm or mg/kg)*	*(ppm or mg/kg)*	*(ppm or mg/kg)*	*(ppm or mg/kg)*	*(ppm or mg/kg)*	*(ppm or mg/kg)*	*(ppm or mg/kg)*	*(ppm or mg/kg)*
08													
09													
10													
11													
12													
13	63,346	38.0											
	304,728	*182.8*											
14	9,169	5.5											
	58,012	*34.8*											
15	110,855	66.5							9.0			21.0	1.3
	443,589	*266.1*							*36.2*			*84.0*	*5.3*
16													
17													
18	12,002	7.2											
	15,003	*9.0*											
19	103,021	61.8	42.4									3.2	.8
	373,408	*224.0*	*153.9*									*11.5*	*2.9*
20	12,002	7.2											
	15,003	*9.0*											
21													
22													
23													
24													
25	198,540	119.1							13.0	4.8		6.3	
	867,007	*520.1*							*56.9*	*21.2*		*27.6*	
26													
27													
	549,110	*329.4*											
28	43,842	26.3							11.1			15.9	2.5
	248,383	*149.0*							*62.8*			*90.2*	*14.1*
29													

TABLE 4-101 COMPOSITION OF FEEDS—

Feed #	Feed Name—Description	Moisture Basis: As Fed or M-F (moisture-free)	Proximate Analysis: Dry Matter	Ash	Crude Fiber	Ether Extract (Fat)	N-Free Extract	Crude Protein	Digestible Protein: Ruminant
			(%)	(%)	(%)	(%)	(%)	(%)	(%)
MINERAL SUPPLEMENTS									
530	Ammonium chloride	As Fed							
		M-F	100.0					160.0	
531	Ammonium phosphate, monobasic	As Fed	97.0	34.5				68.8	
		M-F	100.0	35.6				70.9	
532	Ammonium phosphate, dibasic	As Fed	97.0	34.5				112.4	
		M-F	100.0	35.6				115.9	
533	Bone meal, steamed	As Fed	97.1	77.0	1.4	11.3	0	12.8	8.7
		M-F	100.0	79.3	1.4	11.6	0	13.2	9.0
534	Calcium carbonate	As Fed	99.6						
		M-F	100.0						
535	Calcium phosphate, monobasic, from defluorinated phosphoric acid	As Fed	97.0						
		M-F	100.0						
536	Calcium phosphate, monobasic, from furnace phosphoric acid	As Fed	100.0						
		M-F	100.0						
537	Calcium phosphate, dibasic, from defluorinated phosphoric acid	As Fed	97.0	91.0					
		M-F	100.0	93.8					
538	Calcium phosphate, dibasic, from furnace phosphoric acid	As Fed	97.0	91.0					
		M-F	100.0	93.8					
539	Calcium sulfate, anhydrous (gypsum)	As Fed	84.8						
		M-F	100.0						
540	Colloidal clay (soft rock phosphate)	As Fed							
		M-F	100.0						
541	Curacao phosphate	As Fed	100.0	100.0					
		M-F	100.0	100.0					
542	Defluorinated phosphate from phosphoric acid	As Fed							
		M-F	100.0						
543	Diammonium phosphate	As Fed	97.0	34.5				112.4	
		M-F	100.0	35.6				115.9	
544	Dicalcium phosphate (see Calcium phosphate, dibasic)								
545	Disodium phosphate	As Fed							
		M-F	100.0						
546	Dolomite limestone (see Limestone, magnesium)	As Fed	99.8						
		M-F	100.0						
547	Limestone, grnd	As Fed	99.9	95.9					
		M-F	100.0	96.9					
548	Limestone, magnesium (dolomite)	As Fed	99.8						
		M-F	100.0						
549	Monoammonium phosphate	As Fed	97.0	34.5				68.8	
		M-F	100.0	35.6				70.9	
550	Monocalcium phosphate (see Calcium phosphate, monobasic)								

MINERAL SUPPLEMENTS

Feed #	Macrominerals							Microminerals				
	Calcium (Ca)	Phosphorus (P)	Sodium (Na)	Chlorine (Cl)	Magnesium (Mg)	Potassium (K)	Sulfur (S)	Cobalt (Co)	Copper (Cu)	Iron (Fe)	Manganese (Mn)	Zinc (Zn)
	(%)	*(%)*	*(%)*	*(%)*	*(%)*	*(%)*	*(%)*	*(ppm or mg/kg)*	*(ppm or mg/kg)*	*(%)*	*(ppm or mg/kg)*	*(ppm or mg/kg)*
530												
				66.28								
531	.50	24.00	.06		.45		.7		80.0	1.200	400	300
	.52	*24.74*	*.06*		*.46*		*.7*		*82.0*	*1.237*	*412*	*309*
532	.57	20.00	.04		.45	.01	2.5		91.3	1.200	400	342
	.59	*20.60*	*.04*		*.46*	*.01*	*2.6*		*94.1*	*1.237*	*412*	*353*
533	29.82	12.49	5.53		.32	.18	2.44	.000	11.1	0.085	22	126
	30.71	*12.86*	*5.69*		*.33*	*.19*	*2.51*	*.000*	*11.5*	*0.088*	*23*	*129*
534	37.85	.04	.02	.04	.50	.06	.09		24.0	.034	279	
	38.00	*.04*	*.02*	*.04*	*.50*	*.06*	*.09*		*24.1*	*.034*	*280*	
535	15.91	20.95								.002		
	16.40	*21.60*								*.002*		
536	22.00	23.0		.00					80.0	.002		220.0
	22.00	*23.0*		*.00*					*80.0*	*.002*		*220.0*
537	21.34	18.7			.60	.07					304	
	22.00	*19.3*			*.62*	*.07*					*313*	
538	26.29	18.7			.60	.07					304	
	27.10	*19.3*			*.62*	*.07*					*313*	
539	21.96	.01			2.21		19.96			.170		
	25.90	*.01*			*2.61*		*23.54*			*.201*		
540	17.00	9.00										
541	34.00	15.00										
	34.00	*15.00*										
542	32.00	18.00										
543	.57	20.00	.04	.00	.45	.01	2.5		91.3	1.200	400	342
	.59	*20.60*	*.04*	*.00*	*.46*	*.01*	*2.6*		*94.1*	*1.237*	*412*	*353*
544												
545	21.50	32.0	.00						.001			
546	22.26	.04		.12	9.97	.36				.077		
	22.30	*.04*		*.12*	*9.99*	*.36*				*.077*		
547	33.97	.02	.06	.03	2.06	.11	.04			.349	269.2	
	34.00	*.02*	*.06*	*.03*	*2.06*	*.12*	*.04*			*.350*	*269.6*	
548	22.26	.04		.12	9.97	.36				.077		
	22.30	*.04*		*.12*	*9.99*	*.36*				*.077*		
549	.50	24.00	.06		.45		.7		80.0	1.200	400	300
	.52	*24.74*	*.06*		*.46*		*.7*		*82.0*	*1.237*	*412*	*309*
550												

TABLE 4-101 COMPOSITION OF FEEDS–

Feed #	Feed Name—Description	Moisture Basis: As Fed or M-F (moisture-free)	Proximate Analysis: Dry Matter	Ash	Crude Fiber	Ether Extract (Fat)	N-Free Extract	Crude Protein	Digestible Protein: Ruminant
			(%)	(%)	(%)	(%)	(%)	(%)	(%)
MINERAL SUPPLEMENTS									
551	Monosodium phosphate, anhy	As Fed	87.0						
		M-F	*100.0*						
552	Organic iodide	As Fed							
		M-F	*100.0*						
553	Oyster shells, grnd (flour)	As Fed	99.6	90.2				1.0	
		M-F	*100.0*	*90.6*				*1.0*	
554	Phosphate rock, defluorinated	As Fed	100.0	100.0					
		M-F	*100.0*	*100.0*					
555	Phosphate soft rock (colloidal clay)	As Fed							
		M-F	*100.0*						
556	Phosphoric acid, feed grade (ortho)	As Fed	75.0						
		M-F	*100.0*						
557	Potassium chloride	As Fed	100.0						
		M-F	*100.0*						
558	Potassium iodide	As Fed	92.0						
		M-F	*100.0*						
559	Sodium bicarbonate	As Fed							
		M-F	*100.0*						
560	Sodium chloride	As Fed	100.0						
		M-F	*100.0*						
561	Sodium phosphate, monobasic from furnace phosphoric acid, anhy	As Fed	87.0						
		M-F	*100.0*						
562	Sodium phosphate, dibasic from furnace phosphoric acid	As Fed							
		M-F	*100.0*						
563	Sodium tripolyphosphate	As Fed	96.0						
		M-F	*100.0*						
564	Steamed bone meal	As Fed	97.1	77.0	1.4	11.3	0	12.8	8.7
		M-F	*100.0*	*79.3*	*1.4*	*11.6*	*0*	*13.2*	*9.0*

MINERAL SUPPLEMENTS

Feed #	Macrominerals							Microminerals				
	Calcium (Ca)	Phosphorus (P)	Sodium (Na)	Chlorine (Cl)	Magnesium (Mg)	Potassium (K)	Sulfur (S)	Cobalt (Co)	Copper (Cu)	Iron (Fe)	Manganese (Mn)	Zinc (Zn)
	(%)	(%)	(%)	(%)	(%)	(%)	(%)	(ppm or mg/kg)	(ppm or mg/kg)	(%)	(ppm or mg/kg)	(ppm or mg/kg)
551		22.18	16.53							.001		
		25.50	*19.00*							*.001*		
552										.001		
										.001		
553	37.95	.07	.21	.01	.30	.10				.286	133.6	
	38.00	*.07*	*.21*	*.01*	*.30*	*.10*				*.287*	*134.1*	
554	32.00	16.25	4.00			.09			22	.920	220.0	44.0
	32.00	*16.25*	*4.00*			*.09*			*22*	*.920*	*220.0*	*44.0*
555	17.00	9.00										
556		23.70	.01			.01	.05			.002		
		31.60	*.03*			*.01*	*.07*			*.003*		
557			.01	47.30		50.50				.000		
558			.01	.01		21.67				.000		
			.01	*.01*		*23.56*				*.000*		
559												
			27.36	*.00*		*.01*				*.001*		
560			39.34	60.66								
561		22.18	16.53							.001		
		25.50	*19.00*							*.001*		
562												
		21.50	*32.0*	*.00*						*.001*		
563		24.00	28.80							.004		
		25.00	*30.00*							*.004*		
564	29.82	12.49	5.53		.32	.18	2.44	.000	11.1	0.085	22	126
	30.71	*12.86*	*5.69*		*.33*	*.19*	*2.51*	*.000*	*11.5*	*0.088*	*23*	*129*

TABLE 4-101 COMPOSITION OF FEEDS—

Feed #	Feed Name—Description	Moisture Basis: As Fed or M-F (moisture-free)	Proximate Analysis: Dry Matter	Ash	Crude Fiber	Ether Extract (Fat)	N-Free Extract	Crude Protein	Digestible Protein: Ruminant	Non-Ruminant	Horse Rabbit
VITAMIN SUPPLEMENTS			(%)	(%)	(%)	(%)	(%)	(%)	(%)	(%)	(%)
565	Alfalfa, min 15% protein, dehy *(Medicago sativa)*	As Fed	92.3	9.1	25.5	2.3	39.8	15.6	11.5		10.9
		M-F	*100.0*	*9.8*	*27.7*	*2.5*	*43.1*	*16.9*	*12.4*		*11.8*
566	Alfalfa, min 17% protein, dehy	As Fed	92.7	9.8	24.3	2.5	38.5	17.6	13.1		12.5
		M-F	*100.0*	*10.5*	*26.2*	*2.7*	*41.6*	*19.0*	*14.1*		*13.5*
567	Alfalfa, min 20% protein, dehy	As Fed	92.1	10.6	20.9	2.7	37.9	20.1	15.1		14.5
		M-F	*100.0*	*11.5*	*22.7*	*2.9*	*41.1*	*21.8*	*16.3*		*15.7*
568	Alfalfa, min 22% protein, dehy	As Fed	92.9		18.5			22.5			
		M-F	*100.0*		*19.9*			*24.2*			
569	Alfalfa meal, sun cured, pelleted	As Fed	92.2	10.1	23.6	2.0	39.5	17.0	12.4		12.0
		M-F	*100.0*	*10.9*	*25.6*	*2.2*	*42.9*	*18.4*	*13.4*		*13.0*
570	Alfalfa leaf meal, sun cured	As Fed	91.4	11.2	16.1	1.9	40.8	21.5	16.3		15.7
		M-F	*100.0*	*12.2*	*17.6*	*2.1*	*44.6*	*23.5*	*17.9*		*17.1*
571	Brewers' grains, dried	As Fed	91.0	3.8	14.7	6.6	40.2	25.8	19.1	20.3	19.8
		M-F	*100.0*	*4.2*	*16.1*	*7.2*	*44.2*	*28.3*	*21.0*	*22.4*	*21.8*
572	Buttermilk, condensed	As Fed	29.2	3.6	.1	2.4	12.3	10.8	9.7		
		M-F	*100.0*	*12.5*	*.2*	*8.1*	*42.4*	*36.9*	*33.2*		
573	Cereal grain pasture, immature	As Fed	14.2	1.5	3.5	.4	6.6	2.2	1.6		1.5
		M-F	*100.0*	*10.6*	*24.9*	*2.6*	*46.3*	*15.6*	*11.3*		*10.7*
574	Cod liver oil *(Gadus morrhua)*	As Fed	100.0			63.3					
		M-F	*100.0*			*63.3*					
575	Cod liver oil meal	As Fed	92.5	2.9	.7	28.9	9.6	50.4			
		M-F	*100.0*	*3.1*	*.8*	*31.2*	*10.4*	*54.5*			
576	Crab meal *(Callinectes sapidus)*	As Fed	92.3	40.8	10.5	2.0	7.6	31.4			
		M-F	*100.0*	*44.2*	*11.3*	*2.2*	*8.2*	*34.1*			
577	Distillers' corn grains, solubles, dried *(Zea mays)*	As Fed	92.5	4.6	9.1	10.3	41.4	27.0	13.2		
		M-F	*100.0*	*5.0*	*9.8*	*11.2*	*44.8*	*29.2*	*14.3*		
578	Distillers' grains, solubles, dehy	As Fed	92.1	6.2	3.4	8.9	44.8	28.8			
		M-F	*100.0*	*6.8*	*3.7*	*9.6*	*48.6*	*31.2*			
579	Fish meal, mech-extd	As Fed	88.4	20.2	1.0	5.6	2.2	59.4	52.9	54.7	
		M-F	*100.0*	*22.8*	*1.1*	*6.3*	*2.5*	*67.2*	*59.8*	*61.8*	
580	Fish solubles, condensed	As Fed	54.4	13.1	.6	5.1	0	35.7	31.8	34.3	
		M-F	*100.0*	*24.0*	*1.2*	*9.3*	*0*	*65.7*	*58.5*	*63.1*	
581	Fish solubles, dried	As Fed	93.7	14.8	.5	9.3	3.5	65.5			
		M-F	*100.0*	*15.8*	*.5*	*9.9*	*3.8*	*69.9*			
582	Fish, sardine, solubles, condensed *(Clupea* spp)	As Fed	49.7	10.2	0	9.4	.6	29.5		28.3	
		M-F	*100.0*	*20.5*	*0*	*18.9*	*1.2*	*59.4*		*57.0*	
583	Fish meal, sardine, mech-extd	As Fed	92.9	15.7	.9	5.4	5.8	65.2		63.2	
		M-F	*100.0*	*16.9*	*.9*	*5.8*	*6.3*	*70.1*		*68.0*	
584	Grass pasture, green, immature, high quality	As Fed	30.5	2.9	7.7	1.1	13.5	5.3			
		M-F	*100.0*	*9.4*	*25.4*	*3.5*	*44.2*	*17.5*			
585	Liver meal	As Fed	92.7	6.3	1.4	15.7	2.9	66.5			
		M-F	*100.0*	*6.8*	*1.5*	*17.0*	*3.1*	*71.7*			

TAMIN SUPPLEMENTS

Feed #	TDN Ruminant (%)	TDN Non-Ruminant (%)	TDN Horse Rabbit (%)	DE Ruminant Mcal per lb	DE Ruminant Mcal per kg	DE Non-Ruminant kcal per lb	DE Non-Ruminant kcal per kg	DE Horse Rabbit Mcal per lb	DE Horse Rabbit Mcal per kg	ME Ruminant Mcal per lb	ME Ruminant Mcal per kg	ME Non-Ruminant ME kcal per lb	ME Non-Ruminant ME kcal per kg	ME Non-Ruminant Chicken ME_n kcal per lb	ME Non-Ruminant Chicken ME_n kcal per kg	ME Horse Rabbit Mcal per lb	ME Horse Rabbit Mcal per kg	NE_m Mcal per lb	NE_m Mcal per kg	NE_g Mcal per lb	NE_g Mcal per kg	NE_{lc} Mcal per lb	NE_{lc} Mcal per kg
65	53.2			1.0	2.3					.9	1.9			303	666			.5	1.2	.3	.6	.6	1.4
	57.7			*1.1*	*2.5*					*1.0*	*2.1*			*328*	*722*			*.6*	*1.3*	*.3*	*.7*	*.7*	*1.5*
66	53.4			1.1	2.4					.9	1.9			551	1,213			.5	1.2	.3	.6	.6	1.4
	57.6			*1.1*	*2.5*					*1.0*	*2.1*			*595*	*1,309*			*.6*	*1.3*	*.3*	*.7*	*.7*	*1.5*
67	53.4			1.1	2.4					.9	1.9			720	1,585								
	58.0			*1.2*	*2.6*					*1.0*	*2.1*			*782*	*1,721*								
68														802	1,764								
														863	*1,899*								
69	52.2			1.0	2.3					.9	1.9												
	56.7			*1.1*	*2.5*					*1.0*	*2.1*												
70	57.6			1.1	2.5					.9	2.1												
	62.9			*1.3*	*2.8*					*1.0*	*2.3*												
71	61.7	40.4	47.0	1.2	2.7	809	1,779	1.0	2.1	1.0	2.2	730	1,606	960	2,112	.8	1.7	.6	1.3	.4	.8	.7	1.5
	67.8	*44.4*	*51.6*	*1.4*	*3.0*	*889*	*1,956*	*1.0*	*2.3*	*1.1*	*2.5*	*803*	*1,766*	*1,055*	*2,321*	*.9*	*1.9*	*.6*	*1.4*	*.4*	*.8*	*.8*	*1.7*
72	25.6	27.8		.5	1.1	557	1,225			.4	.9	493	1,085										
	87.8	*95.3*		*1.7*	*3.8*	*1,910*	*4,202*			*1.5*	*3.2*	*1,691*	*3,721*										
73	8.6			.2	.4					.1	.3												
	60.7			*1.2*	*2.7*					*1.0*	*2.2*												
74																							
75	99.4	122.6		2.0	4.4	2,457	5,405			1.6	3.6	2,088	4,594										
	107.4	*132.5*		*2.1*	*4.7*	*2,656*	*5,843*			*1.8*	*3.9*	*2,257*	*4,966*										
76	27.9	31.3		.5	1.2	627	1,380			.5	1.0	559	1,229	827	1,819								
	30.3	*33.9*		*.6*	*1.3*	*680*	*1,495*			*.5*	*1.1*	*605*	*1,332*	*895*	*1,970*								
77	74.8	94.3		1.5	3.3	1,890	4,157			1.2	2.7	1,703	3,746	1,115	2,453			.9	2.0	.6	1.3	1.0	2.3
	80.9	*101.9*		*1.6*	*3.6*	*2,043*	*4,495*			*1.3*	*2.9*	*1,841*	*4,050*	*1,205*	*2,652*			*1.0*	*2.2*	*.6*	*1.4*	*1.1*	*2.5*
78	80.1	83.4		1.5	3.4	1,672	3,679			1.3	2.9	1,500	3,301	1,386	3,049								
	87.0	*90.6*		*1.7*	*3.8*	*1,816*	*3,995*			*1.4*	*3.1*	*1,629*	*3,584*	*1,505*	*3,311*								
79	61.9	63.5		1.2	2.7	1,274	2,802			1.0	2.2	1,050	2,309	1,072	2,359								
	70.1	*71.9*		*1.4*	*3.1*	*1,441*	*3,170*			*1.1*	*2.5*	*1,188*	*2,613*	*1,213*	*2,669*								
80	39.1	43.3		.8	1.7	869	1,911			.6	1.4	718	1,580										
	71.9	*79.7*		*1.5*	*3.2*	*1,597*	*3,514*			*1.2*	*2.6*	*1,321*	*2,907*										
81	71.2	73.3		1.4	3.1	1,469	3,232			1.2	2.6	1,203	2,646	1,295	2,850								
	76.0	*78.2*		*1.5*	*3.4*	*1,568*	*3,449*			*1.3*	*2.8*	*1,284*	*2,824*	*1,382*	*3,041*								
82	39.9	46.5		.8	1.8	932	2,051			.6	1.4	783	1,723										
	80.3	*93.6*		*1.6*	*3.5*	*1,876*	*4,127*			*1.3*	*2.9*	*1,576*	*3,467*										
83	66.4	65.0		1.3	2.9	1,304	2,868			1.1	2.4	1,067	2,347	1,322	2,908								
	71.5	*70.0*		*1.5*	*3.2*	*1,403*	*3,086*			*1.2*	*2.6*	*1,148*	*2,525*	*1,422*	*3,129*								
84	20.3			.4	.9					.3	.7							.2	.5	.1	.3	.3	.6
	66.4			*1.3*	*2.9*					*1.1*	*2.4*							*.7*	*1.6*	*.5*	*1.0*	*.9*	*1.9*
85	82.5	89.7		1.6	3.6	1,798	3,956			1.4	3.0	1,465	3,224										
	89.0	*96.8*		*1.8*	*3.9*	*1,940*	*4,269*			*1.5*	*3.2*	*1,581*	*3,479*										

TABLE 4-101 COMPOSITION OF FEEDS

Feed #	Feed Name—Description	Moisture Basis: As Fed or M-F (moisture-free)	Dry Matter	Macrominerals: Calcium (Ca)	Phosphorus (P)	Sodium (Na)	Chlorine (Cl)	Magnesium (Mg)	Potassium (K)	Sulfur (S)	Microminerals: Cobalt (Co)	Copper (Cu)	Iron (Fe)	Manganese (Mn)	Zinc (Zn)
VITAMIN SUPPLEMENTS			(%)	(%)	(%)	(%)	(%)	(%)	(%)	(%)	(ppm or mg/kg)	(ppm or mg/kg)	(%)	(ppm or mg/kg)	(ppm or mg/kg)
565	Alfalfa, min 15% protein, dehy	As Fed	92.3	1.39	.25		.44		1.17						
		M-F	*100.0*	*1.51*	*.27*		*.48*		*1.26*						
566	Alfalfa, min 17% protein, dehy	As Fed	92.7	1.54	.23		.46				.272	6.8	.033	33.0	
		M-F	*100.0*	*1.66*	*.25*		*.50*				*.293*	*7.4*	*.036*	*35.6*	
567	Alfalfa, min 20% protein, dehy	As Fed	92.1	1.65	.29		.58					15.6	.039	62.8	
		M-F	*100.0*	*1.79*	*.31*		*.63*					*17.0*	*.042*	*68.3*	
568	Alfalfa, min 22% protein, dehy	As Fed	92.9	1.48	.28	.11	.52	.34	2.51	.30		11.1	.045	37.0	20.0
		M-F	*100.0*	*1.59*	*.30*	*.12*	*.56*	*.36*	*2.70*	*.32*		*11.9*	*.048*	*39.8*	*21.5*
569	Alfalfa meal, sun cured, pelleted	As Fed	92.2	1.45	.22	.04	.24	.33	2.15	.25	.230	31.3	.038	48.0	
		M-F	*100.0*	*1.57*	*.24*	*.04*	*.26*	*.36*	*2.33*	*.27*	*.249*	*34.0*	*.041*	*52.0*	
570	Alfalfa leaf meal, sun cured	As Fed	91.4	1.67	.24	.09	.43	.35	2.06		.198	10.5	.036	40.9	
		M-F	*100.0*	*1.83*	*.26*	*.10*	*.47*	*.38*	*2.25*		*.216*	*11.5*	*.039*	*44.8*	
571	Brewers' grains, dried	As Fed	91.0	.27	.48	.26	.18	.14	.09	.30	.061	21.0	.025	37.0	
		M-F	*100.0*	*.30*	*.53*	*.28*	*.19*	*.15*	*.10*	*.33*	*.067*	*23.1*	*.027*	*40.7*	
572	Buttermilk, condensed	As Fed	29.2	.44	.26	.31	.12	.19	.23	.03					
		M-F	*100.0*	*1.51*	*.89*	*1.06*	*.41*	*.65*	*.79*	*.10*					
573	Cereal grain pasture, immature	As Fed	14.2			.02	.01			.01					
		M-F	*100.0*			*.11*	*.10*			*.08*					
574	Cod liver oil	As Fed	100.0												
		M-F													
575	Cod liver oil meal	As Fed	92.5	.16	.69										
		M-F	*100.0*	*.17*	*.75*										
576	Crab meal	As Fed	92.3	14.99	1.57	.94	1.52	.88	.45	.32		32.8	.435	133.9	
		M-F	*100.0*	*16.24*	*1.70*	*1.02*	*1.65*	*.95*	*.49*	*.35*		*35.6*	*.472*	*145.1*	
577	Distillers' corn grains, solubles, dried	As Fed	92.5	.19	.79	.36	.17	.25	.65	.30	.111	61.8	.031	28.5	
		M-F	*100.0*	*.21*	*.85*	*.39*	*.18*	*.27*	*.70*	*.32*	*.120*	*66.8*	*.034*	*30.8*	
578	Distillers' grains, solubles, dehy	As Fed	92.1	.21	1.23	.15		.46	1.84		.196	71.7	.030	64.2	138.1
		M-F	*100.0*	*.22*	*1.33*	*.16*		*.50*	*2.00*		*.213*	*77.9*	*.033*	*69.7*	*150.0*
579	Fish meal, mech-extd	As Fed	88.4	5.48	3.33	1.07	1.21	.21	.39	.24	.106	14.6	.036	22.8	
		M-F	*100.0*	*6.20*	*3.77*	*1.21*	*1.37*	*.24*	*.44*	*.27*	*.119*	*16.5*	*.041*	*25.8*	
580	Fish solubles, condensed	As Fed	54.4	.66	.75	3.29	4.08	.02	1.88	.13	.076	52.0	.037	12.8	41.3
		M-F	*100.0*	*1.21*	*1.39*	*6.06*	*7.51*	*.04*	*3.47*	*.24*	*.140*	*95.6*	*.067*	*23.6*	*76.0*
581	Fish solubles, dried	As Fed	93.7	1.27	1.69										
		M-F	*100.0*	*1.36*	*1.80*										
582	Fish, sardine, solubles, condensed	As Fed	49.7	.14	.83	.18	.28	24.9	.18	.11		25.8	.002	24.9	
		M-F	*100.0*	*.28*	*1.67*	*.36*	*.56*	*50.1*	*.36*	*.22*		*51.9*	*.004*	*50.1*	
583	Fish meal, sardine, mech-extd	As Fed	92.9	4.50	2.61	.18	.41	.10	.33		.183	20.1	.030	22.1	
		M-F	*100.0*	*4.84*	*2.81*	*.19*	*.44*	*.11*	*.35*		*.196*	*21.7*	*.032*	*23.8*	
584	Grass pasture, green, immature, high quality	As Fed	30.5							.20					52.0
		M-F	*100.0*							*.66*					*170.4*
585	Liver meal	As Fed	92.7	.56	1.26						.135	89.4	.063	8.8	
		M-F	*100.0*	*.61*	*1.36*						*.145*	*96.5*	*.068*	*9.5*	

ITAMIN SUPPLEMENTS

Feed #	Fat-Soluble Vitamins				Water-Soluble Vitamins								
	A (1 mg car. = 1,667 IU-A)	Carotene (Pro-vita-min A)	E (α-toco-pherol)	K	B_{12}	Biotin	Choline	Folic Acid (Folacin)	Niacin (Nicotinic Acid)	Panto-thenic Acid	Pyridoxine (B_6)	Riboflavin (B_2)	Thiamin (B_1)
		(ppm or mg/kg)	*(ppm or mg/kg)*	*(ppm or mg/kg)*	*(ppm or mg/kg)*	*(ppm or mg/kg)*	*(ppm or mg/kg)*	*(ppm or mg/kg)*	*(ppm or mg/kg)*	*(ppm or mg/kg)*	*(ppm or mg/kg)*	*(ppm or mg/kg)*	*(ppm or mg/kg)*
565	105,521	63.3		9.90			889		38.7	18.0		12.4	
	114,356	*68.6*		*10.63*			*963*		*41.9*	*19.5*		*13.4*	
566	184,704	110.8	120.0	8.68			1,058	6.31	22.7	28.4		15.7	3.9
	199,206	*119.5*	*129.4*	*9.35*			*1,141*	*6.81*	*24.5*	*30.6*		*17.0*	*4.2*
567	204,374	122.6	140.0	14.70			1,108		37.9	40.6		15.9	6.7
	222,044	*133.2*	*152.0*	*15.78*			*1,203*		*41.2*	*44.1*		*17.2*	*7.3*
568	420,918	252.5	151.0				1,853	3.00	58.8	33.0	7.80	17.4	4.2
	453,091	*271.8*	*162.5*				*1,995*	*3.23*	*63.3*	*35.5*	*8.40*	*18.7*	*4.5*
569	97, 853	58.7											
	106,188	*63.7*											
570	164,700	98.8				.28	895	5.93	53.6	28.0		15.9	4.4
	180,036	*108.0*				*.31*	*979*	*6.48*	*58.6*	*30.6*		*17.4*	*4.9*
571						0	1,559	.22	42.7	8.4	.65	1.5	.7
						0	*1,714*	*.24*	*46.9*	*9.3*	*.72*	*1.7*	*.7*
572												14.5	
												49.8	
573	132,693	79.6											
	934,520	*560.6*											
574													
575					.0009				132.1	46.0	32.8	33.2	18.0
					.0010				*142.8*	*49.8*	*35.5*	*35.9*	*19.5*
576					.4458		2,011		44.2	6.6		5.9	
					.4830		*2,178*		*47.9*	*7.2*		*6.4*	
577	6,335	3.8	40.0			.67	2,814	.82	73.8	12.8	2.22	10.1	2.9
	6,835	*4.1*	*43.2*			*.72*	*3,043*	*.89*	*79.8*	*13.9*	*2.40*	*10.9*	*3.1*
578	1,834	1.1			.0029	2.85	4,254		143.2	25.4	8.67	11.3	6.9
	2,000	*1.2*			*.0031*	*3.09*	*4,619*		*155.5*	*27.5*	*9.42*	*12.3*	*7.5*
579			18.5		.2495		3,510		60.8	8.7	14.14	6.5	1.3
			20.9		*.2823*		*3,972*		*68.8*	*9.8*	*16.00*	*7.4*	*1.4*
580	2,334	1.4			.7111	.17	4,346		182.0	38.2	13.05	15.7	5.9
	4,334	*2.6*			*1.3079*	*.31*	*7,993*		*334.8*	*70.3*	*24.01*	*28.8*	*10.9*
581							5,535		280.1	56.8		17.8	
							5,907		*298.9*	*60.6*		*19.0*	
582					1.0410	.13	3,009		355.8	41.2		16.8	4.0
					2.0946	*.27*	*6,055*		*715.9*	*82.9*		*33.7*	*8.0*
583					.1719		2,983		62.5	9.3		6.0	.4
					.1850		*3,209*		*67.3*	*10.0*		*6.4*	*.5*
584	194,706	116.8											
	638,294	*382.9*											
585					.5019	.02		5.56	205.0	31.0		46.3	.2
					.5416	*.02*		*6.00*	*221.2*	*33.4*		*50.0*	*.2*

TABLE 4-101 COMPOSITION OF FEEDS–

Feed #	Feed Name—Description	Moisture Basis: As Fed or M-F (moisture-free)	Proximate Analysis						Digestible Protein		
			Dry Matter	Ash	Crude Fiber	Ether Extract (Fat)	N-Free Extract	Crude Protein	Ruminant	Non-Ruminant	Horse Rabbit
VITAMIN SUPPLEMENTS			(%)	(%)	(%)	(%)	(%)	(%)	(%)	(%)	(%)
586	Liver & glandular meal	As Fed	93.4	5.7	1.9	15.8	3.9	66.0			
		M-F	*100.0*	*6.2*	*2.0*	*16.9*	*4.2*	*70.7*			
587	Meat solubles, dried	As Fed	90.0	5.7				80.0			
		M-F	*100.0*	*6.3*				*88.9*			
588	Rice bran w/germ, sol-extd *(Oryza sativa)*	As Fed	90.9	13.6	12.0	3.1	47.9	14.3	9.9		10.6
		M-F	*100.0*	*15.0*	*13.2*	*3.4*	*52.7*	*15.7*	*10.9*		*11.7*
589	Rice polishings, dried	As Fed	90.4	6.9	3.2	11.8	56.1	12.5	8.9	10.3	8.9
		M-F	*100.0*	*7.6*	*3.6*	*13.1*	*62.0*	*13.8*	*9.9*	*11.4*	*9.9*
590	Wheat germ meal *(Triticum spp)*	As Fed	90.0	4.3	3.0	10.9	45.6	26.2	24.7		
		M-F	*100.0*	*4.8*	*3.3*	*12.1*	*50.7*	*29.1*	*27.4*		
591	Wheat germ oil	As Fed	100.0								
		M-F	*100.0*								
592	Whey, dried	As Fed	94.0	9.7	0	.8	69.6	13.8	10.6		11.2
		M-F	*100.0*	*10.3*	*0*	*.9*	*74.1*	*14.7*	*11.3*		*11.9*
593	Yeast, brewers', dried *(Saccharomyces cerevisiae)*	As Fed	93.0	6.9	1.5	1.0	39.4	44.3			
		M-F	*100.0*	*7.4*	*1.6*	*1.1*	*42.3*	*47.6*			
594	Yeast, torula, dried *(Torulopsis utilis)*	As Fed	93.0	7.8	2.0	2.5	32.4	48.3	42.7	38.7	
		M-F	*100.0*	*8.4*	*2.2*	*2.7*	*34.8*	*51.9*	*45.9*	*41.6*	

TAMIN SUPPLEMENTS

Feed #	TDN Ruminant	TDN Non-Ruminant	TDN Horse Rabbit	Digestible Energy Ruminant		Digestible Energy Non-Ruminant		Digestible Energy Horse Rabbit		Metabolizable Energy Ruminant		Metabolizable Energy Non-Ruminant ME		Metabolizable Energy Non-Ruminant Chicken ME_n		Metabolizable Energy Horse Rabbit		Net Energy-Ruminant NE_m		Net Energy-Ruminant NE_g		Net Energy-Ruminant NE_{lc}	
	(%)	(%)	(%)	Mcal per lb	Mcal per kg	kcal per lb	kcal per kg	Mcal per lb	Mcal per kg	Mcal per lb	Mcal per kg	kcal per lb	kcal per kg	kcal per lb	kcal per kg	Mcal per lb	Mcal per kg	Mcal per lb	Mcal per kg	Mcal per lb	Mcal per kg	Mcal per lb	Mcal per kg
586	83.4	90.8		1.7	3.7	1,821	4,006			1.4	3.0	1,488	3,273	1,300	2,860								
	89.3	*97.3*		*1.8*	*3.9*	*1,950*	*4,291*			*1.5*	*3.2*	*1,594*	*3,506*	*1,392*	*3,062*								
587																							
588	53.9	54.0		1.1	2.4	1,082	2,381			.9	1.9	1,004	2,209	537	1,181								
	59.3	*59.4*		*1.2*	*2.6*	*1,190*	*2,619*			*1.0*	*2.1*	*1,105*	*2,430*	*590*	*1,299*								
589	81.8	87.0		1.6	3.6	1,744	3,837			1.4	3.0	1,626	3,577	1,300	2,860								
	90.4	*96.2*		*1.8*	*4.0*	*1,928*	*4,242*			*1.5*	*3.3*	*1,798*	*3,955*	*1,438*	*3,164*								
590	86.4	96.2		1.7	3.8	1,928	4,241			1.5	3.2	1,745	3,838	1,334	2,935			1.0	2.2	.6	1.4	1.0	2.3
	96.0	*106.9*		*1.9*	*4.2*	*2,142*	*4,712*			*1.6*	*3.5*	*1,938*	*4,264*	*1,482*	*3,261*			*1.1*	*2.4*	*.7*	*1.6*	*1.2*	*2.6*
591																							
592	82.2	70.5		1.6	3.6	1,414	3,111			1.4	3.0	1,363	2,999	842	1,852								2.2
	87.5	*75.1*		*1.8*	*3.9*	*1,505*	*3,310*			*1.5*	*3.2*	*1,450*	*3,190*	*895*	*1,970*								*2.3*
593	79.7	76.7		1.6	3.5	1,538	3,384			1.3	2.9	1,329	2,923	949	2,087								
	85.7	*82.5*		*1.7*	*3.8*	*1,654*	*3,639*			*1.4*	*3.1*	*1,429*	*3,143*	*1,020*	*2,244*								
594	68.3	65.6		1.4	3.0	1,314	2,890			1.2	2.6	957	2,106	1,102	2,425			.8	1.7	.5	1.2	.9	1.9
	73.5	*70.5*		*1.5*	*3.2*	*1,413*	*3,108*			*1.3*	*2.8*	*1,030*	*2,265*	*1,185*	*2,608*			*.9*	*1.9*	*.5*	*1.2*	*1.0*	*2.1*

TABLE 4-101 COMPOSITION OF FEEDS—

Feed #	Feed Name—Description	Moisture Basis: As Fed or M-F (moisture-free)	Dry Matter	Macrominerals							Microminerals				
				Calcium (Ca)	Phosphorus (P)	Sodium (Na)	Chlorine (Cl)	Magnesium (Mg)	Potassium (K)	Sulfur (S)	Cobalt (Co)	Copper (Cu)	Iron (Fe)	Manganese (Mn)	Zinc (Zn)
VITAMIN SUPPLEMENTS			(%)	(%)	(%)	(%)	(%)	(%)	(%)	(%)	(ppm or mg/kg)	(ppm or mg/kg)	(%)	(ppm or mg/kg)	(ppm or mg/kg)
586	Liver & glandular meal	As Fed	93.4	.66	1.14						.202	97.1	.049	7.3	
		M-F	100.0	.71	1.22						.217	104.0	.052	7.8	
587	Meat solubles, dried	As Fed	90.0	.45	.67										
		M-F	100.0	.50	.74										
588	Rice bran w/germ, sol-extd	As Fed	90.9	.08	1.36										
		M-F	100.0	.09	1.50										
589	Rice polishings, dried	As Fed	90.4	.05	1.24	.11	.13	.65	1.02	.17			.016		
		M-F	100.0	.06	1.37	.12	.14	.72	1.13	.19			.018		
590	Wheat germ meal	As Fed	90.0	.07	1.04	.16		.26	.88	.32	.122	8.8	.011	134.9	123.2
		M-F	100.0	.08	1.16	.18		.29	.98	.35	.135	9.8	.012	149.9	136.9
591	Wheat germ oil	As Fed	100.0												
		M-F	100.0												
592	Whey, dried	As Fed	94.0	.87	.79	.48	.70	.13	1.20	1.05	.109	43.1	.016	4.6	
		M-F	100.0	.93	.84	.51	.74	.14	1.28	1.12	.116	45.9	.017	4.9	
593	Yeast, brewers', dried	As Fed	93.0	.15	1.48	.10		.25	1.48	.49	1.476	24.6	.005	5.9	37.4
		M-F	100.0	.16	1.59	.11		.26	1.59	.53	1.587	26.5	.005	6.3	40.2
594	Yeast, torula, dried	As Fed	93.0	.57	1.68	.01		.13	1.88			13.4	.009	12.7	99.2
		M-F	100.0	.61	1.81	.01		.14	2.02			14.4	.010	13.7	106.7

ITAMIN SUPPLEMENTS

Feed #	Fat-Soluble Vitamins				Water-Soluble Vitamins								
	A (1 mg car. = 1,667 IU-A)	Carotene (Pro-vitamin A)	E (α-tocopherol)	K	B_{12}	Biotin	Choline	Folic Acid (Folacin)	Niacin (Nicotinic Acid)	Pantothenic Acid	Pyridoxine (B_6)	Riboflavin (B_2)	Thiamin (B_1)
		(ppm or mg/kg)	*(ppm or mg/kg)*	*(ppm or mg/kg)*	*(ppm or mg/kg)*	*(ppm or mg/kg)*	*(ppm or mg/kg)*	*(ppm or mg/kg)*	*(ppm or mg/kg)*	*(ppm or mg/kg)*	*(ppm or mg/kg)*	*(ppm or mg/kg)*	*(ppm or mg/kg)*
586									161.6	106.1		40.7	2.6
									173.1	*113.6*		*43.6*	*2.8*
587					.8818								
					.9798								
588							1,080		303.8	23.6		2.6	
							1,188		*334.2*	*26.0*		*2.9*	
589			90.0			.62	1,317		409.4	58.8	27.95	1.8	19.6
			99.6			*.69*	*1,457*		*452.7*	*65.0*	*30.90*	*2.0*	*21.7*
590			133.8				3,010	2.00	47.3	25.8	7.30	5.8	23.5
			148.7				*3,344*	*2.20*	*52.6*	*28.7*	*8.11*	*6.4*	*26.1*
591													
				5,000									
592	500	.3			.0300	.40	1,698	.90	11.2	47.7	2.50	29.9	3.7
	500	*.3*			*.0319*	*.42*	*1,806*	*.96*	*11.9*	*50.7*	*2.66*	*31.8*	*3.9*
593						1.08	4,183	14.76	467.5	110.7	46.75	32.0	110.7
						1.16	*4,497*	*15.87*	*502.6*	*119.0*	*50.26*	*34.4*	*119.0*
594						1.12	2,909	23.25	500.1	82.9	29.48	44.4	6.2
						1.20	*3,129*	*25.00*	*537.8*	*89.1*	*31.70*	*47.7*	*6.7*

TABLE 4-102

AMINO ACID COMPOSITION OF FEEDS
(Feed moisture bases: A-F = As-fed; M-F = Moisture-free)

Feed Name—Description	Moisture Basis: As Fed or M-F (moisture-free)	Dry Matter	Arginine	Cystine	Glycine	Histidine	Isoleucine	Leucine	Lysine	Methionine	Phenylalanine	Serine	Threonine	Tryptophan	Tyrosine	Valine
		(%)	(%)	(%)	(%)	(%)	(%)	(%)	(%)	(%)	(%)	(%)	(%)	(%)	(%)	(%)
ENERGY FEEDS																
Bakery residue, dried	A-F	91.6	.45	.16	.90	.10		.80	.30	.15			.60	.09		
	M-F	*100.0*	*.49*	*.17*	*.98*	*.11*		*.87*	*.33*	*.16*			*.66*	*.10*		
Barley grain	A-F	89.0	.57	.19	.40	.27	.53	.80	.47	.17	.62	.45	.36	.16	.36	.62
	M-F	*100.0*	*.64*	*.21*	*.45*	*.30*	*.60*	*.90*	*.52*	*.20*	*.70*	*.50*	*.40*	*.18*	*.40*	*.70*
Barley grain, Pacific Coast	A-F	89.0	.51	.15	.41				.30	.13				.12		
	M-F	*100.0*	*.57*	*.17*	*.46*				*.34*	*.15*				*.13*		
Beans, navy	A-F	90.0	1.50	.20					1.60	.25				.25		
	M-F	*100.0*	*1.67*	*.22*					*1.78*	*.28*				*.28*		
Beet, sugar, pulp, dried	A-F	91.0	.30	.01		.20	.30	.60	.60	.00	.30		.40	.10	.40	.40
	M-F	*100.0*	*.33*	*.01*		*.22*	*.33*	*.66*	*.66*	*.01*	*.33*		*.44*	*.11*	*.44*	*.44*
Bread, dehy	A-F	86.0	.30	.12					.20	.09				.08		
	M-F	*100.0*	*.35*	*.14*					*.23*	*.10*				*.09*		
Coconut, (Copra) meal, mech-extd	A-F	93.0	2.41	.24	.92	.43	.62	1.26	.71	.33	.84	.99	.68	.20	.47	.92
	M-F	*100.0*	*2.60*	*.26*	*.99*	*.46*	*.67*	*1.35*	*.76*	*.36*	*.90*	*1.07*	*.73*	*.22*	*.51*	*.99*
Coconut (Copra) meal, sol-extd	A-F	92.0	2.44			.35	.82	1.29	.71	.33	.86		.77	.20		1.12
	M-F	*100.0*	*2.65*			*.38*	*.89*	*1.40*	*.77*	*.36*	*.94*		*.84*	*.22*		*1.22*
Corn & cob meal (ear corn)	A-F	87.0	.40	.14	.20				.20	.14				.07		
	M-F	*100.0*	*.46*	*.16*	*.23*				*.23*	*.16*				*.08*		
Corn germ meal	A-F	93.0	1.15	.31				1.64	.86	.33	.77		.86	.29	1.44	1.25
	M-F	*100.0*	*1.24*	*.33*				*1.76*	*.93*	*.36*	*.83*		*.93*	*.31*	*1.55*	*1.35*
Corn gluten feed	A-F	90.4	.80	.20	1.49	.60	1.20	2.60	.80	.30	.90	.80	.80	.20	.90	1.30
	M-F	*100.0*	*.88*	*.22*	*1.64*	*.66*	*1.30*	*2.88*	*.88*	*.33*	*1.00*	*.89*	*.89*	*.22*	*1.00*	*1.44*
Corn gluten meal	A-F	91.0	1.40	.60	1.50	1.00	2.31	7.62	.80	1.00	2.91	1.50	1.40	.20	1.00	2.21
	M-F	*100.0*	*1.54*	*.66*	*1.65*	*1.10*	*2.54*	*8.38*	*.88*	*1.10*	*3.20*	*1.65*	*1.54*	*.23*	*1.10*	*2.43*
Corn grain	A-F	87.0	.35	.09		.17	.43	1.04	.26	.17	.43	.70	.26	.09	.43	.43
	M-F	*100.0*	*.40*	*.10*		*.20*	*.50*	*1.20*	*.30*	*.20*	*.50*	*.80*	*.30*	*.10*	*.50*	*.50*
Corn grain #2, 54 lb/bu	A-F	89.0		.09						.09				.09	.45	.36
	M-F	*100.0*		*.10*						*.10*				*.10*	*.50*	*.40*
Corn grain, high lysine	A-F	92.0	.67	.16					.38	.13				.11		
	M-F	*100.0*	*.73*	*.17*					*.41*	*.14*				*.12*		
Distillers' dried grains, w/o solubles	A-F	92.5	1.01			.51	1.62	2.73	.81	.40	1.11		.81	.20	.91	1.21
	M-F	*100.0*	*1.09*			*.55*	*1.75*	*2.95*	*.87*	*.44*	*1.20*		*.87*	*.22*	*.98*	*1.31*
Distillers' dried corn grain w/solubles	A-F	93.0	1.05	.46	1.10	.70	1.50	2.10	.90	.55	1.50	1.30	1.00	.22	.70	1.50
	M-F	*100.0*	*1.13*	*.49*	*1.18*	*.75*	*1.61*	*2.26*	*.97*	*.59*	*1.61*	*1.40*	*1.07*	*.24*	*.75*	*1.61*
Distillers' rye grain w/solubles, dehy	A-F	90.5	1.00			.70	1.50	2.10	1.00	.40	1.30	1.20	1.10	.30	.50	1.60
	M-F	*100.0*	*1.11*			*.77*	*1.66*	*2.32*	*1.11*	*.44*	*1.44*	*1.33*	*1.22*	*.33*	*.55*	*1.77*
Distillers' dried solubles	A-F	92.1	8.68	.40		.71	1.41	2.82	1.21	.50	1.41	.61	1.11	.20	.91	1.61
	M-F	*100.0*	*9.42*	*.44*		*.77*	*1.53*	*3.07*	*1.31*	*.55*	*1.53*	*.66*	*1.20*	*.22*	*.99*	*1.75*
Grain screenings	A-F	90.3	.68		.59	.30	.52	.98	.48	.14	.64	.57	.46		.32	.63
	M-F	*100.0*	*.75*		*.65*	*.33*	*.58*	*1.08*	*.53*	*.16*	*.71*	*.63*	*.51*		*.35*	*.70*
Hominy feed (corn grits), white	A-F	89.9	.40			.20	.30	.90	.40	.10	.40	.30		.10	.40	.40
	M-F	*100.0*	*.45*			*.22*	*.33*	*1.00*	*.45*	*.11*	*.45*	*.33*		*.11*	*.45*	*.45*
Kafir grains	A-F	90.0	.36			.27	.54	1.62	.27	.18	.63		.45	.18		.63
	M-F	*100.0*	*.40*			*.30*	*.60*	*1.80*	*.30*	*.20*	*.70*		*.50*	*.20*		*.70*
Linseed meal, mech- or sol-extd, 35% protein	A-F	90.6	3.20	.66					1.20	.63				.56		
	M-F	*100.0*	*3.53*	*.73*					*1.33*	*.69*				*.62*		
Manure, poultry, cage, dried	A-F	88.6	.38	.15	1.33	.23	.36	.43	.39	.12	.35	.37	.35	.53	.35	.46
	M-F	*100.0*	*.43*	*.17*	*1.50*	*.26*	*.41*	*.49*	*.44*	*.14*	*.40*	*.42*	*.40*	*.60*	*.40*	*.52*
Manure, poultry, floor, dried	A-F	84.5	.43	.14	2.55	.20	.58	.77	.49	.13	.49	.51	.52		.37	.74
	M-F	*100.0*	*.51*	*.17*	*3.02*	*.24*	*.69*	*.91*	*.58*	*.15*	*.58*	*.60*	*.62*		*.44*	*.88*
Milo grain	A-F	89.0	.36	.18		.27	.53	1.42	.27	.09	.45	.53	.27	.09	.36	.53
	M-F	*100.0*	*.40*	*.20*		*.30*	*.60*	*1.60*	*.30*	*.10*	*.51*	*.60*	*.30*	*.10*	*.40*	*.60*
Oats, grain	A-F	89.0	.57	.15	.49	.09		.34	.36	.18				.12	1.07	
	M-F	*100.0*	*.64*	*.17*	*.55*	*.10*		*.38*	*.40*	*.20*				*.14*	*1.20*	

(Continued)

TABLE 4-102 (Continued)

Feed Name—Description	Moisture Basis: As Fed or M-F (moisture-free)	Dry Matter	Arginine	Cystine	Glycine	Histidine	Isoleucine	Leucine	Lysine	Methionine	Phenylalanine	Serine	Threonine	Tryptophan	Tyrosine	Valine
		(%)	(%)	(%)	(%)	(%)	(%)	(%)	(%)	(%)	(%)	(%)	(%)	(%)	(%)	(%)
)ats, grain, Pacific Coast	A-F	91.2	.60	.17	.40				.40	.13				.12		
	M-F	100.0	.66	.19	.44				.44	.14				.13		
'eanut meal, mech-extd, 45% protein	A-F	92.0							1.26	.58						
	M-F	100.0							1.41	.65						
'eanut meal, sol-extd, 47% protein	A-F	92.0	5.90			1.20	2.00	3.70	2.30	.40	2.70		1.50	.50	1.80	2.80
	M-F	100.0	6.45			1.31	2.19	4.04	2.51	.44	2.95		1.64	.55	1.97	3.06
ice bran	A-F	91.0	.50	.10		.20	.40	.60	.50		.40		.40	.10		.60
	M-F	100.0	.55	.11		.22	.44	.66	.55		.44		.44	.11		.66
ice groats, polished (polished rice)	A-F	89.0	.36	.09	.71	.18	.44	.71	.27	.22	.53		.36	.09	.62	.49
	M-F	100.0	.40	.10	.80	.20	.50	.80	.30	.25	.60		.40	.10	.70	.55
ice polishings, dried	A-F	90.4	.50	.10		.10	.30	.50	.50		.30		.30	.10		
	M-F	100.0	.56	.11		.11	.33	.56	.56		.33		.33	.11		
ye, distillers' grains w/solubles, dehy	A-F	90.5	1.00			.70	1.50	2.10	1.00	.40	1.30	1.20	1.10	.30	.50	1.60
	M-F	100.0	1.11			.77	1.66	2.32	1.11	.44	1.44	1.33	1.22	.33	.55	1.77
ye grain	A-F	89.0	.57	.14		.27	.53	.71	.47	.19	.62	.62	.36	.12	.27	.62
	M-F	100.0	.64	.16		.30	.60	.80	.53	.21	.70	.70	.40	.14	.30	.70
creenings, grain	A-F	90.3	.68		.59	.30	.52	.98	.48	.14	.64	.57	.46		.32	.63
	M-F	100.0	.75		.65	.33	.58	1.08	.53	.16	.71	.63	.51		.35	.70
orghum, grain, all analyses	A-F	89.0	.38	.15	.31	.27	.53	1.42	.29	.12	.44	.53	.27	.10	.35	.53
	M-F	100.0	.43	.17	.35	.30	.60	1.60	.32	.14	.50	.60	.30	.11	.40	.60
orghum, kafir, grain	A-F	90.0	.36			.27	.54	1.62	.27	.18	.63		.45	.18		.63
	M-F	100.0	.40			.30	.60	1.80	.30	.20	.70		.50	.20		.70
orghum, milo, grain	A-F	89.0	.36	.18		.27	.53	1.42	.27	.09	.45	.53	.27	.09	.36	.53
	M-F	100.0	.40	.20		.30	.60	1.60	.30	.10	.51	.60	.30	.10	.40	.60
pelt grain	A-F	89.6	.45			.18	.36	.63	.27	.18	.45		.36	.09		.45
	M-F	100.0	.50			.20	.40	.70	.30	.20	.50		.40	.10		.50
Sugar beet pulp, dried	A-F	91.0	.30	.01		.20	.30	.60	.60	.00	.30		.40	.10	.40	.40
	M-F	100.0	.33	.01		.22	.33	.66	.66	.01	.33		.44	.11	.44	.44
Triticale grain	A-F	90.0	.50	23	.41	.22	.32	.57	.29	.10	.41	.38	.28	.16	.26	.43
	M-F	100.0	.56	.26	.45	.24	.35	.63	.32	.11	.45	.42	.31	.18	.29	.48
Wheat bran	A-F	89.0	1.00	.30	.90	.30	.60	.90	.60	.10	.50		.40	.30	.40	.70
	M-F	100.0	1.12	.34	1.01	.34	.67	1.01	.67	.11	.56		.45	.34	.45	.79
Wheat grain	A-F	89.0	.71	.27	.89	.27	.53	.89	.45	.18	.62		.36		.32	.66
	M-F	100.0	.80	.30	1.00	.30	.60	1.00	.51	.20	.70		.40		.36	.74
Wheat middlings (red dog flour)	A-F	89.0	1.00	.36	.70	.40	.70	1.20	.60	.10	.50	.76	.50	.20	.50	.80
	M-F	100.0	1.12	.41	.79	.45	.79	1.35	.67	.11	.56	.85	.56	.22	.56	.90
Wheat middlings, standard	A-F	90.0	.91	.20	.41	.41	.80	1.21	.71	.18	.70	.80	.60	.20	.41	.80
	M-F	100.0	1.01	.22	.45	.45	.89	1.34	.79	.20	.78	.89	.67	.22	.45	.89
Wheat millrun	A-F	90.0	.89	.20	.59				.59	.28				.20		
	M-F	100.0	.99	.22	.66				.66	.31				.22		
Wheat shorts	A-F	90.0	1.24	.37	.94	.47	.58	1.12	.84	.28	.70	.79	.63	.19	.51	.86
	M-F	100.0	1.38	.41	1.04	.52	.64	1.24	.93	.31	.77	.88	.70	.21	.57	.95
PROTEIN FEEDS																
Barley malt sprouts	A-F	91.9	1.21	.25		.49	1.00	1.51	1.32	.30	.83		.93	.41		1.34
	M-F	100.0	1.32	.28		.53	1.09	1.64	1.43	.33	.91		1.01	.45		1.46
Beans, navy	A-F	90.0	1.50	.20					1.60	.25				.25		
	M-F	100.0	1.67	.22					1.78	.28				.28		
Blood meal	A-F	91.0	3.27	1.43	4.58	3.87	.90	10.10	5.48	1.06	5.28		3.95	1.04		7.06
	M-F	100.0	3.59	1.57	5.03	4.25	.99	11.10	6.02	1.17	5.80		4.34	1.14		7.76
Brewers' grains, dried	A-F	92.0	1.35	.29		.62	1.51	2.58	1.00	.46	1.45		1.04	.35	1.19	1.69
	M-F	100.0	1.47	.32		.68	1.64	2.80	1.09	.50	1.58		1.13	.38	1.30	1.84
Buttermilk, dried	A-F	93.0	1.07	.40	.20	.80	2.16	3.11	2.20	.72	1.42	1.41	1.44	.46	1.01	2.39
	M-F	100.0	1.15	.43	.22	.86	2.32	3.34	2.36	.77	1.53	1.51	1.55	.50	1.08	2.57
Casein, dried	A-F	90.7	3.41	.30	1.50	2.51	5.72	8.62	7.02	2.71	4.61	5.21	3.81	1.00	4.71	6.82
	M-F	100.0	3.76	.33	1.66	2.77	6.31	9.51	7.74	2.99	5.09	5.75	4.20	1.11	5.20	7.52
Coconut (Copra) meal, mech-extd, 20% protein	A-F	92.7	2.41	.24	.89	.41	.60	1.21	.70	.33	.81	.96	.66	.20	.46	.89
	M-F	100.0	2.60	.26	.99	.46	.67	1.35	.76	.36	.90	1.07	.73	.22	.51	.99

(Continued)

TABLE 4-102 (Continued)

Feed Name—Description	Moisture Basis: As Fed or M-F (moisture-free)	Dry Matter	Arginine	Cystine	Glycine	Histidine	Isoleucine	Leucine	Lysine	Methionine	Phenylalanine	Serine	Threonine	Tryptophan	Tyrosine	Valine
		(%)	(%)	(%)	(%)	(%)	(%)	(%)	(%)	(%)	(%)	(%)	(%)	(%)	(%)	(%)
Coconut (Copra) meal, sol-extd, 21% protein	A-F	91.2	2.41			.35	.81	1.28	.70	.33	.86		.77	.20		1.11
	M-F	100.0	2.65			.38	.89	1.40	.77	.36	.94		.84	.22		1.22
Corn distillers' grains, dried	A-F	92.5	1.01			.51	1.62	2.73	.81	.40	1.11		.81	.20	.91	1.21
	M-F	100.0	1.09			.55	1.75	2.95	.87	.44	1.20		.87	.22	.98	1.31
Corn germ meal	A-F	93.0	1.15	.31				1.64	.86	.33	.77		.86	.29	1.44	1.25
	M-F	100.0	1.24	.33				1.76	.93	.36	.83		.93	.31	1.55	1.35
Corn gluten feed	A-F	90.4	.80	.20	1.49	.60	1.20	2.60	.80	.30	.90	.80	.80	.20	.90	1.30
	M-F	100.0	.88	.22	1.64	.66	1.33	2.88	.88	.33	1.00	.89	.89	.22	1.00	1.44
Corn gluten meal	A-F	91.0	1.40	.60	1.50	1.00	2.31	7.62	.80	1.00	2.91	1.50	1.40	.20	1.00	2.21
	M-F	100.0	1.54	.66	1.65	1.10	2.54	8.38	.88	1.10	3.20	1.65	1.54	.23	1.10	2.43
Cottonseed meal, mech-extd, 41% protein	A-F	94.0	4.22	.71	1.67	1.02	1.23	2.15	1.49	.53	2.08	1.59	1.23	.49	1.07	1.79
	M-F	100.0	4.49	.76	1.78	1.09	1.31	2.29	1.59	.56	2.21	1.69	1.31	.52	1.14	1.90
Cottonseed meal, prepress sol-extd, 41% protein	A-F	92.5	4.26	.67	1.59	1.05	1.24	2.25	1.62	.49	2.08	1.62	1.22	.44	1.07	1.75
	M-F	100.0	4.61	.72	1.72	1.13	1.34	2.43	1.75	.53	2.25	1.75	1.32	.48	1.16	1.89
Cottonseed meal, sol-extd, 41% protein	A-F	91.1	4.36	.67	1.61	1.02	1.20	2.17	1.59	.49	2.00	1.57	1.21	.48	1.06	1.64
	M-F	100.0	4.79	.73	1.77	1.12	1.32	2.38	1.74	.54	2.20	1.72	1.33	.53	1.16	1.80
Cottonseed meal, glandless	A-F	92.7							1.76							
	M-F	100.0							1.90							
Distillers' dried grains, w/o solubles	A-F	92.5	1.01			.51	1.62	2.73	.81	.40	1.11		.81	.20	.91	1.21
	M-F	100.0	1.09			.55	1.75	2.95	.87	.44	1.20		.87	.22	.98	1.31
Distillers' dried corn grain w/solubles	A-F	93.0	1.05	.46	1.10	.70	1.50	2.10	.90	.55	1.50	1.30	1.00	.22	.70	1.50
	M-F	100.0	1.13	.49	1.18	.75	1.61	2.26	.97	.59	1.61	1.40	1.07	.24	.75	1.61
Distillers' dried solubles	A-F	92.1	8.68	.40		.71	1.41	2.82	1.21	.50	1.41	.61	1.11	.20	.91	1.61
	M-F	100.0	9.42	.44		.77	1.53	3.07	1.31	.55	1.53	.66	1.20	.22	.99	1.75
Distillers', rye, grains, w/solubles, dehy	A-F	90.5	1.00			.70	1.50	2.10	1.00	.40	1.30	1.20	1.10	.30	.50	1.60
	M-F	100.0	1.11			.77	1.66	2.32	1.11	.44	1.44	1.33	1.22	.33	.55	1.77
Feather meal, hydrolyzed	A-F	93.2	5.82	2.95	6.70	.51	4.25	7.99	1.97	.59	4.76		4.50	.49	2.46	6.54
	M-F	100.0	6.24	3.17	7.19	.55	4.56	8.57	2.11	.63	5.11		4.83	.53	2.64	7.02
Fish, anchovy, meal	A-F	92.0	3.82	.60	3.66	1.60	3.12	5.01	5.09	2.00	2.82	2.41	2.80	.76	2.26	3.51
	M-F	100.0	4.15	.65	3.98	1.74	3.39	5.44	5.53	2.17	3.06	2.62	3.04	.83	2.46	3.81
Fish meal, mech-extd	A-F	88.4	3.73	.57	3.93	1.53	3.64	4.69	5.17	1.72	2.68		2.49	.67	1.91	3.26
	M-F	100.0	4.23	.65	4.44	1.73	4.12	5.31	5.85	1.95	3.03		2.82	.76	2.17	3.68
Fish meal, Menhaden, 60% protein	A-F	91.4	4.09	.63	4.56	1.56	3.01	4.83	5.14	1.89	2.67	2.44	2.69	.72	2.15	3.45
	M-F	100.0	4.48	.68	4.99	1.71	3.29	5.28	5.62	2.07	2.92	2.67	2.94	.78	2.35	3.77
Fish, sardine, meal	A-F	92.9	3.27	1.01	4.58	1.79			5.99	1.99			2.59	.68		4.09
	M-F	100.0	3.52	1.08	4.93	1.93			6.44	2.14			2.79	.73		4.40
Fish, tuna, meal	A-F	93.2	3.74	.45	4.28	1.81	2.36	3.81	4.18	1.45	2.17	2.21	2.30	.59	1.74	2.77
	M-F	100.0	4.01	.48	4.59	1.94	2.53	4.09	4.48	1.55	2.33	2.37	2.47	.63	1.86	2.97
Fish solubles, condensed	A-F	50.1	1.50	.40	3.00				1.60	.80				.30		
	M-F	100.0	2.99	.80	5.99				3.19	1.60				.60		
Fish solubles, dried	A-F	93.7	2.66	.76	5.71	2.66	1.74	2.76	3.06	1.22	1.33	2.05	1.23	.64	.72	1.95
	M-F	100.0	2.84	.81	6.10	2.84	1.86	2.95	3.27	1.30	1.42	2.19	1.31	.69	.77	2.08
Linseed meal, mech- or sol-extd, 35% protein	A-F	90.6	3.19	.66		.59	1.51	1.97	1.20	.62	1.49		1.25	.56		1.84
	M-F	100.0	3.52	.72		.66	1.68	2.19	1.32	.68	1.65		1.39	.61		2.04
Liver meal, animal	A-F	92.7	4.10	.90	5.60	1.50	3.40	5.40	4.80	1.30	2.90	2.50	2.60	.60	1.70	4.20
	M-F	100.0	4.43	.97	6.05	1.62	3.67	5.83	5.18	1.40	3.13	2.70	2.81	.65	1.84	4.54
Meat & bone meal	A-F	93.6	3.56	.60	6.81	.90	1.69	3.09	3.29	.64	1.79		1.79	.29		2.39
	M-F	100.0	3.80	.65	7.28	.96	1.81	3.30	3.52	.69	1.91		1.91	.31		2.55
Meat meal (meat scraps)	A-F	93.5	3.87	.69	8.04	.69	1.39	2.79	3.77	.79	1.57		1.48	.35		2.12
	M-F	100.0	4.14	.74	8.60	.74	1.49	2.99	4.03	.85	1.68		1.58	.38		2.27
Milk, whole, dried	A-F	93.7	.90			.70	1.30	2.50	2.20	.60	1.30		1.00	.40	1.30	1.70
	M-F	100.0	.96			.75	1.39	2.67	2.35	.64	1.39		1.07	.43	1.39	1.81
Milk, skimmed, dried	A-F	94.0	1.20	.50	.20	.90	2.30	3.30	2.80	.80	1.50	1.60	1.40	.40	1.30	2.20
	M-F	100.0	1.28	.53	.21	.96	2.45	3.51	2.98	.85	1.60	1.70	1.49	.42	1.38	2.34
Navy beans	A-F	90.0	1.50	.20					1.60	.25				.25		
	M-F	100.0	1.67	.22					1.78	.28				.28		
Peanut meal, mech-extd, 45% protein	A-F	92.0	3.70			.77	1.55	2.13	1.30	.60	1.78		1.07	.43		1.73
	M-F	100.0	4.02			.84	1.69	2.32	1.41	.65	1.94		1.16	.47		1.88

(Continued)

TABLE 4-102 (Continued)

Feed Name—Description	Moisture Basis: As Fed or M-F (moisture-free)	Dry Matter	Arginine	Cystine	Glycine	Histidine	Isoleucine	Leucine	Lysine	Methionine	Phenylalanine	Serine	Threonine	Tryptophan	Tyrosine	Valine
		(%)	(%)	(%)	(%)	(%)	(%)	(%)	(%)	(%)	(%)	(%)	(%)	(%)	(%)	(%)
Peanut meal, sol-extd, 47% protein	A-F	92.0	5.93	.60		1.21	2.01	3.72	2.30	.40	2.71		1.51	.51	1.81	2.81
	M-F	100.0	6.45	.65		1.31	2.19	4.04	2.50	.43	2.95		1.64	.55	1.97	3.06
Poultry, manure, cage, dried	A-F	88.6	.38	.15	1.33	.23	.36	.43	.39	.12	.35	.37	.35	.53	.35	.46
	M-F	100.0	.43	.17	1.50	.26	.41	.49	.44	.14	.40	.42	.40	.60	.40	.52
Poultry, manure, floor, dried	A-F	84.5	.43	.14	2.55	.20	.58	.77	.49	.13	.49	.51	.52		.37	.74
	M-F	100.0	.51	.17	3.02	.24	.69	.91	.58	.15	.58	.60	.62		.44	.88
Poultry viscera w/feet & heads, rendered	A-F	93.4	3.50	1.00	7.10				3.70	1.00				.45		
	M-F	100.0	3.75	1.07	7.60				3.96	1.07				.48		
Rape seed, Canada, cooked, sol-extd, 40% protein	A-F	92.0	2.42		2.11	1.19	1.58	2.95	2.33	.84	1.67	1.85	1.85	.53	.92	2.11
	M-F	100.0	2.63		2.30	1.29	1.72	3.20	2.54	.91	1.82	2.01	2.01	.57	1.00	2.30
Rape seed meal, sol-extd, 34% protein	A-F	90.3	2.27	.32	1.94	1.13	1.49	2.71	2.18	.77	1.54	1.71	1.71	.56	.86	2.06
	M-F	100.0	2.51	.35	2.15	1.25	1.65	3.00	2.42	.85	1.71	1.89	1.89	.62	.95	2.28
Rye distillers' grains, w/solubles, dehy	A-F	90.5	1.00			.70	1.50	2.10	1.00	.40	1.30	1.20	1.10	.30	.50	1.60
	M-F	100.0	1.11			.77	1.66	2.32	1.11	.44	1.44	1.33	1.22	.33	.55	1.77
Safflower meal, mech-extd, 20% protein	A-F	91.0	1.43	.52	1.46	.40	.77	1.11	.65	.36	1.05		.59	.27		.99
	M-F	100.0	1.57	.57	1.60	.44	.85	1.22	.71	.40	1.16		.65	.30		1.09
Safflower meal w/o hulls, mech-extd, 42% protein	A-F	88.5	3.07	.51	2.56				1.33	.72				.51		
	M-F	100.0	3.47	.58	2.89				1.50	.81				.58		
Sesame seed meal, mech-extd, 44% protein	A-F	93.0	4.76	.58	4.17	.94	1.83	2.96	1.29	1.38	2.03		1.56	.77		2.16
	M-F	100.0	5.12	.63	4.48	1.02	1.97	3.19	1.39	1.49	2.18		1.68	.83		2.33
Shrimp meal	A-F	89.9	2.30	.45		.63	1.42	2.04	2.00	.72	1.39		1.32	.39		1.78
	M-F	100.0	2.56	.50		.70	1.58	2.27	2.23	.80	1.55		1.47	.43		1.98
Soybean flour	A-F	92.3	3.07	.60		.69			4.17	.89	1.79			.99		.30
	M-F	100.0	3.33	.65		.75			4.52	.97	1.94			1.08		.32
Soybean meal, mech-extd 41% protein	A-F	90.0	2.60	.60	2.50	1.10	2.80	3.60	2.70	.80	2.10	2.01	1.70	.60	1.40	2.20
	M-F	100.0	2.89	.67	2.78	1.22	3.11	4.00	3.00	.89	2.33	2.23	1.89	.67	1.56	2.44
Soybean meal, sol-extd, 44% protein	A-F	89.0	3.20	.67	2.10	1.10	2.50	3.40	2.90	.60	2.20	2.59	1.70	.60	1.40	2.40
	M-F	100.0	3.60	.75	2.36	1.23	2.81	3.82	3.26	.67	2.47	2.91	1.91	.67	1.57	2.70
Soybean meal, sol-extd, 49% protein	A-F	89.8	3.80	.80	2.30	1.20	2.60	3.80	3.20	.73	2.70		2.00	.65	2.00	2.70
	M-F	100.0	4.23	.89	2.56	1.34	2.90	4.23	3.56	.81	3.01		2.23	.72	2.23	3.01
Sunflower meal, mech-extd, 41% protein	A-F	92.9	4.20	.80	2.90	1.10	2.40	3.00	2.00	1.60	2.40		1.60	.60		2.40
	M-F	100.0	4.52	.86	3.12	1.18	2.58	3.23	2.15	1.72	2.58		1.72	.64		2.58
Sunflower meal, sol-extd, 44% protein	A-F	93.0	3.50	.70	2.70	1.00	2.10	2.60	1.70	1.50	2.20		1.50	.50		2.30
	M-F	100.0	3.76	.75	2.91	1.07	2.26	2.80	1.83	1.61	2.37		1.61	.54		2.47
Tankage (digester tankage)	A-F	92.0	3.60	.50	6.51	1.90	1.90	5.10	4.00	.80	2.70		2.40	.70		4.20
	M-F	100.0	3.91	.54	7.08	2.07	2.07	5.54	4.34	.87	2.93		2.61	.76		4.57
Whey, dried	A-F	94.0	.34	.30		.20	.90	1.25	1.00	.20	.35	.30	1.29	.17	.20	.69
	M-F	100.0	.36	.32		.21	.96	1.33	1.07	.21	.37	.32	1.38	.18	.21	.74
Yeast, torula, dried	A-F	93.0	2.59	.60	2.56	1.40	2.90	3.50	3.80	.80	3.00		2.60	.50	2.10	2.90
	M-F	100.0	2.79	.65	2.75	1.51	3.12	3.76	4.09	.86	3.23		2.80	.54	2.26	3.12
DRY FORAGES & ROUGHAGES																
Alfalfa hay, all analyses	A-F	91.4	.64			.27	.73	.91	.55	.09	.55		.64	.09		.64
	M-F	100.0	.70			.30	.80	1.00	.60	.10	.60		.70	.10		.70
Alfalfa meal, dehy, min. 15% protein	A-F	92.8	.71	.27	.81	.30	.69	1.10	.59	.19	.81	.71	.60	.19	.40	.71
	M-F	100.0	.76	.29	.87	.32	.74	1.19	.64	.21	.87	.76	.65	.20	.43	.76
Alfalfa meal, dehy, min. 22% protein	A-F	92.9	1.00	.20	1.10	.50	.90	1.70	1.00	.40	1.20	.90	1.00	.60	.80	1.20
	M-F	100.0	1.08	.21	1.18	.54	.97	1.83	1.08	.43	1.29	.97	1.08	.64	.86	1.29
Grass, dehy	A-F	89.0	.99	.19	.72	.46	1.38	1.98	1.06	.31	1.30		.89	.31	.46	1.57
	M-F	100.0	1.11	.21	.81	.52	1.55	2.22	1.19	.35	1.46		1.00	.35	.52	1.76
Manure, poultry, cage, dried	A-F	88.6	.38	.15	1.33	.23	.36		.39	.12	.35		.35			.46
	M-F	100.0	.43	.17	1.50	.26	.41		.44	.14	.40		.40			.52
Manure, poultry, floor, dried	A-F	84.5	.43	.14	2.55	.20	.58	.73	.49	.13	.49	.97	.52	.45	.23	.74
	M-F	100.0	.51	.17	3.02	.24	.69	.86	.58	.15	.58	1.15	.62	.53	.27	.88
SILAGES & HAYLAGES																
Corn stalks (stalklage)	A-F	66.0	.26			.21	.17	.31	.15	.06			.17	.17		.26
	M-F	100.0	.39			.32	.26	.47	.23	.09			.26	.26		.39
PASTURE & RANGE PLANTS																
Rye pasture	A-F	20.6					.35	.54	.29	.06	.23		.47	.04		.29
	M-F	100.0					1.70	2.60	1.40	.30	1.10		2.30	.20		1.40

TABLE 4-103

IODINE CONTENT OF FEEDS

	(moisture basis)	(ppm or mg/kg)
PROTEIN FEEDS		
Crab meal	As Fed *M-F*	.558 .604
Fish, menhaden	As Fed *M-F*	2.004 2.193
DRY FORAGES AND ROUGHAGES		
Alfalfa meal, dehy, min. 15% protein	As Fed *M-F*	.120 .129
Alfalfa meal, dehy, min. 22% protein	As Fed *M-F*	.200 .215
Alsike clover hay	As Fed *M-F*	.160 .183
Bermudagrass hay	As Fed *M-F*	.104 .115
Crimson clover hay	As Fed *M-F*	.059 .060
Lespedeza hay, all analyses	As Fed *M-F*	.145 .159
Redtop hay	As Fed *M-F*	.092 .099
Soybean hay	As Fed *M-F*	.021 .024
MINERAL SUPPLEMENTS		
Bone meal, steamed	As Fed *M-F*	33.199 34.188
Potassium chloride	As Fed *M-F*	 .0
Organic iodide (EDDI)	As Fed *M-F*	800,000
Potassium iodide	As Fed *M-F*	 765,000
Sodium iodide	As Fed *M-F*	 847,000
VITAMIN SUPPLEMENTS		
Alfalfa, min 22% protein, dehy	As Fed *M-F*	.200 *.215*
Crab meal	As Fed *M-F*	.558 .604
Fish solubles, condensed	As Fed *M-F*	1.194 2.196
Fish, sardine, solubles, condensed	As Fed *M-F*	4.934 9.927

TABLE 4-104

SELENIUM CONTENT OF FEEDS

	(moisture basis)	(ppm or mg/kg)
ENERGY FEEDS		
Grain screenings	As Fed *M-F*	.757 .837
Grain screenings, uncleaned	As Fed *M-F*	.805 .874
Oats, white, grain, 34 lb/bu	As Fed *M-F*	.339 .383
Wheat grain, hard red, spring	As Fed *M-F*	.518 .581
Wheat grain, hard red, winter	As Fed *M-F*	.283 .320
Wheat grain, soft red, winter	As Fed *M-F*	.041 .047
Wheat grain, gaines	As Fed *M-F*	.040 .047
Wheat shorts	As Fed *M-F*	.388 .446
Wheat middlings (red dog, flour)	As Fed *M-F*	.299 .342
PROTEIN FEEDS		
Fish, anchovy	As Fed *M-F*	1.383 1.504
Fish, menhaden	As Fed *M-F*	2.224 2.433
Fish, tuna	As Fed *M-F*	4.595 4.928
Poultry, manure, cage, dried	As Fed *M-F*	.466 .526
Wheat germ meal	As Fed *M-F*	.359 .411
DRY FORAGES AND ROUGHAGES		
Alfalfa meal, dehy, min 15% protein	As Fed *M-F*	.501 .540
Alfalfa meal, dehy, min 22% protein	As Fed *M-F*	.540 .581
PASTURE AND RANGE PLANTS		
Barley, grass	As Fed *M-F*	 1.400
MINERAL SUPPLEMENTS		
Calcium carbonate	As Fed *M-F*	.07 .07
Calcium phosphate, monobasic, from furnace phosphoric acid	As Fed *M-F*	.60 .60
Phosphate rock, defluorinated	As Fed *M-F*	.60 .60
VITAMIN SUPPLEMENTS		
Alfalfa, min 22% protein, dehy	As Fed *M-F*	.540 .581
Wheat germ meal	As Fed *M-F*	.359 .411

TABLE 4-105

VITAMIN D CONTENT OF FEEDS

	(moisture basis)	(IU/lb)	(IU/kg)
Alfalfa, hay, early bloom	As Fed M-F	813.5 903.9	1,791.9 1,991.0
Alfalfa, ensiled	As Fed M-F	39.9 131.2	87.8 289.8
Alfalfa, grazed	As Fed M-F	.1 .1	.1 .2
Alfalfa, min 17% protein, dehy	As Fed M-F	199.1 214.7	438.5 473.0
Beet, sugar, pulp dehy	As Fed M-F	250.4 275.2	551.6 606.2
Clover, red, hay	As Fed M-F	761.5 868.3	1,677.3 1,912.5
Cocoa shells	As Fed M-F	1,225.8 1,271.2	2,700.0 2,800.0
Cod-liver oil meal	As Fed M-F	17,978 19,435.7	39,600 42,810
Corn, distillers' grains w/solubles	As Fed M-F	230.2 250.2	507.0 551.1
Corn silage	As Fed M-F	45 182	100 400
FISH OIL Cod liver	As Fed M-F	136,200 136,200	300,000 300,000
Halibut liver	As Fed M-F	4,540,000 4,540,000	10,000,000 10,000,000
Herring	As Fed M-F	40,860 40,860	90,000 90,000
Menhaden	As Fed M-F	34,050 34,050	75,000 75,000
Pilchard	As Fed M-F	27,240 27,240	60,000 60,000
Tuna liver	As Fed M-F	10,442,000 10,442,000	23,000,000 23,000,000
Milk, cattle	As Fed M-F	1.3 11.0	2.9 24.2
Milk, cattle dehy	As Fed M-F	.1 .1	.3 .3
Milk, skimmed, dehy	As Fed M-F	.2 .2	.4 .4
Soybean, hay	As Fed M-F	287.4 322.2	633.0 709.8
Timothy, hay, early bloom	As Fed M-F	629.3 715.6	1,386.2 1,580.6

SECTION 5

PASTURE AND RANGE FORAGES; GREEN CHOP[1]

Contents

	Page
PART I—PASTURE	
Classes of Pasture	588
Pastures for Cattle, Sheep, and Horses	589
Adapted and/or Common Grasses and Legumes of U.S.A.	589
Recommended Cattle, Sheep, and Horse Pastures	591
Area 1, Northern Humid Area, Table 5-1	592
Area 2, Central Humid Area, Table 5-2	596
Area 3, Southern Humid Area, Table 5-3	600
Area 4, Eastern Coastal Area, Table 5-4	606
Area 5, Northern Great Plains Area, Table 5-5	610
Area 6, Southern Great Plains Area, Table 5-6	614
Area 7, Northwest Intermountain Area, Table 5-7	620
Area 8, Southwest Area, Table 5-8	624
Area 9, Northwest Coastal Area, Table 5-9	628
Area 10, California Coastal Area, Table 5-10	634
Pastures for Swine	636
The Corn Belt and North Central States, Table 5-11	638
The South, Table 5-12	640
The Atlantic Coast States, Table 5-13	646
The West, Table 5-14	650
Seeding and Management of Subhumid, Humid, and Irrigated Pastures	652
Establishing a New Pasture	652
Improving or Renovating an Old Pasture	654
Pasture Management	654
Extending the Grazing Season	655
Caring for Pasture and Recreational Areas, and for Grass Along Roads, Lanes, and Bridle Paths	655
PART II—THE WESTERN RANGE	
Range Management Considerations	657
Stocking Rate	658
Season of Use	658
Kind of Livestock	659
Range Grazing Systems	660
Continuous Grazing	660
Rotation Grazing	660
Rotation-Deferred Grazing	660
Rest-Rotation Grazing	661
Range Improvement Methods	661
Conservative Stocking	661
Distribution of Animals on the Range	661
Range Reseeding	662
Grazing Publicly Owned Lands	662
Agencies Administering Public Lands	663
PART III—GREEN CHOP	
Green Chop Considerations	665
Factors Favorable to Green Chop	666
Factors Unfavorable to Green Chop	666

[1]The author gratefully acknowledges the helpful suggestions of the following authorities who reviewed the narrative of this entire section: Mr. Robert E. Williams, Head Range Conservationist, and Mr. B. D. Blakely, Head Agronomist, Soil Conservation Service, USDA; and Prof. Grant Harris, Range Management Specialist, Washington State University, Pullman, Wash.

Also, due acknowledgment is made at the top of each of Tables 5-1 to 5-14 of the authoritative recommendations made therein, especially for this book, by the several competent authorities residing in each of the respective recognized U.S. pasture areas.

Fig. 5-1. All flesh is grass!

Grass is the nation's largest and most remunerative crop, and the cornerstone of successful livestock production.

As the ever-increasing human population of the world consumes a higher proportion of grains and seeds directly, there will be increased reliance on grass for meat, milk, and wool production. In this connection, it is noteworthy that petroleum is not needed to make wool, and that animals do not require fuel to graze the land and recover the energy that is stored in the grass. Noteworthy, too, is the fact that animals are completely recyclable; they produce a new crop each year and perpetuate themselves through their offspring. But it takes thousands of years to create coal, oil, and natural gas; and when they're gone, they're gone forever.

PART I—PASTURE

A pasture is an area of land on which there is a growth of forage that animals may graze. Pasture and grazing land account for 43.9%, or 832 million acres, of continental United States (exclusive of Hawaii and Alaska); and the grass, browse, and other forage produced thereon, provides 44.3% of all U.S. livestock feeds. No method of harvesting has been devised which is as cheap as that which can be accomplished by animals.

CLASSES OF PASTURE

Broadly speaking, all U.S. pastures may be classified as either (1) tame (seeded) pastures, or (2) native pastures. Although no sharp line of demarcation exists between the two groups, tame pastures include those which either receive more than approximately 20 inches of rainfall annually or are irrigated. They are the seeded (cultivated) pastures of the Corn Belt, the South, the East, and the irrigated areas, an smaller and scattered moderate to high rainfall areas throughout the West. The native pastures includ those range pastures which receive less than 2 inches of rainfall annually.

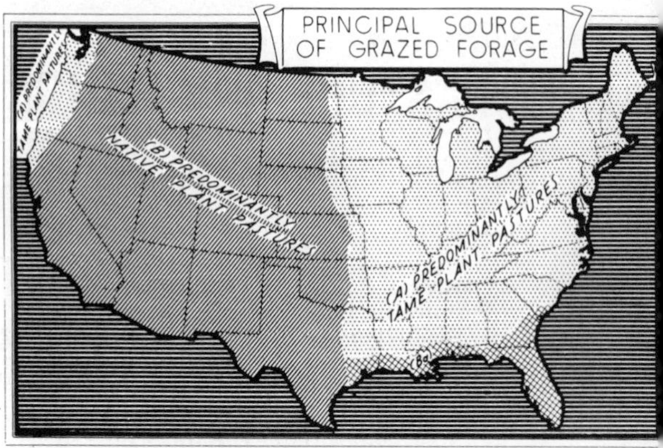

Fig. 5-2. The two major U.S. pasture areas—(A) tame (seeded), ar (B) native (range)—about equally divide the 48 contiguous states in east and west halves. (Courtesy, USDA)

Pasture may be further classified as—

1. **Permanent pastures**—Those which, wit proper care, last for many years. They are most com monly found on land that cannot be used profitabl for cultivated crops, mainly because of topography moisture, or fertility. The vast majority of the farms c the United States have one or more permanent pas tures, and most range areas come under this classifica tion.

2. **Rotation pastures**—Those that are used as part of the established crop rotation. These are seede pastures that are generally used for two to seven year before plowing.

3. **Temporary and supplemental pastures**—Thos that are used for a short period; and they are usuall annuals, such as Sudangrass, sorghum, millet, rye barley, wheat, oats, or rape. They are generall seeded for the purpose of providing supplementa grazing during the season when the permanent or ro tation pastures are relatively unproductive.

Pastures vary greatly in quality, depending o type, soil, growing conditions, and stage of maturity Mature grasses, especially those that are leached an bleached, are low in palatability, digestible energy protein, and carotene, and in some of the minerals Grasses are usually adequate in calcium, magnesium and potassium, but they are apt to be borderline o deficient in phosphorus, and they may be low in som of the trace minerals.

Most grazing areas can be improved by seedin new and better varieties of grasses and legumes, b fertilizing, and by management, including scientifi cally controlled grazing, avoiding overgrazing by bot domestic livestock and wild animals, and supplemental feeding.

PASTURES FOR CATTLE, SHEEP, AND HORSES

Beef cattle, dairy cattle, and sheep compete with each other for many of the grazing areas of the United States. Horses also relish the same kind of pastures, but competition from them is relatively minor. It should be noted, however, that certain plants are poisonous to one or more animal species, but not to others. For example, *Sudan and Sudan hybrids should never be pastured by horses because of the danger of cystitis.*[2]

The economy and importance of pasture for cattle and sheep are attested by the following facts, demonstrated in many experiments and on thousands of successful livestock farms and ranches:

1. **Beef cattle**—It is estimated that 87 percent of the total feed supply of all U.S. beef cattle is derived from forage (see Section 4, Table 4-2); in season, this means pasture. Good pasture alone will produce 200 to 400 pounds of beef per acre annually (in weight of calves weaned, or in added weight of older cattle); superior pastures will do much better.

Generally speaking, cattle can be finished more cheaply on pasture than in the drylot because (a) less labor is required, for the animals do their own harvesting; (b) grass is the cheapest of all roughages; (c) less expensive protein supplement is required; (d) the animals scatter their own manure, thus alleviating hauling it; and (e) fewer buildings and less equipment are necessary.

2. **Dairy cattle**—It is estimated that more than one-half of the nation's milk is produced on pasture. Good pasture alone will provide cows with sufficient nutrients for body maintenance and for the production of about 20 pounds of milk daily; superior pasture will do even better.

Although high-producing dairy cows should be fed grain when on pastures, it is recognized (a) that the better the pasture, the smaller the grain and the protein supplement requirements; and (b) that pastures materially lower the cost of milk production.

3. **Sheep**—It is estimated that 90 percent of the total feed supply of all U.S. sheep is derived from forage (see Table 4-2); for the most part, this means pasture. No other class of farm animals is so well adapted to the utilization of maximum quantities of pasture as sheep. They are unique in that the vast majority of the young are marketed as milk-fed animals directly off grass.

But grass—the nation's largest crop—should not be taken for granted. Again and again, scientists and practical farmers and ranchers have demonstrated that the following desired goals in pasture production are well within the realm of possibility:

- To produce higher yields of palatable and nutritious forage.
- To extend the grazing season from as early in the spring to as late in the fall as possible.
- To provide a fairly uniform supply of feed throughout the entire season.

At the outset, it should be recognized that no one plant embodies all the desirable characteristics necessary to meet the above goals. None of them will grow year-round, or during extremely cold or dry weather. Each of them has a period of peak growth which must be conserved for periods of little growth. Consequently, the progressive stockman will find it desirable (1) to grow more than one species, and (2) to plan pastures for each season of the year. In general, a combination of permanent, rotation, and temporary pastures—accompanied by scientific management—will best achieve these ends.

Adapted and/or Common Grasses and Legumes of U.S.A.

The specific grass or grass-legume mixture will vary from area to area, according to differences in soil, temperature, and rainfall. Fig. 5-3 shows the 10 generally recognized U.S. pasture areas; and Chart 5-1, page 590, shows the best adapted and/or most common grasses and legumes for each of these areas.

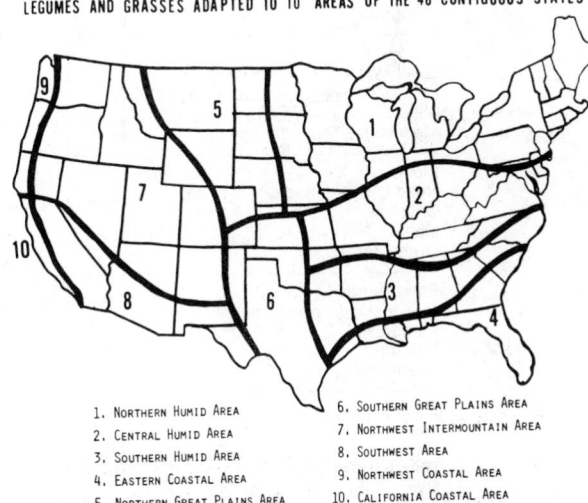

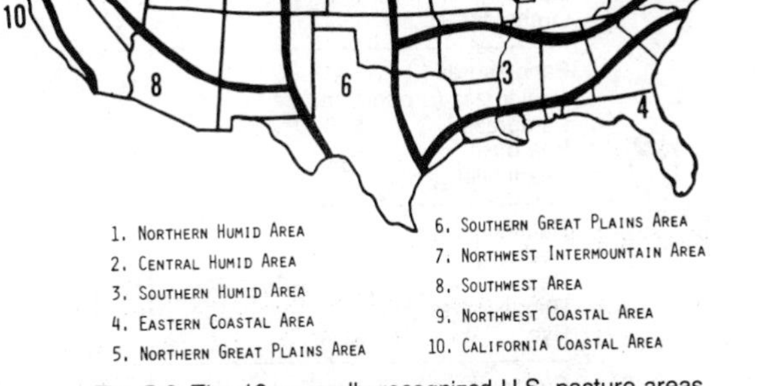

Fig. 5-3. The 10 generally recognized U.S. pasture areas.

In using Fig. 5-3, Chart 5-1, and Tables 5-1 to 5-10, pages 592 to 636, bear in mind that many species of forages have wide geographic adaptation,

[2]Cystitis is a fatal inflammation of the bladder of horses.

but subspecies or varieties often have rather specific adaptation. Thus, alfalfa, for example, is represented by many varieties which give this species adaptation to nearly all states. Variety then, within species, makes many forages adapted to widely varying climate and geographic areas. The county agricultural agent or state agricultural college can furnish recommendations for the area that they serve.

CHART 5-1

ADAPTED GRASSES AND LEGUMES (INCLUDING BROWSE AND FORBS) FOR CATTLE, SHEEP, AND HORSE PASTURES, BY 10 GEOGRAPHICAL AREAS OF THE U.S. (see Fig. 5-3 for geographical areas)

	Areas of the U.S.									
	1	2	3	4	5	6	7	8	9	10
Grasses, shrubs, forbs										
Alfilaria (filaree)								x	x	
Bahiagrass (a paspalum)			x	x		x				
Beardgrass (a bluestem)								x		
Bermudagrass		x	x	x		x		x		x
Bluegrass	x	x			x		x	x		
Bluestem					x	x		x		
Bristlegrass (a millet)								x		
Bromegrass	x				x		x	x		
Buckwheat (wild)								x		
Buffalograss					x	x				
Buffelgrass				x		x				
Canarygrass (including reed)	x	x			x		x		x	
Carib grass				x						
Chamiza (fourwing saltbush)								x		
Cottontop								x		
Curly mesquite (a Hilaria)						x		x		
Dallisgrass (a paspalum)			x	x		x		x		x
Digitgrass				x						
Dropseed						x		x		
Fescue, tall	x	x	x		x	x	x	x	x	x
Foxtail							x		x	
Galleta (a Hilaria)						x		x		
Grama grass					x			x		
Hardinggrass										x
Indiangrass						x				
Indian ricegrass								x		
Indianwheat								x		
Johnsongrass (a sorghum)			x	x		x				
Junegrass					x	x		x		
Kleingrass						x				
Lovegrass						x		x		
Mesquite (vine; a panicum)								x		
Millet	x	x	x	x						
Mormon tea (ephedra, jointfir)								x		
Muhly								x		
Needlegrass (needle-and-thread)					x		x			
Oatgrass									x	
Oats	x	x	x	x		x		x	x	x
Orchardgrass	x	x	x	x	x		x	x	x	x
Pangolagrass			x							
Panicgrass (a panicum)						x		x		
Paragrass				x						
Pea bush								x		
Pearlmillet		x	x	x						
Ratany								x		
Redtop	x						x			
Rescuegrass			x	x						
Rhodesgrass				x						
Rye	x	x	x	x	x	x		x		x
Ryegrass, annual			x	x					x	
Ryegrass, perennial	x								x	
Sacaton								x		
St. Augustine grass			x							
Sorghum-Sudan hybrids		x	x	x		x				

(Continued)

CHART 5-1 (Continued)

	Areas of the U.S.									
	1	2	3	4	5	6	7	8	9	10
Stargrass			x							
Sudangrass	x	x	x	x		x	x	x		x
Switchgrass (a panicum)					x	x				
Three-awn (wiregrass)								x		
Timothy	x	x					x		x	
Tobosa (a Hilaria)						x				
Wheat	x	x	x	x	x	x	x	x	x	x
Wheatgrass					x	x	x	x		
Wild-rye					x	x	x			
Wintergrass, Texas						x				
Winterfat (white sage)								x		
Legumes										
Aeschynomene			x	x						
Alfalfa (lucerne)	x	x	x	x	x	x	x	x	x	x
Alyceclover			x	x						
Austrian winter peas (field pea)			x	x					x	
Black medic (yellow trefoil)			x			x		x		
Bur-clover		x	x							x
Clover, alsike	x	x	x		x		x	x	x	
Clover, arrowleaf			x	x						
Clover, crimson	x	x	x	x						
Clover, Hubam (white sweet clover)								x		
Clover, Ladino	x	x	x	x	x	x	x	x	x	x
Clover, prairie								x		
Clover, red	x	x	x	x	x		x	x	x	
Clover, strawberry					x		x	x		x
Clover, subterranean			x						x	
Clover, white	x	x	x	x	x	x	x	x	x	x
Cowpeas			x	x						
Crown vetch	x									
Hairy indigo				x						
Lespedeza (annual)		x	x	x						
Lespedeza (perennial, sericea)		x	x	x						
Peas (flat)									x	
Soybeans	x	x	x	x						
Sweet clover	x	x		x	x	x	x	x		
Trefoil, birdsfoot	x	x	x	x	x		x	x	x	x
Velvet beans			x	x						
Vetch			x	x	x	x		x		

Recommended Cattle, Sheep, and Horse Pastures

Tables 5-1 to 5-10 contain (1) a list of the most important grasses and legumes, and (2) the authoritative[3] recommendations and facts relative to cattle, sheep, and horse pastures for each of the following 10 generally recognized U.S. grazing areas (see Fig. 5-3):

Table 5-1, Area 1, Northern Humid Area
Table 5-2, Area 2, Central Humid Area
Table 5-3, Area 3, Southern Humid Area
Table 5-4, Area 4, Eastern Coastal Area
Table 5-5, Area 5, Northern Great Plains Area
Table 5-6, Area 6, Southern Great Plains Area
Table 5-7, Area 7, Northwest Intermountain Area
Table 5-8, Area 8, Southwest Area
Table 5-9, Area 9, Northwest Coastal Area
Table 5-10, Area 10, California Coastal Area

The first portion, designated as "Table—A," of each of the respective Tables 5-1 to 5-10, is a summary of the most abundant grasses and legumes comprising the forage in each of the 10 generally recognized U.S. grazing areas shown in Fig. 5-3, page 589.

Where applicable, the pasture recommendations for each of these areas is further broken down with separate tables provided for each into (1) permanent pastures, (2) rotation pastures, and (3) temporary and supplemental pastures in each of the areas. Although these recommendations are listed according to the major soil type and moisture conditions of each area, it is recognized that there are further differences in individual farms and in small areas. For more specific and individual farm recommendations, therefore, the farmer or rancher is urged to seek the advice of local authorities or to write to his state agricultural college.

[3]From competent pasture specialists residing in each of the 10 recognized U.S. pasture areas, with due acknowledgment given.

The following points are pertinent to the recommendations given in Tables 5-1 to 5-10:

1. **Fertilizer rates**—The fertilizer rates given are on a per acre basis and for guide purposes only. Because of the high price of fertilizer, along with concern relative to possible pollution of ground water, fertility rates should be determined by taking the soil test values and adjusting them by a combination o usage and yield desires. Although the practice of soi testing is increasing, from authoritative sources th author has determined that of the nation's currentl chemically fertilized pastures (not including lime) only an estimated 15 to 20 percent are fertilized o the basis of soil tests. *Soil tests are urged.* But fo those who do not soil test, for whatever reason, th

TABLE 5-1A—MOST ABUNDANT GRASSES AND LEGUMES COMPRISING THE FORAGE I

(This table contains the authoritative listing, made especially for this book, c Weeks, Agronomist, University of Massachusetts, Amherst, Mass.; Dr. A Robert R. Seaney, Agronomist, Cornell University, Ithaca, N.Y.; and Dr Park, Penn.)

Grasses—Bluegrass, Kentucky Bromegrass, smooth Canarygrass, reed	Fescue, tall Millets Oats	Orchardgrass Redtop Rye, winter	Sudangrass Timothy Wheat, winter

TABLE 5-1B—GENERAL RECOMMENDATIONS FOR ROTATIONAL PASTURES FOR CATTLE

(This table contains the authoritative recommendations, made especially fo Martin E. Weeks, Agronomist, University of Massachusetts, Amherst, Mass. Robert R. Seaney, Agronomist, Cornell University, Ithaca, N.Y.; and Dr. Joh Penn.)

Soil Type and Moisture Condition	Recommended Seedings (mixtures or pure seedings); Lb of Each and Total Lb/Acre	Seeding Recommendations	
		How to Seed	When to Seed
I. Well-Drained Soils 1. For hay, silage, or green chop	Alfalfa, and .... 10 Smooth bromegrass, or .... 8 Timothy, or .... 4 Late-maturing orchardgrass 4 14-18	Drill or broadcast	Apr. to June or Aug. 1-15
	Alfalfa (alone) 15	Drill or broadcast	Apr. to June or Aug. 1-15
2. All-purpose	Alfalfa .... 6 Red clover .... 3 Ladino clover .... 1 Bromegrass, or .... 8 Timothy, or .... 4 Late-maturing orchardgrass 4 14-18	Drill or broadcast	Apr. to June or Aug. 1-15
3. Pasture	Birdsfoot trefoil, or .... 6 Ladino clover, and .... 2 Bromegrass, or .... 8 Late-maturing orchardgrass .... 4 or Timothy 4 8-14	Drill or broadcast	Apr. to June or Aug. 1-10 (spring seed trefoil, preferably with Timothy)
(For beef cattle)	Crown vetch .... 10 Timothy, or .... 4 Late-maturing orchardgrass .... 4 or Tall fescue 6 14-16	Drill or broadcast	Apr. to June or Aug. 1-10
	Ladino clover .... 1 Tall fescue 10 11	Drill or broadcast	Aug. 15 to Sept. 15 or Jan. 15 to Mar. 15

Footnote on last page of table.

ertilizer guidelines given in Tables 5-1 to 5-10 are etter than no soil test and no guidance at all.

After legumes have been lost from a sward, it is ecommended that they be reestablished in the grass od. If the latter is not feasible, an annual nitrogen pplication at the rate of about 60 pounds of actual itrogen per acre per year should be applied.

Where a center pivot irrigation system is used, nitrogen is usually applied in increments, three to four times during the growing season.

2. **Varieties**—The best guide for the varietal selection of grasses and legumes from the numerous varieties available is the use of certified seed of an adapted variety.

PERMANENT AND/OR ROTATIONAL PASTURES IN AREA 1, NORTHERN HUMID AREA

he following competent specialists residing in the area: Dr. Martin E. R. Schmid, Agronomist, University of Minnesota, St. Paul, Minn.; Dr. John E. Baylor, Agronomist, The Pennsylvania State University, University

egumes—Alfalfa Crown vetch Clover, alsike	Clover, crimson Clover, Ladino Clover, medium red	Clover, white Lespedeza, Korean	Sweet clover Trefoil, birdsfoot

HEEP, AND HORSES IN AREA 1, NORTHERN HUMID AREA

his book, of the following competent specialists residing in the area: Dr. r. A. R. Schmid, Agronomist, University of Minnesota, St. Paul, Minn.; Dr. . Baylor, Agronomist, The Pennsylvania State University, University Park,

Recommended Fertilizer Practices[1] (in lb per acre N, P_2O_5, and K_2O; from left to right, nitrogen, phosphoric acid, and potash, respectively)		Time of Use (or length of grazing season)		Carrying Capacity (animal units/ acre)	Comments
Initial Application at Seeding Time	Annual Maintenance Application After Pasture Is Established	When to Start	When to End		
)-60-60 to 40-120-120 or 0-80-0 + manure (40-80-80 in Northeast)	(No. 1) 0-40-40 to 0-60-180 or (No. 2) 30-60-40 to 80-80-80				Lime to a pH of 6.5 to 7.0. Inoculate alfalfa seed. Alfalfa, brome, and orchardgrass (larger seeds) are best planted on a friable soil surface and firmed by packing. When legume dominates, apply annual fertilizer (No. 1) in part amounts after first and second cuttings. Manure may replace fertilizer as the fall or winter treatment. When grass predominates, apply annual fertilizer (No. 2) in fall or early spring, or part in June after early harvest. If needed, apply borax at 35 lb/acre to establish seeding; then at 10 lb annually for maintenance.
Row 15-45-15	0-70-210				
-60-60 to 40-120-120 or 0-80-0 + manure	(No. 1) 0-40-40 to 0-60-180 (No. 2) 30-60-60 to 80-80-80	May 1-15	Sept. 1-15	1⅓ to 1⅔	(Same as above)
-60-60 to 40-120-120	(No. 1) 0-40-40 to 0-60-180 (No. 2) 30-60-60 to 80-80-80	May 1-15	Sept. 1-15	1⅓ to 1⅔	(Points 1 to 5 same as above) Trefoil should be seeded in the spring, and the seed should be inoculated.
)-60-60 to 40-120-120	(No. 1) 0-40-40 to 0-60-180 30-60-60 to 80-80-80	May 1-15	Sept. 1-15	1⅓ to 1⅔	Do not apply nitrogen (annual maintenance) if there is an adequate stand of legumes.
50-50-50	0-40-120				For winter feed, harvest the first crops as round bales or stacks, left on the ground. Winter graze.

(Continued)

TABLE 5-1

Soil Type and Moisture Condition	Recommended Seedings (mixtures or pure seedings); Lb of Each and Total Lb/Acre	Seeding Recommendations	
		How to Seed	When to Seed
II. Fair to Good, Imperfectly Drained Soils 1. For hay, silage, or green chop	Alfalfa ... 6 Red clover ... 3 Bromegrass, or ... 8 Timothy, or ... 4 Late-maturing orchardgrass ... 4 13-17	Drill or broadcast	Apr. to June or Aug. 1-20
2. All-purpose	Alfalfa ... 6 Red clover ... 3 Ladina clover ... 1 Bromegrass, or ... 8 Timothy, or ... 4 Late-maturing orchardgrass ... 4 14-18	Drill or broadcast	Apr. to June or Aug. 1-15
	Alfalfa ... 6 Birdsfoot trefoil ... 6 Bromegrass, or ... 8 Timothy, or ... 4 Late-maturing orchardgrass ... 4 16-20	Drill or broadcast	Apr. to June or Aug. 1-10
3. Pasture	Red clover ... 3 Ladino ... 2 Bromegrass, or ... 8 Late-maturing orchardgrass or Kentucky bluegrass ... 4 Ryegrass (optional) ... 5 9-18	Drill or broadcast	Feb. to Mar. or Apr. to June or Aug. 1-15
(For beef cattle)	Ladino clover ... 1 Tall fescue ... 10 11	Drill or broadcast	Apr. to June or Aug. 1-20
III. Wet, Poorly Drained Soils 1. For hay, silage, or green chop	Red clover ... 6 Alsike clover ... 3 Timothy, or ... 6 Reed canarygrass ... 6 15	Drill or broadcast	Feb. to Mar. or Apr. to June or Aug. 1-15
2. All-purpose	Red clover ... 4 Alsike clover ... 3 Ladino clover ... 1 Timothy, or ... 6 Reed canarygrass ... 6 14	Drill or broadcast	Feb. to Mar. or Apr. to June or Aug. 1-15
	Birdsfoot trefoil ... 6 Timothy, or ... 4 Reed canarygrass ... 4 10	Drill or broadcast	Feb. to Mar. or Apr. to June or Aug. 1-10

Footnote on last page of table.

—(Continued)

Recommended Fertilizer Practices[1] (in lb per acre N, P_2O_5, and K_2O; from left to right, nitrogen, phosphoric acid, and potash, respectively)		Time of Use (or length of grazing season)		Carrying Capacity (animal units/ acre)	Comments
Initial Application at Seeding Time	Annual Maintenance Application After Pasture Is Established	When to Start	When to End		
0-60-60 to 40-120-120 or 0-80-0 + manure	(No. 1) 0-40-40 to 0-60-180 (No. 2) 30-60-60 to 80-80-80				Lime to a pH of 6.5 to 7.0. Grasses may be fall seeded with small grains, and legumes may be overseeded in Feb.-Mar. period. When legume dominates, apply annual fertilizer (No. 1) in part amounts after first and second cuttings. Manure may replace fertilizer as the fall or winter treatment. When grass predominates, apply annual fertilizer (No. 2) in fall or early spring, or part in June after early harvest.
0-60-60 to 40-120-120 or 0-80-0 + manure	(No. 1) 0-40-40 to 0-60-180 (No. 2) 30-60-60 to 80-80-80	May 1-15	Sept. 1-15	1 to 1½	(Same as preceding)
0-60-60 to 40-120-120 or 0-80-0 + manure	(No. 1) 0-40-40 to 0-60-180 (No. 2) 30-60-60 to 80-80-80	May 1-15	Sept. 1-15	1 to 1½	(Same as preceding plus the following) Trefoil should be seeded in the spring, and the seed should be inoculated.
0-60-60 to 40-120-120 or 0-80-0 + manure	(No. 1) 0-40-40 to 0-60-180 (No. 2) 30-60-60 to 80-80-80	May 1-15	Sept. 1-15	1 to 1½	(Same as preceding plus the following) For pasture use, domestic ryegrass seeded in the spring hastens feed growth that season.
0-60-60 to 40-120-120	(No. 1) 0-40-40 to 0-60-180 (No. 2) 30-60-60 to 80-80-80	Nov.	Apr.	1	Harvest first crop for hay as round bales left in the field. Winter graze new growth and round bales.
0-60-60 to 40-120-120	(No. 1) 0-40-40 to 0-60-180 (No. 2) 30-60-60 to 80-80-80				Lime to a pH of 6.0. When legume predominates, apply annual fertilizer (No. 1), preferably in part amounts after spring and fall harvests. When grass predominates, apply annual fertilizer (No. 2) in fall or early spring, or part in June after harvest.
0-60-60 to 40-120-120	(No. 1) 0-40-40 to 0-60-180 (No. 2) 30-60-60 to 80-80-80	May 1-15	Sept. 1-15	¾ to 1¼	(Same as above)
0-60-60 to 40-120-120	(No. 1) 0-40-40 to 0-60-180 (No. 2) 30-60-60 to 80-80-80	May 1-15	Sept. 1-15	¾ to 1¼	(Same as above plus the following) Trefoil should be seeded in the spring or early summer and the seed should be inoculated.

(Continued)

TABLE 5-1

Soil Type and Moisture Condition	Recommended Seedings (mixtures or pure seedings); Lb of Each and Total Lb/Acre	Seeding Recommendations	
		How to Seed	When to Seed
3. Pasture	Alsike clover ..... 4 Ladino clover ..... 2 Timothy, or ..... 6 Reed canarygrass ..... 6 Ryegrass (optional) 5 12-17	Drill or broadcast	Feb. to Mar. or Apr. to June or Aug. 1-15
(For beef cattle)	Ladino clover ..... 1 Tall fescue 10 11	Drill or broadcast	Apr. to June or Aug. 1-20
4. Pasture, hay, silage, or bedding	Reed canarygrass 12-14	Drill or broadcast	Apr. to May or Aug. 1-25

TABLE 5-1C—GENERAL RECOMMENDATIONS FOR TEMPORARY OR SUPPLEMENTAL PASTURES FO

(This table contains the authoritative recommendations, made especially for this book, o
ronomist, University of Massachusetts, Amherst, Mass.; Dr. A. R. Schmid, Agronomist
University, Ithaca, N.Y.; and Dr. John E. Baylor, Agronomist, The Pennsylvania State

Soil Type and Moisture Condition	Recommended Seedings (mixtures or pure seedings); Lb of Each and Total Lb/Acre	Seeding Recommendations	
		How to Seed	When to Seed
I. Well-Drained Soils	Wheat or rye 50-100	Drill or broadcast	Aug. to Sept.
	Oats (as companion crop) 30-80	Drill or broadcast	Apr. to May
	Oats (alone) 60-100	Drill or broadcast	Aug.
	Sudangrass, or sorghum-Sudangrass 25-35	Drill or broadcast	June
II. Wet, Poorly Drained Soils	Japanese millet 15-20	Drill or broadcast	June
	Sudangrass, or sorghum-Sudangrass ..... 20 Japanese millet 15 35	Drill or broadcast	June

TABLE 5-2A—MOST ABUNDANT GRASSES AND LEGUMES COMPRISING THE FORAGE I

(This table contains the authoritative listing, made especially fo
Howell N. Wheaton, Agronomist, University of Missouri, Columbia, Mo.; Dr
velopment Center, Wooster, Ohio; and Drs. Henry A. Fribourg and Joe D

Grasses—Bermudagrass Bluegrass, Kentucky Bromegrass Canarygrass, reed	Fescue, tall Orchardgrass Pearlmillet	Ryegrasses Sorghum-Sudangrass hybrids	Sudangrass Timothy

-(Continued)

Recommended Fertilizer Practices[1] (in lb per acre N, P_2O_5, and K_2O; from left to right, nitrogen, phosphoric acid, and potash, respectively)		Time of Use (or length of grazing season)		Carry-ing Capac-ity (animal units/ acre)	Comments
Initial Ap-plication at Seeding Time	Annual Maintenance Application After Pas-ture Is Established	When to Start	When to End		
-60-60 to 40-120-120	(No. 1) 0-40-40 to 0-60-180 (No. 2) 30-60-60 to 80-80-80	May 1-15	Sept. 1-15	¾ to 1¼	Lime to a pH of 6.0. When legume predominates, apply annual fertilizer (No. 1), preferably in part amounts after spring and fall harvests. When grass predominates, apply annual fertilizer (No. 2) in fall or early spring, or part in June after harvest. For pasture use, domestic ryegrass added in a spring seeding hastens feed growth that season.
-60-60 to 40-120-120	(No. 1) 0-40-40 to 0-60-180 (No. 2) 30-60-60 to 80-80-80	Nov.	Apr.	1	Harvest first crop for hay as round bales left in the field. Winter graze new growth and round bales.
)-60-60 to 60-60-60	50-0-0 to 110-60-60	May 1-15	Sept. 1-15	1 to 1½	Lime to a pH of 6.0. Most tolerant of extremes of both dry and wet soils. Reed canarygrass develops firm turf and responds to nitrogen fertilization with high yield of moderately profitable feed at an immature stage of growth.

[1]These recommendations serve only as guides. Use soil tests to determine accurately lime and fertilizer needs. Apply fertilizer by the band method when establishing a new eeding.

ATTLE, SHEEP, AND HORSES IN AREA 1, NORTHERN HUMID AREA
e following competent specialists residing in the area: Dr. Martin E. Weeks, Ag-
niversity of Minnesota, St. Paul, Minn.; Dr. Robert R. Seaney, Agronomist, Cornell
niversity, University Park, Penn.)

Recommended Fertilizer Practices[1] (in lb per acre N, P_2O_5, and K_2O; from left to right, nitrogen, phosphoric acid, and potash, respectively). Initial Application at Seeding Time	Time of Use (or length of grazing season)		Carry-ing Capac-ity (animal units/ acre)	Comments
	When to Start	When to End		
40-40-40	Apr. to May 15	May to June	2	About 30 to 45 days grazing. Remove livestock in mid-April if grain is to be harvested. May be grazed in Oct. and Nov. if seeded early.
20-60-60 to 40-120-120	June	July	2	When seeded as companion crop for legume. About 30 days' grazing.
40-40-40	Oct.	Nov.	1½	When seeded alone. About 30 days' grazing.
30-30-30 to 60-60-60	July	Sept.	1 to 2	Can be seeded at 10-day intervals for grazing use.
30-30-30 to 60-60-60	July	Sept.	1 to 2	Can be seeded at 10-day intervals for grazing use.
30-30-30 to 60-60-60	July	Sept.	1 to 2	Can be seeded at 10-day intervals for grazing use.

[1]These recommendations serve only as guides. Use soil tests to determine accurately lime and fertilizer needs. Apply fertilizer by the band method when establishing a new eeding.

ERMANENT AND/OR ROTATIONAL PASTURES IN AREA 2, CENTRAL HUMID AREA

his book, of the following competent specialists residing in the area: Dr.
obert W. Van Keuren, Agronomist, Ohio Agricultural Research and De-
urns, Agronomists, University of Tennessee, Knoxville, Tenn.)

egumes—Alfalfa Birdsfoot trefoil Clover, crimson Clover, Ladino	Clover, medium red Clover, white Lespedeza, Korean or Kobe	Lespedeza, sericea Sweet clover

TABLE 5-2B—GENERAL RECOMMENDATIONS FOR ROTATIONAL PASTURES FOR CATTL

(This table contains the authoritative recommendations, made especially f Howell N. Wheaton, Agronomist, University of Missouri, Columbia, Mo.; D velopment Center, Wooster, Ohio; and Drs. Henry A. Fribourg and Joe [

Soil Type and Moisture Condition	Recommended Seedings (mixtures or pure seedings); Lb of Each and Total Lb/Acre		Seeding Recommendations	
			How to Seed	When to Seed
I. Good, Well-Drained Soils	Orchardgrass, or Bromegrass, and Alfalfa	4-8 8-10 10-15 14-25	Drill or broadcast	Jan. 15 to Apr. 15 or Aug. 15 to Sept. 15
	Orchardgrass, or reed canarygrass Bromegrass Ladino clover	6-8 6 1-2 13-16	Drill or broadcast	Aug. 15 to Sept. 15 or Jan. 15 to Apr. 15
	Timothy Alfalfa	2-4 10-12 12-16	Drill or broadcast	Aug. 15 to Sept. 15 or Jan. 15 to Apr. 15
	Orchardgrass, or Bromegrass, and Med. red clover, or Alfalfa	6-8 8-10 6-8 6-8 12-18	Drill or broadcast	Aug. 15 to Sept. 15 or Jan. 15 to Apr. 15
II. Poor, Well-Drained Soils	Reed canarygrass Ladino, or alsike clover	6-8 1 7-9	Drill or broadcast	Jan. 15 to Mar. 15 Aug. 15 to Sept. 15
	Orchardgrass Korean lespedeza	10 15 25	Drill or broadcast	Aug. 15 to Sept. 15 Jan. 15 to Apr. 15
	Tall fescue, and Med. red clover, or Alfalfa, and La. white or Ladino clover	8-10 3-4 5-6 1-2 12-18	Drill or broadcast	Aug. 15 to Sept. 15 or Jan. 15 to Apr. 15
	Lespedeza, adapted annual	25-30	Drill or broadcast	Mar. 1 to Apr. 15
III. Wet, Poorly Drained Soils	Tall fescue Ladino clover	10-15 1-2 11-17	Drill or broadcast	Aug. 15 to Sept. 15 or Jan. 15 to Apr. 15
	Reed canarygrass Alsike clover Ladino clover	6-8 3 ¼ 9¼-11¼	Drill or broadcast	Aug. 15 to Sept. 15 or Jan. 15 to Apr. 15
	Tall fescue Alsike clover, and La. white or Ladino clover	8-10 2-3 1-2 11-15	Drill or broadcast	Aug. 15 to Sept. 15 or Jan. 15 to Apr. 15

HEEP, AND HORSES IN AREA 2, CENTRAL HUMID AREA

ıis book, of the following competent specialists residing in the area: Dr. obert W. Van Keuren, Agronomist, Ohio Agricultural Research and De- urns, Agronomists, University of Tennessee, Knoxville, Tenn.)

Recommended Fertilizer Practices[1] (in lb per acre N, P_2O_5, and K_2O; from left to right, nitrogen, phosphoric acid, and potash, respectively)		Time of Use (or length of grazing season)		Carrying Capacity (animal units/ acre)	Comments
Initial Application at Seeding Time	Annual Maintenance Application After Pasture Is Established	When to Start	When to End		
30-120-120	0-60-180	Apr. 15	Sept. 15	1 to 1½	Rotation grazing necessary to retain alfalfa stand. Do not apply nitrogen (annual maintenance) if there is an adequate stand of legume.
30-120-120	0-60-60 to 60-60-120	Apr. 15	Sept. 15	1 to 1½	Rotation grazing desirable. Do not apply nitrogen (annual maintenance) if there is an adequate stand of legume. Graze orchardgrass heavily in the spring to keep it acceptable.
30-120-120	0-60-180	Apr. 15	Sept. 15	1 to 1½	Rotation grazing necessary to retain alfalfa stand. Sow timothy at 5 lb/acre in late summer or fall, and 10 lb/acre in spring.
30-120-120	0-60-60 to 60-60-120	Apr. 15	Sept. 15	1 to 1½	Do not apply nitrogen (annual maintenance) if there is an adequate stand of legume.
10-30-30	0-60-60 to 30-60-60	Apr. 15	Oct. 1	1	Do not apply nitrogen (annual maintenance) if there is an adequate stand of legume.
10-30-30	0-60-60 to 30-60-60	Apr. 15	Oct. 1	1	Lespedeza must be spring seeded.
30-120-120	0-60-60 to 60-60-120	Mar.	Nov.	1 to 1½	Do not apply nitrogen (annual maintenance) if there is an adequate stand of legume. Graze tall fescue heavily to keep it acceptable.
0-30-30	0-30-30	May 15	Oct. 1	½	
30-120-120	0-60-60 to 60-60-120	Mar.	Nov.	1 to 1½	For lands which do not long remain under water. Do not apply nitrogen (annual maintenance) if there is an adequate stand of legume.
30-120-120	60-60-60	Apr. 1	Oct. 1	1 to 1½	Do not apply nitrogen (annual maintenance) if there is an adequate stand of legume.
30-100-100 to 30-200-120	40-40-40	Apr. 1	Oct. 1	1 to 1½	For lands which do not long remain under water. Do not apply nitrogen (annual maintenance) if there is an adequate stand of legume.

[1]These recommendations serve only as guides. Use soil tests to determine accurately lime and fertilizer needs. Apply fertilizer by the band method when establishing a new eeding.

TABLE 5-2C—GENERAL RECOMMENDATIONS FOR TEMPORARY OR SUPPLEMENTARY PASTURE

(This table contains the authoritative recommendations, made especially for this boo Agronomist, University of Missouri, Columbia, Mo.; Dr. Robert W. Van Keuren, Ag Drs. Henry A. Fribourg, and Joe D. Burns, Agronomists, University of Tennesse

Soil Type and Moisture Condition	Recommended Seedings (mixtures or pure seedings); Lb of Each and Total Lb/Acre		Seeding Recommendations	
			How to Seed	When to Seed
I. Good, Well-Drained Soils	Wheat Korean lespedeza	90-120 25 115-145	Drill or broadcast	(wheat) Aug. 20 to Sept. 1;[2] (lespedeza) Feb. and Mar.
	Winter barley, or Oats, or Rye	90-135 64-96 56-84	Drill or broadcast	Aug. 15 to Oct. 1
	Lespedeza, annual	25	Drill or broadcast	Feb. and Mar.
	Sudangrass	25-30	Drill or broadcast	May 1 to July 1
	Pearlmillet	20-30	Drill or broadcast	May 1 to July 1
	Soybeans, and Sudangrass, or Pearlmillet	30 15 10 40-45	Drill or broadcast	May 1 to July 1
II. Poor, Well-Drained Soils	Rye	120	Drill or broadcast	Aug. 15 to Oct. 1
	Sudangrass	25-30	Drill or broadcast	May 15 to July 15
	Soybeans	30	Drill or broadcast	May 1 to July 15
	Lespedeza, annual	25-30	Drill or broadcast	Jan. and Feb.[3]
III. Wet, Poorly Drained Soils	Pearlmillet	20-30	Drill or broadcast	May 1 to July 1
	Sudangrass	25-30	Drill or broadcast	May 1 to July 1
	Pearlmillet, or Sudangrass	10 15 25	Drill or broadcast	May 1 to July 1

TABLE 5-3A—MOST ABUNDANT GRASSES AND LEGUMES COMPRISING THE FORAGE

(This table contains the authoritative listing, made especially for this book, land, Agronomist, Auburn University, Auburn, Ala.; and Dr. W.

Grasses—Bahiagrass[1] Bermudagrass	Dallisgrass Fescue, tall	Johnsongrass Orchardgrass	Ryegrass, annual

OR CATTLE, SHEEP, AND HORSES IN AREA 2, CENTRAL HUMID AREA

f the following competent specialists residing in the area: Dr. Howell N. Wheaton, nomist, Ohio Agricultural Research and Development Center, Wooster, Ohio; and noxville, Tenn.)

Recommended Fertilizer Practices[1] (in lb per acre N, P_2O_5, and K_2O; from left to right, nitrogen, phosphoric acid, and potash, respectively). Initial Application at Seeding Time	Time of Use (or length of grazing season)		Carrying Capacity (animal units/ acre)	Comments
	When to Start	When to End		
30-60-60	Oct. 1 Mar. 1 July 1	Dec. 31 Apr. 15 Oct. 1	1 to 2	Lespedeza must be spring seeded. Apply 30 lb of nitrogen as spring top dressing.
30-60-60	Oct. 1 and Mar. 1	Dec. 31 and May 15	1 to 2	Apply 30 lb of nitrogen as spring top dressing.
0-30-30	July 1	Oct. 1	1 to 2	Broadcast on any grain crop.
60-40-40	June 1	Sept. 1	1 to 3	Apply 60 lb of nitrogen after first harvest.
60-40-40 to 120-40-40	June 15	Oct. 1	1 to 3	
60-60-60	June 15	Sept. 15	1 to 3	
60-40-40	Oct. 1 and Mar. 1	Dec. 31 and May 1	1 to 2	Apply 30 lb of nitrogen as a spring dressing.
60-40-40	June 1	Oct. 1	1 to 2	Apply 60 lb of nitrogen after first harvest.
60-60-60	June 15	Sept. 15	1	
0-30-30	July 1	Oct. 1	1 to 2	Annual lespedeza not recommended without cereal or grass except on nontillable land, such as the Ozark region. In latter area, broadcasting usually is the only possible method, and early seeding is recommended as the seed is more likely to be covered by natural processes.
60-40-40	June 15	Oct. 1	1 to 2	Apply 60 lb of nitrogen after first harvest.
60-40-40	June 1	Oct. 1	1 to 3	For soil wet in winter, but dry enough in summer. Apply 60 lb of nitrogen after first harvest.
60-60-60	June 15	Oct. 1	1 to 3	

[1]These recommendations serve only as guides. Use soil tests to determine accurately lime and fertilizer needs. Apply fertilizer by the band method when establishing a new seeding.
[2]At this early date, use Hessian fly resistant wheat only.
[3]On nontillable land.

PERMANENT AND/OR ROTATIONAL PASTURES IN AREA 3, SOUTHERN HUMID AREA

he following competent specialists residing in the area: Dr. Carl S. Hove-Voodhouse, Jr., Agronomist, North Carolina State University, Raleigh, N.C.)

Legumes—Alfalfa Clover, arrowleaf	Clover, crimson Clover, Ladino	Clover, medium red Clover, white	Lespedeza, annual Lespedeza, sericea

[1]Limited to lower Gulf Coast Region.

TABLE 5-3B—GENERAL RECOMMENDATIONS FOR ROTATIONAL PASTURES FOR CATTL

(This table contains the authoritative recommendations, made especially f
S. Hoveland, Agronomist, Auburn University, Auburn, Ala.; and Dr. W. W

Soil Type and Moisture Condition	Recommended Seedings (mixtures or pure seedings); Lb of Each and Total Lb/Acre		Seeding Recommendations	
			How to Seed	When to Seed
I. Upland Soils	Dallisgrass White clover	15 2 17	Drill or broadcast " "	(dallisgrass) Mar. to Apr.; (white clover) Oct. to Nov.
	Tall fescue White or Ladino clover	10 2-3 12-13	Drill or broadcast	Sept. to Oct.
	Coastal Bermudagrass Reseeding crimson clover, or Arrowleaf clover	3,000-5,000 stolons 20 5	Sprigged in Drill or broadcast	Mar. to Apr. 1 Oct. to Nov. Oct. to Nov.
	Orchardgrass, and White or Ladino clover	10 2-4 12-14	Drill or broadcast	Sept. to Oct.
	Bahiagrass Arrowleaf clover Crimson clover	15 5 20 40	Drill or broadcast " " " "	(bahiagrass) Feb. to Apr.; (clover) Oct. to Nov.
II. Bottomland Soils	Dallisgrass Korean lespedeza, and/or Ladino clover, regal	15 15 3 18-33	Drill or broadcast " " " "	(dallisgrass) Mar. to Apr.; (lespedeza) Feb. to Apr.; (clover) Oct. to Nov.
	Tall fescue	10	Drill or broadcast	Sept. to Oct.
	Orchardgrass, or Tall fescue, and Ladino clover	6 5 3 8-9	Drill or broadcast	Sept. to Oct.

HEEP, AND HORSES IN AREA 3, SOUTHERN HUMID AREA

his book, of the following competent specialists residing in the area: Dr. Carl
Voodhouse, Jr., Agronomist, North Carolina State University, Raleigh, N.C.)

Recommended Fertilizer Practices[1] (in lb per acre N, P_2O_5, and K_2O; from left to right, nitrogen, phosphoric acid, and potash, respectively)		Time of Use (or length of grazing season)		Carrying Capacity (animal units/acre)	Comments
Initial Application at Seeding Time	Annual Maintenance Application After Pasture Is Established	When to Start	When to End		
0-60-60 or 0-50-100	0-60-60 or 0-50-100	Mar.	Oct.	3/5	Restricted to clay soils such as Black Belt. Establish dallisgrass in spring, preferably. White clover should be overseeded in fall. White clover should be grazed close in late spring to allow dallisgrass to begin growth. Ladino preferred where clover survives summers. Use La. white clover where reseeding is required.
0-60-60 to 60-60-100	0-60-60 or 0-50-100 to 100-60-60 or 100-25-50	Oct.	June	4/5	Tall fescue will perform satisfactorily under poor drainage conditions. Not adapted to sandy soils of the lower South. Maintenance applications of N will vary depending upon the botanical composition of the mixture.
0-120-100	80-20-40 to 200-50-100	Feb.	Oct.	1	Lime to pH of 6.0. Nitrogen should be applied to the coastal Bermudagrass after it begins to grow. The amount of N required for establishment and maintenance will depend upon level of production desired. May range up to 400 lb N/acre/yr.
0-75-75 or 0-50-100 to 30-120-100	0-50-50 to 100-60-60 or 100-25-50	Oct.	June	4/5	Limited to upper part of region. Maintenance application of N will vary depending upon the botanical composition of the mixture.
0-120-100	0-60-60 to 200-60-60	Feb.	Oct.	1	Limited to lower South. Recommended for areas too sandy and droughty for dallisgrass. Very competitive with legumes.
0-120-100	0-60-60	Mar.	Oct.	1	Lime to pH of 6.0. No P_2O_5 or K_2O needed in Delta Area. Establish dallisgrass and lespedeza, then overseed with clover. Graze clover close in late spring to give dallisgrass and lespedeza opportunity to begin growth.
60-120-100	100-60-50	Nov.	May	1	Lime to pH of 6.0. No P_2O_5 or K_2O needed in Delta Area. Productive on poorly drained soils.
30-120-100 to 60-120-100	0-60-60 to 100-60-60	Nov.	May	1	Lime to pH of 6.0. No P_2O_5 or K_2O needed in Delta Area. Orchardgrass limited to northern part of the region.

[1]These recommendations serve only as guides. Use soil tests to determine accurately lime and fertilizer needs. Apply fertilizer by the band method when establishing a new seeding.

TABLE 5-3C—GENERAL RECOMMENDATIONS FOR TEMPORARY OR SUPPLEMENTAL PASTURE

(This table contains the authoritative recommendations, made especially for thi
land, Agronomist, Auburn University, Auburn, Ala.; and Dr. W. W. Woodhouse

Soil Type and Moisture Condition	Recommended Seedings (mixtures or pure seedings); Lb of Each and Total Lb/Acre		Seeding Recommendations	
			How to Seed	When to Seed
I. All Soils Except as Noted Under Comments	Oats	120	Drill	Sept.
	Wheat or rye	120	Drill	Sept.
	Wheat or rye Arrowleaf clover (yuchi), or Crimson clover	60 5 20 65-80	Drill Broadcast Drill or broadcast	(wheat or rye) Sept. (arrowleaf clover) Sept. (crimson clover) Sept.
	Red clover (adapted variety), or	10	Drill or broadcast	(med. red clover) Sept.
	Kobe or Korean lespedeza, or	25	Drill or broadcast	(lespedeza) Mar. to Apr.
	Hairy vetch	30	Drill or broadcast	(vetch) Sept. to Oct.
	Ryegrass Crimson clover	20 20 40	Drill or broadcast	Sept.
	Rye Ryegrass Arrowleaf clover (yuchi)	60 10 5 75	Drill Broadcast Broadcast	Sept. Sept. Sept.
	Pearlmillet, Sudangrass, or Sorghum-Sudangrass hybrids	30 10 10-30	Drill or broadcast	Apr. to June
	Korean or Kobe lespedeza	25-35	Drill or broadcast	Mar. to Apr.
	Alyceclover	20	Drill or broadcast	May to June
	Sericea lespedeza	15	Drill or broadcast	Mar. to Apr. 10

OR CATTLE, SHEEP, AND HORSES IN AREA 3, SOUTHERN HUMID AREA

ook, of the following competent specialists residing in the area: Dr. Carl S. Hove-r., Agronomist, North Carolina State University, Raleigh, N.C.)

Recommended Fertilizer Practices[1] (in lb per acre N, P_2O_5, and K_2O; from left to right, nitrogen, phosphoric acid, and potash, respectively)		Time of Use (or length of grazing season)		Carrying Capacity (animal units/acre)	Comments
Initial Application at Seeding Time	Annual Maintenance Application After Pasture Is Established	When to Start	When to End		
60-60-50	60-0-0, top dressed in Feb.	Nov.	Apr.	1/2	Earlier fall production than other small grains but less winter hardy than rye or wheat. Oats will perform satisfactorily on fairly wet soils.
60-60-50	60-0-0, top dressed in Feb.	Nov.	Apr.	1/2	Rye is more winter hardy than oats, but it heads earlier in the spring. Rye superior on the sandy soils.
60-60-60	None to 60-0-0	(wheat or rye and arrowleaf clover) Nov. ↔ June; (wheat or rye and crimson clover) Nov. ↔ Apr.	June; Apr.	1/2	Use scarified arrowleaf clover seed. Yuchi arrowleaf clover extends grazing season 2 months longer than crimson clover.
0-60-60	0-60-60	(rye or wheat, and med. red clover) Nov. ↔ July	July		
0-60-60	0-60-60	(rye or wheat and lespedeza) Nov. ↔ Oct.	Oct.		
0-60-60	0-60-60	(rye or wheat and vetch) Nov. ↔ Apr.	Apr.		
60-60-60	None to 60-0-0	Dec.	May	1/2	Crimson clover limited to well-drained soil.
60-60-60	60-0-0	Nov.	June	3/4	This is the best long-season winter annual mixture for well-drained soils in the lower South.
50-50-50	50-0-0 to 100-0-0	June	Sept.	2	Top dress with 30 lb N each month grazed. Seed 10 lb in rows, 25 lb broadcast.
0-60-60		June	Oct.	4/5	Usually seeded in small grain, requires no fertilizer if small grain well fertilized. Lime to pH of 6.5. Most popular in northern half of region.
0-90-50		July	Sept.	4/5	Lime to pH of 6.0. Limited to Gulf Coast Region. Very susceptible to nematodes on sandy soils.
0-40-40 to 0-70-70	0-20-20 to 0-40-60	May of 2nd year	Sept.	3/4	Avoid grazing first year. Use scarified seed. Use herbicide at planting for weed control. Avoid use of N at planting as it encourages weed competition.

[1]These recommendations serve only as guides. Use soil tests to determine accurately lime and fertilizer needs. Apply fertilizer by the band method when establishing a new seeding.

TABLE 5-4A—MOST ABUNDANT GRASSES AND LEGUMES COMPRISING THE FORAGE I

(This table contains the authoritative listing, made especially for this book, c
Agronomist, University of Florida, Gainesville, Fla.; and Dr. Glenn W. Bu

Grasses—Bahiagrass, Argentine Bahiagrass, Paraguay 22 Bahiagrass, Pensacola Bermudagrass, coastal Bermudagrass, coastcross #1	Bermudagrass, Suwannee Bermudagrass, common Buffelgrass Carib grass Dallisgrass	Digitgrass, Pengola Digitgrass, Transvala Digitgrass, slenderstem Johnsongrass Pearlmillet	Stargrass, McCaleb St. Augustine grass, roselawn Sorghum-Sudangrass hybrids Sudangrass

TABLE 5-4B—GENERAL RECOMMENDATIONS FOR PERMANENT PASTURES FOR CATTLE

(This table contains the authoritative recommendations, made especially fc
O. Mott, Agronomist, University of Florida, Gainesville, Fla.; and Dr. Glen

Soil Type and Moisture Condition	Recommended Seedings (mixtures or pure seedings); Lb of Each and Total Lb/Acre	Seeding Recommendations	
		How to Seed	When to Seed
I. Moist, Sandy Soils	Pangolagrass (vegetative) 200-400	Broadcast and cut in, or plant in rows	Mar. to Aug.
	Pensacola bahiagrass 10-15	Drill or broadcast	Anytime
	Argentine bahiagrass 10	Drill or broadcast	Anytime
	Coastal, or coastcross, Bermudagrass (vegetative) 300-500	Sprig in	Mar. to Aug.
	St. Augustine grass (vegetative) 500-1,000	Broadcast or row	Mar. to Aug.
	Dallisgrass 20	Drill	Sept. to Dec.
	Paragrass (vegetative) 500-1,000	Broadcast and cut in	Mar. to Sept.
	Carib (vegetative) 500-1,000	Broadcast and cut in	Mar. to Sept.
	Hairy indigo, alone or with grass 6-8	Drill or broadcast	Apr. to July
	White clover 5	Broadcast	Oct. to Dec.
	White clover with any of first 7 grasses above 5	Broadcast	Oct. to Dec.
	Sweet clover (Hubam) 12-15	Broadcast	Oct. to Dec.
	Sweet clover with any of first 9 grasses above 10	Broadcast	Oct. to Dec.
	Kenland red clover 12	Broadcast	Oct. to Dec.
	Kenland red clover with any of first 9 grasses above 12	Broadcast	Oct. to Dec.
II. Sandy Loams	Pangolagrass (vegetative) 200-400	Broadcast and cut in, or plant in rows	Mar. to Aug.
	Pensacola bahiagrass 10-15	Drill or broadcast	Anytime
	Argentine bahiagrass 10	Drill or broadcast	Anytime
	Coastal or Suwannee Bermudagrass (vegetative) 300-500	Broadcast or row	Mar. to Aug.
	Dallisgrass 20	Drill	Sept. to Dec.
	White clover with any of grasses above .. 5	Broadcast	Oct. to Dec.
	Sweet clover (Hubam) with any of grasses above 10	Broadcast	Oct. to Dec.
	Red clover with any of grasses above ... 12	Broadcast	Oct. to Dec.
	Crimson clover with any of above grasses 20-25	Broadcast	Oct. to Dec.

Footnote on last page of table.

ERMANENT AND/OR ROTATIONAL PASTURES IN AREA 4, EASTERN COASTAL AREA

he following competent specialists residing in the area: Dr. G. O. Mott,
on, Research Geneticist, USDA, Tifton, Ga.)

Legumes—Aeschynomene, Alfalfa, Alyceclover, Austrian winter peas, Clover, arrowleaf, Clover, crimson, Clover, Ladino, Clover, red, Clover, Louisiana red, Clover, white, Cowpeas, Hairy indigo, Lespedeza, common, Kobe, Lupine, Sweet clover, Velvet beans, Vetch

SHEEP, AND HORSES IN AREA 4, EASTERN COASTAL AREA

this book, of the following competent specialists residing in the area: Dr. G.
W. Burton, Research Geneticist, USDA, Tifton, Ga.)

Recommended Fertilizer Practices[1] (in lb per acre N, P_2O_5, and K_2O; from left to right, nitrogen, phosphoric acid, and potash, respectively)		Time of Use (or length of grazing season)		Carrying Capacity (animal units/ acre)	Comments (Use proper inoculant on all legumes)
Initial Application at Seeding Time	Annual Maintenance Application After Pasture Is Established	When to Start	When to End		
30-30-30	100-25-50 to 200-50-100	Mar.	Nov.	1/2 to 1 1/2	Lime to pH of 5.5-6.5; add minor elements where needed. Sometimes grazed year-around.
30-30-30	60-30-30 to 120-60-60	Jan.	Dec.	1/5 to 1	Lime to pH of 5.0-6.5.
30-30-30	60-30-30 to 120-60-60	Mar.	Nov.	1/5 to 1	Lime to pH of 5.0-6.5.
30-30-30	100-25-50 to 200-50-100	Mar.	Nov.	1/2 to 1 1/2	Lime to pH of 6.0-6.5.
0-30-100 + 15 lb CuO	0-60-60 to 0-40-80 + 5 lb CuO	Jan.	Dec.	1/2 to 3	Lime to pH of 5.0-6.5, and add minor elements. Also adapted to organic soils, but omit nitrogen.
30-30-30	40-0-0 to 80-0-0	Jan.	Dec.	1/5 to 1	Lime to pH of 5.5-6.5.
0-30-100 + 15 lb CuO	0-60-60 to 0-40-80 + 5 lb CuO	Mar.	Nov.	1/2 to 3	Lime to pH of 5.5-6.5, and add minor elements. Also adapted to organic soils, but omit nitrogen.
0-30-100 + 15 lb CuO	0-60-60 to 0-40-80 + 5 lb CuO	Mar.	Nov.	1/2 to 4	Lime to pH of 5.5-6.5, and add minor elements. Also adapted to organic soils, but omit nitrogen.
None	None	June	Oct.	2 to 3	Lime to pH of 5.0-6.5. No specific inoculant.
0-60-60	0-50-100	Feb.	June	1 to 3	Lime to pH of 5.5-6.7.
0-60-60	0-50-100	Feb.	June	1 to 3	Top seeded on sod.
0-60-60	0-50-100	Feb.	June	1 to 3	Lime to pH of 6.0-6.7.
0-60-60	0-50-100	Feb.	June	1 to 3	Top seeded on sod.
0-60-60	0-50-100	Feb.	June	1 to 3	Lime to pH of 5.5-6.7.
0-60-60	0-50-100	Feb.	June	1 to 3	Top seeded on sod.
30-30-30	60-15-30 to 120-30-60	Mar.	Nov.	1/2 to 1 1/2	Lime to pH of 5.5-6.5. Sometimes grazed year-around.
30-30-30	60-30-30 to 120-60-60	Jan.	Dec.	1/5 to 1	Lime to pH of 5.0-6.5
30-30-30	60-30-30 to 120-60-60	Mar.	Nov.	1/5 to 1	Lime to pH of 5.0-6.5.
30-30-30	60-30-30 to 120-60-60	Mar.	Nov.	1/2 to 1 1/2	Lime to pH of 6.0-6.5.
30-30-30	60-15-30	Jan.	Dec.	1/5 to 1	Lime to pH of 5.5-6.5.
0-60-60	0-50-100	Feb.	June	1 to 3	Lime to pH of 5.5-6.7.
0-60-60	0-50-100	Feb.	June	1 to 3	Lime to pH of 5.5-6.7.
0-60-60	0-50-100	Feb.	June	1 to 3	Lime to pH of 5.5-6.7.
0-60-60	0-50-100	Jan.	Apr.	1 to 3	Lime to pH of 5.5-6.5.

(Continued)

TABLE 5-4B

Soil Type and Moisture Condition	Recommended Seedings (mixtures or pure seedings); Lb of Each and Total Lb/Acre	Seeding Recommendations	
		How to Seed	When to Seed
II. Sandy Loams (Cont.)	White clover .... 4 Sweet clover 7 11	Drill or broadcast	Oct. to Dec.
	White clover .... 4 Red clover 8 12	Drill or broadcast	Oct. to Dec.
	Hairy indigo, alone or with grass 6-8	Drill or broadcast	Apr. to July

TABLE 5-4C—GENERAL RECOMMENDATIONS FOR TEMPORARY OR SUPPLEMENTAL PASTURES

(This table contains the authoritative recommendations, made especially for this Agronomist, University of Florida, Gainesville, Fla.; and Dr. Glenn W. Burton, Re-

Soil Type and Moisture Condition	Recommended Seedings (mixtures or pure seedings); Lb of Each and Total Lb/Acre	Seeding Recommendations	
		How to Seed	When to Seed
I. Moist, Sandy Soils or Clay, Loam or Clay or Silt Loam Soils	Oats 60-90	Drill	Oct. to Dec.
	Rescuegrass or .... 24-30 Ryegrass 20-30	Broadcast	Oct. 1 to Nov. 1
	Pearlmillet 20-25	Drill	Apr. to July
	Italian or perennial ryegrass 25-30	Broadcast	Oct. to Dec.
	Rye or wheat 60	Drill	Oct. to Dec.
	Oats .... 60 Crimson clover 20 80	Drill or broadcast	Sept. to Nov.
	Pearlmillet .... 20 Hairy indigo 5 25	Drill or broadcast	Apr. to July
	Cowpeas 60-90	Drill	Apr. to June
	Alyceclover 20	Drill	Apr. to June
	Pearlmillet .... 15 Alyceclover 15 30	Drill	Apr. to July
	Soybeans 60-90	Drill or broadcast	May to July
	Sorghum-Sudangrass hybrids, or Sudangrass 8-10 24-30	Drill Broadcast	Mar. 15 to June 30
	Velvet bean 10-15 30-45	Drill Broadcast	Mar. 15 to June 30

—(Continued)

Recommended Fertilizer Practices[1] (in lb per acre N, P_2O_5, and K_2O; from left to right, nitrogen, phosphoric acid, and potash, respectively): Initial Application at Seeding Time	Annual Maintenance Application After Pasture Is Established	Time of Use (or length of grazing season): When to Start	When to End	Carrying Capacity (animal units/acre)	Comments
0-60-60	0-50-100	Feb.	June	1 to 3	Lime to pH of 6.0-6.5.
0-60-60	0-50-100	Feb.	June	1 to 3	Lime to pH of 5.5-6.5.
None	None	June	Oct.	2 to 3	Lime to pH of 5.0-6.5. No specific inoculant.

[1]These recommendations serve only as guides. Use soil tests to determine accurately lime and fertilizer needs. Apply fertilizer by the band method when establishing a new seeding.

FOR CATTLE, SHEEP, AND HORSES IN AREA 4, EASTERN COASTAL AREA

book, of the following competent specialists residing in the area: Dr. G. O. Mott, search Geneticist, USDA, Tifton, Ga.)

Recommended Fertilizer Practices[1] (in lb per acre N, P_2O_5, and K_2O; from left to right, nitrogen, phosphoric acid, and potash, respectively): Initial Application at Seeding Time	Annual Maintenance Application After Pasture Is Established	Time of Use (or length of grazing season): When to Start	When to End	Carrying Capacity (animal units/acre)	Comments
50-50-50	50-0-0, apply Jan. to Feb.	Dec.	Feb. or Apr. (see first comment)	1 to 1½	Winter annual. May be overseeded on permanent pastures or on a prepared seedbed. Stop grazing in Feb. to get grain crop.
20-30-20		Dec.	Apr.	1	Winter annuals. May be overseeded on permanent pastures or on a prepared seedbed.
50-50-50	50- to 100-0-0, apply June to July	May	Sept.	2 to 3	Nitrogen and management are the most critical factors with this crop.
50-50-50	50-0-0, apply Dec. to Jan.	Dec.	Apr.	1 to 1½	
50-50-50	50-0-0, apply Jan. to Feb.	Dec.	Mar. or Apr.	1 to 1½	
50-50-50	50-0-0, apply in Jan.	Dec.	Apr.	2	
50-50-50	30-0-0 to 60-0-0, apply in June	May	Oct.	2 to 3	Indigo extends grazing season. Red clover seeded with cultipacker extends grazing.
0-40-80	None	May	Aug.	1 to 2	Also used for hay.
0-40-80	None	June	Oct.	1 to 2	Also used for hay.
50-50-50	30-0-0 to 60-0-0, apply in June	May	Oct.	2 to 3	Alyceclover extends grazing season.
0-50-100	None	May	Aug.	1 to 2	For clay, loam, or clay or silt loam soils only. Also used for hay.
20-30-20	None	June	Oct.	1 to 1½	This is a summer annual. Also, it may be used for hay.
20-30-20	None	June	Sept.	1 to 1½	This is a summer annual. Also, it may be used for hay.

[1]These recommendations serve only as guides. Use soil tests to determine accurately lime and fertilizer needs. Apply fertilizer by the band method when establishing a new seeding.

TABLE 5-5A—MOST ABUNDANT GRASSES AND LEGUMES COMPRISING THE FORAGE I

(This table contains the authoritative listing, made especially for this book, o
Agronomist, USDA and Montana State University, Bozeman, Mont.; and Dr

Grasses—Bluegrass Bluestem, big[1] Bluestem, little[1] Bromegrass Buffalograss[1]	Canarygrass, reed Fescue Grama grass[1] Junegrass[1] Needlegrass[1]	Orchardgrass Switchgrass[1] Wheatgrass, crested Wheatgrass, intermediate Wheatgrass, pubescent	Wheatgrass, slender Wheatgrass, tall Wheatgrass, Western Wild-rye, Russian

TABLE 5-5B—GENERAL RECOMMENDATIONS FOR PERMANENT PASTURES FOR CATTLE,

(This table contains the authoritative recommendations, made especially for
S. Cooper, Agronomist, USDA and Montana State University, Bozeman,
coln, Neb.)

Soil Type and Moisture Condition	Recommended Seedings (mixtures or pure seedings); Lb of Each and Total Lb/Acre		Seeding Recommendations	
			How to Seed	When to Seed
I. Well-Drained Soils	Crested wheatgrass, or Russian wild-rye, or smooth bromegrass	8-12	Drill	Early fall or early spring
	Crested wheatgrass Smooth bromegrass Alfalfa	4 4 3 11	Drill	Grass in fall and alfalfa in early spring.
	Crested wheatgrass Russian wild-rye Sweet clover	4 4 3 11	Drill	Grass in fall and sweet clover in early spring; or both in spring.
	Crested wheatgrass, or bromegrass, and Slender wheatgrass, or intermediate wheatgrass, and Alfalfa, or Yellow sweet clover, and Russian wild-rye, or green needlegrass, or side-oats grama, or switchgrass, or western wheatgrass, or big bluestem	6-8 6-8 1-3 2-4 0-3 16-23	Drill	Grass in fall and legume in spring; or both in spring. (Switchgrass and big bluestem in spring only.)
II. Wet, Poorly Drained Lands	Reed canarygrass, and Ladino clover or alsike clover	6-8 2 8-10	Drill	Spring
	Reed canarygrass Tall wheatgrass Meadow foxtail Alsike clover	3-4 5-6 2-3 2-3 12-16	Drill	Spring or fall
III. Wet, Poorly Drained Saline Lands	Slender wheatgrass, or tall wheatgrass, and Ladino clover, or strawberry clover	10 2 12	Drill	Spring

ERMANENT AND/OR ROTATIONAL PASTURES IN AREA 5, NORTHERN GREAT PLAINS AREA

he following competent specialists residing in the area: Dr. Clee S. Cooper, V. J. Moline, Agronomist, The University of Nebraska, Lincoln, Neb.)

egumes—Alfalfa Clover, alsike	Clover, Ladino Clover, medium red	Clover, strawberry Clover, white	Sweet clover Trefoil, birdsfoot

[1]Found mostly in native range; seldom seeded.

HEEP, AND HORSES IN AREA 5, NORTHERN GREAT PLAINS AREA

his book, of the following competent specialists residing in the area: Dr. Clee Mont.; and Dr. W. J. Moline, Agronomist, The University of Nebraska, Lin-

Recommended Fertilizer Practices[1] (in lb per acre N, P_2O_5, and K_2O; from left to right, nitrogen, phosphoric acid, and potash, respectively)		Time of Use (or length of grazing season)		Carrying Capacity (animal units/ acre)	Comments
Initial Application at Seeding Time	Annual Maintenance Application After Pasture Is Established	When to Start	When to End		
0-0-0 to 20-0-0	30-0-0 to 60-0-0	May 10-15	July 10-15	¼ to 1	Intermediate wheatgrass may be used when rainfall is above 14 in. annually. Annual fertilizing not usually applied in western part of area but needed in eastern portion. Seed in late summer (Aug. 1 to Sept. 15) in Neb.
0-0-0 to 20-0-0	30-0-0 to 60-0-0	May 10-15	July 10-15	¼ to 1	Fertilizing with nitrogen not necessary when good stands of alfalfa are present. Seed in late summer (Aug. 1 to Sept. 15) in Neb.
0-0-0 to 20-0-0	30-0-0 to 60-0-0	May 10-15	July 10-15	¼ to 1	Alfalfa can be used when rainfall is 14 to 16 in. Annual fertilizing not usually applied. Nitrogen should be applied in eastern portion after sweet clover is gone.
0-0-0 to 20-0-0	30-0-0 to 60-0-0	May 10-15	Aug. 10-15	¼ to 1	Annual fertilizing not usually applied. Nitrogen may be necessary if legumes are not a good stand.
0-0-0 to 20-0-0	30-0-0 to 60-0-0	May 10-15	Oct. 10-15	1½ to 2	Ladino clover not hardy in the North.
0-0-0 to 20-0-0	30-0-0 to 60-0-0	When sod permits	Oct.	1 to 2	May seed in late summer (Aug. 1 to Sept. 15) in Neb.
0-0-0 to 20-0-0	30-0-0 to 60-0-0	May 10-15	Oct. 10-15	1 to 2	Ladino and strawberry clover not hardy in the North. Seed in late summer (Aug. 25 to Sept. 10) in Neb.

[1]These recommendations serve only as guides. Use soil tests to determine accurately lime and fertilizer needs. The heavier fertilizer applications may be used in the higher rainfall areas—18 to 22 inches.

TABLE 5-5C—GENERAL RECOMMENDATIONS FOR ROTATIONAL PASTURES FOR CATTLE,

(This table contains the authoritative recommendations, made especially fo S. Cooper, Agronomist, USDA and Montana State University, Bozeman, coln, Neb.)

Soil Type and Moisture Condition	Recommended Seedings (mixtures or pure seedings); Lb of Each and Total Lb/Acre	Seeding Recommendations	
		How to Seed	When to Seed
I. Irrigated Lands	Smooth bromegrass, or Intermediate wheatgrass ... 6 Orchardgrass ... 5 Alfalfa ... 6 17	Drill	Grass in early spring or fall; alfalfa in spring only.
	Birdsfoot trefoil (Empire) ... 6 Kentucky bluegrass, or Orchardgrass ... 10 16	Drill	Early spring
	Orchardgrass ... 10 Ladino clover ... 3 Med. red clover ... 3 16	Drill	Grass in early spring or fall; alfalfa in spring only.
II. Wet, Poorly Drained Soils	Reed canarygrass ... 8-10 Ladino clover ... 3 11-13	Drill or broadcast	Early spring
	Reed canarygrass ... 4-6 Tall wheatgrass ... 6-8 Redtop ... 2-3 Alsike clover ... 2-3 14-20	Drill	Late spring or fall
III. Wet, Saline Land	Slender wheatgrass, or Western wheatgrass, and ... 6 Tall wheatgrass, and ... 6 Ladino clover, or Strawberry clover ... 2 14	Drill	Early spring

SHEEP, AND HORSES IN AREA 5, NORTHERN GREAT PLAINS AREA

this book, of the following competent specialists residing in the area: Dr. Clee Mont.; and Dr. W. J. Moline, Agronomist, The University of Nebraska, Lin-

Recommended Fertilizer Practices[1] (in lb per acre N, P_2O_5, and K_2O; from left to right, nitrogen, phosphoric acid, and potash, respectively)		Time of Use (or length of grazing season)		Carrying Capacity (animal units/acre)	Comments
Initial Application at Seeding Time	Annual Maintenance Application After Pasture Is Established	When to Start	When to End		
20-40-0	50-30-0 (dry land)	May 10-15	Oct. 1-15	2 to 3, or higher with added fertilizer	Stop pasturing when down to about 4 in. If fertilizer is not added annually, add 100 lb P_2O_5 every 3 years. Seed in late summer (Aug. 1 to Sept. 15) in Neb.
20-40-0	50-30-0	May 10-15	Oct. 1-15	2 to 3	Stop pasturing when down to about 4 in.
20-40-0	50-30-0	May 10-15	Oct. 1-15	2 to 3	This mixture not hardy in the northern part of the area. Stop pasturing when down to about 4 in. If fertilizer is not added annually, add 100 lb P_2O_5 every 3 years.
20-40-0	50-30-0	May 10-15	Oct. 1-15	1 to 2	Meadow foxtail and/or alsike clover may be added if desired. Ladino clover not hardy in northern part of area.
20-40-0	50-30-0	When sod permits	Oct. 1-15	1 to 2	Seed in early spring or late summer in Neb.
20-40-0	50-30-0	May 10-15	Oct. 1-15	1 to 1½	Seed in late summer in Neb.

[1]These recommendations serve only as guides. Use soil tests to determine accurately lime and fertilizer needs.

TABLE 5-5D—GENERAL RECOMMENDATIONS FOR TEMPORARY OR SUPPLEMENTAL PASTURE

(This table contains the authoritative recommendations, made especially for this Agronomist, USDA and Montana State University, Bozeman, Mont.; and Dr. W. J

Soil Type and Moisture Condition	Recommended Seedings (mixtures or pure seedings); Lb of Each and Total Lb/Acre		Seeding Recommendations	
			How to Seed	When to Seed
I. Irrigated Lands	Winter rye	60	Drill	Sept. 1-10
	Winter wheat	60	Drill	Sept. 1-10
	Spring rye	80	Drill	Early spring
	Oats Peas	60 120 180	Drill	Early spring
	Sudangrass	25	Drill	Early summer
II. Nonirrigated (dryland)	Winter rye	50-70	Drill	Sept. 1-10
	Winter wheat	50	Drill	Sept. 10
	Oats	60	Drill	Early spring
	Sudangrass	20-30	Drill	May 20 to June 1
	Spring rye	60	Drill	Early spring

TABLE 5-6A—MOST ABUNDANT GRASSES AND LEGUMES COMPRISING THE FORAGE IN PERMANENT

(This table contains the authoritative listing, made especially for this book, of the following State University, Stillwater, Okla.; and Mr. E. H. McIlvain, Agronomist, USDA, Southern

Grasses—Bermudagrass Bluestem, big Bluestem, little Bluestem, plains Bluestem, sand Bluestem, yellow Buffalograss	Buffelgrass Dropseed, sand Fescue, tall Galleta Grama grass, blue Grama grass, side-oats Indiangrass	Johnsongrass Junegrass Kleingrass Lovegrass, sand Lovegrass, weeping Mesquite, curly Panicgrass, blue	Paspalum, sand Sudangrass Switchgrass Tobosa Wheatgrass Wild-rye, Russian Wintergrass, Texas

OR CATTLE, SHEEP, AND HORSES IN AREA 5, NORTHERN GREAT PLAINS AREA

ook, of the following competent specialists residing in the area: Dr. Clee S. Cooper, Ioline, Agronomist, The University of Nebraska, Lincoln, Neb.)

Recommended Fertilizer Practices[1] (in lb per acre N, P_2O_5, and K_2O; from left to right, nitrogen, phosphoric acid, and potash, respectively). Initial Application at Seeding Time	Time of Use (or length of grazing season)		Carrying Capacity (animal units/acre)	Comments
	When to Start	When to End		
60-0-0	Mar.	Apr.	¾ to 1½	May be grazed to June if grain crop is not harvested. Some winter grazing may be obtained in years when snow cover is not too heavy.
60-0-0	Mar.	Apr.	¾ to 1½	May be grazed to June if grain crop is not harvested. Some winter grazing may be obtained in years when snow cover is not heavy.
60-0-0	May	June	¾ to 1½	
None	May	June	1 to 2	
60-0-0	July	Sept.	1 to 2	
None	Mar.	Apr.	¼ to ¾	May be grazed to June if grain crop is not harvested. Some winter grazing may be obtained in years when snow cover is not too heavy.
None	Mar.	Apr.	¼ to ¾	May be grazed to June if grain crop is not harvested. Some winter grazing may be obtained in years when snow cover is not too heavy. Remove animals at first joint stage.
None	May 25	July 15	½	Depends upon adequate rainfall.
None	June 25-30	Sept. 15	½ to 2	Dependent upon timely summer rains for good start.
None	May	June	¼ to ¾	

[1]These recommendations serve only as guides. Use soil tests to determine accurately lime and fertilizer needs.

AND/OR ROTATIONAL PASTURES IN AREA 6, SOUTHERN GREAT PLAINS AREA

competent specialists residing in the area: Dr. C. M. Taliaferro, Agronomist, Oklahoma Great Plains Field Station, Woodward, Okla.)

Legumes—Alfalfa	Clover, Ladino	Sweet clover	Vetch, hairy

TABLE 5-6B—GENERAL RECOMMENDATIONS FOR PERMANENT PASTURES FOR CATTLE

(This table contains the authoritative recommendations, made especially fo
M. Taliaferro, Agronomist, Oklahoma State University, Stillwater, Okla.; an
Woodward, Okla.)

Soil Type and Moisture Condition	Recommended Seedings (mixtures or pure seedings); Lb of Each and Total Lb/Acre	Seeding Recommendations	
		How to Seed	When to Seed
I. Clay or Silty Soils (15-25 in. rainfall)	Blue grama grass ... 6-8 Side-oats grama grass ... 2-3 Buffalograss ... 1-2 9-13	Drill	Mar. and Apr.
	Blue grama grass ... 6-8 Side-oats grama grass ... 2-3 Buffalograss ... 1-2 Western wheatgrass ... 6-8 15-21	Drill	Mar. and Apr.
	Native bluestem mix ... 15 (big bluestem, little bluestem, switchgrass, Indiangrass, blue grama, side-oats grama)	Drill	Mar. and Apr.
	Buffalograss ... 2-5	Drill	Mar.
II. Sandy Soils (15-25 in. rainfall)	Blue grama grass ... 6-8 Side-oats grama grass ... 2-3 Sand lovegrass ... 1 Switchgrass ... 1-2 10-14	Drill	Mar. and Apr.
	Sand bluestem ... 6 Sand lovegrass ... 1½ Switchgrass ... 1-2 Little bluestem ... 4 13	Drill	Mar. and Apr.
	Weeping lovegrass ... 2-4	Drill	May
III. Sandy or Sandy Loam Soils (above 25 in. rainfall)	Bermudagrass ... 15-20 bu. (sprigs)	Sprig in Drill vetch	Spring Fall

SHEEP, AND HORSES IN AREA 6, SOUTHERN GREAT PLAINS AREA

this book, of the following competent specialists residing in the area: Dr. C.
Mr. E. H. McIlvain, Agronomist, USDA, Southern Great Plains Field Station,

Recommended Fertilizer Practices[1] (in lb per acre N, P_2O_5, and K_2O; from left to right, nitrogen, phosphoric acid, and potash, respectively)		Time of Use (or length of grazing season)		Carrying Capacity (animal units/acre)	Comments
Initial Application at Seeding Time	Annual Maintenance Application After Pasture Is Established	When to Start	When to End		
None	None	Apr. to May	Oct. to Nov.	1/20 to 1/10	Do not overgraze. Good winter pasture if rested during summer. Generally seeded in stubble mulch of Sudangrass, cane, or grain sorghum.
None	None	Oct.	May	1/20 to 1/10	Where western wheatgrass is used, seed early. Defer during summer, graze during winter.
None	None	May 1-10	Sept. 30	1/5	The exact proportions of grasses seeded in the mixture are unimportant, but the locally dominant ones should predominate in any mixtures seeded for permanent native pastures.
	60-0-0, side dressing annually	Apr. 15	Oct. 1	1/3	Should not be grazed until after first of year.
None	None	Apr. to May	Oct. to Nov.	1/20 to 1/10	Do not overgraze. Good winter pasture if rested during summer. Generally seeded in stubble mulch of Sudangrass, cane, or grain sorghum.
None	None	Apr.	Oct.	1/15	
0-40-0 to 30-40-0; or 0-50-0 to 60-50-0	60-20-0 (split N in 2 applications)	Apr. 1	Nov. 1	1/5 to 1/2	Good winter pasture. Use soil tests to determine accurately fertilizer needs.
0-40-0 to 30-40-0; or 0-50-0 to 60-50-0	60-40-0 (split N in 2 applications)	Mar. 15	Nov. 1	1/5 to 1/2	Overseed legume after Bermudagrass is established. Graze lightly during pod-setting to ensure reseeding. Adapted to eastern side of area. Use soil tests to determine accurately lime and fertilizer needs.

[1]These recommendations serve only as guides. Use soil tests to determine accurately lime and fertilizer needs.

TABLE 5-6C—GENERAL RECOMMENDATIONS FOR ROTATIONAL PASTURES FOR CATTLE

(This table contains the authoritative recommendations, made especially fo M. Taliaferro, Agronomist, Oklahoma State University, Stillwater, Okla.; and Woodward, Okla.)

Soil Type and Moisture Condition	Recommended Seedings (mixtures or pure seedings); Lb of Each and Total Lb/Acre	Seeding Recommendations	
		How to Seed	When to Seed
I. Irrigated Lands	Smooth, bromegrass .... 12 Alfalfa 3 15	Drill	Sept. 1
	Bermudagrass, adapted variety (sprigs) .... 15-20 bu. Ladino clover 3	Sprig in Drill	Early summer or fall
	Bermudagrass (sprigs) .... 15-20 bu. Vetch .... 30 Small grain 90	Sprig in Drill	Early summer or fall

TABLE 5-6D—GENERAL RECOMMENDATIONS FOR TEMPORARY OR SUPPLEMENTAL PASTURES

(This table contains the authoritative recommendations, made especially for this Taliaferro, Agronomist, Oklahoma State University, Stillwater, Okla.; and Mr. E. H.

Soil Type and Moisture Condition	Recommended Seedings (mixtures or pure seedings); Lb of Each and Total Lb/Acre	Seeding Recommendations	
		How to Seed	When to Seed
I. Dry Land	Sudangrass 15	Drill	Apr. to July 1
	Johnsongrass .... 15-20 Sweet clover 8-12 23-32	Drill Drill	Apr. to May Fall or Spring
II. Sandy to Loamy Soils	Wheat, rye, barley, or oats .... 30-40 Vetch 10-12 40-52	Drill	Sept. to Oct.
III. Silty Soils	Wheat, barley, or oats 40-60	Drill	Sept. to Oct.
	Elbon or Bonel rye 40-60	Drill	Sept. to Oct.
IV. Irrigated Lands	Wheat, rye, barley, or oats 60	Drill	Sept. 1
	Sudangrass 30	Drill	Apr. 15 to July
	Wheat, rye, barley, or oats .... 40-50 Vetch 10-12 50-62	Drill	Sept. to Oct.
	Elbon or Bonel rye 40-60	Drill	Sept. to Oct.

HEEP, AND HORSES IN AREA 6, SOUTHERN GREAT PLAINS AREA

his book, of the following competent specialists residing in the area: Dr. C.
r. E. H. McIlvain, Agronomist, USDA, Southern Great Plains Field Station,

Recommended Fertilizer Practices[1] (in lb per acre N, P_2O_5, and K_2O; from left to right, nitrogen, phosphoric acid, and potash, respectively)		Time of Use (or length of grazing season)		Carrying Capacity (animal units/acre)	Comments
Initial Application at Seeding Time	Annual Maintenance Application After Pasture Is Established	When to Start	When to End		
60-60-30	60-60-30	Mar.	Dec.	1	Shorter season of use in northern part of region.
200-80-20	200-60-20	Mar.	Sept.	3 to 5	Use adapted variety of Bermudagrass. May be overseeded with small grain and vetch in the fall. More nitrogen may be added, depending on the soil and desired carrying capacity.
200-80-20	200-60-20	Mar.	Sept.	3 to 5	More nitrogen may be added, depending on the soil and desired carrying capacity.

[1]These recommendations serve only as guides. Use soil tests to determine accurately lime and fertilizer needs.

FOR CATTLE, SHEEP, AND HORSES IN AREA 6, SOUTHERN GREAT PLAINS AREA

book, of the following competent specialists residing in the area: Dr. C. M.
McIlvain, Agronomist, USDA, Southern Great Plains Field Station, Woodward, Okla.)

Recommended Fertilizer Practices[1] (in lb per acre N, P_2O_5, and K_2O; from left to right, nitrogen, phosphoric acid, and potash, respectively)		Time of Use (or length of grazing season)		Carrying Capacity (animal units/acre)	Comments
Initial Application at Seeding Time	Annual Maintenance Application After Pasture Is Established	When to Start	When to End		
		June to July	Nov.	1/2 to 2	In western part of area, Sudan is frequently seeded in rows, at the rate of 10 lb/acre. Danger of prussic acid poisoning if drought or frost damaged. In eastern section, 60-30-0 fertilizer may increase carrying capacity to 2 animal units/acre.
0-40-0		Apr. to June	Sept.	1/2	
0-40-0 to 0-50-0, if shown to be needed	0-40-0	Nov.	June	1/5 to 2	
		Nov.	June	1/5 to 1/3	Stop grazing in Mar. to Apr. for grain crop.
		Nov.	June	1/5 to 1/3	Stop grazing in Mar. to Apr. for grain crop.
30-30-0	30-0-0, side dressing	Oct. 15	Apr.	1/3 to 1	Stop grazing in Mar. to Apr. if grain crop is desired.
60-30-0		June 1	Oct. 1	1/2 to 2	If drilled in 40- to 42-in. rows, seed 10 lb/acre.
0-40-0 to 0-50-0, if needed		Nov.	June	1/2 to 2	
		July Nov.	June	1/2 to 2	

[1]These recommendations serve only as guides. Use soil tests to determine accurately lime and fertilizer needs.

TABLE 5-7A—MOST ABUNDANT GRASSES AND LEGUMES COMPRISING THE FORAGE IN

(This table contains the authoritative listing made especially for this book, of Agronomist, Colorado State University, Fort Collins, Colo.; Dr. John L Pullman, Wash.; and Dr. Chester L. Canode, Agronomist, USDA, Pullman,

Grasses—Bluegrass, big Bluegrass, Kentucky Bromegrass Canarygrass, reed	Fescue, tall Meadow foxtail Orchardgrass Redtop	Sudangrass Timothy Wheatgrass, beardless Wheatgrass, crested	Wheatgrass, intermediate and pubescent Wheatgrass, Siberian Wheatgrass, tall Wild-rye grass, Russian

TABLE 5-7B—GENERAL RECOMMENDATIONS FOR PERMANENT PASTURES FOR CATTLE,

(This table contains the authoritative recommendations, made especially for D. Dotzenko, Agronomist, Colorado State University, Fort Collins, Colo.; Dr. Pullman, Wash.; and Dr. Chester L. Canode, Agronomist, USDA, Pullman,

Soil Type and Moisture Condition	Recommended Seedings (mixtures or pure seedings); Lb of Each and Total Lb/Acre	Seeding Recommendations	
		How to Seed[1]	When to Seed
I. Nonirrigated, Well-Drained Soils 1. Under 12-in. rainfall	Any one of the following 3 seeded alone: Crested wheatgrass, or Russian wild-ryegrass, or Beardless wheatgrass 6-8	Drill	Fall, except for Russian wild-rye grass, which should be spring seeded.
2. 12- to 15-in. rainfall	Big bluegrass, or Crested wheatgrass .... 6-8 Alfalfa 5 11-13	Drill	Spring
3. 15- to 18-in. rainfall	Intermediate or pubescent wheatgrass .... 6 Alfalfa 5 11	Drill	Spring
4. Over 18-in. rainfall	Intermediate wheatgrass, or .... 6 Orchardgrass, or .... 4 Smooth bromegrass, and .... 6 Alfalfa 5 9-11	Drill	Spring
II. Nonirrigated, Wet, Poorly Drained Soils; Nonsaline	Meadow foxtail, or .... 5 Fawn fescue, and .... 4 Birdsfoot trefoil or Alsike clover 2 6-7	Drill	Spring
	Reed canarygrass .... 12 Birdsfoot trefoil 2 14	Drill	Spring or fall
III. Nonirrigated, Wet, Poorly Drained Soils; Saline	Fawn fescue .... 8 Strawberry clover 3 11	Drill	Spring
IV. Nonirrigated, Dry, Saline	Tall wheatgrass (Alkar) 8-10	Drill	Spring

ERMANENT AND/OR ROTATIONAL PASTURES IN AREA 7, NORTHWEST INTERMOUNTAIN AREA

he following competent specialists residing in the area: Dr. A. D. Dotzenko, Schwendiman, Agronomist, USDA and Washington State University, Wash.)

Legumes—Alfalfa Clover, alsike	Clover, Ladino Clover, medium red	Clover, strawberry Clover, white	Sweet clover Trefoil, birdsfoot

SHEEP, AND HORSES IN AREA 7, NORTHWEST INTERMOUNTAIN AREA

this book, of the following competent specialists residing in the area: Dr. A. John L. Schwendiman, Agronomist, USDA and Washington State University, Wash.)

Recommended Fertilizer Practices[2] (in lb per acre N, P_2O_5, and K_2O; from left to right, nitrogen, phosphoric acid, and potash, respectively)		Time of Use (or length of grazing season)		Carrying Capacity (animal units/acre)	Comments
Initial Application at Seeding Time	Annual Maintenance Application in lb/acre After Pasture Is Established[3]	When to Start	When to End		
0-0-0 to 20-0-0	None	Apr. 1 to May 1	June 15 to June 30	1/10 to 1/4	Crested wheatgrass most easily established of all grasses in low rainfall areas. Some fall grazing (Sept. 15 to Oct. 15) if rainfall adequate.
0-0-0 to 20-0-0	None	Apr. 1 to May 1	June 15 to June 30	1/7 to 1/3	Crested wheatgrass better adapted to more shallow soils. Some fall grazing (Sept. 15 to Oct. 15) if rainfall adequate. Apply 100 lb gypsum each second year where sulfur is a limiting nutrient for alfalfa.
0-0-0 to 20-0-0	None	Apr. 1 to May 1	June 15 to June 30	1/7 to 1/3	Some fall grazing (Sept. 15 to Oct. 15) if rainfall adequate. Apply 100 lb gypsum each second year where sulfur is a limiting nutrient for alfalfa.
0-0-0 to 20-0-0	None	May 1 to May 15	July 15	1 1/2	Some fall grazing (Sept. 15 to Oct. 15) if rainfall adequate. Apply 100 lb gypsum each second year where sulfur is a limiting nutrient for alfalfa. Molybdenum may be needed or a fertilizer for proper alfalfa growth and nodulation.
20-20-20 or 20-20-0, depending on soil test	None	May	Oct.	3/4 to 1 1/4	
0-100-0 to 0-200-0	30-40-0 to 70-60-0	May	Oct.	3/4 to 1 1/4	
20-20-20 or 20-20-0, depending on soil test	None	May	Oct.	3/4 to 1 1/4	
20-20-20 or 20-20-0, depending on soil test	None	May	Oct.	3/4 to 1 1/2	

[1]Broadcast only when the area is too rough or wet for the proper operation of a drill.
[2]These recommendations serve only as guides. Use soil tests to determine accurately lime and fertilizer needs.
[3]After pasture is established, an annual application of nitrogen is recommended if the legume is absent or constitutes a very small part of the pasture mixture. Under such circumstances, the annual application of nitrogen in pounds per acre should be: In 12- to 18-inch rainfall area, 20 to 30 lb; over 18-inch rainfall area, 40 to 60 lb; and irrigated areas, 80 to 120 lb.

TABLE 5-7C—GENERAL RECOMMENDATIONS FOR ROTATIONAL PASTURES FOR CATTLE

(This table contains the authoritative recommendations, made especially fo D. Dotzenko, Agronomist, Colorado State University, Fort Collins, Colo.; Dr Wash.; and Dr. Chester L. Canode, Agronomist, USDA, Pullman, Wash.)

Soil Type and Moisture Condition	Recommended Seedings (mixtures or pure seedings); Lb of Each and Total Lb/Acre		Seeding Recommendations	
			How to Seed	When to Seed
I. Irrigated, Well-Drained Soils 1. Adequate water season long	Smooth bromegrass, southern strain, or Orchardgrass, late variety, or Tall fescue, and Ladino clover, or Alfalfa	6 4 6 2 3 6-9	Drill	Spring
2. Short water	Intermediate wheatgrass, or Smooth bromegrass, southern strain, or Tall fescue, and Alfalfa	8 6 6 5 11-13	Drill	Spring
II. Irrigated, Poorly Drained Soils; Nonsaline	Meadow foxtail, or Reed canarygrass, or Tall fescue, and Birdsfoot trefoil, or Alsike clover	5 6 4 2 6-7	Drill	Spring
III. Irrigated, Poorly Drained Soils; Saline	Tall wheatgrass, or Tall fescue, and Strawberry clover	8-10 6 3 9-13	Drill	Spring

HEEP, AND HORSES IN AREA 7, NORTHWEST INTERMOUNTAIN AREA

his book, of the following competent specialists residing in the area: Dr. A.
ohn L. Schwendiman, USDA and Washington State University, Pullman,

Recommended Fertilizer Practices[1] (in lb per acre N, P_2O_5, and K_2O; from left to right, nitrogen, phosphoric acid, and potash, respectively)		Time of Use (or length of grazing season)		Carrying Capacity (animal units/acre)	Comments
Initial Application at Seeding Time	Annual Maintenance Application After Pasture Is Established[2]	When to Start	When to End		
0-20-20 or 20-20-0, according to soil test	40-40-0 to 60-60-0; or 40-40-40 to 60-60-60, according to soil test	May	Oct.	2 to 2 1/2	Heavier fertilizing and rotational grazing may increase capacity to as high as 3 animal units/acre.
0-20-20 or 20-20-0, according to soil test	40-40-40 or 40-40-0, according to soil test	May	Oct.	2 to 2 1/2	Heavier fertilizing and rotational grazing may increase capacity to as high as 3 animal units/acre.
0-20-20 or 20-20-0, according to soil test		May	Oct.	3/4 to 1 1/4	May be seeded with companion crop in spring or following small grain harvest. Carrying capacity may increase with irrigation and fertilization.
0-20-20 or 20-20-0, according to soil test		May	Oct.	3/4 to 1 1/4	

[1]These recommendations serve only as guides. Use soil tests to determine accurately lime and fertilizer needs.

[2]After pasture is established, an annual application of nitrogen is recommended if the legume is absent or constitutes a very small part of the pasture mixture. Under such ircumstances, the annual application of nitrogen in pounds per acre should be: In 12- to 18-inch rainfall area, 20 to 30 lb; over 18-inch rainfall area, 40 to 60 lb; and irrigated reas, 80 to 120 lb.

TABLE 5-7D—GENERAL RECOMMENDATIONS FOR TEMPORARY OR SUPPLEMENTAL PASTURES FOR CATTL

(This table contains the authoritative recommendations, made especially for this book, of the fc University, Fort Collins, Colo.; Dr. John L. Schwendiman, Agronomist, USDA and Washingtc Wash.)

Soil Type and Moisture Condition	Recommended Seedings (mixtures or pure seedings); Lb of Each and Total Lb/Acre	Seeding Recommendations	
		How to Seed	When to Seed
I. Irrigated Lands	Rye (Rosen or Abruzzi) 75-100	Drill	Sept.
	Wheat 75-100	Drill	Sept.
	Oats 75-100	Drill	Mar.
	Sudangrass 30-40	Drill	June
II. Wet, Poorly Drained Soils	Rye 75-100	Drill	Sept.
	Wheat 75-100	Drill	Sept.
	Oats 75-100	Drill	Mar.
	Sudangrass 30-40	Drill	June

TABLE 5-8A—MOST ABUNDANT GRASSES AND LEGUMES COMPRISING THE FORAGE

(This table contains the authoritative listing, made especially for this book, Agronomist, University of Arizona, Tucson, Ariz.; Dr. E. H. Jensen, A ronomist, New Mexico State University, Las Cruces, N.M.)

Grasses seeded in irrigated pastures or dryland farms:			
Bermudagrass Bluegrass, Kentucky Bromegrass Dallisgrass	Fescue, tall Lovegrass, weeping Orchardgrass Panicgrass, blue	Ryegrass, annual Ryegrass, perennial Sudangrass	Wheatgrass, crested Wheatgrass, intermediate Wheatgrass, tall or pubescent
Grasses seeded or native in dryland ranges:			
Beardgrass Bluestem Cottontop Curly mesquite Dropseed	Galleta Grama grass Indian ricegrass Junegrass	Lovegrass Muhly Panicgrass Plains bristlegrass	Sacatons, alkali Three-awns Vine-mesquitegrass Wheatgrass, slender or Western

HEEP, AND HORSES IN AREA 7, NORTHWEST INTERMOUNTAIN AREA

owing competent specialists residing in the area: Dr. A. D. Dotzenko, Agronomist, Colorado State tate University, Pullman, Wash.; and Dr. Chester L. Canode, Agronomist, USDA, Pullman,

Recommended Fertilizer Practices[1] (in lb per acre N, P_2O_5, and K_2O; from left to right, nitrogen, phosphoric acid, and potash, respectively). Initial Application at Seeding Time	Time of Use (or length of grazing season)		Carrying Capacity (animal units/ acre)	Comments
	When to Start	When to End		
20-0-0 to 40-0-0	Dec.	Apr.	1/3 to 1/2	Will carry 2 to 3 animal units during lush growth of Apr. to early May.
20-0-0 to 40-0-0	Dec.	Apr.	1/2 to 3/4	May be pastured until fully utilized, June 15 to June 30.
20-0-0 to 40-0-0	Apr. 15 to 30	June 15 to 30	1/3	
0-20-0	July	Sept. 20	2 to 3	High carrying capacity where summers are warm. Responds to nitrogen, but this increases danger of prussic acid poisoning. Not safe to pasture Sudan after frost due to danger of prussic acid poisoning.
20-0-0 to 40-0-0	Dec.	Apr.	1/3 to 1/2	Will carry 2 to 3 animal units during lush growth of Apr. to early May.
20-0-0 to 40-0-0	Dec.	Apr.	1/2 to 3/4	May be pastured until fully utilized, June 15 to June 30.
20-0-0 to 40-0-0	Apr. 15 to 30	June 15 to 30	1/3	
0-20-0			2 to 3	High carrying capacity where summers are warm. Responds to nitrogen, but this increases danger of prussic acid poisoning. Not safe to pasture Sudan after frost due to danger of prussic acid poisoning.

[1]These recommendations serve only as guides. Use soil tests to determine accurately lime and fertilizer needs.

PERMANENT AND/OR ROTATIONAL PASTURES IN AREA 8, SOUTHWEST AREA

the following competent specialists residing in the area: Dr. Robert Dennis, ronomist, University of Nevada, Reno, Nev.; and Dr. Rex D. Pieper, Ag-

Legumes seeded in irrigated pastures or dryland farms:			
Alfalfa Black medic Clover, alsike	Clover, Hubam Clover, Ladino Clover, medium red	Clover, strawberry Clover, white Sweet clovers, biennial	Trefoil, birdsfoot Vetch
Forage plants seeded or native in dryland ranges:			
Alfilaria Buckwheat, brush California coffeeberry	Chamiza False mesquite Indianwheat	Lotus spp. Mormon tea Pea bush	Prairie clover Range ratany Winterfat

TABLE 5-8B—GENERAL RECOMMENDATIONS FOR ROTATIONAL PASTURES FOR CATTLE

(This table contains the authoritative recommendations, made especially fo
Robert Dennis, Agronomist, University of Arizona, Tucson, Ariz.; Dr. E. H
Pieper, Agronomist, New Mexico State University, Las Cruces, N.M.)

Soil Type and Moisture Condition	Recommended Seedings (mixtures or pure seedings); Lb of Each and Total Lb/Acre		Seeding Recommendations	
			How to Seed	When to Seed
I. Irrigated Lands 1. Up to 3,000 ft altitude Warm season	Bermudagrass, NK 37, coastal, or Alicia	 3-5 or 1-2 bales sprigs	Drill Broadcast	Mar. to June Oct. to May
Cool season	Overplant Bermudagrass with oats, wheat or rye, or Ryegrass, or Bur-clover, or Black medic, or Ladino clover	100 25 15 4 2 2-100	Drill	Sept. 15 to Oct. 15
	Tall fescue, Goar, alta, or fawn Alfalfa	12 5 17	Drill	Oct. 15 to Nov. 15 Feb. 15 to Apr. 15
2. 3,000 to 5,500 ft altitude	Orchardgrass, Latar or Potomac, or Tall fescue, alta, fawn, or Goar, or Smooth bromegrass, Lincoln or Manchar, or Meadow bromegrass, Regar, with Alfalfa or Ladino clover	8 12 15 15 3 2 10-18	Drill	Aug. 15 to Oct. 1 or Mar. 15 to May 15
3. 5,500 ft and up	Smooth bromegrass, Lincoln or Manchar, or Meadow bromegrass, Regar, or Tall fescue, alta or fawn, or Orchardgrass, Potomac or Latar, with Alfalfa, or White Dutch clover, or Ladino clover	15 15 8 3 2 2 10-18	Drill	May 1 to June 15 or Aug. 15 to Sept. 1
II. Irrigated Lands With High Water Table	Meadow foxtail, Garrison, or Reed canarygrass, loreed, with Strawberry clover, Salina, or Alsike clover	5 10 3 3 8-13	Drill	Apr. 15 to June 15
III. Irrigated Alkaline or Saline Soils 1. Up to 3,000 ft altitude Warm season	Bermudagrass, NK 37, or coastal	3-5 or 1-2 bales	Drill Drill	Mar. to June Oct. to May
Cool season	Tall wheatgrass, Jose, or Alkar, with narrowleaf birdsfoot trefoil, or Sweet clover, Madrid	10-15 5 5 15-20	Drill Drill Drill	Oct. 15 to Nov. 15 or Feb. 15 to Apr. 15
	Tall fescue, Goar, alta, or fawn, with Narrowleaf birdsfoot trefoil	12 5 17	Drill Drill	Oct. 15 to Nov. 15 Feb. 15 to Apr. 15
2. 3,000 ft and up	Tall wheatgrass, Alkar, or Tall fescue, alta, fawn, or Goar, with Strawberry clover, Salina, or Narrowleaf birdsfoot trefoil	10-15 12 3 5 13-20	Drill Drill Drill Drill	Mar. 15 to June 15 or Aug. 15 to Oct. 1

SHEEP, AND HORSES IN AREA 8, SOUTHWEST AREA

his book, of the following competent specialists residing in the area: Dr. Jensen, Agronomist, University of Nevada, Reno, Nev.; and Dr. Rex D.

Recommended Fertilizer Practices[1] (in lb per acre N, P_2O_5, and K_2O; from left to right, nitrogen, phosphoric acid, and potash, respectively)		Time of Use (or length of grazing season)		Carrying Capacity (animal units/acre)	Comments
Initial Application at Seeding Time	Annual Maintenance Application After Pasture Is Established	When to Start	When to End		
50-0-0	50-0-0 applied in early spring and after each harvest	May	Oct.	2 to 3	Other improved varieties of seed-producing Bermudagrass are also recommended.
50-0-0	50-0-0 applied after each harvest	Nov. 15 to Dec. 15	Continue grazing Bermudagrass throughout summer.	⅓ to ⅔	Late seedings will reduce forage yields. Inoculate legumes with proper bacteria.
50-0-0	None to 50-0-0 applied in fall and after each harvest if alfalfa stand is poor	Jan. 1 to Feb. 15	Dec. 1 to Dec. 31	1 to 2	Grass fields will be low in summer. Use adapted variety of alfalfa.
50-0-0	None to 50-0-0 applied in spring and after each harvest if legume stand is poor	Apr. 15 to May 15	Sept. 15 to Oct. 15	1 to 2	Tall fescue will give most carrying capacity but lowest daily gains. Use adapted variety of alfalfa. Narrowleaf birdsfoot trefoil may be substituted for Ladino.
50-0-0	None to 50-0-0 applied in spring and after each harvest if legume stand is poor	Apr. 15 to May 15	Sept. 1 to Oct. 1	1 to 2	Spring seeding should be made at high elevations or in northern portion of region. Use adapted variety of alfalfa. Plant white clover instead of Ladino in cold areas.
50-0-0	None to 50-0-0 application in spring and after each harvest if legume stand is poor	Apr. 1 to May 15	Sept. 1 to Oct. 15	1 to 2	Time of seeding, length of grazing season, and carrying capacity will vary with altitude and latitude.
50-0-0	50-0-0 applied early in spring and after each harvest	May	Oct.	1 to 2	Other improved varieties of seed-producing Bermudagrass are also recommended.
50-0-0	None to 50-0-0 applied in fall and after each harvest if legume stand is poor	Jan. 1 to Feb. 15	Dec. 1 to Dec. 31	¾ to 1½	Grass yields will be low during summer.
50-0-0	None to 50-0-0 applied in fall and after each harvest if legume stand is poor	Jan. 1 to Feb. 15	Dec. 1 to Dec. 31	¾ to 1½	Forage yields will be low during summer. Inoculate trefoil with proper bacteria. Seed this mixture on moderately saline-sodic soils only.
50-0-0	None to 50-0-0 applied in spring and after each harvest if legume stand is poor	Apr. 15 to May 15	Sept. 15 to Oct. 15	¾ to 1½	Date of seeding and length of grazing period depends upon latitude and altitude. On severe sites substitute Madrid sweet clover (5 lb/acre) for birdsfoot trefoil. For cold areas spring seeding is recommended.

[1]These recommendations serve only as guides. Use soil tests to determine accurately lime and fertilizer needs.

TABLE 5-8C—GENERAL RECOMMENDATIONS FOR TEMPORARY OR SUPPLEMENTAL PASTURE

(This table contains the authoritative recommendations, made especially for thi Agronomist, University of Arizona, Tucson, Ariz.; Dr. E. H. Jensen, Agronomist State University, Las Cruces, N.M.)

Soil Type and Moisture Condition	Recommended Seedings (mixtures or pure seedings); Lb of Each and Total Lb/Acre	Seeding Recommendations	
		How to Seed	When to Seed
I. Irrigated Lands (southern portion of area only)	Sudangrass 25	Drill	Mar. 15 to June 1
	Sorghums 4-6	Drill	June 1
	Sorgos 4-6	Drill	June 1
	Barley 100	Drill	Sept. 15 to Oct. 15
	Oats 100	Drill	Sept. 15 to Oct. 15
	Rye 100	Drill	Sept. 15 to Oct. 15
	Ryegrass .... 8-10 Bur-clover 5 13-15	Drill	Sept. 15
	Hubam clover 10	Drill	Sept. 15
	Med. red clover 4	Drill	Sept. 15
	Sweet clover 4	Drill	Sept. 15

TABLE 5-9A—MOST ABUNDANT GRASSES AND LEGUMES COMPRISING THE FORAGE IN

(This table contains the authoritative listing, made especially for S. McGuire, Agronomist, Oregon State University, Corvallis, Ore.; Dr. W. W. and Dr. Kenneth J. Morrison, Agronomist, Washington State University,

Grasses—Canarygrass, reed Fescue, tall	Meadow foxtail Oatgrass, tall	Orchardgrass Ryegrass (annual and perennial)	Timothy

'OR CATTLE, SHEEP, AND HORSES IN AREA 8, SOUTHWEST AREA

ook, of the following competent specialists residing in the area: Dr. Robert Dennis, Jniversity of Nevada, Reno, Nev.; and Dr. Rex D. Pieper, Agronomist, New Mexico

Recommended Fertilizer Practices[1] (in lb per acre N, P_2O_5, and K_2O; from left to right, nitrogen, phosphoric acid, and potash, respectively)		Time of Use (or length of grazing season)		Carrying Capacity (animal units/ acre)	Comments
Initial Application at Seeding Time	Annual Maintenance Application After Pasture Is Established	When to Start	When to End		
50-0-0	50-0-0 applied after each harvest.	June	Oct.	2 to 3	At 3,000- to 5,000-ft elevations, Sudan is grazed from about June 15 to Sept. 15. Do not graze frozen Sudangrass. Do not graze until plants are 20 in. tall.
35-0-0	35-0-0, applied every 2 mos.	July 1	Nov. 1	1 to 2	Drilled in rows.
35-0-0	35-0-0, applied every 2 mos.	July 1	Nov. 1	1 to 2	Drilled in rows.
30-0-0 to 60-0-0		Oct. 15	Mar. 1	⅓ to ⅔	
30-0-0 to 60-0-0		June 15↔ Oct. 15↔	July 15 June 1	⅓ to ⅔	
30-0-0 to 60-0-0		Apr. 15	June 1	⅓ to ⅔	Use in northern portion of area only.
35-0-0	35-0-0, applied in early spring and every 2 mos.	Jan. 1	June 1	½ to 1	
0-100-0		Oct. 30	June 1	½ to ⅔	Use in southern portion of area only.
0-100-0		Oct. 30	June 1	½ to ⅔	Use only in higher elevations.
0-100-0		Oct. 30	June 1	½ to 1	

[1]These recommendations serve only as guides. Use soil tests to determine accurately lime and fertilizer needs.

PERMANENT AND/OR ROTATIONAL PASTURES IN AREA 9, NORTHWEST COASTAL AREA

this book, of the following competent specialists residing in the area: Dr. W. Heinemann, Animal Scientist, Washington State University, Prosser, Wash.; Pullman, Wash.)

Legumes—Alfalfa Clover, alsike Clover, Ladino	Clover, medium red Clover, subterranean	Clover, white Flat pea	Trefoil, big Trefoil, birdsfoot

TABLE 5-9B—GENERAL RECOMMENDATIONS FOR PERMANENT PASTURES FOR CATTLE

(This table contains the authoritative recommendations, made especially fo
S. McGuire, Agronomist, Oregon State University, Corvallis, Ore.; Dr. W. W
and Dr. Kenneth J. Morrison, Agronomist, Washington State University

Soil Type and Moisture Condition	Recommended Seedings (mixtures or pure seedings); Lb of Each and Total Lb/Acre	Seeding Recommendations	
		How to Seed	When to Seed
I. Uplands	Orchardgrass, and .... 12 White clover, or .... 3 Birdsfoot trefoil 4 15-16	Drill	Spring or fall
	(Droughty Sites) Orchardgrass .... 6 Tall fescue .... 8 Subterranean clover 6 20	Drill	Fall
	Perennial ryegrass .... 15 Subterranean clover 6 21	Drill Broadcast on steep slopes (aerial)	Fall
II. Land Subject to Flooding for Short Periods	Meadow foxtail .... 8 Big trefoil 2 10	Drill	Spring
	Tall fescue, or fawn fescue .... 8 Meadow foxtail .... 6 Big trefoil 2 16	Drill	Spring
III. Land Subject to Flooding for Long Periods	Meadow foxtail .... 8 Big trefoil 2 10	Drill	Spring
	Reed canarygrass .... 8 Big trefoil 2 10	Drill	Spring
IV. Irrigated Lands	Orchardgrass .... 12 Ladino, or New Zealand white clover 2 14	Drill	Spring or fall
	Orchardgrass .... 10 Alfalfa, or .... 5 New Zealand white clover 2 12-15	Drill	Spring or fall

;HEEP, AND HORSES IN AREA 9, NORTHWEST COASTAL AREA

his book, of the following competent specialists residing in the area: Dr. W. leinemann, Animal Scientist, Washington State University, Prosser, Wash.; 'ullman, Wash.)

Recommended Fertilizer Practices[1] (in lb per acre N, P_2O_5, and K_2O; from left to right, nitrogen, phosphoric acid, and potash, respectively)		Time of Use (or length of grazing season)		Carrying Capacity (animal units/ acre)	Comments
Initial Application at Seeding Time	Annual Maintenance Application After Pasture Is Established	When to Start	When to End		
0-60-0	0-60-0	Apr.	Oct.	1	Sulfur, boron, zinc, or potash may be lacking in some areas. Contact the local county extension agent for advice. All permanent pastures may be grazed as early as Apr. in the southern portion of the region.
0-60-0	0-60-0	May Nov.	Oct. Year-round	½	Sulfur, boron, zinc, or potash may be lacking in some areas. Contact the local county extension agent for advice. Use year-round as available and needed.
0-60-0	0-60-0	Nov.	Year-round	½	Use year-round as available and needed.
0-60-0	0-60-0	Apr.	Oct.	2 to 2½	Sulfur, boron, zinc, or potash may be lacking in some areas. Contact the local county extension agent for advice.
0-60-0	0-60-0	Apr.	Oct.	2 to 2½	Sulfur, boron, zinc, or potash may be lacking in some areas. Contact the local county extension agent for advice.
0-60-0	0-60-0	Apr.	Oct.	¾ to 1¼ to 2 to 2½	Sulfur, boron, zinc, or potash may be lacking in some areas. Contact the local county extension agent for advice.
0-60-0	0-60-0	Apr.	Oct.	2 to 2½	Sulfur, boron, zinc, or potash may be lacking in some areas. Contact the local county extension agent for advice.
0-60-0	0-60-0	Apr.	Oct.	2 to 2½	Rotation pasture of this sort is preferred to a permanent pasture. It is usually difficult to preserve the balance of grass and legume necessary for high productivity over 5 or 6 years. Sulfur, boron, zinc, or potash may be lacking in some areas. Contact the local county extension agent for advice.
0-60-0	0-60-0	Apr.	Oct.	2 to 2½	Sulfur, boron, zinc, or potash may be lacking in some areas. Contact the local county extension agent for advice.

[1]These recommendations serve only as guides. Use soil tests to determine accurately lime and fertilizer needs.

TABLE 5-9C—GENERAL RECOMMENDATIONS FOR ROTATIONAL PASTURES FOR CATTL

(This table contains the authoritative recommendations, made especially f S. McGuire, Agronomist, Oregon State University, Corvallis, Ore.; Dr. W. W and Dr. Kenneth J. Morrison, Agronomist, Washington State Universit

Soil Type and Moisture Condition	Recommended Seedings (mixtures or pure seedings); Lb of Each and Total Lb/Acre	Seeding Recommendations	
		How to Seed	When to Seed
I. Moist Bottomland	Perennial ryegrass, or ... 15 Orchardgrass, or ... 12 Tall fescue, and ... 15 Ladino or white clover ... 2 14-17	Drill	Spring
	Orchardgrass, and ... 12 White clover ... 3 15	Drill	Spring
	Orchardgrass, or ... 12 Tall fescue, and ... 15 Birdsfoot trefoil ... 5 17-20	Drill	Spring
	Ryegrass, perennial ... 15 Clover, red or ... 5 Clover, white ... 2 17-20	Drill	Fall
II. Fertile Uplands	Tall fescue, or ... 15 Orchardgrass, and ... 12 White clover ... 3 15-18	Drill	Spring
	Tall fescue, or ... 15 Orchardgrass, and ... 12 Birdsfoot trefoil ... 5 17-20	Drill	Spring
III. Land Subject to Flooding for Short Periods	Meadow foxtail, or ... 8 Tall fescue, and ... 15 Birdsfoot trefoil or ... 5 Big trefoil ... 2 13-20	Drill	Spring
IV. Land Subject to Flooding for Long Periods	Meadow foxtail, or ... 8 Reed canarygrass, and ... 15 Big trefoil ... 5 13-20	Drill	Spring
V. Irrigated Lands	Orchardgrass ... 12 Ladino or New Zealand white clover ... 2 14	Drill	Spring or fall
	Orchardgrass ... 10 Alfalfa, or ... 5 New Zealand white clover ... 2 12-15	Drill	Spring or fall

SHEEP, AND HORSES IN AREA 9, NORTHWEST COASTAL AREA

this book, of the following competent specialists residing in the area: Dr. W. Heinemann, Animal Scientist, Washington State University, Prosser, Wash.; Pullman, Wash.)

Recommended Fertilizer Practices[1] (in lb per acre N, P_2O_5, and K_2O; from left to right, nitrogen, phosphoric acid, and potash, respectively)		Time of Use (or length of grazing season)		Carrying Capacity (animal units/ acre)	Comments
Initial Application at Seeding Time	Annual Maintenance Application After Pasture Is Established	When to Start	When to End		
0-60-0	0-60-0	Apr.	Oct.	2	The Ladino needs adequate moisture to remain productive. All rotation pastures may be grazed as early as Apr. 1 in the southern portion of this region. Sulfur, boron, or potash may be lacking in some areas. Contact the local county extension agent for advice.
0-60-0	0-60-0	Apr.	Oct.	2	Sulfur, boron, or potash may be lacking in some areas. Contact the local county agent for advice. New Zealand white clover has resistance to gray garden slug.
0-60-0	0-60-0	Apr.	Oct.	2	Sulfur, boron, or potash may be lacking in some areas. Contact the local county agent for advice.
0-60-0	0-60-0	Apr.	Oct.	2	This is a short rotation pasture, which will last about 3 years. Excellent yields and good quality.
0-60-0	0-60-0	Apr.	Oct.	½ to ¾	Sulfur, boron, or potash may be lacking in some areas. Contact the local county agent for advice.
0-60-0	0-60-0	Apr.	Oct.	½ to ¾	Sulfur, boron, or potash may be lacking in some areas. Contact the local county agent for advice.
0-60-0	0-60-0	Apr.	Oct.	¾ to 1¼	Spring flooding may reduce time of use. Sulfur, boron, or potash may be lacking in some areas. Contact the local county agent for advice.
0-60-0	0-60-0	Apr.	Oct.	¾ to 1¼	Spring flooding may reduce time of use. Sulfur, boron, or potash may be lacking in some areas. Contact the local county agent for advice.
0-60-0	0-60-0	Apr.	Oct.	2 to 2½	Heavier fertilization and rotation grazing may increase carrying capacity to 3 animal units per acre. Sulfur, boron, or potash may be lacking in some areas. Contact the local county agent for advice.
0-60-0	0-60-0	Apr.	Oct.	2 to 2½	Sulfur, boron, or potash may be lacking in some areas. Contact the local county agent for advice.

[1]These recommendations serve only as guides. Use soil tests to determine accurately lime and fertilizer needs.

TABLE 5-9D—GENERAL RECOMMENDATIONS FOR TEMPORARY OR SUPPLEMENTAL PASTURES FOR CATTLE

(This table contains the authoritative recommendations, made especially for this book, of the fol University, Corvallis, Ore.; Dr. W. W. Heinemann, Animal Scientist, Washington State University Wash.)

Soil Type and Moisture Condition	Recommended Seedings (mixtures or pure seedings); Lb of Each and Total Lb/Acre	Seeding Recommendations	
		How to Seed	When to Seed
I. Moist Bottomland and Uplands	Rye (Rosen or Abruzzi) ... 100	Drill	Sept.
	Rye ... 60 Crimson clover ... 15 75	Drill	Sept.
	Sudangrass ... 20	Drill	May to June
	Ryegrass, annual ... 20	Drill	Sept.
II. Irrigated Lands	Rye (Abruzzi) ... 100	Drill	Aug. to Sept.
	Sudangrass ... 20	Drill	May to June

TABLE 5-10A—MOST ABUNDANT GRASSES AND LEGUMES COMPRISING THE FORAGE IN

(This table contains the authoritative listings, made especially for this book, Clawson, Agronomist, University of California, Davis, Calif.; and Dr. William

Grasses—Bermudagrass Dallisgrass	Fescue, tall Orchardgrass	Ryegrass, annual Ryegrass, perennial	Sudangrass

TABLE 5-10B—GENERAL RECOMMENDATIONS FOR ROTATIONAL PASTURES FOR CATTLE,

(This table contains the authoritative recommendations, made especially for James Clawson, Agronomist, University of California, Davis, Calif.; and Dr.

Soil Type and Moisture Condition	Recommended Seedings (mixtures or pure seedings); Lb of Each and Total Lb/Acre	Seeding Recommendations	
		How to Seed	When to Seed
I. Irrigated Lands 1. For hardpan soils	Strawberry clover, Salina ... 3 Ladino clover ... 2 Ryegrass, annual ... 4 Ryegrass, perennial ... 3 Orchardgrass ... 8 20	Drill or broadcast	Oct. to Nov. or Feb. to Mar.
2. For heavy textured soils	Narrowleaf birdsfoot trefoil ... 3 Strawberry clover, Salina ... 3 Ladino clover ... 2 Ryegrass, annual ... 3 Ryegrass, perennial ... 3 Orchardgrass ... 6 20	Drill or broadcast	Oct. to Nov. or Feb. to Mar.
3. For deep soils	Alfalfa ... 20-25	Drill or broadcast	Oct. to Nov. or Feb. to Mar.
II. Alkaline or Saline Soils	Narrowleaf birdsfoot trefoil ... 5 Ryegrass, annual ... 3 Ryegrass, perennial ... 3 Strawberry clover, Salina ... 3 Tall fescue ... 6 20	Drill or broadcast	Oct. to Nov. or Feb. to Mar.

;HEEP, AND HORSES IN AREA 9, NORTHWEST COASTAL AREA

owing competent specialists residing in the area: Dr. W. S. McGuire, Agronomist, Oregon State
'rosser, Wash.; and Dr. Kenneth J. Morrison, Agronomist, Washington State University, Pullman,

Recommended Fertilizer Practices[1] (in lb per acre N, P_2O_5, and K_2O; from left to right, nitrogen, phosphoric acid, and potash, respectively). Initial Application at Seeding Time	Time of Use (or length of grazing season)		Carrying Capacity (animal units/ acre)	Comments
	When to Start	When to End		
10-0-0 to 60-0-0	Nov.	Apr.	¼ to ⅓	
'0-60-0	Nov.	Apr.	⅓ to ½	Southern half of the region only. Winters too severe for best production of crimson clover in northern half of area 9.
10-0-0 to 60-0-0	July	Oct.	2	Plant when soil is warm. Use annual varieties low in prussic acid.
10-0-0 to 60-0-0	July	Oct.	½	If kept for hay, apply another 60 lb N in early April.
10-0-0 to 60-0-0	Nov.	Apr.	⅓ to ½	
10-0-0 to 60-0-0; apply similar amount after each harvest.	July	Oct.	2	Do not plant until soil is warm.

[1]These recommendations serve only as guides. Use soil tests to determine accurately lime and fertilizer needs.

PERMANENT AND/OR ROTATIONAL PASTURES IN AREA 10, CALIFORNIA COASTAL AREA

of the following competent specialists residing in the area: Dr. W. James
F. Lehman, Agronomist, University of California, Davis, Calif.)

Legumes—Alfalfa Bur-clover, California	Clover, Ladino	Clover, strawberry Clover, subterranean	Trefoil, birdsfoot

SHEEP, AND HORSES IN AREA 10, CALIFORNIA COASTAL AREA

this book, of the following competent specialists residing in the area: Dr. W.
William F. Lehman, Agronomist, University of California, Davis, Calif.)

Recommended Fertilizer Practices[1] (in lb per acre N, P_2O_5, and K_2O; from left to right, nitrogen, phosphoric acid, and potash, respectively). Initial Application at Seeding Time	Time of Use (or length of grazing season)		Carrying Capacity (animal units/ acre)	Comments
	When to Start	When to End		
25-60-0	Mar. 1	Nov. 1	1½	Strawberry clover is less bloat producing than many legumes. Also, it will withstand both alkaline and waterlogged conditions. Adapted to hardpan soils with poor subsoil drainage, generally shallow surface soil, clay or hardpan underneath, medium textured. Nitrogen applied to seeding time. P_2O_5, K_2O, and sulfur applied as needed for specific soils.
25-60-0	Mar. 1	Nov. 1	1¼	Adapted to heavy textured soils with poor surface drainage and subsoil drainage. Deep soils, heavy textured throughout. Nitrogen applied at seeding time. P_2O_5, K_2O, and sulfur applied as needed for specific soils.
25-60-0	Mar. 1	Nov. 1	2	Adapted to deep soils, with good subsoil drainage, generally deep and permeable, medium to light textured. Nitrogen applied at seeding time. P_2O_5, K_2O, and sulfur applied as needed for specific soils.
25-60-0	Mar. 1	Nov. 1	1	Adapted to saline soils with poorly drained subsoil, medium to heavy textured, shallow to deep. Nitrogen applied at seeding time. P_2O_5, K_2O, and sulfur applied as needed for specific soils. Generally speaking, tall fescue is substituted for orchardgrass in areas adapted to cotton.

[1]These recommendations serve only as guides. Use soil tests to determine accurately lime and fertilizer needs.

TABLE 5-10C—GENERAL RECOMMENDATIONS FOR TEMPORARY OR SUPPLEMENTAL PASTURE

(This table contains the authoritative recommendations, made especially for thi
Clawson, Agronomist, University of California, Davis, Calif.; and William F. Lehmar

Soil Type and Moisture Condition	Recommended Seedings (mixtures or pure seedings); Lb of Each and Total Lb/Acre	Seeding Recommendations	
		How to Seed	When to Seed
I. Irrigated Lands	Sudangrass 15-20	Drill	Apr. 15 to July 1
	Ryegrass, annual 20	Drill or broadcast	Sept. to Nov.
	Barley 75-150	Drill	Feb. to Mar.
	Oats 75-150	Drill	Feb. to Mar.

PASTURES FOR SWINE

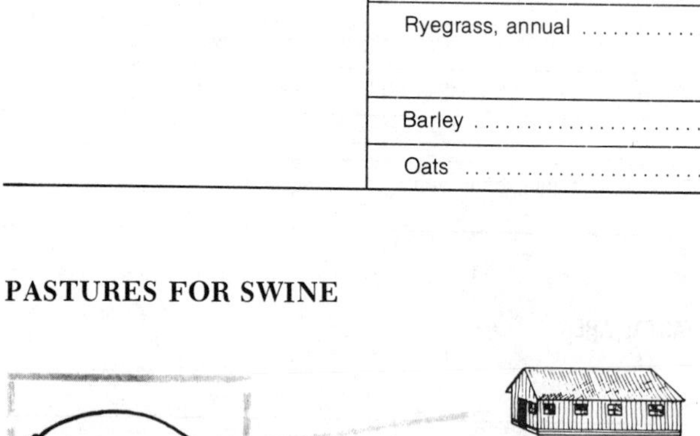

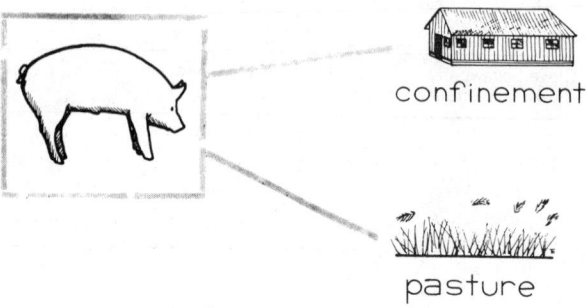

Fig. 5-4. There are now two alternatives for the swine breeding herd—and the wise manager will choose between them.

Prior to 1950, pastures were considered essential for successful swine production. But the importance of pastures for hogs declined with increased knowledge of nutrition and more confinement production.

In recent years, confinement rearing, in which pigs are confined in buildings from birth to market, has increased. But most swine producers follow a program of partial confinement. One survey revealed that 81% of U.S. swine producers confine sows at farrowing time, 65% provide confinement for nurseries, 35% confine growing-finishing pigs, but only 11.3% confine the sow herd.[4] The main reasons given for going to confinement are savings in labor and land. The main problems encountered are high investment in buildings, more disease troubles, rations become more critical, manure disposal and odor control problems are greater, and sow fertility is lowered.

[4]*Hog Farm Management*, Aug. 1973, p. 39.

In season, most producers utilize pasture fo breeding animals. The vast majority of the nation' gestating sows and herd boars are kept in movabl houses, preferably on clean pasture on land that ha been plowed since hogs were last on the area. In an event, pastures for boars and for sows-litters are no obsolete; rather, there now exists two alternatives fo the breeding herd—confinement vs pasture—and th wise manager will choose between them.

Today, only 13.5 percent of U.S. swine feed is derived from roughage (see Table 4-2), including pasture. Yet, often hogs will yield greater returns from a acre of good pasture than any other class of farm animals.

Swine pastures make for a saving in feed costs, i both grain and protein supplements. With properl balanced rations used, the feed saving effecte through the utilization of swine pastures is about a follows:

1. Good pastures will reduce (a) the grain required in producing 100 pounds of pork by 15 to 20 percent, and (b) the protein supplement required in producing 100 pounds of pork by 20 to 50 percent.

2. An acre of good pasture will result in savings of 500 to 1,000 pounds of grain and 300 to 500 pounds of protein supplement.

3. With mature brood sows, good pastures may lower feed costs by 50 percent. With pastures that possess a heavy legume content, sows can be fed 2 pounds less grain and 1/2 pound less protein supplement per head per day during gestation. In fact, sows may get the major portion of their feed from good pastures up to 6 to 8 weeks before farrowing. The condition of the sows is the best guide as to the amount of concentrate feeding necessary.

OR CATTLE, SHEEP, AND HORSES IN AREA 10, CALIFORNIA COASTAL AREA

ook, of the following competent specialists residing in the area: Dr. W. James
gronomist, University of California, Davis, Calif.)

Recommended Fertilizer Practices[1] (in lb per acre N, P_2O_5, and K_2O; from left to right, nitrogen, phosphoric acid, and potash, respectively). Initial Application at Seeding Time	Time of Use (or length of grazing season)		Carrying Capacity (animal units/ acre)	Comments
	When to Start	When to End		
60-0-0 to 100-0-0; additional nitrogen may be applied after each harvest	June	Nov.	3	Sudangrass does best where soil is fertile, weather warm and growing season long. Graze drought- or frost-injured Sudangrass cautiously. Begin grazing when grass is 18 to 24 in. tall. Do not add nitrogen fertilizer where Sudangrass follows a legume or is seeded on dry land.
60-0-0 to 100-0-0; additional nitrogen may be applied after each harvest	Dec.	June	2-3	Excess nitrogen fertilizer can cause nitrogen poisoning.
60-0-0 to 100-0-0	Apr.	June	1 to 2	
60-0-0 to 100-0-0	Apr.	June	1 to 2	

[1]These recommendations serve only as guides. Use soil tests to determine accurately lime and fertilizer needs.

4. When growing-finishing hogs are on high-quality legume pasture, the protein level of the ration can be reduced by 2 percent as compared to confinement feeding. When finishing pigs are on a limited-feeding program, only ½ to ¾ as many animals per acre can be carried as when pigs are full-fed.

5. Good pastures furnish a convenient and economical way to compensate for the protein, mineral, and vitamin deficiencies of grain and other high-energy feeds. This does not infer that protein, mineral, and vitamin supplements need not be added to the ration. Rather, the problem is simplified, and it may be solved at lower cost.

6. Good pastures make for a slight saving in minerals—about 3 pounds of mineral saved per acre of pasture.

Tables 5-11 to 5-14, pages 638 to 652, contain the generally accepted recommendations and facts relative to swine pastures (including crops for hogging down) for each of the 4 major swine production areas of the United States (see Fig. 5-5 for the 4 areas). Logically, the swine pasture recommendations coincide with the 4 well-known swine production areas.

For further and more specific information on swine pasture crops for a particular farm, consult the local county extension agent or vocational agriculture instructor, or write to the state agricultural college.

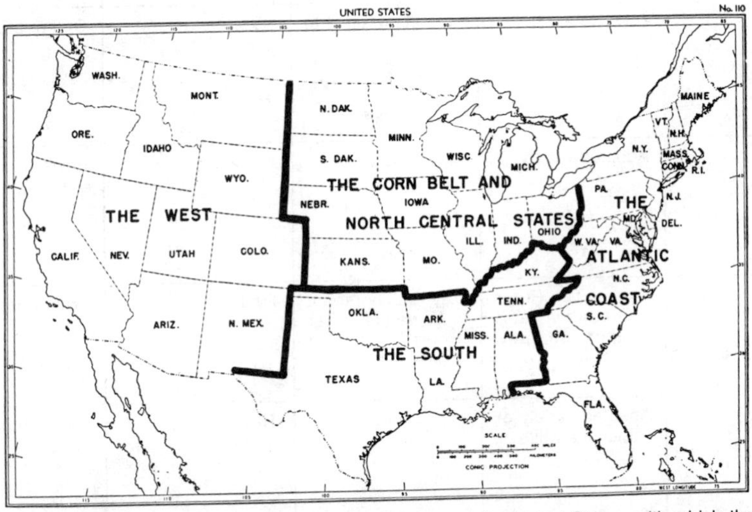

Fig. 5-5. The four major swine production areas of the United States, with which the swine pasture recommendations in Tables 5-11 to 5-14 coincide: (1) Corn Belt and North Central States, (2) the South, (3) the Atlantic Coast, and (4) the West. Some rather characteristic swine feeding practices and pastures are common to each area. (Drawing by R. F. Johnson)

TABLE 5-11—GENERAL RECOMMENDATIONS FOR PLANTING AND GRAZING PASTUR

(This table contains the authoritative recommendations, made especial Dr. A. R. Schmid, Agronomist, University of Minnesota, St. Paul, Minn Mo.; Dr. W. J. Moline, Agronomist, The University of Nebraska, Lincoln search Center, Wooster, Ohio)

Crop	Variety	When to Plant	Seeding Rate/Acre	Method of Planting	Time Elapsing After Planting, Before Pasturing
			(lb)		*(mo)*
Alfalfa	Adapted variety	Early spring or late summer	12-15	Drill or broadcast	4-5
Bromegrass (or preferably brome-grass-legume mixtures)	Achenbach Lincoln Southland Lyon Lancaster	Spring or late summer	Mixture[2] 5-7 lb bromegrass, 10-15 lb alfalfa	Drill or broadcast	4-5
Ladino		Spring or late summer	1-2	Drill or broadcast	Spring of seconc year, but fre quently may b lightly grazec fall of first year.
Oats	Adapted varieties	Early spring	65-130	Drill or broadcast	1½-2
Rape	Biennial type such as dwarf Essex	Mar. to June	4-10	Rows or broadcast	1½-2
Red clover	Adapted variety	Spring (commonly seeded on winter grain in the spring)	8-10	Drill or broadcast	4
Rye	Adapted variety	Fall, from Aug. 1 to Dec. 1	85-170	Drill or broadcast	1-2
Soybeans	Adapted variety	May to July	60-120	Rows or drill	1½-2
Sudangrass or sorghum-Sudangrass hybrids	Sweet Tift Piper Greenleaf and other adapted varieties	May to July (about 2 weeks after corn planting time is best)	20-30	Drill or broadcast	1½-2
Sweet clover	Adapted variety of biennial white or yellow	Early spring	10-15	Drill or broadcast	3-4

Footnotes on last page of table.

ROPS FOR SWINE IN THE CORN BELT AND NORTH CENTRAL STATES

r this book, of the following competent specialists residing in the area: r. Howell N. Wheaton, Agronomist, University of Missouri, Columbia, eb.; and Dr. Robert W. Van Keuren, Agronomist, Ohio Agricultural Re-

Time of Use (or grazing season)		No. Full-fed Growing-Finishing Pigs /Acre[1]	Adaptation and Other Comments
Start	End		
/hen 6-8 in. high	Late fall, but allow 6-8 in. of growth to remain.	20-25	Where soils are sufficiently fertile and not acid. Cannot tolerate poorly drained soils. Provides palatable nutritious forage over a long grazing season, from early spring until late fall. Alfalfa-grass mixtures are frequently used for swine pastures.
pr. 1	Nov. 1	15-20	A palatable perennial pasture crop, sometimes used as a swine pasture. Will endure heavy grazing and tramping. Is not so valuable a swine pasture as alfalfa, where the latter is adapted. Bromegrass-alfalfa mixtures are equal to straight alfalfa for swine pastures.
/hen 4-6 in. high	Late fall, but allow 3-5 in. of growth to remain.	20-25	Where adapted, Ladino clover is superior to alfalfa as a swine pasture. Will last 3-5 years in pasture. Ladino-grass mixtures are frequently used for swine pastures. Winter-hardy varieties now being developed will extend its use even more in the northernmost states.
pring; 6 to 8 wks. after planting	Early summer (4-6 wks. after starting grazing)	10-15	Oats are a good pasture while they last, but pasture period is short.
pring, 6-8 wks. after planting (when 4-6 in. high)	Late fall, because not killed by mild frost.	20-25	Although not a legume, rape is nearly equal to alfalfa as a swine pasture. It is palatable to pigs. It can be grown easily and at low cost. In the northern areas, rape is frequently seeded with one of the cereals (commonly oats) and field peas. If grazed when wet, rape sometimes causes blistering or sunscalding, especially with white hogs or hogs with white markings or spots. Rape is subject to destruction by plant lice, aphids, or flea beetles.
Vhen 6-8 in. high	Late fall, but allow 4-6 in. of growth to remain with first-year red clover.	10-15	Requires well-drained, fairly rich soil that has plenty of lime. Where adapted, ranks next to Ladino and alfalfa as a pasture crop for swine. Later spring growth and lower yield than alfalfa. Red clover-grass mixtures are excellent for swine pastures.
ate fall	Until covered by snow, and again in early spring	12-14	Late summer or early fall seeded rye is best used as a winter and early spring swine pasture for (1) growing-finishing fall pigs, (2) fall bred sows, (3) early farrowed spring pigs, or (4) lactating sows. For desirable winter pasture, rye should be seeded in the late summer, thus allowing for ample growth prior to cold weather. Rye is seldom used for winter pasture in the northern part of the area.
Vhen plants are 6-8 in. high	Until crop is consumed	12-15	For swine pastures, soybeans are less valuable than alfalfa, the clovers, or rape. They should be either grown in rows or drilled with adequate spacing to allow hogs to move through them with a minimum of tramping damage. They require warm weather for rapid growth.
uly 15 (6 in. high)	Late fall, until frost	15-20	Sudangrass is a rank-growing warm-weather annual plant, palatable to swine. Where the growth is normal and not checked by drought or frost, there is little or no danger of prussic acid poisoning. Anyway, swine usually are not affected.
Vhen 4-6 in. high	Oct. 1, with first-year sweet clover	15-20	Where adapted, Ladino, alfalfa, red clover, and rape are preferred to sweet clover as swine pastures, because of their greater palatability and longer grazing season. With biennial sweet clover, the first year's growth is much superior to the second year's growth for swine pasture. May be fitted easily into established cropping systems. Sweet clover-grass mixtures are frequently used for swine pastures. Clipping at intervals may be necessary in order to alleviate coarse woody stems.

(Continued)

TABLE 5-

Crop	Variety	When to Plant	Seeding Rate/Acre	Method of Planting	Time Elapsing After Planting, Before Pasturing
			(lb)		*(mo)*
Winter wheat	Adapted varieties	Spring or fall (fly-free varieties may be seeded in Aug.)	90	Drill or broadcast	1-2
CROPS FOR HOGGING-DOWN					
Corn	Adapted hybrids	Spring	8-16	Rows	3-5
Soybeans	Adapted varieties	May to July	60-120	Rows or drill	4¼-5

TABLE 5-12—GENERAL RECOMMENDATIONS FOR PLANTI

(This table contains the authoritative listin competent specialists residing in the area: D sity, Auburn, Ala.; Dr. E. H. McIlvain, A Station, Woodward, Okla.; Dr. C. M. Taliaferr ter, Okla.; and Dr. Henry A. Fribourg and M Tennessee, Knoxville, Tenn.)

Crop[1]	Variety	When to Plant	Seeding Rate/Acre	Method of Planting	Time Elapsing After Planting, Before Pasturing
			(lb)		*(mo)*
Alfalfa	Adapted variety	Spring or fall	20	Drill or broadcast	4-6
Arrowleaf clover	Yuchi	Sept.	5	Broadcast	2-4
Bermudagrass	Adapted variety	Early spring	12-20 bu sprigs/acre	Bermuda, which is seedless, is propagated by transplanting roots, which are set out about 2 ft apart.	Second year
Cowpeas	Whippoorwill Iron New era Chinese red	May to July	60-180	Rows or drill	1½-2
Crimson clover	Dixie Auburn Chief	Fall	20	Drill or broadcast	2-4
Kudzu		Dec. to Apr.	400-600 roots/acre	New fields of kudzu are established from root runners, from cuttings, or from seedling plants.	May be light grazed secor year, but prefe able to dela grazing un third year.

Footnotes on last page of table.

–(Continued)

Time of Use (or grazing season)		No. Full-fed Growing-Finishing Pigs /Acre[1]	Adaptation and Other Comments
Start	End		
ate fall or spring	June 1	10-12	Some northern Corn Belt swine producers spring seed Peruvian alfalfa with winter wheat. Ordinarily, it furnishes reasonably good grazing after the spring sown winter wheat is gone. Peruvian alfalfa does not live through the winter where temperatures drop much below 10 degrees above zero. Wheat is seldom used for winter pasture in the northern part of the area.
/hen well dented	60 days after turning hogs into the field	An acre of 100 bu corn will provide about 30 days' feed for 30-34 pigs weighing 75-100 lb or for 16-24 pigs weighing 100-200 lb.	Unless damaged (as by hail), crops are seldom hogged-down. The practice was obsoleted by the mechanical corn picker. Unless the price of corn is very low in relation to the price of protein supplements, it will pay to (1) self-feed a supplement where corn alone is available, or (2) hand-feed ¼ to ⅓ lb of supplement per head daily to pigs having access to a supplemental crop grown in or adjacent to the corn.
Vhen the beans are developed	60 days after turning hogs into field	An acre of 35 bu soybeans will produce 375-450 lb of pork.	Swine will make excellent gains in hogging-down soybeans, but soft pork will be produced. Because of the soft pork problem and other limitations, soybeans are usually harvested and sold as a cash crop.

[1]Carrying capacity will vary according to supplemental feeding of swine, thickness of the stand, soil type, soil fertility, rainfall, temperature, etc.
[2]Bromegrass may be sown in mixtures with alfalfa, sweet clover, red clover, or Ladino; singly or in combination.

ND GRAZING PASTURE CROPS FOR SWINE IN THE SOUTH

nade especially for this book, of the following
Carl S. Hoveland, Agronomist, Auburn Univer-
onomist, USDA, Southern Great Plains Field
Agronomist, Oklahoma State University, Stillwa-
Joe D. Burns, Agronomists, The University of

Time of Use (or grazing season)		No. Full-fed Growing-Finishing Pigs /Acre[2]	Adaptation and Other Comments
Start	End		
When 6-8 in. high	Late fall, but allow 6-8 in. of growth to remain.	15-20	Where soils are sufficiently sweet. Cannot tolerate poorly drained soils. It provides palatable nutritious forage over a long grazing season, from early spring until late fall. Alfalfa-grass mixtures are frequently used for swine pastures. Rotational grazing will prevent rapid depletion of the stand.
When 4-6 in. high	June-July	25	Yuchi is productive longer than most varieties.
June 1, or second year	Nov. 1	10-12	Bermudagrass provides a short grazing season for swine, starting late in the spring. Bermudagrass, lespedeza, white clover, and/or Ladino are often seeded together for swine pasture.
When 6-8 in. high	Until crop is consumed	12-15	Where adapted, alfalfa, the clovers, rape, and soybeans are superior to cowpeas as a swine pasture. Cowpeas will succeed under a greater diversity of conditions than soybeans or velvet beans. Sometimes cowpeas are grown with corn or sorghum for hogging-down.
When 4-6 in. high	Apr.-May	20	May be seeded with adapted small grain or ryegrass. Winter annual; a very excellent producer especially with small grain.
After once established, begin pasturing in May or June	Frost or even later	20-25	Kudzu is an excellent plant for erosion control on hillsides and gullies, but it is decreasing as a swine pasture. Avoid overgrazing kudzu.

(Continued)

TABLE 5-1

Crop[1]	Variety	When to Plant	Seeding Rate/Acre	Method of Planting	Time Elapsing After Planting, Before Pasturing
			(lb)		*(mo)*
Ladino or Louisiana giant white	Ladino La. White	Spring or fall	2-3	Drill or broadcast	3-6
Lespedeza	Kobe Korean	Early spring	Kobe: 40 Korean: 25	Drill or broadcast	2-3
Oats	Adapted variety	Spring or fall	90	Drill or broadcast	2-3
Rape	Dwarf Essex	Spring or fall	4-10	Drill or broadcast	1½-2
Red clover	Kenland La. S-1	Fall or spring (commonly seeded on winter grain)	6-10	Drill or broadcast	3-6
Rye	Balbo Bonel Elbon	Late summer or fall	90	Drill or broadcast	2-3
Soybeans	Adapted variety	May to June	60-120	Rows or drill	1½-2
Sudangrass	Sweet Piper Lahoma Green leaf	Mar. to July (about 2 wks. after corn planting time is best)	30-45	Drill or broadcast	1-1½
Vetch (best known as an important winter cover crop in the southeastern states)	Hairy	Mostly fall sown (Sept. and Oct.)	20-30	Drill or broadcast	3-5
Winter wheat	Adapted variety	Fall	90-120	Drill or broadcast	2-3
CROPS FOR HOGGING-DOWN					
Chufas		Apr. to June		Rows	
Corn	Adapted hybrids	Spring	8-16	Rows	3-6

Footnotes on last page of table.

–(Continued)

Time of Use (or grazing season)		No. Full-fed Growing-Finishing Pigs /Acre[2]	Adaptation and Other Comments
Start	End		
Vhen 4-6 in. high	Summer to fall, but allow 3-5 in. of growth to remain.	20-25	Where adapted, Ladino clover is superior to alfalfa as a swine pasture. Will last 3 to 5 years in pasture. Ladino-grass mixtures are frequently used for swine pastures.
Vhen 4-6 in. high	Early fall to frost (Oct.)		An annual which reseeds itself. Will grow on poor acid soils. Less palatable to swine than alfalfa or Ladino. Not ready for early spring but grows in summer.
Vhen 6-8 in. high	Spring (Feb.)	10-15	Spring oats are a good pasture while they last, but pasture period is short. Winter oats provide pasture from Nov. to May.
ummer, fall, or early spring—when 6-8 in. high	Until crop is consumed	20-25	Although not a legume, rape is nearly equal to alfalfa as a swine pasture. It is palatable to pigs. It can be grown easily. Rape is sometimes planted in Sept. or Oct. for use as a late winter and early spring swine pasture (Jan. to May). If grazed when wet, rape sometimes causes blistering or sunscalding, especially with white hogs or hogs with white markings or spots. Rape is subject to destruction by plant lice or aphids.
Vhen 6-8 in. high	Late fall and winter, but allow 4-6 in. of growth to remain with first-year red clover.	10-15	Requires well-drained, fairly rich soil that has medium amounts of lime. Where adapted, ranks next to Ladino and alfalfa as a pasture crop for swine. Later spring growth and lower yield than alfalfa. Red clover-grass mixtures are excellent for swine pastures.
Vhen 6-8 in. high	Spring	15-20	Late summer or early fall seeded rye is best used as a winter and early spring swine pasture, for (1) growing-finishing fall pigs, (2) fall bred sows, or (3) early farrowed spring pigs. For desirable winter pasture, rye should be seeded in late summer, thus allowing for ample growth prior to cold weather.
Vhen plants are 12-24 in. high (summer)	Until crop is consumed	12-15	For swine pastures, soybeans are less valuable than alfalfa, the clovers, or rape. They should be grown either in rows or drilled with adequate spacing to allow hogs to move through them with a minimum of tramping damage. They require warm weather for rapid growth.
pr.-July (12-15 in. high)	Late fall until frost	15-20	Sudangrass is a warm weather annual plant, palatable to swine. Sudangrass-cowpea or Sudangrass-soybean mixtures are frequently used for swine pastures.
an. to Mar.	Apr. to May	12-15	Vetch supplies late winter and early spring swine pasture. For pasture, it is usually seeded with a small grain crop.
Vhen 6-8 in. high	Spring	10-12	Wheat is more palatable to hogs than rye, but it provides less forage per acre. Rye will make earlier pasture than wheat, but wheat will be available later in the spring. Winter wheat can be grazed lightly by hogs in the fall and early spring without lowering the yield of grain.
all (over winter grazing)	Spring		Chufas should be supplemented properly with proteins and minerals. They produce soft pork. Currently, few chufas are grown for swine.
Vhen well dented	60 days after turning hogs into the field	An acre of 100 bu corn will provide about 30 days' feed for 30-34 pigs weighing 75 to 100 lb or for 16-24 pigs weighing 100-200 lb.	The practice of hogging-down decreased with the advent of the mechanical corn picker. Cowpeas, velvet beans, or soybeans are sometimes grown with corn for hogging-down or corn and peanuts are planted in alternate rows; but the yield of corn and pounds of pork produced per acre are lowered. Sometimes pasture is provided adjacent to the cornfield. Unless the price of corn is very low in relation to the price of protein supplement, it will pay to (1) self-feed a supplement where corn alone is available, or (2) hand-feed ¼ to ⅓ lb supplement per head daily to pigs having access to a supplemental crop grown in or adjacent to the corn.

(Continued)

TABLE 5-1

Crop[1]	Variety	When to Plant	Seeding Rate/Acre	Method of Planting	Time Elapsing After Planting, Before Pasturing
			(lb)		*(mo)*
Peanuts	Spanish (for fall hogging off); runner (for winter hogging off)	Late spring	30-50	Rows	
Sorghum	Adapted variety	Spring	6-in. rows; 20-24, broadcast	Rows or broadcast	5-5½
Soybeans	Adapted variety	Late spring	60-90	Rows or drill	5-5½
Sweet potatoes				Rows	
Velvet beans	Florida Georgia Alabama	Spring, after danger of frost	4-12 when planted with corn	Rows with corn are most popular.	4-6

—(Continued)

Time of Use (or grazing season)		No. Full-fed Growing-Finishing Pigs /Acre[2]	Adaptation and Other Comments
Start	End		
Fall or winter, depending upon the variety of peanuts	Late fall or early spring, depending on the variety of peanuts	An acre of 35 bu peanuts will produce 425-625 lb of pork.	Peanuts are frequently planted with corn when the crop is to be hogged off. Provide free access to salt and a calcium supplement. Peanuts will produce soft pork.
When grain is developed	60 days after turning hogs into field	An acre of 60 bu sorghum will produce 600-800 lb of pork.	The grain sorghums are generally grown in areas where climatic and soil conditions are less favorable to corn. Since feeding trials show that the threshed grain sorghums are worth about 90% as much as shelled corn, it is reasonable to assume that, where the acre yields are comparable, the hogging-down value of the 2 crops is of about the same relationship. Cowpeas, velvet beans, or soybeans are sometimes grown with grain sorghum for hogging-down, or grain sorghum and peanuts are planted in alternate rows; but the yield of corn and pounds of pork produced per acre are lowered. Sometimes pasture is provided adjacent to the grain sorghum field. Unless the price of grain sorghum is very low in relation to the price of protein supplement, it will pay to (1) self-feed a supplement where sorghum alone is available, or (2) hand-feed ¼ to ⅓ lb of supplement to pigs having access to a supplemental crop grown in or adjacent to the grain sorghum.
When the beans are developed	60 days after turning hogs into field	An acre of 35 bu soybeans will produce 425-625 lb pork.	Swine will make excellent gains in hogging-down soybeans, but soft pork will be produced. Because of the soft pork problem and other limitations, soybeans are usually harvested and sold as a cash crop.
Late fall	Early spring	An acre of 270 bu sweet potatoes will provide about 30 days' grazing for 10-15 pigs weighing 100-200 lb.	Sweet potatoes can be grown on soil that is sandy and too thin for good corn production. Best results are usually obtained when pigs grazing sweet potatoes are given ⅓ to ½ of a grain ration in addition to a protein supplement of animal or marine origin and a mineral mixture. Generally it is most profitable to sell sweet potatoes as a cash crop, using swine to glean the fields or consume the culls.
Late fall, when beans are mature (Nov.)	Early spring (Feb.)	An acre of 30 bu velvet beans will produce 375-450 lb pork.	Velvet beans are a warm-weather plant. They are usually planted with a supporting plant such as corn, pearlmillet, Japanese sugarcane or sorghum. Swine do not do well on velvet beans alone.

[1]Other pasture crops—generally seeded in mixtures—sometimes used for swine pastures in the South, are: cattail or pearlmillet, Johnsongrass (not seeded, but where present), sweet clover, alta or Kentucky 31 fescue, and dallisgrass.
[2]Carrying capacity will vary according to supplemental feeding of swine, thickness of the stand, soil type, soil fertility, rainfall, temperature, etc.

TABLE 5-13—GENERAL RECOMMENDATIONS FOR PLANTING AND GRAZING PASTURE

(This table contains the authoritative recommendations, made especially for this book, of the following competent specialists sity of Florida, Gainesville, Fla.; Dr. W. W. Woodhouse, Jr., Agronomist, North Carolina State University, Raleigh, N.C.; and

Crop	Variety	When to Plant	Seeding Rate/Acre	Method of Planting	Time Elapsing After Planting, Before Pasturing
			(lb)		*(mo)*
Alfalfa	Adapted variety	Spring or late summer	15-25	Drill or broadcast	4-6
Bermudagrass (for South Atlantic states)	Coastal or Suwannee Coastcross #1 and other strains	Early spring	5-10 lb or sprigs	Coastal Bermuda which is seedless is propagated by transplanting stolons which are set out about 2 ft apart.	2-4
Cowpeas	Whippoorwill Iron New era Clay	May to July	60-180	Rows or drill	1½-2
Crimson clover (for South Atlantic states)	Dixie Auburn Autauga Chief	Fall	12-20	Drill or broadcast	2-3
Kudzu	Adapted strains	Dec. to Apr.	400-600 roots per acre	New fields of kudzu are established from root runners, from cuttings, or from seedling plants.	May be lightly grazed second year, but preferable to delay grazing until third year.
Lespedeza (for Mid- and South Atlantic states)	Climax Kobe Korean Sericea Common	Early spring	20-30; unhulled seed	Drill or broadcast	2-3
Millet	Starr Gahi Tifleaf 1	Apr. to July	15-20	Drill or broadcast	1-1½
Oats	Adapted	Spring, late summer, or fall	48-120	Drill	2-3
Orchardgrass (for North Atlantic states)	Adapted	Spring or fall	6-12	Drill or broadcast	4-5
Rape	Dwarf Essex	March to June in North, fall and winter in South.	4-10	Rows or broadcast	1½-2
Red clover	Adapted	Spring (in North), or fall (in South); commonly seeded on winter grain	10-15	Drill or broadcast	2-3
Rye	Rosen (in North) Balbo (in South) Abruzzi (in South) Gator (in South)	Fall	90-120	Drill	1½-2

Footnote on last page of table.

ROPS FOR SWINE IN THE ATLANTIC COAST STATES

esiding in the area: Dr. Robert R. Seaney, Agronomist, Cornell University, Ithaca, N.Y.; Dr. G. O. Mott, Agronomist, Univer-
r. John E. Baylor, Agronomist, The Pennsylvania State University, University Park, Penn.)

Time of Use (or grazing season) Start	End	No. Full-fed Growing-Finishing Pigs /Acre[2]	Adaptation and Other Comments
When 8-12 in. high	Late summer or fall, but allow 6-8 in. of growth to remain.	15-20	Where soils are sufficiently limed. Cannot tolerate poorly drained soils. It provides palatable nutritious forage over a long grazing season, from early spring until late fall. Alfalfa-grass mixtures are frequently used for swine pastures. Responds to boron in addition to phosphorus and potassium on some soils.
When 6-8 in. high	Nov. 15	10-12	Bermudagrass provides a short grazing season for swine, starting late in the spring. Bermudagrass, lespedeza, white clover, and/or Ladino are often seeded together for swine pasture.
When plants are 4-6 in. high	Until crop is consumed	12-15	Where adapted, alfalfa, the clovers, rape, and soybeans are superior to cowpeas as a swine pasture. Cowpeas will succeed under a greater diversity of conditions than soybeans or velvet beans. Sometimes cowpeas are grown with corn or sorghum for hogging-down.
Jan.	May	10-15	A winter annual, which makes very good growth during winter months. Associates well with cereal grains, and as such is one of the highest producing pasture crops during the winter season.
After once established, begin pasturing in May or June.	Frost or even later	20-25	Kudzu is an excellent plant for erosion control on hillsides and gullies, but it is decreasing as a swine pasture. Avoid overgrazing kudzu.
Late spring or summer	Early fall	10-12	An annual which reseeds itself. Will grow on poor acid soils. Less palatable to swine than alfalfa or Ladino. Not ready for early spring grazing, but excellent in summertime.
10 in.	Until frost	25-30	In North Carolina, millet is preferred to Sudangrass.
Fall (including winter grazing)	Spring	10-15	Oats are a good pasture while they last, but pasture period is short. In the Deep South, oats make an excellent winter pasture for swine.
When 6-8 in. high	Late fall	10-15	Usually seeded in a mixture with a legume. More useful as a pasture in such a mixture. Adapted to medium fertile to fertile soil.
Spring, 6-8 weeks after planting (when 6-10 in. high)	Late fall, because not killed by mild frost.	20-25	Although not a legume, rape is nearly equal to alfalfa as a swine pasture. It is palatable to pigs. It can be grown easily and at low cost. In the northern areas, rape is frequently seeded with one of the cereals (commonly oats) and field peas. If grazed when wet, rape sometimes causes blistering or sunscalding, especially with white hogs or hogs with white markings or spots. Rape is subject to destruction by plant lice, aphids, or flea beetles.
When 6-8 in. high	Late fall, but allow 4-6 in. of growth to remain with first-year red clover. Winter months in South	10-15	Requires well-drained fairly rich soil that has medium amounts of lime. Where adapted, ranks next to Ladino and alfalfa as a pasture crop for swine. Later spring growth and lower yield than alfalfa. Red clover-grass mixtures are excellent for swine pastures.
Fall (including winter grazing)	Spring (including winter grazing)	15-20	Late summer seeded rye is best used as a winter and early spring swine pasture, for (1) growing-finishing fall pigs, (2) fall bred sows, or (3) early farrowed spring pigs. For desirable winter pasture, rye should be seeded in the late summer or fall, thus allowing for ample growth prior to cold weather.

(Continued)

TABLE 5-13

Crop	Variety	When to Plant	Seeding Rate/Acre	Method of Planting	Time Elapsing After Planting, Before Pasturing
			(lb)		*(mo)*
Soybeans	Biloxi Locally adapted varieties	Apr. to July	60-120	Rows or drill	1½-2
Sudangrass	Trudan Sweet Sudax	Apr. to July	15-20	Drill	1-1½
White clover	Louisiana S-1 Nolin white Regal Ladino Tillman Ladino	Spring or fall	2-6	Drill or broadcast	3-5
CROPS FOR HOGGING-DOWN **Corn**	Adapted hybrids	Spring	8-16	Rows	3-5
Chufas		Apr. to June	25-30	Rows 2 ft apart; 12 in. in rows	5
Peanuts	Spanish (for summer and fall hogging-off) Runner (for fall and winter hogging-off)	Late spring	35-50	Rows	3½-4½
Sorghum	Adapted dwarf grain variety	Late spring	8 in rows; 25-35 broadcast	Rows or broadcast	3-4½
Soybeans	Adapted variety	Late spring to early summer	60-120	Rows or drill	4½-5
Sweet potatoes				Rows	
Velvet beans (Southern states only)	Florida Georgia Alabama (or other adapted variety)	Spring after danger of frost	4-12 when planted with corn	Rows with corn are most popular.	6

–(Continued)

Time of Use (or grazing season) Start	End	No. Full-fed Growing-Finishing Pigs /Acre[2]	Adaptation and Other Comments
When plants are 6-8 in. high	Until crop is consumed	12-15	For swine pastures, soybeans are less valuable than alfalfa, the clovers, or rape. They should either be grown in rows or drilled with adequate spacing to allow hogs to move through them with a minimum of tramping damage. They require warm weather for rapid growth.
lay-July (18-24 in. high)	Late fall, until frost	25-30	Sudangrass is a rank-growing warm-weather annual plant, palatable to swine. Where the growth is normal and not checked by drought or frost, there is little or no danger of prussic acid poisoning. Anyway, swine are not usually affected. Millet is replacing Sudan in most sandy soils.
When 6 in. high	Late fall, but allow 3-5 in. of growth to remain. Season long in South.	20-25	Fall seed in Fla., Ga., S.C., and N.C. Where adapted, Ladino clover is superior to alfalfa as a swine pasture. Will last 3 to 5 years in pasture. Ladino-grass mixtures are frequently used for swine pastures, although Ladino alone is preferable. Also, a Ladino-alfalfa mixture is preferred to Ladino alone in certain areas.
When well dented	60 days after turning hogs into the field	An acre of 100 bu corn will provide about 30 days' feed for 30-34 pigs weighing 75-100 lb or for 16-24 pigs weighing 100-200 lb.	The practice of hogging-down corn decreased with the advent of the mechanical corn picker. Cowpeas, velvet beans, or soybeans are sometimes grown with corn for hogging-down, or corn and peanuts are planted in alternate row; but the yield of corn and pounds of pork produced per acre are lowered. Unless the price of corn is very low in relation to the price of protein supplement, it will pay to (1) self-feed a supplement where corn alone is available, or (2) hand-feed ¼ to ⅓ lb of supplement per head daily to pigs having access to a supplemental crop grown in or adjacent to the corn.
all (over winter grazing)	Spring	An acre of chufas will produce about 330 lb of pork.	Currently, few chufas are grown for swine. Chufas should be supplemented properly with proteins and minerals. They produce soft pork. Satisfactory for delayed grazing.
all or winter, depending upon the variety of peanuts	Late fall or early spring, depending on the variety of peanuts	An acre of 25-30 bu peanuts will produce 300-350 lb of pork.	Peanuts are frequently planted with corn when the crop is to be hogged-off. Provide free access to salt and a calcium supplement. Peanuts will produce soft pork. Graze when nuts are mature.
When mature	60 days after turning hogs into field	An acre of 60 bu sorghum will produce 600-800 lb of pork.	The grain sorghums are generally grown in areas where climatic and soil conditions are less favorable to corn. Since feeding trials show that the threshed grain sorghums are worth about 90% as much as shelled corn, it is reasonable to assume that, where the acre yields are comparable, the hogging-down value of the 2 crops is of about the same relationship. Cowpeas, velvet beans, or soybeans are sometimes grown with grain sorghum for hogging-down, or grain sorghum and peanuts are planted in alternate rows; but the yield of sorghum and pounds of pork produced per acre are lowered. Sometimes pasture is provided adjacent to the grain sorghum field. Jnless the price of grain sorghum is very low in relation to the price of protein supplement, it will pay to (1) self-feed a supplement where sorghum alone is available, or (2) hand-feed ¼ to ⅓ lb of supplement per head daily to pigs having access to a supplemental crop grown in or adjacent to the grain sorghum.
When beans are developed	60 days after turning hogs into field	An acre of 35 bu soybeans will produce 425-625 lb pork.	Swine will make excellent gains in hogging-down soybeans, but soft pork will be produced. Because of the soft pork problem and other limitations, soybeans are usually harvested and sold as a cash crop.
Late fall	Early spring	An acre of 270 bu sweet potatoes will provide about 30 days grazing for 10-15 pigs weighing 100-200 lb and produce about 675 lb of pork.	Sweet potatoes can be grown on soil that is sandy and too thin for good corn production. Best results are usually obtained when pigs grazing sweet potatoes are given ⅓ to ½ of a grain ration in addition to a protein supplement of animal or marine origin and a mineral mixture. Generally it is most profitable to sell sweet potatoes as a cash crop, using swine to glean the fields or consume the culls.
Late fall, when beans are mature (Nov.)	Early spring (Feb.)	An acre of 30 bu velvet beans will produce 375-450 lb of pork.	Velvet beans are a warm-weather plant. They are usually planted with a supporting plant, such as corn, pearlmillet, Japanese sugarcane, or sorghum. Swine do not do well on velvet beans alone.

[1]Other pasture crops—generally seeded in mixtures—sometimes used for swine pastures in the southern Atlantic Coast states are: pearlmillet or cattail millet, Johnsongrass (not seeded, but where present), alta or Kentucky 31 fescue, and dallisgrass.

[2]Carrying capacity will vary according to supplemental feeding of swine, thickness of the stand, soil type, soil fertility, rainfall, temperature, etc.

TABLE 5-14—GENERAL RECOMMENDATIONS FOR PLANTIN

(This table contains the authoritative recommen following competent specialists residing in th of Arizona; Dr. W. James Clawson, Agronomist the University of California; Dr. A. D. Dotzenko S. Cooper, Agronomist, USDA and Montan University of Nevada; Dr. Rex D. Pieper, Ag S. McGuire, Agronomist, Oregon State Univer tist, and Dr. Kenneth J. Morrison, Agronomist Chester L. Canode and Dr. John L. Schwendi

Crop	Variety	When to Plant	Seeding Rate/Acre	Method of Planting	Time Elapsing After Planting, Before Pasturing
			(lb)		*(mo)*
Alfalfa	Adapted variety	Spring or late summer	6-20	Drill or broadcast	4-5
Birdsfoot trefoil	Cascade Granger Common narrow-leaf	Spring or late summer	6	Drill or broadcast	Spring of seconc year
Bromegrass (or preferably bromegrass-legume mixtures)	Manchar Lincoln and southern strains	Spring or late summer	Mixture² 8 lb bromegrass, 4-6 lb alfalfa	Drill or broadcast	4-5
Ladino	Pilgrim	Spring or late summer	2-4	Drill or broadcast	Spring of second year, but frequently may be lightly grazed fall of first year.
Peas	White Canadian Alaska	Spring	120	Drill	1-1½
	Austrian Winter peas	Fall	120	Drill	1-1½
Red clover	Kenland Pennscott Lakeland Mammoth	Spring or late summer (commonly seeded on winter grain in the spring)	5-10	Drill or broadcast	4
Rye	Abruzzi or Balbo (northern section) Merced Elbon (southern section)	Late summer or fall	85-170	Drill or broadcast	1½-2
Strawberry clover	Salina	Spring or fall	2-4	Drill or broadcast	Spring of second year, but may be lightly grazed fall of first year.
Sudangrass	Piper Sweet Hybrids	May to July (about 2 weeks after corn planting time is best)	20-30	Drill or broadcast	1-1½

Footnotes on last page of table.

ND GRAZING PASTURE CROPS FOR SWINE IN THE WEST

ations, made especially for this book, of the rea: Dr. Robert Dennis, Agronomist, University nd Dr. William F. Lehman, Agronomist, both of gronomist, Colorado State University; Dr. Clee tate University; Dr. E. H. Jensen, Agronomist, onomist, New Mexico State University; Dr. W. ity; Dr. Wilton W. Heinemann, Animal Scien- oth of Washington State University; and Dr. nan, Agronomists, USDA)

Time of Use (or grazing season)		No. Full-fed Growing-Finishing Pigs /Acre[1]	Adaptation and Other Comments
Start	End		
When 6-8 in. high	Late fall, but allow 6-8 in. of growth to remain.	15-20	Cannot tolerate poorly drained soils. It provides palatable nutritious forage over a long grazing season, from early spring until late fall. Alfalfa-grass mixtures are frequently used for swine pastures.
When 6-8 in. high	Late fall, but allow 4-6 in. of growth to remain with first-year birdsfoot trefoil.	15-18	Well adapted to soils of lower pH and poorer drainage than alfalfa. Long-lived plant, 15 to 20 years. Highly nutritious; second only to Ladino in ease of management. Slow to become established. Will live with bluegrass better than any other legume.
When 6-8 in. high	Late fall, but allow 6-8 in. of growth to remain.	15-20	Bromegrass, a palatable perennial pasture crop, is sometimes used as a swine pasture. It will endure heavy grazing and tramping. It is not as valuable a swine pasture as alfalfa, where the latter is adapted. Bromegrass-alfalfa mixtures are equal to straight alfalfa for swine pastures.
When 4-6 in. high	Late fall, but allow 3-5 in. of growth to remain.	20-25	Where adapted, Ladino clover is superior to alfalfa as a swine pasture. Will last 2-5 years in pasture. Ladino-grass mixtures are sometimes used for swine pastures. Ladino should be used only where season long water is available.
Apr. 20-May 10	Aug.	12-14	For annual crop program in the northern states. Provide an excellent pasture over a short period. Varieties equal in feed value.
When 6-8 in. high	Late fall, but allow 4-6 in. of growth to remain with first year red clover.	10-15	Requires well-drained fairly rich soil that has medium amounts of lime. Where adapted, ranks next to Ladino and alfalfa as a pasture crop for swine. Later spring growth and lower yield than alfalfa. Red clover-grass mixtures are excellent for swine pastures.
Mar. 15	July 1	12-14	Fall seeded rye is best used as a winter and early spring swine pasture, for (1) growing-finishing fall pigs, (2) fall bred sows, or (3) early farrowed spring pigs. For desirable winter pasture, rye should be seeded in the late summer, thus allowing for ample growth prior to cold weather.
When 4-6 in. high	Late fall, but allow 3-5 in. of growth to remain.	20-25	Strawberry clover will stand both alkaline and waterlogged conditions.
May-July (15 in. high)	Late fall, until frost	15-20	Sudangrass is a rank-growing, warm-weather annual plant, palatable to swine. Where the growth is normal and not checked by drought or frost, there is little or no danger of prussic acid poisoning. Anyway, swine are not usually affected.

(Continued)

TABLE 5-14

Crop	Variety	When to Plant	Seeding Rate/Acre	Method of Planting	Time Elapsing After Planting, Before Pasturing
			(lb)		*(mo)*
Sweet clover	White sweet clover Yellow sweet clover Hubam (annual)	Early spring	10-15	Drill or broadcast	3-4
CROPS FOR HOGGING-DOWN					
Peas, ripe field	Any adapted variety	Early spring	120	Drill	3½

SEEDING AND MANAGEMENT OF SUBHUMID, HUMID, AND IRRIGATED PASTURES

This section, and the subsections under it, has reference to those pastures which either receive above approximately 20 inches of rainfall annually or are irrigated. This includes the pastures of the Corn Belt, the South, the East, and the irrigated valleys and smaller and scattered moderate- to high-rainfall areas throughout the West.

Establishing a New Pasture

The following practices are usually adhered to in successfully establishing a new pasture in the subhumid, humid, and irrigated areas:

1. **Adapted varieties and suitable mixtures are selected**—The first requisite of successful pastures is that adapted varieties of grasses and/or legumes shall be selected for the area and for the purposes intended. Tables 5-1 to 5-14 give the general recommendations of competent specialists residing in different areas of the United States. For more specific recommendations for a particular farm or ranch, the stockman should consult such local authorities as the county agent, vocational agriculture instructor, or successful neighbors.

Where grass-legume mixtures are to be grown, a 50-50 mixture is satifactory for most purposes and conditions.

2. **The soil is tested and fertilized**—The soil is tested (see Section 8 of this book for directions) and fertilized (and limed if necessary) according to needs. The three elements required by all grasses and legumes in greatest abundance are nitrogen, phosphorus, and potassium. In addition, where legumes are grown, acid soils need lime. The pH of the soil should be about 6.5. It is best to work lime into the soil considerably in advance of seeding, but commercial fertilizers should be applied at seeding time.

A thin, uniform mulch of barnyard manure is especially valuable in establishing a new seeding.

3. **High-quality seed is purchased**—The seed should be of good quality, of high germination and purity as indicated on the tag, and free of noxious weeds. Also, proof of origin is of prime importance when an imported variety is secured. Certified seed carries more assurance of being high quality than noncertified seed, and gives proof of its origin much as a registration certificate does on a purebred animal.

4. **Scarified legume seed is used**—In the purchase of certain legume seed, it is important that it be scarified, which breaks the seed coat and allows faster moisture penetration—thus assuring quicker and more uniform germination and a better stand the first year.

5. **Legume seed is inoculated**—Since legumes can use nitrogen from the air provided they are inoculated with the proper bacteria, it is important that legume seed be inoculated. The proper inoculant can be determined by referring to Table 5-15.

Inoculant comes in several different forms, usually in a can with directions given thereon. It is important that the seed not be treated more than a few hours before seeding because these nitrogen-fixing bacteria are easily killed by drying, heat, sunlight, or by chemical seed treatment.

6. **A good seedbed is prepared**—A good seedbed is free from weeds, fine-textured, firm and moist.

Weeds are usually destroyed by growing row crops or a small grain the year preceding seeding to pasture and by cultivating frequently following the harvesting of this crop.

There are many different ways in which to prepare a good seedbed. Perhaps as good a method as any consists in (a) plowing as far in advance of seeding as possible, (b) discing, (c) harrowing one or more times to level up the field and smooth down the sur-

-(Continued)

Time of Use (or grazing season)		No. Full-fed Growing-Finishing Pigs /Acre[1]	Adaptation and Other Comments
Start	End		
'hen 6-8 in. high	Sept. with first-year sweet clover	15-20	Where adapted, Ladino, alfalfa, red clover, and rape are preferred to sweet clover as swine pastures, because of their greater palatability and longer grazing season. With biennial sweet clover, the first year's growth is much superior to the second year's growth for swine pasture. May be fitted easily into established cropping systems. Sweet clover-grass mixtures are frequently used for swine pastures. Clipping at intervals may be necessary in order to alleviate coarse woody stems.
urn in any time after pods begin to form.	When crop is consumed	About 400 lb of pork per acre	Adapted to the Northwest. For best results, provide minerals and a limited amount of grain in addition to the peas.

[1]Carrying capacity will vary according to supplemental feeding of swine, thickness of the stand, soil type, soil fertility, rainfall, temperature, etc.
[2]Bromegrass may be sown in mixtures with alfalfa, sweet clover, red clover, or Ladino, singly or in combination. Other adapted grasses can be used with such legumes.

ace, and (d) cultipacking or rolling. A properly pre-ared seedbed should be so firm that one barely eaves a footprint when walking across it; the firmer he better from the standpoint of moisture conserva-ion of small seeds.

7. **The seeding operation is timed and carried out roperly**—The seeding time will vary, being determined primarily by the area and by the species or mixture used (see Tables 5-1 to 5-14).

The actual seeding operation may be (a) by broadcasting, with a whirlwind seeder or by hand; or (b) by drilling, with any one of several types of conventional seeders. Drilling is the preferred method, for it ensures more uniform placement of seed in both depth and amount of seed per acre and results in a more uniform stand.

TABLE 5-15

LEGUME INOCULATION GUIDE

Group No.	Cross-inoculation Group (so-called because all members in a group can be inoculated by the same kind of legume bacteria)
I	For alfalfa, white sweet clover, yellow sweet clover, Hubam, bur-clovers, black medic (yellow trefoil), bitter clover, button clover, and fenugreek.
II	For all trifolium clovers—red, mammoth red, medium red, alsike, white, Ladino, wild white, white Dutch, crimson, hop clovers, cluster, McNeill, strawberry, subterranean, berseem, zigzag, rabbit-foot, Persian, Carolina, and buffalo clover.
III	For garden peas, field peas, Canadian field peas, sweet peas, perennial peas, Austrian winter peas, Tangier peas, all vetches (hairy, Hungarian, common, monantha, purple, Oregon), broad beans, and lentils.
IV	For garden beans (string, snap, wax), navy beans, kidney beans, pinto beans, and scarlet runner beans.
V	For cowpeas (black-eyed peas or black-eyed beans), peanuts, velvet beans, crotalarias, Florida beggarweed, mung beans, jack beans, partridge peas, pigeon peas, tepary beans, tick trefoil, adzuki beans, alyceclover, mat beans, and guar.
VI	For all varieties lespedeza—Japan clover, common lespedeza, Korean, Kobe, Tennessee 76, and sericea.
VII	Lupines (all varieties).
VIII	For all varieties soybeans, including edible soybeans.
IX	The following legumes require special strains of legume bacteria: sesbania, black locust, dalea woods' or clover, crown vetch, garbanzo or chick peas, big trefoil, hairy indigo, serradella, sainfoin, and thornbroom, lotus, or birdsfoot trefoil, cracca, or tephrosia.

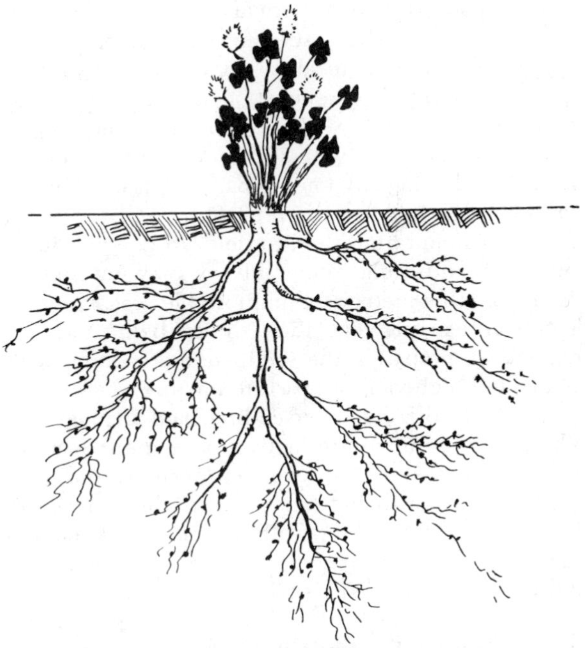

Fig. 5-6. Nodules (small bumps) containing nitrogen-fixing bacteria on a well-inoculated red clover root. (Drawing by R. F. Johnson)

Since most grass and legume seeds are very small, they should not be covered deep. A good rule of thumb is that they should not be covered more than four or five times the width of the seed; usually, this means not more than one-fourth inch.

8. **A companion or nurse crop may or may not be included**—The value of planting a "companion" or nurse crop—usually consisting of annuals—with new seed crops is controversial.

The advantages are (a) it furnishes a crop of value while the new seeding is being established, (b) it lessens erosion, and (c) it reduces the weed population.

The disadvantages are (a) it may retard the growth of the seedlings for whose protection it is grown, and (b) it may rob the new seeding of so much moisture that it kills them during dry spells unless the companion crop is harvested early as pasture, hay, or silage.

Improving or Renovating an Old Pasture

In altogether too many cases, old permanent pastures are merely gymnasiums for livestock. Generally, this condition exists because the least productive areas are used for pastures and because little attention is given to fertility and pasture management.

Permanent pastures in subhumid, humid, or irrigated areas that are run-down may be brought back into production by either of the following methods:

1. **By reseeding without growing a crop in the interim**—Poor, run-down permanent pastures are frequently renovated by reseeding without growing a crop in the interim; in other words, pasture follows pasture. This kind of renovation is designed to increase pasture yields without subjecting the soil to excessive erosion and without keeping the area out of pasture production any longer than necessary. The actual operations involved in renovating will vary from area to area, and from field to field. In general, it involves (a) cultivating (preferably by plowing, but by discing or other methods in unplowable areas) so as to destroy all existing vegetation, (b) fertilizing and liming, and (c) preparing the seedbed and seeding with an adapted high-yielding pasture mixture.

2. **By fertilizing, overseeding, and managing**—Where a fair but unproductive permanent pasture stand exists, pasture improvement or renovation may consist in (a) fertilizing (and liming where needed), (b) seeding (overseeding) with desirable and adapted varieties, and (c) managing in accordance with the outline which follows. Usually the fertilizer and the seed are worked into the soil with a disc and spring-tooth harrow, but a minimum of the existing sod is destroyed.

Pasture Management

Many good pastures have been established only to be lost through careless management. Good pasture management in the subhumid, humid, and irrigated areas involves the following practices:

1. **Controlled grazing**—Nothing contributes more to good pasture management than controlled grazing. At its best, it embraces the following:

a. **Protecting first-year seedings**—First-year seedings should be grazed lightly or not at all in order that they may get a good start. Where practical, instead of grazing, it is preferable to mow a new first seeding about 3 inches above the ground and to utilize it as hay or silage, provided there is sufficient growth to so justify.

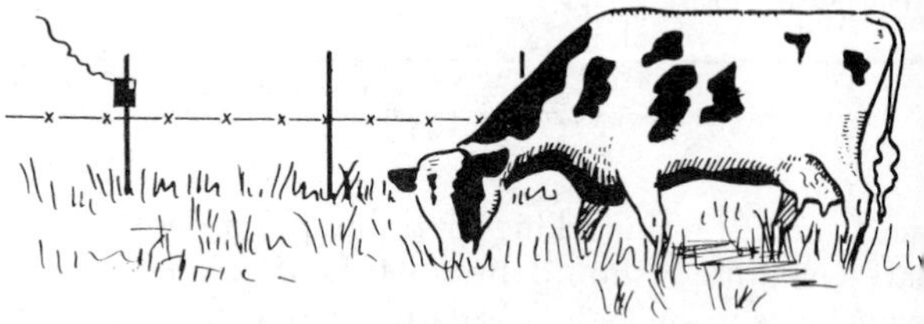

Fig. 5-7. At its best, controlled grazing embraces protection of first year seeding, rotation or alternate grazing, shifting the location of salt, shade and water, deferred spring grazing, and avoiding close late fall grazing, overgrazing, and undergrazing. (Drawing by R. F. Johnson)

b. **Rotation or alternate grazing**—Rotation or alternate grazing is accomplished by dividing a pasture into fields (usually two to four) of approximately equal size, so that one field can be grazed while the others are allowed to make new growth. This results in increased pasture yields, more uniform grazing, and higher quality forage.

Generally speaking, rotation or alternate grazing is (1) more practical and profitable on rotation and supplemental pastures than on permanent pastures, (2) more productive with high-producing dairy cows than with other farm animals, and (3) more beneficial where parasite infestations are heavy—as is usually the case with sheep—than where little or no parasitic problems are involved.

c. **Shifting the location of salt, shade, and water**—Where portable salt containers are used, more uniform grazing and scattering of the droppings may be obtained simply by the practice of shifting the location of the salt to the less grazed areas of the pasture. Where possible and practical, the shade and the water should be shifted likewise.

d. **Deferred spring grazing**—Allow 6 to 8 inches of growth before turning out to pasture in the spring, thus giving grass a needed start. Anyway, the early spring growth of pastures is high in moisture and washy.

e. **Avoiding close late fall grazing**—Pastures

that are closely grazed late in the fall start late in the spring. With most pastures, 3 to 5 inches of growth should be left for winter cover.

f. **Avoiding overgrazing**—Never graze more closely than 2 to 3 inches during the pasture season. Continued close grazing reduces the yield, weakens the plants, allows weeds to invade, and increases soil erosion. The use of temporary and supplemental pastures, such as Sudan, may "spell off" regular pastures through seasons of drought and other pasture shortages and thus alleviate overgrazing.

g. **Avoiding undergrazing**—Undergrazing seeded pastures should also be avoided, because (1) mature forage is unpalatable and of low nutritive value; (2) tall-growing grasses may drive out such low-growing plants as white clover due to shading; and (3) weeds, brush, and coarse grasses are more apt to gain a foothold when the pasture is grazed insufficiently. It is a good rule, therefore, to graze the pasture fairly close at least once each year.

2. **Clipping pastures and controlling weeds**—Pastures should be clipped at such intervals as necessary to control weeds (and brush) and to get rid of uneaten clumps and other unpalatable coarse growth left after incomplete grazing. Pastures that are grazed continuously may be clipped at or just preceding the usual haymaking time; rotated pastures may be clipped at the close of the grazing period. Weeds and brush may also be controlled by chemicals, by burning, etc.

3. **Topdressing**—Like animals, for best results grasses and legumes must be fed properly throughout a lifetime. It is not sufficient that they be fertilized (and limed if necessary) at or prior to seeding time. In addition, in most areas, it is desirable and profitable to topdress pastures with fertilizer annually, and, at less frequent intervals, with reinforced manure and lime (lime to maintain a pH of about 6.5). Such treatments should be based on soil tests, and are usually applied in the spring or fall.

4. **Scattering droppings**—The droppings should be scattered at the end of each grazing season in order to prevent animals from leaving ungrazed clumps and to help them fertilize a larger area. This can best be done by the use of a brush harrow or chain harrow.

5. **Grazing by more than one class of animals**—Grazing by two or more classes of animals makes for more uniform pasture utilization and fewer weeds and parasites, provided the area is not overstocked. Different kinds of livestock have different habits of grazing; they show preference for different plants and graze to different heights.

6. **Irrigating where practical and feasible**—Where irrigation is practical and feasible, it alleviates the necessity of depending on the weather.

EXTENDING THE GRAZING SEASON

In the South and in Hawaii, year-around grazing is a reality on many a successful farm. By careful planning, other areas can approach this desired goal.

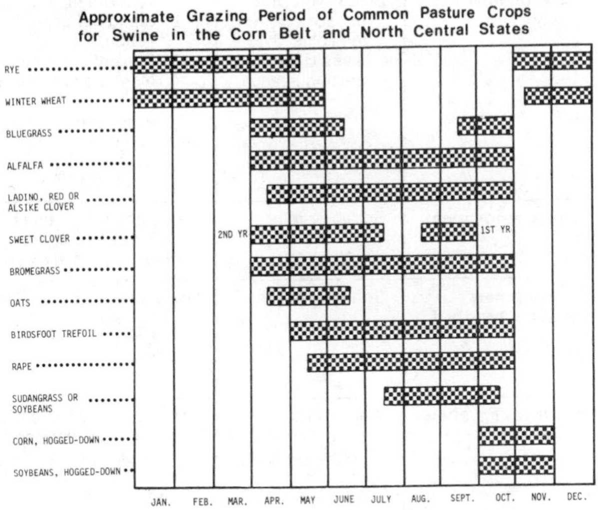

Fig. 5-8. As shown, by selecting the proper combination of crops, year-around grazing can be achieved. For example, 12 months' hog pasture may be obtained by using two crops only, rye and Ladino clover.

Fig. 5-8, based on Table 5-11, illustrates in graphic form the growth period of each of the common swine pasture plants of the Corn Belt and North Central States. As noted, by selecting the proper combination of crops, swine pastures for each month of the year are assured. A similar chart for each area of the country and for each class of animal can be developed from the tables presented in this section.

In addition to lengthening the grazing season through the selection of species, earlier spring pastures can be secured by avoiding grazing too late in the fall and by the application of a nitrogen fertilizer in the fall or early spring. Nitrogen fertilizers will often stimulate the growth of grass so that it will be ready for grazing 10 days to 2 weeks earlier than unfertilized areas.

CARING FOR PASTURE AND RECREATIONAL AREAS, AND FOR GRASS ALONG ROADS, LANES, AND BRIDLE PATHS

Table 5-16, page 656, tells how successful operators maintain, renovate, and seed pastures, and how they care for racetracks, show-rings, and for grass along roads, lanes, and bridle paths. These areas can no longer be taken for granted. They're big and important—and they'll get bigger. Hence, they merit the combined best recommendations of scientists and practical operators.

TABLE 5-16

GUIDE FOR CARING FOR PASTURE AND RECREATIONAL AREAS, AND ALONG ROADS, LANES, AND BRIDLE PATHS

For	When	How to Do It	Comments
Pasture maintenance	Spring and fall Scatter droppings in spring and fall, plus 3 to 4 times during the grazing season.	Use a chain-type tine harrow to— 1. Tear out the old, dead material. 2. Stimulate growth through gentle cultivating action. 3. Prevent a sod-bound condition. 4. Increase moisture penetration. 5. Scatter animal droppings to— a. Help control parasites. b. Fertilize a larger area. c. Prevent animals from leaving ungrazed clumps.	Altogether too many livestock pastures are merely gymnasiums or exercising grounds. This need not be so. Through improved pasture maintenance, stockmen can— 1. Produce higher yields of nutritious forage. 2. Extend the grazing season from early in the spring to late in the fall. 3. Provide a fairly uniform supply of feed throughout the entire season.
Pasture renovation	Spring or fall	Use a chain-type tine harrow to work the fertilizer and seed into the soil, and yet destroy a minimum of the existing sod.	Run-down pastures can be brought back into production without plowing and reseeding.
Preparing new pasture seedbed	Spring or fall	Use a chain-type tine harrow to— 1. Level. 2. Smooth down. 3. Pack.	When properly prepared, a seedbed should be so firm that a person barely leaves a footprint when walking across it. The firmer the better from the standpoint of moisture conservation and small seeds.
Racetracks; show-rings	Whenever the track or ring becomes bedded. Just before the race or show; and between races or show events.	Set a chain-type tine harrow for maximum or light penetration, depending on the condition of the track or ring. Use a chain-type tine harrow as a drag mat to smooth and fill holes.	Good racetracks and show-rings must be firm, yet resilient. Because it's flexible, this harrow can be pulled at good speed, as is necessary between races or show events, and yet do an excellent job of smoothing and filling holes.
Along roads, lanes, and bridle paths	Whenever they become rough or uneven.	Use maximum penetration of chain-type tine harrow to put in shape; then turn harrow over to level and fill up holes.	

PART II—THE WESTERN RANGE[5]

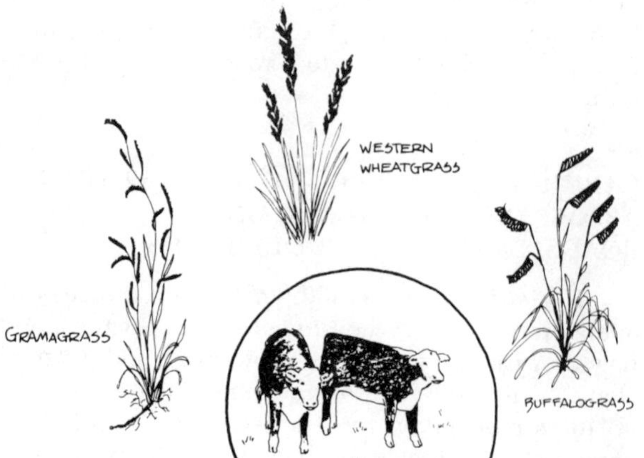

Fig. 5-9. Three common range grasses.

Various geographical divisions are assumed in referring to the western range area—the native pasture area. Sometimes reference is made to the 17 range states, embracing a land area of approximately 1.16 billion acres. At other times, this larger division is broken down, chiefly on the basis of topography, into (1) the Great Plains area (the 6 states of North Dakota, South Dakota, Nebraska, Kansas, Oklahoma, and

[5]The author is very grateful to the following specialists who reviewed this section: Mr. Frank J. Smith, Director of Range Management, Forest Service, U.S. Department of Agriculture, Washington D.C.; Mr. Charles H. Stoddard, Director, Bureau of Land Management, U.S. Department of the Interior, Washington, D.C. (now retired and serving as Resource Consultant, Wolf Springs Forest, Minong, Wisc.); Dr. Leo B. Merrill, In Charge, Texas A&M University Agricultural Research Station at Sonora, Tex.; and Dr. Grant A. Harris, Chairman, Forestry and Range Management, Washington State University, Pullman, Wash.

Texas); and (2) the 11 western states (Arizona, California, Colorado, Idaho, Montana, Nevada, New Mexico, Oregon, Utah, Washington, and Wyoming). In addition to these major and commonly referred to geographical divisions, there are numerous other groupings. These are of importance to the stockman in that they affect the type of management and, to some extent, the kind of animals kept.

Almost half (47.7%) of the land area in the 11 western states is federally owned. Domestic livestock graze on 73% of this area. In 1980, lands in the 11 western states administered by the Bureau of Land Management and the U.S. Forest Service provided grazing all or part of the year for 5,981,980 head of all classes of livestock—cattle, horses, sheep, and goats.

Because of the magnitude of the range livestock industry and the fact that it is a highly specialized type of operation, considerable discussion will be devoted to the range area and the care and management of cattle and sheep in the range method.

The carrying capacity of much of the western range is low, and little of it provides yearlong grazing. Moreover, variation in vegetative types, climate, and topography in the range country is accompanied by great diversity in the seasonal use made of it. As a result, rangelands are usually grazed during different parts of the year, and the herds and flocks migrate with the season, moving to the mountains and higher elevations in summer and returning to the lower ranges in winter.

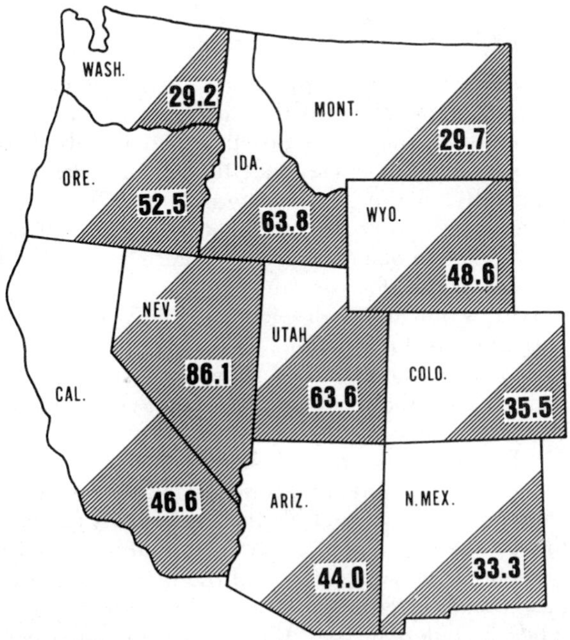

Fig. 5-10. A map showing the 11 western states and the proportion of land in each of these states that is owned by the U.S. Government. (Source: *Public Land Statistics 1980*, U.S. Department of the Interior, Bureau of Land Management, Washington, D.C., p. 9, Table 7)

From the standpoint of vegetation and utilization by livestock, ranges differ from cultivated pastures as follows:

1. **They are less productive**—Generally, their productive capacity is lower. This is as one would expect, for they are largely made up of the residue remaining after the usable agricultural lands have been taken up. Also, plant growth on rangelands frequently is limited by low and undependable rainfall (even drought), short growing seasons, shallow or rocky soil, alkali or salt accumulations, steep topography, etc. Under such conditions, forage plants are usually less resistant to grazing damage than those growing under a more favorable environment.

2. **They are more likely to progress to less palatable plants**—Range vegetation consists of a mixture of native and introduced plants, varying greatly in palatability, nutritive value, and productive ability. Grazing animals select the most palatable plants first. Thus, unless careful management is practiced, the best plants are crowded out through a combination of grazing injury and competition from the ungrazed, low-value plants. Continued poor management can result in good forage plants being almost completely replaced by low value annual, weedy, or shrubby vegetation, or left denuded and subject to severe erosion.

3. **They are more difficult to restore when depleted**—Once a range becomes depleted, it is a slow process to rebuild it. Plowing and drilling are impractical on most rangelands; thus, very often the only feasible way of restoring a range to good condition is to stock it conservatively and manage it well.

4. **They often serve multiple uses**—Rangelands often have other uses in addition to grazing values. Among such uses are: water production, timber production, mineral production, wildlife production, and recreation (camping, hiking, picnicking, etc.).

Thus, many people, in addition to the livestock producer, have an interest in the grazing management practiced on ranges. This is part of the justification given for federal government ownership of large tracts of rangeland.

RANGE MANAGEMENT CONSIDERATIONS

Good range management may be achieved if an inventory or analysis is made of the forage resources and all contributing factors, followed by a sound plan of management based upon the analysis. Consideration should be given to such factors as proper stocking rate, and safe degree of use, season of use, kind of livestock, condition and trend of forage, soil stability, system of use, improvements needed, etc.

Stocking Rate

The key to successful long-term operation of rangeland lies in making (1) a reliable determination of the land that is suitable or adaptable to grazing use over a long period of time; (2) a realistic estimate of grazing capacity for this land; and (3) a flexible stocking rate, even within a single season, followed by (a) application of proper stocking intensity, and (b) frequent observations to determine the effect of the stocking rate upon changes in condition of the forage cover. Too light stocking wastes forage, while too heavy stocking results in a change of forage plant cover from an abundance of valuable forage plants to an abundance of worthless plants.

Of course, the stocking rate for any given unit may vary widely from year to year, and within a given season, depending on the forage production as affected by weather and other factors. For this reason, stocking should either be adjusted to forage yield each year, and within season, or set at a constant rate that will assure a sustained yield of most valuable forage plants (constant stocking at about 25% below average capacity will usually achieve the latter).

Recognition must also be given to the fact that animals do not graze uniformly over a range unit—that certain areas are more attractive to them. Consequently, some areas produce most of the grazed forage, while others may go practically unused. Cattle tend to congregate on fairly level creek bottoms, ridge tops, and around water and shade; whereas sheep, especially if herded, can be moved more uniformly over a unit. But even sheep graze some areas more heavily than others if not herded properly. For the purpose of determining grazing capacity, the key areas—those rather extensive parts of the range which are most heavily grazed—must be given greatest consideration. If preferred or key areas are maintained in good condition, the whole unit will generally remain in good condition. Conversely, if key areas are allowed to deteriorate, the grazing capacity of the whole unit will be endangered.

Grazing capacity determinations are relatively complex and require careful study over a period of several years. They are arrived at most simply and accurately by observing soil stability conditions and changes in plant cover. If the best plants are being destroyed and soil movement is observed, numbers of animals or season of use should be reduced; conversely, if excessive forage remains at the end of the grazing season, numbers should be slowly increased until a balance is struck.

The following rule of thumb, applied to the more heavily grazed key areas, may be used in arriving at the proper stocking rate: "Use half and save half, and the half you save will grow bigger and bigger." The rule refers to half the weight, which is concentrated at the bottom of the plant, and not to half the height. Thus, when the 50% rule of thumb is applied to bluebunch wheatgrass, a common range plant, it means that about 75% of the bunches have been grazed to an average stubble height of about 4 inches and the remaining 25% of the plants left relatively ungrazed.

In arriving at grazing capacity, it is generally wise to seek assistance from qualified range technicians who need to know:

1. The potential of the particular range.
2. The present state of the vegetation as it relates to potential on each site.
3. The alternative methods of changing present conditions to meet management objectives, including such things as flexible stocking, seeding, brush control, fences, watering, and trails.

The commonly used terms for describing range condition are (1) excellent, (2) good, (3) fair, and (4) poor. If the range is covered almost entirely with high-value forage plants which are producing near maximum for the site and has a stable soil, it is classified as being in excellent condition. If the best plants are scarce and the soil is exposed and unstable, it is classified as being in poor condition. Good and fair classifications are intermediate. The trend in condition is also important; if the range condition is improving, the trend is upward, and vice versa. Actually, the range condition reflects the kind of management practiced in the past.

Season of Use

A prime requisite of successful management for both cattle and sheep is that there shall be as nearly year-around grazing as possible and that both the animals and the range shall thrive. In some areas, especially in the southwestern Great Plains area, these conditions are met without necessitating extensive migration of animals. The winter climate is mild, and the native forages cure well on the stalk, thus providing nutritious dry feed at times when green vegetation is not available. Generally speaking, however, most of the cattle and sheep from such areas are marketed via the feeder route rather than as grass-fat slaughter animals.

In general, the most desirable management, both from the standpoint of the animals and the vegetation, consists of the proper seasonal use of the range. Although there is wide variety in the customs and requirements for seasonal use of the range—because of the spread in climate, topography, and vegetative types included in the vast expanse of range country—seasonal-use ranges are usually placed in four major classes: (1) spring-fall, (2) winter, (3) spring-fall-winter, and (4) summer.

Because a range band of sheep can be moved and herded on unenclosed areas with greater ease than a herd of cattle and because investigations in range livestock management have been conducted more extensively with sheep, greater seasonal use of ranges is made with sheep. On the other hand, the more progressive cattlemen are finding ways and means of adopting many of the same methods.

Despite the value of yearlong grazing, it is recognized that the prevalence of severe winters in some parts of the West preclude winter grazing except to a limited degree, and stock must be fed during at least a part of the winter season. Where these conditions prevail, cattle and sheep are usually wintered in the irrigated valleys, close to the feed supply, especially a supply of alfalfa or meadow hay.

Some pertinent points in determining the proper season of use of the range follow:

1. **Elevation**—Generally speaking, vegetative development is delayed 10 to 15 days by each 1,000-foot increase in elevation. Also, severe storms occur later in the spring and earlier in the fall at higher altitudes than at low, desert locations.
2. **Availability of water**—Certain desert areas are so poorly watered that only the occurrence of winter snows makes their use practical.
3. **Early forage "washy"**—Early spring forage is extremely "washy," and may be incapable of supporting stock. Spring grazing should be delayed until the plants are developed enough to meet the nutritive needs of animals.
4. **Soil tramping**—Soil tramping may be serious in early spring. In order to avoid plant damage and soil compaction, grazing should be delayed until the soil is firm.
5. **Poisonous plants grow early**—Most poisonous plants are very early growers and cause their greatest damage when animals are turned out too early. Larkspur, which affects cattle, and death camas, which affects sheep, are two examples. Poisoning losses from these two plants are usually negligible if stock are detained until the best forage plants have made suitable growth.
6. **Winter range should be saved**—If stock are allowed to remain on winter ranges too long after spring growth begins, the next winter's feed will be reduced, because the forage produced on these ranges grows mainly during spring and early summer.

Kind of Livestock

Sheep and cattle share in the utilization of the western range. In fact, some ranges are simultaneously grazed by these two kinds of animals. This dual system of grazing is practical and beneficial provided that the grazing capacity for each is properly adjusted so that the major forage plants are properly used, and that, at intervals, a careful determination is made of condition and trend of soil and forage.

Actually, economic factors—often unrelated to range characteristics—probably have the greatest influence on the selection and popularity of kinds of livestock. The kind which the operator feels will return the greatest net profit is selected, and the choice changes with changing times. Nevertheless, range characteristics may be so specific as to favor one kind of livestock to the point that other kinds would be produced under handicap. Among such range characteristics which should be considered in the choice of a kind of livestock are:

1. **Poisonous plants**—The presence of certain poisonous plant species may limit the use of the range to one kind or another of livestock. Thus, larkspur is a serious menace to cattle, but normally sheep are not affected by it. On the other hand, generally cattle may safely graze lupine-infested ranges, which are sometimes extremely dangerous to sheep. Many other examples of selective poisoning could be cited.
2. **Topography**—Cattle prefer level to gently rolling topography, whereas sheep and goats are better adapted than cattle to steep, rocky, or bushy ranges. The latter seem to have a natural instinct for climbing, and, through the efforts of the herder, they can be encouraged to graze the more difficult terrain. In addition, because of greater ease in herding and moving about on unfenced public domain, sheep are trailed about more than cattle, thus more effectively utilizing seasonal ranges.
3. **Water**—Sheep and goats are better adapted than cattle to more poorly watered ranges, because they can go for longer periods without water. Also, sheep utilize snow as a sole source of water more satisfactorily than cattle; therefore, sheep use range dependent on snow for water more efficiently than cattle.
4. **Vegetative cover**—In general, sheep do not utilize tall-growing grasses so effectively as cattle. Sheep and goats are weed eaters and browsers; and goats probably do better than sheep on a straight diet of browse. Horses are more selective than any other kind of livestock; they prefer grasses, although they will eat small amounts of other kinds of forage. Hogs do best on acorns, pods of certain leguminous shrubs, roots, and other concentrated feeds found on the range only during limited seasons and in certain areas, principally in the Southeast and Southwest.
5. **Predators, insects, and diseases**—Coyotes are serious predators of sheep, but bother cattle very little. Thus, heavy concentrations of coyotes, or other sheep-killing predators, may make sheep raising unprofitable, but not present serious problems to cattle production. The presence of certain insects and diseases may also become factors in the selection of the best suited kind of livestock.

Theoretically, the most efficient use of most ranges can be made by two or more kinds of livestock grazing at the same time; by "common use" or "dual use." The most popular combination is that of cattle and sheep. Destructive grazing often results therefrom, however, because common use requires much more critical grazing management than grazing with one kind of livestock only. This is so primarily because the most popular parts of the range—waterholes, creek bottom meadows, ridge tops, etc.—are the most preferred by all kinds of animals; and, in addition, many of the most valuable forage plants are preferred by both cattle and sheep. As a result, only very careful management can prevent the destruction of these most valuable range areas. In brief, a range unit cannot support its full quota of cattle in addition to its total capacity for sheep. Rather, a studied adjustment should be made to fit the particular unit, based on topography, water distribution, class of forage, and other considerations.

6. **Big game population**—Deer compete more directly with sheep, and elk with cattle.

RANGE GRAZING SYSTEMS

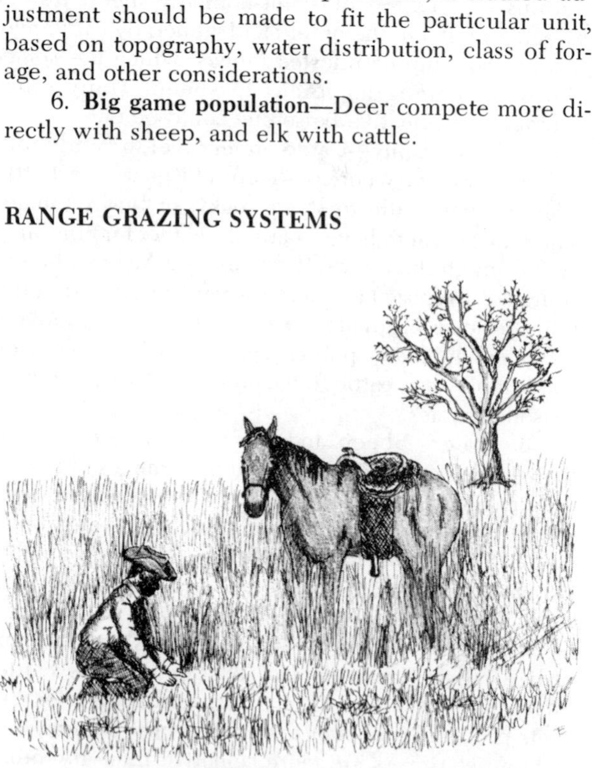

Fig. 5-11. A good grazing system is essential for profitable ranching.

Consciously or unconsciously, ranchers follow one or various combinations of the following grazing systems: (1) continuous grazing, (2) rotation grazing, (3) rotation-deferred grazing, or (4) rest-rotation grazing.

Currently, most progressive ranchers of the northwestern United States use combinations of the four grazing systems, adjusting the system to fit their range needs. In the Great Plains and other regions of good summer rainfall, continuous grazing is most widely used.

Continuous Grazing

Perhaps most ranges of the West are grazed mor or less continuously, although some rest is given ther through the use of seasonal ranges (such as in migra ing to summer ranges in the mountains). Wher continuous grazing is *moderate*, it is perhaps mor suitable and practical than rotation-deferred or res rotation grazing under the following conditions: (1 when the construction of fences or barriers is ver costly, (2) when the important forage species are nc dependent upon reseeding for reproduction, and (3 when seasonal ranges are available and used.

Until and unless more research studies revea that rotation grazing is superior, from the standpoin of both stock and vegetation, continuous grazing wil be followed most extensively.

The Texas Station reports that their studies shov conclusively that, from the standpoint of range im provement and economic return, rest-rotation grazin is superior to continuous grazing.

Rotation Grazing

Rotation grazing is that system in which the graz ing of areas is alternated at intervals throughout the season. A heavy concentration of animals is placed on a given area for a few weeks, after which all the stock are moved on to another area or areas and are finally returned to the first field when the growth is sufficient to withstand another period of grazing. This system is best adapted to the utilization of cultivated pastures in the irrigated valleys of the West or to the humid regions of the United States. However, if a high intensity-low frequency system is followed, it will work on the arid ranges of the West.

Many sheep allotments on national forests are managed on a rotation basis, on a once-over system. Other sheep allotments are used every other year, particularly in high rough country.

Rotation-Deferred Grazing

In rotation-deferred grazing, the range usually is divided into three to five or more units. The grazing on at least one unit is deferred each year until after the seed crop has matured or through a complete growing season or period. The next year a second area is deferred and the grazing on the first area is delayed as late as possible to afford opportunity for the young seedlings to become established. By so treating a new unit each year, the entire area is rested, allowed to reseed itself, and grazed in rotation.

Sometimes rotation-deferred grazing is used as a part of rotation grazing and continuous grazing to improve plant vigor, ensure natural seed production and establishment, or in conjunction with practical range reseeding and brush control.

Rest-Rotation Grazing

This is a relatively new range grazing system. It consists of resting one subunit (range or area), while grazing the others.

Range specialists are agreed that a rest-rotation grazing system has much to offer. However, there are two schools of thought as to the best way in which to apply it, with the controversy centering around the length of the rest period and the intensity of grazing. The two systems are:

- **Rest-rotation grazing, with the alternate resting of each pasture for one year**—Essentially, this system consists of dividing a grazing area into 3 pastures and using them in a 3-year rotation. One pasture is rested while the other 2 are grazed. Here is how it works:

1. Rest one pasture completely for one year, to allow establishment of valuable seedlings. During this rest period, both low-value and high-value plants will recover; but it is hoped that the latter will be favored.
2. Crowd all the livestock on the two remaining, unrested pastures, thereby forcing the grazing of low-value plants and unpopular sites. However, the following differences will prevail in the grazing of these two pastures:

   a. One pasture will be grazed continuously all season.

   b. The other pasture will be allowed to mature seeds before turning stock in, following which it will be grazed heavily so as to tramp the seeds into the ground.

3. Rotate the use of the pastures the next year. Thus, during a given three-year period, each of three pastures will be (a) completely rested one year, (b) grazed continuously all season one year, and (c) allowed to make seed, followed by heavy grazing one year.

- **Rest-rotation grazing, using a high intensity-low frequency system**—This system uses one herd for five or more pastures. Each pasture is grazed for several weeks (but not utilized much over half), then rested for at least four months. Advocates of the high intensive-low frequency system claim that, from the standpoint of utilizing less palatable forage, it is far superior to rest-rotation, with each pasture rested one year.

RANGE IMPROVEMENT METHODS

The warning signals of a range that is on the downgrade and that is in need of improvement are:

1. Desirable forage plants "going out" and being replaced by undesirable ones; the number of young, inferior plant species increasing.
2. Thinning of perennial grass cover, with the grass tufts breaking down and dying; and an increase in annual plants and perennial weeds. The poorer the condition of the range, the more rapidly this process takes place.
3. Weakened vitality of the important forage plants as shown by pale color and reduced height and yield in periods favorable to good growth.
4. Increased soil erosion, by wind and/or by water.
5. Excessive trampling damage.

The above warning signals have appeared in various intensities over part of the western range area of the United States.

There is no quick, easy, and inexpensive method by which poor ranges can be improved. However, one or more of the following methods should be employed.

Conservative Stocking

Usually controlled stocking and natural range reseeding are accomplished by employing one or more of the following practices: (1) rotation-deferred grazing; (2) rest-rotation grazing; (3) a lighter continuous grazing load; or (4) a shorter season of use, such as may be accomplished by using supplemental pastures.

With seeded areas in humid regions, it is generally recommended that pastures be grazed fairly close at least once each year and that the plants not be permitted to reseed. In continuously grazed range areas, however, good management consists in limiting the number of animals to the point where at least one-fourth of the seed from the better forage plants is permitted to ripen annually. This requires frequent and careful inspection to make sure that the better forage species are not progressively being eliminated.

Distribution of Animals on the Range

Next to the proper rate of stocking and proper seasonal use, distribution of the animals on the range is the most important feature in range management. Proper distribution of animals is reflected in more even utilization of the forage. This assignment is more difficult with cattle than with sheep, especially on rough mountainous land. Cattle have a strong tendency to utilize the flatter areas and to congregate around watering places. Also, sheep are usually herded.

Better distribution of animals on the range may be accomplished through (1) fencing; (2) riding the range (or herding); (3) providing sufficient watering places (under ideal conditions, the distance between water in rough country should not exceed ½ mile for cattle or 1 to 2 miles for sheep and goats; in level country, 1 to 1½ miles for cattle or 2 to 3 miles for sheep and goats—water hauling on both cattle and sheep ranges is increasing); (4) systematically locating salt grounds away from watering areas in underused areas and salting at the proper intervals and in the right quantities; (5) building trails into inaccessible parts of the range; and (6) controlling livestock pests such as grubs, and flies, which cause animals to congregate and seek protection.

Range Reseeding

Where improper management and overstocking have seriously reduced the quality of the forage and the grazing capacity of the range, some method of reseeding may be the only logical alternative.

Where considerable of the better forage plants remain, natural regeneration is preferred. The latter is accomplished by managing the grazing season so as to favor the propagation of the remaining desirable native forage and by eliminating low-value brush and other competing vegetation. Often the recovery process can be speeded up by eliminating most undesirable vegetation through the use of herbicides. In Western United States, this process is especially successful on range land dominated by sagebrush and wyethia where a residual stand of native grasses is present as an understory. There are also other weed types which respond well to chemical treatment.

Where most of the desirable plants have been destroyed and the soils are suitable, artificial reseeding is advocated—even though it is expensive and subject to failure.

It is recognized (1) that only a relatively small proportion of the western range can be seeded, (2) that seeding is not a satisfactory substitute for good management practices needed to prevent further destruction of forage plants, (3) that if seeding is necessary, some practice(s) followed in the past has been faulty, and (4) that seeded areas require a high level of management to maintain them.

For successful seeding, the following rules are usually adhered to:

1. Select a site where success can reasonably be achieved. Rainfall, soil, topography, and other site factors must be suitable for seeding; otherwise, the practice is foredoomed.

2. Prepare a firm, weed-free seedbed. Light soils receiving limited rainfall are notoriously difficult to firm into a satisfactory seedbed. Also, seedlings of perennial grasses commonly used in range reseeding are weak competitors with weeds.

3. Select a species or mixture of species adapted to the climate and other features of the area being planted and fitted to the seasonal forage needs of th particular ranch. Keep mixtures simple; single specie are often best.

4. Plant just prior to the season of most dependa ble moisture and growing conditions.

5. Cover the seed, but do not get it too deep. Th best depth for most range grasses is ¼ to ¾ inch depth regulators on drills are advisable. In drier areas deep furrow drills are usually more successful.

6. Protect the seeding from grazing use until th plants have become well established. This may re quire from one to three years—depending on specie planted, weather conditions, etc.

7. Fence the seeding or otherwise provide means of managing it. Without adequate control, ani mals congregate on the new seedings and destro them through overuse.

GRAZING PUBLICLY OWNED LANDS

The ownership of United States land is sum marized in Table 5-17.

TABLE 5-17
OWNERSHIP OF U.S. LAND (48 STATES)[1]

Ownership	Acreage	Percentage of Total
	(million acres)	*(%)*
1. Private ownership	1,311	69.1
2. Indian land	51	2.7
3. Public ownership	535	28.2
(a) Federal	415	21.9
(b) State and local governments	120	6.3

[1]USDA sources.

About 44% of the public lands are in Alaska. Be cause of its remoteness and northern location, lan development has been slow in this state. As a result the federal government still owns over 89% of all th lands in Alaska.

The other 56% of the public lands is located i the 48 contiguous states, but it is not evenly distrib uted across the country. Over 87% of the federal land outside Alaska is in the 11 western states.

Today, in the 11 western public land states, th federal government owns and administers approxi mately 261 million acres on which grazing is allowed At one time or another during the year, domestic cat tle and sheep graze on about half of these publi lands. More of the public lands are used for this pur pose than for any other economic activity. In 1980 lands in the 11 western states administered by th Bureau of Land Management and the U.S. Fores ervice provided grazing all or part of the year for ,981,980 head of all classes of livestock—cattle, orses, sheep, and goats.

gencies Administering Public Lands

Because much of the grazing land that ranchers ely upon to maintain their cattle and sheep enter-rises is built up into operating units by leasing or by btaining use permits from several federal and state gencies, private corporations, and individuals, it is nperative that the owner have a working knowledge f the most important of these agencies. Some range perators are placed in the position of using range ented from as many as six landlords—either private, tate, and/or federal.

The bulk of federal land is administered by the ollowing six agencies: the Bureau of Land Manage-nent, the U.S. Forest Service, the Bureau of Indian ffairs, the Department of Defense, the National Park ervice, and the Bureau of Reclamation. The largest and area from the standpoint of grazing permits and tilization of grazing areas by animals is administered y the first three of these agencies; hence, each of hese three agencies is discussed at this point, fol-owed by pertinent information relative to state and ocal government-owned lands, and railroad-owned ands.

1. **Bureau of Land Management**—The Bureau of and Management of the U.S. Department of the In-erior administers about 46% of all federal lands. Ap-roximately half the lands it manages are in Alaska. The remainder is almost entirely in the 11 western tates.

From the standpoint of the stockman, the most mportant function of the Bureau of Land Manage-nent is its administration of the grazing districts es-ablished under the Taylor Grazing Act of 1934 and of he unreserved public land situated outside of these listricts which are subject to grazing lease under Sec-ion 15 of the Act. This federal act and its amend-nents authorize the withdrawal[6] of public domain rom homestead entry and its organization into graz-ng districts administered by the Department of the Interior. Also, this legislation, as amended, allows the Bureau of Land Management to administer state and privately owned lands under a cooperative arrange-nent.

In 1980, the Bureau of Land Management had 54 listricts, operating in the 11 western states and total-ing 154.8 million acres of public lands. In these dis-tricts, 12,793 operators were granted privileges to graze 3,409,050 head of livestock for an average of about 5 months each year. These operators paid the United States, as grazing fees for this range use, a total of $20,034,523. In addition to livestock use, in 1980, public lands supported, for approximately 5 months of the year, an estimated 1.9 million big game animals, most of which were deer.

In addition to, and outside of, the grazing dis-tricts, in 1980 the Bureau of Land Management supervised 17 million acres of public domain in the western states, most of which was leased to 7,695 stockmen for 1.4 million head of livestock for about 5 months. These operators paid rentals in the amount of $3,479,096 for the use of these lands.

Each district is administered by a District Man-ager, who is a technically trained employee of the Bureau of Land Management. He is responsible to the state bureau office for the proper use, management, and welfare of the public land resources of his dis-trict. In turn, the state office is responsible to the Di-rector's office in Washington, D.C.

Grazing privileges are allocated to individual operators, associations, and corporations on the basis of (a) priority of use; (b) ownership or control of base property dependent on grazing district land for forage during certain seasons of the year, or control of per-manent water needed to graze district land; (c) prox-imity of home ranch to the grazing district; and (d) adequate property to supply the feed needed along with grazing privileges, to maintain throughout the year the livestock permitted on public range. All of these lands are subject to classification and disposal under Sections 7 and 14 of the Taylor Grazing Act, for any higher use or other appropriate purpose. Grazing privileges may, therefore, be cancelled whenever such lands are determined to be more suitable for other purposes.

A fee is charged for grazing privileges. In 1981, the basic fee was equivalent to $2.31 per animal unit month (AUM). An AUM is the equivalent of the graz-ing of a mature cow, two weaner calves, or five sheep for one month.

The Taylor Grazing Act has been responsible for many changes, not all of which have been popular. Some stockmen complain about the loss of their ranges; others tell of increased costs; and there are those who resent government controls, and, above all, the confusion which results from dealing with several agencies. Without doubt, many of these criticisms are justified, and some errors in administration should be rectified; but those who would be fair are agreed that the ranges as a whole have improved under the supervision of the Bureau of Land Management and that further improvements are in the offing.

2. **U.S. Forest Service**—Almost one-fourth of the federal lands are administered by the Forest Service. Over 100 million acres of the national forests are used for grazing under a system of permits issued to local

[6]On May 28, 1954, a bill was signed by President Eisenhower ifting the 142-million-acre limitation on public domain lands that can be included in Taylor Grazing Act districts.

farmers and ranchers by the Forest Service of the U.S. Department of Agriculture. In 1980 about 1,255,234 mature sheep and goats and 1,317,696 mature cattle and horses (mostly cattle), owned by over 16,160 paid permit operators, were grazed on national forests for some part of the year. In addition, there were many calves and lambs for which no fee is charged and additional stock that were grazed under free permits to local settlers. This made an estimated total of 2.8 million domestic animals grazed on the national forest ranges in 1980.

The Forest Service issues 10-year term permits to stockmen who hold preferences and annual permits to those who hold temporary use. Among other things, the permit prescribes the boundaries of the range which they may use, the maximum number of animals allowed, and the season in which grazing is permitted.

Preferences may be acquired through prior use, through a grant, or through purchase of land or livestock, or both, of a user who already holds a preference.

The requisites in order to qualify for a permit are:

a. **Ownership**—The ownership of both the livestock and commensurate ranch property.

b. **Dependency**—The need for forest range in order to round out an operation to obtain proper and practical use of commensurate property.

c. **Commensurability**—The ability of the land to support livestock during the period when not on forest land.

A grazing preference is not a property right. Rather, it is approved for the exclusive use and benefit of a person to whom allowed. Preferences or permits may be revoked in whole or in part for a clearly established violation of the terms of the permit, the regulations upon which it is based, or the instructions of forest officers issued thereunder.

A ranger administers the grazing use on each National Forest Ranger District. Several districts (usually three to six or more) comprise a national forest. A forest supervisor, with his staff, administers the national forest. Several national forests, under the direction of a regional forester and staff, comprise a forest region. The Chief administers the Forest Service from Washington, D. C., under the supervision of the Secretary of Agriculture.

As is true in the administration of Taylor grazing districts, local farmers and ranchers act in an advisory capacity in the allocation of grazing privileges and in details of administration of the national forests. About 840 such livestock associations and advisory boards are recognized and in operation.

Forest Service grazing fees are based on a formula which takes into account livestock prices over the past 10 years, the quality of forage on the allotment, and the cost of ranch operation. In 1981, ave age charges were $2.31 per animal unit month (AUM or $2.31 for a mature cow, two weaner calves, or sheep, for a month.

Although shortcomings exist in the manageme of the national forests, it is generally agreed that the ranges have been vastly improved under the admin tration of the Forest Service. Many of them now a proach the quality that existed in their virgin sta Perhaps the most heated arguments betwe stockmen and the Forest Service arise over the re tive importance attached to the multiple use of b game and other wildlife, recreation, etc. For exampl it was estimated that in 1979 there were 3.3 milli big game animals (75% of which were deer) in t national forests. As would be expected, these wi animals compete with domestic animals for use of t range, thus creating a most difficult problem.

3. **Bureau of Indian Affairs**—Most Indian lanc comprising nearly 51 million acres, are really not pu lic lands. Rather, these lands are held in trust for t benefit or use of the Indians and are merely admini tered by the Bureau of Indian Affairs of the Depa ment of the Interior. Because over 80 percent of I dian lands are in the range area of the West, they a suited primarily to livestock. Thus, it is notewort that the sale of livestock and animal by-products reg larly account for ⅔ of the total Indian agricultural i come. Although the Indians themselves own most the stock grazed on these lands, animals owned nonIndians utilize ¼ of the Indian lands devoted grazing. Provision for such use is handled under lea agreement jointly approved by the Indian owners ar the Bureau of Indian Affairs.

Many of the Indian lands have suffered serio vegetative depletion, but a concerted effort is nc being made to decrease livestock numbers in keepi with available feed supplies and to improve the qua ity of animals produced. However, overstocking co tinues to be a difficult problem on the Navajo, Ho and Papago Reservations.

4. **State and local government-owned lands**— total of 120 million acres are owned by state and loc governments. For the most part, the management these areas is diverse and confused, each state a local government having established different regul tions relative to the lands under its ownership. general, however, such lands are operated on a stip lated lease arrangement. On many such areas, ran depletion has been severe.

5. **Railroad-owned lands**—Recognizing that t main deterrent to rapid settlement and developme of the West was the lack of adequate transportati facilities, the federal government very early encou aged the construction and westward extension of t railroads by means of large grants of land. It was i tended that the railroads should sell or otherwi

ɪtilize these lands in financing their costs of construc-ion. These initial grants, totaling 94,355,739 acres, onsisted of alternate sections extending in a check-rboard fashion for a distance of from 10 to 40 miles n each side of the right-of-way. Today, less than 20 nillion acres of these lands are held by railroads. Many of these holdings are leased to stockmen; but because of inconvenience, past abuses, or other reasons, some of these lands are considered worthless for grazing. In general, railroad lease agreements do not restrict the number of stock to be grazed or the season during which the land may be so used.

PART III—GREEN CHOP

Green chop is fresh herbage that is cut and hopped in the field, then fed to animals in confinement. It is also called zero grazing, or soilage. Green chop minimizes the loss of moisture, color, nutrients, and wastage. Alfalfa, Ladino clover, orchardgrass, bromegrass, grass-legume mixtures, Sudangrass, corn, sorghum, soybeans, and cereal grains are sometimes used in this manner. With tall growing crops, more feed value may be realized from a given area than can be obtained by conventional pasturing. However, green chop requires special equipment and harvesting every day. Also, there are harvesting problems in wet weather.

Most green chop is fed to lactating dairy cows, usually in combination with hay or silage because the total intake tends to be greater. Green chop has increased with herd size, more intensive form of dairying, drylotting of cows, and high grain prices. Also, the use of green chop has been facilitated by the greater mechanization present on modern dairy farms.

Fig. 5-12. Harvesting green chop, or soilage.

GREEN CHOP CONSIDERATIONS

Experiments and experiences indicate that the following points should be thoroughly considered before starting a green chop program:

1. **Daily consumption**—At least 100 pounds of green chop per head per day must be harvested for lactating cows.

2. **Emergency reserve**—A supply of hay and/or silage should be available to meet emergencies caused by machinery breakdowns.

3. **Integrate with hay and silage making**—It is important that a green chop program be integrated with hay and silage making, both from the standpoint of amortizing machinery cost and from the standpoint of feed reserves. Surplus forage at peak growth should be made into silage or hay.

4. **Farm application**—Before attempting this sytem of summer feeding, a careful analysis of the time, machinery, investment, and labor requirements should be made.

5. **Green chop calendar**—By carefully planning a green chop program, it is possible to cut and feed green forage to cows continuously from spring to fall.

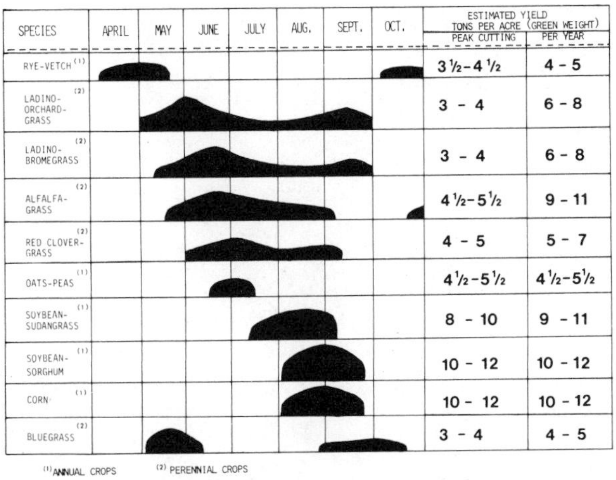
SPECIES
APRIL
MAY
JUNE
JULY
AUG.
SEPT.
OCT.
ESTIMATED YIELD TONS PER ACRE (GREEN WEIGHT)
PEAK CUTTING
PER YEAR
RYE-VETCH (1) 3 ½-4 ½ 4 - 5
LADINO-ORCHARD-GRASS (2) 3 - 4 6 - 8
LADINO-BROMEGRASS (2) 3 - 4 6 - 8
ALFALFA-GRASS (2) 4 ½-5 ½ 9 - 11
RED CLOVER-GRASS (2) 4 - 5 5 - 7
OATS-PEAS (1) 4 ½-5 ½ 4 ½-5 ½
SOYBEAN-SUDANGRASS (1) 8 - 10 9 - 11
SOYBEAN-SORGHUM (1) 10 - 12 10 - 12
CORN (1) 10 - 12 10 - 12
BLUEGRASS (2) 3 - 4 4 - 5
(1) ANNUAL CROPS
(2) PERENNIAL CROPS

Fig. 5-13. A green chop calendar for northeastern United States. (Recommended by Rutgers University, the Soil Conservation Service, and the U.S. Department of Agriculture)

FACTORS FAVORABLE TO GREEN CHOP

Among the factors favorable to green chop are the following:

1. **Increased carrying capacity**—With tall growing crops, 30 to 50 percent greater carrying capacity per acre can be achieved than can be secured from conventional grazing.
2. **It makes it more practical to use distant, small, and scattered pastures**—With large herds, pastures may be so far from the milking parlor as to make it impractical to drive lactating cows back and forth. Likewise, it may not be practical to place animals on small and scattered plots.
3. **It lessens refusal**—Less forage is refused as green chop than when grazed. This is attributed to the fact that it is easier for the caretaker to see leftover forage in a manger than in the field, with the result that feed allowance of green chop is more nearly synchronized with animal needs than carrying capacity of pastures; hence, wastage is lessened with green chop.
4. **It lessens stress due to change**—There is always a certain amount of change where grazing is involved—change from drylot feeds to green grass, and change from pasture to pasture. Even though green chop varies during the season, it usually makes for less change than conventional pasture grazing.
5. **It alleviates the need for fencing, water, and shade**—In areas where green chop is produced, there is no need for fence, water, or shade, because there are no animals therein.

FACTORS UNFAVORABLE TO GREEN CHOP

Like many other good things, there are disadvan- tages to green chop; among them, the following:

1. **Higher harvesting cost**—Green chop require special equipment, and it must be harvested ever day. Hence, it involves more cost for machinery an labor than where animals do their own harvesting.
2. **Green chop makes for late spring start**—Cow can be placed on pasture to graze at least two week earlier than green chopping can commence.
3. **Quantity of green chop varies**—Once gree chopping begins, it is difficult to keep up with it Some of it will likely have to be stored as silage o hay, for later use.
4. **Manure and urine not animal-spread**—Wher green chop is used, manure and urine are not animal spread back on the land. In order to keep green cho production at a high level year after year, either ma nure or chemical fertilizer, or both, must be spread o the land.
5. **Bloat, overeating, toxicity, and hardwar disease**—Bloat can be a problem with green chop especially if it is high in legume. The problem o overeating and toxic plants may be worse with gree chop than with pasture. Toxic plants are chopped an mixed with the feed so that animals cannot pick an choose to avoid them. Likewise, the hardware prob lem may be exaggerated with green chop, becaus pieces of wire may be chopped into bite-sized frag ments.

SECTION 6

HAY AND CROP RESIDUES

Contents

Page

PART I—HAY

History of Hay 669
Magnitude and Importance 669
Kinds of Hay 671
Alfalfa (Lucerne) 672
Cereal Hays 672
Oat Hay 672
Clovers 672
Grass Hays 673
Timothy 673
Lespedeza 673
Soybeans, Cowpeas, Vetch 674
Hay Quality 674
Importance of Hay Quality 674
Characteristics of High-Quality Hay 675
Visual Inspection 675
Federal Grades of Hay 676
Chemical Composition 676
Making Quality Hay 678
Haymaking Systems 683
Long, Loose Hay 683
Chopped Hay 683
Packaged Hay 684
Bales 684
Stacks 684
Cubes (Wafers) 685
Pellets 685
Artificial Drying 685
Storing 687
Additives and Preservatives for Hay 687
Spontaneous Combustion 687
Buying and Selling Hay 688
Hay Sources 688
How Hay Is Priced 688
Futuristic Hay Evaluating and Pricing 688
Freedom from Toxic Residues 691
Hay Shrinkage 692
Hidden Hay Costs 692
Hay Feeding Fundamentals 693

PART II—CROP RESIDUES

Corn Residues 697
Grazing (Pasturing) 697
Stalklage 698
Husklage (Shucklage) 698
Feeding Value of and Supplements for Corn Residues 698
Other Crop Residues 699
Sorghum (Milo) 699
Soybean Refuse 700
Small Grain Refuse 700
Legume and Grass Seed Straws 700
Cottonseed Hulls 700
Treating Crop Residues to Increase Digestibility 700

Fig. 6-1. Haymaking has gone modern! *Left:* The way it used to be done—a backbreaking pitchfork job. *Right:* A pick-up baler with ejector—a bale loader.

Hay is the most important harvested forage fed to livestock, although many other crop residues can be and are utilized. Among the crop residues fed to animals are cornstalks and husklage, sorghum stalks, soybean refuse, small grain straws and chaff, legume and grass seed straws, cottonseed hulls, corncobs, oat hulls, peanut hulls, and a host of others.

The dry forages are all higher in fiber and lower in total digestible nutrients than the concentrates. Hay averages about 28% fiber, and straw approximately 38%, whereas such concentrates as corn and wheat contain only 2 to 3% fiber.

From a feeding standpoint, the following general characteristics of forages are pertinent, although some well-known forages can be cited as exceptions to each characteristic:

1. **Bulk**—They are bulky feeds with a low weight per unit of volume.
2. **Fiber and energy**—They contain more than 18 percent crude fiber, and they are lower in energy than the concentrates.
3. **Digestibility**—They are generally lower in digestibility than the concentrates, due to lignin content.
4. **Minerals**—They are generally higher in calcium, potassium, and trace minerals than most concentrates; but phosphorus content is apt to be moderate to low.
5. **Vitamins**—They are higher in fat-soluble vitamins than most concentrates. Legume hays are good sources of B vitamins.
6. **Protein**—They are variable in protein content. Legume hays may run 20 percent or more crude protein, whereas other forages, such as straws, may have only 3 to 4 percent crude protein.

From an overall nutrition standpoint, forages ma range from very good nutrient sources (such as high quality alfalfa hay) to very poor feeds (such as straws hulls, and some browse). Nevertheless, all of then can be used advantageously, provided (1) they are properly prepared and supplemented, and (2) the feeder uses judgment in selecting the species anc class of animal to which the particular forage is fed Availability, costs, and results should be the deter mining factors in their use, just as the economics o the situation should determine the use of any othe feed ingredient.

PART I—HAY

Hay is forage harvested during the growing period and preserved by drying for subsequent use. I is primarily a cattle, sheep, and horse feed, althougl alfalfa (especially dehydrated alfalfa) may be includec in swine and poultry rations. Average quality hay runs 25 to 35 percent crude fiber and 45 to 55 percen TDN.

Drying, or making hay, is the most commo method of preserving forage for storage, primarily be cause it is relatively easy to handle. It can be stored o transported long, chopped, pelleted, cubed, or pack aged into various types and sizes of bales. Moderi equipment, including conditioners, hastens drying time; and automated systems facilitate handling.

The great capacity and specialized functions o the rumen allow cattle and sheep to use hay, anc other forages, in large amounts. Bacteria and protozo in the rumen break down and make available to the host animal part of the nutrients in cellulose or fi brous material.

In addition to the nutrients that it contains, and to its value in providing feed throughout the year, hay has other values. Dry feed is essential for the proper functioning of the digestive tract; it acts as a stimulant in moving the feed through the intestines, and it maintains the proper conditions in the rumen for the microbial action which plays such a vital role in the digestion of the fibrous portions of feeds. Hay is often used as a supplement to "washy" pastures and succulent silages. Also, it speeds along the development of the rumen function of the young calf, lessens the incidence of ketosis and displaced abomasum in cattle, and prevents a lowering of the fat content of the milk of lactating cows (unless it is finely ground). Also, and most important, good quality hay is a hedge against high concentrate prices, for when the price of such feeds increases disproportionately increased amounts of hay may be fed and concentrates may be decreased, with a higher net return to the producer.

Fig. 6-2. Many serious problems in high-producing dairy herds have been traced to the lack of hay in rations. Increased incidence of ketosis and displaced abomasums have been corrected by adding small amounts of hay to rations.

Despite its several advantages, hay has several shortcomings. It varies in nutrient content and palatability more than any other feed, because of differences in the (1) crops from which it is made, (2) stage of cutting, (3) handling, and (4) weather damage during curing. Not even ruminants can consume enough hay alone to meet the demands of high production; for example, on hay alone, dairy cows will produce only 50 to 70 percent as much milk as they would on a ration consisting of 50 percent concentrates. Also, fiber is poorly digested by monogastric animals, with the result that hay serves primarily as a source of minerals and vitamins for swine and poultry.

An estimated 80 to 85 percent of all hay is fed on the farms or ranches on which it is produced, rather than being purchased. It is important, therefore, that stockmen know how to produce good hay, as well as how to feed it, for most of them determine their own destiny from the standpoint of quality. For this reason, this section covers hay from production to feeding.

HISTORY OF HAY

More than 2,000 years ago, the Roman agricultural writer, Columella, described haymaking as "throwing forage loosely together for a few days to heat and concoct itself and then cool before putting into the mow." But another 20 centuries elapsed before haymaking changed materially. As recently as 1850, it was cut with a scythe; and pitchfork haymaking persisted into the present century. Beginning about 1940, scientists and engineers pooled their efforts to transform forages into high-quality hay. Haymaking went modern, with automated one-man pick-up balers, field choppers, cubing machines, round bales, mechanically compressed stacks, and other modern equipment, replacing the backbreaking, labor-costing methods of old.

Automated haymaking and surplus grains were ushered in together. At the close of World War II, U.S. grain bins bulged. Then, suddenly, in the early 1970s, there were world food shortages. The 20-year grain-feeding binge in the United States began reversing itself. Now, and in the future, more and more grain will be used for direct human consumption. Animals (especially cattle and sheep) will increasingly be "roughage burners."

MAGNITUDE AND IMPORTANCE

Hay is the most important harvested forage for U.S. livestock and ranks third among all livestock feeds, being exceeded only by pasture and corn. The importance of the nation's hay crop is further attested to by the fact that the total area devoted to hay in the United States runs around 60 million acres, the total production averages from 125 to 130 million tons, and the annual crop is worth more than $8 billion. On an air-dry tonnage basis, about 3 times as much hay is produced as silage.

Despite the importance of hay, no other feed crop suffers a higher loss of nutrients from the time it is cut to the time it reaches the manger. During the curing process, the quality and feeding value of hay is decreased rapidly by rain, sun bleaching, raking, handling when too dry, and storing with too much moisture. Studies by the U.S. Department of Agriculture revealed that the following losses accrued in field cured, second-cut alfalfa hay from the time of cutting to the time of feeding: leaves, 35%; dry matter, 20%; and proteins, 29%.

The longer hay remains in the field until it is dry enough to store, the greater the nutrient losses (Fig.

6-3). These losses have been estimated to have a feeding value of more than a billion dollars annually.

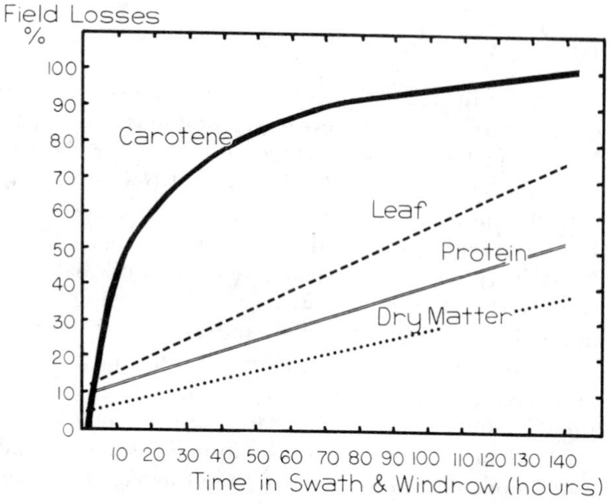

Fig. 6-3. Losses in sun-curing alfalfa hay as related to time in the field to reduce moisture to a safe level for storage. (Adapted by the author from *High Quality Hay*, ARS 22-52, Agricultural Research Service, USDA)

Hay As an Energy Source

Hay is primarily a source of energy for cattle, sheep, and horses. Table 6-1 shows the percentage of total energy (TDN) intake provided to these species by hay and other kinds of feeds.

As shown in Table 6-1, hay is a more important source of energy for dairy cows than for any other class of farm animal. But it is also an important feed source for horses and beef cattle. It is noteworthy, too, that better than one-half the total hay tonnage produced in North America is fed to dairy cattle, while beef cattle consume almost 40 percent of all hay produced. As increasing quantities of concentrates go to feed the world's hungry people, it is expected that stockmen will depend even more on hay to meet a larger percentage of the total feed needs of ruminants.

Comparative Value of Hay

In ruminant rations, hay is primarily a source of energy, but the legumes also serve as a source of protein. For swine and poultry, hay (especially alfalfa) is fed primarily as a source of minerals and vitamins.

Table 6-2 shows the comparative cost of the total digestible nutrients (TDN) and crude protein (CP) of 100 pounds of dry matter from corn grain, corn silage and three different types of alfalfa hay, (high-, medium-, and low-quality).

It is noteworthy that, in terms of energy and protein provided by the different feeds, high-quality alfalfa hay had a value of 83 percent that of corn grain. These calculations show that for supplying energy and protein, high-quality hay is competitive with the other two major livestock feeds.

Hay As a Grain Replacement

In the future, stockmen will increasingly rely upon the ability of ruminants to convert coarse forage, grass, and by-product feeds, along with a minimum of concentrate, into food for human consumption, thereby competing less for humanly edible grains.

Ruminants can make this transition to more forage with ease, for to them it is merely a return to nature. This fact was confirmed by the U.S. Department of Agriculture in a study of all-forage rations for fattening cattle, the results of which are reported in Table 6-3.

As a result of the experiment summarized in Table 6-3, the U.S. Department of Agriculture researchers reported as follows: (1) Beef cattle of an acceptable quality were produced on a pelleted, all-forage ration; (2) steers on an all-forage ration had to

TABLE 6-1

PERCENTAGE OF ENERGY SUPPLIED BY HAY AND OTHER KINDS OF FEEDS[1]

Animal	Hay	Other Harvested Forages	Pasture	All Forage	Concentrates	Total
	(%)					
Dairy cows not in lactation	29.0	5.9	45.7	80.6	19.4	100
Lactating cows	23.1	19.4	19.6	62.1	37.9	100
Horses and mules	18.3	10.2	50.9	79.4	20.6	100
Fattening beef cattle	16.3	8.7	5.2	30.2	69.8	100
Other beef cattle	15.5	4.1	71.7	91.3	8.7	100
Sheep and goats	4.7	3.1	81.8	89.6	10.4	100

[1]Based on USDA data. From paper entitled, "Hay Production, Preservation and Quality," by John E. Baylor, The Pennsylvania State University, *Beef Cattle Science Handbook*, Vol. 13, 1976, p. 199, published by Agriservices Foundation, edited by M. E. Ensminger.

TABLE 6-2

COMPARISON OF COST OF TDN AND CRUDE PROTEIN IN 100 POUNDS OF CORN GRAIN, CORN SILAGE, AND THREE QUALITIES OF ALFALFA HAY (Moisture-free Basis)[1]

	No. 1 Corn Grain	Corn Silage	Alfalfa Hay High-Quality	Alfalfa Hay Medium-Quality	Alfalfa Hay Low-Quality
TDN (%)	94.7	67.7	58.9	57.6	54.7
TDN value[2] ($)	4.92	3.52	3.06	3.00	2.84
Crude protein (%)	10.2	8.2	16.1	14.1	11.9
Crude protein value[3] ($)	1.37	1.10	2.16	1.89	1.59
Total value (TDN + CP) ($)	6.29	4.62	5.22	4.89	4.43
Value compared to corn grain (%)	100	73.4	83.0	77.7	70.4

[1]Composition values for corn grain and corn silage from Section 4, Table 4-101, Feed Composition, of this book; and for alfalfa hay from *Feeding Washington Dairy Cows*, Washington State University Ext. Bull. 486, p. 8.

[2]Corn grain used as standard TDN source. Corn priced at $94/ton, and TDN cost/pound of 5.2 cents.

[3]44% soybean meal used as standard protein source, priced at $118/ton, with cost per pound of crude protein of 13.4 cents/pound.

TABLE 6-3

FEEDLOT PERFORMANCE AND CARCASS EVALUATION OF STEERS FED ALL-FORAGE VS ALL-CONCENTRATE RATIONS[1]

Item	All-Forage Ration[2]	All-Concentrate Ration[2]
Average daily feed intake (lb)	23.3	16.0
(kg)	10.59	7.26
Average daily feed intake in % of body weight	3.23	2.15
Average daily gain (lb)	2.3	2.8
(kg)	1.05	1.27
Feed-gain ratio	10.06	5.71
Average carcass grade	Low Choice	Medium Choice
Dressing percentage (%)	55.4	59.9
Marbling score	Abundant	Abundant
Rib eye area (sq in.)	11.0	10.6
(sq cm)	71.1	68.5
Fat over rib eye (in.)	.37	.67
(mm)	9.4	17.0
Taste panel evaluation[3]	7.6	7.2

[1]Oltjen, R. R., T. S. Rumsey, and P. A. Putnam, "All-Forage Diets for Fattening Beef Cattle," *Journal of Animal Science*, Vol. 32, 1971, pp. 327-333.

[2]Corn grain provided 90% of the all-concentrate ration; pelleted alfalfa provided 98% of the all-forage ration.

[3]Overall desirability rated on a scale of 1 to 9, with 9 being the most desirable.

be fed a month longer than those on the all-concentrate ration; (3) the all-forage steers consumed about 95% as much metabolizable energy and were about 86% as efficient converters of it to body weight gains as were the all-concentrate steers; (4) the forage-fed steers had only 55% as much backfat and 80% as much fat over the rib eye as did the all-concentrate steers; (5) there was a 4.5% difference in dressing percentage in favor of the animals receiving the all-concentrate ration; and (6) at early 1975 prices of the 2 rations (which were approximately equal) and of the beef produced from them, the steers on the all-concentrate ration returned about $30 per head more to the feeder than the all-forage steers, primarily because of their 4.5% higher dressing percentage. Based on this study, the following conclusion may be drawn: Since cattle fed high-roughage rations normally have lower dressing percentages, forages must be cheaper than grain in order for the feeder to obtain the same net return; this situation usually exists relative to pastures, but it doesn't always apply to dry forages.

KINDS OF HAY

Hays are made from a great variety of legumes, grasses, or cereal crops. In terms of total tonnage produced annually, alfalfa accounts for approximately 58 percent of the nation's hay production. Many different kinds of hay make up the other 42 percent of the country's hay supply; among them, cereal hays made from oats, barley, wheat, and rye, and grass hays made from Bermudagrass, prairie grass, redtop, Johnsongrass, orchardgrass, and timothy. Also, more and more farmers are coming to appreciate the flexibility afforded by growing varieties of grasses and legumes that may be used 3 ways: for pasture, for hay, or for silage. With such an arrangement, surplus pasture may be converted into hay, or, if the weather is not favorable for haymaking, the crop can be ensiled. The pasture tables in Section 5 of this book, Tables 5-1 to 5-10, are equally applicable for hay (and silage) crops. This is especially true of the taller growing varieties listed therein.

Whenever feasible, it is recommended that a legume be grown for hay, for the reasons that, in comparison with grasses, legumes are (1) higher in protein, vitamins, and minerals; (2) higher yielding; and (3) nitrogen-fixing when inoculated, because the bacteria (rhizobia) on their roots take free atmospheric nitrogen from the air. However, a mixture of grasses and legumes is often preferred for reasons of palatability, ease in curing, erosion control, and lessening bloat. Also, where horses are involved, a good-quality grass hay may be preferable.

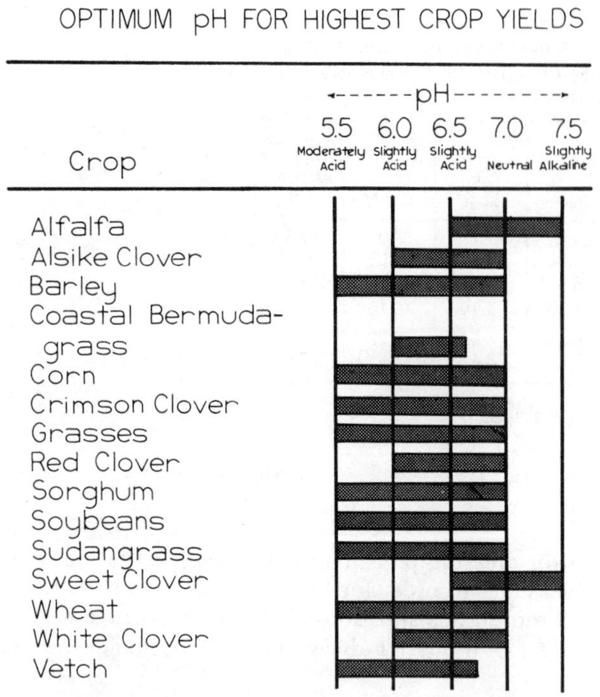

Fig. 6-4. Crops have different pH requirements. Legumes, such as alfalfa, need a nearly neutral soil to promote growth of nodule-forming bacteria.

Alfalfa (Lucerne)

Alfalfa is an important, perennial, leguminous forage plant with trifoliate leaves and bluish-purple flowers. It is grown widely, principally for hay. Alfalfa is capable of surviving dry periods because of its extraordinarily long root system, and it is adapted to widely varying conditions of climate and soil. It yields the highest tonnage per acre and has the highest protein content of the legume hays.

Alfalfa may be grown as a single-species crop or in combination with other legumes and with grasses. Usually, the yield and quality of mixtures are closely related to the amount of alfalfa in the mixture. Alfalfa is high in calcium, protein, and carotene, and in many other minerals and vitamins. It is subject to loss of leaves if not harvested properly; and since the leaves are the most nutritious part of the plant, the feeding value of alfalfa is materially lowered by shattering.

Alfalfa does not do well on acid soils; hence, in many areas it is necessary to include lime as part of the fertilization program. Neither will it thrive where the water table is high, because of its long taproot. Also, it is vulnerable to overirrigation that interferes with the aeration of the soil. When grown on dry land (land without irrigation) and in areas of low annual rainfall, alfalfa may produce only one cutting and a limited yield per season, or perhaps 2 crops or fairly good yield in years with more precipitation. In the warmer irrigated areas of southwestern United States growers may get as many as 8 cuttings each year. Yields per cutting vary from about ½ ton to 3 or more tons per acre. The crop will continue to yield well for 2 to 6 or more years, depending on the location and the care that it receives. It is often used in a short term rotation for just one year in order to improve the fertility and tilth of the soil for a cash crop to be grown the following year.

There are many varieties and strains of alfalfa with variations in the time to maturity, resistance to disease and insects, and feeding value. For individual farm varietal recommendations, the farmer should seek the advice of his county extension agent or soil conservation office, or write to his state agricultural college.

Cereal Hays

Small grains, such as barley, oats, rye, and wheat make satisfactory hay crops if cut when the stems and leaves are still green. They make the most nutritious hay when cut in the soft dough stage. If permitted to develop to a more mature stage, the yield will be higher and the nutritional qualities will be satisfactory as long as the vegetative parts are still green. Cereal grain crops are generally cut for hay (1) when it is necessary to get them off the ground early, such as when they are used as companion crop (nurse crop in establishing grasses or legumes for pasture or hay or (2) when double cropping is planned. In comparison with the legumes and most of the grasses, the cereal hays are lower yielding and not so nutritious. They are generally low in protein, calcium, and carotene; and if permitted to get mature, they are usually higher in fiber than the more common hay crops.

OAT HAY

Oat hay is an excellent feed for horses. It's easy to cure, and horses like it. Early cutting (in the soft dough stage) greatly increases its feeding value, due to the higher protein content. Even though considerable energy is stored in the kernels at maturity, shattering of the grain during harvesting of mature oats results in energy losses and decreased feeding value compared with early cut hay.

Oat hay is low in protein; hence, its feeding value is greatly enhanced when it is fed with alfalfa or some other legume.

Clovers

The clovers are usually grown for hay in combination with grasses. The clover-timothy combination s the most popular. In comparison with alfalfa, clover-timothy mixed hays are lower in protein and not so high in quality. The lower quality is due to the fact that, generally speaking, at cutting time the timothy is at the right stage of maturity whereas the clover is overripe.

Red clover (medium red clover) is the most common of the clovers used as livestock feed. It is grown mostly in the north central and northeastern states and is usually seeded with timothy. It does not have as deep a root system as alfalfa, but it is more penetrating than some of the clovers with shorter growth profiles. Nutrient content and yield per acre are not competitive with alfalfa, either as a hay crop or for other uses as a forage.

Alsike clover is used with or in place of red clover. The two clovers are similar in many ways, but alsike clover has a more shallow root system and can grow on land with a higher water table than either red clover or alfalfa.

Ladino clover (one of the white clovers) has gained in popularity in recent years, more as a pasture plant than as a hay crop. Even though it is not usually considered to be a hay crop, it adds to the quality of hay in mixtures of grasses and legumes. In combination with alfalfa, it provides a lush growth right down to the ground, whereas alfalfa usually loses its lower leaves and becomes more stemmy in this region.

Sweet clover was once quite important as a hay crop, but it has declined in popularity in recent years, partially because of the superior quality and yield of alfalfa with which it competes, and partially because of "sweet clover disease" which it sometimes produces. This poisoning results from a mold which grows in the rank stems of sweet clover before they are dry and produces the chemical, dicoumarol, which prevents blood from clotting. Sweet clover disease can be prevented by more rapid drying of the crop or by administering vitamin K.

Grass Hays

Most tall-growing grasses may be used as hay crops. Timothy, native grasses, orchardgrass, bromegrass, and Johnsongrass are among the grasses which are commonly harvested for hay. Sudangrass and some of the hybrid sorghums, which may be considered as either cereal grains or grasses, are becoming more popular as livestock feed, for both pasture and hay.

The grasses are generally lower in protein and calcium, higher in fiber, and less palatable than the standard legume hays; and, except for Sudangrass and hybrid sorghums, they are not as high yielding as most of the legumes. However, they will grow under more diverse conditions than most legumes and they often appear as the native vegetation in areas which cannot be cultivated. Grass hays may be fed to any class of livestock, but they are improved when mixed with legumes to balance out some of the nutrients in which they are deficient. Grass hays are generally preferred by horsemen because they cause fewer digestive disturbances in equines than do the legumes.

TIMOTHY

Timothy is the preferred hay by most horsemen. Although it may be grown alone, it is commonly seeded in mixtures with medium red or alsike clover.

Timothy is easy to harvest and cure. However, in comparison with hay made from the legumes, it is low in crude protein and minerals, particularly calcium.

As with all other forages, the feeding value of timothy is affected by the stage of growth of the plants at the time of cutting. With increasing maturity, (1) the percentage of crude protein decreases, (2) the percentage of crude fiber increases, (3) the hay becomes less palatable, and (4) the digestibility decreases. However, delaying cutting until timothy has reached the full bloom stage, or later, usually results in the highest yields. When both yield and quality are considered, the best results are obtained when timothy is cut for hay at the early bloom stage.

Lespedeza

Lespedeza is a legume similar to alfalfa, but with the capability of growing on soils too poor for alfalfa. It is especially valuable in the southern states, where it is used both for pasture and hay.

There are several types of lespedeza, most of which originated in the Orient. The annuals have become the most widely grown legume in some sections of the South. If cultured properly, an annual will reseed itself and act more like a perennial than an annual. Among the annuals are common lespedeza, Korean, Kobe, and other varieties which have been developed by plant breeders in this country.

Lespedeza is not a high-yielding plant; but it makes an excellent feed if (1) it is cured without weather damage, and (2) it is fine stemmed and free from foreign material. It is a good forage for most classes of livestock; but it should be harvested when in the early bloom stage in order to reduce the accumulation of lignin and tannins, both of which lower its value as a feed. Where alfalfa can be grown satisfactorily, lespedeza usually cannot compete as a feed crop, either on the basis of its nutritional value or on its yield.

Sericea lespedeza is a perennial legume which grows much taller and coarser than the annual varieties. Like the annuals, it is especially valuable on land that is too poor for alfalfa; it will even outyield the annual varieties on such land. It should be cut a few inches above the base of the stems, because the new growth arises from the shoots, rather than from the crown as does alfalfa. It may be used either as a hay or as a pasture crop, or as a combination of the two.

Soybeans, Cowpeas, Vetch

Soybeans, cowpeas, and vetch, all of which are legumes, are often made into hays. They are not as valuable as alfalfa; generally, they are stemmy and difficult to cure. Nevertheless, if they are cut at the proper stage of maturity and cured without loss of leaves, they make good feed.

It is sometimes advantageous to get a hay crop off the ground early so that a second crop may be seeded. Soybeans, cowpeas, or vetch may be used for this purpose. They may either be seeded alone or in combination with grasses or cereals of similar growth patterns, thereby permitting them to be harvested together—and in time for the second crop to be planted.

Other Potential Hay Crops

In periods of serious drought, many other plants are used for feed. Russian-thistle was used to keep cattle alive during the drought of the 1930s. Even when harvested at a very early stage, it is a better feed when ensiled than when dried for hay. Its abrasive surface is less irritating to the mouths of livestock when softened as silage.

Pigweeds, sow thistles, and other weeds have been used as livestock feeds in emergencies. Even though their yield is low and their nutrient content is poor, they will sustain life.

In using weeds and other unusual crops for livestock feed in times of emergency, it is always advisable to obtain authoritative information relative to toxic substances and the stage at which it is best to harvest them for feeding.

HAY QUALITY

A high-quality hay is one that possesses the physical and chemical characteristics commonly associated with palatability and an abundance of feed nutrients.

Hay quality—the difference between high-, medium-, and low-quality forage—is determined by stage of maturity at cutting time, weathering in the field, and method of harvesting and storing.

Importance of Hay Quality

Hay is feed. Thus, as with any feed, it's the end results from feeding hay—the value as determined by animals—that count. Generally speaking, stockmen recognize that the feeding value of hay varies according to quality. However, it is doubtful that they realize just how much returns in production—in meat, milk, wool, reproduction, and speed and endurance—are affected by quality.

Feeding trials at the Washington Station showed conclusively, the effect of hay quality on milk production. Alfalfa hay at three stages of maturity—prebloom, one-tenth bloom, and full bloom—was fed to lactating cows. The chemical composition of the three different quality hays, the milk yields obtained and the body weight changes of the cows are given in Table 6-4.

TABLE 6-4

EFFECT OF HAY QUALITY ON PERFORMANCE OF LACTATING COWS[1]

	Pre-bloom	1/10 Bloom	Full Bloom
	(%)		
Crude protein	16.1	14.1	11.9
Crude fiber	31.8	35.6	37.4
TDN	58.9	57.6	54.7
	(lb)		
Daily hay consumption per 1,000-lb cow	11.8	9.9	8.0
Daily production of 4% milk per cow	35.3	35.2	32.4
Daily change in body weight per cow	+ 0.52	+ 0.28	− 0.20

[1]*Feeding Washington Dairy Cows*, Washington State University Ext. Bull. 486, p. 8.

Table 6-4 shows that both the protein content and TDN content decrease with hay maturity, whereas the fiber content increases. Also, and most important, the more mature the hay, (1) the lower the hay consumption, (2) the lower the milk production, and (3) the greater the effect on the weight of the cows. Calculations based on this experiment show that if hay cut at full bloom is worth $40 per ton, hay cut at one-tenth bloom is worth $70 per ton.

At the Illinois Station, test cows ate 2.21 lb of early cut hay per 100 lb of body weight compared to 1.6 lb of late cut hay. This meant that cows on late cut hay ate 9 lb less per day than cows on hay cut in the bud stage. To make up this nutritional difference in intake, the cows on the late cut hay required 7 lb more grain daily for each cow. Hence, a 100-cow herd would require 700 lb more grain daily as a penalty for late cutting.

Cornell workers found that quality determines the amount of hay that an animal will consume. Cows consumed 34 percent more of an early cut hay (harvested June 9) than of a late cut hay (harvested July 9). In another New York experiment, cows consumed

.5 pounds more early cut than late cut hay, produced 12 pounds more milk per day, and gained weight. Cows on the late cut hay lost weight.

The Washington, Illinois, and New York studies referred to above confirm what experienced feeders know—many times animals simply cannot consume sufficient amounts of low-quality forage to permit optimum production. For example, 32 pounds of low-quality hay would be required per day to provide sufficient energy for a 1,000-pound beef cow nursing a calf. It is unlikely, however, that a cow would consume this large quantity of poor quality hay.

The Ohio Station compared hay containing 67% TDN to hay containing 61% TDN. The cows on the high-energy hay required approximately 7 pounds less grain daily than the cows on the low-energy hay. When hay constituted 56% of the TDN in the ration, the cows on the low-energy hay needed 15 pounds more grain per day than the cows on the high-energy hay.

The importance of hay quality in beef cattle gains was pointed up in studies conducted by the Tennessee Station (Table 6-5).

TABLE 6-5
COMPARISON OF DIFFERENT QUALITIES OF ALFALFA HAY[1]

	Good Quality	Fair Quality	Poor Quality
Hay:			
Chemical composition (%)			
Dry matter	90.1	89.3	91.5
Crude protein	16.8	14.2	12.5
Crude fiber	26.5	31.6	42.7
Ether extract	1.7	1.3	1.0
NFE	37.6	35.8	30.8
Bomb calorimeter Kcal/gm	5.1	5.3	5.4
Digestibility (%)			
Dry matter	59.5	57.5	44.9
Crude protein	69.1	67.4	54.5
Crude fiber	40.7	40.5	35.9
Ether extract	24.8	26.8	22.5
NFE	65.6	68.2	51.8
Energy	55.2	55.2	40.6
Total digestible nutrients (%)	48.0	47.6	38.6
Caloric value of TDN (kcal/lb)	2,667	2,771	2,569
Cattle:			
No. of steers/treatment	12	12	12
Av. wt. and gain/head (lb)			
Initial wt.	568	576	567
Final wt.	700	681	563
Total gain	132	105	−4
Daily gain	1.85	1.49	−0.06
Av. daily feed (lb)			
Hay, fed	17.2	16.6	13.8
Hay, consumed	17.1	16.5	11.6
Salt	0.06	0.09	0.10

[1]Adapted by the author from data provided by the University of Tennessee, Knoxville, Tenn.

As noted in Table 6-5, the cattle on poor quality alfalfa hay actually lost 0.06 lb per head per day, whereas those of fair quality alfalfa hay gained 1.49 lb daily, and those on good quality alfalfa hay gained 1.85 lb daily.

Thus, experiments and experiences show that, in addition to the low nutrient content that characterizes poor quality hay, a more serious loss may follow from feeding it. Studies show that part of the poor results obtained from feeding low-quality hay can be attributed to its failure to support maximum microflora in the rumen, with the result that the digestibility of the crude fiber suffers. Hand in hand with the decline in microflora activity, roughage consumption goes down. Of course, if animals won't eat feed, it won't do them any good.

Characteristics of High-Quality Hay

High-quality hay is readily consumed and digested by animals, with the nutrients derived therefrom utilized efficiently in carrying out body functions.

The characteristics of high-quality hay, and the importance of hay quality, have long been recognized. The commonly used indicators of hay quality are visual inspection, federal grades, and chemical composition.

VISUAL INSPECTION

Fortunately, hay quality and value can be estimated by certain characteristics. It is important, therefore, that those who grow hay, and those who buy and sell hay, be acquainted with those recognizable characteristics of hay which indicate high palatability and nutrient content. If in doubt, the animals will tell them, for they like and thrive on high-quality hay.

The easily recognized characteristics of hay of high feeding value are:

1. It is made from plants cut at an early stage of maturity (see Table 6-10, page 680), thus assuring the maximum content of protein, minerals, and vitamins, and the highest digestibility.
2. It is leafy, thus giving assurance of high protein content.
3. It is bright green in color, thus indicating proper curing, a high carotene or provitamin A content, and palatability.
4. It is free from foreign material, such as weeds, stubble, etc.
5. It is free from must or mold and dust.
6. It is fine stemmed and pliable—not coarse, stiff and woody.
7. It has a pleasing, fragrant aroma; it "smells" good enough to eat.

FEDERAL GRADES OF HAY

The U.S. Department of Agriculture has used the recognizable characteristics of quality in hay as a basis for determining the federal grades (see Table 6-6 for the grades and grade requirements of three of the largest hay groups). Stockmen who buy or sell hay should familiarize themselves with these federal grades, for they may be used to determine market values and feeding quality.

TABLE 6-6
GRADES AND GRADE REQUIREMENTS OF THREE COMMON KINDS OF HAYS[1]

Kind and Grade of Hay	Leafiness (minimum percent of leaves)	Color (minimum percent of green color)	Maximum Foreign Material
	(%)		
Alfalfa and alfalfa mixed hay:			
U.S. No. 1	40	60	5
U.S. No. 2	25	35	10
U.S. No. 3	10	10	15
Sample grade	(hay which will not grade 1, 2, or 3)		
Timothy and clover hay:			
U.S. No. 1	40	40	10
U.S. No. 2	25	30	15
U.S. No. 3	10	10	20
Sample grade	(hay which will not grade 1, 2, or 3)		
Soybean and soybean mixed hay:			
U.S. No. 1	40	40	10
U.S. No. 2	25	25	15
U.S. No. 3	10	10	20
Sample grade	(hay which will not grade 1, 2, or 3)		

[1]At least 4 grades (1, 2, 3, and sample grade) are provided for each of the following 11 groups of hay:

1. Alfalfa and alfalfa mixed hay
2. Timothy and clover hay
3. Prairie hay
4. Johnson and Johnson mixed hay
5. Grain, wild oat, vetch, and mixed hay
6. Lespedeza and lespedeza mixed hay
7. Soybean and soybean mixed hay
8. Cowpea and cowpea mixed hay
9. Peanut and peanut mixed hay
10. Grass hay
11. Mixed hay

For more detailed information relative to the subject of federal grade of hay, the reader is referred to the *Handbook of Official Hay and Straw Standards*, which is for sale by the Superintendent of Documents, Washington, D.C.

CHEMICAL COMPOSITION

Visual estimates of hay quality are of value and should be used, but often they are unreliable; it's like predicting the eating quality of meat by inspecting animals on foot. Thus, federal grades of hay lack in accuracy, even when made by experts. A Pennsylvania study revealed errors of as much as 5 percent in crude protein and 9 percent in TDN (energy) content of a forage, with evaluations made by trained individuals. Despite this fact, it should be emphasized that visual examination and grading of hay are still needed, especially for (1) weed detection and color, and (2) predicting palatability. Chemical analyses should supplement, but not replace, existing methods of hay grading. Also, it is recognized that any method of determining hay quality by means of chemical analyses is of value only if it is related to feeding value.

Fortunately, research has shown a high relationship between the chemical composition—especially the protein and fiber—of hay and its feeding value for animals. As a result, a growing number of states now have laboratories where, at a nominal charge, a quick check of moisture, protein, and/or crude fiber can be made. As hay matures, protein decreases (pounds of protein per acre decreased from 935 lb in early cut hay to 605 lb in late cut hay, according to a Cornell study) and fiber increases. Likewise, weathering lowers the protein and raises the fiber content since soluble nutrients are washed out by rain and leaves are lost during harvest. It is also noteworthy that palatability seems to be negatively correlated with crude fiber levels—the higher the fiber content, the lower the palatability; this is important, for if the animals won't eat, they can't produce. Cornell investigators found that cows ate 2 lb more of the early cut hay per day than of the late cut hay.

Correct Sampling Necessary

No forage test is any better than the sample taken. Thus, the most important single step in determining the chemical composition of hay is sampling. No matter how accurate the chemical analysis, a poor sampling technique can easily invalidate the results and lead to an erroneous conclusion. For instructions on correct sampling, and to determine if the state has a laboratory for analyzing forage samples, the stockman should see his local county agent or write to his state college of agriculture.

It is difficult to obtain a representative, meaningful sample of forages. At least 12 regular bales should be sampled for a hay analysis. Also, samples of different cuttings should be taken. Conventional bales can best be sampled by use of a special probe which cuts into the bale and draws a sample. Large packages—large round bales and stacks—may be sampled by taking hand-grab samples from inside several bales or stacks, or by actually cutting into the round bale or stack with a large crosscut handsaw.

Hay samples should be placed in a plastic bag or freezer carton; otherwise, the moisture content will not be meaningful.

What Tests To Make

The *proximate feed analysis* is the most common analysis for chemical evaluation of hay. The analyses include (1) moisture (water) or dry matter (DM); (2)

otal (crude) protein (CP or TP = N × 6.25); (3) ether xtract (EE) or fat; (4) ash (mineral salts); (5) crude iber (CF)—the poorly digested carbohydrates; and (6) nitrogen-free extract (NFE)—the more readily digested carbohydrates (calculated rather than measured chemically). It is not always necessary to run all of these values, depending on what use is to be made of the results.

Other analyses which are often useful in evaluating hay are: calcium, phosphorus, carotene, and certain trace minerals. There are times when the amount of vitamins and amino acids in the feed might be useful, particularly when hay is used as a supplement in nonruminant rations. These are costly to run, however, and it may not be economically feasible to run very many of them.

The use of crude fiber as a measure of the poorly digested carbohydrates is quite controversial. In some instances, the digestibility of crude fiber, as determined by actual feeding tests, has been nearly as high as some of the other nutrients. Efforts have been made to refine the fiber measurement, so that it will approximate more closely the undigested carbohydrate portions of feed. Some of the alternative analyses include: acid detergent fiber, alkaline detergent fiber, neutral detergent fiber, lignin, and cell wall constituents. Acid detergent fiber (ADF) is rapidly replacing the old crude fiber test. It is determined by boiling the feed sample in a specially prepared acid detergent solution which leaves a residue consisting chiefly of cellulose and lignin. In using a crude fiber value, one must be aware of its limitations. In general, however, it identifies those feeds high in fiber and serves as a guideline in making judgments in ration formulation.

Laboratories generally do not determine digestibilities of feed nutrients because of the high cost of such procedures. Some may run a simulated rumen digestibility in a flask inoculated with rumen microorganisms. Even though the latter is not very accurate, it provides a useful guideline in feed evaluation. Digestibilities, whether determined or estimated, are necessary to report digestible dry matter, digestible protein, TDN, and other digestible nutrients.

Evaluating Test Results

Test results can best be evaluated by comparing them with some standard. The testing laboratory may provide such information, possibly along with recommendations for applying the test results in balancing rations. For convenience, average crude protein and crude fiber values of some common hay crops are given in Table 6-7.

If a chemically analyzed sample runs higher in protein and lower in fiber than the average figures given in Table 6-7, it means that the sample is better than average quality hay; conversely, if it is lower in protein and higher in fiber, the sample tested is below average quality. Of course, stockmen should not settle for average protein or fiber content, for, on the whole, the vast majority of the U.S. hay crop is of low quality. For this reason, one should strive for the upper figures of the range given in Table 6-7. Certainly, poor quality hay can be fed, and, under certain circumstances, it may even be economical and quite satisfactory—for example, for dry cows. However, when buying poor hay, the purchase price should be lowered accordingly; and the feed analysis should also be used as a basis of balancing the ration. By the same token, it is usually good business to pay a premium for high-quality hay. Some stockmen are very wisely applying an escalator principle to hay purchases. They may pay $2.00 to $3.00 per ton for each 1% of protein above an agreed-upon figure; or they may dock the price by a corresponding amount if the content is lower. For example, if a vendor guarantees to deliver alfalfa with 15% crude protein and it is agreed that a $3.00 per ton premium will be paid for each 1% protein in excess of this figure, a $9.00 per

TABLE 6-7

APPROXIMATE CHEMICAL COMPOSITION (MOISTURE FREE) OF VARIOUS SUN-CURED HAYS

Kind of Hay	Crude Protein		Crude Fiber	
	Average	Range	Average	Range
	(%)			
Alfalfa	15.5	12.0-19.0	29.0	22.0-36.0
Bromegrass	10.5	6.0-15.0	28.0	24.0-31.0
Ladino clover	18.5	16.0-21.5	22.0	18.5-23.0
Red clover	12.0	10.5-18.5	27.0	18.0-34.0
Lespedeza	13.0	11.5-14.5	27.0	22.5-32.5
Oat hay	5.0	4.0- 6.0	28.0	26.0-32.0
Orchardgrass	8.1	6.0-14.0	30.0	26.0-31.0
Soybean	14.5	9.0-16.5	28.0	20.5-41.0
Timothy	6.5	5.5- 9.5	30.0	28.0-31.5
Sudangrass	8.8	6.5-11.0	28.0	26.0-30.5

ton premium would be added for alfalfa running 18% crude protein.

In some cases, hay is also purchased on the basis of moisture content; and the price drops as this moisture content increases—thereby discouraging selling water. For example, if 15% moisture is agreed upon, hay running 18% moisture would be docked 3% in price. Thus, if the base price of hay is $80.00 per ton, the price would be lowered to $77.60 per ton ($80.00 × 3% = $2.40; then, $80.00 − $2.40 = $76.40).

Others use crude fiber in the same manner—paying a premium for hay of low fiber content. Still others apply more complicated formulas, such as the California Hay Evaluation System, to arrive at the dollar value of hay.

Thus, a chemical test provides informed appraisal of hay values. Except for actual feeding trials, it is the most infallible method of evaluating hay quality. With this information at hand, stockmen can do a better job of feeding animals and realizing higher profits. Additionally, an analysis and payment on an incentive quality basis should result in the following benefits: (1) compensate the grower for extra effort in keeping hay quality high; (2) induce more hay growers to produce high-quality hay, rather than just tonnage; and (3) provide livestock producers with better-quality feed, which, in turn, will result in greater feed efficiency. Supply and demand will still regulate the price. But, in the end, hay will be upgraded; it will be cut at an earlier stage of maturity and put up with greater care.

California Hay Evaluation System

University of California workers (Meyer and Lofgren) developed tables, based on just two chemical analyses—dry matter and crude fiber (using a modified AOAC method for the latter), (1) to predict TDN and digestible protein; and (2) to calculate the dollar value of hay on (a) a net energy plus protein value, and (b) a net energy value alone. Because of space limitations, these extensive tables are not reproduced herein. Instead, anyone interested in using the California Hay Evaluation System may obtain the tables, along with instructions on their use, by writing the University of California, at Davis.

MAKING QUALITY HAY

The object of haymaking is to (1) harvest the crop at the optimum stage of maturity which will provide the maximum yield of nutrients per acre without damage to the next crop, and (2) cure properly, which involves lowering the water content of the green herbage from 65 to 85 percent to 20 percent or less.

Hay quality begins with the soil and ends with the manger, with many intermediate factors affecting it along the line. Once forage is cut, opportunities to increase nutrient content are over; from that point on quality can only be preserved.

There is no one best haymaking method or kind of equipment. These must necessarily vary with the size of the operation, the kind of hay, the climate of the area, the individual farm or ranch conditions and buildings, and the available labor and machinery and their cost. Yet, the principles of good haymaking and the objectives sought are the same everywhere.

About 80% of all harvested forage is now baled. Only 10% is stored as loose hay and cubes, and the remaining 10% is stored as hay crop silage.

For convenience in the ensuing discussion, the factors affecting hay quality have been grouped as follows: growing, haying equipment, harvesting, curing, haymaking systems, and storing.

Growing Forage

Growing forage for hay has long been neglected. Average per acre yields are still well under one-half their potential. Little more than 1 acre of hay in 10 is fertilized on a regular basis; and the precious few acres that are fertilized get an average of only 12 to 15 pounds per acre—a paltry amount compared to corn, which receives an average of about 200 pounds of fertilizer per acre.

The steps in growing quality hay are:

1. **Match crop to soil**—Some forage crops will do better on certain soils than others. So, the crop should be fitted to the soil.

2. **Choose quality seed, and proven varieties and mixtures**—Most grass and legume seeds are extremely small and contain very little stored food material. Thus, it is important that good seed be used.

Also, recommended varieties should be selected. If there is a choice, consider simple grass-legume mixtures. Research has shown that in many areas and over a period of years such mixtures are frequently higher yielding and more persistent than legumes grown alone. Mixtures are also easier to harvest and cure as hay.

3. **Lime and fertilize**—To obtain top yields of hay, lime the soil (if needed) to a pH of 7, then use the right kinds and amounts of fertilizer for the species and soil. Legumes, such as alfalfa, need a nearly neutral soil to promote growth of nodule-forming bacteria. Fertilization should be based on soil test.

4. **Get good stand**—A good stand is important. In this connection, it is noteworthy that it takes about one-half million successfully established individual forage plants per acre for a productive stand.

5. **Irrigate where practical**—In arid and semiarid

regions, irrigation is a must for hay production. Whether or not it will pay to irrigate hay crops in humid and semihumid areas will depend on the water supply, irrigation costs, and returns. Census figures show that more than 12 million acres of U.S. hay and pasturelands are irrigated—that's a little less than 1 in 3 of all irrigated acres.

6. **Control insects and diseases**—Forage insects destroy a lot of hay. But they can be controlled by using a proper combination of insecticides, cultural control practices (such as timely cutting), and biological control agents.

Haying Equipment

Mechanization of hay harvesting started with the use of the barn hayfork in 1864, followed by the hay-loader in 1874 and the side-delivery rake in 1893. But the greatest milestone in haymaking was the automatic pick-up baler in 1940.

In addition to saving labor, modern harvesting equipment improves the quality of hay. For example, conditioned hay dries in ⅓ to ½ the time required for nonconditioned hay. Artificial curing of hay, using either natural or heated air, can reduce the field exposure time of hay an additional 20 to 50 percent.

Among the new machines that have been developed for handling hay are the following: the mower-conditioner-windrowers, bale throwers, self-propelled automatic bale wagons, loose hay stackers, large round balers, mechanical stack machines, and wafering or cubing machines.

The largest remaining problems of hay handling are (1) its removal from storage, and (2) its mechanical feeding to animals.

A brief rundown on today's haying tools follows:

1. **Mowers**—Simplicity, high speeds, and greater widths have kept the cutter bar mower popular.

2. **Mower-conditioners**—These combine cutting and conditioning in one operation.

Fig. 6-5. Mower-conditioner.

3. **Windrowers**—Multipurpose self-propelled windrowers provide up to 16 feet of cutting capacity. Most models feature hay conditioners as standard or optional equipment.

Fig. 6-6. Windrower.

4. **Rakes**—Side-delivery rakes are still widely used to make hay crop windrows. Parallel bar rakes, which are most popular today, reduce impact between rake teeth and hay by moving hay from the outside of the swath to the windrow in less than 13 feet of travel.

5. **Balers**—Conventional square balers are still the most popular method of packaging. They come in either twine or wire-tie models.

Large round balers, making bale weights from 900 to 1,500 pounds, are increasing in popularity, especially with cow-calf operators, and even with some dairymen.

6. **Bale handling**—Mechanization took the backache out of bale handling. Today, the following types of sophisticated hay harvesting and handling equipment are available: bale throwers that throw bales where needed and eliminate lifting; bale conveyors for putting bales where desired; automatic bale wagons, which pick up bales from the field, load, transport, and stack them; and stack retrievers for transporting bales far removed from where they will be used or fed.

7. **Stack machines**—These are hydraulically operated machines that compress long hay into stacks, weighing from 1 to 6 tons.

Harvesting at Proper Stage

Whether the crop is a grass or a legume, or a combination, quality is very closely related to the maturity of the plants at the time of harvest. Young, immature plants are high in protein and low in fiber and lignin. As hay crops mature, feeding value goes down and fiber content increases.

Macdonald College of Quebec, Canada, found a 33 percent increase in the coarse stems of red clover and a corresponding decline in the proportion of fine leafy material in a period of 11 days from the time the heads were forming to the early bloom stage. Table 6-8 shows the changes in chemical constituents of alfalfa (a legume) and timothy (a grass) as affected by maturation.

TABLE 6-8

EFFECT OF MATURATION ON HAY QUALITY
(Alfalfa Data from the Nebraska Station;
Timothy Data from the Missouri Station)

Kind of Hay	Crude Protein	Crude Fiber
	(%)	(%)
Alfalfa:		
Prebloom	22.0	25
1/10 bloom	19.2	27
1/2 bloom	18.8	28
Seed stage	14.1	37
Timothy:		
Heading	8.0	31
Full bloom	5.9	34

In Wisconsin, first-growth alfalfa-bromegrass hay was harvested each year for three years at four stages of maturity. Dry matter digestibility, digestible energy, milk production, and animal gains all declined sharply with increasing maturity of the hay (Table 6-9).

TABLE 6-9

NUTRITIVE VALUE OF FIRST-GROWTH ALFALFA-BROMEGRASS HAY HARVESTED OVER A THREE-YEAR PERIOD, AT ARLINGTON, WISCONSIN[1]

Stage of Growth	Dry Matter Digestibility[2]	Digestible Energy[3]	4% Milk Production[4]	Animal Gain[5]
	(%)	(kcal/g)	(lb/day)	(lb/day)
Vegetative	71.4	3.20	45	0.38
First flower	64.6	2.86	29	0.21
Full bloom	58.0	2.54	15	0.15
Green seed pod	55.2	2.43	4	0.05

[1]Adapted by the author from University of Wisconsin data.
[2]Animal digestion trial data, values similar to total digestible nutrients.
[3]Animal energy digestibility × forage gross energy.
[4]Estimated for a 1,200-lb cow fed hay alone.
[5]Lambs fed hay alone, 1962 only.

Like crude protein, many of the other desirable quality factors—including carotene (precursor of vitamin A) and the B vitamins—decrease as plants mature. Vitamin D content is the one exception—it increases as the forage is sun-cured. As proof that alfalfa hay becomes progressively lower in feed value as it gets older, the U.S. Department of Agriculture found that a given acreage of alfalfa cut in early bloom produced 404 lb of butterfat; when cut at the half-bloom stage, 345 lb; and when cut in full bloom, only 331 lb.

Everything considered, there is a loss of about one percent in nutrient value for each day that hay harvest is delayed beyond the late vegetative stage of growth.

Table 6-10 gives guidelines relative to the proper forage-harvesting stage for maximum protein and minimum fiber.

TABLE 6-10

HAY CUTTING GUIDE

Kind of Hay	When to Cut
Alfalfa	Prior to 1/10 bloom, or when new shoots begin to develop from the crown.
Red clover	Early bloom to 1/2 bloom stage.
Alsike clover	Early bloom to 1/2 bloom stage.
Crimson clover	In bloom.
Sweet clover	When blooming begins.
Cowpeas	When pods are 1/2 to fully matured.
Soybeans	When pods are 1/2 to fully matured.
Lespedeza	When in full bloom.
Sericea	When 12-15 inches high.
Ladino clover	In full bloom.
Grasses	Heading out to bloom stage.
Sudangrass	When it begins to head out.
Small grains	When grain is in the milk to soft dough stage.
Grass-legume mixtures	When the legume is at the proper stage of development as described above.

Curing (Drying)

Proper curing ensures that (1) the hay can be stored safely without heating excessively or becoming moldy; and (2) the maximum leafiness, green color, aroma, nutrient value, and palatability shall be retained. To the end that these desired objectives may be achieved, the following information is pertinent:

1. **Moisture content**—Freshly cut forage contains 75 to 80% moisture, whereas the maximum moisture content for safe hay storage is as follows:

   For loose hay—25% moisture.
   For baled hay—20 to 22% moisture (the lower figure for larger bales).
   For chopped hay—18 to 20% moisture.
   For cubes—16 to 17% moisture.

Hay of a higher moisture content than indicated should not be stored because (a) its value may be greatly lowered due to mold or to nutrient losses accompanying fermentations, and (b) of the ever-present danger of spontaneous combustion and a costly fire.

Two rule-of-thumb methods used by farmers in determining when hay is dry enough for storage are:

a. **The twist method**—Twist a wisp of the hay in the hand. If the stems are slightly brittle and there is no evidence of moisture on the twisted stems, the hay can be stored safely.

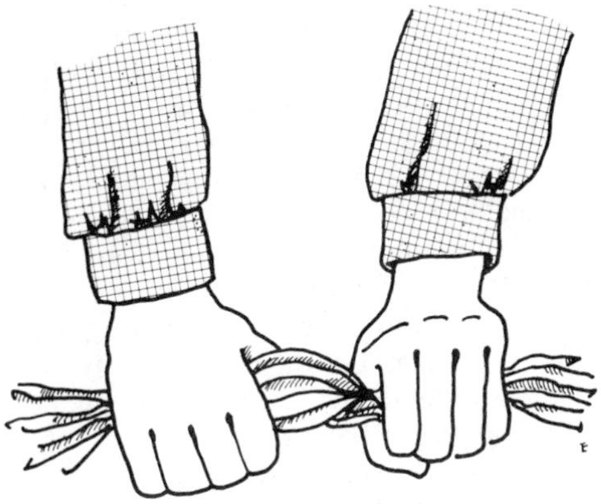

Fig. 6-7. The "twist method" used by farmers in determining when hay is dry enough for storage. If the stems are slightly brittle and there is no evidence of moisture on them when twisted, the hay can be stored safely.

b. **The scrape method**—Scrape the outside of the stems with the finger or thumbnail. If the epidermis can be peeled from the stem, the hay is not sufficiently cured. If the epidermis does not peel off, the hay is usually dry enough to stack or put in the mow.

2. **Shattering losses**—Legume forages contain a larger proportion of leaves than do grasses, but, unfortunately, the fine, thin legume leaves dry out more rapidly than the coarse stems to which they are attached. This results in considerable shattering losses, unless great care is taken. The importance of this condition is readily apparent when it is realized that in alfalfa, for example, 50% of the total weight of the plant is contained in the leaves, but the leaves contain 70% of the protein and 90% of the carotene content of the entire plant.

In field curing hay, losses from leaf shattering range from 2 to 5% for grass hay and 3 to 39% for legume hays, with as much as 15 to 20% for legume hays field cured under the most favorable conditions. Based on extensive experiments with field cured alfalfa hay, the U.S. Department of Agriculture reported that leaf losses averaged 38.5% when none of the hay was wet; 47.3% when the hay was wet by 2 showers; and 74.5% when the hay was wet by 3 showers—and milk production per acre was 19.7% less where cows were fed rain damaged, field cured alfalfa hay in comparison with field cured hay without damage by rain.

3. **Bleaching and fermenting losses**—In general, the carotene or pro-vitamin A content of freshly cured hay is proportional to the greenness. With severe bleaching, more than 90 percent of the vitamin A potency may be destroyed.

Even under the best of conditions, there is an unavoidable loss through fermentation, especially losses in sugars, starch, and carotene. With good weather and proper curing methods, however, these losses will not be excessive.

4. **Leaching losses from rain**—The leaching losses from rain are less severe soon after mowing, but increase in severity with the curing process. Also, repeated showers are more damaging than one heavy rain. Experimental studies have revealed that damaging rains may lower the feeding value of hay by ¼ to ⅓, or even more with severe exposure.

Bad weather can result in both excessive dry matter losses and losses in feeding quality. A U.S. Department of Agriculture study showed, for example, that getting hay in without rain can mean saving in protein of 18 percent. Other losses from rain damage on field-cured hay are shown in Table 6-11.

Losses from weather damage may be reduced (1) by using haymaking equipment that reduces the field drying time, and (2) by understanding and using existing weather aides.

TABLE 6-11

AVERAGE ESTIMATED LOSSES FROM RAIN DAMAGE ON FIELD-CURED HAY[1]

Hay	Losses		
	Field and Storage	Feeding	Total
	(%)	(%)	(%)
All rained on	32.6	8	40.6
50% rained on	28.8	7	35.8
No rain	17.4	5	22.4

[1]From paper entitled "Hay Production, Preservation and Quality," by John E. Baylor, The Pennsylvania State University, *Beef Cattle Science Handbook*, Vol. 13, 1976, p. 199, published by Agriservices Foundation, edited by M. E. Ensminger.

FIELD CURING

As the name would indicate, field curing embraces all curing methods in which hay is cured in the field. The common steps, methods, and equipment used in field curing are:

1. **Cutting and curing in the swath or windrow**—Cutting, followed by curing in the swath or

windrow, is the first step in haymaking, regardless of the subsequent method or type of equipment employed.

Any one of several types of mowers may be used, for all of them are designed to get the hay down. The most important thing is that the hay be cut at the proper stage of maturity (as noted in Table 6-10).

The following points are also pertinent to cutting hay and curing it in the swath or windrow:

a. **Direction of mowing**—It is highly desirable to mow in the same direction as will be traveled in raking and in picking up the hay in subsequent operations.

b. **Time of the day to mow**—Some opinions to the contrary, quality of hay is only slightly affected by either the time of day at which the forage is cut (the proportion of sugar does vary with the time of day), or the presence of dew.

c. **Hay conditioning**—Hay conditioning machinery may be used to speed up the field curing process. These machines are designed so that slow-drying stems are split, cracked, crushed, or broken as they pass through rollers or knives. As a result, the stems dry at about the same rate as the leaves; and there is more uniform curing and less leaf loss. The three types of hay conditioning machinery are:

(1) **The crusher**—It has two flat rollers under spring tension which actually crush stems as the hay passes through.

(2) **The crimper**—It has 2 corrugated rollers which crack the stems every 2 or 3 inches as the hay passes through.

(3) **The flail harvester (or chopper)**—It has loose pointed knives on a horizontal cylinder. These knives chop the hay in about 6-inch lengths and also crack the stems vertically.

Crushers and crimpers pick up the hay from a swath, pass it through a set (or sets) of rollers, and drop it back into the same swath. Usually, they operate directly behind the tractor and condition the hay that was cut the previous round. Some units both mow and condition the hay with the same power transmission system, whereas others are separate pieces of equipment. The latter are cheaper, but they necessitate that the field be gone over twice.

The flail harvester cuts the hay and conditions it in one operation, or it will also condition hay that is in the windrow. It operates somewhat like a silage cutter with a blower tube and downspout which places the hay back on the ground in a windrow.

Although hay conditioners have considerable merit, farmers and ranchers with smaller hay acreages may not be able to justify the added cos of the equipment.

d. **Length of swath curing**—Hay dries mor rapidly in the swath than in the windrow, even i the windrow is small and fluffy. Therefore, i should be left to cure in the swath as long as i possible without damaging it; until the forage i wilted but before there is danger of the leave shattering and of excessive bleaching and loss o carotene. At this stage, the moisture content wil be about 40 percent. The point at which swath curing is sufficient should be carefully determined, because the leaves become dry and brittle, especially on legumes, long before the stems are cured.

No definite period of time for swath curing can be assigned as it will vary according to the tonnage of hay per acre, temperature, sunshine, wind, and atmospheric humidity. Instead, the time should be determined entirely by the condition of the hay in the swath.

Sometimes curing in the swath is speeded up by using swath fluffers; machines which pick up the forage, lift it to a standing position, and then release it to fall loosely onto the stubble—thereby fluffing and serrating the hay.

Where it rains infrequently, sometimes hay is not cured in the swath at all. Under such conditions, it may be cut and windrowed immediately. For the latter purpose, swathers—self-propelled mowers that cut and windrow hay in one simultaneous operation—may be used.

2. **Raking**—After the hay has wilted sufficiently in the swath, but while it is still tough and the leaves will not shatter, it should be windrowed. For this assignment, the side-delivery rake is preferred to the dump rake and to the hay tedder (or kicker). The side-delivery rake rolls hay into fluffy, cylindrical windrows, which allows for good circulation of air; whereas dump rakes produce large windrows which are apt to remain damp underneath and bleach excessively on top, and hay tedders tend to shatter the leaves of legumes. Where the hay crop is exceedingly heavy, the size windrow can be kept desirably small by limiting the width raked into each windrow.

If considerable shattering appears probable, it may be desirable to do the raking early in the morning when the dew makes the hay a bit tough.

Where windrowed hay is rained on, wait until the top half dries out, and then turn it upside down with the side-delivery rake (the use of the tedder for rewindrowing is not recommended because of excessive shattering).

3. **Cocking**—Formerly, well-made cocks, often adorned by hay caps, were considered a necessary part of good haymaking. However, this practice has

greatly decreased, due primarily to higher labor costs and the advent of modern haymaking machinery. Today, the cocking of hay is confined almost entirely to use (a) in hot, arid regions where the leaves shatter if the hay is left in the swath or windrow for any appreciable length of time, and (b) as an emergency measure in order to protect hay when a storm is imminent.

Haymaking Systems

In haymaking, the term "system" refers to a team of processes and machines that does the work from field through feeding, saves crop nutrients, reduces manpower requirements, and eliminates drudgery. When each step is mechanized, it must be matched; otherwise, man and machines end up waiting.

In recent years, automation has had great impact on hay. Some haymaking systems are completely mechanized from field to feeding.

There is no one best haymaking system for all conditions. Nevertheless, all good systems are fast, make handling easy, save labor and nutrients, and increase profits. Baling is the most popular hay-handling system in North America.

LONG, LOOSE HAY

The acreage harvested as long, loose hay has declined sharply in recent years, especially in the humid areas, because (1) of high labor cost, and (2) long hay is too bulky for mechanized feeding. Nevertheless, long, loose hay is still popular in many western areas where specialized handling equipment is used. Moreover, some of the newer systems of handling and self-feeding loose hay show promise.

The two common methods of handling loose hay are:

1. **Loading with hay-loader directly from windrows**—In this method, cured hay is loaded on a truck or wagon directly from the windrow by means of a hay-loader. Usually the hay is then transported to a barn or stack where it is unloaded by fork or sling and mowed away by hand. Sometimes it is chopped into the barn or other storage area.

The hay-loader method requires less investment in equipment than the pick-up baler or field chopper; but, unless the acreage is small, the cost of handling on a per ton basis may be excessive because of the relatively high labor requirement. For these reasons, the use of the hay-loader is usually confined to smaller hay operations.

2. **Hauling cured hay from windrows or cocks with buck rakes, sweep rakes, or sled**—In the West, much of the hay is cured in windrows or cocks and then transported by buck rakes, sweep rakes, or sleds to field stacks where it is stacked by hay stackers or other large mechanical devices. Then after going through a sweat in the stack, the hay is fed out as loose hay, or, if intended for market, it is baled.

Without doubt, this method results in the production of the highest percentage of good quality hay of any known method, primarily because (a) more latitude is permissible in the moisture content when stacking than when baling or chopping, (b) the practice predominates in an area which normally has good haying weather, and (c) the method is prevalent on farms and ranches where haymaking is frequently a major enterprise and where the operators have the know-how to produce good hay.

In a somewhat modified form, this method of haying has spread from the West to other sections of the United States. In the latter areas, where hay is generally grown for winter feed rather than for sale, it is usually transported to barns or sheds from nearby fields by means of a buck rake. It is then unloaded with a hay sling, grapple fork, or blower. Where limited distances and small acreages are involved, this is probably the cheapest method of handling hay, because a minimum of equipment and labor is required.

CHOPPED HAY

Chopped dry hay fits into some feeding systems, particularly in the West.

For safe storage, the moisture of chopped hay should not exceed 18 to 20 percent.

Two common methods of chopping freshly cured hay follow:

1. **Field chopping cured hay directly from the windrow**—In this method, a field chopper gathers the cured hay from the windrow, chops it, and blows it into a truck or trailer. The chopped hay is then blown by a special blower into the barn, stack, or other storage area. This method requires less labor and time than any other method of haymaking except cubing.

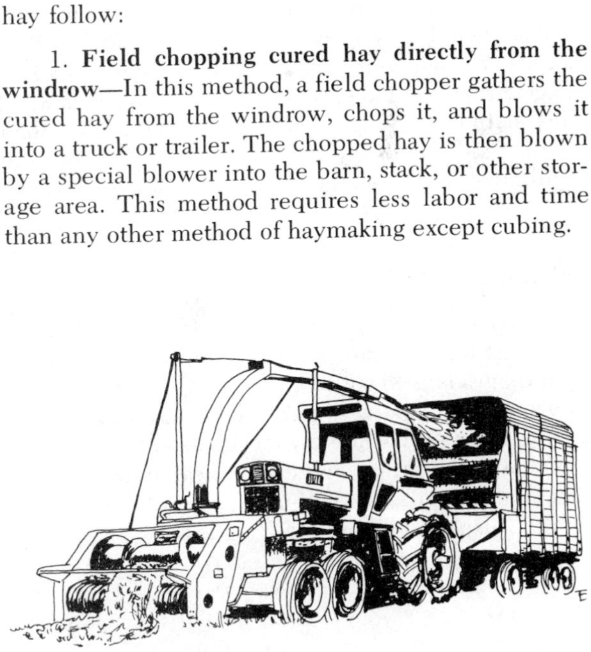

Fig. 6-8. Forage harvester field chopping cured hay directly from the windrow.

Some pertinent facts relative to field chopping cured hay directly from the windrow are:

a. **Equipment cost**—The equipment cost, on a per ton basis—for the chopper, the equipment for hauling, and the blower for unloading—is apt to be high unless a considerable acreage of hay is involved, or the operator can also use this equipment for corn or sorghum silage.

b. **Moisture content**—The moisture content must be lower than is permissible for baling—from 18 to 20 percent.

c. **Convenience in feeding**—Chopped hay is often more convenient to feed.

d. **Feeding value and wastage**—Chopping does not increase the feeding value of the hay, but it may make for less wastage, especially where low-quality hay is involved.

e. **Length of cut**—A 1½- to 2-inch cut is recommended. Finer chopping requires more power, increases the tendency to heat in storage, and makes for dusty and unpalatable feed. Regardless of the length of cut, repeated blower action, first at the chopper and later at the barn, pulverizes the leaves and aggravates the dust nuisance.

f. **Storage space**—Chopped hay requires only ⅓ to ½ as much storage space as long hay. This has the advantage of effecting a saving in space, but the disadvantage of requiring caution in order to prevent overloading mows.

2. **Chopping into the barn or other storage area**—In this method, the cured hay is generally hauled from the windrow or cock to the barn or other storage area where it is chopped by a hay chopper or silage cutter and blown directly into the storage area. This method is slower and requires more labor than where field cured hay is chopped directly from the windrow, but less expensive equipment is necessary.

PACKAGED HAY

Great strides have been made in hay packaging. Round bales and mechanically compressed stacks lend themselves to outdoor storage. Compressing hay into cubes and pellets makes for many advantages; among them, (1) completely mechanized handling; (2) high density, with more economical transportation and storage; (3) easier self-feeding and higher intake by animals; and (4) lower feeding losses.

BALES

The following choices of bales are available:

1. **Conventional square bales**—Square bales are the most common and best-known method of haymaking.

2. **Large round bales**—Many makes and model of large round balers are on the market.

Fig. 6-9. Round bale. It weighs 1,200 to 1,500 pounds, sheds water, and may be left in the field for winter feeding.

STACKS

These are rectangular-shaped (one system makes a circular stack), mechanically pressed haystacks. Long, loose hay is blown into a wagon and pressed down by a hydraulically operated canopy roof. Stacks range in size from 7 to 10 ft wide, 8 to 22 ft long, and 8 to 11 ft high, and they range in weight from 1 to 6 tons.

Fig. 6-10. Loaflike stacked hay; machine made, 2½- to 4-ton size, sheds rain and snow and resists wind damage.

The stack system saves labor and permits nearly the same latitudes as loose hay, with the efficiencies of mechanization.

Limitations of the system are (1) investment costs are high, (2) there is more surface of the stored hay to the weather, (3) heavy rain or snow may seep down through the stacks if they are not formed properly, and (4) they are not efficiently transported over long distances. Nevertheless, the principles involved are good; hence, some of the problems will likely be overcome with more experience.

CUBES (WAFERS)

Engineers are continuing to improve field cubers for forages—machines that can move across hayfields, pick up windrows of forage, and produce dense, high-quality forage cubes or wafers. In the future, this may be the most desirable method of harvesting hay. However, some obstacles must be overcome, including the following:

1. Lowering cost—cubing now costs about $6 to $8 per ton more than baling.
2. Lessening equipment weights, power requirements, and costs.

Fig. 6-11. Hay cuber in operation, making cubes 1¼ inches square by 2 to 3 inches long.

Nevertheless, the benefits derived from field cubing of hay are great. It (1) simplifies haymaking, (2) lessens transportation and storage space—cubes or wafered roughages weigh between 45 and 55 pounds per cubic foot (baled hay density is 25 to 32 lb), (3) reduces labor, (4) makes automatic hay feeding possible, (5) decreases nutrient losses, (6) eliminates dust, and (7) makes for increased feed consumption, gains, and feed efficiency. Also, with cubing, the spread between high- and low-quality roughage is narrowed; that is, within reason the poorer the quality of the roughage, the greater the advantage from cubing or wafering. The latter is so because such preparation assures complete consumption of the roughage.

Hay cubes are of special interest to dairymen because they have the advantages of pellets, without their disadvantage. Like finely ground forage that is pelleted, cubes can be readily automated—they are ready to feed; and, in comparison with long hay, there is less transportation and storage cost. But they will not lower the fat content of the milk, as will pellets when appreciable quantities of them are fed.

New cubing processes have been developed and engineering technology has overcome many of the obstacles that formerly plagued cubing; consequently, field cubing or wafering of forages is increasing.

PELLETS

Pelleted forages are finely ground, then condensed. The advantages of pellets are:

1. Pelleted feeds are less bulky than any other hay package (pelleted roughage requires 1/5 to 1/3 as much space as is required by the same roughage in loose or chopped form), and are easier to store and handle—thus lessening transportation, building, and labor costs. For these reasons, it is particularly advantageous to use pelleted feeds where storage space is limited and feed must be transported considerable distances, conditions which frequently characterize small enterprises.
2. Pelleting prevents animals from selectively refusing ingredients likely to be high in certain dietary essentials; each bite is a balanced feed.
3. Pelleting practically eliminates wastage. Since animals frequently waste up to 20 percent of long hay, less pelleted feed is required. Wastage of conventional feed is highest where low-quality hay is fed and/or feed containers are poorly designed.
4. Pelleting eliminates dustiness and lessens heaves.

The biggest deterrent to increased pelleting at the present time is the difficulty of processing chopped forage coarse enough so that it will not cause digestive disturbances. A minimum of a ¼-inch chop is recommended.

ARTIFICIAL DRYING

The use of forced air, either heated or unheated, for final drying is the most dependable way in which to produce high-quality hay. The application of heat provides faster drying, saves more leaves, and reduces losses of nutrients. Yet, artificial drying is on the decline in the United States because of the added cost in heating the air.

On the average, about 4,000 pounds of water are evaporated from each ton of hay. With a yield of 2 tons per acre, this means removal of approximately 4 tons or 960 gallons of water.

Hay-drying equipment permits handling while hay is green and tough enough to withstand mechanized processes without excessive leaf loss, and it minimizes weather damage. The most common methods of artificial drying are (1) mow curing, (2) artificial dehydrators, and (3) wagon dryers.

Mow Curing

Mow curing, or drying, refers to the practice of curing partially dried hay—either long, chopped, or baled—in barn mows equipped with ventilation systems through which either unheated or heated air is forced. Generally mow dried hay is greener, leafier, and of a higher grade than similar hay that is field cured. This is particularly true in those humid areas which consistently have poor haymaking weather. Of course, during ideal haymaking weather, it is possible to make about as good quality hay by field curing as can be made by mow curing.

Although higher quality hay can generally be produced in mow curing than in field curing, the cost is also greater. For this reason, in areas where poor haymaking weather generally prevails and field curing is hazardous, stockmen may well consider the desirability of making silage.

Artificial Dehydrators

Artificial dehydrating refers to that process in which forage is taken from the field as soon as it is cut (or in some instances after wilting), put through a hay chopper or silage cutter, and dried in large dryers of various types. For the most part, this method of curing is limited to large commercial operations which process early cut alfalfa (or its leaves) and other legume and/or grass crops chiefly as a supplement for swine and poultry. Occasionally, artificially dehydrated hay is produced for other classes of animals, especially in those areas which rarely have good haymaking weather.

Other pertinent facts relative to artificial dehydrating of hay are:

1. **Dryer**—The most popular type of artificial dehydrator in use in this country is one that uses a high initial heat (1,200°F to 1,400°F), and which is usually heated by oil. In a good dryer, the forage does not get sufficiently hot to be injured, primarily because of the cooling effects produced when the water evaporates from the plant tissues, and because the forage is in contact with the hot air for only a few minutes.

Any burning on the leafy portions of the forage indicates that the temperature was too high or that the forage was overdried.

2. **Nutritive value of artificially dehydrated hay**—Generally, artificially dehydrated forage is of high nutritive value, for few leaves are lost and the maximum content of protein, carotene, and riboflavin is retained. But the protein may be slightly less digestible due to the effect of the heat of dehydration, and the vitamin D content is low for the reason that in the curing process the hay is not exposed to the vitamin D-imparting ultraviolet rays of the sunlight.

Dehydrated grass and grass-legume mixtures are generally higher in total digestible nutrients than straight legumes, an important consideration when dehydrated forages are to replace a part of the concentrate ration.

3. **Cost**—Due primarily to high equipment and fuel cost, artificial drying is usually considered practical only in commercial operations where large tonnages are processed. Even then, the artificially dehydrated forage must command a premium price over field cured hay in order to justify the higher investment in equipment, the higher cost in moving the heavier high-moisture forage from field to dryer, and in operating the dehydrator.

The stockman is well justified in paying a premium price for high-quality dehydrated forage for use in swine and poultry rations. Except in special circumstances or as a partial grain replacement, however, it is seldom economical to feed artificially dehydrated hay to cattle or sheep. For ruminants, it is generally more practical to make silage in those areas commonly having poor haying weather. This is especially true where the forage is to be fed out on the farm on which it is produced.

Wagon Dryers

Wagon dryers were developed to reduce the high labor requirements of batch or platform drying. Essentially, they are batch dryers on wheels.

Some wagons are covered; they are known as "covered wagons" because of the outward appearance of the ballooned cover on the wagons during drying. The cover consists of a durable, lightweight material that won't leak water or air. Heat is conveyed under the cover. Other wagons are open and adapted to drying in a shed. In both types, the drying wagons are loaded in the field directly behind the baler, taken to the drying system where they are connected to a main air duct, and dried. After drying, they are taken to the storage place and unloaded. This procedure eliminates the unloading and reloading of wagons required in batch or platform drying.

Generally, baling is started when the hay has dried to 30 to 45 percent moisture.

Although good quality hay can be produced by wagon drying, it requires more labor and handling than the mower systems of haymaking.

Storing

Good hay should never be poorly stored. Naturally, the type of storage will vary from area to area. In the more arid sections where little rainfall comes during the fall and early winter, a good stack of loose or baled hay may provide entirely satisfactory storage. On the other hand, in high-rainfall areas, more expensive waterproof storage should be provided. At and between these two extremes hay may be and is successfully stored in many different ways in different sections of the country.

In the West, a considerable amount of hay is chopped (either at the time of gathering from the windrow or adjacent to the stack) and stack stored. Most such stacks are round, and are built by sliding a snow fence toward the top as the stack is built. Generally these stacks are rounded off at the top and left uncovered. The advantages claimed for stack storage of chopped hay are (1) minimum labor in haymaking, (2) minimum stack storage space and spoilage, and (3) ease of feeding.

Where different kinds and qualities of hay are produced or purchased, each kind and each quality should be stored in such manner that it will be accessible when needed. Otherwise, it may not be convenient to provide for variety in feeding and to feed some of the low-quality along with some of the high-quality hay.

ADDITIVES AND PRESERVATIVES FOR HAY

Farmers in many countries of the world have traditionally added about 20 pounds of salt per ton of new hay, at the time of stacking or putting it into the hay mow, in the belief that the salt would prevent the hay from molding and heating. Carefully controlled experiments have failed to substantiate claims that salt will prevent excess heating or sweating; nor has it prevented spontaneous combustion of hay. However, when salt is used in moderate amounts, it may improve the color, aroma, and palatability of poor-quality hay. It is recognized, too, that much higher levels of salt—quantities sufficiently high to harm animals—may prevent mold.

Over the years a number of products, both liquids and powders, said to preserve hay have become commercially available. While claims have been made that there will be no heating or molding when these materials are used, the results have been highly variable.

Recent studies of several state experiment stations have shown that propionic acid or a combination of propionic and other organic acids can be used successfully to preserve baled hay stored at 30 percent or less moisture. Anhydrous ammonia and ammonium isobutyrate have also been found to be effective in preventing heating and preserving the quality of high-moisture hay. In general, depending on the moisture content of the hay, application rates of 1 to 1½ percent (20 to 30 lb/ton) of the additive are required for effective preservation.

At the time of this writing, researchers are continuing to evaluate chemical hay preservatives and methods of application. But, based on presently available information, the following recommendations appear to be justified:

1. Generally, grass hay baled at less than 25% moisture, and legume hay at 20% moisture or less, will keep satisfactorily without a preservative.
2. For legume hay with between 20 to 25% moisture, apply at least ¼ lb of actual propionic acid per each 50-lb bale. This means that if the commercial hay preservative contains only 20% propionic acid, 1¼ lb of it should be added per each 50-lb bale.
3. For hay running 25 to 30% moisture, apply at least ½ lb of actual propionic acid per 50-lb bale.
4. Acid preservatives should be used only in situations where hay *must* be baled at high moisture (above 20% for legume hay, and above 25% for grass hay).
5. Complete and uniform coverage of the treated hay is essential for successful preservation. While other methods of application may be developed, the most practical method appears to be spraying it onto the hay as it enters the baling chamber.

Spontaneous Combustion

Wet hay ferments and generates heat. Sometimes this results in spontaneous combustion and fire, usually about a month to six weeks after storing. Here are the facts:

1. **Symptoms of heating**—The warning signals are: hay that feels hot to the hands, strong burning odor, and visible vapor.
2. **Temperature of hot hay**—Hot spots may be located by probing the hay with a steel rod. Then the temperature of the hot spots may be tested with a thermometer (a dairy thermometer or other type) attached to a wire and dropped down a pipe. If the hay is over 140°F, it should be checked periodically during the day.

   If the hay is 160°F, it should be checked hourly.

   If the hay is 180°F, there are apt to be fire pockets; and it should be removed from a barn.
3. **Cooling hay**—Hay that is heating may be cooled by discharging through pipes into the hot areas either dry ice or liquid carbon dioxide.
4. **Removing hot hay**—When a fire is imminent and hot hay must be removed, it is important to have plenty of help on hand including the fire department.

Then the hay should be removed cautiously and without wetting unless necessary.

5. **Precautions**—Never walk on hay that is heating—place planks over it, and do not breathe hot and noxious fumes.

BUYING AND SELLING HAY

Historically, most hay has been fed on the farms where it was produced. But this practice is changing. Today, some 15 to 20 percent of the hay produced in the United States is sold off the farm. In 4 states—Arizona, California, New Mexico, and Washington—over 40 percent of the hay produced is sold.

New hay markets are developing. As dairymen, beef cattlemen, cattle feeders, and sheep feeders become more specialized, they prefer to grow animals and rely on other specialists to grow hay. Also, more high-quality hay is needed for expanding horse numbers. Mushroom growers are becoming an important market for hay, too; they require a high-energy hay, although it can be moldy.

Additionally, considerable hay is exported. In order to save space and cut down transportation costs, this generally requires a special hay package—pellets, wafers, or high-density bales.

Hay Sources

The five basic sources of hay in the United States are:

1. **Hay dealers or brokers**—Hay dealers or brokers are important suppliers of hay. In California, 90 percent of the hay is marketed through this channel. For the nation as a whole, dealers and brokers market 5 to 10 percent of the hay produced. They purchase hay from the producer and sell it to the consumer.
2. **Neighbor to neighbor**—This is the oldest market channel for hay, and it is still quite common. This is a disorganized market without any particular pricing structure. Hay is purchased on visual inspection.
3. **Associations of cooperatives**—These organizations vary in size, from very small to very large. The San Joaquin Hay Producers Association in California is one of the largest, grossing over $10 million. Such associations normally purchase hay for cash and store it or move it to the consumer. This is a high-risk operation, because there are no futures markets that enable hedging protection.
4. **Auctions**—Hay auctions are trading centers where hay is sold by public bidding to the buyer who offers the highest price. Hay is sold by visual inspection, and the price is determined by supply and demand.
5. **Contract**—This is an agreement between a hay producer and a stockman to supply hay of a specified quality at a prior agreed-upon price. It assures both the buyer and the seller an orderly market. Such eventualities as weather damaged hay should be covered in the contract.

How Hay Is Priced

Traditionally, most hay has been sold and bought by using the ancient art of bartering. Present official standards and grades are not used extensively in price making. But this situation is changing. Several states now have programs for selling and buying hay on the basis of analysis—especially protein and moisture.

But there are a number of obstacles which must be overcome before the marketing of hay on the basis of analysis will be widely adopted; among them, the following:

1. Keeping the cost of analysis nominal.
2. Developing simple and practical sampling techniques.
3. Reducing the time between sampling and getting the results.
4. Arriving at a satisfactory pricing structure.

With increased marketing of hay, these problems will likely be resolved. Also, there will likely evolve (1) a more standard method of analyzing hay for quality, and (2) a system of pricing consistent with the quality of the product. In the meantime, sellers and buyers will continue to rely on visual inspection.

Futuristic Hay Evaluating and Pricing

Recognizing the problems of hay marketing, the American Forage Grassland Council (AFGC) formed in 1972, a Hay Marketing Task Force, made up of representatives of the AFGC, industry, and the National Hay Association, and charged them with the following responsibilities: (1) to identify hay marketing problems, (2) to determine problem priorities and possible practical solutions, and (3) to develop specific recommendations for action that are within its abilities or the abilities of someone it can stimulate into action.

This committee identified two problem areas: (1) determining the price of hay by some realistic measurement of feed value, and (2) consideration of forage packages to facilitate economically long distance transport when needed and to enable more efficient storage and complete mechanical handling. To date the task force has confined its efforts primarily to problems of measurements of feed value and improved hay standards for marketing hay.

At the outset, the Hay Marketing Task Force pinpointed the need for some scientific measurement of the constituents of forage—such as moisture, crude protein, fiber, energy, TDN, or some combination of these—as the underlying problem in hay marketing.

• **Proposed new hay standards**—The first challenge of the committee was to research and select the best practical methods of chemical analysis suitable for the widest range of hay species, then set up a system of standards with which forages of varying feed value could be classified.

Following careful study, three chemical values were selected as best suited to measure hay quality: crude protein (CP), acid detergent fiber (ADF), and neutral detergent fiber (NDF). Studies showed that ADF is the best predictor of forage dry matter digestibility, while NDF is good for predicting intake, or the amount of a given forage an animal can consume. All three analyses can be run in most forage testing labs.

Using these analyses, the committee developed new proposed hay standards, including five hay grades and one sample grade for all legumes and grasses (Tables 6-12 and 6-13). Included in the standards is a relative feed value rating making it possible to relate directly the feed value of various grades of both legumes and grasses (Table 6-14). (See page 690 for Table 6-13, and see page 691 for Table 6-14.)

The proposed standards also include descriptions of maturity, leafiness, color, foreign material, and freedom from must or mold. These are included primarily to help illustrate what is expected of each forage grade and what external factors might affect feed value. When the standards are adopted and in practice, the visual factors probably will not be too important. The grade will be determined primarily by chemical analysis.

In these proposed standards, legume grade 4 has been established as a base and given a relative feed value of 100. Legume hays fall in grades 1 to 4. But using these values, the best pure grass hay will grade no higher than 2.

Before these standards can be used for the marketing of hay, they must first be approved by the Federal Grain Inspection Service of the U.S. Department of Agriculture.

• **Infrared reflectance analysis**—A second and ex-

TABLE 6-12

PROPOSED MARKET HAY GRADES FOR LEGUMES AND LEGUME-GRASS MIXTURES

Grades	Stage of Maturity International Term	Definition	Physical Description	Typical Chemical Composition[1] CP	ADF	NDF	Relative Feed Value
				(%)	(%)	(%)	
1 Legume hay	Prebloom	Bud to first flower; stage at which stems are beginning to elongate to just before blooming.	40-50% leaves*; green; less than 5% foreign material; free of mold, musty odor, dust, etc.	>19	<31	<40	>140
2 Legume hay	Early bloom	Early to mid-bloom; stage between initiation of bloom and stage in which ½ of the plants are in bloom.	35-45% leaves*; light green to green; less than 10% foreign material; free of mold, musty odor, dust, etc.	17-19	31-35	40-46	124-140
3 Legume hay	Mid-bloom	Mid- to full bloom; stage in which ½ or more of plants are in bloom.	25-40% leaves*; yellow-green to green; less than 15% foreign material; free of mold, musty odor, dust, etc.	13-16	36-41	47-51	101-123
4 Legume hay	Full bloom	Full bloom and beyond.	Less than 30% leaves*; brown to green; less than 20% foreign material; slight musty odor, etc.	<13	>41	>51	100
6 Sample grade** Hay which contains more than a trace of injurious foreign material (toxic or noxious weeds and hardware) or that definitely has objectionable odor or is undercured, heat damaged, hot, wet, musty, moldy, caked, badly broken, badly weathered or stained, extremely overripe, dusty, which is distinctly low quality or contains more than 20% foreign material or more than 20% moisture.							

[1]Chemical analyses expressed on dry matter basis. Chemical concentrations based on research data from NC and NE states and Florida. Dry matter (moisture) concentration can affect market quality. Suggested moisture levels are Grades 1 and 2 <14%, Grade 3 <18%, and Grade 4 <20%. CP = crude protein; ADF = acid detergent fiber; NDF = neutral detergent fiber; relative feed value = digestible dry matter intake.

*Proportion by weight.

**Slight evidence of any factor will lower a lot of hay by one grade.

TABLE 6-13

PROPOSED MARKET HAY GRADES FOR GRASSES AND GRASS-LEGUME MIXTURES

Grades	Stage of Maturity International Term	Definition	Physical Description	Typical Chemical Composition[1] CP[2]	ADF	NDF[3]	Relative Feed Value
				(%)			
2 Grass hay	Prehead	Late vegetative to early boot; stage at which stems are beginning to elongate to just before heading; 2-3 weeks' growth.**	50% or more leaves*; green; less than 5% foreign material; free of mold, musty odor, dust, etc.	>18	<33	<55	124-140
3 Grass hay	Early head	Boot to early head; stage between late boot where inflorescence is just emerging until the stage in which ½ inflorescences are in anthesis; 4-6 weeks' growth.**	40% or more leaves*; light green to green; less than 10% foreign material; free of mold, musty odor, dust, etc.	13-18	33-38	55-60	101-123
4 Grass hay	Head	Head to milk; stage in which ½ or more of inflorescences are in anthesis and the stage in which seeds are well formed but soft and immature; 7-9 weeks' regrowth.**	30% or more leaves*; yellow-green to green; less than 15% foreign material; free of mold, musty odor, dust, etc.	8-12	39-41	61-65	85-100
5 Grass hay	Post head	Dough to seed; stage in which seeds are of doughlike consistency until stage when plants are normally harvested for seed; more than 10 weeks' growth.**	20% or more leaves*; brown to green; less than 20% foreign material; slightly musty odor, dust, etc.	<8	>41	>65	<85
6 Sample grade*** Hay which contains more than a trace of injurious foreign material (toxic or noxious weeds and hardware) or that definitely has objectionable odor or is undercured, heat damaged, hot, wet, musty, moldy, caked, badly broken, badly weathered or stained, overripe, dusty, which is distinctly low quality, or contains more than 20% foreign material or more than 20% moisture.							

[1]Chemical analyses expressed on dry matter basis. Chemical concentrations based on research data from NC and NE states and Florida. Dry matter (moisture) concentration can affect market quality. Suggested moisture levels are Grade 2 <14%, Grade 3 <18%, and Grades 4 and 5 <20%. CP = crude protein; ADF = acid detergent fiber; NDF = neutral detergent fiber; relative feed value = digestible dry matter intake.

[2]Fertilization with nitrogen may increase CP concentration in each grade by up to 40%.

[3]Tropical grasses may have higher NDF concentrations than indicated in this table.

*Proportion by weight.

**For grasses that do not flower or for which flowering is indeterminant.

***Slight evidence of any factor will lower a lot of hay by one grade.

citing development resulting from the efforts of the task force was the discovery of the potential of infrared reflectance analysis as an accurate and rapid means of measuring forage feed quality. When perfected, this technique should make it possible to scan a sample of forage and within minutes read out crude protein, digestible protein, dry matter digestibility, moisture, acid detergent fiber, neutral detergent fiber, and many other measurements of forage feed value.

Researchers are now working on a portable unit which, when perfected, can be placed at strategic locations throughout the United States, such as hay marketing centers, feed stores, and county extension centers. This instrument will collect information from the forage sample, then relay the data to a central computer via telephone. The computer will analyze the data, estimate forage feed value, and relay the information back to the remote terminal keyboard for use in hay pricing or formulating feed rations. The time involved for an analysis by this method should not exceed 2 minutes.

Researchers are confident the system can be developed. The challenge at present is to get the cost of the remote terminal low enough that it can be used commercially.

• **Summary**—At today's feed prices, hay (espe-

TABLE 6-14

TYPICAL COMPOSITION OF DIGESTIBLE DRY MATTER (DDM), DRY MATTER INTAKE (DMI), AND DIGESTIBLE DRY MATTER INTAKE (DDMI) FOR PROPOSED MARKET HAY GRADES DESCRIBED IN TABLES 6-12 and 6-13[1]

	Legume Hays			Grass Hays			
Grade	*In vivo* DDM	DMI gm/W kg$^{.75}$	DDMI gm/W kg$^{.75}$	*In vivo* DDM	gm/W kg$^{.75}$	gm/W kg$^{.75}$	Relative Feed Value[2]
	(%)			(%)			(%)
1	>70	>80	>57	---	---	---	>140
2	66-70	75-80	50-57	>72	>69	>49	124-140
3	58-65	68-74	41-49	62-72	65-69	41-49	101-123
4	<58	<68	<41	55-61	59-64	33-40	85-100
5	---	---	---	<55	<59	<33	<85

[1]Formulas used to calculate relative feed value: legume - DDM = 65.5 + 0.975 ADF% - 0.0277 ADF%^{2}; DMI = 39 + 2.68 NDF% − 0.0410 NDF%^{2}; grasses - DDM = 34.8 + 2.56 ADF% - 0.0491 ADF%^{2}; DMI = 54.8 + 1.22 NDF% - 0.0176 NDF%^{2}; DDMI = $\frac{DDM \times DMI}{100}$; relative feed value = $\frac{DDMI \times 100}{40}$; DDM = *In vivo* dry matter digestibility; DMI = dry matter intake; DDMI = digestible dry matter intake.

[2]Relative feed value is an estimate of overall forage quality. It is calculated from intake and digestibility of dry matter when forages of known composition were fed to sheep. The values are relative, however they are equally appropriate for all classes of livestock. Relative feed value estimates the intake of digestible energy when the forage is the only source of dietary energy and protein.

cially alfalfa) is a profitable crop to grow, both for direct utilization through livestock and as a cash crop, *provided* it is of high quality and is priced at its true value for feed. However, there is general agreement that if hay is to take its rightful place in the marketplace as a major source of protein and energy, many of the traditional selling arts must be replaced by developments of modern science.

But there are a number of obstacles which must be overcome before the marketing of hay on the basis of analysis will be widely adopted; among them, the following:

1. Keeping the cost of analysis nominal.
2. Developing simple and practical sampling techniques.
3. Reducing the time lag between sampling and getting the results. Hay dealers report that much of their hay is purchased one day and sold the next. As a result, they do not have time to take advantage of a laboratory analysis. Therefore, hay is not evaluated properly by traditional chemical analysis unless done so by the producer in advance of selling.
4. Developing standards and grades based on feed value.
5. Arriving at a satisfactory pricing structure.

With increased marketing of hay, these problems will likely be resolved. Also, there will evolve (1) a more standard method of analyzing hay for quality, and (2) a system of pricing consistent with the quality of the product sold. In the meantime, sellers and buyers will continue to rely on visual inspection.

Other developments are necessary if hay is to become a leading cash crop, including:

1. **More markets.** There is need for more markets—such as auctions, dealers, and associations—to provide a ready market outlet for hay growers, and to attract both ample supplies of hay and buyers.
2. **Price protection for the dealer while he has possession of the hay.** There is need for something similar to the futures market that will permit hedging protection.
3. **More hay grower storage facilities.** There is need for additional grower storage so as to spread hay marketing over a longer period and avoid harvest market gluts that depress prices.
4. **Better hay market information.** There is need for hay growers and hay buyers to be better informed relative to supply, demand, and going prices.

Freedom from Toxic Residues

With the emphasis on residues in foods, it is important that hay be free from those residues which are prohibited. If meat or milk (or products derived from them) are found to have residues, the blame cannot be shifted to the hay grower by the livestock producer, unless there is a clear-cut case of fraudulent representation. The best assurance of freedom from such residue rests with the integrity of the hay grower, or the ones who represent him in selling his hay.

Most of the pesticides have "zero" tolerance in foods. The chlorinated hydrocarbons—DDT, chlordane, heptachlor, aldrin, dieldrin, toxaphene, and many others—have been banned for use on feed crops. They have been replaced by the organic phosphates—malathion, parathion, ronnel, and ciodrin—which have a shorter decay period (the period required for their residues to be reduced to zero).

It is important that livestock producers and hay growers be well informed relative to (1) the chemicals which are banned, and (2) the conditions of application for those chemicals which are still permitted. In this way, disastrous financial effects from confiscation of market animals or products can be averted.

Hay Shrinkage

Hay buyers should figure hay shrinkage closely. Here's why: If a ton of hay containing 90% dry matter is bought for $80, 1,800 lb of dry matter have been purchased at this price. However, if the $80 per ton hay contains 80% dry matter, 1,600 lb of dry matter have been purchased for this same price. Purchase of the high-moisture, 80% dry matter hay has resulted in a loss of 200 lb of hay, or 1/9 of the dry matter, worth $8.89 (1/9 × $80). Thus, if the 90% dry matter is worth $80 per ton, then 80% dry matter hay is worth only $71.11 ($80.00 − $8.89) per ton. If 1,000 tons of hay are involved, that's a loss of $8,890 (1,000 × $8.89).

In addition to moisture losses, newly harvested hay may be expected to lose about 5 percent weight from going through the sweat.

Hidden Hay Costs

When buying and feeding hay, all costs and losses—handling, shrinkage and wastage, grinding costs and losses, insurance, interest, and storage—in getting it from the point of purchase to the feed bunk should be taken into consideration. The example that follows underscores this situation.

Which is the best buy—baled alfalfa hay at $75.00 per ton roadside or sun-cured alfalfa pellets at $117.50? Simple arithmetic would indicate that baled hay is the cheaper, by $42.50 per ton. But, what are the facts?

Most researchers report that pellets have a higher feeding value for beef cattle and sheep than long or chopped hay. But they warn that the increased rate and efficiency of gain may not be sufficient to cover the cost of pelleting, running $12 to $16 per ton.

Unfortunately, most people fail to account for all the costs and losses inherent in getting baled hay from roadside to feed bunk. Generally speaking, they do not make allowance for hauling costs, storage, shrinkage and wastage, chopping costs and losses, insurance, interest, and breakdown losses in time and equipment due to hardware in the bales—not to mention the fuss and muss of handling baled or chopped hay.

A comparison of baled alfalfa hay vs alfalfa pellets, based on 1980 figures in the San Joaquin Valley of California, reveals the figures shown in Table 6-15.

TABLE 6-15

HIDDEN HAY COSTS

	Baled Alfalfa Hay	Sun-cured Alfalfa Pellets
	(per ton)	(per ton)
Cost in the feed bunk:		
Price at roadside	$ 75.00	
Hauling and stacking	7.50	
Moisture, shrinkage, wastage, and spoilage	7.50	
Chopping or grinding costs	13.00	
Grinding losses	4.50	
Insurance	3.00	
Interest (6 months)	5.00	
Storage cost	2.00	
Total cost in the feed bunk	$117.50	$117.50
Added returns from pelleting:[1]		
Value of improved rate of gain		5.00
Value of improved feed conversion		15.00
Total		$20.00

[1]Computations by the author, based on feeding trials conducted by the Illinois, California, Washington, and Oklahoma Experiment Stations.

Since it is net returns that count, in each case (bales vs pellets) one must consider (1) cost delivered to the feed bunk, and (2) added returns from pelleting in terms of more gains and/or less feed. In the above example, when all the hidden losses and costs were added to baled hay, the price in the bunk was $117.50 per ton—the same as for sun-cured alfalfa pellets.

Actual feeding trials conducted by Illinois, California, Washington, and Oklahoma Experiment Stations—with roughage constituting 80 to 100 percent of the rations—show that in terms of increased rate and efficiency of gains, each ton of pellets is worth $20 per ton more than a ton of chopped forage. In this example, pellets were the best buy, by $20 per ton—the amount of the added feeding value, since baled and pelleted hay cost the same when delivered to the bunk.

In 1948, the author and his associates at Washington State University, in what appears to be the earliest work ever done on pelleted forage for cattle, reported nearly ¾ pound more daily gains from pelleting over coarsely chopped forage.

In one Illinois Station test, weaned steer calves were fed timothy-alfalfa either (1) long (baled), (2) chopped, or (3) pelleted. Gains were increased 175 percent by pelleting. Even with an allowance of $10 per ton for pelleting, feed cost per pound of gain was 3.7 cents cheaper for calves fed the pellets.

On the average, cattle on high-roughage rations (above 80%) will eat about ⅓ more pellets than long or chopped hay, make about ½- to ¾-lb faster daily gains, and require 200 to 250 lb less feed per 100 lb gain.

Of course, each stockman should apply his own cost figures to baled hay and pellets in order to determine which is the best buy for him. The important thing is that all costs be accounted for—that in computing the cost of baled hay there be added such hidden costs and losses as handling, shrinkage and wastage, grinding costs and losses, insurance, interest, and storage. Additionally, allowance should be made for the added feeding value of the pellets. Also, the age and grade of cattle, other available feeds and prices, and starter vs finishing rations must be considered.

This same method of cost analysis can be applied to hay cubes, or wafers, or to any other alternate method of haymaking.

HAY FEEDING FUNDAMENTALS

Feeding is the end of the line for hay. No matter how carefully it has been grown, harvested, and stored, all that has gone before can be dissipated if it is improperly fed—unless hay feeding fundamentals are observed.

Monogastric animals, including swine and poultry, must eat a large percentage of grains and other concentrates and depend almost entirely on digestive enzymes to break down these compounds. But ruminants, with their four stomach compartments and the help of microorganisms, can subsist largely, or entirely, on bulky, high-fiber forages which, because of their low energy per unit weight of dry matter, must be consumed in large quantities to supply their nutrient needs. The horse, because of his greatly enlarged cecum and large intestine, can utilize quantities of hay intermediate between simple-stomached and ruminant animals.

The economics of the situation—the relative price of forage and grain—calls for greater emphasis on forage accompanied by less grain feeding. With greater quantities of forage incorporated in rations, it is expected that performance—the production of meat and milk—will decrease. However, maximum net returns, rather than just maximum production, will be the primary objective.

Increasingly, forage testing will be used in two ways: (1) to purchase hay on a quality basis, and (2) to balance rations more precisely.

Hay Preparation

Hay is fed as long hay or in processed form. The common methods of processing are: chopping, grinding, cubing, and pelleting.

Considerable hay is chopped in the West, for two reasons: (1) It facilitates handling, and (2) it lessens refusal and waste. Low-quality and coarse forages usually benefit more from chopping than high-quality forages.

Hay is usually finely ground when it is incorporated in mixed swine and poultry rations. Fine grinding is not desirable for ruminants; it results in reduced rumen acetate production and lower milk fat percentage.

Both cubing and pelleting (1) make automatic hay feeding feasible, (2) decrease nutrient losses, and (3) eliminate dust. Also, they narrow the spread between high- and low-quality forage; that is, the poorer the quality of the forage, the greater the advantage from cubing or pelleting. This is so because such preparation assures complete consumption. Also, cubing or pelleting, especially the latter, usually speeds up the passage of forage through the digestive system.

On the average, cattle on high-roughage (above 80% roughage) or all-roughage rations will eat about ⅓ more pellets than long or chopped hay (due to increased density and more rapid passage through the digestive tract), make about ½ to ¾ lb faster daily gains, and require 200 to 250 lb less feed per 100 lb of gain. Also, it is recognized that low-quality roughages are improved most by pelleting.

Cubes offer most of the advantages of pelleted forages, with few of the disadvantages. They alleviate fine grinding, and they facilitate automation in both haymaking and feeding, and they lower milk fat percentage only slightly, if at all.

Complete pelleted rations—in which the hay and grain are combined, then pelleted—are finding an increasing place for horses, and perhaps swine. Among the virtues ascribed to all-pelleted rations are (1) they prevent selective eating—if properly formulated, each mouthful is a balanced diet; (2) they alleviate waste; (3) they eliminate dust (thereby lessening heaves in horses); (4) they lessen labor and equipment; and (5) they lessen storage.

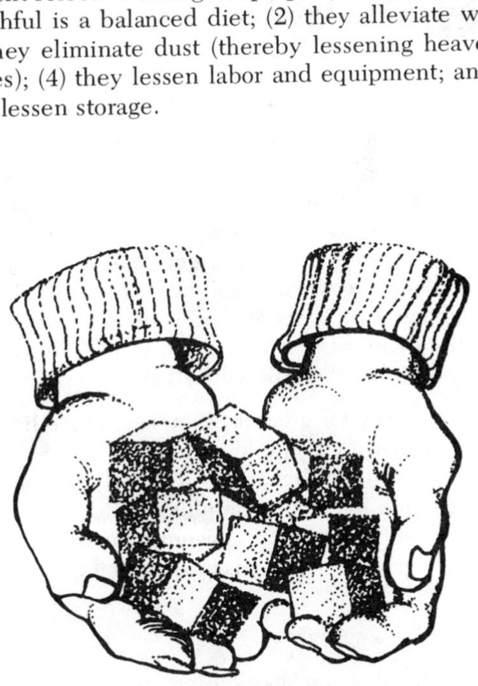

Fig. 6-12. Alfalfa hay cubes—1¼″ square × random length (usually 2″ to 3″ long), with a bulk density of 25 to 32 pounds per cubic foot.

Hay Feeding Systems

Hay may either be (1) self-fed, or (2) limited fed.

Most hay is self-fed. With a manger full of hay in front of them, hay consumption is limited only by the capacity of animals—by the amount that they can hold. That's the reason that ruminants eat more cubes and pellets than long hay of equal quality.

Limited feeding of hay is accomplished either (1) by hand feeding the hay and the concentrate allowances, or (2) by using a complete, mixed ration.

Fig. 6-13. Complete, mixed ration—with the proportion of hay and concentrate controlled—being conveyed by a self-unloading truck into a fenceline bunk.

The vast majority of large cattle and sheep feedlots feed complete rations, in which the quantity of hay is limited. Also, an increasing number of large commercial dairies are switching to complete rations. Most experiments and experiences have not shown any difference between mixed rations and the feeding of roughage and concentrates separately insofar as rate and efficiency of gain are concerned. However, a mixed ration has the following advantages:

1. It makes for greater efficiency in feeding and lessens the sorting at the feed bunk.
2. Where the roughage is relatively unpalatable, a mixed ration forces consumption.
3. When concentrate consumption is to be limited, mixing with the roughage is desirable.

Feeding Hay Packages

Round bales and stacks are another alternative to conventional, 110- to 130-pound, square bales. There is little doubt that in many livestock operations large hay packages can greatly decrease labor and result in similar animal performance. However, special attention needs to be paid to methods of feeding these big hay packages—round bales or compressed stacks; otherwise, waste can easily wipe out any saving in labor.

Under an in-field storage system, the bales or stacks are dropped where they are made, and remain there until needed for fall or winter grazing. Little or no labor is involved in making hay except for mowing, windrowing, and baling; and there is no manure to haul. Cows graze the grass growth which occurs subsequent to baling and consume the bales or stacks in the field. There is little or no labor in feeding. The cows go to the feed, rather than necessitating that the feed be taken to the cows.

The following methods are used in grazing round bales or small stacks, along with regrowth in the field:

1. **Continuous access to all bales or stacks**—In this system, the cows are given continuous access to all the bales or stacks in a field.
2. **Strip grazing bales and stacks**—In this system, the cows are given access only to those bales or stacks which are to be consumed within a given period of time. This is accomplished by using an electric fence and cross-stripping the field containing the bales or stacks. Such a strip-grazing program will increase the number of cow days by at least 35 percent.

A Pennsylvania State University study showed that cows given access to grass regrowth and large round bales during October and November wasted 26 percent of the available standing grass and round bales. It should be noted, however, that considerable hay is always wasted, even under the best circumstances.

Strip grazing will work for both bales and small haystacks. With stacks, it is especially important to limit-feed or strip-graze in order to avoid excess wastage.

3. **Other feeding methods**—Other systems require more investment than in-field grazing of bales or small stacks, but the additional numbers of cattle carried per acre may justify the increased cost. Among such systems are the following:

   a. **Hay packages (large round bales or small stacks) placed in rows and grazed with an electric fence**—Usually, one side of an existing hayfield is used for such a storage and feeding area.

   b. **Portable feeding gates (fences)**—Portable feeding gates (or fences), usually made of metal, may be placed around large bales or stacks. Their main advantages are that the cows cannot gain access to any part of the stack except that exposed to them through the gate, and that the cows cannot trample the forage, since the gate protects the hay. Most feeding gates are designed so that the cows can push them toward the bale or stack as the hay is consumed.

c. **Feeding wagons**—There are several designs and sizes of self-feeding wagons, both commercial and homemade. They give much flexibility; wagons can be easily taken to the area where the round bales are stored, then pulled to various feeding locations.

d. **Three-sided feeder**—Bales may be placed on concrete and enclosed in a three-sided feeder.

The following points are pertinent to feeding arge round bales and stacks:

1. **Frozen ground works best**—In-field grazing of ig hay packages works best on frozen ground, be-ause cows eat more of the hay, and trample less of it han when on soft ground.
2. **Weed spots can result**—Small bales do not eave bare spots following grazing in the field. A sod eadily pervades the space beneath the bales. How-ver, when large bales or small stacks are left in the ield throughout summer, fall, and most of the winter, are, weedy spots are likely to appear.
3. **Dairy cows can use them, too**—Although most f the large bales and stacks are being used by beef ow-calf operators, they can also be used advantage-usly for dairy herds. Hay can be provided to a dairy erd in a three-sided feeder.
4. **Feeding space**—On a per cow basis, 18 to 24 nches of hay feeding space per cow should be al-owed if all animals are to eat at the same time. Six to 0 inches per cow will suffice when hay is available t all times.

BALING WIRE DANGER

Short pieces of baling wire are dangerous to cat-le. If consumed, they are likely to pierce the stomach nd damage the heart, probably killing the animal. Among stockmen, this condition is known as "hardware disease"; among veterinarians, it is more echnically known as *traumatic pericarditis*. Because f this hazard, it is urged (1) that bales of hay and traw be broken by pulling the whole wire off rather han by cutting, and (2) that all used baling wire be arefully folded and placed in a barrel (and later dis-osed of) rather than left on the ground and allowed o get mixed with either the feed or the manure.

Hay Feeding Schedule

In general, animals may be given as much non-egume hay as they will consume, regardless of their revious ration. Thus, feedlot cattle and lambs are usually started with a full feeding of grass hay. Then, as grain is increased, the consumption of hay is de-creased. With legume hay, however, it is necessary that they be gradually accustomed to it; otherwise, looseness and scouring will likely result.

Proportion of Hay to Concentrate

Cattle, sheep, and horses will eat 2 to 3 pounds of hay per 100 pounds of body weight if fed hay alone. Also, it is noteworthy that the higher the quality of the hay, the more of it they will eat, with the result that the grain requirement will be lessened.

The economics of the situation—the comparative price of hay and concentrate—along with the management practices, will determine the proportion of hay to concentrate. Thus, during the period of low grain prices in relation to forage prices—from about 1950 to 1970—it was desirable to feed finishing cattle high-energy rations and to maximize gains. But the grain-fed cattle binge ended with the world grain shortages and high-priced grains of the early 1970s. In the years ahead, with grain becoming more scarce and higher in price than forages, comparatively speaking, more forage and less grain will be fed to finishing cattle, and net returns will be more important than high rate of gain. Cattle and sheep will increasingly be "roughage burners." Stockmen will rely upon the ability of the ruminant to convert coarse forage, grass, and by-product feeds, along with a minimum of concentrate, into palatable and nutritious food for human consumption, thereby competing less for humanly edible grains. Increasingly, the steer and the lamb of tomorrow will be produced on a maximum of milk and grass and a minimum of grain. More and more U.S. cereal grains will be used for human food, just as has been true, historically, in much of the rest of the world.

Ruminants can make the transition to more roughage with ease. For them, it is merely a "return to nature," for they evolved as consumers of forage.

The best buy (hay vs grain) may be determined by calculating the cost per pound of TDN and of protein in the hay and grain being compared. Then, the proportion of hay to concentrate can be varied accordingly. If hay is the best nutrient buy, feed more hay and less grain. On the other hand, if grain is the best buy, feed more grain and less hay.

Ruminants Need Hay

Ruminants need some roughage. Hay fed early in life will develop the calf's rumen and prevent anemia.

Many serious problems in high-producing dairy herds have been traced to lack of hay in the ration; among them, increased incidence of ketosis and displaced abomasums. In addition to these maladies attributable to no-hay rations, it is noteworthy that milk fat percent can be as much as one percent less on an all-silage and concentrate ration.

Different Qualities of Hay May Be Used

The type of ration which will be least costly and result in satisfactory performance will differ according to species, level of performance, reproductive status, age, etc. For example, the nutritive needs of a dry, pregnant beef cow are much lower than those of a high-producing dairy cow. Thus, a low-quality hay may be quite satisfactory for wintering a beef cow without calf at side, whereas high-producing, lactating cows should always have high-quality hay. Also, high-quality hay is important for swine and poultry. Where forage is incorporated in monogastric rations, high-quality dehydrated alfalfa is most commonly used. High-quality hay is also essential for horses.

Hay Waste and Refusal

In a recent Texas Station study involving 10 different hay feeding racks, the cows at the best conventional feeder still wasted 14 percent of their hay. Dairymen have commonly accepted 10 percent refusal as normal.

High-priced feeds and smaller margins are causing stockmen to scrutinize hay losses, and to do something about them. Chopping hay and/or adding molasses will lessen wastage. But feeding high-quality hay is the best way in which to lessen waste and refusal.

Supplementing the Hay Ration

Hay is generally lower in energy and higher in fiber than most grains and concentrates. Legume hays have a high calcium content, but they vary in available phosphorus. If sun-cured properly, they are high in carotene and vitamin D, along with many of the B vitamins. A supplement should supply the nutrients that are most likely to be lacking in the hay; thus, supplements for alfalfa hay should be high in energy and phosphorus and low in fiber. Also, carotene or vitamin A should be provided if the hay has been bleached or turned brown. Salt is lacking in hays and other natural feedstuffs and should be provided as a supplement, along with trace minerals that are deficient in the local area.

Stretching the Hay Supply

Most grains contain 75 to 80 percent TDN (total digestible nutrients), while most medium to good quality hays contain 45 to 50 percent TDN. Hence, as a general rule of thumb, about 5 pounds of grain equal 8 pounds of hay, provided they are of comparable quality; for example, No. 2 hay and No. 2 corn. Thus, it follows that if corn can be bought for $100 per ton, hay should be bought at $62.50 per ton, or less.

If the price of hay is less than five-eighths the price of grain, relatively more of it should be fed whereas, if the price of hay is higher than this, relatively more grain will make for cheaper production.

When hay is scarce and high in price, the hay supply for ruminants and horses can be stretched as follows:

1. Feed only half to two-thirds the normal ration of hay, but be on the alert for digestive disturbances.
2. Replace each 3 pounds of hay deleted with pounds of grain.
3. Make the maximum use of such feeds as cottonseed hulls, corncobs, straw, and grass aftermath in the ration for (a) all but 5 percent of the alfalfa (or other legume) hay of grower rations; and (b) all of the "hottest" finishing ration, adding such supplementary proteins, minerals, and vitamins as necessary to balance the ration.
4. Get finishing cattle and lambs, and animals being fitted for show or sale, on high-concentrate rations as expeditiously as possible. In cattle, eliminate the stocker feeding period—get weaned calves on full feed as quickly as possible.
5. Provide such supplementary proteins, minerals, and vitamins as necessary. This is especially important with gestating-lactating females or young growing animals. This may be accomplished by (a) feeding some legume, either hay or silage, and/or (b) adding suitable protein, mineral, and vitamin supplements. For example, pregnant cows that are in medium to good condition can be wintered satisfactorily on 12 to 20 lb of straw or other low-quality roughage, plus 1 lb of oilseed cake or meal (or equivalent protein supplement), or on straw plus 4 to 5 lb of alfalfa or other legume hay. Unless cows have had good green pasture in the fall, and consequently have a store of vitamin A in their bodies, alfalfa pellets or a vitamin A supplement should be fed. In addition, straw-fed cattle should always have access to a mineral supplement high in calcium.

PART II—CROP RESIDUES

Crop residues are the forages that remain after harvesting a grain crop. Among such crop residues are: cornstalks and husklage, sorghum stalks, soybean refuse, small grain straws and chaff, and legume and grass seed straws. Crop residues must be fed to the right class of animals, and they must be properly supplemented.

As the ever-increasing human population of the world consumes a high proportion of grains and seeds, and their by-products, directly, animals will utilize a maximum of crop residues and a minimum of products suitable for human consumption.

Generally speaking, crop residues may either be grazed, processed as dry feed, or made into silage (see ection 7 for a discussion of crop residue ensilage). 'he important thing to remember is that their relavely low value, in comparison with grains, necessiates low cost harvesting, storing, and feeding. Also, ney must be fed to the right class of animals, and ney must be properly supplemented. Remember, too, hat there is a marked difference between economical wintering and deficient wintering.

CORN RESIDUES

Of all crop residues, the residue of corn is produced in greatest abundance and offers the greatest potential for expansion in cow numbers. In 1980, 73,061,000 acres of corn, yielding 91 bushels per acre, were harvested in this country. For the most part, over and above the grain, 2½ to 3 tons of dry matter produced per acre (40 to 50% of the energy value of the total corn plant) were left to rot in the field. That's 200 million tons of potential cow feed wasted, enough to winter 151.5 million dry pregnant cows consuming an average of 22 pounds of corn refuse per head per day during a 4-month period. Moreover, mature cows are physiologically well adapted to utilizing such roughage. It is noteworthy, too, that when corn residue is used to the maximum as cow feed, acreage which would otherwise be used to pasture the herd is liberated to produce more corn and other crops. Also, there are many other crop residues, which, if properly utilized, could increase the 151.5 million cow figure given above.

Although corn refuse offers tremendous potential as a cow feed, there are difficulties in harvesting and storing it. But science and technology have teamed up and are working ceaselessly away at solving these problems.

Broadly speaking, three alternate methods of salvaging corn refuse are being used: (1) grazing, (2) harvesting and dry feeding, and (3) ensiling (see Section 7 for a discussion of corn residue ensilage); with different ways of accomplishing each. The choice of the method should be determined primarily by (1) cost, (2) the proportion of refuse utilized, and (3) how well it meshes in with other farm enterprises—for example, in some cases the need for fall plowing will necessitate removal of the material from the land and eliminate grazing as an alternative. Costs and the proportion of the refuse salvaged vary widely, as shown in Table 6-16.

Table 6-16 shows that harvesting corn refuse as silage resulted in salvaging the highest yield per acre and at the lowest cost per ton of the five harvesting systems reported, and that the cost of each method of harvesting was lowered as the acreage increased.

Grazing (Pasturing)

This refers to turning the animals directly into the stalk field–the traditional way of utilizing cornstalks. Letting the animals do their own harvesting is the simplest and least expensive method devised for utilizing a crop. However, there is considerable wastage, and it is not possible to prolong the winter feeding period. In an open fall and winter, 2 acres of cornstalks will carry a pregnant cow for 100 to 120 days. But the following problems are associated with this method of harvesting:

1. **Selective grazing**—Cows are selective grazers. They will consume the more palatable portions of corn refuse first, in the following order: corn ears, husks, leaves, and stalk.

TABLE 6-16

HARVESTING COST PER TON OF DRY MATTER OF CORN FORAGE BY VARIOUS HARVESTING SYSTEMS[1]

Harvesting System	Forage Yield/ Acre[2]	Harvesting Cost/Ton[3]		
		80 Acres	160 Acres	240 Acres
	(tons)	($)	($)	($)
Corn silage, with forage harvester	6.0	5.30	(100 Acres) 4.75	---
Corn stalklage, with flail-type forage harvester	1.5	10.85	8.36	7.59
Corn stalklage, with 1 ton stacker	1.5	6.75	5.49	5.20
Corn stalklage, with 3 ton stacker	1.5	8.86	6.48	5.83
Corn husklage, with strawbuncher behind	1.0	6.22	3.21	2.22

[1]Ayres, George E., "Harvesting and Handling Forage for Beef Cows," paper presented at Second Annual Corn Belt Cow-Calf Conference, Ottumwa, Iowa, February 24, 1973.

[2]Tons of dry matter.

[3]Includes cost of all harvesting machines and labor, but does not include storage costs.

2. **Waste**—Only an average of one-third of the stover is actually used, with the amount varying from 15 to 40 percent, depending primarily on weather conditions.

3. **Fencing**—Many cornfields are unfenced; hence, fence must be constructed in order to confine the animals. Also, strip grazing (grazing a part of the field at a time) will improve the utilization of stalks in large fields by making more uniform nutrition available throughout the grazing period. It prevents selective grazing over the entire field, with the result that the animals consume the more palatable portions of the plant first, and leave the bare stalks until last.

The fencing problem may be solved economically with an electric fence.

4. **Snow cover; fall plowing; soil puddling; stock water**—In the northern part of the United States, snow cover prevents grazing for part of the winter and necessitates a reserve feed supply.

Another drawback is that grazing prevents fall plowing, a recommended practice on heavy soils. Also, cattle may puddle and pack the soil, which lowers crop yields.

Frequently, supplying the herd with drinking water is costly. It may necessitate drilling a well or piping or hauling water from a distance.

Stalklage

Stalklage refers to all the residue remaining after harvesting corn with a combine or picker. It may either be stored as dry stalklage or ensiled (see Section 7 for a discussion of stalklage ensilage).

Stalklage is more difficult to collect than husklage; and more expensive, since it involves more equipment and another trip across the field. A number of different machines for harvesting stalklage are being used; among them, forage harvesters, balers, stackers with flail pickups, and choppers and stackers. By operating the machine a few inches above the ground so as to prevent excess soil pickup, a yield of 1 to 3 tons of residue per acre may be obtained, with the moisture content ranging from 20 to 55 percent, depending on the time of harvest. Stacked or baled cornstalks should be at the low end of this moisture range to reduce heating and spoilage.

Cows like dry stover. Self-feeders around a stack make feeding convenient. Leftover material may be used as bedding.

Husklage (Shucklage)

Husklage refers to the forage discharged from the rear of a combine when harvesting corn. It consists of the husks, cobs, and any grain carried over the combine, collected in a wagon or straw buncher pulled behind the combine. This operation minimizes labor and does not slow the grain harvest, because th husklage piles can be dumped at the end of the fiel for supplemental feeding or later pickup by a front end loader and moved to another location for stackin or ensiling. The moisture content of this material wil usually run between 30 and 40 percent, and the yield will be between 1 and 1.5 tons per acre.

The greatest difficulty encountered in feedin husklage dumps at the end of the field is waste. Depending on weather conditions, as much as 50 percen of the material may be wasted. But wastage o husklage dumps can be materially lessened by controlling access to them.

Stacking of husklage has been satisfactory fo some producers.

Ensiling husklage, along with recutting and adding water, results in increased cow consumption and less rejection of cobs.

Since grazing stalk fields is widely practiced in the fall and winter, the feeding of baled, piled, o stacked corn residues in the field permits feeding cows on stalks most of the winter without supplemental feeding of hay or silage. Some molds may develop in the collected material, but usually they are not sufficient to affect either the feed intake or health of mature cows.

Feeding Value of and Supplements for Corn Residues

Table 6-17 lists the daily nutritive requirements of a dry pregnant cow weighing 1,000 pounds. Table 6-18 gives the nutritive composition of air-dry corn stover and husklage.

TABLE 6-17

NUTRITIVE REQUIREMENTS OF A DRY PREGNANT COW (MIDDLE THIRD OF PREGNANCY) WEIGHING 1,000 POUNDS[1]

Dry matter, daily	14.8 lb
TDN, daily	7.9 lb
Crude protein, % of ration	5.9%
Calcium, % of ration	0.18%
Phosphorus, % of ration	0.18%
Vitamin A	19,000 IU

[1]*Nutrient Requirements of Beef Cattle*, Fifth Revised Edition, NRC-National Academy of Sciences, Washington, D.C., 1976, pp. 26, 32.

Not all corn refuse will be of the same composition as Table 6-18; some will be better, some will be poorer. The quality declines with the passing of time following grain harvest; and the more severe the weather, the greater the decline. Also, cultural practices during the growing season may alter corn residue quality.

Studies show that a 1,000-lb cow will eat approx-

TABLE 6-18

ANALYSIS OF AIR-DRY CORN STOVER AND HUSKLAGE[1]

	Corn Stover	Husklage
	(%)	(%)
DN	48	57
rude Protein	4.5	3.4
alcium	0.4	0.02
hosphorus	0.07	0.05
itamin A	---	---

[1]*Cow-Calf Information Roundup*, University of Illinois, 1971, Table 2, p. 10.

mately 18 to 20 lb per day of palatable, air-dry stover, r about 2 lb of air-dry stover per cwt per day. She vill eat slightly larger amounts of husklage. This conumption, along with the information presented in Tables 6-17 and 6-18, suggests that stover and/or usklage rations will meet the daily energy (TDN) eeds of dry pregnant cows, but such rations will be lightly deficient in protein, and low in phosphorus nd vitamin A. Nevertheless, the highest and best use or corn residue is for dry pregnant cows for the eriod following conception to about 30 days before alving.

For corn refuse feeding, mature cows should be n medium to good condition at the start of the winter eeding period; and they should not be permitted to ose over 10 to 15 percent of their weight from fall hrough calving. Heifer weight losses should be nder 5 percent. When weight loss approaches this imit, it's time to feed some grain or silage.

The following additional information is pertinent o the feeding value and supplementation of corn resdues for cattle:

1. **Digestibility**—The components of corn residue rank as follows in digestibility, in descending order: remaining grain, husk, leaf, cob, and stalks.

2. **Energy**—Corn residues provide adequate energy to maintain dry pregnant cows, but they must be supplemented with additional energy when fed to cows nursing calves or to young, growing animals.

3. **Protein**—The crude protein content of corn stover is on the low side for dry pregnant cows. It runs 4.5% (Table 6-18), whereas a dry pregnant cow requires 5.9% crude protein (Table 6-17). Thus, it is recommended that ½ lb per head per day of a 30 to 40% crude protein equivalent (CPE) supplement be provided.

It follows that the protein content of corn refuse is much too low to support either productivity or growth. For example, a 1,000-lb lactating cow requires a daily allowance of 2.0 lb of crude protein. However, a daily consumption of 20 lb of stover will provide only about 0.9 lb—less than half the need.

For nursing cows, the protein deficiency of stover and/or husklage may be corrected by supplementation with the following, on a per head per day basis: 2 pounds of a 40 percent protein supplement, or 6 pounds of a good legume hay. If desired, the protein supplement may be provided in the form of protein blocks, with one block provided for each 15 cows. Where hay is fed, it should be taken to the field, rather than fed in a feedlot, as this will encourage the cows to stay in the field and graze the cornstalks.

4. **Minerals**—Phosphorus should be provided to all cattle fed corn residue. Calcium may be deficient, especially for lactating cows. Also, some of the trace elements may be deficient. Hence, it is recommended that all cattle on high corn refuse have free access to a complete mineral. A mineral mixture with a Ca:P ratio of 1:2 is recommended for gestating cows. Lactating cows might perform better on a 1:1 ratio.

5. **Vitamin A**—Corn residue, along with other crop residues, is deficient in vitamin A. Hence, it must be supplemented. The precalving and postcalving (heavy milking) needs of approximately 19,000 and 36,000 IU per head per day, respectively (NRC-1976), may be met by feeding vitamin A supplement, intramuscular injection of vitamin A solution, or by feeding adequate levels of green-leafy hay.

It is important that corn residue be tailored to match the cow's nutritional needs. This is relatively simple with dry pregnant cows, where supplementation with a high phosphorus mineral and vitamin A will usually suffice. Beginning four to six weeks before calving and continuing through the lactation period, much heavier supplementation is necessary; in addition to phosphorus and vitamin A, protein must be added, and preferably some energy and calcium for nursing cows.

OTHER CROP RESIDUES

A host of crop residues, other than corn residue, can be used for feeding cows. Some of these follow.

Sorghum (Milo)

Cows will make good use of sorghum as a winter feed. It can be grazed or harvested and stored either as dry feed or silage. The sorghum plant stays green late in the fall; hence, good sorghum stover silage can be made without additional water. In comparison with corn residue, sorghum residue (1) is higher in protein content (corn residue averages 4.5% crude protein, whereas sorghum residue averages 6.5%); (2) is less palatable (if given a choice, cows will select corn refuse in preference to sorghum refuse); (3) comprises a lower percentage of the total plant dry matter than corn (40% of the total plant dry matter of sorghum is residue compared with 40 to 50% for corn); and (4) is lower yielding.

After harvesting, sorghum will send up new shoots if moisture permits. The prussic acid content of these shoots may be harmful to grazing animals; hence, cattlemen should be aware of this possible poisoning. These shoots can be grazed safely four to six days after a hard killing frost.

Soybean Refuse

The stems and pods of soybean refuse available for feeding, yield approximately one-fourth ton per acre, with a ratio of stems to pods of about 2:1. The digestibility of stems is low—25 to 35%—due to their high lignin content (18 to 30% for the stalk portion). The digestibility of pods is much higher, ranging from 58 to 63%.

Small Grain Refuse

This refers to (1) straw and (2) tailings—the chaff and grain behind the combine. In the days of binders and threshing machines, straw stacks were commonplace; and they were extensively used for winter cattle feed. With the advent of combines, much of the straw was left to rot in the field. During periods of hay scarcity and high-priced hay, straw is frequently used as either a "hay-stretcher" or "hay-replacer."

Of the common cereal straws, oat straw is the most palatable and nutritious. Barley straw ranks second, and wheat straw is third.

Straw is a bulky feed, and it must be properly supplemented. It is low in protein (wheat straw averages about 3.6% crude protein), low in phosphorus, and low in vitamin A. Dry pregnant cows can be wintered on straw plus a daily allowance of either 5 to 6 pounds of good quality alfalfa hay or 1 to 2 pounds of a 30 to 40% protein supplement, along with free access to a high phosphorus mineral (one containing at least 12% phosphorus). If no legume hay is fed, vitamin A should be fed or injected. When oilseed meals are scarce and high in price, some slow-release nonprotein nitrogen may be used.

The tailings—the chaff and grain behind the combine—are generally used by farmers and ranchers, either as dry feed or mixed with silage.

Legume and Grass Seed Straws

In addition to the cereal straws, other low-cost roughages available in certain sections of the United States are lentil straw, field pea straw, bean straw, clover straw, and bluegrass straw.

Cottonseed Hulls

Cottonseed hulls are one of the most important roughages in the South, especially for cattle. They supply 43.7% TDN, which is about as much as is fu nished by late-cut grass hay or by oat straw. They a low in protein (3.9%)—and practically none of it digestible—low in calcium (.13%), very low in pho phorus (.06%), and lacking in carotene. To corre these deficiencies when fed to dry pregnant cow hulls should be supplemented with a daily allowanc of either (1) 6 pounds of a good quality legume hay, ((2) 2 pounds of a 30 to 40% protein supplement, alon with free access to a complete mineral high in pho phorus, unless a phosphorus-rich supplement such a cottonseed meal is fed. If no legume is fed, vitamin should be fed or injected.

When properly fed, cottonseed hulls are abou equal in quality to fair quality grass hay and are wort more per ton than corn or sorghum stover, straw, c poor hay. Also, they can be fed without furthe processing—there is no chopping; and they are wel liked by cattle, even when fed as the only roughage.

Pelleted hulls are now on the market. In compari son with regular hulls, they are more digestible, re quire less transportation and storage space—becaus of their high density—and are easier to handle.

TREATING CROP RESIDUES TO INCREASE DIGESTIBILITY

Crop residues are inefficiently utilized by ani mals because of the high content and poor digestibil ity of the fibrous fraction. This poor digestibility is re lated to the extent of lignification of the cell wal component of these low-quality forages. Althoug crop residues provide a satisfactory ration for dry ges tating cows, they do not provide sufficient energy fo either young ruminants or lactating cows—they sim ply cannot hold enough of these low-qualit roughages to provide adequate energy. This prompt interest in increasing the digestibility of these cro residues.

There are many approaches to delignifying an increasing the digestibility of crop residues; amon them, treatment with sodium hydroxide, potassiun hydroxide, ammonium hydroxide, and pressurize heating. The potential of such treatments become apparent when it is realized that straw, for example, i only 30 to 40% digestible before treatment. However when pressure heated with water, it becomes 50 t 60% digestible; and digestibility increases to 70 t 80% when sodium hydroxide is added prior to cook ing. By treating corn husklage and milo residue, the Nebraska Station was able to increase their energy value to 90% that of corn silage.

Lowering the cost of treating crop residues to increase digestibility is the primary area which must be researched before these procedures can be applied in practical operations.

SECTION 7

SILAGE AND HAYLAGE; HIGH-MOISTURE GRAIN[1]

Contents

Contents	Page
The Ensiling Process	702
Advantages and Disadvantages of Silage	703
Silo	704
Conventional Upright (Tower) Silos	705
Gastight (Oxygen-Limiting) Silos	706
Pit Silos	706
Horizontal Silos	706
Trench Silos	706
Bunker or Self-Feeder Silos	707
Temporary Silos	708
Silage Storage Losses	708
Kinds of Silage	708
Corn and Sorghum Silage	708
Corn and Sorghum Residue Silage	709
Grass (Hay Crop) Silage	709
Other Silage Crops	709
Combining Crops for Silage	710
Rain-Damaged Hay Silage	710
Frosted Crop Silage	710
Corn or Sorghum Silage Vs Grass Silage	710
Corn Silage with Urea	711
Harvesting Methods and Machinery	711
Characteristics of Good Silage	712
How to Make Good Silage	712
Harvest at Proper Stage of Maturity	712
Cut to Proper Length	713
Control the Moisture Content	713
How to Lower the Moisture Content	713
Conditioning-Wilting	713
Adding Dry Hay or Straw	713
Combining with Corn or Sorghum Silage	713
Adding a Dry Preservative	713
How to Increase the Moisture Content	713
How to Determine the Moisture Content	714
Add Preservatives Only When Needed	715
Some Silage Preservatives	717
Silage Preservative Considerations	717
Silage Preservative Recommendations	717
Fill Rapidly	717
Distribute Forage Uniformly in the Silo	718
Seal or Top-off the Silo	718
Feeding Value and Economy of Silage	718
Silage Pointers	718
Coating the Silo	718
Nutrient Losses in Leakage	718
Exposure to Air	718
Removal of Silage from Silo	718
Moldy Silage	719
Silage for Summer Feeding	719
Effect of Silage on Milk Odor and Flavor	719
Dangerous Silage Gases	719
Haylage (Low-Moisture Silage)	720
High-Moisture Grain	721
Acid Preservation of High-Moisture Grain	

[1]The author gratefully acknowledges the helpful suggestions of the following authorities who reviewed this entire section: Dr. Howard J. Larsen, Department of Dairy Science, The University of Wisconsin, Madison, Wisc.; and Dr. George D. Marx, Dairy Scientist, University of Minnesota, St. Paul, Minn.

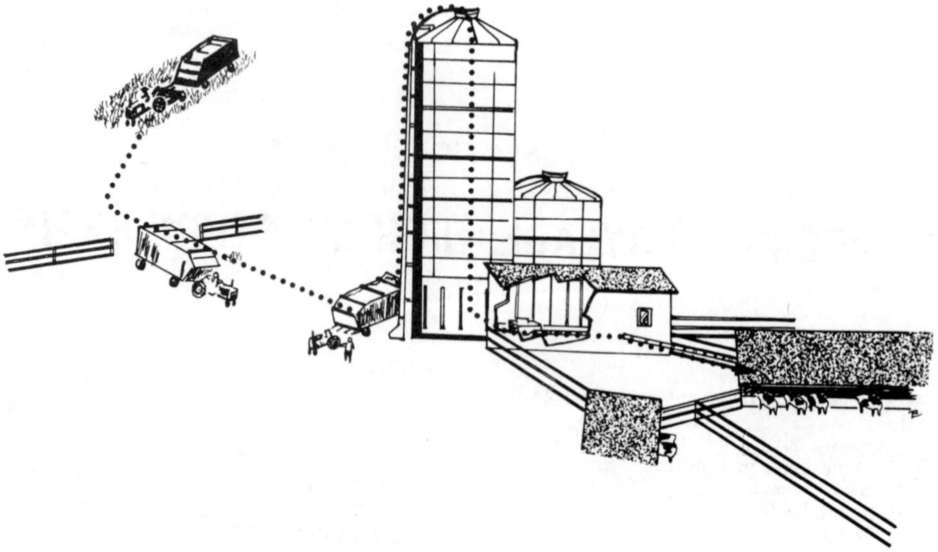

Fig. 7-1. Silage has gone modern! Feeds are harvested at the peak of nutritional value, blown in at structure's top, preserved by oxygen-limiting storage, then fed out the bottom of the silo by an unloader.

Silage may be defined as fermented forage plants. It is a very old method of preserving feed. Columbus found that the Indians used pits or trenches in which to store their grain, and, centuries earlier in the Old World, silos were used as a means of preserving both grain and green forage. The first tower silo built in the United States by white man is said to have been erected by F. Morris in Maryland in 1876.

Silage making is one of the 3 common methods of utilizing forage crops, the other 2 methods being pasturing and haying. Pasturing is the least expensive of the 3 methods, but it is seasonal in nature. In the spring and early summer, forage plants generally grow faster than they can be utilized by normal grazing, and become dormant in cold weather.

The surplus forage produced during the growing season may be preserved for feeding during the winter months and other periods of pasture scarcity by haymaking which, next to grazing, is the most efficient method during dry weather. But weather conditions are not always favorable to haymaking. Ensiling, on the other hand, can be done in inclement weather. Also, it has the added virtues of succulence and of preserving a higher proportion of the nutrients of the plant than can be accomplished in haymaking, although slightly greater cost may be involved than in normal field curing of hay.

Silage is primarily a beef and dairy feed, where it is used as part or the only roughage in the ration. It is also a good sheep feed. Sometimes it is fed to brood sows. Although horses will eat silage, it is not considered a good feed for equines.

The importance of silage in this country is shown by the fact that over 170 million tons are made annually. Further, there is ample evidence that silage making is on the increase. It is estimated that 12,000 new silos are constructed in the United States each year, with tower silos increasing and horizontal silos decreasing.

THE ENSILING PROCESS

The ensiling process refers to the changes which take place when forage or feed with sufficient moisture to cause fermentation is stored in a silo in the absence of air. An understanding of these changes is likely to lead to the production of more high-quality silage.

The entire ensiling process requires two to three weeks, during which time the following aerobic (with air) and anaerobic (without air) activities predominate:

1. **Aerobic activity**—The living plant cells of the forage continue to respire, or breathe, consuming the oxygen of the silage-entrapped air, producing carbon dioxide and water, and releasing energy or heat. Simultaneously, aerobic yeasts and molds thrive and multiply.

During this period, which is very short, the temperature seldom rises to 100°F, provided (a) the materials have been properly prepared, and (b) the right steps for making high-quality silage have been followed.

2. **Anaerobic activity**—When the available oxy-

[2]Estimate by the National Silo Association, Inc., Cedar Rapids, Iowa; in a personal communication of August 2, 1976, to the author.

en of the entrapped air has been consumed, naerobic bacteria—chiefly acid-forming and proteoytic—multiply at a prodigious rate. Simultaneously, he molds and the yeasts die, but continue in a minor vay to function as enzyme systems which produce alohol and other end products.

The combined anaerobic activity produces the ollowing changes: (a) The carbohydrates and sugars especially the sugars) are broken down into lactic cid (the acid in sour milk), some acetic acid (the acid n vinegar), and a small amount of other acids and alohols; (b) small quantities of the proteins are broken lown into ammonia, amino acids, amines, and mides; and (c) the acidity finally reaches a point vhere the bacteria themselves are killed, and the silage-making process is completed. At the last stage, silage in a good silo will remain unchanged for a very ong time—up to 10 to 15 years. However, if it is exosed to air—with the opening of the silo, or through air pockets—yeasts, followed by molds, will again become active.

It is generally known that the above described changes in the ensiling process and the quality of the silage produced can be altered in a number of ways; by moisture content, legume content, length of cut, speed of filling, distribution in the silo, and amount of packing.

Some additional and pertinent facts about the ensiling process follow:

3. **Optimum pH**—The formation of a pH of 3.5 to 4.5 is the key to good wilted silage preservation, because it will prevent the growth of bacteria, including those which cause rotting and putrefaction. Excellent low-moisture silage (45 to 60% moisture) is frequently made in the pH range of 4.0 to 4.5, and even up to 5.0.

4. **Sugar content of forage**—When cut at the proper stage of maturity, corn and sorghum forage possess just the right amount of sugar for the production of good silage. If cut when too immature, the silage will be sour, "sloppy," unpalatable, and lower in feeding value because the plants store nutrients very rapidly as the grain ripens.

Today, high sugar content corn is being grown for silage. It makes very high-quality feed, although it is not superior to conventional corn made into silage.

In making good quality grass silages, either the forage should be wilted below 70 percent moisture content or a preservative should be added. Two theories have been advanced for this necessity; namely, (a) that such forages are low in sugar, and (b) that although sufficient sugars are present, the physical conditions are not suitable for the growth of the desirable acid-producing organisms. Without this precaution, high-moisture or grass-legume silages form a cold fermentation containing high levels of butyric acid, which is undesirable.

5. **Effects of preservatives**—The addition of carbohydrate preservatives, such as molasses and grains, (a) speeds up the formation of lactic and acetic acids, and (b) provides bacteria with a readily available source of energy. On the other hand, the addition of a preservative (a) increases the cost of the silage, and (b) results in certain nutrient losses of the preservative in the ensiling process.

The addition of mineral acids, such as phosphoric acid, (a) has a protein-sparing effect, through inactivating the proteolytic bacteria and their enzymes, and (b) provides the necessary acids for preservation. However, such acids often lower the palatability of the silage.

Ensiled forages at dry matter levels of 20 to 60 percent can be preserved in excellent condition without loss of palatability for up to 10 months by adding 1 percent propionic acid, or any other organic acid such as formic acid and related compounds. But with modern equipment for harvesting and storing, and with our present knowledge of the variables necessary to control and stimulate fermentation, no general recommendations are made for the addition of preservatives to wilted grass silage or haylage. By carefully following the standard recommendations, high-quality silage can be made without the added cost of a preservative.

ADVANTAGES AND DISADVANTAGES OF SILAGE

Some of the *advantages* of silage are:

1. It retains a higher proportion of the nutrients of plants than can be accomplished by haymaking, even if the weather is satisfactory for the latter, chiefly because shattering and bleaching losses are held to a minimum. Thus, grass silage preserves 85% or more of the feed value of the crop, whereas haymaking under the best of conditions will preserve only 80%, and under poor conditions only 50 to 60%.

2. It makes possible the production of the maximum quantity of feed per acre of land and increases the livestock carrying capacity of the farm or ranch. Thus, corn, the chief U.S. silage crop, (a) yields more total digestible nutrients per acre than most other forage crops, and (b) has 30 to 50 percent higher feeding value as silage than when fed as grain and stover.

3. It is feasible to produce a top-quality feed during times of inclement weather when it would be impossible to cure properly the forage crop.

4. It is the most economical form in which the whole stalk of corn or sorghum can be processed and stored.

5. It requires less storage space per pound of dry

matter than dry hay, even when the latter is baled or chopped. A cubic foot of silage contains at least three times as much dry weight of feed as a cubic foot of long hay stored in the mow.

6. It practically eliminates the danger of loss by fire if stored within the recommended moisture range.

7. It is the most satisfactory and economical way in which to preserve a number of by-product feeds.

8. It makes it possible to remove forage crops from the land earlier than would otherwise be possible.

9. It is one of the best methods of controlling the European corn borer since the removal of cornstalks is required in making silage.

10. It helps to control weeds, which are often spread through hay or fodder.

11. It is the cheapest form in which a good succulent winter feed can be provided on most farms and ranches.

12. It is a better source of protein and of certain vitamins, especially carotene, and perhaps some of the unknown factors, than dried forage.

13. It is a very palatable feed and slightly laxative in nature.

14. It makes for less waste, the entire plant being eaten with relish, an important consideration with coarse stemmy forages.

15. It is without a peer from the standpoint of longtime storage, holding the feeding value of protein, carbohydrates, and carotene better than any other method of preservation, and providing a desirable backlog against drought or any other crop failure.

16. It may be completely mechanized as a feeding system, thereby eliminating much labor and time.

17. It offers many advantages over pasture, including (a) no fencing required, (b) approximately one-third more forage from the same acreage, (c) harvesting at optimum maturity, (d) more uniform quality, (e) little or no bloat, and (f) closer observation of animals that are confined to a lot or corral.

Some of the *disadvantages* of silage are:

1. It requires a silo or storage structure and other special equipment, for best results. In comparison with the simpler methods of field curing and storing hay, this is likely to mean higher costs—an important consideration with a small operator.

2. It possesses considerably less vitamin D than sun-cured hay.

3. It necessitates that two to three times as much tonnage be handled as when the same forage is dried for hay, due to the high water content.

4. It incurs an added expenditure when preservatives are necessary.

5. It may be an expensive forage preservation system, especially if a dry forage is fed along with it.

6. It lessens the amount of organic material returned to the soil, which is needed in some soil types.

SILO

Silage may be stored in almost any kind of container. The main requisites of a good silo, regardless of kind, are:

1. That its size be in keeping with the number and kind of animals to be fed daily, the length of the feeding period, and the amount of forage available for ensiling. Directions on how to determine the size silo to build are given in this book under the section on Buildings and Equipment.

2. That it exclude air from the stored material, including entrance of air around the doors of tower silos.

3. That the sidewalls be straight and smooth in order to prevent the formation of air pockets and allow for unimpeded packing.

4. That it be of adequate depth, thus making for better packing and less surface area to total mass exposed.

5. That it be properly reinforced. This point is especially important where direct cut grass silage is made, because it exerts from ½ to 2½ times as much pressure on the walls as does corn silage. Thus, tower silos which were originally built for corn or sorghum silage but which are to be filled with wet grass silage should be either (a) reinforced with extra bands placed around the lower part to strengthen the walls if an inspection reveals that the existing strength is not adequate, or (b) not filled to more than half capacity.

6. That adequate provision be made for the escape of surplus juices, either by a drain or by a gravel bottom.

7. That it be conveniently located and accessible in all kinds of weather, from the standpoint of both filling and feeding.

Silos may be classified according to the five basic methods used for processing forages. Each method is associated with the shape and material of the structure, which also influences the efficiency of preserving the silage. The different shaped structures are also adapted to different methods of filling and unloading. Within each classification there are many variations of each type depending upon the manufacturer.

The kind of silo decided upon and the choice of construction material should be determined primarily by the cost and by the suitability to the particular needs of the farm or ranch.

Silos may be classified as follows:

I. Conventional Upright (Tower) Silos
   1. Concrete stave
   2. Galvanized steel

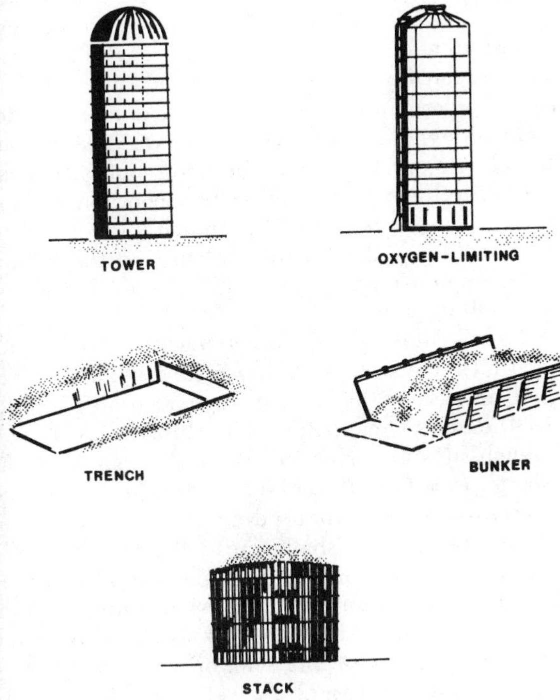

Fig. 7-2. Kinds of silos: (1) tower silo, (2) oxygen-limiting, (3) trench silo, (4) bunker (aboveground level), and (5) enclosed stack silo. The last two are both aboveground temporary silos.

3. Wood stave
4. Monolithic concrete (poured in place)
5. Tile block
6. Brick

II. Gastight (Oxygen-Limiting) Silos
   1. Glass-lined structures
   2. Concrete stave
   3. Galvanized steel
   4. Monolithic concrete

III. Pit Silos

IV. Horizontal Silos
   1. Trench silos (belowground level)
   2. Bunker (aboveground level)

V. Temporary Silos
   1. Enclosed stacks
   2. Open stacks
   3. Modified trench-stack silos
   4. Plastic or polyethylene bag silos

Some pertinent information relative to each main kind of silo is given in the discussion which follows, but it is not within the scope of this book to give detailed silo plans and specifications. The latter may be obtained from local authorities, from silo manufacturers, or by writing to the state agricultural college.

Conventional Upright (Tower) Silos

The upright or tower silo, which is sometimes referred to as the "watch tower of prosperity," is a cylinder built aboveground. Its round shape withstands pressure well and is adapted to good packing.

The tower silo is a permanent farm structure, and, as such, should be constructed to withstand long usage. Although tower silos are usually handy, they are, in their initial cost, generally the most expensive of all types. However, they have the following advantages: (1) durability, (2) minimum top and side spoilage, (3) convenience for feeding during periods of inclement weather, and (4) well adapted to automation (loading and unloading machinery).

Recent developments in construction of tower silos have been made in (1) bottom unloaders, with elimination of doors; (2) center-core unloaders, with elimination of most of the doors; (3) top-unloaders; and (4) large diameter features (24 to 30 feet in diameter). These new developments are also available in gastight (oxygen-limiting) silos.

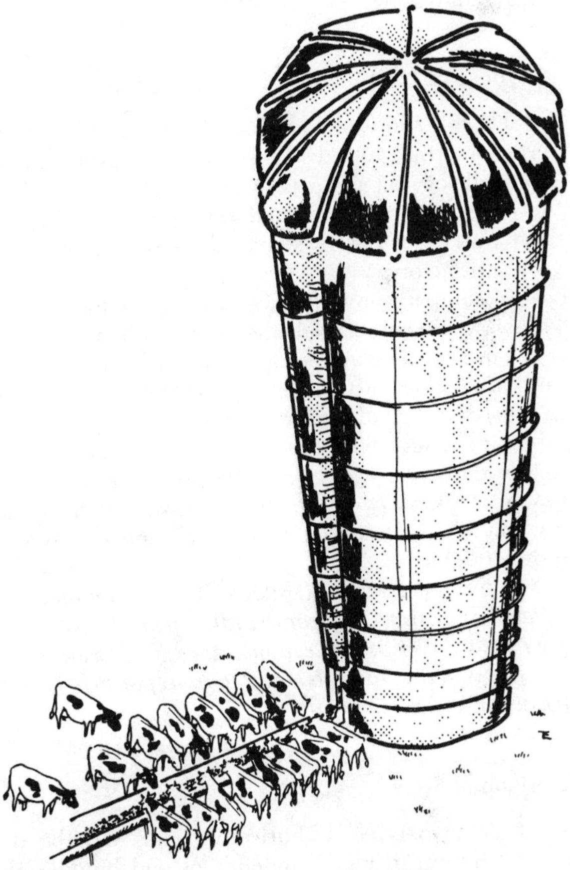

Fig. 7-3. Upright (tower) silo—a crop processing container and storage structure.

Gastight (Oxygen-Limiting) Silos

These silos resemble conventional tower silos, but they are more expensive because of their construction.

Sealed silos are designed for storage of wilted or even overwilted forage with as little as 40 to 55 percent moisture content or for the storage of high-moisture grain containing 20 to 25 percent moisture. These structures may be partly filled on widely separated dates, provided they are sealed between fillings. Packing and tramping of forage is not necessary, although distribution is desirable.

Practically all outside air is kept out of the oxygen-limiting silo, and carbon dioxide formed during fermentation is kept in. A plastic breather bag and a pressure relief valve located in the top of some of these structures compensate for differences in inside and outside pressures without allowing outside air to contact the forage. Before each filling, the plastic breather bag should be checked for holes and flaws, the pressure relief valves should be inspected, the structure should be checked for leaks, door seals should be inspected, and unloaders should be checked for wear in order to prevent malfunctioning during unloading.

Pit Silos

The pit silo is shaped like the tower silo, but inverted into the ground. It resembles a well or cistern. The walls of a pit silo may or may not be lined. Where the water table is low enough that the silo will not fill with water, such as in semiarid areas, the pit silo is very satisfactory.

In comparison with tower silos, pit silos have the following *advantages*: (1) they are never damaged by storm or fire, (2) they require less reinforcing, (3) they minimize silage losses because of not having doors, and (4) they avoid frozen silage. But they have the following *disadvantages*: (1) they are dangerous, due to the frequent presence of suffocating carbon dioxide gas, and (2) considerable work is involved in removing the silage, despite the development of a number of hoist devices.

Before entering a pit silo, it is recommended that a lighted cigarette lighter, candle, or lantern be lowered into the silo. If the flame goes out, assume that the pit is dangerous to enter and replenish it with fresh air before entering.

Horizontal Silos

Only two types of horizontal silos will be discussed herein; namely, trench silos and bunker silos (or horizontal surface silos), both of which may be adapted to self-feeding.

TRENCH SILOS

The trench silo is a horizontal, trench-like stru ture that can be built quickly and at low cost. It most popular in areas where the weather is not t severe and where there is good drainage. The walls a trench silo may or may not be lined, but for makin good silage they should always be smooth; and the may or may not be a floor. A trench silo should b wider at the top than at the bottom, and the botto should slope away from one end in order that exce juices will drain off.

Trench silos have the *advantages* of: (1) low in tial cost; (2) low cost of filling machinery, for a blow is not necessary; (3) relative freedom from freezin and (4) ease of construction. The chief *disadvantage* of trench silos in comparison with tower silos are th (1) larger area to seal, (2) higher spoilage losses, an (3) inconvenience in feeding during inclemen weather. Because of shallowness, the forage shoul be packed very thoroughly in a trench silo by drivin a tractor back and forth over it. When filling is com pleted, the top should be carefully sealed (1) by 3 to inches of limestone or dirt, with or without a seedin of rye or winter wheat; (2) by wet straw, poor-qualit hay, marsh grass, or sawdust; (3) by a mixture of di and straw; (4) by waterproof paper lapped about 1 inches at the joints and covered with dirt or straw; o (5) by polyethylene, plastic, aluminum, or other mate rials.

BUNKER OR SELF-FEEDER SILOS

As a laborsaving measure, some operators ar now constructing horizontal silos aboveground (o slightly recessed)—usually with concrete floors, an side walls of wood, concrete, or other materials—an self-feeding silage to cattle by making use of either feeding fence or an electrified pipe suspended 30 t 48 inches from the floor of the silo.

In the first of these methods, the fence (or gate hurdle, or stanchion) through which the animals pu their heads, is placed at the end (or side) of the sil and moved back as the silage is eaten. Feed waste are reduced if the bottom section of the gate or hurdl is solid and the top slopes away from the silo (abou 7½ inches out at the top, on gates 5 feet high), so tha the animals are forced to eat down toward the botton of the stack.

The principle of the electrified pipe is similar t that of the feeding fence—both are designed to self feed silage without wastage. However, in compariso with the fence, the pipe is much lighter and easier t move, and not so easily broken.

Most of those who have self-fed successfully from bunker silos recommend the following:

1. That there be about 6 inches of space per animal.

2. That the silage not be piled higher than 6 feet.
3. That the silage be loosened in front of the ence or pipe.
4. That there be once-a-day policing (a) to push neaten good silage back up against the stack face, nd (b) to push the spoiled material and manure to the ear of the silo.

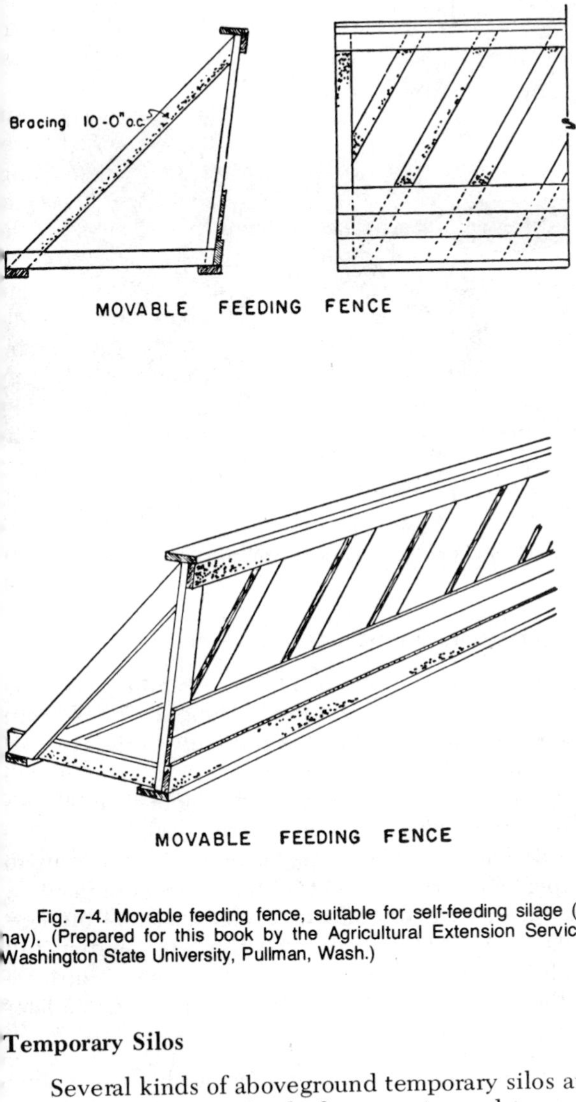

Fig. 7-4. Movable feeding fence, suitable for self-feeding silage (or hay). (Prepared for this book by the Agricultural Extension Service, Washington State University, Pullman, Wash.)

Temporary Silos

Several kinds of aboveground temporary silos are used. Generally, this kind of storage is used to meet emergencies, to supplement permanent silos, or to ensile such by-product feeds as cannery refuse, pea vines, and beet tops or pulp. Aboveground temporary silos are low in cost, can be erected on short notice, require no special foundation, and can be set up on almost any level site convenient for filling and feeding.

The amount of spoilage in aboveground temporary silos can be kept to a minimum by having straight sides, considerable height, proper packing, and protection with fiber-reinforced paper, plastic, or other suitable material. Also, the use of propionic acid, formic acid, or other effective organic acids, applied to harvested materials at the time of chopping, is very effective in reducing spoilage in temporary silos. It is important that stacked material treated with organic acid be covered with plastic to prevent dilution by rain and snow.

The spoilage on the sides of temporary silos will vary from 4 to 20 inches, with greater spoilage in grass silage than in easier-keeping corn or sorghum silage.

Perhaps most aboveground temporary silos can be classed as belonging to one of the following four kinds:

1. **Enclosed stacks**—These are built entirely aboveground, without trenches or holes. They are upright, are generally circular, and are enclosed by snow or picket fences, poles, wooden staves, heavy woven wire, or other materials. Most of them are lined with tar paper, plastic, or tough fiber-reinforced paper made especially for the purpose. Because of the relatively weak walls of these silos, their height should not be greater than twice their diameter unless poles are set at 4 to 6 points around their circumference and tied together at the top.

2. **Open stacks**—These are similar to enclosed stack silos, except that no supports or walls are used. As would be expected, greater spoilage is encountered in the open stack than in the enclosed stack, because of the greater evaporation and spoilage which accompanies the exposed sides. Less spoilage, percentagewise, occurs in stacks of considerable size—stacks that contain 500 to 1,000 tons or more silage.

3. **Modified trench-stack silos**—This silo, which is intermediate between a trench and stack silo, is adapted to areas where the ground-water level is high. It is constructed by excavating a shallow trench 12 to 18 inches deep, by piling the excavated earth on either side of the trench to support the silage and to keep out surface water, by packing silage thoroughly in and over the trench to a height of 10 to 15 feet, and by covering the stack with any one of the materials recommended for covering the trench silo (see trench silo). The modified trench-stack silo is designed to give greater protection and less spoilage than can be accomplished by open or closed stacks. Also, this type of silo is easier to feed from than a trench silo.

4. **Plastic silos**—Plastic films are now available for use as temporary silos. Also, they are used (a) as covers for trench, bunker, and tower silos, and (b) as silo liners. If not punctured, plastic is airtight. Several types of plastic silos can be evacuated with vacuum pumps following filling, thereby lessening the air within.

Among the special advantages attributed to plastic silos are (a) economy ($1.50 to $3.00 per ton of stored silage, depending upon stack size—the greater the size, the more economical); (b) adaptation to small quantities of silage and for out-of-the-way places; and (c) reduction of spoilage to less than 5 percent, providing sealing is complete and puncturing is prevented. But good sealing and nonpuncturing are difficult to accomplish. Present plastic materials are very susceptible to puncture by sharp objects (fingernails, plant stems, implements, animals, etc.). Hence, a minimum of 10 percent loss should be expected.

Gases begin collecting in a plastic silo immediately upon sealing, with the bag reaching its maximum extension in 36 hours.

Silage Storage Losses

Silage storage losses vary widely between kinds of silos and are generally larger than most farmers realize. Estimated (1) average, and (2) range of silage storage losses are given in Table 7-1.

TABLE 7-1
ESTIMATED (1) AVERAGE, AND (2) RANGE OF SILAGE STORAGE LOSSES

Type of Silo	Percent of Loss	
	Average	Range
Gastight upright	5	1-10
Conventional upright	6	2-12
Horizontal (trench)	15	8-25
Open stack	20	12-30

KINDS OF SILAGE

A great variety of crops can be and are made into silage. A rule of thumb is that crops that are palatable and nutritious to animals as pasture, as green feed, or as dry forage also make palatable and nutritious silage. Likewise, crops that are unpalatable and unnutritious as pasture, as green feed, or as dry forage also make unpalatable and unnutritious silage.

Most silage in the United States is made from either corn or sorghum, with corn silage far in the lead—over 15 times as much corn silage as sorghum silage is made. In 1980, 111.1 million tons of corn silage and 7.1 million tons of sorghum silage were produced in the United States. At the present time, it is estimated that 70 percent of the nation's silage is made from corn and sorghum and 30 percent from grasses, legumes, and other feeds. In addition to the kinds of silage already mentioned, silage is made from the small grains, wastes from food processing (sweet corn, green beans, green peas), root crops, and various vegetable residues.

Corn and Sorghum Silage

For the United States as a whole, corn ranks firs in importance as a silage crop. Generally more tota digestible nutrients can be obtained from an acre o corn as silage—which will yield from 5 to 25 tons o forage per acre, with an average of about 7 tons—tha can be obtained from an acre of any other crop. Also corn ensiles easily without the aid of a preservative and keeps almost indefinitely in a good silo.

Fig. 7-5. Corn silage is King of the forages for dairy cattle.

There are three kinds of corn silage; namely,

1. **The whole corn plant**—When at the peak of its nutritive value and right for ensiling, the whole corn plant contains 1½ times the nutrients of the ripened grain that the plant would have yielded. Also, in corn silage made from the whole crop, more than 90 percent of the nutrients produced are saved.
2. **Earcorn silage**—The ensiled ears contain up to 68 percent of the nutrients of the entire corn plant.
3. **Shelled-corn silage**—This consists of the kernels only. At 70% dry matter (30% moisture), shelled-corn silage contains 61 to 66% of the nutrients in the whole crop (also see High-Moisture Grain later in this section).

The sorghums are more dependable and higher yielding than corn in certain areas, particularly in unirrigated, and relatively dry areas, of western and southwestern United States. Also, the sorghums are higher in sugar content than corn forage.

Corn and Sorghum Residue Silage

Corn and sorghum residues—the forages that remain after harvesting a grain crop of corn or sorghum—may be used as cattle feed three ways: (1)

azed, (2) harvested and fed dry, or (3) ensiled and d as silage. The discussion that follows will be limed to ensiling corn and sorghum residue.

When ensiled, cornstalks (stover) produce a roduct known as corn stover silage or cornstalk silage. When stalks are processed as silage, the use of a orage harvester equipped with a screen or a ecutter-blower at the silo is necessary in order to hop the material finely. Fine chopping will ensure ood packing and improve consumption by avoiding electivity.

Where corn stover silage is to be made, the residue should be harvested as soon as possible after the rain is taken off, before the residue loses any moisure. At that time, the grain moisture will generally be nder 30% and the refuse moisture will be above 8%. In an airtight silo, 40 to 45% moisture will suffice. In an unsealed or bunker silo, the moisture content should be 48 to 55% for proper lactic acid formaion. Water may be added at the silo if necessary. As a recaution, some authorities recommend the addition f 56 pounds of corn meal (or other finely ground rain) per ton of corn stover silage, as a means of proiding carbohydrates from which acids will form and ct as a preservative. With husklage, the latter precauion is not necessary since there is sufficient grain emaining in the husk and cob.

The biggest deterrent to harvesting stalklage, in ither dry or ensiled form, is the cost—primarily for he equipment. Rather than own such expensive quipment, which is used for a short period only, cusom harvesting of stalklage is likely cheapest for most perators.

Husklage—the forage discharged from the rear of corn combine, and consisting of the husks, cobs, and ny grain carried over the combine—may also be ensiled. Ensiling husklage, along with recutting and dding water, results in increased cow consumption nd less rejection of cobs.

Like corn, sorghum stover may either be grazed r harvested and stored either as dry feed or silage. Because the sorghum plant stays green late in the fall, good sorghum stover silage can be made without additional water.

Grass (Hay Crop) Silage

Grass (hay crop) silage refers to silage made from ny of the green chops which might otherwise be pasured or dried and made into hay. This includes grasses (such as timothy), legumes (such as alfalfa), grass-legume mixtures, and cereal grains (such as oats).

Tables 5-1 to 5-10, pasture tables (see Section 5, Pasture and Range Forages; Green Chop), are equally applicable for the production of grass silage, because, in practical operations, any adapted grasses and/or legumes may be used three ways: for grazing, for hay, or for silage.

Grass silage can be produced in those areas where the climate is too cool and the growing season too short for corn or sorghum silage.

Generally speaking, a higher percentage of good quality silage is made from corn and sorghum than is made from grass, but this need not be so. Consistently good grass silage can be made provided the forage is (1) cut early, and (2) wilted to 55 to 65 percent. The chief reason why these special precautions are necessary in ensiling grass silage is that they contain a much smaller percentage of fermentable material than does corn or sorghum forage cut at the proper stage.

Other Silage Crops

In the West, Northwest, and North Central states, where the weather is cool and the growing season is short, sunflowers are sometimes grown for silage. Although they yield and ensile well, sunflower silage is neither as palatable nor as nutritious as corn, sorghum, or grass silage. Pound for pound, sunflower silage is about 80 percent as valuable as corn silage.

Throughout the United States, a great array of by-product feeds are ensiled, especially in the less expensive and temporary types of silos. Among such by-products are: grain chaff, pea and bean vines, beet tops and pulp, potatoes, cannery refuse, cull and surplus fruits and vegetables, pulp and trimming wastes from market vegetables and fruits, wet brewers' and distillers' grains, and almond hulls. Sometimes Russian-thistles and other weeds are ensiled.

When potatoes, which contain about 80 percent moisture, are ensiled for cattle, it is recommended either (1) that 20 to 25 lb of dry hay, straw, or chaff be run through the ensilage cutter with each 100 lb of potatoes, or (2) that 1 ton of corn or sorghum silage be chopped with each 500 lb of potatoes. Frozen and sprouted potatoes should not be ensiled. Potato silage intended for swine should be made from cooked or steamed potatoes ensiled alone in a shallow pit or silo.

Either of the above methods recommended for ensiling potatoes for cattle is equally adapted for the preservation of other high-moisture crops, such as apples, beets, pears, tomatoes, cauliflower, kale, and trimming wastes from market vegetables—provided the added forage is in proportion to their respective moisture contents.

Cabbage, rape, and turnips should not be ensiled, as they make unsatisfactory, watery, foul-smelling silage.

Combining Crops for Silage

Sometimes, in order to lower the moisture content, to alleviate the necessity of a preservative, and to assure better quality silage, forages of high sugar content are combined with forages of low sugar content. Thus, excellent silage can be made by mixing 1 ton of sorghum forage with each 3 tons of grass silage material, or a ton of corn forage with each ton of grass forage material (less sorghum forage is necessary than corn forage, because of the higher sugar content of the former).

At times such combination silage crops are even grown together; for example, corn and soybeans, millet or Sudangrass and soybeans, oats and peas, etc.

A major difficulty in combining ensiling crops is that it is almost impossible to synchronize the stage of maturity of different crops so that they reach maximum yield and nutritive level at the same time.

Rain-Damaged Hay Silage

Partly cured hay that has been rained upon may be salvaged as silage (although it will not be of high quality), provided it is finely chopped, distributed evenly, and packed in the silo thoroughly enough to squeeze out the air. It is recommended that it be placed in the bottom of the silo, and, preferably, that alternate loads of a green crop be mixed with it. Otherwise, satisfactory packing can be obtained by putting a few loads of greener-than-ordinary material on top of it.

Frosted Crop Silage

Sometimes corn, sorghum, sunflowers, small grains, beans, and other crops, which may or may not have been intended for silage, are frosted before they reach the silage cutting stage. Such crops may be salvaged as silage. They should be cut at recommended moisture contents and ensiled according to directions. If they are too dry, water should be added.

Corn or Sorghum Silage Vs Grass Silage

Frequently stockmen are confronted with making a choice between corn or sorghum silage and grass silage. Under these circumstances, the following facts are pertinent:

1. Where adapted, corn or sorghum will generally produce a greater tonnage of feed per acre than grass silage.

2. Good quality corn or sorghum silage can be made more consistently and with greater ease than good quality grass silage.

3. Corn or sorghum silage may be more palatable than grass silage, even when the latter is carefully preserved.

4. Grass silage is generally higher in protein and carotene but lower in total digestible nutrients and vitamin D (wilted grass silage is higher in vitamin D than unwilted) than corn or sorghum silage (generally grass silage contains about 90 percent as much TDN as corn silage, but it will equal corn silage in TDN where 150 pounds of grain per ton have been added as a preservative). Thus, grass silage generally requires the addition to the ration of less protein supplement but more total concentrates than corn or sorghum silage. This would indicate that corn or sorghum silage would be slightly preferable to grass silage in high roughage finishing rations for beef cattle and sheep, whereas grass silage would be preferable in high roughage rations for dairy animals and young beef cattle and sheep.

5. Grass silage can be produced in those areas where the climate is too cool and the growing season too short for corn or sorghum silage.

6. The production of grass silage will result in less soil washing than the production of corn or sorghum silage on lands subject to erosion.

7. Grass silage will freeze next to the silo wall more than corn or sorghum silage, especially if it is ensiled when too wet (unwilted).

8. The silo can be kept working full time by using both grass and corn or sorghum silage; ensiling the first cutting of grass silage for summer feeding, and ensiling corn or sorghum silage for winter feeding.

Corn Silage with Urea

Corn silage is relatively low in protein, containing only about 2.3% crude protein on a wet basis (as fed), or 8.3% on a dry basis. Hence, if 10 pounds of urea are added to a ton of corn silage at filling time, it will raise the crude protein content of the silage from 2.3% to 3.7% on a wet basis, or from 8.3% to 13.3% on a dry basis. For most grower cattle and dry cows on straight silage rations, this silage-urea combination will meet the protein requirements.

If urea is added to silage, it must be well mixed. This is usually accomplished if the recommended amount of urea is added by applicator to the forage in the field at the time of harvest.

The practice of adding urea to silage does result in some loss of urea with seepage. Even under ideal conditions, about 10 percent of the urea may be lost; and under average conditions, up to 30 percent of the urea may seep away. Also, the addition of urea to silage makes for less flexibility at feeding time because the same amount of urea is contained in all the silage.

Cattle full fed corn silage (.04% sulfur) and a protein supplement high in urea (no sulfur) may be borderline so far as sulfur intake is concerned. To avoid a possible sulfur deficiency under such circumstances, achieve a nitrogen-sulfur ratio not wider than 15:1.

Where sulfur is needed, it may be provided in any of the following ways:

1. Feed a commercial protein supplement containing additional sulfur.
2. Mix sodium sulfate in the protein at the rate of (a) 16 lb per ton of supplement for a high protein supplement fed at the rate of 1 lb per day, or (b) 10 lb per ton of supplement fed at the rate of 2 lb per day.
3. Mix sodium sulfate in the salt at the rate of 8 percent (8 lb/100 lb salt). This level is based on the assumption that the animals have free access to salt and the intake is around .10 lb of salt per day. If the intake of salt or minerals is below this, the supply of sulfur may be inadequate.

HARVESTING METHODS AND MACHINERY

There is no one best silage-making method or kind of equipment. These must necessarily vary with the kind of silage, the kind of silo, the size of operation, and the available labor and machinery and the cost.

Three principal kinds of machines are used for harvesting silage; namely, field forage harvesters, row-crop binders, and stationary silo fillers.

Fig. 7-6. Harvesting corn for silage with a field forage harvester.

Field forage harvesters, which were first developed around 1936, are more widely used than any other type of equipment for harvesting silage. Also, they tend to be concentrated in those states where the production of silage is most important. With different attachments, field forage harvesters can be used to harvest row crops for silage, grass silage as a standing crop or from the windrow, and hay from the windrow. Also, they can be used to harvest straw and other kinds of forage. Field choppers can even be adapted for grinding and blowing high-moisture cob corn for ensiling. With appropriate attachments, a modern field harvester can be used to harvest all major ensiled crops. Thus, with a minimum of complementary machinery, a forage harvester can be the major piece of equipment in providing a completely ensiled ration for beef cattle, dairy cattle, or sheep. But such equipment is expensive and may or may not be economical where a small operation is involved.

Field chopped forage is generally transported on wagons equipped with mechanical unloading devices or by means of dump trucks. Blowers and conveyors are used in filling both tower and horizontal silos. Frequently, trench silos are filled by dumping over the sides.

The use of row-crop binders reached a peak in 1942, following which they declined. Reduction in the numbers of these machines reflects the increased use of cornpickers, field forage harvesters, and grain combines. In addition to being used to harvest silage, row-crop binders are also used to harvest corn for grain, corn for fodder, and sorghum as bundle feed.

Beginning in 1951, the use of stationary silo fillers declined markedly. Today, few of them are used.

CHARACTERISTICS OF GOOD SILAGE

In order to make good quality silage, stockmen need to know what constitutes silage quality. They need to be acquainted with those recognizable characteristics of silage which indicate high palatability and nutrient content. The easily recognized characteristics of silage of high feeding value are:

1. **Odor**—It has a "clean," rather pleasing acid odor, in contrast to the foul or objectionable odor of poor silage.
2. **Taste**—The taste is pleasing, not bitter or sharp.
3. **Absence of mold and rot**—There is no visible mold, and it is not musty or slimy.
4. **Uniformity**—It is uniform in moisture and color. Generally green or brownish silage is good; tobacco brown or dark brown silage indicates excessive heat; and black silage is rotten and should not be fed.
5. **Animal acceptance**—Animals like and thrive on good silage.

HOW TO MAKE GOOD SILAGE

In addition to using a sound silo of proper size, those who make good silage generally harvest at the proper stage of maturity, cut to proper length, control the moisture content of the forage, add a preservative only when needed, fill rapidly, distribute and tramp the forage in the silo, and seal or top-off the silo. Each of these factors will be discussed.

Harvest at Proper Stage of Maturity

Harvesting at the proper stage of maturity assures the maximum yield and nutrient content.

Fig. 7-7 shows the effect of stage of maturity of the corn plant on total dry matter accumulation.

The "black layer test" can be applied quickly and easily to determine when to harvest corn fo maximum yield and nutrient quality (see Fig. 7-8).

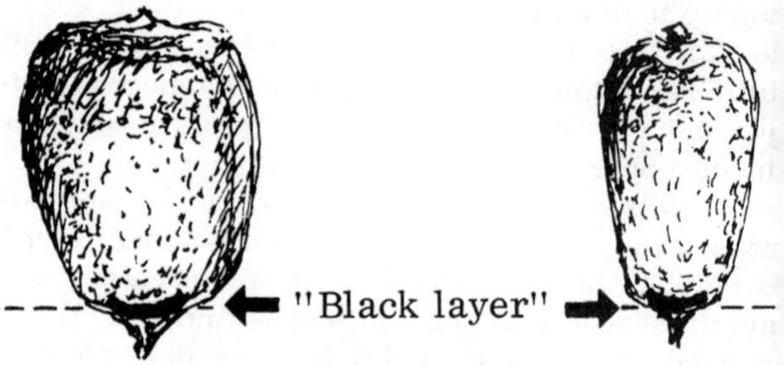

Fig. 7-8. Black layer near the tip of the kernel indicates that the gra is physiologically mature and ready for the silo.

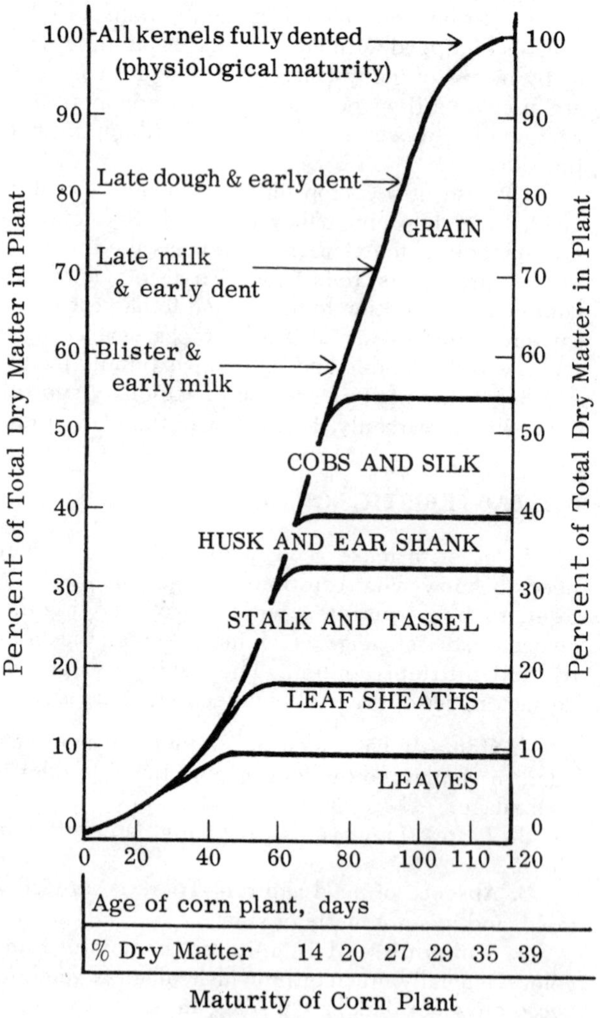

Fig. 7-7. Effect of maturity of corn plant on total dry matter accumulation. (From: Iowa State University Special Report No. 48)

When the grain reaches physiological maturity several layers of cells near the tip of the kernel tur black, forming the "black layer." This layer can be de tected by removing several kernels from the middle o the ear, thence split them lengthwise or just cut of the tip, and look for the black layer near the tip. If th black layer is present, the grain is physiologically ma ture and ready for the silo.

At the black layer stage, usually the grains ar dented and glazed, the lower 4 to 6 leaves of the cor plant are brown, and the plant contains 60 to 67 per cent moisture. It can be cut 3 to 4 weeks past thi stage with very little loss in dry matter or in feedin value.

Sorghum should be cut for silage when the seed are hard.

Grass silage forages (grasses, legumes, and cerea crops) should be cut at the same stage at which they would make the best hay. (See Table 6-10 in Sectior 6, Hay and Crop Residues.)

Cut to Proper Length

The length of the cut sections affects the packin and, hence, the quality of the silage. Also, the prope length of cut varies with the crop and the moistur content. Thus, for corn and sorghum crops, forag harvesters should be set to make a theoretical cut of ¼ to ⅜ inch. If the knives are sharp and set up to the cutter bar, this will result in about 15 percent of the particles being 1 ½ inches and over, and the remainder shorter than this. Such a combination of particle size is necessary for high-quality feed. Grass silage should be more finely chopped than corn or sorghum silage. Also wilted and dry forage and forage with hollow stems should be chopped more finely than forage of high-moisture content, thus permitting more thorough packing and eliminating air pockets.

Control the Moisture Content

Moisture content is one of the most important factors in determining quality of silage. Experimental work and practical experience have indicated that 60 to 67 percent is the best moisture content of a crop to be ensiled. However, low-moisture silage of 40 to 60 percent moisture is now being preserved successfully in (1) oxygen-limiting silos, and (2) tall conventional silos that are properly topped-off with heavy, wet forage or sealed with plastic.

Forage containing more than 60 to 67 percent moisture (1) is heavier and more costly to handle than is necessary; (2) is apt to produce slimy, putrid silage, due to the presence of butyric and other undesirable acids; (3) will have excessive leakage of the juices and some loss of nutrients, except carotene, from the silo; (4) will result in excessive deterioration in the silo walls due to the high acidity; and (5) will exert higher pressure on the silo walls—for the greater the moisture content, the greater the pressure.

If corn and sorghum are harvested at the stage recommended, their moisture content will be right. However, freshly cut grass and/or legume forage contains 75 to 80 percent moisture, which means that for proper ensiling its moisture content must be lowered by 10 to 15 percent.

HOW TO LOWER THE MOISTURE CONTENT

The moisture content of silage material may be lowered by any one or a combination of the following methods: by conditioning and/or wilting, by adding dry hay or straw, by combining with corn or sorghum silage, or by adding a dry preservative of grain, dried molasses, or dried by-products of citrus or beets.

Conditioning-Wilting

This method is particularly applicable to the making of grass silage. Conditioning and/or wilting of grass silage increases the percentage of sugar in the forage, lessens the leakage of juice from the silo, lessens the pressure on the silo walls, and decreases the destructive action of the acids on the silo walls.

The needed 10 to 15 percent reduction in the moisture content of grass silage material can be accomplished by wilting for about two hours on a good drying day and up to one day or longer in slow drying weather.

The combination of conditioning and wilting is the method most commonly followed today. Excellent equipment is available for conditioning.

Excess drying should be avoided, as it will result in the forage becoming too dry for proper ensiling.

Adding Dry Hay or Straw

The moisture content of any wet silage material can be lowered effectively by mixing dry hay or straw with it at the time of filling. Thus, during poor wilting weather, the moisture content of grass forage can be brought within the desired range by adding 5 to 20 percent hay or straw. Also, this is the standard method of lowering the moisture content when it is desired to ensile such high-moisture products as potatoes (see "Other Silage Crops").

Conditioning and wilting is the preferred method of lowering the moisture content of grass silage, rather than adding dry hay or straw.

Combining with Corn or Sorghum Silage

Sometimes the moisture content can be lowered sufficiently merely by mixing high-water content crops with low-water content crops (see sections entitled "Other Silage Crops" and "Combining Crops for Silage"). Simultaneously, usually more desirable bacterial action can be assured by this procedure.

Adding a Dry Preservative

Such dry preservatives as ground grain, corn-and-cob meal, dried molasses, and dried citrus meal, citrus pulp, or beet pulp will reduce the moisture content of freshly cut and unwilted forage, and, in turn, lessen the leakage (or seepage) from the silo.

HOW TO INCREASE THE MOISTURE CONTENT

If the crop is overripe and too dry when cut, or if it becomes overwilted, it will be necessary either to add water to the silo, or, perhaps preferably, to make it into hay.

Drier material may be used for silage by cutting shorter and packing more thoroughly. If necessary, water should be added or the dry material should be mixed with very green, freshly cut material by alternating loads.

HOW TO DETERMINE THE MOISTURE CONTENT

If corn or sorghum forage is harvested at the recommended stage (see section on "Harvest at Proper Stage of Maturity"), its moisture content will be satisfactory for silage making. On the other hand, wilting is always preferable with forages intended for grass silage.

With a little practice, farmers can usually determine when grass forage has wilted sufficiently and when the moisture content is between 60 to 67 percent. Here are some rule-of-thumb methods:

1. **The twist method**—Before chopping, the forage should be so well wilted that the stems may be twisted without breaking, but the limp leaves should show no signs of dryness. This test cannot be used for such coarse crops as sweet clover.

2. **The grab test (or squeeze method)**—This test consists in taking a handful of the chopped forage and giving it a good hard squeeze for about one minute. Then open the hand slowly, note the condition of the ball of forage in the hand, and refer to the following guides:

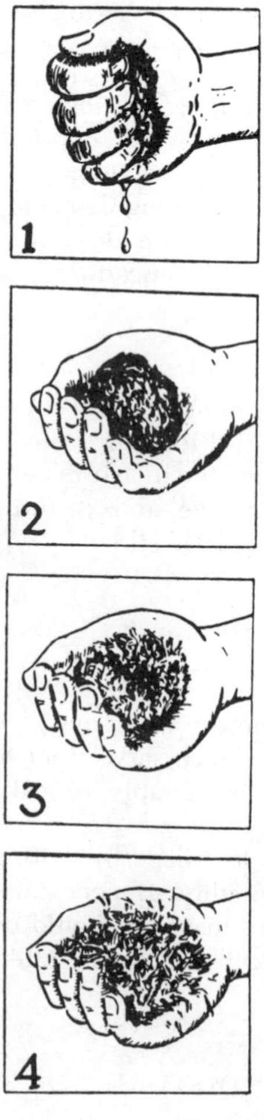

1. Juice runs freely or shows between the fingers. The crop contains 75 to 85% moisture and is too wet to make high-quality silage without treatment. Silages made from crops in this condition will lose large quantities of juice. When possible, wilt these crops. If they must be ensiled without wilting, use an effective chemical preservative (not all of them are effective) or 200 pounds of ground grain per ton of crop.

2. The ball holds its shape—the hand is moist. The crop contains 68 to 75% moisture. Some juices will escape from tower silos. Additional wilting in the field is desirable. Where this is not done, use a chemical preservative or 150 pounds of ground grain per ton of crop, or layer with wilted crops. Odors will be strong without some treatment.

3. The ball expands slowly—no dampness appears on the hand. The crop contains 60 to 67% moisture. This is the best condition for ensiling legumes without treatment.

4. The ball springs out in the opening hand. The crop contains less than 60% moisture. Only very young crops wilted to this condition can be safely ensiled. Others are likely to mold in the silo unless layered with wet crops or placed in gastight silos.

Fig. 7-9. The grab test.

3. **The oven-drying method**—If in doubt or until more experience is obtained, the moisture content of a sample of any kind of forage may be obtained in about an hour's time by the following procedure:

a. Weigh an empty tray on kitchen, bathroom, dairy barn, or other scales.

b. Spread some of the forage in a thin layer o the tray.

c. Weigh the tray and forage.

d. Subtract "a" from "c," in order to obtain th weight of the green forage sample.

e. Place the tray and sample in a preheate oven at 275°F. If the oven doesn't have a vent leave the door slightly open.

f. When the forage seems to be dry, weigh i again. Return it to the oven and reweigh at 5 to 1(minute intervals until the weight is nearly constant. Record final weight of the tray and dry forage.

g. Subtract "a" from "f," in order to obtain the weight of the dry forage sample.

h. Subtract "g" from "d," in order to obtain the weight of the water in the sample.

i. Divide "h" by "d" (and multiply by 100), in order to obtain the percent of moisture in the forage.

4. **Other methods**—The heated-oil method and certain patented devices (as forced air dryers) may be used for moisture determination. However, some of these methods are slow and/or expensive.

Add Preservatives Only When Needed

A number of materials are available to incorporate into silage, with claims made that they will improve the preservation of nutrients, nutritive value, and/or palatability of the silage.

Thorough testing of these materials would necessitate that each of them be used at several levels, with many kinds of silage, with each forage at various moisture contents, and under different storage conditions. Obviously, it would be highly impractical, if not impossible, to carry out such an extensive study. However, there is sufficient understanding of the process of silage formation, the requirements for the preservation of silage nutrients, and the mode of action of the ingredients used in various additives to make sound decisions as to whether they might be economically worthwhile. Also, some experimental testing has been done with certain additives.

In order to be effective, a preservative should serve one or more of the following purposes:

1. Provide fermentable carbohydrates.
2. Furnish additional acids to increase acid conditions.
3. Inhibit undesirable types of bacteria and molds.
4. Reduce the amount of oxygen present, directly or indirectly.
5. Reduce the moisture content of the silage.
6. Absorb some acids which might otherwise be lost in seepage.

SOME SILAGE PRESERVATIVES

Many preservatives have been used, either alone or in combinations. Some of these follow:

• **Molasses (including cane or blackstrap, beet, corn, and citrus molasses)**—Some green forages, such as legumes and certain grasses, are rather low in sugar content. Hence, adding molasses, which is high in sugar, may increase lactic and acetic acid production and improve silage quality and preservation. Also, molasses improves the palatability of silage and increases its nutritive value. For legumes, about 80 lb of molasses per ton are generally used, and for grasses about 40 lb per ton (molasses weighs 12 lb/gal). Additions of much less than these amounts, as an ingredient in mixed preservatives, is of little value. Much of the feeding value of the molasses is retained in the silage under good storage conditions and where there is no seepage loss.

Molasses may be added in either liquid or dehydrated form as the forage enters the blower.

When a grass and/or legume is wilted to 50 to 60 percent moisture content and adequately protected from air, an excellent feed with a good aroma and keeping quality can be secured without the addition of molasses (sugar). Neither is the addition of sugar needed for corn silage.

• **Grain and other feed ingredients**—Silage made from legumes or grasses may be improved under certain conditions by the addition of ground grain (corn, wheat, or barley), ground ear corn, beet pulp, citrus meal, citrus pulp, or other appropriate feed ingredients. The ground material will reduce the moisture content (adding 150 lb of ground grain to a ton of green forage will reduce the moisture content by about 5%); provide additional sugar and starch from which acids may be produced; likely improve palatability; and increase the feeding value of the silage, for almost all of the feeding value of the grain will be retained if a silo which properly excluded air is used.

When green forage is ensiled at a proper moisture content, there is usually no advantage to adding grain for the purpose of preservation or palatability. If the primary concern is to reduce the moisture content of the silage, cheaper materials, such as ground corncobs, cottonseed hulls, oat hulls, or ground hay, may be more appropriate than ground grain.

Ground grain may be added by feeding it into the blower from a properly adjusted hopper attachment.

• **Urea**—Urea has been added to corn or other low-protein forages to improve the protein content of the silage. Generally it is added at the rate of 10 pounds per ton of green material, at which level it increases the protein content by about 1.3 percent. As a rule, there are no decided advantages in this practice as compared to feeding the urea in a supplement with the silage. Further, occasional problems with palatability or excessive urea loss may occur. If urea is added to silage, its cost will not be recovered unless advantage is taken of the higher protein content by reducing protein supplementation of the ration.

• **Limestone**—Limestone (calcium carbonate) may be added at a level of 0.5 to 1.0 percent to corn silage to increase acid production. It neutralizes some of the acids as they are formed, allowing the lactic acid bacteria to perform longer and to produce more acids.

Research has not shown any consistent increase in the nutritive value of silage treated with limestone.

The addition of limestone at ensiling time raises the naturally low calcium content of corn silage—a fact which should be considered when balancing rations. For lactating dairy cows on high alfalfa rations, this may be disadvantageous; but it can be rectified by increasing the phosphorus of the ration. Yet, experiments have shown that dairy cattle do not respond to the addition of limestone to silage. For most beef cattle rations, however, the added calcium is a virtue because (1) when high corn silage rations are fed, the added calcium is needed, and (2) when silage is limited, little calcium is added.

• **Bacteria and mold inhibitors**—Bacitracin and other antibiotics have been used as silage additives. Sodium propionate, and other organic acids, because of their mold inhibiting properties, have also been used as preservatives. Salt has been used because, at an appropriate level, it inhibits certain microorganisms without preventing the action of bacteria which produce the desirable acids. Antibiotics, mold inhibitors, and salt are not essential and are of questionable value in silage fermentation or preservation if air is properly excluded. If air is not excluded, they do little if anything to preserve the silage unless they are added at very high levels.

• **Sodium metabisulfite**—Sulfur dioxide (SO_2, a gas) forced into silage will decrease fermentation and improve carotene preservation. However, this is a complicated process, so sodium metabisulfite ($Na_2S_2O_5$, sometimes called sodium sulfite) is used instead. This salt acts like sulfur dioxide, but is much easier to handle. Eight lb of sodium metabisulfite contain the equivalent of 5 lb of sulfur dioxide. Thus, it is generally recommended that it be applied at the rate of 8 lb per ton of green forage. It should be applied evenly, and the men tramping in the silo should wear air-filter masks as a precautionary measure.

Sodium metabisulfite has no apparent value for corn silage other than reducing carotene losses. Its use with legume forages reduces carotene losses and may improve the odor of the silage. Its effect on palatability is variable. The saving in carotene and the improvement of odor in legume silages are of very little economic value and are outweighed by the cost and inconvenience of sodium metabisulfite application and occasional problems with palatability.

Sodium metabisulfite has been found to reduce the production of toxic gases in silage of high nitrate content. Whether its use for this purpose alone would be of enough value to outweigh its cost and the inconveniences of application is questionable.

• **Bacterial cultures**—Silage preservatives containing cultures of acid-forming bacteria (*Lactobacillus*) are on the market. The basis for including these in the preservative is to provide an inoculum or to increase the numbers of these bacteria and ensure rapid fermentation. There are two schools of thought relative to the value of bacterial cultures as silage preservatives. The advocates claim that these products increase the dry matter, energy, and protein of the silage. Others report that the addition of such "cultures" is unnecessary and of questionable value. Then, the latter add these clinchers: There are always sufficient numbers of these bacteria present on the ensiled material to bring about the proper fermentation. Moreover, the number of live bacteria present in these preparations cannot be guaranteed with accuracy, and the number of bacteria added through such preservatives is insignificant in comparison with numbers already present on the ensiled material.

• **Yeast cultures**—Yeast cultures have also been included in certain silage additives. However, yeasts will sometimes grow in silage without an inoculum being added. When this happens, the silage is of a yeast odor, which is considered undesirable. Yeast does have nutritional value, but because of the small quantity involved in preservatives the contribution in this regard is minimal.

In summary, therefore, there is no good basis for adding yeast cultures to ensiled material.

• **Enzymes**—Cultures of molds, or of molds with other microorganisms, are sometimes added to silage to provide a source of enzymes. It is claimed that these enzymes improve the nutritive value of the silage by increasing its digestibility or digestible nutrient content. Although the enzyme activity of these preparations has not been measured experimentally, no doubt they vary considerably from batch to batch. Further, the quantity of enzymes added by a preservative is insignificant compared to those already present in the silage.

In summary, therefore, there is no reliable evidence that enzymes improve the fermentation or the digestibility, or that they increase the level of any of the nutrients in silage. The only improvement one may expect from such preparations is that which is added by the preparation itself, but the amount generally recommended for use makes this insignificant.

• **Mineral acids**—Mineral acids (hydrochloric acid, sulfuric acid, phosphoric acid) have been used as silage preservatives, almost entirely in Europe, in connection with the ensiling of high-moisture material. These acids substitute for the acids produced by bacterial action. However, they are very corrosive, causing problems in their application, including problems with the silo walls and silage handling equipment. Of the three acids, phosphoric is preferred because (1) it is less corrosive than sulfuric acid or hydrochloric acid, (2) it may enhance the mineral feed value of the silage, and (3) it increases the residual manure value of the silage. But phosphoric acid may introduce a problem of proper calcium-phosphorus ratio. This can result in some abnormal conditions and unsatisfactory performance in the animals to which it is fed.

When mineral acid preservatives are used, it is recommended that ground limestone or some other form of calcium or sodium carbonate be fed to animals at the rate of approximately 1 ounce for each 10 pounds of silage in order to neutralize the acid and prevent any undesirable effect therefrom.

In general, the use of mineral acid preservatives is not considered as desirable as the use of molasses or grain, because (1) they produce sourer and less palatable silage; (2) they may damage clothing, machinery, and/or masonry silo walls, due to their corrosiveness; and (3) they do not add to the nutrient value of the silage except that they may better preserve the carotene.

In general, the use of mineral acids has more disadvantages than advantages.

• **Organic acids**—Propionic acid and other organic acids such as formic acid and related compounds are the newest "family" of preservatives. They are used in a manner similar to mineral acids, but they are much less corrosive and not so difficult to handle, although precautions must be taken.

Organic acids will enhance the preservation of forage without the loss of palatability. Even so, like all preservatives, they cost money; hence, they are recommended only in special cases in making silage. When propionic acid is used, the following guidelines should be observed:

1. Add 1 percent propionic acid to the forage in the field at the time of harvest or at the chopper.
2. Limit the presence of oxygen by using a sound silo.
3. Prevent dilution of organic acid treated silage by rain or snow by covering it with plastic when it is stored outside or in a temporary silo.

It appears that organic acids will find their greatest use in the preservation of high-moisture grain (see section on "High-Moisture Grain").

SILAGE PRESERVATIVE CONSIDERATIONS

A variety of silage preservatives are presently on the market; and, no doubt, new ones will follow. For the most part, these products have been inadequately tested. Yet, a farmer or rancher is often in the position of having to decide whether an additive shall be used. In addition to understanding the silage-forming process and how different additives function, he should consider the following:

1. Preservatives will not substitute for the proper exclusion of air.
2. Preservatives do not produce nutrients in silage.
3. Preservatives that add nutrients to the silage will be partially lost with any spoilage or seepage.
4. The cost of a preservative is usually high in relation to the value of the silage.
5. Chemical analyses are of very limited use in evaluating silage preservatives.

SILAGE PRESERVATIVE RECOMMENDATIONS

When added to silage, the following materials will increase the amount of nutrients it contains:

1. Molasses will increase the total digestible nutrients (TDN, or energy) and may improve fermentation in legumes and certain grasses.
2. Grain or grain by-products will increase protein, total digestible nutrients, and dry matter.
3. Urea will increase the nitrogen (crude protein).
4. Limestone will increase the calcium content.

None of the above products is essential to good silage formation when conditions of moisture and storage are right. Yet, under special circumstances they can be recommended for use. For example, molasses, grain, or grain by-products might be a wise addition to silage when conditions do not allow for proper wilting prior to ensiling, or when an "all-in-one" silage is being made. Urea may be an appropriate addition to an "all-in-one" silage or where increasing the protein content of the silage will simplify its feeding. It is doubtful that there is any justification for adding limestone unless this is a convenient method of calcium supplementation. The economy of most additives of this type depends largely on how well their nutrients are retained in the silage and the use made of them in balancing the rations.

When forages are stored at the proper moisture content, and when air is properly excluded, nutrient losses are low and a good quality silage forms. Additives such as lactic acid bacteria, sodium metabisulfite, mold inhibitors, antibiotics, salt, mold cultures (enzymes), yeast cultures, mineral acids and sodium formate plus sodium nitrite can, therefore, do little if anything to improve the preservation of the silage or its feeding value. When high-moisture material is ensiled, grain (in some cases molasses) is superior to any of these additives. When air is not properly excluded, none of these additives will correct the large fermentation and spoilage losses.

In short, there is no substitute for good management of forage crops for silage, with proper control of such factors as stage of maturity at harvest, harvesting methods, moisture content, fineness of chopping, distribution and packing, and exclusion of air. Generally speaking, manufacturers and salesmen of silage preservatives have recognized this fact and have promoted good silage-making methods—they have been good teachers. Often their counsel on how to make good silage has been more effective than their products.

Fill Rapidly

Once silo filling is started, it should be rapid, so as to avoid spoilage before the silo is filled and sealed. Generally speaking, a silo should be filled in two days or less.

Distribute Forage Uniformly in the Silo

In order to avoid the presence of air pockets and spoilage, it is essential that any kind of chopped forage be distributed uniformly in the silo and that it be packed well. Proper silo distribution is obtained by keeping the material nearly level or slightly higher at the center, whereas added packing is obtained by having one or more persons tramping in the silo. Silo distribution equipment is available for keeping the material in the silo level.

Corn, sorghum, and sunflower silage need not be tramped from the standpoint of preservation and quality of silage, but tramping will result in a somewhat larger tonnage being stored.

Where the forage, regardless of the kind, is harvested at a green, immature stage and cut into short lengths, tramping will not be necessary; but uniform distribution is very important. The only filling precaution under these conditions is to see that the top is carefully leveled and well packed whenever filling is stopped.

Grass silage (especially when wilted), hollow-stemmed forages, and forages that have matured or dried beyond the best silage stage should always be tramped well, especially near the wall.

Mechanical distributors are very helpful, especially in silos of 14-foot or larger diameters.

Seal or Top-off the Silo

Sealing or topping-off is necessary in order to avoid excess spoilage, especially with grass silage, which tends to dry out on the surface and to shrink away from the silo walls. This may be accomplished by carrying out one or more of the following procedures:

1. Leveling off the top and thoroughly tramping the last few feet, especially near the walls.
2. Topping-off the silo with two to three loads of wetter material.
3. Covering the top with plastic cut to fit the silo diameter and turned up against the silo wall a distance of 5 to 8 inches.

FEEDING VALUE AND ECONOMY OF SILAGE

A common rule of thumb is that 3 lb of 70 percent moisture grass silage or 2 lb of 40 percent haylage are equivalent to 1 lb of hay of similar kind and quality; a difference due primarily to the higher water content of silage or haylage. Suggested practical rations for different classes of livestock in which silage is incorporated, usually in combination with hay or some other dry forage, are given in this book in Section 4, Feeding.

Many factors enter into any figures which propose to show the comparative economy of silage vs dry forages; among them, (1) the comparative yield of total digestible nutrients per acre, (2) the cost per ton for preserving and storing, (3) the relative nutrient and feeding value, (4) the distribution of labor, (5) the control of weeds, (6) the kind of haymaking weather, (7) the hazard of curing so much hay without it becoming overripe, (8) the price per ton, and (9) the machinery and efficiency of each method, etc., etc.

SILAGE POINTERS

Some additional pointers which may be of value to the farmer or rancher who is making or feeding silage follow.

Coating the Silo

Since wet grass silage has a somewhat more corrosive action on concrete than does corn or sorghum silage, it may be desirable to apply a protective coating to the inside of concrete silos, whether of solid concrete or of stave construction. The problem is to find an effective, nontoxic, and economical coating. For information on the latest recommendations, the farmer or rancher should contact the local county agent, vocational agricultural instructor, or cement dealer, or write to the state college of agriculture.

Nutrient Losses in Leakage

Seepage losses vary with the moisture content, depth of silage, distribution of the silage, and the amount of nutrients in the seepage. Seepage losses may be as high as 14 percent of the dry matter stored.

The nutrient losses vary, but generally they are in proportion to the run-off. The nutrients lost in seepage from a 100-ton silo may equal the nutrients in ¾ ton or more of hay.

Exposure to Air

Spoilage begins the moment silage is exposed to the air. Therefore, once the silo is opened for use, feed should be removed daily. In the wintertime, a minimum of 1½ inches of silage should be removed daily from a tower silo; in the summertime, 3 inches.

Also, it should be realized that spoilage is likely to occur on the surface of the ensiled material if more than two days elapse between filling periods.

Removal of Silage from Silo

In the past, the common method of feeding silage was by hand. In the present era of bigness and automation, the removal and feeding of silage is being automated.

It is possible to achieve complete push button controlled feeding in an upright silo. With horizontal silos, the silage may be handled with a manure scoop, and sometimes a mechanical unloading wagon or truck—depending on distance from feedlot to silo.

Self-feeding from both tower and horizontal silos can be achieved, but this requires more management on the part of the operator.

Moldy Silage

Moldy silage may be harmful. Any spoilage that causes animals to go off feed, or that upsets the metabolic processes, should not be fed.

Some conditions cause certain molds to produce toxins. The toxins are called *mycotoxins* and the effects of the toxins on animals are called *mycotoxicoses*.

Mature ruminants appear to tolerate higher levels of mycotoxins than young ruminants, monogastric animals, or horses.

One way in which to determine the potential toxicity of moldy silage is to feed it to some less valuable animals for at least two weeks. Observe the animals daily for signs of toxicity—such as reduced gain and going off feed. If no toxic effects are noticed, it is probably safe to feed the suspect silage to other animals. If ill effects are noticed, switch them to other feed immediately and dispose of the suspect silage by spreading it on the land and plowing it under.

Recently, it has been determined that farmers may become afflicted with mycotoxicoses, characterized by severe congestion of the respiratory tract, high fever, and sometimes complete immobility. Thus, prevention of mold growth and protection from molds is essential from the standpoint of the people who feed moldy silage, as well as the animals that consume it.

Silage for Summer Feeding

Some stockmen, especially dairymen, use silage effectively as a summer feed. This practice is especially desirable in those areas where pastures dry up during the hot, dry months. It appears that more and more dairymen will go to year-around silage feeding in a corral.

Effect of Silage on Milk Odor and Flavor

Silage sometimes affects the flavor and odor of milk, especially when ensiled too wet. This effect may be somewhat more pronounced with some silages than with others. The dairyman will do well, therefore, to feed all silages after, rather than before, milking.

Dangerous Silage Gases

Gases formed during fermentation may become hazardous when making and feeding all types of silages unless precautions are observed. The gases are heavier than air and may accumulate near the surface of the silage in pit silos or in tower silos. *Pit silos are always dangerous,* even long after the filling operations. Although tower silos are not free from gas danger, large quantities of gas seldom accumulate unless the doors are put in too far above the silage level.

Farmers have long known about the suffocating effect of carbon dioxide gas formed in silage. It is the most common and the most dangerous of the gases from silage, because it is invisible.

Recently, it also has been recognized that nitrogen dioxide gas is formed by high-nitrate silages and can cause a sometimes fatal disease called nitrogen dioxide pneumonia in man and livestock. Some plants—such as legumes, oats, barley, wheat, corn, sorghums, many pasture grasses, and certain weeds—appear to accumulate especially high concentrations of nitrates during droughts and when grown on high-nitrate soils. When these plants are made into silage, poisonous nitrogen dioxide gas forms until a week or 10 days after filling the silo.

Carbon dioxide gas may be detected by lowering a lighted cigarette lighter, lantern, or candle to the level of the silage. If the flame goes out, the oxygen content of the atmosphere in the silo is dangerously low. Nitrogen dioxide gas can be detected (1) by its yellow or reddish-brown color, or (2) by means of starch-iodide paper (obtained from drug stores or chemical supply houses), which turns blue in the presence of nitrogenous compounds.

Precautions against hazards caused by silage gases include (1) operating the blower for a 15-minute period if it is still connected, (2) swinging a piece of canvas, a tree branch, or a burlap bag vigorously so as to agitate the air and dilute gases that may be present, or (3) taking proper life support equipment when entering an oxygen-limiting, or sealed, silo. Also, adequate provision for ventilation of the silo through the roof is essential.

A victim of silo gas should be moved into fresh air as soon as possible, and artificial respiration should be applied. A physician should be called immediately.

HAYLAGE (Low-Moisture Silage)

Haylage is made from grass and/or legume that is wilted to 40 to 45 percent moisture content before ensiling. Properly made haylage has a pleasant aroma and is a palatable, high-quality feed. Animals usually receive more dry matter and net feed value in haylage than in silage made from the same cut.

Haylage is growing in popularity, especially as a dairy feed. Its nutritive value depends on the stage of the growth of the crop when cut and the percentage of dry matter in it.

Haylage is easy to prepare and preserve in a gas-type silo where air is excluded. But it can be made in a conventional silo provided certain precautions designed to keep out the air are taken.

Haylage is preserved by processes somewhat dif-

Fig. 7-10. Field chopping haylage into covered forage wagon.

ferent from those for wilted or unwilted silage. It must be stored in a silo which can be made as near airtight as possible so that the oxygen present is soon used up and the carbon dioxide that is produced is trapped and held within the silo. These conditions prevent the forage from spoiling by molding, oxidizing, heating, etc. A limited fermentation takes place compared to that occurring in wilted or unwilted silage.

Air exclusion is the key to the success or failure of making low-moisture silage. While rapid air exclusion is important in any silage making procedure, it is doubly important with low-moisture silage. This calls for rigid management of gastight, or oxygen-limiting, silos. Under these conditions, low-moisture silage can be stored satisfactorily with minimum dry matter losses.

Research data from several universities shows that haylage may be stored satisfactorily in conventional silos if they are (1) well constructed, and (2) capped with plastic until feeding is initiated. Haylage stored in this manner generally has the same feeding value as that stored in gastight silos.

HIGH-MOISTURE GRAIN

In recent years, a considerable quantity of high-moisture (22 to 40% moisture) (1) earcorn, (2) shelled corn, and (3) small grains have been ensiled successfully. Such high-moisture grain results when crops are planted late, frost-killed, or harvested when wet. For best results, the silo should be as nearly airtight as it can be made, and the spoilage on top should be kept to a minimum by covering the moist grain with some tight sheet. Ensiling high-moisture grain alleviates costs for drying and risks of molding or heating.

For best results, it is recommended that high-moisture grain containing about 30% moisture be stored and fed. It may deviate from this figure plus or minus 10%, but water should be added when it falls below 26%. Thus, grain not containing 26% moisture should be reconstituted by adding water. Also, a roller mill that will handle high-moisture shelled grain is necessary.

High-moisture ground ear corn can be stored satisfactorily in any upright silo. Many types of hammer mills, forage harvesters with recutter screens, and grinders are available and will do an excellent job in preparing the grain. Roller mills can be used in preparing high-moisture grain for oxygen-limiting silos but not for conventional stave or cement silos. The latter require fine grinding. Cob particles should be no larger than ½ inch, and 90 percent of all kernels should be at least cracked. In conventional silos leveling and packing should be done with care. Distributors should be augmented with hand leveling.

If high-moisture grain is stored in a trench silo instead of an airtight upright silo, it should be ground then firmly packed. Otherwise, spoilage will result.

When high-moisture grain is bought, it should be purchased at a lower figure than dry grain. Likewise a greater quantity of it must be fed. Table 7-2 shows how the ration needs to be changed when high moisture corn is included, as well as the dollar value of corn at different moisture contents; and the computations that follow are pertinent to the use of high moisture corn.

To estimate feed value in percent of dry corn:

Dry corn (15% m) has 85% dry matter.

High-moisture corn (30%) has 70% dry matter.

Therefore, the 30% high-moisture corn has 70/85 × 100 = 82% as much energy feed value.

To estimate dollar value of high-moisture corn:

Dry corn (15% m) costs $2.75/bu.

High-moisture (30%) corn has 82% as much energy feed value.

Therefore, it should cost no more than: $2.75 × .82 = $2.25/bu.

To estimate amount of high-moisture corn to substitute for dry corn:

Total ration mix = 2,000 lb
Dry (15%) corn = 1,500 lb
Other ingredients = 500 lb

How much high-moisture (30%) corn should replace the dry corn?

TABLE 7-2
COMPARATIVE VALUE OF CORN OF DIFFERENT MOISTURE CONTENTS

Moisture	Dry Matter	Lb to Equal 100 Lb Dry Corn	Lb to Equal 1 Bu Dry Shelled Corn	Lb to Equal 1 Bu Dry Ear Corn	Estimated Feed Value of Dry Corn	Estimated Dollar Value of Dry Corn
(%)	(%)	(lb)	(lb)	(lb)	(%)	($)
15	85	100	56	70	100	2.75
20	80	106	60	74	94	2.58
25	75	113	63	79	88	2.42
30	70	121	68	85	82	2.25
35	65	131	73	92	76	2.09
40	60	142	79	99	71	1.95

121 lb of 30% moisture corn = 100 lb dry corn.

Therefore, 1,500 lb × 1.21 = 1,815 lb of 30% moisture corn would be needed to replace the 1,500 lb dry corn.

High-moisture (30%) corn	=	1,815 lb
Other ingredients in mix	=	500 lb
Total ration mix	=	2,315 lb

Thus, due to the moisture, the mix with the high-moisture corn is about 1/7 less nutritious than the dry corn mix, and it is necessary to feed about 1/7 more of this mix than when feeding the dry corn mix.

Acid Preservation of High-Moisture Grain

The acid preservation of high-moisture grain involves the addition of 1 to 1½ percent propionic acid (or a mixture of propionic acid with either acetic acid or formic acid) to high-moisture cereal grain to inhibit mold or spoilage, thereby alleviating artificial drying or the necessity to store in an airtight silo. Table 7-3 gives the recommended level of application of propionic acid to ground ear corn and shelled corn of varying moisture contents and periods of storage.

Guidelines for the use of organic acids from the standpoints of (1) application, (2) storage facility, (3) handling and feeding, and (4) safety precautions follow. Also, the mode of action of organic acids is discussed.

• **Application guidelines—**

1. Check grain for moisture content so selected rate of acid application meets requirement for preservation.
2. Treat the grain immediately after harvesting so as to eliminate heating and mold development.
3. Make sure that all the grain is coated with the acid and that the application rate is correct.
4. Treat outdoors if possible; if not, provide adequate ventilation.
5. After treatment, flush equipment with water or untreated grain to prevent corrosion.
6. Observe safety guidelines at all times.

• **Storage facility guidelines**—Nearly any weatherproofed facility can be used to store treated grains. Among the facilities that may be and are used are the following:

1. Metal bins and buildings coated with acid-resistant paint. (See your building supplier for recommendations.)
2. Wooden bins, cribs, and buildings.
3. Quonset-type buildings, provided grain is not stored on dirt floor. Again, metal should be protected.
4. Concrete silos or bins; may coat to prevent pitting in new concrete. Also, covering concrete floors with plastic is suggested.

TABLE 7-3

AMOUNT OF PROPIONIC ACID RECOMMENDED TO TREAT GROUND EAR AND SHELLED CORN OF VARYING MOISTURE CONTENTS FOR STORAGE PERIODS OF 6, 9, AND 12 MONTHS[1]

% Moisture	% by Weight	Lb/Bu	Lb/Ton	Gal/Ton	Cost/Ton @ 30¢/Lb	Cost/Ton @ 40¢/Lb
6 Months' Storage						
20	.33- .50	.18-.28	6.6-10	.79-1.20	1.98-3.00	2.64-4.00
25	.50- .65	.28-.36	10-13	1.20-1.56	3.00-3.90	4.00-5.20
30	.65- .85	.36-.48	13-17	1.56-2.04	3.90-5.10	5.20-6.80
35-40	.85-1.05	.48-.59	17-21	2.04-2.56	5.10-6.30	6.80-8.40
9 Months' Storage						
20	.40- .60	.22-.34	8-12	.96-1.44	2.40-3.60	3.20- 4.80
25	.60- .85	.34-.48	12-17	1.44-2.04	3.60-5.10	4.80- 6.80
30	.85-1.10	.48-.62	17-22	2.04-2.64	5.10-6.60	6.80- 8.80
35-40	1.10-1.40	.62-.78	22-28	2.64-3.36	6.60-8.40	8.80-11.20
12 Months' Storage						
20	.50- .75	.28-.42	10-15	1.20-1.80	3.00-4.50	4.00- 6.00
25	.75-1.00	.42-.56	15-20	1.80-2.40	4.50-6.00	6.00- 8.80
30	1.00-1.25	.56-.70	20-25	2.40-3.00	6.00-7.50	8.80-10.00
35-40	1.25-1.50	.70-.84	25-30	3.00-3.60	7.50-9.00	10.00-12.00

[1]The lower levels of application have produced acceptable results where good mixing and distribution of the acid was accomplished. The higher application rates are recommended if excellent mixing is not attained.

• **Handling and feeding—**

1. Since the acid is absorbed by the grain, protection is provided after removing from storage. Thus, treated grain can be mixed with dry grains and other feeds.

2. Treated grains can be transported. They have been successfully moved from one silo to another. However, movement from a large diameter to a smaller diameter silo is suggested for better packing.

3. Acid-treated high-moisture corn is readily accepted by cattle. In studies at The University of Wisconsin, dairy cows offered propionic acid-treated high-moisture ground ear corn, consumed more dry matter than those offered similar ensiled corn or dried corn.

• **Safe handling of grain preservatives**—When preservatives are applied, the following safety precautions should be observed by the user:

1. Follow the instructions on the label.

2. Avoid storage of acids with fuels, lubricants, and pesticides. Store acids *only* in original container, tightly closed, with the bungs upright.

3. Use organic acids *only* on grain destined for animal feed.

4. If grain is treated in a building, adequate ventilation must be provided.

5. Acid should not be allowed to come in contact with skin or eyes. Protective gloves, goggles, respirators, or a face shield and protective clothing should be worn when there is a risk of contact.

6. A supply of water must be available to wash away any acid coming in contact with skin or eyes.

7. When contact occurs, drench and remove contaminated clothing immediately.

8. Clean augers with water or untreated grain after treatment to avoid corrosion.

9. Organic acids are flammable so care should be exercised to avoid fire. Keep the acid away from any ignition source and maintain good ventilation in areas when it is being applied.

• **Mode of action of organic acids**—High-moisture shelled corn, ground ear corn, and whole ear corn should be treated promptly after harvesting to avoid heating. When properly applied, propionic acid and other acid-type grain preservatives will kill most molds and related organisms on the outside of the treated grain. As the acid moves into the kernel, it kills the embryo, thus nearly eliminating respiration and enzymatic activity. These actions prevent heating. While the exact mode of action is not known, the acid preservatives continue to inhibit growth of molds. The effect is probably due in part to the lower pH created by the acid; however, not all products which depress pH inhibit mold. To be effective, the treatment must provide continuing protection against mold and other microbial growth. The organic acids previously mentioned have shown reasonable ability to provide extended protection.

SECTION 8

MANAGEMENT

Contents **Page**

Marking or Identifying Animals 724
Weaning 728
Dehorning, Castrating, and Docking Farm Animals 729
Spaying Heifers 734
Bedding Animals 735
Making and Keeping the Soil Productive 736
Barnyard Manure 737
Commercial Fertilizers 740
How to Determine Soil Deficiencies 741
How to Take Soil Samples 742
Soil Fertility Guide 742
Lime Acid Soils 752
Treat Saline and Alkaline Soils 756
Control Pests; Lessen Waste 757
Weed and Brush Control 758
Rodent and Bird Control 759
Controlling and Eliminating Rats 762
Controlling and Eliminating Other Rodents and Birds 765
Some Beef Cattle Management Practices 768
Managing the Beef Bull 768
Managing Beef Cows 769
Calving Two-year-old Heifers 769
Managing Confined (Drylot) Cows 772
Preconditioning Calves 772
Management of Feedlot Cattle 773
Handling Newly Arrived Cattle 773
Dairy Beef 776
Hogs Following Cattle 776
Some Dairy Management Practices 777
Managing the Dairy Bull 777
Managing Dry and Lactating Cows 777
Managing Dairy Calves 779
Managing Replacement Heifers 779
Cow Testing Programs 779
Some Sheep Management Practices 779
Managing the Ram 779
Managing Ewes 780
Some Swine Management Practices 782
Managing the Boar 782
Managing Breeding Swine 783
Managing Sows at Farrowing 783
Feeding Systems for Swine 785
Swine Skills 786
Some Horse Management Practices 786
Normal Breeding and Foaling Seasons 786
Managing the Stallion 787
Managing Mares at Foaling 788
Managing the Newborn Foal 789
Care of the Feet 791
Grooming 792
Transporting Horses 793
Stable Management 793

Management is the art of caring for, handling, or controlling. In a livestock operation, it gives point and purpose to everything else. It can make or break a livestock outfit.

Pertinent facts relative to, along with methods of accomplishing, some important livestock management practices are covered in this section. Management practices of importance to all classes of livestock are presented in the first part of the section, followed by specific class of livestock practices.

MARKING OR IDENTIFYING ANIMALS

The method of marking or identifying animals will vary according to the class of animals and the objectives sought. Thus, some methods of marking are well adapted to one class of animals but not to another; horn brands, for example, may be used on horned cattle only. Also, on the western range, marking or branding is primarily a method of establishing ownership and/or age; whereas in the small herd, particularly in the purebred herd, it is a means of ascertaining ancestry or pedigree. (For breed registry association rules relative to marking or identifying animals, see Section 17 of this book.)

Marking or Identifying Cattle

The method employed to mark cattle should be determined primarily by the objective sought. Fig. 8-1 shows a number of methods of marking or identifying

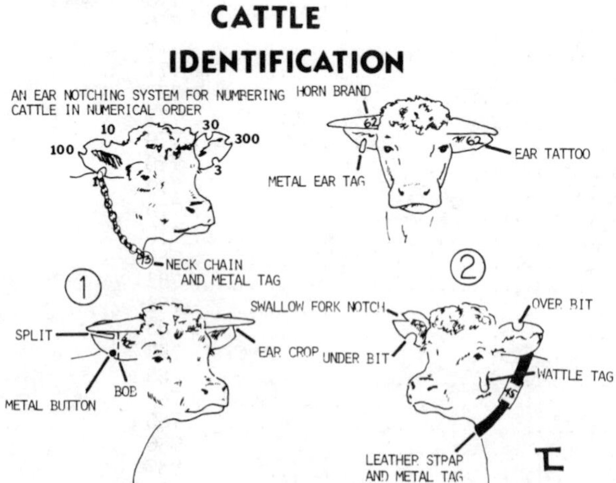

Fig. 8-1. Composite drawings showing a number of methods of cattle identification. It is unlikely that any individual animal will carry more than one or two of these methods of identification.

cattle. A description of each of these and othe methods follows:

1. **Hide brands**—When properly applied, hid brands are permanent. Throughout the range countr the hide brand is recognized as the cattleman' trademark. Most of the western states require th each brand be recorded, as to both type and locatio in order to avoid duplication. When stock are ru close to a state boundary, the same brand may be re corded in two states.

In addition to the regular brand, many ranche identify the age of the females by adding the la number of the year (usually at a different location Thus, heifer calves born in 1984 might be identifie by the number 4. At the end of 10 years, the number are used over again, for there is seldom any difficult in determining ages where there is a 10 year sprea In those states where brands are recorded, thes added numbers or brands must also be approved b and recorded with the registrar of brands.

Hide brands have the disadvantage of being u sightly, and hot iron brands lower the market value the hide. For these reasons, they are not recom mended except when necessary for identificatio purposes. Even then, it is desirable that their size b as small as possible, consistent with serving the pr mary objective of the brand.

The pertinent facts relative to branding are:

a. **Time**—In the range country, the usual pra tice is to brand calves at the same time they a castrated and vaccinated against blackleg.

b. **Location on animal**—The brand is locate on a body area where it may be easily seen an where it will do the least possible damage.[1] Hi and thighs are favorite body areas for brands.

c. **Preparation**—Usually calves are thrown f branding—roped by the hind legs and dragged t the place of branding. Older cattle, however, a restrained in a chute. Some ranchers now pref to use specially designed branding chutes f calves.

d. **Three methods of applying brands are:**

(1) **The hot iron**—To date, this has been th preferred method. In this method, the irons a heated to a temperature that will burn suff ciently deep to make the scab peel but whic will not leave deep scar tissue. The prope temperature of the hot iron is indicated by yellowish color. Branding is accomplished b placing the heated branding iron firmly again the body area which it is desired to mark an

[1]In arriving at both the kind and location of the brand, the own should first check with the brand inspector or the local county age or veterinarian to determine if any part of the animal is reserved f state or federal disease control programs; for example, the cheek the cattle is used for brucellosis reactor identification.

by not allowing it to slip for the few seconds when the hide is burned. The branding iron should be kept free from dirt and adhering hair at all times. Where electricity is available, the electric iron may be used; it keeps an even temperature and, if properly used, makes a clear, uniform brand.

(2) **Freeze marking**—This method, developed by the U.S. Department of Agriculture and Washington State University, makes use of a super-chilled (by dry ice or liquid nitrogen) branding "iron" which is applied to the closely clipped surface for about 20 seconds, following which the hair grows out white. On white cattle, deliberate overbranding (30 seconds or more) will produce a bald brand. When properly done, the method is painless, permanent, and there is no hide damage.

(3) **Branding fluids**—Branding fluids, which are less widely used in making hide brands, consist of caustic material applied by means of a cold iron. Best results are secured if the area is first clipped. In comparison with traditional hot iron branding, the chemical method of producing hide brands is slower; the results are generally less satisfactory, particularly if the operator is inexperienced with the method; and the resulting brand is less permanent.

e. **Characteristics of a good brand**—A good brand is one that is easily read, that is of simple design and yet cannot be easily changed or tampered with, that has no welds or thick points in the iron, and that interferes with the circulation as little as possible. Thick points mean deeper burning and slower healing; whereas small enclosed areas, such as a small "O," will slough out entirely.

2. **Earmarks**—Earmarks are permanent and easily recognized but unsightly. They may be administered with either a sharp knife or a regular ear notcher. Sometimes polled animals are individually identified through ear notches; at other times ear notches are used on commercial ranches to indicate month or season of birth. In such instances a definite value is assigned to each area location. When earmarks are used in commercial operations for purposes of establishing ownership, however, they are uniform and recorded for any given ranch. Some of the more common ownership earmarks are "crops," "swallow forks," "bobs," "overbits," "underbits," and "splits."

3. **Metal or plastic earmarkers (tags and buttons)**—Metal or plastic earmarkers are easily attached, but sometimes they are easily lost. They also frequently rub and scratch the skin, thus making openings for screwworm infestation—an important consideration in the South.

The U.S. Department of Agriculture requires that most cattle two years of age or older be eartagged to identify the animals as to their herd of origin before they are shipped across state lines.

4. **Neck chains or straps**—Neck chains or straps are the most frequently used means of identifying polled cattle. Occasionally, chains or straps may be lost, but this is not particularly serious if the caretaker is on the alert and immediately replaces each one that is lost, without allowing several losses to accumulate before taking action. Neck chains or straps must be adjusted, for young animals grow, or animals change in condition.

5. **Horn brands**—Horn branding for individual identification is commonly used among breeding or sale animals of the horned breeds. Usually horn brands are made by heating small copper numbers with a blow torch or charcoal burner. On mature animals, this method of branding works fairly well, but it cannot be used on young animals while the horns are still growing, unless it is repeated at intervals.

6. **Tattoos**—The purebred beef cattle registry associations require that registered animals be individually tattooed. This method of marking consists of piercing the skin with instruments equipped with needle points which form letters or numbers. This operation is followed by rubbing indelible ink into the freshly pierced area. On dark-skinned animals, tattoos are difficult to read.

It is well to disinfect the tattooing instrument carefully between each operation in order to alleviate the hazard of spreading warts to the pierced area. Warts make it impossible to read the tattoo.

7. **Other identifications**—Other identification marks used on the range include: (1) "buds" formed by making a strip incision through the nose; (2) "wattles" made by cutting down a strip of skin on the jaw bone; and (3) "dewlaps" formed by cutting down a strip of skin on the brisket. New marking devices in different stages of research and development include the following:

a. **Radio transmitter in the second stomach, an electronic device**—The animal swallows a small radio transmitter enclosed in a ¾-inch × 2½-inch plastic capsule, which lodges in the second stomach. From there, it transmits a coded number when signaled by a receiving unit to do so. The transmitter can be retrieved at slaughter and reused.

b. **Implant behind the poll or along backbone, an electronic device**—A ¼-inch cube device, coded to give specific information about the animal, is implanted behind the poll or along the backbone. This device may include such information as birth date, original owner, state of origin, year of implant, and temperature of animal. This method holds great promise as a means of combatting cattle theft.

c. **Laser brand**—On October 28, 1975, the United States Patent Office issued to Washington State University a patent relating to the use of coherent light for humane, permanent identification of animals.

The laser technique makes marks resembling freeze marks or firebrands at speeds up to 30 nanoseconds (a 30 billionth of a second). The marks are less blurred than a firebrand when used to produce firebrandlike marks. When used to produce a freezelike mark, it has the advantage of producing white hair instantaneously. Bald freezelike marks can also be produced but time must be allowed for shedding of hair following lasing. Application is humane in that the speed is faster than the pain reflex.

Table 17-2, Section 17, Marking or Identifying Guide for Registered Beef and Dual-Purpose Cattle, summarizes the pertinent regulations of the beef and dual-purpose cattle registry associations relative to marking or identifying.

Marking or Identifying Sheep and Goats

Sheep operators often find it necessary to mark sheep for one or more of the following reasons:

1. Identification of western sheep on ranges, especially public land, where the brands of different owners may get mixed.
2. Identification with a "buck brand" at breeding time.
3. Identification of ewes and lambs at lambing time.

The need is for a brand which will satisfactorily (1) serve for identification purposes, and (2) scour out.

The common methods of marking or identifying sheep are:

1. **Branding fluids**—There are on the market commercial branding fluids which possess the following desirable features: (a) They will remain on the sheep for a year, and (b) they can be removed from the wool by normal scouring methods.

2. **Marking the ram**—When a number of rams are turned in with a large band of ewes, it is impossible to detect individual rams that may be failing to settle ewes. Moreover, it is quite likely that a different ram will serve the ewe should there be a recurrence of heat, or perhaps more than one ram may serve the ewe at the time of estrus. When only one ram is being used on a small flock, however, it is important to know whether the ewes are getting with lamb. Then, too, with a purebred flock, individual breeding records are rather important.

A breeding record can best be kept by smearing the breast of the ram and the area between his forelegs every day or two with a thick paste which is noninjurious to the wool, or by using a special marking harness on the ram. Then as the ram serves the ewe a mark will be left on her rump.

The color of the paste should be changed every 16 days (the approximate estrous cycle of the ewe) so that one can determine whether ewes that have been bred are returning in heat. For example, during the first 16-day interval, the thick paste used on the ram might well be a mixture of ordinary lubricating oil and yellow ochre; for the second 16-day interval it might be lubricating oil and venetian red; and for the third 16-day interval (if there is still some question about some of the ewes having settled) it might be a paste made by using lubricating oil and lamp black (thus proceeding from light to dark colors).

Naturally, if a good percentage of the ewes are found coming in heat for a second time, the ram should be regarded with suspicion, and perhaps another ram should be obtained. The sterility in some instances may be temporary because of high condition and lack of exercise.

3. **Earmarks**—Identification for sheep can be provided by earmarks, made with either a sharp knife or a regular ear notcher. Such marks are permanent and easily recognized, but unattractive.

Where individual identity is desired, as is necessary in a purebred flock, a definite value is assigned to each area location. With a commercial band, however, the same mark is administered to all animals.

4. **Metal or plastic ear tags**—Most purebred sheep are provided with a metal or plastic ear tag; and sometimes two. Where two tags are used, generally one of them is the individual or flock number assigned by the owner, whereas the other is the individual number assigned by the breed registry association.

Metal or plastic ear tags are easily attached but easily lost. Also, they frequently rub or scratch the skin, thus making openings for screwworm infestation.

5. **Ear tattoos**—These are administered in the same manner as described for cattle.

Table 17-4, Section 17, Marking or Identifying Guide for Registered Sheep and Goats, summarizes the pertinent regulations of the sheep and goat associations relative to marking or identifying.

Marking or Identifying Swine

The common method of marking or identifying swine consists in ear notching the litters. Pigs are generally marked at the same time that the needle teeth are removed. Purebred breeders find it necessary to employ a system of marking so that they may determine the parentage of the individuals for purposes of registration and herd records. Even in the commercial herd, a system of identification is necessary if the gilts are to be selected from the larger and more efficient litters. The ear marks are usually made with a special V-notcher.

The most common notching system, and the one recommended by most swine registry associations, is illustrated in Fig. 8-2.

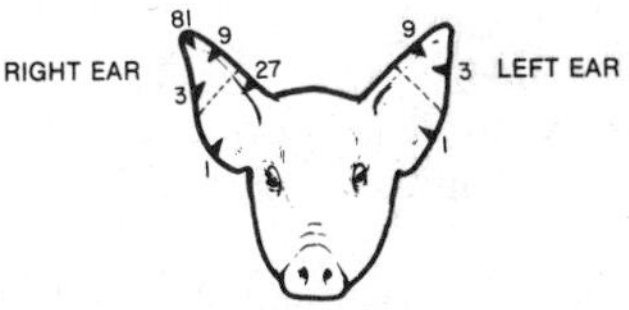

Fig. 8-2. Individual ear notching system. The right ear is used for the litter mark, and the left ear is used for the individual pig number.

Table 17-5, Section 17, Marking or Identifying Guide for Registered Swine, summarizes the pertinent regulations of the swine associations relative to marking or identifying.

Marking or Identifying Horses

The correct identification of Thoroughbred horses racing at the various tracks of the country is important. Thus, one of the duties of a steward, through his horse identification assistant, is that of assuring that each starter in a race is actually the horse named in the entries. This is necessary because only a relatively small percentage of the more prominent racehorses are fondly recognized by sight by the public, the vast majority of racehorses being known only by names and past performances.

To the end that the identity of each horse shall be guaranteed, The Jockey Club requires the following of all horses running at member tracks of the Thoroughbred Racing Association:

1. **Lip tattoo**—The Thoroughbred Racing Protective Bureau (TRPB), Inc., utilizes this method to guarantee to the public the identity of each and every horse running at the tracks of their members. The system consists of tattoo branding, with forgery-proof dyes, The Jockey Club serial number (the registry number) under the upper lip of the horse, with a prefix letter added to denote the age of the horse (see Fig. 8-3). The process is both simple and painless. It is applied by expert crews of the TRPB to two-year-olds as they come to each TRA track.

Lip tattoo equipment and ink similar to that used by the Thoroughbred Racing Protective Bureau is manufactured and sold commercially (distributed by Stone Mfg. & Supply Company, 1212 Kansas Ave., Kansas City, Mo.). Step-by-step instructions for the "do-it-yourselfer" follow:

a. After the digits are placed in the head of the tattoo gun, place the gun head with the digits in a dish of antiseptic (such as Zephiran chloride, available at any drug store).

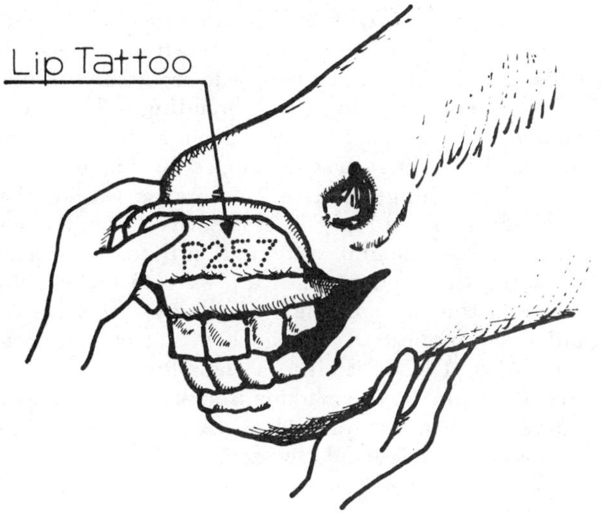

Fig. 8-3. A drawing showing the lip tattoo under the upper lip of a horse. The prefix letter denotes the age of the horse, and the numbers denote The Jockey Club registry number.

b. Roll and hold upper lip back with fingers; do not place anything back of the lip.

c. Wipe upper lip clean with cotton saturated with rubbing alcohol.

d. Shake gun to dry off excess antiseptic.

e. Apply tattoo gun, making sure gun and digits are square with lip. Hold gun rigidly and with sufficient pressure to withstand recoil action of gun.

f. Apply ink and rub into perforations with thumb. Use more ink if bleeding persists. Leave any excess ink on lip.

Initially, the use of the horse lip tattoo was limited to the identification of Thoroughbred racehorses running on member tracks of the Thoroughbred Racing Association, an exclusiveness which The Jockey Club maintained by not making available to others either the tattoo equipment or the formulation of the ink. However, both lip tattoo equipment and ink are presently available through some of the stockmen's supply houses.

2. **Photographs**—This consists of a life-size picture of the night eyes (or chestnuts) of each horse, together with pictures of the sides, front and rear of the animal, showing all natural markings. This corresponds to the human system of fingerprinting employed by the FBI and police departments throughout the world.

Studies have revealed (a) that no two chestnuts are exactly alike, and (b) that from the yearling stage on these chestnuts retain their distinctive size and shape. The chestnuts are photographed, and then classified according to (a) size, and (b) distinctive pattern.

In comparison with the lip-tattoo system, "fingerprinting" horse chestnuts is more costly and necessitates more highly trained people to record and use it.

3. **Freeze marking (cold branding)**—This new method of identifying horses, known as freeze marking (cold branding), was developed by Dr. R. Keith Farrell, Washington State University. Called the Angle System (see Fig. 8-4), it is derived from the ancient Arabic numeral system. It utilizes the basic principle that straight lines are easy to make with crude instruments. It offers simplicity, preciseness, universal application, and good visual communication. Also, it lends itself to a computerized data retrieval system. Freeze marking is now used to identify horses in the United States (Arabians and Appaloosas), New Zealand, Sweden, and Egypt.

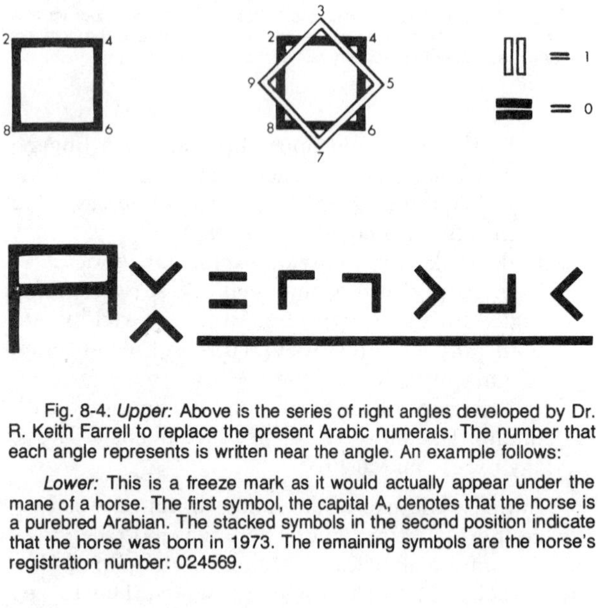

Fig. 8-4. *Upper:* Above is the series of right angles developed by Dr. R. Keith Farrell to replace the present Arabic numerals. The number that each angle represents is written near the angle. An example follows:

Lower: This is a freeze mark as it would actually appear under the mane of a horse. The first symbol, the capital A, denotes that the horse is a purebred Arabian. The stacked symbols in the second position indicate that the horse was born in 1973. The remaining symbols are the horse's registration number: 024569.

Called freeze marking to escape painful associations with the term "branding," the technique utilizes heavy copper stamps, or marking rods, chilled in either liquid nitrogen or dry ice, and 95 percent alcohol. The area to be marked is shaved and scrubbed with a 95 percent alcohol wetting solution to aid in conducting the intense cold and to withdraw body heat.

Placing the copper stamp against the animal's body for 10 to 20 seconds destroys pigment-producing cells (melanocytes) and produces a pigment-free skin area. Hairs growing back in this area will be white. Longer application times result in more balding, a condition necessary for producing legible marks on white or light-colored animals.

A freeze mark that produces white hair causes only minimal changes in the hide and does not seriously impair leather properties. Freeze marks that produce baldness cause some permanent scarring and hide damage. Severe freeze mark damage, however, is minimal compared to fire brand damage.

Freeze marking is more legible than fire branding. Marks are much more distinct, and last just as long. No open wound is produced, which eliminates disease and insect infestations, and freeze marking is relatively painless.

Because some horsemen, particularly those who show horses, object to a visible mark (like Fig. 8-4) the mark is usually placed on the neck under the mane. It is applied approximately 2 inches below the eruption of the mane and about midway between the poll and withers. An area approximately 2 × 7 inches is clipped close to the skin and washed with alcohol. The iron is then applied to the clipped area of the neck.

WEANING

Weaning is the stopping of young animals from suckling their mothers.

Weaning age varies according to species, but normally it is about as follows:

Species	Normal Weaning Age
Calves	6-8 months
Lambs	5 months
Pigs	5-8 weeks
Foals	4-6 months

Currently, there is much interest in early weaning—in weaning animals earlier than the normal ages indicated above. Without doubt, this practice will increase.

If the young are consuming considerable feed at the time of the separation (perhaps by creep feeding), weaning will result in very little disturbance or setback.

The separation should be complete and final, preferably with no opportunity for the young to see or hear their mother again. In no case should the dam be returned to her offspring once the separation has been made. Such practice will only prolong the weaning process and give rise to digestive disorders in the young.

The feed of the dam should be decreased a few days prior to the separation, and should be more bulky for a few days after the removal of the young and until the udder has dried up.

When drying up lactating females, spoiled udders

can be alleviated by adhering to the following procedure:

1. Decrease the ration, and do not feed milk-stimulating feeds just prior to, during, or immediately after weaning—until the udder has dried up.

2. Let "back pressure" in the udder build up. Examine the udders of cows or of mares at intervals, but do not milk them out. If the bag fills up and gets tight, rub an oil preparation (such as camphorated oil or a mixture of lard and spirits of camphor) on it, *but do not milk it out.* At the end of five to seven days, when the bag is soft and flabby, what little secretion remains (perhaps not more than half a cup) may be milked out if so desired.

DEHORNING, CASTRATING, AND DOCKING FARM ANIMALS

Table 8-1, page 731, lists and Figs. 8-5 to 8-9 illustrate the common methods of dehorning, castrating, and docking animals, and gives the directions for accomplishing each. Many stockmen routinely administer these management practices; others call upon a veterinarian for all or part of them. Perhaps the most important thing is that they be done at the proper time.

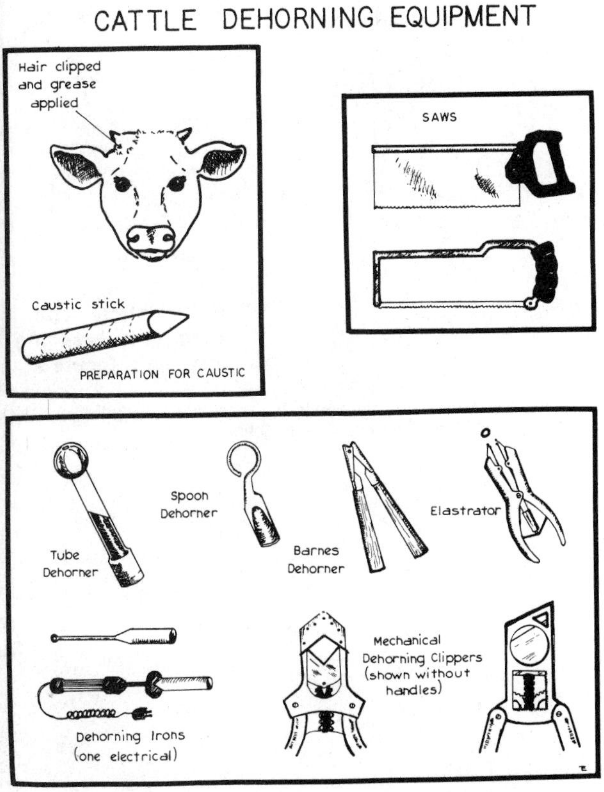

Fig. 8-5. Common instruments used for dehorning cattle.

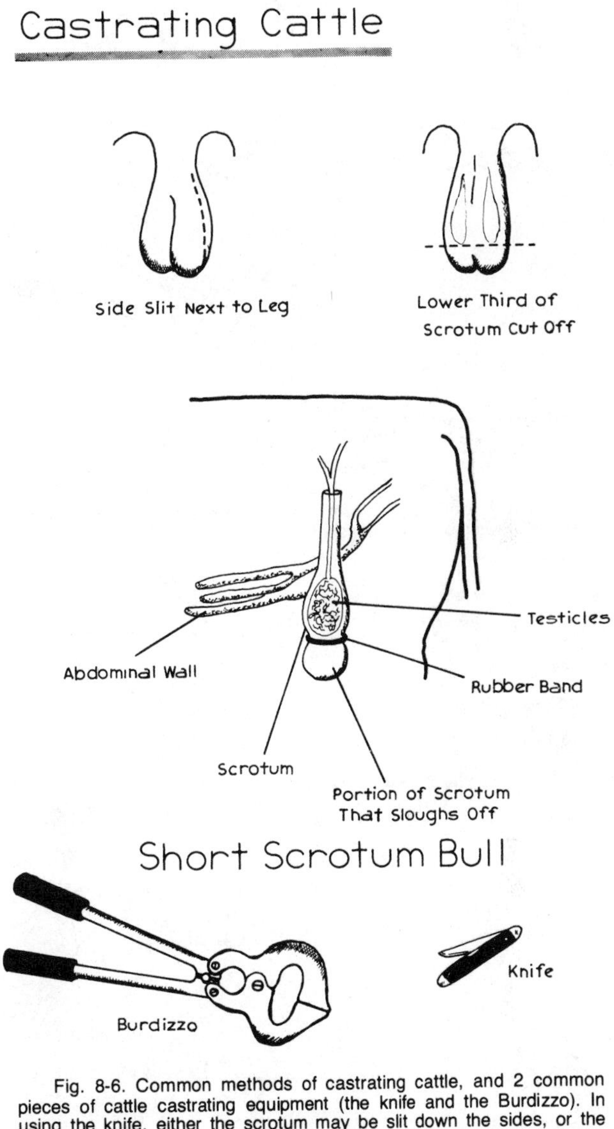

Fig. 8-6. Common methods of castrating cattle, and 2 common pieces of cattle castrating equipment (the knife and the Burdizzo). In using the knife, either the scrotum may be slit down the sides, or the lower third may be removed. With the Burdizzo, first the cord is worked to the side of the scrotum, and then the instrument is clamped on about 1½ to 2 inches above the testicle, where it is held for a few seconds. The short scrotum method of castrating is detailed in Table 8-1. (See Fig. 8-7 for an elastrator.)

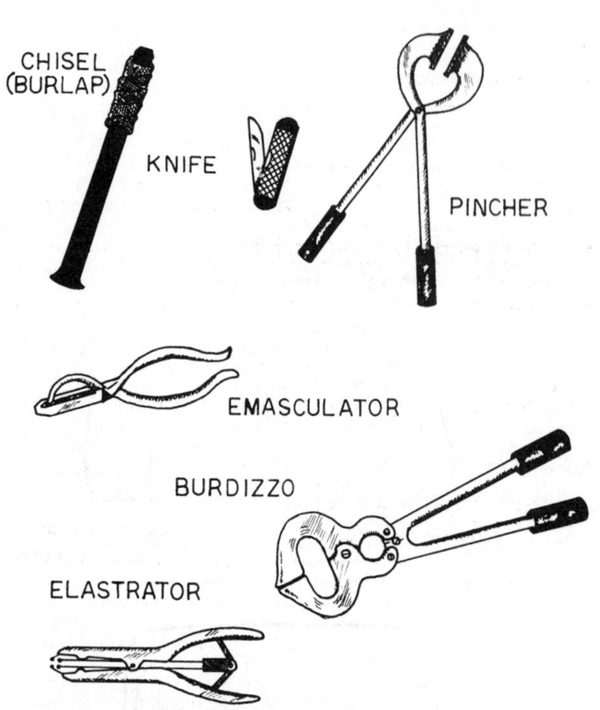

Fig. 8-7. Common instruments used for docking and castrating lambs. (Drawing by R. F. Johnson)

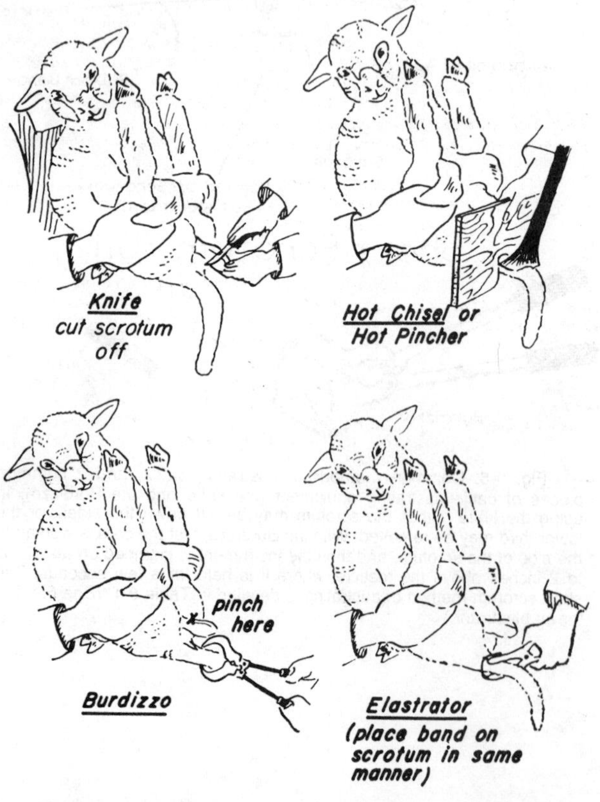

Fig. 8-8. Methods of docking and castrating lambs. (Drawings by R. F. Johnson)

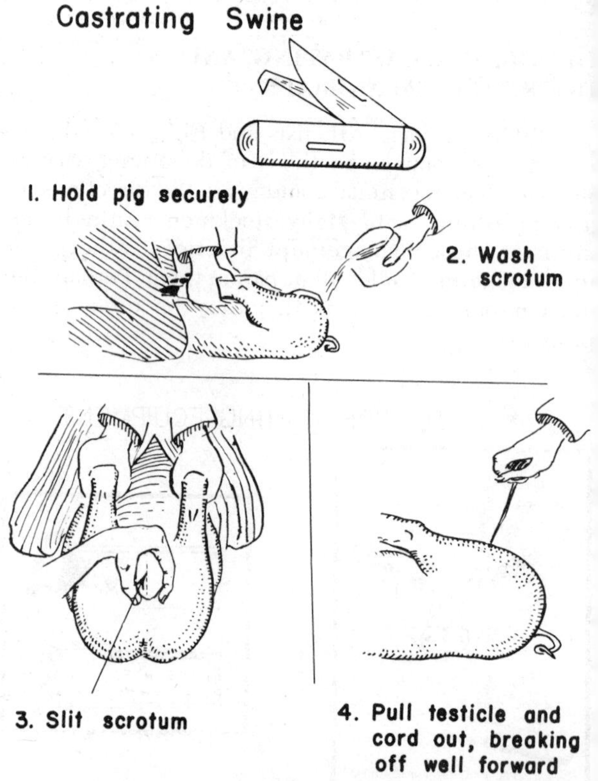

Fig. 8-9. Swine are castrated with a knife, by following the four steps herewith illustrated. (Drawing by R. F. Johnson)

TABLE 8-1

DEHORNING, CASTRATING, AND DOCKING GUIDE

Class of Animal	Skill	Method	Directions	Remarks
Cattle	**Dehorning**	Chemical (dehorning pastes or sticks)	Use caustic potash (potassium hydroxide) or caustic soda (sodium hydroxide) in either paste, stick, or lacquer base form. (The accompanying use of vaseline is not necessary with the lacquer base.) Apply when the calf is 3 to 10 days old. Clip or shear the hair from around the buttons, and surround the area with a ring of heavy grease or vaseline to protect the eyes against the chemical. Rub the chemical over the button until blood appears, protecting the hands while doing so.	Regardless of the method used, dehorning should be done at as young an age as possible, thus lessening weight loss, work, needed equipment, and bruising. About 2 wk are needed following dehorning for yearling animals to regain their original weight. With 2-yr-old or older cattle, the horns are usually tipped. The use of chemicals should be limited to small herds kept under supervision. Following the application of a dehorning paste or stick, keep calves away from their dams for a few hours and out of rain for 24 hr.
		Saws and clippers	Confine or restrain animal to be dehorned in a suitable chute, pinch gate, squeeze pen, or cattle stock. Calves may be handled by throwing, by snubbing them to a fence post, or by tying one side of the body against a strong fence or solid wall. Whatever the instrument used (saws or clippers), remove the horn with about ¼ to ½ in. of the skin around its base.	Clippers are satisfactory for removing the horns of younger cattle, but the hard, brittle horns of mature cattle can best be removed with a saw. Saws and various kinds of shears and clippers are the most widely used method of dehorning in the range country. Electrically operated saws are now available. Less insect trouble is encountered when dehorning is done in the early spring or late fall. If the operation is performed in fly season, apply a fly repellent to the wound.
		Hot iron	The hot-iron method of dehorning consists of the application of a specially designed hot iron to the horns of young calves. Where electricity is available, the electric hot iron may be used; it keeps an even temperature, without getting too hot or too cold.	The hot-iron system of dehorning is bloodless and may be used at any time of the year, but it can be used on young calves only.
		Dehorning spoon and dehorning tube	The steps and directions for using the dehorning tube are as follows: 1. Restrain the calf. 2. Select a tube of proper size to fit over the base of the horn and about ¼ in. of skin all the way around. 3. Place the cutting edge straight down over the horn and then push and twist until the skin has been cut through, making a cut ⅛ to ⅜ in. deep. 4. Hold the tube at about a 45° angle and rapidly turn and shove the cutting edge until the button comes off.	The dehorning spoon (or gouge) is a small instrument with which the horns of young calves can be gouged out. The dehorning tube is a newer instrument than the spoon, and is faster, less tiresome to use, and more certain to avoid regrowth. Either instrument can be used on calves up to 60 days of age. Dehorning tubes come in 4 sizes, varying from ¾ to 1⅛ in. in diameter. Practice cleanliness, and disinfect instruments (except hot irons) between animals in order to lessen infection and disease.
		Elastrator	Use on cattle when horns are 2½ to 6 in. long. Stretch the ring with the elastrator instrument and place over the horn well down into the hairline. Smaller horns drop off in 3 to 6 wk; larger horns may take 2 mo.	The elastrator is an instrument for use in stretching a specially made rubber ring, which may be used in dehorning cattle, as well as in castrating cattle and sheep. Some cattlemen report that they have obtained good results when using the elastrator for dehorning; others report that they have been disappointed. The use of the elastrator is probably the least desirable of the methods herein listed.

(Continued)

TABLE 8-1 (Continued)

Class of Animal	Skill	Method	Directions	Remarks
	Castrating	Slitting scrotum down the sides	Pull one testicle down at a time and hold it firmly to the outside so that the skin of the scrotum is tight over the testicle. With a sharp knife, make an incision on the outside of the scrotum next to the leg. It is important that the incision extend well down to the end of the scrotum to allow for proper drainage and that it extend through both the scrotum and membrane. If desired, the membrane need not be slit; simply remove it along with the testicle. Removal of all, or a substantial portion, of the membrane alleviates the possibility of it collecting blood and forming a clot. Remove the testicles by pulling them out. In older cattle, excessive bleeding may be prevented through severing the partially withdrawn cord by scraping with a knife or by clamping with an emasculator.	Castration is best done when calves are 4 to 1 wk of age. Young animals are usually thrown t be castrated, whereas animals 8 mo of age o older may be more easily operated on in standing position. It is best to perform the operation in the earl spring or late fall to avoid infestation from flies otherwise, a repellent should be applied to th wound. Keep cutting tools and hands clean an sterilized. Where screwworms exist, a fly repellent shoul be applied to the wound and the animal shoul be kept under close observation until th wound has healed over; or a bloodless metho of castration (such as the Burdizzo) should b used. Currently, there is a trend toward castrating a slightly older ages than formerly, primarily (1 to take advantage of the higher gains an greater efficiency of bulls, and (2) to lessen th hazard of urinary calculi.
		Removal of lower end of scrotum	Remove approximately the lower third of the scrotum, exposing the testicles from below. Slit the membrane covering each testicle. If desired, the membrane need not be slit; simply remove it along with the testicle. Removal of all, or a substantial portion, of the membrane alleviates the possibility of it collecting blood and forming a clot. Remove the testicles by pulling them out. Cords may be pulled in calves up to 3 or 4 mo of age; emasculators may be used on older calves and bulls.	Except in small calves, post castration infectio has been associated with castration by removal of the lower end of the scrotum, due to the bottom of the scrotum curling up and stopping drainage. The lateral sides of remaining scrotum may be slit up to the body to help drainage.
		Burdizzo pincers	Throw the animal. Work the cord to the side of the scrotum, and then clamp the Burdizzo on about 1¾ to 2 in. above the testicle, where it is held for a few seconds. Then repeat this operation on the same cord at a location about ¼ in. removed from the first one. Repeat the same procedure on the other testicle.	Burdizzo pincers (named after their inventor, Dr. Burdizzo, and manufactured in Italy) make a "bloodless castration." In using the Burdizzo, it is important that the cord not slip out, that only one cord be clamped at a time, and that there be no interference with the circulation of the blood through the central portion of the scrotum. This method of castration is satisfactory if done properly and by an experienced operator.
		"Russian method"	Make incision and remove spermatozoa-producing tissue of the testicle, but leave intact the sheathing layer that produces testosterone—a growth hormone.	This method of castration, which originated in the U.S.S.R., is now being evaluated in the U.S. The Russians claim that animals castrated in this manner gain like bulls, but have carcasses like steers.
		Elastrator rings	The elastrator works best on young calves under 2 mo of age. Hold the calf in either a sitting or lying position. Press both testicles through the ring and to the lower end of the scrotum, and then release the rubber ring.	The elastrator (developed in New Zealand) is an instrument for use in stretching a specially made rubber ring, which may be placed over the scrotum to castrate young calves.
		Short scrotum bulls	The scrotum is shortened by bringing it through a distended rubber band with an elastrator when the calf is 1 to 3 mo of age. Before the band is released, the testicles are moved near the abdominal wall. The scrotum below the rubber band sloughs off after 3 or 4 wk. Shortening the scrotum requires considerably less time than castration—it can be done in 15 to 30 seconds; and there is no weight loss. As a result of the shortened scrotum, the testicles lie close to the abdominal wall and the animal is sterile. The testicles then develop to about half the weight of those from fertile bulls of the same age and weight.	This patented method of rendering intact males infertile, known as "More-Lean Beef," was developed by the New Mexico Station. A short scrotum animal is really a pseudocryptorchid. The short scrotum treatment does not change either the temperament or the urge of the animal; hence, there will be riding if sexes are not kept separate. The rate and efficiency of gains of short scrotum bulls and intact bulls are about the same, and the carcasses of short scrotum bulls are leaner than the carcasses of steers.

(Continued)

TABLE 8-1 (Continued)

Class of Animal	Skill	Method	Directions	Remarks
heep	**Docking**	Knife or shears	Press the skin toward the body before cutting, leaving loose skin above the cut which will close over the wound. With small flocks, some sheepmen make a practice of tying a string or placing a rubber band around the tail prior to cutting, preventing a loss of blood in this manner. If this is done, the string or band should be removed 3 or 4 hr later. Sever the tail at the place desired.	All lambs should be docked when they are 7 to 14 days of age. Strong lambs may be docked and castrated at the same time, with castrating being done first. Where cutting instruments are used, keep the hands clean and the instruments clean and disinfected. The lamb is usually held with its back to the assistant, who grasps the hind and front legs of the same side in each hand. Sever the tail about an inch from the body as measured on the underside of the tail.
		Hot iron (pincers or chisels)	Do not heat instruments beyond a very dull red color. Protect the lamb's buttocks by placing the tail in a slot in the end of a board or by putting it through a hole in a board. Sever the tail rather quickly, avoiding any more burning than necessary to prevent bleeding.	The use of hot instruments results in less loss of blood and less danger of infection than the use of a knife, but the wound heals more slowly.
		Emasculator	Close the emasculator over the tail at the point at which it is to be severed.	The emasculator crushes a part of the tissue while cutting, thus lessening the loss of blood.
		Burdizzo pincers	Close the jaws over the tail at the point at which it is to be severed. Cut the tail off inside the closed jaws, after the Burdizzo jaws have been closed.	Burdizzo pincers may be used for castrating as well as for docking. This is a bloodless method.
		Elastrator	Draw the tail through the ring, and then release the rubber ring at the point where it is desired to sever it. Where there are scouring or unsanitary conditions, it may be advisable to cut off the tail of the lamb after the ring has been applied.	The elastrator is an instrument for use in stretching a specially made rubber ring, which may be placed around the tail to dock young lambs. The elastrator is a bloodless method of docking. The rubber band cuts off the blood supply, and atrophy follows.
	Castrating	Knife	Grasp the tip of the scrotum, and hold it tight while cutting off the lower third. Draw out the exposed testicles together with the surrounding membranes with either the hands or teeth.	All male lambs not to be left as rams should be castrated when they are 7 to 14 days of age. Strong lambs may be castrated and docked at the same time, with castrating being done first. Where cutting instruments are used, keep the hands clean and the instruments clean and disinfected. The lamb is usually held with his back to the assistant, who grasps the hind and front legs of the same side in each hand.
		Burdizzo	Work the cord to the side of the scrotum, and then clamp the Burdizzo on above the testicle, where it is held for a few seconds. Repeat the same procedure on the other testicle. Complete atrophy of the testicles follows in about 6 wk.	This is a bloodless method in which the testicles are made functionless through destroying their channels of nourishment.
		Elastrator	Press both testicles through the ring and to the lower end of the scrotum and then release the rubber ring.	The elastrator is an instrument for use in stretching a specially made rubber ring, which may be placed around the scrotum to castrate young lambs. The elastrator is a bloodless method of castrating.
		Short scrotum rams	The technique is the same as for bulls; hence, see "short scrotum bulls."	The best time to shorten the scrotum is when lambs are 1 to 3 mo of age.

(Continued)

TABLE 8-1 (Continued)

Class of Animal	Skill	Method	Directions	Remarks
Swine	**Castrating**	Knife	Restrain or hold the animal in a manner in keeping with its age and size and the number of helpers available. A young pig is generally either (1) suspended by its hind legs with the back toward the helper (the helper also clamps his knees against the pig's ribs, near the shoulders), or (2) held on its back on the top of a table (this requires either (a) a castration crate, or rack, or (b) two helpers—one grasping the front legs and the other the rear legs). Large boars are usually snared around the upper jaw and behind the tusks, with the free end of the snare tied to a post; then further restraint is applied by either tying all four legs or by hoisting the hind legs, with the animal castrated in either a standing or lying position. Wash the hands thoroughly with soap and water and rinse with a good disinfectant. If the scrotum is dirty, wash it with soapy water, using a coarse fiber brush. After washing, disinfect the area. Also, disinfect the knife before and between operations. With a sharp knife, slit the scrotum on eacn side (the one-incision method—with the cut made directly between the testicles—is satisfactory, and is preferred by some) as each testicle is presse outward. Extend both cuts well down to allow for proper drainage, and cut deep enough to exter through the scrotum and membrane. (If desired, the membrane need not be slit on young boar simply remove it along with the testicle. However, it is desirable to slit the membrane on old boar rather than break down fibrous attachments of membrane to scrotum.) Pull the cord out (with tension directed backward; otherwise, the inguinal ring may be torn and evi ceration produced)—or break the cord off well forward. Use the emasculator on old boars. In f season, apply an insect repellent to the wound.	Male pigs not intended for breeding purpos should be castrated while they are still suc ling their dams, and far enough in advance weaning to allow healing before being sep rated from the dam (the incision usually hea in 2 to 3 wk). The operation should not be do at the same time that pigs are vaccinate Generally this means that castration should b done within the first 4 wk (some castra 5-day-old pigs); pigs that are weaned at 4 w of age or earlier should not be castrated with 1 wk of the time of weaning. In preparation f the operation, pigs should be kept off feed short time. Boars that are no longer useful in a breedir program may be castrated to remove the bo odor before marketing, which operation known as stagging. By the time the castratic wound has healed (in 3 to 4 wk), the odor us ally disappears enough to allow the stag to b marketed. Pigs with undescended testicles or ruptur (scrotal hernias) should be operated on by veterinarian.

Castrating Horses

Regardless of age or time, castration of colts is best performed by an experienced veterinarian. A colt may be castrated when only a few days old, but most horsemen prefer to delay the operation until the animal is about one year of age. Although there is less real danger to the animal and much less setback with early altering, the practice results in imperfect development of the fore parts. On the other hand, leaving the colt entire for a time will result in more muscular, bold features and better carriage of the fore parts. Therefore, weather and management conditions permitting, the time of altering should be determined by the development of the individual. Thus, underdeveloped colts may be left entire six months or even a year longer than overdeveloped ones. Breeders of Thoroughbred horses usually prefer to have the horses first race as an entire.

There is less danger of infection if colts are castrated in the spring of the year soon after they are turned out on a clean pasture. Naturally, this should be done sufficiently early to avoid hot weather and fly time.

Spaying Heifers

In females, the operation corresponding to castra tion is known as spaying. Under most conditions, de sexing of the heifers is not recommended because: (1 the operation is more complicated and difficult, re quiring a very experienced man; (2) spaying is at tended with more danger than castration; (3) it lower both rate and efficiency of gains; (4) it eliminates th heifers for possible replacement purposes or sale a breeding stock; and (5) experiments and practical op erations with spayed heifers have generally show that the selling price obtained is not sufficientl higher to compensate for the lower and less efficier gains plus the attendant risk of the operation.

A summary of 11 experiments in which spaye heifers and intact, open heifers were compared re vealed that spaying made for 9.9 percent slower rat of gain and increased the feed required per 10 pound gain by 8.5 percent.[2]

Spaying does prevent the possibility of heifer

[2]*Montana Farmer-Stockman*, summary prepared by R. A. Be lows, U.S. Range Livestock Experiment Station, Miles City, Mon Oct. 2, 1975, p. 38.

ecoming pregnant and eliminates the necessity of eparating heifers from bulls or steers.

EDDING ANIMALS

Bedding or litter is used primarily for the purposes of keeping animals clean and comfortable. But edding has the following added values from the tandpoint of the manure:

1. It soaks up the urine which contains about ne-half the total plant food of manure.
2. It makes manure easier to handle.
3. It absorbs plant nutrients, fixing both ammonia nd potash in relatively insoluble forms that protect hem against losses by leaching. This characteristic of edding is especially important in peat moss, but of ittle significance with sawdust and shavings.

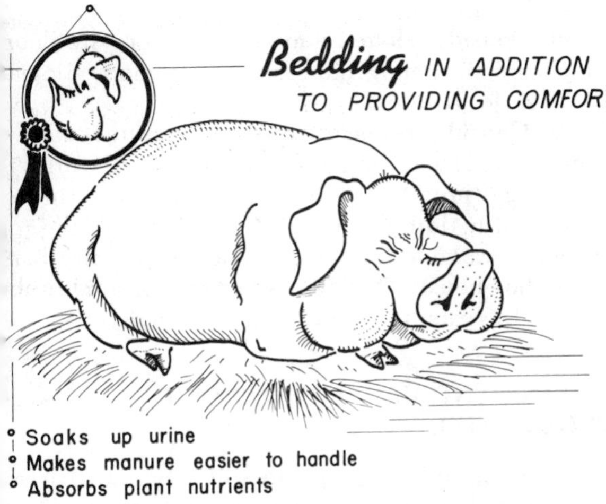
Bedding IN ADDITION
TO PROVIDING COMFORT,
Soaks up urine
Makes manure easier to handle
Absorbs plant nutrients

Fig. 8-10. Bedding is used primarily for the purposes of keeping animals clean and comfortable, but it also soaks up urine, makes manure easier to handle, and absorbs plant nutrients. (Drawing by R. F. Johnson)

Kind and Amount of Bedding

The kind of bedding material selected should be determined primarily by (1) availability and price, (2) absorptive capacity, (3) cleanness (this excludes dirt or dust which might cause odors or stain livestock), (4) ease of handling, (5) ease of cleanup and disposal, (6) nonirritability from dust or components causing allergies, (7) texture or size (for example, material that will not get into the wool of sheep), and (8) fertility value or plant nutrient content. In addition, a desirable bedding should not be excessively coarse, and should remain well in place and not be too readily kicked aside.

Table 8-2 lists some common bedding materials and gives the average water absorptive capacity of each. In addition to these bedding materials, many other products can be and are successfully used for this purpose, including leaves of many kinds, tobacco stalks, buckwheat hulls, processed manure (made by separating solid fibers from the liquid and water-soluble material in animal wastes), and shredded paper.

TABLE 8-2
WATER ABSORPTION OF BEDDING MATERIALS

Material	Lb of Water Absorbed/ Cwt of Air-Dry Bedding
Barley straw	210
Cocoa shells	270
Corn stover (shredded)	250
Corncobs (crushed or ground)	210
Cottonseed hulls	250
Flax straw	260
Hay (mature, chopped)	300
Leaves (broadleaf)	200
(pine needles)	100
Oat hulls	200
Oat straw (long)	280
(chopped)	375
Peanut hulls	250
Peat moss	1,000
Rye straw	210
Sand	25
Sawdust (top-quality pine)	250
(run-of-the-mill hardwood)	150
Sugar cane bagasse	220
Tree bark (dry, fine)	250
(from tanneries)	400
Vermiculite[1]	350
Wheat straw (long)	220
(chopped)	295
Wood chips (top-quality pine)	300
(run-of-the-mill hardwood)	150
Wood shavings (top-quality pine)	200
(run-of-the-mill hardwood)	150

[1]This is a micralike mineral mined chiefly in South Carolina and Montana.

The availability and price per ton of various bedding materials vary from area to area, and from year to year. Thus, in the New England states shavings and sawdust are available, whereas other forms of bedding are scarce, and straws are more plentiful in the central and western states.

Table 8-2 shows that bedding materials differ considerably in their relative capacities to absorb liquid. Other facts of importance relative to certain bedding materials and bedding uses are:

1. **Wood products (sawdust, shavings, tree bark, chips, etc.)**—The suspicion that wood products will hurt the land is rather widespread but unfounded. It is true that shavings and sawdust decompose slowly, but this process can be expedited by the addition of nitrogen fertilizers. Also, when plowed under, they increase soil acidity, but the change is both small and temporary.

Softwood (on a weight basis) is about twice as absorptive as hardwood, and green wood has only 50 percent the absorptive capacity of dried wood.

Wood wastes should not be used for bedding sheep, as the fine material tends to get embedded in the fleece.

2. **Cut straw**—Cut straw will absorb more liquid than long straw; cut oats or wheat straw will take up about 25 percent more water than long straw from comparable material. But there are disadvantages to chopping; chopped straws may be dusty, and they are not suited for bedding sheep because the fine particles get into the fleece.

3. **Bedding for farrowing sows**—Farrowing sows should be bedded lightly with chopped or short material that will not interfere with the movement of the pigs.

4. **Fertility value**—From the standpoint of the value of plant food nutrients per ton of air dry material, peat moss is the most valuable bedding, and wood products the least valuable.

The minimum desirable amount of bedding to use is the amount necessary to absorb completely the liquids in manure. Some helpful guides to the end that this may be accomplished follow:

a. Per 24-hours confinement, the minimum daily bedding requirements, based on uncut wheat or oats straw, of different kinds of livestock are as follows: cow, 9 lb; steers, 7 to 10 lb; sheep, 1 lb; hogs, ½ to 1 lb; and horses, 10 to 15 lb. With other bedding materials these quantities will vary according to their respective absorptive capacities (see Table 8-2, page 735). Also, more than these minimum quantities of bedding may be desirable where cleanliness and comfort of the animal are important. Comfortable animals lie down more and utilize a higher proportion of the energy of the feed for productive purposes (cattle and sheep require 9% less energy when lying down than when standing).

b. Under average conditions, about 500 pounds of bedding are used for each ton of excrement.

Reducing Bedding Needs

In most areas, bedding materials are becoming scarcer and higher in price, primarily because (1) geneticists are breeding plants with shorter straws and stalks, (2) of more competitive and remunerative uses for some of the materials, and (3) the current trend toward more confinement rearing of livestock requires more bedding.

Stockmen may reduce needs and costs as follows:

1. **Collect liquid excrement separately**—Where the liquid excrement is collected separately in a cistern or tank, less bedding is required than where the liquid and solid excrement are kept together.

2. **Chop bedding**—Chopped straw, waste hay, fodder, or cobs will go further and do a better job of keeping animals dry than long materials. Chopped straw, for example, will soak up approximately 25 percent more moisture than long straw.

3. **Use deep-bedding system**—For wintering c tle and sheep, for loose housing of dairy cows, a under certain other conditions, a deep-litter system letting the bedding build up beneath; adding a lig sprinkling of fresh bedding on top at intervals—w keep the animals warm and dry, and save in bedding

4. **Ventilate quarters properly**—Proper ventil tion lowers the humidity and keeps the bedding dr

5. **Feed and water away from sleeping qua ters**—Animals should be fed and watered in areas r moved from their sleeping quarters. With this type arrangement, they defecate less in the sleeping area

6. **Provide exercise area**—Where possible an practical, provide for winter exercise in well-draine dry pastures or corrals, without confining animals t or near their sleeping quarters.

7. **Mound cattle lots**—In open lots, mounds co ered with bedding materials provide drainage an lessen bedding needs.

8. **Consider slotted floors**—Slotted or wire floor which are becoming increasingly common, alleviat the need for bedding.

9. **Consider rubber mats**—Rubber (either solid c foam rubber) bedding replacers (or more correctl speaking, they are bedding-savers, for a limite amount of bedding is usually sprinkled over the to are now available for use in stanchioned dairy barn and in hog houses. The life expectancy of solid rubbe mats is 12 years; for foam rubber, about 4 years.

MAKING AND KEEPING THE SOIL PRODUCTIVE

Making and keeping the soil productive is th very foundation of a successful agriculture, of nationa

Fig. 8-11. In their own way, the American Indians made and kept the soil productive. Maize (corn) was often fertilized by placing a fish in each hill. (Drawing by R. F. Johnson)

prosperity, It has been well said that good soils, good farms and ranches, and good living go hand in hand.

Barnyard Manure

The term manure refers to a mixture of animal excrements (consisting of undigested feeds plus certain body wastes) and bedding.

Animals provide manure for the fields, a fact which was often forgotten during the era when chemical fertilizers were relatively abundant and cheap. One ton of average manure contains 10 lb of nitrogen (N), 5 lb of phosphoric acid (P_2O_5), and 10 lb of potassium (K_2O). At 1983 prices (per pound: N = 25¢, P_2O_5 = 20¢, and K_2O = 10¢), it's worth $4.50 per ton.

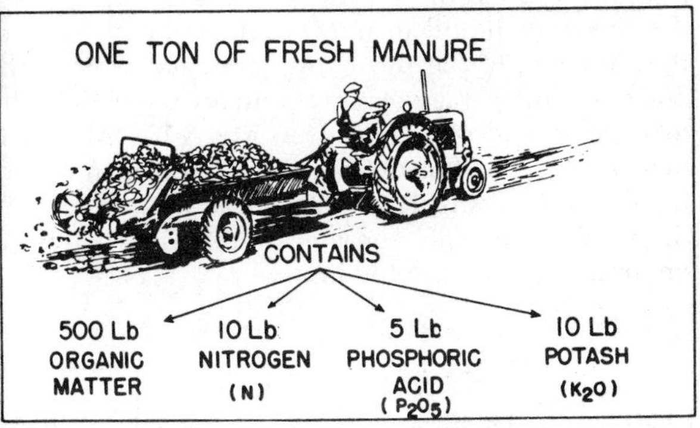

Fig. 8-12. The contents of one ton of average fresh manure.

The energy crisis prompted concern that farmers would not have sufficient chemical fertilizers at reasonable prices in the years ahead. Since nitrogenous fertilizers are oil- and petroleum-based, there is cause for concern. As a result, a growing number of American farmers are returning to organic farming; they are using more manure—the unwanted barnyard centerpiece of the past 40 years, and they are discovering that they are just as good reapers of the land and far better stewards of the soil.

AMOUNT, COMPOSITION, AND VALUE OF MANURE PRODUCED

The quantity, composition, and value of manure produced vary according to species, weight, kind and amount of feed, and kind and amount of bedding. The author's computations are on a fresh manure (exclusive of bedding) basis. Table 8-3 presents data by species per 1,000 pounds liveweight, whereas Table 8-4 gives yearly tonnage and value.

The data in Table 8-3 and Fig. 8-13 are based on animals confined to stalls the year around. Actually, the manure recovered and available to spread where desired is considerably less than indicated because (1) animals are kept on pasture and along roads and

TABLE 8-3

QUANTITY, COMPOSITION, AND VALUE OF FRESH MANURE (FREE OF BEDDING) EXCRETED BY 1,000 POUNDS LIVEWEIGHT OF VARIOUS KINDS OF FARM ANIMALS

(1) Animal	(2) Tons Excreted/Year/1,000 Lb Liveweight[1]	Composition and Value of Manure on a Tonnage Basis[2]						
		(3) Excrement	(4) Lb/ton[3]	(5) Water	(6) N	(7) P_2O_5[4]	(8) K_2O[4]	(9) Value/Ton[5]
				(%)	*(lb)*	*(lb)*	*(lb)*	*($)*
Cow (beef or dairy)	12	Liquid	600	79	11.2	4.6	12.0	4.92
		Solid	1,400					
		Total	2,000					
Steer (finishing cattle)	8.5	Liquid	600	80	14.0	9.2	10.8	6.42
		Solid	1,400					
		Total	2,000					
Sheep	6	Liquid	660	65	28.0	9.6	24.0	11.32
		Solid	1,340					
		Total	2,000					
Swine	16	Liquid	800	75	10.0	6.4	9.1	4.69
		Solid	1,200					
		Total	2,000					
Horse	8	Liquid	400	60	13.8	4.6	14.4	5.81
		Solid	1,600					
		Total	2,000					
Poultry	4.5	Total	2,000	54	31.2	18.4	8.4	12.32

[1]*Manure Is Worth Money—It Deserves Good Care,* University of Illinois Circ. 595, 1953, p. 4.
[2]Columns 5, 6, 7, and 8 from *Farm Manures,* University of Kentucky Circ. 593, 1964, p. 5, Table 2.
[3]From *Reference Material for 1951 Saddle and Sirloin Essay Contest,* compiled by M. E. Ensminger, p. 43: data from *Fertilizers and Crop Production,* by Van Slyke, published by Orange Judd Publishing Co.
[4]P_2O_5 can be converted to phosphorus (P) by dividing the figure given above by 2.29, and K_2 can be converted to potassium (K) by dividing by 1.2.
[5]Calculated on the assumption that nitrogen (N) retails at 25¢, P_2O_5 at 20¢, and K_2O at 10¢ per pound in commercial fertilizers.

lanes much of the year, where the manure is dropped, and (2) losses in weight often run as high as 60 percent when manure is exposed to the weather for a considerable time.

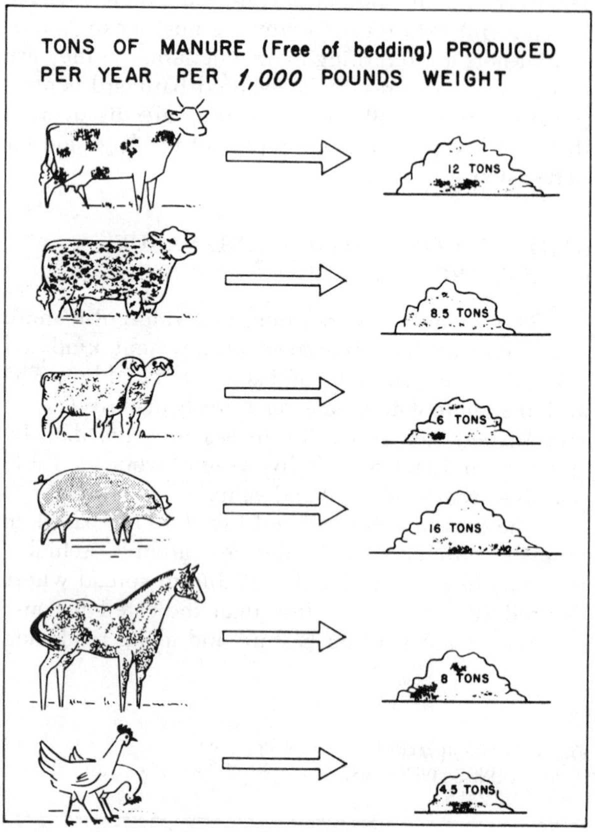

Fig. 8-13. On the average, each class of stall-confined animals produces per year per 1,000 pounds weight the tonnages shown above. (Drawing by R. F. Johnson)

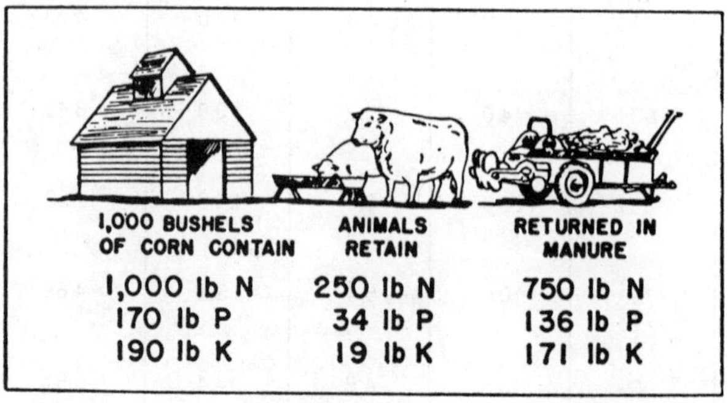

Fig. 8-14. Animals retain about 20 percent of the nutrients in feed; the rest is excreted in manure. (Drawing by Steve Allured)

As shown in Fig. 8-14, about 75% of the nitrogen, 80% of the phosphorus, and 85% of the potassium contained in animal feeds are returned as manure. In addition, about 40% of the organic matter in feeds is excreted as manure. As a rule of thumb, it is commonly estimated that 80% of the total nutrients in feeds are excreted by animals as manure. A ton of fresh barnyard manure has approximately the composition shown in Fig. 8-12.

Naturally, it follows that the manure from well-fed animals is higher in nutrients and worth more than that from poorly fed ones. For example, steer manure produced by fattening cattle liberally fed on nutritious concentrates is more valuable than that produced from cattle wintered on hay.

The urine makes up 20% of the total weight of the excrement of horses, and 40% of that of hogs; these figures represent the 2 extremes in farm animals. Yet the urine, or liquid manure, contains nearly 50% of the nitrogen, 6% of the phosphorus and 60% of the potassium of average manure; roughly one-half of the total plant food of manure (see Fig. 8-15). Also, it is noteworthy that the nutrients in liquid manure are more readily available to plants than the nutrients in the solid excrement. These are the reasons why it is important to conserve the urine.

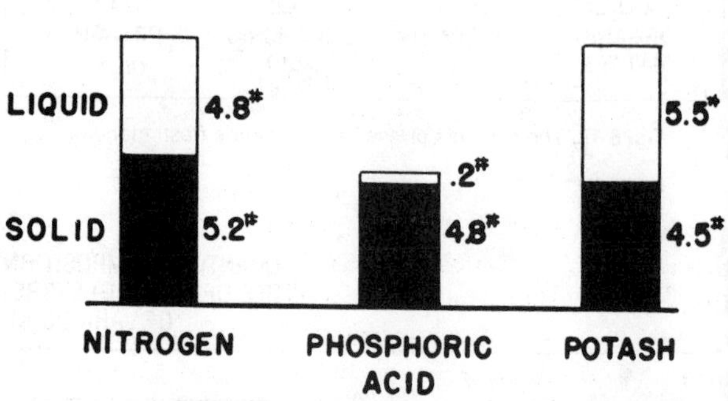

Fig. 8-15. Distribution of plant nutrients between liquid and solid portions of a ton of average farm manure. As noted, the urine contains about half the fertility value of manure. (Drawing by Steve Allured)

The actual monetary value of manure can and should be based on (1) increased crop yields, and (2) equivalent cost of a like amount of commercial fertilizer. Numerous experiments and practical observations have shown the measurable monetary value of manure in increased crop yield. Table 8-3, page 737 (footnote 5), gives the equivalent cost of a like amount of commercial fertilizer.

Currently, we are producing manure (exclusive of bedding) at the rate of 1.5 billion tons annually (see Table 8-4). That is sufficient manure to add nearly 1 ton each year to every acre of the total land area (1.9 billion acres) of the United States.

Based on equivalent fertilizer prices (see Table 8-3, right-hand column) and livestock numbers (Table 8-4), the yearly manure crop is worth $8 billion. That is a potential annual income of $3,333 for each of the nation's 2.4 million farms.

TABLE 8-4

TONNAGE AND VALUE OF MANURE (EXCLUSIVE OF BEDDING) EXCRETED IN 1981 BY U.S. LIVESTOCK[1]

Class of Livestock	Number of Animals on Farms[2]	Average Liveweight	Tons Manure Excreted/Year/ 1,000 Lb Liveweight[3]	Total Manure Production	Total Value of Manure[4]
		(lb)	(tons)	(tons)	($)
Cattle (beef and dairy; including steers)	115,013,000	900	11	1,138,628,700	5,602,053,200
Sheep	12,942,000	100	6	7,765,200	87,902,064
Swine	64,520,000	200	16	206,864,000	968,316,160
Chicken, layers	302,110,000	4.5	4.5	6,117,727	75,370,396
Broilers	4,149,200,000	3.5	4.5	65,349,900	805,110,760
Turkeys	164,871,000	22	4.5	16,322,229	201,089,860
Horses	8,300,000	1,000	8	66,400,000	385,784,000
				1,507,447,756	8,125,626,440

[1]In these computations, no provision was made for animals that died or were slaughtered during the year. Rather, it was assumed that their places were taken by younger animals, and that the population of each species was stable throughout the year.
[2]From USDA, Statistical Reporting Service; and assumed as average throughout the year.
[3]*Manure Is Worth Money—It Deserves Good Care,* University of Illinois Circ. 595, 1953, p. 4.
[4]Computed on the basis of the value per ton given in the right-hand column of Table 8-3. Cattle manure computed at value of $4.92/ton.

Of course, the value of manure cannot be measured alone in terms of increased crop yields and equivalent cost of a like amount of commercial fertilizer. It has additional value for the organic matter which it contains, which almost all soils need, and which farmers and ranchers cannot buy in a sack or tank.

Also, it is noteworthy that, due to the slower availability of its nitrogen and to its contribution to the soil humus, manure produces rather lasting benefits, which may continue for many years. Approximately ½ of the plant nutrients in manure are available to and effective upon the crops in the immediate cycle of the rotation to which the application is made. Of the unused remainder, about ½, in turn, is taken up by the crops in the second cycle of the rotation; ½ of the remainder in the third cycle, etc., etc. Likewise, the continuous use of manure through several rounds of a rotation builds up a backlog which brings additional benefits, and a measurable climb in yield levels.

Stockmen sometimes fail to recognize the value of this barnyard crop because (1) it is produced whether or not it is wanted, and (2) it is available without cost. Most of all, no one is selling it. Whoever heard of a traveling manure salesman?

HOW MUCH MANURE CAN BE APPLIED TO THE LAND

With today's heavy animal concentration in one location, the question is being asked: How much manure can be applied to the land without depressing crop yields, making for salt problems in the soil, making for nitrate problems in feed, contributing excess nitrate to groundwater or surface streams, or violating state regulations?

Based on earlier studies in mid-western United States, before the rise of commercial fertilizers, it would appear that one can apply from 5 to 20 tons of manure per acre, year after year, with benefit.

Heavier applications can be made, but probably should not be repeated every year. With rates higher than 20 tons per annum, there may be excess salt and nitrate buildup. Excess nitrate from manure can pollute streams or groundwater and result in toxic levels of nitrate in crops. Without doubt the maximum rate at which manure can be applied to the land will vary widely according to soil type, rainfall, and temperature.

State regulations differ in limiting the rate of manure application. Missouri draws the line at 30 tons per acre on pasture, and 40 tons per acre on cropland. Indiana limits manure application according to the amount of nitrogen applied, with the maximum limit set at 225 pounds per acre per year. Nebraska requires only one-half acre of land for liquid manure disposal per acre of feedlot, which appears to be the least acreage for manure disposal required by any state.

When a farmer has sufficient land, he should use rates of manure which supply only the nutrients needed by the crop rather than the maximum possible amounts suggested for pollution control.

MANURE USES OTHER THAN AS A FERTILIZER

Recycling manure as a livestock feed is the most promising of the nonfertilizer uses. Various processing methods are being employed; some are even feeding manure without processing. More and more feedlot manure will be either (1) incorporated in a grower ration, or (2) fed to range cattle during periods when range supplementation is beneficial, with the residues distributed over grazing areas where they would have fertilizing value. Further experimentation and Food and Drug Administration approval will be required before the use of manure feeds becomes widespread, but some researchers predict that eventually wastes may supply up to 20 percent of the nation's livestock feed, thereby freeing an equivalent amount of grain for human consumption.

Manure may also serve as a source of energy, which, of course, is not new. The pioneers burned dried bison dung, which they dubbed "buffalo chips," to heat their sod shanties. In this century, methane from manure has been used for power in European farm hamlets when natural gas was hard to get. While the costs of constructing plants to produce energy from manure on a large-scale basis may be high, some energy specialists feel that a prolonged fuel shortage will make such plants economical. India now has about 10,000 anaerobic digestion plants in operation. In 1974, Monfort of Colorado, operator of 2 of the world's largest cattle feedlots, announced that it had granted an option for the construction of a facility to produce 4,000,000 cubic feet of methane gas per day from the 225,000 tons of dry weight manure produced yearly in one of its feedlots. The announcement further stated that the process (anaerobic digestion) reduces the odor associated with manure handling and improves the residue as a fertilizer. The methane, of course, will be usable like natural gas. There is nothing new or mysterious about this process. Sanitary engineers have long known that a family of bacteria produces methane when they ferment organic material under strictly anaerobic conditions. (Grandad called it swamp gas; his city cousin called it sewer gas.) However, it should be added that, due to capital and technical resources needed, for some time to come, the production of methane by anaerobic digestion will likely be limited to municipal or corporate industries. If all animal manure were converted to energy, it has been estimated that it could produce energy equal to 10 percent of the petroleum requirements or 12½ percent of our natural gas requirements.

One researcher has come up with a way in which to combine manure with broken glass to produce bricks, decorative and roofing tiles, wall core material, and garden stones.

Commercial Fertilizers[3]

Valuable as it is, the average livestock farm or ranch simply doesn't produce enough manure to maintain the fertility of its soil. This is so because (1) animals take out about one-fifth of the feed nutrients,[4] (2) even with the most approved methods of handli manure, there are certain additional losses before gets into the soil, (3) barnyard manure is low in pho phorus (but the availability coefficient of the manu is high compared to that of nitrogen), (4) it is not a ways profitable or good business to feed to livesto all of every crop produced, and (5) it is seldom prac cal to use enough purchased feeds to make up for a the plant food deficiencies inevitable in the use homegrown feeds only, due to the forces indicated points 1 and 2 above. In brief, few stockmen b enough feed or save enough manure to maintain th original fertility of the soil.

Thus, the addition of commercial fertilizer green manure crops, and crop residues (such as stra chaff, and stalks), is necessary if the fertility of the so is to be maintained. Every stockman should know few important fertilizer facts so that he can use thes products more profitably.

WHAT'S A FERTILIZER GRADE?

The grade of a fertilizer is the nutrient percen age of the product by weight.

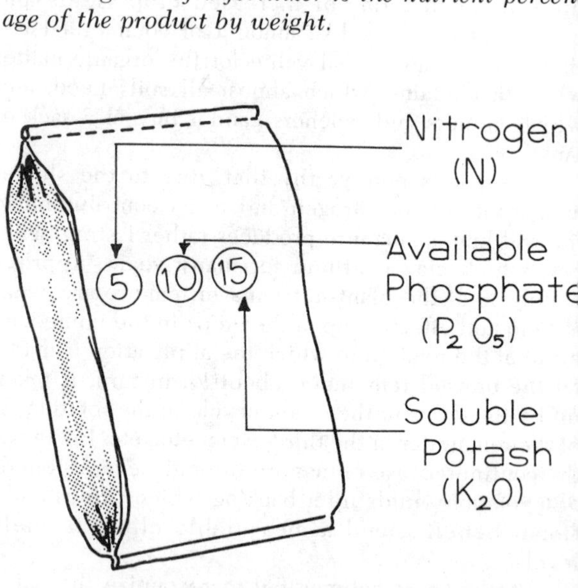

Fig. 8-16. The grade, or analysis, of a fertilizer is always designate by three numbers, which are always listed from left to right.

[3]This section was authoritatively reviewed by Dr. H. M Reisenauer, Professor of Soil Science, University of California, Davi Mr. William C. White, The Fertilizer Institute, Washington, D.C and Mr. Sidney H. Bierly, General Manager, California Fertilizer A sociation, Sacramento, Calif.

[4]The sale of a 1,000-lb steer removes fertility equivalent to 150 l of sodium nitrate (24 lb of N), 100 lb of superphosphate (20 lb P_2O_5 and 50 lb of limestone (47.5 lb $CaCo_3$). (Little potash is removed b animals; most of it is voided in the feces.) In the sale of 10,000 lb c milk—the annual production of a good cow—nitrogen and phosphori acid are removed in amounts equivalent to that found in 300 lb c ammonium sulfate and 200 lb of superphosphate.

Grade is expressed in a set of three numbers, always read from left to right: (1) percent of total nitrogen (N), (2) available phosphate (P_2O_5), and (3) soluble potash (K_2O). Thus, a 5-10-15 grade fertilizer contains 5% N, 10% P_2O_5, and 15% K_2O. The remaining 70% of the product consists of other elements, such as calcium, chlorine, and oxygen. If a nutrient is missing, it is represented by a zero; such as 45-0-0 for urea, 0-44-0 for triple superphosphate, 0-0-60 for potassium chloride (muriate), and 18-46-0 for diammonium phosphate.

HOW TO DETERMINE SOIL DEFICIENCIES

It is recognized that general information relative to the most common plant nutrient deficiencies of soil types or large areas is insufficient, for each farm and even each field presents an individual problem. Thus, in order that fertilizers and lime or gypsum may be used most effectively and profitably, farmers are in need of dependable methods for testing soils, and of other methods of determining plant food deficiencies. Some methods now employed to determine plant food deficiencies in soils are:

1. **Soil tests**—Laboratory soil tests are widely used to determine the supply of available nutrients. Essentially, soil testing is a chemical (or biological) procedure conducted in a laboratory under controlled conditions. For a reliable test, a representative sample must first be taken, for, regardless of how carefully the test itself is conducted, it is no better than the sample which was tested (for instructions on sampling, see the section entitled, "How to Take Soil Samples").

But a soil analysis alone is not enough. The person interpreting the laboratory results and making fertilizer recommendations should also consider (a) the kind of yield responses obtained from field applications of fertilizers in the area, and (b) past soil management practices and future plans. With this information at hand, an experienced person can make intelligent recommendations relative to levels of application of lime and fertilizer elements—nitrogen, phosphorus, potassium, and in some cases other elements.

Today, soil testing is big business; running from a few thousand samples per year in some states to more than a hundred thousand annually in others. Most states have a soil testing service; and, in addition, many of them have county laboratories. Some states do not make any charge for soil testing; others charge a nominal fee. Also, many commercial laboratories and fertilizer companies test soils.

2. **Plant hunger signs (deficiency symptoms)**—Any stockman can recognize when pigs or other animals are hungry. Plants, too, have ways of showing hunger—of showing deficiency symptoms when extreme shortages exist of certain essential elements. Also, characteristic plant symptoms indicate the presence of saline and alkali soils. But some experience is necessary for such recognition. Also, the visible hunger signs in plants may be revealed too late for effective treatment of current crops. The information obtained, however, may be used in preparing for the following season. The common plant hunger signs—color, leaf shapes, growth habits, etc., are shown in Table 8-5, page 744.

3. **Plant analysis**—The fertilizer needs of crops can also be determined through systematic sampling and chemical analysis of plants.

Even though a soil test shows adequate amounts of essential elements, at certain periods of growth the plant may not have a balanced nutrition. For example, wet, dry, cool, or hot weather may cause poor plant growth even when the soil is fertile. A plant analysis aids in diagnosis under the latter conditions. Of course, the very nature of plant analyses would indicate that they cannot be used in determining fertilizer needs at seeding time; but they can be used as guides in making fertilizer applications the following year.

4. **Biological tests**—Most of the early experiments designed to determine fertilizer needs of soils made use of a biological method, and these methods are still used. The field plot is the oldest, as well as the best known, of the biological methods, and this is followed by the greenhouse pot test.

a. **Field plot trials and demonstrations**—It is always desirable to test crop response to fertilizer in actual field tests. Usually such tests are made on a series of plots of equal size by treating the plots in different ways with various kinds and amounts of fertilizer. The benefits of soil treatments are then easily seen.

b. **Greenhouse pot tests**—Essentially these tests consist in filling a number of pots with soil material and in adding various fertilizers. The need for fertilizer is then indicated by the growth of the crop, by the amount of dry matter produced, and/or by analysis of the plant ash.

c. **Other biological tests**—Biological tests, utilizing either plants or microorganisms grown in incubators, may be used to evaluate available levels of several of the nutrient elements. For example, nitrification rate determinations may be used to evaluate the release of available nitrogen by soils.

5. **Portable soil-testing kits**—Many commercial soil-testing kits are now on the market, ranging in price from a few dollars for a simple pH unit to $100 and over for more elaborate apparatus for testing most of the essential elements. Most of these kits have merit and will give desirable results if used on certain

kinds of soils according to specific directions. However, all soil-testing kits have the following inherent weaknesses which must be recognized before they can be used successfully:

a. The difficulty of keeping the glassware clean.

b. The problem of contamination and deterioration of the reagents.

c. The hazard of properly interpreting the results in terms of lime and fertilizer needs.

How to Take Soil Samples[5]

If soil tests for lime and fertilizer needs are to be of value, the samples must be taken properly and the results interpreted correctly. Although the specific instructions relative to how to take such samples will vary somewhat with the laboratory running the test, the following general principles apply (also, see Fig. 8-17):

1. **Number of samples**—Where fields are uniform (from the standpoint of soils, topography, crop growth, and past treatment) one composite topsoil (plow depth) sample from every 10 acres is usually ample. Where fields are not uniform, however, more samples should be taken.

Never mix topsoil and subsoil samples.

2. **Surface litter**—Scrape away surface litter before sampling.

3. **Sampling equipment**—A spade, trowel, or soil auger may be used in taking the samples.

4. **Sampling procedure**—For each sample, take a uniform, but thin, slice down from the surface to the usual plow depth. With soils that are in pastures or lawns, the sample should be limited to the upper 2 to 4 inches. Take about one sample from each acre; then thoroughly mix these samples in a bucket or other container and take one composite sample therefrom.

5. **Dry sample**—Spread each composite sample out on a clean sheet of paper and let dry at room temperature.

6. **Pack**—Remove any stones or roots and pack about a pint of each sample in a soil sample carton, or other suitable container, and forward to the laboratory making the test. Wrap securely and address clearly.

7. **Added information**—Knowledge of past cropping and fertilizer history is important when making a fertilizer recommendation. An "information sheet" is usually provided by the chemical testing laboratory for use in submitting this information. Fill this out, place it in an envelope, stamp it, and secure it to the outside of the soil sample package.

[5]This section was authoritatively reviewed by Dr. A. R. Halvorson, who is in charge of the Soil Testing Laboratory, Washington State University, Pullman, Wash.

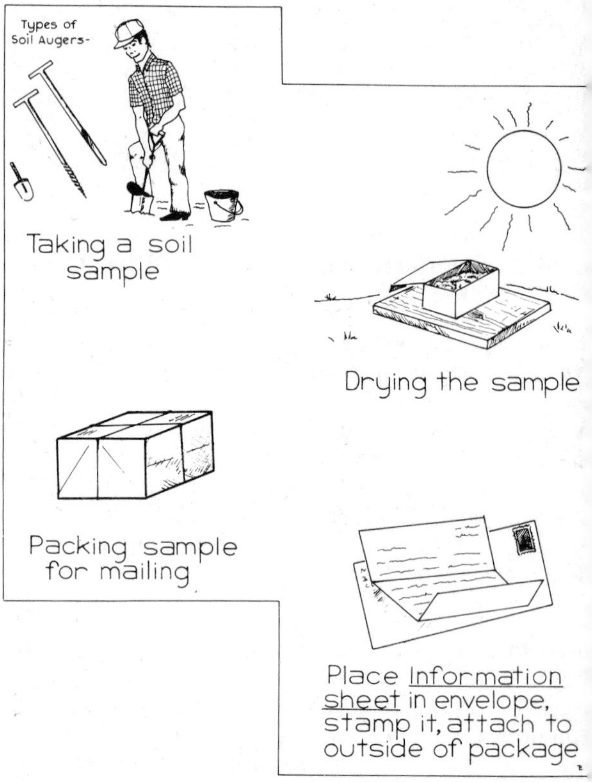

Fig. 8-17. Although some variations in instructions exist between laboratories, the soil sampler is generally admonished to pay particular attention to the four steps illustrated above. (Drawing by R. F. Johnson)

8. **Further information**—Further information on soil sampling and soil testing can be obtained from the county extension agent, Soil Conservation Service, or fertilizer dealer.

9. **Follow instructions**—After receiving your soil test results and recommendations, study them carefully and lime and fertilize in keeping therewith.

Soil Fertility Guide

A list of the essential plant nutrient elements, together with information relative to their respective functions, plant hunger signs, common sources, and other pertinent facts, is presented in Table 8-5, page 744, Soil Fertility Guide.

In addition to the chemical elements listed in Table 8-5, plants require carbon, hydrogen, and oxygen. These three come from the air and water; whereas the other elements must be obtained from the soil.

KIND AND AMOUNT OF COMMERCIAL FERTILIZER TO APPLY

It is impossible to give accurate directions relative to the kind and amount of fertilizer to apply to all soils because, in addition to the hundreds of different soil conditions, consideration should be given to (1) the previous crops, (2) the current crop rotation, (3) the farming system, (4) the price of crops, (5) the price of fertilizers, and (6) the amount and quality of manure and crop residue available.

The era of phosphate formation

Fig. 8-18. Primeval giants, such as rhinoceros (left) and mastodons (right), roamed Florida during the phosphate forming era. (Drawing by R. F. Johnson)

On some soils and under some conditions, "complete fertilizers"—those furnishing nitrogen, phosphate, and potash—will pay best; on others, a two-element fertilizer may be best; and, under still other conditions, a fertilizer containing only one plant nutrient may be best.

The amount of fertilizer to apply should be determined primarily by (1) the level of available nutrient as determined by soil test, (2) crop requirement for the nutrient, and (3) the amount of barnyard manure available. Fig. 8-19 shows that crop "appetites" do vary.[6] Also, it is generally recognized that the grain farmer will need to use about twice as much commercial fertilizer per acre as the livestock farmer who covers his soil with manure every three or four years.

Before purchasing and using fertilizers, the farmer is admonished to confer with his local county agent, vocational agriculture instructor, or fertilizer dealer.

HOW TO DETERMINE THE BEST BUY IN COMMERCIAL FERTILIZERS

Each of the chief plant nutrients—nitrogen, phosphorus,[7] and potassium—has about the same value per pound in any of its respective forms. Usually the fertilizer with the highest composition is the cheapest, but this is not always the case. In addition to cost of nutrients, consideration should be given to the following:

1. **Additional nutrients**—Some fertilizers, such as ammonium sulfate and single superphosphate, contain additional nutrients (in this case sulfur) which may be of vital importance to crop production. In these cases, an additional value can be attached to

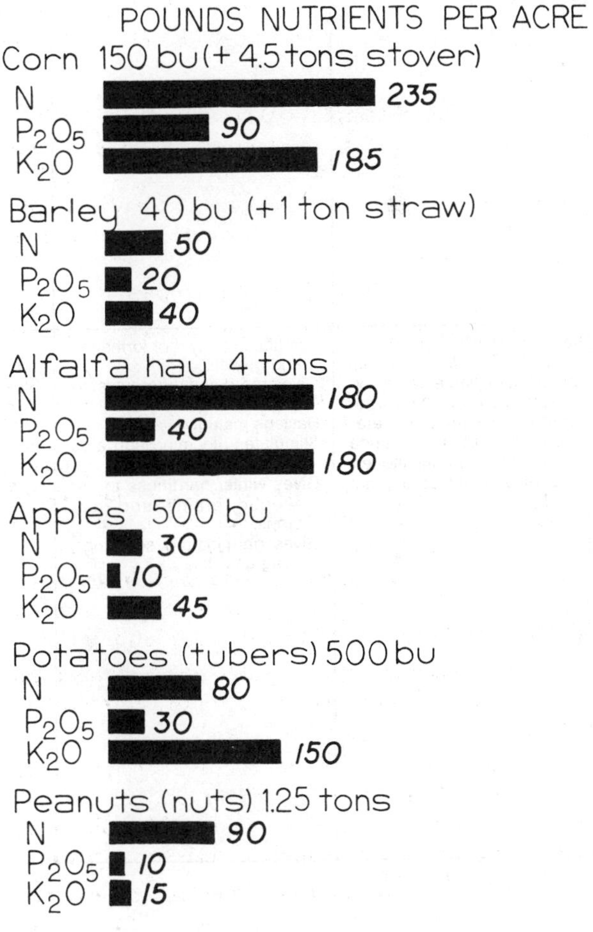

Fig. 8-19. Crop appetites vary. The above bar diagrams show the pounds of nutrients per acre required to produce six important crops. Legumes, such as peanuts and alfalfa, normally get the greater part of their nitrogen from the air. (Bar diagrams based on data from *The Fertilizer Handbook*, published by The Fertilizer Institute, Washington, D.C.)

[6]The amount of fertilizer to apply cannot be based entirely on how much plant food it is estimated that the crop will take out, primarily because crops do not use all plant foods added in fertilizers with the same efficiency. For example, under average conditions, crops probably take up ½ of the nitrogen added and ⅔ of the potash, but only 10 to 20% of the phosphoric acid. Actually, one should add such amounts of fertilizer as will produce maximum economical yields under the specific crop and environmental conditions.

[7]This refers to available P_2O_5, for the method of computation is not necessarily accurate with reference to total P_2O_5.

TABLE 8-5—

Plant Food	Functions in Plants	Hunger Signs	Remarks
Nitrogen (N) (Pure N is a colorless, odorless, inert gas and constitutes about 80% of the air. It must be combined with other elements before it can be put into fertilizers or used as plant food.)	Gives dark green color to plants. Induces rapid growth. Increases yield of leaf, fruit, or seed. Improves quality of leaf crops. Produces rapid growth. Increases protein content of food and feed crops. Feeds soil microorganisms during their decomposition of low-nitrogen organic materials.	A sickly yellowish green color. Slow growth. Drying up or firing of the leaves, which starts at the bottom of the plant and proceeds upward. In plants like corn, grains and grasses, the firing starts at the tip of the bottom leaves and proceeds down the center or along the midrib.	Fortunately, the earth's N supply is unlimited. The air over each acre of land contains 35,000 tons of N which can be utilized by either (1) the synthetic nitrogen fixation process, or (2) properly inoculated leguminous plants which have the ability to take N from the air and convert it into plant food. Today, more than $^2/_5$ of the total N fertilizer is sold to farmers as separate unmixed commercial fertilizers. Nitrogen has about the same value per pound in any of its forms, provided the materials are properly applied. Although plants may absorb nitrogen from a number of compounds, the nitrate form (NO_3) is the most readily utilized. Other forms of nitrogen are converted to nitrate through the activities of soil microorganisms. Nitrate nitrogen may be lost from soils by leaching or denitrification. Ammonia nitrogen is held in soils against leaching by the cation exchange complex; thus, fall application of ammonia fertilizers during periods of reduced microbial activity—when the soil temperature is near freezing—may be advisable. The main objection to organic nitrogen fertilizers is the cost, which is generally 3 to 5 times as much as that supplied by inorganic forms. Most of the fertilizer nitrogen used in the U.S. today is carried by synthetics.
Phosphorus (P) (Pure P is a very reactive substance that bursts into flame when exposed to the air. It must be combined with other elements to curb its violence before it can be put into fertilizers or used as a plant food.)	Stimulates early root formation and growth. Gives rapid and vigorous start to plants. Hastens maturity. Stimulates blooming and aids in seed formation. Gives winter hardiness to fall seeded grains and hay crops. Gives germinating seedlings a big assist.	Purplish leaves, stems, and branches. Slow growth and maturity. In corn, the stalks are small and slender. In small grains, lack of stooling. Low yields of grain, fruit, and seed.	Soils should be tested for phosphorus, with this nutrient applied where needed. Many soils are deficient in P and respond to phosphate fertilization. On the average, the P content of surface soils is only about $^1/_2$ that of N and $^1/_{20}$ that of potassium. Much of the P present in soils is insoluble and unavailable to plants; especially in acid soils. The U.S. possesses $^1/_3$ of the total known world supply of P—about 15 billion tons—in the form of rock phosphate deposits; enough to last this nation an estimated 2,000 years. About 40% of the U.S. supply is in Florida and Tennessee, and 60% is in the Rocky Mountain states of Idaho, Montana, Utah, and Wyoming. In fertilizers, P is guaranteed in the form of available phosphoric acid (P_2O_5). Superphosphate, the principal phosphorus fertilizer material, is produced by treating ground rock phosphate with sulfuric or phosphoric acid or mixtures of the two. The phosphorus in superphosphate is more readily available than that in rock phosphate. Little phosphorus is lost from leaching, but it is lost by erosion.

Footnotes on last page of table.

OIL FERTILITY GUIDE[1]

Source	Average Composition[2]											
	Nitrogen	Available Phosphate	Potash	Calcium	Magnesium	Sulfur	Chlorine	Copper	Manganese	Zinc	Boron	Approximate Calcium Carbonate Equiv.
	%	P_2O_5 %	K_2O %	%	%	%	%	%	%	%	%	lb/ton
mmonia, Anhydrous	82	—	—	—	—	—	—	—	—	—	—	−2,960
mmonia, Aqua	16-25	—	—	—	—	—	—	—	—	—	—	−720 to −1,080
mmonium nitrate	33.5	—	—	—	—	—	—	—	—	.01	—	−1,180
mmonium nit. limestone mixtures	20.5	—	—	7.3	4.4	.4	.4	—	—	—	—	0
mmonium sulfate	21	—	—	.3	—	23.7	.5	.3	—	.1	—	−2,200
mmonium sulfate-nitrate	26	—	—	—	—	15.1	—	—	—	—	—	−1,700
alcium cyanamide	21	—	—	38.5	.06	.3	.2	.02	.04	—	—	+1,260
alcium nitrate	15	—	—	19.4	1.5	.02	.2	—	—	—	—	+ 400
itrogen solutions	21-49	—	—	—	—	—	—	—	—	—	—	−750 to −1,760
odium nitrate	16	—	.2	.1	.05	.07	.4	.07	—	—	.01	+ 580
rea	46	—	—	—	—	—	—	—	—	—	—	−1,680
rea-form	38	—	—	—	—	—	—	—	—	—	—	−1,360
asic slag, open hearth	—	8-12	—	29.0	3.4	.3	—	—	2.2	—	—	+1,000
one meal	2- 4.5	22-28[3]	.2	20-25	.4	.1	.2	—	—	.02	—	+400 to +500
hosphoric acid	—	52-60	—	—	—	—	—	—	—	—	—	−1,000 to −1,400
ock phosphate	—	[4]	—	33.2	.2	.3	.1	—	.03	—	—	+200
uperphosphate, Normal	—	18-20	.2	20.4	.2	11.9	.3	—	—	.01	—	0
uperphosphate, Concentrated	—	42-50	.4	13.6	.3	1.4	—	.01	.01	—	.01	0
uperphosphoric acid	—	69-76	—	—	—	—	—	—	—	—	—	—

(Continued)

TABLE 8-

Plant Food	Functions in Plants	Hunger Signs	Remarks
Potassium (K_2O) (Pure potassium is highly reactive and dangerous to handle. It must be combined with other elements before it can be put into fertilizers or used as plant food.)	Imparts increased vigor and disease resistance to plants. Produces strong, stiff stalks, thus reduces lodging. Increases plumpness of the grain and seed. Essential to the formation and transfer of starches, sugars, and oils. Imparts winter hardiness to legumes and other crops. Essential in stomatal movement and water relations of plants.	Mottling, spotting, streaking, or curling of leaves, starting on the lower levels. Lower leaves scorched or burned on margins and tips. These dead areas may fall out, leaving ragged edges. In corn, grains, and grasses, firing starts at the tip of the leaf and proceeds down from the edge, usually leaving the midrib green. Premature loss of leaves and small, knotty, poorly opened bolls on plants like cotton. Corn and other similar plants falling down prior to maturity due to poor root development.	U.S. soils are far richer in potash than in nitroge or phosphoric acid. Unfortunately, most of the K in the soil is in form that plants cannot readily use and must b supplemented from the fertilizer bag. Soils should be tested for potassium with this nu trient applied where needed. The U.S. possesses known commercial potas deposits sufficient to last for several genera tions. At the present time, principal U.S. com mercial production is from deposits in Calif N. Mex., Utah, and Mich. Potash has about the same value per lb in any c its forms. Therefore, it should be purchased i whatever form it can be obtained at the lowes price. Potassium is usually the cheapest of th three main fertilizer elements. Not as much potash as nitrogen is lost from th soil by leaching; although there may be con siderable K loss from sandy soils.
Calcium (Ca) (The calcium supply is frequently a limiting factor in humid soils or those where rainfall exceeds evaporation. It is the important constituent of lime.)	Promotes early root formation and growth. Improves general plant vigor and stiffness of straw. Influences intake of other plant foods. Neutralizes poisons produced in the plant. Encourages grain and seed production. Increases calcium content of food and feed crops.	Young leaves in terminal bud become "hooked" in appearance and die back at the tips and along the margins. Leaves have wrinkled appearance. In some cases, young leaves remain folded. Light green band along margin of leaves. Short and much-branched roots.	Soils should be tested for acidity before applyin lime or other sources of calcium. Most soils in the humid regions of the U.S. ar "sour" or acid, whereas in arid or semiarid re gions soils are usually alkaline or "sweet." Liming corrects soil acidity and also serves a an economical source of calcium. Legumes respond more to lime than other crops Lime should be spread with a regular lim spreader or an end-gate seeder. Generally, the purer and the finer the limestone the greater the speed of its effect.
Magnesium (Mg) (Magnesium is most likely to be deficient on sandy soils, especially along the Atlantic Coast.)	Is an essential part of chlorophyll, which gives the green color to leaves. Is necessary for formation of sugar from carbon dioxide and water in sunlight. Regulates uptake of other plant foods. Acts as carrier of phosphoric acid in the plant. Promotes the formation of oils and fats. Plays a part in the translocation of starch.	A general loss of green color which starts in the bottom and later moves up the stalk. The veins of the leaf remain green. Cotton leaves often turn a purplish-red color between the green veins. Weak stalks with long branched roots. Definite and sharply defined series of yellowish-green, light yellow, or even deep white streaks throughout entire leaf as with corn. Leaves curve upward along the margins.	Soils may be tested for magnesium.
Sulfur (S) (Sulfur limits crop growth in many areas of the U.S.)	Gives increased root growth. Helps maintain dark green color. Promotes nodule formation on legumes. Stimulates seed production. Encourages more vigorous plant growth.	Young leaves light green in color have even lighter veins. Short, slender stalks. Slow, stunted growth. Spotting of leaves, as with potatoes. Immature fruit, light green in color.	In humid regions, there is a sulfur loss of 20 to 3C lb per acre yearly. Sulfur is added to soils in precipitation and irrigation waters. In industrial areas additions from the atmosphere are sufficient to meet crop needs. The sulfur content of irrigation waters varies widely, but is frequently adequate.

Footnotes on last page of table.

—(Continued)

Source	Nitrogen	Available Phosphate	Potash	Cal-cium	Magne-sium	Sul-fur	Chlo-rine	Cop-per	Manga-nese	Zinc	Boron	Approximate Calcium Carbonate Equiv.
	Average Composition[2]											
	%	P_2O_5 %	K_2O %	←%							→	*lb/ton*
otassium chloride (muriate)	—	—	60-62	.1	.1	—	47.0	—	—	—	.03	0
otassium magnesium sulfate	—	—	22	—	11.2	22.7	1.5	—	—	—	—	0
otassium sulfate	—	—	50	.7	1.2	17.6	2.1	.001	—	—	.002	0

Source	*(Ca %)*
oil amendments:	
Blast furnace slag ($CaSiO_3$)	29.3
Calcitic limestone ($CaCO_3$)	31.7
Dolomitic limestone ($CaCO_3$ + $MgCO_3$)	21.5
Gypsum ($CaSO_4 \cdot 2H_2O$)	22.5
Hydrated lime ($Ca(OH)_2$)	46.1
Marl ($CaCO_3$)	24.0
Precipitated lime (CaO)	60.3
ertilizers:	
Calcium cyanamide ($CaCN_2$)	38.5
Calcium nitrate ($Ca(NO_3)_2$)	19.4
Phosphate rock ($3Ca_3(PO_4)_2 \cdot CaF_2$)	33.1
Superphosphate, normal ($Ca(H_2PO_4)_2$)	20.4
Superphosphate, triple ($Ca(H_2PO_4)_2$)	13.6

Source	*(Mg %)*
olomitic limestone ($MgCO_3$ + $CaCO_3$)	11.4
psom salt ($MgSO_4 \cdot 7H_2O$)	9.6
ieserite, calcined ($MgSO_4 \cdot H_2O$)	18.3
agnesia (MgO)	55.0
otassium—magnesium sulfate ($K_2SO_4 \cdot 2MgSO_4$)	11.2

Source	*(S %)*
mmonium sulfate ($(NH_4)_2SO_4$)	23.7
opperas ($FeSO_4$)	12.0
opper sulfate ($CuSO_4$)	12.8
psom salt ($MgSO_4 \cdot 7H_2O$)	14.0
ypsum ($CaSO_4 \cdot 2H_2O$)	16.8
anganese sulfate ($MnSO_4$)	14.5
ormal superphosphate ($Ca(H_2PO_4)_2$ + $CaSO_4 \cdot 2H_2O$)	11.9
otassium-magnesium sulfate ($K_2SO_4 \cdot 2MgSO_4$)	22.0
eax Micronutrients (MPP)	10.0
ulfur, elemental (S)	30-99.6
ulfur dioxide (SO_2)	50.0
riple superphosphate ($Ca(H_2PO_4)_2$)	1.4
inc sulfate ($ZnSO_4 \cdot H_2O$)	18.0

(Continued)

TABLE 8-

Plant Food	Functions in Plants	Hunger Signs	Remarks
Trace elements: **Boron** (B)	Cell division, viability of pollen grams, fruit formation, carbohydrate and water metabolism, protein synthesis.	Alfalfa—stunted growth and bright yellow leaves; apples—external and internal cork; citrus—heavy fruit shedding and yellowing of leaves; cotton—excessive shedding of flower buds and bolls.	Because of the possibility of toxicity, trace el ments should be added only when one is ce tain they are needed and when the amount r quired is known. *Avoid an overdose. If i doubt, seek expert advice.*
Chlorine (Cl)	Recently demonstrated as essential to plant growth. Chlorine is believed to stimulate the activity of some enzymes and to influence carbohydrate metabolism, the production of chlorophyll, and the water holding capacity of plant tissue.	Chlorine deficiency of field-grown crops has not yet been found in the U.S.	Only small amounts of chlorine are needed. Under certain conditions an excess may be problem.
Cobalt (Co)	Cobalt has been shown to be essential for nitrogen fixation by bacteria in the root nodules of alfalfa. Also, cobalt is considered a needed constituent of the animal protein factor.	Soil supplies are generally adequate.	
Copper (Cu)	Copper is required for chlorophyll formation and several enzyme systems.	Citrus—dieback of twigs, yellow leaves; small grains—twisted leaves, yellowing along leaf margins with "tip burn"; corn—yellowing between leaf veins; vegetables—dieback of leaves.	
Iron (Fe)	Iron is essential for chlorophyll formation and respiration (energy transfer).	Need indicated by pale yellowish color of foliage, in the presence of adequate amounts of nitrogen and on soils that are high in lime or manganese.	
Manganese (Mn)	Necessary for chlorophyll production and for carbohydrate and nitrogen metabolism. Also, high manganese levels reduce iron levels.	Need indicated by pale green to yellow and red colors between green veins of leaves of tomatoes and beets, resinous spots on leaves of citrus, chlorosis of crops such as spinach and soybeans on overlimed soil, and "gray speck" on oats.	
Molybdenum (Mo)	Essential for nitrogen fixation by bacteria in legume nodules and in plant nitrogen metabolism.	Yellow leaves similar to that of nitrogen deficiency.	

Footnotes on last page of table.

—(Continued)

Source	Nitrogen	Available Phosphate	Potash	Calcium	Magnesium	Sulfur	Chlorine	Copper	Manganese	Zinc	Boron	Approximate Calcium Carbonate Equiv.
				Average Composition[2]								
	%	P_2O_5 %	K_2O %	←			%				→	lb/ton
											(B %)	
Boron frits (Frit)											10-17	
Borax ($Na_2B_4O_7 \cdot 10H_2O$)											11	
Boric acid (H_3BO_3)											17	
Sodium pentaborate ($Na_2B_{10}O_{16} \cdot 10H_2O$)											18	
Sodium tetraborate												
Fertilizer borate-46 ($Na_2B_4O_7 \cdot 5H_2O$)											14	
Fertilizer borate-65 ($Na_2B_4O_7$)											20	
Solubor ($Na_2B_{10}O_{16} \cdot 10H_2O$)											20	
							(Cl %)					
Potassium chloride (muriate) Also, many fertilizers carry small amounts of chlorine.							47					
In deficient areas, cobalt is normally supplied in the mineral mix or feed. However, the application of 3 lb of cobalt sulfate per acre is effective on pasture.												
								(Cu %)				
Basic copper sulfate ($CuSO_4 \cdot H_2O$)								13-53				
Copper(ic) ammonium phosphate ($Cu(NH_4)PO_4 \cdot H_2O$)								32				
Copper chelates ($Na_2CuHEDTA$) ($NaCuHEDTA$)								9, 13				
Copper chloride ($CuCl_2$)								16.9				
Copper frits (Frit)								40-50				
Reax Copper (CuMPP)								5-6				
THIS copper (CuMPP)								5				
Copper Silviplex (CuMPPP)								6				
Copper sulfate monohydrate ($CuSO_4 \cdot H_2O$)								35				
Copper sulfate pentahydrate ($CuSO_4 \cdot 5H_2O$)								25				
Cupric oxide (CuO)								75				
Cuprous oxide (Cu_2O)								89				
Rayplex Cu (CuPF)								5-6.7				
											(Fe %)	
Ferrous ammonium phosphate ($Fe(NH_4)PO_4 \cdot H_2O$)											29	
Iron ammonium polyphosphate ($Fe(NH_4)HP_2O_7$)											22	
Iron frits (Frit)											30-40	
Ferric sulfate ($Fe_2(SO_4)3 \cdot 4H_2O$)											23	
Ferrous sulfate ($FeSO_4$) $\cdot 7H_2O$)											19	
Iron chelates (NaFeEDTA or FeHEDTA)											5-15	
Reax Iron (FeMPP)											10-12	
THIS Iron (FeMPP)											5	
Iron Silviplex (FeMPPP)											6	
Rayplex Fe (FePF)											9-9.6	
									(Mn %)			
Manganese chelate (MnEDTA)									12			
Reax Manganese (MnMPP)									10-12			
THIS Manganese (MnPP)									5			
Manganese Silviplex (MnMPPP)									7			
Manganese sulfate ($MnSO_4 \cdot 3H_2O$)									26-28			
Manganese frit (Frit)									35			
Manganous oxide (MnO)									41-68			
Rayplex Mn (MnPF)									8-8.5			
							(Mo %)					
Ammonium molybdate ($(NH_4)_6Mo_7O_{24} \cdot 2H_2O$)							54					
Molybdenum trioxide (MoO_3)							66					
Molybdenum frit (Frit)							30					
Sodium molybdate ($Na_2MoO_4 \cdot 2H_2O$)							39					

(Continued)

TABLE 8-5

Plant Food	Functions in Plants	Hunger Signs	Remarks
Zinc (Zn)	Zinc is involved in enzyme systems essential to protein synthesis and functions in processes related to the development of floral parts, grain and seed production, and in rate of maturity of plant and seed.	Legumes—Small bronze colored spots on older leaves; fruit trees, retarded terminal growth, narrow leaves with yellow tissue between the veins; corn—yellow stripe on either side of midrib.	

[1]Much of the information in columns 2 and 3 of this table was taken, by special permission, from *Our Land and Its Care*, The Fertilizer Institute, 1015 18th St., N.W., Washington, D.C. The principal fertilizer materials and their compositions are from *The Fertilizer Handbook*, 1972, pp. 62-63, published by The Fertilizer Institute, 1015 18th St. Washington, D.C.

[2]Most of the percentages larger than one of N_2 P_2O_5 and K_2O are the usual guarantees. Where more than one grade is sold, the range is indicated by two numbers separated by a dash. The rest of the percentages are averages compiled by A. L. Mehring from many published analyses. A minus sign indicates the number of pounds of calcium carbonate needed to neutralize acid formed when 1 ton of the material is added to the soil. A plus sign indicates basic materials, and a zero physiologically neutral materials.

these fertilizers, over and above their respective nitrogen and phosphorus contents.

2. **Cost of application**—Where both nitrogen and sulfur are needed, it will usually cost the farmer less to purchase and apply ammonium sulfate in one operation than to apply the least costly form of nitrogen, which is usually anhydrous ammonia, and then a sulfur compound in a second operation.

3. **Service and reputation**—In addition to cost as such, consideration should be given to the reputation of the dealer and his services. Often the services rendered are of greater value than the fertilizer.

Table 8-6 shows the method used in determining the best buy of one-nutrient (or simple) fertilizers.

The best buy in mixed fertilizers can be determined on a comparative basis whereby the fertilizer constituents are evaluated according to what they would cost in one-nutrient or simple fertilizers. To this cost is added a justifiable mixing charge. Naturally, mixed fertilizers are somewhat more expensive on a per unit nutrient basis than one-nutrient or simple fertilizers.

Table 8-7 shows the method of determining the best buy in complete or two-element fertilizers. In this particular example, a two-element fertilizer is used, but the same method applies to complete fertilizers, simply include the value of the added potassium.

It will be noted that in Table 8-6, nitrogen and available P_2O_5 were valued at 20¢ and 24¢ per pound, respectively. In this particular case, these figures (20¢ and 24¢) were used because they represented the current costs of N and P_2O_5, respectively, in ammonium nitrate and treble superphosphate—the cheapest sources of each nutrient in one nutrient or simple fertilizers. Also, it will be noted that the cost per ton for mixing was estimated at \$10. Naturally, these figures will vary from area to area and time to time. Simply (1) arrive at the lowest cost of each nutrient in available one-nutrient or simple fertilizers, and (2) add an estimated mixing charge.

In addition to the actual cost of the material, consideration must be given to the cost of transportation, storage, and labor used in applying the fertilizer. These costs may be difficult to evaluate; but if the actual cost of the nutrients from one source is the same as another, farmers will gradually learn to take the one requiring the least labor. The higher-analysis goods require less labor in handling, and fewer stops in applying the material.

WHEN TO APPLY FERTILIZERS

It is important that commercial fertilizers be applied at the proper time. For most annual crops, this means just prior to, or at the time of, seeding; but supplemental applications may be made as either side-dressings or as liquid applications. Pastures and haylands may be fertilized almost any season of the year, but generally greater responses are obtained from early spring applications. Supplemental applications to pasture and hay crops may be made by topdressing or through the sprinkler system.

HOW TO APPLY FERTILIZERS

There is no one best way in which to apply a fertilizer. Several methods are used, and there are advantages to each. The method of application should be determined by the crops, soil, climate, time and rate of application, and kinds of fertilizer and equip-

—(Continued)

Source	Average Composition[2]												Approximate Calcium Carbonate Equiv.
	Nitrogen	Available Phosphate	Potash	Cal- cium	Magne- sium	Sul- fur	Chlo- rine	Cop- per	Manga- nese	Zinc	Boron		
	%	P_2O_5 %	K_2O %	←				%			→		*lb/ton*
										(Zn %)			
Zinc Carbonate ($ZnCO_3$)										52-56			
Zinc chelate ($Na_2ZnEDTA$)										14			
Zinc chelate (NaZnHEDTA)										9			
Zinc oxide (ZnO)										78-80			
Reax Zinc (ZnMPP)										10-12			
THIS Zinc (ZnMPP)										7			
Zinc Silviplex (ZnMPPP)										7			
Zinc sulfate ($ZnSO_4 \cdot H_2O$)										36			
Rayplex Zn (ZnPF)										10			

[3]Total P_2O_5. All of the P_2O_5 in natural organics is considered available.
[4]30-36% total P_2O_5, which is relatively unavailable in some soils.

ment available. Regardless of the method selected, the aim should be to get the fertilizer in the soil where it will do the most good.

Some of the common methods of applying fertilizers are:

1. **Broadcasting**—Consists in spreading fertilizer uniformly over the land by means of a fertilizer distributor. This method is confined almost entirely to permanent or semipermanent pastures and orchards, in which the fertilizer may or may not be worked into the soil. When fertilizer is broadcast on uncultivated crops, the method is commonly called top-dressing.

2. **Side-dressing**—Consists in putting fertilizer along each side of the row after crops are up and growing, using some special fertilizer drill or attachment. Side-dressing is usually confined to the applica-

TABLE 8-6

METHOD OF DETERMINING THE BEST BUY IN ONE-NUTRIENT OR SIMPLE FERTILIZERS

Fertilizer No. and Formula ($N-P_2O_5-K_2O$)	Lb of Nutrient/ Ton of Fertilizer	If the Cost/ Ton Is:	Cost of 1 Lb of Nutrient	Remarks
(#1) 33-0-0	33 × 20 = 660	$132.00	$\frac{\$132.00}{660} = \$0.20$	Thus, the #1 nitrogen fertilizer is the best buy, the N costing 20¢/lb in comparison with 22¢/lb in the #2 fertilizer.
(#2) 21-0-0	21 × 20 = 420	$ 92.40	$\frac{\$92.40}{420} = \$0.22$	
(#3) 0-45-0	45 × 20 = 900	$216.00	$\frac{\$216.00}{900} = \$0.24$	Thus, the #3 phosphorus fertilizer is the best buy, the available P_2O_5 costing 24¢/lb in comparison with 26¢/lb in the #4 fertilizer.
(#4) 0-19-0	19 × 20 = 380	$ 98.80	$\frac{\$98.80}{380} = \$0.26$	

TABLE 8-7

METHOD OF DETERMINING THE BEST BUY IN COMPLETE OR TWO-ELEMENT FERTILIZERS

Fertilizer No. and Formula ($N-P_2O_5-K_2O$)	Lb of Nutrient/ Ton of Fertilizer	Nutrient Value/Ton	Total Value /Ton	Remarks
(#1) 4-12-0	4 × 20 = 80 12 × 20 = 240	80 × $0.20 = $16.00 240 × $0.24 = $57.60 $73.60 Plus $10.00/ton for mixing	$73.60 10.00 $83.60	Thus, the comparable value/ton of these two fertilizers is $83.60 and $284.40, respectively.
(#2) 11-48-0	11 × 20 = 220 48 × 20 = 960	220 × $0.20 = $44.00 960 × $0.24 = $230.40 $274.40 Plus $10.00/ton for mixing	$274.40 10.00 $284.40	Therefore, if fertilizer #1 is selling above $83.60/ton while fertilizer #2 is selling at or below $284.40, the latter is the best buy, and vice versa.

tion of quick-acting nitrogen fertilizers to truck or other crops, where large amounts of readily available fertilizer are desired. Sometimes as many as three side-dressing applications are made during the season.

3. **Drilling with seed**—Consists in applying the fertilizer in the row when the seed is drilled or planted. It is the accepted method of fertilizing small grains and grasses. Drilling should be used only with seed that can tolerate contact with fertilizer, and usually it is restricted to rather light applications.

4. **Banding along the row**—Consists in distributing the fertilizer in bands one or more inches wide on either or both sides of the row while planting, usually at or below the seed level. It differs from normal drilling in that the fertilizer is in close proximity to but not in direct contact with the seed.

5. **Deep drilling**—Consists in placing fertilizer in bands at desired depths. Anhydrous ammonia must be placed 4 to 6 inches deep.

6. **Plowsole of deep furrow**—Consists of placing fertilizer on the bottom of each furrow, a practice which is helpful to plants when the surface of the soil becomes dry during the growing season.

7. **Foliar application**—Consists in spraying dilute solutions of salts of certain metals (copper, manganese, iron, or zinc) when deficiencies of these trace elements show up. Also, the Russians are experimenting with the use of foliar applications of liquid nitrogen to increase the nitrogen content of plants, with the application made a few days ahead of harvesting.

8. **Manure-fertilizer spreading**—Consists in adding fertilizer, usually a phosphate, at the time of spreading manure.

9. **Liquid fertilizer distribution**—Consists in applying fertilizers in solutions either through attachments on planters and cultivators or through injection into irrigation water (either in ditches or in sprinklers) and distribution with the water. Also, liquid fertilizers are sometimes applied at the time of transplanting crops.

Different variations and combinations of these methods can be and are used. For example, part of the fertilizer is sometimes placed on fields by broadcasting, with the remainder drilled in; some may be broadcast and plowed under, with the rest broadcast after plowing. Also, the practice of custom application of fertilizers is increasing. For information on the best method to apply fertilizers, the farmer or rancher should consult the local county agent, vocational agriculture instructor or fertilizer dealer, or write to the state agricultural college.

Lime Acid Soils

Lime may be defined as any compound of calcium or of calcium and magnesium capable of neutralizing soil acidity.

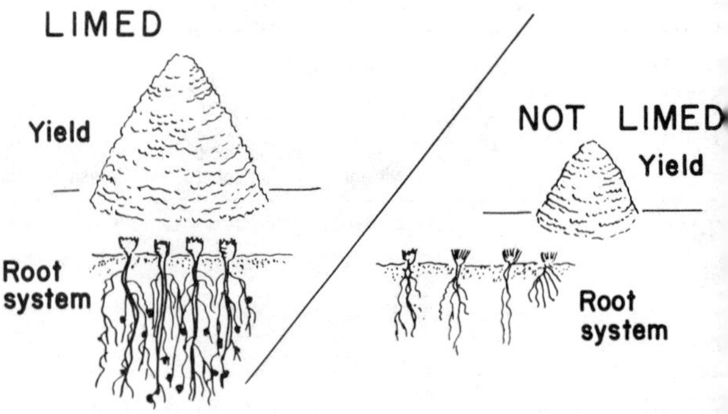

Fig. 8-20. Liming produces a larger crop, and also a larger amount of stubble and root, which remains in the soil to maintain the supply of organic matter. (Drawing by R. F. Johnson)

Some soils originally had, and still have, plenty of lime in them. Others never did have enough of this constituent to produce successfully the best legumes. Still others once had enough lime, but subsequently it has been leached out and removed by crops. As much as 400 pounds of lime per acre may be leached away annually with drainage water, and every crop takes out lime.

It has been estimated that 70 percent of the tillable land in the eastern half of the United States is acid or "sour" as a result of losses by leaching and crop removal of such basic elements as calcium, magnesium, and potassium. In the arid or semiarid regions of the West, the soils are usually alkaline or "sweet."

Liming is the first step in the improvement of most acid soils, and, in addition, it serves as an economical source of calcium (and sometimes magnesium).

FUNCTIONS OF LIME

In addition to serving as a fertilizer material, especially in humid soils, lime is a soil amendment which functions in acid soils as follows:

1. It corrects soil acidity.
2. It supplies calcium for plants and animals, and sometimes magnesium (if the lime is dolomitic).
3. It stimulates soil microbe (bacteria) activity.
4. It speeds the decay of organic matter and the liberation of nitrogen and other plant foods.
5. It makes potassium more efficient in plant nutrition; as a result of liming, plants absorb more cheap calcium and less of the more expensive potassium.
6. It increases availability of molybdenum.
7. It improves crop yields.
8. It reduces the toxicity of iron, manganese, and aluminum.

9. It increases the efficiency of manures and fertilizers.

10. It decreases soil erosion and improves soil structure as a result of increased vigor and density of plants.

THE pH VALUE OF THE SOIL

Soil pH is an excellent indicator of general soil conditions; hence, it should always be determined. However, pH values should be augmented by a lime-requirement determination, as described in the section on "How to Determine Lime Needs."

The pH value of a soil is a convenient method of expressing the degree of acidity or alkalinity. The pH scale is divided into 14 divisions, or pH units numbered from 0 to 14. A soil with a pH of 7.0 is neutral. Soils with pH values below 7.0 are acid or sour, while those above 7.0 are alkaline or sweet. A pH of 5.0 is 10 times more acid than a pH of 6.0, and a pH of 4.0 is 10 times more acid than a pH of 5.0. Thus, a soil having a pH value of 4.0 is 100 times more acid than one having a pH of 6.0. The pH of the majority of soils lies within the range of 4.0 to 8.0. Most soils in the humid region are acid or sour as a result of losses by leaching and crop removal, whereas most soils in arid or semiarid regions are alkaline or "sweet."

It is recognized that the determination of pH does not give a direct measure of the amount of exchangeable acidity, but only the "free" or ionized acidity. Therefore, recommendations for the amount of lime had best be made by a soil technician. Such a person recognizes (1) that soils with considerable clay and organic matter need more lime than do sandy or highly weathered soils of the same pH level; (2) that the type of clay may also be a factor; and (3) that pH is less critical provided all the required nutrients are supplied in the proper amounts and ratios—for example, where molybdenum is needed and added, some crops which were normally considered to grow well only at a pH of 6.5 and above will do equally well on more acid soils.

Soils in the pH range of 6.0 to 8.0 are apt to be more trouble-free than those higher or lower. Values of pH 5.0 or under may be indicative (1) of deficiency or unavailability of such elements as calcium, magnesium, phosphorus, molybdenum, and boron, or (2) of toxic amounts of zinc, manganese, aluminum, nickel, and other elements because of increased solubility.

Values above pH 8.5 indicate (1) the presence of sodium carbonate and/or high exchangeable sodium, (2) the need to treat with gypsum, sulfur, or other acidic materials, and (3) the need to leach out. Values of pH 8.0 to 8.5 often denote the presence of free lime; and in this range and higher the availability of

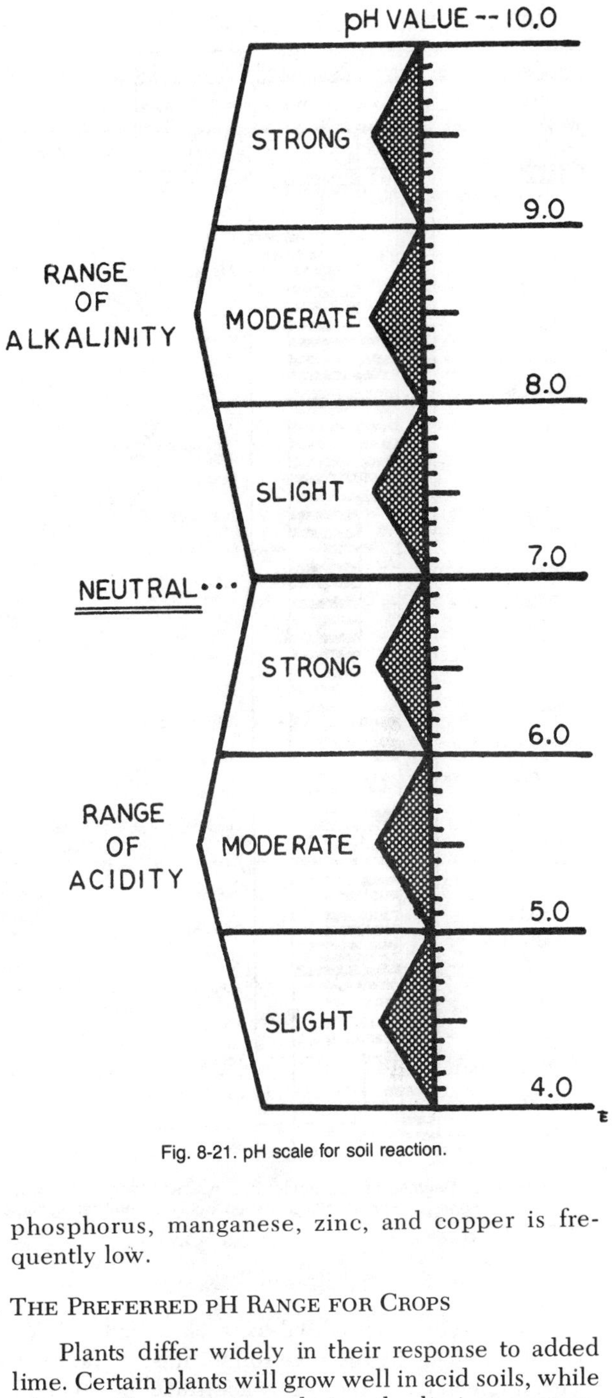

Fig. 8-21. pH scale for soil reaction.

phosphorus, manganese, zinc, and copper is frequently low.

The Preferred pH Range for Crops

Plants differ widely in their response to added lime. Certain plants will grow well in acid soils, while others will not. In considering the liming program, therefore, the type of crop to be grown is of importance (see Fig. 8-22, page 754).

How to Determine Lime Needs

The pH of a soil is a measure of intensity of acidity or alkalinity, but it does not tell how much lime (calcium carbonate) or other liming material to put on the soil.

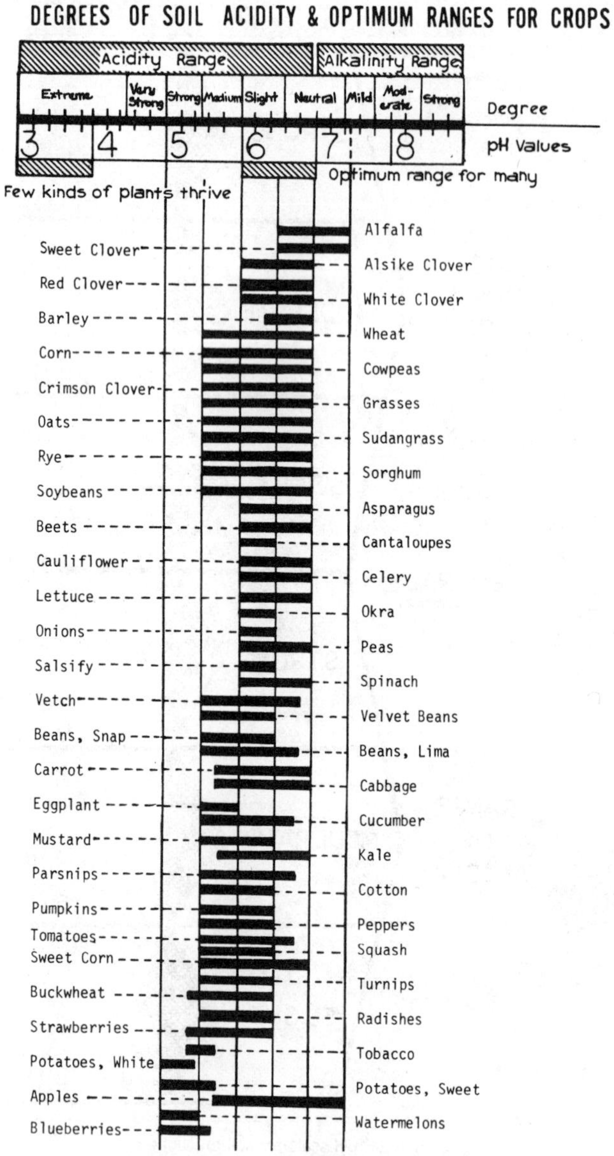

Fig. 8-22. Degrees of soil acidity and optimum ranges for crops. (From: *The Fertilizer Handbook*, published by The Fertilizer Institute, Washington, D.C.)

Wherever it is available, a chemical soil test should be made by the state soil testing laboratory, by a commercial laboratory, or by a fertilizer company. In general, soil testing laboratories use a combination of tests to determine the need for lime and the amount of lime to apply. Many laboratories estimate the amount of acid in a soil sample by measuring the change of pH of a buffered solution—one that resists changes in pH—when the solution is mixed with the soil sample.

Another laboratory method involves a direct titration with calcium hydroxide. The latter determination is the most reliable of the two, but it takes the most time and equipment.

With a chemical soil analysis available, along with background information of the major soil series and types in the area they serve, together with their previous responses to lime, the state laboratory is in the best position to make liming recommendations for the specific farm, crops, and conditions. In the absence of a laboratory service, a soil test kit in the hands of an experienced person will give reasonably satisfactory results.

The following methods may also be used in detecting acid soils and in arriving at lime needs:

1. **The electrometric method—glass electrode**—The glass electrode pH meter is considered the most reliable method of determining the pH values of soils and is a rather reliable method of determining the lime needs.

However, figures obtained by a person unable properly to check and standardize the apparatus may at times be questionable.

2. **The colorimetric (dye) test**—There are a number of different commercial quick test kits for determining pH colorimetrically. Each kit contains directions for making the test and a table for converting pH readings to lime requirements. This method is very simple and easy, but much less accurate than the electrometric method.

3. **Crop growth**—Acid soils are indicated by (a) the repeated failure of such crops as alfalfa and clover, and/or (b) the dominant growth of such plants as sheep sorrell, mayweed, certain mosses, bentgrasses, fescues, timothy, and redtop. Naturally, such observations give no indication as to the quantity of lime needed.

4. **Trial application of lime**—If in doubt relative to lime requirements, the application of lime in a trial strip is recommended. Such a strip may be of any width or length across the field being tested.

As previously indicated, other properties than pH—soil texture, the kind of clay, the content of organic matter, and the amounts and ratios of nutrients—must be considered when estimating the need for lime. Thus, the more clay and humus there is in a soil, the more limestone needed to change the pH.

There may be considerable differences in soil acidity, even within a given field—especially on fields that are not level. In some fields, the tops of the ridges may be alkaline, the slopes acid, and the bottoms alkaline.

Soil samples for testing for lime should be taken as directed in the earlier section entitled, "How to Take Soil Samples."

Kind and Amount of Lime to Use

Nature has provided an abundant supply of liming materials in most parts of the country. In general, the particular product selected should be determined by the following:

1. The comparative cost based on the neutralizing value (N.V.) of one ton of each material (see section on "How to Determine the Best Buy in Lime").
2. The convenience and cost in handling.
3. The relative speed of action in the soil, as determined by fineness, hardness, etc. The rate of reaction of a liming material with the soil is essentially a function of the size of particle; the finer the lime, the more rapid the changes brought about by its application.

   The fineness of lime is measured by screen tests. A standard mesh screen is one in which the number of openings per linear inch corresponds to its screen number. Thus, a 10-mesh screen has 10 openings per linear inch, or 100 openings per square inch. Although the fineness of limestone upon which recommendations are based varies greatly among states, a common guarantee for lime is—

   85 percent through a 15-mesh sieve.
   30 percent through a 100-mesh sieve.
4. The magnesium content of the limestone where this element is needed. When the soil contains less than 50 pounds to the acre of available magnesium, dolomitic limes should be used.

The amount of lime to apply per acre should be determined by (1) the degree of soil acidity, (2) the kind of soil, (3) the kind of lime, (4) the frequency of application, and (5) the kind of crops to be grown.

Under most general farming conditions, the application of lime should be regulated so as to maintain the soil reaction at a pH of about 6.5 or in the neutral range, as determined by a pH test. Generally this will require an initial application of 2 to 5 tons of ground limestone (or equivalent lime) and subsequent applications of about 1 ton every 5 to 10 years thereafter. Smaller quantities should be applied to sandy soils at any one time, but the applications should be made more frequently.

A rule of thumb commonly followed is that it requires about 1 ton of ground limestone (or equivalent lime) per acre to change the soil reaction by one pH unit (say, from 5.5 to 6.5) in a strongly acid sandy loam; loams require approximately 2 tons; clay loams 3 tons; and muck soils 5 tons. If soils are fairly high in organic matter, as they are in the northern states, it will take more lime to accomplish the same change in pH. Conversely, in southern soils, which are lower in organic matter, less lime will be needed to accomplish same pH change.

How to Determine the Best Buy in Lime

Table 8-8 shows how to determine the best buy in some of the common liming materials. As noted, the relative neutralizing power and value (N.V.) of the different forms of lime is usually calculated on the basis of pure calcium carbonate ($CaCO_3$), taken as 100 percent. Also, it is assumed, for purposes of comparison, that each product is 100 percent pure. Actually,

TABLE 8-8
HOW TO DETERMINE THE BEST BUY OF SOME COMMON LIMING MATERIALS

Kind of Lime	Lb Required to Equal the Soil Neutralizing Power of 2,000 Lb of Ground Limestone, Assuming Pure Products in Each Case	Price/Ton That One Could Afford to Pay if Ground Limestone Costs $10.00/Ton	Remarks
Ground limestone ($CaCO_3$)	2,000	$10.00	
Marl (on a dry basis) ($CaCO_3$)	2,000	10.00	
Hydrated lime ($CaOH_2$)	1,480	13.51	Lb for lb, hydrated lime is more valuable than ground limestone because 74 lb of pure calcium hydroxide contains the same amount of calcium and has the same power to neutralize soil acidity as 100 lb of calcium carbonate.
Burned lime (CaO)	1,120	17.86	Lb for lb, burned lime is more valuable than ground limestone because 56 lb of pure calcium oxide contain the same amount of calcium and have the same power to neutralize soil acidity as 100 lb of calcium carbonate.
Oystershells ($CaCO_3$)	2,000	10.00	

most products are not pure, and they may vary considerably within themselves. For example, over the country as a whole, ground limestone, marl, and oystershells probably will not average over about 90 percent calcium carbonate. Most states require that the calcium carbonate equivalent of all liming materials be printed on the bag.

Table 8-8 shows that, based on pure materials in each case, 1,120 pounds of burned lime or 1,480 pounds of hydrated lime are equivalent to 2,000 pounds of agricultural limestone. This means that if ground limestone costs $10.00 per ton, the farmer could afford to pay $13.51 per ton for hydrated lime or $17.86 per ton for burned lime. Like ground limestone, the value of marl and oystershells to correct soil acidity is determined by the percent of calcium carbonate present. Thus, marl or oystershell testing 90 percent has the same neutralizing power as ground limestone of the same test, pound for pound.

When to Apply Lime

Two facts are of paramount importance in determining when to apply lime: (1) It takes some time for lime to dissolve enough really to become effective, and (2) some crops, such as alfalfa and sweet clover, respond more to lime than other crops. Therefore, with a rotation such as CORN-CORN-OATS-CLOVER, it would be best to apply the lime to the crop preceding the seeding of the legume; between the first and second year corn in this case.

Ground limestone can be applied at any time of the year, and without hazard of injury to crops (except potatoes, where lime may make conditions more favorable for scab). Perhaps the only precaution is that it should not be applied when the ground is soft and there is danger of trucks puddling the soil. Burned and hydrated lime should not be applied to growing crops due to their burning effect on plants.

Lime is commonly applied on rough plowed land or after a single harrowing, and incorporated into soils through subsequent cultivation. Also, it is frequently applied as a top-dressing on pastures, lawns, and other grassed areas.

How to Apply Lime

Since damage can be done by getting too much lime in spots, even spreading is important. Also, lime should be mixed with the surface soil.

Generally, lime is spread by the dealer who (1) has a truck with a specially built V-shaped bed and a centrifugal type of spreader, and (2) charges very little more for spreading the lime than for dumping it in one place on the field.

If the farmer spreads his own lime, it should be done with a regular lime spreader or an endgate seeder. Also, it can be spread with manure by putting lime in the gutters behind dairy cows or by putting lime on the loaded manure spreader.

Treat Saline and Alkaline Soils

Saline and alkaline conditions lower the productivity and value of large areas of agricultural land in the United States—an estimated one-fourth of our irrigated land and less extensive acreages of nonirrigated crop and pasturelands.

Saline and alkaline soils are soils that have been harmed by soluble salts, consisting mainly of sodium, calcium, magnesium, chloride, and sulfate, and secondarily of potassium, bicarbonate, carbonate, nitrate, and boron.

Saline soils contain excessive amounts of soluble salts only. Alkaline soils contain excessive absorbed sodium. Because leaching may have occurred previously, alkaline soils do not always contain excess soluble salt.

Salt-affected soils, which occur mostly in arid or semiarid regions, are problem soils that require special remedial measures and management practices.

Salt comes from the water and to a lesser degree from fertilizers. Colorado River water contains about 1¼ tons of soluble salts per acre foot. Salts accumulate in soils unless sufficient water is moved through the profile to carry dissolved salts deep below crop root zones.

• **Handling saline soils**—Excess salinity can injure plants. But some crops are extremely salt tolerant, notably cotton, sugar beets, and barley. Vegetables are relatively sensitive to salinity.

There are no magical chemicals that can tie up salts; however, there are ways to increase water penetration. Some soils may be improved by profile modification, such as chiseling or slip plowing. Organic amendments worked into the soil are helpful. If animal manures are used, the soil-manure mixture should be preirrigated before planting. This will allow for some leaching of the salts contained in manure as well as release of ammonia.

• **Handling alkaline soils**—Gypsum is widely used in reclaiming alkaline soils in western United States. Practically all cropland in California, Arizona, Nevada, and eastern Oregon is alkaline, rather than acidic, on the pH scale. Approximately one million tons of gypsum are incorporated in California soils each year. Gypsum ($CaSo_41{\cdot}2H_2O$) is the common name for calcium sulfate, a mineral used in the fertilizer industry as a source of calcium and sulfur. Another common name is landplaster.

Other amendments (in addition to gypsum) used as correctives for alkaline soils are sulfur, sulfuric acid, and ferric sulfate. Soil correctives for alkaline soils either contain calcium or make the calcium in the soil available for correcting alkaline conditions.

Leaching alkaline soils with irrigation water is an important step in their reclamation. Adequate drainage is needed to remove the excess soluble salts which result from the application of the soil amendment.

Calcium is extremely important in the reclamation of alkaline soils. Where there is an excess of sodium attached to clay particles, it tends to disperse or deflocculate the soil. This makes the particles pack together in such a way that water either permeates very slowly or cannot get through at all.

When gypsum is applied, it supplies calcium. When sulfur containing materials are used, they render the natural calcium in the soils more soluble. In turn, the calcium replaces the excess of absorbed sodium.

The replacement of the sodium on the soil particles by calcium results in an aggregation, or grouping, of the small particles so that there are more large pore spaces in the soil. This improved soil structure permits more rapid water and air penetration and improves soil tilth.

The quantities of soil amendments to be used in the reclamation of alkaline soils is dependent upon a number of factors, the most important of which is the natural calcium in the soil and its relation to the exchangeable sodium present. Other factors include buffering capacity, organic matter content, and the physical characteristics of the soil.

A Program of Soil Management for the Stockman

The following program of soil management is recommended for the livestock farm or ranch:

1. *Feed livestock* for income.
2. *Handle the manure* so as to return the maximum fertility value to the land.
3. *Test soil* and apply fertilizer, lime, or gypsum as needed.
4. *Grow inoculated legumes* to furnish nitrogen, humus, and good livestock feed.
5. *Rotate crops* to ensure good soil structure and keep down weeds.
6. *Keep hilly and rolling ground* in pasture as much as possible.
7. *Utilize all crop residues* by returning them to the soil.

Where to Go for Soil Fertility Help

Any one or all of the following sources may be called upon for counsel and advice on problems of soil fertility:

1. Soil testing laboratories.
2. County extension agent.
3. The state agricultural college.
4. Successful neighboring farmers who have similar conditions.
5. Professional farm managers and consultants.
6. Local fertilizer dealers.
7. Fertilizer trade associations.

CONTROL PESTS; LESSEN WASTE

Waste of food supplies will increasingly nag the consciences and pocketbooks of all people—producers and consumers alike.

Pests cause an estimated 30 percent annual loss in the worldwide potential production of crops, livestock, and forests.[8] Every part of our food, feed, and fiber supply—including marine life, wild and domestic animals, field crops, horticultural crops, and wild plants—is vulnerable to pest attack. Obviously, if these losses could be prevented, or reduced, world food supplies would be increased.

Remember that a worldwide annual loss of 30 percent potential food productivity occurs despite the use of advanced farming technology and mechanized agriculture. Remember, too, that in many of the developing countries losses greatly exceed this figure.

Pests of many kinds attack plants during all stages of their growth, and they attack food and food products after harvest—in storage, during transportation to market, in warehouses, in elevators, in ships, in supermarkets, and in homes after purchase. Here are a few notable pest losses:

Plant diseases and insects—Disease organisms kill plants, cause rotting and blemishing of food products, and reduce crop yields and quality.

Insects devour growing crops, lower yields and quality, and attack grains and other food products in storage and during transport. Also, insects harbor and transmit diseases to plants, animals, and man.

More than 160 bacteria, 250 viruses, and 8,000 fungi are known to cause plant diseases. In the United States alone, approximately 10,000 species of insects are destructive enough to be called "enemies"; and about four-fifths of them are injurious to crops.

• **Weeds**—Weeds reduce yields and quality by competing with crops for water, nutrients, light, and space. Also, they poison livestock, interfere with harvesting, and slow the flow of water for irrigation and drainage. In the United States, some 2,000 species of weeds and brush cost farmers and ranchers an estimated $8 billion annually, with reductions in quantity and quality of crops heading the list.

---

[8]Ennis, W. B., Jr., W. M. Dowler, and W. Klassen, "Crop Production to Increase Food Supplies," *Science*, Vol. 188, No. 4188, May 9, 1975, pp. 593-598. The authors are staff scientists on the National Program Staff, Agricultural Research Service, U.S. Department of Agriculture, Beltsville, Md.

• **Rats**—Each year rats consume feed, damage additional feed, destroy property, and spread disease, for a total cost of $28 per rat. Thus, with an estimated U.S. rat population of 100 million, this means that the nation's yearly keep on rats totals nearly $2.8 billion.

Although they are not as damaging as rats, mice, gophers, and other rodents should also be controlled—and for the same reasons.

• **Birds**—Birds are gluttonous and filthy. They consume and contaminate much feed and spread many diseases. Hence, they should be controlled. In a study of a 12,000-head cattle feedlot in California, University of California researchers found that the 10,000 to 20,000 birds that came to "dinner" ate an average of about 350 pounds of feed each day, for a total of 57,750 pounds during the 5½-month winter season. Iowa cattle feeders figure that Starlings add $3 to $4 to the cost of each steer marketed. Some western feedlot operators estimate that Starling nuisance and feed costs add 2¢ to the cost of each pound of gain.

In addition to the direct losses caused by pests, there are hidden, or indirect, losses: losses in efficiency, and losses in the input of energy involved in crop production—wasted energy. And losses in suffering!

Indeed, science and technology have been the great multipliers. Together, they have upped the ounce to the pound, the pint to the bushel, and the dozen to the gross. Despite this accomplishment, the very existence of man is threatened; not because he cannot produce enough food, but because he has not protected that which he has produced from the ravages of pests—from losses in food and fiber, in lowered efficiency, and in deaths. Like a thief in the night, each year pests steal away billions of dollars in losses to farmers and increased cost to consumers; and for the most part, they go unrectified. Only when millions of people die from starvation are steps taken to control them.

By applying the science and technology that we already have, food and fiber losses can be reduced substantially—perhaps by as much as 30 to 50 percent. The net result would be an increase of 10 to 15 percent in the world food supply, with no new land required. In no other way can the hungry gap be filled so quickly and at so little cost.

Weed and Brush Control[9]

A weed may be defined as a plant (1) growing where it is not wanted and interfering with desired land use, or (2) with a negative economic value within the framework of current land use. Weeds are classified as follows:

1. **Summer annuals**—Seeds produced during the summer and plant dies before winter. Seeds germinate in the spring. Examples are: pigweed, lamb's quarter, and crabgrass.

2. **Winter annuals**—Produce seeds in spring and early summer and plant dies. Seeds germinate in the fall and plants live through the winter. Examples are chickweed, peppergrass, and annual bluegrass.

3. **Biennials**—Seeds germinate in spring and produce leafy plants which lie dormant during the winter. The following season they develop seed stalks which produce flowers and seed. Examples are: wild carrot and bull thistle.

4. **Perennials**—Plants which live longer than two years and produce successive crops of seeds. This group includes woody plants and plants which are killed to the ground each winter and produce new tops the following year. The latter are called herbaceous perennials. Examples are quackgrass, Johnsongrass, Bermudagrass, bindweed, and dandelion.

Fig. 8-23. Some spray from the air; others spray from the ground (Drawing by Steve Allured)

It is estimated that weeds and brush cost U.S

[9]The author gratefully acknowledges the helpful suggestions of the following authorities who reviewed the section on Chemical Weed and Brush Control: Dr. Thomas J. Muzik, Weed Specialist, Department of Agronomy, and Dr. Ben Roche, Professor of Forestry and Range Management, Washington State University, Pullman, Wash; Dr. Joe Antognini, Senior Research Agronomist, and Messrs C. G. Randall and D. F. Dye, Stauffer Chemical Co., Mountain View, Calif., and Mr. Lawrence Southwick, Agricultural Chemical Development, The Dow Chemical Company, Midland, Mich.

The material presented in Table 8-9 is based on factual information believed to be accurate, but it is not guaranteed. Like most manufacturers, the author makes no warranty of any kind, expressed or implied, concerning the use of any product; the buyer assumes all risks of using or handling, whether in accordance with directions or not. Where the instructions and precautions given in Table 8-9 are in disagreement with those of competent local authorities or reputable manufacturers, always follow the recommendations of the latter two agencies.

ırmers and ranchers $8 billion annually; losses accru- ıg from the following:

1. Reductions in quantity and quality of crops, ıcluding deteriorations in pastures and rangelands.
2. Depreciation in land values.
3. Reductions in the quantity and quality of live- tock products through sickness and death of animals om poisonous plants, objectionable odors and flavors f milk, and damage to wool.
4. Increased labor, equipment, rotation, and emical costs.
5. Increased expenditures for cleaning small rain.
6. Harboring insects and diseases, thereby in- reasing depredations and control costs.
7. Clogging irrigation and drainage ditches.

Various methods are effective and should be used ı the control of weeds, including—

1. The use of clean, certified seed that is free of veed seed.
2. Clean cultivation.
3. Mechanical means; mowing and cutting.
4. Rotations.
5. Pasture improvement, especially through ıanagement and renovation.
6. Chemical weed control.

Since most farmers and ranchers are already amiliar with the first four methods, and the fifth one s treated under the pasture section (Section 5) of this ɔook, only chemical weed control will be elaborated ıpon herein. It should be understood, however, that hemical weed control is but a modern tool, designed o reduce or replace cultivation, and that its use may ıot prove profitable, or at least not as profitable as it hould, unless it is accompanied by good manage- nent practices.

Although chemicals have been used for centuries o control weeds, the most marked scientific advances n this field have occurred since about 1944, with the dvent and general acceptance of 2, 4-D. Today, no ɔhase of agricultural production is being subjected to nore rapid and momentous technical changes than hemical weed control. There are more than 700 her- ɔicide formulations and combinations, common ıames, and trade names on the market; and new ones re being added constantly. It is impossible, there- ore, to incorporate in a book of this type all the latest indings and recommendations relative to the mate- ials used in chemical weed control. Instead, the eader is admonished to keep in touch with his local ounty extension agent, or other local authorities, per- aining to the latest developments and recommenda- ions. Table 8-9 is merely presented as a general uide for the use of some chemicals for weed and ɔrush control. Both chemical and trade names are used. Use of trade names is for clarification to explain the use of the chemical involved. Inclusion of a trade name does not imply an endorsement of that particu- lar brand or herbicide; neither does exclusion imply nonapproval.

In addition to the information contained in Table 8-9, the following suggestions are also pertinent in chemical weed control:

1. The successful use of any chemical weed con- trol program depends primarily upon two factors, namely:

   a. Choosing the right weed or herbicide, keep- ing in mind that no one chemical is the answer to every weed problem.

   b. Applying the herbicide properly, according to the directions found on the container and the directions obtained from the county agricultural agent or other local authority or from the state ag- ricultural college.

2. Always consult the container label for safety precautions. Certain chemical weed killers are toxic and present health hazards when swallowed, allowed to contact the skin, or breathed in the form of vapor, spray mist, or dust.
3. Check with the county extension agent, or other authority, for limitations relative to (a) date of use, (b) weather factors, (c) geographic areas, (d) size of area sprayed, (e) method of application, (f) formula- tion of application, (g) quantity of application, and (h) the feed withdrawal period prior to slaughter or milk production.
4. Chemical weed and brush killers may kill or seriously injure many desirable forms of vegetation. Therefore, do not use them where there is danger of drift or spray mist or vapors contacting desirable plants.

Rodent and Bird Control[10]

Without exception, U.S. livestock establishments are plagued by one or more rodents, and/or with birds. Among such pests are rats, mice, English spar- rows, starlings, and pigeons. They should be con- trolled and, if possible, eliminated because they (1) spread diseases and parasties, (2) damage feeds and buildings, and (3) decrease profits.

Rodents and birds can be effectively and effi- ciently eliminated by following the directions given in this section or by engaging the services of profes- sionally trained pest control operators.

[10]The author gratefully acknowledges the helpful suggestions of the following authorities who reviewed this section: Mr. Richard Smith, Fish and Wildlife Service, U.S. Department of the Interior, Washington, D.C.; and Mr. Joe Abrams, Director, Information Office, Wisconsin Alumni Research Foundation, Madison, Wisc.

TABLE 8-9

GUIDE FOR THE CHEMICAL CONTROL OF WEEDS AND BRUSH

Chemical Weed or Brush Killer	Crops and/or Weeds and Brush	Strength and/or Rate of Application	Application Notes
2, 4-D is a growth-regulating substance, and selective brush and weed killer.	Recommended for the control of many woody species, including sagebrush, rabbitbrush, buckbrush, and western snowberry. Also, effective for the control of broad-leaved weeds, such as bindweed, Canada-thistle, cocklebur, tall larkspur, Russian-thistle, waterhemlock, and tansy ragwort. It may be used in cereal crops, in pasture, in lawns, and on rangeland. It is not injurious to most established grasses as normally used. It may be used in granular form in corn.	*Apply according to directions found on the container.*	It is safe to handle. Not poisonous to livestock, but when used poisonous plants, they may become mc palatable and therefore more hazardou Thus, animals should be kept out of pois plant-infested areas for 15 to 30 days f lowing treatment. Apply when plants are actively growing. Avoid spray drift to valuable crops or oth plants. Grass production is usually increased follo ing the application of 2, 4-D, due to t control of weed competition.
2, 4, 5-T is a growth-regulating substance, and selective brush and weed killer.	Recommended for the control of brush, including mesquite, osage orange, prickly pear cactus, wild roses, and wild blackberries, which are resistant to 2, 4-D. Also effective on oak and maple. Recommended for use in killing brush along fencerows, ditch banks, and roadsides. Especially effective on poison ivy and poison oak. It is also recommended for frill and basal application for control of individual trees. It will reduce tall larkspur, but retreatment may be required.	*Apply according to directions found on the container.*	It is safe to handle. Not poisonous to livestock, but when used poisonous plants, they may become mo palatable and therefore more hazardou Thus, animals should be kept out of pois plant-infested areas for 15 to 30 days fc lowing treatment. Spray any time from the first full leaf stage (which time it is most effective) to a shc time before frost. Avoid spray drift to valuable crops or oth plants. If esters are used, it is usually best to u low-volatile products. If tall growth is cut, stumps can be treated control sprouting by using a 1 to 10 so tion of acid equivalent to carrier (oil). Basal or frill treatment can be done at a time of year. Grass production is usually increased follov ing the application.
2, 3, 6-TBA is Benzac or Trysben water-soluble liquid based on amine salt of trichlorobenzoic acid.	For control of broad-leaved weeds and certain woody species; particularly useful for deep-rooted noxious perennials such as bindweed (wild morning glory), Russian knapweed, bur ragweed, and Canada-thistle.	*Apply according to directions found on container.*	Application as directed will result in loss soil productivity for an extended period; used for spot treatment in cropland, treate area should be removed from crop produ tion as residues may result in damage crops.
Ammate (AMS) is a nonselective product used as both a brush and weed killer.	Recommended for the control of hard-to-kill brush, trees, and weeds and as a stump treatment following cutting to prevent resprouting. Effectively controls mixed stands of brush. It may be used to kill brush and weeds along fencerows, on irrigation ditch banks, and drainage ditches; to kill brush in pastures and woodlots; and for spot treatment of brush on croplands or around farm buildings.	*Apply according to directions found on the container.*	Apply any time after brush reaches the fu leaf stage and until foliage begins to di color or until frost, whichever is earlier. Nonvolatile. Safe to use near herbicid sensitive crops. Not poisonous to livestock and nonflamm ble. Apply with a pressure sprayer (grour equipment) so that foliage is thorough wetted. Thoroughly clean equipment after usin Apply light oil to prevent rusting or corr sion. Material is nonselective and will suppre grass temporarily.
Boron compounds are soil sterilants.	Recommended for the control of deep-rooted perennials around storage buildings and tanks, driveways, etc.	*Apply according to directions found on the container.*	Not poisonous to animals when used in orc nary amounts. Heavy applications will sterilize the soil f extended periods which may be an adva tage in certain areas. All of the boron compounds have fir retardant properties.
Carbamates are selective weed killers. 1. 3 Chloro IPC (Osopropyl N 3 Chlorophenyl carbamate).	Effective against most germinating grasses and many broad leaves. Most widespread use has been grass seed crops—particularly alta fescue, bentgrass, ryegrass, and similar crops. May be used on some perennial legume crops such as red clover, alfalfa, and lotus.	*Apply according to directions found on the container.*	Lasts longer in soil than IPC and is genera effective on a wider range of grass a broad-leaved species. Apply in fall. Do not apply on strawberries or peas.

(Continue

TABLE 8-9 (Continued)

Chemical Weed or Brush Killer	Crops and/or Weeds and Brush	Strength and/or Rate of Application	Application Notes
2. IPC Prophan	Used to kill weedy annual grasses in certain perennial crops such as chewing fescue and creeping red fescue. Used to control weedy grasses in certain legume crops such as alfalfa, clover, and lotus. Used to control wild oats in dry peas and certain weedy grasses in strawberry fields.	*Apply according to directions on the container.*	Most effective on cool season grasses; not too satisfactory in the warmer irrigated areas on warm season grasses.
inoseb (DNBP) 1. General Dinitro sprays will kill aboveground vegetation.	Recommended for the control of most annual weeds and grasses, and in killing top growth of perennial weeds and grasses. Recommended for the control of vegetation in areas which are difficult to reach with mechanical equipment; for example, along fencerows, roadsides, ditch banks, in machinery storage areas, etc.	*Apply according to directions found on the container.* The chemical is usually added to diesel or other inexpensive oil and emulsified in water; or it is used to fortify herbicidal oils.	Toxic to livestock. Therefore, both the concentrate and the prepared spray should be kept out of the reach of people and animals. Use sufficient spray to wet the foliage. Best results are secured during warm sunny weather. These weed killers will not sterilize the soil.
2. Selective Dinitro sprays will injure or kill certain plants while others are not affected; usually Dinitro Amine.	Preemergence: 1. Recommended for the control of both broad leaf and grassy weeds in such large seeded crops as corn, beans, potatoes, cotton, peanuts, peas, and soybeans. Postemergence: 1. Recommended for the control of broad-leaved annual weeds in grasses, small grain, corn, alfalfa, clover, peas, and vetch. 2. Ineffective on grasslike weeds beyond the early seedling stage.	*Apply according to directions found on the container.*	Toxic to livestock. Therefore, both the concentrate and the prepared spray should be kept out of the reach of people and animals. Constant exposure to the fumes is harmful. Men doing continuous custom spraying should use respirators. Warm days with temperatures of 65° to 80°F are ideal for spraying. Equipment and clothing should be washed before reuse. Both the crop and the weeds should be sprayed at the proper stage. Consult the directions on the container. Avoid contact with body and clothing because it is a yellow dye.
owpon (dalapon)	Very effective for controlling such perennial grasses as Bermudagrass, Johnsongrass, and quackgrass. Effective on cattails. Used in pasture renovation programs where undesirable species have overtaken pastures.	*Apply according to directions found on the container.*	Dalapon is most effective when used as a foliage spray at 5 to 10 lb per acre, although it is also absorbed through plant roots. For best results, use added wetting agent in sprays of "Dowpon M." It is not poisonous to livestock. There is no long-term soil sterilization, although persistence is greater under dry conditions.
ybar (fenuron) a weed and brush killer, is a pelleted formulation for dry application to control brush or weeds.	Recommended for brush control in pastures, rangeland, fencerows, and drainage ditches. Recommended as soil sterilant in low to medium rainfall areas for control of most weeds around barns, silos, grain storage areas, etc. Used as spot treatment for bindweed control.	*Apply according to directions found on the container.*	Easily applied with simple equipment. Basal application is preferred for treatment of scattered brush (stems more than 4 to 5 feet apart; for more dense stands, grid, or broadcast application is preferred). The higher dosage rates are used on soils high in clay and/or organic matter. See details on product labeling for further directions.
armex (diuron) is used as both a soil sterilant and selective type herbicide in certain crops, such as alfalfa.	As soil sterilant, recommended for control of most weeds around barns, silos, grain storage areas, machinery sheds, and in irrigation and drainage ditches. As a selective herbicide recommended for control of annual weeds in established stands of dormant alfalfa. See label for other crop uses.	*Apply according to directions found on label.*	Low in toxicity to man and animals. Nonflammable. Noncorrosive. Lower rates are used for annual weeds, and higher rates for both annual and perennial weeds. See details on package label for further directions. Can injure shade trees from root absorption, so apply with care.
ilvex is a growth-regulating substance and selective brush and weed killer.	It is effective on about the same brush species as 2, 4, 5-T, except that it is more effective on oaks and maples. Very effective on turf weeds including chickweed. Also, it will reduce tall larkspur, perhaps better than 2, 4, 5-T. Silvex is widely used to control acquatic weeds.	*Apply according to directions found on container.*	It is safe to handle. Silvex is recommended as a foliage application as well as for basal and frill treatment. Avoid spray drift to valuable crops or other plants. Where needed and used in pasture and range spraying, grass production can be increased significantly. Silvex is not poisonous to livestock.

(Continued)

TABLE 8-9 (Continued)

Chemical Weed or Brush Killer	Crops and/or Weeds and Brush	Strength and/or Rate of Application	Application Notes
TCA is an oil sterilant type of grass killer and also selective on certain crops when used at lower rates.	Recommended for the control of quackgrass, Johnsongrass, Bermuda-grass, foxtail, and other undesirable grasses—both annual and perennial.	*Apply according to directions found on the container.*	Not poisonous to livestock, and no fir hazard. TCA will kill many plants with which it come in contact, but it is more toxic to grasse than to broad-leaved plants.
	It may be used in control of local infestations in croplands or for general application on noncroplands. It may be used selectively to control grass in flax, beets, and sugarcane.	*It may also be applied dry.*	TCA will sterilize the soil from 1 to 3 mo c longer (the period of time varying accord ing to rainfall) when used at higher rates For this reason, it is sometimes used t control the grass around machinery lots, o installations, and along roadsides. Thoroughly wash exposed skin and rins equipment after use.
Televar (Monuron) is used as both soil sterilant and selective type herbicide in certain crops.	As soil sterilant recommended for control of most weeds around barns, silos, grain storage areas, machinery sheds and in irrigation and drainage ditches. See package label for crop uses.	*Apply according to directions found on label.*	Low in toxicity to man and animals. Nonflammable. Noncorrosive. As soil sterilant gives prolonged weed con trol and is especially suited for applicatio to medium and heavy soil types and/o areas of light rainfall. See details on package label for further di rections.
Tordon (Picloram)	Recommended for the control of woody species, deep rooted perennial weeds, and annual weeds; on rangeland and pastures, and on right-of-ways.	*Apply according to directions found on the label.* Available in spray and pellet formulations.	Follow label precautions. Not poisonous to livestock. Avoid spray drift off of target areas.

CONTROLLING AND ELIMINATING RATS

Rats are known to have plagued mankind since the beginning of recorded history. They are referred to as the deadliest and most destructive of all animal enemies of man. Other pertinent facts relative to them are:

1. There are an estimated 100,000,000 rats in the United States.

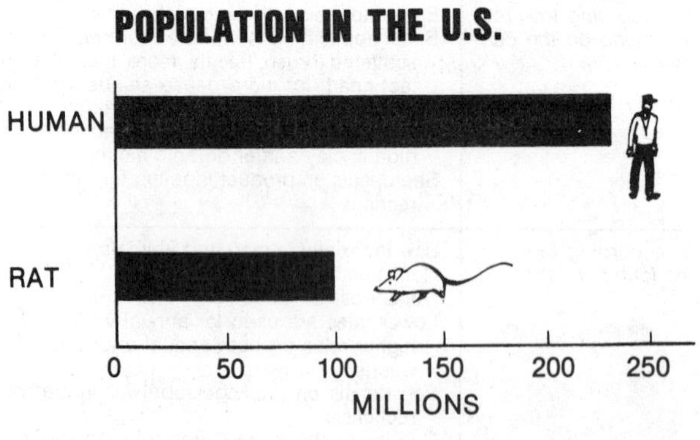

Fig. 8-24. U.S. human and rat populations.

2. Each year, on the average, every rat (a) consumes and damages feed worth $10, and (b) contaminates additional feed, destroys property, and spreads diseases costing another $18. Thus, the yearly keep of each rat amounts to over $28 or $2.8 billion for all U.S. rats.

Fig. 8-25. Rats are costly! Annually, it costs more than $28 to keep rat or nearly $2.8 billion for all U.S. rats.

3. The United Nations Food and Agriculture Organization estimates that rats eat or contaminate 42.5 million tons of the world's grains each year—enough to feed 200 million people.

4. Rats spread many dreaded human and animal diseases—including typhus, bubonic plague, infectious jaundice, rat bite fever, tularemia, and fatal food poisoning of man; and trichinosis, pseudorabies, and various kinds of fleas, lice, mites and several internal parasites of animals.

5. Rats start fires by gnawing through the insulation of wires, kill baby chicks, and weaken building foundations through burrowing under them.

6. Rats become capable of producing young when they are 90 to 120 days of age, produce annually

Fig. 8-26. Rats spread many dreaded human and animal diseases, kill baby chicks, start fires by gnawing through the insulation of wires, and weaken building foundations by burrowing under them. (Drawing by R. F. Johnson)

6 to 10 litters averaging 8 each, and under natural conditions, live to 1 year of age.

WHAT ANTICOAGULANTS[11] ARE AND WHAT THEY DO

Here are the pertinent facts relative to the anticoagulants (warfarin, prolin, pival, fumarin, diphacin, and others that are likely to appear on the market):

1. They are multiple dose poisons; that is, they must be consumed in small quantities over a period of several days before animals develop a fatal hemorrhagic condition.
2. Some of these compounds, e.g., warfarin, are closely related or similar to dicumarol, the "Dr. Jekyll and Mr. Hyde" chemical which (a) prevents blood clots following surgery, and (b) causes sweet clover disease. They kill rats by causing them to bleed to death internally.
3. They are tasteless and odorless. Thus, rodents usually do not develop bait shyness, but continue to eat the poison until they are killed.
4. They may be purchased as (a) a ready-mixed bait containing 0.025 percent active ingredient (some mouse baits contain 0.05 percent active ingredient for faster results), or (b) a concentrate with 0.5 percent active ingredient. They are sold under various brand or trade names at most feed stores, drugstores, hardware stores, and other outlets.
5. It takes from 5 to 14 days of continuous feeding to kill a rat; one dose isn't enough.
6. Most poisoned rats die in their ground burrows or other harbors and are not seen. There may be odors, but this is not apt to be a serious problem.

HOW TO USE ANTICOAGULANTS

For effective results, they should be used according to the following directions:

1. Obtain the ready-mixed material, or thoroughly mix the concentrate according to the directions found on the container.
2. Obtain 5 to 10 pounds of finished bait to control the average rat infestation on a farm. Greater quantities may be necessary where the infestation is heavy.

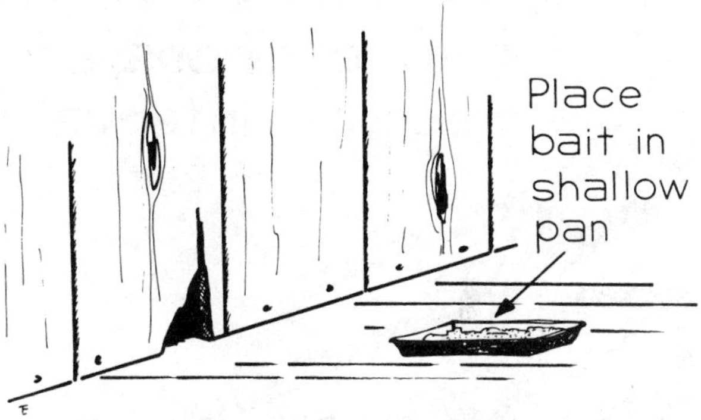

Fig. 8-27. One to two cups of mixed bait should be placed in each shallow open tray, and the trays should then be placed in areas frequented by rats. (Drawing by R. F. Johnson)

3. Place 1 to 2 cups of the mixed bait in each container, and station containers in areas frequented by rats. Shallow open trays make the best containers.
4. Protect bait from children, pets, and farm animals wherever necessary. This can be accomplished by the use of simple bait stations, which may consist of (a) tunnels made by boards about 4 feet long and 10

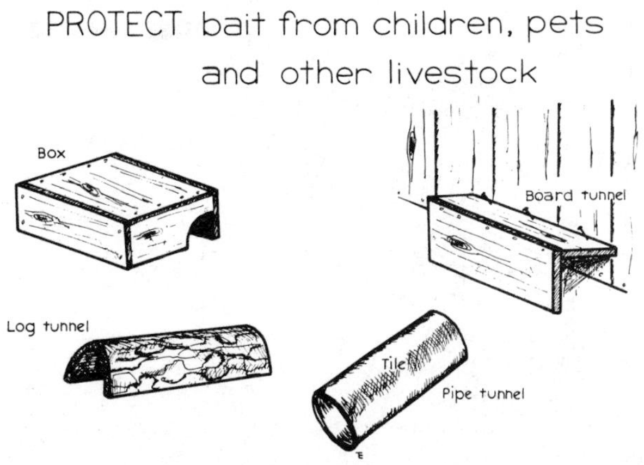

Fig. 8-28. Some common types of bait stations.

[11]Another common rat poison is zinc phosphide. At the present time, however, the anticoagulants are the preferred poison for use on the farm or ranch. In the house, rats and mice should be trapped, especially if small children are present.

to 12 inches wide nailed or hinged to the floors or walls at 45-degree angles, or (b) boxes with openings about 3 inches in diameter. If boxes are used, open bottoms are preferred, for rats (and mice) seem more willing to enter boxes set on floors over which they are used to traveling.

5. It may be helpful to place a can or other container of water in each bait station. Rats drink readily when water is available.

6. Examine the bait stations daily for the first week, and at less frequent intervals (perhaps every other day) after the first heavy feeding period. Never allow a bait container to be empty for more than 24 hours.

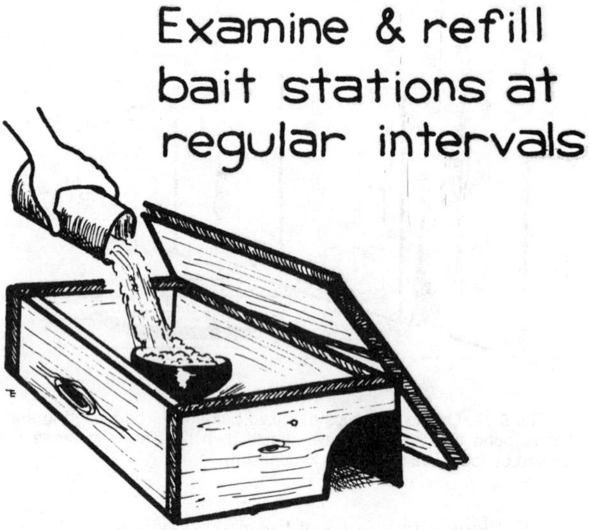
Examine & refill
bait stations at
regular intervals

Fig. 8-29. Bait stations should never be allowed to be empty for more than 24 hours. (Drawing by R. F. Johnson)

Bait for at
least 14 days
JUNE
S M T W T F S
1 2 3 4 5 6
7 8 9 10 11 12 13
14 15

Fig. 8-30. Fourteen days continuous baiting is recommended.

7. Continue baiting for at least 14 days, or for as long as evidence of feeding is observed. Where a heavy population of rats is present in nearby dumps or fields, permanent bait stations should be established and kept in operation.

8. Replace moldy, decomposed, or sour bait.

9. Sift white flour on the floor or ground around bait stations and watch for tracks. If there are no tracks within three nights, move the bait station.

Sift white flour
around bait station

Fig. 8-31. The presence of rats can be detected by sifting white flour around the bait stations.

CAUTIONS IN USING ANTICOAGULANTS

1. Do not let children, cats, dogs, chickens, or farm animals eat the bait. Because anticoagulants are used at a low concentration (.025%), it is not likely that children or farm animals will consume a sufficiently large single dose or enough repeated small doses to cause any ill effects. Yet, it must be realized that it is a poison and that it may sicken or kill man or animals if they eat enough; thus, observe carefully all caution statements appearing on the label.

2. Call a doctor immediately if children are known to have accidentally eaten the bait material. Then follow the directions on the label.

In order to allay needless fears relative to the use of warfarin, it is noteworthy that there is no known case of accidental injury to humans in the United States, due to the use of the product.

3. Consult the veterinarian if farm animals consume baits. Where large quantities have been consumed, the treatment indicated may be very similar to that for humans.

4. Where many rats are to be baited, it is good sanitary practice to pick up and destroy dead rats.

EW RODENTICIDE

A new rodenticide, RH787, marketed under the ade name Vacor, appears promising. It kills rats by nterfering with the metabolism of some of the B omplex vitamins. It differs from the anticoagulants in hat only one feeding is necessary for the rat to obtain lethal dose. However, it is approved only for baiting nside buildings.

LEANING UP PREMISES AND AT-PROOFING BUILDINGS

The old adage that "an ounce of prevention is vorth a pound of cure" still applies, despite the amazng results secured from the use of anticoagulants. remises should be cleaned up and buildings should e rat-proofed.

Cleaning up the premises may well include the ollowing:

1. Burning trash and rubbish piles, and selling unk.
2. Ripping out dark, enclosed places, such as vooden floors, platforms, or unused stairs.
3. Storing lumber, firewood, pipes, posts, and other materials on racks 18 to 24 inches off the ground.
4. Replacing bag storage of grain and feed by rodent proof bulk bins wherever practical, and cleaning up all spilled grain.

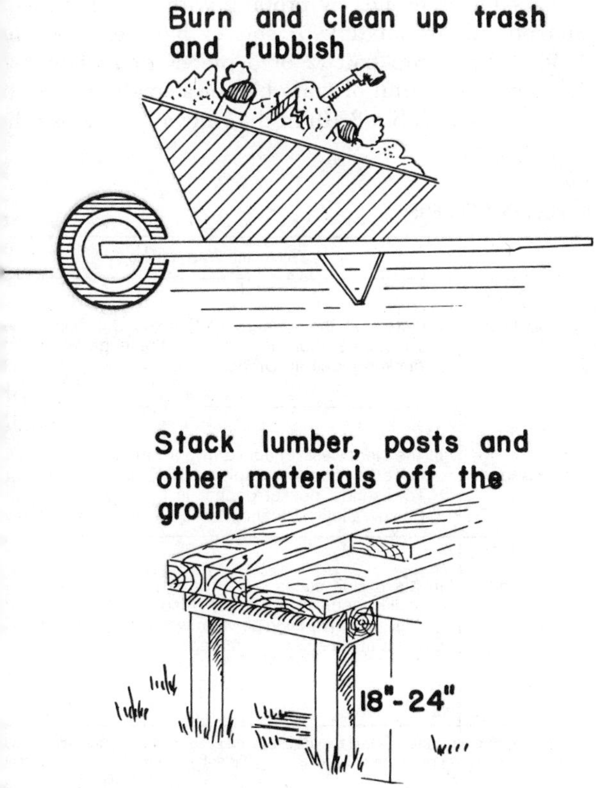

Burn and clean up trash and rubbish
Stack lumber, posts and other materials off the ground
18"-24"

Fig. 8-32. Cleaning up the premises is an important part of any rat-control program. (Drawing by R. F. Johnson)

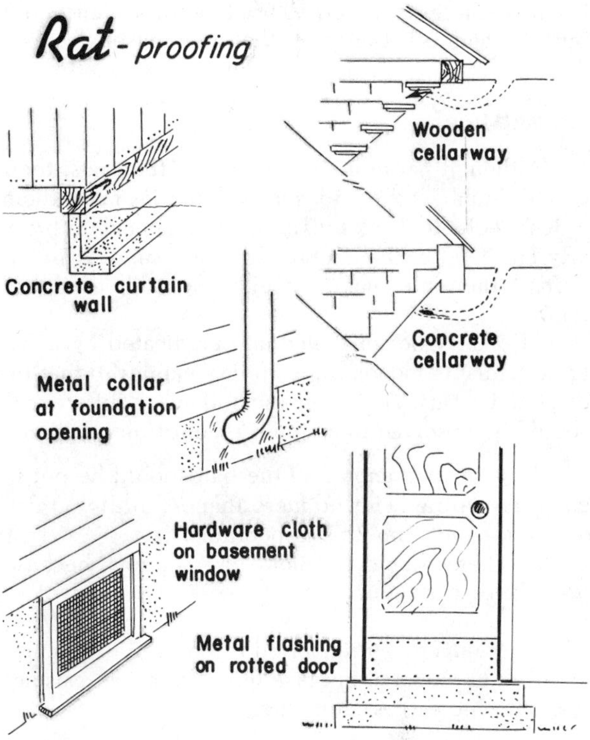

Fig. 8-33. Some methods of rat-proofing buildings. (Drawing by R. F. Johnson)

Buildings may be rat-proofed as follows:

1. By installing a concrete curtain wall around the foundation (24 in. deep, and projecting 12 in. horizontal and toward outside of building). This prevents rats from burrowing under the buildings. In no sense is the curtain wall designed to support the building.
2. By sealing the cellarway with concrete. This will block out rat burrows and make the stairway rat-proof.
3. By closing all foundation openings with metal collars, shields, or masonry.
4. By screening broken basement windows with hardware cloth.
5. By covering gnawed and rotted doors and other parts of the building with metal flashings.

CONTROLLING AND ELIMINATING OTHER RODENTS AND BIRDS

Although they may not be as damaging as rats, it

is, nevertheless, important that other rodents and birds be controlled—and for the same reasons.

HOUSE MICE

Although house mice are smaller than rats, their greater numbers and widespread activities make them at least equally dangerous insofar as damage to property and food contamination are concerned. It is important, therefore, that every effort be made to control them.

Mice may be controlled and eradicated by using anticoagulants indoors in a similar manner to that indicated for rats. However, the following differences should be observed in a mouse eradication program:

1. A smaller amount of the bait should be put in each tray (a tray with no more than ¼ in. lip) at any one time—only 1 to 2 tablespoonfuls.
2. There should be more bait stations, because mice have a smaller home range—often 10 feet or less.
3. A longer feeding period is required in a mouse eradication program, because mice are more resistant than rats and eat less at a time.

POCKET GOPHERS

Pocket gophers, which are characterized by fur-lined cheek pouches in which they carry food, are medium-sized burrowing rodents which live almost entirely underground. There are many species, and they are widely distributed over much of the nor central, southern, and western parts of the Unite States.

Evidence of the presence of pocket gophers is i dicated by (1) many nearly circular mounds thrown u in hay or pasture areas, and (2) young fruit and shad trees eaten off below the ground during the winter.

Pocket gophers can be eradicated either b poisoning or by trapping. Where large and heavily i fested areas are involved, poisoning is fastest an cheapest—especially if a gopher-bait applicator i used. This tractor-mounted unit makes an artifici burrow and meters poisoned grain into it. Five to te acres can be treated per hour. For a demonstration c the mechanical burrow-builder, contact your count extension agent or nearest representative of the Div sion of Animal Damage Control, U.S. Fish an Wildlife Service.

For poisoning smaller areas or rough ground– areas not suited to use of the mechanical applicator– the following procedure is recommended:

1. Select and prepare a poison bait suited to th locality, and species of pocket gopher (see Tabl 8-10). This is important, because pocket gophers var in their tastes.
2. Locate the main runway (10 to 16 inches bac from the mound on the side where the horseshoelik depression is found) by probing with a pipe, trowe sharpened broomstick or shovel handle (see Fig 8-34). Then drop bait (2 or 3 pieces of root bait, level tablespoonful of grain bait or a small handful o leaf bait (see Table 8-10) in the runway and close th

TABLE 8-10

BAIT GUIDE FOR POISONING POCKET GOPHERS

Formula	Where and When to Use
No. 1—Root Bait Cut 1½ lb clean, fresh sweet potatoes or carrots into pieces 1½ to 3 in. long and ½ in. square. Dust ⅛ oz strychnine alkaloid (powdered) evenly over sliced roots with sifter (pepper box). Use while fresh—same day.	Effective in the western half of the U.S., on gen *Thomomys*, but ineffective on the large Willamet pocket gopher in Oregon.
No. 2—Leaf Bait[1] Gather, free from dew or rain, 1¼ lb clean, fresh, green clover or alfalfa leaves. Dust ⅛ oz strychnine alkaloid (powdered) evenly over spread out leaves with sifter (pepper box). Use while fresh—same day.	In the Northwest. Effective in poisoning the large W lamette pocket gopher in Oregon, and th Townsend's pocket gopher in the Snake River ar Boise Valleys in Idaho.
No. 3—Grain Bait[2] Mix well ¾ pt water and ½ oz laundry starch and bring to a boil, while stirring constantly. Cook until the paste is free of lumps. Stir into the paste ¼ pt corn syrup and ½ oz glycerin. Mix in a 1-gal container with 1 oz strychnine alkaloid (powdered) and 1 oz baking soda. Pour hot paste over this mixture while stirring thoroughly. Pour the whole mixture over 12.5 lb wheat (plump kernels), steam-rolled oats, or maize. Stir until kernels are well coated, and then spread to dry.	In the prairie states, effective in poisoning the larg pocket gophers *(Geomys)*. Also effective in poiso ing the genus *Thomomys* in western half of U.S. Grain baits keep well in storage.

[1]One specialist in Idaho, who has had much experience in the eradication of the pocket gopher, recommends that the leaf bait be prepared as follows: (1) gather the top 5 6 in. of alfalfa plants, (2) dip them in water and shake off excess moisture, so that the powder will adhere to the plant, and (3) dip the moist plant in a mixture of equal parts strychnine and saccharine (powder). The sugar makes the poison more palatable.

[2]Strychnine-poisoned grain may be either home-mixed or purchased commercially. Also the U.S. Fish and Wildlife Service sometimes makes mixed products available at nominal price through the office of the county agent (and through the County Agricultural Commissioners in California) or through bait mixing stations when it is not availab commercially.

robed opening with the heel. Poison the entire field t one time.

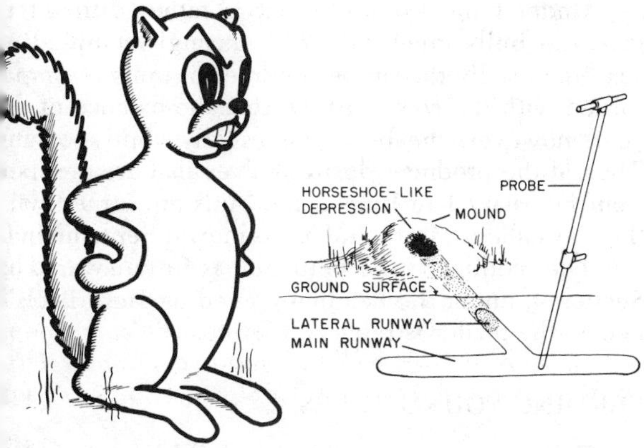

Fig. 8-34. Right way to use a probe to locate the main runway. Drawing by Steve Allured)

3. Drag a harrow down the mounds.

4. If new mounds appear, administer a second oisoning.

In trapping, the following procedure is recomnended:

1. Locate the runway as indicated for poisoning.

2. Open the lateral to the main runway with a arden trowel, or similar instrument, and insert ommercially-made pocket gopher spring traps in the nain runway (see Fig. 8-35).

3. Tie trap with a piece of light wire attached to a take.

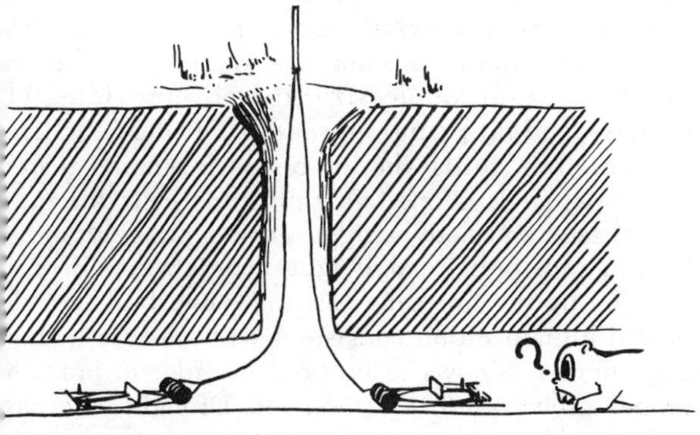

Fig. 8-35. Right way to trap pocket gophers. (Drawing by Steve Alured)

GROUND SQUIRRELS

In certain areas, the population of ground squirels is so high that they destroy as much as 25 percent of the growing and ripening grain. Although these rodents may be trapped or shot, such methods are neither effective nor practical when there are many squirrels. With dense populations, the following control measures are recommended:

1. Poison the squirrels by scattering strychnine-treated grain bait (made as directed in formula No. 3 of Table 8-10, but with steam rolled oats substituted for the wheat and of the following concentrations: for Columbian, Townsend, and Washington ground squirrels, use 1 oz of strychnine to 12 lb of steam rolled oats) on clean, hard surface near the entrance of burrows. One quart of poisoned grain bait is sufficient for 40 to 60 baits of 1 to 2 teaspoonfuls each.

Caution should be exercised to keep strychnine-poisoned grain from animals, poultry, and game birds.

2. Apply the new rodenticide, Ramik Green, according to the manufacturer's directions. Ramik is weather-resistant and is approved for both indoor and outdoor use.

3. Gas the squirrels with calcium cyanide, which may be administered in either the dust or the granular form.

Poisoning and gassing are about equally effective in controlling ground squirrels, but the former is less time-consuming. Further, it may be desirable to use both methods simultaneously, especially where difficulty is encountered in getting squirrels to eat poisoned grain.

ENGLISH SPARROWS, STARLINGS, AND PIGEONS

Most bird contamination around the farm is caused by English sparrows, starlings, and pigeons.

The following methods of control are effective:

1. Keep most such birds out of buildings by placing ½-inch hardware cloth over all windows and other openings, particularly around eaves.

2. Kill the birds by shooting. For best results, (a) induce the birds to feed at certain places, (b) scatter the grain in long, narrow lanes along which shooting may be directed when the birds flock to feed, and (c) use No. 10 shot.

3. Trap the birds. In many localities shooting is impossible, and poisoned baits are impractical or undesirable, or both. In such places, unwanted bird populations may be controlled by the persistent use of traps. Satisfactory traps of many kinds and styles may be homemade at low cost by anyone with a moderate amount of skill. For specifications and information on the locations and operation of various traps for different birds, the farmer and rancher should contact the county agent, vocational agriculture instructor, or write to the state college of agriculture.

4. Poison the birds by the following method: Build a platform 4 to 6 feet high, and during the

winter expose thereon unpoisoned grain for several days. Get the sparrows, starlings, and/or pigeons accustomed to feeding there. Then put out strychnine-poisoned grain (prepared as indicated in Table 8-10 (page 766), except use 1 ounce of strychnine to 8 quarts of grain), using the same kind of grain to which the birds have become accustomed.

5. Use one of the new bird control chemicals (such as "Avi-control" or "Starlicide") according to manufacturer's directions.

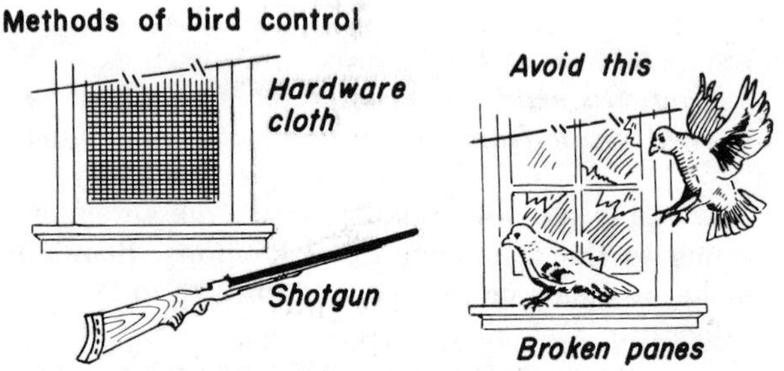

Fig. 8-36. Some common methods of controlling birds around the farm. (Drawing by R. F. Johnson)

SOME BEEF CATTLE MANAGEMENT PRACTICES

Successful cattlemen practice good management. Some beef cattle management practices that are not covered elsewhere in this book follow.

Managing the Beef Bull

Outdoor exercise throughout the year is one of the first essentials in keeping the bull virile and in a thrifty, natural condition. The finest and easiest method of providing such exercise is to arrange for a well-fenced, grassy paddock (about 2 acres is a good size for one bull). Many valuable sires have been ruined through close confinement in a small stall—or more likely yet—through being kept knee deep in mud within a small filthy enclosure. In addition to the valuable exercise obtained in the grassy paddock, the animal gets succulent pasture, an ideal feed for the herd bull.

A satisfactory and inexpensive shelter should be provided for the bull. The most convenient arrangement is to have this within or adjacent to the paddock, so that the bull may run in and out at will. Sufficient storage space for feed, along with materials and equipment for caring for the bull, should be provided in this building. Normally, purebred bulls are kept in separate stalls and enclosures, though some successful purebred breeders regularly run several valuable bulls in one enclosure. Bulls used in commercial herds are usually run together, both on the range and when separated out from the cows. Because of their scuffling and fighting, there is more injury hazard when bulls are handled in a group.

Under range conditions, it is rather difficult to give the bulls much attention during the breeding season. Usually the proper number of bulls is simply turned with the cow herd. During the balance of the year, however, the bulls are usually kept separate. Thus, if the producer desires calves that are dropped from February 1 to June 1, the bulls are turned with the cows about May 1 and are removed September 1.

The feeding of the herd bull is fully covered in Section 4, under the heading, "Feeding Herd Bulls"; hence, the reader is referred thereto.

FEEDING YOUNG BULLS

Following weaning, bulls should be fed and developed sufficiently to show their inherited characteristics, but without excessive finishing. Simultaneously, they should be given plenty of exercise. Overfeeding and lack of exercise are apt to result in infertility, low-quality sperm, and unsound feet and legs.

To achieve proper development, young bulls should gain at least 2½ pounds daily from weaning to 12 to 15 months of age. This will necessitate a daily feed allowance equal to about 2½% of their body weight, with a ration comprised of 50% or more concentrate. From 15 months to 3 years old, they should make a daily gain of 2 to 2¼ pounds and receive a feed allowance equal to 2 to 2¼% of their body weight, with the proportion of roughage increased after the first year.

Without doubt, the least laborious and most convenient management arrangement in handling young bulls consists in allowing a group, not exceeding 10 to 15 head of uniform size and age, the run of a pasture or enclosure of ample size, thereby providing (1) exercise, and (2) pasture in season. Of course, wherever possible, bulls should be performance tested while being developed. Ideally, this calls for individual feed and body weight records, although group feeding plus individual weight records will suffice.

During the breeding season, young bulls should be fed a grain ration consistent with pasture quality and number of cows to be bred in order to promote proper growth and development. Drought, overpasturing, and poor quality pastures are situations in which grain supplementation is particularly needed. Heavy service and poor pasture with no supplemental feeding may shorten the breeding career of a young bull.

After the breeding season, yearling bulls generally need 5 to 6 pounds of grain per day along with good forage.

FEEDING SALE BULLS

Bull sales are generally held in late winter and early spring, at which time mostly yearling and two-year-old bulls are sold. In order to attract buyers, they are usually grain fed from an early age. Most bull buyers—especially commercial cattlemen in rougher range areas—would rather have their new bulls in less than fitted sale condition. They find that such bulls are more fertile and more apt to follow the cows when turned to pasture during the breeding season.

Handling highly conditioned sale bulls during the critical period—after the sale is over, and just ahead of the breeding season—is all-important. Experienced cattlemen "let them down" and yet retain strong, vigorous animals. They do this successfully by (1) providing plenty of exercise, (2) increasing the amount of bulky feeds, such as oats, in the ration, (3) cutting down gradually on the grain allowance, and (4) retaining the succulent feeds and increasing the pasture and hay.

SEMEN AND FERTILITY EVALUATION

Semen quality is based on evaluating the ejaculate for (1) density of concentration, (2) rate of movement, (3) motility, and (4) morphology.

It is advisable to have a semen or fertility evaluation made before buying a bull, especially when purchasing for a single sire herd. The penalty for an infertile multisire herd is not so great. Although a semen test will not provide absolute assurance that the bull will settle the females to which he is exposed, it is a strong indicator. Bulls that are not producing sperm cells, that are producing a high percentage of nonmotile or abnormal sperm cells, and that have infections in the reproductive tract should be avoided. No matter how high the growth rate, and no matter how admirable all the other qualities, a bull that cannot sire a calf crop is useless.

Managing Beef Cows

Reproduction—the production of calves—is the first and most important requisite of the cow-calf system, for if cows fail to produce the cattleman will soon be out of business.

BREEDING AND CALVING SEASONS

The season at which the cows are bred depends primarily on the facilities at hand, taking into consideration the feed supply, pasture, equipment, labor, and weather conditions; whether the cattle are being produced for ordinary commercial or for purebred purposes; and whether they are strictly beef or dual-purpose cattle.

• **Advantages of spring calves**—The production of spring calves has the following advantages:

1. The cows are bred during the most natural breeding season—at a time when they are on pasture, gaining in flesh, and more likely to conceive. The calving percentage is usually higher, therefore, with a system of spring calving.
2. The calves will be in shape to sell directly from the cows in the fall, at which time there is a good demand for feeder calves.
3. If the calves are to be sold as yearlings, one wintering is saved; or if they are to be sold at weaning time, no wintering is required.
4. Because of greater utilization of cheap roughage, dry cows may be wintered more cheaply.
5. Less labor and attention is required in caring for the calves the first winter.
6. Spring calves require less grain and utilize the maximum amount of pasture and forage.

• **Advantages of fall calves**—The production of fall calves has the following advantages:

1. The cows are in better condition at calving time.
2. The cows give more milk for a longer period.
3. The calves make better use of the grass during their first summer.
4. The calves escape flies, screwworms, and heat while they are small.
5. Upon being weaned the following spring, the calves can be placed directly on pasture instead of in a drylot; or, if the desire is to sell, they usually find a ready market ahead of the influx of fall feeder calves from the range area.
6. When the intention is to sell market milk from dual-purpose cows, fall calves are usually best. The greater flow of milk is obtained during the period of highest prices.

CALVING TWO-YEAR-OLD HEIFERS

Unless forage is abundant and cheap, cattlemen can, and should, breed yearling heifers to calve as two-year-olds. But, in doing so, the following practices should be observed:

1. Select the heaviest and highest scoring individual heifers at weaning. Weight at weaning is a means of evaluating the dam's milking ability.
2. Keep heifers separate from older cows.
3. Start with 50% more weaner replacement heifers than needed if it is the intent to maintain the same size herd—with no provision for expansion whatsoever. This means that for every 100 cows in the herd, 20 replacement heifers are actually needed each year in order to maintain the same size herd. (There is about a 20% replacement in each herd each year.)

However, 30 weaner replacement prospects (50% more than actually needed) should be held simply because, based on averages, 10 of them will either die or have to be culled before they replace older cows.

4. Replacement heifers should be fed for gains of approximately 1.0 lb per head per day from weaning to calving. From weaning to mid-pregnancy, 1¼ lb gain per day is about right. From mid-pregnancy until calving, the gain may be lowered to 0.75 lb per day.

5. Select yearlings and coming 2-year-old heifers on the basis of individuality and rate of gain. Also, cull heifers with small pelvic openings; those with large pelvic openings (above 34 sq in., or 220 sq cm) have less calving difficulty. Avoid excessively fat heifers.

6. Breed only well-developed heifers, weighing 600 to 750 lb (depending on breed) at 13 to 14 months of age. Size at breeding is more important than age. Also, some breeds come in heat and mature a little earlier than others.

7. Breed heifers 20 days earlier than the cow herd and restrict the breeding season to 45 days. This gives a short concentrated calving period; therefore, proper attention and help can be given heifers at calving time.

8. "Flush" feed heifers to gain approximately 1.5 lb per head daily beginning 20 days before the start of and continuing through the breeding season.

9. Breed heifers to a bull known to sire small calves at birth.

10. Feed a well-balanced ration, and feed for continuous gain of 0.75 to 1 lb during the pregnancy period; but don't get them too fat.

11. Feed heifers to weigh at least 775 lb by 120 days before calving.

12. Feed heifers to gain 100 to 120 lb from 120 days prior to calving. Heifers should weigh at least 875 lb just before calving and approximately 775 lb shortly after calving.

13. Give heifers special care at calving time. This should include—

a. Providing adequate facilities, including (1) a pull stall, and (2) small pens, each suitable for confining a heifer and her calf for approximately 24 hours of "mothering up."

b. Moving each heifer into the calving area approximately 2 weeks before the expected calving date.

c. Checking heifers for calving at 2-hour intervals.

d. Rendering assistance quickly and expertly when it is needed.

e. Removing heifer and calf from calving area within 24 hours after birth and putting them into a clean, dry pasture or other similar area.

14. Provide superior nutrition—well balanced, and rather liberal—during the lactation period, because a heifer's nutritional requirements double after calving. This requires a good ration—one containing adequate energy and proteins, and fortified with the necessary minerals and vitamins. In season, usually this can be accomplished by keeping these heifers on good pastures, with or without supplemental feeding both during pregnancy and lactation. When good grass is not available—in the winter, early and late, or during droughts—proper feeding must be relied upon.

15. If practical, wean early; at 2 to 6 months of age, rather than the normal 7 months.

16. Run heifers that calved as two-year-olds in a separate herd until after they have had their second calf.

17. Cattlemen are admonished to try calving half of their replacement heifers as two-year-olds to start with—to make sure that they know what is involved before going all out.

Of course, the below-average breeder—the fellow who has lightweight, poorly developed heifers, and who wouldn't think of staying up nights and having cold, numb fingers while serving as nursemaid to a heifer and a newborn calf—should take another year and stick to calving out three-year-olds. Likewise, calving three-year-olds may be practiced in those areas where forage is especially abundant and cheap, while concentrates are scarce and high.

CARE OF COWS AT CALVING

The gestation period of a cow is about 283 days, but it may vary a few days in either direction. The careful and observant caretaker will make definite preparations in ample time. It is especially important that first-calf heifers be watched at calving time, for frequently they will need some assistance. Older cows that habitually have trouble in parturition may well be culled from the herd.

• **Signs of approaching parturition**—Usually, the first sign of approaching parturition is a distended udder, which may be observed some weeks before calving time. Near the end of the gestation period, the content of the udder changes from a watery secretion to a thick, milky colostrum. As parturition approaches, there generally will be a marked shrinkage or falling away of the muscular part of the region of the tailhead and pinbones, together with a noticeable enlargement and swelling of the vulva. The immediate indications that the cow is about to calve are extreme nervousness and uneasiness, separation from the rest of the herd, and muscular exertion and distress.

• **Preparing for calving**—At the time the signs of approaching parturition seem to indicate that the calf ıay be expected within a short time, arrangements ɔr the place of calving should be completed.

During the seasons of the year when the weather ; warm, the most natural and ideal place for calving ; a clean, open pasture away from other livestock. Iogs should not be allowed in the same place with he cow, for they are likely to injure or kill the young alf. They have even been known to injure the cow.

Under pasture conditions, there is decidedly less langer of either infection or mechanical injury to the ·ow and calf. In commercial range operations, it is ·ommon practice to ride the range more frequently at alving time. A better procedure consists in having a mall pasture adjoining headquarters into which ıeavy springing cows are placed a few days before :alving. With the added convenience of such an ar- ·angement, the animals can be given more careful at- ention.

During inclement weather, the cow should be ›laced in a roomy (10 or 12 ft square), well-lighted, vell-ventilated, comfortable box stall or maternity ›en which should first be carefully cleaned, disin- ·ected, and bedded for the occasion.

• **Normal presentation**—Labor pains in a mild orm usually start some hours before actual parturi- :ion. After a time, the water bag appears on the out- side, usually increasing in size until it ruptures from the weight of its own contents. This is followed :losely by the appearance of the amniotic bladder (the second water bag), with the fetus. With the rupture of the second water bag, the straining becomes more violent, and presentation soon follows. Most com- monly in presentation, the front feet come first fol- lowed by the nose which is resting on them, then the shoulders, the middle, the hips, and then the hind legs and feet.

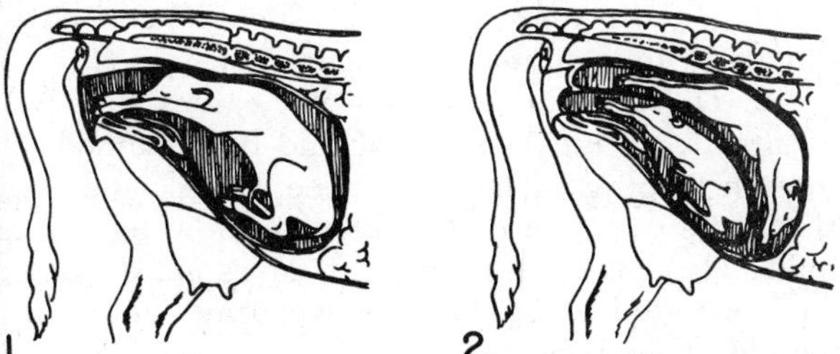

1. Normal Presentation

2. Normal Twin Positions

Fig. 8-37. 1. Normal single presentation; the back of the fetus is directly toward that of the mother, the forelegs are extended toward the vulva, and the head rests between the forelegs. If it is necessary to render assistance, apply ropes above the ankle joints and pull alternately downward on each leg as the cow strains.

2. Normal twin positions. If delivery does not proceed normally, this is a case for a veterinarian.

With posterior presentation (hind feet first), there is likely to be difficulty in calving. Moreover, there is considerably more danger of having the calf suffocate through rupture of the umbilical cord and strangulation.

• **Rendering assistance**—If presentation is normal and within an hour or two after the onset of signs of calving, no assistance will be necessary. On the other hand, if the cow has labored for sometime with little progress or is laboring rather infrequently, it is usually time to give assistance. Such aid will usually consist of fastening small ropes around the pasterns and pulling the young outward and downward as the cow strains. This should be done by an experienced caretaker or a competent veterinarian. It is always well to be reminded that rough, careless, or unsanitary methods at such a time may do more harm than good.

• **The newborn calf**—If parturition has been normal, the cow can usually take care of the newborn calf, and it is best not to interfere. However, in unusual cases, it may be necessary to wipe the mucus from the nostrils to permit breathing; or, more rarely yet, artificial respiration methods must be applied to some calves. This may be done by blowing into the mouth, working the ribs, rubbing the body rather vigorously, and permitting the calf to fall gently. The cow should be permitted to lick the calf dry.

With calves born in sanitary quarters or out on clean pastures, there is little likelihood of navel infection. To lessen the danger of such infection, the navel cord of the newborn calf should be treated at once with a 10 percent solution of tincture of iodine.

A vigorous calf will attempt to rise in about 15 minutes and usually will be nursing in half an hour to an hour. The weaker the calf, the longer the time before it will be able to be up and nursing. Sometimes it may even become necessary to assist the calf by holding it up to the cow's udder.

The colostrum (the milk yielded by the mother for a short period following the birth of the young) is most important for the well-being of the newborn calf. Aside from the difference in chemical composition, compared with later milk, the colostrum contains antibodies which temporarily protect the calf against certain infections, especially those of the digestive tract.

Usually it is best to keep the cow and calf in a small pasture for a few days. After this, they may be turned back with the main herd. Nothing is better for the cow at calving time than plenty of grass, and both the cow and calf will be helped by an abundance of fresh air and sunshine. The cow may deliberately hide the calf for the first few days, and the job may be so thoroughly done as to require considerable cleverness on the part of the caretaker to find it.

• **The afterbirth**—Under normal conditions, the fetal membranes (placenta or afterbirth) are expelled from 3 to 6 hours after parturition. Should they remain as long as 24 hours after calving, competent assistance should be given by an experienced caretaker or a licensed veterinarian. The operation of removing a retained afterbirth requires skill and experience; and, if improperly done, the cow may be made a nonbreeder. Furthermore, before doing this, the fingernails should be trimmed closely; the hands and arms should be thoroughly washed with soap and warm water, disinfected, and then lubricated with petroleum jelly or linseed oil. In no case should a weight be tied to the placenta in an attempt to force removal.

As soon as the afterbirth is ejected, it should be removed and burned or buried in lime, thus preventing the development of bacteria and foul odors. This step is less necessary on the open range, where animals traverse over a wide area.

Managing Confined (Drylot) Cows

An increasing number of beef cows are being confined to a drylot, all or part of the year. Among the management practices peculiar to this type of operation are the following:

1. **Dehorning**—Cows kept in a drylot should either be naturally polled, dehorned, or have their horns tipped. (When the latter is done, leave only a 3- or 4-inch stub.)
2. **Scours**—Calf scours is the bane of the confinement cow-calf operator. A drylot aggravates scours and favors a buildup of the scour problem over the years.

Wherever possible, it is strongly recommended that cows be removed from the drylot immediately before calving and placed in a sizable clean pasture (one that has been idle for a period of time) for calving out, thereby alleviating most, if not all, scouring. Also, a good pasture will stimulate milk flow and make for a good nutritional start in life for calves.

Where calving on clean pasture is not practical, the following precautions against scours should be taken:

a. Clean, disinfect, and bed the maternity stall after each birth.

b. Inject the newborn calf with 1 million IU of vitamin A and vaccinate against calf scours.

c. Limit the feed of heavy producing cows until the calf is 10 days to 2 weeks old.

3. **Consider early weaning**—Weaning calves at two months of age will save feed and result in getting the cows rebred more quickly; hence, early weaning should be considered.
4. **Alternate day feeding**—Cows kept in confinement may be fed every other day without altering performance, thereby effecting a saving in labor.
5. **Dust control**—In dry, windy areas, pens should be equipped with sprinklers for dust control.

SEMICONFINEMENT (or Partial Confinement) COWS

A semiconfinement (or partial confinement) oper ation is one which takes advantage of grazing durin part of the year, such as winter grazing of corn or sor ghum stalks or seasonal grazing of pastures. In addi tion to providing low-cost feed and allowing the ani mals to do their own harvesting, breeding may b timed so that the calves will be dropped on clean pas ture as a means of (1) preventing calf scours, and (2 stimulating milk flow.

Preconditioning Calves

Preconditioning is a way of preparing the calf t withstand the stress and rigors of leaving its mother learning to eat new kinds of feed, and shipping from the farm or ranch to the feedlot. To the cow-calf pro ducer, it is a program of management, nutrition, an immunization.

The steps used in preconditioning may, an should, vary somewhat among areas, farms, an ranches. The important thing is that the program b written down, adhered to rigidly, then certified to b both the owner and the veterinarian. The produce should take the lead in developing such a program but he should seek the counsel of his veterinarian an potential buyers.

Opinions differ rather widely as to what consti tutes properly preconditioned cattle. However, th following preconditioning program is presented wit the hope that the beef producer will use it (1) as yardstick with which to compare his existing program or (2) as a guidepost so that he and his local veterinar ian, and other advisers, may develop a similar an specific program for his own enterprise:

1. **Handle quietly**—Calves should be handle quietly, with a minimum of excitement.
2. **Dehorn and castrate**—All calves that wil eventually go into feedlots should be dehorned (al though tipping of horns is acceptable), and they should be castrated unless they are to be fed out a bulls. There is far less stress if calves are dehorne and castrated well ahead of weaning—about tw months of age is best.
3. **Wean**—Calves should be weaned 30 days ahead of shipment.
4. **Start on feed**—Adjust to feed bunks and wate troughs and start on a ration similar to that which they will get in the feedlot. For the first three days following weaning, calves should have access to loose grass hay. Additionally, they should be started on a ration o about the following composition:

Crude protein, minimum %	12.0
Calcium, %	.5

Phosphorus, %	.3
Vitamin A, IU/lb	5,000
Net energy for production (NEp), Mcal	38
Roughage to concentrate ratio, approx.	40:60

If weaning is totally impractical, calves should be started on a creep feed similar to the above ration.

This type of ration will be very similar to the starting ration that calves will receive when they arrive in the feedlot.

Use medicated feed only on the recommendation of your veterinarian.

5. **Vaccinate**—Vaccinate 2 weeks after weaning. If calves were vaccinated for blackleg, malignant edema, and leptospirosis before 3 months of age, revaccinate. Simultaneously, vaccinate for "red nose" (infectious bovine rhinotracheitis, or IBR), bovine virus diarrhea (BVD), and para-influenza 3 (PI-3). In some instances, clostridial toxoids for types C and D are needed. Follow your veterinarian's advice for vaccination procedures. If a direct sale to a feedlot is involved, the calves should be vaccinated in keeping with the regular program of the feedlot.

6. **Treat for parasites**—At the time of weaning, and prior to shipment, calves should be checked for both internal and external parasites, and treated as necessary. Usually this involves (a) treating for grubs, through either spray, pour-on, or feed; (b) spraying for lice; and (c) checking for worm eggs, and worming if necessary.

7. **Reduce time from farm or ranch to feedlot**—Every effort should be made to reduce the total time between the moment calves leave the farm or ranch and when they arrive at the feedlot.

Where either truck or rail shipments are longer than 36 hours (the 28-hour law governing rail shipments may be extended to 36 hours upon written request of the owner), unload en route for the purpose of giving feed, water, and rest for a period of at least 5 consecutive hours before resuming transportation.

8. **Reduce stress and exposure to infection**—The stress and exposure to infection during the marketing and transportation periods should be reduced to a minimum.

9. **Preconditioning certificate**—It is extremely important that records be kept of all husbandry, nutritional, and medical histories, and that the man who sells the feeder cattle should provide the man receiving them with a written record of all of them. This will help the feedlot operator fit the cattle to his program and minimize costly and unnecessary procedures. A suggested preconditioning certificate is herewith presented as Fig. 8-38, page 774.

Management of Feedlot Cattle

There are many facets of cattle management. Only those that are unique to cattle feedlots will be discussed in the sections that follow.

HEALTH OF FEEDLOT CATTLE

Loss from disease is greater in cattle feedlot operations than in any other type of cattle enterprise. The movement of cattle, stress conditions, methods of purchase, feeding of concentrated feeds, population density, sometimes unsanitary conditions, and the bigness and complexity of the operation all contribute to disease incidence; and disease incidence is directly proportional to population density.

The health of feedlot cattle can be greatly improved through a program involving the following: (1) preconditioning, (2) moving cattle directly from the producer's farm or ranch to the feedlot (fewer than 20% of the cattle now move directly from producer to feedlot), (3) reducing the time between ranch and feedlot, (4) lessening the amount of stress and exposure to infection during marketing and transportation periods, (5) providing the man receiving the cattle with more adequate medical and nutritional history of the cattle, (6) handling of incoming feedlot cattle properly, and (7) diagnosing and treating sick cattle early.

HANDLING NEWLY ARRIVED CATTLE

The most critical period for feeder cattle is the first 21 to 28 days in the feedlot. The following recommendations pertaining to incoming cattle will minimize death losses and maximize performance:

- **Provide clean, dry, comfortable quarters**—Whether it be an open lot or a building, incoming cattle should be provided with clean, dry, comfortable quarters. A dry and comfortable bed for resting is very essential because cattle are tired and have a low resistance to respiratory diseases.
- **Process upon arrival**—The relative merits of processing calves (1) at point of origin, (2) upon arrival at destination, or (3) two to three weeks after arrival are often debated. Experiments show that processing at arrival is best, and that processing at point of origin is preferable to delayed processing.

  When processing, steers and heifers should be separated, as they will feed better that way.
- **Provide clean fresh water**—Give the cattle easy access to clean, fresh water because they are usually dehydrated and thirsty upon arrival and will drink water before they eat feed. Open water tanks are preferable to automatic water bowls because most farm and ranch cattle are accustomed to drinking from tanks or ponds.
- **Provide a palatable ration**—Feeding a palatable ration—one that cattle will start eating soon after they

PRECONDITIONING CERTIFICATE

Date ______________________

Number of Cattle ______________________

Steers ________ Heifers ________ Bulls ________ Breed ______________________ Age ________ Brand ________

PRACTICES, TREATMENT, AND VACCINATION

	Date	Product Brand	Signature of Responsible Person
Castrated			
Dehorned			
Blackleg & Malignant Edema Vaccination			
Shipping Fever Vacc.			
Lepto Vaccine			
IBR Vaccine			
EVD Vaccine			
Other Vaccines			
Weaned			
Wormed			
Grub Treated			
Lice (Treatment) Spray or Dip			
Vitamin A.D.E. Inj.			
Medication, Antibiotics, Sulfa, electrolytes			

Ration during preconditioning period ______________________

Loading Point ______________________

The undersigned hereby declares and certifies that the practices, treatments, and vaccinations indicated above have been carried out and administered to all of the cattle described and identified by this certificate.

Date: ______________ SELLER: ______________
(Signature of owner or authorized representative)

Date: ______________ CATTLE PRODUCER: ______________
(Signature)

Date: ______________ SELLER'S VETERINARIAN: ______________
(Signature)

Fig. 8-38. Preconditioning certificate.

are unloaded in the feedlot—will reduce the incidence of shipping fever and make the cattle recover their weight loss more rapidly.

1. **Roughage**—The best roughage for newly arrived feedlot cattle is *long grass hay,* because it is very similar in composition and taste to the grass to which most feedlot cattle have been accustomed. Thus, cattle will usually eat long grass hay more quickly than any other roughage. In areas where grass hays are not available, or are too expensive to feed, any other nonlegume roughage can be fed, such as

corn silage, sorghum silage, cottonseed hulls, corncobs, or grass-legume hay that contains more grass than legumes. Above all, do not feed high-quality alfalfa hay because it is too laxative and it will cause scouring which will trigger shipping fever. The same may be said relative to alfalfa haylage or alfalfa silage.

Corn silage of approximately 65 percent moisture content is an excellent feed for new cattle. If cattle do not eat the corn silage too well at the outset, the feeder should sprinkle a little grass hay on the top of it to encourage them to start eating.

2. **Concentrate**—Incoming cattle may be fed approximately 4 lb of concentrate per head daily, consisting of 2 lb of grain and 2 lb of protein supplement. The protein supplement should be fortified so as to provide 50,000 IU of vitamin A daily. For heavily stressed cattle, the protein supplement should also contain a high level of antibiotic, or a combination of antibiotic and a bactericidal agent such as sulfamethazine. The following level of antibiotic-sulfamethazine is recommended:

> Feed 350 mg of antibiotic plus 350 mg of sulfamethazine per head daily to newly arrived cattle for a period of 28 days. With the antibiotic-sulfamethazine treatment, shipping fever is practically alleviated.

Do not feed urea for the first 28 days after the cattle arrive. Starvation destroys the ability of the rumen to utilize urea or other nonprotein nitrogen and makes cattle more sensitive to urea toxicity. Therefore, it is not wise to put extra stress on cattle by using urea during this adjustment period.

- **Satisfy mineral hunger**—Incoming cattle are usually hungry for minerals, especially if they have been on dry range forage. Thus, they should have access either to a mineral mixture consisting of two parts of dicalcium phosphate and one part of salt, or to a good commercial mineral.
- **Observe, isolate, and treat sick animals**—Newly arrived cattle should be observed at least twice daily. Sick animals should be removed and treated. Treating sick animals promptly, rather than waiting until tomorrow, may mean the difference between life and death. Animals that show clinical signs of shipping fever—sunken eyes, runny nose, drooling at the mouth, labored breathing, and/or weaving (unsteady gait)—should be isolated in a separate "sick pen" or "hospital."

Rest, fresh water, good feed, proper medication, and TLC (tender loving care) are the cardinal essentials for preventing shipping fever and death losses.

AMOUNT TO FEED; FULL VS LIMITED FEEDING

Feed intake is one of the key factors affecting feedlot performances. The reason for emphasis on high feed intake is that once a sufficient amount of the ration is consumed to meet the maintenance needs of a finishing animal, the remainder is converted to gain with remarkable efficiency. Thus, by adding 4 lb to the daily feed intake of a 600-lb steer, rate of gain may be increased by 1$^1/_{10}$ lb per day. Conversely, poor feed intake results in too high a percentage of the total nutrients being expended for maintenance.

Thus, finishing cattle should receive a maximum ration over and above the maintenance requirements. In general, they will consume daily an amount (on an air-dry basis) equal to 2.5 to 3.0 percent of their liveweight. Feed intake will vary according to the condition of the cattle, the palatability of the feeds, the energy of the ration (in general, animals eat to meet their energy needs), the weather conditions, and the management practices. For example, older and more fleshy cattle consume less feed per hundredweight than do younger animals carrying less condition; thus, mature, overfinished steers will consume feeds in amounts equal to about 1.5 percent of their liveweight, whereas thin steers under 2 years of age will consume fully twice as much feed per unit liveweight.

Limited feeding means just what the term indicates—not giving the animals all they want. Limited feeding generally decreases the rate of gain, adversely affects feed conversion, and increases cost of gains. Under most conditions, cattle should be full fed throughout the finishing period.

MUD PROBLEM

Studies show that mud can reduce cattle gains by as much as 25 to 35 percent. Thus, it is important that the problem be minimized, especially in high rainfall areas. Good drainage is the first essential. This should be assured at the time the feedlot is located and constructed. Mounds, preferably perpendicular to the feed bunk, will provide cattle a dry place on which to lie down. Concrete aprons along the bunk will provide them with solid footing on which to stand and feed. Also, lessening of cattle density during the winter months—fewer animals per lot—is an effective method of controlling the mud problem. Thus, many feedlots plan to feed fewer cattle during the muddy season.

BIRD CONTROL

Nationwide, starlings constitute the major feedlot bird problem. Other feed-consuming bird species commonly identified in feedlots are: brewer blackbirds, redwing blackbirds, and cowbirds.

Some large commercial feedlots estimate their starling population at 100,000 per lot. Some western feedlot operators compute the cost for overwintering each 1,000 starlings at $100.

In addition to feed consumption, birds contaminate feed and spread diseases—to both animals and humans. The starling has been incriminated in the spread of coccidiosis among animals, transmissible gastric enteritis (TGE) in swine, and histoplasmosis in humans.

Recordings of distressed bird calls, carbide cannons, and harassment or killing with guns achieve only partial control. Many chemicals and baits have been tested, and a few have been found to be effective. However, some states do not allow the use of chemicals in bird control. Therefore, before using any chemical, the cattle feeder should check with the appropriate federal, state, and local departments of health.

FLY CONTROL

The housefly is the most common type of fly found around cattle feedlots. It is a scavenger and does not feed on animals, but it does cause irritation and annoyance. Stable flies, which are blood feeders, may also be present in certain areas and certain feedlots.

Effective housefly control requires proper animal waste management and good feedlot sanitation. The basic objective in fly control is to eliminate possible sources of fly development. This can be accomplished by the following steps: (1) Provide proper drainage and avoid wet spots; (2) remove manure immediately after a pen is vacated; and (3) remove manure and spilled feed at important fly-breeding areas such as fence lines, feed bunks, hospital pens, horse pens, truck-washing stations, and receiving and shipping areas. Chemical control should be used in conjunction with the proper waste management techniques, and not as a sole means of control. Residual and space sprays aid in reduction of adult flies; and larvicides may be applied to areas of intense larval development such as manure stockpiles, hospital, and horse pens.

FEEDLOT POLLUTION CONTROL

Pollution control is a most critical factor in site selection and operation of a cattle feedlot. Remoteness from urban development is recommended because of dust and odor. Also, before constructing a cattle feedlot, the owner should familiarize himself with both state and federal regulations. The state regulations can be secured from the state water board. They differ from state to state, but most states require a catch basin (detention pond) sufficient to contain the runoff from a storm of the magnitude of the largest rainfall during a 48-hour period of the most recent 10 years. A feedlot may minimize runoff by locating near the top of the slope and, if necessary, by using diversion embankments to divert runoff from other areas.

Cattle feedlots located near centers of populations are having an increasing number of complaints lodged against them because of manure, dust, and odor. Lawsuits, based on the nuisance law, are being filed against them.

(Also see Section 1, under the heading "Control Pollution.")

DAIRY BEEF

Dairy beef is just what the term implies—it's beef derived from cattle of dairy breeding, or from dairy X beef crossbreds.

Beef from dairy cows, heifers, steers, bulls, and veal calves accounts for (1) about one-fourth of the beef consumed in the United States, and (2) approximately 2.2 percent of farm cash income (amounting to $3.1 billion in 1980).

Economic conditions favor growing and finishing dairy steers and heifers, rather than marketing veal calves. The principles and practices involved in dairy beef growing-finishing programs are similar to those followed in finishing animals of beef breeding of comparable sex, age, quality, and growthiness.

HOGS FOLLOWING CATTLE

Cattle feeders who have a convenient source of feeder pigs, who are not "allergic" to keeping hogs, and whose cattle lots are fenced hog-tight, can add to their net income by having hogs follow cattle. The following hog:cattle ratio is recommended, using 75- to 150-pound pigs:

	If Whole Shelled Corn Is Fed	If Ground or Rolled Grain Is Fed
	(Pig:steer ratio)	*(Pig:steer ratio)*
Calves	1:3	1:5
Yearlings	1:2	1:4
Two-year-olds	1:1½	1:3

For every 50 bushels of whole corn fed to yearling cattle, approximately 50 pounds of pork will be produced. Allowing 50¢ for hogs, and subtracting $10 per pig for protein and other costs, that's $15 per pig.

Because pigs sometimes inflict injury on heifers (injuring the vulva when they are lying down), their use is generally limited to steers.

Sows may be used, but because of their size they may create problems from getting into the feed and water facilities.

SOME DAIRY MANAGEMENT PRACTICES

Successful dairymen pay close attention to the details of management. Some dairy cattle management practices that are not covered elsewhere in this book follow.

Managing the Dairy Bull

Bull calves raised for breeding purposes should be fed and handled much the same as heifers. Older bulls should be kept in thrifty, vigorous condition, but they should not be permitted to get too fat. Mature bulls can be fed the same grain ration as the lactating cows. Depending on the quality of the roughage, usually about ½ pound of grain per 100 pounds of body weight will suffice for the mature bull. Also, individual differences must be considered, for some bulls are easier keepers than others.

In addition to the grain and roughage ration, the bull should have free access to a double compartment mineral box, with ground salt in one side and a suitable mineral mixture in the other.

Managing Dry and Lactating Cows

The management of dry and lactating cows materially affects the efficiency of production and the quantity and quality of the milk produced. Because of the numerous and diverse management practices involved in caring for dry and lactating cows, they are herein presented in summary form:

- Worm cows if they need it especially during the dry period. If worming is done during the lactation period, it should be under the direction of the veterinarian.
- Treat for external parasites if necessary, using approved insecticides and application.
- Develop a written-down herd health program in cooperation with the local veterinarian, then follow it.
- Do not rebreed cows until at least 60 days after calving.
- Strive for a 10-month lactation period, a 60-day dry period, and a 12-month calving interval.
- Examine herd for pregnancy at regular intervals.

MANAGED MILKING

The physiology of the discharge of milk is a delicate process, and it requires the close cooperation of the milker and the cow if it is to be successful. A managed milking program is made up of the following coordinated steps:

1. **Preparing the equipment**—Prior to milking, the equipment to be used in the milking process should be assembled and sanitized. Also, it should be checked and adjusted if necessary.
2. **Preparing the cow**—Under natural conditions, the cow is primed or stimulated by the suckling of the calf. This process can be simulated by washing the cow's teats and udder with warm water (120° to 130°F), then massaging and drying them with a paper towel. Following this process, remove 2 or 3 streams of milk from each quarter into a strip cup (never strip milk onto the floor) and examine for visible evidence of mastitis. Also, this (a) washes out any debris adhering to the end of the teat, and (b) enhances the letdown effect.

   About 45 seconds after the priming stimulus, the udder becomes full and firm (especially in early lactation), and milk occasionally will leak from the teats. This is evidence that the cow has let down her milk and is ready for the next step.
3. **Attaching the teat cup and beginning**—About 1 minute after washing the udder, and not more than 1½ minutes, the teat cups should be attached and milking should begin. Most cows will milk out in 3 to 6 minutes, depending upon the amount of milk and the characteristics of the cow. Also, and most important, each quarter should be milked individually, because some quarters milk out faster than others.
4. **Stripping by machine**—When it is apparent that the cow is about milked out, she should be machine stripped. This consists of pulling down on the teat cups with one hand, and massaging the udder downward with the other. This process should not take over about 20 seconds.
5. **Removing the teat cups**—Both incomplete and overmilking should be avoided. The greatest cause of machine injury is leaving the teat cups on too long. Incomplete milking usually results because one or more quarters are more difficult to milk than the others.

   As soon as the udder is empty, and before the teat cups crawl up, they should be removed, properly and gently. Then, dip the teats with a fresh disinfectant solution (100 ppm idophor or chlorine, or other sanitizing agent). This will remove the milk from the ends of the teats and prevent the invasion of bacteria into the udder. Also, it will avoid attracting flies.

   As soon as the teat cups have been removed from the udder of the cow, they should be cleaned. First, dip them in clean, cold water to remove milk inside the liners, then put them in a clean, warm, approved sanitizing solution. Change the solution after each five to seven cows.

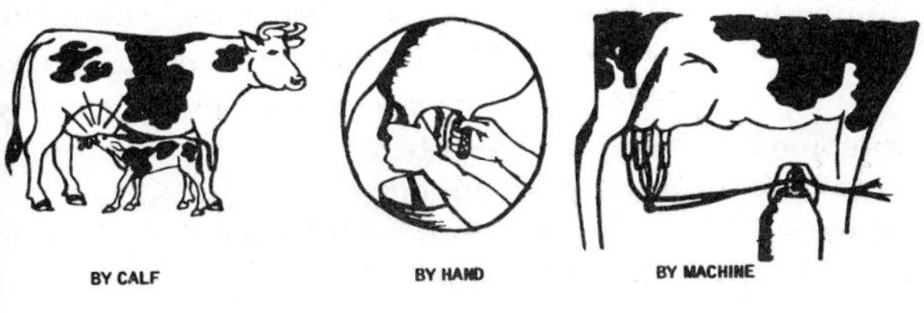

Fig. 8-39. Three ways to milk a cow.

6. **Cleaning up equipment**—After milking the last cow, all milking equipment should be thoroughly cleaned and put away.

7. **Milking time**—The actual milking time per cow will range from 3 to 6 minutes, with an average time of 3½ minutes for cows in mid-lactation. But additional time must be allowed for let-down, adjustments, and interval between cows.

The number of machines one man can manage successfully depends upon the type of barn, the type of milking equipment, the ability of the milker, and the jobs he has other than milking.

One man should handle no more (preferably less) than the following number of units:

Type Milker	Units per Man
Stanchion barns:	
Bucket	2
Pipeline	3
Milking parlor:	
Walk-through	3
Side-opening	3
Herringbone	4

With a 3-inch-line elevated parlor, one man will average 18 to 25 cows per hour. With a 4-inch-line parlor, one man will average 25 to 30 cows per hour. However, additional time must be allowed to bring the cows in from the outside, setting up, cleaning up, as well as milking problem cows.

8. **Milking order**—Cows that have mastitis or a history of chronic mastitis are a source of infection to noninfected cows. Hence, it is well to milk "clean" cows first. A desirable milking order in stanchion barns is:

a. First calf heifers that have been free of mastitis.

b. Older cows that have been free of mastitis.

c. Cows that have a previous history of mastitis, but which no longer show symptoms.

d. Cows with quarters producing abnormal milk.

FEEDS AFFECTING MILK FLAVOR

Consumers want milk to taste like milk—not like silage, grass, or weeds.

Although feeds are not the only cause of milk flavors, they are major contributors. Feed flavors enter the milk through the digestive system, respiratory system, and by direct absorption. Research indicates that most feed flavors are detectable in the milk 20 minutes after the feed is consumed, and that they are usually most pronounced at the end of two hours.

Feed flavors that enter the milk through the respiratory system can usually be detected much sooner than those entering through the digestive system. For example, if a cow breathes air reeking with silage odors, these flavors can be detected in the milk almost immediately. Flavors that are directly absorbed by milk are less common, but they appear if the milk is left exposed for a long enough period.

The following control measures are recommended to alleviate feed flavors:

1. **Avoid sudden change to fresh, lush pasture**—Cows should be shifted from winter feeding, or old pasture, to new and lush pastures on a gradual basis. Also, cows should be taken out of such pastures two to three hours before milking. For the same reasons, freshly cut grass should not be fed immediately before milking.

2. **Control and avoid undesirable weeds**—Many weeds when eaten by cows will impart a strong flavor to milk; among them, wild onions, skunk cabbage, some members of the mustard family, bitterweed, carrot weed, ragweed, and others. With modern weed killers, it is easier to get rid of these weeds than formerly, so they should be eliminated from pasture and hayfields utilized by milk cows.

3. **Silage flavor**—Silage flavor is both common and objectionable. It can be avoided by feeding all silages after milking, never before or during milking. Usually one will be safe if silage is not fed within two to four hours of milking time, but it's safer to feed it shortly after milking. This permits the flavor-causing material to pass through the cow's digestive system before the next milking.

If cows breathe the odor of silage, it will appear as flavor in the milk. Thus, silage should never be left in the mangers or feed alleys. In fact, it is preferable that it be fed in the corral, and not in the area where the cows are being milked.

PESTICIDE RESIDUES

Pesticides are chemicals that are used to kill pests—insects, weeds, and rodents. These products are very necessary for food and milk production. Our abundant supply of wholesome foods would not have been available without their use. Yet, it is important that they be properly used, and that certain precautions be taken. The following points are pertinent to their proper use:

1. **Pesticides that have been associated with milk contamination**—Among the pesticides that have been associated with milk contamination in the past are the following chlorinated hydrocarbons: aldrin, dieldrin, heptachlor, expoxide, DDT and its isomers, toxaphene, and lindane. Of course, other pesticides may become a problem in the future.

2. **How pesticides contaminate milk**—They are absorbed by animal fat. Since milk contains fat, it is one channel through which the animal eliminates pesticides from its body.

3. **Length of time that a contaminated cow may give contaminated milk**—Cases are known where residues have been detected in milk for four to eight months after discontinuing the feeding of contaminated feeds.

4. **Ways that milk may become contaminated**—Milk becomes contaminated (a) by spraying animals with nonrecommended pesticides, (b) by using these materials in back rubbers and vaporizers, (c) by feeding forages and concentrates which have been contaminated with these materials, (d) by allowing cows to drink pesticide-contaminated water, and (e) by using milk utensils that have become contaminated through their use for chores other than handling milk or in milk production. Hence, milk contamination can be prevented by avoiding any of these avenues of contamination.

5. **The meaning of the word "tolerance" as applied to a chemical residue**—A given tolerance is that amount of chemical residue, usually expressed in ppm (parts per million) set by the FDA (Food & Drug Administration), that remains on or in a commodity at harvest and which is at least 100 times less than that amount of the chemical known to be toxic to experimental animals.

A zero tolerance, as applied to chemical residue, means that no amount of the pesticide chemical may remain on or in the raw agricultural commodity when it is offered for shipment. The tolerance level for pesticides in milk is 1.25 ppm on a fat basis. This means that there must be less than 1.25 parts of the pesticide (DDT, for example) to 1 million parts of milk fat figured on a weight basis.

Managing Dairy Calves

One of the most important phases of dairy production is that of managing dairy calves. Statistics show that more than 20 percent of dairy calves die of sickness or disease before reaching maturity.

The recommended practices in managing dairy calves follow, in summary form:

- Dehorn anytime after 10 days of age, using an electric dehorner or caustic potash.
- Remove extra teats before heifers are six months old; cut them off with clean scissors and disinfect area with iodine.
- Check for scours.

Managing Replacement Heifers

Good management of replacement heifers embraces the following principles and practices:

- Separate bulls and heifers before six months of age; do not have over three months' difference in age of animals within a given group.
- Treat for worms when the need is demonstrated.
- Breed heifers at 15 to 18 months of age, but also consider weight and size.
- Accustom bred heifers to milking barn procedure beginning about one month prior to calving.

Cow Testing Programs

Individual cow records are a must in any progressive dairy production program. Dairymen use records as a guide for feeding, for locating, and culling out the least profitable cows, and for maintaining a permanent, detailed record of each cow. Records necessitate that each cow be individually identified, and that there be milk and butterfat production records.

The various testing programs sponsored by federal and state extension services are discussed in Section 3 of this book; hence, the reader is referred thereto. (See "Alternate DHI Testing Plans.")

SOME SHEEP MANAGEMENT PRACTICES

Several sheep management practices that are not covered elsewhere in this book follow.

Managing the Ram

If possible, the ram should be secured considerably in advance of the breeding season, thereby providing an opportunity to become acclimated before being placed in service. In case of show or sale rams, it may also be advisable to remove some of their surplus flesh.

Stud rams are usually kept separate from the ewes except during the breeding season. Their quarters need not be elaborate or expensive. Usually, a dry shelter that will provide protection during times of inclement weather is all that is necessary. Plenty of exercise should be provided at all times.

Rams may subsist largely on pasture and dry roughage. If the pasture has been scanty prior to the breeding season, the rams may be conditioned by feeding a little grain, usually not more than one pound daily. Rams are usually fed some grain when being fitted for show or sale, but it must be remembered that excess fat may actually be harmful from a breeding standpoint.

PREPARING THE RAM FOR MATING

As the weather is usually rather warm at the time of the breeding season, shearing the ram just prior to this will make him more active. This is especially true of old show or sale rams. Where rams are not sheared completely, they should at least have the wool clipped from the neck and from the belly in the region of the penis, for this will result in copulation with greater ease. It is also important to see that the hoofs of the ram are properly trimmed prior to the breeding season.

MARKING THE RAM

When several rams are turned in with a large band of ewes, it is impossible to detect individual rams that may be failing to settle ewes. Moreover, it is quite likely that a different ram will serve the ewe should there be a recurrence of heat, or perhaps more than one ram may serve the ewe at the time of estrus. When only one ram is being used on a small flock, however, it is important to know whether the ewes are getting with lamb. Then, too, with a purebred flock individual breeding records are rather important.

A breeding record can best be kept by using a marking harness (breeding harness), containing a crayon (different colored crayons are available), on the ram, or by smearing the breast of the ram and the area between his forelegs every day or two with a thick paste. Then, as the ram serves the ewe, a mark will be left on her rump. Paint or tar should never be used for this purpose.

The color of the crayon or paste should be changed every 16 days (the approximate estrous cycle of the ewe) so that one can determine whether ewes that have been bred are returning in heat. For example, during the first 16-day interval, the thick paste used on the ram might well be a mixture of ordinary lubricating oil and yellow ochre; for the second 16-day interval, it might be lubricating oil and venetian red; and for the third 16-day interval (if there is still some question about some of the ewes having settled), it might be a paste made by using lubricating oil and lamp black (thus proceeding from light to dark colors).

Naturally, if a good percentage of the ewes are found coming in heat for a second time, the ram should be regarded with suspicion, and perhaps another ram should be obtained. The sterility in some instances may be temporary because of high condition and lack of exercise.

Managing Ewes

Managing ewes at breeding time and at lambing time is extremely important, because it determines the size lamb crop.

TRIMMING AND TAGGING THE EWES FOR MATING

Tagging is the removal of tags or locks of wool and dirt about the dock. It is important that this job be done prior to the breeding season in order to prevent the ewes from befouling themselves and to remove obstacles for the service of the ram.

MANAGING EWES AT LAMBING

There is an appalling lamb death loss of 24.6 per cent from birth to weaning. Most of these losses occur in the first few days of life.

As lambing time approaches, unsheared ewes should be tagged. This consists of shearing the wool from around the udder, flank, and dock. The ewe should also be placed where she has plenty of room away from any jamming or crowding. The grain al-lowance should be materially reduced, but the roughage allowance may be continued if it is certain that it is of good quality and palatable. Careless feeding at this stage is likely to result in milk fever following parturition. At this time, the wool around the udder should be clipped short in order to allow the lamb to find the teats readily. If breeding records have not been kept, the signs of approaching parturition must be relied upon. A nervous, uneasy disposition; a sinking in front of the hips; and fullness of udder are such indications.

• **The lambing pen**—Just before lambing, or immediately thereafter, the ewe should be placed in a lambing pen. These pens are usually 4 feet square and are made by placing together two hinged hurdles which are then set against the walls of the sheep barn. Use of the lambing pen prevents other sheep from trampling on the newborn lamb; eliminates the possibility of the lamb wandering away and becoming chilled; and, through keeping the dam and offspring together, lessens the danger of disowned lambs.

Lambing pens should be clean, dry, well bedded and well ventilated and should be located so as to be free from drafts. During extremely cold weather, addi-tional warmth may be provided for the first few hours after birth by throwing a blanket over the top of the pen.

• **Normal presentation**—A good rule for the shepherd to follow is to be near during parturition but not to disturb the ewe unless she needs help. Normal presentation of the lamb consists of having the forelegs extended with the head lying between them, although some lambs are delivered hind legs first. Even though the lambs are born in clean quarters, tincture of iodine should be applied to the navel soon after birth. The latter precaution may not be necessary when lambs are dropped on an uncontaminated pasture or range, although many range operators report that they have found it necessary to apply iodine to the navel of lambs born on the range as well as in the shed.

• **Taking the lamb**—If the ewe has labored for some time with little progress or is laboring rather in frequently, it is usually time to give assistance. If the lamb is not in the proper position, such assistance consists of inserting the hand and arm in the vulva and turning the lamb so that the forefeet and head are in position to be delivered first. Delivery may then be helped by pulling the young outward and downward as the ewe strains. Before doing this, however, the fingernails should be trimmed closely and the hands and arms should be thoroughly washed with soap and warm water, disinfected, and then lubricated with Vaseline or linseed oil.

• **Chilled and weak lambs**—Lambs arriving during cold weather may become chilled before they have dried. One of the most effective methods of reviving a chilled lamb is to immerse the body, except for the head, in water that is as warm as one's elbow can bear. The lamb should be kept in this for a few minutes and then removed and rubbed vigorously with cloths. It then should be wrapped in an old blanket, a sheepskin, or other heavy material and should be given some warm milk as soon as possible. Another convenient and effective method of drying and warming a chilled lamb consists of putting it into a box containing a light bulb or electric heater.

When strong, healthy ewes have been properly fed and cared for during pregnancy, there will be a minimum of weak lambs. The shepherd should first make certain that the membrane has been removed from the nostrils and that breathing has started. Blowing into the mouth, lifting the body and dropping it a short distance, working the legs, and pressing the sides are artificial methods of starting breathing that may revive lambs that at first appear lifeless.

After breathing has started and the navel cord has been painted with iodine, an attempt should be made to get the lamb to nurse. Quite often even a very weak lamb will nurse the ewe if it is held to the teat. If it refuses to nurse in this manner, some of the colostrum of the ewe should be milked into a sterilized bottle, and the lamb should be fed a few teaspoonfuls each hour by means of the bottle and nipple, until it gains strength.

If the ewe has no milk, an attempt should be made to obtain milk from another ewe that has just lambed, and perhaps in a few hours the normal flow of milk will start.

• **Disowned lambs**—When lambing pens are used, the number of disowned lambs is kept to a minimum. For the most part, disowning of lambs is due to improper feeding during pregnancy or because of a poor milk supply, an inflamed udder, or a maternal instinct that is not sufficiently developed, as is often true in ewes with their first lambs.

For the first few days, a ewe seems to recognize her young by scent or sense of smell. When difficulty is encountered in getting a ewe to own her own lamb or when it is desired to transfer or "graft" a lamb (as may be necessary with the loss of a lamb or when there are twins on an old ewe), deception in the sense of smell is an effective approach. One of the most common practices is to milk some of the ewe's milk on the rump of the lamb and then to smear some of it on the nose of the ewe. Many good shepherds take some of the mucus from the mouth and nose of the newborn lamb and smear it over the nose of the ewe. If these methods fail and the ewe persists in fighting the lamb away, blindfold her so that she cannot see the lamb. As a last resort, and when all other methods have failed, tie a dog in an adjoining pen. Sometimes the latter method will cause latent maternal instincts to rise to a surprising degree.

Occasionally, a ewe will fail to own one of a pair of twin lambs. When this condition exists, about all that can be done is that the shepherd be patient in training the disowned lamb to nurse at the same time as its mate. Both lambs are usually kept from the ewe and turned with her at intervals.

• **The orphan lamb**—A lamb may be orphaned through the death of its mother or because of the inability of the mother to suckle it. The most satisfactory arrangement for the orphan is to provide a foster mother. The good shepherd will try to have every ewe raise a lamb. There may be a ewe that has just lost her lamb or a strong, healthy ewe with just one lamb. When a lamb dies at birth and it is desired to transfer or "graft" another lamb on the ewe, two procedures are common. Sometimes a ewe will accept another lamb provided that the lamb to be adopted is first rubbed with the body of the dead lamb that it is to replace. Though a bit more bothersome, a more effective approach consists of removing the skin from the dead lamb and tying it over the lamb to be adopted. After 2 or 3 days, the skin may be removed gradually, a piece at a time. The latter method is commonly used in the range bands of the West.

When it is impossible to transfer an orphan lamb to another ewe, it may be raised either on cow's milk or on milk replacer. Of course, the problem will be simplified if the lamb has received some colostrum (the first milk) from its mother or from another ewe.

If cow's milk is used, it should not be diluted, because cow's milk is lower in butterfat and total solids than ewe's milk. Milk replacer should be mixed according to the manufacturer's directions.

Both cow's milk and milk replacer should be warmed to 100°F and fed in sterilized bottles. During the first few days, the orphan should be fed about one ounce of milk (or milk replacer) every two hours. Gradually, the quantity may be increased and the intervals spaced further apart.

• **Feed and water after lambing**—Following parturition, the ewe is in a feverish condition and should be handled carefully. She may be watered immediately after lambing, and at frequent intervals thereafter, but she should never be allowed to gorge. It is also a good plan to take the chill off the water before giving it to her. In general, feeds of a bulky and laxative nature should be provided during the first few days. A mixture of equal parts of oats and wheat bran may be fed in very limited quantities, with all the hay that can be consumed. Heavy grain feeding at this time may cause udder trouble in the ewe and digestive disturbances in the lamb. The feed may be gradually increased until the ewe is on full feed in about a week.

• **Examination of the udder**—During the first two days following lambing, the udder should be examined night and morning. Sometimes a lamb will nurse one side only. If all the milk is not being taken by the lamb, the udder should be milked out and the ration lessened accordingly. If the udder becomes swollen and feverish, it should be milked out, bathed with warm water, and then dried. Following this, it should be painted with tincture of iodine. This treatment should be repeated once or twice daily, as necessary. Lambs should not be allowed to suckle their mothers when the udder is in such a condition. It is also a good plan to isolate the affected ewes from the rest of the flock.

SOME SWINE MANAGEMENT PRACTICES

Some management practices unique to swine, and not covered elsewhere in this book, follow.

Confinement Vs Pasture

In recent years, confinement rearing, in which pigs are confined in buildings from birth to market, has increased. But most swine producers follow a program of partial confinement. A 1973 survey revealed that 81% of U.S. swine producers confine sows at farrowing time, 65% provide confinement for nurseries, 35% confine growing-finishing pigs, but only 11.3% confine the sow herd.[12] The main reasons given for going to confinement are savings in labor and land. The main problems encountered are high investment in buildings, more disease troubles, rations become more critical, manure disposal and odor control problems are greater, and sow fertility is lowered.

In season, most producers utilize pastures for breeding animals. The vast majority of the nation's gestating sows and herd boars are kept in movable houses, preferably on clean pasture on land that has been plowed since hogs were last on the area. In any event, pastures for sows and litters are not obsolete; rather, there now exists two alternatives for the breeding herd—confinement vs pasture—and the wise manager will choose between them.

[12]*Hog Farm Management*, Aug. 1973, p. 39.

Grouping Hogs

Grouping, along with separating hogs by sexes, ages, and sizes, is important. The following practices are generally advocated by successful producers:

1. **Gilts to be retained for the breeding herd**—They should be separated from market hogs at four to five months of age.
2. **Pregnant gilts and sows**—They should be kept separate during the gestation period, unless they are self-fed a bulky ration.
3. **Boars of different ages**—Junior and mature boars should not be run together. Boars of the same age or size can be run together during the off-breeding season.
4. **Adjusting size of litter**—Where possible, the size of litters should be adjusted to the number of functioning teats and the nursing ability of the sow. Transferring pigs from sow to sow should be done as early as possible; three to four days after farrowing is usually the maximum length of time that this can be done, unless the odor of the pigs is masked, when it may be possible to transfer at a later date.
5. **Running sows and litters together**—Pigs should be about 2 weeks old before placing sows and litters together, although small groups may be put together as early as 1 week. The age difference between such litters should not be more than 1 week in a central farrowing house or 2 weeks on pasture. Not more than 4 sows and litters should be grouped together in a central farrowing house; and not more than 6 on pasture.
6. **Creep feeding**—A maximum of 40 pigs per creep may be allowed.
7. **Early weaning**—In early weaning, not over 10 pigs should be placed together up to 3 weeks of age; 20 may be placed together at 3 to 4 weeks of age; and 25 at 5 weeks of age.
8. **Pigs of different weights**—Growing-finishing pigs of varying weights should not be run together. It is recommended that the range in weight should not exceed 20 percent above or below the average.

Managing the Boar

Proper care and management of the herd boar is most essential for successful swine production. Too frequently the boar is looked upon as a necessary evil and is neglected. Under such conditions, he is usually confined to a small, filthy pen—a typical pigsty—exercise is discouraged; and the feeding practices are anything but intelligent.

• **Feed, shelter, and exercise**—Outdoor exercise throughout the year is one of the first essentials in keeping the boar in a thrifty condition and virile. This may be accomplished by providing a well-fenced pasture. Even then, the herdsman may find it necessary to walk old boars or boars that are being fitted for the shows. In addition to the valuable exercise that is obtained in the pasture lot, green succulent pasture furnishes valuable nutrients for the herd boar. The amount of feed provided should be such as to keep the boar in a thrifty, vigorous condition at all times. He should be neither overfat nor in a thin, run-down condition. The concentrate allowance should be varied with the age, development, and temperament of the individual; breeding demands; roughage consumed, etc. Feeding the boar is more fully covered in Section 4 of this book; hence, the reader is referred thereto.

A satisfactory but inexpensive shelter should be provided for the boar, and he should be allowed to run in and out at choice.

Boars of the same age or size can be run together during the off-breeding season, but boars of different ages should not be kept together.

• **Ranting**—Some boars pace back and forth along the fence, often chopping their jaws and slobbering. Such action is called ranting. Young boars that take to excessive ranting may go off feed, become "shieldy," and fail to develop properly. Although this condition will not affect their breeding ability, it is undesirable from the standpoint of appearance. Isolation from other boars or from the sow herd is usually an effective means of quieting such boars. Should the boar remain off feed, placing a barrow or a bred sow in the pen with him will help to get him back on feed.

Managing Breeding Swine

Management, more than any other factor, determines how well swine reproduce and survive, and how nearly they perform to their genetic potential.

NORMAL BREEDING AND FARROWING SEASONS

The season in which the sows are bred and the question of raising 1 or 2 litters a year vs multiple farrowing (scheduling breeding so that the litters arrive throughout the year, rather than once or twice per year as is the case in the conventional 1- or 2-litter systems), depend primarily on the facilities at hand. The location of the producer (particularly the weather conditions in the area), availability and price of feeds, condition and growth of the sows, equipment for handling pigs during the winter months, available labor, and the type of production (purebred or commercial) should be taken into consideration. No positive advice can be given, therefore, for any and all conditions. Sows will breed any time of the year; but, as in other farm animals, the conception rate is much higher during those seasons when the temperature is moderate and the nutritive conditions are good. For the country as a whole, spring pigs are preferred, as is shown by the size of the spring pig crop in comparison with the fall pig crop.

No one expects the seasonal pattern of hog production to be completely eliminated, but, because of the several recognized advantages of multiple farrowing to both the processor and the producer, it is likely that it will increase sufficiently to make for a lessening of some of the market gluts of the past.

BREEDING PRACTICES

The following breeding practices are recommended:

1. **Breeding following early weaning**—When weaning under two weeks of age, breed sows on the second heat period after weaning. When weaning at three weeks or older, it is satisfactory to breed sows on the first heat period following weaning.
2. **First service of boars**—Whenever practical, it is recommended that boars be allowed to serve females outside the breeding herd (some market hogs) prior to serving those in the breeding herd.

Managing Sows at Farrowing

It has been conservatively estimated that from 30 to 35% of the pigs farrowed never reach weaning age, and an additional loss of 5 to 10% occurs after weaning. This means that only 60% of the pig crop reaches market age.

The careful and observant herdsman realizes the importance of having everything in readiness for farrowing time. If the pregnant sows have been so fed and managed as to give birth to a crop of strong, vigorous pigs, the next problem is that of saving the pigs at farrowing time. Good management will give a powerful assist to this end.

• **Signs of approaching parturition**—The immediate indications that the sow is about to farrow are extreme nervousness and uneasiness, an enlarged vulva, and a possible mucous discharge. She usually makes a nest for her young, and milk is present in the teats.

• **Preparation for farrowing**—About three or four days prior to farrowing, the sow should be isolated from the rest of the herd. It is important, however, that moderate exercise be continued while the animal is in the farrowing quarters.

• **Sanitary measures**—Before being moved into the farrowing quarters, the sow should be thoroughly scrubbed with soap and warm water, especially in the region of the sides, udder, and undersurface of the body. This removes adhering parasite eggs (especially the eggs of the common round worm) and disease germs.

The house should be thoroughly cleaned to reduce possible infection. This may be done by scrubbing the walls and floors with boiling-hot lye water made by using one can of lye to 15 gallons of water. If the farrowing house has dirt floors, the top 2 or 3 inches of soil should be replaced by an equal quantity of clean clay soil. The sow should then be placed in her new quarters.

• **The quarters**—Hogs are sensitive to extremes of heat and cold and require more protection than any other class of farm animals. This is especially true at the time of parturition. It is recommended that the farrowing house temperature be maintained at 60° to 70°F, and that it not go below 40°F or above 85°F. Along with this temperature, there should be adequate ventilation at all times. In cold areas and during the winter months, use heat lamps or pig brooders when the farrowing house temperature falls below 65°F (see Section 9).

The main requirements for satisfactory housing are that the quarters be dry, sanitary, and well ventilated and that they provide good protection from heat, cold, and winds.

• **The guard rail**—A guard rail around the farrowing pen is an effective means of preventing sows from crushing their pigs. The importance of this simple protective measure may be emphasized best by pointing out that approximately one-half of the young pig losses are accounted for by those pigs that are overlaid by their mothers. The rail should be raised 8 to 10 inches from the floor and should be 8 to 12 inches from the walls. It may be constructed of two-by-fours, two-by-sixes, or strong poles or steel pipe.

• **Bedding**—The farrowing quarters should be lightly bedded with clean, fresh material. Any good absorbent that is not too long and coarse is satisfactory. Wheat, barley, rye, or oat straw; short or chopped hay; ground corn cobs; peanut hulls; cottonseed hulls; shredded corn fodder; and shavings are most commonly used.

• **The attendant**—The herdsman should be on the job, especially during times of inclement weather. It may be necessary to free the newborn pigs from the enveloping membrane and to help them reach the mother's teat. In cold weather the young should be dried off and other precautions taken to avoid chilling.

If the sow has labored for some time with little progress or is laboring rather infrequently, assistance should be given. This usually consists of inserting the hand and arm in the vulva and gently correcting the condition preventing delivery. Before doing this, the fingernails should be trimmed closely, and the hands and arm should be thoroughly washed with soap and warm water, disinfected, and then lubricated with petroleum jelly or linseed oil.

As soon as the afterbirth is expelled, it should be removed from the pen and burned or buried in lime. This prevents the sow from eating the afterbirth and prevents the development of bacteria and foul odors. Many good swine producers are convinced that eating the afterbirth encourages the development of the pig-eating vice. Dead pigs should be removed for the same reason.

It is also well to work over the bedding; remove wet, stained, or soiled bedding and provide clean, fresh material.

• **Chilled and weak pigs**—Pigs arriving during cold weather are easily chilled. Under such conditions, it may be advisable to take the pigs from the mother as they are born and to place them in a half-barrel or basket lined with straw or rags. In extremely cold weather, a few hot bricks or a jug of warm water (properly wrapped to prevent burns) may be placed in the barrel or basket; or the pigs may be taken to a warm room until they are dry and active.

One of the most effective methods of reviving a chilled pig is to immerse the body, except the head, in water as warm as one's elbow can bear. The pigs should be kept in this for a few minutes, then removed and rubbed vigorously with cloths.

• **Orphan pigs**—Pigs may be orphaned either through sickness or death of their mother. In either event, the most satisfactory arrangement for the orphans is to provide a foster mother. When it is impossible to transfer the pigs to another sow, they may be raised on cow's milk or milk replacer. The problem will be simplified if the pigs have received a small amount of colostrum (the first milk) from their mother.

If cow's milk is used, it is preferable that it be from a low-testing cow. Do not add cream or sugar; however, skim milk powder, at the rate of a tablespoonful to a pint of fluid milk may be added, if available. Milk replacer should be mixed according to the directions found on the container. The first 2 or 3 days the orphans should be fed regularly every 2 hours and the milk should be at 100°F. Thereafter, the intervals may be spaced farther apart. All utensils (pan feeding or a bottle and nipple may be used) should be clean and sterilized.

Orphan pigs should be started on a prestarter or starter ration when they are one week old. Also, a source of iron should be provided (in keeping with instructions given in Section 4).

• **Artificial heat**—During times of inclement weather, artificial heat usually must be provided, especially for pigs farrowed in northern United States. Most large central hog houses are equipped with a heating unit for use in winter farrowing, designed to maintain the temperature at 60° to 70°F.

Individual houses may be insulated by banking with straw and other insulating materials. Then a lantern or oil burner may be suspended from the top of the house. It must be remembered, however, that here is considerable fire hazard with this practice. The electric pig brooder is a much safer heating unit for either the central or the movable hog house. The principles involved are identical to those of the electric chick brooder.

Feeding Systems for Swine

The choice of the swine feeding system(s) and the choice of the ration(s) must go hand in hand. For example, if the grain and the protein supplements are to be self-fed in separate feeders or compartments, it is important that they be of equal palatability; otherwise, pigs will consume too much of one and too little of the other. A listing and discussion of each of the common feeding systems follows.

• **Complete self-fed rations**—The trend is toward the use of complete self-fed rations for growing-finishing pigs, because, in comparison with free-choice of ingredient feeding, they (1) lend themselves better to automation, (2) provide better control of nutrient intake, and (3) result in faster gains than free-choice of ingredient feeding.

Complete rations may be formulated either by "building from the ground up" (by adding each ingredient, one by one), or by mixing a balanced protein supplement with ground grain.

A survey made by the University of Illinois revealed that the most-used ration by swine producers was prepared by mixing ground grain and a commercial formula supplement (a protein, mineral, and vitamin supplement).

Where producers do their own mixing, they favor simplified rations. Fortunately, a simple ration of corn and soybean meal, fortified with minerals and vitamins, will generally give as good results as a more complex ration consisting of many ingredients. Of course, with large volume buying and computerized formulations, commercial feed companies can use more complex rations (with many ingredients) advantageously, especially from the standpoints of enhancing palatability and balancing amino acids, minerals, and vitamins.

• **On-floor feeding**—On-floor feeding is particularly suited to the controlled feeding of growing-finishing swine or the breeding herd. Feeding in the sleeping area encourages cleanliness, since pigs are less inclined to dung where they eat. Feed wastage is reduced to a minimum when the animals do not have more feed available than they will consume at one eating. Even though automated, restricted feeding requires close attention, because the daily feed intake of pigs is affected by weather.

• **Self-fed free-choice**—Shelled corn and protein supplements may be fed separately and free-choice. Generally, pigs fed free-choice rations in separate feeders or compartments will not make as uniform or as fast gains as pigs fed a complete mixed ration. The free-choice system requires more supervision, as the palatability of the grain or the protein supplement may vary and the pigs will then overeat or undereat the supplement or the grain. There is very little, if any, difference in economy of gain between feeding a free-choice or a complete ground mixed ration.

• **Liquid feeding**—Liquid feeding usually involves mixing predetermined amounts of feed and water prior to, or at the time of, feeding. When properly used, this method can practically eliminate feed dust in the feeding area and minimize wastage. Ratios of feed and water can be varied to produce a free-flowing liquid or a thick paste. In some cases, feed is automatically dropped into the water in the feed trough. Research has shown no difference in rate of gain or feed per pound of gain in pigs full fed on liquid or dry feeds. Neither does liquid feeding have any effect on dressing percentage, carcass measurements, or carcass quality.

• **Limit feeding**—With gestating sows, limit feeding to 4 to 6 pounds per head daily is a must in order to keep them from getting too fat. Overly fat sows have difficulty in farrowing and give birth to weak or dead pigs. With growing-finishing pigs, it is a way in which to increase slightly the proportion of lean to fat in the carcass. A discussion of limit feeding of (1) gilts and sows, and (2) growing-finishing pigs follows.

1. **Gilts and sows**—Replacement gilts should be started on a limited feeding program at 180 to 200 pounds, and all gestating sows and gilts should be limit fed. Limit feeding may be accomplished by any one of the following methods:

a. **By feeding bulky, fibrous feed,** like silage, haylage, or alfalfa, with such feed constituting at least one-third of the ration. Actually, this is a way in which to lower the energy content of the ration. Although bulky feeds will hold the weight down, they usually do not lower feed cost.

b. **By interval feeding,** in which gilts or sows are turned to self-feeders for 2 to 8 hours every second or third day. Under this system, gilts will usually eat around 12 lb of feed at a time (or an average of 4 lb per day) and older sows will consume around 15 lb (or an average of 5 lb per day). The amount of feed consumed in interval feeding may be controlled either (1) by varying the interval, from every other day to twice a week, (2) by varying the length of time that the gilts and sows are left on the self-feeders (from 2 to 8 hours), or (3) by hand feeding.

Experiments and experiences show that reproductive performance is the same with either interval feeding or daily hand feeding a limited amount. Turning sows to self-feeders at intervals requires less labor than daily hand feeding, but it does result in greater stress on fences and equipment.

c. **By group hand feeding** a limited ration to several sows. This is apt to result in the "bossy" sows getting too much and the "timid" sows getting too little. This problem can be partially alleviated by feeding over a large area.

d. **By individual feeding** in either individual pens or in tie stalls, tethered by a neck collar or belt.

2. **Growing-finishing pigs**—Sometimes growing-finishing pigs are limit fed in order to produce leaner carcasses. Usually, it is started when pigs weigh around 100 pounds and feed is limited to about 85 to 95 percent of what pigs of comparable age consume when self-fed. Limit feeding of market hogs results in slower gains, increased labor, and more mechanization. Thus, unless sufficient premium is paid for the modestly leaner carcasses, it cannot be justified.

• **Pelleted rations**—The use of a complete pelleted ration for growing-finishing hogs will increase the average daily gain by 2 to 5 percent and improve the feed efficiency by approximately 5 to 10 percent. Thus, when a complete ration is purchased, buying a pelleted feed may be more economical than buying a meal. But the advantage of pelleting will usually not be sufficient to offset the cost of hauling grain to the mill and having a pelleted ration made. Also, pellet machines are costly; hence, the purchase of such equipment cannot be justified with the volume of feed handled by most swine producers.

Swine Skills

Among the essential swine skills which the swine producer must perform are the following:

1. **Clipping the boar's tusks**—The common procedure in preparation for removing the tusks consists of drawing a strong rope over the upper jaw and tying

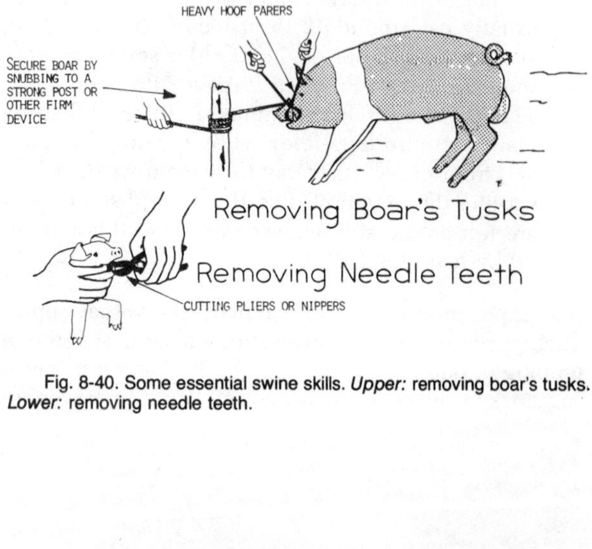

Fig. 8-40. Some essential swine skills. *Upper:* removing boar's tusks. *Lower:* removing needle teeth.

the other end securely to a post or other object. As the animal pulls back and the mouth opens, the tusks may be cut with a bolt clipper.

2. **Removing the needle teeth**—Newborn pigs have eight small, tusklike teeth (so-called needle or black teeth), two on each side of both the upper and lower jaw, which are usually clipped by means of pliers or special forceps, thereby alleviating the likely possibility of the pigs inflicting injury upon the sow's udder or on each other. In removing the teeth, care should be taken to avoid injury to the jaw or gums, for injuries may provide an opening for germs; for this reason only the tips of needle teeth should be clipped.

3. **Ringing**—When rooting starts, the herd should be "ringed"; and this applies to all hogs past weaning age. Older animals can be restrained by a rope or instrument placed around the snout, whereas young pigs can be held.

Many types of rings can be and are used, but the fishhook type is most common. Rings (generally 1 to 3 rings) are usually placed in the snout, just back of the cartilage but away from the bone; although some producers prefer to use a ring that is placed through the septum (the partition of the nose). Others cut the cartilage on top of the snout, but this causes a rather severe setback and should be practiced with caution.

SOME HORSE MANAGEMENT PRACTICES

Horse management practices vary between areas and individual horsemen. In general, however, the principles of good management are the same everywhere, and they apply whether one horse or many horses are involved. Some horse management practices that are not covered elsewhere in this book follow.

Normal Breeding and Foaling Seasons

The most natural breeding season for the mare is in the spring of the year. Usually mares are gaining flesh at this time; the heat period is more evident, and they are more likely to conceive. Furthermore, the spring-born foal may be dropped on pasture—with less danger of infection and with an abundance of exercise, fresh air, and sunshine to aid in his development; and there will be good, green, succulent pasture for the mare. Such conditions are ideal.

Also, it must be remembered that the showman will want to give consideration to having the foals dropped at such time that they may be exhibited to the best advantage. The same applies to the person who desires to sell well-developed yearlings of the light horse breeds or to race two-year-olds. It is noteworthy, however, that the percentage of barren mares that conceive at an early breeding is markedly lower than is obtained later in the season. Nevertheless, some mares do conceive early in the year, and even a small percentage is advantageous to some breeders.

Managing the Stallion

Although certain general recommendations can be made, it should be remembered that each stallion should be studied as an individual, and his care, feeding, exercise, and handling should be varied accordingly.

• **Quarters for the stallion**—The most convenient arrangement for the stallion is a roomy box stall which opens directly into a 2- or 3-acre pasture paddock, preferably separated from the other horses by a double fence. A paddock fence made of heavy lumber is safest. The stall door opening into such a paddock may be left open except during extremely cold weather; this will give the stallion plenty of fresh air, sunshine, and additional exercise.

• **Feeding the stallion**—The feed and water requirements of the stallion are adequately discussed in Section 4. In addition to this, it may be well to reemphasize that, in season, clean lush pastures produced on fertile soils are excellent for the stallion. Grass is the horse's most natural feed, and it is a rich source of vitamins that are so necessary for vigor and reproduction. Perhaps the ideal arrangement in providing pasture for the stallion is to give him access to a well-sodded paddock.

• **Exercise for the stallion**—Most horsemen feel that regular, daily exercise for the stallion is important. Certainly, it is one of the best ways in which to keep a horse in a thrifty, natural condition. It has also been assumed that forced exercise is of importance in improving semen quality. However, recent studies with dairy bulls cast considerable doubt on the relationship of exercise to fertility. For example, in one study involving dairy bulls used in artificial insemination, 8 bulls which were force exercised were compared with a like number which were kept in box or tie stalls, without forced exercise. The exercised group showed a nonreturn rate of 63.8 percent, whereas the bulls that were not exercised showed a nonreturn rate of 65 percent; hence, the bulls without exercise were actually a little more fertile than the exercised ones.[13] This points up the need for well-controlled experiments on the importance of exercise of the stallion on semen quality.

Stallions of the light horse breeds are most generally exercised under saddle or hitched to a cart. Thus, Standardbred stallions are usually jogged 3 to 5 miles daily while drawing a cart. Thoroughbred stallions and saddle stock stallions of all other breeds are best exercised under saddle for from 30 minutes to 1 hour daily, especially during the breeding season. Exercise should not be hurried or hard; the walk and the trot are the best gaits to use for this purpose. After the stallion is exercised, he should be rubbed down and cooled off before he is put up, especially if he is hot. Better yet, the ride should be so regulated at the end that the horse will be brought in cool, in which case he can be brushed off and turned into his corral.

Frequently, in light horses, bad feet exclude exercise on roads, and faulty tendons exclude exercise under the saddle. Under such conditions, one may have to depend upon (1) exercise taken voluntarily by the stallion in a large paddock, (2) longeing or exercising on a 30- to 40-foot rope, or (3) leading.

Longeing should be limited to a walk and a trot; and, if possible, the stallion should be worked on both hands; that is, made to circle both to the right and to the left. It is also best that this type of exercise be administered within an enclosure. Two precautions in longeing are: (1) do not longe a horse when the footing is slippery, and (2) do not pull the animal in such manner as to make him pivot too sharply with the hazard of breaking a leg.

Leading is a satisfactory form of exercise for some stallions if it is not practical to ride them. In leading, a bridle should always be used—never a halter—and one should keep away from other horses and be careful that the horse being ridden is not a kicker.

Where several stallions are exercised, a properly installed mechanical exerciser driven by an electric motor may be used as a means of lessening labor. It is similar to the merry-go-round type of equipment used to exercise dairy bulls.

The objection to relying upon paddock exercises alone is that the exercise cannot be regulated, especially during inclement weather. Some animals may take too much exercise and others too little. Moreover, merely running in the paddock will seldom, if ever, properly condition any stallion. Nevertheless, a 2- to 3-acre grassy paddock should always be provided, even for horses that are regularly exercised. Stallions that are worked should be turned out at night and on idle days.

• **Grooming the stallion**—Proper grooming of the stallion is necessary, not only to make the horse more attractive in appearance, but to assist exercise in maintaining the best of health and condition. Grooming serves to keep the functions of the skin active. It should be thorough, with special care taken to keep all parts of the body clean and free from any foulness, but not so rough nor so severe as to cause irritation either of the skin or the temper.

---

[13]*Physiology of Reproduction and Artificial Insemination of Cattle*, W. H. Freeman and Co. Publishers, 1961, p. 625.

Managing Mares at Foaling

The period of parturition is one of the most critical stages in the life of the mare. Through carelessness or ignorance, all of the advantages gained in selecting genetically desirable and healthy parent stock and in providing the very best of environmental and nutritional conditions through gestation can be quickly dissipated at this time.

• **Work and exercise**—Saddle or light-harness mares should be exercised moderately in the accustomed manner. If they are not used, other gentle exercise, such as leading, should be provided. This is especially important if they have not been accustomed to being on pasture and if it is desired to avoid any abrupt changes in feeding at this time.

• **Signs of approaching parturition**—Usually the first sign of approaching parturition is a distended udder, which may be observed 2 to 6 weeks before foaling time. About 7 to 10 days before the arrival, there will generally be a marked shrinkage or falling away of the muscular parts of the top of the buttocks near the tailhead and a falling of the abdomen. Although the udder may have filled out previously, the teats seldom fill out to the ends more than 4 to 6 days before foaling; and the wax on the ends of the nipples generally is not present until within 2 to 4 days before parturition. About this time the vulva becomes full and loose. As foaling time draws nearer, milk will drop from the teats; and the mare will show restlessness, break into a sweat, urinate frequently, lie down and get up, etc. It should be remembered, however, that there are times when all signs fail and a foal may be dropped when least expected. Therefore, it is well to be prepared as much as 30 days in advance of the expected foaling time.

• **Preparation for foaling**—When signs of approaching parturition seem to indicate that the foal may be expected within a week or 10 days, arrangements for the place of foaling should be completed. Thus, the mare will become accustomed to the new surroundings before the time arrives.

During the spring, summer, and fall months when the weather is warm, the most natural and ideal place for foaling is a clean, open pasture away from other livestock. Under these conditions, there is decidedly less danger of either infection or mechanical injury to the mare and foal. Of course, in following this practice, it is important that the ground be dry and warm. Small paddocks or lots that are unclean and foul with droppings are unsatisfactory and may cause such infectious troubles as navel ill.

During inclement weather, the mare should be placed in a roomy, well-lighted, well-ventilated, comfortable, quiet box stall which should first be carefully cleaned, disinfected, and bedded for the occasion. It is best that the mare be stabled therein at nights a week or 10 days before foaling so that she may become accustomed to the new surroundings. The foaling stall should be at least 12 feet square and free from any low mangers, hay racks, or other obstructions that might cause injury to either the mare or the foal. After the foaling stall has been thoroughly cleaned, it should be disinfected to reduce possible infection. This may be done by scrubbing with boiling hot lye water, made by using 8 ounces of lye to 20 gallons of water (one-half this strength of solution should be used in scrubbing mangers and grain boxes). The floors should then be sprinkled with air-slaked lime. Plenty of clean, fresh bedding should be provided at all times.

A foaling stall somewhat away from other horses and with a smooth, well-packed clay floor is to be preferred. The clay floor may be slightly more difficult to keep smooth and sanitary than concrete or other such surface materials, but there is less danger to the mare and the newborn foal from slipping and falling; and it is decidedly better for the hoofs.

• **Feed at foaling time**—Shortly before foaling, it is usually best to decrease the grain allowance slightly and to make more liberal use of light and laxative feeds, especially wheat bran. If there are any signs of constipation, a wet bran mash should be provided.

• **The attendant**—A good rule for the attendant is to *be near but not in sight*. The presence of the attendant may prevent possible injury to the mare and foal; and, when necessary, he may aid the mare or call a veterinarian.

• **Parturition**—The first actual indication of foaling is the rupture of the outer fetal membrane, followed by the escape of a large amount of fluid. This is commonly referred to as the rupture of the "water bag." The inner membrane surrounding the foal appears next, and labor then becomes more marked.

With normal presentation, a mare foals rapidly, usually not taking more than 15 to 30 minutes. Usually, when the labor pains are at their height, the mare will be down, and it is in this position that the foal is generally born, while the mare is lying on her side with all legs stretched out.

In normal presentation, the front feet, with heels down, come first, followed by the nose which is resting on them, then the shoulders, the middle (with the back up), the hips, and then the hind legs and feet. If the presentation is other than normal, a veterinarian should be summoned at once, for there is great danger that the foal will smother if its birth is delayed. If the feet are presented with the bottoms up, it is a good indication that they are the hind ones, and there is likely to be difficulty.

If after reasonable time and effort have been expended a mare appears to be making no progress in parturition, it is advisable that an examination be made and assistance be rendered before the animal has completely exhausted her strength in futile efforts at expulsion. In rendering any such assistance, the following cardinal features should exist: (1) cleanliness; (2) quietness; (3) gentleness; (4) perseverance; and (5) knowledge, skills, and experience.

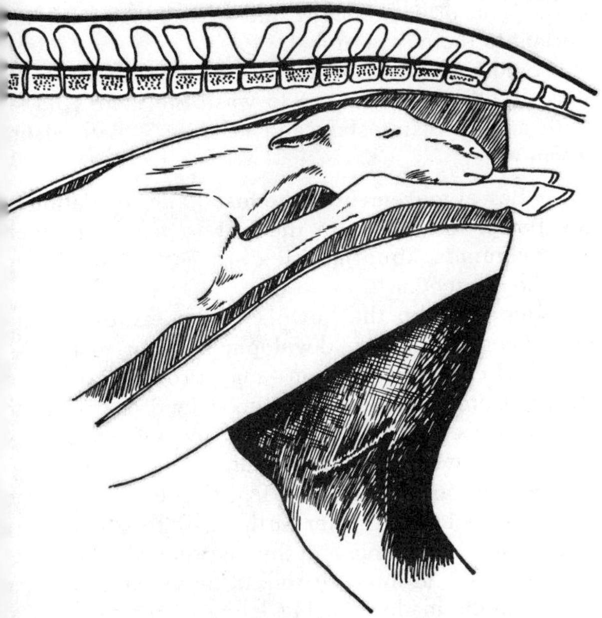

Fig. 8-41. Normal presentation. The back of the fetus is toward the back of the mother, the forelegs are extended toward the vulva with the heels down and the nose rests between the forelegs.

When parturition is unduly delayed or retarded, the fetus often dies from twists or knots in the umbilical cord, or from remaining too long in the passage. In either case, there may be stoppage of fetal circulation or lack of oxygen for the fetus, or both.

If foaling has been normal, the attendant should enter the stable to make certain that the foal is breathing and that the membrane has been removed from its mouth and nostrils. If the foal fails to breathe immediately, artificial respiration should be applied. This may be done by blowing into the mouth of the foal, working the ribs, rubbing the body vigorously and permitting the foal to fall around. Then after the navel has been treated, the mare and foal should be left to lie and rest quietly as long as possible so that they may gain strength.

• **The afterbirth**—If the afterbirth is not expelled soon after the mare gets up following foaling, it should either be tied up in a knot or tied to the tail of the mare. This should be done so that the foal or mare will not step on it, thereby increasing the danger of inflammation of the uterus and foal founder in the mare. Usually the afterbirth will be expelled within 1 to 6 hours after foaling. If it is retained for a longer period, or if lameness is evident, the mare should be blanketed, and an experienced veterinarian should be called. Retained afterbirth often causes laminitis, which is recognized by lameness in the mare. This is usually treated by feeding easily digested feed for a period of 36 hours and by applying cold applications to the mare's feet until the condition is relieved.

To prevent development of bacteria and foul odors, the afterbirth should be removed from the stall and burned or buried in lime as soon as possible.

• **Cleaning the stall**—Once the foal and mare are up, the stall should be cleaned. Wet, stained, or soiled bedding should be removed. The floor should be sprinkled with lime; and clean, fresh bedding should be provided. Such sanitary measures will be of great help in preventing the most common type of joint ill.

If the weather is extremely cold and the mare hot and sweaty, she should be rubbed down, dried, and blanketed soon after getting on her feet.

• **Feed and water after foaling**—Following foaling, the mare usually is somewhat hot and feverish. She should be given small quantities of lukewarm water at intervals, but she should never be allowed to gorge. It is also well to feed lightly and with laxative feeds for the first few days. The very first feed might well be a wet bran mash with a few oats or a little oat meal soaked in warm water. About one-half the usual amount should be fed. Usually, for the first week, no better grain ration can be provided than bran and oats. The quantity of feed given should be governed by the milk flow, the demands of the foal, and the appetite and condition of the mare. Usually the mare can be back on full feed within a week or 10 days after foaling.

• **Observation**—The good horseman will be ever alert to discover difficulties before it is too late. If the mare has much temperature (normal for the horse is about 101°F), something is wrong and the veterinarian should be called. As a precautionary measure, many good horsemen take the mare's temperature a day or two after foaling. Any discharge from the vulva should be regarded with suspicion.

Managing the Newborn Foal

Immediately after the foal has arrived and breathing has started, it should be thoroughly rubbed and dried with warm towels. Then it should be placed in one corner of the stall on clean, fresh straw. Usually the mare will be less restless if this corner is in the direction of her head. The eyes of a newborn foal should be protected from a bright light.

• **Navel cord**—To reduce the danger of navel infection (which causes a disease known as joint ill or navel ill) the navel cord of the newborn foal should be treated at once with a solution of tincture of iodine (or

Metaphen or Merthiolate may be used). This may be done by placing the end of the cord in a wide-mouthed bottle nearly full of tincture of iodine while pressing the bottle firmly against the abdomen. This is best done with the foal lying down. The cord should then be dusted with a good antiseptic powder. Dusting with the powder should be continued daily until the stump dries up and drops off and the scar heals, usually in three or four days. If an antiseptic powder is not available, air-slaked lime may be used. Any foreign matter that accumulates on the navel should be pressed out, and a disinfectant should be applied.

If left alone, the navel cord of the newborn foal usually breaks within 2 to 4 inches from the belly. Under such conditions, no cutting is necessary. However, if it does not break, it should be severed about 2 inches from the belly with clean, dull shears or it may be scraped into with a knife. Never cut diagonally across. A torn or broken blood vessel will bleed very little, whereas one that is cut directly across may bleed excessively. If severing of the cord is resorted to, it should be immediately treated with iodine.

• **The colostrum**—The colostrum is the milk that is secreted by the dam for the first few days following parturition.

The strong, healthy foal will usually be up on its feet and ready to nurse within 30 minutes to 2 hours after birth. Occasionally, however, a big awkward foal will need a little assistance and guidance during its first time to nurse. The stubborn foal should be coaxed to the mare's teats (forcing is useless). This may be done by backing the mare up on additional bedding in one corner of the stall and coaxing the foal with a bottle and nipple. The attendant may hold the bottle while standing on the opposite side of the mare from the foal. The very weak foal should be given the mare's first milk even if it must be drawn in a bottle and fed by nipple for a time or two. Sometimes these weak individuals will nurse the mare if steadied by the attendant.

Aside from the difference in chemical composition, the colostrum (the milk yielded by the mother for a short period following the birth of the young) seems to have the following functions:

1. It contains antibodies that temporarily protect the foal against certain infections, especially those of the digestive tract. Because newborn foals are unable to produce antibodies for some time after birth, they must acquire preformed antibodies through colostrum, which is especially high in immune lactoglobulins. To be effective in protection against disease, however, colostrum must be ingested within a few hours after birth, preferably within 15 to 30 minutes, because gut closure occurs about 24 to 30 hours after birth. Subsequently, the foal digests these larg molecular weight proteins, with the loss of their im munization properties.

2. It serves as a natural purgative, removing feca matter that has accumulated in the digestive tract.

This, therefore, explains why mares should not b milked out prior to foaling and why colostrum is im portant to the newborn foal.

Before allowing the foal to nurse for the first time it is usually good practice to wash the mare's udde with a mild disinfectant and to rinse it with clear warm water.

• **Bowel movement of the foal**—The regulation c the bowel movement in the foal is very importan Two common abnormalities are constipation an diarrhea or scours.

Impaction in the bowels of the excrement ac cumulated during the development prior to birth— material called meconium—may prove fatal if nc handled promptly. Usually a good feed of colostrur will cause elimination, but not always—especiall when foals are from stall-fed mares.

Bowel movement of the foal should be observe within 4 to 12 hours after birth. If by this time the has been no discharge and the foal seems rather slug gish and fails to nurse, it should be given an enema This may be made by using 1 to 2 quarts of water a blood heat, to which a little glycerin has been addec or warm, soapy water is quite satisfactory. The solu tion may be injected with a baby syringe (one havin about a 3-inch nipple) or a tube and can. This trea ment may be repeated as often as necessary until th normal yellow feces appear.

Diarrhea or scours in foals may be associated wit infectious diseases or may be caused by unclean su roundings. Any of the following conditions may brin on diarrhea: contaminated udder or teats; nonremova of fecal matter from the digestive tract; fretfulness c temperature above normal in the mare; an excess c feed affecting the quality of the mare's milk; colc damp bed; or continued exposure to cold rains. A treatment is not always successful, the best practice i to avoid the undesirable conditions.

Some foals scour during the foal heat of the mare which occurs between the seventh and ninth day fo lowing foaling.

Diarrhea is caused by an irritant in the digestiv tract that should be removed if recovery is to be ex pected. Only in exceptional cases should an astrin gent be given with the idea of checking the diarrhea and such treatment should be prescribed by the vete inarian.

If the foal is scouring, the ration of the mar should be reduced, and a part of her milk should b taken away by milking her out at intervals.

Care of the Feet

The important points in the care of a horse's feet are to keep them clean, prevent them from drying out, trim them so they retain proper shape and length, and shoe them correctly when shoes are needed.

Each day, the feet of horses that are shod, stabled, or used should be cleaned and inspected for loose shoes and thrush. Thrush is a disease of the foot, caused by a necrotic fungus, and characterized by a pungent odor. It causes a deterioration of tissues in the cleft of the frog or in the junction between the frog and bars. This disease produces lameness and, if not treated, can be serious.

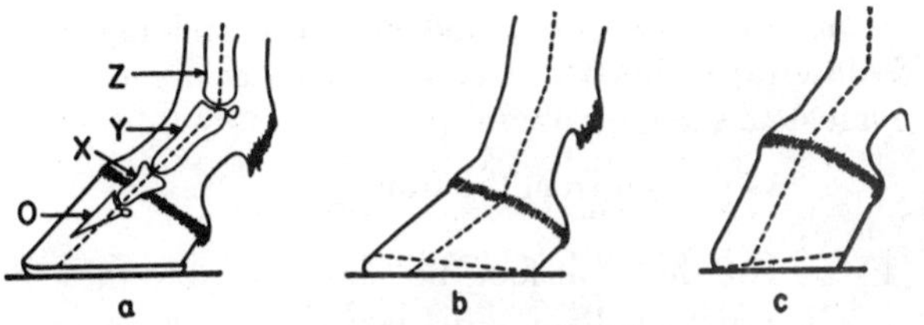

Fig. 8-42. (a) Properly trimmed hoof with normal foot axis: O—coffin bone; X—short pastern bone; Y—long pastern bone; Z—cannon bone. (b) Toe too long, which breaks the foot axis backward. Horizontal dotted line shows how hoof should be trimmed to restore normal posture. (c) Heel too long, which breaks the foot axis forward. Horizontal dotted line shows how trimming will restore the correct posture.

• **Proper stance; correcting common faults**—Before trimming the feet or shoeing a horse, it is important to know what constitutes both proper and faulty conformation (see Section 11).

Fig. 8-42 shows the proper posture of the hoof and incorrect postures caused by hoofs grown too long in either toe or heel. The slope is considered normal when the toe of the hoof and the pastern have the same direction. This angle should always be kept in mind and changed only as a corrective measure. If it should become necessary to correct uneven wear of the hoof, the correction should be made gradually over a period of several trimmings.

Prior to the trimming of the feet, the horse should be inspected while standing squarely on a level area—preferably a hard surface. Then it should be seen in action, both at the walk and the trot.

The hoofs should be trimmed every month or six weeks, whether the animal is shod or not. If shoes are left on too long, the hoofs grow out of proportion. This may throw the horse off balance and place extra stress upon the tendons. Hence, the hoofs should always be kept at the proper length and the correct posture. They should be trimmed near the level of the sole; otherwise, they will split off if the horse remains unshod. The frog should be trimmed carefully, with only ragged edges removed that allow the filth to accumulate in the crevices, and the sole should be trimmed sparingly, if at all. The wall of the hoof should never be rasped.

Table 8-11 shows the common faults and how to correct them through proper trimming.

TABLE 8-11
COMMON FOOT FAULTS, AND HOW TO CORRECT THEM

Fault	How It Looks	How to Trim
Splayfoot	Front toes turned out, heels turned in.	Trim the outer half of the foot.
Pigeon-toed	Front toes turned in, heels turned out—the opposite of splayfoot.	Trim the inner half of foot more heavily; leave the outer half relatively long.
Quarter crack	Vertical crack on side of hoof.	Keep the hoof moist, shorten the toes, and use a corrective shoe.
Cocked ankles	Standing bent forward on fetlocks—most frequently the hind ones.	Lower the heels to correct. However, raising the heels makes for more immediate horse comfort.
Contracted heels	Close at the heels.	Lower the heels and allow the frog to carry more of the weight, which tends to spread the heels apart.

• How to recognize good and faulty shoeing—The following checklist may be used as a means to evaluate a shoeing job:

1. **As viewed from the front—**

Yes No
- ☐ ☐ Are the front feet the same size, the toes the same length, and the heels the same height?
- ☐ ☐ Is the foot in balance in relation to the leg?
- ☐ ☐ Is the foot directly under the leg, is the axis of the foot in prolongation to the axis of the upper leg bones, and is the weight of the body equally distributed over the feet?

2. **As viewed from the side—**

Yes No
- ☐ ☐ Does the axis of the foot coincide with the axis of the pastern?
- ☐ ☐ Does the slope of the wall of the hoof parallel the slope of the pastern?
- ☐ ☐ Has the lower outer border of the wall been rasped?
- ☐ ☐ Does the conformation of the foot and the type of shoe warrant the rasping done?

3. **As the height and strength of nailing are inspected closely—**

Yes No
- ☐ ☐ Do the nails come out of the wall at the proper height and in sound horn?
- ☐ ☐ Are the nails driven to a greater height in the wall than necessary?
- ☐ ☐ Is the size of the nail used best suited for the size and condition of the foot and the weight of the shoe?
- ☐ ☐ Are the clinches of sufficient thickness to ensure strength?
- ☐ ☐ Are the clinches smooth and not projecting?

4. **As the shoe is scrutinized—**

Yes No
- ☐ ☐ Is the toe of the shoe fitted with sufficient fullness to give lateral support to the foot when breaking over and leaving the ground?
- ☐ ☐ Are the branches of the shoe from the bend of the quarter to the heel fitted fuller than the outline of the wall to provide for expansion of the foot and normal growth of horn?
- ☐ ☐ Are the heels of the shoe of sufficient length and width to cover the buttresses?
- ☐ ☐ Are the heels finished without sharp edges?
- ☐ ☐ Does the shoe rest evenly on the bearing surface of the hoof, covering the lower border of the wall, white line, and buttresses?
- ☐ ☐ Is the shoe concaved so that it does not rest upon the horny sole?
- ☐ ☐ Are the nail heads properly seated?
- ☐ ☐ Is the shoe the correct size for the foot?
- ☐ ☐ Will the weight of the shoe provide reasonable wear and protection to the foot?
- ☐ ☐ Have the ragged particles of the horny frog been removed?

• Care of the foal's feet—Foals may become unsound of limb when the wear and tear is not equally distributed due to an unshapely hoof. On the other hand, faulty limbs may be helped or even corrected by regular and persistent trimming. Such practice also tends to educate the foal and to make shoeing easier at maturity. If the foal is run on pasture, trimming of the feet may be necessary long before weaning time. A good practice is to check the feet regularly every month or six weeks and to trim a small amount each time if trimming is needed, rather than trim too much at any one time. Tendons should not receive undue strain by careless trimming of the feet. Usually, only the outer rim should be trimmed, though sometimes it is necessary to cut down the heel or frog or to shorten the toes. The necessary trimming may be done with the rasp, farrier's knife, and nippers (using the rasp for the most part).

Before the feet are trimmed, the foal should be inspected first while standing squarely on a hard surface. Then it should be seen in action, both at the walk and the trot.

• Treatment of dry hoofs—When hoofs become dry and brittle, they sometimes split and cause lameness. The frogs lose their elasticity and are no longer effective shock absorbers. If the dryness is prolonged, the frogs shrink and the heels contract.

Dry hoofs usually can be prevented by keeping the ground wet around the watering tank, attaching wet burlap sacks around the hoofs, or applying a hoof dressing. Several satisfactory commercial hoof dressings are on the market. Also, a good homemade product may be made as follows:

6 parts fish oil (cod liver oil)
1 part pine tar oil
1 part Creolin
2 parts glycerin

The above mix should be stirred well before using and applied daily. If fish oil is not available, raw linseed oil may be substituted.

Grooming

Proper grooming is necessary to keep a horse attractive and help maintain his good health and condition. Grooming cleans the hair, helps keep the skin functioning naturally, lessens skin diseases and parasites, and improves the condition and fitness of the muscles.

Grooming should be rapid and thorough but not so severe that it makes the horse nervous or irritates his skin. Horses that are kept in stables or small corrals should be groomed thoroughly at least once a day. When horses are worked or exercised, they should be groomed both before and after the work or exercise.

Wet or sweating animals should be handled as follows:

1. Remove the tack as fast as possible, wipe it off, and put it away.
2. Remove excess water from the horse with a sweat scraper and then rub him briskly with a grooming or drying cloth to dry his coat partially.
3. Cover the horse with a blanket and walk him until he is cool.
4. Allow the horse to drink two or three swallows of water every few minutes while he is cooling and drying.

To assure that the horse is groomed thoroughly and that no body parts are missed, follow a definite order of grooming. This may vary according to individual preference.

Grooming may be checked by rubbing the fingertips against the natural lay of the hair. If the coat and skin are not clean, the fingers will get dirty and gray lines will show on the coat where the fingers passed. The cleanliness of the ears, face, eyes, nostrils, lips, sheath, and dock can be determined by inspection.

Transporting Horses

Horses are transported via trailer, van, truck, rail, boat, and plane. Today, transportation by motor (trailer, van, or truck) is most common because of the distinct advantage of door-to-door movement. Regardless of the method, however, the objectives are the same: to move them safely, with the maximum of comfort, and as economically as possible. To this end, selection of the equipment is the first requisite. But equipment alone, no matter how good, will not suffice.

The trip must be preceded by proper preparation including conditioning of horses; and horses must receive proper care, including smooth movement, en route.

In summary form, the requisites of good transportation, with special emphasis on motor transportation, follow:

1. Provide good footing.
2. Drive carefully.
3. Make nurse stops.
4. Provide proper ventilation.
5. Teach horses to load early in life.
6. Provide health certificate and statement of ownership.
7. Schedule properly.
8. Have the horses relaxed.
9. Clean and disinfect public conveyance.
10. Have a competent caretaker accompany horses.
11. Use shanks except on stallions.
12. Feed lightly.
13. Water liberally.
14. Pad the stalls.
15. Take along tools and supplies.
16. Check shoes, blankets, and bandages.
17. Be calm when loading and unloading.
18. Control insects.

Trailers, vans, and trucks have the very great advantage of being able to load from in front of one stable and unload in front of another.

The trailer is usually a 1- or 2-horse unit, which is drawn behind a car or truck. Generally speaking, this method of transportation is best adapted to short distances—less than 500 miles. Horses are trailered to shows, races, endurance rides, breeding establishments, to new owners, from one work area to another on the range.

The van or vanlike trailer is a common and satisfactory method of transportation where three to eight horses are involved. There is hardly any limit to the kinds of vans, ranging from rather simple to very palatial pieces of equipment.

Rail and boat shipments are seldom used anymore. Where valuable horses are involved, they have given way to the greater speed and flexibility of plane transportation.

Stable Management

The following stable management practices are recommended:

1. Remove the top layer of clay floors yearly; replace with fresh clay, and level and tamp. Also, keep the stable floor higher than the surrounding area, thereby making for dryness.
2. Keep stalls well lighted.
3. Use properly constructed hayracks to lessen waste and contamination of hay, with the possible exception of maternity stalls.
4. Scrub concentrate containers at such intervals as necessary, and after feeding a wet mash.
5. Work over bedding daily, removing excrement and wet, stained or soiled material, and provide fresh bedding.
6. Practice rigid stable sanitation to prevent fecal contamination of feed and water.
7. Lead foals when taking them from the stall to the paddock and back, as a way in which to further their training.
8. Restrict the ration when horses are idle, and provide either a wet bran mash the evening before an idle day or turn idle horses to pasture.
9. Provide proper ventilation at all times—by means of open doors, windows that open inwardly from the top, or stall partitions slatted at the top.
10. Keep stables in repair at all times, so as to lessen injury hazards.

SECTION 9

BUILDINGS AND EQUIPMENT

Contents **Page**

Location of Farm or Ranch Headquarters 796
Farmstead Arrangement 797
Layout of Livestock Operations 798
Requisites of Livestock Buildings 798
Space Requirements of Buildings and Equipment 802
Environmentally Controlled Buildings 808
Heat Production of Animals 809
Vapor Production of Animals 810
Recommended Environmental Conditions of Animals 810
Naturally Ventilated Buildings 810
Roofs 812
Floors for Stalls or Stables 812
Solid Floors 812
Slotted Floors 812
Paved Lots and Feeding Floors 813
Manure Management 814
Manure Production and Storage 814
Manure Handling 814
Requisites of Livestock Equipment 814
Pollution Control 815
Guidelines Relative to Facility and Equipment Costs 816
How to Determine the Size Barn to Build 816
How to Determine the Size Silo to Build 817
Beef Cattle Buildings and Equipment 820
Beef Cattle Barns 820
Beef Cattle Sheds 820
Facilities for Finishing Cattle 822
Beef Cattle Equipment 827
Dairy Cattle Buildings and Equipment 830
Stall Barns 830
Loose Housing 830
Manure Management 833
Dairy Cattle Equipment 835
Sheep Buildings and Equipment 836
Sheep Finishing Facilities and Equipment 836
Confinement of Sheep on Slotted Floors 836
Sheep Equipment 838
Swine Buildings and Equipment 839
Space, Temperature, and Grouping of Early Weaned Pigs 839
Swine Buildings and Systems 839
Portable Houses 840
Slotted Floors 840
Swine Equipment 842
Horse Buildings and Equipment 843
Horse Barns 843
Shades 846
Show-Ring 846
Horse Equipment 846
Concrete Structures for the Farm and Ranch 849
Fences for Livestock 852
Paint for the Farm and Ranch 860
The Farm Pond 865

Properly designed, constructed, and arranged buildings and equipment are an asset to the farm or ranch. They increase animal production, make for labor and feed efficiency, conserve crops and manure, provide comfort for man and beast, and add to the beauty of the farm landscape. In serving these purposes, it is not necessary that buildings and equipment be either elaborate or expensive, because the amount of barn rent that an animal can pay is limited. There is an old saying in the livestock industry that, "Fancy white houses and big red barns won't put fat on animals. You just need feed and water." Yet, a stockman needs adequate facilities. In addition to protecting animals from the elements, barns and shelters should be evaluated from the standpoints of storage of feed, saving of labor, and handling of manure.

Fences should be considered. Repairing fencing or adding new fencing is expensive. Also, it should be kept in mind that the life of the fence depends to a great extent on the life of the posts, and the stability of the corner posts.

Increasingly, the U.S. livestock industry has moved to more and more confinement—to production in limited quarters, with or without shelter, but with no pasture. The reasons: (1) it lessens land cost, (2) it minimizes management time, (3) it minimizes labor, (4) it maximizes control of animals, (5) it maximizes genetic potential of animals, and (6) it maximizes feed conversion. These motivating forces will be accentuated in the future, with the result that confinement production will continue to increase. The trend and acceptance is toward more closed, insulated, mechanically ventilated, environmentally controlled structures—especially for poultry, swine, and dairy enterprises. But the shift to confinement structures and high-density production operations has introduced new problems and accentuated old ones, especially in the areas of animal behavior, manure management, optimum environment, and flexibility.

Family happiness is important to the success of any livestock venture. Thus, the home should come in for its share of consideration, also. If it is not completely modern so far as plumbing, heating, and lighting are concerned, it will need to be made so as soon as possible.

In this section, special attention is given to space requirements, environmental control, types of floors, handling manure, and laborsaving buildings and equipment. The effect of man-made environment on animal behavior is accorded special and rather complete treatment in Section 1.

Because of variations in climatic conditions, sizes and types of enterprises, and systems of management, no attempt will be made herein to present detailed building and equipment plans and specifications. Rather, it is proposed merely to convey suggestions regarding some of the desirable features of buildings and equipment in use for each class of farm animals in various parts of the country. For detailed plans and specifications for a particular locality, the stockman should (1) study successful buildings and equipment on neighboring farms or ranches; (2) consult the local county agricultural agent, vocational agriculture instructor, or lumber dealer; and/or (3) write to the state college of agriculture.

LOCATION OF FARM OR RANCH HEADQUARTERS

The headquarters or farmstead is the base of all farm or ranch operations. It is important, therefore that it be located and planned correctly for efficient profitable, and pleasant operation.

Since most farms have been established for several years, there are relatively few opportunities for selecting entirely new sites for headquarters on unimproved land. However, as buildings become obsolete or need replacement, as the system of farming changes, or as farms are combined into larger units opportunities do arise for relocation of the headquarters area. The changing picture in agriculture also leads to the need for revising and improving existing headquarters. The same principles which apply in locating and planning headquarters on unimproved land also apply to improving existing headquarters as buildings are replaced or added, and should be a part of a master plan. Likewise, when appraising the desirability of an existing headquarters, these same points should be considered. These factors are:

1. **Central location**—From the standpoint of management, the most convenient location for the headquarters is near the center of the unit or in the middle of the long side. Normally, a natural farm control center of this type makes for (a) the best accessibility to fields, for both equipment and stock, and (b) the most desirable overall visual supervision of operations. However, if the farm or ranch is deep, the location of the headquarters near the center of the unit may require a private all-weather lane or road leading from the highway to the headquarters, at added cost for construction and maintenance.

2. **Water supply**—Water must be available and plentiful. The availability of electricity and automatically operated pumps makes it possible to locate the headquarters farther from the source of water supply when it is desirable to do so to obtain other advantages. In irrigated areas, it is also desirable to be near an irrigation turnout.

3. **Roads**—It is preferable that the headquarters be located near an all-weather road or highway that is well maintained, but one should avoid having the farmstead on both sides of a road. Normally, a location along an all-weather road has better access to electric

and telephone lines, the school bus, mail, religious and recreational facilities, and other services. Also, in irrigated areas, the irrigation turnout is usually near an all-weather road. When along dirt and gravel roads, the headquarters should be placed far enough away, taking into consideration the direction of the prevailing winds, to keep the dust from becoming a nuisance.

4. **Telephone and electricity**—If possible, the headquarters should be near well-maintained telephone and electric lines. Farming or ranching is a business, and it is difficult to conduct any kind of business without access to a telephone. Likewise, electricity is essential for the operation of most modern utilities and automated equipment.

5. **Service facilities**—The farm or ranch should have convenient access to an established mail route, a school bus from a good school, delivery services (milk, laundry, bread, etc.), the church of preference, and recreational facilities of interest.

6. **Size and shape of area**—The area for the headquarters should be of adequate size, usually from two to five acres, and nearly square in shape. With further mechanization, smaller farmsteads are now needed than in the era when horses were extensively used. In general, the tendency is to have too large a farmstead; this adds to maintenance and weed-control costs and keeps valuable land out of production.

7. **Topography**—The topography should be high and level with no abrupt slopes. A relatively level area requires less site preparation, thus lowering building costs.

8. **Drainage**—The soil should be porous and the slope gentle, for this makes for dry corrals and lots. Animal health is much more easily maintained when the yards and buildings are well drained; and the work of the caretaker is more pleasant and easier under such conditions.

9. **Erosion control**—On those farms or ranches in which soil erosion is a problem, it is desirable that the headquarters and fields be located to permit contour farming.

10. **Vegetation; windbreaks**—Natural shade, trees for windbreaks, and a well-sodded area are valuable attributes. If a natural windbreak (hills or trees) is not available, wind protection may be provided by planting trees, or by utilizing the buildings themselves as windbreaks to protect yards, lots, and other open areas.

11. **Soil fertility**—A good fertile soil is desirable for the garden and yard.

12. **View**—A scenic view, especially from the living side of the house, makes for a "heap of living." Also, a location near the top of a slope may permit a commanding view of the rest of the farm.

The consideration of the above points in locating, planning, and improving the farm or ranch headquarters can add materially to the convenience, comfort, and pleasure of the operator, and to the profitable operation of the enterprise.

FARMSTEAD ARRANGEMENT

The arrangement—which means the location and orientation of individual buildings within the site—should make for ease of use, economy of labor and cost in operation, and attractiveness. In general, for conservation of space and time, the barn and other service buildings should be located around a central court and should be so arranged that most of them can be seen from the house.

In planning a new farm or ranch headquarters or in altering an old one, buildings, fences, lots, trees, etc., should be added according to an established master plan; for, once constructed, buildings are difficult and expensive to move.

In arriving at the best arrangement, the farmstead cannot and should not be modeled after one popular pattern. Instead, consideration should be given to the following pertinent points:

1. **The house location comes first**—As the farm or ranch house is the headquarters or office of the farm business as well as a home, its location is of greatest importance in farmstead arrangement. The ranch house should be located: (a) on a high area which is well drained away from the house and will command a view of other buildings as well as one or more scenic views; (b) where it is easily accessible; (c) in the direction of the prevailing winds, sheltered from strong winter winds but not blocked from summer breezes; (d) to obtain the maximum of sunlight in the north and a minimum of sunlight in the south; (e) with access to either the front or back door, but with best access to the front door; (f) where there is adequate yard which can be well landscaped; and (g) nearest the road, but generally not closer than 100 to 150 feet. (If the road is especially dusty or noisy, this distance should be increased.)

2. **Orientation**—Fortunately, the farm or ranch headquarters need not be oriented with the compass. Although in general the farmstead plan will be developed to present the front to the road, most buildings can be turned, quarter-turned, or reversed, as may be necessary to take advantage of the prevailing winds, sunlight, view, etc. In general, livestock barns are placed with the long axis north and south, whereas, when possible, livestock sheds in northern areas are faced to secure direct sunlight and yet to face away from the direction of the prevailing winds. In the South, sheds are usually oriented for maximum shade and storm protection.

3. **Direction of wind**—The house should be located on the windward side of the headquarters, with

special consideration given to summer winds. Swine barns should be located at the greatest distance from the other farm buildings, especially from the house and dairy barn, and so that the prevailing winds will not blow from the swine barn to the house. Unless hills form a natural windbreak, it is desirable to arrange suitable tree plantings for this purpose or to locate the buildings so that they can be used to shelter open lots. Usually, a tree windbreak is located 75 to 150 feet from the buildings to be protected, with three to seven rows of trees 20 to 75 feet wide.

4. **Efficiency**—The buildings should be located so as to require a minimum of walking when doing the chores. This means that those buildings in which the most time is spent—such as the dairy barn, poultry house, and machine shop—should be closest to the house, that accessory buildings should be located close to those buildings which they service, and that the buildings should be near enough together to permit efficiency of labor without making a fire hazard. Likewise, animal barns should be convenient to feedlots and pastures.

5. **Corrals and lots**—The buildings and their adjacent corrals and lots should be arranged so that the buildings are accessible without walking through feedlots and corrals.

6. **Fire protection**—Farm buildings should be far enough apart so that fire will not spread easily from one building to another. In general, this means at least 100 feet apart in the case of large buildings. In acquiring added fire protection through spacing buildings farther apart, one should avoid extreme distances that will mean inefficiency in operation; fire insurance is probably cheaper than labor.

7. **Appearance**—Careful attention to the headquarters arrangement can add to the attractiveness of the entire unit. Manure piles and unsightly objects should not be visible from the main highway or house; shrubbery and trees should be planted to screen unsightly objects; fences and buildings should be repaired and painted regularly; and yards, driveways, and corrals should be kept free of rubbish, scattered farm machinery, etc.

8. **Gates and lanes**—The adoption of larger machinery has necessitated wider gates and lanes than have been commonplace in the past. Often the wider lanes can serve as pasture as well as roadway.

9. **Expansion**—Provision should be made for easy expansion of the farmstead. Many times buildings can be expanded in size by extending their length, provided no other buildings or utilities interfere.

LAYOUT OF LIVESTOCK OPERATIONS

Prior to starting construction, the stockman may avoid much subsequent difficulty and expense by first doing some paper and pencil planning. He should first decide on the specific kind, or kinds, of livestock and the size of the enterprise. Then, he should sketch out the buildings and equipment required to meet these needs in the most efficient and economical manner. In particular, the preliminary layout of livestock operations should include the following:

1. **The management system**—The management system will greatly affect the kind, size, and amount of buildings and equipment. For example, the swine producer must first decide whether he will (a) farrow out pigs, or buy feeders; (b) farrow twice a year, or multiple farrow—if he is going to maintain a breeding herd; (c) rely on multipurpose buildings, or have special buildings (for example, a special house for farrowing and a special house for finishing); (d) buy commercial, ready-mixed feeds, or use a maximum of homegrown feeds; or (e) follow confinement rearing or pasture rearing—or a combination of the two.

2. **Plans for the flow of all materials**—The stockman should develop detailed plans for the flow of all materials, with primary consideration given to maximum automation and minimum labor and expense. These plans should include provision for (a) delivering the proper feed to the animals at the desired time and place, (b) providing a sanitary water supply, (c) delivering and distributing bedding, (d) removing manure, and (e) marketing animals. All these considerations, and more, enter into the handling of materials to, within, and from buildings.

The above information, constituting the layout of operations, should first be put on paper in sketch form, by the producer. From this, the architect and/or engineer can design, or recommend for purchase, buildings and equipment which most effectively and economically meet the production requirements of the specific enterprise.

REQUISITES OF LIVESTOCK BUILDINGS

When planning to build, the owner should first pose to, and ask of, himself the searching question "Why is this structure to be built?" Livestock buildings should not be monuments. Instead, they should be production tools; they should contribute to the farm or ranch operations, and they should not only pay for themselves but they should pay satisfactory returns on the investment.

Each farm or ranch is different, and the type and size of buildings will vary accordingly. Among the factors determining the type and size of buildings are: (1) kind and fertility of the soil, (2) available markets, (3) size of farm, (4) tenant or owner operation, (5) kind and amount of livestock and crops to be grown, (6) personal preference, (7) climatic conditions of the region, and (8) storage requirements. Thus, the specific requisites of animal buildings will vary according to the needs of the region, state, community, and individual farm or ranch.

Before starting construction, there should be a complete plan. This step should precede arranging financing and starting site preparation and construction; and it should be observed whether you (a) do the construction yourself, (b) have a contractor build it, (c) buy a site-erected manufactured building, or (d) buy a prefabricated building. Much of the needed planning information may be obtained from this book. Additional planning information, along with working drawings for building construction, are available from manufacturers and dealers. Also, standard plans prepared by your university or the U.S. Department of Agriculture may be purchased from your Extension Agricultural Engineer or from the following:

Midwest Plan Service
Iowa State University
Ames, Iowa 50011

Certain general requirements of animal buildings should always be considered; and it is with these that the ensuing discussion will deal. Once buildings are constructed, there is a practical limit to the changes that can be made in remodeling. Consequently, it is most important that very careful consideration be given to the following requisites:

1. **Environmental control**—Man achieves environmental control through clothing, vacationing in resort areas, air-conditioned homes and cars, etc. But most animals are produced under the environmental (weather, etc.) conditions peculiar to the area, with only minor modifications thereof. However, limited basic research has shown that animals are more efficient—that they make more rapid gains and require less feed—if raised under conditions of ideal temperature and humidity. The primary reason for having buildings, therefore, is to modify the environment. Proper barns and other shelters and shades, insulation, ventilation, heating, and air conditioning (in some cases) can be used to approach the environment that we wish. Eventually, many hogs will likely be produced under controlled environmental conditions, in air-conditioned houses that can be heated or cooled as desired. Controlled environment will also be extended more and more to cattle and sheep feedlots; for example, summer temperatures will be moderated through shades, sprinklers, and air-circulating equipment. Also, we shall give increasing attention to other stresses such as space requirements—to the number of animals that can be run together, as affected by class, age, size, and sex. Naturally, the investment in environmental control facilities must be balanced against the expected increased returns; and there is a point beyond which further expenditures for environmental control will not increase returns sufficient to justify the added cost. This point of diminishing returns will differ between sections of the country, quality of the animals (more cost can be justified for valuable purebreds), and operators; and climate, labor costs, feed costs, market prices, and other factors all enter into the picture.

2. **Properly insulated**—Proper insulation, which slows up the passage of heat, is desirable in the summer as well as in the winter. In the winter, it helps to hold the heat within the building, while in the summer it tends to prevent the heat from entering the structure.

In addition to being properly insulated, farm buildings need a vapor barrier (material which has a low rate of vapor transmission) to resist the entrance of moisture into the wall.

3. **Well ventilated**—Ventilation refers to the changing of air—the replacement of foul air with fresh air. Livestock barns and shelters should be well ventilated, but care must be taken to avoid direct drafts. Animals cannot do well in poorly ventilated stalls or sheds. If a choice must be made between warmth and ventilation, secure the latter. The primary purposes of ventilation are to remove excess moisture and foul odors. A desirable ventilation system accomplishes these objectives with a minimum temperature variation inside the building.

Ventilation may be secured by various systems. The simplest method usually consists of one or more of the following: (a) open shed, (b) open doors, (c) windows that open inwardly from the top, or (d) building or stall partitions left slatted or open at the top. A more complete method of ventilating tight buildings consists of a system of intake and outtake flues operated on the basis of either gravity or forced ventilation. Whatever the system, proper ventilation is one in which the foul air is drawn off and harmful humidity conditions are eliminated without excess heat loss or creation or drafts.

4. **Keep proper humidity**—Humidity refers to the amount of moisture in the air. The air inside animal buildings picks up moisture from respiration and excrement. When the humidity of a building is high, the evaporation from the surface of the body practically ceases. This causes a sensation of warmth in the summer and coldness in the winter. When there is a

difference of several degrees between inside and outside temperatures, the moisture condenses on the cold surfaces. In freezing weather, this condensation forms frost. Condensation on the walls or other surfaces is evidence of unsatisfactory moisture conditions. Such condensation is objectionable because it is harmful for animals to go from a moist, warm barn into the cold outside air and because the excess moisture causes the structure to decay or deteriorate.

Experimental studies indicate that a 1,000-lb cow breathes into the air approximately 10 lb of moisture per day. Thus, from a herd of 42 cows there would be given off 420 lb or about 50 gallons of water per day. This amount of moisture must be removed daily in order to keep the barn free from dampness, thus necessitating proper ventilation.

Strawlofts in sheds or one-story barns and haymows in two-story barns reduce moisture condensation and frosting on the ceiling.

5. **Reasonable construction and maintenance cost**—The word "reasonable" is used herein because low cost buildings of durable construction are a thing of the past.

If buildings are to be practical and economical, there must be reasonable construction and maintenance costs. In the first place, they cannot be built if the necessary capital is not available. Secondly, livestock barns and shelters must be paid for out of the enterprises they house, if desirable economic relationships are to be maintained. In considering building cost, the following additional points are pertinent: (a) low initial cost of buildings that do not serve satisfactorily the intended needs may result in lower profits, just as low-cost breeding stock may not prove profitable; (b) annual maintenance cost must be considered along with initial cost; (c) cost should be computed on the basis of per unit production rather than total dollars for the building—for example, if the yearly cost of a set of swine buildings (including amortization, interest on investment, taxes, insurance, and maintenance) is $1,000, the building cost per pound of pork produced is 10 cents if 10,000 pounds of pork are produced, whereas it is only 1 cent per pound of pork produced if 100,000 pounds of pork are produced; (d) consideration must be given to those intangible and often difficult to measure, but nevertheless real, savings in animal deaths, feed efficiency, and labor.

In addition to the practical aspects, however, certain intangible values accrue from having good buildings, such as the pride and satisfaction derived in caring for animals under such circumstances, the influence that it may have on the children—in causing them to want to stay on the farm or ranch, the reduced hazards, and, in the case of purebred herds and flocks, the advertising value that accompanies an attractive set of buildings.

6. **Flexible design; multiple-use versus specialized, single-use buildings**—In order that buildings may be adapted to changes in a changing world they should be as flexible in design as possible. This needed flexibility can best be achieved by constructing a one-story building with the maximum of movable partitions and equipment. Also, flexibility is one of the recognized virtues of open sheds—they may be used to accommodate cattle, sheep, hogs, or horses; or they may serve for machinery, feed, or bedding storage.

Flexibility of use is especially important on tenant-operated farms, because a change in tenants is usually accompanied by a change in type of operations; and, after all, nearly one-half of all farms in the United States are tenant operated.

At the present time, it is impossible to predict the kind and extent of changes that will come with further mechanization and by such developments as pelleted forages. It is obvious, however, that they will materially affect conventional storage and feeding practices, and that one will not be able to take full advantage of these and other developments unless maximum flexibility in building design exists.

On the other hand, it is recognized that highly specialized operations, of which there will be more in the future, favor specialized, single-use buildings rather than flexible, multiuse structures. For example, more and more large, highly specialized swine producers are constructing specially designed farrowing houses and specially designed finishing houses, rather than multiple-use structures in which pigs are both farrowed and finished. This is being done because the more highly specialized, but less flexible, buildings make for greater efficiency of production on the part of both the animal and the farm operator. In recognition that specialized, single-use type buildings cannot be easily remodeled and converted to other uses, should the future economy make their present use unprofitable, many operators are very wisely amortizing these inflexible structures over a shorter than normal period of time.

7. **Reduce labor**—Because of the high cost of labor, now and in the future, labor requirements must be held to a minimum. To this end, and to the extent that it is practical, all possible laborsaving considerations and devices should be a part of building construction.

8. **Utility value**—Buildings should have utility value; that is, they should be designed so that they will best serve their intended purpose. Generally, it may be said that the chief purpose or function of livestock buildings is to make it possible for animals to convert plants into more nutritious products. They should be constructed with this and other utility purposes in mind. To accomplish this, it is important that the man who is to use the building—the farmer or

rancher—should have the most to say about the design, provided that the design is well planned.

9. **Protect newborn animals**—Young calves, lambs, pigs, and foals are "animal babies" and must be protected as such. Suitable buildings make it possible to save more newborn animals.

10. **Attractiveness**—Any structure that has utility value and is erected in good proportions and in harmony with the natural surroundings will have aesthetic value and add to the value of the farm or ranch and the general enjoyment of rural living. Attractive buildings also add materially to the sale value of farm or ranch property.

11. **Durability**—Livestock barns and shelters should be adequately durable to stand firm against wind and weather (including snow load) and to last for a sufficient span of time without excessive maintenance cost, but they need not be so expensively or permanently constructed that there is danger of their becoming obsolete before being worn out. Thus, it is noteworthy that, with the mechanization of farms and ranches, many a horse barn throughout the country is either idle or has been remodeled, after a fashion, for the storage of the tractor.

Income tax authorities recognize the need for replacing farm buildings with new and modern structures. Thus, a barn depreciated at the rate of 5 percent annually is entirely written off the books in 20 years. In general, farmers and ranchers assume that the useful life of a building will average about 40 years.

12. **Dryness**—Buildings should be constructed to assist in providing a reasonably dry bed for animals. This means that the barn or shelter should be located on high ground with drainage away from the building and that, except in dry areas, the structure should be provided with eave troughs and downspouts which empty into a tile line draining away from the building. Proper ventilation and direct sunlight also aid materially in providing dry bedding. In open sheds, it is important that there be sufficient depth to assure dryness.

13. **Well lighted**—Proper lighting is essential for visibility and the convenience of the caretaker. A well-lighted building may be obtained in either of four ways: (a) through an open shed or door arrangement, (b) by providing adequate windows or plastic materials designed to let light through, (c) by artificial lights, or (d) through a combination of the previous three ways as is usually the case.

14. **Direct sunlight**—Direct sunlight possesses disinfecting properties, and arrangements should be made to obtain its benefits when practical. With modern chemical disinfectants, however, sunlight is not as necessary as formerly. An open shed or door arrangement is the best means of securing direct sunlight.

15. **Sanitary**—Sanitation is essential for disease prevention and parasite control. This means that livestock barns should be constructed so that they may be easily cleaned, thoroughly disinfected, and free from vermin. Sanitation is also promoted structurally by providing for direct sunlight and elimination of moisture. Smooth walls and hard-surfaced floors are the most satisfactory from this standpoint. It must be recognized, however, that hard-surfaced floors are hard on animals, especially cattle, sheep, and horses.

16. **Easily cleaned; manure handling**—A barn or shelter which is arranged so that it may be easily cleaned is more likely to be kept in a sanitary condition. Construction that will permit the operation of a power-operated manure loader, a tractor with a blade on the front or back, or a mechanical gutter cleaner, in addition to such things as smooth walls and floors, are definite assets from the standpoint of ease in cleaning.

Large drive-through doors and adequate ceiling height are essential for the operation of a laborsaving manure loader. In general, a minimum ceiling height of nine feet is recommended when it is planned to use a power-operated manure loader. (For further information relative to manure handling, see discussion headed "Manure Handling," page 814, and Section 8, Management.)

17. **Convenient**—Livestock buildings should be constructed to furnish the greatest possible convenience, which means fewer steps and man-hours. This means that attention should be given to the most convenient arrangement for feeding, watering, and bedding the animals and for cleaning of barns. Among the most important points to be considered are the locations of the silo, water tanks, feed rooms or bins, hay chutes, driveways, feed alleys, stalls, doors, and windows and the arrangement for removing manure with a minimum of labor. Greater convenience usually means that one man can and will care for more animals, and this means greater profits. It is significant to note that an hour saved each day is equivalent to eliminating a full month's work in a year.

18. **Adequate space for animals**—Livestock barns and shelters should be of adequate size to accommodate the existing herds and flocks, expected young stock, and any contemplated early expansion in operations. (See Tables 9-1 to 9-5.)

19. **Adequate space for feed and bedding storage**—In arriving at the space needs for feeds and bedding storage, consideration should be given to: (a) the crop productivity of the farm, (b) the size of animal population to be maintained, (c) the management practices, and (d) the length of the winter feeding period (see Feeding Guides in Section 4 and Table 9-7). The proper storage of hay or grain, in a barn that keeps out rain, means more tons of sound, properly cured roughage or grain for feed.

20. **Minimum fire risk**—The principal causes of fire losses in farm buildings are smoking, spontaneous combustion (usually from heat generated by moist hay

or forage), lightning, defective flues, sparks on roof, faulty wiring, ignition caused by accidents with equipment, trash fires, gasoline, and matches. The structural safeguards against fire include tile-lined flues, spark arrestors on chimneys, fire-resistant or fireproof shingles, masonry or metal construction, approved wiring that is inspected at intervals, approved installations of equipment, lightning protection in those areas where thunderstorms are frequent and intense, and "no smoking" signs which are observed. In addition to these precautions, exit doors should be provided so that animals may be removed from the building quickly in case of fire.

21. **Safety**—In the construction of livestock buildings, it is important that safety features be given consideration. For example, doors and ceilings should have adequate height; doors should be of sufficient width; door sills should be low or omitted; and paved lots should not be dangerously smooth and slick.

22. **Rodent control**—Livestock barns, and especially the feed storage parts thereof, should be constructed to provide the maximum protection against rats and mice. This is important both from the standpoint of feed conservation and disease prevention. New buildings may be rat proofed as follows:

a. By installing a concrete curtain wall around the foundation. This prevents rats from burrowing under the buildings.

b. By using hardware cloth around feed storage bins, or by using metal lining or metal bins.

23. **Surrounded by suitable corrals or lots**—Animal buildings should be provided with adjacent, well-drained, safe, durably fenced, and attractive corrals or lots. In small lots and in muddy areas, pavement is recommended. If the entire corral or lot is not surfaced, it will be helpful at least to pave a strip 15 feet wide in front of feed racks and around watering tanks. It is recommended that small corrals or lots be fenced with planks or poles. For larger yards, heavy wire fencing—sometimes with the wire doubled—and heavy posts set 8 to 10 feet apart may be used. Barbed wire fences are hazardous when used to confine animals in small yards; many an animal has been injured and made useless through such an arrangement.

24. **Adapted to present and future needs**—In building a livestock barn or shelter, the farmer or rancher should give consideration to present needs and future plans, the present and potential production, and the various uses for which the barn is needed now and in the future. Thus, in constructing a barn, it is desirable that it be of a size that will meet the existing conditions and be so built as to permit the erection of additions without disturbing the convenient arrangement or without tearing down any part of the structure.

25. **Protect animal health**—Animal barns and shelters should provide healthful indoor living conditions for the occupants. This is most important, for healthy animals are the profitable and efficient ones.

Space Requirements of Buildings and Equipment

One of the first and frequently one of the most difficult problems confronting the stockman who

TABLE 9-1—SPACE REQUIREMENTS OF BUILDINGS

Class, Age, and Size of Animal	Barn or Shed		Shades		Feedlots[1]		Hay or Silage	
	Floor Area per Animal	Height of Celing[3]	Shade per Animal	Shade Height	Area If Ordinary Dirt Lot	Area If Paved Lot	Length per Animal[4]	Width If Feeds from 1 Side
	(sq ft)	*(ft)*	*(sq ft)*	*(ft)*	*(sq ft)*	*(sq ft)*	*(in.)*	*(in.)*
Cows, 2 years or over	40-50	8½-10	30-40	10-12	300[5]	50-100	24-30	30
Yearling finishing cattle	Solid floor: 30-40 Slotted floor: 20-25	"	25-35	"	125-200[5]	30-50	20	"
Calves, 350 to 500 lb	20-30	"	15-25	"	100-175[5]	20-50	18	"
Cows in maternity stall	100-120	"	35-40	"	1-2 acre pasture paddock	—	30	—
Herd bulls	100-150	"	35-45	"	"	—	"	30

[1]Allow slope of ¼ to ½ in./ft in paved lots, and ½ in. or more in dirt lots (depending on soil and climate conditions).
[2]Feed bunks should be about 8 in. deep for calves and 12 in. for older cattle.
[3]Minimum ceiling height of 9 ft necessary where a power-operated manure loader is to be used.
[4]With liberal grain or other concentrate feeding, half the recommended space given herein. With bunker or self-feeder silos, allow 6 in./animal.
[5]More space is desirable under some soil and climatic conditions.

wishes to construct a building or item of equipment is that of arriving at the proper size or dimensions. Tables 9-1 to 9-7, pages 802 to 808, contain some conservative average figures which, it is hoped, will prove helpful. In general, less space than indicated may jeopardize the health and well-being of the animals; whereas more space may make the buildings and equipment more expensive than necessary.

Where of comparable age and size and where handled similarly, the space requirements of dairy cattle are like those given for beef cattle in Table 9-1. But since dairy cattle are kept for milk whereas beef cattle are kept for meat, there are differences in their management and space requirements. Thus, the special space requirements of dairy cattle—for dry and lactating cows, dairy calves, and replacement heifers—are given later in this section under the heading "Dairy Cattle Buildings and Equipment."

The space requirements of most swine are given in Table 9-3. The space requirements and other requisites of early weaned pigs are given in Table 9-21, page 839.

RECOMMENDED MINIMUM WIDTH OF SERVICE PASSAGES

In general, the requirements for service passages are similar, regardless of the kind of animals. Accordingly, the suggestions contained in Table 9-6, page 808, are equally applicable to cattle, sheep, swine, and horse barns.

STORAGE SPACE REQUIREMENTS FOR FEED AND BEDDING

The space requirements for feed storage for the livestock enterprise—whether it be for cattle, hogs, sheep, or horses, or, as is more frequently the case, a combination of these—vary so widely that it is difficult to provide a suggested method of calculating space requirements applicable to such diverse conditions. The amount of feed to be stored depends primarily upon: (1) length of pasture season, (2) method of feeding and management, (3) kind of feed, (4) climate, and (5) the proportion of feeds produced on the farm or ranch in comparison with those purchased. Normally, the storage capacity should be sufficient to handle all feed grain and silage grown on the farm and to hold purchased supplies. Forage and bedding may or may not be stored under cover. In those areas where weather conditions permit, hay and straw are frequently stacked in the fields or near the barns in loose, baled, or chopped form. Sometimes poled framed sheds or a cheap cover of waterproof paper or wild grass is used for protection. Other forms of low-cost storage include temporary upright silos, trench silos, temporary grain bins, and open-wall buildings for hay.

Table 9-7, page 808, gives the storage space requirements for feed and bedding. This information may be helpful to the individual operator who desires to compute the barn space required for a specific livestock enterprise. This table also provides a convenient means of estimating the amount of feed or bedding in storage.

AND EQUIPMENT FOR BEEF CATTLE

Manger, or Rack			Feed Bunk or Trough for Hand-Feeding Grain[2]				Self-Feeder	Water	
Width If Feeds from 2 Sides	Width If Attached Side of Barn	Height at Throat	Length per Animal	Width If Feeds from 1 Side	Width If Feeds from 2 Sides	Height at Throat	Trough Length If Feeder Is Kept Filled	Water per Animal per Day	Water Trough
	(in.)	(in.)	(in.)	(in.)	(in.)	(in.)	(in.)	(gal)	
48-60	30	24	24-30	18-30	48	24	6-12 per animal	12	Allow 1 linear ft of open water tank space for each 10 cattle; or one automatic watering bowl for each 25 cattle.
"	"	20	18-24	18	"	22	6-9	10	
"	"	20	18	"	"	18	6-8	8	A satisfactory water temperature range in winter is 40-45°F; in summer, 60-80°F.
—	—	26	30	—	—	30	9-12	15	
36-40	30	"	"	30	36-40	"	"	"	

Remarks:
Animals with horns require about 1 linear ft more manger or trough space per animal than the figures given in this table. Movable hayracks or feed bunks are usually 12 to 16 ft in length.
Provide a paved area of at least 10 ft around waterers, feed bunks, and roughage racks.
For specifications on slotted floors see "Confinement Feeding; Slotted Floors," later in this section.
Re: Water per animal per day. A minimum of 20 gal/day is needed for continuous flow to keep the water clean, and to keep it from freezing in the winter months.

TABLE 9-2—SPACE REQUIREMENTS O

Class, Age, and Size of Animal	Barn or Shed			Feedlot		Shades	
	Floor Area per Animal	Height of Ceiling	Window Space (not including open sheds)	Area if Ordinary Dirt Lot	Area if Paved Lot	Area per Animal	Height
	(sq ft)	*(ft)*	*(sq ft)*	*(sq ft)*	*(sq ft)*	*(sq ft)*	*(ft)*
Dry ewes	16	8½-10	1 sq ft window space per 35 sq ft floor space	16-20	16	10-12	8-10
Ewes with lambs	20	"	"	30	20	14	8-10
Stud rams	20-30	"	"	30-60	25	15	8-10
Feeder lambs	6	"	"	25-30	16	6-8	8-10

[1]For self-feeding silage, allow 5 linear in. of manger or rack space for mature sheep and 4 in. for feeder lambs.

Remarks: Wide barn doors are needed to prevent crowding and possible injury to pregnant ewes. Doors at least 8 ft wide are preferable. For specifications of slotted floors, see section entitled, "Confinement of Sheep on Slotted Floors."

TABLE 9-3—SPACE REQUIREMENTS OI
(See Table 9-21, page 839, fo

Age and Size of Animal	Swine Buildings					Shades		Pasture or Feeding Floor	
	Inside Sleeping Space or Shelter per Animal[1]	Height of Ceiling[2]	Height of Pen Partition	Hog Door Height	Hog Door Width	Shade per Animal	Shade Height	Good Pasture	Paved Feeding Floor in Addition to Sleeping Space, When Confined, per Animal
	(sq ft)	*(ft)*	*(in.)*	*(in.)*	*(in.)*	*(sq ft)*	*(ft)*	*(animals/acre)*	*(sq ft)*
Sows before farrowing:									
Gilts	15-17	7-8	36	36	24	17	4-6	10-12	15-20
Mature sows	18-20	"	"	"	"	20	"	8-10	"
Sows with pigs:									
Gilts	48	"	"	"	"	50	"	6-8	48
Mature sows	64	"	"	"	"	60	"	6-8	64
Herd boars	15-20	"	48	"	"	15-20	"	¼ acre/boar	15-20
Growing-finishing swine:									
(1) weaning[6] to 75 lb	5-6[7,8]	"	30	"	"	6	"	20 on full-feed; 10-15 limited feed	6-8[9]
(2) 75 lb to 125 lb	6-7[7,8]	"	33	"	"	7	"	"	7-9[9]
(3) 125 lb to market	8-10[7,8]	"	36	"	"	10	"	"	8-10[9]

[1]Space requirements are less with slotted floors (see section on slotted floors under "Buildings and Equipment for Swine").
[2]Ceiling heights in excess of 7 to 8 ft make for cold hog houses in the northern half of the United States.
[3]For example, a 6-ft feeder open on both sides has 12 linear ft of feeding space.
[4]The drinking water should not fall below 35° to 40°F during the winter.
[5]With creep provided for pigs in addition.

UILDINGS AND EQUIPMENT FOR SHEEP

Hay or Silage Manger, or Rack (for hand-feeding)				Feed Trough (for grain or roots; hand-feeding)				Self-feeder (for concentrate or roughage)	Water
Length per Animal[1]	Width if Feeds from 1 Side	Width if Feeds from Both Sides	Height at Throat	Length per Animal	Width if Feeds from 1 Side	Width if Feeds from 2 Sides	Height at Throat	Trough Length if Feeder is Kept Filled	Water per Animal per Day
(in.)	*(in.)*	*(in.)*	*(in.)*	*(in.)*	*(in.)*	*(in.)*	*(in.)*	*(in.)*	*(gal)*
12	14-16	20-24	12-15	12	14-16	20-24	10-15	6 (when salt is used as a governor, 3" will suffice).	2
"	"	"	"	"	"	"	"	—	3
"	"	"	"	20-24	"	"	"	—	3
"	12-14	18-22	10-12	12	"	12	8-12	3" for conc. alone; 4" for complete ration, or hay, or silage.	1

rovide a paved area of at least 5 ft around waterers, feed bunks, roughage racks, and entrances to sheds.
rovide water space as follows: 1 linear foot of open tank per 10 head of sheep, or one automatic bowl per 15 head. Maintain water temperature above 35°F in winter and below 5°F in summer.

UILDINGS AND EQUIPMENT FOR SWINE
arly Weaned Pigs)

Feeding Equipment				Watering Equipment[4]			
Self-feeder Space (animals per linear ft, or per hole)		Percent of Total Self-feeder Space Given to Protein Supplement		Feed Trough Space per Animal for Hand-feeding	Water Trough Space per Animal for Hand-feeding	Automatic Watering Cups (two openings considered 2 cups)	Comments
Dry-Lot	Pasture	Dry-Lot	Pasture				
(no. animals/linear ft)[3]	*(no. animals/linear ft)*[3]	*(% total feeder space)*	*(% total feeder space)*	*(linear ft/animal)*	*(linear ft/animal)*		
3	3	15	10-15	1½	1½	1 cup/12 gilts	When alfalfa hay is fed in rack, allow 4 sows per linear ft.
2	4	"	"	2	2	1 cup/10 sows	
1[5]	1[5]	"	"	1½[5]	2	1 cup/4 sows	For the pig creep, provide a minimum of 1' of feeder space per 5 pigs, see that the edge of the feeder trough does not exceed 4" above the ground floor, and do not allow more than 40 pigs per creep.
1[5]	1[5]	"	"	1½[5]	2	1 cup/4 sows	
1	1	"	"	2	2	1 cup/2 boars	
4	4-5	25	20-25	¾	¾	1 cup/20 pigs	
3	3-4	20	15-20	1	1	"	When salt or mineral is fed free-choice provide 3 linear ft of mineral box space or 3 self-feeder holes per 100 pigs.
3	3-4	15	10-15	1¼	1¼	"	

[6]For early weaning (under 5 to 6 weeks) space requirements, see Table 9-21.
[7]Over and above the sleeping space given herein, pigs that are confined from weaning to market should be provided the feeding floor space recommended in the column neaded "Paved feeding floor in addition to sleeping space, when confined, per animal."
[8]The larger area in the summertime.
[9]The larger area when fed from troughs; the smaller area is adequate where self-feeders are used.

TABLE 9-4
SPACE REQUIREMENTS OF BUILDINGS AND EQUIPMENT FOR HORSES

Kinds, Uses, and Purposes	Recommended Plan	Box Stalls or Shed Areas				Tie Stalls
		Size	Height of Ceiling	Height of Doors	Width of Doors	Size
Smaller horse establishments: Horse barns for pleasure horses, ponies, and/or raising a few foals.	12′ × 12′ stalls in a row; combination tack-feed room for 1 and 2 stall units; separate tack and feed rooms for 3 or more stall units. Generally, not more than a month's supply of feed is stored at a time. Use of all-pelleted rations (hay and grain combined) lessens feed storage space requirements.	Horses: 12′ × 12′ Ponies:[1] 10′ × 10′	8′-9′	8′	4′	5′ wide; 9′-12′ long
Larger horse breeding establishments: The following specially designed buildings may be provided for different purposes:						
BROODMARE AND FOALING BARN	A rectangular building either (1) with a central aisle, and a row of stalls along each side, or (2) of the "island" type, with 2 rows of stalls, back to back, surrounded by an alley or runway. Ample quarters for storage of hay, bedding, and grain. A record or office room, toilet facilities, hot water supply, veterinary supply room, and tack room are usually an integral part of a broodmare barn.	12′ × 12′ to 16′ × 16′	9′	8′	4′	—
STALLION BARN	Quarters for one or more stallions, with or without feed storage. A small tack and equipment room. Stallion paddocks, at least 300′ on a side, adjacent to or in close proximity.	14′ × 14′	9′	8′	4′	—
BARREN MARE BARN	An open shed or rectangular building, with a combination rack and trough down the center or along the wall. Storage space for ample hay, grain, and bedding.	150 sq ft per animal	9′	8′	4′	—
WEANLING; YEARLING QUARTERS	Open shed or stalls. The same type of building is adapted to both weanlings and yearlings, but different ages and sex groups should be kept separate. When stalls are used, two weanlings or two yearlings may be placed together.	10′ × 10′	9′	8′	4′	—
BREEDING SHED	A large, roofed enclosure with a high ceiling; laboratory for the veterinarian, hot water facilities, and stalls for preparing mares for breeding and holding foals.	24′ × 24′	15′-20′	8′	9′	—
ISOLATION (QUARANTINE) QUARTERS	Small barn, with feed and water facilities and adjacent paddock; for occupancy by new or sick animals.	12′ × 12′	9′	8′	4′	—
For riding academies and training and boarding stables	Either (1) stalls constructed back to back in the center of the barn, with an indoor ring around the outside; (2) stalls around the outside and a ring in the center; or (3) stalls on either side of a hallway or alleyway, and an outdoor ring.	12′ × 12′	9′	8′	4′	5′ wide; 9′-12′ long

[1]Even for ponies, a 12′ × 12′ stall is recommended since (1) it costs little more than a 10′ × 10′, and (2) it affords more flexibility—it can be used for bigger horses when and if the occasion demands.

TABLE 9-5

KIND, SIZE, AND LOCATION OF FEED AND WATER EQUIPMENT FOR HORSES

Equipment for	Kind of Equipment	Materials and Design	Sizes for		In Stall		In Corral		Remarks
			Horses	Ponies	Location	Height	Location	Height	
Concentrates	Pail; tub.	Metal, plastic, or rubber; usually with screw eyes, hooks, or snaps for suspending.	16-20 qt	14-16 qt	Front of stall.	⅔ height of animal at withers; or 38"-42" for horses, and 28"-32" for ponies.	Along fence-line.	Same height as in stall.	For sanitary reasons, removable concentrate containers are preferable, so that they can be taken out and easily and frequently cleaned, which is especially important after feeding a wet mash.
	Box.	Wood.	Width 12"-16" Length 24"-30" Depth 8"-10"	Width 10"-12" Length 20"-24" Depth 6"-8"					If desired, a pie-shaped metal pan set in a wooden shelf can be mounted in a front corner of the stall and pivoted in such manner that it can be pulled outward for filling and cleaning, then returned into the stall and locked in place.
Hay	Stall rack.	Metal, fiber, or plastic.	25-30 lb	10-15 lb	Corner of stall; in trailer or van.	Bottom of rack same height as horse or pony at withers.			Hay racks (1) alleviate contaminated hay and lessen parasitic infestation, and (2) lessen pawing and waste. Racks should open at bottom so that dirt, chaff, and trash may be removed or will fall out. For stallions and broodmares, always use high racks to alleviate injury hazards.
	Manger.	Wood.	Width 30" Length 24"-30"	Width 20" Length 20"	Front or corner of stall.	30"-42" for horses; 20"-24" for ponies.			
	Corral rack.	Wood.	Large enough to provide one day's supply of hay for intended number of horses.				In fenceline if it feeds from one side only. On high ground if it feeds from both sides.	Top of rack may be 1' to 2' higher than horse at withers.	Corral hayracks that feed from both sides should be portable.
Mineral	Box.	Wood.			Corner of stall.	Same height as concentrate box.	Fence corner.	⅔ height of horse at withers.	If mineral container is stationed in the open—in a corral, or in a pasture—it should be protected from wind and rain. Mineral containers should have 2 compartments—one for mineral mix, and the other for salt.
	Self-feeder.	Metal or wood.				"	Fence corner.		
Water	Stall, automatic.	Metal; 1 cup or 2 cups.			Front corner of stall.	24"-30"			The daily water requirements are: mature horse, 12 gal; foals to 2-yr.-olds, 6-8 gal; and ponies, 6-8 gal. In colder areas, waterers should be heated and equipped with thermostatic controls. A satisfactory water temperature range in the winter is 40°-45°F; in summer, 60°-80°F. Watering facilities should be designed so as to facilitate draining and cleaning. Locate water facilities proper distance from feed containers; otherwise, horses will (1) carry feed to the waterer, or (2) slobber water in the concentrate container. A 20" × 30" automatic waterer will accommodate about 25 horses; and a two-cup waterer will serve 12 head. Check automatic waterers daily.
	Corral, automatic.				In fence corner of line fence.	24"-30"			
	Pail.	Metal, plastic, or rubber.			Front of stall.	⅔ height of withers; or 38"-42" for horses, and 28"-32" for ponies.			
	Tank.	Concrete; steel.					Set in fence so that there are no protruding corners; or paint white out in corral or pasture.	30"-36"	Allow 1 linear ft of tank to each 5 horses. Tanks should be equipped with a float valve, which should be protected.

TABLE 9-6

RECOMMENDED MINIMUM WIDTH FOR SERVICE PASSAGES

Kind of Passage	Use	Minimum Width
Feed alley	For feed cart	4′-0″
Driveway	For wagon, spreader, or truck	9′-0″
Doors and gate	Drive-through	9′-0″
Doors and gate	Stall door or paddock	4′-0″

TABLE 9-7

STORAGE SPACE REQUIREMENTS FOR FEED AND BEDDING[1]

Kind of Feed or Bedding	Pounds per Cubic Foot	Cubic Feet per Ton	Pounds per Bushel of Grain
Hay-straw:			
1. Loose			
Alfalfa	4.4-4.0	450-500	
Nonlegume	4.4-3.3	450-600	
Straw	3.0-2.0	670-1,000	
2. Baled			
Alfalfa	10.0-6.0	200-330	
Nonlegume	8.0-6.0	250-330	
Straw	5.0-4.0	400-500	
3. Chopped			
Alfalfa	7.0-5.5	285-360	
Nonlegume	6.7-5.0	300-400	
Straw	8.0-5.7	250-350	
Corn:			
15½% moisture:			
Shelled	44.8		56.0
Ear	28.0		70.0
Shelled, ground	38.0		48.0
Ear, ground	36.0		45.0
30% moisture:			
Shelled	54.0		67.5
Ear, ground	35.8		89.6
Barley, 15% moisture	38.4		48.0
Ground	28.0		37.0
Flax, 11% moisture	44.8		56.0
Oats, 16% moisture	25.6		32.0
Ground	18.0		23.0
Rye, 16% moisture	44.8		56.0
Ground	38.0		48.0
Sorghum grain, 15% moisture	44.8		56.0
Soybeans, 14% moisture	48.0		60.0
Wheat, 14% moisture	48.0		60.0
Ground	43.0		50.0

[1]Adapted by the author from *Beef Housing and Equipment Handbook*, Midwest Plan Service, Iowa State University, Ames, Iowa, 1968, p. 62.

Environmentally Controlled Buildings

Environment may be defined as all the conditions, circumstances, and influences surrounding and affecting the growth, development, and production of a living thing. In animals, this includes the air temperature, relative humidity, air velocity, wet bedding, dust, light, ammonia buildup, odors, and space requirements. Control or modification of these factors offers possibilities for improving animal performance. There is still much to be learned about environmental control, but the gap between awareness and application is becoming smaller.

The *critical temperature* is that temperature at which the heat created by digestion and body metabolism just equals that which the animal dissipates by convection, evaporation, radiation, and conduction. The *comfort zone* is the range in temperature within which the animal may perform with little or no discomfort. At temperatures below the comfort zone additional nutrients need to be converted to heat to keep the body warm; and at temperatures above the comfort zone, nutrients are needed to help keep the animal cool. The *optimum temperature* is the temperature at which the animal responds most favorably as determined or measured by maximum rate of gain or production, feed efficiency, and/or production.

The critical temperature and comfort zone vary with different species, ages, breeds, and the physiological and productive status of animals. The species differences result primarily from the kinds of thermoregulatory mechanism provided by nature such as type of coat (hair, wool, feathers), sweat glands, etc. Thus, hogs, which have a light coat of hair, are very sensitive to extremes of heat and cold. On the other hand, nature gave cattle an assist through growing more hair for winter and shedding hair for summer, with the result that they can withstand higher and lower temperatures than hogs.

The temperature varies according to age, too. For example, the comfort zone of newborn lambs is 75 to 80°F, whereas the comfort zone of mature sheep is 4[illegible] to 75°F.

There are also breed differences, which make it possible to select animals well adapted to specific environments. For example, the Santa Gertrudis breed of cattle, which evolved from a Brahma X Shorthorn cross, is intermediate between its parent breeds in heat tolerance. The Jersey and Brown Swiss breeds of dairy cattle will maintain their milk production in hot summer temperatures better than Holsteins or other dairy breeds.

Animals that consume large quantities of roughage produce more heat during digestion; hence they have a different critical temperature than the same animals on a high-concentrate ration. Because of this, experienced cattle feeders decrease the roughage and increase the concentrate of finishing cattle during the hot summer months.

Stresses at both high and low temperatures are increased with high humidity. The cooling effect of evaporating sweat is minimized and the respired air has less of a cooling effect. As humidity of the air increases, discomfort at any temperature, and nutrient utilization, decrease proportionately.

Air movement (wind) results in body heat being

emoved at a more rapid rate than when there is no vind. In warm weather, air movement may make the nimal more comfortable; but in cold weather, it adds o the stress of temperature. At low temperatures, the utrients required to maintain body temperature are ncreased as the wind velocity increases. In addition o the wind, a drafty condition where the wind passes hrough small openings directly onto some portion or ll of the animal body will usually be more detrimen- al to comfort and nutrient utilization than the wind tself.

Stockmen were little concerned with the effect of nvironment on animals as long as they roamed pas- ures and ranges. Space requirements, wet bedding, mmonia buildup, odors, and manure disposal were o problem. But the concentration of animals into maller spaces changed all this. With the shift to con- inement structures and high-density production op- rations, building design became more critical.

Environmentally controlled buildings are costly o construct, but they make for the ultimate in animal omfort, health, and efficiency of feed utilization. Also, they lend themselves to automation, which re- ults in a saving in labor; and, because of minimizing pace requirements, they effect a saving in land cost. Today, environmental control is rather common in poultry and swine housing, and it is on the increase with other classes of livestock.

Before an environmental system can be designed for animals, it is important to know their (1) heat pro- duction, (2) vapor production, and (3) space require- ments. This information is as pertinent to designing livestock buildings as nutrient requirements are to balancing ration.

HEAT PRODUCTION OF ANIMALS

The heat production of animals is given in Table 9-8.

Table 9-8 may be used as a guide, but in doing so, consideration should be given to the fact that heat production varies with age, body weight, ration, breed, activity, house temperature, and humidity at high temperatures. As noted, Table 9-8 gives both total "heat production" and "sensible heat produc- tion." "Total heat production" includes both sensible heat and latent heat combined. Latent heat refers to the energy involved in a change of state and cannot be measured with a thermometer; evaporation of water or respired moisture from the lungs are exam- ples. Sensible heat is that portion of the total heat,

TABLE 9-8

HEAT PRODUCTION OF ANIMALS[1]

Heat Source	Unit		Heat Production, Btu/hr			Heat Production, Kcal/hr		
			Temperature	Total	Sensible	Temperature	Total	Sensible
	(lb)	*(kg)*	*(°F)*			*(°C)*		
Cow	1,000	453.6	40	3,600	2,640	4	907.2	665.3
			70	3,000	1,550	21	756.0	390.6
Calves (6-10 months)			60	780	660	16	196.6	166.3
			80	720	420	27	181.4	105.8
Sheep	100	45.4	*0.039 in. fleece length:*			*0.1 cm fleece length:*		
			45	560	500	7	141.1	126.0
			70	320	245	21	80.6	61.7
			3.937 in. fleece length:			*10.0 cm fleece length:*		
			45	245	185	7	61.7	46.6
			70	260	125	21	65.5	31.5
Hog:								
Sow & litter (3 weeks after farrowing)	400	181.4	—	2,000	1,000	—	504.0	252.0
Fattening	200	90.7	35	860	740	2	216.7	186.5
			70	610	435	21	153.7	109.6
Horse			70	1,800-2,500[2]	—	21	453.6-630	—

[1]Adapted by the author from 1974 *Agricultural Engineers Yearbook*, St. Joseph, Mich., ASAE Data Sheet D-249.2, p. 424, except for horse. Data for horse from *Farm Buildings*, by John C. Wooley, McGraw-Hill Book Company, Inc., 1946, p. 140, Table 24.

[2]Armsby and Kriss, in a paper entitled, "Some Fundamentals of Stable Ventilation," published in the *Journal of Agricultural Research*, Vol. 21, June 1921, p. 343, list the total heat output as follows: a 1,000-lb horse, 1,500 Btu per hour; a 1,500-lb horse, 2,450 Btu per hour.

measurable with a thermometer, that can be used for warming air, compensating for building losses, etc. Heat is measured in British thermal units (Btu). One Btu is the amount of heat required to raise the temperature of 1 pound of water 1°F.

VAPOR PRODUCTION OF ANIMALS

Animals give off moisture during normal respiration; and the higher the temperature, the greater the moisture. This moisture should be removed from buildings through the ventilation system. Most building designers govern the amount of winter ventilation by the need for moisture removal. Also, cognizance is taken of the fact that moisture removal in the winter is lower than in the summer; hence, less air is needed. However, lack of heat makes moisture removal more difficult in the wintertime. Table 9-9 gives the information necessary for determining the approximate amount of moisture to be removed.

temperature control, which requires moving more ai than in the winter.

RECOMMENDED ENVIRONMENTAL CONDITIONS FOR ANIMALS

The comfort of animals (or man) is a function o temperature, humidity, and air movement. Likewise the heat loss from animals is a function of these three items.

The prime function of the winter ventilation system is to control moisture, whereas the summer ventilation system is primarily for temperature control. I air in livestock barns is supplied at a rate sufficient to control moisture—that is, to keep the inside relative humidity in winter below 75 percent—then this wil usually provide the needed fresh air, help suppress odors, and prevent an ammonia buildup.

Temperature, humidity, and ventilation recommendations for different classes of livestock are given in Table 9-10. This table will be helpful in obtaining

TABLE 9-9
VAPOR PRODUCTION OF ANIMALS[1]

Vapor Source	Unit		Temperature		Vapor Production		Vapor Production	
	(lb)	*(kg)*	*(°F)*	*(°C)*	*(lb/hr)*	*(Btu/hr)*	*(kg/hr)*	*(Kcal/hr)*
Cow	1,000	453.6	40	4	0.92	960	0.42	241.9
			70	21	1.38	1,450	0.63	365.4
Calves (6-10 months)			60	16	0.11	120	0.05	30.2
			80	27	0.29	300	0.13	75.6
Sheep	100	45.4			*0.039 in. fleece length:*		*0.1 cm fleece length:*	
			45	7	0.06	60	0.03	15.1
			70	21	0.07	75	0.03	18.9
					3.937 in. fleece length:		*10.0 cm fleece length:*	
			45	7	0.06	60	0.03	15.1
			70	21	0.13	135	0.06	34.0
Hog: Sow & litter (3 weeks after farrowing)	400	181.4			0.97	1,020	0.44	257.0
Fattening	200	90.7	35	2	0.11	120	0.05	30.2
			70	21	0.16	175	0.07	44.1
Horse			70	21	0.729	—	0.33	—

[1]Adapted by the author from 1974 *Agricultural Engineers Yearbook*, St. Joseph, Mich., ASAE Data Sheet, D-249.2, p. 424, except for horse. Data for horse from *Farm Buildings*, by John C. Wooley, McGraw-Hill Book Company, Inc., 1946, p. 141, Table 25.

Since ventilation also involves a transfer of heat, it is important to conserve heat in the building to maintain desired temperatures and reduce the need for supplemental heat. In a well-insulated building, mature animals may produce sufficient heat to provide a desirable balance between heat and moisture; but young animals will usually require supplemental heat. The major requirement of summer ventilation is

a satisfactory environment in confinement livestock buildings, which require careful planning and design.

Naturally Ventilated Buildings

Stockmen continue to have interest in naturally ventilated buildings, as well as in environmentally controlled buildings, primarily because of their signif

TABLE 9-10

RECOMMENDED ENVIRONMENTAL CONDITIONS FOR ANIMALS

Class of Animal	Temperature				Acceptable Humidity	Commonly Used Ventilation Rates[1]					Drinking Water			
	Comfort Zone		Optimum			Basis	Winter[2]		Summer		Winter		Summer	
	(°F)	(°C)	(°F)	(°C)	(%)		(cfm)	(m^3/min.)	(cfm)	(m^3/min.)	(°F)	(°C)	(°F)	(°C)
Beef cow	40-70	5-21	50-60	10-15	50-75	1,000 lb (or 454 kg)	100	2.8	200	5.7	50	10	60-75	15-24
Steer, enclosed bldg. on slotted floor	40-70	5-21	50-60	10-15	50-75	1,000 lb (or 454 kg)	100	2.1-2.3	200	14.2	50	10	60-75	15-24
Dairy cow	40-70	5-21	50-60	10-15	50-75	1,000 lb (or 454 kg)	100	2.8	200	5.7	50	10	60-75	15-24
Dairy calves	50-75	10-24	65	17		per 100 lb (45 kg)	10		25					
Sheep:														
Ewe	45-75	7-24	55	13	50-75		20-25	.6-.7	40-50	1.1-1.4	40-45	5-8	60-75	15-24
Feeder lamb	40-70	5-21	50-60	10-15	50-75		15	.3	30	.65	40-45	5-8	60-75	15-24
Newborn lamb	75-80	24-27												
Swine:														
Sow, farrowing house	60-70	15-20	65	17	60-85	Sow and litter	80	1.4	210	2.8	50	10	60-75	15-24
Newborn pigs (brooder area)	80-90	27-32	85	29	60-85									
Growing-finishing hogs	60-65	15-17	60	15	60-85	125 lb (or 57 kg)	15	.7	75	2.1	50	10	60-75	15-24
Horse	45-75	7-24	55	13	50-75	1,000 lb (or 454 kg)	60	1.7	160	4.5	40-45	5-8	60-75	15-24
Newborn foal	75-80	24-27												
Poultry:														
Layers	50-75	10-24	55-70	13-20	50-75	per bird	2		5		50	10	60-75	15-24
Broilers	85-95	21-27	70	24	50-75	per lb body weight	½		1		50	10	60-75	15-24
Turkeys	95-100 (beginning poults)	35-38				per lb body weight	½		1		50	10	60-75	15-24

[1]Generally two different ventilating systems are provided; one for winter, and an additional one for summer. Hence, as shown in Table 9-10, the winter ventilating system in a beef cow barn should be designed to provide 100 cfm (cubic feet/minute) for each 1,000-pound cow. Then, the summer system should be designed to provide an added 100 cfm, thereby providing a total of 200 cfm for summer ventilation.

In practice, in many buildings, added summer ventilation is provided by opening (1) barn doors, and (2) high-up hinged walls.

[2]Provide approximately ¼ the winter rate continuously for moisture removal.

icantly lower construction and operating costs. Because no attempt is made to regulate temperature, the costs of heavy insulation, tight fitting doors and windows, and a mechanical ventilation system are averted. Of course, buildings which for management reasons must be maintained at temperatures above winter levels are not suited to natural ventilations (e.g., farrowing barns or dairy stanchion barns).

Naturally ventilated buildings can be successfully used for most livestock housing; among them, (1) free-stall housing for dairy cattle; (2) loafing or bedded pack barns for dairy, beef, or sheep; (3) swine-finishing buildings; and (4) calf barns.

Naturally ventilated buildings are mainly a shell to protect animals from rain and snow, and to protect the building contents (grain, hay, etc.). Winter inside temperatures will often be within 3 to 10°F of outside temperatures. Thus, such buildings are often referred to as cold confinement livestock buildings.

A naturally ventilated building has a continuous opening at the high point (normally the ridge) of the building for air exhaust and continuous openings or inlets along the long sidewalls of the building for fresh air. The size of these openings is based on rules of thumb or experience. Air entering along the sidewalls (normally under the eaves) of the building is warmed by the heat from the animals in the building and picks up moisture as it rises toward the ridge. The continuous open ridge allows this warm, moist air to escape, thus completing the air exchange process.

During the warm weather, the building should serve mainly to keep rain out and act as a sunshade. Large continuous openings in the sidewalls allow summer breezes to blow through the building.

Typical naturally ventilated buildings can be divided into two types as follows:

1. **Open front**—These buildings have one long side completely open at least one-half the height of the sidewall. The open side faces away from the direction of prevailing winter winds, normally to the south or southeast.

2. **Enclosed**—These buildings have all sides closed but provide continuous eave openings and large doors or vent panels for summer conditions. Enclosed naturally ventilated buildings offer more protection from wind and precipitation than open front buildings.

Roofs

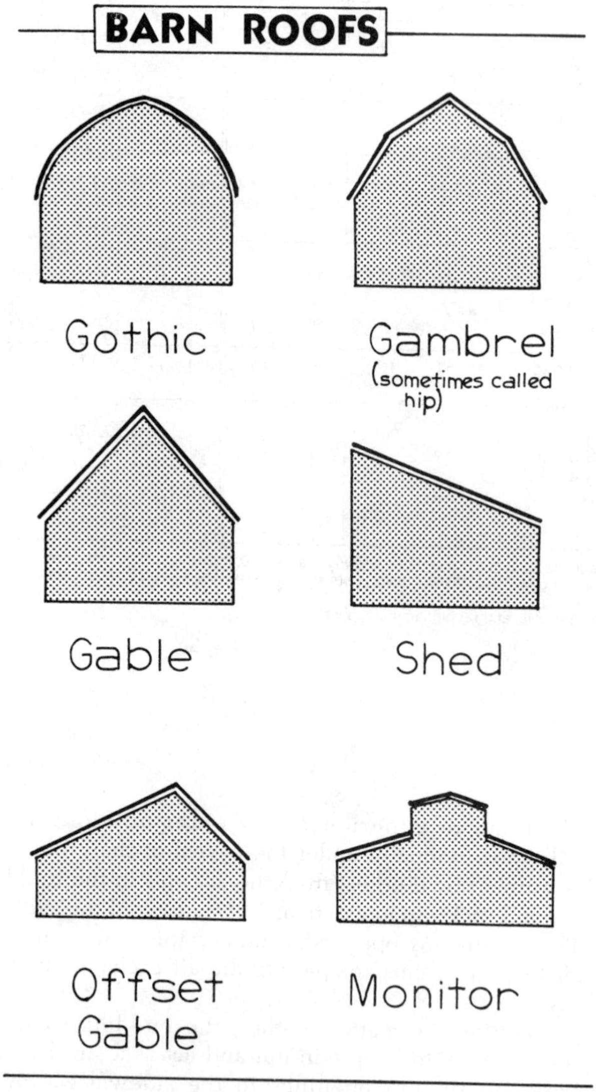

Fig. 9-1. Six roof shapes.

Six roof shapes are shown above. The three most widely used styles are shed, gable, and offset gable.

The shape, style, slope, and type of roof construction selected should be based upon the function to be served, economy of construction, strength, and appearance.

On permanent buildings, the roofing should last 15 to 20 years without replacement. Among the more durable roofings are: cedar shingles, cement-asbestos shingles, asphalt shingles, steel sheets, and aluminum.

Floors for Stalls or Stables

After considering both the advantages and disadvantages of the many types of flooring materials, most practical stockmen are agreed that, under average conditions and where a solid floor is desired, concrete is the most satisfactory flooring for central hog house and that clay is the most satisfactory flooring for cattle sheep, and horse barns or shelters.

Most stockmen feel that a perfect flooring material has not yet been developed, as each of the existing types has certain disadvantages. Rough wooden floors furnish good traction for animals and are warm to lie upon; but they are absorbent and unsanitary. They also lack durability and often harbor rats and other rodents. Concrete floors are durable, impervious, easily drained, and sanitary; but they are rigid and without resilient qualities, are slippery when wet and are cold to lie upon. Clay floors are noiseless and springy, and afford a firm natural footing unless wet; but they are difficult to keep clean and level.

SOLID FLOORS

Solid floors for animals may be and are constructed of numerous materials—including clay, clay with a concrete border, plank, concrete, concrete with board surfacing, cork brick, creosoted wooden blocks, cinders, or various combinations of these materials. Regardless of the type of flooring material, for a good dry bed there should be a combination of surface and subsurface drainage, together with a cover provided by a suitable absorbent litter.

SLOTTED FLOORS

Slotted floors are floors with slots through which the feces and urine pass to a storage area below or nearby. Such floors are not new; they have been used in Europe for over 200 years. More and more slotted floors are being used for swine and poultry in this country, and there is increased interest in using them for cattle and sheep.

The main *advantages* of slotted floors are (1) they facilitate automation and save labor; (2) they lessen or eliminate bedding; (3) they facilitate handling of manure; (4) they necessitate less space per animal; (5) they require less land; (6) they increase sanitation; (7) they lessen mud, dust, odor, and fly problems; and (8) they lessen pollution.

The chief *disadvantages* of slotted floors are (1) higher initial cost than conventional solid floors, (2) less flexibility in the use of the building, (3) any spilled feed is lost through the slots, (4) animals raised on slotted floors resist being driven over a solid floor, and (5) environmental conditions become more critical.

Slats may be made of wood, concrete, or metal (steel or aluminum). Table 9-11 summarizes pertinent facts pertaining to each type of material.

The design of concrete slats is important (see Fig. 9-2 and Table 9-12 for recommendations).

TABLE 9-11
MATERIALS FOR FLOOR SLATS

Material	Expected Life	Advantages	Disadvantages	Comments
Wood	2-4 years	Low initial cost.	Difficult to maintain spacing. Not too durable.	Make from hardwood, like oak.
Concrete	10-15 years	Long life. May be homemade.	Quality control difficult when poured on site.	
Metal (steel or aluminum)	4-8 years	Easily cleaned.	High cost.	Special erosion-resistant metal now available.
Flattened expanded steel mesh		Low cost.	Will not take too much weight.	Satisfactory for young lambs and for young pigs to 50 lb weight.

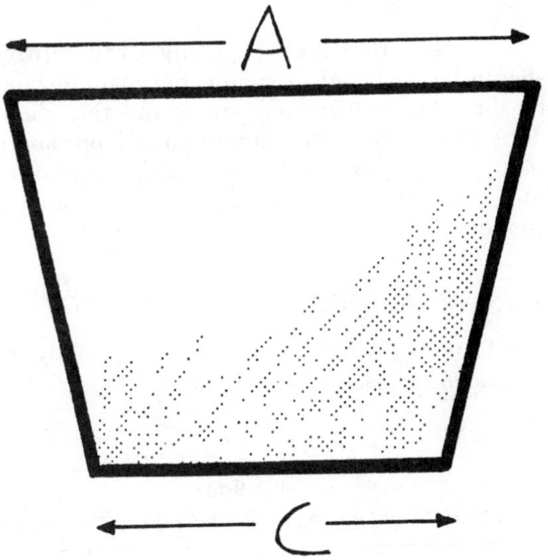

Fig. 9-2. Concrete slat. (see Table 9-12 for recommended A-B-C dimensions)

(2) less waste in feed and bedding; (3) greater conservation of manure; (4) greater sanitation and fewer diseases; and (5) more animal comfort, which means greater gains and production.

TABLE 9-12
CONCRETE SLAT DIMENSIONS[1]

Class of Animal	Length	See Fig. 9-2 A	B	C	Space Slats	Floor Space
	(ft)	*(in.)*	*(in.)*	*(in.)*		
Cattle (beef and dairy): Designed for approx. 250 lb per linear foot live load	6	6	6	3	*Calves:* 1¼" apart	For 1,000 lb: 20-25 sq ft
	8	6	6	3	*Steers,* weaning to market: 1½" to 1¾"	
	10	6	7½	3		
Hogs and sheep: Designed for approx. 100 lb per linear foot live load	4	4	3½	3	Space slats either a uniform ⅜" apart or ¾" to 1" apart. Spaces between ⅜" and ¾" are not recommended because pigs' legs may get caught. Space slats 1" apart behind the sow to improve cleaning.	*Dry ewe:* 6 sq ft; *Ewe & lamb:* 8 sq ft; *Feeder lamb:* 4 sq ft; *Sows:* 8 sq ft; *Sow and litter:* 20-30 sq ft; *Growing-finishing:* 1. To 75 lb: 5 sq ft 2. 75-125 lb: 6 sq ft 3. 125 lb market: 8 sq ft
	6	4	4	3		
	8	5	4½	4		
	10	5	5	4		
	12	5	5½	4		

[1]For concrete slats, use a 7½ bag mix with ¾" maximum aggregate.

Paved Lots and Feeding Floors

In those areas in which barnyard mud is a usual winter problem, a concrete lot floor for beef and dairy cattle, hogs, and sheep may constitute one of the most profitable improvements. A properly constructed paved feedlot results in (1) a saving in time and labor;

The paved feedlot should be located where it will make for convenience in feeding and watering operations, will be sheltered from the prevailing winds, and will be exposed to the sun. In most areas, these conditions are met by locating the paved lot adjacent to the building housing the animals and on the south or east side thereof.

Manure Management

Modern livestock buildings and equipment should be designed to handle the manure produced by the animals that they serve; and this should be done efficiently, with a minimum of labor and pollution, so as to retrieve the maximum value of the manure, and make for maximum animal sanitation and comfort.

MANURE PRODUCTION AND STORAGE

Table 9-13 shows the approximate daily manure production of each class of animal.

TABLE 9-13

APPROXIMATE DAILY MANURE PRODUCTION, WITHOUT BEDDING[1]

Animal	Cu Ft/Day Solids and Liquids[2]	Gallons/ Day[3]
1,000-lb cow	1½	11
1,000-lb steer	1	7½
10 head of sheep	½	4
10 head of hogs:		
50 lb	⅔	5
100 lb	1⅓	10
150 lb	2¼	17
200 lb	2¾	20½
250 lb	3½	26
1,000-lb horse	¾	5½

[1]Adapted by the author from *Michigan State University Circular Bull. 231.*
[2]There are about 34 cu ft in a ton of manure.
[3]One cu ft = 7½ gal.

Manure may be stored in a separate tank or it may be left to accumulate in a pit under slotted floors.

Storage capacity can be computed as follows:

Storage capacity = No. of animals × daily manure production × desired storage time in days + extra water.

Example: 80 cows (1,000 lb each) × 1½ cu ft × 120 days = 14,400 cu ft; 7½ gal × 14,400 cu ft = 108,000-gal capacity.

Extra water must often be added to liquefy the wastes. Thus, if the manure is to be pumped, 1/5 to 3/5 of the storage volume may be needed for the extra water. For irrigation, there should be about 95 percent water and 5 percent manure. Water should be kept to a minimum if the manure is to be field spread with a tank wagon.

Generally three to six months' storage capacity is desirable.

Manure Gases

When stored inside a building, gases from liquid wastes create a hazard and undesirable odors. Most (95% or more) of the gas produced by manure decomposition is methane, ammonia, hydrogen sulfide, and carbon dioxide. Several have undesirable odors or possible animal toxicity, and some promote corrosion of equipment. Table 9-14 gives some properties of the more abundant gases.

TABLE 9-14

PROPERTIES OF THE MORE ABUNDANT MANURE GASES[1]

	Gas	Weight Air = 1	Physiologic Effect	Other Properties
CH_4	Methane	½	Anesthetic	Odorless, explosive
NH_3	Ammonia	⅔	Irritant	Strong odor, corrosive
H_2S	Hydrogen sulfide	1+	Poison	Rotten-egg odor, corrosive
CO_2	Carbon dioxide	1⅓	Asphyxiant	Odorless, mildly corrosive

[1]*Beef Housing and Equipment Handbook*, Midwest Plan Service, Iowa State University, Ames, Iowa, 1968, p. 10.

Animals and people can be killed (asphyxiated), because methane and carbon dioxide displace oxygen.

Most gas problems occur when manure is agitated or when ventilation fans fail.

No one should enter a storage tank, unless (1) the space over the wastes is first ventilated with a fan, (2) another person is standing by to give assistance if needed, and (3) wearing self-contained breathing equipment—the kind used for fire fighting or scuba diving.

It is important that maximum building ventilation be provided when agitating or pumping wastes from a pit. Also, an alarm system (loud bell) to warn of power failures in tightly enclosed buildings is important, because there can be a rapid buildup of gases when forced ventilation ceases.

MANURE HANDLING

Modern handling of manure involves maximum automation and a minimum loss of nutrients. Among the methods being used, with varying degrees of success, are slotted floors emptying or pumping into irrigation systems; storage vats; spreaders (including those designed to handle liquids alone or liquids and solids together); dehydration; power loaders; conveyers; industrial-type vacuums; lagoons; and oxidation ditches. Actually, there is no one best manure management system for all situations; rather, it is a matter of design and using that system which will be most practical for a particular set of conditions.

REQUISITES OF LIVESTOCK EQUIPMENT

Generally speaking, *livestock equipment refers to*

structures other than barns or shelters used in the care and management of animals. Much of this equipment is portable.

When contemplating equipment, as when planning to build a self-feeder, the operator should first ask himself if it is needed. Why is it being purchased or built? Unless it makes a contribution to the farm or ranch operations and efficiency, and unless it is needed and will pay for itself, it cannot be justified.

Much livestock equipment had best be purchased, rather than homemade. When buying, look for rugged construction, simple operation, and durable materials and finishes. When making equipment yourself, use a good plan and follow the recommendations as to types and grades of materials. Working drawings of many items of equipment have been prepared by the universities and the U.S. Department of Agriculture. These may be purchased from your Extension Agricultural Engineer or from the following regional planning service:

Midwest Plan Service
Iowa State University
Ames, Iowa 50011

The size and design of livestock equipment may differ; that is, not all hay racks or self-feeders, for example, are the same. Yet there are certain fundamentals of livestock equipment that are similar regardless of the kind of equipment, the design, or the size. These requisites are:

1. **Utility value**—Equipment should be useful, practical, and efficient.
2. **Simple construction**—As much livestock equipment is homemade, simple construction is essential.
3. **Durable**—Livestock equipment receives hard and heavy use. Thus, it should be strongly and durably built.
4. **Dependable**—Livestock equipment should be dependable, so that it will function without getting out of order. Overly complicated equipment sometimes requires more of the operator's time than if it were not available.
5. **Low annual cost and upkeep**—Like animal barns and shelters, livestock equipment must be paid for out of animal profits. It is important, therefore, that it have a low annual and upkeep cost. Because of lower maintenance cost and longer years of usefulness, it may be cheaper in the long run to pay a higher initial cost for more durable and substantial equipment than to purchase whatever is cheapest.
6. **Movable**—Equipment that is used away from buildings should be movable. Bunks, racks, and self-feeders should be built on skids so that they may be moved from one location to another.
7. **Accessible**—Stationary or less-portable equipment, such as stocks and dipping vats, should be readily accessible.
8. **Save feed**—Much grain and hay may be saved when fed in properly constructed equipment. When such equipment is used, animals clean up grain and hay and do not throw or root it out of the feeder.
9. **Reduce labor**—Modern equipment—such as large self-feeders and large hay racks that require infrequent filling, and feed carts and farrowing crates—reduces the labor required in feeding operations.
10. **Conserve manure**—Feeding equipment may be located in either the barn or feed yard, from whence the accumulated manure is hauled directly to the fields or is placed in the manure pit. Or, by carefully moving the feeding facilities about in the fields, the fertility may be scattered where it is most needed.

Automation

Automation is a coined word meaning the mechanical handling of materials. Stockmen automate to lessen labor and cut costs.

Modern equipment has practically eliminated the pitchfork, bucket, and basket. Such chores as feeding, watering, bedding, and barn cleaning have been, or are being, mechanized. Stockmen are using more self-unloading trucks and trailers, self-feeders, feed bunk augers and belts, laborsaving grain- and forage-processing equipment (producing pellets, cubes, or wafers, etc.), automatic waterers, and manure disposal units. Automation of the livestock industry will increase.

Scales

Scales are a valuable piece of equipment for the modern stock farm or ranch; for they make it possible to determine weights of animals on production-testing studies, to secure the accurate rate of gains of animals being finished, to sell animals on the farm or ranch on a weight basis, and to buy and sell feed on a weight basis. For greatest usefulness, scales should be so arranged that a pen may be set up quickly when weighing mature animals or may be removed when weighing feed.

A convenient place for the farm scales is in the farm court, next to the corrals or feedlot. In this location, the scale is convenient for weighing livestock or loads of feed and supplies.

POLLUTION CONTROL

Broadly speaking, the environmentalists have voiced the following major concerns pertaining to animal pollution:

1. That excess nitrates may cause a disorder in human babies, commonly known as "blue babies."
2. That the oxygen demand of runoff sludge may suffocate fish.
3. That odors and dust are a nuisance.

A general discussion of "manure" appears in this book in Section 8, Management; and a general discussion of pollution is presented in Section 1, Animal Behavior and Environment. Thus, the reader is referred to these two sections for a more complete discussion of the subject of "pollution control."

GUIDELINES RELATIVE TO FACILITY AND EQUIPMENT COSTS

Overinvestment is a rather common mistake. Stockmen are prone to invest more in land and improvements than reasonably can be expected to make a satisfactory return. Cattle feedlot operators frequently invest too much in feed mills and equipment; and sometimes small cattle feeders fail to recognize that it may cost half as much to mechanize to feed 500 head as it costs to mechanize to feed 2,000 head.

In order to lessen overinvestment by the uninformed, guidelines are useful. Here are some:

1. **Guideline No. 1—The break-even point on how much you can afford to invest in equipment to replace hired labor can be arrived at by the following formula:**

$$\frac{\text{Annual saving in hired labor from new equipment}}{.15} = \text{amount you can afford to invest}$$

Example:

If saving in hired labor costs is $10,000 per year, this becomes—

$$\frac{\$10{,}000}{.15} = \$66{,}667\text{, the break-even point on new equipment.}$$

Since labor costs are going up faster than machinery and equipment costs, it may be good business to exceed this limitation under some circumstances. Nevertheless, the break-even point, $66,667 in this case, is probably the maximum expenditure that can be economically justified at the time.

2. **Guideline No. 2—The break-even point on new facility-equipment costs is five times the annual salary of each person replaced.**

Assuming an annual cost plus operation of power machinery and equipment equal to 20% of new cost, the break-even point to justify replacement of one hired man is as follows:

If annual cost of one hired man is—	The break-even point on new investment is—
$ 8,000 (20%) × 5	$40,000
10,000 (20%) × 5	50,000
12,000 (20%) × 5	60,000

In the above figures, it is assumed that the productivity of men at different salaries is the same, which may or may not be the case.

Example:

Assume that the new cost of added equipment comes to $10,000, that the annual cost is 20% of this amount, and that the new equipment would save 2 hours of labor per day for 6 months of the year. Here's how to figure the value of labor to justify an expenditure of $10,000 for this item:

$10,000 (new cost) × 20% = $2,000, which is the annual ownership use cost.

$2,000 ÷ 360 hours (labor saved) = $5.56/hour.

So, if labor costs less than $5.56/hour, you probably shouldn't buy the new item.

HOW TO DETERMINE THE SIZE BARN TO BUILD

The length and depth of the barn (its size) may be varied according to needs. The size barn to build for any given farm or ranch may be determined as follows:

1. Estimate the number and kind of animals to be quartered and compute the total animal space requirements from Tables 9-1, 9-2, 9-3, and 9-4.
2. Compute the yearly feed requirements of the animals to be fed and quartered by referring to the Feeding Guides in Section 4 giving consideration to the length of the pasture season (unless animals are to be raised entirely in confinement) and the quantity and quality of the grass.
3. Estimate the farm production of feeds and bedding to be stored in the barn. In most operations this should coincide reasonably close to the total animal requirement (point No. 2), but there may be circumstances where the feed and bedding storage requirements are more or less than the animal feed requirements.
4. Estimate the total tonnage of feed and bedding to be stored by correlating the animal feed needs and

the farm or ranch production (correlate the results of points 2 and 3). Then determine the total storage space requirements for feed and bedding from Table 9-7, page 808.

5. Determine the size of barn to build from the total animal space requirements and the total yearly feed and bedding storage requirements (points 1 and 4).

HOW TO DETERMINE THE SIZE SILO TO BUILD

The size of silo to build should be determined by needs. With tower type and pit silos, this means (1) that the diameter should be determined by quantity of silage to be fed daily, and (2) that the height (depth in a pit silo) should be determined by the length of the silage feeding period. Similar consideration should be accorded with trench silos.

Size of Tower Silo

If the diameter is too great, the silage will be exposed too long before it is fed; and, unless a quantity is thrown away each day, spoiled silage will be fed.

The minimum recommended rate of removal of silage varies with the temperature. In most sections of the United States, it is desirable that a minimum of 1½ inches of silage be removed from tower silos daily during the winter feeding period, with the quantity increased to a minimum of 3 inches when summer feeding is practiced. Of course, the total daily silage consumption on any given farm or ranch will be determined by (1) the class and size of animals, (2) the number of animals, and (3) the rate of silage feeding. Some suggestions on how much silage to feed cattle, sheep, and horses are found in the Feeding Guides given for each of these species in Section 4, Feeding.

Silo height should be determined primarily by the length of the intended feeding period. In general, however, the height should not be less than twice, nor more than three and one-half times the diameter. The greater the depth, the greater the unit capacity. Extreme height is to be avoided because (1) of the excessive power required to elevate the cut silage material, and (2) of the heavier construction material required. Also, it is noteworthy that, with silos of the larger diameters, more labor is required in carrying the silage to the silo door for removal.

Table 9-15 may be used as a guide in computing the proper diameter of tower silo for any given farm or ranch.

Fig. 9-3, page 818, shows capacities of tower silos of different heights and diameters. It is based on well-eared corn silage harvested in the early dent stage, cut in ¼-inch lengths, well-tramped when filled, and with the silo refilled once after settling for a day.

Fig. 9-3 can be adapted for corn silage of different stages of maturity and grain content, and for other kinds of silage, by applying the rules of thumb given in Table 9-16, page 819.

The following example will serve to illustrate how to determine the size tower silo to build:

> Over a period of years, a farmer plans to winter 34 head of 425-pound stocker calves on a ration of corn silage and protein supplement. There is a 240-day wintering period. No increase in the herd is planned. What size tower silo should he build?

The answer is obtained as follows:

1. First, here are the silage requirements:

   a. Section 4, Table 4-26, page 288, indicates that 425-pound stocker calves on a ration of corn silage and protein supplement should receive about 30 pounds of silage per head per day.

   b. 34 × 30 = 1,020 pounds of silage required daily for the 30 calves.

   c. 1,020 × 240 = 244,800 pounds or 122.4 tons, of silage required for the 240-day wintering period for the 34 calves.

2. Next, here is the size silo to build:

   a. Table 9-15 shows that in order to remove 1,005 pounds of silage daily (which is only slightly less than the 1,020 lb needed daily),

TABLE 9-15

MAXIMUM DIAMETER OF TOWER SILO TO BUILD IF SILAGE IS TO BE KEPT FRESH

Inches of Silage Removed Daily	Total Silage Removed Daily with an Inside Silo Diameter of:					
	10 Feet	12 Feet	14 Feet	16 Feet	18 Feet	20 Feet
	(lb)	*(lb)*	*(lb)*	*(lb)*	*(lb)*	*(lb)*
Summer: 3 in. daily will remove[1]	786	1,312	1,539	2,010	2,545	3,142
Winter: 1½ in. daily will remove[1]	393	656	770	1,005	1,272	1,571

[1]The pounds listed in each of the columns to the right are approximations based on an average constant weight of 40 lb of silage per cu ft.

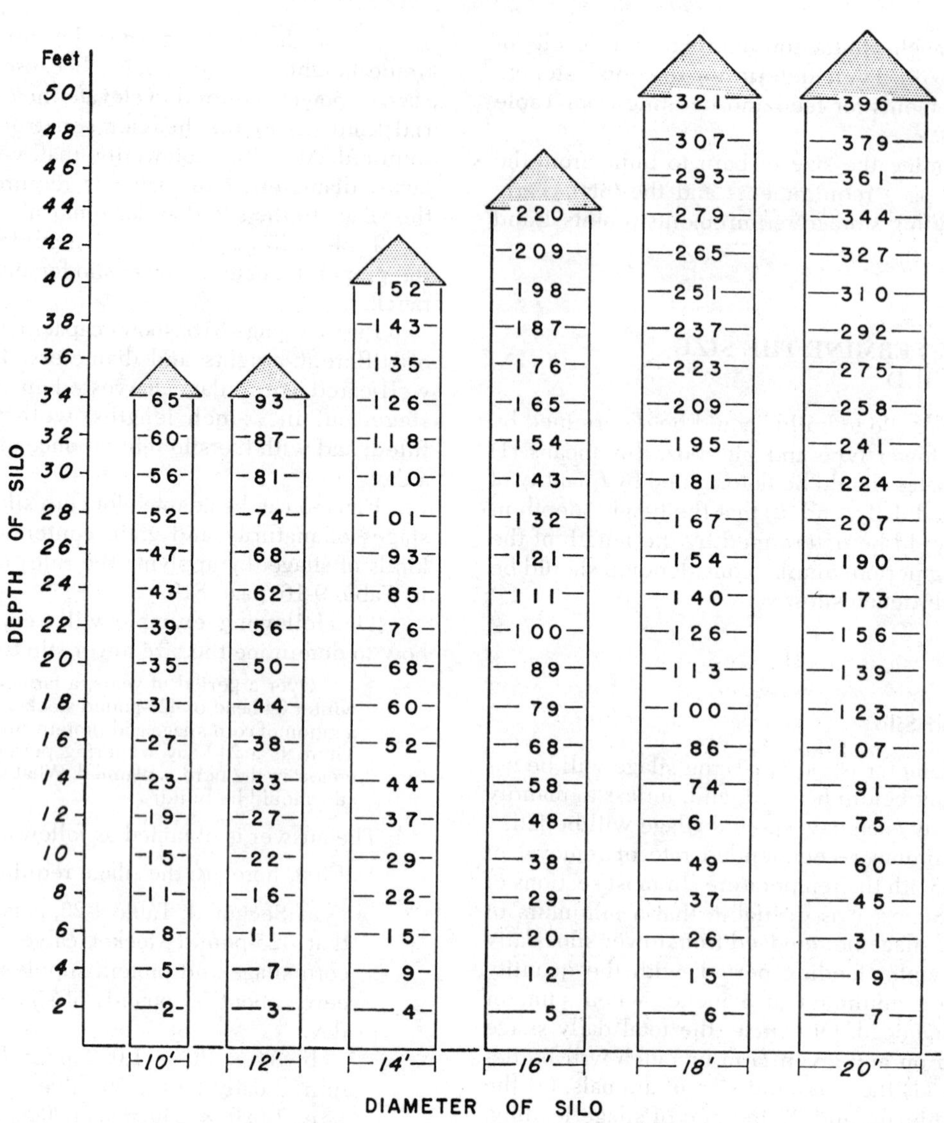

Fig. 9-3. Capacity in tons of settled corn silage in tower silos of varying sizes (based on data reported in USDA Circ. 603). See Table 20-4 of this book for tabular material. (Drawing by R. F. Johnson)

with 1½ inches removed from the top of the silo each day, the diameter of the silo should not be greater than 16 feet.

b. Fig. 9-3 can now be used as a guide in determining both the proper height (or depth) and diameter of the silo. Fig. 9-3 shows that a silo 16 ft in diameter and 27 ft high will hold 127 tons of silage, which would allow for 4.6 tons spoilage in excess of the required 122.4 tons. However, the height of a silo should not be less than twice the diameter. It appears best, therefore, to plan on a 14-ft diameter silo. As noted in Fig. 9-3, 34 ft of settled silage in a 14-ft diameter silo will provide 126 tons of silage, which would allow for 3.6 tons spoilage in excess of the required 122.4 tons. To allow for settling, an additional 4 to 6 ft should be added to the height, thus making a 38- to 40-ft height.

c. The size silo to build to meet the needs outlined in this example, therefore, is one that is 14 feet in diameter and 38 to 40 feet high.

TABLE 9-16

EFFECT OF KIND OF SILAGE ON WEIGHT

Kind of Silage	Changes to Be Made in the Number of Tons Shown in Fig. 9-3
1. For corn silage ensiled when less mature than usual	Add 5 to 10%
2. For corn ensiled when dry or overripe	Deduct 5 to 10%
3. For corn very rich in grain	Add 5 to 10%
4. For corn with very little grain	Deduct 5 to 10%
5. For sorghum silage	Use the same weights as used for corn silage of comparable grain and maturity.
6. For sunflower silage	Add 5 to 10%
7. For grass silage	Add 10 to 15%[1]

[1]For this reason, a stronger structure is necessary where grass silage is stored.

Size of Trench Silo

As in an upright silo, the cross-sectional area of a trench silo should be determined by the quantity of silage to be fed daily. The length is determined by the number of days of the silage feeding period. The only difference is that generally greater allowance for spoilage is made in the case of trench silos, though this factor varies rather widely.

Under most conditions, it is recommended that a minimum 4-inch slice be fed daily from the face (from the top to the bottom of the trench) of a trench silo during the winter months, with a somewhat thicker slice preferable during the summer months.

The dimensions, areas, and capacities given in Table 9-17 are based on the assumption that the silage weighs 35 pounds per cubic foot,[1] which is an average figure for corn or sorghum silage. Thus, a trench silo 8 ft deep, 6 ft wide at the bottom, and 10 ft wide at the top has a cross-sectional area of 64 ft. This size silo will hold 747 pounds of silage for each 4-inch slice, or 2,240 pounds of silage for each 1-ft slice, or 112 tons in a trench 100 ft long.

For illustrative purposes, let us use the same example and silage requirements as were used in the section on "Size of Tower Silo," but this time determine the size trench silo to build rather than the size tower silo. Briefly, the requirements are for 1,020 pounds of silage daily for a 240-day wintering period. As noted in Table 9-17, one day's feed or 1,020 pounds of silage (1,062 lb to be exact) can be obtained in each 4-in. slice of a trench silo 8 ft wide at the bottom, 14 ft 8 in. wide at the top, and 8 ft deep; or a 91 square foot cross-sectional area. The cross-sectional area should not be larger than this if a 4-in. slice is to be removed daily in order to alleviate spoilage.

In order to obtain a 240-day feed supply, the filled trench must be 80 feet long (240 by ⅓—the ⅓ representing ⅓ ft or 4 in.).

The size trench silo to build to meet the specified needs, therefore, is one that is 8 ft wide at the bottom, 14 ft 8 in. wide at the top, 8 ft deep, and 80 ft long. In

TABLE 9-17

DIMENSIONS, CROSS-SECTION AREA OF TRENCH SILO, AND WEIGHT OF SILAGE IN 4-INCH SLICE AND PER LINEAL FOOT[1]

Side Slope per Foot of Depth (inches)	Depth	Bottom Width	Top Width		Cross-Sectional Area	Weight of Silage: 4-Inch Slice	Weight of Silage: 1-Foot Slice
	(ft)	(ft)	(ft)	(in.)	(sq ft)	(lb)	(lb)
3	4	5	7	0	24	280	840
4	4	6	8	8	29	338	1,015
5	4	7	10	4	33	385	1,155
3	6	6	9	0	45	525	1,575
4	6	7	11	0	54	630	1,890
5	6	8	13	0	63	735	2,205
3	8	6	10	0	64	747	2,240
4	8	7	12	4	77	898	2,695
5	8	8	14	8	91	1,062	3,185
3	10	6	11	0	85	992	2,975
4	10	8	14	8	113	1,318	3,955
5	10	10	18	4	142	1,657	4,970

[1]*Silos, Types and Construction*, USDA, Farmers' Bulletin No. 1280, p. 55.

[1]Because the silage in trench silos is generally not so deep and well-packed as the silage in tower silos, an average figure of 35 lb per cubic foot is used herein for trench silos and 40 lb for upright silos. With all types of silos—including aboveground and belowground types—the weight of a cubic foot of silage varies with the kind and maturity of material, moisture content, length of cut, rate of filling, and depth of the silo. Corn silage harvested when about 74% of the grain has passed the milk stage and containing approximately 70% moisture is considered average silage. Volume for volume, sorghum silage weighs about the same as corn silage. Grass or grass-legume silage is 10 to 15% heavier than corn silage.

order to take care of spoilage and to provide a measure of safety, it is recommended that the actual length be from 85 to 90 ft.

About 8 feet for a trench silo is the most economical depth from the standpoint of cost and feeding. Of course, in filling it is desirable to pile silage 3 feet higher over the center of the trench and round it off. This provides for settlement.

BEEF CATTLE BUILDINGS AND EQUIPMENT

The economical production of beef cattle in most sections of the United States depends largely upon the investment in practical, durable, and convenient buildings and equipment, as well as upon the care, feeding, and management of the herd. As would be expected in a country so large and diverse as the United States, there are wide differences in the system of beef production. In a broad general way, a major difference in management exists between the farm herd method and the range cattle method. In addition, further management differences exist within each area according to whether the enterprise is commercial or purebred, whether it is a cow-calf proposition or devoted to one of the many methods of growing stockers and feeders or cattle for finishing, or whether it is a combination of two or more of these systems of beef production. Climatic differences also vary, all the way from nearly year-round grazing in the deep South to a long winter-feeding period in the northern part of the United States. Then, too, the size of the herd may vary all the way from a few animals up to an operation involving many thousands of head. Finally, there is the matter of availability of materials and labor and individual preferences.

Beef cattle are not so sensitive to extremes in temperature—heat and cold—as are dairy cattle or swine. In fact, mature beef animals will withstand extremely cold weather if kept dry.

Beef cattle shelters are of two kinds, natural and artificial. The former includes hills and valleys, timber, and other natural windbreaks. The artificial shelters include those man-made structures (solid fences, stacks, barns, and sheds) designed to protect cattle against the elements—heat, cold, wind, rains, and snows. It is with beef cattle barns and sheds that this discussion will deal.

In addition to protecting the animals during severe cold and stormy weather and at winter calving time, these structures should (1) provide a reasonably dry bed for the animals, (2) simplify feeding and management, (3) provide storage for feed and bedding when necessary, and (4) protect young calves.

Although the discussion describes beef cattle barns and sheds, it must be recognized that on small farms, which have a limited number of beef cattle, the animals are usually housed in a general-purpose barn or shed or in extensions to other barns rather than in separate and specially designed beef cattle structures.

Beef Cattle Barns

Barns are more substantial structures than sheds and provide more complete protection for stock in the colder areas. In addition to housing the animals, such structures usually provide adequate facilities for all of the roughage and bedding needed during the winter season and for a considerable proportion of the concentrates. Stalls, pens, and storerooms may also be included—additions which are especially important where a breeding herd is to be served. In general, beef cattle barns effect a saving of labor and time in feeding, and save feed.

The type of barn is determined by the kind of stock and the method of handling and management. Fig. 9-4 shows a pole barn.

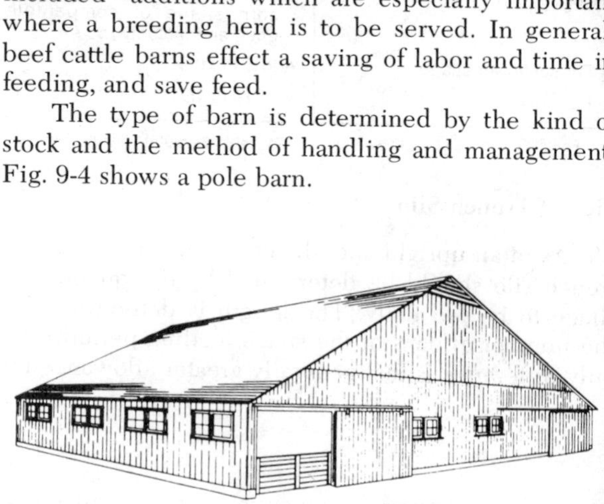

Fig. 9-4. Enclosed beef barn, pole type construction. (From: *Handbook of Building Plans*, p. 137. Courtesy, Midwest Plan Service, Iowa State University, Ames, Iowa)

Beef Cattle Sheds

Sheds are the most versatile and widely used beef cattle shelters throughout the United States. They are used for cattle in the feedlot, as a range shelter for dry cows with calves, and for housing young stock. They usually open to the south or east, preferably opposite to the direction of the prevailing winds and toward the sun. They are enclosed on the ends and sides. Sometimes the front is partially closed, and in severe weather drop-doors may be used. The latter arrangement is especially desirable when the ceiling height is sufficient to accommodate a power manure loader.

So that the bedding be kept reasonably dry, it is important that sheds be located on high, well-drained ground; that eave troughs and downspouts drain into suitable tile lines, or surface drains; and that the structures have sufficient width to prevent rain and snow from blowing to the back end. Sheds should be a minimum of 24 ft in depth, front to back, with depths up to 36 ft preferable. As a height of $8\frac{1}{2}$ ft is necessary to accommodate some power-operated manure loaders, when this type of equipment is to be used in the shed, a minimum ceiling height of 9 ft is recommended. The extra 6 inches allow for the accumulation of manure. Lower ceiling heights are satisfactory when it is intended to use a blade or pitchfork in cleaning the building.

The length of the shed can be varied according to needed capacity. Likewise, the shape may be either a single long shed or in the form of an L or T. The long arrangement permits more corral space. When an open shed is contemplated, thought should be given to feed storage and feeding problems.

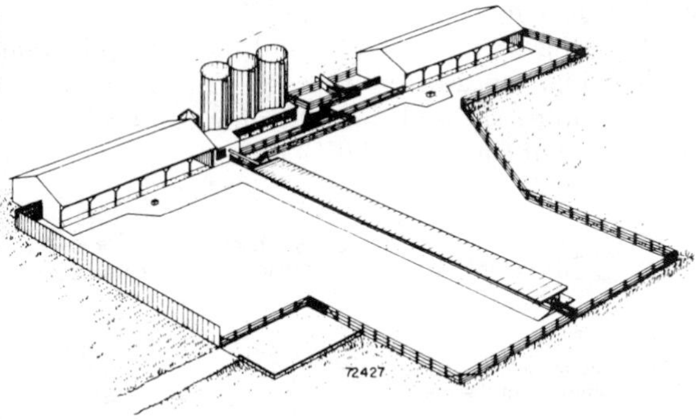

Fig. 9-5. Beef cattle layout, with open sheds and outside feeding. (From: *Handbook of Building Plans*, p. 133. Courtesy, Midwest Plan Service, Iowa State University, Ames, Iowa)

Facilities for Drylotting Beef Cows

Facilities for drylotting beef cows must be adequate and functional, but they need not be elaborate. The requirements change somewhat according to the stage of the reproductive cycle and the area. For example, dry pregnant cows require less space than cows suckling calves. Dry and well-drained areas require less space than high-rainfall, poorly drained areas. Cold areas require shelter, and hot areas require shade. However, the basic facility needs, subject to some adaptation, for a confinement cow-calf operation are as follows:

• **Site**—Pollution is a most critical factor in site selection of a confinement cow-calf operation. Remoteness from urban development is recommended because of dust and odor. Also, before constructing facilities, the owner should familiarize himself with both state and Federal regulations, then comply therewith.

• **Corral space**—Under average conditions, and an unsurfaced lot, it is recommended that 300 sq ft of corral space be provided per cow-calf. More or less space may be provided, depending primarily on rainfall, drainage, dust control, pollution control, etc. Up to 400 sq ft may be required in a wet, muddy area. The Arizona Station provided 355 sq ft per cow and calf; the Texas Station provided 200 sq ft for a dry pregnant cow and 300 sq ft per cow-calf after calving; and a successful commercial operator near Toronto, Canada, provides 2,700 sq ft of corral space, or $1/16$ acre, per cow for his 400-head operation.

In addition to corral space, provision must be made for feed storage, alleys, and working areas. So, under average conditions the total space requirements for confinement (drylot) beef cows is on the order of one acre for 60 to 65 cows.

• **Fencing**—The specifications for a desirable fence are 54 to 60 in. high; posts of 3- to 4-in. diameter pipe, or treated wood with 5-in. top diameter, set in concrete and spaced 8 to 12 feet apart; enclosed (except for the feed bunk area) by 4 strands of $3/8$-in. steel cable spaced at intervals (from ground up) of 18 in., 12 in., 12 in., and 13 in. Four strands of the cable should be placed above the feed bunk, with these spaced equal distances apart. The cable above the bunks should be so spaced as to keep a cow from butting a calf into the trough. Pipe or wooden rails may be used in place of steel cable if desired.

The above arrangement will confine the cows, but not the calves. Thirty-nine-inch woven wire, with number 9 top and bottom wires and 6-inch mesh, should be placed around the entire corral. Also, calves must be kept out of the feed bunks.

• **Cows per pen**—Except at calving time, 50 to 100 cows may be run together in one group.

• **Shelter and shade**—Where winters are severe and snowfall is heavy, a shelter should be provided. This is especially important for newborn calves. A high and dry, deep, pole-type shed, opening away from the direction of prevailing winds is excellent. In areas with mild winters, a windbreak may be provided by hills, trees, or board fences.

In hot climates, a shade 12 feet or more high, oriented north-south (so that the sun will shine under the shade early in the morning and late in the evening), should be provided, with approximately 40 square feet of area per cow-calf.

• **Bunk and concrete apron**—Feed bunks, which form part of the pen fence, should provide 24 to 30 in. of space per cow. Bunks should be constructed of concrete with the outside (alley side) of the bunk 22 to 36 in. high, and the inside (the pen side) 22 to 24 in. high. The bottom of the bunk should be rounded and about 18 in. wide.

An 8-foot concrete apron (platform), 4 to 6 inches thick, should extend from the feed bunk into the pen to provide solid footing for the cattle. The apron should have a slope of one inch per foot, which will make it nearly self-cleaning.

• **Water troughs**—Confinement cows should have access to water at all times. One linear foot of trough space for each eight cows is sufficient. There should be sufficient water supply to provide 20 gallons per head daily. Shallow, low capacity troughs are preferable, since frequent drainage is necessary to keep them clean. Continuous-flow troughs are excellent and help to keep clean water. Also, in most areas, continuous-flow troughs will not freeze during the winter months; hence, heat is not required. Heated and insulated troughs are necessary in cold areas.

• **Maternity stalls**—Where cows calve in confinement, maternity stalls 100 to 120 square feet in size should be provided for occupancy during calving and continuing for 1 to 2 days thereafter. This is particularly true of first-calf heifers. When several heifers calve the same day in a large corral, they frequently claim the wrong calf, or no calf at all—with the result that they end up as dogies. Thus, it is best that a heifer and her calf be confined to a maternity stall until they pair up.

When calving during the winter months in cold areas, the maternity stalls should be in a barn or shed; and heat lamps should be provided. In warm areas, uncovered pens will suffice. All maternity stalls should be cleaned, disinfected, and freshly bedded for each birth.

• **Calf creep**—Calves should be fed separately from the cows, in a creep.

• **Hospital area**—A special hospital area should be provided, with individual pens in which to treat and isolate sick animals.

• **Loading and unloading facilities, working chutes, squeeze, and scales**—An area should be provided for receiving and shipping cows, working animals, and weighing. The size and facilities of this area will be determined by the number of cows.

Facilities for Finishing Cattle (Also see "Beef Cattle Equipment")

Cattle-feeding facilities and equipment are a manufacturing plant, wherein animate objects (cattle) convert feed into beef. Hence, they merit the same level of competence in planning and design as any other sophisticated manufacturing plant.

Some preliminary feedlot planning considerations follow:

1. Decide on the type of facilities: (a) feedlot (open pen), (b) cold confinement, or (c) warm confinement.
2. Decide on the number of cattle and the feed and storage requirements with provision for expansion.
3. Determine the justifiable investment in cattle-feeding facilities.
4. Select the facilities, equipment, and arrangement that best fit the management program you have chosen; for example, (a) fenceline bunks and a central feed-processing plant, (b) upright storage with distributors and bunks, or (c) self-feeders.
5. Design a system that is practical, laborsaving, environmentally suitable for economical gains of cattle, and attractive.

An open lot without shelter is the cheapest type of feedlot construction of any. In the Southern Plains area, where the weather is mild and shelters are unnecessary, investment costs range from $40 to $60 per head of capacity.[2]

Housing increases costs, and the more elaborate the housing, the greater the cost. University of Minnesota studies showed the following costs relative to three types of cattle feedlot facilities.[3]

Type of Facility	Sq Ft per Animal	Cost per Animal Capacity
		($)
Open shed	20	105.00
Cold confinement	17	172.50
Warm confinement (heated)	17	255.00

OPEN PEN FEEDLOT

An open pen feedlot is, as indicated by the name, a lot in which the cattle are in the open—usually it is without shelter (except for such natural protection as may be afforded by trees, hills, or wind fences, or perhaps a roof over the feed bunks or shades).

• **Location**—In the present day and age, pollution control is the first and most important consideration in locating a cattle feedlot. The location should avoid (1) neighbors complaining about odors, flies, and dust, and (2) pollution of surface and underground water. Also, feedlots should be located on a well-drained site, with area available for expansion. Whenever possible, they should be built on a slope, preferably at the top of it. There should be a minimum amount of runoff from areas above lots (a diversion terrace can be used if necessary); and there should be ample space below feedlots for necessary water pollution control measures. Also, feedlots should be located

[2]Estimates by the author.

[3]*Feedlot Management*, Nov. 1972, p. 44; with estimated 50% increase in construction cost from 1972 to 1982 added by the author.

where there is ample space for expansion, if and when desired. Of course, the space requirements will vary. But minimum space requirements for an open feedlot—including lots, mill, office, etc.—are approximately 8/10 acre per 100 head or 7 acres per 1,000 head. In order to allow for expansion to double this size, it is recommended that there be 1.4 acres per 100 head, or 12 acres per 1,000 head.

• **Layout**—Prior to starting construction, anyone contemplating a feedlot may avoid much subsequent difficulty and expense by doing some paper and pencil planning at the outset. First, decide on the size of the enterprise and the management system. Then, establish traffic routes for animals, feed, cleaning equipment, supply trucks, etc. Next, sketch out the facilities and equipment required to meet these needs in the most efficient and economical manner, including pens, mill, scales, office, etc. Where the area permits, the ideal feedlot should be U-shaped, with the following arrangement: the facilities for receiving and loading out cattle, scales, milling feed, office, and equipment barn should be located near the center of the U. Pen facilities should be located on the three closed sides of the U. The open end of the U should be connected to a public road. Also, trench silos should be located at the mouth of the U. The mouth of the U should be kept open in order to allow for the flow of livestock, feed, and visitors. The smallest pens should be located as near the feed mill as possible, in order to minimize travel time in feeding. Feed alleys should parallel the legs and closed end of the U, forming a semicircle around the feed mill area. The corners of the pens should be rounded to allow feed trucks to turn at all intersections.

The above information, constituting the layout of operations, should first be put on paper in sketch form, by the cattleman. From this, the engineer or consultant can design facilities and equipment which most effectively and economically meet the production requirements of the specific enterprise.

• **Pens**—Pens are the working end of the business; they have much to do with the well-being and performance of the cattle. Hence, their design is most important; and the more severe the climate, the more important the design becomes. Consideration should be given to the following points when designing cattle pens:

1. **Drainage**—To supplement the natural drainage, surface grading should be done prior to starting construction. A grade of 4 to 6 percent should be established. Excessive slopes (above 10%) should be avoided, as they make for difficult footing and are subject to erosion.

2. **Mounds**—Mounds of dirt in each pen will provide a drier resting area for cattle. They should be 6 to 8 feet high, with a 4:1 or 5:1 (horizontal to vertical) slope on the sides. The top of the mound should be fairly narrow (about 10 feet wide) and crowned for good drainage. The size of the mound will vary with the size of the lot and the number of cattle for which it is intended that it shall provide a rest area. Each mound should be large enough so that the cattle can rest on the upper half of it; 10 to 15 square feet of mound per animal will accomplish this, although not all animals will use mounds. The orientation of earth mounds is unimportant, so long as they do not block feedlot drainage. The mound should be built parallel with the general lot drainage to assure that liquids can readily drain from the mound area.

Mounds will require some maintenance. A logical time for this work is when manure is cleaned from the pen.

3. **Surface**—When cattle are fed in outside pens during favorable weather, the type of pen surface (concrete vs dirt) is probably of little importance. Where good drainage cannot be provided, concrete surfaced pens should be considered.

4. **Size**—Climate and the amount of paving are the main factors determining pen size. It will vary anywhere from a minimum of 30 sq ft per animal with a surface lot and open housing to 400 sq ft per head for an unsurfaced lot in a wet, muddy area. With an open, dirt lot, 75 sq ft of pen space per head is adequate in a dry climate, whereas up to 400 sq ft per animal may be required in a wet climate. On the average, a pen space allowance of 125 to 200 sq ft per head is recommended if the lot is unpaved.

Custom feedlots require a variety of pen sizes in order to accommodate each customer's cattle in separate pens. In custom feedlots, the majority of pens with a capacity for 120 head appear to be desirable, because of customer convenience. From the standpoint of trucking requirements, pens should be sized to hold multiples of 60 head. In most feedlots, where some custom feeding is done, it is recommended that feedlot sizes vary from 120-head capacity to 300 head.

5. **Shape**—Pens should be rectangular in shape, preferably with rounded corners to allow the feed truck to turn at all intersections. The depth of the pen will depend upon the length of the bunk; there must be sufficient bunk space to accommodate the cattle in any given pen; then there should be sufficient depth to provide the number of square feet per animal intended.

6. **Fences**—The specifications for the most desirable feedlot are: 54 to 60 in. high; posts of 3- to 4-in. diameter pipe, or treated wood with 5-in. top diameter, set in concrete and space 8 to 12 feet apart; enclosed (except for the feed bunk area) by 4 strands of 3/8-in. steel cable spaced at intervals (from ground up) of 18 in., 12 in., 12 in., and 13 in. Three strands of the cable should be placed above the feed bunk, with

these spaced equal distances apart. Steel or aluminum pipe, wood, or wire may be used in place of steel cable if desired.

Working corral fences should be higher and stronger than ordinary feedlot fences. They may be constructed with treated wood posts, 6 in. top diameter, spaced 8 feet apart and set in concrete, enclosed by 2-in. dimension lumber; or 4-in. diameter pipe and ⅞-in. sucker rod or equivalent may be used.

7. **Feed bunks and aprons**—Six to nine inches of bunk space per head are adequate when cattle have access to feed at all times. The bunk forms part of the fence for the pen. It should be constructed of concrete with the outside (alley side) of the bunk 22 to 36 in. high, and the inside (the pen side) 18 to 22 in. high. The bottom of the bunk should be rounded and about 18 in. wide.

A 6- to 8-foot concrete apron (platform), 4 to 6 inches thick should extend from the feed bunk into the pen to provide solid footing for the cattle. The apron should have a minimum slope of ½- to 1-inch per foot (a 1-in. per ft slope will be nearly self-cleaning). A lip at the low end of the concrete apron, to divert the water along the length of the slab and out the pen, rather than merely off the slab area and into the pen dirt area, will alleviate the low muddy spot that usually develops at the juncture of the concrete and dirt.

8. **Water troughs**—Feedlot cattle should have access to clean fresh water at all times. One linear foot of trough space for each 10 head of cattle is sufficient. Shallow, low capacity troughs are preferable, since frequent drainage is necessary to keep them clean. There should be sufficient water supply to provide 20 gallons per head per day. Continuous-flow troughs are excellent and help to keep clean water.

9. **Windbreaks and shades**—In those sections of the country where snow and cold winds are a problem, windbreaks may be provided by hills, trees, or board fences. Where board fences are used, they are generally constructed of 1-inch lumber and are 7 feet high.

In hot climates, shades should be provided. Shades should be 12 feet or more high; provide 20 to 25 square feet per animal; and be oriented north and south, so that the sun will shine under the shade early in the morning and late in the evening. Shades may be run east-west in the hot deserts of the southwest to take fullest advantage of the cooler north sky.

Fans, sprinklers, and other cooling devices increase the effectiveness of shades.

10. **Back scratchers**—Some feedlots provide back scratchers in their feeding pens. A back scratcher is a horizontal suspended arm with a burlap-type cloth attached at cattle back level. This device is so arranged that the cloth is always saturated with insect repellent. The repellent is placed on the animal's back through contact with the scratcher, thereby preventing flies and other insects from molesting the cattle.

11. **Gates**—Gates should be of the same height as the fences. The length of most gates should be coordinated with the width of alleyways and crowding areas. Thus, gates along a 12-foot alley should be 12 feet long and capable of swinging either way.

• **Alleys**—Two types of alleys are common to most feedlots—feed alleys and drive alleys. In feedlots below approximately 5,000-head capacity, feed and drive alleys should be combined, for cost reasons. In larger feedlots, it is recommended that feed alleys be separated from drive alleys, so that feeding and cattle movement can be done without excessive interference. Feed alleys, and combined feed and drive alleys, should be at least 20 feet wide, so as to permit the passing of trucks. Working and drive alleys, which do not handle truck traffic, should be 12 feet wide.

• **Cattle loading and unloading facilities**—The facilities for loading and unloading cattle are the connecting link between the corrals and the various types of vehicles used to transport cattle in and out. They should include a truck dock (with loading chute), scales, working alley, and holding pens. These facilities should be located centrally so that a smooth flow of cattle trucks in and out can be maintained. However, they should be somewhat removed from the feed-mill area, in order to separate the traffic flow of feed and cattle and alleviate congestion of traffic.

• **Scales**—Smaller feedlots can use a combination cattle and commodity scale. With larger feedlots, however, it is recommended that there be separate scales for each of these uses.

The scale should have sufficient capacity to meet the largest anticipated volume demand, without the necessity of dividing a lot of cattle and weighing in two or more drafts. In the larger feedlots, this calls for a scale approximately 10 feet wide and 60 feet long, with a capacity of 100,000 pounds. To avoid exceeding the weighing capacity of a scale, provide one square foot or less of platform space for each 110 pounds of rated capacity.

The scale should be equipped with a ticket printer, in order to verify weight records and avoid human error in market transactions.

• **Cattle processing facilities**—A separate area should be provided where cattle can be branded, dehorned, castrated, vaccinated, and sprayed or dipped. This should include crowding areas and a curved chute 18 to 30 feet long to aid in the movement of the cattle to the squeeze chute. Generally cattle will work best in a chute constructed with a modest curvature; a sharp curve may spook them. A manually operated squeeze chute is satisfactory for a small feedlot, but larger feedlots should use a hydraulic squeeze chute for reasons of efficiency and ease of operation.

• **Spraying pen: dipping vat**—Periodic and regular treatment for the control of lice, grubs, summer flies, and other parasites is essential to good management. For this purpose, a spray pen or a dipping vat will be needed.

A spray pen should not be over 15 feet wide, and it should have solid sides, good drainage, and a rough-paved or gravel surface. This type of pen will keep the animals close enough to the spray nozzle to give good spray penetration and reduce drift of the spray materials.

Dipping vats are increasing in popularity. A properly constructed system of chutes and pens with a dipping vat can provide fast, positive control. A metal vat can be purchased or a concrete vat can be built. The vat is usually 28 to 32 feet long, plus an entry chute at one end and a drip pen on the other.

• **Hospital areas**—Hospital pens are for holding sick or injured animals. From 2 to 5 percent of the feedlot area should be allocated for this purpose in lots feeding mostly yearling cattle. Lots feeding calves will need more intensive care areas. Each hospital area should be equipped with a squeeze chute, refrigerator, running water, medicine, equipment storage, and feed facilities for the sick animals. Two or three small pens at the hospital for cattle at various stages of recovery are recommended. In areas subject to severe weather, shelter should be provided for hospitalized animals.

• **Feed mill; feed delivery equipment**—Mill type will be determined by the number of cattle to be fed and the feeds to be used. Generally speaking, it is recommended that feedlots with fewer than 5,000 head of cattle use a self-mixing, self-unloading truck, especially if milling of the grain is not necessary immediately prior to feeding. For lots above 5,000-head capacity, a mixing mill and self-unloading feed truck are more economical. Mill type will be determined by the feeds that are to be used. If whole corn is fed, a minimum of mill facilities is needed. Where sorghum grain is the primary feed, steam-flaking equipment or other similar processing equipment, must be considered. Steam-flake type mills, along with storage facilities, make for added costs of from $17 to $20 per head of capacity.

Mill capacity is also important. For a 5,000-head feedlot, the mill should be able to process 20 tons of feed per hour; for a 10,000-head feedlot, it should be able to process 35 to 40 tons per hour. Modern mills automatically mix the feed, with operation from a central control panel.

Some small to medium-sized feedlots use conveyors (auger or belt-type) to move feed from the mill to the cattle. Unless the mill and the cattle are in close proximity (not more than 1,000 to 1,200 ft apart), conveyors are likely to be more costly than trucks. Also, when it comes to changing the feed formula of cattle, they are less flexible than trucks. Confinement systems, with a great concentration of cattle and short runs, make conveyors more practical.

• **Office and parking**—Since the office is the headquarters of the cattle feedlot business, its location is of importance. It should be located near the main access road in order to minimize nonessential traffic in the feedlot proper. In most feedlots, the truck scales are adjacent to the office, and weighings are made in the office.

• **Equipment storage and repair building**—Most commercial feedlots have a separate building in which they store trucks, tractors, silage loaders, spare parts, and miscellaneous supplies. Also, this building is used for repair work.

• **Lights**—Feedlot lights serve three purposes: (1) They have a calming and quieting effect on the cattle, although there is no conclusive experimental work to support the claim that lights will improve rate of gain or feed efficiency; (2) they prevent prowlers and pilfering; and (3) they make for convenience in working or loading cattle at night. Mercury vapor lights, equipped with photo eyes (which turn on automatically), are recommended.

• **Pollution control**—Pollution control is a most critical factor in site selection and operation of a cattle feedlot. Remoteness from urban development is recommended because of dust and odor. Also, before constructing a cattle feedlot, the owner should familiarize himself with both state and Federal regulations.

Various methods of handling manure are being studied, including recycling it for feed, the production of methane gas, the production of garden fertilizer, etc. Yet, today, and perhaps for sometime to come, the vast majority of manure will be spread on farmland for use as a fertilizer. Thus, it is imperative that adequate farmland be available, and that a suitable location be made for stockpiling waste until it is used.

• **Manure cleaning equipment**—Pens are generally cleaned after the cattle have been removed and before a new lot is brought in.

The most common manure cleaning equipment consists of a front-end loader attached to a tractor and a dump truck. The front-end loader is a scoop 4 to 5 feet wide, and 15 inches deep. The truck is a 4-wheel dump. Typically, feedlots with 1,000-head capacity use one tractor-loader and one truck; 5,000-head capacity lots use one tractor and two trucks; and 10,000-head capacity lots use two tractor-loaders and four trucks. The number of trucks needed is dependent on the distance that the manure must be hauled. The average load hauled from the pens is 7,000 pounds.

Feedlots with large pens sometimes use self-loading wheel scrapers.

CONFINEMENT FEEDING; SLOTTED FLOORS

Currently, there is much interest in cattle confinement feeding and slotted floors. The main deterrent is cost; construction costs vary with type of structure and may range up to $300 per steer space.

Confinement cattle feeding refers to feeding in limited quarters, generally 20 to 25 square feet per yearling animal, which is about ⅛ the space normally allotted to a yearling in an unsurfaced lot and ⅓ that of a paved lot. The confinement is usually under roof on slotted floors.

Slotted floors are floors with slots through which the feces and urine pass to a storage area immediately below or nearby.

Interest in confinement feeding and slotted floors was ushered in for the purposes of (1) automating and saving labor; (2) cutting down on bedding and facilitating manure handling; (3) lessening mud, dust, odor, and fly problems; (4) increasing gains and saving feed; (5) lessening land requirements; and (6) lessening pollution.

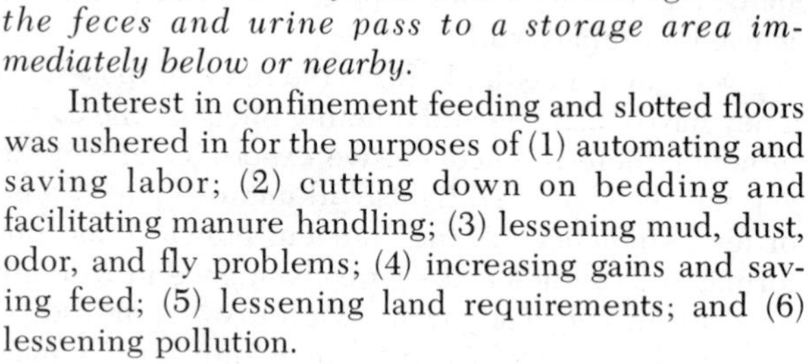

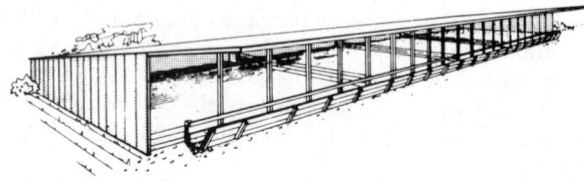

Fig. 9-6. Cold confinement (open-front), 34-foot shed roof, slotted floor, fenceline bunk. (From: *Handbook of Building Plans*, p. 153. Courtesy, Midwest Plan Service, Iowa State University, Ames, Iowa)

Research has shown conclusively that cattle fed during the winter months in cold areas gain faster and more efficiently if they are sheltered. However, as pointed out earlier under the section headed "Open Pen Feedlot," the per head cost is much higher for confined or sheltered cattle. Thus, the decision on whether or not cattle confinement can be justified, even in the northern part of the United States, should be determined by economics. Will the cattle in confinement quarters gain sufficiently more rapidly and efficiently to justify the added cost? Of course, manure disposal and pollution control should also be considered. Also, a choice may be made between cold confinement and warm confinement.

• **Cold Confinement**[4]—Cold confinement refers to a more or less open shed for confining cattle; hence, winter temperatures therein are within a few degrees of outdoor temperatures. Open sheds should be faced away from the direction of the prevailing winds. Additionally, doors or other openings in the closed walls should be provided for summer ventilation.

• **Warm Confinement**[5]—Warm confinement refers to a confinement building for cattle which is sufficiently insulated and ventilated to maintain inside winter conditions above 35°F in severe weather, and in the range of 50° to 60°F most of the time.

• **Design Requirements**—Based on information and experiences presently available, the following figures may be used as guides:

1. *Floor space*—Allot 15 to 30 square feet per animal exclusive of the bunk and alley, with an average of 20 to 25 square feet for a 1,000-pound animal.
2. *Animals per pen*—25 to 100 head per pen, with 25 to 30 being most common.
3. *Bunk*—Allow 6 to 18 inches of linear bunk space per animal, with the amount of feeding space determined by frequency of feeding and size of animal.
4. *Waterers*—Locate one waterer per 25 head at the back (opposite feed bunk) of each pen, preferably in the center.
5. *Slats*—Reinforced concrete, steel, or aluminum may be used. Most concrete slats are 5 to 6 in. wide across the top, 6 to 7 in. deep, tapered to 3 to 4 in. wide at the bottom, and placed so as to provide a slot width of 1½ to 1¾ in.
6. *Manure production and storage*—Manure production will vary with size of animal and *kind of feed*, but it will be approximately as follows:

Animal	Cu Ft/Day Solids and Liquid	%Water	Gallons/Day
1,000-lb steer	1-1½	80-90	7½ - 10¾

Here is how to determine how much manure will need to be stored:

Storage capacity = Number of animals × daily manure production × desired storage time (days) + extra water.

A rule of thumb is that when the pit occupies the entire area beneath the cattle, it will fill at a rate of 8 to 10 inches per month.

[4]The terms "cold confinement" and "warm confinement" refer to winter conditions. Without mechanical cooling, both systems are "warm" during the summer months.

[5]Ibid.

Beef Cattle Equipment

There is hardly any limit to the number of different articles of beef cattle equipment, and the design of each. Figs. 9-7 through 9-16, pages 827 to 829, suggest designs of those articles that are most common. Suitable equipment saves feed and labor, conserves manure, and makes for increased production. Detailed plans and specifications for such equipment can usually be obtained through the local county agricultural agent, vocational agriculture instructor, lumber dealer, or through writing to the college of agriculture in the state.

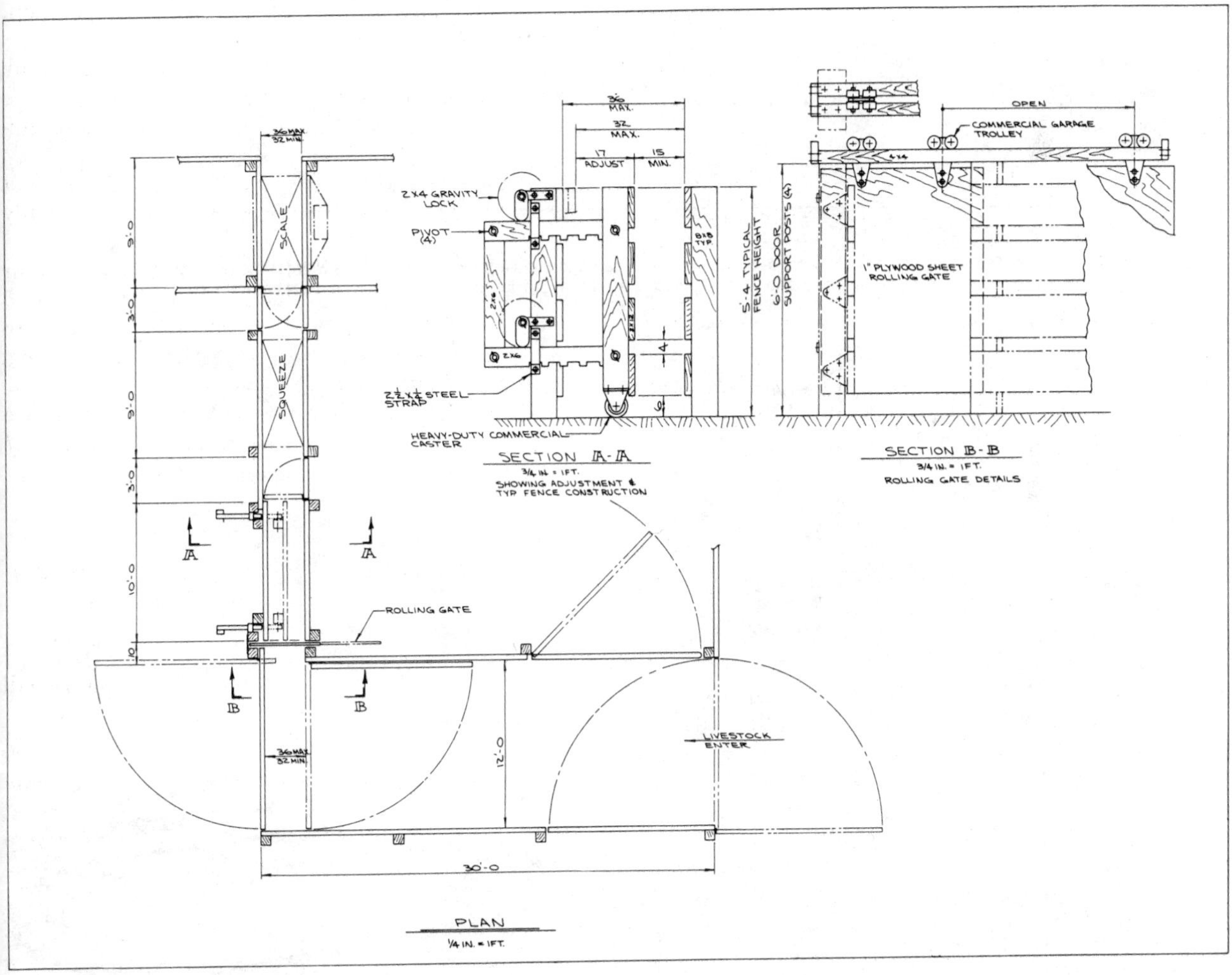

Fig. 9-7. An excellent chute, squeeze, and scale arrangement which can be adapted to any cattle corral (new or old). As shown, it features (1) a very short approach chute, with vertical sides that are adjustable for width, and (2) a gate located so that animals have no choice other than to go into the chute. The squeeze should (1) have a stanchion-type head gate, and (2) be designed so that the sides close in on the back of the animal. With this arrangement, animals will stand still without struggling. Where the narrow portion of the squeeze is at the bottom, squeezing tends to lift the animal off its feet and causes it to fight. (Drawing prepared under the direction of Dr. A. T. Ralston, Department of Animal Science, Oregon State University)

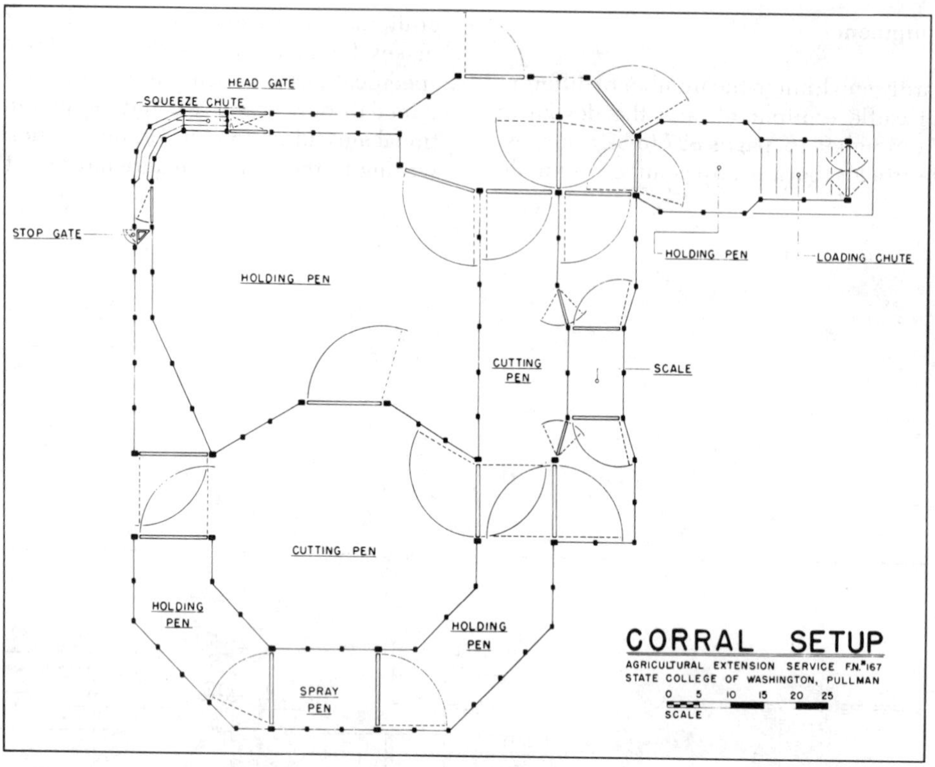

Fig. 9-8. Convenient cattle corrals. Some variations can and should be made in keeping with (1) the number of animals to be worked, (2) the size and topography of the area available, and (3) the facilities adjacent thereto.

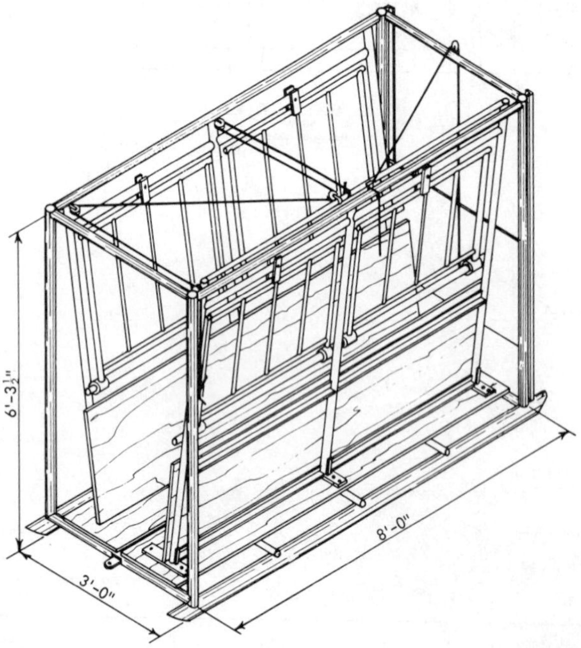

Fig. 9-9. A workable cattle squeeze. It may be more practical to buy a commercially built squeeze than to build one. (From: *Beef Housing and Equipment Handbook*, p. 41. Courtesy, Midwest Plan Service, Iowa State University, Ames, Iowa)

Fig. 9-10. A concrete water tank. When located in the fenceline, it will serve two corrals or pastures; and the top fence boards will keep animals out of the tank. Usually the water level is controlled by means of a float.

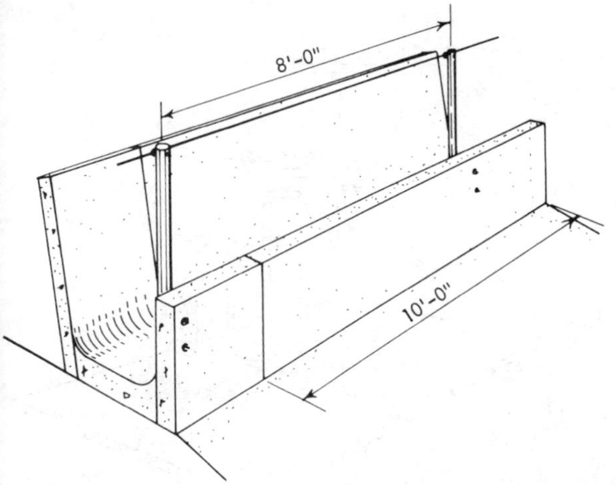

Fig. 9-11. Concrete fenceline bunk. Slope apron 1 inch per foot away from bunk. (Make front of bunk flush with apron [or fill opening], so manure cannot accumulate under the bunk.) (From: *Beef Housing and Equipment Handbook*, p. 52. Courtesy, Midwest Plan Service, Iowa State University, Ames, Iowa)

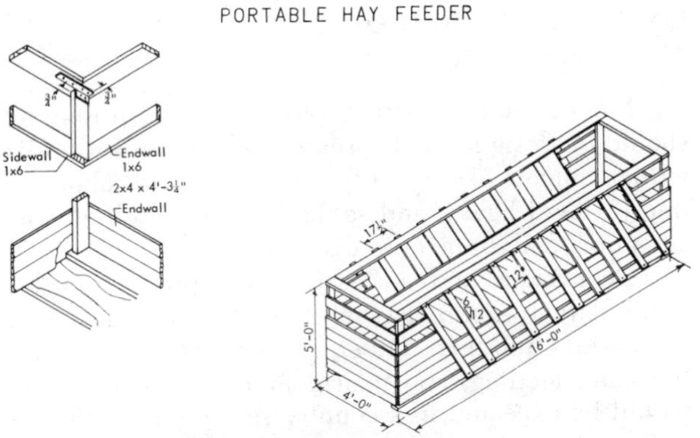

Fig. 9-14. Portable hay feeder. With this type of feeder, cattle work down from the top. Hence, there is less wastage of the leaves and finer particles than with overhead racks. (From: *Beef Housing and Equipment Handbook*, p. 56. Courtesy, Midwest Plan Service, Iowa State University, Ames, Iowa)

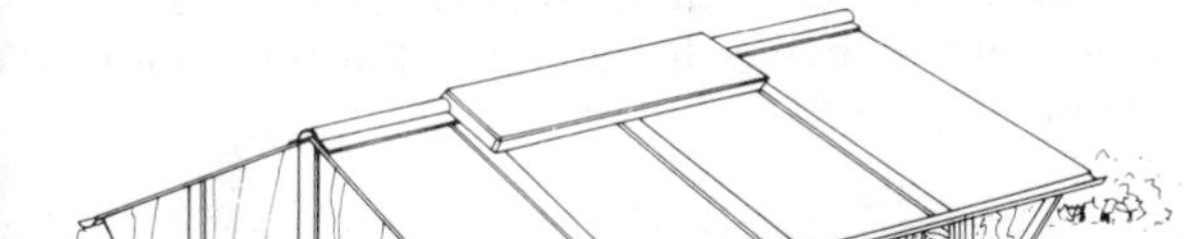

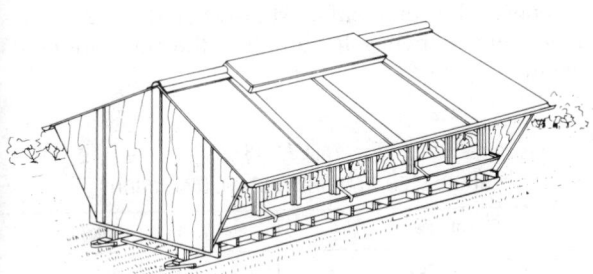

Fig. 9-12. A self-feeder for cattle. Note the door in the roof for filling and the runners for moving. (From: *Beef Housing and Equipment Handbook*, p. 58. Courtesy, Midwest Plan Service, Iowa State University, Ames, Iowa)

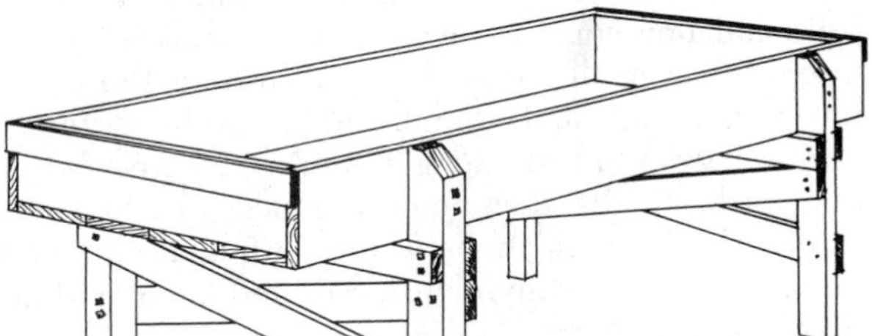

Fig. 9-15. Cattle feed bunk. This type of feed bunk is usually placed out in the open and used for feeding silage and/or grain.

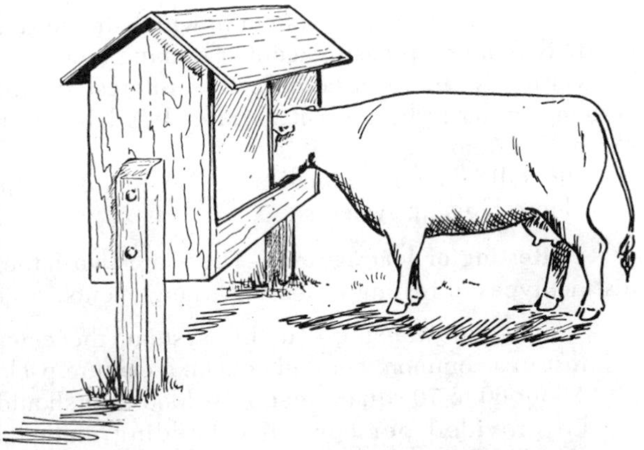

Fig. 9-13. A salt-mineral feeder for cattle. Note that it is a two-compartment arrangement, and that the feeder is enclosed except for the feeding side.

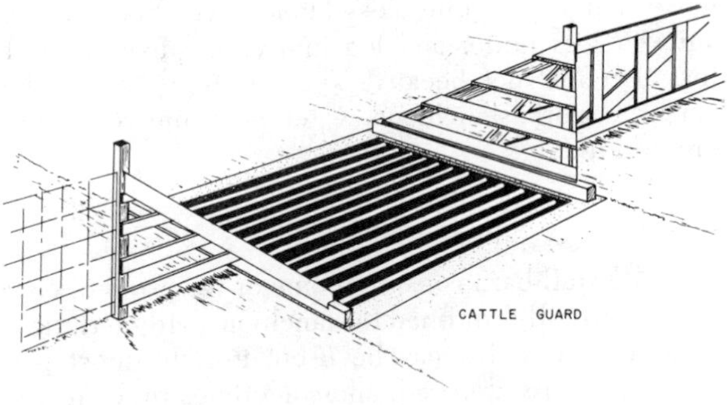

Fig. 9-16. *Cattle guards* may be set in a fence to permit convenient passage of automobiles and trucks but deter cattle, hogs, sheep, and most horses.

DAIRY CATTLE BUILDINGS AND EQUIPMENT

Modern dairy cattle buildings and equipment should be designed to facilitate (1) freedom and individual comfort of cows, (2) automation and laborsaving, (3) herd health and sanitation, (4) bedding conservation, and (5) manure disposal.

Economical and successful dairy production depends largely upon the investment in practical and convenient buildings and equipment, as well as upon the care, feeding, and management of the herd. As would be expected in a country so wide and diverse as the United States, there are wide differences in the systems and facilities of dairy production. A major difference exists between (1) the family-owned and operated farm herd, which usually relies on pastures in season and produces its own winter roughage; and (2) the large, commercial type, drylot operation. In addition, facilities differ according to whether the operation is devoted to producing milk or raising heifer replacements, or a combination of both. Further facility differences exist between a strictly commercial enterprise and a purebred herd. Climate also makes for differences; from the animals being out in the open much of the time in the South and in California, to the need for warmth and a long winter feeding period in the North. Finally, there are differences due to meeting the health regulations of the particular area and market, the availability of materials and labor, and individual preferences.

Two basic facility systems are used for lactating cows: (1) stall barns, and (2) loose housing. Thus, the starting point in designing lactating cow facilities is to select the system.

Many arguments are heard relative to the merits of each system. Experiments show that, with proper care and management, similar milk production can be achieved with either system. The stall barn allows the cows to be displayed to greater advantage, which is particularly important in the purebred herd; and, generally speaking, the cows are observed more frequently than in loose housing. On the other hand, loose housing requires less labor, saves bedding, results in less udder and leg injury, and usually costs less to construct because such items as expensive concrete work, stanchions, water cups, and ventilation are omitted.

Stall Barns

The stall barn consists of one or two rows of cows that are usually confined to stanchions, although tie or comfort-type stalls may be used. For the most part, concrete floors are used; and sometimes they are covered with rubber mats. The floor slopes into a gutter, which is usually 16 in. wide and 8 in. deep on the stall side and 6 in. deep on the alley side.

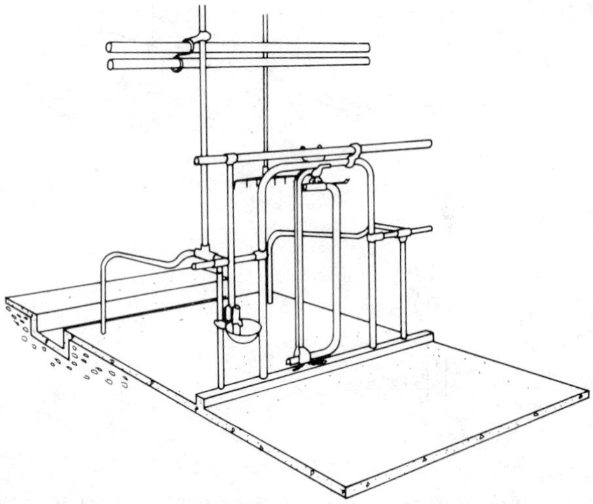

Fig. 9-17. Stall barn stanchion stall. (From: *Dairy Housing and Equipment Handbook*, p. 4. Courtesy, Midwest Plan Service, Iowa State University, Ames, Iowa)

Table 9-18 gives suggested stanchion-stall dimensions. Tie or comfort stalls should be 2 inches wider and 4 inches longer than the measurements given.

TABLE 9-18

SUGGESTED STANCHION-STALL DIMENSIONS

Size of Cow	Stall Width	Stall Length
Small cow (Jersey)	3′6″ to 3′10″	4′6″ to 4′10″
Medium cow (Guernsey, Milking Shorthorn, Red Poll)	4′0″ to 4′3″	5′0″ to 5′4″
Large cow (Holstein, Brown Swiss)	4′4″ to 4′6″	5′4″ to 5′8″

Loose Housing

Loose housing is that system in which the herd is handled on a group basis except at milking time.

Currently, there is considerable interest in this system, primarily because it requires less labor than the stall system.

The following functional areas are involved in most loose housing systems:

1. **Resting or loafing area**—There are two rather distinct types of resting or loafing arrangements:

a. *Group housing*—In this system, the cows rest in a common, bedded area on a manure pack. About 60 to 70 square feet of bedded area should be provided per cow. The bedding material should be deep to begin with; then each day the droppings should be removed and 10 to 12 pounds of fresh bedding per cow added.

The reported advantage of group housing over individual stall housing is that the cost per cow is less.

b. *Freestall housing*—This is a modification of the group housing system. It consists of individual open stalls, which are bedded. For small breeds, use 4- × 7-foot stalls; for large breeds, use 4- × 7½-foot stalls.

The reported advantages of freestall housing over group housing are: bedding costs may be reduced by as much as 75 percent; less labor is required to bed the cows; the cows are cleaner, with the result that less cow-washing time is required; and less space per cow is necessary.

2. **Calf and maternity quarters**—The author recommends the use of individual calf stalls for the first six weeks, followed by the use of group pens thereafter. Generally, one individual pen is provided for each 10 cows. Calf pens should be dry and well ventilated.

Also, maternity and isolation stalls should be provided. It is recommended that 100 to 120 square feet be allowed per stall, with one such stall per 20 milk cows in the herd.

3. **Feeding area**—Separate feeding and bedded areas are preferred. Also, more and more large dairies are emulating cattle fattening operations as a labor-saving device, by feeding complete rations in fence-line feeders, with the herd separated into two or more production groups.

4. **Exercise yard**—This is usually a paved area which serves as a place for exercise. Approximately 100 square feet per cow should be allowed. It should be designed so that it can be cleaned daily by scraping.

5. **Holding area**—This area is for the purpose of confining cows in preparation for milking. It should be paved, easy to clean daily, and funneled to the parlor. About 15 to 20 square feet per cow should be allowed.

6. **Milking parlors**—Loose housing systems lend themselves particularly well to the use of milking parlors. Although there is a wide range in choices in milking parlors, the three most common ones are:

a. *Tandem-type*—Where the cows stand "Indian file" in line and broadside to the operator's pit.

b. *Herringbone*—Where the cows stand in groups at an angle to the operator's pit—like herringbone (see Fig. 9-18, page 832).

c. *Rotary (or Carousel)*—Where the parlor, which may incorporate either tandem or herringbone principle, rotates around the operator.

Fig. 9-18 shows diagrams of different milking parlor arrangements, whereas Table 9-19 shows the capacities of milking parlors.

TABLE 9-19

MILKING PARLOR CAPACITIES[1]

Parlor Size	Cows/Hr	Cows/Man-hr	Milk Lb/Man-hr
Double 3 HB	34	29	440
Double 4 HB	41	39	870
Double 6 HB	58	30	640
Double 8 HB	71	35	720
Double 10 HB	92	44	1,040
Double 2 SO	41	41	882
Double 3 SO	51	47	1,031
Rotary (turnstyle)	96	48	—
Polygon	134	67	1,566

[1]From *Dairy Housing and Equipment Handbook*, p. 22, published by Midwest Plan Service, Iowa State University, Ames, Iowa. Data from a field survey.
HB = herringbone parlor. SO = side opening parlor.

Facilities and Equipment for Dry and Lactating Cows

Pertinent information relative to lots, housing, and feed and water facilities of dry and lactating cows follows:

• **Lot and Housing Facilities—**

1. Provide the following minimum lot space per head:

Kind of lot	Sq. Ft/Animal
All paved	100
Paved and dirt	150
Dirt	200

2. Slope paved lots ¼ to ½ inch/foot, and slope dirt lots ½ inch or more per foot.

3. Pave (rough finish) a 15- to 20-foot area around waterers, feed bunks and racks, and entrances to sheds.

4. Open sheds may be used under a loose housing system. Free stall housing saves bedding, keeps the cows cleaner, and saves labor. For small breeds, use 7- × 4-foot stalls; for large breeds, 7½- × 4-foot stalls.

5. Keep the temperature in stall barns 40°F or more; and in milking parlors 50° to 60°F during the winter.

6. Provide for proper ventilation—changing of air—in cold climates and during the winter months, with care taken to avoid direct drafts and coldness. A fan system installed according to manufacturer's directions is best. For each 1,000 pounds of animal weight in a well-insulated barn, the fan should be capable of removing a minimum of 100 cubic feet per minute (cfm).

7. Bed all cows except in dry climates.

8. Provide for efficient manure disposal, selecting the system best adapted for the particular dairy and area.

• **Feed and Water Facilities—**

1. Provide 24 to 30 inches per head of manger space for roughage feeding.

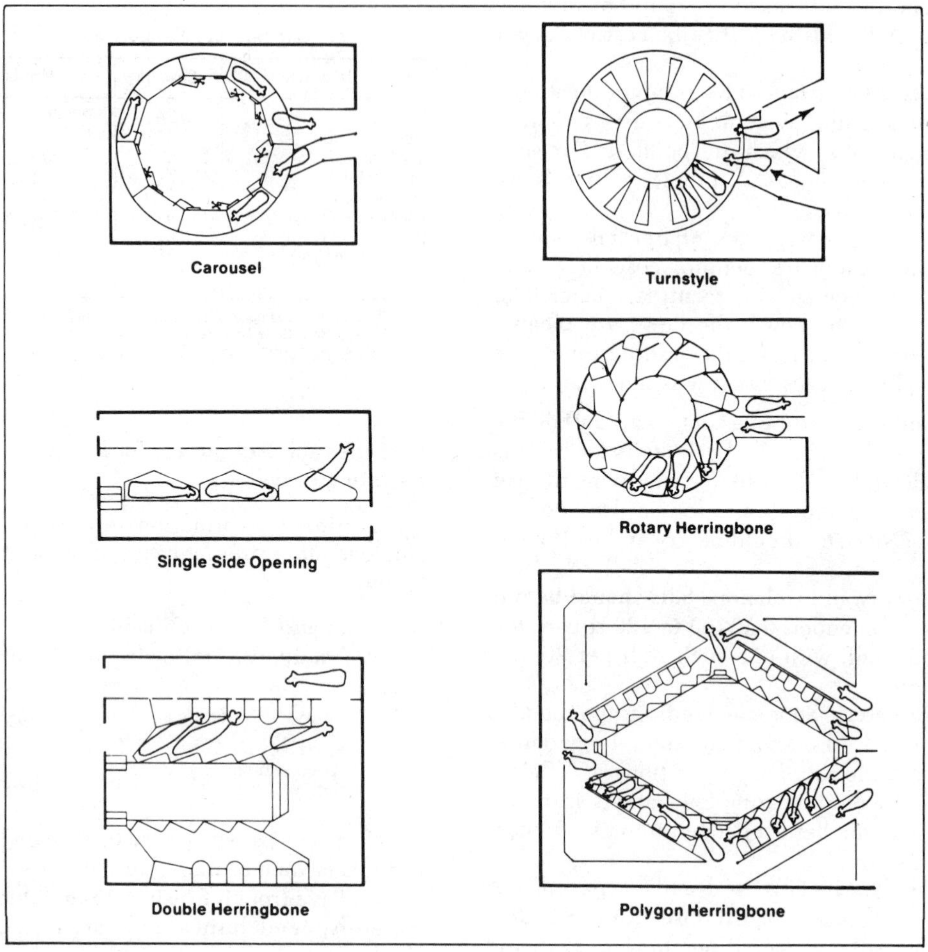

Fig. 9-18. Diagrams of different milking parlors. (From: *Dairy Housing and Equipment Handbook*, p. 23. Courtesy, Midwest Plan Service, Iowa State University, Ames, Iowa)

2. Make bunks and roughage racks 24 to 30 inches wide when cattle are fed from one side; 36 inches wide when feeding from both sides.
3. Provide adequate water space: (a) 1 linear foot of open tank per 8 to 10 head, or (b) one automatic bowl per 15 head.
4. Keep water temperature within the range of 35° to 80°F; warm it to 50°F in the winter.

MILKING EQUIPMENT

Basically, there are two types of milking equipment: (1) the bucket system, and (2) the pipeline system.

In the bucket system, the milk is received directly into a nearby vacuumized portable bucket, which may be either of two types: (1) floor type, or (2) suspended type.

Conventional pipeline systems use a rigid heat-resistant glass or stainless sanitary pipe for carrying vacuum from the milk receiver to the individual milking units, and for carrying the milk from the units to the receiver. Pipeline milkers may be used in any of the following types of facilities: (1) stanchion barn, (2) herringbone milking parlor, (3) side-opening milking parlor, (4) walk-through milking parlor, or (5) rotary parlor.

Regardless of make, the mechanical milking systems can be separated broadly into four major parts: (1) vacuum supply, (2) milk flow, (3) pulsation, and (4) milking unit.

There are two characteristics of milk which make it ideal for the development of bacteria: (1) It is a well-balanced food in which bacteria thrive, and (2) as it comes from the cow, the temperature is ideal for bacterial growth. For these reasons, milk must be cooled to at least 50°F (preferably to 40°F) as soon as possible in order to inhibit bacterial growth.

Milk may be handled by either of two systems: the can system, or the bulk system. Until 1939, when the bulk system was first introduced in California, all milk was handled in cans. Today, the trend is to more and more bulk tanks. Although the initial cost is greater than where cans are used, the greater returns over a period of time, justify the expense. Further, many dairymen are facing the situation of being forced into going to bulk tanks if they are to retain a market outlet.

Generally speaking, the following advantages accrue to the use of bulk tanks, in comparison with cans: (1) a saving in labor, (2) less loss in milk, (3) alleviating 10-gallon cans, (4) higher butterfat tests (due to butterfat being left on lids of cans), (5) a saving in hauling costs, and (6) a premium paid by the plant.

Facilities and Equipment for Dairy Calves

Pertinent information relative to facilities and equipment for dairy calves follows:

• Housing—

1. House calves separately (in individual pens or tie stalls)—from birth until at least one week after milk or milk substitute is discontinued. Thereafter, they may be raised in groups, with (a) a maximum of 10 head per group (preferably 6 to 8 per group), and (b) a maximum age difference of 2 months between calves.
2. Provide a minimum of 24 square feet pen space for individual calves; 2½- × 4-foot tie stalls; 30 square feet for calves in groups without outside runs.
3. Solid partitions between individual pens reduce drafts and chilling. Front of pens should be wire or slatted.
4. Preferred pen temperature is within the range of 50° to 75°F.

• Feed and Water—

1. Feed boxes for individual calves should be 8 × 10 × 6 in. deep, and they should be removable so as to facilitate cleaning. Troughs for group feeding should be 10 wide and 6 in. deep, with 2 linear feet per calf; provide stanchions. Top of feed containers should be 20 in. from the floor, and feed containers should be located in corner of pen away from water.
2. Automatic drinking cups are preferred for both individual quarters and group pens (one cup 5 to 8 calves). Where pails or tanks are used, keep them clean. Top of drinking cups for calves should be 20 inches from the floor.
3. Always locate water facilities at a corner away from the feed.

Facilities and Equipment for Replacement Heifers

Pertinent information relative to facilities and equipment for replacement heifers follows:

• Lots and Housing—

1. Provide the following lot space per head:

Kind of Lot	Sq Ft/Animal
All paved	50-75
Paved and dirt	75-100
Dirt	100-150

2. Slope paved (rough finish) lots ¼ to ½ inch/foot, and dirt lots ½ inch or more per foot.
3. Pave (rough finish) a 15- to 20-foot area around waterers, feed bunks and racks, and entrances to sheds.
4. Provide an open shed; allow 20 to 30 square feet per head for small cattle, and 30 to 40 square feet per head for large cattle. Bed sheds as needed.
5. Provide artificial shade in hot climates if natural shade is not available. Allow 20 to 30 square feet per animal, and build shade 8 to 10 feet high.

• Feed and Water Facilities—

1. Provide feed bunks that are 24 to 30 in. above the ground (to top of bunk; with height determined by size of cattle); 8 to 12 in. deep (12 in. deep for silage); and 18 to 24 in. wide when feeding from one side, 36 in. wide when feeding from 2 sides.
2. Allow the following amount of feeder space per head:

	Grain	Roughage
Small cattle	12″	18″
Large cattle	18″	24″

3. Allow 1 linear foot of water tank space for each 10 animals and one automatic watering bowl for each 25 animals. Water temperature may range from 35° to 80°F; warming to 50°F in the winter is desirable.

Manure Management

Planned manure management is an important part of a modern dairy production program. The collection, transport, storage, and land application of manure must be compatible with sanitary milk production, housing systems, and pollution control. Likewise, manure should be handled so as to retain its highest value as a fertilizer.

Prior to construction, any proposed waste management system should be approved by the appropriate regulatory, public health, and milk market officials.

Manure can be handled in either of two ways:

- **As a solid**, which runs 20 to 30 percent solids, and which refers to feces plus bedding, or feces after liquid separation.
- **As a liquid**, which may be up to 15 percent solids, and which refers to feces, urine, and sometimes dilution water.

Manure can be hauled (1) daily and spread on available land, usually as a solid; or (2) from storage, either as a solid or liquid, and spread on cropland at a convenient time.

The average daily dairy cattle production of manure is shown in Table 9-20.

TABLE 9-20
DAILY DAIRY CATTLE MANURE PRODUCTION, SOLIDS AND LIQUIDS[1]
(Manure at 87.3% water and 62 lb/cu ft density.)

Animal Size	Total Manure Production			Nutrient Content N	P	K
(lb)	(lb/day)	(cu ft/day)	(gal/day)	(lb/day)		
150	12	0.19	1.5	0.06	0.010	0.04
250	20	0.33	2.4	0.10	0.020	0.07
500	41	0.66	5.0	0.20	0.036	0.14
1,000	82	1.32	9.9	0.41	0.073	0.27
1,400	115	1.85	13.9	0.57	0.102	0.38

[1]From *Dairy Housing and Equipment Handbook*, p. 37. (Courtesy, Midwest Plan Service, Iowa State University, Ames, Iowa)

SOLID MANURE

Manure can be handled as a solid if it is mixed with bedding or if the liquids are allowed to evaporate or drain away.

Manure from a stall barn is usually loaded directly into a spreader with a barn cleaner and spread on the land daily.

Where long-term storage is planned, provision should be made for a period of 180 days or more. To calculate the size of storage area, multiply the number of 1,000-pound cow units by 2.5 cubic feet per day (this figure includes bedding), then multiply by the number of days of storage desired.

Storage by stacking works best with manure containing bedding. It is well suited for use with stall barns and up to 80 cows in the herd. The investment in facilities is usually lower than with liquid storage systems. Moving stacked manure to the land requires a manure loader, a spreader, and a tractor.

LIQUID MANURE

Yearlong liquid storage of manure is a practical goal for dairymen. It permits incorporating manure into the soil at the best time to preserve fertilizer value. Also, milkhouse and parlor wastes can go into a liquid manure storage.

In comparison with storing solid manure, storage of liquid manure is generally more costly. Also, odors can be a problem, especially when agitating and spreading; and labor requirements may interfere with field work.

Several types of storage are used for liquid manure: storage tank under the barn, outside-below ground storage tank, earthen storage basin, and above ground storage (silo).

Liquid manure is moved to storage by dropping through slots into a storage area below, automatic floor scrapers, tractor scrapers, barn cleaners, or by pumping from a hopper or tank into the storage structure.

SLOTTED ALLEYS

The scraping of alleys takes time and effort, and it can easily be neglected during busy seasons, resulting in undesirable conditions.

Slotted alleys eliminate the labor and cost of scraping and the cost of scraping equipment, because wastes pass directly through the slots into the storage area below. Manure does not build up on the floor. As a result, cows' feet remain comparatively clean, and cows track little manure into the milking parlor from the slotted holding area.

BARNYARD RUNOFF CONTROL

Precipitation that falls on or flows across manure-covered areas or manure stacks can cause severe pollution to streams, lakes, or ponds. This runoff must be kept from reaching usable private or public waters. Local regulations usually govern runoff control systems.

Fig. 9-19 shows a two-unit collection system; (1) a

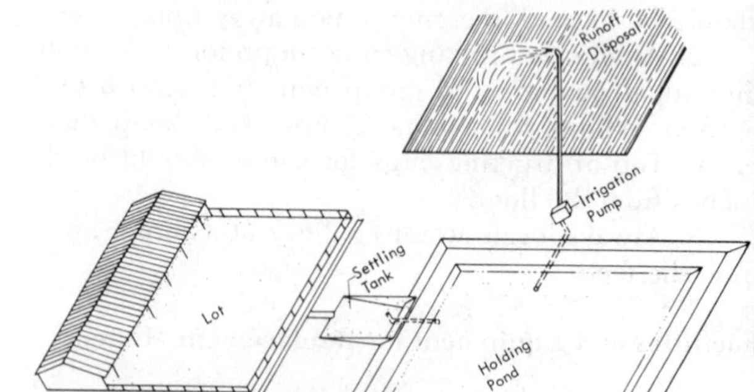

Fig. 9-19. Runoff control system. (From: *Dairy Housing and Equipment Handbook*, p. 48. Courtesy, Midwest Plan Service, Iowa State University, Ames, Iowa)

settling basin to remove most of the solids, and (2) detention pond to hold the water.

The best method of applying liquid manure to the land is through an irrigation system. Most irrigation systems can handle fluid waste with up to 4 percent solids, which are typical of lot runoff and effluent from a lagoon or milkhouse.

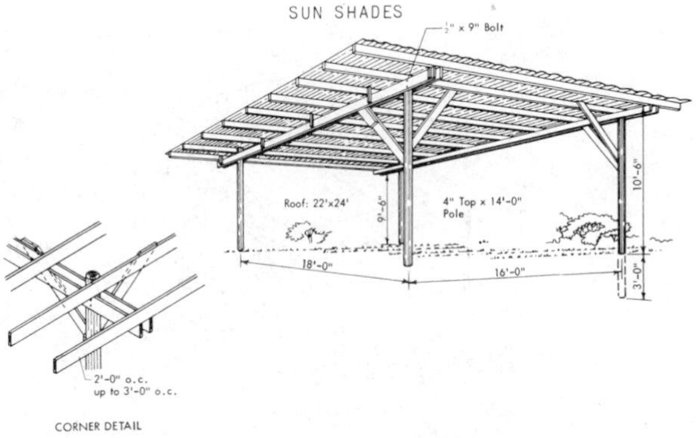

Fig. 9-21. Sun shade. Allow 20 to 25 sq ft per animal. Paint top side of roof white and underside of roof black. Orient row of shades north and south so sun will shine under shade early morning and late evening. (From: *Dairy Housing and Equipment Handbook*, p. 93. Courtesy, Midwest Plan Service, Iowa State University, Ames, Iowa)

Dairy Cattle Equipment

There are numerous articles of dairy cattle equipment, and many designs of each. Figs. 9-20 through 9-23 show some of the most common articles and the usual designs.

Detailed plans and specifications for dairy cattle equipment can usually be obtained through the local county agricultural agent, vocational agriculture instructor, lumber dealer, or through writing to the college of agriculture in the state.

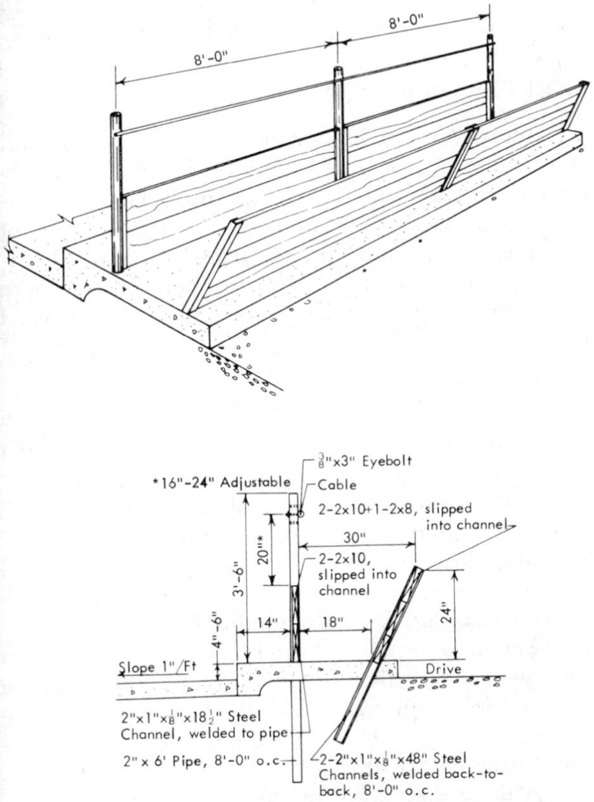

Fig. 9-20. Fenceline bunk. (From: *Dairy Housing and Equipment Handbook*, p. 55. Courtesy, Midwest Plan Service, Iowa State University, Ames, Iowa)

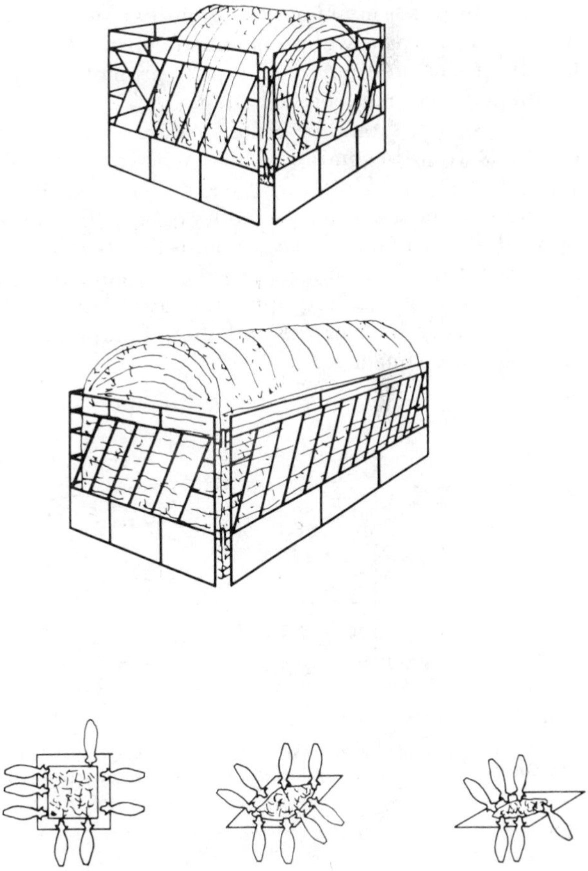

Pinned corner connections allow panels to collapse, thus providing complete hay cleanup.

Fig. 9-22. Slant bar feeder panels for stacks. (From: *Dairy Housing and Equipment Handbook*, p. 61. Courtesy, Midwest Plan Service, Iowa State University, Ames, Iowa)

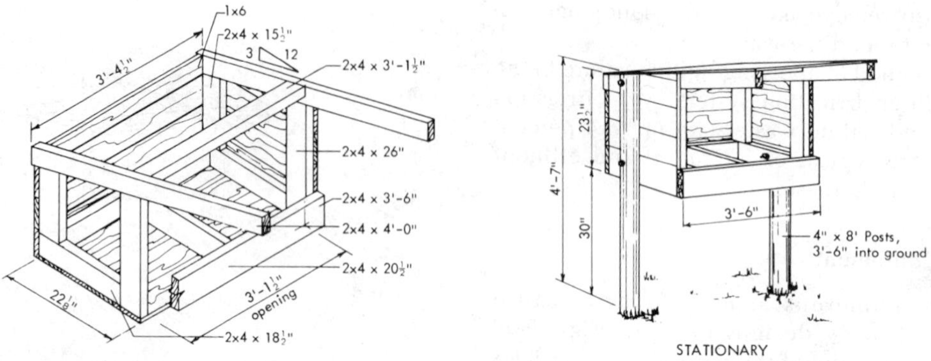

Fig. 9-23. Stationary mineral feeder. (From: *Dairy Housing and Equipment Handbook*, p. 67. Courtesy, Midwest Plan Service, Iowa State University, Ames, Iowa)

SHEEP BUILDINGS AND EQUIPMENT

Sheep do not require expensive or elaborate buildings and equipment, but this statement should not be construed to mean that the facilities for the sheep enterprise should not be carefully planned. On the contrary, it pays well to plan and construct sheep buildings and equipment that will promote sheep health and conserve feed and labor.

The shelter should be of such nature as to protect the flock from becoming soaked with rain or wet snow. Dry snow or bitter cold has no harmful effect, and up until lambing time, a shelter open to the south on well-drained ground may be entirely satisfactory.

Except for the smaller space requirements, many sheep buildings and equipment closely resemble those used by beef cattle, and their functions and requisites are similar.

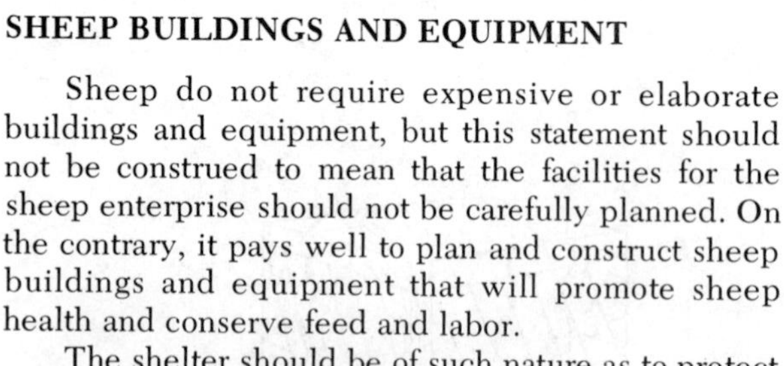

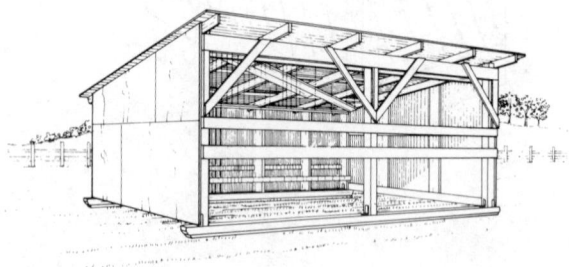

Fig. 9-24. Movable sheep shelter. (From: *Sheep Handbook Housing and Equipment*, p. 54. Courtesy, Midwest Plan Service, Iowa State University, Ames, Iowa)

Sheep Finishing Facilities and Equipment

Drylot feeding refers to feeding under restricted conditions. This may either be (1) open-yard feeding, or (2) shelter or barn feeding.

OPEN-YARD FEEDING

Open-yard feeding is the common method of finishing lambs in the irrigated area of the West, though a few eastern lamb-feeding operations are in open yards. In this system, equipment costs are kept to a minimum—the facilities merely consisting of an enclosed and well-drained yard which may or may not have a natural or constructed windbreak, and the necessary feed bunks. Open-yard feeding is often used by large operators who feed thousands of lambs.

SHELTER OR BARN FEEDING

Because of inclement weather in the fall and early winter, many of the lamb-feeding operations in the central and eastern states are in drylots which afford shelter. In some instances, the lambs are kept under cover without an exercising lot. These barns may consist of anything from an open shed to more costly and elaborate structures, including slotted floors.

Confinement of Sheep on Slotted Floors

Today, there is much interest in confinement sheep production and slotted floors, among both ewe-lamb producers and lamb feeders. For both groups, it offers new hope for eliminating internal parasites, lessening labor, increasing environmental control, lessening space requirements, increasing gains and saving feed, eliminating bedding, and lessening mud, odor, and fly problems. Additionally, for the ewe-lamb producer, it fits in with early weaning and multiple lambing; and for the lamb feeder, it provides a way in which to achieve greater animal comfort during hot weather.

Without doubt, the interest of sheepmen in slotted floors has been accentuated by the extent and success of slotted floors in poultry, swine, and cattle production.

The shelter above the slotted floor may range all the way from a mere shade over the top of the floor to a completely enclosed, environmentally controlled building; or it may be somewhere between those two extremes—for example, a shed open to one side.

• **Design requirements for slotted floors for sheep**—Unfortunately, there is limited experimental work on which to base design recommendations for slotted floors for sheep. Consequently, the recommendations given in this section are based on producer experiences and such research as is available. Some of them may, and likely will, change as further knowledge becomes available. Nevertheless, the recommendations that follow may serve as useful guides and stimulate further studies:

1. **Floor space**—Allow 8-10 sq ft per dry ewe, 10-12 sq ft for a ewe and a lamb, and 4-5 sq ft for a feeder lamb.
2. **Animals per pen**—Not more than 100 pregnant ewes, or 50 ewes and lambs, or 500 feeder lambs in each group.
3. **Bunk space**—Allow 12 in. per ewe. For feeder lambs, allow 6 in. per head for twice-a-day feeding, and 3 in. per head for self-feeding.

With slotted floors, hay should always be chopped or ground. Long hay may pile up on the slotted floors and prevent manure from dropping through, thereby providing a place for internal parasite development and causing the sheep to befoul themselves.

4. **Waterers**—Locate one automatic waterer at the back (opposite the feed bunk) of each pen. In cold areas, waterers should be equipped with electric heating. Provide one automatic waterer for 15 ewes or 20 feeder lambs. With an open tank, allow one foot per 10 ewes or 15 feeder lambs.
5. **Slats**—Slats are usually made from wood, concrete, or metal. Wood slats are less costly than concrete or metal, but they are also less durable; and, if they're green, they will warp and expand. Slats should be strong enough to support a 200-pound concentrated load at the center.

Slat width and spacing (slot width) are governed by animal comfort and cleaning efficiency. Narrow slats and too wide spacing can cause injury to the feet. On the other hand, wide slats and narrow openings result in floors that are not completely self-cleaning.

Most sheepmen (including both ewe-lamb producers and lamb feeders) seem to favor ¾ to ⅞ inch spacing (slots) between 2-inch slats.

Access to the pit should be provided through the slotted floor if the manure is handled as a liquid and a manure pump is used. For this purpose, either a steel grid or removable slats may be used.

• **Manure production and storage**—Sheep manure may be handled either as a liquid or a solid. Because sheep feces are rather dry, in comparison with the feces of cattle and hogs, sheep manure lends itself to handling as a solid.[6] On the other hand, handling it as a liquid may, in some operations, offer certain advantages from the standpoint of automation and laborsaving. In either case, a storage area beneath the floor is required.

The quantity of manure produced varies according to the size of animal, kind and amount of feed, and amount of bedding; but it will be approximately as follows for each 1,000 pounds of sheep:

Cu Ft/Day Solids and Liquid	Percent Water	Gallons/ Day
0.6	70	4.5

When sheep manure is handled as a liquid (rather than as a solid), extra water will need to be added to liquify the wastes. From $^1/_5$ to $^3/_5$ of the storage volume may be needed for extra water if the manure is to be pumped. For irrigation, there should be about 95 percent water and 5 percent manure.

There are about 34 cubic feet in a ton of manure.

Of course, the total manure storage capacity will depend on the frequency of cleaning. Here is how to determine how much manure will need to be stored:

Storage Capacity = no. of sheep × daily manure production × desired storage time (days) + extra water if handled as a liquid.

Pits range up to 10 feet deep. With finishing lambs, a 5-foot pit, cleaned every 100 days, will suffice.

Where manure is to be handled as a liquid, storage tank dimensions and proportions should follow the recommendations of the manure agitator manufacturer. Access to the pit is usually provided from the slotted floor, via either a steel grid or removable slats.

Where manure is handled as a solid, the height of the storage area is determined by two factors: (1) frequency of cleaning, and (2) method of cleaning. The manure may be removed by means of a tractor-mounted loader or a scraper. Less working height is required where the building is arranged so floor sections may be removed. But the removing of floor sections does require labor. So, consideration should be given to having greater floor height, with access to the pit via doors or removable panels opening from the ends or sides.

Regardless of whether sheep manure is handled as a liquid or a solid, the area from the floor to the

[6]Fresh sheep manure is about 14% lower in moisture than cow manure.

ground must be completely enclosed to prevent drafts.

Sheep Equipment

Figs. 9-25 to 9-31 show some of the most common articles of sheep equipment, along with the usual designs.

Detailed plans and specifications for sheep equipment can usually be obtained through the local county agricultural agent, vocational agriculture instructor, lumber dealer, or through writing to the college of agriculture in the state.

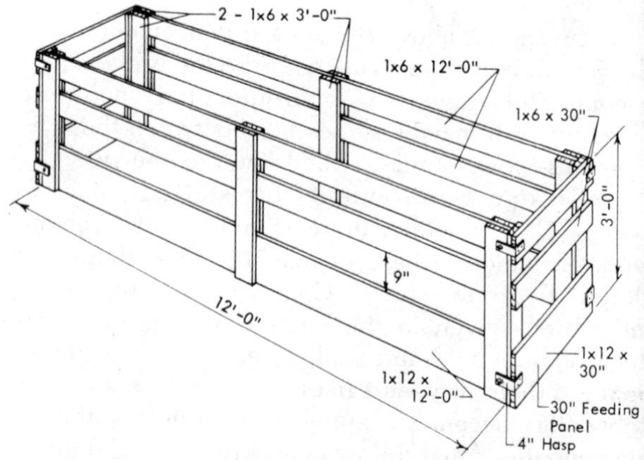

Fig. 9-27. Hay feeder. (From: *Sheep Handbook Housing and Equipment*, p. 41. Courtesy, Midwest Plan Service, Iowa State University, Ames, Iowa)

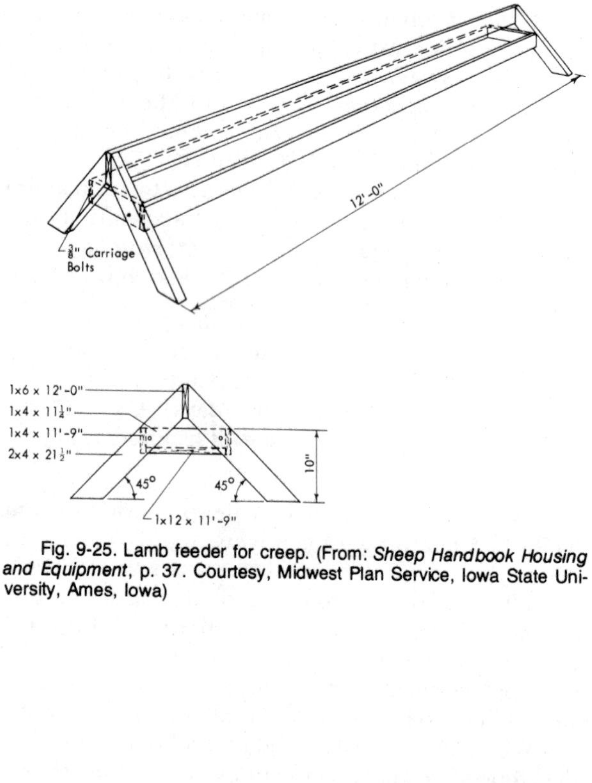

Fig. 9-25. Lamb feeder for creep. (From: *Sheep Handbook Housing and Equipment*, p. 37. Courtesy, Midwest Plan Service, Iowa State University, Ames, Iowa)

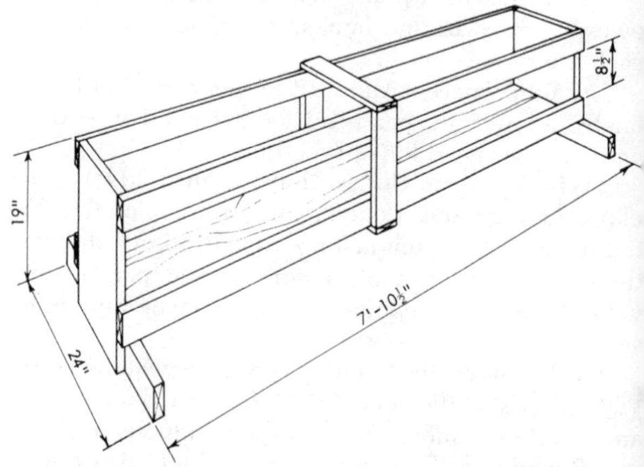

Fig. 9-28. Lamb grain bunk. (From: *Sheep Handbook Housing and Equipment*, p. 38. Courtesy, Midwest Plan Service, Iowa State University, Ames, Iowa)

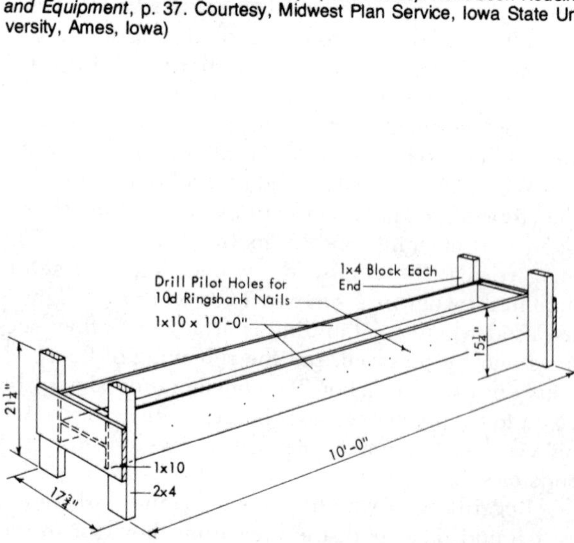

Fig. 9-26. Grain bunk for ewes. (From: *Sheep Handbook Housing and Equipment*, p. 37. Courtesy, Midwest Plan Service, Iowa State University, Ames, Iowa)

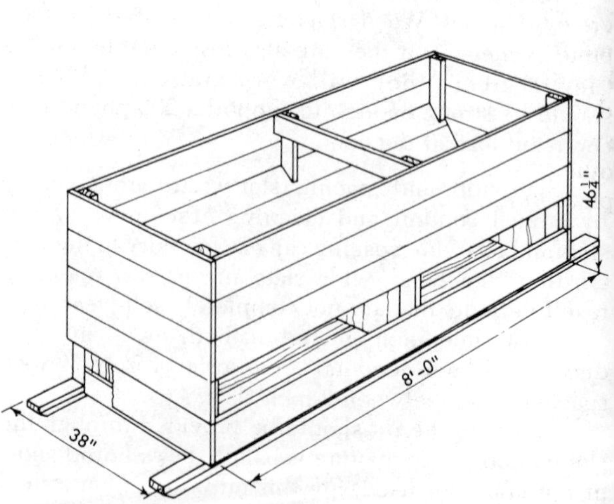

Fig. 9-29. Self-feeder for feeding grain or pellets to finishing lambs. (From: *Sheep Handbook Housing and Equipment*, p. 35. Courtesy, Midwest Plan Service, Iowa State University, Ames, Iowa)

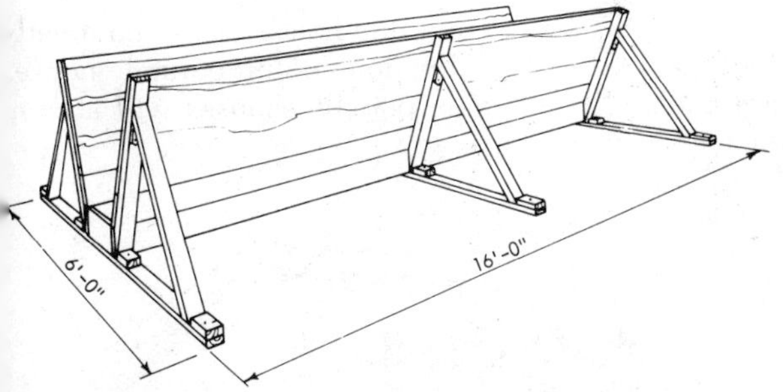

Fig. 9-30. Foot bath. (From: *Sheep Handbook Housing and Equipment*, p. 60. Courtesy, Midwest Plan Service, Iowa State University, Ames, Iowa)

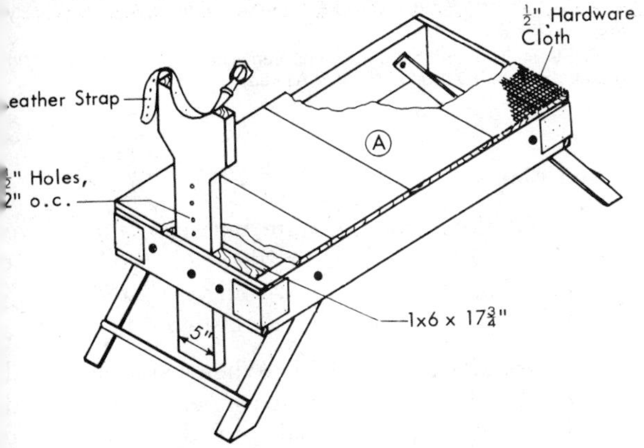

Fig. 9-31. Blocking stand. (From: *Sheep Handbook Housing and Equipment*, p. 65. Courtesy, Midwest Plan Service, Iowa State University, Ames, Iowa)

SWINE BUILDINGS AND EQUIPMENT

Properly designed swine buildings and equipment should provide for quartering, feeding, and handling hogs in accordance with recommended production practices. The functions and requisites of swine buildings and equipment differ from those for other classes of livestock in that increased emphasis is placed on (1) temperature control—because hogs are so sensitive to extremes of heat and cold; and (2) ventilation, sanitation, and manure disposal—because swine are confined more (and confinement rearing is increasing) than other four-footed animals.

Space, Temperature, and Grouping of Early Weaned Pigs

For early weaned pigs (pigs weaned under six weeks), warm, dry, well-ventilated, draft-free housing is essential, and supplemental heat (such as a heat lamp) and special feeders and waterers are recommended; at its best, this entails an extra building—a nursery. Also, the conditions listed in Table 9-21 should prevail.

Swine Buildings and Systems

Fig. 9-32 illustrates swine movement through a sequence of three stages of production. As indicated, a swine building does not function alone; rather, it is one part of a system. A brief description of one-stage, two-stage, and three-stage production follows:

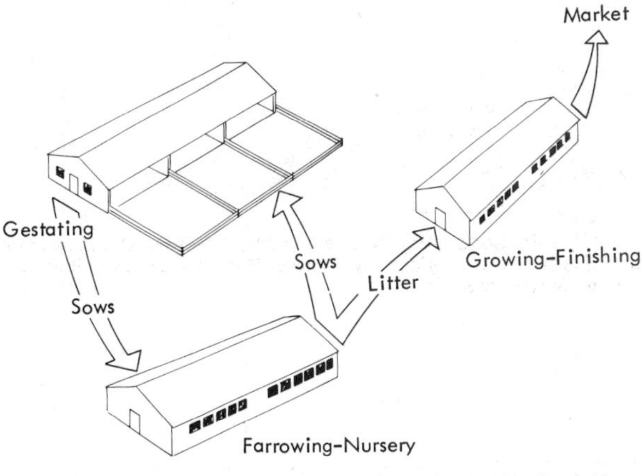

Fig. 9-32. A swine building system. (From: *Swine Handbook Housing and Equipment*, p. 4. Courtesy, Midwest Plan Service, Iowa State University, Ames, Iowa)

TABLE 9-21

RECOMMENDED CONDITIONS FOR EARLY WEANING

Conditions	Age in Weeks				
	1	2	3	4	5
Minimum pig wt., lb	5	9	12	15	21
Farrowing house temperature, °F	75-80	75-80	74	70	70
Minimum floor space per pig, sq ft	4	4	4	5	5
Minimum no. of pigs per linear ft of feeder space	5	5	4	4	4
Maximum no. of pigs per linear ft of water space	12	12	12	10	10
Maximum no. of pigs of uniform size per group	10	10	10	20	25

● **One-stage production**—Farrow-to-finish production, all in one pen of one house, is no longer common. Feeder pig production (which does not involve finishing by the breeder) is a one-stage system. Normally, about 4 litters per year are sold from the pens in which they are farrowed, at 40- to 60-pound weights.

● **Two-stage production**—Under this system, pigs are farrowed, nursed, weaned, and started in one pen to about 60 pounds, or 12 weeks of age. Then they are moved to a finishing unit for the next 12 weeks. Usually, 3 to 4 litters per year are raised this way.

● **Three-stage production**—The three-stage system is conducted about as follows: (1) pigs farrowed in stalls of a farrowing-nursery unit, where they are held until weaning; (2) started or grown in pens with supplemental heat from about 25 to 100 pounds; and (3) finished to market weight in still another unit.

Portable Houses

Portable houses were originally designed to accommodate one sow and her litter. In recent years, however, the size of movable houses has been greatly increased. Consequently, double-unit portable houses are common, and some portable houses will accommodate as many as six sows.

Fig. 9-34. The portable double-unit Sunshine Hog House. Note folding doors in front. (Drawing by Steve Allured)

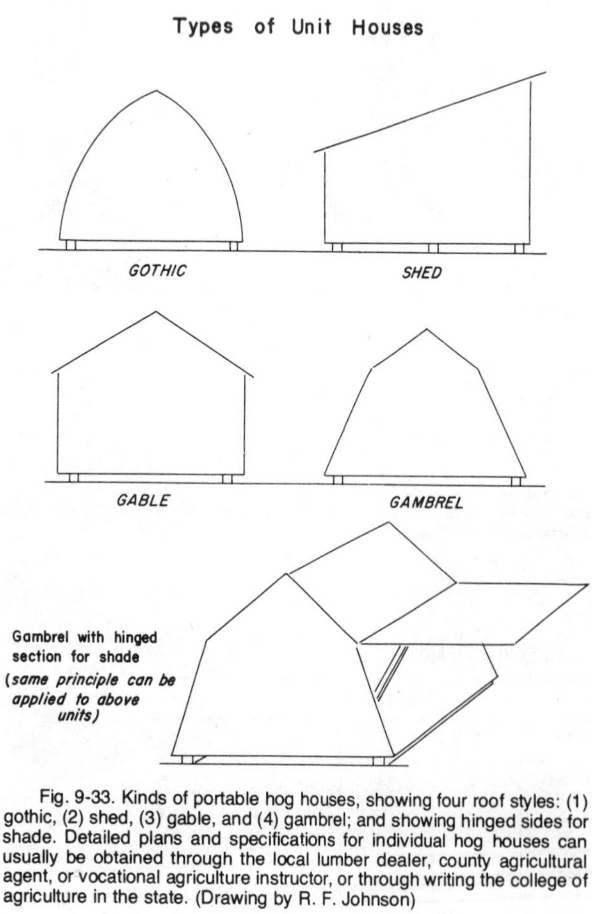

Fig. 9-33. Kinds of portable hog houses, showing four roof styles: (1) gothic, (2) shed, (3) gable, and (4) gambrel; and showing hinged sides for shade. Detailed plans and specifications for individual hog houses can usually be obtained through the local lumber dealer, county agricultural agent, or vocational agriculture instructor, or through writing the college of agriculture in the state. (Drawing by R. F. Johnson)

Slotted Floors

Slotted floors for hogs are not new; they have been used in Europe for over 200 years. Slotted floor buildings should be warm and well ventilated.

The main *advantages* of using slotted floors are:

1. **Less space is needed**—When using a slotted floor, only half as much floor space per hog is needed as when using a conventional solid floor. Thus, one can either (a) place twice as many hogs in a building with slotted floors as in a conventional building, or (b) construct a building of one-half the size.
2. **Bedding is eliminated**—Bedding is expensive; that is, when all costs are computed—initial purchase price, storage space, handling, spreading it in the pens, and removal of the straw with the manure.
3. **Manure handling is reduced**—Conventional solid floors require frequent removal of manure. With a slotted floor, manure may be handled every month, or every six months, depending on the type and size of the pit under the floor.

The chief *disadvantages* of slotted floors are:

1. **There is a higher initial cost.**
2. **There is less flexibility in the use of the building.**
3. **Any spilled feed may be lost through the slots.**
4. **Pigs raised on slotted floors resist being driven over a solid floor.**
5. **Environmental conditions become more critical.**

● **Slotted floor construction**—The main features that control the successful operation of a slotted floor system for swine are:

1. **The storage pit under the slats**—A concrete-

ned pit is located under the slat floors. Depth of the it varies with the method of manure disposal. If the ıanure is to be drained to a lagoon, the pit usually ınges from 1½ to 2 feet deep. If 6 months' storage is esired, the pit should be 5 to 6 feet deep. The floor f the pit should slope slightly (about 1 inch per 25 eet) to the cleanout location.

2. **Slats**—Slats are usually made from concrete, teel, or wood. Wood slats are less costly than con- rete or steel, but also less durable. Hardwood slats of ak, elm, hickory, or maple may last two to five years.

Slat width and spacing (slot width) are governed y size of hogs and cleaning efficiency. Narrow slats re usually more effective in farrowing and nursery nits, but too wide spacing of narrow slats can cause ıjury to the feet and legs of finishing hogs. On the ther hand, wide slats and narrow openings result in oors that are not completely self-cleaning.

The recommended slot widths are given in Table -22.

Slats stay cleaner if they are run at right angles to ıe pigs' major traffic pattern.

TABLE 9-22

RECOMMENDED SLOT WIDTH (SLOTTED FLOORS)

Age or Weight	Slot Width
Newborn pigs[1]	⅜" and 1"
25-40 lb	1"
40 lb to market, and farrowing	1"

[1]Cover openings with plywood, sheet metal, or mesh during farrowing. Use 1 in. lots behind sow; use ⅜ in. slots elsewhere.

• **Partially slotted floors**—A solid section permits imited feeding on the floor to help keep pens clean, r permits under-floor heat in a farrowing or nursery rea. Where a partially slotted floor is used, the ratio f slotted to solid floor should be about 3 to 1 or 4 to , the objective being to limit the space on the solid concrete floor so that there is just enough room for the pigs to lie down.

The following plans and practices are pertinent to he success of a partially slotted floor arrangement:

1. Place the feeder in one end of the pen, and the waterer in the other on slats.
2. Use a continuous feeder along one wall; this rrangement will alleviate dunging along the wall.
3. Avoid long pens; the distance from the feeder o the slots shouldn't be more than 12 feet. A hog will eldom move more than 12 feet before defecating.

Hog Manure

Handling manure has become a major problem for producers who raise large numbers of hogs in confinement. Careful waste management is necessary in order to—

1. Maintain good swine health through sanitary facilities.
2. Avoid pollution of air and water.
3. Comply with local, state, and federal regulations.

On a 100-lb liveweight basis, the approximate daily production of manure is: ⅛ cu ft, 1.0 gal., or 7.5 lb. The average density of manure is 59 lb/cu ft.

Table 9-23 shows the approximate daily manure production.

TABLE 9-23

APPROXIMATE DAILY MANURE PRODUCTION, FREE OF BEDDING[1]

Class Weight	Waste Production			
	Liquids and Solids		Wet Solids Only	
(lb)	*(cu ft)*	*(gal)*	*(cu ft)*	*(lb)*
Pigs:				
40	.06	.5	.04	2.4
100	.13	1.0	.1	5.9
150	.21	1.7	.15	8.8
210	.30	2.2	.2	12.0
Sows and boars:				
300	.43	3	.3	17.5
500	.71	5	.5	30
Sow and litter:	.55	4	.5	30

From: *Swine Handbook Housing and Equipment*, p. 33, Table 6. (Courtesy, Midwest Plan Service, Iowa State University, Ames, Iowa)

Generally swine producers have a choice in the method of handling wastes. Manure from bedded areas and drained solid floors is usually handled as a solid. Wastes from unbedded floors and some lots are semisolid. Wastes under slotted floors are usually liquid.

• **As a solid**—Solid manure results from catching and holding excrement in bedding, or by allowing the liquids to run off, leaving the solids to be handled separately.

Handling solid manure requires (1) solid floors that can be bedded or drained; (2) a minimum of equipment; and (3) an area on which to spread the solids.

Where manure is to be handled as a solid, it is recommended—

1. That sloping floors be installed, waterers be located where manure accumulation is desired, and pens be kept full of pigs.
2. That manure be hauled directly to fields when possible, but that it not be spread on frozen ground.
3. That any necessary stockpiling be done with care; where it will be convenient to load it on a

spreader, where it will not pollute water, and where water can be diverted away from it.

• **As a liquid**—Swine producers generally handle manure as a liquid for one or more of the following reasons:

1. It usually requires minimal time and labor, because it can be stored in tanks or lagoons until spread.
2. Disposal of liquid manure can be postponed to fit field schedules, soil conditions, and expected rainfall, provided the storage unit is adequate.
3. Objectionable odors, fly problems, and unsightliness are minimized when wastes are stored as a liquid in a covered storage.

Handling liquid manure requires:

1. Scrapers, gutters, slotted floors, or drains to move the wastes into storage.
2. A watertight storage unit to which water can be added.
3. Pumps, agitators, and augers to stir and remove the liquid manure.
4. A tank truck or wagon, and land on which to spread the manure.

• **Lagoons—A lagoon is a body of water or pond into which liquid manure is discharged, where it is digested by bacterial action.** Most of the fertility value of the manure is wasted, but the savings in equipment and labor may offset the loss. The use of a lagoon gets the excrement out of the building or off the floor and out of the way, but does not conserve its fertility value.

• **Oxidation ditch—An oxidation ditch is a storage unit, oxygenated to promote aerobic bacterial digestion.** Although an oxidation ditch may control odor, the effluent is still a potential pollutant.

• **Settling basin—A settling (or debris) basin is a separating and holding unit.** It is usually a part of the surface runoff control system. Liquids enter the basin where they slow down and drop their undissolved solids. The liquids are drained off, leaving the solids to dry for removal and field spreading. Settling basins are usually smaller and much shallower than a lagoon—it is intended that they dry out.

• **Consider fish farming**—In China, some of the communes use hog manure in fish farming.

In China, such hog-fish combinations (1) provide pollution control of pig manure, and (2) recycle and use feed twice—first through hogs, and second through fish. The hogs are conveniently located, adjacent to the fish ponds. So, they flush the manure from the pig pens directly into the fish ponds.

Our major concerns about the hog-fish combination are: (1) that the excess nitrates and nitrites from manure may change the plant life of the water so that it may be undesirable for fish, and (2) that oxyge starvation of the fish may ensue. It appears that tl Chinese have achieved a proper "balance" and a leviated these hazards.

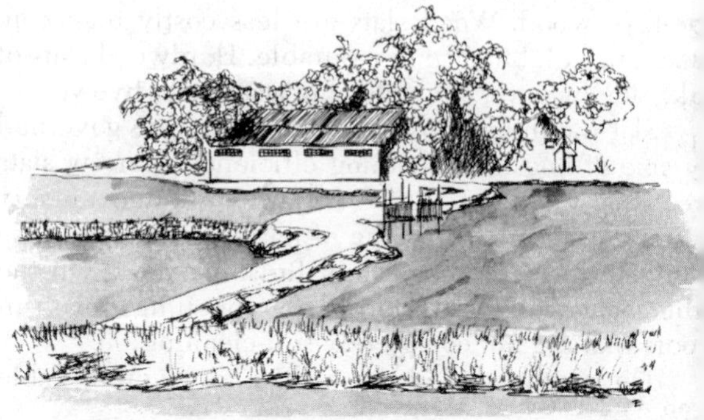

Fig. 9-35. A hog-fish combination in China, on Kwang Li Peopl Commune, Kaw Yao County, Kwang Tung Province, where, in 1971, th produced 49,700 pigs and had a fish catch of 863,000 lb. The fish we fed entirely, and solely, on pig manure flushed from the hog barns direc into the fish ponds, plus grass clippings from the pond bunks.

Swine Equipment

Figs. 9-36 to 9-40 show some of the most commc articles of swine equipment, along with the usual d signs.

Detailed plans and specifications for swir equipment can usually be obtained through the loc county agricultural agent, vocational agriculture i structor, lumber dealer, or through writing to the cc lege of agriculture of the state.

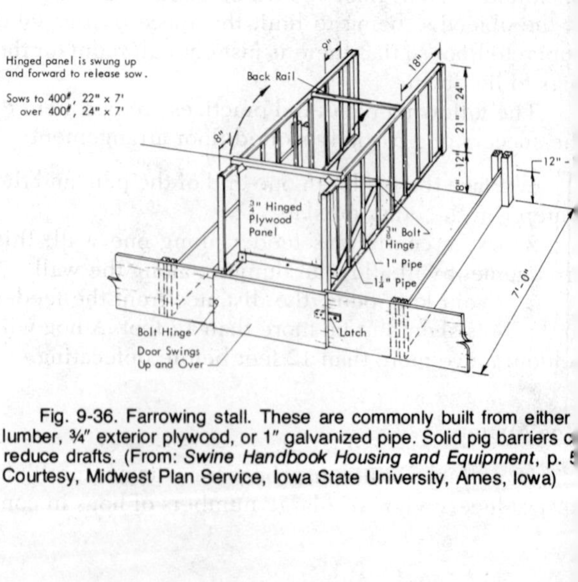

Fig. 9-36. Farrowing stall. These are commonly built from either lumber, ¾" exterior plywood, or 1" galvanized pipe. Solid pig barriers c reduce drafts. (From: *Swine Handbook Housing and Equipment*, p. Courtesy, Midwest Plan Service, Iowa State University, Ames, Iowa)

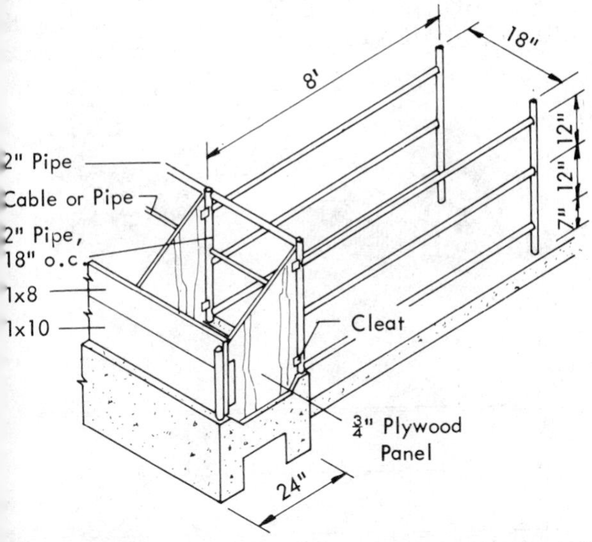

Fig. 9-37. Feeding stall. These stalls should be about 18″ wide and 8′ long. Long narrow stalls discourage the fast eater from bothering slower eaters. If shorter stalls are used, a gate or other device should be provided to keep the sows in the stalls until all of them have finished eating. (From: *Swine Handbook Housing and Equipment*, p. 62. Courtesy, Midwest Plan Service, Iowa State University, Ames, Iowa)

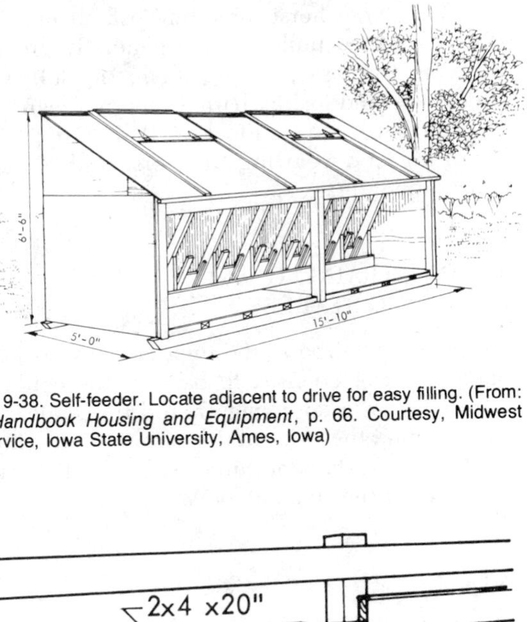

Fig. 9-38. Self-feeder. Locate adjacent to drive for easy filling. (From: *Swine Handbook Housing and Equipment*, p. 66. Courtesy, Midwest Plan Service, Iowa State University, Ames, Iowa)

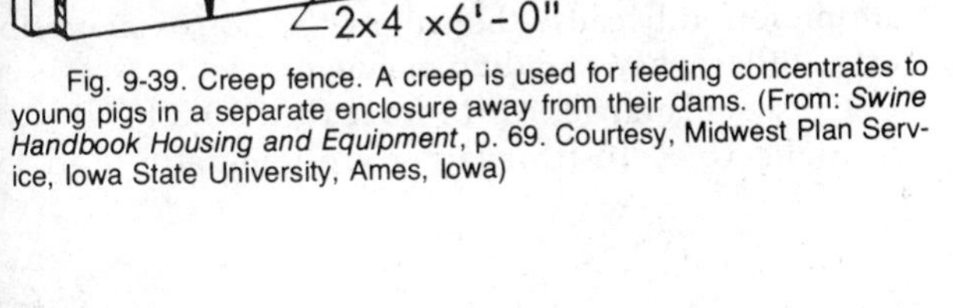

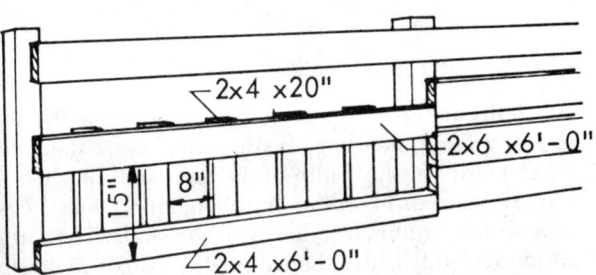

Fig. 9-39. Creep fence. A creep is used for feeding concentrates to young pigs in a separate enclosure away from their dams. (From: *Swine Handbook Housing and Equipment*, p. 69. Courtesy, Midwest Plan Service, Iowa State University, Ames, Iowa)

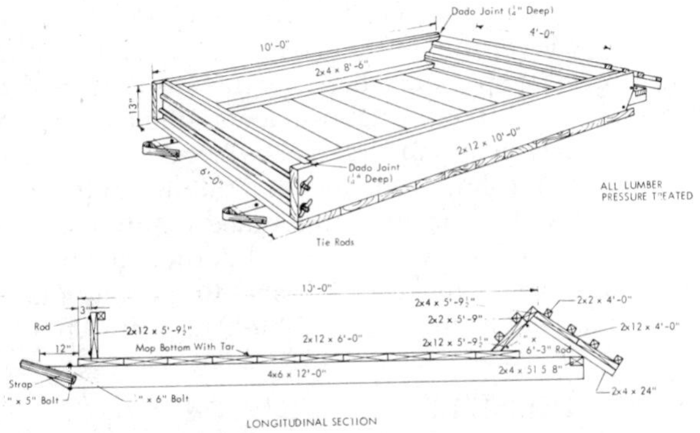

Fig. 9-40. Hog wallow. Allow 1 wallow of the dimensions shown for about 8 market hogs. Either move the wallow frequently to avoid mudholes, or provide a concrete platform sloped to drain outside the lot. Fill wallows only 2″ × 3″ deep to minimize slopping. (From: *Swine Handbook Housing and Equipment*, p. 75. Courtesy, Midwest Plan Service, Iowa State University, Ames, Iowa)

HORSE BUILDINGS AND EQUIPMENT

Adequate and well-designed buildings and equipment make horse care easier and add to the personal satisfaction of the owner and caretaker.

Two of the most important considerations in planning horse facilities are health and safety—not only of the horses, but also of the people who either come in contact with the animals and facilities or who live nearby. Other important considerations include sound construction, laborsaving conveniences, and building style in harmony with the surroundings.

The same basic principles are involved when selecting and equipping a barn to stable only one or two animals as for a larger number.

It is recognized that the design of both the buildings and equipment used for light horses is likely to be dominated by the fads and fancies of the owner. Figs. 9-41 to 9-51 suggest designs of some buildings and equipment for light horses.

Horse Barns

The horse barn, whether large or small, should be well planned, durable, and attractive. Basically, its purposes are to provide an environment that protects the horses from temperature extremes, keeps them dry and out of the wind, eliminates drafts, provides fresh air in both winter and summer, and protects them from injury.

Ample space should be provided for the well-being of the horses, and for the convenience, safety, and enjoyment of the people who care for and use them.

The needs for housing horses and storing materials vary according to the intended use of the building. Broadly speaking, horse barns are designed to serve (1) small horse establishments—the owner with one to a few head, (2) large horse breeding establishments, or (3) riding academies and training and boarding stables. A summary of the space requirements of buildings for horses is presented earlier in this section under the heading "Space Requirements of Buildings and Equipment," Table 9-4, page 806.

SMALL HORSE ESTABLISHMENTS

When 1 or 2 riding horses or ponies are kept, they are usually stabled close to the house, which makes for greater convenience in their care and use. In most cases, box stalls are built in a row and provision is made for limited feed, bedding, and tack storage; usually a combination feed and tack room for units with 1 to 2 stalls, and separate feed and tack rooms with 3 or more stalls. Fig. 9-41 shows an attractive small barn for 2 horses; a barn which was designed by the author for the U.S. Department of Agriculture.

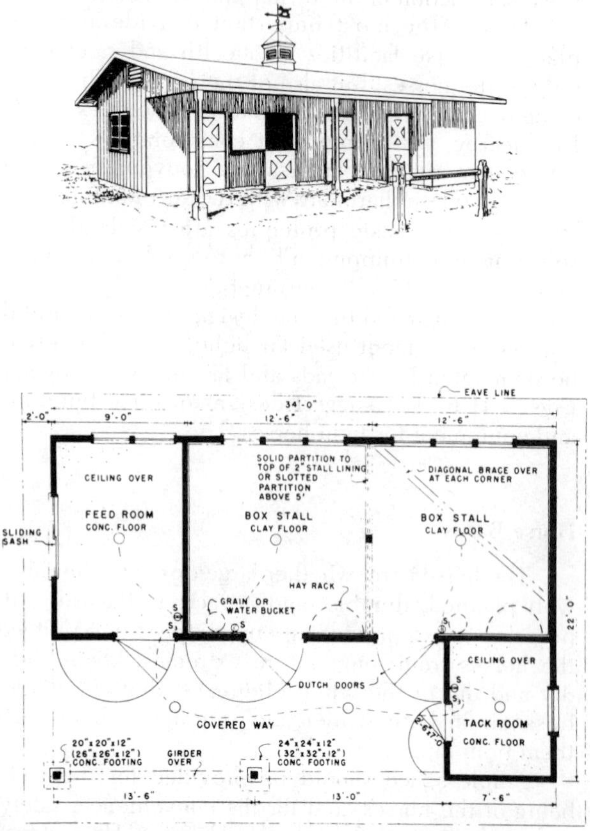

Fig. 9-41. Riding horse barn above and floor plan below. Barn has two box stalls, a feed room, and a tack room.

LARGE HORSE BREEDING ESTABLISHMENTS

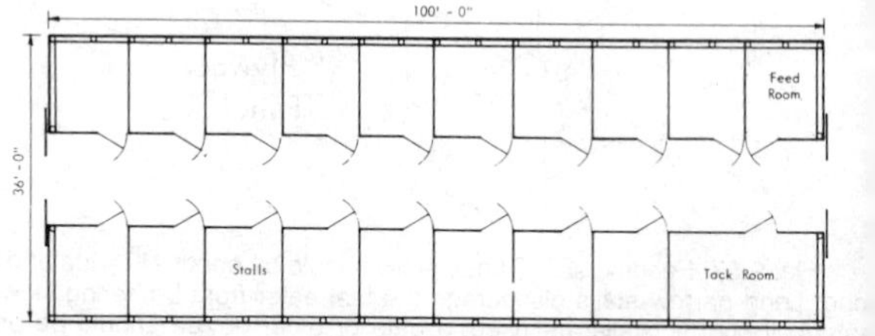

Fig. 9-42. A 17-stall horse barn; 36′ × 100′ in size, with a 20′ × 2 tack room, a 10′ × 12′ feed room, and a 12′ alley. (From: *Horse Han book Housing and Equipment*, p. 56. Courtesy, Midwest Plan Servic Iowa State University, Ames, Iowa)

With large horse breeding establishments, spe cially designed buildings are generally provided fo different purposes; among them, the following: (1 broodmare and foaling barn, (2) barren mare barn, (3 stallion barn and paddocks, (4) breeding shed, (5 weanling and yearling quarters, and (6) isolatio (quarantine) quarters.

RIDING, TRAINING, AND BOARDING STABLES

For this purpose, the quarters may consist of (1 stalls constructed back to back in the center of th barn with an indoor ring around the stalls, (2) stall built around the sides of the barn with the ring in th center, or (3) stalls on either side of a hallway or a leyway and the ring outdoors.

STALLS

Stalls are of two general types: (1) box stalls; an (2) tie, straight, standing, or slip stalls. As tie stalls di fer primarily in the width of the area and their use i less common in breeding establishments, the discus sion will be confined to loose or box stalls. The latte are preferred because they allow the horses more lib erty, either when standing or lying down.

Box stalls range in size from 10 × 10 ft to 16 × 1 ft. Stalls 16 × 16 ft or larger are generally used fo foaling mares. Tie stalls are usually 5 ft wide and 9 to 12 ft long. The length of a tie stall should be measured from the front of the manger or grain box to the

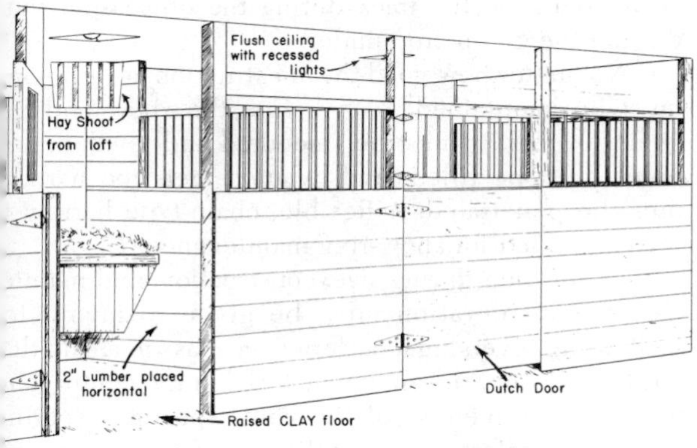

Fig. 9-43. A satisfactory type of box stall for horses.

rear of the stall partition. (See section headed, "Space Requirements of Buildings and Equipment" for stable specifications, Table 9-4, page 806.)

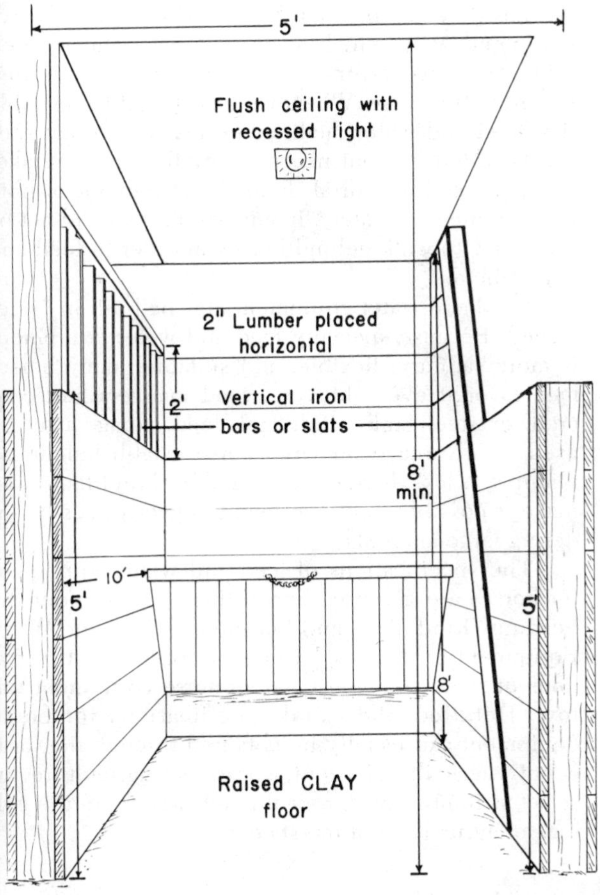

Fig. 9-44. A satisfactory type of tie stall for horses.

Adequate quarters for a horse should be (1) ample in size and height for the particular type of animal; (2) properly finished and without projections; (3) dry with good footing; (4) equipped with suitable doors; (5) provided with ample windows for proper lighting; (6) well ventilated; (7) cool in summer and warm in winter; (8) equipped with suitable mangers, grain containers, watering facilities, and mineral boxes; and (9) easy to keep clean.

STALL FLOOR

A raised clay floor covered with a good absorbent bedding, with proper drainage away from the building, is the most satisfactory flooring for horse stables. Clay floors are noiseless and springy, keep the hoofs moist, and afford firm natural footing unless wet; but they are difficult to keep clean and level. To lessen the latter problems, the top layer should be removed each year, replaced with fresh clay, and leveled. Also, a semicircular concrete apron extending into each stall at the doorway will prevent horses from digging a hole in a clay floor at this point. This arrangement is particularly desirable in barns for yearlings, as they are likely to fret around the door.

Rough wooden floors furnish good traction for animals and are warm to lie upon; but they are absorbent and unsanitary, they often harbor rats and other rodents, and they lack durability.

Concrete, asphalt, or brick floors are durable, impervious to moisture, easily cleaned, and sanitary; but they are rigid and without resilient qualities, slippery when wet, hazardous to horses, and cold to lie upon. It is noteworthy that concrete and asphalt, generously covered with bedding, are widely used for stable floors throughout eastern and western Europe.

There is great need for an improved stall floor covering material for horses—one which will lessen (1) the amount of bedding needed, and (2) the labor and drudgery of cleaning.

TACK ROOM

A tack room is an essential part of any barn. With one or two stall units, a combination tack and feed room is usually used, for practical reasons. On large establishments, the tack room is frequently the showplace of the stable. As such, the owner takes great pride in its equipment and arrangement. Also, depending upon the use of the horses, the tack room takes on an air and personality that represents the horses in the stalls.

Generally, the tack rooms in stables where there are American Saddle Horses, harness horses, or hunters are rather formal. In the vernacular of the "horsey set," they maintain the "Boston touch." On the other hand, where western-style riding prevails, generally the formal tack rooms have been replaced by more practical, simple rooms.

Tack rooms should be floored, rodent-proof and bird-proof, and ceiled over.

Shades

A shade, either trees or man-made, should be provided for horses that are in the hot sun.

The most satisfactory man-made horse shades are (1) oriented with a north-south placement, (2) at least 12 to 15 feet in height (in addition to being cooler, high shades allow a mounted rider to pass under), and (3) open all around.

Show-Ring

There are no standard specifications relative to size, type of construction, and maintenance of show-rings. Yet, all the better rings meet certain basics.

The National Horse Shows Association recommends the following ring sizes: indoor ring, 110 × 220 ft; outdoor ring, 120 × 240 ft. It is recognized, however, that many good show-rings are either smaller or larger than these dimensions.

In addition to ring size, consideration must be given to proper footing—to achieving resilience, yet firmness and freedom from dust. With an outdoor ring, establishing proper drainage and constructing a good track base are requisite to all-weather use. Drainage is usually secured by (1) locating the ring so that it is high, with the runoff away from it, and (2) installing a perforated steel pipe (with the perforations toward the bottom side), or drainage tile, underneath the track if necessary.

Resilience, with firmness, is usually secured by mixing organic matter with dirt or sand. For example, the entire ring at the Spanish Riding School is covered with a mixture of ⅔ sawdust and ⅓ sand, which is sprinkled at intervals to keep the dust down.

In many indoor rings of the United States, 6 to 8 in. of tanbark on a dirt base are used. Unless tanbark is wetted down at frequent intervals, it tends to pulverize and give poor footing. Others mix shavings and/or sawdust with dirt or sand to obtain a covering of 18 to 24 in. of the material. One good ring with which the author is familiar was prepared by laying down 9 in. of wood shavings, 2 in. of sawdust, and 4 in. of sand—all of which were mixed together, then oiled. Still others add a bit of salt, because it holds moisture when wetted down, thereby minimizin dust.

For outdoor rings, needed organic matter for re silience is sometimes secured by seeding rye, or othe small grain, on the track during the off-season, the discing the green crop under.

No matter how good the construction, a show-rin must be maintained, both before the show and be tween events. It must be smoothed and leveled, an holes must be filled; and, when it gets too hard, i must be penetrated. A flexible, chain-type harrow i recommended for show-ring maintenance.

In addition to ring size, construction, and mainte nance, consideration must be given to layout fo facilitating reversing a performance class in a ring tha has turf or other decorative material in the center; an to the attractiveness of the ring; spectator seatin capacity, comfort, and visibility; nearby parking; an handling the crowd.

Horse Equipment

Although the design of horse equipment is likel to be dominated by the fads and fancies of the owne the basic needs are merely for simple but effectiv equipment with which to provide hay, concentrates minerals, and water—without waste, and withou hazard to the horse. Whenever possible, it is desirabl that feed and water facilities be located so that the can be filled without necessitating that the caretake enter the stall or corral, from the standpoint of bot convenience and safety. In any event, it should not b necessary to walk behind horses in order to feed an water them.

Feed and water equipment may be built in or de tached. Because specialty feed and water equipmen is more sanitary, flexible, and suitable, many horse men favor it over old-style wood mangers and con crete or steel tanks. Bulk-tank feed storage may b used to advantage on large horse establishments t eliminate sacks, lessen rodent and bird problems, an make it possible to obtain feed at lower prices by or dering large amounts.

The specifications of feed and water equipmen for horses are given earlier in this section under th heading "Kind, Size, and Location of Feed and Wate Equipment for Horses," Table 9-5, page 807. Illustra tions of some common types of horse equipment fol low. Detailed plans and specifications for hors equipment can usually be obtained through the loca county agricultural agent, vocational agriculture in structor, lumber dealer, or through writing to the col lege of agriculture of the state.

Fig. 9-45. Hay manger (corner) and mineral box (to right of manger). Note door opening from the feed room directly over the hay manger; a convenient arrangement permitting feeding the hay without entering the stall. (Drawing by Steve Allured)

Fig. 9-47. Horse watering tank set in the fence so that there are no protruding corners that might cause injury.

Fig. 9-46. A feed bucket. (Drawing by Steve Allured)

Fig. 9-48. A type of automatic waterer, of which there are several commercially manufactured ones. This equipment may be placed to one side of the manger in a stall.

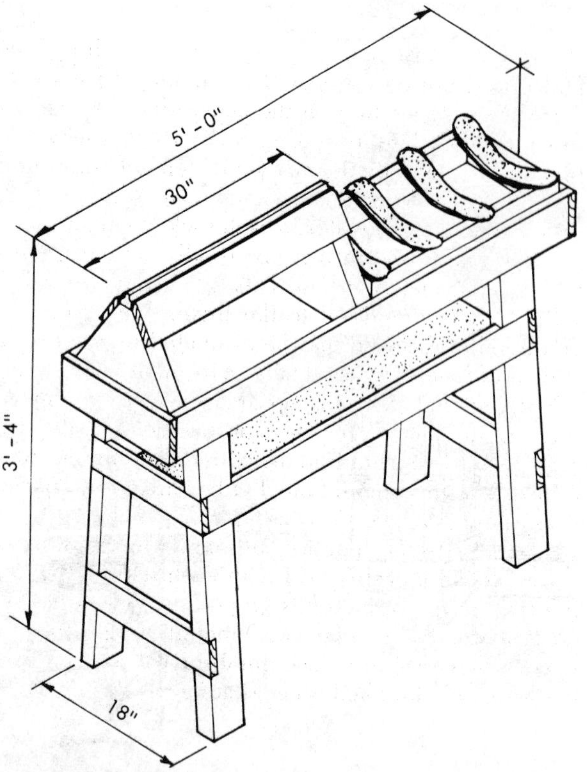

Fig. 9-49. Saddle cleaning rack. (From: *Horse Handbook Housing and Equipment*, p. 30. Courtesy, Midwest Plan Service, Iowa State University, Ames, Iowa)

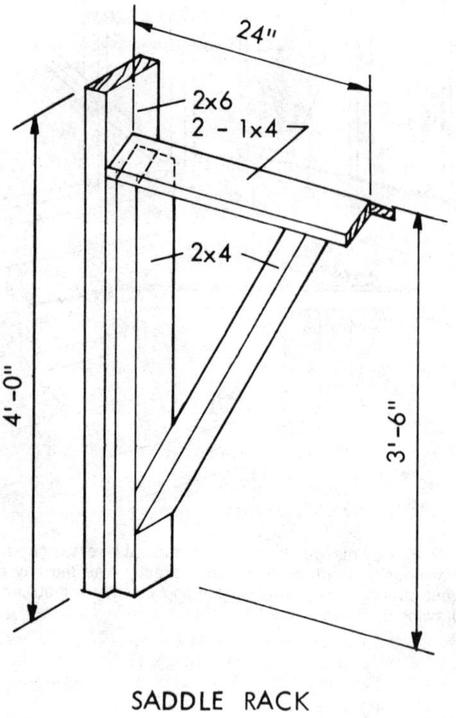

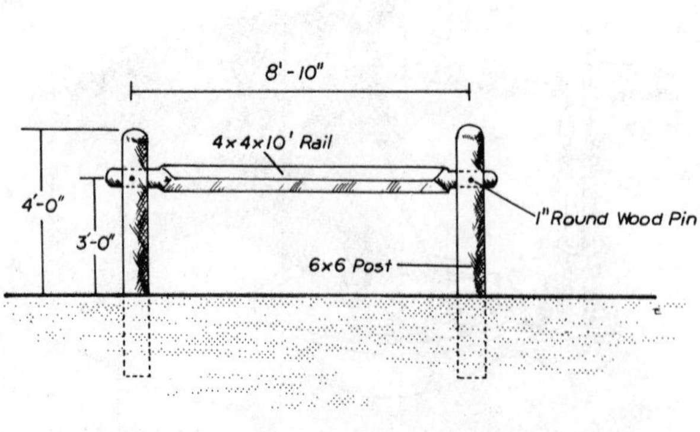

Fig. 9-50. Hitching rail. After shaping and prior to installing, the rail and posts should be treated with a preservative.

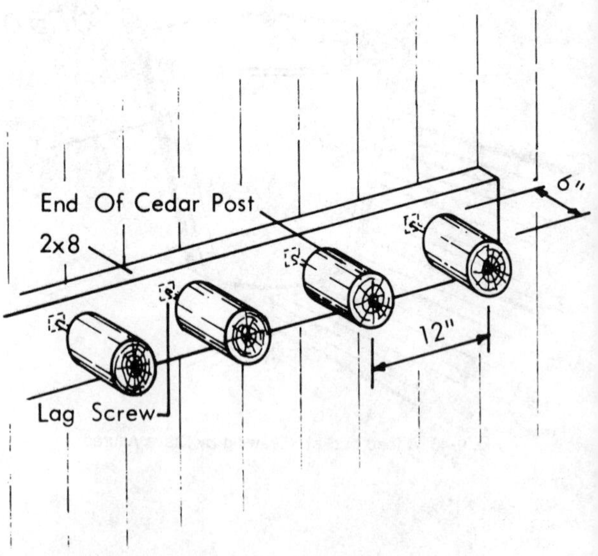

Fig. 9-51. Tack room equipment. Top: saddle racks. Bottom: bridle racks. (From: *Horse Handbook Housing and Equipment*, p. 29. Courtesy, Midwest Plan Service, Iowa State University, Ames, Iowa)

CONCRETE STRUCTURES FOR THE FARM AND RANCH[7]

Progressive stockmen are making increased use of concrete. Although they usually either contract the job or hire experienced help where major structures are involved, most farmers and ranchers do their own concrete work on the smaller and simpler jobs.

Selecting the Mix for the Job

Concrete mixes are often designated by volume proportion, such as 1:2¼:3; the figures from left to right referring to the proportions or parts of cement, sand, and gravel—all measured by the same unit of volume. Such designation is incomplete, however, because it leaves unspecified the amount of one of the most important ingredients—water. In fact, the amount of water and the amount of cement determine the quality of the concrete. Aggregates such as sand and gravel are only fillers. The amounts of these to use varies, depending on the consistency of the mix desired. Therefore, in mixing concrete, it is essential to measure carefully the amount of cement and water. For most farm jobs, the recommended water-cement ratio is 6 gallons of water to each sack of cement.

The sand used in the mix usually contains moisture, which means that less than 6 gallons of water will actually be added to the mixer. The moisture condition of the sand may be evaluated by squeezing it in your hand. Dry sand falls apart; moist sand forms a ball; and wet sand glistens and leaves excess moisture on the hand. After the sand has been evaluated, refer to Table 9-24 to find the correct amount of water to use. Use this amount only! Other mixes for special uses are also shown in Table 9-24.

For half-sack batches, use half the amount of water shown in Table 9-24 with each half-sack of cement. Always maintain this proportion of water to cement and vary the aggregate amounts to get the desired mix workability. The durability of the concrete decreases if additional water is added.

[7]In the preparation of this section, invaluable assistance was given by the Portland Cement Association, Skokie, Ill. Also, the Portland Cement Association provided much of the material presented herein.

TABLE 9-24

WATER-CEMENT GUIDE; AND SUGGESTED TRIAL MIXES

Type of Job	Total Amount of Water to Each Sack of Cement	Maximum Size Aggregate	Amount of Water to Use at the Mixer per Sack of Cement When Sand Is— Damp[1]	Wet[2] (average)	Very Wet[3]	Suggested Mixture for 1-sack Trial Batches[4] Cement Sacks	Aggregates Fine	Coarse
	(gal)	*(in.)*	*(gal)*	*(gal)*	*(gal)*	*(cu ft)*	*(cu ft)*	*(cu ft)*
Concrete subject to severe wear, weather, or weak acid and alkali solutions; such as milk coolers, tanks, creamery floors, etc.	5	¾	4½	4	3½	1	2	2¼
Most farm concrete jobs, such as floors, steps, walks, driveways, septic tanks, dairy and swine barn floors, silos, manure pits, water tanks, basement walls, etc.	6 6	1 1½	5½ 5½	5 5	4¼ 4¼	1 1	2¼ 2½	3 3½
Concrete in thick sections and not subject to freezing, such as thick footings, foundations, walls, engine bases, etc.	7	—	6¼	5½	4¾	1	3	4

[1]Damp describes sand that will fall apart after being squeezed in the palm of the hand.
[2]Wet describes sand that will ball in the hand when squeezed but leaves no moisture on the palm.
[3]Very wet describes sand that has been subjected to a recent rain or recently pumped. Balling a sample in the hand will leave moisture on the palm and the sand-gravel glistens in the light.
[4]Mix proportions will vary slightly depending on gradation of aggregate.

How to Estimate Amount of Concrete Needed

When a farmer or rancher is doing his own concrete work, he needs to know how to figure quantities of materials needed. With this information at hand, he can purchase adequate materials, and yet avoid excesses and wastage.

For convenience in estimating material needs, the following guides are presented:

Table 9-24, *Water-Cement Guide; and Suggested Trial Mixes*. This shows the recommended water-cement ratio for use with sand of varying amounts of moisture, and gives suggested trial mixes.

Table 9-25, *Guide for Estimating Quantity of Concrete Needed for 100 Sq Ft of Slabs of Various Thicknesses*. This is a convenient table for estimating quantities for paved corrals, barn floors, sidewalks, etc.

TABLE 9-25

GUIDE FOR ESTIMATING QUANTITY OF CONCRETE NEEDED FOR 100 SQ FT OF SLABS OF VARIOUS THICKNESSES[1]

Thickness of Concrete	Amount of Concrete
(in.)	*(cu yd)*[2]
3	0.92
4	1.24
5	1.56
6	1.85
8	2.46
10	3.08
12	3.70

[1]In the figures presented, no allowance is made for waste and variables in work. To provide for these, it is good practice to increase material quantities by 5 to 10%.

[2]If it is desired to convert cubic yards to cubic feet, simply multiply by 27.

Table 9-26, *Guide for Estimating Amounts of Materials Needed for a Cubic Yard of Concrete*. This table may be used as a buying guide.

The following information may also be helpful in figuring materials in concrete:

Twenty-seven cu ft equals 1 cu yd.
One sack of cement equals 1 cu ft, and weighs 94 lb.
One ton of sand contains about 0.82 cu yd or 22 cu ft.
One ton of gravel contains about ¾ cu yd or 20 cu ft.
One gal of water weighs 8.33 lb.
One cu ft of water weighs 62.4 lb.

The unit of measure for concrete is the cubic yard, which contains 27 cubic feet. To determine the amount of concrete needed, find the volume in cubic feet of the area to be concreted and divide this figure by 27. The following formula can be used to determine the amount of concrete needed for any square or rectangular area:

$$\frac{\text{Width, ft} \times \text{Length, ft} \times \text{Thickness, ft}^{8}}{27} = \text{Cubic yards}$$

For example, a 4-inch thick floor for a 30 × 90-foot building would require:

$$\frac{30 \times 90 \times 0.33}{27} = 33.00 \text{ cubic yards of concrete}$$

The amount of concrete determined by the above formula does not allow for waste or slight variations in concrete thickness. An additional 5 to 10 percent will be needed to cover waste and other unforeseen factors.

[8]The thickness dimension must be changed to feet or parts of a foot. The decimal part of a foot was used in the example. However, the fractional part may be used instead. Table 9-27 gives both the fractional and decimal parts of a foot for several common thickness dimensions.

TABLE 9-26

GUIDE FOR ESTIMATING AMOUNTS OF MATERIALS NEEDED FOR A CUBIC YARD OF CONCRETE

Suggested mixture for 1-sack trial batches[1]			Materials per cu yd of concrete				
	Aggregates			Aggregates			
Cement sacks	Fine	Coarse	Cement	Fine		Coarse	
(cu ft)	*(cu ft)*	*(cu ft)*	*(sacks)*	*(cu ft)*	*(lb)*	*(cu ft)*	*(lb)*
With ¾-in. maximum size aggregate							
1	2	2¼	7¾	17	1550	19½	1950
With 1-in. maximum size aggregate							
1	2¼	3	6¼	15.5	1400	21	2100
With 1½-in. maximum size aggregate							
1	2½	3½	6	16.5	1500	23	2300
With 1½-in. maximum size aggregate							
1	3	4	5	16.5	1500	22	2200

[1]Mix proportions will vary slightly depending on gradation of aggregate.

How to Determine the Amount of Materials to Order

Table 9-26 gives the number of sacks of cement and the amount of aggregate (in cubic feet and in pounds) needed to produce a cubic yard of concrete for water-cement ratios of 5, 6, and 7 gallons of water per sack of cement (with the amount of water per sack of cement determined by the type of job and the dampness of the sand, as tabulated in Table 9-24). Since it is generally impossible to recover all of the aggregate, a 10 percent allowance has been included to cover normal wastage.

TABLE 9-27

HOW TO CHANGE THICKNESS IN INCHES TO FRACTIONS AND DECIMAL PARTS OF A FOOT FOR USE IN CALCULATING QUANTITIES OF CONCRETE

Inches	Fractional part of foot	Decimal part of foot
4	4/12 or 1/3	0.33
5	5/12	0.42
6	6/12 or 1/2	0.50
7	7/12	0.58
8	8/12 or 2/3	0.67
10	10/12 or 5/6	0.83
12	1	1.00

How to Make Quality Concrete

High-quality concrete is made by (1) selecting suitable materials (cement, aggregate, and water); (2) thoroughly mixing them together in the right proportions; and (3) correctly placing, finishing, and curing the resulting mix. To this end, the following simple rules should be followed:

1. **Water**—The water used in mixing concrete should be clean enough to drink.
2. **Sand**—The sand should be clean, hard, and well graded.
3. **Gravel**—The gravel (or rock or slag) should be clean and hard and range in size below the maximum specified for the particular kind of work.
4. **Cement**—The cement should be free from hard lumps due to dampness (lumps which do not readily pulverize when squeezed in one's hand).
5. **Measuring**—Measure water and cement for each mix. It is especially important that no more water be used than is indicated in Table 9-24.

Water may be measured in a pail marked off in gallons and half gallons. Cement is measured by shovelfuls or by dividing the sack in the desired proportions.

6. **Mixing**—For machine-mixing, allow one to two minutes after all materials are in the mixer.

For hand-mixing, proceed as follows: (a) place the sand on a watertight mixing platform; (b) spread the cement evenly over the sand and turn the two materials with a shovel until they are thoroughly mixed, as evidenced by the uniform color; (c) spread the mixture out evenly, add the gravel, and again mix thoroughly; and (d) form a hollow in the material, slowly add the measured quantity of water, and mix until every particle has been completely covered with cement paste. Hold rigidly to the predetermined amounts of water and cement; then vary the amounts of sand and gravel according to the consistency of the mix desired.

7. **Ready-mixed concrete**—When ordering ready-mixed concrete for average farm jobs, specify a mix based on Table 9-28.
8. **Workable mix**—Table 9-24 gives the proportions of water to cement to use. Aggregates are used until the mix is workable (one which is smooth and plastic, and will place and finish well). For the first trial batch, use the proportions of sand to gravel shown in Table 9-24. If this batch is too stony, use more sand and less gravel in the next batch. If it is oversanded, reduce the amount of sand and increase the amount of gravel until the desired consistency is obtained. Remember: always adjust the mix by varying the amounts of aggregates.
9. **Forms**—Forms, constructed to conform to the desired shape, should be rigid, tight, and well-braced. Then they should be oiled so that they can be easily removed.
10. **Placing**—Freshly mixed concrete should be placed in 6- to 12-inch layers into the forms and then tamped, spaded, or vibrated enough to eliminate air pockets.
11. **Construction joints**—Where there is some delay in completing a concrete job (even from one day's run to another), (a) roughen the surface with a stiff broom before it hardens, and (b) wet the surface and cover it with a layer of cement mortar ½ inch thick before adding concrete.
12. **Cold weather**—During cold weather, the water, sand, and gravel should be heated before mixing, and the new concrete should be protected from freezing for at least three days. Do not deposit concrete on frozen ground or in forms containing frost or ice.
13. **Finishing**—Newly placed concrete should first be leveled off with a strikeboard or wood float and then allowed to stiffen before finishing in final form. For a gritty, nonskid floor (such as is best for barn floors and driveways), finish with a wood float; for a still rougher floor (such as is best for concrete corrals), finish with a broom. For a smooth, dense surface (as is best for feed mangers, poultry house floors, and basement floors), finish with a steel trowel.
14. **Curing**—For proper curing, concrete needs moisture, and should be protected from drying out for at least 5 days. Depending on the type of structure, cover with canvas, burlap, earth, or straw, and keep wet for the required time.
15. **Reinforcing**—Where an important structure is

involved, the advice of a competent architect or engineer should be sought in the design and installation of reinforcing. It is best to use regular reinforcing rods or mesh. Never use scrap iron or rusty fence wire for reinforcement.

16. **Dry floors**—For a dry concrete floor (such as is desired in hog houses, grain storage buildings, etc.), (a) select a well-drained site; (b) provide a 6- to 12-inch well-tamped fill of gravel or crushed rock; (c) place a moisture barrier of 55-pound asphalt roll roofing, tough waterproof paper, or 6 mil plastic film over it, lapping the points 6 inches; and (d) place the concrete slab.

TABLE 9-28

GUIDE FOR ORDERING READY-MIXED CONCRETE

WHEN ORDERING CONCRETE FOR					
FLAT WORK (using 1½-in. maximum size aggregate)			FORMED WORK (using ¾-in. maximum size aggregate)		
Severe Exposure	Normal Exposure	Mild Exposure	Severe Exposure	Normal Exposure	Mild Exposure
Garbage feeding floors, floors in dairy plants	Paved barnyards, floors for farm buildings, sidewalks	Footings, concrete improvements in mild climates	Mangers for silage feeding, manure pits	Reinforced concrete walls, beams, tanks, foundations	Concrete improvements in mild climates
SPECIFY cement content: minimum number of sacks per cu yd of concrete					
7	6	5	7¾	6½	5½
water content (includes water contained in aggregates): maximum gal per sack of cement					
5	6	7	5	6	7
GET medium-consistency concrete (3-in. slump)					

FENCES FOR LIVESTOCK[9]

Confinement of animals is not new. The pioneers planted trees and shrubs, such as osage orange, or built rail or stone fences to hold their livestock. Then, in 1873, Joseph F. Glidden of DeKalb, Illinois, invented barbed wire, or the "devil's rope," as it was dubbed by those who considered the barbs inhumane or who opposed its use because it marked the beginning of the end of the open range and free grazing.

Good fences (1) maintain farm boundaries, (2) make livestock operations possible, (3) reduce losses to both animals and crops, (4) increase land values, (5) promote better relationships between neighbors, (6) lessen accidents from animals getting on roads, and (7) add to the attractiveness and distinctiveness of the premises.

Fence Specifications

Except for corrals, woven and/or barbed wire fences are used for most classes of livestock. Frequently, special materials are used in horse fences. Fence construction specifications for different classes of livestock, and different materials, are given in Tables 9-29, 9-30, and 9-31.

Wire Fences

The discussion which follows will be limited primarily to wire fencing, although it is recognized that such materials as rails, poles, boards, stone, and hedge have a place and are used under certain circumstances. Also, where there is a heavy concentration of animals, such as in corrals and in feed yards, there is need for a more rigid type of fencing material than wire. Moreover, certain fencing materials have more artistic appeal than others.

[9]In the preparation of this section, the author had the benefit of the authoritative review and helpful suggestions of the following: Professor Henry Giese, Department of Agricultural Engineering, Iowa State University, Ames, Iowa; Dr. Richard E. Phillips, Extension Agricultural Engineer, University of Missouri, Columbia, Mo.; Mr. Harold L. Coons, Keystone Steel and Wire, Peoria, Ill.; and Dr. Donald W. Bates, Extension Agricultural Engineer, University of Minnesota, St. Paul, Minn.

TABLE 9-29

WOVEN WIRE FENCE CHART

Kind of Stock	Recommended Woven Wire Height	Recommended Weight of Stay Wire	Recommended Mesh or Spacing Between Stays	Recommended Number of Strands of Barbed Wire to Add to Woven Wire[1]	Comments
	(in.)	(gauge)	(in.)		
Cattle	47, 48, or 55	9 or 11	12	1 strand 2″ to 3″ above top of woven wire, with points 4″ or 5″ apart, to prevent animals from breaking down woven wire.	Also satisfactory for all farm animals, except young pigs. Fences for cattle feedlots should be constructed of wood, cable, pipe, or other strong material, and should be 60″ high.
Sheep	32	11 or 12½	12	2 strands on top.	Sheep fences should total 39″ in height. Twelve-in. mesh is best for sheep as they will not get their heads caught if they attempt to reach through. With a heavy concentration of feeder lambs, use wooden fence 39″ high.
Swine	26, 32, or 39	9 or 11	6	1 strand on bottom.	Barbed wire on bottom prevents rooting under.
Horses (also see Table 9-31, Horse Fence Chart)	55 or 58	9 or 11	12	1 strand on top; with points 4″ or 5″ apart.	Also satisfactory for all farm animals except young pigs. Cyclone, wood, pole, or other durable and attractive materials are usually used around the headquarters.
All farm animals	26 or	9 or 11	6	3 strands on top; 1 strand on bottom.	
	32	9 or 11	6	2 strands on top; 1 strand on bottom.	

[1]The American Society of Agricultural Engineers' standard for barbed wire calls for 4 in. spacing with 2-point wire and 5 in. spacing with 4-point wire.

TABLE 9-30

BARBED WIRE FENCE CHART[1]

Kind of Stock	Recommended Number of Points	Recommended Spacing Between Points	Recommended Weight of Strands	Recommended No. of Lines of Barbed Wire to Install	Comments
		(in.)	(gauge)		
Cattle or horses; in farm pastures	2 or 4	4 or 5	12½	4-5	Two-point barbs are 4 in. apart; 4-point are 5 in. apart.
Cattle or horses; on the range	2 or 4	4 or 5	12½	2 or 4	Not all animals will be restrained by 2 or 3 strands.
Sheep	Barbed wire is not considered suitable for sheep because it tears the fleece.				
Swine	2 or 4	4 or 5	12½	6	A 6-strand barbed wire fence for swine may cost more to build and maintain than a woven wire fence.

[1]The American Society of Agricultural Engineers' standard for barbed wire calls for 4 in. spacing with 2-point wire and 5 in. spacing with 4-point wire.

HOW TO SELECT WIRE

The kind of wire to purchase should be determined primarily by the class of animals to be confined.

Tables 9-29, 9-30, and 9-31 are suggested guides.

Twenty-six- or thirty-two-inch woven wire, with or without barbed wire, is frequently used as movable fence for hogs, especially when hogging down crops.

The following additional points are pertinent in the selection of wire:

1. **Styles of woven wire**—The standard styles of woven wire fences are designated by numbers as 958, 1155, 849, 1047, 741, 939, 832, and 726.

The first one or two digits represent the number of line (horizontal) wires; the last two, the height in inches; i.e., 1155 has 11 horizontal wires and is 55 in. in height. Each style can be obtained in either (a) 12-in. spacing of stays (or mesh), or (b) 6-in. spacing of stays. Also, a special 2-in. mesh is available for horses.

2. **Mesh**—Generally, a close-spaced fence with

TABLE 9-31

HORSE FENCE CHART

Material	Material Specifications		Construction Details			Comments
	Post	Line Fence or Rails	Fence Height	Number and Spacing Rails or Mesh of Wire	Distance Between Post on Centers	
Steel or Aluminum Rail	7½′ 7½′ 8½′	10′ or 20′ rail 10′ or 20′ rail 10′ or 20′ rail	60″ 60″ 72″	3 rails; 20″ centers 4 rails; 15″ centers 4 rails; 18″ centers	10′ 10′ 10′	Because of the strength of most metal rails, fewer rails and posts are necessary than where wood is used.
Board	7½′, 4″-8″ diameter 8½′, 4″-8″ diameter	2″ × 6″, or 2″ × 8″ 2″ × 6″, or 2″ × 8″	60″ 72″	4 boards 5 boards	8′ 8′	
Poles	7½′, 4″-8″ diameter 8½′, 4″-8″ diameter	4″-6″ diameter 4″-6″ diameter	60″ 72″	4 poles 5 poles	8′ 8′	
Woven Wire	7½′, 4″-8″ diameter	9 or 11 gauge stay wire	55″-58″	12″ mesh	12′	Woven wire is satisfactory for larger areas where the concentration of animals is not too great. But it is not recommended for corrals, paddocks, or small pastures. Use 1 or 2 strands of barbed wire (with points 3″ to 4″ apart) on top.
	7′-8′ Heavy channel steel	11 or 12½ gauge	60″, 72″, or 84″	2″ mesh, 4″ horizontal	5′4″	Manufactured (Keystone) in 16′ panels. The 2″ mesh prevents a hoof from going through and alleviates leg damage.
	7′-8′ Heavy channel steel	16 guage	60″ or 72″	2″ mesh, 2″ horizontal	5′4″	Manufactured (Keystone) in 16′ panels.
	7′-8′ Heavy channel steel	¼″ steel	72″	8″ mesh, 2″ horizontal at bottom 8″ mesh, 4″ horizontal at top	5′4″	Manfactured (Keystone) in 16′ panels. This is welded fabric from ¼″ steel, not woven style.

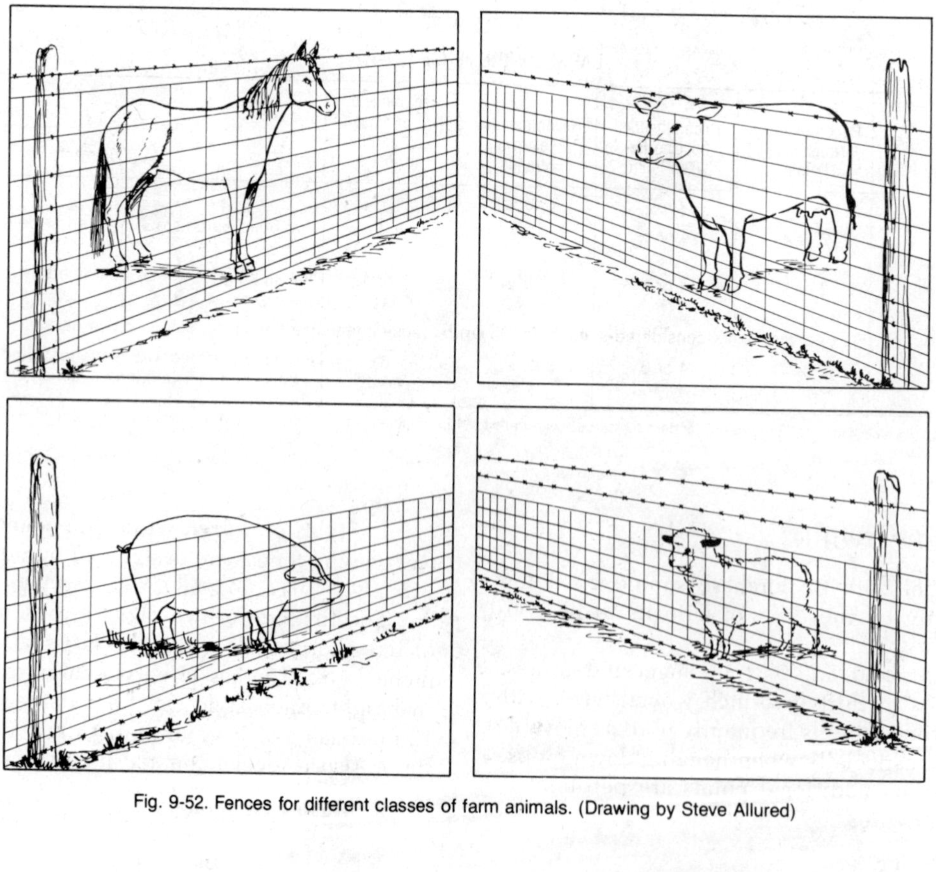

Fig. 9-52. Fences for different classes of farm animals. (Drawing by Steve Allured)

ay or vertical wires 6 in. apart (6-in. mesh) will give etter service than a wide-spaced (12-in. mesh) fence. However, some fence manufacturers believe that a 2-in. spacing with a No. 9 wire is superior to a 6-in. pacing with No. 11 filler wire (about the same mount of material is involved in each case).

3. **Weight of wire**—A fence made of heavier weight wires will usually last longer and prove heaper than one made of light wires. Heavier or arger size wire is designated by a smaller gauge umber. Thus, No. 9 gauge wire is heavier and larger han No. 11 gauge. Woven wire fencing comes in Nos. , 11, 12½, and 16 gauges—which refers to the gauge f the wires other than the top and bottom wires. Heavy barbed wire is 12½ gauge. But there is a ighter, high tensile barbed wire which comes in 14 to 6 gauge.

Heavier or larger wire than normal should be used in those areas subject to (a) salty air from the ocean, (b) smoke from industries of close proximity, which may give off chemical fumes into the atmosphere, (c) rapid temperature changes, or (d) overflow or flood.

Likewise, heavier wire than normal should be used in fencing (a) small areas, (b) where a dense concentration of animals is involved, and (c) where animals have already learned to get out.

4. **Styles of barbed wire**—Styles of barbed wire differ in the shape and number of the points of the barb, and the spacing of the barbs on the line wires. The two-point barbs are commonly spaced 4 inches apart while four-point barbs are generally spaced 5 inches apart. Since any style is satisfactory, selection is a matter of personal preference.

5. **Standard size rolls or spools**—Woven wire comes in 20 and 40 rod rolls; barbed wire in 80 rod spools.

6. **Wire coating**—The kind and amount of coating on wire definitely affects its lasting qualities. Galvanized coating is most commonly used to protect wire from corrosion. Coatings are specified as Class I, Class II, and Class III. The higher the class number, the greater the coating thickness and performance.

HOW TO SELECT POSTS

Three kinds of material are commonly used for fence posts: wood, metal, and concrete. The selection of the particular kind of posts should be determined by (1) the availability and cost of each, (2) the length of service desired (posts should last as long as the fencing material attached to it, or the maintenance cost may be too high), (3) the kind and amount of livestock to be confined, and (4) the cost of installation.

Wood posts—Osage orange, black locust, chestnut, red cedar, black walnut, mulberry, and catalpa—each with an average life of 15 to 30 years without treatment—are the most durable wood posts, but they are not available in all sections. Untreated posts of the other and less durable woods will last 3 to 8 years only, but they are satisfactory if properly butt treated (to 6 to 8 in. above the ground line) with a good wood preservative.

The proper size of wood posts varies considerably with the strength and durability of the species used. In general, however, large posts last longer than small ones. Satisfactory line posts of osage orange or of other woods that have been pressure treated may be as small as 2½ in. in diameter; whereas line posts of other woods should be 4 to 8 in. in diameter at the smaller end. Split posts should be a minimum of 5 in. in diameter. Line posts are generally 7 to 8 feet in length, depending on the height of the fence to be constructed.

Wood corner, end, and gate posts should be substantial, usually not less than 8 inches in diameter. Also, they should be long enough so that they can be set in the ground to a depth of at least 36 inches.

Metal posts—Metal posts (made of steel or wrought iron) last longer, require less storage space when not in use, require less labor in setting than wood posts, and are fire resistant. Also, they may give protection against lightning by grounding the current. However, such protection is questionable in dry weather or in areas with a low water table. Metal posts are usually higher in price than wood posts.

Metal line posts are made in different styles and cross sections. Heavier studded "T" or "Y" section posts are most popular for livestock, although lighter channel posts may be used for temporary and movable fences. Line posts are available in lengths of 5 to 8 feet in increments of 6 inches. Metal corner, end, and gate posts are commonly made from angle sections, and come in 7 to 9 feet lengths.

Concrete posts—When properly made, concrete posts give excellent service over many years. In general, however, they are expensive.

WHAT TO USE IN TREATING POSTS

The less durable types of fence posts will last about five times longer when treated than when untreated. This affects yearly savings in two ways: (1) in the costs of posts, and (2) in the labor involved in fence construction.

Although the relative durability of posts does not materially affect initial fencing costs, the length of life of the posts is the greatest single factor in determining the cost of a fence on an annual basis.

Some common preservatives are: creosote, pentachlorophenol, zinc chloride, and chromated zinc chloride. Creosote and "penta" should be used only on dry seasoned posts; the others are effective on green posts with the bark left on. Preservatives should always be used in keeping with the directions of the manufacturer.

Pressure treating is preferable because it forces the preservative to the center of the post—leaving none of the wood untreated.

HOW TO MEASURE AND FIGURE A FENCING JOB

In addition to deciding upon the kind of materials to use, the farmer or rancher must reach a decision relative to (1) the quantity of wire, and (2) the number and dimensions of corner, end, and gate posts, and of line posts. The following pointers may be helpful to this end:

1. **Area to be fenced**—The shape of the field has much to do with the quantity of fencing required (less fencing being required to fence the same acreage in an oblong field). Thus, the best way to determine how much fencing is needed is to measure the plot of ground on all sides and add these measurements together; to this add an additional small quantity for wrapping around corner, end, and gate posts.

2. **Corner, end, and gate posts**—The corner, end, and gate posts should be (a) long enough so that they can be set 3½ to 4 feet deep; (b) heavy and well braced; and (c) used at each corner or end, with two such posts used on each side of gates.

Similar heavy posts should be set (a) at intervals of 30 to 40 rods apart where there is a long, straight stretch of fence, and (b) at the top and bottom of each slope if the ground is hilly.

3. **Line posts**—The line posts for field and pasture fence should be spaced about 12 to 16½ feet apart (wider spacing is usually used on western ranges, because less wire is held up), whereas in a corral fence they should not be spaced more than about 8 to 12 feet apart.

Driven line posts should be long enough so that they can be driven into the ground to a depth of about 2 feet, while tubular posts should be long enough so that they can be set 30 inches deep—preferably in concrete.

4. **Gates**—Normally gates are widths of 8, 10, 12, 14, or 16 feet, and heights of 48, 50, or 55 inches. It is always well to secure a wide enough gate to accommodate the machinery or portable buildings common to the farm or ranch.

Generally gates are subjected to more wear than any other part of the fence. In addition to performing the functions of a fence, they must be made to open and close in a convenient manner. Simple swinging gates made of steel or wood are most popular.

Gates should be well-braced and durable.

5. **Costs**—Fence costs vary widely with types of construction and prevailing prices of materials and labor. Wire, posts, and labor are the largest items of costs.

HOW TO ERECT A WOVEN WIRE FENCE[10]

The following steps and pointers should be adhered to in the construction of a good fence:

1. **Preparing the fence line**—After the fence line has been laid out, the ground area over which the wire will be stretched should be cleared of all rocks, stumps, and irregularities which might hinder the proper stretching of the fencing with the bottom of the fence close to the ground and parallel to the ground line.

2. **Making concrete footings**—If concrete footings are desired, prepare mix according to the directions given in the section on "Concrete Structures for the Farm and Ranch." Let the concrete set a minimum of seven days before stretching the fence.

3. **Setting and bracing corner, end, and gate posts in concrete**—The principal requisite of a neat appearing and long lasting fence is corner, end, and gate posts that are firmly set and well braced. These posts are the foundation of the fence, and will take most of the strain of the fence, as well as the whole strain of stretching. In short, they are the most important part of the fence.

Steel corner, end, and gate posts and their braces are usually set in concrete as illustrated in Fig. 9-53. Note (a) that the concrete extends about 6 inches below the post; (b) that it is a little higher next to the post than to the outside, thus causing water to flow away from it in order to alleviate freezing and chipping off the concrete; and (c) that the concrete of the brace runs so that its long way is in the direction of the brace.

Wood corner, end, and gate posts are usually set in slightly larger concrete blocks than steel posts. Where it is not convenient to set them in concrete they may be set with heavy stones tamped well around them (mixing dirt between the stones to make a firm job).

Another difference is in bracing. With wood posts, a double span brace arrangement is preferable to a single span anchor. The double span (see Fig. 9-54) is cheap and simple, and takes less time and labor to set than the single span with necessary anchoring.

[10]Where considerable fencing is to be done, the farmer or rancher should also consider the use of a versatile, laborsaving fence building machine. These machines drive posts, dispense wire, and stretch fence in one operation. They will handle woven wire, barbed wire, or a combination of both.

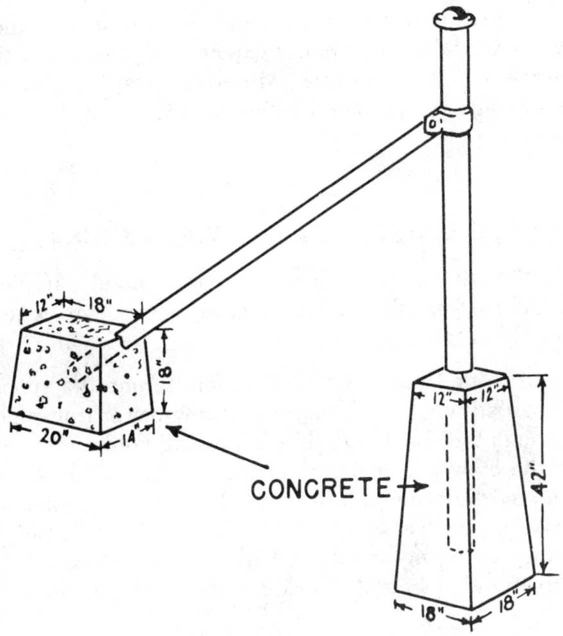

Fig. 9-53. A steel end post and brace set in concrete. (Drawing by Steve Allured)

4. **Locating line posts**—After the corner or end posts are located, line fence posts are best located by (a) measuring off along the line fence and placing a small stick at each interval or spacing decided upon, and (b) lining-up the holes through sighting between the two corner or end posts as a helper moves a straight pole into line as directed (with a white cloth around the top).

5. **Setting line posts**—Drive steel posts with a post driver or sliding sleeve to a depth of about 24 inches if the soil is solid, or to a depth of 30 to 36 inches if the ground is loose or if there is a sag in the ground along the fence line.

Wooden line posts may be set by either (a) digging holes and setting them to a depth of at least 30 inches with the soil tamped solidly around them, or (b) sharpening the lower ends and driving them into the ground with a maul to a depth of about 24 inches.

Never drive the post deeper than the height of the fence being constructed.

In cases where there is quite a pronounced sag in the ground or where the ground is soft, one or two of the line posts at the bottom of the sag should be set in concrete.

6. **Grounding for lightning protection**—In areas subjected to frequent thunderstorms, it is good practice to ground the fence line by setting steel posts or pipes at intervals of about every 40 rods in those areas where animals commonly congregate.

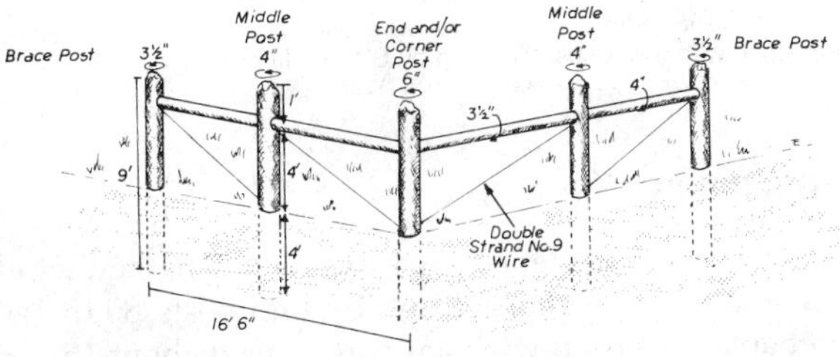

Fig. 9-54. This type of double span corner arrangement is preferable to a single span anchor where wood posts are used.

How to Stretch the Fence

Proper execution of the following steps will result in a tightly stretched fence:

1. **Attach the fencing properly to end post or corner post**—This consists of (a) cutting out one or two of the stay wires on the end of the roll in order to fasten the fencing to the end post easily, (b) passing the line or horizontal wires of the fence around the post and wrapping the ends of these wires firmly around the corresponding line wires of the fence (making sure (1) that the stay wire is straight up and down, parallel with the end post; otherwise, all the stay wires in the entire length of fence will be out of line, and (2) that the line wires coming up to the end posts are straight in line with the fence line, so that stretching will not have a tendency to twist the end post around in the ground).

2. **Unroll the fence**—The fence should be unrolled so that (a) it will be flat upon the ground with its bottom edge close to the bottom of the line posts, and (b) it extends beyond the next end or corner post.

3. **Attach chain-type stretchers and stretch**—Fence stretchers should be fastened to the end post, and the clamp bar of the stretchers should be attached to the wire a few feet from this last mentioned post. The clamp bar should be attached *firmly*, just back of and parallel with one of the stay wires of the fence, to make for even stretching. Make sure that the chains on the stretchers are straight. A single-jack stretcher is satisfactory for 26- or 32-inch woven wire, but a double-jack stretcher should be used for higher fences.

Gradually take up the slack with the stretchers. But before any tension is actually applied to the fence, stop and go down the fence line, setting up the fence along the line posts by hand. If a long stretch of fence is involved, a few temporary fastenings at intervals to the line posts may be made with wire.

Then stretch the fence until the tension curves are about ½ to ⅔ their normal size. Overstretching

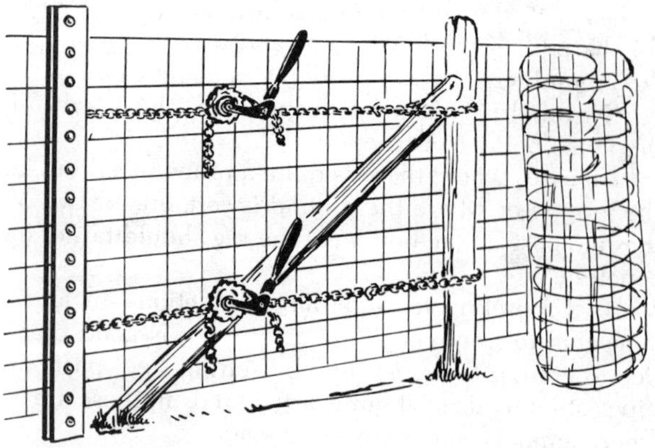

Fig. 9-55. Stretching woven wire with chain-type stretchers. (Drawing by Steve Allured)

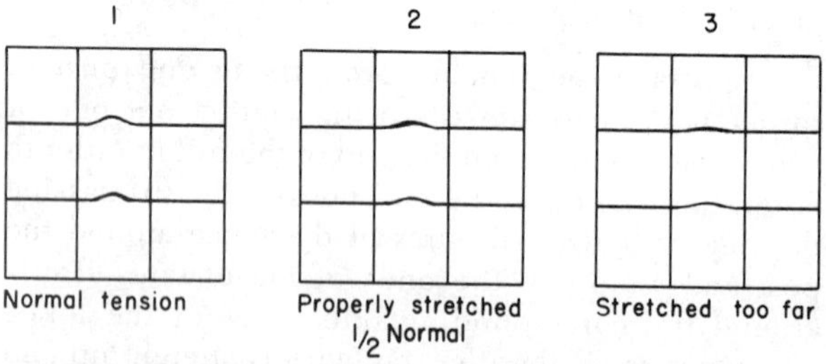

Fig. 9-56. Fence should be stretched properly, as indicated by the tension curves. When stretched properly, the tension curves are about ½ to ⅔ their normal size. (Drawing by Steve Allured)

removes these tension curves and injures the fence. The "springs" of the fence expand and contract as temperatures change and when the fence gets a hard blow.

4. **Secure to end posts**—When the stretching is tight enough, (a) cut off the fence sufficiently beyond the end post so that the line wires may be wrapped around it, and (b) take up the slack in each line wire between the stretcher and the end post and wrap each of them around the post and the corresponding line wires of the fence. Start with the bottom wire first.

Then release the fence stretchers.

5. **Fasten to line posts**—Following stretching, the fence should be fastened to the line posts. Wire fencing can be fastened to wooden posts with regular fence staples using 1-inch staples for hard woods and 1½-inch staples for soft woods. Less splitting of the wood will occur if the staples are driven in diagonally to the grain. Do not drive staples tightly against the wire.

Steel posts usually have self-fastening clips on them, into which the line wires can be inserted and closed.

Line wires should be loose enough so that the wire can slide a bit when it expands or contracts with changes of temperature. Moreover, staples driven tightly against the wire may weaken it and cause it to break.

HOW TO ERECT A BARBED WIRE FENCE

Erecting a barbed wire fence is relatively simple. It differs from the construction of woven wire fence chiefly as follows:

1. **Unrolling**—Two men generally unroll a spool of barbed wire by (a) running a crow bar or iron pipe through the spool, and (b) each taking hold of an end of the crow bar or pipe and walking along the fence line as the wire unrolls. With large fencing jobs, it may be practical to use the tractor for the purpose of either unrolling or rolling barbed wire; using a reel driven either from the tractor wheel or the engine.

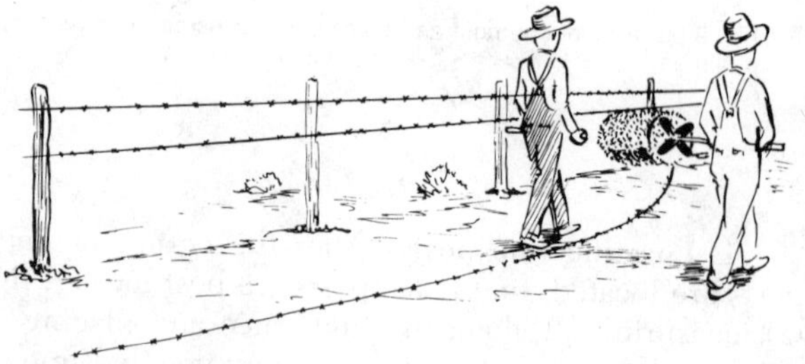

Fig. 9-57. Unrolling barbed wire. (Drawing by Steve Allured)

2. **Stretching**—Barbed wire can be stretched by using a tackle block stretcher equipped with wire grips.

Fig. 9-58. Stretching barbed wire with a tackle block stretcher equipped with wire grips. (Drawing by Steve Allured)

SUSPENSION FENCES

A suspension fence consists of 3 to 5 barbed wires stretched tightly and held up by line posts 80 to 120 feet apart. Twisted wire stays are placed about 16 feet apart. Suspension fences are in use in southwestern United States and in Australia. They cost about one-

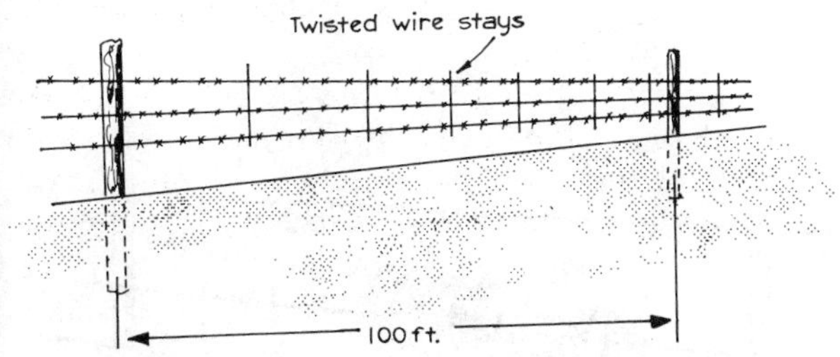

Fig. 9-59. Suspension fences save money—and work. The line posts are spaced 80 to 120 ft apart, and the twisted wire stays are about 16 ft apart.

half as much as conventional woven-wire or barbed wire fences.

A suspension fence sways in the breeze, with the result that cattle are spooked by the movement. When they attempt to rub on it, the wires "give" and they back away. Suspension fences require less expensive upkeep than conventional fences. If the bulls fight through it, the wire springs back into position after they run over or under it.

Suspension fences should work as well with sheep as with cattle.

Suspension fences are best adapted to even terrain. Whenever slope direction changes, a support post should be set. The fence needs either wooden or steel line posts, at 80- to 120-foot intervals, and it must be securely braced at the corners and every one-quarter mile. Wire is stretched taut so that there is not more than 3 inches of sag between the line posts.

The wire should be attached to wooden posts with L-shaped, deformed, shanked staples, long U-shaped staples or a piece of 18 to 20 gauge metal strip ½- by 1½-inch placed over the wire and held with a six penny nail on each end. Regular wire clips should be used on steel posts. The wires must move freely under the holders. Don't let the wire stays touch the ground.

A mile of four-wire fence with stays 16 feet apart takes 16 rolls of wire, 53 line posts, corner and stretch posts, 300 spiral wire stays, metal strips, nails, brace wire, and about 108 man-hours of labor.

Electric Fences

Where a temporary enclosure is desired or where existing fences need bolstering from roguish or breachy animals, it may be desirable to install an electric fence, which can be done at minimum cost.

The following points are pertinent in the construction of an electric fence:

1. **Safety**—If an electric fence is to be installed and used, (a) necessary safety precautions against accidents to both persons and animals should be taken, and (b) the farmer or rancher should first check into the regulations of his own state relative to the installation and use of electric fences. *Remember that an electric fence can be dangerous.* Fence controllers should be purchased from a reliable manufacturer; homemade controllers are dangerous and should not be used.

2. **Charger**—The charger should be safe and effective (purchase one made by a reputable manufacturer). There are four types of chargers: (a) *the battery charger,* which uses a 6-volt hot shot battery; (b) *the inductive discharge system,* in which the current is fed to an interrupter device called a circuit breaker or chopper which energizes a current limiting transformer; (c) *the capacitor discharge system,* in which the power line is rectified to direct current and the current is stowed in the capacitor; and (d) *the continuous current type,* in which a transformer regulates the flow of current from the powerline to the fence.

3. **Wire height**—As a rule of thumb, the correct wire height for an electric fence is about three-fourths the height of the animal; with two wires provided for sheep and swine. Following are average fence heights above the ground for different animals:

Cattle—30 to 40 inches
Calves—12 to 18 inches
Sheep, two wires—One wire 8 to 10 inches and the other 16 to 18 inches
Swine, two wires—One wire 6 to 8 inches and the other 14 to 16 inches
Horses—30 to 40 inches
Mixed Livestock, three wires—8, 12, and 32 inches.

4. **Posts**—Either plastic or steel posts may be used for electric fencing. Corner posts should be as firmly set and well braced as required for any non-electric fence so as to stand the pull necessary to stretch the wire tight. Line posts (a) need only be heavy enough to support the wire and withstand the elements, and (b) may be spaced 40 to 50 feet apart for horses and cattle, and 25 to 40 feet apart for sheep and swine.

5. **Wire**—New 4 point 12½ gauge barbed wire is preferred, because the barbs will penetrate the hair of animals and touch the skin, but smooth wire can be used satisfactorily. Rusty wire should never be used, because rust is an insulator.

6. **Insulators**—Wire should be fastened to the posts by insulators and should not come into direct contact with posts, weeds, or the ground, unless plastic or fiberglass posts are used (plastic and fiberglass do not require insulators). Inexpensive solid glass, porcelain, or plastic insulators should be used, rather than old rubber or necks of bottles.

7. **Grounding**—One lead from the controller

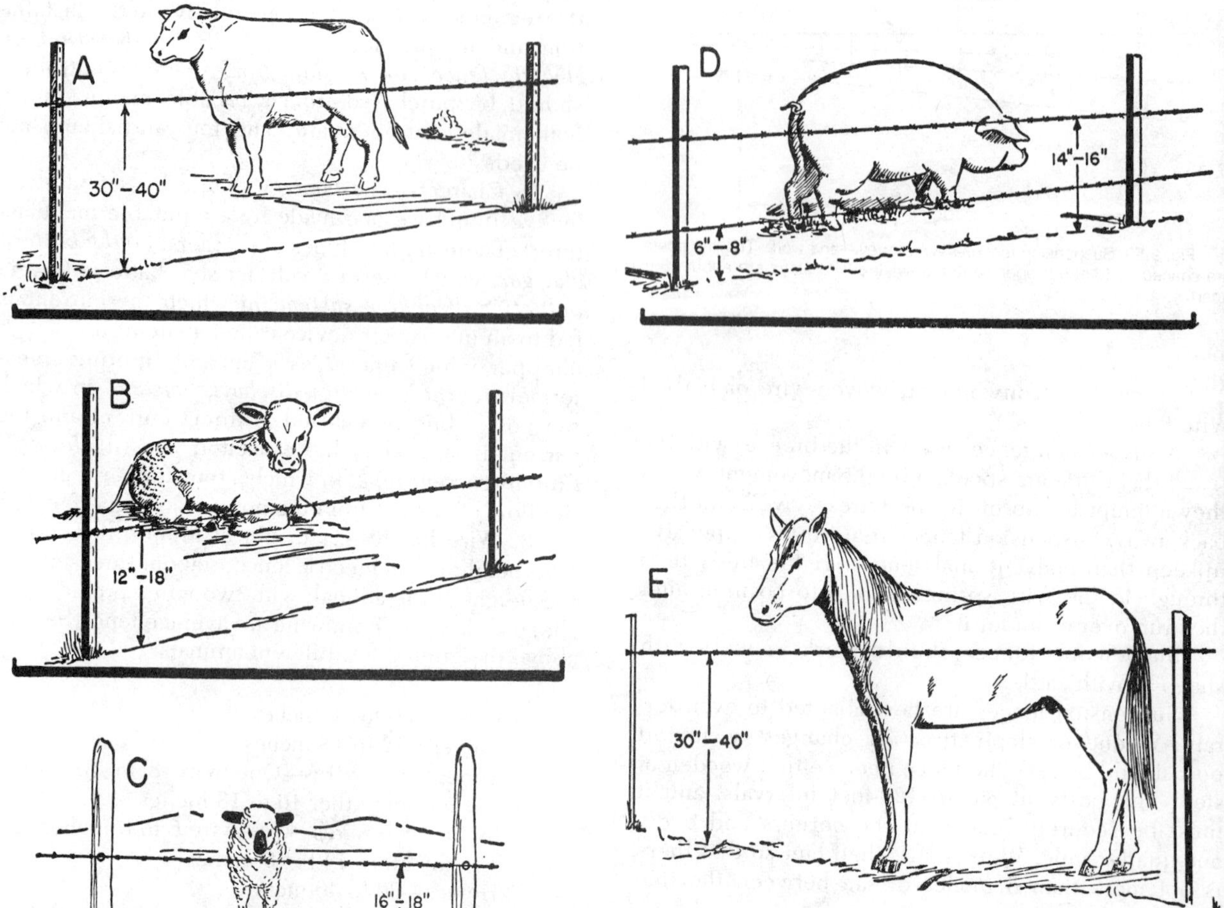

Fig. 9-60. Recommended number of wires and spacing for electric fences for each class of animals; A, cattle; B, calves; C, sheep; D, swine; and E, horses.

should be grounded to a pipe driven into the moist earth. *An electric fence should never be grounded to a water pipe, because it could carry lightning directly to connecting buildings.* A lightning arrestor should be installed on the ground wire.

Dog-Proof Fences for Sheep

Sheep corrals should be fenced with dog-proof fencing. Although such fencing is difficult to construct and expensive, it will pay dividends in the protection afforded.

A satisfactory dog-proof fence should be a minimum of 6 feet in height (preferably 7 ft), and the wire should have diamond mesh rather than the usual horizontal type. Also, it is preferable that the wire and bracings should be on the inside of the posts. To keep dogs from digging under the fence, a single strand of barbed wire should be placed on the surface of the ground on the outside of the posts; this arrangement is far more effective than placing wire in the ground directly below the woven wire.

PAINT FOR THE FARM AND RANCH[11]

The chief purposes of painting farm buildings and equipment are: (1) to preserve them from the effects of the weather, and (2) to add to their attractiveness. In addition, interior painting, such as is done in most homes, makes walls and ceilings more sanitary and dark rooms lighter.

[11]In the preparation of this section, the author had the benefit of the authoritative review and helpful suggestions of the following: Mr. Robert S. Taub, Staff Scientist, The Sherwin-Williams Company, Cleveland, Ohio.

How to Estimate Quantities of Paint Needed

The chief factors determining the quantities of paint needed for a given building are: (1) the square feet of surface area to be painted, and (2) the kind of coating material, the character of surface to be covered, and the number of coats to be applied.

To find the number of square feet of surface area on a house or barn (or other farm building), proceed as follows:

1. Measure the distance around the building and multiply it by the height up to the eaves. This will give the total side area of the building, not including the gables.
2. Calculate the gable area by multiplying the height of the triangle or gable by half its width.
3. Do not deduct for doors, windows, or other openings, as they will compensate for surface variation and waste.

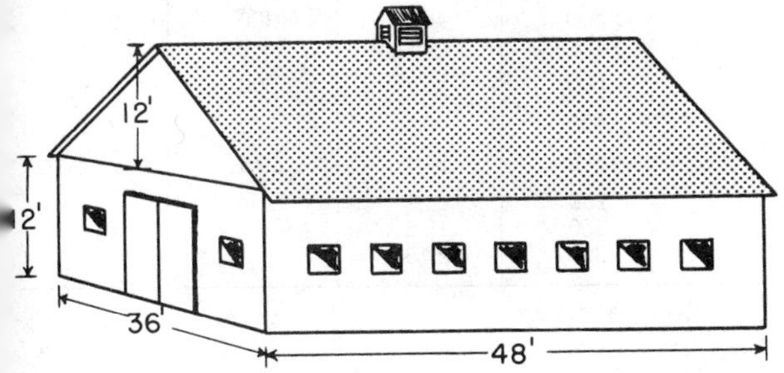

Fig. 9-61. How much paint do I need for this barn? The answer to this question is given in the accompanying narrative. (Drawing by Steve Allured)

4. Estimate the trim requirements on the basis of one tenth, or 10 percent, of the body paint requirements.

For purposes of illustrating how to figure quantities of materials needed, the area of the barn shown in Fig. 9-61 is herewith calculated:

1. Total side area (distance around × height)
   168 ft × 12 ft = 2,016 sq ft
2. Total gable area (each gable = height × ½ width)
   12 ft × 18 ft = 216 sq ft for one gable
   216 × 2 = 432 sq ft for two gables
3. Total area
   2,016 + 432 = 2,448 sq ft

Once the square feet of surface area to be painted has been determined, the number of gallons of paint required for the job can be obtained by referring to Table 9-32, page 862, Guide for Estimating Quantities of Paint Needed. Simply divide the total square feet to be painted by the number of square feet a gallon of paint will cover. Thus, in the case of the barn shown in Fig. 9-61, having a total area of 2,448 square feet and assuming a smooth—previously painted or primed wood surface, the number of gallons of paint is:

2,448 ÷ 400 = 6.1 gal (roughly 6 gal) paint required for 1 coat

2,448 ÷ 200 = 12.2 gal (roughly 12 gal) body paint required for 2 coats

10 × 10% = 1 gal (approximately) of trim required.

Of course, the figures given in Table 9-32 are conservative estimates for painting under average conditions. As noted in Table 9-32, the quantity of paint is affected by (1) the kind of coating materials, (2) the character of surface to be painted, and (3) the number of coats of paint to be applied. Additional factors affecting the quantity of paint are: (4) the consistency of the paint, (5) the thickness of the applied paint film, (6) the temperature, (7) the manner of application, and (8) the experience of the painter.

Whitewashes

Many different formulas are used for whitewashes. Perhaps a whitewash made according to the directions which follow is as good as any, for it is white and it sticks well:

1. Mix ½ bushel (4 gal or 38 lb) quicklime with boiling water, adding the water slowly and stirring constantly until a thin paste results.
2. Add 1 gallon salt to the lime paste and stir thoroughly.
3. Add water to bring the whitewash to the proper consistency for spraying or painting.
4. Just before using, add to each pailful (about 3 gal) a handful of cement and a teaspoonful of ultramarine bluing. The cement makes the whitewash stick to any surface, and the bluing gives it a snow-white appearance.

Paint Pointers

Observation of the following pointers will ensure more durability and satisfaction in farm and ranch paint jobs:

1. **Kind and quality of paint**—The first requisite in planning to paint is to select the right kind of paint. A specific kind of paint, enamel, or varnish is essential for each type of surface to be painted (see Table 9-32).

The second requisite is that the paint shall be of high quality, for the labor cost is the same regardless

TABLE 9-32
GUIDE FOR ESTIMATING QUANTITIES OF PAINT NEEDED

Coating Materials	Character of Surface to Be Painted	Square Area Covered by One Gallon			
		1 Coat		2 Coats	
		(sq ft)	*(sq m)*	*(sq ft)*	*(sq m)*
Asphalt-asbestos roof cement	Smooth	75	7.0		
Asphalt roof paint	Smooth	175	16.3		
	Rough	125	11.6		
Latex wall paint (interior)	Smooth (prime plaster with solvent based wall primer)	400	37.2		
	Rough—previously painted stipple or textured walls	150-200	13.9-18.6		
Solvent based (interior)	Smooth (prime dry wall with latex primer)	350	32.5		
Latex flat or gloss (exterior)	Smooth—previously painted or primed wood	400	37.2	200	18.6
	Wood shakes	200	18.6	100	9.3
	Smooth sound masonry	300	27.9	125-175	11.6-16.3
	Rough masonry (stucco, brick)	175-250	16.3-23.2	100-150	9.3-13.9
	Filled aggregate block	250-350	23.2-32.5	125-200	11.6-18.6
Solvent based gloss or flat (exterior)	Smooth—previously painted or primed wood	350-400	32.5-37.2		
Solvent based stain	Smooth wood	500	46.5	300	27.9
	Rough siding or shakes	125-175	11.6-16.3	75-100	7.0-9.3
Latex based stain (exterior)	Smooth wood	500	46.5	300	27.9
	Rough siding or shakes	125-175	11.6-16.3	75-100	7.0-9.3
Shellac (interior)	Smooth wood	600	55.7	300	27.9
Spar varnish (exterior)	Smooth wood	500	46.5	275	25.5
Varnish (interior)	Smooth wood	550	51.1	275	25.5
Whitewash	Wood	250	23.2		
	Brick	200	18.6		
	Plaster	300	27.9		

of the quality, and the cost of paint represents only about one-fourth the total cost of a painting job.

2. **Ready-mixed vs home-mixed paints**—Most farmers and ranchers use ready-mixed paints, primarily because it is more convenient to use them. Also, most of them contain materials not available to home mixing, which impart desirable properties to the paint; they are more uniform; exact quantities may be purchased; and there is a wide choice of colors.

3. **Choice of color**—Although choice of color should remain a matter of personal preference, farm folks should be aware:

a. that red, orange, and yellow are warm colors, which attract attention

b. that cream, buff, peach, and light tan are also warm colors, but they blend into the surroundings and are pleasing;

c. that green, blue, and violet are cool;

d. that light colors give an impression of greater spaciousness and dark shades tend to shrink areas;

e. that farm buildings can be made to belong to the landscape through the selection of the proper shades of both body paint and trim;

Fig. 9-62. Farm buildings can be made to belong to the landscape through the selection of the proper shades of both body paint and trim.

f. that light colors reflect more illumination than dark ones, an important consideration in interior painting; and

g. that the color of the finished job will appear somewhat deeper than that of the color chip (chart).

4. **When to paint**—In all cases, follow the manufacturer's directions. However, the most common recommendations on when to paint are as follows:

a. **New wood buildings**—Apply a priming coat of paint as soon as the weather permits, and the finishing coat or coats as soon as possible after the

Fig. 9-63. The surface tells you when it is time to repaint! Repaint efore checking, cracking, or excess chalking is evident.

primer is thoroughly dry—usually within 24 hours in good drying weather.

b. **Metal**—Gutters, roofing, and water tanks fabricated with galvanized iron should be primed with special galvanized iron primers if solvent based topcoats are used. Latex paints may be applied directly to new, clean galvanized surfaces without a special primer. Use two coats on metal.

Iron and steel surfaces should be painted before rusting occurs and the first coat should be a rust resisting metal primer.

c. **Repainting**—Usually outside painting will last from 3 to 5 years. Repaint before old finish starts to check or crack, thus preventing moisture from getting in back of the paint film. The better grades of modern type paints wear by chalking, thus eliminating this checking or cracking characteristic.

5. **Preparation of surface**—Proper preparation of ne surface prior to painting consists in doing the folowing:

a. Remove all dirt, plaster, or other similar substances by wiping with a cloth or brushing with a stiff brush.

b. Remove all cracked, peeling loose paint, and chalk. Treat mildewed surfaces with a solution made by mixing approximately 1 quart of household bleach to 3 quarts of water, with the concentration of the bleach varying according to the severity of the mildew. Scrub solution on with a stiff bristle brush. Allow to remain on the surface 30 minutes. Rinse thoroughly; use rubber gloves, goggles, and protective clothing when applying.

c. Smooth rough places with abrasive paper.

d. Touch up all knots, sappy streaks, and bare spots with a good exterior undercoater.

e. Putty, with a putty of a color to match that of the finish, all nail holes and cracks. Where a two- or three-coat job is involved, it is best to putty after the first coat.

f. If grease is present on metals that are to be painted, first clean with paint thinner and scrape and brush free of all dirt, rust, and loose paint.

For masonry, proper preparation of the surface consists of doing the following:

a. **When repainting masonry**—Clean off mildew and any excess chalk; remove loose and peeling paint; spot prime bare spots with a masonry conditioner; and apply latex or solvent based topcoats. Porous or sand blasted stucco or masonry should first be treated with masonry conditioner prior to topcoating with latex paints.

b. **New masonry**—Apply two coats of latex if in sound condition. Observe the following:

(1) Fill aggregate block with a block filler prior to painting.

(2) Either remove form-saving oils or other surface treatments on concrete, or allow to weather two years before repainting.

(3) Allow glazed shingles to weather two years before painting.

(4) Allow new brick to weather at least one year before painting. Also, make certain that the mortar joints are sound.

All undereave and protected areas should be washed with detergent and water, rinsed thoroughly, and allowed to dry before painting. This will help prevent peeling caused by invisible salt deposits that are not removed by normal rain washing. Sand lightly any remaining glossy areas and remove dust prior to painting.

6. **Good paint weather**—Good paint weather is clear and dry, with a temperature of 60° to 80°F. Do not paint when insects are numerous and there is considerable dust. Paint can be applied at any time in heated buildings.

7. **Moisture**—Moisture, which is indicated when paint blisters and peels, ruins more paint jobs than any other single factor. The moisture problem may exist because either the wood is damp at the time of

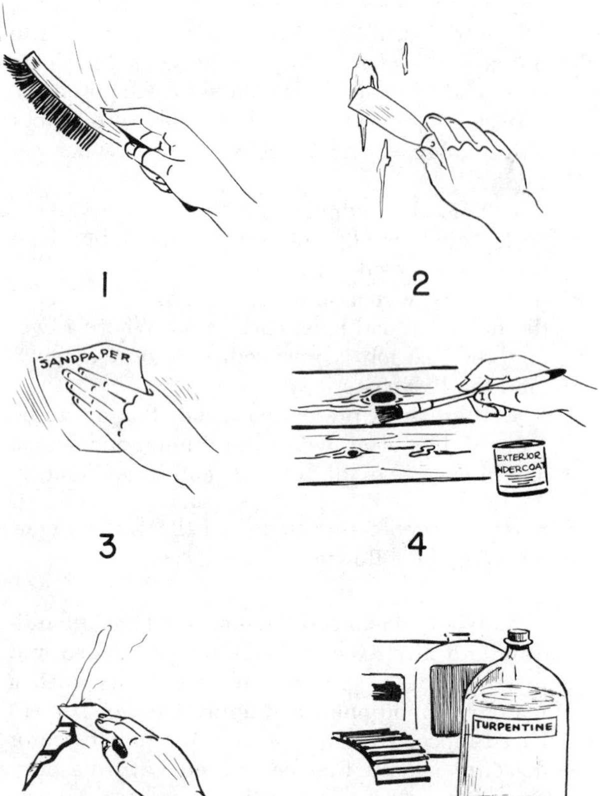

Fig. 9-64. Steps in the preparation of the surface for painting: (1) brush (or wipe with a cloth) to remove all dirt, plaster, or other similar substances; (2) remove cracked, peeling, and loose paint; (3) smooth rough places with sandpaper; (4) touch up knots, sappy streaks, and bare spots with a good exterior undercoater; (5) putty all nail holes and cracks; and (6) if grease is present on metals, first clean with paint thinner or turpentine.

painting or moisture is coming from within the building itself.

Before painting, ample time must be allowed for all surfaces to dry out after being subjected to snow, rain, or even heavy dew or frost. Longer time is required for unpainted wood, since such surfaces are more absorbent.

There are on the market good commercial moisture registering instruments that will measure accurately the moisture content in substrates to be painted.

8. **Application of paint**—In the actual painting operation proceed as follows:

a. Follow the manufacturer's directions which are printed on the containers.

b. Stir the paint thoroughly before and while using, because all paints settle.

c. Use a good primer and sealer on new wood.

Fig. 9-65. Stir the paint thoroughly before and while using, becaus all paints settle.

Paint as soon as possible after erection to preven peeling.

d. Do not thin ready-mixed latex or solven based paints unless label instructions permit.

e. Do not spread latex paints too thinly. Generally, they should be applied at the rate of abou 1 gallon per 400 square feet.

f. Consider the temperature at the time of ap plication. When using latex paints, make certai that the air temperature and the temperature c the substrate are more than 50°F. When usin solvent based paints, apply when the temperatur is 65° to 85°F.

9. **Number of coats**—For new work or for ol work where the paint has been completely remove three coats are recommended. Two coats are recommended for a repaint job.

10. **The paint brush**—The following pointers ar important relative to paint brushes:

a. Buy a good brush. For solvent based paint brushes made from carefully selected Chines hog bristles are best. For latex paints, brushe made from nylon or other synthetic bristle ar recommended.

b. Use a brush of the proper size; a large on for a wall, and a small one for tight corners.

c. Clean and care for the brush as follows:

(1) When a hog bristle brush is to be use the next day, wash it in a paint thinner or i turpentine and suspend it by a wire throug the handle in a container of raw linseed oil s that the bristles do not touch the botton Never leave a hog bristle brush in paint or i water. When using nylon brushes with late based paints, wash them out with deterger and water and suspend them in water.

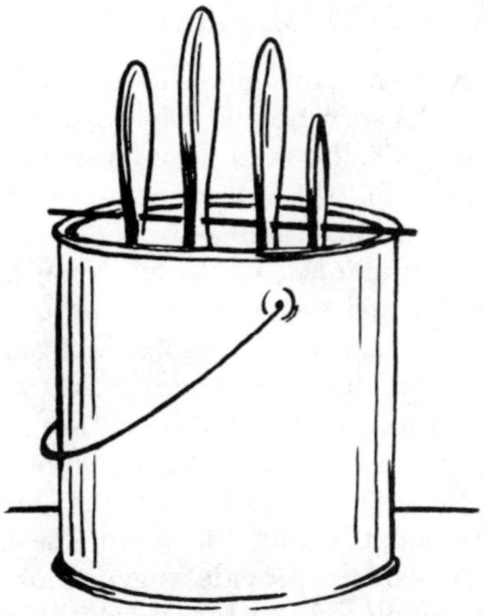

Fig. 9-66. Brushes that are to be used the next day. Hog bristle brushes should be suspended by a wire through the handle in a container of raw linseed oil so that the bristles do not touch the bottom. Synthetic bristle brushes should be suspended in like manner, but in water.

(2) When a paint brush is to be put away for several weeks, (a) clean out all the paint—use a paint thinner on hog bristle brushes used on solvent based paints, and use water on synthetic bristles; (b) wash with a detergent and rinse thoroughly; (c) straighten the bristles with a comb; and (d) wrap carefully with a fairly porous paper (like brown wrapping paper) to preserve the shape of the bristles and allow to dry in a warm place.

11. **Spray painting**—The practice of spray painting farm buildings is increasing. Because many farmers have air compressors and because spraying is less costly than brushing, this trend will increase. Spraying will give very satisfactory results if properly done; however, when painting new or badly weathered wood, it is necessary to brush the priming coat in order to work the paint into the pores of the wood for proper sealing.

12. **Painting over creosote**—There is no sure way of preventing creosote from bleeding through paint films. A good drying exterior undercoater used as a first coat performs as well as anything. However, it should first be tried out on a small test area before painting the entire structure. Apply it on both weathered and unweathered sides and check for any bleeding through. If small amounts of bleeding do occur, dark colors result in less tendency for it to show.

13. **Painting over whitewash**—Some paints are recommended as satisfactory for painting over whitewashed surfaces. However, only whitewash should be applied over previously whitewashed surfaces unless they are first thoroughly wire brushed to remove all the remaining whitewash coating.

14. **Paint removers**—Paint may be removed by the use of commercial or homemade paint removers, by blow torch, by scrapers, by wire brush, and by sandpaper.

15. **Removing paint from glass**—Paint spots on glass may be removed by one of the following methods: (a) by scraping with a razor blade, provided the spots are small; or (b) by using a paint and varnish remover.

16. **Removing paint from hands**—Paint is injurious to the skin and should not be allowed to remain on the hands for any great length of time. It may be removed by washing the hands with kerosene or by rubbing them with pure lard. Painters often rub linseed oil into their hands before painting.

17. **Paint is poisonous**—*Lead-containing paints are poisonous to farm animals, especially to cattle, and more particularly to calves.* Thus, such animals should not be allowed access to freshly painted surfaces and to discarded paint buckets and other containers.

For safety, it is recommended that a lead-free (zinc oxide and titanium base) paint or whitewash be used on the interior of stables and on corral fences for calves.

When painting inside buildings, the windows should be open in order to obtain air circulation and to keep inhalation of gases to a minimum.

18. **Paint is combustible**—Because paints, varnishes, and oils are combustible, (a) they should be stored at safe distances from heat and open flames, and (b) paint-soaked rags should be disposed of promptly.

THE FARM POND

It is estimated that about 50,000 farm ponds are developed annually in the United States; half of them by financial assistance provided by federal and state governments (the usual assistance to qualifying farmers amounts to 50 to 60 percent of the total cost of the pond). It is further estimated that there are approximately one million ponds on American farms. When properly planned and built, ponds are generally successful in regions where a tight subsoil prevails but unsuccessful in those areas where the subsoil is open or abortive.

A well-planned pond contributes to greater farm earnings and may pay for itself in a short time if less expensive sources of water cannot be developed. In addition, it may provide a number of incidental benefits and pleasures for the farm family. Among the multiple uses that may be made of farm ponds are the following:

1. Water supply for livestock.

2. Irrigation of crops.
3. Fire protection.
4. Food and recreation. A pond may be stocked with fish, and it will attract birds and wildlife. Also, it may provide the family with a suitable swimming and skating area.
5. Control of gullies and flooding of farm land by impounding peak runoff water.
6. Domestic use. If special care is taken to protect the water from contamination, ponds can be used for watering gardens and flushing toilets; and, with chemical purification or boiling (all surface water is considered contaminated according to public health standards), pond water can be used for household purposes.

How to Build a Pond

There are four major types of ponds in common use: (1) dug-out ponds fed by groundwater, (2) ponds fed by surface runoff, (3) ponds fed by springs or creeks, and (4) off-stream storage ponds. The discussion that follows pertains primarily to the latter three types, which normally require a small earth dam to impound the water.

With a little know-how and technical assistance, satisfactory farm ponds can be built by using a farm tractor, scraper, and plow. However, the use of heavy equipment, which can usually be rented or contracted, is faster and often less expensive. Only a minimum of materials need be purchased.

The ensuing discussion will suggest the general procedure to follow in building a dam and the resulting reservoir in most localities.

PRELIMINARY CONSIDERATIONS

The first step is to secure the counsel and advic of the Soil Conservation Service or other local au thorities. Specifically, reliable information should b secured on the following points: (1) the suitability o the soil, for holding water and dam construction; (2 the water requirements; (3) the adequacy of the wate supply; (4) the existing state laws pertaining to im pounded water; and (5) the probable costs and bene fits. These, and similar considerations, will indicat whether the proposal is feasible and sound. It is als wise to see what experience the neighbors have ha relative to the above and other points. If the decisio is then reached to build the pond, these same au thorities can usually provide suggestions relative t (1) design details, (2) construction, (3) periodic in spections during construction, and (4) operation an maintenance.

DESIGNING

Sound design of even a small earth-fill dam i usually so involved that competent engineering as sistance is necessary. This can usually be obtaine from agricultural engineers in the Soil Conservatio Service or in other local or private organization These specialists will usually make surveys and plan and evolve construction blueprints similar to Fig 9-67, 9-68, and 9-69 prior to starting constructio However, a well-informed owner will be better abl to assist the engineer in making certain major desig decisions, and be more competent in carrying out th pond construction. To the latter end, the farmer c rancher should be familiar with such information a follows.

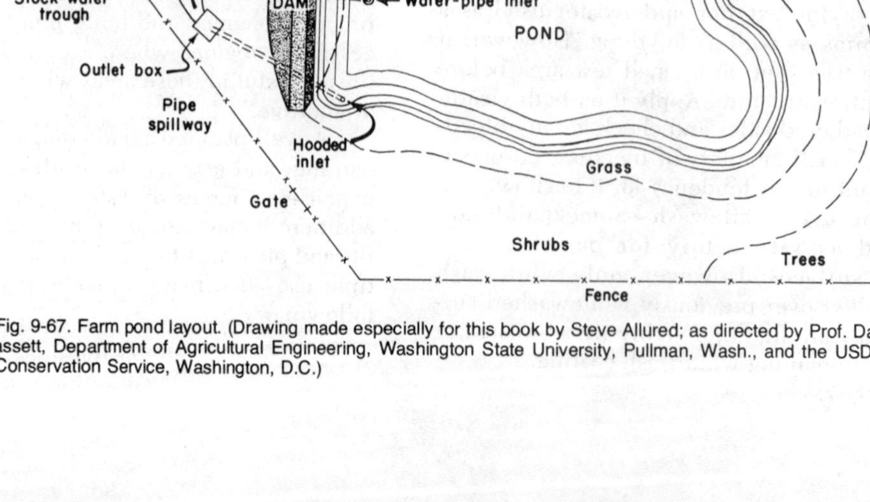

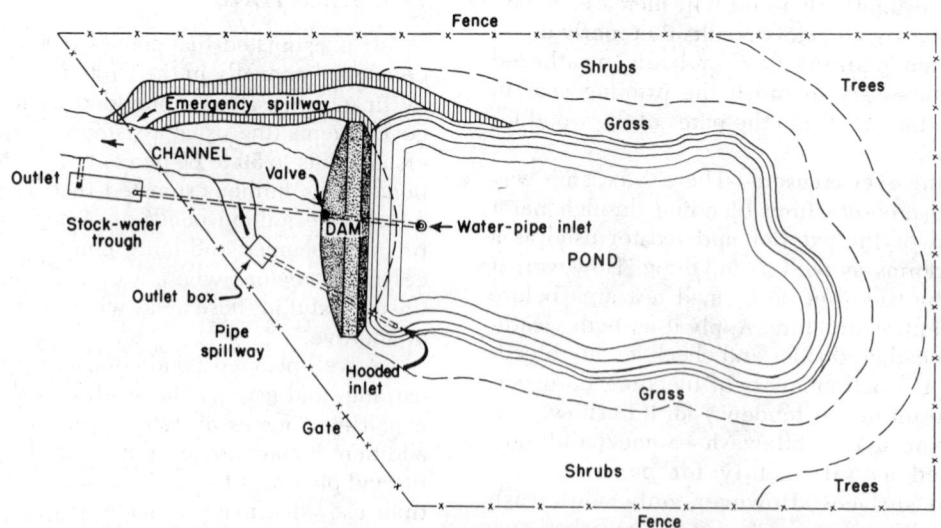

Fig. 9-67. Farm pond layout. (Drawing made especially for this book by Steve Allured; as directed by Prof. Day L. Bassett, Department of Agricultural Engineering, Washington State University, Pullman, Wash., and the USDA Soil Conservation Service, Washington, D.C.)

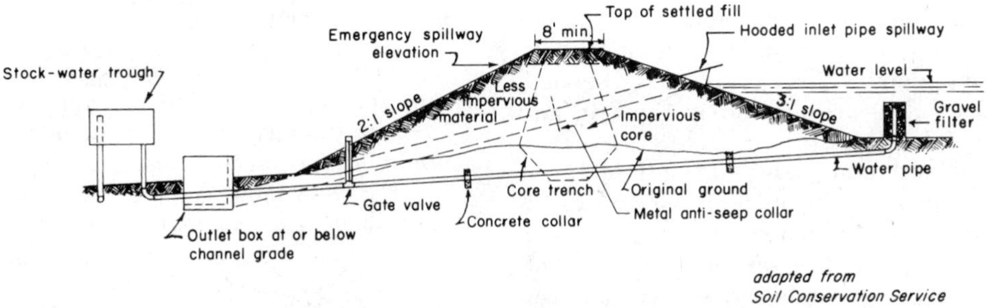

Fig. 9-68. Cross section of fill showing certain fill dimensions, core trench, hooded inlet pipe spillway, emergency spillway elevation, and water pipe system. (Drawing made especially for this book by Steve Allured; as directed by Prof. Day L. Bassett, Department of Agricultural Engineering, Washington State University, Pullman, Wash., and the USDA Soil Conservation Service, Washington, D.C.)

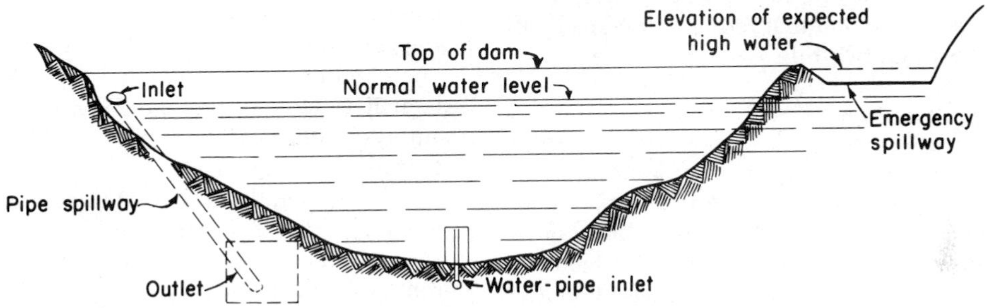

Fig. 9-69. Profile through axis of dam, showing the relative elevations of the emergency spillway, pipe spillway, and top of dam. (Drawing made especially for this book by Steve Allured; as directed by Prof. Day L. Bassett, Department of Agricultural Engineering, Washington State University, Pullman, Wash., and the USDA Soil Conservation Service, Washington, D.C.)

Survey of the Area

A topographic survey of the watershed and the pond site should first be made by a trained technician. A general soil survey of the area is also necessary and can be noted on the topographic map, together with important features such as vegetative cover, farm unit boundaries, and building sites and roads. From these surveys, plans showing the detailed dimensions of the dam and other features may be made.

Location

The proper pond site is extremely important. Here are the most important considerations relative thereto:

1. **Convenience**—First and foremost, the pond should be located where it will conveniently serve the purposes for which it is intended.

2. **Size of watershed**—The drainage area should be large enough to keep water in the pond during dry periods, but not so large as to flood the pond and cause heavy spillway flow during heavy rains.

3. **Cover of watershed**—Whenever possible, the pond should be located to receive the drainage from a watershed that is covered with permanent grass or trees. Where this is not possible, the drainage area should be protected by terracing, strip cropping, or other applicable soil conservation measures. Direct drainage from eroding cultivated land will likely result in the rapid filling of the pond with silt. If barnyards are in the drainage area, diversion terraces should be constructed to lead the runoff from the yards around the pond.

If the pond is to be used primarily as a water supply, it should not be located in a deep gully, as it will likely fill rapidly with silt unless an adequate program of gully stabilization is adopted.

4. **Depth of water**—In warm climates the depth of water at the deepest part should be at least 6 to 8 ft; in cold climates, depths of 10 to 12 ft are needed. Depths of 12 to 15 ft are needed in the arid and semiarid parts of the country.

5. **Type of soil material**—The soil material should be suitable; as a foundation for the dam, as fill material for the earth embankment, and as basin material to prevent excessive seepage from the bottom of the pond. Clay or sandy clay subsoils are usually best for holding water.

6. **Spillway possibilities**—The most economical spillway is a wide, flat channel, built at either end of the dam in undisturbed soil.

7. **Cost**—Everything else being equal, the most desirable location from the standpoint of minimum construction cost is the one that requires the smallest dam, such as between two hills.

Size

A pond should have sufficient storage capacity to meet the intended needs. The size pond to build on a given farm should be determined by four factors; namely, (1) the volume of water needed, based on average daily requirements; (2) the annual rainfall; (3) the kind of vegetation and terrain; and (4) the evaporation and seepage, which vary according to geographic location but which may increase the design capacity by 50 to 100 percent.

If there is not sufficient volume of water available to fill a large enough reservoir to meet all the intended needs, the size pond should then be limited to the volume of water actually available.

For general planning purposes, the quantity of water, in cubic feet, that can be stored in a given pond may be estimated from the following formula:

Volume (cubic feet) = $^4/_{10}$ the greatest depth (ft) × the surface area of the pond (sq ft).

Foundation Requirements

Improper soil foundation conditions can cause great difficulty in operation and maintenance, and even the failure of the dam. Thus, great care should be given to foundation examination and subsequent dam design.

Knowledge of surface, subsurface, and bedrock conditions is necessary in the design of an earth embankment. Small earth dams may be built on a wide range of foundations provided prior investigation has been made to determine the conditions, and the design is prepared accordingly. For farm ponds, these investigations may be accomplished by trained technicians using soil augers or shovels.

Embankment Design

The most common mistake is to make the banks too steep and the dam too narrow. Although the specific dimensions of the dam will vary, certain minimum specifications are necessary for safe design.

Fig. 9-68 shows a typical cross section of a small earth dam with commonly accepted features. A 3 to 1 slope on the water side of the dam and a 2 to 1 slope on the lower side is recommended. A width of 8 ft at the top of dams up to 10 ft in height is recommended with the minimum top width being 12 ft if it is to be used as a roadway. The settled dam crest should be at least 1½ ft above the level of the water in the spillway during design discharge (allowing about 15% for the normal settlement of a new dam). Provision should be made for a core trench for the purpose of preventing seepage water from flowing beneath the dam. Also, where rather porous material must be used for the dam, it is well to provide for a core wall, which is simply a continuation of the core trench backfill to the top of the dam.

If possible, the design of the embankment should be based upon the economical use of available materials adjacent to the site. Pervious material for the front and back faces is normally not difficult to locate. If satisfactory core materials high in clay content are unavailable, or must be hauled some distance, consideration should be given to designing a core wall of concrete or steel. (However, core walls of concrete or steel are rarely economically practical for farm ponds unless the need for water is critical.)

Seepage Beneath the Dam

Under certain conditions, the designer may wish to install a toe drain to collect the water which would seep along the contact plane between the fill and the foundation, thereby preventing undermining and subsequent failure of the dam.

Normally, toe drainage is not necessary for the dams of farm ponds unless the embankment rests on solid rock, hardpan, or an impervious clay. Also, it is recognized that most structures can be satisfactorily tied into hardpan or impervious clay foundations if they are first properly cored and scarified.

Spillway and Water Supply Pipes

Spillway design is important; many pond failures are due to inadequate spillway capacity or protection.

All ponds that may receive flood runoff should be equipped with an emergency spillway of proper width and depth that will safely bypass peak runoff from the heaviest rains or snow melt which exceed the temporary storage capacity of the reservoir. This emergency spillway is normally a vegetated waterway (although masonry material is sometimes used) located in a depression in the rim of the impounding area or around one end of the dam. A vegetated emergency spillway should never be placed over the fill in the dam.

Since vegetation will not survive under continued submergence, it is necessary to provide a mechanical spillway to handle minor flood flows and the base flow moving through the pond. This mechanical spillway usually consists of a pipe installed in or beneath the fill equipped with appropriate entrance and exit protection devices, antiseep collars, and valves. The inlet to the mechanical spillway is usually set one foot below the emergency spillway.

In order to remove water from the pond for the intended uses—for example, for watering stock—it may be necessary to install water pipes through or beneath the fill to appropriate installations below the dam. These water pipes should be equipped with appropriate filters, valves, antiseep collars, and exit protection devices.

Fig. 9-68, page 867, illustrates the use of the hooded inlet pipe spillway, the vegetated emergency spillway, and the water pipe. Either of two types of mechanical spillways may be used: (1) the hooded pipe inlet spillway, which is most common, or (2) the drainage-type spillway. The latter type spillway permits a pond to be completely drained, a desirable feature for fish propagation, mosquito control, or desilting and cleaning.

CONSTRUCTING THE DAM

When design details for the particular structure are completed and presented in drawings such as Figs. 9-67, 9-68, and 9-69, construction may go forward step by step as follows:

1. **Mark the pond waterline and dam site**—Usually these boundaries are first staked out, including the centerline of the dam and the toe of its slopes.
2. **Clear the site**—Both the dam site and the area to be covered by water, including ground at least 10 feet back from the shoreline, should be cleared of all trees, stumps, brush, large stones, and other objects.
3. **Remove the topsoil from the dam site**—The topsoil from the dam site should be removed, taking it down to a good, firm foundation. This topsoil may be used to cover the face of the dam and as a seedbed in the emergency spillway.
4. **Plow the dam site**—The dam site should be plowed lengthwise to obtain a better bond between the foundation and fill materials.
5. **Construct the core trench**—Construct the core trench as follows:

   a. Dig a trench about 4 feet wide (or wide enough to accommodate the digging equipment) and at least 2 feet deep (or deeper if necessary to reach a fairly impervious material) lengthwise along the centerline of the dam site.

   b. Backfill the trench with the most impervious material available, preferably well-packed moist clay or sandy clay soil in a moist condition.

   c. Where rather porous material is used for the dam, continue the core trench to the top of the dam as the fill progresses, compacting the material with a sheepsfoot roller if possible; making what is known as a core wall.

6. **Install all outlet water pipes**—Outlet pipes for livestock water, irrigation, fire protection, and other farmstead purposes should be installed before the fill is placed in the embankment. This is accomplished by digging (chiefly with a tractor and plow and then finishing with hand shovels) a trench or trenches across the dam site at the desired points. If the lines are to be used the year-around, they should be installed below the frost level; otherwise, they should be equipped with shut-off and drain valves to prevent freezing.

   The size pipe used will depend upon the volume of water required. Usually 1½-inch galvanized pipe is used for stock, with the pipe leading to a tank of the proper size located below the dam. The stock tank should be equipped with an ordinary float valve to control the flow of water and a drain plug to facilitate cleaning.

   The inlet end of each pipe should be equipped with a filter to avoid possible plugging of the line. Suitable filters may be either purchased or homemade. A satisfactory filter may be made as follows:

   a. Perforate a piece of a supply pipe about 3 feet long by drilling ¼-inch holes 2 inches apart through the pipe wall.

   b. Install the perforated pipe in the pond vertically, and extend it up with nonperforated pipe to the desired position in the pond.

   c. Place a drum or other container, filled with gravel or crushed stone and sand, about the drilled portion of the inlet pipe.

7. **Install toe drainage or foundation drainage if specified**—Where seepage-handling facilities have been included in the design, they should be installed in accordance with the design specifications before the fill is placed in the embankment. If drainage system detail is not available in the design, satisfactory drains may be constructed as follows:

   a. Install 6-in. perforated pipe in a trench a minimum of 18 in. square, parallel to the downstream toe of the fill and extending the length of the dam and approximately 10 to 15 feet from the toe in the foundation soil. The perforated pipe is (1) connected to a lead-out pipe at its lowest elevation to deliver the collected seepage a safe distance downstream from the toe of the dam, and (2) installed with the perforation on the bottom side and resting on approximately 6 in. of clean sand and gravel mixture.

b. Cover the pipe completely with a mixture of sand and gravel before the dirt fill material is place over it.

8. **Install the mechanical or pipe spillway**—Mechanical spillways usually employ one of two basic designs: the hooded inlet pipe spillway through the fill as shown in Fig. 9-68, or a horizontal pipe installed beneath the fill with a vertical entrance section and a valved opening at the bottom through which the pond can be drained. If any portion of the spillway is to be installed in the foundation material, it should be placed before the embankment fill is made. If it is designed to go through the embankment at a given elevation, it should be installed when the fill has been completed to a slightly higher elevation.

If a hooded inlet is to be used, its construction details should be available in the design or may be obtained from Soil Conservation Service engineers.

A satisfactory pipe spillway of the second type mentioned above can be constructed, using for the horizontal portion, rigid pipe such as concrete, asbestos cement, or metal 6 inches or larger in diameter. If concrete pipe is used, unequal settlement of the embankment will tend to crack the joints. For this reason the joints in concrete pipe should not be mortared or made rigid, but instead the pipe should be supported on concrete cradles and joints sealed with a vulcanized endless rubber gasket. Regardless of the material used, antiseep collars should be placed around the culvert in such size and number as will increase the length of the seepage line by approximately 20 percent of the length of the pipe.

The vertical section of this type spillway may be constructed when conditions will allow its practical installation. It should have a cross-sectional area at least 1½ times as large as that of the horizontal pipe and should be at least 4 times as deep as the diameter of the pipe. The inlet may be built of concrete, of concrete blocks, or of pipe.

A baffle may be attached to the mouth of the inlet to prevent whirlpool action. This baffle may be made of corrugated metal sheets or other durable material bolted to the inlet. It should extend a minimum of 2 feet from each side of the inlet and protrude 12 inches below and 36 inches above the lip of the inlet.

Where needed to protect human life or where debris will be a problem, a rigid debris guard that cannot easily be removed should be installed around the inlet. A simple and economical debris guard can be constructed by making a frame using four vertical posts and horizontal cross braces across the top of the posts, completely enclosing the sides and the top of the frame with woven wire. The sides of the debris guard should be at least 2 feet horizontally from the sides of the inlet pipe and should extend up to the design depth of flow.

The outlet end of the spillway pipe and the toe of the earth fill should be protected against erosion. This may be done by extending the outlet pipe a minimum of eight feet beyond the toe of the slope or by building a stone or concrete apron or spreader at the outlet end.

Where the pond is to be stocked with fish, the pipe spillway should be designed so as to permit the complete drainage from time to time. If a hooded inlet pipe spillway is used as shown in Fig. 9-68, some other provision may be necessary to drain the pond.

9. **Build up the earth fill**—After all installations have been made, the excavation of the pond and the placing of the earth fill for the dam can be completed. This may be accomplished with (a) an ordinary farm tractor, plow, and scraper of appropriate size, or (b) heavy equipment.

The fill of the dam should be built up in uniform layers about 6 inches deep, with each layer well compacted. If the fill material is dry, it should be sprinkled with water in order to obtain proper compaction. A sheepsfoot roller or other compaction device is recommended.

To obtain the proper slope to the sides of the dam and to bring the fill to the proper level, it is best to check, from time to time, with (a) a surveyor's level and rod, or (b) a carpenter's level with a template.

10. **Construct the emergency spillway**—This spillway (also known as an auxiliary or overflow spillway) is necessary in order to protect the dam against heavy rains and flood water. It will normally be installed around one end of the dam. A broad, flat-bottomed channel with a covering of sod or of masonry suitable to control erosion may be used. The dimensions and slope of this channel are critical and should be established by the design engineer.

11. **Seed the banks**—In order to prevent silting, immediately following completion of the pond, all areas up to the water's edge should be smoothed and seeded to adapted grass or grass-legume mixtures. Steep banks that are not accessible to a drill should be seeded by hand. Mulching the exposed surfaces with straw, or preferably with manure, will help hold the soil until grass is established.

The presence of water-fowl and other forms of wildlife can also be encouraged by planting trees and berry bushes near the pond. However, avoid planting trees and bushes on the dam, as their roots may weaken the structure. The herbaceous stock should not be planted so close to the pond bank as to prevent mowing the strip immediately adjacent to the water's edge. Both sides of the earth dam should be mowed periodically as a rank growth of vegetation invites rodents which will burrow into the embankment and may cause failures or loss of water.

Fence Animals Out

Animals should always be fenced away from the pond and dam site in order to ensure clean, clear water and to safeguard them against drinking water contaminated with their own feces. The fence should keep animals away from the upstream sides of the pond and away from the dam.

Install a Springboard or Drifting Raft if Swimming Is Desired

If swimming is desired, a support for a springboard may be anchored to the floor of the pond at the time of construction; otherwise, a drifting raft may be anchored in the center of the pond.

Stock with Fish if Desired

Clear water ponds stocked with fish provide both recreation and food. A well-managed pond will yield upwards of 150 pounds of fish per acre each year. In addition, a good supply of fish will lessen the mosquito menace.

Information on fertilizing the pond and on the proper kind of fish for stocking can be obtained from state and federal fish hatcheries, from the state agricultural college, or from the U. S. Department of Agriculture.

Maintenance

If properly designed and constructed, a small earth dam and resulting pond should require a minimum amount of maintenance. However, like any other farm structure, the dam and the pond should be given regular inspection and maintenance as required. Floating debris, rodent infestation, and shore weeds should be removed to protect the earth fill and spillway. Eroded areas in the spillway and around the discharge pipe should be repaired immediately. Submerged weeds and algae should be controlled by chemical means without injury to fish or livestock if the treatment conforms to the recommendations of the state agricultural colleges or federal fish and wildlife authorities. Under no conditions should trees or heavy vegetation be permitted to grow on the embankment.

SECTION 10

ANIMAL HEALTH, DISEASE PREVENTION, AND PARASITE CONTROL[1]

By

Dr. Robert F. Behlow, DVM, Professor and Extension Veterinarian,
North Carolina State University, Raleigh, North Carolina

and

Dr. M. E. Ensminger, Ph.D., Distinguished Professor, University of Wisconsin–River Falls;
Adjunct Professor, California State University–Fresno; and
Collaborator, U.S. Department of Agriculture

Contents **Page**

Signs of Good Health 874
Signs of Ill Health 875
Genetic Resistance to Disease 875
Animal Diseases 876
Relative Susceptibility of Animals and Man to TB Bacilli 916
Hardware Disease 916
Liver Abscesses 916
Animal Parasites 917
Choice of Wormer 917
Forms of Insecticides 917
Recommended Compounds for Control of Internal Parasites, by Animal Species, Table 10-4 918
Application of Insecticides 919
Use Insecticides Safely 919
Parasites and Their Control 919
Cattle Parasites and Their Control, Table 10-5 920
Sheep and Goat Parasites and Their Control, Table 10-6 930
Swine Parasites and Their Control, Table 10-7 940
Horse Parasites and Their Control, Table 10-8 950
Mites and Their Control, Table 10-9 958
Poisonous Plants 958
Common Poisonous Plants 961
Chemical Poisoning 978
Disinfectants 984
General Animal Sanitation and Disease Prevention 986
A General Program of Animal Health, Disease Prevention, and Parasite Control 987
A Program of Beef Cattle Health, Disease Prevention, and Parasite Control 988
Cattle Feedlot Disease and Parasite Control Program 990
A Program of Dairy Cattle Health, Disease Prevention, and Parasite Control 991
A Program of Sheep Health, Disease Prevention, and Parasite Control 994
A Program of Swine Health, Disease Prevention, and Parasite Control 996
Specific Pathogen-Free (SPF) Pigs 999
A Program of Horse Health, Disease Prevention, and Parasite Control 1000
Regulations Relative to Disease Control 1001
Food and Drug Administration (FDA) 1002
U.S. Department of Agriculture (USDA) 1002
State Veterinarians, Sanitary Commissions, and Boards 1003
Interstate Requirements with Respect to Animal Health 1003
Quarantine 1003
Foreign Disease Protection 1003
Federal Quarantine Center 1004
Indemnity Payments 1004
Federal Indemnity Payments 1004
State Indemnity Payments 1006

[1]The material presented in this section is based on factual information believed to be accurate, but is not guaranteed. Where the instructions and precautions given herein are in disagreement with those of competent local authorities or reputable manufacturers, always follow the latter two.

Each year, stockmen suffer staggering losses from diseases and parasites—internal and external. Death takes a tremendous toll. Even greater economic losses—hidden losses—result from failure to reproduce living young, and from losses due to retarded growth and poor feed efficiency, carcass condemnations and decreases in meat quality, and labor and drug costs. Also, considerable cost is involved in keeping out diseases that do not exist in the United States; and quarantine of a diseased area may cause depreciation of land values or even restrict whole agricultural programs. Additionally, and most important, some 200 different types of infectious and parasitic diseases can be transmitted from animals to human beings; among them, such dreaded diseases as brucellosis (undulant fever), leptospirosis, anthrax, Q fever, rabies, trichinosis, tuberculosis, and tularemia. Thus, rigid meat and milk inspection is necessary for the protection of human health. This is added expense which the producer, processor, and consumer must share, and which adds substantially to the cost of food and fiber.

It has been conservatively estimated that annual U.S. losses from the more important diseases, parasites, and pests of livestock and poultry aggregate $10 billion.[2]

Studies, including extensive surveys made by the author, reveal that American farmers and ranchers suffer the following appalling losses:

1. Twelve percent of the cows that are bred never calve.

2. Calf losses from birth to weaning run 6%.

3. Ten percent of all calves (beef and dairy combined) are afflicted by calf scours, and 18% of all dairy calves so afflicted die.

4. Cattle feedlot losses on calves run about 2.0%; and on yearlings, about 1.4%.

5. About 1.5 million head of cattle die in feedlots each year, at an estimated loss of over $750 million.

6. Sterility and delayed breeding in dairy cattle make for an estimated yearly loss of $60 per cow, or a national total of $650 million.

7. Dairy herds average 10% breeding difficulty at any one time and 1.85 services per conception.

8. Retained placenta occurs in about 10% of the parturitions of dairy cattle.

9. Nearly 40% of all dairy cows have some form of mastitis, which causes a yearly loss of $225 per afflicted cow, according to the National Mastitis Council.

10. Five percent of the ewes bred never lamb.

11. Twenty-five percent of the lambs born die between birth and weaning.

12. About 2.5% of lambs on finishing rations die.

13. Fifteen percent of the sows bred never farrow.

14. Thirty percent of the pigs born die between birth and weaning.

15. Horsemen are spending millions on needless concoctions.

16. One-half of all pregnant mares either abort or produce weak foals.

17. We produce only a 50% foal crop, which means that two mares are kept a whole year to produce one foal.

18. Six percent of the foals born die between birth and weaning.

It is not intended that this book shall serve as a source of home remedies. Rather, the enlightened stockman will institute a program designed to assure herd health, disease prevention, and parasite control. When animal disease troubles are encountered, he will not attempt to diagnose or treat but will call upon his local veterinarian in exactly the same manner as he calls upon the family doctor when human ill health is encountered. But a well-enlightened producer will (1) be in a better position to institute a program designed to assure herd health, (2) more readily recognize any serious outbreak of disease and promptly call a veterinarian, (3) prevent unnecessary suffering of sick animals, (4) be better qualified to assist the veterinarian in administering treatment, and (5) be more competent in carrying out a program designed to bring the disease under control with a minimum spread of the infection.

SIGNS OF GOOD HEALTH

In order that stockmen may know when animal disease strikes, they must first know the signs of good health, any departure from which constitutes a warning of trouble. Some of the signs of good health are:

1. **Contentment**—Healthy animals appear contented; the cow will stretch on rising, the sheep will stand or lie quietly, the pig will curl his tail, and the horse will look completely unworried when resting.

2. **Alertness**—Healthy animals are alert and bright eyed and will prick their ears up on the slightest provocation.

3. **Eating with relish, and cudding by ruminants**—In healthy animals, the appetite is good and the feed is attacked with relish (as indicated by

---

[2]The author arrived at the $10 billion figure in two ways:

1. From listing prepared by the associate deans and directors of veterinary research programs for the Council of Deans, Association of American Veterinary Medical Colleges; based on information available on disease losses as of February 1, 1981, and including cattle (beef and dairy), sheep, swine, poultry, horses, and fish. This listing showed animal diseases losses of $10 billion per year.

2. *Losses in Agriculture*, Ag. Hdbk. No. 291, Agricultural Research Service, USDA, 1965, pp. 73-82, Tales 26-32. This report estimated annual U.S. losses from the more important diseases, parasites, and pests of livestock and poultry at $2.8 billion for the period of 1951-60; during this 1951-60 period livestock and poultry had an average farm value of $14 billion. The author updated the $2.8 billion figure by applying 1981 livestock and poultry numbers and values ($60.8 billion), and allowing for 1/6 fewer annual losses in 1981 than during the period 1951-60, due to improved animal health. So, 14:2.8 :: 60.8:X. Then, X = $12 billion. Then, 1/6 (or .1667%) × 12 = $2 billion. So, $12 billion − $2 billion = $10 billion.

eagerness to get to the trough, wagging the tail, etc.). In cattle and sheep, cudding is a sure sign of good health, and is one of the first things to disappear in sickness.

4. **Sleek coat and pliable and elastic skin**—A sleek, oily coat and a pliable and elastic skin characterize healthy animals. When the hair coat loses its luster and the skin becomes dry, scurfy, and hidebound, there is usually trouble.

5. **Bright eyes and pink eye membranes**—In healthy animals, the eyes are bright and the membranes—which can be seen when the lower lid is pulled down—are whitish pink in color and moist.

6. **Normal feces and urine**—The consistency of the feces varies with the diet; for example, when animals are first turned on lush grass, they will be loose. Also, the consistency and dryness of the feces vary between species, but they should be firm and not dry. And there should not be large quantities of undigested feed. The urine should be clear. Both the feces and urine should be passed without effort, and should be free from blood, mucus, or pus.

7. **Normal temperature, pulse rate, and breathing rate**—Table 10-1 gives the normal temperature, pulse rate, and breathing rate of farm animals. In general, any marked and persistent deviations from these normals may be looked upon as a sign of animal ill health.

TABLE 10-1

NORMAL TEMPERATURE, PULSE RATE, AND BREATHING RATE OF FARM ANIMALS

Animal	Normal Rectal Temperature		Normal Pulse Rate	Normal Breathing Rate
	Average	Range		
	(°F)	(°F)	(min)	(min)
Cattle	101.5	100.4-102.8	60-70	10-30
Sheep	102.3	100.9-103.8	70-80	12-20
Goats	103.8	101.7-105.3	70-80	12-20
Swine	102.6	102 -103.6	60-80	8-13
Horses	100.5	99 -100.8	32-44	8-16
Poultry	106.0	105 -107	200-400	15-36

Every stockman should provide himself with an animal thermometer, which is heavier and more rugged than the ordinary human thermometer. The temperature is measured by inserting the thermometer full length in the rectum, where it should be left a minimum of three minutes. Prior to inserting the thermometer, a long string should be tied to the end to aid in retrieving it.

In general, infectious diseases are ushered in with a rise in body temperature, but it must be remembered that body temperature is affected by stable or outside temperature, exercise, excitement, etc.

The pulse rate indicates the rapidity of the heart action. The pulse of different farm animals is taken at the following body areas: cattle, either on the outside of the jaw just above its lower border, on the soft place immediately above the inner dewclaw, or just above the hock joint; sheep and swine, on the inside of the thigh where the femoral artery comes in close proximity to the skin; and horse, either at the margin of the jaw where an artery winds around from the inner side, at the inside of the elbow, or under the tail. It should be remembered that the younger, the smaller and the more nervous the animal, the higher the pulse rate. Also, the pulse rate increases with exercise, excitement, digestion, and high outside temperature.

The breathing rate can be determined by placing the hand on the flank, by observing the rise and fall of the flanks, or, in the winter, by watching the breath condensate in coming from the nostrils. Rapid breathing due to recent exercise or excitement should not be confused with disease.

SIGNS OF ILL HEALTH

Most sicknesses are ushered in by one or more departures from the signs of good health. They're foretold by signs of poor health, by indicators that tell the expert caretaker that all is not well—that tell him that his animals will go off feed tomorrow, and that prompt him to do something about it today.

Among the signs of animal ill health are: lack of appetite—the animal does not eat or graze normally; listlessness; droopy ears; sunken eyes; humped-up appearance; abnormal dung—either very hard or watery dung suggest an upset in the water balance or some intestinal disturbance following infection; abnormal urine—repeated attempts to urinate without success or off-colored urine should be cause for suspicion; abnormal discharges from the nose, mouth, and eyes, or a swelling under the jaw; unusual posture—such as standing with the head down or extreme nervousness; persistent rubbing or licking; dull hair coat and dry, scurfy, hidebound skin; pale, red, or purple mucous membranes lining the eyes and gums; reluctance to move or unusual movements; higher than normal temperature; labored breathing—increased rate and depth; altered social behavior such as leaving the herd and going off alone; and sudden drop in production—weight gains, milk, wool, or work.

GENETIC RESISTANCE TO DISEASE

Genetics as a tool for eliminating or controlling certain diseases holds promise. In this area, plant breeders have led the way. In 1905, it was discovered that certain varieties of wheat were more resistant to mycotic stem rust than others, thereby laying the foundation for important advances in the knowledge of genetic resistance to disease. Subsequently, scientists have evolved with many varieties of plants showing genetic resistance to disease. Evidence that similar genetic resistance to disease hold for animals has been demonstrated by experiences and experiments. For example, Brahman cattle are more resistant to certain parasites, notably Texas fever, than the British breeds.

The application of genetics to disease control in animals presents greater problems than in plants; it's more expensive and time-consuming. Also, to be of greatest practical value, it would be necessary to develop strains or breeds of animals that are genetically resistant to several diseases. Nevertheless, the stakes are high and this approach is worthy of greater attention than it has received in the past.

ANIMAL DISEASES

Table 10-2 is a summary of the common nonnutritional diseases and ailments affecting animals. (The nutritional diseases and ailments of animals are covered in Table 4-100 of Section 4.)

TABLE 10-2—ANIMAL

Disease	Species Affected	Cause	Symptoms and Signs (or age group most affected)
Actinomycosis, and **Actinobacillosis** (see Lumpy jaw)			
African sleeping sickness (African trypanosomiasis)—a chronic infectious disease caused by protozoa.	All animals and man.	Flagellates (protozoa) of the genus *Trypanosoma*. The trypanosomes live in the blood of their host, where they multiply and release poisonous byproducts of metabolism.	In man or domestic animals, they invade the nervous system, causing lethargy and finally death. The trypanosomes are spread from host to host by bloodsucking tsetse flies. When a tsetse fly withdraws blood from an infected animal, trypanosomes are sucked into its intestine, where they multiply and undergo developmental changes. They migrate to the fly's salivary glands, in which they further develop and multiply. The fly can then transfer the trypanosomes via saliva into a vertebrate host.
Anaplasmosis (see Table 10-5, Cattle Parasites and Their Control)			
Anthrax (splenic fever, charbon)—an acute, infectious disease. 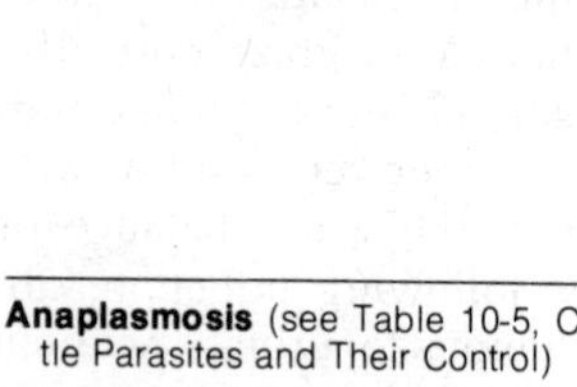 Fig. 10-1	All warm-blooded animals and man.	*Bacillus anthracis*, a large, rod-shaped organism.	History of sudden death. Sick animals are feverish, excitable, and later depressed. They carry head low and lag behind herd. Respiration is rapid. There are swellings over the body, especially around the neck region. Milk secretion may turn bloody or cease entirely, and there may be a bloody discharge from all body openings. Cattle are most susceptible. Most frequent in mature animals on summer pasture.

Footnote on last page of table.

DISEASES AND HEALTH[1]

Distribution and Losses Caused By	Treatment	Control and Eradication	Prevention	Remarks
The disease is confined to Africa, wherever tsetse flies (the vectors) occur. It makes large areas of Africa uninhabitable for man. If trypanosomiasis were brought under control, the Savannah pastures of the tsetse fly-infested area would carry a cattle population of 120 million head, almost equal to the total cattle population of the U.S.	Treatments have been disappointing.	It is difficult to control, because many wild animals serve as a reservoir of trypanosomes.	Prevention is dependent upon control of the tsetse fly vector by means of insect repellants, insecticides, and brush clearing.	This organism has a life cycle similar to that of the malarial parasite. Like malaria, it can be spread from man to man, but unlike malaria, it can also be carried from animals to man.
General throughout the world in so-called anthrax districts.	Early treatment with massive doses of penicillin may be effective.	Quarantine infected herds, and withhold all milk and other products from the market until the danger of disease transmission is past. All carcasses and contaminated material should be burned completely or buried deeply and covered with quicklime, preferably on the spot. Vaccinate all exposed but healthy animals, rotate pastures, and initiate a rigid sanitation program. Spray affected and normal animals to avoid fly transmission of infection.	In infected areas, vaccination should be repeated each year, usually in the spring; and there should be adequate fly control by spraying animals during the insect season. At least 9 types of biologics (serums, bacterins, and vaccines) are now available and used in anthrax prevention. The choice of the bacterin should be left to the veterinarian or livestock sanitary officials. Prevention of anthrax in man depends on (1) eradication of the disease in animals; (2) elimination of industrial infections (tanneries, woolen mills, and factories utilizing animal hair); and (3) early diagnosis and prompt treatment of infected cases.	The farmer or rancher should never open the carcass of a dead animal suspected of having died from anthrax; instead, the veterinarian should be summoned at the first sign of an outbreak. Control measures should be carried out under the supervision of a veterinarian. The bacillus that causes anthrax can survive for years in a spore stage, resisting all destructive agents.

(Continued)

TABLE 10-

Disease	Species Affected	Cause	Symptoms and Signs (or age group most affected)
Arthritis (see Navel infection)			
Atrophic rhinitis	Swine (apparently the disease in swine is not related to atrophic rhinitis in man).	*Bordetella bronchiseptica* and other bacteria. Also, a calcium-phosphorus imbalance or a calcium deficiency in growing pigs has been shown to produce similar, if not identical, lesions.	Persistent sneezing, which be comes more pronounced as th pigs grow older, is the firs symptom. At 4 to 8 wks. of age the snout begins to show wrir kles, and the snout may bulg and thicken. At 8 to 16 wks. c age, the snout and face ma twist to one side. Affected pig become rough all over, an make slow and inefficien gains. Nose bleeding is ofte seen. Actual death may be du to pneumonia. Young pigs under 60 to 80 l weight are most susceptible.
Babesiasis (see Equine piroplasmosis, Table 10-8, Horse Parasites and Their Control)			
Bacillary hemoglobinuria (or red water disease)—an acute, infectious disease.	Cattle. Sheep to a lesser degree.	*Clostridium hemolyticum*, an anaerobic bacterium.	Death in 2 days or less. Appetite rumination, and milk flow suddenly cease. High temperature and rapid breathing. Both urine and feces usually blood tinged. All ages affected, but most losses occur in cows over 1 yr. of age on summer pasture.
Bighead (or swellhead, photosensitization)	Mostly sheep and cattle, although it occurs in horses, swine, and goats.	Any of the following plants: little leaf horsebush *(Tetradymia glabrata)*, spineless horsebush *(T. canescens)*, agave *(Agave lecheguilla)*, sachuiste *(Nolina texana)*, smartweed *(Polygonum persicaria)*, St. Johnswort *(Hypericum perforatum)*, buckwheat, and wet alsike clover.	Affects skin of white or light-colored animals. Swelling of the ears, eyelids, anc lips, accompanied by intense itching and seepage of fluic through the skin.
Black disease 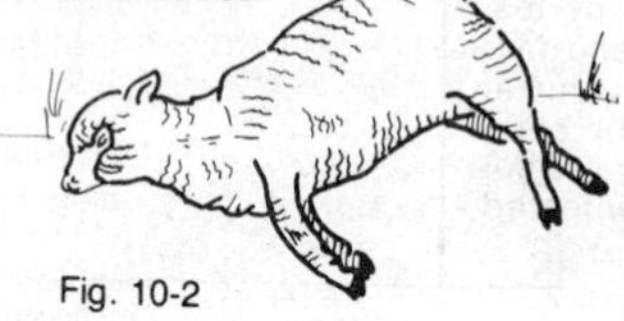 Fig. 10-2	Sheep.	A spore-bearing anaerobe, *Clostridium novyi*. Immature liver flukes *(Fasciola hepatica)*, which are present in the bile ducts, provide site for *C. novyi* action.	Rapid death with no distinc symptoms. Usually affects adult sheep tha are on pastures harboring fluke-bearing snails, and strikes during the summer and fall.

—(Continued)

Distribution and Losses Caused By	Treatment	Control and Eradication	Prevention	Remarks
The disease is quite widespread in the U.S. and has been reported in other countries. It is estimated that 5 to 10% of the slaughtered swine coming from major swine-producing areas of the U.S. have turbinate atrophy.	*B. bronchiseptica* is sensitive to the sulfonamide drugs. The 2 most commonly used ones are: 1. Sulfamethiazine medication in the feed at the level of 100 to 450 g of sulfamethiazine per ton of complete ration. 2. Sodium sulfathiazole administered in the drinking water, at the level of ⅓ to ½ g per gallon of water. Young animals should be treated for 5 weeks, whereas older animals need to be treated only 4 weeks.	The following control plans are effective: 1. Isolate bred females in separate lots and never allow contact with any other swine except their offspring until they are culled. Keep individual litters separate until a month after removal of the sow at weaning time. Then select and isolate new breeding stock from those litters which show no evidence of symptoms. 2. Allow the pigs to nurse the sow one or a few times (to obtain colostrum), and then remove and raise them as orphaned pigs; but never allow them to get near the head of their dam. 3. Obtain specific pathogen-free breeding stock. 4. Procure the offspring by hysterotomy and raise them in as nearly sterile an environment as possible. 5. Catch the pigs at birth on a sterile cloth, with subsequent removal to an isolated area. When considering the 4th and 5th methods given above, caution must be given that hand-rearing baby pigs that have received no colostrum is beset by many difficulties.	Select breeding stock from herds known to be clean, and isolate for a period of 30 days or until their litters have been weaned without showing symptoms. Use clean farrowing quarters. If feeder pigs are purchased, select animals above 60-80 lb in wt, as they are less susceptible. Separate different age groups. A bacterin licensed by the USDA became available in 1977.	Atrophic rhinitis is not the same as bull nose. No simple test is available to check for carrier swine.
It is found in poorly drained areas and is possibly associated with liver fluke damage. It has been reported on the poorly drained swampy pastures and meadows of Sierra Nevada and on the Coast Range Mountains of Calif., Nev., and Ore. Also in the Ellensburg Valley of Wash. and in La., Fla., Mont., Ida., and Tex., and in the Andes of Chile. A mortality up to 100% occurs in untreated cases.	Antiserum from hyperimmunized animals, in conjunction with antibiotics.	Remove cattle from areas of known infection. Vaccinate at 6-month intervals in heavily infected areas.	Vaccination about 2 wks. prior to the time of the previous annual outbreak is the best practical control measure. Also, the destruction of snails (by draining stagnant water and using bluestone) may aid in prevention.	It should be understood that bloody urine (red water) may also be one of the symptoms in such conditions and diseases as lack of phosphorus, cattle tick fever, leptospirosis, plant poisoning, and anthrax. Therefore, one should not be misled by this symptom alone.
Worldwide.	No antidotes have been discovered. Keep animals in shade during day and let them graze at night. Soothing and protective preparations may be applied to affected skin to allay itching and control infection; and the veterinarian may inject antihistamines.	Where practical, eradicate the causative plant(s). Practice good range or pasture management, so that good forage is available and animals have less tendency to eat weeds.	Keep animals away from areas where causative plants are located. On the range, feed a supplement when trailing animals through such plant-infested areas.	Condition results when animals become hypersensitive to sunlight following ingestion of certain substances, usually plants, containing sensitizing agents that render the tissue cells abnormally sensitive to light.
Throughout the U.S., wherever fluke-bearing snails abound.	The disease progresses so rapidly that treatment is not successful.	Eliminate the fluke. See discussion of Liver fluke in Table 10-6, Sheep and Goat Parasites and Their Control. Vaccinate with a toxoid animals exposed to fluke-bearing snails.	Destroy fluke-bearing snails. Vaccinate with a toxoid annually before going on snail-infested pastures (a sheep vaccinated 2 successive years is very resistant).	The bacteria gain entrance to the sheep when they are infested by the flukes. Consult with the veterinarian for differential diagnosis and control. Vaccination is the most practical means of control.

(Continued)

TABLE 10-

Disease	Species Affected	Cause	Symptoms and Signs (or age group most affected)
Blackleg (or black quarter, quarter ill, emphysematous gangrene)—a very infectious, highly fatal disease.  Fig. 10-3	Cattle, and less frequently sheep and goats.	*Clostridium chauvoei*, an anaerobic bacterium.	**Cattle:** Lameness, and swellings over the neck, shoulder, flanks thighs, and breast, which crackle under pressure. High fever, loss of appetite, and severe depression. Death usually occurs within 2 days after onset of symptoms. Most frequently seen in cattle ranging from 4 months to 2 yrs. of age, but may occur in older animals. **Sheep:** Swellings are most frequently in the region of a recent wound; from shearing, castrating, docking, bruising from fighting, or from parturition. In sheep, all ages may be affected.
Blue bag (see Mastitis)			
Bluetongue[2] (at first called "Sore muzzle" in the U.S.)—an infectious but noncontagious virus disease.	Sheep. Cattle are sometimes mildly affected.	A virus, transmitted by insects of the *Culicoides* sp.	A blue tongue, high temperature (104°-107°F), depression and loss of appetite, rapid and extreme loss of weight, reddened mucous membrane of the mouth which turns purplish or blue in color, frothing of the saliva, formation of lip ulcers, offensive odor, discharge from the eyes and nose, weakness, appearance of a red band at the top of the animal's hoof, lameness (and in extreme cases the hoof may slough off), and loss of wool. When the disease first appeared in the U.S., it was erroneously diagnosed as "sore mouth."
Bovine pulmonary emphysema (also known as cow asthma, panters, lungers, fog fever, skyline fever, summer pneumonia, green grass poisoning, and grunts)	Cattle.	Unknown.	The disease is characterized by rapid and labored breathing (the animal forces air from the lungs, and may grunt with each breath). Affected animals may breathe through and froth at the mouth. The temperature remains normal or only slightly elevated, and the appetite is good.
Bovine virus diarrhea (BVD, mucosal-disease complex)	Cattle. Deer.	Virus.	Characterized by diarrhea and dehydration. Incubation period of 7 to 9 days following exposure to the virus. High temperature (104°-107°F) for 2 to 5 days. Nasal discharge, rapid breathing, depression. The animal goes partially off feed. Some animals have a prompt recovery; in other cases, signs persist accompanied by a nasal discharge and diarrhea. Sometimes blood flecks occur in the feces. Cough, eye lesions, and lameness may affect 10% of the herd. Abortions may occur 3 to 6 weeks after the infection. A marked loss in milk production occurs in lactating cows.

Footnote on last page of table.

—(Continued)

Distribution and Losses Caused By	Treatment	Control and Eradication	Prevention	Remarks
Widespread, especially in the western range states. Infected territories are referred to as "hot areas." Few recoveries.	Massive doses of antibiotics will sometimes save an animal, provided it is given during the early stages of the disease. But a good immunization program is the key to preventing losses from blackleg.	Burn or bury carcasses. Vaccinate all healthy animals. Eradication of blackleg from pastures is difficult, if not impossible.	In areas where blackleg is known to have existed, vaccinate all cattle at 3 to 4 mos. of age. In endemic areas, the first vaccination should be given at one month of age. Since young calves may not develop lasting immunity, all animals under 4 mos. of age when vaccinated in the spring should be revaccinated in the fall. In areas where the disease is prevalent, some cattle owners vaccinate all new additions to the herd. In "hot areas," vaccinate sheep 2 to 4 wks. before shearing, castrating, and docking.	Vaccination in cattle usually effective for a period of from 9 to 12 months, and a natural resistance tends to develop when the animal is about 2 yrs. of age. In "hot areas," it may be necessary to vaccinate at 3 to 4 mos. of age and annually thereafter until 3 yrs. of age. In sheep, vaccination is used as a routine measure in spore-infested areas, and also as a preventive at the beginning of an outbreak.
Bluetongue has been known in South Africa since 1876. Also, it has appeared in Cyprus and Palestine. It was first definitely diagnosed in the U.S. in 1953, although it probably had been present in Texas since 1948 where it was known as "sore muzzle." In the U.S., found only in the Southwest. In South Africa, mortality rates usually do not exceed 10%, but have run as high as 70%. So far, the disease has proved less virulent in the U.S. (from 5-30% of stricken sheep may die), and the mortality rate has been considerably lower. A severe economic loss results from the reduction in condition and wool.	Methods of treatment tried to date have been of little value. Good nursing will save some affected animals. Secondary complications may be treated by the veterinarian with appropriate antibiotics and sulfonamides. May also be necessary to treat or prevent secondary screwworm infections of the lesions in the Southwest.	Banning the inshipment of sheep from infected areas will help, but it must be remembered that gnats and other insects that transmit the disease are carried about by automobiles, trucks, and planes.	Vaccination with vaccines (a modified virus vaccine that is grown and attenuated in chick embryo culture) is successful against the presently existing U.S. strain of the virus. Vaccinate all ewes and rams at shearing time each year, and all replacement lambs at 3½ mos. of age. Do not vaccinate ewes during the first 2 months of pregnancy. Otherwise, a large number of "crazy lambs" may be born. Night housing of sheep on high land beyond the flying range of the vectors also protects against the exposure of the sheep to the bites of virus-infected midges.	It is not transmitted by contact, but it is transmitted by insects—such as sand flies. Bluetongue becomes prevalent in summer and stops abruptly after the first hard frost. Rams are more susceptible than ewes. A sheep that has had bluetongue becomes immune for life against that strain, but not against other strains. Suspected cases of bluetongue should be diagnosed by the local veterinarian and then reported to state or federal officials.
Widely distributed, especially where cattle are moved from dry range to green mountain or irrigated pastures in the fall or late summer.	When bovine pulmonary emphysema strikes, call a veterinarian. Atropine and antihistamines are useful in treating the disease.		Prevention consists in avoiding sudden changes from dry or poor pasture to immature, green feed. Some dry hay should be fed while making the transition.	
This disease is widespread in the U.S. The chief loss is economic, caused by loss in weight, poor growth—even after recovery, loss of milk production, and from abortions. Mortality is low, rarely exceeding 5%.	Antibiotics or sulfonamides effectively combat the secondary bacterial invaders that accompany the disease.	Practice good sanitation and disease preventive measures; isolate sick animals.	Avoid contact with affected animals, and do not use contaminated feed and water. Isolate new animals for 30 days. A modified live-virus vaccine is available. One vaccination should last a lifetime. Do not vaccinate pregnant cows because of possible abortions and birth defects, and do not vaccinate calves under 6 months of age because it may be ineffective due to the temporary immunity from colostrum of immune cows. Vaccinate cattle on farms or feedlots where this disease is a constant problem.	The New York virus diarrhea and the Indiana virus diarrhea are the same virus. Serologic surveys have been made (New York and Florida), showing that between 50 and 63% of the cattle had been infected by the time they reached adulthood. Diarrhea is not a good name for this disease since not all animals exhibit the diarrhea.

(Continued)

TABLE 10-2

Disease	Species Affected	Cause	Symptoms and Signs (or age group most affected)
Brucellosis (Bang's disease, undulant fever, malta fever)—a hidden disease; one of the most serious and widespread affecting the livestock industry.	Cattle. Sheep (rarely). Goats. Swine. Man. The *suis*, *abortus*, and *melitensis* strains are seen in horses, and both the *suis* and *melitensis* strains are seen in cattle, but the incidence is less frequent than *abortus*. Man is susceptible to all 3 types.	*Brucella abortus*. *B. suis*. *B. melitensis*.	The act of abortion is the most characteristic symptom (especially in cattle), although not all animals that abort are affected and not all affected animals abort. In cattle, the typical symptoms are (1) abortion in the last third of pregnancy, (2) retained afterbirth, (3) several services per conception, and (4) uterine infections. There are no marked symptoms in goats. In swine, abortion and sterility are not so common as in cattle; infection may cause swollen joints and lameness, and swelling or atrophy of the testes, epididymus, and prostate in the male. In man, the disease is characterized by chills, headache, severe night sweats, fever, and extreme weakness.

Fig. 10-4

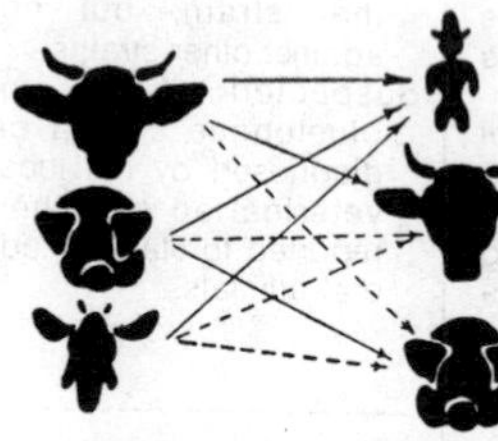

SOURCES OF INFECTION
Dotted lines indicate sometimes a source

Fig. 10-5

—(Continued)

Distribution and Losses Caused By	Treatment	Control and Eradication	Prevention	Remarks
Worldwide. It is the most important U.S. animal-human disease, and there is great economic loss in fewer animal offspring, in breeding and parturition trouble, in lowered milk production, etc. For the U.S. as a whole, fewer than 1% of all cattle tested (including both beef and dairy animals) react. It is rather common in goats, but rare in sheep.	Since there is no successful animal treatment, farmers and ranchers should not waste valuable time and money on so-called cures that are advocated by fraudulent operators. In man, the recommended treatment is the antibiotic Auremoycin administered for at least 3 weeks.	The nationwide cooperative federal-state brucellosis eradication program has been very successful in reducing the incidence of bovine brucellosis in the U.S. The program consists of bood testing and certifying brucellosis-free herds and areas. The certification progresses from an individual herd, thence to an area or county, thence to a state. In 1974, 30 states were certified as brucellosis free and 20 more had modified certified status. **Cattle:** Two principles are involved: (1) finding infected animals and eliminating them from the herd, and (2) vaccination. In heavily infected herds where valuable animals are involved, the test-and-slaughter plan is not practical. In such herds, vaccination with Strain 19 at 2 to 10 months of age is recommended. Currently, a new French vaccine, Strain H-38, is being tested in the U.S. It uses a killed brucellosis organism, whereas Strain 19 uses a live organism. Strain H-38 offers new hope for brucellosis eradication, but it is still in the experimental stage. In lightly infected herds, blood testing and removal of reactors is recommended. If there is danger of exposure, calfhood vaccination should be used as a protective measure. A federal-state cooperative plan for the control and eradication of brucellosis is in progress in the U.S. under this program. *Certified* herds are those that are free of the disease; *Modified Certified Areas* are areas that, as a result of complete testing, are considered nearly free of the disease; *Certified Brucellosis-free* are former Modified Certified Areas in which continued testing indicates that the disease has been completely eradicated. **Goats:** Blood testing and the elimination of reactors is recommended. It is claimed that strains of *B. melitensis* used as a bacterin (killed vaccine) and as a vaccine induce a high degree of immunity in sheep and goats. **Swine:** Several plans for eradication of brucellosis in swine are followed. With an infected commercial herd, it is recommended that the entire herd be sold for slaughter. With a valuable purebred herd, blood testing and slaughter is recommended, with the separation of the pigs from the sows if there are several reactors. Vaccination of swine has not been successful and is not recommended.	Buy replacement animals that are free of the disease and that are from herds known to be free of the disease. Divert or fence off drainage from infected areas. Animals that are purchased or that are shown should be isolated for 30 days and tested before adding to the herd. Avoid visiting infected farms or premises, as the germs may be brought home on shoes or clothing. For the same reason feeds should not be bought from such farms, and one should be aware of used feed bags. Do not use calfhood vaccination unless (1) there is a disease problem in the herd or in an adjoining herd, or (2) it is so required in order to ship cattle into certain states. However, in problem areas vaccinate with Strain 19; dairy heifer calves should be vaccinated at 2 to 6 months of age and beef heifers at 2 to 10 months of age.	Brucellosis derives its name from a British Army surgeon, Sir David Bruce, who, in 1887, discovered the bacteria later named *Brucella melitensis*. In cattle, it is called Bang's disease after a Danish veterinarian, who isolated *B. abortus* in 1896. Pasteurizing milk and cooking meat make these foods safe for human consumption. There is ample evidence that boars transmit the disease. Bulls are less apt to do so. The following tests are used for diagnosis of the disease in cattle: 1. Agglutination test, of which there are two common methods: a. The *tube, or "slow" method*—In which a blood sample is taken from the jugular vein; the blood is allowed to clot and the serum to separate; and the serum is mixed in small test tubes with a suspension of specially selected strain of *B. abortus*. Complete agglutination in dilutions of 1:100 and higher are positive. b. *The plate or rapid test*—This is a rapid agglutination test which is done on a glass slide or plate. The antigen consists of specially selected strains of *B. abortus* stained with gentian violet and brilliant green. 2. *Milk ring test*—This is a modification of the agglutination test which is done with milk. The test involves mixing the antigen with fresh milk. The test depends on the fact that clumps of agglutinated organisms are carried to the surface by rising fat globules. A positive test is indicated by a purple cream layer with white milk below. The milk ring test is a highly efficient and accurate screening test for locating infected dairy herds. 3. *Card test*—This test involves the use of a disposable card on which blood serum or plasma is mixed with buffered whole-cell suspension of *B. abortus* (antigen), which reacts (agglutinates) with antibodies in the blood serum of animals infected with brucellosis. The agglutination test is also used for diagnosing swine brucellosis. A specially prepared *Brucella* antigen is added to swine blood serum at definite dilution rates. If brucellosis is present, clumping, or agglutination, occurs in the test sample.

(Continued)

TABLE 10-2

Disease	Species Affected	Cause	Symptoms and Signs (or age group most affected)
Calf diphtheria—an acute, infectious disease.  Fig. 10-6	Cattle.	*Spherophorus necrophorus*, a soil organism; the same organism that is often found in foot rot.	Difficulty in breathing, eating, and drinking. Drooling. Yellowish crumbling masses and patches of dead tissue (diphtheritic membranes) on the borders of the tongue, adjacent to the molar teeth, and in the throat. It is largely confined to housed suckling calves and young feedlot cattle.
Calf scours (infectious diarrhea, or white scours)—an acute, contagious, and often rapidly fatal disease.	Cattle, especially young or newborn calves, under 2 days of age.	Both viruses and bacteria. A reo-virus is responsible for calf scours in calves from 12 hours to 5 days of age. A second virus, the corona virus, is responsible for scours in calves 5 to 10 days of age.	Calf scours can vary from a mild to a severe disease. In the mild form, the main symptom is softer than normal feces. The severely affected calf initially appears depressed and has a lack of appetite. Then begins a severe diarrhea which consists of yellowish, foul smelling, watery or foamy feces. These calves can have a rough hair coat, sunken eyes, and appear emaciated. In very acute cases, death can occur before diarrhea is observed; however, death usually occurs 2 to 3 days after the onset of diarrhea. Some degree of associated pneumonia occurs more frequently in stabled dairy calves than in beef calves on the range.
Caseous lymphadenitis (or pseudotuberculosis, lungers disease)—a chronic fatal disease of sheep.	Sheep.	The bacillus, *Corynebacterium pseudotuberculosis (C. ovis,* old name).	The first noticeable symptom is rapid breathing after exercise. Breathing becomes progressively more difficult, and respiration more rapid. In the final stage, affected animals show heavy flank breathing even when at rest. Early in the course of the disease, animals cough occasionally; later, there are often prolonged, spasmodic coughing spells. There is no rise in body temperature, unless an acute secondary lung infection sets in, and there is no loss of appetite. Affected sheep lose flesh and become extremely emaciated and weak. Eventually death occurs. Most losses occur in sheep 4 years old or older.

Footnote on last page of table.

—(Continued)

Distribution and Losses Caused By	Treatment	Control and Eradication	Prevention	Remarks
The disease occurs throughout the U.S. If untreated, the mortality is high. Mild cases may recover if treatment is given in early stages.	Treatment consists in using sulfa drugs or broad spectrum antibiotics. The local application of a proteolytic enzyme for removal of the dead tissue is indicated.	Segregate sick animals, and clean and disinfect all quarters. Check all animals daily. The services of a veterinarian should be sought in handling an outbreak of calf diphtheria.	Practice good management and sanitation, including cleaned and disinfected maternity stalls.	Calf diphtheria has no relationship to the diphtheria of humans.
Scours is the cause of more calf deaths than all other diseases combined. It is estimated that 10% of all calves in the U.S. are affected by the disease, and that 8% of beef calves and 18% of dairy calves so affected die.[3] It is further estimated that calf scours costs the cattle industry $200 million annually.[4]	Treatment of severely affected diarrheic calves should include: discontinuing feeding milk for 24 to 48 hours; giving fluids orally and by injection to combat dehydration; administering gastrointestinal protectants; and giving antibiotics orally and by injection. The choice of the antibiotic should be made by the veterinarian, for in many areas the bacteria associated with calf diarrhea are resistant to many of the available drugs.	To keep the disease away from the herd, one must prevent primary infection of the newborn. This rests on strict sanitary measures and isolation. The disease can be introduced by adding calves or adult animals from another herd. Calf diarrhea frequently occurs when a newly assembled herd begins to calve.	The most effective preventive measure of calf scours is a new modified live virus vaccine, developed by the University of Nebraska and released in 1973 by Norden Laboratories under the name Scourvax-Reo (after the reo-virus that the University of Nebraska identified as the primary cause of scours in young calves). Studies show that the product is about 80% effective in preventing the early, or neonatal, form of the disease that scientists have found is caused by a reo-virus. The vaccine is not 100% effective because it was developed to counteract a specific virus—reo-virus—while there are a number of other agents that can also cause diarrhea. Scourvax-Reo is sprayed into the back of the calf's mouth in a 4.0-ml dose 12 to 24 hours after birth. The modified live virus is incapable of causing disease, but produces an immune response when given orally to a susceptible calf.	If the virus causing calf diarrhea in a herd can be identified, vaccination of the cows before parturition or of the calves at birth with that specific agent will reduce the incidence and severity of the disease. Vaccination of cows before parturition with bacterins has been less successful because of the variation between strains of bacteria.
Occurs mostly in range bands, but rarely in farm flocks. Affected animals do not recover.	There is no specific treatment. Vaccines are ineffective. Good nursing will prolong life, but the desirability of doing so is questionable.	Animals showing symptoms should be removed and slaughtered. In badly infected flocks, shear lambs first, thereby alleviating the possibility of spreading the disease through shearing wounds.	In buying stock, particularly old ewes, be careful not to introduce this disease. Dip shears in disinfectant between sheep, and treat all shearing wounds with tincture of iodine.	Although it is largely speculation, it is thought that infections may occur through (1) docking, castrating, and shearing wounds, and (2) inhalation of dust. Many extensively infected animals that are in excellent condition are seen at slaughter.

(Continued)

TABLE 10-2

Disease	Species Affected	Cause	Symptoms and Signs (or age group most affected)
Cholera, hog—a highly contagious disease.  Fig. 10-7	Swine.	A filtrable virus.	Sudden onset, fever, loss of appetite, and weakness. Exhibit a scissorlike gait, and drink considerable water. Belly may be purplish-red in color. There is constipation alternating with diarrhea, and coughing is often evident. Disease is often confused with erysipelas. Pigs from immune sows are resistant to infection for several weeks after birth.
Circling disease (or listerellosis, encephalitis, or listeriosis)—an infectious disease.  Fig. 10-8	Cattle. Sheep. Goats. Also reported in swine, in horses, and in man.	*Listeria monocytogenes*, a bacterial infection.	Depression, staggering, circling, and strange awkward movements. Cows may abort. Positive diagnosis can be made only by laboratory examination of the brain.
Contagious ecthyma (see Sore mouth)			
Cowpox—an acute infectious virus disease of the udder and teats. 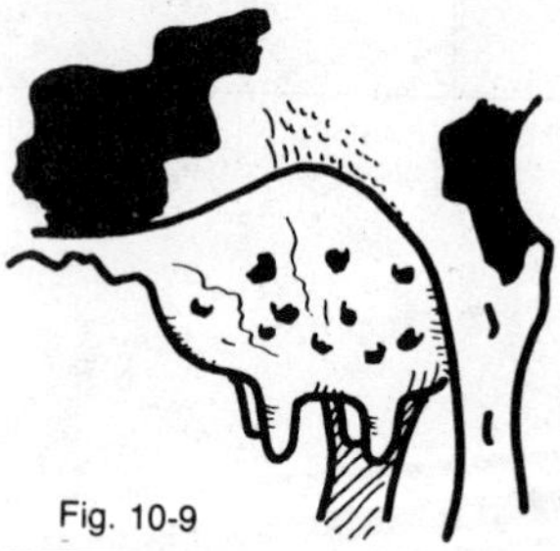 Fig. 10-9	Cattle. Man. (Transmissible from milkers to cows, and vice versa.)	A virus.	Reddish, painful spots appear on the udder; in 24 to 48 hours, these change to yellowish-white blisters. They then enter the pus stage, and in 10 to 12 days develop into scabs. In rare cases the pox eruptions may appear on the body or limbs. Occurs mostly in dairy herds.

—(Continued)

Distribution and Losses Caused By	Treatment	Control and Eradication	Prevention	Remarks
Cholera occurs wherever hogs are grown throughout the world, but the disease is more common in areas where large numbers of swine are kept and in garbage-feeding sections. It was first found in the U.S. in 1833. The disease is characterized by 95 to 100% morbidity, and almost as high mortality. On January 31, 1978, the U.S. Secretary of Agriculture declared the U.S. free of hog cholera.	No known treatment after the disease has developed, but serum (or hog cholera antibody concentrate) is often beneficial when administered in the incubation stage.	Any new outbreak of hog cholera in the U.S. is followed by immediate slaughter of the infected herd (with indemnity), decontamination of the premises, and thorough investigation to locate the source of the infection.	The passage, in 1961, of Public Law 87-209 for hog cholera eradication in the U.S. marked the beginning of the end of the disease in this country. But this drastic step was not without precedent. Both Canada and Great Britain had used the slaughter method to stamp out hog cholera. Federal funds were made available in 1962, and the following 4-phase program was initiated soon thereafter, with the objective of establishing a hog cholera-free swine population: *Phase I. Preparation*—Educating; surveys to determine incidence, to improve and standardize diagnostic systems, and to promote disease reporting; and enforcement of garbage cooking. *Phase II. Reduction of Incidence*—Quarantine of infected and exposed pigs; and stopping intrastate movement of infected pigs. *Phase III. Elimination of Outbreaks*—Depopulation (with indemnity) of infected premises, and increasing the control of biologics used. *Phase IV. Protection Against Reinfection*—Prompt enforcement of all procedures in Phase III and 21-day segregation of all swine imported from outside the state, with restrictions on the importation of pigs from states having endemic hog cholera. The program has been highly successful. Sporadic outbreaks of hog cholera have occurred, and will likely continue to occur. But we must remember that no country has ever eliminated hog cholera through the use of a vaccine. We must remember, too, that even if complete control by vaccination could be achieved, those countries demanding hog cholera-free pork would not likely accept pork if antibodies to hog cholera vaccination could be detected in the exporter's swine.	It is noteworthy that prior to the National Hog Eradication Program, hog cholera caused an estimated average annual loss of $2,945,000 per year.
Widespread.	Various sulfa derivatives, alone and in combination with antibiotics, have shown beneficial results if given early.	Isolate animals with circling disease. Move unaffected animals to clean premises. Use caution in handling infected tissues as humans are susceptible to the disease.	No commercial vaccine is available in the U.S. The following program may aid in preventing the disease: 1. Do not store silage in a silo that is in poor repair. 2. Do not feed silage from the top layer of an upright silo. 3. Never feed moldy or spoiled silage. 4. Provide clean, dry quarters during inclement weather. 5. Provide clean drinking water. 6. Control parasites. 7. Avoid stress.	Incidence ranges from 1 to 7% in an infected herd of cattle. The mortality rate of affected animals is extremely high. Silage samples can be submitted to a diagnostic laboratory to determine if *Listeria* are present.
Throughout the U.S. It is not considered a serious disease.	There is no treatment that will destroy the virus or shorten the duration of the disease. A 3% orthophenylphenate solution applied as a wash to the udder and teats or Whitfield's ointment will tend to ease tissue sensitivity and soften the skin.	Isolate pox-infected cows and arrange for their care by a separate attendant. Have the milker wash his hands thoroughly as he goes from cow to cow. Destroy flies and insects. Vaccination may stop spread.	Avoid contact with infected animals or milkers recently vaccinated against smallpox. Isolate new additions to the herd for 2 weeks. Milk diseased cows last.	As a rule, one attack of cowpox confers a lasting immunity. Milkers that have been recently vaccinated against smallpox (with the cowpox vaccine) may introduce the disease.

(Continued)

TABLE 10-2

Disease	Species Affected	Cause	Symptoms and Signs (or age group most affected)
Distemper (or strangles)—a widespread contagious disease.  Fig. 10-10	Horses. Mules.	*Streptococcus equi*, a bacterium.	Depression, loss of appetite, high fever, and discharge of pus from the nose. By the 3rd or 4th day of the disease, the glands under the jaw start to enlarge, become sensitive, and eventually break open and discharge pus. A cough is present. Any age but most common in young stock.
Dysentery, swine (or bloody scours, vibrionic dysentery, hemorrhagic dysentery)—an acute, infectious disease.	Swine.	*Vibrio coli*, a bacterium; and possibly other unknown causes.	The most characteristic symptom of swine dysentery is a profuse bloody diarrhea. Sometimes the feces are black instead of bloody and contain shreds of tissue. Most affected animals go off feed, and there is a moderate rise in temperature. Some pigs die suddenly after a couple of days of illness, whereas others linger on for two weeks or longer. On autopsy or postmortem, the large intestine is found to be inflamed and bloody.
Edema disease (or enterotoxemia, gut edema, gastric edema, and edema of the bowel)	Swine (young pigs).	Associated with the *E. coli* syndrome.	Usually affects most thrifty, rapid-growing pigs, 6 to 16 weeks of age; commonly ushered in by high temperature and swollen eyelids; constipation, inability to eat, and a staggering gait may be observed; affected pigs may display nervous symptoms such as fits or convulsions; as the disease progresses, the hog becomes completely paralyzed; death usually follows in from a few hours to 2 to 3 days.
Encephalomyelitis (see Equine encephalomyelitis)			
Enteritis (or swine enteritis)—a general term which includes at least 3 separate enteritic diseases.	Swine.	The cause of each of the 3 enteritic conditions is given, along with the symptoms and signs, in the column to the right.	The symptoms of each of the enteritic diseases are: *Acute salmonellosis*—An acute enteritis and septicemia, usually fatal. *Infectious enteritis*—An acute enteritis and gastritis, due to *S. cholerae-suis*. Some cases progress rapidly and terminate in death; others become chronic, with the affected pigs becoming stunted and unthrifty. *Necrotic enteritis*—Due to (1) a B vitamin deficiency and *S. cholerae-suis*, or (2) a similar condition—fibrino necrotic enteritis—caused by *S. cholerae-suis* and *Spherophorus necrophorus* (more common now), unsanitary conditions, etc. In necrotic enteritis, there is an inflammation (enteritis) of the large intestine in particular which leads to the development of areas of "necrotic" or dead tissue in the linings of the intestines; hence, the name necrotic enteritis or "necro" as it is often called.

—(Continued)

Distribution and Losses Caused By	Treatment	Control and Eradication	Prevention	Remarks
Worldwide. Death loss is very low.	Good nursing is the most important treatment. This includes clean, fresh water, good feed, and shelter with uniform temperature away from drafts. The veterinarian may prescribe one of the sulfas and/or antibiotics, or both. Early treatment is of the utmost importance in distemper.	Put affected animals in strict quarantine. Clean and disinfect contaminated quarters and premises.	Vaccination is the best preventive for strangles. Previously unvaccinated horses should receive the vaccine in three 10-ml doses at weekly intervals. One "booster" shot is sufficient for horses that are known to have had strangles. Vaccination should be on a yearly basis with one "booster" dose. The vaccine should not be given to newborn foals.	Affected animals are usually immune for the remainder of life.
Coast to coast, but it is most common in the Corn Belt, where the swine population is densest. Outbreaks of the disease are usually associated with animals that pass through central markets or public auctions. The death rate may vary from less than 10% to more than 90% of the herd, with an average of about 25% unless treatment is effective.	Some of the antibiotics, arsenicals, sulfonamides, and quinoxalines administered by the veterinarian and in keeping with the manufacturer's directions, may reduce death losses. Good management and nursing will help. Milk seems to aid recovery. There is a tendency for relapses following treatment. Because infected swine eat very little, if at all, medication through the drinking water is essential.	In case of an outbreak, sick animals should be removed from the healthy ones and a rigid program of sanitation initiated.	Avoid public stockyards and auction rings, isolate newly acquired animals, and practice rigid sanitation.	Some animals that have gone through an outbreak remain carriers although they may appear to be healthy. When such carriers are introduced into a herd, the signs may not appear in the contact animals until the end of several weeks or months.
It appears to be increasing in the U.S. The disease is usually fatal to young pigs.	Treatment of edema disease has not been highly successful. Treatment is aimed at reducing the factors believed to predispose increase of *E. coli* and avoiding sudden changes in management. Some producers feel that they lessen the incidence of the disease when there is an outbreak by administering an antibiotic.	The edema disease treatment and management practice that has earned the best rating involves withholding (for 24 hours) or sharp reduction in the amount of feed for a short but variable length of time, and avoiding stresses.	There is no justification, based on present knowledge of this disease, to warrant consideration of quarantine measures. Avoid abrupt feed changes and great stress.	
Wherever hogs are raised, especially under crowded and filthy conditions.	Consult the veterinarian. Appropriate sulfa drugs, B vitamins, and/or antibiotics may be indicated.	Segregate sick animals and move well animals to clean quarters. Clean and disinfect contaminated premises.	Good sanitation and management. Quarantine incoming swine 3 weeks before introduction to the herd.	It is no longer considered adequate to diagnose the condition as enteritis; it is now necessary to separate and identify the agents which may damage the intestinal tracts of swine. Internal parasites may cause enteritis, particularly in the southern states.

(Continued)

TABLE 10-2

Disease	Species Affected	Cause	Symptoms and Signs (or age group most affected)
Enterotoxemia (or overeating disease, pulpy kidney disease)—an acute disease of sheep.  Fig. 10-11	Sheep.	*Clostridium perfringens*, Type D, an anaerobic bacterium.	Loss of appetite, sluggishness, diarrhea, staggering blindly about, and convulsions. Usually the course of the disease is acute, with affected animals dying within a few hours. Affects sheep of all ages in a high state of nutrition—on a lush feed of grain, milk, or grass.
Epizootic bovine abortion (see Foothill abortion)			
Equine abortion (virus abortion, equine arteritis, rhinopneumonitis, bacterial abortion, etc.)—premature expulsion of the fetus. Fig. 10-12	Horses.	Causes of abortion in mares may be grouped into two types: (1) infectious agents, such as viruses, bacteria, and fungi; and (2) noninfectious abortions, such as twinning, hormonal deficiencies, congenital anomalies, and miscellaneous causes.	Expulsion of the fetus at any period prior to the time that the foal can survive out of the uterus. Rhinopneumonitis is a mild, usually nonfatal disease of the upper respiratory tract, commonly seen in young horses in the fall or winter. It is characterized by a cough, a nasal discharge, a loss of appetite, and a temperature of 102° to 105° F. As the disease progresses, the temperature returns to normal, but the nasal discharge and cough may persist for several weeks. In older horses, the disease may be so mild as to go unnoticed. Most abortions due to this virus occur between the eighth and eleventh months of gestation, although they may occur as early as the fifth month. Sometimes the foal is born alive at term, but dies at 2 to 3 days of age due to infection by the virus. The virus of *equine arteritis* may also cause abortion. It produces more obvious signs of illness than equine rhinopneumonitis, including discharges from the eyes and nose, fever (102-106° F), and filling (edema) of the limbs. A laboratory examination is necessary conclusively to establish the presence of the specific virus. Up to 50% of affected pregnant mares may abort. Bacterial infection is a common cause of abortion in mares. Several species of bacteria have been incriminated. *Salmonella abortus equi*, which was formerly responsible for abortion storms, still occurs sporadically. However, the most common cause of bacterial abortion at the present time is organisms of the streptococci group. Other bacteria frequently cultured from aborted feti include *E. coli*, *Klebsiella*, and *Staphylococci*. They may cause abortion at any stage of pregnancy. But, generally *Streptococci* cause abortion during the first 5 months, whereas *E. coli* are more apt to cause abortion during the last half of pregnancy. Bacterial abortion is often characterized by retention of the placenta, as well as by metritis or inflammation of the uterus. Fungi do not attack the fetus directly; rather they cause degeneration of the placenta so that the fetus has insufficient nourishment. For this reason, the aborted fetus is often small and only a fraction of the normal weight for its gestational age. If abortion does not occur, the foal may be carried to full term and be born in a reasonably vigorous, but undersized and undernourished state. Most mycotic abortions occur during the second half of pregnancy. There is no vaccine.

—(Continued)

Distribution and Losses Caused By	Treatment	Control and Eradication	Prevention	Remarks
Wherever lambs are finished out. Overeating disease is responsible for the largest number of death losses in lambs in western feedlots. In unvaccinated feedlot lambs, minimum losses of 1% may be expected. In explosive outbreaks, losses may range from 10 to 40%.	Once the disease has developed, treatment is unsuccessful. Great difficulty is encountered in getting affected lambs back on feed.	Control of explosive outbreaks late in the feeding period consists of the following: 1. Reduce the concentrate allowance by 50% for one week or longer. 2. Market all lambs carrying adequate condition for slaughter. 3. Vaccinate the remaining lambs with bacterin or toxoid and gradually return to full feed. 4. Consider the use of Type D antitoxin (which is expensive) to stop the losses. It will confer temporary immunity (2-3 weeks), following which a long-lasting immunity may be established by vaccinating with bacterin or toxoid.	Make a gradual change from range to feedlot conditions. Precondition lambs by placing them on hay and concentrates before weaning. Vaccinate lambs with either a bacterin or toxoid soon after their arrival in the feedlot, provided they are in good condition and not wet. Allow at least 10 days after vaccination for immunity to develop. Sometimes, revaccination with the bacterin or toxoid (a booster shot) is required 2 to 4 weeks following the first vaccination. Young lamb losses during the first 6 weeks of life may be prevented by vaccinating the pregnant ewes. Ewes that have not been vaccinated previously should be vaccinated twice: 2 to 4 weeks apart, with the second vaccination being given 2 to 4 weeks prior to lambing. Thereafter, an annual booster shot should be given 2 to 4 weeks prior to lambing.	This disease affects the biggest and fastest-growing lambs. The feeling persists among some feeders that a higher than usual incidence of enterotoxemia occurs in feedlot lambs on high-silage rations.
Incidence highest in areas where the greatest number of foals are produced. It is estimated that, for the U.S. as a whole, ⅓ of all pregnant mares either abort or produce weak, infected foals.	Isolate mares that have aborted, and accord them good feed and care.	Quarantine animals that have aborted. Burn or bury the bedding and fetus. Disinfect contaminated premises. Isolate newly introduced animals.	Preventive measures embrace avoidance of all possible causes. It begins with mating only healthy mares to healthy stallions and with being scrupulously clean at the time of breeding. New horses should always be isolated as a preventative measure, and aborting mares should be quarantined. Where abortions have occurred in the broodmare band, the special cause in the matter of feed, water, exposure to injuries, overwork, lack of exercise, and so forth may often be identified and removed. Avoid constipation, diarrhea, indigestion, bloating, violent purgatives or other potent medicines—including administering cortisones in late pregnancy, painful operations, and slippery roads. The following points are pertinent in controlling abortion in a band of broodmares: 1. Prevent rhinopneumonitis by following a planned immunization program under the direction of a veterinarian. 2. Prevent equine arteritis in areas where the disease is a problem by administering the vaccine. 3. Prevent *Salmonella* abortion on premises known to be contaminated by vaccinating all pregnant mares each year. 4. Control and prevent bacterial abortion by mating only healthy mares to healthy stallions and observing scrupulous cleanliness at the time of service and examination. Suture mares where necessary. 5. Keep broodmares healthy and in good flesh, and feed a ration that contains all the essential elements of nutrition.	Sanitation and herd health are important factors in lessening the incidence of abortions, regardless of kind. Consult the local veterinarian whenever abortion occurs. Cattle abortion is not a factor in producing abortion in mares.

(Continued)

TABLE 10-2

Disease	Species Affected	Cause	Symptoms and Signs (or age group most affected)
Equine encephalomyelitis (or sleeping sickness)—a virus, epizootic (epidemic) disease transmitted by insects. 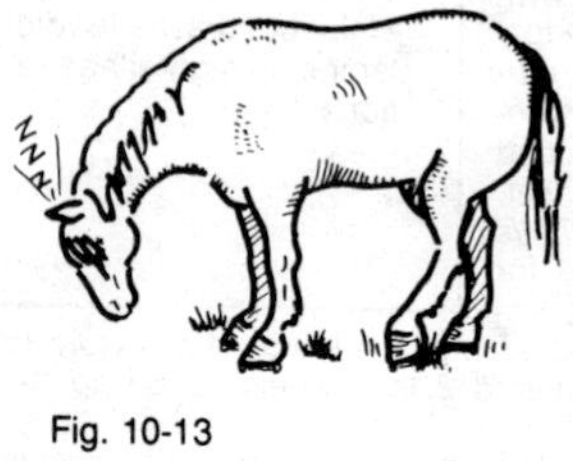 Fig. 10-13	Horses. Mules. Man. Birds (chickens, pheasants, etc.). Wild rodents.	The disease is caused by several distinct viruses. The 3 most active types in the U.S. are: Eastern equine encephalomyelitis, Western equine encephalomyelitis, and Venezuelan equine encephalomyelitis.	In early stages, the animal walks aimlessly about, crashing into objects. Later it may appear sleepy, standing with a depressed head. Grinding of the teeth, inability to swallow, paralysis of the lips, and blindness may be noted. Paralysis may cause the animal to go down. If affected animal does not recover, death occurs in 2 to 4 days.
Equine infectious anemia (E.I.A. or swamp fever)—an infectious virus disease.	Horses. Mules.	Virus.	Symptoms vary, but some of the following are usually seen: high and intermittent fever, depression, stiffness and weakness—especially in the hindquarters, anemia, jaundice, edema and swelling of the lower body and legs, unthriftiness, and loss of condition and weight—even though the appetite remains good. Most affected animals die within 2 to 4 weeks.
Equine infectious metritis (Contagious equine metritis)—a contagious venereal disease of horses.	Horses.	A gram-negative cocco-bacillus grown under micro-aerophilic conditions. Thus, the causative organism will grow with a limited oxygen supply. The causative organism may be spread at the time of mating or by means of contaminated instruments.	The disease is highly contagious. A profuse pussy discharge from the vulva 3 to 5 days after the mares have been covered, smearing the buttocks, and matting the tail; low conception; and some mares have early abortions. There are no clinical signs of the disease in the stallion. Thus, one cannot tell whether or not a stallion has the disease by examining him.

—(Continued)

Distribution and Losses Caused By	Treatment	Control and Eradication	Prevention	Remarks
Since 1930, the Eastern and Western types of the disease have assumed alarming proportions in the U.S. Then, in 1971, Venezuelan equine encephalomyelitis first occurred in the U.S., when an outbreak was reported in Texas. Generally speaking, mortality from the Western type does not exceed 50%, whereas that from the Eastern and Venezuelan types is 90% or higher.	Treatment is not very effective, because of the rapid course of the disease. Since the Western type progresses more slowly and results in a lower mortality rate than the Eastern and Venezuelan types, it lends itself to more supportive treatment. Good nursing is perhaps the best and most important treatment. The maintenance of fluid and electrolyte balance is recommended. No specific therapeutic agent is known to influence the course of the disease.	Prompt disposal of all infected carcasses; destruction, if possible of insect breeding grounds; and discouragement of movement of animals from an epizootic area to a clean one.	Prevention entails vaccination of all horses against the 3 separate strains. Vaccination against Eastern and Western encephalomyelitis involves 2 injections, 7 to 14 days apart, given annually in April or May. Protection against Venezuelan equine encephalomyelitis requires one injection only, given annually. Also, there is a new killed virus vaccine which can be used to vaccinate simultaneously against all 3 types—Eastern, Western, and Venezuelan. A veterinarian should administer the vaccine.	Birds and wild rodents are natural disease hosts for Western type. Mosquitoes (*Culex tarsalis*) transmit the disease. Man and horse infections are incidental, out of species infections. The public health aspects of this disease are unrelated to horse infections.
It was first reported in France in 1843, and it has existed in different sections of the U.S. for at least 60 years. The USDA reported that of the horses tested for equine infectious anemia in 1974, 9,089 or 2.56% tested positive.	No successful treatment known.	After horse owners test and eliminate all infected animals from their herds, the Coggins Test can be used to protect their stock from reinfection, (1) by buying horses only after they have been tested and found free from the disease, (2) by not allowing untested horses to be stabled or pastured with their own, and (3) by not taking their horses to any assembly point (show, sale, racetrack, trail ride, etc.) where prior testing is not required.	Apply the "Coggins Test" for diagnosing E.I.A. Repeat the test to confirm all positive reactions. Positive reactors are identified with an "A" in a visible brand or lip tattoo, which stands for anemia. Animals so branded are quarantined and cannot be moved except for slaughter or approved research purposes, In order to prevent bringing in the disease, in 1976 the USDA amended the import regulations to require that imported horses pass the Coggins Test to assure that they are free of equine infectious anemia.	Infected horses may be virus carriers for years and represent a source of danger for susceptible horses. Beginning in 1977, several states modified or repealed their Coggins testing regulations, in response to industry pressure on the following bases: 1. The Coggins Test is adequate and valuable to confirm a diagnosis of E.I.A., but it should not be used indiscriminately as a screening test on apparently healthy animals. 2. None of the healthy positive reactors have been shown to have transmitted the disease. 3. The positive horse may never have been sick. 4. The old gelding which grazes across the road likely never has and never will be tested.
The disease has been reported in France, Great Britain, Ireland, and Australia. Infection may prevent conception, cause the fertilized egg (and embryo) to die and be aborted, damage the placenta (thus putting the newborn foal at risk), or have no noticeable effect. Among the mares covered on the affected stud farms of Newmarket, England, in 1977, about 30% showed clinical signs of the disease; and the losses for the year were estimated at $30 million.	Antibiotics are the common treatment of mares; administered (1) locally into the uterus, or (2) systemically by intravenous, intramuscular, or subcutaneous injection. Stallions are given antibiotics systemically and washed and cleaned thoroughly.	Never allow infected stallions to breed mares, and never use contaminated instruments. Require a disease (health) history of each mare sent to a stud for breeding, and have all the mares examined by a veterinarian upon their arrival at a stud farm for breeding. In 1977, the USDA banned the importation of breeding stock from affected countries. Young animals and geldings were not included in the ban.	The disease may be prevented by artificial insemination, using fresh semen treated with an extender containing an antibiotic. Of course, many breed registries require natural service, although they will accept reinforcement by A.I.	Mares and stallions that are infected one year will breed successfully the next year.

(Continued)

TABLE 10-2

Disease	Species Affected	Cause	Symptoms and Signs (or age group most affected)
Equine influenza	Horses. Mules.	Influenza is caused by any one of a group of related viruses.	All age horses are susceptible to these viruses; however, young horses, 1 to 3 years of age, are most susceptible. Older animals are more resistant due to previous exposure to these viruses. Symptoms develop 2 to 10 days after exposure. Onset is marked by rapidly rising temperature which may reach 106° F. The fever persists 2 to 10 days. Other signs include loss of appetite, extreme weakness and depression, rapid breathing, a dry cough, and a watery discharge from the eyes and nostrils which is later followed by a white- to yellow-colored nasal discharge.
Erysipelas, swine—an acute or chronic infectious disease.  Fig. 10-14	Swine, but also reported in sheep, rabbits, turkeys, and man.	*Erysipelothrix insidiosa (rhusiopathiae)*, a bacterium.	Three forms: 1. In acute septicemic form, it resembles hog cholera. Affected animals show a high fever and frequently there is edema of the nose (which causes the animals to breathe with a snoring sound) and of the ears and limbs. There may be purplish patches under the belly similar to those described for hog cholera. 2. In the diamond-skin form, which is a subacute form, there are reddish rectangular plaques in the skin; and there may be a partial sloughing off of the ears and tail. 3. In the chronic form, the knees and hocks are generally swollen and stiff. Most serious in pigs 3 to 12 months of age, but hogs of all ages are susceptible.
Fescue foot (fescue toxicity)	Cattle. Also reported in sheep in Australia.	The exact cause of this condition is still unknown. It is either a mycotoxin on the fescue or some change in the fescue plant itself that makes it toxic under some conditions.	There are variations in the severity of symptoms in cattle on toxic pastures. Some animals show no apparent lameness, whereas others show varying degrees of sloughing (necrosis) on the ends of their tails. During the summer, cattle grazing toxic pastures show a poor growth rate, increased temperatures, and increased pulse and respiratory rates. The only complaint that cattlemen make is the fact that the cattle are not doing as well as in previous years. In some herds, the weaning weights decline for 2 or 3 years before cattlemen realize that they have a problem.
Foot-and-mouth disease—a highly contagious disease.  Fig. 10-15	All cloven-footed animals, but mainly cattle, sheep, and swine. Man is mildly susceptible.	Small filtrable viruses, of at least 6 types, all of which are immunologically distinct from one another.	Water blisters in the mouth (snout in hogs), on the skin between and around the claws of the hoof, and on the teats and udder. Fever is another symptom.

—(Continued)

Distribution and Losses Caused By	Treatment	Control and Eradication	Prevention	Remarks
Widespread throughout the world. It frequently appears where a number of horses are assembled, such as racetracks, sales, and shows. Death rate is low, but economic loss is high. It interrupts training, racing, and showing schedules, and it may force the withdrawal of animals from sales.	Treatment should be handled by the veterinarian. Avoid exercise during period of elevated temperature. The early use of antibiotics and/or sulfa drugs may prevent some of the complicated secondary conditions.	Avoid transmission of the virus through contaminated feed, bedding, water, buckets, brooms, on the clothing and hands of attendants, and on transportation facilities.	Vaccination with a killed virus using 2 doses, with the second injection given 4 to 12 weeks after the first. For continued protection, each vaccinated animal must receive a booster annually, or at any time demanded by exposure or epizootic conditions. All new animals should be isolated for 3 weeks. Any sick animals should be quarantined.	
Throughout the U.S. Death losses are high in the acute form, ranging from 50 to 75%.	Serum provides satisfactory treatment if given early enough, especially in conjunction with penicillin.	Isolate all sick animals and examine the herd daily for new cases. Clean and disinfect contaminated premises.	On infected farms, administer one of the following products: 1. Erysipelas vaccine (avirulent), available through your veterinarian; 2. Erysipelas bacterin; or 3. Oral erysipelas vaccine, in the water.	The most unpredictable and one of the most important diseases of swine. Veterinary aid is necessary for an accurate diagnosis.
Most cases of fescue toxicity occur among cattle that graze pure stands of fescue during late fall and winter; and most toxic stands of fescue pasture are several years old. Fescue toxicity is more prevalent in animals suffering from malnutrition or parasitism.	No medication is effective for cattle with fescue foot. In severe cases where sloughing has occurred, the animal should be destroyed for humane reasons. Cattle usually recover completely if they are removed from fescue pasture or fescue hay and are given other feed or pasture as soon as the first signs of the disease appears.	Avoid fescue pastures that are toxic. But this may not be practical because fescue is a valuable pasture grass in certain areas.	Proper management of fescue pastures is the best way to prevent fescue toxicity. Toxic pastures should be renovated and some legume should be seeded with the fescue. It requires good pasture management, along with fertilization, to maintain a good fescue-legume pasture.	
Scattered in different countries of the world. The disease is not present in the U.S., but there have been at least 9 outbreaks, each of which was stamped out. The last outbreak occurred in 1929. Mortality of adult animals is not high, but there is great economic loss due to the decreased usefulness and productivity of affected animals.	Treatment is not satisfactory.	In the case of U.S. outbreak, the quarantine of the area, and the slaughter and burial of all infected and exposed animals, with the owners paid indemnities. To date, vaccines have not been used in the U.S. because they have not been regarded as favorable to rapid, complete eradication of the infection.	Quarantine at ports of entry; assistance with eradication in neighboring countries. Neither live animals nor fresh or frozen meats can be imported from any country in which it has been determined that foot-and-mouth disease exists (meat imports from these countries must be canned or fully cured). Vaccines containing one or more immuno-types of the virus are manufactured in Europe and Argentina. The immunity produced from such vaccines lasts for only 3 to 6 months; hence, animals must be vaccinated 3 to 4 times per year. Recently, promising studies have been made with live vaccines prepared with attenuated strains.	In Sept., 1946, foot-and-mouth disease appeared in Mexico. From that date until Sept. 1, 1952, except for a 9-month period (from Sept. 1, 1952, to May 23, 1953), the U.S. Secretary of Agriculture closed the Mexican border to importations of most livestock and meat products. In Feb., 1952, foot-and-mouth disease appeared in Canada. From that date to March 1, 1953, the Canadian Border was closed to imports of virtually all livestock and meat products.

(Continued)

TABLE 10-2

Disease	Species Affected	Cause	Symptoms and Signs (or age group most affected)
Foothill abortion (epizootic bovine abortion, EBA)	Cattle.	Virus (psittacoid virus) The soft-bodied pajaroello tick, *Ornithodoros coriaceus* is the vector.	Cows may abort when about 3 to 6 months pregnant. Some calves stillborn and weak.
Foot rot (or foul foot)—an infectious disease.  Fig. 10-16	Cattle. Sheep. Goats.	Perhaps more than one causative agent or agents. The soil organism, *Spherophorus necrophorus* is most frequently recovered from cases of foot rot in cattle, and *Fusiformis nodosus* in conjunction with *Spirochaeta penortha* in sheep.	Lameness is usually the first symptom. In the early stages, there is a reddening and swelling of the skin just above the hoof, between the toes, or in the bulb of the heel. If not arrested, pus may be discharged and there may be a characteristic foul odor. Later the joint cavities may be involved, and the animal may show fever and depression, lose weight, and even die. In sheep, foot rot rarely involves the coronary band or extends above the top of the hoof.
Garget (see Mastitis)			
Glanders (or farcy)—an acute or chronic infectious disease.  Fig. 10-17	Horses. Mules. Donkeys. It can be transmitted to other animals and man through close contact.	*Malleomyces mallei*, a bacterium.	Chronic form is most often observed in the horse—affecting the lungs, skin, or nasal passages. There may be a nasal discharge which later becomes pus, and/or nodules and ulcers in the skin. With the lung type there is generally loss in condition, lack of endurance, bleeding and mucous discharge from the nose, and coughing. The skin of extremities may develop ulcers that exude a honeylike tenaceous discharge. Acute form is seen more in mules and donkeys. Death usually occurs in a week after many or all of the symptoms noted above have been seen. Prevalent in areas where the horse is still relied upon for transportation and work. The disease has largely disappeared from the mechanized areas of the world (including U.S.) due to decline in horse population and lack of contact for spread.

—(Continued)

Distribution and Losses Caused By	Treatment	Control and Eradication	Prevention	Remarks
The abortion rate frequently reaches 65%.			Move cattle out of tick-infested area (dryland brush area) during the 3- to 6-months gestation period. 2 g/head/day of chlortetracycline in the feed will prevent foothill abortion.	Aborting animals usually are immune and should be retained in the herd.
It is a potential hazard wherever cattle, sheep, or goats are kept, especially in wet, muddy areas. Foot rot seldom causes death, but infected animals generally lose weight and, if lactating, produce less milk. Young animals become stunted, and males may be rendered useless at breeding time. Sheepmen often get discouraged and dispose of entire flocks because of this disease.	Place in a clean, dry place. **Cattle:** If necessary trim away the affected part of the foot. Also, check for and, if necessary, eliminate foreign bodies in or around the hoof. The veterinarian may use sulfonamide or antibiotic therapy—accompanied by cleaning, disinfecting, and packing the affected area. **Sheep:** Treat as follows: 1. Examine every foot of every sheep. Trim each foot showing infection, removing enough of the horn of the hoof thoroughly to expose all diseased tissue. 2. Walk all sheep through suitable disinfectant solution and move to clean ground. The 2 most widely used disinfectants are (1) formaldehyde, 10%, and (2) copper sulfate, 20%. Repeat foot bath at weekly intervals until foot rot disappears; then continue at 2-week intervals for another 2 months. Two weeks after initial antiseptic treatment, examine feet of each sheep a second time, to detect and trim infections overlooked first time or developed subsequently. Treatment with sulfadiazine, given daily for 20 days, has proven fairly effective.	Segregate infected animals in dry places, and clean and disinfect contaminated areas. Add organic iodide to either the salt or the grain ration according to directions. For prevention in cattle, the usual recommendation is to use 50 mg of EDDI per head daily on a continuous basis; for treatment, once cattle have foot rot, the usual recommendation is 500 mg per head daily for 2 to 3 weeks. **Sheep:** After trimming, and treatment in a foot bath, place sheep in a clean, dry pasture. One that has not been used for 30 days would be considered clean.	Drain muddy pastures and remove sharp stones; segregate new animals. Add organic iodide to either the salt or the grain ration according to directions. Recommended preventive measures to avoid entrance of infection into a herd or flock are: 1. Purchase foundation and replacement animals from a known clean source. 2. If animals come from a questionable or unknown source, pass through a public market, or are transported by public conveyance, (a) trim their feet on arrival, (b) walk them through a recommended disinfectant solution, and (c) isolate them for 1 month. **Sheep:** Allow land previously pastured by sheep to remain idle 4 weeks before turning other sheep on it. *Note:* Recent experimental studies cast considerable doubt on the use of EDDI to prevent foot rot. Thus, more experimental work is needed.	Cross infections of foot rot between cattle and sheep do not occur. A sheep with foot rot may spread infection up to 3 years, but contaminated land loses its ability to infect within 3 weeks. Since copper sulfate is corrosive for most metals, it should be prepared in earthenware or wooden containers. For foot bath, use a suitable container filled to a depth of 2 in. with the antiseptic solution of choice. Stand affected animals in solution for 2 minutes; walk normal appearing animals through it. A 10% formaldehyde solution may be made by mixing 1 gal of 38% formaldehyde and 9 gal of water. A 20% copper sulfate solution may be made by mixing 1⅔ lb of copper sulfate per gallon of water. Visibly affected sheep should be kept standing in the solution 5 to 10 minutes.
		Apply the "mallein test" to suspected or exposed animals. Destroy infected animals, and clean and disinfect contaminated equipment and premises.	Avoid inhalation or ingestion of the causative organism. Do not use public watering places. No method of immunization is available.	Glanders is no longer present in the U.S. To avoid horses with glanders being brought into the U.S. from infected areas, they must be tested prior to entry.

(Continued)

TABLE 10-2

Disease	Species Affected	Cause	Symptoms and Signs (or age group most affected)
Gut edema (see Edema disease)			
Hemorrhagic septicemia (see Shipping fever)			
Infectious atrophic rhinitis (see Atrophic rhinitis)			
Infectious bovine rhinotracheitis (IBR or red nose)	Cattle.	Virus.	Affected animals go off feed and lose weight; generally cough; may show pain in swallowing; usually slobber and show a nasal discharge; breathe rapidly, with difficulty, and in severe cases through the mouth; show severe inflammation of the nostrils, trachea, and windpipe; have a high fever, 104° to 107° F; and may remain sick for as long as a week. When the disease breaks out, 25 to 100% of the animals are affected. Death loss rarely exceeds 5%. Although IBR is usually thought of as a respiratory disease, it may cause inflammation of the eyes and/or vagina. Also, it may cause abortion.
Infectious embolic meningo-encephalitis (thromboembolic meningo-encephalitis)	Cattle.	A hemophiluslike gram-negative bacterium. Further investigation is needed.	Feedlot cattle, in fall and winter months. Affects both sexes, usually animals 1 to 2 years of age. Characterized by incoordination, coma, sometimes blindness, and always fever (near 107° F). Death usually follows in 2 to 4 days. Positive diagnosis can be made upon autopsy, by the inflamed areas of infection observed in the brain.
Influenza (see Swine influenza)			
Johne's disease (chronic bacterial dysentery, or paratuberculosis)—a chronic, incurable, infectious disease. 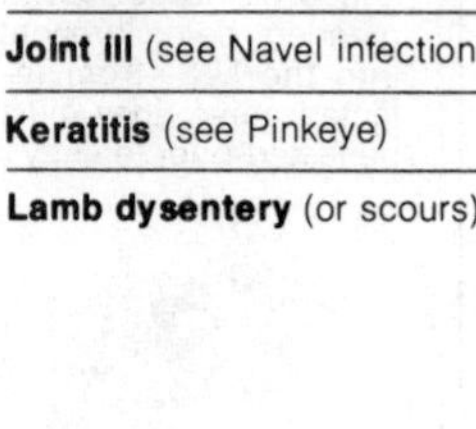 Fig. 10-18	Cattle. Sheep and goats, sometimes. Swine and horses, rarely.	*Mycobacterium paratuberculosis*, a bacterium.	Loss of flesh, and intermittent diarrhea and constipation—with the former becoming more prevalent. Affected animals may retain a good appetite and normal temperature. The feces are watery but contain no blood and have a normal odor. The disease is almost always fatal, but with the animal living from a month to 2 years. Upon autopsy, the thickening of the infected part of the intestines, covered by slimy discharge, is all that is evident.
Joint ill (see Navel infection)			
Keratitis (see Pinkeye)			
Lamb dysentery (or scours)	Sheep.	Infection with virulent strains of *Echerichia coli*, with cold, wet, unsanitary conditions as predisposing factors.	Affects lambs first few days after birth; seldom occurs after first week. Diseased lambs are weak, depressed, and do not care to suckle; profuse diarrhea (scours) that may be tinged with blood; usually temperature rises and the lamb may become gaunt or bloat; and death may follow in a few hours.

—(Continued)

Distribution and Losses Caused By	Treatment	Control and Eradication	Prevention	Remarks
Throughout the U.S. The main economic losses are in poor growth, loss of weight, loss of milk production, abortions, and loss of time and cost of drugs. Death losses rarely exceed 5%.	No known treatment, but sulfonamides and antibiotics effectively combat the secondary bacterial invaders that accompany the disease.	Practice good sanitation and disease preventive measures; isolate sick animals.	Infectious bovine rhinotracheitis can be prevented by the use of a vaccine, of which there are two types. The modified live virus vaccine provides lasting immunity, but it should not be used on pregnant cows or on calves under 6 months of age. Killed virus vaccines must be repeated.	IBR was first found in a Colorado feedlot in 1950.
Western region of the U.S. Only 1 to 2 cases develop in a lot at a time, but 10% of cattle may be affected before the disease runs its course.	If identified early, treatment with one of the broad-spectrum antibiotics may help affected animals.	If the situation is serious enough, (1) change diet (to at least 50% roughage) and pens, or (2) scatter animals out on pasture.	Application of the usual sanitary measures may help prevent the disease. Also, for the first 28 days after arrival, fortify the ration with 350 mg of Aureomycin plus 350 mg of sulfamethiazine per head per day.	Polioencephalomalacia, which also affects feedlot cattle and causes incoordination, may be confused with infectious embolic meningoencephalitis; but fever is rarely associated with polioencephalomalacia.
Widespread; in practically every country where cattle are raised on a large scale. Apparently, it is increasing in the U.S.	No satisfactory treatment is known.	If infection strikes, have herd tested with "Johnin" at intervals of 3 to 6 mos., remove reactors, disinfect quarters, and isolate young stock from mature animals. The Johnin test, as with many other tests, is not entirely accurate, as some affected animals fail to react to it. Difficult to eradicate from a herd.	Effective prevention is accomplished by keeping the herd away from infected animals. Purchase new or replacement animals from reputable breeders and disease-free herds.	This is a chronic incurable infectious disease. It resembles tuberculosis in many respects. The disease seems to involve calfhood exposure, with no evidence of infection until 6 to 18 months later.
Occurs in sheep throughout the world. Lambs born on the range are less susceptible than those born in a lambing shed. The mortality is very high.	There is no highly effective treatment, once lambs are affected. Various drugs such as sulfonamides, antibiotics, antidiarrheals, and anthelmintics may save a few. There is no lamb dysentery vaccine available for routine prevention. An antitoxin (*C. perfringens* B, C, D) is sometimes used for treatment when clostridial scours is suspected.	Practice rigid sanitation; isolate diseased animals and move healthy animals to clean quarters or ground; keep lambs clean, warm, and dry.	Hold lambing ewes on dry, well-drained areas, and keep the lambs clean, warm, and dry.	

(Continued)

TABLE 10-2

Disease	Species Affected	Cause	Symptoms and Signs (or age group most affected)
Leptospirosis	Cattle. Sheep. Goats. Swine. Horses. Dogs. Foxes. Rats and other rodents. Humans (the disease is transferable between species).	Several species of corkscrew-shaped organisms of the spirochete group. *Leptospira pomona* primarily affects cattle and swine.	In most herds, leptospirosis is a mild disease. However, the symptoms may vary from herd to herd. In general, the symptoms noted in cattle are (1) high fever (103° to 107° F), (2) poor appetite, (3) abortion anytime, (4) bloody urine, (5) anemia, and (6) ropy milk. All ages of cattle, and both sexes, are affected (including steers). **Swine:** Leptospirosis is usually characterized by abortion, pigs born dead or weak, and unthrifty market hogs. **Equine:** Leptospirosis is characterized by fever, inappetence, mild depression, and occasionally jaundice. **Human:** Leptospirosis is characterized by abrupt onset of fever with chills, headache, vomiting, and pains in the extremities, joints, and muscles.
Listerellosis (see Circling disease)			
Liver necrosis of lamb (or umbilical necrobacillosis, necrotic hepatitis)—a form of hepatitis affecting young lambs.			
Lockjaw (see Tetanus)			
Lumpy jaw and wooden tongue—2 infectious, chronic diseases.  Fig. 10-19	Cattle. Swine. Horses. Man.	*Actinomyces bovis* causes lumpy jaw. *Actinobacillus lignieresei* causes wooden tongue.	Lumpy jaw is usually confined to the bones of the lower jaw, although the upper jaw and nasal bones may be involved. Affected bones become enlarged, spongy, and filled with creamy pus. Inflamed cauliflower masses of tissue may spread out and appear on the surface, discharging a pus of foul odor. The teeth may become loosened. Wooden tongue attacks chiefly the tissue in the throat area of cattle, but it is also seen in the tongue, stomachs, lungs, and lymph glands. Usually it first makes its appearance as a small swelling under the skin in the infected area. The enlargements usually break open and discharge pus. If the tongue is involved, it will increase in size and hardness. There will be constant drooling and the animal will lose weight due to impaired eating. Lumpy jaw and wooden tongue occur most frequently in young cattle during the period of changing teeth.

—(Continued)

Distribution and Losses Caused By	Treatment	Control and Eradication	Prevention	Remarks
Leptospirosis was first observed in man in 1915-1916, in dogs in 1931, and in cattle in 1934. It was first reported in cattle and swine in the U.S. in 1944 and 1952, respectively, although it has been found in dogs in the U.S. since 1939. Bovine leptospirosis has been reported in Europe, Australia, and the U.S. Surveys indicate this disease is widespread in the cattle population of the U.S. as well as in many other parts of the world. Losses due to this disease amount to over $100 million annually in the U.S. Mortality is low in most outbreaks; however, in young calves, it may be high. The main losses are from poor growth in beef cattle and loss of milk production in dairy cows. If it were not for abortions, this disease would go undetected in many herds.	Treatment of animals, which should be prescribed by a veterinarian, may include blood transfusions, administration of selected antibiotics, and good care. Antibiotics give fairly good results if cases are treated promptly. It appears that selected antibiotics must be used to eliminate shedders. High levels of the tetracycline drugs (400-500 g per ton) can be used in swine complete feeds for 14 days to help remove the carrier phase. In human leptospirosis, the M.D. should be consulted relative to treatment.	The disease is spread by infective urine; therefore, spread animals out over a large area; avoid congestion in a corral or barn. Fence off waterholes or ponds or slow running streams. Isolate sick animals or new additions to the herd. Discard milk from diseased cows. Clean and disinfect the barns; exterminate rodents. Administer leptospirosis vaccine to all cows and sows on problem farms each year. Keep different classes of livestock separated, because leptospirosis can be spread from one species to another.	Vaccinate susceptible animals annually if disease is present in area. Purchase clean animals, isolate for 30 days and retest. Vaccination of people with a suspension of killed leptospires has been employed and reported successful in several countries.	Carrier animals—animals that have had leptospirosis and survived—may spread the infection by shedding spirochetes in the urine. The infected urine may then either (1) be breathed as a mist in cow barns, or (2) contaminate feed and/or water and thus spread the infection. Breeding bulls can transmit this disease to cows. It is known that recovered cattle can remain carriers for up to 3 months and swine can remain carriers up to 1 year. Leptospirosis is mainly a warm-weather disease. The spirochetes seldom survive for more than 30 days outside the animal. Stagnant water favors their survival.
A most serious malady of range sheep, although it occurs in farm flocks. The mortality is high.	No treatment is effective.	Clean and disinfect old, filth-laden barns, sheds, and corrals.	Sanitation is the key to prevention. Lambs born on the range or clean pasture are seldom affected. Clean and disinfect lambing quarters prior to use. Treat navel cord of newborn lambs with tincture of iodine.	In sheep, the germ which causes liver necrosis in lambs also causes complications in sore mouth and plays a role in foot rot. Another condition attributed to complications of necrobacillosis in sheep is a form of venereal disease which affects the sheath and penis of the ram and the vulva of the ewe.
Throughout the U.S.	The veterinarian may, under certain conditions, (1) administer a water solution of an iodine salt of sodium or potassium, (2) prescribe an antibiotic, (3) resort to surgery, or (4) use X-ray therapy. Treatment of lumpy jaw is not very satisfactory, but most cases of wooden tongue yield readily to treatment. Sometimes treatment with organic iodine (EDDI) is effective. Add to the ration 250 to 500 mg/head/day for 2 to 3 weeks. Superficial abscesses should be opened, drained, and swabbed with tincture of iodine.	Segregate and treat or eliminate infected animals, and, if practical, do not feed materials having sharp awns (foxtail, barley, rye, bearded wheat, etc.)	Since the organisms causing lumpy jaw and wooden tongue gain entrance to the body through injuries or abrasions, the only prevention consists in not feeding such feeds as foxtail, barley, rye, and bearded wheat—unless there is no practical alternative. Under some conditions, organic iodine appears to be effective in the prevention of lumpy jaw in cattle. For prevention, add to the ration 50 mg of ethylenediamine dihydriodide (EDDI)/head daily.	These 2 different and distinct microorganisms produce similar chronic diseases affecting mainly the head of cattle—hence, the name "big head." The same fungus that causes lumpy jaw occasionally attacks the udder of sows where it is characterized by many small abscesses filled with calcified granules. On rare occasions, the lumpy jaw organism has also been found in the fistulous withers of the horse in conjunction with *Brucella* organisms. The condition may spread throughout the body, resulting in emaciation and a condemned carcass upon slaughter.

(Continued)

TABLE 10-

Disease	Species Affected	Cause	Symptoms and Signs (or age group most affected)
Malignant edema (gas gangrene)	Cattle. Sheep. Goats. Swine. Horses.	Malignant edema is caused by *Clostridium septicum* and related bacteria.	This is an acute infectious, bı noncontagious, disease charac terized by gangrene and em physema around a wound. The affected animal goes off feec breathes rapidly, and is prc foundly depressed. A swellin forms around the wound. gaseous and malodorous flui exudes from the wound. In ac vanced stages of the disease the animal is prostrated an often disoriented. There may may not be a rise in tempera ture. Death occurs after course of 12 to 48 hours. Th mortality rate is high.
Mastitis (or garget, blue bag in sheep)—an infectious inflammation or irritation in the udder which interferes with the normal flow of milk and/or its quality. 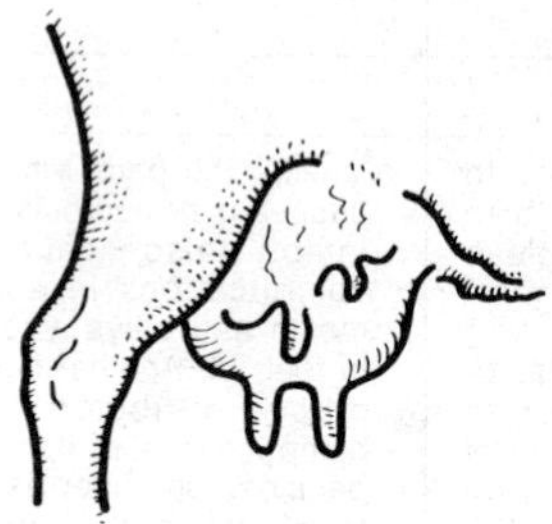 Fig. 10-20	Cattle. Sheep (known as blue bag). Goats. Swine.	Mastitis may be either infectious or noninfectious. Infectious mastitis, resulting from the invasion of bacteria in the gland, may be from several different types of bacteria. Over 95% of all cases are caused by the following species of streptococci and staphylococci: *Streptococcus agalactiae*, *Streptococcus dysgalactiae*, *Streptococcus uberis*, and *Staphylococcus aureus*. Infection with any of these organisms is usually chronic, with flare-ups occurring at regular intervals. No amount of drugs given to cows today can prevent another attack next month under the same conditions. Noninfectious mastitis is the result of injury, chilling, bruising, or rough or improper milking.	In acute mastitis, the udder is hc very hard, and tender. The an mal will have an increase i temperature, refuse to eat, "los its cud," have dull eyes and rough coat. Because of th soreness of the udder, the an mal stands in an awkward pos tion, moves about and lie down with reluctance and diff culty. Milk will be reduce greatly, and may becom lumpy or watery. Abcesses ma appear on the udder. Deat often occurs in untreated acut mastitis. In chronic mastitis, the on symptom which may be note is that the milk will be thick c lumpy. **Sheep:** Mastitis of sheep, if ur treated, usually results i chronic discharging teats c gangrene (blue bag) and deatl Affected ewes tend to separat out from the flock or band Generally only one side c udder is affected. It is hot an painful; and, in early stages secretion from teat is (1) thi and reddish due to hemor rhage, or (2) thick, creamy, an yellowish in color. Usually ac companied by rise in tempera ture, 104° to 106° F. Followin gangrene (when the udder be comes cold and turns dark i color), death may follow; o with recovery, a portion of th udder may slough off.
Metritis (or inflammation of the uterus)—an inflammation of the uterus usually caused by various types of bacteria.	Cattle. Horses. Sheep. Swine.	Various types of bacteria. Laceration at the time of birth or retention of the afterbirth are the principal predisposing causes.	Usually develops soon after ani mal has given birth. Symptom include: a foul smelling dis charge from the vulva that be comes thick and yellow o white, and finally brownish o bloodstained, and chilling, hig temperature, rapid breathing marked thirst, loss of appetite and lowered milk production. Pressure on the right flank may produce pain. Animal may lie down and refuse to get up. Affected animal may die in 1 to 2 days; or the acute infection may develop into a chronic form, producing sterility.

—(Continued)

Distribution and Losses Caused By	Treatment	Control and Eradication	Prevention	Remarks
The incidence in a single herd may be high following castration, dehorning, or accidental wounds.	In the early stages of the disease, treatment with massive doses of antibiotics may be effective.		Since malignant edema is associated with contamination of wounds, the disease can be partially prevented by minimizing wounds and by castrating and dehorning under hygienic conditions. Vaccination of young cattle with a vaccine containing *C. septicum* (for malignant edema) along with *C. chauvoe* (for blackleg) at the time of the blackleg vaccination(s) will give some protection against malignant edema. Also, antibiotics may be administered 4 to 5 days following surgery.	
Worldwide. Mastitis takes heavier toll from the dairy industry than any other single disease. Losses are chiefly in decreased milk production and poor quality milk. A 1981 survey conducted by *Hoard's Dairyman* showed that 87% of the respondents encountered mastitis during the year. The listing prepared by the associate deans and directors of veterinary research programs for the Council of Deans, Association of American Veterinary Medical Colleges, showed estimated mastitis losses of $368 million per year, based on losses as of February 1, 1981. Mastitis causes a yearly loss of $225 per afflicted cow, according to the National Mastitis Council.	Consult a veterinarian. Mastitis is usually treated by the intramammary injection of antibiotics, sulfa drugs, nitrofurans, or combinations of these drugs. Acute cases should also be treated systemically by the veterinarian. Local application of hot packs and udder massage will increase circulation in chronic cases; however, hot, acute, painful glands should be treated with cold applications. **Sheep:** Antibiotics or sulfas are the treatments of choice. If treatment is started early enough, the udder may return to normal milk production. Surgical removal of gangrenous half of udder is not successful; thus, ewes not capable of producing milk should be marketed.	Although mastitis is usually apparent, it may be a "hidden" disease. Therefore, several different tests have been developed for detecting the presence of the causative microorganisms in lactating cows; among them, (1) *screening tests*, or *presumptive tests*, made either at the side of the cow or at the bulk tank, of which the California Mastitis Test (CMT) is the most widely used one; and (2) *specific laboratory tests* designed to detect the causative organism. A reasonable goal, based on using CMT test, is to have at least 75% of the bucket milk samples score negative or trace; less than 75% negative (−) and trace (T) bucket readings indicates a milking management problem. On an individual quarter basis, 90% of the samples scoring negative or trace indicates a well-managed herd.	Milk all diseased cows last. Use strip cup before milking. Promptly remove new cases from the milking line and segregate. Wash udders with clean individual towels placed in chlorine solution (with a strength of 200 ppm); then wring the towel dry and wipe. Milk properly, in a regular, rapid, and thorough manner. Dip the ends of the teats in a chlorine solution after milking (chlorine with a strength of 200 ppm). Before milking each cow, wash hands with soap and water, disinfect them with chlorine solution, and wipe them dry, preferably with paper towels. Do not milk on the floor. Do not permit wet hand milking. Before sweeping the floor, sprinkle it with lime or superphosphate. Use plenty of straw bedding for each cow.	Although mastitis in sows and in ewes is somewhat different than that found in milk cows, the same general principles in preventive and control measures are applicable. The methods of treatment, control and eradication, and prevention of mastitis are the same for both milk cows and milk goats.
Throughout the U.S.	Treatment should be left to the veterinarian. Most cases are treated by the introduction (in solution or tablets) of an antibiotic or sulfa into the uterus.	Avoid constipation in sows since it predisposes to metritis and mastitis. (See Prevention.)	Alleviate as many of the predisposing factors as possible, including bruises and tears while giving birth, exposure to wet and cold, and the actual introduction of disease-causing bacteria during delivery or the manual removal of the afterbirth. Provide clean stalls, and, if assistance becomes necessary, first disinfect the hands and arms as well as the external genitals.	

(Continued)

TABLE 10-2

Disease	Species Affected	Cause	Symptoms and Signs (or age group most affected)
Mucosal disease (see Bovine virus diarrhea)			
Mycoplasmal pneumonia (see Virus pneumonia)			
Navel infection (or joint ill, navel ill actinobacillosis, arthritis in lambs)—an infectious disease of newborn animals.  Fig. 10-21	Horses. Cattle. Sheep: in lambs it is commonly called arthritis. It affects newborn foals, calves, and lambs.	Several kinds of bacteria are involved.	Loss of appetite, swelling, soreness and stiffness in the joints and general listlessness. Umbilical swelling and discharge. Two forms occur in lambs: (1) suppurative (pus-forming), caused by pus-producing bacteria and marked by swelling of and pus in the affected joints and (2) nonsuppurative, caused by the swine erysipelas organism, in which most lambs recover within a month without treatment, but some develop a chronic lameness.
Necrotic enteritis (see Enteritis)			
Pinkeye (or keratitis)—an infectious eye ailment. 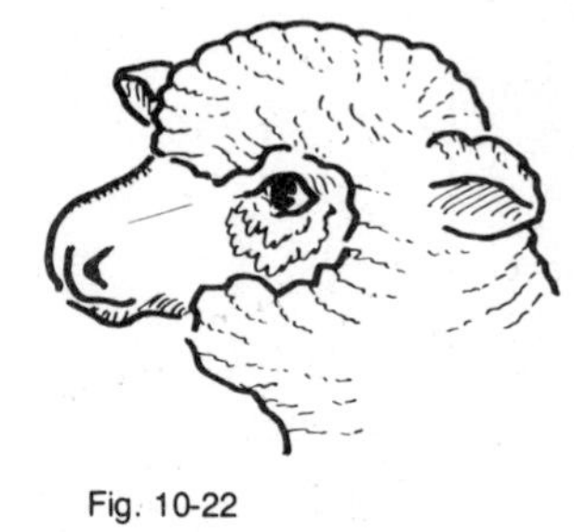 Fig. 10-22	Cattle. Sheep. Goats.	The most common bacterial form of the disease is caused by *Moraxella bovis (Hemophilus bovis)*. Infection with this bacteria can be started with vitamin A deficiency, injuries, dust, insects, or strong sunlight. Also, a virus form of pinkeye is caused by the "red nose" or infectious bovine rhinotracheitis (IBR) virus.	Liberal flow of tears and tendency to keep the eyes closed; redness and swelling of the lining membrane of the eyelids and sometimes of the visible part of the eye; and there may be discharge of pus and ulcers of the cornea. If unchecked, blindness may follow. In viral pinkeye, the eyeball itself is only slightly affected. IBR mainly affects the eyelids and the tissues surrounding the eyes. It causes a severe swelling of the lining of the lids.
Pneumonia—an inflammation of the lungs.	All animals.	Causes are numerous, including (1) many microorganisms, (2) a number of different viruses, (3) inhalation of water or medicines given by untrained persons as a drench, and (4) changeable weather during the spring and fall.	Ushered in by a chill, followed by elevated temperature. Quick, shallow respiration, discharge from the nostrils and perhaps eyes, and a cough. Legs wide apart, drop in milk production, loss of appetite and constipation. Crackling noises with breathing, and gasping for breath.

—(Continued)

Distribution and Losses Caused By	Treatment	Control and Eradication	Prevention	Remarks
Throughout the U.S. About 50% of the infected foals die and many that survive have deformed joints. In calves and lambs, the mortality is not so high. In suppurative arthritis in lambs, the mortality rate is high and chances of recovery poor.	The veterinarian may give a blood transfusion from the dam to the offspring, or in other cases he may administer a sulfa drug, an antibiotic, a serum, or a bacterin.	(See Prevention.)	Practice sanitation and hygiene at mating and parturition. Feed iodized salt to pregnant mares in iodine-deficient areas. Soon after birth, treat the navel cord of newborn animals with iodine.	Navel infection in calves and lambs occurs less frequently than in foals. Providing clean quarters for the newborn and painting the navel cord with iodine constitute the best preventive measures for all classes of livestock.
The disease is widespread throughout the U.S., especially among range and feedlot cattle. Pinkeye is encountered in nearly half of U.S. beef cattle herds and affects 3% of all beef cattle. Deaths are rare. One record showed that affected steers gained an average of 50 lb less during the grazing season than those not affected.	The most common treatment for bacterial pinkeye is the application of antibiotics or sulfa drugs to the affected eye as ointments, powders, or sprays; preferably, with treatment made twice daily. Recovery is speeded up by keeping the infected animals in a dark barn. A commercially produced protective eye patch is now available. It completely covers the infected eye, holding the medication in place, protects the eye from insects and bright sunlight, and reduces the work and expense of handling and isolation. Held in place by a special adhesive, the eye patch drops off and decomposes after about 7 to 10 days. Treatment of IBR conjunctivitis is seldom of value, although antibiotics sometimes help reduce the secondary bacterial infection.	Isolate affected animals, control insects, provide good nutrition—including adequate vitamin A, and control IBR conjunctivitis by a vaccination program.	Prevention of bacterial pinkeye consists in the following: controlling face flies and other insects that feed around the eyes; good nutrition, including adequate vitamin A; and isolation of affected animals. IBR conjunctivitis may be prevented by proper vaccination of animals prior to onset of the disease. The herd should not be vaccinated once the disease appears; nor should pregnant cows be vaccinated. Affected animals should be isolated.	Pinkeye is highly infectious and may spread rapidly through a herd, producing drop in production or occasional blindness in some animals.
Nationwide. If untreated, 50 to 75% of affected animals die. Pneumonia causes 1/5 of all nonnutritional mortality in U.S. beef cattle.	Place sick animals in quiet, clean quarters away from drafts, and give easily digested nutritious feeds. Sulfonamides and antibiotics are effective in the treatment of most acute pneumonias. But they are ineffective against virus pneumonia, except for keeping down secondary bacterial pathogens.	Segregate sick animals. Pigs are subject to epizootic pneumonia, known as virus pneumonia, VPP, or infectious pneumonia. It is caused by a *Mycoplasma*. Two methods have been used for eradication of virus pig pneumomia: (1) specific pathogen-free (SPF) pigs, and (2) farrowing old sows in isolation and ascertaining that the pigs are free of VPP before adding them to the herd.	Provide good hygienic surroundings and practice good, sound husbandry.	Changeable weather during the spring and fall and drafty, damp barns are conducive to pneumonia. Always try to establish what organism is causing the condition. Many terminal bacterial pneumonias start from virus diseases.

(Continued)

TABLE 10-

Disease	Species Affected	Cause	Symptoms and Signs (or age group most affected)
Pox, swine (or *Variola suilla*)—an acute infectious disease.	Swine.	There are 2 types of swine pox, each caused by a different type of virus.	Swine pox is characterized b small red spots, which appea over large parts of the body especially on the ears, neck undersurface of the body, an inside of the thighs. These spots grow rapidly and reach the size of a dime. A hard nodule de velops in the center of each. Several days later, small pea-sized vesi cles (blisters) develop; at first these contain a clear fluid, but later the contents become puslike. Soon these blisters dry up, leaving dar brown scabs, which fall off. Preceding the skin changes noted above some animals show fever, chills, and refusal to feed.
Pseudorabies (PRV) (Aujeszky's Disease, mad itch)	Most species of domestic and wild animals. Swine are the natural hosts and chief reservoir of the disease.	Virus of the Herpesvirus group.	Baby pigs may die with few, i any, clinical signs evidenced However, death is usually pre ceded by fever which may ex ceed 105°F, dullness, loss o appetite, vomiting, weakness incoordination, and convul sions. In pigs less than 3 week old, death losses frequently approach 100%. After 3 weeks of age, the signs and death losses decrease. In adult pigs, the signs may be very mild and include fever, off feed coughing, sneezing, vomiting, diarrhea, constipation, convulsions itching, middle ear infections, and blindness. Sows infected in middle pregnancy may abort mummified fetuses. Sow infected late in pregnancy often abort or give birth to weak, shaker, o stillborn pigs. Piglets infected prior to birth usually die within 2 days. In animals other than swine, the disease is characterized by intens itching, self-mutilation, convulsions, and death.
Pulpy kidney disease (see Enterotoxemia)			
Q fever (nine mile fever)	Cattle. Goats. Sheep. Man.	The causative organism of Q fever is *Coxiella burneti*. Ticks are the most important vector. Man may acquire Q fever through the inhalation of contaminated dust (including tick feces). However, most persons become infected through exposure to livestock, or through the *ingestion* of their products (raw milk or meat of infected animals).	This disease is classified as rickettsial disease, although th mode of infection to man differ from that of other infections i this group. In man, the disease manifests it self by acute onset, chills prostration, and fever. Head ache is pronounced in mos cases. The fever is continuou and lasts from a few days to or 3 weeks. It resembles in fluenza.
Rabies (or hydrophobia, madness)—an acute infectious disease. 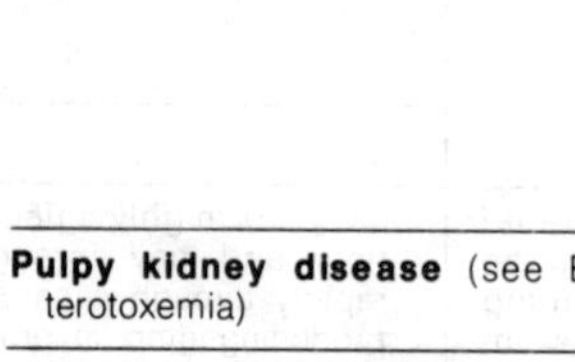 Fig. 10-23	All warm-blooded animals and man.	A filtrable virus which is usually carried into a bite wound by a rabid animal.	Disease manifests itself in forms; (1) the furious form, an (2) the dumb form. In early stages of the furious form there is loss of appetite, cessa tion in milk secretion, anxiety restlessness, and a change i disposition. Next there is mad ness, excitability, loud bellow ing, pawing of the ground, ina bility to swallow, and violen butting of the head. At thi stage, the animal is ver dangerous, attacking and biting itself or other animals and man Posterior paralysis strikes o the fourth or fifth day, followe by a coma and death on abou the sixth day.

—(Continued)

Distribution and Losses Caused By	Treatment	Control and Eradication	Prevention	Remarks
Widely distributed in the Midwest, where it is an important disease of young pigs. Very few pigs die.	No treatment is known. Good management and nursing will help.	Vaccination against both viruses is possible, but is not recommended for 2 reasons: (1) The disease has no great economic importance, and (2) the use of a live vaccine would introduce more virus in an infected environment.	The disease appears to be transmitted primarily by lice, and less often by other insects and by contact. Therefore, lice control is the best preventive measure.	Swine which recover from the disease are immune to further attacks from the specific type of virus that caused the disease, but not to attacks from the other virus. One type of virus is related to that causing pox in various other species of animals, but the other is not.
Widespread and of considerable economic importance in midwestern U.S. The disease is widespread in Europe, where it causes heavy losses.	Drugs and feed additives are not effective for control or treatment of pseudorabies.	The following control measures are recommended for the protection of herds: 1. Dispose of dead pigs properly (burning or burying). 2. Buy tested breeding stock and isolate them for 30 days. 3. If you raise breeding stock, do not buy feeder pigs. 4. Get feeder pigs from a farm that has not had the disease. 5. Keep visitors away from swine premises. 6. Keep stray dogs, cats, and wildlife off the premises. 7. Keep swine and cattle separate. 8. Isolate show stock for 30 days after the fair is over.	A modified live virus vaccine for pseudorabies is available. But state regulatory officials must authorize vaccine usage. Breeding stock should be vaccinated twice per year prior to breeding. Pigs from unvaccinated sows may be vaccinated after 3 days of age. Pigs from immunized sows should be vaccinated when 3 to 8 weeks old.	Pseudorabies is not related to rabies. The disease was first recognized as an infectious disease of cattle and dogs in Hungary by Aujeszky in 1902. The continued increase and severity of the disease may be due to rearing more pigs in large confinement units or by the cessation of cholera vaccination. A serum test is the most practical herd test.
Q fever was first recognized in Australia in 1935. The disease has been identified in cattle in some 35 states within the U.S. and therefore is recognized as endemic. A recent Ohio study indicates that the disease is more prevalent among large than among small dairy herds.	Treatment with an antibiotic may reduce the duration of fever and illness, but the response to antibiotic therapy usually is not dramatic.	A vaccine against Q fever is now available.	Avoid ticks, take care in aiding animals through parturition, and pasteurize milk properly. Since milk-borne transmission of Q fever has occurred, pasteurization temperatures have been elevated slightly to ensure killing of the causative organism. Vaccination appears to have some value in the control of the disease in both livestock and people.	The Armed Forces have studied this disease from the standpoint of "biological warfare." The atomic bomb dropped on Nagasaki weighed about 5 tons and killed an estimated 80,000 people; but only a fraction of an ounce of chick embryo tissue inoculated with *C. burneti* could, if properly placed, infect a billion persons and would likely be fatal to as many as 10 million.
Less than 10% of the cases appear in cattle, horses, swine, or sheep.	After the disease is fully developed, there is no known treatment. Where animals are bitten or exposed to rabies, see the veterinarian.	Persons bitten by a rabid animal should immediately report to the family doctor who usually administers Semple-type vaccine, although irradiated vaccines are used to some extent. With severe bites, especially those around the head, antiserum is particularly indicated. Complete eradication would be difficult to achieve because of the reservoir of infection in wild animals.	Immunize all dogs, and regulate the licensing, quarantine, and transportation of dogs. Vaccinate in areas where rabies is present. Several new rabies vaccines are available, and others are being developed and tested experimentally. Thus, the choice of a rabies vaccine for animals should be made by the veterinarian. Control wild carnivores and bats.	Rabies is generally transmitted to farm animals by dogs, or by wild animals (skunks, foxes, etc.). Where people are bitten or exposed to rabies, they should see their local doctor. He may use a vaccine made of (1) killed virus, nervous tissue origin, or (2) killed virus, duck embryo origin. Also, new and promising vaccines have been developed and are being tested experimentally.

(Continued)

TABLE 10-

Disease	Species Affected	Cause	Symptoms and Signs (or age group most affected)
Red water disease (see Bacillary hemoglobinuria)			
Rhinopneumonitis (see Equine abortion)			
Scrapie—a virus disease.	Sheep.	The cause and method of spread of scrapie are not fully understood. It has been proven that scrapie is an inoculable disease and that the transmissible agent can be passed in series, indefinitely, from sheep to sheep in filtrates prepared from the tissues of infected sheep. This is characteristic of a virus, but the transmissible agent can withstand boiling and exposure to concentrations of chemical agents which inactivate all known viruses. Moreover, no evidence of the development of antibodies has been obtained and the lesions of nerve cell degeneration do not resemble those usually associated with a virus infection. Thus, the virus theory appears to be untenable.	Despite intense itching, whic acccounts for the name scrapi (affected sheep rub themselve against fences and other ob jects to relieve the itching, an in doing so scrape off wool), a uncoordinated gait is consid ered the most reliabl symptom. Scrapie affects the nervous sys tem of sheep. Animals are rest less and excitable, walk un steadily, suffer from thirst, ar weak, become paralyzed, an die. The appetite remains goo and there is no rise in tempera ture. Scrapie seldom appear in animals under 18 months c age and is usually fatal after course of some weeks o months.
Shipping fever (or hemorrhagic septicemia)—one of a group of infectious diseases.  Fig. 10-24	Cattle. Sheep, especially lambs.	Shipping fever is caused from multiple infection due to the interaction of viruses and bacteria, accentuated by environmental conditions creating physical tension or stress. Change in weather and feed, overcrowding, hard driving, lack of rest, and improper shelter all help usher in the disease.	Develops rapidly and lasts for week or less. Usually high tem perature, discharge from th eyes and nose, a hacking cough, difficulty in breathing and there may be a swelling i the region of the neck. Some times there is diarrhea. In ver acute forms, animals may di without showing symptoms. Young animals most susceptible but animals of all ages af fected.
Sore mouth (or contagious ecthyma)—a highly contagious virus disease.  Fig. 10-25	Sheep. Goats. Man.	A specific virus, often complicated by the necrosis organism, *Spherophorus necrophorus*.	Infected lambs and kids refus feed and appear depressec Small vesicles appear on th lips, gums, and tongue, caus ing these parts to be red an swollen. Vesicles break an form sores that bleed easil and become encrusted with scab. The sores may becom infected or may spread to th teats, udder, and feet (jus above the coronet) of th mother. This is especially a disease c lambs and kids.
Strangles (see Distemper)			

—(Continued)

Distribution and Losses Caused By	Treatment	Control and Eradication	Prevention	Remarks
The disease has long been known in France and Britain, but was not reported in Canada until 1938 and in the U.S. until 1947.	No cure is known.	Because the means of natural transmission are still unknown, no adequate control measures are available. In the U.S., control has been by quarantine, and by destruction of infected flocks, with indemnity payments (federal and some states) made to the owner.		Suspected cases of scrapie should be diagnosed by the local veterinarian and then reported to state or federal officials. In the U.S., an eradication program is in progress, in an attempt to prevent scrapie from becoming established in this country.
Worldwide, especially among thin and poorly nourished young animals that are subjected to shipment by rail or truck during bad weather. It is a serious problem to both shippers and receivers of animals. Death losses may be high in untreated cases.	Isolate affected animals. Treatment should be handled by a veterinarian. In the early stages, the use of antibiotics and/or sulfonamides will control the accompanying bacterial infection.	Institute good feeding and management.	**Cattle:** As a preventative measure, one should eliminate as many as possible of the predisposing factors that lower the animal's vitality. Also, newly purchased animals should be isolated for 2 to 3 weeks before being placed in the herd. Several vaccines, both modified and inactivated, are available. Vaccination should be done 3 to 4 weeks before exposure. Where cattle have been subjected to great stress—long shipment, extensive handling, and/or exposure to severe weather conditions—it is recommended that they be handled as follows: Adult cattle should be given long grass or oat hay, rolled oats, and/or wheat bran during the first week; newly weaned calves can be given a calf starter ration in addition; all animals should have access to plenty of clear fresh water at all times. For the first 28 days after arrival, fortify the ration with 350 mg of Aureomycin plus 350 mg of sulfamethiazine per head per day. Newly arrived cattle should also receive 50,000 IU of vitamin A per head daily and have free access to a good mineral mixture. **Sheep:** Reducing stress and adding high levels of antibiotics or sulfas to the feed during susceptible periods may be of some value in reducing the incidence of pneumonia.	The name hemorrhagic septicemia should be discarded because this name describes neither the disease nor its cause.
Throughout the U.S. The mortality (from secondary causes) is low, but of economic importance in that it results in unthriftiness and loss of weight.	Infected lips, mouths, and nostrils should be treated by applying an ointment containing a broad-spectrum antibiotic to the lesions.	If the disease is in serious proportion, immunization with a vaccine may be necessary in order to bring it under control. Isolate, if practical, infected animals and prevent secondary invaders.	General sanitation. A specific vaccine, similar in nature to smallpox vaccine, will also produce an immunity. The latter should be used either (1) at the time of docking and castrating, or (2) at least 10 days before shipping feeder lambs in areas in which the disease is known to be very prevalent. But do not vaccinate noninfected sheep on noninfected ground because this will infect the premises.	If young lambs or kids go off feed for an extended time, the udder of the dam may become caked, predisposing blue bag or mastitis. Sore mouth is often complicated by the necrosis organism and by screwworms. Scabs from sore mouth lesions remain infective for several years and carry infection over from year to year, on the premises.

(Continued)

TABLE 10-2

Disease	Species Affected	Cause	Symptoms and Signs (or age group most affected)
Streptococci infection (or streptococcic septicemia)	Swine.	Bacteria of the streptococcus group.	It is often confused with hog cholera, because, like cholera, it is characterized by sudden onset and death within 12 to 18 hours. Affected animals show weakness, prostration, and high temperature. Diarrhea and bloody urine may occur. It often localizes in the joints, causing a chronic arthritis.
Swine dysentery (see Dysentery, swine)			
Swine erysipelas (see Erysipelas, swine)			
Swine influenza (or hog flu)	Swine.	Type A influenza virus, which is similar to the virus that caused the worldwide flu epidemic of 1918 in humans. Indeed, there is substantial evidence that flu was first introduced to swine at that time—that this is a classic example of a human disease that was transmitted to animals.	Makes appearance suddenly. High fever, loss of appetite, a cough, and discharge from the eyes and nose are seen. Animals reluctant to move, but may sit up like dogs to facilitate breathing.
Tetanus (or lockjaw)—chiefly a wound infection disease; see Fig. 10-26.  Fig. 10-26	Chiefly in horses (and other equines) and man, but occurs in cattle, swine, sheep, goats.	A powerful toxin (more than 100 times as toxic as strychnine) liberated by the bacterium *Clostridium tetani*, an anaerobe.	Usually associated with a wound. First sign of tetanus is a stiffness about the head. The animal often chews slowly and weakly and swallows awkwardly. Third or inner eyelid protrudes over forward surface of eyeball. With the slightest movement or noise, animal shows violent spasms. Usually remains standing until close to death. All ages susceptible.
Transmissible gastroenteritis (TGE)—destructive, contagious disease among young pigs.	Swine.	A virus.	It is characterized by marked scouring in all cases, and by vomiting in some cases. Usually, the body temperature remains normal. The stomach and intestines are inflamed. Once the disease strikes, it spreads rapidly; the entire herd may become noticeably affected in 2 to 3 days.

—(Continued)

Distribution and Losses Caused By	Treatment	Control and Eradication	Prevention	Remarks
It appears to be fairly common in swine.	Sulfa drugs and antibiotics, administered by the veterinarian, are commonly used in treatment.	Good management with emphasis on sanitation.	Antibiotics in the feed will help prevent this condition. Proper treatment of the navel with iodine at birth is an essential management practice.	All ages of swine are susceptible. It occurs under both good and poor sanitary and management conditions.
Throughout the U.S Mortality seldom exceeds 2%. The principal loss from swine flu is the lingering debility, which results in the animals not making economic gains.	Provide warm, dry, clean quarters and minimum rations. Antibiotics and sulfonamides may be used on a herd basis to control various bacterial invaders. Expectorants aid if respiration is difficult.	Correct any faulty feeding and management with the herd. Avoid bringing in new animals.	Use dry, clean hog lots that are rotated, thus breaking up the life cycle of the lungworm (the lungworm harbors the virus which, in the presence of the bacteria, causes swine influenza).	This disease is precipitated by sharp cold or windy days, particularly during the spring and fall. There is serologic evidence of cross infection of humans and swine with the type A influenza virus.
Worldwide, but in the U.S. occurs most frequently in the South. Death occurs in over 80% of the affected cases.	Place the animal under the care of a veterinarian and keep it quiet. Good nursing is important. Tetanus antitoxin may be helpful if administered at the time of the first symptoms.	(See Prevention.)	Immunity against tetanus can be obtained through inoculation with either toxoid or antitoxin. Toxoid is an injection of neutralized tetanus toxin to stimulate the horse to build its own antibodies. Antitoxin is a concentrated serum with tetanus toxin antibodies taken from another horse and administered as a preventative measure following wounds, surgery, or foaling. Active immunization is achieved through 2 injections of tetanus toxoid at 2- to 4-week intervals, followed by annual booster injections. If an immunized horse is wounded 2 months or more following such immunization, it is recommended that the veterinarian administer another toxoid injection at that time. If a horse not previously immunized is wounded, it is recommended that the veterinarian administer antitoxin, which will give passive protection for up to 2 weeks.	On some "hot" premises, all surgery should be accompanied with tetanus antitoxin.
There is a high mortality (90-100%) in pigs less than 3 to 4 weeks old, but the death loss in older swine is low.	No effective treatment is known, but good feeding and management help. Antibiotics or sulfonamides may minimize secondary bacterial complications.	Move and scatter the sows that have not farrowed. Continuous farrowing should be avoided, because it tends to perpetuate the disease. Isolation of new additions to the herd.	The most effective preventive measure consists in exposing the sows to the disease before they farrow. They will then develop resistance or protection, and the pigs will get the antibodies from the milk. However, this procedure should be used only where it is inevitable that sows will be exposed at farrowing time and where there is no danger of spreading the disease to neighboring herds.	The incubation period may be only 18 hours.

(Continued)

TABLE 10-2

Disease	Species Affected	Cause	Symptoms and Signs (or age group most affected)
Tuberculosis—a chronic infectious disease. (Fig. 10-27 shows a positive reaction to the TB test.) 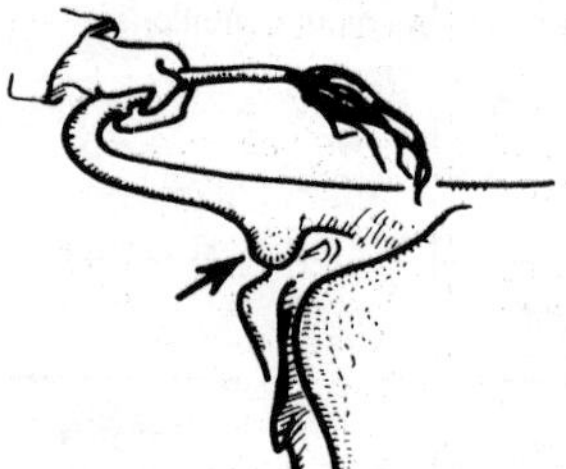 Fig. 10-27	All animals and man. Tuberculosis in sheep and goats is rare and of chronic character.	*Mycobacterium tuberculosis*, of which there are 3 kinds: (1) the human, (2) the bovine, and (3) the avian (bird) types.	Animals usually get tuberculosis of the lungs and lymph nodes, although in poultry the liver, spleen, and intestines are chiefly affected. In cows, the udder sometimes becomes infected and swollen in chronic cases. Many times infected animals show no outward physical signs of the disease. There may be loss in weight, swelling of joints, and a chronic cough and labored breathing. Other seats of infection are genitals, central nervous system, and the digestive system. **Sheep:** Manifest few symptoms; observed on post-mortem. Coughing is prominent in goats, but not in sheep.
Tularemia	Cattle. Sheep. Horses. Swine. Man. Cats. Chipmunks. Dogs. Hamsters. Muskrats. Beavers. Coyotes. Foxes. Mice. Rats. Rabbits. Deer. Opossums. Squirrels. Guinea pigs, and many other wild mammals.	It is caused by *Francisella tularensis* (closely related to the plague bacillus). The chief vectors of tularemia are ticks (especially the dog tick and the wood tick) and bloodsucking flies (especially the deerfly and horsefly), but it may also be spread by fleas, certain mosquitoes, and lice. Infected animals that die may contaminate water and thereby spread the infection to sheep and perhaps other species.	In man the disease is characterized by headache, chills, fever, and vomiting, accompanied by irregular fever, which lasts for several weeks.
Vaginitis (or granular venereal disease, granular vaginitis)—an infectious disease which localizes in the cow's vulva and on the penis and prepuce of the bull, causing an inflammation of varying intensity.	Cattle.	Unknown.	Tissue of the vagina is reddish, roughened, and granular in appearance. Infected animals are usually difficult breeders.
Vesicular exanthema	Almost exclusively swine, but horses are slightly susceptible. Man is not affected.	A virus. In order to distinguish vesicular exanthema from foot-and-mouth disease, tests may be made with horses and cattle. Also guinea pigs may be infected with foot-and-mouth disease but are resistant to the virus of vesicular exanthema.	Similar to foot-and-mouth disease. Small vesicles (like water blisters) appear around the head, particularly on the snout, nose, or lips. Also, these blisters appear on the feet where the hair and the horny part of the hoof meet, on the ball of the foot, on the dewclaws, between the toes, and on the udder and teats of nursing sows. Affected animals go lame, have a high temperature, and usually go off feed for 3 to 4 days.

—(Continued)

Distribution and Losses Caused By	Treatment	Control and Eradication	Prevention	Remarks
Worldwide. The incidence of tuberculosis in the U.S. is steadily declining.	In humans, tuberculosis can be arrested by hospitalization and complete rest, but in animals this method of treatment is neither effective nor practical. Also, no known medical treatment is effective with animals.	**Cattle:** Periodic testing and removal of reactors is the only effective method of control. The following is an effective control program for animals: Disposing of tubercular swine, cattle, and poultry. Applying strict sanitation, and Rotating feedlots and pastures. **Sheep:** Testing with avian tuberculin may be of assistance.	Removal and supervised slaughter of reactor animals, and pasteurization of milk and creamery by-products. Fig. 10-27 shows a positive reaction to the intradermic (into the skin) tuberculin test in a cow. This reaction indicated the presence of TB. Avoid pasturing or housing cattle and swine with chickens.	The relative susceptibility of man and animals to the 3 different kinds of tuberculosis germs is shown in Table 10-3. All states will accept for entry the following: (1) accredited herds which have been tested for TB within the past 12 months, or (2) cattle which have had individual negative tests within the past 30 days. Cattle for export must be tested for tuberculosis and found free of the disease within 90 days of shipment.
This disease of mammals was first described in 1911 in Tulare County, Calif. (from whence came the name), as a "plaguelike disease of rodents." In 1919 it was described as "deerfly fever." The disease has been found throughout the U.S. Also, it has been reported in the U.S.S.R. and Japan.	Streptomycin is very effective in the treatment of this disease.	Avoid picking up sick, easily caught, or dead rabbits. Wear rubber gloves when handling wild game.	Cooking readily renders the infected tissues safe for human consumption. Vaccines are effective.	Primary modes of transmission to man include bites from infected ticks or other arthropods, the handling of infected animals, and the inhalation of dust or vapor containing *F. tularensis*.
Throughout the U.S. Losses are in terms of lower percentage calf crop and decreased milk production.	Treatment consists in sexual rest. The condition will clear up by itself in several weeks. The veterinarian may place antibiotics in the uterus and vagina, or he may treat by injecting antibiotics or sulfa drugs.	(See Prevention.) Artificial insemination aids in control.	Purchase clean animals from clean herds, and avoid the use of bulls that have been exposed to the injection. In problem herds, vaccination 30 to 60 days prior to the breeding season should be considered.	It is believed that the infection is commonly transmitted by the bull at the time of service, but this is not the only means of transmission since virgin heifers may be infected.
The course of the disease is 1 to 2 weeks; the mortality is low. Pregnant sows often abort and nursing sows fall off noticeably in milk production.	No treatment is known. Good nursing will help.	Rigid quarantine of affected areas (applicable to both herds and pork products) and the destruction of affected herds appear to constitute the best control and eradication. The virus that causes vesicular exanthema is readily killed when exposed to a 2% sodium hydroxide (lye) solution or a 4% sodium carbonate (soda ash) solution. Cooking garbage breaks the chain of infection from infected meat to susceptible swine. Thus, this is the principal means of control.	Avoid the introduction of hogs from infected areas. The disease is transmitted by contact with or consumption of feed and water contaminated with the virus. Uncooked garbage that contains pork trimmings has often been the source of the virus.	Hogs are resistant after recovery, but only for a few months. No immunizing agents are available. Usually symptoms appear 24 to 48 hours after contact with the virus.

(Continued)

TABLE 10-2

Disease	Species Affected	Cause	Symptoms and Signs (or age group most affected)
Vibriosis (vibrionic abortion, vibrio fetus)	Sheep. Cattle.	The bacterium *Vibrio fetus*.	Many abortions among ewes, especially during the last 4 to 6 weeks of pregnancy, with lambs stillborn. Infected cattle herds are characterized by (1) abortions in the middle third of pregnancy, (2) several services per conception, and (3) irregular heat periods. For diagnosis, laboratory methods must be used.
Virus pneumonia (or VPP, mycoplasmal pneumonia of swine; also see Pneumonia)	Swine.	A small cocobacillary organism, *Mycoplasma hypopneumonia*. Three different swine diseases are caused by *Mycoplasma* spp. infection: mycoplasmal pneumonia, mycoplasmal arthritis, and mycoplasmal polyserositis arthritis.	Coughing is the first and most characteristic symptom. It generally begins 10 to 16 days following exposure to diseased animals, and it may persist indefinitely. Coughing is most marked when pigs come out to feed in the morning or following vigorous exercise. Diarrhea usually occurs when pigs first begin to cough, but it lasts only 2 to 3 days. Temperatures are only slightly elevated, and even when they reach 105° F, the pigs do not look sick. In general, affected pigs eat well, but gains are slow and feed utilization poor.
Warts—infectious; caused by a virus. 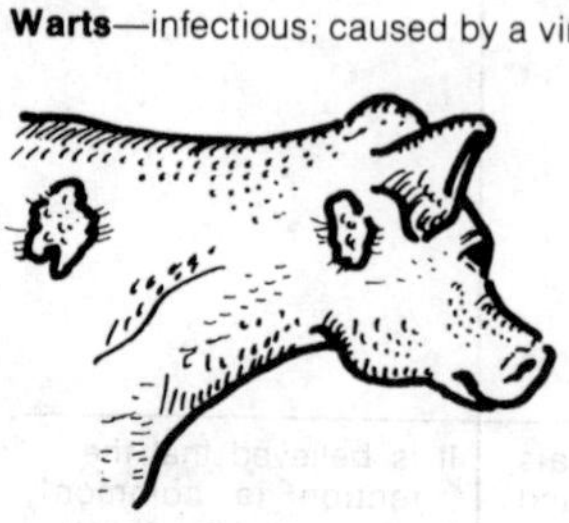 Fig. 10-28	Cattle and other animals and man.	Virus (apparently each class of animals is attacked by a specific wart virus).	Growths on the skin varying from very small to large pendulous growths weighing several pounds. Warts may appear anywhere on the body, especially on the teats or around the head.
White cattle scours (see Calf scours)			
Winter dysentery, bovine (or winter scours)—an acute infectious disease of stabled cattle.	Cattle.	An organism, *Vibrio jejuni*, and perhaps a virus.	The period of incubation is extremely short, varying from 3 to 5 days. A profuse watery diarrhea is the main symptom. Often feces are dark brown in color, and tend to become darker, when intestinal hemorrhages occur. Usually the temperature remains normal and the appetite remains unchanged. Calves and young animals are least susceptible, but animals of all ages are affected. The seasonal incidence of the disease, the age and number of animals affected, together with the suddenness of the onset, are helpful in arriving at a correct diagnosis.
Wooden tongue (see Lumpy jaw)			

[1]The illustrations for this table were prepared by R. F. Johnson.
[2]Dr. T. Moll, DVM, College of Veterinary Medicine, Washington State University, very kindly reviewed the summary herewith presented relative to this disease.
[3]*Better Beef Business*, April 1973.
[4]*Successful Farming*, May 1973.

—(Continued)

Distribution and Losses Caused By	Treatment	Control and Eradication	Prevention	Remarks
Vibriosis occurs in sheep and cattle throughout the U.S. Reports indicate that the distribution is worldwide. In affected flocks, abortions may be up to 80%. In general, ewes show little clinical evidence of infection.	There is no treatment for sheep. Cattle are treated by injecting drugs into the uterus and/or by allowing sexual rest.	Females that have aborted should be isolated and all aborted fetuses and membranes should be burned. Control of vibriosis in sheep or cattle requires sound management, strict sanitation, early diagnosis, and proper vaccination.	A vaccine is available. Repeat annually. Avoid contact with diseased animals, and contaminated feed, water, and materials. Artificial insemination is a rapid and practical method of stopping infection from cow to cow.	Ewes that abort one year may raise perfectly healthy lambs in succeeding years. Some probably remain as "carriers." Transmission in sheep appears to be from consuming infected feed and water, whereas in cattle it is from an infected bull.
It is estimated that virus pig pneumonia afflicts ½ the herds in the U.S. and lowers efficiency by 20%.	Broad-spectrum antibiotics or sulfonamides will lessen the bacterial complications that accompany virus pig pneumonia, although they will not clear the primary pneumonia. Good nursing care will help.	The disease is continued in infected herds by contact of young pigs with carrier sows. The most infallible control consists of a disease-free pig program.	Avoid the purchase of animals from infected herds; purchase "disease-free pigs."	If management is poor, many pigs develop serious pneumonia due to secondary bacterial infection of the lungs. The age 14 to 26 weeks seems to be the most critical for such secondary lung complications. Virus pneumonia may well be the world's most important swine disease.
Throughout the U.S. They are a nuisance, but their presence does not normally interfere with the animal's health. However, they damage the hide, making the leather weak in the area where the wart is found. Also, in the South, animals with warts are more subject to screwworm infection.	Soften with oil for several days, then tie off the growth with thread or snip it off with sterile scissors and paint the stump with tincture of iodine; The wart vaccine helps in some cases; or the veterinarians may resort to surgical removal of extremely large warts.	(See Prevention.)	Segregate "warty" cattle. Clean and disinfect all exposed pens, stables, chutes, and rubbing posts. Milk "warty" cows last.	Wart cases usually recover spontaneously after 4 to 8 months, and the warts drop off.
Through the central, eastern, and southern states. It causes few death losses, but afflicted animals lose in condition; and, in lactating animals, there is a sharp reduction in milk flow.	The veterinarian may administer antibiotics or an intestinal antiseptic. For the relief of dehydration, I.V. injections of physiological salt solution, accompanied by glucose, are useful.	Separate the initial case(s) from the herd and practice rigid sanitation.	Isolate replacement animals. Separate from the herd any animal suffering from an acute attack of dysentery.	It is a disease of stabled cattle, both dairy and beef, most frequently occurring between the months of November and March.

TABLE 10-3
RELATIVE SUSCEPTIBILITY OF ANIMALS AND MAN TO THREE DIFFERENT KINDS OF TUBERCULOSIS BACILLI

Species	Susceptibility to Three Kinds of Tuberculosis Bacilli			Comments
	Human Type	Bovine Type	Avian (bird) Type	
Humans	Susceptible	Moderately susceptible	Questionable	Pathogenicity of avian type for humans is practically nil.
Cattle	Slightly susceptible	Susceptible	Slightly susceptible	
Swine	Moderately susceptible	Susceptible	Susceptible	Ninety percent of all swine cases are due to the avian type.
Chickens	Resistant	Resistant	Very susceptible	Chickens only have the avian type.
Horses and mules	Relatively resistant	Moderately susceptible	Relatively resistant	Rarely seen in these animals in the U.S.
Sheep	Fairly resistant	Susceptible	Susceptible	Rarely seen in these animals.
Goats	Marked resistance	Highly susceptible	Susceptible	Rarely seen in these animals in the U.S.
Dogs	Susceptible	Susceptible	Resistant	Highly resistant.
Cats	Quite resistant	Susceptible	Quite resistant	Usually obtained from milk of tubercular cows.

Relative Susceptibility of Animals and Man to TB Bacilli

There are three kinds of tuberculosis bacilli—the human, the bovine, and the avian (bird) types. Practically every species of animal is subject to one or more of the three kinds, as shown in Table 10-3.

Hardware Disease

The term "hardware disease" (traumatic gastritis) is used to describe the condition that results from swallowing foreign materials, usually metal (nails, wire, screws, pins, etc.). Cattle are involved more than other classes of animals. In most cases, the metal is found only in the reticulum (second stomach).

Nearly 7,000 cattle are condemned each year by the Federal Meat Inspection Service as unfit for food because of hardware disease. Clinical reports indicate that the problem is increasing due to the use of more chopped feeds and more contamination. Sharp objects will injure the lining of the stomach and cause infection and inflammation, a condition known as traumatic gastritis.

Hardware disease is a problem in cattle because of their eating habits and stomach arrangement. The usual source of metals is the feed. The animals eat rapidly and are not able to sort foreign objects from their feed.

• **Prevention**—Avoid foreign objects getting into the feed through good management. Also, install strong magnets, in keeping with the manufacturer's directions, (1) at the outlets of mechanical silo unloaders, and (2) in feed processing equipment.

• **Symptoms**—The most common symptoms are: loss of appetite and digestive disturbance; slow and stiff movement and arched back; elbows that bow outward; decreased rumen movement and chewing; possible diarrhea; tendency to stand with the front feet elevated so as to lessen the pressure of the viscera on the inflamed area; rise in body temperature; and swellings under the jaw, at the brisket, and at the hock joints. Bulls may be reluctant to mate.

• **Treatment**—Powerful magnets may be permanently placed in the cow's second stomach, for the purpose of holding objects that have not penetrated the stomach wall. However, the only sure cure for traumatic gastritis is veterinary surgery. Surgery will be successful only if performed before the condition has progressed to the point that damage has been done to the heart or other organs.

Liver Abscesses

Abscesses, as indicated by the name, are single or multiple abscesses on the liver, observed at slaughter. Usually the abscess consists of a central mass of necrotic liver surrounded by pus and a wall of connective tissue. At slaughter, most livers affected with abscesses are condemned for human food.

In some lots of cattle on high concentrate rations, as high as 75 percent of the livers have been condemned. On the average, however, it will probably run between 5 and 10 percent. Since the liver of a

,000-pound steer weighs approximately 11 pounds, its condemnation represents a considerable monetary loss. The loss from reduced feed efficiency and gains may be even greater.

• **Cause**—The direct cause of most bovine liver abscesses is *Spherophorus necrophorus*, the same bacteria which causes foot rot. This organism, which is ever present in ruminal contents, penetrates the covering epithelium through points of injury, discontinuity, and necrosis. Factors back of rumenitis include (1) rapid rate of change from a diet of roughage to one high in concentrate, and (2) fattening with a diet containing more than 25 percent concentrate.

• **Symptoms and signs**—Liver abscesses generally go undetected until cattle are slaughtered. However, reduced feed intake and gains near the end of the feeding period may be indicative.

• **Prevention and treatment**—Liver abscesses can be greatly reduced, but not entirely eliminated, by continuous feeding during the fattening process of the commonly used antibiotics (Aureomycin, Terramycin, Bacitracin, or Tylan). Also, the incidence of liver abscesses can be reduced by the gradual change from a ration of high roughage to a ration of high concentrate.

ANIMAL PARASITES[3]

Animals are attacked by a wide variety of internal and external parasites, the prevention and control of which is one of the quickest, cheapest, and most dependable methods of increasing production with no extra animals, no additional feed, and little more labor. This is important, for, after all, the farmer or rancher bears the brunt of this reduced meat, milk, and wool production, wasted feed, and damaged hides. It is hoped that the discussion that follows may be helpful in (1) preventing the propagation of parasites, and (2) causing the destruction of parasites through the use of the most effective wormer or insecticide.

Choice of Wormer

Knowing what internal parasites are present within an animal is the first requisite to the choice of the proper drug, or anthelmintic. Since no one drug is appropriate or economical for all conditions, the next requisite is to select the right one; the one which, when used according to directions, will be most effective and produce a minimum of side effects on the animal treated. So, coupled with knowledge of the kind of parasites present, an individual assessment of each animal is necessary. Among the factors to consider are age, pregnancy, other illnesses and medications, and the method by which the drug is to be administered. Some drugs characteristically put animals off performance for several days after treatment, whereas others have less tendency to do so. Some drugs are unnecessarily harsh or expensive for the problem at hand, whereas a safe inexpensive alternative would be equally suitable.

Each livestock establishment should, in cooperation with the local veterinarian and/or other advisor, evolve with a parasite control program and schedule. It is recommended that several different wormers be used, and that they be rotated. Also, a schedule of treatments should be prepared, based on knowledge of the life cycles of the various parasites.

Table 10-4, page 918, lists the common chemical compounds for the control of internal parasites, by species. Although this is a valuable guide, it is recognized that wormers are constantly being improved, and that new ones are becoming available. So, the stockman should consult his local veterinarian relative to the choice of drug to use on his animals at the time.

Forms of Insecticides

Insecticides for use on animals may be purchased in several forms. The most common are emulsifiable concentrates, dusts, wettable powders, and oil solutions. When treating animals, be sure to use only insecticide formulations that are prepared specifically for livestock.

• **Emulsifiable concentrates (EC)**—Emulsifiable concentrates, which are probably the most common type of formulation, are solutions of insecticides in petroleum oils or other solvents. An emulsifier has been added so that the solution will mix well with water. On occasion, usually after extended storage, an EC may separate into its various parts; in that case, it should be discarded. An emulsion may also separate if it is allowed to stand after the concentrate has been added to the water; periodic agitation will help prevent this.

---

[3]This section was authoritatively reviewed by the following: Dr. A. C. Todd, Ph.D., Department of Veterinary Science, University of Wisconsin, Madison, Wisc.; Dr. George T. Edds, DVM, Ph.D., Department of Veterinary Science, University of Florida, Gainesville, Fla.; Dr. Jack Dunlap, DVM, Parasitologist, Washington State University, Pullman, Wash.; Dr. William P. Johnson, DVM, American Cyanamid Co., Princeton, N.J.; Dr. O. H. Siegmund, DVM, Merck Sharp and Dohme, Rahway, N.J.; Dr. W. J. Smith, DVM, Pitman-Moore, Inc., Washington Crossing, N.J.; Dr. John P. Sepesi, DVM, Shell Chemical Company, San Ramon, Calif.; and Dr. R. J. Boisvenue, Ph.D., Eli Lilly Company, Greenfield, Ind.

The use of trade names of wormers and insecticides in this section does not imply endorsement, nor is any criticism implied of similar products not named; rather, it is recognition of the fact that stockmen, and those who counsel with them, are generally more familiar with the trade names than the generic names.

TABLE 10-4
RECOMMENDED COMPOUNDS FOR CONTROL OF INTERNAL PARASITES, BY ANIMAL SPECIES[1]

Wormer	Trade Name—Manufacturer	Cattle			Sheep									Swine								Horses								Wormer
		Coccidiosis	Gastrointestinal Nematode Worm	Lungworm	Brown Stomach Worm	Coccidiosis	Cooperias	Hookworm	Lungworm	Nodular Worm	Stomach Worm	Trichostrongyles	Whipworm	Ascarids	Coccidiosis	Kidney Worm	Lungworm	Nodular Worm	Stomach Worm	Threadworm	Whipworm	Ascarids	Bots	Pinworm	Stomach Worm	Strongyles, Large	Strongyles, Small	Tapeworm	Threadworm	
Amprolium[2]	Amprol (Merck)	x				x									x															Amprolium[2]
Cambendazole	Camvet																					x		x					x	Cambendazole
Carbon disulfide																						x	x		x					Carbon disulfide
Coumaphos	Co-Ral Baymix (Bayvet Corp.)		x		x		x	x		x	x	x	x																	Coumaphos
Dichlorvos	Atgard Equigard (Shell) Equigel													x				x	x		x	x	x	x		x	x			Dichlorvos
Dithiazanine iodide and piperazine citrate	Dizan (Elanco)																					x		x			x			Dithiazanine iodide and piperazine citrate
Haloxon	Luxon Loxon		x		x		x	x		x	x	x																		Haloxon
Hygromycin B	Hygromix (Elanco)													x				x			x									Hygromycin B
Lead arsenate	Bi-forma (Texas Pheno Co.)																											x		Lead arsenate
Levamisole	Tramisol (Am. Cyanamid)		x	x	x		x	x	x	x	x	x	x	x		x	x	x	x	x	x	x			x	x	x			Levamisole
Levamisole-piperazine																								x	x				x	Levamisole-piperazine
Mebendazole	Telmin (Pitman-Moore)																					x		x		x	x			Mebendazole
Parbendazole	(Helmatac)		x		x		x	x		x	x																			Parbendazole
Phenothiazine	(Du Pont)				x		x	x		x	x															x	x			Phenothiazine
Phenothiazine, low level	Pheno-Sweet (Farnam)																									x	x			Phenothiazine, low level
Phenothiazine-piperazine	Pheno-Pip (Haver-Lockhart)																					x				x	x			Phenothiazine-piperazine
Phenothiazine-trichlorfon	Equiverm (Texas Pheno Co.)																					x	x	x		x	x			Phenothiazine-trichlorfon
Piperazine	Wonder Wormer (Farnam)													x								x					x			Piperazine
Piperazine-carbon disulfide	Parvex (Upjohn)																	x				x					x			Piperazine-carbon disulfide
Piperazine-carbon disulfide-phenothiazine	Parvex-Pheno (Upjohn)																					x				x	x			Piperazine-carbon disulfide-phenothiazine
Pyrantel tartrate	Banminth (Pfizer)													x				x				x		x		x	x			Pyrantel tartrate
Tetramisole	Tetramizole Bayer 9051								x																					Tetramisole
Thiabendazole	Thibenzole Equizole (Merck)		x		x		x	x		x	x	x								x		x		x		x	x		x	Thiabendazole
Thiabendazole-piperazine	Equizole-A (Ft. Dodge)																					x		x		x	x		x	Thiabendazole-piperazine
Thiabendazole-trichlorfon	Equivet-14 (Farnam)																					x	x	x		x	x		x	Thiabendazole-trichlorfon
Trichlorfon	Anthon (Chemagro) Bot-X (Farnam) Dyrex (Ft. Dodge)												x									x	x	x						Trichlorfon
Trichlorfon-phenothiazine-piperazine	Dyrex T.F. (Ft. Dodge)																					x	x	x		x	x			Trichlorfon-phenothiazine-piperazine

[1]The products listed have 90% efficacy or more. This list is not complete. Inclusion of trade names does not imply endorsement.
[2]In the U.S., it is permitted in feed only as an aid in the control of coccidiosis of chickens and turkeys.

• **Dusts**—Dusts are applied directly to animals in the dry form and cannot be used as sprays.

• **Wettable powders**—Wettable powders are also dry, but the addition of a dispersing and wetting agent allows them to be suspended in water for application to cattle. Continuous agitation of the mixture is important when treating with wettable powders.

• **Oil solutions**—Oil solutions are insecticides dissolved in oil; no emulsifier is added. These materials are usually ready for use and should not be added to water.

Application of Insecticides

The availability of an insecticide and the type of application(s) for which it was formulated are of prime importance, but the treatment of a herd or flock is dictated pretty much by the animal species, number of animals, available handling facilities, time or season, the target pest, management practices, and cost. The common methods of insecticide application are (1) spraying, (2) dipping, (3) back rubbers, (4) pour-on, and (5) feed additives.

Use Insecticides Safely

Certain basic precautions must be observed when insecticides are to be used because, used improperly, they can be injurious to man, domestic animals, wildlife, and beneficial insects. Follow the directions and heed all the precautions on the labels.

• **Selecting insecticides**—Always select the formulation and insecticide labeled for the purpose for which it is to be used.

• **Storing insecticides**—Always store insecticides in original containers. Never transfer them to unlabeled containers or to food or beverage containers. Store insecticides in a dry place out of reach of children, animals, or unauthorized persons.

• **Disposing of empty containers and unused insecticides**—Properly and promptly dispose of all empty insecticide containers. Do not reuse. Break and bury glass containers. Chop holes in, crush, and bury metal containers. Bury containers and unused insecticides at least 18 inches deep in the soil in a sanitary landfill or dump, or dump in a leveled isolated place where water supplies will not be contaminated. Check with local authorities to determine specific procedures for your area.

• **Mixing and handling**—Mix and prepare insecticides in the open or in a well-ventilated place. Wear rubber gloves and clean dry clothing (respirator device may be necessary with some products). If any insecticide is spilled on you or your clothing, wash with soap and water immediately and change clothing. Avoid prolonged inhalation. Do not smoke, eat, or drink when mixing and handling insecticides.

• **Applying**—Use only amounts recommended. Apply at the correct time to avoid unlawful residues in meat. Avoid treating animals younger than specified on the label. Avoid retreating more often than label restrictions. Avoid drift on nearby crops, pastures, livestock, or other nontarget areas. Avoid prolonged contact will all insecticides. Do not eat, drink, or smoke until all operations have ceased and hands and face are thoroughly washed. Change and launder clothing after each day's work.

• **In case of an emergency**—If you accidentally swallow an insecticide, induce vomiting by taking one tablespoonful of salt in a glass of water. Repeat if necessary. Call a doctor.

• **Withdrawal**—After spraying, dipping, or dusting your animals with pesticides, observe the prescribed number of days interval between the last treatment and slaughter. Refer to the container labels for this information.

Parasites and Their Control

A summary, by species, of the common animal parasites and their control is given in the following tables:

Table 10-5, Cattle Parasites and Their Control, page 920.

Table 10-6, Sheep and Goat Parasites and Their Control, page 930.

Table 10-7, Swine Parasites and Their Control, page 940.

Table 10-8, Horse Parasites and Their Control, page 950.

Table 10-9, Mites and Their Control, page 958.

Few dosages for the control of internal parasites are given; instead, users are admonished to *follow the directions on the label*. Also, few insecticides are suggested for the control of external parasites because of (1) the diversity of environments and management practices under which they occur, (2) the varying restrictions on the use of insecticides from area to area, and (3) the fact that registered uses of insecticides change from time to time. Information about what is available and registered for use in a specific area can be obtained from the county agent, extension entomologist, or agricultural consultant.

The insecticide recommendations given for horses are for animals not used for human food. Where horses are to be slaughtered for human food, the tolerance levels and withdrawal periods given on the manufacturer's label should be followed with care.

TABLE 10-5—CATTLE PARASITES

Parasite	Species Affected	Symptoms and Signs of Affected Animals (or Damage Inflicted)
Anaplasmosis—is an infectious disease caused by a minute parasite, *Anaplasma marginale*, which invades the red blood cells. See Fig. 10-29, which shows greatly enlarged red blood cells from an animal with anaplasmosis. The black dots near the margins of the cells are *Anaplasma marginale*. 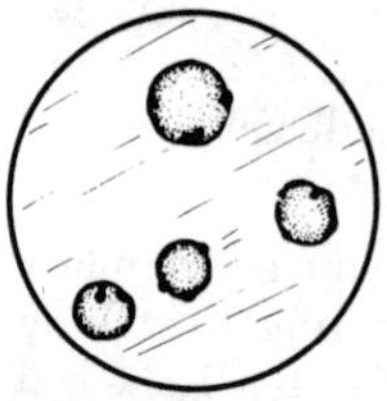 Fig. 10-29	Cattle, especially adults.	Calves usually have the mild type, either not manifesting any symptoms or simply becoming "dumpy" for a few days and then apparently recovering, though their blood remains the permanent abode of the parasite. Mature animals develop severe anemia and usually show a rapid, pounding heart action, labored and difficult breathing, dry muzzle, marked depression, tremors of the muscles, loss of appetite, and marked reduction in milk flow. Also, the eyes and other mucous membranes and the skin may become yellow, there may be depraved appetite, and sick animals may show brain symptoms and an inclination to fight. Urine is normal color. Blood is thin and watery. In severe cases, death may follow in one to a few days.
Blowfly—the blowfly group consists of a number of species of flies all of which breed in animal flesh.  Fig. 10-30	All farm animals.	Infected wounds and soiled hair. Maggots spread over the body, feeding on the skin surface, producing severe irritation, and destroying the ability of the skin to function. Infected animals rapidly become weak, fevered, and unthrifty.
Bovine trichomoniasis—is a protozoan venereal disease of cattle caused by *Trichomonas foetus* (see Fig. 10-31), which are one-celled, microscopic in size, and capable of movement. 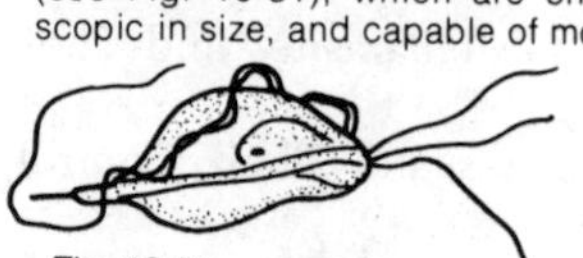 Fig. 10-31	Cattle.	Infected bulls do not usually exhibit visible symptoms, but they are the source of spread from cow to cow. Infected cows are characterized by (1) abortions in the first third of pregnancy, (2) uterine infections, (3) irregular heat periods, and (4) several services per conception. Diagnosis can be confirmed microscopically.
Cattle tick fever (or Texas fever, splenic fever)—is an infectious protozoan disease of cattle caused by a one-celled protozoa called *Babesia bigemina*, which depends upon the tick for transmission and survival.  Fig. 10-32	Cattle, especially adults.	High temperature, rapid breathing, enlarged spleen, engorged liver, pale and yellow membranes, and red to black urine.
Coccidiosis—is a parasitic disease caused by protozoan organisms known as coccidia.	Cattle. Sheep. Goats. Pet stock. Poultry. Each class of animals harbors its own species of coccidia; thus, there is no cross infection between animals.	Diarrhea and bloody feces, and pronounced unthriftiness and weakness.

AND THEIR CONTROL

Distribution and Losses Caused By	Treatment	Prevention and Control	Remarks
Throughout the world, especially in warm climates. From 25 to 60% of infected animals may die. The mortality rate may vary from 2.5% to 50-60%.	Chlortetracycline (Aureomycin) or tetracycline (Terramycin) is effective as an early treatment, and as a means of eliminating carriers when used at high levels. Good care and nursing and keeping animals quiet will help. Blood transfusions are effective.	Immunize cattle by administering 2 doses of anaplasmosis vaccine at least 4 weeks apart. Consult with your veterinarian relative to vaccination or elimination of carrier programs. Spray cattle and buildings to ward off biting insects that spread the disease. Sterilize surgical instruments, dehorners, and needles. Market animals that have recovered from the disease, except in endemic areas.	All animals that recover remain permanent carriers of the parasite. Anaplasmosis is transmitted by ticks, horseflies, mosquitoes, and probably other biting insects, and by such mechanical agencies as needles and surgical and dehorning instruments. Usually a summer disease.
Widespread, but present greatest problem in Pacific Northwest, and in southern and southwestern states. Death losses not excessive, but production is lowered.	When animals become infested with blowfly maggots, their wounds should be treated twice weekly with a smear, dust, or pressurized spray of the proper insecticide, used according to manufacturer's directions.	Eliminate the blowfly by: Destroying dead animals by burning or deep burial. The use of traps, poisoned baits, and electrical screens. Using repellents, such as pine tar oil.	Damage inflicted is similar to that caused by screwworms, and the treatment for both types of maggots is the same.
Throughout the U.S. Economic loss in beef cattle is primarily due to the low percentage calf crop in infected herds. In dairy cattle, delayed return to milk production lowers profits.	Dimetridazole is effective in treating infected bulls. Infected cows should be rested for a minimum of 3 months and then test-bred by artificial insemination to a known clean bull.	Use bulls free of the infection. Use artificial insemination. If practical, sell infected animals for slaughter or allow 90 days of sexual rest. Exercise great precaution in introducing new animals in the herd, in breeding outside cows, and in taking cows outside the herd for breeding purposes.	The infected bull is the source of the infection. A Univ. of Calif. scientist reports that bulls may be successfully treated by the administration of (1) sodium iodide given intravenously, and (2) bovoflavin salves applied to the penis. The compound Berenil has been shown to be relatively effective as a treatment.
Confined to the Gulf Coast area. Death occurs in about 10% of the chronic and 90% of the acute cases. Infected young animals are stunted, mature animals are emaciated, and the milk flow of infected dairy animals is greatly reduced. Death losses are much higher in countries where anaplasmosis also occurs in cattle.	Successful treatment of sick animals depends upon early recognition of the disease and prompt treatment. Agents traditionally used include trypan blue, trypaflavine, and quinuronium sulfate.	Avoid contact with the cattle fever tick, the only natural agent by which cattle tick fever is transmitted. Treat animals at regular intervals with a suitable insecticide. If practical, render pastures tick-free by excluding host animals (cattle, horses, and mules) for 8 to 10 mos., thus starving the ticks.	The protozoa invade the red blood cells of cattle. The parasite is transmitted to cattle by ticks and is carried over in ticks by egg transmission. Although immune, recovered animals are permanent carriers of the disease. In infected areas, native cattle are either immune or only slightly affected. With the exception of a few possible areas along the Gulf Coast, the cattle fever tick has been eradicated—thus controlling tick fever.
Worldwide. There is lowered gain and production in infected animals, along with some death losses. Most severe in calves.	Amprolium (Amprol) is reported to be effective. But, in the U.S., it is permitted in feed only as an aid in the control of coccidiosis of chickens and turkeys.	Avoid feed and water contaminated with the protozoa that causes the disease. Segregate affected animals. Remove and properly dispose of manure and contaminated bedding daily. Drain low, wet areas. Keep animals in a sunny, dry place.	In the oocyst stage, the parasite may resist freezing and certain disinfectants and may remain viable outside the body for months, but it is readily destroyed by direct sunlight or complete drying. Most cattle outbreaks are among dairy calves and feeder calves.

(Continued)

TABLE 10-5

Parasite	Species Affected	Symptoms and Signs of Affected Animals (or Damage Inflicted)
Flies—Several species of flies attack or annoy cattle. The most common ones follow: 1. **Face fly** *(Musca autumnalis)*-was first found in this country in New York in 1953. It is a close relative of and similar in appearance to the housefly. 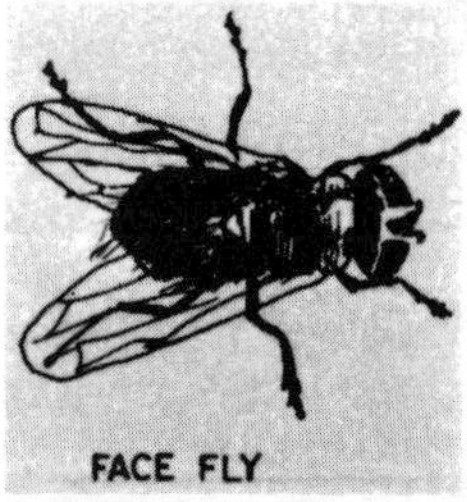 Fig. 10-33	The face fly is primarily a pest of cattle, although it also attacks horses and sheep.	The face fly does not bite, but its habit of clustering around the eyes, mouth, and nostrils is extremely annoying to animals, interfering with their vision and breathing, and preventing normal grazing. Large populations force animals to leave pastures and seek relief in wooded areas and shelters.
2. **Horn fly** *(Haematobia irritans)*. This fly, which is about one-half the size of an ordinary housefly, is one of the most numerous and worst annoyances of cattle. 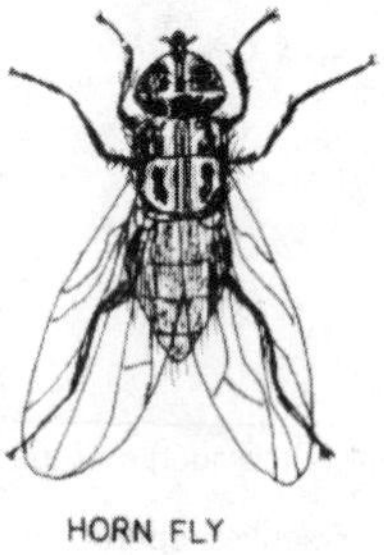 Fig. 10-34 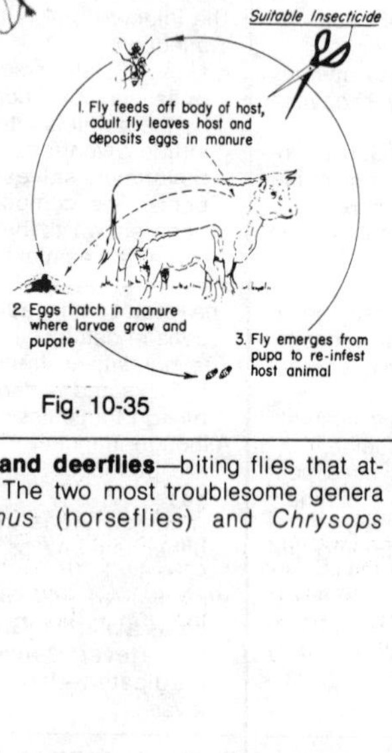 Fig. 10-35	Cattle, almost exclusively.	Tormented cattle often refuse to graze during the day and seek protection by hiding in dark buildings, brush, or tall grass. Heavily infested cattle may also have rough, sore skin, and suffer an inevitable loss in condition.
3. **Horseflies and deerflies**—biting flies that attack cattle. The two most troublesome genera are: *Tabanus* (horseflies) and *Chrysops* (deerflies)	Cattle. Horses.	The bite from the slashing mouthparts of these insects is very painful, and animals try to dislodge the fly with their tail or tongue or by stamping their feet. Heavily attacked animals stop grazing and tend to bunch together or seek shelter. Severe outbreaks can seriously affect weight gain.
4. **Houseflies** *(Musca domestica)*—are nonbiting flies that are common around barns and lots.	Cattle. Horses.	Although houseflies are nonbiting, they cause serious economic losses through annoyance of cattle and by disease transmission. Also, they create public health problems.

—(Continued)

Distribution and Losses Caused By	Treatment	Prevention and Control	Remarks
The face fly is more troublesome and causes greater losses than the horn fly.	Regular spraying or dusting, or by self-treatment with dust bags or back rubbers.	Prevention consists in scattering or removing fresh cow manure. Coumaphos in a complete feed. See manufacturer's label for dosage recommendations.	When cattle enter a barn or darkened area, the fly leaves the animal's face and rests on fence posts, gates, sides of barns, etc. The adult fly hibernates in attics and other protected places during the winter. The face fly lays its eggs in fresh manure, where the larvae develop.
Throughout the U.S. Losses inflicted by horn flies include lowered gains, and lowered milk production.	The horn fly can be very effectively controlled in the adult stage by treating cattle. When sprays are used, only a small deposit of insecticide on the hair is sufficient. Also, complete coverage is not necessary because horn flies move about on the animal sufficiently to come in contact with insecticide deposited on almost any part of the animal's body. Thus, protection of beef cattle from attacks of horn flies can be accomplished by use of sprays, dusts, dust bags, dips, back rubbers, and pour-ons, used according to manufacturer's directions.	In small pastures and where it is practical, spread fresh droppings with a spring-tooth harrow in order to hasten their drying. At frequent intervals, haul out cattle manure around barns, and spread thinly on the land, preferably with a manure spreader.	The horn fly is often found resting at the base of the horn, hence the name.
Tabanids are found in all parts of the U.S., and large numbers may be expected wherever there are extended areas of permanently wet, undeveloped land and a mild climate. Generally, horseflies are more of a problem to livestock than deerflies, but deerflies are often extremely annoying in the coastal areas of the South and the mountain areas of the West.		No really satisfactory method exists of controlling horseflies or deerflies. If possible, avoid pasturing cattle near swampy wooded areas when these flies are numerous. Also, sheltering animals is often beneficial since tabanids do not ordinarily enter enclosures.	Horseflies are also implicated in disease transmission because their habit of feeding on one animal and immediately attacking another can result in the direct mechanical transfer of pathogenic organisms that live in blood.
Houseflies become numerous both inside and outside barns and farm buildings. Perhaps they are the most abundant insect pest of feedlots. Houseflies are annoying to cattle and people, and they can spread human and animal diseases.	Several insecticides in spray or bait forms may be used to control adult flies in barns.	Insecticides alone will not control houseflies. Adequate sanitary measures, including proper disposition or handling of manure, are necessary to eliminate fly breeding areas. Spread manure thinly in fields so fly eggs and larvae will be killed by drying and heat.	Houseflies breed in manure, garbage, and decaying vegetable matter. The eggs hatch after an incubation period of 12 to 36 hours. The larvae feed on the organic medium and grow to full size in 6 to 11 days.

(Continued)

TABLE 10-5

Parasite	Species Affected	Symptoms and Signs of Affected Animals (or Damage Inflicted)
5. **Stablefly** *(Stomoxys calcitrans)*, which is about the size of a housefly, is usually found in the vicinity of animals. See Fig. 10-37 for the life history and habits of the stablefly.	All farm animals and man.	Fly-fighting and restlessness. In seeking natural protection, animals frequently resort to mudholes, brush, etc.
Gastrointestinal worms, including many species. These may be found in the stomach, small intestine, and colon.	Cattle. Sheep. Goats. Poultry. Horses. Swine. Dogs. Cats.	The symptoms are not specific, but infested animals generally show loss of weight, anemia, and/or diarrhea.
Grubs (warbles, heel fly)—cattle grubs are the maggot stage of honeybeelike insects known as heel flies, warble flies, or gadflies. In the U.S. there are 2 species of grubs with similar habits; namely, the common cattle grub and the northern cattle grub.	Cattle.	Attack of heel fly, in spring or early summer, causes cattle to run madly with their tails high over their backs in an attempt to escape. Grub (larva) in the back, usually from December to May, causes a conspicuous swelling.

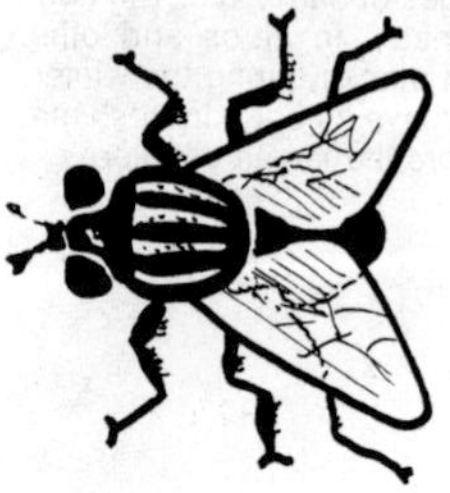

Fig. 10-36

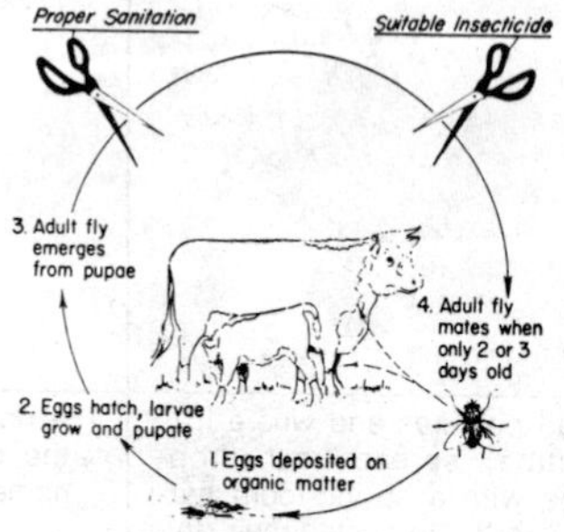

Fig. 10-37

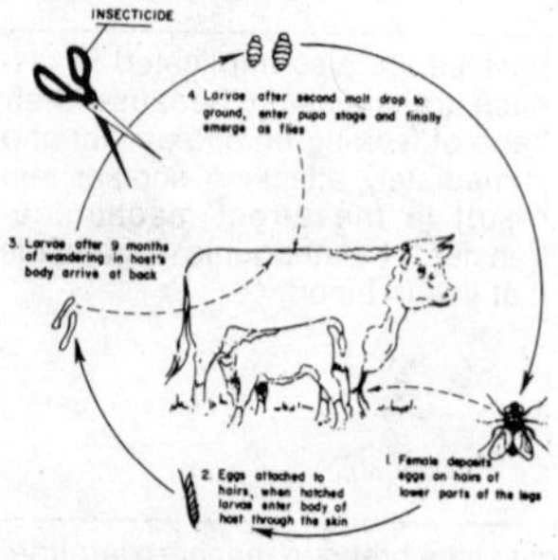

Fig. 10-38

Fig. 10-39

—(Continued)

Distribution and Losses Caused By	Treatment	Prevention and Control	Remarks
All temperate regions of the world. Losses include: Decreased gains in beef cattle. Lowered milk production in dairy cattle. Possible transmission of certain diseases and parasites. Gains and/or milk production may be lowered by as much as 50% in seasons when the number of flies becomes large.	With stableflies, insecticides should be used as a supplement to good sanitation, rather than as the principal method of control, because alone they may not do a satisfactory job. Residual sprays are the most effective method of treatment. They should be applied inside and outside barns and other farm structures where stableflies rest. The spray may be applied by spray gun or by fogging or misting devices. Care should be taken to prevent contamination of feed and drinking water, and animals should be removed from the area during the spraying. Application of insecticides to the cattle may afford only temporary relief. The preferred method of application is spraying since the insecticide should be applied to the legs and lower body of the animal.	Control of stableflies by direct application of insecticides to cattle is usually not satisfactory. They are best controlled by sanitation and by application of insecticides to the resting surfaces. Sanitation is the most effective method of controlling stableflies in such areas as feedlots and barnyards because it breaks the life cycle by removing the breeding sites. Barnyards and feedlots should be well drained; manure and decaying organic matter should be removed from inside and outside buildings and disposed of weekly or more often, if possible, by spreading it out to dry (this kills developing larvae). If manure cannot be spread, it should be placed in compact piles where the surface will dry quickly and become unattractive to females.	As a blood-feeding fly, the stablefly is implicated in carrying disease, and high populations cause reduced weight gains in cattle.
One or more species are found in most areas throughout the U.S. The losses are in terms of lowered feed efficiency caused by disturbed digestion, lowered meat and milk production, and some death losses.	Therapeutic doses before breeding, after calving, and when grazing is at its height of one of the wormers listed in Table 10-4, Recommended Compounds for Control of Internal Parasites, by Species. Use drug of choice according to manufacturer's directions. Based on Table 10-4, the cattleman and his veterinarian can (1) rotate drugs to prevent parasitic resistance, and (2) select the method of administration easiest to follow (although varying according to drug, they may be given as a drench, bolus, feed or mineral mix, paste, or injection). Phenothiazine at the rate of 1 gram per day mixed in the salt affects the reproduction of parasites not removed by treatment.	Rotate pastures. Segregate calves from mature animals. Avoid overstocking or overgrazing pastures since the infective larvae are mainly on the bottom inch of grass. Cross-graze with cattle and horses. Keep feeders and waterers sanitary.	In areas of constant exposure, routine treatment is recommended. Consult with your veterinarian relative to a program.
Throughout the U.S. The following kinds of losses are incurred: Decreased gains or milk production, mechanical injury, and even death. Carcass damage. Shock to animals.	Applying a systemic insecticide to cattle as soon as possible after the activity of the heel flies ceases since these insecticides kill the young larvae in the animal's body. When the grubs are near the back or located in the back, treatments are less effective, and the possible side effects are more likely. Side effects may also occur when there is concentration of grubs in the gullet or spinal cord of treated cattle. A single treatment with a systemic insecticide should give excellent control of cattle grubs. For the correct timing, each owner is advised to check with his local county agent or consultant. Systemics may be administered as sprays, dips, pour-ons, or as feed additives. Never use more than one systemic insecticide at a time, and always use a systemic in keeping with manufacturer's directions.	Effective and complete eradication necessitates area campaigns; farm by farm, county by county, and state by state. Coumaphos in a complete feed. See manufacturer's label for dosage recommendations.	The cattle grub or heel fly is probably the most destructive insect attacking beef and dairy animals.

(Continued)

TABLE 10-5

Parasite	Species Affected	Symptoms and Signs of Affected Animals (or Damage Inflicted)
Lice—small, flattened, wingless insect parasites of which there are several species, most of which are specific for a particular class of animal. 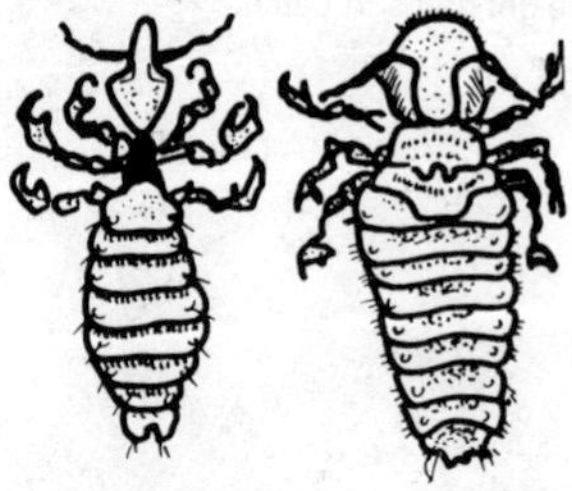 Fig. 10-40	Cattle (with other species for other classes of animals).	Intense irritation, restlessness, and loss of condition. There may be severe itching and the animal may be seen scratching, rubbing, and gnawing at the skin. The hair may be rough, thin, and lack luster; and scabs may be evident. Lice are apt to be most plentiful around the root of the tail, on the inside of the thighs, over the ankle region and along the neck and shoulders. One type sucks blood and may cause the animal to become anemic.
Liver fluke (or liver rot, *Fasciola hepatica)*—a flattened, leaflike brown worm, usually an inch long (see Fig. 10-41).  Fig. 10-41	Cattle. Sheep. Goats and other animals.	Anemia, as indicated by pale mucous membranes, digestive disturbances, loss of weight, and general weakness. Severe liver damage, with liver condemned and not permitted for use as human food.
Lungworm *(Dictyocaulus viviparus)*—white, threadlike worms 1½ to 3 in. long, found in the trachea and bronchi of cattle.	Cattle, especially calves. Sheep. Goats. Swine. Cats.	Symptoms include coughing, labored breathing, loss of appetite, unthriftiness, and intermittent diarrhea. Death may follow, probably from suffocation or pneumonia.
Measles *(Cysticercosis, or measly beef)*—a parasitic disease of cattle, cysticercosis is an invasion of the musculature and viscera by larvae. *Cysticercus bovis*, or *Taenia saginata*, the beef tapeworm of man. Cattle are the intermediate host and people the definite host.	Cattle.	Most cases of beef measles produce few signs in live cattle. In the carcass, the cysticerci (cysts) are readily discernible. Carcasses that are excessively infested are unsatisfactory for food and should be condemned. If only a few cysticerci are found, the entire carcass is frozen sufficiently long to ensure that the cysticerci are killed.
Mites—are very small parasites that produce mange (scabies, scab, itch).	Cattle. Each class of animals has its own species of mange mites, except in the case of the sarcoptic mite which is transferable between classes and to man.	Marked irritation, itching, and scratching, crusting over of the skin, accompanied by formation of thick, tough, wrinkled skin.

—(Continued)

<table>
<tr><th>Distribution and Losses Caused By</th><th>Treatment</th><th>Prevention and Control</th><th>Remarks</th></tr>
<tr><td>Widespread.
Lice retard growth, lower milk production, and produce unthriftiness.</td><td>For effective control, all members of the herd must be treated simultaneously at intervals, and this is especially necessary during the fall months about the time they are placed in winter quarters. Cattle should be inspected for lice periodically throughout the winter and spring and retreated when necessary. Insecticides applied by spraying or dipping are the most effective against lice, but some control may be obtained by dusting.</td><td>Because of the close contact of cattle during the winter months, it is practically impossible to keep them from becoming infested with lice.</td><td>Lice show up most commonly in winter and on ill-nourished and neglected animals.</td></tr>
<tr><td>Worldwide, wherever there are low-lying wet areas and suitable snails.
Lowered gains and milk production and feed inefficiency are the chief losses. In addition, vast quantities of liver are condemned each year at the time of slaughter.
In 1973, a total of 1,300,000 livers were condemned.</td><td>Under the direction of a veterinarian, administer one of the following wormers: carbon tetrachloride, hexachloroethane, or hexachlorophene. Follow manufacturer's label for dosage recommendations.</td><td>Drainage or avoidance of wet pastures.
Where relatively small snail-infested areas are involved, it may be practical to destroy the snail (carrier of liver fluke), preferably in the spring season, through—
Applying 3 to 6 lb of copper sulfate (bluestone or blue vitrol) per acre of grassland, mixing and applying the small quantity of copper sulfate with a suitable carrier (such as a mixture of 1 part of the copper sulfate to 4 to 8 parts of either sand or lime); and
Treating ponds or sloughs with 1 part of copper sulfate to 500,000 parts of water.</td><td>When copper sulfate is used in the dilutions indicated, it is not injurious to grasses and will not poison farm animals, but it may kill fish.
Snail-infested pastures should not be used for making hay. Copper sulfate solutions are not curative for infected animals.</td></tr>
<tr><td>Throughout the U.S. particularly in association with wet pastures.
Along coastal areas and around the Great Lakes.</td><td>Levamisole (Tramisol) given as a drench, bolus, injection or in the feed is effective against lungworms. Like all wormers, it should be given according to the manufacturer's directions.</td><td>Practice rigid sanitation.
Do not spread infested manure on pastures.
Where practical, segregate calves from older animals.
Keep calves on a good ration.
Utilize dry pastures, if possible.</td><td>Replacement stock from herds with a history of lungworm infection should be (1) received with great caution, and (2) treated with levamisole.
Adults often gain immunity.</td></tr>
<tr><td rowspan="2">Beef measles is worldwide. However, the incidence is highest in Africa, the Middle East, Asia, and South America. In the U.S., the measles problem is largely confined to the Southwest.
At the time of slaughter, losses result from extensive trimming and prolonged storing of mildly infected carcasses and from condemning heavily infected carcasses.</td><td>No effective treatment for bovine measles is known.</td><td>Humans are the sole host for the adult tapeworm, and cattle are the only intermediate host. No other animals are involved in the life cycle. Thus, control consists of disposing of human excrement in such manner that it cannot come in contact with cattle.</td><td rowspan="2">The name is a misnomer, for it has no relationship, and little resemblance, to human measles. In humans, the disease is caused by a virus; in cattle, the disease is caused by a tapeworm cyst which lodges and grows in the muscle tissue.</td></tr>
<tr><td colspan="2">In endemic areas, workers employed in and around feedlots, cattle pastures, and dairies should be medically examined for beef tapeworm parasitism, and infested individuals should be treated for removal of the worms. Sanitary latrines should be provided for caretakers, and they should be forbidden to defecate in feedlots or pastures where cattle feed. At slaughter, cysticecus-infested meat should be disposed of in a manner which avoids the inclusion of viable cysticerci in human food. Meats should be thoroughly cooked to destroy viable cysticerci.</td></tr>
<tr><td>Widespread.
Mites retard growth, lower milk production and gains, and produce unthriftiness. Also, the skin is made less valuable for leather.</td><td>Mites can be controlled by spraying or dipping infested animals with suitable insecticidal solutions, and by quarantine of affected herds. Beef cattle may be treated with coumaphos, lime-sulfur, prolate, or toxaphene. Lime-sulfur is the only pesticide permitted for the control of mites on dairy cattle.</td><td>Avoid contact with diseased animals or infested premises.
Scabies is a reportable disease in the U.S. So in the case of an outbreak, contact the local veterinarian or livestock sanitary official.
Control by spraying or dipping infested animals with suitable insecticides, and quarantine affected herds.</td><td>There are 2 chief forms of mange: sarcoptic mange (caused by burrowing mites), and psoroptic mange (caused by mites that bite the skin and suck blood but do not burrow).
The disease appears to spread most rapidly during the winter months and among young and poorly nourished animals.</td></tr>
</table>

(Continued)

TABLE 10-5

Parasite	Species Affected	Symptoms and Signs of Affected Animals (or Damage Inflicted)
Mosquitoes—particularly species of the genera *Aedes, Psorophora*, and *Culex*, are a severe nuisance to cattle in many areas.	All farm animals.	Mosquitoes may occur in such abundance that cattle refuse to graze. Instead, they bunch together or stand neck deep in water to protect themselves from attack. Moreover, mosquitoes will annoy cattle day and night, so they can cause serious losses in meat production—or even death in extreme cases. Also, they may be disease carriers.
Ringworm—a contagious disease of the outer layer of skin caused by certain microscopic molds or fungi. 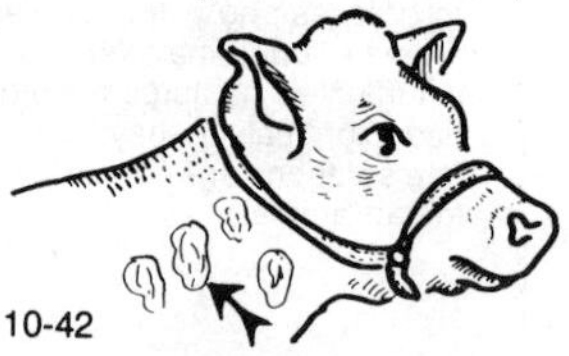 Fig. 10-42	All animals and man.	Round, scaly areas almost devoid of hair appear mainly in the vicinity of the eyes, ears, side of the neck, or the root of the tail. Mild itching usually accompanies the disease.
Screwworm—the screwworm fly raises its maggots in the open wounds of animals. 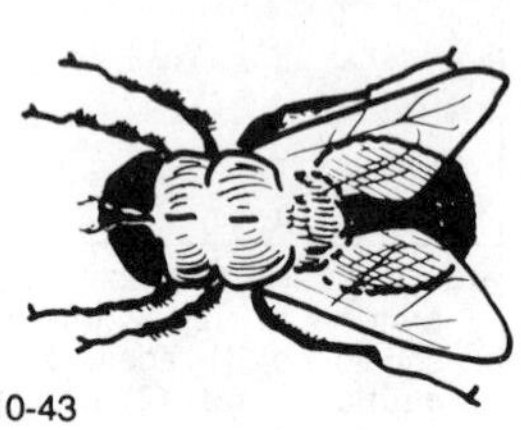 Fig. 10-43	All farm animals.	Loss of appetite and condition, and listlessness due to infested wounds.
Ticks—the lone star tick (Amblyomma americanum), the Gulf Coast tick *(A. maculatum)*, the Rocky Mountain wood tick *(Dermacentor andersoni)*, the Pacific Coast tick *(D. occidentalis)*, and the American dog tick *(D. variabilis)* are 3-host species that attack cattle during the summer months. The black-legged tick *(Ixodes scapularus)* is a 3-host tick that is common in late winter and early spring. The winter tick *(D. albipictus)* is a 1-host species found on cattle and horses in the fall and winter. In addition, larvae and nymphs of the so-called "spinose" ear tick *(Otobius megnini)*, a 1-host species, attach deep in the ears of cattle and feed there for several months. 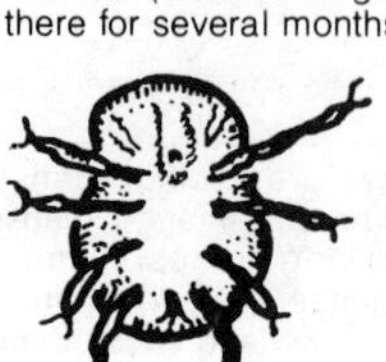 Fig. 10-44	Cattle.	Generally speaking, injury to cattle from tick parasitism varies directly with numbers of parasites. Ticks feed exclusively on blood. Thus, when several hundred ticks feed, the host becomes anemic, unthrifty, and loses weight. In addition, some female ticks generate a paralyzing toxin. The spinose ear tick, commonly called the "ear tick," takes up residence along the inner surfaces of the ear and in the external ear canals, where it is extremely annoying. Cattle heavily parasitized by spinose ear ticks droop their heads, rub and shake their ears, and turn the head to one side.
Warbles (see Grubs)		

—(Continued)

Distribution and Losses Caused By	Treatment	Prevention and Control	Remarks
Mosquitoes are rather widely distributed, but they are most numerous in the southeastern U.S., especially in swampy regions that have permanent pools of water or that are exposed to frequent flooding. Sometimes they kill cattle, although this is rare.		Mosquitoes can be controlled in several ways: (1) by elimination of breeding places, through providing fills, ditches, impoundments, improved irrigation methods, and other means of water manipulation; (2) by chemical destruction of larvae, by treating the relatively restricted breeding areas with proper larvicides; and (3) by chemical destruction of adults. Elimination of breeding sites is by far the most satisfactory and effective method of control. However, either this method or chemical destruction of larvae may not be economically practical if the breeding area is extensive. When the latter is the case, control can best be accomplished through group action, such as mosquito abatement districts. The cattle producer can achieve some control by fogging or spraying the pasture or rangeland, and some relief can be obtained by spraying the animals with insecticides.	Almost all female mosquitoes must take a blood meal before they can lay eggs. (The males do not suck blood, but feed on nectar and other plant juices.) Eggs are laid singly or in rafts on the surface of the water or on the ground in depressions that are flooded by tidal waters, seepage, overflow, or rainwater. The larvae and pupae are aquatic.
Throughout the U.S. It is unsightly and affected animals may experience considerable discomfort, but actual economic losses are not too great.	Clip the hair from the affected areas, remove scabs with a brush and mild soap. Paint affected areas with tincture of iodine or salicylic acid and alcohol (1 part in 10) every 3 days until cleared up. Certain proprietary remedies available only from veterinarians have proved very effective in treatment.	Isolate affected animals. Disinfect everything that has been in contact with infested animals, including curry combs and brushes. Practice strict sanitation.	Though ringworm may appear among animals on pasture, it is far more prevalent as a stable disease. It is usually a winter disease, with recovery the following summer after the animals are turned to pasture.
Mostly in the southern and southwestern states, in which areas it formerly caused 50% of the normal annual livestock losses.	When maggots (larvae) are found in an animal, they should be removed and sent to the proper authorities for identification, and the animal should be treated with a proper insecticide or smear. Additional treatment or control measures will be supervised by inspection personnel.	Prevention in infested areas consists mainly of keeping animal wounds to a minimum and of protecting those that do materialize. In 1958, the USDA initiated an eradication program. Screwworm larvae were reared on artificial media. Two days before fly emergence, the pupae were exposed to gamma irradiation at a dosage which caused sexual sterility but no other deleterious effects. Sterile flies were distributed over the entire screwworm infested region in sufficient quantity to outnumber the native flies, at an average rate of 400 males per square mile per week. The female mates only once and, therefore, when mated with a sterile male does not reproduce. There was a decline in the native population each generation until the native males were so outnumbered by sterile males that no fertile matings occurred and the native flies were eliminated. This program has virtually eliminated all the losses caused by screwworms in the U.S. Unfortunately, the states bordering on Mexico are periodically reinfested by mated female flies from Mexico. For this reason, permanent elimination of screwworms from the U.S. by the sterile-male technique cannot be hoped for until they are also eradicated in Mexico.	Benzene (not benzine) or chloroform may be used to kill the maggots. Screwworms sometimes penetrate the dimple in front of a cow's udder.
Ticks are widely distributed, especially throughout the southern part of the U.S.; but they are usually seasonal in their activities. Ticks suck blood. They cause economic losses by transmitting diseases; by restlessness, anemia, and inefficient feed utilization; and by necessitating expensive treatments. Among the diseases transmitted to or produced in cattle by ticks are Texas fever, anaplasmosis, Q fever, tick paralysis, and piroplasmosis.	Because most species of ticks, except the ear tick, attach to the external surfaces of cattle, dipping and spraying are the most effective methods of control; however, dusts may be used. To treat animals for ear ticks, the chemical should be applied into the ears of the cattle.		

TABLE 10-6—SHEEP AND GOAT

Parasite	Species Affected	Symptoms and Signs of Affected Animals (or Damage Inflicted)
Bankrupt worms (see Trichostrongyles)		
Bladder worms (or larval tapeworms)—see the following: Gid tapeworm, Hydatid, Sheep measles, and thin-necked bladder worm.		
Blowfly (wool maggot)—the blowfly group consists of a number of species of flies, all of which breed in necrotic animal flesh and in exudates, and the wool maggot that attacks soiled fleece areas (Fig. 10-45 shows blowfly maggots). 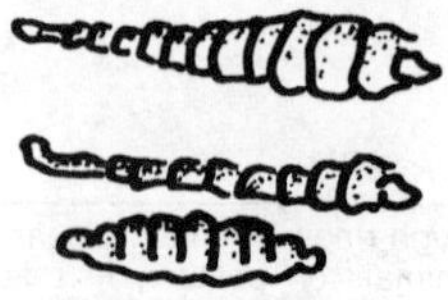 Fig. 10-45	All farm animals.	Infested wounds and soiled fleece. Maggots spread over the body, feeding on the skin surface, producing severe irritation, and destroying the ability of the skin to function. Infested animals rapidly become weak, fevered, and unthrifty.
Brown stomach worm *(Ostertagia circumcincta and O. trifurcata)*—brown hairlike worms about ½ inch long. 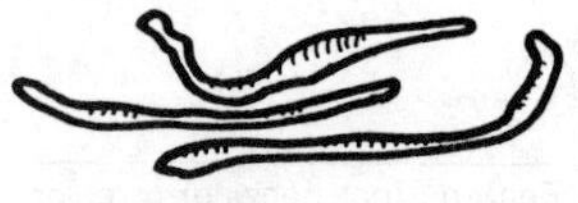 Fig. 10-46	Sheep. Goats. Also occurs in cattle, but with different species of worms.	It is difficult to attribute any specific symptoms to brown stomach worms, because infestations with them are usually accompanied by infestations with the common stomach worm and other parasites.
Coccidiosis—is a parasitic disease caused by protozoan organisms known as coccidia (see Fig. 10-47 for the life history and habits of coccidia). 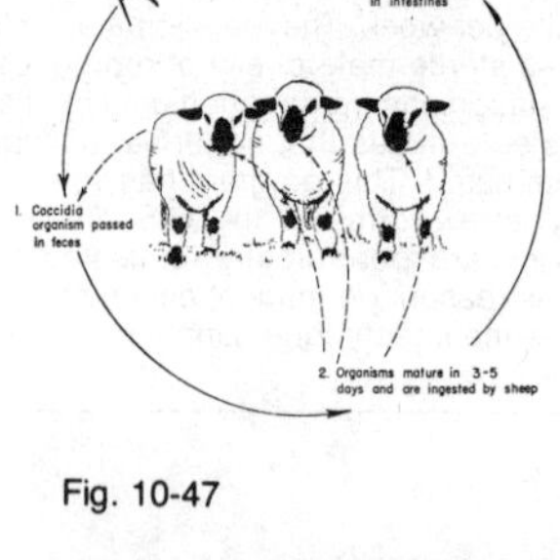 Fig. 10-47	Sheep. Cattle. Goats. Pet stock. Poultry. Each class of animals harbors its own species of coccidia; thus, there is no cross infection between animals.	Diarrhea and bloody feces, and pronounced unthriftiness and weakness.
Cooperias (including 4 species: *C. curticei, C. oncophora, C. punctata,* and *C. pectinata*)—the 4 species of parasites classed as Cooperias are small hairlike worms less than ⅓ inch in length.	Sheep. Goats. Cattle.	No specific symptoms, but affected animals exhibit diarrhea (scours), depression, loss of appetite, loss of weight, and retarded growth.
Flies—the common species of flies are fully discussed in Table 10-5, Cattle Parasites and Their Control; hence, the reader is referred thereto.		

PARASITES AND THEIR CONTROL

Distribution and Losses Caused By	Treatment	Prevention and Control	Remarks
Widespread, but present greatest problem in Pacific Northwest, and in South and southwestern states. Death losses not excessive, but production is lowered.	Coumaphos (Co-Ral). Dioxathion. Ronnel. *Follow directions on label.*	Eliminate the blowfly and decrease the susceptibility of animals to infestation by— Destroying dead animals by burning or deep burial; Using traps, poisoned baits, and electrical screens; Using repellents, such as pine tar oil; and Docking lambs and tagging sheep at intervals.	Most blowfly damage is limited to the wool maggot fly.
More prevalent in western U.S.	See Table 10-4, Recommended Compounds for Control of Internal Parasites, by Animal Species. Use drug of choice according to manufacturer's directions.	The recommended measures for prevention and control are the same as those listed for the stomach worm; hence, see those listed under the latter.	See the remarks listed under the stomach worm, which are also applicable here.
Worldwide. There are lowered gains and production, and frequently high mortality in feedlot lambs.	Amprolium (Amprol). Diphenthane 70. Furacin compounds. Sulfonamides. *Follow directions on label.* In the U.S., amprolium is permitted in feed only as an aid in the control of coccidiosis of chickens and turkeys.	*Feedlot lambs:* In feedlot lambs, where the disease is most prevalent, good management and natural resistance are important. To these ends, move feeders into feedlots with a minimum of stress and shrink, allow for plenty of space, keep water and feed troughs free from fecal pellets, maintain dry lots and bedding, start animals on grain feed gradually, and segregate affected animals if practical. The same principles and practices apply to other sheep, when trouble is encountered. Also, when practical, drain or fence low, wet areas; and keep animals in a sunny, dry place.	In the oocyst stage, the parasite resists many disinfectants and may remain viable outside the body for months, but it is readily destroyed by direct sunlight or complete drying. Some sheepmen protect feeder lambs from coccidiosis by using a feed mixture containing sulfa drugs (such as sulfaguanidine) which check the growth of parasites. Medicated feed may also be fed to lambs when they have to be kept closely confined with ewes.
Widely distributed, but damage is not heavy except in an occasional flock with excessive infection.	See Table 10-4, Recommended Compounds for Control of Internal Parasites, by Animal Species. Use drug of choice according to manufacturer's directions.	The recommended measures for prevention and control are identical to those listed for the stomach worm; hence, see those listed under the latter.	See the remarks listed under the stomach worm, which are also applicable here.

(Continued)

TABLE 10-6

Parasite	Species Affected	Symptoms and Signs of Affected Animals (or Damage Inflicted)
Gid tapeworm *(Coenurus cerebralis)*—the larval form of one of the 4 species of bladder worm or tapeworm, found in dogs and related carnivores, which also affect sheep and goats (see Fig. 10-48). 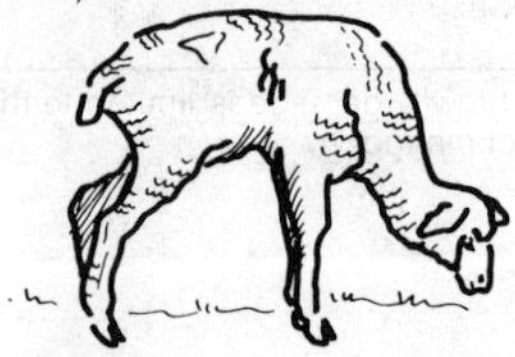 Fig. 10-48	Sheep. Goats. Pet stock.	Disease known as coenurosis, the symptoms of which are defects in vision and disturbances in movements. Affected animals may stumble, run into objects, walk with the head high or in circles and there may be at least a partial paralysis of the hindquarters.
Grub-in-the-head *(see Sheep bots)*		
Head bots or nose bots *(see Sheep bots)*		
Hookworm *(Bunostomum trigonocephalum)*—a white, thin worm varying from ½ to 1 in. in length (see Fig. 10-49).  Fig. 10-49	Sheep. Goats. Also occurs in cattle, but with different species of worms.	Anemia, edema, and unthriftiness; similar to the symptoms of the common stomach worm.
Hydatid *(Echinococcus granulosus)*—the larval form of one of the 4 species of bladder worm or tapeworm, found in dogs and related carnivores, which also affect sheep and goats (Fig. 10-50 shows adult hydatid).  Fig. 10-50	Sheep. Goats. Cattle. Swine. Horses. Man.	Ordinarily, no specific or distinctive symptoms are observed, but in heavy infestations there may be shallow respiration, emaciation and weakness.
Keds (see Sheep keds)		
Lice—small flattened, wingless insect parasites of which there are 2 groups; sucking lice and biting lice. Biting lice, *Damalinia ovis*, are most common but least harmful. The sucking species are (1) body lice *(Linognathus ovillus* and *L. africanus)*, and (2) the foot louse *(L. pedalis)*. Sucking lice are larger than biting lice. Body lice are found anywhere on body where wool is dense, while foot lice are usually found on legs below knees and hocks.  Fig. 10-51	Sheep, with other species for other classes of animals.	Intense irritation, restlessness, and loss of condition. There may be severe itching and the animal may be seen scratching, rubbing, and gnawing at the skin. The wool may be matted and lack luster; and scab may be evident.

—(Continued)

Distribution and Losses Caused By	Treatment	Prevention and Control	Remarks
Spotted and rare over the U.S.	No known treatment. Surgery is successful in some cases.	Elimination of stray dogs. Examination, and proper worm treatment when necessary, of all dogs that may come in contact with sheep and goats. Proper disposal of all carcasses of infested animals.	In afflicted sheep and goats, cysts containing the larvae *(Coenurus cerebralis)* of the tapeworm eggs voided by dogs or other carnivorous animals are found on the brain and spinal cord.
Widespread over North America, with heaviest U.S. infestations in the southern states.	See Table 10-4, Recommended Compounds for Control of Internal Parasites, by Animal Species. Use drug of choice according to manufacturer's directions.	The recommended measures for prevention and control of the hookworm are identical to those of the stomach worm; hence, refer to the latter.	See the remarks listed under the stomach worm, which are also applicable here.
Sparsely scattered over the U.S., but not great numbers of sheep and goats seriously affected. Few deaths occur.	No known treatment.	Eliminate stray dogs. Examination, and proper worm treatment when necessary, of all dogs that may come in contact with sheep and goats. Proper disposal of all carcasses of infested animals.	
Lice are not as common on sheep as on other domestic animals, but they do occur occasionally. Lice retard growth, lower wool production, and produce unthriftiness.	In fall or spring after shearing apply one of the following insecticides in keeping with the manufacturer's label: Carbaryl Ciodrin Coumaphos (Co-Ral) Diazinon Dioxathion Malathion Methoxychlor Ronnel (Korlan) Rotenone Toxaphene	For effective control, all sheep should be treated simultaneously.	Lice show up most commonly in winter and on ill-nourished and neglected animals. Spray at 400 lb pressure.

(Continued)

TABLE 10-6

Parasite	Species Affected	Symptoms and Signs of Affected Animals (or Damage Inflicted)
Liver fluke *(Fasciola hepatica)*—a flattened, leaflike brown worm, usually about an inch long. See Fig. 10-52 for the life history and habits of the liver fluke.	Sheep. Goats. Cattle, and other animals.	With heavy fluke infestations, death may occur without any definite symptoms being evident. Sometimes there may be a distinct pot-bellied condition caused by the escape of fluids into the body cavity through damage to the liver.

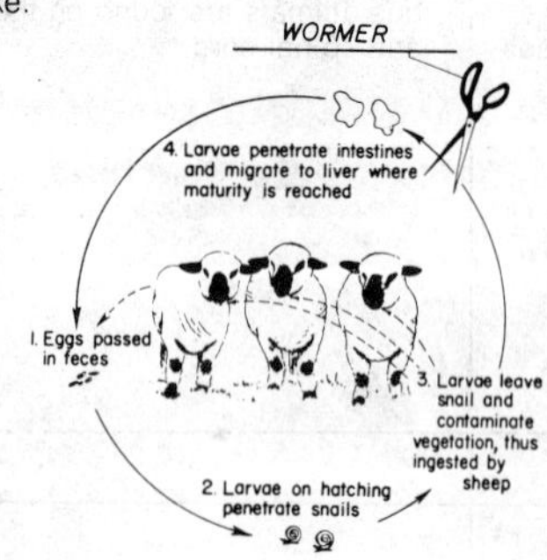

Fig. 10-52

Parasite	Species Affected	Symptoms and Signs of Affected Animals (or Damage Inflicted)
Lungworms (including the thread lungworm, *Dictyocaulus filaria*, and the hair lungworm, *Muellerius capillaris)*—thread lungworms are white and up to 4 in. long, whereas hair lungworms are thinner and much shorter. See Fig. 10-53 for the life history and habits of the lungworm.	Sheep. Goats. Also occurs in cattle, but with different species of worms.	No specific symptoms are associated with infestations of hair lungworms. The thread lungworm is especially damaging to lambs and kids. First symptom is coughing, usually accompanied by a dirty, pussy discharge from the nostrils. Other symptoms include rapid and difficult breathing, unthriftiness, loss of appetite, and lowered head and extended neck. Death may follow in 2 to 3 mos., probably from infection or pneumonia.

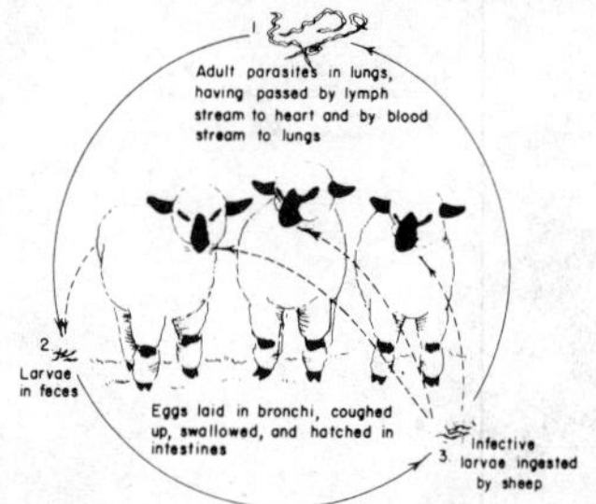

Fig. 10-53

Parasite	Species Affected	Symptoms and Signs of Affected Animals (or Damage Inflicted)
Mange (see Mites)		
Mites (or sheep scab, scabies, mange)—are very small parasites that produce mange of which there are 2 forms: (1) sarcoptic mange, caused by burrowing mites; and (2) psoroptic mange, caused by mites that bite the skin but do not burrow. Psoroptic or common scab is the most important form of sheep scabies in the U.S.	Sheep. Each class of animals has its own species of mange mites, except in the case of sarcoptic mite which is transferable between classes and to man.	Marked irritation, itching, and scratching. Crusting over of the skin, accompanied or followed by the loss of wool and the formation of thick crusts or scabs on the skin; hence, the name scabies. The disease appears to spread most rapidly among young and poorly nourished animals.
Nodular worm *(Oesophagostonum columbianum)*—a white worm about ⅝ in. long, found in the cecum and colon of sheep and goats. See Fig. 10-54 for the life history and habits of the nodular worm.	Sheep. Goats. Also occurs in cattle and swine, but with a different species of worms in each.	Unthriftiness, reduced fleece and mutton yields, and death losses. In general, the symptoms accompanying severe nodular worm infestation are not unlike those of general parasitism.

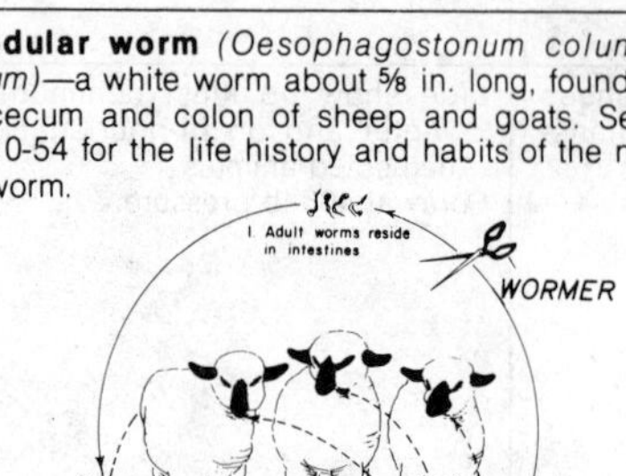

Fig. 10-54

—(Continued)

Distribution and Losses Caused By	Treatment	Prevention and Control	Remarks
Worldwide. In the U.S. it is most common in areas where low-lying, wet pastures and the intermediate snail hosts abound. Results in lowered meat and wool production. Produces many "fluky livers" or "rotten livers," which are condemned in packinghouses each year. Also death losses.	Under the direction of a veterinarian, administer one of the following anthelmintics: carbon tetrachloride, hexachloroethane, or hexachlorophene. If the diet is low in calcium, reinforce it with a calcium supplement.	Drainage or avoidance of wet pastures. Where relatively small snail-infested areas are involved, it may be practical to destroy the snail (carrier of liver fluke), preferably in the spring season, through the following: 1. Applying 3 to 6 lb of copper sulfate (bluestone or blue vitrol) per acre of grassland, mixing and applying the small quantity of copper sulfate with a suitable carrier (such as a mixture of 1 part of the copper sulfate to 4 to 8 parts of either sand or lime). 2. Treating ponds or sloughs with 1 part of copper sulfate to 500,000 parts of water.	When copper sulfate is used in the dilutions indicated, it is not injurious to grasses and will not poison farm animals, but it may kill fish. Snail-infested pastures should not be used for making hay. Infested sheep are usually treated in Nov. or Dec., with 2 or even 3 treatments at 4- to 6-week intervals. The fluke, may also carry *Clostridium novyi*, an anaerobic bacterium which causes black disease.
Widely distributed throughout the world. Death losses occasionally occur.	Levamisole (Tramisol), which is used in treatment of lungworm in cattle, is said to be effective in the treatment of sheep lungworm, also. Tetramisole (Bayer 9051). Treatment should be followed by good feeding and nursing.	Remove animals from infested ground and place them on dry pastures where clean water is available. Drain or avoid low, wet pastures. Do not spread fresh manure of infested sheep and goats on pastures used by these animals. Since old animals may be carriers without showing symptoms, weaned lambs and kids should be kept away from mature animals whenever practical to do so.	
Widespread. It is easily transmitted by contact from one sheep to another, and it spreads very rapidly after being introduced in a flock. Mites retard growth and gains, lower wool production, and produce unthriftiness. Also, the pelt is made less valuable.	Apply one of the following insecticides in keeping with the manufacturer's label: Lime-sulfur Toxaphene Control by spraying or dipping infested animals with suitable insecticides, and quarantine affected flocks or bands.	Avoid contact with diseased animals or infested premises. Scabies is a reportable disease in the U.S. So, in case of an outbreak, contact the local veterinarian or livestock sanitary official.	The disease appears to spread most rapidly during the winter months.
Worldwide. Mostly in central, eastern, and southern U.S., with only limited damage. In addition to lowered fleece and mutton yields and to death losses, the presence of nodular worm in slaughtered animals causes all or a considerable portion of the large and small intestine to be unfit for surgical sutures (catgut) or for casings. It is estimated that the latter 2 losses total half a million dollars annually.	See Table 10-4, Recommended Compounds for Control of Internal Parasites, by Animal Species. Use drug of choice according to manufacturer's directions.	The recommended measures for prevention and control of nodular worm are identical to those of the stomach worm; so, refer to the latter.	See the remarks listed under the stomach worm, which are also applicable here.

(Continued)

TABLE 10-

Parasite	Species Affected	Symptoms and Signs of Affected Animals (or Damage Inflicted)
Ringworm—a contagious disease of the outer layers of skin caused by a microscopic mold or fungus of the genus *Trichophyton*. 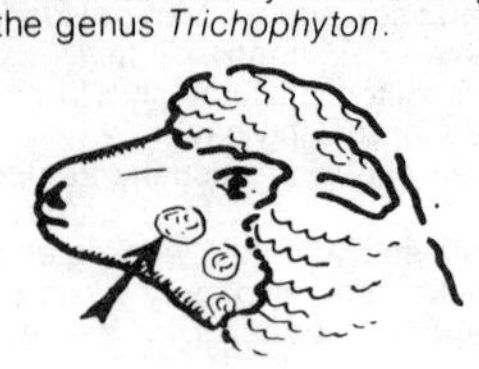 Fig. 10-55	All animals and man, but ringworm in sheep is seldom observed.	Round, scaly areas almost devoid of hair (or wool) ap pear mainly in the vicinity of the eyes, ears, side of th neck, or the root of the tail. Mild itching usually accompanies the disease.
Scab (see Mites)		
Screwworm—the screwworm fly raises its maggots in the open wounds of animals, where they live on live flesh. Fig. 10-56 shows a screwworm maggot.  Fig. 10-56	All farm animals.	Loss of appetite and condition, and lowered thrift an vigor.
Sheep bots (*Oestrus ovis*, nasal botfly)—this condition, commonly called grub-in-the-head, is due to a beelike fly about the size of the common horsefly. See Fig. 10-57 for the life history and habits of the sheep nasal fly. 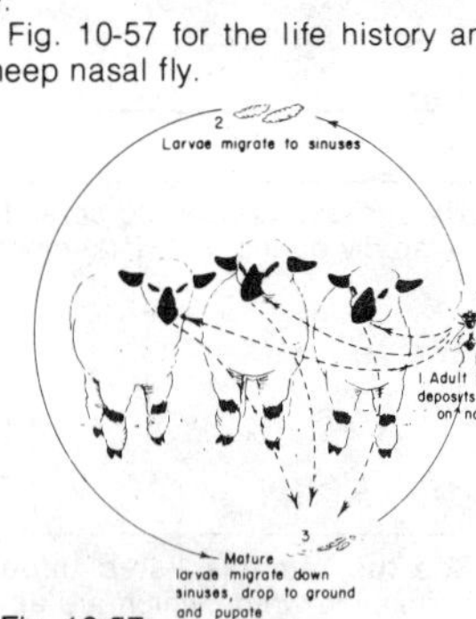 Fig. 10-57	Sheep.	When the flies attempt to deposit their larvae around th nostrils of sheep, the animals cease to feed, becom restless, press their noses against other sheep, and/c seek shelter. Grub infestation results in a snotty nose, and there ma be difficult breathing and frequent sneezing.
Sheep keds (or sheep ticks, *Melophagus ovinus*)—a hairy, bloodsucking fly without wings, which ranges up to ¼ in. in length. 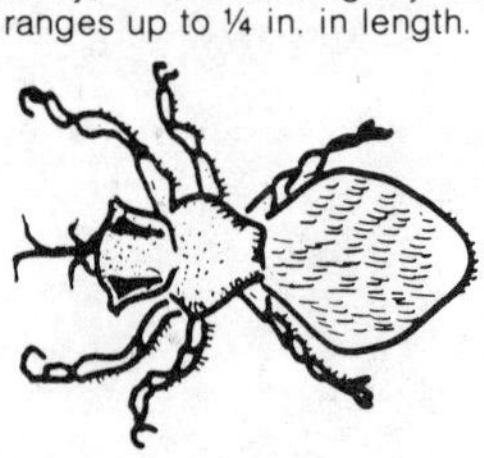 Fig. 10-58	Sheep. Goats.	Marked reduction in condition, anemia, biting, an scratching, and loss of and damage to wool.

—(Continued)

Distribution and Losses Caused By	Treatment	Prevention and Control	Remarks
Ringworm of sheep is of little economic importance in the U.S.	Clip the hair or wool from the affected areas and paint sores with tincture of iodine every 3 days until cleared up.	Isolate affected animals. Disinfect everything that has been in contact with infested animals, including cards and brushes. Practice strict sanitation.	Although ringworm may appear among animals on pastures, it is far more prevalent as a stable disease among livestock during the winter.
Mostly in the southern or southwestern states, in which areas it formerly caused 50% of the normal annual livestock losses.	Apply one of the following insecticides: Coumaphos (Co-Ral) Ronnel (Korlan)	Keep animal wounds to a minimum. Schedule branding, castrating, docking, and other stock operations that necessarily produce wounds during the winter season or early spring when the flies are least abundant and active. Kill all possible maggots during the winter and spring months. If possible, keep wounded and infested animals in a screened, fly-proof area. Screwworm infestation is lessened by castrating with Burdizzo pincers, by eliminating sources of mechanical injury, and by having newborn animals arrive at the season of least fly activity.	Some stockmen apply pine-tar oil to recent wounds and to the navels of newborn animals to repel flies. Benzene (not benzine) or chloroform may be used to kill the maggots. The federal program of release of sterile male flies has been very successful in screwworm eradication in southern U.S. Screwworms have been eradicated from the U.S. except for sporadic outbreaks in south Texas near Mexico.
Worldwide. Although death losses are rare, there is loss in condition, both at the time the fly attacks and while the larvae are in the nasal passage.	Ruelene. Follow directions on label.	The following measures, during the sheep nasal fly season, may lessen, but not eliminate, sheep nasal fly: With a small flock and where practical, keep sheep in a darkened barn during the day. Apply pine-tar oil about the nostrils of the sheep every few days, thus repelling some of the flies. But the latter is not too satisfactory.	
Throughout the U.S., but especially prevalent in the northern states. Losses include retarded growth of young animals, loss in condition of mature animals, and damage to the fleece. Fine wool breeds of sheep and Angora goats are not seriously affected.	Treatment after shearing as soon as the shear cuts heal (including unshorn lambs) with any one of the following insecticides, used according to the manufacturer's label: Coumaphos (Co-Ral) Diazinon Dieldrin Dioxathion (Delnav) Malathion Pyrethrins + synergist Ronnel (Korlan) Rotenone Toxaphene	Prevention and control involves spraying or dipping all sheep as soon as the cuts heal up following shearing.	Poorly housed and poorly fed animals are most likely to suffer from sheep ticks.

(Continued)

TABLE 10-

Parasite	Species Affected	Symptoms and Signs of Affected Animals (or Damage Inflicted)
Sheep measles *(Cysticercus ovis)*—the larval form of one of the 4 species of bladder worm or tapeworm, found in dogs and related carnivores, which also affect sheep and goats. Fig. 10-59 shows *C. ovis*.  Fig. 10-59	Sheep. Goats.	No specific symptoms have been attributed to infestatio with sheep measles.
Small stomach and intestinal worms (see Trichostrongyles).		
Stomach worms *(Haemonchus contortus,* or twisted stomach worm, common stomach worm)—is the most destructive parasite of sheep and goats. Worms are ¾ to 1½ in. long, about the size of a horse hair in diameter, and the live females are striped like a barber pole (see Fig. 10-60 for the life history and habits of the stomach worm). 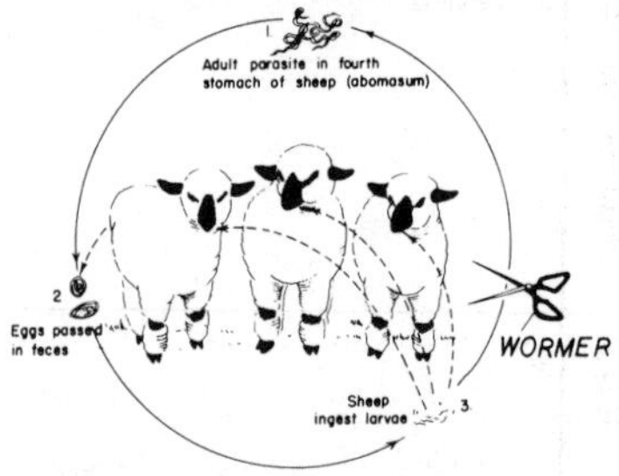 Fig. 10-60	Sheep. Goats. Cattle, although cattle are usually not seriously harmed.	No specific symptoms, because (1) sheep are seldor infested with one kind of parasite only, and (2) identi cal symptoms may be exhibited in cases of infestatio by other parasites. Sheep heavily infested with para sites become unthrifty, listless, thin, and weak. Th membranes of the eyes, nose, and mouth becom pale, and there may be diarrhea, loss of wool, and watery swelling under the lower jaw and along the ab domen. Lambs and kids are more seriously affected than olde animals.
Tapeworms (including the common species: *Monieza expansa, M. benedeni,* and the fringed tapeworm, *Thysanosoma actinoides)*—Common tapeworms are long, flat, ribbonlike worms which range up to a width of ¾ in. and a length of 20 ft. The fringed tapeworm differs in that it is smaller and it has fringed segments (see Fig. 10-61 for the life history and habits of the tapeworm). 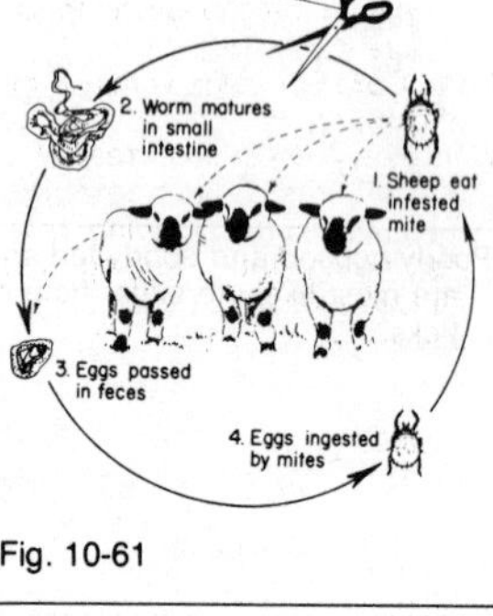 Fig. 10-61	Sheep. Goats.	Common tapeworms do not produce any marked o specific symptoms. However, the fringed tapeworn may cause death of the host through blocking the cys tic duct, gallbladder, and the ducts of the liver an pancreas. Infected animals usually have normal appetites.

—(Continued)

Distribution and Losses Caused By	Treatment	Prevention and Control	Remarks
Worldwide. In the U.S., they are most prevalent in the West. Few death losses due to sheep measles. Chief economic loss occurs at slaughter, because infested carcasses are trimmed or condemned according to the degree of infestation.	There is no known treatment for the removal of the parasite.	Elimination of stray dogs. Examination, and proper worm treatment when necessary, of all dogs that may come in contact with sheep and goats. Proper disposal of all carcasses of infested animals.	The parasite is not transmissible to man; the removal or condemnation of affected carcasses is done to assure clean, wholesome meat.
Mortality from stomach worms is high in sheep; however, in cattle, mortality is low but economic losses may be large. The incidence is greatest in warm, wet areas of the U.S.	See Table 10-4, Recommended Compounds for Control of Internal Parasites, by Animal Species. Use drug of choice according to manufacturer's directions.	Rotate pastures, changing the flock to clean, fresh pastures at about 2-week intervals. But, in the event of an outbreak of stomach worms, treat the sheep before turning them on to new pastures. Horses and hogs, which are not harmed by the stomach worm, may be rotated with sheep and goats. Lambs can be kept free from stomach worms by confining them to a drylot (without grass or weeds) or to a barn.	See manufacturer's label for proper drug withdrawal prior to slaughter for food.
Worldwide. Lambs are more susceptible than older sheep. Common tapeworms are not an important factor in sheep raising, because they are practically harmless. Only under poor nutritional conditions will they increase unthriftiness. The fringed tapeworm is confined largely to the range bands of western U.S., whereas the other species occur mostly in the central, eastern, and southern states. The fringed tapeworm may cause death, and in the packing plant livers infected with this parasite (that is, when the worms are found) are condemned as unfit for human consumption.	Dichlorophen (Diphenthane 70, Teniathane). Lead arsenate, given in bolus doses of ½ gram for adult sheep. It is generally available in commercial preparations as a mixture with phenothiazine or thiabendazole for the control of roundworms and tapeworms with one drench.	So little is known about the life history of the fringed tapeworm (for example, its life cycle is not known with certainty) that it is impossible to make intelligent recommendations relative to control measures. Rotating pastures will help.	The following directions and precautions should be observed in administering treatments: 1. The entire flock should be treated at the same time. 2. Overdosing should be avoided because toxic reactions may result. 3. Following treatment, the animals should be moved to fresh pasture if possible. Do not drench ewes within 2 months of lambing. With heavy infections, treat at 20- to 30-day intervals, beginning immediately after shearing time and continuing until freezing weather. Fall treatment is especially important.

(Continued)

TABLE 10-6

Parasite	Species Affected	Symptoms and Signs of Affected Animals (or Damage Inflicted)
Thin-necked bladder worm *(Cysticercus tenuicollis)*—the larval form of one of the 4 species of bladder worm or tapeworm, found in dogs and related carnivores, which also affect sheep and goats (see Fig. 10-62).  Fig. 10-62	Sheep. Goats.	Usually there are no external symptoms, and since light infestations are the rule, no attention is called to the parasite. Thin-necked bladder worms burrow into the liver and/or thin membranes of the abdominal cavity of sheep and goats, producing tissue damage.
Trichostrongyles (small stomach and intestinal worms; including 4 species; *Trichostrongylus axei, T. colubriformis, T. vitrinus,* and *T. capricola)*—small hairlike worms less than ⅓ in. in length.	Sheep. Goats. Cattle.	Severe unthriftiness, diarrhea, and anemia.
Twisted stomach worm (see Stomach worm)		
Whipworm *(Trichuris ovis)*—a white worm 1½ to 2 in. in length, with slender anterior and heavy posterior portions that resemble the lash and handle, respectively, of a whip.	Sheep. Goats.	No well-defined symptoms, though persistent bloody diarrhea and unthriftiness may be evident.
Wool maggot (see Blowfly)		

TABLE 10-7—SWINE PARASITES

Parasite	Species Affected	Symptoms and Signs of Affected Animals (or Damage Inflicted)
Ascarids (or large intestinal roundworm, *Ascaris lumbricoides)*—yellowish or pinkish worms, 8 to 15 in. long, almost the size of a lead pencil (see Fig. 10-63 for the life history and habits of ascarids). 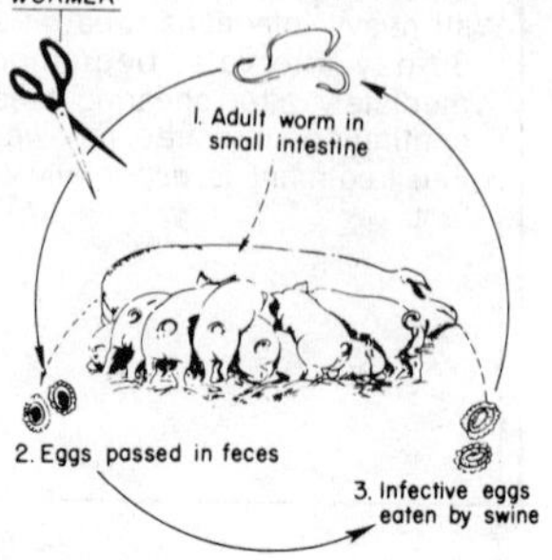 Fig. 10-63	Swine.	Young pigs become unthrifty and stunted, and there is usually coughing, "thumpy" breathing, and there may be a yellow color to the mucous membrane due to blockage of the bile ducts. Principal damage is produced by migrating larvae which produce liver damage and lung lesions resulting in verminous pneumonia.

—(Continued)

Distribution and Losses Caused By	Treatment	Prevention and Control	Remarks
Worldwide.	No known treatment.	Elimination of stray dogs. Examination and proper worm treatment, when necessary, of all dogs that must come in contact with sheep and goats. Proper disposal of all carcasses of infested animals.	
Death losses are not unusual, particularly when animals are on scant rations.	See Table 10-4, Recommended Compounds for Control of Internal Parasites, by Animal Species. Use drug of choice according to manufacturer's directions.	Although not as effective, the recommended measures for prevention and control of trichostrongyles are the same as those of the stomach worm; refer to the latter.	See manufacturer's label for proper drug withdrawal prior to slaughter for food.
Worldwide. Damage can be severe in specific flocks with numerous death losses.	See Table 10-4, Recommended Compounds for Control of Internal Parasites, by Animal Species. Use drug of choice according to manufacturer's directions.	Clean pastures and rotation grazing are the key to prevention and control.	

AND THEIR CONTROL

Distribution and Losses Caused By	Treatment	Prevention and Control	Remarks
Worldwide. Losses include stunted growth, uneconomical gains, and sometimes death in young animals.	See Table 10-4, Recommended Compounds for Control of Internal Parasites, by Animal Species. Use drug of choice according to manufacturer's directions.	Prevention consists in keeping the young pigs away from infection.	See manufacturer's label for proper drug withdrawal prior to slaughter for food.

(Continued)

TABLE 10-7

Parasite	Species Affected	Symptoms and Signs of Affected Animals (or Damage Inflicted)
Blowfly—the blowfly group consists of a number of species of flies all of which breed in animal flesh.	All farm animals.	Infested wounds and soiled hair. Maggots spread over the body, feeding on the skin surface, producing severe irritation, and destroying the ability of the skin to function. Infested animals rapidly become weak, fevered, and unthrifty.
Coccidiosis—a parasitic disease caused by protozoan organisms known as coccidia (see Fig. 10-64 for the life history and habits of coccidia). 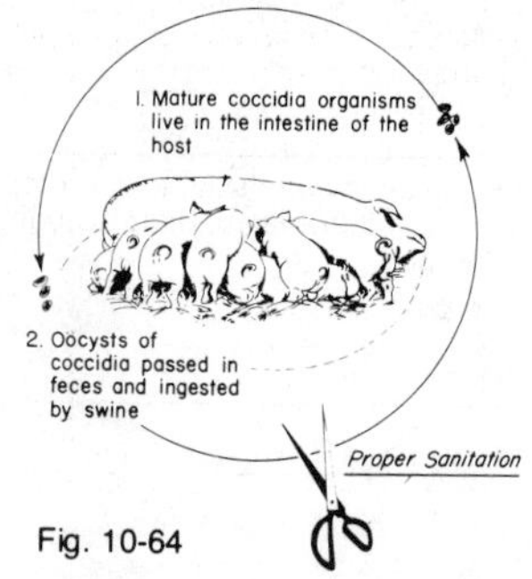 Fig. 10-64	Swine. Cattle. Sheep. Goats. Pet stock. Poultry. Each class of animals harbors its own species of coccidia; thus, there is no cross infection between animals.	Diarrhea and bloody feces, and pronounced unthriftiness and weakness.
Flies—the common species of flies are fully discussed in Table 10-5, Cattle Parasites and Their Control; hence, the reader is referred thereto.		
Kidney worm *(Stephanurus dentatus)*—a thick-bodied black-and-white worm up to 2 in. in length (see Fig. 10-65 for the life history and habits of the kidney worm). 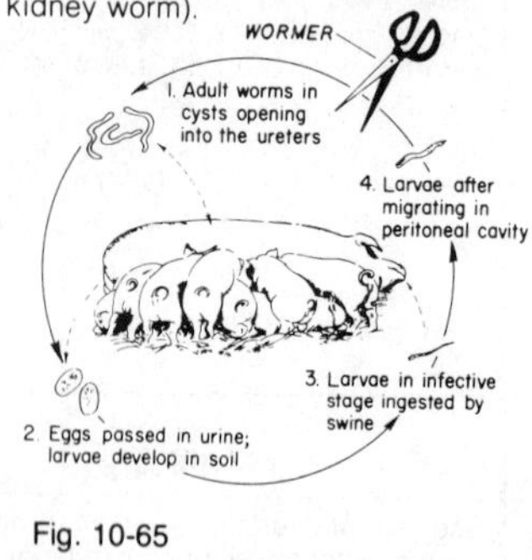 Fig. 10-65	Swine and cattle, but not so damaging in the latter.	No specific symptoms can be attributed to kidney worm infestation. The growth rate is markedly retarded and frequently pus is discharged in the urine. The parasite may also seriously affect the ability of the sow to produce young.
Large intestinal roundworm (see Ascarids)		
Lice—small, flattened, wingless insect parasites of which there are several species, most of which are specific for a particular class of animals. 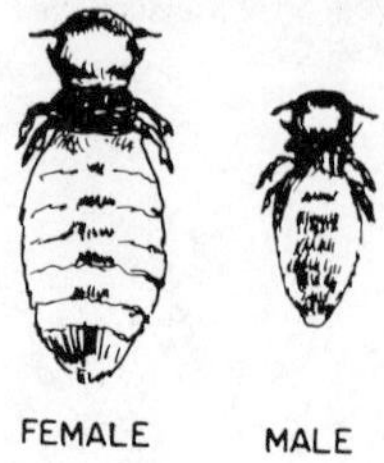 Fig. 10-66	Swine, with other species for other classes of animals.	Intense irritation, restlessness, and loss of condition. There may be severe itching and the animal may be seen scratching and rubbing. The hair may be rough, thin, and lack luster; and scabs may be evident. Lice are apt to be most plentiful around the root of the tail, on the inside of the thighs, and around the neck and ears.

—(Continued)

Distribution and Losses Caused By	Treatment	Prevention and Control	Remarks
Widespread, but present greatest problem in Pacific Northwest and in South and southwestern states. Death losses not excessive, but production is lowered.	Treat wounds twice weekly with a smear, dust, or pressurized spray of the proper insecticide, used according to manufacturer's directions.	Eliminate the blowfly by: 1. Destroying dead animals by burning or deep burial. 2. The use of traps, poisoned baits, and electrical screens. 3. Using repellents, such as pine-tar oil.	Damage inflicted is similar to that caused by screwworms, and the treatment for both types of maggots is the same.
Worldwide. Death losses are rare, but there is lowered gain and production in infected animals.	Good nursing will help. Sulfamerazine, sulfamethazine, or sulfaquinoxaline. Amprolium (Amprol) is reported to be effective. But, in the U.S., it is permitted in feed only as an aid in the control of cocciciosis of chickens and turkeys.	Avoid feed and water contaminated with the protozoa that causes the disease. Segregate affected animals. Remove and properly dispose of manure and contaminated bedding daily. Drain low, wet areas. Keep animals in a sunny, dry place.	In the oocyst stage, the parasite resists low temperatures and disinfectants and may remain viable outside the body for months, but it is readily destroyed by sunlight or complete drying.
Mostly in southern U.S. where it is one of the most damaging worm parasites affecting swine. Losses incurred include (1) inefficient gains, (2) upon slaughter, severe trimming or even condemnation of the damaged carcass, and (3) liver condemnations.	Levamisole (Tramisol) in the drinking water or feed is highly effective in the treatment and control of kidney worms.	The gilt-only method was first proved at the Coastal Plain Experiment Station, Tifton, Ga. This method is based on the fact that the kidney worm may take as long as a year to reach the egg-laying stage. With this method, gilts are bred only once; then at weaning time, they are sent to slaughter before mature kidney worms develop. By using only young breeding stock for 4 farrowing seasons (2 years), the parasites are eliminated. Consult your veterinarian for a control program based on using levamisole.	Carcass and liver losses at slaughter are ultimately borne by the swine producer in the form of lowered market prices. See manufacturer's label for proper withdrawal prior to slaughter for food.
Widespread. Lice retard growth, lower milk production, and produce unthriftiness.	Crotoxyphos (Ciodrin) Coumaphos (Co-Ral; do not use on pigs before weaning). Dioxathion (Delnav) Malathion Methoxychlor Ronnel (Korlan) Toxaphene Follow labels for mixing directions, application, and safety precautions.	Because of the close contact of swine during the winter months, it is practically impossible to keep them from becoming infested with lice. For effective control, all swine should be treated simultaneously at intervals, especially in the fall about the time they are placed in winter quarters.	Lice show up most commonly in winter and on ill-nourished and neglected animals. Lice are capable of transmitting the virus of swine pox.

(Continued)

TABLE 10-7

Parasite	Species Affected	Symptoms and Signs of Affected Animals (or Damage Inflicted)
Lungworms *(Metastrongylus* spp*)*—worms thread-like in diameter, 1 to 1½ in. long, white or brownish in color, found in the air passages (see Fig. 10-67 for the life history and habits of the lungworm). 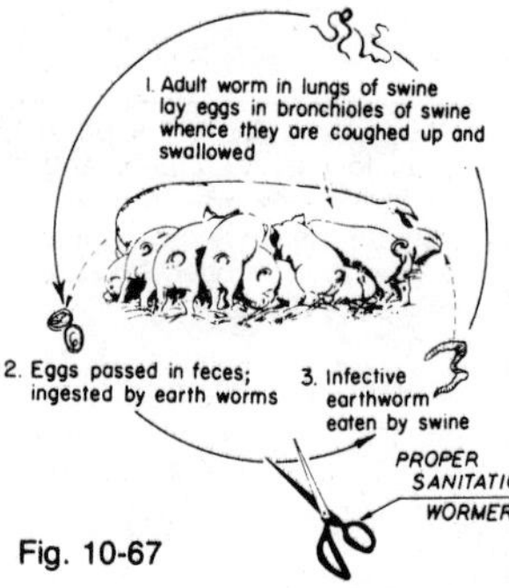 Fig. 10-67	Swine. Also occurs in cattle, sheep, and goats, but with different species of worms in each.	Pigs heavily infected with lungworms become unthrifty, stunted, and are subject to spasmodic coughing.
Mange (see Mites)		
Mites—are very small parasites that produce mange (scabies, scab, itch).	Swine. Each class of animals has its own species of mange mites, except in the case of the sarcoptic mite which is transferable between classes and to man.	Marked irritation, itching, and scratching. Crusting over of the skin, accompanied by formation of thick, tough, wrinkled skin.
Nodular worm—of which 4 species occur in swine, all of which are slender, whitish to grayish in color, and ⅓ to ½ in. in length (see Fig. 10-68 for the life history and habits of the nodular worm). 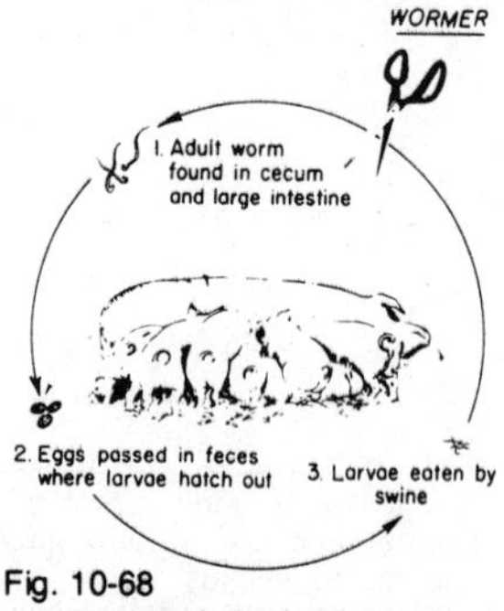 Fig. 10-68	Swine.	No specific symptoms. Weakness, anemia, emaciation, diarrhea, and general unthriftiness occur.
Ringworm—a contagious disease of the outer layers of skin caused by certain microscopic molds or fungi.  Fig. 10-69	All animals and man.	Round, scaly areas almost devoid of hair appear mainly in the vicinity of the eyes, ears, side of the neck, or the root of the tail. Mild itching usually accompanies the disease.

—(Continued)

Distribution and Losses Caused By	Treatment	Prevention and Control	Remarks
Throughout the U.S., but heaviest infestation of swine occurs in southeastern states. In addition to the lowered growth and feed efficiency, there is evidence to indicate that lungworms may be instrumental in the spread of swine influenza.	Levamisole (Tramisol) in drinking water or feed is 90 to 100% effective. See manufacturer's label for dosage recommendations.	Keep hogs away from those areas where earthworms are likely to abound. Remove manure piles and trash, and drain low places. Ring the snout. Routinely treat with levamisole.	See manufacturer's label for proper drug withdrawal prior to slaughter for food.
Widespread. Mites retard growth, lower gains, and produce unthriftiness.	Lime-sulfur. Malathion. Toxaphene. Follow container label for mixing directions, application, and safety precautions.	Avoid contact with diseased animals or infested premises. Scabies is a reportable disease in the U.S. So, in case of an outbreak, contact the local veterinarian or livestock sanitary official. Control by spraying or dipping infested animals with suitable insecticides, and quarantine affected herds.	There are two chief forms of mange; sarcoptic mange (caused by burrowing mites), and psoroptic mange (caused by mites that bite the skin and suck blood but do not burrow). The disease appears to spread most rapidly during the winter months and among young and poorly nourished animals.
Widely distributed over U.S., but damage is heaviest in southeastern states. In addition to the usual lack of thrift, the intestines of severely infested animals are not suited for either sausage casings or food (chitterlings).	See Table 10-4, Recommended Compounds for Control of Internal Parasites, by Animal Species. Use drug of choice according to manufacturer's directions.	A strict program of swine sanitation, accompanied by pasture rotation, constitutes a successful and practical preventive measure.	Dichlorvos and levamisole are broad-spectrum wormers; hence, they control ascarids and whipworms, in addition to nodular worms.
Throughout the U.S. It is unsightly and affected animals may experience considerable discomfort, but actual economic losses are not too great.	Clip the hair from the affected areas, remove scabs with a brush and mild soap. Paint affected areas with tincture of iodine or salicylic acid and alcohol (1 part in 10) every 3 days until cleared up.	Isolate affected animals. Disinfect everything that has been in contact with infested animals, including brushes. Practice strict sanitation.	Though ringworms may appear among animals on pasture, it is far more prevalent as a stable disease.

(Continued)

TABLE 10-7

Parasite	Species Affected	Symptoms and Signs of Affected Animals (or Damage Inflicted)
Screwworm *(Cochliomyia hominivorax)*—the maggots of the screwworm fly develop in the open wounds of animals.	All farm animals.	Loss of appetite, unthriftiness, and lessened activity.
Stomach worms—of which 3 species infest swine. *Ascarops strongylina* and *Physocephalus sexalatus*, commonly known as "thick stomach worms," are reddish in color and up to an inch long. *Hyostrongylus rubidus*, commonly known as the "red stomach worm," is reddish, small, delicate, slender, and about 1/5 in. in length (see Fig. 10-70 for the life history and habits of the stomach worm).	Swine. Also occurs in cattle, sheep, and goats, but with a different species of worms in each.	Unthriftiness and marked loss of appetite.

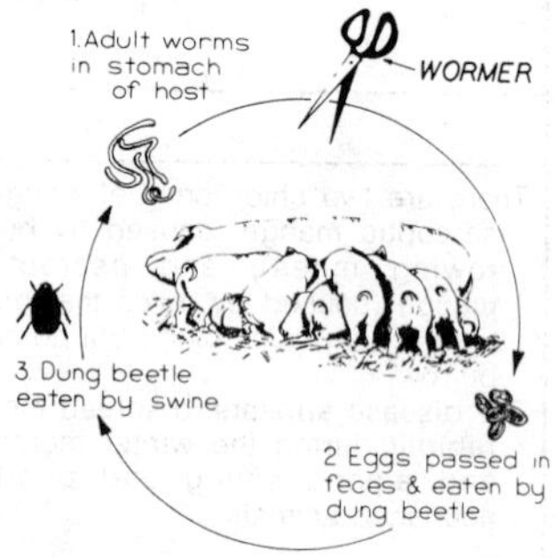

Fig. 10-70

Parasite	Species Affected	Symptoms and Signs of Affected Animals (or Damage Inflicted)
Thorn-headed worms *(Macracanthorhynchus hirudinaceus)*—white to bluish worms, cylindrical to flat, up to the size of a lead pencil, with rows of hooks which it uses for attachment purposes (see Fig. 10-71 for the life history and habits of the thorn-headed worm).	Swine.	No specific symptoms, although swine infested with thorn-headed worms exhibit the general unthriftiness commonly associated with parasites. Digestive disturbance accompanies severe cases.

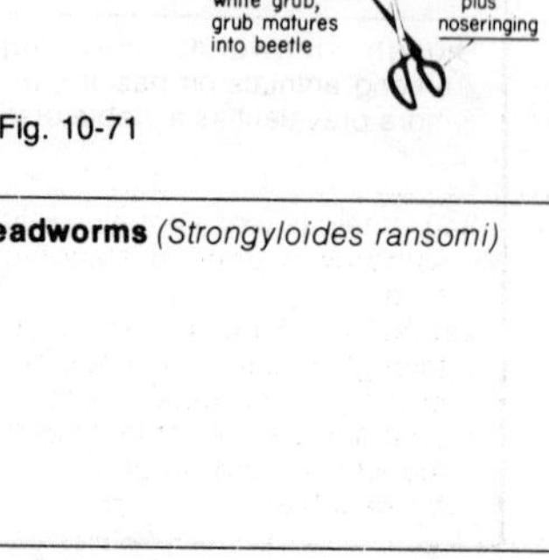

Fig. 10-71

Parasite	Species Affected	Symptoms and Signs of Affected Animals (or Damage Inflicted)
Threadworms *(Strongyloides ransomi)*	Swine. Also occurs in cattle, but with different species of worms.	In heavy infestations, it is a serious disease causing scours, anemia, and severe weight and death loss, particularly in young pigs. Light infections may show no symptoms.

—(Continued)

Distribution and Losses Caused By	Treatment	Prevention and Control	Remarks
Mostly in the southern and southwestern states; in which areas it formerly caused 50% of the normal annual livestock losses.	When maggots (larvae) are found in an animal, they should be removed and sent to the proper authorities for identification, and the animal should be treated with the proper insecticide or smear. Additional treatment or control measures will be supervised by inspection personnel.	Keep animal wounds to a minimum. Schedule stock operations that necessarily produce wounds during the winter season when the flies are least abundant and active. Kill all possible maggots during the winter and spring months. If possible, keep wounded and infested animals in a screened, fly-proof area.	Benzene (not benzine) or chloroform may be used to kill the maggots. The screwworm eradication program, in which sterile males were used, has practically eliminated the screwworm in the U.S.
Widespread throughout the U.S. Losses are chiefly in stunted growth and waste of feed.	Dichlorvos (Atgard). Levamisole (Tramisol) is the treatment of choice. See manufacturer's label for dosage recommendations.	Preventive measures for the control of stomach worms are similar to those advocated for the control of ascarids; hence, the reader is referred thereto.	On farms where a good parasite control program is practiced, this parasite is no problem.
Common in southern U.S. Losses include slow growth, inefficient feed utilization, death losses, and damaged intestines that are unfit for sausage casings.	No known drug treatment is entirely satisfactory for removing thorn-headed worms.	Keep pigs from feeding in areas where they might obtain the white grub of the June bug, the intermediate host. Sanitation, clean ground, and nose-ringing are effective preventive measures.	
It occurs mainly in the southern and southeastern states.	Levamisole (Tramisol) in the drinking water or feed. Thiabendazole (Thibenzole). See manufacturer's label for directions.	Good sanitation, pasture rotation, and frequent worming.	See manufacturer's label for proper withdrawal prior to slaughter for food. Larval transmission may occur through sow's milk, frequently resulting in serious infections in young pigs. For this reason, infected sows should be disposed of by slaughter.

(Continued)

TABLE 10-7

Parasite	Species Affected	Symptoms and Signs of Affected Animals (or Damage Inflicted)
Trichinosis *(Trichinella spiralis)*—a parasitic disease of man contracted largely by consuming infested pork, eaten raw, or imperfectly cooked (see Fig. 10-72 for the life history and habits of trichina). 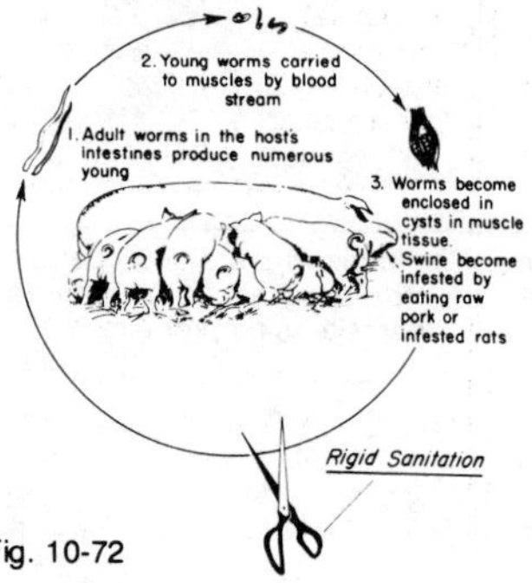 Fig. 10-72	Swine. Man.	No specific symptoms in hogs, even when the parasite is present in the muscle tissue, its usual abode.
Whipworm *(Trichuris suis)*—1½ to 2 in. in length, with slender anterior and heavy posterior portions that resemble the lash and handle, respectively, of a whip (see Fig. 10-73 for the life history and habits of the whipworm). 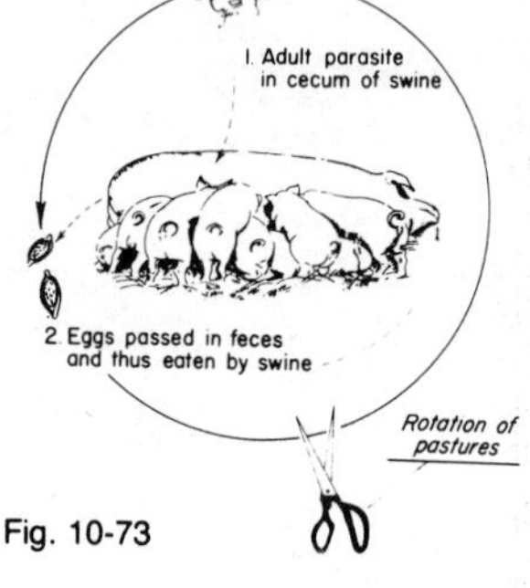 Fig. 10-73	Swine.	Infected animals may develop a diarrhea, and in heavy infections the diarrhea becomes bloody. In massive infections, growth may be noticeably retarded, and the animal may become weak and finally die.

—(Continued)

Distribution and Losses Caused By	Treatment	Prevention and Control	Remarks
Old studies (conducted prior to current garbage-cooking laws) showed (1) less than 1% of pork from grain-fed hogs infected with trichinosis, and (2) 5 to 6% infection of pork in hogs fed uncooked garbage.	There is no practical treatment for infected hogs. Infected humans should be under the care of an M.D.	Prevention of trichinosis in man may be obtained by: 1. Thoroughly cooking all pork at a temperature of 137°F before it is consumed; or 2. Freezing pork for a continuous period of not less than 20 days at a temperature not higher than 5°F. Trichinosis in swine may be lessened by: 1. Destruction of all rats on the farm; 2. Proper carcass disposal of hogs and other animals that die on the farm; and 3. Cooking all garbage and offal from slaughterhouses.	In man, the disease is usually accompanied by a fever, digestive disturbances, swelling of infected muscles, and severe muscular pain (in the breathing muscles as well as others). Microscopic examination of pork is the only way in which to detect the presence of trichina, but such a method is regarded as impractical in meat inspection procedure.
Widely scattered throughout the U.S., but heaviest infestation in southeastern states. Losses include slow gains, feed inefficiency, and some deaths.	Dichlorvos (Atgard). Hygromycin B (Hygromix). Levamisole (Tramisol). Use drug of choice according to manufacturer's directions.	Clean, well-drained pastures; rotation grazing; and plenty of sunlight are the key to the prevention and control of the whipworm.	The whipworm is increasing in the Corn Belt and in the Southeast.

TABLE 10-8—HORSE PARASITES

Parasite	Species Affected	Symptoms and Signs of Affected Animals (or Damage Inflicted)
Ascarids, *Parascaris equorum* (or white worm, large roundworm)—the female varies from 6 to 14 in. in length and the male from 5 to 13 in. When full grown, both are about the diameter of a lead pencil. Fig. 10-74 shows the life history and habits of the ascarid, *Parascaris equorum*. 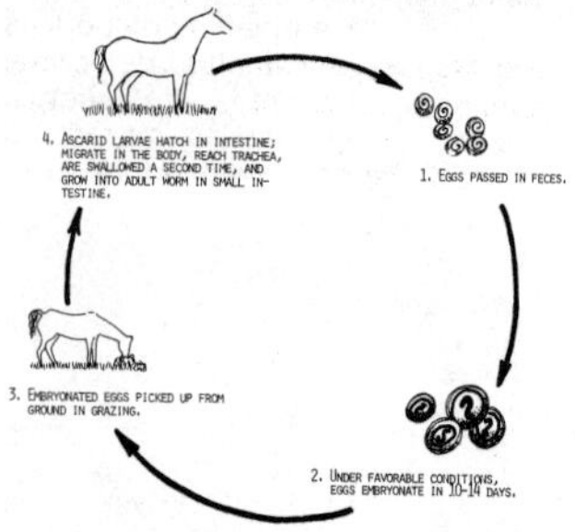 Fig. 10-74	Horses. Mules. Zebras.	The injury produced by ascarids covers a wide range, from light infections producing moderate effects to heavy infections which may be the essential cause of death. Death is usually due to a ruptured intestine. Serious lung damage caused by migrating ascarid larvae may result in pneumonia. More common, and probably more important, is a retarded or impaired growth and development manifested by potbellies, rough hair coats, and digestive disturbances. Especially affect foals and young animals, but are rarely important in horses over 5 years of age; older animals develop acquired immunity from earlier infections.
Blowfly—the blowfly group consists of a number of species of flies, all of which breed in animal flesh.	All farm animals.	Infested wounds and soiled hair. Maggots spread over the body, feeding on the skin surface, producing severe irritation, and destroying the ability of the skin to function. Infested animals rapidly become weak, fevered, and unthrifty.
Bots—three species of horse botflies are pests of horses in the U.S.: the common horse bot or nit fly *(Gastrophilus intestinalis)*, the throat bot or chin fly *(G. nasalis)*, and the nose bot or nose fly *(G. hemorrhoidalis)*. 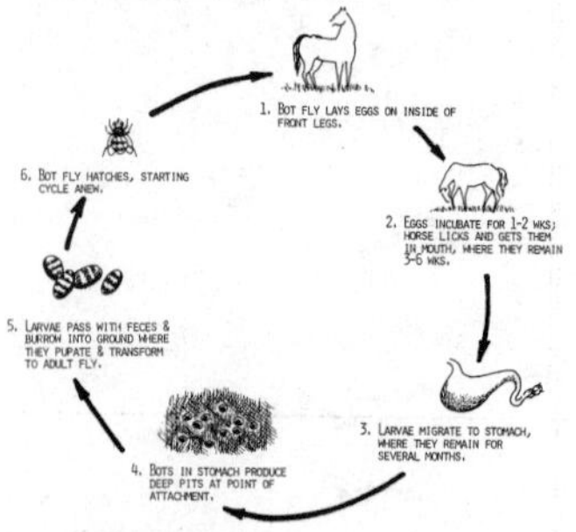 Fig. 10-75	Horses. Mules. Asses. Zebras.	Animals attacked by the botfly may toss their heads in the air, strike the ground with their front feet, and rub their noses on each other or on any convenient object. Infected animals may show frequent digestive upsets and even colic, lowered vitality and emaciation, and reduced work output.
Dourine—a chronic venereal disease, *Trypanosoma equiperdum*; a protozoa. Spread mostly through mating. 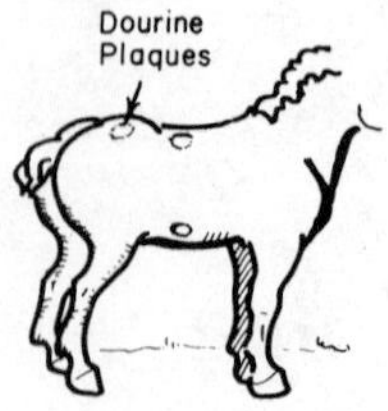 Fig. 10-76	Horses. Asses.	Redness and swelling of the reproductive organs of both the mare and stallion. Frequent urination, and increased sexual excitement in both sexes. A pussy discharge may be noted. Firm, round, flat swellings (dollar plaques) eventually appear on the body and neck. In advanced stages, there may be paralysis of the face, knuckling of the joints of the hind limbs, and dragging of the feet.

AND THEIR CONTROL

Distribution and Losses Caused By	Treatment	Prevention and Control	Remarks
Throughout the U.S. Presence of ascarids results in: Loss of feed through feeding worms. Lowered work efficiency (including performance on the track and in the show-ring). Retarded growth of young animals. Lowered breeding efficiency. Death in severe infestations.	See Table 10-4, Recommended Compounds for Control of Internal Parasites, by Animal Species. Use drug of choice according to manufacturer's directions. In addition to selecting the particular drug(s) for ascarid control, the horseman should set up a definite treatment schedule, then follow it. The advice of the veterinarian should be sought on both points. Also, to preclude the possibility that worms may become resistant to a drug that is used continuously, the veterinarian may recommend a rotation of drugs.	Keep foaling barn and paddocks clean. Store manure in pit 2 to 3 weeks. Provide clean feed and water. Place young foals on clean pasture. Worm all mares 4 to 6 weeks before foaling. Worm foals and yearlings on a regular basis.	Foals usually first acquire ascarid infection from contaminated stalls and paddocks. Foals should be treated early in life, before the ascarids have a chance to mature and become large enough to block the intestine. As a precaution, mares should not be treated closer than 30 days before foaling or within 14 days after foaling.
Widespread, but present greatest problem in Pacific Northwest, and in South and southwestern states. Death losses not excessive, but production is lowered.	When horses become infested with blowfly maggots, their wounds should be treated twice weekly with a smear, dust, or pressurized spray of the proper insecticide, used according to manufacturer's directions.	Eliminate the blowfly by: 1. Destroying dead animals by burning or deep burial. 2. The use of traps, poisoned baits, and electrified screens. 3. Using repellents, such as pine-tar oil.	
Worldwide. Presence of bots results in: Loss of feed through feeding worms. Itching and loss of tail hair due to rubbing. Lowered work efficiency. Retarded growth of young animals. Lowered breeding efficiency. Death in severe infestations.	See Table 10-4, Recommended Compounds for Control of Internal Parasites, by Animal Species. Use drug of choice according to manufacturer's directions. In the late fall or early winter at least one month after the first killing frost, administer one of the recommended drugs.	Frequent grooming, washing, and clipping. Prevention of reinfestation is best assured through community campaigns in which all horses within the area are treated. Fly nets and nose covers offer some relief from the attacks of adult botflies. Thirty days prior to worming, the eggs of the botfly which may be clinging to the body should be destroyed by (1) clipping the hair, and/or (2) washing with warm water at 120°F. The insides of the knees and the fetlocks especially should be treated in this manner.	As a precaution, mares should not be treated closer than 30 days before foaling or within 14 days after foaling.
Worldwide, but now rare in the U.S.	No successful treatment known.	Destroy all infected animals. Avoid mating with an infected animal, and apply modern hygiene.	An official test of the blood serum may be obtained from the USDA, on request.

(Continued)

TABLE 10-8

Parasite	Species Affected	Symptoms and Signs of Affected Animals (or Damage Inflicted)
Equine piroplasmosis *(Babesiasis)*—the disease is caused by either of two protozoa, *Babesia caballi* or *B. equi*, which invade the red blood cells.	Horses. Donkeys. Mules. Zebras.	Similar to equine infectious anemia (or Swamp fever). A positive diagnosis is made by demonstrating the protozoa in the red blood cells or by antigen-antibody serum tests. Clinical signs include fever (103-106° F), anemia, icterus, depression, thirst, lacrimation, and swelling of the eyelids. Constipation and colic may occur. The urine is yellow to reddish in color. The incubation period is 1 to 3 weeks.
Flies and mosquitoes—flies and mosquitoes are usually classed as follows: 1. *Biting flies and mosquitoes*—this includes horseflies, deerflies, stable flies, horn flies, black flies, biting midges, and mosquitoes. 2. *Nonbiting flies*—include the face fly and housefly. 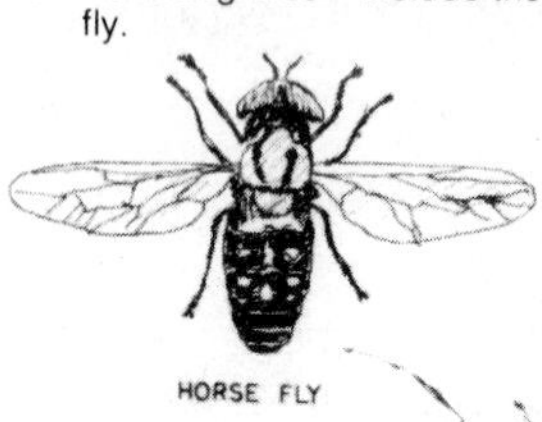 Fig. 10-77 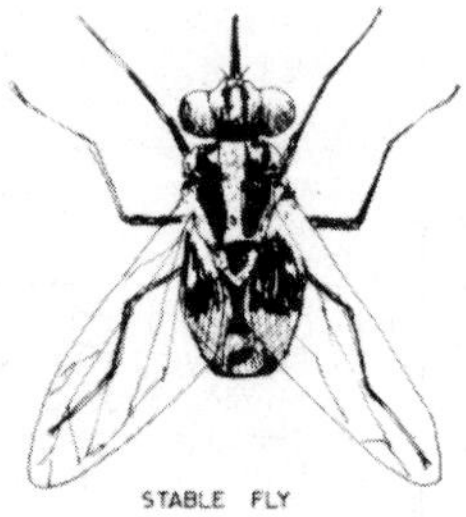 Fig. 10-78 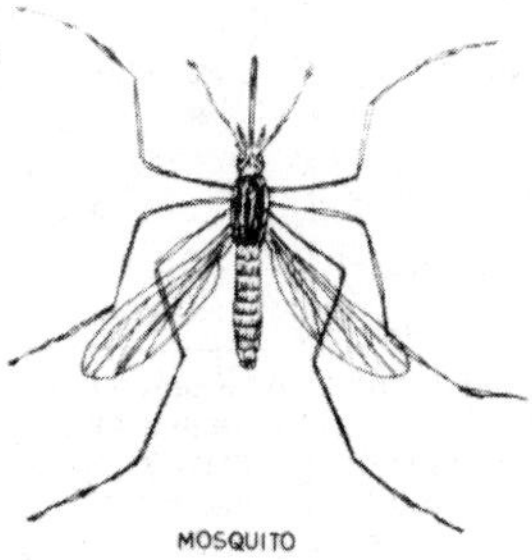 Fig. 10-79	Horses.	They lower the vitality of horses, mar the hair coat and skin, produce a general unthrifty condition, lower performance, and make for hazards when riding or using horses. Also, they may temporarily or permanently impair the development of foals and young stock.
Intestinal threadworms (see Threadworms)		
Lice—small, flattened, wingless insect parasites of which there are several species, most of which are specific for a particular class of animals. 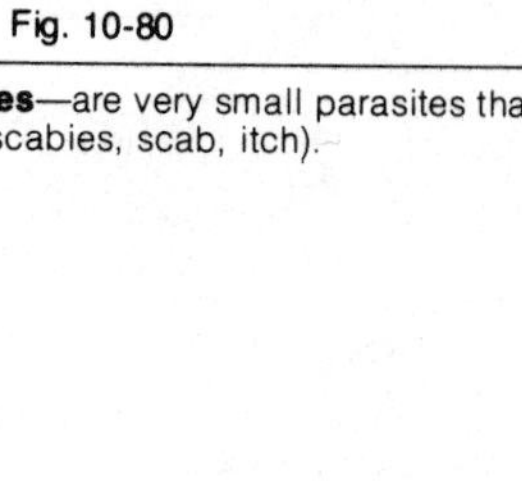 Fig. 10-80	Horses. Mules, with other species for other classes of animals.	Intense irritation, restlessness, and loss of condition. There may be severe itching and the animal may be seen scratching, rubbing, and gnawing at the skin. The hair may be rough, thin, and lack luster; and scabs may be evident. Lice are apt to be most plentiful around the root of the tail, on the inside of the thighs, over the fetlock region, and along the neck and shoulders.
Mites—are very small parasites that produce mange (scabies, scab, itch).	Horses. Mules. Each species of animals has its own species of mange mites, except in the case of the sarcoptic mite, which is transferable between classes and to man.	Marked irritation, itching, and scratching. Crusting over of the skin, accompanied by formation of a thick, tough, wrinkled skin.

—(Continued)

Distribution and Losses Caused By	Treatment	Prevention and Control	Remarks
Worldwide. First diagnosed in the U.S. in 1961, in Florida.	A new and promising treatment, developed by the University of Florida, consists in 2 successive treatments with Diampron, 4 mg/lb body weight, given intramuscularly 48 hours apart. This will eliminate the positive carrier state.	Tick control is the most effective approach. The tropical horse tick, *Dermacentor nitens*, is the vector in the U.S. The disease may also be spread by the vampire bat.	Where equine piroplasmosis is suspected, call the veterinarian. The carrier state persists from 10 months to 4 years.
Wherever there are horses. Flies and mosquitoes are probably the most important insect pests of horses.	*Biting flies and mosquitoes;* Use an insecticide, of which there are several, according to manufacturer's directions. Treat manure piles and buildings for fly control; and treat wet areas that harbor mosquitoes. Fly repellents containing pyrethrins which last 4 to 8 hours after application have been developed for the control of horseflies and deerflies on horses. *Nonbiting flies:* Face fly control is difficult to achieve. Pyrethrin repellents applied to the horse's face and head, will repel nonbiting flies for 8 to 12 hours. A mask or net can be made and attached to the halter so that its movements protect the horse's eyes from face flies when on pasture. Residual sprays, when applied to the sunny surfaces of barns, shelters, and fences where face flies congregate reduce populations.	*Biting flies and mosquitoes:* Sanitation—the destruction of the breeding areas of the nests—is the key to the control of biting flies and mosquitoes. Do not allow manure or other breeding areas to accumulate. Spread manure in fields (to dry) every day or two. Control horseflies, deerflies, and mosquitoes by filling low spots in corrals or paddocks and draining all water-holding areas. *Nonbiting flies:* Sanitation is the most efficient method of reducing populations of houseflies. Sanitation may be additionally important if the horses are located near an urban area, in order to avoid complaints from neighbors. Residual sprays will eliminate many houseflies. Also, houseflies are attracted to baits (insecticides mixed with sugar or other attractive material), which are effective housefly killers.	Flies and mosquitoes can be the vector (carrier) of serious diseases.
Widespread. Lice retard growth, lower work efficiency, and produce unthriftiness.	Periodic application, according to the directions on the label, of one of the following insecticides: Coumaphos (Co-Ral) Crotoxyphos Dioxathion Malathion Rotenone	Because of the close contact of horses during the winter months, it is practically impossible to keep them from becoming infested with lice. For effective control, all horses should be treated simultaneously at intervals, especially in the fall about the time they are placed in winter quarters.	Lice show up most commonly in winter and on ill-nourished and neglected animals. Although rarely dipped, horses may be so treated for lice, using any one of the mixtures and procedures recommended for dipping cattle.
Widespread. Mites retard growth, lower work efficiency, and produce unthriftiness.	Any of the following insecticides, used as a spray or dip, will meet the federal regulations for the control of mites of horses: lime-sulfur (2% "sulfide sulfur"), lindane (0.05-0.06%), or toxaphene (0.5-0.6%); using 2 applications of each insecticide 10-14 days apart.	Avoid contact with diseased animals or infested premises. In case of an outbreak, contact the local veterinarian or livestock sanitation official. Control by spraying or dipping infested animals with suitable insecticides, and quarantine affected herds.	There are 2 chief forms of mange: sarcoptic mange (caused by burrowing mites), and psoroptic mange (caused by mites that bite the skin and suck blood but do not burrow). The disease appears to spread most rapidly during the winter months and among young and poorly nourished animals. Although rarely dipped, horses may be so treated if it is practical and convenient to do so.

(Continued)

TABLE 10-8

<table>
<tr><th>Parasite</th><th>Species Affected</th><th>Symptoms and Signs of Affected Animals (or Damage Inflicted)</th></tr>
<tr><td>Pinworms (or rectal worms)—of which 2 species are frequently found in horses. Oxyuris equi are whitish worms with long, slender tails, whereas Probstmyria vivipara are so small as to be scarcely visible to the naked eye (see Fig. 10-81 for the life history and habits of the pinworm).

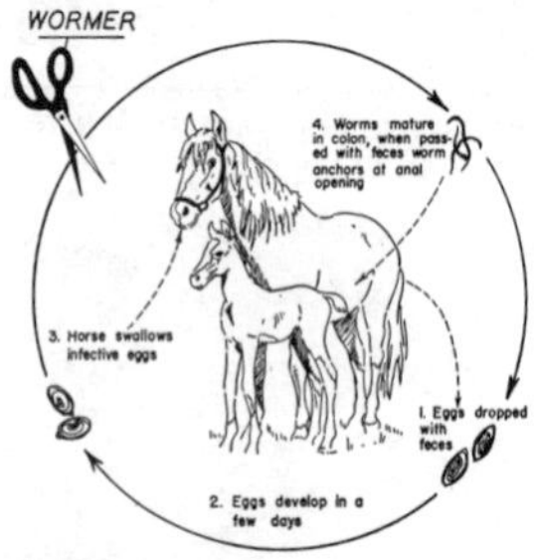

Fig. 10-81</td><td>Horses.
Also occurs in man, but with different species of worms.</td><td>Irritation of the anus and tail rubbing. Heavy infections may also cause digestive disturbances and produce anemia.
The large pinworm is most damaging to the horse.</td></tr>
<tr><td>Ringworm—a contagious disease of the outer layers of skin caused by certain microscopic molds of fungi.

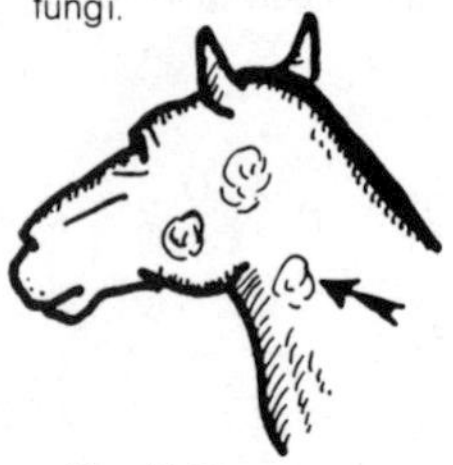

Fig. 10-82</td><td>All animals and man.</td><td>Round, scaly areas almost devoid of hair appear mainly in the vicinity of the eyes, ears, side of the neck, or the root of the tail.
Mild itching usually accompanies the disease.</td></tr>
<tr><td>Screwworm—the screwworm fly raises its maggots in the living flesh of animals.

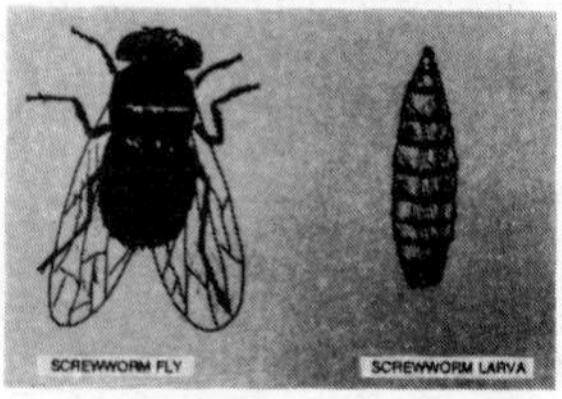

Fig. 10-83</td><td>All farm animals.</td><td>Loss of appetite, unthriftiness, and lowered activity.</td></tr>
<tr><td>Stomach worms—a group (3 kinds are important in horses) of parasitic worms that produce an inflammation of the stomach (see Fig. 10-84 for the life history and habits of the stomach worm).

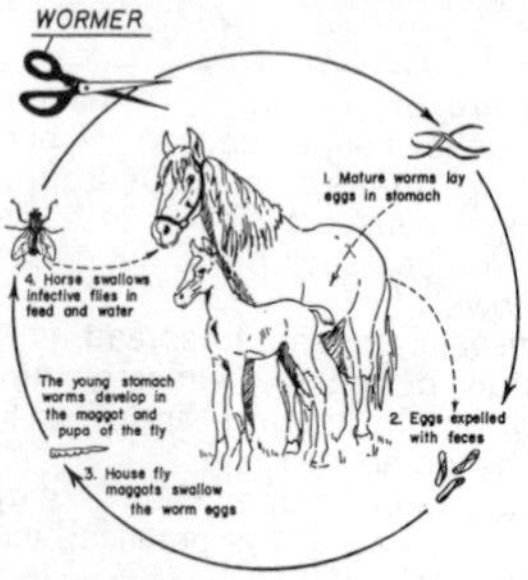

Fig. 10-84</td><td>Horses.
Trichostrongylus axei is also a common parasite of cattle, sheep, and a number of other animals.</td><td>Loss of condition.
Severe gastritis.
Summer sores, a skin disease.</td></tr>
</table>

—(Continued)

Distribution and Losses Caused By	Treatment	Prevention and Control	Remarks
Throughout the U.S.	See Table 10-4, Recommended Compounds for Control of Internal Parasites, by Animal Species. Use drug of choice according to manufacturer's directions.	Sanitation and keeping animals separated from their own excrement.	
Throughout the U.S. It is unsightly and affected animals may experience considerable discomfort, but actual economic losses are not too great.	Clip the hair from the affected areas, remove scabs with a brush and mild soap. Paint affected areas with tincture of iodine or salicylic acid and alcohol (one part in ten) every 3 days until cleared up. Copper napthenate or dichlorphene is also effective.	Isolate affected animals. Disinfect everything that has been in contact with infested animals, including curry combs and brushes. Practice strict sanitation.	Though ringworm may appear among animals on pasture, it is far more prevalent as a stable disease.
Mostly in the southern and southwestern states, in which areas it formerly caused 50% of the normal annual livestock losses.	When maggots (larvae) infest the flesh of an animal, a sample of the larvae should be sent to proper authorities for identification, and the animal should be treated with a proper insecticide. Additional treatment or control measures will be supervised by inspection personnel if the larvae are screwworms. The application of dusts of coumaphos (Co-Ral) and ronnel (Korlan) will provide relief.	Keep animal wounds to a minimum. The screwworm eradication program, by sterilization, has been very effective. This consists in sterilizing male screwworms, in the pupal stages with gamma rays. Male screwworms mate repeatedly, but females mate only once. Thus, when a female mates with a sterilized male, only infertile eggs are laid. The release of millions of sterilized males has led to the near eradication of screwworms from most of the U.S.	Benzene (not benzine) or chloroform may be used to kill the maggots. Screwworms sometimes infest the prepuce of geldings.
Throughout the U.S. Wasted feed and lowered efficiency are the chief losses.	Carbon disulfide. Levamisole-piperazine. Both of the above treatments are effective when used according to manufacturer's directions.	Houseflies and stable flies are the vectors. Hence, the control of flies is the best method of preventing and controlling stomach worms.	Infection of *T. axei*. Also occurs in ruminants. Dilute formaldehyde and astringents are commonly used in the treatment of summer sores.

(Continued)

TABLE 10-8

Parasite	Species Affected	Symptoms and Signs of Affected Animals (or Damage Inflicted)
Strongyles—of which there are about 60 species; 3 are large (up to 2 in. in length) and the rest small (some scarcely visible to the naked eye). The large strongyles are variously referred to as bloodworms *(Strongylus vulgaris)*, palisade worms, sclerostomes and red worms (see Fig. 10-85 for the life history and habits of *S. vulgaris*). 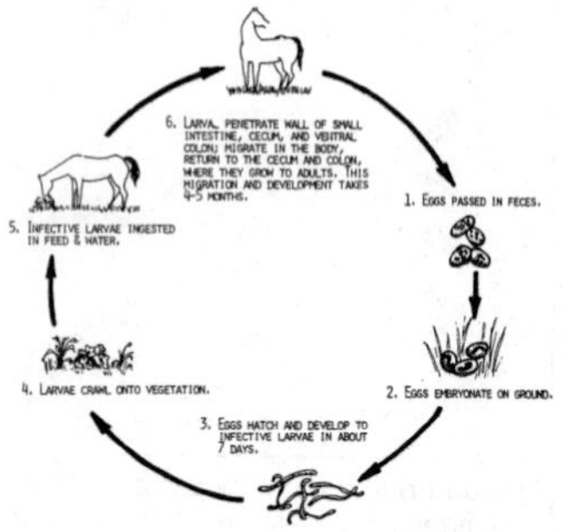 Fig. 10-85	Horses. Mules.	Lack of appetite, anemia, progressive emaciation, a rough hair coat, sunken eyes, digestive disturbances including colic, a tucked-up appearance, and sometimes posterior paralysis and death. Collectively these symptoms indicate the disease known as strongylosis. Harmful effects greatest with younger animals. One species of large strongyles *(S. vulgaris)* may permanently damage an intestinal blood vessel wall, resulting in death at any age. Also, they may cause severe colic, which may terminate in death.
Summer sores *(Cataneous habronemiasis)*—caused by larval form of *Habronema* spp (large stomach worm) in skin. Also known as Jack sores or Bursatti.	Horses. Mules. Jacks.	Unsightly and uncomfortable skin lesions of various sizes.
Tapeworms—of which 3 species are of economic importance in the horse. *Anoplocephala perfoliata* is most common and most damaging.	Horses.	Heavy infections may cause digestive disturbances, loss in weight, and anemia.
Threadworms *(Strongyloides westeri)*	Horses.	Diarrhea in foals.
Ticks 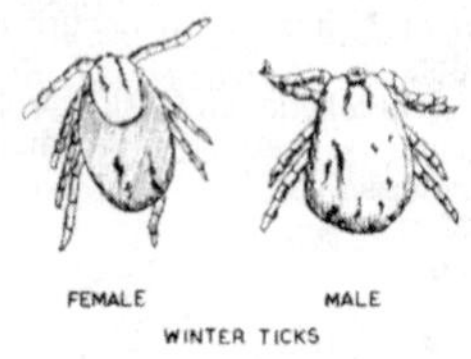 Fig. 10-86	Horses.	Ticks reduce the vitality of horses through constant irritation and loss of blood. Massive infestations may cause anemia, loss of weight, and even death. "Head heaviness" is often associated with massive infestations of ear ticks. Other losses may result from the simple presence of the ticks on the animal, a factor called "tick worry."

—(Continued)

Distribution and Losses Caused By	Treatment	Prevention and Control	Remarks
Throughout the U.S., wherever horses and mules are pastured. Presence of strongyles results in: Loss of feed through feeding worms. Lowered work efficiency (including on the track and in the show-ring). Retarded growth of young animals. Lowered breeding efficiency. Death in severe infestations.	See Table 10-4, Recommended Compounds for Control of Internal Parasites, by Animal Species. Use drug of choice according to manufacturer's directions.	Gather up manure daily from pastures and barns and store it in a pit for 2 to 3 weeks. Avoid moist pasture and overstocking. Rotate pastures.	Give treatment to both sexes and all ages. Strongyles are not transmissible to ruminants or swine. Heavily infected animals may have one or more of the 3 species of large strongyles along with 10 to 12 species of small strongyles.
Occurs throughout the U.S.	Dilute formaldehyde or ronnel and astringents are commonly used in the treatment of summer sores. Surgical removal or cauterization of the sores may be resorted to.	Good fly control, as flies are the vector.	Veterinary diagnosis is sometimes required to differentiate this lesion from "proud flesh" and ringworm.
Throughout the northern part of the U.S. Losses are primarily in wasted feed and retarded growth.	Treatment not normally suggested, primarily because only light infections are encountered. Lead arsenate (Bi-forma), used according to manufacturer's directions.	Sanitation, good husbandry, and proper manure disposal.	
Very common in foals.	See Table 10-4, Recommended Compounds for Control of Internal Parasites, by Animal Species. Use drug of choice according to manufacturer's directions.		Threadworms are self-limiting. They disappear by the time the foals are 6 months of age.
Ticks are particularly prevalent on horses in the southern and western parts of the U.S.		Because most species of ticks, except the ear tick and the tropical horse tick, attach to the external surfaces of horses, an application of the recommended insecticide by spray or wipe-on will give effective control. Ear ticks and tropical horse ticks should be treated by applying the chemical into the ears of the horses. Since horses are often confined to rather small areas, treatments of the premises may also help control heavy infestations of ticks. Recommended insecticides for control of ticks (except ear tick) are: crotoxyphos, coumaphos (Co-Ral), dioxathion, malathion, and pyrethrins. The recommended insecticides for control of ear ticks are: lindane and ronnel (Korlan).	Ticks are important to horsemen because they may transmit diseases such as equine piroplasmosis (carried by *Anocentor nitens*) or cattle fever (carried by the *Boophilus* species). Also, most of the ticks mentioned may be vectors of anaplasmosis, and several species can cause tick paralysis in hosts.

TABLE 10-9—MITES

(Livestock owners who suspect scabies [mites] should immediately contact

Animal and Insect	Insecticide	Tolerance (ppm in fat unless otherwise indicated)	Min. Days from Last Application to Slaughter or Use of Milk
Cattle (beef)			
Mites 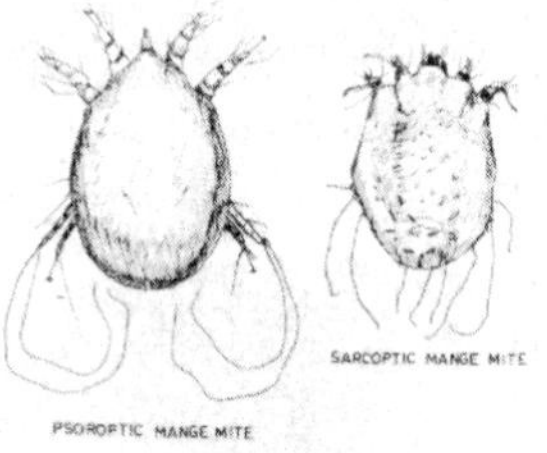 Fig. 10-87	Lime-sulfur[2]		
	Coumaphos		0
	Prolate		21
	Toxaphene	7	28
Cattle (dairy)			
Mites	Lime-sulfur		
Sheep and Goats			
Mites	Lime-sulfur		
	Toxaphene	7	28
Swine			
Mites	Lime-sulfur		
	Toxaphene	28	30(S)
	Malathion	4	
Horses			
Mites	Lime-sulfur		
	Lindane		
	Toxaphene		

[1]This table was authoritatively reviewed by a staff member of the federal regulatory division charged with the control of scabies; namely, Dr. Glen O. Schubert, Chief Staff Veterinarian, Bacterial and Parasitic Diseases, Animal and Plant Health Inspection Service, USDA, Hyattsville, Md.

POISONOUS PLANTS[4]

Poisonous plants have been known to man since time immemorial. Biblical literature alludes to the poisonous properties of certain plants, and history records that hemlock (a poison made from the plant from which it takes its name) was administered by the Greeks to Socrates and other state prisoners.

No section of the United States is entirely free of poisonous plants, for there are hundreds of them. But the heaviest livestock losses from them occur on the western ranges because (1) there has been less cultivation and destruction of poisonous plants in range areas, and (2) the frequent overgrazing on some of the western ranges has resulted in the elimination of some of the more nutritious and desirable plants, and these have been replaced by increased numbers of the less desirable and poisonous species. It is estimated that poisonous plants account for 8 to 10 percent of all range animal losses each year; and even more in some areas.

[4]In the preparation of this section, the author had the benefit of the authoritative review and suggestions of the Director of the Poisonous Plant Research Laboratory, USDA, Logan, Utah.

Diagnosis of Plant Poisoning

The diagnosis of plant poisoning in animals is not an easy or precise procedure. Any case of sudden illness or death with no apparent cause is commonly considered to be a poisoning. This may not always be correct. When large numbers of animals are suddenly

AND THEIR CONTROL[1]

their local veterinarian and/or state or federal regulatory official for advice.)

Formulation and Strength	Amount of Formulation per Animal (unless otherwise indicated)	Where and When to Apply	Remarks
2% "sulfide sulfur" spray or dip[3]	Wet animal thoroughly.	Make 2 or more applications of lime-sulfur at intervals of 10 or 14 days.	Lime-sulfur has been the time-honored treatment for scabies since the USDA approved it years ago.
0.3%	Do	Do	
0.2-0.25%	Do	Do	
0.5-0.6% spray or dip	Do		
2% spray or dip[3]	Wet animal thoroughly; treat for one minute. (These instructions apply to all animals and treatments.)	2 applications 10-14 days apart.	Lactating dairy cows are usually sprayed. Lime-sulfur is the only pesticide permitted for dairy cattle.
2% "sulfide sulfur"[3]	Wet animal thoroughly.	Make two applications 10-14 days apart.	Sheep and goats should be dipped, rather than sprayed.
0.05-0.6% dip	Wet animal thoroughly.	Do	
2% "sulfide sulfur" spray or dip[3]	Amount will vary with size of animal and amount of hair.	Make two applications 10-14 days apart.	When spread, it takes 2-4 qt/animal, depending on the size.
0.05-0.06% spray or dip	Do 1 gal. 50 to 57% emulsifiable conc. plus 1 lb of nonfoaming detergent/100 gal of water.	Do	
2% "sulfide sulfur" spray or dip[3]	Dosage will vary with size of animal and amount of hair.	Make 2 applications 10-14 days apart.	
0.05-0.06% spray or dip	Do	Do	
0.5-0.6% spray or dip	Do	Do	

[2]Certain proprietary brands may be used. However, lime-sulfur may be homemade by mixing 12 lb of unslaked lime (or 16 lb of commercial hydrated lime; not airslaked lime) and 24 lb of flowers of sulfur or sulfur flour to 100 gal of water.

[3]Lime-sulfur and nicotine dips must be heated and maintained at 95° to 105°F.

affected, however, a suspicion of poisoning is justified until it has been proven otherwise.

Symptoms or signs induced by eating poisonous plants may include (1) sudden death; (2) transitory illness; (3) general body weakness; (4) disturbance of the central nervous, vascular, and endocrine systems; (5) photosensitization; (6) frequent urination; (7) diarrhea; (8) bloating; (9) chronic debilitation and death; (10) embryonic death; (11) fetal death; (12) abortion; (13) extensive liver necrosis and/or cirrhosis; (14) edema and/or abdominal dropsy; (15) tumor growths in tissues; (16) congenital deformities; (17) metabolic deficiencies; and (18) physical injury.

No general set of symptoms and signs per se irrefutably provides all the information necessary to make a diagnosis of plant poisonings. Nevertheless, a careful description of the toxic signs coupled with information pertaining to available plants provides a meaningful basis for a tentative diagnosis. Additional information essential to a poisonous plant diagnosis includes (1) type of feed, site grazed, and availability of water; (2) identification and relative abundance of all poisonous plants available to animals; (3) amount and stage of growth of the various poisonous plants being grazed; (4) the toxicity and palatability of the plants in relation to their stage of growth; (5) time from eating the plants until onset of toxic signs; (6) species, age, and sex of animals affected; (7) clinical signs of toxic reactions; (8) chemical analysis of plants; and (9) a careful evaluation of all the information relative to the etiology of the disease.

Why Animals Eat Poisonous Plants

A frequently asked question is: Why do animals eat poisonous plants? The answer is not simple, but among the reasons are the following: (1) total lack of sufficient palatable forage—the animals are hungry; (2) decrease in palatability and nutrients of mature, weathered range grasses, with the result that poisonous plants become more appealing, comparatively speaking; (3) insufficient spring grass; (4) rain, melting snow, and heavy dew may enhance the palatability of some poisonous plants; and (5) going without water too long, which results in a reduction in feed intake, then, after watering, they develop a ravenous appetite and eat anything in sight—including less palatable poisonous plants.

Poisonous plants vary in palatability—between species, and within species, and at different stages of growth. For example, poison hemlock is never palatable and is eaten only as a last resort—when palatable forage is not available. Locoweed and black nightshade are eaten at any stage of growth or when mixed with hay. Others, such as lupines, horsebrush, and death camas may be eaten only at certain stages of growth. Still others, such as milk vetch, larkspur, and halogeton, are highly palatable to livestock at any and all times, with the result that if they're present animals will seek them out and there will be losses. Then, too, certain plants are poisonous to cattle but not to sheep (and vice versa), as shown in Table 10-10.

TABLE 10-10

TYPE OF RANGE ANIMAL SUSCEPTIBLE TO POISONOUS PLANTS AT DEFINITE SEASONS

Poisonous to Cattle	Time of Year	Poisonous to Sheep	Time of Year	Poisonous to Cattle & Sheep	Time of Year
Low larkspur	Spring	Death camas	Spring	Broomweed	Spring and summer
Oak	Spring	Greasewood	Fall	Chokecherry	Spring
Tall larkspur	Early summer and early fall	Horsebrush	Spring	Copperweed	Summer
Timber milk vetch	Spring	Rubberweed	Summer	Desert parsley	Spring
Water hemlock	Spring	Sneezeweed	Summer	Halogeton	All year
				Loco	Spring
				Lupine	Summer and fall
				Milkweeds	Summer
				Veratrum	Summer

Preventing Losses from Poisonous Plants

With poisonous plants, the emphasis should be on prevention of losses rather than on treatment, no matter how successful the latter. The following are effective preventative measures:

1. **Follow good pasture or range management** in order to improve the quality of the pasture or range. Plant poisoning is nature's sign of a "sick" pasture or range, usually resulting from misuse. When a sufficient supply of desirable forage is available, poisonous plants may not be eaten, for they are usually less palatable. On the other hand, when overgrazing reduces the available supply of the more palatable and safe vegetation, animals may, through sheer hunger, consume the toxic plants.

2. **Know the poisonous plants common to the area.** This can usually be accomplished through (a) studying drawings, photographs, and/or descriptions; (b) checking with local authorities; or (c) sending two or three fresh whole plants (if possible, include the roots, stems, leaves, and flowers) to the state agricultural college—first wrapping the plants in several thicknesses of moist paper.

By knowing the poisonous plants common to the area, it will be possible—

a. To avoid areas heavily infested with poisonous plants which, due to animal concentration and overgrazing, usually include waterholes, salt grounds, bed grounds, and trails.

b. To control and eradicate the poisonous plants effectively, by mechanical or chemical means (as recommended by local authorities) or by fencing off.

c. To recognize more surely and readily the particular kind of plant poisoning when it strikes, for time is important.

d. To know what first aid, if any, to apply, especially when death is imminent or where a veterinarian is not readily available.

e. To graze with a class of livestock not harmed by the particular poisonous plant or plants, where this is possible. Many plants seriously poisonous to one kind of livestock are not poisonous to another, at least under practical conditions.

f. To shift the grazing season to a time when the plant is not dangerous, where this is possible.

That is, some plants are poisonous at certain seasons of the year, but comparatively harmless at other seasons.

g. To avoid cutting poison-infested meadows for hay when it is known that the dried cured plant is poisonous. Some plants are poisonous in either green or dry form, whereas others are harmless when dry. When poisonous plants (or seeds) become mixed with hay (or grain), it is difficult for animals to separate the safe from the toxic material.

3. **Know the symptoms that generally indicate plant poisoning**, thus making for early action.

4. **Avoid turning to pasture in very early spring.** Nature has ordained most poisonous plants as early growers—earlier than the desirable forage. For this reason, as well as from the standpoint of desirable pasture management, animals should not be turned to pasture in the early spring before the usual forage has become plentiful.

5. **Provide supplemental feed during droughts, after plants become mature, and after early frost.** Otherwise, hungry animals may eat poisonous plants in an effort to survive.

6. **Avoid turning very hungry animals where there are poisonous plants**, especially those that have been in corrals for branding, etc.; that have been recently shipped or trailed long distances; or that have been wintered on dry forage. First feed the animals to satisfy their hunger or allow a fill on an area known to be free from poisonous plants.

7. **Avoid driving animals too fast when trailing.** On long drives, either allow them to graze along the way or stop frequently and provide supplemental feed.

8. **Remove promptly all animals from infested areas when plant poisoning strikes.** Hopefully, this will check further losses.

9. **Treat promptly, preferably by a veterinarian.**

Treatment of Plant-Poisoned Animals

Unfortunately, plant-poisoned animals are not generally discovered in sufficient time to prevent loss. Thus, prevention is decidedly superior to treatment.

When trouble is encountered, the owner or caretaker should *promptly* call a veterinarian. In the meantime, the animal should be (1) placed where adequate care and treatment can be given, (2) protected from excessive heat and cold, and (3) allowed to eat only feeds known to be safe.

The veterinarian may determine the kind of poisonous plant involved (1) by observing the symptoms, and/or (2) by finding out exactly what poisonous plant was eaten through looking over the pasture and/or hay and identifying leaves or other plant parts found in the animal's digestive tract at the time of autopsy.

It is to be emphasized, however, that many poisoned animals that would have recovered had they been left undisturbed, have been killed by attempts to administer home remedies by well-meaning but untrained persons.

Common Poisonous Plants

The list of poisonous plants is very extensive. Nevertheless, both the stockman and the veterinarian should have a working knowledge of the principal poisonous species in the area in which they operate. Table 10-11, page 962, lists, alphabetically by common name, the most prevalent poisonous native plants affecting U.S. livestock and gives a pertinent summary relative to each of them. There are many duplications of names. Moreover, local names are not always in agreement with the more common names.

Surprising as it may seem, plants do not readily fall into poisonous and nonpoisonous groups. Some are poisonous only at certain seasons of the year and under specific conditions.

TABLE 10-11—SOME IMPORTANT

Fig. 10-88

Fig. 10-89

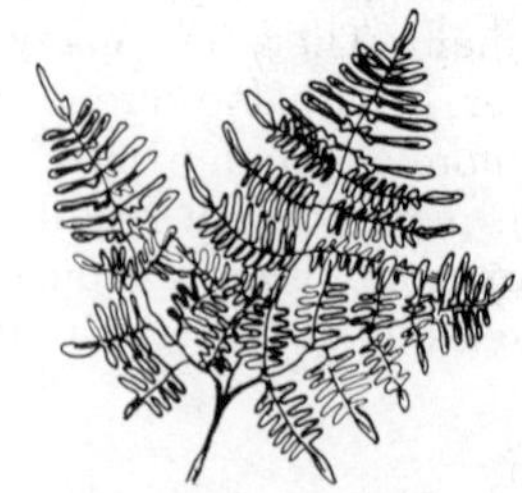

Fig. 10-90

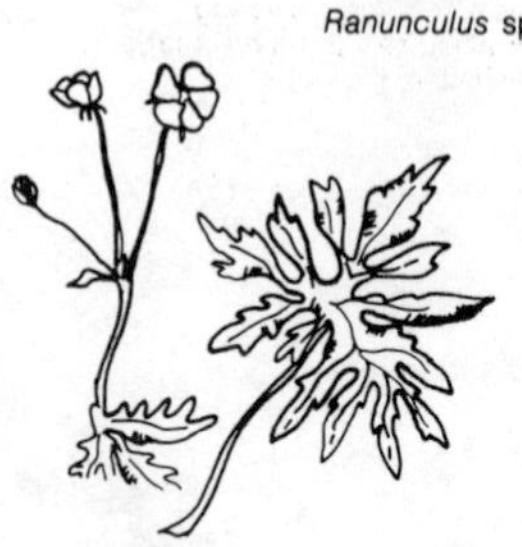

Fig. 10-91

Common and Scientific Name of Plant	Description of Plant	Where It Grows	Species Affected	Parts of Plant That Usually Cause Poisoning
Arrowgrass (goosegrass, sourgrass) *Triglochin maritima; Triglochin palustris*	A perennial which resembles grass except the leaves are thicker and circular. Six to 12 in. tall, with stems half rounded.	Alkaline marshes and wet areas throughout the U.S.	Cattle. Sheep.	Leaves, either green or dry (hay).
Bitterweed (see Rubberweed) *Hymenoxys odorata*	Annual; bright yellow flowers; pungent odor and bitter taste.	On misused land.	Sheep, although other animals may be poisoned.	
Bitterweed (see Sneezeweed)				
Bracken fern (brake fern, eagle fern) *Pteridium aquilinum*	Perennial ferns with black root stocks; wide, broadly triangular 3-part fronds 2 to 4 ft long on 3-ft stems.	Thickets and wooded hills throughout the U.S.	Cattle. Horses. Sheep, to a lesser extent.	The fronds; either green or dry (hay).
Buttercup *Ranunculus* sp	A perennial, 16 to 32 in. high, with yellow flowers.	Widely distributed in wet pastures and meadows.	All animals are susceptible, but cattle are most often poisoned.	Green parts of plant (dried plants in hay are harmless).

Footnote on last page of table.

POISONOUS PLANTS OF THE U.S.[1]

Conditions Under Which Poisoning Usually Occurs	Toxic Symptoms	Treatment	Prevention	Remarks
Especially in dry seasons and after first fall frost.	Nervousness, abnormal breathing, trembling or jerking of muscles, blue coloration of lining of mouth, spasms or convulsions, respiratory failure.	**Sheep:** Intraperitoneal injection of 1 g of sodium nitrite and 2 g of sodium thiosulfate in a 20% solution of water. **Cattle:** Double the dose given above for sheep.	Keep animals off this forage after drought or frost. Eradication of plant not practical.	Prussic acid in the leaves is the toxic principle.
Overgrazed range and dry years; winter to summer.	Depression, vomiting, salivation, and general weakness. There may be a green salivary discharge.	Remove from infected area and give good quality feed and water. No medical treatment is known.	Keep away from plant. Do not overgraze. Grub out plants. Treat with herbicide.	
Grazing fern-infested areas in dry periods of late summer when other succulent feed is scarce; or when hay contains dry ferns.	**Ruminants**: High fever, loss of appetite, depression, difficult breathing, excess salivation, nasal and rectal bleeding, bloody urine, hemorrhaged mucous membranes. **Horses**: Loss of weight and condition, progressive lack of coordination, marked depression, crouched stance, arched back, twitching muscles, general body weakness, weak but fast pulse, unable to stand, convulsions, or spasms.	**Ruminants**: Few recover. **Horses**: Intravenous injection of thiamin hydrochloride.	Provide alternate forage or suitable feed supplement to reduce intake. Eradication of plant may be practical.	This is a cumulative poison. Swine consume root stocks without injury. Thoroughly cooked young shoots are sometimes used for human consumption.
Animals allowed to graze buttercup-infested pastures consume the plant along with other succulent forage.	Poisoned cows may give milk which is bitter or reddish in color. Severe poisoning causes diarrhea, twitching of the ears and lips, difficult breathing, and eventually convulsions.	Call a veterinarian promptly. Tannin may be helpful, but treatment is largely symptomatic.	Do not permit animals to graze buttercup-infested pastures when other forage is scant and dry. Mow buttercups each year before seeding thus lessening their spreading.	Buttercups are most poisonous just before flowering. Milk acquires a flavor that persists in the butter.

(Continued)

TABLE 10-11

	Common and Scientific Name of Plant	Description of Plant	Where It Grows	Species Affected	Parts of Plant That Usually Cause Poisoning
 Fig. 10-92	**Chokecherry** *Prunus virginiana*	A shrub or small tree with smooth dark bark; green oval leaves with small-toothed edges; white flowers in umbrella clusters; and red fruit.	Common along fence rows, in thickets and in abandoned fields over a wide range of soil conditions in the U.S.	Sheep. Cattle, sometimes.	Wilted leaves.
 Fig. 10-93	**Cocklebur** *Xanthium* sp	The cocklebur has large, grapevine-shaped leaves which are stiff and hairy, hairy stems, an olive-shaped bur with spines. It is generally less than 3 ft tall.	In fields and waste areas of eastern U.S., and in low wet places of western U.S.	Swine are most often poisoned, but cattle and sheep are affected.	Seedlings at the 2-leaf stage.
 Fig. 10-94	**Copperweed** *Oxytenia acerosa*	Perennial, 3 to 5 ft tall; shrublike, but stems die back to root crown annually; leaves divided into 3 to 5 long, narrow segments; flowers appear in late summer, and are in small numerous heads and orange-yellow color.	Dry alkali flats southern Colo., Utah, and N.M. to southern Calif.	Cattle. Sheep.	
 Fig. 10-95	**Death camas** (camas, poison sage, swampgrass, alkali-grass, poison onion) *Zigadenus paniculatus*	Perennial herbs resembling an onion except the bulb is odorless and the leaves are flat. Produces white flowers in early spring; 4 to 18 in. tall.	Drouthy, coarse-textured soils on hills, and on locally wet meadows.	Sheep. Sometimes cattle and horses.	All parts of the plant, either green or dried.

—(Continued)

Conditions Under Which Poisoning Usually Occurs	Toxic Symptoms	Treatment	Prevention	Remarks
Consumption of wilted leaves during drought, or branches that have been cut or broken.	Nervousness, abnormal breathing, trembling or jerking muscles, blue color of lining of mouth, spasms or convulsions, respiratory failure.	Same as treatment for arrowgrass. Death is rapid, so treatment may not be effective.	Keep animals away from chokecherry when they are hungry or thirsty.	Prussic acid is the toxic principle.
Consumption of seedlings in spring or early summer.	Prostration, rapid and weak pulse, low temperature, and vomiting. Death, if it occurs, within 24 hrs. Inflamed stomach.	Emergency treatment consists in giving the animal—by mouth or through a stomach tube—such fatty substances as raw linseed oil, mineral oil, lard, cream, or whole milk.	Keep animals, especially swine, out of cocklebur-infested pastures during late spring and early summer when cocklebur seeds are sprouting. Lessen cocklebur infestation by mowing or chopping plants before seeding.	The cocklebur seedling—at which stage the poisoning hazard is greatest—is very different from the mature plant. The young seedlings are attached underground to the easily recognized burs from which they sprout.
On fall trails, where other feed is scarce.	Loss of appetite, depression, weakness, coma.	None known.	Provide alternate forage. May be practical to eradicate plants.	
Spring, especially very early spring.	Rapid breathing, excessive salivation, nausea, weakness and staggering, convulsions, coma, death.	None known.	Spray plants with 2,4-D in early stages of growth.	Swine are thought to be immune from death camas poisoning.

(Continued)

TABLE 10-11

Common and Scientific Name of Plant	Description of Plant	Where It Grows	Species Affected	Parts of Plant That Usually Cause Poisoning
 Fig. 10-96 **Greasewood** (chico) *Sarcobatus vermiculatus*	An erect but much branched, thorny shrub, 2 to 5 ft tall, high in water content, with fleshy, bright green leaves and gray bark on the older stems.	Drouthy, saline and saline-alkali soils of western U.S.	Cattle. Sheep.	Leaves.
 Fig. 10-97 **Halogeton** *Halogeton glomeratus*	An annual characterized by a little hooked spine on tip of leaf, high water content in the spring, and seed covered with white scalelike wings. Resembles Russian thistle.	Along roadsides, and on overgrazed ranges in desert areas of the West.	Sheep. Cattle, sometimes.	Green or dry tops.
 Fig. 10-98 **Henbane** *Hyoscyamus niger*	Annual or biennial herb; stems much branched and clammy—hairy; disagreeable odor.	Drouthy, coarse-textured soils in western U.S. and Canada.	Cattle. Sheep. Horses.	All parts of the plant, especially the seeds.
 Fig. 10-99 **Horsebrush** (spring rabbit brush; coal-oil brush) *Tetradymia* sp	Shrub 2 to 4 ft high with yellow flowers in spring. Spiny, silvery-white leaves.	Intermountain region, mostly dry semideserts.	Sheep.	Leaves and twigs, especially before flowering.

—(Continued)

Conditions Under Which Poisoning Usually Occurs	Toxic Symptoms	Treatment	Prevention	Remarks
Most cases of poisoning occur in early spring or when very hungry animals are allowed to feed exclusively on greasewood.	**Early signs:** Dullness, loss of appetite, lowering of head, reluctance to follow herd, irregular gait. **Advanced signs:** Drooling and frothing (white) at mouth, nasal discharge, progressive weakening, rapid and shallow breathing, coma, death.	None effective.	Provide supplemental feed.	Greasewood is regarded as a valuable forage plant for winter and early spring grazing by sheep, but it should never form their exclusive diet. Salts of oxalic acid are the poisons, as they are in halogeton.
In late fall or winter, when sheep first get on winter range and prior to leaching out poisons by snow or rain.	**Early signs:** Dullness, loss of appetite, lowering of head, reluctance to follow herd, irregular gait. **Advanced signs:** Drooling and frothing (white) at mouth, nasal discharge, progressive weakening, rapid and shallow breathing, coma, death.	None effective.	Provide supplemental feed. Good management.	Halogeton has spread over many of the ranges of the West. It has come into and made large areas of intermountain ranges almost useless for grazing. Salts of oxalic acid are the poisons, as they are in greasewood.
Where other vegetation is scarce and animals are hungry.	Thirst, convulsions, vomiting, loss of voluntary motion, difficult breathing, irregular pulse, stupor, and death.		Avoid pasturing hungry animals in henbane-infested areas in which other vegetation is scarce.	
Spring trail on depleted range, although not always.	Depression; weakness; swelling of head, neck, ears, eyelids, and nose; heavy, drooping ears that hang straight down; loss of milk in nursing ewes; peeling of skin and wool from head, ears, and back.	Move animals to shade and hand feed hay and water for a few days.	Avoid grazing plants. Eradication of plants not practical.	Both plants (little-leaf horsebrush and spineless horsebrush) belong to the sunflower family.

(Continued)

TABLE 10-11

Fig. 10-100

Fig. 10-101

Fig. 10-102

Fig. 10-103

Common and Scientific Name of Plant	Description of Plant	Where It Grows	Species Affected	Parts of Plant That Usually Cause Poisoning
Indian hemp (hemp dogbane) *Apocynum cannabinum*	Branching perennial, 1½ to 5 ft high. Stems contain a milky juice. Leaves opposite, oblong in shape, with smooth margins. Flowers greenish-white, borne in clusters at the ends of the stems and branches. Fruit a long, slender pod containing many seeds that bear tufts of "floss."	Widely distributed throughout the U.S. and Canada, in open places, often in coarse soil and along streams.	Cattle. Sheep. Horses.	All parts of the plant are poisonous, either fresh or dried in hay.
Jimmyweed (rayless goldenrod) *Haplopappus heterophyllus*	A low-growing half-shrub with erect stems arising from the woody crown to a height of 2 to 4 ft. The leaves are resinous, narrow, and alternate, and may be toothed along the margins. Stems bear flat-topped clusters of yellow flowers.	Colo., Tex., N.M., and Ariz., particularly in moist areas.	All species.	Leaves, either green or dry; hence, jimmyweed is poisonous at all times.
Larkspur—of which there are a number of species. *Delphinium* sp	Most perennial; deeply indented leaves in 3-5-7's. In the spring spurred, usually blue flowers are clustered along tops of stems. Resemble cultivated larkspur or delphinium.	Plains and in moist areas under aspen groves and along streams on mountain ranges of western U.S.	Cattle. Rarely, sheep and horses.	All parts of the green plant.
Laurels *Kalmia* sp	Long, lace-shaped, leathery, evergreen leaves; pink and white flowers bunched in clusters among leaves; 2 to 10 ft high.	Moist woods, swamps, mountains throughout U.S.	Sheep especially, but all animals.	Leaves.

—(Continued)

Conditions Under Which Poisoning Usually Occurs	Toxic Symptoms	Treatment	Prevention	Remarks
Animals find Indian hemp distasteful; hence, only very hungry animals will eat it.	Increased body temperature; hard rapid pulse; vomiting; dilation of pupils; blue coloration of lining of mouth and nostrils; progressive weakening with marked irregular gait; coma; convulsions; marked, labored breathing.	Mineral oil drench and lots of water. Heart stimulants.	Avoid grazing. Eradicate plants with 2,4-D.	
Inadequate feed. Generally on ranges in poor condition.	The disease is known as "trembles," because affected animals tremble, which is especially noticeable about the nose, hips, and shoulders. Animals become weak and eventually die.	The intravenous administration of glucose and calcium gluconate is helpful.	Keep animals from infested areas. Grub out jimmyweed or treat with herbicide.	Human beings may be afflicted through consuming milk or butter from poisoned cows. Jimmyweed belongs to the sunflower family.
Consumption in early spring, especially if other forage is scarce.	Nervousness; staggering and falling; nausea; excessive salivation; frequent swallowing; twitching muscles; bloating; rapid, irregular heart action; respiratory paralysis.	No treatment for poisoning. Place head of animal uphill to avoid bloat.	Graze larkspur-infested ranges with sheep. Kill plants with 2,4,5-T, or Silvex.	If the animal is badly bloated, the use of a stomach tube or a trocar may be necessary.
Overgrazed pastures, especially in winter and spring.	Frothing at mouth, nasal secretions, paralysis, death.	Laxatives, demulcents, and nerve stimulants may help. But no treatment is very successful.	Avoid laurel patches.	Of the laurels, *Leucothoe davisiae* is the most important. Meat is poisoned.

(Continued)

TABLE 10-11

Common and Scientific Name of Plant	Description of Plant	Where It Grows	Species Affected	Parts of Plant That Usually Cause Poisoning
Locoweed (loco)—of which there are about 100 species, some of which are not poisonous. *Oxytropis* sp 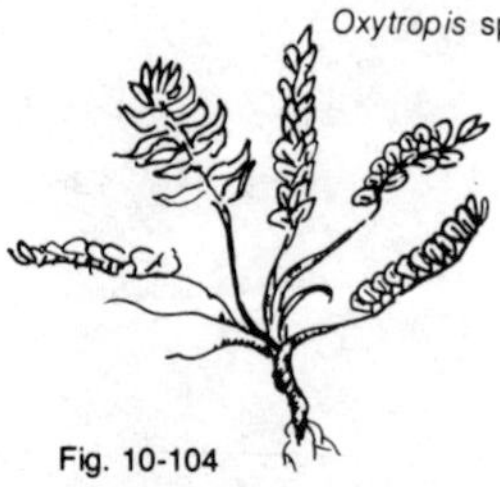 Fig. 10-104	Perennial herbs with erect or spreading stems. Flowers and stems resemble garden pea, but much smaller.	Plains and some mountain valleys, western half of U.S.	Cattle. Sheep. Horses.	Both green and dry plants are poisonous.
Lupines (wild bean, blue pea, sundial, Quaker's bonnets, Indian beans)—not all lupines are poisonous. *Lupinus* sp  Fig. 10-105	Annual and perennial herbs, and a few shrubs. Leaves divided into 2 to 9 parts, attached to a single point like spokes of a wheel. Spikes of blue sweet pea-shaped flowers in early summer.	Throughout U.S., under a wide range of soil and moisture conditions.	Cattle. Sheep. Sometimes other species.	Especially the seeds, but the leaves may also cause poisoning. Both green and dry plants are dangerous.
Milkweed—of which there are several species. *Asclepias* sp 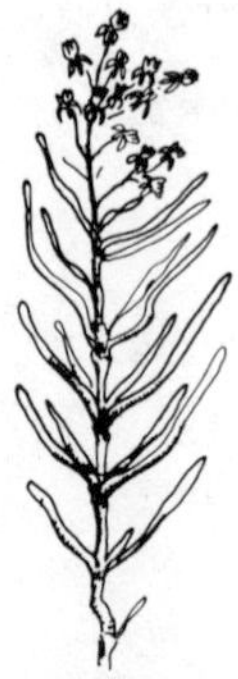 Fig. 10-106	Produces pods with many silk-covered seeds. Plants secrete milk from stem and leaf when broken.	Widely distributed throughout the U.S.	Cattle. Sheep. Horses.	Green plants or dry hay.
Nightshade (silverleaf nightshade, cutleaf nightshade) *Solanum elaeagnifolium* *Solanum triflorum* 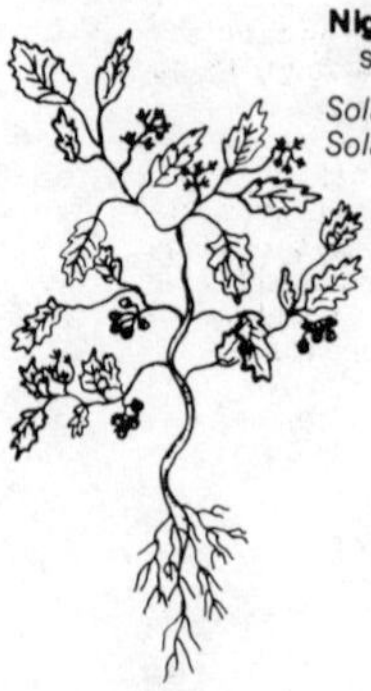 Fig. 10-107	Annual; small (½ to 2 ft tall), bushy plant with small star-shaped white flowers borne in clusters and green or dark blue-black berries.	Common on coarse-textured soils in disturbed areas and in open woods throughout the U.S. In pea fields in the Northwest.	Cattle. Sheep. Goats. Swine. Horses.	Unripe fruit and leaves.

—(Continued)

Conditions Under Which Poisoning Usually Occurs	Toxic Symptoms	Treatment	Prevention	Remarks
Grazing on ranges when more palatable forage is not available or feeding hay containing locoweed. Locoweeds frequently begin growth earlier than grasses, so they are especially dangerous in the spring.	Loss of flesh; irregular gait; loss of sense of direction; nervousness; weakness; withdrawal from herd or flock; loss of muscular control; violent actions when disturbed.	None known.	Keep good forage available when on infested range. Kill plants with 2,4-D.	"Loco" is a word of Spanish extraction, meaning crazy.
Late summer and fall.	Rough, dry hair coat; nervousness; depression; reluctance to move; difficulty in breathing; twitching leg muscles; loss of muscular control; frothing at mouth; convulsions; coma; death.	None known.	Supplemental feeding when on infested ranges. Keep animals from dense stands. Avoid eating, especially in seed stage.	
If grass is scant or dry, hungry animals may eat the bitter tasting milkweed.	Loss of muscular control; staggering and falling; violent spasms; bloating; rapid, weak pulse; difficulty in breathing; respiratory paralysis.	None known.	Supplemental feeding when on infested ranges. Avoid milkweed in hay. Kill plants with 2,4,5-T.	
Whenever animals feed in areas where nightshade is abundant.	Two types of effects may be observed: (1) nervous effects—apathy, drowsiness, salivation, labored breathing, trembling, progressive weakness or paralysis, prostration, and unconsciousness; or (2) gastrointestinal irritation—nausea, abdominal pain, vomiting, diarrhea, and sometimes blood.	None known.	Grub out the plants and remove them from the area. Kill plants with 2,4-D or 2,4,5-T.	Mature cattle and horses rarely eat enough to be seriously affected. Where too great concentrations are not involved, nightshade-containing forage may be used as silage without hazard provided the nightshade is cut at an early stage of maturity.

(Continued)

TABLE 10-11

Common and Scientific Name of Plant	Description of Plant	Where It Grows	Species Affected	Parts of Plant That Usually Cause Poisoning
Oak (shinnery oak; scrub oak) *Quercus* sp 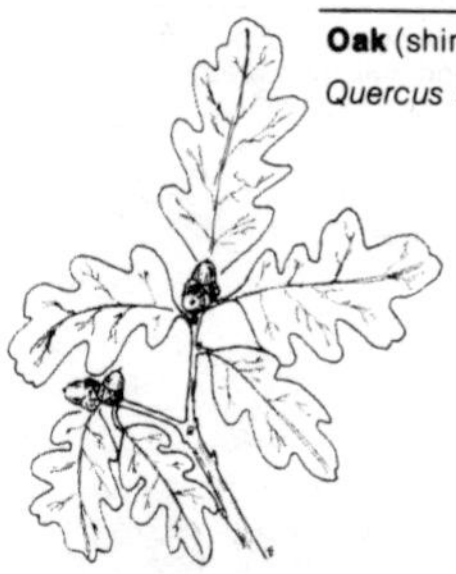 Fig. 10-108	Shrubs and trees best recognized by (1) their fruit, the acorn, and (2) their leathery leaves with wavy margins.	In thickets throughout the central, southern, and western states.	Cattle.	Leaves and leaf buds. Especially dangerous if early bud growth is blackened by a late frost in spring.
Oleander *Nerium oleander* sp  Fig. 10-109	Ornamental evergreen shrub or small tree, 5 to 25 ft tall. Showy (red, pink, or white) flowers; shiny leathery leaves, and long slender twin pods.	In warm areas.	Most animal species and man.	All parts, especially leaves, fresh or dry.
Pine needles (ponderosa pine; western yellow pine; jack pine) *Pinus ponderosa* 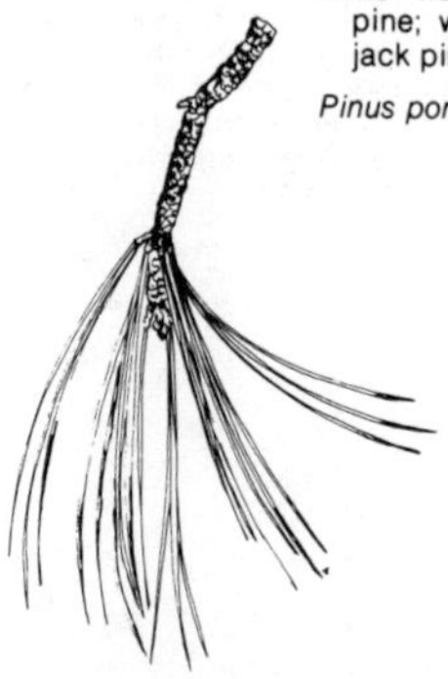 Fig. 10-110	A long-lived tree growing 50 to 150 ft tall. Needles are usually in groups of 3. Seeds are in prickly cones.	At moderate elevations east of the Cascade Mountains, from California to British Columbia. Also, in S.D., Wyo., Mont., Ariz., and N.M.	Cattle.	Green pine needles or slash.
Poison hemlock (poison parsley) *Conium maculatum* 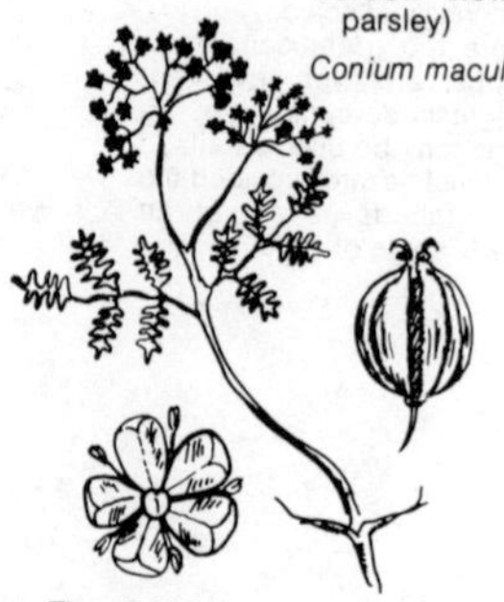 Fig. 10-111	Biennial; purple spotted, parsleylike leaves; parsniplike roots; showy, umbrella-clustered flowers; disagreeable odor; 3 to 6 ft tall.	Dry waste areas and about farm buildings, especially in eastern U.S. and on the Pacific Coast.	Cattle. Sheep. Horses. Other.	Roots, tops, and seeds.

—(Continued)

Conditions Under Which Poisoning Usually Occurs	Toxic Symptoms	Treatment	Prevention	Remarks
Animals turned into woodlots in early spring before grass becomes abundant.	Gaunt, tucked-up appearance; constipation, frequently followed by profuse diarrhea; weakness; tendency to remain near water; reluctance to follow the herd; emaciation; mucus in droppings; dark-colored urine; collapse.	Remove from herd. Drench with mineral oil. Provide feed and water.	Conservative grazing. Supplemental feeding. Kill plants with 2,4,5-T.	Among animals exposed to oak poisoning, only a few are actually poisoned.
Usually consumed by animals as clippings carelessly discarded.	Severe gastroenteritis followed by cardiac response, increased pulse rate, cold extremities, dilation of pupils, discoloration of mouth, sweating, abdominal pain, nausea, vomiting, weakness, bloody feces.		Avoid grazing in pastures where oleander is close. Oleander is used as an ornamental plant more than as a range plant.	
Cattle will eat the pine needles anytime. But the danger period from the standpoint of abortion is greatest the latter part of gestation.	Pregnant cows, free of brucellosis, abort, especially during the last 3 months of pregnancy; excessive hemorrhage; retained placenta; septic metritis, often followed by peritonitis. Calf may be born normal, but weak, if cow affected near end of pregnancy.	None known.	Keep pregnant cows away from yellow pines if possible. Cows will eat needles even when well fed.	It is suspected that the high turpentine content of yellow pine needles actually causes the abortion, for there is evidence that turpentine can cause abortion in the human female.
Generally in the spring when the herbage is green.	Nervous trembling; salivation; lack of coordination; bloating; dilation of pupils; rapid, weak pulse; blue color of mouth lining; respiratory paralysis; coma.	Large doses of mineral oil.	Destroy plants with Silvex before they bud.	Poison hemlock is the same poisonous plant that history records as having been administered by the Greeks to Socrates and other state prisoners.

(Continued)

TABLE 10-11

Common and Scientific Name of Plant	Description of Plant	Where It Grows	Species Affected	Parts of Plant That Usually Cause Poisoning
Rayless goldenrod (see Jimmyweed)				
Rubberweed *Hymenoxys* sp  Fig. 10-112	Perennial, with small yellow asterlike flowers and aromatic leaves.	Dry mountains, Colo., Utah, Ariz., and N.M.	All; but especially sheep.	
Snakeweed (broom snakeweed; broomweed) *Gutierrezia* sp  Fig. 10-113	Low, perennial half-shrubs growing 1 to 2 ft tall. Many branched and quite resinous. Leaves are linear, entire, and alternately arranged. Yellow flowers in small composite heads.	Dry range and desert of the West, from Mexico to Idaho.	Cattle. Sheep. Goats.	Leaves, during early development.
Sneezeweed (staggerwort, swamp sunflower) *Helenium* sp  Fig. 10-114	A perennial with large yellow to orange heads that resemble the sunflower except that they have yellow centers.	Widespread in wet meadows and pastures, and along ditches and streams in eastern U.S. and Canada, and on moist slopes and meadows in western U.S.	Sheep. Sometimes, cattle, horses, mules.	Flowers are most dangerous, but all parts of the plant are somewhat poisonous.
Spring parsley (wild carrot) *Cymopterus watsonii*  Fig. 10-115			Cattle. Sheep.	

—(Continued)

Conditions Under Which Poisoning Usually Occurs	Toxic Symptoms	Treatment	Prevention	Remarks
Spring and fall, in overgrazed areas.	Loss of appetite, cessation of rumination, frothing, green nasal discharge.			It contains a certain amount of rubber.
Occurs most frequently on overgrazed ranges.	Listlessness, off feed, rough hair coat, diarrhea or constipation, mucus in feces, blood in urine, nasal discharge, crusting and peeling of muzzle. Abortion, retained placenta, vulvar swelling, death.	Remove from feed at first sign of toxicity.	Avoid close grazing of infested range. Supplemental feeding while on infested range.	
Eating flowers in late summer and early fall.	Depression, weakness, lack of coordination, irregular pulse, coughing, shortness of breath, chronic vomiting and spewing, frothing at mouth, bloating, convulsions.	Remove from infested range. Mature plants more toxic than young plants.	Provide mineral supplement. Control plants with 2,4-D.	
	Mature animals: Sunburn and blistering of exposed areas (nostrils, muzzle, udder, teats, genital organs); painful to nurse young, which may starve unless otherwise fed. **Lambs and calves:** Frantic attempts to nurse; loss of weight.	Remove to shade. Animals recover when they stop eating plants.	Provide other feeds on infested range.	

(Continued)

TABLE 10-11

Common and Scientific Name of Plant	Description of Plant	Where It Grows	Species Affected	Parts of Plant That Usually Cause Poisoning
St. Johnswort (goatweed, Klamath weed) *Hypericum perforatum*	An erect, much-branched perennial herb, 1 to 2½ ft high, with slender stems and elliptic to oblong, small, oppositely set leaves, and deep-yellow, black-dotted flowers.	Fields, waste places, and hills across northern half of U.S.	All animals with white skin areas.	Leaves.
Tansy ragwort (stinking Willie, Baughlan) *Senecio* sp	A biennial or perennial, strongly scented herb with simple stems, 8 in. to 3 ft high. The whole plant has a strong, unpleasant odor when crushed.	Roadsides, waste places, and sometimes pastures; especially in eastern U.S. and in the Pacific Northwest.	Cattle. Horses.	Green or dried tops.
Timber milk vetch *Astragalus miser*	It flowers in June and July. Flowers resemble sweet peas and vary in color from creamy white to shades of violet. Produces seeds with slender pods.	Elevations of 6,000 to 11,000 ft in the Rocky Mountain states. Also in the Black Hills of South Dakota.	Cattle. Sometimes sheep, goats, and horses.	Any part of green plants, from emergence until they dry up in late summer or are killed by frost.

Fig. 10-116

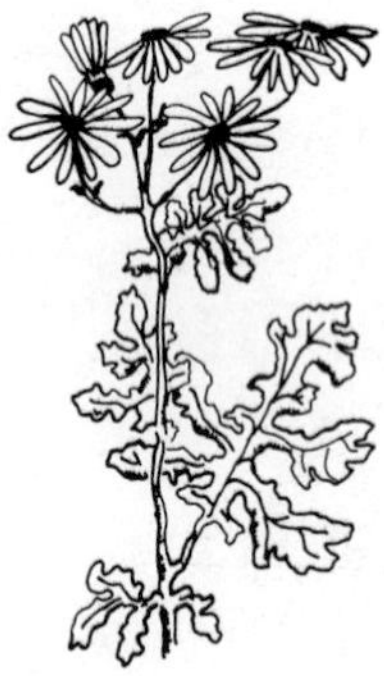

Fig. 10-117

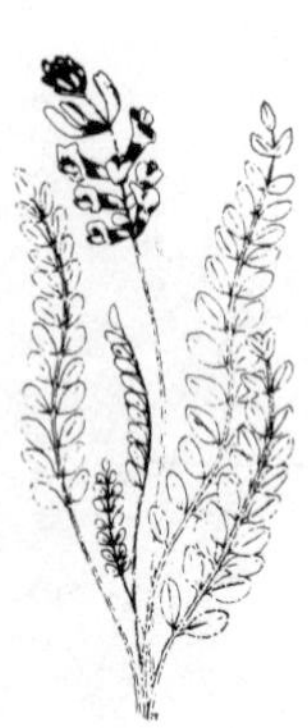

Fig. 10-118

—(Continued)

Conditions Under Which Poisoning Usually Occurs	Toxic Symptoms	Treatment	Prevention	Remarks
Feeding on considerable of the plant and being in bright sunlight.	Restlessness, scratching head, crouching diarrhea, rapid pulse, increased temperature, sunburn of white skin areas, blisters, matted wool or hair, swollen eyelids, clouded eyes, convulsions.	Move to shade. Treat with healing oil. Provide feed and water.	Avoid grazing infested areas. Kill plants with herbicides and by biological means.	Except when other forages are scarce, animals will not eat enough to cause pronounced symptoms. The symptoms produced are due to light sensitization. Lecheguilla and beargrass in the Southwest, horsebrush in the West, and buckwheat in the East and Southeast produce similar symptoms.
Only where feed is scarce, for tansy is bitter and unpalatable.	Weakness, a nervous, staggering gait, and postmortem shows liver lesions.		Avoid grazing hungry animals in tansy-infested areas where other forage is sparse. Destroy the plants before they bear seed.	
Even when other forage is available, cattle readily eat milk vetch. Sheep graze it sparingly unless they have no other forage.	Nervousness; frequent urination; irregular gait; muscular weakness; inability to stand; rapid, weak pulse; white coloration of lining of mouth; coma; paralysis of respiratory tract. If eaten in early pregnancy, it causes deformities in the offspring.	None known.	Control plants by treating before bloom with Silvex or 2,4,5-T.	

(Continued)

TABLE 10-11

Fig. 10-119

Fig. 10-120

Fig. 10-121

Common and Scientific Name of Plant	Description of Plant	Where It Grows	Species Affected	Parts of Plant That Usually Cause Poisoning
Veratrum (false hellebore, skunk cabbage) *Veratrum californicum*	The plant reaches a height of 6 to 8 ft. It is a robust perennial of the lily family. Leaves may measure 9 to 12 in. long and 3 to 6 in. wide. It has cream-colored flowers, which grow in clusters. Seedpods turn black as they ripen.	Western U.S., on moist, open meadows and hillsides at elevations of 5,000 to 11,000 ft.	Cattle. Sheep. Goats.	Green leaves and plant tops, from the time it starts to grow until after it is killed by frost.
Water hemlock—of which there are several species. *Cicuta* sp	A perennial, with parsley-like leaves; hollow, jointed stems and hollow pithy rootstock and roots. The flowers are borne in umbrella clusters, and the stems are streaked with purple ridges. Oil has pungent odor. Two to 6 ft tall.	Wet meadows, pastures and floodplains of western and eastern U.S.; generally absent in the plains states.	All animals and man.	Tubers only. (The leaves and fruit, either green or dried in hay, can be eaten without danger.)
White snakeroot *Eupatorium rugosum*	Perennial, 1 to 5 ft tall, with erect stems. Leaves opposite, oval with pointed tips and sharply toothed edges; the blade strongly 3-ribbed, the upper surface dull and the lower shiny. Flowers showy, snow-white in open terminal clusters, blooming in late summer.	From eastern Canada to Saskatchewan, and south to Tex., La., Ga., and Va.	Cattle. Horses.	All parts of plant, fresh or dry.

[1]The illustrations for this table were prepared by R. F. Johnson and Toby Escola.

CHEMICAL POISONING

In the everyday pursuits of modern agriculture, more and more chemicals that may, under certain conditions, be poisonous to animals are being used. Thus, great care should be exercised in handling these products; the labels on the containers should be read and heeded carefully, and partly used packages and empty containers should not be left where animals have access to them.

When chemical poisoning happens, it can be both devastating and perplexing. Usually, the causative agent can be diagnosed after an investigation of the environment and the feed. However, few poisons can

—(Continued)

Conditions Under Which Poisoning Usually Occurs	Toxic Symptoms	Treatment	Prevention	Remarks
Sheep and goats readily eat it. Cattle will eat it if other forage is scarce.	Excessive salivation; frothing; general weakness; irregular gait; vomiting; paralysis of legs; fast, irregular heartbeat; slow, shallow breathing; coma and convulsions; deformed young at birth.	None effective. Keep in shade. Intravenous injection of sodium pentabarbital.	Keep animals away during breeding season. Control plants with Silvex.	It is sometimes called wild corn and corn cabbage.
Especially in early spring when, in the process of eating the young leaves and stems, the animal may pull the tubers out of the ground and eat them.	Muscle twitching, rapid pulse, rapid breathing, tremors, convulsions, dilation of pupils, excessive salivation, frothing of mouth, coma, rapid death.	None known.	Keep animals away from water hemlock and provide good quality forage. Control plants with 2,4-D or 2,4,5-T.	
Snakeroot is moderately distasteful to animals. It may be eaten along with hay. On pasture or range, it is not normally eaten unless other vegetation is scarce.	**Cattle:** Listlessness, constipation, violent trembling, breath with acetone odor. **Horses:** Sluggishness, marked depression, incoordination—especially hind parts, and inability to swallow.	Purgatives will help eliminate the drug. Stimulants and calcium gluconate are helpful.	Keep animals away from infested areas.	

be diagnosed with certainty by clinical symptoms alone; yet a chemical analysis of stomach contents is costly, impracticable, and inadvisable unless there is sufficient evidence to justify a laboratory determination for a particular poison.

When trouble is encountered, the owner or caretaker should promptly call a veterinarian. In the meantime, the animal(s) should be (1) placed where adequate care and treatment can be given, (2) protected from excessive heat and cold, and (3) allowed to eat only feeds known to be safe.

Table 10-12, page 980, lists the most common chemical poisons and presents pertinent facts pertaining to each.

TABLE 10-12—SOME POTENTIALLY

Poison	Source	Species Affected	Symptoms and Signs
Arsenic (As)	Arsenic used to control insects and weeds, and to defoliate crops. Accumulation of arsenic in soils may sharply decrease crop growth and yields, but it is not a hazard to animals or humans that eat plants grown in these fields, provided they do not eat the foliage of recently sprayed plants.	All farm animals.	The onset is sudden; characterized by groaning, restlessness, and rapid breathing. Death in 3 to 4 hours, or, if less material is consumed, in a few weeks. Autopsy reveals severe hemorrhagic inflammation of stomach and intestines, with perhaps areas of erosion on mucous membranes.
Copper sulfate ($CuSo_4 \cdot 5H_20$)	Overdosing with copper sulfate (bluestone), or the use of too concentrated solution, in treating for parasites.	All farm animals.	Abdominal pains, vomiting, and diarrhea. Autopsy shows stomach and intestines intensely inflamed and stomach lining coated with copper sulfate.
Ergot (a parasitic fungus)  Fig. 10-122	It replaces the seed in the heads of grasses and cereal grains, in which it appears as a purplish-black, hard banana-shaped dense mass from ¼ to ¾ inch long. Most common in rye, wild rye, bromegrass, and dallisgrass.	Cattle. Sheep. Horses. Man.	Acute ergot poisoning, caused by large quantities eaten at one time, may produce paralysis of the limbs and tongue, disturbance of the gastrointestinal tract, and abortion. Chronic poisoning produces gangrene of the extremities, with subsequent sloughing off of hooves, ears, and tail. Delirium, spasms, and paralysis may occur before death.
Fluorine (F) (fluorosis) 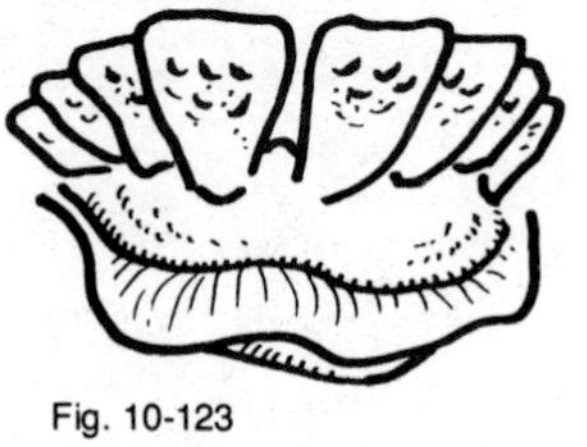 Fig. 10-123	Ingesting excessive quantities of fluorine through either the feed or water.	All farm animals, poultry, and man.	Abnormal teeth (especially mottled enamel) and bones, stiffness of joints, loss of appetite, emaciation, reduction in milk flow, diarrhea, and salt hunger.
Lead (Pb)	Lead is discharged into the air from auto exhaust fumes and other sources. Lead pollution of feed and food crops as a result of lead being deposited on the leaves and other edible portions of the plant by direct fallout. Inhaling airborne lead. Lead may get into feed or food and water from contact with lead pipes, utensils, or discharged storage batteries.	All farm animals; but cattle and sheep are especially susceptible.	Symptoms develop rapidly in young animals, but slowly in mature animals. Feces may become very dark gray and be tinged with blood. Salivation, champing of the jaws, frenzy, blindness, convulsions, coma, and death. Mature animals usually have diarrhea and show incoordination, especially in the hind limbs, and prostration.
Mercury (Hg)	Mercury is discharged into air and water from industrial operations and is used in herbicide and fungicide treatments. Consumption of seed grains treated with fungicides that contain mercury, for the control of fungus diseases of oats, wheat, barley, and flax. Mercury poisoning has occurred where mercury from industrial plants has been discharged into water and then accumulated in fish and shellfish.	All farm animals, but especially cattle and hogs.	Gastrointestinal, renal and nervous disturbances; but impossible, on basis of symptoms, to differentiate mercury from other poisons. Case history of animals consuming mercury-treated grains should be considered strong circumstantial evidence.

POISONOUS ELEMENTS

Distribution and Losses Caused By	Prevention	Treatment	Remarks
Arsenic has long been a leading cause of chemical poisoning.	Keep animals away from arsenic.	Handled by the veterinarian. If caught in time, first remove the material from the animal. Sodium thiosulfate may be used, and supportive treatment may be indicated.	
	Never give copper sulfate in a concentration greater than 2% or in a dose of more than 4 fluid ounces of a 1% solution.		
Ergot is found throughout the world.	Never feed heavily ergot-infested hay or grain.	If noticed in time, stricken animals may recover if put on good feed. Tannin used as a drench is an antidote, and sedations, such as chloral hydrate, may be given to nervous animals.	
The water in parts of Ark., Calif., S.C., and Tex. has been reported to contain excess fluorine. Occasionally, throughout the U.S. high-fluorine phosphates are used in mineral mixtures.	Avoid the use of feeds, water, or mineral supplements containing excessive fluorine.	Any damage may be permanent, but animals which have not developed severe symptoms may be helped to some extent, if the source of excess fluorine is eliminated.	Fluorine is a cumulative poison. 100 ppm (0.01%) fluorine of the total dry ration is the borderline in toxicity for cattle, sheep, and pigs. At levels of 25 to 100 ppm, some mottling of the teeth may occur over periods of 3 to 5 years. In breeding animals, whose usefulness exceeds 3 to 5 years, the permissible level is 50 ppm of the total dry ration. Not more than 65 to 100 ppm fluorine should be present in dry matter of rations when rock phosphate is fed.
Rather extensive, because of the wide use of lead preparations in agriculture.	Avoid sources of lead.	If damage to tissue has been extensive, treatment is of little value; in any event it should be handled by a veterinarian. The best chemical antidote is protein (milk, eggs, blood serum).	Lead poisoning can be diagnosed positively by analyzing the blood tissue for lead content. It is a cumulative poison.
When, through ignorance or negligence, mercury-treated grain is fed to animals.	Do not feed livestock seed grains treated with a mercury-containing fungicide. Surplus of treated grain should be burned and the ash buried deep in the ground.	Treatment is not too satisfactory. The best antidote is protein (milk, egg, blood serum).	Ultimate diagnosis depends upon demonstrating the presence of mercury in the tissues, especially in the kidneys and liver. Food and Drug Administration prohibits use of mercury-treated grain for feed or food. Mercury is a cumulative poison.

(Continued)

TABLE 10-12

Poison	Source	Species Affected	Symptoms and Signs
Mycotoxins (toxin-producing molds; e.g., *Aspergillus flavus*, *Penicillium cyclopium*, *P. islandicum*, and *P. palitans*)	Aflatoxin (most studied of the group) associated with peanuts, brazil nuts, silage, corn and most other cereals, hay, and grasses. The mold can produce toxic compounds on virtually any food (even synthetic) that will support growth. While aflatoxin appears to cause most of the problem, it is not the only mycotoxin to be feared. Other mycotoxins are being studied.	Turkeys. Ducklings. Pheasants. Trout. Cattle. Swine. Man. In all species, the young are far more susceptible than mature animals. Generally, ruminants appear to tolerate higher levels of mycotoxins and longer periods of intake than simple-stomached animals.	*The toxic symptoms from continued (long-term) intake of aflatoxin are:* **Animal** / **Aflatoxin Level[1]** *(ppb)* / **Symptoms** **Beef cattle:** 450 lb / 700 / Liver damage 1,000 / Reduced growth and feed efficiency **Dairy cattle:** Calves (milk fed) / 200 / Fatal Lactating cows / 20 / Drop in milk yield; Aflatoxin secreted in milk **Sheep** / 1,750 / Reduced fertility **Swine:** 50 lb / 280 / Reduced growth and feed efficiency 80 lb / 450 / Liver damage 615 } 810 } / Reduced growth and feed efficiency Breeding herd / 450-1,500 / Abortions, dead pigs at birth
Nitrate-nitrite poisoning (oat hay poisoning, cornstalk poisoning)	Consuming high-nitrate feeds—feeds with a high-nitrate content due to nitrate fertilization, drought, etc. Eating nitrate or nitrite fertilizer, or drinking pond water containing same.	Cattle, sheep, and horses; especially cattle.	Accelerated respiration and pulse rate; diarrhea; frequent urination; loss of appetite; general weakness, trembling, and a staggering gait; frothing from the mouth; lowered milk production; abortion; blue color of the mucous membrane, muzzle, and udder due to lack of oxygen; and death in 4½ to 9 hr after eating lethal doses of nitrate.
Pitch (clay pigeon poisoning)	Expended clay pigeons; roofing material, certain types of tar paper, and plumbers' pitch.	Pitch poisoning is confined almost entirely to hogs.	An acute, highly fatal disease, characterized, clinically, by depression, and pathologically by striking liver lesions.
Protein (protein poisoning is nonexistent)	Some opinions to the contrary, there is no proof that heavy feeding of high-protein feeds is harmful, provided the ration is balanced out in all other respects and the animal's kidneys are normal and healthy.	None.	Some horses appear to be allergic to certain proteins or to excesses of specific amino acids, as a result of which they may develop "protein bumps."
Salt poisoning (NaCl—sodium chloride)	Brine from cured meats; wet salt. Where large amounts of brine or salt have been mixed in hog slop. When excess salt is fed following salt starvation. When salt is improperly used to govern self-feeding of concentrate.	All farm animals, but swine and sheep most frequently affected.	Sudden onset—1 to 2 hours after ingesting salt; extreme nervousness; muscle twitching and fine tremors; much weaving, wobbling, staggering, and circling; blindness; weakness; normal temperature, rapid but weak pulse, and very rapid and shallow breathing; diarrhea; death from a few hours up to 48 hours. Convulsions seldom occur, except in pigs.
Selenium poisoning (Se) (alkali disease)  Fig. 10-124	Consumption of plants grown on soils containing selenium.	All farm animals and man.	Loss of hair from the mane and tail in horses, from the tail in cattle, and a general loss of hair in swine. In severe cases, the hoofs slough off, lameness occurs, food consumption decreases, and death may occur by starvation.

[1]FDA regulations do not permit grain or feed containing more than 20 parts per billion of aflatoxin to be fed to animals.

—(Continued)

Distribution and Losses Caused By	Prevention	Treatment	Remarks
Widely distributed throughout the world. In addition to the effect of mycotoxins on the animal's health, milk and eggs are contaminated by the residues or mycotoxins, or their metabolic products.	The prime cause of aflatoxin is moisture; hence, proper harvesting, drying, and storage are important factors in lessening contamination and toxin production. Propionic and acetic acids will inhibit mold growth; hence, their use in preserving high-moisture grains is encouraged.	Remove the source of the mold. Animals suffering from molds frequently respond to vitamin B injections. Iron therapy may be helpful, since hemorrhaging is a frequent problem.	Certain molds produce toxins, or mycotoxins. Aflatoxin has been clearly shown to be a carcinogen (tumor producing). Ultraviolet irradiation and anhydrous ammonia under pressure will reduce the toxicity of aflatoxins and, if continued long enough, will deactivate them entirely. Not all toxins are harmful. For example, zearalenol is being commercially produced as a growth promotant hormone for cattle.
Excessive nitrate content of feeds is an increasingly important cause of poisoning in farm animals, due primarily to more and more high nitrogen fertilization. Nitrate poisoning may be lowered by (1) feeding high levels of carbohydrates or energy feeds (grain or molasses) and vitamin A, (2) feeding limited amounts of high-nitrate forage, (3) alternating or mixing high- and low-nitrate forages, and (4) ensiling forages high in nitrates, since fermentation reduces some of the nitrates to gas.	Regard any amount of nitrate nitrogen over 0.5% of the total ration (moisture-free basis) as a potential source of trouble. When in doubt, have the feed analyzed.	A 4% solution of methylene blue (in a 5% glucose or a 1.8% sodium sulfate solution) administered by a veterinarian intravenously at the rate of 100 cc/1,000 lb liveweight. Yeast culture fed as follows: 1/4 lb daily for cattle, and 1/10 lb per day for lambs.	Nitrate does not appear to cause the actual toxicity. During digestion, the nitrate is reduced to nitrite, a far more toxic form (10 to 15 times more toxic than nitrates). In cows and sheep, this conversion takes place in the rumen (paunch); in horses, in the cecum.
Wherever there is pitch.	Do not allow animals access to pitch-containing products.	No known treatment.	Pastures containing clay pigeons are dangerous for years; deaths having been reported 35 years after area was used for trapshooting.
High-protein diets are sometimes fed (1) in the South where cottonseed meal may be cheaper than low-protein feeds, and (2) in the West where alfalfa is usually abundant and cheap.			In a high-protein ration, amino acids left over after the protein requirements have been met are deaminated or broken down in the body, in which process a part of each amino acid is turned into energy, and the remainder is excreted via the kidneys. Generally speaking, high-protein feeds are more expensive than those high in carbohydrates or fats, with the result that there is the temptation to feed too little of them.
Salt poisoning is relatively rare.	If animals have not had salt for a long time, they should first be hand-fed salt, gradually increasing daily allowance until they leave a little in the mineral box; then self-feed.	Provide large quantities of fresh water to affected animals. Those unable to drink should be given water via stomach tube, by the veterinarian.	Indians and pioneers handed down many legendary stories about huge numbers of wild animals that killed themselves simply by gorging at a newly found salt lick after having been salt-starved for long periods of time.
In certain regions of western U.S.—especially certain areas in S.D., Mont., Wyo., Neb., Kan., and perhaps areas in other states in the Great Plains and Rocky Mountains. Also, in Canada.	Abandon areas where soils contain selenium, because crops produced on such soils constitute a menace to both animals and man.	Although arsenic has been shown to counteract the effects of selenium toxicity, there appears to be no practical method of treating other than removal of animals from affected areas.	Chronic cases occur when animals consume feeds containing 8.5 ppm of selenium over an extended period; acute cases occur on 500 to 1,000 ppm. The toxic level of selenium is in the range of 2.27 to 4.54 mg/lb of feed.

DISINFECTANTS

A disinfectant is a bactericidal or microbicidal agent that frees from infection (usually a chemical agent which destroys disease germs or other microorganisms, or inactivates viruses).

The high concentration of animals and continuous use of modern livestock buildings often results in a condition referred to as disease buildup. As disease-producing organisms—viruses, bacteria, fungi, and parasite eggs—accumulate in the environment, disease problems can become more severe and be transmitted to each succeeding group of animals raised on the same premises. Under these circumstances, cleaning and disinfection become extremely important in breaking the life cycle. Also, in the case of a disease outbreak, the premises must be disinfected.

Under ordinary conditions, proper cleaning of barns removes most of the microorganisms, along with the filth, thus eliminating the necessity of disinfection.

Effective disinfection depends on five things:

1. Thorough cleaning before application.
2. The phenol coefficient of the disinfectant, which indicates the killing strength of a disinfectant as compared to phenol (carbolic acid). It is determined by a standard laboratory test in which the typhoid fever germ often is used as the test organism.
3. The dilution at which the disinfectant is used.
4. The temperatures; most disinfectants are much more effective if applied hot.
5. Thoroughness of application, and time of exposure.

Disinfection must in all cases be preceded by a very thorough cleaning, for organic matter serves to protect disease germs and otherwise interferes with the activity of the disinfecting agent.

Sunlight possesses disinfecting properties, but it is variable and superficial in its action. Heat and some of the chemical disinfectants are more effective.

The application of heat by steam, by hot water, by burning, or by boiling is an effective method of disinfection. In many cases, however, it may not be practical to use heat.

A good disinfectant should (1) have the power to kill disease-producing organisms, (2) remain stable in the presence of organic matter (manure, hair, soil), (3) dissolve readily in water and remain in solution, (4) be nontoxic to animals and humans, (5) penetrate organic matter rapidly, (6) remove dirt and grease, and (7) be economical to use.

The number of available disinfectants is large because the ideal universally applicable disinfectant does not exist. Table 10-13 gives a summary of the limitations, usefulness, and strength of some common disinfectants.

When using a disinfectant, *always read and follow the manufacturer's directions.*

TABLE 10-13
DISINFECTANT GUIDE

Kind of Disinfectant	Usefulness	Strength	Limitations and Comments
Alcohol (ethyl-ethanol, isopropyl, methanol)	Primarily as skin disinfectants and for emergency purposes on instruments.	70% alcohol—the content usually found in rubbing alcohol.	They are too costly for general disinfection. They are ineffective against bacterial spores.
Boric acid[1]	As a wash for eyes, and other sensitive parts of the body.	1 oz in 1 pt water (about 6% solution).	It is a weak antiseptic. It may cause harm to the nervous system if absorbed into the body in large amounts. For this and other reasons, antibiotic solutions and saline solutions are fast replacing it.
Chlorines (sodium hypochlorite, chlormine-T)	Used for dairy equipment and as deodorants. They will kill all kinds of bacteria, fungi, and viruses, providing the concentration is sufficiently high.	Generally used at about 200 ppm for dairy equipment and as a deodorant.	They are corrosive to metals and neutralized by organic materials. Not effective against TB organisms and spores.
Cresols (many commercial products available)	A generally reliable class of disinfectant. Effective against brucellosis, shipping fever, swine erysipelas, and tuberculosis.	Cresol is usually used as a 2 to 4% solution (1 cup to 2 gal of water makes a 4% solution).	Cannot be used where odor may be absorbed, and, therefore, not suited for use around milk and meat.
Formaldehyde (gaseous disinfectant)	Formaldehyde will kill anthrax spores, TB organisms, and animal viruses in a 1 to 2% solution. It is often used to disinfect buildings following a disease outbreak. A 1 to 2% solution may be used as a foot bath to control foot rot.	As a liquid disinfectant, it is usually used as a 1 to 2% solution. As a gaseous disinfectant (fumigant), use 1½ lb potassium permanganate plus 3 pt of formaldehyde. Also, gas may be released by heating paraformaldehyde.	It has a disagreeable odor, destroys living tissue, and can be extremely poisonous. The bactericidal effectiveness of the gas is dependent upon having the proper relative humidity (above 75%) and temperature (above 30° C and preferably near 60° C).

Footnote at end of table.

(Continued)

TABLE 10-13—(Continued)

Kind of Disinfectant	Usefulness	Strength	Limitations and Comments
Heat (by steam, hot water, burning, or boiling)	In the burning of rubbish or articles of little value, and in disposing of infected body discharges. The steam "Jenny" is effective for disinfection (example: poultry equipment) if *properly employed,* particularly if used in conjunction with a phenolic germicide.	10 minutes' exposure to boiling water is usually sufficient.	Exposure to boiling water will destroy all ordinary disease germs but sometimes fails to kill the spores of such diseases as anthrax and tetanus. Moist heat is preferred to dry heat, and steam under pressure is the most effective. Heat may be impractical or too expensive.
Iodine[1] (tincture)	Extensively used as skin disinfectant, for minor cuts and bruises.	Generally used as tincture of iodine, either 2% or 7%.	Never cover with a bandage. Clean skin before applying iodine. It is corrosive to metals.
Iodophor (tamed iodine)	Primarily used for dairy utensils. Effective against all bacteria (both gramnegative and grampositive), fungi, and most viruses.	Usually used as disinfectants at concentrations of 50 to 75 ppm titratable iodine, and as sanitizers at levels of 12.5 to 25 ppm. At 12.5 ppm titratable iodine, they can be used as an antiseptic in drinking water.	They are inhibited in their activity by organic matter. They are quite expensive.
Lime (quicklime; burnt lime; calcium oxide)	As a deodorant when sprinkled on manure and animal discharges; or as a disinfectant when sprinkled on the floor or used as a newly made "milk of lime" or as a whitewash.	Use as a dust; as "milk of lime"; or as a whitewash, but *use fresh.*	Not effective against anthrax or tetanus spores. Wear goggles when adding water to quicklime.
Lye (sodium hydroxide; caustic soda)	On concrete floors; in milk houses because there is no odor; against microorganisms of brucellosis and the viruses of foot-and-mouth disease, and vesicular exanthema. In strong solution (5%), effective against anthrax and blackleg.	Lye is usually used as either a 2% or 5% solution. To prepare a 2% solution, add 1 can of lye to 5 gal of water. To prepare a 5% solution, add 1 can lye to 2 gal water. A 2% solution will destroy the organisms causing foot-and-mouth disease, but a 5% solution is necessary to destroy the spore of anthrax.	Damages fabrics, aluminum, and painted surfaces. Be careful, for it will burn the hands and face. Not effective against organism of TB or Johne's disease. Lye solutions are most effective when used hot. **Diluted vinegar can be used to neutralize lye.**
Lysol (the brand name of a product of cresol plus soap)	For disinfecting surgical instruments and instruments used in dehorning, castrating, and tattooing. Useful as a skin disinfectant before surgery, and for use on the hands before castrating.	0.5 to 2.0%.	Has a disagreeable odor. Does not mix well with hard water. Less costly than phenol.
Phenol (carbolic acid): 1. Phenolics—coal tar derivatives 2. Synthetic phenols	They are ideal general-purpose disinfectants. Effective and inexpensive. They're very resistant to the inhibiting effects of organic residue; hence, they are suitable for barn disinfection, and foot and wheel dip-baths.	Both phenolics (coal tar) and synthetic phenols vary widely in efficacy from one compound to another. So, note and follow manufacturer's directions. Generally used in a 5% solution.	They are corrosive, and they're toxic to animals and humans. Ineffective on fungi and viruses. Effective against all bacteria and fungi, including TB organism.
Quaternary ammonium compounds (QAC)	Very water soluble, ultrarapid kill rate, effective deodorizing properties, and moderately priced. Good detergent characteristics and harmless to skin.	Follow manufacturer's directions.	They can corrode metal. Adversely affected by organic matter. Not very potent in combatting viruses. Not effective against TB organisms and spores. Not effective against anthrax and tetanus.
Sal soda	It may be used in place of lye against certain diseases.	10½% solution (13½ oz to 1 gal water).	
Sal soda and soda ash (or sodium carbonate)	They may be used in place of lye against foot-and-mouth disease and vesicular exanthema.	4% solution (1 lb to 3 gal water). Most effective in hot solution.	Commonly used as cleaning agents, but have disinfectant properties, especially when used as a hot solution.
Soap	Its power to kill germs is very limited. Greatest usefulness is in cleansing and dissolving coatings from various surfaces, including the skin, prior to application of a good disinfectant.	As commercially prepared.	Although indispensable for sanitizing surfaces, soaps should not be used as disinfectants. They are not regularly effective; staphylococci and the organisms which cause diarrheal diseases are resistant.

[1]Sometimes loosely classed as a disinfectant but actually an antiseptic and practically useful only on living tissue.

GENERAL ANIMAL SANITATION AND DISEASE PREVENTION

In order to reduce the possibility of disease, it is important that there be adopted certain management practices relative to the environment of the animal; among them, proper ventilation, housing, manure disposal, pasture rotation, and carcass disposal.

Ventilation

The need for ventilation is not as great for the animal as it is for human beings, for most of the animal's life is spent out of doors where plenty of fresh air is available. Ventilation is significant only when animals are housed in crowded quarters.

Ventilation is the act of causing the movement of air through buildings with the objective of supplanting foul air with fresh air containing needed oxygen. Contrary to common opinion, when a feeling of discomfort is noticed, it is the result of oxygen starvation rather than carbon dioxide poisoning.

The amount of moisture in the air is important. When improper ventilation prevents proper evaporation, the moisture content of the air increases. If humidity rises too high, interfering with heat elimination, heat stroke may ensue. Moist air generally is a more favorable medium for the existence of microorganisms, thus lending itself well to the transmission of contagious diseases. When one animal is infected with a contagious disease and is closely housed with others, an epidemic will usually follow. The air may also pick up various noxious gases, such as ammonia from decomposing urine, which may cause irritation to the sensitive membranes of the mouth, eyes, nose, and respiratory tract.

Ventilation is measured in cubic feet per minute (cfm). The required ventilation differs according to species of animal, size of animal, and outside temperature. (Also see Section 9 of this book.)

Housing

Proper drainage and dryness, adequate space, and good lighting are some of the requirements for good housing. In addition, animal quarters must be of such construction as to facilitate proper cleaning, disinfection, and maintenance of sanitary conditions. This includes suitable floors, adequate waste disposal, and proper absorbent bedding. Further discussion of the requisites of livestock buildings is found in Section 9.

Manure Disposal

Situations that compel animals to live in close contact with their own body excreta are most injurious to physical well-being. Urine, feces, exhalations, and nose and mouth discharges may often contain disease-producing agents, and furnish an ideal medium for the growth of microorganisms. Stockmen are fully aware of the miraculous recovery many animals undergo when taken from small, unsanitary enclosures to good, clean pastures.

The importance of removing excrement frequently from the immediate surroundings (enclosures, barns, and loafing sheds) cannot be stressed too much. The method of disposal of solid and liquid manure is also very important. As this manure may contain a variety of parasites and eggs, proper disposal offers an excellent opportunity for breaking the life cycles of these parasites. On the other hand, if left in an accessible place for animals, manure can be a rich, never-ending source of disease and parasitism.

In order to ensure the killing of many harmful parasites, one may store manure (for two weeks to a month) so that the heat generated will cause their death. It should be stored in a covered concrete pit and located far enough away from the buildings to prevent contamination. These enclosures should be inaccessible to all animals. Spraying manure pits with a suitable insecticide will inhibit fly development. If the manure is believed to be free of specific infectious microorganisms (for example, tuberculosis, brucellosis, and blackleg), it may be spread daily on arable land containing no animals. Here the purifying elements—such as rain, sunshine, soil, and vegetable processes—will tend to render the manure sanitary. Food and water should always be protected from contamination by manure.

It is also important that the manure spreader should not move from areas where commercial, sick, and/or isolated animals are held into areas where breeding animals are kept.

Pasture Rotation

Pasture rotation provides a very practical method of control of many diseases and parasites. Permanent pastures used by one species of animal may be regarded as highly dangerous for profitable endeavors. A method by which land areas for pasturage are systematically changed periodically to crop production is recommended.

As many parasites (including bacteria) are often specific for a certain host (for example, bots of horses affect no other animal), frequently pastures may be rotated between different species.

Carcass Disposal

In the disposal of carcasses, it is a safe rule to assume that all of them are a source of some infection and subsequently to adopt the proper sanitary precautions.

The most sanitary method of destroying a carcass is to burn it, preferably at the site of death in order to prevent the contamination of surrounding ground. A trench of sufficient size should be prepared, a fire built, and the animal placed on top so that it will be consumed in its entirety. An incinerator may be used for poultry.

The most common method of large animal carcass disposal is by burial. So that this method will be effective, the carcass should be buried deep and covered with quicklime. The top of the carcass should be at least 4 feet below the surface of the ground and in soil from which there is no danger of contamination by drainage. Burial should not be near a flowing stream, for this will only serve to spread the disease downstream.

Near large centers of population, rendering plants will take carcasses, and they afford the easiest method of disposal.

When an animal dies, it is recommended that a veterinarian be called immediately to perform a postmortem examination. This is done in an attempt to study the abnormal conditions present and to determine the cause of death. It is never safe for one who is uninformed about specific disease lesions to open an animal carcass. Such practice may not only serve to spread a very highly contagious disease but may also expose the operator to a dangerous infection.

It is also unsafe to feed the carcass to other animals. Such procedure may cause the animal consuming it to become sick, or it may serve only to spread the disease.

Fig. 10-125

A GENERAL PROGRAM OF ANIMAL HEALTH, DISEASE PREVENTION, AND PARASITE CONTROL

Consciously or unconsciously, most stockmen follow a program of animal health. But, unfortunately, few give sufficient thought to it; and fewer yet write it down and adhere rigidly to it. As a result, it is often a hit and miss proposition and little attention is given to disease prevention until it is too late—until some disease or parasite takes a heavy toll. At such times, one becomes well aware that "an ounce of prevention is worth a pound of cure."

Because livestock and poultry diseases cost an estimated 10 billion dollars annually, it behooves each stockman to develop and follow a written-down program of animal health, disease prevention, and parasite control. The owner himself should take the lead in developing such a program; but he should mobilize the best wit, wisdom and judgment of others—including his local veterinarian, and perhaps the county agent, the vocational agriculture instructor, and/or successful neighbors. Also, it must be recognized that cumulative changes in this program will need to be made from time to time; in light of new problems, new biologics, etc.

Any general program of animal health, disease prevention, and parasite control should embrace certain management practices relative to the environment of the animal; among them, proper ventilation, housing, manure disposal, pasture rotation, and carcass disposal (see section headed "General Animal Sanitation and Disease Prevention"). Additionally, the following general program is applicable to any one or all classes of farm animals:

1. **Provide good housing and ventilation**—Although housing and close confinement predispose animals to disease, it is often very necessary; to facilitate handling, to combat the elements, and/or to furnish protection when young are arriving.

Experimental studies indicate that a 1,000-lb cow breathes into the air approximately 10 lb of moisture per day. Thus, from a herd of 42 cows there would be given off 420 lb, or about 50 gallons of water per day. This amount of moisture must be removed daily in order to keep the barn free from dampness.

When ventilation is poor and there is a difference of several degrees between inside and outside temperature, the moisture condensation forms frost. Thus, condensation on the walls or other surfaces of barns, along with wet animals, gives evidence of unsatisfactory moisture conditions. Such condensation is objectionable because it is harmful for animals to go from a moist, warm barn into the cold outside air and because the excess moisture causes the structure to decay or deteriorate. Thus, in no case should warmth of a building be obtained at the cost of poor ventilation.

Ventilation may be secured by various systems. The simplest method usually consists of one or more of the following: (a) open shed, (b) open doors, (c) windows that open inwardly from the top, or (d) building or stall partitions left slatted or open at the top. A more complete method of ventilating tight buildings consists of a system of intake and outtake flues operated on the basis of either gravity or forced ventilation. Whatever the system, proper ventilation is one in which the foul air is drawn off and harmful humidity conditions are eliminated without excess heat loss or creation of drafts.

2. **Keep barn idle for one month or longer**—All barns should be emptied of animals for a minimum period of one month each year, thus permitting thorough cleaning, disinfecting, and drying out.

3. **Provide suitable feed containers**—Avoid feeding off the ground because of the hazard of spreading diseases and parasites.

4. **Control and exterminate rodents and birds**—Without exception, U.S. livestock establishments are plagued by one or more rodents and/or with birds. Among such pests are rats, mice, English sparrows, and pigeons. They should be controlled, and eliminated if possible, because they (a) spread diseases and parasites, (b) damage feeds and buildings, and (c) decrease profits. (For pointers on this subject, see discussion in Section 8 of this book.)

5. **Isolate new animals**—Strictly from a disease prevention standpoint, there is much merit in maintaining a closed herd and not bringing in new animals; but this is not always practical. When new animals must be added to the herd or flock, (a) secure a health certificate signed by the local veterinarian, and (b) isolate them in separate quarters—in a separate barn, lot, or pasture—for a minimum period of three weeks and arrange for their feed and care by a separate caretaker.

Thoroughly clean and disinfect the isolation stall after each animal(s) is removed and before a new animal(s) is placed therein. Disinfect with a hot 3 percent lye solution, followed by the use of another recommended disinfectant (recommended by your local veterinarian), such as Lysol or sodium orthophenylphenate.

6. **Restrict commercial stock trucks**—Do not permit commercial stock trucks to drive on the premises unless they have been thoroughly disinfected.

7. **Use caution in showing**—Despite all of their many virtues, there is a disease hazard in showing at livestock shows.

8. **Use disinfectants**—A disinfectant is defined as any biological, physical, or chemical agent capable of exerting changes in environment unfavorable for the continued survival of microorganisms.

Under ordinary conditions, proper cleaning of barns removes most of the microorganisms, along with the filth, thus eliminating the necessity of disinfection. In case of a disease outbreak, however, the premises must be disinfected. Also, it is desirable that an adequate foot disinfection program be maintained at all times for all visitors entering barns (see Table 10-13, page 984, for proper disinfectant).

9. **Call the veterinarian**—Effective animal health programs call for full cooperation between the producer and the veterinarian, with the former calling upon the latter in exactly the same manner as well-informed people call upon the family doctor when human ill health is encountered.

A PROGRAM OF BEEF CATTLE HEALTH, DISEASE PREVENTION, AND PARASITE CONTROL

Although the exact program will and should vary according to the specific conditions existing on each

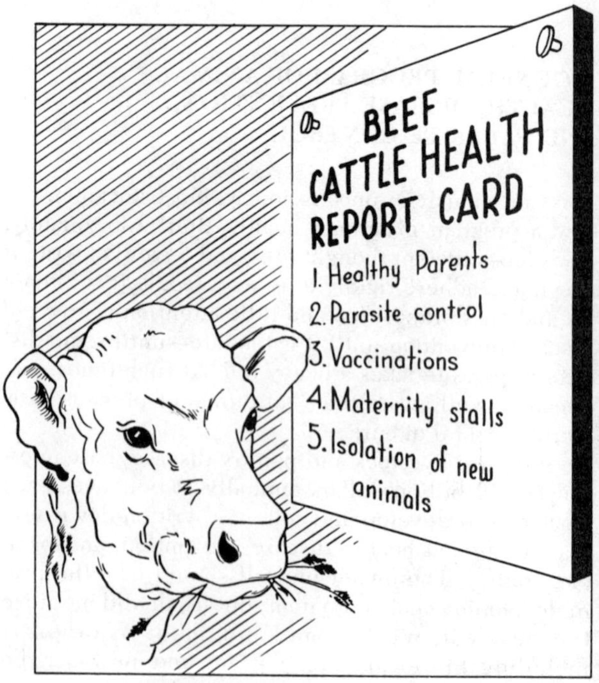

Fig. 10-126

individual farm or ranch, the basic principles will remain the same. With this thought in mind, the following program of beef cattle health, disease prevention, and parasite control is presented with the hope that the beef cattle producer will use it (1) as a yardstick with which to compare his existing program, and (2) as a guidepost so that he and his local veterinarian, and other advisors, may develop a similar and specific program for his own enterprise.

I. General Beef Cattle Program

1. Breed only healthy cows to healthy bulls.

2. Avoid either an overfat or a thin, emaciated condition in all breeding animals.

3. Flush cows by providing more lush pasture or by feeding grain so that they gain approximately 1.5 pounds per head daily beginning 20 days before the start and continuing throughout the breeding season. This practice will cause more cows to come into heat, to breed early in the season, and to conceive at first service.

4. Provide plenty of exercise for bulls and pregnant cows, preferably by allowing them to graze in well-fenced pastures in which plenty of shade and water are available.

5. Keep lots and corrals well drained and as dry as practical to prevent breeding places for foot rot, other diseases, and parasites. Fence cattle out of pasture mudholes for the same reason.

6. If possible, divert drainage from adjacent infected premises and avoid across-the-fence contact with the neighbor's cattle unless they are definitely disease free. Do not visit farms where infectious diseases exist, as the germs may be brought home on shoes, clothing, or vehicles. For the same reason, feeds should not be bought from such farms, and one should beware of used feed bags.

7. If rented pastures must be used, avoid areas on which cattle have overwintered; and, preferably, use only those rented pastures that have not had cattle on them for one year or that have been plowed in the interim.

8. Test the entire herd for tuberculosis and brucellosis each fall or at the time they are brought in from pasture and placed in winter quarters.

9. Eliminate the breeding ground of parasites as far as practical and use the proper insecticide or anthelmintic for their control.

10. Keep commercial cattle—such as stocker and feeder cattle, and finishing cattle—in isolated areas away from breeding animals.

11. Have all cows checked for pregnancy 45 to 60 days after the breeding season is over. When problems are encountered, immediately consult a veterinarian.

12. When disease troubles are encountered, isolate affected animals and follow the instructions and prescribed treatment of a veterinarian.

II. New Stock

1. Vaccinate calves against blackleg and malignant edema in areas that are endemic for these diseases.

2. Isolate newly acquired animals for a minimum of three weeks, during which time they should be cared for by a separate caretaker.

3. While in isolation, test all newly acquired breeding animals for tuberculosis, brucellosis, leptospirosis, anaplasmosis, vibriosis, and Johne's disease; first, however, make every reasonable effort to ascertain that they come from herds which are known to be free from these and other diseases.

4. Spray newly acquired animals for lice control; and check them for internal parasites, and treat where indicated.

5. When possible, it is preferable to purchase virgin heifers and bulls, from a disease control standpoint. Isolate "tried" (nonvirgin) bulls for a period of 3 weeks, and then turn them with a limited number of virgin heifers; observe these heifers for 30 to 60 days after breeding, as an aid in preventing the introduction of breeding diseases.

6. Thoroughly clean and disinfect the isolation stall after each animal(s) is removed and before a new animal(s) is placed therein.

III. Calving Time

1. When weather conditions permit, allow parturient cows to calve in a clean, uncontaminated, open pasture. During inclement weather, place the cows in isolated, roomy, light, well-ventilated maternity stalls—which should first be carefully cleaned, thoroughly disinfected, and provided with clean bedding for the occasion. After calving, all wet, stained, or soiled bedding should be removed and the floor sprinkled with lime; the afterbirth should be burned or buried deep in lime; and, if there has been trouble, the cows should be kept isolated until all discharges have ceased.

2. Unless the calves are born on a clean pasture away from possible infection, treat the navel cord of each newborn animal with tincture of iodine.

3. See that the newborn calf gets colostrum milk as soon as possible. But bear in mind that the antibodies of colostrum depend upon the dam's disease history, either directly or through vaccinations.

IV. **Suckling Calves**

1. If the baby calves are confined to stalls, scrub stalls thoroughly twice each week with warm soap solution and disinfect the walls and feed bunks and/or mangers.

2. Vaccinate calves with blackleg and malignant edema bacterin at 2 to 3 months of age (at 1 month of age in endemic areas).

Preconditioning (also see Section 8, Management)

Preconditioning is the schedule of practices used in preparing feeder calves to withstand the stress of leaving their mothers, shipping, and adapting to feedlot conditions. It consists of good management, along with immunological practices and treatment for parasites.

Changed environment; excitement of sorting, loading, and shipping; long periods without feed and/or water; movement through one or more assembly points; change of feed; and exposure to disease—all add up to fatigue, stress, shrink, and *lowered disease resistance.*

The steps used in preconditioning may, and should, vary somewhat among areas, farms, and ranches. But generally it involves the following:

1. Handling calves quietly, with a minimum of excitement.
2. Dehorning and castrating well ahead of weaning time—about two months of age is best.
3. Weaning calves 30 days ahead of shipment.
4. Starting calves on a ration that will be similar to the starting ration that they will receive when they arrive in the feedlot.
5. Vaccinating in keeping with the advice of your veterinarian.
6. Treating for parasites, internal and/or external, before weaning.
7. Reducing time, stress, and exposure to infection from the farm or ranch to the feedlot.
8. Providing a preconditioning certificate, which is a record of all husbandry, nutritional, and medical histories provided by the seller of feeder cattle to the buyer. (See Section 8, Management, for preconditioning certificate.)

Cattle Feedlot Disease and Parasite Control Program[5]

Cattle feeders should have and follow a written-down cattle feedlot health program, developed in cooperation with the feedlot veterinarian. The following outline will serve as a useful guide for this purpose.

1. Process (dehorn, castrate, and vaccinate) incoming cattle soon after they are unloaded from the truck, rather than wait. At that time, also inject with vitamin A and implant with any approved growth stimulant intended to use.

   Newly arrived cattle are usually vaccinated against bovine virus diarrhea, red nose, shipping fever, and lepto; and given high injections of vitamin A if they are stressed (250,000 to 1,000,000 IU, depending on size of cattle and degree of stressing).

2. For the first 28 days after arrival, fortify the natural protein supplement (do not feed incoming cattle urea) with the following per head per day: 350 mg of Aureomycin plus 350 mg of sulfamethiazine (Aureo-S-700, which is a combination of an antibiotic and the bactericidal agent sulfamethiazine) *plus* 50,000 IU of vitamin A.
3. Administer plenty of TLC (tender loving care) to the sick.
4. Treat sick animals three times in the first 24 hours. It will reduce repeats and chronic illness.
5. Keep hospital chutes and pens clean and well bedded.
6. Disinfect chutes, tanks and syringes, and balling guns to prevent spread of disease.
7. Seek professional advice if there is no response to medication within 48 hours.
8. Autopsy all dead animals for an accurate diagnosis. If one specific problem is causing a lot of deaths, seek prevention.
9. Don't change rations too fast.
10. Have all members of the health team—detection, treatment, and convalescence—performing at top level.
11. Control parasites:

    a. **Internal parasites**—Worm calves, if necessary, with one of the following drugs, used according to manufacturer's directions: Phenothiazine, thiabendazole (Thibenzole or T.B.Z.), or tetramisole (Tramisol).

    Treat only (1) if animals appear to be heavily parasitized or are from areas where previous experience has shown that they are heavily parasitized, or (2) if 300 or more eggs per gram (epg) of dry feces are found.

    b. **External parasites**—Treat for external parasites (commonly lice, grubs, flies, and ticks), if necessary, using a recommended insecticide and following the manufacturer's instructions on the label or container.

[5]This section was authoritatively reviewed by Dr. Donald R. Mackey, DVM, cattle feedlot veterinarian, Greeley, Colo.

c. **Control of flies around the feedlots and feedmill**—Prevent flies by starting the following program early in the season: sanitation, prevention of breeding areas, and use of residual sprays and mists before the fly population builds up. The aerial application of such materials as naled (Dibrom) or malathion will generally reduce the adult population for short periods of time. The following residual sprays and baits applied to fly resting areas such as fences, barns, sides of fence bunks, shades, and other structures (not on animals) give some adult fly control: diazinon, dichlorvos (DDVP), dimethoate (Cygon), fenthion (Baytex), methoxychlor, ronnel (Korlan), and trichlorfon (Dipterex). Before using any chemical in a control program directed toward either the adult fly or the larvae, read the label carefully to determine the usage and restrictions of the material.

• **Care of sick feedlot cattle**—Proper care of sick cattle necessitates two things: (1) suitable hospital facilities, and (2) prompt and correct diagnosis and treatment.

Diagnosis and treatment are very important in the health program of any cattle feedlot. The pen checkers and hospital technician cannot completely operate this phase of the program alone. The yard foreman must supervise it, although he has far too many responsibilities to be completely responsible for the program. The consulting veterinarian must establish general policies and treat unusual and difficult cases, although he cannot examine every animal as it enters the hospital, simply because the health program costs would be too high. Thus, diagnosis and treatment is really a team approach. Each person involved should contribute to the program according to his responsibilities and abilities, thereby assuring the most effective and economical health program.

Diagnosis is the art or act of recognizing disease from its symptoms. Thus, correct diagnosis assumes that those responsible for the cattle feedlot health program recognize what is "normal." Then, any deviation from the normal should be reason for concern and further study. Practical experience in both the normal and abnormal is essential in arriving at a correct diagnosis.

Treatment should be left in the hands of the veterinarian. He should give instructions for the proper use of drugs and biologicals—the administration, dosage levels, indications, and contraindications of a given product, and signs and lesions of specific diseases. It is his responsibility to keep management current on the latest findings regarding research, new products, and disease conditions.

A PROGRAM OF DAIRY CATTLE HEALTH, DISEASE PREVENTION, AND PARASITE CONTROL

Although the exact herd health program will vary according to the specific conditions existing on each individual dairy farm, the basic principles will remain the same. Accordingly, the following program of dairy cattle health, disease prevention, and parasite control is presented with the hope that the dairyman will use it (1) as a yardstick with which to compare his existing program, or (2) as a guide so that he and his local veterinarian may develop a similar and specific program for his enterprise.

I. General Dairy Cattle Health

The following general, overall health program should be observed on the dairy farm:

1. Provide plenty of exercise for pregnant cows, preferably by allowing them to graze in well-fenced pastures in which plenty of shade and water are available.

DAIRY CATTLE HEALTH REPORT CARD

1. Milk Production Record
2. Reproductive Record
3. Mastitis Screening Test
4. Blood Tests
5. Vaccinations
6. Parasite Control
7. Isolation

Fig. 10-127. Criteria of dairy cattle health.

2. Keep lots and corrals well drained and as dry as practical, to prevent breeding places for foot rot and parasites. Fence cattle out of pasture mudholes and ponds for the same reason.

3. If possible, divert drainage from adjacent farms and avoid common line fences which permit direct contact with the neighbor's cattle. Do not visit farms where infectious diseases exist, as the organisms may be brought home on shoes, clothing, or vehicles. For the same reason, feeds should not be bought from such farms, and one should not use reused feed bags.

4. If rented pastures must be used, avoid areas on which cattle have overwintered; and, preferably, use only those rented pastures that have not had cattle on them for one year or that have been plowed and renovated in the interim.

5. Follow a sound feeding program to prevent ketosis, milk fever, and nutritional diseases.

6. Prevent hardware disease by keeping wire, nails, and loose metals away from cows.

7. Test annually for tuberculosis, unless you are participating in a "back tagging program" which may permit less frequent testing.

The "back tagging program," instituted by the cooperative federal-state governments, calls for the identification of all cull cows that go to slaughter. Under this program, all dairy and beef herds are constantly monitored for tuberculosis and other infectious and contagious diseases.

8. Test all additions to the herd for brucellosis, using the blood test. Also, test for leptospirosis simultaneously, as the same blood sample can be used for both tests.

The "milk ring test," which is run on all Grade A dairies every three months, will serve as a screening test to monitor the herd against brucellosis.

9. Worm all calves and replacement heifers with a suitable vermifuge, as young animals are susceptible to parasite damage.

10. Use artificial insemination, unless (a) AI is not available, or (b) there are compelling reasons to use a bull in natural service in a particular breeding program.

11. Lessen calving difficulties of first calf heifers by (a) feeding replacement heifers liberally, and (b) breeding when they reach adequate size to a bull whose calves are small at birth.

12. When disease problems strike, isolate affected animals and obtain the services of a veterinarian. Remember that early diagnosis is the key to prompt treatment and response.

13. Isolate for three weeks all animals entering the herd from shows or sales.

II. Infectious Disease Control

Effective vaccines exist for the control of the following infectious diseases of dairy cattle:

Anaplasmosis	Malignant edema
Anthrax	Rabies
Bacillary hemoglobinuria	Rhinotracheitis
Blackleg	Shipping fever
Brucellosis (calfhood vaccination)	Staphylococcus mastitis
Calf scours	Tetanus
Enterotoxemia	Vibriosis
Leptospirosis	Virus diarrhea
	Warts

No effective vaccines exist for the control of the following infectious diseases of dairy cattle:

Actinobacillosis (wooden tongue)	Listeriosis
Actinomycosis (lumpy jaw)	Malignant catarrhal fever
Calf pneumonia	Mastitis
Cowpox	Mycotic infections
Foot rot	Pyelonephritis
Infectious keratitis (pinkeye)	Ringworm
Johne's disease	Tuberculosis
	Winter dysentery

The dairyman and his veterinarian should develop a vaccination program, then follow it routinely. Such a program will vary from herd to herd, depending upon (1) the location of the farm, and (2) the disease situation on the farm and in the immediate area. Remember that the loss of just one animal, or the lowered milk production that usually accompanies a disease outbreak, will often exceed the cost of vaccinating the entire herd.

Also, some vaccinations can be used to control certain diseases once they have been accurately diagnosed.

III. Reproductive Disease Control

Reproduction is one of the most important parts of a dairy health program. Each day that a cow is "open" beyond 100 days following freshening, costs the dairymen $.50 to $1.00 per day. The following reproductive disease control program is recommended:

1. **Records**—Check all open cows for heat twice daily, and record heat periods. Cows remain in heat for only 14 to 16 hours. Hence, for highest conception, those found in heat in the morning should be bred that evening, and those found in heat in the evening should be bred the next morning.

Also, keep accurate and complete breeding and calving records.

2. **Pregnancy examination**—A veterinarian or experienced technician should make a pregnancy examination anytime 35 days following service. Also, in problem herds, the veterinarian may perform an

involution checkup 30 to 50 days after calving to determine if the ovaries and uterus are normal.

3. **Retained placenta; other genital abnormalities**—Cows with retained placenta should be examined and treated by the veterinarian 1 to 3 days following calving. Likewise, prompt treatment should be given for any abnormality of the genital organs, such as heat cycles of abnormal length or cystic ovaries.

IV. Modern Milking

A good milking program will lessen disease—especially mastitis, the bane of all dairymen—and increase profits from 5 to 25 percent. From a herd health standpoint, a modern milking program should embrace the following:

1. **Udder and teat protection**—Most mastitis is predisposed by injuries to the udder and teats. The best injury-preventive measure consists in using liberal amounts of clean, dry bedding.

2. **Sanitation**—Clean udders and teats are essential to the production of clean milk of high quality. This calls for (a) adequate space and a clean barn; (b) clean, dry corrals (free from low, wet, muddy, or swampy areas); and (c) clean, sterile milking equipment.

3. **Proper milking procedure**—First, the cow must be stimulated to release (let down) her milk. This can be accomplished by the following:

a. Keeping the cow content in a quiet, peaceful atmosphere. Also, and most important, the milker should love his work; otherwise, it is almost impossible to have this type of atmosphere.

b. Keeping regular milking hours.

c. Washing the teats and lower udder, which both stimulates and cleanses.

After stimulation, rapid and complete milking must follow immediately. Within about 45 seconds following the stimulation, pressure builds up in the cow's udder due to the action of the hormone oxytocin, which is released by the pituitary gland. This hormone action lasts for about 5 to 6 minutes. It causes the muscles to contract around the milk cells. So, within 1 minute, and not more than 1½ minutes, following the stimulation, attach the teat cups and begin milking.

Machine strip each teat when it is nearly collapsed and before milk flow has stopped. This is accomplished by pulling down gently on the teat cups and massaging each quarter with a gentle downward motion. Do not prolong machine stripping as it will cause teat injury; about 20 seconds per cow is sufficient.

4. **Milking teat dip**—Use a postmilking teat dip (100 ppm iodophor or chlorine, or other sanitizing agent) to lessen the incidence of new infections during lactation.

5. **Milking machines repair**—Keep milking machines—the most used machines on the dairy farm—in good operating condition at all times. Have them checked and serviced regularly; the vacuum and pulsations are of the greatest importance.

6. **Mastitis tests**—Quality milk cannot be produced if there is infectious mastitis in the herd. Clinical mastitis milk is unfit for human consumption because of (a) abnormal physical characteristics (flaky, lumpy, stringy, watery, or bloody); (b) high bacterial count; (c) poor flavor—usually salty; and (d) possible drug residues.

Dairymen should be familiar with the following mastitis screening tests, and use one or more of them:

a. **The strip cup test**—It should be used prior to each milking to spot such abnormalities as flakes, clots, pus, or watery milk.

b. **The California Mastitis Test (CMT)**—This is a method for detecting subclinical infections, which in the past usually went undetected until the more advanced stages of infection were reached.

The CMT test should be conducted on the "foremilk" of each cow once monthly, and the results for each animal should be recorded.

A positive relationship exists between the bacteriological status of the udder quarters and the intensity of the CMT test. Milk from uninfected quarters will yield a low test score; milk from infected quarters will yield a high test score.

c. **Wisconsin Mastitis Test (WMT)**—This is a screening test for somatic cells in bulk milk tanks. Dairy plants and health departments routinely run this test to check on the quality of milk that is supplied. Counts above 1,500,000 cells indicate a serious mastitis problem. Such milk is unsuitable for interstate shipment. Moreover, if 3 out of 5 tests are above the maximum, the milk plant will refuse the milk.

7. **Mastitis treatment**—If there are more than 8 cases of mastitis per 100 cows per month, your losses are excessive and something should be done about it.

Owners of problem herds should seek the help of a veterinarian. He should make an accurate diagnosis and isolate the causative organism. This eliminates the guesswork, permits specific treatment, and keeps the permanent udder damage to a minimum.

Clinical mastitis at freshening can be reduced to a minimum by treating cows with high CMT test scores during the dry period. The efficacy of treatment is higher during this period, because the gland is resting and damaged tissue may be repaired before freshening. Also, there is no milk loss due to antibiotic or drug residues.

V. **Calving Time**

1. When weather conditions permit, allow parturient cows to calve in a clean, uncontaminated, open pasture. During inclement weather or when pasture is not available, place the cows in isolated, roomy, light, well-ventilated maternity stalls—which should first be carefully cleaned and thoroughly disinfected (following the mixing directions printed on the container) and provided with clean bedding for the occasion. After calving, all wet, stained, or soiled bedding should be removed and the floor sprinkled with lime; the afterbirth should be burned or buried deep in lime; and, if there has been trouble, the cows should be kept isolated until all discharges have ceased.

2. Unless the calves are born on a clean pasture away from possible infection, treat the navel cord of each newborn animal with tincture of iodine.

VI. **Suckling Calves**

1. If the baby calves are confined to stalls, scrub stalls thoroughly twice each week with warm soap solution and disinfect the walls and feed bunks and/or mangers.

2. If you are in a brucellosis area, vaccinate all dairy heifer replacements against brucellosis at two through six months of age, observing the state regulations.

3. Vaccinate calves with blackleg and malignant edema bacterin at about two months of age and again at weaning time if these diseases are prevalent in the area.

VII. **New Stock**

1. Vaccinate incoming calves against blackleg and malignant edema in areas that are endemic for these diseases.

2. Isolate newly acquired animals for a minimum of three weeks, during which time they should be cared for by a separate caretaker.

3. While in isolation, test all newly acquired breeding animals for tuberculosis, brucellosis, leptospirosis, anaplasmosis, and Johne's disease; first, however, make every reasonable effort to ascertain that they come from herds which are known to be free from these and other diseases.

4. Spray newly acquired animals for lice control; and check them for internal parasites, and treat where indicated.

5. When possible, it is preferable to purchase virgin heifers from a disease control standpoint.

6. Thoroughly clean and disinfect the isolation stall after each animal(s) is removed and before a new animal(s) is placed therein. Disinfect with a hot three percent lye solution, followed by the use of another recommended disinfectant (recommended by your local veterinarian), such as one of the phenolic germicides (see Table 10-13, page 985).

VIII. **Sound Practices**

The following practices are valuable adjuncts to the herd health program:

1. **Health records**—Each cow should be individually identified and have an individual health record. The latter should show all vaccinations, tests, past diseases, and treatments. This will assist the veterinarian in establishing a more accurate diagnosis when illness is encountered.

2. **Routine tests**—Routine tests should be conducted for brucellosis, tuberculosis, leptospirosis, and occasionally other diseases.

3. **Dehorning**—Dehorning prevents many injuries. Calves should be dehorned at an early age, when it is both easier and safer.

4. **Removal of extra teats**—Removal of extra teats imparts a better appearance to the udder and lessens some udder disease problems.

5. **Care of feet**—Sore feet markedly reduce production. So, the feet should be examined routinely, and trimmed and treated when necessary; thereby preventing many cases of lameness and foot disease.

6. **Parasite control**—Make routine examinations for both internal and external parasites, and treat as necessary. Parasites result in lowered feed efficiency and poor growth of young stock.

A PROGRAM OF SHEEP HEALTH, DISEASE PREVENTION, AND PARASITE CONTROL[6]

Among practical sheep raisers, the importance of disease prevention and parasite control is well known. The following program of sheep health, disease prevention, and parasite control is presented with the hope that the sheep producer will use it (1) as a yardstick with which to compare his existing program, and (2) as a guidepost so that he, with his local veterinarian and other advisors, may develop a similar and specific program for his own enterprise.

I. **General Sheep Program**

1. Limit visitors, and have them disinfect or change their shoes. Salesmen, rendering truck personnel, shearers, sheep buyers, neighbors, and other visitors can unknowingly carry bacteria or other infectious agents on their persons, clothing, and footwear. For this reason, (a) unnecessary visitors should be kept out of the sheep premises; and (b) before entering the premises, visitors should be required either to step into a shallow metal vat containing a foam rubber

[6]This section was authoritatively reviewed by the following sheep specialists: Dr. Robert M. Jordan, Department of Animal Science, University of Minnesota, St. Paul, Minn.; and Dr. J. B. Outhouse, Purdue University, Lafayette, Ind.

Fig. 10-128

mat immersed in a 5 percent creosol solution, or to put on clean rubber boots or plastic shoe covers provided for them.

2. Clean and disinfect the sheep barn (or sheep quarters) thoroughly at least once per year. First remove all manure. Then, either (a) scrub down the pens, walls, floors, and feeders with a lye solution (1 lb of lye to 15 to 20 gal of water), or (b) apply a high-pressurized water sprayer and steam cleaner. Following cleaning and disinfecting, it is highly desirable that the barn be opened for air circulation and left empty for 3 to 6 weeks.

3. Sprinkle a light application of superphosphate (.0 - 46 - 0) to fresh bedding; to keep the moisture down, to reduce ammonia fumes, and to aid in foot rot control.

4. Avoid overcrowding; provide 12 to 15 square feet of floor space for each mature ewe of the larger breeds. Provide proper ventilation and keep the quarters clean and dry at all times.

5. Keep the feed and water troughs clean.

6. Force the flock to take plenty of exercise.

7. Control foot rot. Trim the feet of all ewes at shearing time and examine for evidence of foot rot. Isolate and treat lame sheep by trimming, followed by walking them through a foot bath at weekly intervals until foot rot disappears.

8. In known bluetongue areas, vaccinate all sheep prior to the insect season, observing the following precautions: (a) Do not vaccinate ewes during the first 2 to 3 months of pregnancy, for it will likely result in abnormal lambs; and (b) delay until 4 months of age vaccinating lambs that are suckling immune dams, because the ewes' milk will impair the active immunity produced by the vaccine.

9. Rotate the flock on clean, well-drained pastures at about two-week intervals.

10. Eliminate stray dogs, and examine and administer, when necessary, the proper worm treatment for the removal of tapeworms to all dogs that must come in contact with sheep, because the cyst stage of tapeworms may occur in sheep.

11. Give any animal showing evidence of infestation with internal or external parasites prompt and modern treatment, provided that a known and satisfactory treatment exists. Preferably, and when feasible, the parasite should be eliminated by breaking its life cycle before it enters the sheep.

12. Control most internal parasites through the following program:

a. Two weeks before turning to pasture, drench each adult sheep, with the wormer of choice.

b. During the pasture season, worm sheep at such intervals as necessary.

c. At the end of the pasture season, again individually drench with the wormer of choice. Use different wormers in rotation, to lessen the hazard of a strain of stomach worm becoming resistant to a wormer.

13. Spray all sheep, including the lambs, with an approved insecticide for "tick" and wool maggot (blowfly) control, in from 10 to 30 days after shearing and again in July and August. An exception should be made in the case of very thin ewes or young lambs for which spraying should be either delayed or of lowered concentration.

14. Tag all ewes prior to the breeding season.

15. Keep commercial feeder lambs in isolated areas away from breeding animals. When such lambs are placed in the feedlot, vaccinate them against enterotoxemia (overeating disease). If sore mouth (contagious ecthyma) has been present on the premises, vaccinate against it, also.

16. In the case of sick or dead sheep, promptly consult with the veterinarian with the view of diagnosing and instituting proper control measures if necessary.

II. Lambing Time

1. Before lambing, trim all "tags" from the udder and dock. If a warm barn is available, shear the ewes before lambing. Shorn ewes will take better care of their lambs because they will seek shelter, rather than stand outside in the snow or in a chilling rain. Also, more shorn ewes can be housed in the same space, and lambs can nurse shorn ewes better than ewes with a heavy fleece.

2. Clean and disinfect the lambing quarters thoroughly, well in advance of the lambing season.

3. Keep the holding pens as clean and as dry as possible, by applying lime as indicated and by changing the bedding frequently.

4. Treat the navel of the newborn lamb with 7 percent tincture of iodine.

5. In the spring, vaccinate lambs against "overeating disease"; and if sore mouth has been present on the premises, vaccinate against it, also.

III. **New Stock**

1. Avoid the inshipment of sheep from areas known to be infected with such contagious diseases as bluetongue and scrapie.

2. Do not make any additions to the ewe flock during the last three months of pregnancy.

3. Isolate, for a period of 30 days, all sheep, including show sheep, which are brought from the outside to be added to the flock. Also, take the following precautions with such new animals:

a. On arrival, treat for parasites, both external and internal.

b. Vaccinate for overeating disease; and if sore mouth has been present on the premises, vaccinate against it, also.

A PROGRAM OF SWINE HEALTH, DISEASE PREVENTION, AND PARASITE CONTROL[7]

Successful swine production necessitates the application of health-conserving, disease-prevention, and parasite-control measures to the breeding, feeding, and management of the herd. By nature, swine possess clean habits, if only they are given an opportunity. In altogether too many cases, however, the pig is placed in crowded conditions, old hog lots, and filthy quarters. Such conditions favor the attack by the common diseases and parasites of swine.

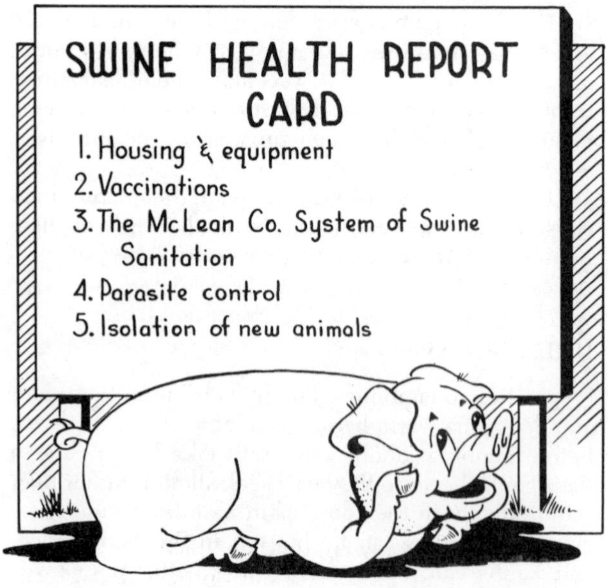

Fig. 10-129

[7]This section was authoritatively reviewed by the following swine specialists: Dr. Robert H. Grummer, Meat and Animal Science Department, The University of Wisconsin, Madison, Wisc.; Dr. Richard F. Wilson, Animal Science Department, The Ohio State University, Columbus, Ohio; and Dr. E. R. Barrick, Department of Animal Science, North Carolina State University, Raleigh, N.C.

The following program of swine health, disease prevention, and parasite control is presented with the hope that the swine producer will use it (1) as a yardstick with which to compare his existing program, and (2) as a guide so that he, with his local veterinarian and other advisors, may develop a similar and specific program for his own enterprise.

I. **General Swine Health Program**

The basic goal of swine sanitation is that there be healthy animals raised in clean, dry quarters that are well ventilated, with plenty of clean water (water fountains should be drained and cleaned frequently). At its best, this involves observance of the following rules:

1. Plan the entire physical plant layout for efficient, rapid, and adequate cleaning, and locate it so as to allow natural drainage away from buildings and feeding floors.

2. Provide proper ventilation and temperature control in facilities so that they will be warm and dry.

3. Separate farrowing and nursery area at least 100 feet from other buildings, with all traffic away from area.

4. Provide isolation quarters at least 300 feet from other swine buildings; and so that drainage will be away from, rather than toward, any swine lots, pastures, or buildings.

5. Remove soiled bedding and body discharges frequently—daily, if necessary.

6. Provide for adequate disposal of manure. Do not spread hog manure on hog pastures.

7. Clean and disinfect all buildings and equipment immediately after each period of use.

8. Allow buildings to remain idle three days to three weeks before bringing in a new group of hogs, according to specific needs and depending on how well cleaning and disinfecting are carried out.

9. Do not allow other hog producers and visitors in pens or on pastures with the herd without their first changing their outer clothing and washing footwear in a disinfectant solution; and do not allow equipment and dogs from other farms and delivery vehicles to enter pens or pastures with hogs. Post buildings and lots with signs requesting compliance with these rules.

10. Dispose of dead animals (a) through licensed, properly equipped rendering trucks; (b) by deep burial—at least 6 feet, with carcass covered with lime before dirt is replaced; or (c) by complete burning.

11. Dispose of contaminated bedding by (a) burial

or burning; (b) complete soaking of entire mass with 3 percent solution of USP cresol compound, with chloride of lime—30 percent available chlorine—or with formaldehyde; or (c) spreading on cropland.

II. The Breeding Herd

1. Maintain a closed herd insofar as is possible. Do not allow breeding stock to come into contact, either directly or indirectly, with hogs from other herds. Start with breeding stock that is apparently free of infectious diseases. Do not allow poultry to mingle with breeding stock.

2. Buy breeding stock from as few herds as possible (one is best). Insist on evidence of freedom from disease in the herds from which you buy stock. Such evidence may include general disease certification programs, brucellosis validation or specific pathogen-free certification, as well as observation of the herd by the buyer or his veterinarian. As a further safeguard, keep newly purchased breeding stock isolated for at least six weeks. During this period, observe them for symptoms of disease, and blood test for brucellosis.

3. Validate the herd for brucellosis by blood testing all animals six months of age or over. One test is adequate to certify the herd, provided all hogs are tested and found negative. Then, recertify the herd annually thereafter, by passing a single negative test on the entire herd. If brucellosis is found, select and follow the appropriate eradication plan as given in Table 10-2 of this Section under "Brucellosis."

4. Vaccinate gilts or sows against leptospirosis at the time of purchase. Revaccinate all sows annually in areas where the disease is known to be present. Confer with your veterinarian.

5. Vaccinate females and boars for erysipelas if recommended by the veterinarian. Sows can be vaccinated during pregnancy, to within a month to three weeks before farrowing. This increases the antibody level of the sow's milk so that the pigs probably will not need to be vaccinated during the first few weeks of life; vaccination of pigs should be delayed since a high antibody level in the sow's milk may prevent development of satisfactory immunity by the pigs. The advice of a veterinarian should be sought on erysipelas control programs in herds where the disease has been a problem.

6. Where there is exposure to transmissible gastroenteritis (TGE), upon the advice of your veterinarian vaccinate sows twice—six weeks and two weeks prior to farrowing.

7. Control lice, mange, and other external parasites on sows and boars (see Table 10-7, Swine Parasites and Their Control, for effective insecticides). Since low-grade, unnoticed mange may be present on sows in the summer, it is good insurance routinely to treat every sow in the fall or winter during the last six weeks before farrowing.

8. Treat sows for ascarids (large roundworms) two weeks prior to breeding and two weeks prior to farrowing, using the drug of choice according to the manufacturer's directions (see Table 10-7, Swine Parasites and Their Control, "Ascarids," for list of drugs).

9. House sows in well-ventilated, draft-free buildings. Bedding should be dry at all times.

10. Group sows according to age and weight, with no more than 20 to 25 sows to a group.

11. Arrange housing and feeding facilities to ensure maximum exercise. However, do not force sows or boars to travel great distances over rough, frozen ground or on ice.

12. Provide adequate shade for the breeding herd during summer months. Access to a sanitary wallow or a sprinkler may increase the number of live pigs farrowed, but hogs should be kept out of filth such as old wallows that are used year after year.

13. Isolate and treat sows showing signs of flu or pneumonia. Do not breed sows that are suffering from flu or pneumonia; be certain they have recovered before breeding.

14. Wash sows with warm water and soap and rinse with mild antiseptic solution before moving them into the farrowing area. When washing the sow, particular care should be taken to remove the small plug of dirt from the end of each teat. Also, sows may be sprayed for mange and lice at this time.

15. If sows have been rather generously fed during gestation, reduce the allowance and use a laxative feed after they are placed in farrowing stalls. The amount necessary to satisfy them will vary, but should not be more than approximately 30 to 50 percent of normal daily intake during the latter part of gestation.

Where sows have been fed 4 lb per head per day until 30 days before farrowing, and then 4 to 6 lb daily from this time until they are placed in farrowing stalls, no reduction in allowance is necessary until farrowing. Give little or no feed on day of farrowing, then feed 4 lb the first day after farrowing.

16. Inject sows with antibiotics before and after farrowing to prevent uterine infections following farrowing, provided the veterinarian so recommends.

III. Farrowing to Weaning

The following program will materially lessen baby pig losses:

1. Provide farrowing quarters that are warm, dry, and free of drafts, as it will help prevent scours and other baby pig diseases. Scrape, clean, and disinfect the farrowing house between farrowing periods. Having the farrowing house idle for three weeks between farrowing periods helps to check the buildup of infectious organisms. Farrowing stalls, guardrails, or other structural design should be used to save pigs from

being laid on by sows. Use a small amount of bedding such as coarsely ground corncobs, chopped straw, or wood shavings. Keep bedding clean and dry.

2. Be on hand when sows farrow, or at least check on them every few hours. Remove the newborn pigs from the surrounding membrane and clean the mucus from their noses and mouths. Placing the pigs under a brooder or heat lamp may save some of them from being laid on or chilled.

3. Help weak or chilled pigs to nurse, as pigs should receive colostrum soon after birth.

4. Provide heat lamps or other supplemental heat in the creep area for baby pigs, to maintain temperatures of 85° to 90°F for the first few days following birth. Suspend the heat lamp from 24 to 28 inches above the bedding and follow other rules for safe use of heat lamps.

5. Even up litters and provide for orphan pigs by switching among sows farrowing within 48 hours of each other. But first the pigs should be allowed to nurse, and runt pigs should be destroyed, before extra pigs in a litter are transfered to another sow. Attempts to transfer pigs to a sow that has farrowed more than 3 or 4 days beforehand are seldom successful, since teats that are not nursed dry up. Masking the body odor of the pigs by spraying the litter and the transferred pigs with a disinfectant may be helpful in changing older pigs.

6. Sever the navel cord, leaving ½ to 1 inch. This can be done by grasping it between the thumb and finger of each hand and pulling until it parts. The cord can be cut, but this may result in excessive bleeding. Disinfect navel with 7 percent tincture of iodine solution to protect against infections that enter through the navel.

7. Clip "needle teeth" soon after birth to prevent cuts on the sow's udder or injury to pigs' noses when they fight. Cut the tip off the tooth with a pair of sharp sidecutter pliers. Be careful not to crush the teeth, injure the gum, or leave jagged edges, or you may do more harm than good.

8. Prevent nutritional anemia (iron deficiency) in pigs kept on wood or concrete floors for two weeks or more after farrowing. This can be accomplished by—

a. Injecting a soluble iron compound (containing 150 mg of iron) at 1 to 3 days of age. Repeat at 2 to 3 weeks if necessary. Be sure needles and syringes are boiled for 25 to 30 minutes to avoid abscess formation.

b. Placing clean, hog manure-free sod in each farrowing pen several times a week.

c. Giving iron pills or liquids to pigs at weekly intervals.

d. Using iron-fortified baby pig feeds in the creep area.

9. Castrate male pigs early—at 3 days to 2 weeks of age. This reduces shock and the possibility of infection. Care should be taken to make the incisions low enough to drain properly. Sterilize instruments used in castration before use by boiling for 15 minutes. Keep them clean by placing the instrument in a disinfectant solution before and after each pig is castrated. On farms where tetanus (lockjaw) is a problem, keep pigs confined to clean quarters until healed.

10. Provide clean, fresh drinking water.

11. Where disease conditions warrant, feed antibiotics in the creep feed until the pigs reach 75 to 100 pounds.

12. Vaccinate pigs for erysipelas if this disease is in the herd, or if it is prevalent in your community. Follow the advice of your veterinarian in the choice of the vaccine and in the vaccination schedule.

13. Treat pigs for ascarids (large roundworm), using according to manufacturer's directions one of the products listed in Table 10-7, Swine Parasites and Their Control, "Ascarids."

14. Protect suckling pigs from lice and mange by treating the sows and boars. Do not spray pigs before they are 6 weeks old. Treating the bedding with ronnel (Korlan) at the rate of ½ pound per 100 square feet will elimiate lice, but it has not been shown to be effective against mange.

IV. Weaning to Market

1. Provide dry, well-bedded, draft-free sleeping quarters that are well ventilated. Pigs put in buildings that are cold and drafty are susceptible to diseases such as pneumonia, and they usually make poor daily gains. Extra bedding will help reduce the stress of cold weather. In hot weather, adequate shade is necessary. A mist spray system or wallow helps reduce the stress of hot weather, but hogs should be kept out of stagnant pools or old wallows that are used year after year.

2. Separate pigs into groups according to size, rather than age. Keep groups as uniform as possible with regard to size. Generally, it is not advisable to have more than 40 feeder pigs in a group.

3. Feed well-balanced rations, properly fortified. Keep clean drinking water available at all times. Provide one fountain for each 30 head and no more than 4 pigs per feeder hole.

4. Control internal parasites, especially ascarids (roundworms), through sanitation, pasture rotation, and use of drugs. The McLean County system of rotating pastures and sanitation remains the best method of preventing infection by worms of pigs on pasture. For pigs in confinement, thorough and frequent cleaning of buildings and floors is the best control program.

Worm routinely if autopsy or fecal examinations reveal adult worms or worm eggs, using dichlorvos, piperazine, or levamisole hydrochloride according to manufacturer's directions.

5. Follow a louse and mange control program routinely. A spray or dip is best, and malathion or toxaphene, used according to manufacturer's directions, are best for control of both lice and mange.

Small groups of animals should be sprayed at one time, and complete coverage of the body is necessary to control the parasites. Retreatment in 10 to 14 days is recommended. If weather does not permit the use of a spray or dip, dusting powders can be used. Pigs to be treated properly must be confined. If temperatures are extreme, simply hold pigs in confinement until they are dry.

6. Isolate sick hogs immediately, and notify the veterinarian.

V. Purchased Feeder Pigs

Normally, death losses from the time of delivery of feeder pigs to your farm until market time should not exceed 2 percent. About half these losses occur during the first week after arrival.

Feeder pigs that are in good condition and purchased from a reliable source (preferably a single source), then handled as described below, will have the best chance of getting off to a good start with minimum death losses:

1. Abide by the legal health regulations of your state. These laws are for your protection. Vaccinations and other health certificates should be furnished by the seller and demanded by the buyer.
2. Avoid unnecessary handling, watch the pigs closely during this critical period, and work with your veterinarian to maintain optimum herd health.
3. Keep newly arrived feeder pigs isolated from other hogs for at least three weeks.
4. Avoid contact between feeder pigs and breeding stock, because feeder pigs can sometimes be carriers of diseases which can be spread to other hogs on the farm. Wear different outer clothing and boots when working with two sets of pigs to avoid spreading disease.
5. Provide warm, dry, draft-free, disinfected, well-bedded quarters.
6. Allow ample space for feeding, watering, and sleeping; provide one feeder space at the feeder for each 5 pigs, one waterer for each 30 pigs, and 4 to 6 square feet of sleeping space per pig.
7. Feed a low-protein (10% or less) ration for the first 3 to 4 days.
8. Use high antibiotic level for the first three days, preferably in the drinking water.
9. Dust bedding for control of lice and mange before pigs arrive. Do not spray pigs sooner than five days after arrival or when weather is unfavorable.
10. Do not worm or castrate pigs for at least 10 days after arrival.
11. Keep pigs of different sizes separated; otherwise, the bigger ones will crowd the smaller ones away from feed and water.
12. When signs of trouble appear, call your veterinarian.

Specific Pathogen-Free (SPF) Pigs

Specific pathogen-free pigs are pigs that are free of disease at birth. Pathogen-free pigs are obtained from their dam 2 to 4 days prematurely by hysterectomy, caesarotomy, or hysterotomy. Also, pigs may be caught at natural birth in sterile canvas bags, in sterile basins, or on sterile canvas towels.

Hysterectomy means removal of the womb. With this technique, the pigs are freed without transversing the birth canal, thereby eliminating any chance of the pigs becoming infected while passing through the birth canal. Although hysterectomy represented the ultimate in disease control, it had one major disadvantage. Many laboratories could not comply with meat inspection regulations; hence, they experienced difficulty in marketing the carcass of the sow. As a result, this forced commercial laboratories to obtain the pigs by Caesarean section (C-section), with methods developed to keep the newborn pigs separated from the contaminated environment of the sow. Fig. 10-130 diagrammatically presents the SPF approach to elimination of chronic diseases from the national swine population.

This system, which is licensed, embraces the following provisions: (1) obtaining pigs by a surgical process (Caesarean section) 2 to 4 days before normal birth; (2) rearing pigs in individual isolation until one week old, and in groups of 8 to 12 until 4 weeks old; (3) rearing pigs in groups of 10 to 20 from 4 weeks old to maturity, on farms from which all other swine have been removed, to which no new stock is introduced, and where the producer avoids contact with other swine; (4) resuming normal birth of SPF pigs on these clean farms; and (5) restocking other "clean" farms—farms that have no swine or only SPF swine, and on which the owner avoids contact with other swine.

Primary SPF herds are those originating from surgically derived stock and maintained in strict isolation. Any new blood lines added to the herd must also be obtained by surgical means. These primary herds are used to supply secondary multiplying herds, which in turn supply breeding stock to commercial swine producers.

The SPF method is drastic and costly. However, at the present time, it is the only means whereby atrophic rhinitis, mycoplasma pneumonia (MP), and swine dysentery can be controlled and eradicated. In addition to these diseases, SPF herds must be validated brucellosis-free, leptospirosis-free, and with no evidence of lice or mange.

The National SPF Accrediting Agency, Inc., Conrad, Iowa 50621 is responsible for supervising the SPF program and for issuing an Accreditation Certificate to those who qualify.

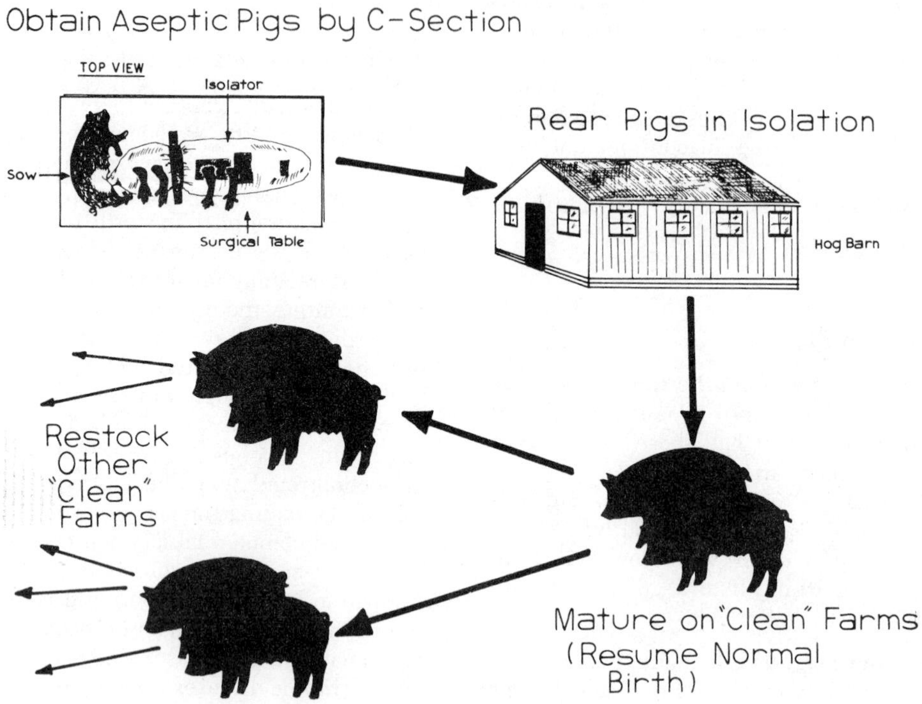

Fig. 10-130. Diagrammatic outline of swine repopulation method.

A PROGRAM OF HORSE HEALTH, DISEASE PREVENTION, AND PARASITE CONTROL

In addition to following a program embracing superior breeding, sound management, and scientific feeding, the good horseman will adhere to a strict sanitation and disease-prevention program designed to protect the health of his animals. Although the exact program will differ from farm to farm, the basic principles will remain the same. With this in mind, the following program of horse health, disease prevention, and parasite control is presented with the hope that the horseman will use it (1) as a yardstick with which to compare his existing program, and (2) as a guidepost so that he and his local veterinarian, and other advisors, may develop a similar and specific program for his own enterprise.

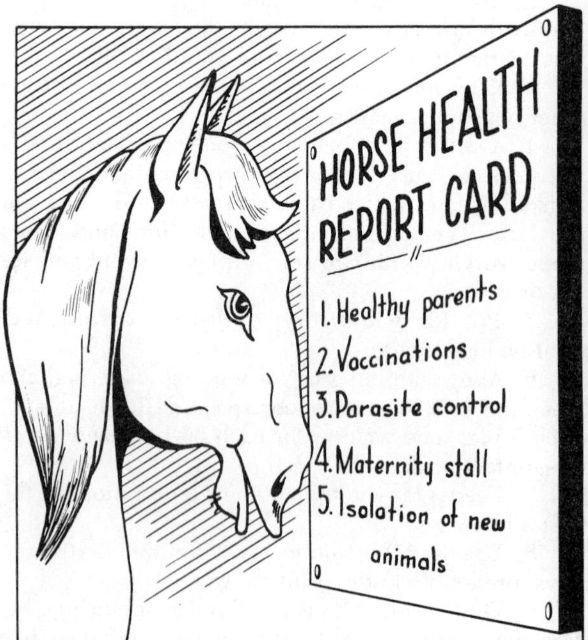

Fig. 10-131

I. General Horse Health Program

The following health program is recommended for all horses:

1. Have on hand first aid supplies, and know when and how to use them in case of accident or sudden illness.

2. Vaccinate against the most common diseases (Table 10-2).

3. Avoid public feeding and watering facilities.

4. When signs of infectious disease are encountered, promptly isolate affected animals, provide them with separate water and feed containers, and follow the instructions and prescribed treatment of the veterinarian.

5. Prevent or control parasites by adhering to the following program:

a. Provide good sanitary practices and a high level of nutrition.

b. Have adequate acreage; use temporary seeded pasture rather than permanent pasture, and practice rotation grazing.

c. Pasture young stock on clean pastures, never allowing them to graze on an infested area unless the area has been either plowed or left idle for a year in the interim.

d. Do not spread fresh horse manure on pastures grazed by horses; either store the manure in a suitable pit for at least two weeks, or spread it on fields that are to be plowed and cropped.

e. When small fields or paddocks must be used, pick up the droppings at frequent intervals.

f. Keep pastures mowed and harrowed (use a chain harrow).

g. Prevent fecal contamination of feed and water.

h. Follow a worming program and schedule to control internal parasites.

i. When external parasites are present, apply the proper insecticide.

j. If cattle are on the farm, alternate the use of pastures between cattle and horses, since horse parasites will die in cattle.

k. Avoid overgrazing because there are more parasites on the bottom inch of the grass.

II. Breeding and Foaling Health Program

The following health program is recommended where horses are bred and foals are produced:

1. Mate only healthy mares to healthy stallions and observe scrupulous cleanliness at the time of service and examination. Never breed a mare from which there is a discharge.

2. Provide plenty of exercise for the stallion and the pregnant mare, either in harness or under the saddle or in roaming over a large pasture in which plenty of shade and water are available.

3. During the spring and fall months when the weather is warm, allow the mare to foal in a clean, open pasture, away from other livestock. During inclement weather, place the mare in a roomy, well-lighted, well-ventilated box stall—which first should be cleaned carefully, disinfected thoroughly with a lye solution (made by adding 1 can of lye to 12 to 15 gal of water), and provided with clean straw (not shavings) for the occasion. After foaling, all wet, stained, or soiled bedding should be removed and the floor lightly dusted with lime (excessive lime is irritating to the eyes and nasal passages of foals). The afterbirth should be examined for completeness, and after ascertaining that all of it has been discharged, it should be burned or buried in lime; and the mare should be kept isolated until all discharges have ceased.

4. To lessen the danger of navel infection, promptly treat the navel cord of the newborn foal with tincture of iodine.

5. As a precaution against foaling diseases and other infectious troubles, the veterinarian may administer antibiotics to both the mare and foal on the day of foaling.

III. Health Program for New Horses and Visiting Mares

The following health program is recommended where new horses and visiting mares are brought into the herd:

1. Isolate new animals for a period of three weeks before adding them to the herd. During this period, the veterinarian may (a) administer sleeping sickness vaccine (in season) and tetanus toxoid, (b) make a thorough general and parasitic examination, and (c) give a genital examination of breeding animals, and treat where necessary.

2. Require that mares brought in for breeding be accompanied by a health certificate issued by a veterinarian. Beware of mares that have had trouble in foaling or have lost foals.

3. If feasible, cover visiting mares near their own isolation quarters, using tack and equipment that is not interchanged with that used for mares kept on the establishment.

REGULATIONS RELATIVE TO DISEASE CONTROL

Certain animal diseases are so devastating that no individual farmer or rancher could long protect his herds and flocks against their invasion. Moreover, where human health is involved, the problem is much too important to be entrusted to individual action. In the United States, therefore, certain regulatory activities in animal disease control are under the supervision of various federal and state organizations. Federally, this responsibility is entrusted to the U.S. Department of Health and Human Services and the U.S. Department of Agriculture.

Food and Drug Administration (FDA)

In 1906, the U.S. Congress enacted the Pure Food and Drug Law. Concurrently, the Federal Meat Inspection Act was passed. Both laws became effective in 1907. The U.S. Food and Drug Administration (FDA) was established as a separate unit of the U.S. Department of Agriculture in 1927. Then, in 1940, the FDA was transferred from the USDA to the Federal Security Agency, presently, the U.S. Department of Health and Human Services.

The FDA is charged with the responsibility of safeguarding American consumers against injury, unsanitary food, and fraud. It also protects industry against unscrupulous competition. It inspects and analyzes samples and conducts independent research on such things as toxicity (using laboratory animals), disappearance curves for pesticides, and long-range effects of drugs.

U.S. PUBLIC HEALTH SERVICE (USPHS)

This section of the Department of Health and Human Services is concerned with the prevention and treatment of disease. It works in the areas of vector control, pollution control, and control of communicable diseases of man. A part of this important complex is the National Institute of Health (NIH), which was formed in 1930, and which is composed of the following nine sister institutes: the National Cancer Institute, the National Heart Institute, the National Institute of Allergy and Infectious Diseases, the National Institute of Arthritis and Metabolic Diseases, the National Institute of Dental Research, the National Institute of Mental Health, the National Institute of Neurological Diseases and Blindness (including multiple sclerosis, epilepsy, cerebral palsy, and blindness), the National Institute of Child Health and Human Development, and the National Institute of General Medical Science. In addition to its own research program, the USPHS provides grants for health related research at many universities and research institutes in the United States.

U.S. Department of Agriculture (USDA)

The following four divisions of the U.S. Department of Agriculture have primary responsibilities in the area of animal and human health:

1. **The Animal and Plant Health Inspection Service**—This division is charged with maintaining the wholesomeness and safety of meats processed in packing plants that ship meat and meat products, including poultry and poultry products, interstate. Veterinarians and other trained personnel make the inspections. Its purpose is to protect consumers against infected meats and fraudulent and unsanitary preparation of meat products. The inspection first consists of an examination of the live animal so that any unfit beast may be removed and disposed of properly. Secondly, the carcasses and internal organs are inspected for any abnormalities of animals carrying infectious diseases. Centers of infection sources may be located, thus assisting the livestock owners in the vicinity.

2. **The Labeling and Registration Section**—This section in the USDA has responsibility for the proper labeling and safe use of pesticides. Manufacturers of pesticides must present new products with their proposed labels for approval before they are authorized to sell them. The label must indicate, as a minimum, the following: the name of the product; the active and inactive ingredients, together with the percentage of each, in the formulation; the pest(s) controlled; directions for use—including the method and rate of application; any restrictions to be observed in application and handling; and an antidote—if known.

It is the responsibility of the FDA, however, to set legal tolerances for pesticides on or in raw agricultural products. Also, it sets the "safe" interval between last application of the insecticide and the time of harvest of the crop or the slaughter of the animal.

Thus, through the cooperative supervision of the USDA and the FDA, both the pesticide user and the consumer of the product are safeguarded.

3. **Veterinary Services**—This division of the USDA is responsible for programs to control and eradicate (if possible) certain diseases of livestock; e.g., brucellosis, tuberculosis, scabies, and hog cholera. It does the following things: conducts nationwide federal-state cooperative programs for the control and eradication of animal diseases; suppresses spread of disease through control of interstate and international movements of livestock; keeps informed of the overall disease situation nationally and internationally, administers laws to ensure humane treatment of livestock and certain laboratory animals; collects and disseminates information on morbidity and mortality; and provides training for USDA employees and others in related government agencies.

4. **Stockyards Inspection**—With the advent of large public markets, public stockyards inspection was initiated. This is an addition to the regular inspection performed on animals by meat inspectors prior to slaughter. Among the principal diseases for which inspections are made are: anthrax, scabies of cattle and sheep, tick or splenetic fever, hog cholera, and erysipelas of swine.

Not only are the incoming shipments of livestock

inspected, but a reinspection is made of outgoing shipments. Tests for tuberculosis and brucellosis are accomplished, and dipping for scabies is performed before shipments are allowed to return to farms and ranches.

State Veterinarians, Sanitary Commissions, and Boards

Most states have state veterinarians, or comparable officials, who direct the livestock sanitary and regulatory programs within their respective states. Stockmen may secure the regulations applicable to the state in which they reside by writing to their State Department of Agriculture.

Interstate Requirements with Respect to Animal Health[8]

Health requirements governing the interstate movement of livestock are issued by both state and federal animal health agencies. Federal regulations are contained in Title 9 of the Code of Federal Regulations, the principal provisions of which are:

Part 71 (General Requirements)—Prohibits or restricts the interstate movement of animals affected with or exposed to a communicable disease.

Part 72 (Texas [Splenetic] Fever in Cattle)—Describes areas under quarantine and requires dipping in a permitted dip prior to movement from an infected area.

Part 73 (Scabies in Cattle)—Affected cattle are required to be dipped in a permitted dip prior to interstate movement.

Part 75 (Communicable Diseases in Horses)—Reactors to the test for equine infectious anemia are required to be identified and shipped interstate for slaughter.

Part 76 (Hog Cholera and Other Communicable Swine Diseases)—Several conditions under which interstate movement may be made are set forth, depending upon the status of the swine and the state of origin with respect to hog cholera and the feeding of garbage.

Part 77 (Tuberculosis in Cattle)—Requires the branding of reactors and tuberculin testing of cattle which originate in areas which are not modified accredited.

Part 78 (Brucellosis)—Requires that reactors be identified and moved for slaughter; exposed cattle must be identified and may be moved for slaughter or to a quarantined feedlot. Other cattle must be tested, or vaccinated, and so certified, depending upon the status of the state and herd of origin.

Part 83 (Screwworms)—Describes the Screwworm Control Zone, the areas of recurring infestation, and the inspection and treatment requirements for movement from these areas.

In addition to the federal interstate regulations, each of the states has requirements for the entry of livestock. Generally, these requirements include compliance with interstate regulations. States usually require a certificate of health or a permit, or both, and additional testing requirements depending upon the class of livestock involved.

Detailed information can be obtained from state or federal animal health officials, or from accredited veterinarians in all states. Shippers are urged to obtain such information prior to making interstate shipments of livestock.

Quarantine

Many highly infectious diseases are prevented by quarantine from (1) gaining a foothold in this country, or (2) spreading. *By quarantine is meant (1) segregation and confinement of one or more animals in the smallest possible area to prevent any direct or indirect contact with animals not so restrained; or (2) regulating movement of animals at points of entry.*

When an infectious disease outbreak occurs, drastic quarantine must be imposed to restrict movement out of an area or within areas. The type of quarantine varies from one involving a mere physical examination and movement under proper certification to the complete prohibition against the movement of animals, produce, vehicles, and even human beings.

Foreign Disease Protection

Distance no longer provides a buffer against the invasion of foreign diseases. More than 90 percent of animals imported into the United States arrive by air. A jet plane can outpace the development of clinical signs of disease in an animal that has been exposed to infection just prior to shipment. This prompts great concern for epizootic diseases capable of crippling or destroying entire livestock populations. Such diseases still exist in Asia, Africa, and Latin America; among them, dreaded diseases such as rinderpest, contagious bovine pleuropneumonia, foot-and-mouth disease, African horse sickness, African swine fever, Newcastle disease, fowl plague, trypanosomiasis, East Coast fever, and piroplasmosis.

Until 1875, the importation of livestock into the United States was free and easy. But, that year the United States prohibited the importation of cattle and hides from Spain, where foot-and-mouth disease was rampant. By 1880, European countries were refusing to buy cattle or beef from the United States, for fear of

[8]This section was authoritatively reviewed by Dr. F. W. Hansen, Jr., DVM, Animal and Plant Health Inspection Service, USDA, Hyattsville, Md.

getting contagious bovine pleuropneumonia. Then, in 1884, Congress established the Bureau of Animal Industry in the U.S. Department of Agriculture and gave the Secretary of Agriculture authority to enforce quarantine laws.

Today, there are stations at several entry points, where inspectors of the USDA's Veterinary Service Division inspect all animals and poultry to be imported to the United States. If no communicable diseases are found, the animals may be quarantined for a period of time, during which time they are treated for external parasites and subjected to various tests—e.g., horses are tested for glanders and equine infectious anemia, and cattle are tested for brucellosis. At the end of the quarantine period, if no communicable diseases are found, they are released to the purchaser.

The Veterinary Services Division is also charged with the responsibility of safeguarding against diseases introduced by the importation of zoo animals into this country. Wild animals brought into this country must undergo an extensive quarantine period abroad, followed by a further quarantine period at the animal quarantine station at Clifton, New Jersey. Moreover, they are allowed to go only to certain approved zoos, where the zoo animals are isolated from domestic livestock and where proper measures are taken to dispose of waste to prevent the spread of diseases.

Federal Quarantine Center

A Federal Quarantine Center was authorized in Public Law 91-239, signed by the President on May 6, 1970; and a 16.1-acre site for the Center was selected at Fleming Key, near Key West, Florida.

The Quarantine Center is designed to hold some 400 head of cattle, or other species in equivalent numbers, at one time, for a five-month quarantine period. This maximum security station enables American livestock producers to import breeding animals from all parts of the world, while at the same time safeguarding our domestic herds and flocks from such diseases as foot-and-mouth disease, rinderpest, piroplasmosis, and others.

Indemnity Payments[9]

Where certain animal diseases are involved, the stockman can obtain financial assistance in eradication programs through federal and state sources.

Note well: Both federal and state indemnity payments are subject to change. So, for current regulations, the stockman should contact his local veterinarian or State Department of Agriculture.

[9]This section, including both "Federal Indemnity Payments" and "State Indemnity Payments," was authoritatively reviewed by Dr. F. J. Mulhern, DVM, Administrator, Animal and Plant Health Inspection Service, USDA, Hyattsville, Md.

FEDERAL INDEMNITY PAYMENTS

Information relative to indemnities paid to owners by the federal government for animals disposed of as a result of outbreaks of certain diseases is given in Chapter I, Subchapter B, Title 9 of the Code of Federal Regulations, a summary of which follows:

• **Brucellosis and tuberculosis**—The indemnity payments to owners by the federal government where brucellosis and tuberculosis are involved change from time to time. But the pertinent regulations that existed when this book was written follow:

1. **Brucellosis**—

a. **Affected cattle**—Owners of cattle destroyed which are affected with brucellosis may be paid an indemnity by the Department not to exceed $50 for any grade animal or $100 for any registered animal, except in Alaska, Hawaii, Puerto Rico, and the Virgin Islands where no payment for any animal destroyed shall exceed $100.

b. **Herd depopulation**—The Deputy Administrator may authorize the payment of federal indemnity to owners whose cattle are destroyed because of brucellosis not to exceed $50 for any nonregistered animal or $100 for any registered animal which (1) has been found to be exposed; and (2) is a part of a known infected herd, the destruction of which will, in the opinion of the Deputy Administrator, contribute to the brucellosis eradication program.

c. **Exposed cattle**—The Deputy Administrator may authorize the payment of federal indemnity to owners for cattle destroyed because of brucellosis not to exceed $50 for any nonregistered animal or $100 for any registered animal which has been found to be exposed by reason of previous association with brucellosis affected cattle, the destruction of which will, in the opinion of the Deputy Administrator, contribute to the brucellosis eradication program.

2. **Tuberculosis**—

a. **Affected cattle**—The Department may pay owners an indemnity for cattle affected with tuberculosis not to exceed 90 percent of the difference between the appraised value of each animal so destroyed and the net salvage received by the owner; *provided* (1) that no such payment may exceed $350 for each animal, and (2) that any joint state-federal indemnity payment, plus salvage, does not exceed the appraised value of each animal.

Also, the following regulations apply specifically to tuberculosis: The Deputy Administrator may authorize the payment of indemnity to owners of cattle which are destroyed because of tuberculosis not to exceed $100 for any grade animal or $200 for any purebred animal which (1) has been found to be exposed, and (2) is a part of a known infected herd, the destruction of which will, in the opinion of the Deputy Administrator, contribute to the tuberculosis eradication program; *provided* that the joint state-federal indemnity payments plus salvage do not exceed the appraised value of the animals.

b. **Appraisals of cattle destroyed because of tuberculosis**—Cattle to be destroyed because of tuberculosis shall be appraised by Veterinary Services or state representative, with due consideration given to their breeding value as well as to their dairy or meat value. Where purebreds are involved, the owner shall either (1) see that the animals are accompanied by their registration papers at time of appraisal; or (2) be granted a reasonable time, by the veterinarian in charge, in which to present papers. Veterinary Services may decline to accept any appraisal that appears to be unreasonable or out of proportion to the market value of cattle of like quality.

3. **Marking (or identifying) and slaughtering brucellosis and tuberculosis reactors**—Prior to marketing, the cattle must be marked or identified as follows:

a. Brucellosis reactor cattle must be branded with a 2- to 3-inch-high letter "B" on the left jaw, and tagged with a metal federal or state reactor tag in the left ear.

b. Tuberculosis reactor cattle must be branded with a 2- to 3-inch-high letter "T" on the left jaw, and tagged with a metal federal or state reactor tag in the left ear.

c. The cattle on which indemnity payments are made must be slaughtered within 15 days after the appraisal is made, unless an extension of time is granted by Veterinary Services.

4. **Example**—For purposes of illustrating the federal indemnity payments where tuberculosis of cattle is involved, let us assume the following:

a. That a cow reacted to a test for tuberculosis.

b. That the designated appraiser valued this breeding cow at $850.

c. That the properly branded and eartagged cow was sent directly to slaughter where she netted the owner $150, after deducting certain marketing costs.

What indemnity payments may the owner expect?

a. The difference between the appraised value and the net market price is $850 − $150 = $700.

b. Since federal payments irrespective of state payments are computed at 90% of the difference between appraised value of the cow and salvage after deducting certain marketing costs, therefore 90% of $700 = $630.

c. However, the maximum federal payment is limited to $350 for either a registered or grade cow.

d. Therefore, the federal indemnity payment in this case would be $350.

e. Thus, the money the stockman may receive from this cow valued at $850 is $150 in salvage, $350 federal indemnity, and a varying amount of state indemnity based upon the particular state's regulations.

• **Foot-and-mouth disease, pleuropneumonia, rinderpest, and other contagious or infectious animal diseases which constitute an emergency and threaten the livestock industry of the country**—Under Title 9, Part 53, of the U.S. Code of Federal Regulations, the Secretary of Agriculture of the U.S. Department of Agriculture may declare a national emergency due to the existence of foot-and-mouth disease, rinderpest, contagious pleuropneumonia, or any other communicable disease of livestock and poultry which threatens the livestock industry of the country. Upon agreement with state authorities to enforce quarantine restrictions and orders and to participate equally in the payment of indemnities:

1. The federal government will pay to the owner 50% (and in the case of exotic Newcastle disease up to 100%) of the expenses of purchase, destruction, and disposition of animals and materials required to be destroyed because of being contaminated by or exposed to such disease. The appraisal of animals shall be based on the fair market value and shall be determined by the meat, egg production, dairy, or breeding value of such animals. (In the case of hog cholera, under Part 56, of the U.S. Code of Federal Regulations, the federal government can pay 100% of the appraised value up to a maximum of $300 per head for purebred, inbred, hybrid, or breeding swine, and a maximum of $150 per head for all other swine.) In the case of grade swine which are affected with or exposed to hog cholera, only females shall be eligible for appraisal based on breeding value and no such appraisal shall exceed 3 times the animals' meat or feeding value. Animals may be appraised in groups providing (a) they are the same species and type; and (b) where appraisal is by the head, each animal in the group is the same value per head, or where appraisal is by the pound, each animal in the group is the same value per pound.

2. The federal government will pay to the owner 50% (and in the case of exotic Newcastle disease and hog cholera up to 100%) of the expenses of purchase and destruction of contaminated materials (parts of barns or other structures, equipment, feed, bedding, clothing, and other items) that must be destroyed be-

cause they cannot be properly cleaned and disinfected.

3. The appraised value of animals and materials must be established by either (a) Veterinary Services (VS) and a state representative jointly, or (b) a VS representative alone, provided the state authorities approve.

4. Animals affected by or exposed to disease shall be destroyed promptly after appraisal and disposed of by burial or burning, unless otherwise specifically authorized by Veterinary Services of the Animal and Plant Health Inspection Service of the U.S. Department of Agriculture.

In order to reduce the cost of eradicating emergency disease to the stockman and to the state and federal governments, it is essential that suspicious cases be promptly reported. If such a disease is suspected, a report should promptly be made to your practicing veterinarian and to state and federal animal health officials.

STATE INDEMNITY PAYMENTS

Although subject to change as the laws of the respective states change, Table 10-14 summarizes the information relative to state indemnity payments as they existed at the time this book was written. It is suggested that each stockman secure the regulations applicable to the state in which he resides by writing to his State Department of Agriculture.

TABLE 10-14

SUMMARY OF STATE INDEMNITY PAYMENTS FOR REACTORS[1, 2]

Disease	No State Indemnity Paid	States that Pay More than Federal Maximum	States that Pay Less than Federal Maximum	States that Pay Amounts Equaling Federal Maximum
Brucellosis	Ind. Okla. Kan. Tenn. Miss. Tex. Mont. Va. Nev. W. Va. N.M. Wyo. N.D. V.I. Ohio	Calif. N.J. Conn. N.Y. Hawaii Ore. Iowa Penn. Mich. R.I. N.H. Wisc.	Ariz. Me. Ark. Mass. Del. Mont. Fla. Neb. Ga. N.C. Ida. S.D. Ill. Utah Ky. Wash. P.R.	Ala. Alaska Colo. La. Md. Mo. S.C.
Tuberculosis	Ariz. Okla. Ky. W. Va. La. Wyo. Minn. V.I. Miss.	N.J. Penn. R.I.	Ala. Kan. N.D. Ariz. Me. Ohio Calif. Mass. Ore. Conn. Mich. S.D. Del. Mo. Tenn. Fla. Mont. Tex. Ga. Neb. Utah Hawaii Nev. Vt. Ida. N.H. Va. Ill. N.M. Wash. Ind. N.Y. P.R. Iowa N.C.	Alaska Colo. Md. Minn. Wisc.

[1]Foot-and-mouth disease and other animal diseases which threaten the livestock industry, as covered in CFR Title 9, Chapter 1, Subchapter B, Part 53.

[2]Most states have provision for the expenditure of emergency funds in the event of an outbreak of such foreign animal diseases, and other funds would be forthcoming through special action of their respective state legislatures.

SECTION 11

SELECTING AND JUDGING LIVESTOCK

Contents **Page**

Bases of Selection 1007
Qualifications of a Good Judge 1007
Do's and Don'ts for Contest Judges 1008
How to Use the Judging Guides, Tables 11-1 Through 11-5 1008
Scorecard Judging 1008
Judging Beef Cattle 1010
Breeding Beef Cattle Scorecard 1013
Steer Scorecard 1014
Judging Dairy Cattle 1015
Dairy Cow Scorecard 1018
Judging Sheep 1019
Breeding Sheep Scorecard 1022
Market Lamb Scorecard 1022
Judging Swine 1023
Breeding Swine Scorecard 1026
Market Barrow Scorecard 1027
Judging Horses 1028
Horse Scorecard 1031
Determining the Age of Animals by the Teeth 1032
Determining the Age of Cattle by the Teeth 1032
Determining the Age of Sheep and Goats by the Teeth 1032
Determining the Age of Swine by the Teeth 1032
Determining the Age of Horses by the Teeth 1032
Comparative Animal Ages 1041

The great livestock shows throughout the land have exerted a powerful influence in molding animal types. At the same time, producers of meat animals are ever aware of market demands as influenced by consumer preferences, dairymen recognize that the main function of dairy cattle is the production of milk, and producers of light horses are cognizant of the importance of performance. It is realized, however, that only a comparatively few animals on farms and ranches are subjected annually to the scrutiny of experienced show-ring judges or market specialists. Rather, the vast majority of purebred animals and practically all commercial animals are evaluated by practical stockmen—men who conduct their own buying and selling operations. In general, these men are intensely practical; no animal meets with their favor unless it has utility value. Such stockmen have no interest in the so-called breed fancy points, and they may not be able to express fluently their reasons for selecting certain animals while culling others. But successful stockmen become quite deft in their evaluations. They are generally good judges of livestock.

BASES OF SELECTION

There are four bases of selection; namely, (1) selection based on type or individuality, (2) selection based on pedigree, (3) selection based on show-ring winnings, and (4) selection based on production testing. Only the first one—commonly called judging—will be enlarged upon in the sections which follow.

QUALIFICATIONS OF A GOOD JUDGE

The essential qualifications which a good judge of any class of stock must possess are:

1. **Knowledge of the parts of an animal**—This consists of mastering the language that describes and locates the different parts of an animal.

2. **A clearly defined ideal or standard of perfection**—The successful livestock judge must know what he is looking for.

3. **Keen observation and sound judgment**—The good judge possesses the ability to observe both good conformation and defects, and to weigh and evaluate

the relative importance of the various good and bad features.

4. **Honesty and courage**—The good judge of any class of livestock must possess honesty and courage, whether it be in making a show-ring placing or in conducting a breeding and marketing program. For example, it often requires considerable courage to place a class of animals without regard to (a) placings in previous shows, (b) ownership, and (c) public applause. It may even take greater courage and honesty with oneself to discard from the herd a costly animal whose progeny has failed to measure up.

5. **Logical procedure in evaluating**—There is always great danger of the beginner making too close an inspection; he often gets "so close to the trees that he fails to see the forest." Good judging procedure consists of the following three separate steps: (a) observing at a distance (20 to 30 feet) and securing a panoramic view where several animals are involved, (b) using close inspection (and handling cattle and sheep), and (c) moving the animal in order to observe action. Also, it is important that a logical method be used in viewing an animal from all directions, as for example (a) side view, (b) rear view, and (c) front view; thus avoiding overlooking anything and making it easier to retain the observations that are made.

6. **Tact**—In discussing either (a) a show-ring class, or (b) animals on a stockman's farm or ranch, it is important that the judge be tactful. The owner is likely to resent any remarks which indicate that his animal is inferior.

Do's and Don'ts for Contest Judges

Members of 4-H Clubs, FFA students, college judging classes, and other prospective livestock judges should first become thoroughly familiar with the six qualifications of a good judge as outlined in the section above. Next, they should observe the following do's and don'ts:

1. **Do's:**

a. Make certain how the class is numbered, and keep the numbers straight.

b. Get a clear picture of the class and of each individual animal in mind, so that they will be remembered when giving reasons.

c. Keep in a position of vantage where the class can be seen at all times; usually this means some distance away rather than too close.

d. Make placings on the basis of the big things.

e. Make certain that the card is filled out completely and correctly, and that the correct numbers are kept in mind.

f. If permissible, make concise notes that will assist in recalling each individual in the class; record such things as distinctive color markings, outstanding faults, etc.

g. When giving reasons, use good poise and look the judge in the eye.

h. Talk reasons clearly, and with conviction and confidence.

i. Give reasons in logical sequence; give the major reason first.

j. Use breeding terms in a breeding class and market terms in a market class; and use terms appropriate to the class of animals (for example, round for beef cattle, mammary system for dairy cows, leg for sheep, ham for hogs, and croup for horses).

k. Use comparative and descriptive terms in giving reasons. Avoid such vague terms as "good," "better," and "best."

l. Concede or grant good points and faults, regardless of the placing of the animal.

2. **Don'ts:**

a. Don't act on hunches; if the first placing is arrived at after due consideration and in a logical manner, stick to it.

b. Don't place animals on the basis of small relatively unimportant characters.

c. Don't destroy self-confidence and self-respect by discussing the class with others before giving reasons.

d. Don't pay attention to what you overhear others say about a class; be an independent judge.

e. Don't give wordy and meaningless reasons.

f. Don't bluff; if you don't know the answer to a question, say so.

How to Use the Judging Guides, Tables 11-1 Through 11-5

Tables 11-1 through 11-5 are handy guides for judging beef cattle, dairy cattle, sheep, swine, and horses. It is suggested that the beginner use these as follows:

1. Examine animals in the order indicated; namely, (a) side views, (b) rear view, (c) front view, and (d) handling (etc., etc.).

2. Study the points listed under "ideal type," and know "the common faults."

3. Rank or place the animals according to their consistent rating on all points, especially the most important ones, or if you prefer, use the scorecard method which follows.

SCORECARD JUDGING

A scorecard is a listing of the different parts of an animal, with a numerical value assigned to each part according to its relative importance. It is a standard of excellence. The use of the scorecard involves studying each part, then assigning a score to each.

Different methods of scoring individual animals have evolved. All of them are based on visual appraisal. This point bears emphasis because stockmen and students often get the erroneous impression that, just because some visual scoring system (scoring systems based on visual appearance, in contrast to actual weights, measurements, etc.) is recommended for or used in conjunction with a production testing program, it must be more accurate than all other scoring systems. This isn't true. All are visual methods, and the score resulting from the use of any of them is no better than the person making it. Some method of selecting all animals by score, preferably on a systematic and written down basis, is the important thing.

The following new and modern scorecards are presented in the sections that follow:

Fig. 11-11, Breeding Beef Cattle Scorecard, page 1013
Fig. 11-12, Steer Scorecard, page 1014[1]
Fig. 11-23, Dairy Cow Scorecard, page 1018[2]
Fig. 11-34, Breeding Sheep Scorecard, page 1022[1]
Fig. 11-35, Market Lamb Scorecard, page 1022[1]
Fig. 11-46, Breeding Swine Scorecard, page 1026[1]
Fig. 11-47, Market Barrow Scorecard, page 1027[1]
Fig. 11-61, Horse Scorecard, page 1031

As noted, the scorecard gives each of several traits a value, which total 100 for a perfect score.

A scorecard is a valuable teaching aid for beginners. It systematizes judging and avoids any part of the animal being overlooked. However, a scorecard has the following limitations: (1) It is not adapted to evaluating a great number of animals, or to comparative or show-ring judging, because of the time involved in using it; (2) a near worthless animal may score quite high—for example, an animal that is so structurally unsound that it can hardly walk may have a rather high total score; (3) it evaluates each part of an animal, rather than the system—the skeletal system, the muscle system, etc.; (4) it is based almost entirely on consumer needs (for example, on the end product—meat, in meat animals); and (5) it accords precious little consideration as to whether, or how, an animal can be changed better to conform to the needs and desires of man.

[1]Adapted by the author from *Animal Science and Industry Laboratory Manual,* by Able, Bill V., Dell M. Allen, Robert H. Hines, and Miles McKee, Kansas State University, Kendall/Hunt Publishing Company, Dubuque, Iowa.

[2]Adapted by the author from unified scorecard prepared by The Purebred Dairy Cattle Association, and approved by The American Dairy Science Association.

JUDGING BEEF CATTLE

The parts of a beef animal are shown in Fig. 11-1, whereas Table 11-1 is a judging guide for beef cattle. Also, two beef cattle scorecards follow; one for breeding beef cattle, the other for steers (see pages 1013 and 1014).

Beef type and size have changed through the years. Today, ideal beef animals are growthy and well muscled—they're meat-type. Such animals gain rapidly and possess carcass excellence. They are longer, less deep bodied, trimmer through the middle, more upstanding, and less smooth than the type previously in vogue. Meat-type animals show muscling when viewed from any angle, plus necessary minimum outside finish when ready for slaughter. Muscling is indicated by bulges and creases; hence, great smoothness is no longer a virtue.

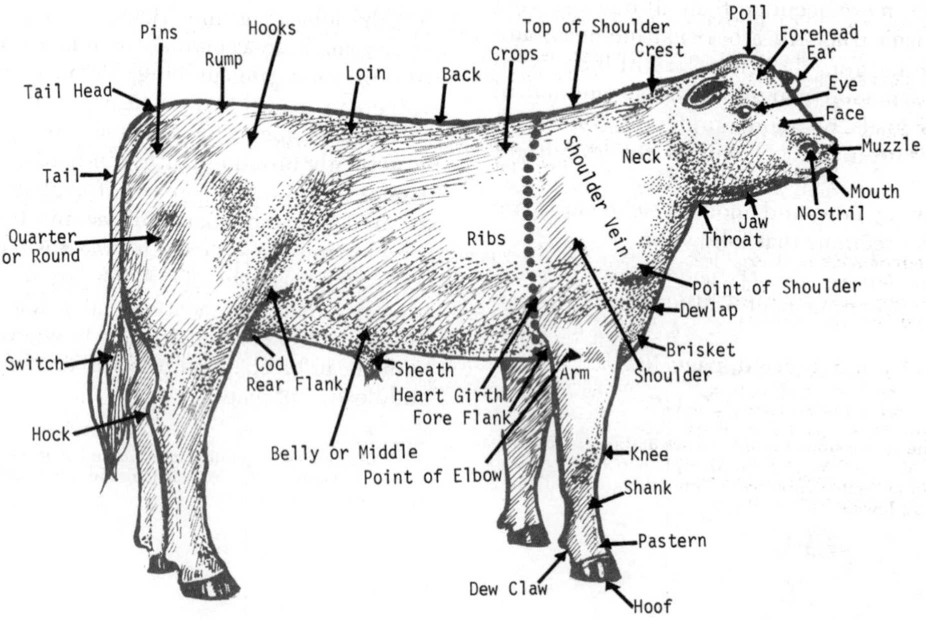

Fig. 11-1. Parts of a steer. The first step in preparation for judging beef cattle consists in mastering the language that describes and locates the different parts of an animal.

TABLE 11-1
JUDGING GUIDE FOR BEEF CATTLE

Procedure for Examining, and What to Look for	Ideal Type	Common Faults
Side view:		
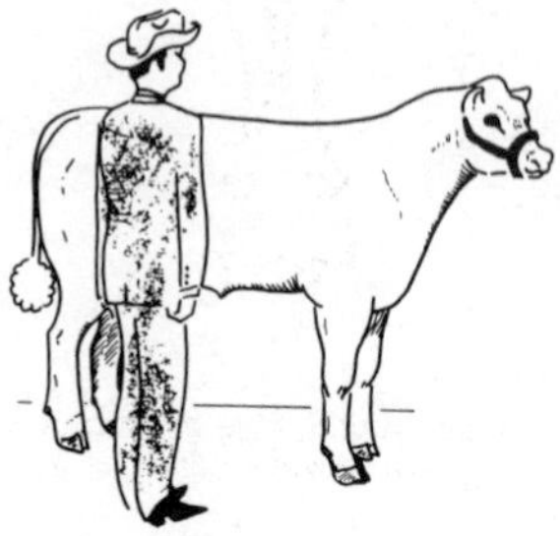 Fig. 11-2	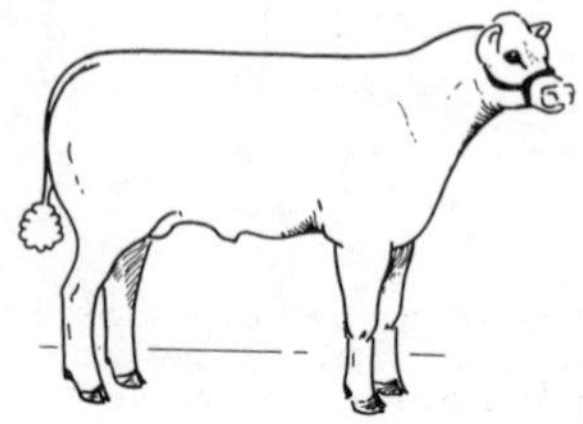 Fig. 11-3	 Fig. 11-4
1. **Reproductive efficiency** (not applicable to market animals)—Femininity in females, and masculinity in males; giving evidence of reproductive efficiency—which is the first	1. Very feminine females; long bodied, lean, smooth muscled; refined, feminine head; lean shoulder, and hindquarters; and a good, functional udder.	1. Females with steery appearance—coarse front; protruding brisket; bristly hair on neck and top of shoulders; rounded hindquarters; and fat deposits over body.

(Continued)

TABLE 11-1 (Continued)

Procedure for Examining, and What to Look for	Ideal Type	Common Faults
Reproductive efficiency (continued) requisite of beef cattle kept for breeding purposes.	Very masculine bull. Muscles well developed and clearly defined; and well developed testicles of equal size.	Bulls lacking masculinity; underdeveloped crest; muscles lacking development and not clearly defined; testicles small, imbalanced, or with one carried high; scrotum that is twisted or filled with fat.
2. **Muscling**—Bulginess of muscles in those areas least affected by fatness—the arm, forearm, gaskin, and stifle; movement and bulge of muscles as animal walks.	2. Heavily muscled in arm, forearm, gaskin, and stifle; muscles that move and bulge as the animal walks.	2. Lacking muscling. Light arm, forearm, gaskin, and stifle. Little movement or bulging of muscles as animal walks.
3. **Size**—Indicated by height to top of shoulders and length from nose to tailhead. Young breeding animals and steers that are long, tall, and not excessively fat—indications that they will continue to grow.	3. Adequate size, as evidenced by height to top of shoulders and length from nose to tailhead. Growthy young animals are long, tall, and not excessively fat.	3. Undersized; small and dumpy. Bulls showing signs of early sexual maturity; they are not likely to make continued rapid growth and reach large mature size.
4. **Freedom from waste**—Trimness in both breeding and slaughter cattle.	4. Very trim and free from waste; not too fat; and with no loose hide on the throat, dewlap, brisket, fore flank, navel or sheath, or twist.	4. Loose hide that is filled, or will fill, with fat. Excessively fat breeding cattle, accompanied by lowered reproduction. Excessively fat slaughter cattle, with reduced carcass value.
*5. **Structural soundness**—Straight legs, that are properly set; feet that are large and properly shaped; and hocks and knee joints that are clean and free from puffiness.	5. Legs that are straight, true, and squarely set; feet that are large, wide, and deep at the heel, with toes of equal size and shape that point straight ahead; and hocks and knee joints that are correctively set and clean.	5. Sickle-hocked, post-legged, back at the knees (calf-kneed), over at the knees (buck-kneed), or puffiness or swelling of knee or hock joints.
*6. **Breed type**—Characteristics true to breed, as distinguished by color and markings, shape of head, presence or absence of horns (and shape of horns if present), set of ears, body shape, and size. Commercial cattlemen can disregard this trait. Likewise, it is unimportant in steers.	6. True-to-breed characteristics; in color and markings; body shape, and size.	6. Breed characteristic associated with undesirable traits.
Rear view: 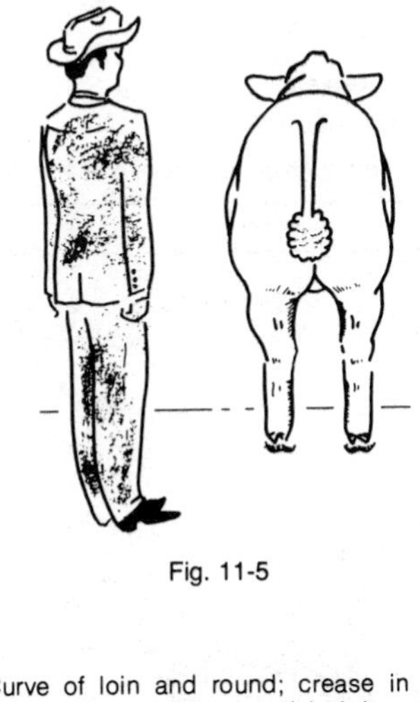 Fig. 11-5	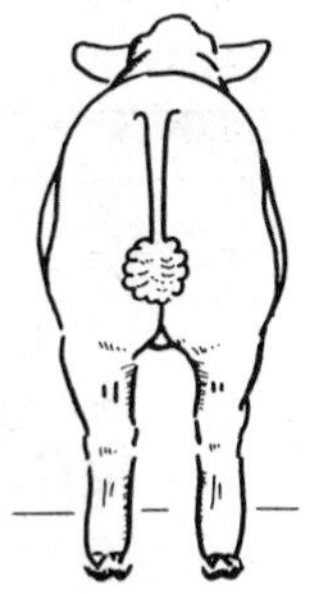 Fig. 11-6	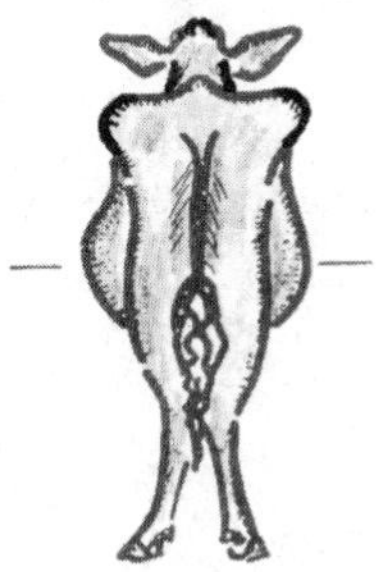 Fig. 11-7
1. Curve of loin and round; crease in thigh; groove down topline, and bulging of loin eye on each side of backbone.	1. Well curved loin and round; marked crease in thigh; well-defined groove down topline, with loin eye bulging on each side of backbone.	1. Flat loin and round; little or no crease in the thigh or groove down topline.
*2. Straightness and set to hind legs. Size and shape of hind feet. Cleanness of hock.	2. Hind legs straight and squarely set. Large feet that are wide and deep at the heels, with toes of equal size that point straight ahead; hock correctly set and clean.	2. Cow-hocked; puffiness and swelling of hock joints. Small feet.

(Continued)

TABLE 11-1 (Continued)

Procedure for Examining, and What to Look for	Ideal Type	Common Faults
Front view:		
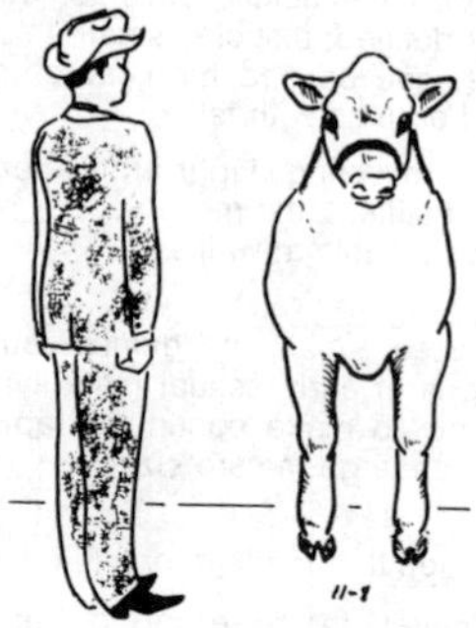 Fig. 11-8	 Fig. 11-9	 Fig. 11-10
*1. Shapeliness of head.	1. A shapely head, true to breed type—as indicated by presence or absence of horns (shape of horns if present), and set of ears.	1. A plain head.
*2. Sex character.	2. Females show femininity about the head and front end—they have lean cheeks, jaw, neck, brisket, and shoulders. Bulls show masculinity; they are on the "look," with head up and ears cocked, and they have a well-developed crest.	2. Cows lacking femininity; bulls lacking masculinity.
3. Brisket.	3. A neat, trim brisket.	3. Heavy and wasty in the brisket.
4. Width of chest.	4. A wide chest.	4. A narrow chest.
*5. Set to front legs; size and shape of front feet.	5. Correctly set front legs, with front feet of proper size and shape.	5. Crooked front legs; puffiness and swelling of the knee joints; curled toes.
Handling:		
1. Muscling.	1. Heavily muscled, meaty.	1. Light muscled.
2. Quality of hide and mellowness.	2. A loose, pliable, mellow hide.	2. Coarse hided and hard.
3. Finish.	3. Desirable finish.	3. Lacking finish; or overdone, soft and flabby.

*Not as important in market steers.

	Perfect Score	ANIMAL No. 1	No. 2	No. 3	No. 4	Etc.
REPRODUCTIVE EFFICIENCY: **Highly fertile female**—Feminine—long body, lean, smooth muscled; refined, feminine head; lean cheek, jaw, neck, brisket, shoulder, and hindquarters; and a good functional udder (or promise of udder development in a heifer). **Avoid lowly fertile female**—Steery appearance—coarse, heavy front, masculine rather than feminine; protruding brisket; bristly hair on neck and top of shoulders; rounded hindquarters; and fat deposits on the face, brisket, shoulders, hips, rump, pins, below the vulva, and in front of the udder. **Highly fertile bull**—Masculine—"he's on the look," with head up and ears cocked; well-developed crest; muscles well developed and clearly defined, especially in the regions of the neck, loin, and thigh; and well-developed genitalia, with testicles of equal size and well defined, and a proper neck to the scrotum. **Avoid lowly fertile bull**—Lacking masculinity—ears not alert; undeveloped crest; muscles lacking development and not clearly defined; testicles small, unbalanced, or with one carried high; scrotum that is twisted or filled with fat.	20					
MUSCLING: **Well muscled**—Bulging in those areas least affected by fatness—the arm, forearm, gaskin, and stifle muscles move and bulge as animal walks. Look for curved loin and round; crease in thigh; well-defined groove down topline, with loin eye bulging on each side of backbone. Look for calves with long, smooth muscling, indicating continued growth. Since muscling is a masculine trait, it is more important in bulls and steers than in heifers. *Avoid* coarse shoulders in breeding cattle, because it is usually associated with calving problems.	20					
SIZE: **Adequate size**—As indicated by height to top of shoulders and length from nose to tailhead. Young breeding animals and steers should be long, tall, and not excessively fat—indications that they will continue to grow. *Avoid* bulls showing signs of early sexual maturity; they are not likely to make continued rapid growth and reach large mature size.	15					
FREEDOM FROM WASTE: **Freedom from waste; trimness**—In both breeding and slaughter cattle. Excessively fat breeding cattle usually have lowered reproduction. Excessively fat slaughter cattle have reduced carcass value. *Avoid* loose hide that is filled, or will fill, with fat. Look for loose hide on the throat, dewlap, brisket, fore flank, navel or sheath, and twist. Look for fat over back ribs, point of shoulder, and a long backbone; since no muscle should be found at these places, if you feel something, it is fat.	15					
STRUCTURAL SOUNDNESS: **Structurally sound**—Legs straight, true, and squarely set; feet large, wide, and deep at the heel, with toes of equal size and shape that point straight ahead; hock and knee joints correctly set and clean. *Avoid* sickle-hocked, post-legged, back at the knees (calf-kneed), over at the knees (buck-kneed), or puffiness or swelling of knee or hock joints.	15					
BREED TYPE: **Characteristics true to breed**—Breed distinguished by color and markings, shape of head, presence or absence of horns (and shape of horns if present), set of ears, body shape, and size. *Avoid* breed characteristics associated with undesirable traits. Commercial cattlemen can disregard this trait. Likewise, it is unimportant in steers.	15					
TOTAL	100					

Fig. 11-11. Breeding Beef Cattle Scorecard.

	Perfect Score	ANIMAL No. 1	No. 2	No. 3	No. 4	Etc.
CONFORMATION:	60					
General appearance—(10 points)						
Muscular, thick, legs set wide apart, stylish, well balanced, large framed, adequate size for age. (10)						
Hindquarters—(29 points)						
Loin—meaty, thick, full deep loin edge. (10)						
Rump—long, level, full, and square. (7)						
Quarter—long, deep, thick, bulging, meaty. (10)						
Legs—correct, set wide apart. (2)						
Forequarters—(16 points)						
Back—thick, muscular, strong. (7)						
Ribs—bold spring, deep forerib. (3)						
Shoulders—smooth, muscular. (3)						
Crops—full. (0.5)						
Neck—clean, balanced. (0.5)						
Brisket—trim, neat dewlap. (1)						
Legs—correct, set wide apart. (1)						
Middle—(5 points)						
Stretchy, trim, straight underline. (5)						
Finish—(35 points)	35					
Amount and uniformity of finish—over the back, loins, ribs, rump, and shoulder. (30)						
Trim in flanks, cod, and along underline. (5)						
Quality—(5 points)	5					
Trimness of head, hide, fineness of hair, ample bone.						
TOTAL	100					

Fig. 11-12. Steer Scorecard.

JUDGING DAIRY CATTLE

The main function of the dairy cow is to produce milk. Her appearance is not always indicative of her productive ability. However, a cow's appearance does tell us about her potential and her wearing ability.

The parts of a dairy cow are shown in Fig. 11-13, whereas Table 11-2 is a judging guide for dairy cattle. Also, a scorecard for dairy cattle follows (see page 1018).

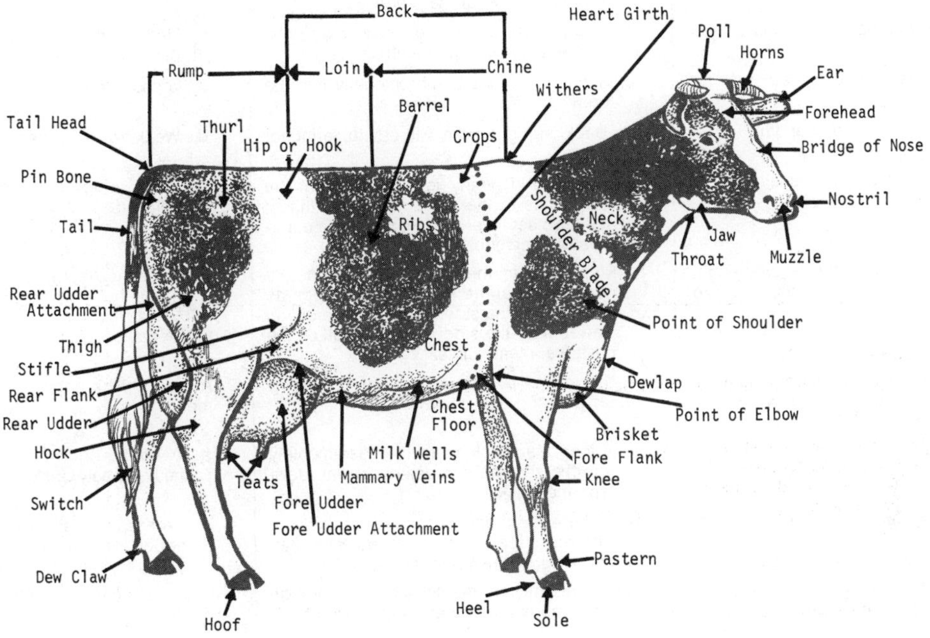

Fig. 11-13. Parts of a dairy cow.

TABLE 11-2
JUDGING GUIDE FOR DAIRY CATTLE

Procedure for Examining, and What to Look for	Ideal Type	Common Faults
Side view:		
 Fig. 11-14	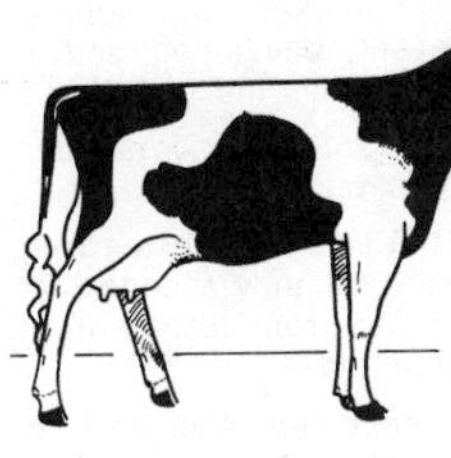 Fig. 11-15	 Fig. 11-16
1. **General appearance**—Style and attractiveness, size, relative length and depth throughout, straightness of back and levelness of rump, attachment and shape of udder, and quality as denoted by fineness of hair, smoothness of joints, fineness of bone, and absence of coarseness throughout. At the walk, observe the style and carriage, straightness of the legs, and the blending of the parts.	1. Great style and beauty.	1. Plain; lacking in attractiveness and in symmetry and balance.

(Continued)

TABLE 11-2 (Continued)

Procedure for Examining, and What to Look for	Ideal Type	Common Faults
Side view: (continued)		
a. **Breed characteristics**—Breed type and size.	a. True to the breed as evidenced by the conformation of the head and color markings. Adequate size for the breed.	a. Lacking in breed character. Undersized.
b. **Head**—Relative proportion to body; clean cutness.	b. Head in proportion to body, moderate length, clean-cut, and alert.	b. Head either too big or too little for the body; plain headed.
c. **Shoulder blades**—Smoothness.	c. Shoulder blades blend smoothly into the body.	c. Loose or out at the shoulders.
d. **Back**—Straightness and strength.	d. Back straight from withers to tailhead; strong.	d. Weak, sway backed; high tailhead.
e. **Rump**—Length, width, and levelness.	e. A long and nearly level rump, from the hip bones to the pin bones; high thurls; tailhead set level with the back bone and free of coarseness.	e. Short, sloping rump; low thurls; high tailhead.
f. **Legs and feet**—Set to the legs, strength of pasterns, cleanness of joints, and size of feet.	f. Proper set to the legs; hind legs nearly perpendicular from hock to pastern as viewed from the side; front legs straight as viewed from the side; standing well on the pasterns; clean joints.	f. Sickle hocks (hind legs set too far forward); back at the knees (calf-kneed); over at the knees (buck-kneed); down on the pastern; puffy around the joints.
2. **Dairy character**—Evidence of milking ability; angularity; openness, without weakness; not coarse.	2. A triangular body; loose, without being weak; and showing quality.	2. A round, thick body; and sluggish temperament.
a. **Neck**—Length, leanness, smoothness, cleanness about the throat, dewlap, and brisket.	a. A long lean neck, which blends smoothly into the shoulders; clean-cut throat, dewlap, and brisket.	a. A short, thick neck; open shoulders; heavy, wasty brisket.
b. **Withers, ribs, flanks, thighs**—Shape of withers; width between ribs, and shape and length of ribs; shape of thigh.	b. Withers sharp and free of excessive flesh; ribs wide apart, and rib bones wide, flat, and long; thighs rather thin and flat.	b. Excessive flesh over the withers; close ribbed, and rib bones narrow, round, and short; and full, fleshy thighs.
3. **Body capacity**—Relatively large in proportion to size of animal; ample capacity, strength and vigor.	3. Great capacity, primarily achieved through length of body rather than extreme depth.	3. Lacking capacity; short bodied; and lacking depth.
a. **Barrel**—Support of barrel; length and depth.	a. Strongly supported; long, with adequate depth; well sprung ribs; greater depth of barrel toward rear.	a. Lacking capacity; short bodied; and lacking depth in the middle.
b. **Heart girth**—Spring of forerib; fullness of crops; width of chest floor.	b. Well sprung forerib; full crops; wide chest floor.	b. Lacking spring of forerib; slack in the crops; and narrow chest floor.
4. **Mammary system**—Attachment; balance; capacity; texture.	4. Udder strongly attached, well balanced, capacious, and fine textured.	4. Udder broken away from the body; different size halves and quarters; small; coarse.
a. **Udder**—Shape, attachment, texture.	a. Symmetrical; moderately long, wide, and deep; strongly attached; noticeable but not deep division between right and left halves; division between front and rear quarters not marked, and right and left quarters evenly developed; soft, pliable, and collapsed after milking.	a. Quarters not evenly developed; deeply cut between quarters or halves; excessive depth of udder; meaty textured; weakly attached.
b. **Fore udder**—Length, width, depth, and attachment.	b. Moderate length, uniform width from front to rear, and strongly attached.	b. Fore udder not extended well forward, lacking width, and weakly attached.
c. **Rear udder**—Height, width, shape, and attachment.	c. The rear udder should extend high up, and be wide, slightly rounded, of fairly uniform width from top to floor, and strongly attached.	c. Not attached high, and narrow, flat, not uniform in width from top to floor, and weakly attached.
d. **Teats**—Size and placement.	d. Teats of convenient size and squarely placed.	d. Teats too large or too small, and poorly placed.
e. **Mammary veins**—Size, length, crookedness, branching.	e. The milk veins are large, long, crooked, and branching; the milk wells are large and numerous; and the veins on the udder are large, crooked, and numerous.	e. Small milk veins and wells.

(Continued)

TABLE 11-2 (Continued)

Procedure for Examining, and What to Look for	Ideal Type	Common Faults
Rear view:		
 Fig. 11-17	 Fig. 11-18	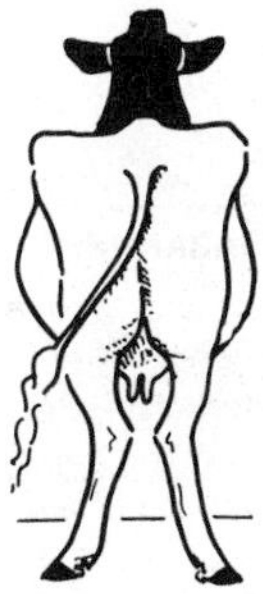 Fig. 11-19
1. Width over back.	1. Vertebra well defined.	1. Vertebra not well defined; meaty.
2. Width over loins and rump.	2. Wide over the loins and rump.	2. Narrow over the loin and rump.
3. Width between hip and pin bones.	3. Wide between the hip and pin bones.	3. Narrow between the hip and pin bones.
4. Height and width of thurls. The width between hips and pins and the levelness of rump are believed to be associated with the size and shape of the udder.	4. Thurls high and wide apart.	4. Low thurls.
5. Straightness of hind legs.	5. Hind legs straight as viewed from the rear.	5. Cow hocked; puffiness and swelling of hock joints.
6. Size and shape of hind feet.	6. Feet short, compact, and well rounded with deep heel and level sole.	6. Small feet; toes not of equal size.
7. Rear udder attachment.	7. Rear udder that extends high and is wide and strongly attached.	7. Rear udder not extended high up; weakly attached.
Front view:		
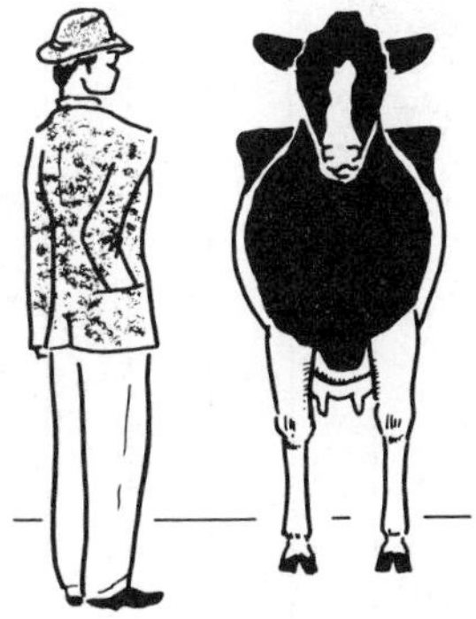 Fig. 11-20	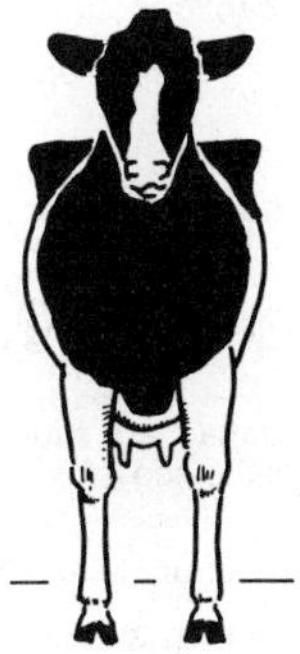 Fig. 11-21	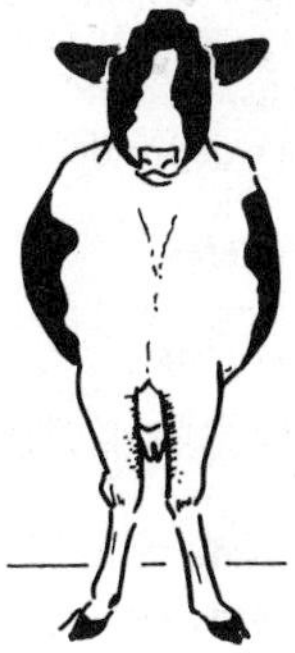 Fig. 11-22
1. Shapeliness of head.	1. A shapely head, with a broad muzzle, large nostrils, and a strong jaw.	1. A plain head.
2. Sex character.	2. Cows show femininity about the head and front end; bulls show masculinity and have a well-developed crest.	2. Cows lacking femininity; bulls lacking masculinity.
3. **Neck**—Length, leanness; blending of neck and shoulders; throat, dewlap, and brisket.	3. Neck long, lean, and blended smoothly into the shoulders, with clean-cut throat, dewlap, and brisket.	3. Short, thick neck; neck not blending smoothly into shoulders; leathery and wasty about the throat, dewlap, and brisket.
4. **Chest**—Width of floor.	4. Wide chest floor, indicating constitution and chest capacity.	4. A narrow chest.
5. **Front legs and feet**—Set to front legs; size and shape of front feet.	5. Forelegs medium in length, straight, wide apart, and squarely placed.	5. Crooked front legs; puffiness and swelling at knee joints; curled toes.

	Perfect Score	ANIMAL No. 1	No. 2	No. 3	No. 4	Etc.
GENERAL APPEARANCE:	30					
Attractive individuality with femininity, vigor, stretch, scale, harmonious blending of all parts and impressive style and carriage. All parts of a cow should be considered in evaluating a cow's general appearance. (10)						
Breed characteristics—Breed type and size.						
Head—Clean-cut, proportionate to body; broad muzzle with large, open nostrils; strong jaws; large, bright eyes; forehead, broad and moderately dished; bridge of nose straight; ears medium size and alertly carried.						
Shoulder blades—Set smoothly and tightly against the body.						
Back—Straight and strong; loin, broad and nearly level. (10)						
Rump—Long, wide, and nearly level from hook bones to pin bones: clean-cut and free from patchiness; thurls, high and wide apart; tailhead, set level with back line and free from coarseness; tail, slender.						
Legs and feet—bone flat and strong, pasterns short and strong, hocks cleanly moulded. Feet, short and compact and well rounded with deep heel and level sole. Forelegs medium in length, straight, wide apart, and squarely placed. Hind legs, nearly perpendicular from hock to pastern, from the side view, and straight from the rear view. (10)						
DAIRY CHARACTER:	20					
Evidence of milking ability, angularity, and general openness, without weakness; freedom from coarseness, giving due regard to period of lactation.						
Neck—Long, lean, and blending smoothly into the shoulders, clean-cut throat, dewlap, and brisket.						
Withers—Sharp.						
Ribs—Wide apart, rib bones wide, flat, and long.						
Flanks—Deep and refined. (20)						
Thighs—Incurving to flat, and wide apart from the rear view, providing ample room for the udder and its rear attachment.						
Skin—Loose, and pliable.						
BODY CAPACITY:	20					
Relatively large in proportion to size of animal, providing ample capacity, strength, and vigor.						
Barrel—Strongly supported, long and deep; ribs highly and widely sprung; depth and width of barrel tending to increase toward rear. (10)						
Heart girth—Large and deep, with well-sprung foreribs blending into the shoulders; full crops; full at elbows; wide chest floor. (10)						
MAMMARY SYSTEM:	30					
A strongly attached, well balanced, capacious udder of fine texture indicating heavy production and long period of usefulness.						
Udder—Symmetrical, moderately long, wide, and deep, strongly attached, showing moderate cleavage between halves, no quartering on sides; soft, pliable, and well collapsed after milking; quarters evenly balanced. (10)						
Fore udder—Moderate length, uniform width from front to rear and strongly attached. (6)						
Rear udder—High, wide, slightly rounded, fairly uniform width from top to floor, and strongly attached. (7)						
Teats—Uniform size, of medium length and diameter, cylindrical, squarely placed under each quarter, plumb, and well spaced from side and rear views. (5)						
Mammary veins—Large, long, tortuous, branching "Because of the natural undeveloped mammary system in heifer calves and yearlings, less emphasis is placed on mammary system and more on general appearance, dairy character, and body capacity. A slight to serious discrimination applies to overdeveloped, fatty udders in heifer calves and yearlings." (2)						
TOTAL	100					

Fig. 11-23. Dairy Cow Scorecard.

JUDGING SHEEP

Basically, the description of body parts and the desirable meat cuts of sheep and beef cattle, are similar. Sheep simply come in a smaller package. However, sheep are covered with wool, which makes it important that they be handled in order to be sure of their conformation. Also, the fleece should be considered in breeding classes.

The parts of a sheep are shown in Fig. 11-24, whereas Table 11-3 is a judging guide for sheep. Also, two scorecards for sheep follow; one for breeding sheep, and the other for market lambs (see page 1022).

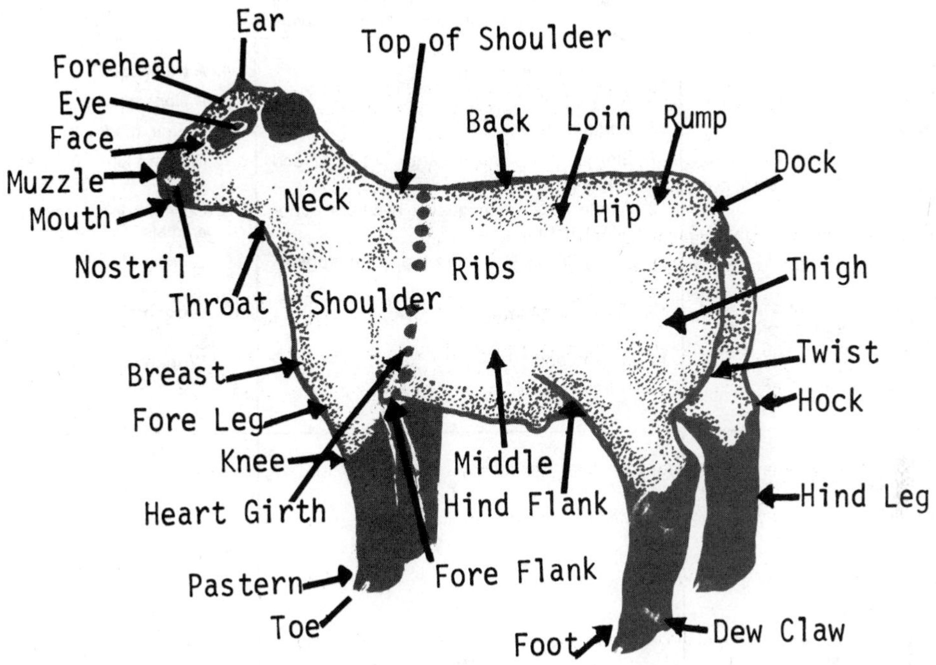

Fig. 11-24. Parts of a sheep. The first step in preparation for judging sheep consists in mastering the language that describes and locates the different parts of an animal.

TABLE 11-3
JUDGING GUIDE FOR SHEEP

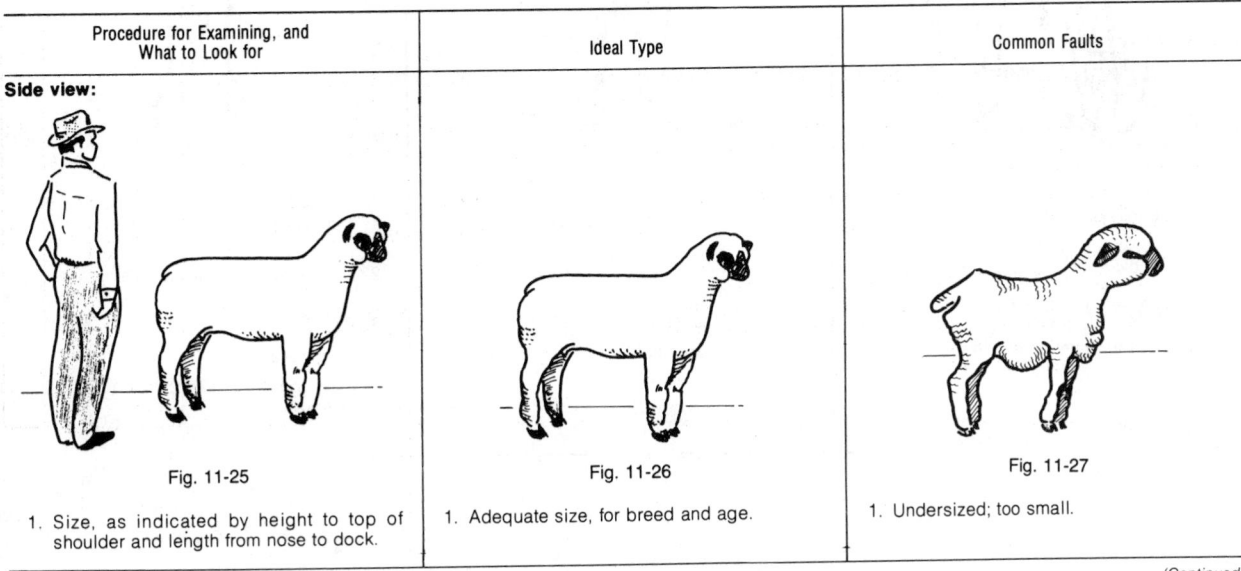

Procedure for Examining, and What to Look for	Ideal Type	Common Faults
Side view:  Fig. 11-25	 Fig. 11-26	 Fig. 11-27
1. Size, as indicated by height to top of shoulder and length from nose to dock.	1. Adequate size, for breed and age.	1. Undersized; too small.

(Continued)

TABLE 11-3 (Continued)

Procedure for Examining, and What to Look for	Ideal Type	Common Faults
Side view: (continued)		
2. Balance and symmetry.	2. Balanced and symmetrical.	2. Lacking in balance and symmetry.
3. Stretch, with variation according to breed and age.	3. Breeds vary; some are more stretchy than others. Young animals should be growthy—they should be long, tall, and not excessively fat.	3. Dumpy lambs—which are not likely to make continued rapid growth or reach large mature size.
4. Strength of back.	4. A strong, straight back.	4. A weak back.
5. Levelness of rump.	5. A level rump.	5. A sloping rump.
6. Trimness of underline.	6. A trim underline.	6. High in the flanks.
*7. Straightness of legs and strength of pasterns.	7. Straight, true, and squarely set legs, and strong pasterns.	7. Crooked legs, and weak pasterns.
8. Size of bone.	8. Ample bone, with quality (wethers rather fine boned).	8. Coarse boned, lacking quality (or breeding animals too fine boned).
9. Style.	9. Plenty of style; a pleasing, alert appearance.	9. Lacking in style.
*10. Breed type (markings, shape of head, and fleece characteristics true to the breed).	10. Showing plenty of breed type.	10. Lacking breed type.
11. Freedom from wrinkles.	11. Smooth bodied.	11. Wrinkles along the neck, especially in fine-wool breeds.
(* at end of table)		
Rear view:		
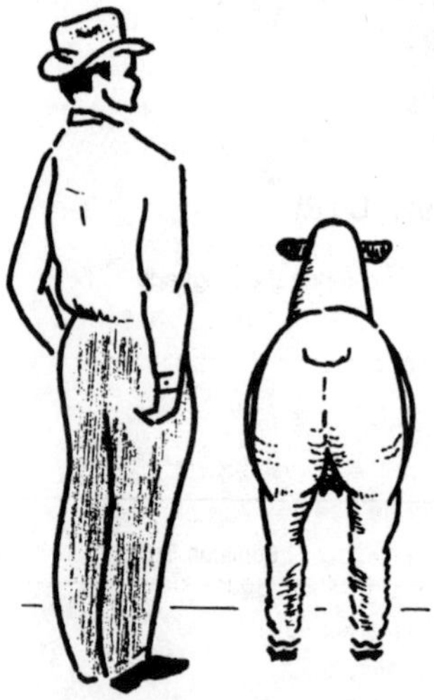 Fig. 11-28	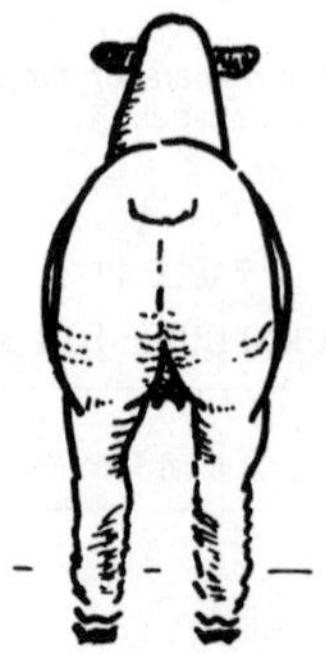 Fig. 11-29	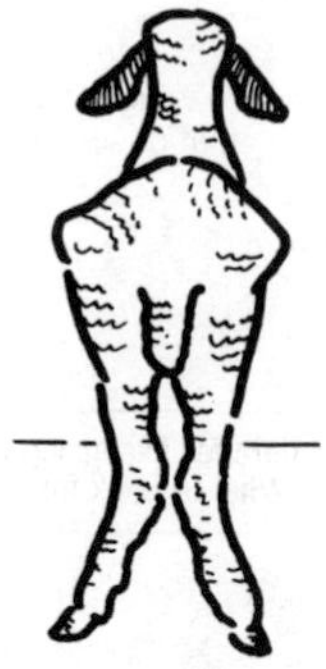 Fig. 11-30
1. Uniformity of width from front to rear.	1. Uniformly wide from front to rear.	1. Narrow bodied.
2. Curve over back, loin, and round.	2. Well covered or rounded over back, loin, and rump.	2. Narrow over the back and loin.
3. Trimness of middle.	3. Trim in the middle.	3. Paunchy.
4. Depth and plumpness of leg.	4. Deep, plump leg.	4. A light leg.
*5. Set of hind legs	5. Legs set wide apart.	5. Legs too close together; cow hocked; or bow legged.
(* at end of table)		

(Continued)

TABLE 11-3 (Continued)

Procedure for Examining, and What to Look for	Ideal Type	Common Faults
Front view:  Fig. 11-31	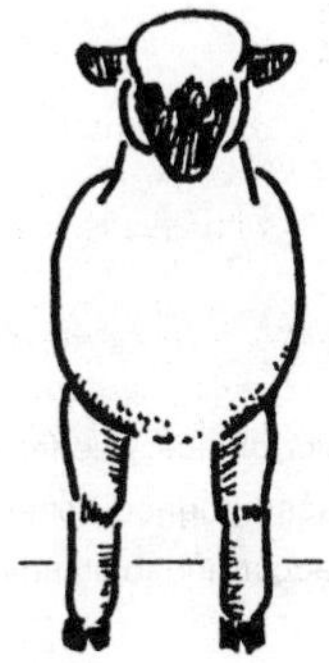 Fig. 11-32	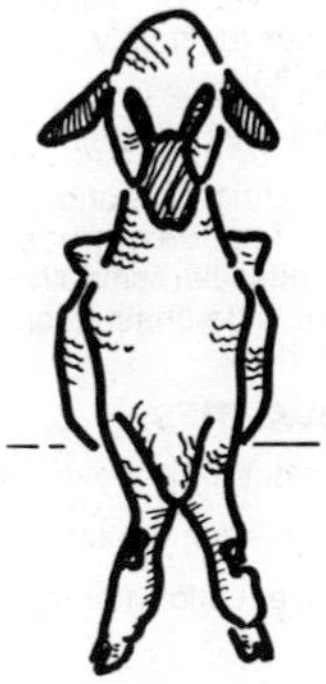 Fig. 11-33
*1. Shapeliness of head.	1. Head shape true to breed.	1. A plain head.
*2. Sex character.	2. Ewes show femininity; rams masculinity.	2. Ewes lacking femininity; rams lacking masculinity.
3. Brisket.	3. A neat, trim brisket.	3. Heavy in the brisket; and "apron" or folds of skin.
4. Width of chest.	4. A wide chest.	4. A narrow chest.
5. Set of front legs. (at end of table)	5. Correctly set front legs.	5. Crooked front legs.
Handling and fleece:		
1. Muscling.	1. Well muscled.	1. Lacking muscling.
2. Smoothness over the shoulders.	2. Smoothly laid in shoulders.	2. Open in the shoulders.
3. Fullness of fore rib.	3. Well-sprung fore rib.	3. Pinched in the heart girth.
4. Rounding of back, loin, and rump.	4. Rounded over the back, loin, and rump.	4. Narrow over the back and loin; and not carrying width out to the dock.
5. Depth and plumpness of leg.	5. A deep, plump leg.	5. A light leg.
†6. Finish.	6. Desirable finish.	6. Lacking finish.
*7. Quality of fleece.	7. A long, dense, clean fleece, with uniformly good quality throughout.	7. A coarse, open fleece, lacking quality and uniformity; kempy black fibers.

*Not as important in market wethers as in breeding animals.
†Not important in breeding animals.

	Perfect Score	ANIMAL No. 1	No. 2	No. 3	No. 4	Etc.
CONFORMATION:	73					
General appearance—(25 points)						
Size and scale—big for age, roomy, heavy bone. (15)						
Type—straight lined, balanced, deep ribbed, long, stylish. (10)						
Hindquarters—(26 points)						
Leg—muscular, plump, thick, deep. (9)						
Rump—long, level, full, square dock. (7)						
Loin—wide, strong, meaty. (9)						
Twist—deep, full. (1)						
Forequarters—(22 points)						
Back—wide, straight, strong. (8)						
Ribs—bold spring, deep ribbed. (6)						
Shoulders—muscular, smooth. (4)						
Chest—deep, wide chest floor. (3)						
Neck—short, thick. (1)						
BREEDING QUALITIES:	27					
Head—clean-cut, bright eyes, feminine or masculine, proper color of face, free from wool blindness. (5)						
Underpinning—strong pasterns, legs correctly and squarely placed, rugged bone. (15)						
Fleece—dense, uniform crimp, long staple, pink skin, fineness according to standard of the breed, free from black fiber. (7)						
TOTAL	100					

Fig. 11-34. Breeding Sheep Scorecard.

	Perfect Score	ANIMAL No. 1	No. 2	No. 3	No. 4	Etc.
CONFORMATION:	55					
General appearance—(10 points)						
Straight top and underline, muscular, thick, legs set wide apart, stylish, well balanced, adequate size for age.						
Hindquarters—(26 points)						
Legs—straight, set wide apart. (2)						
Twist—clean, muscular. (1)						
Leg—meaty, plump, long, deep, thick. (8)						
Rump—long, level, thickly muscled. (6)						
Loin—meaty, thick, deep loin edge, straight. (9)						
Forequarters—(13 points)						
Back—thick, straight. (6)						
Ribs—bold spring, deep forerib. (3)						
Shoulders—muscular, smooth. (2)						
Neck—short, thick. (1)						
Breast—wide, deep chest floor, trim. (0.5)						
Legs—straight, set wide apart. (0.5)						
Middle—(6 points)						
Middle—trim, free from wastiness.						
FINISH:	40					
Uniformly covered with the correct amount of finish over back, ribs, loin, rump. (30)						
Covering over shoulder, dock. (5)						
Trim in flanks, cod, etc. (5)						
QUALITY:	5					
Smooth pelt, head trim and refined, ample bone.						
TOTAL	100					

Fig. 11-35. Market Lamb Scorecard.

JUDGING SWINE

The desired meat-type hog yields a high percentage of lean meat with a minimum amount of fat. It is heavily muscled, correct in finish, firm, well-balanced, and has adequate length of side.

The parts of a hog are shown in Fig. 11-36, whereas Table 11-4 is a judging guide for swine. Also, two swine scorecards follow; one for breeding swine, and the other for market barrows (see pages 1026 and 1027).

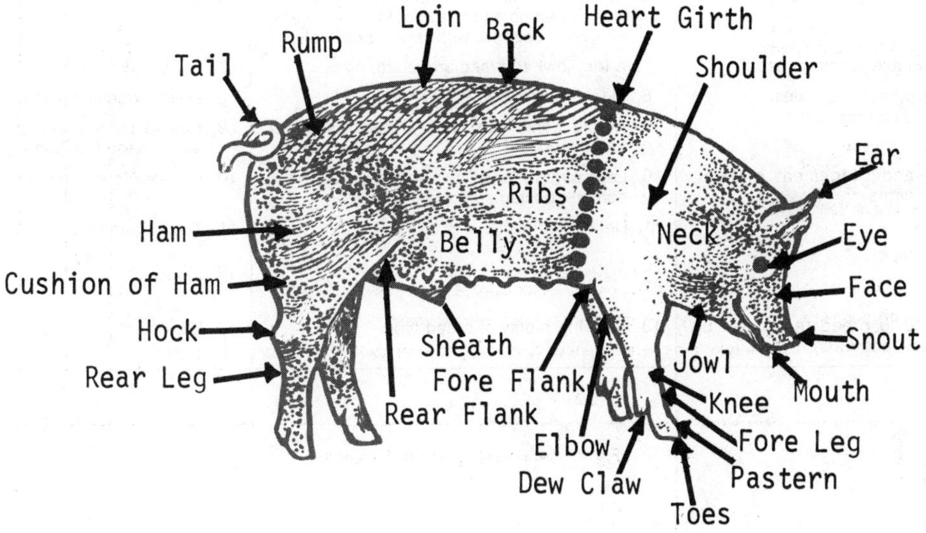

Fig. 11-36. Parts of a hog. The first step in preparation for judging hogs consists in mastering the language that describes and locates the different parts of the animal.

TABLE 11-4

JUDGING GUIDE FOR SWINE

Procedure for Examining, and What to Look for	Ideal Type	Common Faults
Side view:		
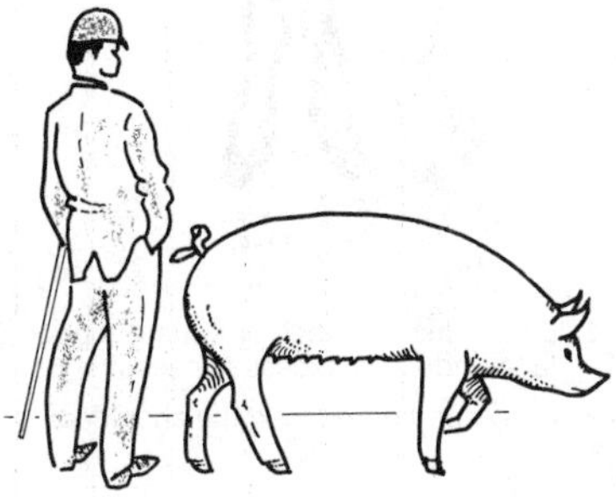 Fig. 11-37	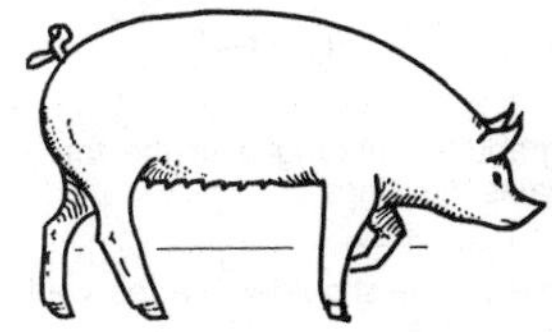 Fig. 11-38	 Fig. 11-39
1. Size.	1. Adequate size.	1. Undersized; too small.
2. Balance and symmetry.	2. Balanced and symmetrical, with all parts in the right proportion.	2. Lacking in balance and symmetry, such as long necked, and slack framed.
3. Length of side from a point in the center of the ham to the forepart of the shoulder. (This corresponds to measuring from the front of the aitch bone to the front of the rib in the carcass.)	3. A long side. Meat-type barrows should be 29 inches or longer at 200-pound weight.	3. A short side.

(Continued)

TABLE 11-4 (Continued)

Procedure for Examining, and What to Look for	Ideal Type	Common Faults
Side view: (continued)		
4. Depth of body.	4. Moderate depth of body, and moderately deep in the fore and rear flanks.	4. A shallow body; high in the flanks.
5. Topline.	5. A moderate and evenly arched back; high tail setting.	5. Low back of the shoulders; steep in the rump; low tail setting.
6. Underline; *teats.	6. Underline is firm, trim, and free from wrinkles. Breeding animals should have at least 12 evenly spaced, well-developed teats.	6. Wasty middle. Blind or inverted nipples.
7. Trimness of jowl and length of neck.	7. A trim jowl and medium length neck.	7. Heavy jowl and short, thick neck.
8. Shoulders; smoothness, wrinkles.	8. Muscular and free from wrinkles.	8. Rough, open shoulders; wrinkles.
9. Size of bone.	9. Ample bone with quality (barrows rather fine boned).	9. Coarse bone, lacking quality (or breeding animals too fine boned).
10. Straightness and placement of legs; *strength of pasterns.	10. Legs straight, true, and squarely set on the corners; pasterns strong.	10. Crooked legs; weak pasterns.
†11. Finish.	11. Desirable finish.	11. Too fat and lardy.
12. Style.	12. Plenty of style.	12. Lacking in style.
*13. Breed type true to breed (as shown by color, head, face, and ears).	13. Showing plenty of breed type.	13. Lacking breed type.
Rear view:		
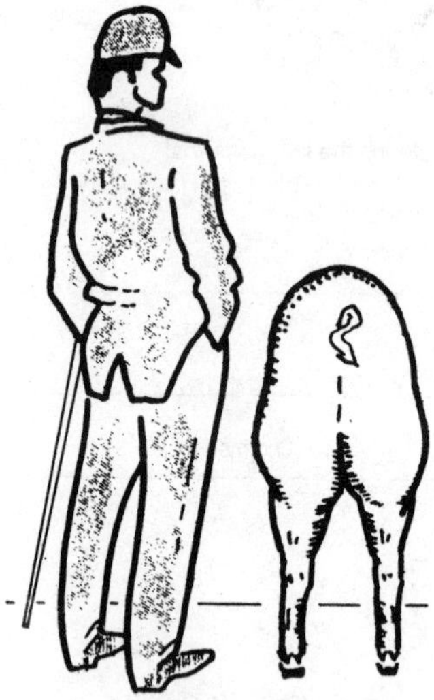 Fig. 11-40	 Fig. 11-41	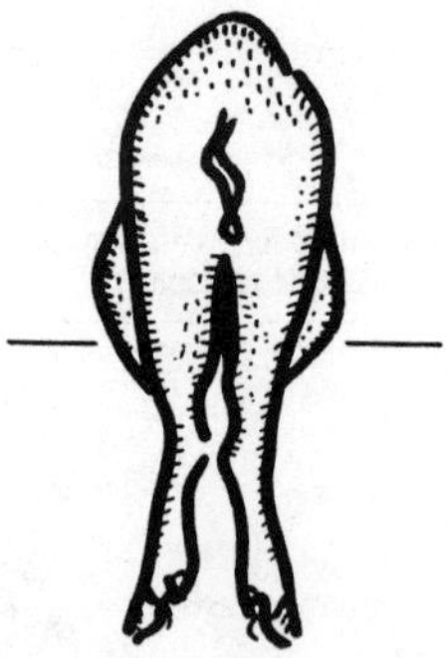 Fig. 11-42
1. Width over the top.	1. A gradual rounding over the top which indicates meatiness.	1. Either "fish backed" or square topped (the latter indicates excess fatness).
2. Shoulder muscling.	2. Shoulder slightly bulging with the space between the shoulder blades well filled in with muscle.	2. Heavy and/or rough in the shoulders.
3. Width and fullness of loin.	3. Wide, strong loin.	3. Narrow and pinched over the loin.
4. Rump.	4. Long, with a gradual slope toward a high set tail. Slightly rounded from side to side over the top, with no sign of excessive fatness.	4. A steep rump and low tail setting, which cut down on the size of the ham.
5. Plumpness, fullness, trimness, and firmness of ham; length of shank.	5. Deep, thick, slightly bulging, and firm ham; meated well down to hocks.	5. A light ham; long shank.
*6. Set of hind legs.	6. Legs set well apart.	6. Crooked hind legs.

(Continued)

TABLE 11-4 (Continued)

Procedure for Examining, and What to Look for	Ideal Type	Common Faults
Front view:		
 Fig. 11-43	 Fig. 11-44	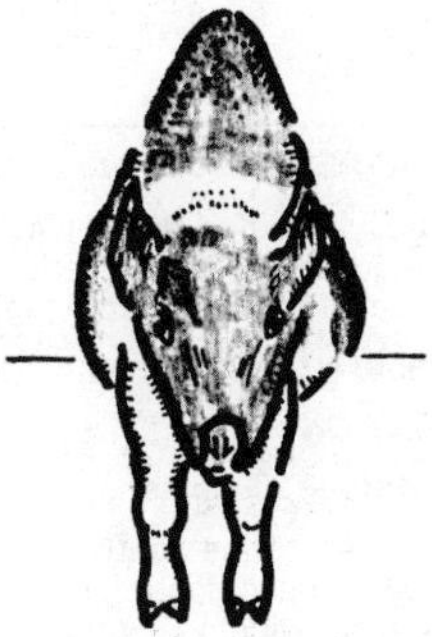 Fig. 11-45
*1. Shapeliness and trimness of head.	1. Head of medium length, wide between the eyes, and trim. The face, ears, and nose of breeding animals true to breed characteristics.	1. A plain head.
*2. Sex character.	2. Females show femininity; males show masculinity.	2. Females lacking femininity; males lacking masculinity.
*3. Set of front legs.	3. Correctly set front legs.	3. Crooked front legs.

*Not as important in market barrows as in breeding animals.
†Not important in breeding animals.

	Perfect Score	ANIMAL No. 1	No. 2	No. 3	No. 4	Etc.
GENERAL APPEARANCE:	25					
Type—heavy muscled, lean, trim, firm, smooth, long bodied, ham and rump should be wider than rest of body, moderately deep forerib, uniformly arched top, well balanced and stylish with a high degree of development in the valuable region of the hindquarters, same standards as for the market barrow. (12)						
Size—ample size, scale, and ruggedness for age. (13)						
CONFORMATION:	45					
Hindquarters—(26 points)						
Hind legs—set wide apart, out on the corners giving an indication of abundant muscling. (3)						
Ham—wide, deep, long, full, firm, meaty, deep in the seam. (9)						
Rump—long, wide, uniformly turned, high tail setting, meaty. (6)						
Loin—muscular turn, long, lean. (8)						
Forequarters—(14 points)						
Back—muscular turn, long, lean, uniformly arched, full spring of rib. (7)						
Shoulders—smooth, muscular, free from fatty creases and wrinkles, no evidence of fat deposits at the elbow. (4.5)						
Head—clean-cut, trim, firm jowl. (2)						
Neck—short. (0.5)						
Middle—(5 points)						
Deep, roomy middle but not loose or wasty, belly trim and firm, deep ribbed.						
BREEDING QUALITIES:	30					
Underpinning—(13 points)						
Legs—straight as viewed from side, front, and rear, squarely set under corners of body, strong, straight toes. (8)						
Pasterns—strong, short, straight, but not buckled over. (1)						
Action—free, easy, unhindered walk, not stiff or "peggy." (4)						
Mammary system—(12 points)						
Six (6) sound nipples on a side, nipples prominent and evenly spaced; no evidence of inverted or blind teats.						
Breed character—(5 points)						
Head—varies with the breed, wide between the eyes, sows feminine and boars masculine. (3)						
Ears—relatively small and refined, not large and coarse thereby hindering vision; breeds having erect ears should show no tendency for drooping; breeds having drooping ears should show no tendency for being erect. (2)						
TOTAL	100					

Fig. 11-46. Breeding Swine Scorecard.

	Perfect Score	ANIMAL No. 1	No. 2	No. 3	No. 4	Etc.
CONFORMATION:	60					
General appearance—(12 points)						
Heavy muscled, lean, trim, firm, long bodied, ham and rump should be wider than the rest of the body, uniformly arched top, well balanced and stylish with a high degree of development in the valuable region of the ham and loin, adequate size for age.						
Hindquarters—(28 points)						
Hind legs—set wide apart, out on the corners, giving an indication of abundant inner muscling, straight. (2)						
Ham—wide, deep, long, full, firm, meaty, deep in the seam. (10)						
Rump—long, wide, uniformly turned, high tail setting, meaty. (7)						
Loin—muscular turn, lean, long, uniformly arched. (9)						
Forequarters—(14 points)						
Back—muscular turn, long, lean, uniformly arched, uniform width, full spring of rib. (7)						
Shoulders—smooth, muscular. (5)						
Neck—short. (0.5)						
Head—clean-cut, refined, trim firm jowl. (1.5)						
Middle—(6 points)						
Side—long, moderately deep forerib, smooth, free from wrinkles. (3)						
Underline (belly)—trim, firm, no evidence of looseness or wastiness. (3)						
FINISH:	36					
Ham—firm and free from wrinkles at the base, firm in the crotch. (5)						
Rump—no evidence of a counter sunk tail setting. (5)						
Back and loin—lean, meaty turn, evidence of abundant muscling accompanied by minimum amount of backfat. (8)						
Shoulders—firm and smooth, free from fatty creases and wrinkles, no evidence of fat deposit at the elbow. (8)						
Jowl—trim and firm. (5)						
Belly—trim and firm. (5)						
QUALITY:	4					
Smooth throughout, not creased or wrinkled. (2)						
Bone—ample substance of bone, definitely not fine but not overly coarse either. (2)						
TOTAL	100					

Fig. 11-47. Market Barrow Scorecard.

JUDGING HORSES

Horses are judged on the basis of body conformation and performance. A horse must conform to the specific type that is needed for the function he is to perform; and, additionally, he should conform to the characteristics of the breed that he represents.

The parts of a horse are shown in Fig. 11-48, whereas Table 11-5 is a judging guide for light horses. Although light horses only are covered in this section, the same methods and principles apply to judging draft horses and mules. Also, a horse scorecard follows (see page 1031).

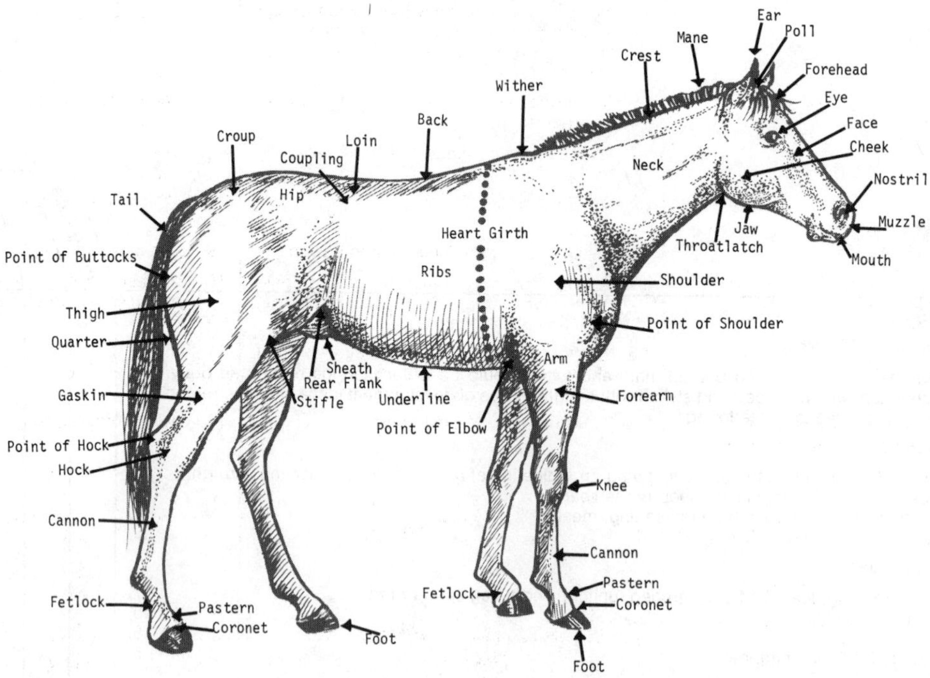

Fig. 11-48. Parts of a horse. The first step in preparation for judging horses consists in mastering the language that describes and locates the different parts of the animal.

TABLE 11-5

JUDGING GUIDE FOR LIGHT HORSES[1]

Procedure for Examining, and What to Look for	Ideal Type	Common Faults
Side view:		
 Fig. 11-49	 Fig. 11-50	 Fig. 11-51
1. Style and beauty.	1. High carriage of head, active ears, alert disposition and beauty of comformation.	1. Lacking style and beauty.
2. Balance and symmetry.	2. All parts well developed and nicely blended together.	2. Lacking in balance and symmetry.
3. Neck.	3. Fairly long neck; carried high; clean-cut about the throat latch; with head well set on.	3. A short, thick neck; ewe-necked.

(Continued)

TABLE 11-5 (Continued)

Procedure for Examining, and What to Look for	Ideal Type	Common Faults
Side view: (continued)		
4. Shoulders.	4. Sloping shoulders (about a 45° angle).	4. Straight in the shoulders.
5. Topline.	5. A short, strong back and loin, with a long, nicely turned and heavily muscled croup, and a high, well-set tail; withers clearly defined and of the same height as the high point of croup.	5. Sway backed; steep croup.
6. Coupling.	6. A short coupling as denoted by the last rib being close to the hip.	6. Long in the coupling.
7. Middle.	7. Ample middle due to long, well-sprung ribs.	7. Lacking middle.
8. Rear flank.	8. Well let down in the rear flank.	8. High cut rear flank or "wasp waisted."
9. Arm, forearm, and gaskin.	9. Well-muscled arm, forearm, and gaskin.	9. Light-muscled arm, forearm, and gaskin.
10. Legs, feet, and pasterns.	10. Straight, true, and squarely set legs; pasterns sloping about 45°; hoofs large, dense, and wide at the heels.	10. Crooked legs; straight pasterns; hoofs small, contracted at the heels, and shelly.
11. Quality.	11. Plenty of quality, as denoted by clean, flat bone, well-defined joints and tendons, refined head and ears, and fine skin and hair.	11. Lacking quality.
12. Breed type (size, color, shape of body and head, and action true to the breed represented).	12. Showing plenty of breed type.	12. Lacking breed type.
Rear view:		
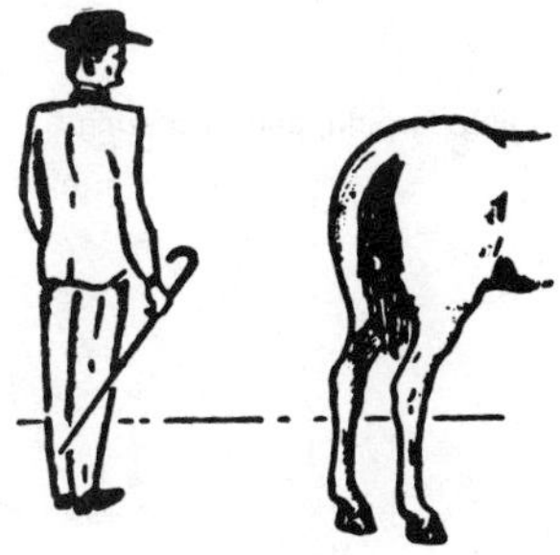 Fig. 11-52	 Fig. 11-53	 Fig. 11-54
1. Width of croup and through rear quarters.	1. Wide and muscular over the croup and through the rear quarters.	1. Lacking width over the croup and muscling through the rear quarters.
2. Set to the hind legs.	2. Straight, true, and squarely set.	2. Crooked hind legs.
Front view:		
 Fig. 11-55	 Fig. 11-56	 Fig. 11-57
1. Head.	1. Head well proportioned to rest of body, refined, clean-cut, with chiseled appearance; broad, full forehead with great width between the eyes; jaw broad and strongly muscled; ears medium sized, well carried, and attractive.	1. Plain headed; weak jaw.

(Continued)

TABLE 11-5 (Continued)

Procedure for Examining, and What to Look for	Ideal Type	Common Faults
Front view: (continued)		
2. Sex character.	2. Refinement and femininity in the brood mare; boldness and masculinity in the stallion.	2. Mares lacking femininity; stallions lacking masculinity.
3. Chest capacity.	3. A deep, wide chest.	3. A narrow chest.
4. Set to the front legs.	4. Straight, true, and squarely set.	4. Crooked front legs.
Soundness:		
1. Soundness, and freedom from defects in conformation that may predispose unsoundness.	1. Sound, and free from blemishes.	1. Unsound; blemished (wire cuts, capped hocks, etc.)
Action[2]:		
1. At the walk.  Fig. 11-58	1. Easy, prompt, balanced; a long step, with each foot carried forward in a straight line; feet lifted clear of the ground.	1. A short step, with feet not lifted clear of the ground.
2. At the trot.  Fig. 11-59	2. Rapid, straight, elastic trot, with the joints well flexed.	2. Winging, forging, and interfering.
3. At the canter.  Fig. 11-60	3. Slow, collected canter, which is readily executed on either lead.	3. Fast and extended.

[1]The illustrations for this table were prepared by R. F. Johnson.

[2]The 3 most common gaits are given herein. Five-gaited horses must perform 2 additional gaits. In judging, (1) every horse should be observed at each intended gait, and (2) trained horses should be examined when performing at the use for which they are intended.

	Perfect Score	ANIMAL No. 1	No. 2	No. 3	No. 4	Etc.
BREED TYPE:	15					
Animals should possess the distinctive characteristics of breed represented, including— **Color** **Height at maturity** **Weight at maturity**						
FORM:	35					
Style and beauty—Attractive, good carriage, alert, refined, symmetrical, and all parts nicely blended together.						
Body—Nicely turned; long well-sprung ribs; heavily muscled.						
Back and loin—Short and strong, wide, well muscled, and short coupled.						
Croup—Long, level, wide, muscular, with a high-set tail.						
Rear quarters—Deep and muscular.						
Gaskin—Heavily muscled.						
Withers—Prominent, and of the same height as the high point of the croup.						
Shoulders—Deep, well laid in, and sloping about a 45° angle.						
Chest—Fairly wide, deep, and full.						
Arm and forearm—Well muscled.						
FEET AND LEGS:	15					
Legs—Correct position and set when viewed from front, side, and rear.						
Pasterns—Long, and sloping at about a 45° angle.						
Feet—In proportion to size of horse, good shape, wide and deep at heels, dense texture of hoof.						
Hocks—Deep, clean-cut, and well supported.						
Knees—Broad, tapered gradually into cannon.						
Cannons—Clean, flat, with tendons well defined.						
HEAD AND NECK:	10					
Alertly carried, showing style and character.						
Head—Well proportioned to rest of body, refined, clean-cut, with chiseled appearance; broad, full forehead with great width between the eyes; ears medium sized, well carried, and attractive; eyes large and prominent.						
Neck—Long, nicely arched, clean-cut about the throat-latch, with head well set on, gracefully carried.						
QUALITY:	10					
Clean, flat bone; well-defined and clean joints and tendons and fine skin and hair.						
ACTION:	15					
Walk—Easy, springy, prompt, balanced, a long step, with each foot carried forward in a straight line; feet lifted clear of the ground.						
Trot—Prompt, straight, elastic, balanced, with hocks carried closely, and high flection of knees and hocks.						
DISCRIMINATION:						
Any abnormality that affects the serviceability of the horse.						
DISQUALIFICATION:						
In keeping with breed registry or show regulations.						
TOTAL	100					

Fig. 11-61. Horse Scorecard.

DETERMINING THE AGE OF ANIMALS BY THE TEETH

The life-span of farm animals is relatively short, and their productiveness or usefulness declines with advancing years. The age of animals, therefore, is of practical importance to the breeder, the seller, and the buyer.

The approximate age of animals can be determined by the teeth as described and illustrated herewith. There is nothing mysterious about this procedure. It is simply a matter of noting the time of appearance and the degree of wear of the temporary and permanent teeth. The temporary or milk teeth are readily distinguished from the permanent ones by their smaller size and whiter color.

It should be realized, however, that theoretical knowledge is not sufficient and that anyone who would become proficient must also have practical experience. The best way to learn how to recognize age in any class of farm animals is to examine the teeth of individuals of known ages.

Determining the Age of Cattle by the Teeth

At maturity cattle have 32 teeth, of which 8 are incisors in the lower jaw. The 2 central incisors are known as pinchers; the next 2 are called first intermediates; the third pair is called second intermediates or laterals; and the outer pair is known as the corners. There are no upper incisor teeth; only the thick, hard dental pad.

Table 11-6, page 1033, illustrates and describes how one may determine the age of cattle by the teeth.

The dental requirements for show steers are given in Section 12—Fitting and Showing Livestock, under "Age and Show Classification of Beef Cattle."

Determining the Age of Sheep and Goats by the Teeth

Mature sheep and goats have 32 teeth, of which 24 are molars and 8 are incisors. As in cattle, all incisors are in the lower jaw. Also, as in cattle, the 2 central incisor teeth are called pinchers; the adjoining ones, first intermediates; the third pair, second intermediates; and the outer ones, corners. There are no tusks.

Table 11-7, page 1034, illustrates and describes how one may determine the age of sheep by the teeth.

Determining the Age of Swine by the Teeth

Mature swine have 44 teeth. Of these, 12 are front teeth or incisors, 4 are tusks or tushes, 4 are premolars, and 24 are molars (with half of each kind of teeth found in each jaw). As in the horse, the incisors are grouped in 3 pairs in each jaw and are called centrals, intermediates, and corners in accordance with their relative locations. The tusks are more prominent in the male than in the female.

Table 11-8, page 1035, illustrates and describes how one may determine the age of swine by the teeth.

Determining the Age of Horses by the Teeth

The mature male horse has a total of 40 teeth,[3] whereas the young animal, whether male or female, has 24. These are as listed on page 1037, in Table 11-9.

As the tushes are usually not present in the mare, the mature female may be considered as having a total of 36[3] teeth rather than 40 as in the male.

---

[3]Quite commonly, a small, pointed tooth, known as a "wolf tooth," may appear in front of each first molar tooth in the upper jaw, thus increasing the total number of teeth to 42 in the male and 38 in the female. Less frequently, there are 2 more "wolf teeth" in the lower jaw, which increases the total number of teeth in the male and female to 44 and 40, respectively.

TABLE 11-6

GUIDE TO DETERMINING THE AGE OF CATTLE BY THE TEETH[1]

Drawing of Teeth	Age of Animal	Description of Teeth
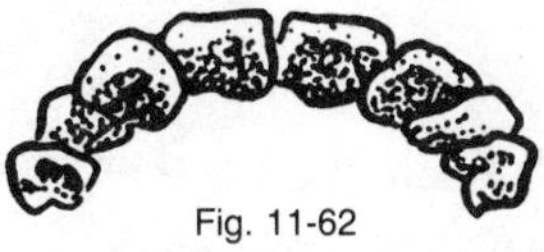 Fig. 11-62	At birth to 1 month	Two or more of the temporary incisor teeth present. Within first month, entire 8 temporary incisors appear.
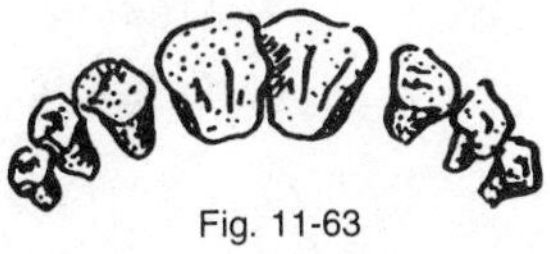 Fig. 11-63	2 years	As a long-yearling, the central pair of temporary incisor teeth or pinchers is replaced by the permanent pinchers. At 2 years, the central permanent incisors attain full development.
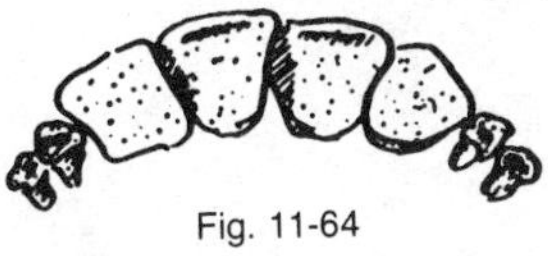 Fig. 11-64	2½ years	Permanent 1st intermediates, one on each side of the pinchers, are cut. Usually these are fully developed at 3 years.
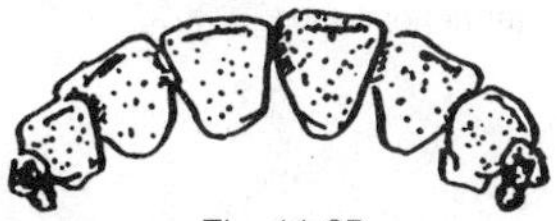 Fig. 11-65	3½ years	The 2nd intermediates or laterals are cut. They are on a level with the 1st intermediates and begin to wear at 4 years.
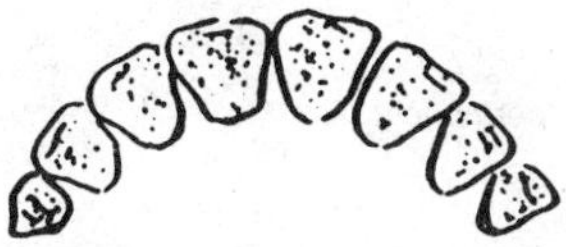 Fig. 11-66	4½ years	The corner teeth are replaced. At 5 years the animal usually has the full complement of incisors with the corners fully developed.
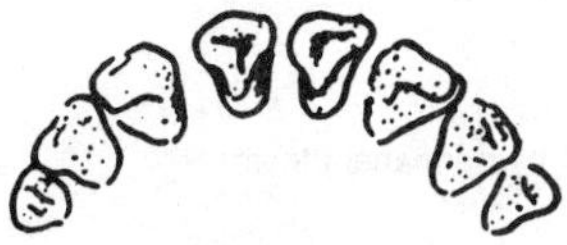 Fig. 11-67	5 or 6 years	The permanent pinchers are leveled, both pairs of intermediates are partially leveled, and the corner incisors show wear.
 Fig. 11-68	7 to 10 years	At 7 or 8 years, the pinchers show noticeable wear; at 8 or 9 years, the middle pairs show noticeable wear; and at 10 years, the corner teeth show noticeable wear.
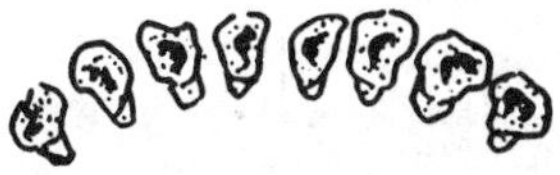 Fig. 11-69	12 years	After the animal passes the 6th year, the arch gradually loses its rounded contour and becomes nearly straight by the 12th year. In the meantime, the teeth gradually become triangular in shape, distinctly separated, and show progressive wearing to stubs. These conditions become more marked with increasing age.

[1]The illustrations for this table were prepared by R. F. Johnson. See Section 12—Fitting and Showing Livestock—for the dental requirements for show steers.

TABLE 11-7

GUIDE TO DETERMINING THE AGE OF SHEEP BY THE TEETH[1]

Drawing of Teeth	Age of Animal	Description of Teeth
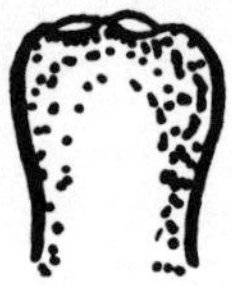 Fig. 11-70	Newborn lamb	None of the teeth may be present, although sometimes the 2 pinchers and also the 2 first intermediates are pressing through the gums or even are cut through.
 Fig. 11-71	3 months	A full set of completely developed temporary incisor teeth present.
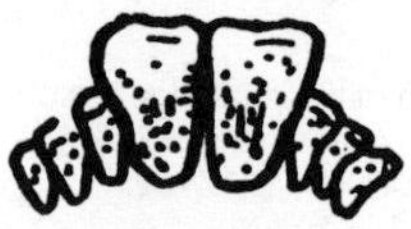 Fig. 11-72	12 to 15 months	Temporary pinchers are replaced by the 2 permanent ones.
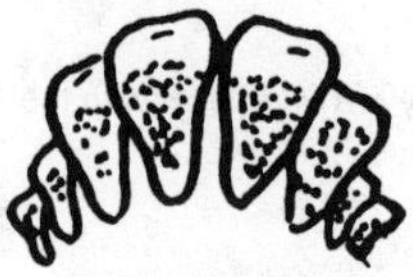 Fig. 11-73	2 years	First temporary intermediates replaced by permanent teeth.
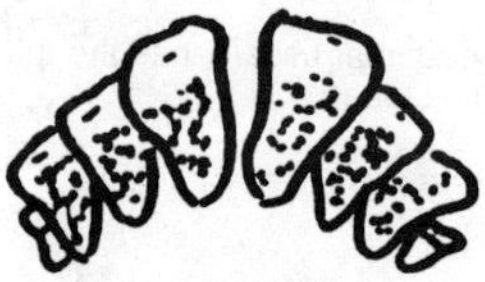 Fig. 11-74	3 years	Second temporary intermediates replaced by permanent teeth.
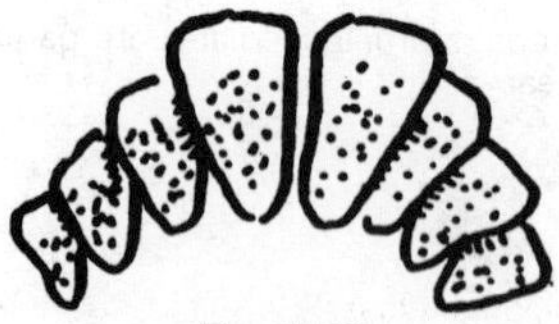 Fig. 11-75	4 years	Two temporary corner incisors replaced by permanent teeth. Animal has "full-mouth."
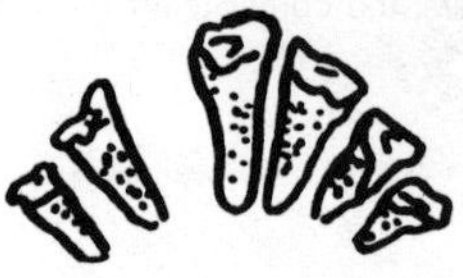 Fig. 11-76	After 4 years	After the sheep has a solid mouth (at 4 years), it is impossible to tell the exact age. With more advanced age, the teeth merely wear down and spread apart, and the degree of wearing and spreading is an indication of age. The normal number of teeth may be retained until 8 or 9 years, but often some are lost after about the 5th or 6th year, resulting in a "broken mouth." When most of the teeth have disappeared, animals are known as "gummers."

[1]The illustrations for this table were prepared by R. F. Johnson.

TABLE 11-8

GUIDE TO DETERMINING THE AGE OF SWINE BY THE TEETH[1]

Drawing of Teeth	Age of Animal	Description of Teeth
 Fig. 11-77	At birth	Eight sharply pointed, so-called "needle teeth," consisting of 2 tusks and 2 corner incisors on each jaw. These teeth are commonly cut off in order to avoid discomfort and injury to the nursing sow.
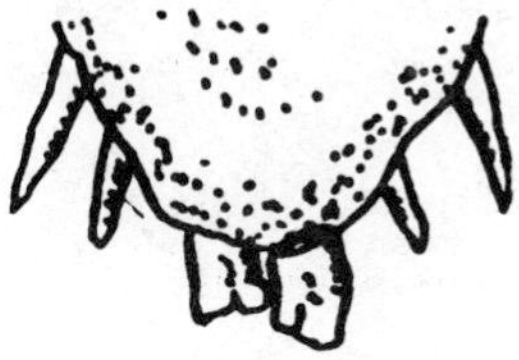 Fig. 11-78	4 to 5 weeks	Central temporary incisors appear, 2 in the upper and 2 in the lower jaw.
 Fig. 11-79	6 to 12 weeks	At 6 to 8 weeks, 2 temporary intermediate incisors appear. At 12 weeks, central incisors fully grown.
 Fig. 11-80	6 months	Temporary corner incisors are shed, and the permanent corners appear.
 Fig. 11-81	9 months	The permanent tusks take the place of the temporary tusks.
 Fig. 11-82	12 months	Central permanent incisors replace the temporary centrals.

(Continued)

Footnote on last page of table.

TABLE 11-8 (Continued)

Drawing of Teeth	Age of Animal	Description of Teeth
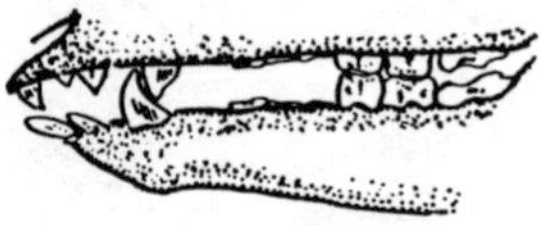 Fig. 11-83	12 to 15 months	First 3 temporary molars, on each side of upper and lower jaw, shed and replaced by permanent molars.
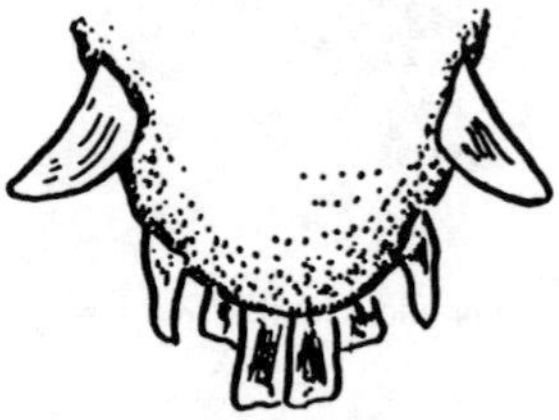 Fig. 11-84	18 to 20 months	At 18 months, temporary intermediate incisors shed and replaced by permanent ones. At 20 months, permanent intermediate incisors in line with centrals.
 Fig. 11-85	Above 20 months	Past 20 months, the age of hogs cannot be estimated with certainty. At 2 years, however, the incisors, including the intermediates, will show wear, and the 6th or last molars (one upper and one lower on each side) will be fully up and about to come in contact. Beyond this stage, advance in age is merely indicated by progressive wear of the teeth.

[1]The illustrations for this table were prepared by R. F. Johnson.

TABLE 11-9

NUMBER AND TYPES OF HORSE TEETH

Number of Teeth of Mature Animal	Number of Teeth of Young Animal	Types of Teeth
24	12	Molars or grinders.
12	12	Incisors or front teeth (the 2 central incisors are known as centrals or nippers; the next 2—one on each side of the nippers—are called intermediates or middles; and the last—or outer pair—the corners).
4	None	Tushes or pointed teeth. These are located between the incisors and the molars in the male.

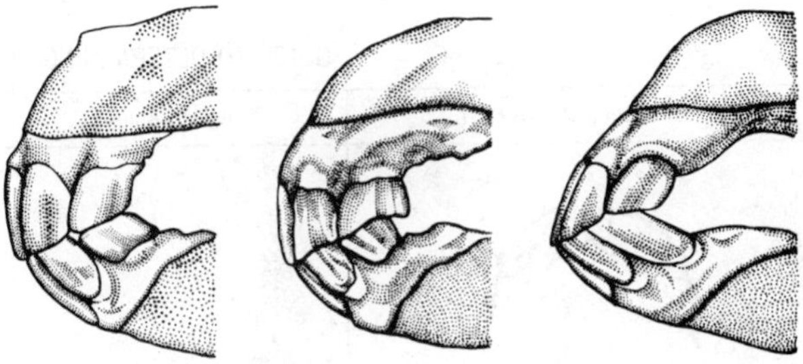

Fig. 11-86. Side view of 5-, 7-, and 20-year-old horse mouth. Note that as the horse advances in age the teeth change from nearly perpendicular to slanting sharply toward the front. (Drawing by R. F. Johnson)

The permanent incisor teeth of young horses five to seven years of age are elliptical or long from side to side; whereas when the animal becomes older, these teeth become triangular, with the apex of the triangle pointed upward. As the animal advances still more in age, the teeth become more slanting. Instead of curving to approach a right angle with the jaws, they slant outward.

From 5 to 12 years of age the wearing surface of the cups is the most reliable indication of age. At fairly regular intervals, according to age, the cups disappear with wear.

Table 11-10 illustrates and describes how one may determine the age of horses by the teeth.

After 12 years of age, even the most experienced horsemen cannot accurately determine the age of an animal. It is known, however, that with more advanced age the teeth change from oval to triangular and that they project or slant forward more and more each year.

It must also be realized that the environment of the animal can very materially affect the wear on the teeth, often making it impossible to determine accurately the age of animals. For example, the teeth of horses raised in a dry, sandy area will show more than normal wear. Thus, the 5-year-old western horse may have a 6- or even 8-year-old mouth. The unnatural wear resulting in the teeth of cribbers or animals with parrot mouth or undershot jaw also makes it difficult to estimate age.

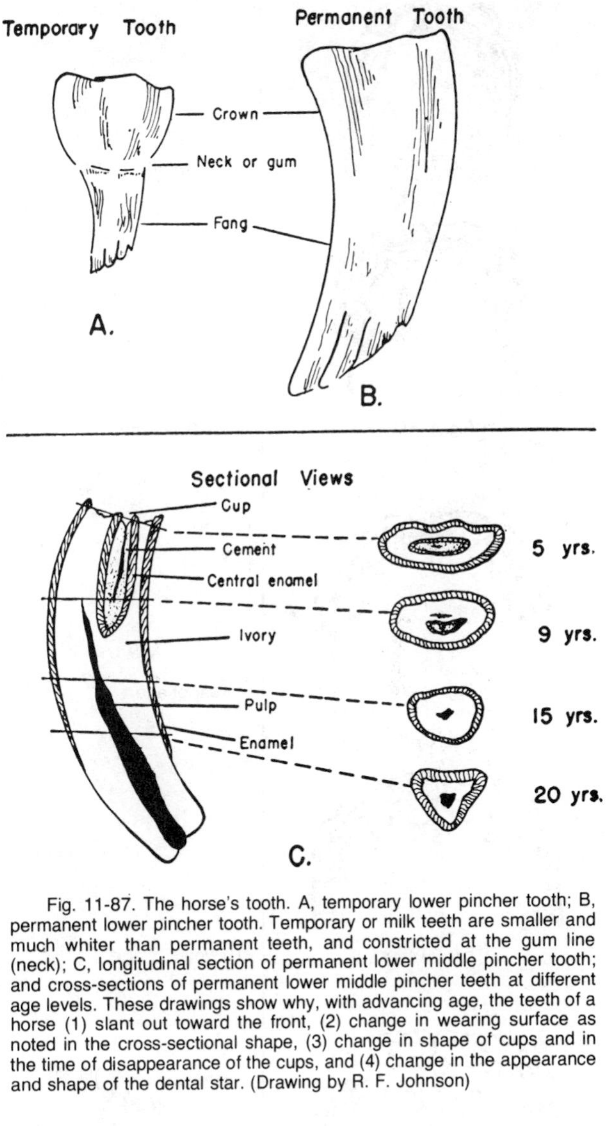

Fig. 11-87. The horse's tooth. A, temporary lower pincher tooth; B, permanent lower pincher tooth. Temporary or milk teeth are smaller and much whiter than permanent teeth, and constricted at the gum line (neck); C, longitudinal section of permanent lower middle pincher tooth; and cross-sections of permanent lower middle pincher teeth at different age levels. These drawings show why, with advancing age, the teeth of a horse (1) slant out toward the front, (2) change in wearing surface as noted in the cross-sectional shape, (3) change in shape of cups and in the time of disappearance of the cups, and (4) change in the appearance and shape of the dental star. (Drawing by R. F. Johnson)

TABLE 11-10

GUIDE TO DETERMINING THE AGE OF HORSES BY THE TEETH[1]

Drawing of Teeth	Age of Animal	Description of Teeth	
 Fig. 11-88	At birth or before 10 days of age	First or central upper and lower incisors appear.	Appearance of temporary teeth
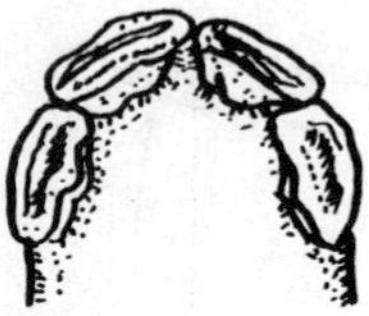 Fig. 11-89	4 to 6 weeks of age	Second or intermediate upper and lower incisors appear.	
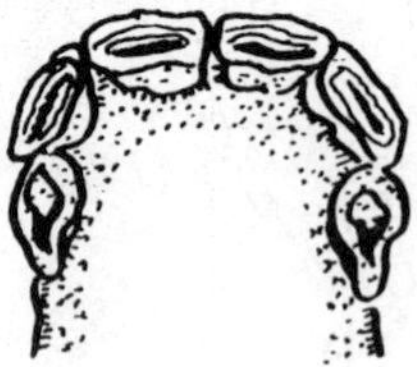 Fig. 11-90	6 to 10 months	Third or corner upper and lower incisors appear.	
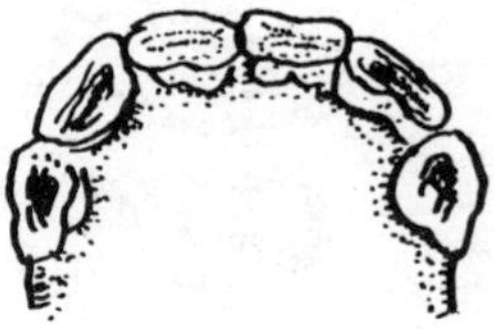 Fig. 11-91	1 year of age	Crowns of central incisors show wear.	Wear of temporary teeth
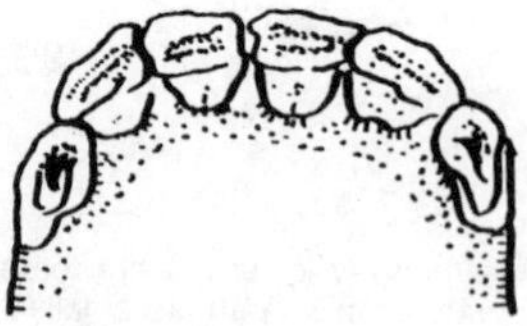 Fig. 11-92	1½ years of age	Intermediate incisors show wear.	
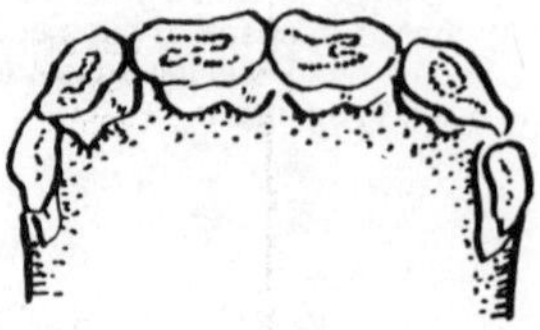 Fig. 11-93	2 years of age	All temporary incisors show wear.	

Footnote on last page of table.

(Continued)

TABLE 11-10 (Continued)

Drawing of Teeth	Age of Animal	Description of Teeth	
 Fig. 11-94	2½ years of age	First or central incisors appear.	Appearance of permanent teeth
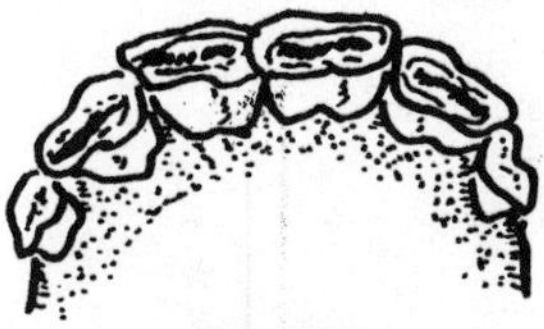 Fig. 11-95	3½ years of age	Second or intermediate incisors appear.	
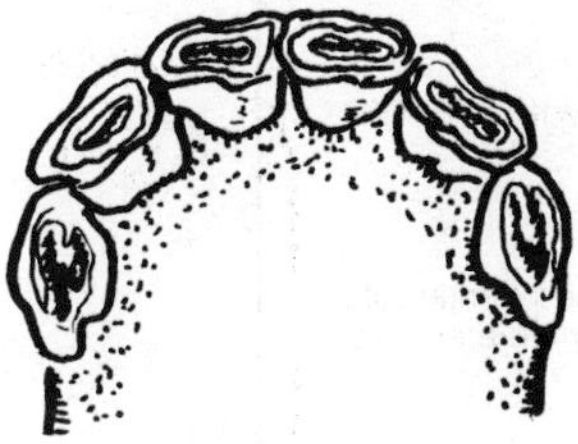 Fig. 11-96	4½ years of age	Third or corner incisors appear.	
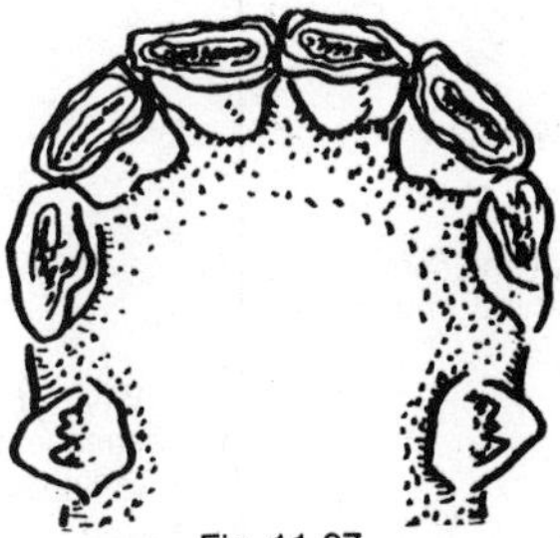 Fig. 11-97	4 to 5 years of age (in male)	Canines appear.	
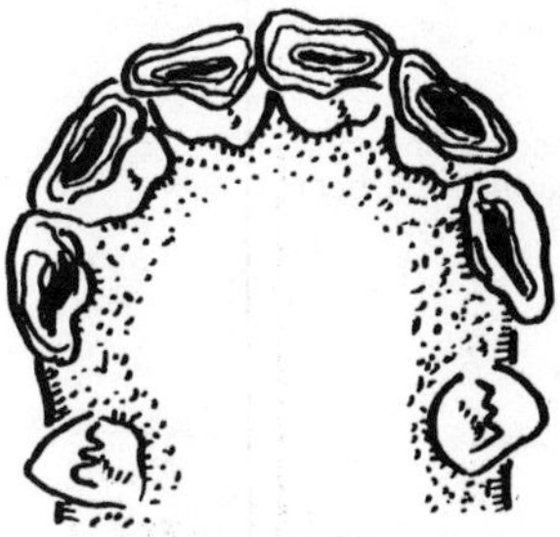 Fig. 11-98	5 years of age	Cups in all incisors.	Wear of permanent teeth
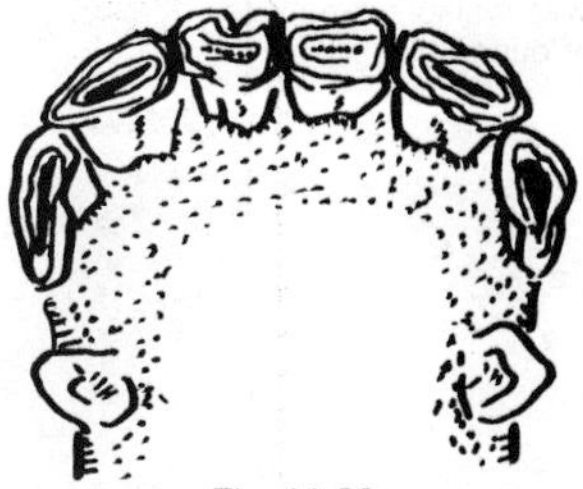 Fig. 11-99	6 years of age	Cups worn out of lower central incisors.	

(Continued)

TABLE 11-10 (Continued)

Drawing of Teeth	Age of Animal	Description of Teeth	
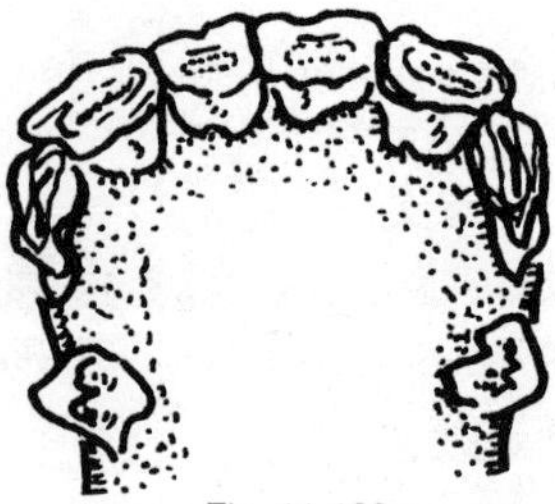 Fig. 11-100	7 years of age	Cups also worn out of lower intermediate incisors.	
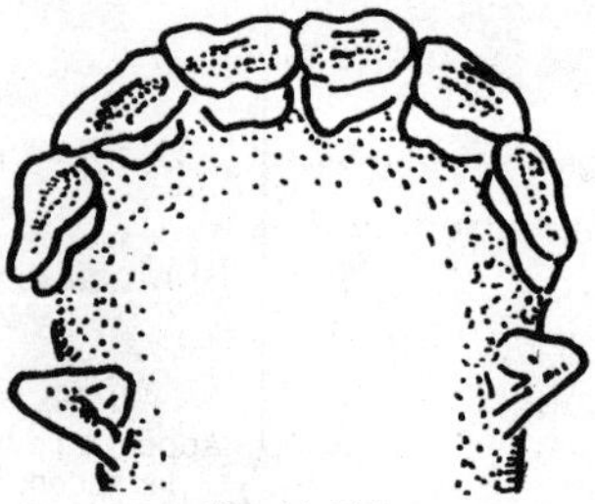 Fig. 11-101	8 years of age	Cups worn out of all lower incisors, and dental "star" appears on lower central and intermediate pairs.	
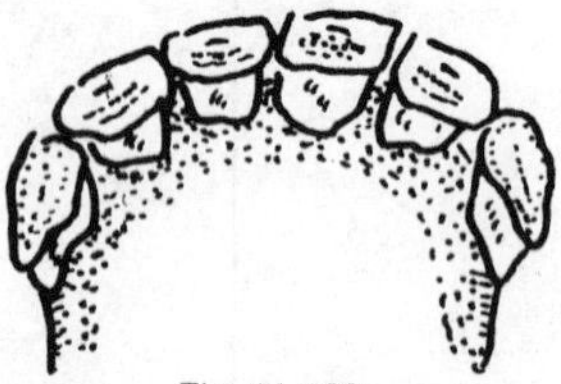 Fig. 11-102	9 years of age	Cups also worn out of upper central incisors, and dental "star" appears on upper central and intermediate pairs.	Wear of permanent teeth
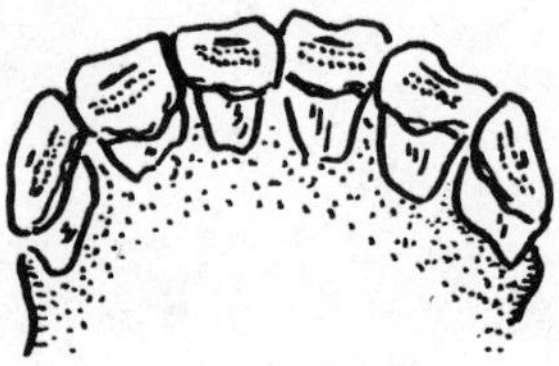 Fig. 11-103	10 years of age	Cups also worn out of upper intermediate incisors, and dental "star" is present on all incisors, both upper and lower.	
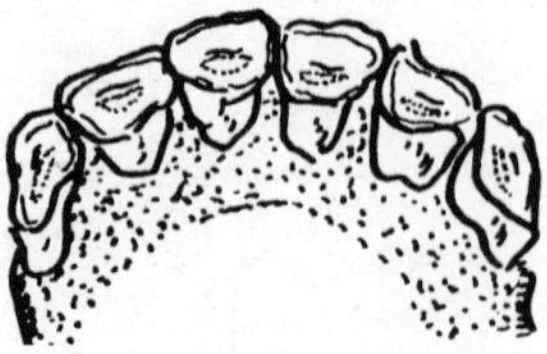 Fig. 11-104	11 years of age	Cups worn out of all upper and lower incisors, and dental "star" approaches center of cups.	
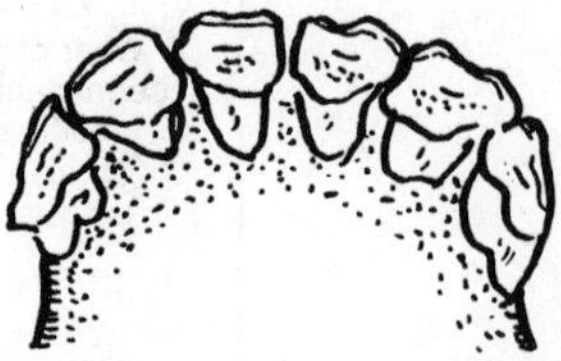 Fig. 11-105	12 years of age	No cups. "Smooth mouthed."	

[1]The illustrations for this table were prepared by R. F. Johnson.

Comparative Animal Ages

Many "oldest animal" ages are claimed, but few of them are substantiated. Certainly, if the record ages claimed by many human centenarians are undocumented and unacceptable, the asserted records of many lower animals are even more suspect. There is need, therefore, for some mathematical method of relating the ages of other animals to that of man. Most people use a direct ratio, or rule of thumb. For example, a dog's age is commonly assumed to be one-seventh that of a man; hence, a dog 7 years old would be equivalent in age to a man 49. But since the oldest dog on record lived to the age of 34, according to the 1- to 7-year theory, a man could live to 34 × 7, or 238 years. This, of course, is more than twice the highest reliably recorded human age. The fallacy of the rule-of-thumb method (the 1- to 7-ratio in the case of dog to man) is that it fails to take into consideration *both* (1) the average age at maturity, and (2) the average age at death.

Mr. David P. Willoughby developed an ingenious, but relatively simple, "prediction formula," which takes into consideration both the time required to reach maturity and the maximum age recorded.[4]

As shown in the first two columns of Table 11-11, he assumes (based on information available to him) for each species (1) the average age at maturity, and (2) the average age at death.

In Willoughby's formula, the average age at death is B, and the average age at maturity is A. Then the ratio of the maximum potential age, C, to B is C/B = 0.32 B/A + 0.6.

Using Willoughby's formula, and securing from Table 11-11 the average age of a horse at death (22 years) and the average age of a horse at maturity (5 years), the maximum predicted age of a horse is—

$$C/22 = [0.32\ (22/5) + 0.6], \text{ or}$$
$$C = 22 \times [0.32\ (22/5) + 0.6] = 44 \text{ years.}$$

By applying this formula to each species listed in Table 11-11, the maximum predicted ages shown in the right-hand column were obtained. Although recognizing that the Willoughby formula makes use of 2 assumed values (average age at maturity, and average age at death), in the opinion of the author of this book it is a more accurate way to predict maximum age than the direct ratio method (1 to 7 dog to man ratio, for example).

TABLE 11-11

COMPARATIVE AGES, IN YEARS, OF MAN AND SOME LOWER ANIMALS[1]

Species	Av. Age at Maturity	Av. Age at Death	Maximum Age	
			Reported	Predicted
(All males)	*(Years)*	*(Years)*	*(Years)*	*(Years)*
Man	21	70	115+	117
Camel	8	31?	40+	57
Cat	1	10	35	38
Cow	4	19	30+	40
Dog	1.5	11.5	34	35
Elephant, Asiatic	25	60	77+	82
Hog	1.0	8.0	13	25
Horse	5	22	62	44
Mouse	.182	2.2	6	9.8
Sheep	1.5	9	15+	22.7

[1]Adapted by the author. Also, the author added swine.

[4]Mr. Willoughby is Honorary Associate in Vertebrate Paleontology at the Los Angeles County Museum of Natural History. Table 11-11 reproduced herein is taken from *Natural History*, Vol. LXXVIII, No. 10, published by The American Museum of Natural History, New York, N.Y., December, 1969, pp. 56-59.

SECTION 12

FITTING AND SHOWING LIVESTOCK

Contents

Contents	Page
Showing Beef Cattle	1043
Training, Grooming, and Showing Beef Cattle	1043
Age and Show Classification of Beef Cattle	1047
Showing Dairy Cattle	1048
Training, Grooming, and Showing Dairy Cattle	1048
Age and Show Classification of Dairy Cattle	1050
Showing Sheep	1051
Training, Grooming, and Showing Sheep	1051
Age and Show Classification of Sheep	1053
Showing Swine	1054
Training, Grooming, and Showing Swine	1054
Age and Show Classification of Swine	1056
Showing Horses	1056
Training, Grooming, and Showing Horses	1056
Age and Show Classification of Horses	1058
Shipping to the Fair	1059

There is no higher achievement than that of breeding and fitting a champion; an animal representing an ideal which has been produced through intelligent breeding and then fitted to the height of perfection. Also, it is realized that most of the advertising value of the show-ring accrues to those who exhibit winners, thus behooving the exhibitor to select, fit, and show his animals to the best possible advantage.

The selection of the prospective show animals is the first and most important assignment. Some pointers relative to type or individuality are given in Section 11, Selecting and Judging Livestock.

The second requisite for successful showing is to fit the animals to the peak of perfection. Directions on feeding show animals and suggested fitting rations are contained in Section 4, Feeding.

Assuming that show animals have been carefully selected and properly fed, there yet remains the assignment of parading before the judge. In order to present a pleasing appearance in the show-ring, animals must be well trained, thoroughly groomed, and properly shown. Competition is keen, and often the winner will be selected by a very narrow margin. Close attention to details may, therefore, be a determining factor in the decisions.

Also, most successful showmen attempt (1) to select animals as old as possible within the respective age classifications, so that they may show to the best possible advantage, and (2) to fill as many show classes as possible, thus increasing their chances of winning. In order to meet these requisites, the showman must be fully versed relative to the usual show classifications.

SHOWING BEEF CATTLE

Pertinent information relative to showing beef cattle is summarized in the two sections which follow.

Training, Grooming, and Showing Beef Cattle

Table 12-1 is an illustrated guide for training, grooming, and showing beef cattle.

TABLE 12-1

TRAINING, GROOMING, AND SHOWING GUIDE FOR BEEF CATTLE

The Essential Steps in Training, Grooming, and Showing	Why	How
1. Gentling and posing  Fig. 12-1	Such schooling makes it possible for the judge to see the animal at its best.	The logical steps in gentling and posing prospective show cattle are: Petting and brushing. Accustoming to rope halter. Leading with rope halter; walking forward from the left side and holding the rope (or halter strap) neatly coiled in the right hand. Teaching the animal to stand or pose properly; standing squarely on all 4 feet, with the back straight and the head held on a level with the top of the back. Accustoming to the show halter in advance of the show.
2. Trimming, cleaning, and oiling the feet  Fig. 12-2	So that the animal will stand squarely and walk properly, and that the feet will be clean and attractive.	The feet should be trimmed whenever they need it. With some calves, it may be necessary to administer the first trimming as early as 3 months of age, and to repeat the trimming at intervals of 3 to 4 weeks. Proper trimming, cleaning, and oiling the feet may be accomplished as follows: Either secure the animal in a set of stocks or throw him. Square up the soles and sides of the feet, and cut back the toes; using a chisel, nippers, farrier's knife, and/or rasp. Learn to trim feet correctly by observing and by trimming under the direction of an expert. Before entering the ring, thoroughly clean and oil the hoofs and dew claws. Never trim the feet nearer than a week, and preferably 2 weeks, prior to the show, thus giving the animal ample time to get used to the trimming and avoiding possible tenderness at show time.
3. Training and polishing the horns 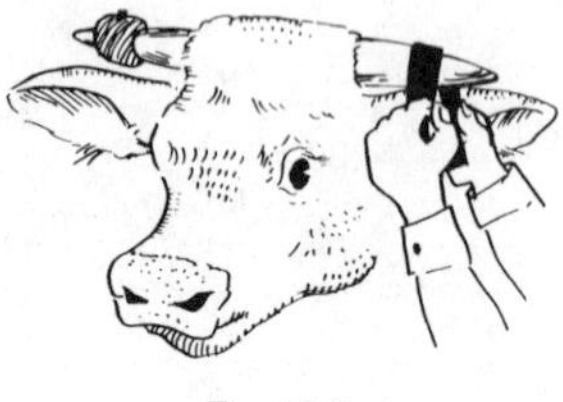 Fig. 12-3	A well-curved set of horns will command the admiration of the judge, but poorly shaped horns will give the head a coarse, unattractive appearance.	Proper training and polishing of the horns may be accomplished as follows: When the horns are 3 to 4 inches long, apply ½-pound weights for such time and at such intervals as necessary to obtain the desired shape. Shorten extremely long horns by cutting them back. They can be cut back 2 to 3 inches; removing a half inch at a time at no more frequent intervals than a month or 6 weeks. As a rule, most of the black tip can be removed without harming the sensitive part. Smooth the horns down a week or two before the show, using a sharp knife, a rasp, or a steel scraper; then give the final touches with sandpaper, fine emery cloth, steel wool, or a flannel cloth and emery dust. Apply a polish, such as (1) a paste made by mixing olive oil or sweet oil with pumice stone or tripoli, or (2) glycerine, linseed oil, or mineral oil. Then rub briskly with a flannel cloth.
4. Grooming 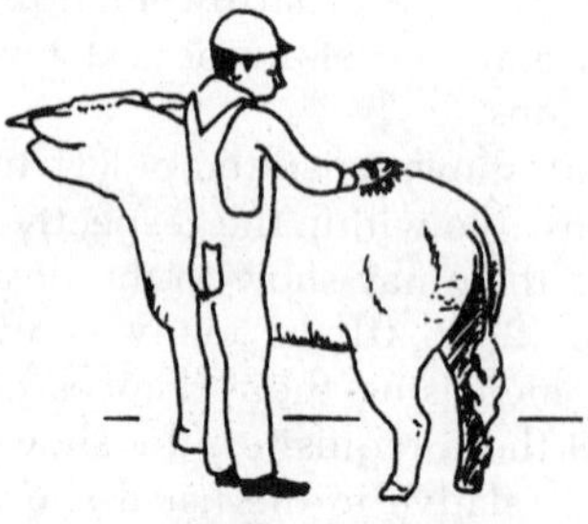 Fig. 12-4	Vigorous brushing (1) stimulates the circulation in the hide, keeping it in a loose pliable condition, (2) brings out the natural oil in the hair, and (3) removes dandruff, dirt, and dead hair.	With a brush and woolen cloth; use a curry comb sparingly. Brush long-haired animals downward and then upward in the opposite direction of the lay of the hair in order to make it loose and fluffy.
5. Blanketing  Fig. 12-5	It helps to keep the animal clean, alleviates annoyance from flies, gives the hair a more glossy appearance, and helps to keep the hide mellow and pliable. However, the use of a blanket may cause excessive sweating and loss of hair; thus, most cattlemen limit blanketing to those animals on which the hide needs softening.	Either blanket with a commercially made blanket, or make a cheap but satisfactory one out of burlap.

(Continued)

TABLE 12-1 (Continued)

The Essential Steps in Training, Grooming, and Showing	Why	How
6. Clipping 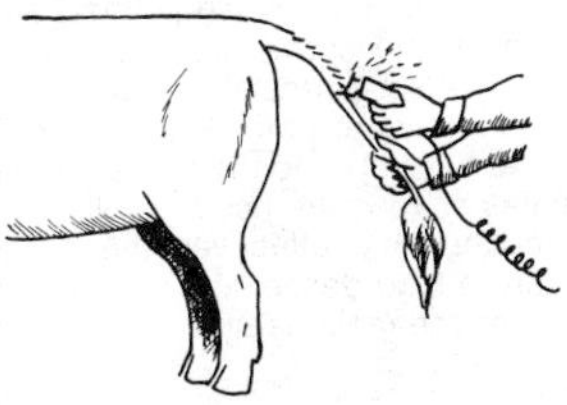 Fig. 12-6	The tail is clipped in order to show the fullness of the twist and the thickness or beefiness of the hindquarters. The head of polled animals is clipped so that it will appear cleaner cut, more shapely, and the poll more clearly defined. Not all animals are clipped.	Clipping is best done with electric clippers, although hand clippers may be used. In general, clipping of the tail should begin above the switch even with the point where the fullness of the twist begins to fail, and should extend to the tail head where it is tapered off so that the tail head blends nicely with the rump. The heads of polled or dehorned animals should be clipped from a point just back of the jawbone and ½ inch behind the ears; leaving the eyelashes and the hair on the nose. With Angus, do not clip either the inside or the outside of the ears; with polled or dehorned steers, clip the outside of the ears but leave the inside untrimmed. Clipping should be done 6 or 7 days before the show so that the clipped hair will lose its stubby appearance. Not all animals are clipped. Usually custom decrees the following: **Steers:** 1. All tails clipped. 2. Heads clipped if naturally polled or dehorned (clip the outside of the ears, but leave the inside untrimmed). 3. Horned heads not clipped. **Breeding cattle:** 1. Angus—Heads (but do not clip either the inside or the outside of the ears) and tails clipped. 2. Shorthorns (both horned and polled)—Neither tails nor heads clipped. 3. Herefords (both horned and polled)—Tails clipped; heads not clipped.
7. Washing  Fig. 12-7	Frequent washing keeps the animal clean; stimulates a heavy growth of loose, fluffy hair; and keeps the skin smooth and mellow.	Beginning 4 to 6 weeks before the show, wash the animal regularly once each week. Since washing usually makes an animal appear gaunt for a time, it is usually best not to wash within 10 to 12 hours of the show. In preparation for washing, (1) brush thoroughly with a stiff brush in order to remove as much dirt and dandruff as possible, and (2) secure the animal with a chain about the neck instead of a rope, but make sure that this chain can be quickly and easily released in case of trouble. Wash the animal carefully by the following procedure: 1. Prepare a bucket of soapy solution by mixing 1 to 2 cups of good concentrated liquid coconut oil shampoo in 1½ gallons of warm water (or high grade castile soap may be used). 2. Wet the animal thoroughly all over (except in the ears; either hold the ear down or shut it off with the hand as each side of the head is washed). 3. Soap the animal by dipping a rice brush into the bucket of soap solution, and by scrubbing until a good thick lather covers all parts. Rinse off all the soap with care. Any soap left may tend to cause dandruff. Remove surplus water from all parts of the animal by using the back of a Scotch comb (or a scraper). If desired (and this step is a matter of preference), spray lightly with a solution made by adding one tablespoonful of creosote to one gallon of water. The dip helps to hold the curl. Some cattlemen eliminate this step, because they feel dip causes dandruff. Curl the animal as described in the next step.
8. Curling 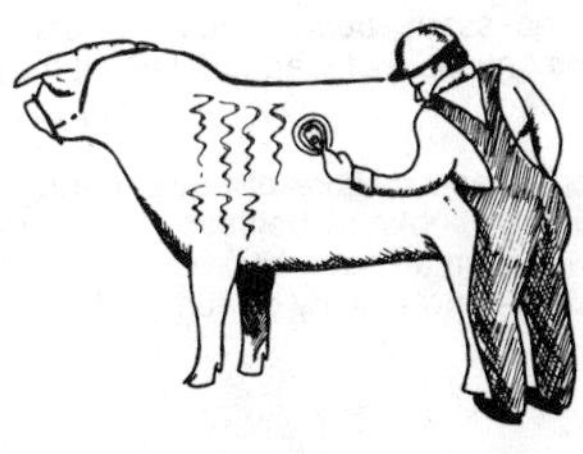 Fig. 12-8	In order to emphasize the strong points, and to soften and minimize any roughness or weakness in conformation.	Experienced showmen vary the method or type of curling according to the individuality of the animal; giving consideration to the length of hair and to whether it is straight or naturally curly and to the conformation and condition of the animal. Also, there are some differences between breeds; for example, in Angus breeding classes it is common practice to curl the hair in the regions of the neck, the forepart of the shoulders and on the rear part of the thighs—leaving the hair on the other parts of the body uncurled. The most skillful showmen are able to produce a more natural effect rather than something that is quite artificial in appearance. Regardless of the type of curl desired, in preparation therefor the animal is always either (1) washed, rinsed, and scraped free of surplus water—as described under washing; or (2) dry-brushed, wetted down all over (preferably by means of a hand sprayer, although satisfactory wetting down may be accomplished by dipping a stiff brush into the water or creosote solution and by brushing the hair smoothly against the animal) with water alone or a

(Continued)

TABLE 12-1 (Continued)

The Essential Steps in Training, Grooming, and Showing	Why	How
		solution made by adding one tablespoonful of creosote to each gallon of water, and scraped free of any surplus water. Naturally, the latter procedure is followed when a complete washing is not necessary or desired, as is frequently the case when animals are curled daily or twice daily when on the show circuit or when curled 1 to 2 hours before showing. Also, animals are frequently wetted down without washing when they are being curled 1 to 2 times daily, as is customary 7 to 10 days prior to the first show. In general, the following distinct types of curls are used:

The fluffy curl—Which is really a misnomer, for it is not a curl in the true sense of the word. Instead, the object is to produce a hair effect very much like that of a fluffy teddy bear or a fluffy ball of fur. When the art is plied by a master showman to a beast which possesses a long dense coat of hair, but which does not possess a tight natural curl (for example, most Shorthorns are straight-haired), the fluffy curl produces the most natural effect of all methods.

Step by step, the fluffy curl is produced as follows:

1. Either wash or wet down the hair of the animal as previously described.
2. Part the hair (with a Scotch comb) from back of the poll to a point just back of the withers, and comb (with a Scotch comb) the moist hair in the direction of its natural lay (from back of the withers toward the tail head along the top line; downward on the rest of the body) on all body parts except the head and rear end.
3. Beginning along the top of the animal, make waves ½ in. to 1 in. apart (and about 2 in. wide) by drawing the back tip of a round curry comb in the direction of the natural lay of the hair in a short, wavy manner. Make waves on all body parts except the head, rear end, and legs. Proceed to step No. 4, thus allowing the necessary 2 to 3 minutes for the curl to set before applying step No. 5.
4. Comb (with a Scotch comb) and/or brush the hair on the head forward and toward the muzzle on horned animals (and polled animals whose heads have not been clipped), outward from the median line at the top of the neck and withers, and outward from the median line of the rear end; thus accentuating the width in these regions. Brush the hair on the ears toward the head.
5. In order to tip back the ends of the hairs, while the coat is drying, brush it lightly (with a dry rice brush) opposite the direction of its natural lay (upward on all body parts except the top line; forward on the top line from the tail head to a point just back of the withers). Such brushing should begin at the tops of the hoofs and extend to the top of the animal; except in steers, where fineness of bone is desired, it is usually best not to fluff out the hair below the knees and hocks.

As the hair becomes fairly dry (the drying time varying with the weather and density of hair), discontinue brushing and draw (with a Scotch comb) it opposite the direction of its natural lay.

After the hair of straight-haired animals is well trained, one may start this (step No. 5) with the Scotch comb, without first brushing as indicated above. But the procedure as indicated should always be followed with animals that are naturally curly-haired.

6. Continue to comb the dry hair outward and upward (with a large, coarse, hard rubber comb).

The wave curl—Which is sometimes referred to as the Hereford curl, because it is especially popular with Hereford showmen. This type of curl may be produced on any animal which possesses a rather tight natural curl, such as characterizes most Herefords, whereas the fluffy curl is produced on straight-haired animals.

Step by step, the wave curl is produced in exactly the same manner as previously outlined for the fluffy curl. Thus, the reader is referred to the previous section. The chief difference is the end result obtained from applying step No. 5; namely, in brushing and combing the hair in the opposite direction of its natural lay. In straight-haired animals (such as most Shorthorns), the application of this step in tipping the hair back—thus producing the fluffy curl; whereas in naturally curly-haired animals (such as most Herefords), the application of this step results in denser, tighter curls—thus producing the wave curl.

In producing the fluffy curl and the wave curl, one further difference may be observed. In the wave curl, step No. 6 may be either omitted or applied lightly depending on how tight a curl is desired.

The parallel curl—In which parallel lines spaced about one inch apart are marked along the body.

Step by step, the parallel curl is produced as follows:

1. Either wash or wet down the hair of the animal as previously described.
2. Part the hair (with a Scotch comb) along the center top line from back of the poll to the tail, and comb (with a Scotch comb) the moist hair downward on all parts except the head and rear end.
3. Mark off (with a "liner") parallel lines about one inch apart along the sides of the body, from in front of the shoulders to the rear edge of the thigh and from the point where the back and the sides blend to the knees and hocks. Either line both sides from rear to front or vice versa.
4. From the tops of the hoofs to the point where the back and sides blend, draw (with a Scotch comb) the hair upward, and then in order to curl the ends of the hairs, brush upward with a dry brush while it is drying. With steers, where fineness of bone is desired, it is usually best to comb the hair downward below the knees and hocks.
5. Comb (with a Scotch comb) the hair on the head forward and toward the muzzle on horned animals, from the median line across the withers from the center top line to where the back and sides blend, and from the median line of the rear end, thus accentuating the width in these regions.

The diamond curl—In which the animal is first lined (with a liner) in the shape of diamonds is sometimes used. Like the parallel curl, this method is easily mastered by the amateur and is valuable in training the hair in early stages. However, like the parallel curl, it leaves marks that still show on the animal after further brushing, combing, and drying, and produces a distinctly artificial effect instead of the much sought natural effect.

9. Oiling  Fig. 12-9	To impart "bloom" to the animal.	An animal in full bloom doesn't necessarily need artificial oil on its hair. However, if the hair seems unusually dry and lifeless, it may be desirable to apply oil, provided it is limited to a light application properly applied. Either a commercial oil or a homemade mixture of equal parts of glycerine, sweet oil, and rubbing alcohol may be used. After curling, apply lightly and evenly with a small hand sprayer or woolen cloth, and then brush the animal. Avoid a shiny or gummy finish.

(Continued)

TABLE 12-1 (Continued)

The Essential Steps in Training, Grooming, and Showing	Why	How
10. Cleaning and fluffing the switch 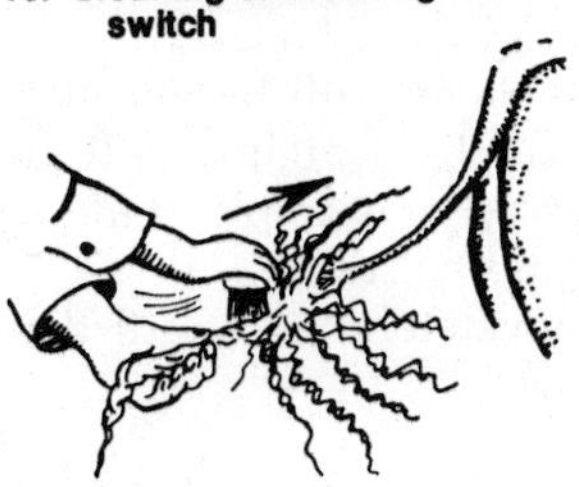 Fig. 12-10	In order to enhance attractiveness.	The evening before the show, wash the switch thoroughly with soap and warm water. If it is white, add a little blueing to the wash water to brighten the hair. Rinse the switch in alum water (the alum will make it fluffy). While the hair is damp, braid the switch into about six 3-strand braids, and tie all of them together at the bottom. About an hour before the show, unbraid and fluff by brushing the hair upward a few strands at a time.
11. Showing  Fig. 12-11	In order to win.	The following guiding principles are adhered to by successful showmen: Train the animal long before entering the ring. Have the animal carefully groomed and ready for the parade before the judge. Dress neatly for the occasion. Enter the ring promptly when the class is called. Lead the animal from the left side (walking near the left shoulder), with the halter strap neatly coiled in the right hand. Pose the animal correctly so that it stands on all four feet with the back held perfectly straight and the head on a level with the top of the back. Keep one eye on the judge and the other on the animal. Keep calm and collected. Work in close partnership with the animal. Be courteous and respect the rights of other exhibitors. Be a good sport; win without bragging and lose without squealing.

Age and Show Classification of Beef Cattle

Distinct and separate show classifications are provided for breeding animals and for fat steers. Also, there is a breakdown of several classes within each category. These follow:

• **Breeding beef cattle**—Beef cattle show classes vary according to breed. Most breed associations make available to livestock shows, and to breeders, a "Standard Show Classification" setting forth their recommended classes and ages. The Standard Classification of the American Hereford Association is given in Table 12-2. Some breed differences exist; hence, the showman should secure from the breed registry and/or show(s) the classification for the breed that he is exhibiting.

Table 12-2 shows that there are no classes for breeding females older than spring yearlings. Older female classes have been eliminated from the classification because it is believed that the place for females

TABLE 12-2
TYPICAL SHOW CLASSIFICATION FOR BREEDING BEEF CATTLE

Class	Age of Bulls	Age of Females
Junior calves	Born after January 1 of the current show year. In shows between January 1 and May 1, this class for bulls is divided into spring calves, and those born after April 1.	Same as bulls.
Winter calves	Born between November 1 and December 31 of the previous year.	Same as bulls.
Senior calves	Born between September 1 and October 31 of the previous year.	Same as bulls.
Summer yearlings	Born between May 1 and August 31 of the previous year.	Same as bulls.
Spring yearlings	Born between March 1 and April 30 of the previous year.	Same as bulls.
Junior yearlings	Born between January 1 and February 28 of the previous year.	No class.
Senior yearlings	Born between September 1 and December 31, 2 years prior to the show.	No class.
Two-year-olds	Born between March 1 and August 31, 2 years prior to the show.	No class.

of such age should be at home where they can start production rather than continue under fitted show condition with the possible impairment in future productivity and possible loss to the breed.

It should be noted that there is a trend in the major cattle shows throughout the country to have breeding classes with a maximum two- to three-month age span per class.

In addition to providing for individual classifications for each sex as shown in Table 12-2, the major shows also make provision for championships and for various group classes. Since these differ somewhat between both shows and breeds, no attempt will be made to list them herein. Instead, the showman is admonished to study the premium list of the show or shows in which he plans to exhibit. Entries must be made for both individual and group classifications, but no entries for championship classes are required.

• **Steers**—The steer classification varies considerably; with some shows following age divisions, others following weight divisions, and still others following a combination of both age and weight. Each system has its advocates.

The National Western Stock Show of Denver, Colorado, provides the following Market Steer rules and classifications:

1. **Teeth and weight**—Steers may have permanent teeth (middle incisors) up, but not in wear. Any steer weighing less than 950 pounds or over 1,350 pounds will be ineligible to show.

2. **Carcass information**—All steers must be sold through the National Western Stock Show for immediate slaughter. Carcass information will be obtained from the top five placings in each class.

3. **Breeds and divisions**—Steers will be put into classes (a) by breeds, and (b) by weights, with a maximum of 30 steers per class. (For example, if there are 150 steers entered of a breed, there will be 5 classes.) Thus, the class is not determined until the steers are weighed at the show.

SHOWING DAIRY CATTLE

Showing is an important part of the purebred dairy business. It provides an excellent means of publicizing individual herds and of evaluating and comparing breeding stock. Additionally, a number of breed programs are furthered through dairy cattle shows; and many members of 4-H Clubs and Future Farmers of America (FFA) acquire and build interest in dairy cattle through exhibiting animals and participating in judging contests at dairy cattle shows.

Training, Grooming, and Showing Dairy Cattle

Table 12-3 is an illustrated guide for training, grooming, and showing dairy cattle.

TABLE 12-3

TRAINING, GROOMING, AND SHOWING GUIDE FOR DAIRY CATTLE

The Essential Steps in Training, Grooming, and Showing	Why	How
1. Selecting and fitting	Selection of the prospective show animal is the first and most important assignment. Unless the right kind of animal is selected, no amount of fitting and showing can make a champion. The second requisite is proper fitting so as to enhance the attractiveness of the animal, without excessive fatness.	Select show animals and begin filling at least 2 months ahead of the show. Separate show animals from the rest of the herd. Animals with excess condition will be placed down in the show because of lack of dairy character. So, if the animal is carrying excess condition (coarse at the withers, patchy over the pinbones, throaty, or fat), place her on a low maintenance ration. If the animal is thin and in poor condition, feed extra grain daily. Some young animals will grow faster than others, so carefully observe their growth pattern. Feed plenty of hay since this will develop capacity and body depth. The grain mixture can be any home-grown grain (corn, oats, barley, etc.) plus minerals.
2. Brushing daily; blanketing	Brush to encourage the hair to lie flat and appear smooth and sleek. Blanket to keep the animal clean.	With a brush and a woolen cloth. Use a curry comb sparingly. Use a thick blanket, either a commercial or a homemade one.
3. Clipping 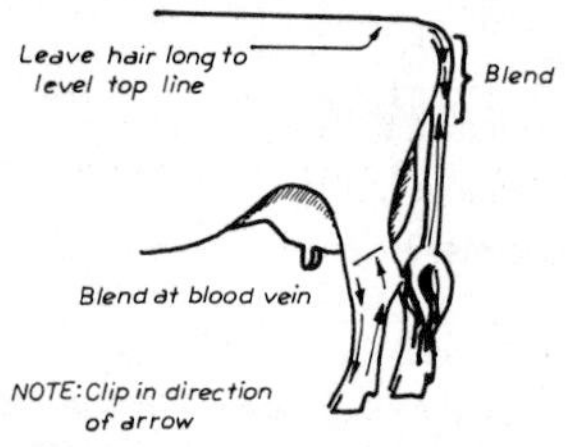 Fig. 12-12. How to clip legs and tail.	To accentuate quality and dairy character. Clip the *tail* to accentuate the switch and make the tail appear slender. Clip the *legs* to accentuate quality of bone and correct stance. Clip the *udder and belly* to bring out veining and show quality. Clip the *head and neck* to impart a clean-cut appearance.	Do not clip the entire animal (1) unless it has an extremely rough hair coat, has not lost the winter hair coat, has stained areas, or shows excess sun bleaching; or (2) if the show is less than 2 months away. Clip the animal on the tail, legs, udder-belly, head, and neck. Start at the top of the switch and clip up the tail against the grain of the hair to within 4 to 5 inches of the tailhead. Clip the tailhead with the grain of the hair (referred to as blending) being careful not to call attention to defects in the rump region. If there is a high area, clip closely. If a low point exists, leave the hair. Clip the side and back of the rear legs from the hock down closely against the grain of the hair. Take advantage of natural lines and attempt to correct the legs by carefully removing hair. The blood vein in the hock region makes an excellent point for blending. The

(Continued)

TABLE 12-3 (Continued)

The Essential Steps in Training, Grooming, and Showing	Why	How
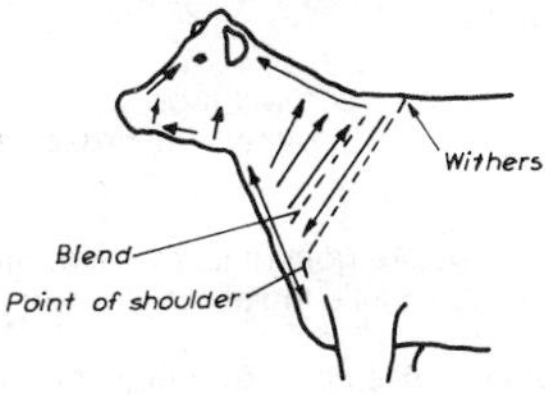 Fig. 12-13. How to clip the head and neck.		back of the legs should be clipped in a straight line to the point of the pinbones. Clip the entire udder. But do not clip the belly of calves or heifers since it makes the animals appear shallow bodied. Extremely long, woolly hair can be removed by holding the clipper away from the body when clipping. The belly on cows should be clipped only enough to show milk veins to advantage. Clip the head and neck closely against the grain of the hair. Clip the area forward of a line formed by the point of the shoulders and the front of the withers. The natural crease formed by the neck and shoulders can be used to blend the long hair. Care should be exercised when clipping about the head because animals are sensitive to the clipper and may throw the head. Hair should be removed from inside the ears. If the animal will not allow the clipper, use scissors to clip the long hair in the ears. Be careful with the scissors! At intervals, lubricate the clipper by dipping the blades in a shallow widemouthed can of light oil or kerosene to remove dirt, dust, and minimize wear and dulling. When a good job of clipping has been done, no lines are visible where blending has occurred.
4. Trimming hooves 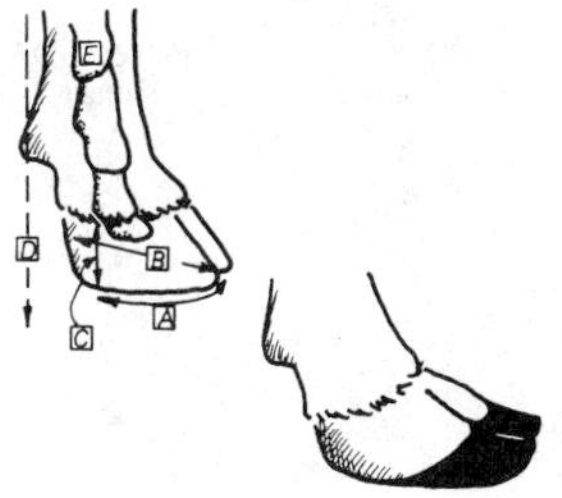 Fig. 12-14. Top: normal foot. Bottom: foot that needs attention.	So that the animal will stand squarely and walk properly, and that the feet will be clean and attractive. Long toes increase chances of foot rot, punctures from the sole, and broken or cracked toes. Each of these leads to lameness in the animal. The bottom or sole of the foot should be trimmed so that the animal stands squarely on her feet with the wall (outside shell) supporting her weight.	The normal foot should appear as shown in Fig. 12-14, top; it should be well rounded (see A), have short toes (see B), and be deep at the heel (see C). The proper leg formation puts the leg directly under the weight it is to support (see D). Be familiar with the bone structure in the foot, and see that the weight of the animal is carried properly by that structure (see E). The bottom drawing of Fig. 12-14 shows a foot that needs attention. The dark area shows the amount of growth that needs to be removed. When the toes are too long, it causes the animal to carry too much weight in an unfavorable position; too much weight is on the heel and not enough on the toes. If hooves are too long, shorten them with a hoof trimmer, nipper, or a chisel and mallet. A hoof knife (tool with a U-shaped tip) will remove excess growth and is safer than other tools. The foot can be smoothed with a rasp. Wear leather gloves to protect your hands. The following methods can be used to work on the feet: Restrain the animal in a set of stocks or a restraining table, or throw it. Work on one foot at a time. Proceed with care to avoid injuring yourself or your animal.
5. Washing	To make and keep the animal clean.	Use a mild soap (detergent can cause skin irritation) and scrub with a coarse brush. Avoid getting water in the ears and slowly accustom the animal to water (avoid cold water and high pressure hoses). Remove all soap with a good rinse and extra water with a squeegee. Blanket the animal. Avoid washing more than once a week since it removes the animal's natural oils and coarsens the hair. Wash the stained areas and switch frequently until clean. Check the poll and head area since dirt accumulates there quickly. Remove wax and dirt from inside the ears with a clean cloth dampened with rubbing alcohol.
6. Oiling	To impart "bloom" to the animal. Oil is not needed unless the hair is unusually dry and lifeless.	Several good commercial oils are on the market. However, a homemade oil can be prepared by mixing equal parts of glycerine, sweet oil, and rubbing alcohol. If the animal is oiled, use it only on dark hair because it may yellow white hair.
7. Training	Because practice makes for perfection, and proper training is necessary if the animal is to be shown to advantage.	Practice well ahead of the show. The showman and his animal must work as a team, each knowing what to expect of the other. Get the animal used to the halter that you will be using in the show. Train it to walk slowly one half step at a time in a clockwise direction. Always keep the animal between you and the judge. The lead strap should be on the left side of the animal loosely coiled in one hand (not wrapped around trapping your hand). The other hand should grasp the halter next to the head of your animal for control. You should be able to walk forward and backward with your animal under complete control and in view of the judge at all times. When you stop your animal, her feet should be positioned correctly as follows:

Calves and heifers: The front feet should be parallel or straight across from each other. The hind foot *nearest* the judge should be one half step BACK. Usually this is the right rear leg (the heifer appears longer and more stretchy this way).

Cows: Again the front feet are parallel, but the hind foot *nearest* the judge is one half step FORWARD (this is normally the right hind leg). This allows the judge to see the rear and fore udder attachments.

All movements and positioning of the animal and her feet should be done with halter commands (not with your feet or body pushing). Allow 2 to 4 feet between you and the animal ahead of you when circling. When you are called in by the judge, line up closely to the animal next to you (less chance of the judge placing another animal above you). Never stop your animal with her front feet in a hole or going downhill since it makes her look smaller.

(Continued)

TABLE 12-3 (Continued)

The Essential Steps in Training, Grooming, and Showing	Why	How
8. At the show	In order to win.	At the show, you should know the answers to the following questions: 1. What time does the show begin? 2. What breeds will be shown first and when will it be your turn? 3. Are all health, registration, and entry papers in order and checked in?

The morning of the show, follow the same feeding, watering, bedding, and grooming routine established at home. Just before going into the ring, give her a final drink of water, but watch her sides (good spring of rib, but not rounded). If your animal dislikes the water, add a little molasses to cover up chlorine, mineral, and other tastes. You may do this at home to get your animal used to the molasses-tasting water.

Be prompt and ready to go into the show-ring. Watch the judge at all times and follow directions closely. Present your animal to best advantage, moving slowly, and keeping her between you and the judge. Once you are called in to line up, move smartly, but do not run. Don't cut in front of someone who is placed above you. Above all, exhibit good sportsmanship and a positive attitude. Be a modest winner and a gracious loser.

Age and Show Classification of Dairy Cattle

The show classifications for 1978 fairs arranged in order of judging for a two-day program, recommended by The Purebred Dairy Cattle Association, Peterborough, New Hampshire, follows.

Note well: Each year, the age and show classification is updated. For example, in 1984 the class 1 reads: Bull calf, born after June 30, 1983, and over four months of age.

Class	First Day	No. of Places	Percentage Money
Class 1.	Bull Calf, born after June 30, 1977, and over four months of age.	8	5
Class 2.	Yearling Bull, born after June 30, 1976, and before July 1, 1977.	8	5
Class 3.	Champion Bull, winners of Classes 1 and 2.		
Class 4.	Heifer Calf, born after June 30, 1977, and over four months of age.	16	8
Class 5.	Junior Yearling Heifer (not in milk), born after December 31, 1976, and before July 1, 1977.	10	6
Class 6.	Senior Yearling Heifer (not in milk), born after June 30, 1976, and before January 1, 1977; Senior Yearling Heifers that have freshened show in Two-Year-Old Cow Class.	12	7
Class 7.	Junior Champion Female, winners of Classes 4, 5, and 6 (Rosette).		
Class 8.	Reserve Junior Champion Female, except Junior Champion, winners of Classes 4, 5, and 6 and second place winner in Junior Champion's Open Single Class (Rosette).		
Class 9.	Junior Get-of-Sire. Group to consist of four animals under two years of age, none of which has freshened; either sex, the get of one sire; not more than two can be bulls. Sire must be named and each exhibitor is limited to one entry sired by the same bull. At least three animals must have been bred by exhibitor. Animals may be owned by one or more exhibitors.	5	4
	Second Day		
Class 10.	Two-Year-Old Cow, born after June 30, 1975, and before July 1, 1976.	17*	11
Class 11.	Three-Year-Old Cow, born after June 30, 1974, and before July 1, 1975.	15*	9
Class 12.	Four-Year-Old Cow, born after June 30, 1973, and before July 1, 1974.	12*	8
Class 13.	Aged Cow, five years old and over, born before July 1, 1973.	20*	13
Class 14.	Senior Champion Female, winners of Classes 10, 11, 12, and 13 (Rosette).		
Class 15.	Reserve Senior Champion Female, except Senior Champion, winners of Classes 10, 11, 12, and 13, and second place winner in Senior Champion's Open Single Class (Rosette).		
Class 16.	Grand Champion Female, winners of Classes 7 and 14 (Rosette).		
Class 17.	Reserve Grand Champion Female, except Grand Champion, winners of Classes 7, 8, 14, and 15 (Rosette)		
Class 18.	Best Five Head	10	24
		TOTAL	100%
Class 19.	Premier Sire. Recognition to be awarded to the owner, or last recorded owner of the sire whose progeny accumulates most points on not less than four and not more than eight progeny in the open single classes. This will not be a lead out class. The management will calculate the points for every sire having four or more progeny exhibited in the show. The scale of points will be the same as that for determining Premier Breeder.		
Class 20.	Premier Exhibitor's Award—The Exhibitor winning the most points on not to exceed eight animals owned and exhibited by himself in the open single classes shall be designated the Premier Exhibitor. The scale of points will be the same as that for determining Premier Breeder.		
Class 21.	Premier Breeder's Award—The Breeder winning the most points on not to exceed eight animals in the open single classes, exhibited by himself and/or other exhibitors shall be designated the Premier Breeder. The point system for determining Premier Breeder is as follows:		

Placings	1	2	3	4	5	6	7	8	9	10
Points Senior Females	20	18	16	14	12	10	8	6	4	2
Points Bulls and Junior Females	10	9	8	7	6	5	4	3	2	1

* BEST UDDER AWARD

It is recommended that in each of the milking cow classes the judge select the 1st and 2nd best uddered cows in each class. This selection shall be made on the basis of udder alone and a milk out will be at the discretion of the judge. It is recommended that the two cows in each class be given special recognition just prior to the final placing of their respective class. It is not recommended that the best uddered cows from each of the classes compete for an overall best udder award of the show.

SHOWING SHEEP

Pertinent information relative to showing sheep is summarized in the two sections which follow.

Training, Grooming, and Showing Sheep

Table 12-4 is an illustrated guide for training, grooming, and showing sheep.

TABLE 12-4
TRAINING, GROOMING, AND SHOWING GUIDE FOR SHEEP

The Essential Steps in Training, Grooming, and Showing	Why	How
1. Gentling and posing  Fig. 12-15	In order that the judge may see the animal at its best.	Hold the sheep by grasping the wool lightly under the chin with the left hand; standing or squatting to the left and front of the sheep. Train the animal to stand with the legs squarely placed, the back straight, and the head held erect.
2. Trimming the feet  Fig. 12-16	To keep the sheep strong and straight on the pasterns.	Trim at approximately 2-month intervals, with the final trimming given 2 to 4 weeks in advance of the show. Hold the animal in a sitting position and trim hoofs with a sharp pocket knife or with ordinary pruning shears.
3. Blanketing breeding sheep  Fig. 12-17	To keep the fleece clean, to "condition" it, and to keep it compact.	By using either (1) a ready-made blanket, or (2) a homemade blanket made from heavy cotton sacks.
4. Deciding how and what to trim  Fig. 12-18	In order to enhance the attractiveness of the individual through accentuating its strong points and covering up its weaknesses.	Carefully study each individual animal as nature made it, and visualize how it can and should appear after trimming.
5. Securing animal on trimming stand 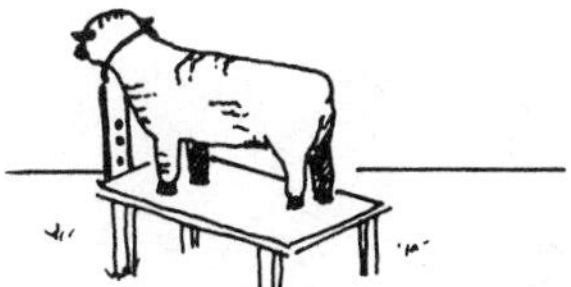 Fig. 12-19	The trimming stand is the most satisfactory arrangement for holding sheep for trimming, but animals may be held fast by another person or secured by means of a rope halter or yoke.	The trimming stand, which may be homemade, is a platform on which the animal is stood and which has an arrangement at one end for securing the head in approximate show position.

(Continued)

TABLE 12-4 (Continued)

The Essential Steps in Training, Grooming, and Showing	Why	How
6. Giving first "rough" blocking  Fig. 12-20	To make the top level and to make the animal appear as attractive as possible. Also in wethers, to make them firm to the touch when handled over the back and loin.	The time to give the first rough blocking is affected by (1) the age of the sheep, (2) whether animals are being exhibited in breeding or fat classes, (3) the rules of the show, (4) the shearing date, and (5) the length, density, and condition of the fleece. Generally speaking, however, the first rough blocking is scheduled about as follows: For yearling or mature sheep, at least 6 weeks before the show, and preferably much earlier. For lambs, at least 2 to 3 weeks before the show, and preferably much earlier; although trimming of lambs is later than with mature sheep because of their shorter wool. For wethers, very early in the season for they keep cooler and feed better with a short fleece. In preparation for blocking, the fleece is (1) dampened by hand-spraying and then by brushing on it a dip solution prepared by adding 2 tablespoonfuls of creosol dip to a gallon of water, and (2) vigorously combed with a circular-type curry comb or a Scotch comb. This develops the desired "face" on the fleece. **The first rough blocking consists of flattening down the back (by cutting not shorter than ¾ inch in breeding sheep and from ¼ to ½ inch in wethers) and removing a little wool from such other parts of the body, especially the sides, as necessary in order to impart a blocky appearance.** In the case of wethers, trimming should be close along the topline in order to make them firmer to the touch. Do not clip small dung locks from around the dock or britch; wash them out with a solution of soap and warm water to which 2 tablespoonsful of creosol dip has been added to each gallon of water.
7. Giving the second trimming 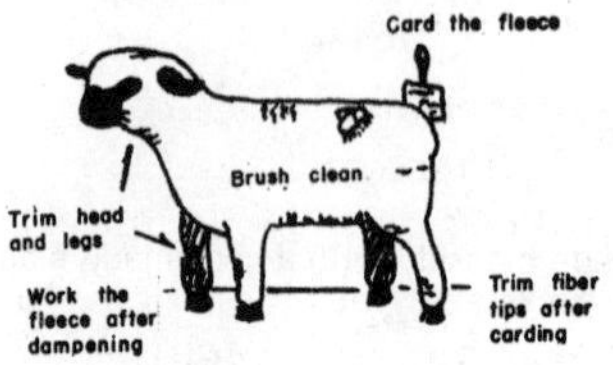 Fig. 12-21	To impart the desired appearance and to result in a blending together of all body parts.	The second trimming, which is usually given within 2 to 3 weeks of the show, consists of the following procedure: Dampening the fleece, as outlined under the first rough blocking. Brushing and carding in order (1) to get the fibers separated and parallel to each other, and (2) to give increased fullness to certain regions. Trimming with sharp shears as desired.
8. Giving the third and final trimming  Fig. 12-22	To bring the blocking and trimming operations closer to perfection.	The third and final trimming, which is administered immediately before sheep are sent to the show, consists of dampening the fleece (as described under first rough blocking) and of alternate brushing, carding, and trimming with sharp shears.
9. Giving preshow touching up 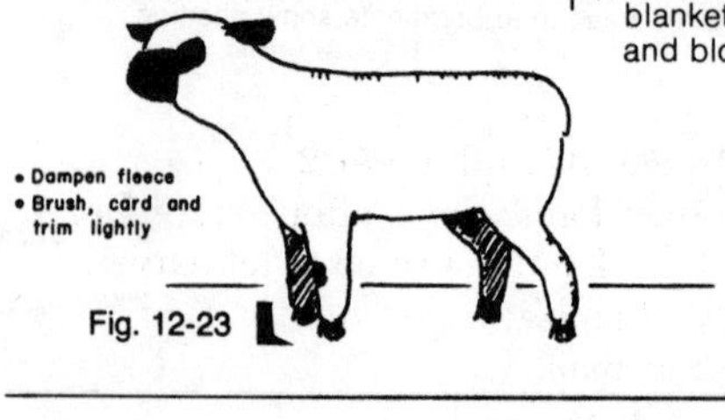 Fig. 12-23	To remove any marks or irregularities in the fleece caused by shipment or blanketing, and to impart freshness and bloom.	Dampen the fleece (as described under rough blocking) and go over lightly, with the brush, card, and shears.

(Continued)

TABLE 12-4 (Continued)

The Essential Steps in Training, Grooming, and Showing	Why	How
10. Showing  Fig. 12-24	In order to win.	The guiding principles in showing sheep are: Touch up the animal well in advance of show time, and be prompt in parading before the judge when the class is called. Dress neatly for the occasion. Without being an "eager beaver," get a favorable position in the line. Pose the animal correctly; standing squarely on its feet with the head up and back well supported. The showman should station (either standing or squatting) himself to the left and front of the animal and should hold his animal by grasping the wool lightly under the chin with the left hand. His right hand is kept free for use in keeping the sheep set up properly. Push against the animal's breast with your knee in order to get him to bolster his back when the judge is handling him. When the judge walks to the front, step aside so that he may obtain a good head and front view of the animal. Keep calm and collected; radiate confidence and experience. Keep one eye on the judge. When the judge requests that the position in the line be changed, move into the new position from behind the line. Be courteous to other exhibitors and the judge. Be a gracious winner or an equally good loser.

The procedure given in Table 12-4 is particularly applicable to mutton-type sheep. With animals of the long-wool and the fine-wool breeds, the following differences should be observed:

• **The Long-wool Breeds (Lincoln, Cotswolds, and Leicesters):**

1. There is a tendency to accentuate the length of fleece through stubble shearing and allowing more than 12 months growth before showing.
2. The fleece is usually oiled or dressed, most commonly with wool fat thinned down by heating, and rubbed on with a brush or cloth.
3. Blankets are not used immediately after oiling or dressing, for they may cause sweating and fleece discoloration.
4. The fleece is neither carded nor blocked. Shears are used only for trimming about the dock.

• **The Fine-wool Breeds (Rambouillet and Merino):**

1. As with long-wool sheep, there is a tendency to accentuate the length of fleece through stubble shearing and allowing more than 12 months growth before showing.
2. The fleece is never blocked with shears.
3. The fleece is usually oiled, most commonly with wool fat, three to four weeks before the show; and a second light application may be given a week before the show if necessary. The dressing is usually patted on the surface and not rubbed in.
4. Sometimes the fleece is colored. One such product is made by mixing 1 tablespoonful of burnt amber and 2 tablespoons of lampblack to each 1½ pints of dressing. It should be noted, however, that the practices of artificial oiling and coloring are now being discouraged or even outlawed in many shows.
5. Blankets are always used sparingly on fine-wool sheep, because their presence may cause excess sweating and fleece discoloration.

Age and Show Classification of Sheep

In most sheep shows there are two main classifications: breeding sheep and market sheep. Sometimes there is an added classification for feeder sheep.

1. **Breeding sheep**—The breeding classes are for purebreds only. Usually the number of different breeds for which classifications are indicated and the amount of premium money allotted to each varies according to the popularity of the breed within the area and the premium money support accorded by the breed associations. At most fairs, a base date of September 1 is used for breeding classes, although a few use a base date of January 1. Most 4-H and FFA shows use the latter.

With the base date of September 1, lambs must have been dropped on or after September 1 of the year preceding the one in which they are shown; and yearlings must have been born on or after September 1 of the second preceding year. Aged animals include all those older than the yearling group.

The following sheep show classification is used at most of the major sheep shows, with some minor variations such as a pair of lambs or yearlings versus a pen of 3.

Class	
1	Ram, 1 year old and under 2.
2	Senior Ram Lamb, born after September 1.
3	Junior Ram Lamb, born after January 1.
4	Pair Ram Lambs (or pen of 3).
5	Champion Ram.
6	Reserve Champion Ram.

7 Ewe, 1 year old and under 2.
8 Senior Ewe Lamb, born after September 1.
9 Junior Ewe Lamb, born after January 1.
10 Pair of yearling ewes (or pen of 3).
11 Champion Ewe.
12 Reserve Champion Ewe.
13 Exhibitors Flock (Ram any age, 2 yearling ewes, 2 ewe lambs).

2. **Wethers**—The wether classifications followed at two major Midwest state fairs follow:

a. **Iowa State Fair**—Separate classes for each of the major breeds, plus a class for crossbreds. A minimum weight of 80 pounds required. In addition to the class winners, Iowa State Fair has a Grand Champion Market Lamb and a Reserve Grand Champion Lamb.

b. **Ohio State Fair**—Ohio provides separate breed classes, plus a class for grades or crossbreds. However, they draw the minimum eligible weight at 85 pounds, except for wethers of the Montdale and Southdown breeds, where it is lowered to 80 pounds.

SHOWING SWINE

Pertinent information relative to showing swine is summarized in the two sections which follow.

Training, Grooming, and Showing Swine

Table 12-5 is an illustrated guide for training, grooming, and showing swine.

TABLE 12-5

TRAINING, GROOMING, AND SHOWING GUIDE FOR SWINE

The Essential Steps in Training, Grooming, and Showing	Why	How
1. Gentling and posing 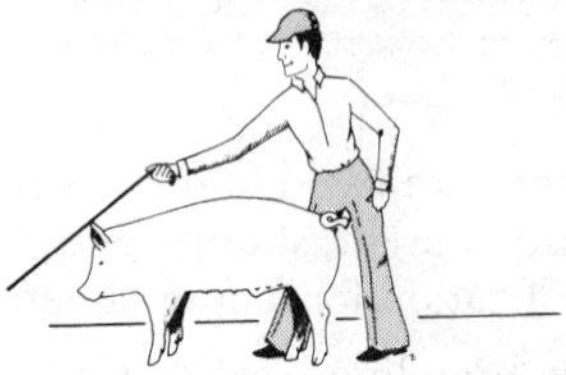 Fig. 12-25	So that the judge may see the animal to best advantage.	The pig should be gentled by handling, without becoming a pet. It should be trained to respond to either the cane or the whip but not both; so that it can be brought to a stop and the proper pose when desired, and that it can be guided by merely placing the cane or whip alongside the head.
2. Trimming the feet 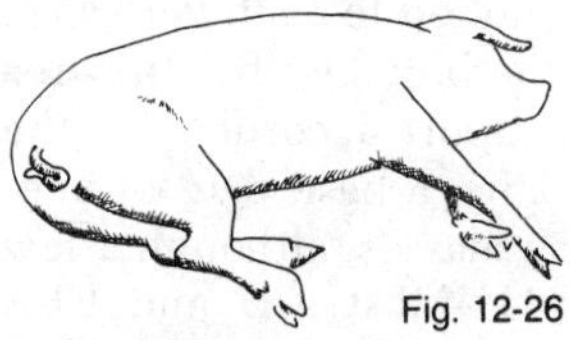 Fig. 12-26	In order that the animal may stand squarely and walk properly and that the pasterns will appear straight and strong.	First, get the animal to lie on its side, which can usually be accomplished by merely stroking its belly. Then trim the toes to the proper length and square up the bottoms by using a small rasp and a knife. Also shorten and dress down the dew claws. The toes should be trimmed regularly, with some animals as often as every 6 weeks. Do not trim within 2 weeks prior to the show.
3. Removing tusks and ring  Fig. 12-27	Tusks in boars over 1 year of age are dangerous, from the standpoint of both the handlers and other animals. Rings may make the nose sore and the animal irritable and somewhat unmanageable.	To remove tusks, tie the animal to a post with a strong rope, draw up the upper jaw, and use a bolt clipper. To remove rings, tie the animal as for the removal of tusks, and cut the rings with wire pliers or nippers.

(Continued)

TABLE 12-5 (Continued)

The Essential Steps in Training, Grooming, and Showing	Why	How
4. Clipping 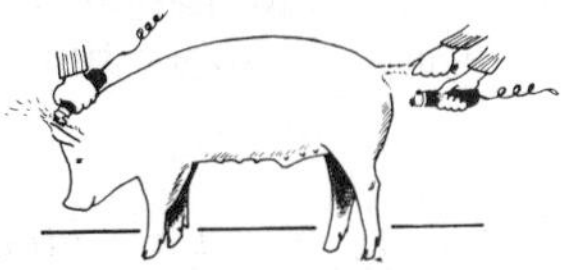 Fig. 12-28	To enhance the trimness of the animal, and to show off the teats of a gilt.	With hand or electric clippers, a few days before the show. Usually, the clipping operation includes the following body areas: 1. The ears, both inside and outside. 2. The tail, extending from above the switch to the tailhead. 3. The head and jowl, from which the long hairs only are clipped. 4. The belly of gilts.
5. Washing  Fig. 12-29	To keep the animal clean, to make the skin smooth and mellow, and to assist in shedding the coat of older animals.	Wet the skin and hair thoroughly with lukewarm water, rub the hair with tar soap until a suds is formed, and then work the suds into the hide with the hands and brush. Add a small amount of blueing to the water when washing white hogs. Following washing, rinse off all traces of soap and place the animal in a clean pen to dry.
6. Oiling colored hogs  Fig. 12-30	In order to soften the skin and hair and to give the necessary bloom to the coat.	Apply with a cloth or brush any clear, light vegetable oil. Paraffin oil mixed with rubbing alcohol is very satisfactory. Usually a light application of oil is given the night before the show, and the animal is rubbed with a lightly oiled rag just before entering the ring.
7. Powdering white pigs  Fig. 12-31	In order to make white hogs or white spots on dark hogs appear clean and attractive.	The animal should first be thoroughly washed and allowed to dry. Before entering the ring, white hogs or white spots on dark hogs are usually powdered with talcum powder or corn starch.
8. Showing  Fig. 12-32	In order to win.	The guiding principles adhered to by most successful swine showmen are: Train the animal long before entering the ring. Have the animal carefully groomed and ready for the parade before the judge. Dress neatly for the occasion. Enter the ring promptly when the class is called. Do not crowd the judge but keep your animal in a position of vantage at all times. Avoid being smothered by the mob. Keep in the open. Keep one eye on the judge and the other on the pig. Center your attention entirely on showing the hog. The animal may be under the observation of the judge when you least suspect it. Keep calm, confident, and collected. Remember that the nervous showman creates an unfavorable impression. Work in close partnership with the animal. Be courteous and respect the rights of other exhibitors. Do not allow the hog to bite or fight other animals. Be a good sport. Win without bragging and lose without squealing.

Age and Show Classification of Swine

Classifications for swine breeding classes are based on age. Barrow classifications are usually based on weight rather than age.

1. **Breeding swine**—The classifications for breeding swine vary somewhat from fair to fair. The classifications shown in Table 12-6 are used at the Iowa State Fair and the Ohio State Fair, two of the great Corn Belt swine shows.

TABLE 12-6

SHOW CLASSIFICATION GUIDE FOR BREEDING SWINE

Class	Iowa State Fair	Ohio State Fair
Junior Yearling	Farrowed on or after January 1 and before July 1 of the preceding year.	Farrowed on or after January 1 and before June 30 of the year preceding the show.
Senior Pigs	(No Senior Pig Class.)	Farrowed on or after July 1 and before December 31 of the year prior to the show.
Fall Pigs	Farrowed between July 1 and November 30 of the previous year.	(No Fall Pig Class.)
December Through March Pigs	Farrowed between December 1 of the previous year and March 31 of the year shown.	(None.)
January Pigs	Farrowed on or after January 1 and before February 1 of the year shown.	Same as Iowa State Fair.
February Pigs	Farrowed on or after February 1 and before March 1 of the year shown.	Same as Iowa State Fair.
March Pigs	Farrowed on or after March 1 of the year shown.	Same as Iowa State Fair.
Senior Champion	Winners of junior yearling and fall classes.	First prize junior yearling animals and first prize senior pigs.
Junior Champion	Winners of December through March classes.	Winners of January through March classes.
Grand Champion	Selected from Senior and Junior Champions.	Same as Iowa State Fair.
Reserve Grand Champion	(None.)	The remaining champion and the second place animal from the class in which the Grand Champion was selected.
Certified Pair	Two littermates (2 boars, 2 gilts, or 1 of each) farrowed on or after January 1 of the previous year shown, from a certified litter.	To consist of 1 boar and 1 gilt farrowed on or after January 1 of the current year, from a certified litter.

Show regulations differ relative to the stipulations of the groups; thus, the exhibitor should study the rules of the particular show in which he desires to exhibit.

2. **Barrows**—The barrow classifications of the Ohio State Fair follow:

a. All barrows must have been farrowed on or after February 1 of the year shown.

b. By breeds (with similar classifications for each of the major breeds, and for crossbreds):

(1) Barrow, 190 to 205 pounds.
(2) Barrow, 210 to 225 pounds.
(3) Barrow, 230 to 240 pounds.
(4) Champion Barrow, by breeds.
(5) Reserve Champion Barrow, by breeds.
(6) Grand Champion Barrow over all breeds.
(7) Reserve Grand Champion Barrow over all breeds.

The weight divisions of different barrow shows vary considerably, thus making it imperative that the exhibitor study these carefully. In fact, the breeding program should be planned with this information in mind, because animals farrowed at the proper time may be better fitted and may reach the proper bloom for different weight divisions, if the breeding and feeding programs are properly synchronized.

SHOWING HORSES

Horses are shown either (1) in hand (in breeding classes), wearing either a halter or bridle; or (2) in performance classes.

The performance classes for horses are so many and varied that it is not practical to describe them in this book. Instead, the reader is referred to the official Rule Book of the American Horse Shows Association and to the rules printed in the programs of the local horse shows.

Training, Grooming, and Showing Horses

Table 12-7 is an illustrated guide for training, grooming, and showing horses.

TABLE 12-7

TRAINING, GROOMING, AND SHOWING GUIDE FOR HORSES

The Essential Steps in Training, Grooming, and Showing	Why	How
1. Training	To reach a high degree of proficiency for the intended use—pleasure riding, racing, jumping, or whatnot.	There are many different ways in which to train a horse. Although the approach may differ, most good trainers schedule the schooling of the horse as follows: 1. **Training the foal**—Put a halter on the foal when it is 10 to 14 days old. Teach it to lead and stand properly. 2. **Training the yearling**—Teach the yearling the meaning of "whoa" and his name; to stand when hobbled; and to get used to the blanket and saddle. 3. **Training at 18 months old**—At 18 months of age, teach the young horse to drive, turn, stop, and back up—by using plowlines; to flex his neck and set his head; to respond to the bosal, and to get used to leg pressure. 4. **Training the 2-year-old**—Train the 2-year-old to respond to the aids (legs, hands and reins, and voice); to back; and if a western horse, train him to pivot and to make a sliding stop.
2. Grooming	Proper grooming is necessary to (1) make and keep the horse attractive, and (2) maintain good health and condition. Grooming cleans the hair, keeps the skin functioning naturally, lessens skin diseases and parasites, and improves the condition and fitness of the muscles.	To assure that the horse will be groomed thoroughly and that no body parts will be missed, follow a definite order. This may differ according to individual preference, but the following procedure is most common: 1. Clean out the feet. 2. Groom the body. 3. Brush the head, comb and brush the mane and tail. 4. Wipe with the grooming cloth. 5. Check the grooming. 6. Wash and disinfect grooming equipment.
3. Clipping and shearing	It makes horses look sharp and feel sharp. Show-ring custom decrees certain breed differences in haircuts and hairdos.	When clipping and shearing, proceed as follows: 1. **Protect the ears**—by placing a wad of cotton in them, to cut down on the noise from clippers and prevent hair from falling into the ears. 2. **Clip long hairs**—by removing long hairs from about the head, the inside of the ears, on the jaw, and around the jetlocks.

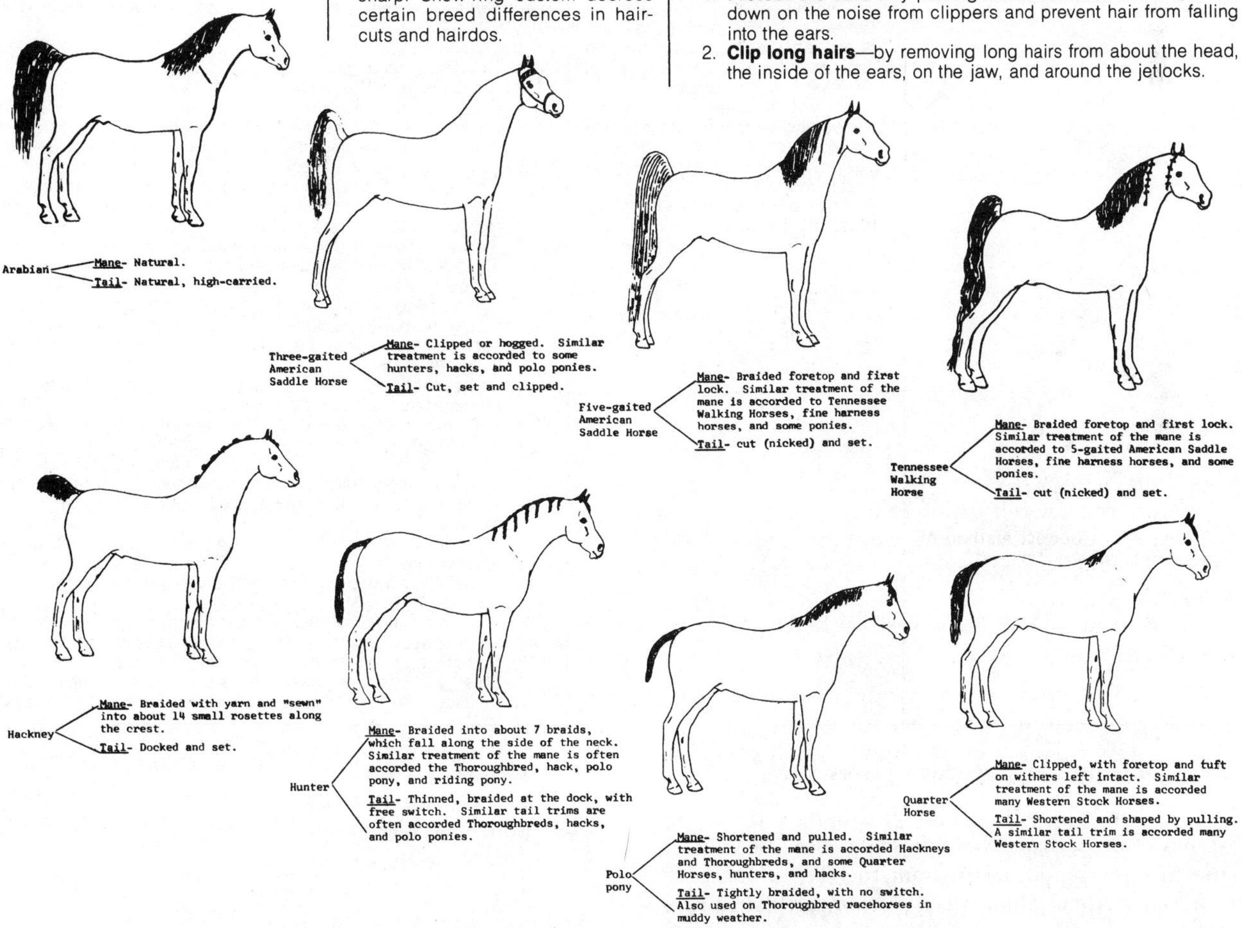
Arabian
Mane- Natural.
Tail- Natural, high-carried.
Three-gaited American Saddle Horse
Mane- Clipped or hogged. Similar treatment is accorded to some hunters, hacks, and polo ponies.
Tail- Cut, set and clipped.
Five-gaited American Saddle Horse
Mane- Braided foretop and first lock. Similar treatment of the mane is accorded to Tennessee Walking Horses, fine harness horses, and some ponies.
Tail- cut (nicked) and set.
Tennessee Walking Horse
Mane- Braided foretop and first lock. Similar treatment of the mane is accorded to 5-gaited American Saddle Horses, fine harness horses, and some ponies.
Tail- cut (nicked) and set.
Hackney
Mane- Braided with yarn and "sewn" into about 14 small rosettes along the crest.
Tail- Docked and set.
Hunter
Mane- Braided into about 7 braids, which fall along the side of the neck. Similar treatment of the mane is often accorded the Thoroughbred, hack, polo pony, and riding pony.
Tail- Thinned, braided at the dock, with free switch. Similar tail trims are often accorded Thoroughbreds, hacks, and polo ponies.
Polo pony
Mane- Shortened and pulled. Similar treatment of the mane is accorded Hackneys and Thoroughbreds, and some Quarter Horses, hunters, and hacks.
Tail- Tightly braided, with no switch. Also used on Thoroughbred racehorses in muddy weather.
Quarter Horse
Mane- Clipped, with foretop and tuft on withers left intact. Similar treatment of the mane is accorded many Western Stock Horses.
Tail- Shortened and shaped by pulling. A similar tail trim is accorded many Western Stock Horses.

Fig. 12-33. Common haircuts and hairstyles for different breeds and uses of horses.

(Continued)

TABLE 12-7 (Continued)

The Essential Steps in Training, Grooming, and Showing	Why	How
4. Washing (shampooing)	Washing (1) cleans the animal—it removes the dirt, stains, and sweat that cannot be removed by grooming; (2) makes for a fine hair coat with a good sheen; and (3) keeps the skin smooth and mellow.	In preparation for shampooing, (1) groom the horse carefully, (2) secure the animal for washing either by having someone hold him by the shank or by tying, and (3) have shampoo concentrate, warm water, buckets, and sponges available. To assure that the horse will be washed thoroughly and that no body part will be missed, follow a definite order. After shampooing, rinse the horse with warm water, using either a bucket and sponge or a hose (if the horse is used to the latter). Then complete the washing operation as follows: 1. Scrape with a "sweat scraper" held snugly against the hair to remove excess water, using long sweeping strokes, except do not scrape the head and legs. 2. Dry with a clean dry sponge or coarse towel, squeezing it out at intervals. 3. Blanket the horse and walk him until he is completely dry. 4. Apply a coat dressing if desired.
5. Coat dressing	To achieve the all-important "bloom" or eye appeal in show, parade, and sale animals. A coat dressing will not take the place of the natural conditioning of the horse, which can be achieved only through proper feeding, health, grooming, and shampooing.	Proper grooming should always precede the use of coat dressing. Coat dressing is best applied by means of a heavy cloth (preferably terry cloth). Moisten the rag with the dressing and rub the coat vigorously in the direction of the natural lay of the hair; then brush to bring out the bloom. Coat dressing should always be used following washing, and for show, parade, or sale. It is best to apply a heavier application of coat dressing 12 to 24 hours ahead of the event, then go over the horse with a lightly dressed rag just ahead.
6. Showing  Fig. 12-34. Correct method of leading when showing "in hand."	Horse shows provide entertainment for spectators, and recreation, sport, and competition for the exhibitors. Also, the show-ring has been, and will continue to be, an important medium for getting horses and people together in one place and at one time to compare, design, and engineer the most desirable models. Also, horse shows stimulate improved breeding, for winning horses (and their relatives bring good prices).	The following practices are recommended for showing in hand, or at halter: 1. Train the horse early. 2. Groom the horse thoroughly. 3. Dress neatly for the show. 4. Enter the ring promptly and in tandem when the class is called. Line up at the location indicated by the ringmaster or judge unless directed to continue around the ring in tandem. 5. Stand the horse squarely on all 4 feet with the forefeet on higher ground than the hind feet if possible. The standing position of the horse should vary according to the breed. For example, Arabians are not stretched, but American Saddlers are trained to stand with their front legs straight under them and their hind legs stretched behind them. Other breeds generally stand in a slightly stretched position, somewhat intermediate between these two examples. When standing and facing the horse, hold the lead strap or rope in the left hand 10 to 12 inches from the halter ring. Try to make the horse keep his head up. 6. Unless the judge directs otherwise, the horse should first be shown at the walk and then at the trot. 7. Keep the horse posed at all times; keep one eye on the judge and the other on the horse. 8. When the judge signals to change positions, the exhibitor should back the horse out of line, or if there is room, turn him to the rear of the line and approach the new position from behind. 9. Try to keep the horse from kicking when he is close to the other horses. 10. Keep calm; a nervous showman creates an unfavorable impression. 11. Work in close partnership with the horse. 12. Be courteous and respect the rights of other exhibitors. 13. Do not stand between the judge and the horse. 14. Be a good sport; win without bragging and lose without complaining.

Age and Show Classification of Horses

In hand (breeding) classes differ by breeds and shows. But perhaps the following classifications for the Morgan breed, taken from the *Rule Book of the American Horse Shows Association*, are fairly representative:

Weanling colts
Yearling colts
Two-year-old colts
Three-year-old stallions
Four-year-old stallions
Five-year-old and over stallions
Sire and Get class (stallion to be shown with 2 to 4 of his get)
Get of sire (2 to 4 get to be shown)
Weanling fillies
Yearling fillies

Two-year-old fillies
Three-year-old mares
Four-year-old mares
Five-year-old and over mares
Broodmare and foal
Dam and produce (mares to be shown with two or more produce)
Produce of Dam (2 to 4 produce to be shown)
Show Champion
Reserve Champion
Also, in hand classes may be provided for geldings

Performance classes for horses are so numerous and varied that it is not practical to describe them here. Instead, the showman should refer to the official rule book of the American Horse Shows Association (which is published annually) and to the rules printed in the programs of local horse shows.

SHIPPING TO THE FAIR

Today, most show animals are shipped via either truck or air. Very few are transported via rail nowadays.

Regardless of the method of shipping, it is important that the following details receive consideration:

1. **Schedule properly**—Schedule the transportation so that animals will arrive within the limitations imposed by the show, and at least two or three days in advance of the date that they vie for awards.

2. **Clean and disinfect any public conveyance**—Before using, thoroughly clean and disinfect any type of public conveyance.

3. **Place feed and supplies on a deck**—If space is at a premium, place the feed supply, bedding, and show equipment on a deck or platform in the truck.

For cattle, a deck 5½ to 6 feet above the floor will allow for air circulation and placing younger cattle thereunder; for sheep and swine, a clearance of 5 feet is adequate.

4. **Bed properly**—For cattle and sheep and for hogs in cool weather, the floor should first be sanded so that the animals will not slip, and then covered with long, clean, bright straw.

During warm weather, properly wetted sand makes the best bedding for swine. In extremely hot weather, it may also be necessary to air-condition the conveyance.

The floor of vehicles used for transporting horses should be covered with coconut matting made for the purpose, rubber mats, or sand covered with straw.

5. **Load animals properly**—The loading arrangement will vary between classes of animals. Herewith are the preferred arrangements:

a. **For cattle**—In transporting by truck, cattle are generally stood crosswise of the truck, with the largest animal near the cab and tied facing to one side. The direction of facing the remaining animals is alternated; the second animal is faced in the opposite direction from the first, and so on. In trucking it is usually best to tie cattle fairly short and near enough together so that they will not lie down.

b. **For sheep**—Provide suitable and necessary partitions for separate penning of the sexes and for separating out the rams that are not accustomed to running with each other. Do not overcrowd; allow sufficient space for the sheep to bed down in comfort.

c. **For swine**—Provide suitable and necessary partitions for separate penning of animals of different ages and sexes. Show hogs should not be crowded.

d. **For horses**—Most horses are transported via trailer or van. The trailer is usually a 1-or 2-horse unit, in which the horse(s) faces toward the direction of travel. Vans are commonly used where 3 to 8 horses are transported, with the horses either (1) stabled abreast and facing the direction of travel, or (2) stabled crosswise, with the alternating animals facing in opposition direction.

6. **Take feed along**—When mixed feeds are used, as is usually the case in fitting rations, a supply adequate for the entire trip should be taken along. This will reduce the hazard of animals going off feed because of feed changes.

7. **Feed limited rations**—Limit show animals to half feed at the last feed before loading out and while in transit. A heavy "fill" is likely to result in digestive disturbances and overheating in warm weather.

8. **Handle animals quietly**—In transit, the animals should be handled quietly and should not be allowed to become hot or to be in a draft.

SECTION 13

MARKETING LIVESTOCK AND MILK

Contents **Page**

PART I—LIVESTOCK MARKETING

Methods of Marketing Livestock 1062
Preparing and Shipping Livestock 1062
  How to Prevent Shipping Bruises, Crippling, and Death Losses 1063
  Number of Animals in a Truck or in a Railroad Car 1063
  Kind of Bedding to Use for Animals in Transit 1064
  Shrinkage in Marketing Animals 1064
    Shrinkage Tables 1064
Market Classes and Grades of Livestock 1066
  Factors Determining Market Grades 1066
  Market Classes and Grades of Cattle 1066
  Market Classes and Grades of Sheep 1066
  Market Classes and Grades of Hogs 1066
Some Livestock Marketing Considerations 1069
  Cyclical Trends in Market Livestock 1069
  Seasonal Changes in Market Livestock 1069
  Dockage 1070
  Livestock Marketing Costs 1070
  Meat Check-off 1070
Parity and Parity Prices in Farm Animals 1070

PART II—MARKETING MILK

Farm Production and Handling of Milk 1071
  How Milk Is Sold by Dairymen 1072
Market Channels for Milk and Dairy Products 1072
How Milk Is Priced and Regulated 1072
  Federal Milk Marketing Orders 1072
  State Milk Control 1073
  Cooperatives 1073
  Other Regulatory Programs 1073
    Sanitary Regulations 1073
    Standards and Grades 1073
    State Trade Practice Laws 1074
  Methods of Pricing or Paying for Fluid Milk 1074
  The Milk Price Support Program 1075

Marketing is the end of the line. From the producer's standpoint, it is that part which gives point and purpose, and profit or loss, to all that has gone before. Market receipts constitute the only source of reimbursement to the producer for his work.

In the past, the producer of meat animals or milk could be successful if he knew how to breed, feed, and manage his stocks. Today, this is not enough; preconsidered, if not prearranged, markets are essential.

In Part I, discussion is limited to the marketing of four-footed meat animals—cattle, sheep, and hogs. Because of its distinct and different market channels and procedures, milk is treated in a separate section, Part II.

PART I—LIVESTOCK MARKETING

Market day is the producer's payday—hence, it is the most important single day of operation to him.

Livestock marketing embraces those operations beginning with loading animals out on the farm, ranch, or feedlot and extending until they are sold to go into processing channels.

METHODS OF MARKETING LIVESTOCK

The producer of livestock is confronted with the perplexing problem of where and how to market his animals. Usually there is a choice of market outlets, and the one selected often varies with different species of livestock and among sections of the country. The methods of marketing also differ between slaughter and feeder animals, and all of these differ from the marketing of purebreds.

Prior to the advent of terminal public markets in 1865, country selling accounted for virtually all sales of livestock. But sales of livestock in the country declined with the growth of terminal markets, until the latter method reached its peak at the time of World War I. Country selling was reactivated by the large nationwide packers beginning about 1920, in order to meet the increased buying competition of the small interior packers. The decline in the proportion of all livestock moving through terminal public markets was largely accounted for by the growth in country selling until the late 1930s and by the growth of both country selling and auctions since.

In 1980, meat-packers (812 of them) purchased their animals through the following channels:[1]

	Cattle (%)	Calves (%)	Sheep (%)	Hogs (%)
Direct, country dealers, etc.	77.0	42.8	79.6	76.6
Auctions	14.6	51.2	12.9	9.9
Terminal markets	8.4	6.0	7.5	13.5

Of course, it is generally recognized that these figures are continuing to shift, with terminal market sales decreasing and with auction and country sales increasing.

Most U.S. livestock are marketed through four channels—terminals, auctions, direct, or carcass grade and weight basis. But there are other methods, including country commission firms, local markets, and concentration yards; order buyers, local plants and retailers; cooperative shipping associations and cooperative selling associations; telephone auctions, telephone direct selling, teletype auctions, television auctions; and selling on consignment (custom method). It should be realized, however, that there is duplication in the listing. For example, order buyers operate at public stockyards and auctions, as well as conduct free-lance country operations.

The choice of a market outlet should be determined strictly by the net returns from the sale of livestock; effective selling and net returns are more important than selling costs.

[1]*Packers and Stockyards Résumé*, Vol. 19, No. 5, USDA, March, 1982.

PREPARING AND SHIPPING LIVESTOCK

Improper handling of livestock immediately prior to and during shipment may result in excess shrinkage; high death, bruise, and crippling losses; disappointing sales; and dissatisfied buyers. Unfortunately, many stockmen who do a superb job of producing animals dissipate all the good things that have gone before by doing a poor job of preparing and shipping to market. Generally speaking, such omissions are due to lack of know-how, rather than any deliberate attempt to take advantage of anyone. Even if the sale is consummated prior to delivery, negligence at shipping time will make for a dissatisfied customer. Buyers soon learn what to expect from various producers and place their bids accordingly.

In addition to the important specific considerations covered in the sections which follow, these general considerations should be accorded in preparing livestock for shipment and in transporting them to market:

1. **Select the best-suited method of transportation**—truck or rail.
2. **Feed and water properly prior to loading**—Withhold grain feeding of all classes of livestock 12 hours before loading (omit one feed). Cattle and sheep may be allowed free choice to dry, well-cured grass hay up to loading time, but they should not be allowed access to water within 2 to 3 hours of shipment.
3. **Keep animals quiet**—Remember that "easy does it."
4. **Comply with the requirements for health certificates, permits, and brand inspection**—This is necessary where interstate shipments are involved.
5. **Comply with the federal 28-hour law[2] in rail shipments**—This prohibits transporting livestock by rail for a longer period than 28 consecutive hours without unloading, feeding, watering, and resting for 5 consecutive hours before resuming transportation. On request of the owner, the period can be extended to 36 hours.
6. **Feed or graze cattle or sheep in transit if advantageous**—This refers to a provision of railroads whereby stockmen may be granted permission to graze or finish animals for a period up to 12 months, at some intermediate stop between their point of origin and the market to which they will be consigned at the end of the finishing period.
7. **Use partitions in the truck or car when necessary**—It is important to separate species, sexes, and age groups.
8. **Avoid shipping during extremes in weather**—It is best not to ship when it is very cold or very hot.

[2]No such law applies to truck transportation of animals.

How to Prevent Shipping Bruises, Crippling, and Death Losses

Losses from bruising, crippling, and death that occur during the marketing process represent a part of the cost of marketing livestock; and, indirectly, the producer foots most of the bill. It is estimated that these losses have a monetary value of $61 million annually, a staggering sum indeed.[3]

The following precautions are suggested (in addition to those already covered under the main heading, "Preparing and Shipping Livestock") as a means of reducing livestock marketing losses from bruises, crippling, and death:

1. Dehorn cattle, preferably when young.
2. Remove projecting nails, splinters, and broken boards from feed containers and fences.
3. Keep feedlot free from old machinery, trash, and any obstacle that may bruise.
4. Remove protruding nails, bolts, or any sharp objects in truck or car.
5. Bed properly (see Table 13-3).
6. Use good loading chutes; not too steep.
7. With two or more decks, have upper deck(s) high enough to prevent back bruises on animals below.
8. Use partitions (a) in cars and trucks that are not fully loaded, to keep animals closer together; and (b) in very long trucks, to keep animals from crowding from one location to another.
9. In rail shipments, place "bull board" in position and secure before car door is closed on loaded cattle.
10. Drive trucks carefully. Slow down on sharp turns and avoid sudden stops.
11. Inspect load enroute to prevent trampling of animals that may be down. If an animal goes down, get it back on its feet immediately.
12. Back truck slowly and squarely against unloading dock.
13. Unload slowly. Do not drop animals from upper to lower deck; use cleated inclines.
14. Never lift sheep by the wool.

All of these precautions are simple to apply; yet all are violated every day of the year.

In a nationwide survey involving 775,000 hogs and 163,000 cattle, Livestock Conservation Institute found that 8.5 percent of all market hogs and 6.4 percent of all market cattle showed unmistakable and costly carcass bruises.

[3]Estimates by Livestock Conservation Institute, in a letter to the author of this book.

Fig. 13-1. Don't kick hogs! It costs an estimated $5 for each such act. Handlers should take time when loading hogs. Of course, it helps if the handlers know more than the hogs, but this isn't necessary if they have more time. (Drawing by R. F. Johnson)

Number of Animals in a Truck or in a Railroad Car

Overcrowding of market animals causes heavy losses. Sometimes a truck or a railroad car is overloaded in an attempt to effect a saving in hauling charges. More frequently, however, it is simply the result of not knowing space requirements.

Normally railroad cars are either 36 or 40 feet in length, but truck beds are variable in size. The size of the truck or car, and the class and size of animals, determine the number of head that should be loaded therein. For comfort in shipping, the truck or car should be loaded heavily enough so that the animals stand close together, but both overloading and underloading should be avoided. Tables 13-1 and 13-2 give some indication as to the number of market animals that may be loaded in a truck or railroad car.

TABLE 13-1

NUMBER OF ANIMALS FOR SAFE LOADING IN A TRUCK[1]

Length of Truck Floor	Kind and Weight of Animals		
	Cattle	Lambs	Hogs and Calves
(ft)	*(1,000 lb)* *(454 kg)*	*(100 lb)* *(45.4 kg)*	*(225 lb)* *(102 kg)*
8 (2.4 m)	4	20	16
12 (3.7 m)	7	31	24
15 (4.6 m)	9	40	30
20 (6.1 m)	12	54	40
24 (7.3 m)	15	65	48

[1]Recommendations of Livestock Conservation Institute, Oak Brook, Ill.

TABLE 13-2

ANIMALS PER RAILROAD CAR[1]

Car Size	Kind and Weight of Animals		
	Cattle	Lambs	Hogs
	(1,000 lb) *(454 kg)*	*(100 lb)* *(45.4 kg)*	*(225 lb)* *(102 kg)*
36-foot car (11 m)	26	110	73
40-foot car (12 m)	28	125	81

[1]Recommendations of Western Weighing and Inspection Bureau, Chicago, Ill. In loading double-deck cars, the upper deck should contain 10% fewer than the lower deck.

Kind of Bedding to Use for Animals in Transit

Among the several factors affecting livestock losses, perhaps none is more important than proper bedding and footing in transit. This applies to both truck and rail shipments.

Footing, such as sand, is required at all times of the year, to prevent the truck or car floor from becoming wet and slick, thus predisposing to injury of animals by slipping or falling. Bedding, such as straw, is recommended for warmth in the shipment of calves, sheep, goats, and swine during extremely cold weather, and as cushioning for dairy cows, breeding stock, or other animals loaded lightly enough to permit their lying down. Recommended kinds and amounts of bedding and footing materials are given in Table 13-3.

TABLE 13-3

GUIDE RELATIVE TO BEDDING AND FOOTING MATERIAL WHEN TRANSPORTING LIVESTOCK[1,2,3]

Class of Livestock	Kind of Bedding for Moderate or Warm Weather; above 50°F	Kind of Bedding for Cool or Cold Weather; below 50°F
Cattle	Sand, 2 inches	Sand; for calves use sand covered with straw
Sheep and goats	Sand	Sand covered with straw
Swine	Sand, ½ inch to 2 inches[4]	Sand covered with straw
Horses and mules	Sand	Sand

[1]Straw or other suitable bedding (covered over sand) should be used for protection and cushioning breeding stock that are loaded lightly enough to permit their lying down in the car or truck.

[2]Sand should be clean and medium-fine, and free from brick, stones, coarse gravel, dirt, or dust.

[3]Fine cinders may be used as footing for cattle, horses, and mules, but not for sheep or hogs. They are picked up by and damage the wool of sheep, and they damage hog casings.

[4]In hot weather, wet sand down before loading and while enroute. Drench hogs when necessary, but never apply water to the backs of hot hogs—it may kill them.

Shrinkage in Marketing Animals

The shrinkage (or drift) is the weight loss encountered from the time animals leave the feedlot until they are weighed over the scales at the market. Thus, if a steer weighed 1,000 lb at the feedlot and had a market weight of 970 lb, the shrinkage would be 30 lb or 3.0 percent. Shrink is usually expressed in terms of percentage. Most of this weight loss is due to excretion, or in the form of feces and urine and the moisture in the expired air. On the other hand, there is some tissue shrinkage, which results from metabolic or breakdown changes.

In many of the large cattle feedlots of the West, it is common practice to give slaughter cattle buyers a 4 or 5 percent pencil shrink. However, in so doing, it is understood that the cattle will be sorted at least 2 days ahead of marketing, that they will be given their regular feed the night before, that they will be allowed free access to water overnight, and that they will be weighed over the owner's scales on the morning that they go to market.

The most important factors affecting shrinkage are:

1. **The fill**—Naturally, the larger the fill animals take upon their arrival at the market, the smaller the shrinkage.

2. **Time and distance in transit**—The longer the animals are in transit and the greater the distance, the higher the total shrinkage. Also, the shrink takes place at a rapid rate during the first part of the haul, and then decreases as time in transit progresses.

3. **Truck vs rail transportation**—Based on practical experience and observation, most stockmen are of the opinion (a) that truck shipments result in less shrinkage than rail shipments for short hauls, and (b) that rail shipments result in less shrinkage than truck shipments for long hauls.

4. **Season**—Extremes in temperature, either very hot or very cold weather, result in higher shrinkage. Shrink is at a minimum between 20° and 60°F.

5. **Age and weight**—Young animals of all species shrink proportionally more than older animals because of their lower carcass yield caused by less body fat and greater amount of fill in proportion to liveweight. For this reason feeder cattle shrink about 25 percent more than finished cattle.

6. **Overloading and underloading**—Overloading always results in abnormally high shrinkage. Unless animals are partitioned off properly, underloading will also result in excess shrinkage.

7. **Rough ride, abnormal feeding, and mixed loads**—Each of these factors will increase shrinkage.

On the average, the following shrinkage is obtained on market animals:

a. Cattle .... from 3 to 6 percent.
b. Sheep .... from 6 to 10 percent.
c. Hogs .... from 1 to 2 percent.

SHRINKAGE TABLES

Both sellers and buyers must give consideration to shrinkage. For example, if a buyer offers $40.00 per

TABLE 13-4

EVALUATING A BID WITH A PENCIL SHRINK DEDUCTED

Bid	2%	3%	4%	6%	8%
$49.00	$48.02	$47.53	$47.04	$46.06	$45.08
48.00	47.04	46.56	46.08	45.12	44.16
47.00	46.06	45.59	45.12	44.18	43.24
46.00	45.08	44.62	44.16	43.24	42.32
45.00	44.10	43.65	43.20	42.30	41.40
44.00	43.12	42.68	42.24	41.36	40.48
43.00	42.14	41.71	41.28	40.42	39.56
42.00	41.16	40.74	40.32	39.48	38.64
41.00	40.18	39.77	39.36	38.54	37.72
40.00	39.20	38.80	38.40	37.60	36.80
39.00	38.22	37.83	37.44	36.66	35.88
38.00	37.24	36.86	36.48	35.72	34.96
37.00	36.26	35.89	35.52	34.78	34.04
36.00	35.28	34.92	34.56	33.84	33.12
35.00	34.30	33.95	33.60	32.90	32.20
34.00	33.32	32.98	32.64	31.96	31.28
33.00	32.34	32.01	31.68	31.02	30.36
32.00	31.36	31.04	30.72	30.08	29.44
31.00	30.38	30.07	29.76	29.14	28.52
30.00	29.40	29.10	28.80	28.20	27.60
29.00	28.42	28.13	27.84	27.26	26.68
28.00	27.44	27.16	26.88	26.32	25.76
27.00	26.46	26.19	25.92	25.38	24.84
26.00	25.48	25.22	24.96	24.44	23.92
25.00	24.50	24.25	24.00	23.50	23.00
24.00	23.52	23.28	23.04	22.56	22.08
23.00	22.54	22.31	22.08	21.62	21.16

TABLE 13-5

PRICE NEEDED TO COMPENSATE FOR SHRINKAGE

Asking	2%	3%	4%	6%	8%
$50.00	$51.02	$51.55	$52.08	$53.19	$54.35
49.00	50.00	50.52	51.04	52.13	53.26
48.00	48.97	49.48	50.00	51.06	52.17
47.00	47.96	48.45	48.96	50.00	51.09
46.00	46.94	47.42	47.92	48.94	50.00
45.00	45.92	46.39	46.88	47.87	48.91
44.00	44.90	45.36	45.83	46.81	47.83
43.00	43.88	44.33	45.97	45.74	46.74
42.00	42.86	43.30	43.75	44.68	45.65
41.00	41.84	42.27	42.71	43.62	44.57
40.00	40.82	41.24	41.67	42.55	43.48
39.00	39.80	40.21	40.63	41.49	42.39
38.00	38.78	39.18	39.58	40.43	41.30
37.00	37.76	38.14	38.54	39.36	40.22
36.00	36.73	37.11	37.50	38.30	39.13
35.00	35.71	36.08	36.46	37.23	38.04
34.00	34.69	35.05	35.42	36.17	36.96
33.00	33.67	34.02	34.37	35.11	35.87
32.00	32.66	32.99	33.33	34.04	34.78
31.00	31.63	31.96	32.29	32.98	33.70
30.00	30.61	30.93	31.25	31.91	32.61
29.00	29.59	29.88	30.21	30.85	31.52
28.00	28.57	28.87	29.17	29.74	30.43
27.00	27.55	27.84	28.12	28.72	29.35
26.00	26.53	26.80	27.08	27.66	28.26
25.00	25.51	25.77	26.04	26.60	27.17
24.00	24.49	24.74	25.00	25.53	26.09
23.00	23.47	23.71	23.96	24.47	25.00

hundred with a 4 percent shrink allowance, the producer will want to know how much he will receive. The answer can be quickly and easily obtained from Table 13-4, as follows: Look at $40.00 under column 1, headed "bid." Go across to column 4, headed "4%." As shown, the producer will receive $38.40 for his cattle.

If the producer has decided that $40.00 is his minimum asking price, he may refuse the offer in the example above and ship his cattle to market. Then he will wish to know how much he will have to receive in order to compensate for shrinkage. The answer can be quickly and easily obtained from Table 13-5, as follows: Look at $40.00 under column 1, headed "Asking"; then read under the proper column to the right. Thus, if the animals shrink 4 percent during marketing, the price will have to be $41.67 in order to compensate for shrinkage. Of course, this price compensates for shrinkage only. It does not consider transportation, for example, even though it is an important cost of marketing livestock.

MARKET CLASSES AND GRADES OF LIVESTOCK

Broadly speaking, *the market class is the use to which animals are put*, whereas *the market grade is a measure of how well the animal fulfills the requirements for the class.* More accurately, however, the market class is determined by all of those factors affecting the use and value of the animal, except the final grade. Thus, in cattle, the market class is determined by whether the animals are cattle or calves; by the general use to which the animals are put (slaughter cattle, stocker and feeder cattle, milkers and springers, vealers, slaughter calves, and stocker and feeder calves); by the sex (steers, heifers, cows, bulls), by the age (yearling or two-year-old or over steers), and by weight.

Grading livestock is the act of sorting, dividing, or designating animals of similar classes and grades. The grade is the final subdivision in the classification process. It indicates the relative degree of excellence of an animal or group of animals. When grading is properly and expertly done, each individual of a specific class and grade group is quite similar to other individuals in that group, regardless of whether the animals are in the same pen or in separate markets hundreds of miles removed from each other.

Factors Determining Market Grades

Market grades are determined by attributes associated with market preferences and valuation. But the relevant attributes differ widely among species—in numbers, in range of variability, and in ease of objective measurement. Among species grade differences are the following:

1. Eight grades are used to cover the range in quality of steer and heifer carcasses, in comparison with five for lambs and hogs.
2. In addition to quality grades, five separate yield, or cutability grades (i.e., yield of boneless, closely trimmed retail cuts) are used in conjunction with quality grades in both beef and lamb.
3. Pork grades incorporate yield and quality considerations without separate quality and yield designations.

Quality grades in beef carcasses are based on the palatability-indicating characteristics of the lean. Cutability, or yield, grades of beef are determined primarily by objective measurements. All graded beef is dual graded—it is graded for both quality and yield.

Pork carcass grades rely heavily on objective measurements to determine expected combined yield of lean cuts, used in conjunction with a subjective determination of "acceptable" or "unacceptable" quality of the lean.

Thus, drawing up standards or specifications for a system of grades is a complex problem. Some of the attributes upon which grades are based can be evaluated directly; others must be evaluated indirectly, through indicators. For example, the yield of lean cuts of pork is related to (or indicated by) backfat thickness, carcass weight, and carcass length.

Market Classes and Grades of Cattle

The generally accepted market classes and grades of live cattle are summarized in Table 13-6. The first five divisions and subdivisions include those factors that determine the class of the animal or the use to which it will be put. The grades indicate how well the cattle fulfill the requirements to which they are put.

Market Classes and Grades of Sheep

The market classes and grades of sheep (see Table 13-7) follow closely the pattern for the classes and grades of cattle and swine. One notable difference is that a sizable number of sheep are sold as breeders. For the most part, this class is made up of mature western ewes that are sold to country buyers for the purpose of producing one to two more crops of lambs before again being returned to the market. Usually such ewes can be acquired at a lower cost than ewe lambs. Another difference between sheep and other species is found in the fact that one feeder class, namely the shearers, is based on wool value as well as adaptability for further feeding.

Market Classes and Grades of Hogs

The market classes and grades of swine were developed in much the same manner as the classifications of cattle were developed and brought into use. They also serve much the same purpose. Swine classes and grades do differ from those used in cattle and sheep in that (1) there are no age divisions by years (e.g., cattle are classified as yearling and two-year-old and over), and (2) rarely are hogs of any kind purchased on the market for use as breeding animals. As in the classification of market cattle, the class of market hogs indicates the use to which the animals are best adapted, whereas the grade indicates the degree of perfection within the class.

The market classes and grades of hogs are summarized in Table 13-8.

TABLE 13-6
THE MARKET CLASSES AND QUALITY GRADES OF CATTLE

Cattle or Calves	Use Selection	Sex Classes	Age	Weight Divisions Wt. (Group)	(lb)	Commonly Used Quality Grades[1]
Cattle	Slaughter cattle[1]	Steers	Yearlings	Light Medium Heavy	750 down 750-950 950 up	Prime, Choice, Good, Standard, Utility, Cutter, Canner
			2-year-old and over	Light Medium Heavy	1,100 down 1,100-1,300 1,300 up	Prime, Choice, Good, Standard, Commercial, Utility, Cutter, Canner
		Heifers	Yearlings	Light Medium Heavy	750 down 750-900 900 up	Prime, Choice, Good, Standard, Utility, Cutter, Canner
			2-year-old and over	Light Medium Heavy	900 down 900-1,050 1,050 up	Prime, Choice, Good, Standard, Commercial, Utility, Cutter, Canner
		Cows	All ages	All weights		Choice, Good, Standard, Commercial, Utility, Cutter, Canner
		Bullocks	24 mo. and under	All weights		Prime, Choice, Good, Standard, Utility
		Bulls		All weights		None (yield graded only)
	Feeder cattle	Steers	Yearlings	Light Medium Heavy Mixed		Prime, Choice, Good, Standard, Utility, Inferior
			2-year-old and over	Light Medium		Prime, Choice, Good, Standard, Commercial, Utility, Inferior
		Heifers	Yearlings	Light Medium Heavy Mixed		Prime, Choice, Good, Standard, Utility, Inferior
			2-year-old and over	Light Medium Heavy Mixed		Prime, Choice, Good, Standard, Commercial, Utility, Inferior
		Cows	All ages	All weights		Choice, Good, Standard, Commercial, Utility, Inferior
		Bullocks	24 mo. and under	All weights		Prime, Choice, Good, Standard, Utility, Inferior
		Bulls	24 mo. and over	All weights		None
	Milkers and springers	Cows (milkers or springers)	All ages	All weights		None
Calves	Vealers	No sex class (Sex characteristics of no importance at this age)	Under 3 mo.	Light Medium Heavy	110 down 110-180 180 up	Prime, Choice, Good, Standard, Utility
	Slaughter calves	Steers Heifers Bulls	3 mo. to 1 year	Light Medium Heavy	200 down 200-300 300 up	Prime, Choice, Good, Standard, Utility
	Feeder calves	Steers Heifers Bulls	Usually 6 mo. to 1 year	Light Medium Heavy Mixed		Prime, Choice, Good, Standard, Utility, Inferior

[1]In addition to the quality grades, there are the following yield grades for all slaughter cattle, except bulls: Yield Grade 1, Yield Grade 2, Yield Grade 3, Yield Grade 4, and Yield Grade 5; with Yield Grade 1 representing the highest cutability, and Yield Grade 5 the lowest. Thus, slaughter cattle are graded for both quality and yield grade.

TABLE 13-7

THE MARKET CLASSES AND QUALITY GRADES OF SHEEP

Sheep or Lambs	Use Selection	Sex Classes	Age	Weight Division	(Pounds)	(Kilograms)	Commonly Used Grades
Sheep	Slaughter sheep	Ewes	Yearling	Light	90 down	40.9 down	Prime, Choice, Good, Utility[1]
				Medium	90 to 100	40.9-45.4	
				Heavy	100 up	45.4 up	
			Mature (2-year-old or older)	Light	120 down	54.5 down	Choice, Good, Utility, Cull[1]
				Medium	120-140	54.5-63.6	
				Heavy	140 up	63.6 up	
		Wethers	Yearling	Light	100 down	45.4 down	Prime, Choice, Good, Utility[1]
				Medium	100-110	45.4-49.9	
				Heavy	110 up	49.9 up	
			Mature (2-year-old or older)	Light	115 down	52.2 down	Choice, Good, Utility, Cull[1]
				Medium	115-130	52.2-59.0	
				Heavy	130 up	59.0 up	
		Rams	Yearling	All weights			Prime, Choice, Good, Utility[1]
			Mature (2-year-old or older)	All weights			Choice, Good, Utility, Cull[1]
	Feeder sheep	Ewes and wethers	Yearlings	All weights			Fancy, Choice, Good, Medium, Cull
		Ewes	Mature (2-year-old or older)	All weights			Choice, Good, Medium, Cull
	Breeding sheep	Ewes (rams occasionally purchased as breeders, but not listed in market reports)	Yearlings, 2-, 3-, or 4-year-olds and older	All weights			Fancy, Choice, Good, Medium, Cull
Lambs	Slaughter lambs	Ewes, wethers, and rams	Hothouse lambs	60 down			Prime, Choice, Good, Utility[1]
		Ewes, wethers, and rams	Spring lambs	Light	70 down	31.8 down	Prime, Choice, Good, Utility[1]
				Medium	70-90	31.8-40.9	
				Heavy	90 up	40.9 up	
		Ewes, wethers, and rams	Lambs	Light	75 down	34.0 down	Prime, Choice, Good, Utility[1]
				Medium	75-95	34.0-43.1	
				Heavy	95 up	43.1 up	
	Feeder lambs	Ewes and wethers	All ages	All weights			Fancy, Choice, Good, Medium, Cull
	Shearer lambs	Ewes and wethers	All ages	All weights			Choice, Good, Medium

[1]In addition to the above quality grades, there are five yield grades applicable to all lamb and mutton carcasses, denoted by numbers 1 through 5, with the yield grade 1 representing the highest degree of cutability. Thus, slaughter sheep and lambs may be graded for (1) quality alone, (2) yield grade alone, or (3) both quality and yield grades.

TABLE 13-8

THE MARKET CLASSES AND QUALITY GRADES OF HOGS

Hogs or Pigs	Use Selection	Sex Class	Weight Divisions (lb)	(lb, continued)	(kg)	(kg, continued)	Commonly Used Grades
Hogs	Slaughter hogs	Barrows and gilts (often called butcher hogs)	120-140	240-270	55-64	109-123	U.S. No. 1,
			140-160	270-300	64-73	123-136	U.S. No. 2,
			160-180	300-330	73-82	136-150	U.S. No. 3,
			180-200	330-360	82-91	150-163	U.S. No. 4,
			200-220	360-400	91-100	163-182	U.S. Utility.
			220-240	400 lbs up	100-109	182 up	
		Sows (or packing sows)	270-300	400-450	123-136	182-204	U.S. No. 1,
			300-330	450-500	136-150	204-227	U.S. No. 2,
			330-360	500-600	150-163	227-272	U.S. No. 3,
			360-400	600 lbs up	163-182	272 up	Medium
							Bull
		Stags	All weights				Ungraded
		Boars	All weights				Ungraded
	Feeder hogs	Barrows and gilts	120-140		55-64		U.S. No. 1, U.S. No. 2,
			140-160		64-73		U.S. No. 3, U.S. No. 4,
			160-180		73-82		U.S. Utility, Cull.
Pigs	Slaughter pigs	Barrows, gilts, and boars	Under 30		13.6		Ungraded
			30-60		13.6-27.2		
		Barrows and gilts	60-80		27.2-36.3		Ungraded
			80-100		36.3-45.4		
			100-120		45.4-54.5		
	Feeder pigs	Barrows and gilts	80-100		36.3-45.4		U.S. No. 1, U.S. No. 2,
			100-120		45.4-54.5		U.S. No. 3, U.S. No. 4,
							U.S. Utility, Cull.

SOME LIVESTOCK MARKETING CONSIDERATIONS

Enlightened and shrewd marketing practices generally characterize the successful livestock enterprise. Among the considerations of importance in marketing animals are those which follow.

Cyclical Trends in Market Livestock

The price cycle as it applies to livestock may be defined as that period of time during which the price for a certain kind of livestock advances from a low point to a high point and then declines to a low point again. In reality, it is a change in animal numbers that represents the stockman's response to prices. Although there is considerable variation in the length of the cycle within any given class of stock, currently, the price cycle of the different classes of animals is about as follows: hogs, 4 years; sheep, 9 to 10 years; and cattle, 10 years.

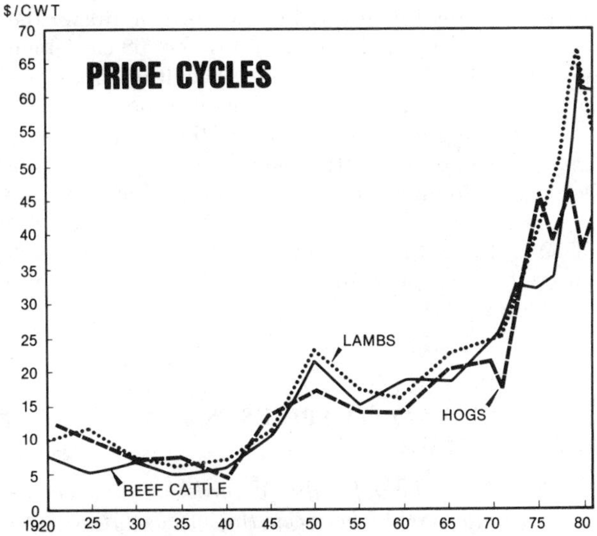

Fig. 13-2. Average price received by U.S. farmers for each class of livestock, 1920-1981. In general this shows that, currently, the price cycle of each class of animals is approximately as follows: hogs, 4 years; sheep, 9 to 10 years; and cattle, 10 years. (Source: USDA; AMS; Livestock, Meat, Grain and Seed Division)

The species cycles are a direct reflection of the rapidity with which the numbers of each class of farm animals can be shifted under practical conditions to meet consumer meat demands. Litter-bearing and early producing swine can be increased in numbers much more rapidly than either sheep or cattle. When market hog prices are favorable, established swine enterprises are expanded, and new herds are founded, so that about every four years, on the average, the market is glutted and prices fall, only to rise again because too few hogs are being produced to take care of the demand for meats.

It is noteworthy that cattle cycles were formerly 15 years or longer, but that they have shortened to about 10 years, due to the earlier maturity of modern cattle and the marketing of cattle at younger ages.

Normal cycles are disturbed by droughts, wars, general periods of depression or inflation, and federal controls.

Seasonal Changes in Market Livestock

As would be expected, (1) seasons of high market prices are generally associated with light marketings, and seasons of low market prices with heavy marketings, and (2) the seasons of high and low market prices vary with different classes of livestock. It must be realized, however, that the normal seasons of high

TABLE 13-9

WHEN TO BUY AND WHEN TO SELL FARM ANIMALS

Kind and Class of Animal	Lowest Prices	Highest Prices
Slaughter steers,		
Prime	Nov., Dec.	July, Aug.
Choice	Nov., Dec.	July, Aug.
Good	Nov., Dec.	July, Aug.
Choice slaughter heifers	Oct. - Dec.	July, Aug.
Cows (slaughter)	Nov., Dec., Jan.	Mar. - June
Feeder steers	Nov., Dec., Jan.	July, Aug.
Choice slaughter lambs	Jan., Feb.	April, June
Feeder lambs	July, Aug., Sept.	March, Apr., May
Slaughter barrows and gilts	April - May	July, Aug.
Sows (slaughter)	April - May	July, Aug., Sept.
Feeder pigs	June	March - Apr., Aug., Sept.

and low market prices may be changed by such factors as (1) federal farm programs and controls, (2) business conditions and general price levels, (3) feed supplies and weather conditions, (4) wars, etc.

In recent years, seasonal patterns have not been as reliable as they used to be. Year-round finishing of cattle and lambs in large commercial feedlots and year-round farrowing of sows in confinement have made for more uniform marketing throughout the year and lessened seasonality in livestock marketing. Thus, when arriving at livestock forecasts and marketing advice, proper reservation should be exercised in considering seasonal patterns. Anyway, it is not always wise to plan production to hit the highest market, for sometimes that would push up production costs more than enough to offset the gains from higher prices. Nevertheless, a careful study of normal seasonal prices will serve as a useful guide (see Table 13-9).

Dockage

Dockage refers to deductions made in the liveweight of market animals because of excessive dressing losses, or because part of the product is of low quality. Some common dockages on livestock markets are:

1. **Piggy sows**—Usually docked 40 pounds, but it may range from 0 to 50 pounds, depending on the market.
2. **Stags (hogs)**—Usually docked 70 pounds, but it may range from 40 to 80 pounds, depending on the market.
3. **Cattle with lumpy jaws**—Usually bought subject to the amount of wastage.

Livestock Marketing Costs

Stockmen need to be acquainted with livestock marketing costs. Although commission and yardage rates vary widely (1) according to size of consignment, and (2) between markets, Table 13-10 summarizes the estimated average charges throughout the country.

TABLE 13-10

ESTIMATED AVERAGE LIVESTOCK MARKETING COSTS OF TERMINALS VS AUCTIONS MARKETS[1]

Type of Market Outlet	Rate per Head			
	Cattle	Calves	Hogs	Sheep and Lambs
	($)	($)	($)	($)
Terminals	6.45	4.25	2.35	1.60
Auctions	7.30	4.80	2.40	1.80

[1]Estimates made by the author.

Meat Check-off

Several beef, lamb, and pork check-off programs are in operation. Each program is under different sponsorship; some nationwide, others statewide. They are characterized by little correlation between them and varying degrees of effectiveness.

In 1976, the National Pork Producers Council increased its voluntary check-off from 5 cents to 10 cents per market hog. It was projected that this would generate approximately $3 million per year, from about 65 percent of all hogs slaughtered.

On May 28, 1976, legislation officially known as the Beef Research and Consumer Education Act was signed into law by President Gerald R. Ford. Its stated purpose:

> To establish, finance, and carry out a coordinated program of research, producer and consumer education, and promotion to improve, maintain, and develop markets for cattle, beef and beef products, and to provide an adequate, steady supply of high-quality beef and beef products readily available to consumers at reasonable prices.

The bill was modeled after current cotton and egg check-off programs. It was based on a value added formula at the contribution rate of 0.3 of 1%. In other words, when Seller "A" sells an animal to Buyer "B" for $200, he actually receives $199.40 ($200.00 minus 0.3%). If Buyer "B" then fattens the animal and sells to Buyer "C" for $400.00, he actually receives $398.80. Not until the final sale before slaughter is the money actually collected and sent to the Beef Board, which will then determine exactly how to spend it.

After the first year of operation, the assessment rate could be set anywhere from 0.1% to 0.3% of the sale price of the cattle sold, and the assessment raised above 0.5% only with the approval of the producers in a new referendum.

The entire program was voluntary, so at any step in the selling process the seller could request his contribution back.

It was projected that the program would generate $30 to $40 million annually once it was fully operational. The Beef Board—which was to administer the funds—was supposed to be composed of 68 cattlemen appointed by the Secretary of Agriculture from recommendations made by cattle organizations.

According to the provisions of the Beef Research and Consumer Education Act, before the program could go into effect at least ⅔ of the cattlemen voting in a referendum had to approve it. In the referendum of 1977, fewer than the required ⅔ of the cattlemen voted for the program; hence, it was not implemented.

PARITY AND PARITY PRICES IN FARM ANIMALS

Parity may be defined as a yardstick for measuring the relationship between the prices farmers receive for the products they sell and the prices they pay for the things they buy, including interest, taxes, and farm wage rates.

The price of a farm commodity may be said to be at parity when a given unit has the same purchasing power that it had in the base period. Thus, if a swine producer received parity price per hundred weight for hogs, he should be able to take the money and buy as much with it as he could back in 1910 to 1914, the base period. If the price of things farmers buy for their production program doubles, the parity price of farm commodities also doubles, to keep in line or equal parity.

It is important that the stockman be informed relative to parity prices because most farm price support legislation is based on some percentage of parity. To this end, the following facts are presented:

1. **The old parity formula**—The base period 1910 to 1914 was long used in figuring the parity price of most farm products. This period was selected originally because it was felt that price relationships between farm and nonfarm products were fair to both farmers and nonfarmers during this period.

Prior to 1950, the parity formula based on 1910 to 1914 was used for the following purposes:

a. To determine the relationship between the prices farmers receive and the prices farmers pay, and

b. To determine the relationships among the parity prices of individual commodities.

Obviously, the weakness in using a fixed and historic base period, such as the years 1910 to 1914, is that it gets more and more out of date with each passing year. For example, production-cost relationships have been materially affected since 1910 to 1914 by the passing of horse and mule power and their replacement by tractors and trucks, by the development of hybrid corn, and by the greatly increased acreage of soybeans, etc.

2. **The new parity formula which became effective January 1, 1950**—In 1948 and again in 1949, Congress passed acts which revised and brought up to date the parity formula, called the "new formula." Although over-all price relationships between what farmers get and what they pay are still based on the years 1910 to 1914, the parity prices of individual farm commodities are figured by a new formula that reflects recent price relationships. The latter is accomplished by using the latest 10-year averages. This is an improvement, because it helps reflect changes in price relationships that have taken place in farming since 1910 to 1914.

The new parity formula lowers parity prices for many crops and raises them for some livestock and livestock products.

3. **Basic and nonbasic commodities**—The basic commodities are: wheat, corn, cotton, rice, peanuts, and tobacco. All other farm commodities are considered nonbasics.

4. **How to figure parity by the new formula**—Step by step, parity prices under the new formula are computed as follows:

a. Secure the average price of the commodity during the past 10 years.

b. Divide this price by the average of the index of price received (1910 to 1914 = 100) for the same period.

c. Multiply this figure by the current index of prices paid for commodities and services, interest, taxes, and wage rates.

5. **How to use parity**—Parity may be used in a number of ways:

a. It can be used to determine how farmers, on the average, are faring on the basis of what they sell as it relates to what they buy; or

b. It can be applied to income, to provide a farmer with the relationship of his level of living compared to other occupations; or

c. It can be applied to a single agricultural commodity (like corn, wheat, soybeans, cattle, hogs, or vegetables—singly, or as a group) to see how it compares in its relationship with parity and with other commodities.

6. **Where can current parity price figures be obtained?**—Current parity price figures for each commodity are reported monthly by the U.S. Department of Agriculture, Washington, D.C.

PART II—MARKETING MILK

Marketing milk is big business. In 1981, United States farm sales of milk totaled 132.6 billion pounds, which dairymen sold at an average price of $13.80 per cwt, for total cash receipts of $17.7 billion.

In our present system, the marketing of milk and dairy products is handled largely by specialists, usually under a multitude of regulations and controls. However, a successful milk producer must understand milk markets and the factors affecting them if he is to take full advantage of his opportunities.

FARM PRODUCTION AND HANDLING OF MILK

Satisfactory milk marketing necessitates one basic ingredient—quality milk; and, ultimately, this means more income for the dairyman.

The difference in price between Grade A milk and the lower grades is considerable. But it goes beyond this; quality can mean increased consumer demand.

Buyers, consumers, and health departments all have a distinct interest in the quality of milk marketed and used for manufacturing.

Quality milk can be produced only when dairymen pay special attention to a number of factors; among them, herd health, the layout and structure of the barn and milk house, clean cows, care of the utensils, cooling and storage of milk, and transportation of milk to market.

How Milk Is Sold by Dairymen

Dairymen sell most of their product in the form of whole milk. At one time, they marketed a considerable amount of their product as farm-separated cream, but the proportion of this product has been declining. In 1976, farm-separated cream was equivalent to only 0.2 percent of total marketing, compared to 38 percent in 1940.

Annual whole milk production in the United States increased by only 16 billion pounds from 1950 to 1981. However, during this same period milk used on the farms where produced decreased dramatically. In 1950, 15.6% of the milk was used on farms where produced, compared with only 1.7% in 1981. Obviously, the point has been reached where little additional marketing of milk by farmers can be expected through decreased use on the farm. Future increases in milk marketings will have to come from increased production.

MARKET CHANNELS FOR MILK AND DAIRY PRODUCTS

Milk moves from the farm to the consumer in the following three stages:

1. Assembly and transportation from farms to processing plants.
2. Processing and packaging or manufacturing into various dairy products.
3. Distribution of packaged milk and manufactured milk products to consumers.

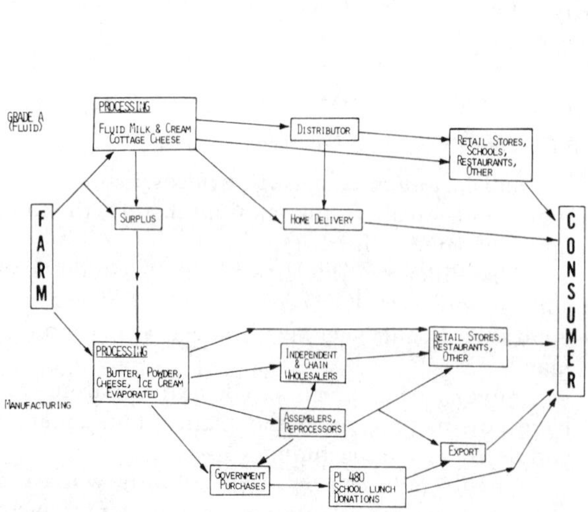

Fig. 13-3. Marketing channels for milk and dairy products. (From *Organization and Competition in the Dairy Industry*, Tech. Study No. 3, National Commission on Food Marketing, p. 17, Fig. 7.)

Also, dairymen market their milk as (1) Grade A milk, or (2) manufacturing milk (Grade B).

The market channels for milk and dairy products are shown in Fig. 13-3.

HOW MILK IS PRICED AND REGULATED

Chaotic conditions in milk marketing, resulting from the breakdown of private controls, and the serious economic plight of farmers during the depression years of the early 1930s, brought requests from organized producers and distributors for government control. Out of this evolved two forms of government controls—those established by the federal government and those established by state governments; both were designed to bring more stability into the marketing of milk. Today, federal and state agencies affect, directly or indirectly, the pricing of all milk marketed by dairy farmers in the United States. It has been estimated that more than two-thirds of all milk eligible for fluid markets is affected by milk orders.

Federal Milk Marketing Orders

Federal milk marketing orders are established and administered by the Secretary of Agriculture under acts of Congress passed in 1933 and 1937. They are legal instruments, and they are very complex. However, stated in simple terms, they are designed to stabilize the marketing of fluid milk and to assist farmers in negotiating with distributors for the sale of their milk. Prices paid to farmers are controlled, but there is no direct control of retail prices.

Federal orders are not concerned with sanitary regulations. These are administered by state and local health authorities.

On January 1, 1981, there were 48 different federal milk market orders, each with a market administrator and provision for setting minimum farm prices and regulating transactions between farmers and milk dealers in their area. In 1980, slightly more than two-thirds of all milk marketed in the country and 80% of the Grade A milk were covered by federal orders.

Prices in other Grade A markets are influenced by prices established under federal orders or state control programs. Additionally, dairy support programs directly affect the prices of both manufacturing grade milk marketed by farmers and the milk farmers sell as farm-separated cream.

Those desiring further information or clarification regarding milk marketing orders should obtain the following U.S. Department of Agriculture publications:

USDA-CMS—"*The Federal Milk Marketing Order Program*," Marketing Bull. No. 27, April, 1968.

USDA-AMS—*"Questions and Answers on Federal Milk Marketing,"* AMS - 559, March, 1975.

For information about a specific milk marketing order, a copy of the actual order of interest is recommended.

State Milk Control

In 1981, sixteen states had milk control programs that also included some authority over milk prices.

In setting minimum farm prices, state control agencies often operate in a manner similar to federal milk orders. Classified pricing principles are used and prices are set for a particular market and not necessarily for the whole state.

State orders establish prices at various stages along the way—farm, wholesale, retail, as well as regulating trade practices. It is noteworthy, however, that fewer and fewer state milk commissions set retail milk prices. The foes of retail pricing point out that on the average, retail price setting results in a lower price to the farmer than where retail prices are let alone.

Because of their inability to cope with out-of-state milk, state milk controls will likely decline in importance in the future; they will be replaced by federal milk orders.

Cooperatives

The practice of dealing separately with a large number of producers led to dissatisfaction in a number of cases. To rectify this situation, cooperatives were organized. These cooperative associations are of two general types:

1. Bargaining associations which do not handle any milk, but make all business arrangements.
2. Associations which process and distribute milk or assemble it for fluid use.

About 75% of the total deliveries of milk to plants and dealers in the United States is handled by cooperatives.

Other Regulatory Programs

Because of the essential nature of milk, plus the fact that it is easily contaminated and a favorable medium for bacterial growth, it is inevitable that numerous regulatory programs have evolved around it—federal, state, and local, some having been designed to control prices and assure a reasonably uniform flow of milk, and others for sanitary reasons.

SANITARY REGULATIONS

The sanitation of milk and dairy products is assured by the enforcement of sanitary regulations by federal, state, and local authorities.

All major cities and states have sanitary regulations governing the production, transportation, processing, and delivery of milk. Unfortunately, from area to area, there are a bewildering number of different regulations, with the result that milk going to more than one city market is often subjected to duplication and confusion in inspection. Also, sanitary and health regulations have sometimes been used as barriers to keep milk out of a certain area for competitive reasons.

In 1923, the U.S. Public Health Service (USPHS) established an Office of Milk Investigations, and, in 1924, the USPHS published its first Grade A pasteurized milk ordinance. Subsequently, this regulation has been revised several times.

Producers are issued permits allowing them to ship Grade A milk. The permit is revoked if either the bacteria count of raw milk exceeds 100,000 per milliliter or the cooling temperature exceeds 40°F in three of the last five samples.

The standard plate count of Grade A pasteurized milk may not exceed 20,000 per milliliter nor the coliform count 10 per milliliter in three of the last five samples or the processor's permit will be revoked.

The Food and Drug Administration (FDA) is charged with inspecting dairy products and processing plants for contamination and adulteration.

In an attempt to improve the quality of milk, on April 6, 1967, a National Conference of Interstate Milk Shippers passed a resolution calling for milk to contain not more than 1,500,000 leucocytes per milliliter on herd milk (leucocytes in milk are an indirect measure of mastitis in cows).

STANDARDS AND GRADES

In most states, milk is graded as follows:

Grade A
Grade B, or manufacturing grade
Grade C, or reject.

Grade A milk must meet high-quality standards, and must be produced, processed, and handled in an approved manner in approved facilities and equipment. Producers are issued permits allowing them to ship Grade A milk. But the permit is revoked if either the bacteria count of raw milk exceeds 100,000 per milliliter or the cooling temperature exceeds 40°F in 3 of the last 5 samples.

As with Grade A milk, each state adopts and enforces its own regulations to control milk used for the manufacture of milk products. The standards are neither as stringent, nor as uniform, for Grade B milk as for Grade A milk.

The U.S. Department of Agriculture has established grades for butter, cheese, and nonfat dry milk (see Section 14, Meat and Milk, Table 14-11).

STATE TRADE PRACTICE LAWS

For more than 20 years, there has been considerable concern about competitive practices in the sale of fluid milk products and ice cream. Among the unfair trade practices sometimes observed or suspected in the marketing of milk are: discriminatory price cutting, secret rebates, loans, advertising rebates, furnishing and servicing equipment, and the giving of gifts and free signs.

In 1976, 37 states had laws concerned with the marketing and/or pricing of milk and dairy products. Without doubt, more states will enact dairy fair trade practice laws, and this approach will be used by the state as a substitute for complete milk control.

Methods of Pricing or Paying for Fluid Milk

Economists refer to the different systems of paying for milk as "price plans." These plans, which in actual practice generally involve two or more plans—for example, pricing based on (1) class, (2) grade, and (3) base-surplus—are:

1. **Flat price plan**—This was the common method up to World War I. The milk producer was paid a uniform price for all milk sold regardless of quality or the use made of it.

2. **Use classification plan**—Most marketing orders established two use classes—Class I and Class II.

Class I milk generally includes milk used in fluid form such as whole fluid milk, or milk for creamed drinks which must be made from milk approved by local health authorities. Generally speaking, Class I prices are 35 to 100 percent higher than Class II prices.

Class II milk usually includes milk in excess of fluid needs, which is used to make manufactured dairy products—primarily butter, nonfat dry milk, and cheese.

On some markets, a further division is made, primarily for milk going into cottage cheese, with the result that there are three classes of milk—Class I, Class II, and Class III.

3. **Blend price**—When dealers buy according to classification prices, they may pay producers a blend price. The blend is an average of class prices weighted by the volume of milk in each class, usually quoted at a specific point and for a specific test of milk.

4. **Quality grade plan**—Frequently, the terms Grade A and Grade B (usually called Manufacturing Grade Milk) are encountered in milk marketing. Although there may be some local variations in their use, Grade A usually refers to milk produced under conditions which make it acceptable for fluid use in a given market. Grade B often refers to milk produced under conditions which do not make it acceptable for fluid milk use—it's manufacturing milk.

The production of Grade A milk relative to that of Grade B milk has been increasing in recent years (see Fig. 13-4). In 1981, farmers sold almost 98.3 out of every 100 pounds of milk they produced. Of the milk which they sold, 85% of it was eligible for the fluid market as Grade A; 15% of it was manufactured milk, or Grade B; and 1.1% of it was sold directly to consumers.

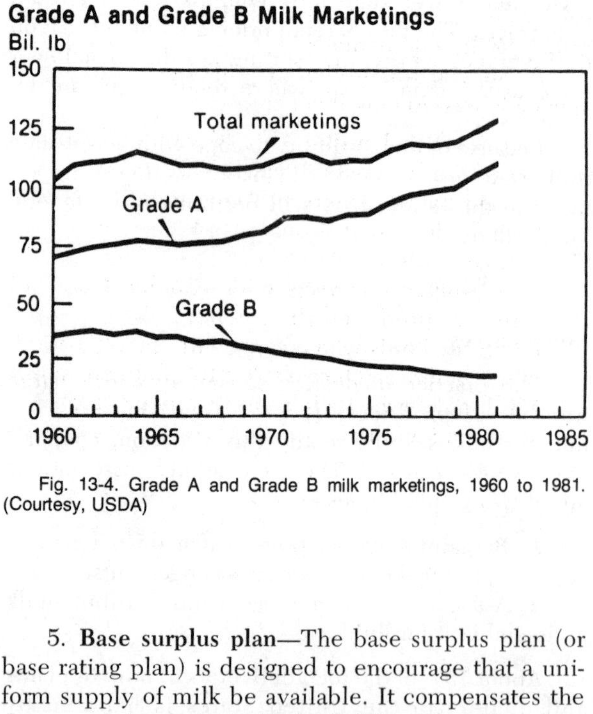

Fig. 13-4. Grade A and Grade B milk marketings, 1960 to 1981. (Courtesy, USDA)

5. **Base surplus plan**—The base surplus plan (or base rating plan) is designed to encourage that a uniform supply of milk be available. It compensates the dairyman who maintains a high fall production, when more milk is needed. The base period is established during the lowest production months, usually over a period of three to six months. Then, a dairyman's base is established by the average amount of milk delivered during the base period. His base may be modified from time to time.

6. **Butterfat test price plan**—The butterfat test of milk affects the price. The common practice is to establish a price for 100 pounds of milk of a specified butterfat test. Usually 3.5% butterfat is the basis for pricing, although several markets have established

their base at as high as 4.0% butterfat. Then a price differential (per point or 0.1%) is set up for milk testing above or below this amount.

7. **Solids-not-fat price plan**—Today, the emphasis on the food value of milk is shifting from fat content to the other solids, especially protein. This is feasible because tests for solids-not-fat have been devised, and these are proving practical for field use. It is anticipated that this system of pricing milk will expand in the future.

On the average, whole milk contains about 2¼ lb of solids-not-fat for each pound of milk fat. Thus, milk testing 4 percent butterfat contains approximately 9 lb of solids-not-fat, to a total of 13 lb of solids per hundredweight.

8. **Gallon or quart plan**—Occasionally, a producer supplies milk to a distributor on a per gallon or per quart basis. Since average milk weighs 2.15 lb to the quart and 8.6 lb to the gallon, 100 lb of milk would be equivalent to 46.5 quarts or 11.6 gallons. Thus, one can easily compute the possible returns from selling milk by different methods.

9. **Special milk**—Certain milks are sold under special labels. Among them are:

a. **Certified milk**—This is milk that is produced under special sanitary conditions prescribed by the American Association of Medical Commissioners. It is sold at a higher price than ordinary milk.

b. **Golden Guernsey milk**—Golden Guernsey milk is produced by owners of purebred Guernsey herds who comply with the regulations of the American Guernsey Cattle Club. Such milk is sold under the trade name "Golden Guernsey," at a premium price.

c. **All-Jersey milk**—This is produced by registered Jersey herds whose owners comply with the regulations of the American Jersey Cattle Club. It is sold at a premium price under the trademark of "All Jersey."

The Milk Price Support Program

Some of the price support programs pertaining to surpluses since World War II had their origin in wartime programs designed to increase production. Following the war, the demand for dairy products for military and foreign use declined sharply. Thereupon, the Agricultural Act of 1948 extended the price support authorization at 90% of parity for milk and butterfat; and, beginning the following year, the Agricultural Act of 1949 authorized and directed the Secretary of Agriculture to support manufacturing milk prices to producers at between 75 and 90% of parity. Although the Act has been amended several times since, it still provides the authority for the milk price support program. In 1981, the law was amended so as to require that the Secretary of Agriculture support the price of milk at between 70 and 90% of parity, adjusted annually. In 1981, U.S. farmers received $13.80 per cwt for all milk sold to plants; with a price differentiation of $13.96 per cwt for fluid (Grade A) milk, and $12.75 per cwt for manufacturing (Grade B) milk.

In several of the postwar years, the government has made substantial purchases of dairy products—through CCC and other purchase programs—to support prices at announced levels.

SECTION 14

MEAT AND MILK[1]

Contents **Page**

PART I—MEAT

Meat Inspection 1078
Federal Meat Inspection 1078
State Meat Inspection 1079
Meat Grading 1079
Federal Grades of Meats 1080
Beef and Veal 1080
The Dressing Percentage of Cattle and Calves 1081
Beef and Veal Cuts and How to Cook Them 1082
Lamb 1085
The Dressing Percentage of Lambs 1085
Lamb Cuts and How to Cook Them 1086
Pork 1087
The Dressing Percentage of Hogs 1087
Pork Cuts and How to Cook Them 1088
Meat Preservation 1089
Freezing Meats 1089
Directions for Selecting, Preparing, and Freezing Meats 1090
How Long Can Meat Be Frozen? 1091
How to Thaw and Cook Frozen Meats 1092
Methods of Curing Pork on the Farm 1092
Methods of Cooking Meats 1094
Packinghouse By-products from Slaughter 1095
Meat Import Law (Public Law 88-482) 1096

PART II—MILK

Produce Quality Milk 1098
Physiological Factors Affecting Amount and Composition of Milk 1098
Environmental Factors Affecting Amount and Composition of Milk 1099
Mastitis 1100
Milk Flavor 1101
Beware of Pesticide Residues 1102
Sanitary Regulations 1102
Standards and Grades of Milk and Milk Products 1103
Uses of Milk 1103

Perhaps most people consume meat and milk simply because they like them. They derive a rich enjoyment and satisfaction therefrom. For flavor, variety, and appetite appeal, they are unsurpassed.

But animal products are far more than just very tempting and delicious foods. From a nutrition standpoint, they contain certain essentials of an adequate diet. This is important, for how we live and how long we live are determined in large part by our diet.

It is estimated that the average American gets the percentages of his food nutrients shown in Fig. 14-2 from animal products. Foods of animal origin (meat, milk, and their various by-products) are especially important in the American diet; they provide nearly all of the vitamin B-12, 2/3 of the total protein, about

Fig. 14-1. Meat on the table.

[1]The author is very grateful to the following person who reviewed Part I—Meats of this section and made many helpful suggestions for its improvement: Mr. John C. Pierce, Director, Livestock Division, Agricultural Marketing Service, USDA.

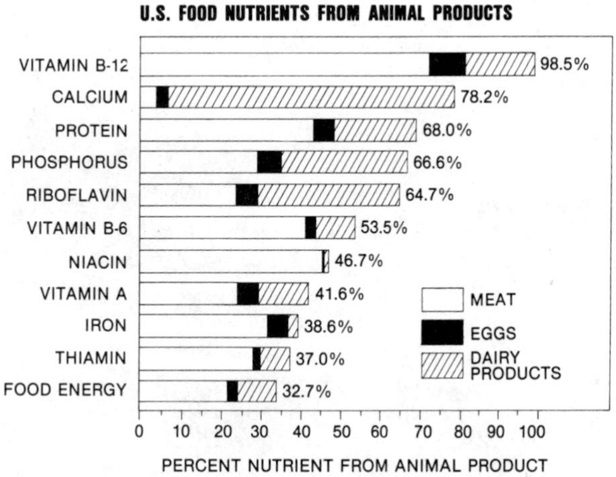

Fig. 14-2. Percentage of food nutrients contributed by animal products of the total nutrient supply in the United States in 1980.

1/3 of the total energy, nearly 4/5 of the calcium, 2/3 of the phosphorus, and significant amounts of the other minerals and vitamins needed in the human diet.

Although this section is devoted primarily to the final animal products—meat and milk—it must be remembered that the top grades of these important food constituents represent the culmination of years of progressive breeding, the best in nutrition, vigilant sanitation and disease prevention, superior care and management, and modern marketing, processing, and distribution. Thus, the efficient availability of the highest quality meat and milk is dependent upon the well-coordinated operation of the whole field of animal science. Much effort and years of progress have gone into the production of meat and milk.

PART I—MEAT

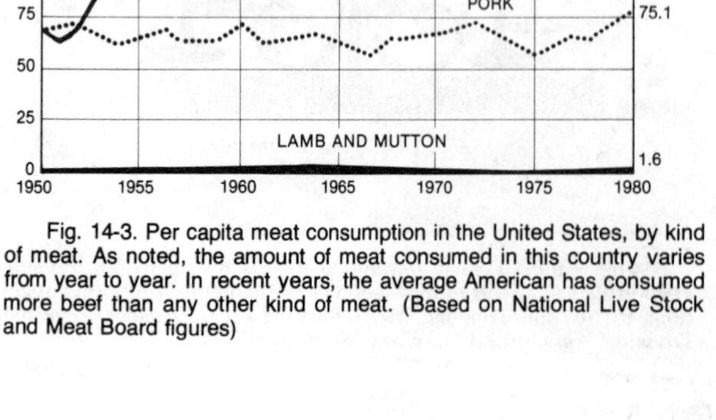

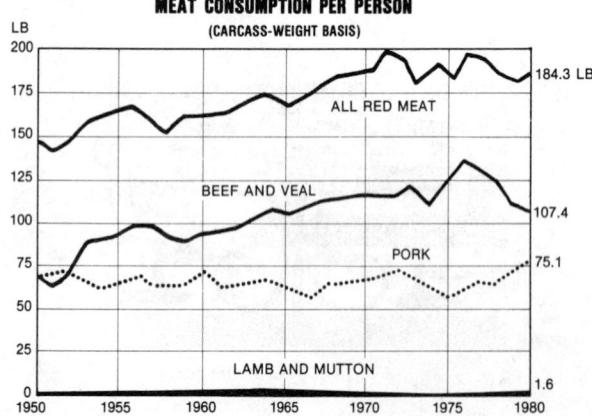

Fig. 14-3. Per capita meat consumption in the United States, by kind of meat. As noted, the amount of meat consumed in this country varies from year to year. In recent years, the average American has consumed more beef than any other kind of meat. (Based on National Live Stock and Meat Board figures)

Part I contains the latest information relative to the processing and preparation of meat for the table.

MEAT INSPECTION

With the growth of the far-flung livestock-marketing system and meat slaughtering, processing, and distribution, it soon became apparent that federal and state supervision was necessary to protect the public health. The various services rendered, legislative acts, and state and federal organizations carrying out these functions will be discussed briefly.

Federal Meat Inspection

The federal government requires supervision of establishments which slaughter, pack, render, and prepare meats and meat products for interstate shipment and foreign export; it is the responsibility of the respective states to have and enforce legislation governing the slaughtering, packaging, and handling of meats shipped intrastate, but state standards cannot be lower than federal levels. The meat inspection laws do not apply to farm slaughter for home consumption, although all states require inspection if the meat is sold.

The meat inspection service of the U.S. Department of Agriculture was inaugurated, and is maintained, under the Meat Inspection Act of June 30, 1906. This act was updated and strengthened by the Wholesome Meat Act of December 15, 1967. The latter statute (1) requires that state standards be at least to the levels applied to meat sent across state lines; and (2) assures consumers that all meat sold in the United States is inspected either by the federal government or by a state program of equal standards. The Animal and Plant Health Inspection Service of the U.S. Department of Agriculture is charged with the responsibility of meat inspection.

The purposes of meat inspection are (1) to safeguard the public by eliminating diseased or otherwise unwholesome meat from the food supply, (2) to enforce the sanitary preparation of meat and meat products, (3) to guard against the use of harmful ingredients, and (4) to prevent the use of false or misleading names or statements on labels. Personnel for carrying out the provisions of the act are of two types: professional or veterinary inspectors who are graduates of accredited veterinary colleges, and nonprofessional food inspectors who are required to pass a Civil Service examination. The inspections consist of the following two types:

1. Antemortem (before death) inspection is made in the pens or as the animals move from the scales after weighing. The inspection is performed to detect evidence of disease or any abnormal condition that

would indicate a disease. Suspects are provided with a metal ear tag bearing the notation "U.S. Suspect No. . . . ," and are given special postmortem scrutiny. If in the antemortem examination there is definite and conclusive evidence that the animal is not fit for human consumption, it is "condemned," and no further postmortem examination is necessary.

2. Postmortem (after death) inspection is made at the time of slaughter and includes a careful examination of the carcass and viscera (internal organs). All good carcasses are stamped "U.S. Inspected and Passed," whereas the inedible carcasses are stamped "U.S. Inspected and Condemned." The latter are sent to the rendering tanks, the products of which are not used for human food.

In addition to the antemortem and postmortem inspections referred to, the government meat inspectors have the power to refuse the application of the mark of inspection to meat products produced in a plant that is not sanitary. All parts of the plant and its equipment must be maintained in a sanitary condition at all times. In addition, plant employees must wear clean, washable garments, and suitable lavatory facilities must be provided for hand washing.

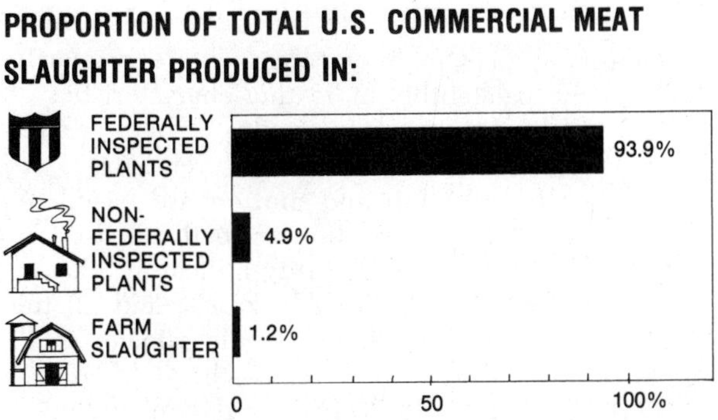

Fig. 14-4. Proportion of total U.S. meat slaughter produced in (1) federally inspected plants, (2) nonfederally inspected plants, and (3) farm slaughter, 1981. (Source: USDA, Economic Research Service)

Meat inspection regulations require the condemnation of all or affected portions of carcasses of animals with various disease conditions, including pneumonia, peritonitis, abscesses and pyemia, uremia, tetanus, rabies, anthrax, tuberculosis, various neoplasms (cancer), arthritis, actinobacillosis, and many others.

Most of the larger meat-packers are under federal inspection; hence, they are allowed to ship interstate. Fig. 14-4 shows the proportion of the total United States meat slaughter that was produced in (1) federally inspected plants, (2) nonfederally inspected plants, and (3) farm slaughter, in 1981.

State Meat Inspection

States have varying legislation governing the slaughtering and further processing of meats produced for intrastate commerce. However, the Wholesome Meat Act of 1967 requires that the state standards be equal to the federal standards. Inspection that was often formerly conducted under local ordinances is now conducted, with one exception, by state employees. The one exception is the city of Chicago, where city employees conduct inspection under the overall supervision of the state.

The Wholesome Meat Act gave the states the option of either conducting their own inspection service, or turning the responsibility over to the federal government. The federal government pays up to half the costs for states running their own inspection service. In most of these states, the service is administered by the State Department of Agriculture. Quite frequently, they simply apply the federal regulations.

MEAT GRADING

Practically all packers and retailers identify their higher grades of meats with individual, alluring private brands; either alone, or in combination with federal grades. The company's reputation depends upon consistent standards of quality for all meats that carry its brands. The brand names are also effectively used in advertising campaigns.

Federal grading of beef was first started as a special service to United States Steamship Lines in 1923; and on February 10, 1925, the 68th Congress passed an act setting up a federal meat grading service. But

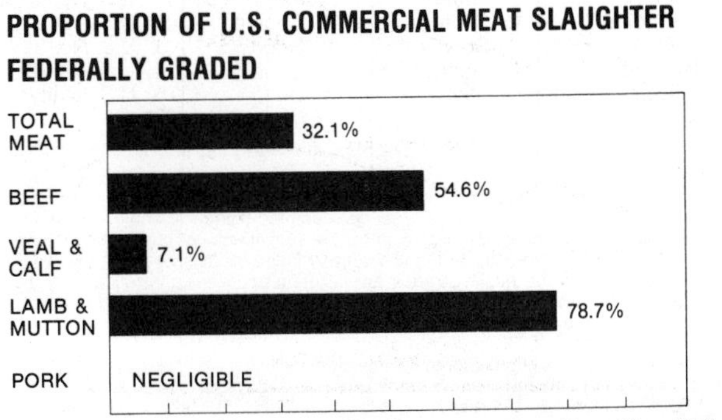

Fig. 14-5. Proportion of U.S. commercial meat production: federally graded in 1981. (Data provided by USDA, Livestock, Meat, Grain, and Seed Division, Agricultural Marketing Service)

commercial meat grading was not inaugurated until 1927. The service is on a voluntary basis, with a charge of $16.20 per hour made therefor ($19.20/hr. on Saturday and Sunday).

Federal Grades of Meats

The grade of meat may be defined as a measure of its degree of excellence based on quality, or eating characteristics of the meat, and the yield, or total proportion of primal cuts. Naturally, the attributes upon which the grades are based vary between species. Nevertheless, it is intended that the specifications for each grade shall be sufficiently definite to make for uniform grades throughout the country and from season to season, and that on-hook grades shall be correlated with on-foot grades.

Both producers and consumers should know the federal grades of meats and have a reasonably clear understanding of the specifications of each grade. From the standpoint of producers this is important, for meat over the block is the ultimate objective. From the standpoint of consumers, especially the housewife who buys most of the meat, this is important, because (1) in these days of self-service, prepackaged meats there is less opportunity to secure the counsel and advice of the meat cutter when making purchases, and (2) the average consumer is not the best judge of the quality of the various kinds of meats on display in the meat counter.

Federally graded meats are so stamped (with an edible vegetable dye) that the grade will appear on the retail cuts as well as on the carcass and wholesale cuts. These are summarized in Table 14-1.

TABLE 14-1

FEDERAL GRADES OF MEATS BY CLASSES[1]

Beef[2]	Veal	Mutton and Lamb[2]	Pork
1. Prime[3]	1. Prime	1. Prime[4]	1. U.S. No. 1
2. Choice	2. Choice	2. Choice	2. U.S. No. 2
3. Good	3. Good	3. Good	3. U.S. No. 3
4. Standard	4. Standard	4. Utility	4. U.S. No. 4
5. Commercial	5. Utility	5. Cull[5]	5. U.S. Utility
6. Utility			
7. Cutter			
8. Canner			

[1]In rolling meat, the letters U.S. precede each federal grade name. This is important as only government-graded meat can be so marked. For convenience, however, the letters U.S. are not used in this table or in the discussion which follows.

[2]In addition to the quality grades given herein, there are the following yield grades of beef and lamb (and mutton) carcasses: Yield Grade 1, Yield Grade 2, Yield Grade 3, Yield Grade 4, and Yield Grade 5. Thus, beef and lamb carcasses are graded for both quality and yield grade.

[3]Cow beef is not eligible for the Prime grade.

[4]Limited to lamb and yearling carcasses.

[5]Limited to mutton carcasses.

The word "quality" implies superiority. It follows that quality grades indicate the relative superiority of carcasses or cuts in palatability characteristics. In turn, palatability is associated with tenderness, juiciness, and flavor. The major quality-indicating characteristics are (1) color and texture of bone, which indicates the age of the animal; (2) firmness; (3) texture; and (4) marbling—the intermixing of fat among the muscle fibers.

Yield grades indicate the *quantity* of meat—the amount of retail, consumer-ready, ready-to-cook, or edible meat that a carcass contains. Yield grade is not to be confused with dressing percent. Dressing percent refers to the amount of carcass from a live animal, whereas yield grade refers to the amount of edible product from the carcass.

The attributes which determine meat grades differ widely among species. Among species grade differences are the following:

1. Eight grades are used to cover the range in quality of steer and heifer carcasses, in comparison with only five for veal, lamb, and pork.
2. In addition to quality grades, five separate yield, or cutability grades (i.e., yield of boneless, closely trimmed retail cuts) are used in conjunction with quality grades in both beef and lamb.
3. Pork grades incorporate yield and quality considerations without separate quality and yield designations.

Quality grades in beef carcasses are primarily based on the palatability-indicating characteristics of the lean. Cutability, or yield, grades of beef are determined primarily by objective measurements.

The grades of lamb and mutton are based on separate evaluations very similar to those used in beef; namely, (1) the palatability-indicating characteristics of the lean—the *quality grade*; and (2) the indicated percent of trimmed, boneless major retail cuts to be derived from the carcass—the *yield grade*.

Pork carcass grades rely heavily on objective measurements to determine expected combined yield of lean cuts, used in conjunction with a subjective determination of "acceptable" or "unacceptable" quality of the lean.

Some of the attributes upon which grades are based can be evaluated directly; others must be evaluated indirectly, through indicators. For example, the yield of lean cuts of pork is related to (or indicated by) backfat thickness, carcass weight, and carcass length.

BEEF AND VEAL

Beef exceeds pork in farm production but not in processing. Considerably less beef than pork is cured. The vast majority of it is either consumed fresh or frozen. With some minor variations, veal is slaughtered and processed in the same manner as beef.

The Dressing Percentage of Cattle and Calves

Cattle are not all beef, and beef is not all steak! It is important, therefore, that those who slaughter animals and those who purchase wholesale or retail cuts know the approximate (1) percentage yield of chilled carcass in relation to the weight of the animal on foot, and (2) yield of different retail cuts.

Figure 14-6 illustrates these points. As noted, an

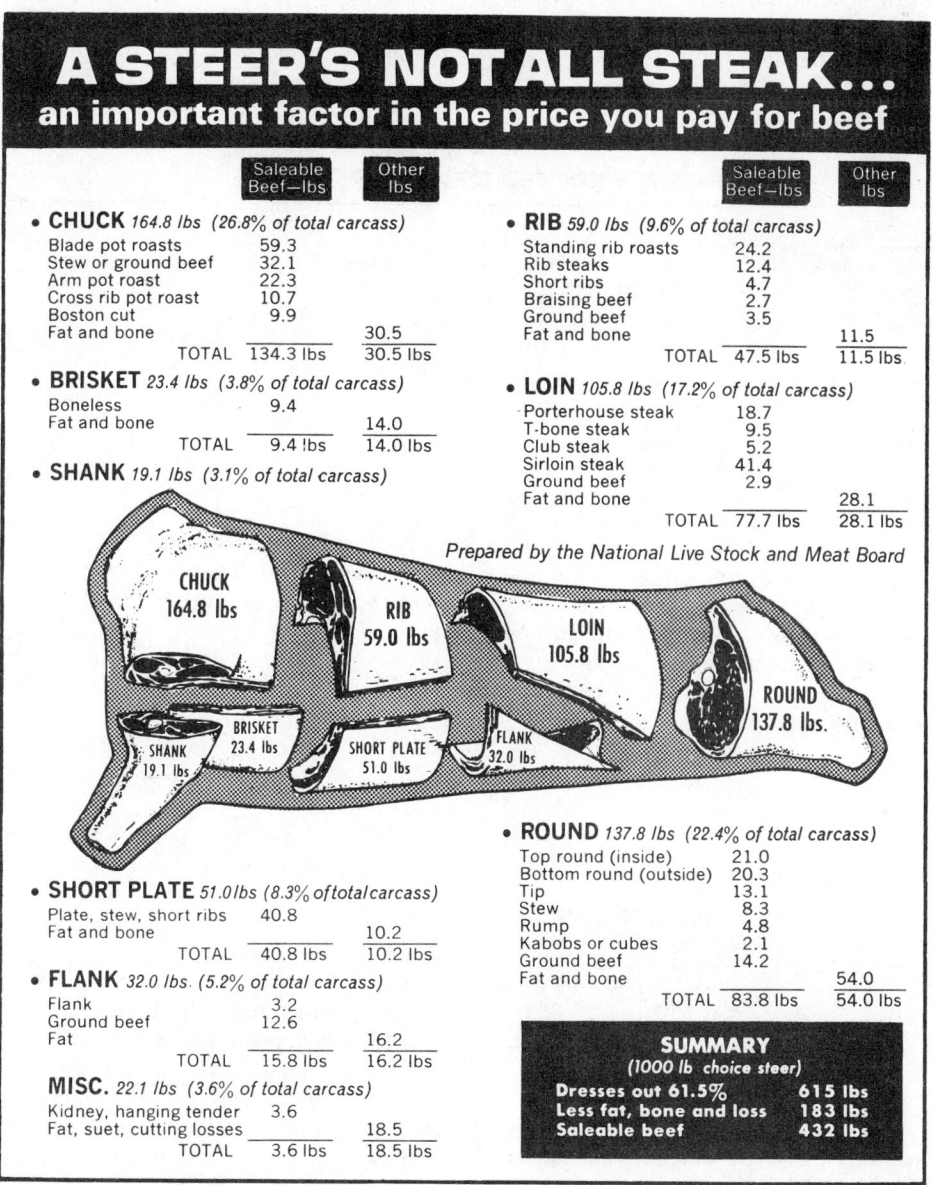

A STEER'S NOT ALL STEAK...
an important factor in the price you pay for beef

• **CHUCK** *164.8 lbs (26.8% of total carcass)*

	Saleable Beef—lbs	Other lbs
Blade pot roasts	59.3	
Stew or ground beef	32.1	
Arm pot roast	22.3	
Cross rib pot roast	10.7	
Boston cut	9.9	
Fat and bone		30.5
TOTAL	134.3 lbs	30.5 lbs

• **BRISKET** *23.4 lbs (3.8% of total carcass)*

	Saleable Beef—lbs	Other lbs
Boneless	9.4	
Fat and bone		14.0
TOTAL	9.4 lbs	14.0 lbs

• **SHANK** *19.1 lbs (3.1% of total carcass)*

• **RIB** *59.0 lbs (9.6% of total carcass)*

	Saleable Beef—lbs	Other lbs
Standing rib roasts	24.2	
Rib steaks	12.4	
Short ribs	4.7	
Braising beef	2.7	
Ground beef	3.5	
Fat and bone		11.5
TOTAL	47.5 lbs	11.5 lbs

• **LOIN** *105.8 lbs (17.2% of total carcass)*

	Saleable Beef—lbs	Other lbs
Porterhouse steak	18.7	
T-bone steak	9.5	
Club steak	5.2	
Sirloin steak	41.4	
Ground beef	2.9	
Fat and bone		28.1
TOTAL	77.7 lbs	28.1 lbs

Prepared by the National Live Stock and Meat Board

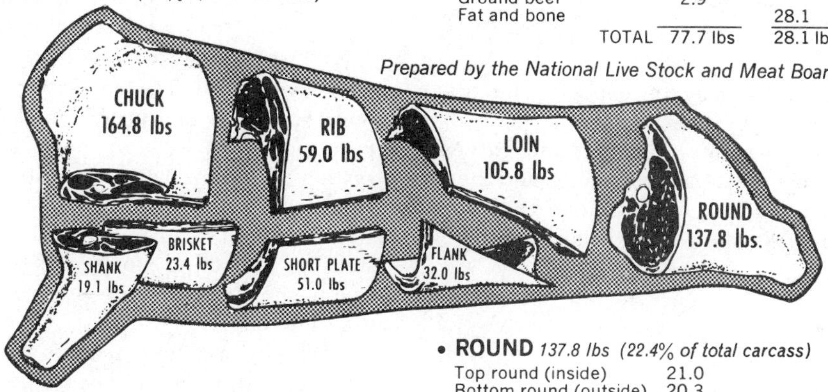

• **SHORT PLATE** *51.0 lbs (8.3% of total carcass)*

	Saleable Beef—lbs	Other lbs
Plate, stew, short ribs	40.8	
Fat and bone		10.2
TOTAL	40.8 lbs	10.2 lbs

• **FLANK** *32.0 lbs. (5.2% of total carcass)*

	Saleable Beef—lbs	Other lbs
Flank	3.2	
Ground beef	12.6	
Fat		16.2
TOTAL	15.8 lbs	16.2 lbs

MISC. *22.1 lbs (3.6% of total carcass)*

	Saleable Beef—lbs	Other lbs
Kidney, hanging tender	3.6	
Fat, suet, cutting losses		18.5
TOTAL	3.6 lbs	18.5 lbs

• **ROUND** *137.8 lbs (22.4% of total carcass)*

	Saleable Beef—lbs	Other lbs
Top round (inside)	21.0	
Bottom round (outside)	20.3	
Tip	13.1	
Stew	8.3	
Rump	4.8	
Kabobs or cubes	2.1	
Ground beef	14.2	
Fat and bone		54.0
TOTAL	83.8 lbs	54.0 lbs

SUMMARY
(1000 lb choice steer)

Dresses out 61.5%	615 lbs
Less fat, bone and loss	183 lbs
Saleable beef	432 lbs

Supply and Demand are not the only factors in the price you pay for beef. For instance, today's modern-type 1,000 lb choice steer produces an approximate 615 lb carcass which the packer sells to a retailer who trims away 183 lbs of fat, bone and waste . . . ending up with only 432 lbs of beef that he cuts, wraps and sells to customers.

Of that a surprisingly small amount is steak and a much larger quantity is roasts as shown in the chart above. Retail stores put a higher price on steak and a lower price on pot-roasts and ground beef so that they sell it all . . . not end up with only less-in-demand cuts like pot-roasts and short ribs left in the cooler.

Fig. 14-6. Cattle are not all beef, and beef is not all steak! This shows the approximate (1) percentage yield of chilled carcass in relation to the weight of the animal on foot, and (2) yield of different retail cuts. (Courtesy, National Live Stock and Meat Board, Chicago, Ill.)

average 1,000-pound steer will yield about a 615-pound carcass or 432 retail cuts,[2] only 23.4 pounds of which will be porterhouse, T bone, and club steaks.

The chief factors determining the dressing percentage of cattle are (1) the amount of fill, (2) the finish or degree of fatness, (3) the general quality and refinement (refinement of head, bone, hide, etc.), and (4) the size of udder. The better grades of steers have the highest dressing percentage, with thin Canner cows showing the lowest yield. Table 14-2 gives the dressing percentages that may be expected for different grades of cattle and calves.

TABLE 14-2

DRESSING PERCENT OF CATTLE AND CALVES, BY GRADE[1]

Cattle			Calves and Vealers		
Grade	Range	Average	Grade	Range	Average
Prime	61-66	62	Prime	61-66	63
Choice	58-64	60	Choice	57-63	59
Good	57-61	59	Good	55-59	57
Standard	55-60	57	Standard	52-57	55
Commercial	54-62	57	Utility	47-54	51
Utility	49-57	53	Cull	40-48	46
Cutter	45-54	49			
Canner	40-48	45			

[1]From USDA sources.

The average liveweights of cattle and calves dressed by commercial meat-packing plants, and their percentage yields in meats, for the year 1980 are given in Table 14-3. As shown, cattle average 59% and calves 60%.

TABLE 14-3

AVERAGE LIVEWEIGHT, CARCASS YIELD, AND DRESSING PERCENTAGES OF ALL CATTLE AND CALVES COMMERCIALLY SLAUGHTERED IN THE UNITED STATES IN 1980[1]

	Average Liveweight		Dressed Weight		Dressing Percentage
	(lb)	*(kg)*	*(lb)*	*(kg)*	*(%)*
Cattle	1,071	487	634	288	59
Calves	250	111	149	67	60

[1]Source: *Livestock and Meat Statistics, Supplement for 1980*, USDA, Statistical Bulletin No. 522, pp. 86 and 107, Tables 92 and 114.

[2]The loss of 183 pounds between the carcass and the cuts (from 615 to 432 pounds) is due to the removal of bones, normal shrinkage, fat, and meat trimmings which are always removed.

Beef and Veal Cuts and How to Cook Them

There was a time when each area of the United States had its traditional cuts of beef and veal. However, increased central processing and boxed beef prompted the need for greater uniformity in cutting and labeling, among both packers and retailers. Out of this need arose a new nationwide standardized identification-labeling system, coordinated by the National Live Stock and Meat Board and adopted by the industry in 1973. The names for various cuts of beef, pork, and lamb sold in U.S. food stores were reduced from more than 1,000 to about 300. As a result of this system of uniform labeling, a rib eye steak is a rib eye steak—not a Delmonico steak at one place, a filet steak someplace else, or a Spencer steak or beauty steak in still other stores, depending on where you live in the United States—or even where you shop in the same city.

Whether a beef carcass is cut up in the home or by an expert, it should always be cut across the grain of the muscle tissue, and the thick cuts should be separated from the thin cuts and the tender cuts from the less tender cuts.

In order to buy and/or process beef and veal wisely, and to make the best use of each part of the carcass, the consumer should be familiar with the types of cuts and how each should be processed.

Every grade and cut of meat can be made tender and palatable provided it is cooked by the proper method. Also, it is important that meat be cooked at low to moderate temperatures, usually between 300° and 325°F for roasting. At this temperature, it cooks slowly, and as a result is juicier, shrinks less, and has a better flavor than when cooked at high temperatures.

Fig. 14-7, page 1083, shows the wholesale and retail cuts of beef and gives the recommended method(s) of cooking each. Fig. 14-8, page 1084, presents similar information for veal.

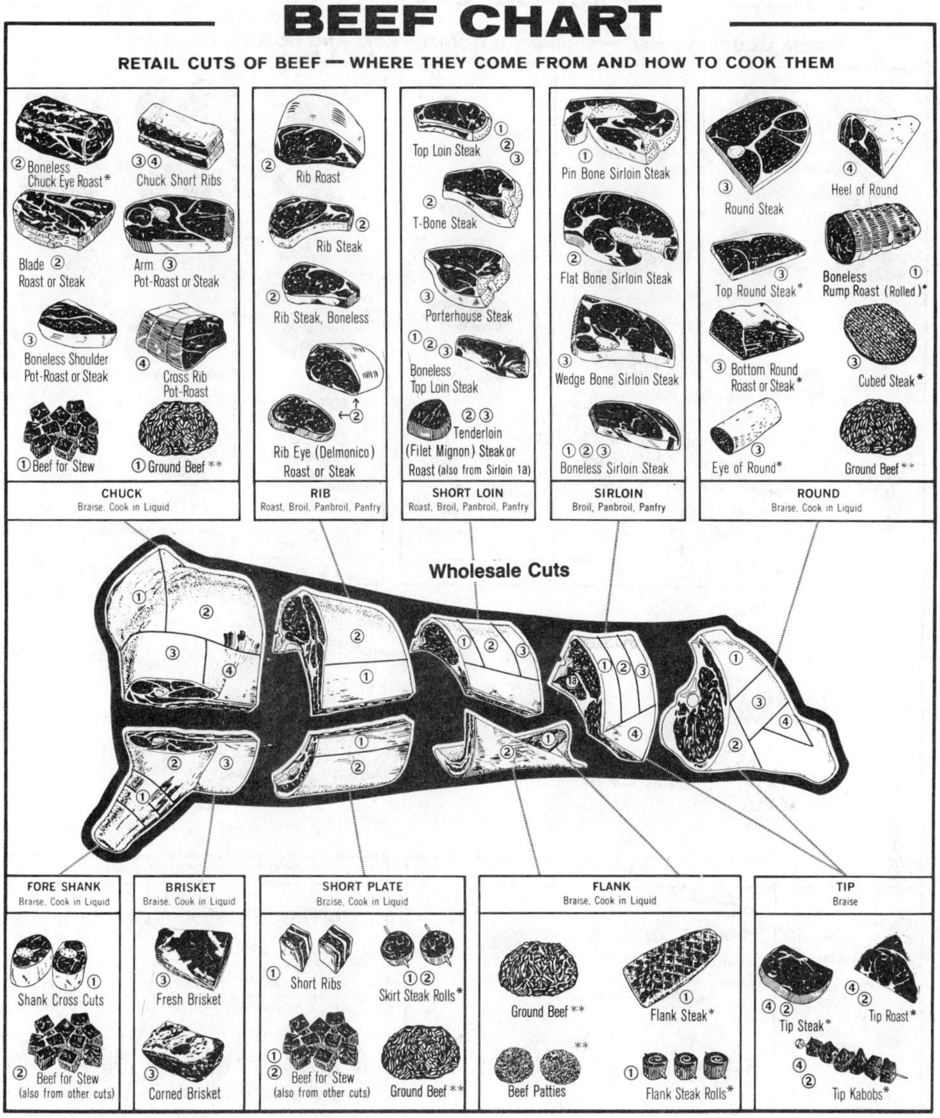

*May be Roasted, Broiled, Panbroiled or Panfried from high quality beef.
**May be Roasted (Baked), Broiled, Panbroiled or Panfried.

Fig. 14-7. The wholesale and retail cuts of beef, and the recommended method(s) of cooking each. (Courtesy, National Live Stock and Meat Board, Chicago, Ill.)

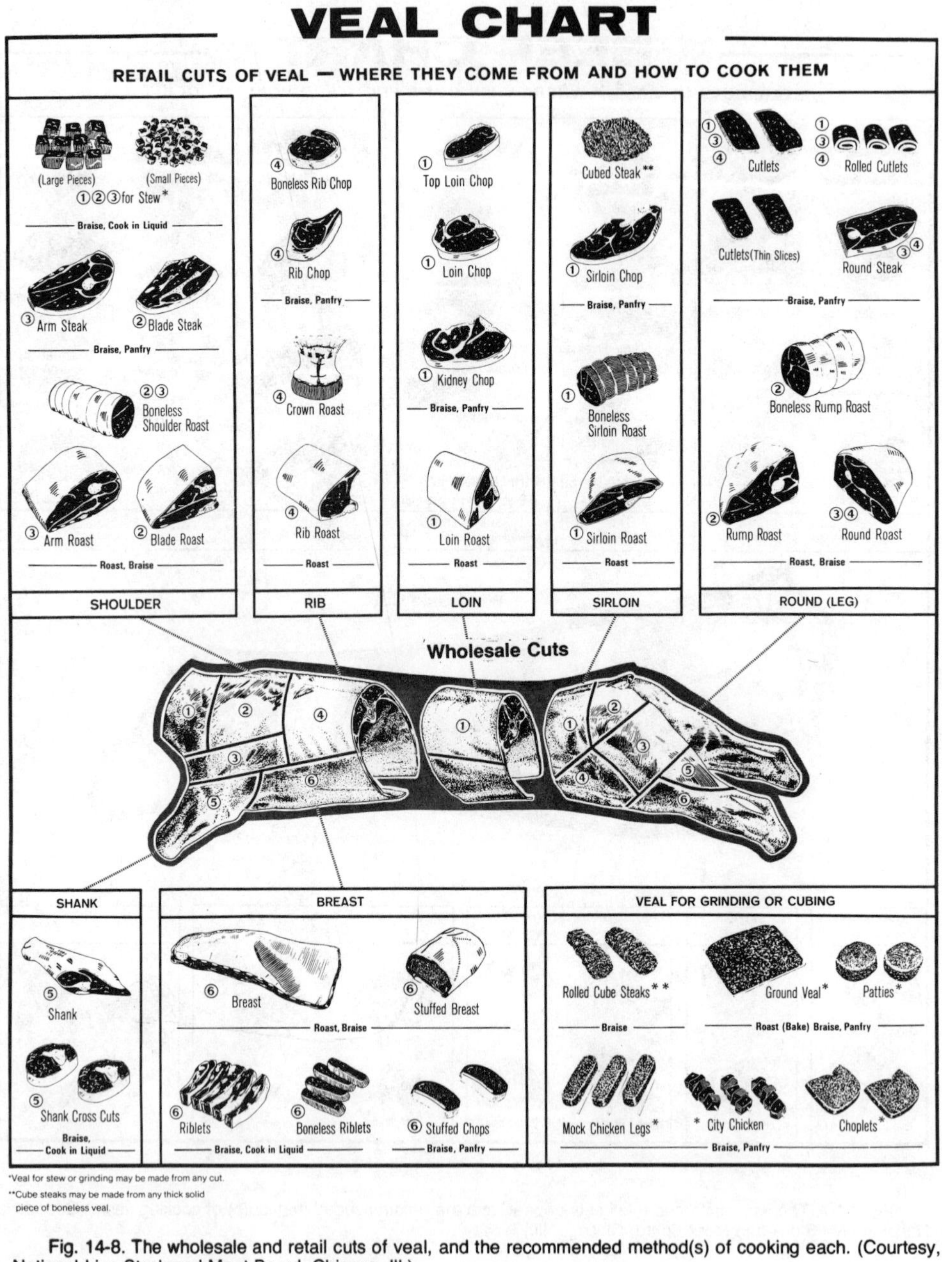

Fig. 14-8. The wholesale and retail cuts of veal, and the recommended method(s) of cooking each. (Courtesy, National Live Stock and Meat Board, Chicago, Ill.)

LAMB

Most of the sheep meat eaten in this country is lamb. Accordingly, the discussion which follows will be limited to lamb, and mutton will not be covered.

Even though lamb constitutes a smaller proportion of the total U.S. meat supply than any other class of meat, these facts are noteworthy: (1) lamb, like other meats, is easily digested, and, therefore, is widely used in the diet of convalescents, (2) there is less religious prejudice against lamb and mutton than any other meat except fish, and (3) fewer lamb and mutton carcasses (percentagewise) are condemned by meat inspectors than any other class of livestock.

The Dressing Percentage of Lambs

TABLE 14-4

DRESSING PERCENT OF LAMBS AND SHEEP (MUTTON) BY GRADE[1]

Lambs (wooled)			Sheep (excludes yearlings)		
Grade	Range	Average	Grade	Range	Average
Prime	49-55	52	Choice	49-54	52
Choice	47-52	50	Good	47-52	49
Good	45-49	47	Utility	44-48	46
Utility	43-47	45	Cull	40-46	43
Cull	40-45	42			

[1]From USDA sources.

Table 14-4 gives the dressing percentages that may be expected from the different grades of sheep and lambs. As would be expected, the highest dressing percentage is obtained when animals are slaughtered following shearing. Lambs of the mutton breeds yield a somewhat higher percentage of carcass than those of the so-called wool breeds.

The average liveweight of sheep and lambs dressed by federally inspected meat-packing plants and their percentage yield in meat for the year 1980 was as shown in Table 14-5.

TABLE 14-5

AVERAGE LIVEWEIGHT, CARCASS YIELD, AND DRESSING PERCENTAGES OF ALL SHEEP AND LAMBS COMMERCIALLY SLAUGHTERED IN THE U.S. IN 1980[1]

	Average Liveweight		Average Dressed Weight		Average Dressing Percentage
	(lb)	*(kg)*	*(lb)*	*(kg)*	*(%)*
Sheep and lambs	112	50.9	55	25.0	49.1

[1]Source: *Livestock and Meat Statistics, Supplement for 1980*, USDA, Statistical Bulletin No. 522, pp. 86 and 107, Tables 92 and 114.

Lamb Cuts and How to Cook Them

The two major wholesale cuts of lamb are the (1) hindsaddle, and (2) foresaddle. The division into hindsaddle and foresaddle is made between the twelfth and thirteenth ribs, with one pair of ribs remaining on the hindsaddle. Each of these two larger cuts comprises about 50 percent of the carcass weight.

The hindsaddle is further subdivided into the leg and loin. The foresaddle is subdivided into the shoulder, rack (rib), fore shank, and breast.

Fig. 14-9 shows the common wholesale and retail cuts of lamb, and how to cook the retail cuts.

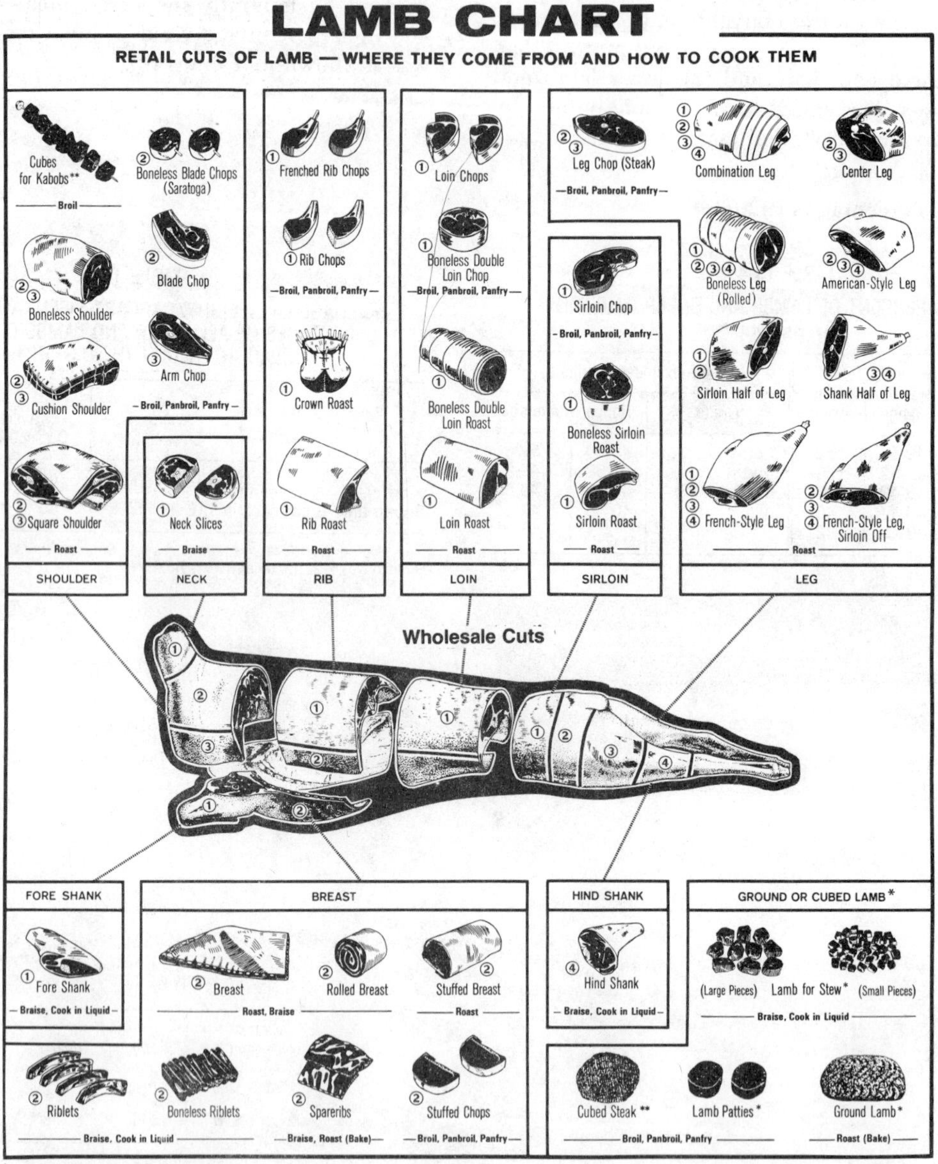

* Lamb for stew or grinding may be made from any cut.
**Kabobs or cube steaks may be made from any thick solid piece of boneless Lamb.

Fig. 14-9. Wholesale and retail cuts of lamb—where they come from, and how to cook them. (Courtesy, National Live Stock and Meat Board, Chicago, Ill.)

PORK

Generally, pork is cheaper than beef or lamb, and the lower grades are more tender and palatable than the comparable lower grades of other meats. Also, more hogs are farm slaughtered than any other class of animals.

The Dressing Percentage of Hogs

Pigs are not all pork chops! It is important, therefore, that those who slaughter hogs and the consumer know the approximate (1) percentage yield of chilled carcass in relation to the weight of the animal on foot, and (2) yield of different retail cuts. Fig. 14-10 illustrates these points.[3] As noted, on a liveweight basis only about 3⅓ percent of a pig is center cut pork chops.

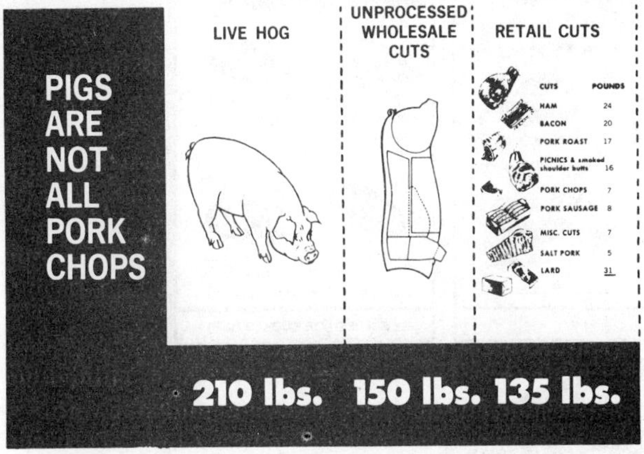

Fig. 14-10. Pigs are not all pork chops! This shows the approximate (1) percentage yield of chilled carcass in relation to the weight of the animal on foot, and (2) yield of different retail cuts. (Courtesy, American Meat Institute)

Because hogs have a smaller digestive capacity, fill is less important in determining their dressing percentage than is the case with cattle. The degree of finish and the style of dressing are the important factors affecting dressing precentage in hogs. U.S. No. 1 hogs dressed packer style (with head, leaf fat, and kidneys removed) dress about 70 percent, whereas hogs dressed shipper style (head left on, and leaf fat and kidneys in) dress 4 to 8 percent higher.

[3]There is a loss of about 15 pounds (from 150 to 135 pounds) in processing a whole pork carcass into retail cuts. This is due to the inevitable losses in boning, curing, smoking, trimming, etc.

Table 14-6 gives the approximate dressing percentages that may be expected from the different grades of barrows and gilts. It is generally recognized that fat, lardy-type hogs give a higher dressing percentage than can be obtained with meat- or bacon-type animals. Because lard frequently sells at a lower price than is paid for hogs on foot, an excess yield of lard very obviously represents an economic waste of feed in producing the animals and is undesirable from the standpoint of the processor. Accordingly, attaching great importance to the projected dressing percentage of hogs is outmoded. The more progressive buyers are now focusing their attention on the cutout value of the carcass, especially on the maximum yield of the more sought primal cuts of high quality.

TABLE 14-6

APPROXIMATE DRESSING PERCENT OF BARROWS AND GILTS, BY GRADE[1]

Grade	Range	Average
U.S. No. 1	68-72	70
U.S. No. 2	69-73	71
U.S. No. 3	70-74	72
U.S. No. 4	71-75	73
Utility	67-71	69

[1]From USDA sources.

Hogs have a relatively smaller barrel and chest cavity than cattle and sheep. In addition, they are dressed with their skin and shanks on. Consequently, they dress higher than other classes of slaughter animals.

The average liveweight of hogs, dressed packer style by federally inspected meat-packing plants, and their percentage yield in meat for the year 1980 were as shown in Table 14-7.

TABLE 14-7

AVERAGE LIVEWEIGHT, CARCASS YIELD, AND DRESSING PERCENTAGES OF ALL HOGS COMMERCIALLY SLAUGHTERED IN THE U.S. IN 1980[1]

	Average Liveweight		Average Dressing Weight		Dressing
	(lb)	*(kg)*	*(lb)*	*(kg)*	*(%)*
Hogs	242	110	172	78	71.07

[1]Source: *Livestock and Meat Statistics, Supplement for 1980*, USDA, Statistical Bulletin No. 522, pp. 86 and 107, Tables 92 and 114.

Pork Cuts and How to Cook Them

A minimum of 24 hours chilling at temperatures ranging from 33° to 38°F is necessary to remove properly the animal heat and give the carcasses sufficient firmness to make possible a neat job of cutting. After chilling, the carcasses are brought to the cutting floor where they are reduced to the wholesale cuts.

The method of cutting varies somewhat according to the value of lard and the relative demand for different cuts. Despite some variation, the most common

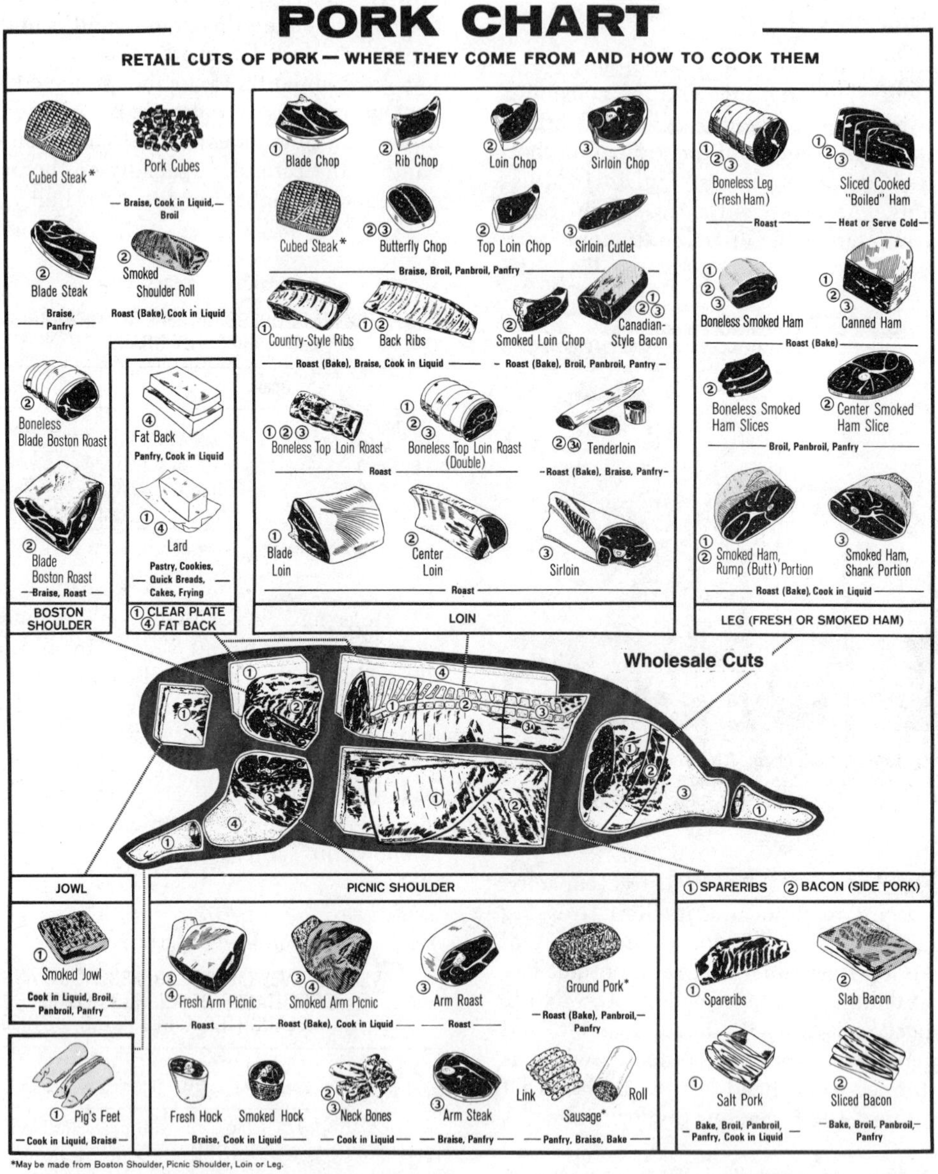

Fig. 14-11. Wholesale and retail cuts of pork—where they come from, and how to cook them. (Courtesy, National Live Stock and Meat Board, Chicago, Ill.)

wholesale cuts of pork are ham, bacon, loin, picnic shoulder, Boston butt, jowl, spareribs, and feet.

Market hogs weighing from 180 to 200 pounds will yield from 45 to 55 percent of their liveweight in the 4 primal cuts: the ham, loin, picnic shoulder, and Boston butt. However, because of the relatively higher value per pound of these cuts, they make up three-quarters of the value of the entire carcass.

MEAT PRESERVATION

From time immemorial, one of man's major food problems has been that of preserving meats over a period of time and in a condition suitable as a food. Fundamentally, meat preservation is a matter of controlling putrefactive bacterial action. Various methods of preserving meats have been practiced through the ages, the most common of which are (1) drying, (2) smoking, (3) salting, (4) freezing, (5) canning, and (6) making into sausages.

Fig. 14-12. Indians drying and smoking venison.

In this country, meat curing is largely confined to pork, primarily because of the keeping qualities and palatability of cured pork products. Considerable beef is dried or corned and some lamb and veal are cured, but none of these is of such great importance as cured pork.

No phase of meat preservation and merchandising has received greater interest in recent years than frozen meats. For this reason, further discussion is devoted to this subject in the section that follows.

Fig. 14-13. The freezer temperature should be 0°F or lower.

Freezing Meats

Freezing is not a new method of meat preservation, for in arctic regions meats have been frozen since time immemorial. But special freezing methods are important recent developments. The rapid growth of freezer lockers, and the increased popularity of food freezers, have made this method of meat preservation available to homes throughout America.

When properly prepared and stored, frozen meats resemble fresh meats in appearance, flavor, appetite appeal, and food value; and they furnish a welcome diversion to the familiar stocks of canned, salted, and cellar stored food.

Among the reasons for the increased interest in frozen meats in recent years are the following:

1. Uniform supplies of meats can be available throughout the year by freezing. This is particularly advantageous for such products as lamb, which is highly seasonable.

2. The consumer is accepting frozen foods in greater quantities, especially frozen prepared foods, fruits, and vegetables. Hence, consumer confidence in frozen food has improved; and with it, frozen meat is becoming more acceptable.

3. Home freezers are in widespread use. With frozen food space available, there is a tendency to fill it up by purchasing greater quantities of food in frozen form.

4. Improved quality control is possible in frozen meat merchandising. Cutting and packaging can be done at scheduled times and under strict supervision, rather than as dictated by the need to refill the fresh meat display case.

5. Central fabricating (cutting) is facilitated by frozen meat merchandising, thereby eliminating or greatly reducing the need for in-store cutting.

6. Improved utilization of the entire carcass results from frozen meat merchandising. Cuts do not depreciate in retail display cases as is true in fresh meat merchandising; and shrinkage and spoilage are practically eliminated. As a result, the retailer is able to merchandise all of the cuts to the best advantage.

7. No case of botulism from frozen foods is on record, and three weeks of zero storage will kill any trichinae (the parasites that cause trichinosis) in pork.

DIRECTIONS FOR SELECTING, PREPARING, AND FREEZING MEATS

The following simple directions will assure frozen meats of high quality:

1. **Selection**—Whether selecting live animals for slaughter or purchasing wholesale cuts, it is important that the following requisites be met:

a. **Healthy animals**—Meat should come from healthy aniamls. Every stockman knows, better than words can describe, how a sleek coat of hair, a bright eye, and vigor indicate animal health. In the case of carcasses or wholesale cuts that are slaughtered in either federal or state inspected plants, their respective required-by-law inspections give full assurance of the wholesomeness and cleanliness of the product.

b. **Desirable weights**—Meat should come from animals (or cuts) of a weight that will give the size steaks, chops, and roasts desired. Popular weights are: steers and heifers, 600 to 1,000 lb; calves 150 to 250 lb; lambs, 70 to 100 lb; hogs, 180 to 225 lb.

c. **Adequate but not excessive finish**—Meat should have adequate finish, without being excessively fat and wasty. Good and Choice grades of beef and lamb are most popular.

d. **High quality**—Meat should be of high quality; for freezing will retain quality, but it cannot improve it.

2. **Slaughtering**—Whether slaughtering is done on the farm, at the locker plant, or at the packing plant, a good, clean job of dressing is important. To accomplish this, animals intended for slaughter should (a) be held off feed for at least 24 hours before slaughter, (b) have access to fresh water at all times, (c) not be overheated or excited, since animals in this condition do not bleed well, and the meat will not keep as long, and (d) be protected from unnecessary bruising, which results in bloody spots on the carcass that must be trimmed out and wasted.

3. **Chilling**—Immediately after slaughter, carcass meat should be hung in a clean place where the temperature will neither be freezing nor above 36°F, preferably at a chill room temperature of 32° to 34°F. Chilling removes body heat (the temperature of freshly slaughtered meat is around 100°F) and arrests the destructive growth of bacteria, molds, and yeasts which multiply rapidly at temperatures around 70°F, but slowly at temperatures between 32° and 34°F. Improper chilling (too low temperature) may ruin the flavor of the meat or even render it unfit for use.

Be sure that meat is not exposed to odors or contaminations of any kind before or during chilling as warm flesh is receptive to musty odors, paints, and disinfectants.

4. **Aging meat**—Pork and veal should be cut, packaged, and frozen as soon as they are thoroughly chilled; usually pork is cut within 48 hours after slaughter. This is especially important with pork because it has a tendency to become rancid rather quickly.

Beef and lamb improve in flavor and tenderness if allowed to hang in temperatures ranging from 34° to 38°F. During this time, the enzymes in the meat carry on a self-digestion of the connective tissues, which is a tenderizing process called aging or ripening.

However, aging of meat that is subsequently frozen does not appreciably enhance its tenderness. This is so because freezing within itself is a tenderizing process due to cellular breakdown. Besides, aging results in more rancidity and a shorter storage life. Therefore, it may be concluded that even beef and lamb should not be aged unless they are to be consumed immediately following storage in the freezer.

When beef and lamb are to be aged, length of the aging period should be determined by the kind, size, quality, and finish of the meat. In general, however, 7 to 10 days appears to be sufficiently long for beef of average finish, and 5 to 7 days for lamb. Beef and lamb carcasses that are not well protected with fat should not be aged over 5 days.

The inexperienced person should seek the counsel of an expert in determining whether to age or how long to age a particular carcass or wholesale cut.

5. **Cut meat to family preferences**—The size of roasts, number and preferred thickness of steaks, amount of ground meat, proportion of fat in the sausage, and the size package should be determined by the size family and their preferences.

Boning meat (except for steaks and chops) is recommended because as much as 25 percent less storage space is required, and the danger of the bones puncturing the wrapping is eliminated.

When having meat cut by an expert, one should also plan ahead for company meals. Fancy cuts, such as double pork chops with a pocket for dressing, crown roast of lamb, and other fancy cuts can be prepared.

6. **Wrapping**—The quality of frozen meat is determined to a considerable extent by the proper selection of wrapping paper and the method of wrapping. Improper wrapping results in "drying out" or "freezer burn" of meat, which affects both the appearance and the flavor of the frozen product.

Numerous types of papers, bags, and cartons may be and are successfully used in wrapping meats for storage. The perfect wrapper should be moisture and vaporproof; easy to fold, wrap, and handle; tough enough to resist tearing; capable of receiving ink or China pencil labeling; and inexpensive. It should be added that, since no existing wrapper or container possesses all of these qualities, each processor must select the one or ones best suited to his conditions. *Regular butcher paper, ordinary waxed paper, or grocery bags should never be used.*

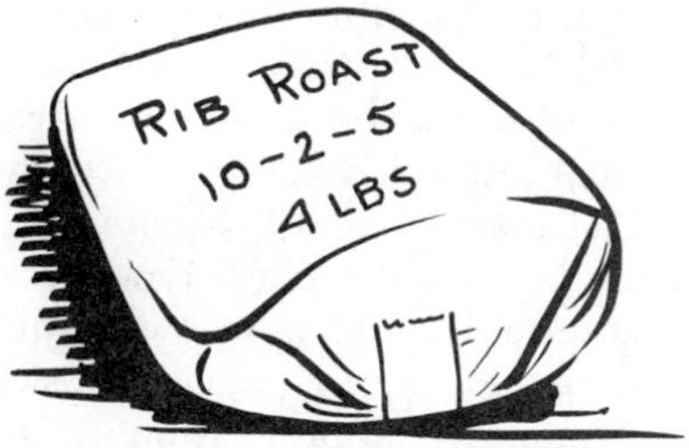

Fig. 14-14. A package of meat properly labeled. (Drawing by Steve Allured)

Some additional meat wrapping tips are:

a. **Keep cool**—Keep meat cool both during and after wrapping.

b. **Separate cuts**—Place two sheets of waxed, or otherwise waterproofed, paper between steaks, chops, or hamburger patties that are to be wrapped together. This will (1) permit separating them without first thawing the entire package, and (2) facilitate broiling or pan frying from the frozen state.

c. **Package for each meal**—In general, each package should contain the meat needed for a single meal, but if the portions are separated as described (in point b.), one or more steaks, chops, or hamburgers can be removed and the package resealed and placed in the freezer.

d. **Label properly**—Label each package with a special waterproof ink or China marking pencil in order to be able to locate with certainty the desired kind of meat weeks and months later. The label should give the kind and cut of meat enclosed, the number of possible servings (including weight if convenient), and the date of packaging.

e. **Freeze immediately**—Start freezing immediately following wrapping and labeling. If it is not practical to place the packages in the freezer at once, put them in a household refrigerator until they can be removed to the freezer.

f. **Avoid refreezing if possible**—Refreezing results in loss of juices when rethawed, thereby affecting juiciness and flavor.

7. **Freezer temperatures**—The freezer temperature should be 0°F or lower.

8. **Seasoning**—In general, meats intended for freezing should not be salted. On the other hand, spices appear to enhance the keeping qualities of meats.

9. **Curing**—It is not advisable to freeze cured meats because the salt therein favors the development of rancidity upon freezing. When freezing of cured meat is necessary, the storage time should not exceed 60 days.

HOW LONG CAN MEAT BE FROZEN?[4]

This is a controversial subject. It depends upon many factors, such as proper slaughtering, chilling, aging, wrapping, slicing, and the speed with which meat is frozen after wrapping. On the average, however, it appears that meat may be safely stored at 0°F for the periods shown in Table 14-8.

TABLE 14-8
STORAGE GUIDE FOR FROZEN MEATS

Product	Storage Period	Comments
	(month)	
Beef	6 to 12	
Fresh pork	4 to 6	
Lamb and veal	6 to 9	
Ground beef and lamb	3 to 4	
Ground pork	4 to 6	
Sausage	3	
Liver, heart, etc.	3 to 4	
Ready-cooked beef stew, without potatoes	6	**To prepare:** Remove from package. Heat one hour on low heat in covered saucepan, or in casserole for one hour in oven at 325°F.
Baked ham	3 to 4	**To prepare:** Thaw; then heat 15 minutes per pound in 350°F oven.
Ready-cooked meat loaf	6	**To prepare:** Heat for 1 to 1½ hours at 350°F or thaw at room temperature and serve cold.

[4]An interesting historical sidelight on the length of time that meat may be preserved by freezing is found in the discovery of elephants, once hunted by primitive man, frozen in ice cliffs along the coast of Siberia. Perfectly preserved in their airtight tombs, these elephants have been kept in frozen storage for unnumbered thousands of years.

HOW TO THAW AND COOK FROZEN MEATS

Frozen meat may be either (1) removed from frozen storage, unwrapped, and cooked from the frozen state, or (2) left wrapped and thawed before being cooked.

The following pointers are pertinent to this subject:

1. **Thawing time**—The following rules of thumb may be used in estimating the thawing time:

In a refrigerator .... 5 to 10 hours per pound.
Room temperature ...... 2 to 3 hours per pound.
Before a fan .... 1 to 1½ hours per pound.

Unless meat is thawed slowly, there will be excessive loss of juices and the cooked product will be dry and tasteless.

2. **Increased roasting time from frozen state**—In roasting meat from the frozen state, add an additional 20 minutes per pound of meat. Better yet, if this procedure is followed, (a) remove the frozen meat from the freezer, (b) place it in the oven, and (c) when it is completely thawed, place a meat thermometer in the thickest part to determine desired doneness.

3. **Increased broiling time from frozen state**—In broiling steaks from the frozen state, add an additional 20 to 30 minutes.

4. **Cooking time**—The cooking time for thawed meat is the same as for fresh meat.

5. **Ground meats and heart**—Such meats are usually thawed completely before cooking.

Methods of Curing Pork on the Farm

Farm meat curing other than freezing is largely confined to pork, primarily because of the keeping qualities and palatability of cured pork products. Accordingly, the discussion in this section will be limited to pork only.

The secret of pork curing is to use good sound meat, the correct curing method and formula, clean containers, and to be fortunate enough to secure cool curing weather.

The primary meat curing ingredients and the functions of each are:

1. **Salt**—It preserves by inhibiting or retarding the bacteria that cause spoilage. In excess quantities, salt makes meat less palatable.

2. **Sugar**—Sugar is used to offset some of the harshness of salt. A combination of 7 pounds of salt and 3 pounds of white or brown sugar is a basic mixture.

3. **Saltpeter** (nitrate of potash)—It is used to enhance the bright red color of the lean meat. Also, and most important, nitrite in meat ensures against the development of botulism.

4. **Commercial cures**—In addition to salt, or salt and sugar, commercial cures frequently contain spices and flavorings to impart characteristic flavor, appearance, and/or aroma.

NOTE WELL: There is considerable concern over nitrite-cured bacon, which produces low levels of nitrosamines when cooked at high temperatures, because nitrosamines have been found to be carcinogenic in rats. An expert panel appointed by the U.S. Secretary of Agriculture has recommended that nitrates be eliminated entirely from meat cures, that the initial use of nitrite be limited to 156 ppm, and that the residual nitrite be restricted according to the class of product from 50 ppm in sterile canned meats up to 125 ppm in cooked sausage types like frankfurters. The desire and intent is to lower the levels of nitrite so cooking does not produce nitrosamines, yet maintain enough nitrite to prevent botulism.

Table 14-9 is a Pork Curing Guide for curing pork on the farm.

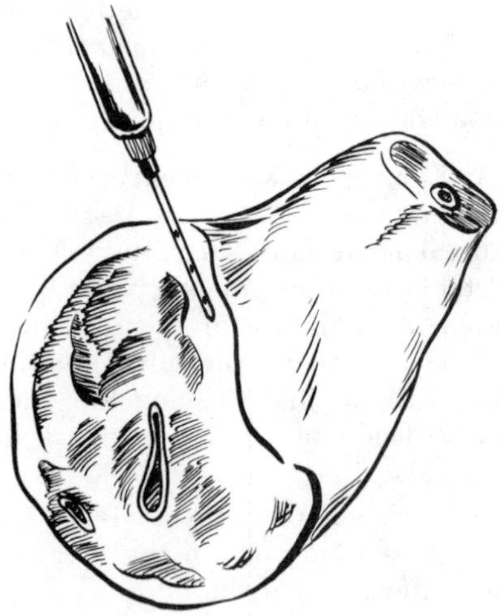

Fig. 14-15. Tissue pumping a ham. (Drawing by Steve Allured)

TABLE 14-9

PORK CURING GUIDE

Curing Method and Formula	Kinds of Cuts to Which Adapted	Curing Directions	Comments
1. **Dry-salting:** Salt only	Backs, sides, and other cuts of pork; especially heavy fat backs and bacons.	Rub and sprinkle over the cuts 7 to 10 pounds of salt per 100 pounds of meat. Pile meat closely in layers and resalt at intervals. Let cure for 3 to 4 weeks. A curing temperature of about 40°F is preferable.	Dry-salting was formerly the most common method of preserving farm pork, but, for the most part, it has now been replaced by the dry-cure and sweet-pickle methods.
2. **Dry-cure (or dry-sugar cure):** Mixture per 100 lb meat: 7 lb table salt 3 lb sugar (brown, or granulated sugar)	Ham, bacon, and shoulders.	**For bacon,** (1) rub on 1 ounce of the dry-cure mix per pound of bacon (sprinkling any surplus of this amount on the rib side), and (2) let cure 7 days per inch of thickness (thus a side of bacon 2 inches thick would be left in cure 14 days). **For hams,** (1) rub on 1 ounce (if hams weigh over 20 pounds use 1¼ to 1½ oz) of the dry-cure mix per pound of ham, applying in 3 rubbings at 3 to 5 day intervals, and (2) let cure 7 days per inch of thickness (as measured directly back of the aitch bone). **For shoulders,** (1) rub on 1 ounce of the dry-cure mix per pound of shoulder, applying in 2 rubbings at 3 to 5 day intervals, and (2) let cure 7 days per inch of thickness. Meats in cure should be placed on a clean shelf, on a table, or in a barrel with a drain at the bottom so the juice can drain off.	The dry-cure produces more palatable products than dry-salting. In comparison with the sweet-pickle cure, the dry-cure is more rapid, requires less equipment, results in a higher shrink, and gives a stronger cure. It is the preferred cure in warm areas. Curing should not be attempted where the temperature is over 50°F because spoilage is apt to take place before the cure can penetrate the ham or shoulder.
3. **Immersed in brine method (or sweet-pickle cure):** Mixture per 100 lb meat: 7 lb table salt 3 lb sugar (brown or granulated sugar) 5 gal water (8 lb/gal, boiling water or unboiled, cold water—depending on its purity) As noted, the ingredients are the same in the dry-cure and in the sweet-pickle cure; the difference being that in the dry-cure the mixture is applied directly to the meat while in the brine cure the mixture is dissolved in water and the meat submerged in it.	Hams, bacon, and shoulders.	Weigh out the ingredients of the formula (water may be measured), and stir until thoroughly mixed. Pack the meat skin side down into either a crock or a clean, well-soaked odorless, hardwood barrel. Place the thicker and heavier cuts in the bottom of the container. Pour cool pickle or brine over the meat until the liquid just covers it. Four to five gallons of pickle will cover 100 pounds of meat. Place a weight, such as a lid, on top of the meat to hold it under the liquid, and weight it down with a clean stone or other suitable object. Let the meat cure in the pickle 11 days per inch of thickness. Thus the thinner and lighter cuts on the top of the container must be removed first. When large vats are used, overhauling (meaning the rehandling or repacking) of the meat is recommended in order to permit a more uniform distribution of the pickle, but this is not necessary with the small quantities cured on most farms. If the brine should spoil (as evidenced by cloudiness), remove all the cuts, wash them with cold water, and place them in a new brine with ⅔ of the original ingredients.	A combination of salt and water is called a brine or pickle; and with the addition of sugar, it is known as a sweet-pickle cure. In comparison with the dry-cure, the sweet-pickle cure results in a milder flavor and gives less shrinkage. For best results, the curing-room temperature should not rise above 40°F; temperatures above 50°F are too high for safe pickle curing. A salometer or salinometer is a ballasted, glass vacuum tube graduated in degrees, which is sometimes used for testing the strength or salinity of pickle. A test of about 75°F is preferred under most conditions for storage without refrigeration; 65°F where refrigerated storage is available.

(Continued)

TABLE 14-9 (Continued)

Curing Method and Formula	Kinds of Cuts to Which Adapted	Curing Directions	Comments
4. **Pumping, followed by either dry-cure or sweet-pickle cure:** By means of a plunger-type syringe, sweet-pickle cure (see formula under "sweet-pickle cure") is injected into the ham or shoulder and then the ham or shoulder is subjected to either the dry-cure or sweet-pickle cure for the balance of the curing.	Hams.	Mix the sweet-pickle formula as directed under "sweet-pickle cure." Inject into the center of the ham an amount of sweet-pickle cure equivalent to 8 to 10 percent of the weight of the cut. Following pumping, complete the curing process by either the dry-cure or sweet-pickle cure methods, but *lessen the curing time by ⅓.*	Pumping hastens the curing process in the center of the cut and lessens spoilage. Meat-packers use a modified pumping method for hams, known as the artery cure. In this process, the sweet-pickle is injected into the femoral artery under 25 pounds of pressure, and the subsequent dry-cure or sweet-pickle cure is limited to 2 weeks. The advantages of the artery cure are speed and uniform flavor.
5. **Smoking:** Hickory, oak, hard maple, and apple are favorite smoke woods. Hardwood sawdust is also satisfactory. Two methods of smoking are: 1. *For light mahogany smoke,* smoke 24 to 48 hours at a temperature of approximately 125° to 135°F. 2. *For meat that is stored for summer use,* smoke at a temperature of 80° to 100°F at intervals of approximately 5 to 10 days, over a period of several weeks.	Hams, bacon, and shoulders.	If the meat has been either dry-cured or sweet-pickle cured, prior to smoking, it should be (1) soaked from ½ to 3 hours in cold water (using the longer soaking period for the heavier cuts), (2) scrubbed with a clean stiff brush, and (3) hung to dry overnight in the smokehouse. This removes the excess salt on the outside and alleviates the formation of salt streaks. Hang the cuts so that they will not touch each other, since this will cause streaking. After meat is smoked, many people like to season it heavily with black pepper.	Smoking is a common practice in the home-curing of pork, and most packer dry-cured and sweet-pickled cured pork cuts are smoked, as well as many items of sausage. Smoking adds flavor, makes for a more desirable appearance, and improves the keeping qualities of meats. The higher the temperature, the greater the absorption of smoke and the darker the color. After the smoked meat has cooled, it should be carefully wrapped with heavy parchment paper, put into muslin bags, and hung in a dry, dark, cool, well-ventilated place.

METHODS OF COOKING MEATS

Every grade and cut of meat can be made tender and palatable provided it is cooked by the proper method. Also, it is important that meat be cooked at low temperature, usually between 300° and 350°F. At this temperature, it cooks slowly, and as a result it is juicier, shrinks less, and is better flavored than when cooked at high temperatures.

The method used in meat cookery depends on the nature of the cut to which it is applied. In general, the types of meat cookery may be summarized as follows:

1. **Dry-heat cooking**—Dry-heat cooking is used in preparing the more tender cuts; those that contain little connective tissue. The common methods of cooking by dry heat are: (a) roasting, (b) broiling, and (c) panbroiling (see Fig. 14-16).

a. **How to roast:**

(1) Season with salt and pepper, if desired.

(2) Place fat side up on rack in open roasting pan.

(3) Insert thermometer.

(4) Roast in oven at 300° to 350°F.

(5) Do not add water, nor cover, nor baste.

(6) Roast until the meat thermometer registers rare, medium, or well-done, as desired.

For best results, a meat thermometer should be used to test the doneness of roasts

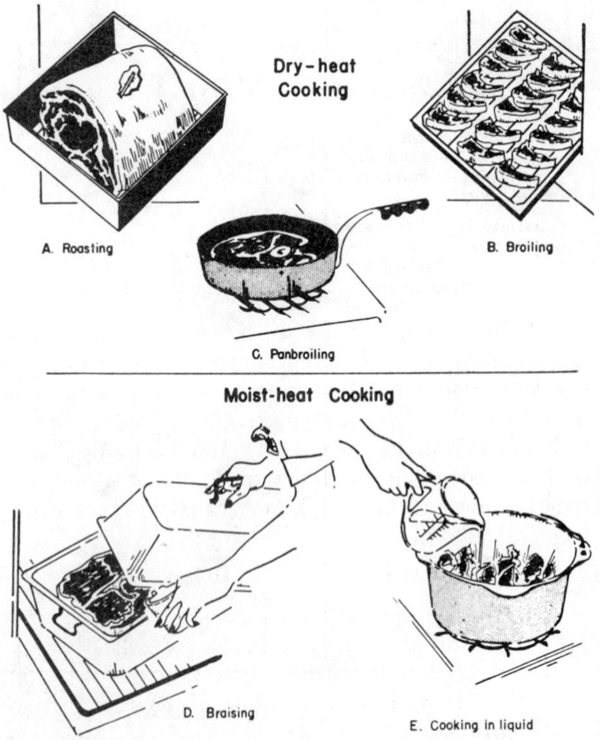

Fig. 14-16. Common methods of meat cookery. *Dry-heat cooking*: A, roasting; B, broiling; and C, panbroiling. *Moist-heat cooking*: D, braising; and E, cooking in liquid. (Drawing by R. F. Johnson)

(and also for thick steaks and chops). It takes the guess work out of meat cooking. Allowing a certain number of minutes to the pound is not always accurate; for example, rolled roasts take longer to cook than ones with bones.

The thermometer is inserted into the cut of meat so that the bulb reaches the center of the largest muscle, and so that it is not in contact with fat or bone. Naturally, frozen roasts need to be partially thawed before the thermometer is inserted, or a metal skewer or ice pick will have to be employed in order to make a hole in frozen meat.

As the oven heat penetrates, the temperature at the center of the meat gradually rises and is registered on the thermometer. The meat can be cooked as desired—rare, medium, or well-done, except for pork which should always be cooked well-done (160° to 170°F for cured pork; 185°F for fresh pork).

b. **How to broil:**

(1) Set the oven regulator for broiling.

(2) Place meat on the rack of the broiled pan and cook 2 to 5 inches from heat.

(3) Broil until the top of meat is brown.

(4) Season with salt and pepper.

(5) Turn the meat and brown the other side.

(6) Season and serve at once.

c. **How to panbroil:**

(1) Place meat in a heavy, uncovered frying pan. Cook slowly.

(2) Do not add fat or water.

(3) Turn at intervals to ensure even cooking.

(4) As fat accumulates, pour it off.

(5) Brown meat on both sides.

(6) Do not overcook. Season.

2. **Moist-heat cooking**—Moist-heat cooking is generally used in preparing the less tender cuts, those containing more connective tissues that require moist heat to soften them and make them tender. In this type of cooking, the meat is surrounded by hot liquid or by steam. The common methods of moist-heat cooking are: (a) braising, and (b) cooking in liquid (see Fig. 14-16).

a. **How to braise:**

(1) Brown the meat on all sides in a small amount of hot fat in a heavy utensil.

(2) Season with salt and pepper.

(3) Add small amount of liquid, if necessary.

(4) Cover tightly.

(5) Cook at simmering temperature, without boiling, until tender.

b. **How to cook in liquid (large cuts and stews):**

(1) Brown on all sides in hot fat, if desired.

(2) Season with salt and pepper, if desired.

(3) Cover with water and cover kettle tightly.

(4) Cook slowly (simmer but not boil) until done.

(5) Add vegetables just long enough before serving to be cooked, if desired.

PACKINGHOUSE BY-PRODUCTS FROM SLAUGHTER

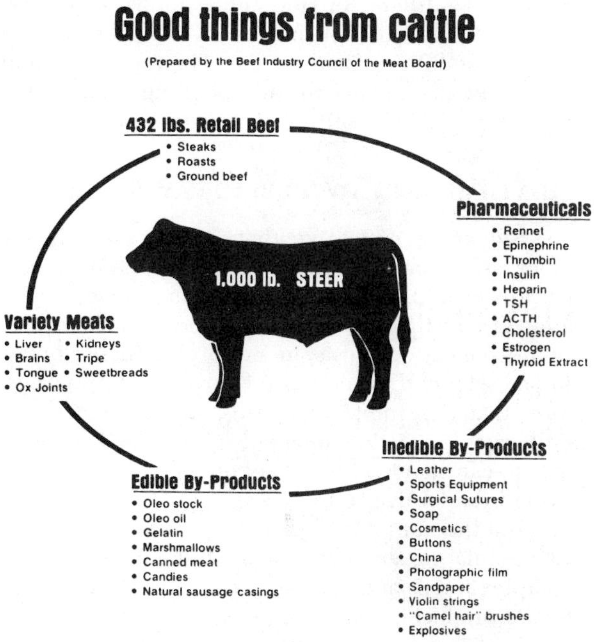

Fig. 14-17. Many good things come from cattle in addition to about 432 pounds of steaks, roasts, and hamburger normally yielded by a 1,000-pound steer. Several of these products are shown in the above figure. (Courtesy, National Live Stock and Meat Board, Chicago, Ill.)

The meat or flesh of animals is the primary object of slaughtering. The numerous other products are obtained incidentally. Thus, all products other than the carcass meat are designated as by-products, even though many of them are wholesome and highly nutritious articles of the human diet.

The relative value of carcass meat and by-products varies, both according to the class of livestock and from year to year. It is estimated that packers retreive the following percentages of the live cost of different classes of slaughter animals from the value of the by-products: 1,000-lb Choice steer, 8.21%; 100-lb Choice lamb, 13.63%; and 235-lb hog,

10.59%. Sheep pelts and cattle hides alone account for 73 and 40%, respectively, of the value of lamb and beef by-products.[5]

In contrast to the four early-day by-products—hide, wool, tallow, and tongue—modern cattle slaughter alone produces approximately 80 by-products which have a great variety of uses. Although many of the by-products from cattle, sheep, and hogs are utilized in a like manner, there are a few special products which are peculiar to the class of animals (e.g., wool and "catgut" from sheep).

The complete utilization of by-products is one of the chief reasons why large packers are able to compete so successfully with local butchers. Were it not for this conversion of waste material into salable form, the price of meat would be much higher than under existing conditions. In fact, under normal conditions, the wholesale value of the carcass is about the same as the cost of the animal on foot. The returns from the sale of by-products cover all operating costs and return a reasonable profit.

MEAT IMPORT LAW (PUBLIC LAW 88-482)

Stockmen recognize that, because of cheaper labor (and often cheaper feed supplies), farmers in the surplus meat producing countries can produce meat at a lower cost than the American producer. Transportation distances and costs are not prohibitive in obtaining meat from these countries. It would appear, therefore, that only protective walls—tariffs, quotas, and embargo legislation enacted by the U.S. Government—can stand in the way of increased meat competition from foreign sources.

But the question of regulating the supply of meat, like regulating any other commodity, raises extremely complex issues and makes for conflicts of interest between producers and consumers.

In the early 1960s, cattlemen became deeply concerned over rising imports of beef. Instead of holding at the historical 1% to 5% level of U.S. beef production, imports spiraled to more than 10% of U.S. production. A number of factors caused the increased imports, the most significant of which was that the U.S. market was the most lucrative outlet in the world. Although the U.S. State Department favored voluntary agreements with major exporters, American cattlemen insisted on more stringent regulations. Finally, Congress enacted the Meat Import Law (Public Law 88-482) in August, 1964, to become effective January 1, 1965. This bill provided for import quotas, based on a formula, for fresh, chilled, and frozen beef, veal, mutton (not lamb), and goat meat—including both carcass and boneless meat. It did not include lamb or canned meats. The law was for the purpose of limiting annual imports of the specified meats to a level comparable to the designated base period 1959-1963, with an adjustment, or "growth factor," based on changes in domestic production relative to the base period.

The base quota was established as the average annual quantity imported during the base period (1959-1963), which was 725,400,000 pounds, or 4.6% of domestic production during those years. Each year, the growth factor is determined by calculating the percentage by which the estimated U.S. commercial production of the specified meats in the current calendar year and the 2 preceding years (i.e., a 3-year average) exceeds (or falls short of) the average annual U.S. production during the base period. Thus, to determine the growth factor at the beginning of a given year, it is necessary to estimate production for that year. Then, the calculated growth factor is multiplied times the base quantity to determine the amount of increase (or decrease) in the base. This increase (or decrease) added to the base gives an adjusted base quota. With U.S. production on the increase, the annual adjustment has always been upward. The Act allows a 10% leeway above the adjusted base quota before quotas are applied to individual countries. Thus, a "quota trigger point" is determined at 110% of the adjusted base quota.

The Secretary of Agriculture is required to estimate at the beginning of each quarter year the quantity of prospective imports. If the estimated quantity of prospective imports exceeds the trigger point, the President is required to invoke a quota on imports of these meats. In case quotas are imposed, the total import quota is allocated among the countries from whom the United States is importing on the basis of shares supplied by those countries during a representative period.

The law does contain provision under which the President is empowered to suspend or increase quotas when he deems it in the best interest of the nation to do so because (1) of overriding economic or national security interest, (2) the supply of meat is inadequate to meet domestic demand at reasonable prices, or (3) international agreements have been entered into which will have the same effect as the Act.

It appears that voluntary agreements will be the chief means of controlling future imports. But the fact that a law exists may have considerable psychological effect on negotiations.

The current quotas and tariffs of live animals and meats are given in Table 14-10.

[5]Figures provided by Dr. James L. Pearson, Deputy Director, Economic Research Service, USDA, Washington, D.C., in a personal communication to the author.

TABLE 14-10

U.S. QUOTAS AND TARIFFS OF LIVE ANIMALS AND MEATS[1]

Import Item	Quotas (no head/ year)[2]	Tariff (or duty)		
		1[3]	LDDC[4]	2[5]
Animals for breeding[6]	None	Free		Free
Cattle:				
Cattle weighing:				
under 200 lb	200,000	1.0¢/lb		2.5¢/lb
between 200 and 700 lb		1.0¢/lb		2.5¢/lb
Dairy cattle weighing:				
over 700 lb		Free		3.0¢/lb
Other cattle[7]	400,000	1.0¢/lb		3.0¢/lb
Beef and Veal (fresh, chilled, or frozen)	(See footnote 8)	2.0¢/lb		6.0¢/lb
Sheep:				
Live sheep		Free		$3/head
Mutton		2.0¢/lb	1.5¢/lb	5.0¢/lb
Lamb		0.5¢/lb		7.0¢/lb
Goats:				
Live goats		$1.50/head		$3/head
Goat meat		1.2¢/lb	Free	5.0¢/lb
Swine:				
Live hogs		Free		2.0¢/lb
Pork		Free		2.5¢/lb
Horses:				
Valued under $150/head		Free		$30/head
Valued over $150/head		Free		20% ad valorem tax

[1]*Tariff Schedules of the United States Annotated (1983)*, USITC Publication 1317, United States International Trade Commission, Washington, D.C. 20436.
[2]Includes Canada, Mexico, and all other countries.
[3]Products of Canada and all other coutries not designated LDDC or 2.
[4]Products of Least Developed Developing Countries.
[5]Products of communist countries.
[6]Must be purebreds of a recognized breed and registered in a recognized registry book.
[7]For not over 400,000 head entered in the 12-month period beginning April 1, in any year, of which not over 120,000 shall be entered in any quarter beginning April 1, October 1, or January 1.
[8]Legislation of August 1964 established a basic limit of 725.4 million lb plus an added factor based on U.S. production.

PART II—MILK

Fig. 14-18. Milk is good—and good for you.

The first food that nature provides for all young mammals, including man, is milk. Long before recorded history, man found that milk was good—and good for him. As a result, he augmented the secretion of women's mammary glands by domesticating milk-producing animals and selecting them for higher production for his own use. For the most part, this included the cow, whose importance in milk production is attested to by her well-earned designation as, "the foster mother of the human race."

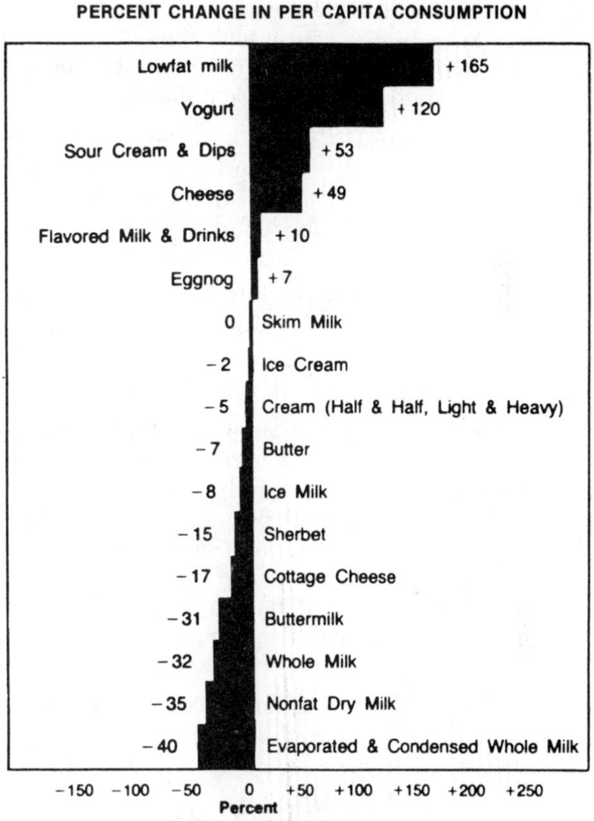

Fig. 14-19. Change in United States per capita consumption of milk and dairy products 1971 to 1981, based on sales. Bars to the left of the vertical line represent decreases; bars to the right represent increases. (From: *Milk Facts*, Milk Industry Foundation, Washington, D.C., 1982, p. 14)

Records exist of cows being milked as early as 9000 B.C. The Bible contains many references to milk; one of the most remembered of which is found in Exodus 3:8—"milk and honey." Also, Sanskrit writings, thousands of years old, relate that milk was one of the most essential of all goods. Five centuries before the time of Christ, Hippocrates, generally recognized as the Father of Medicine, recorded for posterity his evaluation of the nutritional value of milk in the statement, "Milk is the most nearly perfect food."

PRODUCE QUALITY MILK

Consumers and health departments have a distinct interest in the quality of milk.

Quality milk can be produced only when the dairyman pays special attention to a number of factors:

1. **Health of the herd**—The herd should be free from diseases that might be spread to human beings through the milk. Bacteria in milk coming from cows must be eliminated. Mastitis is the most important herd health problem at the present time.

2. **Clean animals**—The milker should clean the flanks and udders of cows just prior to milking to prevent dirt from getting into the milk. Clean floors and bedding and a well-drained yard make the cleaning job easier.

3. **Clean equipment**—All milking equipment should be kept as clean and free from bacteria as possible. Bacteria grow in cracks and rough spots on equipment if it is not washed properly.

4. **Cool and store milk properly**—Proper cooling and storage of milk on the dairy farm require facilities which will cool the milk promptly from the in-the-pail temperature of about 90°F down to 40°F, and then hold it at that temperature until it is collected. Bacteria will reproduce (divide) once every 30 minutes in 70 to 90 degree temperature; thus, in 12 hours, one bacterium can reproduce 16 million. Cooling will control this growth.

5. **Keep barn and milk house clean**—The milking barn should be clean and should have a concrete floor. Barn odors may be eliminated by having a building well ventilated.

A milk room is important to the convenience of the operator, and an aid to the production of high-quality milk.

6. **Control flies**—Fly control measures are important to dairymen. Flies add to the bacterial count of milk; cases are on record of flies carrying as many as 1,250,000,000 bacteria. They can carry typhoid, dysentery, and other contagious diseases.

Breeding places for flies, such as manure piles and mudholes, should be eliminated.

7. **Control bacteria**—In summary, here is how the bacterial count in milk can be kept down:

a. Rinse the utensils and equipment with hot water after cleaning so they dry off quickly.

b. Remove all milkstone[6] from the equipment, as bacteria must have food.

c. Cool the milk as quickly as possible to 40° F, as bacteria like high temperatures.

d. Wash and sanitize with proper cleaning and sterilizing material.

e. Have a well-lighted barn and milk house, as bacteria like darkness rather than light.

Physiological Factors Affecting Amount and Composition of Milk

The variation in the butterfat composition of milk at the plant has puzzled dairymen. And since the fat content of the milk has a bearing on the paycheck, it's an economic factor, too.

A number of physiological factors affect the amount and composition of milk:

1. **Breed and individual inheritance**—Variation in the ability of cows to produce total milk, fat, and solids-not-fat is an inherited characteristic. There is both a breed difference and an individual difference. In general, total milk production decreases and butterfat content increases by breeds in the following order: Holstein, Brown Swiss, Ayrshire, Guernsey, and Jersey.

Within the Holstein breed, a range in butterfat from 2.6 to 6.0 percent has been reported; and within the Jersey breed, from 3.3 to 8.4 percent. Similar variation between breeds and individuals exists in total milk production.

2. **Stage of lactation**—The greatest variation in the composition of milk takes place immediately following parturition, within the first five days after freshening. The secretory product known as colostrum, found in the udder at the time of calving and produced for a short time thereafter, is not milk as such. It contains more globulins, vitamins A and D, iron, calcium, magnesium, chlorine, and phosphorus than does milk; but it contains less lactose and potassium than milk.

Total milk production generally increases for the first month following freshening, then it decreases gradually thereafter. Conversely, the butterfat test is usually higher toward the end of the lactation period than soon after freshening.

3. **Persistency**—This refers to the level at which milk production is maintained as lactation progresses. Generally speaking, following the peak lactation period, about a month after freshening, the total milk production each month is approximately 90 percent of that of the previous month.

[6]Milkstone is a complex mixture of milk and water minerals with entrapped fat, protein, soil particles, and microorganisms, plus cleaner and sanitizer residues. This film adheres tightly to the surface of milk-handling equipment and requires special acid treatment for removal.

4. **Estrus; pregnancy**—Milk and butterfat production may fluctuate, usually downward, on the day of or the day following the heat period. Pregnancy seems to have little effect on milk composition. However, beginning about the fifth month of pregnancy, total production of gestating cows declines more rapidly than that of nonpregnant cows. It has been estimated that the energy requirement of the fetus are equivalent to about 400 to 600 pounds of milk.

5. **Calving interval**—Research indicates that it is most profitable for cows to calve at 12-month, rather than longer, intervals. With an 8-week dry period, this means a lactation period of 10 months.

6. **First- and last-drawn milk**—The percentage of fat in last-drawn milk is higher than that in first-drawn milk. The reasons for this are not known.

7. **Age**—The age of a cow has a definite effect on production. Most cows reach maturity and maximum milk production at about 6 years of age, following which there is a decline in production. Records indicate that cows produce approximately 25 percent more milk at maturity than they do as 2-year-olds. Also, after passing their prime—after 6 years of age—butterfat gradually decreases with advancing age.

New "age adjustment factors" have been developed to standardize 305-day lactation records to a mature equivalent basis and to minimize environmental variation due to month of the year in which the record began. These factors remove, with considerable accuracy, recent environmental effects from age and month of calving in individual breeds and regions. A complete list of adjustment factors for milk and fat by breeds, by regions (and for the United States, by month of calving) and by age, is contained in the following report: *USDA-DHIA Factors for Standardizing* 305-Day Lactation Records for Age and Month of Calving, ARS-NE-40, U.S. Department of Agriculture, September, 1974.

8. **Size**—Within a breed, large cows usually produce more milk than small cows. However, according to Brody of the Missouri Station, for each 100-pound increase in body weight, production increases only 70 percent of the proportional increase in body size.

Environmental Factors Affecting Amount and Composition of Milk

All animals, including dairy animals, are the result of two forces—heredity and environment. Because of this, the maximum development of dairy cattle characteristics of economic importance—particularly total milk production—cannot be achieved unless there are optimum conditions of environment. Among the environmental factors affecting amount and composition of milk are the following:

1. **Feed**—If milk cows are not fed, or if they do not eat, they will not produce. There are a number of ways in which feed may affect the quantity and/or composition of milk. Among them—

a. **Underfeeding**—By underfeeding we usually refer to not providing sufficient energy. The degree of milk reduction therefrom is related to the extent of underfeeding and the length of time it exists.

b. **Challenge or lead feeding in early lactation**—One of the most critical periods for proper feeding is immediately following freshening. It is very difficult for high-producing cows to consume enough feed to supply the energy needs for production at this time. As a result, most cows lose weight during this period. The current system of increasing the concentrates, beginning 2 to 3 weeks before freshening until the cow is consuming 1.0 to 1.5 pounds of concentrate per hundred pounds body weight at calving, is known as challenge or lead feeding. In this system, after freshening, cows are fed to their inherited capacity for milk production as determined by profitability; at the point where the added milk produced does not pay for the added feed, it is time to discontinue further feed increases.

c. **Deficiency of nutrient(s)**—A deficiency of any essential nutrient required by the cow will lower milk production and feed efficiency, rather than make for significant changes in the composition of milk.

d. **Some feed ingredients and rations influence milk composition**—Some feeds reduce the fat percentage of milk. Among such feeds are: cod-liver oil and other fish oils, certain pasturages (especially lush spring pastures), and pearl millet. Also, fine grinding of forage, too small an amount of roughage, or heated starch will lower the butterfat content of milk. On the other hand, such feeds as whole cottonseed, soybeans, and coconut oil result in an increase in the fat content of milk.

The amounts of fat-soluble vitamins A, D, and E in milk are influenced by the amounts of these particular vitamins in the ration, and in the case of vitamin D, exposure to sunlight is a factor, also.

2. **Length of dry period**—A dry period of approximately 60 days is recommended following each lactation period. This is important because it permits the cow's body to store up reserves so as to meet the rigorous demand of the next lactation, and it permits proper involution and conditioning of the udder. A short dry period usually results in lower milk production.

3. **Condition at calving time**—Cows that are in a thin, run-down condition at calving time produce less milk than cows in good condition. Excessive condition will also lower milk production after freshening,

but it should be added that this seldom happens in good producing dairy cows.

Cows in good flesh at calving time have been observed to start their lactation with 25 percent more milk production than those calving in poor condition. Generous feeding of thin cows following freshening may eliminate some of this difference, but it is questionable that thin, high-producing cows can ever consume enough to catch up.

4. **Frequency of milking**—Frequency of milking does result in more total milk produced. Cows milked 3 times a day consistently produce more milk than those milked twice a day, and cows milked 4 times a day produce more milk than those milked 3 times daily. It has also been observed that cows milked more frequently are more persistent in their production throughout the lactation; that is, milk production declines less rapidly as lactation progresses. Of course, a decision as to whether or not it pays to milk more than twice daily will depend on whether the additional milk more than covers the added labor and other costs of obtaining it. In a limited number of herds, managed for intensive production, three daily milkings have been possible.

Frequency of milking has no effect on butterfat percentage.

5. **Irregular feeding and milking**—Unequal interval between milkings affects both quantity and composition of milk; more milk of slightly lower fat content is obtained following the longer intervals.

6. **Change of milkers**—High-producing dairy cows are under stress, with the result that they are usually very sensitive to any changes, including that of the caretaker. Creating a pleasant, quiet, and comfortable environment causes a cow to perform more efficiently.

7. **Environmental temperature; season**—Butterfat percentage of milk varies with the season, being higher in the fall and winter and lower in the spring and summer. It may vary up and down seasonally by an average of .3 to .5 percent. Solids-not-fat also show a seasonable variation, with the low point in the spring and summer. The reasons for these changes are not known; it may be due to temperature and humidity, changes in body weight, or kinds and amounts of feeds may be reflected.

Severe weather conditions usually decrease the amount of total milk produced and may influence the fat test either up or down. Temperatures above 85°F greatly affect cows and the situation is accentuated when high temperatures are accompanied by high humidity.

It is also noteworthy that cows calving in the fall months consistently produce more than those calving at other times of the year. Cows calving in the spring produce the least. This difference may be as much as 10 to 15 percent. This phenomenon may be due in part to temperature; but more than likely available feeds including spring pastures to which fall-calving cows respond so well, may be a factor.

8. **Day-to-day variation**—Research has shown that day-to-day butterfat tests vary from 0.1 to 2.0 percent.

9. **Disease**—Disease does affect milk secretion, in both total production and composition, with the degree of the effect determined by the kind and severity of disease. Mastitis will, for example, lower both the total production of milk and the composition thereof.

10. **Drugs**—Many types of drugs have been used in an effort to increase milk production and affect its composition. Most of them have no effect, so it is questionable that they can be used on a practical basis.

When added to the feed at certain levels, thyroprotein (thyroxine) stimulates the cow to produce more milk of a higher percentage of fat. However, to be effective, it must be added at a specific time during the lactation period and cows must be fed more when they are receiving the drug.

Oxytocin will, on a temporary basis, increase yields of both milk and fat. This is because it permits greater release of milk from the udder. But it must be administered just after each milking in order to get the residual milk which makes its administration both expensive and time-consuming. Hence, it is not considered a practical procedure.

11. **Prepartum milking**—Prepartum milking is the practice of milking cows 10 days to 2 weeks before they are due to freshen. Those who follow this practice usually do so because they believe it will lessen congestion and swelling of the udder and belly of the cow. Among some, the feeling also persists that it will lessen the incidence of both mastitis and udder edema. It is known that prepartum milking will result in cows producing normal milk at the time of freshening, rather than colostrum. Thus, where prepartum milking is done, it is necessary to save (freeze) the early milk in order to have colostrum available for the newborn calf.

Mastitis (also see Section 10)

Mastitis refers to an inflammation of the udder. The term *mastitis* is from the Greek word *mastos*, for breast, and *itis* refers to inflammation of.

The listing prepared by the associate deans and directors of veterinary research programs for the Council of Deans, Association of American Veterinary Medical Colleges, showed estimated mastitis losses of $368 million per year, based on losses as of February 1, 1981.

It has been said that dairymen themselves are responsible directly, or indirectly, for 90 percent of their mastitis troubles; however, most dairymen blame their milking machine. The three main routes through which mastitis comes are: (1) dirty, or poorly adjusted, milking equipment; (2) poor milking practices; and (3) injuries to cows because of their surroundings.

Several species or groups of microorganisms may cause mastitis, but over 95 percent of all cases are caused by the following species of streptococci and staphylococci: *Streptococcus agalactiae, Streptococcus dysgalactiae, Streptococcus uberis,* and *Staphylococcus aureus.* Infection with any of these organisms is usually chronic, with flare-ups occurring at regular intervals. No amount of drugs given to cows today can prevent another attack next month under the same conditions.

Although mastitis is usually apparent, it may be a "hidden" disease. Therefore, several different tests have been developed for detecting the presence of the causative microorganisms; among them, (1) *screening test,* or *presumptive tests,* made either at the side of the cow or at the bulk tank, of which the California Mastitis Test (CMT) is the most widely used one, and (2) *specific laboratory tests* designed to detect the causative organism. A reasonable goal, based on using the CMT test, is to have at least 75% of the bucket milk samples score negative or trace; less than 75% negative (-) and trace (T) bucket readings indicates a milking management problem. On an individual quarter basis, 90% of the samples scoring negative or trace indicates a well-managed herd.

Through the years, many different kinds of drug therapy have been used—including dyes, chemicals, sulfas, antibiotics, and nitrofurans. Many times such drugs have been effective, at least temporarily; in any event, acute cases of mastitis should be treated by a veterinarian.

In summary, it may be said that the dairymen themselves are unwittingly setting the stage for mastitis flare-ups in their herds, by providing the ideal conditions—poor milking practices, poor milking equipment, and improper surroundings. By rectifying these shortcomings—through managed milking and sanitation—dairymen can reduce or eliminate mastitis.

Milk Flavor

Most consumers base the quality of any product on its flavor; and milk is no exception. They want milk that "tastes good." The flavors most often found in milk, and their cause and prevention, are:

1. **Feed and weed flavors**—Consumers do not want milk that tastes like silage, grass, or weeds.

2. **Oxidized flavors**—This has been described as a cardboard flavor. Some causes of oxidized flavor are (a) metallic contamination from copper and iron, which may be alleviated by using stainless steel; (b) exposure to sunlight or just daylight; (c) foaming; and (d) drylot feeding. Feeding vitamin E to the milking herd will reduce or eliminate oxidized flavors.

3. **Rancid flavors**—This flavor is caused by a breakdown of the butterfat which releases strong-flavored acids. This action is caused by the enzyme lipase, which is present in all milk. The primary causes of rancid milk are (a) stripper cows (those well advanced in lactation); (b) excessive agitation of milk, due to high lifts and sharp turns in pipeline milking; and (c) slow cooling with foaming.

Although feeds are not the only cause of milk flavors, they are major contributors. Feed flavors enter the milk through the digestive system, respiratory system, and by direct absorption. Research indicates that most feed flavors are detectable in the milk 20 minutes after the feed is consumed, and that they are usually most pronounced at the end of 2 hours.

Feed flavors that enter the milk through the respiratory system can usually be detected much sooner than those entering through the digestive system. For example, if a cow breathes air reeking with silage odors, these flavors can be detected in the milk almost immediately. Flavors that are directly absorbed by milk are less common, but they appear if the milk is left exposed for a long enough period.

The following control measures are recommended to alleviate feed flavors:

1. **Avoid sudden change to fresh, lush pasture**—Cows should be shifted from winter feeding, or old pasture, to new and lush pastures on a gradual basis. Also, cows should be taken out of such pastures two to three hours before milking. For the same reasons, freshly cut grass should not be fed immediately before milking.

2. **Control and avoid undesirable weeds**—Many weeds when eaten by cows will impart a strong flavor to milk; among them, are wild onions, skunk cabbage, some members of the mustard family, bitterweed, carrot weed, ragweed, and others. It is easier to get rid of these weeds today than formerly, so they should be eliminated from pasture and hayfields utilized by milk cows.

3. **Silage flavor**—Silage flavor is both common and objectionable. It can be avoided by feeding all silages after milking, never before or during milking. Usually one will be safe if silage is not fed within two to four hours of milking time, but it's safer to feed it shortly after milking. This permits the flavor-causing material to pass through the cow's digestive system before the next milking.

If cows breathe the odor of silage, it will appear as flavor in the milk. Thus, silage should never be left in the mangers or feed alleys. In fact, it is preferable that it be fed in the corral, and not in the area where the cows are being milked.

4. **Barny**—This flavor(s) is caused by dirty stables, poor ventilation, unclean milking, and unclean cows—all of which can be alleviated.

5. **Salty**—This flavor, which masks the slightly sweet flavor of milk, is caused by mastitis, stripper cows, or certain individual cows. Milk from cows that have mastitis, or from strippers, should not be marketed.

6. **Malty**—Malty flavor is primarily due to high bacteria count. The remedy is to keep bacteria out of milk as much as possible, and to prevent growth of those that do get into it. Clean and cold milk will practically eliminate malty flavor. Also, milk handlers should pick up all the milk and not leave any of it in the farm bulk tank.

7. **High-acid, sour milk**—This is due to very high bacterial count. In these days of mechanical refrigeration, there is no excuse for sour milk; simply cool it as rapidly as possible from the 90°F temperature of the milk pail to 40°F.

8. **Unnatural or foreign**—This refers to flavors that come from medicinal agents and disinfectants. The control of such off-flavors consists in (a) handling medicines and disinfectants so that the flavor or odor from them will not get into the milk, and (b) using chemical sanitizers only in the concentrations indicated by the directions. Do not market milk from drug-treated cows for at least 72 hours after last treatment, or longer if so prescribed on the drug label or by the veterinarian.

For good-tasting milk, the dairyman should keep it clean, keep it cold, feed silage after milking (not before), use good quality feed, and not ship milk from problem cows.

Beware of Pesticide Residues

Pesticides are chemicals that are used to kill pests—insects, weeds, and rodents. These products are very necessary for food and milk production. Our abundant supply of wholesome foods would not have been available without their use. Yet, it is important that they be properly used, and that certain precautions be taken. The following points are pertinent to their proper use:

1. **Pesticides that have been associated with milk contamination**—Among the pesticides that have been associated with milk contamination in the past are the following chlorinated hydrocarbons: aldrin, dieldrin, heptachlor, expoxide, DDT and its isomers, toxaphene, and lindane. Of course, other pesticides may become a problem in the future, for pesticides change—some old ones are banned, and some new ones develop.

2. **How pesticides contaminate milk**—They are absorbed by animal fat. Since milk contains fat, it is one channel through which the animal eliminates pesticides from its body.

3. **Length of time that a contaminated cow may give contaminated milk**—Cases are known where residues have been detected in milk for four to eight months after discontinuing the feeding of contaminated feeds.

4. **Ways that milk may become contaminated**—Milk becomes contaminated (a) by spraying animals with nonrecommended pesticides, (b) by using these materials in backrubbers and vaporizers, (c) by feeding forages and concentrates which have been contaminated with these materials, (d) by allowing cows to drink pesticide-contaminated water, and (e) by using milk utensils that have become contaminated through their use for chores other than handling milk or in milk production. Hence, milk contamination can be prevented by avoiding any of these avenues of contamination.

5. **The meaning of the word "tolerance" as applied to a chemical residue**—A given tolerance is that amount of chemical residue, usually expressed in ppm (parts per million) set by the FDA (Food and Drug Administration) that remain on or in a commodity at harvest and which is at least 100 times less than that amount of the chemical known to be toxic to experimental animals.

A zero tolerance, as applied to chemical residue, means that no amount of the pesticide chemical may remain on or in the raw agricultural commodity when it is offered for shipment. The tolerance level for pesticides in milk is 1.25 ppm on a fat basis. This means that there must be less than 1.25 parts of the pesticide to one million parts of milk fat figured on a weight basis.

SANITARY REGULATIONS

The sanitation of milk and dairy products is assured by the enforcement of sanitary regulations by federal, state, and local authorities.

All major cities and states have sanitary regulations governing the production, transportation, processing, and delivery of milk. Unfortunately, from area to area, there are a bewildering number of different regulations, with the result that milk going to more than one city market is often subjected to duplication and confusion in inspection. Also, sanitary and health regulations have sometimes been used as barriers to keep milk out of a certain area for competitive reasons.

In 1923, the U.S. Public Health Service (USPHS) established an Office of Milk Investigations, and, in 1924, the USPHS published its first Grade A pasteurized milk ordinance. Subsequently, this regulation has been revised several times.

The U.S. Public Health Service evolved with the

"Grade 'A' Pasteurized Milk Ordinance—1965 Recommendations of the USPHS" as a means of alleviating duplication of effort and confusion. Among its provisions are that, on delivery, milk shall not contain over 100,000 bacteria per milliliter, and it should have a maximum temperature of 50°F. This is, without a doubt, the most complete set of sanitary milk rules available, and it is being adopted by more and more regulatory officials.

The Food and Drug Administration (FDA) is charged with inspecting dairy products and processing plants for contamination and adulteration.

In an attempt to improve the quality of milk, on April 6, 1967, a National Conference of Interstate Milk Shippers passed a resolution calling for milk to contain not more than 1,500,000 leucocytes per milliliter on herd milk (leucocytes in milk are an indirect measure of mastitis in cows).

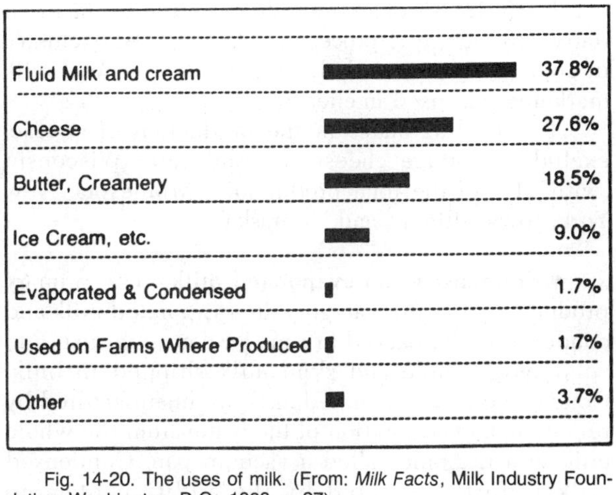

Fig. 14-20. The uses of milk. (From: *Milk Facts*, Milk Industry Foundation, Washington, D.C., 1982, p. 27)

STANDARDS AND GRADES OF MILK AND MILK PRODUCTS

TABLE 14-11

SELECTED DAIRY PRODUCTS GRADED BY USDA

Product	Volume	Share of U.S. Production
	(million lb)	*(%)*
Butter	665	76
Cheese	74	4
Nonfat dry milk	491	56

The U.S. Department of Agriculture has responsibility for the development of standards and grades for milk and dairy products. Milk is graded as Certified, Class I (Grade A), or Class II (Grade B).

The major dairy products for which the USDA has established grades, and the proportion graded are shown in Table 14-11.

It is expected that more and more dairy products will be federally graded.

USES OF MILK

In 1981, 37.87% of the milk sold by United States farmers was used in fresh milk, the rest, 62.13%, was used as manufactured dairy products. The per capita consumption of fluid whole milk totaled 139 pounds in 1981. Fluid milk is retailed as pasteurized milk, homogenized milk, fortified milk (vitamin D), skim milk, flavored milk (whole milk with flavor added), or flavored milk drink (skim milk with flavor added).

A few pertinent points relative to each of the manufacturing products follow:

• **Cream**—Cream is made by concentrating the fat portion of milk. This is done by passing milk through a cream separator. In commerce, whipping cream contains about 40% fat; coffee or table cream, 18 to 20%; and half-and-half, 12%.

• **Ice cream and similar frozen desserts**—Today, 99 percent of all frozen desserts in the United States consist of ice cream, ice milk, sherbet, and mellorine (made with a vegetable fat base). Other frozen desserts include frozen custard, frosted malted milk, artificially sweetened ice cream and ice milk, and water ices.

• **Nonfat dry milk**—Among the dried products produced from milk are nonfat dried milk, for both human food and animal feed; dried whey, for both human food and animal feed; dried whole milk; and dried buttermilk. Of these, the production of nonfat dry milk is by far the most important. In 1981, 1,306 million pounds of nonfat dried milk were produced in the United States.

Nonfat dried milk has many uses, principally as an ingredient in other dairy and food products, although its use in the home has grown considerably in recent years. Despite its wide variety of uses nonfat dry milk has been in surplus much of the time.

• **Cheese**—Cheese is made by (1) exposing milk to specific bacterial fermentation, or (2) treating with enzymes, or both methods, to coagulate some of the proteins.

Milk can be, and is, processed into many different varieties of cheese. Some are made from whole milk, others from milk that has had part of the fat removed, and still others from skim milk. American types of cheese (Cheddar, Colby, washed curd, stirred curd, Monterey, and Jack) make up 62 percent of the nation's cheese output. The most important variety produced from skim milk is cottage cheese. Other important types of cheese are Italian (mostly soft varieties), Swiss, Muenster, brick, blue, and processed cheese.

In 1981, 46% of all the milk used in manufactured dairy products was processed into cheese (exclusive of cottage cheese), and about 27.6% of the total milk marketed was used in cheese.

The leading states in the production of cheese, excluding cottage cheese, by rank, are: Wisconsin (with 41% of the total production), Minnesota, New York, Iowa, Illinois, and Nebraska.

• **Condensed and evaporated milk**—The primary products within this category are evaporated milk and condensed milk packed in cans for consumer use, and condensed whole and skim milk shipped in bulk. Condensed and evaporated milk are manufactured by removing a major portion of the water from the whole milk in a machine called a vacuum pan. Condensed milk is further treated by the addition of large amounts of sugar.

Candy manufacturers, especially bakers and ice cream processors, are large users of condensed milk.

The production of evaporated and condensed whole milk is declining. In 1945, 10.8% of milk marketings of the United States were used in evaporated and condensed milk, compared with only 1.7% in 1981.

• **Butter**—Butter is made from cream. In 1981, 18.5% of the milk supply was separated into cream for the making of butter. As marketed, it consists of about 80% milk fat. The remainder is water, salt, and traces of other substances.

The per capita consumption of margarine surpassed butter in 1957. In 1979, the per capita consumption of butter was 4.5 pounds, whereas the per capita consumption of margarine was 11.5 pounds.

Minnesota is the leading butter-producing state; Wisconsin ranks second.

SECTION 15

WOOL AND MOHAIR[1]

Contents **Page**

Classes of Wool 1105
Combing or Staple Wool 1106
French Combing Wool 1107
Clothing Wool 1107
Carpet Wool 1107
Wool Grading 1107
Grades of Wool 1107
The Blood System 1107
Numerical System 1108
Worsted Spinning Count 1108
Wool Sorting 1108
How to Produce Quality Wool 1108
Chemical and Laser Shearing 1108
Classes and Grades of Mohair 1109
The National Wool Act 1110

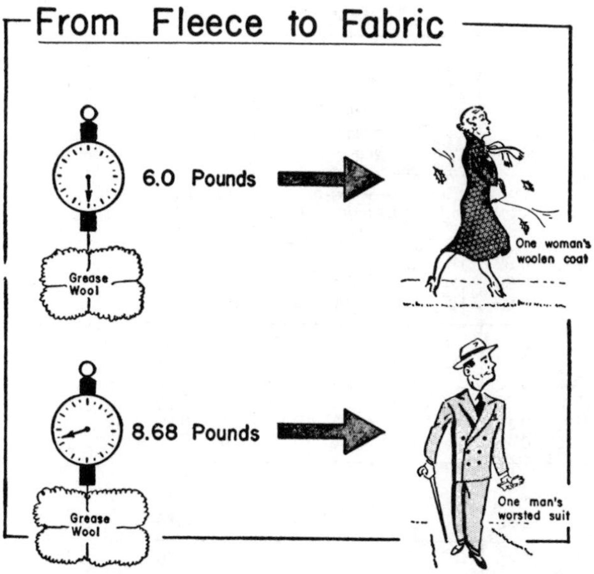

Fig. 15-1. Wool, from fleece to fabric. It requires 6 pounds of grease wool (nearly ¾ of an average fleece) to make one woman's woolen coat. It requires 8.68 pounds of grease wool (slightly more than one fleece) to make one man's worsted suit.

It is important that sheep and goat producers, and those who counsel with them, be familiar with the classes and grades of wool and mohair. With this information at hand, producers can more intelligently (1) place their production in line with market demands, and (2) market their products. To this end, a rather complete discussion follows relative to the classes and grades of wool and mohair.

CLASSES OF WOOL

The wool trade recognizes two major classes of wool—"apparel wool" and "carpet wool." In 1980, 93 percent of the wool consumed in the United States was apparel wool, and 7 percent was carpet wool. As the names imply, most apparel wools are those suitable for manufacture into yarns and fabrics for human clothing, whereas most wools of the carpet class are used in making floor covering.

Apparel wools are further classified according to use as (1) combing wool or staple wool—the long-fibered wools within the class; (2) French combing wool—the wools of intermediate length; and (3) clothing wool—the short-fibered wools. Although these three classes are based largely on length of fiber,

[1]Much information and many suggestions for improvement of this section were received from Dr. Clair E. Terrill, Staff Scientist, Agricultural Research Service, USDA, Beltsville, Md., and from Mr. Harry C. Reals, Jr., In Charge, Livestock Division Wool and Mohair Laboratory, Standardization Branch, USDA, Denver, Colo.

other factors—such as supply and demand, fiber diameter, purity, condition, etc.—are important in determining the use made of wools. Thus, many wools used by the woolen industry are longer than some used in worsted manufacture; and a considerable amount of wool classed as clothing is used in the worsted industry. In general, however, the manufacturer can realize the greatest profit by utilizing apparel wools according to their best adaptation as indicated by the three classes. Further, carpet wool is not suited for use as apparel wool.

Combing or Staple Wool

Combing or staple wools are usually referred to as the highest priced and best wool obtained from sheep. A 64s combing wool should be 3 inches or more in length, with the length varying according to grade as shown in Table 15-1. By and large, combing wools are used for making worsted fabrics. They take their name from the fact that one of the main processes in worsted manufacturing is the combing operation, which separates the long fibers from the short ones. The long fibers are used to make worsted cloths, and the short fibers (called noil) are used in the making of woolen cloths. In the former, the fibers are laid parallel to each other; whereas in the latter, the shorter wool fibers that are used in making woolen cloths and felts are laid in every direction—in fact, the more mixing, the better in woolens. These differences are of importance to the consumer. Among other things, they explain why "worsted suits" hold their press better than "woolen suits."

In the United States, wools with sufficiently long fibers and otherwise adapted to the making of worsted cloth are commonly combed on the Noble or Bradford comb.

Prior to World War II, approximately 70% of the apparel wool used in the United States was processed on the worsted system of manufacture. However, since that time there has been a gradual decline in the use of worsteds and an increase of woolens. In 1980, 50.3% of the mill consumption of wool was processed on the worsted system and 49.7% on the woolen system.

TABLE 15-1
COMPARATIVE WOOL CHART

	Standard Specifications[1]			Length Classes[2]							
Old Blood Grade	Numerical Count Grade	Limits for Average Fiber Diameter (Microns)[3]	Variability Limit for Standard Deviation Maximum (Microns)	Combing Wool		French Combing Wool				Clothing Wool	
				over		*from* →		*to*		*under*	
				(in.)	*(cm)*	*(in.)*	*(cm)*	*(in.)*	*(cm)*	*(in.)*	*(cm)*
Fine	Finer than 80s	Under 17.70	3.59	—	—	—	—	—	—	—	—
Fine	80s	17.70-19.14	4.09	2¾	6.99	1¼	3.81	2¾	6.35	1¼	3.81
Fine	70s	19.15-20.59	4.59	2¾	6.99	1¼	3.81	2¾	6.35	1¼	3.81
Fine	64s[4]	20.60-22.04	5.19	2¾[4]	7.62	1¼	3.81	2¾	6.35	1¼	3.81
½ Blood	62s	22.05-23.49	5.89	3	8.89	1½	3.81	3	7.62	1½	3.81
½ Blood	60s	23.50-24.94	6.49	3	8.89	1½	3.81	3	7.62	1½	3.81
⅜ Blood	58s	24.95-26.39	7.09	3¼	8.89	2	6.35	3¼	8.89	2	6.35
⅜ Blood	56s	26.40-27.84	7.59	3¼	8.89	2	6.35	3¼	8.89	2	6.35
¼ Blood	54s	27.85-29.29	8.19	3½	10.16	2½	6.35	3½	10.16	2½	10.16
¼ Blood	50s	29.30-30.99	8.69	3½	10.16	2½	6.35	3½	10.16	2½	10.16
Low ¼	48s	31.00-32.69	9.09	4	11.43	—	—	—	—	4	11.43
Low ¼	46s	32.70-34.39	9.59	4	11.43	—	—	—	—	4	11.43
Common	44s	34.40-36.19	10.09	5	12.70	—	—	—	—	5	12.70
Braid	40s	36.20-38.09	10.69	5	12.70	—	—	—	—	5	12.70
Braid	36s	38.10-40.20	11.19	5	12.70	—	—	—	—	—	—
Braid	Coarser than 36s	Over 40.20	—	—	—	—	—	—	—	—	—

[1]Standards for grades of wool as published by the U.S. Department of Agriculture, August 20, 1965, Federal Register (7 CFR Part 31). These standards became effective January 1, 1966.

[2]There are no USDA official lengths for the different classes. The lengths given herein (which are unstretched staple length) are in keeping with trade practices and were provided for use in this book by The Livestock Division Wool Laboratory, Standardization Branch, USDA, Denver, Colo.

[3]A micron is 1/25,400 of an inch.

[4]Grade 64s, 2¾ inch staple is par value on the wool futures contract.

NOTE: Common and Braid are not classified according to length because these wools are practically always of combing length. Carpet wool includes all those not suited to the three classes listed.

French Combing Wool

French combing wools are in between the combing wools and the clothing wools in length. These wools are manufactured on the French or Heilman comb, which is designed to use wools and still produce worsted fabrics. Thus, the French system utilizes much wool that is not long enough for manufacture on the regular worsted system known as Noble combing. This system of combing is becoming more popular.

Clothing Wool

Clothing wool is the name usually given to the shortest wool. This wool is too short to be manufactured on the worsted system, but it can be used successfully on the woolen system. Although longer fibers can be used in making woolens, they are usually more expensive than short fibers and hence are reserved for making worsteds which usually sell at a slightly higher price than woolens. The term "clothing wool," however, does not mean that the wool is suitable only for fabrics to be made into clothing. This type wool is also used to make felts.

Carpet Wool

Carpet wools, which are usually the coarsest wools, are of low quality because they (1) contain mixtures of very coarse, hairy fibers, and (2) vary markedly in fiber length. The chief requisite of carpet wool is resilience, the quality that makes it resistant to matting down and to wear under the constant scuffing of passing feet. Most of this wool comes from long-wooled sheep and from sheep that show lack of breeding. Most carpet wools are imported because U.S. flocks, except the sheep kept by Navajo Indians, have been improved to the point where the vast majority of wool grades as apparel wool. Carpet wools come chiefly from New Zealand, the United Kingdom, and Argentina.

WOOL GRADING

Wool grading is based primarily on fiber diameter or fineness, but consideration is also given to length. Many manufacturers desire wool of certain finenesses only. This means that the wool must be separated at the warehouse and like fleeces must be piled by themselves. This process is called wool grading, and it is done by a highly trained wool man.

A graded pile or bale of wool does not infer that all the wool therein is of one diameter. This is so because any single fleece of wool as it comes from the sheep may possess several different grades. Thus, a 60/62s combing wool simply means that the greater part of the wool on the fleece is of that fineness and length. The manufacturer knows that some wool in these fleeces, especially on the shoulder part of the fleece, will be finer; and also that some wool, as on the britch, will be considerably coarser. Because of this, a further separation, known as sorting follows. The ability to grade wool, which is acquired only with considerable experience, requires a keen sense of sight combined with the sense of touch and rare good judgment.

Grades of Wool

There are many factors which enter into the value of grease wool, but among the most important are diameter, length, and clean wool fiber present.

The average diameter of fiber, and the limits for the variation in diameter for the various grades, are shown in Table 15-1. Maximum limits to the variation allowed for each grade are expressed by the statistical term "standard deviation." In application, if there is too much variation in fiber diameter, the wool is assigned to the next coarser grade. Wool can be separated roughly, after a little experience, into three broad market grades according to its diameter: (1) fine wool, (2) medium wool, and (3) coarse or braid wool. More accurately speaking, however, there are two distinct methods of grading wool according to diameter with several grades in each. The older method is called the blood system; the newer method is called the numerical count. A comparison of these two systems is contained in Table 15-1.

An experienced wool man determines the grade of wool by the senses of sight and touch; however, for use in more objective grade determination and for arbitration purposes, where there may be a dispute as to grade before final settlement, there is a prescribed scientific test. A copy of this test may be obtained from the USDA, Standardization Branch, Wool Laboratory, Denver, Colorado, which explains microprojector equipment recommended and also sampling and testing procedures. Testers and research workers may also use calipers, or photographic or air-flow equipment for grade determination.

The grades of wool produced vary widely between areas. Thus, about 75 percent of the wool produced in Texas, New Mexico, Arizona, and Nevada grades fine and ½ blood, whereas ⅜ blood and ¼ blood wool predominate in the North Atlantic, East North Central, West North Central, and South Atlantic States.

THE BLOOD SYSTEM

The blood system divides all wool, from finest to coarsest, into six market grades. These are: (1) Fine, (2) ½ blood, (3) ⅜ blood, (4) ¼ blood, (5) low ¼ blood, and (6) Common and Braid. Originally, these fractional "blood" names denoted the amount of Merino blood in the sheep producing the wool. At the present time these names indicate wool of a certain fiber diameter only and have no connection whatsoever with the amount of Merino blood in the sheep. The blood grades, therefore, are merely trade names identifying the different grades of wool, without relationship to the breeding of the sheep, and are rapidly being replaced with the numerical count.

NUMERICAL SYSTEM

The numerical system divides domestic wool into 14 grades, and each grade is designated by a number. The numbers range from 80s for the finest of wool down to 36s for the coarsest. This method gives more grades, and thus finer divisions can be made; and this is more satisfactory to the wool dealers and manufacturers. Table 15-1 shows the correlation between the two grade systems.

Worsted Spinning Count

Theoretically, the wool-quality-number system is based on the number of hanks of yarn (each hank representing 560 yards) that can be spun from one pound of such scoured wool in the form of top. Wool of 50s quality, therefore, should spin 50 × 560 yards per pound of top, if spun to the maximum on the worsted system of manufacture. Unfortunately, this is not always true; the lower grades will not spin up to their number. Moreover, it is noteworthy that, in actual practice, wools are rarely spun to their maximum limit. Furthermore, spinning count is not determined by diameter alone; such factors as fiber length, moisture conditions, and the skill of the workmen influence the count that may be spun. It may be concluded, therefore, that neither the blood system nor the numerical count denotes accurately what it is supposed to indicate according to derivation.

WOOL SORTING

Sorting is the operation of taking an individual fleece, untying the twine, opening the fleece, and separating it into the various grades that were grown on the different body areas. This operation is usually done in the mill, but occasionally it is done in a warehouse. The reason for this is that a mill knows exactly what qualities of wool it wishes to put into a fabric. The object of sorting is to obtain large lots of wool that are very uniform in fiber diameter, length, strength, and other characteristics. It is easy for an inexperienced person to distinguish a very fine wool from a very coarse wool, but it takes considerable training to be able to separate two consecutive grades, such as 56s from 58s. Sorting is always done on the grease wool. The dusting and scouring operations open up the fleece into small pieces and homogenize it so that sorting of scoured wool is impossible. Sorting is necessary on wool if a uniform worsted with a certain spinning count is desired. The thoroughness of the sorting varies according to the type of fabric to be made from the wool. If the wool is not to be spun to the maximum count and if uniformity is not too important, then only a superficial sorting is necessary.

HOW TO PRODUCE QUALITY WOOL

If U.S. sheep producers are to survive the inroads of imports and synthetic fibers, it is imperative that they market a higher quality product—one that does not require unnecessary processing expenditures in the textile mills.

Despite remarkable improvement in wool wrought through improved breeding, handlers, buyers, and processors of the domestic wool clip are in general agreement that the overall quality of the nation's wool clip has declined in recent years, and that the primary reason for this decline is the generally poor preparation of the clip. The most common explanations or excuses back of this are: (1) carelessness and indifference on the part of sheep producers, and (2) lack of skilled and dedicated shearers. Whatever the cause, all are agreed that a change must be made if the domestic wool clip is to meet its increasing competition from imported wool and fabrics and from man-made fibers.

Observance of the following wool production and handling practices will result in marketing a higher quality product:

1. Producing superior fleeces through (a) feeding properly, (b) protecting the on-the-back fleece from foreign material, and (c) tagging sheep at intervals.
2. Using a scourable branding material, where branding or identifying is necessary.
3. Shearing in a clean place.
4. Using skilled shearers.
5. Packing properly.
6. Shipping in clean trucks or cars, and keeping the wool bags dry.

Chemical and Laser Shearing

High shearing cost and a scarcity of sheep shearers have spurred interest in finding an easier and less costly way of removing the fleece from sheep. Two experimental approaches appear promising; one involving the use of a chemical, the other the use of a laser beam.

- **Chemical shearing**—The chemical approach involves the use of the chemical, Cyclophosphamide

(CPA). This synthetic drug, discovered several years ago by German scientists doing cancer research, was observed to cause patients' hair to fall out.

When the CPA pill is given to sheep orally (with a balling gun), it temporarily stops cell growth, constricting the fiber at the skinline and causing it to break. From 7 to 12 days after administering the drug, sheep may be sheared with the bare hands. They can be stripped as naked as the human body—they can be sheared without nicks, second cuts, or shearing skill.

To date, no harmful side effects have been reported from the use of CPA, including its use on pregnant ewes. However, even if FDA approval is forthcoming, the following problems will deter any widescale use of chemical shearing:

1. The necessity of handling sheep twice—when administering the pill and when defleecing 7 to 12 days later.
2. The susceptibility of "bald" sheep to sunburn or cold.
3. The loss of wool by rubbing on post and brush.
4. The variability of different body areas in response to the chemical; the wool on the back, shoulders, and sides is usually removed rather easily, whereas the wool from around the face and legs comes off with difficulty.
5. Variations in dosage levels, apparently due to individual and body weight differences.
6. Variations between animals in the time interval required from dosing to defleecing.
7. The regrowth of wool in Suffolks may come in black.

Despite the problems enumerated above, chemical shearing merits further study. The stakes are high, especially for the small flock owner who has difficulty in obtaining shearers.

- **Laser beam shearing**—A group of Australians, headed by a former sheep shearer, has patented a laser beam for shearing. These ingenious inventors reasoned that lasers, which had already been used to cut woolen cloth and steel, could be adapted to shearing sheep. The laser actually severs the wool by burning. However, the Australian developers feel that it can be governed so that it will selectively cut only wool—that it can be designed so that it will automatically switch off when the beam strikes tissue or any other material differing in density from wool.

CLASSES AND GRADES OF MOHAIR

Kid hair is finest and is especially sought by mills. The fleeces from adults—especially bucks and old wethers—are the coarsest; and that from yearlings is intermediate between the other classes. These classes can be recognized by the grower and should be packed separately at shearing time. In addition, those fleeces that are extremely coarse, weak, and shorter than 6 inches or those having an excess of kemp, burs, or other foreign matter should be kept separate from clean, strong fleeces of desirable length and fineness.

TABLE 15-2

SPECIFICATIONS FOR THE OFFICIAL GRADES OF GREASE MOHAIR

	Fiber Diameter		
	Limits for Average (microns)	Maximum Standard Deviation (microns)	Approximate Number of Fiber Measurements[1]
Finer than 40s	Under 23.01	7.2	1,000
40s	23.01-25.00	7.6	1,000
36s	25.01-27.00	8.0	1,200
32s	27.01-29.00	8.4	1,200
30s	29.01-31.00	8.8	1,400
28s	31.01-33.00	9.2	1,400
26s	33.01-35.00	9.6	1,600
24s	35.01-37.00	10.0	1,600
22s	37.01-39.00	10.5	1,800
20s	39.01-41.00	11.0	2,200
18s	41.01-43.00	11.5	2,200
Coarser than 18s	43.01 and over		2,600

[1]The number of fibers to measure for each test shall be the number needed to attain confidence limits of the mean within ± 0.40 micron at a probability of 95%. Measurement of the approximate number of fibers for the grades listed above may serve as a guide to meet the required confidence limits. The numbers indicated are based on mohair matchings.

The current U.S. Department of Agriculture grade standards (1) for grease mohair, and (2) for mohair top are based on average fiber diameter (fineness) and fiber diameter dispersion. Grease mohair refers to the fleece as it comes from Angora goats, and before processing; mohair top is the processed fiber obtained after raw mohair has been scoured, carded, and combed.

As with wool, the grades of mohair are based primarily on the presumed spinning count obtainable on the Bradford system (or the number of 560-yard hanks to the pound). In practice, fineness is associated with softness and is recognized by the experienced touch when handled between the thumb and fingers.

Mohair has certain physical and chemical properties which are basic to its commercial value as a textile fiber. The average fiber diameter is the major consideration as this characteristic determines, to a large degree, the type of fabric or product for which the mohair may be used. Other characteristics affecting grease mohair value are its length, yield of clean mohair, strength, luster, color, and character. Grease mohair standards (grades) became effective August 1, 1971, and mohair top grades became effective January 1, 1973.

The official grades of grease mohair and the specifications of each are given in Table 15-2.

THE NATIONAL WOOL ACT

Through passage, and subsequent extension, of the National Wool Act (first passed in 1954), Congress recognized wool as an essential and strategic commodity which is not produced in the United States in sufficient quantity to meet domestic needs.

The incentive payments are financed from the duties collected on the imports of wool. Also, the Act authorized an industry self-help program for the purpose of developing and conducting advertising and sales promotion programs for lamb and wool.

In the 1980 marketing year, the incentive price per pound of wool was 123 cents. For mohair, it was 290.3 cents per pound. Pulled wool, taken from sheep slaughtered for meat, was supported at a level comparable to shorn wool.

In order to secure the most benefit from this Act, the sheepman should (1) sell for the highest price possible, and (2) obtain complete sale records. For example, let us assume that the national average wool price is 88.1 cents per pound. To bring the national average price of 88.1 cents to the incentive level of 123 cents, each producer's price would need to be increased by 39.6% (123 − 88.1 = 34.9; then 34.9 ÷ 88.1 = 39.6%). Therefore, if you sell 1,000 pounds of wool for 88.1 cents per pound, you will get (1) $881 from the buyer, and (2) $348.88 (39.6% more) from the Department of Agriculture, making a total of $1,229.88.

But if you sell your wool for 130 cents per pound, instead of 88.1 cents, the story is as follows: You will get (1) $1,300 from the buyer, and (2) $514.80 (39.6% of $1,300) from the Department of Agriculture, making a total return of $1,814.80. This shows how the returns on 1,000 pounds of wool could be increased by $584.92 through careful marketing.

SECTION 16

LAW ON THE LIVESTOCK FARM[1]

Contents **Page**

Leases and Leasing 1112
Choosing the Right Farm Operator 1112
How to Find a Good Tenant 1112
Landlord To Do's 1112
Lease Objectives 1113
Minimum Legal Essentials of a Written Lease 1113
Provisions in Addition to Minimum Legal Essentials 1113
Desirable Procedure in Setting Up Lease Terms 1113
Verbal Leases Vs Written Leases 1113
One-Year Vs Longtime Leases 1114
Types of Farm Leases 1114
Additional Pertinent Points Relative to Leases 1115
Lease Forms and Assistance 1115
Other Legal Documents 1115
Bill of Sale 1115
Chattel Mortgages 1115
Uniform Commercial Code (U.C.C.) 1116
Syndicated Animals (Partnership) 1116
Legal Aspects of Fencing 1116
Animals on Highways 1117
Trespass by Animals 1117
Handling Estrays (Strays) 1118
Agisters 1118
Sheep-killing Dogs 1118
Horse Protection Act (Soring Horses) 1119
Persons Injured by Animals 1119
Brands and Brand Inspection 1119
Livestock Operations Which Must Be Licensed 1120
Liability 1120
Workmen's Compensation Acts 1120
Social Security Law 1121
Guides to Keeping Out of Legal Difficulties 1122

"Ignorance of the law excuses no one".

Fig. 16-1. (Drawing by Steve Allured)

The primary function of laws is protection. Also, they provide certain guides for dealing with others.

Observance of the principles herewith outlined will help avoid legal difficulties and losses. When the stockman is contemplating business deals involving questions about which there are uncertainties, however, he should consult a reliable attorney. Observance of the latter point is important, for "ignorance of the law excuses no one." Also, it is generally recognized (1) that attorney fees are cheaper than a court trial, and (2) that, when differences of opinion exist, it is usually less expensive and more satisfactory to compromise than to take the question to court.

---

[1]In the preparation of this section, the author had the authoritative review of the following: Prof. H. W. Hannah, Professor Agricultural Law, University of Illinois, Urbana, Ill.; and Mr. R. A. Hensel, lawyer-stockman, Waterville, Wash.

LEASES AND LEASING[2]

Leases are the legal agreements by which landowners and farm operators do business. Since about 1½ million farmers rent land, a total of 3 million people (landowners plus farm operators) face basic leasing problems.

A lease may be oral or written, long or short, adapted or not adapted; and, as a written instrument, it may be valid or invalid. In short, as a means of aiding the parties to it, a lease may be good, bad, or indifferent. Good written farm leases can lead to better understanding and closer cooperation between landowners and tenants.

Choosing the Right Farm Operator

Most of the problems are solved when the right operator is selected. Although there are no exact formulae, most of the better tenants possess the following characteristics:

1. Honesty.
2. Willingness to cooperate in a successful farm management program.
3. Thorough knowledge of the proper care of all crops and livestock to be included in the farm business.
4. Ability, health, energy, and initiative to do good work in proper season.
5. Sufficient equipment and financial backing to operate the farm efficiently, unless these are furnished by the owner.
6. Favorable attitude toward adoption of new methods and practices, as rapidly as their merit is established.
7. Interest in soil conservation.
8. Interest in preventing the introduction and spread of weeds.
9. Pride and interest in the farm and the community life.
10. Willingness to make minor repairs to buildings and farm.
11. Neatness about the farm and their person.
12. Spouses who are interested in farm life and the farm operation.

HOW TO FIND A GOOD TENANT

Usually a good tenant can be found in the locality of the farm through (1) letting the word out via the present tenant (provided he is a good one), (2) inquiring at local elevators, lumber yards, county extension office, etc., and (3) advertising in local papers. It is seldom practical to move a tenant very far because farming conditions vary between communities.

Thorough investigation of a prospective tenant is essential. To this end, the landlord should—

1. Check the references which the prospective tenant gives, keeping in mind that these were selected by him.
2. Investigate through his (the landlord's) friends.
3. Make inquiry of local businessmen as to the prospective tenant's methods of doing business.
4. Check the county records for the tenant's mortgage activity, and investigate his credit contacts to determine his financial status.

Landlord To Do's

In order to assure good relationships, the landlord should observe the following to do's:

1. Be as sure as humanly possible that a satisfactory man has been picked before offering him a lease.
2. Make certain that the combination of the farm, the plan of operation, and the rental terms will assure a reasonable income for both parties.
3. Prepare a legally binding written lease, which binds both the landlord and the tenant. A one-sided lease will start the tenant out with a suspicious attitude.
4. Introduce the tenant to the farm.
5. Put the tenant at ease in business relationships.
6. Help the tenant by passing on new ideas and suggestions on better ways to do things.
7. Be a cooperator—not a boss.
8. Treat the tenant as an honest person; this point should have been investigated thoroughly before renting to him. However, if he proves to be dishonest, get rid of him as quickly and as gracefully as possible.
9. Remember that the tenant is a human being with pride, problems, and faults the same as yourself.
10. Be prompt and on time in all your business relationships.
11. Arrange for tenant settlements, preferably monthly, and be present to work them out at the agreed-upon time.
12. Arrange your business prior to contacting the tenant so it can be transacted promptly and efficiently.
13. Be cordial, but do not overdo visiting; a good tenant is a busy man and does not want to be unduly delayed in his work.
14. Be firm, but fair, in all your business transactions.
15. Give written instructions, retaining a copy for your own file.

---

[2]In the preparation of this section, the author benefited from the review of Doane Agricultural Service, Inc., St. Louis, Mo.

Lease Objectives

A good lease should meet the following objectives:

1. It should provide for a fair division of income and expenses between landlord and tenant.
2. It should provide for a system of farming that will maintain production and improvement.
3. It should, when possible, give assurance to a good tenant that his lease will be continued.
4. It should be simple, yet it should state in full the important agreements between the two partners.
5. It may provide a purchase option to the tenant in the event the owner wishes to sell.

Minimum Legal Essentials of a Written Lease

To be legal, a written lease must meet the following requirements:

1. It must give the name and address of both parties.
2. It must give the legal description of the property leased.
3. It must state the beginning and ending dates.
4. It must give the divisions of crops and designate the place of delivery.
5. It must state cash provisions, if any, and give the time and place of payment of same.
6. It must provide for possession.
7. It must be signed by both parties; preferably by both husband and wife.
8. It must be notarized if so required by state law.

Provisions in Addition to Minimum Legal Essentials

A desirable farm lease contains more than the minimum legal essentials. Usually, it is to the mutual best interest of both the landlord and the tenant that the following points be considered in written agreements:

1. Landlord's right of entry to maintain improvements, and to plan and inspect farm operations.
2. Storage space for the rental share of crops.
3. Prohibiting subletting without owner's consent.
4. Yielding possession at the end of the lease period.
5. Restricting work of tenant off the farm.
6. Tenant's guarantee of workmanship.
7. Upkeep of fences, buildings, trees, and natural resources.
8. Keeping the premises in repair.
9. Interest on deferred rents.
10. Penalty in the case of failure or default of the tenant.
11. Landlord's lien on crops and livestock.
12. Responsibility for costs in case of disputes.
13. Agreement is binding on the heirs and assigns.
14. Agreement for renewals.
15. Some details of management. The crop rotation, and soil and livestock management plans should be worked out and made a part of the lease. They should not be so rigid as to be unworkable; nor should they be so general as to preclude the best use of land and improvement.
16. Reimbursement for capital improvements performed by operator, and provision for removal of portable improvements provided by the tenant.
17. Adjusting cash rental to a changed price level.
18. Control of weeds, hauling of manure, etc.
19. Kind of records that will be kept.
20. Special agreement not covered by normal practice; such as provisions for the use of legume crops, homegrown lumber, wood, gravel, and other items of special nature.
21. Information on the rights of third parties; hunting, use of water, rights of way, etc.
22. Provision for arbitration in case of disagreement.

Desirable Procedure in Setting Up Lease Terms

The following provisions add to the protection and to the ease of living up to the lease terms:

1. Set payments to fall due when the tenant will have cash; when crops or livestock normally are marketed.
2. Check all leases, notes, and chattel mortgages handled in the transaction, to be sure they are all in the same name.
3. In some states, it is advisable to have leases acknowledged before a notary public and filed.
4. Make leasing arrangements early in the season.
5. If possession is doubtful, lease subject to getting possession.
6. Use possession notice, and follow the law applicable to the state involved.
7. Do not allow any person to occupy real estate without a written lease.

Verbal Leases Vs Written Leases

Although verbal leases are legal in many states, most oral leases protect neither party and are often the cause of disagreement.

Written leases have the following advantages:

1. They serve as a basis of settlement in case of dispute.

2. They serve as a memorandum to prevent disputes, because both parties can refer to the written terms.

3. They serve as a basis for minor changes when needed.

4. They give assurance that both parties will consider all phases of the contract.

5. They protect the heirs in case of death of a principal.

6. They serve as a history of the farm's operation.

7. They make the terms of rental definite.

8. They give an opportunity to provide for a reasonable notice period to terminate lease.

One-Year Vs Longtime Leases

Generally speaking, one-year leases are preferable to longtime leases for the following reasons:

1. They can be renewed annually when there is no disagreement.

2. They can be adjusted from year to year to care for changing conditions.

In one-year leases, the tenant can and should be protected by clauses agreeing to pay for the following improvements made by him:

1. Soil improvements that extend beyond the year concerned.

2. Permanent grass or legume seedings, for which he contributed expenses, but did not benefit.

3. Buildings, or major repairs to buildings, not normally his expenditure.

Types of Farm Leases

The most common types of farm leases are:

1. **Cash lease**—This is a good type of lease for (a) the small farm or where the landlord lives at a distance, and (b) a tenant who has adequate livestock, equipment, and working capital. It encourages livestock farming because all of the crop can easily be fed on the farm. Also, it is simple, with little chance for controversy.

There are two types of cash leases: (a) that type in which a fixed rent per acre is agreed upon when the lease is drawn, and (b) that type in which the rent is adjusted to prices of farm products which prevail during the lease year. Under the second plan, the landlord bears part of the risk of price changes; however, it is difficult to keep cash rent in line with farm product prices. If product prices are used as a basis for rent changes, the products, markets, and dates should be specified.

The landlord may prefer a cash lease because (a) the amount paid is definite, and (b) it requires less supervision by the owner. On the other hand, it may not always be desirable from the standpoint of the landlord because (a) it generally makes for lower income, (b) it gives him less control of the farm, and (c) it is difficult to collect rent if crops fail.

The tenant may prefer a cash lease because (a) it will make for more profit if he is a successful manager, (b) it makes for more independence in the operation, and (c) it makes for more profit in the good years.

2. **Crop share lease**—In this type, crops are shared and cash is paid for pasture and/or hay land. This is the most used lease in grain areas.

Crop share leases are adapted to (a) areas where land is good and nearly all tillable, and (b) young tenants and those with less capital. However, they lend little encouragement to livestock farming.

The landlord may prefer a crop share lease because (a) the rental more nearly equals the value of the land, and (b) there is more opportunity for supervision of the farm.

The tenant may prefer a crop share lease because (a) there is less risk, especially because of crop failures, (b) it requires less capital, and (c) the landlord is more inclined to improve the farm and increase productivity.

3. **The livestock share lease**—This type of lease fits the tenant who wants to raise livestock, but cannot finance a program. It is especially suited where tenant and landlord get along well and where the landlord can make a good contribution in management.

In order for this type of lease to work best, the landlord should live close to the farm, and either give it his personal attention or arrange for adequate management help such as can be provided through a professional farm management service.

The landlord may prefer a livestock share lease because (a) it encourages more livestock and more manure, (b) low-quality crops can be utilized more easily, (c) he retains an active interest in management, and (d) it generally makes for more profit.

The tenant may prefer a livestock share lease because (a) the risk is less since rent is based on net income on the farm, (b) it requires less tenant capital, (c) the landlord is more willing to make improvements, and (d) he can gain experience from the guidance of a successful owner.

4. **Manager operator or partnership**—This includes a number of leases in which the landowner normally furnishes all the capital, while the manager furnishes only the labor, but shares in the gross returns, usually on a basis of a stipulated percentage—normally 35 to 40 percent. Among such leases, are the various father-and-son business agreements.

With this type of lease, generally the owner supplies the equipment, livestock, capital, and general supervision; and the young man handles most of the work and the day-to-day management decisions.

Among the basic requirements for this type of lease are separate living quarters for each of the families, a farm business large enough to support two families, ability to get along together, a good set of records, and revision of the agreement from time to time.

After a few years under this type of lease, usually the operator prefers, and is able, to switch to a regular crop share, livestock share, or cash lease.

Except for a father-and-son agreement, this type of lease is normally not too satisfactory and should be avoided because (a) it is rather hazardous as the operator has only labor at stake, and (b) disputes arise over the system of accounting.

Additional Pertinent Points Relative to Leases

Herewith are additional pertinent points relative to leases:

1. **One-year leases**—A tenant with a one-year lease should have it renewed at the end of the year; otherwise, he may find (a) that the landlord can require him to move on less notice than that established in the original lease, or (b) that the issues which the original lease covered are not now covered. Without renewal of the lease in writing, the lessee becomes a year-to-year tenant.

2. **Notice to vacate land**—Under a written agreement, a tenant is entitled only to the period of notice specified in his lease; unless the laws of the state stipulate otherwise. If the tenant does not have a written agreement, or if the written lease has expired and the lessee is holding the land as a year-to-year tenant, the period of notice will be variable according to state law.

It is in the best interest of both landlord and tenant that the written lease provide for a period of notice, and that it contain an automatic renewal clause in the case of no notification.

3. **Landlord's lien**—Most states provide for a landlord's lien, which gives the owner of the land a claim against the crops of his tenant for rent. These vary somewhat from state to state.

4. **Landlord's right to harvest crops**—Poor health, discouragement, poor crops, or better immediate prospects elsewhere sometimes lead a tenant to move before the term for which he has rented a farm has expired. Occasionally, a tenant even leaves before harvest without having made any arrangements to take care of growing crops. Under such circumstances, usually the state law (a) protects the landlord's right to harvest crops, and (b) permits the tenant later to redeem crops matured and harvested by the landlord provided he first pays any rent due and other reasonable compensation to the lessor.

5. **Tenant's right to take removable fixtures**—Most state laws permit the tenant to take with him removable improvements that are either (a) brought with him, or (b) constructed at his own expense; usually these include hog houses, brooder houses, temporary fences, and cribs. Generally, a fixture is considered removable when the parties intend it to be and when it can be removed without undue injury to the land or other buildings. Actually, these matters should be fully covered in the lease or in a subsequent written agreement with the landlord. The lease or agreement should either (a) allow the tenant to remove the structure(s), or (b) provide for payment of a fair value when, and if, the tenant leaves.

In order to be entitled to this privilege, however, usually the law states that the tenant must meet the following three stipulations:

a. He must not owe the landlord back rent; otherwise, the landlord may hold the improvements.

b. He must have put the improvements on the land.

c. He must remove them before his term expires.

Lease Forms and Assistance

Several good lease forms are available. Usually, they can be obtained from county agents, state colleges of agriculture, or professional farm managers.

The person who has not had experience writing a lease form will need some help. Since a good lease must be both agriculturally and legally sound, it is well to call upon both an agriculturist and an attorney.

OTHER LEGAL DOCUMENTS

In addition to leases, some other legal documents are: Bill of Sale, Chattel Mortgages, Uniform Commercial Code (U.C.C.), and Syndicated Animals (partnerships).

Bill of Sale

This is a document given by a seller to a buyer, establishing the fact that legal title has passed hands to the new owner; and by the bill of sale the seller warrants that he is the owner or agent for the owner and has the lawful right to sell the property, that the property is free of encumbrance, and that he will defend the buyer's title. Such bill of sale should be received in the purchase of livestock and equipment. In many states, the title certificate replaces a bill of sale for all licensed vehicles such as trucks and automobiles. Also, in certain states the brand slip comprises the identification of animals with a bill of sale.

Chattel Mortgages

Stockmen frequently use short-term credit to finance livestock (especially feeder cattle and lambs), crops, and equipment and supplies. The chattel mortgage is the most common form of security for such loans. It allows the debtor (mortgagor) possession and use of mortgaged chattels (personal property) and at the same time gives to the seller or lender (mortgagee) a security interest superior to the rights of purchasers or transferees of the mortgagor, regardless of the actual knowledge of such persons concerning encumbrances on the chattel. It enables farmers to obtain loans because it gives the lender a high type of security.

The following facts about chattel mortgages are important:

1. They must be recorded within the time and in the manner provided by state law; they must be prepared and signed in accordance with state law; and they become effective when not renewed or foreclosed within the time specified by law.
2. They must contain an accurate description of the property mortgaged.
3. They usually (in most states) attach to progeny born of mortgaged females.
4. They do not attach to crops until the seed is in the ground unless the state law provides otherwise.
5. They are good on feed, seed, or fertilizer if the same mortgage includes animals to which the feed is fed, or the crop on which the seed and fertilizer are used.
6. They prohibit the sale of mortgaged chattels without the consent of the mortgagee. If they are sold without consent, the buyer does not acquire title and the mortgagor is subject to penalty. The mortgagee then may recover the property from the purchaser or the purchaser's transferee, and may recover damages from the mortgagor. Also, many states stipulate that it is a crime to dispose wrongfully of mortgaged property.
7. They differ from state to state. Accordingly, the appropriate state law must be studied thoroughly by those whose business requires the use of chattel mortgages.

Uniform Commercial Code (U.C.C.)

In recent years, most states have adopted the Uniform Commercial Code. It allows such security instruments as chattel mortgages and conditioned sale contracts to be replaced by a single security agreement that creates a security interest in personal property. The old instruments may be used, but their effect will be to create a security interest.

In those states having the Uniform Commercial Code, the form required by the particular state must be filed in compliance with the state law.

The Uniform Commercial Code can be used to integrate all the financial needs of the farmer in one transaction. The approval of future advances and provision for after-acquired personal property as collateral will give flexibility to farm production financing.

Syndicated Animals (Partnership)

Like livestock leases and father-son farming plans, syndicated animals are a form of "farm partnership." Such animals may be owned by two or more partners. In a syndication, liability is mutually shared; that is, all parties are mutually or fully responsible for the liabilities of the syndicate. However, the syndicate is not liable for personal, as distinct from syndicate, debts. Syndicate agreements may be written in practically any form desired by the partners. For example, the services of a syndicated sire and the expenses for his keep may be shared according to shares (investment) just as contributions to any other partnership in the form of capital, labor, and management may be shared in any proportions.

LEGAL ASPECTS OF FENCING

Most states have laws pertaining to boundary fences. In some states, however, the fencing regulations are left largely to the counties and townships. Although these laws vary greatly from state to state and are subject to frequent change, the following conditions usually prevail:

1. **Inside fences**—No stipulations are made relative to the materials used on inside fences (fences other than between boundaries).
2. **Boundary fences**—Usually state laws require every landowner to enclose his land with a fence tight enough and strong enough to turn livestock.

Some states deny the landowner any damages for trespass of livestock if he does have his land properly fenced; whereas other states permit collection of damages for trespass by animals even though the landowner suffering the damages does not have his own land fenced.

It is a rather common point of law that the condition of the fence at the point where the stock passes over or through in trespassing determines whether it is a suitable fence. The fact that it is not high enough at some other point, or that someone left a gate open on the other side of the farm, has nothing to do with the case. Thus, the argument is settled solely by the condition of the fence at the place where the animals went through, and not by its condition at any other place.

Although state laws vary rather widely, the predominant decisions of state courts on various situations involving livestock and fences are as follows:

a. When the livestock owner has good fences, is not aware that his animals habitually break out, has not been negligent, and makes an immediate attempt to get them back when they do break out, he is not liable for damage caused by them.

b. When animals break through an adjoining owner's part of a division fence, and such fence is not good, the owner of the animals cannot be held liable for damages inflicted by their trespass.

c. The owner of animals may be held liable for damages inflicted by their trespass provided—

(1) His animals are known to be in the habit of breaking out, regardless of how good the fences may be.

(2) His fences are not good.

(3) He has caused their trespass through negligence, such as by leaving a gate open, or by stampeding animals until they break out.

3. **Misdemeanor**—In some states, if anyone willfully or negligently (a) leaves open or tears down a gate provided for the convenience of the public, (b) tears down a fence on another person's property, or (c) allows livestock to run at large, the act is classed as a misdemeanor. Upon conviction, such person is subject to fine or imprisonment, or both.

4. **Responsibility for division fences**—The responsibility of neighbors for the construction and maintenance of boundary or division fences is the subject of frequent controversy.

Where the partition fence extends north and south, custom decrees that the owner whose land is east of the fence must build the north half, and the owner whose land lies on the west side of the fence must build the south half. Where the partition fence extends east and west, the owner whose land lies north of the fence builds the west half and the owner whose land lies south of the fence builds the east half. A simple customary rule regarding the apportionment of a division fence is one which gives a responsibility to the landowner for that portion of the fence which is on his right as he stands on his own property and faces the fence. Where landowners agree to some other division, the agreement should be put in writing, acknowledged before a Notary Public, and recorded by the County Recorder.

5. **Railroad fences**—Railroad companies are generally required by state laws to construct and maintain fences along their rights-of-way, provided the land adjacent thereto is otherwise enclosed. Also, they are required to maintain at road crossings cattle guards sufficient to prevent stock from getting on the railroad. When such fences and guards are not constructed and maintained properly, the railroad company may be liable for all damages which may be done by its trains.

6. **Check state laws**—Before constructing a boundary fence, the farmer or rancher should first examine with care the state laws in order to make certain that he is complying with the legal requirements.

Many states have legal fence statutes which settle the question as to what kind of division fence must be built.

ANIMALS ON HIGHWAYS

Sometimes farm animals get out on highways. If, under these circumstances, a user of the highway runs into a loose animal and is injured and/or has his vehicle damaged, he frequently tries to collect damage from the owner of the animal. Although state, county, and/or township laws vary, and it is not possible to predict with accuracy what damages, if any, may be recovered in particular instances, the following general rules apply:

1. If a farmer or rancher is negligent in maintaining his fences and allows his animals to get on the road, he can be held liable for damage resulting to persons using the highway.
2. If a farmer or rancher has good fences that are well maintained, but has one or more animals which he knows are in the habit of breaking out, he may be held liable for damages caused by such animals.
3. If animals get on to the highway, despite the facts that there are both good fences and no animals that are known habitually to get out, the owner may be held liable for any damage inflicted provided he knew that the animals were out and made no reasonable effort to get them in.
4. If the farmer or rancher is not negligent in any way, he may or may not be judged liable for the damage inflicted by his animals, depending on the state law and other circumstances.
5. If a farmer is driving animals along or across a highway, he is not likely to suffer liability for any damages unless it can be proved that he was negligent. Stock-crossing signs usually increase the caution exercised by motorists, but do not excuse a farmer from exercising due care.
6. In some states, laws provide that a farmer may, under the supervision of and with varying amounts of assistance from highway authorities, construct an underpass for his animals and for general farm use.

TRESPASS BY ANIMALS

Livestock owners who do not use reasonable care in restraining animals may be held liable for damages caused by their trespassing. Among the kinds of damages for which the courts have held that the owner may be responsible are:

1. The destruction of growing crops.
2. The transmitting of disease.
3. The breeding of females by trespassing sires.

The amount of damages in such cases is based on

the difference in value to the owner between the actual progeny and the intended progeny. Damage may be considered where the female is a registered purebred and the culprit is a scrub.

Generally state laws stipulate that the owner of land on which animals are trespassing may do anything reasonable to terminate the trespass, including the following:

1. Drive them back to the place from whence they came.
2. Call the owner and ask him to get them.
3. Confine, feed, and care for the animals until the owner comes and takes them; collect costs for same.

HANDLING ESTRAYS (STRAYS)

An estray is usually defined as a domestic animal (not including dogs or cats) of unknown ownership running at large. In some states, poultry at large may be regarded as other straying animals and may be taken up to prevent damage.

Although there is no uniformity in the state laws governing the handling of estrays (strays), some of the more common provisions are:

1. That either (a) landowners, or (b) local authorities may confine such animals and care for them.
2. That following confinement of estrays, a reasonable attempt must be made to locate the owner. Some laws specify public posting and the giving of notice in local papers.
3. That the taker-up is entitled to make reasonable use of estrays while they are in his custody; for example, work a horse or milk a cow.
4. That upon coming for estrays, the owner must satisfy the claims of the taker-up for feed, housing, care, and other costs.
5. That if the owner does not claim his animals, they either (a) become the property of the taker-up, or (b) must be sold at public auction, with reimbursement made to the taker-up for expenses incurred and with the balance turned in as county funds.

AGISTERS

The term agister is taken from English law. *It refers to one who pastures, feeds, or cares for the livestock of another for hire.*

Most states possess lien laws in favor of agisters. Such laws generally provide that agisters shall have a claim against the animals for agreed or reasonable charges, and that such claim may be enforced by retention and sale of enough animals to satisfy it. Unless the law so states, it is usually assumed that the lien expires when possession of the animals terminates.

In order to be entitled to lien consideration, the agister must keep his premises properly fenced, take reasonable precautions against injury to animals, and provide suitable feed, water, and shelter. Also, in case of death or injury loss due to neglect, such as from feeding poisoned grain, the agister is liable if he knows or should have known of the circumstances.

Although interpretations have varied, generally the courts have established that the following rules shall prevail:

1. **That the agister must have the animals in his possession**—To be entitled to a statutory lien, the person claiming it must have the animals in his charge and under his control. Thus, commercial feed companies are not entitled to a lien on the basis that they supply feed on credit.
2. **That there must be a signed or implied agreement**—To be entitled to a lien, there must have been an agreement, either signed or implied, covering their pasture, feed, and/or care.
3. **That a security agreement shall take precedence**—A security agreement takes precedence over an agister's lien unless—

   a. The mortgagee consents to an agreement whereby persons other than the mortgagor shall feed and care for animals, or
   b. The mortgage is executed while animals are under agistment.

4. **That the agister must give the owner written notice of sale and publicize same**—After unsuccessfully requesting reasonable compensation and while still in possession of the animals, the agister must give the owner written notice of the time and place of sale as required by the particular lien statute. Also, there must be due publication of notice. The animals may then be sold, with the agister retaining the amount which he claims and paying the balance, if any, to the owner.

SHEEP-KILLING DOGS

In many areas, one of the greatest causes for discouragement in sheep production is the problem of sheep-killing dogs. The problem is accentuated by the failure of dog owners to recognize that even the most lovable pet may roam the countryside at night, molesting and killing sheep and other domestic animals.

Unfortunately, from state to state, there is wide variation in laws pertaining to sheep-killing dogs. Few such laws are entirely satisfactory from the standpoint of the sheep owner, and most of them are not aggressively enforced. Most such state laws provide for one or more of the following forms of legal protection against dogs:

1. **That dogs must be licensed**—This provision

has two objectives; namely (a) to eliminate those dogs which the owner does not consider worthy of a license fee, and (b) to build up a county indemnity fund for payment to animal owners who suffer damage from dogs. Usually the maximum indemnity payment for various kinds of animals is stipulated by law, and claims must be presented through the township supervisor or other designated official.

2. **That it is a misdemeanor to allow a dog known to possess harmful tendencies to run at large**—Some state laws make it a misdemeanor to keep such a dog unless it is confined or chained. Also, these states usually make the owner an insurer for any and all damages inflicted by such a dog.

3. **That animal-molesting dogs may be killed**—Some state laws allow the owner of domestic animals the right to pursue and kill dogs not accompanied by their owners when they are discovered in the act of killing, wounding, or chasing domestic animals.

4. **That animal-molesting dogs may be poisoned**—Some state laws allow a sheep owner to put out poison for dogs on his own premises, provided it is done with reasonable care and good intentions.

5. **That damages may be collected from the dog's owner**—In some states, the law provides that the owner of animals killed or injured by dogs has a right of action against the dog's owner for all damages caused by the dog.

Such laws as the above have done some good, but—regardless of the printed law—it is generally recognized that sheep-killing dogs make for much ill-feeling and that any damages collected seldom cover the actual losses. Under these circumstances, the best protection for a flock owner still consists of a dog-proof corral for lotting at night.

HORSE PROTECTION ACT (SORING HORSES)

Soring is the use of painful methods and devices to enhance a horse's gait in the show-ring. It evolved as a means of producing a fast, flashy gait in Tennessee Walkers, for the show-ring. Soring is the practice of using caustic liquid, commonly called "scooter juice," along with chains or shackles, to make a walking horse's front ankles sore. This process, combined with feet seven or more inches long, heavy shoes, and some drastic training, creates the desired show-ring gait.

A true running walk is executed at a speed of 6 to 8 miles per hour and will not exceed 10 miles per hour. It is done with economy of effort to both the horse and the rider and is not very showy. In an effort to increase the speed to 15 to 18 miles per hour and obtain high action in front, yet keep the gait from being classed as a rack, horses are sometimes sored by means of blisters, chains, and whatnot, so that they scoot their hind feet far under them in order to keep the weight off their sore front feet. The soreness, along with accompanying long toes and heavy shoes (secured by bands over the feet, in addition to nails), cause the horse to pick his front feet up very high as he leaps through the air. Actually the fast, artificial gait that results more nearly resembles a rack than a running walk.

The Horse Protection Act, making it illegal to show or exhibit sored horses or to conduct a horse show in which sored horses are allowed to participate, was passed by the U.S. Congress and signed into law in 1970. But soring persisted, finally culminating in the Horse Protection Act Amendment of 1976, with stricter provisions. The act as amended defines soring as any practice that causes a horse to suffer physical pain, distress, or lameness while walking, trotting, or otherwise moving.

The U.S. Department of Agriculture, Animal and Plant Health Inspection Service, is charged with enforcing the Act.

A first offender, violating the provisions of the Act is subject to a civil penalty up to $2,000. Criminal penalties carry a maximum fine of $3,000 or a year in jail, or both. Second offenders may be fined up to $5,000 or 2 years in jail, or both.

PERSONS INJURED BY ANIMALS

The owner of farm animals may be held liable for personal injuries caused by them under the following circumstances:

1. When he negligently allows or causes them to commit the injury.

2. When he is aware that he owns a vicious animal, and when such an animal inflicts injury upon someone who was not acting negligently.

It is a common-law rule that "a dog is entitled to one bite," and that after that the owner may be liable for injury to others. However, some states have passed laws removing this protection and holding the owner liable for the first attack.

BRANDS AND BRAND INSPECTION

In the range country, brands are used as a means of determining the ownership of animals and of lessening theft. To meet these needs, each of the western states has laws governing the recording and inspection of brands and the transfer of branded animals. These laws generally contain the following provisions:

1. **Recorded brands**—Ranchers are required to register any brand they use, and, after its approval by the Registrar of Brands of the state agency in charge, to use that specific brand on their livestock—usually

on cattle, sheep, horses, and mules: (For information relative to types of identification, the reader is referred to Section 8, Management, under the heading, "Marking or Identifying Animals."

2. **Bill of sale**—When animals are sold, a bill of sale or other written evidence of transfer must be signed by the seller and given to the purchaser.

3. **Local brand inspectors**—Local brand inspectors, usually under the supervision of the state department or commissioner of agriculture, inspect all animals leaving their district to determine if any are being sold by a person other than the rightful owner.

4. **Inspection of hides**—Frequently there is provision for an inspection of hides at slaughter houses as a further means of disclosing theft and wrongful sale.

5. **Slaughtering in remote places**—Usually the slaughtering of animals in remote places is prohibited, for purposes of lessening theft.

6. **Penalties**—Violations, especially theft and effacing or changing of brands, are subject to severe penalties.

Stockmen operating in those states which have brand laws should become thoroughly familiar with the provisions thereof, and should recognize that law enforcement against rustlers and thieves can only be as good as the existing brands and brand inspection program. In case of suspected theft, the first question that the sheriff is prone to ask is "what brand did the lost animal have?" Unbranded range animals are an open invitation to thieves, and in the case of loss, make for a cold reception from law enforcement officials, for they can be of little help unless there is positive animal identification.

LIVESTOCK OPERATIONS WHICH MUST BE LICENSED

Although state laws vary, the following livestock and livestock operations are generally subject to license and regulation:

Auctioneers
Auction sale ring operations
Bull lessors
Commission merchants handling meat, livestock, and livestock products
Dealers of livestock
Dead animal disposal
Dogs
Feed dealers
Horseshoers
Meat dealers
Meat and produce peddlers
Pet dealers
Poultry dealers
Public carriers of livestock
Racetrack operators
Rendering plant operation
Sires for public service
Slaughter house operation
Stockyards (public) operation
Traders (itinerant) of horses and mules
Veterinarians
Weighing (public) of livestock

LIABILITY

Most farmers are in such financial position that they are vulnerable to damage suits. Moreover, the number of damage suits arising each year is increasing at an almost alarming rate, and astronomical damages are being claimed. Studies reveal that about 95 percent of the court cases involving injury result in damages being awarded.

Comprehensive personal liability insurance protects a farm operator who is sued for alleged damages suffered from an accident involving his property or family. The kinds of situations from which a claim might arise are quite broad, including suits for injuries caused by animals, equipment, or personal acts.

Both workmen's compensation insurance and employer's liability insurance protect farmers against claims or court awards resulting from injury to hired help. Workmen's compensation usually costs slightly more than straight employer's liability insurance, but it carries more benefits to the worker. An injured employee must prove negligence by his employer before the company will pay a claim under employer's liability insurance, whereas workmen's compensation benefits are established by state law, and settlements are made by the insurance company without regard to who was negligent in causing the injury. Conditions governing participation in workmen's compensation insurance vary among the states.

WORKMEN'S COMPENSATION ACTS

The workmen's compensation acts are laws making industrial employers responsible for injuries to their employees and laying down certain conditions with respect to liability insurance. Under certain conditions, the courts have ruled that those laws also apply to farm workers.

Farmers who employ three or more workers, even if only part time, and do not carry workmen's compensation insurance, may be liable for judgment for damages if an accident injures a workman.

Workmen's compensation insurance, available to farmers at moderate cost, protects the employer against all claims for damages arising from injuries to an employee. Farmers can obtain full information on this insurance, as well as help in preparing the application, from the local county agricultural agent.

SOCIAL SECURITY LAW[3]

The pertinent provisions of the present Social Security Law as it pertains to farmers and ranchers are:

1. **Who is covered**—The law covers all (a) self-employed farmers whose net earnings from farm or nonfarm operations are $400 or more annually, and (b) farm labor and domestic help on farms (including cooks and housekeepers) provided (1) they are paid $150 or more in cash wages in a calendar year, by any one farm operator, or (2) they work for one farmer for 20 days or more in a year for cash wages figured on an hourly, daily, weekly, or monthly basis. Agricultural workers admitted to the United States on a temporary basis from any foreign country are not eligible.

2. **Amount paid in**—

a. A self-employed farmer reports his earnings and pays the social security self-employment tax at the time he files his annual tax return with the Internal Revenue Service.

He may report his actual net earnings or an amount under an optional method. If his gross income from farming is $2,400 or less, he may report his actual net (if $400 or more) or two-thirds of his gross; if his gross income is more than $2,400 and his net farm income is less than $1,600, he may report for social security (but not for income tax purposes) either his actual net or $1,600. If gross income is over $2,400 and the actual net is over $1,600, actual net earnings must be reported.

In 1982, a self-employed person paid 9.35% on net earnings up to $32,400. The tax rate will eventually go to 10.75% in 1990. The taxable base will continue to increase as earning levels rise.

b. The self-employed farmer is also responsible for reporting the wage of his farm laborers and any domestic help he may have. In 1982, he should have deducted 6.7% of each employee's pay up to $32,400. He adds 6.7% of his own to this and pays the total (13.4%) to the Internal Revenue Service.

3. **What are the benefits?**—Depending on creditable earnings, the benefits are approximately as follows:

a. For a retired farmer (65 in 1982), up to $729 per month; or up to 20% less if he chooses to take benefits between 62 and 65.

b. For a retired farmer and his wife (both 65 in 1982), up to $1,093 per month.

c. For a widow or widower, surviving child, or surviving dependent parent, up to $768 per month.

d. For a farmer under 65 who is suffering from a severe disability which has lasted, or is expected to last, 12 calendar months, or to result in death, up to $768 per month, provided he has had at least 5 years of work under social security in the 10-year period just before the disability began. (A worker who becomes disabled before 31 needs fewer work credits—in some cases as little as 1½ years; a worker who becomes disabled at 43 or older needs more than 5 years of credit.) Also, his eligible dependents can get the same benefits as the dependents of a farmer retired at 65.

e. Besides monthly payments, an eligible survivor can receive a lump-sum death payment of $255 on the record of the deceased worker.

f. Medicare offers both hospital insurance and medical insurance under social security for most people 65 and over and for some disabled people under 65. Older people eligible for monthly social security benefits have hospital insurance automatically. Or those who are not eligible for monthly benefits can get it by paying a monthly premium ($113) starting July, 1982. People who have been entitled to disability checks for 24 or more months, and insured workers and their dependents who need dialysis treatment or a kidney transplant because of permanent kidney failure, also have this protection.

People who are covered under hospital insurance have medical insurance automatically unless they state they don't want it. The premium for this coverage is $12.20 a month through July, 1983.

The Social Security Administration administers another program called Supplemental Security Income. It is financed from general revenues rather than from social security taxes. It pays monthly checks to people in financial need who are 65 or older, blind, or disabled. More information about this program is available at any social security office.

The number on his social security card is very important to the farm operator as well as to the hired farm worker. It identifies the individual's social security record and is the key to future benefit payments. It is important, therefore, that a person's social security number is on the social security reports for both the self-employed farmer and the agricultural worker.

Those who expect to draw social security payments later should check with the Social Security Administration every three years, especially if they change jobs frequently, to make sure that their rec-

---

[3]In the preparation of this section, the author had the benefit of the review and suggestions of Mr. John Percy, Director, Office of Information, Social Security Administration, Department of Health & Human Services, Baltimore, Md. 21235.

ords are in order and that their correct earnings are credited to their individual social security records.

For a social security card—either a new card or a duplicate of one that has been lost—or for more information about retirement, survivors, and disability insurance, Medicare health insurance, or Supplemental Security Income, get in touch with the nearest social security office.

GUIDES TO KEEPING OUT OF LEGAL DIFFICULTIES

Herewith are some guides to keeping out of legal difficulties:

1. **Use written contracts**—Use written contracts instead of verbal contracts whenever possible, because there is less opportunity for dispute later.
2. **Pay for an option**—An option or promise to leave an offer open should always be secured by a small payment; otherwise, the agreement may be revoked at pleasure.
3. **Require surrender of a note**—Upon paying a note, require its surrender. Otherwise, it may be sold and you may be required to pay it again.
4. **Give adequate warning when lending a treacherous animal**—If a treacherous animal is lent to a neighbor, he should be warned of these traits; otherwise, the owner may be held liable for any harm or damage that the animal may inflict.
5. **Consider trees on boundary lines as joint property**—Trees standing on boundary lines are the property of both owners, and their disposal must be by mutual agreement. Also, one cannot legally claim fruit from a tree standing upon another man's property even though the branches extend over the boundary.
6. **Be aware of auto passenger responsibility**—If the owner of an auto offers a pedestrian (or hitchhiker) a ride, he may be liable for any injury to him because of careless driving, defective equipment, or any action whereby an accident results.
7. **Pay money only to an authorized agent**—Never pay money to an agent unless you know he is authorized to make collections. When payment is made, be certain to secure a signed receipt.
8. **Pay by check**—Pay debts and bills by check; then there is written proof of payment.
9. **Secure adequate protection through insurance**[4]—In these times of high court judgments, it is imperative that the stockman have adequate insurance protection. Without such protection, or without substantial wealth, he is at the mercy of the claims-conscious public. A judgment could put him out of business unless he had adequate insurance to cover that judgment.

   Most stockmen strive to keep their fences in proper repair, their equipment in satisfactory order, their employees properly educated about the hazards of the occupation in which they are engaged; and to handle their entire operation in a safe and sane manner. Yet, accidents do happen, and, when that time comes, an insurance policy is the answer to the financial part of the problem.

   Recently, the insurance industry developed liability policies designed specifically for the farmer and stockman. These policies are blanket-type liability policies designed to cover the farmer's legal liability arising out of his operation of the farm or ranch. Some of the general provisions covered by such policies are:

   a. Liability for bodily injury or property damage to employees or guests.
   b. Medical aid where the policyholder is liable.
   c. Property damage as a result of breachy animals.
   d. Liability for accidents on highways and public roads caused by animals.
   e. Bodily injury and property damage liability for personal acts of the stockman and his family.

   Such policies, being tailored for the actual needs of farmers and ranchers, are quite flexible and can be written to suit each individual's particular needs.

   The important thing is that the stockman should take advantage of adequate liability coverage, which can be obtained at little added cost, for the time is past when one can have a secure feeling with a $10,000 policy. The high cost of claim settlement and the increased amount of jury verdicts make it desirable that limits of liability be increased. There should be adequate coverage to assure the stockman protection when and if needed and to keep him in business.
10. **Have a will made**—Most important of all, the farmer or rancher should have a will that covers his property and disposes of it in keeping with his wishes. (See Section 2, Business Aspects, "Wills.")

---

[4]In the preparation of this section, the author was assisted by the following insurance agents: Mr. Charles I. Palmerton, Special Agent for the Northwestern Mutual Insurance Co.; and Mr. Ivan R. Sayles, Sayles Insurance Co., Pullman, Wash.

SECTION 17

BREED REGISTRY ASSOCIATIONS

Contents **Page**

Breed Registries 1123
Breed Registry Associations, Table 17-1 1124
Association Rules Relative to Marking or Identifying Animals 1123
Marking or Identifying Guide for Registered Beef and Dual-purpose Cattle, Table 17-2 1132
Marking or Identifying Guide for Registered Dairy Cattle, Table 17-3 1134
Marking or Identifying Guide for Registered Sheep and Goats, Table 17-4 1135
Marking or Identifying Guide for Registered Swine, Table 17-5 1137
Marking or Identifying Guide for Registered Horses and Ponies, Table 17-6 1138
Association Rules Relative to Artificial Insemination 1123
A.I. Rules of Beef and Dual-purpose Cattle Associations, Table 17-7 1139
A.I. Rules of Dairy Cattle Associations, Table 17-8 1141
A.I. Rules of Sheep and Goat Associations, Table 17-9 1142
A.I. Rules of Swine Associations, Table 17-10 1143
A.I. Rules of Horse Associations, Table 17-11 1144

BREED REGISTRIES

Fig. 17-1. Breed registries record the lineage.

A breed registry association consists of a group of breeders banded together for the purposes of: (1) recording the lineage of their animals, (2) protecting the purity of the breed, (3) encouraging further improvement of the breed, and (4) promoting the interest of the breed. A list of the breed registry associations is given in Table 17-1, page 1124.

Association Rules Relative to Marking or Identifying Animals

Tables 17-2 to 17-6, pages 1132–1138, summarize the pertinent regulations of the breed associations—beef and dual-purpose cattle, dairy, sheep and goats, swine, and horses—relative to marking or identifying registered animals.

Association Rules Relative to Artificial Insemination

Most of the cattle and horse registry associations have rules with which there must be compliance if animals produced artificially are to be eligible for registry. Fewer of the sheep, goat, and swine registry associations have rules governing the registry of animals produced through artificial insemination, but most of them are receptive to the idea.

Tables 17-7 to 17-11, pages 1139–1147, summarize the pertinent regulations relative to the registration of animals produced by artificial insemination.

TABLE 17-1
BREED REGISTRY ASSOCIATIONS

Class of Animal	Breed	Association and Address
Beef and dual-purpose cattle	**Angus**	American Angus Assn. 3201 Frederick Blvd. St. Joseph, Mo. 64501
	Ankina	Ankina Breeders 5803 Oakes Road Clayton, Ohio 45315
	Ankole-Watusi	Ankole-Watusi International Registry Star Route 45 Hebron, N. Dak. 58638
	Barzona	Barzona Breeders Assn. of America P.O. Box 631 Prescott, Ariz. 86302
	Beefalo	American Beefalo Assn., Inc. 116 Executive Park Louisville, Ky. 40207
	Beef Friesian	Beef Friesian Society 210 Livestock Exchange Bldg. 4701 Marion Street Denver, Colo. 80216
	Beefmaster	Beefmaster Breeders Universal Suite 350, G.P.M. South Tower 800 N.W. Loop 410 San Antonio, Tex. 78216
		Foundation Beefmaster Assn. Livestock Exchange Bldg. Suite 200 4701 Marion Street Denver, Colo. 80216
		National Beefmaster Assn. 817 Sinclair Building Fort Worth, Tex. 76102
	Belted Galloway	Belted Galloway Society, Inc. P.O. Box 5 Summitville, Ohio 43962
	Blonde d'Aquitaine	American Blonde d'Aquitaine Assn. Route B, Box 230 Grandview, Ida. 83624
	Braford	International Braford Assn., Inc. P.O. Box 1030 Ft. Pierce, Fla. 33450
	Brahman	American Brahman Breeders Assn. 1313 La Concha Lane Houston, Tex. 77054
	Brangus	International Brangus Breeders Assn., Inc. 9500 Tioga Drive San Antonio, Tex. 78230
	Charolais	American-International Charolais Assn. 1610 Old Spanish Trail Houston, Tex. 77054
	Chianina	American Chianina Assn. P.O. Box 890 Platte City, Mo. 64079
	Devon	Devon Cattle Assn., Inc. P.O. Box 628 Uvalde, Tex. 78801
	Dexter	American Dexter Cattle Assn. 707 W. Water Street Decorah, Iowa 52101
	Galloway	American Galloway Breeders Assn. Route 1, Box 106A Athol, Ida. 83801
		Galloway Cattle Society of America Hennepin, Ill. 61327

(Continued)

TABLE 17-1 (Continued)

Class of Animal	Breed	Association and Address
Beef and dual-purpose cattle (continued)	**Gelbvieh**	American Gelbvieh Assn. 5001 National Western Drive Denver, Colo. 80216
	Hereford	American Hereford Assn., The 715 Hereford Drive Kansas City, Mo. 64105
	Limousin	North American Limousin Foundation 100 Livestock Exchange Bldg. 4701 Marion Street Denver, Colo. 80216
	Maine-Anjou	American Maine-Anjou Assn. 564 Livestock Exchange Bldg. 1600 Genesee Street Kansas City, Mo. 64102
	Marchigiana	American International Marchigiana Society P.O. Box 342 Lindale, Tex. 75551
	Milking Shorthorn	American Milking Shorthorn Society 313 S. Glenstone Springfield, Mo. 65802
	Murray Grey	American Murray Grey Assn. 1222 North 27th Street Billings, Mont. 59107
	Normande	American Normande Assn. P.O. Box 350 Kearney, Mo. 64060
	Pinzgauer	American Pinzgauer Assn. P.O. Box 1003 Norman, Okla. 73070
		Canadian Pinzgauer Assn. No. 108 Stockmans Centre 2116 27th Avenue N.E. Calgary, Alberta, Canada T2E 7A6
	Polled Hereford	American Polled Hereford Assn. 4700 East 63rd Street Kansas City, Mo. 64130
	Polled Shorthorn	American Polled Shorthorn Society 8288 Hascall Street Omaha, Neb. 68124
	Ranger	Ranger Cattle Company Box 21300, North Pecos Station Denver, Colo. 80221
	Red Angus	Red Angus Assn. of America 4201 I-35 North Denton, Tex. 76201
	Red Brangus	American Red Brangus Assn. P.O. Box 1326 Austin, Tex. 78767
	Red Poll	American Red Poll Assn. Box 35519 Louisville, Ky. 40232
	Romagnola	Canadian Romark Assn. Box 177 Jarvie, Alberta, Canada T0G 1H0
	Salers	American Salers Assn. P.O. Box 30 Weiser, Ida. 83672
	Santa Gertrudis	Santa Gertrudis Breeders International P.O. Box 1257 Kingsville, Tex. 78363
	Scotch Highland	American Scotch Highland Breeders' Assn. P.O. Box 81 Remer, Minn. 56672
	Shorthorn	American Shorthorn Assn. 8288 Hascall Street Omaha, Neb. 68124

(Continued)

TABLE 17-1 (Continued)

Class of Animal	Breed	Association and Address
Beef and dual-purpose cattle (continued)	**Simmental**	American Simmental Assn. 1 Simmental Way Bozeman, Mont. 59715
	South Devon	North American South Devon Assn. P.O. Box 68 Lynnville, Iowa 50153
		International South Devon Assn. P.O. Box 68 Lynnville, Iowa 50153
	Sussex	Sussex Cattle Assn. of America P.O. Drawer AA Refugio, Tex. 78377
	Tarentaise	American Tarentaise Assn. 123 Airport Road Ames, Iowa 50010
	Texas Longhorn	Texas Longhorn Breeders Assn. of America 3701 Airport Freeway Fort Worth, Tex. 76111
	Welsh Black	United States Welsh Black Cattle Assn. Route 1 Wahkon, Minn. 56386
Dairy cattle	**Ayrshire**	Ayrshire Breeders' Assn. 2 Union Street Brandon, Vt. 05733
	Brown Swiss	Brown Swiss Cattle Breeders' Assn., The Box 1038 Beloit, Wisc. 53511
	Dutch Belted	Dutch Belted Cattle Assn. of America, Inc. P.O. Box 358 Venus, Fla. 33960
	Guernsey	American Guernsey Cattle Club, The 2105J S. Hamilton Road Columbus, Ohio 43227
	Holstein-Friesian	Holstein-Friesian Assn. of America P.O. Box 808 Brattleboro, Vt. 05301
	Illawarra	International Illawarra Assn. 1722JJ S. Glenstone Avenue Springfield, Mo. 65804
	Jersey	American Jersey Cattle Club, The P.O. Box 27310 Columbus, Ohio 43227
	Milking Shorthorn	American Milking Shorthorn Society 1722JJ South Glenstone Avenue Springfield, Mo. 65804
Sheep	**Cheviot**	American Cheviot Sheep Society R.R. #1, Box 100 Clarks Hill, Ind. 47930
	Columbia	Columbia Sheep Breeders Assn. of America P.O. Box 272 Upper Sandusky, Ohio 43351
	Cormo	American Cormo Sheep Assn. 18106 Woodgate Road Montrose, Colo. 81401
	Corriedale	American Corriedale Assn., Inc. Box 29C Seneca, Ill. 61360
	Cotswold	American Cotswold Record Assn. 282 Meaderboro Road Rochester, N.H. 03867
	Debouillet	Debouillet Sheep Breeders Assn. 300 S. Kentucky Roswell, N.M. 88201

(Continued)

TABLE 17-1 (Continued)

Class of Animal	Breed	Association and Address
Sheep (continued)	**Delaine Merino**	American & Delaine Merino Record Assn. 1193 Township Road 346 Nova, Ohio 44859
		Black Top & National Delaine Merino Sheep Assn. 290 Beech Street Muse, Penn. 15350
		Texas Delaine Sheep Assn. Route 1 Burnet, Tex. 78611
	Dorset	Continental Dorset Club, Inc. P.O. Box 577 Hudson, Iowa 50644
	Finnsheep	Finnsheep Breeders Assn., Inc. P.O. Box 34303 Indianapolis, Ind. 46234
	Hampshire	American Hampshire Sheep Assn. Box 345 Ashland, Mo. 65010
	Lincoln	National Lincoln Sheep Breeders' Assn. R.R. #6, Box 24 Decatur, Ill. 62521
	Montadale	Montadale Sheep Breeders' Assn., Inc. P.O. Box 44300 Indianapolis, Ind. 46244
	Natural Colored	Natural Colored Wool Growers Assn. Route 2, Box 2382 Davis, Calif. 95616
	North American Clan Forest	North American Clan Forest Assn. High Meadow Farm Ferryville, Wisc. 54628
	North County Cheviot	American North County Cheviot Sheep Assn. 717 Fall Creek Road Longview, Wash. 98632
	Oxford	American Oxford Down Record Assn. Route 4 Ottawa, Ill. 61350
	Panama	American Panama Registry Assn. Route 1 Jerome, Ida. 83338
	Polypay	American Polypay Sheep Assn. 1934 East Rua Bronco Sandy, Utah 84092
	Rambouillet	American Rambouillet Sheep Breeders' Assn., The 2709 Sherwood Way San Angelo, Tex. 76901
	Romney	American Romney Breeders Assn. 4375 N.E. Weslinn Drive Corvallis, Ore. 97333
	Shropshire	American Shropshire Registry Assn., Inc. P.O. Box 1970 Monticello, Ill. 61856
	Southdown	American Southdown Breeders' Assn. Rt. 4, Box 14B Bellefonte, Penn. 16283
	Suffolk	American Suffolk Sheep Society 55 East 100 North Logan, Utah 84321
		National Suffolk Sheep Assn. P.O. Box 324 Columbia, Mo. 65201
	Targhee	U.S. Targhee Sheep Assn. P.O. Box 40 Absarokee, Mont. 59001
	Tunis	National Tunis Sheep Registry R.D. 1 Wayland, N.Y. 14572

(Continued)

TABLE 17-1 (Continued)

Class of Animal	Breed	Association and Address
Goats	**Angora**	American Angora Goat Breeders' Assn. P.O. Box 195 Rocksprings, Tex. 78880
Dairy goats	**All breeds**	American Dairy Goat Assn. P.O. Box 865 Spindale, N.C. 28160
		American Goat Society, Inc. Route 2, Box 112 DeLeon, Tex. 76444
		International Dairy Goat Registry, Inc. Rt. 2, Box 365 Dublin, Tex. 76446
		National Pygmy Goat Assn. Fern Avenue Amesbury, Mass. 01913
Swine	**Berkshire**	American Berkshire Assn. 601 W. Monroe Street Springfield, Ill. 62704
	Chester White	Chester White Swine Record Assn. P.O. Box 228 Rochester, Ind. 46975
	Duroc	United Duroc Swine Registry 1803 W. Detweiller Drive Peoria, Ill. 61615
	Hampshire	Hampshire Swine Registry 1111 Main Street Peoria, Ill. 61606
	Hereford	National Hereford Hog Record Assn. Route 1, Box 37 Flandreau, S.D. 57028
	Landrace	American Landrace Assn., Inc. Box 647 Lebanon, Ind. 46052
	Poland China	Poland China Record Assn. 368 West Douglas, Box B Knoxville, Ill. 61448
	Spotted	National Spotted Swine Record, Inc. 110 W. Main Street Bainbridge, Ind. 46105
	Tamworth	Tamworth Swine Assn. 2656 Horner Road Winchester, Ohio 45697
	Yorkshire	American Yorkshire Club, Inc. Box 2417 West Lafayette, Ind. 47906
Light horses	**American Bashkir Curly**	American Bashkir Curly Registry P.O. Box 453 Ely, Nev. 89301
	American Creme Horse	Worldwide Horse Registry for American White and American Creme Box 79 Crabtree, Ore. 97335
	American Fox Trotting Horse	American Fox Trotting Horse Breed Assn., Inc. P.O. Box 666 Marshfield, Mo. 65706
	American Mustang	American Mustang Assn., Inc. P.O. Box 338 Yucaipa, Calif. 92399
	American Paint Horse	American Paint Horse Assn. P.O. Box 18519 Fort Worth, Tex. 76118
	American Part-Blooded	American Part-Blooded Horse Registry 4120 S.E. River Drive Portland, Ore. 97222

(Continued)

TABLE 17-1 (Continued)

Class of Animal	Breed	Association and Address
Light horses (continued)	**American Saddlebred Horse**	American Saddlebred Horse Assn. 929 South Fourth Street Louisville, Ky. 40203
	American White Horse	Worldwide Horse Registry for American White and American Creme Box 79 Crabtree, Ore. 97335
	Andalusian	American Andalusian Assn. P.O. Box 1290 Silver City, N.M. 88061
	Appaloosa	Appaloosa Horse Club, Inc. P.O. Box 8403 Moscow, Ida. 83843
	Arabian	Arabian Horse Registry of America, Inc. 3435 South Yosemite Denver, Colo. 80231
	Buckskin	American Buckskin Registry Assn., Inc. P.O. Box 1125 Anderson, Calif. 96007
		International Buckskin Horse Assn., Inc. P.O. Box 357 St. John, Ind. 46373
	Chickasaw	Chickasaw Horse Assn., Inc., The P.O. Box 607 Love Valley, N.C. 28677
		National Chickasaw Horse Assn. Route 2 Clarinda, Iowa 51232
	Cleveland Bay	Cleveland Bay Assn. of America Box 182 Hopewell, N.J. 08525
	Galiceno	Galiceno Horse Breeders Assn., Inc. 111 East Elm Street Tyler, Tex. 75702
	Hackney	American Hackney Horse Society P.O. Box 174 Pittsfield, Ill. 62363
	Half-Arabian and Anglo-Arabian	International Arabian Horse Assn. 224 E. Olive Avenue Burbank, Calif. 91503
	Half-bred Thoroughbred	American Remount Assn., Inc. P.O. Box 1066 Perris, Calif. 92370
	Hanoverian	American Hanoverian Society, The 809 West 106th Street Carmel, Ind. 46032
	Hungarian Horse	Hungarian Horse Assn. Bitterroot Stock Farm Hamilton, Mont. 59840
	Lipizzan	Lipizzan Assn. of America Woolworth Tower New York, N.Y. 10279
	Missouri Fox Trotting Horse	Missouri Fox Trotting Horse Breed Assn., Inc. P.O. Box 637 Ava, Mo. 65608
	Morab	Morab Horse Registry of America P.O. Box 143 Clovis, Calif. 93612
	Morgan	American Morgan Horse Assn., Inc. Box 1 Westmoreland, N.Y. 13490
	Norwegian Fjord	Norwegian Fjord Assn. of North America 29645 N. Callahan Road Round Lake, Ill. 60073
	Palomino	Palomino Horse Assn., Inc., The P.O. Box 324 Jefferson City, Mo. 65102

(Continued)

TABLE 17-1 (Continued)

Class of Animal	Breed	Association and Address
Light horses (continued)	**Palomino (continued)**	Palomino Horse Breeders of America P.O. Box 249 Mineral Wells, Tex. 76067
	Paso Fino	American Paso Fino Horse Assn., Inc. 907 Penn Avenue Pittsburgh, Penn. 15221
		Paso Fino Owners and Breeders Assn., Inc. P.O. Box 764 Columbus, N.C. 28722
	Peruvian Paso	American Assn. of Owners & Breeders of Peruvian Paso Horses P.O. Box 2035 California City, Calif. 93505
		Peruvian Paso Horse Registry of North America P.O. Box 816 Guerneville, Calif. 95446
	Pinto Horse	Pinto Horse Assn. of America, Inc. 7525 Mission Gorge Road, Suite C San Diego, Calif. 92120
	Quarter Horse	American Quarter Horse Assn. 2736 W. 10th Street Amarillo, Tex. 79168
		Standard Quarter Horse Assn. 4390 Fenton, #206 Denver, Colo. 80212
	Rangerbred	Colorado Ranger Horse Assn., Inc. 7023 Eden Mill Road Woodbine, Md. 21797
	Spanish Barb	Spanish Barb Breeders Assn. P.O. Box 7479 Colorado Springs, Colo. 80907
	Standardbred	United States Trotting Assn. 750 Michigan Avenue Columbus, Ohio 43215
		National Trotting & Pacing Assn., Inc. 575 Broadway Hanover, Penn. 17331
	Tennessee Walking Horse	Tennessee Walking Horse Breeders' and Exhibitors' Assn. of America P.O. Box 286 Lewisburg, Tenn. 37091
	Thoroughbred	Jockey Club, The 380 Park Avenue New York, N.Y. 10017
	Ysabella	Ysabella Saddle Horse Assn., Inc. c/o Prairie Edge Farm Route 3 Williamsport, Ind. 47993
Ponies	**American Gotland Horse**	American Gotland Horse Assn. P.O. Box 263 Jenks, Okla. 74037
	American Walking Pony	American Walking Pony Assn. Route 5, Box 88 Upper River Road Macon, Ga. 31201
	Connemara Pony	American Connemara Pony Society R.D. 1 Hoshiekon Farm Goshen, Conn. 06756
	Miniature Horse	International Miniature Horse Registry Box 907 Palos Verdes, Calif. 90274
	National Appaloosa Pony	National Appaloosa Pony, Inc. Box 206 Gaston, Ind. 47342
	Pony of the Americas	Pony of the Americas Club, Inc. P.O. Box 1447 Mason City, Iowa 50401

(Continued)

TABLE 17-1 (Continued)

Class of Animal	Breed	Association and Address
Ponies (continued)	**Shetland Pony**	American Shetland Pony Club P.O. Box 435 Fowler, Ind. 47944
	Welara	American Welara Pony Society Box 401 Yucca Valley, Calif. 92284
	Welsh Cob	Welsh Cob Society of America 225 Head of the Bay Road Buzzards Bay, Mass. 02532
	Welsh Pony	Welsh Pony Society of America Box 2977 Winchester, Va. 22601
All horses and half-breds	Any and all colors and types of horses (including animals not eligible for registry, eligible but not registered, or registered in existing associations) including both light and draft horses.	National Recording Office Box 79 Crabtree, Ore. 97335
	Half-bred Thoroughbreds: Section 1: **The American Remount Half-Thoroughbred**—Must have one Thoroughbred parent registered in the American (Jockey Club) Stud Book.	American Remount Assn., Inc. (Half-Thoroughbred Registry) P.O. Box 1066 Perris, Calif. 92370 (Formerly the Half-Bred Stud Book operated by the American Remount Association, but now a privately owned registry.)
	Section 2: **The American Remount Anglo**—Must have one Thoroughbred parent registered in the American (Jockey Club) Stud Book and the other parent registered in the Stud Book of a recognized breed. Section 3: **The American Remount Thoroughbred Kind**—Must have one Thoroughbred parent of a recognized Foreign Registry or must have both parents registered in the American (Jockey Club) Stud Book but be ineligible for registry in the American (Jockey Club) Stud Book. Section 4: **The American Remount Hunter-Jumper**—Must be a minimum of 36 months of age; be performance certified by an approved Equine Practitioner, a Master of Fox Hounds, an Official of the American Horse Show Association, or a Steward of the American Remount Association; and be ineligible for registry in the American (Jockey Club) Stud Book. Section 5: **The American Remount Polo Pony**—Must be performance certified by an approved Equine Practitioner, a 5-goal rated player, an Officer of the U.S. Polo Association, or a Steward of the American Remount Association; and be ineligible for registry in the American (Jockey Club) Stud Book. Section 6: **The American Remount Endurance Horse**—Must be performance certified by an approved Equine Practitioner, an Official of the American Horse Show Association, or a Steward of the American Remount Association; and be ineligible for registry in the American (Jockey Club) Stud Book. Section 7: **The American Remount Record**—This is an Identification Certificate issued to a horse that has apparent Thoroughbred ancestry but is not otherwise eligible for registry in the American (Jockey Club) Stud Book or any other recognized Stud Book.	
	Half-bred Arabian: 1. **Anglo-Arabs** must carry not more than ¾ and not less than ¼ Arabian blood. May be one of the following: a. By Thoroughbred stallions and out of registered Arabian mares; b. By registered Arabian stallions and out of registered Thoroughbred mares; c. By registered Thoroughbred or Arabian stallions and out of registered Anglo-Arab mares; or d. By Anglo-Arab stallions and out of registered Anglo-Arab mares, registered Thoroughbred mares, or registered Arabian mares. 2. **Half-Arabians** are by registered Arabian stallions and out of mares that are not registered Thoroughbreds or Arabians.	International Arabian Horse Assn. 224 East Olive Avenue Burbank, Calif. 91503
	Half-bred, grade, and crossbred horses involving—American Saddle Horse, Appaloosa, Hackney, Morgan, Quarter Horse, Standardbred, Tennessee Walking Horse, Welsh Pony, and certain other breeds.	American Part-Blooded Horse Registry 4120 Southeast River Drive Portland, Ore. 97222

(Continued)

TABLE 17-1 (Continued)

Class of Animal	Breed	Association and Address
Draft horses	**American Cream**	American Cream Horse Assn. Route 1, Box 88 Hubbard, Iowa 50122
	Belgian	Belgian Draft Horse Corporation of America P.O. Box 335 Wabash, Ind. 46992
	Clydesdale	Clydesdale Breeders Assn. of the United States Route 1, Box 131 Pecatonica, Ill. 61063
	Percheron	Percheron Horse Assn. of America P.O. Box 141 Fredericktown, Ohio 43019
	Shire	American Shire Horse Assn. 14410 High Bridge Road Monroe, Wash. 98272
	Suffolk	American Suffolk Horse Assn., Inc. Route 1, Box 212 Ledbetter, Tex. 78946
Jacks, donkeys, and mules	**Jack and Jennet**	Standard Jack and Jennet Registry of America Sulphur Run Farm Route 1, Box 194 Elk Run, Ky. 42733
	Miniature Donkey	Miniature Donkey Registry of the United States, Inc. 1108 Jackson Street Omaha, Neb. 68102
	Donkey and Mule	American Donkey and Mule Society, Inc. Route 5, Box 65 Denton, Tex. 68102

TABLE 17-2

MARKING OR IDENTIFYING GUIDE FOR REGISTERED BEEF AND DUAL-PURPOSE CATTLE

Breed	Association Rules Relative to Marking
Angus	Each animal, for which application for registry is submitted, must be tattooed alike in both ears. Each breeder may devise his own tattooing system, using a series of numbers or letters, or a combination of numbers and letters. Tattoo marks are limited to 4 units in each ear, and only standard numerals or letters are acceptable. Each animal of the same sex to be registered by any one breeder must be tattooed differently.
Barzona	The holding brand and private herd number must be fire or freeze branded on an animal before it may be registered as a Barzona. Before a calf is weaned, it must have an ear tattoo if the private herd number brand has not been applied.
Beef Friesian	The Association prefers that the tattooed herd prefix be in the right ear and the individual identification tattoo number in the left ear. If an ear tag is also used, this tag should be in the right ear.
Beefmaster	**Foundation Beefmaster Association:** The Association accepts any permanent marking system, such as fire brand or tattoo. The Association does not specify any particular type of marking for certified animals, but each animal and each herd must carry a prefix name. In order that each Beefmaster may be permanently identified with the breeder thereof, the breeder must use a prefix name, such as "Jones Beefmaster," "Smith Beefmaster," etc., to designate his cattle. Thus, in a unique way, the responsibility for the continued improvement of the breed is placed squarely upon the individual breeder.
Belted Galloway	A tattoo in either or both ears is required. Also, breeders are encouraged to use individual herd identifications such as ear tags or chains.
Braford	Animals are identified with the owner's fire brand, plus numbers, plus one bar for cattle with 2 generations of Braford x Braford matings and 2 bars for 3 generations.
Braham	A holding brand (a symbol, letter, combination of letters and/or symbols, numerals, replica of some object, etc., to denote ownership or breeder) and private herd number (both branded by fire) are required on a calf before it may be registered.
Brangus	Each animal for which application is submitted must be fire branded on the body with the owner's holding brand and a private herd number, and the year brand. The application for registration or enrollment must show where this brand is located on the body. The Association suggests starting with the number "1" on the private herd number and numbering consecutively. The holding brand is any mark, initial, or number, or combination of all 3 which the breeder chooses to use, and which is approved by the breed registry (IBBA).

(Continued)

TABLE 17-2 (Continued)

Breed	Association Rules Relative to Marking
Charolais	Private herd number (tattoo or fire brand) and holding brand (tattoo or fire brand) required. The breeder may devise his own system, using a series of numbers or letters, or a combination of numbers and letters.
Chianina	All animals must be individually tattooed.
Devon	Each breeder is assigned a herd tattoo code of 3 letters, which must be applied to the right ear. The individual herd number plus the letter code for the year of birth must be tattooed in the left ear.
Dexter	Tattoo in either or both ears.
Galloway	Must have ear tattoo.
Gelbvieh	Each calf must be permanently identified with ear tattoo, freeze brand, or fire brand, showing breeder's 3-letter herd prefix, a 3 or 4 digit number, and the international year code letter suffix. Example: RFR 3136 E; where RFR stands for Rocky Ford Ranch, the 3136 is the number within the individual herd; and the E the 1973 year code.
Hereford	Each animal, for which application for registry is submitted, must be tattooed in one ear. Tattooing in both ears is recommended. The Association recommends (1) starting with number "1" in each ear and proceeding upward in regular order to 999; (2) preferably, limiting numbers to 3 digits; and (3) using one or more means of identification in addition to the tattoo (horn brands, neck chains, or ear tags).
Limousin	All animals must have the first owner's herd prefix in one ear (4 letters). Each individual tattoo number must also have a letter at the end indicating the year of birth. The herd prefix may be in one ear and the calf's herd number and year code in the opposite ear.
Maine-Anjou	The permanent identification of an animal must consist of 3 parts: (1) breeder letters—a set of 3 or 4 letters to identify the first owner; (2) animal number—which is an individual identification; and (3) year letter—which represents the year the animal was born.
Marchigiana	All registered animals must have either an ear tattoo or a brand. The Association prefers that breeders also use herd letters, which they will reserve on request. The Association uses the tattoo as part of the animal's registered name; for example, XYZ Miss Letargo 338.
Murray Grey	Each animal must be tattooed in the left ear. If member so desires, a corresponding tattoo may also be placed in the right ear. The international year and letter designations are required.
Pinzgauer	Any three numbers, or letters, and one additional letter tattoo character must be inserted in either ear of each animal prior to registration. The last character of the tattoo must be the international letter denoting the year of birth.
Polled Hereford	Each animal, for which application for registry is submitted, must be identified by a permanent tattoo. The Association recommends 2 tattoos; a herd code in one ear, and numerals indicating the calf's number and a letter indicating the year of birth in the other. Each animal of the same sex registered by any one breeder should be tattooed differently.
Polled Shorthorn	Each animal must have a tattoo number in its ear. No two animals of the same sex in a herd may be tattooed identically.
Red Angus	Calves must be tattooed, not exceeding 4 digits in one ear, as follows: Right ear to have owner's assigned (by Sec.-Treas.) letters and last digit of year of animal's date of birth. Left ear to have herd identification number of owner's own system, but the animal must be definitely identified and without duplication in the herd. Special symbols, diagonals, brands, bars, joined letters, etc., cannot be recorded.
Red Brangus	Permanently identified by fire brand, including the owner's or breeder's brand and the animal's private herd number.
Red Poll	Breeders and first owners select their own tattoo marking system. However, identifying numbers must be tattooed in both ears, but either the same or different marks may be used in each ear (if different marks are used, they must be so specified on application for registry). Animals of the same sex and near the same age must be tattooed with a different number, but animals of different sex and widely different age may have the same number.
Salers	Calf must have individual tattoo number prior to registration. Only alphabetical letters and/or numbers may be used. Tags may be used to augment tattoos, but they cannot be used to replace them.
Santa Gertrudis	Prior to classification, each animal must be numbered by fire brand or freeze brand so that it can be individually identified. Two or more animals of the same sex may not bear the same number in a given herd for a minimum period of 10 years.
Scotch Highland	Owners must submit herd designation, which must be tattooed in ear or branded on animal; and herd designation must have prior approval in registry office so that no two breeders use the same designation. The Association recommends (1) that the left ear carry the 2 letters for herd designation, followed by the year (thus, a calf in 1976 on the Double X ranch might be marked XX76), and (2) that individual animal numbers either appear in the right ear or be branded on the animal.
Shorthorn	Each animal, for which application for registry is submitted, must be tattooed with a number in one ear. A letter or initial may or may not precede the number. The application for registry must show whether the tattoo appears in the calf's right or left ear. Duplication of numbers in the same sex and herd is not permissible.
Simmental	Each animal to have a private herd number (brand or tattoo), which shall include the International Year/Letter Designation.
South Devon	Each animal must be tattooed or branded.
Sussex	A designation mark comprised of 3 letters will be allocated by the Association for the exclusive use of each breeder in tattooing calves. A "year letter" denotes the year of birth. Calving season is also indicated by numbers.

(Continued)

TABLE 17-2 (Continued)

Breed	Association Rules Relative to Marking
Tarentaise	Each animal for which registration is applied must be tattooed with a breeder's member or nonmember number (assigned by the Association), the year letter designate of the animal's birth (international year/letter designation), and the individual animal's identification number. The Association suggests that this breeder number be tattooed either in front or above the year letter designate and identification number. Member and nonmember numbers assigned by the A.T.A. office are six (6) digit numbers, and the breeder member numbers always begin with 0 and nonmember numbers begin with a 9. The breeder member number is the last three (3) digits of the number assigned, and the nonmember number is the last four (4) digits preceded by the 9 (5 digits in all).
Texas Longhorn	Fire or acid brands, showing ownership and private herd number are required.
Welsh Blacks	All purebred animals must be ear tattooed. Tattoo shall be in the ear (left or right) as designated by the Association Secretary and shall include the following: herd letters, assigned by the Secretary; and the herd numbers of the animal, along with the designated year letter to indicate the year of birth. In grading up animals, a brand and an identification number may be used or the animal must be ear tattooed with both the herd letters and an identification number as in the purebred book.

TABLE 17-3

MARKING OR IDENTIFYING GUIDE FOR REGISTERED DAIRY CATTLE

Breed	Association Rules Relative to Marking
Ayrshire	All calves must be tattooed before leaving individual pens or ties. Both ears may be used and the letters and numbers in the ears must be stated on the application for registration. Tattoos must include at least one letter and one number. Duplicate tattoos are not allowed in the same herd. The letters "I," "O," "Q," and "V" may be used only if accompanied by one or more other letters in the same ear. Tattoos may not exceed a total of 5 letters and numbers per line in each ear. The number shall be followed by the year letter designated by the Association.
Brown Swiss	Each animal must be tattooed in ear with indelible ink with such letters and numbers as the owner may select. No two animals in the same herd, of the same sex, shall have the same number. Both ears may be used. If only one ear is used, it is recommended that it be the left ear. All calves must be tattooed before leaving individual pens or ties.
Guernsey	The animal must be plainly tattooed in the ear with indelible ink or paste before application for registration is made. Both ears may be used, but it is not required. The tattoo may consist of either a series of numbers or a combination of letters and numbers selected by the owner but may not exceed a total of 6 numbers and letters. No two animals in the same herd and of the same sex can have the same tattoo. Vaccination tattoos are not acceptable as identification for registration.
Holstein-Friesian	Ear tag identification may be used but is not considered official for registration by the Association.
Jersey	All calves must be tattooed before leaving individual pens or ties. Both ears may be used and the letters and numbers in the ears must be stated on the application for registration. Tattoos must include at least one letter and one number. The letters "I," "O," "Q," and "V" may be used only if accompanied by an additional letter in the same ear. Tattoos may not exceed a total of 7 letters and numbers in each ear. The Association recommends that calves be tattooed in both ears with the same tattoo.
Milking Shorthorn	Milking Shorthorn cattle cannot be registered unless they have been tattooed in the ear with an individual identification number. An identification letter or initial may precede the number if desired. The application for registry of a calf must show whether the tattoo number appears in the right or left ear. Duplication of numbers for calves of same sex in the same herd is not permissible. Use of initial letters without a number is not sufficient—a number is required.

TABLE 17-4

MARKING OR IDENTIFYING GUIDE FOR REGISTERED SHEEP AND GOATS

Breed	Association Rules Relative to Marking
Cheviot	All sheep for which application is made for registry must be marked with either (1) a metallic label in the ear, or (2) tattoo numbers. The metal label or tattoo must bear the breeder's or applicant's name or initials thereon. Numbers shall not be duplicated. Should any sheep be labeled with another breeder's label, it shall not be changed. If a label is lost, it must be promptly replaced with a duplicate of the original. Association ear labels will be furnished by the secretary as part of the registration charges, but there is a specific charge for duplicate labels.
Columbia	At the time sheep are inspected for registration, the inspector must ear tag (with a tag provided by the Columbia Sheep Breeders' Association of America) and tattoo all recorded sheep in the left ear (the ear tag and tattoo numbers being identical). Sheep rejected by the inspector must be identified by placing "0000" (4 zeros) in the left ear with a tattoo. Flock number must be placed in the right ear.
Corriedale	Individual flock tag with assigned designation of name or initials. Either metal or plastic tags accepted. The registration number tag is not required but may also be used at the breeder's option.
Cotswold	Breeder must use private ear labels with name and number, which must be given when application is being made for registry. The Association provides a metal ear tag for the registry number.
Debouillet	All sheep must be ear tagged by the inspector, preferably in the right ear. Ear tag identifies the flock number (each breeder having his own flock number), the year of birth, whether permanent registry, and the individual number.
Delaine Merino	**American and Delaine-Merino Record Assn.:** The breeder's ear tag—which must carry the name of the breeder and the flock number of the sheep, with no duplication of flock numbers—is the identity of the sheep under registration. **Black-Top Delaine Merino Sheep Breeders' Assn.:** Each animal admitted to record shall have a number assigned to it, which shall correspond with the number of the ear tag furnished by the Secretary.
Dorset	Sheep offered for registration must bear (in either ear) the owner's private flock record tag with number, and this number must be given on the application form. When approved for registration, an Association tag is issued with certificate of register. The latter tag must be inserted in the other ear. Tattoo marking is accepted, but the Association tag must be used also.
Finnsheep	Breeder's tag, carrying the name and initials of the owner of the dam when lamb was dropped, plus the lamb's individual number and tattoo, placed in left ear. Association number and ear tags will be furnished by Association and shall be placed with tattoo in right ear.
Hampshire	When approved for registration, an individual ear tag (with Association numbers) for each animal is forwarded to the owner. This tag must be inserted either in place of the private tag or in addition thereto. The Association recommends the use of both tags, thereby providing identity if one tag is lost.
Lincoln	A tattoo or metal ear tag may be used for the private or flock number. The Association provides a loop-style metal ear tag for the registry number.
Montadale	The Montadale Association provides a standard ear tag for each animal accepted for registry. Each breeder is required to purchase flock tags with name, address, and consecutive numbers.
Oxford	Breeder must use private ear number, which must be given on the pedigree when making application for registry. Association ear tag number is furnished when animal is registered.
Panama	At the time the sheep are inspected for registration, the inspector must tattoo in the right ear all sheep that are accepted. The tattoo shall consist of the breed insignia followed by the flock number of the respective flock. Also, at the time of inspection, a small V-shaped notch is cut into the extreme tip of the right ear of all animals accepted for registration.
Rambouillet	The Association requires the use of, but does not furnish, metal or plastic ear tag containing a name or initial and a number, as identification. Some breeders use duplicate tags, one for each ear. Prior to registry, each breeder must file with the secretary a sample ear tag identical to those in use in his flock. The Association recommends, but does not require, supplementary means of identification, such as tattoos.
Romney	When applications are approved, the secretary shall assign the record number and furnish a metal ear tag with numbers corresponding to that shown on the certificate of registration, which tag shall be placed in the left ear of the sheep (or if preferred, the number may be tattooed in the left ear). The right ear should be used for the breeder's tattoo or tag.
Shropshire	Prior to recording, each sheep must wear a permanent identification in the ear. This may be either a tag or tattoo, and it must show the number and the applicant's name or initials thereon. No number shall be duplicated; hence, no 2 sheep can be registered with the same number. The Association assigns a registry number and furnishes an ear tag bearing numbers corresponding to those shown on the certificate of registry. This tag must be inserted in the ear of the sheep, promptly. Thus, all registered Shropshires should be wearing 2 ear tags (one in each ear). Should either tag get lost, it should be replaced with a duplicate of the original.
Southdown	Each animal shall wear a label in the ear. This tag shall show a flock number and the name or initial of the breeder, owner, or applicant for registration.

(Continued)

TABLE 17-4 (Continued)

Breed	Association Rules Relative to Marking
Suffolk	**American Suffolk Sheep Society:** All registered sheep and those eligible for registration must be positively identified at all times by means of ear tag, ear notch, tattoo, Wilmark, or some other method. **National Suffolk Sheep Association:** 1. Each sheep must be ear marked with the breeder's individual identification and number, by means of an ear tag, a tattoo, or ear notches. 2. Also, the Association furnishes a tag carrying the animal's registration number, which must be inserted in the animal's ear in addition to the breeder's private identification.
Targhee	The "breeder's identification" of all sheep consists of a flock ear tag inserted in the left ear at the time of birth. This tag must bear the breeder's name and year of birth. The "flock registration" of sheep passing inspection requires that the right ear shall carry (1) the breeder assigned tattoo number, and (2) the Association symbol tag. The "stud registration" of sheep passing inspection requires that the right ear shall carry (1) the tattooed Association individual number corresponding to the ear tag number, and (2) the Association serially numbered ear tag.
Tunis	When approved for registration, an individual ear tag (with Association numbers) for each animal is forwarded to the owner. This tag must be inserted either in place of the private tag or in addition thereto.
Angora goats	In order to be eligible for registration, goats must have any 2 of the following 3 marks of identification, the choice of the 2 marks being optional with the owner: 1. A metal or plastic ear tag with the owner's initial and private number. 2. The owner's private number either tattooed or burned in one ear. 3. Notched according to the system shown.  Fig. 17-2. One of the accepted methods for marking registered Angora goats. (Drawing by R. F. Johnson)
Dairy goats	**American Goat Society, Inc.:** The American Goat Society, Inc., reports that "Tattooing is practiced by breeders. Herd letters are assigned by the record association on request." **American Dairy Goat Assn.:** Herd letters are assigned for use in right ear. The left ear should carry an individual identifying tattoo; i.e., 1980 should be designated "A," and the first 3 kids of that year would be tattooed as follows: "A-1," "A-2," and "A-3."

TABLE 17-5

MARKING OR IDENTIFYING GUIDE FOR REGISTERED SWINE

Breed	Association Rules Relative to Marking
American Landrace	The Association accepts either ear notching or tattoo, or both. The Association recommends, but does not require, the Universal Ear Notching System shown in Fig. 17-3. 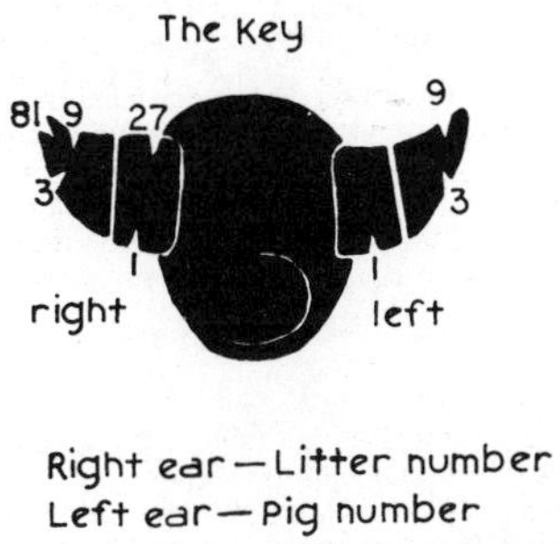 Fig. 17-3. Universal Ear Notching System (also known as the 1-3-9-27 system) used, or recommended, by most swine registry associations. Right ear is used for litter mark, and left ear is used for individual pig number. Up to 161 litters can be marked with this system.
Berkshire	The Association requires that all litters and each individual pig be identified according to the Universal Ear Notching System shown in Fig. 17-3, prior to one week of age.
Chester White	The Association recommends the Universal Ear Notching System shown in Fig. 17-3.
Duroc	The Association requires the Universal Ear Notching System shown in Fig. 17-3.
Hampshire	The Association requires the Universal Ear Notching System shown in Fig. 17-3.
Hereford	Ear Notches by the National Hereford Hog Record Association required. 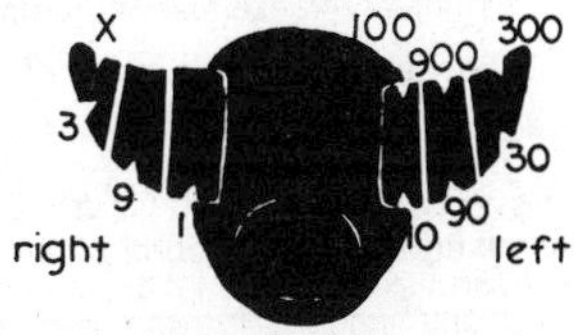 Fig. 17-4. Shows all the notches used in this system. Of course, no one pig would have all the notches indicated. Provision is made for individual notching of ears and numbering the pigs individually from 0 to 999. The left ear carries the litter number and the right ear carries the individual pig number. For example, pig number 118 carries the litter marks 100 and 10, and the individual mark within the litter is 8; thus his number is 118. The system provides for the identification of the litters by 10s, starting with litter 0 (no marks) and continuing to 990. The pigs within each litter are marked individually starting with 0 (no marks) and running to 9. This takes care of 10 pigs. If there are more than 10 pigs in the litter, the others are marked by means of a notch called ×.
Inbred breeds (registered by Inbred Livestock Association)	The Association requires that all pigs be marked according to the system shown in Fig. 17-4.
OIC	Animals must be ear notched and the position of the notch indicated on pedigree when sent for record. All pigs in the same litter must receive and bear the same ear mark, except by permission for experimental purposes.
Poland China	The Association requires the Universal Ear Notching System shown in Fig. 17-3.
Spotted	All pigs must be ear marked within 10 days of birth, and each litter must be marked differently. The Association suggests the Universal System shown in Fig. 17-3.
Tamworth	The Association requires the Universal Ear Notching System shown in Fig. 17-3.
Yorkshire	The Association requires that individual notches be placed in the ear of each pig no later than one week of age; and the Association requires the Universal Ear Notching System shown in Fig. 17-3.

TABLE 17-6

MARKING OR IDENTIFYING GUIDE FOR REGISTERED HORSES AND PONIES

Breed	Association Rules Relative to Marking
American Creme	Tattoo or natural marks required for registry.
American Gotland Horse	Foals tattooed in upper lip; with number, or with number and letter; in sequence in breeder's herd. Tattoo becomes a permanent part of registration record.
American White Horse	Tattoo or natural marks required for registry.
Appaloosa	Appaloosas that race must have their registration number tattooed in the upper lip.
Arabian	The Association sponsors a voluntary freeze marking program.
Quarter Horse	Racing Quarter Horses must be tattooed on the inside of the upper lip.
Shetland Pony	Lip tattooing is strongly recommended but not required.
Standardbred	Horse racing at extended parimutuel meetings must be lip tattooed.
Thoroughbred	Lip tattoo on the inside of the upper lip required of race horses.

TABLE 17-7

A.I. RULES OF BEEF AND DUAL-PURPOSE CATTLE ASSOCIATIONS

Breeds	Registry Association	Pertinent Rules or Attitude of Each Registry Association Relative to Artificial Insemination
Angus	American Angus Assn. 3201 Frederick Blvd. St. Joseph, Mo. 64501	The owner of the dam must obtain one A.I. Service Certificate from the owner of the sire for each calf to be registered. Unlimited number of certificates available. When applying for A.I. Service Certificates, the sire owner must certify that to the best of his knowledge the bull has or has not transmitted any of the following genetic defects: red coat color, dwarfism, osteopetrosis, or double muscling. Any such defect will be listed on the A.I. Service Certificate, and upon verification, such information will be made available to any member upon request. Calves may be registered after sire's death, without time limit, if his death and semen inventory are reported to the Association within 90 days of death.
Barzona	Barzona Breeders Assn. of America P.O. Box 631 Prescott, Ariz. 86302	Animals resulting from A.I. eligible for registration, provided both parents are registered in the Association and the sire has been blood typed.
Beef Friesian	Beef Friesian Society 210 Livestock Exchange Bldg. 4701 Marion Street Denver, Colo. 80216	Calves resulting from A.I. service eligible for registration provided a Semen Sale Report has been filed by the seller of the semen. These reports are due twice annually from all sellers of Beef Friesian semen.
Beefmaster	Beefmaster Breeders Universal G.P.M. Tower South, Suite 350 800 N.W. Loop 410 San Antonio, Tex. 78216	No distinction is made between A.I. and natural service.
	Foundation Beefmaster Assn. Livestock Exchange Bldg. Suite 200 4701 Marion Street Denver, Colo. 80216	There are no A.I. rules relative to certified Beefmasters. However, in the BBU upgrading program, affidavits are required to the effect that semen from certified Beefmaster sires was used A.I. on listed base females.
	National Beefmaster Assn. 817 Sinclair Bldg. Fort Worth, Tex. 76102	
Belted Galloway	Belted Galloway Society, Inc. P.O. Box 5 Summitville, Ohio 43962	A.I. accepted provided the owner of the bull from which the semen was secured signs a service permit.
Blonde d'Aquitaine	American Blonde d'Aquitaine Assn. Route B, Box 230 Grandview, Ida. 83624	Open A.I. policy. The owner of the dam obtains all necessary forms from the Associaton and completes them. Percentage calves can be recorded in a grading-up program.
Braford	International Braford Assn. P.O. Box 1030 Fort Pierce, Fla. 33450	A.I. is permissible, but discouraged.
Brahman	American Brahman Breeders Assn. 1313 La Concha Lane Houston, Tex. 77054	Animals resulting from A.I. eligible for registration provided there has been compliance with the rules of the Association. So, those intending to breed a Brahman cow A.I. should request the complete rules, *A.I. in Brahman cattle*, from the breed association office.
Brangus	International Brangus Breeders Assn. 9500 Tioga Drive San Antonio, Tex. 78230	A.I.-produced calves are eligible for registration provided (1) bulls from out-of-herd are blood typed, and (2) an A.I. service certificate is obtained, executed as directed, and returned with the application for registry. Calves may be registered after sire's death if his death and semen inventory are reported to the Association within 90 days of death.
Charolais and Charbray	American-International Charolais Assn. 1610 Old Spanish Trail Houston, Tex. 77054	A.I.-produced calves from out-of-herd accepted for registration provided the breeder has complied with the regulations governing A.I. A.I.-produced calves within a herd, where the breeder is the owner of both the sire and dam at the time of service, are accepted the same as progeny from natural service.
Chianina	American Chianina Assn. P.O. Box 890 Platte City, Mo. 64079	The requirements for registry of A.I.-sired calves is the same as for calves sired by natural service. A blood type record of all sires used in A.I. must be on file with the Chianina Association.
Devon	Devon Cattle Assn. P.O. Box 628 Uvalde, Tex. 78801	Progeny of A.I. are registered without restriction provided (1) sire is licensed for A.I. by the Association, and (2) one of the following is furnished: (a) a statment from the sire owner identifying the dam and specifying the date she was bred, or (b) a breeding certificate from the inseminator identifying the licensed bull, the cow, and the breeding date.
Dexter	American Dexter Cattle Assn. 707 W. Water Street Decorah, Iowa 52101	Semen must be from a registered Dexter bull that is approved by the Association.
Galloway	American Galloway Breeders Assn. Route 1, Box 106A Athol, Ida. 83801	Open A.I. rules. The owner of the dam obtains an application for registration from the Association and has it signed by the owner of the sire. Calves may be registered after the death of their sire if his death and semen inventory are reported within 30 days of death.

Footnote on last page of table.

(Continued)

TABLE 17-7 (Continued)

Breeds	Registry Association	Pertinent Rules or Attitude of Each Registry Association Relative to Artificial Insemination
Galloway (continued)	Galloway Cattle Society of America Hennepin, Ill. 61327	
Gelbvieh	American Gelbvieh Assn. 5001 National Western Drive Denver, Colo. 80216	Open A.I. policy. The owner of the dam obtains all necessary forms from the Association and completes them. Percentage calves can be recorded in a grading-up program.
Hereford	American Hereford Assn. 715 Hereford Drive Kansas City, Mo. 64105	The following 2 basic requirements are added to the normal rules of eligibility for registering an A.I.-produced calf: 1. The person registering the calf must be a part owner of the service bull on the breeding date; and there is a limit of four owners per bull. 2. An A.I. blood typing permit must have been previously issued on the service bull by the Association.
Limousin	North American Limousin Assn. 100 Livestock Exchange Bldg. 4701 Marion Street Denver, Colo. 80216	Open A.I. policy. The owner of the dam must obtain a standard PDCA[1] type Breeding Receipt completed and signed by the inseminator. This inseminator must have a signature card on file with the Association. The Breeding Receipt must be attached to the application for registration or recordation. Percentage calves can be recorded in a grading-up program. Calves may be registered after sire's death if his death and semen inventory are reported to the Association within 90 days of death.
Maine-Anjou	American Maine-Anjou Assn. 564 Livestock Exchange Bldg. 1600 Genesee Street Kansas City, Mo. 64102	Open A.I. policy. The only requirement is that bulls used in A.I. be blood typed.
Marchigiana	American International Marchigiana Society P.O. Box 342 Lindale, Tex. 75551	Open A.I. policy. All nonmembers registering A.I.-sired calves are required to blood type them in order to be certain of heritage.
Murray Grey	American Murray Grey Assn. 1222 North 27th Street Billings, Mont. 59107	The registration requirements for A.I. and natural service are the same. The breeder certifies as to the sire semen used. However, a blood type record of all sires in A.I. must be on file with the Association.
Normande	American Normande Assn. P.O. Box 350 Kearney, Mo. 64060	The registration requirements are the same for animals sired by A.I. or natural service. However, a blood type record of all sires used in A.I. must be on file with the Association.
Polled Hereford	American Polled Hereford Assn. 4700 East 63rd Street Kansas City, Mo. 64130	When it is desired to use a bull A.I., the owner must have him blood-typed and obtain from the Association a permit to breed A.I. The owner of the dam must obtain one A.I. Service Certificate from the owner of the sire for each calf to be registered. This must be completed and attached to the application for registration. The number of A.I. Service Certificates issued is limited; hence, it is wise to obtain the certificate before breeding the cow. The owner of a bull must report his death to the Association within 30 days, and provide an inventory of semen. Various restrictions apply as to the period of time that semen may be used following the death of a bull. But there are no restrictions on the use of semen from a superior sire after death. The registration requirements are the same for animals sired by A.I. or natural service. However, a blood type record of all sires used in A.I. must be on file with the Association.
Polled Shorthorn	American Shorthorn Assn. 8288 Hascall Street Omaha, Neb. 68124	Calves sired by A.I. are eligible for registration provided the following rules are complied with: The sire has been blood-typed and a copy of the report is on file with the Association; and an A.I. Service Certificate properly executed accompanies the application for registry. The semen of a dead bull may be used indefinitely.
Red Angus	Red Angus Assn. of America 4201 I-35 North Denton, Tex. 76201	An A.I. Bull Permit must first be completed and signed by the owner of the sire and the owner of the dam, and filed with the Association. The owner of the dam then purchases a photostatic copy of this permit from the Association, and attaches it to the application for registration. Calves can be registered for 15 years after the death of their sire if the sire's death and semen inventory were reported within 90 days of his death.
Red Brangus	American Red Brangus Assn. Box 1326 Austin, Tex. 78767	The owner of the dam must also be the owner of the sire or one of not more than three owners of the sire. All bulls used in A.I. must be blood-typed, with a record of the blood type filed with the registry Association. Eligible calves may be registered for 15 years following the death of a sire.
Red Poll	American Red Poll Assn. Box 35519 Louisville, Ky. 40232	The owner of the dam must obtain a standard PDCA[1] type Breeding Receipt completed and signed by the inseminator. This inseminator must have a signature card on file with the Association. The Breeding Receipt must be attached to the application for registration. Semen can be used after the death of the bull.
Salers	American Salers Assn. P.O. Box 30 Weiser, Ida. 83672	Open A.I.

Footnote on last page of table.

(Continued)

TABLE 17-7 (Continued)

Breeds	Registry Association	Pertinent Rules or Attitude of Each Registry Association Relative to Artificial Insemination
Santa Gertrudis	Santa Gertrudis Breeders International P.O. Box 1257 Kingsville, Tex. 78363	All bulls which are the source of semen for out-of-herd A.I. must be blood-typed, with a record of the blood type filed with the Association. The owner of the dam must obtain one A.I. Service Certificate from the owner of the sire for each calf to be registered. This must be completed and attached to the application for registration. The number of A.I. Service Certificates issued is limited so it's wise to obtain the certificate before breeding the cow. Percentage females can be recorded in a grading-up program without an A.I. Service Certificate. Frozen semen can be used after the death of the sire if his death and semen inventory are reported to the Association within 90 days of his death.
Scotch Highland	American Scotch Highland Breeders' Assn. P.O. Box 81 Remer, Minn. 56672	The owner of the dam must obtain a Breeders' Certificate from the Association which must be completed and signed by the inseminator on the date of service. This inseminator must have a signature card on file with the Association. One copy of this form must be sent immediately to the Association. One copy is retained by the owner of the dam and is used as the "application for registration" of the resulting calf. A.I. calves must be registered before one year of age.
Shorthorn	American Shorthorn Assn. 8288 Hascall Street Omaha, Neb. 68124	Calves sired by A.I. are eligible for registration provided the following rules are complied with: The sire has been blood-typed and a copy of the report is on file with the Association; and an A.I. Service Certificate properly executed accompanies the application for registry. The semen of a dead bull may be used indefinitely.
Simmental	American Simmental Assn. 1 Simmental Way Bozeman, Mont. 59715	Open A.I. policy. The only requisite is that the bull must be blood-typed before any of his A.I. resulting calves can be registered. Percentage calves can be recorded in a grading-up program.
South Devon	North American South Devon Assn. P.O. Box 68 Lynnville, Iowa 50153	Open A.I. policy. The owner of the dam obtains all necessary forms from the Association and completes them. Percentage calves can be recorded in a grading-up program.
	International South Devon Assn. P.O. Box 68 Lynnville, Iowa 50153	Calves resulting from A.I. or ova transplant accepted for registry. Part ownership of the sire not required. Blood type of sire must be filed with registry.
Sussex	Sussex Cattle Assn. of America, P.O. Drawer AA, Refugio, Tex. 78377; with actual registration in the Sussex Cattle Society, 12 Lonsdale Gardens, Tunbridge Wells, Kent, England	Sussex cattle produced by means of A.I. are eligible for registration. Semen must be from a Sussex bull registered with either the American Sussex Cattle Association or the English Herd Book, or both. List of bulls used by A.I. must be filed with the Association by owner upon commencing A.I. To register an A.I. calf, a statement must be attached to the application giving the following information: identity of the approved bull, identity of the cow, date of breeding, and certification by the A.I. technician who inseminated the cow.
Tarentaise	American Tarentaise Assn. 123 Airport Road Ames, Iowa 50010	Open A.I. policy. A blood type record of all sires in A.I. must be on file with the Association. For Tarentaise raised in the U.S. or Canada, it is recommended (1) that a sire used in A.I. have a yearling weight ratio of 110 or above, and (2) that any Tarentaise cross bull used in an upgrading program have a weaning weight ratio of 110 or above.
Texas Longhorn	Texas Longhorn Breeders Assn. of America 3701 Airport Freeway Fort Worth, Tex. 76111	A.I.-produced calves are eligible for registration provided (1) semen is furnished by a reputable breeder, breeding service, or inseminator, and taken from approved bulls; (2) the cow bred by A.I. is properly identified; and (3) the application for registration is accompanied by a breeding receipt, signed by the inseminator, giving certain specified information.
Welsh Black	United States Welsh Black Cattle Assn. Route 1 Wahkon, Minn. 56386	Open A.I. policy. The inseminator must record each insemination on an approved A.I. breeding record form at the time of insemination; and a copy of this record form must accompany each application for registry.

[1]PDCA is the abbreviation for Purebred Dairy Cattle Association.

TABLE 17-8

A.I. RULES OF DAIRY CATTLE ASSOCIATIONS

Breed	Registry Association	Pertinent Rules or Attitude of Each Registry Association Relative to Artificial Insemination
Ayrshire	Ayrshire Breeders' Assn. 2 Union Street Brandon, Vt. 05733	Compliance with the "Requirements Governing Artificial Insemination of Purebred Dairy Cattle," adopted by The Purebred Dairy Cattle Association and The National Association of Animal Breeders. Copy may be obtained from each dairy breed registry association.
Brown Swiss	Brown Swiss Cattle Breeders' Assn. Box 1038 Beloit, Wisc. 53511	Compliance with the "Requirements Governing Artificial Insemination of Purebred Dairy Cattle," adopted by The Purebred Dairy Cattle Association and The National Association of Animal Breeders. Copy may be obtained from each dairy breed registry association.

(Continued)

TABLE 17-8 (Continued)

Breed	Registry Association	Pertinent Rules or Attitude of Each Registry Association Relative to Artificial Insemination
Guernsey	The American Guernsey Cattle Club 2105J S. Hamilton Road Columbus, Ohio 43227	Compliance with the "Requirements Governing Artificial Insemination of Purebred Dairy Cattle," adopted by The Purebred Dairy Cattle Association and The National Association of Animal Breeders. Copy may be obtained from each dairy breed registry association.
Holstein-Friesian	Holstein-Friesian Assn. of America Box 808 Brattleboro, Vt. 05301	Compliance with the "Requirements Governing Artificial Insemination of Purebred Dairy Cattle," adopted by The Purebred Dairy Cattle Association and The National Association of Animal Breeders. Copy may be obtained from each dairy breed registry association.
Jersey	The American Jersey Cattle Club P.O. Box 27310 Columbus, Ohio 43227	Compliance with the "Requirements Governing Artificial Insemination of Purebred Dairy Cattle," adopted by The Purebred Dairy Cattle Association and The National Association of Animal Breeders. Copy may be obtained from each dairy breed registry association.
Milking Shorthorn	American Milking Shorthorn Society 1722JJ S. Glenstone Avenue Springfield, Mo. 65804	The American Milking Shorthorn Society does not certify technicians. But a breeding receipt is required where a cow is serviced by an affiliated technician, and a receipt of purchase is required where semen is purchased. No breeding receipt is required where the owner of the dam is the owner of the bull from which semen was obtained.

TABLE 17-9

A.I. RULES OF SHEEP AND GOAT ASSOCIATIONS

Breed	Registry Association	Pertinent Rules or Attitude of Each Registry Association Relative to Artificial Insemination
Cheviot	American Cheviot Sheep Society, Inc. R.R. #1, Box 100 Clarks Hill, Ind. 47930	Accepted if certified to by veterinarian or A.I. technician and accompanied by $10 registration fee.
Columbia	Columbia Sheep Breeders Assn. of America P.O. Box 272 Upper Sandusky, Ohio 43351	No rules at present.
Corriedale	American Corriedale Assn., Inc. Box 29C Seneca, Ill. 61360	No rules. Breeders who wish to use A.I. should write to the registry for prior approval.
Debouillet	Debouillet Sheep Breeders Assn. 300 S. Kentucky Roswell, N.M. 88201	No rules.
Delaine Merino	The American and Delaine Merino Record Assn. 1193 Township Road 346 Nova, Ohio 44859	No rules at present, but the Association is favorable.
Dorset	The Continental Dorset Club P.O. Box 577 Hudson, Iowa 50644	No rules. To 1976, no A.I.-sired lambs accepted.
Finnsheep	Finnsheep Breeders Assn., Inc. P.O. Box 34303 Indianapolis, Ind. 46234	No restrictions on the use of A.I. Pedigree identity same as required for natural service. Sheep produced by either A.I. or natural service are so identified on the records of the Association.
Hampshire	The American Hampshire Sheep Assn. Box 345 Ashland, Mo. 65010	No printed rules. But the Association is receptive.
Lincoln	National Lincoln Sheep Breeders' Assn. R.R. #6, Box 24 Decatur, Ill. 62521	No rules.

(Continued)

TABLE 17-9 (Continued)

Breed	Registry Association	Pertinent Rules or Attitude of Each Registry Association Relative to Artificial Insemination
Montadale	Montadale Sheep Breeders' Assn., Inc. P.O. Box 44300 Indianapolis, Ind. 46244	No rules.
Oxford	American Oxford Down Assn. Route 4 Ottawa, Ill. 61350	No rules. But receptive to A.I. registration.
Rambouillet	American Rambouillet Sheep Breeders' Assn. 2709 Sherwood Way San Angelo, Tex. 76901	Accepted. Require proof that semen was taken from the particular ram.
Romney	American Romney Breeders Assn. 4375 N.E. Weslinn Drive Corvallis, Ore. 97333	No rules.
Shropshire	American Shropshire Registry Assn. P.O. Box 1970 Monticello, Ill. 61856	No rules.
Southdown	American Southdown Breeders' Assn. Route 4, Box 14B Bellefonte, Penn. 16283	No printed rules. But Secretary states that the Association has accepted limited number of A.I.-sired sheep certified to by collector and inseminator and accompanied by $15 fee.
Suffolk	American Suffolk Sheep Society 55 East 100 North Logan, Utah 84321	Breeder must furnish sire's pedigree; owner must sign application for registry.
	National Suffolk Sheep Assn. P.O. Box 324 Columbia, Mo. 65201	Registry of A.I.-produced lambs is limited to owner of ram and ewe at time of service.
Tunis	National Tunis Sheep Registry R.D. 1 Wayland, N.Y. 14572	No rules.
Angora Goats	American Angora Goat Breeders Assn. Rocksprings, Tex. 78880	No rules.
Dairy Goats	American Dairy Goat Assn. P.O. Box 865 Spindale, N.C. 28160	Accepted provided application for registration is accompanied by (1) Certificate of Collection of Semen signed by the owner or lessee of buck or owner of semen, and (2) Certification of Artificial Insemination signed by the owner of doe or owner's agent, and provided each doe inseminated is individually tattooed and record of her A.I. is sent to the ADGA.
	American Goat Society, Inc. Route 2, Box 112 DeLeon, Tex. 76444	Accepted provided (1) sire is approved by official classifier, (2) there is proof of breeding service and identification of female, and (3) reliable record of frozen semen ampule is presented.

TABLE 17-10

A.I. RULES OF SWINE ASSOCIATIONS

Breed	Registry Association	Pertinent Rules or Attitude of Each Registry Association Relative to Artificial Insemination
Berkshire	American Berkshire Assn. 601 West Monroe Street Springfield, Ill. 62704	Accepted provided an A.I. certificate is signed and dated by both the inseminator and the owner of the boar.
Chester White	Chester White Swine Record Assn. Box 228 Rochester, Ind. 46975	Accepted. If the owner of the sow being bred is not the owner of the boar, a breeding certificate must be obtained from the owner of the boar.
Hampshire	Hampshire Swine Registry 1111 Main Street Peoria, Ill. 61606	Accepted provided a special A.I. breeding certificate is executed and properly signed by the owner of the boar, the receiver of the semen, the owner of the sow, and the inseminator. Charges are the same as for natural service.

(Continued)

TABLE 17-10 (Continued)

Breed	Registry Association	Pertinent Rules or Attitude of Each Registry Association Relative to Artificial Insemination
Hereford	National Hereford Hog Record Assn. Route 1, Box 37 Flandreau, S.D. 57028	No rules.
Poland China	Poland China Record Assn. 368 W. Douglas, Box B Knoxville, Ill. 61448	Accepted. But litter entry form, or application for registry of first pig out of litter, must be accompanied by (1) properly executed National Association of Swine Records approved A.I. form, and (2) $10 fee.
Spotted	National Spotted Swine Record, Inc. 110 West Main Street Bainbridge, Ind. 46105	A.I.-sired pigs accepted. Affidavit required where litter is sired by a boar in an A.I. stud.
Tamworth	Tamworth Swine Assn. 2656 Horner Road Winchester, Ohio 45697	No rules.
Yorkshire	American Yorkshire Club, Inc. Box 2417 West Lafayette, Ind. 47906	A $5/litter registration fee (plus charges for individual pedigrees at the regular rate) is charged on A.I.-sired litters (1) if sire is owned jointly by two or more breeders, (2) if sire is not owned by owner of dam at time of service, or (3) if semen is sold, bought, or used from another breeder or commercial provider.

TABLE 17-11

A.I. RULES OF HORSE ASSOCIATIONS

Breed	Registry Association	Present Rules or Attitude of Each Registry Association Relative to Artificial Insemination
Light horses and ponies:		
American Bashkir Curly	American Bashkir Curly Registry P.O. Box 453 Ely, Nev. 89301	No rules relative to A.I. to date.
American Creme Horse	Worldwide Horse Registry for American White and American Creme Box 79 Crabtree, Ore. 97335	A.I. discouraged, but accepted.
American Gotland Horse	American Gotland Horse Assn. P.O. Box 263 Jenks, Okla. 74037	No rules.
American Mustang	American Mustang Assn., Inc. P.O. Box 338 Yucaipa, Calif. 92399	Accepted provided certified authentication is provided by the attending veterinarian of both mare and the stallion.
American Paint Horse	American Paint Horse Assn. P.O. Box 18519 Fort Worth, Tex. 76118	Accepted only if A.I. used (1) within 24 hours of collection, and (2) on premises of collection.
American Part-Blooded	American Part-Blooded Horse Registry 4120 S.E. River Drive Portland, Ore. 97222	Accepted provided customary proof of breeding is furnished.
American Saddlebred Horse	American Saddlebred Horse Assn. 929 S. Fourth Street Louisville, Ky. 40203	Accepted provided A.I. takes place (1) on premises where stallion is standing, and (2) in presence of owner or party authorized to sign certificate of breeding.
American Walking Pony	American Walking Pony Assn. Route 5, Box 88 Upper River Road Macon, Ga. 31201	Accepted only if (1) stallion and mare are on same premises and insemination is done by licensed veterinarian, or (2) same veterinarian collects, transports, and inseminates if sire and dam are on different premises.
American White Horse	Worldwide Horse Registry for American White and American Creme Box 79 Crabtree, Ore. 97335	A.I. discouraged, but accepted.

(Continued)

TABLE 17-11 (Continued)

Breed	Registry Association	Present Rules or Attitude of Each Registry Association Relative to Artificial Insemination
Andalusian	American Andalusian Assn. P.O. Box 1290 Silver City, N.M. 88061	The breed registry reports: "Not practiced."
Appaloosa	Appaloosa Horse Club, Inc. P.O. Box 8403 Moscow, Ida. 83843	Accepted provided (1) accompanied by natural service in same heat period, and (2) semen used only where stallion is standing.
Arabian	Arabian Horse Registry of America, Inc. 3435 South Yosemite Denver, Colo. 80231	Accepted provided (1) stallion licensed by Arabian Horse Registry of America for A.I., (2) both collection and insemination take place on same premises, and (3) semen not stored longer than 48 hours.
Buckskin	American Buckskin Registry Assn., Inc. P.O. Box 1125 Anderson, Calif. 96007	**American Buckskin Registry Assn., Inc:** Accepted provided (1) prior notice of intent to use the stallion A.I. is given the Association, and (2) semen is used on premises of collection immediately following collection. **International Buckskin Horse Assn., Inc:** Accepted provided (1) registry is notified in advance of intent to use A.I., (2) collecting, storing, and inseminating performed by approved practitioner, and (3) letters A.I. incorporated in Stud Book and on certificate.
Chickasaw	Chickasaw Horse Assn., Inc., The P.O. Box 607 Love Valley, N.C. 28677 National Chickasaw Horse Assn. Route 2 Clarinda, Iowa 51232	No rules.
Connemara Pony	American Connemara Pony Society R.D. 1 Hoshiekon Farm Goshen, Conn. 06756	Accepted provided (1) fresh semen is used, and (2) signed letters are on file from stallion owner, mare owner, the veterinarian who collected the semen, and the veterinarian who did the inseminating.
Galiceno	Galiceno Horse Breeders Assn., Inc. 111 E. Elm Street Tyler, Tex. 75702	Accepted, without rules or restrictions.
Hackney	American Hackney Horse Society P.O. Box 174 Pittsfield, Ill. 62363	Accepted provided insemination takes place on premises where stallion is standing and in presence of owner or party authorized to sign certificate of breeding for stallion.
Half-Arabian and Anglo-Arabian	International Arabian Horse Assn. 224 E. Olive Avenue Burbank, Calif. 91503	Accepted provided (1) stallion licensed by Arabian Horse Registry for A.I., and (2) registration certificate is stamped to show animal was A.I. produced.
Lipizzan	Lipizzan Assn. of America Woolworth Tower New York, N.Y. 10279	No rules.
Missouri Fox Trotting Horse	Missouri Fox Trotting Horse Breed Assn., Inc. P.O. Box 637 Ava, Mo. 65608	No rules. No horses registered by A.I. to date.
Morab	Morab Horse Registry of America P.O. Box 143 Clovis, Calif. 93612	Accepted provided (1) stallion owner notifies the registry, annually, of intent to use A.I., and (2) insemination is done immediately following collection.
Morgan	American Morgan Horse Assn., Inc. Box 1 Westmoreland, N.Y. 13490	Accepted only if (1) collection and insemination by licensed veterinarian, (2) insemination immediately following collection, and (3) both collection and insemination are on same premises, or (4) need for medical or safety reasons.
National Appaloosa Pony	National Appaloosa Pony, Inc. Box 206 Gaston, Ind. 47342	No rules.
Palomino	Palomino Horse Assn., Inc., The P.O. Box 324 Jefferson City, Mo. 65102 Palomino Horse Breeders of America P.O. Box 249 Mineral Wells, Tex. 76067	**Palomino Horse Assn., Inc.:** Accepted only if semen used at time and place of collection. **Palomino Horse Breeders of America:** Accepted provided there is a properly signed breeder's certificate, but they do not advocate.

(Continued)

TABLE 17-11 (Continued)

Breed	Registry Association	Present Rules or Attitude of Each Registry Association Relative to Artificial Insemination
Paso Fino	American Paso Fino Horse Assn., Inc. 907 Penn Avenue Pittsburgh, Penn. 15221	
	Paso Fino Owners and Breeders Assn., Inc. P.O. Box 764 Columbus, N.C. 28722	Each request handled on individual basis. But a veterinarian must be in attendance at time of both collection and insemination.
Peruvian Paso	American Association of Owners and Breeders of Peruvian Paso Horses P.O. Box 2035 California City, Calif. 93505	**American Association of Owners and Breeders of Peruvian Paso Horses:** A.I. accepted only (1) in case of injury to stallion, and (2) if stallion and mare are on same premises.
	Peruvian Paso Horse Registry of North America P.O. Box 816 Guerneville, Calif. 95446	**Peruvian Paso Horse Registry of N.A.:** Accepted only (1) deemed necessary (in writing) by veterinarian, (2) stallion and mare are at same location, and (3) stallion serves mare naturally once each heat period.
Pinto	Pinto Horse Assn. of America, Inc. 7525 Mission Gorge Road, Suite C San Diego, Calif. 92120	Accepted provided (1) intent of each A.I. breeding requested in letter to registrar; (2) a veterinarian (not necessarily the same one for each step) certifies to collection of semen, insemination of mare, and birth of foal; and (3) blood type evidence of parentage, along with foaling date, is furnished.
Pony of the Americas	Pony of the Americas Club, Inc. P.O. Box 1447 Mason City, Iowa 50401	A.I. on the farm where the stallion is located is approved. But mailing of semen is not allowed.
Quarter Horse	American Quarter Horse Assn. 2736 W. 10th Street Amarillo, Tex. 79168	A.I. not permitted, unless insemination (1) immediately follows collection, and (2) is at the place or premises of collection.
	Standard Quarter Horse Assn., Inc. 4390 Fenton, #206 Denver, Colo. 80212	
Rangerbred	Colorado Ranger Horse Assn., Inc. 7023 Eden Mill Road Woodbine, Md. 21797	No rules. No request to use A.I.
Shetland Pony	American Shetland Pony Club P.O. Box 435 Fowler, Ind. 47944	Accepted only if (1) sire and/or dam incapable of natural service, (2) sire and dam owned by same person, and (3) owner retains dam until she foals.
Spanish Barb	Spanish Barb Breeders Assn. P.O. Box 7479 Colorado Springs, Colo. 80907	No rules. Do not advocate A.I.
Standardbred	United States Trotting Assn. 750 Michigan Avenue Columbus, Ohio 43215	Accepted provided (1) fresh semen is used (frozen or dessicated not permitted), and (2) insemination takes place on same day and same premises where semen was produced.
Tennessee Walking Horse	Tennessee Walking Horse Breeders' and Exhibitors' Assn. of America P.O. Box 286 Lewisburg, Tenn. 37091	Accepted provided insemination (1) is done on premises where stallion is standing, and (2) takes place in presence of owner or party authorized to sign certificate of breeding for the stallion used.
Thoroughbred	Jockey Club, The 380 Park Avenue New York, N.Y. 10017	Natural service only. But the immediate A.I. reinforcement of the stallion's service with a portion of the ejaculate produced by the stallion during such cover is permitted.
Trakehner	American Trakehner Assn., Inc. P.O. Box 268 Norman, Okla. 73070	Accepted provided (1) there is prior approval by the Association, (2) semen is used immediately following collection and at the place or premises of collection, and (3) insemination is under supervision of licensed veterinarian.
Welsh Pony	Welsh Pony Society of America Box 2977 Winchester, Va. 22601	Foals produced by A.I. not accepted for registry.

(Continued)

TABLE 17-11 (Continued)

Breed	Registry Association	Present Rules or Attitude of Each Registry Association Relative to Artificial Insemination
Draft horses:		
Belgian	Belgian Draft Horse Corporation of America P.O. Box 335 Wabash, Ind. 46992	Accepted provided stallion and mare were on the same farm at the time mare was bred.
Clydesdale	Clydesdale Breeders Assn. of the United States Route 1, Box 131 Pecatonica, Ill. 61063	No rules.
Percheron	Percheron Horse Assn. of America P.O. Box 141 Fredericktown, Ohio 43019	Accepted provided (1) semen is obtained from member of Percheron Horse Association of America, or from reputable A.I. establishment; (2) blood type of stallion is on file with the Association; (3) authentication is furnished by owner of stallion, owner of mare, and veterinarian who implanted; and (4) application for registry of A.I.-produced foal is filed before June 1 of the year following date of foaling.
Shire	American Shire Horse Assn. of America 14410 High Bridge Road Monroe, Wash. 98272	Will accept. No rules.
Suffolk	American Suffolk Horse Assn., Inc. Route 1, Box 212 Ledbetter, Tex. 78946	No rules, but favorable toward A.I.
Jacks and donkeys:		
Donkeys	American Donkey and Mule Society, Inc. Route 5, Box 65 Denton, Tex. 68102	No rules.
	Miniature Donkey Registry of the United States, Inc. 1108 Jackson Street Omaha, Neb. 68102	No rules.
Jacks and Jennets	Standard Jack and Jennet Registry of America Sulphur Run Farm Route 1, Box 194 Elk Run, Ky. 42733	Eligible for registration. No stipulations.

SECTION 18

AGRICULTURAL MAGAZINES

Fig. 18-1. Breed magazines publish news items and promote breeds.

The agricultural magazines publish news items and informative articles of special interest to stockmen. Also, many of them employ field representatives whose chief duty it is to assist in the buying and selling of animals.

Table 18-1 lists agricultural magazines and classes them as follows:

1. General agricultural magazines.
2. General livestock magazines.
3. Class of livestock magazines; general or by breeds.

No claim is made that either all or the best magazines are included in Table 18-1. Rather, it is hoped that stockmen will find this list helpful as they choose one or more magazines.

TABLE 18-1
AGRICULTURAL MAGAZINES

Devoted To	Breed (Where Applicable)	Publication	Address
General agriculture		Agri-Finance	5520 Touhy Avenue, Suite G Skokie, Ill. 60076
		Big Farmer	Big Farmer, Inc. 131 Lincoln Highway Frankfort, Ill. 60423
		California Farmer	83 Stevenson Street San Francisco, Calif. 94105
		Capper's Weekly	616 Jefferson Topeka, Kan. 66607
		Country Guide	1760 Ellice Avenue Winnipeg, Manitoba, Canada R3H 0B6
		Country Journal	P. O. Box 870 Manchester Center, Vt. 05255
		Doane's Agricultural Report	800 Manchester Road St. Louis, Mo. 63144
		Farm Journal	230 West Washington Square Philadelphia, Penn. 19105

(Continued)

TABLE 18-1 (Continued)

Devoted To	Breed (Where Applicable)	Publication	Address
General agriculture (continued)		Farmer's Digest	Box 363 Brookfield, Wisc. 53005
		Farmland News	P. O. Box 7305 Kansas City, Mo. 64116
		Grange News	3104 Western Avenue Seattle, Wash. 98121
		MFC News, The	Box 500 Madison, Miss. 39110
		National 4-H News	7100 Connecticut Avenue Chevy Chase, Md. 20815
		National Future Farmer, The	P. O. Box 15160 Alexandria, Va. 22309
		Northwest Farm Paper Unit	Box 2160 Spokane, Wash. 99210
		Ohio Farmer, The	1350 W. Fifth Avenue Columbus, Ohio 43212
		Progressive Farmer	3737 Nobel Avenue Dallas, Tex. 76204
		Successful Farming	1716 Locust Des Moines, Iowa 50336
		Texas Farm and Ranch News	118 W. Nakoma San Antonio, Tex. 78216
		Today's Farmer	201 S. Seventh Street Columbia, Mo. 65201
		West Texas Livestock Weekly	Box 3306 San Angelo, Tex. 76901
General livestock		Advanced Animal Breeder	Box 1033 Columbia, Mo. 65201
		Animal Nutrition and Health	Sandstone Building Mount Morris, Ill. 61054
		California Livestock News	3382 El Camino Avenue #6 Sacramento, Calif. 95814
		Drovers Journal, The	P. O. Box 2939 Shawnee Mission, Kan. 66201
		Feed Management	Sandstone Building Mount Morris, Ill. 61054
		Feedstuffs	Box 1289 Minneapolis, Minn. 55440
		Journal of Animal Science	American Society of Animal Science 309 W. Clark Street Champaign, Ill. 61820
		Livestock Breeder Journal	Box 4264 Macon, Ga. 31208
		Meat Industry	P. O. Box 72 Mill Valley, Calif. 94942
		National Provisioner, The	15 W. Huron Street Chicago, Ill. 60610
		New Mexico Stockman	P. O. Box 7127 Albuquerque, N.M. 87194
		Professional Nutritionist, The	2226 Clay Street San Francisco, Calif. 94115
		Record Stockman, The	4877A Packinghouse Road, Suite 201 Denver, Colo. 80216
		Western Livestock Journal	Nelson R. Crow Publishing, Inc. P. O. Drawer 17F Denver, Colo. 80216
		Western Livestock Reporter	P. O. Box 30758 Billings, Mont. 59107

(Continued)

TABLE 18-1 (Continued)

Devoted To	Breed (Where Applicable)	Publication	Address
Beef and dual-purpose cattle	**General (covers all breeds)**	Arkansas Cattle Business	5400 Murray Little Rock, Ark. 72209
		Beef	1999 Shepard Road St. Paul, Minn. 55116
		Beef Week	P. O. Box 4264 Macon, Ga. 31208
		Better Beef Business	Box 7386 Louisville, Ky. 40207
		Calf News	18345 Ventura Blvd., Suite 303 Tarzana, Calif. 91356
		California Cattleman	P. O. Box 1684 Auburn, Calif. 95603
		Cattleman, The	1301 W. Seventh Street Fort Worth, Tex. 76102
		Cattlemen	1760 Ellice Avenue Winnipeg, Manitoba, Canada R3H OB6
		El Ganadero Internacional	11201 Morning Court San Antonio, Tex. 78213
		Feedlot Management	Box 67 Minneapolis, Minn. 55440
		Florida Cattleman & Livestock Journal	P. O. Box 1403 Kissimmee, Fla. 32741
		Gulf Coast Cattleman	11201 Morning Court San Antonio, Tex. 78213
		Ideal Beef Memo	Route 1, Box 79 Huxley, Iowa 50124
		Oregon Cattleman	1000 N.E. Multnomah Street Portland, Ore. 97232
		Southern Beef Producer	P. O. Box 843 Franklin, Tenn. 37064
		Stock Show, The	The Stock Show Publishing Co. P. O. Box 601 Albany, Tex. 76430
		Washington Cattleman, The	P. O. Box 96 Ellensburg, Wash. 98926
	Beef breed publications:		
	Angus	Angus Journal	Frederick & Brookside St. Joseph, Mo. 64501
		Western States Angus News	Box 30299 Portland, Ore. 97230
	Beefmaster	The Beefmaster Cowman	11201 Morning Court San Antonio, Tex. 78213
	Brahman	Brahman Journal, The	P. O. Box 220 Eddy, Tex. 76524
	Brangus	Brangus Journal	9500 Tioga Drive San Antonio, Tex. 78230
	Charolais	Charolais Journal	1610 Old Spanish Trail Houston, Tex. 77054
	Chianina	American Chianina Journal	P. O. Box 890 Platte City, Mo. 64079
	Hereford	American Hereford Journal, The	Box 4059 Kansas City, Mo. 64101
		Canadian Hereford Digest	5160 Skyline Way N.E. Calgary, Alberta, Canada T2E 6V1
		Texas Hereford	4609 Airport Freeway Fort Worth, Tex. 76117

(Continued)

TABLE 18-1 (Continued)

Devoted To	Breed (Where Applicable)	Publication	Address
Beef and dual-purpose cattle (continued)	**Limousin**	International Limousin Journal	P. O. Box 2205 Fort Collins, Colo. 80522
	Maine-Anjou	Maine-Anjou International	334 Ninth Avenue Calgary, Alberta, Canada T2E 0V6
	Murray Grey	Murray Grey News	P. O. Box 30085 Billings, Mont. 59107
	Polled Hereford	Polled Hereford World	4700 E. 63rd Street Kansas City, Mo. 64130
	Red Angus	American Red Angus	4201 I-35 North Denton, Tex. 76201
	Red Poll	Red Poll News	Box 35519 Louisville, Ky. 40232
	Santa Gertrudis	Santa Gertrudis Journal, The	P. O. Box 2386 Fort Worth, Tex. 76101
	Shorthorn	Shorthorn Country	8288 Hascall Street Omaha, Neb. 68124
	Simmental	North American Simmental	P. O. Box 878 Stephensville, Tex. 76401
		Simmental Country	13 - 4101 19th Street N.E. Calgary, Alberta, Canada T2E 7C4
		Simmental Shield Update	P. O. Box 511 Lindsborg, Kan. 67456
	Welsh Black	Welsh Black Cattle World	Route 1 Wahkon, Minn. 56386
Dairy cattle	**General (covers several breeds)**	Dairy Contact	Suite 214 11802 - 124th Street Edmonton, Alberta, Canada T5L 0M3
		Dairy Herd Management	P. O. Box 67 Minneapolis, Minn. 55440
		Dairyman, The	P. O. Box 819 Corona, Calif. 91720
		Dairymen's Digest	Box 5040 Arlington, Tex. 76005
		Dairynews	831 James Street Syracuse, N.Y. 13203
		Hoard's Dairyman	28 Milwaukee Avenue W. Fort Atkinson, Wisc. 53538
		Pennmarva Magazine	1717 Guynn Oak Avenue Baltimore, Md. 21207
	Dairy breed publications:		
	Ayrshire	Ayrshire Digest	2 Union Street Brandon, Vt. 05733
		Canadian Ayrshire Review	1160 Carling Avenue Ottawa, Ontario, Canada K1Z 7K6
	Brown Swiss	Brown Swiss Bulletin, The	P. O. Box 1038 Beloit, Wisc. 53511
	Guernsey	Guernsey Breeders Journal	Box 27410 Columbus, Ohio 43227
	Holstein	California Holstein News	1177 West Hedges Fresno, Calif. 93728
		Holstein Journal	335 Lesmill Road Don Mills, Ontario, Canada M3B 2V1

(Continued)

TABLE 18-1 (Continued)

Devoted To	Breed (Where Applicable)	Publication	Address
Dairy cattle (continued)	**Holstein (continued)**	Holstein World	P.O. Box 288 Sandy Creek, N.Y. 13145
		Texas Holstein News	Route 1 Buda, Tex. 78610
	Jersey	Canadian Jersey Breeder	343 Waterloo Avenue Guelph, Ontario, Canada N1H 3K1
		Jersey Journal	Box 27310 Columbus, Ohio 43227
	Milking Shorthorn	Journal of the Milking Shorthorn and Illawarra Breeds	1722 JJ S. Glenstone Avenue Springfield, Mo. 65804
Sheep	**General (covers all breeds)**	Montana Wool Grower	P. O. Box 1693 Helena, Mont. 59601
		National Wool Grower	600 Crandall Bldg. Salt Lake City, Utah 84101
		Ranch Magazine	Box 2678 San Angelo, Tex. 76902
		Sheep Breeder and Sheepman, The	P. O. Box 796 Columbia, Mo. 65201
		Shepherd Magazine, The	R.D. 1, Box 67 Sheffield, Mass. 01257
		Speaking of "Columbias"	P. O. Box 272 Upper Sandusky, Ohio 43351
Goats	**General (covers all breeds of milk goats)**	Better Goat Keeping	Harvard, Mass. 01451
		Dairy Goat Journal	P. O. Box 1808 Scottsdale, Ariz. 85252
Swine	**General (covers all breeds)**	Hog Farm Management	P. O. Box 67 Minneapolis, Minn. 55440
		National Hog Farmer	1999 Shepard Road St. Paul, Minn. 55116
		Nebraska Pork Talk	Box 487 Madison, Neb. 68748
		Pig Farming	Fenton House, Wharfedale Road Ipswich, Suffolk, England IP1 4LG
		Pork	P. O. Box 2939 Shawnee Mission, Kan. 66201
		Southern Hog Producer	P. O. Box 843 Franklin, Tenn. 37064
	Swine breed publications:		
	Berkshire	Berkshire News, The	601 W. Monroe Street Springfield, Ill. 62704
	Chester White	Chester White Journal	P. O. Box 228 Rochester, Ind. 46975
	Duroc	Duroc News	1803 W. Detweiller Drive Peoria, Ill. 61615
	Hampshire	American Hampshire Herdsman	1111 Main Street Peoria, Ill. 61606
	Landrace	American Landrace, The	Box 647 Lebanon, Ind. 46052
	Poland China	Poland China World, The	Box B Knoxville, Ill. 61448
	Spotted	Spotted News	110 W. Main Street Bainbridge, Ind. 46105

(Continued)

TABLE 18-1 (Continued)

Devoted To	Breed (Where Applicable)	Publication	Address
Swine (continued)	**Tamworth**	Tamworth News	2656 Horner Road Winchester, Ohio 45697
	Yorkshire	Yorkshire Journal	Box 2417 West Lafayette, Ind. 47906
Horses	**General (covers several breeds)**	American Farriers Journal	P. O. Box L Harvard, Mass. 01451
		American Horseman	Box 12186 Fort Worth, Tex. 76116
		Bluegrass Horseman, The	Box 389 Lexington, Ky. 40501
		Bridle and Bit	Star Route 2 Box 680 Cave Creek, Ariz. 85331
		California Horseman's News	P.O. Box 474 San Marcos, Calif. 92069
		California Horse Review	3203 Orange Grove Avenue North Highlands, Calif. 95660
		Canadian Rider	491 Book Road West Ancaster, Ontario, Canada L9G 3L3
		Capital Horseman	14405 W. 52nd Avenue Arvada, Colo. 80002
		Chronicle of the Horse, The	P. O. Box 46 Middleburg, Va. 22117
		Corinthian, The	10077-C Yonge Street Richmond Hill, Ontario, Canada L4C 1T7
		Corral, The	P. O. Box 110 New London, Ohio 44851
		Cuttin' Hoss Chatter, The	P. O. Box 12155 Fort Worth, Tex. 76116
		Equestrian Trails	P. O. Box 44135 Sylmar, Calif. 91342
		Equus	656 Quince Orchard Road Gaithersburg, Md. 20760
		Horse and Horseman	Box HH Capistrano Beach, Calif. 92624
		Horse & Rider	P. O. Box 555 Temecula, Calif. 92390
		Horse Play	Box 545 Gaithersburg, Md. 20760
		Horse Show	598 Madison Avenue New York, N.Y. 10022
		Horse World	Box 1007 Shelbyville, Tenn. 37160
		Horseman	Box 10973 Houston, Tex. 77292
		Horseman	1485 W. Front Street Lincroft, N.J. 07738
		Horseman's Review	Box 12 Monroe Center, Ill. 61052
		Horsemen's Corral	P. O. Box 110 New London, Ohio 44851
		Horsemens Gazette, The	Box 202 Badger, Minn. 56715
		Horsemen's Journal, The	Suite 317 6000 Executive Blvd. Rockville, Md. 20852

(Continued)

TABLE 18-1 (Continued)

Devoted To	Breed (Where Applicable)	Publication	Address
Horses (continued)		Horsemen's Yankee Pedlar	872 Southbridge Street Auburn, Mass. 01501
		Horses—All	Box 550 Nanton, Alberta, Canada T0L 1R0
		Horses Unlimited	Box 10530 Gladstone, Mo. 64118
		Lariat, The	12675 S.W. First Street Beaverton, Ore. 97005
		National Horse Journal, The	P. O. Box 927, Station F Toronto 5, Ontario, Canada
		National Horseman, The	11603 Shelbyville Road Middletown, Ky. 40243
		Northeast Horseman	P. O. Box 131 Hampden, Me. 04444
		Practical Horseman	Gum Tree Center Unionville, Penn. 19375
		Shining Mountain Sentinel	P. O. Box 15 Brady, Mont. 59416
		Southern Horseman, The	P. O. Box 5735 Meridian, Miss. 39301
		Tack 'N Togs	P. O. Box 67 Minneapolis, Minn. 55440
		Turf and Sport Digest	511 Oakland Avenue Baltimore, Md. 21212
		Washington State Horsemen Canter, The	P. O. Box "X" Kirkland, Wash. 98033
		Western Horseman Magazine	P. O. Box 7980 Colorado Springs, Colo. 80933
	Horse breed publications:		
	Appaloosa	Appaloosa News	Box 8403 Moscow, Ida. 83843
		Appaloosa World	P. O. Box 1035 Daytona Beach, Fla. 32019
		Appy, The	7427 Southwood Drive Mentor, Ohio 44060
	Arabian	Arabian Horse Journal, The	P. O. Box 260 Mt. Airy, Md. 21771
		Arabian Horse Times, The	R.R. #3 Waseca, Minn. 56093
		Arabian Horse World	2650 E. Bayshore Road Palo Alto, Calif. 94303
	Belgian	Belgian Review (annual)	P. O. Box 335 Wabash, Ind. 46992
	Connemara	Connemara, The (annual)	R.D. 1, Hoshiekon Farm Goshen, Conn. 06756
	Hackney	Hackney Journal, The	Box 4333 New Windsor, N.Y. 12550
	Morgan	Morgan Horse, The	Box 1 Westmoreland, N.Y. 13490
	Mustang	American Mustang World	28751 10th Street Lake Elsinore, Calif. 92330
	Paint Horse	Paint Horse Journal, The	P. O. Box 18519 Fort Worth, Tex. 76118
	Palomino	Palomino Horses	P. O. Box 249 Mineral Wells, Tex. 76067

(Continued)

TABLE 18-1 (Continued)

Devoted To	Breed (Where Applicable)	Publication	Address
Horses (continued)	**Peruvian Paso**	Peruvian Horse World Review	P. O. Box 1807 Birmingham, Ala. 35201
	Pinto	Pinto Horse International	P. O. Box 125 Lake Orion, Mich. 48035
	Pony of the Americas	Pony of America	P. O. Box 1447 Mason City, Iowa 50401
	Quarter Horse	All Western Quarter Horse News	P. O. Box 3266 Logan, Utah 84321
		Intermountain Quarter Horse	P. O. Box 217 Sandy, Utah 84091
		Quarter Horse Digest	Route 2, Box 14 Gann Valley, S.D. 57341
		Quarter Horse Journal	Box 9105 Amarillo, Tex. 79105
		Speedhorse	P. O. Box 1000 Norman, Okla. 73070
	Rangerbred	Rangerbred News, The	18232 Lawndale Homewood, Ill. 60430
	Saddlebred	Bluegrass Horseman, The	P. O. Box 389 Lexington, Ky. 40501
		Saddle Horse Report, The	P. O. Box 1007 Shelbyville, Tenn. 37160
		Saddle Horse West	21282 S. Horton Road West Linn, Ore. 97068
	Shetland Pony (also Hackney, Welsh, and Miniature)	Pony Journal, The	P. O. Box 435 Fowler, Ind. 47944
	Standardbred	Harness Horse	P. O. Box 1831 Harrisburg, Penn. 17105
		Hoof Beats	750 Michigan Avenue Columbus, Ohio 43215
	Tennessee Walking Horse	Voice of the Tennessee Walking Horse	Box 286 Lewisburg, Tenn. 37091
		Walking Horse Report	P. O. Box 1007 Shelbyville, Tenn. 37160
	Thoroughbred	Arizona Thoroughbred, The	P. O. Box 35055 Phoenix, Ariz. 85069
		Backstretch, The	19363 James Couzens Highway Detroit, Mich. 48235
		Blood Horse	Box 4038 Lexington, Ky. 40504
		British Columbia Thoroughbred	4023 E. Hastings Street Barnaby, British Columbia, Canada V5C 2J1
		Florida Horse	Box 2106 Ocala, Fla. 32678
		Maryland Horse, The	P. O. Box 427 Timonium, Md. 21093
		Oregon Horse, The	P. O. Box 17248 Portland, Ore. 97217
		Thoroughbred of California, The	201 Colorado Place Arcadia, Calif. 91006
		Thoroughbred Record, The	P. O. Box 4240 Lexington, Ky. 40544
		Turf and Sport Digest	511 Oakland Avenue Baltimore, Md. 21212
		Washington Horse, The	P. O. Box 88258 Seattle, Wash. 98188

(Continued)

TABLE 18-1 (Continued)

Devoted To	Breed (Where Applicable)	Publication	Address
Horses (continued)	**Trakehner**	American Trakehner, The	Box 7 Arlington Heights, Ill. 60006
	Welsh Pony	Welsh Pony World	4531 Dexter Street N.W. Washington, D.C. 20007

SECTION 19

WHERE TO GO FOR HELP[1]

Contents Page

Colleges of Agriculture in the U.S.A. and Canada 1160
Organizations and Agencies Serving the Livestock Industry 1161
Agricultural Council of America 1161
Agricultural Stabilization and Conservation 1161
Agriservices Foundation 1161
American Farm Bureau Federation 1161
American Feed Manufacturers Association 1161
American Horse Council, Inc. (AHC) 1162
American Meat Institute (AMI) 1162
American Sheep Producers Council, Inc. 1162
American Society of Agricultural Consultants 1162
American Society of Animal Science 1162
American Veterinary Medical Association 1162
Canada Department of Agriculture 1162
Canadian Society of Animal Science 1162
County Agricultural Agent 1163
Dairy Herd Improvement Association (DHIA) 1163
Dairy Herd Improvement Registry (DHIR) 1163
Farmers' Educational and Cooperative Union of America 1163
Fertilizer Institute 1163
4-H Clubs 1163
Future Farmers of America (FFA) 1163
Livestock Conservation Institute 1163
Livestock Marketing Association 1164
National Academy of Sciences 1164
National Association of Animal Breeders (NAAB) 1164
National Cattlemen's Association (NCA) 1164
National Council of Farmer Cooperatives 1164
National Cutting Horse Association 1164
National Farmers' Organization (NFO) 1164
National Farmers Union 1164
National Grange 1165
National Independent Meat Packers Association, The (NIMPA) 1165
National Live Stock and Meat Board 1165
National Live Stock Producers Association 1165
National Livestock Exchange 1165
National Silo Association, Inc. 1165
National Swine Repopulation Association 1165
National Wool Growers Association 1165
Soil Conservation Service 1166
State Departments of Agriculture 1166
U.S. Department of Agriculture 1166
United States Animal Health Association 1166
Veterans Administration 1166
Vocational Agriculture Instructor 1166
Western States Meat Association (WSMA) 1166
Poison Information Centers 1166

[1]In the preparation of this section on Where to Go for Help, the author had the benefit of the authoritative review and suggestions of the Information Division, USDA, Washington, D.C.

Large manufacturing companies can and do hire their own specialists. For them, it is good business, for they know that they can sell more of their products if they are good than if they are poor.

Small farmers and ranchers, on the other hand, cannot afford to hire their own specialists, despite the fact that modern farming is fully as complex as manufacturing. This is so chiefly because of the smaller size of farms. There are, for example, about 2.7 million U.S. farms and ranches, whereas only four manufacturers make over nine-tenths of the nation's automobiles.

Under these circumstances the farmer needs to know where to go for help—the help of specialists provided by various government agencies (federal, state, and county), and by commercial companies and trade associations. This section is presented for this purpose.

COLLEGES OF AGRICULTURE IN THE U.S.A. AND CANADA

U.S. stockmen can obtain a list of available bulletins and circulars, and other information regarding livestock, by writing to (1) their State Agricultural College (Land-Grant Institution), or (2) the Superintendent of Documents, Washington, D.C.; or by going to the local County Extension Office (Farm Advisor) of the county in which they reside. Canadian stockmen may write to the Department of Agriculture of their province or to their provincial university. A list of U.S. Land-Grant Institutions and Canadian Provincial Universities follows:

State	Address
Alabama	School of Agriculture, Auburn University, Auburn, Ala. 36830
Alaska	Department of Agriculture, University of Alaska, Fairbanks, Alaska 99701
Arizona	College of Agriculture, University of Arizona, Tuscon, Ariz. 85721
Arkansas	Division of Agriculture, University of Arkansas, Fayetteville, Ark. 72701
California	College of Agricultural and Environmental Sciences, University of California, Davis, Calif. 95616
Colorado	College of Agricultural Sciences, Colorado State University, Fort Collins, Colo. 80521
Connecticut	College of Agriculture and Natural Resources, University of Connecticut, Storrs, Conn. 06268
Delaware	College of Agricultural Sciences, University of Delaware, Newark, Del. 19711
Florida	College of Agriculture, University of Florida, Gainesville, Fla. 32611
Georgia	College of Agriculture, University of Georgia, Athens, Ga. 30601
Hawaii	College of Tropical Agriculture, University of Hawaii, Honolulu, Hawaii 96822
Idaho	College of Agriculture, University of Idaho, Moscow, Ida. 83843
Illinois	College of Agriculture, University of Illinois, Urbana, Ill. 61801
Indiana	School of Agriculture, Purdue University, Lafayette, Ind. 47907
Iowa	College of Agriculture, Iowa State University, Ames, Iowa 50010
Kansas	College of Agriculture, Kansas State University, Manhattan, Kan. 66506
Kentucky	College of Agriculture, University of Kentucky, Lexington, Ky. 40506
Louisiana	College of Agriculture, Louisiana State University and A & M College, University Station, Baton Rouge, La. 70803
Maine	College of Life Sciences and Agriculture, University of Maine, Orono, Me. 04473
Maryland	College of Agriculture, University of Maryland, College Park, Md. 20742
Massachusetts	College of Food and Natural Resources, University of Massachusetts, Amherst, Mass. 01002
Michigan	College of Agriculture and Natural Resources, Michigan State University, East Lansing, Mich. 48823
Minnesota	College of Agriculture, University of Minnesota, St. Paul, Minn. 55101
Mississippi	College of Agriculture, Mississippi State University, State College, Miss. 39762
Missouri	College of Agriculture, University of Missouri, Columbia, Mo. 65201
Montana	College of Agriculture, Montana State University, Bozeman, Mont. 59715
Nebraska	College of Agriculture, University of Nebraska, Lincoln, Neb. 68503
Nevada	The Max C. Fleischmann College of Agriculture, University of Nevada, Reno, Nev. 89507
New Hampshire	College of Life Sciences and Agriculture, University of New Hampshire, Durham, N.H. 03824
New Jersey	College of Agriculture and Environmental Science, Rutgers University, New Brunswick, N.J. 08903
New Mexico	College of Agriculture and Home Economics, New Mexico State University, Las Cruces, N.M. 88003
New York	New York State College of Agriculture, Cornell University, Ithaca, N.Y. 14850
North Carolina	School of Agriculture, North Carolina State University, Raleigh, N.C. 27607
North Dakota	College of Agriculture, North Dakota State University, State University Station, Fargo, N.D. 58102
Ohio	College of Agriculture and Home Economics, The Ohio State University, Columbus, Ohio 43210
Oklahoma	College of Agriculture and Applied Science, Oklahoma State University, Stillwater, Okla. 74074
Oregon	School of Agriculture, Oregon State University, Corvallis, Ore. 97331
Pennsylvania	College of Agriculture, The Pennsylvania State University, University Park, Penn. 16802
Puerto Rico	College of Agricultural Sciences, University of Puerto Rico, Mayaguez, Puerto Rico 00708
Rhode Island	College of Resource Development, University of Rhode Island, Kingston, R.I. 02881
South Carolina	College of Agricultural Sciences, Clemson University, Clemson, S.C. 29631
South Dakota	College of Agriculture and Biological Sciences, South Dakota State University, Brookings, S.D. 57006

(Continued)

State	Address
Tennessee	College of Agriculture, University of Tennessee, P. O. Box 1071, Knoxville, Tenn. 37901
Texas	College of Agriculture, Texas A & M University, College Station, Tex. 77843
Utah	College of Agriculture, Utah State University, Logan, Utah 84321
Vermont	College of Agriculture, University of Vermont, Burlington, Vt. 05401
Virginia	College of Agriculture, Virginia Polytechnic Institute, Blacksburg, Va. 24061
Washington	College of Agriculture, Washington State University, Pullman, Wash. 99163
West Virginia	College of Agriculture and Forestry, West Virginia University, Morgantown, W.Va. 26506
Wisconsin	College of Agricultural and Life Sciences, University of Wisconsin, Madison, Wisc. 53706
Wyoming	College of Agriculture, University of Wyoming, University Station, P. O. Box 3354, Laramie, Wyo. 82070
In Canada	
Alberta	University of Alberta, Edmonton, Alberta T6H 3K6
British Columbia	University of British Columbia, Vancouver, British Columbia V6T 1W5
Manitoba	University of Manitoba, Winnipeg, Manitoba R3T 2N2
New Brunswick	University of New Brunswick, Fredericton, New Brunswick E3B 4Z7
Ontario	University of Guelph, Guelph, Ontario N1G 2W1
Quebec	Faculty d'Agriculture, L' Universite' Laval, Quebec City, Macdonald College of McGill University, Montreal, Quebec H9X 1C0
Saskatchewan	University of Saskatchewan, Saskatoon, Saskatchewan S7N 0W0

ORGANIZATIONS AND AGENCIES SERVING THE LIVESTOCK INDUSTRY

Listed herein are various organizations and agencies whose purpose is to assist farmers, ranchers, and related industries. Some of these are nationwide and are affiliated with separate regional or state organizations of similar kind and interest, and, in turn, some of the latter are organized into district or county groups.

Agricultural Council of America

1625 "I" Street, N.W.
Washington, D.C. 20006

The Agricultural Council of America was formed in 1973, spearheaded by the late Congressman Jerry Litton, of Missouri. The main purpose of the organization is to bring together all the diverse segments of agriculture as an organized voice to communicate with the urban consumer.

Agricultural Stabilization and Conservation

The U.S. Department of Agriculture maintains in each county (usually at the county seat) an office of the Agricultural Stabilization and Conservation Committee (ASC), which is directed by a committee of local farmers. They can provide both cost-sharing and technical assistance to stockmen who desire (1) to improve the vegetative cover on their pasture land by means of artificial reseeding or by control of competitive shrubs; (2) to improve water penetration on soil by means of furrowing, chiseling, ripping, scarifying, or listing; or (3) to get better distribution of grazing by constructing wells, developing springs or seeps, installing pipelines for livestock water, or constructing permanent-type fences.

Cost sharing by the ASC usually amounts to their paying approximately 50 percent of the cost involved in performing the practice and providing the necessary engineering and other technical assistance required.

Agriservices Foundation

648 West Sierra Avenue
P.O. Box 429
Clovis, California 93613

Agriservices Foundation is a nonprofit foundation serving world agriculture whose dedicated purposes are to foster and support programs of education, research, and development, which will contribute toward wider and more effective application of science and technology to the practice of agriculture, for the benefit of mankind.

Agriservices Foundation conducts the following:

1. Programs in world food, hunger, and malnutrition.

2. Study-tours abroad, involving scientific and cultural exchange.

American Farm Bureau Federation

225 Touhy Avenue
Park Ridge, Illinois 60068

The American Farm Bureau was formed in 1920, as the farm bureau of a city chamber of commerce. Today, it is national in scope, with organizations on both the state and county level. It maintains an active educational and legislative program.

American Feed Manufacturers Association

1701 N. Ft. Myer Drive
Arlington, Virginia 22209

The American Feed Manufacturers Association is a nationwide organization of feed manufacturers banded together primarily (1) to improve the quality and to promote the use of commercial feeds, (2) to encourage high standards on the part of its members, and (3) to protect the best interests of the feed manufacturer and the stockman in legislative programs.

American Horse Council, Inc. (AHC)

1700 "K" Street, N.W., Suite 300
Washington, D.C. 20006

The American Horse Council, which was organized in 1969, is the only national trade association dedicated to protecting and promoting the horse industry through a united effort, encompassing all breeds, functions, and horse related activity. AHC represents 2 million horsemen through individual membership and over 80 member organizations.

The Council's activities and accomplishments include: following legislation, especially tax matters affecting horsemen; obtaining federal equine research funds; studying and solving racing problems; pressing for legislation to prohibit interstate off-track wagering; encouraging the networks to televise more horse events; studying land use and ordinance regulations affecting horsemen; working for increased trails for pleasure riders; publishing a brochure on careers in the horse industry; and publishing a Horse Industry Directory.

American Meat Institute (AMI)

P.O. Box 3556
Washington, D.C. 20007

The American Meat Institute was founded in 1906, although it has changed its name along the way.

Today, it has a membership of 990, consisting of meat packers, processors, sausage manufacturers, meat suppliers, canners, and other related industries.

The AMI issues the following publications: *Meatfacts* (annual); *Financial Facts About the Meat Packing Industry* (annual); *Newsletter* (irregular); and *Weekly Report* (weekly).

With a staff of 35, the American Meat Institute protects and furthers the interests of its members through legislative matters, interpreting government rules and regulations that affect the industry, increasing the efficiency of processing and distributing meats, disseminating material about the industry to government leaders and consumers, and improving the quality and promoting the consumption of meats.

American Sheep Producers Council, Inc.

200 Clayton Street
Denver, Colorado 80206

The American Sheep Producers Council was formed on September 12, 1955, under authority granted in Section 708 of the National Wool Act of 1954.

The incentive payments are financed from duties collected on the imports of wool and wool products. Also, the Act authorized an industry self-help program for the purpose of developing and conducting advertising and sales promotion programs for lamb and wool. Each sheep producer contributes from his incentive payments 1 cent per pound for wool and 5 cents per 100 pounds of lamb sold. This provides a promotion and advertising fund.

American Society of Agricultural Consultants

8301 Greensboro Drive
Suite 470
McLean, Virginia 22102

The American Society of Agricultural Consultants is a professional association whose members are located throughout the U.S. and Canada. It was organized December 10, 1963, at Fresno, California, with the author of this book as its first president. Members of the Society must meet rigid standards based on their experience, training, knowledge, performance, and ability to render independent decisions. The agricultural consultant is the person who is regularly called upon to display one of the most valuable of all commodities—judgment.

American Society of Animal Science

425 Illinois Building
113 N. Neil Street
Champaign, Illinois 61820

The American Society of Animal Science, which was founded in 1908, is a society of persons with interest in animal science and livestock production.

American Veterinary Medical Association

930 North Meacham Road
Schaumburg, Illinois 60172

This is a professional organization of veterinarians and others who are interested in animal health.

Canada Department of Agriculture

Manyberries, Alberta, Canada

Canadian farmers and ranchers can receive assistance and publications by writing to the Canada Department of Agriculture at the address given above.

Canadian Society of Animal Science

Suite 907 — 151 Slater Street
Ottawa, Ontario, Canada K1P 5H4

This is a Canadian society of persons with interest in animal science and livestock production.

County Agricultural Agent

The county agricultural agent (or farm advisor), usually located in the County Court House, is qualified to give the latest findings of the state agricultural college, and the USDA, and to help in applying these findings to the individual farm or ranch.

Dairy Herd Improvement Association (DHIA)

This program, first adopted in 1926, is the most complete of all dairy production and record plans. More than half of the cows in the United States on production test are on this program. Both registered and grade cows can be enrolled.

State and local Dairy Herd Improvement Associations (DHIA) conduct the program among dairymen, working through the Cooperative Extension Service in cooperation with the Federal Extension Service and the Animal Science Research Service of the USDA.

In this program, a supervisor or tester, employed by the local or state testing association, visits the herd one day each month. He identifies all cows in the herd, and he weighs and takes representative samples of the milk from all animals in the herd for 2 consecutive milkings (3 milkings on herds on 3-times-daily milkings). He then combines the milk samples and tests them for butterfat or sends them to a central testing laboratory. Records are obtained on an individual cow basis on monthly and accumulative records for milk and fat (the latter in pounds and percentage); amount and cost of feed, and income over feed cost; breeding dates, calving dates, dry dates, and other factors affecting productivity; and in some testing associations, somatic cells or the California Mastitis Test (CMT) is made as an aid in monitoring udder health.

The above information is fed into a computer, programmed to provide monthly summaries on (1) individual cows, and (2) the herd; and this information is sent to each dairyman.

Additional reports are provided in most states. For example, in addition to the Monthly Report, they may provide the following: Lactation Record Report (a twice annual report of lactation records completed by the members of the herd during the previous six-month period), Reproduction Management Report, Individual Cow Record, Meritorious Lifetime Production Certificate, Herd Production Certificate, and Calf Record Report.

Dairy Herd Improvement Registry (DHIR)

This is the Standard DHIA record *plus* added requirements to satisfy the needs of breed associations. Among the latter are surprise tests, made when the milk production of certain cows exceeds the breed average or another specified amount. Only registered dairy cows are eligible for DHIR records. The production records of herds enrolled in DHIR are mailed to the respective breed registries for official recording.

Farmers' Educational and Cooperative Union of America

P.O. Box 2251
Denver, Colorado 80201

The Farmers' Union also has organizations on both the state and county level. It maintains an active educational and legislative program.

Fertilizer Institute

1015 Eighteenth Street, N.W.
Washington, D.C. 20036

This is a fertilizer trade association. Among its services, it disseminates valuable information on soil fertility.

4-H Clubs

This is an organization of farm boys and girls, 9 through 19 years of age, under the sponsorship of the agricultural extension service of the land grant colleges. This organization is for the purpose of developing knowledge, leadership, and ability through agriculture, home economics, and community service projects. For further information about 4-H clubs, contact the local county agent.

Future Farmers of America (FFA)

This is an organization of farm boys and girls enrolled in vocational agriculture. For further information about the FFA, contact the local vocational agriculture instructor.

Livestock Conservation Institute

1100 Jorie Boulevard, Suite 143
Oak Brook, Illinois 60621

Livestock Conservation Institute is a nonprofit organization which was founded in 1916. It serves as a clearinghouse for all sectors of the livestock and meat industry in sponsoring research and educational programs designed to eradicate diseases among livestock and reduce the losses from parasites and bruises and other injuries incurred in handling and shipping livestock.

It is estimated that the program of Livestock Conservation saves the livestock and meat industry $642 million annually by lessening losses that can result from deaths, diseases, injuries, and parasites.

Livestock Marketing Association

301 East Armour Boulevard
Kansas City, Missouri 64111

This is a trade association of livestock auction markets; auction markets are trading centers where animals are sold by public bidding to the buyer who offers the highest price per hundredweight or per head.

National Academy of Sciences

National Research Council
2101 Constitution Avenue, N.W.
Washington, D.C. 20418

The National Research Council is a division of the National Academy of Sciences. The Academy was established in 1916 to promote the effective utilization of scientific and technical resources. The NRC, which is a private, nonprofit organization of scientists, publishes bulletins periodically giving the nutrient requirements of domestic animals. Copies of the NRC requirements may be purchased at a nominal charge from the Academy at the address given above.

National Association of Animal Breeders (NAAB)

P.O. Box 1033
Columbia, Missouri 65201

The NAAB is the national organization representing all of the artificial insemination organizations in the United States, along with associate members from many other countries.

National Cattlemen's Association (NCA)

5420 South Quebec Street
P.O. Box 3469
Englewood, Colorado 80155

The National Cattlemen's Association, which had its beginning in 1898 in Denver, Colorado, is a nonprofit trade association representing approximately 280,000 cattlemen. It is the national spokesman for all segments of the beef industry, including both cow-calf men and feeders.

The NCA's principal services include government affairs, public information, management education, and marketing information. Also, the Association sponsors a special market analysis service, known as Cattle-Fax, and a special truck clearinghouse service aimed at improving truck transportation and efficiency, known as Trans Fax.

The NCA has offices at three locations. Its headquarters are in Englewood, Colorado. Additionally, it has a branch office in Omaha, Nebraska, and a Government Affairs Office at 1015 National Press Building, Washington, D.C.

National Council of Farmer Cooperatives

1129 20th Street, N.W.
Washington, D.C. 20036

This is a national federation dedicated to the promotion of the interests of farmer cooperatives through its influence on various governmental and other agencies. The Council provides an avenue through which cooperatives are advised of current economic, technological, legal, and other developments. Also, it provides a forum through which better understanding among cooperatives can be attained.

National Cutting Horse Association

P.O. Box 12155
Fort Worth, Texas 76116

The National Cutting Horse Association was formed to promote and encourage the showing of cutting horses in the contest arena. To this end, the Association developed a standard method under which all cutting horse contests are conducted in an equitable manner and standardized the rules for judging the event. Today, the National Cutting Horse Association approves some 750 open and championship shows each year. It is the only cutting horse association operating throughout the entire United States.

National Farmers' Organization (NFO)

720 Davis Avenue
Corning, Iowa 50841

This is the youngest (1965) and smallest of the four major farm organizations. It is highly market-oriented, and it frequently withholds farm products from markets during times of depressed prices.

National Farmers Union

The National Farmers Union was formed in Texas in 1902, making it the second oldest national organization. It concentrates on action programs to preserve and strengthen family farms which it considers vital to a representative democracy. It gives strong support to cooperatives. But it now devotes most of its effort to legislative projects.

National Grange

1616 "H" Street, N.W.
Washington, D.C. 20006

The National Grange (Patrons of Husbandry) was founded in 1868. It had its origin just after the Civil War in an effort to unite farmers, both North and South, in a common cause for farm betterment.

Today, it combines social, educational, and economic objectives. It places emphasis on moral and spiritual idealism. Its program provides activities for men, women, and youth.

National Independent Meat Packers Association, The (NIMPA)

734 15th Street, N.W.
Washington, D.C. 20005

The National Independent Meat Packers Association was established in 1942 to "protect and further the legitimate interests of the independent segment of the meat-packing industry." This objective is accomplished by representing members in legislative matters; interpreting government rules and regulations; disseminating material to government leaders and consumers; providing meetings for members to review new ideas; and communicating with members on a continuing basis.

National Live Stock and Meat Board

444 N. Michigan Avenue
Chicago, Illinois 60611

The National Live Stock and Meat Board was organized in 1921. It is the only service organization working for the country's total livestock and meat industry. Funds from the industry support the Meat Board's consumer marketing, research, education, and promotion programs. The "core" program of the Board pertains to beef, pork, lamb, veal, and processed meats.

The percentage of dollars allocated by species programs in based on the percentage of income which the Board receives from each species.

All Meat Board revenue, with the exception of income from the sale of materials, comes from livestock producers, feeders, and meat-packers. Producer/feeder contributions are on a per head basis when livestock are marketed through participating livestock marketing agencies and packer livestock stations. Additionally, some packers contribute on a per-head-slaughtered basis; and special funding for special projects comes from various industry groups.

Producer-feeder representatives comprise the majority of the Board of Directors, but the Directorate also has representation from the other segments of the industry—marketing, packing-processing, meat retailing, and food service.

National Live Stock Producers Association

307 Livestock Exchange Building
Denver, Colorado 80216

The National Live Stock Producers Association was formed on May 10, 1930. It is the nation's largest livestock marketing association, handling over 10 million head of livestock valued at $1.5 billion annually for some 300,000 farmers and ranchers. It provides cooperative marketing services to livestock farmers and ranchers on nearly 150 markets throughout the livestock producing areas of the United States. The Association also has six regional credit affiliates which make livestock loans.

National Livestock Exchange

104 S. Muskego Avenue
Milwaukee, Wisconsin 53233

This is a trade association of livestock commission firms and others, operating on terminal livestock markets.

National Silo Association, Inc.

Box 247
Cedar Falls, Iowa 50613

The National Silo Association is dedicated to the advancement of ensiled feeding programs and the automated equipment that makes such feeding programs possible. Among its services, it has a marketing committee and it distributes a number of publications (free, or at nominal cost) of value to meat and milk producers in feeding programs.

National Swine Repopulation Association

SPF Accreditation
P.O. Box 4405
Lincoln, Nebraska 68504

The National Swine Repopulation Association is responsible for supervising the Specific Pathogen-Free (SPF) swine program and for issuing an Accreditation Certificate to those who qualify.

National Wool Growers Association

600 Crandall Boulevard
Salt Lake City, Utah 84101

The National Wool Growers Association is a trade association representing the sheep and wool growers of the United States.

Soil Conservation Service

The Soil Conservation Service is part of the U.S. Department of Agriculture. Technicians are specialists in soil and water management. They are qualified to determine contour lines and to make farm plans. In organized Soil Conservation districts, facilities are available for classifying land according to its capabilities for use and resistance to erosion damage. From this information, a complete conservation program can be worked out for each farm or ranch, to be applied in large or small steps as desired.

State Departments of Agriculture

Each state has a State Department of Agriculture, which renders various valuable services, including livestock sanitary and regulatory work. With the exceptions noted below, all of these are located at the state capitals of the respective states. The three exceptions are:

Maryland: State Department of Agriculture, College Park, Maryland
Nevada: State Department of Agriculture, Reno, Nevada
New Mexico: State Department of Agriculture, State College, New Mexico

U.S. Department of Agriculture

Washington, D.C. 20250

The U.S. Department of Agriculture has a competent staff of specialists serving all the various branches of the nation's agriculture. Also, it operates a Research Center, with headquarters at Beltsville, Maryland, and numerous other laboratories and offices throughout the country.

A list of U.S. Department of Agriculture publications can be obtained by directing a request to the Office of Information, U.S. Department of Agriculture, Washington, D.C. 20250. Many of these publications are also available at the offices of county agricultural agents or through U.S. congressmen and senators.

United States Animal Health Association (formerly U.S. Live Stock Sanitary Association)

2810 Buford Road
Richmond, Virginia 23235

The primary responsibility of the United States Animal Health Association is to establish uniform methods and rules for the control of brucellosis.

Veterans Administration

810 Vermont Avenue, N.W.
Washington, D.C. 20005

For information relative to veterans' rights, veterans should contact the nearest local veterans' office or write to the address given above.

Vocational Agriculture Instructor

Many high schools have vocational agriculture departments staffed with one or more instructors. They teach good farming practices. Their students are members of the Future Farmers of America. Like the county agent, instructors are usually acquainted with the local modifications which may be desirable in order to obtain the best results from the general recommendations made in this book.

Western States Meat Association (WSMA)

88 First Street
San Francisco, California 94105

The Western States Meat Association was organized in 1946, with 48 charter members. Today, it has 700 members.

The WSMA furthers the interest of and serves as the spokesman for its members—most of which are in the West. It specializes in serving its members through (1) disseminating news and technical data of importance to the industry, (2) interpreting government regulations, and (3) keeping them abreast of legislative matters in Washington, D.C. The Association provides professional know-how, on its staff or on a consultant basis, in the areas of public relations, lobbying, publicity, and research. Also, it maintains a labor library, a traffic manager, and a legal staff.

POISON INFORMATION CENTERS

With the large number of chemical sprays, dusts, and gases now on the market for use in agriculture, accidents may arise because of operators being careless in their use. Also, there is always the hazard that a child may eat or drink something that may be harmful. Centers have been established in various parts of the country where doctors can obtain prompt and up-to-date information on treatment of such cases, if desired.

Local medical doctors have information relative to the Poison Information Centers of their area, along with some of the names of their directors, telephone numbers, and street numbers. When calling any of these centers, one should ask for the "Poison Information Center." If this information cannot be obtained locally, call the U.S. Public Health Service at Atlanta, Ga.; or Wenatchee, Wash.

SECTION 20

WEIGHTS AND MEASURES

Contents

Page
Weights and Measures of Common Feeds 1167
Estimating Weight of Grain in a Bin 1168
Estimating Weight of Hay in a Stack or in a Barn 1168
Estimating Weight of Silage in a Silo 1171
Estimating Animal Weights 1172
Animal Units 1174
Land Area 1175
Legal Description of Land 1175
Land Measures 1175
How to Determine Acreage 1176
The Metric System 1176

Weights and measures are the standards employed in measuring weights, quantities, and volumes. Even among primitive people, such standards are necessary; and with the growing complexity of life they become of greater and greater importance.

The United States and a few other countries use standards that belong to the *customary*, or English, system of measurement. This system evolved in England from older measurement standards, beginning about the 1200s. All other countries—including England—now use a system of measurements called the *metric system*, which was created in France in the 1790s. The United States is now in the process of converting to the metric system.

Weights and measures form one of the most important parts of modern agriculture. This section contains pertinent information relative to the most common standards used by U.S. stockmen.

WEIGHTS AND MEASURES OF COMMON FEEDS

In calculating rations and mixing concentrates, it is usually necessary to use weights rather than measures. However, in practical feeding operations it is often more convenient for the farmer or rancher to measure the concentrates. Table 20-1 will serve as a guide in feeding by measure.

TABLE 20-1
WEIGHTS AND MEASURES OF COMMON FEEDS

Feed	Approximate Weight	
	Lb per Quart	Lb per Bushel
Alfalfa meal	0.6	19
Barley	1.5	48
Beet pulp (dried)	0.6	19
Brewers' grain (dried)	0.6	19
Buckwheat	1.6	50
Buckwheat bran	1.0	29
Corn, husked ear	—	70
Corn, cracked	1.6	50
Corn, shelled	1.8	56
Corn meal	1.6	50
Corn-and-cob meal	1.4	45
Cottonseed meal	1.5	48
Cowpeas	1.9	60
Distillers' grain (dried)	0.6	19
Fish meal	1.0	35
Gluten feed	1.3	42
Linseed meal (old process)	1.1	35
Linseed meal (new process)	0.9	29
Meat scrap	1.3	42
Milo (grain sorghum)	1.7	56
Molasses feed	0.8	26
Oats	1.0	32
Oats, ground	0.7	22
Oat middlings	1.5	48
Peanut meal	1.0	32
Rice bran	0.8	26
Rye	1.7	56
Sorghum (grain)	1.7	56
Soybeans	1.8	60
Tankage	1.6	51
Velvet beans, shelled	1.8	60
Wheat	1.9	60
Wheat bran	0.5	16
Wheat middlings, standard	0.8	26
Wheat screenings	1.0	32

ESTIMATING WEIGHT OF GRAIN IN A BIN

Sometimes stockmen need to estimate the weight of grain in storage. Such estimates are difficult to make because of differences in moisture content, depth of material stored, and other factors. However, the following procedure will enable one to figure feed quantities fairly closely.

1. **Corn (shelled) or small grain in rectangular cribs or bins**—Multiply the width by the length by the average depth (all in feet) and multiply by 0.8 to get the number of bushels (multiplying by 0.8 is the same as dividing by 1¼, the number of cubic feet in a bushel).

2. **Ear corn in rectangular cribs or bins**—Multiply the width by the length by the average depth (all in feet) and multiply by 0.4 to get the number of bushels (multiplying by 0.4 is the same as dividing by 2½, the number of cubic feet in a bushel of ear corn).

3. **Round bins or cribs**—To find the cubic feet in a cylindrical bin, multiply the squared radius by 3.1416 by the depth.

Thus, the volume of a round bin 20 feet in diameter and 10 feet deep is determined as follows:

a. The radius is half the diameter, or 10 feet
b. $10 \times 10 = 100$
c. $100 \times 3.1416 = 314.16$
d. $314.16 \times 10 = 3{,}141.6$ cubic feet
e. Where shelled corn or small grain is involved, one should multiply $3{,}141.6 \times 0.8$, which equals 2,513.28 bushels of grain that it would hold if full.
f. Where ear corn is involved, one should multiply $3{,}141.6 \times 0.4$ which equals 1,256.64 bushels of ear corn that it would hold if full.

ESTIMATING WEIGHT OF HAY IN A STACK OR IN A BARN

Stockmen and hay dealers frequently buy and sell large quantities of hay in the stack or in the barn. This practice is especially prevalent in the western and Great Plains states, where cattle and sheep are brought into the valleys to be wintered on hay bought from valley hay producers. Under such circumstances, the weight of hay is usually estimated, because (1) no scales are available, and/or (2) it is impractical to weigh the hay due to the time, labor, and wastage involved. In many such instances, the hay is fed directly from the stack or barn, in racks arranged about it. Under these and other circumstances, there is need for a simple and reasonably accurate method of estimating the weight of hay in a stack or in a barn.

In order to estimate the tonnage of hay in a stack or in a barn, it is necessary (1) to compute the volume of hay, and (2) to know the number of cubic feet per ton of hay. Table 20-2 gives the latter information.

TABLE 20-2

CUBIC FEET PER TON OF HAY

Feed	Settled 1 to 2 Mo.	Settled over 3 Mo.
	(cu ft)	*(cu ft)*
Alfalfa	485	470
Clover	512	500
Hay, baled (closely stacked)	150-200	150-200
Hay, chopped	225	210
Straw, baled	200	200
Straw, loose	1,000	600-1,000
Timothy	640	625
Wild hay	600	450

In using Table 20-2, it should be recognized that many factors—other than kind of hay, form (loose, chopped, or baled), and period of settling—affect the density of hay in a stack or in a barn, including (1) moisture content at haying time, and (2) texture and foreign material.

It is relatively simple to compute the volume of hay in a mow, but it is more difficult to determine the volume of a stack. Although different rules or formulas may be and are used, the following are recommended by the U.S. Department of Agriculture:[1]

1. **Volume of hay in barns**—Multiply the width by the length by the height, all in feet, and divide by the cubic feet per ton as given in Table 20-2.

2. **Volume of hay in oblong stacks**—Three types of oblong stacks are common, as shown in Fig. 20-1.

The volume of each type of oblong stack may be determined as follows:

a. For low, round-topped stacks—
$(0.52 \times 0) - (0.44 \times W) \times W \times L$
b. For high, round-topped stacks—
$(0.52 \times 0) - (0.46 \times W) \times W \times L$
c. For square, flat-topped stacks—
$(0.56 \times 0) - (0.55 \times W) \times W \times L$

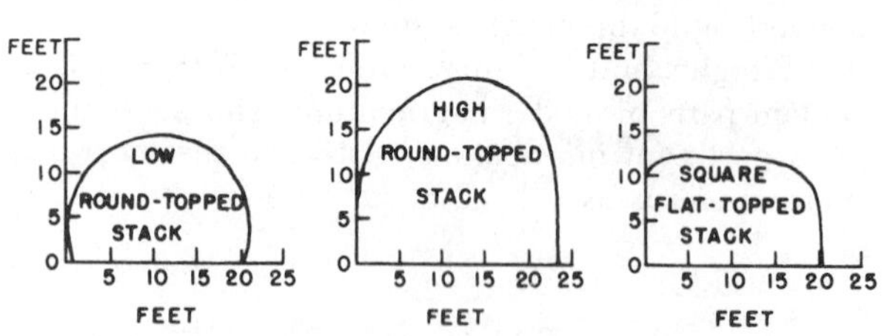

Fig. 20-1. Three column types of oblong stacks.

In these formulas "0" is the "over" or "overthrow," which is the distance in feet from the ground on one side of the stack, up and over the stack and

[1]*Measuring Hay in Stacks*, USDA Leaflet No. 72.

down to the ground on the other side; W is the width; and L is the length.

The application of this formula will be illustrated by the following example:

Example—It is desired to estimate the amount of alfalfa hay in a low, round-topped type of oblong stack that has settled 4 months. The stack is 20 feet wide, 30 feet long, and has an over of 40 feet.

The answer is secured as follows:

a. Volume = (0.52 × 40) − (0.44 × 20) × 20 × 30 = 7,200 cubic feet.

b. Table 20-2 shows that there are 470 cubic feet per ton of settled alfalfa.

c. 7,200 ÷ 470 = 15 tons of hay.

TABLE 20-3

VOLUME OF ROUND STACKS OF HAY OF SPECIFIED DIMENSIONS[1]

(Volume figures given to the nearest 5)

Circumference	Indicated Volume in Cubic Feet When the Over Is—												
	25 feet	26 feet	27 feet	28 feet	29 feet	30 feet	31 feet	32 feet	33 feet	34 feet	35 feet	36 feet	37 feet
(feet)													
45	825	960	1,090										
46	840	975	1,105	1,235									
47	855	990	1,120	1,250	1,385	1,505							
48	870	1,005	1,135	1,265	1,400	1,525	1,650	1,785					
49	885	1,020	1,150	1,285	1,420	1,540	1,670	1,805	1,935				
50	900	1,035	1,165	1,300	1,435	1,560	1,690	1,825	1,955	2,090	2,215		
51	915	1,050	1,180	1,315	1,450	1,580	1,710	1,845	1,980	2,110	2,240	2,370	2,495
52	930	1,065	1,200	1,330	1,465	1,600	1,730	1,865	2,000	2,130	2,265	2,400	2,530
53	945	1,080	1,215	1,345	1,485	1,615	1,750	1,880	2,020	2,155	2,290	2,430	2,560
54	960	1,095	1,230	1,360	1,500	1,630	1,770	1,900	2,040	2,180	2,320	2,460	2,595
55	975	1,110	1,245	1,380	1,515	1,650	1,790	1,920	2,065	2,205	2,345	2,490	2,630
56	990	1,125	1,260	1,395	1,530	1,665	1,810	1,940	2,085	2,230	2,375	2,520	2,660
57	1,005	1,140	1,275	1,410	1,550	1,685	1,830	1,960	2,105	2,250	2,400	2,545	2,695
58	1,020	1,155	1,290	1,435	1,565	1,705	1,850	1,980	2,125	2,275	2,425	2,575	2,725
59	1,035	1,170	1,310	1,450	1,580	1,720	1,865	2,000	2,150	2,300	2,455	2,605	2,755
60	1,050	1,185	1,325	1,465	1,600	1,740	1,885	2,020	2,170	2,325	2,480	2,635	2,790
61	1,065	1,200	1,340	1,485	1,615	1,760	1,905	2,040	2,195	2,345	2,510	2,665	2,825
62	1,080	1,215	1,355	1,500	1,635	1,775	1,925	2,060	2,215	2,365	2,535	2,695	2,855
63	1,095	1,230	1,370	1,515	1,655	1,795	1,945	2,080	2,235	2,390	2,560	2,725	2,890
64	1,110	1,245	1,385	1,530	1,670	1,810	1,960	2,100	2,260	2,415	2,585	2,755	2,920
65	1,125	1,260	1,400	1,545	1,685	1,830	1,980	2,120	2,280	2,440	2,615	2,780	2,950
66	1,140	1,275	1,420	1,560	1,705	1,850	2,000	2,140	2,300	2,465	2,640	2,810	2,985
67	1,155	1,290	1,435	1,575	1,720	1,865	2,020	2,160	2,325	2,485	2,665	2,840	3,015
68	1,170	1,305	1,450	1,595	1,740	1,885	2,040	2,180	2,345	2,510	2,690	2,870	3,050
69	1,185	1,320	1,465	1,610	1,755	1,905	2,055	2,200	2,365	2,530	2,715	2,900	3,080
70	1,200	1,335	1,480	1,625	1,770	1,925	2,075	2,220	2,385	2,555	2,745	2,930	3,115
71	1,215	1,350	1,495	1,640	1,790	1,940	2,095	2,240	2,405	2,580	2,770	2,960	3,145
72	1,230	1,365	1,515	1,660	1,805	1,960	2,115	2,260	2,430	2,605	2,795	2,990	3,175
73	1,245	1,380	1,530	1,675	1,820	1,975	2,135	2,280	2,450	2,625	2,825	3,015	3,210
74	1,260	1,395	1,545	1,690	1,840	1,995	2,150	2,300	2,470	2,650	2,850	3,045	3,245
75		1,410	1,560	1,705	1,855	2,010	2,170	2,320	2,495	2,675	2,875	3,075	3,275
76		1,425	1,575	1,725	1,870	2,030	2,190	2,340	2,515	2,695	2,905	3,105	3,310
77			1,590	1,740	1,890	2,050	2,210	2,360	2,540	2,720	2,930	3,135	3,340
78			1,605	1,755	1,905	2,070	2,230	2,380	2,560	2,745	2,955	3,165	3,375
79				1,775	1,925	2,090	2,250	2,400	2,580	2,765	2,980	3,195	3,405
80				1,790	1,945	2,105	2,270	2,420	2,605	2,790	3,010	3,225	3,440
81				1,805	1,960	2,125	2,285	2,440	2,625	2,815	3,035	3,255	3,470
82				1,820	1,975	2,145	2,305	2,460	2,645	2,835	3,060	3,280	3,500
83					1,995	2,160	2,325	2,480	2,665	2,860	3,090	3,310	3,535
84						2,180	2,345	2,500	2,690	2,880	3,115	3,340	3,570
85								2,520	2,710	2,905	3,140	3,370	3,600
86									2,735	2,930	3,170	3,400	3,635
87											3,195	3,430	3,665
88												3,460	3,700
89												3,490	3,730
90													3,765
91													
92													
93													
94													
95													
96													
97													
98													

(Continued) *(Continued)*

Footnote on last page of table.

TABLE 20-3 (Continued)

Circumference	Indicated Volume in Cubic Feet When the Over Is—												
	38 feet	39 feet	40 feet	41 feet	42 feet	43 feet	44 feet	45 feet	46 feet	47 feet	48 feet	49 feet	50 feet
(feet)													
45													
46													
47													
48													
49													
50													
51													
52	2,665	2,795											
53	2,700	2,835	2,975										
54	2,735	2,875	3,015	3,160									
55	2,770	2,915	3,060	3,210	3,360	3,505							
56	2,805	2,955	3,105	3,255	3,415	3,565	3,720						
57	2,845	2,995	3,150	3,305	3,465	3,625	3,785	3,940					
58	2,880	3,035	3,195	3,350	3,515	3,680	3,850	4,010	4,175				
59	2,915	3,075	3,235	3,400	3,570	3,740	3,915	4,080	4,245	4,415			
60	2,950	3,115	3,280	3,445	3,625	3,795	3,975	4,150	4,320	4,490	4,670		
61	2,985	3,155	3,325	3,495	3,675	3,855	4,040	4,215	4,390	4,570	4,750	4,925	
62	3,020	3,195	3,365	3,540	3,730	3,915	4,105	4,285	4,465	4,650	4,830	5,015	5,200
63	3,055	3,235	3,410	3,585	3,780	3,970	4,165	4,355	4,540	4,730	4,910	5,105	5,295
64	3,090	3,275	3,455	3,635	3,835	4,030	4,230	4,425	4,615	4,805	4,995	5,195	5,390
65	3,125	3,315	3,495	3,680	3,885	4,085	4,290	4,490	4,690	4,885	5,075	5,285	5,485
66	3,160	3,355	3,540	3,730	3,935	4,145	4,355	4,560	4,760	4,960	5,160	5,370	5,580
67	3,195	3,395	3,585	3,780	3,990	4,205	4,420	4,630	4,830	5,040	5,245	5,460	5,670
68	3,230	3,430	3,630	3,825	4,045	4,265	4,485	4,695	4,900	5,120	5,330	5,550	5,765
69	3,265	3,470	3,670	3,875	4,095	4,320	4,545	4,760	4,970	5,195	5,415	5,640	5,860
70	3,300	3,510	3,715	3,920	4,150	4,375	4,610	4,825	5,045	5,275	5,495	5,730	5,955
71	3,335	3,550	3,760	3,970	4,205	4,435	4,670	4,895	5,120	5,355	5,580	5,820	6,050
72	3,375	3,590	3,805	4,015	4,255	4,495	4,735	4,965	5,195	5,435	5,665	5,910	6,145
73	3,410	3,630	3,845	4,065	4,310	4,550	4,795	5,030	5,270	5,515	5,750	6,000	6,240
74	3,445	3,665	3,890	4,110	4,360	4,610	4,855	5,095	5,340	5,595	5,835	6,090	6,335
75	3,480	3,705	3,935	4,160	4,415	4,670	4,915	5,165	5,415	5,675	5,915	6,180	6,430
76	3,515	3,745	3,975	4,205	4,465	4,725	4,980	5,235	5,490	5,750	6,000	6,270	6,525
77	3,550	3,785	4,020	4,250	4,520	4,785	5,045	5,305	5,560	5,830	6,085	6,355	6,620
78	3,585	3,825	4,065	4,300	4,570	4,840	5,105	5,370	5,635	5,910	6,170	6,445	6,715
79	3,620	3,865	4,105	4,345	4,625	4,895	5,170	5,440	5,710	5,990	6,255	6,535	6,810
80	3,655	3,905	4,150	4,395	4,675	4,955	5,235	5,510	5,785	6,070	6,340	6,625	6,905
81	3,690	3,945	4,195	4,440	4,730	5,010	5,295	5,575	5,855	6,145	6,425	6,715	7,000
82	3,725	3,985	4,240	4,490	4,785	5,070	5,360	5,645	5,930	6,225	6,510	6,800	7,090
83	3,760	4,025	4,280	4,535	4,830	5,130	5,425	5,715	6,005	6,305	6,595	6,890	7,185
84	3,795	4,065	4,325	4,580	4,885	5,190	5,485	5,785	6,080	6,385	6,675	6,980	7,280
85	3,830	4,105	4,365	4,630	4,935	5,245	5,550	5,850	6,155	6,465	6,760	7,070	7,375
86	3,865	4,145	4,410	4,675	4,990	5,300	5,615	5,920	6,230	6,545	6,845	7,160	7,470
87	3,900	4,185	4,455	4,725	5,040	5,360	5,680	5,990	6,300	6,620	6,930	7,250	7,565
88	3,940	4,220	4,500	4,770	5,090	5,420	5,745	6,060	6,375	6,700	7,015	7,340	7,660
89	3,975	4,260	4,540	4,815	5,145	5,475	5,805	6,125	6,450	6,780	7,100	7,430	7.755
90	4,010	4,300	4,585	4,860	5,200	5,535	5,865	6,195	6,525	6,860	7,185	7,520	7,845
91	4,045	4,340	4,630	4,910	5,250	5,595	5,930	6,265	6,600	6,940	7,270	7,605	7,940
92	4,080	4,380	4,670	4,955	5,305	5,650	5,995	6,335	6,675	7,020	7,355	7,695	8,035
93		4,420	4,715	5,005	5,360	5,710	6,055	6,400	6,750	7,095	7,440	7,785	8,130
94		4,460	4,760	5,050	5,410	5,765	6,120	6,470	6,825	7.175	7,525	7,875	8,225
95			4,805	5,100	5,465	5,825	6,180	6,540	6,895	7,255	7.610	7,965	8,320
96				5,150	5,515	5,885	6,245	6,610	6,970	7,335	7,695	8,055	8,415
97				5,195	5,570	5,945	6,310	6,680	7,045	7,415	7,780	8,145	8,510
98					5,625	6,000	6,370	6,750	7,120	7,495	7,865	8,285	8,605

[1]From USDA Leaflet No. 72, Table 1, p. 5.

3. **Volume of hay in round stacks**—The rules or formulas used for oblong stacks do not apply to round stacks. But Table 20-3 gives the volume of round stacks when the circumference is between 45 and 98 feet and the over between 25 and 50 feet.

The volume of stacks having circumferences or overs greater or less than those given in Table 20-3 can be calculated by using the following formula:

$$\text{Volume} = (0.04 \times 0) - (0.012 \times C) \times C^2$$

In this formula, C equals the circumference or distance around the stack at the ground, and 0 equals the over or distance from the ground on one side over the peak to the ground on the other side (usually it is best to take two over measurements at right angles to each other, and to average them).

Thus, the computation of the volume of a large round stack may be illustrated by the following example:

Example—It is desired to determine the amount of alfalfa hay in a round stack that is 100 feet in circumference and has an average over of 60 feet.

The answer is secured as follows:

a. Volume = (0.04 × 60) − (0.012 × 100) × $(100)^2$ = 12,000 cubic feet.

b. Table 20-2 shows that there are 470 cubic feet per ton of settled alfalfa.

c. 12,000 ÷ 470 = 25.5 tons of hay.

ESTIMATING WEIGHT OF SILAGE IN A SILO

Sometimes, either (1) for inventory purposes, or (2) for purposes of buying or selling, stockmen need to estimate the amount of silage remaining in a silo after part of it has been fed out. For a tower-type silo, this may be done by referring to Table 20-4 and following the directions given herein.

TABLE 20-4

GUIDE FOR ESTIMATING AMOUNT OF SILAGE IN A TOWER-TYPE SILO[1]

Depth of Settled Silage	Total Quantity of Settled Silage, from the Top to the Depth Indicated, in Silos Having a Diameter of:					
	10 feet	12 feet	14 feet	16 feet	18 feet	20 feet
(feet)	*(tons)*	*(tons)*	*(tons)*	*(tons)*	*(tons)*	*(tons)*
1	1	1	1	2	2	3
2	2	3	4	5	6	7
3	3	5	6	8	10	13
4	5	7	9	12	15	19
5	6	9	12	16	20	25
6	8	11	15	20	26	31
7	10	14	19	25	31	38
8	11	16	22	29	37	45
9	13	19	26	34	43	53
10	15	22	29	38	49	60
11	17	24	33	43	55	67
12	19	27	37	48	61	75
13	21	30	41	53	67	83
14	23	33	44	58	74	91
15	25	36	48	63	80	99
16	27	38	52	68	86	107
17	29	41	56	73	93	115
18	31	44	60	79	100	123
19	33	47	64	84	106	131
20	35	50	68	89	113	139
21	37	53	72	94	120	148
22	39	56	76	100	126	156
23	41	59	81	105	133	164
24	43	62	85	111	140	173
25	45	65	89	116	147	181
26	47	68	93	121	154	190
27	50	71	97	127	161	198
28	52	74	101	132	167	207
29	54	77	105	138	174	215
30	56	81	110	143	181	224
31	58	84	114	149	188	232
32	60	87	118	154	195	241
33	62	90	122	160	202	249
34	65	93	126	165	209	258
35	67	96	131	171	216	267
36	—	—	135	176	223	275
37	—	—	139	182	230	284
38	—	—	143	187	237	292
39	—	—	148	193	244	301
40	—	—	152	198	251	310
41	—	—	—	204	258	318
42	—	—	—	209	265	327
43	—	—	—	215	272	335
44	—	—	—	220	279	344
45	—	—	—	226	286	353
46	—	—	—	—	293	361
47	—	—	—	—	300	370
48	—	—	—	—	307	379
49	—	—	—	—	314	387
50	—	—	—	—	321	396

[1]From USDA Circular No. 603, by J. B. Shepherd and J. E. Woodward; with all decimals rounded off to the nearest whole number. This tabular material was used as a basis for Fig. 9-3 under Buildings and Equipment.

Table 20-4, page 1171, is for well-eared corn silage harvested in the early dent stage, cut in ¼-inch lengths, well-tramped when filled, and with the silo refilled once after settling for a day. The depth indicated in the left-hand column is the actual depth of the settled silage and not the height of the silo. As noted, silage is more compact and heavier as the depth increases.

Table 20-4, page 1171, can be adapted for corn silage of different stages of maturity and grain content, and for other kinds of silages, simply by applying the following rules of thumb:

Kind of Silage	Changes to Be Made in the Number of Tons Shown in Table 20-4
1. For corn silage ensiled when less mature than usual	Add 5 to 10%
2. For corn ensiled when dry or overripe	Deduct 5 to 10%
3. For corn very rich in grain	Add 5 to 10%
4. For corn with very little grain .	Deduct 5 to 10%
5. For sorghum silage	Use the same weights as used for corn silage of comparable grain and maturity.
6. For sunflower silage	Add 5 to 10%
7. For grass silage	Add 10 to 15%

To estimate the amount of average corn silage remaining in a tower silo after part of it has been fed out, proceed as follows:

1. Estimate the actual depth of silage left in the silo.

2. Estimate the original total depth of silage in the silo after settling 30 days.

3. Determine the feet of silage removed by subtracting the depth of silage left (1, above) from the original depth of silage in the silo (2, above).

4. Using Table 20-4, determine the original tonnage of silage contained in the silo.

5. Using Table 20-4, determine the amount of silage removed.

6. Determine the tonnage of silage remaining by subtracting the amount of silage removed (5, above) from the original tonnage (4, above).

As an example, let us assume that 10 ft of well-eared corn silage harvested in the early dent stage is left in a silo having a diameter of 14 ft and that, after settling, the entire depth of silage was 40 ft before feeding started. What tonnage of silage remains?

The answer is obtained as follows:

a. Estimated depth of silage remaining = 10 feet

b. Estimated original depth of settled silage = 40 feet

c. Feet of silage removed 40 – 10 = 30 feet

d. Original tonnage of silage before feeding as determined from Table 20-4; for a 14-ft diameter silo, 40 ft deep = 152 tons

e. Amount of silage removed as determined from Table 20-4; for a 14-ft diameter silo from which 30 ft has been removed = 110 tons

f. Estimated tonnage of silage remaining; 152 – 110 = 42 tons

To estimate the amount of corn or sorghum silage in a trench silo (whether it is full or partly used), multiply the average width in feet by the depth in feet to get the cross-sectional area, then multiply by the length to get the volume, and finally multiply the volume by 35 (the average weight of one cubic foot of corn silage in a trench silo) in order to obtain the pounds of silage. For example, the amount of silage in a trench silo 8 ft wide at the bottom, 12 ft wide at the top, 8 ft deep and 40 ft long is computed as follows:

1. $\frac{8 + 12 = 20}{2}$ = 10 feet; average width

2. 10 × 8 = 80 square feet; cross-sectional area

3. 80 × 40 = 3,200 cubic feet; volume of the silo

4. 3,200 × 35 = 112,000 pounds, or 56 tons capacity.

ESTIMATING ANIMAL WEIGHTS

Feeders who finish large numbers of animals have scales in their feedyards for use in determining in-weights, out-weights, and interim weight gains of animals while they are on feed. Likewise, both purebred and commercial breeders usually have scales. However, those with only one animal, or a few head—such as 4-H Club and FFA members, and part-time farmers—may not have scales. As a result, rations cannot be accurately evaluated, rate of gain cannot be calculated, and an animal's "weight readiness" for a livestock show or for market cannot be determined. Under such circumstances, a simple but reasonably accurate method of estimating body weight is very useful. Fortunately, animal weights may be determined with reasonable accuracy by taking two body measurements (length and heart girth), then applying a certain formula.

- **Estimating weight of beef cattle—**

Here is how to do it:

Step 1—Measure the circumference (heart girth), from a point slightly behind the shoulder blade, thence down over the forefribs and under the body, behind the elbow (distance C of Fig. 20-2).

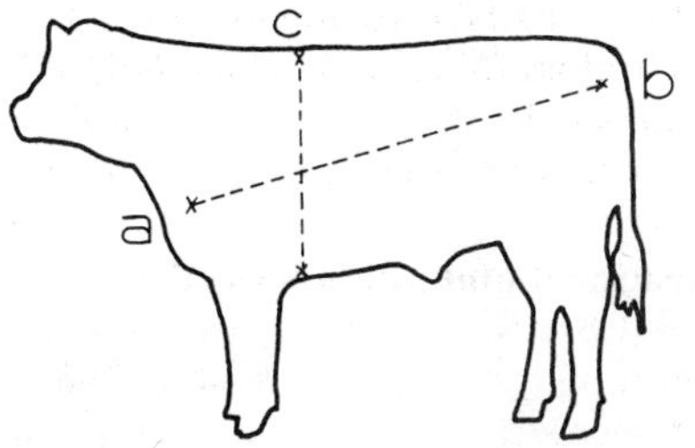

Fig. 20-2. How and where to measure beef cattle.

Step 2—Measure the length of body, from the point of the shoulder to the point of the rump (pinbone), in inches (distance A-B of Fig. 20-2).

Step 3—Take the values obtained in Steps 1 and 2 and apply the following formula to calculate body weight:

Heart girth × heart girth × body length ÷ 300 = weight in pounds.

Example of a beef animal—

Assume that the heart girth measures 76 inches and the body length, 66 inches. How much does the animal weigh?

$76 \times 76 = 5{,}776$
$5{,}776 \times 66 = 381{,}216$
$381{,}216 \div 300 = 1{,}270$ pounds

• Estimating weight of dairy heifers—

Weight for age is important in dairy heifers from the standpoint of determining the growth progress made in herd replacements.

Table 20-5 shows the weight and heart girth measurements of dairy calves or heifers at monthly intervals up to 21 months of age. If the dairyman does not have scales, he can measure the heart girth with a tape (see Fig. 20-3) and use Table 20-5, to estimate weight within 95 percent accuracy.

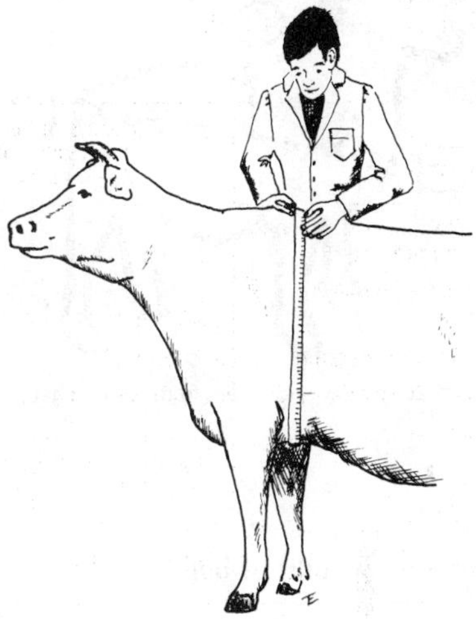

Fig. 20-3. How to tape measure a heifer.

TABLE 20-5

NORMAL HEART GIRTH MEASUREMENT AND WEIGHT OF CALVES AND HEIFERS DURING THE GROWING PERIOD[1]

Age in Months	Brown Swiss, Holstein		Ayrshire, Milking Shorthorn, Red Poll		Guernsey		Jersey	
	(in.)	(lb)	(in.)	(lb)	(in.)	(lb)	(in.)	(lb)
Birth	31	96	29½	72	29	66	24½	56
1	33½	118	32	98	31½	90	29½	72
2	37	161	35½	132	34½	122	32½	102
3	40¼	213	38¾	179	38	164	35½	138
4	43½	272	42¾	236	41¼	217	38¼	181
5	47	335	45½	291	44¼	265	41½	228
6	50	396	48¼	340	47	304	44½	277
7	52½	455	51¼	408	49¾	362	47¼	325
8	54¾	508	53	447	51¾	410	49¾	369
9	57	559	55	485	53¾	448	51¾	409
10	58¾	609	57	526	55	486	53¼	446
11	60½	658	58	563	56¾	521	55	481
12	62½	714	59	583	58¼	549	56½	520
13	63¼	740	60¾	630	59¼	587	57½	540
14	64¼	774	62	666	60½	615	58½	565
15	65¼	805	63	703	61¾	640	59	585
16	66¼	841	64	731	62½	674	59¾	611
17	67¼	874	65¼	758	63½	696	60½	635
18	68½	912	66	781	65	727	61½	660
19	69¼	946	66½	813	65½	752	62½	687
20	70½	985	67½	841	66¼	780	63	712
21	71½	1,025	68½	885	67½	816	64	740

[1]Adapted by the author from the following sources: Body weights for Holsteins and Jerseys from USDA Tech. Bull. 1098 and 1099. Heart girth measurements for these weights taken from Research Bull. 194 (1960), Neb. Ag. Exp. Sta. Weights and heart girth measurements for Ayrshires and Guernseys calculated from data furnished by Professor H. P. Davis, University of Nebraska.

• **Estimating weight of sheep and goats—**

The weight of sheep and goats is estimated in the same way as beef cattle; hence, it involves making the measurements and applying the formula given for beef cattle. There is one important precaution, however; with unshorn sheep, be sure to part, or compress, the wool to ensure an accurate heart girth measurement.

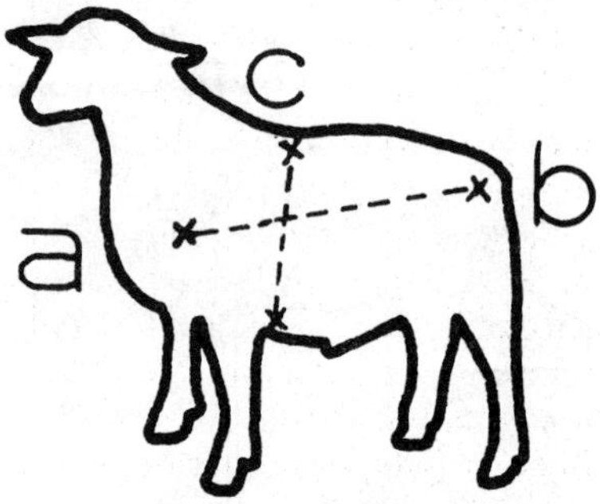

Fig. 20-4. How and where to measure sheep.

• **Estimating weight of hogs—**

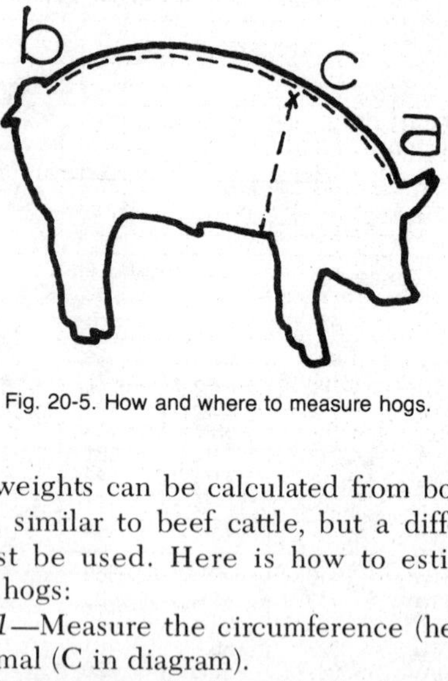

Fig. 20-5. How and where to measure hogs.

Hog weights can be calculated from body measurements, similar to beef cattle, but a different formula must be used. Here is how to estimate the weight of hogs:

Step 1—Measure the circumference (heart girth) of the animal (C in diagram).

Step 2—Measure the length of body (A-B in Fig. 20-5). With the animal standing or restrained in the position shown in Fig. 20-5, measure the distance from the poll (between the ears), over the backbone, to the base of the tail.

Step 3—Apply the following formula:

Heart girth × heart girth × length ÷ 400 = weight in pounds

Note: For hogs weighing less than 150 lb, add 7 lb to the weight figure obtained from the formula. For animals weighing 151 to 400 lb, no adjustment is necessary.

• **Estimating weight of horses—**

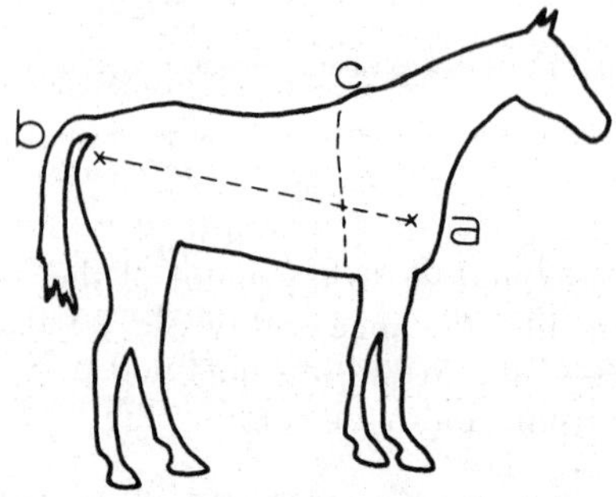

Fig. 20-6. How and where to measure horses.

It is easy to estimate the weight of a horse; and tests have shown that the results obtained this way are accurate within 3 percent of actual scale weight. This procedure is as follows:

Step 1—Measure the circumference (heart girth) of the body in inches (C in diagram).

Step 2—Measure the length of body from the point of the shoulder to the point of croup (A-B in the diagram).

Step 3—Apply the following formula to calculate the weight of the horse:

Heart girth × heart girth × length ÷ 300 + 50 lb = weight of horse

Example: Assume that the heart girth is 70 inches and the body length is 65 inches. How much does the horse weigh?

70″ × 70″ × 65″ ÷ 300 + 50 lb = weight
4,900 × 65 = 318,500
318,500 ÷ 300 = 1,061 lb
1,061 + 50 = 1,111 lb body weight

ANIMAL UNITS

An animal unit is a common animal denominator, based on feed consumption. It is assumed that one mature cow represents an animal unit. Then, the comparative (to a mature cow) feed consumption of other age groups or classes of animals determines the proportion of an animal unit which they represent. For example, it is generally estimated that the ration of one mature cow will feed 5 mature ewes, or that 5 mature ewes equal 1.0 animal unit.

The original concept of an animal unit included a weight stipulation—an animal unit referred to a

1,000-lb cow, with or without a calf at side. Unfortunately, in recent years, the 1,000-lb qualification has been dropped. Certainly, there is a wide difference in the daily feed requirements of a 900-lb range cow and of a 1,500-lb exotic cow. Both will consume dry matter on a daily basis at a level equivalent to about 2 percent of their body weight.

Hence, a 1,500-lb cow will consume 50 percent more feed than a 1,000-lb cow.

Also, the period of time to be grazed has an effect on the total carrying capacity. For example, if an animal is carried for one month only, it will take one-twelfth of the total feed required to carry the same animal one year. For this reason, the term "animal unit months" is becoming increasingly important. So, in addition to the weight factor, the time factor has a distinct bearing on the ultimate carrying capacity of a tract of land.

Table 20-6 gives the animal units of different classes and ages of livestock.

TABLE 20-6
ANIMAL UNITS

Type of Livestock	Animal Units
Cattle:	
Cow, with or without unweaned calf at side, or heifer 2 years old or older	1.0
Bull, 2 years old or older	1.3
Young cattle, 1 to 2 years	0.8
Weaned calves to yearlings	0.6
Horses:	
Horse, mature	1.3
Horse, yearling	1.0
Weanling colt or filly	0.75
Sheep:	
5 mature ewes, with or without unweaned lambs at side	1.0
5 rams, 2 years old or over	1.3
5 yearlings	0.8
5 weaned lambs to yearlings	0.6
Swine:	
Sow	0.4
Boar	0.5
Pigs to 200 pounds	0.2
Chickens:	
75 layers or breeders	1.0
325 replacement pullets to 6 months of age	1.0
650 8-week-old broilers	1.0
Turkeys:	
35 breeders	1.0
40 turkeys raised to maturity	1.0
75 turkeys to 6 months of age	1.0

LAND AREA

Stockmen should be familiar with some of the basic facts about land area—legal description, measures, and how to determine acreage—regardless of whether they are owners or renters. This section is designed for this purpose.

Legal Description of Land

The federal government land survey established townships 6 miles square with the boundaries due north and south and due east and west. Townships are located north or south of standard parallels and east and west of prime meridians.

Each township is further divided into 36 square miles or sections, numbered from the northeast corner back and forth, with section 36 in the southeast corner (see Fig. 20-7). The sections are further divided into quarters, the quarter sections into quarters of 40 acres, etc. (See Fig. 20-8, page 1176.) By this method, every parcel of land can be accurately located and identified for legal purposes. For example, the description of a certain 40-acre area may read: The south half of the west half of the southwest quarter of section 1 in township 24, north of range 7 west.

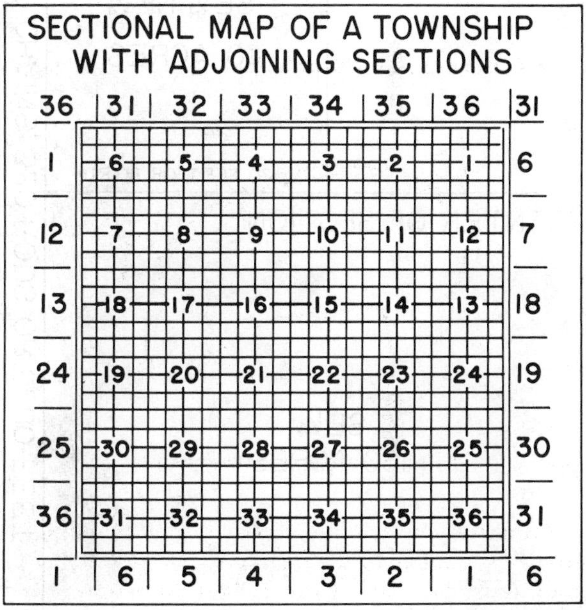

Fig. 20-7. A township, divided into its 36 sections. (Drawing by Steve Allured)

Land Measures

The U.S. Government Land Measures are:

A township—36 sections, each a mile square.

A section—640 acres.

A quarter section—Half a mile square, 160 acres.

An eighth section—Half a mile long and a quarter of a mile wide, 80 acres.

A sixteenth section—A quarter of a mile square, 40 acres.

Other commonly used land measures are:

A rod—16½ feet.

A chain—66 feet, 4 rods, or 100 links each 7.92 inches long.

A mile—320 rods, 80 chains, or 5,280 feet.

A square rod—272¼ square feet.

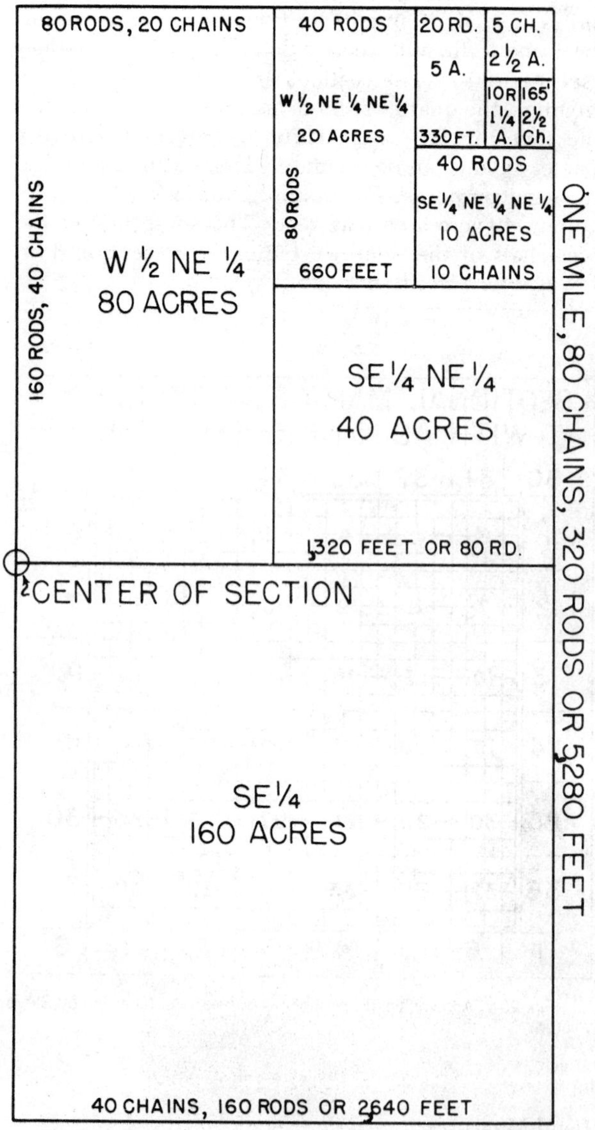

Fig. 20-8. A half section, divided into component units. (Drawing by Steve Allured)

How to Determine Acreage

One acre of land measures 160 square rods, or 4,840 square yards, or 43,560 square feet. Thus, in order to find the number of acres in a piece of land, simply multiply the length by the width, and divide by 160 or 4,840 or 43,560, respectively, as determined by the unit used in computing the area. When the opposite sides of an area are unequal, add them and take half the sum for the mean length or width.

For purposes of illustration, let us determine the acreage of an area measuring 500 feet × 400 feet.

This is done as follows:

1. 500 × 400 = 200,000 sq. ft.
2. 200,000 ÷ 43,560 = 4.59 acres

THE METRIC SYSTEM[2]

Increasingly, stockmen and those who counsel with them need to use the metric system. Hence, they need to have a working knowledge of it, along with conversion tables.

The basic metric units are: the *meter* (length/distance), the *gram* (weight), and the *liter* (capacity). The units are then expanded in multiples of ten or made smaller by 1/10. The prefixes, which are used in the same way with all basic metric units, follow:

"milli-"	=	1/1000
"centi-"	=	1/100
"*deci-*"	=	1/10
"*deca-*"	=	10
"hecto-"	=	100
"kilo-"	=	1,000

Conversion factors: grams to pounds, divide by 454; percent to grams/ton, divide by 11 and move decimal 5 places to right; ppm to percent, move decimal 4 places to left.

Table 20-7 is a conversion table, from metric system to U.S. customary system, and vice versa.

[2]For additional conversion factors, or for greater accuracy, see *Misc. Pub.* 233, the National Bureau of Standards.

TABLE 20-7
WEIGHTS AND MEASURES

LENGTH

Unit	Is Equal To	
Metric System		**(U.S. Customary)**
1 millimicron ($m\mu$)	.000000001 m	.000000039 in.
1 micron (μ)	.000001 m	.000039 in.
1 millimeter (mm)	.001 m	.0394 in.
1 centimeter (cm)	.01 m	.3937 in.
1 decimeter (dm)	.1 m	3.937 in.
1 meter (m)	1 m	39.37 in.; 3.281 ft; 1.094 yd
1 hectometer (hm)	100 m	328 ft, 1 in.; 19.8838 rd
1 kilometer (km)	1,000 m	3,280 ft, 10 in.; 0.621 mi
U.S. Customary System:		**(Metric)**
1 inch (in.)		25 mm; 2.54 cm
1 hand*	4 in.	
1 foot (ft)	12 in.	30.48 cm; .305 m
1 yard (yd)	3 ft	.914 m
1 fathom** (fath)	6.08 ft	1.829 m
1 rod (rd), pole, or perch	16½ ft; 5½ yd	5.029 m
1 furlong (fur.)	220 yd; 40 rd	201.168 m
1 mile (mi)	5,280 ft; 1,760 yd; 320 rd; 8 fur.	1,609.35 m; 1.609 km
1 knot or nautical mile	6,080 ft; 1.15 land miles	
1 league (land)	3 mi (land)	
1 league (nautical)	3 mi (nautical)	

*Used in measuring height of horses.
**Used in measuring depth at sea.

CONVERSIONS (to make opposite conversion, simply divide by the number given instead of multiplying).

To Change	To	Multiply by
inches	centimeters	2.54
feet	meters	.305
meters	inches	39.37
miles	kilometers	1.609
kilometers	miles	.621

(Continued)

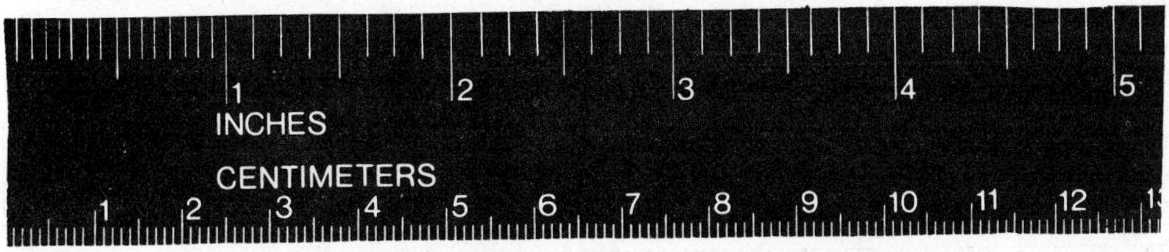

Fig. 20-9 Inches-centimeter scale for direct conversion and reading.

SURFACE OR AREA

Unit	Is Equal To	
Metric System:		**(U.S. Customary)**
1 square millimeter (mm²)	.000001 m²	.00155 in.²
1 square centimeter (cm²)	.001 m²	.155 in.²
1 square decimeter (dm²)	.01 m²	15.50 in.²
1 square meter (m²)	1 centare (ca)	1,550 in.²; 10.76 ft²; 1.196 yd²
1 are (a)	100 m²	119.6 yd²
1 hectare (ha)	10,000 m²	2.47 acres
1 square kilometer (km²)	1,000,000 m²	247.1 acres; .386 mi²
U.S. Customary System:		**(Metric)**
1 square inch (in.²)	1 in. × 1 in.	6.452 cm²
1 square foot (ft²)	144 in.²	.093 m²
1 square yard (yd²)	1,296 in.²; 9 ft²	.836 m²
1 square rod (rd²)	272.25 ft²; 30.25 yd²	25.29 m²
1 rood	40 rd²	10.117 a
1 acre	43,560 ft²; 4,480 yd²; 160 rd²; 4 roods	4,046.87 m² 0.405 ha
1 square mile (mi²)	640 acres	2.59 km² or 259.0 ha
1 township	36 sections; 6 miles square	

CONVERSIONS (to make the opposite conversion, simply divide by the number given instead of multiplying).

To Change	To	Multiply by
square inches	square centimeters	6.452
square centimeters	square inches	.155
square yards	square meters	.836
square meters	square yards	1.196
square miles	square kilometers	2.59

VOLUME

Unit	Is Equal To			
Liquid and dry: **Metric System:**		**(U.S. Customary)** (liquid)		(dry)
1 milliliter (ml)	.001 l	.271 dram (fl)		.061 in.³
1 centiliter (cl)	.01 l	.338 oz (fl)		.610 in.³
1 deciliter (dl)	.1 l	3.38 oz (fl)		
1 liter (l)	1,000 cc	1.057 qt or 0.2642 gal (fl)		.908 qt
1 hectoliter (hl)	100 l	26.418 gal		2.838 bu
1 kiloliter (kl)	1,000 l	264.18 gal		1,308 yd³
U.S. Customary System: **Liquid:**		**(Ounces)**	**(Cubic Inches)**	**(Metric)**
1 teaspoon (t)	60 drops	1/6		5 ml
1 dessert spoon	2 t			
1 tablespoon (T)	3 t	½		15 ml
1 fl oz		1	1.805	29.57 ml
1 gill (gi)	½ c	4	7.22	118.29 ml
1 cup (c)	16 T	8	14.44	236.58 ml; .24 l
1 pint (pt)	2 c	16	28.88	.47 l
1 quart (qt)	2 pt	32	57.75	.95 l
1 gallon (gal)	4 qt	8.34 lb	231	3.79 l
1 barrel (bbl)	31½ gal			
1 hogshead (hhd)	2 bbl			
Dry:				
1 pint (pt)	½ qt		33.6	.55 l
1 quart (qt)	2 pt		67.20	1.10 l

(Continued)

Dry: (continued)			
1 peck (pk)	8 qt	537.61	8.81 l
1 bushel (bu)	4 pk	2,150.42	35.24 l
Solid: **Metric System:**			
1 cubic millimeter (mm^3)	.001 cc		
1 cubic centimeter (cc)	1,000 mm^3	.061	
1 cubic decimeter (dm^3)	1,000 cc	61.023	
1 cubic meter (m^3)	1,000 dm^3	35.315 $ft;^3$ 1.308 yd^3	
U.S. Customary System:		**(Metric)**	
1 cubic inch ($in.^3$)			16.387 cc
1 board foot (fbm)	144 $in.^3$		2,359.8 cc
1 cubic foot (ft^3)	1,728 $in.^3$		.028 m^3
1 cubic yard (yd^3)	27 ft^3		.765 m^3
1 cord	128 ft^3		3.625 m^3

CONVERSIONS (to make opposite conversion, simply divide by the number given instead of multiplying).

To Change	To	Multiply by
ounces (fluid)	cubic centimeters	29.57
cubic centimeters	ounces (fluid)	.034
quarts	liters	.946
liters	quarts	1.057
cubic inches	cubic centimeters	16.387
cubic centimeters	cubic inches	.061
cubic yards	cubic meters	.765
cubic meters	cubic yards	1.308

WEIGHT

Unit	Is Equal To	
Metric System:		**(U.S. Customary)**
1 microgram (mcg)	.001 mg	
1 milligram (mg)	.001 g	.015 gr
1 centigram (cg)	.01 g	.154 gr
1 decigram (dg)	.1 g	1.543 gr
1 gram (g)	1,000 mg	.035 oz
1 dekagram (dkg)	10 g	5.648 dr
1 hectogram (hg)	100 g	3.527 oz
1 kilogram (kg)	1,000 g	35.274 oz; 2.205 lb
1 ton	1,000 kg	2,204.6 lb; 1.102 (short) ton; 0.984 (long) ton
U.S. Customary System:		**(Metric)**
1 grain (gr)	.038 dr	64.8 mg; .065 g
1 dram (dr)	.063 oz	1.772 g
1 ounce (oz)	16 dr	28.35 g
1 pound (lb)	16 oz	453.6 g; 0.454 kg
1 hundredweight (cwt)	100 lb	45.36 kg
1 ton (short)	2,000 lb	907.18 kg; 0.907 (metric) ton
1 ton (long)	2,200 lb	1,016.05 kg; 1.016 (metric) ton
U.S. Customary System:		
1 part per million (ppm)	1 microgram/gram 1 mg/l 1 mg/kg	.454 mg/lb .907 g/ton .0001 % .013 oz/gal
1 percent (%) (1 part in 100 parts)	10,000 ppm 10 g/l	1.28 oz/gal 8 lb/100 gal

(Continued)

Weight (continued)

CONVERSIONS (to make opposite conversion, simply divide by the number given instead of multiplying).

To Change	To	Multiply by
grains	milligrams	64.799
ounces (dry)	grams	28.35
pounds	grams	454
pounds (dry)	kilograms	.454
kilograms	pounds	2.2
milligrams/pound	parts/million	2.2
parts/million	grams/ton	.907
grams/ton	parts/million	1.1
milligrams/pound	grams/ton	2
grams/ton	milligrams/pound	.5
grams/pound	grams/ton	2,000
grams/ton	grams/pound	.0005
grams/ton	pounds/ton	.0022
pounds/ton	grams/ton	453.59
grams/ton	percent	.00011
percent	grams/ton	9,072
parts/million	percent	move decimal 4 places to left

WEIGHTS AND MEASURES PER UNIT

Unit	Is Equal To
Volume per unit area:	
1 liter/hectare	0.107 gal/acre
1 gallon/acre	9.354 l/ha
Weight per unit area:	
1 kilogram/centimeter^2	14.22 lb/in.^{2}
1 kilogram/hectare	0.892 lb/acre
1 pound/square inch	0.0703 kg/cm^2
1 pound/acre	1.121 kg/ha
Area per unit weight:	
1 centimeter^2/kilogram	0.0703 in.^{2}/lb
1 square inch/pound	14.22 cm^2/kg

TEMPERATURE

One centigrade (C) degree is $^1/_{100}$ the difference between the temperature of melting ice and that of water boiling at standard atmospheric pressure. *One centigrade degree equals 1.8°F.*

One Fahrenheit (F.) degree is $^1/_{180}$ of the difference between the temperature of melting ice and that of water boiling at standard atmospheric pressure. *One Fahrenheit degree equals 0.556°C.*

To Change	To	Do This
degrees centigrade	degrees Fahrenheit	multiply by $^9/_5$ and add 32
degrees Fahrenheit	degrees centigrade	subtract 32, then multiply by $^5/_9$

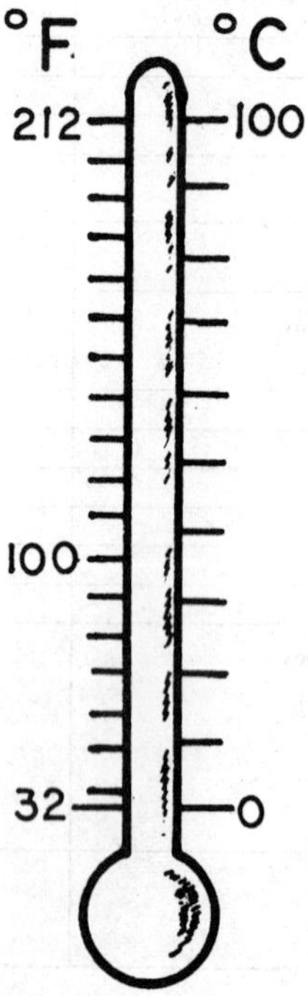

Fig. 20-10. Fahrenheit-centigrade scale for direct conversion and reading.

CALORIES

One calorie is the amount of heat required to raise one gram of water, one degree centigrade at sea level.

Unit	Is Equal To
1 kcal	1000 cal
1 mcal	1000 kcal
1 therm	1000 kcal or 1 mcal

Index

A

Page

Accounts, farm ... 65
Acetonemia ... 438
Acidosis ... 438
Acreage determination ... 1176
Actinobacillosis ... 876
Actinomycosis ... 876
Adaptation
  breeding for ... 17
  quick ... 18
Additives ... 225
  withdrawal period ... 284
Adjustment factors
  dairy ... 147
  sheep ... 156
Aerobic activity ... 702
African sleeping sickness ... 876
Age
  comparative animal ... 1041
  determination ... 1032
    cattle ... 1033
    horses ... 1038
    sheep ... 1034
    swine ... 1035
Age and show classifications
  beef cattle ... 1047
  dairy cattle ... 1050
  horses ... 1058
  sheep ... 1053
  swine ... 1056
Agisters ... 1118
Agonistic behavior ... 6
Agricultural
  chemicals ... 27
  colleges ... 27, 1160
  Council of America ... 1161
  Stabilization and Conservation ... 1161
Agriservices Foundation ... 1161
Air dry ... 454
A.I. rules ... 1139-1147
Alcohol, disinfectant ... 984
Alfalfa ... 672
Alkaline soils ... 756
Allelomimetic behavior ... 7
Alternate AM-PM Test ... 146
American
  Baskir Curly horse ... 206
  Creme Horse ... 206
  Farm Bureau Federation ... 1161
  Feed Manufacturers Association ... 1161
  Gotland Horse ... 206
  Horse Council ... 1162
  Jack Ass ... 212
  La Mancha goats ... 203
  Landrace swine ... 204
  Meat Institute ... 1162
  Merino sheep ... 200
  Mustang ... 206
  Saddle Horse ... 206
  Sheep Producers Council ... 1162
  Society of Agricultural Consultants ... 1162
  Society of Animal Science ... 1162
  Veterinary Medical Association ... 1162

Page

  Walking Pony ... 206
  White Horse ... 207
Amino acid composition of feeds ... 580–583
Amino acids ... 227
  essential ... 227
  nonessential ... 227
Anaerobic activity ... 702
Anaplasmosis ... 876, 920
Andalusian horses ... 207
Anemia ... 438
Angora goats ... 203
Angus
  cattle ... 190
  Hereford cross ... 168
  Red ... 196
Animal(s)
  behavior ... 2
    abnormal ... 17
    applied ... 17
    controlling ... 20
    normal ... 15
  choosing class ... 46, 47
  communication ... 14
  companionship ... 20
  diseases (*see Table 10-2, p. 876*)
  environment ... 21, 811
  estrays ... 1118
  health ... 15, 873
    interstate requirements ... 1003
    program ... 987
  heat production ... 809
  highways, on ... 1117
  parasites ... 917
  persons injured by ... 1119
  protein ... 219
  sanitation ... 986
  sight ... 16
  sleep ... 16
  trespass ... 1117
  units ... 1174, 1175
  vapor production ... 810
  weight estimating ... 1172
Anthelmintics ... 226, 917, 918
Anthrax ... 876
Antibacterial products ... 226
Antibiotics ... 225
  beef cattle ... 284
  dairy cattle ... 342
  horses ... 423
  sheep ... 364
  swine ... 394
Anticoagulants for rats ... 763
Antioxidants ... 226
Aphosphorosis ... 438
Appaloosa horses ... 207
Appraisal, farm ... 41
Appropriative water rights ... 44
Arabian horses ... 207
Arrowgrass ... 962
Arsenicals ... 395
Arsenic poisoning ... 980
Arthritis ... 878

Page

Artificial
  dehydrators ... 686
  drying hay ... 685
  insemination ... 180
    advantages ... 181
    costs ... 188
    electric stimulation ... 183
    female, of ... 186
    limitations ... 182
    preparation of equipment ... 182
    rectal massage ... 183
    registration rules ... 1139-1147
    semen ... 183
      frozen ... 184
    time of ... 185
    vagina ... 183
  lighting ... 23
Ascarids ... 940, 950
  horses ... 950
  swine ... 940
Ascorbic acid ... 233
As-fed ... 454
Asses and mules
  American Jack ... 212
  donkeys ... 212
  Miniature Mediterranean Donkeys ... 212
  mules ... 212
Associations
  breed registry ... 1123
  marking rules ... 1123
Atrophic rhinitis ... 878
Automation ... 815
Ayrshire cattle ... 199
Azoturia ... 440

B

Babesiasis ... 878
Baby pig shakes ... 440
Bacillary hemoglobinuria ... 878
Back fat measuring—swine ... 159
Backgrounding beef cattle ... 316
Balanced rations ... 252, 258
  method of balancing ... 253
    computer ... 257
    net energy ... 256
    square ... 254
    trial-and-error ... 255
Baling wire danger ... 695
Bankrupt worms ... 930
Banks ... 52, 53
  for cooperatives ... 53
Barn size ... 816
Banyard manure (*see* Manure)
Barzona cattle ... 190
Bedding ... 735
  absorbent ability ... 735
  amount ... 735
  eating ... 20
  kind in transit ... 1064
  kinds ... 735
  reducing needs ... 736
  storage space requirements ... 803

Page

- Beef
  - cooking 1082–1084
  - cuts 1081–1083
  - futures trading 97
  - grades 1080
- Beef cattle
  - additives 281, 284, 314
  - age determination 1033
  - A.I. rules 1139
  - Angus 190
  - anthelmintics 917, 918
  - antibiotics 225, 284
  - artificial insemination 187
    - rules 1139
  - backgrounding 316
  - Barzona 190
  - beef and veal 1080
    - cooking 1082–1084
    - cuts 1081, 1084
    - grades 1080
  - Beef Friesian 190
  - Beefmasters 190
  - Belted Galloway 191
  - bloat 440
  - Blonde d'Aquitaine 191
  - Braford 191
  - Brahman 191
  - brands 724
  - Brangus 191
  - breed(s) 190
    - registries 1124, 1125
  - breeding season 769
  - brood cows 295
  - buildings 820–826
    - space requirements 802
  - bull(s)
    - age and service of 123
    - feeding 306, 768, 769
    - finishing 321
    - management 768
  - calves
    - early weaned 301
    - feeding 301
    - orphan 301
    - preconditioning 772
      - certificate 303, 774
  - calving
    - care at 770–772
    - normal 771
    - season 769
    - two-year-olds 769
  - capital needs 46
  - carcass evaluation 143
  - castrating 729, 732
    - equipment 729
  - characters
    - economically important 135
    - heritability 135
  - Charbray 191
  - Charolais 191
  - Chianina 192
  - commercial herd performance testing 145
  - compensatory growth 308
  - computers 73
  - confinement (drylot cows) 772
  - contract feeding 79
  - cows
    - gestation period 122
    - hormonal heat control 177
  - creep feeding 302
  - crop residues 297
  - crossbreeding 168
  - dairy beef 776
  - dehorning 729, 731
    - equipment 729
  - Devon 192
  - Dexter 192
  - dressing percent 1082
  - energy 271
    - deficiency symptoms 272
  - equipment 827–829
  - fattening (*see* finishing)
  - feed(s) 309–311
    - preparation 251, 314
    - substitution table 290–294
    - urea 311
  - feeder's margin 78
  - feeding 260
    - breeding cattle 295
    - grain fed 310
    - guide 282
    - sale cattle 322
    - salt-feed mixtures 300
    - schedule 317
    - short scrotum bulls 321, 729
    - show cattle 322
    - stockers 306–308
    - winter 296
  - fertility evaluation 769
  - finishing (fattening) 308–322
    - amount to feed 318, 775
    - bird control 775
    - break-even prices 76
    - business aspects 75
    - capital needs 49
    - computers in 75
    - condominiums 85
    - confinement facilities 826
    - contract 79, 84
    - co-op feedlots 83
    - cost of 50
    - custom feeding 79
    - dairy beef 776
    - diseases 876
    - feed, amount 775
    - feedlot facilities 822–825
    - fly control 776
    - full vs limited 318, 775
    - futures trading 97
    - health 773
      - program 990
    - hogs following cattle 776
    - incentive basis 63
    - losses 874
    - management 773–776
    - margin 78
    - new arrivals 316, 773
    - over-finishing 319
    - pasture 320–322
    - pollution control 776
    - profit indicators 70
    - record forms 86
  - Fleckvieh 192
  - futures trading 97
  - Galloway 192
  - Gelbvieh 192
  - gestation
    - period 122
    - table 122
  - grain fed 310
  - grooming 1043
  - growth stimulants 303
  - Hays Converter 193
  - health program 988
  - heat (estrus)
    - duration of 122
    - table 122
  - Hereford 193
  - heritability of traits 135
  - hogs following 776
  - identifying 724
  - implants 281, 284, 314
  - import quotas 1097
    - tariff 1097
  - incentive basis 59
  - Indu Brazil (Zebu) 193
  - insemination 187
  - judging 1010
  - Limousin 193
  - Lincoln Red 193
  - liquid supplements 300
  - liveweight average 1082
  - losses 874
  - Maine Anjou 194
  - management practices 768–776
  - Marchigiana 194
  - market classes and grades 1067
  - marking 724, 1132
  - minerals 272, 274–279, 313
  - mites 958
  - most probable producing ability 136
  - multiple births 132
  - Murray Grey 194
  - Normande 194
  - Norwegian Red 195
  - nutrient requirements 260–271
  - parasites (*see Table 10-5, p. 920*) 918
  - parts of steer 1010
  - parturition signs 770
  - pastures (and ranges) 298, 589–636
    - nutrient deficiencies of 299
    - protein blocks 300
    - range cubes 299
    - supplements 299
    - winter 297, 298
  - performance testing 139
  - Piedmont 195
  - Pinzgauer 195
  - Polled Hereford 195
  - Polled Shorthorn 196
  - preconditioning 303, 990
    - certificate 774
  - production testing 134
    - forms 136
    - requisites 140
  - profit indicators 69
  - protein 272
    - deficiency symptoms 272
  - puberty, age of 122
  - range cubes 299
  - range feeding (*see* Pastures)
  - Ranger 196
  - ration(s) 288, 296
    - confinement cows 300
    - cows nursing calves 297
    - dry pregnant cows 296
    - net energy formulation 315
    - replacement heifers 305
    - sale—fitting 323
    - show—fitting 323
    - stockers 307
  - ratios 136
  - record forms 136, 137–139
  - Red
    - Angus 196
    - Brangus 196
    - Poll 196
  - registry associations 1123
  - replacement heifers 304–306
  - reproductive failure, nutritional 296
  - Russian castrates 321
  - salt-feed mixtures 300
  - salt requirements 274
  - Santa Gertrudis 197
  - scorecard
    - breeding cattle 1013
    - steer 1014
  - Scotch Highland 197
  - semen evaluation 769
  - Shorthorn 197
  - show classifications 1047
  - showing 1043
  - Simmental 197
  - slotted floors 813
  - South Devon 197
  - Sussex 198
  - Tarentaise 198
  - Texas Longhorn 198
  - traits, heritability of 135
  - vitamins 280–283, 313
  - water 281, 314
  - weight estimating 1172
  - Welsh Black 198
  - winter feeding 296
  - wormers 917, 918
- Beef Friesian cattle 190

Page

Beefmaster cattle 190
Behavior, animal 2, 4
abnormal 17
agonistic 6
allelomimetic 7
applied 17
care-giving 8
care-seeking 8
companionship 20
complex learning 3
controlling 20
eliminative 9
genetic 2
gregarious 7
ingestive 9
investigative 10
normal 15
sexual 10
shelter-seeking 11
simple learning 2
Behavioral systems 4
Belgian horses 211
Belted Galloway cattle 191
Berkshire swine 204
Bighead 878
Bill of sale 1115
Bird control 758–762, 768, 775
Bitterweed 962
Black disease 878
Black-faced Highland sheep 202
Black leg 880
Bladder worms 930
Bloat 226, 440
Blonde d'Aquitane cattle 191
Blood typing 189
Blowfly 920, 930, 942, 950
Blue bag 880
Bluetongue 880
Boar, age and service of 123
Boarding agreement 94, 96
Bolting feed 20
Bonuses 60
Boric acid 984
Borrower's credit factors 56
Bots 950
Bovine pulmonary emphysema 880
trichomoniasis 920
virus diarrhea 880
Bracken fern 962
Braford cattle 191
Brahman cattle 191
Branding 724
Brands 724, 1119
inspection 1119
recorded 1119
Brangus cattle 191
Red 196
Beathing rate, normal 875
Breed(s) 190
beef cattle 190
dairy cattle 199
goats 203
horses 206
sheep 200
swine 204
Breeding
for adaptation 17
record forms 135
Breed magazines 1149
Breed registry associations 1123, 1132
marking rules 1123
Brown stomach worm 930
Brown Swiss cattle 199
Brucellosis 882, 1004
Bruises, market 1063
Brush
chemical control 760–762
control 758–762
Buckskin horses 207
Budgets 71
Buffalo X cattle hybrids 171
Buildings 795

Page

barn size 816
beef cattle 820
concrete structures 849–852
costs 816
dairy cattle 830
environmental control 22, 808–811
farmstead arrangement 797
heat production, animal 809
horses 843
location 797
moisture production, animal 810
paint 860
rat-proofing 765
requisites 798
roofs 812
service passage width 808
silo size 817
slotted floors 812
beef cattle 802
dairy cattle 830
horses 806
sheep 804
swine 804, 839
space requirements 802–808
beef cattle 802
dairy cattle 830
horses 806
sheep 804
swine 804, 839
storage space 808
ventilation 810, 811
Bull
age and service of 123
cost vs A.I. 188
crossbred 171
Burdizzo 729
Bureau of Indian Affairs 664
Bureau of Land Management 663
grazing fees 663
Business 31
accounts 65
analyzing 69
budgets 71
capital 46
cattle feedlots 75
computers 73
corporation 35
Subchapter S 35
tax-option 35
credit 52
estate planning 115
futures trading 97
inheritance 116
insurance 119
manager 58
organization 32
types 33
partnership 34
general 34
limited 34, 35
proprietorship 33
records 65, 69
taxes 106
Buttercup 962
Butter, graded 1103, 1104
B vitamins 232
By-product(s)
feeds 219
packinghouse 1095
value 1095

C

Calcium 228
-phosphorus ratio 228
supplements 229
Calf
diphtheria 884
scours 884
Calorie system 244
Calves, dressing percentage of 1081
Calving induced 180

Page

Canada Department of Agriculture 1162
Canadian Society of Animal Science 1162
Capital
cattle feedlot 49
cow-calf 46
gain 109
needs 46
Ca-P ratio 228
Carcass
disposal 986
evaluation, beef cattle 143
Care-giving 8
Care-seeking 8
Carotene conversion 231
Carpet wool 1107
Caseous lymphadenitis 884
Castrating 730–734
Cattle
finishing (*see* Beef cattle finishing)
tick fever 920
Central money markets 54
Central testing stations
beef cattle 142
sheep 151
swine 161
Cereal
grain refuse 700
hay 672
Certified
boar 162
litter 162
mating 163
Characters
economically important 135
heritability 135
Charbray cattle 191
Charlais cattle 191
Chattel mortgages 1115
Cheese, graded 1103
Chelated trace minerals 230
Chemical(s)
analysis, feed 241
brush control 760–762
communication 15
poisoning 980
weed control 760–762
Chester White swine 204
Cheviot sheep 200
Chianina cattle 192
Chickasaw horses 207
Chin-ball marker 176
Chlorines 984
Chokecherry 964
Cholera, hog 886
Circling disease 886
Cleveland Bay horses 207
Clipping pastures 655
Clothing wool 1107
Clover hay 672
Clydesdale horses 212
CO_2 danger in silos 719
Coccidiosis 920, 930, 942
Cocklebur 964
Colic 440
Colleges of Agriculture 1160
Color in Shorthorn cattle 122
Colostrum 220
Columbia sheep 202
Combing wool 1106
Combustion, spontaneous in hay 687
Commercial
feeds 259
fertilizers 740
Common stomach worm 938
Communication
animal 14
chemical 15
sound 14
visual displays 15
Companionship, animal 20
Comparative slaughter 246
Compensatory growth of cattle 308

Page

Competitive Livestock Marketing
Association 1163
Complementary 167
Complex learning 3
Computer ration balancing 257
Computers 69, 73
Concrete 849, 852
Conditioning, operant 3
Condominium cattle feedlots 85
Confinement
cows 300
facilities 826
Connemara Pony 208
Conner Prairie swine 204
Contagious ecthyma 886
Continuous grazing 660
Contract(s)
cow rental 88
feeding 79, 84
futures 100–104
heifer replacement 88
horse 91
partnership 118
stocker 87
Cooking 1094
beef 1082
frozen meats 1092
lamb 1086
methods 1094
pork 1088
veal 1082
Co-op
feedlots 83
milk marketing 1073
Cooperias 930
Copper sulfate poisoning 980
Copperweed 964
Corn
high-lysine 221
high-moisture 221
residues 697–699
grazing 697
harvesting 697
silage 708
stalk poisoning 446
with urea 710
Corn-hog ratio 408
Cornstalks, grazing 697
Corporation 35
tax-option 35
Corral design 18
Corriedale sheep 202
Cotswold sheep 201
Cottonseed hulls 700
County agricultural agent 1163
Cow(s)
gestation period 122
pools 89
pox 886
pregnancy testing 131
rental contracts 88
Cow-calf
capital needs 46
computers in 74
incentive basis 62
investment 48
profit indicators 70
returns 47, 48
Cowpea hay 674
Cream 1103
Credit 39, 52
factors considered 55, 56
sources 52
types of 52
Creep feeding
beef cattle 302
horses 432
sheep 374
swine 405
Cresols 984
Cribber 17
Crippling, market 1063

Page

Crooked calf disease 440
Crop residues 29, 219, 297, 667, 696
treating 700
Crossbred bulls 171
Crossbreeding
Angus X Hereford 168
bull selection 170
complementary crosses 167
dairy cattle 172
horses 174
hybrids vs purebreds 168
sheep 172
swine 173
Cross, trihybrid 123
Crude protein content of feeds 226, 455
Cubing forages 249, 685
Curing meat 1092–1094
Custom feeding 79
contract 79, 84
Cutting hay 679–682
Cyclical market trends 1069

D

Dairy cattle
A.I. rules 1141
anthelmintics 917, 918
Ayrshire 199
breeds 199
registries 1126
Brown Swiss 199
buildings 830–834
loose housing 830
space requirements 830
stall barns 830
stanchions 830
bull management 777
calf management 779
challenge feeding 346
contracts 88
dairy beef 776
dry cow management 777
Dutch Belted 199
energy 324
equipment 835
milking 832
fat-corrected milk 347
feed(s) 342
preparation 251, 343
sodium bicarbonate 343
sodium propionate 343
substitution table 290, 347
thyroprotein 342
urea 342
feeding 324
pasture 347
sale 350
show 350
Guernsey 199
health program 991
Holstein-Friesian 199
Illaware 199
Jersey 199
judging 1015
lactating cow management 777
losses 874
management 776–779
milking 777, 778
pesticide residues 778
marking 1134
milk fever 330, 331
milking
modern 993
parlors 831, 832
minerals 329–337
mites 958
nutrient requirements 324–328
parasites (*see Table 10-5, p. 920*) 918
parts of cow 1015
pastures 589–636
protein 324
rations
dry cows 347

Page

lactating cows 343–347
thumb feeding rules 345
replacement heifers 349
managing 779
normal growth 350
sale feeding 350
scorecard 1018
show feeding 350
vitamins 329, 338–341
water 342
weight estimating 1173
wormers 917, 918
Dairy Herd Improvement Association 1163
Dairy Herd Improvement Registry 1163
Death camas 964
Death losses
disease 874
market 1063
Debouillet sheep 200
Deeded land 43
Deerflies 922
Dehorning
cattle 729
methods of, by polled bulls 123–125
Dehydrators, artificial 686
Delaine-Merino sheep 200
Depreciation 72
schedules 111, 113, 114
Devon cattle 192
Dexter cattle 192
DHIA testing 146
DHIR testing 146
DHI testing 145
Digestion trial 242
Digestive systems, types of 215, 216
Dipping animals 919
Disease(s) (*see Table 10-2, p. 876*)
control regulations 1001
foreign protection 1003
genetic resistance 875
indemnity payments 1004–1006
losses 874
prevention 873, 986
transmitted to humans 874
Disease-free pigs 999
Disinfectants 984
guide 984
Distemper 888
Dockage 1070
Docking 729
Dog-proof fences 860
Dominance 12
Donkeys 212
Dorset sheep 200
Double muscling 127
Dourine 950
Dressing percentage of
calves 1082
cattle 1082
hogs 1087
lambs 1085
Drugs 226
Dryers, wagon 686
Dry matter 454
Duroc swine 204
Dutch Belted cattle 199
Dwarfism 123–127
Dysentery, swine 888

E

Easements 45
Eating bedding 20
Edema disease 888
Elimination, manure 19
Eliminative behavior 9
Encephalomyelitis 888
Energy 226
calorie system 244
conservation 28
digestible 244
expressing 243
feeds 454

Page
gross ... 244
measuring of ... 243
metabolizable ... 245
net energy system ... 245
comparative slaughter ... 246
determining ... 245
TDN system ... 243
terms ... 243
utilization ... 245
Ensiling process ... 702
Enteritis ... 888
Enterotoxemia ... 890
Enterprise accounts ... 65, 73
Environment
animal ... 21, 811
effects of grazing ... 25
percent of change due to ... 135
Environmental control ... 22, 808–811
Epizootic bovine abortion (EBA) ... 890
Equine
abortion ... 890
encephalomyelitis ... 892
infectious anemia ... 892
infectious metritis ... 892
influenza ... 894
piroplasmosis ... 952
Equipment ... 795
beef cattle ... 827–829
costs ... 51, 816
dairy cattle ... 835
horses ... 846
requisites of ... 814, 815
sheep ... 838
swine ... 842
Ergot poisoning ... 980
Erysipelas ... 894
Estate planning ... 115
Estrays ... 1118
Ewe
gestation period ... 124
pregnancy testing ... 131

F

Face fly ... 922
Fair, shipping to ... 1059
Farm
account book ... 68
accounts ... 65
acquiring of ... 36
appraisal ... 41
contract ... 42
credit system ... 53
financing ... 39
headquarters location ... 796
layout of operations ... 798
loans ... 40
purchase ... 37
records ... 65
rent ... 37
value of ... 38
Farmers' Educational and Cooperative Union of America ... 1163
Farmers' Home Administration ... 52, 53
Farmstead arrangement ... 797
Fats, feeding ... 220
Federal grades of meat ... 1066
Federal Intermediate Credit banks ... 52
Federal land ... 662
Federal Land Banks ... 52, 53
Feed(s)
additives ... 225
by-product ... 219
classes ... 215–217
commercial ... 259
composition tables ... 455–585
crude protein content of ... 226, 455
definition ... 215
energy content ... 455
environmental effect ... 22
evaluation charts ... 238
high-cellulose treatment ... 250
home-mixed ... 259

Page
laws ... 260
moisture content ... 454
preparation ... 248, 251
storage space requirements ... 808
substitution tables ... 258
beef cattle ... 290–294
dairy cattle ... 290–294
horses ... 428
sheep ... 368–370
swine ... 401
TDN calculations ... 243
weights and measures ... 1167
Feeder's margin ... 78
Feeding
beef cattle ... 260–322
custom ... 79
dairy cattle ... 324–350
horses ... 409–435
sheep ... 351–376
standards ... 250
swine ... 377–408
trials ... 247
Feedlots (*see* Beef cattle finishing)
co-op ... 83
record forms ... 86
Feedstuffs ... 215, 234
best buy ... 234
chemical analysis ... 241
energy value ... 243
evaluation of ... 234
chart method ... 238–241
physical ... 234
tabular method ... 235–238
proximate analysis ... 455
Female, insemination of ... 186
Fence(s) ... 852–860
dog-proof ... 860
electric ... 859
legal aspects of ... 1116
posts ... 855
suspension ... 858
Fertilizer(s) ... 740
amount ... 743
application ... 743, 750
best buy, determination of ... 743
commercial ... 740
grade ... 740
Institute ... 1163
kind of ... 743
Fertilizing pastures ... 572
Fescue foot ... 894
FFA ... 1163
Field
cubing hay ... 685
curing hay ... 681
wafering hay ... 685
Finishing cattle (*see* Beef cattle finishing)
Finnsheep ... 200
Fish pond ... 865
Fitting for show and sale
beef cattle ... 322
dairy cattle ... 350
horses ... 435
sheep ... 376
swine ... 408
Fleckvieh cattle ... 192
Flies ... 922, 930, 942, 952
deerflies ... 922
face ... 922
horn ... 922
horseflies ... 922
houseflies ... 922
stableflies ... 924
Floors ... 812, 813
slotted ... 812, 813
Fluorine
maximum ... 229
poisoning ... 440, 980
Flushing ... 122, 371, 404
Food and Drug Administration ... 1002
Foot-and-mouth disease ... 894, 1005
Foothill abortion ... 896

Page
Foot rot ... 896
Forage
characteristics ... 668
cubing ... 249, 685
preservation methods ... 668
wafering ... 685
Formaldehyde ... 984
Founder ... 442
4-H Clubs ... 1163
Freemartin heifers ... 133
Freeze marking ... 725
Freezing meat ... 1089
French Alpine goats ... 203
French combing wool ... 1107
Future Farmers of America ... 1163
Futures trading livestock ... 97
basis ... 99
contracts ... 100–104
glossary of terms ... 104
hedging ... 99
speculating ... 99

G

Galiceno horses ... 208
Galloway cattle ... 192
Garget ... 896
Gas leases ... 45
Gastrointestinal worms ... 924
Gelbvieh cattle ... 192
Genetic resistance to diseases ... 875
Gestation
period ... 122
table ... 124
Gid tapeworm ... 932
Glanders ... 896
Goats
A.I. rules ... 1143
American La Mancha ... 203
Angora ... 203
breeds ... 203
French Alpine ... 203
marking ... 1136
Nubian ... 203
registry associations ... 1127
Rock Alpine ... 203
Saanen ... 203
Swiss Alpine ... 203
Toggenburg ... 203
Goiter ... 442
Gophers, pocket ... 766
Government land ... 43
Grades
meat ... 1080
milk ... 1074, 1103
Grading
meat ... 1079
milk ... 1073, 1103
Grain
acid-treated ... 222
high-moisture ... 221
processing ... 248
weight in bin ... 1168
Grange ... 1165
Grass
hay ... 673
silage ... 709
tetany ... 442
Grazing
corn residues ... 697
environmental effects of ... 25
fees or charges ... 663
permits ... 663
public-owned ... 662
regions—10 U.S. ... 589
season, extending ... 655
systems ... 654, 660
continuous ... 660
controlled ... 654
deferred ... 654
rest-rotation ... 661
rotation ... 654, 660
rotation-deferred ... 660

Page

Greasewood ... 966
Green chop ... 665
Gregarious behavior ... 7
Grooming
 beef cattle ... 1043
 dairy cattle ... 1048
 horses ... 1056
 sheep ... 1051
 swine ... 1054
Grubs, cattle ... 924
Grubs-in-the-head ... 932
Guernsey cattle ... 199
Gut edema ... 898

H

Habituation ... 2
Hackney horses ... 208
Halogeton ... 966
Hampshire
 sheep ... 200
 swine ... 204
Hanoverian horses ... 208
Hardware disease ... 916
Hay ... 667
 additives ... 687
 advantages ... 669
 artificial dehydrators ... 686
 bales ... 684
 buying ... 688
 California evaluation method ... 678
 characteristics ... 675
 chemical composition ... 676–678
 chopped ... 683
 combustion ... 687
 costs ... 692
 crop silage ... 709
 cubic feet per ton ... 1168
 cubing ... 685
 curing ... 680–683
 cutting ... 679
  time ... 680
 definition ... 668
 dehydrators ... 686
 disadvantages ... 669
 dryer, wagon ... 686
 energy source, as ... 670
 equipment ... 679
 evaluating ... 688
 feeding ... 693
 grades, federal ... 676, 689
 grain replacement ... 670
 growing ... 678
 history ... 669
 importance ... 668
 kinds ... 671
 loose ... 683
 losses ... 669, 670, 681, 696
 magnitude ... 669
 methods of making ... 681–686
  artificial drying ... 685
  field curing ... 681
 moisture content ... 680
 mow curing ... 686
 mowing ... 681
 packaged ... 684
 pellets ... 685
 preservatives ... 687
 pricing ... 688
 production ... 674–678, 696
 protein source ... 670
 raking ... 682
 ruminant needs for ... 695
 sampling ... 676
 selling ... 688
 shrinkage ... 692
 sources ... 688
 spontaneous combustion ... 687
 stacks ... 684
 standards ... 689
 storing ... 687
 stretching supply ... 696
 supplementing ... 696
 toxic residues ... 691
 wafering ... 685
 wagon dryers ... 686
 weight in stack ... 1168
Haylage ... 454
Hays Converter cattle ... 193
Head bots or nose bots ... 932
Headquarters location ... 796
Health
 animal ... 15, 873
 environmental effect ... 23
 programs ... 987
 signs ... 874, 875
Heat
 disinfectant ... 985
 production, animal ... 809
Heat (estrus)
 detection methods ... 176
 duration of ... 122
 hormonal control ... 177
 interval of ... 122
Heaves ... 442
Hedging ... 99
Hemorrhagic septicemia ... 898
Henbane ... 966
Heredity, percent change due to ... 135
Hereford
 Angus cross ... 168
 cattle ... 193
 swine ... 204
Heterosis ... 164
High-lysine corn ... 221
High-moisture grain ... 221
 acid-treated ... 222, 248
Hog(s) (*see* Swine)
 -corn ratio ... 408
 following cattle ... 776
 market
  classes ... 1068
  grades ... 1068
 price, seasonal ... 1069
Holstein-Friesian cattle ... 199
Hookworms ... 932
Hormonal control of heat ... 177
Hormones ... 225
Horn
 brands ... 725
 fly ... 922
Horse(s)
 age determination ... 1038
 A.I. rules ... 1144
 American
  Bashkir Curly ... 206
  Creme Horse ... 206
  Gotland Horse ... 206
  Mustang ... 206
  Saddle Horse ... 206
  Walking Pony ... 206
  White Horse ... 206
 Andalusian ... 207
 anthelmintics ... 917, 918
 antibiotics ... 423
 Appaloosa ... 207
 Arabian ... 207
 artificial insemination ... 188
  rules ... 1144
 Belgian ... 211
 blood typing ... 189
 bloom feeds ... 423
 boarding agreement ... 94, 96
 breed(s) ... 206
  registries ... 1128–1132
 Buckskin ... 207
 buildings ... 843
  space requirements ... 806
 castrating ... 734
 Chickasaw ... 207
 Cleveland Bay ... 207
 clipping ... 1057
 Clydesdale ... 211
 Connemara Pony ... 208
 contracts ... 91
 crossbreeding ... 174
 diseases (*see Table 10-2, p. 876*)
 equipment ... 846
 feed(s) ... 419, 422–424
  bran mash ... 423
  lysine ... 424
  preparation ... 251, 424
  substitution tables ... 428–431
  urea ... 422
 feeding
  breeding animals ... 428
  broodmare ... 428
  creep ... 432
  foals ... 432
  guide ... 424–427
  pleasure horses ... 433
  racehorses ... 434
  sale, for ... 435
  show, for ... 435
  stallions ... 428
  urea ... 422
  young equines ... 428, 432
 fences ... 854
 freeze marking ... 728
 energy ... 413
 Galiceno ... 208
 gestation ... 122
  period ... 122
  table ... 124
 Hackney ... 208
 Hanoverian ... 208
 health program ... 1000
 heat ... 122
  duration ... 122
  interval ... 122
 heritability of performance ... 163
 Hungarian Horse ... 208
 identifying ... 727
 judging ... 1028
 Lipizzan ... 208
 lip tattoo ... 727
 losses ... 874
 marking ... 727, 1138
 minerals ... 414–417
  allowances ... 408, 413
 Missouri Fox Trotting Horse ... 208
 mites ... 958
 Morab ... 208
 Morgan ... 209
 Morocco Spotted Horse ... 209
 National Appaloosa Pony ... 209
 nutrient requirements ... 409–418
 Paint Horse ... 209
 Palomino ... 209
 parasites (*see Table 10-8, p. 951*) ... 918
 parts of ... 1028
 Paso Fino ... 209
 pastures ... 589–636
 Percheron ... 211
 Peruvian Paso Horse ... 209
 Pinto Horse ... 210
 Pony of the Americas ... 210
 protection act ... 1119
 protein ... 414
  poisoning ... 414
 puberty, age of ... 122
 Quarter Horse ... 210
 Rangerbred ... 210
 rations for ... 429
 record forms ... 164–166
 registry associations ... 1128
 scorecard ... 1031
 Shetland Pony ... 210
 Shire ... 211
 show-ring ... 846
 Spanish Barb ... 210
 Spanish Mustang ... 210
 stallion
  age and service of ... 123
  breeding contract ... 94, 95
  exercise ... 787
  feeding ... 787

Page
grooming 787
management 787
quarters 787
syndicate agreement 93
syndicated 91
training 19
treats 423
vitamins 418, 419
water 418
Standardbred 211
Suffolk 211
teeth 1037
Tennessee Walking Horse 211
Thoroughbred 211
Trakehner 211
weight estimating 1174
Welsh Pony 211
wormers 917, 918
Ysabella 211
Horsebrush 966
Horseflies 922
Houseflies 922
Housing 986
Hungarian Horse 208
Husklage 698
Hybrid vigor 164
vs purebreds 168
Hydatid 932
Hydroponics 248
Hypoglycemia 442

I

Identifying animals 724, 1132–1138
Illawara cattle 199
Implants 225
Import meat law 1096
quotas 1097
tariff 1097
Imprinting 3
Incentive basis 59–65
Income, net 72
Indemnity payments 1004–1006
Index, selection 175
Indian
hemp 968
lands 662
Indu-Brazil cattle 193
Infectious atrophic rhinitis 898
Infectious bovine rhinotracheitis 898
Infectious embolic meningo encephalitis 898
Influenza 898
Ingestive behavior 9
Inheritance horns 123
Insecticides 917–919
Insect losses 757
Insemination, artificial 180
advantages 181
beef cattle 187
dairy cattle 187
female, of 186
horses 188
limitations of 182
registration association rules 1123
semen 183
sheep 187
swine 187
time of 185
Insight learning 4
Insurance 52, 119
Interest 57
Interspecies relationship 14
Interstate health requirements 1003
Intestinal threadworms 952
Intrastate offering 54
Inventory, annual 66
Investigative behavior 10
Investment credit 111, 114
Iodine
content of feeds 584
deficiency 442
disinfectant 985
Iodophor 985

J

Page
Jersey cattle 199
Jimmyweed 968
Job description 59
Johne's disease 898
Joint-ill 898
Judge, qualifications of 1007
Judging 1007
beef cattle 1010
contest 1008
dairy cattle 1015
horses 1028
scorecard 1008
sheep 1019
swine 1023

K

Karakul sheep 202
Keds 932
Keratitis 898
Ketosis 444
Kidney worm 942

L

Lacombe swine 205
La Mancha, American, goats 203
Lamb(s) (*see* Sheep)
cooking 1086
cuts 1086
dysentery 898
grades 1066, 1068, 1080
Land
acreage determination 1176
legal description 1175, 1176
measures 1175
ownership 662–665
public 662
agencies 663, 664
Landrace swine, American 204
Large intestinal roundworm 942
Larkspur 968
Laurels 968
Law 1111
bill of sale 1115
chattel mortgages 1115
legal documents 1122
syndicated animals 1116
Lead poisoning 980
Leader-follower 13
Lean meter 161
Learning
complex 3
insight 4
simple 2
Lease(s) 1112–1115
cash 37, 1114
crop-share 1114
farm 37
gas 45
livestock-share 37, 1114
oil 45
Leased land 43
Legal difficulty guides 1122
Legume inoculation 653
Leicester sheep 202
Lenders, credit factors 55
Leptospirosis 900
Lespedeza hay 673
Liability 1120
Lice 926, 932, 942, 952
Licensed livestock operations 1120
Lighting, artificial 23
Lime 752
amount to use 755
application 756
best buy determination 755
disinfectant 985
functions 752
kinds of 755
needs determination 753
Limousin cattle 193
Page
Lincoln
Red cattle 193
sheep 202
Lipizzan horses 208
Liquid supplements 250
Listerellosis 900
Liver
abscesses 916
fluke 926, 934
necrosis of lambs 900
Livestock
associations 1123
Conservation Institute 1163
insurance 119
licensed operations 1120
marketing 1061
methods 1062
shipping 1062
Loading chute design 18
Loans, farm 40
Lockjaw 900
Locoweed 970
Lucerne (alfalfa) 672
Lumpy jaw 900
Lungworm 926, 934, 944
Lupines 970
Lye 985
Lysol 985

M

Magazines 1149
Maine-Anjou cattle 194
Malignant edema 902
Management 723
Manager 58
Managra swine 205
Man-animal relationship 14
Manganese deficiency 444
Mange 934, 944
Manure 26, 737–740, 833
amount applied 739
amount produced 737–739, 814, 834, 841
animal elimination 19
beef cattle 825
composition 737
dairy cattle 833
disposal 986
equipment 825
fertilizer 27
gasses 814
handling 814, 833, 841
pollution control 26
sheep 837
storage 814
swine 840
uses 739
value 737
Marchigiana cattle 194
Mare(s)
gestation period 122
pregnancy testing 132
Margin
feeding 78
price 78
Market
bruises 1063
classes of livestock 1066
cattle 1067
hogs 1068
sheep 1068
crippling 1063
cyclical trends 1069
death losses 1063
grades of livestock 1066
seasonal changes 1069
Marketing
costs 1070
dockage 1070
feeder pig values 1069
livestock 1061
meat check-off 1070
methods 1062

Page

milk ... 1061, 1071
shipping livestock ... 1062
shrinkage ... 1064
Marking animals ... 724, 1132–1138
Mastitis ... 902, 993
Mating table ... 123
Measles ... 926
Measures
feeds of ... 1167
weights and ... 1167
Meat ... 1077
certification, swine ... 162
check-off ... 1070
consumption ... 1078
cooking ... 1094
curing ... 1092–1094
diet contribution ... 1078
freezing ... 1089–1092
grades
beef ... 1082
federal ... 1080
lamb ... 1085
pork ... 1087
veal ... 1082
grading ... 1079
import law ... 1096
inspection ... 1078, 1079
preservation ... 1089
Mercury poisoning ... 980
Metabolism trial ... 242
Metric measures ... 1176–1180
Metritis ... 902
Micotoxins ... 982
Milk ... 1077
adjustment factors ... 147
All-Jersey ... 1075
certified ... 1075
composition ... 1098
condensed ... 1104
consumption per capita ... 1097
cooperatives ... 1073
diet contribution ... 1078
environmental factors ... 1099
evaporated ... 1104
federal control ... 1072
fever ... 330, 444
flavor ... 1101
Golden Guernsey ... 1075
graded ... 1103
grades ... 1074, 1103
handling ... 1071
market channels ... 1072
marketing ... 1061, 1071
orders ... 1072
regulatory programs ... 1073
state laws ... 1074
mastitis ... 1100
only record ... 146
pesticide residues ... 1102
pricing ... 1072, 1074
price supports ... 1075
production ... 1071
quality ... 1098
replacer ... 222
sanitary regulations ... 1073, 1102
standards ... 1073
state control ... 1073
uses ... 1103
weight per gallon ... 1075
Milking
modern ... 993
Shorthorn cattle ... 194
Milkweed ... 970
Mineral(s) ... 227
beef cattle ... 272
chelated ... 230
dairy cattle ... 329
horses ... 415
major or macro ... 227
micro ... 227
rights ... 44
sheep ... 355

Page

supplements ... 230, 454, 568–571
swine ... 383
trace minerals ... 227
Miniature Mediterranean Donkeys ... 212
Missouri Fox Trotting Horse ... 208
Mites ... 926, 934, 944, 952, 958, 959
cattle (beef and dairy) ... 926, 958
control ... 958
dairy ... 958
goats ... 934, 958
horses ... 952, 958
sheep ... 934, 958
swine ... 944, 958
Mohair ... 1105
classes ... 1109
grades ... 1109
Moisture
content, feeds ... 454
-free ... 454
production, animals ... 810
Molasses ... 223
silage preservative ... 715
Molybdenum toxicity ... 444
Montadale sheep ... 201
Moon blindness ... 446
Morab horses ... 208
Morocco Spotted Horse ... 209
Mortgages, farm ... 41
Mosquitoes ... 928, 952
Most probable producing ability ... 136
Mouse control ... 766
Mower
crimper ... 682
crusher ... 682
Mowing hay ... 682
Mucosal disease ... 904
Mules ... 212
Multiple births, cattle ... 132
Murray Grey cattle ... 194
Muscular hypertrophy ... 127
Mutton grades ... 1080
Mycoplasma pneumonia ... 904

N

National
Academy of Sciences ... 1164
Association of Animal Breeders ... 1164
Cattlemen's Association ... 1164
Council of Farmer Cooperatives ... 1164
Cutting Horse Association ... 1164
Farmers' Organization ... 1164
Farmers' Union ... 1164
Grange ... 1165
Independent Meat Packers Association ... 1165
Live Stock and Meat Board ... 1165
Livestock Exchange ... 1165
Live Stock Producers Association ... 1165
Silo Association ... 1165
Swine Repopulation Association ... 1165
Wool Act ... 1110
Wool Growers Association ... 1165
Navel infection ... 904
Necrotic enteritis ... 904
Needle teeth ... 786
Nightshade ... 970
Nitrate-nitrite poisoning ... 982
Nitrate poisoning ... 446
Nodular worm ... 934, 944
Nonprotein nitrogen ... 223
slow-release ... 224
Normande cattle ... 194
North Country Cheviot sheep ... 201
Norwegian Red cattle ... 195
Nubian goats ... 203
Nutrient(s) ... 226
beef cattle ... 260–281
dairy cattle ... 324–342
horses ... 409–418
needs ... 226–233
sheep ... 351–364
swine ... 377–392

Page

Nutritional diseases ... 436–453
Nutrition, environmental effect ... 22

O

Oak (poison) ... 972
Oat hay ... 672
poisoning ... 446
OIC swine ... 205
Oil leases ... 45
Oils ... 220
Oleander ... 972
Operant conditioning ... 3
Organization chart ... 59
Osteomalacia ... 446
Ova transplantation ... 179
Overeating disease (enterotoxemia) ... 890
Owner-Sampler Records ... 146
Oxford sheep ... 201

P

Packinghouse by-products ... 1095
Paint ... 860–865
Paint Horse ... 209
Palatability of feed ... 247
Palomino horses ... 209
Panama sheep ... 202
Parakeratosis ... 446
Parasites (*see table numbers in alphabetized listing below*) ... 917
beef cattle (Table 10-5) ... 920
control ... 873
dairy cattle (Table 10-5) ... 920
horses (Table 10-8) ... 950
losses ... 874
sheep (Table 10-6) ... 930
swine (Table 10-7) ... 940
Parity ... 1070
Partnership ... 34, 35
contract ... 118
general ... 34
limited ... 34
Paso Fino horses ... 209
Pasture(s) ... 588
adapted plants ... 590
areas ... 589
cattle, for ... 589–636
finishing ... 320–322
classes of
permanent ... 588
rotation ... 588
supplemental ... 588
temporary ... 588
clipping ... 655
establishing ... 652
fertilizer rates ... 592
forages ... 587
horses, for ... 589–636
improving ... 654
irrigated ... 652, 655
management ... 652, 654
plants ... 454
regions, 10 U.S. ... 589
renovating ... 654
rotation ... 986
scattering droppings ... 655
seeding ... 652
sheep, for ... 589–636
swine, for ... 636–653
four U.S. areas ... 637
topdressing ... 655
winter ... 297, 298
Pelleting
complete rations ... 250
roughages ... 249
Percheron horses ... 212
Performance testing (*see* Production testing)
Periodic ophthalmia ... 446
Peruvian Paso Horse ... 209
Pest control ... 19, 757
Phenol ... 985
Phosphorus ... 228

Page

supplements ... 229
Photosynthesis ... 29
pH range of soil ... 753
Piedmont cattle ... 195
Pigeon control ... 767
Pine needle abortion ... 972
Pinkeye ... 904
Pinto Horse ... 210
Pinworms ... 954
Pinzgauer cattle ... 195
Pitch poisoning ... 982
Plant
diseases ... 757
hunger signs ... 744–751
Pneumonia ... 904
Pocket gophers ... 766
Poison hemlock ... 972
Poison Information Centers ... 1166
Poisonous elements ... 980–983
common ... 961
diagnosing ... 958
loss prevention ... 960
plants ... 659, 979, 985
treatment ... 961
why animals eat ... 959
Poland China swine ... 205
Polioencephalomalacia ... 448
Polled
bulls for dehorning ... 123
Hereford cattle ... 195
Shorthorn cattle ... 196
Pollution
control ... 24, 815, 825
feedlot ... 776
laws ... 24
regulations ... 24
Ponds ... 865–871
Pony of the Americas ... 210
Pork
cooking ... 1088
curing ... 1092–1094
cuts ... 1088
grades ... 1080
Potato silage ... 709
Pox, swine ... 906
Preconditioning calves ... 303, 772, 990
certificate ... 774
Predators ... 659
Predicted difference ... 150
Pregnancy
disease ... 448
testing cows ... 131
Preservatives
hay ... 687
silage ... 713–717
Production Credit Association ... 52, 53
Production testing ... 134
beef cattle ... 134, 139
dairy cattle ... 145
horses ... 163
sheep ... 151
registry associations, by ... 157
swine ... 157
record associations, by ... 162
Profit indicators
cow-calf ... 70
feedlot ... 70
Progeny testing (*see* Production testing)
Progestagens ... 178
Progesterone ... 284
Property line ... 45
Proprietorship ... 33
Prostaglandins ... 178
Protein ... 226
animal ... 219
feeds ... 454
plant ... 219, 448
poisoning ... 414, 982
single cell ... 224
supplements ... 218
Proved sires, dairy ... 149
Pseudorabies ... 906

Page

Puberty, age of ... 122
Public offerings ... 54
Pulpy kidney disease ... 890, 906
Pulse rate, normal ... 875

Q

Q-fever ... 906
Quarantine ... 1003
Quarter Horse ... 210
Quaternary ammonium compounds (QAC) ... 985

R

Rabies ... 906
Racehorse feeding ... 434
Railroad car, number animals in ... 1063
Railroad-owned lands ... 664
Raking hay ... 682
Ram
age and service of ... 123
hand mating ... 123
Rambouillet sheep ... 200
Ranch
acquiring ... 36
financing ... 39
headquarters ... 796
value ... 38
Range
cubes ... 299
distribution of animals ... 661
forages ... 587
grazing systems
continuous ... 660
rest-rotation ... 661
rotation ... 661
rotation-deferred ... 660
improvement ... 661
management ... 657
plants ... 454
reseeding ... 662
season of use ... 658
stocking rate ... 658, 661
western ... 656
Rangerbred horses ... 210
Ranger cattle ... 196
Rat control ... 758, 762–765
Rations
balanced ... 252
balancing method ... 253–258
computer ... 257
net energy ... 256
square ... 254
trial-and-error ... 255
Ratios, beef cattle ... 136
Rayless goldenrod ... 974
Reasoning ... 4
Record(s)
farm ... 65
books ... 68
forms
beef cattle ... 136, 137–139
cattle feedlot ... 86
dairy cattle ... 147
horses ... 164–166
sheep ... 152, 154, 155
swine ... 158–160
Recreational areas ... 655
Red
Angus cattle ... 196
Brangus cattle ... 196
Poll cattle ... 196
water disease ... 908
Registry associations ... 1123–1132
Regulation A offering ... 54
Relationships
interspecies ... 14
man-animal ... 14
Rent, farm ... 37
Repeatability ... 150
Reproductive failure, nutritional,
beef cattle ... 296
Rest-rotation grazing ... 661

Page

Rhinopneumonitis ... 908
Rickets ... 448
Ringworm ... 928, 936, 944, 954
Riparian water right ... 43
Risk, farm ... 45
Rock Alpine goats ... 203
Rodent control ... 759, 762–767
Romney sheep ... 202
Roofs ... 812
Rotation-deferred grazing ... 660
Rotation grazing ... 654, 660
Roughages
definition ... 216
percentage ... 217
processing ... 249
Rubberweed ... 974

S

Saanen goats ... 203
Salers cattle ... 196
Saline soils ... 756
Sal soda ... 985
Sal soda and soda ash ... 985
Salt
deficiency ... 448
poisoning ... 448, 982
sick ... 448
Sanitary commissions ... 1003
Sanitation, animal ... 986
Santa Gertrudis cattle ... 197
Scab ... 936
Scales ... 815
Scorecard
beef cattle breeding ... 1013
dairy cattle ... 1018
horses ... 1031
judging ... 1008
personnel ... 64
sheep
breeding ... 1022
market lamb ... 1022
steer ... 1014
swine
breeding ... 1026
market barrow ... 1027
Scotch Highland cattle ... 197
Scrapie ... 908
Screwworm ... 928, 936, 946, 954
Seasonal market changes ... 1069
SEC registered offering ... 54
Selecting ... 1007
Selection
bases ... 1007
systems ... 175
index ... 175
minimum culling standards ... 175
tandem ... 175
Selenium
content of feeds ... 584
poisoning ... 450, 982
Semen
collection and handling ... 182
dilutors ... 184
drugs for ... 184
extenders ... 184, 185
packaging ... 186
shipment of ... 184
sperm concentration ... 183
storage of ... 184
thawing frozen ... 187
volume ... 183
Sex determination ... 130
Sexual behavior ... 10
Sheep
additives ... 364
adjustment factors ... 156
age determination ... 1034
A.I. rules ... 1142
American Merino ... 200
anthelmintics ... 917, 918
artificial insemination ... 187
rules ... 1142

Page

Black-faced Highland 202
bots 936
breeds 200
building 836
    space requirements 804
castrating 730, 733
central test station 151
characters
    economically important 152
    heritability 152
Cheviot 200
Columbia 202
Corriedale 202
Cotswold 201
creep feeding 374
cross breeding 172
Debouillet 200
Delaine Merino 200
diseases (*see Table 10-2, p. 876*)
docking 730, 733
    equipment 838
Dorset 200
dressing percent 1085
ewes-management 780
farm testing 151
feed
    preparation 251, 365
    substitution table 368–370
feeding
    breeding animals 371
    ewes 371
    farm flock 372
    growing-finishing lambs 373
    rams 373
    range band 372
    sale, for 376
    show, for 376
finishing lambs 375
    drylot 375
    field 375
Finnsheep 200
flushing 371
gestation
    period 122
    table 122
grades 1068
Hampshire 200
health program 994
heat
    duration 122
    interval 122
heritability of traits 152
identifying 726
implants 361
insemination 187
judging 1019
Karakul 202
keds 936
killing dogs 1118
lamb(s)
    chilled 781
    cooking 1086
    cuts 1086
    disowned 781
    grades 1080
    orphaned 781
lambing
    management 779
    pen 780
Leicester 202
Lincoln 202
liveweight average 1085
losses 874
market
    classes 1068
    grades 1068
marking 726, 1135
    ram 726, 780
measles 938
minerals 355–361
mites 958

Page

Montadale 201
North Country Cheviot 201
nutrient needs 351–364
Oxford 201
Panama 202
parasites (*see Table 10-6, p. 930*) 918
parts of 1019
pastures 589–636
performance test 152
production testing 151, 156, 157
    form 152
    registry associations, by 157
progeny testing rams 156
protein 351
puberty, age of 122
ram
    age and service of 123
    mating 123
Rambouillet 200
range
    forage deficiencies 372
    supplements 372
rations 367
record form 152, 154, 155
registry associations 1126, 1127
Romney 202
scorecard
    breeding sheep 1022
    market lambs 1022
Shropshire 201
slotted floors 836
Southdown 201
Suffolk 201
Targhee 202
traits, heritability 152
Tunis 201
vitamins 355, 360–363
water 364
weaning early 373
weight estimating 1174
wormers 917, 918
Shelter-seeking behavior 11
Shetland Pony 210
Shipping
    bedding 1064
    bruises 1063
    crippling 1063
    death losses 1063
    fever 908
    livestock 1062
    railroad car capacity 1063
    shrinkage 1064
    to fair 1059
    truck capacity 1063
    28-hour law 1062
Shire horses 212
Shorthorn
    cattle 197
    color inheritance 122
Show classification
    beef cattle 1047
    dairy cattle 1050
    horses 1058
    sheep 1053
    swine 1056
Showing 1043
    beef cattle 1043
    dairy cattle 1048
    grooming 1043, 1048, 1051, 1054, 1056
    horses 1056
    sheep 1051
    shipping to fair 1059
    swine 1054
Show-ring, horse 846
Shrinkage in marketing 1064
Shropshire sheep 201
Shucklage 698
Sight, animal 16
Silage 454, 701
    advantages 703
    characteristics 711

Page

    conditioning 713
    corn 708
        husklage 709
        residue 708
        stalklage 709
        stover 709
        urea 710
    cutting length 712
    definition 702
    disadvantages 704
    earcorn 708
    economy 718
    ensiling process 702
    feeding
        fence 707
        value 718
    frosted crop 710
    gases 718
    grass 709
    harvesting
        machinery 711
        methods 711
        stage 712
    hay crop 709
    haylage 701, 719
    high-moisture grain 701, 709, 720
        acid preservation 721
        storage 721
    importance 702
    kinds 708
    low-moisture 719
    making 712
        distribute uniformly 717
        fill rapidly 717
        seal off 718
    milk odor and flavor 719
    moisture content 713
        determining 713
    moldy 718
    potato 709
    preservatives 703, 713–717
        bacteria inhibitors 715
        bacterial cultures 716
        enzymes 716
        grain 715
        limestone 715
        mineral acids 716
        molasses 715
        mold inhibitors 715
        organic acids 716
        sodium metabisulfite 715
        urea 715
        yeast culture 716
    rain-damaged hay 710
    shelled corn 708
    sorghum 708
        residue 708
    storage losses 708
    sunflower 709
    weight estimating 1171
    weight in silo 819
    wilting 713
Silos 704
    bunker 706
    coating 718
    conventional upright 705
    enclosed stack 707
    gastight 706
    horizontal 706
    kinds 704–708
    modified trench stack 707
    nutrient losses 708, 718
    open stack 707
    pit 706
    plastic 707
    self-feeder 706
    size 817–819
    temporary 707
    tower 817
    trench 706, 819
    upright 705, 817

Page

Simmental cattle ... 197
Sleep, animal ... 16
Slotted floors ... 812, 813
  cattle (beef and dairy) ... 813
  sheep ... 813, 836
  swine ... 813, 840
Small stomach and intestinal worms ... 938, 940
Snakeweed ... 974
Sneezeweed ... 974
Soap ... 985
Social
  order ... 12
  relationship ... 12
Socialization ... 3
Social Security Law ... 1121
Soda ash ... 985
Sodium
  bentonite ... 226
  carbonate ... 226
  metabisulfite ... 715
Soft pork ... 408
Soil
  acidity ... 754
  Conservation Service ... 1166
  deficiencies ... 741
  fertility ... 736–757
  guide ... 742, 744–751
  liming ... 752
  management ... 757
  pH value ... 753
  samples ... 742
  tests ... 741
Soil Conservation Service ... 1166
Sonoray ... 161
Sore mouth ... 908
Sorghum
  residues ... 699
  silage ... 708
Soring horses ... 1119
Sound communication ... 14
South Devon cattle ... 197
Southdown sheep ... 201
Sow(s)
  gestation period ... 122
  pregnancy testing ... 131
Soybean
  hay ... 674
  refuse ... 700
Spanish-Barb horse ... 210
Spanish-Mustang horse ... 210
Sparrow, English ... 767
Specific pathogen-free pigs ... 999
Speculation ... 99
Sperm concentration ... 183
Spotted swine ... 205
Spraying pen ... 825
Spring parsley ... 974
Squirrels, ground ... 767
Stablefly ... 924
Stalklage ... 698
Stallion
  age and service of ... 123
  breeding contract ... 94, 95
  mating ... 123
Standardbred horse ... 211
Staple wool ... 1106
Starlings ... 767
State
  Colleges of Agriculture ... 1160
  Departments of Agriculture ... 1166
  Veterinarians ... 1003
Stiff lamb disease ... 450
St. Johnswort ... 976
Stocker(s)
  contracts ... 87
  feeding ... 306–308
Stomach worms ... 938, 946, 954
Strangles ... 908
Straws ... 700
Strays ... 1118
Streptococci infection ... 910

Page

Stress ... 23
  environmental ... 23
Strongyles ... 956
Suffolk
  horses ... 212
  sheep ... 201
Summer sores ... 956
Sunflower silage ... 709
Superovulation ... 179
Sussex cattle ... 198
Sweet clover disease ... 450
Swine
  additives ... 394–396
  age, determination ... 1035
  A.I. rules ... 1143
  American Landrace ... 204
  anthelmintics ... 917, 918
  antibiotics ... 394, 395
  arsenicals ... 395
  artificial insemination of ... 187
  backfat measuring ... 161
  Berkshire ... 204
  boar
    age and service of ... 123
    management ... 782
    mating ... 123
  breed(s) ... 204
    registries ... 1128
  buildings ... 839
    space requirements ... 804, 839
      early weaned pigs ... 839
  castrating ... 730, 734
  central testing stations ... 161
  characters
    economically important ... 157
    heritability ... 157
  Chester White ... 204
  chilled pigs ... 784
  confinement ... 784
  Conner Prairie ... 204
  corn-hog ratio ... 408
  creep feeding ... 405
  crossbreeding ... 173
  diseases (*see Table 10-2, p. 876*)
  dressing percent ... 1087
  Duroc ... 204
  dysentery ... 910
  energy ... 377
  equipment ... 842
  erysipelas ... 910
  farrowing ... 783
    artificial heat ... 784
  feed(s) ... 392, 394
    preparation ... 251, 396
    substitution table ... 401–404
  feeder pigs ... 999
  feeding
    boars ... 406
    breeding gilts ... 404
    breeding swine ... 401
    brood sows ... 404
    growing-finishing ... 406
    hand ... 786
    limited ... 785
    liquid ... 785
    orphans ... 405
    requirements ... 407
    sale, for ... 408
    sex effect ... 407
    show, for ... 408
    systems ... 785
  flushing ... 404
  gestation
    period ... 122
    table ... 122
  grouping hogs ... 782
  guard rail ... 784
  Hampshire ... 204
  health program ... 996
  heat
    duration ... 122

Page

    interval of ... 122
  Hereford ... 204
  heritability of traits ... 157
  identifying ... 726
  influenza ... 910
  insemination of ... 910
  judging ... 1023
  Lacombe ... 205
  lean meter ... 161
  liveweight average ... 1087
  losses ... 874
  management ... 782–786
    breeding season ... 783
    breeding swine ... 783
    skills ... 786
  Managra ... 205
  marking ... 726, 1137
  meat certification ... 162
  milk replacer ... 395
  minerals ... 383–387
  mites ... 958
  needle teeth ... 786
  nutrient
    allowances ... 377–382
    requirements ... 377–392
  OIC ... 205
  parasites (*see Table 10-7, p. 940*) ... 918
  parts of hog ... 1023
  pastures ... 636–653, 782
  Poland China ... 205
  pork
    cooking ... 1088
    curing ... 1092–1094
    cuts ... 1088
    grades ... 1080
  pox ... 906
  probing ... 160
  production testing ... 157, 162
    form ... 158
  protein ... 380
  puberty, age of ... 122
  rations ... 396–400
    complete ... 785
    pelleted ... 786
    self-feeding ... 785
  record form ... 158–160
  registry associations ... 1127
  ringing ... 786
  scorecard ... 1062
    breeding swine ... 1026
    market barrow ... 1027
  show classification ... 1056
  showing ... 1054
  skills ... 786
  slotted floors ... 813, 840
  soft pork ... 408
  sonoray ... 161
  specific pathogen-free ... 999
  Spotted ... 205
  Tamworth ... 205
  testing ... 157
    meat certification ... 162
  traits, heritability of ... 157
  tusks, clipping of ... 786
  vitamins ... 388–393
  water ... 392
  weight estimating ... 1174
  Wessex Saddleback ... 205
  wormers ... 917, 918
  Yorkshire ... 205
Swiss Alpine goats ... 203
Syndicate(d)
  agreement ... 93
  animals ... 1116
  horses ... 91

T

Tail biting ... 20
Tamworth swine ... 205
Tandem selection ... 175

Page

Tansy ragwort 976
Tapeworms 938, 956
Tarentaise cattle 198
Targhee sheep 202
Tax 106
accrual basis 108
capital gain 109
cash basis 108
depreciation 111, 113
estate 115
gift 117
investment credit 110, 113
loss limitations 115
planning 113
pointers 106
shelters 114
Taylor Grazing Act 663
TDN calculation 243
Temperature
building 808, 811
normal animal 875
Tenant 1112
Tennessee Walking Horse 211
Tetanus 910
Texas Longhorn cattle 198
Thin-necked bladder worm 940
Thorn-headed worm 946
Thoroughbred horses 211
Threadworm 946, 956
Ticks 928, 956
Timber 45
milk vetch 976
Timothy hay 673
Toggenburg goats 203
Training animals
early 18
horses 19
Trakehner horses 211
Tranquilizers 226
Transmissible gastroenteritis (TGE) 910
Trespass animals 1117
Trichinosis 948
Trichostrongyles 940
Trihybrid cross 123
Truck, number of animals in 1063
Trusts 118
Tuberculosis 912, 916, 1004
susceptibility to 916
Tularemia 912
Tunis sheep 201
Twisted stomach worm 940

U

Unidentified factors 233
Uniform Commercial Code 1116

Page

Urea 223, 311
corn silage-urea supplements 710, 715
Urinary calculi 450
U.S.
Animal Health Association 1166
Department of Agriculture 1002, 1166
DHIA sire summaries 149
Forest Service 663
grazing fees 664
Public Health Service 1002

V

Vagina, artificial 183
Vaginitis 912
Vapor production, animals 810
Veal
cooking 1082–1084
cuts 1081–1084
grades 1080
Ventilation 810, 811, 986
Veratum 978
Vesicular exanthema 912
Vetch hay 674
Veterans Administration 1166
Veterinarians, state 1003
Vibriosis 914
Virus pneumonia 914
Visual display communication 15
Vitamin(s) 230
A 231
deficiency 452
B 231
beef cattle, for 230
C 233
D 228
content of foods 585
dairy cattle, for 329
E 232
fat-soluble 230
horses, for 418
imbalances 233
K 232
sheep, for 355
supplements 233, 454, 572–579
swine, for 388
water-soluble 230
Vocational agriculture instructors 1166

W

Wafering forages 685
Wagon dryers 686
Warbles 928
Warts 914
Water 233

Page

rights 43
appropriative 44
riparian 43
Water hemlock 978
Weaning 728
Weather effect on animals 22
Weed(s)
chemical control 760–762
control 757–762
Weigh-A-Day-A-Month 146
Weights 1167
feeds 1167
measures 1167
Welsh
Black cattle 198
Pony 211
Wessex Saddleback 205
Western range 656
Western States Meat Packers Association 1166
Whipworm
sheep 940
swine 948
White
cattle scours 914
muscle disease 452
snakeroot 978
Whitewashes 861
Wills 118
Winter dysentery, bovine 914
Wood chewing 20
Wooden tongue 900, 914
Wool 1105
blood system 1107
carpet 1107
clothing 1107
combing 1106
French combing 1107
grades 1106–1108
maggot 940
National Wool Act 1110
numerical system 1108
quality production 1108
shearing 1108
shorting 1108
staple wool 1106
worsted spinning count 1108
Workmen's Compensation Acts 1120
Wormers 917, 918

Y

Yorkshire swine 205
Ysabella horse 211

Z

Zoning animals 28